GEOMORPHOLOGY

Cover illustration by Mateo Gutiérrez

GEOMORPHOLOGY

Mateo Gutiérrez
Departamento de Ciencias de la Tierra,
Facultad de Ciencias, Universidad de Zaragoza,
Zaragoza, Spain

English translation by Patricia Bobeck
Geotechnical Translations,
Austin, TX, USA

CRC Press
Taylor & Francis Group
Boca Raton London New York Leiden

CRC Press is an imprint of the
Taylor & Francis Group, an **informa** business
A BALKEMA BOOK

CRC Press/Balkema is an imprint of the Taylor & Francis Group, an informa business

Typeset by V Publishing Solutions (P) Ltd, Chennai, India

Printed and bound in The Netherlands by PrintSupport4U, Meppel

Published by: CRC Press/Balkema
P.O. Box 447, 2300 AK Leiden, The Netherlands
e-mail: Pub.NL@taylorandfrancis.com
www.crcpress.com – www.taylorandfrancis.com

Library of Congress Cataloging-in-Publication Data

Applied for

ISBN: 978-0-415-59533-9 (Hbk)
ISBN: 978-0-203-09331-3 (eBook)

To my grandchildren

CONTENTS

PREFACE

The success of the book ***Geomorfología Climática*** (2001) and its slightly amplified English translation ***Climatic Geomorphology*** (2005) motivated us to write a book (***Geomorfología***, 2008) of a more general nature with a broader focus on the landforms and the processes that shaped them. The study of these morphologies, some of which formed during past geological periods under environmental conditions very different from those of today, makes it possible to reconstruct the evolution of relief and to infer environmental changes that have involved geological media, the climate, or human activity.

A major advance of Geomorphology in recent decades is the development of techniques that make it possible to quantify morphogenetic processes and rates at which forms change under different environmental conditions. The development of Geochronology, or absolute dating methods, is helping us correct the limitations of relative dating that have prevailed in Geomorphology for many years. The ability to assign numerical ages to both landforms and deposits opens up multiple possibilities for reconstructing the evolution of relief, making correlations, calculating rates, and estimating recurrence periods.

A theme of major concern facing people today is the possible warming of the planet due to the release of greenhouse gases into the environment. Investigations conducted by the scientific community show that this temperature increase is at least partially anthropogenic. Given this more-than-probable cause and effect relationship, the most sensible and prudent path is to design and apply mitigation measures to alleviate this heating that can negatively affect both the natural environment and human society. The information that Geomorphology can provide on the recent past (Historical Geomorphology) may be very useful in making predictions on the activity of these potentially dangerous processes in the future and on the possible effects of environmental changes.

The aim of this book is to provide a general vision of the multiple aspects of Geomorphology and to provide a methodological foundation to approach the study of various branches of geomorphology. To this end, the book contains a basic bibliography (around 3600 references) that can be used for future research. In addition, applied aspects of Geomorphology are covered at the end of each chapter to provide knowledge of the activities of geomorphologists in the professional world. The book Geomorphology (2013) constitutes an extensively and enlarged version of the book Geomorfologia (2008) published in Spanish language by Pearson-Prentice Hall.

I would like to express my appreciation to the many researchers with whom I have collaborated in the preparation of this book: Doctors A. Cendrero, P.G. Silva, F.J. Gracia, J. Remondo, A. Pérez-González, F. Gutiérrez, C. Zazo, G. Benito, J.L. Goy, J. López-Martínez, A. Gómez Ortiz, A. Martín-Serrano, J.L. Peña, J.M. García-Ruiz, J. Rodríguez, C. Martí, B. Valero and C. Sancho, some for their valuable reviews and relevant observations, others for providing excellent photographs that have significantly enriched the book, and finally others who have supplied unpublished data. The following have also provided very good photographic documents that have enriched the book: Dr. A. Gómez-Sal, Dr. Belén Leranoz, P. Lucha, J. Guerrero, and E. Pueyo. To them I express my sincere thanks.

My most effusive thanks to Cinta Marín, member of our research team, who has transcribed most of the book with accuracy and enthusiasm.

Finally, thank you to Nieves for her help, tolerance, and infinite understanding. Our son Francisco has always provided help and collaboration for which I express my sincere gratitude.

September 2012

1

GEOMORPHOLOGY

1.1 Introduction

The word "geomorphology" is derived from the Greek roots *geo* (Earth), *morphos* (shape), and *logos* (study). Thus, geomorphology is the study of the shape of the Earth's surface. Schumm (1991) defines geomorphology as the science of phenomena on or near the earth's surface and the interactions between different types of materials and processes, including solids, liquids and gases. The effects of human activity are often crucial. Some authors consider geomorphology to be one of the most difficult physical sciences (Linton, 1964). Some authors restrict geomorphology to the study of the characteristics of sub-aerial topography but others extend it to the morphology of the ocean floors. Geomorphology also includes the investigation of landforms on other planets; this science is called planetary or extraterrestrial geomorphology (Greeley, 1985; Baker, 1981a; 1993, 2001, 2004a).

Endogenetic and exogenetic forces shape the topography of the earth's surface. Endogenetic forces create high elevations and deep depressions primarily by vertical movement (Teixell *et al.*, 2009). Exogenetic forces continuously denude and wear down the existing topography. This constant battle, which occurs at many scales, has continued throughout the Earth's history. The processes of external dynamics can be categorized as weathering-erosion, transport and sedimentation. The results are groups of erosional and depositional landforms that typically show specific characteristics related to the processes operating in various morphogenetic environments as the various climatic belts oscillate between the Poles and the Equator, which explains the enigmatic relict shapes in many regions (Cotton, 1958). The energy necessary for these processes comes from various sources. Solar radiation reaches the Earth's surface and is partially transformed into heat, which is the principal source of meteorological processes. These processes control weathering, pedology, the development of landform, as well as plant and animal life. In addition to solar radiation, gravitational energy causes sediment transport and mass movements on hillsides, etc. Endogenetic energy creates most of the topography of the Earth's surface (Büdel, 1968, 1977).

Geomorphology has traditionally used medium-scale studies to analyze geomorphic events that over time have shaped the present-day landscape. Some authors call this historical geomorphology (Chorley, 1978). The analysis of active processes on a more detailed scale and the study of temporal variation of landforms have recently led to a field known as quantitative geomorphology (Chorley, 1978) or process geomorphology (Hart, 1986; Thorn, 1988). As Chorley *et al.* (1984) point out, historical studies are based on reading backwards whereas studies of process geomorphology are directed toward prediction. These are the two prevailing tendencies in geomorphology today. In some countries the study of present-day processes has been emphasized to the partial neglect of traditional studies of landscape evolution over long periods of time but knowledge of how microforms develop can provide a valuable understanding of the long history that has created present-day landforms. We can compare this situation with that of geotectonics where the study of microstructures has led to a deeper understanding of the structure of cordilleras (Gutiérrez, 1990). Geomorphology plays a fundamental role in the control of many ecosystem processes, and in turn, ecosystems can have a profound influence on geomorphic processes and landforms (Renschler *et al.*, 2007).

These present-day tendencies in geomorphology, like those in other branches of science, follow thinking and understanding that have evolved over centuries. Geomorphology was born as a branch of science when naturalists stopped describing landforms and began explaining them. More than a century ago, geomorphology was known as orography and hydrography (Tricart, 1965). McGee and Powell of the United States may have been the first to use the term "geomorphology" in the 1880s (Hart, 1986) although Tinkler (1985) claims Laumann (1858) used it for the first time in Germany (Roglic, 1972).

1.2 History of geomorphology

1.2.1 Geomorphology prior to the 20th century

The origins of geomorphology, like many branches of science, are nebulous. They go back to observations and interpretations made by philosophers and thinkers prior to the Christian era, especially to the Greeks. The excellent book by Chorley *et al.* (1984) contains a profound and detailed analysis of the history of geomorphology. The term "geomorphology" had not been coined but the first and fundamental ideas on this subject were observations made as the natural sciences developed.

As Holmes (1965) pointed out, the Greek philosophers made important discoveries in the fields of mathematics and astronomy but they considered the phenomena of the natural world to be manifestations of the powers of mythical deities. In the Mediterranean region, many of these gods were personifications of various aspects of the natural world (Poseidon, Zeus, Pluto, etc.). These beliefs persisted in the development of geomorphic thought until the 18th century (the "teleological period" according to Chorley, 1978).

In spite of these general convictions, some philosophers such as Thales of Miletus (624–565 BC) broke from these constraints. Thales considered the activities of the natural world to be normal acts that could be studied by observation and reason. Also, Anaxagoras (500–428 BC) pointed out that the Earth revolves so slowly compared to our existence that its motions go unnoticed (Cailleux, 1961). He also pointed out that water is raised by the Sun and later falls as rain. At the same time, Plato (429–347 BC) stated that there was an immense cavern inside the Earth filled with water in continuous movement that was the source of all rivers and the place to which all water returned (Adams, 1938). This belief may have been based on observations of karst topography in Greece. In addition, Aristotle (348–322 BC) contributed excellent ideas to geomorphic thought including the hydrologic cycle and river erosion as the agents of landform change.

An example is the Natural History written by Pliny the Elder (23–79 BC), which is a compilation of the work of Greek authors with additional superstitions and legends (Cailleux, 1961). As a result of these works, few advances were made for many centuries.

The general paralysis in the development of science until the 16th century was also evident in the field of geomorphology. During this millennium and a half, the Arabs brought the works of the Greeks to Spain. Jewish scholars translated Arabic manuscripts into Hebrew and the Toledo School translated them into Latin (Cailleux, 1961). Thus during this period knowledge was transmitted primarily in monasteries and in some cases by the patronage of sovereign powers. This is how Avicenna (980–1037) became familiar with the work of Aristotle and later formulated a hypothesis on the desert landscapes of Arabia. Avicenna thought that mountains could be created by the elevation of the ground or by the effects of runoff and wind (Thornbury, 1954).

During this period, the reigning ideas were catastrophist or cataclysmic, encouraged in part by Christian religious thought and the continuing argument for the Great Flood as the explanation of numerous natural facts (Ellenberger, 1988). These beliefs led to dogmatism such as that expressed in 1654 by Dr. J. Lightfoot, Vice Chancellor of the University of Cambridge, who affirmed that the sky and the Earth, center and circumstance, were formed at one time by the Trinity on October 26, 2004 BC at 9 o'clock in the morning (King, 1976).

The ideas of the catastrophists dominated until the 19th century even though the genius Leonardo da Vinci (1452–1519) had expressed important considerations on the origin of mountains (a precursor to the idea of isostasy) in his writings. He pointed out that salt is produced from the evaporation of seawater and that this process shows the slowness of geomorphic processes and the importance of fluvial erosion. Lamentably, his work was not published until the end of the 18th century so his influence on the scientific thought of the period was negligible (Ellenberger, 1988). As a result, catastrophism was the dominant doctrine of this period.

During the 17th, 18th, and part of the 19th centuries, the most significant advances resulted from applied engineering studies, primarily in hydrology. Many of these engineers were French, such as Perrault, de Chezy, Surrel, Guetthard, and Desmarest, who were seeking alternatives to many of the non-functional academic concepts of the day. Zittel (1901) pointed out that the years between 1790 and 1820 constitute "The Heroic Age of Geology."

Catastrophist theories found their rebuttal in the uniformitarian ideas expressed at the end of the 18th century. Hutton (1726–1797) first introduced the principle of uniformitarianism in his *Theory of the Earth* (1788) and Lyell (1797–1875) developed it in his book *Principles of Geology* (1830), which has been reprinted many times. These Scottish geologists established the basics of the modern science of geology as well as the fundamentals of geomorphic reasoning. The uniformitarian theory can be summed up in a single sentence: "the present is key to the past" (Geikie, 1905). This means that different landforms have originated from processes similar to those occurring today and that these processes have operated over long time periods. The way we see the world is the total opposite of catastrophism. Lyell described how mountains erode as a result of weathering and fluvial transport processes, how the transported materials are later deposited in sedimentary basins, and how new mountains arise in these areas.

In 1802 Playfair published his book *Illustrations of Hutton's Theory of the Earth* in which he claimed that valleys are proportional to the size of the tributaries they encompass in areas of uniform lithology and structure and that valleys are subject to widespread fluvial erosion. This is known in geomorphology as Playfair's Law.

But the catastrophists, including the geological clergymen Buckland and Sedgwick, argued that only a universal flood could explain glacial erratics and hanging alpine valleys (Beckinsale and Chorley, 1968). The gradual acceptance of Lyell's ideas finally put an end to the diluvialists.

In 1840 Louis Agassiz (1807–1873) published his book *Studies of Glaciers* incorporating the ideas of predecessors such as Scheuchzer, Martel, de Saussure, and especially Charpentier and put an end to catastrophism. In the book

Agassiz expressed the idea of an ice age and its importance in valley erosion and sedimentation (Rémy, 2007). Ramsay (1862), President of the Geological Society of London, developed these ideas further, and uniformitarianism became firmly established in the field of geology (Price, 1973).

1.2.2 Models of landscape evolution

Davis, Penck, and King proposed the existing models of landscape evolution at different times during the first half of the 20th century. The models of Davis and King are related to specific climate types whereas Penck's model assigns more importance to tectonic aspects than to climate. Basic knowledge of the scientific biographies of these investigators will aid in the understanding of the content, origin, and criticism of each of these alternatives.

William Morris Davis (1850–1934), an outstanding figure in geomorphology, was a professor of geology at Harvard University. His life and work are the subjects of numerous analytical studies, some of them extensive and of great depth (Chorley *et al.*, 1973; King and Schumm, 1980). Although Davis' most fundamental work was originally published in 1906 and later published as a book (Davis, 1954), this prolific investigator's scientific production spanned the years of 1880 until 1938, the publication date of his last article.

Professor Davis graduated from Harvard University. After a three-year stay in Cordoba (Argentina) and a trip around the world, he began teaching at Harvard in 1875. His investigatory work is amazingly abundant, is always written in a clear and brilliant style, and is accompanied by splendid illustrations that make his works most attractive (Higgins, 1975; King, 1976); at the same time he provided excellent tools for the teaching of geomorphology. His great virtue was his ability to synthesize diverse knowledge. Some authors consider him the founder of geomorphology (Tricart, 1965).

His greatest contribution is the model of the geographic cycle, also called the erosional cycle, the normal cycle, or the humid cycle by other authors (Higgins, 1975), which is considered to be the first paradigm in geomorphology (Hart, 1986). It was the first generally accepted theory of landform development (Thorn, 1988). No geomorphic theory has been more influential (Tinkler, 1985). He published his basic thesis in 1889 in the paper on *The Rivers and Valleys of Pennsylvania*, and followed it ten years later with the definitive version: *The Geographic Cycle*.

Davis' idea is that leveled topography may undergo abrupt uplift and be subject to prolonged erosion under conditions of crustal stability. The initial topography goes through stages called youth, maturity, and old age which are defined by different characteristics. The final result is the creation of a **peneplain** with some residual relief projecting above it (*Fig. 1.1*); this residual relief corresponds to the initial planate relief. Renewed uplift initiates a new cycle. In his explanation of a cycle, he analyzes the development of fluvial flow, slopes, and interruptions of the ideal cycle that reactivate erosion as a function of a new base level.

Criticism of this work comes in part from the fact that the model was established in wet temperate zones, which he considered "normal." As late as 1900 Davis considered glacial landforms and arid topography to be "accidents" in this cycle. In a 1905 paper, he established a new geographical cycle for arid climates and in 1906, a cycle for glacial climates (pre-glacial, optimum, and postglacial stages). Although this work generated widespread criticism (Reynaud, 1971) to which Davis responded in other papers, his concept of the erosional cycle was widely accepted. Higgins (1975) points out twelve reasons for this acceptance, with arguments as simple as its simplicity, applicability, rationality, etc.

In addition, Davis' work covered broad fields of geomorphology including glacial, coastal, arid and volcanic landscapes. His relationships with European scientists were significant, especially with German colleagues (Alfred and Walter Penck) whom he visited frequently and for long periods.

His concept of landscape cyclicity became firmly established among investigators of the period and was applied to many fields of geomorphology, leading to karst cycles (*Fig. 1.2*), and marine, glacial, savanna, and periglacial erosional cycles (Higgins, 1975; Tinkler, 1985; Hart, 1986).

Davis' ideas were well received in France and among English-speaking scientists but not in Germany or the Soviet Union. His ideas went practically unnoticed in the eastern and northern European countries. French investigators such as Lapparent and Baulig were fervent supporters, and especially Martonne, whose three-volume book *Treatise on Physical Geography* appeared in 1925–1927. Because of his book, Martonne exercised considerable influence on many geomorphologists. In 1913, Martonne was the first to use the term climatic geomorphology (Stoddart, 1969a; Beckinsale *et al.*, 1991).

Davis had many supporters and numerous critics. Some geomorphologists have gone so far as to hold Davis responsible for setting the science of geomorphology back fifty years (Hart, 1986). Davis was an "organizer" of existing knowledge who used his creative imagination to invent a logical outline but he did not consider the role of vegetative cover (Tricart, 1956). To Tricart, the term "normal erosion" does not seem adequate; he asks, "Normal, compared to what?" He criticized Davis for lack of a method (Tricart, 1965) and for the use of deductive approximation, which is contrary to the inductive scientific method. Strahler (1950) considers Davis' approach to be superficial and inadequate because it leads to a completely qualitative treatment that is inappropriate for dynamic erosional processes (Strahler, 1950a). Other significant criticism had to do with processes

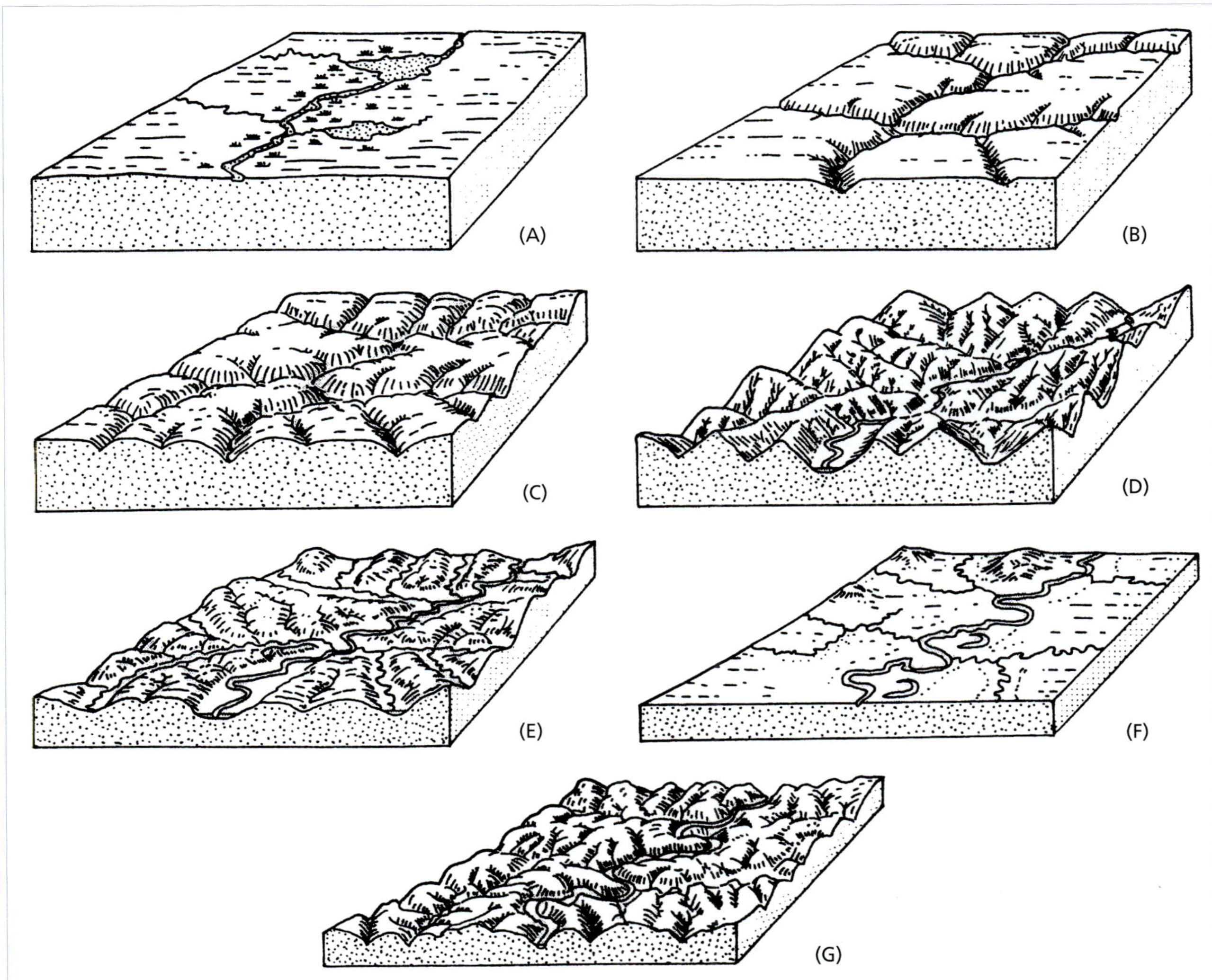

Figure 1.1 Davis erosional cycle. (A) Initial stage of relief. (B and C) Youth stage. (D and E) Maturity stage. (F) Old stage with development of peneplain and some monadnocks. (G) Uplift and beginning of a new erosional cycle. (Drawing by Raisz in Strahler, 1965).

that occur in different climates and produce different types of landforms. In Germany, the cradle of climatic geomorphology (von Richtofen, Machatschek, Passarge and Troll are the most significant authors), Davis' model was fiercely denounced because it failed to consider climate-related processes. Thus, Passarge (1931) states: "The forces that create landscapes are climatic in origin.... Vegetation, for example, depends solely on climate but at the same time, vegetative cover partially but decisively affects landform and sedimentation." Later, in the 1950s, many French investigators abandoned Davis' concepts and defended the principles of climatic geomorphology (Cholley, Dresch, Birot, Tricart and Cailleux).

Considering the foregoing, it is clear that geomorphologists initially were receptive to Davis' ideas but that this acceptance later declined. However, none of his numerous critics offered an alternative model that overcame the defects of his normal erosional cycle.

Walther Penck (1888–1923) was the only geomorphologist to publish an alternative model, which he included in his book entitled *Morphological Analysis* (a title suggested by his father). This important geomorphologist, son of the famous glaciologist Albrecht Penck, obtained his doctoral degree in Heidelberg in 1910, and thereafter spent two years in the Andes working for the geological service of Argentina. His experience in central European geology plus his knowledge of the Andean region, an arid area with significant Quaternary tectonics, considerably influenced the development of his theories. In addition, his fundamentally geological training was augmented by the transmission of geomorphic knowledge from his father. He was a professor in Constantinople from 1915 to 1918, later

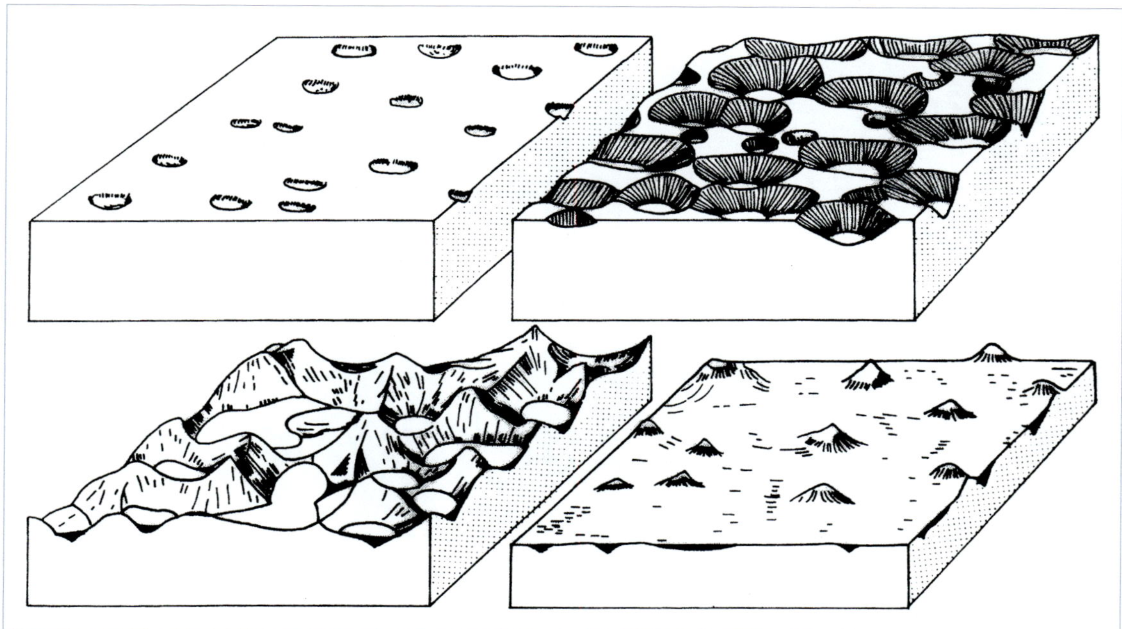

Figure 1.2 Karst cycle (Grund, 1914).

worked at the university in Leipzig, and was working at the Geological Institute at the time of his death. He wrote *Morphological Analysis* while suffering from cancer, which caused his death at the age of 35 (Bremer, 1983). His father published the book in 1924. The book is a geologist's view of geomorphology in which he seeks to interpret diastrophic history (Thornbury, 1954). In this sense, we can consider it a precursor to the geomorphic method in neotectonic investigations.

Penck based his theory on the study of correlative deposits in areas adjacent to uplifted blocks of the Andes and the Alps (Chorley *et al.*, 1973) and deduced from his analysis that relief is generated by slow initial uplift followed by accelerated uplift, later followed by a period of declining uplift, ending in stability. Davis, on the contrary, started with leveled topography that was rapidly uplifted, followed by a prolonged period of crustal stability. William Morris Davis published Walther Penck's ideas in English in an article that appeared in 1932, eight years after their original publication. Penck's model begins with a primary surface (*Primärrümpf)* that is uplifted and eroded, while at the same time the primary surface develops other surfaces that are contained within each other, giving rise to piedmont steps, *Piedmonttreppen* (*Fig. 1.3*). Each resulting step acts as a local independent base level. Relief is determined by the uplift rate (endogenetic forces) and is opposed by the intensity of fluvial erosion (exogenetic forces). When uplift declines, fluvial erosion predominates and later, a terminal leveled surface (*Endrumpf*) develops. Along with this development, Penck also established a theory on slope evolution that is linked to the stages of crustal movements they illustrate (*Fig. 1.4*) (Davis, 1932).

Penck's ideas were not well accepted in Germany (Bremer, 1983). Eight years later, Davis popularized Penck's ideas by publishing a critical and inaccurate version of them (Chorley, *et al.*, 1984). This incorrect interpretation may well have been the result of Penck's prose, which was difficult to understand. A 1953 English translation of the book led to a significant resurgence of Penck's ideas.

Lester Charles King (1907–1989) later developed a third model of topographic evolution. This London-born South African geologist was trained by C.A. Cotton, a geomorphologist from New Zealand and a follower of Davis, who himself wrote three well-known books (1942–1944) of the period. King tried unsuccessfully to apply Davis' concepts to the territory of South Africa. King was also influenced by Penck's ideas, and by the geomorphology of his field area, South Africa, a landscape of vast erosional surfaces bounded by steep scarps.

King first expressed his ideas in a 1953 article entitled *Canons of Landscape Evolution* and developed them more fully later in his 1962 book *The Morphology of the Earth*. His model is cyclical, like that of Davis. Like Davis, he begins with rapid diastrophism followed by a long period of tectonic calm with the development of extensive pediments and the eventual appearance of a **pediplain**. Residual landforms of variable morphology populate this vast leveled surface including mesas in tabular regions and *bornhardts* in granitic areas. King considered the "normal" morphoclimatic system to be

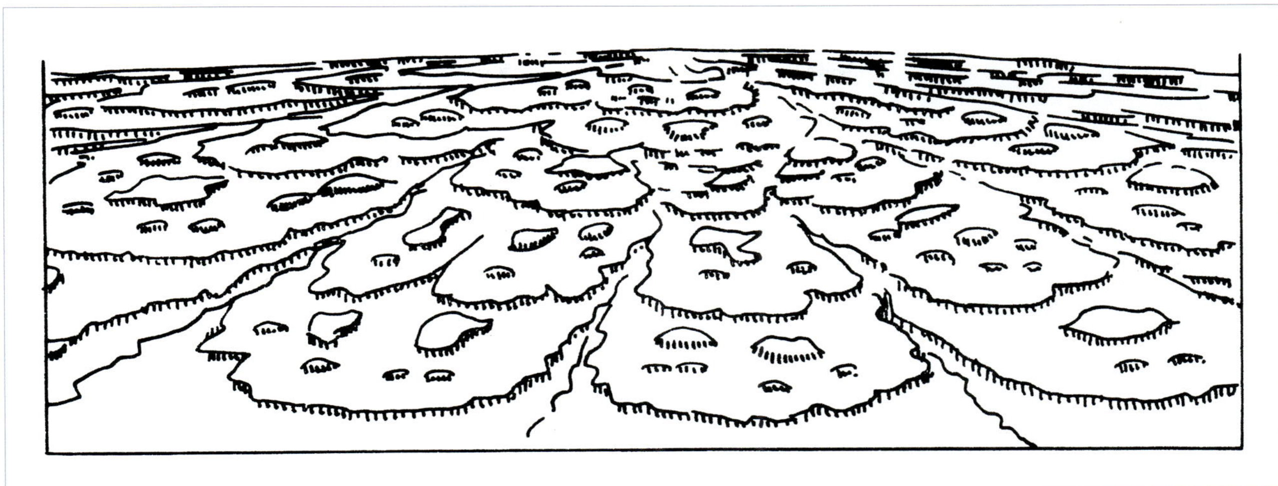

Figure 1.3 Davis' vision of Penck's *piedmonttreppen* on an expansion dome (Davis, 1932).

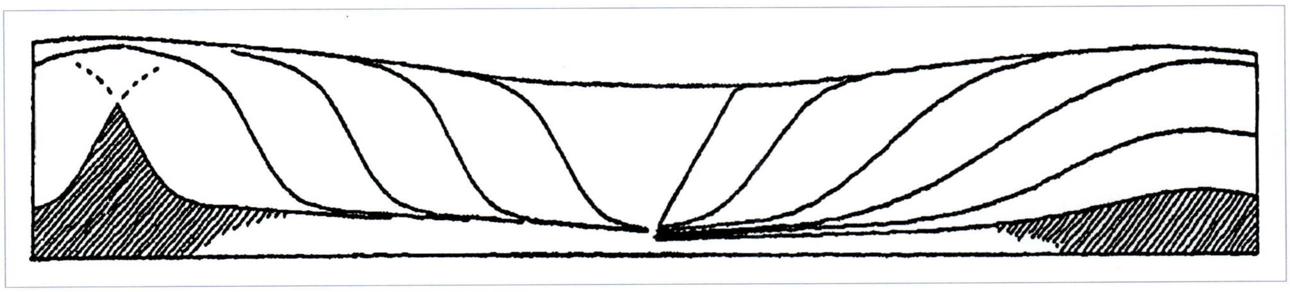

Figure 1.4 Enlargement of valleys and changes in slope profile during development. Right, stages of youth to old age according to W.M. Davis. Left, regression parallel to an escarpment, according to W. Penck (According to Davis, 1932).

an arid area or a savanna with alternating wet-dry seasons. Renewed uplift initiates a new cycle in which the preceding pediplain is eroded by slope retreat in distal areas as new pediments are created. Successive cycles form stepped pediplains on a continental scale, an arrangement similar to *piedmonttreppen*. The mechanism of piedmont formation is more accessible to our observations and more comprehensible than peneplain formation (Jahn, 1954).

King's model is based on parallel scarp retreat and selected ideas from the models of Davis and Penck (Higgins, 1975; Thorn, 1988), as discussed above. King's ideas did not produce broad discussions on the models of Davis and Penck because he published his theory during a time when geomorphology was convulsed by the advent of process analysis and quantification (Higgins, 1975).

Another of King's contributions that generated significant discussion was the concept of *cymatogeny* (King, 1961), which he introduced in his 1962 book. He related this term to the concepts of orogenesis and epeirogenesis. Cymatogeny is defined as regional crustal arching with thousands of meters of vertical uplift. The idea seems to have been inspired by Cloos' (1939) *Hebung*. For King, cymatogenic movements dominate in the crust, an idea that is contrary to plate tectonics but consistent with his belief in an expanding Earth (King, 1983).

Today the models of Davis, Penck, and King are the three primary approaches to landscape evolution. The ideas of Davis and King are cyclic, with clear climatic influences, whereas Penck's assume a continual struggle between erosion and the various uplift rates. These models are summarized in *Figure 1.5*.

1.2.3 Geomorphology in the first half of the 20th century

An important goal of geomorphology is to analyze the relationship between the Earth's surface and crustal changes. Having reviewed some ideas on this relationship, we will

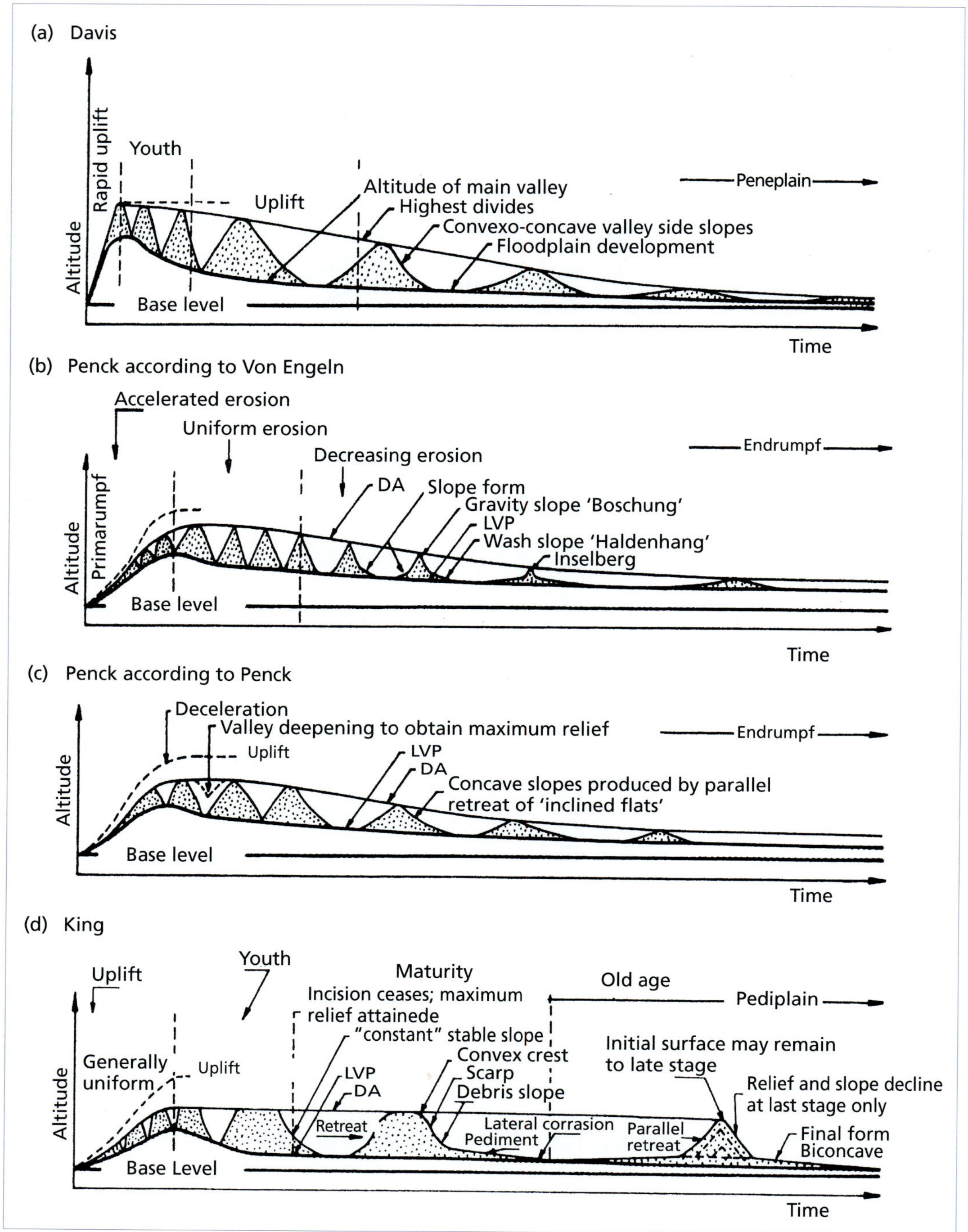

Figure 1.5 Cyclical models of the evolution of landforms related to height and time, with a fixed base level. Also shown are schematic slope profiles (Thornes and Brunsden, 1977).

now examine in detail the major research on the role of tectonics in geomorphology from 1900 until the discovery of plate tectonics in the 1960s (Beckinsale and Chorley, 1968, 1991). We will discuss the influences of crustal and eustatic changes during this period. Chapter 2 of this book will be a brief analysis of plate tectonics and its relationship to geomorphology. Climatic geomorphology and climate change will be discussed in later chapters of this book.

Isostasy is an ancient term; the idea of hydrostatic balance in the Earth's crust was first noted in the middle of the 18th century. It was the target of research during the 19th century when Airy (1855) and Pratt (1859) (*Fig. 2.2*) significantly developed the concept. In his 1890 monograph on Lake Bonneville, Gilbert included a section on the hypothesis of loading and unloading in surface deformation (*Fig. 21.28*). During the first half of the 20th century many investigators were loyal supporters of isostatic compensation while others harshly criticized it.

Epeirogenesis is another idea associated with mountain uplift. In his Lake Bonneville study, Gilbert stated that widespread convex structures might be responsible for the creation of continents. He called them epeirogenies to distinguish them from orogenies. Epeirogenies are examples of diastrophism. These convex structures correspond to the vertical deformations that King (1962, 1983) called cymatogenies.

Climate change caused by solar energy, the gravitational effects of the crust, and eustatic changes were major topics of geomorphic thought from the end of the 19th century until the mid 20th century (Beckinsale and Chorley, 1991). In 1909, Penck and Brückner established the various glacial periods: Günz, Mindel, Riss and Würm. This classification had a major impact on European geomorphologists. In northern Europe the following terms were used, from early to modern: Menap, Elster, Saale, and Weichse, and in North America, in the same order: Nebraska, Kansas, Illinois, and Wisconsin (Lamb, 1977).

During this period various investigators made numerous attempts to estimate the chronology of these ice ages and interglacials. Antevs introduced dendrochronology in 1928. De Geer studied varves in 1500 locations over a period 30 years and published his results in various papers until 1940. Their work was the first to shed light on the chronology of the Late Pleistocene and Holocene.

The first half of the 20th century saw major advances in the understanding of paleoclimates, particularly those of the Tertiary and Quaternary (Beckinsale and Chorley, 1991). In the area of geomorphic investigation, Büdel's (1963) development of morphoclimatology and climatomorphologic geomorphology included the study of paleoforms or inherited forms.

Eduard Suess wrote a major three-volume work entitled *The Face of the Earth* (1883–1908), which was translated into English, French, and Spanish. Suess revealed the evidence for continental transgressions and regressions and concluded that they occurred essentially simultaneously throughout the world. This led him to conclude that continental areas had been stable throughout geologic time, except for local movements. He showed that positive eustatic movements or transgressions took long periods of time and were interrupted by negative movements of shorter duration. He analyzed the comparison between the ideas of the neptunists who believed in the absolute predominance of sea level movements in Earth's history and the ideas of the tectonicists who claimed predominance of vertical activity on the Earth's surface.

In 1894 Albrecht Penck published his book *Morphologie del Erdoberfläche* in which he supported the eustatic theory and established an even greater number of crustal movements than Suess. He considered glacial eustasy to be an obvious reality. As a result, geomorphologists understood that the effects of general base level changes were reflected in the resulting landforms (Chorley, 1963).

In 1834 Charles Darwin hypothesized that slow subsidence of the volcanic peak on the Isla de Morea (Tahiti) caused the reef to grow vertically and the atoll to form. Dana supported this theory in 1872 in his book *Corals and the Coral Islands*. In 1915 eustasy returned to the problem of coral reefs through the work of Daly and his theory on glacial control of coral reefs.

After the mid 1930s, the supporters of diastrophic ideas confronted the eustatic theory and invoked vertical movements as the causes of transgressions and regressions (Beckinsale and Chorley, 1991).

The culmination of the eustatic theory in geomorphology occurred in a 1928 publication by the French geomorphologist Baulig: *Le Plateau Central de la France.* He recognized different erosional surfaces and even indicated that their origin was sub-aerial. He also pointed out that some of these surfaces must have been beveled as a result of higher sea levels.

1.2.4 Process geomorphology

Geomorphology, like other branches of science, proceeds from simple description to quantification. For this reason geomorphology uses methods from other scientific areas, primarily physics. During the 1940s and 1950s some investigators applied the techniques of physics to glaciers and littoral, fluvial, and eolian environments but it was only at the beginning of the second half of this century that a revolution occurred in the development of process geomorphology. Geomorphologists are now more concerned with process than with landforms (Barry, 1997). Some authors consider this the most important theme of modern geomorphology (Hart, 1986). A 1980 paper estimated that 75% of geomorphic investigation in Great Britain was process analysis at a detailed scale (Gardner, 1983). This implies that geomorphologists no longer study the geomorphic evolution of large morphodynamic units. With the development of chronological techniques, we are now seeing a renaissance of historical geomorphology (Summerfield, 2005). The future of geomorphology must be supported by the process analysis and also anthropogenic morphogenesis (Tricart, 1956).

Many geomorphologists say this revolution began when Strahler published papers (1950, 1952) that introduced dynamic or process geomorphology (Higgins, 1975). For Strahler (1950), Gilbert's 1914 paper *The Transportation of*

Debris by Running Water promoted the quantitative dynamic method of geomorphic study. Others consider Leopold and his collaborators, whose papers are summarized in the book *Fluvial Processes in Geomorphology* (Leopold *et al.*, 1964), to be the real pioneers and catalysts of geomorphic process analysis.

Process analysis involves a process-response system in which the process is the creative agent and the response is the resulting landform. Process analysis may include a variety of techniques depending on the diversity of the environments in which they occur (Goudie, 1981a). Many of these techniques are recent inventions and in many cases craftsmen manufacture the equipment. As Hart (1986) says, "necessity is the mother of invention." Process investigations are often difficult because of the complex instrumentation of the system, the numerous interacting processes, and an incomplete knowledge of the numerous mechanisms (Baker and Twidale, 1991). This area of study has led to a reduction of temporal and spatial scales. Thus, for example, processes have been investigated in micro-basins and on individual segments of a slope. In recent decades, great strides have been made in this field to the extent that it is now one of the basic areas of geomorphology. These advances offer hope for increased understanding of landform evolution. Today geomorphologists have reported a wide variety of erosion rates depending on lithology and climate. Saunders and Young (1983) and Goudie (1995) have published compilations of these rates. On steep glacial slopes, the rates are 1–5 mm/year; in temperate zones, 0.01–0.1 mm/year; in semiarid climates 0.1–1 mm/year; in gully zones 1–10 mm/year; and in steep slopes of the tropical jungle 0.1–1 mm/year.

Extrapolation of the collected data to larger regions is difficult. Peeters *et al.* (2008) obtained adequate data by applying it on the scale of a millennium; they claim that measurements of present-day rates may be highly relevant in interpreting relief evolution over the long term.

Process analysis, in both field and laboratory, involves statistical techniques to adequately process the collected data. This effort led to *quantitative geomorphology*. Hydrologic engineer Horton (1945) pioneered this area of geomorphology with his work on the morphometrics of drainage basins. Two of his followers are Strahler (1952, 1956, 1964), and Schumm (1956). The description of drainage basins and channel networks had been a qualitative science but Horton transformed it into a quantitative science that provided numerical data of practical value (Strahler, 1964). The development of quantification has been made possible by computers that can rapidly process large amounts of data.

Geomorphologists use models to deal with the complexity of geomorphic processes (Kirkby, 1987), which has created another field of modern geomorphology. Scale models can partially reproduce reality (the use of wind tunnels or flumes for fluvial erosion). An example of an analog model is the simulation of kaolin by the ice deformation. The third type is a mathematical model, such as those of Ahnert (1987a) and Kirkby (1994), the best-known investigators today. However, even the latter two types of models cannot encompass the total complexity of the interactions between processes (Baker and Twidale, 1991).

1.2.5 Reconstruction of geomorphic history: Geochronology

For many years, geology used **relative chronology** to provide an understanding of a sequence of events but it provided no information on the duration of events. Today, **geochronologic techniques** can provide the ages of some events and this possibility has radically changed our research (Summerfield, 2005a, b; Bishop, 2007). These techniques do not negate the need for traditional relative chronology but they can be used to determine a precise age of the traditional geological time units (Tricart, 1965).

Pollen analysis, the study of the pollen variation in a column or boring, can be used to investigate climate oscillations primarily in the study of Quaternary deposits. At the beginning of the 20th century, Blytt and Sernander established a column of climatic periods ordered from most recent to oldest. For the Postglacial (that is, the last 10,000 years), the periods are: Subatlantic, Suboreal, Atlantic, Boreal and Preboreal. The Late Glacial is subdivided into the Upper Dryas, Allerod, Lower Dryas, and Bolling. The Pleniglacial is subdivided into the Dryas (the oldest) and the Arctic. **Prehistoric** material requires another type of dating.

Dating is an essential part of landscape reconstruction and of chronology, which is an essential component of geology. In addition, numerical dating of rocks makes it possible to calculate weathering and erosion rates (Oguchi *et al.*, 1999). *Table 3.1* shows the most common methods of absolute dating.

The study of recent global changes has promoted the advancement of reconstruction techniques. The principal project of the IGBP (International Geosphere-Biosphere Programme) is called Past Global Changes (PAGES). It includes the following time categories: (1) the last 2,000 years of Earth history, and (2) the last few hundred thousand years. Slaymaker (2000), among others, describes geomorphic contributions to environmental change during these epochs.

The availability of powerful computers and the ability to acquire large data sets may improve precision and confidence in **geomorphic prediction** although there are no guarantees (Wilcock and Iverson, 2003). Much work remains to be done to define the statistical properties of the landscape with the goal of providing a secure basis for prediction in geomorphology (Church, 2003).

1.3 Geomorphic systems

In his 1952 paper on the dynamic bases of geomorphology Strahler proposed that the development of a geomorphic system is complete when landforms and processes are related as dynamic systems and the transformation of mass and energy is a function of time. Many geomorphic processes operate in defined systems that can be isolated for analysis. In his 1980 book, Strahler defined a geomorphic system as a set of interrelated and interconnected elements that 1) are assumed to exist in the real world, and 2) have unique characteristics that humans can measure, describe, and analyze. Geomorphic systems have been complicated by biotic evolution that affects the soil (Twidale, 2003) and by human presence that has also modified the natural system (Clowes and Comfort, 1982). Various disciples of Strahler (Kennedy, 2004a), such as Chorley and Schumm, have described both the theoretical and practical processes of the general systems theory.

According to Chorley and Kennedy (1971), systems can be considered open, closed, or isolated. In an **isolated system,** matter and energy are neither imported nor exported. In a **closed system,** energy is transferred but matter is not. Finally, in an **open system,** energy can be interchanged and matter can move (Hugget, 2003). In all geomorphic systems, there are **endogenetic or internal processes (tectonics, volcanism)** and **exogenetic or external (geomorphic) processes** that are caused primarily by climatic forces (Scheidegger, 1991). Briefly, endogenetic processes create relief and exogenetic processes erode landscapes. The interplay between the two processes is the development of geomorphology. Davis' erosional cycle is part of a closed system whereas an open system comes from general systems theory associated with process analysis (Kennedy, 2004a).

An example is the open fluvial system (Schumm, 1977), which is composed of three areas: Zone 1, the sediment source area, the drainage basin where sediments are produced; Zone 2, a transference zone; and Zone 3, a sedimentation area. The dominant controls in Zone 1 are climate, diastrophism, and soil use whereas in Zone 3 the controls are base level and diastrophism.

1.4 Geomorphic concepts

The capacity of a geomorphic system can be expressed in two ways: first, by the sediment removal rate and second, by the energy consumed in maintaining or transforming the system. This leads to the concept of **equilibrium,** which can be expressed only in the direction of change (Chorley *et al.*, 1984). Chorley and Kennedy (1971) distinguish eight types of equilibrium in geomorphology (*Fig 1.6*). (1) Static equilibrium

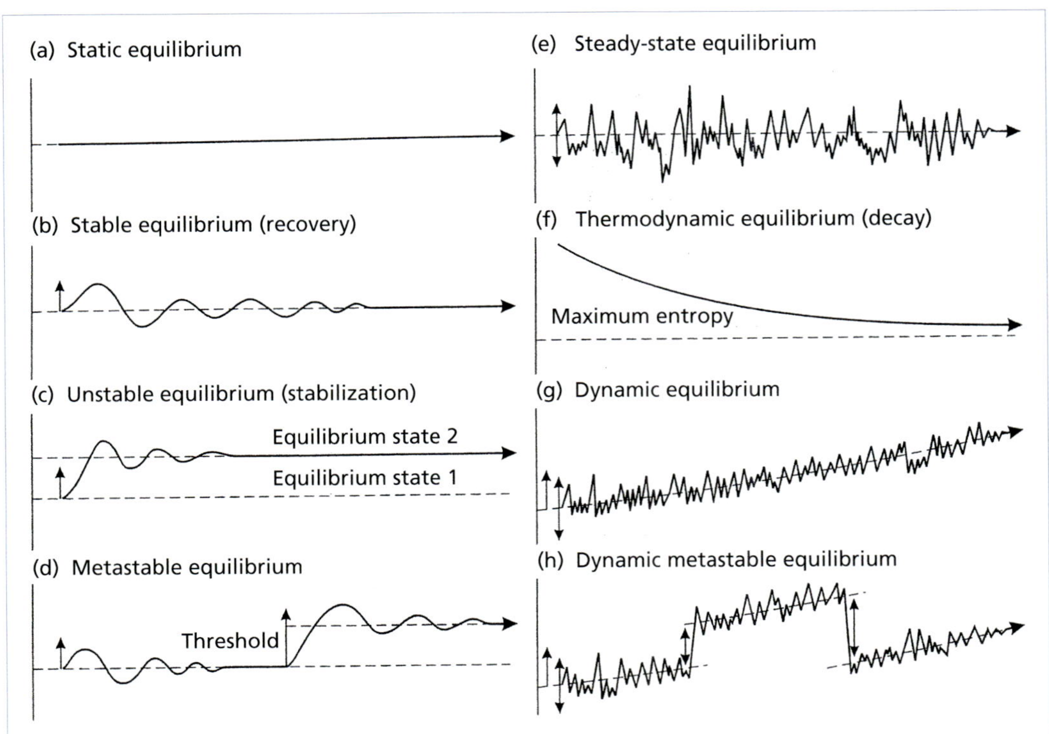

Figure 1.6 Types of equilibrium in geomorphology (Chorley and Kennedy, 1971).

describes the condition of a system where no change is taking place over time. (2) Stable equilibrium describes a condition that tends to revert to a primitive state after a small disturbance. (3) Unstable equilibrium is the condition where a small disturbance forces the system to a new equilibrium state where it stabilizes. (4) Metastable equilibrium is the condition of a system that has crossed an external or internal threshold of the system. (5) A stationary system exists when a system constantly fluctuates around a middle equilibrium state. (6) Thermodynamic equilibrium is the tendency in some systems toward a state of maximum entropy. (7) Dynamic equilibrium consists of a set of fluctuations around a middle state that change in a certain direction. (8) Metastable dynamic equilibrium combines the dynamic and metastable tendencies with fluctuations around a tendency that crosses thresholds.

Many topographic landforms are in dynamic equilibrium (Hack, 1960) where processes are operating but the active elements of the system are in equilibrium among themselves. A particular change, for example, a climate change, can cause the system to exceed a **geomorphic threshold** (Schumm, 1973, 1979), creating instability in the environment. These thresholds can be abrupt or gradual (Begin and Schumm, 1984) and can operate over any time interval, from minutes to millions of years. Geomorphic thresholds can be triggered by extrinsic or intrinsic causes. An example of the extrinsic type is an abrupt change that wipes out vegetation, leading to intense erosion. An intrinsic threshold operates within a system, such as when the continuous addition of particles to a slope reaches a critical threshold of instability. These geomorphic thresholds can sometimes be defined quantitatively (Leopold and Wolman, 1957) *(Fig. 1.7)* and can be used as a basis for the recognition of potentially unstable landforms (Schumm, 1979, 1991). On the other hand, because connections within a geomorphic system can be highly complicated, introduction of an external change may propagate throughout the system in a complex way (Schumm's 1973 **complex response**). The identification of an intrinsic threshold has practical applications. Brunsden (1980, 1990, 2004a) analyzes a set of important concepts related to a geomorphic system response to predominant processes. He points out that the system has **reaction time** to the stimuli, followed by a **relaxation time** necessary for the establishment of new equilibrium conditions. The sum of the reaction and relaxation times is the response time. The period of time that a geomorphic system remains in a state of equilibrium is the **persistence time** *(Fig. 1.8)* (Bull, 1991). As a result, we can have systems with various thresholds, short or long persistence times, and highly variable response times. The latter indicate the sensitivity of the system to change (Brunsden, 1980, 1990; Wright, 1984).

Instability and **geomorphic change** are due to the intervention of extrinsic factors introduced into the system and to the internal properties of systems (intrinsic factors). Conditions of intrinsic instability can result from the fact that they operate near intrinsic thresholds (for example, semiarid erosional systems); high-energy topographic conditions in which slopes approach erosional thresholds (for example, areas of energetic tectonic uplift and/or rapid fluvial incision) and finally, high-energy climatic environments (for example, monsoons and hurricanes). High-energy systems involve major sediment movements that correspond to significant

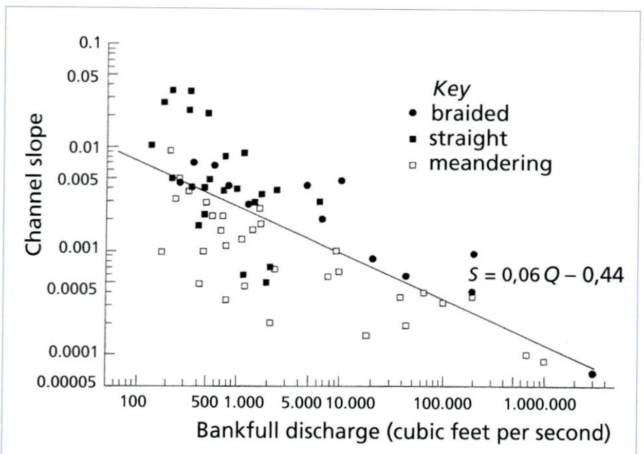

Figure 1.7 Relationship between flow and slope. The line separates meandering and braided channels (Leopold and Wolman, 1957).

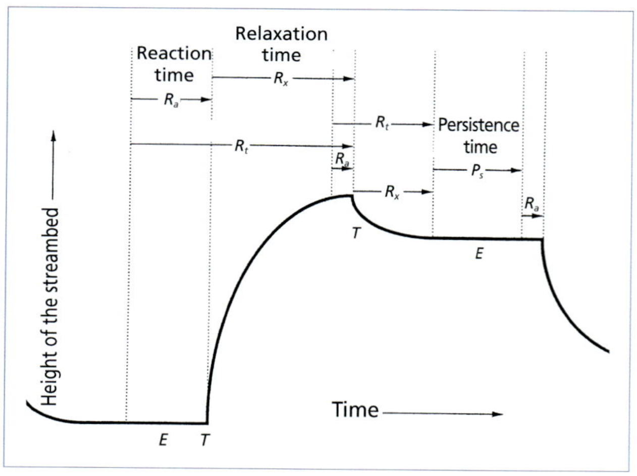

Figure 1.8 The changes in height of the streambed produced by aggradation and degradation are used as references to indicate the components of response time. R_t is the response time, which is the sum of the reaction time R_a and the relaxation time (R_x). P_s is the persistence time under new equilibrium conditions. *T* and *E* are threshold and equilibrium conditions, respectively (Bull, 1991).

rates of intrinsic geomorphic change. This instability can be studied on different time scales from human-induced instabilities to climate change and Quaternary tectonic movements. They can also be extended to spatial scales (Harvey, 2007).

Very little is known about the resistance of natural systems to geomorphic change. An interesting field of investigation is the study of landform sensitivity to change. "The **sensitivity** of a landscape to change can be expressed as the probability that a specific change in the controls of a system produces a measurable, recognizable and persistent response" (Brunsden and Thornes, 1979). Sensitivity can vary in space and time and also according to scale (Thornes and Brunsden, 1977). Geomorphic systems covering small areas can quickly reach equilibrium with new conditions, shortening the time necessary for analysis. Temporal modifications can be either slow and gradual or sudden and catastrophic. Landforms generally respond slowly to climate change because they require longer time periods for analysis (Brunsden, 2001). To understand changes, either fast or slow, in the development of landforms, a researcher must know to the maximum extent possible the magnitudes of the processes and their temporal record. All this helps us understand the sensitivity of geomorphic systems (Allison and Thomas, 1993). Some landforms have high sensitivity to external impulses and respond rapidly to predominant processes, such as fluvial channels. They can be called unstable or fragile (Thomas, 2001). The response of geomorphic systems to climate change probably occurs through changes in vegetative cover; landscapes appear to be very sensitive to high intensity events (Thomas, 2004). Insensitive or stable systems respond slowly or are resistant to change, such as inter-fluvial zones or the plains of Africa and Australia. Insensitive areas are usually landforms composed of resistant lithologies whereas small changes can occur more rapidly in areas in sensitive zones such as gullied areas (Brunsden and Thornes, 1979).

In desert areas, a relatively inactive system may undergo more topographic change than a similar system in other climatic zones. This shows that climatic sensitivity of landforms in arid zones is significant because less relaxation time is necessary to reflect changes in landforms prior to the application of climatic stimulus (Shaw and Thomas, 1993). Trudgill (1976) calls landforms with short relaxation times **labile landforms** *(Fig. 1.9)*. Labile landforms are rapidly modified by extreme events or have very little resistance to change. Their response is almost parallel to climate change. In contrast, **resistant landforms** are subject to weak processes or have a high resistance to change.

It has been assumed that climate change is gradual and the hydrological response is likewise gradual. However, Knox (1972), in his work in southwestern Wisconsin (United States) indicates that climate change may be abrupt and as a result, may trigger a response that, although brief, can run contrary to the soil loss curves of Langbein and Schumm (1958). Knox also points out that an increase in intense and sustained precipitation would erode a channel and increase the slope gradient *(Fig. 1.10C)*. As a result, during this short period soil loss would increase significantly and would later decrease *(Fig. 1.10D)*. However, these fluctuations are difficult to distinguish from the effects of major floods that have a long return period (Chorley *et al*, 1984).

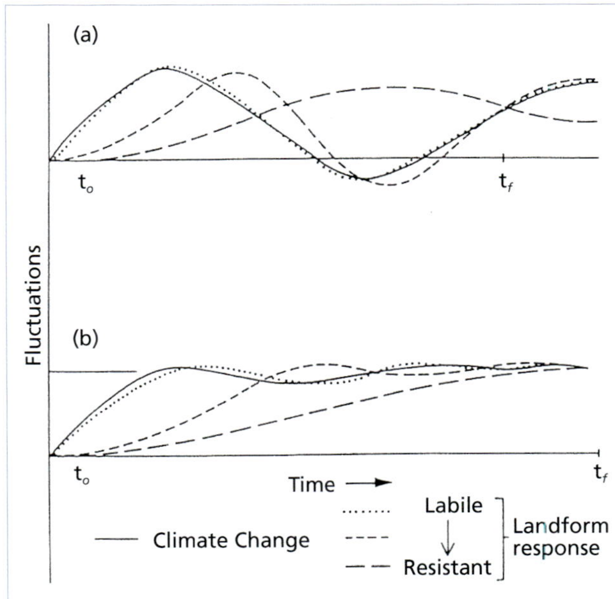

Figure 1.9 Response of labile and resistant landforms through time. A: corresponds to cyclic climate change and B: represents a fluctuating type that later stabilizes (Trudgill, 1976).

1.5 Temporal and spatial scales

In the first decades of the 20th century during the heyday of William Morris Davis's ideas, the problem of scale was simple because the cyclical concept divided landforms into young, mature, and old (Cailleux and Tricart, 1956). However, progress in scientific investigation soon invalidated these divisions. This has led to the establishment of areas of different sizes in geomorphology and these classifications are instruments we use to frame our investigations. In addition, the sizes are practical, as shown by their use in the preparation of geomorphic maps. We can classify landforms according to their dimensions or by the processes that created them (Cailleux and Tricart, 1956). Geomorphology addresses problems that occur over a wide range of temporal and spatial scales (Cox, 2007).

The explanation of the meandering course of a small-scale drainage network involves reasoning that is quite different from that needed to understand the *point bar* of a meander in a large river (Schumm, 1985a, 1991). Similar explanations can be applied to the time dimension but the time problem is more complicated because time cannot be seen and as a result, it is difficult to define. Time is a measure of change (Schumm, 1991). Because experimental studies are conducted in the field over a period of only a few years, great caution must be taken in extrapolating their results to geomorphic systems that operate over much longer time periods (Schumm, 1985a).

The spatial scale implicitly addresses the intrinsic characteristics of structure and the activity of the processes that do geomorphic work. At some spatial scales certain variables become dominant. In addition, processes are not homogeneous at different scales (Chorley *et al.*, 1984).

Table 1.1 shows a classification of morphological characteristics according to the size (Tricart, 1965). This illustration contains material from Tricart's earlier work (1952b) and also from the work of Cailleux and Tricart (1956). The classification is divided into eight orders. The spatial hierarchy of landforms in Chorley *et al.* (1984) is very similar to Tricart's (1965). The classification shown in Table 1.1 is self-explanatory. It ranges from landforms that can be observed from space (Short and Blair, 1986) down to small rugosities that occur on individual rocks.

Baker's (2004b) diagram, adapted from Carey (1962), shows the scales of geotectonic phenomena and is aplicable to landform changes that occur over a wide range of temporal and spatial scales (Fig. 1.11). Temporal and magnitude intervals related to these events can be classified as mega-, meso-, micro- and non- events Table 1.2 (Schumm, 1985a).

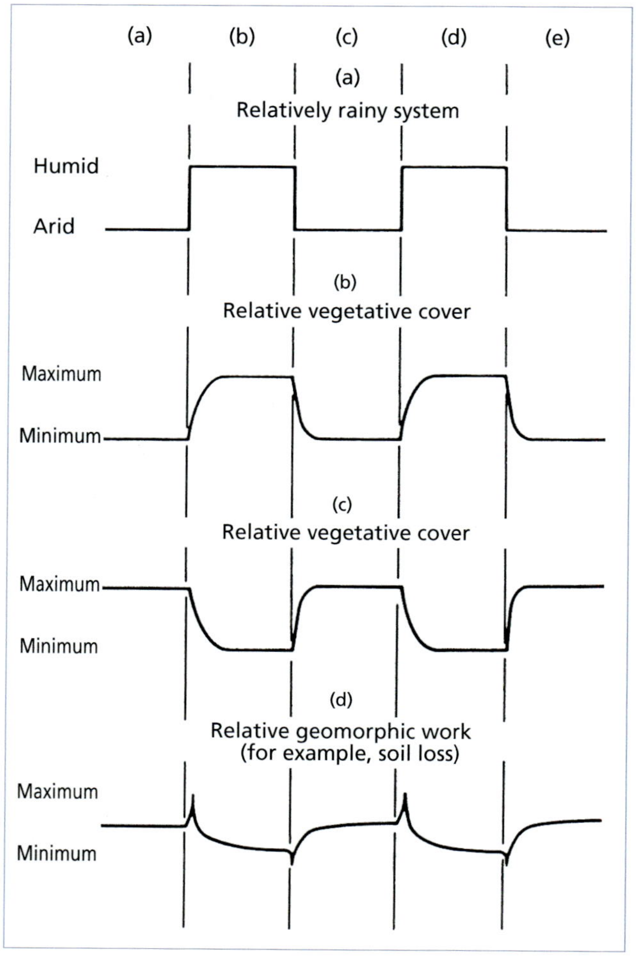

Figure 1.10 Responses of vegetation and landforms to abrupt climate change. The curves are most appropriate for regions with moderate annual precipitation between 250 and 1520 mm (Knox, 1972).

1.6 Applied geomorphology

Applied geomorphology is the application of geomorphic knowledge to research. It proposes solutions to resolve resource, management, and environmental planning problems (Jones, 1980). A simpler definition is Huggett's (2003) statement that applied geomorphology is the study of interaction between human beings and landscapes and landforms.

We could say that applied geomorphology began in the 18th century when French engineers with a naturalist spirit

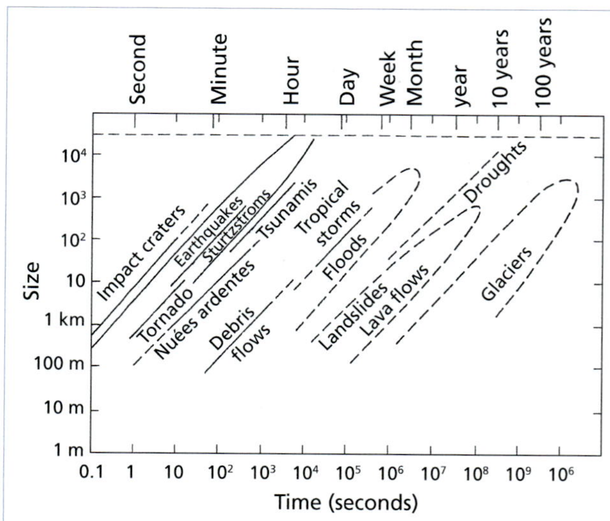

Figure 1.11 Temporal and spatial scales for various geomorphic processes (Baker, 2004b; modified by Carey, 1962).

Table 1.1 **SPATIAL CLASSIFICATION OF MORPHOLOGIES (TRICART, 1965)**

Order	Surface area units (Km^2)	Unit characteristics, with examples	Equivalent climate units	Genetic mechanisms that control relief	Persistence time
I	10^7	Continents, ocean basins, (global configuration)	Large areal systems controlled by astronomical factors	Differentiation of the Earth's crust into sial and sima	10^9 years
II	10^6	Large structural units (Scandinavian shield, Tethys, Congo basin)	Large climatic types (interference of geographical factors with astronomical factors)	Crustal movements, as in the formation of geosynclines. Climate influences on dissection	10^8 years
III	10^4	Principal structural units (Paris basin, Jura, Massif Central)	Subdivisions of climatic types, but with minor erosion importance	Tectonic units that are linked to paleogeography: Erosion rates influenced by lithology	10^7 years
IV	10^2	Basic tectonic units: mountain massifs, horsts, grabens	Regional climates with geographical influences, especially in mountainous areas	Influenced primarily by tectonic factors and secondly by lithology	10^7 years
Isostatic Compensation Threshold					
V	10	Tectonic events: Anticline, syncline, *mont,* valley, etc.	Local climates, influenced by the shape of the topography; sunny, shady, effects of altitude	Predominance of lithology and tectostatics Structural influences	10^6–10^7 years
VI	10^{-2}	Landforms: crests, terraces, cirques, moraines, terminal moraines, alluvial fans	Mesoclimate directly associated to the landform (for example, nivation niches)	Predominance of lithology and tectostatics Influenced by lithology	10^4 years
VII	10^{-6}	Microforms: solifluction lobes, polygonal soils, nebkhas, badlands, gullies	Microclimate, directly associated with morphologies (for example, karren)	Predominance of lithology and tectostatics Influenced by lithology	10^2 years
VIII	10^{-8}	Microscopic: for example, details of corrosion and polish, etc.	Microenvironments	Interference of dynamics and the rock texture	

Table 1.2	VARIABLE SIGNIFICANCE OF GEOLOGIC EVENTS THROUGH TIME (SCHUMM, 1985a)							
Magnitude	**Time scale**							
Related to the event	**1 day**	**1 year**	**10 years**	**10^2 years**	**10^3 years**	**10^5 years**	**10^6 years**	**10^8 years**
Mega-event	Flow or local landslide	Gully	Meander cutoff	Volcanic eruption	Terrace formation	Continental glaciation	Large folds and faults	Mountain construction
Meso-event	Rill	Flow or local landslide	Gully	Meander cutoff	Volcanic eruption	Terrace formation	Continental glaciation	Large folds and faults
Micro-event	Movement of sand grains	Rill	Flow or local landslide	Gully	Meander cutoff	Volcanic eruption	Terrace formation	Continental glaciation
Non event	—	Movement of sand grains	Rill	Flow or local landslide	Gully	Meander cutoff	Volcanic eruption	Terrace formation

undertook the construction of public works (Tricart, 1962). The advent of process geomorphology in the 1950s began a new focus on a closer connection between practical aspects and geomorphology. Today, process quantification values enable geomorphologists to conduct predictive studies that are both useful and necessary in selecting locations for high-level waste repositories, for example.

The applications of geomorphology can be divided into two broad classes (Chorley *et al.*, 1984): 1) Man as a geomorphic agent in terms of the effects of his actions, without regard to geomorphic processes and landforms. 2) Geomorphology as an aid to evaluate site selection and engineering construction and planning. The objectives of applied geomorphology are obviously numerous and have wide application.

There are countless published articles and symposia on applied geomorphology; the majority focus on environmental geomorphology. Books published by Hails (1977), Tricart (1978), Verstappen (1983), Fookes and Vaughan (1986), Allison (2002), and Fookes *et al.* (2005) are compendia of the various branches of applied geomorphology written by geomorphologists who are specialists in each subject. Slaymaker (2000) has published a book on the effects of global warming on geomorphic systems, which will undoubtedly be valuable for predictions.

Geomorphic maps are precise graphical representations of the topographic landforms of an area. Numerous past and present-day influences (St-Onge, 1968) have contributed to the formation of landforms. Earth scientists generally have a keen interest in map preparation (pedology, geology, and botany). Geomorphic maps are a basic document that can be revised to satisfy the concrete requirements of their application (Peña, 1997a). The practical uses of geomorphic cartography are presented in Tricart (1962, 1965, 1972, 1979), Tricart and Demek (1972), Panizza (1972, 1988), Piotrowski *et al.* (1972). Geomorphic maps have numerous applications: agriculture and the forest sector, civil engineering (linear projects, dam construction) (Tricart and Demek, 1972), and mineral exploration (Piotrowski *et al.*, 1972).

"A geomorphic cartographic system can be defined as a series of elements organized in a legend established on the basis of methodological criteria and applied in the preparation of a geomorphic map" (Peña, 1997b). Geomorphic characteristics vary from country to country and as a result, different geomorphic legends are available, which can be adjusted as needed. Another important criterion is the selection of an appropriate scale to satisfy the objectives of the map.

The International Geographical Union has proposed uniform scales for the preparation of geomorphic maps for large territories. They have prepared a Geomorphic Map of Europe at a scale of 1:2,500,000 in 16 maps with a legend in four languages (Bashenina *et al.*, 1968, 1971). The Geomorphic Map of Spain at a scale of 1:1,000,000 (Martín-Serrano, 2005) was presented at the recent Sixth International Conference on Geomorphology (Zaragoza, 2005). The map was accompanied by a report on the map's objectives, design, and contents. The book by Demek and Embleton (1978) and the comments of Van Zuidam (1982) contain a good analysis and discussion of the problems of medium-scale geomorphic maps. An example of one such map is the Geomorphic Map of Aragon, at a scale of 1:300,000 (Peña *et al.*, 2002). For practical purposes, scales of 1:25,000 to 1:100,000 should be used but concrete problems may require more detailed maps. Dumitrashko and Scholz (1978) discuss this concept in their paper on large-scale maps.

Different geomorphic legends have been proposed for different countries because of landform diversity. The standard was established for the detailed map of France (Tricart, 1972). Serrat (1976) translated this legend into Spanish and Spanish geomorphologists have used it with variations. Martín-Serrano *et al.* (2004) produced the legend for the Geomorphic Map of Spain, scale 1:50,000. A Geomorphic Cartographic System for the I.T.C (International Institute for Aerospace Survey and Earth Sciences) has been prepared, submitted for revisions, and presented in English, French and Spanish (Verstappen and Van Zuidam, 1968, 1991; Verstappen, 1970). Poland has a long tradition of detailed geomorphic maps (Klimaszewski, 1956, 1963; Jurasz and Niewiarowski, 1962). Bashenina *et al.* (1960) prepared the legend for Russian geomorphic maps at scales of 1:25,000 and 1:50,000. The first legend for Czech maps was prepared in 1963 and the work of Demek (1972) and Czudek (1973a) has contributed significantly to geomorphic cartography in Czechoslovakia. Both German republics have created geomorphic maps: Leser and Stäblein (1985) produced a legend for a 1:25,000 geomorphic map in the GMK Schwerpunktprogramm. In addition, Panizza (1972) proposed a legend for detailed geomorphic maps of Italy.

Spain has prepared legends for geomorphic maps at a scale of 1:50,000 (Martín-Serrano *et al.*, 2004). For the last few years, all official geologic maps at a scale of 1:50,000 have been published with a color geomorphic map at the same scale. In addition, many other countries are planning detailed legends.

Photo-interpretation today uses images not produced by photographic cameras. For this reason, the term "image interpretation" is more appropriate than "photo-interpretation," which does not cover the entire field of what is actually involved in the interpretation. The term "remote sensing" focuses on recording methods and related technology (Verstappen, 1977).

In this book we propose to analyze applied problems in different environments.

1.7 Structure of geomorphology

In a 1973 article on the history and compartmentalization of geomorphology, Jennings proposed a structure of the discipline (*Fig. 1.12*) based on his personal experiences. In his diagram, he indicates techniques and related studies that can be applied to different regions or domains of geomorphology

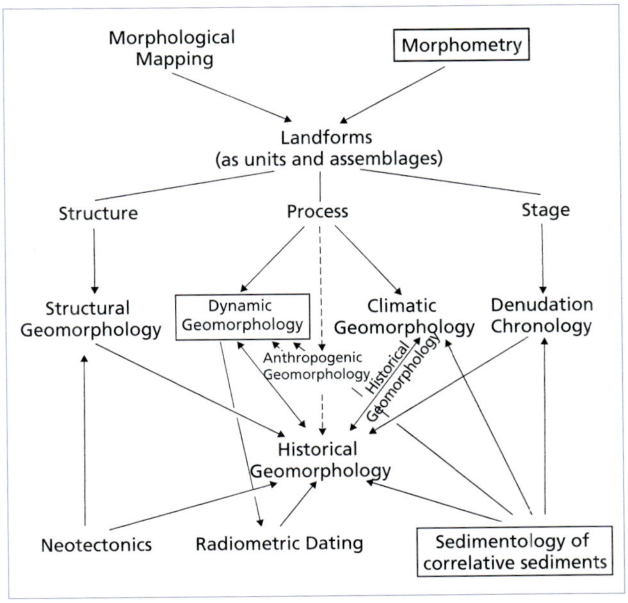

Figure 1.12 Structure of geomorphology (Jennings, 1973).

such as fluvial, coastal, glacial, karst, etc. He also points out subjects that are affected by the mathematical revolution and in part by the new geomorphology.

Tricart and Cailleux (1965) propose a division of three large groups: Climatic or zonal geomorphology, structural geomorphology, and zonal or pluri-zonal geomorphology, which some authors use (Panizza, 1992).

Returning to *Fig. 1.12*, the framework of geomorphology includes structural, dynamic, and climatic geomorphologies and the chronology of denudation. All these contribute to historical geomorphology. Landforms can be mapped and analyzed by morphometrics to determine their static and dynamic structural influences. Process analysis, sedimentology of correlative sediments, and radiometric dating all converge in historical geomorphology. In addition, climate and anthropogenic change appear in Professor Jennings' (1973) structure of geomorphology. Although it is just one version of the field of geomorphology, we believe it provides good coverage of the most significant fields of geomorphology.

2

STRUCTURAL GEOMORPHOLOGY

2.1 Introduction

Structural geomorphology is the study of landforms that owe their character to the lithologies and structures of the constituent rocks. These landforms are known as **structural** (Twidale, 1971; Migon, 2004) or **tectostatic landforms** (Tricart, 1968). Structural geomorphology also includes landforms formed by endogenetic processes without the intervention of denudation. These forms are called **tectonic** (Twidale, 1971; Goudie, 2004c) or **tectodynamic** (Tricart, 1968) **landforms**.

In the geomorphic sense, "structure" includes not only rock type (lithology) and arrangement of strata (stratigraphy and tectonics) but also active and continuous changes in these properties (Twidale, 1971). Structural geomorphology research considers erosion as a constant and neither specifies its mechanisms nor analyzes its processes; it is used only to understand the interaction of erosion and tectodynamics and to indicate how structure influences the type of landform (Tricart, 1968).

Exogenetic forces acting on rock structure produce structural landforms of various magnitudes. Rocks have varying resistance to weathering and erosional processes. Furthermore, external dynamics can create specific landforms when they act on certain rock types. In fact, books have been written to describe the wide variety of landforms in granitic areas. Some important ones are those of Twidale (1982b) and Twidale and Vidal Romaní (2005). In limestone and evaporitic terrains, dissolution creates a wide variety of morphologies called karst. The study of these rocks is an applied science that includes hydrogeology, ore deposits, and geotechnical engineering, among others. Karst is the subject of numerous books (for example, Sweeting, 1972). Sandstone formations also have specific landforms (Young and Young, 1992; Young, 2004; Härtel *et al.*, 2007).

Tectonic geomorphology is the study of the constant struggle between tectonic processes that generate topography and surface processes that slowly denude and rejuvenate the landscape (Burbank and Anderson, 2001). Today, many geomorphologists study small-scale geomorphic processes; they have not focused on the geomorphic advances resulting from the discovery of plate tectonics and large-scale mountain building (Summerfield, 1991; Keller and Pinter, 1996; Burbank and Anderson, 2001).

Tectonic activity generates Earth features including fault scarps, folds, mantle plumes, and ring structures, among others. These features are the result of various types of tectonic activity including diastrophism, epeirogeny, and isostasy (Goudie, 2004c).

2.2 Internal structure of the Earth

Our knowledge of the Earth's interior comes primarily from the study of seismic waves. The propagation of these waves through the Earth make it possible to calculate travel times at different places and to understand the large geospheres that exist in its interior.

Seismic events or earthquakes originate at a point within the Earth called the **focus** or **hypocenter,** which is normally located at a depth of up to 700 km. The vertical projection of the focus onto the surface is the **epicenter**. The size of an earthquake is measured by the energy liberated, which can be measured on the **Richter magnitude scale** (Bullen and Bolt, 1985).

The energy liberated by an earthquake is propagated by elastic waves called seismic waves. The fastest ones, **P waves** (primary), arrive first; they are also called **longitudinal** or **compressive waves**. **S waves** (secondary), also known as **shear** or **transverse waves,** travel only through solids. They do not travel through liquids or molten rocks because these materials lack rigidity. Seismic waves that are transmitted along the Earth's surface are called **surface waves.** Longitudinal and transverse waves are sometimes called precursors because they arrive before surface waves. *Figure 2.1* shows that the Earth's interior is clearly heterogeneous. This heterogeneity is shown by discontinuities that subdivide the Earth's interior. Mohorovicic discovered the first discontinuity while studying seismic waves generated by the 1909 Croatian earthquake. The **Mohorovicic discontinuity**, or Moho, separates the crust from the Earth's mantle. In 1925, investigators discovered the **Conrad discontinuity** in the crust in the Austrian Alps. Older petrologists adopted Conrad's model, which recognized two layers within the Earth: **SIAL**, an upper layer rich in silica and aluminum and **SIMA**, a layer rich in silica and magnesium. Seismic wave velocity increases abruptly at the Moho and in continental and oceanic environments. A **low velocity zone** (LVZ) has been recognized at a depth of 80 to 300 km. Velocity clearly increases in the **transition zone** between the depths of 400 and 670 km that separates the upper mantle from the lower mantle.

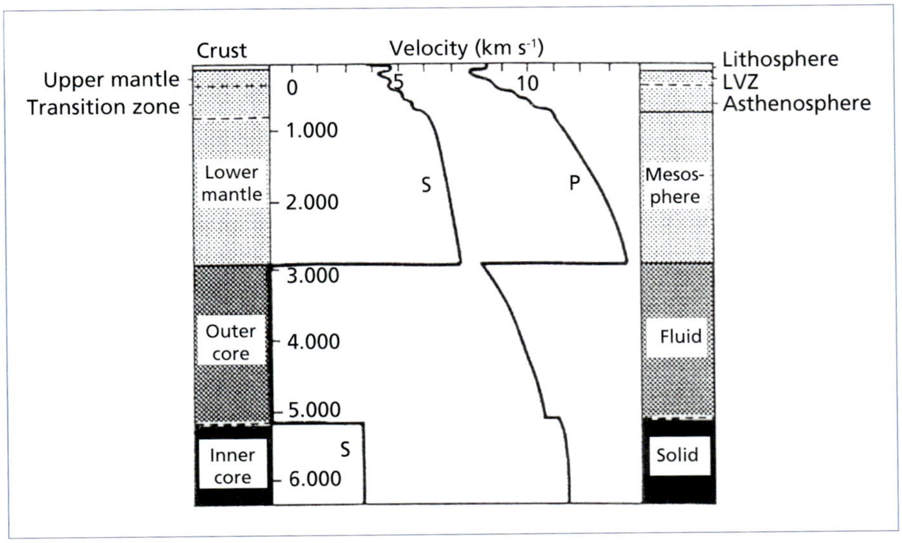

Figure 2.1 Variation of P and S wave velocities with depth in the Earth. (Based on data by Hart *et al.*, 1977) and its correlation with compositional and geologic characteristics (Kearey and Vine, 1996).

In the lower mantle, seismic wave velocity increases gradually to a depth of 2891 km, the **Gutenberg discontinuity**, which separates the mantle from the core. P wave velocity decreases abruptly in the core and S waves are not transmitted in the core, indicating that this portion of the Earth must be in a molten state. P waves increase in velocity at a depth of 5150 km, the boundary between the outer core and the inner core. Some investigators call this boundary the **Lehman discontinuity** (Fowler, 2005).

2.3 Composition of the Earth

Drilling has provided direct knowledge of the composition of the upper part of the crust whereas information on greater depths has been obtained indirectly. Propagation of seismic waves provides numerical and reliable data on the internal composition of the Earth.

The average composition of the upper part of the continental crust is that of granodiorite and quartz-diorite. To obtain mineralogical and chemical data on the lower crust, we use studies of stable mineral assemblages of dry basaltic rocks at pressures greater than 30 kilobars and temperatures between 1000 and 1250°C (Green and Ringwood, 1967). In a "dry" state, this composition is similar to high-pressure acidic and intermediate rocks. If the lower crust is "wet," the basaltic rocks would be amphibolites. At any rate, the lower crust likely has a complex composition with significant regional variations.

Oceanic crust is 6 to 7 km thick. It can be divided into three layers based on wave propagation velocities (Bott, 1982). The top layer is composed of sedimentary materials; the second layer is made up of basalt and dikes of the same composition, and the third is gabbro and metagabbro.

The mantle is thought to be chemically homogeneous and composed of silicates. Because mineralogy and silicate structure change with depth, scientists have proposed a transition zone between the upper and lower mantle at a depth of 400 to 670 km. The composition of the mantle has been inferred from three sources: the study of seismic wave propagation, rock residuum enclosed in basalt at the surface, and investigations of meteorites.

Because oceanic crust is composed of basalt generated in the mantle, researchers think the upper mantle is made of peridotite or eclogite (Harrison and Bonatti, 1981). The seismic velocity of these rocks is similar to that of the upper mantle. For various mineralogical and petrophysical reasons, the composition of the upper mantle is considered to be similar to that of peridotite. The overall composition of the mantle has been estimated on the basis of meteorite research and experimental data. Ringwood (1974) has suggested that the most likely composition is a mixture of about 25% basalt and 75% dunite (a very olivine-rich rock), which Ringwood calls pyrolite.

Figure 2.1 shows the low velocity zone that indicates the presence of molten material. The low velocity zone of the upper mantle is very important for plate tectonics because it is a layer of low viscosity over which the lithosphere and asthenosphere move.

The conditions in the lower mantle are such that no common rock exists there although a few minerals such as corundum (Al_2O_3), periclase (MgO), and rutile (TiO_2) can adjust to these conditions. It seems evident that the mantle is heterogeneous and that the lower mantle must differ mineralogically from the upper mantle. The difference may be in chemical (addition of Fe) or mineralogical composition (polymorphism); the latter seems more likely.

The core begins to appear at a depth of 2891 km at the Gutenberg discontinuity; S waves are not transmitted between the depths of 2891 and 5150 km due to the core's molten state. The interval between 2890 and 5150 km is considered to be the outer core. The inner core is theoretically thought to have a solid-metallic state because it transmits S waves. Compositionally the core appears to be similar to iron meteorites, which are composed of iron with small quantities of nickel and other metals. The high density of the core suggests that it must contain elements much heavier than those in the mantle. Since iron is the only heavy element that appears to be abundant in the solar system, it is thought to be an important constituent of the core. It appears that the core contains about 4% nickel and the remainder iron. The inner core has a seismic velocity and density that suggest a composition of pure iron.

2.4 Lithosphere and asthenosphere

The **lithosphere** is composed of the crust and the upper solid-state mantle *(Fig. 2.1)*, which is defined seismologically as the rigid outermost layer of the Earth. There are two types of lithosphere: **oceanic lithosphere** whose upper part is made up of the thin oceanic crust, and **continental lithosphere** whose upper zone is composed of thick continental crust. The thickness of the lithosphere is highly variable. The boundary of its lower layer with the **asthenosphere** is gradual although it is assumed to be thermal and to coincide with the 1333°C isograd at which olivine, the most abundant mineral in the Earth's mantle, begins to melt. The asthenosphere is a much weaker layer; when subjected to force, it reacts by slowly flowing. The asthenosphere is not molten because S waves propagate through it. Even so, it does contain small quantities of molten material. The relative plasticity of the asthenosphere is primarily due to the effect of its elevated temperature (>1333°C) on the rheology of mantle material.

The asthenosphere is thought to extend to a maximum depth of about 700 km. The properties of the underlying layer are not well known. Seismic waves traveling through it do not display anomalies *(Fig. 2.1)* and indicate that this layer has high resistance. It is called the **mesosphere**; it constitutes the remainder of the lower mantle to the Gutenberg Discontinuity (beginning of the core).

2.5 Isostasy

The theory of **isostasy** provides a basis for the understanding of vertical crustal movements. The rigid **lithosphere** can be considered to "float" on top of the underlying **asthenosphere**. To reach equilibrium, the **lithosphere** adjusts vertically depending on its thickness and density. The American geologist Dutton introduced the term **isostasy** in 1889 to describe this equilibrium state. Two models explain the equilibrium. **Pratt's model** of 1859 shows that on top of an equilibrium surface the volume of the crustal block has its own density, with a larger

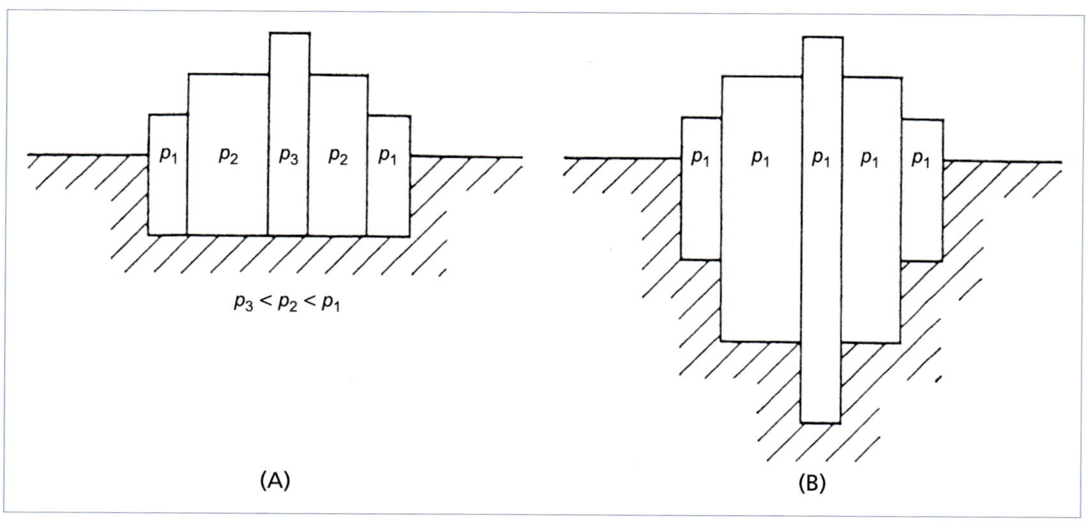

Figure 2.2 Schematic representation of isostatic equilibrium according to (A) Pratt and (B) Airy *(Jacobs, Russell, and Wilson, 1959)*.

volume having a lower density. **Airy's model** of 1855 proposes a better approach to the problem. He estimates that the blocks have the same density but differ in thickness and float on a heavier substrate (Fig. 2.2). The fundamental difference between the two models is the location of what is called **the isostatic compensation level**, which has a constant depth in Pratt's model but is variable in Airy's model, meaning that it is at greater depth under continents than under oceans.

This means that the higher the crust floats (as in cordilleras and altiplanos) the deeper is its inflection into the upper mantle. Thrusting and folding of strata that were initially horizontal thicken the crust and lift its surface isostatically.

When an additional load is added, such as a significant accumulation of ice, the crust sinks to greater depth and when the ice accumulation melts, the upper mantle rises. This process is called **glacio-isostatic rebound**. The most characteristic examples are the melting of the Fennoscandian and Laurentide ice sheets.

2.6 Precursors of global tectonics

The greatest revolution in the earth sciences, plate tectonics, occurred in the 1960s. This new theory pulled together diverse aspects of geological history from the last hundred million years.

Plate tectonics had predecessors throughout past centuries; numerous books and articles discussed its arguments and viewpoints and provide an account of the evolution of these ideas (Coulomb, 1969; Takeuchi, Uyeda and Kanamori, 1970; Rupke, 1970; Tarling and Tarling, 1971; Hallam, 1973; Vine, 1977; Franckel, 1988; Kearey and Vine, 1996 and Jordá, 1998).

Francis Bacon's book *Novum Organum* published in 1620 and *François Placet*'s book *La Corruption du grand et petit monde* called attention to the similarity of the Atlantic coastlines of Africa and South America and the possibility that the two continents were once united. Placet thought the universal flood was the cause of the separation.

During the 17th and 18th centuries, catastrophist ideas reigned in the majority of sciences; in geology they invoked the Great Flood as the fundamental cause of the opening and separation of the Atlantic Ocean. Lilienthal expressed this opinion in 1756 followed by Alexander von Humboldt in 1845 although the latter also attributed geological similarities between the southern continents. In his 1858 book *La création et ses mystères*, Antonio Snider-Pellegrini discussed the similarities between Carboniferous plant fossils in America and Europe and proposed that the continents had been part of a single land mass. At the beginning of the 20th century, the Austrian geologist Eduard Suess, in his five-volume work entitled Das Antlitz der Erde (1883–1909) *The Face of the Earth*, noted a close correspondence between geological formations in the southern hemisphere that had been united in a single continent called **Gondwana**.

During the 19th century the catastrophist or cataclysmist ideas that had survived for centuries slowly gave way to new explanations of the uniformitarian theory expounded by the Scottish geologists James Hutton and Charles Lyell. But it was an American physicist named Taylor in 1910 and even more so a German meteorologist named Wegener (1915) who collected numerous data in favor of **drift** and discarded catastrophism by proposing a mechanism of slow **continental drift**. Wegener, in his book *Die Entstehung der Kontinente und Ozeane/The Origin of Continents and Oceans*, proposed that all continents were united into a supercontinent he called **Pangaea** prior to the beginning of the Mesozoic, 250 Ma (Fig. 2.3).

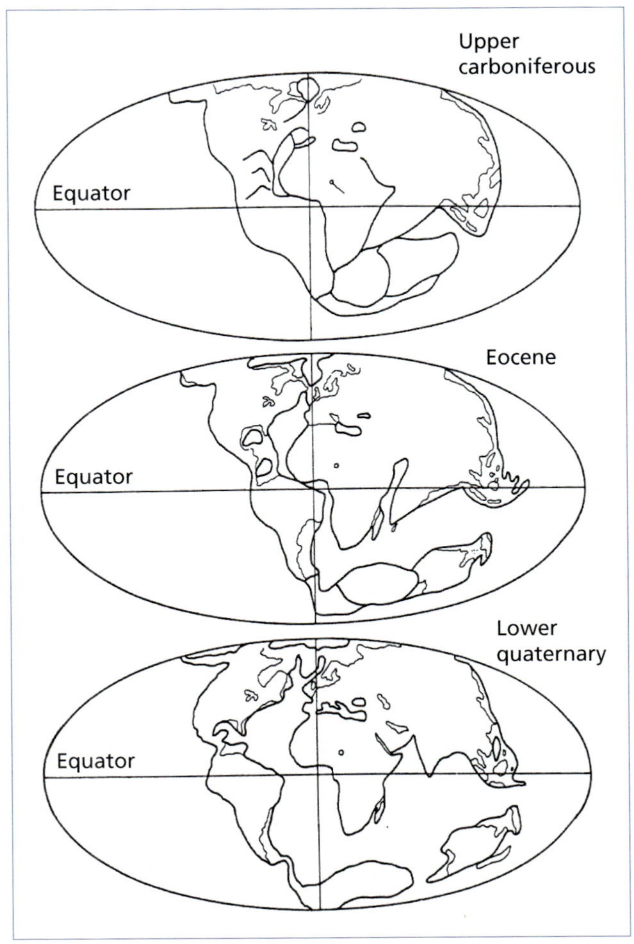

Figure 2.3 Reconstruction of the position of the continents from the late Carboniferous to the early Quaternary (Wegener, 1915).

His argument was that the coastlines fit together like a puzzle. He added the similarity between geological formations and fossils present on both sides of the Atlantic Ocean and paleoclimatic arguments such as tillites and glacial striations that provide evidence of glaciation. He also reconstructed climatic zones from the study of geological formations of different ages.

Wegener's work initially produced a major impact on geologists and geophysicists throughout the world although severe detractors attacked the idea as too fanciful. Sir Harold Jeffreys (1929) pointed out that accepting Wegener's idea required an understanding of the drift mechanism. He objected that the crust and underlying mantle are too rigid to allow such large movements. In addition, investigators at that time questioned how, based on their understanding that the Earth was several million years old, drift occurred between the Paleozoic and the Mesozoic. There were many other objections and the theory was essentially rejected in the 1930s.

Holmes (1928) pointed out that elevated temperatures decreased the resistance of materials and that, as a result, this fact contradicted Jeffreys' ideas. Holmes proposed that the continents moved by thermal convection current caused by heat from radioactive disintegration.

During and after the Second World War more data was collected and ideas that had been abandoned were again reviewed to provide new offshoots to the theory of continental drift.

The study of rock magnetism or **paleomagnetism** made great strides at the end of the 1940s and generated new and important data on the continental movement between the Mesozoic and today. These advances came from geophysical, geological, and topographic investigations conducted in the oceans by Scripps Institution of Oceanography in California and the Lamont Doherty Geological Observatory of Columbia University in New York.

At the end of the 1950s scientists discovered a 60,000-km-long chain of submarine mountains or **oceanic ridges** related to volcanism and earthquakes (*Fig. 2.4*). Hess (1962) collected the existing data and published the **hypothesis of seafloor spreading**, a name Dietz had proposed in 1961; Hess pointed out that the sea floor was moving. Convection (*Fig. 2.5*) forces material from the depths of the earth along the axis of ridges and the sea floor extends and disappears into trenches along the continental margins. The mechanism is similar to a conveyor belt.

The theory of seafloor spreading found strong support in a publication by Vine and Matthews (1963). By studying

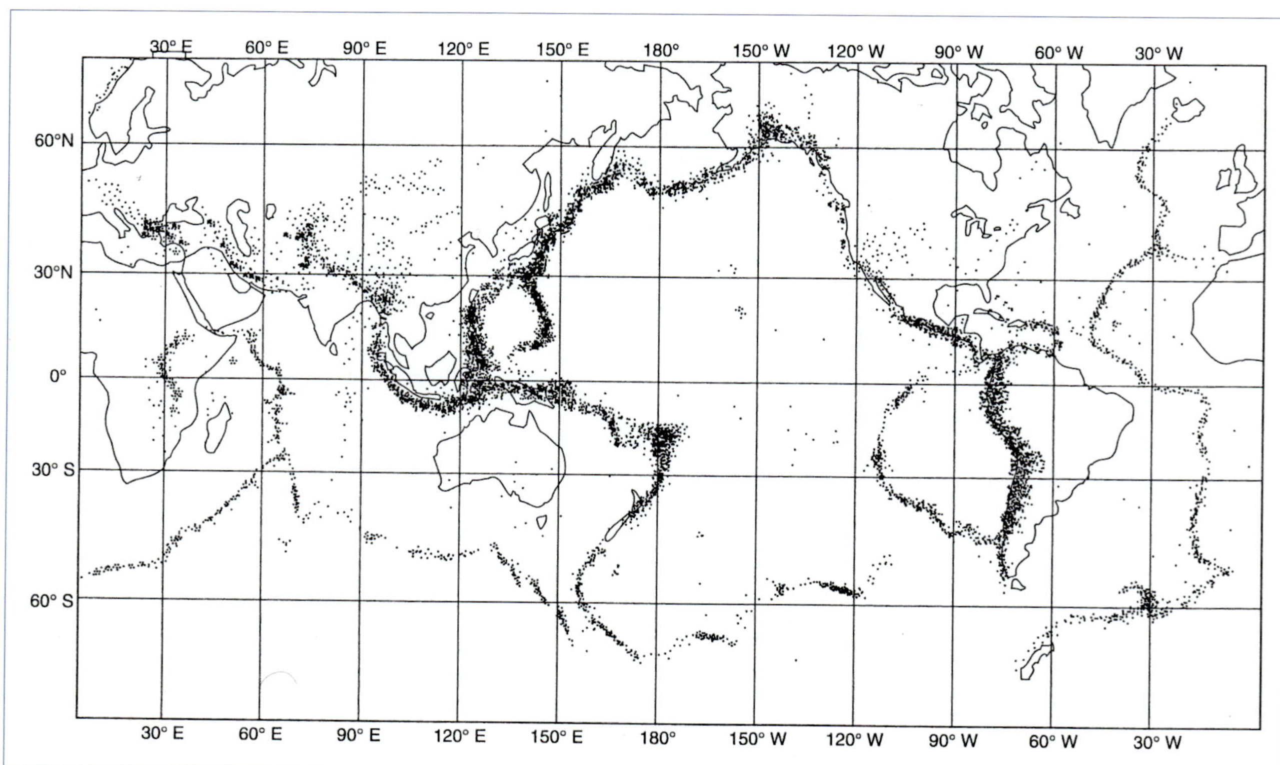

Figure 2.4 Global distribution of high magnitude (>5) epicenters for the period of 1963–1988 (National Earthquake Information Center, in Keller and Pinter, 1996).

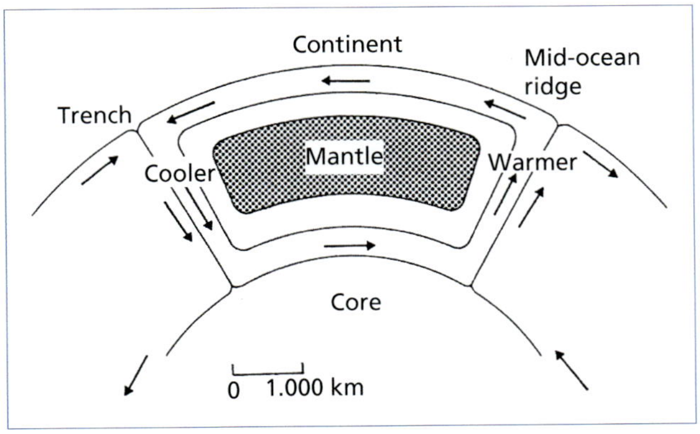

Figure 2.5 The concept of seafloor spreading, according to Hess (1962).

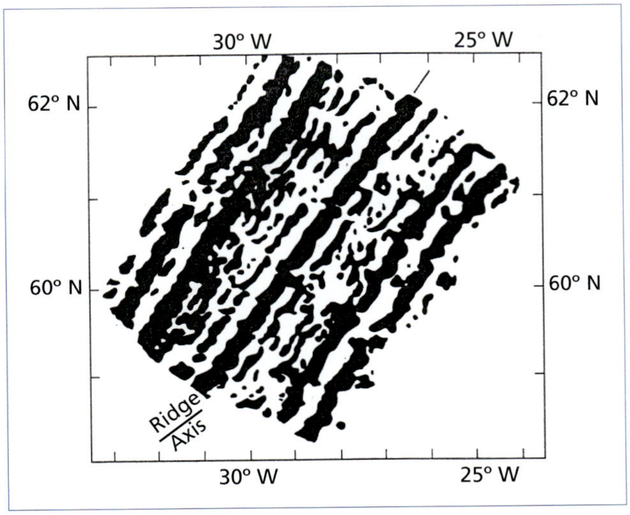

Figure 2.6 Banded system of parallel anomalies on the mid Atlantic ridge (Heirtzler *et al.*, 1966).

magnetic anomalies of the ridges, they discovered banding parallel to the ridge axis; they explained this by magnetization of the oceanic crust in alternating bands of normal and reverse polarity. The alternating bands showed that the magnetic field had reversed its direction various times in the geologic past and the bands had recorded the history of the reversals (Heitzler *et al.*, 1966) (*Fig. 2.6*). By assigning absolute ages to the basalts underlying the think sedimentary layer, Cox, Doell, and Dalrymple (1964) were able to reconstruct a history of magnetization directions for the last 4.5 Ma. Episodes of normal and/or reverse polarity, called **epochs**, can last hundreds of thousands of years. In addition, intervals of sudden polarity changes, called **events**, can last 100–200,000 years. The material that rises from the mantle via convection currents cools, becomes magnetized, and fossilizes its polarity. The geologic periods are named: Brunhes (normal polarity), 0–0.7 Ma; Matuyama (reversed polarity) 0.7–2.4 Ma; Gauss (normal polarity) 2.4–3.5 Ma; and Gilbert (reversed polarity) >3.5 Ma. The periods of magnetic polarity last about a million years, from which scientists calculate extension velocities of several centimeters per year (*Fig. 2.7*) and different velocities for different ocean ridges. This information was used to make a map of the age of the ocean floor (Muller *et al.*, 1997) (*Fig. 2.8*). Opdyke *et al.* (1966) conducted a series of measurements of remanent magnetism of the sediments on the floor of the Antarctic Ocean. The sediment column recorded the history of the reversals in the Earth's magnetic field (Menard, 1971). This discovery opened the way for magnetic stratigraphy of oceanic sediments (Grandstein *et al.*, 2004).

Wilson (1965) was an important precursor of the theory of plate tectonics with his discovery of a new type of fault, called a **transform fault** (*Fig. 2.9*), which

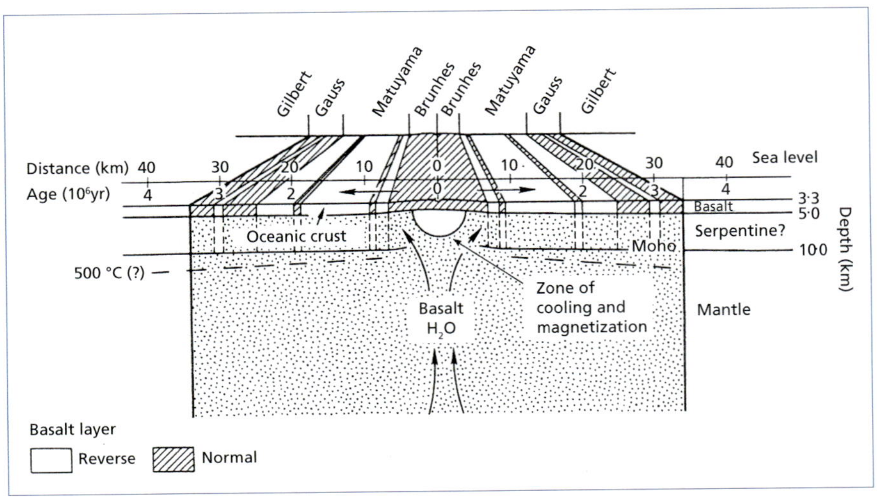

Figure 2.7 Schematic diagram illustrating the mechanism for generation of positive (normal) and negative (reverse) anomalies on both sides of the mid-ocean ridge crest (Vine, 1966).

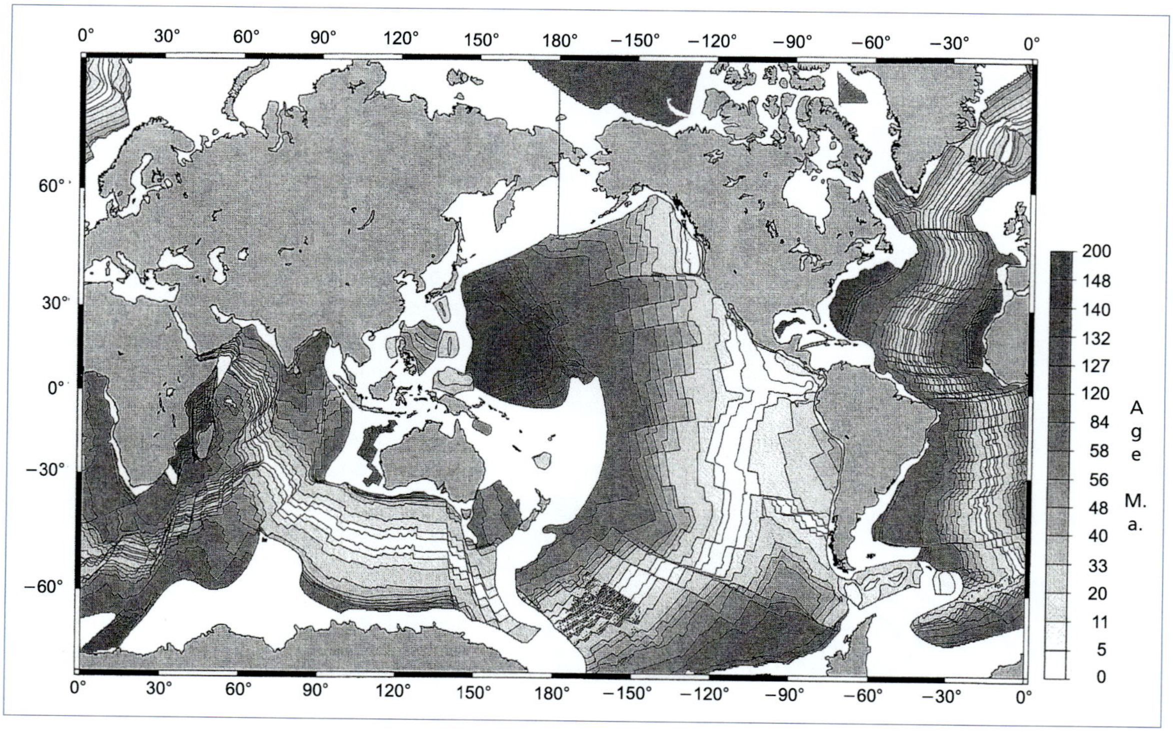

Figure 2.8 Isochron map of the ocean floor (Muller *et al*, 1997).

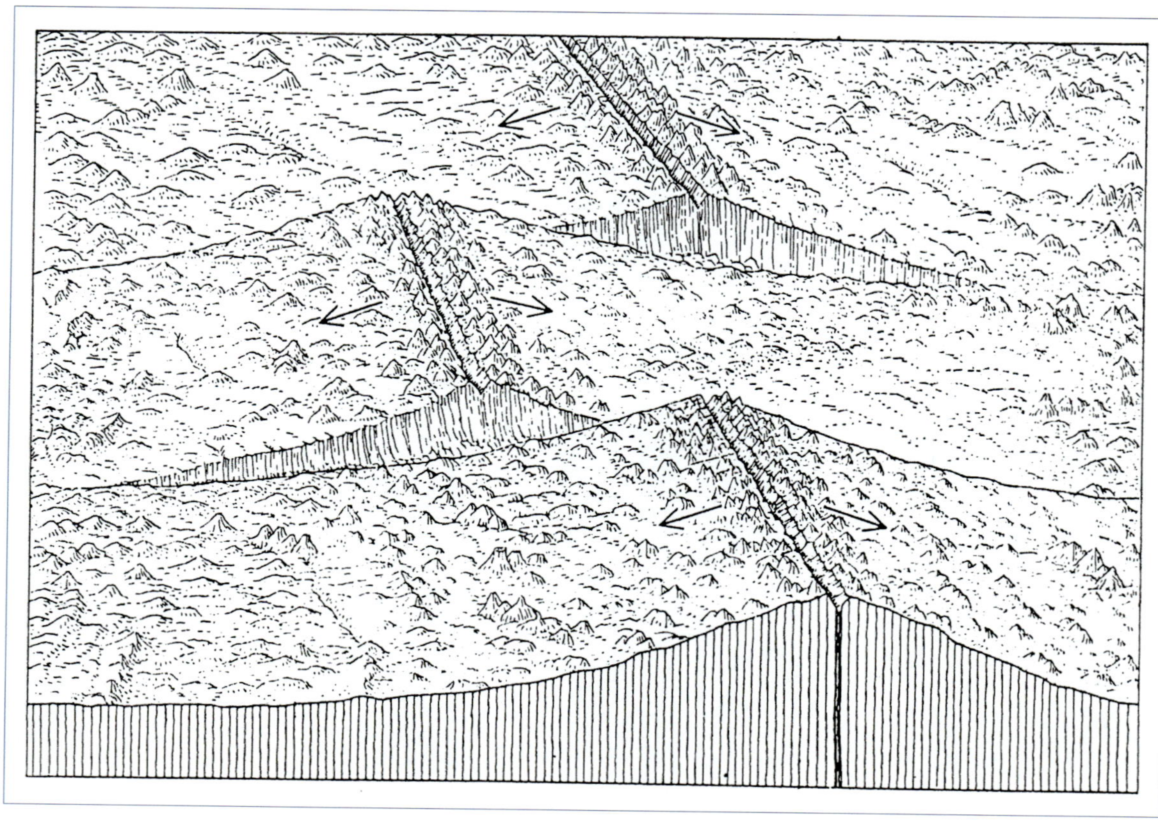

Figure 2.9 Transform faults between two segments of displaced ridges (Menard, 1971).

compartmentalized ridges into a series of segments. These are horizontal displacement faults that terminate abruptly at both ends and should not be confused with strike-slip faults.

McKenzie and Parker (1967), Morgan (1968), and Le Pichon (1968) reviewed previously proposed ideas to determine their overall agreement with the new data on the system of ridges, trenches, transform faults, magnetic anomalies, and movement directions associated with earthquakes. They concluded that there was total agreement and that the seafloor spreading theory seemed to be absolutely indisputable (Hallam, 1973).

As a result of all these discoveries, the earth sciences were in the midst of a great scientific revolution, as J. Tuzo Wilson pointed out, similar to that of astronomy during to the acceptance of the ideas of Copernicus and Galileo, similar to chemistry during to the introduction of the ideas of atoms and molecules, similar to biology during the acceptance of evolution, and finally, similar to physics during the acceptance of quantum mechanics.

2.7 Plate tectonics

The plate tectonics model divides the outer part of the earth into a number of thin rigid plates that move relative to each other. The plates move at the rate of a few centimeters per year. A majority of all earthquakes, volcanic eruptions, trenches, and cordilleras occur at plate boundaries (Fowler, 2005) (*Fig. 2.10*). Plates range in size from 10^6 to 10^8 km^2 and are from 70 km thick beneath oceans to 151 km thick below continents. They are classified as seven large plates (10^8 km^2) (the Eurasia and Pacific plates, for example) and eight intermediate plates (10^6–10^7 km^2) (the Arabian and Caribbean plates, for example) (Fowler, 2005) (*Fig. 2.10*). In addition there are twenty smaller plates (10^5 to 10^6 km^2) (Turcotte and Schubert, 2002).

There are three types of plate boundaries (Isacks *et al.*, 1968) (*Fig. 2.11*):

a. **Divergent boundaries**, areas of tension where new crust is formed by the injection of basaltic magma;

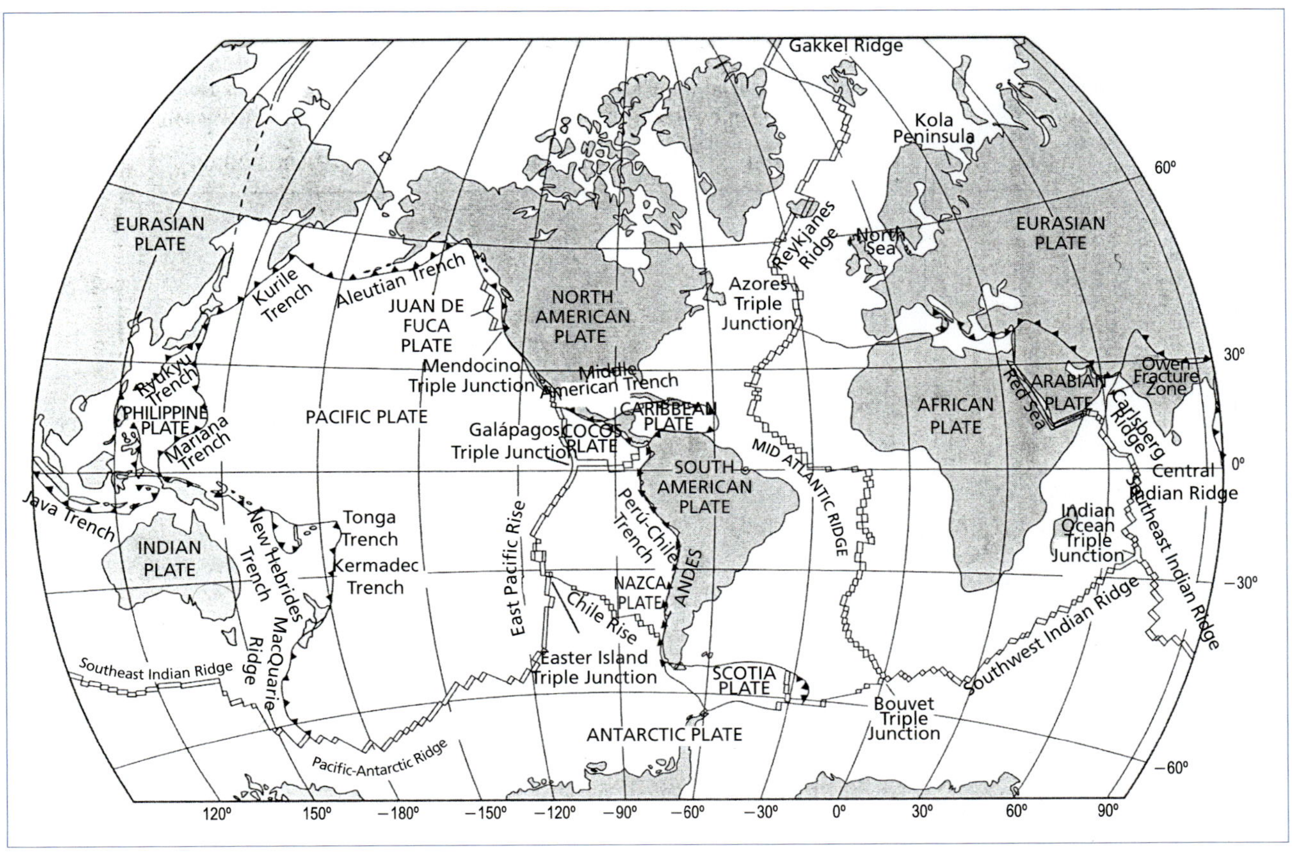

Figure 2.10 Principal tectonic plates, mid-ocean ridges, trenches, and transform faults Fowler, 2005).

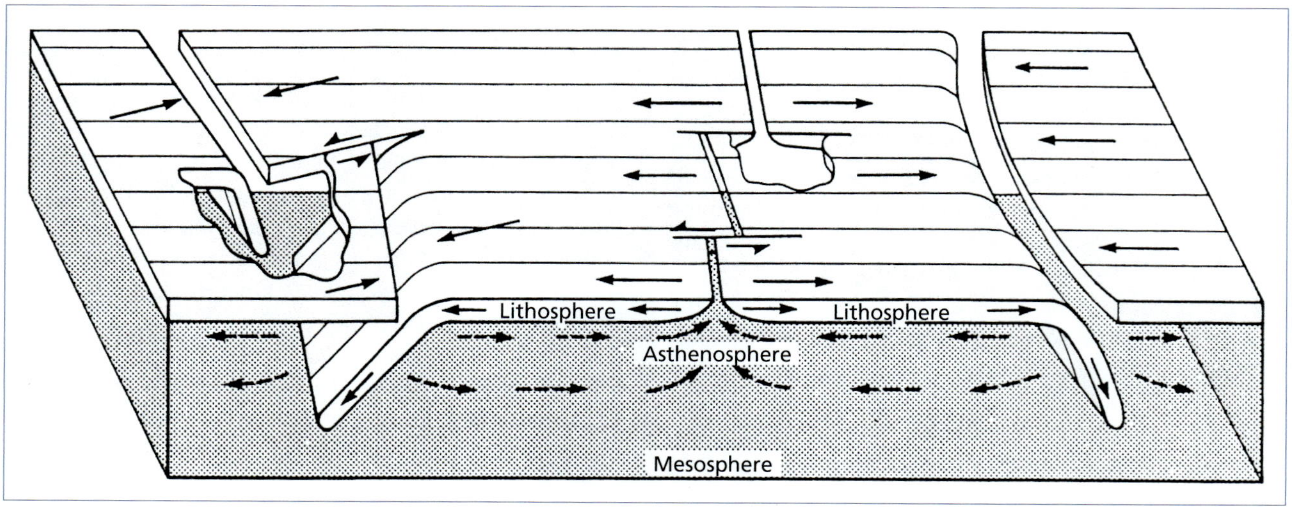

Figure 2.11 Block diagram summarizing the principal characteristics of tectonic plates. Arrows in the lithosphere indicate relative movements. Arrows in the asthenosphere represent complementary flow in the mantle (Isacks, *et al*, 1968).

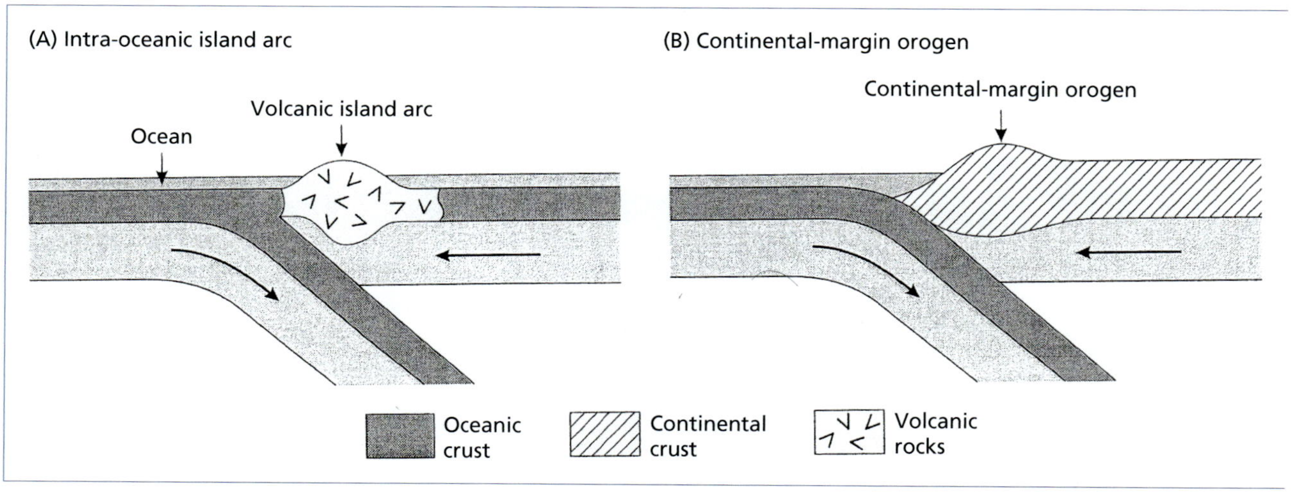

Figure 2.12 Two types of margins. (A) Intra-oceanic island arc, formed when an oceanic plate is subducted under another oceanic plate. This is common in the western Pacific Ocean. (B) Continental-margin orogen, created when an oceanic plate is subducted under a continental plate (example, the Andes) (Summerfield, 1991).

these are located along ocean ridges where the two plates separated by the ridges are moving apart.

A *continental rift* is defined as an elongated depression where the lithosphere has deformed as a result of extension. The most characteristic examples are the *rift valleys* of eastern Africa, the Rhine graben, and Lake Baikal. The African rifts began in the Miocene-Pliocene (Ollier, 1981) and active faulting has occurred after the formation of large erosional surfaces. The rifts begin by *updoming* and extension.

b. **Convergent margins** occur in areas of convergence where lithospheric plates are destroyed as they descend into the interior of the asthenosphere at a dip of about 45°. In the asthenosphere, the descending plate encounters a *Benioff-Wadati zone* where it is destroyed by melting. These processes generate igneous activity in island arcs, trenches, or recent volcanic cordilleras (*Fig. 2.12*). Seismic activity can reach a depth of about 660 km but this is not the end of subduction. Along the curve, starting at the Benioff-Wadati zone, forces are extensional but they are compressional in deeper zones. The collision of two continents results in an *intercontinental collision orogeny* (*Fig. 2.13*). An example is the collision

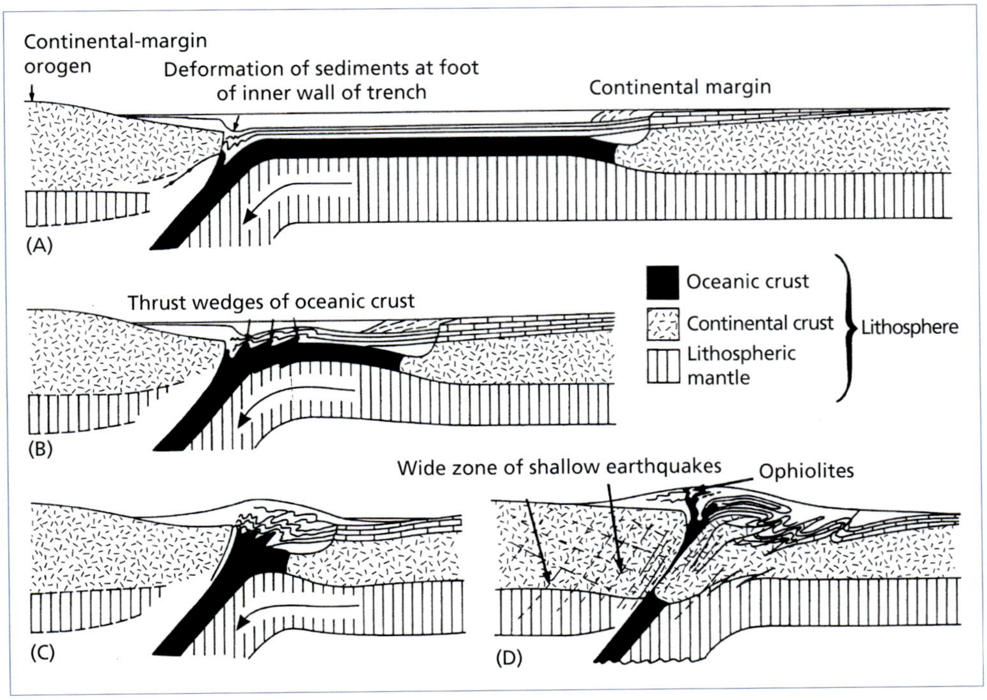

Figure 2.13 Schematic representation of the development of an intercontinental collision orogeny (Dewey and Bird, 1970; modified in Summerfield, 1991).

between India and Asia to form the Himalayas. The subduction zone changes into a suture zone showing where the continents are joined. Ophiolites, gabbros and other rocks that make up the majority of subducted oceanic lithosphere can be found in suture zones. The thickening of the rising continental crust creates high relief. Continental collisions can produce immense horizontal forces. It is believed that the original crust of the Himalayas has been shorted 300 km or more. Deformation of the crust occurs as both brittle and ductile mechanisms. In the upper crust brittle behavior causes significant thrusting or thrust sheets that form large imbricated structures whereas ductile mechanism causes recumbent folds.

c. **Transform zones** Some rigid plates slide against each other along transform faults. The ocean ridge system is made up of a series of segments separated by transform faults. The segments are perpendicular to the direction of extension, and transform faults are parallel to it. Transform faults are passive plate boundaries where adjacent plates slide past each other; they are not involved in extensional or compressional iterations with the other plate.

Wilson (1966) proposed that continental drift was cyclical because the oceans open and close cyclically. This periodicity is known as **Wilson's cycle** (Dewey and Burke, 1974), named to recognize the contribution of J. Tuzo Wilson to the study of plate tectonics.

Figure 2.14 (Turcotte and Schubert, 2002) illustrates the stages of development in the Wilson cycle. *Figure 2.14A* shows the breakup of a continent, which occurs in *rift valleys* such as in East Africa. The second stage is the formation of seafloor spreading with the creation of an oceanic ridge and the rise of molten material from the mantle to generate oceanic crust *(Fig. 2.14B)*. The Red Sea is an example of this stage of the development. Continuing extension produces an ocean basin *(Fig. 2.14C)* such as the Atlantic Ocean, for example.

Continental-oceanic boundaries are called **passive continental margins** or **active continental margins**. Subduction occurs at active continental margins. As the ocean floor of the continental margin grows, the lithosphere becomes thicker and denser. The lithosphere becomes unstable and sinks, becoming a trench, and subduction begins *(Fig. 2.14D)*. Oceanic trenches form adjacent to continents. If the rate of subduction exceeds seafloor spreading, the ocean decreases in size. In time, the oceanic ridge will be subducted *(Fig. 2.14E)*. The remainder of the ridge will also undergo subduction and the continents will collide *(Fig. 2.14F)*.

We can conclude that the theory of plate tectonics explains the principal characteristics of the continents and oceans as well as fundamental smaller features. Mechanisms of global diastrophism are also responsible for small-scale systems, which are important in geomorphology.

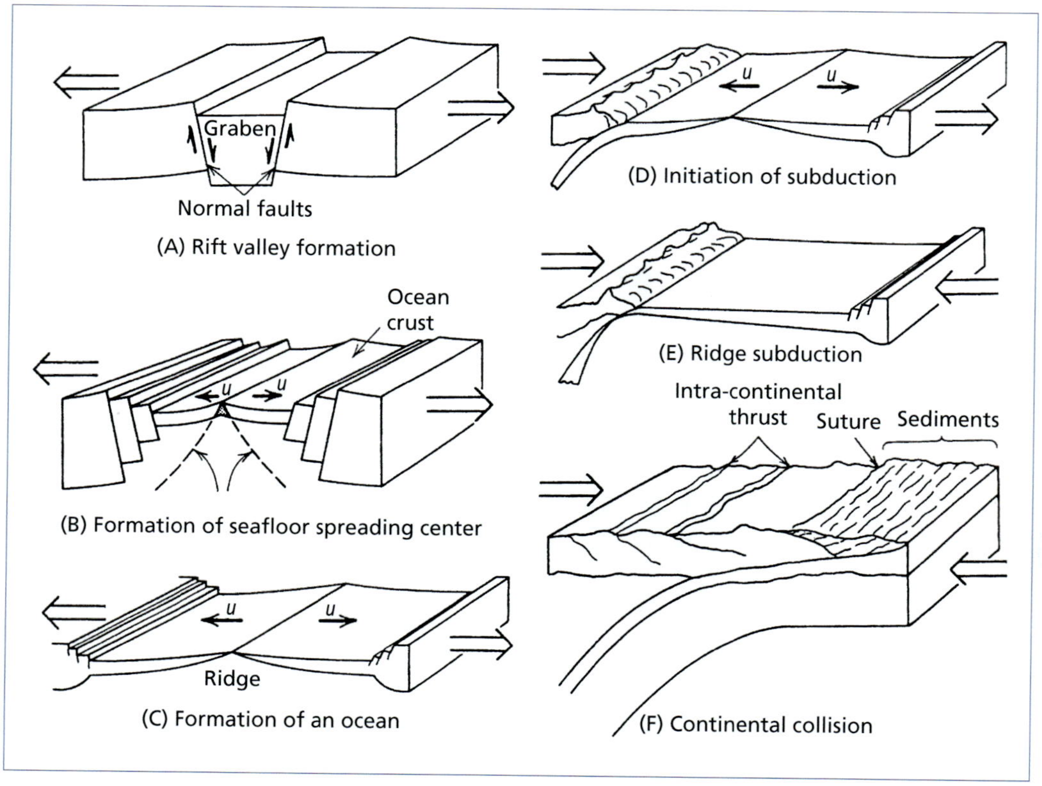

Figure 2.14 Stages in the development of the Wilson cycle (Turcotte and Schubert, 2002).

2.8 Structural landforms

In previous sections we have briefly analyzed the structure of the Earth and its overall structure in terms of plate tectonics. At the beginning of this chapter we defined structural landforms as those that owe their forms to the interaction between various erosional processes and the lithology and structure of rocks.

In the case of rocks with horizontal stratification such as those in the large Spanish Tertiary-age continental basins, exogenetic processes produce differential relief consisting of broad **platforms** and **mesas** (*Fig. 2.15*). Perhaps the most spectacular examples of this residual topography are found in Monument Valley (Arizona) and along the Colorado River (*Fig. 2.16*), which are excellent examples of **stair-step topography**. Mesa morphology, an example of residual topography, requires the presence of a resistant layer at the top of the rock column. Caliche and laterites can also form this resistant layer. The continuous erosion of mesas leads to the disappearance of the upper resistant bed and the mesa becomes a **butte** (*Fig. 2.18*). The term **cuesta** (*Fig. 2.17* and *2.23B*), a Spanish word, is used for a landform created by differential erosion in a series of monoclines with a shallow dip. In hard strata, a scarp forms; its side is called the escarpment of the cuesta and the opposite slope is called the **backslope**. The shape of a cuesta depends on the thickness of the resistant bed at the top of the face. If it is thick, erosion will not cut away much of the face but if the resistant bed is thin and easily eroded, the face will be highly sinuous (Derruau, 1965, Peulvast and Vanney, 2001) (*Fig. 2.19*). When beds are steeply inclined, the resulting landform is called a **hogback** (*Fig. 2.20* and *2.23B*). If the strata dip almost vertically, the landform is called a **razorback** (*Fig. 2.21* and *2.23B*).

Various authors have classified monoclines by establishing limits on the inclination of beds as shown in *Table 2.1*. In addition to the variety of names, this table shows substantial changes in the number of degrees used to establish geomorphic terms. Streams that cut orthogonally to the direction of the strata produce a triangular-faced landform called a **flatiron** (*Fig. 2.22*). John Wesley Powell proposed a genetic classification of rivers in his 1875 book *Exploration of the Colorado River of the West*. William Morris Davis also proposed a classification in his 1889 work, *The Rivers of Northern New Jersey with notes on the classification of rivers in general*. Rivers that flow in the direction of the dip of the strata are called **consequent** or **cataclinal.** Baulig (1938) call streams that flow opposite to the dip of strata **resequent**

Figure 2.15 Mesa composed of sandstones. Jodhpur (India).

Figure 2.16 Platforms of the Colorado and stair-step relief created by the downcutting of the Colorado River (Utah, United States). Photo F. Gutiérrez.

Figure 2.17 Cuesta composed of sandstones. Mula Basin (Murcia Province). Sub-Betic.

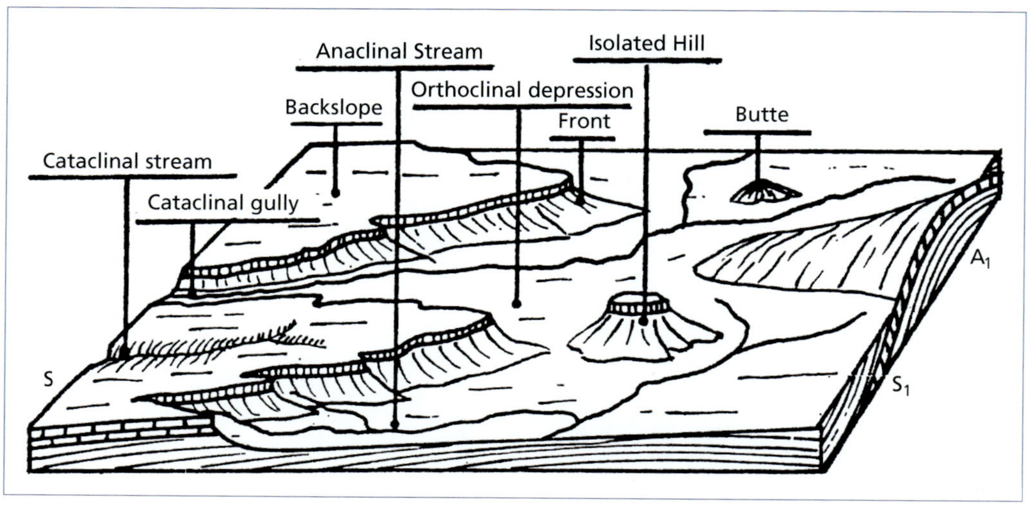

Figure 2.18 Diagram of cuesta topography. Cataclinal or consequent streams and anaclinal streams are present (Viers, 1974).

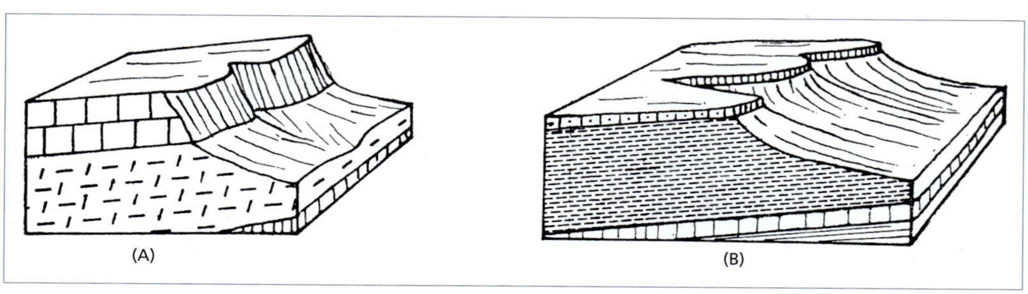

Figure 2.19 Block diagrams of cuestas. (A) Steep front and slight backcutting in thick limestones. (B) Sinuous front with thin sandstone scarps (Viers, 1974).

Figure 2.20 Hogbacks and razorbacks in sands and gypsum. Barbastro Anticline (Province of Lérida, Spain).

Figure 2.21 Group of razorbacks in Cretaceous limestones in the Exterior Sierras of the Pyrenees. Salinas de Jaca (Province of Huesca, Spain).

Table 2.1 **DIFFERENT GEOMORPHIC TERMINOLOGY IN MONOCLINAL SERIES**

Birot (1959)	Simonett (1968)	Viers (1974)
Cuesta < 10°	Cuesta 0.5–5°	Cuesta 1–10°
Crest > 10°	Homoclinal ridges 5–40°	Crest, Outcrop 10–30°
Razorback > 45°	Hogback > 45°	Hogback 30–70° Razorback 70–90°

or **anaclinal** or **subsequent**. A stream that flows parallel to the direction of the layers is called **obsequent** or **subsequent** or **orthoclinal**. These terms come from the work of Baulig (1938), Birot (1959), Derruau (1965) and Viers (1974), Klein (1993), and Peulvast and Vanney (2001). Ollier (1981), Huggett (2003), and the Encyclopedia of Geomorphology (Goudie, 2004) contain a simpler nomenclature for streams in a monoclinal series. Their nomenclature includes a **strike valley**. The stream in a strike valley connects at right angles to **dip streams** and **anti-dip streams** (*Fig. 2.23A*).

The evolution of folded relief results in simple structures such as a set of anticlines and synclines of similar amplitude, which is the congruent style in the nomenclature of Germanic or Saxon tectonic styles (Stille, 1924). In geomorphology these simple folds evolve to form various landforms (*Fig. 2.24*). Areas of higher elevation that have anticlinal structures are called **monts** in the terminology of the Jura. Lower elevation areas that are synclines are called **valleys**. The combination is called **congruent topography**. When water erosion affects the hinge areas of an anticline, a **combe** (from Latin *cymba*) results. A **combe** opens up parallel to two **crests**. A similar stream that incises into the flank of an anticline is called a **ruz**. A stream that entrenches and crosses an anticlinal structure perpendicularly is a narrow pass called a **cluse** (*Fig. 2.25*). This assemblage is called **Jurassian relief**. When the erosion of the growing combe removes the resistant bed of the anticline, it creates **inverted relief** (*Fig. 2.26*), which includes a **breached anticline** and, in a culminating position, a **hanging syncline** (*Fig. 2.27*). If the region remains tectonically stable for a long time, the folded structures are eroded away, leaving **planate relief** (*Fig. 2.28*). This terminology, based on names from the Jura region, comes from the work of Martonne (1909), Birot (1959), Derruau (1965), Viers (1974), Ahnert (1996), and Peulvast and Vanney (2001). If planate relief undergoes generalized uplift, the drainage network erodes downward into the terrain and the resistant layers are then located in an elevated position, leaving all the

Figure 2.22 Flatirons in limestone and marls of the Muschelkalk. Highway between Monterde and Nuévalos (Iberian Cordillera).

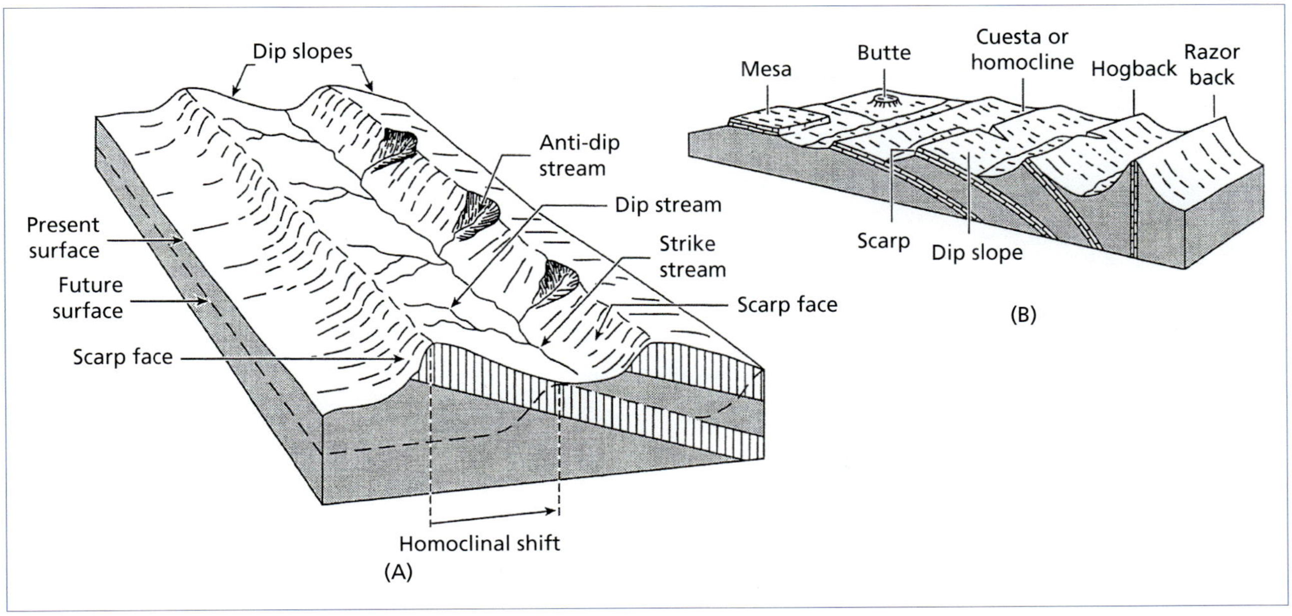

Figure 2.23 Monoclinal series and its relationship to topography. (A) Cuestas and their relationship to drainage. (B) Landforms associated with strata of increasingly greater dip (Goudie, 2004b).

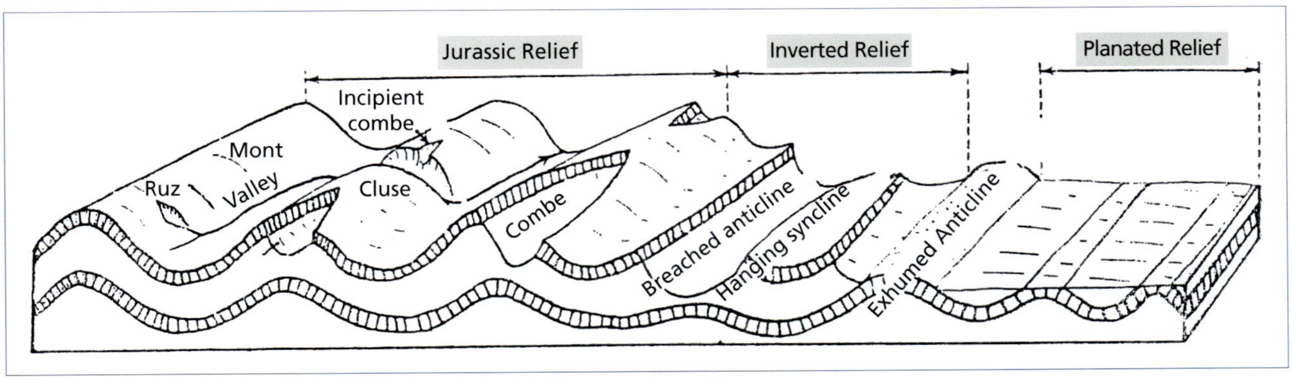

Figure 2.24 Evolution of folded topography (Derruau, 1965).

topmost zones at equal elevation. This type of morphology is known as **Appalachian relief**.

The landforms we have just described are typical of simple congruent folds. However, folding structures are generally more complex so these names are not always used or other terms may be used. Salt diapirs have a distinctive morphologic expression with a notable scarp and a deep depression with a large waterfall, such as the Orduña salt dome (Basque-Cantabrian Mountains, Spain).

A stream is called **antecedent** when it develops prior to tectonic uplift and maintains its direction during uplift. A stream may incise into a horizontal formation that discordantly overlies a folded assemblage. Reaching the surface of the unconformity, the stream may erode through it, flowing through the most easily erodible beds. This process is called **superimposition**.

Faults are fractures with differential movement (*Fig. 2.29C*). Both **normal** and **reverse faults** can have scarps, called **fault scarps,** at the margins of the uplifted and downdropped blocks (Bull, 2007). The uplifted block of a normal fault undergoes intense fluvial erosion and **trapezoidal fault facets** (Derruau, 1965; Zuchiewicz, 2004), which may occur on the fault scarp, may evolve and turn into **triangular fault facets** (*Fig. 2.30* and *Fig. 2.31*) (Bull, 2007). These facets may have a throw of a few hundred meters as a result of repeated fault movement. Alluvial fans often develop at the base of fault scarps, as in the Pre-Cordillera of the Andes and Death Valley (California) (*Fig. 2.32*). They are often preserved in arid

Figure 2.25 Cluse of the Guadalupe River that perpendicularly cuts Cretaceous limestone razorbacks. Aliaga (Province of Teruel, Spain). Central-eastern Iberian Cordillera.

Figure 2.26 Breached anticline in Cretaceous formations, Makhtesh Ramon. Negev Desert. Israel.

Figure 2.27 Hanging syncline in upper Cretaceous formations. Cuevas San Clemente (Province of Burgos, Spain). Northern Iberian Cordillera.

Figure. 2.28 Topography in upper Cretaceous limestones created by fluvial erosion. Nogueruelas (Province of Teruel, Spain). Nogueruelas. Eastern Iberian Cordillera. Photo B. Leranoz.

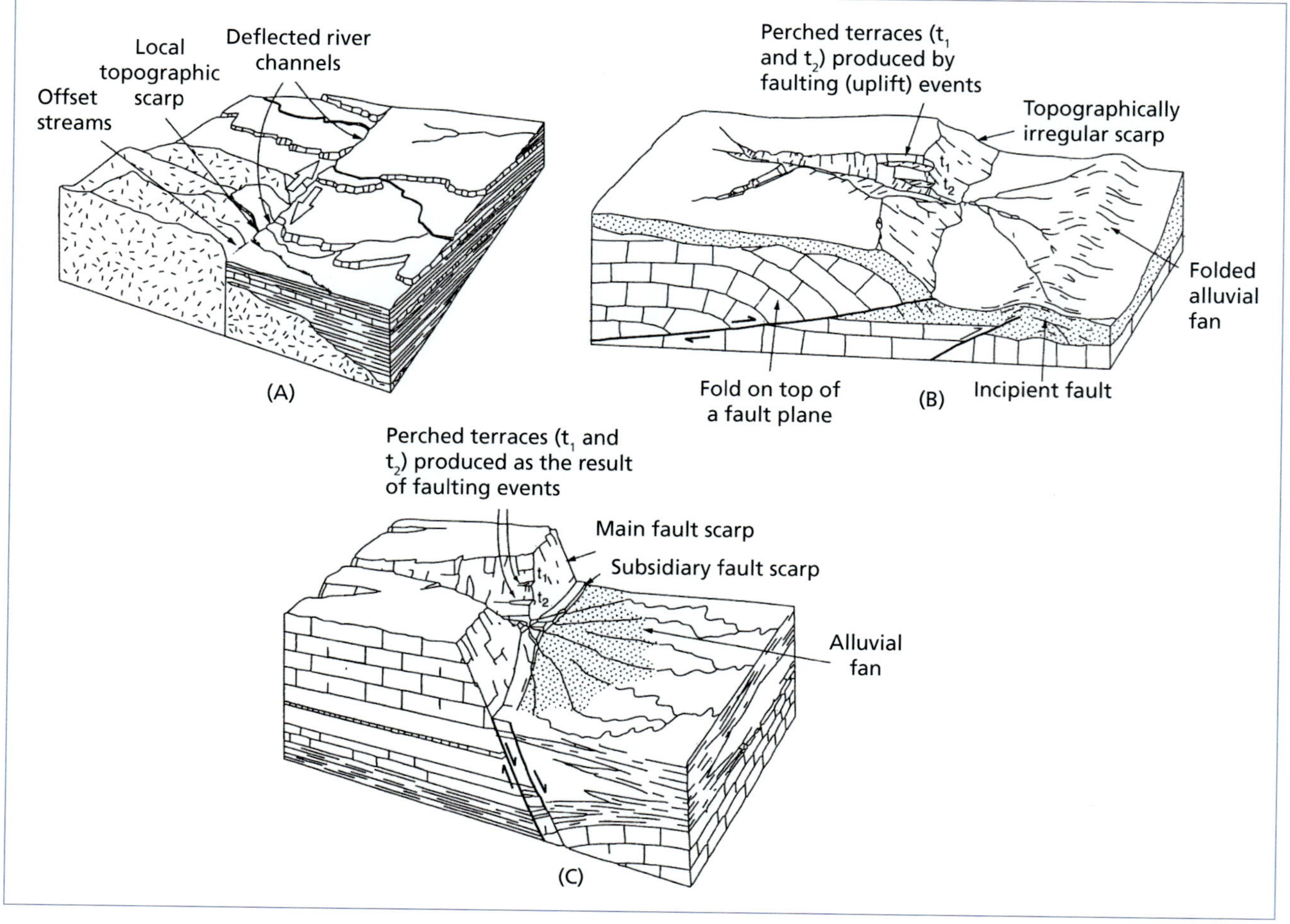

Figure 2.29 Block diagrams showing the topographic expressions of various types of faults. (A) Strike-slip faults showing local topographic scarp, offset fluvial channels, and displaced streams. (B) Low angle reverse fault with a fold over the fault plane. The topography shows an irregular scarp, hanging terraces (t_1 and t_2) produced by fault activity, uplift, and a folded alluvial fan. (C) Normal fault showing main and secondary fault scarps, hanging terraces (t_1 and t_2) produced by uplift and an alluvial fan (Ramsay and Huber, 1987).

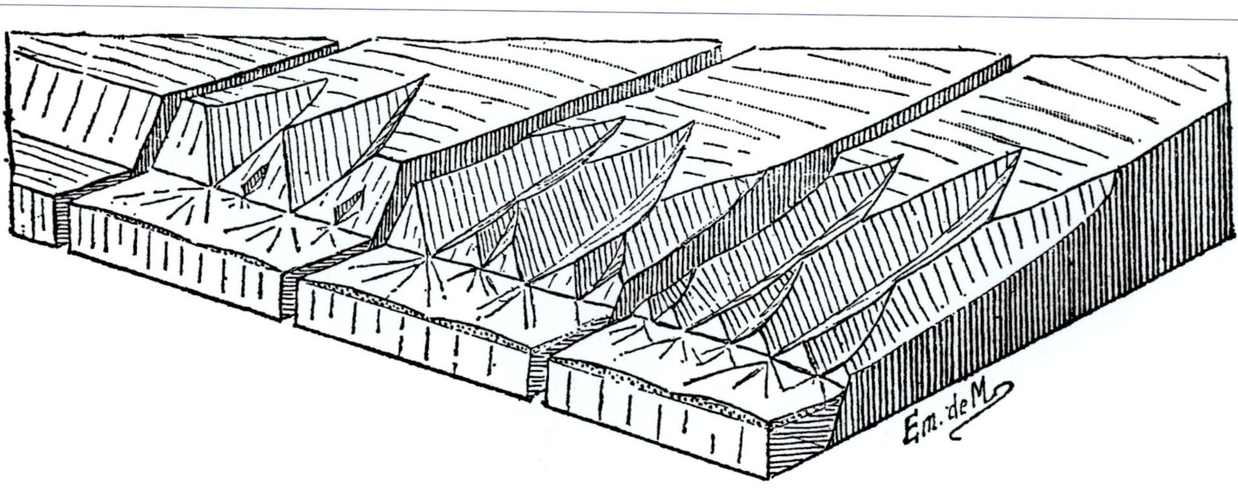

Figure 2.30 Formation of facets indicating a fault (from Martonne, 1909).

Figure 2.31 Scarp and triangular fault facets S. Luis Valley. Río Grande. Sangre de Cristo Mountains. Colorado, United States. Photo F. Gutiérrez.

Figure 2.32 Coalescing alluvial fans and, in the foreground, a playa. Photo taken from Panamint Range. Death Valley (California). Photo F. Gutiérrez.

zones even though the evolutionary status of a fault scarp depends on the relationship between the uplift rate and the intensity of the erosional processes acting on it. In this context, hanging valleys can develop if uplift exceeds downcutting (Ahnert, 1996). Facets can develop by marine abrasion, in glacial valleys, in recently rejuvenated fluvial valleys, and during the erosion of volcanic cones (*planezes*). Normal faults can occur as a set of en echelon faults such as those on the scarp of the Sierra Palomera (Jiloca half graben, Teruel, Spain) and in its northern (*Fig. 2.39*) and southern extensions. Fault scarps disappear when denudation exceeds uplift. When an old fault zone erodes, a **fault line scarp** may develop. Davis (1913) defined two types of fault line scarps. The first is a **resequent** scarp where the entire surface has been degraded except for a resistant formation on the uplifted block, so that the scarp approximates the trace of the fault. The second, an **obsequent** fault line scarp, occurs when the existing rocks on the downdropped block are more resistant than those on the uplifted block. In this case, erosion is more intense on the downdropped block and as a result the scarp is located on the side opposite from the original, producing **inverted relief**.

High angle **reverse faults** are found in areas of crustal shortening where cordilleras develop, as at the boundary of convergent plates associated with subduction. These faults affect pediments and marine and fluvial terraces, creating small scarps. They can occur as a defined fault scarp. The front of a low angle fault may develop a scarp if it is composed of resistant rocks (*Fig. 2.29B*). Due to its low angle, the trace of the fault plane is sinuous and streams may create deep entrants.

Strike-slip faults (*Fig. 2.29A*) are generally highly complex. Streams may flow across them but some faults, such as the San Andres, cross highways and other human structures. Strike-slip faults can be hundreds of kilometers long such as the San Andres fault system, the Alpine Fault in New Zealand, and the Great Glen Fault in Scotland. All of them are areas of intense seismic activity. The Tanna fault (*Fig. 2.33*) in the northern part of the Izu Peninsula of Japan has one kilometer of left-lateral displacement accompanied by vertical

Figure 2.33 The Tanna fault, on the Izu Peninsula (Japan). The direction of movement is shown by markers, and at the back, a roof covers a trench.

displacement. The most recent activity occurred during the Kita-Izu earthquake of 1930. Its recurrence interval over the last 6,000 years is estimated to be 700–1000 years (The Tanna Fault Trenching Research Group, 1983). On the Iberian Peninsula the most important strike-slip fault is the Lorca-Alhama Fault or the Alhama de Murcia Fault (Bousquet, 1979). This fault crosses almost 100 km of the Murcia region of Spain in a SW-NE direction from Huéral-Overa (Almería) to north of the city of Murcia near Orihuela (Alicante). Its left-lateral movement has created numerous examples of offset drainage channels, shutter ridges, and pressure ridges occur along its trace (Silva *et al.*, 1992a, b, c, 1993; Baena *et al.*, 1994). This fault, which has been active during the Quaternary, is known for significant prehistoric, historic, and instrumental seismic activity (Lorca earthquakes of 1579, 1674 and 1818; Silva, 1994).

The classic paper of Wesson *et al.* (1975) describes strike-slip fault landforms (*Fig. 2.34*). Among the landforms are **linear valleys** that flow along major faults, **scarps** resulting from strike-slip faults with a vertical component of movement, **benches** consisting of elevated, slightly planed landforms that may be tilted or convex, **springs** (thermal or not) that often occur along the fault plane, **offset drainage channels** indicating the relative direction of displacement, **sag ponds** related to the relative doming of the fault zone, **linear ridges**, which are small arched areas produced by compression between multiple traces in the fault zone, and **shutter ridges** that form when a fault moves a ridge and places it in front of a stream that may move it noticeably (sometimes in the opposite direction of the fault movement).

The largest structurally controlled landforms are **rift valleys**; like ocean ridges they are associated with extensional forces. *Rift valleys* occur along the axes of oceanic ridges; they are bounded by large normal faults. Smaller *rift valleys* also occur on continents (*Fig. 2.35*). The best-known example is the East Africa rift valley system, which extends more than 6,000 km from the lower reach of the Zambezi River to the Sea of Galilee (*Fig. 2.36*). Its width ranges from 30 to 90 km and its scarp height ranges from 400 to 2000 m. The deepest parts are occupied by numerous lakes of varying sizes, some of them below sea level (Tanganyika, Dead Sea). The African *rifts* are accompanied by basaltic volcanism that extends to the margins. Numerous volcanic cones occur in the bottom of the *rift valley*. The large stratovolcanoes of Mt. Kenya (5,200 m) and Kilimanjaro (5,895 m), the highest points on the African continent, are related to volcanism in the Eastern or Gregory Rift Valley. The normal faults are accompanied by another system of synthetic fault planes (dipping in the same direction as the normal fault) and antithetic fault planes (dipping in the opposite direction) that appear en echelon to the major fault plane (*Fig. 2.37*). Very few geological sections include the two faults that form the boundaries of a tectonic depression. Studies conducted using seismic stratigraphy show that fault systems are highly complex. These investigations show that *rifts* are asymmetrical and are in fact half grabens. Research done on the entire

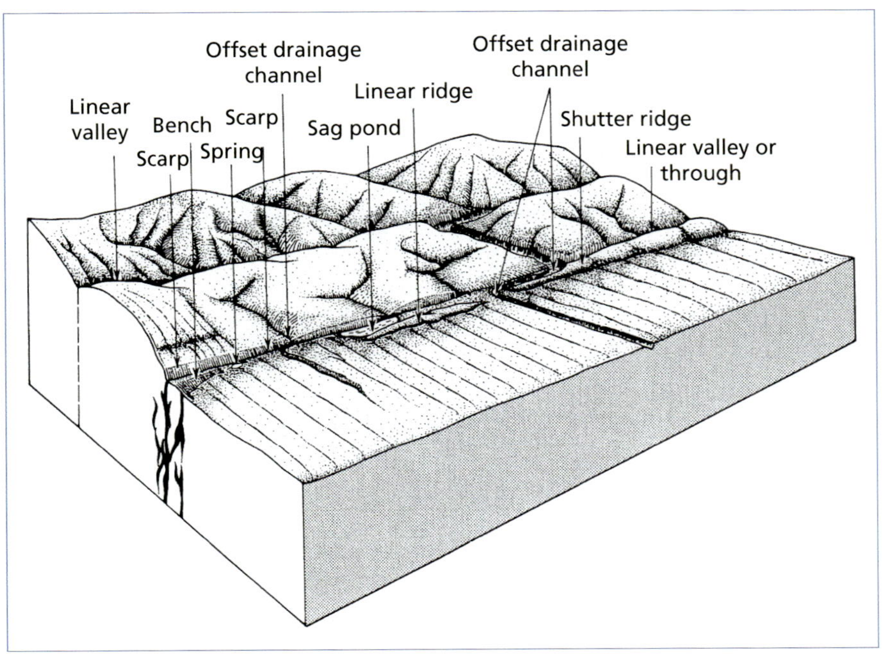

Figure 2.34 Block diagram showing the various landforms associated with recently active strike-slip faults (Wesson *et al.*, 1975).

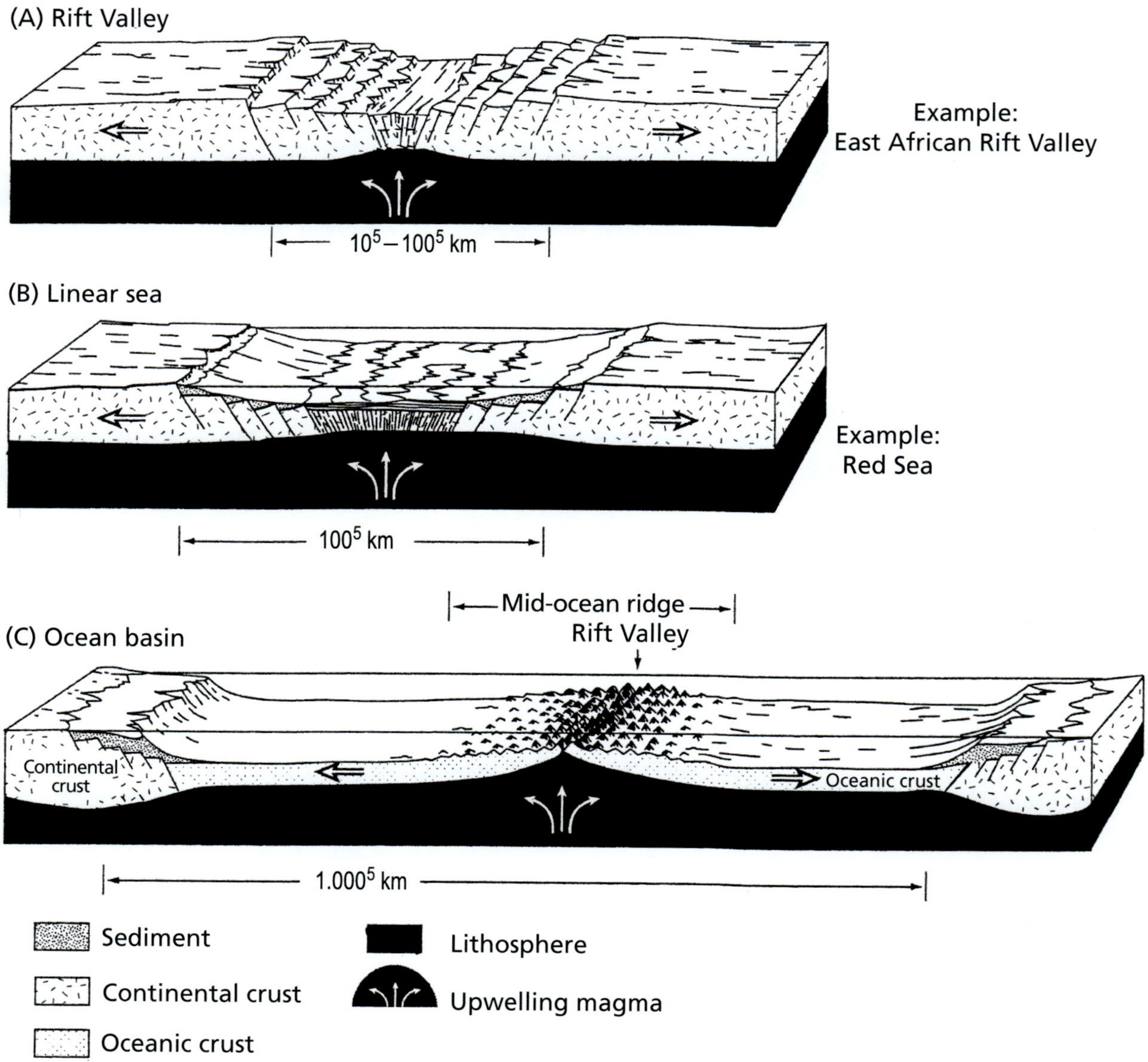

Figure. 2.35 Landforms produced when the crust separates as a result of extension. *Rifting* and extension of the ocean floor creates *rift valleys*, linear seas, and ocean basins (Lutgens and Tarbuck, 1992).

Rift Valley indicates that the faults normally occur on one side and are discontinuous (Summerfield, 1991). Researchers think that extension and crustal thinning, conditions that promote volcanism, create *rift valleys*. This idea encompasses traditional ideas about arching and fracturing at the crest and volcanism (Cloos, 1939). The arching causes tilting of the various ages of erosional surfaces that have developed on the African continent (*Fig. 2.38*).

In addition to the East Africa Rift Valley, other smaller rifts have been recognized such as the Rhine and Baikal grabens. The Rhine graben, 600 km long and about 70 km wide, is one of the best-described examples in the world. Its northern part separates the structural uplifts of the Black Forest and the Vosges. Basaltic volcanism is present in the Rhine graben. The Baikal graben is about 2500 km long and 50 to 60 km wide. Lake Baikal is located in the graben but unlike other rifts, it contains no volcanic rocks. The fault system present in grabens is generally asymmetrical; one flank is steep and the other is gentle, as in the Jiloca half graben (Iberian Cordillera) (*Fig. 2.39*).

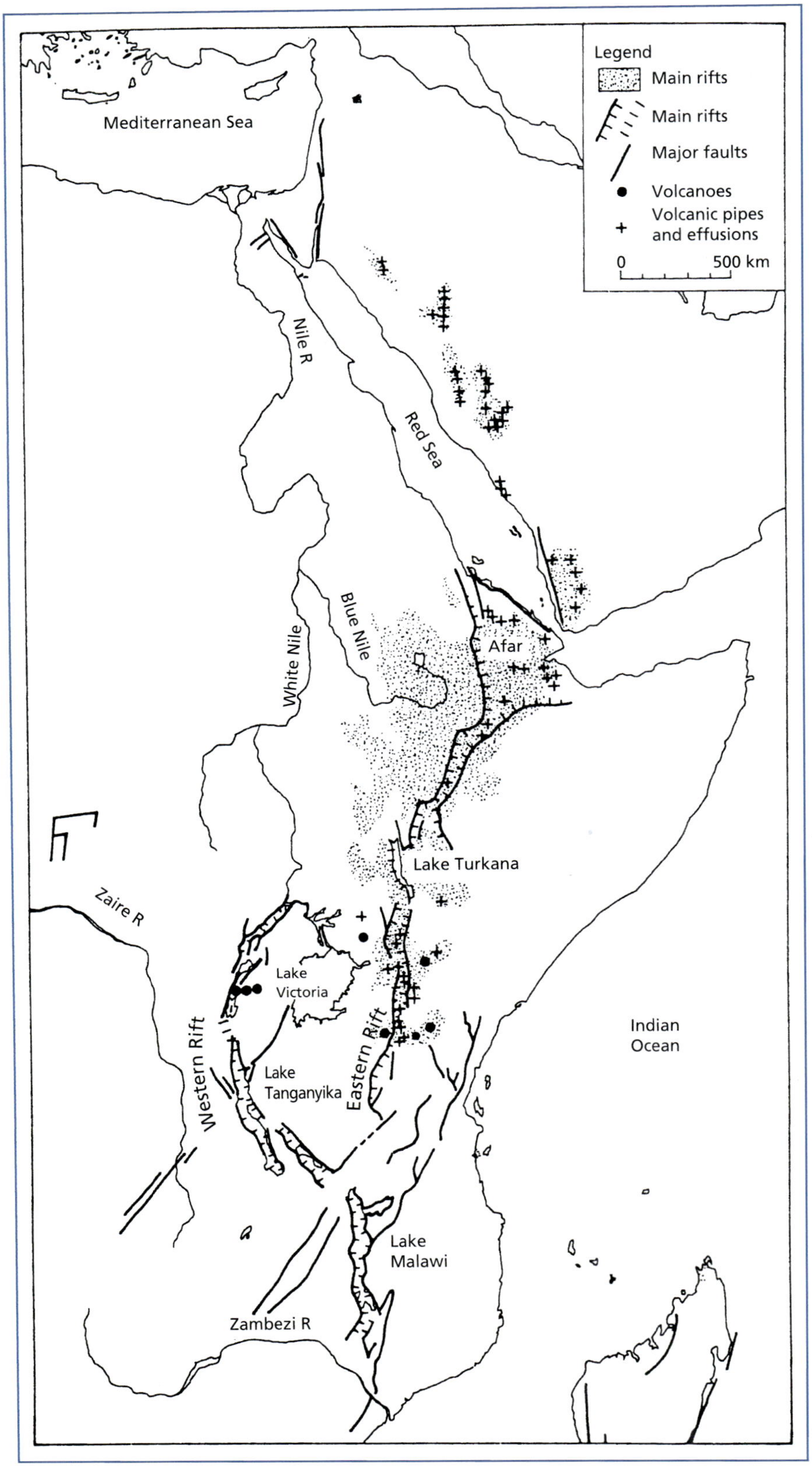

Figure 2.36 Location of the East African Rift Valley system and associated volcanic rocks (Selby, 1985).

Figure 2.37 A step that is slightly tilted toward the main fault. In the distance is the Eastern or Gregory Rift Valley. Longonot (Kenya).

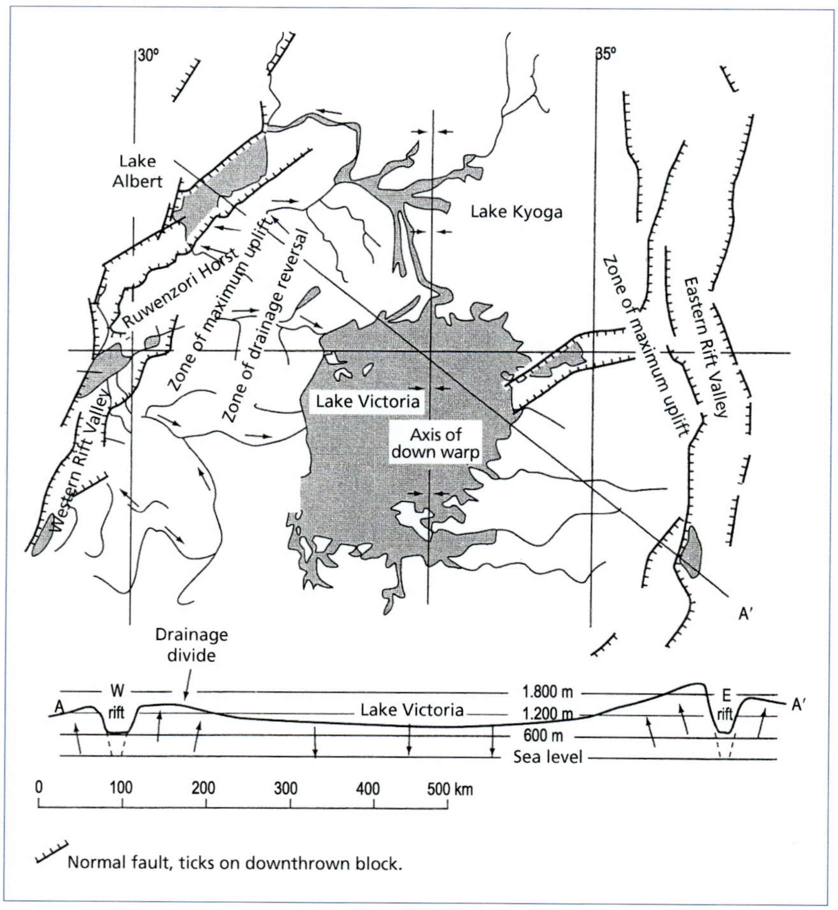

Figure 2.38 On the East African plateau, arching occurs around the rifts, Lake Victoria occupies wide negative doming, and various drainage directions are present (Bloom, 1978).

Figure 2.39 North of the Polje-Jiloca half graben from Sierra Palomera (Iberian Cordillera, Spain). Photo F. Gutiérrez.

2.9 Pseudo-structural landforms

Some landforms seem to be associated with endogenetic forces but in fact are the result of **non-tectonic processes** operating on a thin surficial layer of the Earth's crust (Twidale, 1971, 1976a). Mechanisms that produce these landforms include clay overloads, the force of gravity, and deformation associated with volumetric changes of the surface layers. To this list, we should add impacts by extra-terrestrial bodies. Many of these landforms can lead to interpretations that erroneously point to tectonic mechanisms. Australian investigators have described pseudo-structural landforms in their books (Hills, 1963; Twidale, 1971, 1976a; Ollier, 1981).

Cracks or holes in strata of resistant rocks can be caused by lateral extension (*Fig. 2.40*). Surface deformation occurs in areas where competent formations overlie clay-rich materials. The most typical example is the Northamptonshire area of England (Hollinworth *et al.*, 1944); here streams are eroding limestones and sands that overlie clays. The clays flow toward the valley, causing subsidence along the margins of the *plateau* and creating a broad arch or ridge in the interfluve, called a **camber** (*Fig. 2.41A*). The weight of the overlying rocks causes the clays to flow toward the valley, causing **bulges** (*Fig. 2.41B*) with dips contrary to those of the valley so that the bulges resemble an anticline. The axis of these **valley anticlines** (*Fig. 2.41C*) is subparallel to the stream, and may extend 35 km following the sinuosity of the Colorado River (Potter and McGill, 1978), for example.

The Zagros Mountains (Iran and Iraq) are composed of a thick pre-Albian limestone forming the core of folds of similar amplitude. Overlying them are marls, sandstones, and silts that terminate with a lower Miocene series containing anhydrite (gypsum at the surface). The region is subject to significant fluvial erosion and streams generally follow the structural strike. Various dislocations that are unlike the general style are superimposed on these structures. These dislocations result from downcutting by the drainage network, which has caused instabilities on the flanks of folds (Harrison and Falcon, 1934, 1936). The following landforms are recognized: **knee folds** and structures of **recumbent and cascade folds** (*Fig. 2.42*). In knee folds erosion of the upper flank has created an overturned flank called a **flap**. On the flanks of some

Figure 2.40 Lateral spreading in sandstones on top of Miocene clays. Gabarda Rock Pillars (Province of Huesca, Spain). Ebro Depression.

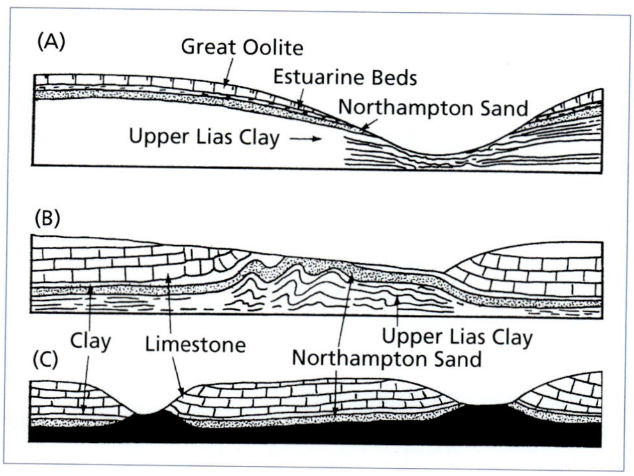

Figure 2.41 Northhamptonshire surface structures. (A) Camber (B) and (C) Valley bulges. (Hollingsworth *et al.*, 1944).

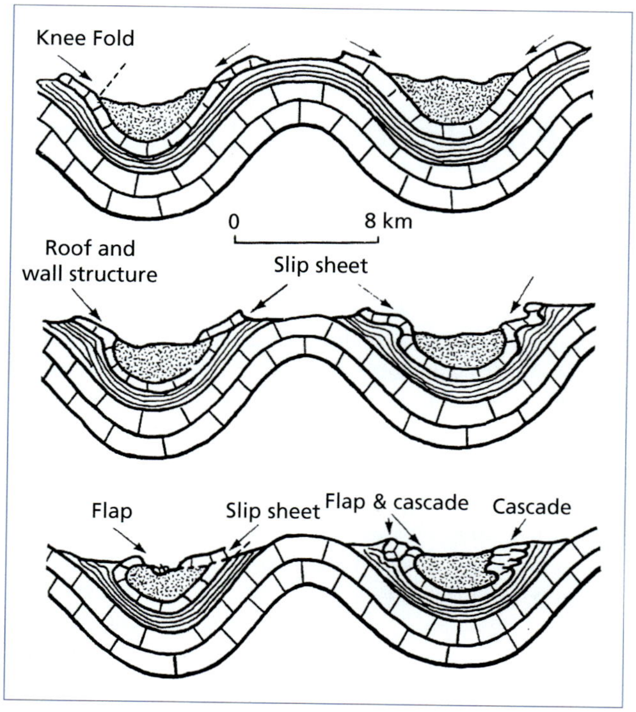

Figure 2.42 Gravity collapse structures. The diagrams show successive stages in the development of a *flap* (Harrison and Falcon, 1934).

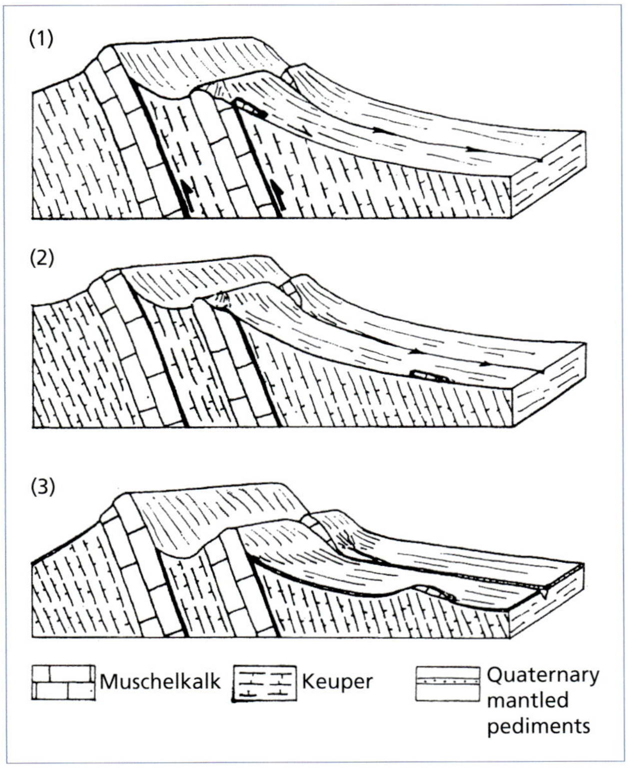

Figure 2.43 Stages of progressive block displacement in the Villel region (Province of Teruel, Spain) (Gutiérrez and Peña, 1979).

of the folds a **slip sheet** has occurred. All these dislocations are **gravity-collapse structures** (Harrison and Falcon, 1934, 1936). The presence of plastic clays contributes decisively to the development of these structures.

An example of a slip sheet is found in the Villel area (Province of Teruel, Spain) where blocks of Muschelkalk dolomite slid over paleoslopes of gypsiferous Keuper clays and are hanging about 70 meters above the bed of the Turia River (Gutiérrez and Peña, 1979c) *(Fig. 2.43)*. These morphologic characteristics resemble the interpretations expressed by many tectonicists on the occurrence of landslides where the influence of morphology is decisive (Lutaud's pre-tectonic facilities, and Glangeaud's concept of morphotectonics, see Auboin, 1961). This type of landslide also brings to mind Goguel's (1965) interpretation of Montmeyan thrusting.

In many parts of the world, craters bear no relationship to volcanic fields. These craters are well preserved in arid areas but lakes may fill them in wetter zones. These closed depressions, which have an uplifted rim and sometimes have central protuberances, have been interpreted as meteorite **impact craters** *(Figs. 2.44* and *2.45)* although some authors (Bucher, 1963, Goguel, 1963) think they originate from gaseous intrusions. The presence of cohesite and stishovite, varieties of high temperature and pressure quartz, in the crater area indicate a high-impact origin. An addition, the presence of meteorite remains confirms the impact origin (Twidale, 1976a). In the next chapters we will analyze other types of

Figure 2.44 Impact crater. Meteor Crater (Arizona). Photo F. Gutiérrez.

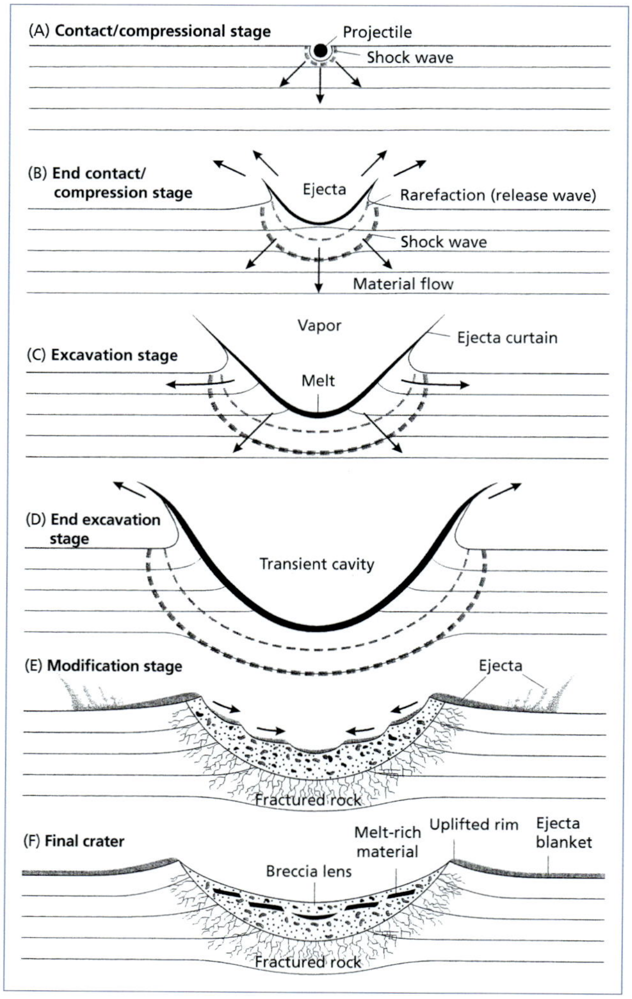

Figure 2.45 Formation of an impact crater. Cross sections showing the progression (*French, 1998*).

non-tectonic surface structures including structures caused by the passage of water and ice and vice versa (wedges, pingos, patterned ground, cryokarst, etc.), expansion phenomena, subsidence resulting from evaporite dissolution, contraction in saline lakes, subsidence caused by mine works and fluid pumping, among others.

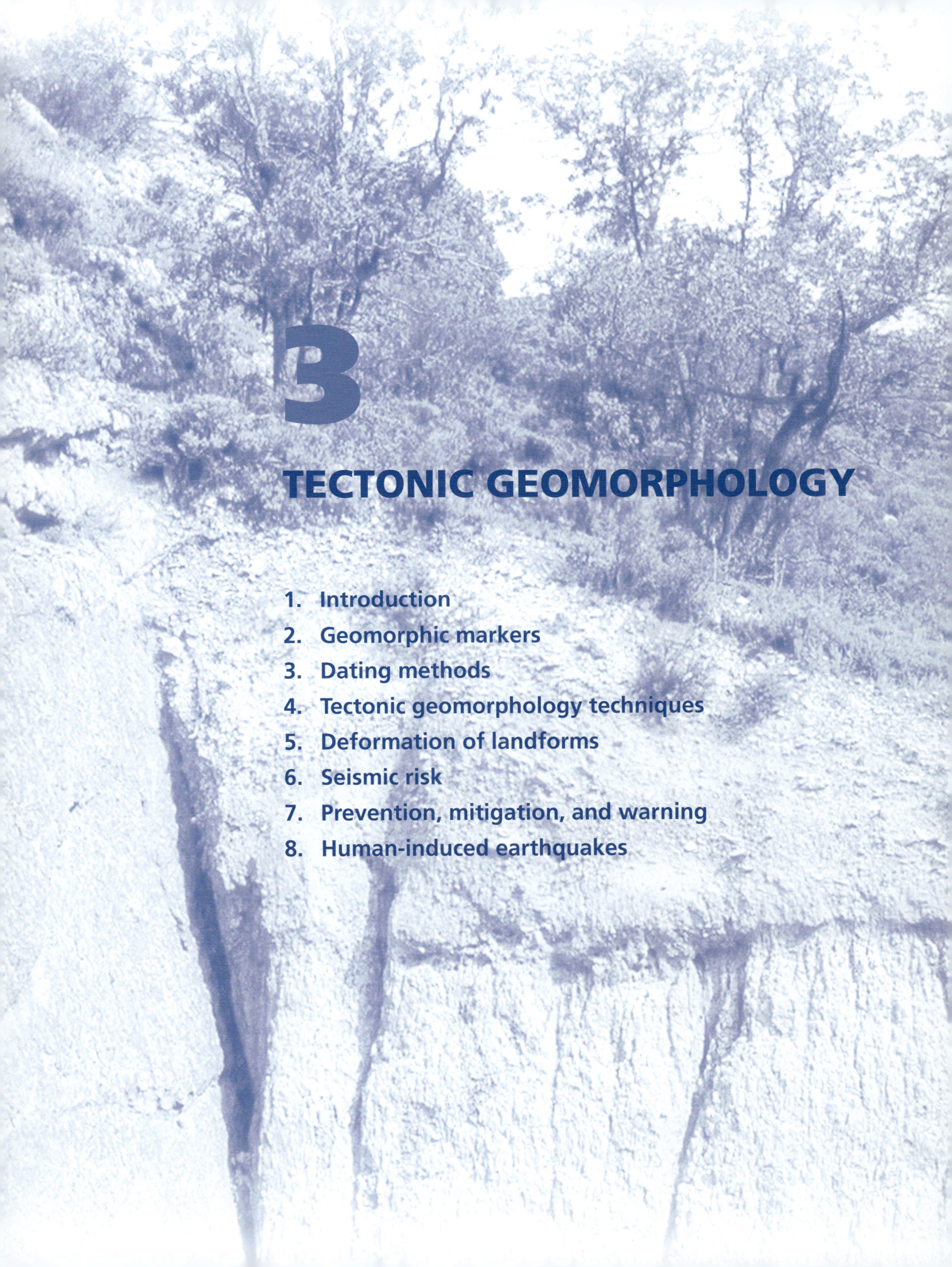

3

TECTONIC GEOMORPHOLOGY

3.1 Introduction

The relief of the Earth's surface is created by the interaction of endogenetic and exogenetic processes. Endogenetic forces create large morphostructural units (Büdel, 1968, 1977) primarily by vertical movements whereas exogenetic forces continuously wear down the existing topography. This battle has persisted relentlessly on multiple scales throughout Earth history (Gutiérrez, 1990). As previously discussed, **tectonic geomorphology** is the study of tectonic processes *(Fig. 3.1)* that create relief and that counteract surface processes that seek to erode them (Yeats *et al.*, 1997; Burbank and Anderson, 2001; Goudie, 2004c).

Keller and Pinter (1996) define tectonic geomorphology in two ways: (1) the study of forms produced by tectonic processes, and (2) the application of geomorphic principles to resolve tectonic problems. For these authors tectonic geomorphology is synonymous with **active tectonics** *(Fig. 3.2)* although these researchers consider active tectonics to be processes that deform the Earth's crust over a human lifetime. Active tectonics is fundamentally the study of earthquakes (McCalpin and Nelson, 1996). These authors define **paleoseismology** as the study of prehistoric earthquakes with a focus on their location, age, and **magnitude**. Paleoseismology differs from **neotectonics**, a term proposed by Obruchev in 1948 (Mescherikov, 1968) for a discipline that studies Neogene and Quaternary crustal movements and the quasi-instantaneous deformation of landforms and sediments during earthquakes (Allen, 1986). For Fairbridge (1981), *neotectonics is the study of movements that create topography*. Caputo and Pavlides (2008) analyze the methods and application of earthquake geology and Caputo and Helly (2008) describe the various disciplines that investigate the history of earthquakes.

Figure 3.1 Normal fault in Jurassic materials and a colluvial sequence. At the contact between the deposits there is a fault throw of 50 cms and slickensides with various microstructures along the fault. Photo taken in 1981. The colluvium has been dated absolutely with ^{14}C: one to 2 m of the surface provided an age of 43,070 ± 1200 BP and another showed an age of >48,500 BP, exceeding the scope of the method (Gutierrez *et al.*, 2005b). Rubielos de la Cérida (Provincia de Teruel). Central Iberian Cordillera.

Figure 3.2 Active fault with slickensides, striations, and grooves. Mula-Kale (Turkey). Photo: F. Gutiérrez.

Prior to 1970 the evaluation of earthquake risk in industrialized countries such as the United States and the USSR was based solely on the historical earthquake record (McCalpin and Nelson, 1996). Nowadays most countries that have seismically active faults conduct regional paleoseismic data studies and analyze seismic risk for these locations (Algermissen *et al.*, 1982). Paleoseismology adds to historical and instrumental seismic records by characterizing and dating large prehistoric earthquakes (Vittori *et al.*, 1991).

The characteristics of erosional or depositional landforms generated during an earthquake are called **co-seismic** in contrast to **aseismic** features, which are formed by erosional and sedimentary processes. Paleoseismologists are limited to studying earthquakes that deform surfaces that are clearly recognizable. These earthquakes are called **morphogenetic earthquakes**. A geological structure genetically related to earthquakes that preserves the stratigraphic record is called a **seismite**.

Tectonic activity, which generally works contrary to gravity, releases energy that creates relief. Tectonic activity gets its energy from a small part of the horizontal movement of lithospheric plates that make up the rigid exterior of the planet. Some of this energy comes from heat produced by the disintegration of radioisotopes and phase changes in the Earth's interior (Burbank and Anderson, 2001). In summary, we know that millions of years of horizontal movement at plate margins are transformed into vertical movement at convergent margins (in other words, mountain systems); they are no more than the sum of individual displacements caused by earthquakes over long periods of time that in turn lead to processes of isostatic elevation.

The study of tectonic geomorphology requires an understanding of geomorphic and tectonic processes. A tectonic geomorphologist must be familiar with the classical theories of landform evolution proposed by Davis, Penck, and King, as discussed previously. Likewise, an understanding of the geomorphic concepts of ridges, complex response, and dynamic equilibrium (Keller and Pinter, 1996) is also essential. These theories and concepts were analyzed in a preceding chapter. A simple example explains why an understanding of these geomorphic concepts is necessary. An earthquake involves a sudden liberation of forces that may change landforms instantly and produce fault scarps. With a long recurrence interval between earthquakes, geomorphic processes can slowly erode the original fault scarp before the next earthquake occurs. In general, the preservation of evidence for past earthquakes depends on the interaction between geomorphic processes prior to deformation (Bull, 1991). When the magnitude of deformation exceeds the intensity of the geomorphic processes, seismic landscapes are created.

The primary effect of an earthquake is abrupt movement; this ground shaking can cause buildings (*Fig. 3.3*) (destruction of Sodom and Gomorra, 4350 BC), bridges, dams, tunnels, and oil and gas pipelines to collapse (Bolt, 1999). The

Figure 3.3 Buildings partially destroyed by the Pugili earthquake of April 1996. Cotopaxi region (Ecuador). Photo F. Gutiérrez.

secondary effects of earthquakes include **liquefaction** (earthquake of Niigata, 1964), which is defined as the transformation of a water-saturated granular material from a solid state to a liquid. The shaking that accompanies earthquakes causes numerous **landslides** (Huascarán, Andes of Peru, 1962). Earthquakes can rupture gas lines and cause significant **fires** such as the fires that occurred after the 1906 San Francisco earthquake. Oceanic earthquakes can cause vertical landslides on the ocean floor that produce to seismic marine waves or **tsunamis**. Waves up to one meter tall with a wavelength of 100 km can move 800 km/hour; when these waves reach the coast they become waves more than 20 m high with velocities of 60 km/hour that are obviously capable of causing significant catastrophes (Malaga, 1689; Lisbon, 1755; Hawaii, 1946; Sumatra, 2004). Earthquakes also cause **floods** by breaching or collapsing dams (Earthen dam in Sheffield, California, 1925.) It has been estimated that a large earthquake in a densely populated area of southern California could kill several thousand people and cause damages amounting to a trillion dollars (Keller and Pinter, 1996). These data show the need to study tectonic geomorphology to obtain a better understanding of the evidence of various types of seismic processes (fault ruptures, secondary effects) on the landscape. Once scientists understand the mechanisms that generate these landforms, they can evaluate risk and alert society.

Tectonic geomorphology is closely related to other branches of science including seismology, Quaternary paleoclimatology, geochronology, structural geology, stratigraphy, geodesy, and geomorphology (Burbank and Anderson, 2001). Integration of all these disciplines will hopefully lead to significant advances in tectonic geomorphology.

3.2 Geomorphic markers

Geomorphic markers are features of the landscape that can be used to analyze evidence of deformation. Fluvial and marine terraces are good examples. The markers must have a known planar or linear geometry prior to tectonic activity (reference surface) that makes it possible to determine the magnitude of the deformation that occurred during folding or fracturing. To ascertain the magnitude of deformation, the duration and amount of deformation must be known (Burbank and Anderson, 2001).

3.2.1 Planar geomorphic markers

Cross sections of **fluvial terraces** are frequently used to study deformation caused by faulting and folding generated by tectonics or climate change. We can use the climate record to estimate the age of a terrace if we know the climate record for the area over time, and if the terrace was created by climate. If we do not have a climate record or if we are uncertain whether the terrace was created by climate, we have to date the marker (terrace). The surfaces of **alluvial fans** are similar to, but more inclined than, those of fluvial terraces. They are also variable and their cross section is slightly concave. Even so, they can be used as markers because their average slope can be used (Avouac *et al.*, 1993). The average slope of **pediments** and **cover pediments** can also be used as markers. In the area of La Laja near the city of San Juan (Argentina) recent 30° reverse faults to the west on a covered slope create small fault scarps along the trace. Small fault scarps are characteristic of the coated alluvial fan surfaces just northwest of the Totana locality (Murcia) that are associated with the Lorca-Alhama Fault (Silva, 1994; Baena *et al.*, 1994).

Marine terraces, beaches and **coastlines** can also serve as planar markers. Marine terraces are formed by sea level variations over long periods of time. Changes in sea level or eustatic changes were significant during the Pleistocene. They reached a 150 m low about 18,000 years ago (Fairbanks, 1989). The constant attack of waves along the coastline creates erosional marine terraces or abrasion platforms that dip about one degree toward the sea. Sea level variation can be determined from the magnitude of tectonic uplift. Elevated marine terraces are used to identify uplift parallel to the coast and spatial variations of crustal movement (*Fig. 3.4*). Constructed marine terraces can form when conditions are favorable for the growth of corals or coralline algae. Due to rapid erosion and accelerated coral growth, terraces can form in several thousand years. On the other hand, sea level has not changed significantly over the last few thousand years and as a result, abrasion platforms and coral terraces can be found along numerous coasts. Spanish coasts, particularly the Almería coast, the Strait of Gibraltar, and the Baleares provide magnificent examples of uplifted marine terraces that are segmented by normal faults (Zazo *et al.*, 1999b).

In some **lakes** such as Lake Bonneville in Utah, the precursor to the Great Salt Lake, numerous erosional platforms up to several hundred meters wide formed during the Pleistocene. These platforms have been used as markers to analyze the isostatic rebound that followed the contraction of the lake (Gilbert, 1890). Marine and lacustrine **deltas** can be used as markers because they provide clear evidence of ancient water levels and they are larger than marine or lacustrine terraces.

Erosional surfaces that formed during long periods of crustal stability can also be used as markers. A structural contour map shows their elevations as domes and broad convex structures as ductile deformation of the erosional surfaces and faults of greater or lesser extent, half grabens and grabens as fragile deformations (*Fig. 3.5*) (Simón, 1983). Many investigators in the *rift valleys* of East Africa use extensive erosional surfaces as markers for analyzing deformation of the rift valley during a period of arching. Likewise, we can use exhumed surfaces or surfaces formed at an angular unconformity as smaller-scale markers.

Figure 3.4 Marine platforms created by continuous uplift since the beginning of the Quaternary. Southern Peru. Photo C. Zazo.

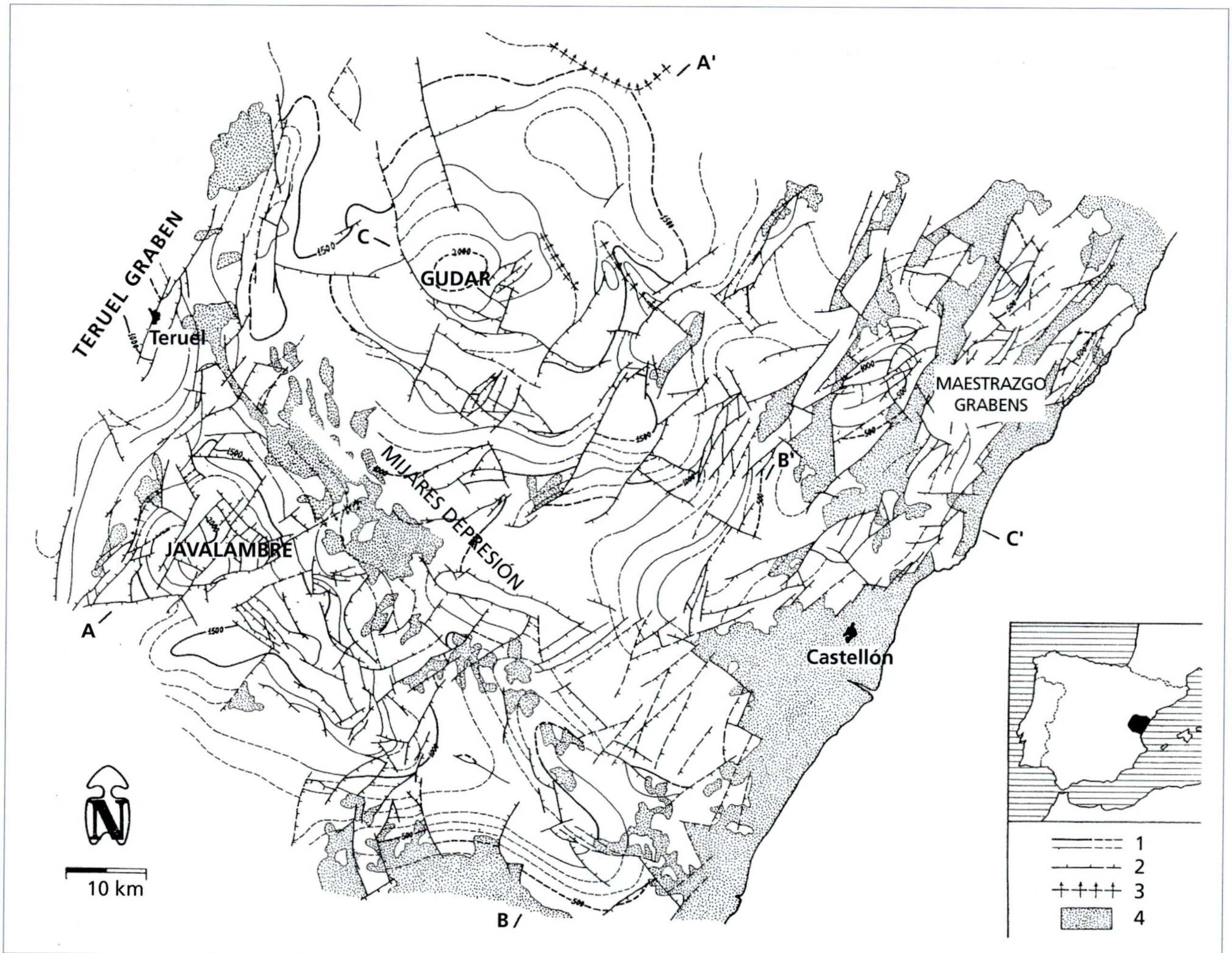

Figure 3.5 Map of structural contours of the fundamental deformed pediplain (Miocene-Pliocene) in the Central-Eastern Iberian Cordillera. 1) Isohipses of the pediplain, 100 m equidistant. 2) Faults presumed to be active during Plio-Quaternary extension. 3) Flexures. 4) Upper Pliocene-Quaternary (Simon, 1983).

3.2.2 Linear geomorphic markers

Planar markers generally occupy large areas. Linear geomorphic markers, although smaller in size, and those created by man can also be used to determine deformation (Burbank and Anderson, 2001).

Riverbeds can be displaced by strike-slip faults and lateral displacements can be measured. Strike-slip movements can decapitate fluvial systems and channels may end abruptly at a fault. Movement along the San Andres Fault has caused numerous channels to flow into the fault. The Carboneras Fault (Almeria, Spain) and the Lorca-Alhama Fault (Murcia) are excellent examples of displaced riverbeds (Goy and Zazo, 1984; Silva, 1994; Baena *et al.*, 1994).

Glacial ridges made up of glacial drift material, and **ridges of rock glaciers** can be used as linear markers. Lateral displacement generated by strike-slip faults can be easily measured.

Likewise, **debris flow levees** located along channel margins can be used as linear markers. Stake lines, railroad tracks, paths, painted lines on streets and highways, and even vehicle tracks documenting co-seismic displacements that occurred during recent earthquakes can also be used (Burbank and Anderson, 2001).

To use this list of features as markers, a geomorphologist must understand the nature of the geomorphic markers used and, to the extent possible, know the age of the markers because deformation rate calculations depend on the age of the displaced markers.

3.3 Dating methods

Dating is perhaps the most important task in estimating tectonic activity but it is often the most difficult part of a tectonic study. Without a chronology it is impossible to calculate the rate of tectonic processes (Keller and Pinter, 1996). Until the advent of radioactive dating methods, researchers used relative dating methods. Today numerous methods of dating are available for the study of tectonic activity. These methods can provide absolute ages or relative dates. An example of a relative date is the following: Surface B is older than C, and A is the most recent. Absolute dating methods have errors inherent to each method. *Table 3.1* shows a list of the absolute dating methods most widely used in tectonic geomorphology and the temporal ranges for which they can be used. Numerous books are now available to explain the various dating methods (Mahaney, 1984; Pierce, 1986; Rutter and Catto, 1995; Noller *et al.*, 2000; Burbank and Anderson, 2001; Banner, 2004).

3.3.1 Relative dating methods

One of the relative dating methods is based on the **seismic velocity of clasts**, which varies as a function of fissure density. The number of cracks in a clast controls the propagation rate of seismic waves. This is a semi-quantitative method (Crook, 1986). A similar technique is based on the theory that spheroidally weathered rocks can be reliably compared to less altered rock. Seismic velocity can be measured in weathering profiles that contain spheroidally weathered rocks and less altered rocks. This technique has been used to produce recent chronologies in glacial deposits. Another relative dating technique involves the measurement of **hydration rings in obsidian**, a glassy mineral that hydrates at outcrops. Hydration rings, measured in microns, grow at a rate that depends on the composition of the obsidian, the surface temperature, and ring growth rate. This technique needs to be calibrated to surfaces of known age in order to provide quantification.

Table 3.1 **ABSOLUTE DATING METHODS (BURBANK AND ANDERSON, 2001)**

Method	Useful interval	Materials
Radioisotopic		
^{14}C	35 Ka	Wood, carbonate shells
U/Th	10–350 Ka	(corals and speleothems)
Thermoluminescence (TL)	30–300 Ka	Quartz-silt
Optically stimulated luminescence (OSL)	0–300 Ka	Quartz-silt
Cosmogenic		
In situ ^{10}Be, ^{26}Al	3–4 Ma	Quartz
He, Ne	Unlimited	Olivine, quartz
^{36}Cl	0–4 Ma	
Chemical		
Tephrochronology	0-several Ma	Volcanic ash
Amino acid racemisation	0–300 Ka; the interval depends on temperature	Carbonate shells
Paleomagnetic		
Identification of reversals	0–700 Ka	Fine sediments, volcanic flows
Secular variations	0–700 Ka	Fine sediments
Biological		
Dendrochronology	10 Ka, depending upon existence of a local master chronology	Wood

The **soil development** method is a qualitative technique based on carbonate and clay accumulations that thicken over time (Birkeland, 1984). In arid regions, high temperatures and the loss of CO_2 in solution causes carbonates to precipitate and accumulate in soils. The total quantity of carbonate in the soil, along with coated clasts present at the base of the clastics, can be used to determine the relative ages of surfaces. The terrace sequence in the Río Cinca (Spain) includes a thick layer of carbonates in older layers that is not present in younger layers (Sancho and Meléndez, 1992).

Lichenometry is a method commonly used in geomorphic investigations to determine the ages of moraines and fallen blocks. The method uses the diameters of certain lichens present in a deposit to obtain an approximate age of the deposit (Innes, 1985; Chueca, 1994). The most commonly used lichen is the genus *Rhizocarpon* whose diameter is thought to increase linearly over time. To obtain more precise ages, local calibration is needed because lichen growth rate is a function of climate. Some investigators think lithology and the lichen's orientation to the sun also affect the growth rate. Like all techniques this method has limitations and uncertainties. Researchers think it is possible to produce detailed chronologies for geomorphic surfaces that are younger than 500 years old.

3.2.2 Absolute dating methods

Table 3.1 shows the most common methods of absolute dating. **Dendrochronology** is a method that uses measurements of the thickness of growth rings of some species of trees. Tree rings grow rapidly in wet seasons and summers, creating wide layers of wood separated by thin layers that record dry seasons and winters. Dendrochronology is based on a standard series of growth rings obtained from a certain set of trees (Stuiver, 1970; Shroder, 1980). Dendrochronology studies have provided the basis for the calibration of the ^{14}C method and have also contributed to studies of rockslides, volcanic flows, and earthquakes (Burbank and Anderson, 2001).

Varve chronology is a dating method based on the number of varves in a sedimentary deposit. A varve is a pair of sedimentary layers deposited in a proglacial lake. A pair consists of a light-colored band of fine sand and a dark-colored band of mud. The banding is produced by melt water that feeds the lakes. Rapid sedimentation that occurs in the springtime produces the light-colored layer and slow flocculation that occurs later in the year produces the dark band. Each varve corresponds to a year. Varve chronology was first used along the margins of the Scandinavian and Laurentide ice sheets.

The **radiocarbon** method is the most commonly used dating technique. It is also the least expensive and the most reliable. Standard radiocarbon analysis can determine dates up to 35,000 years in the past. Acceleration mass spectroscopy (AMS) is a variation of ^{14}C analysis that increases the temporal range of the dating to 50,000 years, produces fewer errors, and requires only a small amount of sample (3 mg of charcoal). The method is based on the measurement of the relative abundance of ^{14}C compared to the stable carbon isotope (^{12}C). To determine absolute ages, ^{14}C dates must be calibrated (Stuiver and Kra, 1986; Blackwell *et al.*, 2006).

The **uranium series** is the basis of a technique that uses the disintegration of uranium radioisotopes to provide a numerical date. Two disintegration series are used: ^{238}U and ^{235}U. Both transition series end with a lead isotope. This method can date material that ranges from 10^4 to 10^6 years old. It can be used to date inorganic and biogenic carbonates; however, the dating of shells can be complicated by later enrichment in carbonate.

Luminescence dating is used for deposits that exceed the maximum age that can be measured by the ^{14}C method. Luminescence, used for sedimentary environments, is based on an estimate of the time that a sediment was exposed to sunlight and a measurement of the radiation received. A related method that uses luminescence produced in a laboratory by the addition of heat is called **thermoluminescense** (TL). A method that uses added light is called **optically stimulated luminescense** (OSL). Because OSL has many of the same advantages as TL, it is preferable to use OSL (Aitken, 1998).

The **cosmogenic** method is a new technique that allows scientists to date the surfaces of rock outcrops and alluvial deposits that have been continuously exposed to cosmic radiation since they formed. This method is evolving rapidly. It measures ^{10}Be, ^{26}Al, ^{36}Cl and other isotopes that were formed when cosmic rays interacted with the Earth's surface. Since these isotopes accumulate in measurable quantities on soil surfaces, alluvial deposits, and rocky surfaces, their concentration is a potential measure of the minimum exposure time of the soil or rock (Bierman, 1994).

Tephrochronology, the study of tephra or pyroclastics, allows scientists to correlate the stratigraphy of previously dated layers of tephra over long distances (Westgate and Naeser, 1998). Although a variety of techniques can be used for dating, the most popular is the **$^{40}K/^{40}Ar$** method, which is based on the disintegration of potassium to argon. This method can be used to date rocks up to 10^5 years old.

The **amino acid racemization** method is based on proteins present in the organic material of skeletons and shells. These proteins are composed of amino acids that over time undergo chemical reactions. After an organism dies, physical and chemical processes break down the amino acids and the degree of alteration can be used as a clock to provide an absolute age of the organism (Kaufman *et al.*, 1992). The advantages of this method include its low cost and the fact that it requires only a small amount (2 mg) of sample.

3.4 Tectonic geomorphology techniques

Tectonic geomorphologists use two methods. They use geomorphic techniques to determine the paleoseismic characteristics of **landforms** (King, 1966; Goudie, 1981a) and use this information to prepare paleoseismic landform maps. The second method involves the use of the **stratigraphic record**; geomorphologists use stratigraphic techniques to map paleoseismic deformation and geophysical methods to find buried faults. They also use trenches to obtain data on paleoearthquakes. To understand the paleoseismology on an area, both regional and local scales need to be considered (McCalpin, 1996a).

Field investigations of active tectonics should begin at a regional scale (thousands of km^2), proceed to a local scale (a few km^2), and end at a site scale (from 1 hectare to several m^2). Many countries provide regional geological maps than can be used as the basis for initial studies. Local and site scale investigations should be done for the specific area indicated.

3.4.1 Cartography of paleoseismic regions

The first step is a survey of surface geology maps to identify areas of recent deformation. A photogeologic study can aid considerably in the preparation of these maps. The process should include photogrammetric measurements with stereoscopic pairs. Satellite images showing faults and lineaments can be very useful (*Fig. 3.5*) (Arlegui and Simón, 2000).

Paleoseismic features can be mapped as structures that deform geological materials or as deformed landforms (McCalpin, 1996a). It is important to conduct a detailed survey of erosional and depositional landforms with emphasis on those that have undergone deformation. Some investigators have proposed legends for morphotectonic maps (Goy *et al.*, 1988, 1991; Baena *et al.*, 1993). As McCalpin (1996a) states, these authors have prepared an excellent system of cartographic units for neotectonic studies. *Figure 3.7* shows a portion of a proposed legend listing 87 geomorphic symbols related to neotectonics.

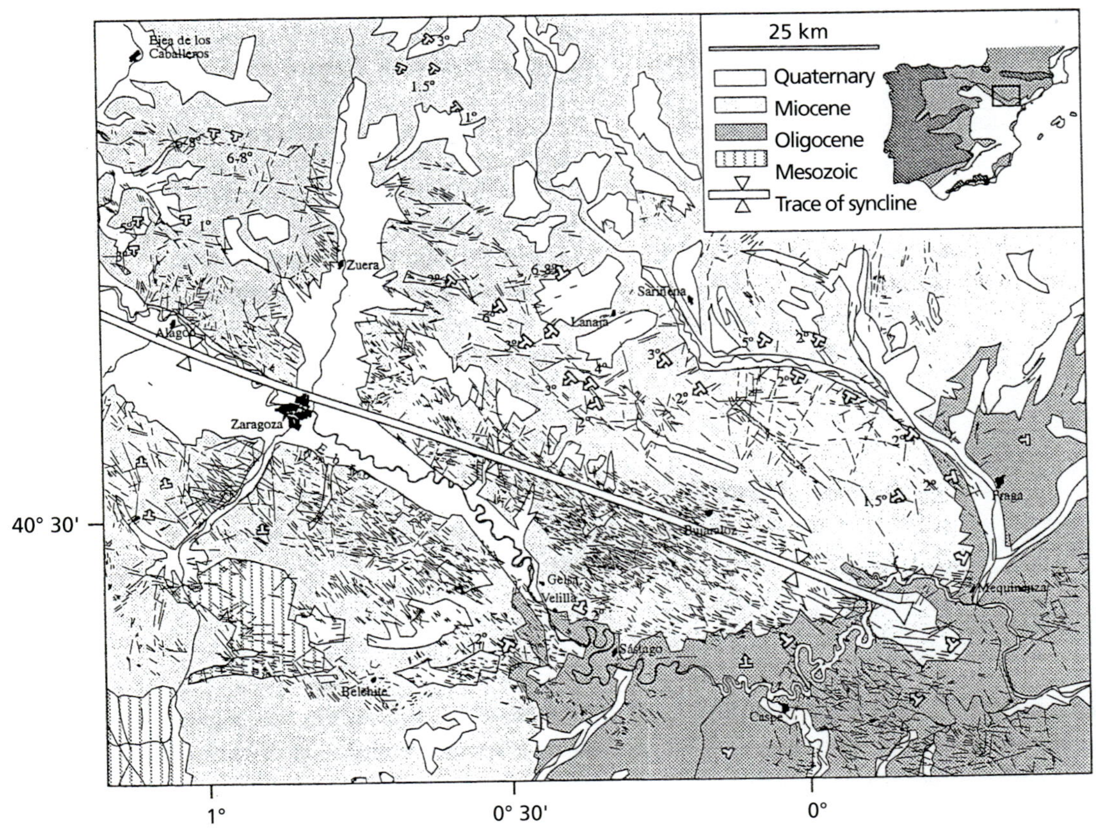

Figure 3.6 Geological-structural diagram of the central sector of the Ebro Basin. The lineaments were mapped from Landsat images and field studies (Arlegui and Simon, 2000).

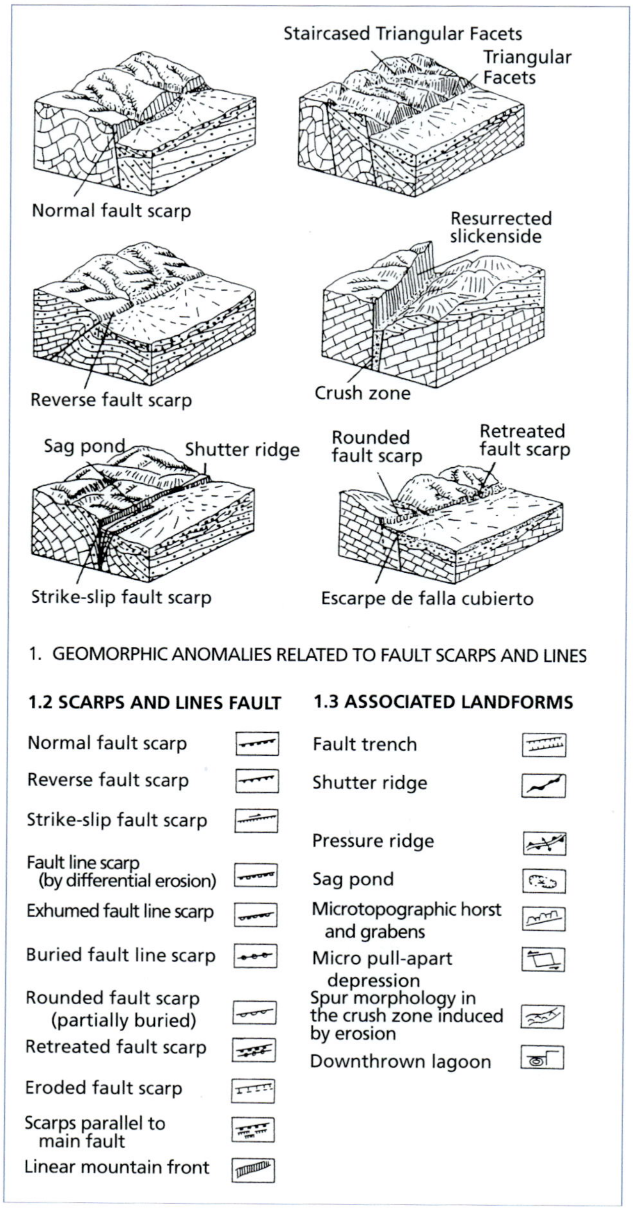

Figure 3.7 Block diagrams of neotectonic deformation and geomorphic mapping symbols. *Top*: Block diagrams that show scarps and fault lines. *Below*: Geomorphic mapping symbols related to scarps and fault lines (Goy *et al.*, 1988, 1991).

Topographic profiles made with detailed maps at a scale of 1:1000 are a simple method to determine fault throws. For this purpose, these profiles should be made perpendicular to the fault scarp. Topographic profiles provide the vertical displacement of dip-slip faults and they can sometimes also provide important data on the age of the fault (Buckman and Anderson, 1979). Terraces created by fluvial and coastal erosion are geomorphic reference levels that can be used for longitudinal and perpendicular topographic profiles to find deformation-related anomalies.

3.4.2 Cartography of paleoseismic stratigraphy

We can use stratigraphy to recognize paleoseismicity in displacement and angular unconformities. We can use secondary evidence such as clastic dikes and soft-sediment deformation. Shallow borings can be done in many locations and deeper borings may penetrate cemented gravels. Paleoseismology is primarily used to correlate calculations of net or accumulated displacement in a fault zone.

A fundamental technique for the study of deformed areas is the excavation of trenches (*Figs. 3.8* and *3.9*). Trench studies (*Fig. 3.10*) provide valuable data on displacement and the recurrence of paleoearthquakes. The trenches should be dug perpendicular to the direction of the fault for faults with a vertical movement and parallel to the fault direction for strike-slip faults. When movement is both strike-slip and dip-slip, several trenches are needed to determine the three-dimensional displacement components. These trenches are used primarily to study stratigraphy and the deformation resulting from paleoearthquake activity. Backhoes are used to dig trenches and operators take care to ensure the stability of the walls to prevent collapses, which have occurred during investigations (McCalpin, 1996b). These studies include analysis of soil type and development, which are indicators of the time necessary for soil formation. In addition, it is essential to find layers with datable material (carbon, organic remains, quartz and feldspar sands, etc.) to date the seismic events that may be recorded in the stratigraphy of the trench walls.

3.4.3 Geophysical techniques in paleoseismology

Geophysical techniques are used to study subsurface geology in fault zones and to detect buried faults that have no morphological expression. The methods used must have a resolution of less than 10 m and must reflect distinct discontinuities present in the subsurface. **Reflection and refraction seismic methods** can detect faults and show the arrangement and thickness of strata in the subsurface but they do not provide data on the age and magnitude of paleoearthquakes. **Ground penetrating radar** (GPR) provides images of the subsurface similar to images produced by reflection seismic. The similarity is due to reflections of transmitted waves received at the surface even though the frequency of reflection seismic is 100 Hz and that of radar is 80–300 MHz. GPR investigations were used to analyze the fault scarp at

Figure 3.8 Trench for the study of recent deformation. Nahal Shehoset (Southern Arava). Western margin of the Dead Sea. Negev Desert Israel.

Figure 3.9 Trench in a compressional area. Ragged Mountain (Alaska). Photo F. Gutiérrez.

Figure 3.10 Plastic syn-sedimentary deformation in a trench in the Ragged Mountains (Alaska). Photo F. Gutiérrez.

Ventas de Zafarraya (Granada); this scarp was activated in 1884 during the most recent large earthquake on the Iberian Peninsula, which occurred at Arenas del Rey with an intensity of XMSK (Reicherter *et al.*, 2003). **Aeromagnetic** surveys can be used to locate areas of intense fault deformation. **Gravimetric methods** can express differences in gravitational attraction on two sides of a fault that has vertical throw. Neither method is commonly used in paleoseismology. Recently **geoelectric exploration** has become popular, especially electrical tomography, a method that produces 2-D pseudo-sections of the fault environment. A combination of seismic, electrical, and aeromagnetic methods was used to analyze a fault scarp in Sencelles (Mallorca) related to the 1851 Palma de Mallorca earthquake (Silva *et al.*, 2001).

3.4.4 Archeological techniques

Archeology provides methods for dating geological materials. Ceramic fragments within material that has moved, such as a slope deposit, can provide a relative age ("*post quem*"). The basis of this dating method is the fact that a slope deposit is later than the age of the youngest ceramic contained in the deposit. Geoarcheology includes the study of geophysics, geochemistry, and Quaternary studies, including geomorphic aspects (Beach *et al.*, 2008).

A three-dimensional stratigraphic study with an age obtained by archeological methods and thermoluminescence has made it possible to create 4-D (four dimensional) models (Brown, 2008).

Ruins of a known shape can supply maximum and minimum ages and also information on the geometry of deformation (Vita-Finzi, 1986; Azañón *et al.*, 2004; Justo *et al.*, 2008). By using several types of ceramics for dating, archeologists can reach a precision of several decades. Part of the Great Wall of China, constructed 400 years ago, has been displaced by an oblique-slip fault with 1.45 m of left-lateral movement and 0.95 m of vertical displacement (Deng *et al.*, 1984). In Spain the only archeological data related to seismic deformation comes from the Roman ruins of Baelo Claudia (Bolonia, Cádiz), which is thought to have been affected by two earthquakes between 40–60 AD and 350–395 AD (Silva *et al.*, 2005 Fig. 3.11).

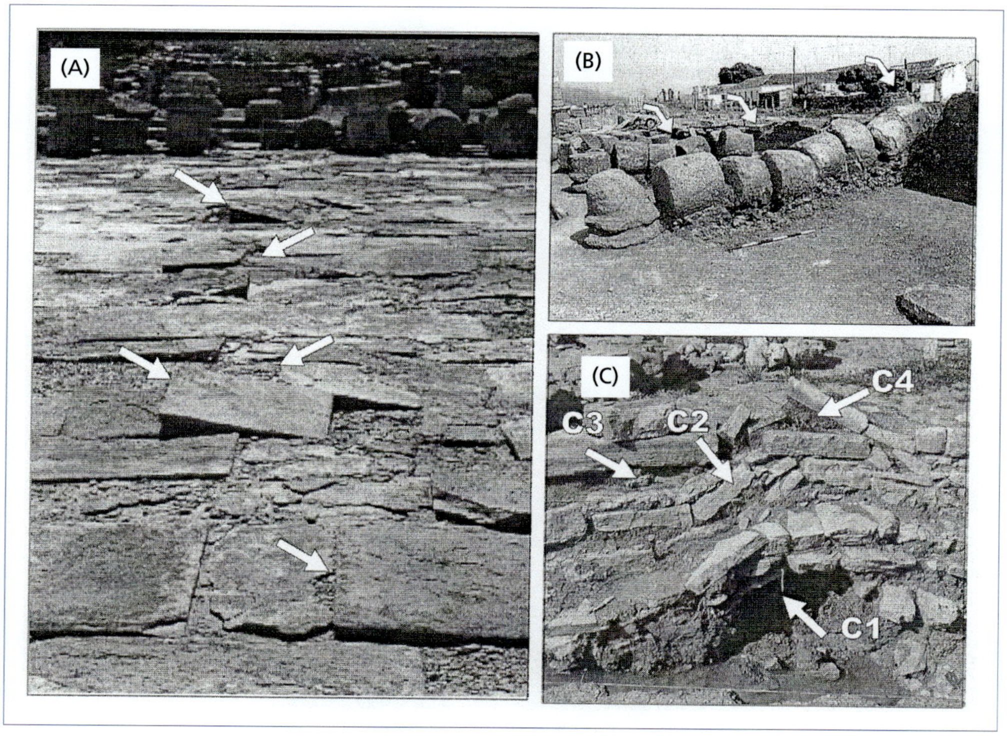

Figure 3.11 (A) Deformed flagstones in the *Forum* of *Baelo Claudia*. The arrows indicate the uplifted flagstones. (B) Columns collapsed to the south of the Basilica (from Sillières, 1997). (C) Remains of the northwestern corner of the *Forum* folded and pushed upward. C1: Folded Roman pavement 40–60 AD; C2: Foundation stone of a collapsed house 40–60 AD; C3: Horizon demolished 40–60 AD; C4: Vertical crack under a house affected by a reverse fault, 350–395 AD (Silva *et al.*, 2005).

3.4.5 Geodetic techniques

Geodesy is the science of the Earth's shape, including its gravimetric shape and its surface changes. Due to plate movements and gravity, the Earth's surface is in constant motion on a regional scale. Geodesy requires the exact position of geographical points. Geodetic surveys repeated before and after an earthquake can provide current information on the geodetic history of deformation.

For most people, the Earth's relief is relatively stable in the long term. However, the precision of some geodetic measurements is so precise that it is possible to detect long-term deformation, especially in tectonically active regions. Geodetic measurements require a network of **control points**.

The tectonic applications of geodesy can be categorized as short term and long term methods. **Near field techniques** focus on deformation on a scale of meters to several tens of kilometers whereas **far field methods** are used for regional and global deformation (Burbank and Anderson, 2001). (For example, a geodetic survey between London and New York can detect the extension rate of the Atlantic Ocean).

Traditional **triangulation networks** are based on the configuration of a triangular system located along a particular fault or region. The points are generally several kilometers apart although they can be as much as 50 km apart *(Fig. 3.12)* (Jordan and Minster, 1988). Repeated measurements of orientations and longitudes between stations show small fault movements. Triangulation is primarily used to obtain the horizontal displacement rate and **precision leveling** can determine small amounts of vertical displacement, for example, less than one mm per km *(Fig. 3.13)* (Yoshikawa, 1970). In addition, near-field techniques make it possible to measure horizontal displacements along strike-slip faults using the **alienation group** technique. This method uses lines of needles hammered into the ground perpendicular to the fault direction and spaced at least 100 m apart. Since the position of the needles is known, it is easy to measure later displacement. This technique also indicates aseismic segments.

The use of **marigraphs** from one of 517 stations around the world (Emery and Autrey, 1991) can indicate changes in sea level height. Glacial melting and the expansion and contraction of the oceanic water column causes sea level to vary on the scale of a decade. At the present, sea level appears to be rising approximately 1.0 mm/year (Mörner, 2004). In addition, changes in salinity, ocean temperature, atmospheric pressure, and ocean currents cause a marked variation in sea level. An oceanographic correction is made to the marigraph record of each place. The resulting data are interpreted as relative sea level changes due to vertical movements of the land. Increases in post-seismic deformation may continue several years after a principal earthquake. For whatever reason, it is difficult to predict the effects of global eustatic changes (Pirazzoli, 2005). The continuous tidal record may be useful in the study of deformation between seismic events even though the system of elevation and subsidence can be interpreted in several ways.

3.4.6 Other techniques

To measure the movement of lithospheric plates, we need techniques that cover larger areas, distances of hundred or thousands of kilometers. **Very Long Baseline Interferometry** (VLBI) is a technique developed in the 1960s that has evolved into the study of real time tectonics (Ryan and Ma, 1998). It measures points on the Earth's surface at a precision of less than 1 cm by observing quasars, the most energetic and distant particles in the universe. Radiotelescopes around the world can track the positions of known quasars and control the signals they emit. This technique uses two radiotelescopes located on contiguous lithospheric plates. The VLBI stations use data from the geometry and retardation time of light rays emitted billions of years ago by these extragalactic objects to provide data on the type and rate of deformation of the Earth's surface (Molnar and Gilbson, 1994).

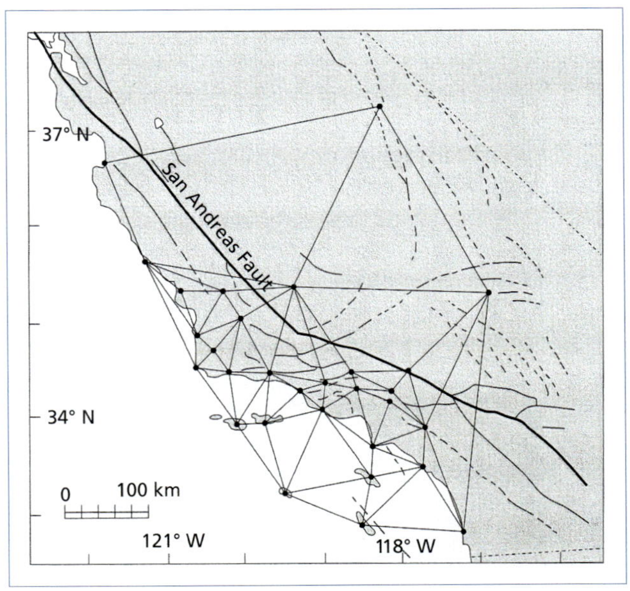

Figure 3.12 Geodetic network for the study of deformation west of the San Andreas Fault along the California Coast Ranges and the continental borderland. Since movement between the Pacific and North American plates is not limited to the San Andreas Fault, but rather is distributed along numerous faults, this dense network was designed to determine exactly where the movement takes place (Jordan and Minster, 1988).

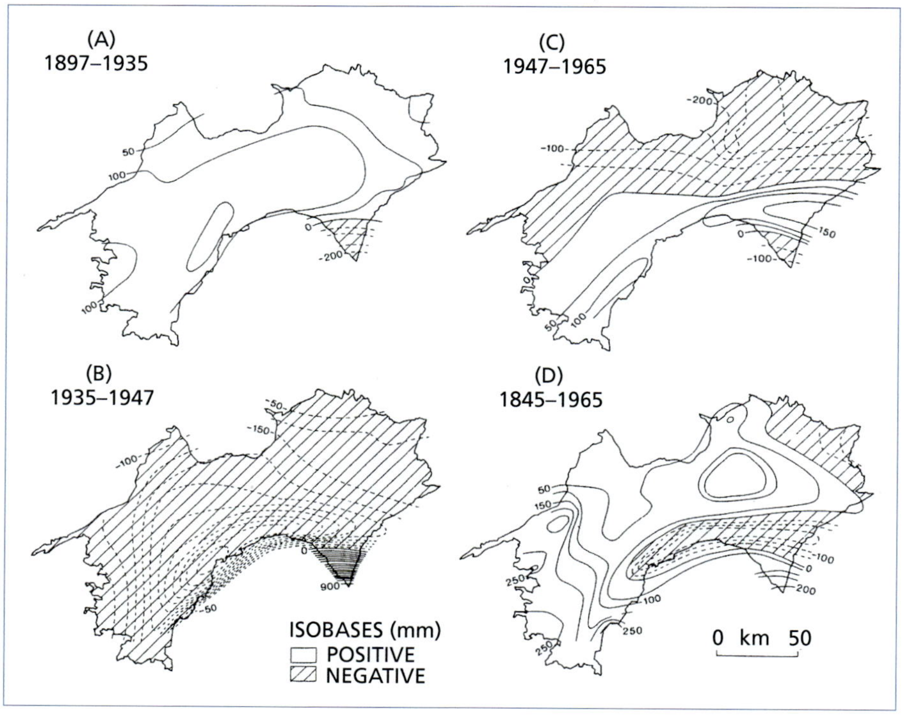

Figure 3.13 Vertical movements of Shikoku Island (Japan) during three time periods (A-C) and for all of period D (1845–1965). The values are based partially on a precise levelling measured in 50 mm isobases. Positive and negative changes in vertical movement can be observed. The effect of the 1946 earthquake is clearly shown (Yoshikawa, 1970).

A recent discovery that is even more promising is the **Global Positioning System (GPS).** The GPS technique can obtain the most accurate location possible in time by using its constellation of 24 satellites that circle the Earth at a height of 20,000 km and send signals to the Earth's surface (Leick, 1990). This technology can calculate the location of a point on the Earth's surface with an accuracy of less than 1 cm. The receivers are portable and the most compact models are about 25 cm in length. Their cost has come down significantly, leading to increasing scientific, commercial, and home use of GPS. Geologists in Southern California have used GPS to analyze deformation along the plate boundary, thus improving the VLBI data (*Fig. 3.14*) (Feigl *et al.*, 1993). The number of permanent GPS stations has increased significantly in tectonically active regions of the world in recent years.

Another new geodetic technique for measuring displacement of the ground over large areas is called **radar interferometry**. **Synthetic aperture radar** (SAR) conducted by satellites at a height of 785 km transmits signals along westward trajectories that form a 23° angle with the vertical. Scientists use the return signal to calculate and record the distance from the satellite to Earth. The resulting representation of Earth movement is called a SAR interferograph. The big advantage of this method is the large area it covers although its resolution is less than GPS. A SAR image covers a surface 60 km by 60 km, more than 300,000 *pixels* that measure an area 100 by 100 m (Massonnet *et al.*, 1994).

In conclusion, over the last two decades measurement techniques have ushered in a new era in high-precision geodetics. Today it is possible to detect small velocity differences between nearby crustal blocks and even inside individual blocks and to better define the relative movement of lithospheric plates via GPS and VLBI observations. GPS networks provide records of seismic deformation almost in real time and radar interferometry provides a detailed view of co-seismic and inter-seismic deformation on a regional scale (Burbank and Anderson, 2001).

3.5 Deformation of landforms

To understand the active tectonics of a region a geomorphologist also needs a precise understanding of the various erosional and depositional morphologies that have formed in this region. Geomorphology provides primary data on the presence of recent deformation. It is difficult to conceive of

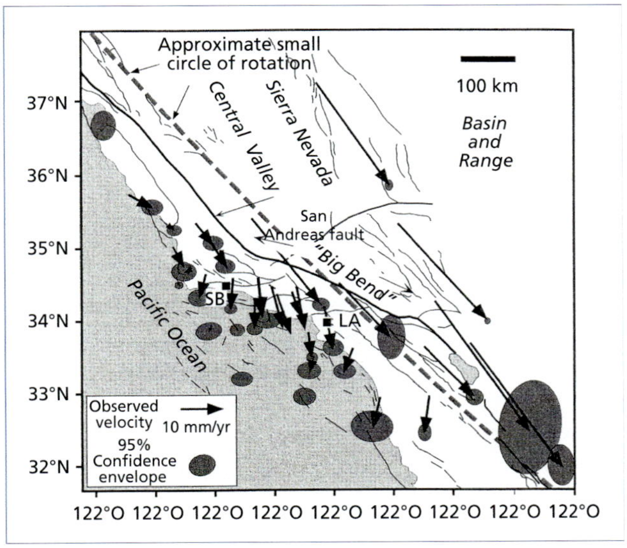

Figure 3.14 Deformation rates in Southern California. The largest displacement vectors east of the San Andreas Fault are parallel to the fault. To the west of Santa Barbara (SB) and Los Angeles (LA) there are significant displacements that are at angles of 30–90° to the direction of the San Andreas fault and that accommodate folds and reverse faults. This variation in directions also indicates the differential rotation of crustal blocks in the region. All this shows that the plate boundary is not concentrated along the San Andreas Fault, but rather is represented by a diffuse zone of deformation that is several hundred kilometers wide (Feigl *et al.*, 1993).

an adequate paleoseismological study that does not incorporate geomorphic principles and foundations.

3.5.1 Geomorphic indices of tectonic activity

Geomorphologists use quantitative measurements to objectively compare different landforms and calculate various parameters (**geomorphic indices**) that can be used to identify the level of tectonic activity. These indices can be used to study a large area rapidly by incorporating data from topographic maps, aerial photographs, and digital terrain models (Keller and Pinter, 1996). These indices can be used to assign **tectonic activity classes** (Bull, 1978; Silva *et al.*, 2003; El-Hamdouni *et al.*, 2008).

One of the indices is **drainage basin asymmetry,** which describes network geometry both quantitatively and quantitatively. A network formed in a region with active tectonic deformation has a different configuration and geometry. Drainage basin asymmetry has been used to detect tilting

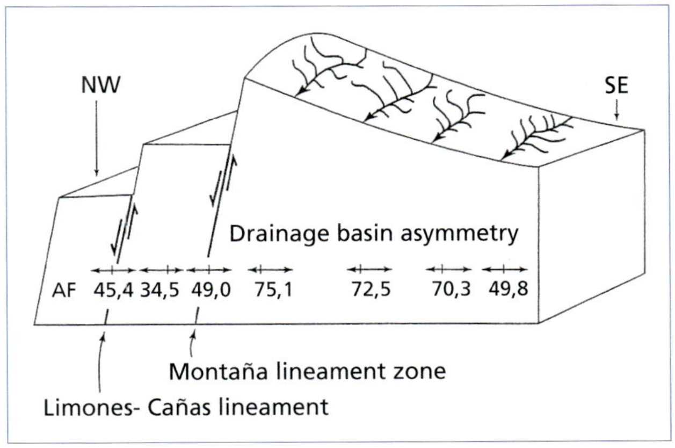

Figure 3.15 Asymmetry factors for the Nicoya Peninsula. They indicate that the alignment of the Montaña tilts to the southeast (Hare and Gardner, 1985).

in large drainage basins (*Fig. 3.15*) (Hare and Gardner, 1985). The **Asymmetry Factor** (*AF*) is defined as:

$$AF = 100\ (A_r/A_t)$$

In which A_r is the area of the basin on the right of the principal branch (looking downstream) and A_t is the total area of the drainage basin. A value of 50 indicates stability whereas deviations from 50 suggest tilting. Another proposed index for basin asymmetry is the **Transverse Topographical Symmetry Factor** (*T*) (Cox, 1994).

$$T = D_a/D_d$$

Here D_a is the distance from the midline of the drainage basin to the midline of the belt of active meanders and D_d is the distance from the midline of the basin to the watershed boundary of the basin. In a symmetrical basin $T = 0$ and asymmetry increases as T approaches 1. The T values are calculated for different valleys. The *FA* method can quickly indicate tilting.

The **Thalweg Gradient Index** (*SL*) was proposed by Hack (1973) and is defined as:

$$SL = (\Delta H/\Delta L)L$$

Here, $\Delta H/\Delta L$ is the slope of the channel or the gradient of the river section; L is the length of the channel from its midpoint to its highest point. The *SL* index is related to the energy of water flow and is sensitive to changes in gradient; therefore it can be used estimate its relationship to tectonic activity and net changes in the longitudinal profile of the river although the latter may also be related to significant lithological contrasts. High index values indicate longitudinal profiles with steep slopes; this may indicate that they have not yet reached an equilibrium profile, possibly as a result of active tectonics.

The Mountain Front Sinuosity Index (Bull, 1977a, 1978) has the following mathematical expression:

$$S_{mf} = L_{mf}/L_s$$

In this equation, L_{mf} is the length of the mountain front at the slope break along the foot of the mountain and L_s is the length of the front measured as a straight line (Fig. 3.16). This method examines fronts associated with tectonic activity and recent uplift, which are relatively straight and have low S_{mf} values. If uplift decreases or ceases, processes of fluvial erosion produce a sinuous mountain front with S_{mf} values that are much higher than 1. Detailed topographic maps and aerial photographs are used to calculate this parameter (Bull and McFadden, 1977a).

Another geomorphic index of tectonic activity proposed by Bull (1977a, 1978) is the **Valley Width/Height Ratio**, which can be expressed as:

$$V_f = 2V_{fw}/[(E_{ld} - E_{sc}) + (E_{rd} - E_{sc})]$$

Where V_f is the width/height ratio of the valley, V_{fw} is the width of the valley bottom, E_{ld} and E_{rd} are the watershed boundary elevations on the left and right margins, respectively, and E_{sc} is the elevation of the valley bottom (Fig. 3.17). These parameters should be calculated at a certain distance from the mountain front for each valley. Low values of V_f indicate deep valleys with rivers that are actively entrenching, which are generally associated with uplift.

Each of these indices provides a relative classification of tectonic activity that is useful in reconnaissance work. Applying an index to a specific region provides more relevant results. In fact, a geomorphologist can develop a **classification of relative tectonic activity** like that of Bull (1977a, 1978, 2007, 2009) for the southwestern United States or that of Silva *et al.*, (2003) for the southeastern part of the Iberian Peninsula. Areas are normally classified as very active, moderately active, and inactive. This classification is a good tool for selecting areas for future field study.

S_{mf} and V_f indices can be assigned to various mountain fronts to classify their relative tectonic activity, as defined by Bull (1977a, 1978), Bull and McFadden (1977), Rockwell *et al.*, (1984) and Silva *et al.*, (2003). These authors establish five classes of mountain fronts with different grades of tectonic activity. The principal characteristics and relative elevation rates of these classes are the following:

The most active mountain fronts (**Class 1**) have low S_{mf} values between 1.0 and 1.6; medium V_f values of less than 1 are associated with tectonic elevation rates between 5 and 1 m/Ka. **Class 2** is characterized by moderate tectonics, S_{mf} index values between 1.4 and 3.0, and V_f values of about 1; these are associated with elevation rates between

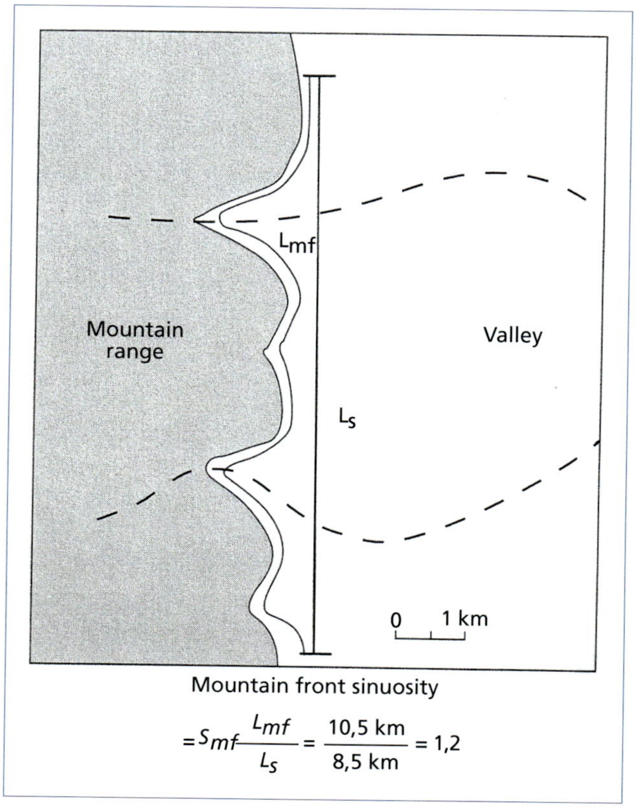

Figure 3.16 Diagram indicating the parameters used in the calculation of the Mountain Front Sinuosity Index and values obtained for the Argus Range (Keller and Pinter, 1996).

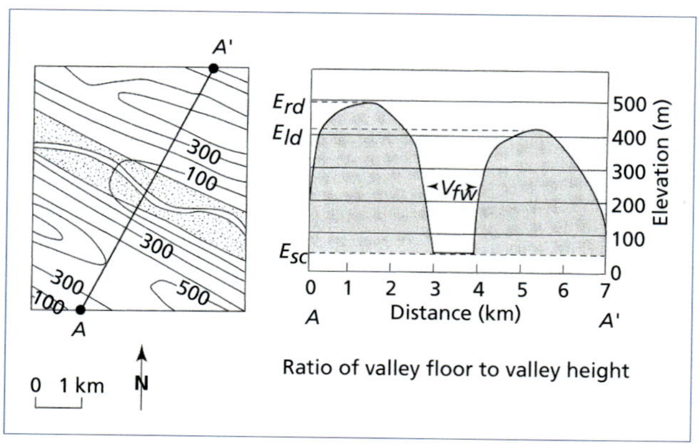

Figure 3.17 Idealized diagram showing the width/height ratio of the valley. The left and right are shown looking downstream (Keller and Pinter, 1996).

1.0 and 0.5 m/Ka. **Class 3** is characterized by weak tectonics, with S_{mf} values between 1.8 and 5.0 and average V_f values greater than 1. **Classes 4** and **5** are mountain fronts that are tectonically inactive and/or controlled by fluvial erosion. In these classes, S_{mf} values are greater than 3.0 and V_f values are about 2; uplift values are insignificant, generally less than 0.005 m/Ka.

This classification is based on a study of mountain fronts in the southwestern United States. As expected, the geodynamics of the Betic Cordilleras are different; their elevation values are less than those in the United States. This fact has led to a classification of mountain fronts in the eastern Betics, specifically in the Guadalentin tectonic depression (Silva *et al.*, 1992b), and production of detailed geomorphic maps (*Fig. 3.18*) with S_{mf} and V_f indices and tectonic activity classes for 16 mountain fronts (Silva, 1994; Silva *et al.*, 2003) (*Table 3.2*). The majority of these fronts have very low values of V_f (<1) and S_{mf} (<2), which puts them in Bull's (1978) Class 1 but there is no superposition of alluvial fans in the proximal zone. Active sedimentation is migrating toward the distal zone, which suggests these mountain fronts belong in Class 2 or 3 (Silva and Harvey, 1991; Silva *et al.*, 1992a and b; Silva *et al.*, 1993; Silva, 1994 and Silva *et al.*, 2003).

3.5.2 The fluvial system and tectonic geomorphology

This section deals with tectonic effects on rivers and their deposits; these are landforms that are very susceptible to change (Schumm, 1977). Tectonic effects can be easily recognized in consolidated rocks where the drainage system is a function of the differential resistance the rocks. The simplest examples are found in various types of drainage networks: rectangular and trellis networks, for example (Schumm *et al.*, 2000).

Geomorphologists use various parameters to describe river shapes and processes and the balance between them in a fluvial system tells us that rivers are very sensitive to any type of change (Keller and Pinter, 1996). Quaternary climate changes and their effects on landforms are very obvious (Bull, 1991; Gutiérrez, 2005). Anthropic activity has considerably affected fluvial aggradation and degradation processes (Goudie, 1981b; Wilkinson, 2005). In addition, as knowledge of tectonic geomorphology increases, geomorphologists can analyze the role of tectonic activity in fluvial system processes and landforms (Adams, 1980; Ouchi, 1985).

Drainage systems adapt to changes of the surface slope; therefore, they hold a valuable record of information

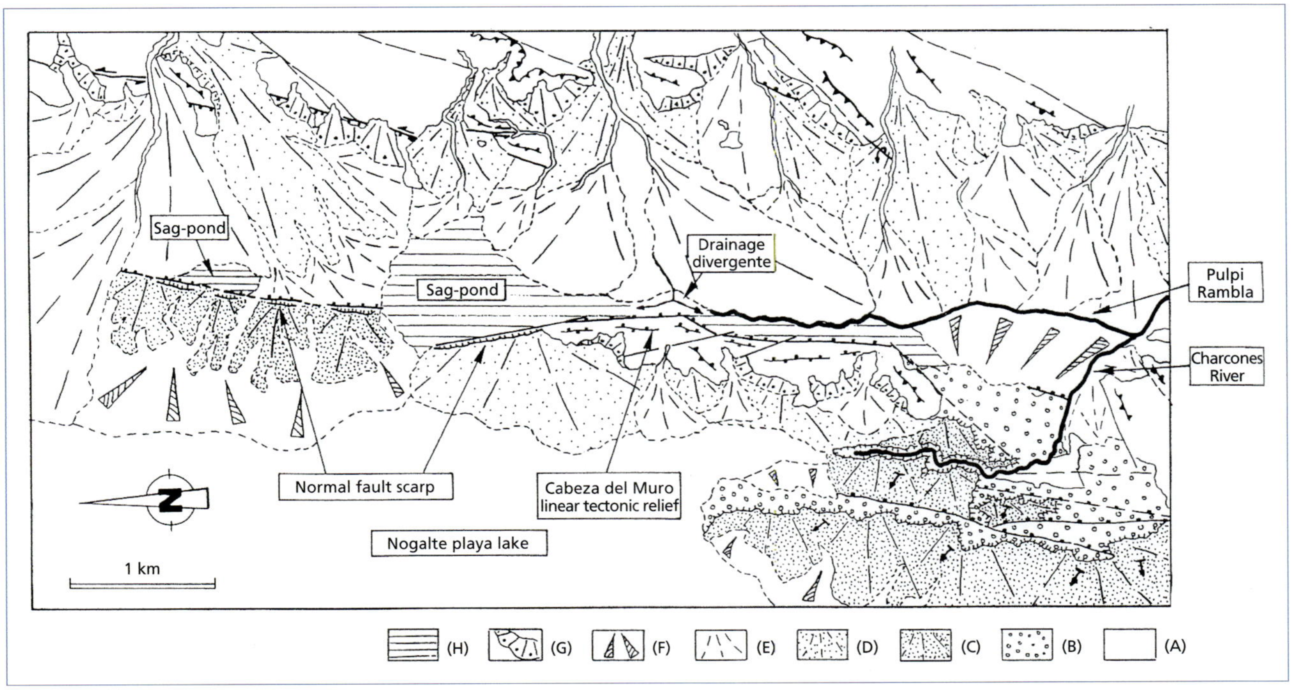

Figure 3.18 Morphotectonic diagram of the Palomares fault zone between Rambla del Aljibejo and the Cuesta of Escarihuela (the connection zone with the Pulpí corridor). The most significant tectonic landforms and geomorphic anomalies are shown. A: Betic substrate; B: Plio-Pleistocene conglomerates; C: 1st Depositional Phase; D: 2nd Depositional Phase; E and F: 3rd Depositional Phase, glacis-type border (F); G: Piedmont Deposits (Upper Pleistocene-Holocene), H: Endorheic zones (*playa-lakes* and *sag-ponds*) (Silva, 1994).

Table 3.2 PRINCIPAL GEOMORPHIC AND STRUCTURAL CHARACTERISTICS OF MOUNTAIN FRONT ACTIVITY CLASSES. GUADALENTÍN DEPRESSION (MURCIA) (SILVA *ET AL*., 2003)

Tectonic class	Principal morphometric and geomorphic features and seismic behavior	Associated uplift rates
Class 1: Active tectonics	(1) Linear mountain fronts: $S_{mf} < 1.5$. (2) V-shaped rambla valleys ($V_f < 0.6$) characterized by active incision. (3) *Overlapped* to *onlapped* fan bodies evolving by distal aggradation. (4) Steep and relatively reduced (in size) fan systems on piedmont areas. (5) Fan and inter-fan channels disconnected from axial fluvial systems. (6) Well-developed and preserved assemblage of tectonic landforms such as pressure ridges, benches, drainage offsets (strike-slip) and triangular facets (normal). (7) Isolated rupture surfaces (<2 km length) with associated vertical slip < 1.4 m. (8) Evidence of multiple paleoseismic events with minimum theoretical recurrence intervals of 1750 to 2600 years. (9) Isolated rare strong historical events (>VIII MSK), but common VIII-VII MSK events with recurrence periods of 100 to 250 years. Some of them entirely aseismic for the last 650 years. (10) Instrumental seismicity down to 4 mb, but exceptionally has reached 5 mb.	≥0.08 m/ka
Class 2: Moderate tectonics	(1) Irregular mountain fronts: S_{mf} values between 1.8 and 2.30. (2) U-shaped rambla valleys (V_f: 0.3–0.80) characterized by valley floor aggradation. (3) *Onlapped* fan bodies evolving by very limited distal aggradation or distal downcutting and headcut development. (4) Gentle slope and relatively extensive sedimentary areas, culminated by caliches in the proximal zones. (5) Fan channels partially connected to axial fluvial systems but the inter-fan channels totally connected, causing dissection of the fan bodies. (6) Degraded and/or buried tectonic forms. (7) No observed surface ruptures. (8) No available data on paleoseismic events, but expected. (9) Rare historic VII-VI MSK events, but common V MSK events randomly distributed through time. (10) Instrumental seismicity down to 3.0 mb.	0.07–0.03 m/ka
Class 3: Inactive	(1) High sinuosity mountain fronts ($S_{mf} \geq 2.8$) characterized by the development of engulfed areas and presence of isolated relief on the piedmont zone. (2) Wide valley floors ($V_{mf} \geq 0.08$) controlled mainly by typical fluvial dynamics and characterized by the occurrence of valley widening and the development of fluvial terraces. (3) Through-dissected fan bodies, which are mainly inherited landforms subjected to surface reworking and extensive caliche development. (4) Extensive, indurated, and highly dissected pediment areas. (5) Ancient fan and inter-fan channels totally connected to axial fluvial systems and integrated in the regional drainage network. (6) Highly degraded and/or buried tectonic landforms, mostly isolated. (7) No observed surface ruptures. (8) Record of rare paleoseismic events, with recurrence intervals of 25,000 to 50,000 years. (9) Exceptional historic V-IV MSK events, but commonly aseismic. (10) Commonly aseismic, but isolated events down to 2.5 mb are randomly recorded.	<0.03 m/ka

Described topics are listed in the following order: (1) S_{mf} index. (2) V_f index. (3) Alluvial fan development. (4) Extension and slope of pediment areas. (5) Characteristics of fan channels and their integration in the regional drainage network. (6) Occurrence and preservation of tectonic landforms. (7) Occurrence and record of surface ruptures. (8) Paleoseismic data. (9) Historical seismicity data. (10) Instrumental seismicity data.

on the evolution of faults and folds (Ollier, 1981). Earthquake-induced changes to fluvial systems that result directly from deformation that occurs during an earthquake can threaten people who live along the riverbanks. Slow deformation of the valley bottom can affect channel stability and pose risks to man-made structures (Schumm, 1986). Changes that occur gradually over thousands of years are more common than simultaneous deformation. These changes highlight areas of active deformation (Schumm, 1986).

Earthquakes and fault movements may cause large rivers that flow across floodplains to undergo **avulsion**, the abandonment of one channel for another. A classic example is the Indus River of Pakistan where in 1819 a large earthquake caused an uplift 6 meters high, 16 km wide, and 80 km long. The earthquake totally blocked the eastern margin of the Indus River and flow was not re-established in the channel until 1828 (Holmes, 1968).

A river may reach an **equilibrium stage** in which active factors cause equilibrium to move in a direction that tends to absorb the effect of change and to re-establish equilibrium (MacKin, 1948; Leopold and Maddock, 1953). This implies that the effects of deformation are propagated upstream and downstream. Numerous rivers have adjusted to tectonic activity (Melton, 1959).

A geomorphologist must know the magnitude of deformation to determine whether it is rapid enough to produce recent adjustment of the channel (Schumm *et al.*, 2000). Another type of deformation is non seismic, which can be described as progressive uplift and subsidence that operates over a specific period of time (*Fig. 3.19*) (United States Department of Commerce, 1972; in Schumm *et al.*, 2000).

3.5.2.1 Uplift, subsidence, and earthquakes in fluvial systems

Numerous cases have been studied, mainly in the United States, in which the fluvial system provides independent evidence to show that geodetic measurements are valid and deformation is real (Adams, 1980). The reconstruction of longitudinal profiles of rivers can be used to show repeated uplift over a long time period. These reconstructions involve erosional and depositional terraces. The primary geomorphic indications of tectonic activity may be landforms created by long-term continuous deformation, which modifies the drainage network and deforms the set of terraces. These characteristics contrast with more subtle river changes (Schumm *et al.*, 2000). Tilting can produce changes in an entire drainage network. In addition, continuous tilting can cause lateral erosion and the development of an asymmetric valley. The formation of large lakes in sedimentary basins may be partially due to tectonic activity. Lakes Titicaca and Uyuni, at elevations of about 4000 m in the Altiplano of

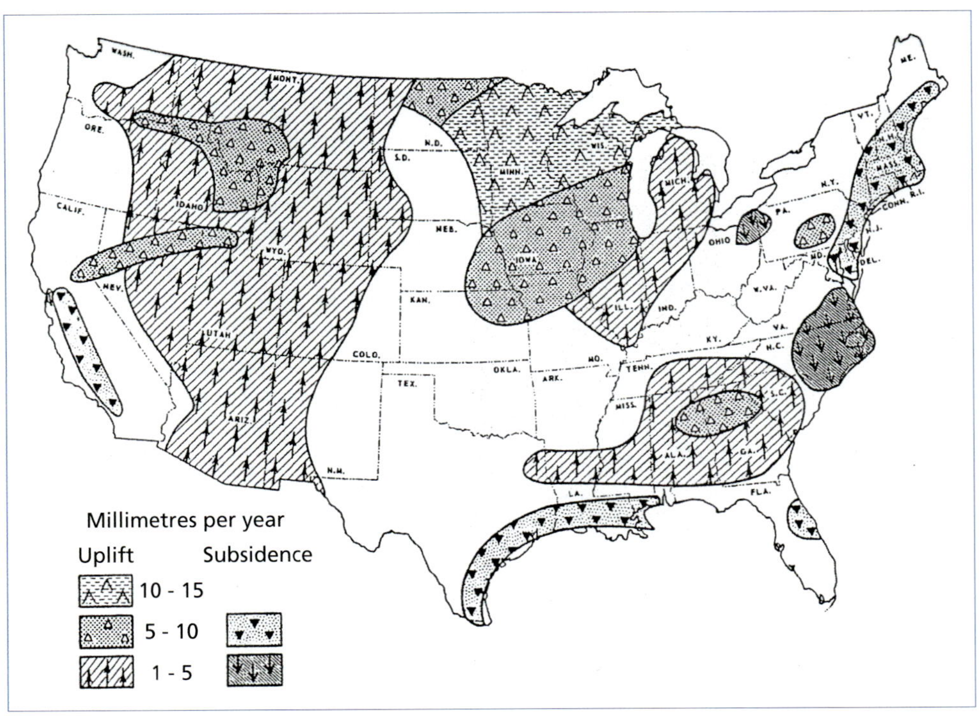

Figure 3.19 Vertical movements obtained by precise levelling by the National Geodetic Survey (United States Department of Commerce, 1972).

Bolivia, are associated with plate convergence and uplift (Schumm *et al.*, 2000).

Riverbeds along the Gulf Coast of the United States show that big rivers adjust to deformation more rapidly than small rivers, including uplift that affects flooding. Floods are much more frequent on the upstream side of uplifts than on their downstream side (*Fig. 3.20*).

The New Madrid earthquakes affected the Mississippi River in significant way and the effects persist 200 years later. Profiles of the maximum channel level (*bankfull stage*) in 1880 and 1915 show a transverse uplift (Lake County Uplift) that is shown by various convexities at the time of measurement. This convexity can be seen over about 40 miles (Russ, 1982) (*Fig. 3.21*). Profiles constructed downstream in 1880, 1976, and 1988 show that the 1880 profile is the lowest of all of them, which demonstrates that the river has eroded. The New Madrid convexity observed in the 1880 profile is not visible in the 1976 and 1988 profiles. The steepest part of the 1976 and 1988 profiles is located downstream of New Madrid. It shows a drop of 1.2 m over 3.2 km. This may reflect the exhumation of a fault scarp in the channel of the Mississippi River. The drop is located 8 km upstream of a fault where a waterfall formed after the earthquake of February 7, 1812 (Russ, 1982).

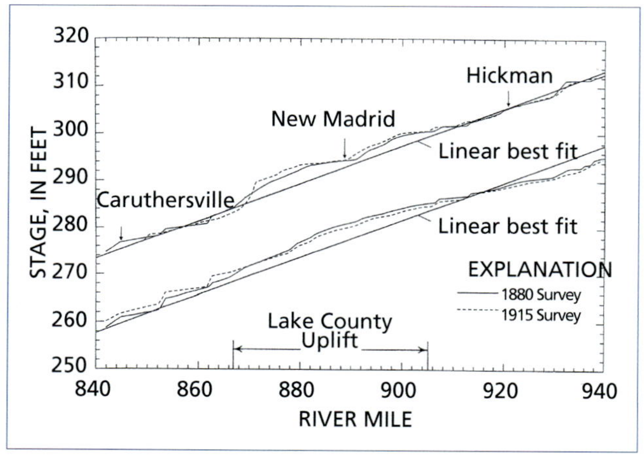

Figure 3.21 Profiles of the maximum channel level (top) and medium stages (below) of the Mississippi River and the location of uplift in Lake County (Russ, 1982, in Schumm *et al.*, 2000).

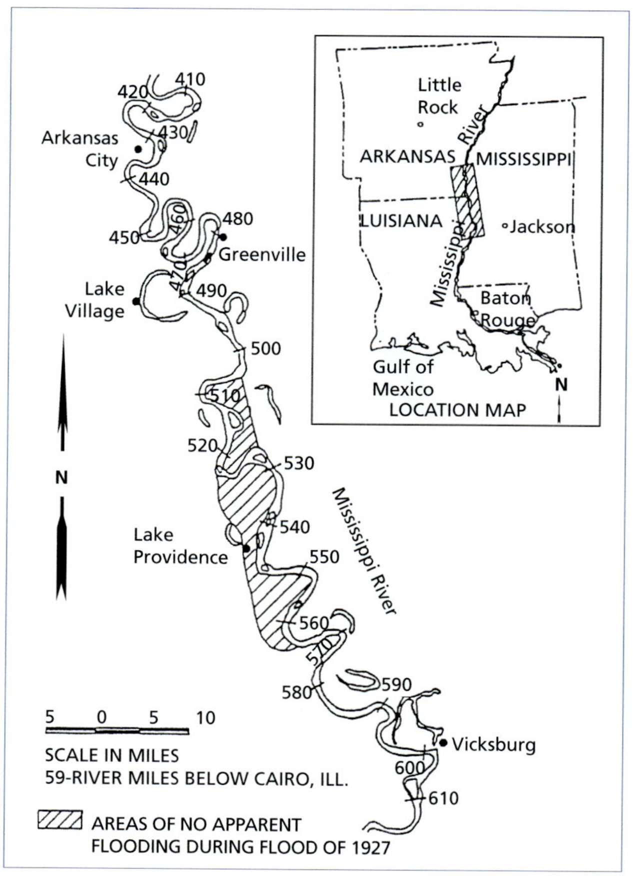

Figure 3.20 Map of the Mississippi River between Arkansas City and Vicksburg indicating areas that were not inundated during the 1927 flood (Schumm *et al.*, 2000).

3.5.2.2 Deformation in colluvial and fluvial deposits

In tectonically active zones, slope profiles can cut faults, producing a step (*Fig. 3.1*). This gradient change causes colluvium to fill the step, a process that continues until the scarp disappears. The **wedge conceptual model** is used in the study of colluvial deposits to interpret the history of a fault (Wallace, 1977). Japanese researchers (Okada *et al.*, 1989) have given the name "D structure" to a fault-and-colluvium unit because it resembles the letter D. After the faulting, detritus fills the step, creating a wedge shape (*Fig. 3.22*). The photograph shows ceramic fragments at the top of the wedge deposits. Because the ceramics postdate the middle Bronze and predate the Iberian epoch, they must have been produced between 1200 and 500 BC. We can conclude that significant tectonic activity occurred after the period of 1200 to 500 BC (Burillo *et al.*, 1985).

If tectonic activity is followed by a prolonged period of stability, the fault scarp disappears, the slope reaches equilibrium, and soil may develop on the scarp (Gutiérrez *et al.*, 1983, Gutiérrez and Peña, 1994a). The soil may in turn be disturbed by a new stage of tectonic activity. A new colluvial wedge may form. Multiple faulting events produce a series of colluvial wedges that stack up on the down-dropped block of the fault. These wedges, which may be separated by a soil

Figure 3.22 Wedge-shaped fault-colluvium contact. Aguatón (Province of Teruel). Central-eastern Iberian Cordillera.

layer, represent stages of sedimentation that occur after fault movement (*Fig. 3.23*) (McCalpin, 1996b). This model has been applied to numerous faults in the western United States to identify up to four paleoseismic events (Forman *et al.*, 1991). A **facies model** has been proposed for sedimentation at the base of a normal fault scarp (Nelson, 1992). This model establishes **detrital facies** made up of clasts and soil fragments, mostly from weathering processes on the slope above the fault plane. As the scarp fills, *slopewash* and *creep* occur on the slope and colluvium that is deposited is finer grained, has better stratification and sorting, and is richer in organic material. These deposits are called **wash facies**.

In areas of normal faults, angular unconformities are present and associated with tilting due to drag on the fault plane. In tilted areas, the dip of the strata may increase with increasing age and at increasing depth (McCalpin, 1996a).

Along the Concud normal fault, near the village of Concud and the city of Teruel (Spain), Quaternary sediments outcropping on the down-dropped block show evidence of several events of seismic movement simultaneous to faulting (Gutiérrez *et al.*, 2005) (*Fig. 3.24*). The oldest event is indicated by faults that displace the QU2 unit but not the overlying QU3 unit. Deformation simultaneous with faulting occurred between 62 and 71.6 Ka. It took several more recent events (events, 2, 3, and 4) to fill three fissures that extend into unit QU3. These fissure fillings are bounded by well-defined surfaces and show very clear differences of texture and color. The two oldest contain large blocks that fell from terrace deposits as rock falls. These events must have occurred after 62 Ka. If four large seismic events occurred over 71.7 Ka, the recurrence interval on this fault is less than 18 Ka. Obviously the meager chronological

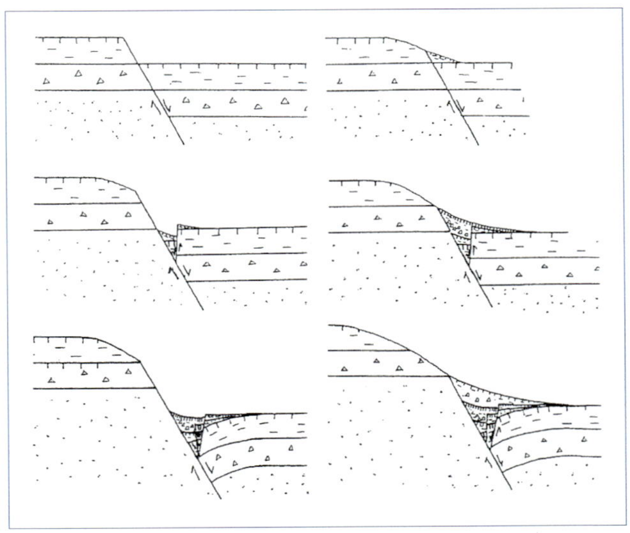

Figure 3.23 Diagrams indicating how colluvial wedges can vary as a result of successive fault displacements. The vertical marks indicate soils (McCalpin, 1996a).

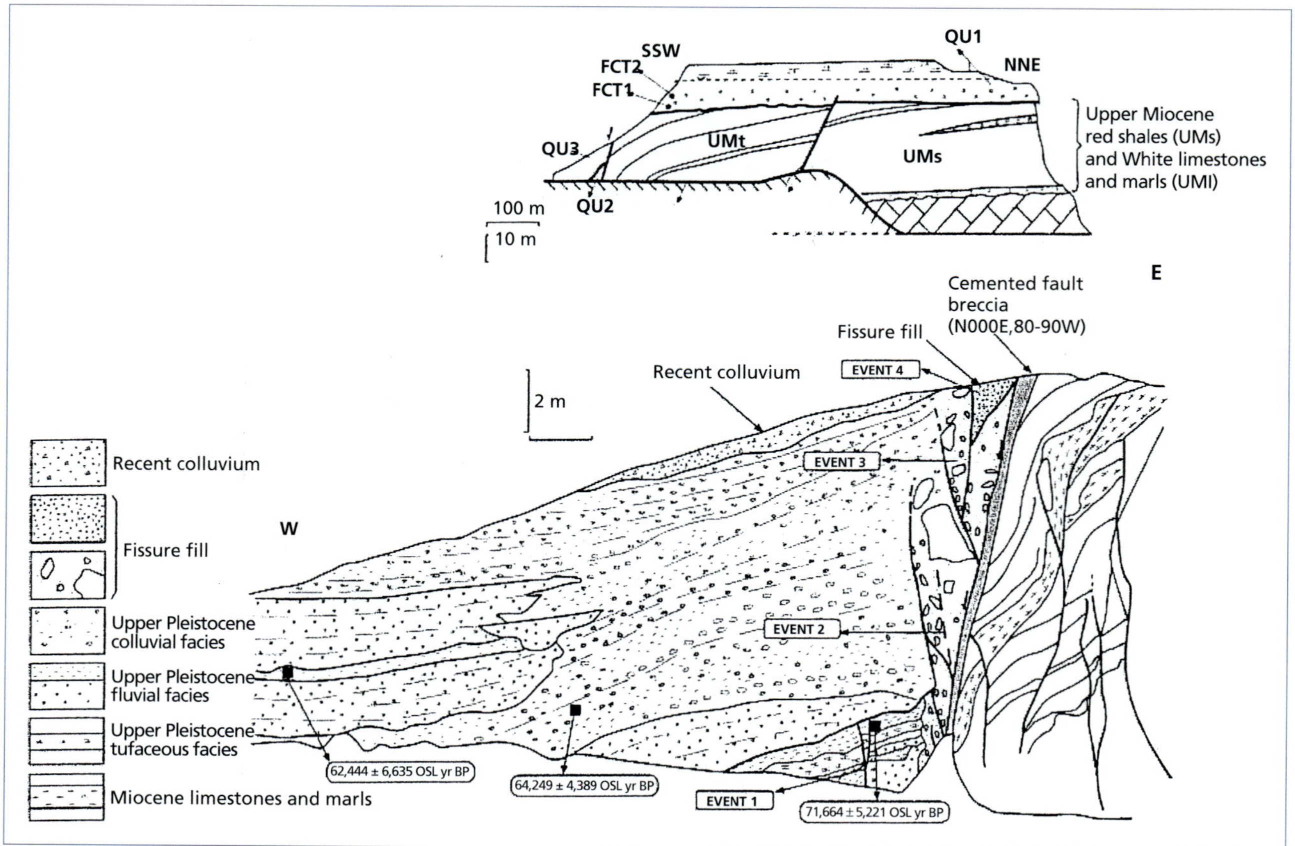

Figure 3.24 Cross sections of the Concud Fault in the area of the Estación de los Baños (Teruel) and paleoseismic interpretation of Pleistocene sediments (Gutierrez *et al.*, 2005).

information on the paleoseismic history of the Concud Fault is not sufficient for us to draw conclusions on the precise frequency of large earthquakes. Supposing that the fault that ruptured during the earthquake is about 23 km long, we can use the empirical ratio of Wells and Coppersmith (1994) to anticipate earthquakes with magnitudes around 6.6 along this structure.

In the eastern area of the Jiloca half graben, a set of faults indicates recent activity (Capote *et al.*, 1981, Gutiérrez *et al.*, 1983a and b; Gutiérrez *et al.*, 2005) (*Fig. 3.25*). In the Rubielos de Cérida region numerous carbonate outcrops are compartmentalized by N-S valley fill. These elongate depressions have been interpreted as tectonic valleys (*Fig. 3.26*) (Gutiérrez *et al.*, 1983a) generated by Quaternary activity along a dominant system of normal synthetic faults and accessory normal antithetical faults (Capote *et al.*, 1981). The latter are oblique to the principal NW-SE faults (Palomera Fault, *Fig. 2.39*), which define the eastern margin of the Jiloca half graben. These narrow grabens are less than 4 km long and 100 to 800 m wide. At the margin of tectonic valleys, the profile of the rocky slope generally shows a small scarp at the contact with the valley fill, usually not more than a meter high.

A quarry located about 2 km northwest of Rubielos de Cérida (Spain) was excavated in 1980 (*Fig. 3.1*), exposing the location of a fault plane with carbonate-cemented fault breccia. Slickensides, striations, and grooves with relatively long wavelengths are present at the surface. The fault has displaced 50 cm of calcareous regolith present on the slope of Jurassic material and on top of the colluvium. It is not a fresh scarp but the slope changes its gradient from about 22° to 30°. The colluvial sequence is cemented and dips 10° practically parallel to the topographic surface. This unit of stratified detrital material is interpreted as *sheetwash* from the Jurassic slope. There are no wedges or other features that would indicate episodes of movement simultaneous with seismic activity.

At this site, fragments of charcoal were collected from 2.1 m and 7.9 m below the surface. ^{14}C dating has shown that the most recent sample is 43,070 ± 1200 BP and the oldest is > 48,500 BP, beyond the scope of the method (Gutiérrez *et al.*, 2005). Using the age of the first

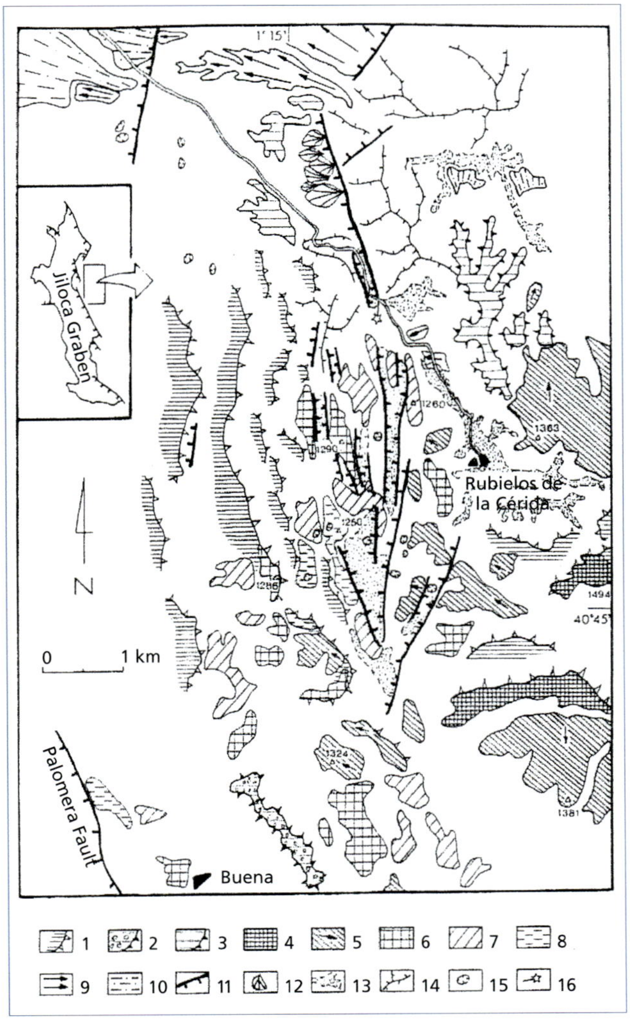

Figure 3.25 Geomorphic map of tectonic valleys in the Rubielos de la Cérida, located on the northeastern margin of the Jiloca semi-graben. 1: Mesozoic limestone cuesta fronts. 2: Paleogene conglomerate mesas. 3: Miocene sand and gravel mesas. 4: Miocene planation surface. 5: Fundamental surface of the Iberian cordillera, 6, 7 and 8: Upper, middle, and lower erosional surfaces. 9 and 10: Levels of lower and upper glacis. 11: Fault scarp. 12: Alluvial fan. 13: Valley floor. 14: Watershed divide. 15: Doline (Gutiérrez *et al.*. 1983a).

sample, we obtain a vertical displacement velocity of 0.05 to 0.07 mm/year. This number suggests slow movement along the fault and leads us to believe that earthquakes of magnitudes greater than 5.6 are not to be expected (Wells and Coppersmith, 1994; Anderson *et al.*, 1996; Villamayor and Berryman, 1999).

Until now we have analyzed vertical movement but in a river significant lateral movements of a channel can give rise to **asymmetrical valleys**. This asymmetry may be related to tectonic influences.

Active tilting and movement toward the fault can affect riverbeds in half grabens *(Fig. 3.27)* (Leeder and Gawthorpe, 1987). Leeder and Alexander (1987) state that the floodplain of the Madison River contains oxbow remains that have a concave orientation toward the present position of the river *(Fig. 3.28)*. According to dating that has been done, the total amount of river migration since 7000 years BP is 1600 m, indicating an average displacement of 23 cm/year. According to these authors, the drift is due to tilting; this agrees with the concavity toward the northeast in the abandoned meanders.

The lower reach of the Indus River in Pakistan is made up of numerous channel systems and wide floodplains over which ancient riverbeds flowed. Rivers change from braided to anastomosing or meandering when they cross various active structures *(Fig. 3.29)*. The river has avulsed numerous times across the floodplain in recent times. The zones where avulsion is most frequent are near significant tectonic landforms (Jorgensen *et al.*, 1993).

Fluvial terraces are the morphologies most often used to detect and measure tectonic activity during the Pleistocene. These terraces are flat and thus they easily preserve fault scarps. A fault that cuts a terrace is obviously more recent than the terrace *(Fig. 3.30)*; on the contrary, a fault that does not cut the terrace postdates the terrace. When doming occurs *(Fig. 3.31)*, sedimentation occurs in low-lying areas and rivers downcut in elevated areas. Sedimentation and entrenching are processes that keep the profile in equilibrium. The morphology of deformed terraces shows the nature of the deformation. For this reason, longitudinal profiles of terraces have been used to detect deformation (Bullard and Lettis, 1993). Deformed lake terraces can also indicate isostatic and tectonic activity. The most basic example is the deformation of the Lake Bonneville terraces in Utah (Crittenden, 1963).

Mountain fronts in arid and semiarid zones show the development of **alluvial fans**. High values of S_{mf} and V_f for the mountain front uplift rate indicate relatively low uplift rates for mountain front activity (Keller and Pinter, 1996). The situation is more complex when the alluvial fan is tilted. If the fan tilts farther than the threshold, the fan is confined and a new fan segment forms farther from the mountain front *(Fig. 3.32)*. A study of alluvial fans in Death Valley shows that tilting produces segmented fans (Hooke, 1972).

3.5.2.3 Deformation in littoral and lacustrine environments

Littoral landforms are used to study tectonic deformation because sea level is a reference level that can be used to

Figure 3.26 Tectonic valley in which the upper Quaternary fill is in normal fault contact with Jurassic formations. The step that can be observed along most of the fault trace corresponds to a fault throw of less than a meter. The Rubielos de la Cérida region (Province of Teruel), northeastern margin of the Jiloca semi-graben. Central-eastern Iberian Cordillera.

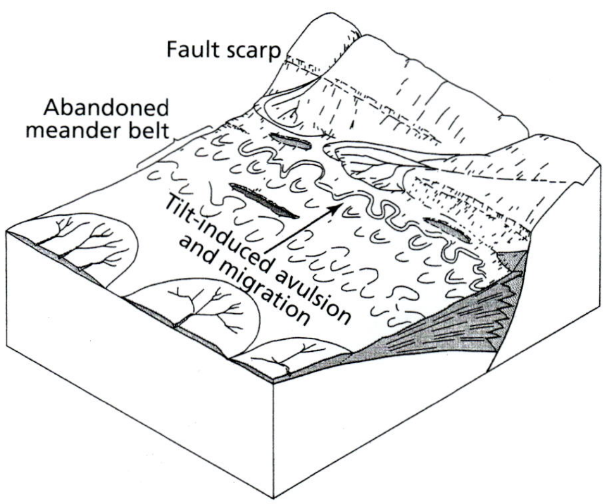

Figure 3.27 Geomorphology of a half graben. The active channel has been displaced by tilting toward the scarp of the active fault (Leeder and Gawthorpe, 1987).

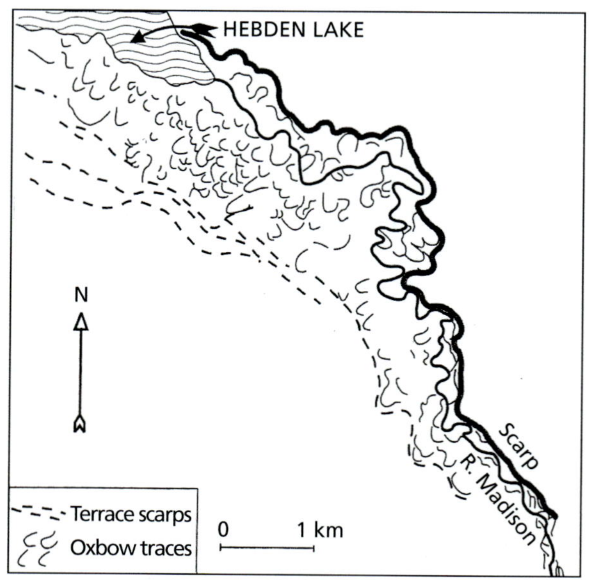

Figure 3.28 Belt of asymmetrical meanders and lateral drift of the Madison River south of Lake Hebden (modified by Schumm *et al.*, (2000) from the figure of Leeder and Alexander, 1987).

determine the age of these landforms (Keller and Pinter, 1996; Bull, 2007; Nelson. 2007). On a global scale, sea level has remained essentially constant for about 6000 years; this makes it possible calculate displacement with respect to sea level.

Along coasts of where immersion is limited by **coral reefs**, sea level rises and reefs are eroded by waves (Chappell, 1974). The shape of the resulting platforms depends on the reef geometry, its orientation to the waves, and the time between successive uplift events. The amount of uplift during successive seismic events can be easily obtained from the

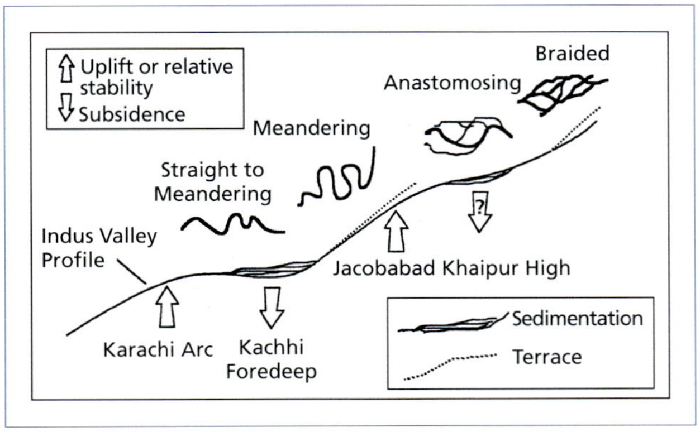

Figure 3.29 Diagram showing the relationship between tectonics and the profile of the Indus Valley and channel types. Changes in slope are exaggerated in the diagram (Jorgensen *et al.*, 1993).

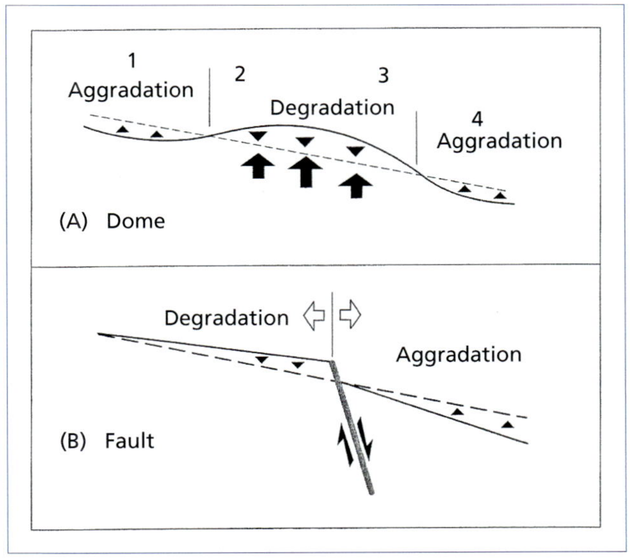

Figure 3.31 Degradation and aggradation zones associated with doming and faulting (Schumm *et al.*, 2000).

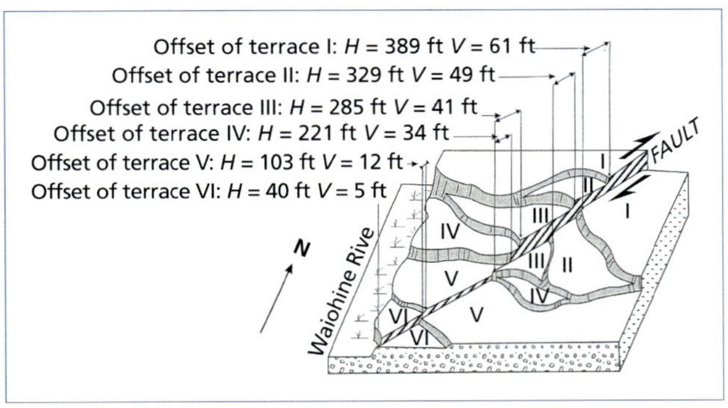

Figure 3.30 Block diagram of the displaced fluvial terraces at the mouth of the Waiohine Gorge, New Zealand. (I – Oldest terrace; VI – Youngest terrace) (H – number of feet of horizontal movement on the fault, V = number of feet of horizontal movement). Length of the fault in the diagram is 0.8 Km (Stevens, 1974).

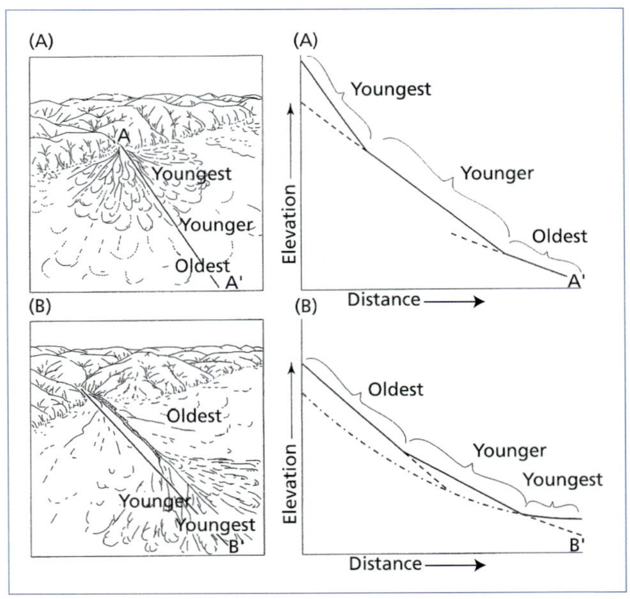

Figure 3.32 Morphology of alluvial fans. (A) The segments of the fan are located near the mountain front and are associated with active uplift. (B) The youngest segments are more distant from the mountain front and are more related to erosion of the mountain block than to uplift (Bull, 1977b).

height of benches and notches (*Figs. 3.33* and *3.34*) compared to analogous features formed along the present day high tide line. The presence of corals colonizing the benches and notches provides a rapid method of dating (Burbank and Anderson, 2001).

Erosional coasts have an **abrasion platform** that has a flat surface and a cliff on the landward side. Coastal uplift creates a hanging abrasion platform or plateau (called rasa in Spanish) (Hernandez-Pacheco, 1950). The platform surface may have detrital sediments and fossils that can be used for age dating. Successive pulses of uplift create a stairway of platforms, such as the well-developed stairway on the Spanish Cantabrian coast, Almería, the Huon Peninsula (New Guinea), and in southern Peru (*Fig. 3.4*).

Faults frequently cut **marine terraces**. When this happens, the fault movement postdates terrace formation. This morphology can be correlated over long distances. Variations in terrace height indicate modifications of the uplift

Figure 3.33 Beach probably uplifted by the 27 March 1964 earthquake. Valdez Fjord (Alaska). Photo F. Gutiérrez.

Figure 3.34 Uplifted notch, possibly due to the 27 March 1964 earthquake. Valdez Fjord (Alaska). Photo F. Gutiérrez.

rate. Along the Atlantic coast of the United States, one single level terrace ranged in height from less than 50 m to about 100 m, reflecting postglacial rebound that has migrated northward over the last 18,000 years (Winker and Howard, 1977). Along the Estrecho coast of Gibraltar from Cádiz to Málaga in Spain, the altimetric distribution of marine terraces of isotopic stages 5e and 5c shows symmetrical doming of the coastline with a maximum of +19 m in Tarifa and minimum of

+1 m in the Cádiz and Málaga areas (Zazo *et al.*, 1999b). In addition, these same terraces have different elevations at the intersection of important faults and the coastline, with up to 6 m of vertical displacement, indicating that they have been active during the last 95,000 years.

A preliminary study was recently conducted in preparation for the construction of the underwater Gibraltar Tunnel to unite Europe and Africa (Silva *et al.*, 2006). The criterion for evaluating fault activity has been seismicity, surface faults, geomorphology of fault traces, identification of paleoseismic fractures, and evaluation of uplift rates on coastal segments. The study included detailed cartography (*Fig. 3.35*) that shows that the NE-SW Cabo de Gracia strike-slip fault has been active during at least the last 128 Ka and that it probably caused moderate events affecting the Roman city of Baelo Claudia (*Fig. 3.11*).

Deformation studies also analyze the position of lake coastlines after isostatic rebound, such as studies done at Lake Bonneville (Crittenden, 1963) and the Lahontan lakes (Hanks and Wallace, 1985). In tectonically active zones such as the Dead Sea, Holocene fan-delta deposits appear to be highly deformed (*Fig. 3.36*) (Galit *et al.*, 1995) and synthetic and antithetic fault systems can be observed.

3.6 Seismic risk

Earthquakes occur most commonly along the margins of lithospheric plates (*Fig. 2.4*) in both extensional zones and convergent areas. Because convergent areas generate more energy than divergent areas, the most catastrophic earthquakes are located along intercontinental collision zones such as the Eurasian region and in areas where continental and oceanic lithospheric plates are converging, such as the Circum Pacific belt.

Figure 3.37 shows the factors that constitute **seismic risk** (Panizza, 1991; Castaldini and Panizza, 2005). **Seismic hazard** is associated with earthquake characteristics (magnitude,

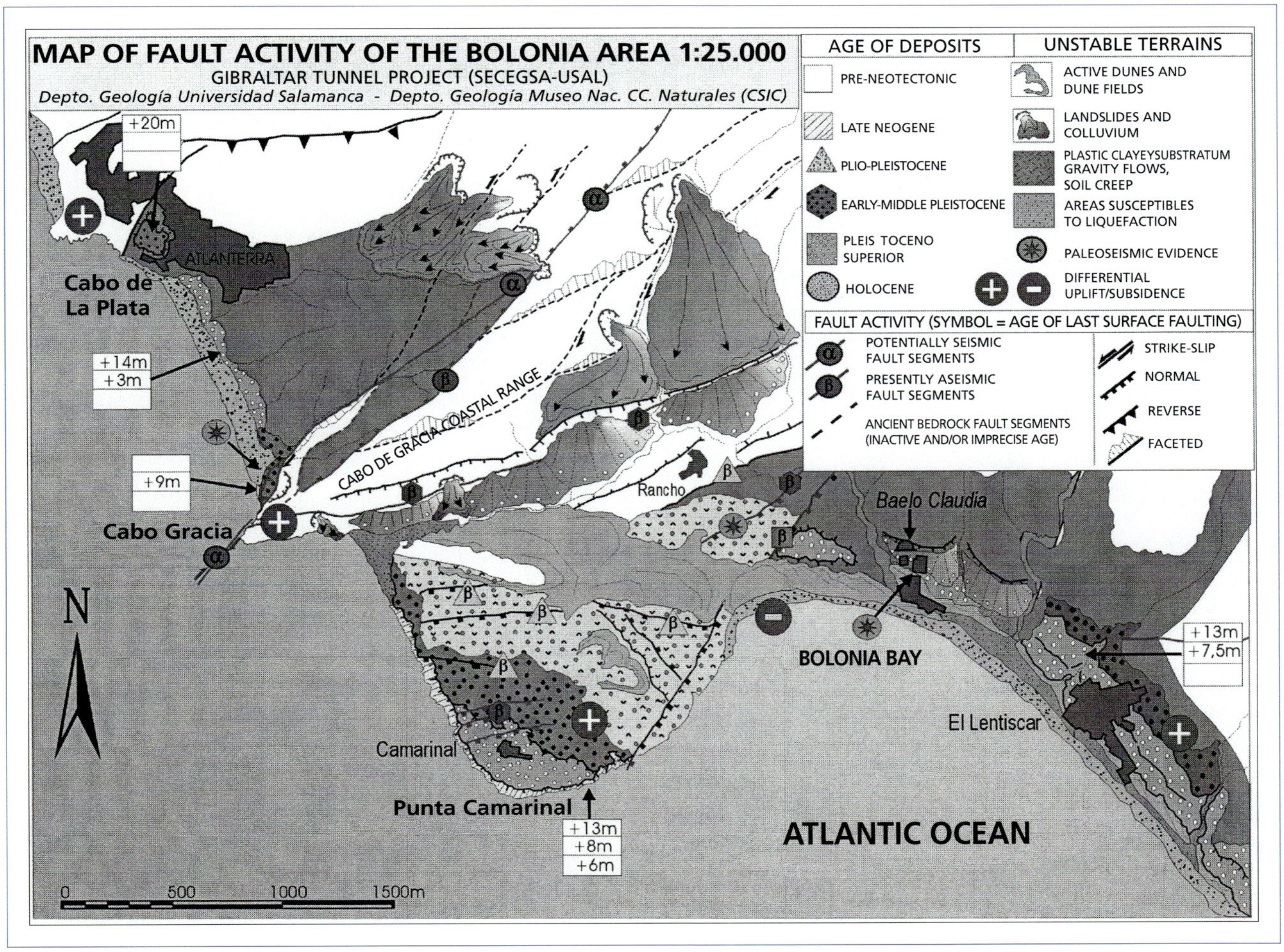

Figure 3.35 Map of active faults in the Bolonia area (Province of Cádiz, Spain). The elevation data indicates shows the actual height of the deposits above sea level, which relate to oxygen isotopic substages: 5e (roof), 5c (middle) and 5a (wall) (Silva *et al.*, 2006).

Figure 3.36 Rambla valley incising into Holocene deposits of a fan delta, deformed by a group of normal faults. Nahal Darga, Dead Sea (Israel).

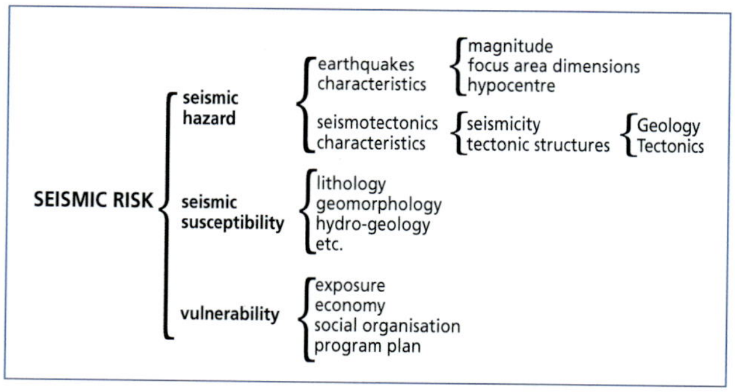

Figure 3.37 Factors involved in the definition of seismic risk (Panizza, 1991).

ground acceleration, hypocenter, mechanism, etc.) and with seismotectonics (seismicity, geology, and tectonics). Seismic sensitivity refers to geological, morphological, and hydrologic factors. The most important geomorphic characteristics affecting seismic sensitivity are slope angle, nature of sediments, landforms, degraded slopes, paleo-landslides and subterranean cavities. Regional vulnerability includes all of the following factors related to a town: construction, infrastructure, economic activity, social organization, and potential programs for expansion to a particular territory (Panizza, 1988).

Seismic hazard includes the size and frequency of earthquakes and associated temblors whereas seismic risk involves death and property damage caused by earthquakes. The product of seismic hazard and vulnerability constitutes seismic risk, which does not depend exclusively on earthquake characteristics (hazards) but also takes into account the vulnerability of a community and its property to earthquakes, landslides, and tsunamis generated by the quake.

Earthquakes are described by their **intensity** on the Mercalli scale, modified by Wood and Neumann (1931) (MME Intensity, *Table* 3.3). In Europe the MSK (Medeved, *et al.*, 1965) scale has been used since 1964. The MSK scale is essentially equivalent to the MME scale and to a more recent scale called the European Macroseismic Scale (EMS Scale), which is revised every 6 years. Intensity is a subjective measure of the strength of an earthquake. Twelve intensity classes have been established based on effects to people and construction. The intensity of an earthquake depends on the distance to the epicenter, acceleration, period, duration, amplitude of seismic waves, ground type, groundwater level, and the type and quality of human structures. It is based on observations and interviews (Costa and Baker, 1981). **Magnitude** is an experimentally determined objective measurement that determines the amount of energy liberated by an earthquake. The most frequent classification of magnitude is the Richter Scale (*Table 3.4*), which was initially calibrated in the southwestern United States by scientists using a torsion seismograph (Wood-Anderson). Since the middle of the 20th

Table 3.3 **MERCALLI EARTHQUAKE INTENSITY SCALE AND EQUIVALENCE TO THE RICHTER SCALE (BRYANT, 1991)**

Scale	Intensity	Description of effects	Maximum acceleration in mm/sec^2	Equivalents Richter scale
I	Instrumental	Detected only on seismographs.	<10	
II	Very weak	Felt by some people.	<25	
III	Weak	Felt by everyone; like a large rumbling.	<50	<4.2
IV	Slight	Felt by people walking; objects fall from bookshelves.	<100	
V	Slightly to strong	Sleepers awaken; church bells ring.	<250	<4.8
VI	Strong	Trees sway; suspended objects swing; objects fall off shelves.	<500	<5.4
VII	Very strong	Mild alarm, walls crack, plaster falls.	<1000	<6.1
VIII	Destructive	Uncontrollable vehicle movement; chimneys fall and masonry fractures; damage to poorly constructed buildings.	<2500	
IX	Ruinous	Some houses collapse; ground cracks; pipes break open.	<5000	<6.9
X	Disastrous	Great number of cracks; many buildings destroyed, liquefaction and extensive displacement.	<7500	<7.3
XI	Very disastrous	Collapse of the majority of buildings and bridges; roads, railways, pipelines and cables destroyed; general triggering of other hazards.	<9800	<8.1
XII	Catastrophic	Total destruction; trees knocked down; uplifted and undulating ground.	>9800	>8.1

century, it has been generalized and applied to all regions of the world. Richter understood that amplitude decreased with increasing distance to the earthquake (Yeats *et al*., 1997). The Richter Scale spans more than nine orders of magnitude. The relationship between magnitude and intensity is shown in Tables 3.5. More than a million earthquakes occur every year. The number of earthquakes of different magnitudes that occurs each year is the **frequency**, shown in Table 3.5. **Large earthquakes** may have a magnitude of 8 or more, **major earthquakes** are between 7.0 and 7.9, and **strong earthquakes** range from 6.0 to 6.9 The largest recorded earthquake, a magnitude of 9.5, occurred in northern Chile in 1960. It killed 5,700 people, destroyed 58,000 houses, and left two million people homeless (McGuire *et al*., 2004). Chapron *et al*., (2006), conducted an investigation of marine basins and calculated that seismic movement contemporary with the 1960 Chile earthquake dislodged 187 × 10^6 m^3 of marine sediments in northern Patagonia.

Tectonic forces deform the Earth elastically and crustal blocks store the energy of deformation. These forces are liberated during movement of blocks along a fault, creating or increasing fault movement. The most widely accepted theory of earthquakes is **elastic rebound**, which the American geologist Henry F. Reid proposed in 1906 after the San Francisco earthquake. Earthquakes occur when deformation stored in the fault exceeds rock strength and friction and the elastic deformation stored in the fault is liberated into the crust and upper mantle. The **hypocenter** is the place where the energy is transformed into elastic or seismic waves. Its projection to the Earth's surface is called the **epicenter**.

Surface vibrations generated by seismic wave energy create seismic hazard. When these surface vibrations affect humans and their property, they are called seismic risk. During the 20th century earthquakes have killed about two million people. Table 3.6 shows historically documented catastrophic earthquakes that have killed 50,000 people (McGuire *et al*., 2004). Yeats has compiled an appendix (Yeats *et al*., 1997) of historical earthquakes of all sizes with dates, magnitudes, name, location, type and fault name and references. The CRED (*Centre for Research on the Epidemiology of Disasters*) estimates that during the 1990s, earthquakes killed 100,000 people and affected the lives of a total of 14 million. The CRED also calculates that earthquakes have caused economic losses of $215 billion, second only to floods. Other dangerous phenomena associated with earthquakes include

Table 3.4 **ENERGY EQUIVALENTS OF EARTHQUAKES COMPARED TO THE RICHTER SCALE. SOURCE: CALIFORNIA DIVISION OF MINES AND GEOLOGY (in COSTA AND BAKER, 1981)**

Earthquake magnitude	TNT equivalent	Example
1.0	0.17 Kg	
1.5	0.9 Kg	
2.0	5.9 Kg	
2.5	28 Kg	
3.0	179 Kg	
3.5	450 Kg	
4.0	5.5 Metric Tons	
4.5	29 Metric Tons	Denver, CO, 1965
5.0	181 Metric Tons	
5.3	455 Metric Tons	San Francisco, 1957
5.5	910 Metric Tons	
6.0	5.7×10^3 Metric Tons	
6.3	14.4×10^3 Metric Tons	Long Beach, CA, 1933
6.5	28.7×10^3 Metric Tons	San Fernando, CA, 1971
7.0	181×10^3 Metric Tons	
7.1	228×10^3 Metric Tons	El Centro, CA, 1940
7.5	910×10^3 Metric Tons	
7.7	$1{,}811 \times 10^3$ Metric Tons	Kern County, CA, 1962
8.0	$5{,}706 \times 10^3$ Metric Tons	
8.2	$11{,}421 \times 10^3$ Metric Tons	San Francisco, CA, 1906
8.5	$28{,}711 \times 10^3$ Metric Tons	Anchorage, Alaska, 1964
9.0	$181{,}999 \times 10^3$ Metric Tons	

Table 3.5 **APPROXIMATE RELATIONSHIP BETWEEN EARTHQUAKE MAGNITUDE AND FREQUENCY (UNITED STATES GEOLOGICAL SURVEY, 1974)**

Magnitude	Expected annual incidence	Felt area (km^2)	Distance felt (km)	Intensity (Modified mercalli scale)
3.0–3.9	49,000	1,940	25	II-III
4.0–4.9	6,200	8,850	50	IV-V
5.0–5.9	800	38,850	110	VI-VII
6.0–6.9	120	165,350	200	VII-VIII
7.0–7.9	18	518,000	400	IX-X
8.0–8.9	1	2,072,000	725	XI-XII

Table 3.6 **CATASTROPHIC EARTHQUAKES > 50,000 HISTORICALLY DOCUMENTED FATALITIES (MCGUIRE, 2004)**			
Location	**Year**	**Fatalities**	**Magnitude, commentary**
Ardabil, Irán	893	150,000	
Halab (Aleppo), Syria	1138	230,000	7.5
Sicilia, Asia Minor	1268	60,000	
Gulf of Chihli (Bo Hai) China	1290	100,000	
Hongtong, Shanxi, China	1303	200,000	8.0
Hua Xian, Shaanxi, China	1556	830,000	~8, many fatalities in the loess
Samaxi (Shemakha), Azerbaijan	1667	80,000	
Sicily, Italy	1693	60,000	
Tabriz, Iran	1727	77,000	
Lisbon (Portugal)	1755	70,000	8.7, and a large tsunami
Iran	1780	200,000	
Calabria, Italy	1783	50,000	
Messina, Italy	1908	70,000–100,000	7.2, also fatalities from a tsunami
Haiyuan, Ningxia, China	1920	200,000	8.6, major fractures and landslides
Kanto, Tokyo, Japan	1923	143,000	7.9, large fire in Tokyo
Tsinghai (Qinghai), China	1927	200,000	7.9, large fractures. 40.912 fatalities
Gansu (Kansu), China	1932	70,000	7.6
Quetta, Pakistan	1935	30,000–60,000	7.5, Quetta almost completely destroyed
Ashgabat, Turkmenistan	1948	110,000	7.3
Northern Peru	1970	66,000	7.9, Mt. Huascaran, large rockslide, flooding, 530,000 dollars in damage
Tangshan, Hebei, China	1976	255,000 (official)	7.5, there may have been 655,000 fatalities
Iran	1990	50,000	7.7, landslides

tremors, ground shaking, liquefaction, landslides, tsunamis, and fire. All of these are considered seismic risk.

Iberia is a small tectonic plate located between Eurasia and Africa at the western end of the Alpine belt. The Alpine orogeny formed the Pyrenees between Eurasia and Iberia and the Betics between Iberia and Africa. The Iberian Peninsula is located in the interior of a mobile plate where active plates have moderate displacement rates (<1 mm/year) and resulting earthquakes have a recurrence interval on the order of 10^3–10^5 years (Santanach, 2001).

Table 3.7 shows historical earthquakes with intensities greater than IX that have occurred from the 14th century to the present. The most destructive earthquake in Spain was the Arenas del Rey (Province of Granada) quake; it occurred on December 25, 1884, killing 750 to 900 people, injuring 2,000, and damaging more than 2,000 houses (Fouqué *et al.*, 1889; López Arroyo *et al.*, 1981; and others). The catastrophic 1829 earthquake of Torrevieja (Province of Alicante, Betic Cordilleras) killed 400 people and destroyed several towns (Rodriguez de la Torre, 1884). Destructive earthquakes have collapsed buildings, cracked the ground, liquefied sediments, and triggered mass movements and other effects.

The distribution of seismic activity in Spain is concentrated in the Betic cordilleras in the southern and southeastern parts of the country (Andalucia, Murcia, and Alicante), and in the Pyrenees and the Catalan coastal chains. Except for deep earthquakes in the Province of Granada, hypocenters are at or near the surface (Capote, 1997).

Figure 3.38 displays major historical Spanish earthquakes on a neotectonic sketch of the Iberian Peninsula. The figure shows epicenter locations and intensities of the 15 most significant earthquakes, according to the author. *Figure 3.38*

Table 3.7 **DESTRUCTIVE HISTORICAL EARTHQUAKES IN SPAIN WITH INTENSITY IX OR GREATER AT THE EPICENTER SINCE THE 14TH CENTURY (ACCORDING TO CAPOTE, 1997; MARTÍNEZ, 2001; SANTANACH, 2001)**

Year	Epicenter	Intensity	Magnitude	Fatalities/commentary
1373	1. Viella (Lérida)	VIII-IX		Rock movement, towers and castles collapsed, unknown number of victims. Damage in Barcelona (grade VI).
1396	2. Tabernes (Valencia)	IX		Fortresses and bridges collapsed, hundreds of houses destroyed, landslides, cracks, and changes in springs.
1427	3. Olot (Gerona)	VIII-IX		
1427	4. Amer (Gerona)	VIII-IX		
1428	5. Queralbs (Gerona)	IX-X		800 fatalities. Significant damage to churches and castles, many dwellings damaged, large cracks in the ground.
1431	6. Atarfe (Granada)	IX		Towers and mosques come apart; part of the Alhambra wall fell.
1504	7. Carmona (Sevilla)	IX		100 fatalities. Destruction of walls, collapse of temple cupolas, many houses damaged, landslides, cracks, change in water regimes. Liquefaction in Seville.
1518	8. Vera (Almería)	IX-X		165 fatalities. Destruction of the entire city. It was rebuilt in another place. Damage to fortifications.
1522	9. Almería	IX		More than 2500 fatalities. Almeria completely destroyed, devastated 80 villages, destroyed harbor, tsunami, considered as large as the earthquake of Lisbon.
1531	10. Baza (Granada)	IX		310 fatalities. Baza and Benamaurel (Granada) practically destroyed.
1645	11. Alcoy (Alicante)	IX		Entire houses collapsed, likewise in Muro de Alcoy.
1680	12. Alhaurín el Grande (Málaga)	IX	6.8–7.4	In Málaga only, there were 200 fatalities and 250 wounded. Destruction of the fortress, 852 houses destroyed, and 1250 damaged. Variation in the flow of springs.
1748	13. Montesa (Valencia)	IX		38 fatalities. Destroyed the towns of Montesa, Sellent, and Estubeny.
1755	14. SW Cabo de San Vicente	VIII		61 fatalities due to earthquakes and 1214 drowned by tsunami waves. Damages in the center and south of the Peninsula. Present day value of losses 763 million Euros.
1804	15. Dalías (Almería)	IX		150–200 fatalities. Churches and fortress towers collapsed, several hundred houses destroyed, aftershocks for seven months.
1806	16. Pinos Puente (Granada)	VIII-IX		12 fatalities, 94 homes destroyed, 1110 damaged, and 118 undamaged. Cracks in the ground.
1829	17. Torrevieja (Alicante)	X	6.9	399 fatalities, 388 injured. About 2900 houses destroyed and more than 2000 damaged. Months of aftershocks, cracks in the ground, and change in water flow.
1884	18. Arenas de Rey (Granada)	IX	6.5–6.7	750–900 fatalities, about 2000 injured. About 1000 houses destroyed and 17,000 damaged. Landslides, fissures, liquefaction, water modifications, strong aftershocks for a year.
1956	19. Albolote (Granada)	VIII	5.0	11 fatalities, 7 houses destroyed, 50 damaged, 300 uninhabitable, 350 were cracked, 4 of the fatalities were caused by landslides.
1969	20. SW Cabo de San Vicente		7.3	4 deaths were caused by panic, slight tsunami. The felt intensity was VII. Some damage in the Province of Huelva.

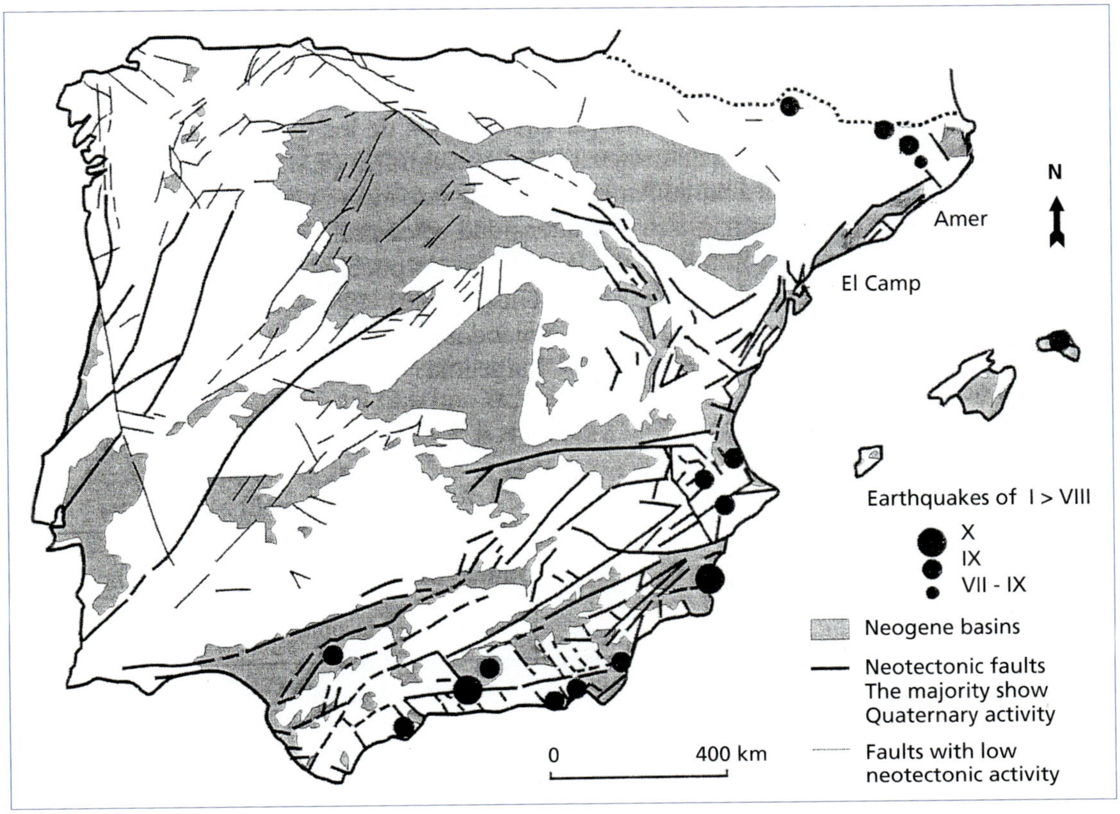

Figure 3.38 Major historical earthquakes shown on a neotectonic diagram of the Iberian Peninsula. (Santanach, 2001).

indicates that the 15 highest intensity earthquakes occurred along particular seismogenic fault (Santanach, 2001). Seismicity studies classify Spain as a country with moderate seismic activity (Hernández-Enrile *et al.*, 2001).

Many deaths and much of the damage caused by earthquakes result from the **vulnerability of buildings.** Causes of building vulnerability include: building walls that are too tall or too thin for their height; concrete that may contain insufficient or bad cement; towers and chimneys constructed on low buildings that destroy small structures when they collapse; narrow streets that block the fall of materials; and others. Buildings and other structures vibrate at different frequencies depending on their composition, shape, and the size. When buildings vibrate at the same frequency, the oscillations of the structures reinforce each other, producing more damage (Costa and Baker, 1981, McGuire *et al.*, 2004). In Spain the standards for earthquake resistant construction have been codified in Norms of Earthquake Resistant Construction NCSE-94 and NCSE-02. An appendix of the publication also provides seismic hazard maps based on horizontal ground acceleration and codified for each Spanish town (AE-88, 1996).

Compressive and extensional forces cause ground shaking that cracks and separates the ground; these horizontal and vertical movements, which can measure several meters, destroy building foundations. The 1906 San Francisco earthquake caused 6.4 m of horizontal displacement. The 1899 Yakutat Bay (Alaska) earthquake caused 15 m of vertical movement (Plafker *et al.*, 1978). Unconsolidated materials vibrate more than rocks. For this reason, buildings located on mud and poorly compacted artificial fill, in swampy areas, and on fluvial sediments (*Fig. 3.39*) were the most severely damaged in the 1906 San Francisco earthquake. Intensity was weak on bedrock (Borcherdt, 1975).

Vibration and shaking during an earthquake cause unconsolidated sediments to compact. The San Francisco earthquake caused a volume decrease of 0.6 m in some places, causing **subsidence**. The 1964 Alaska earthquake caused 1.8 m of subsidence. The process of "seismic" compaction and the resulting differential settling can pose grave dangers for structures (Costa and Baker, 1981).

Seismic ground tremors can reorient particles within unconsolidated sediments, and with increased pressure, can convert a solid granular material into a fluid. This process is called **liquefaction**. This change in state of the ground leads to physical behavior that is dangerous for buildings. During the 1964 magnitude 7.5 earthquake in Niigata (Japan),

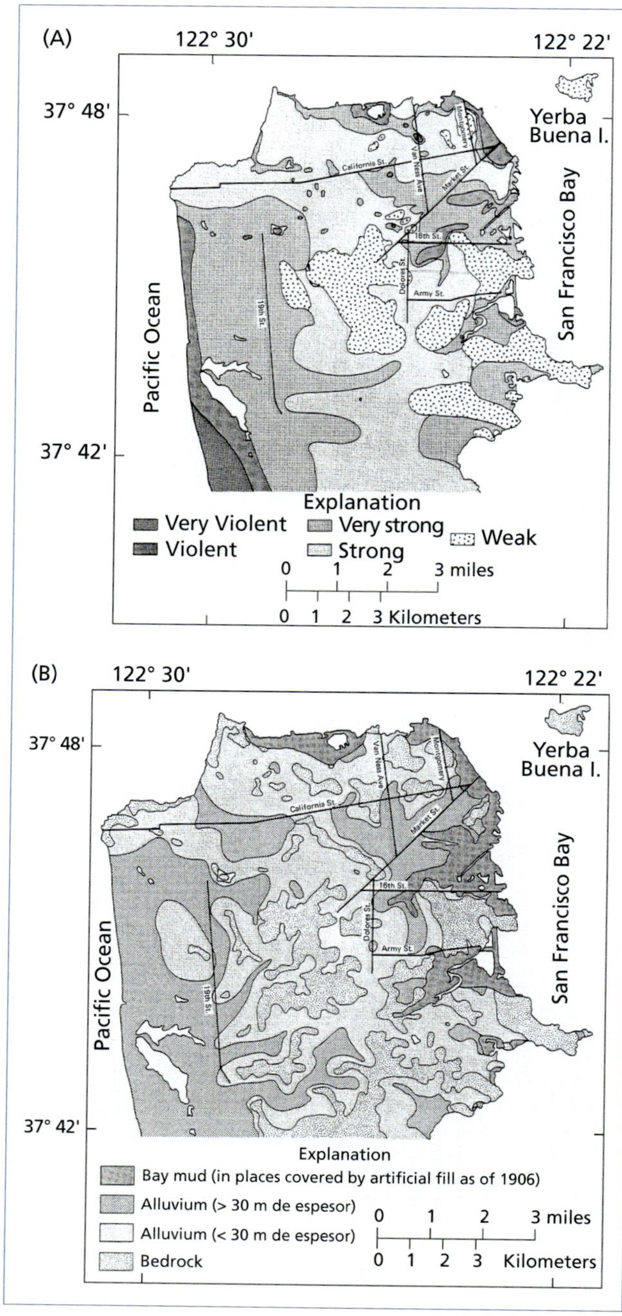

Figure 3.39 Map of apparent intensities of the 1906 San Francisco earthquake. Zones of maximum seismic intensity correspond to areas of silt and alluvium; weak intensity corresponds to areas of bedrock outcrop (Borcherdt, 1975).

liquefaction caused no structural damage to multi-story buildings but caused them to tilt. In the 1989 Loma Prieta earthquake, a magnitude 7.1 quake related to the San Andreas Fault caused significant damage due to liquefaction. The same occurred during the 2001 magnitude 6.8 Nisqually earthquake that affected structures in Seattle (McGuire *et al.*, 2004). Perucca and Moreiras (2006), in their study on

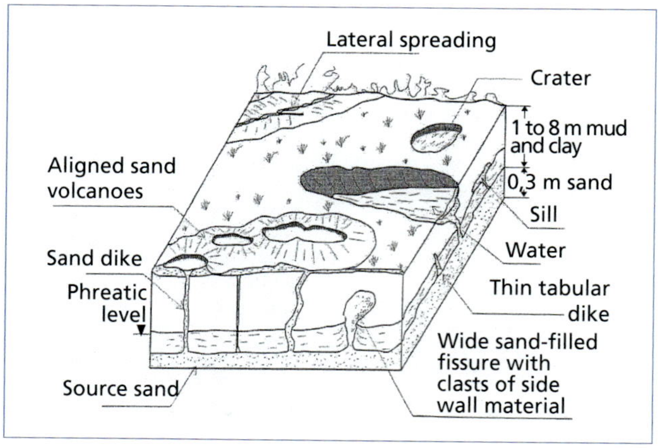

Figure 3.40 Block diagram showing dikes cutting the strata of mud and clay, craters, lateral extension, and sand volcanoes (Perucca and Moreiras, 2006).

liquefaction in the San Juan and Mendoza provinces of western Argentina, show that the provinces have been affected by at least nine destructive earthquakes with magnitudes ≥6.3 between 1861 and 1997. These earthquakes caused liquefaction that created extensive damage. They show that the distance between liquefaction and the epicenter increases with increased earthquake magnitude. They also describe lesser effects caused by liquefaction. Visible effects of liquefaction include sand craters, sand volcanoes (small craters produced by water surging at high pressure), sand lenses or dikes, piping of water and sand boiling, soil collapse, convoluted stratification and lamination, and subaqueous landslides (Perucca and Moreiras, 2006) (*Fig. 3.40*).

Seismic activity can substantially modify poorly compacted sediments and increase the risk to structures. Objects subjected to strong vibration respond with **resonance** to a natural frequency or period. For example, ten-story buildings have a natural period of one second and 20-story buildings have a natural period of two seconds. If the period of a 10-story building is greater than one, the building will respond in a dangerous manner. This phenomenon occurred with catastrophic consequences during the 1985 magnitude-8 earthquake in Mexico City, a city founded by the Aztecs on an island located in the middle of Lake Texcoco. The city had expanded extensively onto drained areas on top of unconsolidated lacustrine sediments. About 60% of the destroyed or seriously damaged buildings had 6 to 12 stories. The area of firm ground, called the hilly area, had no collapse. Most of the collapsed or damaged buildings were relatively modern (from the 1960s and 1970s) buildings that had been designed and constructed according to the earthquake-resistant norms applicable at the time (Meli, 2002; McGuire, 2004).

Landslides will be covered in the chapter on slopes and mass movements. Landslides caused by seismic activity can be catastrophic, such as the 1962 Nevados del Huascarán (Peru) landslide (Plafker and Erickson, 1978; Oliver-Smith,

2002). Landslides can also result from medium or low magnitude earthquakes in mountainous areas with high relief.

Fires that break out after an earthquake are a significant secondary threat that can have devastating effects. After the 1906 San Francisco earthquake, wooden houses caught fire. Because water lines had ruptured, the fires burned uncontrolled for three days and destroyed 28,000 buildings. Fire accounted for 80% of the earthquake damage. After the Great Kanto earthquake that hit Tokyo in 1923, fire destroyed half a million buildings and killed about 100,000 of the total 143,000 who died from the quake (McGuire *et al.*, 2004).

Tsunamis resulting from seismic activity will be discussed in the chapter on littoral geomorphology. **Dam** or **dike breaks** release massive amounts of water in a short time period and cause significant flooding that may result in loss of life and property damage. Jansen (1980) estimates that about 2,000 dams have failed since the 12th century. In addition to seismic activity and dam failure, causes of flooding include overflowing of dams, structural failure due to *piping* (earthen dams), and rupture due to cementation problems (Singh, 1996). A 1953 storm in the North Sea generated high waves that broke dikes in Holland, causing many of the 1,835 fatalities that occurred during that event (McGuire *et al.*, 2004). In 1963, 2000 people died in the town of Longarone (Italy) when gigantic translational landslides on the southern margin of the Vaiont reservoir caused water to overflow the dam. A dam built in the Strait of Rientes en Lorca (Province of Murcia, Spain) between 1755 and 1789 was the highest dam of its time when it was completed. The reservoir filled for the first time in February 1802 and in April, the dam broke, creating a flood that destroyed part of the city of Lorca and killed 600 people. The exact cause of the failure is not known (Singh, 1996). A highly detailed list of 85 dam failures, with names, locations, geometric characteristics of the dam, failure and duration, can be found in Singh (1996).

3.7 Prevention, mitigation, and warning

Physical methods that minimize the impact of seismic hazard are fundamental in reducing seismic risk. The Bam (Iran) earthquake of December 2003 that killed 26,000 people shows the importance of construction materials used in houses. As seismic engineers and seismologists say, "Buildings kill people, earthquakes don't." Because the houses in Bam were made of adobe, most were destroyed during the earthquake because adobe structures collapse even during low-magnitude earthquakes (McGuire *et al.*, 2004).

This fact has led authorities to establish earthquake-resistant construction norms. Damages may be acceptable but not loss of human life. Approximately 75% of earthquake deaths in the 20th century are attributed to the collapse of buildings that were either poorly constructed, constructed of inadequate materials, or built according to poorly applied earthquake-resistant design. Horizontal vibrations are propagated through buildings by their foundations. Objects such as bookcases inside buildings can move or fall so one preventive measure is to fasten this type of furniture to the floor. The way to mitigate seismic tremors is to build structures according to earthquake-resistant construction norms (Costa and Baker, 1981; Bolt, 1999; McGuire *et al.*, 2004).

The study of seismic activity includes applied aspects, especially in the areas of prevention and seismic mitigation. "The lack of complete knowledge of the physical mechanism introduces epistemological uncertainty into the study, which makes it difficult to make seismic predictions in the short term and immediate interval (months and days)" (Mézcua, 2002). The principal sources of seismic data in Spain are the Seismological Data Base of the National Geographic Institute, the General Catalog of Isoseismic Lines (Mézcua, 1982), Seismicity of the Iberomagrebi Area (Mézcua and Martínez, 1983), and more recently, the Catalog of Historical Earthquakes on the Iberian Peninsula (Martínez Solares and Mezcua, 2002).

Studies in Spain have focused on research into historical seismicity or levels of seismic hazard indicated by available data. Another research trend attempts to establish the probability of future events of specific intensity or magnitude (Cendrero *et al.*, 1997). Historical seismicity has been summarized in the preceding paragraphs. Munuera (1969) and Roca and Udías (1976) discuss research on probability and Muñoz (1983) has written an extensive treatise on seismic risk. Chávez *et al.* (1999) simulated the loss of life resulting from the 1428 AD earthquake in Queralbs (La Candelaria), which had an intensity of 9 or 10. They calculate that about 1000 people were killed, 1000 buildings destroyed, and about 140,000 people left homeless. Monetary losses totaled the equivalent of 4.8 billion Euros. A study of the economic and social impacts of seismic risk in Spain shows that, on the basis of maximum seismic risk, potential losses due to earthquakes could be 33.1% of the total losses for the period of 1986–2016, or equivalent to 157 billion Euros.

Mitigation of damages caused by earthquakes is a continuing task for countries affected by seismic activity. Mitigation requires protective strategies including seismological research, application of earthquake-resistant norms, community awareness of protection, action plans in the event of a seismic catastrophe, and appropriate insurance legislation (Martín, 2002). To this list we should add research into paleoseismicity (Capote and Martínez, 2001; Vicente *et al.*, 2001).

Earthquake-resistant norms are perhaps the most suitable instrument for preventing and mitigating earthquake damage to human structures. These norms should be revised and updated periodically to incorporate new advances in seismology and seismic engineering (Blázquez, 2001). The organization for preventive regulation is called the Permanent Commission on Earthquake-Resistant Norms (CPNS). General norms are published in a volume entitled Disaster

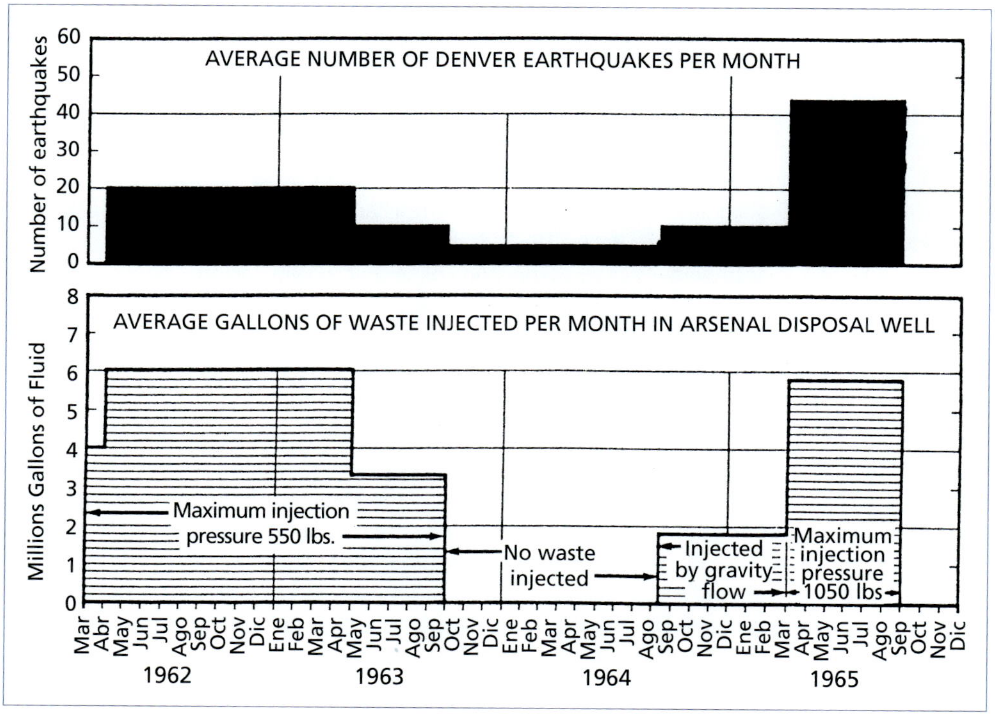

Figure 3.41 Relationship between earthquake frequency and injection of contaminated water (Evans, 1966).

Prevention and Mitigation (UN, 1984). The Civil Protection Agency has Disaster Relief Agencies that are very active following a disaster (Dueñas, 2001). The Instituto Geografico Nacional has established an Earthquake Alarm System that identifies and locates seismic events on the Iberian Peninsula in real time.

Immense effort has been devoted to the reduction of seismic risk by connecting digital seismographs to emergency centers. This **alarm** system makes it possible to respond rapidly before and after large earthquakes. In the United States, data on earthquakes with magnitude greater than 3.5 are sent electronically via satellite, internet, or telephone to federal and international civil defense agencies. This information can also be transmitted to dams, power plants, and railroads (Bolt, 1999).

3.8 Human-induced earthquakes

Between April and November 1962, Denver experienced more than 700 earthquakes. The largest had a magnitude of 4.3 and none of them caused any damage. Even so, the public was concerned; some people even purchased earthquake insurance (Evans, 1966).

Prior to 1962, Denver had not recorded an earthquake since 1882. Epicenters of the 1962–65 earthquakes were near the Rocky Mountain Arsenal well, located at a distance of about 10 km from Denver. The hypocenters were located along a fault plane. The United States Geological Survey conducted seismic studies and determined that the epicenters were even closer to the well than previously thought.

In November 1962 researchers suggested a link between the earthquakes and the injection of contaminated wastewater at the Rocky Mountain Arsenal northeast of Denver. The wastewater was injected into highly fractured and weathered schists and gneisses at a depth of 4,000 m. Sixteen million liters (4.2 million gallons) were injected in March 1962 and the earthquakes started shortly thereafter (*Fig. 3.41*). The well was plugged on February 20, 1966.

Precambrian rocks, such as those below Denver, are impermeable and the fluid injected into them sealed their fractures. High-pressure fluids reduce the resistance to friction. If fluid pressure is high enough, it can cause blocks to move with the result that the elastic relaxation energy is released as an earthquake.

In a case in Inglewood (Los Angeles, United States) fluid injection into an oilfield to increase pressure for enhanced petroleum production was suspected of causing a 1963 earthquake that fractured a wall of the Baldwin Hills dam (Strahler and Strahler, 1974; p. 221–224). Water passed over the damaged dam and caused a flood that inundated numerous houses in the area and caused five deaths. The correlation between fault activity and fluid injection has been clearly demonstrated. Increased fluid pressure decreases cohesion and internal friction on the fault plane and can activate faults.

4

VOLCANIC GEOMORPHOLOGY

4.1 Introduction

Volcanic eruptions are the most powerful manifestation of the energy present in the Earth's interior. Of the approximately 60,000 volcanoes in the world, about 50 erupt above sea level every year, potentially affecting the lives and property of millions of people. The human race is linked to volcanic eruptions in another way. Paleoanthropologists have provided absolute dating of volcanic materials in the East Africa Rift Valley (Tanzania, Kenya, and Ethiopia) that are related to the first hominids (3.7 Ma). Volcanic ash is also known to preserve fossils in perfect condition (Sigurdson, 2000).

Active volcanoes are classified according to their activity, which is of great importance for towns located in these areas. Tsuya and Morimoto (1963) established a classification that includes active volcanoes, which erupt every few years; dormant volcanoes, which erupt once every ten to one hundred years; and finally, extinct volcanoes. It is difficult to differentiate between extinct and dormant volcanoes.

We can define a volcano as *"the surface expression of internal thermal processes of a planet or satellite manifested by the emission of solid, liquid or gaseous products at the surface"* (Francis, 1993). A volcano contains a vent or chimney that connects it to molten material called **magma**. Magma is molten rock that contains gases; magma is not synonymous with **lava**. Magma usually contains a melted silicate phase, a gaseous phase (mostly water), carbon dioxide, and one or more solid phases (rock fragments and mineral crystals). During an eruption, partial degasification converts magma to lava. The material erupted on the surface usually builds a **cone** composed of lava and/or fragmentary material.

Gases are the most important phase in the mechanism of volcanic eruptions. Eruptions can be fluid lavas that liberate gases slowly or highly viscous material that erupts with difficulty at the surface. Expansion of gases causes violent explosions that generate particulate material called pyroclastics. Most of the world's volcanoes erupt a mixture of lavas and pyroclastics (*Fig. 4.1*).

Volcanic terrain occupies only about 5% of the Earth's land surface; the percentage is much higher on the ocean floor. The ocean floor is composed of numerous submarine volcanic accumulations frequently covered by a thin layer of sedimentary deposits.

Figure 4.1 The Teide volcano, 3,718 m (Tenerife, Canary Islands) composed of lava and pyroclastics, from the Roques de García. Photograph taken in 1968.

Volcanism involves a flow of energy and material that causes the planet to gradually lose energy. The heat necessary for volcanism comes from the radioactive disintegration of uranium (^{235}U and ^{238}U), potassium (^{40}K), thorium (^{232}Th), and other radioactive elements in the Earth's interior. Each isotope produces a different amount of heat and has a specific half-life. Melting is limited to a depth of about 200 km in the crust and upper parts of the mantle. Because the molten material is less dense than its source area, it rises to the surface. At the surface of the crust, rocks are usually less dense than magma but the magma is pushed to the surface by the internal pressure of the dissolved gases that expand as the confining pressure decreases (similar to the "champagne bottle" effect).

Volcanoes are located above areas of partial melting inside the Earth. A melt that reaches the surface with a high concentration of dissolved volatiles causes a highly explosive eruption. In other cases, the magma can be held at depth where it cools slowly, creating plutonic rocks in which gravitational settling of crystals, assimilation processes, and mixture and re-melting of crustal rocks can occur (Sparks, 1993).

4.2 Plate tectonics and volcanism

Volcanoes occur in narrow bands related to lithospheric plates (*Fig. 4.2*). Plate tectonics result from the planet's loss of energy (Perfit and Davidson, 2000). The Earth has been losing energy since its formation. Convection is the principal mechanism of energy transfer within the planet. Material heats up at the base of the mantle at a depth of about 2900 km and rises due to its lesser density and plastic behavior. It moves upward until it reaches the **lithosphere**, which is rigid and brittle and covers the entire world. The lithosphere includes the Earth's crust and the upper part of the mantle. It is 70–120 km thick and is broken up into numerous fragments called the **lithospheric plates**. Below the lithosphere is the **asthenosphere**, a layer less rigid than the lithosphere that contains areas of partial melting or material near the melting point. The asthenosphere is the foundation on which the lithospheric plates move (*Fig. 4.3*). The plates encircle

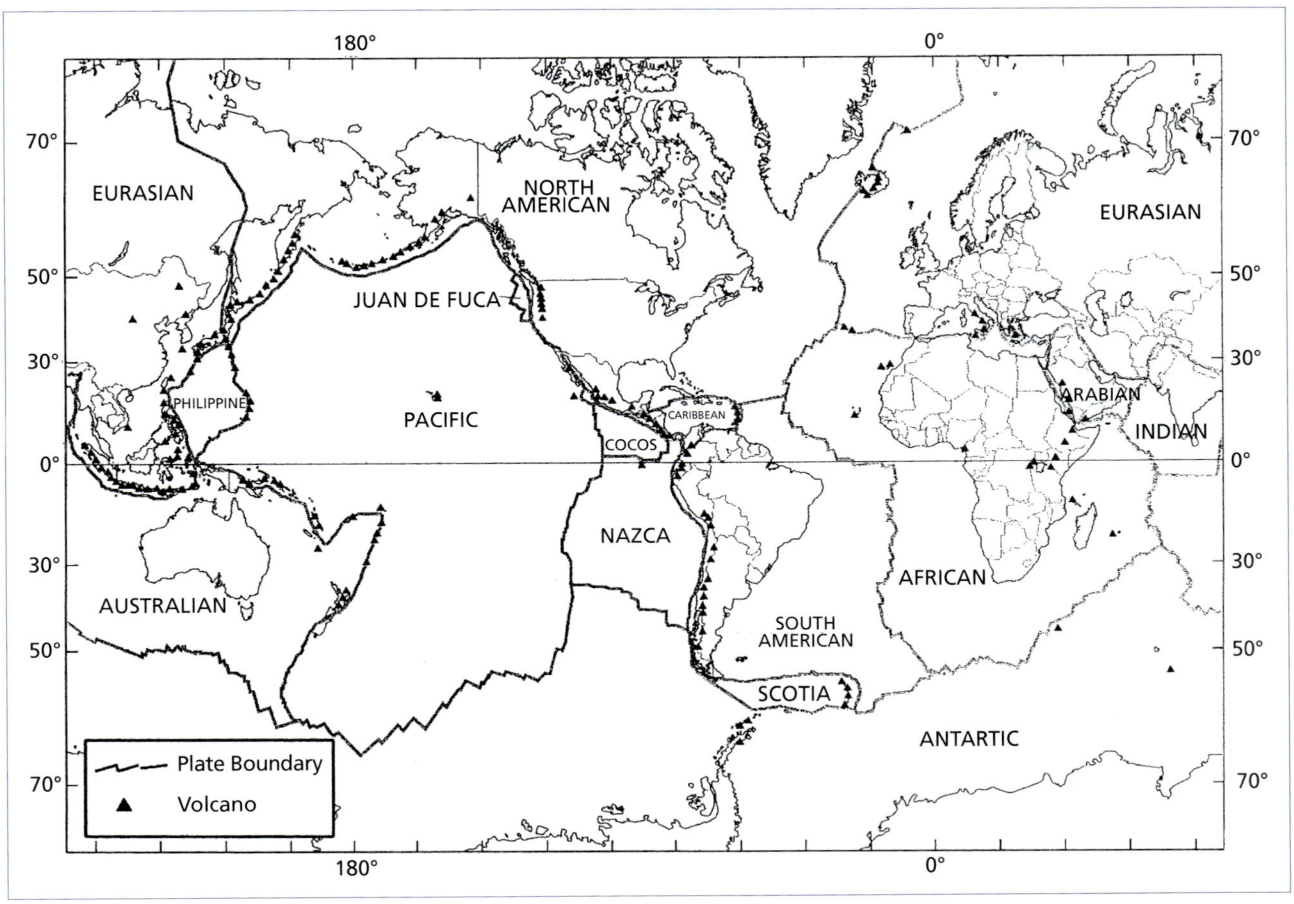

Figure 4.2 World map showing the principal active volcanoes and plate boundaries. (United States Geological Survey, in Perfit and Davidson, 2000).

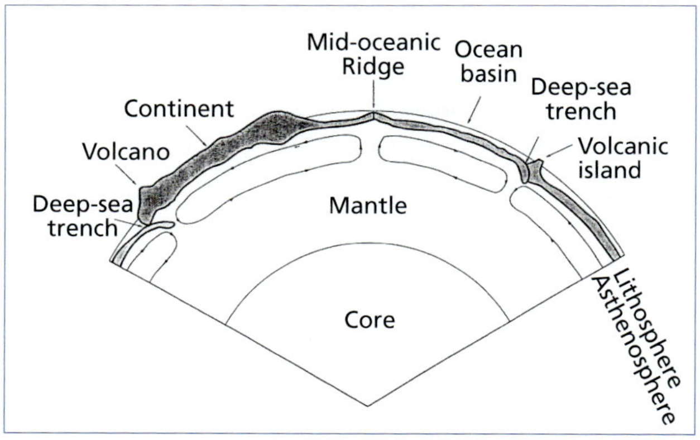

Figure 4.3 Schematic section of the Earth showing the lithosphere, highly exaggerated, and the asthenosphere and some terms used in plate tectonics (Glass, 1982).

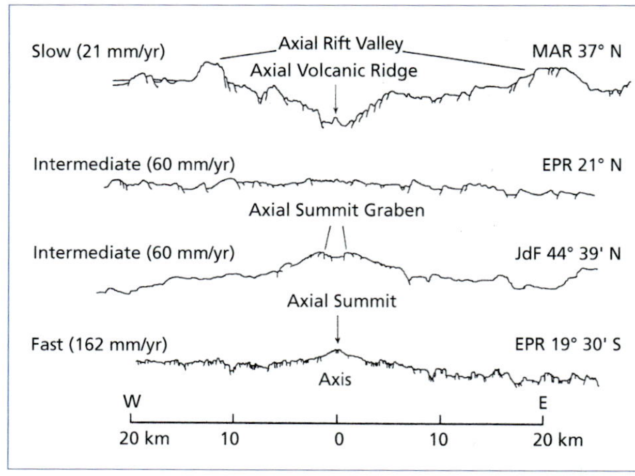

Figure 4.4 Bathymetric profiles of mid-ocean ridges, with different extensional velocities. Higher velocity profiles correspond to the southeast Pacific and lower velocities are those of the northern mid-Atlantic ridge. (Perfit and Davidson, 2000).

the planet and are fragmented into different configurations shaped by their margins, which are classified as **divergent, convergent,** and **transform.**

Divergent or constructive margins are mid-ocean ridges that generate crust and the continental rift valleys in East Africa. Convergent or destructive boundaries are highly complex geologic regions where one plate sinks below the other along what is called the Benioff plane. Transform or conservative boundaries slide horizontally against each other; they are characterized by faults or fracture zones (Perfit and Davidson, 2000).

Mid-ocean ridges are divergent zones where oceanic crust composed of basic rocks is created. These ridges are characterized by high heat flow, weak doming of the lithospheric mantle, and thin crust (Schmincke, 2004). Most of the ridges are submarine, except the subaerial outcrops of the mid-Atlantic ridges in Iceland and Jan Mayen (Norway). Surface studies and direct investigations using submersibles (Chester, 1993) provide valuable data on how ridges function. Central rift fissures, pillow lavas, dikes, and hydrothermal activity occur during this volcanic activity. The entire set of ridges constitutes a range of mountains 70,000 km in length; it is without doubt the longest mountain range on Earth. The center of this range is displaced by transform faults; studies of its cross sections show that the rate of extension (sea-floor spreading) varies along the ridge (Perfit and Davidson, 2000) (*Fig. 4.4*). From north to south, volcanic islands linked to the Mid Atlantic ridge are Jan Mayen, Iceland, Santa Elena, Tristán de Cunha, and Bouvet.

Intra-plate rift-valley-type volcanoes are located primarily in East Africa (*Fig. 4.5*). The rift includes a group of wide lakes located primarily between grabens (Turkana, Nakuru (*Fig. 4.6*), Alberto, Kivu, Tanganika, Malawi). Two large

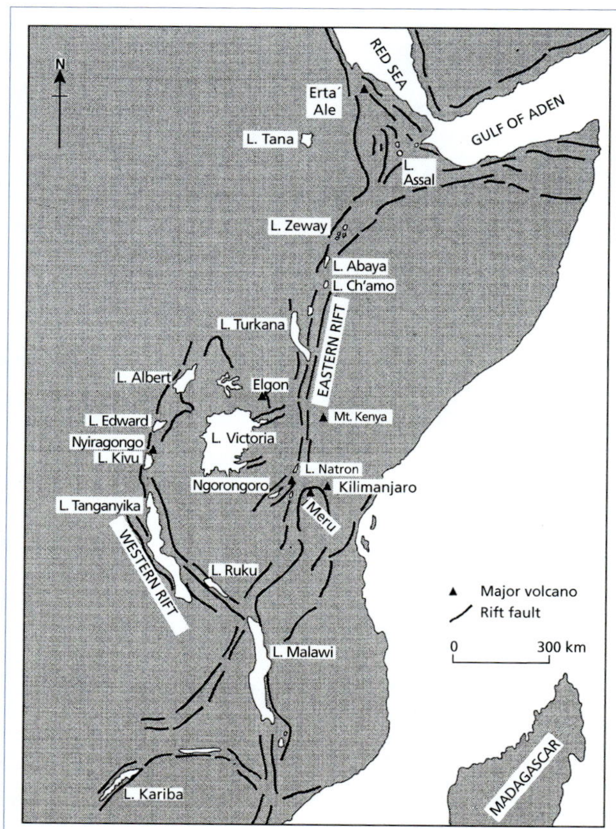

Figure 4.5 Principal lakes, rift valleys, and volcanoes of the East African Rift system (Francis, 1993).

Figure 4.6 Lake Nakuru, Rift Valley Gregory. Kenya.

African volcanoes related to rift valley dynamics are located to the east; they are Mount Kenya (5194 m) and Kilimanjaro (5894 m) and they are located about 100 km from the rift margin (Nyamweru, 1980).

At convergent or destructive margins, one plate sinks beneath another in a **subduction zone,** which is characterized by deep oceanic trenches and volcanic chains above the thrusting plate. It is thought that 80% of active volcanoes are found in subduction zones (Clapperton, 1977). The circum-Pacific volcanic belt has been called the "*Ring of Fire*" to indicate the frequency and danger of eruptions. The "andesite line" indicates the predominant composition of the magmas (Holmes, 1965) and marks the boundaries of the Ring of Fire. Even so, magma composition along this belt is variable and not all eruptions are violent. In plate tectonics the name "volcanic orogenic series" has been proposed for convergent margins (Ringwood, 1974; Perfit and Davidson, 2000) (*Fig. 4.7*). Volcanoes are accompanied by earthquakes to a depth of 600 km in the Benioff zone. The amount of lithosphere recycled into the mantle in a subduction zone is equal to the volume created along divergent plate margins.

The Pacific Ocean margin has a large number of volcanoes, many of them active: Kamchatka, Japan, the Philippines, Celebes, New Guinea, New Hebrides, Samoa in the Western Pacific and the Aleutians, the Cascade range in Canada and the United States, Paricutin, Irazu, the Andes, and Mount Burney in the Eastern Pacific.

Intra-plate volcanism owes its activity to **mantle plumes**, places where magma rises as hot currents or "plumes" (Sparks, 1993); these hot spots form volcanoes at the Earth's surface (Wilson, 1993). This type of volcano can be found on both continental and oceanic plates (Fig. 4.8). The most characteristic example of oceanic intra-plate volcanism is the Hawaiian Islands; they are located thousands of kilometers from the closest plate boundary. The Azores, Ascension, Canary, and Cape Verde islands are all located far from oceanic rifts, and are interpreted by many authors to have formed from mantle plumes. Research indicates that hot spots occur in the mantle and ascend from there to the surface.

Mantle plumes produce a wide variety of rocks on continental intra-plates. The volcanic massifs of the central Sahara, for example, reach heights of 3000 m (Hoggar in

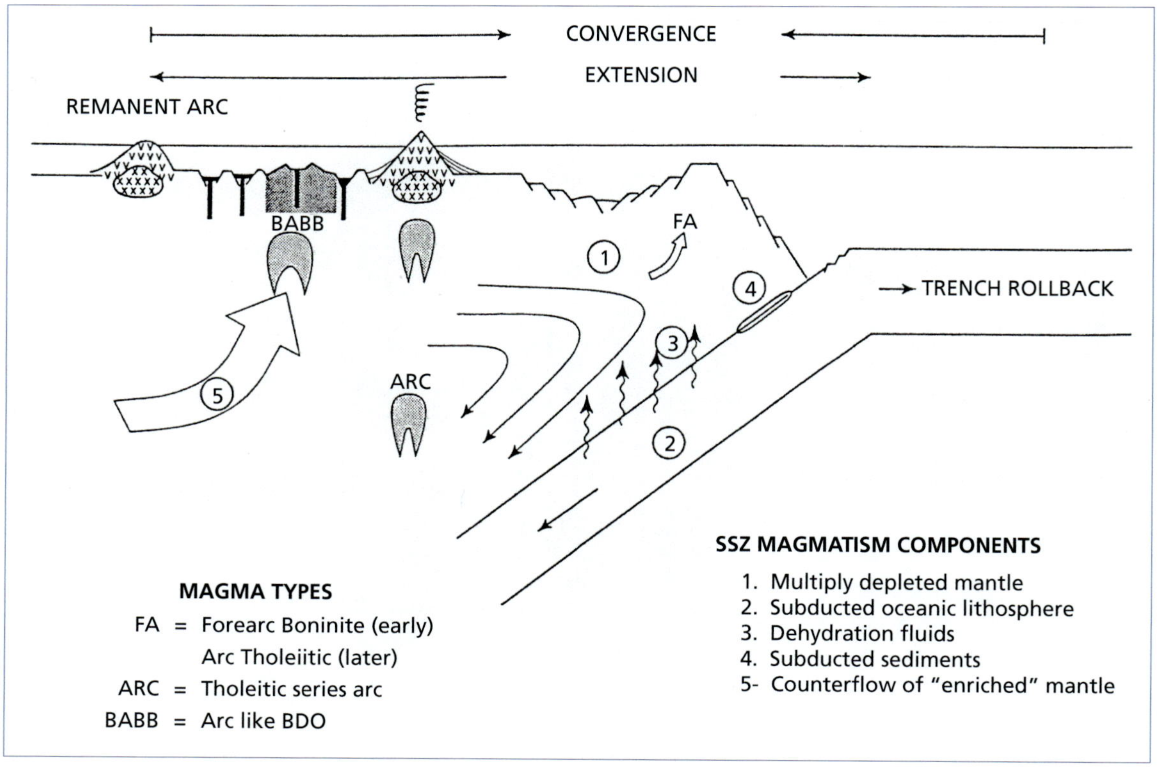

Figure 4.7 Schematic section of an intra-oceanic convergent margin showing the principal tectonic and volcanic characteristics. The arrows and shaded areas represent areas where magma is forming in various tectonic environments below the fore-arc, active arc, and back-arc. Components that may contribute to magmatism in the supra-subduction zone (SSZ) numbered 1 to 5. SSZ = supra-subduction zone. BZPA = Back-arc basalt zone. (Hawkins, 1995).

Algeria and Tibesti in Chad). This Tertiary volcanism occurred on continental crust and caused partial melting and assimilation of continental rocks accompanied by eruptions of rocks with high silica and alkaline contents.

Basaltic magmas that erupt lava onto continents cause **flood basalts**, which have been active over long periods of earth history. Researchers paid little attention to them until the publication of a special volume devoted to them (Macdougall, 1988). These basalts are predominantly tholeiitic although in Decca (northern India) they are alkaline and generally Cretaceous in age. Flood basalts occur throughout the world and can cover thousands or hundreds of thousands of square kilometers as in Paraná, Brazil, where they are Jurassic-Cretaceous in age and occupy an area of 750,000 km^2 (*Fig. 4.9*). The Columbia River (northwestern United States), Karroo (South Africa), Siberia, and other outcrops cover smaller surface areas. These piles of lavas and pyroclastics, which are sometimes several thousands of meters thick, can fill paleo-topography. Differential erosion produces stepped slopes (traps).

4.3 Magma and volcanic activity

Silicate magmas are chemically complex because they contain silicate groups that incorporate numerous elements. Three factors influence the melting and solidification temperatures of magma: composition, pressure, and volatile content. Melting temperatures are lower for more acidic rock compositions (more silica and fewer iron and magnesium silicates). Magma at high pressure (greater depth) melts at higher temperatures than magma at low pressure. Melting temperature decreases as the volatile content increases (Francis, 1993).

Magma can begin to crystallize in the Earth's interior with the formation of phenocrysts, which are minerals that have a high melting point (olivine and pyroxenes in basaltic magmas). Volatiles, which can be measured by a variety of techniques, include: H_2O, SO_2, CO_2, H_2, Cl, F, CO, and others. They play a dominant role in the eruption of magma. For example, basaltic magma at the oceanic ridges contains less than 0.5% water whereas rhyolitic magma has 4–5%.

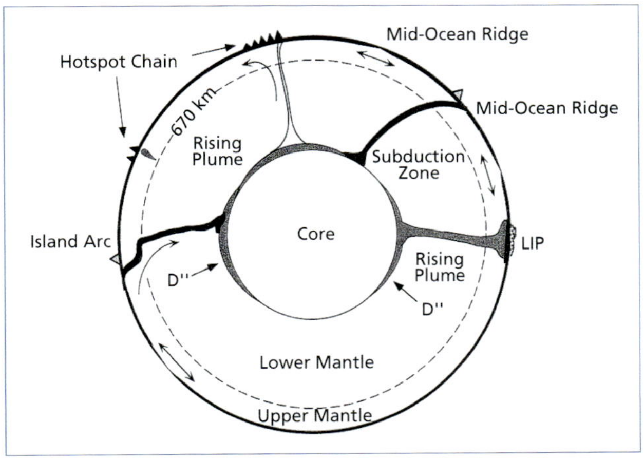

Figure 4.8 Diagram showing the hypothesis of oceanic lithosphere recycled in the mantle and the origin of mantle plumes from the seismic D" layer at the mantle-core boundary. This model is based on tomography images of the interior of the Earth, from seismic studies. The characteristics and depths are without scale. The thickness of layer D" is highly exaggerated. Large igneous provinces (LIP) can form from voluminous volcanism associated with the collision of the head of an ascending plume over the base of the lithosphere. Chains of hot spots can result from shallow plumes from the 670 km discontinuity or from the continuous rise of deep plumes from the base of the mantle (Perfit and Davidson, 2000).

Figure 4.9 Flood basalts. Paraná basalts. Iguazú Falls.

Volatiles, which have low density, concentrate toward the roof of the magma chamber and erupt first (Decker and Decker, 1998).

Viscosity is an important concept in volcanology; it describes the "slowness" of a fluid or its resistance to moving. Magmas become more viscous as they cool; for example, rhyolitic magma becomes eight times more viscous as it cools from 1300°C to 600°C. Basaltic magmas solidify at about 1000°C and their viscosity decrease is very significant (Chester, 1993).

Volcanic activity is classified according to **explosivity**, which is the percentage of particulate material compared to the total volcanic product erupted. When calculated for a volcanic zone, it can be used to compare one volcanic region to another (Rittmann, 1962). Thus, Indonesia's explosivity index is 99%, Southern Italy's is 40%, Iceland's is 39%, and the Central Pacific Ocean's index is 3%. Newhall and Self (1982) propose another explosivity index; it is based on the ratio between total volume of products erupted and the height of the eruptive column (*Table 4.1*).

4.4 Types of volcanic activity

Volcanic eruptions are difficult to classify because their characteristics change as they erupt. Some characteristics last only a few days but others can continue for months or years. In addition, the type of activity can vary over time, resulting in complex processes that must be simplified for classification.

One simple classification is a division into **central, areal,** and **fissural** eruptions. Central volcanism refers to an eruption that ascends through a vent or chimney. Some central-type volcanoes, like Fujiyama, have erupted their materials through summit craters. Others, like Etna, have numerous effusion vents on their flanks, called boccas or satellite vents (Chester *et al.*, 1985). Areal volcanism, also known as multi-orifice volcanism, is characterized by the lack of localized eruption centers situated at concrete points for long periods. Their eruptions are lavas and pyroclastics that pile up to form part of a group of small volcanic edifices. Individual volcanic structures are usually small, generally less than 450 m high.

Table 4.1 **SIZE AND FREQUENCY OF VOLCANIC ERUPTIONS. IEV = EXPLOSIVITY INDEX. AS THE IEV INCREASES, THE ERUPTION GOES FROM A STROMBOLIAN TYPE TO A PLINIAN (NEWHALL AND SELF, 1982; SIMKIN AND SIEBERT, 1994)**

EI	Volume of pyroclastic fall m^3 (km^3), with examples	Height in km of the eruption column	General description	Expected frequency
0	Less than 10^4	Less than 0.1	Not explosive	Varies per year
1	10^4–10^6	0.1–1	Small (Strombolian)	Varies per year
2	10^6–10^7	1–5	Moderate (Strombolian)	10 each year
3	10^7–10^8 Nevado del Ruiz, 1985	3–15	Moderate-large (Sub-Plinian)	2 each year
4	10^8–10^9 (10^{-1}–1) Galunngung, 1982	10–25	Large (Plinian)	4 each decade
5	10^9–10^{10} (1–10) Mt St. Helens, 1980	More than 25	Very Large (Plinian)	5 each century
6	10^{10}–10^{11} (10–10^2) Krakatoa, 1883	More than 25; climate effects	Huge (Plinian)	2 every 300 years
7	10^{11}–10^{12} (10^2–10^3) Tambora, 1815	More than 25; climate effects	Colossal (Ultra-Plinian)	1 every 1000 years
8	Mayor de 10^{12} (more than 10^3) Toba, 74,000 years ago	More than 25; climate effects	Super-eruption (Ultra-Plinian)	1 every 100,000 years

Fissural eruptions occur when magma fills long cracks prior to a surface eruption (Rittmann, 1962) (*Fig. 4.10*). These eruptions are not confined to a particular area but begin at a fissure and form curtains of basaltic eruptions that may diminish to individual eruptive centers connected to the fissure (Iceland and Hawaii). When the eruption ceases, the rock solidifies in the fissure and forms a dike.

Apart from central, areal, and fissural types of eruption, the type or types of eruptive activity are controlled by composition, temperature, and volatile content; these factors determine magma viscosity and explosivity. Violent explosions can also result when magma of any volatile content comes into contact with water present in the crust, producing **hydrovolcanic** or phreatic eruptions (for example, in areas near the sea or significant aquifers). Intensely violent eruptions often have a hydrovolcanic component.

4.5 Principal types of volcanic eruptions

As we have seen, factors that determine eruptions can be highly variable and can change over time. Holmes (1944) proposed a classification based on the degree of eruptive violence (*Fig. 4.11*). However, scientists have observed that a volcano erupts differently over the course of an eruption, which indicates the complexity of volcanic eruptions and the simplification implied in this type of classification.

4.5.1 Icelandic eruptions

In Icelandic eruptions, basalt flows out along fractures created by the extension of the mid-oceanic rift (*Fig. 4.12*) as the North American and Euro-Asian plates diverge. The volcanoes are shown in cross section (*Fig. 4.11*). Not all fissures (locally called *gjä*) become conduits for surface eruptions (*Fig. 4.13*) although magma fills them at depth. The fissures are usually short-lived; they begin to develop as basaltic eruptions with some gas content and the lava flows rapidly, creating lava plains. The eruption finally grows weaker and slowly stops. The most famous volcano in Iceland is Lakagigar; its eruption in 1783 was the most significant and calamitous in history. Active volcanism continues, as shown by the eruption of Helgaffell in 1973.

Figure 4.10 Volcanic cones from fissural eruptions (Rittmann, 1962).

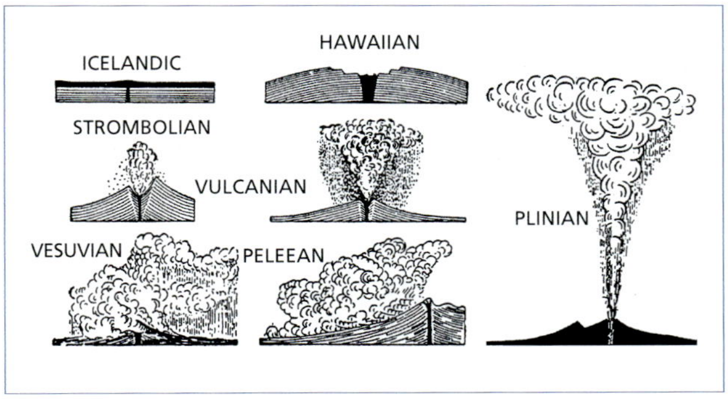

Figure 4.11 Principal types of volcanic eruptions (Holmes, 1944).

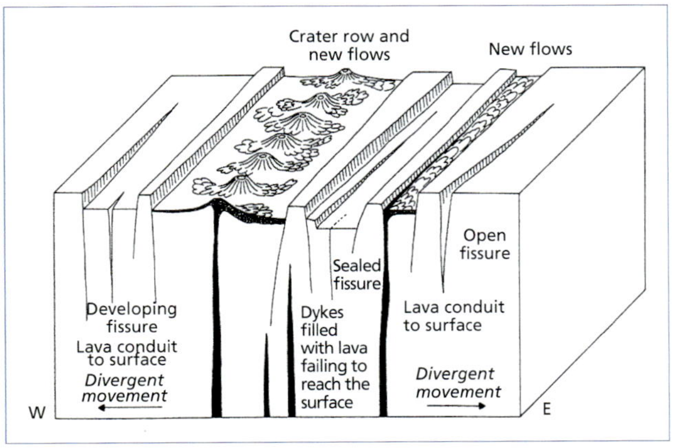

Figure 4.12 Typical island rift and fracture set (no scale) (Scarth, 1994).

The flows from this eruption reached the edge of the town of Heimaey (*Fig. 4.14*) and affected part of the port of Vestmannaeyjar, Iceland's most important fishing port (Bullard, 1976). Fissure eruptions on the island of São Jorge in the Azores are another example of Icelandic eruptions.

A series of flows of this type over a long period of time in past geologic epochs produced *flood basalts* that today occupy broad areas. They were described earlier in this book (Section 4.2).

4.5.2 Hawaiian eruptions

The Hawaiian Islands, in the center of the Pacific Ocean, were built by basaltic eruptions. These islands are also known for fissure eruptions although the dominant type of eruption is lava from a central dome. The islands were formed by repeated accumulations of lava that initially formed submarine mounts (*seamounts*) and later emerged at the surface, creating an archipelago. The eruptions are of low-viscosity basalt accompanied by small explosions due to the minor volatile content. The flows pour over the surface, reaching areas far from the eruptive center, creating topographies with dips of 3–5°. The most important volcanoes are Mauna Loa and Kilauea. Mauna Loa has had 39 recognized eruptions since 1930 and Kilauea less than 60.

A peculiarity of the Hawaiian Islands is the presence of low-viscosity **lava lakes** that persist for long periods and attract many tourists. The lakes are nested in craters and calderas. The Halemaumau crater in the Kilauea caldera was the

Figure 4.13 Crack or distension passage Mid-Atlantic Ridge. Pingvellir (Iceland). Photo F. Gutierrez.

Figure 4.14 Basalt flows from the 1973 eruption affecting the port and houses of the town of Heimaey (Vestmann Islands, Iceland). Photo. F. Gutiérrez.

site of a lava lake for a period of time. Lava lakes can exist in volcanoes in other geodynamic contexts. For example, the lava lake in the Nyiragongo (Zaire) volcano (Tazief, 1979) occupied a crater 1200 m in diameter and 1000 m deep until it emptied rapidly on 10 January 1977 through a system of parallel fissures on the flank of the volcano. Approximately 10 million m^3 of lava flowed out in about an hour at speeds of 100 km/hour with fronts that were two meters high. This event killed between 50 and 100 people and caused significant damage to agriculture (Carracedo, 1988; Fischer *et al.*, 1997).

4.5.3 Strombolian eruptions

The name comes from the volcano on the island of Stromboli, north of Sicily. This volcano is active intermittently and for short durations with moderate explosions that emit pyroclastics to heights of hundreds of meters and lavas that are predominantly basaltic. The explosions are of similar intensity and their cones generally do not exceed a height of 250 m. Stromboli is a stratovolcano that has grown about 3000 m above the ocean floor by the accumulation of lava and pyroclastics. Large lateral collapses that occur on its eastern flank extend to the sea (Romagnoli *et al.*, 2009).

Paricutin (Mexico) is another example of a Strombolian volcano, as is Etna (3,345 m., Sicily), which erupted ten times between 1974 and 1993. Etna has numerous vents on the flanks of the volcano and volcanic deposits alternate between lava and pyroclastics. The 1993 eruption damaged houses built on its flanks (*Fig. 4.15*). The Pacaya (Guatemala) volcano is a Strombolian volcano that is more active than Etna (Francis, 1993).

4.5.4 Vulcanian eruptions

These eruptions owe their name to the volcano on the island of Vulcano, located on the Lipari Islands north of Sicily. They are more highly explosive than the Strombolian type. The particulate matter consists of ash, lapilli, and bombs whereas flows are rhyolitic, trachytic, or andesitic in composition, characteristic of high viscosity magmas. Eruptions of columns of pyroclastics and gases are usually violent and can rise to 10–20 km. Eruptions are intermittent with long periods of repose, on the order of a century, and active phases usually last several months. The lava erupted usually has a limited extent.

The Irazú volcano (Costa Rica) (Fig. 4.22) was very active between 1963 and 1965; it did not produce lava eruptions but it did erupt significant ash. In 1968 it erupted particulate material that reached velocities of 600 m/sec along ballistic trajectories (Morrissey and Martin, 2000). Other volcanoes that erupt similarly are the Sakurajima and Asama volcanoes, both in Japan, and the Ngauruhoe volcano (New Zealand), which erupted in 1975 with explosions of andesitic magma.

4.5.5 Vesuvian eruptions

This type of activity is more violent than Strombolian and Vulcanian eruptions (Holmes, 1944; Ollier, 1969a; Bullard,

Figure 4.15 Houses affected by the flows from the 1993 eruption of the Etna volcano. (Sicily, Italy) Photo F.J. Gracia.

1976). The name comes from Vesuvius, the volcano located in the Bay of Naples. A catastrophic eruption of Vesuvius buried the cities of Pompeii and Herculaneum on 24 August in 79 AD (*Fig. 4.16*). Pompeii was a city of 20,000 inhabitants located 20 km southwest of the volcano. The two-day eruption killed the majority of its inhabitants. Pliny the Younger, who survived the eruption, left a detailed description of it. Eruptions of large columns of gases and pyroclastics caused these materials to fall like rain ("Plinian rain") that accumulated to a thickness of about 5 meters and practically buried both cities, which were located on the flanks of the volcano. This almost instantaneous burial of the cities preserved them along with numerous objects of everyday life. Archeological excavations of the cities began in the 15th century and continue today. These studies have provided a wealth of knowledge on Roman life at the time (*Fig. 4.17*).

4.5.6 Peleean eruptions

This eruptive classification is named for Mont Pelée (Island of Martinique) (*Fig. 4.18*), a volcano located in the center of the arc of the Lesser Antilles. The city of Saint Pierre, the principal city on the island, had 28,000 inhabitants; it was located 7 km south of the volcano. The eruption of 7 May 1902 is famous; it was described in great detail by the French petrologist Lacroix (1904) who studied it while it was active.

Prior to its destructive eruption Mont Pelée (1,397 m) had begun its activity with tremors, phreatic eruptions, and small ash clouds. On 8 May 1902 an explosive eruption from the lower slope ejected a **nuée ardente** which completely destroyed the city of St. Pierre in two or three minutes. Nuée ardentes contain solid particles (large blocks mixed with fine powder) that slide downslope by gravity (Fig. 4.19).

The high temperature gases rise, carry large quantities of dust with them, and form a turbulent cloud up to a height of 30 km. In the case of Mont Pelée, the cloud reached a velocity of 200–500 km/hour and a temperature between 200 and 450°C (Scarth, 1994). The Mont Pelée Museum contains exhibits of partially melted and fused scissors and bundles of nails that provide an approximate temperature of the nuée ardente, which is somewhat higher than the melting point of iron. On 20 May 1902 a second nuée ardente hit St. Pierre.

In November of 1902, a large pillar of solidified lava began to rise above the crater, pushed by the pressure of gases in the magma. It grew about 10 meters a day and reached a height of 310 meters by May 1903 (Francis, 1993). This plug was destroyed by a later eruption in 1929.

4.5.7 Plinian eruptions

This eruption style is named after Pliny the Younger, who saw the eruption of Vesuvius in the year 79 AD and described the volcanic activity in detail. These eruptions are the most violent of all volcanic activity, but they are not very frequent (Williams and Self, 1983). Some recent examples are: Mount St. Helens, United States, (1980), El Chinchón, Mexico,

Figure 4.16 Vesuvius in 2004, in the foreground, ruins of Pompeii where the volcanic ash that buried it has been excavated.

Figure 4.17 Sleeping man. Ruins of Pompeii preserved under a layer of ash emitted by the eruption of Vesuvius in 79 AD.

Figure 4.18 The crest of Mount Pelée covered with clouds and the town of Saint Pierre at the base (Martinique Island, Lesser Antilles).

Figure 4.19 The ruins provide testimony of the catastrophic eruption of Mount Pelée on 8 May 1902, which killed about 25,000 people.

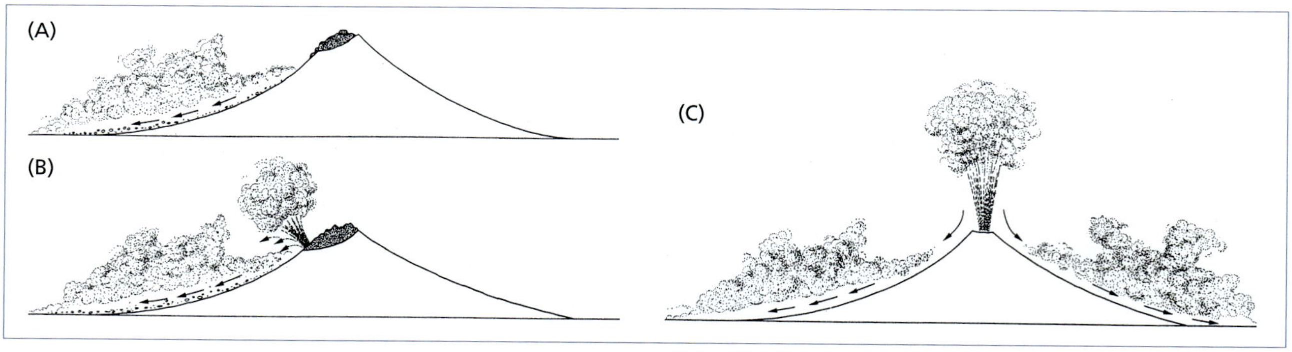

Figure 4.20 Three generation mechanisms for nuées ardentes. (A) Simple gravitational collapse of a dome or flow on a volcano. (B) Explosive fracture of a growing lava dome. (C) Collapse of an eruption column (modified from Francis, 1993).

(1982), and Pinatubo, Philippines, (1991). In this type of activity, the volcano expels fine ash and pumice of an acid composition into the atmosphere in columns that can reach a height of 45 km before the material falls "like rain." The volcano emits lava at the end of the eruption. The ash can be deposited hundreds of kilometers from the volcano and the finest powder can move into stratospheric currents and encircle the globe. The powder that erupted from the Krakatoa caldera (1883) has been found in many places throughout the world. The active period of these eruptions is usually a few days. Some eruptions of this type may have larger dimensions than the preceding types; these are called ultra-Plinian (Walker, 1980).

4.6 Nuées ardentes and pyroclastic flows

In contrast to pyroclastic rains of the Plinian type, nuées ardentes and pyroclastic flows move with a significant horizontal component. These are dense gaseous suspensions that contain solid particles of various sizes and drops of molten lava at temperatures of hundreds of degrees. They can circulate at velocities of hundreds of km per hour (*Fig. 4.20*).

They can result from the collapse of domes or pillars of highly viscous gas-rich lava (as in Merapi, in Indonesia, or Mont Pelée in the French Antilles) that flow downslope, conforming to the topography. They have generated pyroclastic deltas in the Caribbean Sea. They can also result from the collapse of eruptive columns and form ring-shaped *base surges* similar to the rings produced at the base of the "mushrooms" of atomic explosions. These fringes or basal surges move radially from the eruptive column and can be very destructive, such as the surge at La Soufrière on the island of St. Vincent (Lesser Antilles).

4.7 Hydrovolcanic eruptions

These eruptions occur when incandescent volcanic materials come in contact with water in its various forms, such as when the eruption reaches the sea, a lake, or groundwater. The stark temperature contrast vaporizes the water almost instantaneously and the vapor expands abruptly, causing a violent explosion.

4.7.1 Sub-aqueous eruptions

Many volcanic eruptions occur beneath the sea. Marine water exerts high pressure on volcanic vents and causes pyroclastic explosions to occur at shallow depths. As they cool, basaltic lavas solidify as **pillow lavas**, which look like a solid crust. Continuous eruptions over long time periods create islands at mid-ocean ridges such as the island of Surtsey south of Iceland, which formed between 1963 and 1967, and island arcs such as Kick'en Jenny Island south of the Lesser Antilles. Other examples include intra-plate islands such as Hawaii. Rapid solidification of lava in subaqueous environments creates volcanic glass (solidified lava that is amorphous but not crystallized) known as **hyaloclastite**. This glass frequently occurs in the spaces that remain between the pillows that form pillow lava. Hyaloclastite also forms at the bottom of lakes and below glaciers (Chester, 1993) (*Fig. 4.21*).

This type of eruption can occur in lakes within craters such as the one inside the crater of the Irazú volcano (Costa Rica, 3432 m) (*Fig. 4.22*). In these cases, eruptions are smaller and less dangerous to humans. Some of these volcanoes, such as Poas in Costa Rica (*Fig. 4.26*), emit sulfurous gases and the water in the lake can have a pH of less than 1. With tropical rains, sulfur-rich eruptions produce acid rains.

Figure 4.21 Hyaloclastite layers Montaña Amarilla. Tenerife.

Figure 4.22 Lake within the crater of the Irazú volcano (3,432 m). Costa Rica. Photo F. Gutiérrez.

4.7.2 Phreatic eruptions

When a vent interacts with water present in the phreatic zone, a violent explosion occurs that releases vapor and fragments of surrounding rock. Phreatomagmatic eruptions also expel magma; in some cases the proportion of particles exceeds 90%. Fine ash is also picked up and transported by the wind. The Eifel district (eastern Germany) is home to volcanic morphologies called **maars**, which are large closed depressions and craters between 100 and 3000 m wide and 10 to 500 m deep. Seventy-two of them are spread over the region; lakes fill their interiors. Standing out on the general land surface, they are composed primarily of a ring of fragments and lavas with slopes that dip 10–15° toward the outside of the crater and more steeply to the interior. They are of hydrovolcanic phreatomagmatic origin, formed when basaltic magma came into contact with groundwater in river valleys and depressions. The different morphologies of the maars result from a combination of explosive activity and subsidence (Sheridan and Wohletz, 1981 and 1983).

Studies of core samples from the maar of Lago Grande of Montichio, east of Naples (Watts *et al.*, 1996) have led to good paleoclimatic results. The same is hoped for borings that have been installed to a depth of 142.4 meters in the Fuentillejo maar, Lake of the Posadillas, in the Calatrava volcanic field (Ciudad Real, Spain) (Vegas *et al.*, 2004; Martín-Serrano *et al.*, 2009).

4.7.3 Sub-glacial eruptions

These eruptions take place beneath an ice cap; they cause partial melting of the ice, which can cause significant flooding. When the ice cap is thin, eruptions beneath the ice can melt the roof and the eruption becomes a sub-aerial eruption. When the ice is thick, the melting creates an internal lake and the eruption is sub-aqueous, producing *pillow lavas* that pile up, potentially causing pyroclastic explosions (volcanic breccias and hyaloclastites, Bennet *et al.*, 2009) in the shallow lake due to the interaction of the rising magma and the water. The resulting overpressure can destroy the ice dome. Successive eruptions of the sub-aerial type cause volcanic material to pile up and take on the form of planate hills or **tuyas** (*Fig. 4.23*) (Mathews, 1947; Smellie, 2007) as they are called in British Columbia.

The Vatnajökull (Iceland) ice cap lies atop the Grimsvotn volcano and normally drains down its flanks. When an eruption occurs, the ice melts and causes enormous floods called **jökulhaups**, with discharge almost as large as the Amazon. They have great erosive and transport power and are the primary cause of sand and gravel plains known as **sandar.**

4.8 Rock avalanches and lahars

Some eruptions cause gigantic explosions that blow away part of the volcanic edifice, like the eruption of Mount St. Helens (northwestern United States) in May of 1980. These rock **avalanches** are characterized by low fragmentation, high accumulation, and moderate dispersion. A volcano that has generated even larger avalanches is Socompa, located in the Atacama Desert of Chile. The amphitheater created by Mount St. Helens has an angle of 30° whereas those of Socompa reach 70° (Francis *et al.*, 1985). The hyperarid climate of the area has preserved the deposits, making it an excellent place to study avalanche deposits.

These collapses can be caused by magmatic eruptions or by non-magmatic explosions such as phreatic eruptions. They can also result from "cold" explosions of a seismic origin.

A **lahar** is a different type of avalanche. The word lahar comes from Java; some researchers use the names "*debris flow*" or "*mud flow*" for these deposits. Lahars are composed of volcanic ash and water and may contain

Figure 4.23 Table mountain (tuya) created by subglacial eruptions. Mývatk (Iceland). Photo. F. Gutiérrez.

various types of materials, including large blocks (*Fig. 4.24*) (Tazieff and Derruau, 1990). The water can be rain or snowmelt created by the eruption as occurred in 1985 at the Nevado del Ruiz volcano in Colombia. Depending on the water content and the slope it follows, a lahar can reach or exceed a speed of 100 km/hour. Lahars from the slopes of the Nevado del Ruiz volcano reached speeds of 60 km/hour, buried the city of Armero, and killed more than 22,000 people (Voight, 1990). The inhabitants could have been saved by a simple alarm system. In recent times, lahars from the eruption of the Soufrière volcano (Island of Guadaloupe) on 8 July 1976 (*Fig. 4.25*) were not very destructive, but those from the Nevado del Ruiz and Mount St. Helens were notable for the long distances they traveled down the Lagunillas, North Fork Toutle, and South Fork Toutle rivers, respectively.

Figure 4.24 Lahar outcrops showing their heterogeneity. St. Pierre (Island of Martinique, Lesser Antilles).

Figure 4.25 The La Soufrière volcano in May 1994. Island of Guadaloupe. (Lesser Antilles).

Figure 4.26 Fumaroles at the Poas volcano (Costa Rica). Photo F. Gutiérrez.

4.9 Gaseous or hydrothermal eruptions

Volcanic gases can also escape through fissures and vents to form fumaroles, solfataras, and boiling muds, unaccompanied by explosions or lava eruptions. On 21 August 1986 Lake Nyos (Cameroon) released carbon dioxide (called mofeta) that killed 1742 people. The origin of the gas has generated wide discussion.

Fumaroles are composed of gases and water vapor at high temperatures that escape to the atmosphere though vents generally smaller than 10 cm (*Fig. 4.26*). Fumaroles can reach temperatures of 1000°C and often exceed 100°C. They are common; Yellowstone National Park has thousands of fumaroles. Near Naples (Italy) **solfataras** with more than 29 active vents release hot gases of carbon dioxide, sulfur dioxide, and hydrogen sulfide in addition to mud and water vapor. The sulfur reacts with water to produce sulfuric acid, which significantly increases the acidity of the water.

Geysers are jets of boiling water that spurt periodically. The name comes from the Geysir region of Iceland (*Fig. 4.27*). Geyser vents are surrounded by a margin of stepped rings formed by the precipitation of dissolved silica (geyserite). Areas known for geyser activity are Yellowstone National Park, Iceland, and Waiotapu (New Zealand).

Numerous countries have exploited hydrothermal activity for heating: Larderello (Italy), Sao Miguel (The Azores), Reykjavik (Iceland), China, Philippines, Japan, United States, among others.

Figure 4.27 Geyser with a border of geyserite. Strokkurgeysir (Iceland). Photo F. Gutiérrez.

4.10 Lava flows

When liquid magma reaches the surface, it forms lava flows that follow the topography under the influence of gravity. The flow type, velocity, and area covered by the lava depend on their chemical composition, volatile content, degree of crystallization, temperature, and interrelated variables that determine viscosity. The viscosity, together with the volume erupted and the relief over which they flow, determine the velocity that the lava can attain. Individual lava flows can flow distances ranging from several tens of meters to tens of kilometers. In some cases they flow like rivers through narrow reaches; if the volume is large enough they can fossilize landforms (*Fig. 4.28*). Flows are active for periods ranging from hours to weeks; in some cases, such as Hawaii or during the 1730–36 Lanzarote eruption, a volcanic edifice can produce flows for years.

Two large lava groups produce flows of different types. Acid lavas, which are rich in silica, have somewhat lower temperatures and higher viscosity. As a result, they flow slowly and are limited in their extent. Basic lavas (especially basaltic) are silica-poor and have higher temperature and lower viscosity; as a result they can flow tens of kilometers per hour and travel distances of kilometers.

Due to these differences, the behavioral differences of lava can provide a wide diversity of surface structures, many of which reveal the high viscosity and deformable nature of these "solid-liquid" materials.

The low thermal conductivity of lavas sometimes causes an external crust to solidify while the interior lava continues to flow through it as a higher-temperature liquid; incandescent lava be seen through cracks in the crust.

4.10.1 Basalt flows

Basaltic lava flows to the surface at a temperature of about 1200°C. The flow can move 25 to 75 km/hour but the speed decreases as volatiles are released and when the flow comes into contact with the cooler atmosphere. When this happens, the flow cools at the surface and a set of cracks form. The viscosity of the internal lava increases and it flows more slowly, like the front of the flow, and the internal part takes several days to solidify. The flow may assume a linguoid shape; its edges cool and the lava channelizes on the inside, leaving sub-vertical ridge-shaped walls called **levees** (*Fig. 4.29*).

Two principal types of basaltic lava are called by their Hawaiian names, **aa** and **pahoehoe.** The former is the most common type of lava; its rough and spiny surface is difficult and dangerous to walk on. Large fields of aa lava are known as **malpais** in the Canary Islands (*Fig. 4.30*). One variety is block lava, which is characterized by individual fragments that measure 0.25 to 1 m. Part of the town and port of Garrachico

Figure 4.28 Undulating front of lava flows moving toward the Cañadas complex. Llanos de Ucanca. Island of Tenerife. (Canary Islands).

town (Tenerife, Canary Islands) were buried by block lavas during the 1706 eruption (*Fig. 4.31*) that started at the pre-littoral cliff and reached the coast (*Fig. 4.32*) (Romero, 1990).

Pahoehoe lavas are the least viscous of all lavas and as a result they create fluid structures. Unlike aa lavas, pahoehoe surfaces are smooth and glossy. Gas escaping from the flow generates numerous bubbles that create shallow vesicles on the surface. The lava can flow in individual cylindrical bodies that flow on top of each other, creating "sausage" lava (*Fig. 4.33*). Ropy lava, named for its interlaced cable appearance, is one of the best examples of pahoehoe lava. Pahoehoe lavas are so plastic that they can flow around tree trunks, carbonize them, and leave a mold as a vestige (Bullard, 1976).

Often the difference between aa and pahoehoe lava is subtle; it has been noted that pahoehoe lava can transform into aa (Schminke, 2004). It is thought that pahoehoe lavas originate from low viscosity basaltic magmas whereas aa lavas develop from a wider compositional range. In addition, Rowland and Walker (1990) indicate that aa lavas form when eruptions rates are 5 to 10 cubic meters per second.

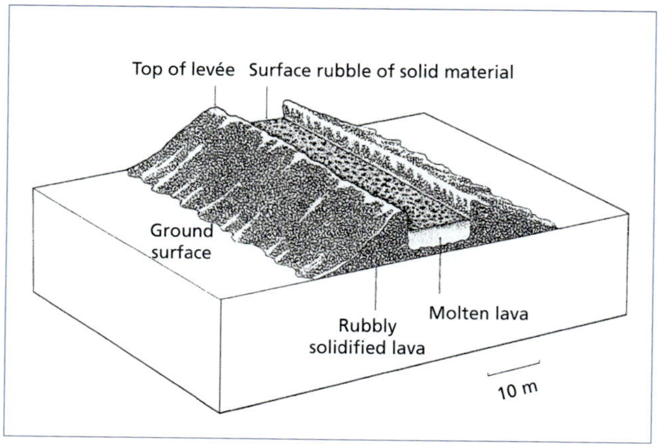

Figure 4.29 Block diagram showing the origin of levées (Francis, 1993).

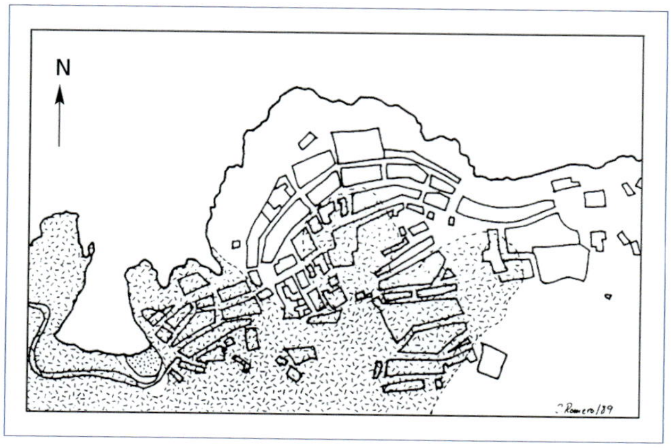

Figure 4.31 Most probable path of flows from the eruption of 5 May to 13 June of 1706 City of Garachico, Island of Tenerife (Canary Islands) (Romero, 1990).

Figure 4.30 Extensive field of aa lavas (malpais). Timanfaya. Island of Lanzarote (Canary Islands).

Figure 4.32 Subrounded block lavas on the slopes dominating the port and city of Garachico. Island of Tenerife. (Canary Islands).

Figure 4.33 "Sausage" lavas from the edge of the Teide Roques of García. Island of Tenerife (Canary Islands).

We have already indicated that the internal flow of basalt lava loses velocity as the thickness of the solidified crust increases and the basalt flows as if in a tube. If the strength of the roof of the lava tube is exceeded as the interior empties, the roof collapses, making it possible for us to observe the trajectory and length of the flow. These morphologies are called **jameos** on the Canary Islands *(Fig. 4.34)* (Bravo, 1964a). Successive collapses of the cap make it possible to follow the trace of the flow. By observing the walls of the cavern, we sometimes find **stalactities** and **stalagmites** at the base, formed by the dripping of the solidified lava; stalagmites are more rare than stalactites.

Lava in lava tubes is highly confined and shows diverse structures formed as a result of this pressure. The most frequent are surface bulges, called **tumuli**, which are generally several meters tall and several tens of meters in diameter *(Fig. 4.35)*. They typically have radial cracking as a result of the vertical push of the melted flow. The thickness can decrease significantly by drainage; as a result the surface lava can collapse, creating **ridges**. Tumuli can also result from the pressure of the underlying lava *(Fig. 4.36)* (Green and Short, 1971, Bullard, 1976). When molten lava breaks through the surface crust in places, gas pushes the lava to erupt and create small cones, called **hornitos** in Spanish *(Fig. 4.37)*. This term is used for cones that are no taller than 10 m; larger ones are called **adventitious cones** (Ollier, 1969a). Some lavas can solidify during aerial effusion creating **pearls** and **Pelé's hair**, named after the Hawaiian goddess of fire. When semisolid lava erupts through cracks or vents, it scratches the walls, creating **striated lava** *(Fig. 4.38)*.

Contracting lava generates fractures, which are generally hexagonal in shape although they can also have four, five, seven, or eight faces. Cooling cracks propagate toward the center of the flow, creating a columnar structure *(Fig. 4.39)*. Columnar basalts usually do not exceed a height of 10 m or a thickness of 60 cm. Excellent examples are the Giants Causeway in Antrim (Northern Ireland). The Organ Rock on the northern coast of Gomera (Canary Islands) is another magnificent example of this type of structure, even though it is not a basalt flow but rather a trachyphonolitic plug (Bravo, 1964b, Cendrero, 1970, 1971; Yanes, 1990). Fluvial scarps with well-developed columnar basalt can be seen at Castelfullit de la Roca (Province of Gerona, Spain). Columnar jointing is very characteristic of, but not exclusive to, large basalt flows.

Lava forms numerous landforms when it flows into the sea, lakes, and marshes. **Lava deltas** commonly occur along island coasts where lavas enter the sea, cool, and move seaward. The cooling can also be very rapid; it can cause violent explosions that emit large amounts of water and generate pseudo-craters (Thoraninsson, 1953b). Flows can extend onto the littoral, creating coastal **platforms of piled-up lava** such as the Costa del Silencio on the Island of Tenerife (Canary Islands, Spain) *(Fig. 4.40)*

Figure 4.34 Jameo that is 8 km long showing partial collapse of a large flow. Jameo de los Almacenes. Basalt from the Corona volcano. Island of Lanzarote (Canary Islands).

Figure 4.35 Tumulus with radial cracking. Roques de García. Island of Tenerife (Canary Islands).

Figure 4.36 Central Ridge. At the bottom of the Cañadas complex. Roques de García. Island of Tenerife (Canary Islands).

Figure 4.37 Hornito called Mantle of the Virgin. National Park of Timanfaya. Island of Lanzarote (Canary Islands).

Figure 4.38 Striated lava from the eruption of the Chinyero volcano in 1909. Island of Tenerife (Canary Islands). Photo F. Gutiérrez.

Figure 4.39 Columnar basalt structure Agulo, Island of Gomera (Canary Islands).

Figure 4.40 Littoral platform from lava filling. Costa del Silencio. Island of Tenerife (Canary Islands).

4.10.2 Acid lava flows

Siliceous lava and some intermediate lavas behave differently because of their higher viscosity. These lavas are rhyolitic, dacitic, and andesitic in composition. They erupt at about 900°C. They are more viscous than basaltic lavas because of their lower temperature and their higher SiO_2 and Al_2O_3 content and because they cool more quickly. Siliceous lavas travel more slowly than basaltic lava, usually not moving more than 10 km/hour. As a result, siliceous lavas flow only a short distance, generally not more than 1–2 km and they are much thicker than basaltic lavas. Acidic lava eruptions cover a smaller area than basaltic effusions.

Andesitic lavas are not very different from basaltic lavas but they have higher viscosity; as a result they move slower and remain close to the effusive vent. Block lavas (*Fig. 4.32*) are typical of andesites. The fronts of the block flows are steep; they can reach heights of 100 m. They are composed of piles of blocks that bang against each other. Lava tubes rarely occur in andesites. **Dacitic lavas** are considerably more viscous than andesites. They move very slowly and form very thick extrusions. Eruption of dacitic lava can be explosive; they often create pyroclastic cones. Because dacitic lava flows are very thick, they create **lava domes** instead of flows. Domes grow in a complex manner. In the Andes, the Chileans call the flat ones *tortas*. *Coulées* are intermediate forms between domes and lava flows, and as they flow they create morphologies similar to glacial *ogives*. Another volcanic form is the **plug** that rises up a vent like a piston. Plugs that push and bulge the topographic surface are known as **cryptodomes. Rhyolitic lavas** have high silica content and a high volatile content, which cause explosive eruptions of abundant pyroclastics. Few historic eruptions of rhyolites are known. These lavas form extrusions like dacites and are frequently found in craters formed later by large explosive eruptions. A special type of lava commonly present in rhyolitic flows is **obsidian**, a dark colored or black glass. It lacks bubbles and breaks with a conchoidal fracture, with edges that can be used for cutting. For this reason it was widely used by many prehistoric cultures to make knives, arrowheads, and other items. It has been and is still used today for jewelry because of its intense shine, which is known as "mirror of the Incas."

4.11 Pyroclastics

Explosive eruptions that emit large amounts of pyroclastics are the most powerful and destructive type of volcanic activity. These eruptions eject pyroclastic material into columns as pyroclastic flows. Recent eruptions of Mount St. Helens (in the northwestern United States) and Mount Pinatubo (Philippines) attest to their enormous destructive power.

Magma is subject to high pressure at depth and this pressure keeps the gases dissolved in the magma. As magma approaches the surface or when a fracture provides a pathway for magma to reach the surface, pressure in the magma decreases rapidly and the gases come out of solution. Bubbles form and gas pressure increases inside the magma, which propels it toward the surface (like bubbles in a bottle of champagne). This activity can cause explosions that fragment the magma and surrounding rocks (Cashman *et al.*, 2000).

Many volcanoes emit more pyroclastics than lava. Most magmas produce pyroclastics but the largest volumes are produced by magmas of acidic or intermediate composition and by phreatic eruptions. The characteristics of pyroclastic deposits depend on sedimentation mechanisms among other factors (Wilson and Houghton, 2000). For many years pyroclastics did not interest scientists but in recent decades more research has been focused on this topic (Fisher and Schmincke, 1984, Sparks *et al.*, 1997). The three principal types of pyroclastic eruptions are **pyroclastic falls**, **pyroclastic surges**, and **pyroclastic flows** (Sparks and Walker, 1973).

4.11.1 Pyroclastic fall deposits

These are deposits that fall to the Earth's surface from a column of gas and fragmentary material that has risen into the atmosphere after an eruption. This material is also called *tephra* (Froese *et al.*, 2008), a term coined by Aristotle. Its thickness can be used to calculate its volume if the topography prior to the eruption is known; this information can be used to prepare isopach maps (*Fig. 4.41*) (Sigurdson *et al.*, 1985). It is also important to know the grain sizes present and their proportions (*Fig. 4.42*). The grain sizes of pyroclastics are:

Ash . . . . <4 mm
Lapilli. . . . . 4–32 mm
Blocks and bombs . . . . >32 mm

Volcanic bombs may have a nucleus composed of country rock or peridotite; they may follow ballistic trajectories and frequently rotate as they travel along their trajectories (*Fig. 4.43*).

In addition to the parameters previously discussed (thickness, size, and sorting), vesicles that may be present are also important; these are small cavities formed by the expansion of gas bubbles during solidification of the magma. The volume of vesicles varies greatly from one eruption to another.

Pyroclastic particles are sorted by weight and size during their transport and fall. The finest particles travel the longest distances. At some locations, they are sorted by grain size with the largest fragments at the base of the deposit and with particle sizes fining toward the top, showing that the upper particles fell more slowly. These deposits may contain

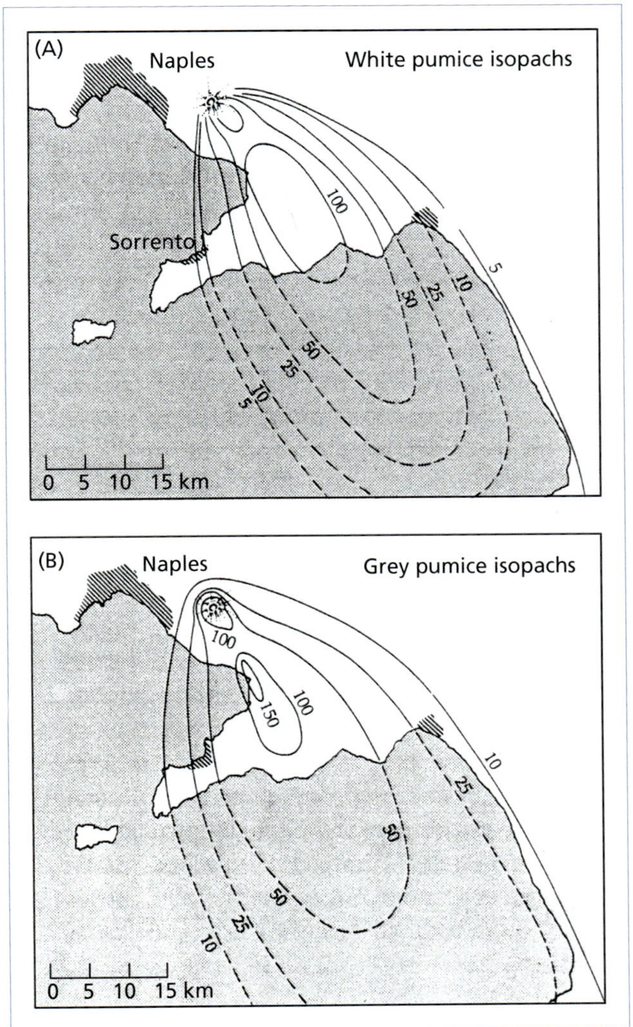

Figure 4.41 Isopach map of the Plinian eruption of Vesuvius showing the spreading caused by the wind. The isopachs are in centimeters. (A) It shows the lower white pumice and (B) the gray pumice. (Sigurdsson *et al*, 1985).

Figure 4.42 Layers of ash and basaltic lapilli with some intercalated bombs Km 31.850 from the highway between El Portillo and La Laguna. Island of Tenerife (Canary Islands). Photograph taken in 1968.

cross stratification *(Fig. 4.44)* that shows the action of strong winds or volcanic explosions. The cross bedding generated by explosions may show radial distribution as a result of expansion caused by the effusive vent.

Pyroclastic fall materials can be cemented or welded; these rocks that are called **agglomerates or volcanic breccia** when formed by coarse particles and **tuffs** when the particles are fine.

4.11.2 Pyroclastic surges

Pyroclastic surges are generated by the lateral expulsion of fragments or an ash-filled cloud flowing across the ground (Wilson, 1993). The explosion and the effect of gravity move this dense hot mixture of particles and gas. The speed and particle concentration can fluctuate abruptly during their movement. Several types of pyroclastic surges have been described: "dry" surges have temperatures higher than 100°C because the water vapor in them condenses and the pyroclastics have no cohesion. When the temperature is less than 100°C, they are called "wet" and the pyroclastics are cohesive; these are characteristic of phreatomagmatic or magmatic eruptions that contain only volatiles (Valentine and Fisher, 2000).

Pyroclastic surges can be generated by various mechanisms. The collapse of the eruptive mixture in the

Figure 4.43 Volcanic bomb. Notice the folds produced by the block rotating in the atmosphere while in the plastic state. Eruption of the Chinyero volcano in 1909. Island of Tenerife (Canary Islands).

Figure 4.44 Cross bedding in pyroclastics. Near the Gourbeque Volcanic Observatory. Island of Guadaloupe. (Lesser Antilles).

vertical eruption column is known as a *base surge*. Another mechanism is a rapid expansion (decompression) or *blast* into the atmosphere of a mixture of eruptive pyroclastics. An excellent example is the explosion on the slope of the Mount St. Helens volcano on 18 May 1980 (Valentine and Fisher, 2000).

4.11.3 Pyroclastic flows

These pyroclastics move laterally as flows of poorly sorted, high density volcanic material at temperatures of hundreds of degrees. They can move at speeds that exceed 100 km/hour and travel distances of more than 100 km. Pyroclastic flow deposits can reach volumes as large as 100 km^3. The morphology of the deposit and its thickness reflect the topography. The term **ignimbrite** is used for deposits that result from the deposition of pyroclastic flows in which pumice is the major component. They contain lenticular fragments resulting from the solidification of parts that were liquid at the time of flow.

Pyroclastic flows have high energy. When an eruptive column collapses, the potential energy in the pyroclastic flow is transformed into kinetic energy that expands laterally. Movement of the pyroclastic flow is more like a liquid than a mass of solid fragments. Pyroclastic flows have low viscosity and move as a high-velocity fluid, including over slightly dipping slopes (Francis, 1993). Because pyroclastic flows move rapidly they do not cool very much during flow and for this reason, they still contain a large amount of heat when they stop. The high temperature causes glass particles to weld together and to the matrix. Ignimbrite eruptions are rare. Research has shown that separate ignimbrite flow units can be distinguished (Smith, 1960, Sparks *et al.*, 1973). Some large ignimbrite outcrops have been formed by two or three flows. The most significant examples of pyroclastic flow eruptions are Los Alamos (New Mexico), the eruption of Tambora (Indonesia) in April 1815, the nuées ardentes from the eruption of Mont Pelée (Island of Martinique, Lesser Antilles) on 8 May 1902, that of the Valley of Ten Thousand Chimneys (Alaska) of June 1912, and the eruption of Mount St. Helens (northwestern United States) on 18 May 1980.

4.12 Volcano morphology

The shape of the volcanic edifice is related to the composition of the lava it erupts and consequently to the type of activity it has demonstrated. In the study volcanic shapes, it is helpful to differentiate between volcanoes that erupt basic and acidic lavas.

4.12.1 Volcanoes that erupt basic lava

We have already seen that basic lavas are very fluid, their eruptions are rapid, and the stacking of these flows creates volcanoes with almost flat slopes. Shield volcanoes have very gentle convex slopes with dips less than 5°. The most significant are found on the Mauna Loa volcano, which rises 10,000 m above the sea floor to an elevation of 4,000 m above sea level. The Kilauea volcano is located on the flank of the Mauna Loa volcano. Other morphologies are lava domes and lava cones, with greater convexity than shield lavas. The difference between them is very subtle (Ollier, 1969a).

4.12.2 Volcanoes that erupt acidic lava

We have previously mentioned that acidic lavas are very viscous and for this reason have difficulty getting through the eruptive vent. This causes over-pressuring that can cause violent explosions.

When lava leaves the vent, it expands and forms a convex dome also called a **cumulo-dome** or **mamelon** (*Fig. 4.45*) (Ollier, 1969a). Their diameters are much wider

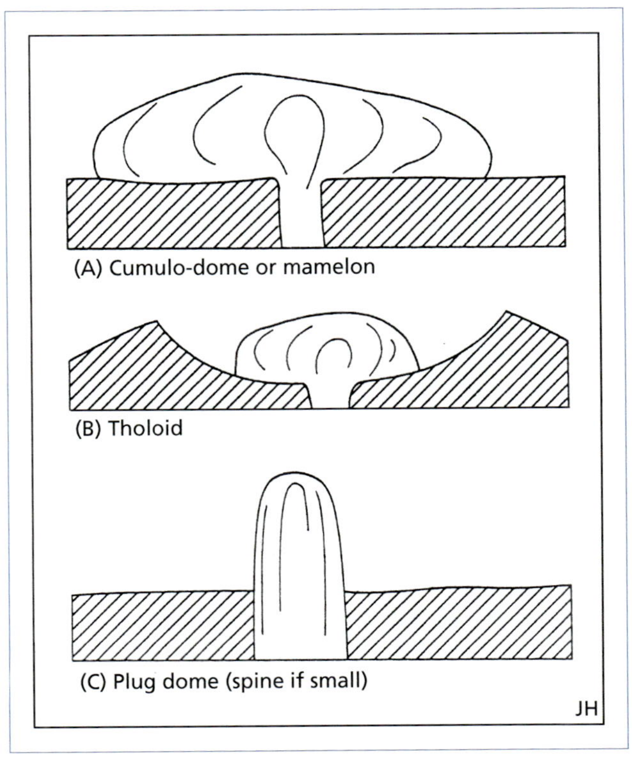

Figure 4.45 Different types of acid lava volcanic edifices. (Ollier, 1969a).

Figure 4.46 Dome on the eastern flank of Mont Pelée. Island of Martinique. (Lesser Antilles).

than the outlet vents. As the viscous lava cools, it becomes a layer. The surfaces crack during growth (Williams and McBirney, 1979). Domes (*Fig. 4.46*) are constructed of rhyolitic, dacitic, or trachitic lavas that erupt at temperatures of 750–900°C. The andesitic dome of the La Soufrière volcano (Island of Guadaloupe) (*Fig. 4.25*) is a good example. The summit is a tortuous morphology of dome folds, spires, and lakes. The last eruption occurred on 8 July 1976.

The morphology of a cumulo-dome or mamelon found inside the crater of a large volcano is called a **nested dome** or **tholoid** (*Fig. 4.45B* and *4.47*). Volcanoes with these domes in the crater can be very active, such as the nested domes of St. Lucia and La Soufrière of St. Vincent (Lesser Antilles) and Poas volcano (Costa Rica) (*Fig. 4.47*). Sometimes the domal structure can be pushed upward by underlying magma, leaving a lateral border that separates it from the rest of the volcano. This mechanism seems to explain the present structure at the summit of Mont Pelée.

Plug domes or **spines** are semi-solidified masses of lava that fill the volcanic vent. This lava can be pushed like a piston by the underlying magma. The most characteristic example of this was the spine of Mont Pelée during the 1902–1903 eruption. The Pitones del Corbet (*Fig. 4.48*) (Island of Martinique) are thought to be plugs of this type. They are made of columns of sub-vertical lavas that can reach heights of several hundred meters. These plugs can grow rapidly and are linked to explosions. They end up being destroyed, like the Aguja of Mont Pelée, which was knocked down in the 1929 eruption.

4.13 Pyroclastic cones

These common forms are normally produced during Strombolian eruptions. They result from moderate volcanic explosions of basalts or andesites that contain intermediate amounts of gas. At the top, a circular or elliptical crater develops that is surrounded by a poorly sorted mixture of fragments that forms a border around the outlet vent. The accumulations are ash and lapilli with dispersed bombs and blocks along with some intercalated lavas. Volcanic cones can reach heights of several hundred meters (*Fig. 4.49*) and they form very rapidly. In 1952, on the island of San Benedicto (Mexico), a cone grew 300 m in twelve days. They are generally grouped together and usually occur as fissure effusions as in the region of the Puys in Auvergne (France). The eruption of Lanzarote during the 18th century and more recent eruptions of Etna have also occurred along fissures. The eruption of Teneguía (La Palma, Canary Islands, 1971) created a pyroclastic cone more than 100 m tall. The morphology of ash cones can change if the wind changes direction and/or intensity or if the explosions vary in magnitude.

Figure 4.47 Nested dome and fumaroles in the crater of the Poas volcano (Costa Rica). Photo F. Gutiérrez.

Figure 4.48 Pitones del Corbet. Island of Martinique (Lesser Antilles).

Figure 4.49 Ash cones and in the foreground, vineyards that have been excavated to protect them from the dry winds. These vineyards were planted over an extensive area of the aa lavas produced by the volcanic eruption of 1730 and 1736. La Garia, Lanzarote (Canary Islands).

4.14 Stratovolcanoes

Large volcanoes like Teide, 3,718 m (Spain) (*Fig. 4.1*), Vesuvius, 1,277 m (Italy), Popocatepetl, 5,452 m (Mexico), and Lincacábor, 5,916 m (Chile) (*Fig. 4.50*), among others, occur in isolation and constitute the world's most spectacular morphologies. These stratovolcanoes are formed by successive eruptions of lavas and pyroclastics. Because of their complexity and long evolution, they are also called **composite** or **polygenetic** volcanoes (Schmincke, 2004). The absolute height of these large volcanic edifices is highly variable, because, as previously discussed, what is truly significant compared to their size is the height of the effusion point at the beginning of the formation of the future stratovolcano.

These polygenetic volcanoes have formed over thousands of years, normally by central or poly-orifice type eruptions on the flanks such as Etna or the Teide-Pico Viejo complex (Carracedo, 2006). Eruptive stages alternate with periods of quiescence during which the volcano may be subjected to erosional processes and external dynamics. When another eruption occurs, it flows down gullies generated by prior fluvial erosion, modifying the riverbeds in an obvious way. Most stratovolcanoes are located around the Pacific Ocean in the Pacific "Ring of Fire." Active stratovolcanoes are relatively young morphologies and many have not been adequately dated. They are estimated to be between 100,000 and a million years old. For example, Mont Pelée is 35,000 years old and its basement is 300,000 years old.

4.15 Calderas

Some volcanoes have deep circular or elliptical depressions at the summit (*Figs. 4.22*, *4.47*, and *4.51*). These depressions are much larger than normal craters, giving the impression that a large explosion has blown away the roof of the cone. These large depressions, normally created by collapse (Williams, 1941), are known by the Spanish word **caldera** (*Table 4.2*). The word caldera is often used in place names in the Canary Islands. The term comes from the Caldera de Taburiente on the Island of La Palma (Canaries), which is not a caldera in the genetic sense of the term; in the Canary Islands, most calderas are scars from large landslides (Ancochea *et al.*, 1994). The largest known caldera is the Yellowstone caldera, which measures 60 km in diameter (Lipman, 2000). Caldera walls are usually sub-vertical and the bottom is situated at a depth of several hundred meters like the Caldera of Santorini (Greece) (*Fig. 4.52*), which formed 3,630 years BP and where an accumulation of pyroclastics and flows can be seen in the

Figure 4.50 Lincacábor volcano (5916 m). It towers over the Salar of the Atacama (Chile). Photo J. Rodríguez Vidal.

Figure 4.51 Crater of the pyroclastic volcano Omuro, Izu Peninsula (Japan).

Table 4.2 SOME WELL-DOCUMENTED CALDERA ERUPTIONS (LITMAN, 2000)

Date	Caldera Name	Location	Edifice; Structure	Topographic diameter of the caldera (km)	Volume of the eruption (km^3)	Geometry of the collapse
1991	Pinatubo	Philippines	Stratocone	2.5	4–5	Funnel?
1968	Fernandina	Galapagos	Shield basalt	5 × 6	0.1	Trap; compound
1912	Katmai	United States	Stratocone	2.5 × 4	12	Funnel?
1883	Krakatoa	Indonesia	Dispersed stratocones	8	10	Funnel?
1750–1790?	Kilauea	Hawaii	Shield basalt	3 × 5	¿?	Plate
1.4 ka	Rabaul	Papua New Guinea	Disperse volcano	10 × 15	11	Plate
1.8 ka	Taupo	New Zealand	Rift/scattered calderas	35	35	Bottom of a sack?[a] compound
3.6 ka	Santorini	Mediterranean	Scattered volcanoes	7 × 10	25	Compound
7.7 ka	Crater Lake	United States	Scattered volcanoes	8 × 10	55	Plate
22 ka	Aira	Japan	Caldera scattered	18 × 22	300	Funnel?[a]
35 ka	Campi Flegrei	Italy	Volcano scattered	13	80	Plate
75 ka	Toba	Indonesia	Dacitic arc fault zone	30 × 80	1500	Plate (R)[b]
95 ka	La Primavera	Mexico	Scattered volcano	11	20	Plate
600 ka	Yellowstone	United States	Caldera scattered	60	1000–2000	Plate (R)
760 ka	Long Valley	United States	Older Volcanoes	15 × 30	600	Plate (R)
1.1 Ma	Valles	United States	Scattered volcano	20 × 22	300	Plate (R)
2.2 Ma	Cerro Galán	Argentina	Scattered volcano	25 × 35	2000	Plate (R)
2.8 Ma	Chegem	Russia	Collision zone	15 × 20	>300	Plate?
5.6–3.8 Ma	Dorobu	Japan	Scattered caldera	6 × 13	¿?	In pieces, compound
25.9 Ma	Lake City	United States	Scattered caldera	15 × 20	200–500	Plate (R)
26.6 Ma	Creede	United States	Scattered caldera	24	>500	Plate (R)
27.6 Ma	Silverton	United States	Scattered caldera	15 × 20	50–100	Trap
27.8 Ma	La Garita	United States	Scattered volcano	35 × 75	5000	Multiple plates? (R)
73 Ma	Tucson	United States	Andes arc	25	>500?	Trap
Ordovician	Scafell	England	Volcanic arc	>15	¿?	In pieces, (compound?)

[a] Can also be interpreted as a subsiding plate.
[b] (R) post-collapse resurgence of the bottom of the caldera.

Figure 4.52 Vertical walls below the town of Santorini, composed of accumulations of pyroclastics and lavas. Santorini Caldera (Greece).

Figure 4.53 Subsidence Caldera On the left, the eruptions take place through annular fractures. In the second from the left, the eruptions continue through a central vent after the formation of the caldera (Bailey and Maule, 1960).

vertical walls. In the center of the caldera is the island of Nea Kameni; it was created by later effusions of aa lava.

The principal structural and morphological elements of calderas are clear borders, an internal wall, fringe faults (may or may not be present), a structural floor, inter-caldera fill (ignimbrites and sliding of the walls), and an underlying magma chamber. Knowledge of these elements provides a basis for the discussion of structural and subsidence processes (Lipman, 2000).

Many calderas were formed during significant eruptions of large amounts of pyroclastic material, which led to rapid emptying of the magma chamber and collapse of its roof. Large depressions formed by this process are called **subsidence calderas** or **Glencoe-type calderas** (Scotland) (*Fig. 4.53*) (Francis, 1993); they resemble a piston subsiding in a cylinder. **Annular dikes** are the geologic expression of the cylinder.

Another model comes from the 16 August 1883 eruption of the Krakatoa volcano located at 6° south latitude

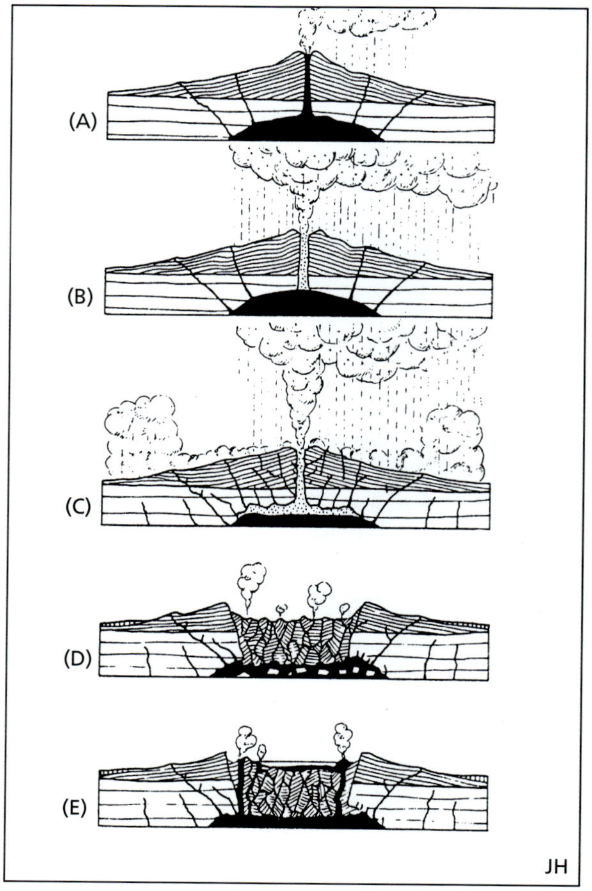

Figure 4.54 Stages in the formation of the caldera at Crater Lake, Oregon: (A) Small explosions. (B) Increase in the violence of the activity. (C) Intense moment of Plinian and Peleean eruptions. The magma sinks below the roof of the magma chamber. (D) Collapse of the cone on the magma chamber. (E) New eruptions at the base of the caldera, especially near the edge (Williams, 1941).

(Sonda Strait, Indonesia) where a large collapse converted the island into three small islets concentric to the place of maximum subsidence where two-thirds of the land disappeared. The eruption began with the effusion of large quantities of pyroclastics (*Fig. 4.54*). Ashes from this eruption have been identified in the two large polar icecaps (Porter, 1981, Schmidt, *et al.*, 2004). The moment of maximum activity was a series of explosions (Plinian and Peleean eruptions) that sent ash into the stratosphere. The magma chamber later sank rapidly and magma later rose near the margin of the caldera. These collapses generated **tsunamis** that killed 36,000 people (Neall, 1996; Begét, 2000). In 1927 a new volcano (Anak Krakatau) appeared in the center of the caldera as a result of a submarine eruption that formed the island. This model is called the **Krakatoa-type caldera** or **Crater Lake type caldera**. The latter name comes from the Crater Lake caldera in Oregon (United States).

Siliceous calderas form groups that make up **caldera complexes** as in the San Juan Mountains of Colorado (United States) (Steven and Lipman, 1976). This area contains 18 known calderas formed between 22 and 30 million years ago. The calderas are superimposed on each other and some appear nested to the extent that it is difficult to determine where one begins and the other ends.

Another type is the **basaltic shield caldera**, such as volcanoes in the Hawaiian Islands, Iceland, Reunion Islands, and the Galapagos. They are produced when the summit collapses like a piston, followed by the eruption of large quantities of lava, which causes the roof of the magma chamber to lose support and sink. These calderas are less deep than the types described previously. Basaltic lavas are found in the interior of the caldera as in the Menengai Caldera (Kenya), which is 12 km in diameter, 485 m deep, and covers an area of 98.5 km^2 (*Fig. 4.55*).

As some volcanoes evolve, collapses or large landslides on the volcanic flanks have occasionally been interpreted as calderas such as Las Cañadas (Tenerife) (Ancochea *et al.*, 1990, 2000; Watts and Masson, 1995; Martí *et al.*, 1994) and Taburiente (La Palma) (Ancochea *et al.*, 1994).

4.16 Volcanic landforms resulting from erosion

In spite of the constructive and destructive character of volcano dynamics, erosion gradually wears down the Earth's surface. A totally extinct volcano can become a planar erosional surface over time. Weathering processes that affect volcanic rocks are identical to those that attack minerals in other rocks, even though the processes and their intensity may vary in different morphoclimatic zones (Gutiérrez, 2001, 2005).

Rivers that drain lava fields erode valleys and modify the landscape. Fluvial action that continues for a long period can carve deep gorges such as the Columbia River Gorge (northwestern United States), which was created by the river flowing over basaltic plateaus. The gorge has reached the substrate, revealing the paleostratigraphy that existed prior to the accumulation of the lavas. On the islands of La Gomera, La Palma, and Grand Canary, erosion has created immense **gullies** hundreds of meters deep. Cendrero and Díaz de Terán (1985) conducted a quantitative study of the erosion on the Island of La Gomera (Canaries) and showed the average depth of the gullies to be 272 m and average erosion to be 0.204 km^3/km^2.

Figure 4.55 Menengai caldera with basaltic lavas in the interior. Center of Gregory Rift Valley. Nakuru (Kenya).

Fluid lavas can flow through the fluvial network and modify the drainage. Significant volumes of lava can impound lakes. When lava flow ceases, rocks under the lava may be more easily eroded than the lava. The undermining of a river along with other processes of external dynamics generates relief in which the higher parts are resistant lavas (such as basalt) and as a result, produces **inversion of relief** *(Fig. 4.56)*.

Pyroclastic cones are sometimes so porous that water totally infiltrates into them and as a result, water does not erode them. Clay from weathering fills the pores in old cones and causes gullying on the slopes. Lava cones erode more easily than pyroclastic cones. Near the crater there is less erosion because the receiving basin is very small; maximum erosion occurs in the middle areas of the slope; in the lower portions, there is less erosion because of the lesser slope. Alluvial fans are deposited in distal areas. Interfluves between gullies are composed of sharp ridges. The combination of gullies and ridges is called **parasol ribbing**, which does not often remain intact for a long time. In stratovolcanoes and old massifs there is more structural control by more resistant lava layers, which may give rise to valleys with scarped sides *(Fig. 4.57)*.

When main gullies are active in a volcanic cone, they become wider and deeper. In the lower part, the valleys are spaced and develop large triangular faces. These faces are called **planèzes**. Continuous erosion of the volcano ends by wearing down the cone until it becomes an assemblage of

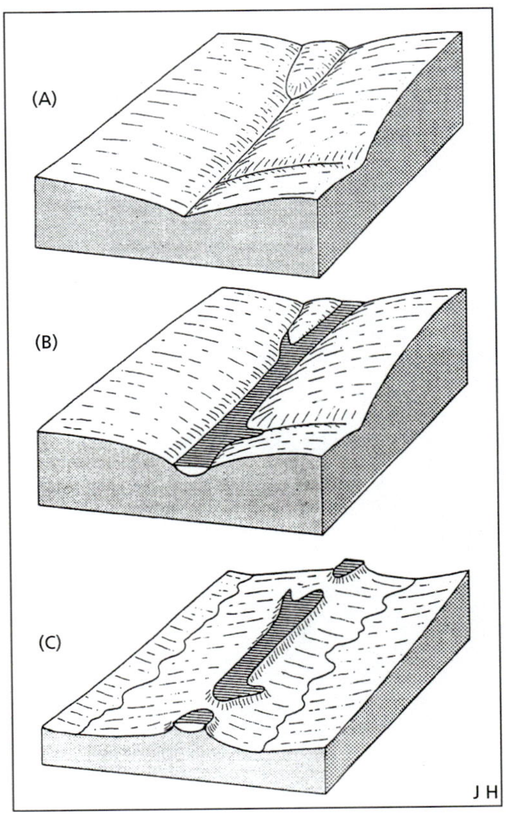

Figure 4.56 Inversion of relief. (Ollier, 1969a).

Figure 4.57 Valleys with steep sides. Taganaga valley, Anaga massif. Tenerife (Canary Islands).

Figure 4.58 Stages of a volcanic eruption: intact volcano, stages of planeze, residual volcano, and volcanic skeleton (Ollier, 1969a).

dikes and plugs that represent the **skeleton of the volcano** (*Fig. 4.58*) (Ollier, 1969a). This skeleton is composed of chimneys that are vents through which the volcanic materials that built the cone were erupted. An example is the *puys* of the Auvergne (France). This evolution is complicated when the volcano reactivates and lava flows down the gully beds.

Large stratovolcanoes such as Teide may have areas with very distinct climates. In the northern part of Tenerife, precipitation is abundant whereas the south has an arid climate; this causes different erosional processes and resulting forms. Periglacial landforms are present above 2,000 m (Criado, 2006).

4.17 Volcanic risk and prediction

Volcanic activity is one of the most catastrophic natural phenomena (*Fig. 4.59A, B*) Tilling, 1993, 2005). The world has about 700 potentially dangerous volcanoes; each year about 50 volcanoes erupt with major or minor risk of catastrophe (ICOG, 1997). There is a reliable record of the activity of large volcanoes located on land. However, medium intensity eruptions are not well known. They do not cause big disasters; an example is the medium magnitude eruptions of the Popocatépetl volcano (1992–1999) (Gomez and Zamorano, 2002).

Catastrophic historical eruptions are known to have occurred at Vesuvius (79 AD), Tambora (1815), Mont Pelée (1902), Mount St. Helens (1980), Nevado del Ruiz (1985),

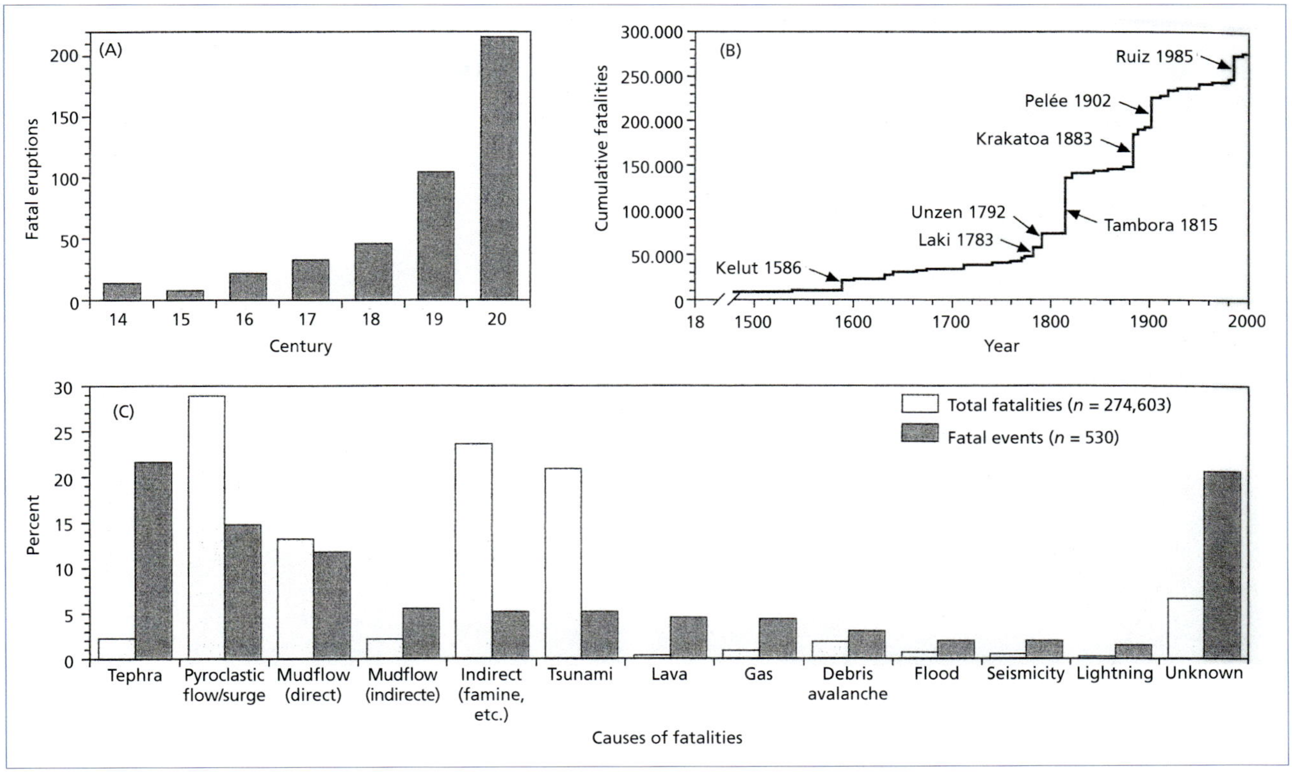

Figure 4.59 (A) Volcanic activity has killed 274,000 people since the 16th century and possibly many more. (B) A large part of the fatalities occurred during lethal events. (C) The majority of fatalities have resulted from pyroclastic flows, lahars, ash falls, or tsunamis (modified from Simkin *et al.*, 2001).

and Pinatubo (1991) (*Table 4.3*). However, many large eruptions do not kill people or cause property damage because the eruptions take place in unpopulated areas such as the Katmai volcano in Alaska in 1912 (Hubd and Inbar, 2002). On the other hand, as population has increased, the number of disasters has also increased; in other words, geologic risk has intensified (Carracedo, 1988). Volcanic soils are very fertile and consequently are highly desirable for agriculture. This is the principal reason for high population density on the flanks of many volcanoes (Schminke, 2004 and Thouret, 2004). Some volcanic eruptions produce devastating lahars that travel tens of kilometers and destroy and fossilize everything in their paths (Fuster, 1992). In addition, ash eruptions affect the stratosphere; ash blocks solar radiation and decreases the temperature of the Earth's atmosphere. For example, volcanic aerosols from the 15 June 1991 eruption of Pinatubo (Philippines) reduced temperatures in the northern hemisphere by 0.5 to 1°C (Self, 2005) and up to 2°C below the normal temperature during the summer of 1992 (Robock, 2002).

In recent years numerous investigations have been done on the effects of volcanic activity on the Earth's surface temperature (Grove, 2004; Solomina *et al.*, 2008). Dendrochronology research (*Fig. 4.60*) (LaMarche and Hirschboeck, 1984) suggests that a relationship between cold events in tree rings and volcanic activity in the last 500 years. A comparison of the data from several of these authors with the *Dust Veil Index, DVI* of Lamb (1970), which measures the impact of volcanic eruptions on the atmosphere, shows a concordance with the nineteen principal volcanic events that occurred during the period of 1500 to 1963. For ten of these 19 eruptions, notable cold events occurred one or two years after the eruption. However, for the other nine, no relationship was established.

Bryson and Goodman (1980) show that following most eruptions recorded between 1945 and 1970, the optical depth of aerosols (measured as the content of the aerosol H_2SO_4) is double the normal value, suggesting increased injection into the stratosphere. These circumstances point to a temperature decrease for these decades. We know that glaciers advanced globally during the decades of the 1960s and 1970s, which supports the preceding discussion. Periods of advance during the Little Ice Age may be due to the same cause (Grove, 2004).

Materials ejected into the stratosphere have different $\mathbf{H_2SO_4}$ contents. The 1963 eruptions of the Agung (Indonesia) volcano and the Fuego volcano (Philippines) in 1974 had

Table 4.3 **PRINCIPAL VOLCANIC CATASTROPHES, WITH MORE THAN 400 FATALITIES (1783–2003). (MODIFIED FROM SIMKIN AND SIEBER, 1996)**

Region, Volcano	Date	Impact
Vesuvius, Italy	1794	60% of the Torre del Greco was buried by flows. Damage in Ottaviano, San Giussepe, and Osma. 400 dead.
Ritter Island, Melanesia	1888	Volcanic avalanche triggers tsunamis. About 3,000 dead.
Tambora, Indonesia	1815	Pyroclastic flows and tsunamis killed 10,000. Resulting starvation killed another 82,000.
Galunggung, Indonesia	1822	Plinian eruption destroyed 114 towns and killed 4,000 persons.
Ruang, Indonesia	1871	Tsunami produced by the collapse of a lava dome killed 400 people.
Krakatoa, Indonesia	1883	Plinian eruption and tsunamis killed 36,400 people.
Kelut, Indonesia	1919	Lahars and possibly pyroclastic flows destroyed 104 towns and killed more than 5,000 people.
Merapi, Indonesia	1930	Pyroclastic flows destroyed 13 villages and killed 1,369 people.
Pinatubo, Philippines	1991	Plinian eruption that killed about 350 people; subsequent illnesses claimed 450 lives. 65,000 people were evacuated.
Unzen, Japan	1792	Collapse of the flank of a volcano and tsunamis killed 14,500 people.
Mount St. Helens, USA	1980	Plinian eruption claimed 57 lives and damages were on the order of a billion dollars.
El Chichón, Mexico	1982	Pyroclastic flows, ash fall, and lahars killed 1,600 people.
Nevado del Ruiz, Colombia	1845	Lahars caused the death of 1,000 people.
Nevado del Ruiz, Colombia	1985	Lahars destroyed Armero and killed 23,000. Damages were estimated at 7.7 billion dollars.
Mont Pelée, Martinique	1902	Pyroclastic flows destroyed St. Pierre and killed about 30,000 people.
Soufriere, St. Vincent	1902	Pyroclastic flows killed 1,700 people.
Lakagigar, Iceland	1783–84	The major historic eruption produced 13–14 km^3 of lava. Gases destroyed harvests and 9,350 people died of starvation.

higher SO_2 and HCl contents than Krakatoa (Indonesia), Santa María (Costa Rica), and Mount St. Helens (United States), which emitted larger quantities of siliceous material (Rampino and Self, 1982). The principal atmospheric effect of eruptions is the conversion of volcanic sulfur to sulfuric acid, which is the dominant aerosol. This aerosol cools the lower troposphere by reflecting solar radiation. From a climatic point of view, the amount of sulfur ejected into the stratosphere is more significant than the magnitude of the eruption (Grove, 2004). This may explain the lack of cold stages following some of the previously discussed eruptions, according to the record of LaMarche and Hirschboeck (1984).

We can determine the sulfur content of the atmosphere by analyzing samples of cores drilled into ice caps. The sulfate content increases significantly after an eruption. This content can be determined from acidity variations measured by electrical conductivity. Numerous eruptions between 9,000 and 110,000 years ago were larger than that of Tambora (1815). The distribution of volcanic eruptions and the values of optical depth calculated by the GISP2 ice core (central Greenland), clearly suggest that volcanism has played an important role in climate history over the last two millennia. Periods of intense volcanic activity from the end of the 16th century until the beginning of the 20th century coincide with periods of low temperatures and glacial advance (Zielinski, 1995).

Pyroclastic flows, ash falls, lahars, and tsunamis are responsible for 90% of the deaths that occur as a direct result of volcanic activity (*Fig. 4.59C*) (Simkin *et al.*, 2001). Lava moves slowly and causes little damage because it

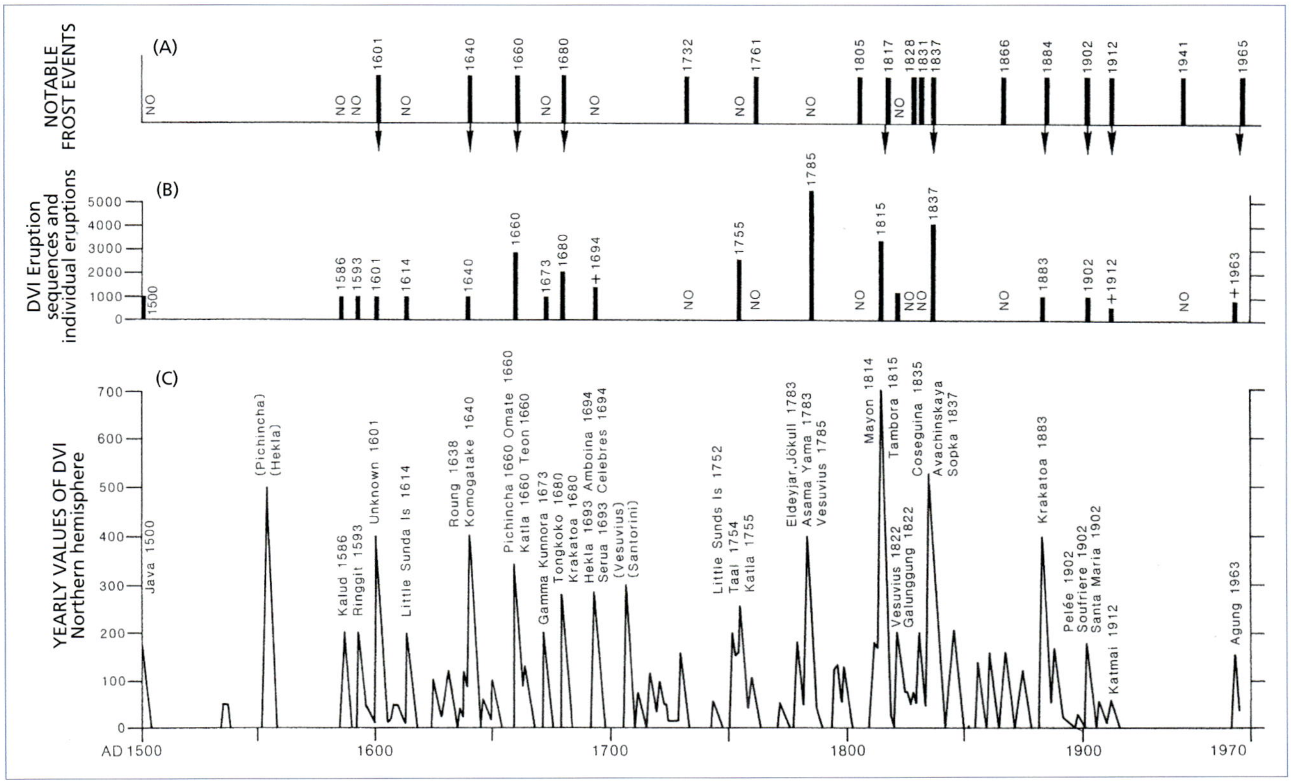

Figure 4.60 Rings showing cold stages in pines in California and the western United States and principal volcanic eruptions: (A) Years with cold stage rings; the arrows indicate associated eruptions. "NO" signifies the absence of principal cold events at the time of a large eruption. (B) Dust Veil Index (DVI) and dates of eruption sequences or individual eruptions. "NO" indicates the apparent absence of large eruptions corresponding to events indicated in A. (C) DVI of each year, with numbers of volcanoes and dates of the principal eruptions (LaMarche and Hirschboeck, 1984).

affects only static structures. Pyroclastic flows are deadly because they reach temperatures of 500–600°C and speeds up to 400 km/hour. They can kill by burning and suffocation and can destroy buildings (*Fig. 4.19*) (McGuire *et al.*, 2004). The Chinchón or Chichonal volcano (Chiapas, southeastern Mexico) has erupted for 7,700 years. At the end of March and beginning of April in 1982, it began erupting ash clouds up to 24 km high. This was the most destructive eruption in Mexico in historic times. Between 1,500 and 2,000 people died in various towns that had been built on the margins of the volcano. The deaths were primarily the result of pyroclastic flows (Espinola *et al.*, 2002).

Eruptive columns contain a mixture of gases, pulverized magma, and solid fragments that reached heights of tens of kilometers. The fragmentary material consists of **tephra,** a term that includes ash, lapilli, volcanic blocks and bombs expelled by the volcano. Tephra can cover areas of tens of kilometers around a volcanic edifice. Ash can form a 30 cm-thick layer at a distance of more than 100 km from the volcano. Ash reduces visibility and is dangerous for people and animals since breathing the fine particles can cause silicosis. It also affects machinery, contaminates water, destroys fish farms, and creates problems for air navigation (Tilling, 2005). Ash deposited on buildings makes them damp, increases their weight, and can cause roofs to collapse.

When a volcano ruptures and collapses, large **debris avalanches** can occur, which are very dangerous. The north flank of the Mount St. Helens volcano (Washington State, northwestern United States) collapsed spectacularly during the eruption of 18 May 1980 (*Fig. 4.61*) (Hickson, 1990). The Unzen volcano (Japan) caused a huge landslide in 1792 and generated a tsunami that killed 14,000 people (McGuire, 2004).

The only area of Spain that has experienced active volcanism since the end of the 15th century (when the islands were occupied by Castile) is the Canary Islands. Eruptions have occurred on the islands of La Palma, Hierro, Tenerife, and Lanzarote (*Fig. 4.62*) (Cendrero *et al.*, 1997) during historic times. In general, volcanic eruptions in the Canary Islands are not very dangerous because the eruptions have low viscosity

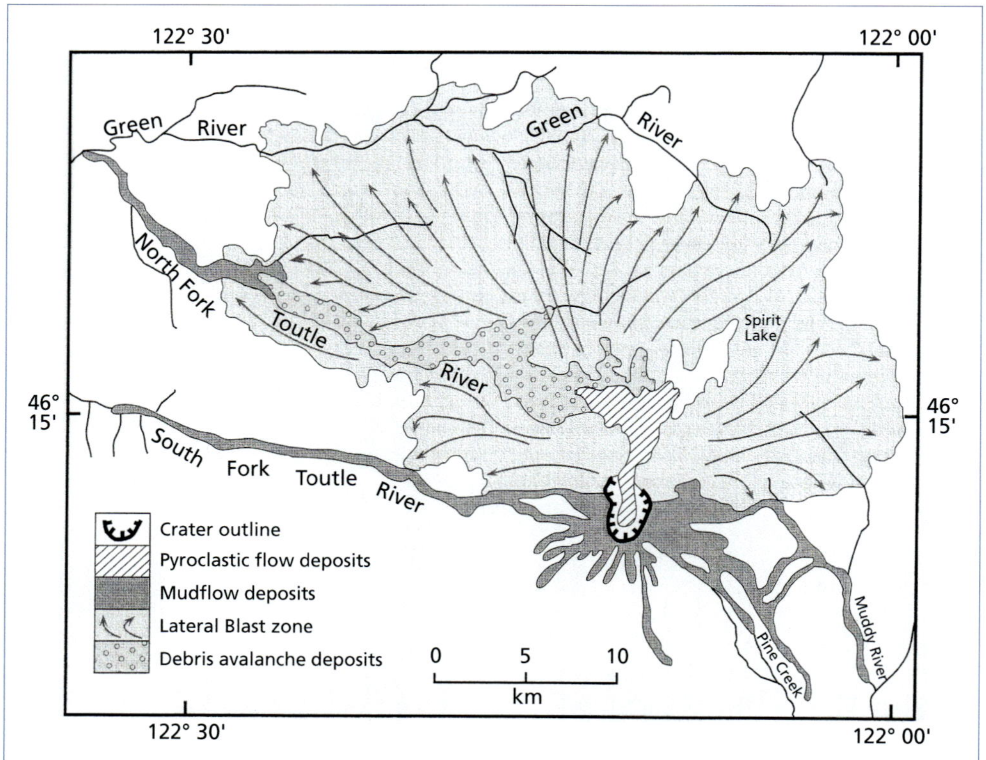

Figure 4.61 Geological scheme of the eruption of 18 May 1980 of the Mount St. Helens volcano. Deposits of pyroclastic rain are not indicated (Tilling, 2005).

and the magmas are low-explosivity alkaline basalt. One of the most significant risks is the possibility of large landslides (Cendrero *et al.*, 1997; Masson *et al.*, 2002; Ancochea, 2004; Paris *et al.*, 2005) *(Fig. 4.62)*. The first edifice of Las Cañadas (Canary Islands) had four episodes of sliding between 2.5 million years and 150,000 years ago (Ancochea *et al.*, 2000; Cantagrel *et al.*, 1999). At La Palma, at least three collapse episodes occurred between 1.2 million years and 700,000 years ago (Ancochea *et al.*, 1994). On the island of El Hierro, three episodes have been identified, one 300,000 years ago, a second <80,000 years ago, and a third that has not been dated (Fúster *et al.*, 1993). Enormous energy was liberated by these slides. It is calculated that the Las Cañadas slide moved about 80 km^3 of rock, which represents about $5x10^{18}$ joules (Cendrero and Dramis, 1996).

The cause most often cited to explain these large slides (McGuire, 2004) is groundwater, which destabilizes the volcanic edifice especially if the magma encounters it near the surface. McGuire (2004) points out that the Shasta volcanoes (California) and Popocatépetl (Mexico) have been the locations of 25–30 km^3 slides. When they occur on islands, these slides can generate catastrophic tsunamis.

In recent decades, vigilance and eruption prevention research have increased with the installation of new and complex instruments that aid in the understanding of the mechanisms of volcanic eruptions (Araña and Ortiz, 1993). Due to the practical impossibility of studying all active and potentially dangerous volcanoes, efforts have been concentrated on 170 of them (Ancochea and Barrera, 2002).

Volcanologists today are capable of predicting with good accuracy "where" and "when" an eruption in a known and monitored area will take place. Even so, the magnitude and characteristics of an eruption cannot be predicted, especially those of the explosive type (Araña and Ortiz, 1993).

Researchers need the best possible geologic understanding of the volcanic areas they investigate. They need to understand the time factor to determine the life of the volcano, types of magma, eruptive mechanisms, and the geological criteria used in the evaluation of volcanic risk, which is inferred from an analysis of the preceding considerations (Araña, 1988).

Monitoring active and dangerous volcanoes is mostly done at various volcano observatories *(Fig. 4.63)*. Rising magma produces indicators that can be measured and that become more frequent and intense closer in time to the eruption. Magma produces earthquakes, deforms the surface of a volcano, and produces changes in gases emitted by fumaroles. On occasion, it heat ups the surface (Scarth, 1994).

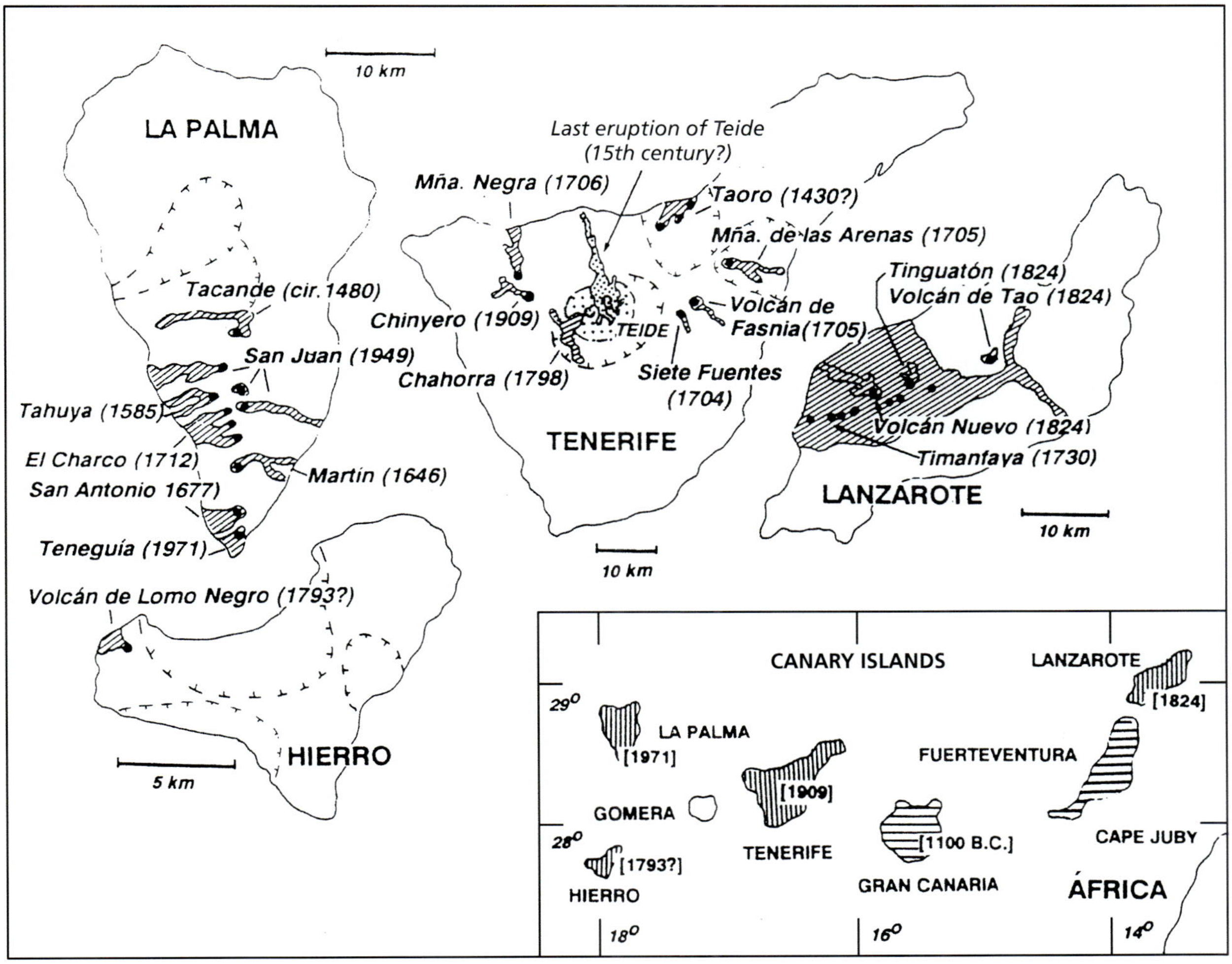

Figure 4.62 Distribution of historical eruptions in the Canaries. In the lower frame: the vertical line indicates islands with historical eruptions; the horizontal line indicates islands with prehistoric activity; in white, no Quaternary activity. Areas of volcanic activity and ages of activity are noted. Caldera-like structures formed by large landslides are found in Hierro, La Palma, and Tenerife (Cendrero *et al.*, 1997).

Instrumental techniques that are used, listed in decreasing order of efficiency, are the following: (Araña, 1988, Martí and Folch, 2005):

a) Geodesy: networks, leveling, clinometry.
b) Geochemistry of gases and thermometry.
c) Geophysics: geo-electricity and geo-magnetism.
d) Geophysics: seismology.
e) Geodesy: gravimetrics.

It is sometimes difficult to approach dangerous volcanoes and in this event, remote sensors are used to monitor them (transmission of data by artificial satellites). Pre-eruptive and eruptive theoretical models are also used (Marti and Folch, 2005).

Geodetic methods are used to analyze deformation of the ground due to volcanic activity. Surveying and triangulation are the standard methods for observing deformation; inclinometers and extensiometers are used for these measurements (Yokohama, 1993).

Changes in the volume and types of gases can be good indicators of an imminent eruption. Modifications in volume or composition may be related to changes in equilibrium conditions in the magma chamber. The most significant changes are increases in the amounts of HCl, HF, and SO_2 and an increase in the S/Cl ratio (Noguchi and Kamiya, 1963). In principle, the surface temperature of the volcano seems to be significant. However, the low thermal conductivity of rocks means that a significant temperature increase at depth

Figure 4.63 The Mont Pelée volcanological observatory (Martinique, Lesser Antilles) located on the flank of the volcano, where complete instrumentation has been installed to monitor the volcano.

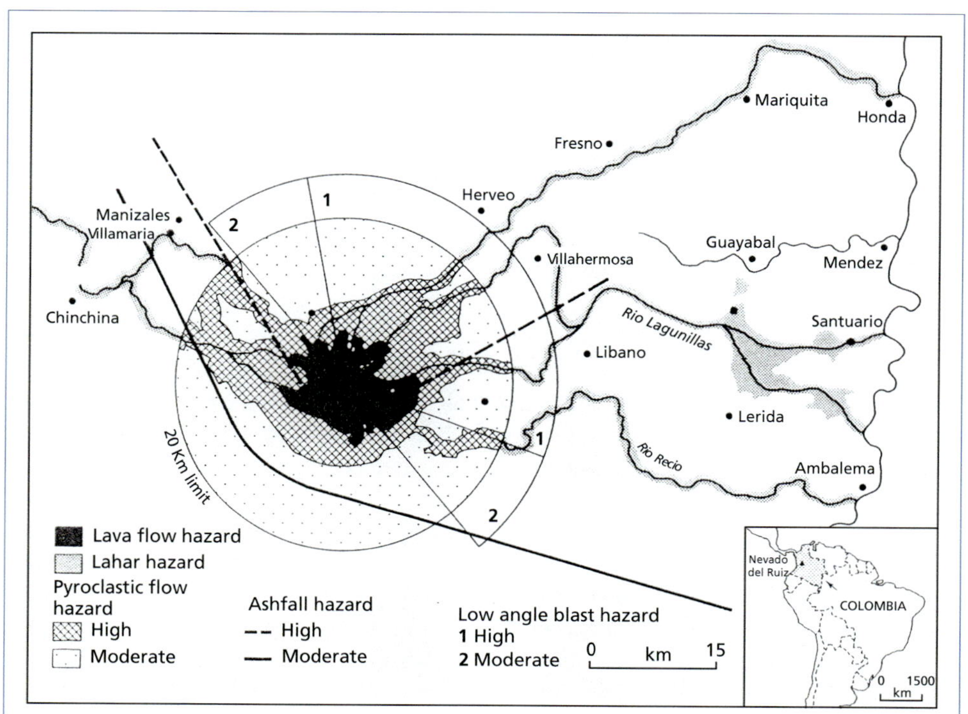

Figure 4.64 Map of volcanic risk for the Nevado del Ruiz (Colombia) (Parra and Cepeda, 1990).

may not be detected at the surface. The monitoring of volcanic activity by geoelectric and geomagnetic methods has been a rapidly developing field in recent years.

Seismology is not very effective in predicting eruptions (Araña, 1988). Only if complete seismic records are available, including interpretations and responses collected over several years in an area, is it possible to predict an eruption in a plausible manner (Schminke, 2004). Gravimetry can provide information on the deep structure of a volcano but like other techniques used to monitor volcanoes, it is controversial.

This research on volcanic characteristics and history makes it possible to construct volcanic risk maps (Parra and Cepeda, 1990) *(Fig. 4.64)*. These maps show areas subject to possible risk. There are also maps of thematic risk (lahars, flows, etc.) and the superpositions of these make it possible to produce summary maps on the risks of a particular volcano. By using these maps, people can take measures to evacuate high danger zones. However, the repeated use of alarms can sometimes desensitize people to risk and cause them to interpret all alarms as false.

Authorities can order evacuations of areas of high volcanic risk. At other times, when flows are relatively slow *(Fig. 4.15)*, authorities can modify the path of the flow as has been done during recent eruptions of Etna. During the 1992 eruption authorities diverted a flow that was headed for the town of Zaffarena (Barberi *et al.*, 1993). Another problem occurs when volcanologists disagree about volcanic prediction and evacuation. In the 1976 eruption of the volcano Soufrière (Guadaloupe, Lesser Antilles) *(Fig. 4.65)* one research team thought that the symptoms indicated an imminent eruption whereas another research team totally disagreed with these predictions (Tazieff and Derruau, 1990).

The dispute arose from the ambiguity of various indicators used to predict the imminence of eruptions (or a violent phase of an ongoing eruption). The rise of magma toward the surface can deform the volcanic edifice, increase the temperature of the gases and the content of the magmatic components in them, increase seismic activity, and cause changes in the electric and magnetic fields. Even so, the fact that magma is rising does not necessarily mean it will reach the surface and erupt because it can sometimes stop at depth. With our present level of knowledge of volcanic prevention, it is not possible to discriminate between these two possibilities.

Figure 4.65 Streets of Basse Terre. At the base of the La Soufrière volcano (Island of Guadalupe, Lesser Antilles).

5

WEATHERING AND RESULTING LANDFORMS

1. Introduction
2. Types of weathering processes
3. Resultant weathering forms
4. Practical aspects of weathering

5.1 Introduction

Weathering is the fundamental process that shapes landforms. In the geological cycle it is the precursor to erosion, transport, and sedimentation. Weathering is also the principal agent in karst formation. It is also essential to soil and regolith development and in the alteration of monumental rocks (Turkington *et al.*, 2005).

Weathering is the response of materials in equilibrium in the lithosphere to conditions of contact or near contact with the atmosphere, hydrosphere, and perhaps most importantly, the biosphere (Reiche, 1950). The biosphere is significant in many ways. Winkler (1973) provides a similar definition similar to Reiche's. Most rocks have been subjected to high pressures and temperatures in the absence of air and water. When they crop out at the surface, they come to equilibrium with conditions at the lithosphere-atmosphere contact: one atmosphere of pressure, lower temperatures than their previous environments, and the presence of air and water (Carroll, 1970). Mechanical weathering processes break up rocks and chemical activity alters many types of minerals. Biological action affects rocks by breaking them up both physically and biochemically and forming a surface weathering film. Weathering promotes erosional processes; the lack of weathering considerably hinders denudation. Weathering products restructure themselves and form authigenic minerals and rocks that may serve as paleoclimatic indicators (Birkeland, 1984; Migon and Lidmar-Bergström, 2002; Bremmen, 2002).

Reiche's definition makes it clear that climate plays a fundamental role in weathering processes. These circumstances are clearly reflected in the diagrams of Peltier (1950) and Wilson (1969) where they relate weathering processes to variations of precipitation and annual median temperatures (*Figs. 12.1* and *12.3*).

The fundamental process that weathers rocks in cold and polar deserts is freezing—thawing (Ugolini, 1986), which is most effective in environments where median temperatures hover around 0°C. In these polar regions, low temperatures affect not only the speed of chemical reactions but also the state of the water. Low temperatures slow down dissolution except in carbonate rocks. Also, low temperatures directly or indirectly affect the colonization, growth, and decomposition of biological components. However, some characteristics of these climates promote physical and chemical processes: duration of solar radiation during the summer, increased ultraviolet radiation, elevated temperature gradients, wet-dry and freeze-thaw cycles, etc. We can conclude that weak weathering in polar regions is the result of adverse climatic factors accompanied by reduced biomass production, slower decomposition of organic material, and decreased leaching and chelate formation.

The weathering of rocks in arid zones is primarily the result of wide daily and seasonal variations of temperature and relative humidity. Moreover, rock outcrops are more common in deserts than in other climatic zones and the lithologic and structural characteristics of outcrops affect the processes and resultant morphologies (*Fig. 5.1*). In these desert environments, physical weathering plays a more important role than biological and chemical weathering. High temperatures and the lack of water result in sparse animal life and vegetation and as a result, reduced biological weathering (Bierman and Caffee, 2001; Viles, 2005). The moisture deficit also decreases the occurrence and intensity of chemical weathering because water is needed to attack crystalline structures and mobilize dissolved ions. Processes such as salt weathering, however, are physical-chemical in nature, which implies the dissolution and precipitation of highly soluble salts that can break up a rock (Cooke *et al.*, 1993). Furthermore, in arid zones iron oxides form by capillary action and silica and alumina mobilize in alkaline media (Smith, 1994). All these factors indicate that chemical weathering processes should predominate in deserts but chemical weathering is actually more common in humid environments. In arid regions, we find landforms and deposits that are inherited from past climatic conditions that may have been very different; these landforms include alteration profiles, laterites, sinkholes, karst cavities, and so forth. They all formed in previous climates that were more humid and where chemical weathering was more aggressive.

Weathering processes cause granular disintegration, flaking, and splitting. Fluvial and eolian erosional processes mobilize loose particles and leave an imprint on diverse rock morphologies as a result of differential weathering processes. These forms are usually small in size and are characterized by the presence of shallow pits in the rock surface (Goudie, 1989a).

In humid tropical zones, the dominant processes are chemical weathering with maximum leaching as a result of higher rainfall. In addition, weathering in these zones is more intense and deeper because of high temperatures that increase the rate of chemical reactions. Although dense vegetation supplies a large amount of organic matter to the surface layer, bacteria consumes most of it, and only a small amount of humus remains in the soil. High temperature and rainfall conditions promote the growth of flora and fauna.

The activities of various animal species play an important role in the mobilization of surface material and also in the construction of small mounds (*Fig. 5.2*). Climate is certainly one of the most significant factors in the weathering process. Water from precipitation is the most important reagent in weathering processes; the amount of water is a fundamental factor in the intensity and type of weathering. In humid tropical regions water attacks crystalline structures, removes bases (Na^+, K^+, Ca^{2+} and Mg^{2+}) and partially removes silica. Water also affects biochemical reaction rates in plants. Temperature

Figure 5.1 Sandstone weathering controlled by rock structure. Paleochannels of Miocene Calcareous sandstone of the Ebro Depression, Alcañiz, Teruel Province, Spain.

Figure 5.2 Small mounds built by crabs for shelter. Fort de France (Island of Martinique, Lesser Antilles).

has direct and indirect effects on weathering processes. High temperatures generally speed up chemical reactions. According to Van't Hoff's law, for each 10°C of temperature increase, chemical reaction rates double or triple (Tricart and Cailleux, 1965).

Weathering processes, especially chemical weathering, produce highly weathered rock. Rock decomposition results from high precipitation and temperatures and the response of vegetation to these factors. Natural water in these areas can have pH values ranging from 4 to 6; this acidity may be due to silicate hydrolysis (Curtis, 1976). The liberation of biogenic carbon dioxide provides a large amount of bicarbonate to water. Humification processes produce humic and fulvic acids that remove cations such as organo-metallic compounds (Thomas, 1994a). These processes form thick weathering profiles, which will be analyzed later.

Duchafour (1979) distinguishes between geochemical weathering, which occurs in the absence of organic material, and biochemical weathering, which is defined by the presence of organic material. **Geochemical weathering** rapidly produces simple crystalline forms by generating free oxides (oxy-hydroxides) or complex forms such as authigenic clays. In contrast, **biochemical weathering** involves slow processes of crystallization and authigenesis. Crystallization of iron oxides and authigenesis of clays at the expense of gels slow down in the presence of organic material. In contrast, the transformation of micaceous minerals is faster when organic and complexing material is present (Robert *et al.*, 1979).

5.2 Types of weathering processes

Weathering processes can be divided into three types: physical (or mechanical), biological, and chemical, and all of them can act simultaneously. Physical weathering breaks up the rock by different mechanical mechanisms (*Fig. 5.3*). Most forces applied to geological materials are external although they can also build up inside the rock.

5.2.1 Physical weathering

5.2.1.1 Sheeting

Many rocks either originate or spend long periods at considerable depths in the lithosphere and are thus subjected to high confining pressures. When uplifted to the surface, these rocks undergo pressure release as a result of unloading; this release generates joint-like fractures that run parallel to the topography, a process known as **sheeting** (*Fig. 5.4*) (Gilbert, 1904, Birot *et al.*, 1968). Sheeting is known by many names (Twidale and Vidal Romani, 2005) and can be recognized by the presence of arched slabs up to 10 m thick (*Fig. 5.5*). Their

Figure 5.3 Physical weathering. Sinai Desert (Egypt).

minimum thickness is 0.2 m. Sheeting cuts through structures in the substrate and therefore postdates the consolidation of the affected rock. It occurs in most climates. Joint spacing increases with depth and is more easily observed in massive rocks such as granites, conglomerates, and sandstones. The difficulty of observing this type of jointing in non-massive rocks is probably due to the existence of old bedding planes that distribute expansion stress (Ollier, 1965). Sheeting can be explained by processes of thermal expansion, chemical weathering, and relaxation due to pressure release, and here

Figure 5.4 Sheeting developed in granites of the Yosemite Valley, California. The sheets are parallel to the ground surface. Photo by J. López-Martínez.

Figure 5.5 Sheeting at the top of Half Dome, Yosemite Valley (California). The effect of sheeting on slopes can be seen in the background. Photo by J. López-Martínez.

it is considered to be topographical jointing (Lewis, 1954; Chapman, 1956; Birot *et al.*, 1968). These explanations are exogenetic in nature. Endogenetic explanations attribute sheeting to forces developed during the emplacement of the granite prior to cooling; others relate it to diapiric intrusions of gneissic domes. Other authors suggest that sheeting is a consequence of lateral compression (Twidale, 1973, 1982b, Vidal Romaní *et al.*, 1995; Vidal Romaní and Twidale, 1998; Twidale and Vidal Romaní, 2005). As will be discussed later, jointing is very important in landform development in tropical cratonic areas (Birot, 1960, 1968). Measurement of in-situ forces has shown that the Australian continent is undergoing horizontal compression (Twidale and Vidal Romaní, 2005). These post-orogenic states can now be recognized in the Earth's crust (Müller, 1964a). Holzhausen (1989) experimental work indicates that lateral compression of partially confined blocks causes upward deformation trajectories (*Fig. 5.6*).

5.2.1.2 Gelifraction

Gelifraction, also called **congelifraction** or **frost shattering**, is the most important weathering process in periglacial zones. As liquid water freezes, its volume increases approximately nine percent. Rock that contains water breaks when the stresses exerted by the interstitial water exceed the rock's tensile strength. If water freezes in a confined space, it can transmit high pressures to the host rock. However, freezing of water in cracks generally does not occur under totally confined conditions because the water may ooze out of the rock, in which case only part of the expansion is exerted on the rock (Ollier, 1969b). Sparse vegetation cover and rock outcrops play an important role in gelifraction in polar environments. Vegetation provides protection while wind abrasion and desiccation of rock outcrops prevents the retention of weathering products (Ugolini, 1986). As a result, weathering is a surface phenomenon in polar regions (Strakhov, 1967) (*Fig. 12.4*). Frost expansion causes a rock to fracture, produces angular clasts (also called cryoclasts), or causes granular disintegration of the rock. Mechanical fragmentation of rocks is an important geomorphic agent as demonstrated by block fields that develop on planate surfaces and talus accumulations on slopes in mountainous areas. Gelifraction is most effective in springtime when snowmelt water penetrates into cracks and subsequently freezes. On the southern slopes of the Sierra Nevada in the Betic Cordillera of Spain, frost shattering reaches its maximum intensity during pre- and post-snow periods (Gómez Ortiz and Salvador Franch, 1998). The continuous production of cryoclasts causes headward erosion of rock slopes. According to field measurements, retreat rates range from 0.3 mm/year in Spitzbergen (Jahn, 1976) to 2.5 mm/year in the Swiss Alps (Barsch, 1977a).

Gelifraction occurs with great intensity on the upper parts of stratovolcanoes that have developed on the plains of central-eastern Africa, such as Mount Kenya and Kilimanjaro. It is also common in the high Andean peaks of Chile, Peru, Ecuador, and the Nevados of Colombia.

The effectiveness of gelifraction depends on rock type and water content. Generally, slates disintegrate more easily than igneous crystalline rocks (Birot *et al.*, 1968). This process is largely controlled by the structural characteristics of the rock especially the density of discontinuity planes in the rock mass (*Fig. 5.7*). Porosity is particularly important in sedimentary rocks; rocks with porosity less than 6% do not undergo

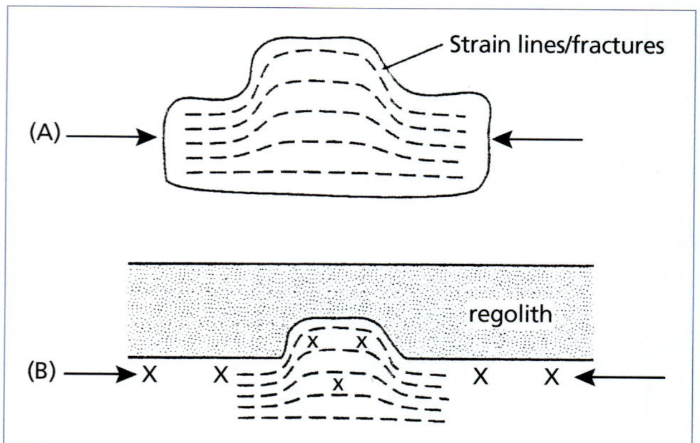

Figure 5.6 Response to compression of a partially confined block: (A) laboratory experiment (Holzhausen, 1989); (B) result of compression of a partially confined block under actual field conditions (in Twidale and Vidal Romani, 2005).

Figure 5.7 Rounded clasts with discontinuity planes affected by gelifraction. James Ross Island, east of the Antarctic Peninsula (Photo by A. Martín-Serrano).

gelifraction and highly porous rocks are not affected by gelifraction (Lautridou and Ozovf, 1982).

Numerous experiments have been done on cryogenic weathering of different rock types under variable conditions (Tricart, 1956a). Experiments performed on schists indicate that increased freezing intensity results in increased fragmentation due to the fatigue effect (Lautridou, 1988) *(Fig. 5.8)*. Disintegration by gelifraction increases as the number of freeze-thaw cycles increases; it decreases over time.

5.2.1.3 Insolation weathering

Temperature variations can cause rocks to dilate and contract and ultimately to break apart. The heat source may be solar radiation or fires, which are relatively frequent in semi-arid environments (Ollier and Ash, 1983). In the first case, the process of physical disintegration is called **insolation weathering** and in the second it is called **thermal shock** (Yatsu, 1988; Dorn, 2003). Rock surfaces in the desert are exposed to high temperatures, for example, 82°C in the Sudan (Goudie, 1989a), and to diurnal oscillations of 54°C as in the Atacama Desert (Keller, 1946). Insolation weathering is sometimes seen in outcrops in the humid tropics where temperatures can sometimes reach 50°C. As a result, the rocks are subject to thermal expansion and contraction. Wetting and drying cycles, which can also be a significant weathering factor, also affect these outcrops.

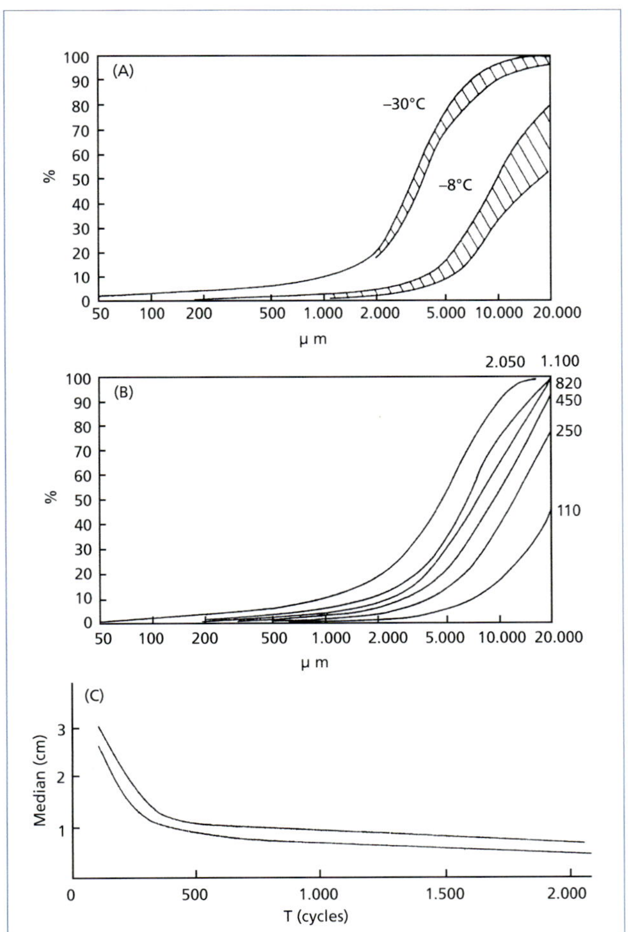

Figure 5.8 Examples of the grain-size distribution of gelifracts produced from Precambrian schists in Normandy. (A) Influence of the frost intensity. (B) Changes in grain size with time (indicated by the number of freezing cycles at −8°C). (C) Reduction with time of the median particle size produced by gelifraction (frost cycles at −8°C) for examples B and F (modified from Lautridou, 1988).

Most rocks contain a variety of minerals that have different degrees of thermal expansion depending on their color and structure. Dark rocks and minerals absorb more heat than light ones. In addition, some minerals expand more in one direction than in another due to their crystalline structure. Expansion causes compressive forces inside the rock and contraction causes tensile forces. These extensional and contraction differences can generate micro-fractures and cause granular disintegration (Birot, 1968). However, rocks are poor heat conductors and significant temperature differences can occur between the surface and the interior of a rock *(Fig. 5.9)*; this also causes significant expansion and contraction that can lead to exfoliation (Birot *et al.*, 1968; Warke and Smith, 1994; Smith and Warke, 1997). Outcropping rocks or blocks broken along parallel fractures (also called cleaved boulders) are interpreted as having been produced by insolation weathering (Ollier, 1983) although other investigators consider the cause to be salt weathering. The sharp noises or cracking that explorers heard at night in the silence of the desert are thought to be rocks breaking as they cooled during the night.

Blackwelder (1925, 1933) and Griggs (1936) conducted classic experiments on heating and cooling cycles of different types of rocks in the absence of water; these experiments showed no rock cracking after a number of cycles equivalent to 244 years of daily weathering. But when they cooled the rock with water instead of in dry air, micro-fractures occurred after a number of cycles equivalent to two and a half years. Humidity is of course present in all deserts as fog, night dew, and sometimes as storms that rapidly cool rocks and cause them to contract. Another factor to keep in mind regarding these early experiments is that rocks gradually weaken over long periods; they are affected by continuous expansion and contraction cycles that lead to rock fatigue and eventual rupture (Griggs, 1936; Aires-Barros *et al.*, 1975 and 1977; Gomez-Heras *et al.*, 2006). Moreover, we must remember that rock failure is not the result of a single mechanism but that generally two or more processes alternate or act in unison (Robinson and Williams, 1994). Temperatures

Figure 5.9 Lichen colony on calcareous Oligocene sandstone and affected by intense exfoliation. Ebro Depression, Albelda, Huesca Province, Spain.

decrease exponentially with depth and the largest decrease occurs within the first centimeters (Peel, 1974; Smith, 1986) (*Fig. 5.10*). Laboratory experiments by Gomez-Heras *et al.* (2006) show that surface temperature differences of various minerals can cause granular disintegration by fatigue. Also, crystal size is an important control on temperature variation of surface minerals. Finally, temperature differences can directly control rock fracture (Gomez-Heras *et al.*, 2006).

Using fire to simulate rock weathering, Goudie *et al.* (1992) found that the disintegration process of a material depends on the rock's physical characteristics (texture, mineralogy and type of cement), in other words, the intrinsic properties of the rock (Allison and Goudie, 1994).

5.2.1.4 Weathering by wetting and drying

Rocks in desert environments are exposed to numerous wetting and drying cycles that destroy the integrity of the rock. The ability of water to alter rocks is shown by durability studies conducted on different rock types (Ollier, 1984). Researchers plunged rocks into water for a day and then dried them in air the next day. Disintegration showed up as surface exfoliation and rock fractures mainly along existing joint planes. The first exfoliations occurred after five cycles and cracking began after ten cycles. Disintegration was complete after 180 cycles. Polarity of water molecules attracted to the mineral surfaces is thought to cause exfoliation. This polarity produces an expansive force that separates the confined surfaces. This mechanism was verified in laboratory experiments where rocks were subjected to the activity of high and low polarity liquids. High polarity liquids caused significant disruptive effects whereas the low polarity types caused no disintegration (Anon, 1966).

Rainfall, dew, and fog can all moisten rock surfaces in deserts. Even dew, which is common in some deserts, can cause a daily wetting and drying cycle (Cooke *et al.*, 1993). Water penetration into a rock depends on pore size and geometry; drying results from evaporation that is controlled by temperature and wind speed. Wetting-and-drying weathering is more effective with more frequent wetting and drying cycles. For this reason rock outcrops located on sunny slopes are more intensely desiccated than rocks in shady areas. They will experience a higher number of wetting and drying cycles during the same time period and the weathering will be more intense. The Tertiary sandstones of the Ebro Basin (Sancho and Benito, 1990) illustrate this fact; the ashlars of historical churches and buildings of the region such as the medieval churches of Torres del Bayo (Gracia, 1985) also corroborate this observation. Both sandstones are intensely altered where

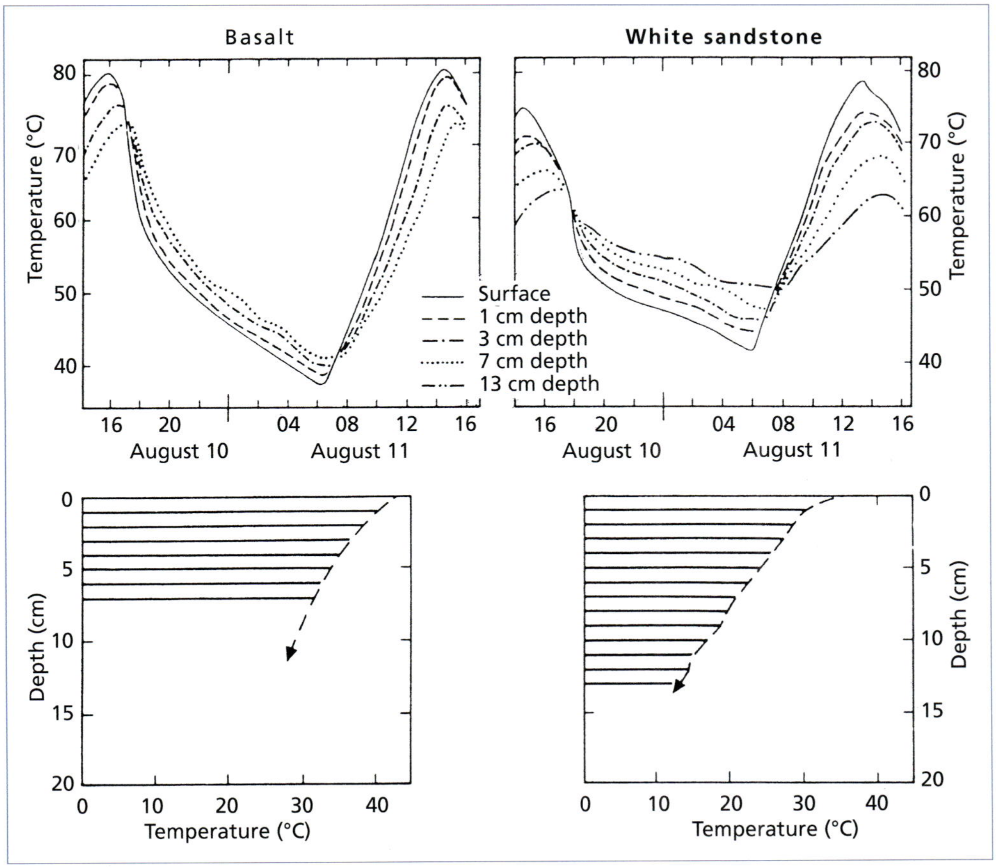

Figure 5.10 Temperature variations of basalt and white sandstone at the surface and at different depths in the Tibesti (central Sahara) (Peel, 1974).

exposed on the side that receives the most sun but weathering is practically nonexistent on the shady side. In the more arid environments of the Sinai desert, however, the observations are the opposite. The reason for this contradiction is not known and should be studied; it may be related to the annual amount, distribution, and type of precipitation.

5.2.1.5 Salt weathering

In deserts, extensive accumulations of highly soluble salts are common in lacustrine depressions or as surface coverings on rocks and sediments; the accumulations occur as light-colored and white efflorescences (Mabbutt, 1977). The precipitation of salts in porous spaces in rocks and later modifications to physical and chemical processes affecting the deposited salts cause rocks to expand. The expansion creates disruptive forces that may disintegrate the rock. This group of processes is more effective in desert and coastal areas than in moderate temperature zones. Salt weathering may be important in polar areas and on Mars (Malin, 1974). Alveoles, tafoni, and caves have been discovered in coarse-grained rocks in Antarctica (Wellman and Wilson, 1965). Examples of salt weathering in the Arctic have also been described (Watts, 1983).

In recent decades investigators (Doornkamp and Ibrahim, 1990) have used laboratory experiments to study the activity and consequences of salt weathering (Evans, 1969–70; Goudie *et al.*, 1970; Goudie, 1974; Cooke, 1979, 1981; Williams and Robinson, 1981; McGreevy, 1982; Sperling and Cooke, 1985; Fahey, 1986; Goudie and Viles, 1995, 1997). This research focuses on environmental conditions and materials affected by salt weathering. Researchers also experimented with different saline solutions to determine various salt weathering intensities, investigated effects and liberated particles, and sought to determine the various processes involved in salt weathering. Cooke *et al.* (1993) contains an excellent synthesis of salt weathering. Salt weathering has also been studied qualitatively to analyze the origin of various micromorphologies. This research will be discussed in the section on forms that result from weathering. Considerable advances have been made in the applied field because salt activity damages highways, buildings, and engineering structures, mainly as a result of reactions between cement and gypsum, which cause expansion and subsequent cracking (Goudie and Viles, 1997).

5.2.1.5.1 The effect of climate

Broad diurnal and seasonal variations of temperature and relative humidity and the presence of dry and sporadic winds promote salt weathering in deserts (Goudie, 2004d). These temperature and humidity variations are not temporary. They occur either on the surface or close to it, causing spatial variations of salt weathering. High temperatures cause saline solutions to evaporate and eventually precipitate. Salts also precipitate from saturated solutions as they cool. Humidity increases can form hydrous salts; temperature increases cause thermal expansion of saline crystals. These crystal growth, hydration, and expansion processes cause significant volumetric changes, leading to disruptive forces in the rocks (Cooke *et al.*, 1993).

Climatic variations affect water capillarity (Cooke *et al.*, 1982) (Fig. 5.11). Depending on the size of the constituent particles and the degree of cementation of the rock, the capillary fringe can reach a height of 3 m in very arid environments. Its upper limit is determined by accumulations of salt crystals; when the capillary front reaches the surface, it produces saline efflorescences (Fig. 5.12).

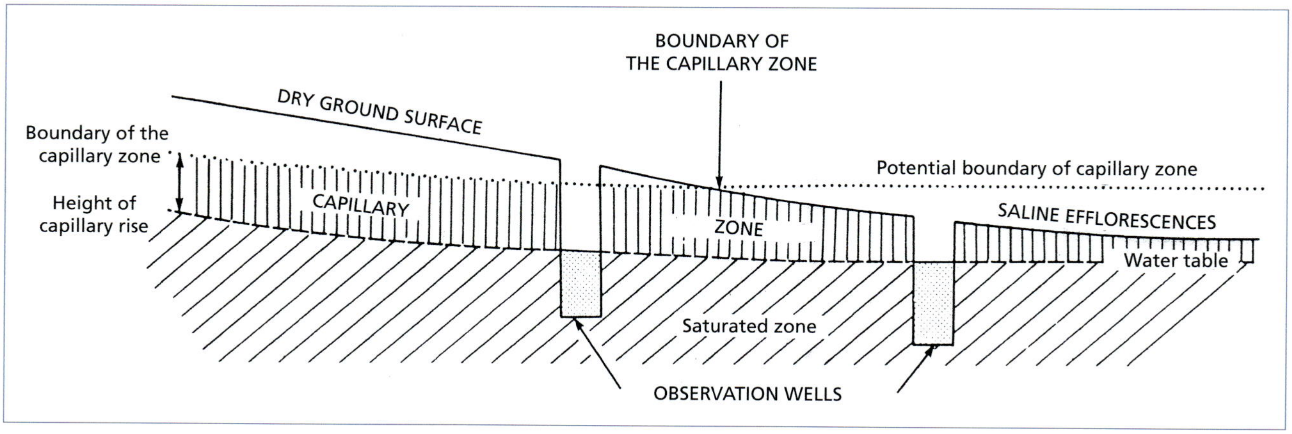

Figure 5.11 Characteristics of the capillary zone (Cooke *et al.*, 1982).

Figure 5.12 Salt efflorescences in Miocene calcareous sandstones of the Ebro Depression, El Tormillo, Huesca Province, Spain.

5.2.1.5.2 Affected rocks

Salt weathering depends on the transmissivity of the rock and the capacity of salt-laden water to infiltrate it. Transmissivity depends on the physical properties of the rock, including effective and trapped porosity, porosity distribution, adsorption capacity, penetration capillarity, desorption, hydric expansion, and resistance to traction, in addition to the types of clay minerals present in the rock.

Tests on the durability of different rock types show that sandstones and limestones are less resistant to salt weathering than igneous and metamorphic rocks. Salt weathering is most effective on porous rocks with high adsorption and water loss capacity, such as poorly cemented sandstones (Mabbutt, 1977).

5.2.1.5.3 Salts: types, origin, and concentration zones

The most common salts involved in salt weathering are chlorides, sulfates, carbonates, and nitrates of sodium, calcium, magnesium, and potassium. Calcium carbonate, $CaCO_3$, is most common in semi-arid zones, sulfates predominate in arid areas, and chlorides are also quite common in coastal deserts (Cooke *et al.*, 1993). Crystallization requires alternating wetting and drying stages; the higher the number of cycles, the greater the intensity of salt weathering. The effectiveness of weathering depends on salt solubility, hydration capacity, and crystallization strength. Some salts, such as mirabilite ($Na_2SO_4.10H_2O$), for example, are more disruptive than freezing water. Laboratory experiments conducted on different saline solutions indicate that sodium and magnesium sulfates are the most effective agents (*Fig. 5.13*) for disintegrating rocks by salt weathering (Goudie *et al.*, 1970). However, some salts increase significantly in volume as they hydrate; as a result they generate important forces that help disintegrate the rock. Common hydrations that occur in nature are the change of thenardite (Na_2SO_4) to mirabilite ($Na_2SO_4.10H_2O$) and anhydrite ($CaSO_4$) to gypsum ($SO_4.2H_2O$). The anhydrite-gypsum transformation increases the rock volume by about 39 percent (Jauzein, 1974).

The type of salt present in a region depends upon the source area lithology, which obviously varies from region to region. In Australia, NaCl is the predominant salt and in the salt flats of the Atacama Desert, the dominant salts are NaCl and sodium and calcium sulfates. In the endorheic area of the Ebro Depression of Spain, the most common salts are sodium chlorides and sulfates (Pueyo, 1978–79) and in some salt lakes of mid-eastern Africa, the preponderant salt is Na_2CO_3 (Natron Lake).

Salts come primarily from ions liberated during the chemical weathering of rocks; the salts are dissolved by runoff and subterranean flow. In littoral zones, aquifer overpumping can cause salt-water intrusion and increase the salinity of the extracted water. Salt is also mobilized by aqueous and eolian erosion of salt deposits. Salts can also come from volcanic dust or gases. In littoral deserts, marine aerosols blowing in from the sea are commonly an important source of salts (Goudie and Viles, 1997). In these areas, the salt concentration gradually decreases towards inland areas (Goudie, 1989b). In Aftout, Mauritania, and in the Senegal delta, waves strike littoral bars and generate water drops that evaporate and create small salt crystals; the crystals float in suspension, forming a whitish fog that the wind pushes toward the continent. Later the rain may wash the salt crystals to the ground. More often, however, the crystals are caught up in the higher relative humidity that is present at night; these hygroscopic crystals attract water that sticks to them and as a result of this enlargement they fall to the ground (Tricart, 1969).

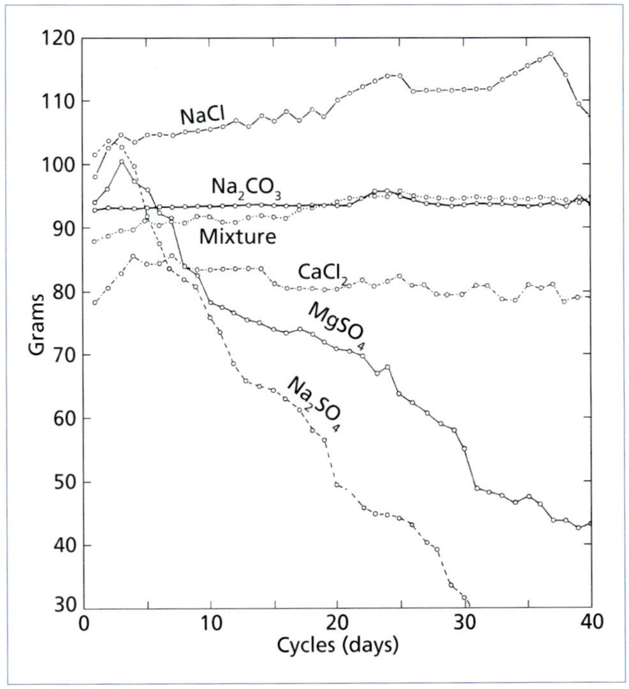

Figure 5.13 Weight variations of samples from the Arden sandstone, after being subjected to salt weathering tests with different types of salts (Goudie *et al.*, 1970).

Salts from diverse sources accumulate in a variety of situations (Cooke *et al.*, 1993). In littoral deserts (Peru-Chile, Namibia, and Western Australia) ocean salt contributes 150 kg of salt/hectare/year. In tidal flats of high salinity seas such as the Red Sea, the Persian Gulf, and the Mediterranean, significant saline sedimentation takes place in littoral sabkhas. In continental interiors, evaporation of lacustrine water that has accumulated as runoff and from groundwater and eolian contributions can generate salt lakes. The different solubilities of salts control their precipitation and result in a zonation showing the presence of chlorides in the interior, sulfates toward the middle with tumuli, and less soluble carbonates along the exterior (*Figs. 5.14* and *17.58*).

5.2.1.5.4 Physicochemical processes

The mechanisms involved in salt weathering are chemical and physical in nature. They include crystallization, hydration, and thermal expansion of highly soluble salts in confined and semi-confined spaces such as pores and cracks in rocks.

Of these three mechanisms, **crystal growth** is the most important (Goudie, 1974); as saline solutions become saturated, precipitates and salt crystals form in pore spaces (Winkler and Singer, 1972). These crystals can continue to grow between the walls that confine them if there is a film of saline solution at the crystal/rock interface. Crystallization generates significant pressures that are transmitted to the rock (Evans, 1969–70). Evaporation, cooling, or mixing of solutions with a common ion can cause precipitation. Some salt solubility (Na_2SO_4, Na_2CO_3, $NaNO_3$, and $MgSO_4$) diminishes at lower temperatures, which may cause precipitation during nightly cooling (Goudie, 1989a). In general, salt crystal growth predominates in areas where wetting and drying cycles are common and in zones of concentrated salt, such as the borders of saline lakes and the interiors of fluvial channels.

Hydration occurs when a salt adsorbs water into its crystalline structure. This mechanism increases volume and causes disruptive effects in confined zones such as granular disintegration and cracking of rocks. Different salts have different hydration pressures. Hydration of Na_2CO_3 and Na_2SO_4, for example, increases their volume as much as 300 percent (Goudie, 1977b). The initial precipitation in the interstices of rocks is in the form of a hydrated salt but due to high ambient temperatures during the day, salts convert to anhydrous phases. At night, when temperature decreases and humidity increases, the salts can re-hydrate. Hydration appears to be primarily a function of temperature and humidity changes that occur in deserts (Cooke *et al.*, 1993). Adsorption and dehydration of crystalline structures can recur several times; the pressures generated by these repeated volume changes and the disintegration effects on the rocks might be quite significant (Goudie and Viles, 1997).

According to laboratory experiments, **thermal expansion** of salt crystals is less effective than hydration (Goudie, 1974) in breaking up a rock. Expansion by insolation warming depends on the thermal expansion of each salt, which is usually higher than the thermal expansion of the minerals in a rock (Cooke and Smalley, 1968) and higher than the thermal oscillations recorded for various deserts. Experiments have been done at temperatures higher than those that affect desert rock surfaces (*Fig. 5.15*) and it is assumed that salt volumetric expansion by insolation heating behaves similarly (Cooke and Smalley, 1968) to experiments. Expan-

Figure 5.14 Zone of sulfate precipitation with selenite domes. In the background, central area of chloride precipitation. El Melah Sabkha. Zarzis, Tunis.

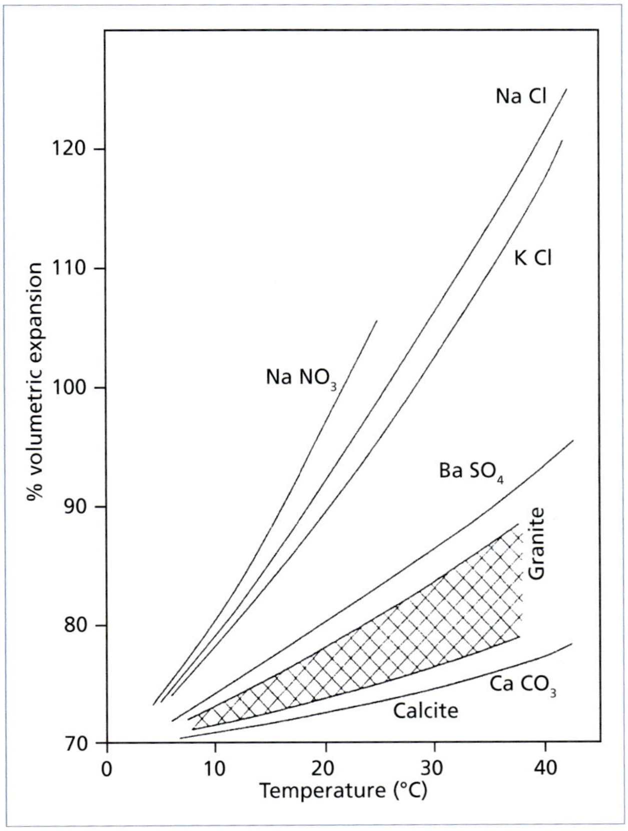

Figure 5.15 Relationship between temperature and volumetric expansion for different salts and granite (Cooke and Smalley, 1968).

sion of salts in pores and fissures of rocks that are close to the Earth's surface creates forces that can crack the rock. Studies conducted in the middle latitudes of the northern hemisphere show that rocks on sunny slopes have a higher degree of weathering than those on shaded slopes. This difference is attributed to the thermal expansion of NaCl (Johannessen *et al.*, 1982).

5.2.2 Biological weathering

Some authors define biological weathering in more precise terms. According to Viles (1988), **biogeomorphology** is a unique term for describing the role of living beings in the decomposition of rocks and the creation of micro-morphologies. Other investigators distinguish between the action of fauna, **Zoogeomorphology** (Butler, 1995) and the activity of flora, **Phytogeomorphology** (Howard and Mitchel, 1955).

Bacteria, fungi, algae, and lichens are abundant in desert environments, mainly due to the redistributing action of the wind. These microorganisms, of which there are numerous species, cause the weathering of rock and soil although their significance in the overall weathering process is not well understood. Environmental conditions and rock or soil characteristics determine the type of communities that will exist at the surface and, therefore, the characteristics of biological weathering (Viles, 1995).

Whatever the climatic conditions, large numbers of living creatures are found on the soil surface, according to Paul and Clark (1989). These microorganisms include: 10^7 bacteria, 10^6 actinomicetes, 10^6 fungi, and 10^4 algae. They are most abundant at the surface and their number decreases rapidly with depth. Heterotrophic **bacteria** oxidize and mineralize organic matter by means of weak and simple organic acids whereas autotrophic bacteria oxidize sulfur to sulfides and nitrogen compounds. Bacteria that are very abundant in humid mountain areas can reduce or oxidize iron and manganese while others such as *Galionella* and *Leptothrix* proliferate in swampy areas (Robert 1993). **Algae**, fungi, and lichens are pioneering plants that colonize rocks, including both acidic crystalline rocks and carbonates. Algae grow on wet rocks and **lichens** are well adapted to dryness. They cause physical weathering by wetting and drying (Viles, 2004).

These organisms grow both on the surface of rocks and soils and inside them. They can grow below the surface where they develop biocrusts *(Fig. 5.16)*, often made of bands of different colors that can measure up to 4 cm thick (Thomas, 1988). Some microorganisms colonize existing cracks and other attach firmly to the surface.

This biological activity, which results from life processes or secreted products, can substantially change surface features through physical and chemical processes (Ollier, 1983; Cooke *et al.*, 1993). Bacteria and lichens expand and contract depending on water content variations; these changes cause the rock to disintegrate by granular disintegration and exfoliation, both at the surface and in confined spaces. Secretion of organic acids can initiate chemical processes that attack rock constituents and create chelates that aid in the export of low-solubility cations. Some lichens are even capable of dissolving quartz (Hallbauer and Jahns, 1977); the zone they alter can extend to a depth of 1–5 mm (Viles and Pentecost, 1994). Carbon dioxide from microbiological activity helps dissolve carbonate surfaces and create biological micro-karren with tiny furrows and voids (Ollier, 1983). Some of these microorganisms also play a fundamental role in the creation of desert varnish.

Another significant aspect of biological activity is related to aqueous and eolian erosion (Thomas, 1988). Plant cover is known to affect sediment mobilization and surface biocrusts also considerably reduce erosion *(Fig. 5.16)*. When eolian activity slows down, dunes are able to stabilize (Veste, 1995) *(Fig. 5.17)*. When infiltration capacity decreases, rilling occurs, even though it is difficult to erode the biocrust cementation *(Fig. 5.18)*.

Figure 5.16 Biocrust developed on Miocene clays and surface colonization by lichens. Observe the lichen's resistance to erosion. Experimental station with 342 mm of average annual precipitation in the Lanaja Ebro Depression, Huesca Province, Spain.

The pressure of growing roots acts like a wedge to rupture rock and cause them to disintegrate (Mitchell, 1988). The amount of carbon dioxide available to solutions is also important; CO_2 comes from gas exchange or respiration of flora and fauna present in the soil and from the decomposition of organic matter. Fauna plays an important role in transferring and mixing materials and in moving regolith to other areas. In the humid tropics, termites and worms do this work; they may be responsible for sorting the upper layer of the soil. **Termites**, of which there are several thousand species, have a high degree of social organization. They transport clay-sized particles to use in building **termitaria**, which they construct with secreted saliva. The termite population of the Ivory Coast has been calculated at 9 million/ha; their number decreases with latitude and altitude (Goudie, 1988). Termitaria can be small mounds *(Fig. 5.19)* or towers (cathedral-termitaria of Tricart, 1974) *(Fig. 5.20)*; they can reach a height of 9 m and have a density of 1000/ha. Large termitaria are built by only a few species, whereas numerous species construct smaller termite mounds. Growth rates are highly variable, from 25 to 500 mm/yr. (Goudie, 1995). After termites abandon the termitaria, it erodes and is finally destroyed. Although the termitaria material is much stronger than the underlying soil, it will erode, normally in several decades (Goudie, 1988). Termites also build small filiform corridors on the soil and vegetation; these corridors are composed of cemented soil particles. These accumulations give us an idea of the amount of material termites can remove. Calculations done by several authors in Nigeria, Senegal, and Kenya show the accumulation rate to range between 300 and 1050 kg/ha/yr. (Goudie, 1988).

Terrestrial **worms** also build constructions similar to those of termites although they are smaller in size, seldom reaching 30 cm in height. Earthworms excavate burrows to depths of 70 cm, thereby increasing soil infiltration rates. These facts show the importance of worms as agents of removal and mixing of soil components. The amount of material removed is estimated at 5–25 mm/year, greater than the quantity moved by termites (Goudie, 1988).

Some crustaceans bore into the soil to construct their shelters, such as in Fort de France (Island of Martinique, Lesser Antilles) *(Fig. 5.2)*. Other vertebrates do the same type of work (rabbits, moles, etc.). Some birds produce guano accumulations and others construct mud nests on the walls of rocky overhangs or perforate soft materials such as sand and silt (bee-eating birds, for example), causing significant bioturbation.

All these weathering processes are relatively important but chemical weathering plays the most significant role in the disintegration of minerals and rocks in humid tropical zones.

Figure 5.17 Biocrust developed on a system of longitudinal dunes affected by rilling. Nizzana Experimental Station (Negev Desert, Israel).

5.2.3 Chemical weathering

Chemical weathering results from a change in chemical environment (Loughnan, 1969). Rocks formed under conditions of high pressure and temperature and in the absence of air become unstable when they come into contact with the atmosphere. Water, oxygen, and carbon dioxide attack rock minerals through exothermic reactions. The soluble products are removed by dissolution, leaving a residue rich in less soluble components. This residue forms authigenic compounds, mainly clay minerals and iron and aluminum oxyhydroxides or sesquioxides, which are in relatively stable equilibrium with environmental conditions.

Methods for studying chemical weathering in rocks are highly diverse. Laboratory experiments have supplied a disparate set of data containing large gaps such as the rate of dissolution reactions. Muxart and Birot (1977) studied the various regoliths developed in different climates.

Many questions remain unanswered: weathering rates, kinetic mechanisms, the nature of secondary products, and the importance of chemical weathering in the evolution of landforms (Colman and Dethier, 1986).

Jenny (1950) proposed a disintegration mechanism for mineral structures. Following the Pauling Rules, the sum of the positive charges in a crystal must be equal to the sum of negative charges but on crystal surfaces valences are not compensated, resulting in an unstable equilibrium. When crystals come in contact with water, OH^- ions join to the existing cations and the H^+ ions bond with oxygen and other negative ions. The H^+ ion, with its small ionic radius, can occupy cation sites on the crystal surface. Cations are then released as hydroxides. In addition, the loss of H^+ ions modifies the pH of the solution.

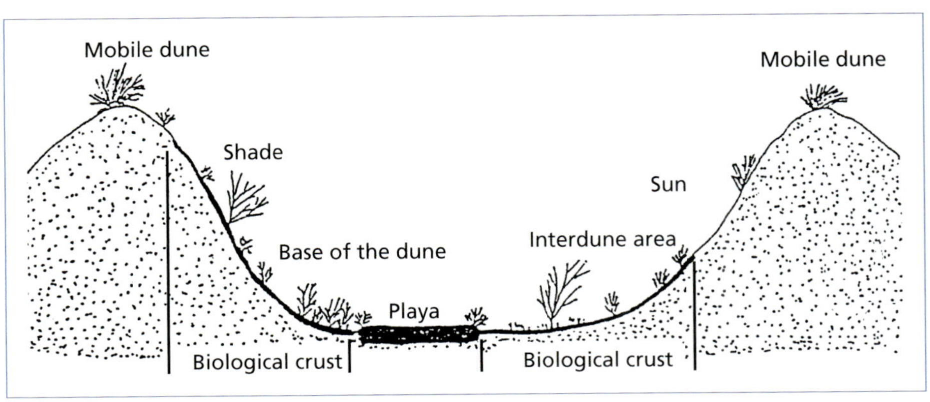

Figure 5.18 Cross-section of a system of longitudinal dunes at the Nizzana Experimental Station (Negev Desert, Israel). (Veste, 1995).

Figure 5.19 Termite mounds in the Masai Mara savanna of Kenya.

Figure 5.20 Inactive termite tower of about 6 m high in the forested savanna of Eastern Bouaké, Ivory Coast.

5.2.3.1 Weathering of silicates

Silicates are the most common minerals in rocks. Mineral species on the Earth's surface have the following approximate abundances: feldspars (30%), quartz (28%), clay minerals and micas (18%), calcite and dolomite (9%), iron oxides (4%), pyroxenes and amphiboles (1%) and other minerals (10%) (Leopold *et al.*, 1964). Cations released during weathering reactions may be easily washed away or may remain as a residue. This is due to physicochemical variations of the solution. This section contains a brief discussion of these factors and their effect on the mobility and precipitation of the most frequent silicate ions.

According to geochemical studies, fresh water **pH** ranges from 4 to 9 (Mason, 1958). *Figure 5.21* shows the solubility of some of the most common compounds as a function of pH (Loughnan, 1969). Sodium and potassium are soluble at all pH values. Calcium and magnesium hydroxides are soluble at the pH of natural waters. However, titanium and aluminum oxides and ferric hydroxide are insoluble in these intervals. Only ferrous and titanium hydroxides and calcium carbonates are affected by the pH of natural waters. Silica solubility is low although an alkaline pH will wash it away relatively easily. The silica solubility curve refers to amorphous silica and not to quartz, which is twenty times less soluble than amorphous silica (Carroll, 1970). Even so, there are environments where acidity and alkalinity are high. For example, aluminum can be leached in the form of aluminates in very basic playa-lake environments. In soils with large quantities of organic matter, pH may be lower than 4, which is low enough to remove aluminum. In addition, sulfur oxidation causes water to become highly acidic.

The oxidation-reduction potential (**Eh**) has an obvious effect on the solubility of polyvalent cations such as Fe and Ti. Chemical elements in the oxidation state are in equilibrium with the environment. Because solubility varies as a function of different valence states, the redox potential affects the mobility of a chemical element. *Figure 5.21* shows the solubility of trivalent and divalent iron. Ferric hydroxide is soluble only at a pH lower than 3 whereas ferrous hydroxide is soluble up to a pH of 8. Hence, the redox potential value is essential to the mobility of iron. The principal factors affecting Eh in weathering environments are the availability of atmospheric oxygen and the presence or absence of organic matter (Loughnan, 1969). Oxidation is an exothermic reaction that takes place above the zone of saturation. Below the water table, the lack of oxygen results in a reducing environment. For this reason, iron in this environment occurs in its ferrous state and is related to the gley horizon, which forms in humid areas that lack drainage. Organic matter is a powerful reducing agent because it oxidizes rapidly and produces carbon dioxide.

Potassium is a highly soluble cation although it is usually retained in the structure of illite-type clay minerals. Nevertheless, under conditions of high precipitation it can be dissolved without leaving any traces in the system.

Chelation consists of the joining of a cation and a complexing agent (chelate) such as EDTA. The result is a ring-shaped organometallic compound with the cation in the middle (Bland and Rolls, 1998). Cations fixed by chelates can be divalent or trivalent and these organometallic complexes are highly soluble. They are created in nature by humic and fulvic acids produced during plant decomposition and root secretion (Keller, 1957).

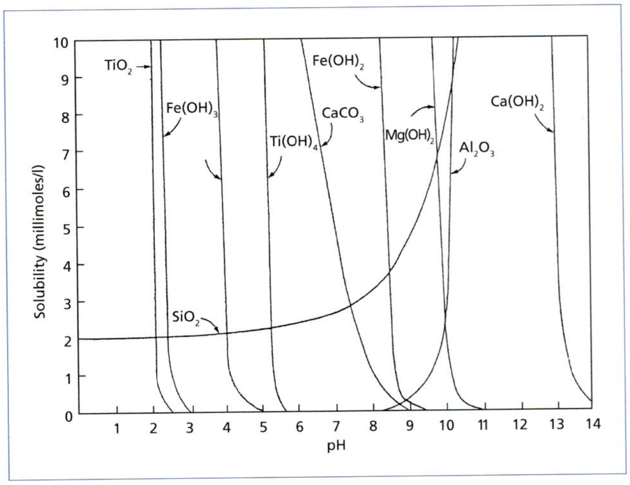

Figure 5.21 Solubility compared to pH for compounds released by chemical weathering (compiled from several authors by Loughnan, 1969).

Ion exchange is the most important process in chemical weathering (Carroll, 1970). It consists of the reaction between ions in solution and those present in minerals. Clay minerals are the main agents responsible for the exchange capacity in rock weathering. If a mineral is not electrically neutral, a cation will adhere to the clay surface and neutralize the negative charge. Al^{3+} frequently replaces Si^{4+} in this process. Each clay mineral has a different ion exchange capacity; minerals of the montmorillonite group have the highest ion-exchange capacity.

Hydration is the addition of water molecules to a mineral structure. When hydrated, iron oxides transform into iron oxyhydroxides. Water is also easily introduced into the crystalline structure of expansive clays. The transformation of anhydrite ($CaSO_4$) into gypsum ($CaSO_4 \cdot 2H_2O$) also involves a considerable increase in volume.

The most important factor controlling the disintegration rate of crystalline structures and resulting products is the quantity of **solution water**. This water removes soluble cations resulting from the hydrolysis process. The amount of dissolution depends on the water volume in contact with mineral surfaces and the solubility of the mineral. Significant *precipitation* occurring over long periods can cause resistant minerals such as quartz to disappear by dissolution. The Weipa bauxites of North Queensland, Australia, were formed from leaching of a substratum of kaolinitic sandstones that contained 90 percent quartz and 10 percent kaolinite. Monsoon precipitation has formed bauxites with 5 percent quartz and small quantities of kaolinite (Loughnan and Bayliss, 1961).

Research has determined the sequence by which the most common cations are lost. Polynov (1937) compared the compositional percentage of dissolved matter in rivers to the composition of drainage basins containing igneous rocks. He obtained the following order of loss: $Ca^{2+} > Na^+ > Mg^{2+} > K^+ > SiO_2 > Fe_2O_3 > Al_2O_3$. Other studies indicate that the order depends on rock type (Reiche, 1950; Carroll, 1970).

Crystalline structure and atomic bond strengths play an important role in mineral weathering. During magmatic crystallization, minerals that form first have the highest melting points. The sequence is represented by Bowen's crystallization series (1928). Goldich (1938) worked out the weathering sequence for the most common silicates from a detailed study of the weathering of igneous and metamorphic rocks (*Fig. 5.22*); his sequence is the inverse of Bowen's because the most easily weathered minerals are the first to crystallize in a silicate melt. The most stable minerals crystallize later in the cooling sequence and have a higher number of more stable covalent bonds.

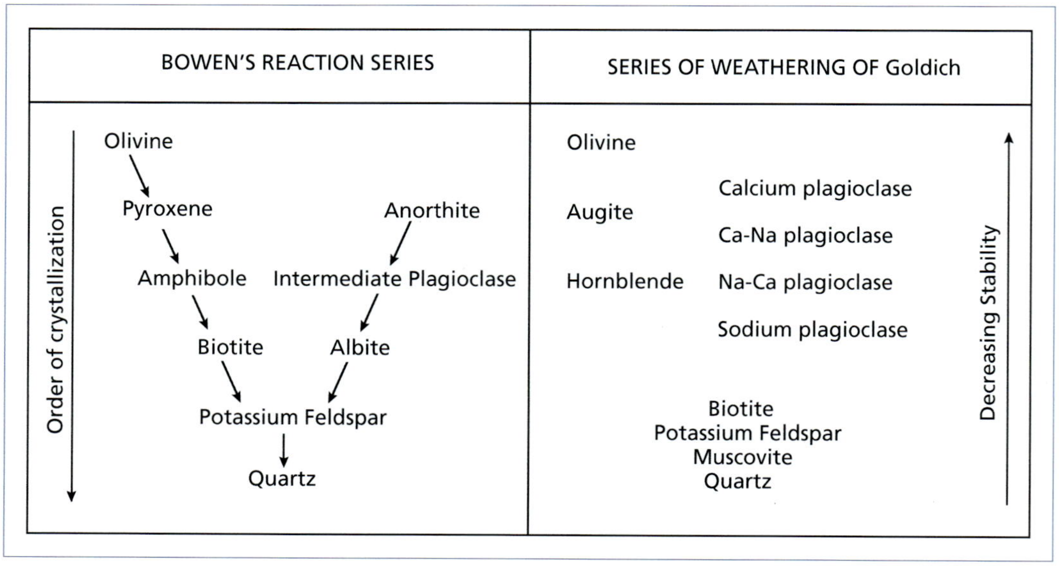

Figure 5.22 Bowen's reaction series and Goldich's weathering sequence. (Adapted by Ollier and Pain, 1996).

5.2.3.2 Degrees of weathering and weathering products

The extent of weathering varies and is measured on the basis of alteration products. Researchers have developed several indices to categorize weathering products (Taylor and Eggleton, 2001). The indices are based on qualitative scales of friability (Ollier, 1965; 1969b), on mechanical properties of the regolith (Lumb, 1962; 1983), and on the chemistry and mineralogy of weathering profiles (Grant, 1969; Sueoka, 1988; Geological Society, 1990). As an example, Grant (1969) called the ratio between the most mobile elements (Na, K, Ca, and Mg) and residual clay minerals "the abrasion pH." The lower the index, the more intense the weathering. Sueoka (1988) proposed a chemical weathering index, which consists of the sum of the molecular weight percentages of Al, Fe, and Ti oxides plus H_2O divided by the percentages of the remaining elements (alkalis, alkaline earths, and silica). Thomas (1994a) and Migon (2006) include other indices in the degree of weathering.

Weathering destroys crystalline structures and releases cations in the process. Highly soluble cations are leached by water while others that remain in the residue are restructured as authigenic minerals (Bland and Rolls, 1998). The most common are clay minerals, quartz, and oxyhydroxides of iron and aluminum, which form lateritic hardpans. These changes are accompanied a marked decrease in resistance to erosion and changes in porosity, permeability, and water content (Lee and Fookes, 2005).

Clay minerals are hydrated silicates of aluminum, iron, and magnesium ordered into sheets; for this reason they are called 'phyllosilicates.' Tetrahedral sheets are formed by Si surrounded by four O ions whereas octahedral layers consist of Al, Mg, or Fe cations at the center surrounded by six O or OH ions. These two basic sheets can be grouped in three different ways. First, a tetrahedral sheet adjacent to an octahedral sheet forms a structure of the 1:1 type. Clay minerals of this type constitute the kaolinite-halloysite group. Secondly, an octahedral sheet can be sandwiched between two tetrahedral sheets. This arrangement is characteristic of 2:1 phyllosilicates, which includes the smectite group (montmorillonite, beidellite, and nontronite) and illites. The third type consists of an octahedral sheet placed between adjacent 2:1 sheets; this type makes up the 2:1:1 clay minerals, and chlorite is the most common example.

The formation of clay minerals is complex because of the wide variety of weathering environments. Clays can also result from the weathering of non-phyllosilicate minerals. These processes release cations, silicon, and alumina that reorganize to form clay minerals. When the altered mineral is a phyllosilicate, weathering occurs mainly in the solid state (Birkeland, 1984).

Studies by Sherman (1952) on basalt weathering in Hawaii indicate an increase in clay percentage with precipitation and changes in the mineralogy of the rock. *Figure 5.23* shows that montmorillonite, a mineral typical of low precipitation regimes, transforms into kaolinite under conditions of intense leaching. When precipitation is high, the soil loses silica, and iron and aluminum oxyhydroxides are formed. Sherman also indicates that during dry periods, the alteration residue becomes enriched in iron and titanium oxides.

When precipitation occurs throughout the entire year, the main products are aluminum oxyhydroxides.

Considering the dominant weathering products, several weathering zones can be identified in the humid tropics (Strakhov, 1967) (*Fig. 12.4*). Pedro (1966, 1968) made the following classification:

1. **Bisiallitization zone** where clay minerals of the 2:1 type (montmorillonite, $4SiO_2 . Al_2O_3 . H_2O$) are present; these are areas with less than 500 mm precipitation.
2. **Monosiallitization zone** where kaolinite ($2SiO_2 . Al_2O_3 . 2H_2O$) is the predominant mineral in the regolith; these are zones with precipitation of 500–1200/1500 mm.
3. **Allitization zone** where gibbsite ($Al(OH)_3$) occurs with kaolinite; these are regions where precipitation is higher than 1500 mm.

Sanches Furtado (1968) has shown that kaolinite is the most abundant mineral in the humid tropics. He also established regional differences based on predominant minerals and precipitation. *Figure 5.24* shows a map of the clay mineralogy in alluvial soils of different climatic regions in Sri Lanka (Herath and Grishaw, 1971). Jenny (1941) analyzed the role of temperature in the generation of clay minerals. He shows that the clay content of regoliths in basic rocks is a direct function of temperature. At an average annual temperature of 10°C, clays make up 15 percent of the soil while at 16°C the clay fraction rises to 50 percent.

Duchaufour (1977) established three main types of alteration in the tropics to broadly characterize different climatic zones. According to this author, **fersiallitic soils** predominate in areas with precipitation from 500 to more than 1000 mm and temperatures between 13° and 20°C. These soils also contain smectites and the process of desilication begins to generate kaolinite in them; iron is present as oxyhydroxides with nontronite, an iron variety of smectite. **Ferruginous soils** are formed in regions with higher temperatures and precipitation and the majority of authigenic clays belong to the kaolinite group. **Ferrallitic soils** are the terminal phase of the evolution and alteration of soils; the primary minerals in them have been completely hydrolyzed, except quartz, and they have released Fe and Al oxyhydroxides. They develop at average annual temperatures higher than 25°C and precipitation greater than 1500 mm.

In general, the type and amount of clay minerals vary with depth although some profiles show no substantial changes. These variations must be related to the intensity of leaching, which removes many ions and silicon down to significant depths. This type of modification is quite common. In this way, surficial gibbsite changes to kaolinite at depth, or, if kaolinite occurs at the surface, it transitions to montmorillonite (Loughnan, 1969). These sequences may be in equilibrium with environmental conditions although conditions may change. If the soil profile is eroded montmo-

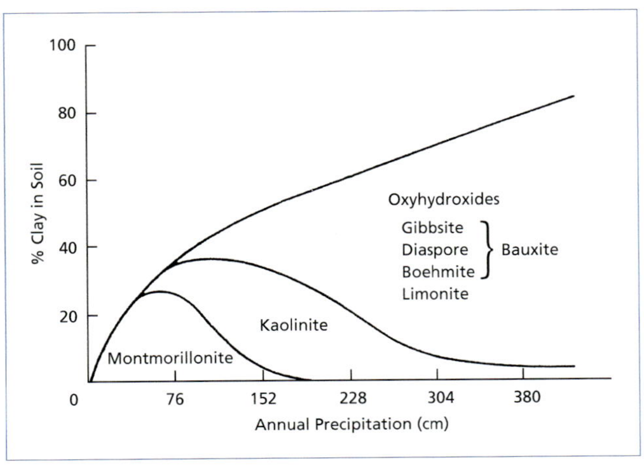

Figure 5.23 Progressive development of clay types in Hawaiian soils, under continuously wet climate conditions (after Sherman, 1952).

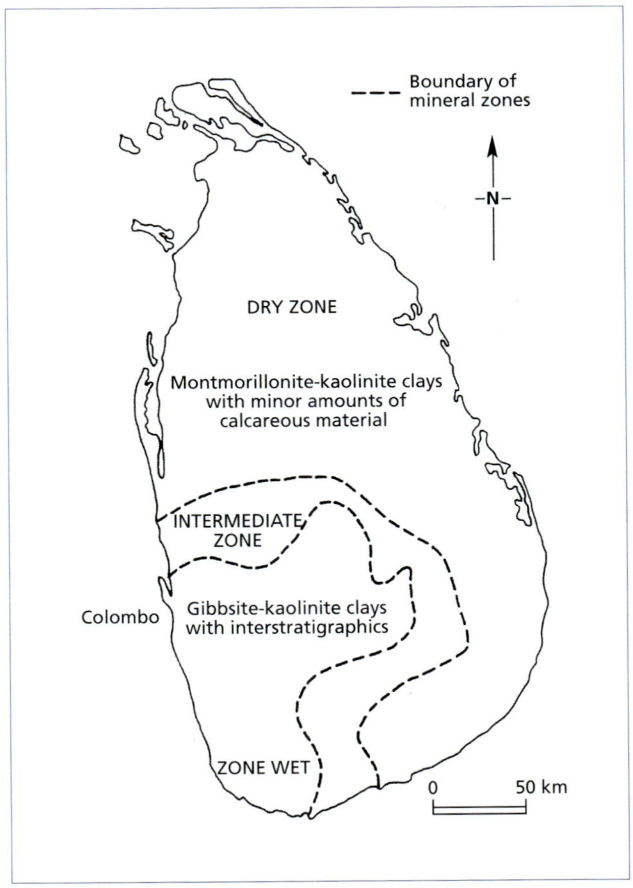

Figure 5.24 Clay mineralogy of alluvial soils in various climatic regions of Sri Lanka (Herath and Grimshaw, 1971).

rillonite may crop out at the surface but in an aggressive climate montmorillonite will weather to kaolinite at the surface (Fig. 5.25). The surface may slowly deflate due to denudation; if this process is in balance with the rate of kaolinite generation, the profile changes its mineralogical makeup (Birkeland, 1984).

Formation and transformation of clay minerals in the weathering profile are slow processes. Hence, the study of clay minerals can be useful for paleoclimatic reconstructions (Birkeland, 1974). As previously indicated, montmorillonite forms in zones of low precipitation whereas kaolinite and halloysite form in regions of high leaching. If we find kaolinite in soils in an arid climate, we can reason that the climate of the area was humid and tropical in the past. On the other hand, when an arid zone that has montmorillonite in the soil profile undergoes a climate change towards high leaching conditions, the montmorillonite generated under that climate transforms into kaolinite. Thus, we can recognize humid to arid climatic changes but not the contrary (Pedro *et al.*, 1969; Singer, 1979/1980). Because mineral transformations are slow, these changes occur only if climate change has taken place over a long period. If the climate change occurred recently, no detectable modifications occur. In volcanoes located on humid tropical zones, pedogenesis is relatively fast although it is frequently interrupted by new volcanic eruptions that bury and fossilize the soils. Under these circumstances the endogenous exhalative activity is more rapid than pedogenesis and any climatic change that occurs would not be recorded in the soil (Fig. 5.26).

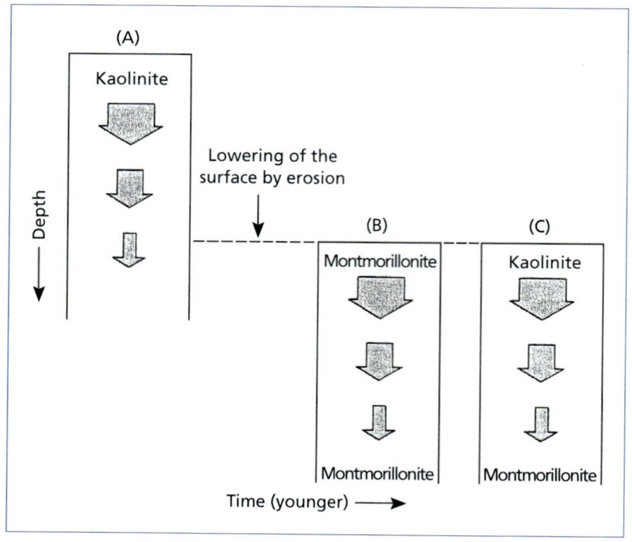

Figure 5.25 Vertical distribution of clay minerals as a function of leaching conditions within the soil. (A) Leaching conditions favor the formation of kaolinite and montmorillonite at different levels in the soil profile. (B) Rapid lowering of the surface by erosion results in montmorillonite outcropping in a soil environment of high leaching. (C) With time, the montmorillonite at the surface becomes desilicified and alters to kaolinite. In the figure, the wider arrows indicate more leaching (after Birkeland, 1984).

5.2.3.3 Weathering profile

Landforms may be covered with a thin layer of weathered material or blanketed by a thick alteration mantle of which the physical and chemical characteristics are significant for landform evolution (Gerrard, 1988).

Altered *in-situ* rock is called **saprolite** although the term **regolith** is more commonly used. The word regolith comes from the Greek word *regos*, which means cover. Regolith also includes weathered rocks that may have undergone thin mass movements (creep) or plant and animal activity (Ollier, 1988a). French researchers call this material **alterite**. Sandy material resulting from weathering is known as **grus** or **lem**. It can originate near the surface or at great depths and is essentially azonal (Migon and Thomàs, 2002).

The limit between unweathered rock and altered material can be sharp, as in granites, or diffuse, as in limestones and slates. Ruxton and Berry (1957) call this limit the **basal weathering surface** although this term has been criticized due to its static implications. Because alteration moves toward deeper areas, it is a dynamic boundary, and therefore it seems preferable to call it a **weathering front** (Mabbutt, 1961a).

5.2.3.4 Weathering depth

The downward development of regolith depends on factors such as climate (temperature and precipitation), biology (vegetation), geomorphology (surface stability), hydrology, geology (rock type and joint density), and chronology (climatic and tectonic changes) (Faniran and Jeje, 1983; Thomas, 1994a).

The weathering depth depends on the balance between weathering intensity and surface erosion, which is closely related to topography. In the tropics, low surface runoff erosion (Fournier, 1960; Douglas, 1969) promotes weathering. Weathered material is easily eroded on slopes but erosion is limited on gentle slopes or flat land surfaces. As a consequence, maximum weathering depths will occur in planar areas where the climate is suitable for rapid weathering, where the topography is smooth, and where vegetation hinders runoff. Erosion is minimal in these zones (Ollier, 1969b).

Within the weathering profile, the maximum intensity occurs in the zone of aeration. In the saturated zone, processes such as reduction, hydrolysis, and chemical substitution are also active although the rate of these reactions is slower (Gerrard, 1988). These circumstances led Lelong (1966) to distinguish between an upper alteration zone where water flows easily and promotes processes of hydrolysis and a lower alteration zone

Figure 5.26 Reddish paleosoils buried by volcanic ash in the Gregory Rift Valley, Kabernet, Kenya.

where percolating water flows slowly through joints. All these processes help explain the great thickness of altered rocks.

Geological research in humid tropical zones focuses on outcrops of unaltered rock. However, applied studies of the regolith (for civil engineering, mining, agronomy, and other purposes) require an understanding of the surface extent of rock outcrops and the depth of the regolith. Mechanical drilling and geophysical methods are used to obtain this information (Nocera and Matano, 2002). Quarries (*Fig. 5.27*), open-pit mines, and excavations for road construction are excellent places to observe the depth of weathering. Ollier (1965) recorded depths of 150 m in some tunnel drillings for a hydroelectric project in Victoria, Australia. Thomas (1965) indicated a depth of 100 m in Nigeria and Ollier (1960) found the same in Uganda. In South America, Nagel (1962) and Feininger (1971) found depths of more than 100 m on metasedimentary ridges in Brazil and on quartz diorites in Colombia, respectively. *Table 2.9* of Migon (2006) shows the thickness of alteration layers in present-day and ancient granitoids. The data contain discrepancies.

Deep weathering profiles are relatively common in arid and semiarid zones of Africa (Thomas, 1994a) and Australia (Mabbutt, 1965b) and in temperate regions of the U.S.A. (Kaye, 1967) and Europe (Demek, 1965; Gutiérrez and Rodríguez, 1978; Molina and Blanco, 1980). These weathering profiles obviously did not originate under the climatic conditions that prevail today in these places. Many authors think these profiles must have originated under humid tropical climates where chemical activity was more intense. Thus, the presence of thick weathering profiles in areas far from the humid tropics is a good indicator of climate change.

Thomas (1994a) has analyzed the spatial distribution of regoliths. The influence of rock structure is apparent because it affects weathering depth. For example, the density of discontinuity planes in the rock mass controls the extent of weathering. Maximum weathering depth is related to maximum discontinuity plane density (Thomas, 1966, 1994a, 2005) (*Fig. 5.28*), which may correspond to areas where fracture systems intersect. In this case, the weathering front is highly irregular and bears no relation to the topographical profile. However, the deepest weathering develops on fluvial divides. This may be due to fluvial erosion of a previous thick weathering profile (*Fig. 5.29*) or to faster weathering on the divides (Thomas, 1994a).

5.2.3.5 Weathering zones

The weathering profile is not an unstructured regolith. On the contrary, sets of distinctive features can be seen within the profile, especially in granitic rocks. Studies by Ruxton and Berry (1957, 1961) and Berry and Ruxton (1959) on granitic rocks in Hong Kong were the first to establish zonation within weathering profiles. Wilhelmy (1958) studied similar topics in eastern Asia, Africa, Central and South America, the Iberian

Figure 5.27 Weathering profile observable on a granite quarry. Note the ferralitic zone in the upper part, thinning towards the valley. Aparecida, Sao Paulo, Brazil.

Peninsula, and Corsica. Mabbutt (1961b) analyzed weathering profiles in western Australia. Within a weathering profile, angularity and the number of fresh rock blocks increase with depth. In some cases jointed blocks have an envelope composed of shells of residual materials and a corestone that may remain totally fresh and coherent. This phenomenon is known as **spheroidal weathering** (*Figs. 5.30* and *5.31*). Geologists think spheroidal weathering results from the outward migration of elements from within the rock (Ollier, 1969b). The upper parts of the profile are more easily eroded down to the zone where blocks predominate. Classification into zones is based on the degree of alteration of the granite rock and the presence of unaltered blocks at the base of the profile (*Fig. 5.32*). The previously cited authors have all applied these criteria to establish divisions in the weathering profile. Gerrard (1988) made a broad compilation of weathering profile classifications devised by different authors. Ollier (1969b) proposed the most widely used zonation (Gerrard, 1988):

1. Soil.
2. Structureless regolith.
3. Regolith that preserves the rock structure (*Fig. 5.33*).
4. Regolith with rounded blocks.
5. Regolith with parallelepiped-shaped blocks in close proximity to each other (*Fig. 5.34*).
6. Fresh rock.

Obviously, the zonation of weathering profiles on sedimentary, metamorphic, and volcanic rocks is highly variable although the Ollier profile type is also applicable. Rock fissuring plays a fundamental role in all of them. Other profile divisions can be based upon chemical, mineralogical, hydrological, or engineering criteria, depending on the objectives (Ollier, 1988a).

5.2.3.6 Weathering rates

Chemical weathering processes operate in all climatic zones but the weathering rate is much higher in the humid tropics (Muxart and Birot, 1977). In studying these processes, it is important to quantify rock weathering.

Various methods are used for quantification. Quantitative data can be obtained from the alteration of building and ornamental stone because the age of installation is known. Numerous authors have done this type of research in different climates (Ollier, 1988a). Approximate data can also be obtained by accelerated aging research conducted

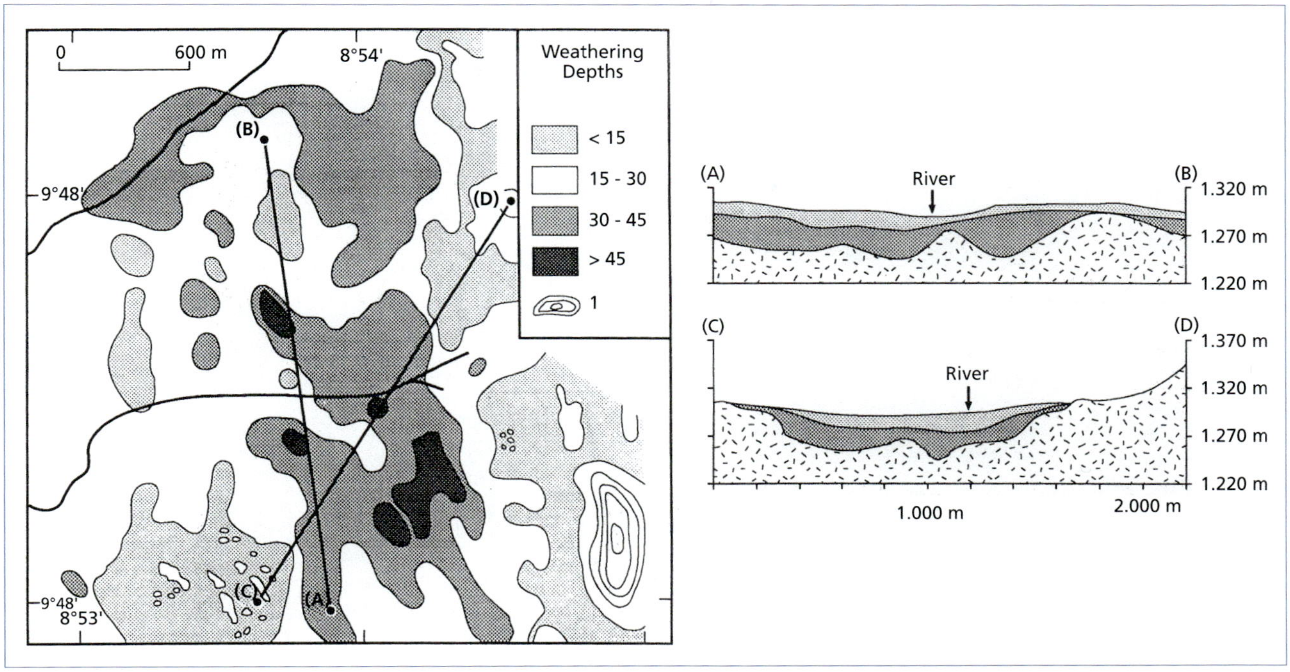

Figure 5.28 Depth of weathering distribution, in meters, in fine-grained biotite granite, near Jos, northern Nigeria. 1 Outcrops of unweathered granite at 15-m contour intervals. Saprolite is shown on the cross-sections (A-B, C-D) (after Thomas, 1966).

Figure 5.29 Thick weathering profile incised by the Mathioya River, Kenya.

in the laboratory. The recorded values are highest at the beginning of the weathering process. Laboratory weathering rates have been estimated to be about 200–400 times higher than rates that actually occur in the field (Swodoba-Colberg and Drever, 1993). Nevertheless, one of the most common methods is based in the quantification of solid and dissolved discharge measured in experimental plots that are a few tens of meters long. The chemical composition of the weathering residue has also been studied (Nahon, 1991). The experimental period is variable but normally ranges from 1 to

Figure 5.30 Spheroidal weathering of volcanic rocks. Fort de France, Martinique Island, Lesser Antilles.

Figure 5.31 Spheroidal weathering in basic lavas of the Gregory Rift Valley, Kabernet, Kenya.

5 years. In the humid tropical savanna zones of Cameroon, removal rates of 1–2 mm/ka of surficial material have been observed (Birot, 1978). In the jungle of the Ivory Coast, the observed rate was 1.2 mm/ka (Roose and Lelong, 1976). In Burkina Faso, Senegal, and Ivory Coast values range from 1 to 8 mm/ka (compiled by Goudie, 1995). Another method consists of studying the bedload, suspended load, and dissolved load discharges from small basins. A study of three experimental basins in New England of western Australia showed that dissolved load is ten times greater than suspended load and that bedload is virtually absent (Zakaria, 1977; in Ollier, 1969a). Denudation values of 14.4 mm/ka for basalts and 4.5 mm/ka for granites have been measured (Ollier, 1969a). Data from Rhodesia indicate that the rate of surface lowering in granites is 15 mm/ka in rainy zones and 6 mm/ka in drier zones (Owens and Watson, 1979). In southeastern Brazil, Vasconcelos *et al.* (1992) propose $^{40}K/^{40}Ar$ and $^{40}Ar/^{39}Ar$ ages of 10.1–5.6 Ma for deep regoliths (>20 m). Erosion values measured on ultrabasic rocks in New Caledonia are about 28 mm/ka (Trescases, 1973, in Muxart and Birot, 1977). A compilation of alteration rates by Thomas (1994a) provided an estimate of 2–50 mm/ka. However, confusion remains regarding the age of a weathering profile and the rate of saprolite formation. In temperate humid areas such as the Appalachians, rates of saprolite formation are 20 m per million years for the medium term (Cleaves *et al.*, 1970. In tropical humid zones, 50 m of saprolite have formed in the last million years (Thomas, 1994a). Finally, long-term weathering rates can be calculated from the absolute dating of geologic materials (pyroclasts, lavas, etc.). However, great

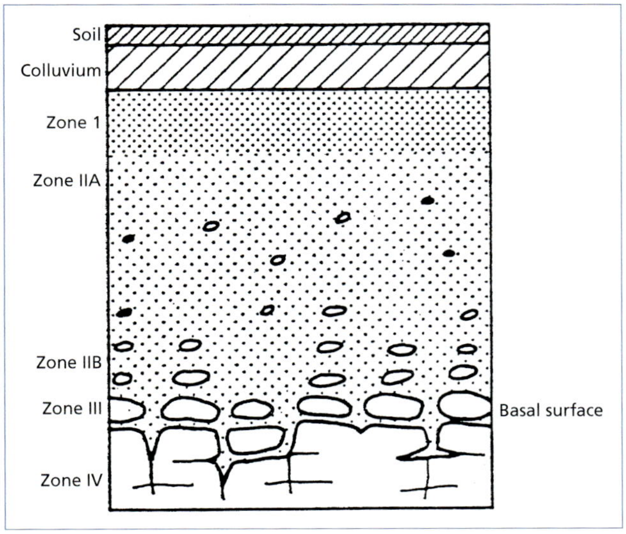

Figure 5.32 Characteristic weathering profile in jointed granite (modified from Ruxton and Berry, 1961).

Figure 5.33 Weathering profile on granite. Primary and *schlieren* structures can be seen together with spheroidal weathering. The photograph was taken on a quarry where altered granite is exploited for construction in La Alberca, Salamanca, Spain.

Figure 5.34 Weathering profile showing the transition between the rounded and the angular block zones. Spanish Central System, Villacastín – Ávila road.

uncertainty remains about the time-weathering relationship, which normally is not linear (Viles, 2001).

Values obtained from experimental methods indicate that most regoliths formed during the Quaternary because average weathering rates are about 20 m per million years. Birot (1978) gave the following values for the alteration of 1 m of granite: 50,000 years in a rainforest and 70,000 years in a savanna. All these estimates are similar to those proposed by Thomas (1994a). When considering these long time periods, we must remember that climate changes may have occurred during the development of the profiles. In dry periods, the weathering rate slows but denudation increases.

A cosmogenic dating technique that uses weathering rings has recently been developed (Gorgon and Dorn, 2005). This method, which uses weathering rings (WR), has numerous applications in geomorphology and related sciences.

5.3 Resultant weathering forms

Rock surfaces in deserts can be affected by differential weathering, a process that creates different types of **microforms**. These landform features occur at the base of vertical walls that contain caves or shelters, on rocky slopes where **alveoles** and **tafoni** occur, and on horizontal or low-gradient surfaces that contain **gnammas** or weathering pits. All these forms can be found at the same rock outcrop; they commonly occur in deserts but they can also be found in other climates (Martini, 1978; Twidale and Corbin, 1963). They develop primarily on isogranular igneous (granite, rhyolite) and sedimentary (sandstone) rocks but they can also be found in conglomerates, gneisses and porphyries (Goudie and Viles, 1997; Goudie, 2004e).

The basal portion of a rock outcrop has a high water content due in part to the rise of water by capillary action. For this reason, weathering is more intense at the base than in the upper part of an outcrop where the rock is drier and weathering is minimal (Mabbutt, 1977). These factors explain **basal undercutting** of rock walls, which is more likely to occur in shady areas and create **caves** or **shelters** (Howard and Selby, 1994). When undercutting affects isolated hills or inselbergs, it can create mushroom-shaped monoliths. Subsurface weathering and later exhumation can also generate this landform; an excellent example is the Eyre Peninsula of southern Australia where there are spectacular overhangs (Twidale, 1962).

Rocky walls of moderate or considerable steepness are often marked by rounded voids that may occasionally completely cover the rock surface. They also occur on loose blocks on a slope. If the hollows are about a centimeter in diameter, they are called **alveoles** (*Fig. 5.35*); when grouped together they resemble a honeycomb. Their origin is unknown; they have been attributed to eolian erosion, exfoliation, frost action, and salt weathering. They are common in coastal environments (Fig. 11.15) (Mustoe, 1982). Mottershead (1994) assigns a minimum weathering rate of 1 mm/year to alveoles in littoral environments.

Larger versions of alveoles, measuring decimeters up to a few meters, are called **tafoni**, an Italian word (*Fig. 5.36*). They usually occur in groups and are circular or elliptical in shape; detritus deposited by wind, rain, or runoff often covers the base of the void (Mellor *et al.*, 1997). Tafoni can grow and interconnect with others; they can also connect with other tafoni as they grow toward the interior of the rock. Like alveoles, they are sometimes oriented along planes of weakness in the rock, which may be imperceptible. The rock surface usually has exfoliation; saline efflorescences often occur inside the tafoni. Again, their origin is not well understood (Twidale and Vidal Romani, 2005). Some authors make a distinction between wall tafoni and basal tafoni related to basal undercutting processes (Smith, 1978; Bradley *et al.*, 1978). Most researchers agree that the dominant process involves the near-surface movement of saline water, which is similar to salt weathering and wetting-and-drying cycle weathering. Other authors cite insolation weathering, frost action, dissolution of carbonated cements, and eolian erosion (Evans, 1969–70; Mustoe, 1982; Young, 1987; Robinson and Williams, 1992). In addition to these subaerial processes, weathering beneath the soil surface and later exhumation can create tafoni that appear as a result of erosion (Dragovich, 1969). Tafoni are rare in extremely arid environments because the lack of water limits the weathering processes.

Tafoni formed along the coast of Corsica, Elba, and Cerdeña can be distinguished from those that occur by cavern weathering (tafoni) and smaller ones (alveoles). Flaking of granites is similar, except in halite crystals, which are common in areas of flaking and granular disintegration. Biotites can be impregnated by halite and subject to crystal growth forces that cause expansion. Thermal expansion by insolation is three times greater in sodium chloride than in other constituent minerals. The biotites break into small fragments that fill small voids, which suggests that physical weathering is more significant than chemical weathering; it produces concave ablation zones characteristic of the largest tafoni (Frenzel, 1989).

Other researchers have analyzed tafoni morphology (Calvin and Cailleux, 1962; Rodríguez and Navascués, 1982; Sancho and Gutiérrez, 1990; Sancho and Benito, 1990). The latter authors claim that tafoni developed in Oligocene sandstones of the Ebro Depression of Spain under semi-arid conditions and that the development depended on slope surface, textural factors, feldspar weathering, and wetting-and-drying and salt weathering. In addition, it appears that tafoni development requires tens of millions of years (Norwick and Dexter, 2002).

Figure 5.35 Alveoles in siliceous sandstone in Buntsandstein, Bezas (Province of Teruel, Spain), Iberian Cordillera.

Figure 5.36 Tafoni on Oligocene calcareous sandstone in the Ebro Depression. The tafoni, which are variable in size, appear to have a similar orientation. Albelda, Huesca Province, Spain.

Figure 5.37 Development of tafoni on Ayers Rock (Uluru) slopes, Australia, Photo P. Lucha.

Small closed depressions called **gnammas** or **weathering pits** develop on the surfaces of horizontal or low-slope outcrops (*Fig. 5.37*); these have numerous local names. They usually occur in groups and they most often have diameters in the range of decimeters although centimeter- and meter-size pits are also known (*Fig. 5.38*); their depth is a function of their diameter. Their plan view shapes vary although circular and elliptic morphologies predominate. Some gnammas have

Figure 5.38 Gnamma in hole, 2 m in diameter, photographed from a helicopter. The gnamma is in Buntsandstein sandstones, Bezas, Sierra de Albarracin (Province of Teruel, Spain), Iberian Cordillera. Photo B. Leranoz.

highly irregular borders, a consequence of the coalescence of individuals. Their cross sections are highly variable (Gutiérrez and Ibáñez, 1979) (*Fig. 5.39*); based on their cross sections they can be classified (Twidale and Corbin, 1963) into three types. The depth of **deep gnammas** or **pits** exceeds their diameter (*Fig. 5.38*). A second type is **shallow gnammas** or **pans**, when the depth is less than their diameter (*Fig. 5.40*); overhanging sidewalls are common in shallow gnammas. The third type, **armchair gnammas**, develop on gentle slopes and have triangular cross sections; they are an intermediate step in tafoni morphologies. All three types result from differential weathering along joints and especially along the intersection of weakness planes although in some cases no structural control can be observed. They can also develop beneath the soil or below an alteration mantle where later denudation is necessary to expose them. Exfoliation, granular disintegration, and, in some cases, abundant lichen colonies can play a role in their formation. The dominant processes are wetting and drying and salt weathering. In the case of calcareous sandstones, carbonate dissolution is significant (Gutiérrez and Ibáñez, 1979). If sandstones are siliceous, quartz leaching takes place at a pH higher than eight (Benito *et al.*, 1993a). Once a closed depression has formed, water may accumulate inside it and chemical weathering takes over. The particles inside the gnamma may be removed by eolian deflation or by overflow when the depression fills up with water. Some outcrops may have small overflow channels eroded into the rock (*Fig. 5.41*). The origin of the overhanging sidewall in some gnammas may be due to the continuous presence of water in the bottom, which concentrates the weathering in this area. When gnammas and tafoni coexist in the same outcrop they may coalesce and form a gnamma shaped like a buttonhole without a background. Headward erosion of a fluvial network may occasionally capture a gnamma.

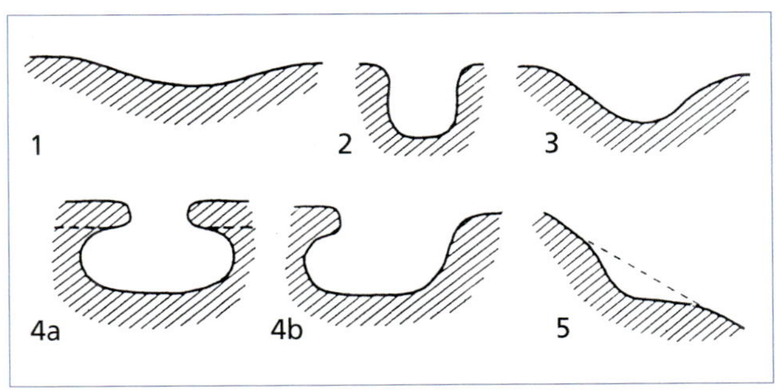

Figure 5.39 Gnamma cross-sections in the Alcañiz region, Ebro Depression. Teruel Province, Spain. 1: Gnamma with smooth borders. 2: Gnamma in hole or pit. 3: Gnamma pan. 4a: Gnamma in with overhanging sidewall. 4b: Asymmetrical gnamma. 5: Gnamma in armchair (Gutiérrez and Ibañez, 1979).

Figure 5.40 Gnamma pan with overhanging sides and a small overflow channel weathered into granite. Monte Louro (Coruña, NW Spain). This region and its splendid small-scale features of granitic landforms are located in an area of high precipitation near the coast.

Rillenkarren can develop in rocks that are neither carbonate nor evaporite; they are found in all climates, but primarily in humid climates. They are also called **weathering flutes** (*Fig. 5.42*). They are most common on rocky surfaces with high dip angles (Migon, 2006) and they may form as a result of weathering, granular disintegration, and water erosion of the resulting detritus (Birot, 1958, Czudek *et al.*, 1964). Williams and Robinson (1994) think that the wetting and drying process plays a significant role. However, Twidale (1982b) thinks they are subsurface in origin. The presence of weathering flutes in various menhirs destroys the hypothesis of a Tertiary origin for these forms (Czervinski and Migon, 1993).

Another feature that develops primarily in granitic rocks is **polygonal cracking** (*Fig. 5.43*), also called hieroglyphics (Twidale and Vidal Romani, 2005). Polygonal cracking occurs in all climatic environments. The polygons range in diameter from 5 to 30 cm (Migon, 2006); they have cracks of up to 5 cm long on the sides, which may extend to a depth of several centimeters (Twidale and Vidal Romani, 2005). Polygonal cracking develops primarily on isogranular and medium grained rocks. Planes of shear dislocation have been invoked as their origin (Vidal Romani, 1991); another hypothesis is based on weathering and erosion focused on large blocks or platforms.

5.4 Practical Aspects of Weathering

5.4.1 Weathering processes and products

Weathering in desert areas is mainly driven by wide variations in temperature and humidity. Temperature fluctuations cause insolation weathering and gelifraction occurs at high altitudes. Humidity variations cause wetting-and-drying weathering and salt weathering. The lack of water impedes chemical weathering and biotic weathering is mostly limited to the physical and chemical degradation caused by algae and lichens.

Wetting-and-drying cycles cause rocks to disintegrate by flaking and cracking (Ollier, 1969b). This type of weathering is an important erosional agent in desert areas where it causes considerable breakage of rocks (Ollier, 1977). In these areas, water circulating in the pores of rocks carries highly soluble salts, especially chlorides and sulfates. Evaporation causes precipitation of these salts, which can be even more disruptive than gelifraction. This process, which is known as salt weathering, acts in concert with wetting and drying.

Figure 5.41 Gnamma on siliceous sandstones in Buntsandstein with overflow channel and significant colonization by lichens.

Figure 5.42 Weathering flutes developed in a granitic block. Cascada del Xallas. Ezaro (Province of la Coruña, NW Spain).

Figure 5.43 Intense polygonal cracking in granite. Sierra de Xures (Province of Orense, NW Spain).

Salt weathering involves crystal growth, volumetric expansion, and salt hydration (Cooke, 1986; Cooke *et al.*, 1993). It is the most important weathering process in desert areas (Doornkamp and Ibrahim, 1990); it can disintegrate rocks into silt-size particles. Evidence of this process can be seen along the margins of salt lakes, in ephemeral channels, and in places where the capillary fringe intersects the soil surface.

The builders of ancient civilizations understood very well that salts adversely affected rock (Ollier, 1977). The Great Pyramid of Cheops (*Fig. 5.44*), for example, was originally covered with a protective layer of resistant limestone slabs. About 1000 years ago, part of the covering was destroyed and the pyramid's core began to undergo the effects of weathering (Emery, 1960). Natural materials continue to be used in construction today but many modern structures are made of concrete and engineers study its behavioral response to weathering.

The effectiveness of salt weathering (Cooke *et al.*, 1982) depends on environmental conditions such as daily fluctuations in temperature and humidity (which cause the crystallization or hydration of salts) and fluctuations of the capillary front. The properties of the affected materials, especially porosity and pore distribution, are equally important. Finally, different salts cause different effects: sodium and magnesium sulfate are the most aggressive (*Fig. 5.13*) (Goudie *et al.*, 1970; Goudie, 1974; Cooke, 1979).

The most common and serious problems related to engineering structures (Cooke, 1986; Goudie, 1994a; Goudie and Viles, 1997; Goudie, 1998) result from fluctuations of the **capillary front**, which can affect building foundations, roads, canals, oil pipelines, and other structures. Strict control over construction materials is therefore fundamental. Only materials that have a low salt content and a high resistance to salt weathering should be used. Laboratory analysis of their durability is recommended. Buildings can show the height of the capillary front in an obvious way. In arid areas of Bahrain, Dubai, and Egypt, the front may reach a height of 4 m but it usually ranges from two to three meters (Cooke *et al.*, 1982). However, in the semi-arid area of Alcañiz in the Ebro Basin of Spain, the capillary front has reached a height of 5 m (*Fig. 5.45*) at the Calatrava Castle. Soluble salts precipitated as a result of capillary action have also been observed on the entrance door to the Calatayud (*Fig. 5.46*). The salts most harmful to concrete are chlorides and sulfates of sodium, calcium, and magnesium; contact with these salts can cause concrete to expand and disintegrate. The use of sulfate-resistant cements is highly recommended (Mehta, 1983).

Highways should not be built in areas where soils are wet as a result of capillary action (*Fig. 5.47*). In such places, salt weathering can cause cracking, voids, crumbling, and disintegration (Cooke *et al.*, 1982). One way to assess the risk of salt weathering is to measure the electrical conductivity of water in the vadose zone and make maps that show salinity differences (*Fig. 5.48*).

Laterite, an erosion-resistant rock formed by weathering, has many uses. It has been used for the construction of temples in Cambodia, for the San Marcos Spanish Castle in Saint Augustine (Florida), and for numerous roads in tropical areas, such as the weather-resistant roads of Gabon (Persons, 1970).

For engineering purposes, we differentiate between laterite, which is an irreversibly hardened rock, and lateritic soil, which softens when moistened and hardens when dried (Persons, 1984). Lateritic soils are easily extracted with axes and pickaxes in quarries. The extracted bricks can be dried in the open air where they harden after at least two years (McFarlane, 1976). These bricks are made today in Agra (India). Their use as a construction material has declined in most tropical countries but persists in areas where laterite is the only resistant rock. Numerous remains of ancient civiliza-

Figure 5.44 The upper part of the Great Pyramid of Egypt is covered by unweathered limestone slabs. The underlying stone, however, shows marked weathering. Photo J.L. Peña.

Figure 5.45 An undulating capillary front in the sandstone slabs of the Calatravas Castle (18th century), reaching a height of 5 m above ground level. The most significant weathering is seen in the building blocks. Alcañiz, Province of Teruel, Ebro Depression, Spain.

tions have been preserved where springs, breakwaters, moats, sewers, and so forth were constructed of such materials and have functioned for hundreds of years (Persons, 1970).

Hardened laterite bricks have been used for wall construction, revetments, small dams, sewers, channels, and pavements. Its pleasant texture and strength make this rock an attractive material for architecture and engineering (Persons, 1970, 1984).

Laterite is also used in the construction of road and airport subgrades. Laterite is commonly used in the streets

Figure 5.46 Precipitation of high solubility salts by rising capillary front. The foundation blocks of limestone show less deterioration than the bricks. Calatayud (Province of Zaragoza) Iberian Cordillera, Spain.

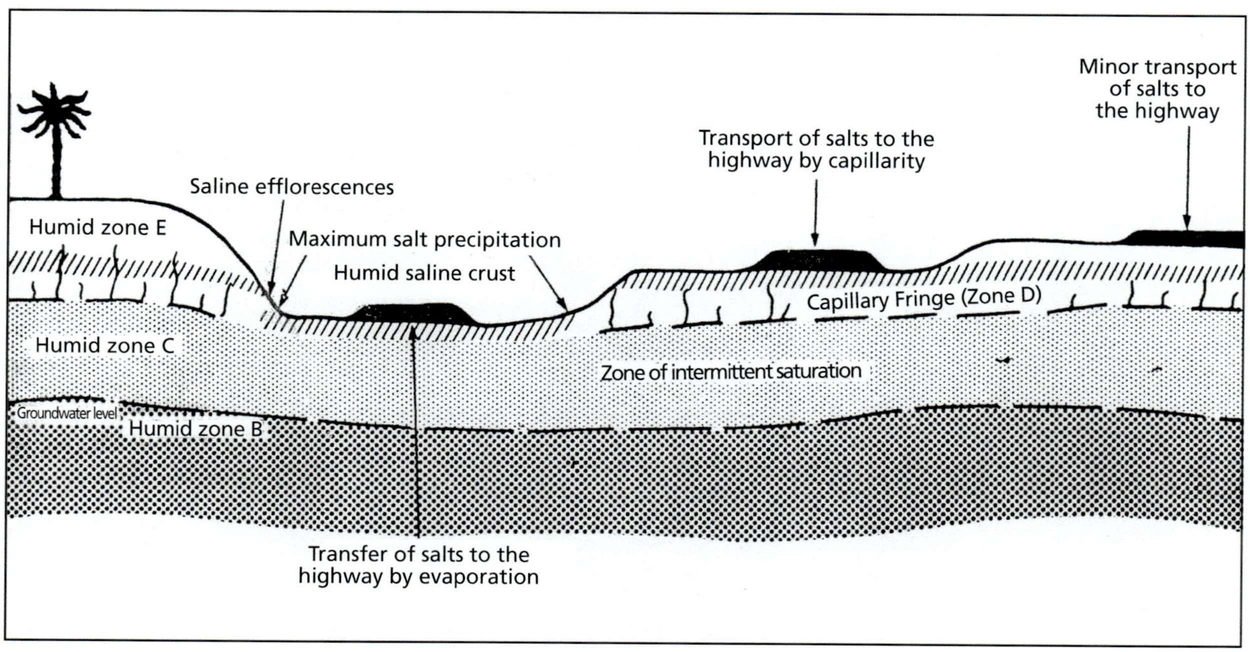

Figure 5.47 Schematic section showing the relevance of soil moisture zones to road construction (after Fookes and French, 1977, in Cooke *et al.*, 1982).).

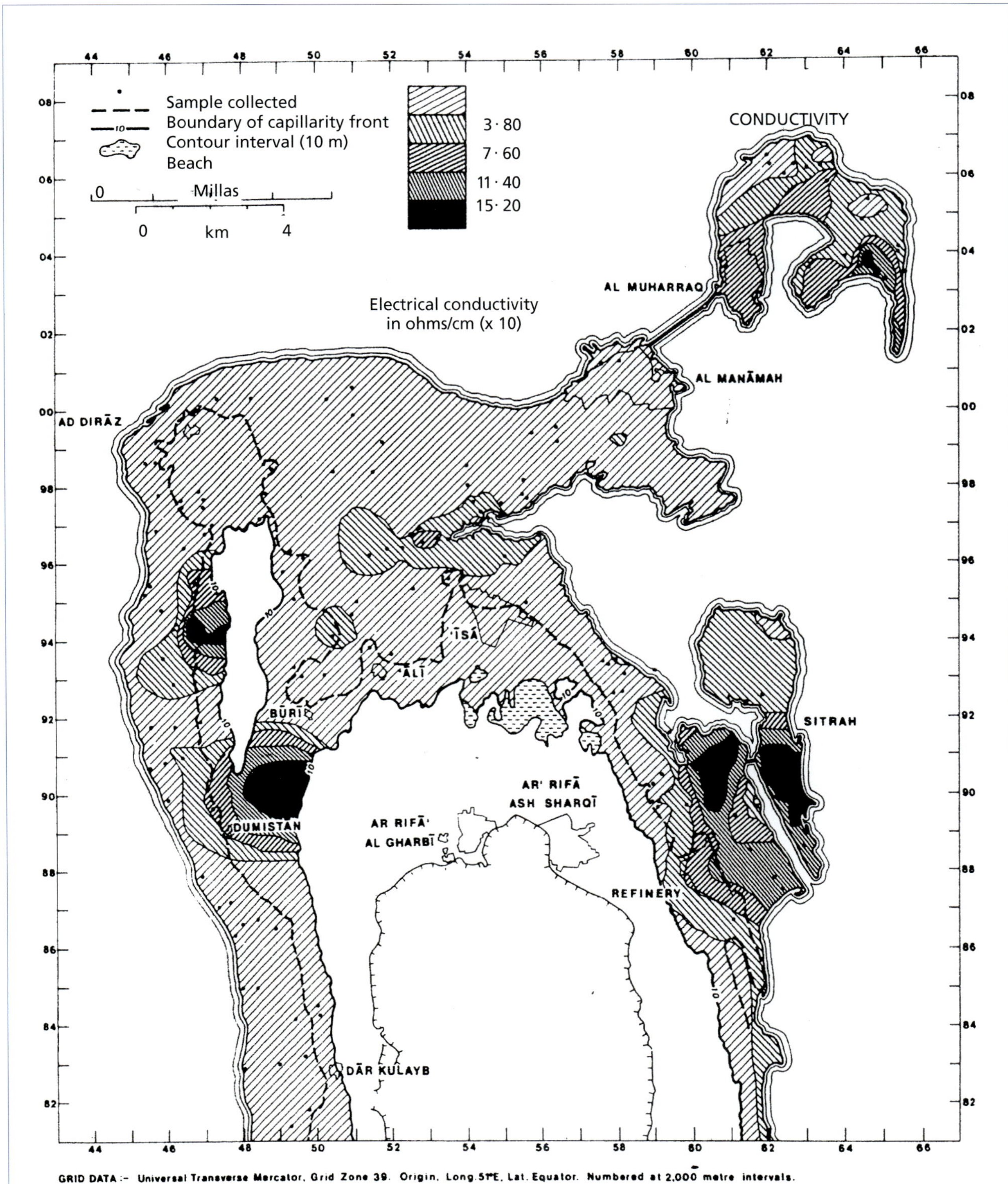

Figure 5.48 Aggressive ground conditions in northern Bahrain revealed by the electrical conductivity of groundwater (after Power and Water, 1976; in Cooke *et al.*, 1982).

and radial roads of Lagos (Nigeria) because of its strength and the proximity of laterite quarries. The construction technique is as follows (*Fig. 5.49*): oblong blocks of laterite are placed on an excavated base. Voids are filled with crushed laterite that ranges in size from 0.1 to 10 mm in Brazil and the various countries of Gulf of Guinea (Persons, 1970; 1984). It is later compacted and a 5-cm layer of crushed laterite is added. In the humid tropics, road subgrade must have good drainage to prevent deformation of the bearing surface. Some lateritic soils are sensitive to moisture and lose strength at high humidity. For this reason, the lateritic soil must be ventilated for a period of time before it is crushed.

Rock fill dams built with ground and compacted laterites are very strong, partly because laterite particles cement together to produce an impermeable surface. Where laterite is unavailable, dams are filled with various rocks that remain permeable after compaction. Additional work is necessary to make these dams impermeable (Persons, 1970; 1984).

5.4.2 Supergene mineral ores and placers

Many mineral deposits form as a result of weathering processes. Metallic elements are easily removed during weathering; they can be transported to the sea or precipitate in the weathering profile. When concentrated they can become a mineral ore of economic interest. This process is known as **supergene enrichment**, which is very important for sulfide ores (Ollier, 1969b; Ollier and Pain, 1996).

Supergene ores are composed of several zones (*Fig. 5.50*). The top of the weathering profile is the oxidized zone. The oxidized zone is divided into two parts: a lower zone of oxides that is more enriched towards the base, and a leached zone where residual iron minerals may concentrate to form a caprock or *gossan*, which is an excellent prospecting indication. Beneath the oxidized zone is a sulfide enrichment zone formed by weathering of the overlying material and precipitation of the weathered elements below the water table under reducing conditions. The boundary between the oxidized and sulfide zones corresponds to the water level at the time of their formation. A slow lowering of the water table favors the development of a thick layer of supergene sulfides. Lowering of the water table can result from climatic and tectonic changes or drainage modifications caused by volcanic eruptions. Due to these lowering processes, the supergene sulfide zone, which varies in thickness, may be located at a depth of tens of meters to hundred of meters and may be even deeper than 600 m in places (Ollier, 1969b; Ollier and Pain, 1996).

Laterites can form exploitable ores of iron, aluminum, nickel, manganese, and gold, among others. Within the alteration profile, metallic concentrations increase with depth; aluminum oxyhydroxides predominate at the top. Metals that have highly charged ions are transported within the regolith as small amorphous particles that later dehydrate and crystallize (Kühnel, 1987). Particles move towards the bottom of the profile and crystallize later as shown by Aleva (1991) for nickel-bearing laterites and by Nahon *et al.* (1985) for manganese-bearing laterites.

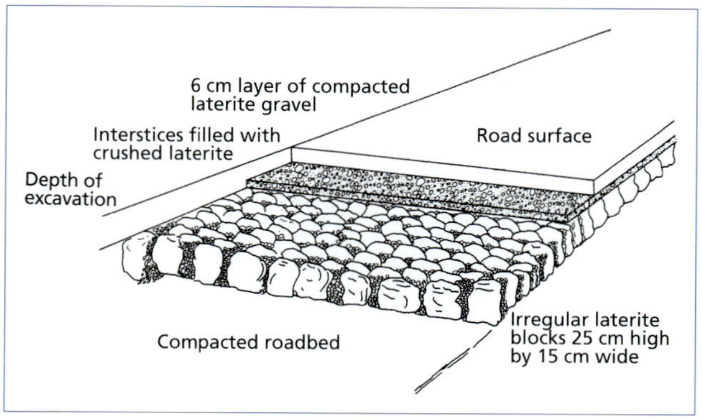

Figure 5.49 Diagram showing laterite used as road subgrade (after Persons, 1970).

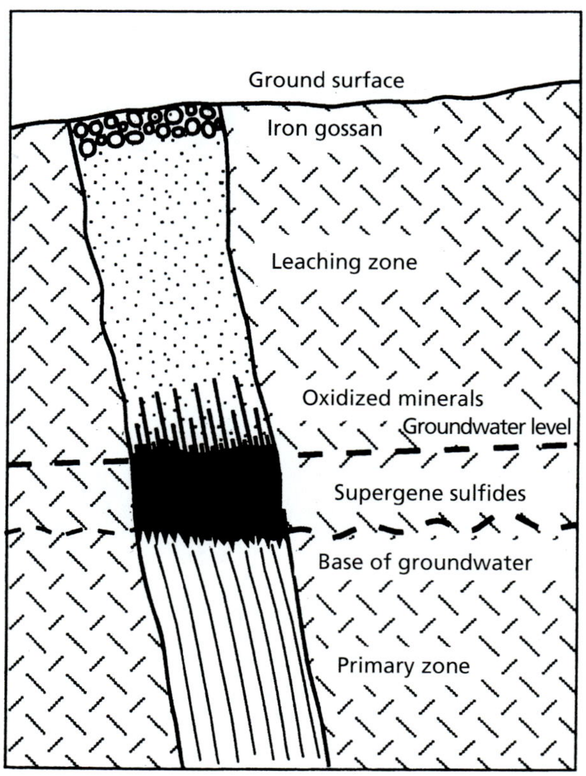

Figure 5.50 Different zones in an ore body formed by supergene enrichment (Ollier and Pain, 1996).

Gold from auriferous laterites is very pure (>95% Au). It precipitates in cracks between quartz grains and also around iron oxide nodules (Thomas, 1994a). It is thought that gold is released from the sulfide crystalline structure under very low pH conditions. It later dissolves as a sulfate and as it moves deeper into the profile, pH and Eh conditions change, causing it to precipitate near the phreatic level (Wilson, 1984). The activity of continental brines has also been considered because when they are introduced into the profile, chlorine waters become highly acidic. Gold solubilizes as a chloride and, under reducing conditions, the increased Fe^{2+} content causes gold to precipitate as small very pure crystals (Mann, 1998). In addition, scientists have pointed out the role of humic acids in the mobilization of colloidal gold.

Numerous iron deposits are related to **karstification**. The iron is mainly limonite, hematite, goethite, and siderite, which are weathering residues of limestones and dolomites. These occur as discrete accumulations that fill fractures or karstic depressions of varying sizes. Contacts between the carbonate rock and iron are highly irregular. A carbonate massif usually develops underground caves as a result of phreatic flow. Spain has several iron ore deposits related to carbonate karst processes: Alquife (Grenada), Sierra Menera (Teruel), Cabárceno (Santander), and Cerro del Hierro (Iron Hill, Seville). They all developed in carbonate rock of different ages. The iron ore of Sierra Menera is the interstratal type and karstification of Ashgillian mixed iron/magnesium carbonates produced secondary iron minerals (*Fig. 5.51*) (Fernández-Nieto *et al*., 1981).

Approximately 10 percent of the world's aluminum production comes from karstic bauxites (Caribbean region and Southeast Asia). These bauxites occur mainly as fill in karst depressions that slowly become deeper as the bauxite deposits grow. They form in warm climates in rocks dating from the Cambrian through the Quaternary (Bardossy, 1981).

Tin-bearing alluvium penetrates the endokarst in the Malayan tower karst where it fills voids and fissures. This is an important tin ore created by early karstification processes and later placer in filling (Sweeting, 1972).

Placers originate by mechanical concentration resulting from gravity separation of heavy minerals (denser than quartz), either by water or wind (Bateman, 1950). For this process to occur, minerals must be released from the altered rock and concentrated by the previously discussed methods. Most placers form in fluvial and littoral environments. The most important minerals mined in this way are gold, diamonds, ilmenite, rutile, cassiterite, zircon, monazite, and gemstones. Although placers are widely distributed throughout the world (*Fig. 5.52*), humid and sub-humid tropical environments provide optimum conditions for their formation (Thomas, 1994a). Most placers result from present-day (*Fig. 5.53*) or

Figure 5.51 Exploitation of iron oxy-hydroxides produced by carbonate karstification. Cantera Corral, Sierra Menera. Ojos Negros, Teruel, Iberian Cordillera. Spain.

recent geomorphic processes such as the diamond placers of the Central African Republic and cassiterite placers in Indonesia. Other deposits formed in past geological periods such as the Tertiary gold-bearing placers of Las Médulas (León province, Spain), the Proterozoic deposits of the Witwatersrand (South Africa), and the Triassic diamond-bearing placers of Swaziland (Sutherland, 1984; 1985).

Quaternary climate fluctuations have produced terraces in fluvial systems that can be mined for heavy minerals of economic interest and that simultaneously can be transported to lower terrace levels. Fluvial channel beds undergo continuous reworking and channel migration. As a result, coarse gravel sedimentation promotes the trapping of small diamonds such as those in Sierra Leone and Ghana (Thomas and Thorp, 1993). In Indonesia and Malaysia, cassiterite grains are stripped from weathering profiles on granites and andesites and form placer deposits in colluvium, alluvial fans, and fillings of endokarstic cavities (Aleva, 1985; Gupta *et al.*, 1987). A wide variety of economic minerals such as ilmenite, zircon, and monazite can concentrate in beach placers. Waves and littoral drift currents may concentrate them. In addition, wind removal of beach sand can create eolian placers on littoral dunes.

Commercial mineral extraction can cause significant environmental change, which is usually tolerated by governments in developing countries. Much of the exploitation of alluvial cassiterite placers in Southeast Asia occurs near the coast. Miners dredge and pump riverbeds to exploit these deposits. This operation causes extreme water turbulence. In addition, a significant part of the suspended load is dumped into the sea where it threatens the life of coral reefs. Another example of the consequences of mining practices is the Madeira Basin (southwestern Amazon River) where dredged gold is amalgamated with mercury; between 5 and 30 percent of the mercury is lost in the river. This pollution has serious effects on the fauna living in the river basin (Reading *et al.*, 1995).

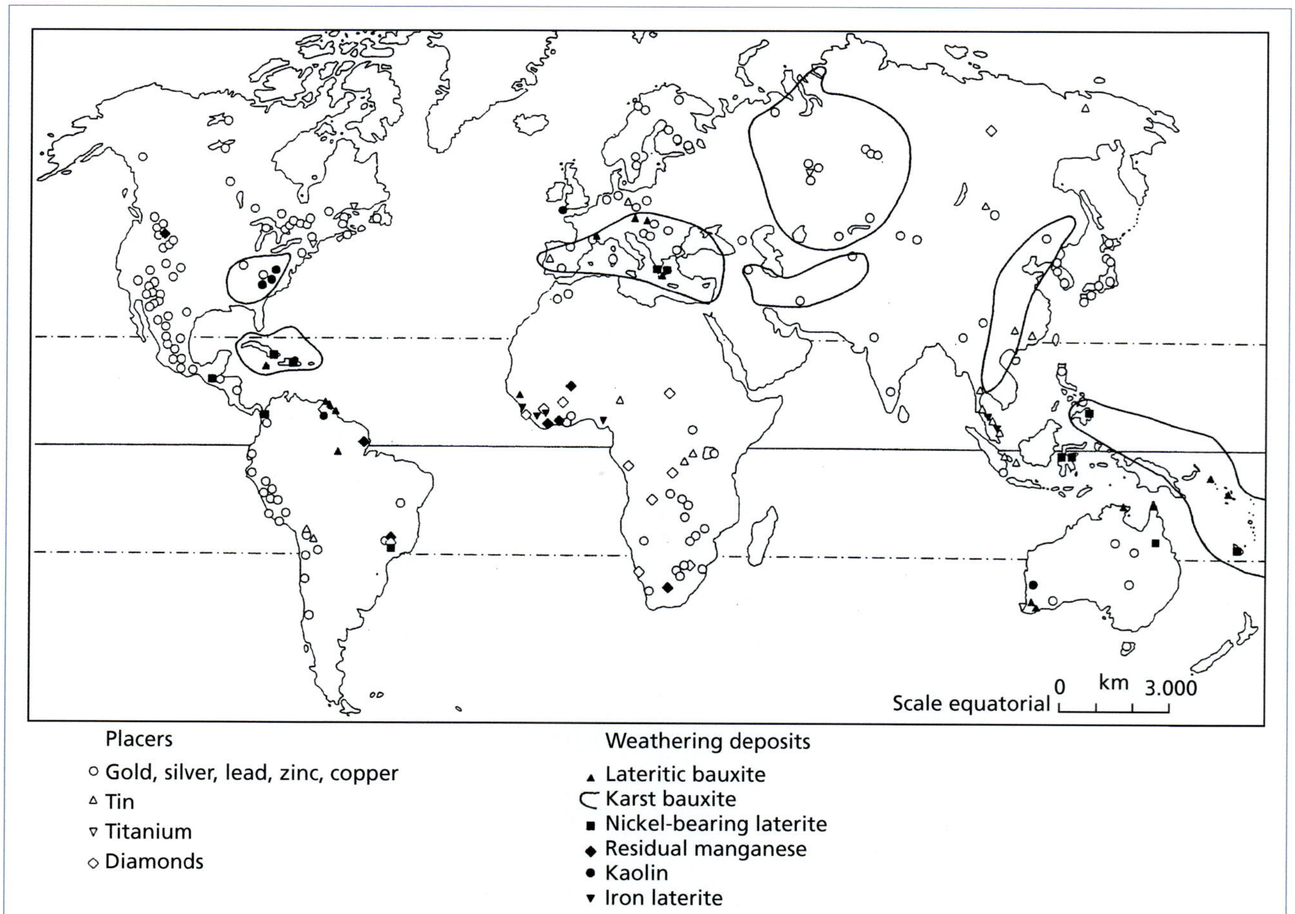

Figure 5.52 Major weathering-generated mineral deposits and placers formed under humid tropical conditions (Reading *et al.*, 1995).

Figure 5.53 Prospecting a gold placer using trays and a suction dredge. Nzi River tributary. Bouaké, Ivory Coast.

5.4.3 Regolith characteristics: Implications for engineering geology

A wide variety of rocks crop out in tropical areas although granites and other chemically resistant quartz-rich rocks are the most abundant. For this reason, most researchers have focused their work on granitic rocks. After weathering, the resulting regolith contains two main particle sizes: non-altered quartz grains and the clay fraction that results from the weathering of feldspar and micas. The regolith has a porosity of up to 60 percent, high permeability, and angles of internal friction between 20° and 30° (Dearman and Shibakova, 1989). Heavy rains can saturate the regolith and catastrophic landslides can occur. The base of the weathering profile contains unaltered blocks of varying sizes. These blocks can cause problems in road excavation and in slope stabilization (Douglas, 1986).

Weathering depth is quite variable. Under high precipitation conditions, the predominant clay minerals in the saprolite are kaolinite and halloysite. Montmorillonite occurs in transition zones to semiarid regions or in poorly drained areas. High clay contents contribute to steep slope instability and promote landslides during periods of heavy rain (Thomas, 1986).

In this chapter we have analyzed weathering profiles and their subdivision into zones (Ollier, 1969b) based on grus structure and composition and fresh rock morphology. The same criteria have been adapted to classify alteration profiles for engineering purposes (Dearman *et al.*, 1978). The Tropical Soil Committee of the International Society of Soil Mechanics and Foundation Engineering (ISSMFE, 1985) used these principles to develop a classification suitable for engineering projects and especially useful for estimating loading capacity for foundation purposes (Dearman and Shibakova, 1989). This classification appears in *Table 5.1*. Fookes (1997, 2005) contains a graphic representation of the classification of various types of weathering *(Fig. 5.54)*. The six different grades closely resemble the zones proposed by Ollier. We should point out that civil engineers and geological engineers study the stability and durability of the weathering mantle and the strength of materials (Ehler, 2005).

A three-dimensional understanding of the regolith is of prime importance for construction of engineering projects. Engineers drill and use geophysical prospecting methods to obtain this information (Ollier and Pain, 1996). Seismic refraction profiling provides good results regarding the depth of the regolith but does not clearly define the various zones (Kesel, 1977a). The hammer seismic technique is the most suitable method for calculating regolith thicknesses less than 30 m. It has been used to distinguish regolith from colluvium on slopes and to determine their evolution (Mills, 1990). The electric method makes it possible to identify layers of varying electrical resistivity and their thicknesses within the regolith. Both land and airborne electromagnetic devices make it possible to produce three-dimensional regolith maps from their vertical conductivity distribution (Street and Anderson, 1993). Ground penetrating radar techniques (GPR) are used to conduct three-dimensional mapping to depths of about 40 m (Mellet, 1990).

5.4.4 Weathering of monuments

People have for a long time noticed signs of deterioration in rocks used to build the world's monuments; according to Winkler (1973), visual observation indicates that the pace of physical deterioration has increased alarmingly and practically exponentially over the last 80 years (Ordaz *et al.*, 1984).

The deterioration of monuments is a topic that concerns town councils, communities, and international organizations. It has been called *rock sickness* or more dramatically, *rock cancer*. It has reached significant proportions in some architectural monuments, buildings, and cultural icons of human patrimony (Esbert *et al.*, 1983). Monuments involve a group of specific factors related to anthropogenic activity including the type of construction (presence of mortar and cement), composition of the urban atmosphere, and rainfall (Robert, 1993).

To safeguard our cultural heritage, UNESCO has published a book (Stovel, 1998) on risk prevention for the world's patrimony. Published by ICEROM (International Centre for the Study of the Conservation and Restoration of Cultural Property) the book focuses on cultural wealth protection, risk prevention, improved patrimony management, administration, and case studies.

Preservation of patrimony is a part of **sustainable development**; it is a principal foundation of the definition

Table 5.1 **CLASSIFICATION OF RESIDUAL TROPICAL SOILS BY DEGREES OF WEATHERING (ISSMFE, 1985)**

Grade	Degree of decomposition	Field recognition	Engineering properties
VI	Soil	No recognizable rock texture; surface layer contains humus and plant roots.	Unsuitable for important foundations. Unstable on slopes when cover is destroyed.
V	Completely Weathered	Rock completely decomposed by weathering in place but texture still recognisable. In types of granitic origin, original feldspars completely decomposed to clay minerals. Cannot be recovered as cores by ordinary rotary drilling methods.	Can be excavated by hand or ripping without use of explosives. Unsuitable for foundations of concrete dams or large structures. May be suitable for foundations of earth dams and for fill. Unstable in high cuttings at steep angles. Requires erosion protection.
IV	Highly Weathered	Rock so weakened by weathering that fairly large pieces can be broken and crumble in the hands. Sometimes recovered as core by careful rotary drilling. Stained by limonite. Less than 50% rock.	Similar to grade V. Unlikely to be suitable for foundations of concrete dams. Erratic presence of boulders makes it an unreliable foundation stratum for large structures.
III	Moderately Weathered	Considerably weathered throughout. Possessing some strength—large pieces cannot be broken. Often limonite-stained. 50 to 90% rock.	Excavated with difficulty without use of explosives. Mostly crushed under bulldozer tracks. Suitable for foundations of small concrete structures and rockfill dams. May be suitable for semipervious fill. Stability in cuttings depends on structural features, especially joint attitudes.
II	Slightly Weathered	Distinctly weathered through much of the rock fabric with slight limonite staining. Some decomposed feldspar in granites. Strength approaching that of fresh rock. More than 90% rock.	Requires explosives for excavation. Suitable for concrete dam foundations. Highly permeable through open joints. Often more permeable than the zones above or below. Questionable as concrete aggregate.
I	Fresh Rock	Fresh rock may have some limonite stained joints immediately beneath weathered rock.	Staining indicates water percolation along joints; individual pieces may be loosened by blasting or stress relief and support may be required in tunnels and shafts.

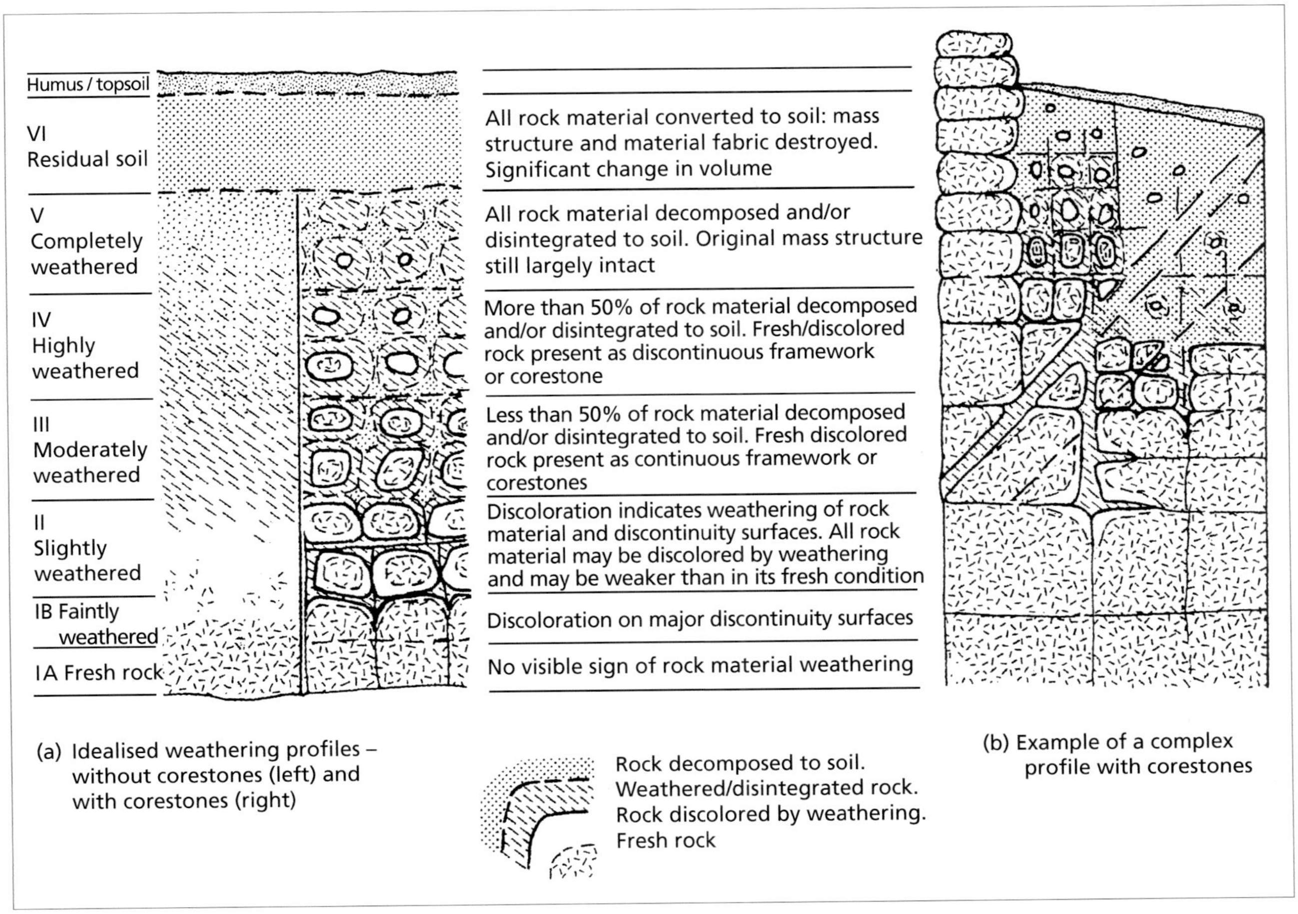

Figure 5.54 Classification of degrees of tropical weathering (Fookes, 1997).

of the political, economic, environmental, and social future at global, national, and regional scales (Panizza and Piacente, 2003). Release of gases, primarily from industrial activities that began during the Industrial Revolution of the 18th century, has accelerated the alteration of rock monuments. In addition to weathering, atmospheric pollution unleashes a chain of mechanisms including surface sedimentation (Prada *et al.*, 1996) that make it important to understand precisely the type of atmospheric particles transported and deposited by the wind (Esbert *et al.*, 2001).

As a result, preservationists need thorough investigations on preservation and restoration to confront the problem. One serious problem caused by the cleaning of monuments is the disappearance of patinas from the rocks. Patinas accumulate over long periods of time as a result of physical, chemical, and biological processes that harden the rock or other construction material. Because of this, cleaning the limestone rock of the Gothic-style Cathedral of Burgos has become highly controversial. Cleaning has removed the patinas and turned the Cathedral from gray to white. Also, the cleaning the bricks of the mudejar towers in Teruel has changed their old pinkish hue to rosy tones.

The application of commercial products to monuments for purposes of preservation has to involve appropriate scientific planning because the applications can be counterproductive and can even accelerate the process they are designed to stop (Esbert *et al.*, 1983, 1997).

The first step in the preservation of rock monuments is a study of the mineralogy and petrophysical properties of the rock and its chemical and biological behavior. This should be included in a detailed "de-visu" study of the various forms of alteration present in the materials (Sanjurjo and Alves, 2006). Using knowledge of meteorological parameters, scientists should conduct studies of alteration and accelerated aging. This detailed investigation should lead to best practices for the preservation of the monument. The excellent volume of Esbert *et al.* (1997) is a compilation of petrographic techniques, physical properties, specific tests, and intervention techniques.

The origin of the weathering flutes in the granites of the Neolithic menhirs (5500–7000 BP) at Carnac, Britain, has been known for a long time. The Egyptians covered the 144-m-high Great Pyramid of Cheops (4700 BC) (*Fig. 5.46*) with limestone flagstones to prevent weathering of the complex. Because limestone alters very slowly in hyperarid areas, over time the limestone flagstones were removed for inclusion in other structures. More recently, scientists have conducted experiments on weathering in natural environments; Hilger (1897) did an experiment in which he separated samples of limestone, sandstone, and micaceous schist into sections 1 to 2 cm long and exposed them to the atmosphere for 17 years. He observed the fragments at intervals and noted alteration characteristics and particles that had come loose. At the end of the long period, he observed that the weathered samples were altered both physically and chemically. Limestone was most affected by chemical attack, sandstone underwent the most physical alteration, and micaceous schist suffered from both weathering and chemical physical and chemical weathering.

In an urban environment, limestone weathers more rapidly than it does in the surrounding rural environment. This was recognized by Archibald Geikie who studied weathering of gravestones in and around Edinburgh (Scotland) (Huggett, 2003). Weathering rates of marble in Dirham (England) were calculated and the values obtained were 2 microns/year in rural environments and 10 microns/year in industrial urban environments (Attewell and Taylor, 1988). At St. Paul's Cathedral in London the addition of sulfates from acid rain pushed the rate up to 0.62 microns/year (Sharp *et al.*, 1982).

Attempts have also been made to study weathering rates by observing the degree of alteration in clasts deposited in river environments that are modern terrace sequences. To do this, the age of the accumulation levels must be known; for the middle valley of Ródano, Bornand (1979) estimates the age at 2.5 million years. The clasts that make up the terraces are primarily limestone, granite, and quartzite. Overlying the terraces is a chronosequence of soils with an increasing degree of evolution with age. Beginning decarbonation with T_2. These are fersialitic soils that culminate as ultisols on the oldest terrace. The sequence shows the rapid and total dissolution of limestone clasts and the gradual disappearance of crystalline clasts. He concludes by establishing a first evolutionary phase that is fersialitic in nature (500,000–800,000 years) in which iron is concentrated and a second phase that involves unreddening and degradation of red soils (ultisol phase of more than a million years). As Millot (1979) points out, in this case there are no available ages of alteration other than stratigraphic methods. Woodward *et al.* (1994) conducted a similar study on soil profiles developed on fluvial terraces in northwestern Greece for which no dates were available.

Durability is the opposite of alterability. It indicates the capacity of a material to resist aggressive action of alteration agents. Determining the durability of a rocky material is a highly complex task because of the number of extrinsic and intrinsic factors involved in the process and the various types of deterioration that can occur (Esbert *et al.*, 1997).

Scientists have attempted to determine weathering rates by knowing the age of the rock; they have observed flagstones of different rock types from quarries located near a cemetery. This methodology has the advantage of a precise age and weathering caused by different natural processes but the method is limited to a time span of 900 years (See Ollier, 1969b).

5.4.4.1 Weathering of sandstones in various climatic environments

The rocks most commonly used for monument construction are granites (Segovia Aqueduct, Escorial Monastery, among others), limestones (Cathedrals of Burgos and Leon, among others), and sandstones. Sandstones have been used throughout history in the construction of various types of monuments primarily because they are easy to work. Sandstones have diverse mineralogical and textural characteristics; matrix, porosity, and degree and type of cementation differ widely within the group. In this section we shall try to provide a brief analysis of sandstone weathering in different climates and to consider the intrinsic characteristics of the various types of sandstones to the extent possible. We shall consider only those monuments subjected to non-destructive atmospheric processes. For this reason, we will exclude structures built with siliceous sandstones in Belfast (Ireland), which are affected by high degrees of anthropogenic pollution and the resulting development of black crust (Smith, Magee, and Whalley, 1994).

The monumental structures built along the margins of the Nile River in **Egypt** during the period of the three pharaonic dynasties, approximately 4800–2950 BP, attract our attention because of the large volume of rock that the Egyptians moved and because the structures are extremely well preserved. The monuments were constructed primarily of two types of rock: Nubian Sandstone (Jurassic-Cretaceous) in horizontal strata and lenses between clay layers and pink Pre-Cambrian granite from the Aswan region. Average rainfall in the area is 2–10 mm/year and winds are from the north (Goudie, 2002). Because water is scarce, natural alteration is minor in spite of the age of the monuments. Exfoliation and flaking of timbers (Temple of Luxor) and bas-reliefs have been observed. In the latter, flaking generally exceeds a centimeter and may coincide with the maximum penetration of a chisel. In places where the phreatic level is close to the surface, capillary action and the deposition of white salts can be seen along the front. Even so, the primary deterioration of the monuments is due to anthropogenic activity: the destruction of numerous reliefs by hammering, very possibly for religious reasons. The Sphinxes of the Temple of Amon in Karnak, the Colossus of Memnon (that represents Amenofis III), the imposing statue of Ramsés II at the entrance to the Temple of Luxor, and the rubble of pink granite 18 m tall from the Statue of Ramsés II in Thebes are clear evidence of man's destruction of this magnificent patrimony.

Construction of the Great Aswan Dam upstream on the Nile was completed in 1970; it created Lake Nasser and drowned a large number of monuments. International aid was focused on saving monuments by relocating them to the edge of the Nile River (Abu Simbel) or moving them to other countries with very different climates. The Egyptian obelisk that had remained intact in the Aswan region has deteriorated in the 30 years since it was moved to the United States and installed in the hot and humid climate of New York (Pedro, 1993).

The province of **Rajhastan (India)**, adjacent to Pakistan has numerous buildings (forts, palaces, and other structures) built of Jurassic sandstone and decorated with magnificent sculptures (*Fig. 5.55*). The most important forts are those of Jodhpur and Jaisalmer; the latter is located in the Thar Desert a few kilometers from the Pakistani border, where temperature extremes range from 3 to 50°C and rainfall is about 150 mm, with most rains coming during the summer monsoon (Dhir *et al.*, 1992).

Some people consider Jaisalmer to be one of the most beautiful small cities in the world, calling it the "pearl of the desert." The city was once a stopping place for caravans from the Far East and Delhi. Within its enormous 5-km-long wall is a museum-city of buildings dating from the 12th century including temples, palaces, and exquisite mansions with marvelously sculpted balconies, all of them constructed of golden-toned siliceous Jurassic sandstone. The monuments have not weathered at all because the environment is arid. We have not observed destruction caused by man.

The city of **Salamanca (Spain)** has an average annual temperature of 12.4°C, annual rainfall of 360 mm, and freezes about 30 days per year; it is a characteristic example of the use of a single type of rock to construct numerous architectural monuments (most of them from the 16th century and others from the 17th and 18th centuries); the rock continues to be used today for building facades. The golden rock of Salamanca is mid-upper Eocene sandstone from the quarries of Villamayor near Salamanca. Various research teams have studied the sandstone and have collaborated to produce the volume "Study of Alteration and Treatment of the Villamayor Rock" (Arribas *et al.*, 1984; Hernández *et al.*, 1984; Ordaz *et al.*, 1984; Vicente *et al.*, 1984; Hernández *et al.*, 1984; Oteo, 1984).

The sandstone is a heterometric arkose. The rock most commonly used in construction has grain diameters between 0.125 and 0.24 mm (Arribas *et al.*, 1984), has a

Figure 5.55 Twelfth-century palace constructed of sandstone. Jaisalmer (India).

white-yellowish-grayish color (Rock Color Chart, 1979), and contains grains of quartz, feldspars, kaolinite, illite, smectites, palygorskite, and accessory minerals. The presence of swelling minerals such as smectites and palygorskite causes volumetric expansion of the crystalline structure. Consequently, wetting and drying processes play an important role in the weathering of the rock (Arribas *et al.*, 1984).

The Villamayor sandstone has numerous pores that allow water to enter. Various teams studied the physical properties of the rocks. We will describe the work done by Ordaz *et al.* (1984), which follows the methodology proposed by Belikov *et al.* (1967) and the International Society for Rock Mechanics (1972). Table 5.2 shows the calculations for physical porosity. The sandstone has high porosity (30–50%). Effective porosity is 96%, which suggests that most pores are interconnected. Absorption capacity is very rapid and as a result, communication between pores happens very quickly. Desorption or water loss by evaporation in the first twenty-four hours was 71.76%. Capillary measurements show a high suction capacity of water. The capillary front rose a minimum height of 7 cm in the first twenty-four hours of testing.

Ordaz and his colleagues also conducted weathering and accelerated aging tests and determined that the primary concern is water. Table 5.3 shows the loss of material by disaggregation in the various layers during accelerated aging. In the test using wetting-drying cycles, the average loss was 3.2% after 30 cycles. Losses for freeze-thaw cycles indicate that gelifraction causes a loss of 5.7% of solid material after 30 cycles on average. Finally, the salt crystallization test using a solution of 15% SO_4 Na_2 causes significant disruption after 10 cycles with the production of 20.8% by weight of loose material. This disintegration is due in large part to the transformation of SO_4 Na_2 (anhydrite) to mirabilite (SO_4 $Na_2 \cdot 10H_2O$), a process that increases volume by 400%.

The goal of these weathering tests was to determine the ability of the material to resist alteration or, in other words, determine its future durability (Ordaz *et al.*, 1984). These results show that water, liquid or solid, causes most of the weathering because of the high porosity of the arkose. Also, the samples that had the highest weight loss are those that contained a larger amount of swelling clays.

Salts, like humidity, are the most effective elements in the deterioration of sandstones. Chemical analyses of fresh rocks in the quarry do not indicate the presence of Cl^-, SO_4^{2-}, and $NO3^-$. However, high solubility salts can be incorporated into sandstone during construction from soil or air pollution (Torraca, 1981). However, because the samples analyzed did not contain SO_4^{2-}, S^{2-} or Cl^-, it is thought that environmental pollution does not play a role.

Another aspect of sandstone weathering is biological activity, shown by the presence of fungus, algae, and lichens (Arribas *et al.*, 1984). The study of altered surfaces may support data on the solubility processes of some chemical species (Hernandez *et al.*, 1984; Ordaz *et al.*, 1984). Observation of the degree of weathering in buildings in the city of Salamanca reveals the presence of alveolization, flaking, exfoliation, flutes, disaggregation (crumbling and loss of cement), and discoloration, crusts, efflorescences and patinas of various types (Hernández *et al.*, 1984; Ordaz, 1984).

In considering methods to remediate the weathering, Vicente *et al.*, (1984) indicate the most appropriate treatments to be the use of synthetic resins (polyvinyl acetate, plexiglas, etc.) or synthetic resins without solvents. However, they claim that "laboratory experiments should be done on

Table 5.2 PHYSICAL PROPERTIES OF THE "PIEDRA DE VILLAMAYOR" (ORDAZ *ET AL.*, 1984)

	Average
Specific weight of mineral grains (ρg)	2,660 Kg/m^3
Apparent Density of dry rock (ρd)	1,760 Kg/m^3
Effective porosity (n_0)	32.9%
Total porosity (n)	33.8%
Degree of pore filling (Sr)	95.9%
Natural water content (w)	17.0%
Saturated water content (i_s)	18.4%
Void index (I_v)	12.6%
Water absorption (W)	14.5%
Swelling coefficient (εs)	2.5×10^{-3}
Coefficient of thermal expansion (α_1)	9.0 m/m °C
Coefficient of capillarity (C)	120 Kg/ $m^2 \cdot min^{0.5}$
Coefficient of water penetration (A)	5.6×10^{-3} m/ $min^{0.5}$

Table 5.3 LOSS OF MATERIAL BY DISINTEGRATION DURING ALTERATION TESTS ON THE "PIEDRA DE VILLAMAYOR" (ORDAZ *ET AL.*, 1984)

Test	N° of cycles	Percentage of material loss by weight
Wetting-drying	30	0.8–6.4% (average 3.2%)
Freezing	30	3.0–9.5% (average 5.7%)
Salt crystallization	10	15.1–29.7% (average 20.8%)

the recommended products in order to determine the most suitable treatment to protect and consolidate the Villamayor sandstone." However, Arribas *et al.* (1984) conducted tests on resins and silicones and rejected the use of the latter because of the temperature changes that render them useless.

The **San Juan de la Peña Monastery** (Province of Huesca) located in the Pyrenean External Sierras, is a symbol of the reign of Aragón (Spain); it is housed in a significant shelter constructed of Miocene conglomerates (*Fig. 5.56*). The temperature ranges from −10° to 32°C and average annual precipitation is 915 mm. The monastery was begun in the 8th century; major parts of the construction date from the 10th and 11th centuries and the cloister was constructed in the 12th century. This monastery was built for the pantheons of the Kings of Aragon and the nobles; since then it has undergone numerous re-incarnations and fires. The Romanesque cloister dating from the 12th century and the Pantheon of Kings, constructed of fine-grained Oligocene sandstones, were studied by the Department of Culture of the General Deputation of Aragon and by the Faculty of Petrology and Geomorphology of the Universities of Oviedo and Zaragoza, respectively (Esbert *et al.*, 1987). The fundamental concern was the state of weathering of the cloister capitals (*Fig. 5.57*). Two samples were analyzed, a fine-grained sample of 100 μm and another of 200 μm, that was less representative of ornamentation. The fine-grained sandstone is composed of quartz (42%), calcite (3%), phyllosilicates (10%), plagioclase (3%), and dolomite (3%).

The physical properties of the fine-grained sample were studied according to the methodologies of Belikov *et al.*, (1987); free absorption of water (more than 90% in the first 4 hours; absorption by capillary action (weight increase as a function of time), capillary penetration (height rise as a function of time) and evaporation curve (*Fig. 5.58*). Porosity and porosity size distribution made it possible to determine the free or effective porosity (9.6%), trapped porosity (6.2%) and the average pore throat radius (0.25 μm) (*Fig. 5.59*) by injecting mercury into the rock (Pellerin, 1980). The linear expansion caused by water absorption was re-obtained by a calculation of free linear swelling according to the standards of the I.S.R.M. (1972). The value, which is very small, occurs in the first hours of the test. The detailed study of the cloister showed that only the capitals located on the west side (*Fig. 5.56*) were well preserved and that their reliefs relate a series of biblical facts. Successive rebuilding and fires in the cloister over the years explain the movement, the reliefs truncated by impacts, poor preservation practices, etc. The tim-

Figure 5.56 Twelfth-century cloister of the San Juan de la Peña Monastery in the Pyrenean External Sierras (Province of Huesca) located in the shade of shelter built of massive Miocene conglomerates, at an elevation of 1200 m. In the distance, the San Voto Chapel that dates from the middle of the 17th century.

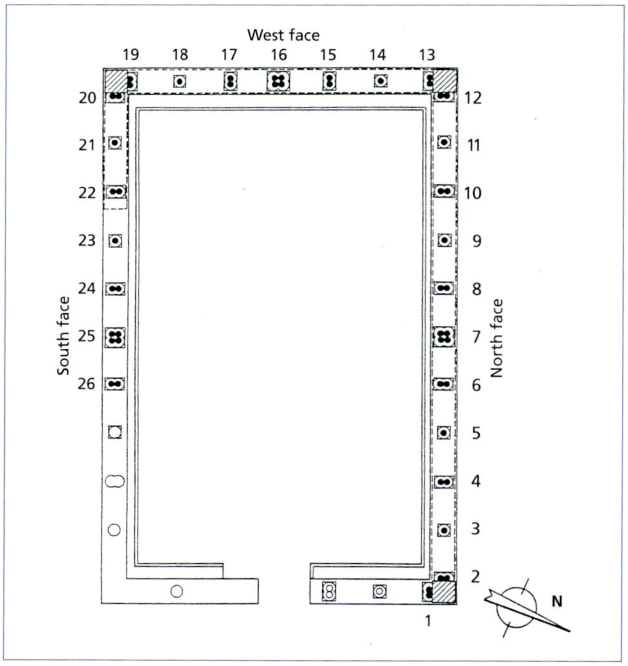

Figure 5.57 Diagram of the Cloister of San Juan de la Peña (Esbert *et al.*, 1987).

bers are mostly replacements except for column number 25, which has a high degree of alveolization.

The capitals have evidently undergone uneven weathering; they are less deteriorated on the western side (from 13 to 19). The presence of dark and pink patinas may reflect the application of ancient pigments. In addition to these patinas, flaking, exfoliation, granular and other common disintegration of mechanical origin are present, such as fissures and flaking, which possibly occurred as workers were transporting them to preserve them.

X-ray diffraction and scanning electron microscopy analysis of the alteration products revealed calcite crystals and various organisms with various ornamentations. Extraction of salts (*Fig. 5.60*) and wet analysis by spectroscopic, chromatographic, and plasma induction techniques show a high percentage of Cl^- and NO_3, which may be responsible for the activity of these organisms.

Tests of accelerated aging conducted with cubic 5-cm pressure gauges indicate that salt weathering has a considerable influence on weight loss (9.77% after 10 cycles) whereas the weight decrease is small in freeze-thaw cycles (0.36% after 15 cycles).

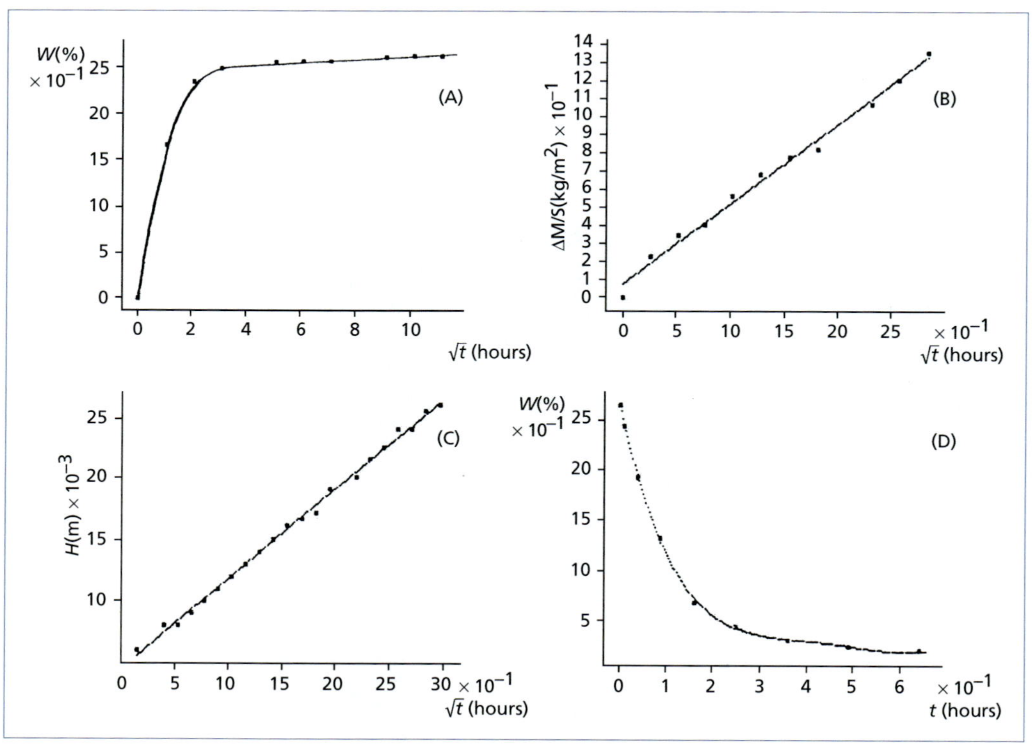

Figure 5.58 Porosity characterization curves all done on 100-μm Sample A, which is light orange gray color in color. (A) Absorption of free water by total immersion (W- water content as a function of time). (B) Capillary absorption (increment of weight per surface unit (ΔM/s). (C) Capillary penetration (height reached as a function of time √t). (D) Evaporation curve (W water content as a function of time in hours) (Esbert *et al.*, 1987).

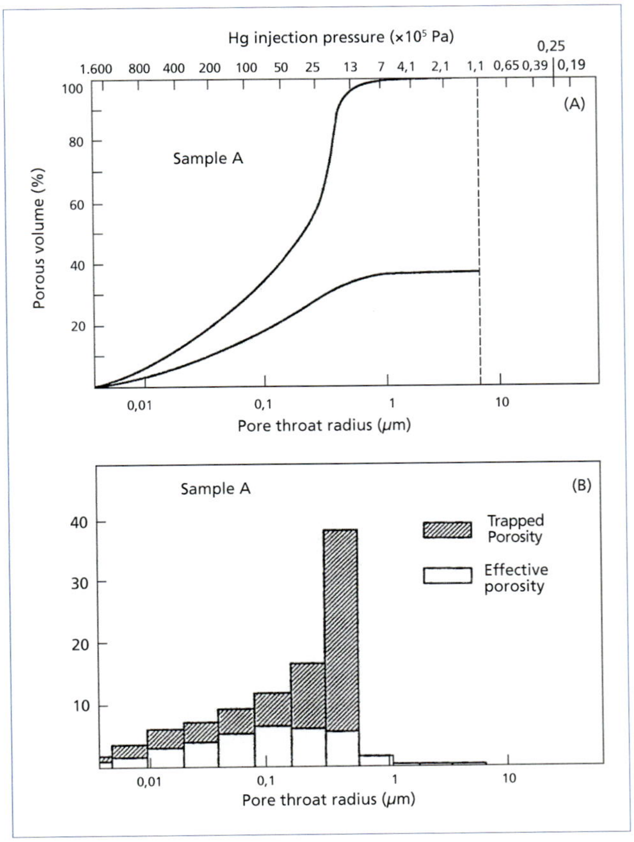

Figure 5.59 Porosimetric curves: pore throat radius distribution. (A) The upper curve corresponds to the first injection and the lower curve to the second. The difference between them indicates trapped porosity. (B) Histogram of the variation of porous volume as a function of pore throat radius (Esbert *et al.*, 1987).

Figure 5.60 Salt extraction on a capital of the San Juan de la Peña Cloister.

Characterization and testing show that varied causes and mechanisms of weathering: the most important is anthropogenic activity linked to various historical reconstructions that have caused partial and total losses of ogees and capitals. Moisture in the sandstones and its texture determine water absorption and other physical properties. Conversion of pore water to ice causes deterioration of the rock. A large number of microorganisms were observed but their activity is thought to have a minor effect on total deterioration.

The monuments of **Angkor** in the northwestern part of Cambodia form a complex of sandstone temples built between the 9th and 13th centuries during the splendid period of the Jemer Empire. The area has an average temperature is 26°C and it receives 1400 mm of rainfall. Average humidity during the dry season is 80% (Delvert, 1968). The sandstone is gray in color and is covered with whitish coating as a result of alteration. Capillary action is the primary cause of weathering; it causes alveolization and tafoni shapes at the base of the temples (up to 5–10 cm). Dense tropical vegetation has invaded many monuments and roots penetrate into existing joints. Lichens can be observed attacking the rock although the efficiency of this process is not known. Chemical analyses indicate very little difference between altered and fresh sandstones. The pink sandstones, which are the most resistant, are cemented by kaolinite whereas gray sandstones are cemented by chlorite.

We can conclude that in humid tropical countries with 1500 mm of rainfall, chemical action is not very active and is not responsible for weathering. Hydration of the rock and the resulting disaggregation cause alteration.

Analysis of the alteration of sandstone monuments in different climates shows that monuments built in arid to hyperarid areas (Jaisalmer and Egypt) are well preserved due to the lack of rainfall and the absence of soil water.

The Villamayor sandstone has high porosity and capillarity. The water that moves through the rock pores destroys primarily during wetting and drying cycles and especially by salt weathering, which causes the most loss of sandstone according to the aging tests. The presence of swelling clays in the matrix causes volumetric expansion of the crystal lattice and consequent swelling. In this subhumid environment, water plays a double role; it penetrates into pores and causes expansive clays to swell.

In the Cloister of San Juan de la Peña, the most significant cause of the deterioration of capitals is anthropogenic, as the fractures show, perhaps due to transport during successive fires or other reconstructions. Even so, the capitals along the western side are in an excellent state of preservation. It is important to protect the cloisters from the direct action of the rain because rain impact on them has almost completely obliterated the reliefs (for example, Romanesque capitals of the Cloister of Santo Domingo of Silos, Burgos). After a thorough study of the capitals of the Monastery of San Juan de la Peña, the Administration decided that no action was necessary.

Finally, the temples of Angkor, which are located on a large plain surrounded by water-filled trenches, are subjected to all the processes of water activity in rock pores. Even so, chemical analyses of fresh and altered rock do not reveal very different features because chemical weathering is not a major process.

In conclusion, we should mention that sandstones are rocks that are easy to sculpt and they are very easily weathered, particularly in areas where water plays a significant role. Sandstones are also a group of rocks with varied composition and different petrophysical properties, making it difficult to draw general conclusions.

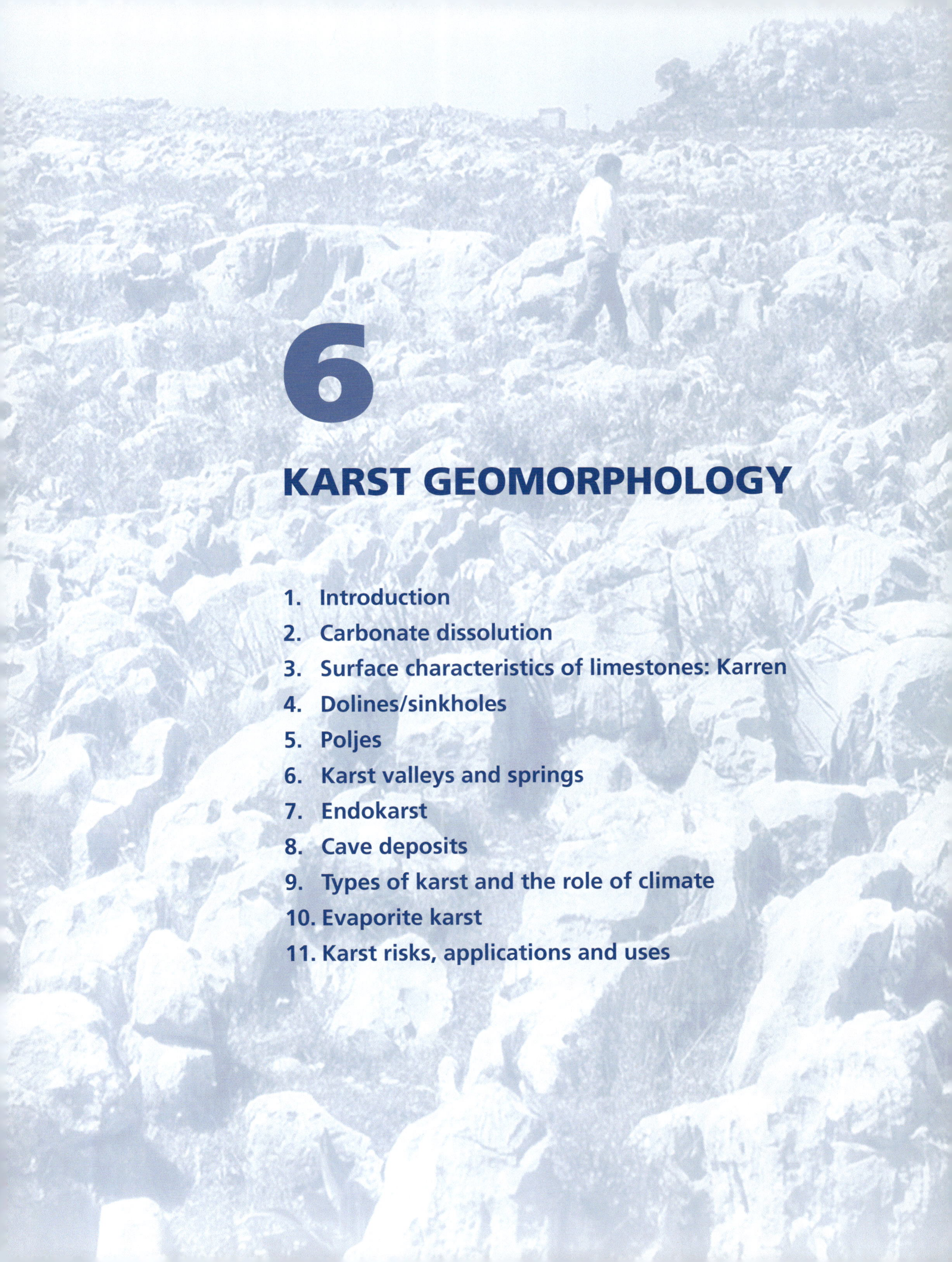

6

KARST GEOMORPHOLOGY

1. Introduction
2. Carbonate dissolution
3. Surface characteristics of limestones: Karren
4. Dolines/sinkholes
5. Poljes
6. Karst valleys and springs
7. Endokarst
8. Cave deposits
9. Types of karst and the role of climate
10. Evaporite karst
11. Karst risks, applications and uses

6.1 Introduction

Karst landforms result from the dissolution of rock by water. In karst areas, the combination of highly soluble rocks and well-developed secondary porosity determines the hydrology and the various morphologies (Ford and Williams, 1989; Ford, 2004). According to White (1988) the characteristic landforms in karst regions are:

- Closed depressions of variable sizes
- Poorly organized surface drainage
- Caves and subterranean drainage systems

Closed depressions *(Fig. 6.1)* come in various sizes. Poljes can extend tens of kilometers along their major axis. They are not connected to each other; they are known as **exokarst**, which is characterized by chaotic and disorganized landforms (Martonne, 1924). Poorly organized surface drainage is the centripetal type of karst; this type of drainage is totally different from the organized drainage present in non-karst rocks. Caves and underground drainage, the third characteristic feature of karst terrain, results from vertical drainage, which creates underground circulation called **endokarst**.

The word "karst" is the Germanic meaning of the term **kras**, the name of the western area of Slovenia that borders Italy where the limestone ground surface lacks vegetation (Kranjc, 2004). The geomorphology schools of Vienna conducted the first karst studies in this area. Jovan Cvijiv (1893) published an extraordinary and extensive paper in German in which he clearly and simply defined most karst landforms. As a result, most terms used to describe karst come from the German language.

Karst has an extensive terminology that includes terms used in different areas. For example, the scientific word "doline" has various names in different parts of Spain: "torca" (generally used in both Castilles), "celada" (Sierra de Albarracín, Iberian Cordillera), and "cuencón (Sierra de Javalambre, Iberian Cordillera). Karst landforms have always attracted man's attention and thus the diverse terminology is not surprising.

Figure 6.1 Aerial view of funnel-shaped sinkhole in Jurassic limestone. Villar del Cobo, Province of Teruel. Iberian Cordillera, Spain. Photo B. Leranoz.

In addition to carbonate rocks, karst occurs in outcrops composed of highly water-soluble minerals such as gypsum, its anhydrous phase (anhydrite), and chlorides, of which halite is the principal example. These minerals, called evaporites, form primarily by evaporation; they crop out in small percentages globally but their dissolution rate is much higher than that of carbonates, which causes rapid morphogenesis. As a result, the interaction of human activity and evaporite dissolution can lead to numerous and significant geologic risks.

Karst has been known for ages; the Greek thinkers tried to explain large rivers flowing from calcareous massifs (surgences) and the sudden disappearance of large flows (blind valleys). In spite of numerous contributions to the literature, in modern history Cvijic's paper (1893) is the foundation on which all future investigations of karst geomorphology will be based. Progress in this research has generally led to various vicissitudes of geomorphology. Thus, Grund (1914) introduced the Davisian cyclical concept to karst *(Fig. 1.2)* and Cvijic (1918) re-iterated it. Karst researchers have consistently emphasized the importance of climate in the development of karst landforms. At the Karst Symposium in Frankfurt participants emphasized that each climate has its own karst development (Lehmann, 1954).

Since 1950, **speleology** has contributed significantly to karst studies because it involves visits to karst cavities, raises questions about their origin, and has led to age determinations of carbonate deposits through the study of oxygen isotopes and U/Th dating (Sweeting, 1981). More recently, the application of AMS (Accelerated Mass Spectrometry) to measure cosmogenic isotopes in rock samples has provided another source of information (Shopov, 2004).

Carbonate formations are more common in recent periods of Earth history. Paleokarst has been recognized in stratigraphic columns by the presence of erosional and depositional karst landforms in environments where they are unlikely to have originated. Karst landforms in environments that have less than 250–300 mm of average annual precipitation are indicators of **paleokarst** (Sweeting, 1972). The study of relict karstification is one of the major contributions of karst to paleoclimatology.

Our knowledge of karst terrain development has enhanced the neotectonic investigations of mountain chains and their influence on morphogenesis. In coastal zones, positive eustatic change causes karstification to stop and leads to aggradation and fossilization of these landforms; however, negative movements of the base level cause the landforms to rejuvenate (Osborne, 2004).

These considerations indicate that while the fundamental karst process is dissolution, soluble rock masses are subject to additional extrinsic factors (De Waele et al., 2009) that affect their evolution. For this reason, investigations of karst are by nature multidisciplinary. To obtain the maximum knowledge possible, a karst massif must be studied in three dimensions with the goal of deducing complex relations between surface-generated morphologies and those originating at depth. This investigation must take into account environmental variation (climatic, tectonic, and anthropogenic) that has affected the karst during its temporal evolution.

Natural morphologies that develop by processes other than dissolution are called **pseudokarst** (Twidale, 1982b; Halliday, 2004). Examples include rills found on granitic slopes and cavities (Vaqueiro *et al.*, 2006) far from alteration profiles (Vidal Romaní and Gracia, 1987) and siliceous speleothems (Vidal Romaní *et al.*, 2003; Sanjurjo *et al.*, 2003). Cavities in quartzites have been studied in Minas Gerais (Brazil) by Corrêa Neto (2000), in Gran Savana (Venezuela) by Piccini (1995), and in South Africa and Mozambique by Martini (2000). Pseudokarst forms include tafoni and gnammas *(Fig. 5.41)*, which occur in various climates and rock types (Bryan, 1928; Renault, 1953; Gutiérrez and Ibáñez, 1979; Young and Young, 1992) and subsurface conduits and closed depressions that occur in clastic rocks as a result of the process of *piping* *(Fig. 16.54)*. This category also includes lava tubes (*jameos*) and accompanying collapse structures *(Fig. 4.34)* and stalactites of solidified volcanic lavas. Pseudokarst morphologies also include those formed by melting of permafrost (thermokarst or cryokarst) such as melt lakes and associated collapse structures *(Fig. 15.59)*, to be discussed in the chapter on periglacial geomorphology.

The distribution of carbonate terrain karst areas is shown in *Figure 6.2* (Ford and Williams, 1989). Ice caps aside, carbonate formations make up approximately 12% of the Earth's emergent land surface. They are most abundant in the northern hemisphere. Twenty-five percent of the world's population lives in karst areas and uses karst water. This fact is important in terms of the possible interference it may cause in karst areas.

A distinction can be made among karst types on the basis of different variables (cover, lithology, tectonics, physiography, hydrology, among others) (Quinlan, 1967). From this broad typology, we will review several widely used terms. Karst landforms can originate in areas devoid of soil and vegetation; in this case they are called **bare karst**. Karstifiable landforms can be found beneath a cover of material often made up of loose, transported sediments. This type is called **covered** or **mantled karst**. When karstification occurs between non-karstified formations, it is called **interstratal karst**. If the cover material is eroded from covered or interstratal karst, then karstifiable rocks can crop out. This type of karst is called **exhumed karst**. Karst researchers also distinguish between **active karst** and **relict karst** (Ford, 2004).

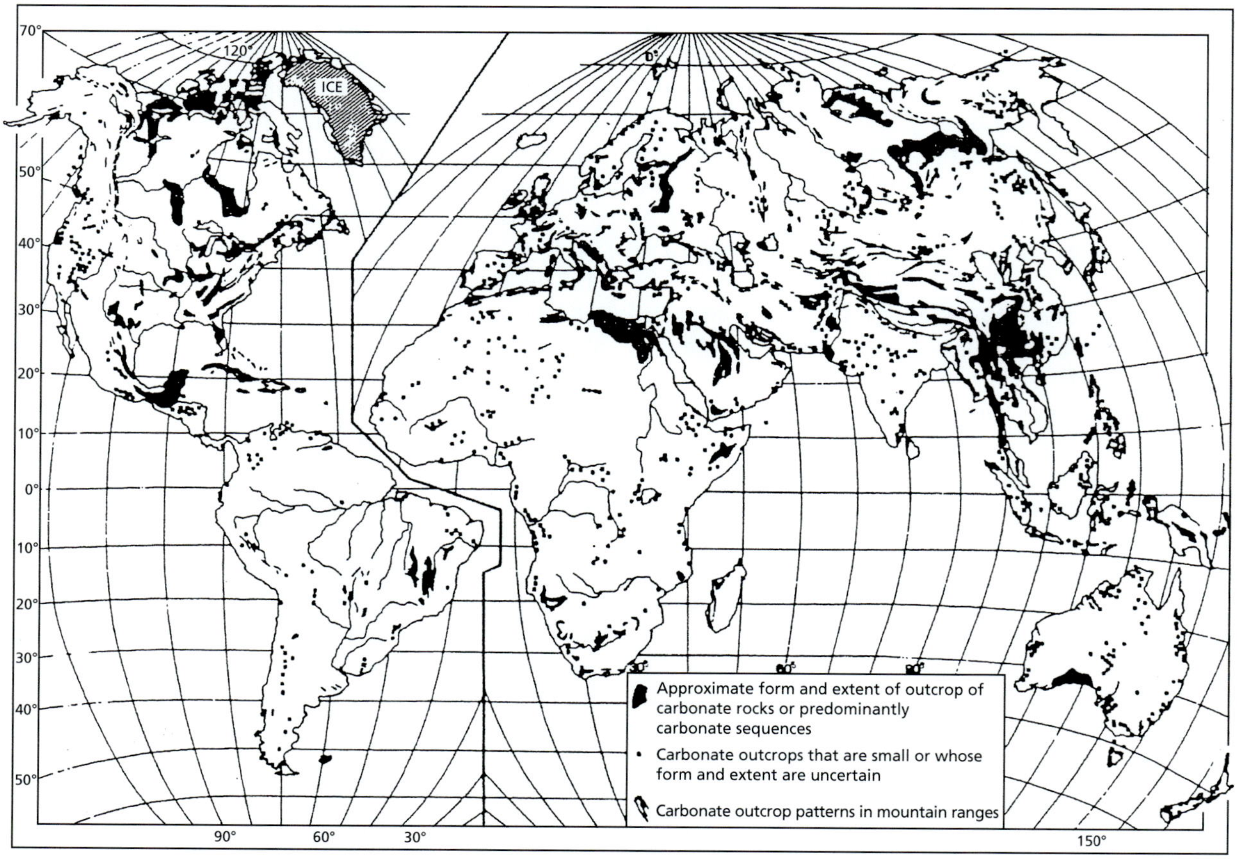

Figure 6.2 Global distribution of carbonate rock outcrops. Karst is present in most areas and is also present below non-karstifiable rocks (Ford and Williams, 1989).

6.2 Carbonate dissolution

Dissolution is a process by which a rock or part of it combines with water to form a solution or group of solutes. Solubility varies widely among minerals; gibbsite is virtually insoluble (0.001 mg/l at pH = 7) but halite is highly soluble (360,000 mg/l at pH = 7) (Gunn, 2004a).

Dissolution or corrosion (Bögli, 1980) of carbonate rocks is a reversible chemical reaction that leads to the formation of carbonate karst in relatively pure limestones and dolomites.

Calcium carbonate, as calcite or aragonite, is not very soluble in pure water. However, carbon dioxide gas is present in the atmosphere and soil and is dissolved in rainwater, groundwater, rivers, lakes, and oceans. Dissolved in water, carbon dioxide takes the form of H_2CO_3, carbonic acid, which is the driving force behind carbonate dissolution and karst generation (Dreybrodt, 1988, 2004).

The following reactions show the dissolution of carbonates (calcite and dolomite):

$$CO_2 + H_2O \leftrightarrow CO_3H^- + H^+$$

$$CaCO_3 + CO_2 + H_2O \leftrightarrow Ca^{2+} + 2HCO_3^-$$

$$CaMg\,(CO_3)_2 + 2CO_2 + 2H_2O \leftrightarrow Ca^{2+} + Mg^{2+} + 4HCO_3^-$$

Dissolution is caused by the addition of H^+, which comes from the dissociation of H_2CO_3. The reaction produces a solution of only 14 mg/l (as $CaCO_3$) at 25°C. The addition of H^+ from other acids (such as the oxidation of sulfides that produces sulfuric acid, for example) increases the solubility.

The solubility of carbon dioxide is a function of temperature. For a normal atmosphere with 0.035% of carbon dioxide per unit volume of dry components at sea level, the maximum dissolution is 1.00 at 0°C, 0.70 at 10°C, 0.44 at 25°C, etc. (Hutchinson, 1957). As a result, the quantity of limestone that can be dissolved at saturation equilibrium per unit volume of water is directly proportional to the partial pressure of carbon dioxide in the air that is in contact with water and inversely proportional to the water temperature (Henry's Law). Calcite solubility decreases with increased temperature at an approximate rate of 1.3% per degree centigrade (Gunn, 2004a).

A significant quantity of carbon dioxide is present in soils due to the respiration of plant roots and bacterial disintegration of organic matter, both of which increase with temperature. CO_2 moves or is stored in larger quantities in sandy soils where it may constitute 31% of the volume (Drake, 1984). Water dissolves carbon dioxide and moves it between pores. In soils, the concentration of carbon dioxide is often 1–2%, about 30–60 times higher than its content in the normal atmosphere; in humus-rich soils with minor ventilation the concentration of CO_2 may reach 25%, which is about 700 times its normal value in the atmosphere (Trombe, 1952). The CO_2 content of air in caves is highly variable although in cavities with good aeration its concentration is similar to that of the atmosphere. However, air in snow contains about 0.1% CO_2 because the smaller molecules of oxygen and nitrogen disseminate far from pores faster than the larger CO_2, molecules (see Sweeting, 1972).

Karst forms generated by biological activity (biokarst, Viles, 1984) tend to be small in size. This activity contributes to weathering and carbonate precipitation. Many organisms can dissolve limestone by secretion of organic acids and carbon dioxide. Along intertidal limestone coasts, considerable biochemical and biophysical erosion can occur as a result of burrowing or lithophagous mollusks, for example. Lichens exert intense biophysical activity in response to wetting and drying; their activity aids in the enlargement of solution pans.

Figure 6.3 shows the amount of carbonate dissolved as a function of carbon dioxide concentration. Because CO_2 solubility is a function of temperature, it follows different curves for different temperatures. These curves indicate the maximum amount of $CaCO_3$ that can be dissolved in carbon dioxide. These equilibrium saturation curves separate fields of oversaturation from fields of undersaturation or ones that contain aggressive water. Dissolution corresponding to a point on the 30°C saturation curve can cool (toward C) where it will dissolve more calcite; if heated (toward W) it will precipitate calcium carbonate; if it comes in contact with air that has a lower CO_2 concentration (toward L) it will lose some CO_2 and will precipitate calcium carbonate, and finally, if the dissolution encounters air with a higher CO_2 concentration (toward G) it will incorporate it and dissolve more calcite. It can take hours or even days for these reactions to reach equilibrium because some of them are slow (Jennings, 1985).

When calcite reaches saturation equilibrium, dissolution stops. Even so, if the saturated water at the surface or

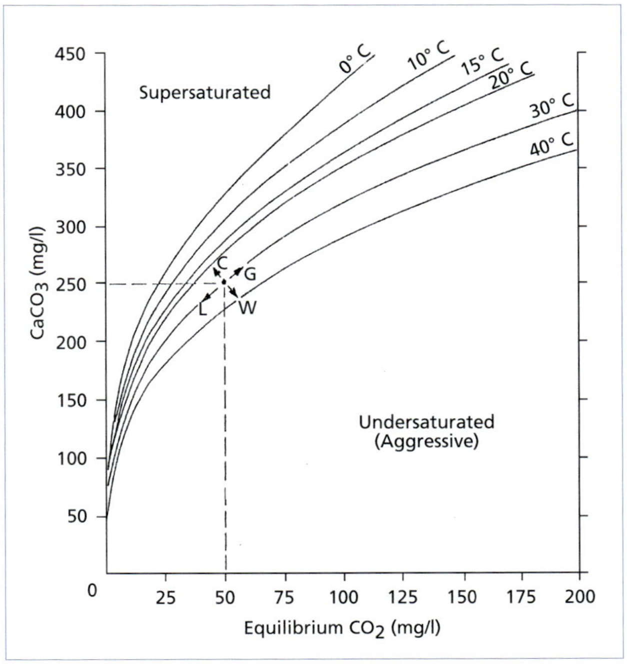

Figure 6.3 Equilibrium saturation curves for calcium carbonate at different temperatures, as a function of dissolved carbon dioxide concentration (Trombe, 1952).

at a shallow depth descends and cools (toward C) it becomes aggressive water. This process is called **corrosion by cooling** (Bögli, 1964). This process is active in the upper part, that is, in the vadose zones of the endokarst. It occurs in zones where there are wide daily temperature variations and clearly defined seasons. The process takes place during the day until midnight and during the hot season (Bögli, 1980).

Water in saturation equilibrium can become corrosive if it mixes with water that has a different calcium carbonate concentration (**corrosion by mixture**) (Bögli, 1964). The convex side of the saturation curve (*Fig. 6.4*) is an oversaturated field and the concave side is an undersaturated field of aggressive waters. Any straight line connecting two points anywhere on the saturation curve passes into the zone of aggressive water. Any mixture of saturated waters W_1 and W_2 produces aggressive water (M1:1 for an equal mixture). A mixture of saturated waters W_3 and W_4 remains oversaturated. The mixture of W_2 saturated and W_5 aggressive are aggressive in any proportion. M1:3 (one saturated part of W_1 and 3 oversaturated parts of W_4) remains oversaturated but M3:1, the opposite mixture remains aggressive.

Dolomite is a calcium and magnesium carbonate [$MgCa(CO_3)_2$] that occurs in calcareous rocks along with calcite ($CaCO_3$). In nature we have rocks that are intermediate between pure calcite (composed of $CaCO_3$), and

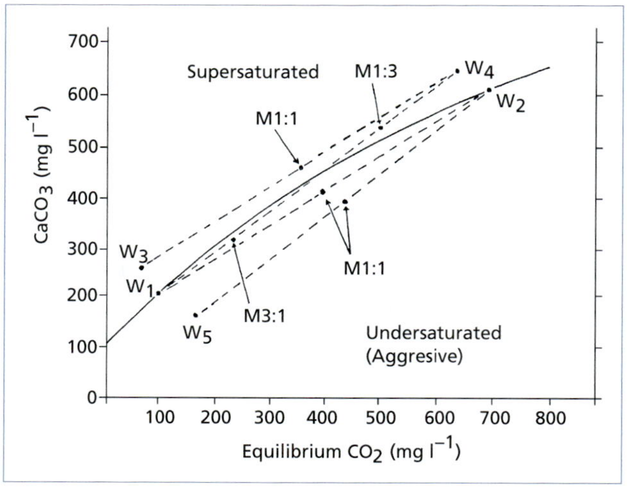

Figure 6.4 Corrosion by mixing, illustrated by several examples (Jennings, 1985).

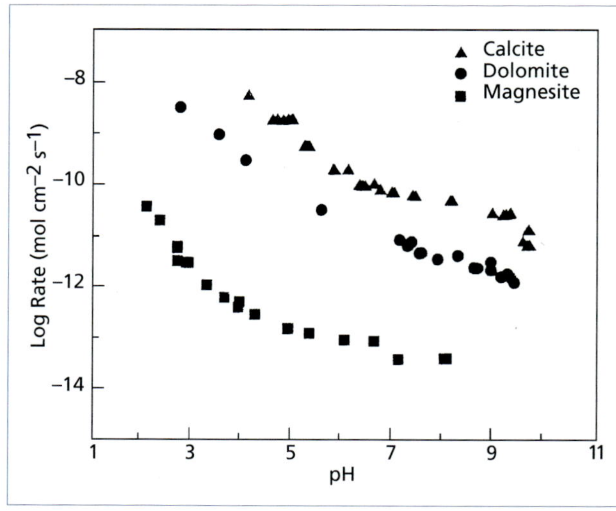

Figure 6.5 Dissolution rates of calcite, dolomite, and magnesite as a function of pH (Chou ***et al.***, 1989).

dolomite (made of only [$MgCa(CO_3)_2$]. They differ in terms of percentages; for example, dolomitic limestone is 50–90% calcite and 10–50% dolomite. Carbonate rocks contain two stable minerals: calcite and dolomite; aragonite is unstable and converts to calcite.

Busenberg and Plumer (1982) studied dolomite dissolution under conditions of partial pressures of 0–1 and temperatures of 1.5 to 65°C. Initially dolomite dissolves in quantities of 5 to 6 mol$\mu m^{-2}s^{-1}$ (at partial pressure of 0.93 atm) but this dissolution later decreases exponentially to very low levels. As a result, Busenberg and Plumer (1982) show that dolomite dissolution can take place in two steps. The first step is the reaction of $CaCO_3$ and H^+ and the second step of the same reaction involves the $MgCO_3$ component, which is controlled by the reaction rate. Chou *et al.* (1989) show that dolomite dissolution, as a function of pH, is somewhat lower than that of calcite, but the values are much higher than those obtained for magnesite (*Fig. 6.5*).

The common ion effect is a phenomenon in which an ion from the dissolution of a mineral from another source affects the amount of mineral that can be dissolved. The common ion effect causes calcite and dolomite solubility to decrease in waters that contain dissolved gypsum (Ford and Williams, 1989).

The quantification of **limestone dissolution** depends on the surface slope, annual precipitation, rainfall intensity, and the prevailing wind direction (Szunyogh, 2005). Reliable data (Douglas, 1968) can be obtained only through a long series of observations. Various methods have been used to obtain limestone dissolution rates.

Pedestals in limestone erratic blocks can be used to determine dissolution rates because they form when the underlying limestone is protected from dissolution. The height of a pedestal provides a dissolution rate if the age of ice retreat is known (Dreybrodt, 1988). Jennings (1985) summarizes the known data on lowering rates calculated by this technique; the rates range from 10 to 42 mm/ka.

A **microerosion meter** has been used to measure short time intervals on outcropping rocks, both on the surface and in cavities (Trudgill *et al.*, 1981). Microerosion measurements collected for ten years in various karst areas and environments in Australia show surface reduction values of 0.005 mm/year; for places with a subsoil or a fluvial bed, the measurements reach a maximum of 0.14 mm/year (Smith *et al.,* 1989). Microerosion measurements on the surface of an erosional platform are 1.486 mm/year and those obtained on soil inside rocky outcrops are 0.093 mm/year over the medium term (Stephenson and Finlayson, 2009). However, this technique has some limitations (Spate *et al.*, 1985). The values obtained depend on climate; higher numbers are observed in humid tropical climates (Jennings, 1985).

Various authors have used different methods to measure corrosion in caves. One of the most common methods involves **circular limestone disks** from a specific area. They are dried, weighed, and are actively dissolved in the cavity; after a period of time, the weight loss can be determined. Trudgill (1976) obtained highly disparate values because the disks are associated with soils with different acidities; numbers obtained at the surface are much lower.

Corbel's (1959) **hydrochemical method** consists of measuring the amount of limestone transported annually by a watercourse and comparing this amount to the surface of the limestone terrain in the drainage basin to determine a quantity in m^3/year/km^2 or a value of the surface lowering in mm/ka. The most outstanding conclusions provided by this method are the fact that the highest values of karst corrosion come from the snowy regions of the Arctic and limestones in mountain chains; temperate areas have much lower values.

Erosion rates in cavities have been investigated by determining the uranium ages of **speleothems** (Gascogne, 1981). For example, if the age of the base of the speleothem is 100,000 years at a height of 2 m above the base level of the current in a passage of the cavity, the average lowering has been 2 mm/ky (Dreybrodt, 1988).

6.3 Surface characteristics of limestones: Karren

Rock surfaces of outcropping limestone are sculpted by small grooves or holes, called karren in the international karst literature, *lapiés* in French, and *lapiaz* in Spanish. Karren can also develop in calcarenites with 90% $CaCO_3$ (Gutiérrez and Ibáñez, 1979). These features, which have widely varying morphologies, are the most noticeable characteristics of dissolution processes in karst areas. They usually occur in groups as **karren fields** *(Fig. 6.6)* Karren can develop from water flowing over the limestone outcrop; in this case it is called **free karren**. If the dissolution occurs in the soil-limestone interval, it is called **covered** or **semi-free karren**. If the soil is denuded and exhumed, the same processes that form free karren can affect it. As a result, karren fields can be complex.

The first studies of karren were done during the second half of the 19th century in Mediterranean and Alpine karst (Ginés, 2004). Karren have been studied in the laboratory and in the field; the major impulse of these studies was the decade of the 1960s with the work of Bögli (1960) in Switzerland, Bauer in Austria, and Miötke (1968) in the Picos de Europa, Spain. An important meeting on karren landforms was held in Palma de Mallorca, Balearic Islands (Fornos and Ginés, 1996).

Researchers have described numerous types of karren generated by climatic and anthropogenic processes (Cvijic, 1924a). Multiple karren classifications have been proposed. Bögli's classification (1980) is based on the difference between free and covered karren. White's classification (1988), slightly modified by Salomon (2000), is based on size, free or covered karren, and slope and shape. In their classification, Ford and Williams (1989) include plan-view circular shapes, linear shapes with structural control, linear shapes controlled by hydrodynamics, and polygenetic shapes. Here we will use

Figure 6.6 Karren field showing kluftkarren and hohlkarren that formed under soil (covered karren) and was later exhumed. Sierra de Cabra. Córdoba. Betic Cordilleras, Spain.

Table 6.1 **TABLE OF SIMPLE KARREN TYPES (SWEETING, 1972)**

Type	Average size	Free or covered surface	Horizontal or inclined surfaces	Sharp or smoothed crests
Rillenkarren	1–2 cm deep, up to 50 cm long	Free	Inclined	Sharp
Trittkarren	3–50 cm high 20–100 m wide	Free	Horizontal	Sharp
Rinnenkarren	50 cm deep up to 20 m long	Free and half-free Half-free	Inclined	Sharp crests, sometimes slightly rounded bases.
Spitzkarren	50 cm wide 50 cm deep	Free	Inclined crests and roof-like surfaces	Sharp
Meanderkarren	50 cm deep up to 20 cm long	Free and half-free	Only slightly inclined	Sharp crests, slightly rounded bases
Rundkarren	12–50 cm deep up to 15 cm long	Covered	Inclined	Smoothed
Kamenitzas Solution Basins	Few cm to more than 3 m in diameter up to 50 cm deep	Both free and covered according to type	Horizontal	Sides: – fretted when free karren – smooth when covered Bases normally smoothed
Kluftkarren	Few cm to 4 m deep up to 4 m wide	Free and covered	Development along joints	Sides: – sharp when free – smooth when covered Bases normally smoothed
Hohlkarren (or *Mohrkarren*)	60 cm–1 m deep 50 cm wide	Covered—formed under peat	Inclined or slightly inclined	Smoothed sides and bases
Deckenkarren	Normally a few mm or cm deep	Covered—formed by direct action of plants	Inclined and horizontal	Smoothed but can be sharp

Sweeting's classification (1972) as shown in *Table 6.1*. This author weights the factors affecting karren formation.

Sweeting assigns great importance to **the nature of chemical reactions**. Bögli (1960) distinguishes four phases; the first is the period of maximum dissolution activity, which corresponds to the moment when the CO_2-charged water comes in contact with the limestone. In later phases, dissolution decreases gradually, reaching a minimum after 60 hours. In phases two and three, free karren are produced (rillenkarren, trittkarren, rinnenkarren and spitzkarren). In the case of covered karren, dissolution takes place in the presence of biogenic CO_2, which produces hohlkarren, rundkarren, and deckenkarren. The remaining types of karren in *Table 6.1* can form as free or covered karren.

Since free karren result from surface dissolution of limestone, the **type and quantity of precipitation** are important factors in karren formation. Snow can contain a large amount of CO_2 whereas drizzle does not seem to be very capable of producing free karren (Miötke, 1968). Tropical rainfall, which is abundant and intense, creates large-scale karren. In arid climates, the characteristics of surface dissolution are **vermiculations** that form on limestone outcrops in small, disconnected grooves that are 1–2 mm deep (*Fig. 6.7*)

The role of the **limestone type**, including its porosity (among other factors) in karren formation is poorly understood. Some limestones, such as the Triassic *schrattenkalk* (karren limestone) of the Alps (Sweeting, 1972) show abundant karren formation. However, karren does not usually develop on chalk, which is very porous, or on finely stratified limestones.

Other factors are the **slope of the outcrop and structure**. Inclination affects karren formation. For example **rillenkarren** (*Fig. 6.8*) develop on 60–80° slopes. The factors that affect its evolution are rainfall intensity, size of water drops, water temperature, and viscosity. Electron microscopy has shown that mechanical mobilization of small limestone particles is one of the principal processes. The removal of particles increases the impact of water drops and these effects are enhanced by algal activity (Fiol *et al.*, 1969). **Rinnenkarren** (*Fig. 6.9*) follow their presence on medium surface slopes.

Figure 6.7 Vermiculations, karren formed on outcropping limestones in arid zones. Avdat, Negev Desert. Israel.

Figure 6.8 Rillenkarren (karren with rills and sharp ridges) formed on the upper Cretaceous Cañones limestone formation. Piedra de San Martín, Pyrenees of Navarra, Spain.

Figure 6.9 Rinnenkarren (solution flutes), walls of structural karren and small rillenkarren formed in Cretaceous limestone. Piedra de San Martín, Pirineo de Navarra, Spain.

Spitzkarren also form on surface outcrops; they are usually three-dimensional pinnacles that are 0.5–30 m tall and up to several meters wide. The pyramidal facets are subvertical. They develop in the humid tropics. Excellent examples are found in Madagascar and Shilin, Lunan County, Yunnan province of China (Ginés, 2004). They are called "subjacent karst" by some authors. **Trittkarren**, **kamenitzas (solution pans),** and **meanderkarren** form in flat-lying areas or areas with very low slopes. Trittkarren form horseshoe-shaped steps. In Texas, kamenitzas are called by their Spanish name, **tinajitas** (Birot, 1966). In jointed limestones, maximum dissolution occurs along these fissures, creating **kluftkarren** (*Fig. 21.8*). These open joints are called *grikes* in England and *cutters* in the United States. Dissolution can create features of highly variable widths and depths. When these features are 2–4 m wide, tens of meters long and 1–5 m deep, they are called **bogaz** (Cvijic, 1893) (*Fig. 6.10*). In Puerto Rico these solution corridors are known as **zanjón** (Monroe, 1964a).

Karren that originate beneath the soil dissolve slowly and biogenic CO_2 plays an important role. The shapes are rounded and soft to the touch, different from free karren whose morphologies are sharp, rough, and cutting to the touch. Karren covered by soil include **rundkarren** and **hohlkarren** (*Fig. 6.11*). This last type of karren is known as *roches perforées* in the French literature. **Deckenkarren** (*Fig. 6.12*) also form under these circumstances when exhumation makes it possible to see rounded karren.

Seawater and some marine organisms (Pholas, Patellas, etc.) in contact with carbonate rock produce **coastal karren**, which perforate the rock. Other karren are characterized by their rugosity and variety of shapes. The chapter on coastal geomorphology will include a discussion of coastal karren.

We have discussed the importance of temperate, arid, and wet tropical climates on karren but there are also well-developed karren fields in areas subject to the action of snow. Cold melt water produces **limestone pavements** or Bögli's **corrosion tables** as in Yorkshire (England) where glacial striations in carboniferous limestone rapidly disappear as a result of dissolution. The pavement is composed of well-developed structural karren with grikes surrounding clints or limestone outcrops (Sweeting, 1966). The relationship between the inclination of the limestone and topography is important in pavement development; they do not form at dips greater than 45° (Williams, 1966).

6.4 Dolines/sinkholes

Cvijic (1893) introduced the Slovenian word doline into the scientific literature; the English-speaking world also uses the word *sinkhole*. Dolines are closed depressions, circular

Figure 6.10 Bogaz with rinnenkarren on its walls. Cretaceous Cañones limestone. Piedra de San Martín. Pyrenees of Navarra.

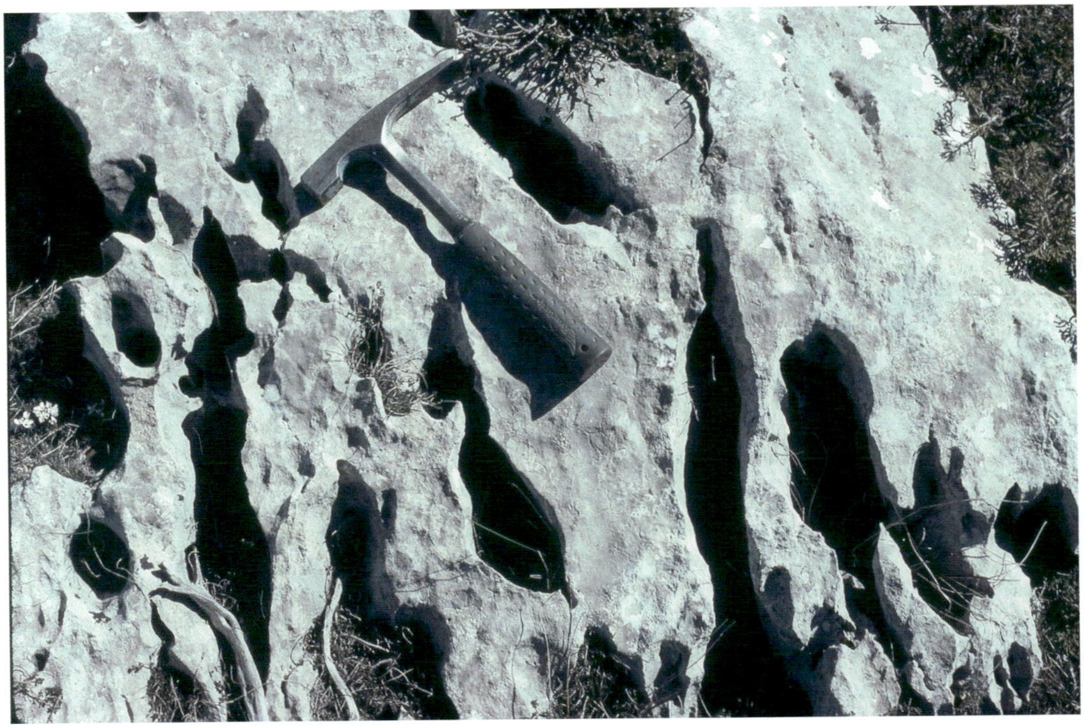

Figure 6.11 Hohlkarren, the most common type of covered karren, formed on Liassic limestone. Sierra de Javalambre, Province of Teruel. Iberian Cordillera, Spain.

Figure 6.12 Deckenkarren that contain hohlkarren. The limestone is Miocene. Bañolas-Besalú. Province of Gerona, NE Spain.

or ellipsoidal in shape in plan view, ranging from meters to a kilometer in diameter and a few meters to more than a hundred meters in depth. In some areas of the world, sinkhole density is 450 per km^2 (Cramer, 1941). Edges of sinkholes range from gentle to steep. They can occur in isolation but are more frequently grouped in fields of sinkholes (*Fig. 6.13*).

Cvijic (1893) classifies sinkholes on the basis of their morphology. **Bowl sinkholes**, which are larger in diameter than in depth, have gentle or sharp boundaries (*Fig. 6.14*). **Funnel-shaped sinkholes** have a diameter two or three times their depth and a floor covered with detritus (*Fig. 6.1* and *6.15*). **Well-shaped sinkholes**, whose diameter is smaller than the depth, have sub-vertical walls and a floor that may provide access to an underground gallery (*Fig. 6.16*). This classification is simple but it is difficult to categorize some sinkholes into one of these three types. **Uvalas**, which result from the connection of two or more sinkholes, generally have flat floors. Another morphological sinkhole variety, which occurs in tropical cone karst, is star-shaped; it is called a cockpit, after the Cockpit region of Jamaica (*Fig. 6.17*).

Cramer (1941) made the first genetic distinction between solution sinkholes and collapse sinkholes. Classifications proposed later have grouped both of these as solution sinkholes and consider them to have formed by this process. Authors differ regarding the distinctions within the group of collapse sinkholes. In addition, some sinkholes form as a result of subsidence. Because most sinkholes form as a result of multiple processes, they are called polygenetic sinkholes (Ford and Williams, 1989).

The processes that generate sinkholes are dissolution, sagging, suffosion, and collapse. The combination of these processes produces seven basic types of sinkholes (Gutiérrez *et al.*, 2007) (*Fig. 6.18*).

6.4.1 Solution sinkholes

When water penetrates into joints and fractures, it dissolves calcite and creates solutes and insoluble material that infiltrate into the karst mass. This action enlarges fracture planes and causes a loss of volume, which is accompanied by settling and lowering of the surface; this process creates a solution sinkhole. The decalcification clay and the material transported into the closed depression remain in the sinkhole. These depressions can range in diameter from 1 to 1,000 m and may be shaped like a bowl (*Fig. 6.14*). Small sinkholes often form in the deposits that cover the floor. If a preferential fracture orientation exists, the sinkholes align along this prevalent direction.

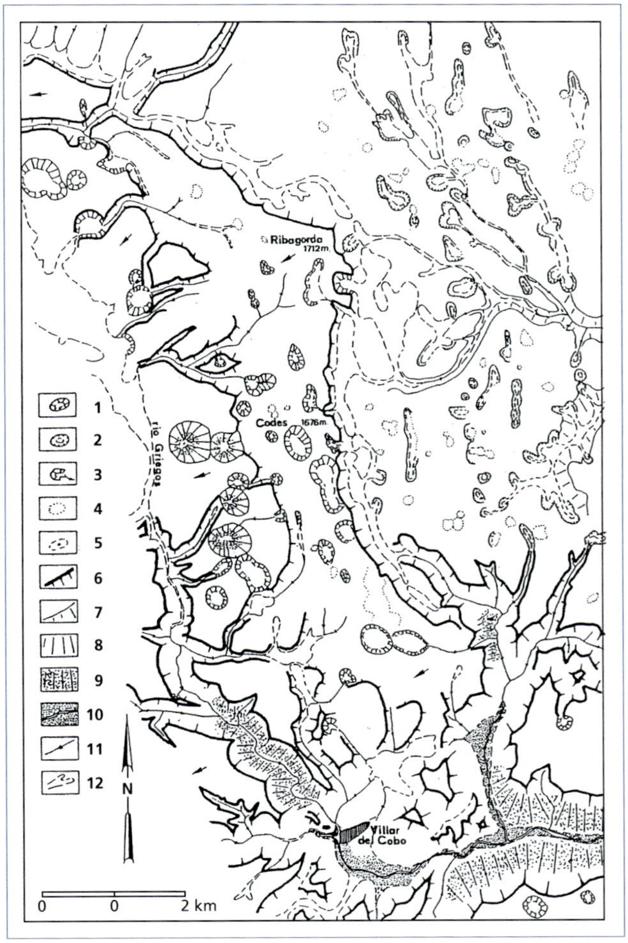

Figure 6.13 Group of dolines on Lias and Dogger carbonates in the Villar del Cobo region (Province of Teruel). Iberian Cordillera, Spain. 1: Funnel-shaped sinkhole 2: Bowl-shaped sinkhole 3: Captured doline 4: Degraded doline floors. 5: Swallow hole. 6: Structural scarps. 7: Non-structural scarps. 8: Regularized slopes. 9: Grèzes litées. 10: Alluvial floors. 11: Linear incision gullies. 12: Flat bottom gullies. (Gutiérrez and Pena, 1979a).

Asymmetrically shaped sinkholes form under a variety of circumstances. If the rock strata dip moderately, the up-dip side of the sinkhole is steeper than the lower side (Sweeting, 1972). Under snowy conditions, the leeward side of the sinkhole is gentle and the other side is steep because more snow accumulates on the leeward side. Also, soil cover contains more biogenic CO_2 that causes more dissolution (Bögli, 1980; Jennings, 1985).

In areas with significant snowfall, sinkholes form as a result of the activity of CO_2 present in the snow; the snow gradually deepens and creates a well-shaped sinkhole or **kotlici** or a **nival doline** (*Fig. 6.19*). Cvijic (1893) calls them *schneedolinen*. They have been observed in the Julian Alps, Pyrenees, South Betics of Malaga (Spain), and the Atlas (Sweeting, 1972). Snow may remain in the bottom of deep sinkholes until the summer and may even persist all summer.

The presence of trees inside sinkholes promotes growth of sinkholes due to mechanical weathering by the root system and increased biogenic CO_2. Trees are obviously not present above the tree line; also sinkholes are not found in the Alps (Corbel, 1957). In forest areas a temperature inversion causes cold air to accumulate in the bottom of deep sinkholes; as a result the tree line is found in the lower areas of the sinkhole (Gèze, 1965).

Terra rossa may accumulate inside sinkholes as result of the decalcification of impure limestone; these accumulations do not exceed a thickness of 4 m. Surface runoff can also deposit alluvial material, especially in large sinkholes. In central Europe, loess and periglacial deposit fillings are common (Sweeting, 1972). In glaciated areas, a sinkhole floor may be covered by a blanket of glacial till, which in turn can be covered by small sinkholes as a response to the dissolution of the limestone substrate (Miötke, 1968).

Small sinkholes can form in a few thousand years and it is quite possible that larger solution sinkholes form over periods of hundreds of thousands of years. Once formed, they can persist for millions of years (Williams, 2004a, b) especially if they are located on large erosional and depositional planation surfaces. When large sinkholes occur in environments with frequent freeze-thaw cycles, the resulting gelifracts produce pebbles and clast deposits on the doline floor, which can be partially absorbed by underlying galleries, as in the Villar del Cobo sinkhole field (*Figs. 6.1* and *6.13*). Gelifract production accompanies slope retreat and increases sinkhole diameter as seen in sections of detritus in flat bottom valleys along sinkhole margins that have been eroded by the retreating sinkhole. When all the interfluves of the sinkholes are small, the topography looks like an egg-carton; this type of landform is called **polygonal karst** (Williams, 1971). This type of karst occurs in temperate latitudes and in the wet tropics.

Well-developed solution sinkholes require significant precipitation and flat land surfaces that lack fluvial hierarchy and where, as a result, most water penetrates into the interior of the karst mass. In addition, highly jointed limestones promote infiltration (Sweeting, 1972).

Limestone and evaporite dissolution can occur under a soil film; maximum dissolution takes place in rocks that are closest to the surface. This zone is called **epikarst** (Klimchouk, 2000a) or the **subcutaneous zone** (Williams, 1983) (*Fig. 6.20*). Epikarst thickness is difficult to define but can be estimated at 15–30 m (Waltham *et al.*, 2005).

Figure 6.14 Bowl-shaped doline, dissolution dolines in other classifications, formed on Lias limestone with clay floor due to decalcification. Sierra de Javalambre. Iberian Cordillera, Spain.

Figure 6.15 Funnel-shaped doline on mid-Jurassic limestone. In the distance, the Tremedal Paleozoic massif. Sierra de Albarracín. Iberian Cordillera, Spain.

Figure 6.16 Well-shaped sinkhole, collapse sinkhole in other classifications, formed in jointed Dogger limestone. The floor of the doline connects with an underground cave. Pozondón. Iberian Cordillera, Spain.

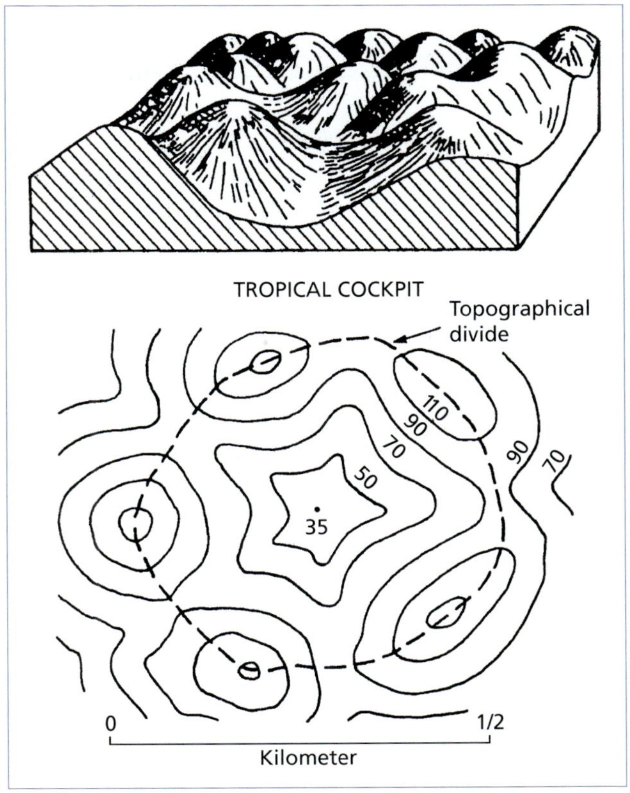

Figure 6.17 Block diagram showing conical karst and star-shaped elevation contours in the cockpit doline (Williams, 1969).

6.4.2 Collapse sinkholes

Several similar genetic sinkhole classifications have been published in recent years (Williams, 2004a, b; Beck, 2005; Waltham *et al.*, 2005). These classifications distinguish two main groups: sinkholes that form as the surface lowers due to dissolution and those that reformed by internal erosion and deformation processes related to subsurface karstification. The first type was discussed in the previous section on solution sinkholes. The second category includes four types classified according to the type of material affected by gravitational movement (rock, insoluble substrate, and unconsolidated cover) and by subsidence (collapse or suffosion). Rocks and insoluble substrates can undergo brittle collapse whereas cover deposits are affected by collapse and suffosion. The second category is more significant in engineering applications (Sowers, 1996; Waltham *et al.*, 2005; among others).

Collapse structures are formed by the following factors: (1) high joint density that greatly decreases the mechanical strength of roof rock in cavities (Sweeting, 1972; Bögli, 1980; Jennings, 1985; White, 1988; Ford and Williams, 1989, 2007; Loucks, 1999; White and White, 1969, 2000; Williams, 2004a, b); (2) absence of holes wide enough to initiate flexure; and (3) vadose conditions that cause the roof of the cavity to have a higher effective weight (White and White, 1969, 2000; White, 1988; Ford and Williams, 1989).

Deviation of gravitational forces around a cavity creates a zone of tension, the top of which is affected by an arched compression zone (Waltham *et al.*, 2005). This zone of tension generates dome-shaped fracture planes and arched roofs. Tharp (1995) studied this fracture mechanism. Roof collapse causes an accumulation of blocks on the floor. As the collapse progresses upward, the assemblage of fractured blocks creates a **trans-stratal breccia pipe** (also called **collapse chimney**, organ pipes, among others) (Andrejchuk and Klimchouk, 2002; Klimchouk and Andrejchuk, 1996a, 2005). This chimney may be several hundred meters high (Johnson,

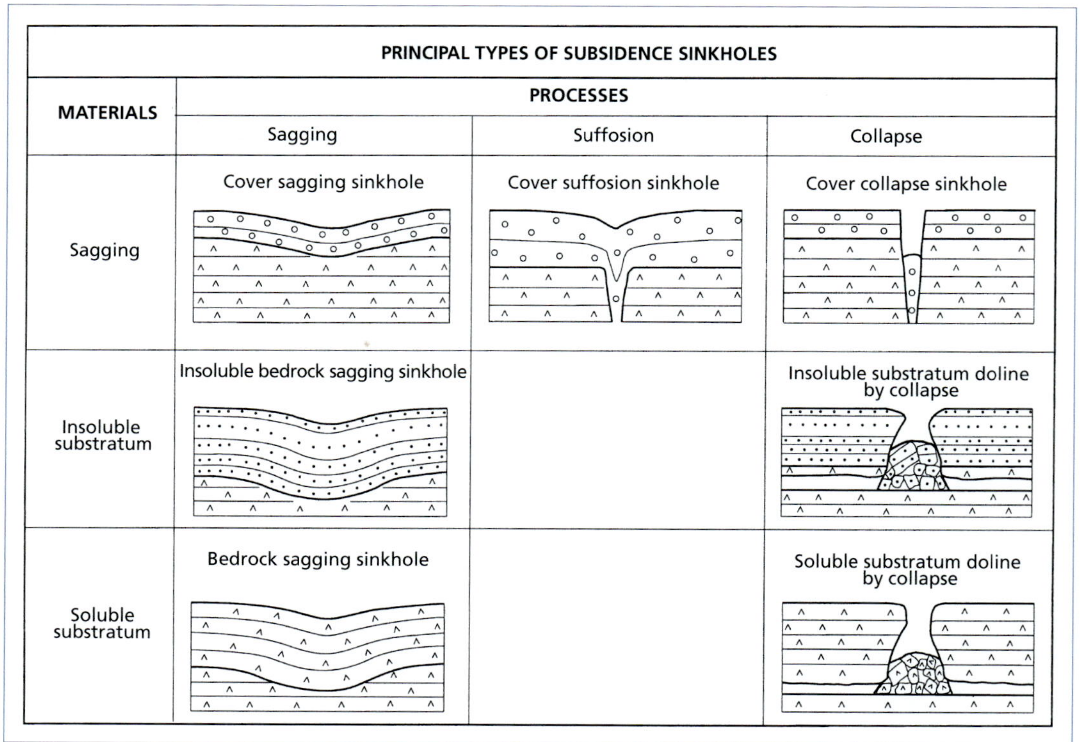

Figure 6.18 Principal types of subsidence sinkholes. These do not include dissolution dolines formed by erosional lowering of the ground surface (Gutiérrez ***et al.***, 2007).

Figure 6.19 ***Kotlicis*** or snow dolines. Cotiella. Interior Sierras of the Pyrenees.

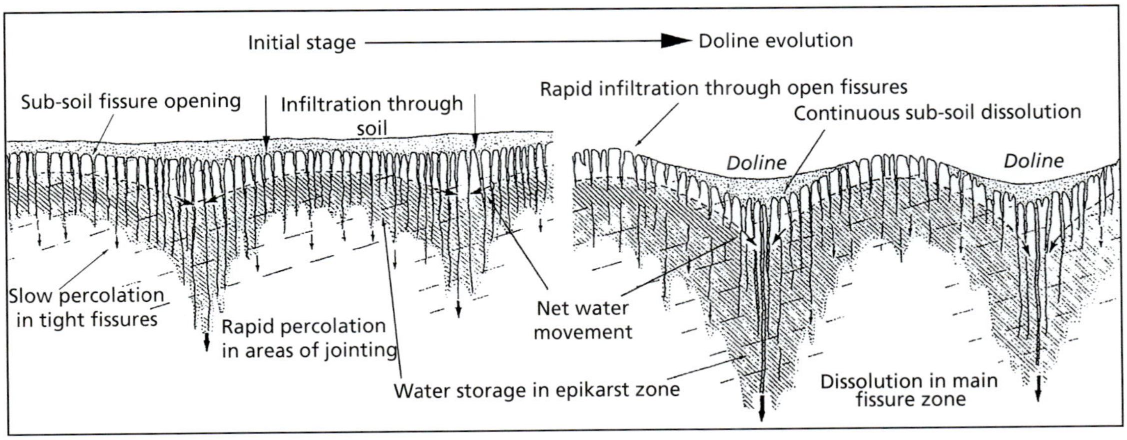

Figure 6.20 Initiation and development of solution sinkholes in areas with increased infiltration and significant tectonic jointing (Williams, 1983).

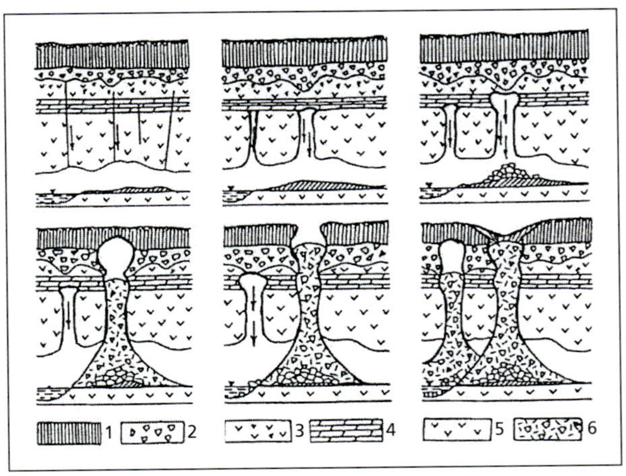

Figure 6.21 Formation of sinkholes by gravity/filtration mechanism, which produces vertical dissolution conduits. 1: Silt, 2: Collapse breccia horizon, 3: Gypsum, 4: Limestone/dolomite, 5: Gypsum/anhydrite, 6: Collapsed material (Andrejchuk and Klimchouk, 2002).

1989; Ford, 1997; Lu and Cooper, 1997; and Warren, 1999) (*Fig. 6.21*). Because collapse breccias are preferential sites for water penetration, they are rapidly karstified and can be transformed into a mass of corroded clasts glued together in a matrix of karst residuum and finally an insoluble mass devoid of clasts (Kerans, 1988; Loucks, 1999; Warren. 1999, 2007).

Sinkholes generated by upward propagation (ascending chimney, *stoping*) of the roof in a rocky cavity due to fracture processes above dissolution voids are called **soluble substratum doline by collapse** (*Fig. 6.16*) (Gutiérrez and Peña, 1979b) or **insoluble substratum doline by collapse** (*Fig. 6.22*), depending on whether the cavity has

Figure 6.22 San Pedro shaft. Insoluble substratum doline by collapse. This type of collapse affects lower Cretaceous marls and limestones. This example is 100 m deep, 90 m in diameter and there is a 20 m deep lake on the floor. Oliete, Province of Teruel. Iberian Cordillera, Spain.

Figure 6.23 Cover subsidence doline. The cover is unconcolidated material and the karst substrate may be Jurassic limestone. In the distance is the Tremedal Paleozoic massif. Orihuela del Tremedal, Province of Teruel. Iberian Cordillera, Spain.

crossed karst or non-karst lithologies, respectively (Gutiérrez and Cooper, 2002; Guerrero *et al.*, 2004, 2006). Both types of sinkholes have vertical margins and vertical dimensions on the order of meters or tens of meters; they both appear suddenly. However, these types of sinkholes are not common (Beck, 2005; Waltham *et al.*, 2005) primarily due to the formation of stable domes (Bögli, 1980; Waltham *et al.*, 2005) and/or because broken blocks underneath them support the roofs (Andrejchuk and Klimchouk, 2002; Klimchouk and Andrejchuk, 2005). **Cover sagging sinkholes** (*Fig. 6.23*), also called **alluvial dolines** and **cover suffosion sinkholes,** result from upward migration (suffosion, *ravelling*) of cover material across dissolution holes (Ward, 2004); they behave in a ductile manner. These sinkholes generally have a bowl or funnel shape (Clayton, 1966) and can be hundreds of meters in diameter (Guerrero *et al.*, 2004, 2006; Gutiérrez *et al.*, 2007). **Cover collapse sinkholes** (*dropouts*) form by upward propagation of cavities that have collapsed across a cohesive and brittle, but unconsolidated, cover. These dolines develop rapidly; their margins are vertical or overhanging. Cover collapse sinkholes are sometimes difficult to distinguish from cover suffosion sinkholes because they tend to fill up and take on a conical or bowl shape. Finally, the classifications include the term **buried sinkhole**, used for sinkholes filled with sediment. Compaction of the filling can lead to compaction sinkholes (Waltham *et al.*, 2005).

With the exception of solution sinkholes, this new classification (Gutiérrez *et al.*, 2007) follows Beck's (2005) methodology of sinkhole systematics, which uses two terms to define each principal sinkhole type: the first describes the material affected by gravitational movement (cover, soluble or insoluble substrate) while the second indicates the principal process involved: sagging, suffosion, and collapse (*Fig. 6.18*). In practice, more than one type of material and various processes can be involved in sinkhole formation (*Figs. 6.24* and *6.25*).

Cenotes are collapse sinkholes with flooded floors that are common in the Yucatán peninsula (Mexico) (Gerstenhauer, 1968; Day, 2004b). In southeastern Australia (Marker, 1976), they appear to be inundated by glacio-eustatic rise as shown by the abundance of submerged vadose speleothems. They are sub-circular in shape with steep walls that extend below the water table to depths that can exceed 60 m. They may be associated with a significant network of submerged cavities (Beddows, 2004). Water in cenotes was vital to the Mayan people because surface water flow is not present in the Yucatán (White, 1988).

Figure 6.24 Cover and bedrock saline sinkhole formed in the lower Miocene gypsum formations in Zaragoza. Mediana de Aragón, Province of Zaragoza. Ebro Depression, Spain.

Figure 6.25 Collapsed and covered doline, probably by collapse of substrate in Miocene gypsum. Villalba Baja, Province of Teruel. Teruel Semi-graben, Iberian Cordillera, Spain.

6.5 Poljes

Poljes are large closed depressions that are the most noticeable and problematic karst feature in temperate regions (Büdel, 1977); they have also been identified in tropical zones. In most Slavic languages, the word polje means a field that can be cultivated (Sweeting, 1972). In Italian poljes are called *campo* or *piano*; in France, *plans*; and in Cuba, *hojos* (Ford and Williams, 1989, 2007). In Spain they are known as *plá* and *navas* in the Iberian Cordillera, and in Navarra, as *raso*. These depressions are common in the Balkan region (Dinarides) where they have been classified into several types (Milanovic, 1981; Mijatovic, 1984). They may form part of tectonic grabens that are filled by great thicknesses of Tertiary deposits (more than 2,000 m). In these cases, which are characteristic of the highest areas of Dinaric karst, the floors of the depressions are mechanically flattened. However, poljes found at lower elevations and near the Adriatic Sea are located on top of carbonate formations where planation surfaces develop as a result of stepped karst corrosion. Poljes have flat floors, with slopes that usually dip 30° or more; they range in size from 0.5 km^2 to more than 500 km^2 (Sauro, 2005). Poljes are usually named after the town where they are located or a nearby town. If fracturing controls their shape, the polje may be elongate, up to 70 km, such as the Livno polje (Sweeting, 1972). In recent decades numerous poljes have been discovered in the Iberian Cordillera (see Gracia *et al.*, 2002); they are generally open (captured by the fluvial network), occur in Mesozoic carbonate formations, and are associated with stepped karstic corrosion surfaces. Most of them occupy a large surface; neotectonic fracturing may control their formation in some cases, such as the poljes of Vistabella (Gutiérrez *et al.*, 1982), Gallocanta (Gracia *et al.*, 2002) (*Fig. 6.26*), and Jiloca (Gracia *et al.*, 2003). Others contain no observable control linked to brittle tectonics (Gutiérrez and Valverde, 1994, among others). Some authors use gravimetry to emphasize the importance of tectonics in polje formation (Gams, 2005). Others suggest they are tectonic *pull-apart* depressions (Vrabec, 1994).

Rivers may drain poljes but more commonly poljes are dispersed across several hydrologic basins. Some poljes remain dry (*Fig. 6.27*) whereas others are periodically flooded (*Figs. 6.28* and *6.29*). Intense rainfall flooded the Zafarraya polje (Lhenaff, 1968) in 1996, greatly increased flow in the Arroyo de la Madre, which drains the polje, and exceeded the infiltration capacity of the ponors (López-Chicano *et al.*, 2002). Poljes situated at higher elevations in the Dinarides are dry but lower ones are frequently flooded (Sweeting, 1972). Karst poljes flood regularly during cold and wet periods (from October to April in Dinaric Karst) and dry up in summer (Bonacci, 1987, 2004). Poljes act as depositional areas for calcareous and non-calcareous detritus; their floors are composed of alluvium transported by the internal fluvial network and by alluvial fans that form at the contact between the slope and the flat floor. In high elevation areas in the Dinárides and Appenines poljes may contain glacial and periglacial deposits. Morphologies called **hums** often stand out on top of the flat floor of the polje.

Surface runoff can flood poljes but springs (karst resurgences) located at the margin of the polje more often cause the flooding. Flooding also occurs when the water table rises and intersects the surface. In addition, poljes drain underground into deep voids called **swallow holes** or **ponors**. When the floor of the polje is covered by alluvium, bedrock collapse or caprock collapse sinkholes (Sweeting's, (1972) alluvial dolines) may form and allow water to infiltrate. These dolines change constantly because they fill with alluvium and, in turn, create other dolines. During flooding, swallow holes become springs or intermittent springs, called **estavelles** in France.

Karstic depressions generated as poljes can be captured by the fluvial system and acquire an open hydrologic nature (Sweeting, 1972; Nicod, 1972; Roglic, 1974; Ristic, 1976; Jennings, 1985; Bonacci, 1987; Ford and Williams, 1989, 2007). However, during the long evolution of a polje, which is characterized by surface lowering due to karstification, there may be non-kartifiable formations that paralyze or abort their evolution (sterile poljes of Nicod, 1979). These circumstances frequently affect the Guadazaón poljes (Gutiérrez and Valverde, 1994) and the poljes of Layna (Gracia *et al.*, 1996) where the upper Cretaceous carbonates are karstified until the water reaches kaolinite-bearing sands (Utrillas Formation) or the descending Jurassic limestones contact the impermeable Keuper substrate.

Gams (1978) points out that although numerous definitions of polje have been proposed, much uncertainty remains about the exact nature of a polje. He maintains that a polje must be a closed depression with steep sides and must have a flat floor, karst drainage, and a minimum floor width of 400 m floor, which Cvijic (1893) sets at a kilometer. Gams proposes a classification of five polje types based on geomorphic and hydrodynamic characteristics. Ford and Williams (1989) think the number can be reduced to three types (*Fig. 6.30*): **border poljes**, which develop in an area where the water table fluctuates in non-karstic rocks and extends throughout limestones; **structural poljes**, associated with grabens and half grabens and characterized by an elongate morphology; and **base-level poljes** which are named for the water table and are found where the regional epiphreatic zone intersects the surface.

Poljes are highly varied and karst processes alone cannot explain their development; thus, they are polygenetic

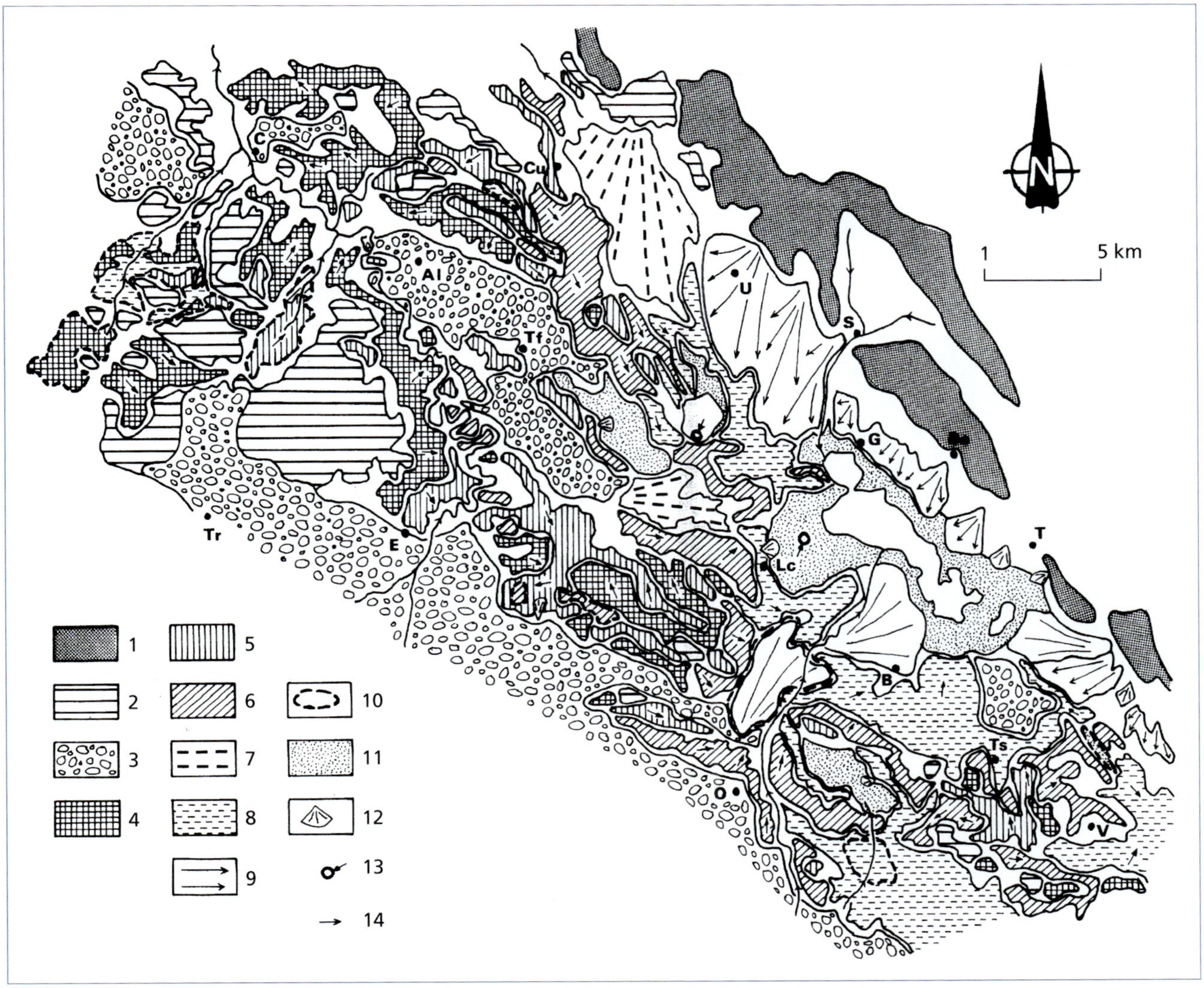

Figure 6.26 Geomorphic map of the Gallocanta polje. 1: Residual topography. 2: Neogene erosional surface. 3: Tertiary deposits. 4: Erosional surface C1. 5: Erosional surface C2. 6: Erosional surface C3. 7: Pediment G3. 8: Erosional surface C4. 9: Pediment G4. 10: Floor of hanging and/or captured polje. 11: Present-day polje floor. 12: Alluvial fan. 13: Swallow hole. 14: Dominant topographic slope. The letters represent towns. The residual Paleozoic relief is in fault contact with the polje surfaces formed in Mesozoic formations and the mechanical contact appears to be fossilized by the cover pediment (Gracia ***et al.***, 2002).

landforms resulting from various processes (Sauro, 2005). Two of the most debated aspects in the discipline of karst are the factors and processes involved in their origin and the development of stepped erosional surfaces (Gospodaric and Habic, 1979).

A thin cover of alluvium with a noticeably flat topography generally covers the limestone substrate of the polje floor. The origin of these depressions is related to the lowering of the corrosion surface of the ground. The flat topography results from corrosion processes that take place under the alluvial cover (levelling by corrosion, cryptocorrosion according to Nicod, 1976); the corrosion processes are controlled by the position of the water table (Birot, 1949a; Pfeffer, 1973; White, 1988; Ford and Williams, 1989, 2007; Bonacci, 2004). The polje floor may be lowered by erosion and clastic sediments may be removed by sinkholes that cover the rocky substrate. Deepening of the weathering front and fluvial erosion of the topographic surface are similar to the generation of etch-surfaces on crystalline rocks (Doppelten Einebnugsflächen) (Büdel, 1957). For this reason, some authors apply this theory to polje development (Büdel, 1977; Thomas, 1989a, b).

The polje floor continues to deepen until the topography reaches the level of frequent flooding (epiphreatic

Figure 6.27 Minde Polje, formed in a semi-graben. Lusitanica Graben. Portugal.

Figure 6.28 Zafarraya Polje during a dry period. Sierra Gorda, Sub Betics. Province of Granada, Spain. Photo J. Rodríguez Vidal.

Figure 6.29 Zafarraya Polje, flooded during the winter of 1997. Sierra Gorda, Sub Betics. Province of Granada, Spain. Photo J. Rodríguez Vidal.

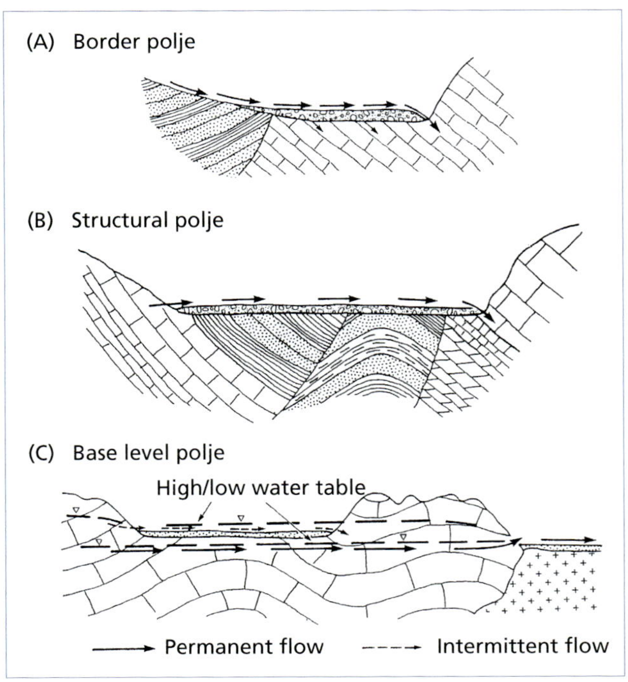

Figure 6.30 Fundamental types of poljes (according to Gams, 1978; summarized in Ford and Williams, 1989).

zone). When this happens, corrosion processes expand the polje floor by retreat of the marginal slopes (Trudgill, 1985), which corresponds to the border corrosion of Roglic (1952), which is controlled by the position of the water table (Ford and Williams, 1989, 2007). The relative drop of the water table increases the hydraulic gradient, causing the polje floor to become deeper as a result of vertical dissolution.

Variations in the relative position of the polje floor and the water table may be due to neotectonic activity and climate changes. Climate change toward wetter (biostasy) or arid (rexistasy) (Erhart, 1955) could cause the phreatic level to rise or fall, respectively. The position of the water table has probably also been affected by entrenchment of the fluvial network during the Quaternary. The genetic interpretation lacks the support of absolute dating of corrosion surfaces (Gracia *et al.*, 2003).

Stepping of the erosional surface has been recognized in poljes of the Iberian Cordillera: seven in the Sierra de Albarracín (Peña *et al.*, 1989), in the La Layna Polje (Gracia *et al.*, 1996) (*Fig. 6.31*) and in the Gallocanta (Gracia *et al.*, 2002). The elevation difference between the highest erosional surface and the polje floor gives an approximate measurement of the depth of the depression created by corrosion lowering (Gracia *et al.*, 2003).

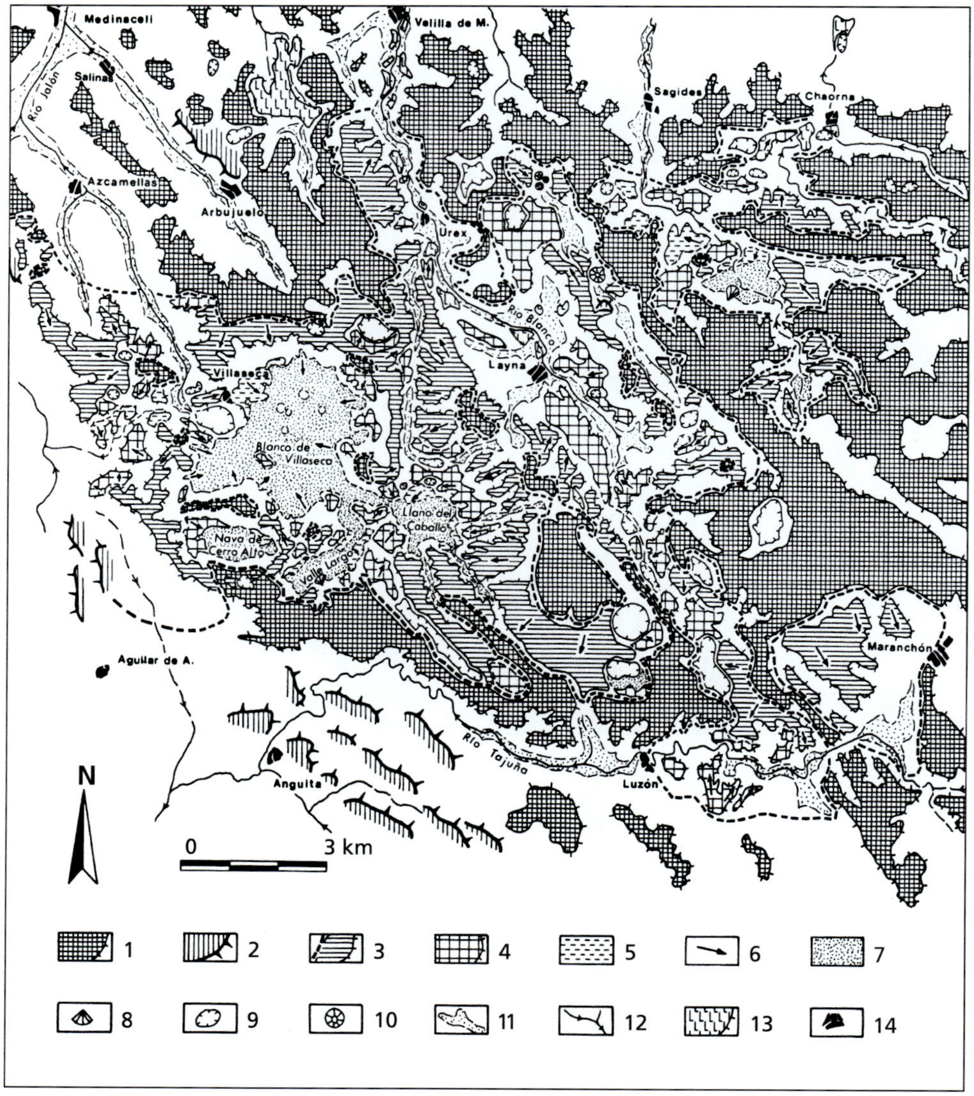

Figure 6.31 Geomorphic map of the central sector of the Layna polje system. 1: Neogene erosional surfaces. 2: Structural relief on Mesozoic materials. 3: External margin of the poljes and upper erosional surface. 4: Intermediate corrosion surface. 5: Lower corrosion surface. 6: Fill material on polje floors. 7: Dominant topographic slope. 8: Alluvial fan. 9: Bowl-shaped doline, closed depression in general. 10: Funnel-shaped sinkhole. 11: Flat-floored valley 12: Principal fluvial network. 13: Travertine edifice. 14: Town nucleus (Gracia ***et al.***, 2002).

Poljes are of great interest for several reasons. For one, they enable us to establish possible chronological relationships between the beginning of polje generation and regional erosional surfaces. They also provide valuable hydrological information and reconstruction of paleohydrological evolution of karst systems as poljes evolve. Finally, karst planation surfaces that fringe the polje floors can be used as markers for detecting neotectonic deformation (Gracia *et al*., 2002).

6.6 Karst valleys and springs

Karst is characterized by vertical subterranean drainage as a result of rapid, high infiltration into outcropping limestone and in the soil. Gunn (1981) shows that underlying caves are supplied through joints (vadose flow) that respond to precipitation within days to weeks. Flows through vertical shafts have the highest infiltration values. Surface runoff rarely

Figure 6.32 Arbayún Gorge or Foz, formed by the Irati River. Pyrenees of Navarra.

reaches rivers. Lateral flow across saturated and unsaturated soils is very limited. The behavior of covered karst (Jennings, 1985) depends on the nature of the soil or surface deposits. Prolonged erosion of weathering profiles composed of clays and iron hydroxides can produce "terra rossa" in the Mediterranean area. When doline fill persists over a long period, hydromorphic soils can form.

Cvijic (1893) introduced the term karst valley. He distinguished four types: pocket valleys, blind valleys, semi-blind valleys, and dry valleys. Sweeting (1972) added allogenic valleys to this classification. As Roglic (1964) points out, the term karst valley is inconsistent with true karst, which absorbs all rainfall. However, karst areas are present on surfaces that are dissected by dense fluvial networks (Gunn, 2004b).

Allogenous valleys are narrow and have steep sides, forming canyons *(Fig. 6.32)* (Ehlen and Wohl, 2002) that sometimes develop as a result of fracturing. These valleys, which proceed from the exterior of the karst mass, outlive the karst mostly because of their high discharge. These allogenous valleys are incised by fluvial erosion and karst corrosion; in layered limestones they may generate **potholes.** When massive limestones are fractured, rivers follow the fracture trace and remain confined within closely spaced vertical walls, such as the gorges of Yecla, Province of Burgos (Northern Spain), and Pitarque, Province of Teruel (Iberian Cordillera, Spain). Gorges may occasionally form as a result of the collapse of karst cavities. Part of the limestone surface may persist as a natural bridge *(Fig. 6.33)*.

$CaCO_3$ deposition may cause allogenous valleys to produce travertine or **calcareous tufa**. Processes responsible for this transformation are degasification of CO_2 in rapids and waterfalls, evaporation in arid areas, and algal activity (Pentecost, 2005; Viles and Pentecost, 1994). The travertine may have stem facies, moss, and stromatolitic facies. Rivers may cut down through them and reveal the travertine thickness and structure *(Fig. 6.34)*. Accretion of carbonate deposits may occasionally form an **accretionary cascade** (Pentecost, 2005) *(Fig. 6.35)*. Travertine dams that form by lateral accretion can obstruct lakes (travertine dams) such as the Lagos de Plitvice in Croatia (Roglič, 1940) or the Lagunas de Ruidera in Albacete, Spain (Ordoñez *et al.*, 1983) *(Fig. 6.36)*.

Some rivers gradually lose their water into the alluvium until they dry up (Sweeting, 1972). The flow is more commonly lost at a point, creating **blind valley**, where the river's total load flows into the interior of the karst mass. Cavities fed by these rivers increase in size by mechanical and karst corrosion. In **semi-blind valleys**, water is generally absorbed at a point and disappears but during flooding, water flows farther along from the normal disappearance point until it is completely lost. As a result, in a semi-blind

Figure 6.33 Natural karst bridge. Puentedey, (Northern Spain) Province of Burgos.

Figure 6.34 Karstified travertine located above the bed of the Júcar River. Uña, Province of Cuenca. Iberian Cordillera, Spain.

Figure 6.35 Waterfall built of travertine, which today appears to be karstified with presence of a cave. Priego, Province of Córdoba. Betic Cordilleras, Spain.

Figure 6.36 Ruidera lakes dammed by a collar of travertine. Albacete (South Iberian Cordillera, Spain). Photo J.L. Peña.

valley, the upper part flows to the normal loss point, and in the lower part water only flows during periods of flooding. A limestone boulder marks the end of a blind valley and the river disappears into a cave at its base. Blind valleys are found in the upper part of calcareous masses.

Pocket valleys (in Spanish, **valles en saco** or **valles de cabecera escarpada**; in French *vallés reculées*) are the opposite of blind valleys, since they are characterized by a karst resurgence at the base of the carbonate complex. They generally form high discharge springs that make up the headwaters of rivers such as the Nacedero del Urederra at the foot of the Raso (polje) of Urbasa (Navarre). Pocket valleys have steep walls and an abrupt cliff at the headwaters.

Dry valleys lack fluvial channels in their beds and are found primarily at the headwaters of fluvial systems. They are common in many karst areas. Dry valleys in limestones have steep slopes and flat floors. These valleys are favorable for sinkhole development because water dissolves the limestone by fissuring into these semi-confined zones (Sweeting, 1972) *(Fig. 6.37)*. They can form in several ways: (1)

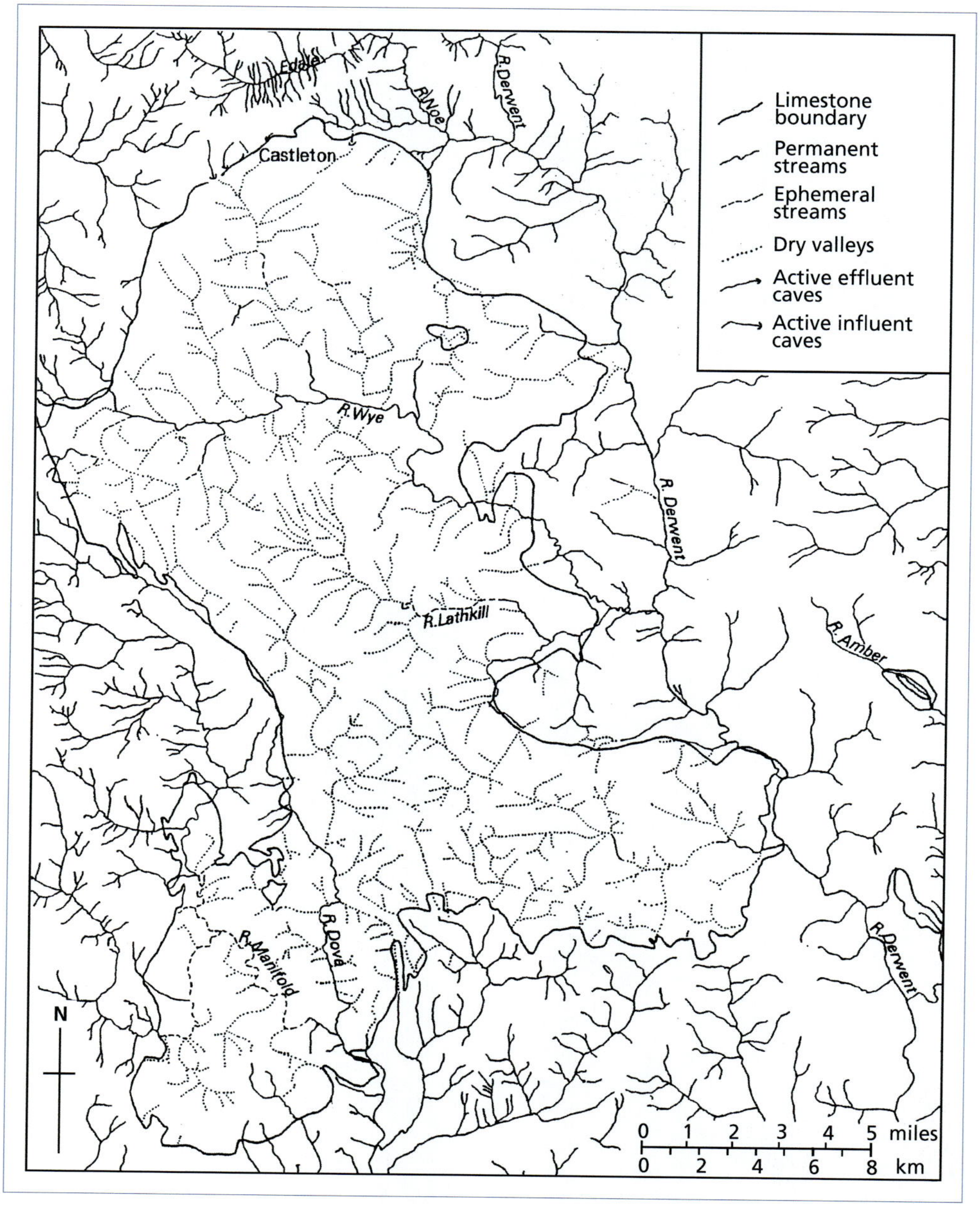

Figure 6.37 System of dry valleys in Carboniferous limestones in Derbyshire (Warwick, 1964).

different climates in the past that had higher precipitation or permafrost, (2) superposition of non-karst strata together with karstification of drainage, and (3) lowering of the water table due to uplift and incision of the principal valleys or scarp retreat (Gunn, 2004b).

Karst springs are places at the surface where water emerges from endokarst. Discharges range from a few mm^3/sec to values in excess of 20 m^3/sec. The karst spring with the highest known flow is Tobio in Papua New Guinea, which discharges >100 m^3/sec. In general, karst springs are found in low topographic positions (Smart and Worthington, 2004a).

Karst springs are categorized as two principal types: those in which the water surges to the surface by free flow and those that flow by forced or artesian flow. The latter are called vauclusian springs. Free-flow springs are similar to those in other non-karst areas although karst flow tends to be much larger. Springs commonly flow from caves (pocket valleys) but they sometimes flow from impermeable layers intercalated between the limestone series (Sweeting, 1972).

Vauclusian springs *(Fig. 6.38)* flow from outcropping or covered limestones under pressure and forced flow such as the Fontaine de Vaucluse in Avignon (France); its flow ranges from 8 to 150 m^3/sec (Trombe, 1974). The water comes from large reservoirs through a highly inclined conduit. An **intermittent spring** contains irregular conduits and siphons *(Fig. 6.38)*. As the figure shows, above a certain water level, the water flows. When the siphon ceases, the spring does not flow (Trombe, 1952).

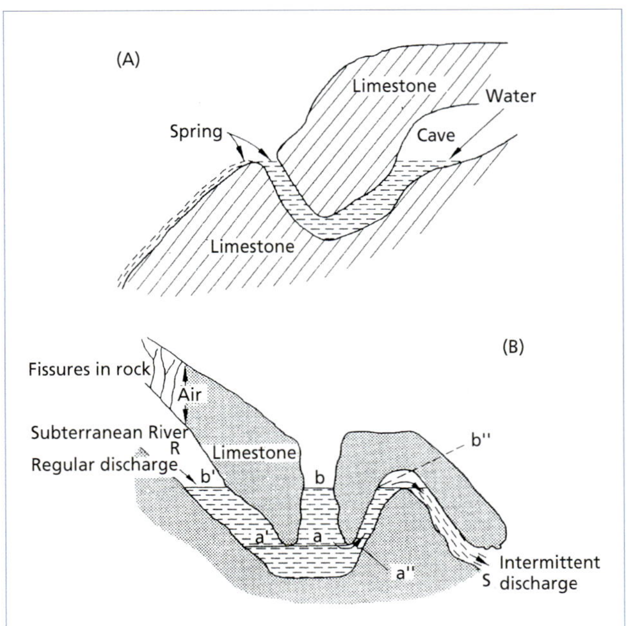

Figure 6.38 (A) Vauclusian spring. (B) Intermittent spring (Trombe, 1952).

Scientists can determine the origin of karst spring water and its karstic or autochthonous origin by its $CaCO_3$ content, which requires knowledge of its subterranean circulation. To understand circulation, **tracers** added to the water at specific points are used to determine the outputs. Numerous types of tracers have been used to study subterranean flow; they include natural, chemical, radioactive isotopes, stable isotopes (Gèze, 1965). The most novel technique is the use of quantitative fluorometry, which involve the use of WT rodamine and LT fluorescein, among others (Ford and Williams, 1989). The majority of karst systems are dendritic; that is, they have various tributaries that feed the principal flow. However, parallel systems have also been identified. As a result, residence time and pathways vary in subterranean flow systems (Smart and Worthington, 2004b). The largest subterranean river is the Trebinjcica (Herzegovina), which has five successive subterranean pathways with a total length of 100 km. It finally flows into the Ombla de Ragusa (Trombe, 1974).

6.7 Endokarst

Endokarst is subterranean karst where erosion occurs in thick formations of lithified soluble rocks. Caves are surface openings that provide access to underground cavities that are longer or deeper than caves. The study of these cavities is a separate discipline called **speleology**, which can be a hobby or a science (Gèze, 1965). Understanding the karstification process requires a basic knowledge of the characteristics, evolution, and origin of endokarst landforms. Numerous genetic classifications of karst cavities have been proposed (see Klimchouk, 2004a) but none of them adequately describes all types of cavities. The most common and important cavities are created by dissolution. Ford and Williams (2007) and Klimchouk (2009) distinguish three genetic types of cave dissolution: 1) syngenetic/cogenetic (coastal and oceanic) in rocks that have high porosity and permeability, 2) hypogenic and confined, primarily where water penetrates into the soluble formation from the bottom, and 3) hypergenic (epigenic), not confined, where water recharge comes from the surface. However, processes other than karst can generate other types of cavities, such as volcanic, glacial, littoral, and suffosion cavities, among others. (White and Culver, 2005).

Dissolution cavities are widely distributed in limestone, gypsum, and, to a lesser extent, in chlorides. In addition to dissolution, the dominant process, other processes such as erosion and those resulting from gravity rupture also generate cavities. The cavities or caves generally do not exceed 1 km in length and 100 m in depth. The largest cave in the world is Mammoth Cave in Kentucky, United States, which is 563 km long. The deepest is the Gouffre Mirobda (–1,733 m) in the Alps near Chamonix, France. The largest subterranean cave, in both surface and volume is the Sarawak Chamber (Malaysia), with a surface of 160,000 m^2 and a volume of 12 million m^3, (Klimchouk, 2004a).

Cave morphology is highly variable. The principal endokarst landforms, galleries and chambers, may be wider and taller overall than adjacent galleries. Galleries may be straight, sinuous, horizontal, inclined, and vertical. Vertical galleries are called shafts. Caves can have a variety of shapes: a single conduit, branched, anastomosing, etc. Dissolution cavities are controlled primarily by hydrogeologic factors, recharge type, and flow (Palmer, 1991, 2000).

Several factors are known to affect cave formation. An experienced researcher can infer significant data on the stages of landform formation from the study of endokarst. These episodes also affect the surface. As a result, from a geomorphic point of view, the value of endokarst landform study is the preservation of stages of the erosional history of an area, especially those that do not remain at the surface (Sweeting, 1972).

6.7.1 Principal factors in cave formation

The factors that affect endokarst landforms are the following, according to Sweeting (1972).

6.7.1.1 Limestone type and structure

Limestone is normally stratified and jointed; bedding planes and joints are the primary channels for water penetration. Also, cavities frequently form in contact with impure layers in limestone. The degree of fissuring is of great importance in terms of hydraulic conductivity. Cave shapes are influenced partially by the inclination of stratification and the degree of jointing (Gillieson, 1996). In horizontal limestones, caves are short and wide. Caves that originate along fracture planes tend to be tall, narrow, and sinuous.

6.7.1.2 Type and amount of water flow

In a cavity, dissolution can be caused by **phreatic flow** (in French, *eau forcé*) in which the entire cave is full of water, or by **vadose flow** (in French, *eau libre*). Vadose flow occurs above the water table whereas saturated flow occurs below the water table. Intermediate flow, also called **paraphreatic** or **epiphreatic,** occurs when water alternates between saturated and vadose flow. Saturated and forced flows have circular or elliptical cross sections because these shapes offer the least resistance to infiltrating water (*Fig. 6.39*). Conduit width does not vary much in homogeneous rock

Figure 6.39 Circular cross section of phreatic flow in an outcropping cave where present-day flow is vadose. Resurgence of the Gato Cave. Río Guadiaro. Hundidero-Gato Karst System (Delannoy, 1998). Grazalema Sierra. Province of Cádiz, Betic Cordilleras, Spain.

but has an elliptical shape in the presence of bedding and vertical planes (*Fig. 6.40B, C*). If saturated flow becomes vadose flow, the free water incises into the floor of the cavity (*Fig. 6.40H, I*).

Bretz (1942) described a set of landforms created by saturated flow in which he clearly distinguishes the various morphologies. These forms are described in Sweeting (1972), Bögli (1980), Jennings (1985), Williams (1983), Ford and Williams (1989, 2007), and Lumberg (2005). These authors distinguish **alveolar** or **honeycombed shapes** that occur on walls and roofs and form complicated systems of small centimeter-sized voids; they result from differential solution. **Pockets** are larger than alveolar shapes; their depth is greater than their width. **Anastomosing** shapes are twisted

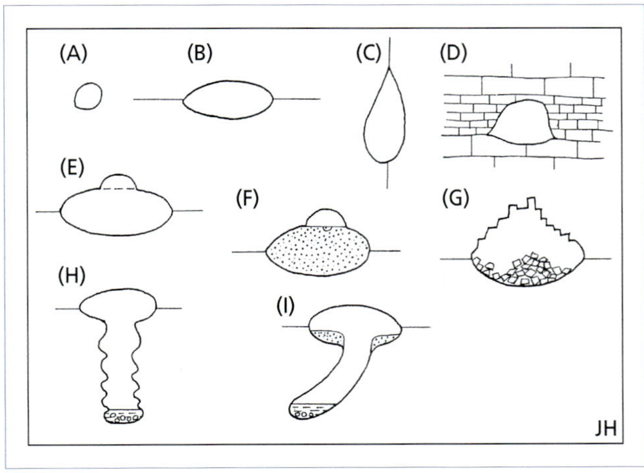

Figure 6.40 Cross-sections of caves, according to Renault (1958), Bögli (1956) and Ollier and Tratman (1969) (compiled by Jennings, 1985). Phreatic conduits: (A) circular in massive rock; (B) elliptical in horizontal stratification; (C) elliptical in vertical fissure; (D) in a group of more soluble layers; (E) half-tube roof due to bubbles; (F) aggradation to form the half-tube roof by dissolution; (G) modified by collapse in vadose flow; (H) incision of the floor of a phreatic conduit; (I) meandering incision in a phreatic conduit.

cavities that are intensely interconnected. They are about a centimeter in size and occur on bedding planes. **Pendants** are rock pillars between anastomosing channels; they can develop along bedding planes and joints with anastomosing shapes. **Pockets of reverse dissolution**, which can form on roofs and walls, form only under saturated conditions. Many occur at points where water converges; they are thought to form as a result of corrosion by the water mixture. The depth and diameter of these latter forms range widely from several centimeters to 20-m chimneys. **Scallops**, which are shaped like shells, are found on roofs and walls; they form as a result of saturated flow (*Fig. 6.41*). They are generally uniform in size and occur in groups. The steeper slope of the shell occurs downstream, and thus can be used to indicate the direction of water flow. According to Curl (1966a) the length of the shell is inversely proportional to the velocity and viscosity of the fluid. Bretz (1942) thought they are a product of dissolution, but Bögli (1980) believes they are formed by sand-bearing water.

6.7.1.3 Physiographic and climatic factors

It has been observed that complex systems in branched caves are characteristic of thin limestone sequences and

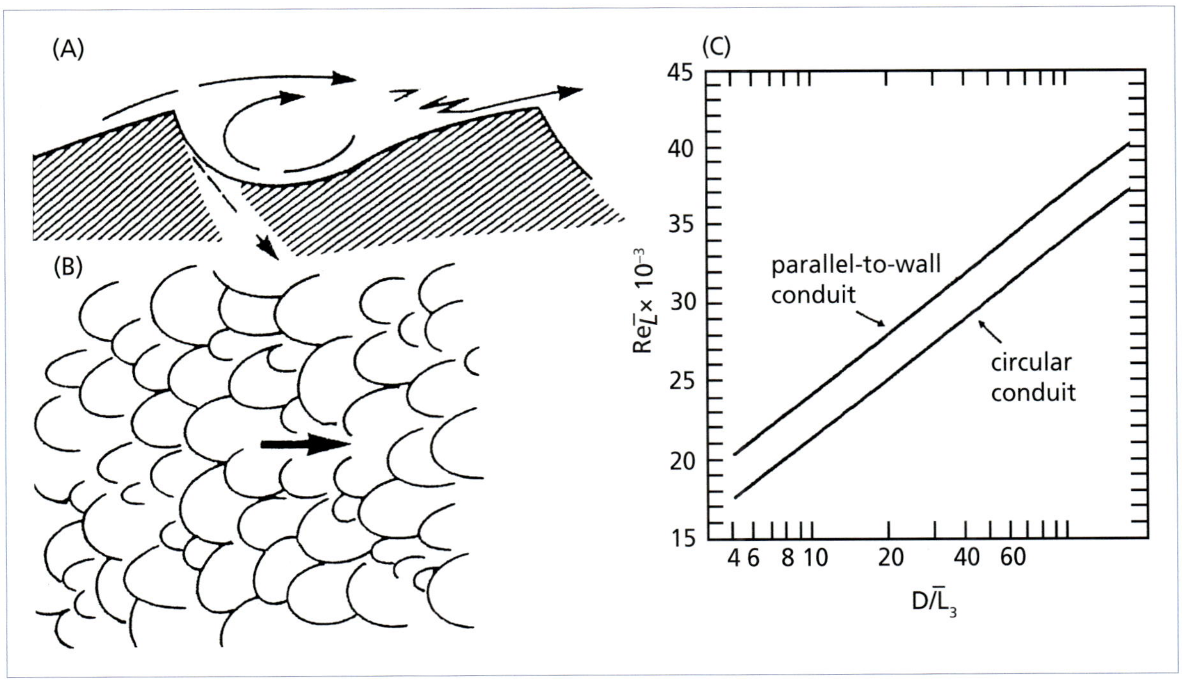

Figure 6.41 Shells or scallops (A) Section perpendicular the shell: (1) Flow of the current. (2) turbulent eddies = location of maximum dissolution. (3) diffusion, mixture, and separation (4) trajectory of the most inclined side where later dissolution occurs (the shell migrates downstream). (B) Characteristics of a fully developed shell system. (C). Relationship between the Reynolds number and the quotient of width or diameter (D) and length of the shell (Curl, 1974).

surface karst whereas deep karst is typical of thick limestone formations, like the Dynaric Karst. Relative height and existing relief are important factors in cave development.

Carbonate rocks have bedding planes that allow water to penetrate. When dissolution begins, the holes are small (<10 μm to 1 mm) and they gradually enlarge by dissolution. Caves can be predominantly saturated or vadose in morphology.

Many models have been proposed to explain endokarst evolution. Ford and Williams (1989) propose a four-stage model of **cave evolution** (*Fig. 6.42*), which is easy to understand in steeply dipping limestone. In stage 1 (bathyphreatic caves), the density of penetrable fractures is very low and the piezometric surface is stable. In stage 2, (phreatic caves with multiple curls) caves form when there are significantly more fractures. Caves with a mixture of shorter and shallow curls and an almost horizontal water table are characteristic of stage 3, with intense fissuring and less strength (caves with a mixture of phreatic components and leveled water table). In other words, the cave has a mixture of curls and passages that gently level toward the water table. In stage 4, (ideal water table cave) the fissuring is so intense and the strength so low that the cave develops along the water table.

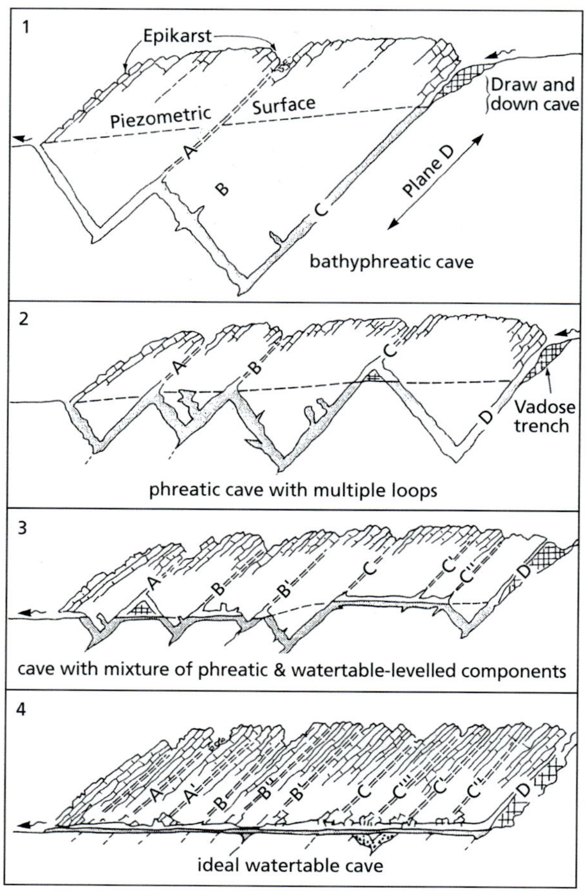

Figure 6.42 Four-stage model showing the difference between the geometry of an unconfined phreatic cave system and phreatic level caves (Ford and Williams, 1989).

Ford (1965) analyzed the temporal evolution of the Mendip Caves, in which sea level is the base level. He proposed a multiphase model as a function of eustatic variations and sea level lowering, in which cavities deepen and others remain hanging and disconnected. A relationship between valley development and cave formation has been demonstrated in many areas such as the Demanova Caves, Slovakia (Droppa, 1966).

Climate affects cave formation through temperature and rainfall variations. Endokarst is more common in humid areas than in dry areas. The presence of karst caves in arid zones shows that the climate was wetter in the past. Low temperatures promote CO_2 dissolution and, as a result, enhance endokarst development. However, the process of frost shattering affects limestone in periglacial climates; caves formed in wet climates are destroyed during periglacial stages. The foregoing indicates that precipitation and temperature fluctuations throughout the Quaternary have considerably affected endokarst development.

6.7.2 Some cave types

Curl (1996b) proposes a karstification index based on the number of caves present in an area. Sweeting (1972) considers this index an indication of fluvial activity in a limestone terrain. Cavities are difficult to classify even though tens of thousands of caves have been explored. Ford and Williams (1989) classify them in two large groups: one group based on their internal characteristics and the other on their relationship with external factors. Each group can be subdivided into several types. In addition, we do not know the depth of the deepest cavities because it is still not possible to dive into water-filled caves deeper than about 200 m. Other caves have very narrow passages that speleologists cannot enter.

Perhaps the most satisfactory classification is based on flow type (Sweeting, 1972). An example of a phreatic cave is Carlsbad Caverns (western United States) (*Fig. 6.43*). It is 330 m deep and occurs in thick limestone. Its origin is unknown; a large part of the dissolution took place in phreatic flow and is not related to an existing river. Bretz (1942) cites numerous examples of phreatic caves.

Many explored caves are the vadose type. The most famous is in the area of Postojna, in classic karst. The Postojna Cave and its tributaries are 17.6 km long but when associated caves are added, its length increases to 50 km. Its present day landforms indicate gravitational water flow, and it is generally more than 10 m wide and high. Rivers in caves adapt to geologic structures. Numerous examples of this type of cave have been described (Sweeting, 1972).

Vertical caves or shafts are cylindrical cavities with grooves on the walls; they occur along fractures or faults. The Yorkshire caves (England) are of this type. In addition to its numerous conduits, **Mammoth Cave** contains shafts developed in the limestone that underlies sandstones. In the western Pyrenees, significant karst has developed in Cretaceous formations where shafts (Fig. 6.44) have developed to depths of several hundred meters (López Martínez, 1986). This area is known as the Karst Massif of the Piedra de San Martín or Larra Karst.

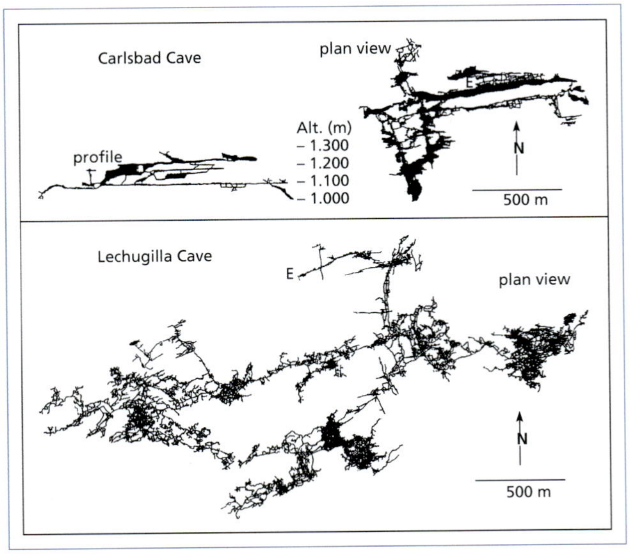

Figure 6.43 Plan view of the Carlsbad Cavern and Lechuguilla Cave systems, both at the same scale (Palmer, 2004).

6.8 Cave deposits

Clastic, chemical, and organic sedimentation occurs in active caves, even while they are undergoing significant erosion. For this reason, caves are excellent traps for sediments. Caves may contain fossils or datable artifacts; absolute dating can be done by isotopic methods. ^{14}C, amino acid racemization, and thermoluminescence can also be used. Cave deposits are classified as **autochthonous** and **allochthonous** (Sweeting, 1972; Ford and Williams, 1989).

It is important to distinguish between **clastic** deposits that form as materials enter the cave and those that originate within the cave. The former, which are much more complex, often contain archeological remains. Interior deposits may lack archeological material but can provide valuable information on the genesis and evolution of the caves (Farrant, 2004). Caves are sometimes full of sediments (Fig. 6.45). The most common autochthonous sediment results from roof collapse (Fig. 6.46). Most clastic sediments in caves are allochthonous, meaning that they originated outside the cave and were transported into it. Most processes, structures, and deposits found in caves are identical to those found in surface fluvial environments. These sequences contain repeated episodes of downcutting and fill, collapse, subsidence, and

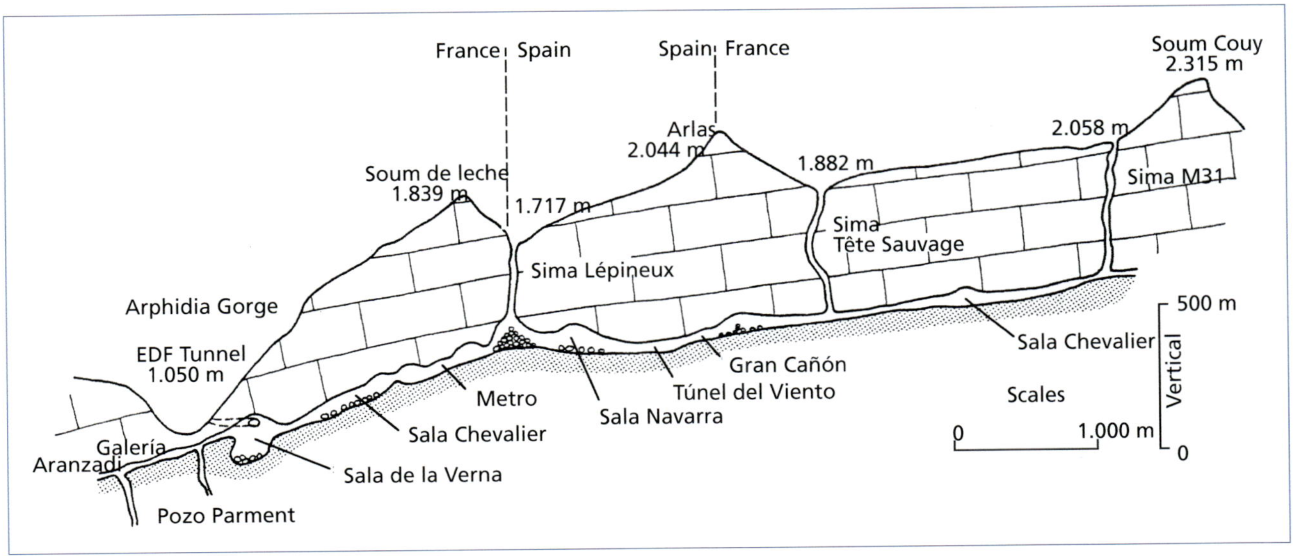

Figure 6.44 Schematic profile of the Piedra de San Martín system. Western Pyrenees (Lopez-Martinez, 1986).

Figure 6.45 Karst cave filled mainly with sandy-argillaceous material. Ancião. Lusitánica Graben Portugal.

Figure 6.46 Roof collapse in an endokarst cave. Wind Cave, Malaysia. Photo F. Gutiérrez.

re-sedimentation that can create complex stratigraphy. Sediments may become fluid and create sliding beds that penetrate far into the interior of caves (Gillieson, 1996).

Paleomagnetism can be used to date clastic sediments. Parés and Pérez-González (1995) describe the excavation of clastic deposits in the Gran Dolina (Atapuerca, Burgos). They discovered numerous human remains that had been deposited more than 780,000 years ago near the Matuyama-Brunhes boundary.

Speleothems intercalated in clastic deposits can be dated by the uranium series. Recent advances in the cosmogenic isotope technique provide another method of dating clastic cave sediments. Research on fluvial sediments in the upper levels of Mammoth Cave has determined incision rates of the Green River over the last 3.5 million years (Granger *et al.*, 2001).

Calcite precipitation produces spectacular decorations inside caves. $CaCO_3$ is deposited by sheets of water in laminar flow and by precipitation of water films during turbulent flow (Dreybrodt, 2005). Calcite is the principal secondary precipitate; it produces a landform called a **speleothem**. Carbonates can also be Mg-rich and hydrated. Determination of calcite speleothem growth rates is difficult due to complex environmental changes. Even so, rates obtained by ^{14}C dating range between 0.01 and 0.26 mm/year; those obtained by the uranium series range from 0.002 to 0.35 (based on various authors, Ford and Williams, 1989). Calcite grows by enlargement and coalescence of small crystals that twin along a certain axis. In caves, aragonite crystallizes as bunches of needles, called **antodites** or cave flowers.

Cave breccia is a calcareous deposit often found at the entrance to Mediterranean caves; it is composed of chaotic angular limestone blocks perhaps related to cold stages when gelifraction created these deposits (Butzer, 1965). Abundant limestone breccias within caves are produced by roof collapse.

Speleothems form by dripping and by water flow. **Stalagmites** (*Fig. 6.47*) have a simple shape, growing from the point of water supply toward the sides and the roof. Researchers can use computers to simulate their growth (Dreybrodt, 2005). Stalagmites are of three types (Franke, 1965): 1) those that have uniform diameter and grow toward the roof of the cavity, created by water of uniform concentration, 2) stalagmites with thick terrace morphology, and 3) the most common shape, a cone.

Flowstone (in French, *gours*) is deposited from a highly uniform flow that accretes on cave walls. Flowstone is most common on the floor or on slightly dipping walls; it can be several meters thick (Ford and Williams, 1989).

Stalactites, the most common type of speleothem (*Fig. 6.49*), form from water that drips from the roof. As the saturated or near-saturated water enters the cave, it loses CO_2 as it comes to equilibrium with the carbon dioxide concentration of the cave air. In dry caves, evaporation is an important factor in this process. As the drop emerges, the stalactite forms a calcite ring; this process slowly forms a cylindrical hole. As the cylinder grows, the water moves through the central conduit. If the water flows rapidly and does not have enough time for all the $CaCO_3$ to be deposited, the water drop falls and creates a **stalagmite**. Stalagmites can connect with stalactites to form columns (*Fig. 6.47*). Stalactites can also be found as draperies, which form from circulating water; they have no central conduit. Draperies are translucent or banded with different colors.

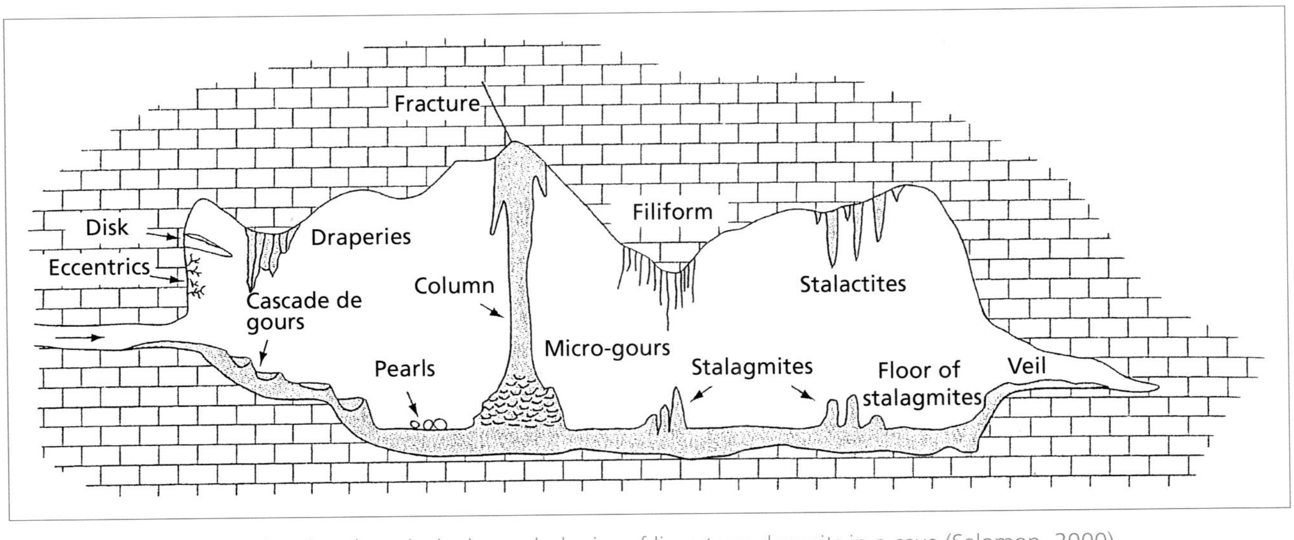

Figure 6.47 Diagram showing the principal morphologies of limestone deposits in a cave (Salomon, 2000).

Figure 6.48 Gour en Deer Cave. Malaysia. Photo F. Gutiérrez.

Gours (*Figs. 6.47* and *6.48*) are small pools with margins made of calcite dams, which give them a stepped appearance. Gours form as water flows over the dams in fine layers and creates eddies that promote ventilation, thus facilitating precipitation of calcium carbonate, which is near saturation (Géze, 1965).

Eccentrics (or **helictites**) (*Fig. 6.47*), which form from infiltrating water, make beautiful ornamentations. They have a sinuous morphology, and grow from a central orifice. Because infiltration is very slow, eccentrics do not form water droplets and thus gravity does not affect their shape. Precipitation occurs around the central orifice; each new layer forms as a cone that is connected to the next one, and as a result, the cones are distorted. Some eccentrics form below water and are different from subaerial eccentrics because their central conduit is not a capillary tube. Davis (2005) thinks that eccentrics grow through the interaction of near-surface groundwater and thermal fluids that rise via a system of fractures as a result of the common ion effect.

Cave pearls (*Fig. 6.47*) are concentric accretions of calcite that form around a foreign nucleus such as a grain of sand. They grow in groups and require gentle oscillation to cause spherical accretion; they range in size from 0.2 to 15 mm.

Calcite ponds are deposits that form on the surface of ponded water in very dry caves. The ponds of calcite crystals

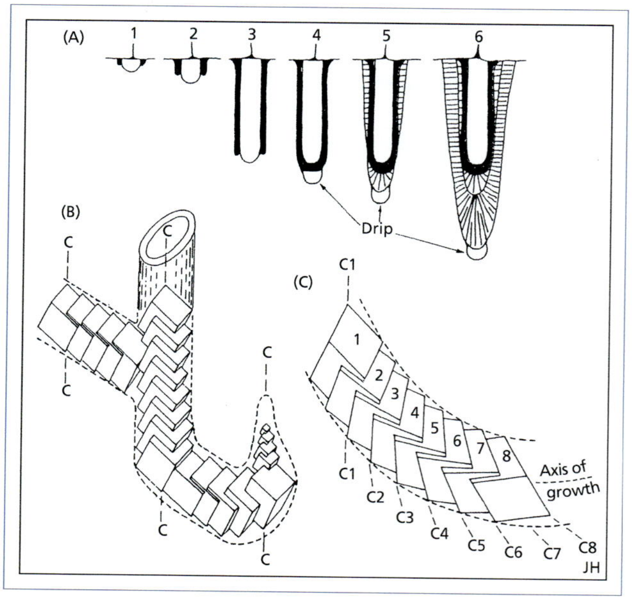

Figure 6.49 (A) Development of a stalactite that is occluded by capillarity and grows in the shape of a cone; (B) crystalline structure of an eccentric with abrupt changes in the direction of growth, and (C) curved eccentric produced by a gradual change in the direction of axis **c** (in Jennings, 1971).

are several tens of millimeters thick and are maintained by surface tension. Another type, acicular calcite crystals, form underwater and grow upward from the floor. Microorganisms can also generate deposits such as **milk** or *rock milk*, which is a white paste that turns to powder when it dries.

6.9 Types of karst and the role of climate

One of the most commonly used karst classifications is the one proposed by Cvijic (1924b). Cvijic's categories include the following types. **Holokarst** is a carbonate mass into which all rainwater infiltrates; it has no surface water flow. **Merokarst** involves impure limestones; dissolution and fluvial erosion are important processes. **Fluviokarst** is traversed by allogenous rivers that compartmentalize the karst mass.

Other authors such as Sweeting (1972) propose a classification based on climate and process: holokarst (true karst), fluviokarst, glacial karst (nival karst), tropical karst and arid and semi-arid karst. Jennings (1985) sets up nine types of karst based on climate.

One of the major debates among geomorphologists is the influence of climate on the development of karst landforms. Many researchers have considered these landforms to be a zonal phenomenon. In his study of karst in Java, Lehmann (1936) recognized that cone karst (*kegelkarst*) occurs specifically in the wet tropics. At the 1953 Karst Symposium in Frankfurt, Lehmann (1954) showed that each climate has its own type of karst development. Even so, a classification must consider the carbonate rock type, thickness, and degree of fissuration (Verstappen, 1964). It is difficult to imagine that the conic and tower shapes of tropical karst form only in carbonate rocks of similar thickness and the same degree of fracturing.

Most geomorphologists agree that different morphologies are a function of climate differences. But karst researchers have been unable to explain the causes of these contrasts. Climatogenic geomorphology, which may have arrived at the limit of its contribution, considers tropical karst as erosional plains with isolated hills (Ford and Williams, 1989). Other authors minimize the climatic parameter (Fenelon, 1974; Gams, 1992; Drogue and Bidaux, 1992; among others). Some researchers emphasize the nature of the limestone, its structure, vegetation, and soils. Salomon says that two conditions are necessary: average temperature higher than 17–18°C and average annual rainfall higher than 1000 mm. If these conditions are not present, typical tropical landforms do not form. These climatic conditions affect both soil and vegetation, and as a result, the production of biogenic CO_2 and organic acids.

Table 6.2 **AVERAGE MAXIMUM DISSOLUTION POTENTIAL (MDP) PER CLIMATE TYPE (GOMBERT, 1995)**

Climate Type	MDP mm/ky	Surface (%)	Number of Stations
Desert	3.4	23	20
Mountain	50.9	5	12
Polar	42.5	15	12
Cold	29.4	19	35
Cold Temperate	36.6	6	38
Hot Temperate	31.9	4	23
Mediterranean	88	2	26
Subtropical	21	9	24
Tropical	92.7	8	25
Equatorial	88.2	9	28
TOTAL/AVERAGE	40.6	100	243

Table 6.2 (Gombert, 1995) shows the maximum dissolution potential (MDP) in mm/ky for different climates. The MDP essentially takes rainfall, average annual temperature, and CO_2 production (aggressivity of the water) into account. The author applies his analysis to 243 representative meteorological stations. He uses them to infer that karst zonation although it varies according to the criteria used. The climate factor is the main one, followed by lithology and degree of fracturing (Salomon, 2005).

Scientists began to study **cold region karst** in the second half of the 20th century. The presence of intense hydrodynamic phenomena and the importance of chemical denudation became known during this period (Pulina, 2005). These findings are based on the work of the Polish scientist Pulina in eastern Siberia during the 1960s and studies of other permafrost-affected regions where chemical corrosion is as important as in temperate regions (Pulina, 1974; Muxart and Birot, 1977).

On continents, cold zones include alpine-type mountains in diverse latitudes and polar zones. Phenomena similar to karst also develop in alpine and subpolar glaciers. The knowledge of glacial caves goes back to the end of the 19th century to the research of the Frenchmen Martel and Forel.

Large karst landforms are rare in polar and subpolar regions. Karst in these areas are old fossil forms (paleokarst) that have evolved and been transformed; glaciers can preserve, inhibit, or destroy karst forms (Ford, 1983). A small-scale karst feature that forms under tundra vegetation is poorly preserved by the intense influx of the frost. Occasionally, groups of gypsum collapse sinkholes are found (Eraso *et al.*, 1995). They are related to areas with subterranean karst in gypsum or limestone that is affected by thermal waters. Small and medium karst features develop better in subpolar climates, as observed along the Scandinavian coasts.

Karst relief is very well developed in permafrost regions. Many closed depressions (thaw lakes, alas, etc.) will be discussed in the chapter on periglacial region geomorphology. They result from thermokarst or melting of permafrost. Even so, there are many additional karst landforms such as those described by Pulina and Trzcinski (1996), Eraso *et al.*, (1995) and Pulina (2005).

Karst relief in the ablation zone of glaciers differs from karst landforms in polar regions and permafrost. The landforms are similar to those in classic karst (Dinarides) but the growth rate is different (Pulina, 2005).

We should consider the effect of glaciokarstic processes on the formation of polar karst landforms and on endokarst in limestone masses. Barrère (1964) discusses the interaction of glacial and karst processes in the Pyrenees. Glacial cirques have been transformed by dissolution into closed depressions called cirque-dolines. Miötke (1968) described the same landforms in the Picos de Europa, Spain. Many closed depressions were formed from clastic detritus during glacial and proglacial sedimentation (Ford, 1979). In addition, the deepest shafts in the world are found in the Alpine domain (Salomon, 2000). Kotlici occur in this same environment when there is no permafrost and snowfall is seasonal. There are 900 dolines/km^2 in the Alps (Bauer, 1962).

Groundwater circulation is well developed in endokarst. Karst groundwater is present under polar glaciers and under permafrost. On Arctic islands, "thermal windows" in the permafrost provide an abundant water supply to aquifers. In spite of good circulation of polar water in karst regions, no caves of recent origin have been discovered. Entrances to old caves have been plugged by ice. However caves are a normal phenomenon in subpolar glaciers (Pulina, 2005).

Due to the low intensity of biochemical processes in Arctic areas, the amount of biogenic CO_2 produced is very small (Smith *et al.,* 1982). As a result, according to Pulina (2005), chemical dissolution rates are about 20 mm/ky less than the rate indicated by Gombert (1995). However, in specific geographical and geological situations (glacial melt waters or presence of juvenile CO_2), chemical dissolution can be intense and comparable to European karst of medium latitudes.

Karst in temperate regions is climatically equivalent to **holokarst** or Cvijic's (1924b) true karst. The most representative example known is Dynaric Karst (Roglic, 1974); a fringe of it of which extends 600 km along the Adriatic Coast to Albania and 100 km inward to the interior. It is covered with closed depressions; the surface of Montenegro is 64% covered with dolines. The carbonate massif is massive pure limestone that is 4,000 meters thick. The mountains are 2,000 meters high and rainfall in some places exceeds 2500 mm/year. There are no surface water flows; water circulates vertically. Horizontal caves are not typical (Sweeting, 1972).

Fluviokarst occurs in temperate regions (Cvijic, 1918, 1925; Roglic, 1960; Sweeting, 1972). It is characterized by the combined activity of fluvial and karstic processes. Most carbonate territory of central and Western Europe has fluviokarstic characteristics. In these areas, allogenous rivers compartmentalize the karst masses. The limestone thickness and outcrop surface is smaller because subterranean flow is not very well developed, as in holokarst, due to the fact that it is controlled by fluvial base levels. Significant mechanical erosion by fluvial channels is added to corrosion. Gorges, sinkholes, poljes, and blind, pocket, and dry valleys are common (Sweeting, 1992). Numerous caves form at the contact with an impermeable level. Sinkholes reach their major development on planar surfaces where drainage hierarchies are rare due to the flatness of these surfaces. Detritus from the decarbonation of impure limestones partially covers the surfaces of this type of karst.

Arid zones are defined by less than 500 mm of annual rainfall (Grove, 1977). Because there is little or no vegetation, **karst in arid regions** has very low CO_2 content, which inhibits karst development when rainfall is less than 250–300 mm/year (Sweeting, 1972). As a result, if we find carbonate karst morphologies in areas that have these rainfall amounts, we must interpret these landforms as having formed in periods of greater rainfall than the present. In arid or hyperarid environments, where rainfall is less than 200 mm/year limestone dissolution is practically negligible; it is limited to surface micro-dissolution, called vermiculation *(Fig. 6.7)*.

The lack of rainfall also means that waters are charged with soluble ions and that they precipitate by evaporation in very disparate environments. In addition, the pH in desert areas is very alkaline. The most important precipitation process is evaporation, which results in the loss of CO_2; as a result, the solution becomes concentrated and $CaCO_3$ precipitates (Pentecost, 2005).

Relative humidity has been measured in caves in arid and semiarid zones; it ranges from 38 to 100% (Jennings, 1983). In the Jerusalem region, cave stalagmites formed in endokarst that was wetter than the current one (Frumkin *et al.*, 1992).

Salomon (2005) cites an example of karst on the Nullarbor Plain (Australia) (Lowry and Jennings, 1974) where rainfall is 400 mm in the coastal area and 150 mm in the interior; karst has developed on Eocene and Miocene limestones. Large karst depressions and numerous kilometer-long caves are present. The Nullarbor limestones are porous and mechanically weak but the excellent development of caliches at the surface increases their mechanical strength with respect to collapse (Williams, 1978). Another example is the Adar Plateau (central Niger) where dolines and uvalas occur, some of them appearing to have been created by collapse. However, the present-day aridity of the Nigerian climate does not support recent genesis of these depressions so most likely they are inherited from wetter paleoclimatic periods.

The most significant differences between **karst in tropical regions** and karst types in other climates are related to a) more rapid dissolution of carbonates in tropical climates because of the higher biogenic CO_2 content (Birot, 1954, 1966; Ford and Williams, 1989), b) intense gullying of slopes as a result of heavy rainfall, which creates irregular closed depressions, and c) high evaporation, which causes very hard calcium carbonate to precipitate on steep slopes, which is often cited as the principal cause of their high inclination (Sweeting, 1972). These landforms develop primarily in tropical areas where limestone crops out in mountain chains: Greater Antilles, some regions that ring the Caribbean Sea, Southwest Asia, Indonesia, New Guinea and some African areas.

Tropical karst landforms are generally divided into two types: **kegelkarst** (**cone karst**) and **turmkarst** (**tower karst**). The distinction is based on the shape of the hills. In the future the classification may be based on the shape of closed depressions, which seems logical (Sweeting, 1972). Balazs (1973) proposed a classification of tropical karst based principally on a height/width ratio. He distinguished three types: doline karst, cone karst, and tower karst. Hills of cone karst are 30–120 m tall and occur with a density of 15–30 hills/km^2 whereas tower karst ranges from 100 to 300 m high and has a density of 5–10 hills/km^2.

Various factors determine the occurrence of these landforms. The type of limestone appears to play an important role. Cone karst hills in Jamaica occur in highly jointed limestones (Sweeting, 1958, 1972) whereas the towers of Puerto Rico (Monroe, 1964b) and those of Sarawak (Malaysia) (Wilford and Wall, 1965) develop in crystalline limestone with intercalated soft layers. The direction and dip of the layers affects landform development; the hills elongate in one direction, like the depressions that occur at their feet (dolines, uvalas, poljes, and valleys). In addition, fracture planes also affect the development of this landform.

There is no clear relationship between total precipitation and the type of tropical karst that develops. Verstappen (1964) found that precipitation of at least 1500 mm was necessary for the formation of cone karst in New Guinea. The cone karst region of Shuicheng (China) has average annual rainfall of 1230 mm (Sweeting, 1995); in the karst tower area of Guilin, the most spectacular of China, annual rainfall measures 1870 mm. To make comparisons, it is also necessary to know the seasonality and intensity of precipitation and the temperature because altitudinal variations clearly affect the formation of cone karst.

The network of open joints that controls surface drainage also promotes water infiltration (Salomon, 2005). In addition to allogenous rivers, which usually have significant flow, the drainage network is composed of dry valleys, blind valleys that disappear into sinkholes, and pocket valleys that grow from a resurgence. Accompanying surface forms are free and covered karren, dolines of various types, uvalas, and poljes. Rapid and high infiltration of water into the interior of the karst mass creates an important system of galleries that are predominantly horizontal; they contain a long and complex history of the development of tropical karst (Barbary *et al.*, 1991; Sweeting, 1995). These cavities form more easily in areas where the water table fluctuates, which is where greater dissolution occurs (Jennings, 1985).

Cone karst is characterized by numerous conical and hemispheroidal hills (*Fig. 6.17*) separated by closed depressions called **cockpits** in Jamaica (Sweeting, 1958). In Java, the density of these hills is 30 cones/km^2 (Lehmann, 1936). Cone height varies throughout the world; the highest occur in Jamaica where they may be 100–130 m high and up to 500 m in diameter (Day, 2004a). The slopes are inclined between 40° and 60°. Cones may be symmetrical or asymmetrical; they are controlled by the dip of the layers or by fracturing. The slopes are frequently covered with a highly resistant limestone crust that can be 5–10 m thick, as in Puerto Rico (Monroe, 1966). Cone karst forms after short storms in which the CO_2-rich water dissolves, evaporates and precipitates $CaCO_3$, especially on windward oriented slopes. The crest of the cones may host elongate depressions up to 100 m in length, called zanjón in Puerto Rico, which form as a result of dissolution and enlargement along a fracture plane; for this reason it is considered a structural karren.

Cockpits are star-shaped closed depressions that develop at the foot of cones; they are separated from each other by saddles. This landform results from intense tropical storms that make gullies on the slopes. Morphometric research on cone karst in New Guinea seems to confirm this interpretation (Williams, 1972). Williams calls it polygonal karst after the polygons resulting from their morphometric analysis. Sometimes cockpits are oriented along fracture lines. When several cockpits coalesce, they create a **glade** (Sweeting, 1958, 1972), a Jamaican word equivalent to an uvala in temperate climate karst. These glades may have an irregular floor with small basins.

The origin of cone karst is not well understood but the ideas of Lehmann (1936), originally developed for the karst of Java (*Fig. 6.50*), have been partially accepted. The process begins with a flat surface that bulges due to endogenetic

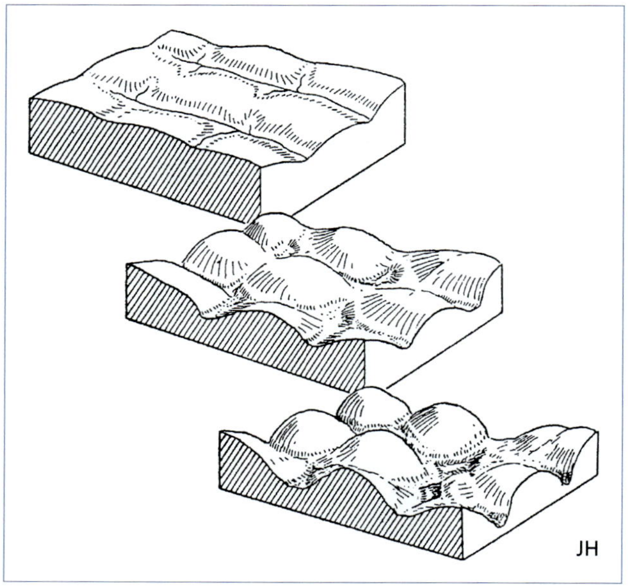

Figure 6.50 Cone karst evolution in Goenoeng Sewoe, Java (Lehman, 1936).

activity. This variation of relief rejuvenates drainage, as shown by the strong incision of gullies in the limestones. During a later stage dismemberment occurs, with closed depression and cone formation because chemical dissolution has a greater effect than fluvial erosion. This hypothesis is accepted primarily for the later stages of the evolution. It is possible, as Sweeting (1972) points out, that better knowledge of the depressions may provide new data to help resolve the formation of cone karst.

Karst towers up to 200 m high and 500 m in diameter (Day and Tang, 2004), also called pinnacle karst and in the Spanish-speaking Greater Antilles, mogotes karst or *(Fig. 6.51)* or pepinos. They occur in the Caribbean, Indonesia, New Guinea, and especially in Southeast Asia *(Fig. 6.52)*. In southern China and northern Vietnam karst towers cover an area of 600,000 km^2 (Sweeting, 1972). Karst towers are steep-sided hills, sometimes isolated and sometimes in groups, that can reach heights of 300 m. A river, lake, or a flood plain usually surrounds the towers.

There are two principal types of karst towers (Sweeting, 1972). The first consists of flood plains that

Figure 6.51 Mogotes in karst in the Sierra de los Organos, the westernmost sierras of Cuba. The plain is part of the Viñales polje. Photo J.L. Peña.

overlie limestones from which the carbonate towers protrude (Fig. 6.53). The other type includes towers that stand above plains of non-carbonate rocks, such as the granite in the center of Jamaica (Sweeting, 1958, 1972) and Malaysia (Jennings, 1985). The karst towers of Guilin (China) are the most spectacular examples; they occur on top of a 2600-m-thick formation of mid-and-upper Devonian and upper Carboniferous limestone (Sweeting, 1995). There are two varieties (Yuang, 1987): *fengling*, which is more extensive, consists of isolated towers with steep sides that protrude from flat limestone surfaces covered with Quaternary alluvium. *Fengcong*, the other type, are composed of an assemblage of grouped towers that are elevated above a rocky basement. The shape and degree of inclination of the slopes depend on the type of limestone. Sharp isolated peaks form in crystalline limestones and dome shapes occur in dolomites. In the Guilin area, statistical research on the towers shows that their average height is 74 m and their density is 1.23–1.59 towers/km^2.

Figure 6.52 Tower Karst. Mulu. Malaysia.

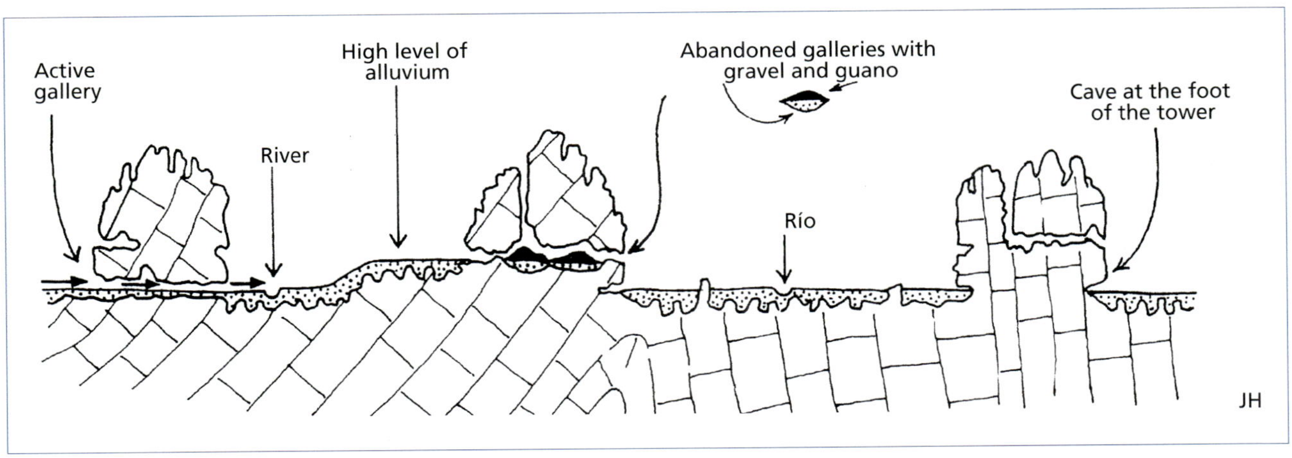

Figure 6.53 Schematic section of tower karst in the Kinta valley, Malaysia (Jennings, 1985).

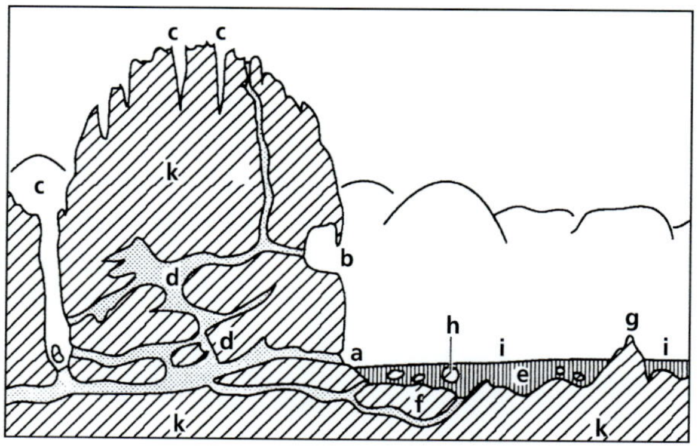

Figure 6.54 Morphological characteristics of tower karst of the Sierra de los Organos (Cuba). a: active cave at the base of a tower; b: abandoned caves with speleothems; c: deep karstified fractures; d: vadose caves related to old plain levels; e: alluviated valley floor; f: sub-fluvial phreatic caves; g: limestone promontories; h: loose limestone blocks; i: surface of a frequently flooded plain; k: limestone tower (Lehmann, 1955).

The *fengcong* towers are located in elevated areas where the inclination of the slope ranges from 30°to 70° depending on the rock type. Depressions of up to several kilometers in length occur between peaks. The slopes of these depressions are steep and are interpreted as ancient river valleys (Sweeting, 1995).

In general, fracturing controls karst towers; it affects these monoliths by developing numerous bogaz and very deep corridors (Fig. 6.54). Sinkholes with steep sides have also been observed; these are possibly the result of cavity collapse. $CaCO_3$ precipitation occurs on the surface of the towers, creating calcareous crusts 1–2 m thick, which are highly resistant to erosion; these were first studied in Puerto Rico (Monroe, 1964b, 1966). This crust protects the tower from karst corrosion because the slow evolution of the slope enables these landforms to persist.

Lakes, swampy zones, and floodplains develop near the feet of the *fengling* type of tower. Intense geomorphic activity occurs at the contact with the base of the tower. Researchers have observed horizontal grooves hundreds of meters long near the foot of the scarp, indicating old lacustrine and fluvial levels. Overhanging slopes, which usually collapse, are associated with the grooves. But the most common feature is the presence of caves at the tower-plain contact with vadose or groundwater springs that accelerate limestone dissolution in the area and steepen the slopes as a result. This basal incision zone may be emphasized by fluvial erosion. If this activity ceases, the angle of the basal segment of the slope diminishes (Gerstenhauer, 1960) and the tower degrades.

In the Guilin karst (south China), caves commonly form in crystalline limestones. They are generally horizontal galleries with numerous large collapse chambers of different heights. Three levels of caves can be seen at $\pm$ 0, 7–15 and 25–35 m above the water level throughout the Guilin region, and other gallery systems have formed at 50, 90, 150, and 300 m. The lowest caves are the most recent and those located at higher elevations are older. This shows that the towers grew as the plain was lowered, primarily by karst corrosion. The development is thought to have begun in the upper Pliocene. The gallery at 90 m contains human remains from the mid Pleistocene (*Gigantopithecus*). Numerous absolute age dates are available, primarily done by U/Th and ^{14}C. Paleomagnetic research indicates that the sediments in one cave are at least 900,000 years old (Sweeting, 1995).

The plains originate by flattening as limestones are dissolved below alluvial sediments or lacustrine or swampy zones. This flattened surface is controlled by oscillations of the water table (Ford and Williams, 1989). The plains compare to marginal plains or margins of polje floors in temperate regions. Thomas (1994a) and Twidale (2004) consider their evolution to be similar to that of an etch plain where the lowering of the surface of the plain is accompanied by tower growth.

6.10 Evaporite karst

Evaporite deposits form in seas, lagoons, and interior lakes where evaporation exceeds surface and underground runoff. A concentrated solution can precipitate as a function of its solubility product and can form evaporites such as gypsum, anhydrite, and calcite (Klimchouk, 2004b). Gypsum and anhydrite cover large expanses of the Ukraine, the United States, and Europe as diapirs, stratified deposits (interstratal karst as in the Ukraine), and beneath an alluvial mantle (covered karst, Zaragoza, Spain). Halite karst can be seen only in very arid regions when salt diapirs pierce the surface, as at Mount Sedom, Dead Sea (Israel). Research on karst processes in evaporites has increased since the 1970s (Klimchouk, 2004b). Evaporite dissolution is much simpler than carbonate dissolution processes. Dissolution depends on the solubility coefficient, the interface between the solid and dissolved phase, contact time, and temperature (Jakucs, 1977).

The principal factors governing karst processes in evaporites are discussed in Gutiérrez and Gutiérrez (1998), Klimchouk (2000b), Jeschke *et al.*, (2001), Gutiérrez *et al.* (2006), Guerrero (2008) and Lucha (2009). The factors are: (1) composition of evaporites and adjacent aquifers (lithology and mineralogy); (2) structure and texture of soluble rocks and adjacent aquifers; (3) amount of flowing water in contact with evaporites and their physico-chemical properties (including saturation index and temperature); (4) flow regime and groundwater conditions (laminar or turbulent, phreatic or vadose); and (5) variation of the water table (or piezometric level).

6.10.1 Evaporite dissolution

Halite (NaCl) solubility is very high and is not very temperature dependent. Its solubility is $6.7 \cdot 10^{-1}$ mmol cm^{-2} s^{-1} (Alkattan *et al.*, 1997). The most common evaporite is gypsum ($CaSO_4 \cdot 2H_2O$), which dissolves rapidly at rates between $1.5 \cdot 10^{-5}$ and $5 \cdot 10^{-5}$ mmol cm^{-2} s^{-1} (Navas, 1990), whereas the value for anhydrite ($CaSO_4$) is $6 \cdot 10^{-6}$ mmol cm^{-2} s^{-1}. When anhydrite comes in contact with water, two molecules of water bond with the crystalline structure to form gypsum. The process is accompanied by a volume increase of 1.557 times (Biese, 1931). This increase can significantly deform the rock structure and can fold and fracture the rock. The temperature-based solution curves of gypsum and anhydrite are bell-shaped with the maximum dissolution of gypsum at 40°C and at 58°C at 1 atm but gypsum loses its crystallization water as it is transformed into anhydrite (Blount and Dickson, 1973). Halite solubility increases with temperature and the common ion effect decreases its solubility (Frumkin, 2000a).

6.10.2 Karst landforms in evaporites

Evaporite deposits form in seas, lagoons, and endorheic lakes when water evaporates and precipitates its salts as a solubility product. The most common evaporites are gypsum, anhydrite, and halite and other associated salts. Halite is difficult to find in outcrop because of its high solubility; it can be found only in highly arid environments. It crops out in regions like Death Valley, the Dead Sea, in Qinghai (Tibetan plateau), in the Canadian islands in the high Arctic (Ford and Williams, 1989), or in open pit salt mines. Salt in these arid environments forms karren, sinkholes, shafts, and endokarst caves.

The best-studied example of salt dissolution is Mount Sedom near the Dead Sea, where precipitation has been recorded at about 50 mm/year (Frumkin, 1994, 2004; Frumkin and Ford, 1995). The salts form part of an elongate diapir (11 by 1.5 km) composed of upper Cenozoic salts; the diapir is growing at a rate of 6–7 mm/year (Frumkin, 2000a). As the salt is exposed to the surface, chlorides are dissolved, leaving a tabular mountain 50 m thick composed of anhydrite, dolomite, and clastic sediments with a nucleus dominated by chloride salts (Frumkin, 2000a). On the surface of the cap,

Figure 6.55 Entrance to a halite karst cave showing collapses and blocks on the floor. Mount Sedom. Dead Sea. Israel.

collapse sinkholes and shafts have developed, some of which are connected to a significant network of subterranean galleries (*Fig. 6.55*). The collapse of salt and gypsum are comparable to that of limestones but in salt and gypsum, collapses occur with less mechanical strength. Breccia pipes are related to structures formed by deep dissolution (Waltham, 1989; Waltham *et al.*, 2005). The caves are Holocene (Frumkin *et al.*, 1991) but the oldest have been gradually destroyed by dissolution and collapse. The cave network is about 90 km long (Frumkin, 2000b). The outcropping chlorides have spectacular karren, sometimes covered with desert dust.

The salt diapir of Cardona (northeastern Spain) also has splendid karren (*Fig. 6.56*). This diapir, with a surface area of 0.9 km^2, is the largest outcrop of salt in Western Europe. The saline rocks are cut by a meander of the Cardener River. Endokarst is well developed and the best-known cave is Forat

Figure 6.56 Rillenkarren in halite with fine intercalations of clay. Salt mines of Cardona, Cataluña, Spain. Photo F. Gutiérrez.

Micó, discovered in 1967, which is 600 m long (Cardona, 1989–90; Cardona and Viver, 2002; Gutiérrez *et al.*, 2004, 2006). On the basis of an inventory of 178 sinkholes, Lucha *et al.*, (2006b) estimated the minimum probability of formation is 4.7–8 dolines/km^2/year.

The Djelfa salt dome (Algeria) (*Fig. 6.57*) is located in an arid zone on a platform at about 1000 m; the platform has numerous sinkhole fields with a high density of well-shaped sinkholes and resurgences near the contact between the salt and the surrounding rock (Salomon, 2000).

Evaporite speleothems are controlled by evaporation. Halite speleothems are much less abundant than those of sulfates (Forti and Hill, 2004). Evaporite speleothems are similar in shape to calcite speleothems but their structure is generally fibrous and filamentous. Evaporite stalactites deviate in the direction of the air current where maximum evaporation occurs (Frumkin and Forti, 1997).

Outcropping karst and karst covered in gypsum have shapes similar to karst in carbonate rocks (Waltham *et al.*, 2005). Gypsum outcrops (*Fig. 6.58*) persist only in arid climates; dry flat-floor valleys shape the fluvial system. Karren that form on gypsum have the form of rillenkarren and knupkarren (*Fig. 6.59*), although kamenitzas (tinajas) and klufkarren are also present (White, 1988; Macaluso and Sauro, 1996).

The size of solution sinkholes in gypsum range from 100 m in diameter and a few meters deep (Sauro, 1996); they forming closed subsidence depressions (bedrock sagging sinkhole). Sinkholes are not common in the extensive outcrops of Miocene gypsum around Zaragoza (Spain) but where gypsum is covered by alluvial material in small valleys or the floodplain of the Ebro River, underlying karstification is significant, with sinkhole development through processes of sagging, suffosion, and collapse (*Fig. 6.60*). Collapse sinkholes are common in the Ebro Depression (Gutiérrez and Gutiérrez, 1998) but southeast of Calatayud there are enormous dolines that measure 4.4 and 12 km^2 in which 200 m of Miocene cover has subsided (Gutiérrez, 1996).

Dissolution of the evaporitic substrate under the alluvium of the Jalón River in Calatayud and the Ebro and Gállego rivers during the Quaternary has caused significant subsidence of the alluvium, which may be more than 100 m thick in terrace deposits where significant angular unconformities can be seen (*Fig. 6.61*) (Gutiérrez and Gutiérrez, 1998; Benito *et al.*, 1998a, 2000; Gutiérrez *et al.*, 2001, 2006; Guerrero *et al.*, 2004, 2006). Also, where the gypsum of the Barbastro-Balaguer Anticline crosses the Noguera-Ribagorzana River (Spain), terraces are up to 110 m thick (Lucha *et al.*, 2008a).

Collapse sinkholes are common in the interstratal karst of the Ukraine (*Fig. 6.62*). The Iren River basin (Pre-Urals,

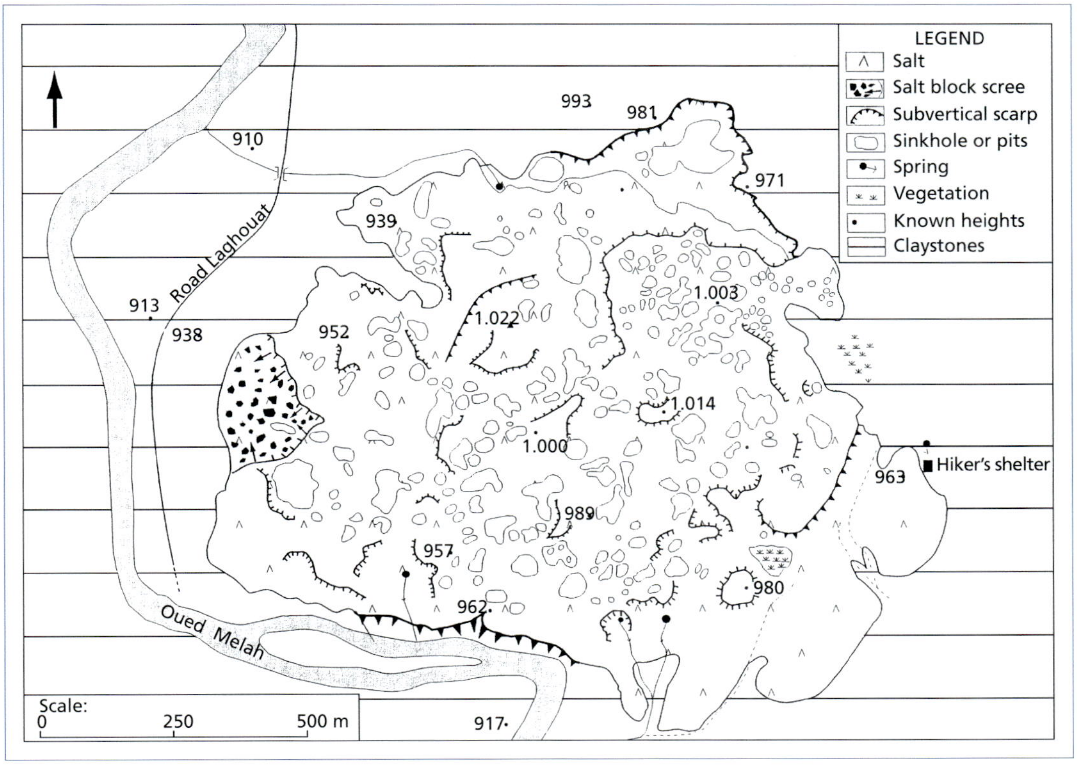

Figure 6.57 Geomorphic diagram of the Rocher de Djelfa (Algeria). (Salomon, 2000).

Figure 6.58 Outcropping Miocene gypsum affected by a fracture network. Near Calatayud, Province of Zaragoza. Calatayud graben, Iberian Cordillera (Spain).

Figure 6.59 Rillenkarren and knupkarren in a secondary gypsum alabaster nodule. Bujaraloz, Province of Zaragoza. Ebro Depression (Spain).

Figure 6.60 Cover collapse sinkhole that provides entry to a cave. It has formed above near-surface compacted gypsum-bearing silts. La Puebla de Alfindén, Province of Zaragoza. Ebro Depression (Spain).

Figure 6.61 Alluvial unconformity within an alluvial sequence that has thickened as a result of subsidence of the Jalón River. Calatayud, Province of Zaragoza (Spain).

Figure 6.62 Collapse sinkhole in interstratal karst in Miocene gypsum. Dankiutsy Collapse. Ukraine. Photo F. Gutiérrez.

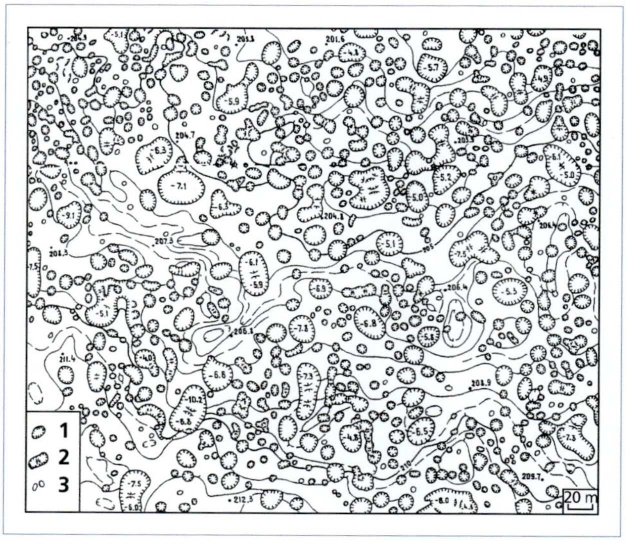

Figure 6.63 Typical landscape of a gypsum karst area in the Iren River basin, Pre-Urals, Russia (Klimchouk and Andrejchuk, 1996a).

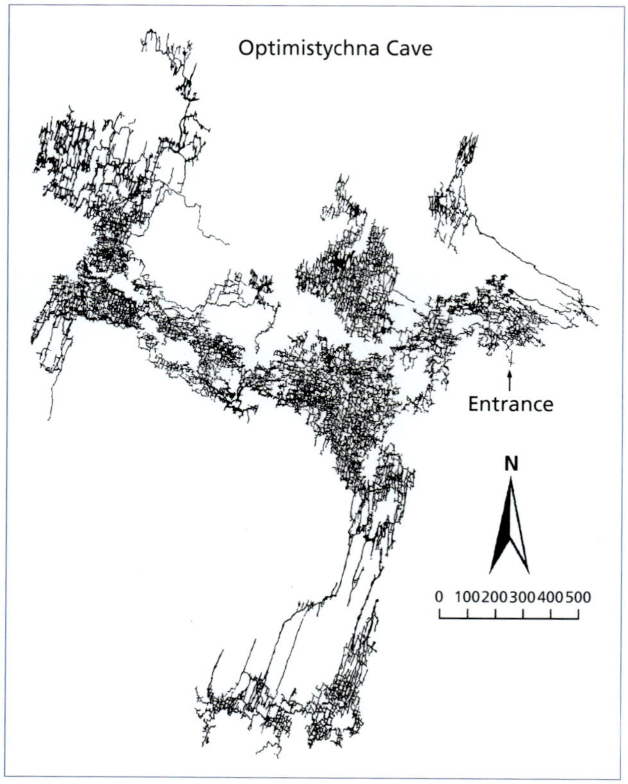

Figure 6.64 Map of Optimistychna caves (Ukraine), formed in gypsum. This is the largest gypsum endokarst in the world (Speleological Club of Aviv, in Klimchouk, 2005).

Russia) is a typical example of an interstratal karst landform with a thin permeable cover (*Fig. 6.63*). The sinkholes have evolved through a variety of mechanisms: as simple collapse caves, as vertical trough structures, and others by suffosion. The oldest dolines are fossilized and stabilized. Some dolines are reactivated and form new landforms often superimposed on older ones. The resulting doline fields are polygenetic and can be highly complex. Dolines can cover 30% of the area (Klimchuk and Andrejchuk, 1996a).

Interstratal karst of the Sorbas (Province of Almería, Spain) occurs in an intermontane basin of the Betic Cordilleras; it is composed of a sequence of Messinian gypsum and marls. Precipitation is less than 250 mm and the average annual temperature is 18°C. There are about 100 dolines in the 10 km^2 area (Calaforra and Pulido-Bosch, 1996, 1997; Calaforra, 1998). In the Ripon area, North Yorkshire, England, in a small 6 km^2 area of interstratal karst developed in Permian materials, dolines range up to 80 m in diameter and 30 m deep, formed by collapse of a system of caves in gypsum (Cooper, 1988; Cooper, 1995).

Karst gypsum caves develop in formations of different ages in various parts of the world (Klimchouk, 2000b). Reaction kinetics are the principal differences between carbonate and evaporite dissolution processes. Gypsum dissolution is primarily controlled by diffusion across the boundary layer. Caves form only when gradients are high, enlargements are significant, and/or when fracture conditions occur (Klimchouk, 2004c). Under unconfined phreatic conditions, caves develop when the initial widths of flow trajectories are large enough to cause fracture under a specific hydraulic gradient. Artesian phreatic conditions are generally not favorable to speleogenesis, especially for gypsum.

The best-known and most extensive system of gypsum caves is in Ukraine where eleven caves exceed 2 km in length. Most are located north of the Dniester River. Some of the caves have been made accessible by gypsum quarries. Caves in the region are authentic labyrinths developed along highly vertical fissures that connect with extensive networks at different elevations. They have an average density of 164 km/km^2; this ratio is the quotient of cave lengths within a specific area. The Optimitichescaja Cave is the world's longest gypsum cave (*Fig. 6.64*) and the world's second longest karst cave (Klimchouk, 2005).

Gypsum speleothems have morphologies that are obviously different from those of calcite because their different genetic mechanism involves oversaturation due to evaporation. Gypsum stalactites are contorted and branched because these speleothems deform in the presence of constant winds (Forti, 1996). Gypsum speleothems and other sulfate minerals have a wide variety of shapes depending on growth rate and the water source area. White (2005) describes 1) bulk crystals (gypsum in the soil, dentate, and acicular); 2) gypsum-bearing crusts (granular and fibrous); and 3) fibrous gypsum (gypsum flowers, angel hair, and cave cotton). Stalactites are common in Italian caves but stalagmites are not. In arid areas, stalagmites are more common than stalactites (Calaforra *et al.*, 1992). Because evaporation is the dominant mechanism in gypsum caves, many speleothem types common in calcite caves, such as eccentrics, are not found in gypsum endokarst. Climate has an almost predominant role in both the formation of chemical deposits in gypsum environments and in their preservation (Forti, 1996).

6.11 Karst risks, applications and uses

Karst processes and landforms can pose multiple risks that generally require specific investigative methods due to the unusual and highly variable characteristics of karst environments (Gutiérrez, 2010). Research on karst applications has increased in recent decades (LaMoreaux *et al.*, 1997; Durán, 1988, 2002; Durán and Burillo, 1989). When karst processes affect humans and their activities, loss of life and property can occur. Karst risk includes loss of sheep and cows into shafts in the Pyrenees, flooding of agricultural land in poljes, and significant water losses from dams and reservoirs. Sinkholes are the most significant karst risk (Gutiérrez, 2010). More than 6,500 sinkholes have formed in the eastern United States since 1950 (Newton, 1986).

The percentage of the Earth's surface subject to karstification is 12% (Ford and Williams, 1989). This fact emphasizes the importance of karst research. To analyze karst risk, scientists must identify, investigate, predict, and mitigate (Gutiérrez *et al.*, 2008a). In addition, authorities need to interview local people who can provide data to help estimate the probability of karst risk (Benito and Gutiérrez, 1988).

Galve *et al.*, (2009b,d) prepared a detailed sinkhole map of a 40.8 km^2 area in a western sector of the city of Zaragoza (northeastern Spain). In recent decades, 70% of the sinkhole area has been covered with anthropogenic sediments, causing 137 ha of karst-created marshes to disappear. This research used aerial photographs taken over time. The investigation indicates that preventive planning should be done on geomorphic maps to avoid significant property damage.

Karstification affects **construction** in many places throughout the world. Numerous collapse sinkholes have occurred in carbonate terrain in Pennsylvania (Foose, 1953), Alabama (Powell and Lamoreaux, 1969), and especially Florida (Sinclair, 1982; among others). Most of them result from groundwater pumping. Williams and Vineyard (1976) conducted a sinkhole inventory from which they concluded that half of sinkholes are caused by human activity. Small collapse sinkholes and subsidence dolines of larger sizes can cause significant property damage (White, 1988). When the layer of insoluble material that overlies limestones or evaporites near the surface is relatively thick, it is often not capable of bearing new loads added by construction. In this case, deep foundations must be laid on top of competent rock (*driller piers*) (Waltham *et al.*, 2005). Micro driller piers can also be used (Sowers, 1996). When a collapse sinkhole forms suddenly, it can break gas and water pipelines, seriously damage electric conduits, and crack buildings and cause them to subside (Friend, 2002).

Building damage due to evaporite dissolution is significant; one example occurred in Ripon, North Yorkshire, (Cooper, 1988, 1995, 1998; Cooper and Waltham, 1999). The town of Calatayud, Zaragoza (Spain) sits on a fan composed primarily of gypsum particles (Gutiérrez, 1998); gypsum dissolution has caused buildings to tilt (*Fig. 6.65*). Old churches develop cracks and may lean to one side (*Fig. 6.66*). On 10 November 2003, a sinkhole collapsed at the entrance of a building and significantly affected its foundation (*Fig. 6.67*). The tenants were evicted and about a year later, the building was demolished. In the town of Orihuela del Tremedal, Province of Teruel (Spain), numerous severely fractured houses had to be demolished. The cause of karst risk in this area is the dissolution of the Keuper evaporites; as they subside they rupture the water supply system, which in turns promotes additional dissolution. Karstification of evaporites has fractured homes in the newly created city of Puilatos, Zaragoza (Spain). The houses were not yet inhabited and the town was completely demolished. More unfortunate was the loss of 34 people by the sudden collapse of karst dolomite in Far West Rand in South Africa (Bezuidenhout and Enslin, 1970).

Gutiérrez *et al.*, (2009) studied an active sinkhole about 100 m long in the city of Zaragoza (northeastern Spain). The active subsidence of this depression, including a sudden 5-m collapse, led to the abandonment of a factory during the 1990s. The sinkhole developed as a result of the karstification of halite and glauberite. Ground penetrating radar (GPR) profiles and electrical resistivity studies have provided structural information about the subsurface. Dating by AMS and OSL make it possible to determine the regional history. The information obtained indicates that the progressive deformation that has affected the building will continue.

Sudden collapses can occur instantaneously, as illustrated by the case of Pedreguer (Province of Alicante, Spain), where on the night of 2 August 1982 an orange grove subsided 73 m; the trigger appears to have been water extraction (Garay, 1986). On 17 January 1984, the Rio Piedra riverbed collapsed instantaneously; its waters fed numerous waterfalls in the Monasterio de Piedra Natural Park. As a result of the collapse, the waterfalls ceased to flow and water did

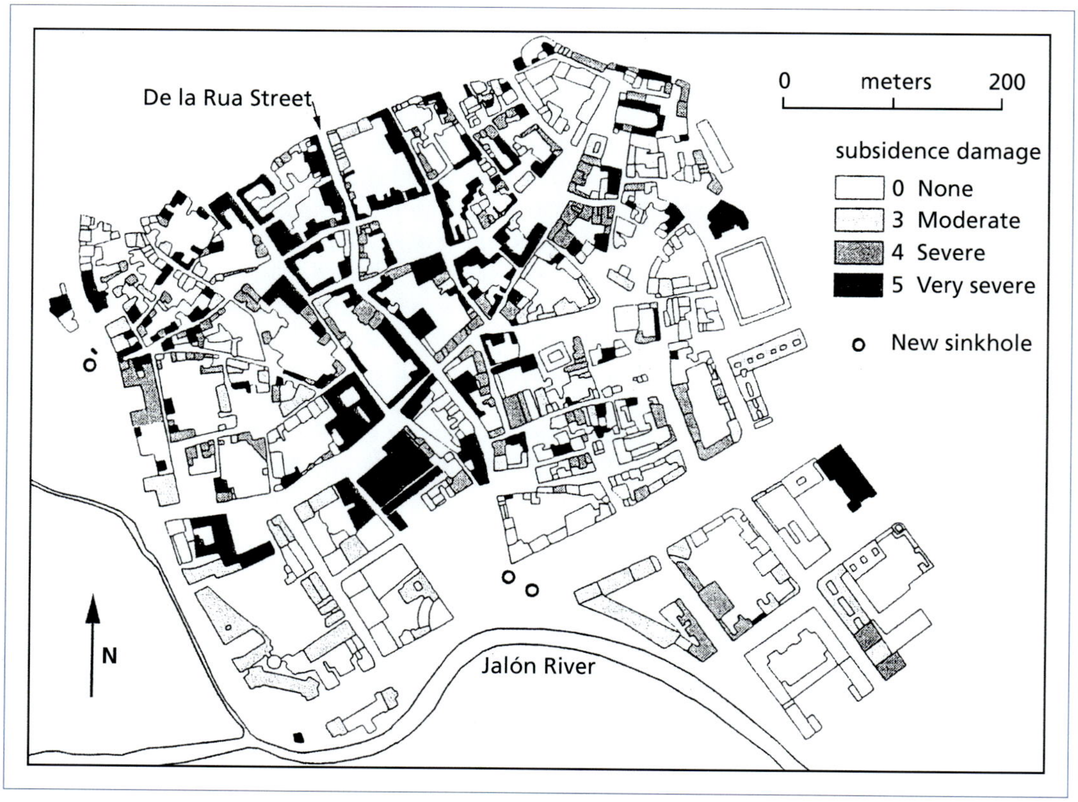

Figure 6.65 Distribution of damage from building subsidence in Calatayud, Spain, due to dissolution and dolines in underlying gypsum (Gutiérrez and Cooper, 2002).

not reach fish hatchery or the city of Nuévalos (Province of Zaragoza, Spain).

Linear constructions (roads, railroads, irrigation canals) are often affected where they cross areas of soluble rocks (*Fig. 6.68*). Risk is much greater when evaporites are involved because they dissolve more rapidly than carbonates (Benson and Kaufmann, 2001). On the outskirts of Zaragoza, the Logroño Highway A-68 is affected by ductile subsidence that is causing the highway and nearby factories to settle. The subsidence is the result of dissolution of evaporites that underlie the fluvial gravels of the Ebro River flood plain (*Fig. 6.70*). This active subsidence has caused a depressed cone that is periodically filled with asphaltic gravel (*Fig. 6.71*). In March 2003 the Pina de Ebro area of the high-speed rail line from Zaragoza to Barcelona was affected by a collapse sinkhole several meters deep; the collapse left the railroad tracks up in the air, seated on ballast and terrace gravels above the underlying karstifiable gypsum (Guerrero *et al.*, 2004). Dissolution of chloride salts in Cardona (Barcelona, Spain) causes numerous collapses and flexures that affect highways in spectacular ways (*Fig. 6.69*) (Lucha, 2008b). The presence of holes, which are the most significant risk in highway construction work, can be inferred in part by geologic mapping aided by geophysical methods (electrical, radar, seismic, tomography, and microgravity) (Pueyo-Anchuela *et al.*, 2009a, b). Construction of a 50 km section of highway across the classic karst region of Slovenia exposed 300 caves (Knez and Slabe, 2004).

In 1954 the Violada Canal, which forms part of a large irrigation network in the Ebro Depression (Spain), was affected by collapse that measured 25 m in diameter (Riba and Llamas, 1962). Unlined canals such as the Imperial Canal of Aragón provide water to underlying gypsum, thus activating the karst process and creating dolines. The relationship between sinkhole density and proximity to the irrigation canal network in the area of Villamayor (Zaragoza) illustrates these circumstances. The maximum sinkhole concentration occurs near irrigation canals and especially at their confluence (*Fig. 6.72*) (Benito and Gutiérrez, 1988).

To determine zones likely to contain sinkholes along linear infrastructures, Guerrero *et al.*, (2008) used a methodology based on an analysis of subsidence characteristics, responses in adjacent slopes, and the distribution of karst depressions. The method was applied to 24 km of high-speed railway between Madrid and Barcelona. The proposed zonation may be used as a basis for future investigations or for the application of mitigation measures.

Figure 6.66 Inclined tower. San Pedro de los Francos. Calatayud, Zaragoza (Spain).

Figure 6.67 Apartment building ("Casa Azul") and the sinkhole that appeared at the entrance on 10 November 2003. The building was later demolished. Photo F. Gutiérrez.

Large **dams** are constructed to mitigate floods, generate hydroelectric energy, and provide water to towns and for irrigation and recreation (Milanovic, 2004). Dams are constructed in narrow places, often of concrete or earth materials. When a reservoir behind a dam is filled, hydrostatic pressure increases and a new groundwater level is established along the valley walls. The principal karst problem is the presence of caves, which can drain water both under the dam and the bottom and walls of the reservoir. In cases of water loss, cement can be injected through boreholes (Bruce, 2003) although this solution is costly. The Montejaque dam, Sierra de Libar (Málaga) was built in the 1920s in karst limestone terrain but it never held water (*Fig. 6.73*) because it is located in the Hundidero area where water is lost into a blind valley and an offshoot of the Gato cave forming the Hundidero-Gato system (*Fig. 6.39*). Dissolution processes increase considerably in karst evaporites and the resulting problems are much more difficult to resolve (James, 1992).

In populated areas sinkholes are often used as waste dumps (*Fig. 6.74*), which can contaminate karst water and cause chemical and/or biological water **pollution**. Karst aquifers have high transmissivity (Paukstys and Narbutas, 1996). Pollution comes from fertilizers, petroleum, and radioactive substances (Klimchouk and Andrejchuk, 1996b). Due to the complexity of karst systems, investigation of waste disposal sites is a concern. Evaporite regions are particularly unfavorable sites for waste disposal (Ford and Williams, 1989, 2007).

Active karst terrain makes good water reservoirs and the **exploitation of water** can have significant risk, as shown by ductile or brittle subsidence. Sinkholes are connected with high-transmissivity karst aquifers and cave systems are water storage areas (Milanovic, 2000; Johnson, 2006; Gutiérrez *et al.*, 2008b). In some places, water exploitation can affect the safety of sensitive structures such as radioactive waste repositories in New Mexico (LaMoreaux *et al.*, 1997;

Figure 6.68 Collapse sinkhole at the edge of Highway A-68 between Zaragoza and Logroño. Photo F. Gutiérrez.

Figure 6.69 Highway affected by rapid dissolution of chloride salts from the Cardona Diapir, Barcelona (Spain). Photo F. Gutiérrez.

Hill, 2003). Karst aquifers are an important reserve of mineral water in the United States; in 1985, 25 million m^3 of water was consumed per day (Peck *et al.*, 1988).

Agriculture (Gams *et al.*, 1993) can be affected by cover collapse sinkholes, which form constantly in active gypsum karst in the cultivated fields of Villamayor (Province of Zaragoza, Spain). Farmers are well aware that sinkholes can form after large storms. These sinkholes can be as large as 15 m in diameter and more than 10 m deep; they may make a loud noise and can occasionally hurl particles outward. The sinkholes may later fill rapidly with insoluble material. In other cases, sinkholes form as a result of irrigation (Benito and Gutiérrez, 1988). Fertilizers normally used in agriculture can pollute karst aquifers. Poljes are flat areas that are usually intensely cultivated but they flood periodically, as does the Zafarraya polje, (*Fig. 6.29*) and cause irreparable danger to agriculture. A similar problem resulted from the flooding of the tectonic polje of Cetinje, Montenegro (Bonacci and Zivaljevic, 1993). Karstification occurs in upper Triassic dolomitic limestone and dolomites in Montenegro. Precipitation in the area ranges from 2,300 to 5,000 mm, with an average of 3,500, the highest in Europe. Intense karstification has destroyed the surface fluvial network; groundwater supplies the polje. The Cetinje polje and the city of Cetinje, located in a hum, were flooded on 16–20 February 1986. Ponors were converted to estavelles and agriculture and patrimony in the lower part of the city were significantly damaged. To mediate flooding of the poljes, evacuation tunnels were drilled and artificial outlets were created (Nicod, 1972).

Figure 6.70 Flexure caused by karst subsidence of gypsum that underlies fluvial gravels. Highway A-68, Zaragoza-Logroño. Photo F. Gutiérrez.

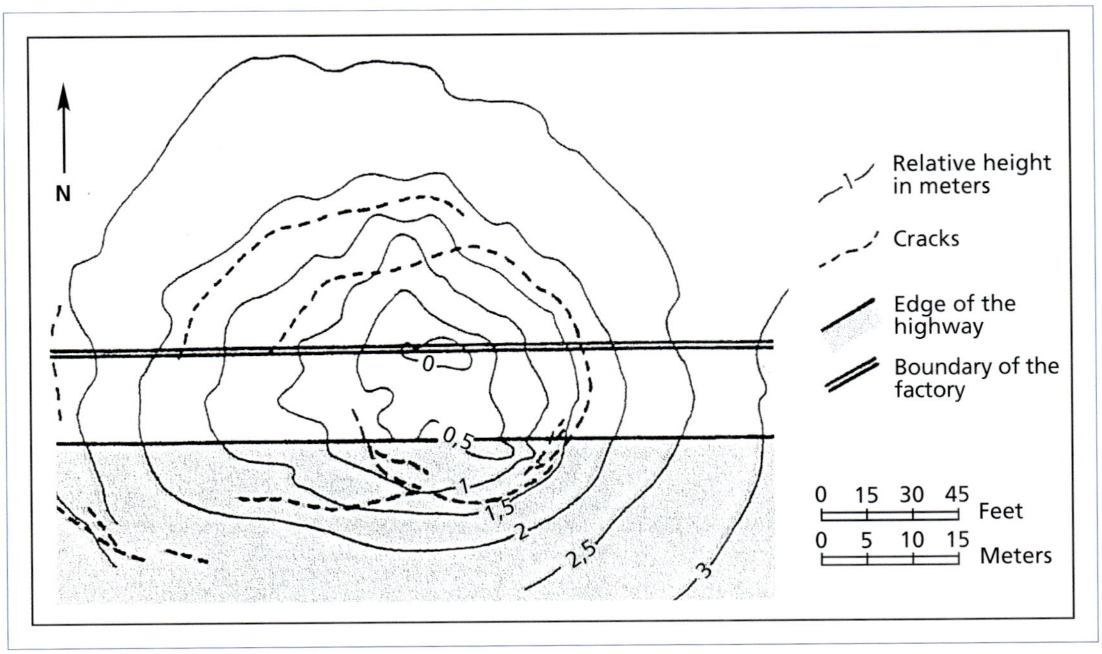

Figure 6.71 Subsidence cone affecting Highway A-68, Zaragoza-Logroño, and a factory. This cone has been filled with several meters of asphalt. The subsidence rate is calculated at 25 cm/year (González, 1970, in Benito and Gutiérrez, 1988).

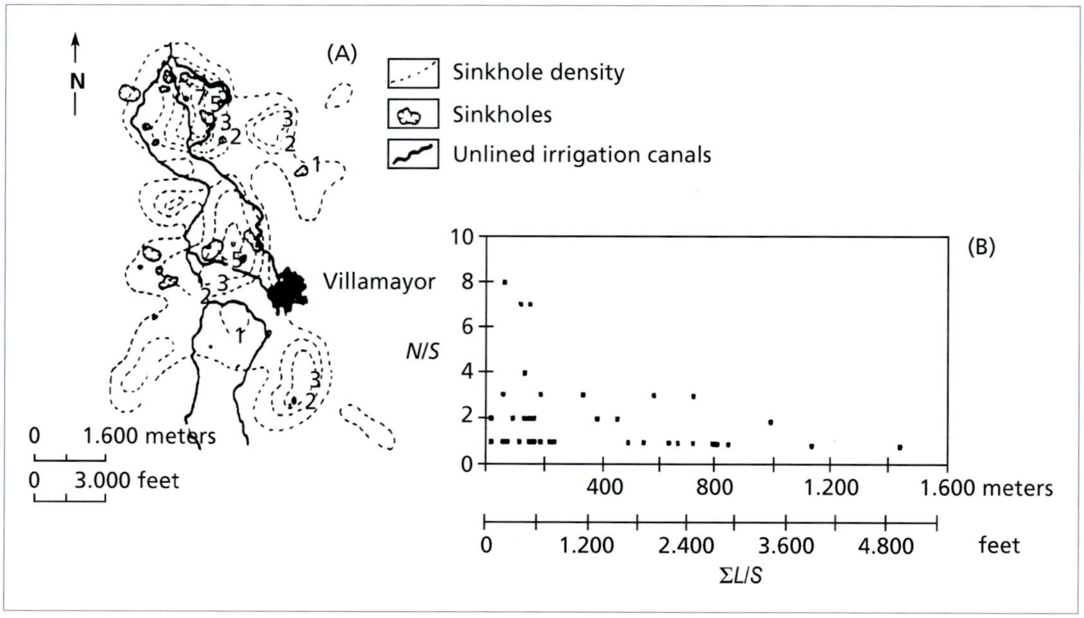

Figure 6.72 (A) relationship between sinkhole density and unlined irrigation channel network. (B) relationship between sinkhole density and distance to the closest irrigation channel. N: number of sinkholes. S: surface unit. ΣL: sum of distances from dolines per surface unit, compared to the closest irrigation channel (Benito and Gutiérrez, 1988).

Figure 6.73 Montejaque arc-shaped dam, which never held water. Sierra de Libar, Province of Málaga. Spain. Photo F. Gutiérrez.

Figure 6.74 Trash disposal in a funnel-shaped sinkhole, in Oligocene gypsum. Alcampel, Province of Lérida (Spain).

Prior to the 1940s, water was used in rotary drilling for petroleum but it dissolved soluble rocks. Now drilling companies use a fluid that does not dissolve salt (Johnson, 2005). Today it is known that high karst porosity plays a significant role in hydrocarbon migration and storage (Lowe, 2000).

In some cases, groundwater can cause problems in underground **mining** as it does in open pit mines; these difficulties can affect production. For this reason, it is imperative to conduct hydrogeological studies to understand the role of water (Fernández-Rubio, 1976, 1989; Fernández-Rubio and Benavente, 1984). Mining can be done in karst, carbonate, and evaporite rocks (Klimchouk and Andrejchuk, 1996b). In the Sierra Menera (Teruel and Guadalajara), iron oxy-hydroxide ore formed in interstratal karst by karstification of mixed carbonates that contained iron and magnesium; the residuum of these carbonates is the ore (Fernández-Nieto *et al.*, 1981) *(Fig. 5.52)*. There are numerous examples of karstification playing a similar role in metallic ore development (Lowe, 2000). Gypsum and chloride mines are exploited in subterranean galleries or quarries; this mining can cause impacts if the extraction is done above a saturated zone. However, if water is also extracted, the impact can be much more complex and significant, as has happened in western Ukraine (Klimchouk and Andrejchuk, 1996b). These authors show that floods have occurred in the Tamansky region (Russia). Also, collapses are common during underground evaporite mining. Salt mining is also done by injecting water through boreholes into the salt, converting it into brine and pumping it out. This practice creates numerous subsidence sinkholes and damages houses in Cabezón de la Sal (Province of Burgos, Spain).

During the mining of iron ore formed by the karstification of underlying carbonates, as in the Ashgilian dolomites of the Sierra Menera, the Cambrian dolomites of Cerro del Hierro (Province of Sevilla, Spain), and the Aptian dolomites of Cabárceno (Cantabria, Spain), the outcrops are exhumed (exploited) at irregular karst contacts, and they are frequently acicular shaped *(Fig. 6.75)*, similar to the exhumed karst needles of Yunan (China). Today these abandoned mines at Cabárceno have been converted into a zoo to take advantage of the irregular landscape and the beautiful needles.

Man and caves have enjoyed a constant relationship throughout history (Hoyos and Jordá, 1989). The discovery of human remains in cave fill is testimony to this link (Finlayson *et al.*, 2006) The most important example is the paleo-anthropological excavation of Atapuerca, Province of Burgos, Spain *(Fig. 6.76)* (Bahn, 2004). The age of the hominids and artifacts at Atapuerca TD6 is early Pleistocene. This level (TD6) is below the Mayutama-Brunhes boundary, about 780,000 years (Parés and Pérez-González, 1995; Carbonell *et al.*, 1995). Humans have merely followed the example of

Figure 6.75 Morphology of karst needles in Aptian dolomites. These landforms are the contact between karstified carbonate rocks exhumed by the mining of iron oxyhydroxides. Cabárceno, Cantabria (Spain).

Figure 6.76 Paleoanthopological excavation in fossil karst in Cretaceous limestones. Relleno de la Gran Dolina (TD) (Atapuerca, Burgos, Spain).

animals; each time conditions became more rigorous, they have sought refuge in caves (Nicod, 1972; Trombe, 1974). The profusion of cave paintings (Altamira, Cantabria) and engravings attest to this relationship, as does the establishment of religious cults in caves (Languedoc, Crete, Bahía-Brazil, Cacahuamilpa-Mexico, southwest Asia, etc.). Caves have also been used for defensive purposes (medieval caves in central Europe) (Gillieson, 1996).

Cave **tourism** is highly diverse: observations of paintings and engravings, sport speleology, and tourist exploitation of caves (in Spain: Nerja, Aracena, Artá, Drach, and others) (López-Martínez and Durán, 1989; Rivas *et al.*, 2004). During the IV International Speleology Congress in 1965, the inaugural session was held in Postojna Cave. In addition, the Symposium of the International Association of Volcanology and Chemistry of the Earth's Interior (Canary Islands, 1968) held a scientific session in a lava tube cave (hameo) in Lanzarote. Bernard Gèze, the President of the International Union of Speleology at that time, led the session. Other caves have been the site of boat trips through endokarst lakes and musical programs (Nerja concerts) for the enjoyment of the special resonance of these karst salons. In France, about 6 million people per year frequent the underground world (Salomon, 2000). Each year, about 20 million people visit tourist caves. Two million people visit Mammoth Cave (Kentucky) every year. There are about 650 tourist caves in the world. Carlsbad Cavern (New Mexico) contains more than 1400 light bulbs,

65 km of electrical cable, 6 km of paths, 70 signs and 50 drains (Gillieson, 1996).

Anthropogenic environmental impacts in karst caves have been documented by Williams (1993): ecological impoverishment, degradation and erosion of soil, cave sedimentation, deterioration of water quality, destruction of landforms, drying up of springs, marine water intrusion, and flooded karst systems. To evaluate these impacts, Liñán *et al.* (2004) used environmental monitoring experiments and continuous observation to conduct a study of seven Spanish tourist caves. This work shows the need for and the usefulness of this type of research in the management and preservation of karst cavities.

Another challenge is the delimitation of karst areas for purposes of protection. At the Lunan Stone Forest Park (exhumed karst in Yunan, China) boundaries were defined by the use of a Spatial Decision Support System (SDSS) which rapidly identified the seven square kilometers of the national park, which was created in 1982 (Zhang *et al.*, 2005). This karst is called by its Mandarin name "shiling" which means stone forest.

Environmental impacts are more common in karst terrains than in others. Development and exploitation by man affects the karst hydrologic regime; they can unleash a series of catastrophes, and as a consequence lead to **legal action**. There are many examples of regulation and litigation related to karst areas (LaMoreaux *et al.*, 1997).

Florida has developed insurance for karst risk to cover personal and property damage as a result of rapid changes in physical and hydrologic characteristics of karst. For this reason, vulnerability and sensitivity maps are needed to describe the cause and effect relationship and provide an adequate description of the geology and hydrology of karst areas (LaMoreaux *et al.*, 1997).

Galve *et al.*, (2009b, c) have developed a method to quantitatively estimate the possibility of doline production (spatial probability) and risk (spatio-temporal probability) for a 50 km^2 surface in the karst evaporites of the Ebro valley. The high susceptibility model has been transformed into a risk map by taking into account the frequency between 2005 and 2006. This model can be used as an objective basis for the application of mitigation, prevention, or correctional measures.

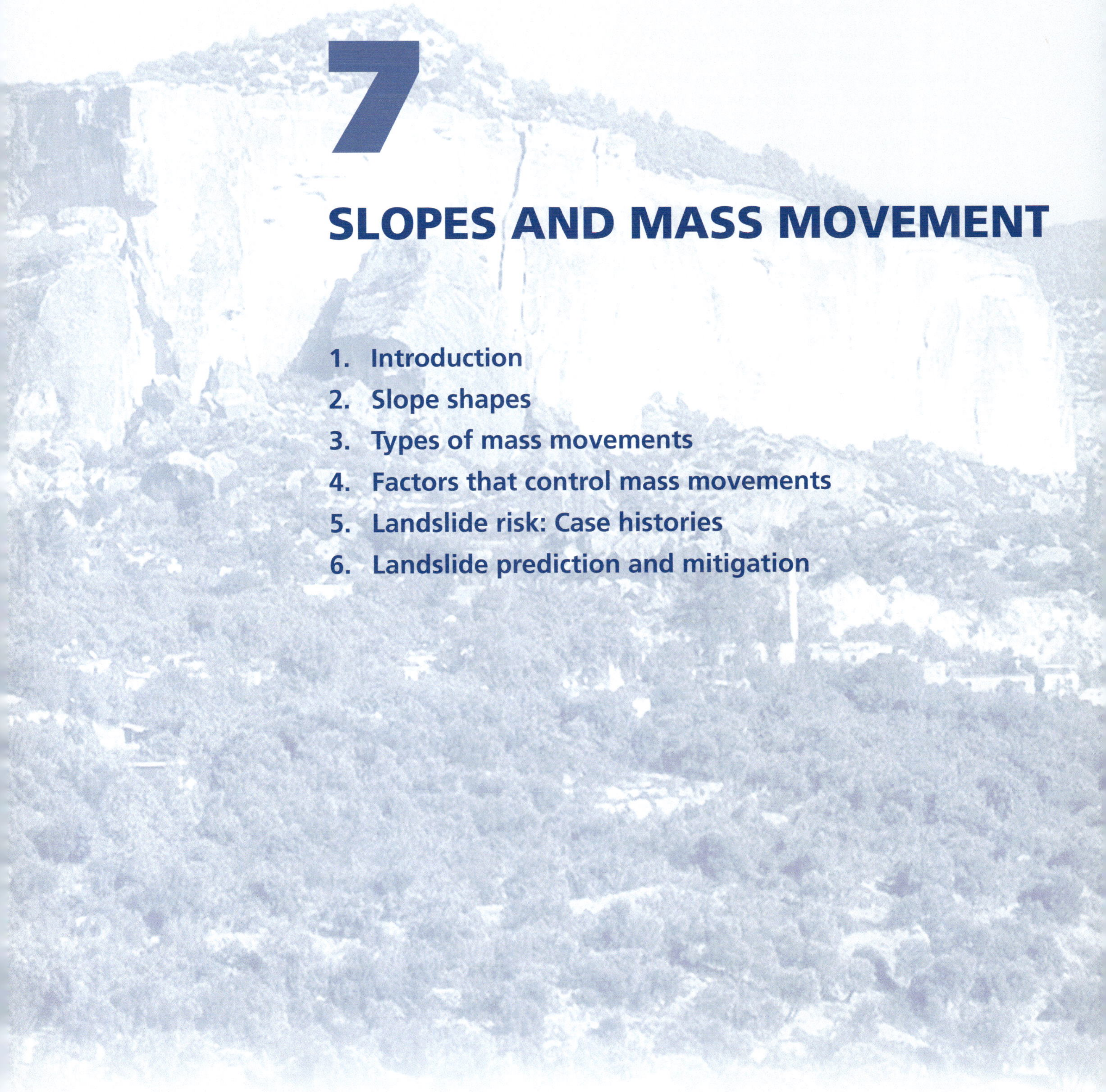

7

SLOPES AND MASS MOVEMENT

1. Introduction
2. Slope shapes
3. Types of mass movements
4. Factors that control mass movements
5. Landslide risk: Case histories
6. Landslide prediction and mitigation

7.1 Introduction

Slopes cover most of the Earth's surface (Young, 1972) and thus are a fundamental component of relief. The study of their shape, significance, and evolution is one of the basic subjects of geomorphology (Douglas, 1977). Comparatively little research was done on slopes until the last five decades. However, research on slope processes has now proliferated in both experimental field plots and the laboratory, models have been developed to assess danger, and procedures have been established for analyzing slope stability. Even so, the research is difficult due to the complexity of the phenomenon because the mechanisms at work are not known in a detailed way (Carson and Kirkby, 1972). In addition, the features used to recognize landslides are rapidly covered by vegetation in humid climates *(Fig. 7.47)*, making it difficult to identify them. The principal exceptions to this comment about slopes are found on the extensive plain that makes up the Colorado Plateau, the páramos of Castile (Spain), the hamadas of the Sahara, erosional surfaces of the Brazilian, African, and Australian shield, and the wide alluvial plains along the Indus-Ganges and the Po.

Early geomorphologists gave special emphasis to the long-term slope evolution. According to Davis and as discussed in various publications, particularly Davis' (1899) first theory of general evolution of relief, slope evolution consisted of a gradual decrease of slope angles in valleys. Penck's (1924) morphological analysis system showed that if fluvial excavation accelerates, the resulting slope profile is convex whereas if downcutting is slow, the slope is concave. King (1953) proposed a third theory of landscape evolution based on slope research by Wood (1942). King proposed that slopes retreat in a parallel manner, a theory that Powell had already published in his 1875 Exploration of the Colorado River of the West, although it had its origin in the work of Fischer (1868, 1872).

Researchers have occasionally discussed the contribution of landslides to landscape evolution. Cendrero and Dramis (1996) propose a ratio between the rate of landslide movement and the temporal magnitude of the lowering of the ground surface. Analysis of available data from various European regions shows that in some cases mass movements are the principal process of topographic evolution.

Strahler (1950b) proposes a slope classification based on the result of a frequency distribution analysis and field studies of slopes in numerous locations. He classifies them as three types. **Highly cohesive slopes,** those composed of clays or strong massive rocks such as granite, schist, or gneiss, have angles of 40° and 50° and are affected by landslides; rivers undercut the bases of their slopes. The second type, **slopes in repose**, which are called block-controlled slopes by Bryan (1922) and gravity slopes by Meyerhoff (1940), have angles of 30°–35°. They are composed of large clasts controlled by the angle of repose. Laboratory research on the angle of repose of fragmentary materials shows characteristics similar to valley slopes (Van Burkolow, 1945). The third type, **slopes reduced by wash and creep**, called rain-washed slopes by Bryan (1922) and wash slopes by Meyerhoff (1940) correspond to the *haldenhänge* of Penck (1924). The activity of water and creep reduces slope inclination to less than the angle of repose.

The morphogenetic forces acting on a slope are perpendicular and parallel components (Jahn, 1954). The perpendicular component includes weathering processes and pedogenesis, which tend to form a weathering film. The parallel component refers to the displacement of material by gravity (landslides, gullying, solifluction, creep, etc.).

As a result, the two components are interdependent and the evolution of a slope is a function of the magnitude of both. Jahn (1954) calls the relationship between the two **the denudation balance**; Tricart (1957) prefers the term **morphogenetic balance**. Schumm and Chorley (1966) use similar reasoning in their work on the Colorado Plateau; they call the quotient between these two components the **weathering relationship**.

When the perpendicular component exceeds the parallel, the regolith deepens and does not move totally downslope. On the contrary, if the parallel component exceeds the perpendicular, the regolith is rapidly denuded and the slope stabilizes when the denudation reaches coherent rock. If it is soft, badlands form. If the relationship is equal to 1, it indicates that the amount of altered material is equal to the evacuated material. This notion corresponds to Hack's (1960) **dynamic equilibrium**.

This morphogenetic balance depends on the inclination of the slope, the nature of the rock, and climate (Tricart, 1957). If the gradient is significant, more detritus will be exported and the slope gradient will increase. Coherent rocks require prior weathering (perpendicular component). Climate intervenes directly or as a result of vegetation and affects both components.

In addition, at the present, "we find ourselves in an unprecedented situation in the four thousand five hundred million years of life on the planet, that for the first time there is a species (man) with the capacity to affect the various natural systems on a planetary level with a qualitative and quantitative importance that could equal and even exceed these natural agents" (Cendrero, 2003).

We need to understand the processes of slope development for many of man's activities including agriculture, forest practices, construction of linear projects (highways, channels, and railroads), and construction of buildings; geomorphologists and engineers do this fundamental work (Dunne and Leopold, 1978; Clark and Small, 1982).

Slopes are composed of two types of materials: rock and soil, which have different mechanical properties and evolve in different ways. Rocks are consolidated and coherent materials that are affected by planes of weakness (stratification, schistosity, and joints) that weaken the rock and favor fragmentation and disaggregation. Soils (regolith, alterite, surface deposits) are unconsolidated materials with low mechanical strength and high porosity.

Mass movements (*Fig. 7.1*) are processes that move materials by gravity; they can be dangerous and even disastrous when they affect human life and property. They are of concern to geotechnical and geologic engineers (Bromhead, 1986). Geomechanics and related disciplines provide an understanding of landslides from an "internal" perspective whereas geomorphology contributes an "external" perspective (Crozier, 1986).

Landslides pose a risk when they threaten human life and property. This risk has increased during the last century due to population increase and the increased area used for resource extraction; these factors have sometimes caused people to use areas of potential risk (Crozier, 1986). In addition, other activities such as emigration toward large urban areas and the construction of numerous engineering works that modify the landscape and cause slope instability generate landslides that can cause injuries and fatalities, property loss, and environmental damage. Landslides pose significant danger in tropical areas because of the presence of a thick regolith and high precipitation in these climates. These conditions favor surface and deep mobilization of the altered layer, which occurs most often with storm rainfall. For this reason, it is necessary to conduct detailed geomorphic cartography and install reconnaissance boreholes (Tricart, 1974b).

According to the calculations of Ayala (1994), since the year 1000 about 300,000 landslides throughout the world have each killed at least 100 people. In Spain the most significant disaster took place in Azagra (Navarre) where a landslide on a gypsum cliff killed more than 100 (Faci *et al.*, 1988). In Spain landslides cause losses on the order of 180 million euros and several deaths every year (Suarez and Regueiro, 1997). Cendrero *et al.* (1997) consider mass movements in Spain to be the third largest geomorphologic risk; they estimate losses for the period of 1986–2016 to be 7.6 billion dollars (Ayala *et al.*, 1987a).

Figure 7.1 Mass movements on slopes in the Appenines. Italy.

The San Francisco Bay, California, geologic map showed one landslide at the beginning of the 20th century. In 1970, the map showed 1200 landslides; by 1980 the number had risen to 70,000 mapped landslides. This is due in part to the strong pressure that humans place on the environment (Brabb, 1989). In the United States, annual damages from landslides were estimated at 1.5 billion dollars in 1985 (Schuster and Fleming, 1996) and Bromhead (1986) puts the number of deaths at about 25 people per year.

7.2 Slope shapes

A slope shape that shows the dip of the slope is a **slope profile** (Douglas, 1977). Slope angles and distances can be measured on a section perpendicular to the slope. A series of irregularities is generally observed, such as concavities and convexities that depend primarily on the lithology of the substrate and processes that affect slope morphology; these factors play a significant role in the morphoclimatic environment *(Fig. 7.2)* (Kirkby, 1976; Toy, 1977).

Geomorphologists characterize a slope by analyzing its topographical profile and assigning to it one or more elementary geometrical shapes. These geometric shapes were established by Savigear (1952, 1956) and later expanded and quantified by Young (1964). Young's work combines Savigear's convex, concave, and rectilinear elements into units, segments, and elements of parts of a slope. The morphology of a profile is analyzed in this way. The techniques used are compiled in King (1966), Goudie (1981a), and Gardiner and Dackombe (1983). Another classification proposed by Dalrymple *et al.* (1968) divides the slope into nine units based on shape and pedogenic processes. Slope measurements make it possible to distinguish nine possible shapes of slope units *(Fig. 7.3)* (Ruhe, 1975).

For **slope cartography**, Young (1972) distinguishes morphological maps, slope angle maps, and genetic morphological maps. Morphological maps are drawn at a detailed scale to determine the surface units of the slope. This technique, which has been developed by numerous authors, is based on the assumption that relief can be divided into uniform morphological units bordered by morphological discontinuities *(Fig. 7.4)* (Curtis *et al.*, 1965). There are two types of morphological slope angle maps: isoclinal maps that show the slope angle at a point; they are generally made

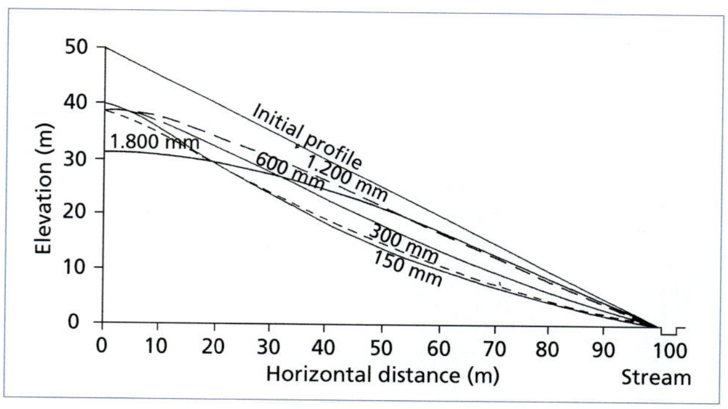

Figure 7.2 Slope profiles simulated for different annual precipitation, assuming 100,000 years of evolution (Kirkby, 1976).

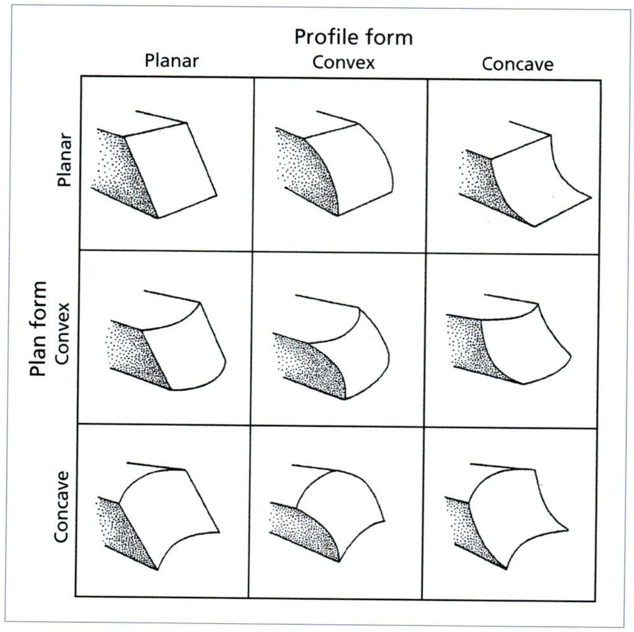

Figure 7.3 Nine possible shapes of slope units. (Based on Ruhe, 1975; classified by Parsons, 1988).

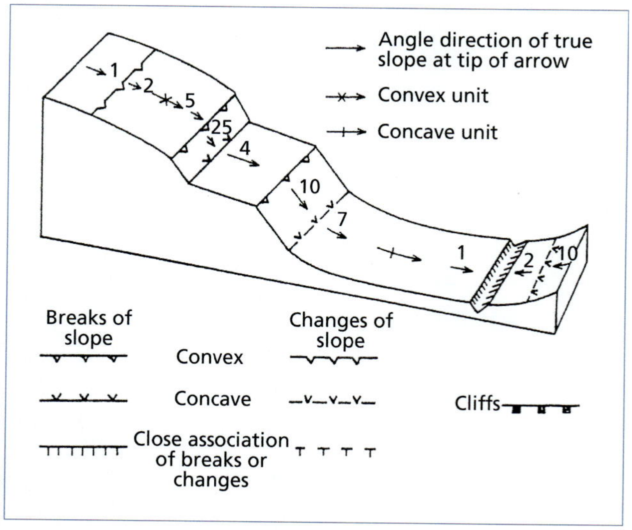

Figure 7.4 Symbol used in geomorphologic cartography (Curtis *et al.*, 1965).

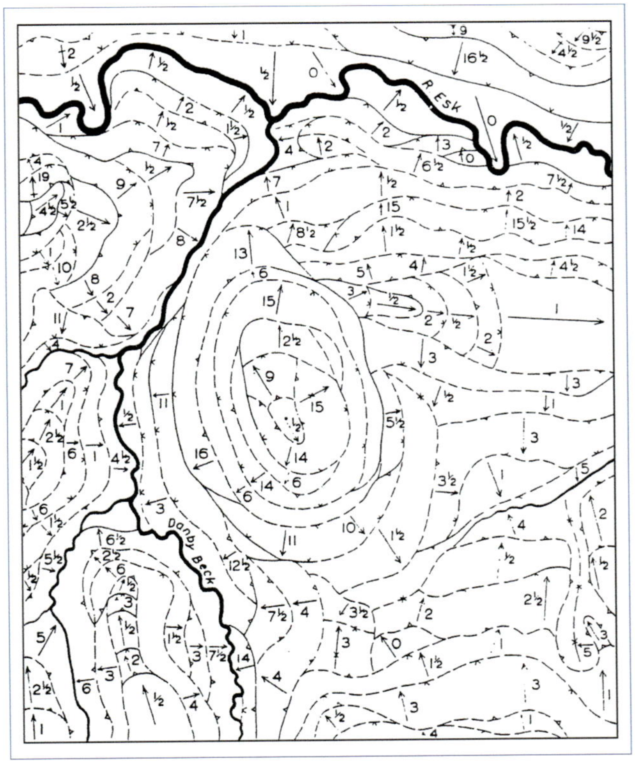

Figure 7.5 Morphological map of part of Eksdale, Yorkshire, England (Gregory and Brown, 1966).

from contour maps (Gregory and Brown, 1966) (*Fig. 7.5*). The other type, called maps of average slope, uses contours to indicate average slope angles over distances of a kilometer. The third type, genetic geomorphic maps, are most commonly used in geomorphic cartography. Landforms are shown according to their possible origin, taking into account present and past processes and the age of the landform. Numerous legends have been proposed for maps at different scales including those of Tricart (1965, 1972), Verstappen and van Zuidam (1968, 1991), St-Onge (1968), Demek (1972), Panizza (1972), Demek and Embleton (1978), Barsch and Liedke (1985), and Peña (1997). These legends generally follow morphological differences present in various countries; for this reason geomorphologists develop systems or legends appropriate for the needs of particular countries (Parsons, 1988). Structural shapes predominate in the preparation of small-scale geomorphic maps whereas climatic landforms are used in large-scale geomorphic mapping.

7.3 Types of mass movements

The different types of mass movements show a wide variety of processes and factors that offer unlimited possibilities for classification (Hansen, 1984a). Mass movements can be classified by the kinematic type of movement, the type of material moved, their type of activity, rate of movement, and stage of development, among others (Cruden and Varnes, 1996). The proliferation of **classifications** has produced a voluminous scientific terminology. A classification suitable for one region may be ambiguous if used in another. In some regions landslides are very rare whereas in others they are so common that they are a significant factor in landform formation (Zaruba and Mencl, 1969, 1982) (*Fig. 7.1*). Different generic terms used to designate mass movements become the title of the classification as discussed by Crozier (1986): *slope failure* (Ward, 1945), *mass wasting* (Yatsu, 1966), *mass movement* (Hutchinson, 1988), *landslides* (Varnes, 1958), and *slope movement* (Varnes, 1978).

Other classifications in addition to those discussed above were popular when they were published. Terzaghi's classification (1943) is based on the physical properties of the affected rocks. Sharpe (1938) classified mass movements as a function of the material moved and the type and rate of movement. Varnes (1958), using criteria similar to Sharpe's, proposed a classification that he later improved considerably (Varnes, 1978). *Figure 7.6* shows the various types of movement (falls, landslides, and flows) that Varnes distinguished. Zaruba and Mencl (1969) base their classification on the nature of the affected rocks and types of movement. The geotechnical classification of Skempton and Hutchinson (1969) uses soil fabric and pore fluid pressure as primary principles. Finally, Crozier's morphometric classification (1975) is based on morphometric indices and groups of processes. The European classification (EPOCH, 1993) is based on type of movement and affected materials. Brunsden's (1993) classification is based on processes that "cause" landslides. There is no consensus on the best mass movement classification system to use because all existing classifications have their limitations. For this reason it is best to use existing terminology to classify and at the same time describe in detail the movements being studied. Today, the most consistent classifications are those of Varnes (1978), Hutchinson (1988), and Cruden and Varnes (1996).

Because various authors use different terminology, it is helpful to first define the term **landslide**, which Cruden (1991) defines as "the movement of a mass of rock, detritus, or earth downward on a slope." The term landslide is usually used in its broad sense to designate almost all varieties of mass movements that occur on slopes, including such processes as falls, topples, or flows in which there is no or very little displacement (Varnes, 1984). The term landslide generally excludes **creep**. Bryan (1922) defines mass movement as a "slow downslope movement of soil and detritus, caused by weathering. Creep is primarily a result of gravity, but is promoted by the presence of water, alternation of

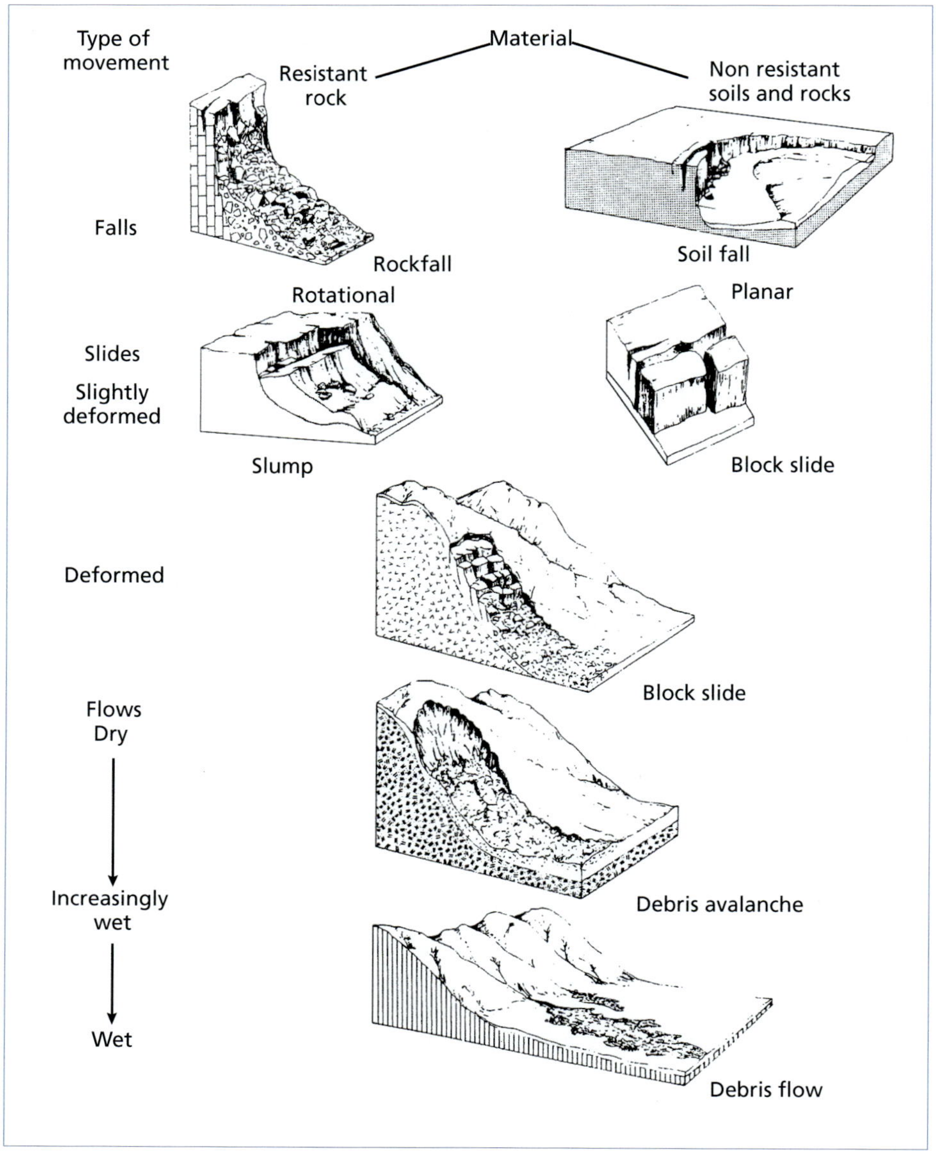

Figure 7.6 Major types of mass movements, according to the classification by Varnes (1958).

wetting and drying, freezing and thawing, growth of roots and the resulting disintegration of rock, and by burrowing of animals." For Zaruba and Mencl (1982), creep (*Fig. 7.7*) is slow deformation, much too small to produce shear rupture as a landslide does. Creep, on the contrary, is confined to the surface of the slope; its velocity decreases with depth. Creep characteristically bends strata and other objects; it is recognized by bending (*Fig. 7.8*). Creep is more intense in periglacial zones (frost creep). In contrast, landslides are rapid movements of earth materials separated from the underlying stable part of the slope by an obvious plane. The term mass movement may be more precise to generically designate these processes because the modifier "mass" establishes the difference from other processes in which transport is not massive (Brunsden, 1984).

It is usually possible to identify various parts and characteristic features of a landslide although their formation depends on the type of mass movement. Varnes' (1958) diagram of a rotational landslide with flow is highly representative (*Fig. 7.9*). This figure proposes a series of easily understood terms. *Figure 7.10* partially reflects the characteristics of the landslide in the preceding figure.

In our discussion of the various types of mass movements, we will not adhere to any classification because we

Figure 7.7 *Creep* in Silurian slates. Sierra de Aracena. Province of Huelva. Southwest Spain.

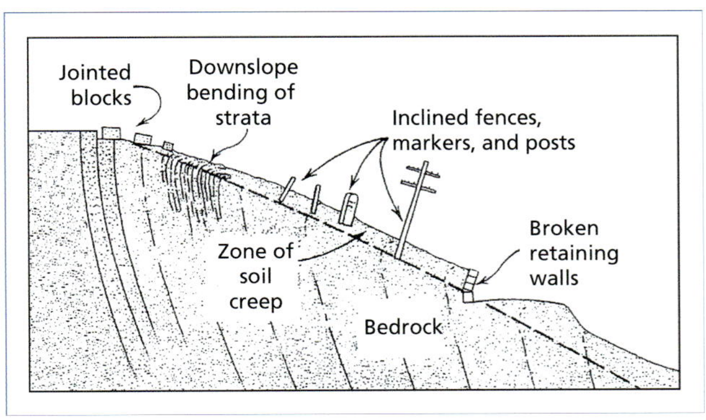

Figure 7.8 *Creep* affecting the upper part of a slope. Various objects show evidence of deformation (Sharpe, 1938).

consider it important to first understand the type of rupture, mechanisms, and causes (Dikau *et al.*, 1996a) of each mass movement and in this way distinguish and investigate them, and then later evaluate slope stability and apply appropriate corrective measures. In addition, it is difficult to categorize a landslide as a particular type. In general, landslides are complex; their complexity increases when various types of materials are involved simultaneously in a single landslide (Glade and Crozier, 2005).

7.3.1 Falls and rockfalls

These are generally defined as a rock mass that breaks loose from a steep slope (cliff, cut) normally with a small cutting surface (Ayala *et al.*, 1987b). The materials fall freely if the slope is sub-vertical (*Fig. 7.11*) and the particles saltate when the slope has a smaller angle. On slopes inclined less than 45°, particles roll (Cruden and Varnes, 1996). Loose blocks usually break into fragments when they hit the slope. In addition, falls can usually be subdivided into rock, detritus, and soil because the mechanics of these processes is similar. The initial rupture surface can be planar, wedge-shaped, stepped, or vertical (Flageollet and Weber, 1996).

Causes vary; one of the most significant is the formation of ice in joints, which increases volume by 9% and can lead to enlargement of the joints. Successive cycles of freezing and thawing cause the rock to break. Intense rains can also be a significant triggering factor. Another very common factor is slope undercutting caused by fluvial (*Fig. 7.12*) and marine (*Fig. 11.16*) erosion. A good example is Niagara Falls where large blocks of massive dolomite have piled up at the base of the cliff (Bromhead, 1986). Seismic shaking can also be a triggering factor as shown by the Huascarán (Perú) landslide (*Table 7.3*) (*Figs. 7.49 and 7.50*). Some towns are subject to significant landslide risk because of steep slopes and fracturing of rocks in a mountain climate; an example is Andorra

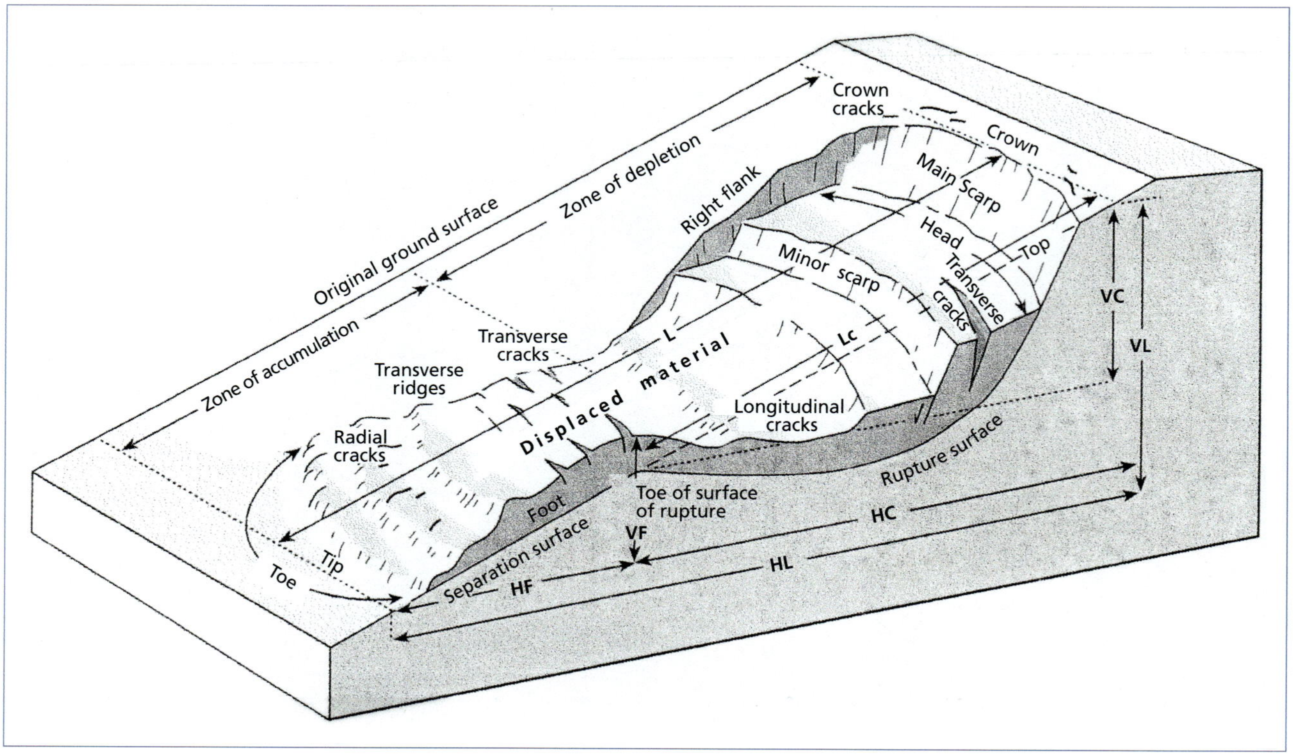

Figure 7.9 Principal characteristics of a slump-earthflow. Note: H is the horizontal distance and V is the vertical distance in the various parts of the slide. (Modified from Varnes, 1958).

Figure 7.10 Combination slide and flow, located at the foot of the building of the Department of Geology of the Faculty of Sciences of Santander (Spain). Photo obtained by a member of the department in 1979.

Figure 7.11 Rockslide. Götsu Canyon. Turkey. Photo F. Gutiérrez.

Figure 7.12 Rockslide caused by fluvial undermining of the Nahal Zin. Israel.

la Vella (central Pyrenees), which is located in a narrow valley (Copons *et al*., 2005).

7.3.2 Topples

A topple is a rotation of a rock or soil mass toward the outside of a slope along a point or axis below the center of gravity of the displaced mass (Cruden and Varnes, 1996) Topples are ruptures that occur in materials in vertical structures; they form from planes of weakness in the rock (*Fig. 7.13*). The separation of columns can occur along one or more surfaces and decompression is usually an essential requirement. A topple can mobilize up to a billion cubic meters of rock and soil (Dikau *et al*., 1996b).

Various types of topples have been identified (Goodman and Bray, 1976; Cruden and Varnes, 1996): a flexure topple commonly occurs in slate, phyllite, and schist (*Fig. 7.14A*). A block topple has closely spaced jointed columns; they are typical of massive limestone and sandstone and columnar volcanic rocks. A *chevron topple* (*Fig. 7.14B*) shows a change in dip along the fracture surface. Finally, flexed *sagging block topples* (*Fig. 7.14C*) are characterized by sagging falsely continuous wide columns affected by movement along numerous joints.

Factors that cause topples are freezing and thawing cycles, saline weathering in sandstones, and dissolution in limestones. These weathering processes weaken the rock, deepen cracks, and promote basal undercutting. In addition, rock topples occur on big scarps whereas topples in detritus occur on small scarps (Dikau *et al*., 1996b; Dikau, 2004).

Figure 7.13 Topple in sandstones, called Nefertiti Green River. Gray Canyon. Utah. United States.

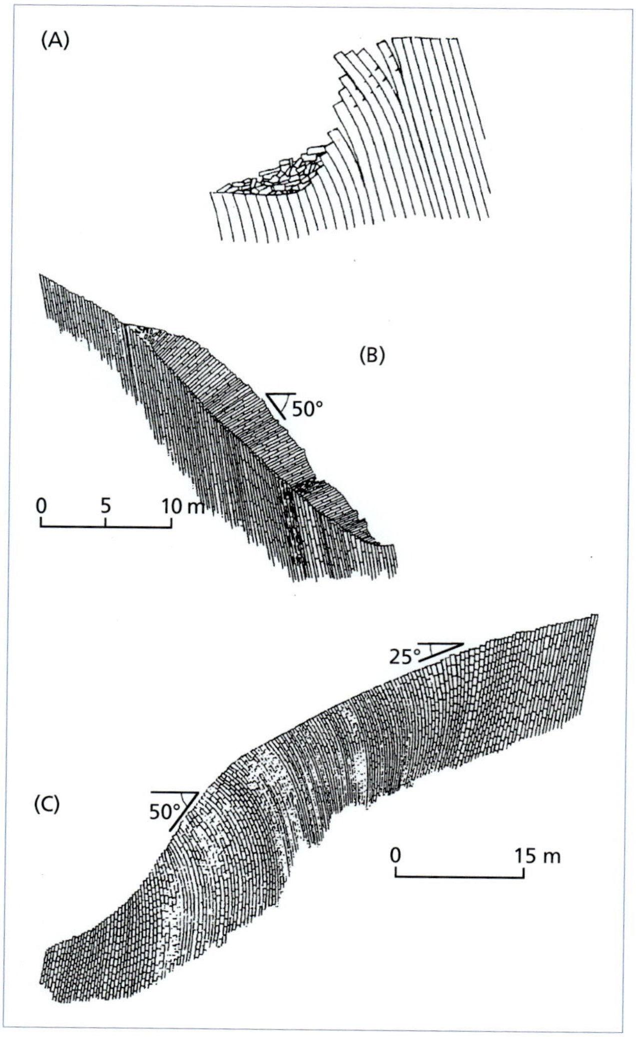

Figure 7.14 Three types of topples. (A) *flexure:* the cracks indicate topple by tension; the fallen blocks indicate that the movement is complex (rock topple and rock slide). (B) *Chevron:* topple of multiple blocks; the surface of the chevron hinge can develop on a landslide rupture surface forming a complex slide involving topple and sliding. (C) *by flexed sagging blocks* (Cruden and Varnes, 1996).

7.3.3 Landslides

"A **landslide** is a slope movement of a soil or rock mass that occurs primarily on a rupture surface or over narrow zones of intense shear deformation" (Cruden and Varnes, 1996). Frequently the first signs of earth movement are surface cracks; a landslide is often triggered along these cracks. However, this is a progressive movement in which the displaced mass slips past the original rupture surface on the original earthen surface, which is thus a separation surface (Varnes, 1978) (*Fig. 7.9*). Once started, sliding is visible on scarps; striated flanks indicate the direction of movement (Ayala *et al.*, 1987b). Landslides can be classified in terms of the degree of activity: active (with present-day movement), dormant or latent (with movement during the last year, but able to be reactivated) and relict or stabilized (incapable of being reactivated) (Zaruba and Mencl, 1969; Cruden and Varnes, 1996; and Crozier, 2004a). Zaruba and Mencl (1982) classify landslides as contemporaneous or fossil according to their age; they also classify them as initial, advanced, or completed landslides.

7.3.3.1 Rotational landslides (*slumps*)

These movements are "rather rotational around an axis parallel to the topographic contours of the slope; this implies shear displacement (sliding) along a concave surface, which is visible or can be easily recognized" (Varnes, 1978). The

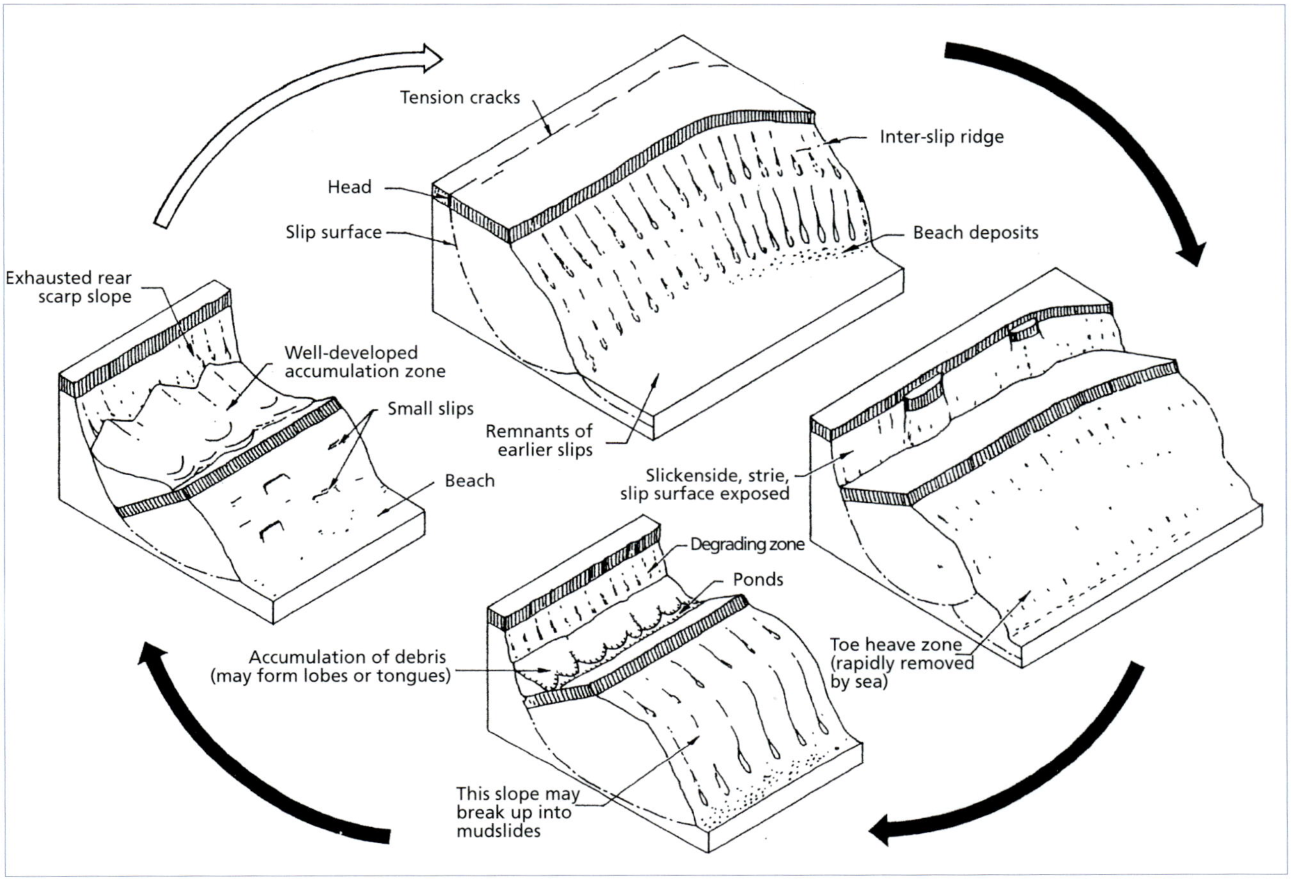

Figure 7.15 Block diagrams illustrating the development cycle of rotational landslides originating along a sea cliff (Bromhead, 1979).

shear surface may be circular or spoon-shaped. Rotational landslides can occur in rocks, detritus, and soils. They show a small amount of internal deformation in the displaced material, which distinguishes them from flows (Buma and van Asch, 1996a).

After a fracture occurs, the sliding mass begins to rotate and can break up into numerous blocks from the transverse cracks that have formed *(Fig. 7.9)*. At the source of the slide, blocks tilt away from the slope *(Fig. 7.15)* (Bromhead, 1979); in these locations, lakes may form that later fill and evolve into marshy areas that contain peat. Striations can be seen on the slipping surface; exposed cracks are concentric in plan view and concave in the direction of movement. On scarps, the impermeable material can develop rills and gullies. In the frontal zone of detritus or soil slides, lobes and radial tension crack systems can form.

Rotational landslides form in a wide variety of materials with very different geotechnical properties. In general, rotational landslides produce hummocky surfaces and drainage network anomalies. According to Varnes (1958), rock slumps move at velocities ranging from a few centimeters per year to several meters per month whereas soil slumps can reach velocities of up to 3 m/second. Their sizes range from a few meters to large complexes covering several hectares *(Fig. 7.16)*.

Conditions that promote the occurrence of rotational slides (Buma and van Asch, 1996) are slopes of highly fractured blocks (rock slumps), slopes with thick regolith or moraine deposits (soils, detritus), non-vegetated slopes, or marly or shale slopes (Selby, 1985), and rocky substrates with horizontal stratification (Young, 1972; Gracia, 1985). To these conditions we should add basal undercutting by rivers and wave action (Varnes, 1978; Bromhead 1979) and anthropogenic activities such as excavation and construction. Earthquakes, explosions, sudden overloading, and the rise of water levels from rain or rapid snowmelt can also promote landslides.

Multiple rotational landslides can occur in two or more landslide units *(Fig. 7.17)*. Their morphology is complex. At the head, movement begins with the formation of tension

Figure 7.16 Large rotational landslide. Wolf Breckenridge. Rocky Mountains. Colorado. United States Photo F. Gutiérrez.

cracks due to decompression. These open cracks gradually enlarge and in turn the mass slowly moves downward; a stepped morphology is formed by the creation of a new crack. Displacement is greatest at the center of the moving mass and during its slide the mass rotates around its shear surface. Its velocity is not dangerous because the rotation is slow.

In addition to causes listed for simple rotational slides, Buma and van Asch (1996) point out that multiple slides can form from the widening of a simple slide, which is usually caused by erosion of basal materials.

Gutiérrez *et al.*, (1994) present an example of this process in the one-hundred-meter gypsum scarp of Alfajarín near Zaragoza (Spain). They show that the scarp originated by northward lateral migration of the Ebro River. Unloading of the mass generated open and vertical joints parallel to the scarp (*Fig. 7.17* and *7.18*). Water penetrating into the fractures dissolved the gypsum and widened the cracks all the way to the impermeable argillaceous base. Karstification of basal gypsum led to subsidence of the blocks. These individual blocks slid in a rotational way (presumably in a retrogressive way) and created closed depressions.

Terracettes are regularly spaced small earthen dams made up of small rotational slides on grassy slopes where the fault throw is generally on the order of a decimeter. Some authors think they result from the trampling of cattle (Higgins, 1982). They occur most often on steep soil slopes (Hutchinson, 1968). Selby (1993) claims "they are the most

Figure 7.17 Multiple rotational landslide in gypsum-bearing material. Alfajarín. Province of Zaragoza. Spain.

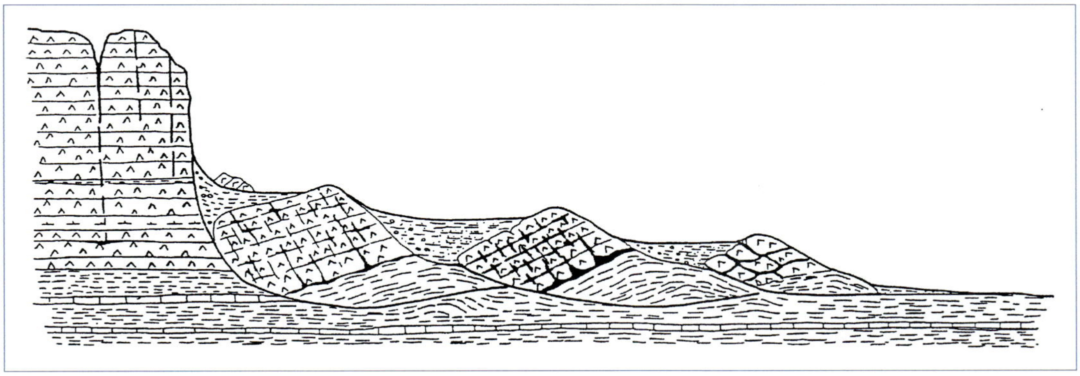

Figure 7.18 Multiple rotational landslides in gypsum where the clay-rich base facilitates sliding. Observe the Quaternary fill in areas of reverse slope and a block that has slid on top of the fill. Alfajarín scarp. Province of Zaragoza. Spain. (Gutiérrez *et al.*, 1994).

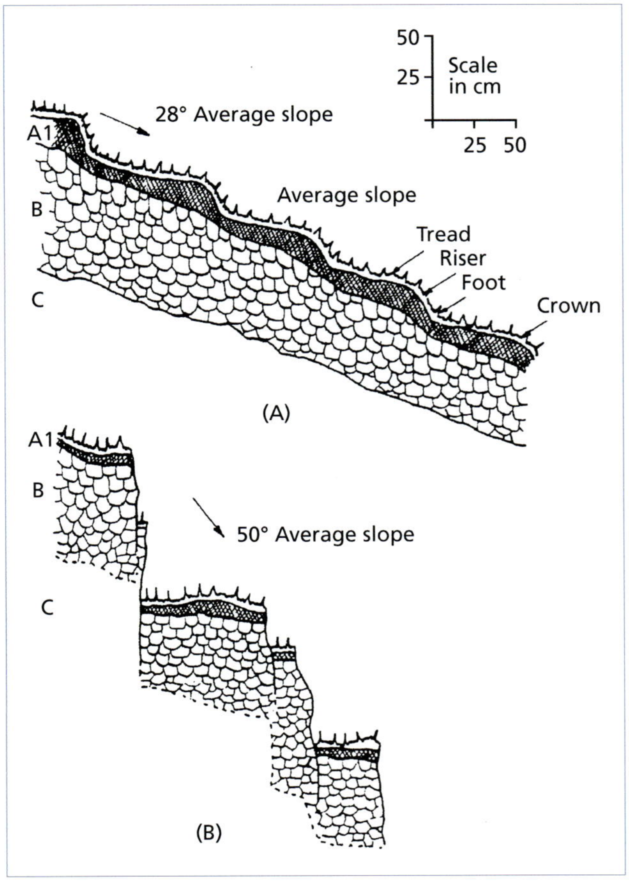

Figure 7.19 Diagrams of low and high angle *terracettes* (Selby, 1993).

prominent surface characteristics attributed to soil creep." He distinguishes two types of terracettes (*Fig. 7.19*): those that occur on low angle slopes where velocities are slow and where grass covers all slopes whereas on higher inclination slopes, velocity and cracking increase at the front of each step; they can also cause small landslides. According to Vincent and Clarke (1980), low-angle terracettes shaped like the steps of a staircase are associated with high liquid limits, low plastic limits, and low clay content. Where the steps are wider, soils have relatively low liquid limits and high plastic limits.

7.3.3.2 Translational landslides

In translational landslides, the mass progresses outward and downward along a generally flat or slightly undulating planar surface and the rotational component is minimal (Varnes, 1978). The sliding mass displaces on top of the earthen surface. The slide occurs most commonly along discontinuity planes in the rock (stratification, schistosity, faults, and joints) and in deposits with obvious variations in shear strength (Dikau *et al.*, 1996a). Translational landslides can occur in rocks, detritus, and soils.

Block slides (*Fig. 7.20*) indicate block movement along a plane, which is usually a bedding plane; they are accompanied by small rotational slides. In the area where the block slide begins, a small graben usually develops and forms a depression. The movement of the block slide occurs by impulses and the velocity of large slides increases with heavy precipitation (Ibsen *et al.*, 1996a). The fundamental causes are an abrupt change of rock type with marked differences in shear strength. Basal undercutting and heavy rainfall are the principal triggering agents for these processes.

Rockslides are typical on mountain slopes and rocky outcrops where the angle of the discontinuity planes in the rock is approximately equal to that of the slope. A good example of this type of rockslide is the Flims (Switzerland) slide that moved 12 km^3 of rock (Heim, 1932). The slope is made up of marly limestone of Malm age dipping 7–12° toward the Rhine. The head of the slide is 1400 m above the river. Slide movement was rapid; the rocky mass rose 150 m

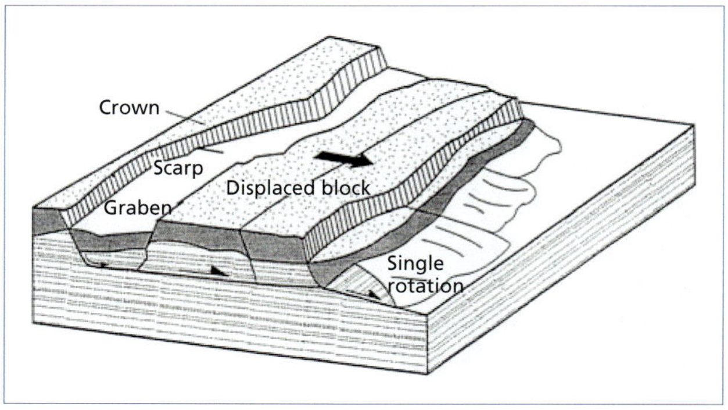

Figure 7.20 Block diagram of a typical block landslide with rupture along an approximately flat surface and with slight rotational landslide (GSL, 1987; in Ibsen *et al.*, 1996a).

up the opposite slope (*Fig. 7.21*). During the Riss-Würm interglacial it generated a 15 km long lake that later emptied. Its age, determined by ^{36}Cl and ^{10}Be, is between 4,900+250 a. and 15,440+1,480 a. (Ivy-Ochs *et al.*, 2009). Another example, larger and more devastating, is the Vaiont rockslide of 9 October 1963, in which the rockslide mass, calculated at 240 million m^3 (*Figs. 7.22, 7.56* and *7.57*) of crushed rock, moved at a speed of 20–25 m/second (Müller, 1964b; Kiersch, 1964, 1983). The movement was controlled by structural discontinuities with well-defined scarps at the head and on the flanks. Another translational rockslide occurred in the Andes of Bolivia (*Fig. 7.23*) where rocks slid along stratification and cut off the highway between Cochabamba and Santa Cruz. To restore highway communications, a tunnel was constructed below the landslide. Costa and Schuster (1988)

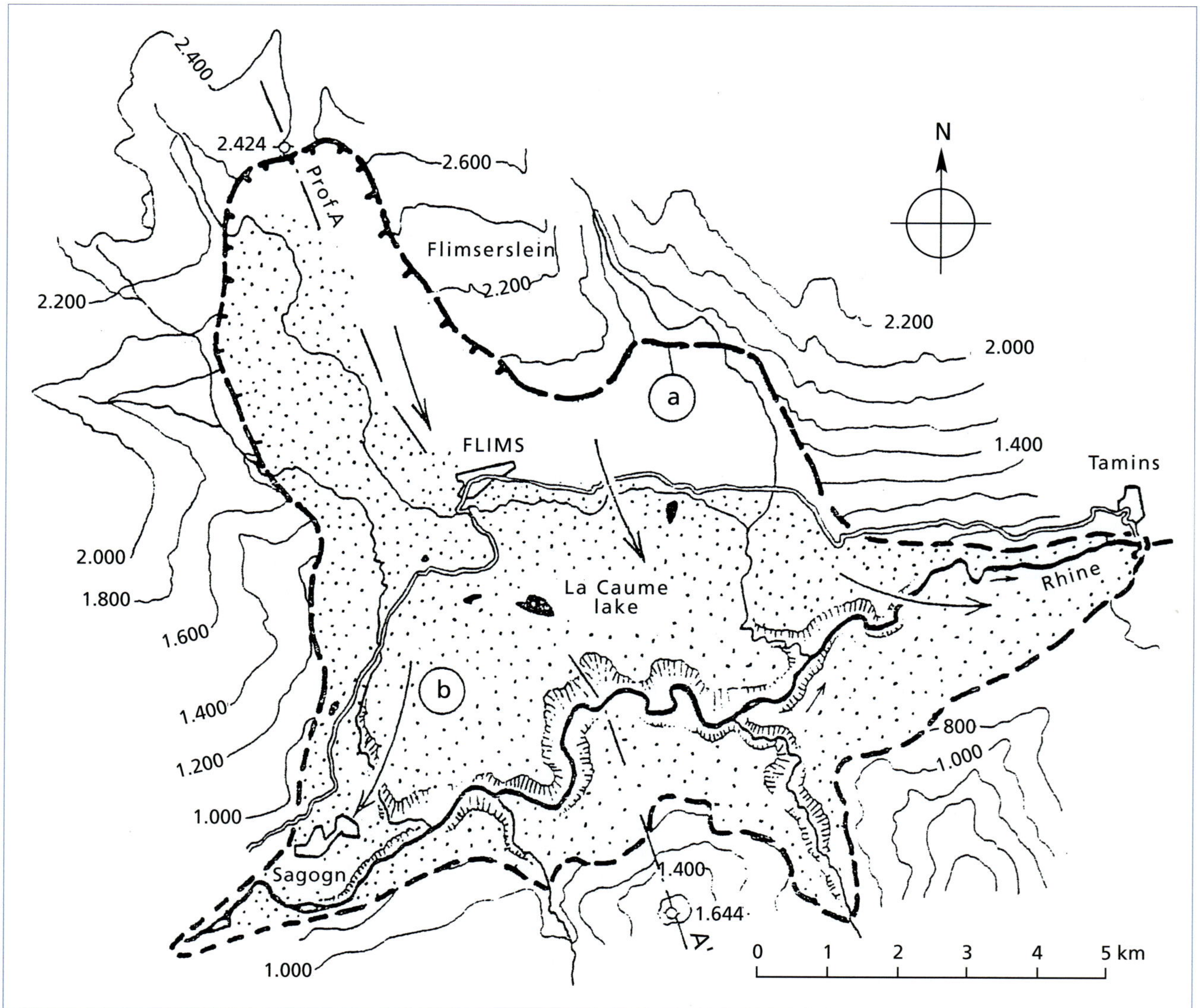

Figure 7.21 Rockslide, near Flims (Switzerland). (A) Boundary of the slide area. (B) Jurassic limestone that slid (Heim, 1932).

Figure 7.22 Translational landslide of the Vaiont rocks. Dolomitic Alps. Italy.

Figure 7.23 Rocky translational landslide that affected the Cochabamba to Santa Cruz highway (central Bolivia). The risk of new landslides was handled by boring a tunnel. Photo M.C. Maldonado.

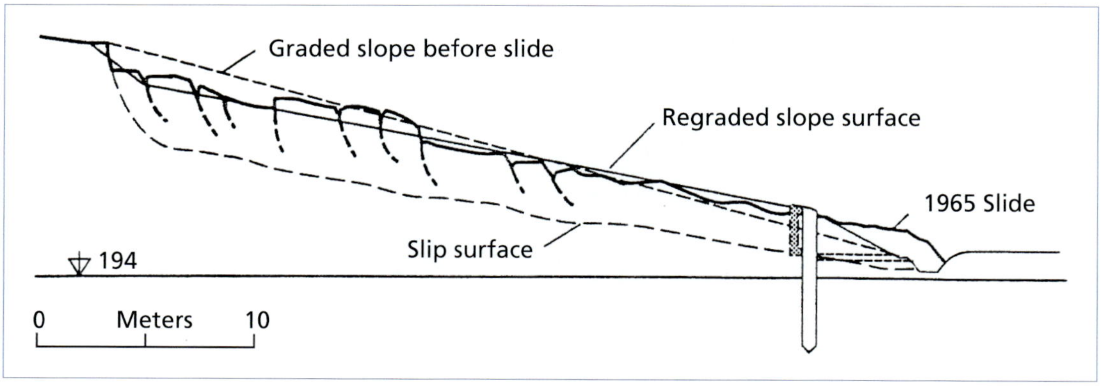

Figure 7.24 Plate landslide triggered by clearing for construction of a railroad. Prior and later stages of the landslide had been indicated East Slovakia (Zaruba and Mencl, 1969).

analyzed in detail the natural dams created by this slide, their types, and causes. The mechanism of rocky landslide movement is difficult to identify because many factors intervene. In general, they have high velocities, higher than rotational slides (Sorriso-Valvo and Gullà, 1996).

Soil and detritus slides, also called *slab slides,* have many other names (Ibsen *et al.*, 1996b). They are common in weathered soils especially in clay-rich regoliths. They move by near-surface shear. The sliding angle is related to the degree and depth of weathering. When the sliding surface is irregular, **grabens** and rotated blocks occur (*Fig. 7.24*) (Zaruba and Mencl, 1969). Slab slides are characterized by their surface occurrence; they are much longer and wider than the thickness of the slide material. At the head is an arc-shaped vertical crack; these slides are generally triggered by heavy rain. As the permafrost melts, the active layer or mollisol moves (Ibsen *et al.*, 1996b).

Debris slides are ruptures of unconsolidated material that break into small pieces as they advance (Varnes, 1978). A debris slide characteristically has shallow depth and a high ratio of length to width (Hutchinson, 1988). (*Fig.7.25*). The affected materials are usually colluvium or alteration mantles. Rupture often occurs along the rock-regolith contact (Corominas, 1996). They travel faster and farther with increased dip and less so with higher clay content (Hutchinson, 1988). Factors that trigger debris slides are earthquakes and intense rainfall that increases pore water pressure and reduces shear strength (Caine, 1980). Debris slides can cause debris avalanches (Guadagno *et al.*, 2005) and *debris flows* (Corominas *et al.*, 1996).

Mudslides or mudflows (in Europe) and earthflows (in America) are a type of mass movement in which clay, silt, or very fine sand move primarily by sliding over shear surfaces with relatively slow movement, creating lobate or elongate landforms (Brunsden, 1984).

Figure 7.25 Debris slide on steep slopes in the Central Cordillera of the Andes that affected the railroad line between San Lorenzo and Ibarra. Ecuador. Photo F. Gutiérrez.

Figure 7.26 Mudflow showing its trajectory and accumulation lobe. Funes. Dolomitic Alps. Italy.

Mudflows are made up of three parts: source area, trajectory, and accumulation lobe or zone (*Fig. 7.26*). The source area is usually shaped like an amphitheater. The length of the trajectory depends on the slope and its length. A lobate accumulation zone develops at the base of the slope; it has a low angle and radial cracks (Brunsden and Ibsen, 1996).

A landslide type that can also be included in this flow category is the slump-earthflow (Sharpe, 1938) (*Fig. 7.9*), which has three distinguishable parts. A slump-earthflow can also be considered a complex landslide. Mudflow movement is slow; it ranges from 1 to 25 m/year although it can reach much higher speeds. Mudflows can be caused by intense rainfall and rapid snow melting, as occurred on the highway between La Puebla de Valverde to Camarena de la Sierra (Province of Teruel, Spain), where a *slump-earthflow* occurred on top of the Keuper clays in March 1977, isolating the town of Camarena de la Sierra for several days. As a result, another highway was built to prevent the isolation of the town. Mudflows often occur in winter in temperate zones. For example, small mudflows occur on north-facing shaly slopes in the *badlands* of Las Bardenas Reales (Navarre, Spain) (*Fig. 17.25*) because of winter rills associated with weak storm precipitation; the slopes are eroded or destroyed by concentrated gullying during periods of storm rainfall (Gutiérrez *et al.*, 1995). Mudflows can also be initiated by the melting of permafrost (*Fig. 15.14*).

7.3.4 Lateral spreading

Terzaghi and Peck (1948) introduced the term *spread* to describe sudden movements of sand and silt that contain interstitial water. The rupture surface is not a rapid shear surface. The spread can result from liquefaction or flow (and extrusion) of softer material (Cruden and Varnes, 1996). On the other hand, Dikau *et al.* (1996a) and Dikau (2004) contend that the expression "lateral spreading" is primarily used to describe lateral displacement of a rock or soil mass over underlying softer material where the basal shearing surface is not well defined.

Some authors claim that the phenomenon called *sackung* is a type of lateral spreading (Jahn, 1964).

7.3.4.1 Lateral spreading in rocks

Lateral spreading in rocks is a lateral displacement of either homogeneous rock masses or cohesive rocks that lie on top of ductile materials (Pasuto and Soldati, 1996; Borgatti and Soldati, 2005). First defined by Jahn (1964), lateral spreading in homogeneous rocks is generally brittle behavior that lacks a well-defined surface or zone of plastic deformation. It is characterized by double ridges, depressions (trenches or grabens) and up-slope oriented scarps (*Fig. 7.27A*) (Pasuto

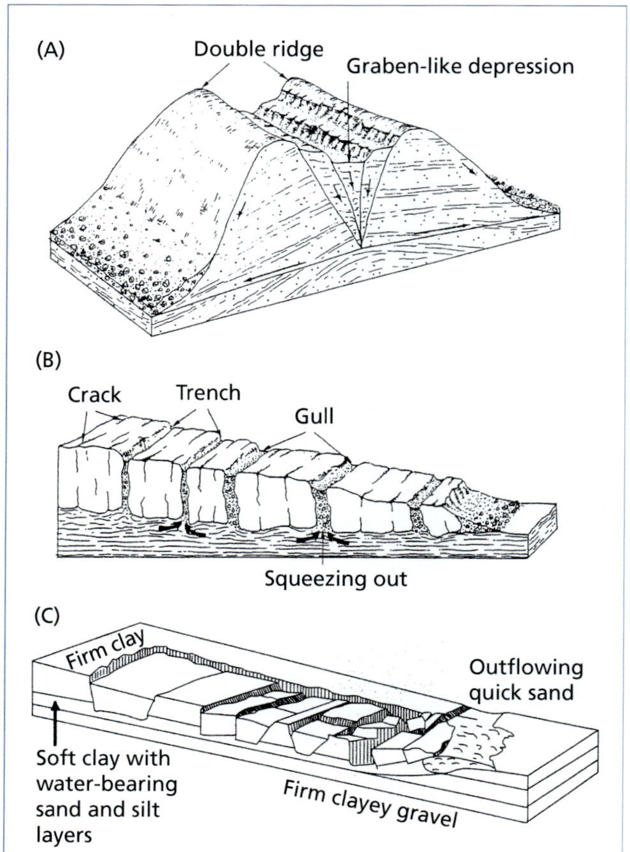

Figure 7.27 Block diagrams indicating (A) lateral extension in a homogeneous rock mass, (B) extensions of brittle cohesive formations on top of ductile materials (Pasuto and Soldati, 1996), and (C) lateral extension with ruptures in fine sand and silt (Varnes, 1978).

and Soldati; Borgatti and Soldati, 2005). The origin of lateral spreading in rocks is not well understood although it may result from earthquakes.

These movements may extend for many kilometers from scarped margins of structural platforms. McGill and Stromquist (1979) have shown the presence of trenches 600 m wide, 300 m deep, and 20 km long along the eastern zone of Cataract Canyon on the Colorado River (USA).

Another type of lateral spreading affects cohesive brittle formations that overlie ductile units, generally as a result of deformation from the extrusion of underlying material (*Fig. 7.27B*) (Pasuto and Soldati, 1996). This produces grabens, gullies, trenches, pseudo-karst depressions, and bulges of ductile material toward the valley. This spreading occurs slowly, starting with cracks that open gradually; in the final stages, blocks slip down the slope (Cancelli *et al.*, 1987). When the process affects numerous blocks, it is considered a landslide caused by the extrusion of ductile rocks (Zaruba and Mencl, 1982). Visco-elastic behavior and deformation resulting in small landslides may mobilize ductile material by extrusion during lateral extension. These partial landslides may coalesce to form a well-defined landslide surface. Zaruba and Mencl (1982) call this type of slope movement "block slides;" it is caused by extrusion of ductile rocks.

Phenomena such as *cambering* and *bulging* (Varnes, 1978; Cruden and Varnes, 1996) can be considered in this category. They are discussed in Section 2.9 of this book.

7.3.4.2 Lateral spreading in soils and detritus

Lateral spreading in soils and detritus can be defined as the settling of soil blocks on top of a layer of sensitive soil or by progressive rupture of the entire sliding mass (Buma and van Asch, 1996b). Deformation of the ground that accompanies this extension can cause loss of life and significant damage to buildings, linear constructions, and other structures.

Quick clays are found in coastal environments located near Quaternary ice margins (Alaska, eastern Canada, and the fjords of Norway and Sweden). They can undergo significant spreading because dissolution of sodium from the cohesive clays converts them to dispersed clays that move easily. Landslides of quick clays begin in areas of low slope and retreat upslope. In the first stage, a rotational landslide occurs along the sliding surface in a layer of plastic clays. The loss of horizontal support triggers rapid retreat upslope and toward the flanks.

7.3.4.3 Sackung

The term was introduced by Zischinsky (1966) to describe surface manifestations of *creep* deep in slopes with laminated rocky substrates. A *sackung* (plural *sackungen*) is a German word for a landform that typically has some scarps on the upper part of the slope (also called anti-slope scarps by Bull, 2009); these scarps are opposite and antithetical and are associated with linear graben-like depressions. Small grabens, which are parallel to topographic contours, fill with detritus and ephemeral pools (*Figs. 7.28* and *7.29*) (Gutiérrez-Santolalla *et al.*, 2005). Sackungen are 15–300 m long and 1–9 meters high. In some cases, the scarp is concave and appears to be the source of a landslide (Bordonau and Vilaplana, 1986).

Various mechanisms have been proposed to explain these scarps on the upper part of slopes. Tabor (1971) thinks differential erosion is the cause. Jahn (1964) suggests that upper-slope scarps result from a combination of gravity-induced spreading and erosion in the area of tension cracks (*Fig. 7.30A*). Other authors suggest that the cause

Figure 7.28 *Sackung* with a small filled "graben" in which a trench is being dug to study sediments and obtain absolute ages. Glacial valley of Estós, near Benasque, central Pyrenees. Spain. Photo F. Gutiérrez.

Figure 7.29 *Sackung* in Ragged Mountain (Alaska). Photo F. Gutiérrez.

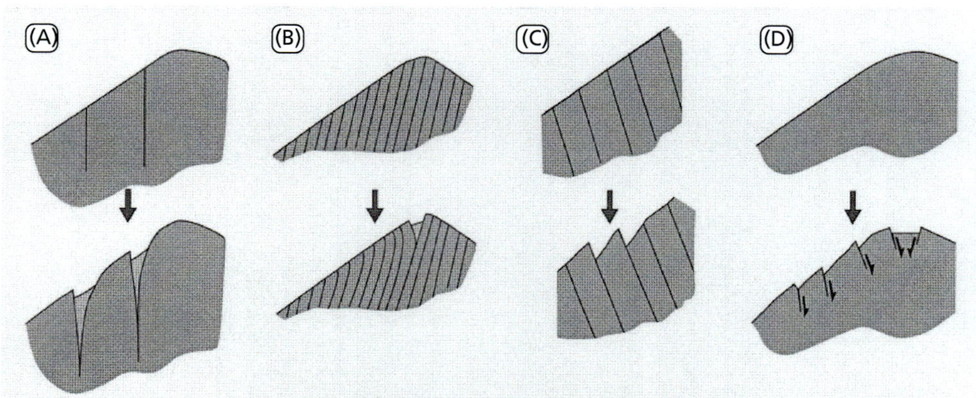

Figure 7.30 Diagrams showing lateral-extension mechanisms that various authors have proposed for the creation of scarps on the upper parts of slopes (Gutiérrez-Santolalla *et al.*, 2005).

of *sackungen* is downslope sagging in intensely laminated rocks (*Fig. 7.30B*) (Zischinsky, (1966) among others). Bovis (1982) suggests that *sackungen* are produced by flexural topples defined by high-dip joints combined with erosion on the upper slope, which create depressions (*Fig. 7.30C*). The most accepted interpretation attributes upper-slope scarps to displacement of highly dipping rupture planes (*Fig. 7.30D*) (McCalpin and Irvine, 1995; Pasuto and Soldati, 1996; Bisci *et al.*, 1996; among others). It is quite possible, however, that the geomorphic characteristics of sackungen are a case of morphological convergence or equifinality (Gutiérrez-Santolalla, *et al.*, 2005).

Various factors have been proposed to explain the lateral spreading that produces *sackungen*: loss of load in glacial valleys where a glacier is retreating (Agliardi *et al.*, 2001); extensional forces on the upper slope (Radbruch-Hall, 1978); earthquake shaking (McCalpin, 1999); co-seismic displacement along deep tectonic faults (McCalpin, 1999), and subsidence due to evaporite dissolution (Gutiérrez, 1998; Calvo *et al.*, 1999). Detailed studies indicate that sackungen pose no risk (Thompson *et al.*, 1997). As a result, it is not clear that *sackungen* are a necessary precondition for catastrophic collapses (Hewitt *et al.*, 2008).

There is little data on the age of *sackungen*. Work done by Gutiérrez-Santolalla *et al.* (2005) shows that the oldest dates for the *sackung* trench on the slope of the Valle de Vallibierna is 5900 calibrated years BP and 6790 calibrated years BP for the Estós valley. This chronology is quite different from the date of the beginning of glacial retreat and as a result, it appears unlikely that *sackungen* are generated by loss of slope loading. McCalpin and Irvine (1995) also found a similar age difference between sackungen age and glacial retreat age in the southern Rocky Mountains. As a result, it is speculative to establish a relationship between *sackungen* and climate change (Gutiérrez-Santolalla, *et al.*, 2005). It seems most probable that seismic activity may be one of the causes (McCalpin, 1999) although paleoseismic investigations would be necessary to corroborate this theory (Jibson, 1996).

A trench excavated in the glacial valley of El Ubagol (central Spanish Pyrenees) shows three displacements based on the stratigraphy of the colluvial wedge and the faults that affect them. The first caused a *sackung* (16.9 ka) and the other two events have ages of 15–8.3 and 8.3–5 Ka. *Sackung* kinematics are thought to be controlled by seismic activity. These features are archives for large paleo-earthquakes (Gutiérrez *et al.*, 2008).

7.3.5 Flows

Flows are mass movements that are fluidized by water or air. They have more internal deformation than a landslide. Flow movements occur on top of a large number of small shear surfaces or because the water content of the mass is so high that the flow behaves as a liquid (Bromhead, 1986).

Flows can be subdivided into various types depending on the material they carry: **debris flows**, **soil flows**, and **rock flows** (Dikau *et al.*, 1996a; Dikau, 2004). Debris flows move rock fragments, blocks, and clasts in a matrix of sand with low clay content. Earth flows are composed of sands, silts, and clays that are water saturated or liquefied and that behave as a fluid (mudflow) although they may occur in sands that lack clays (*grain flows*).

The rheology of the material and the landforms that result from earth and detritus flows are similar; there is a gradual transition between the two extremes. For a flow to be considered a *mudflow*, the material moved must contain more than 50% fine particles.

Unlike the flow types previously described, which occur on top of unconsolidated materials, another type affects consolidated (cohesive) rocks; these are *rock flows* (Varnes, 1978), which are generally associated with deep gravitational deformations (*sackung*).

Debris flows are flows of sediments composed of a mixture of coarse fragments cemented in a matrix of fine particles; they contain water and air in their interior (Johnson, 1970; Innes, 1983; Costa, 1984, 1988a; Johnson and Rodine, 1984; Takahashi, 1991; Coussot and Meunier, 1996; Corominas *et al.*, 1996; Dikau, 2004; Coe *et al.*, 2008). Debris flows occur in most morphoclimatic zones; they can move long distances and are very destructive. The source of the material is slope and alteration deposits. Mountains can also supply sediments from areas composed of alluvial deposits associated with earlier stages of fluvial activity. The water necessary to generate a debris flow usually comes from high intensity storms or from rapid thawing or rainfall on top of a blanket of snow on slopes. These circumstances cause rapid runoff that can mix with unsorted sediments containing abundant argillaceous particles to produce debris flows. These flows behave as viscous fluids to increase velocity and decrease shear resistance (Johnson, 1970). The viscosity of debris flows varies considerably; when they have high viscosity and are voluminous, they erode channels and produce well-defined levees along the overflow margins of the channel. On the contrary, if viscosity is very high, they move along meadow hillsides where they do little erosion and create very small levees (Selby, 1993). All this creates a wave of wet detritus that moves by gravity along the channel (*Fig. 7.31*). The morphology of debris flows is variable although common features are present. These flow deposits are commonly lobate in the frontal zone and undulatory in the anterior parts, which results from successive waves of material (*Fig. 7.32*). Debris flows have lateral and central deposits that reach the frontal slope (Johnson and Rodine, 1984). Debris flows incise considerably into the drainage channel and can overflow, creating marginal deposits (levees). The internal structure (*Fig. 7.33*) is massive with large clasts distributed haphazardly in a fine-grained matrix although local sand or silt lenses can be found that have been deposited by a more fluid flow between successive episodes of debris flows. These stratified deposits, called transition deposits, can form under transitory conditions between a phase of debris and another fluvial phase,

Figure 7.31 Debris flow with lobes emplaced on top of a large talus slope. Ibón de Plan. Aragon Pyrenees Photo P. Lucha.

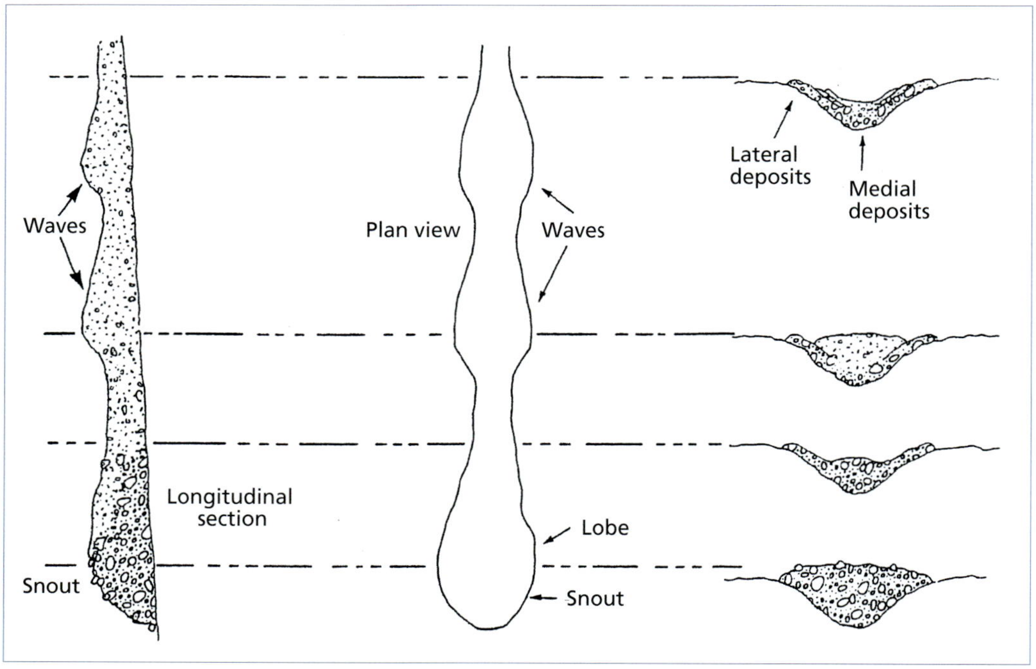

Figure 7.32 Diagram of a lobate debris flow showing waves of detritus and lateral and medial deposits (Johnson and Rodine, 1984).

Figure 7.33 Fault contact between debris flow deposits to the left of the photograph and stratified fluvial sediments on the right. Pleistocene alluvial fans in the Rio Blanco. Precordillera of the Andes Province of Mendoza. Argentina.

frequently at the end of the debris flow (Wells and Harvey, 1987). The latter are more common in the proximal zones of alluvial fans in desert environments (Godt and Coe, 2007).

Two mechanisms have been suggested for the origin of debris flows. The most widely accepted theory proposes that the addition of rainwater and snowmelt transforms detritus landslides from abrupt slopes (27° to 56°) into debris flows; thus from a morphological point of view, the head of a debris flow is usually a debris slide. This soaking of a loose deposit decreases friction between particles and promotes the downslope movement of the debris flow (Campbell, 1975; Costa, 1988a). Another suggested mechanism is that rapidly flowing water encounters a loose deposit and mixes with it to produce a debris flow (Johnson, 1970; Johnson and Rodine, 1984). These authors call this type of initiation of the process "the water hose effect." To produce debris flows by any mechanism, high intensity precipitation is required, which may cause flash floods. These flash floods are not common and the recurrence interval for these deposits is estimated at 300 to 10,000 years (Costa, 1988a); they occur locally.

Debris flows are triggered by intense rainfall on abundant colluvial material; they move through channels and on top of alluvial fans (Johnson and Rodine, 1984) and are finally deposited in low slope areas, in many cases contributing to the head of alluvial fans. The distinction between debris flows and fluvial deposits is not always easy because continuous gradation exists between them (Hooke, 1987). Both result from channelized fluids with different viscosities and a significant capacity for transport and erosion.

Along their reach debris flows remove vegetation, cover highways, canals, and railroads, carry off cars, destroy structures, and cover farm fields (Hooke, 1987). In argillaceous areas in semiarid zones, mudflows can also be channelized by the network of rills and gullies, creating significant risks.

Debris flows are very common in semi-arid zones of the world; numerous researchers have studied them, particularly in California. For example, numerous debris flows have occurred in the Canada Valley of Los Angeles County including flows in 1934, 1938, and from 1962 to 1971 that caused 63 fatalities and significant property losses (see Johnson and Rodine, 1984).

A 1988 earthquake near the city of Gissar in southern Tajikistan near the border with Afghanistan triggered numerous landslides on loess hills saturated with irrigation water. Seismic acceleration caused liquefaction of these materials, generating mudflows that affected the area to a distance of 2 km and moved a total of about 20 million m^3, burying houses on the outskirts of the city (Ishihara, 1989).

In some areas, urban development extends to the heads of fans, which causes larger detritus and mud flows because the urban substrate propagates flow and prevents water losses because it is impermeable (Rantz, 1970).

Lahars are debris flows of pyroclastic material on the flanks of volcanoes that are explosive in nature. Lahars range in size from volumes less than 0.1 million m^3 to collapses of volcanic edifices (>3.5 billion m^3) (Neall, 2004). They are one of the most devastating volcanic hazards because they can cause numerous fatalities and destroy structures located in their paths. Neall (1996) claims that lahars have caused 64,000 fatalities throughout history.

Debris flows have been the subject of experimental field studies (Iverson *et al.*, 2005; Stock and Dietricht, 2006) and modeling studies (Berti and Simoni, 2005).

7.3.6 Complex mass movements

Slope movements often involve one or more types of movement that have already been discussed. Complex mass movements are evident in different parts of the mass moved or in distinct developmental stages of the movements (Varnes, 1978). Thus it is common for a mass to begin to move as a particular type of rupture and evolve later to another type of movement.

Many examples, some of them catastrophic like that of Huascarán (Peru) or Aberfan (United Kingdom), illustrate the most common types of complex mass movements.

7.3.6.1 Rock avalanches

A rock avalanche, also called *sturzstrom* by Hsu (1975), can be defined as "a mass movement, primarily fluidized by air, that rapidly mobilizes large quantities of rock derived from the disintegration of loose rocks and that moves under the influence of gravity" (Seijmonsbergen *et al.*, 2005) (*Figs. 7.34* and *7.35*). The volume of a rock avalanche in the zone of accumulation can exceed 10 million m^3 and it can cover a surface greater than 0.1 km^2 (Angeli *et al.*, 1996). Table 1 from Hewitt *et al.*, (2008) contains an inventory of large

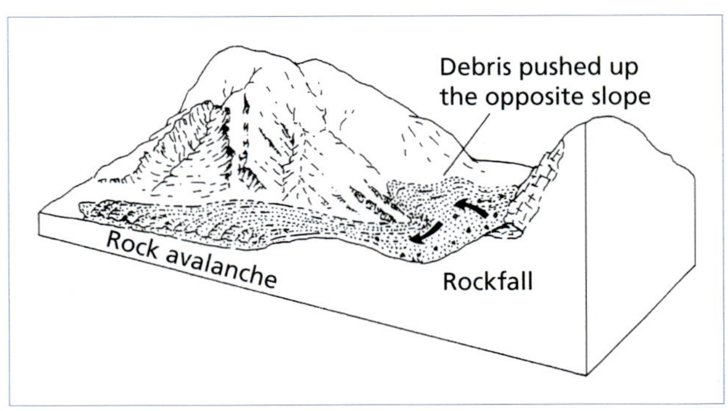

Figure 7.34 Block diagram of a rock avalanche (Angeli *et al.*, 1996).

Figure 7.35 Steep slopes in principal dolomite, landslides, detritus cone, and rock avalanche. Passo de Giou (2236 m). Dolomitic Alps, Italy.

(>10^6 m^3) catastrophic rock avalanches in various mountainous areas of the world.

Velocities of large rock avalanches exceed 90 km/hour and some blocks can travel 350 km/hour. They can travel a distance of up to 10–30 km and climb several hundred meters on an opposite slope. Occasionally a rock avalanche becomes channelized and deposits its load at the foot of a mountain in the shape of a fan (*Fig. 7.36*). Large avalanches form primarily under high mountains with steep slopes that are undercut by fluvial and glacial processes, which create sub-vertical scarps or cliffs where falls of large rocky masses can occur (Selby, 1993).

Mechanical analysis of rock avalanches begins with the initial rupture that causes the landslide and then the subsequent displacement (Angeli *et al.*, 1996). Numerous hypotheses have been proposed to explain the origin of rock avalanches but none seems totally satisfactory. In addition to undercutting, ice retreat after the last glaciation may have caused decompression of mountain slopes (Panizza, 1973) but the principal difficulty is explaining the reasons for the movement of large volumes of rock and the kinematic change that occurs after the initial rupture (Angeli *et al.*, 1996).

An example of their magnitude is the rock avalanche that covered the 11th century Romanesque hermitage of San Adrián Sasabe (Huesca Pyrenees) to the top of the central nave during the 15th century.

7.3.6.2 Flow slides

Flow slides are structural collapses of slope material that is momentarily fluidized and mobilized; these slides can reach high velocities and move considerable distances (Ibsen *et al.*, 1996c, Dikau, 2004) (*Figs. 7.37* and *7.38*). The flow slide is composed of poorly compacted material that loses its cohesion as the mass becomes fluidized. The fluid can be air or water and thus the dominant mechanism is fluidization or liquefaction. The flow slide is a complex mass movement that begins as a landslide and rapidly fluidizes as it progresses and turns into a flow. Its name indicates that it has its own morphological features characteristic of both landslide and flow. When the landslide ends, the subsequent flow ceases shortly afterward (Ibsen *et al.*, 1996c; Dikau, 2004). In general the duration of the flow is usually a few minutes although the front of the flow slide may remobilize for several days afterward, as occurred in Jupille (Belgium) (Calamber and Dantine, 1964). Flow slides occur in materials that have very little cohesion such as glacial tills and coal tipples, as shown by the disastrous flow slide that occurred in Aberfan (United Kingdom) in October 1966 and the ash in Jupille (Belgium) in February 1961. There are also dumps such as Portiño (Coruña, Spain) where in September of 2002 (*Fig. 7.39*) a large landslide occurred and caused major environmental impact. Bishop (1973) conducted a detailed study of this type of flow slide.

Figure 7.36 Channel and fan related to a rock avalanche. Parque Natural di Dolomiti di Sesto. Italy.

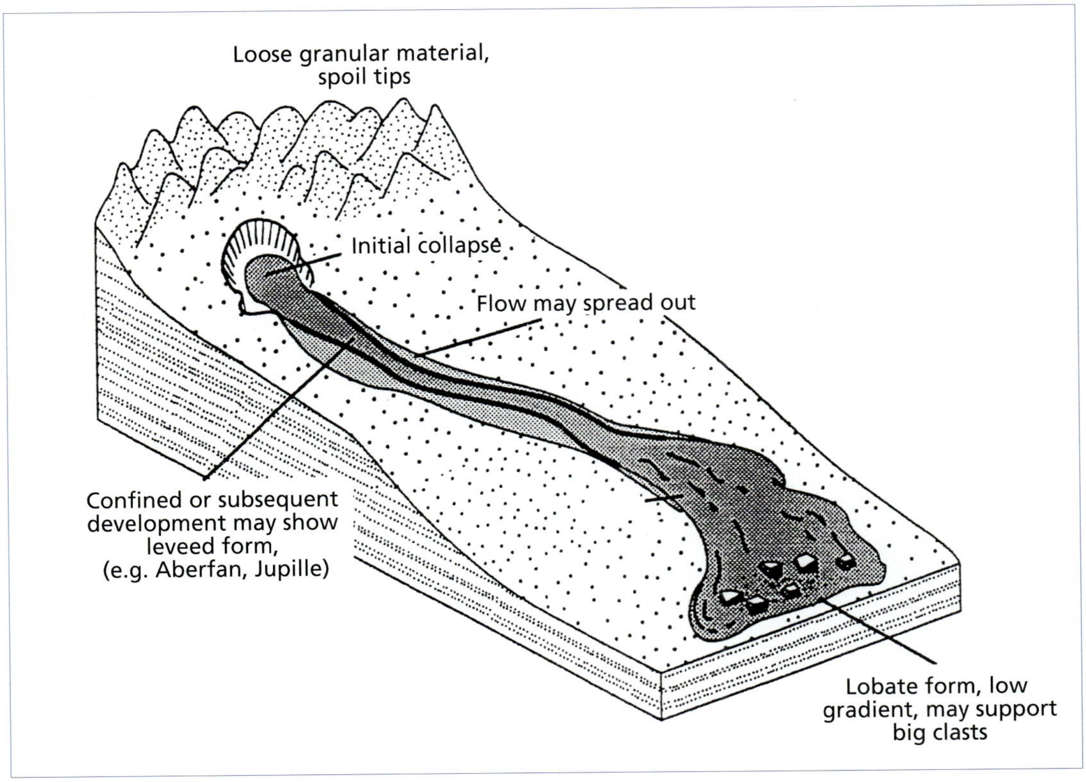

Figure 7.37 Block diagram of a typical flow slide (Ibsen, *et al.*, 1996c).

Figure 7.38 Flow slide in a moraine belt composed of loose heterometric material of glacial *till*. Barranco de Remascaro. Benasque. Aragón Pyrenees. Photo F. Gutiérrez.

Figure 7.39 Flow slide in a dump. Portiño (Coruña). Spain. Photo F. Gutiérrez.

The category of complex mass movements includes the reactivated landslide of La Valette in March of 1988 (Fig. 7.40) in the Alps of Haute-Provence; it occurred on the right side of the Ubaye River in the basin of the Valette River. The principal scarp at the head is 300 m tall and is associated with a rotational landslide. A volume of 6 million m^3 of black Jurassic clays and glacial till flowed from this point along a sliding plane 30 m deep.

The causes of these complex mass movements may be vibrations, seismic temblors, intense rain, rapid overloading, and mobilization of lateral support (Ibsen *et al.*, 1996c). These high-energy flow slides are capable of transporting large blocks. As a result, they have caused numerous human fatalities and material damage to infrastructures, including modifications of river flows.

Figure 7.40 La Valette flow slide. Alpes of Haute Provence. France.

7.3.6.3 Large gravitational landslides

In addition to landslides along predetermined surfaces based on Zaruba and Mencl's (1982) classification, such as the Quaternary landslide of Flims (Switzerland), other ancient rock movements have occurred along bedding planes or on shales, phyllites, and micaschists. These large-scale deformations on mountain slopes, known as *gravitational landslides*, are complex movements. They occur in tectonically elevated sedimentary complexes that contain folded structures. Elevated layer complexes slide downslope with the help of materials that have a low coefficient of friction. These landslides may be very large (Zaruba and Mencl, 1982).

Gravitational landslides related to thrust sheets in the European Alpine mountains have been described in the area of tectonics. Giannini (1951) studied one example in the Apennines where Oligocene stratified clays and sandstones slid on top of Malm clays (Fig. 7.41). Deep gravitational deformation with various stages of evolution has been identified in the Dolomites (Alps) of Cortina d'Ampezzo (Italy). In the upper part of the slide, lateral spreading was accompanied by a block landslide and in the lower part, lateral spreading evolved into block landslides. In contrast falls, topples and small displacements occurred at the base (Pasuto *et al.*, 1994).

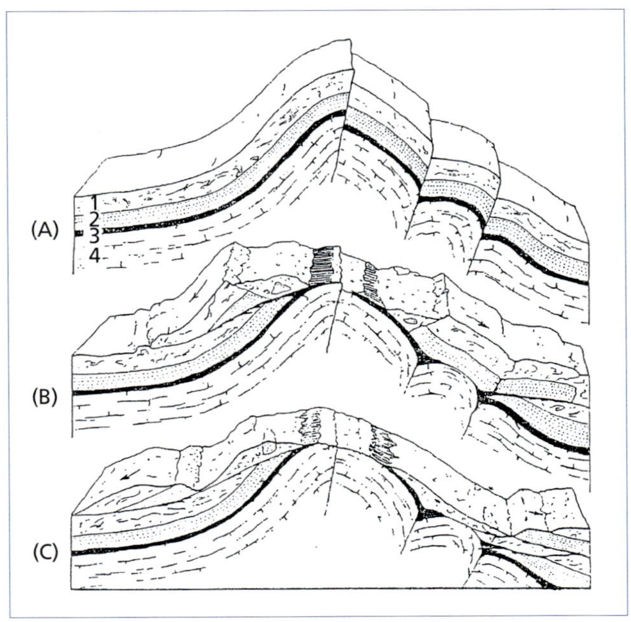

Figure 7.41 Development of a gravitational landslide: 1-stratified clays, 2-sandstones, 3-red clays, 4-siliceous limestones. (A) Uplift of a faulted fold, (B) Gravitational landslide of the apical part of the fold along the clay layer, (C) Final stage after partial denudation (Giannini, 1951).

In the Sierra del Montsec (central pre-Pyrenees), researchers have studied landslides affected by the pre-Oligocene erosional surface and other older large-scale landslides that occurred during the Garumnian and early Eocene. Quaternary landslides have been active until the present. Rosell and Linares (2001) indicate that mass movements occur in thrust fronts that may be synchronous with or later than emplacement of the thrust. These mass movements occur when the relief is most energetic. Thus the front of the Montsec thrust sheet contains mass movements that have been mapped and ordered chronologically. They are related to a set of rotational landslides that appear stepped throughout their temporal evolution (Fig. 7.42). Rosell and Linares think that climate was a triggering factor of these slides.

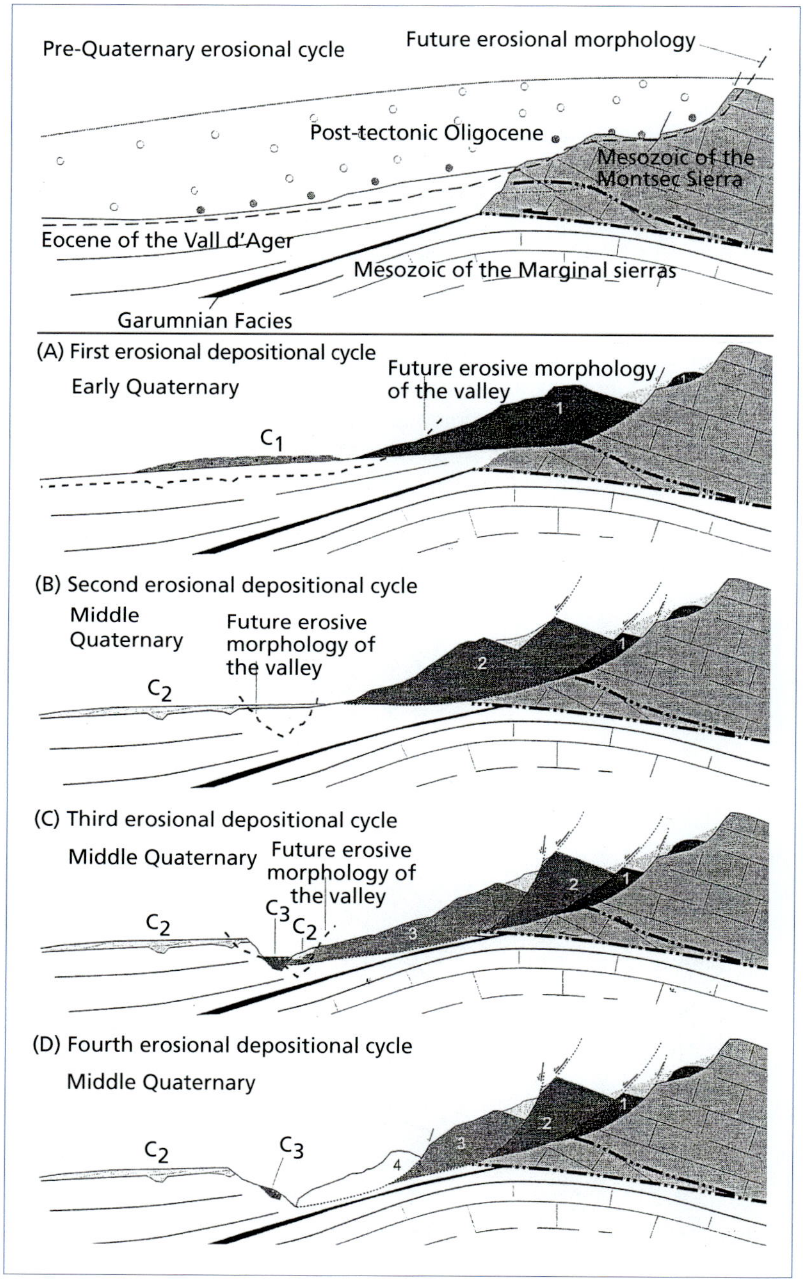

Figure 7.42 Evolutionary model proposed for gravitational landslides in the Sierra del Monsec. It shows the temporal succession of four erosional-depositional cycles, according to the Quaternary record of the area (Rosell and Linares, 2001).

A study of deep gravitational deformation in the mountains of Crimea (Ukraine) by Pánek *et al.* (2009) indicates that extensional discontinuities along with karstification occurred >300 ky BP whereas the evolution of the principal scarp at the top began approximately 110 ky BP.

7.4 Factors that control mass movements

The study of mass movements requires an understanding of factors that affect slope stability and trigger mass movements (Ayala *et al.*, 1987b; Phillips, 2005).

Slope materials are subject to shear forces that move materials downslope; an opposing force, shear strength, counteracts movement. For this reason, a **safety factor** can be calculated, which is the ratio of strength/shear force. On a static slope, strength is greater than shear force and the safety factor is greater than one (Crozier, 2004b). Instability is determined not only by the slope's margin of stability but also by external destabilizing forces that affect it (Crozier, 1986).

Mass movements are intimately linked to the geological cycle. Epeirogenesis, weathering, and erosion are long-term slope processes that establish the conditions that trigger mass movements.

Selby (1982) established two types of slopes based on the intensity of dynamic and static processes. The first are slopes where weathering is less that the potential rate of material removal. The second are slopes where the transport rate does not remove all the regolith, allowing it to accumulate on the slope. An intermediate would be the point where the two processes are in equilibrium.

Overall and over the long term, slope processes are determined by fluvial incision. Valleys widen principally as a result of slope processes whereas the stream at the base of the valley incises vertically and transports material downstream (Cotton *et al.*, 1996). This long-term process is responsible for many of the landforms we see today.

For the purpose of adequately understanding the causes and development of instability, we can distinguish several factors (Crozier, 1986, 2004a). 1) **Preconditioning factors** are static inherent factors that not only affect the stability margin but also act as catalysts for the more effective operation of other dynamic destabilizing factors that may catalyze mass movement by external causes (undermining, rainstorms). 2) **Preparatory factors** are dynamic because by definition they predispose the slope to move; that is, they are factors that make a slope susceptible to movement but do not initiate it (examples are weathering, climate change, tectonic uplift, deforestation, human activity). 3) **Triggering factors** initiate movement and the slope becomes unstable (intense rains, seismic shaking, and undermining). 4) **Sustaining factors** control the duration, shape, and rate of movement on an unstable active slope.

In some regions mass movements are not common whereas in others they are so abundant that they constitute a crucial factor in landform formation *(Fig. 7.25)*. For example, on the mountainous island of Taiwan, where high intensity tropical precipitation occurs, 7810 landslides were identified in an area of 100 km^2 from 1965 to 1977 (Lee, 1981). However, the majority of slopes are stable most of the time.

Numerous causes of landslides have been documented; they include geological, morphological, physical, and anthropogenic (Alexander, 1992; Cruden and Varnes, 1996; Hutchinson, 1988; Croista and Clagne, 2009) although only one can be the trigger (Varnes, 1958). In some cases, landslides can occur without an obvious triggering cause; various factors can contribute, such as weathering that gradually causes the slope to have weakened shear strength (Wieczorek, 1996).

Geologic causes result from the fundamental characteristics of the material that constitutes the slope; these factors are lithology, mineralogical and textural composition of the material, and its structure (Borgatti and Soldati, 2005; Gutiérrez *et al.*, 2005; Bull, 2009). Together they determine geomechanical behavior. Tectonically active areas are susceptible to landslide triggering (Bull, 2009). In addition, the geologic history of the area is important (Ballantyne, 2002).

Rocks and soils contain many planes of weakness that significantly reduce the shear strength of the mass located above unaltered material. These discontinuities also play a fundamental role in groundwater movement (Henscher, 1987). Henscher classifies common geotechnical discontinuities in all types of rocks and soils and distinguishes between tectonic joints, faults, sheeting, and lithological discontinuities. When the discontinuities have angles similar to those of the slopes, it is easy to trigger translational landslides, as in flysch, for example.

In granitic terrains, the most important characteristics affecting the mechanical behavior of regolith are hereditary discontinuities, texture, and types of clay minerals. Logically, the degree of weathering and the structure of the saprolite are also important because slopes covered by weathered blocks evolve by sliding. *Grus* has high porosity and low shear strength; when it becomes saturated, it looses cohesion. Generally, in the regolith of granitic rocks, the sliding plane is located at depth near the contact between fresh rock and the regolith (Tricart, 1974a). Mass movement is generally rotational (Durgin, 1977). This type of landslide predominates south of Minas Gerais (Brazil) in the dome-shaped landform of the "*meias naranjas*" (Tricart, 1974a). The presence of relict discontinuities in the regolith including a basal weathering surface and old landslide planes facilitates the inflow of water, which reduces shear strength and potentially triggers translational landslides. Examples have been cited on the Sierra de la Costa in Venezuela (Garner, 1966).

Kaolinites present in regolith in the humid tropics have a low contraction limit and they do not crack during dry periods. However, the presence of clay minerals of the 2:1 type (montmorillonite) cause significant retraction, generating fissures that allow rapid infiltration and fluidization, which favors landslides (Zaruba and Mencl, 1982). In addition, kaolinitic saprolites have high plasticity and liquid limits whereas these limits are lower in saprolites with montmorillonite. However, significant rain can cause material to reach its liquid limit and trigger numerous landslides of various types; an example is the landslide that occurred in La Fría on the piedmont in the Venezuelan Andes after rains of 899 mm in the months of February and March 1972 (Tricart, 1974a). When slopes are significantly inclined, dissection predominates and clay minerals present are of the 2:1 type because kaolinite requires a long period of formation. The presence of significant quantities of swelling clays accentuates the uplift process and also decreases the shear strength of the slope material.

It is clear that topographic energy, geometry (height, length, and shape) and the exposure of the slope are the principal **geomorphic causes**. A common activity is the modification of slope geometry, which may result from natural causes such as fluvial and littoral undermining, dissolution, piping, etc., or from anthropogenic causes such as clearing done for engineering projects (Brunsden, 1979; Miller and Dunne, 1996; Borgatti and Soldati, 2005). Rejuvenation by river incision forms convex slopes that become unstable; they cause gravitational deformation and in some cases, catastrophic landslides during intense storms (Chigira, 2009).

Landslides are not common on slopes that are only slightly inclined. In the humid African tropics research by various authors (Thomas, 1994a) shows that slope instability occurs above inclinations of 26–28° although these values can differ substantially in other regions of the world. According to various researchers, most landslides occur on slopes that dip between 28 and 45°. When slopes are steep, numerous landslides occur, especially in relation to intense rainfall (*Fig. 7.25*). The geologic history of the slope must also be considered; we can find old relict or fossil landslides that provide valuable data for the interpretation of active landslides (Dramis and Sorriso-Valvo, 1994, Rosell and Linares, 2001).

Weathering of rocks is the result of a combination of physical, chemical, and biological processes that lead to their disintegration. Rocks and soils are affected by external influences and atmospheric variations that cause instabilities that lower shear strength and cohesion of materials (Crozier, 1986; Taylor and Cripps, 1987).

A set of variables governs weathering and soil formation; it includes climate, biological activity, topography (which controls drainage), parent rock, and the time necessary for disintegration and soil formation (Jenny, 1980). One of the principal effects of weathering is grain-size change, which generally leads to a distribution based on particle size or a multimodal distribution, depending on the parent rock. Each type of grain-size distribution is related to a specific change in the strength of the regolith and its subsequent behavior (Brunsden, 1979).

Of all climatic parameters, the most important, is without doubt **precipitation** because it is the principal contributor to slope rupture. It can act through 1) an increase of pore pressure, which reduces the material's strength because it hydrates clays; 2) the weight of snow or rain, as a weathering agent; and 3) basal undercutting when rivers rise (Selby, 1993), which decreases or eliminates support at the slope base and increases the shear stress of the materials (Ayala *et al.*, 1987b).

Water that penetrates into the regolith moves toward a zone of lower potential gradient under both saturated and unsaturated conditions. Under saturated conditions all pores of the regolith are full of water, which is common in the lower parts of the weathering deposit and after an intense rainfall. In general unsaturated flow, which predominates, is characterized by the presence of water only in small pores and by slow flow (Selby, 1993). Increased pore water pressure changes the consistency of the regolith, which simultaneously decreases cohesion and internal friction and promotes landslide movement. A rapid increase in pore water pressure can cause sudden liquefaction especially in fine sands and silts (Zaruba and Mencl, 1969).

According to numerous authors surface landslides in soils and weathered rocks commonly occur on dipping slopes during the most intense moments of a rainstorm. Intensity and duration thresholds must be exceeded to trigger these mass movements (Wieczorek, 1996). Campbell's (1975) research in the Santa Monica Mountains of southern California shows that precipitation above 6.34 mm/hr can trigger surface landslides that can turn into dangerous debris flows. Kumar and Hasegawa (2008) found that of the 677 landslides that occurred in Nepal between 1951 and 2006, only 193 were associated with rain. When daily precipitation exceeds 144 mm, the risk of landslides on Himalayan slopes is high. Kumar and Hasegawa consider precipitation to play the most important role in triggering landslides; they point out that ruptures during daily precipitation and accumulated precipitation show the degree to which previous precipitation affects landslide processes in the Himalaya.

Precipitation often causes catastrophic landslides and floods. From 1 to 4 October 1968 several states in northeastern India received significant monsoon rains that caused landslides and flooding that partially or completely buried the cities of Darjeeling and Jalpaiguri. About 1000 persons died and large property losses occurred especially to oil pipelines, highways, railroads, and dwellings (Davis, 1992). This combined action of flooding and landslides also occurred in Honduras and Nicaragua and other Central American countries between 29 October and 3 November 1998 as a result of rains related to Hurricane Mitch (Hubp and Inbar, 2002; Rubiera, 2002; and Baños, 2002). About 85% of the land was flooded, more than 10,000 people were killed, 15,000 went missing, and major damage affected infrastructures, dwellings, and agriculture. A landslide triggered by the Casitas volcano (Nicaragua) (*Fig. 7.44*) was channelized by the edges of the slope and the constituent pyroclastics were converted into a debris flow by the addition of water. The debris flow spread into agricultural land at the foot of the volcano and buried numerous settlements, killing about 1,500 people. In the distal zone, the thickness of the debris flow was estimated at 2 m (personal communication with Prof. A. Cendrero).

The landslides are a periodic erosional phenomenon that can be catastrophic in some areas (Guerra and Favis-Mortlock, 2002). For example, the region between Sao Paulo and Río de Janeiro is grooved by the steep Serras do Mar and Mantiqueira, which are separated by the Paraiba Valley. Catastrophic landslides occurred here in 1942, 1956, 1966, 1967, and 1988. As a result of the 1942 flooding and landslides, the government established measures related to construction, forest preservation, and soil use. In March 1956, heavy rains fell on the city of Santos with maximum intensities of 250 mm in 10 hours that triggered numerous rotational and translational landslides (Pichler, 1958). In January 1966, 472 mm of rain fell on Río de Janeiro in 72 hours, causing numerous mass movements including block slides, mudflows, and rotational landslides (Tricart, 1974a). This disaster killed more than 1000 people. The most spectacular landslide episode took place in the scarped areas of the Serra des Araras, a part of the Serra do Mar 50 km west of Río de Janeiro. On the 22 and 23 of March 1967 about 275 mm of rain fell with intensities of 114 mm/hour, accompanied by intense thunder. These conditions triggered more than 10,000 landslides in a 180-km^2 area, debris slides, debris avalanches and mudflows that affected water and electrical supplies and highway transportation, cutting off the Sao Paulo-Rio de Janeiro highway. About 1700 people died and property losses were estimated at 1.6 billion dollars (Jones, 1973). In February 1988, debris avalanches in Río de Janeiro killed 200 people and left 20,000 people homeless (Smith and de Sánchez, 1992).

The geomorphic literature describes various rain-induced catastrophic landslide events of different magnitudes. One of the largest occurred on 29 October 1959 in the city of Minatillan, southeast of Mexico City, where massive nighttime mudflows buried numerous homes and killed 800 people as they slept. In the Grand Riviere du Nord (Haiti) on 13 and 14 November 1963, intense tropical rainstorms

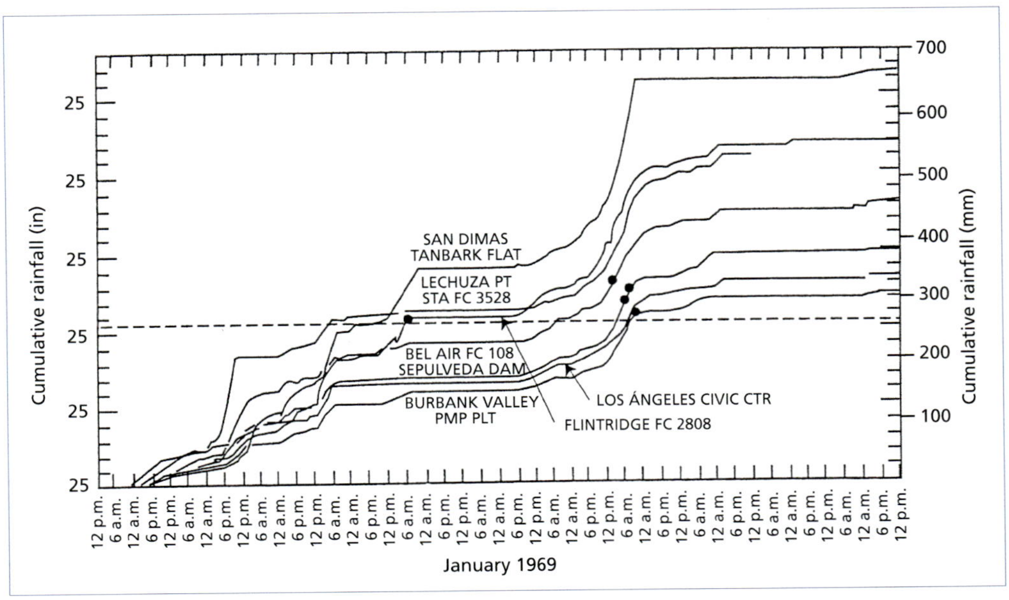

Figure 7.43 Cumulative curves of precipitation records of Santa Monica and the San Gabriel Mountains, Southern California. Each curve corresponds to the station indicated in the figure. The thick points indicate known debris flows. The more inclined parts of the cumulative curves indicate higher precipitation intensity (modified from Campbell, 1975).

Figure 7.44 Landslide on the Casitas volcano and the subsequent debris flow triggered by rains from Hurricane Mitch, at the end of October and first of November 1998. The debris flow materials buried about 1,500 people Northwest of the city of León. Maribios volcanic chain. Nicaragua. Photo A. Gómez Sal.

caused significant mass movements of mud and rock that affected numerous towns. These events killed 500 locals and tourists (Davis, 1992).

In the eastern Pyrenees, high intensity short-duration rainfall has triggered many debris flows and surface landslides on colluvium and altered rocks. Large ruptures have occurred during rains lasting one or two days with rainfall intensities exceeding 200 mm in 24 hours. Moderate intensity rainfall can trigger or reactivate mudflows and rotational and translational landslides in clay formations. The analysis of historical documents and dendrochronology shows that landslides have increased significantly since 1959 (Corominas and Moya, 1996).

Increased runoff due to rainfall or rapid snowmelt causes small debris flows. On the lower parts of slopes and in small streambeds, large storms increase the sediment content of runoff and cause large debris flows (hyperconcentrated flows) (Florsheim *et al.*, 1991).

By studying the sequence of processes triggered by heavy rain in northern Italy, Luino (2005) recognized a first step in which debris slides form on steep slopes and mud-debris flows occur in basins smaller than 20 km^2. With higher precipitation, in basins up to 2000 km^2, mass landslides with volumes of up to two million cubic meters begin to form that can affect the rocky substrate. Finally, in basins larger than 2000 km^2, runoff causes basal undermining of the slopes; after several days of heavy precipitation, deep landslides form that can move significant masses of rock.

Although there is a clear cause and effect relationship between heavy precipitation and the triggering of surface landslides, the role that the rainfall plays in the movement of deep landslides is not so obvious. Many authors have pointed out a relationship but the analysis of the relationship is complex; it requires an understanding of prior rainfall and of the hydrogeological behavior of the slope.

Brand *et al.*, (1985) conducted research on the relationship between precipitation and landslides in Hong Kong; they found that if rainfall intensity exceeds 70 mm/hour, significant landslides occur; when the intensity does not exceed 100 mm/day, landslides are rare. These authors also emphasize the role of prior precipitation because it controls the development of surface or deep landslides.

High intensity precipitation is not common in the humid tropics (Lal *et al.*, 1981). Deep landslides are clearly influenced by immediately preceding precipitation or other long duration rainfall (Thomas, 1994a). In March 1974 it rained 742 mm in 16 days in Tubarau (southeastern Brazil); 240 mm of the rainfall was recorded on the last day. The catastrophic consequences included numerous landslides and floods; one landslide killed 25 people (Bigarella and Becker, 1975).

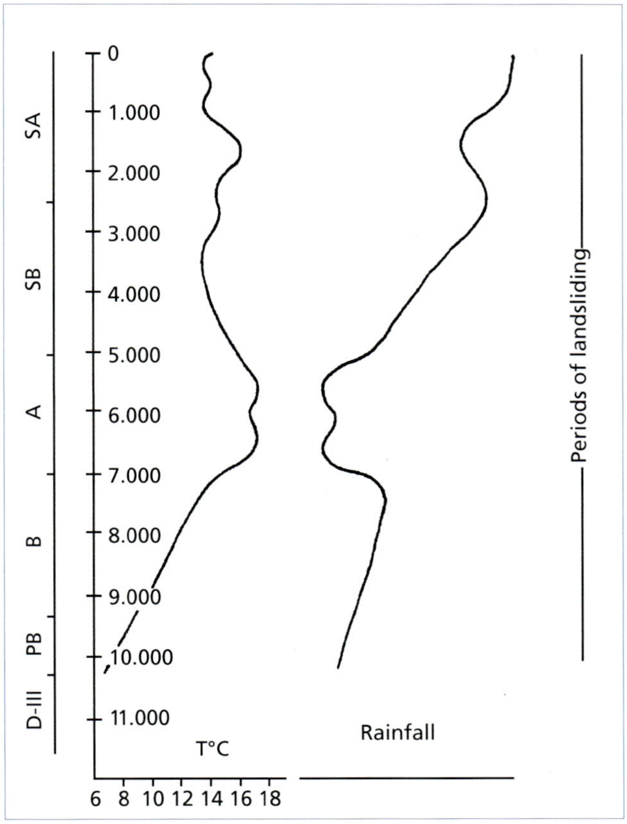

Figure 7.45 Changes of temperature and precipitation during the Holocene in Cantabria (Spain) and attempted correlation with periods of greater landslide occurrence. D-III = Dryas III; PB = Preboreal; B = Boreal; A = Atlantic; SB = Sub-boreal; SA = Subatlantic (González *et al.*, 1996).

Rapid **snowmelt** produced by sudden heating or by rainfall on snow adds water to slopes and increases pore water pressure, which can trigger landslides. There are many examples in the literature that analyze these relationships.

A widely discussed topic is the possible relationship between **climate change** and landslides (González, 1995; González *et al.*, 1996, 1999; Remondo, 2001; Remondo *et al.*, 2005a and b; Soldati *et al.*, 2004). A reconstruction of climate change during the Holocene in the Cantabria region (Spain) (Salas, 1993, 1995) based on pollen records, isotopic dating, and other techniques has shown a correlation between the occurrence of mass movements and implied climate changes for the periods of 8.500–7.500 BP, 3.000–2.500 BP and 500–300 BP. All these are periods of higher than normal precipitation (*Fig. 7.45*) (González, 1995; González *et al.*, 1996). On the other hand, Remondo (2001) and Remondo *et al.*, (2003a, 2005a, b) indicate that increased landslide frequency in recent decades in the mountains in northern Spain is not related to climatic parameters or seismic activity but to socioeconomic indicators of human activity. In addition, their detailed analysis of landslide frequency over a period of 43 years produced and validated sensitivity models and created landslide risk maps for use in quantitative predictions (Remondo *et al.*, 2005b). Remondo *et al.*, (2003b) that established a procedure for validating landslide sensitivity maps; they maintain that validation is essential in determining the predictive value of these susceptibility maps.

The rapid decrease of **water level** on a slope can trigger landslides in earthen dams, along coastlines and on lakeshores, reservoirs, and rivers. This abrupt decrease can be seen in a streambed during a flood (*Fig. 7.46*) in the sudden water level drop in a reservoir or a sea level drop after a storm tide.

Research on the role of **vegetation** in slope stability has increased significantly since the 1960s. Prior to that time, many people did not consider it an important factor but it has now been demonstrated to be quite significant under certain circumstances especially for surface landslides (Greenway, 1987). For more than a century, trees and bushes have been planted on slopes to control erosion and improve stability. However, in a study of 700 landslides triggered by an intense storm in Hong Kong, So (1971) determined that landslides occurred more commonly on vegetated slopes than on bare slopes. In tropical areas, a landslide surface is rapidly colonized by vegetation (*Fig. 7.47*).

Greenway (1987) distinguished between hydrologic and mechanical mechanisms in his work on vegetation and slope stability. Among the most important is the beneficial role of interception, which reduces infiltration by absorption and evaporation losses. In contrast, plant roots and stems increase infiltration capacity and surface flow; this mechanism may promote landslides on highly inclined slopes with surface soils as in Brazil (de Ploey and Cruz, 1979). Roots also take water out of the soil, thus lowering the pore water pressure. One adverse factor is the loss of soil moisture, which can lead to cracking and a high infiltration capacity. Soil reinforcement by roots is an important mechanical factor that increases shear strength. The type, depth of penetration, extent, and strength of the various roots that are present should be taken into account.

Tsukamoto and Kusakabe (1984) studied vegetation effects on slopes in Japan; they classified slopes into four types (*Fig. 7.48*). Type A is a thin soil where roots do not penetrate into the rocky substrate; the soil-rock interface may act as a landslide plane. Type B is similar to type A, except that the roots penetrate into the substrate. Type C slopes have thicker regolith and a transition layer in which soil density and shear strength increase with depth. Roots penetrate into this layer and provide greater slope stability. Finally, type D has a thick regolith; the roots have very little mechanical influence on stability as long as they remain in the regolith. Under the

Figure 7.46 Rotational landslide due to basal undermining as a result of a 2003 flood on the Rio Ebro, upstream of Zaragoza (Spain) Photo J. Guerrero.

Figure 7.47 Plucking scar of a landslide colonized by several months of vegetation. Saint Denis, Island of Martinique (Lesser Antilles).

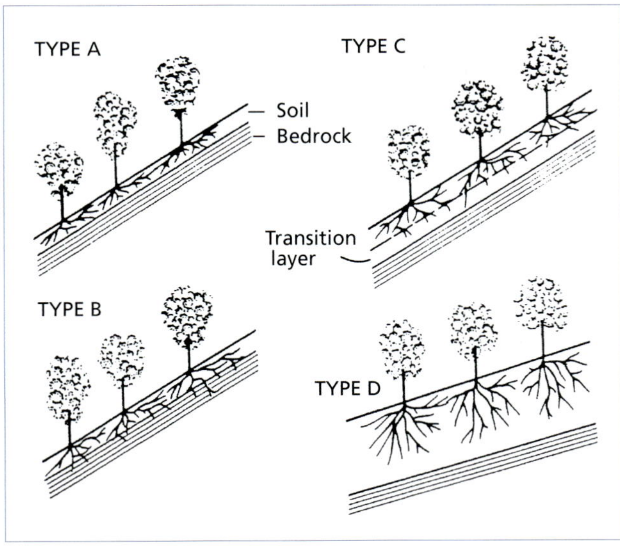

Figure 7.48 Slope classification on the basis of root anchoring and strength (Tsukamoto and Kusakabe, 1984).

root zone, deep landslides can occur. However, roots reduce the susceptibility to erosion. The weight of trees depends on the species, diameter, height, and separation. Overloading of a slope increases the shear force and pore water pressure. In general if the slope angle is small, overloading increases stability. Vegetation can transmit wind-derived destabilizing forces to the soil. Strong winds can pull out trees especially those with shallow root systems. Finally, landslides often occur on deforested slopes as observed in many parts of the world. As mentioned above, vegetation plays a complex role in slope stability; it promotes both slope stability and instability.

Alpine orogenic belts are subject to sporadic **earthquakes** that affect slope stability (Panizza, 1991). Earthquakes cause a gradational increase in shear force on slopes. The primary effect is ground acceleration. The additional force is equal to the mass of the landslide (potential) multiplied by seismic acceleration, which may be 0.5 times greater than gravitational acceleration. Because this force acts in various directions, it can initiate movement by increasing destabilizing forces. Landslides triggered by earthquakes occur near the crests of peaks where susceptibility to landslides is greatest (Veder, 1981). Earthquake-induced acceleration also indirectly affects stability through its effects on material cohesion. Ground vibration can cause some materials to lose their shear strength. The most spectacular example of this effect is liquefaction.

Liquefaction, which affects silts and loose sands, is a process in which the soil becomes saturated with water and behaves like a fluid. Simonett (1967) studied the frequency of earthquake-induced landslides in Papua New Guinea and showed that their number decreases logarithmically with distance from the epicenter. As earthquake magnitude increases, so do the number of landslides and their dimensions. Seismic activity often causes rockfalls, and rock and soil landslides (Crozier, 1986). In addition, traffic, machinery, explosive, and wind vibrations can accelerate slope movements.

Numerous examples of earthquake-induced landslides have been described. The San Fernando (California) earthquake of 1971 triggered about 1000 mass movements. In Friuli (Italy) the earthquake of 6 May 1976 reactivated mass movements and triggered numerous new landslides. One of the most disastrous landslides in the world occurred in Huascarán (Peru). The Cordillera Blanca of Peru has numerous glaciated peaks that rise to more than 6000 m, of which Huascarán (6654 m) is the tallest (*Figs. 7.49* and *7.50*). Two large ice and rock avalanches have flowed into the nearby Santa River valley. The first avalanche killed 4000 people in 1962; the second and larger was triggered by a magnitude 7.7 earthquake on 31 May 1970. This slide moved 50 million cubic meters of ice, rock, debris, and water that moved at velocities of about 280 km/hour and covered 16 km of the valley floor in three minutes. It buried the town of Yungay and killed 18,000 to 20,000 people (*Fig. 7.51*). This avalanche, the largest in recorded history, should be considered an ice and rock avalanche because of the height of the fall, velocity, and volume (Plafker and Erickson, 1978; Evans *et al.*, 2009). On 8 October 2005, the Kashmir earthquake triggered several thousand earthquakes of the sliding and translational landslide type. Debris avalanches (>8 million m^3) buried several towns and dammed up rivers, forming lakes (Owen *et al.*, 2008).

Earthquakes can also have hydrologic consequences. The 1847 Zenkovjic (Japan) earthquake killed 10,000 people and destroyed 20,000 to 30,000 houses. It is famous for blocking the Sai River and for the secondary disasters caused by dam collapses in the valleys. More than 40,000 fracture surfaces were counted. The largest occurred on the slope of the Kouzou mountain where a large quantity of sediments accumulated in the Sai River, creating a reservoir more than 40 km long. Three weeks after the earthquake, the dam collapsed due to water from heavy storm rainfall. Water stored by the Sai River flowed downstream and killed 100 people (Terakawa and Matsuo, 1996).

We know that our increased population and technical capacity to interfere with the environment have considerably affected the processes that affect the Earth's surface (Cendrero, 1992). **Human activity** is obvious in mining operations and civil engineering projects (Ayala *et al.*, 1987b). **Excavation** modifies slope profiles along linear works and creates instabilities. **Blasting** causes vibrations that can act as catalysts for mass movements; in general, blasting decreases stability. **Overloading** is the result of increasing weight;

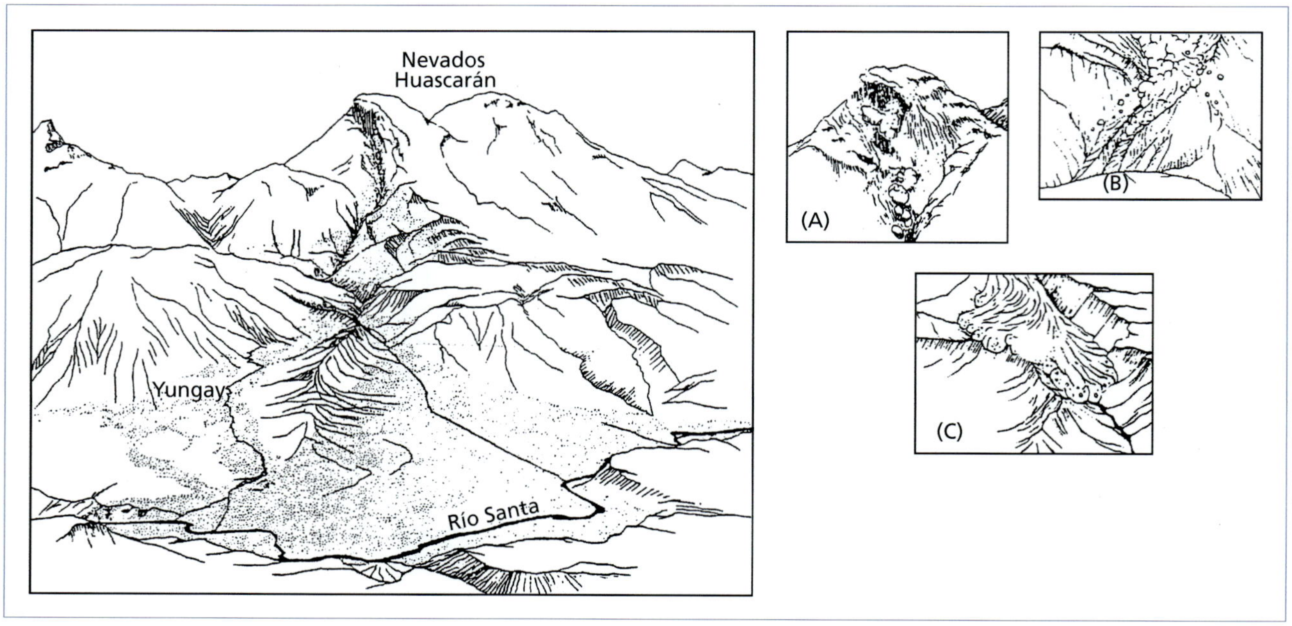

Figure 7.49 Diagram of a rock avalanche (in stippled pattern) near Huascarán, Peruvian Andes, 31 May 1970. (A) Landslide (B) Avalanche movement near terminal moraine belts. (C) Division of the avalanche into lobes at Yungay and Ranrahirca (modified from Plafker and Erickson, 1978).

Figure 7.50 El Huascarán seen from the Rio Santa valley. The catastrophic avalanches of 1962 and 1970 began on the rocky wall of the north peak (left). Cordillera Blanca, Perú. Photo J. López-Martínez.

Figure 7.51 Location of the town of Yungay (Cordillera Blanca, Perú) that was buried by the avalanche catastrophe that occurred in 1970. Photo J. López-Martínez.

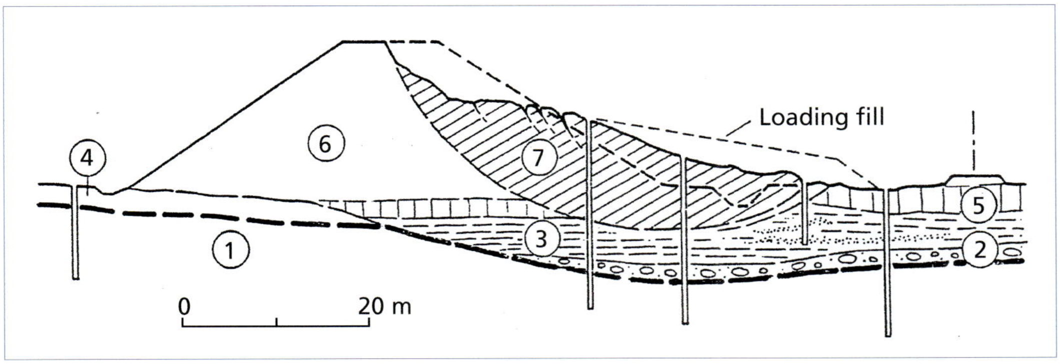

Figure 7.52 Collapse of a railroad embankment near Podlesin, Bohemia, (Czechoslovakia). 1- Permian shales and sandstones 2- Sandy gravels 3-Clay-sand alluvium. 4- Loess. 5- Silty alluvium. 6- Embankment 7- Sliding mass (Zaruba and Mencl, 1982).

it results from the construction of embankments, which often partially collapse. These embankments are constructed in valleys filled with alluvial deposits, which usually contain a large amount of water and can collapse easily (Zaruba and Mencl, 1982). Veder (1981) shows that in the spring of 1969 settling began to occur along a section of the Oberpullendorf (Austria) to Rattelsdorf (Germany) railroad, requiring the use of highly complex measures to stabilize the earthworks. Zaruba and Mencl (1982) discuss another example (*Fig. 7.52*) in Podelsin, near Slany (Czechoslovakia). An embankment was built on top of Permian shales and sandstones and also partially on top of argillaceous-sandy alluvium overlying sandy gravels full of water at low pressure. The substrate extruded progressively after a period of intense rain.

Mining activity (*Figs. 5.52* and *19.2*) includes quarry, cutting, and underground mining exploitation. The first two are excavations, which have already been discussed. Mining in underground tunnels causes subsidence due to the loss of volume; when wooden beams supporting the tunnels deteriorate and lose their capacity to hold up tunnel roofs, collapses can occur, especially in tunnels close to the surface. A good example is the underground coalmines in the eastern part of the United States where collapses occur directly above tunnels and indicate the traces of tunnels (Gray and Bruhn, 1984).

The neighborhood of La Corona (Ejea de Los Caballeros, Zaragoza, Spain) provides an example of how human activities can trigger landslides. This neighborhood is located on top of a scarp created by the downcutting and lateral migration of the Arba de Luesia River. The scarp, composed of Miocene shales and a cover of Quaternary fluvial gravel (Gutiérrez *et al.*, 1994), is the location of numerous mass movements that affect various buildings in the area (*Fig. 7.53*). Research on the landslides indicates that water from a leaking sewer system is the triggering factor. This water rapidly reduces the mechanical strength of the Miocene shales, which are affected by a dense network of discontinuities and characterized by dispersive and expansive behavior (*Fig. 7.54*).

Parsons (1988) enumerated three principal consequences of anthropogenic activity on slopes: 1) generation of artificial slopes, 2) extractive industries that create artificial pathways, and 3) changes in soil use. The first two causes have been discussed previously.

Generally human-induced **changes in soil use** increase soil erosion, runoff, and the basin's drainage area (Strahler, 1956b). While dating debris flow deposits in Scotland using lichenometry, Innes (1983) reached the conclusion that the effects of fire and overgrazing and the resulting modification of soil use have been responsible for increased debris flows over the last 250 years.

Investigators have identified two landslides in the Cornisa Cantábrica (Spain) that occurred during prehistoric-historical times (González *et al.*, 1996, 1999) and increased landslide frequency during recent decades (Remondo, 2001) that seems to bear no relationship to precipitation regimes; instead these events appear to involve a direct and general

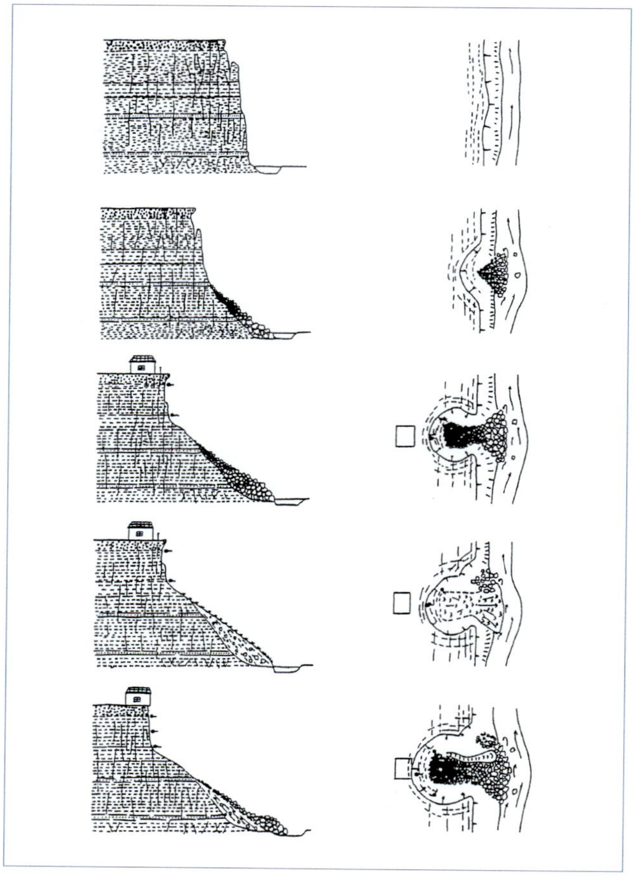

Figure 7.54 Evolutionary diagram of mass movements that affect the Barrio de la Corona (Ejea de los Caballeros). Province of Zaragoza. Spain. (Gutiérrez *et al.*, 1994).

Figure 7.53 Complex landslide movement and block fall that affects the slide scar beneath houses of the Barrio de la Corona (Ejea de los Caballeros). Province of Zaragoza. Spain.

anthropogenic impact on the Earth as a result of the growing influence of human activities related to economic, technological, and population development (Remondo *et al.*, 2005a; Cendrero *et al.*, 2006). This type of geomorphic response to human impacts on surface processes can be a manifestation of "global geomorphic change" independent of climate change (Cendrero, 2003).

In tropical areas, anthropogenic activity primarily affects weathering deposits because desiccation cracks that form during dry periods promote water infiltration during rainy periods. These processes are thought to have partially triggered multiple landslides in the broad deforested area of the Serra do Mar between Sao Paulo and Río de Janeiro (Tricart, 1974a). The lack of tree mass is accompanied by the lack of cohesive power or root capacity in the regolith, the lack of substrate anchoring, and the loss of water adsorption. In addition, shear strength decreases on slopes. These factors trigger landslides during intense precipitation. For example, a 100.7 mm rainstorm that lasted three hours in the Uluguru Mountains (Tanzania) on 23 February 1970 generated 1000 new surface landslides and reactivated other previously existing ones (Temple and Rapp, 1972). However, research by So (1971) questions the role of vegetation as a fundamental factor in landslide development because he claims that 35% of the 700 June 1966 Hong Kong landslides occurred in forested areas, which covered only 8.4% of the region.

7.5 Landslide risk: Case histories

In the area of environmental geomorphology we can categorize problems into two groups: Materials-Resources and Processes-Risks (Cendrero, 1980). Regarding the latter, "it is not surprising that prevention and correction of natural hazards has become a more important global preoccupation in recent years for various countries and numerous international organizations" (Cendrero, 1997).

Landslides are the most common geomorphic phenomenon and the one that occurs most extensively throughout the world (McGuire *et al.*, 2004). Cruden and Varnes (1996) propose a scale of mass movement velocity related to destructive capacity. They point out that landslides can be **catastrophic** if they advance at high speeds and destroy obstacles they encounter in their path. **Intermediate** landslides are slower although they can also cause loss of human life and property. Finally, **sluggish** landslides cause property damage but threats to human life are very rare (*Table 7.1*). Even so, the destructive capacity of a mass movement depends not only on its velocity but also on the mass moved. In addition, the magnitude of a landslide is due to its mass and velocity. Another aspect to consider is the distance that the mass moves.

Table 7.1 CLASSIFICATION OF VELOCITY OF MOVEMENT (MODIFIED FROM CRUDEN AND VARNES, 1996)

Speed Class	Description	Typical Velocity (m/s)	Destructive Impact
Catastrophic			
7	Extremely fast	>5 (~5–50 m/s)	Catastrophically violent. Escape unlikely, many fatalities, buildings destroyed.
Intermediate			
6	Very fast	5.10^{-2}–5 (~m/min–m/s)	Some lives lost. Difficult to escape. Buildings destroyed.
5	Fast	5.10^{-4}–5.10^{-2} (~m/hr–m/min)	Escape possible. Structures destroyed.
Sluggish			
4	Moderate	5.10^{-6}–5.10^{-4} (~m/wk–m/hr)	Some structures can be temporarily maintained.
3	Slow	5.10^{-8}–5.10^{-6} (~m/yr–m/wk)	Damaged structures can be repaired during movement.
2	Very slow	5.10^{-10}–5.10^{-8} (~m/century–m/yr)	Some permanent structures undamaged by movement.
1	Extremely slow	<5.10–^{10}(<m/century)	Movements perceptible only with instruments. Construction possible with precautions.

Natural phenomena are not risks in themselves but become risks when people suffer damages or losses (Selby, 1993) as a result of being exposed to them. It is important to understand what we mean by hazard, exposure, vulnerability and risk (Varnes, 1984; Panizza 1988; Crozier and Glade, 2005; among others). **Hazard** or **threat**, (in Spanish America), is the probability that a natural or anthropogenic phenomenon will occur in a territory in a certain time interval. For example, it is the possibility of a landslide or earthquake occurring in a specific place. **Vulnerability** is the potential degree of losses resulting from the occurrence of a phenomenon of a given magnitude. **Exposure** refers to the elements threatened: population, property, or economic activities exposed to a given threat. **Risk** is the number of lives lost, persons injured, property damaged, and economic activity disturbed as a result of a natural phenomenon. Consequently, risk is the product of hazard multiplied by vulnerability and exposure.

Because landslides are often triggered by other dangerous events such as earthquakes and volcanic eruptions, their impact is undervalued (McGuire *et al.*, 2004). *Table 7.2* shows the principal natural disasters resulting from landslides that caused more than 1000 fatalities (Crozier and Glade, 2005). The available numbers show that between 1990 and 1999, 8658 deaths were attributed to landslides and avalanches, which is an underestimate. Schuster's (1996) research on the 25 most catastrophic landslides of the 20th century is a significant contribution because of its extensive descriptions of various landslides. *Table 7.3* shows a 250-year history of the principal types of landslides, dates, and resulting damages. Landslides occur primarily in mountain belts where seismic activity and volcanic eruptions are significant (McGuire *et al.*, 2004).

This chapter describes numerous landslides and in some cases, their causes. We will analyze some "classic" landslides in more detail because they have been well studied and can thus offer the reader an idea of landslide consequences and the lessons we can gain from their analysis.

The **Rossberg Landslide** (Switzerland), also called the Goldau, occurred on 2 September 1806 (*Table 7.3*). This mass movement was the very fast-moving rock avalanche type (Turner and Jayaprakash, 1996). It occurred in the Rossberg Massif, moved 10 to 20 million m^3 of rock, buried the small town of Goldau, destroyed 300 houses, and killed 457 people (Eisbacher and Clague, 1984). Part of the landslide material was deposited in the Lauerzer See, generating 20-m high waves that damaged towns along the lakeshore. Heim (1932) was the first to study this landslide.

The **Elm Landslide** (Switzerland) occurred on 11 September 1881 (*Table 7.3*) and was studied in detail by Bus and Heim (1881) and Heim (1882, 1932). Hsu (1978) has written an excellent review of Heim's work and these landslides in particular. The Elm landslide is a rock avalanche, similar to the Rossberg. The mass movement was triggered by artificial undermining due to mining of shales from a quarry located at the foot of the vertical slope. An estimated 10 million m^3 of rock fell to the quarry floor from a height of 470 m with velocities of 80 m/sec, killing 115 people. This description and data come from Varnes (1978) and Turner and Jayaprakash (1996).

The **Lower Gros Ventre Landslide** is the most important one in the recorded history in the United States (Voight, 1978). It occurred in the Gros Ventre Valley of northwestern Wyoming in the Teton National Forest on 23 June 1925. In about three minutes, 40 million m^3 of weathered sandstone, limestone, sand, silt, and water-saturated clay debris broke loose, slid, and ran up on the opposite slope of the valley. The slide created a 70 m high dam that impounded the Gros Ventre River (*Fig. 7.55*). There are multiple causes of this mass movement, which slid from a height of 640 m. One cause was the approximately 20° dip of strata toward the valley; the non-homogeneous, leached, weathered, and very porous rock mass had little shear strength. In addition, prior to sliding rain and snowmelt, several magnitude 3.5 earthquakes had affected the area, which may have liquefied the sands. All these are triggering factors. Based on its characteristics, the landslide is classified as a complex type; within it, the debris flow was visco-plastic.

The **Vaiont Landslide** (Dolomite Alps, Italy) occurred on 9 October 1963 and is considered the worst dam disaster in history. Numerous researchers have described it (Muller, 1964b; Varnes, 1978; Costa and Baker, 1981; Kiersch, 1964, 1983; Bromhead, 1986; and others). Kiersch's work is the most significant and extensive. The Vaiont dam is located in a limestone syncline that is karstified in layers with upper Jurassic and lower Cretaceous marls dipping 45° toward the valley. It was a glacial valley and the canyon was formed by postglacial fluvial incision. Existing joints and fractures were partially attributed to postglacial decompression.

At 265.5 m, the arc-shaped Vaiont dam was the tallest in the world at the time. It was not damaged in spite of the fact that it was subject to translational landslides carrying 4 million tons of rock (*Figs. 7.22, 7.56* and *7.57*). The landslide that caused the catastrophe was 1.8 km long and 1.6 km wide and had a volume of 240 million m^3. It moved at 25–30 m/second. The sliding material completely filled the reservoir and in some places reached a height of 150 m above the maximum level of the reservoir. It produced a wave of air, water, and rock that rose 240 m above this level on the opposite slope; the expansive wave caused slight damage to the town of Casso. The wave overtopped the dam by 100 m and reached a height of 70 m at the confluence with the Piave Valley about 2 km downstream of the dam where it wiped out the town of Longarone and affected other nearby towns

Table 7.2 **MAJOR NATURAL LANDSLIDE DISASTERS THAT HAVE CAUSED MORE THAN 1000 FATALITIES. ALL DATA ARE BASED ON SOURCES LISTED IN THE COLUMN ENTITLED "SOURCE." THE TYPE OF PROCESS NAMED WHERE IDENTIFIABLE FROM THE SOURCE; N.I. = NOT IDENTIFIABLE (BASED ON CROZIER AND GLADE, 2005)**

Event	Trigger	Date	Location, Region, Country	Consequences	Source
Landslide	N.I.	24.11.1248	Mont Granier, Savoie, France	1,500 to 5,000 dead	Flageollet (1989)
Landslide	N.I.	1310	Zhigui, Hubei, China	3466 dead	Tianchi (1989)
Landslide	Earthquake	24.11.1604	Arica, Chile	>1000 dead	Nussbaumer (1998)
Landslide	Earthquake	19.06.1718	Gansu Province, China	40,000 families buried	Tianchi (1989)
Landslide	Earthquake	10.10.1786	Kangding-Louding, Sechuan, China	100,000 dead	Tianchi (1989)
Lahar	Volc. eruption	19.02.1845	Volcán Nevado del Ruiz, Colombia	1,000 dead	Nussbaumer (1998)
Landslide	N.I.	09.1857	Montem./ Basilic., Italia	5,000 dead	Nussbaumer (1998)
Lahar	Volc. eruption	1919	Java, Indonesia	5,110 dead 104 towns destroyed	Brand (1989)
Landslide	Earthquake	16.12.1920	Kansu, China	>200,000 dead	Bell (1999) Nussbaumer (1998)
Landslide	Earthquake	1920	Haiyuan, China	100,000 dead	Tianchi (1989)
Debris flow	N.I.	14.12.1941	Huaraz, Peru	4000–6000 dead 1/4 of Huaraz destroyed	Erickson *et al.* (1989) Nussbaumer (1998)
Landslide	Earthquake	1949	Khait, Tajikistan	12.000 dead	Alexander (1995)
Landslide	Earthquake	15.08.1950	Asma, India	Aprox. 30,000 dead	Nussbaumer (1998)
Landslide	Precipitation	1958	Shizuoka, Japan	1,094 dead 19,754 buildings destroyed	Oyagi (1989)
Landslide	N.I.	29.10.1959	Minatilan, Mexico	5.000 dead	Nussbaumer (1998)
Debris flow	N.I.	10.01.1962	Nevados, Mt. Huascarán, Peru	4.000 dead Pueblo of Yungay destroyed	Erickson *et al.* (1989) Nussbaumer (1998)
Landslide	N.I.	10.10.1963	Vaiont Dam in the Piave Valley, Italia	2.600 dead Towns destroyed	Kiersch (1964, 1983) Müller (1964)
Landslide	Precipitation	1967	Sierra das Araras, Brasil	1.700 dead	Erickson *et al.* (1989)
Lahar	Earthquake	31.05.1970	Ancash, Yungaytal, Peru	66.794 dead	Alexander (1995) Nussbaumer (1998)
Landslide	N.I.	20.09.1973	Choloma, Honduras	2.800 dead	Nussbaumer (1998)
Lahar	Volc. eruption	13.11.1985	Nevado del Ruiz, Colombia	>25.000 dead	Nussbaumer (1998)
Landslide	Earthquake	23.01.1989	Tajikistan	Up to 10,000 dead 2 towns destroyed	Nussbaumer (1998)
Landslide	N.I.	07.06.1993	Nepal	3.000 dead	Nussbaumer (1998)
Debris flow	Precipitation	12.1999	Venezuela	30.000 dead 400.000 homeless	Larsen *et al.* (2001)

Table 7.3 NOTABLE LANDSLIDE DISASTERS, 1750–2003 (MCGUIRE ET AL., 2004)

Landslide	Date	Impact
Valtellina, Italy	1762	A rock avalanche of perhaps 10^6 m^3 killed 115–200 people in the villages of Gero and Barcone
Sichuan Province, China	1786	Earthquake triggered landslide that temporarily dammed the Dadu river. When overtopped and broken after 10 days, the subsequent flood killed about 100,000 people
Calabria, Italy	1786	Earthquake-triggered landslides killed about 50,000 across the region
Goldau, Switzerland	1806	Stürzstrom of 35 million m^3 from Rossberg Massif killed 457 people
Elm, Switzerland	1881	Slope destabilized by mining collapsed and destroyed the village of Elm, killing 115 people. The volume of the landslide was 10 million m^3
Frank, Alberta, Canada	1903	30 million m^3 collapsed from Turtle Mountain, burying the outskirts of Frank and claiming more than 75 victims
Sarez, Tajikistan	1911	Earthquake triggered a 2.9 km^3 Stürzstrom that destroyed the village of Sarez
Gansu Province, China	1920	Earthquake-induced landslides claimed at least 100,000 lives
Alma-Ata, Kazakhstan	1921	Debris flows killed 500 in Alma-Ata
Sichuan Province, China	1933	Earthquake-induced landslides killed 6,800 directly and dammed river; another 2,800 killed by flood when dam failed
Kobe, Japan	1938	Rain-induced debris flows killed 505 people and destroyed 100,000 homes in Kobe and its neighbourhood
Kure, Japan	1945	Landslides killed 1,154 and destroyed almost 2,000 homes
Tadzhikistan	1949	Magnitude 7.5 earthquake-triggered regional landslides, which killed 12,000–20,000 people
Arita River, Japan	1953	Landslides killed 460 and destroyed about 4,800 homes
Kanogawa River, Japan	1958	Landslides killed 1,094 and destroyed nearly 20,000 homes
Central Chile	1960	Landslides triggered by Magnitude 9.5 earthquake on 22 May killed more than 200 across Central Chile
Nevados Huascarán, Peru	1962	A Stürzstrom of about 13 million m^3 destroyed nine villages and towns, including most of Ranrahirca (a neighbour to Yungay) and killed 4,000 people (see also 1970)
Vaiont, Italy	1963	Catastrophic collapse induced by filling a reservoir killed more than 2,000 people and wiped out Longarone and satellite hamfets. The volume of the landslide was 200 million m^3
Nevados Huascarán, Peru	1970	Earthquake-triggered Stürzstrom killed 18,000 in Yungay (15,000 deaths) and Ranrahirca. The volume of the landslide was 50–100 million m^3 (see also 1962)
Mayunmarca, Peru	1974	About 1 km^3 buried Mayunmarca and outlying farms, killing 400 people
Puerto Rico	1985	Landslides following storms killed between 129 and 300 people and destroyed more than 100 homes
Nambija, Ecuador	1993	Slopes destabilized by historical goldmines collapsed onto Guayaquil, killing more than 300 people
Sarno, Italy	1998	16 mudflows killed 160 and caused US$ 35 million damage
Central America	1998	Hurricane Mitch triggered about 1 million landslides, killing more than 10,000 and causing damage of US$ 5,000 million. Honduras, Nicaragua, Belize and El Salvador were all affected
Venezuela	1999	Landslides killed about 30,000 along north coast of country and caused damage of US$ 10,000 million
San Salvador, El Salvador	2001	An earthquake-induced mudslide destroyed 400 homes and killed about 1,000 people in San Salvador. The volume of the landslide was about 100,000 m^3

Figure 7.55 Complex landslide of the Lower Gross Ventre (23-June-1925). Northwest Wyoming. Teton National Forest. United States. Photo F. Gutiérrez.

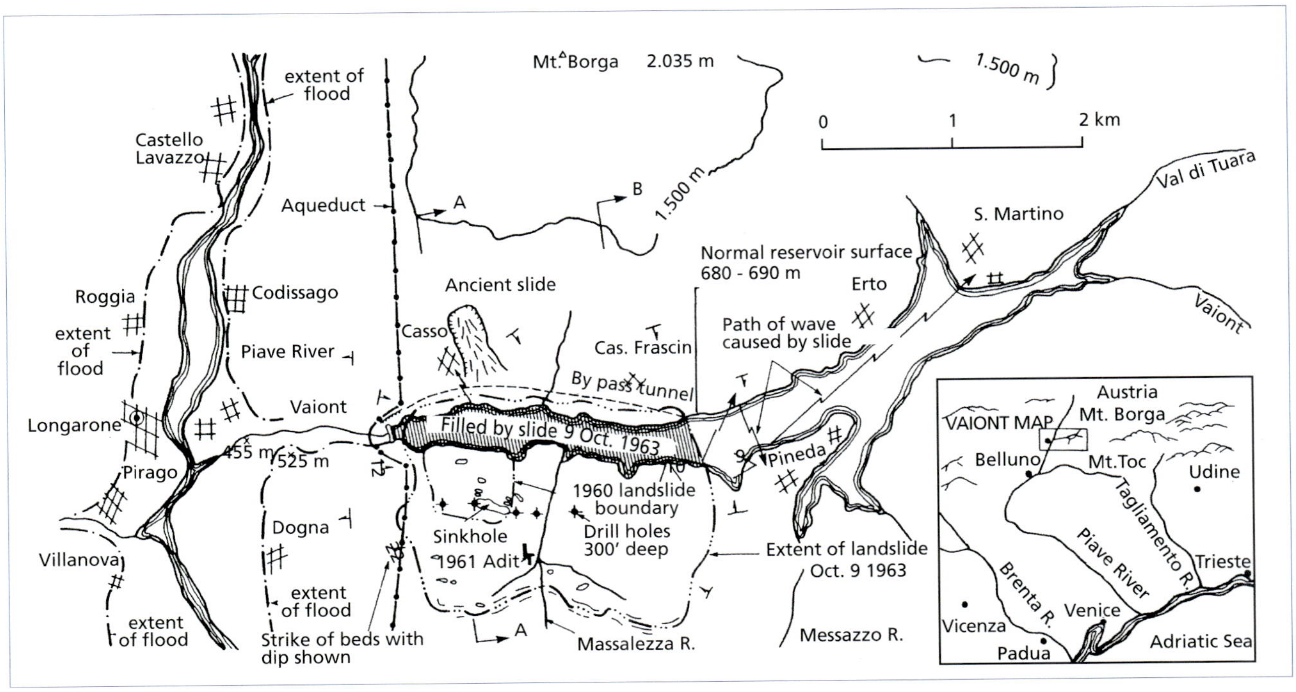

Figure 7.56 Map of the area of the Vaiont dam and the valley of the Piave River that shows the geographical features, boundaries of the slide, and destructive flood waves (Kiersch, 1964).

Figure 7.57 Translational landslide, fill, and dam. Vaiont valley. Dolomitic Alps. Italy.

(Pirago and Villanova). Approximately 2000 people lost their lives. The catastrophe lasted 7 minutes.

The landslide also caused the dammed water to rise to the tail of the reservoir although this was not catastrophic. The landslide also generated numerous earth temblors that were recorded by seismological observatories in various European cities. On 1 October, eight days before the slide, herbivore animals had sensed the danger and moved away. The mayor of Casso had issued an order to evacuate the slopes. Today, the slide mass is gullied. The dam, completed at the beginning of 1963, had a capacity of 150 million m^3. After the required fill tests were done, the dam was filled and reached its capacity of 680 m in April. The rise in water level in the reservoir created an induced phreatic level, which increased the hydrostatic pressure. Heavy rains started on 28 September and continued until 9 October. The reservoir level rose due to surface runoff to about 700 m, about 30 m below the top of the dam. Two months before the collapse, geodetic measurements showed that movement had gradually accelerated and three days before the disaster there was a strong impulse that measured more than 14 cm/day (Kilburn and Petley, 2003). On 2 September, officials tried to lower the reservoir level but the supply of runoff and subterranean flow exceeded the amount that could be drained. There were 120 million m^3 of water stored in the reservoir at the time of the disaster.

Several factors controlled the landslide. 1) The 45° inclination of the strata toward the valley facilitated the translational landslide. 2) The low shear strength of the marls. 3) Endokarst cavities. 4) The joint system that weakened cohesion. 5) Heavy rains. 6) Swelling minerals in clays. 7) Slow gravitational movement, detected in 1960. In that year, a 700,000-km3 landslide has occurred on the southern margin near the dam.

The 1966 **Aberfan Landslide** (Wales) has been studied in detail by Bishop *et al.* (1969) and Bishop (1973) and analyzed by various authors (Bolt *et al.*, 1975; Bromhead, 1986; among others). The landslide occurred on 21 October 1966 in tipple no. 7 of a coalmine. It destroyed houses in the city of Aberfan and a school where 116 children and 5 teachers died; the total death toll was 144 people. This type of landslide had occurred in the Welsh mining basin since 1916. The main movement occurred in two or three minutes; the flow destroyed an electrical cable, blocked 180 m of highway, affected the railroad and a canal, and also dammed up the Taff River. The flow mobilized 180,000 tons of debris, which moved from a height of 435 m, the top of the dump. The cause was high pore pressure in the debris, which reached the point of liquefaction in the lower part of the dump. The upper part was not saturated and a fraction of it moved as if floating on the liquefied material. The thickness of the front of the flow was about 9 m.

These large events are the "tip of the iceberg" of slope processes because they are real disasters that have captured human attention. However, as stated previously, mass movements are one of the most widespread geomorphic processes in the world.

Landslides in Spain are highly varied because they affect rock formations of different ages, morphologies, and climatic environments (Cendrero *et al.*, 1997). Most mass movements occur in the Pyrenees (*Table 7.4*) although they also occur in other areas of marked relief (*Fig. 7.58*) (Suárez and Regueiro, 1997). Suárez and Regueiro analyze in detail the importance and extent of slope movements in Spain.

In the United States (*Fig. 7.59*) (Hays, 1981) landslides most commonly occur in the Appalachians, the Rocky Mountains, and along coastal cordilleras. Altogether about one seventh of the surface of the United States is prone to landslides. The Mississippi Valley and the plateaus south of the Great Plains also have a low but significant landslide risk.

Table 7.4 SOME CATASTROPHIC MAJOR NATURAL LANDSLIDES IN SPAIN (CENDRERO ET AL., 1997)

Location	Date	Type of movement	Damage	Volume (m^3)
Alcoy, Alicante	Dec. 1620	–	City partly destroyed, earthquake triggered	–
Inza, Navarra	Dec. 1714–Apr. 1715	Mudflow	Village destroyed	10^6
Biniarroi, Mallorca	Mar. 1721	Rotational slide	–	–
Corbera, Valencia	Nov. 1783	–	–	–
Azagra, Navarra	July 1874	Rockfall	91 deaths	–
Puigcercós, Lleida	Jan.1882	Translational slide	Village partly destroyed	3×10^5
Bono, Lleida	Oct. 1937	Debris avalanche	River dammed	10^5
Rocabruna, Gerona	Oct. 1940	Debris flow	6 deaths	–
Rosiana, Gran Canaria	Feb. 1956	Translational slide	Village destroyed	5×10^5
Puebla de Arenosa, Castellón	Oct. 1957	Earthflow	Cracks in buildings	–
Senet, Lleida	Aug. 1963	Debris avalanche	River dammed	5×10^4
Benasque, Huesca	Aug. 1963	Debris avalanche	Road destroyed	–
Alcoy, Alicante	Dec. 1964	Rotational slide	Buildings affected	–
Tudela de Veguin, Asturias	1975	Earthflow	Cracks in buildings	9×10^5
Carege, Lleida	Nov. 1982	Rotational slide	Road & bridge destroyed	–
La Guingueta, Lleida	Nov. 1982	Debris avalanche	Village isolated	–
Capdella, Lleida	Nov. 1982	Debris avalanche	3 deaths	–
Pont de Bar, Lleida	Nov. 1982	Translational & complex slide	Village & road destroyed	10^7
La Coma, Lleida	Nov. 1982	Mudflow	–	2.5×10^5
Gósol, Lleida	Nov. 1982	Translational slide & mudflow	–	10^6
Sta Cruz de Moya, Cuenca	Apr. 1984	Translational slide	–	10^5
Olivares, Granada	Apr.1986	Mudflow	Village affected	3.6×10^6

Figures 7.58 and *7.59* show that the most significant and numerous landslides occur in cordillera zones with high relief. This reasoning can be extended to other countries of the world.

7.6 Landslide prediction and mitigation

Today man exerts such strong pressure on the Earth's surface that he is considered a principal morphological agent (Gutiérrez, 2001, 2005, 2007). This situation is exacerbated by population increase and a wide variety of human structures that trigger additional landslide processes. As a result, more people die and property damage increases (Cendrero, 2003).

To this anthropogenic activity we must add the possible effects of global climate change (Corominas, 2006) and surface manifestations of the endogenetic activity of our planet. This set of environmental changes makes prediction difficult. In doing this type of work, it is useful to invoke the Principle of Uniformitarianism but apply it in reverse: "the past and the present are the key to predicting the future" (Cendrero, 2003). In other words, if we study past and present landslides, we will have good tools for prediction.

A good example is the landslide study of Guipúzcoa (Spain) conducted over a 50 year interval by Remondo (2001)

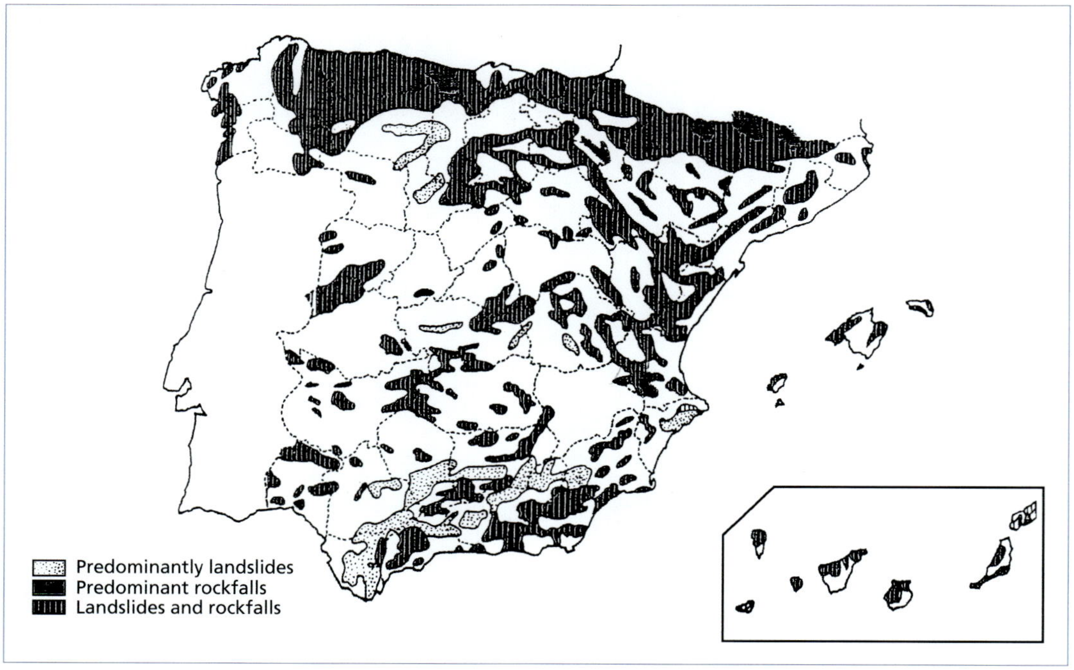

Figure 7.58 Map of Spain indicating the areas of high landslide frequency. (Map created by M. Ferrer, in Suárez and Regueiro, 1997).

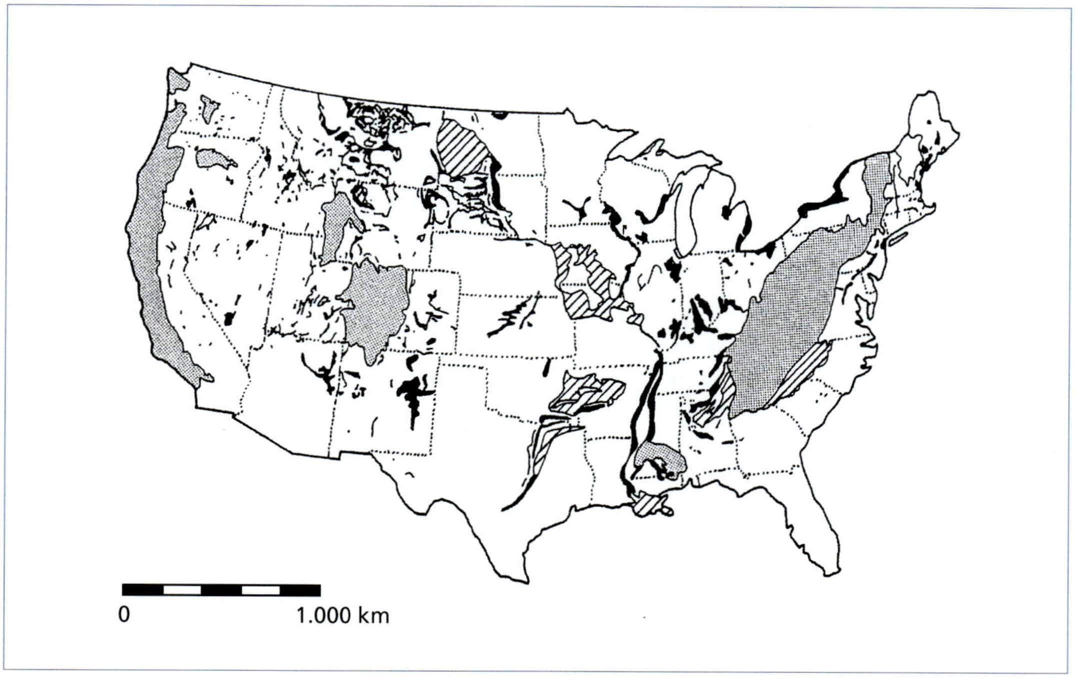

Figure 7.59 Distribution of landslides in the United States. Points Risk of large landslides. Lines: Risk of moderate landslides. Black: Local landslides of variable severity. (Hays, 1981).

and Remondo *et al.* (2003a, b, 2005a, b). Remondo and his colleagues used aerial photographs from seven flights between 1954 and 1997 and an inventory of 2,500 landslides inferred from the interpretation of aerial photographs and fieldwork. They concluded that in little more than 40 years the frequency and rate of the process had increased by an order of magnitude and that it follows a quasi-exponential pattern. Comparative analysis of the frequencies and rates cited and a detailed study of precipitation show that variations of these factors cannot explain the marked increase in frequencies and rates. Nor does seismicity during this period justify this large increase. The most plausible explanation for increased landslides is human activity. These researchers analyzed the variation of the raw interior product (RIP) from 1950 to the present, related to the rate of mobilization. The RIP is an indicator of motive force (increase in the number of persons and their economic and technological capacity). Remondo and colleagues found a good relationship between the latter and denudation rates. The global RIP can be expected to continue to increase in the near future and the mobilization rates will continue to rise as well (Cendrero, 2003).

Two types of measures reduce risk (UNDRO, 1991): structural (physical intervention) and non-structural (planning, legislation, alarm systems, evacuation plans, etc.). Among non-structural methods, susceptibility, hazard or threat, and risk maps are useful tools for mitigating the consequences of potentially hazardous processes because they provide suitable information for better regulation and management of territory and for the development of emergency and civil protection plans.

The development of hazard and risk map techniques has been difficult largely due to the complexity and diversity of mass movements. In spite of the fact that slope stability and unstable slope calculation techniques (the engineering perspective) have improved greatly, they are not viable for areal studies because of their high cost and the amount of time required for observational detail (Hutchinson, 1995). For this reason, regional cartographic analysis must begin with a different (geomorphic) focus. Procedures based on geomorphic analysis include direct geomorphic evaluation based on the criteria of experts and heuristic and statistical (or probabilistic) procedures based on a consideration of controlling factors. The recent development of spatial data analysis techniques, together with the generalized use of Geographic Information Systems have opened up new possibilities for the application of these methods, making them quantitative and easier to validate. Hansen (1984b), Hartlen and Viberg (1988), Gee (1991) and Chacón *et al.*, (2006), among others, have revised landslide risk cartography methods. Even so, most existing maps analyze only susceptibility or the propensity of the ground to slide; few examples quantitatively express landslide hazards and, less so, risk (Remondo *et al.*, 2005b).

Regardless of the method used to map landslide hazard or risk, it is necessary to evaluate the resulting map. For this purpose, Remondo (2003b, 2005b) did validation tests on landslide maps produced between the 1950s and 1980s by comparing them with landslides that occurred during the 1980s and 1990s. The results obtained have been practically equal in both cases.

To conduct prediction and mitigation work on landslides the mechanical behavior of rock and soils must be known. This information can be obtained from field investigations involving aerial photographs, geodetic surveys, geological and geophysical investigations, borings, measurements of penetration strength, piezometric measurements, measurements of deformation at various depths, electrical potential of soils, pH, and redox. Laboratory research complements field studies and can determine the following: water content, soil type, compression tests, determinations of cohesion and the angle of internal friction, X-rays and other parameters. An analysis of the various field and laboratory techniques can be found in Veder (1981).

Numerous authors study landslide mitigation and stabilization (Veder, 1981; Zaruba and Mencl, 1982; Ayala *et al.*, 1987b; Flageollet, 1989; Hotz and Schuster, 1996; Wyllie and Norrish, 1996; Laín, 2002; Rogers, 2002; Panizza, 2005a, b; Borgatti and Soldati, 2005). This work involves a more in depth analysis of the most common techniques used to adequately address landslide mitigation.

To reduce or eliminate geomorphic hazard, geomorphic processes must be modified or protective practices must be put in place (Panizza, 2005a). In the first case, the cause must be known (for example, eliminating surface runoff and penetration into the soil) and in this way effective cohesion of detrital material can be increased. In the second case, slope protection structures or similar structures such as retaining walls can be built (Panizza, 2005b). There are numerous techniques for mitigating the development of recent landslides (Rogers, 1992).

A common method used to stabilize slopes consists of evacuating water through surface or underground **drainage**. Drainage trenches or ditches can be used to conduct water to the external areas of the landslide. Subsurface drainage can be achieved through the use of horizontal drains (Krohn, 1992) (*Fig. 7.60*), vertical wells, drainage galleries and trenches with draining fill.

Forest repopulation in potential landslide areas is a common practice that contributes to slope stability by increased reinforcement by roots and interception of rainfall and evapotranspiration, which reduces pore pressure.

Seeding for vegetative cover on top of slopes promotes stabilization and diminishes surface erosion. Plant roots also absorb water and increase shear strength.

Geotextiles (*Fig. 7.61*) are used as a support or filtration base and for drainage in civil engineering projects.

Figure 7.60 Horizontal drains in the Lacadele landslide. Cortina d'Ampezzo (Italia).

Figure 7.61 Geotextile installed at the head of a landslide to prevent water infiltration into the sliding mass. Note the rapid vegetative colonization of the principal scarp. Saint Denis. Island of Martinique. Lesser Antilles. Photo F. Gutiérrez.

Figure 7.62 Metallic grid on a highway slope; loose fragments can be seen at the base. Dolomitic Alps (Italy).

The photograph shows them being used to prevent infiltration of water into the sliding mass.

Methods of surface correction include the use of **metallic mesh** in rockfall areas to prevent linear projects especially highways and other structures from being damaged by impacts and to retain loose fragments on the upper part of the slope (*Fig. 7.62*).

Gunite and sprayed-on concrete can be used to increase slope stability. Both are concrete mortars of a specific size that are pressure sprayed onto a surface (*Fig. 7.63*). These surface treatment methods prevent weathering but are primarily used to prevent the fall of fragments. Slopes can be reinforced by the use of **rivets**, which are steel bars installed into several meter-long drill-holes; they are used to staple a rock mass and prevent landslides. This type of anchor is also used on slopes with an unstable surface regolith. Normally each of them is attached to cover a 1 to 6 m^2 of slope. **Walls** are concrete structures or gabions (*Fig. 7.64*) that are sometimes used to stabilize landslides or to prevent the fall of cobbles onto linear constructions. There are three types of walls (Jiménez-Salas *et al.*, 1981) (*Fig. 7.65*): **retaining walls**, which are constructed separate from the ground and later filled in; **containment walls**, which are built by prior excavation and followed by the construction of a wall that holds up the unstable slope; and **coating walls**, used as erosion and weathering protection and to provide stabilizing weight.

Figure 7.63 Slope stabilization by a false tunnel at its base. The corrections done consist of a structure of concrete armed with anchors, sprayed-on concrete applied to the top of the slope, and installation of rivets. Dolomitic Alps (Italy).

Figure 7.64 Support wall constructed of gabions and concrete to prevent fragments from falling on the highway. Carretera de Villar del Cobo a Tramascastilla (Province of Teruel). Photo F. Gutiérrez.

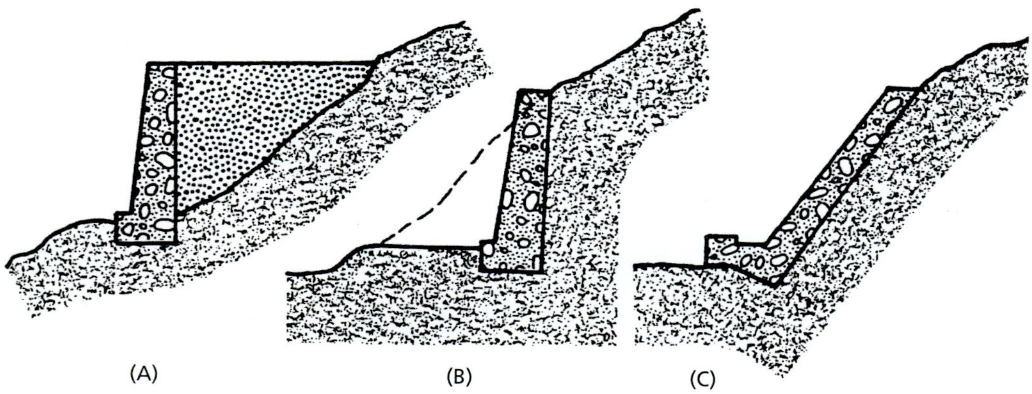

Figure 7.65 Types of protective walls: (A) Retaining. (B) Containment. (C) Coating (Jiménez-Salas *et al.*, 1981).

Another slope stabilization technique involves **cutting off the top**, which has the effect of reducing the weight of the upper part of the slope and in this way reducing the forces that destabilize the slope. Excavated material often accumulates at the foot of the slope, creating a **heel of dirt** for the purpose of increasing slope strength. Another technique to improve slope stability is the construction of intermediate **berms** (or terraces) in the slope profile. They are used in slopes near highway infrastructures. In the case of mining slopes, berms are absolutely necessary for exploitation; they are usually 20 meters wide and 10–15 m high (*Fig. 5.52*).

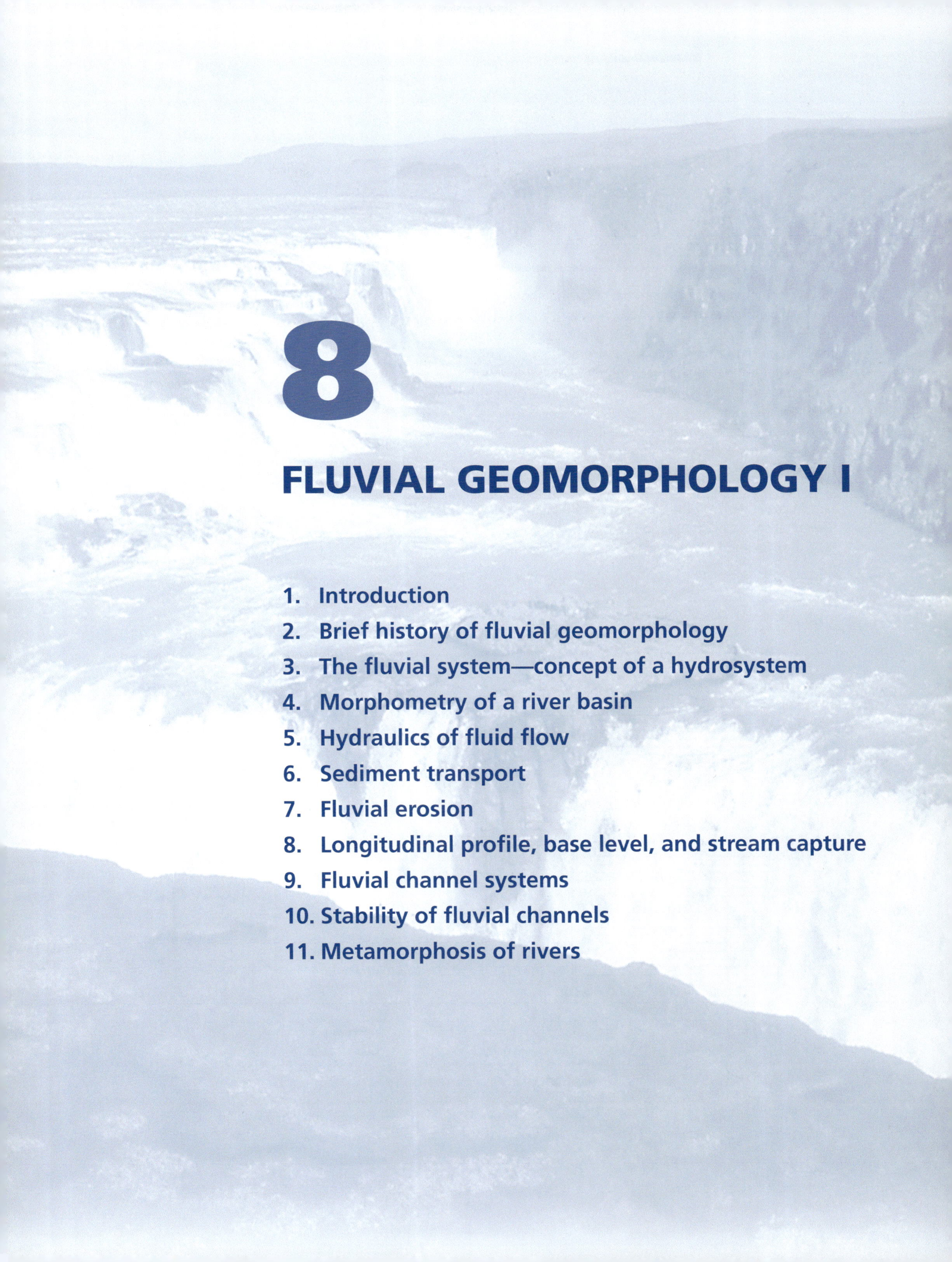

8

FLUVIAL GEOMORPHOLOGY I

1. Introduction
2. Brief history of fluvial geomorphology
3. The fluvial system—concept of a hydrosystem
4. Morphometry of a river basin
5. Hydraulics of fluid flow
6. Sediment transport
7. Fluvial erosion
8. Longitudinal profile, base level, and stream capture
9. Fluvial channel systems
10. Stability of fluvial channels
11. Metamorphosis of rivers

8.1 Introduction

Fluvial geomorphology has many definitions; Gregory (2004a) has compiled many of them. According to Richards (1987), the fundamental goal of fluvial geomorphology is to explain the relationships among the physical processes of channel flow on a moving bed, the mechanism of flow-induced sediment transport, and alluvial channel landforms created by sediment transport.

Rivers are essentially erosional and transport agents that carry water and sediment from continents to the oceans. In spite of the fact that <0.005% of continental water is found in rivers at any given time, water flow is one of the most important forces at work on the Earth's surface (Knighton, 1998). Rivers transport about 19 billion tons of material every year, 80% as solids and 20% as dissolved material (Meybeck, 1979; Milliman and Meade, 1983; Walling, 1987). During floods, the amount of water and sediment carried by some rivers is even higher. For example, the highly destructive Mississippi River flood of 1973 reached Saint Louis with a discharge of 24,120 m^3/sec but fell short of the maximum recorded flood on the Mississippi, which exceeded 56,640 m^3/sec. Even so, this flood is less than the average flow of the Amazon, the largest river in the world (*Fig. 8.1*) (Chorley *et al.*, 1984).

A river is a body of water that flows in a channel. The characteristics of water flow are the domain of hydraulic engineering whereas the dimensions and channel systems are geomorphic subjects (Chorley *et al.*, 1984). Geomorphologists face a challenge in determining channel characteristics because in most cases these features are at least partially covered with water. During dry periods and in arid and semi-arid regions, there is no flow in channels most of the year. Channels in arid zones are ephemeral, which means that water

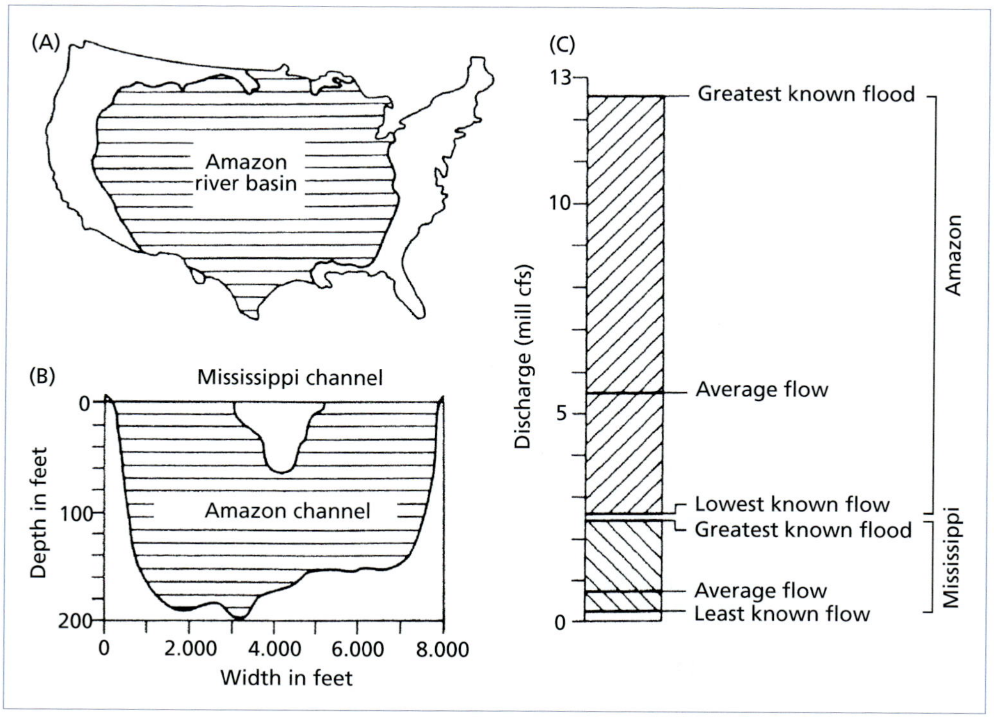

Figure 8.1 The Amazon and Mississippi Rivers. (A) The Amazon River basin covers 6 million km^2; it size is equivalent to 3/4 of the United States. (B) Comparison of cross sections of the Amazon River in Obidos (about 700 km from its mouth) and that of the Mississippi in Vicksburg (about 500 km from the delta). (C) Comparison of discharges of the Amazon River in Obidos and the Mississippi in Vicksburg (Chorley *et al.*, 1984).

flows during periods of rainfall; perennial rivers in humid regions flow year-round. Intermittent rivers flow seasonally, at least one month a year.

The most important geologic erosional process on the Earth's surface is water that runs off the land (Morisawa, 1968). Water creates most landscapes and forms flood plains where many of our cities are built. Fluvial action generates steep slopes that can cause landslides as their instability increases. Glacial melt water flows down braided channels and transports glacial materials downstream. Exorheic rivers deliver water to oceans but those that flow though endorheic areas deposit their loads in interior basins. Finally, underground currents move particles under the Earth's surface (Morisawa, 1985).

8.2 Brief history of fluvial geomorphology

Fluvial geomorphology developed along with the science of geomorphology. During the 17th and 18th centuries, hydraulic engineers such as Perrault, Surrell, Guettard, Desmarest, and De Saussure studied the role of fluvial action in degradation and landscape formation in spite of the diluvialist ideas that prevailed during this time. These ideas were gradually rejected by the work of Scottish geologists Hutton (18th century) and Lyell (19th century) who finally put an end to theory of catastrophism, which had prevailed for many centuries. As scientific geology developed, geomorphology followed.

At the end of the 19th century, North American geologists Dutton and Powell provided the most interesting advances with their documentation of the erosional activity of rivers in canyon formation. To Powell we owe the concept of the base level of erosion; he considered the existence of local base levels and the sea as the general base level. He also classified rivers as consequent, antecedent, and superimposed. At the end of the 19th and beginning of the 20th century, the North American geologist Gilbert contributed his ideas on processes of fluid mechanics, capacity and competence, the concept of grade, and the interdependence of variables within a fluvial system (Morisawa, 1985).

Williams Morris Davis established an erosional cycle (youth, maturity, old age) in 1899; his ideas endured until the advent of quantitative methods in fluvial systems (Morisawa, 1968) in the middle of the 20th century. A treatise that had major impact was *Fluvial Processes in Geomorphology* (Leopold *et al.*, 1964), a collection of the investigations of its authors that provided the research basis for present-day fluvial systems. The following decades saw the publication of numerous books on fluvial geomorphology. Gregory (2004a) discusses the fundamental characteristics of fluvial systems: dynamics and morphology (Morisawa, 1968); form and process (Richards, 1982; Morisawa, 1985; Knighton, 1984, 1998); rivers and landscape (Petts and Foster, 1985); and the fluvial system (Schumm, 1977). All these books document research progress on fluvial processes and resulting landforms. Along with the quantification of drainage basins initiated by Horton (1945), which was compiled and amplified by Zavoianu (1985), numerous laboratory experiments on fluvial geomorphology have been conducted (Schumm *et al.*, 1987). Fluvial systems have been subjected to climate change and tectonic deformation in the past (Schumm *et al.*, 2000; Schumm, 2005), which can trigger or promote paleo-inundations or other events (Baker and Bunker, 1985; Benito and Thorndycraft, 2004a and b).

Knowledge of river behavior and prediction are quite important today and as a result of the knowledge obtained by quantification techniques, we are able to make predictions on fluvial system behavior resulting from environmental variations.

One aspect of fluvial geomorphology that can create numerous environmental problems is resource exploitation (Briersley and Fryirs, 2005). Human beings require water to live. Today overpopulation (more than seven billion inhabitants on the planet) is the most serious problem confronting humanity because we need water from the environment and we are applying intense pressure to obtain it. Rivers provide water for drinking and irrigation, to generate hydroelectric energy, and to cool nuclear power plants. Rivers are exploited as placers and used for navigation and sports; in addition, landforms created by water provide beautiful tourist destinations. Rivers are also used for wastewater disposal. In addition, fluvial ecosystems are of vital importance for particular habitats of flora and fauna. On the other hand, water creates the risk of drought, pollution, and flooding, which can negatively affect human life and property. Man has settled on and exploited the resources of fluvial flood plains and areas of sedimentation in alluvial fans, and without knowing it, has placed himself in danger.

8.3 The fluvial system—concept of a hydrosystem

In the last three decades, the concept of a system (Chorley, 1962) has been used extensively in environmental sciences to link physical, chemical, and biological processes (Piégay and Schumm, 2003). The concept has had a major influence on fluvial geomorphology (Chorley and Kennedy, 1971) and has been integrated into other related disciplines such as ecology, hydrology, human geography, and river management (Hack, 1960; Stoddart, 1965a; Schumm, 1977; Dunne and Leopold, 1978; Petts and Amoros, 1996).

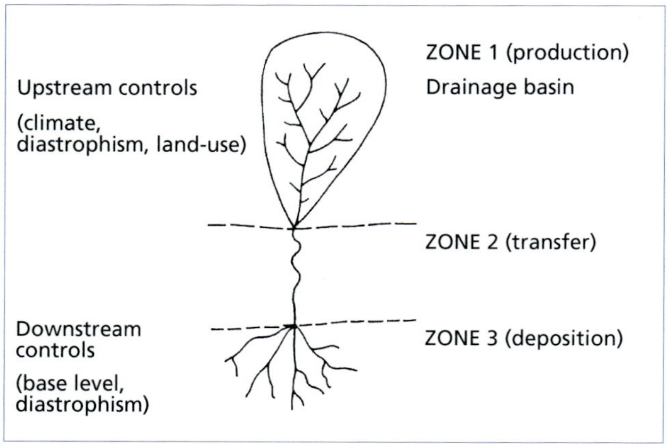

Figure 8.2 Idealized fluvial system (Schumm, 1977).

A system can be defined as a significant combination of things that form a complex assemblage with connections, interrelations, and transfers of energy and matter between them. In the case of a fluvial system this includes not only fluvial channels but also a group of drainage networks, areas of alluvial fan and delta sedimentation, and also runoff and slope sediments (Piégay and Schumm, 2003).

A fluvial system changes over time due to erosional processes and sedimentation; it also responds to climate changes, base level modifications, Quaternary tectonics, and human activities (Park, 1981). This variability makes it difficult to make predictions about a fluvial system.

Figure 8.2 shows an idealized diagram of a **fluvial system** (Schumm, 1977) that is composed of three zones. The uppermost area of the basin (Zone 1) is the area of runoff and sediment production. Zone 2 is the transfer zone and Zone 3 is the area of sedimentation (alluvial fans, deltas, flood plains). The division may seem artificial because rivers transport, erode, and deposit in all zones; however, a single process predominates in each area.

A temporal scale must be established for defining the components of a fluvial system because the level of variables and their overlap change according to the scale used (*Table 8.1*). Time, geology, and climate are independent variables whereas vegetation, relief, paleo-hydrology, and valley dimensions are dependent variables. Over geologic time we have little or no knowledge of dependent variables such as those related to channel dimensions; these variables are classified as indeterminate (Schumm and Lichty, 1965).

Independent morphological variables adjust their response depending on the interaction of variables that reflect the independent hydrological and sedimentological aspects of a river basin. In addition, they define the force exerted by the flowing fluid and the resistance to morphological change of the channel bed materials. According to Richards (1982), the principal independent variables are the current energy, sediment size and its distribution, and the shape of gravel-sized particles. All of these reflect current energy and sediment properties. **Dependent morphological variables** are channel geometry, which is three-dimensional, cross-section, shape in plan view, and longitudinal profile properties; these factors, which constitute the complete morphology, are intimately interrelated.

Table 8.1 **THE STATUS OF DRAINAGE BASIN VARIABLES DURING TIME SPANS OF DECREASING DURATION (SCHUMM AND LICHTY, 1965)**

Drainage basin variables	Status of variables during designated time spans		
	Geological	**Modern**	**Present day**
1. Time	Independent	Not relevant	Not relevant
2. Geology (lithology, structure)	Independent	Independent	Independent
3. Climate	Independent	Independent	Independent
4. Vegetation (type and density)	Dependent	Independent	Independent
5. Relief	Dependent	Independent	Independent
6. Paleohydrology (long term water discharge and sediments)	Dependent	Independent	Independent
7. Dimensions of the valley (width, depth, and slope)	Dependent	Independent	Independent
8. Average water discharge and sediments	Indeterminate	Independent	Independent
9. Morphology of the channel (width, depth, slope, shape, and system)	Indeterminate	Dependent	Independent
10. Observed discharge of water and sediments	Indeterminate	Indeterminate	Dependent
11. Observed flow characteristics (depth, velocity, turbulence, etc.)	Indeterminate	Indeterminate	Dependent

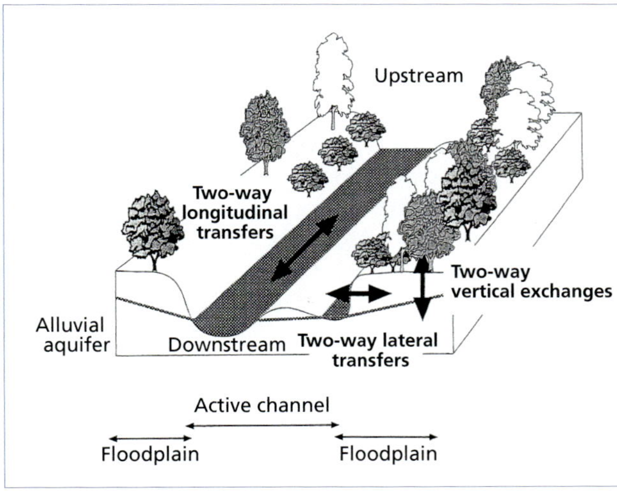

Figure 8.3 The hydrosystem, a complex system with three bidirectional axes: upstream and downstream, channel margins, surface and underground environments (Piégay and Schumm, 2003).

Today geomorphologists conduct fluvial investigations with other environmental scientists; they consider geomorphic processes in relation to biological processes and human activity. The concept of a **hydrosystem** provides a framework in which to evaluate these interactions (Roux, 1982; Amoros and Petts, 1993). A hydrosystem can be defined as a three dimensional system (*Fig. 8.3*) in which the longitudinal (upstream and downstream), lateral (channel margins), and vertical/surface (underground) components transfer energy, matter, and biota. As a result, the system depends on the dynamic interactions of the hydrological, geomorphic, and biological processes that act on its three dimensions over a long period of time (Piégay and Schumm, 2003).

The longitudinal dimension is defined by the relations of upstream and downstream current. For example, an upstream channel is affected by changes in sediment supply, which can cause aggradation or accretion. Downstream, a fall in the base level causes erosion at increasingly high elevations and causes the rocky riverbed to crop out. The lateral dimension is highly complex due to channel changes or metamorphosis. Thus the Ubaye River, a tributary of the Durante in the French Alps, is a braided channel where it flows over marly formations but it gradually becomes a meandering channel and later a straight channel where it flows over more resistant rocks (Piégay *et al.*, 2000). Channel behavior includes floodplain architecture, and accordingly, its biological diversity. The vertical dimension causes degradation or aggradation of the channel; these changes can cause biological and chemical changes in the floodplain. For example, the downcutting of the channel causes a drop in the phreatic level, which clearly affects the flora.

The hydrosystem concept can be considered as an extension of the river system concept (Schumm, 1977), which is applied to large rivers with well-developed floodplains. It involves geomorphic and also biological and chemical processes. Whereas the fluvial system accentuates the temporal and longitudinal dimensions, the hydrosystem concept emphasizes lateral and vertical dimensions, which are more important in large flood plains and have a greater influence on underground alluvial water storage, ecological richness, and regeneration of vegetation on the banks (Piégay and Schumm, 2003).

8.4 Morphometry of a river basin

A river or drainage basin is the entire area that provides surface runoff to the river (Gregory and Walling, 1973). A drainage basin is a well-defined topographic and hydrologic entity; it is a fundamental geomorphic unit (Chorley, 1969). It expresses the interaction of the strength of surface topography and the flow of material and energy (Zavoianu, 1985). To understand basin morphology and process-landform relationships it is necessary to study the configuration of the river basin. This study must be quantitative; numerous methods have been proposed to accomplish this. We will limit ourselves to a study of the most important characteristics that define a drainage basin.

To calculate the **basin area** a geomorphologist defines watershed divides and uses this information to calculate the total area with the aid of graph paper and a planimeter. The idea of ranking rivers in a basin was originated by Horton (1945) with the establishment of a river hierarchy (*Fig. 8.4*). The **order** is a basic property of river networks because it is related to the relative discharge of a channel segment. Of the various ordering systems the most commonly used is Strahler's (1952); he defines a first-order segment as a river segment that does not have a tributary and that flows from its origin. The second order segment originates at the intersection of two first-order segments, a third order results from the confluence of two second-order segments, and so on. The order does not increase when a segment of a given order intersects a segment of lesser order. According to Strahler's method, all fourth-order basins (*Fig. 8.4*) are considered similar whereas in Shreve's system (1975) the magnitude of a channel segment is defined as the total number of tributaries that supply it. For algebraic questions, Scheidegger (1965) uses only even numbers. The magnitude obtained is closely related to the proportion of the total area of the basin.

A commonly used property is the **bifurcation ratio** which was proposed by Horton (1932); it is defined as the ratio of the number of rivers of order n to the number of rivers of a higher order $(n + 1)$: $Rb = N_n/N_{n+1}$. The study of numerous river systems confirms the principle that in a climatic region, if the lithology and status development are uniform, the bifurcation ratio tends to remain constant from one order to the next. Values of this ratio range from 3 to 5 and are characteristic of river systems. However, values of 10 or more are found in very broad basins with alternating outcrops of hard and soft rock (Strahler, 1964). With the entire network ordered and the values plotted on semilog paper, Horton (1945) established that the number of segments of successively lower orders in a given basin tends to form a geometrical progression that begins with the unique segment of higher order and grows according to a constant bifurcation ratio. This definition is known as the **law of stream numbers.** For example, if the bifurcation ratio is 3 and the principal river is of the sixth order, the number of segments will be 1, 3, 9, 27, 81 and 243.

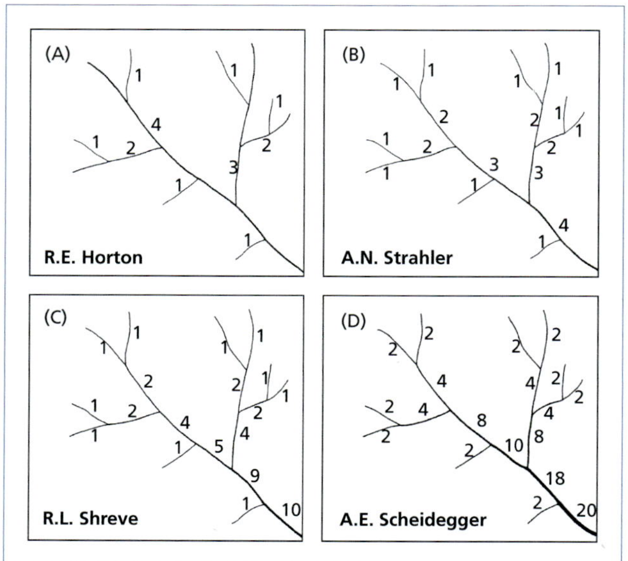

Figure 8.4 Methods of ordering stream and river segments (Gregory and Walling, 1973).

Drainage network density is an important characteristic of a river basin because it reflects geomorphic, lithological, climatic, vegetative, and anthopogenic controls. Drainage network density (Horton, 1932) consists of the ratio between the total length of all reaches of all orders and the basin area: $D = \Sigma l/A$. This reflects the balance between erosional forces and the resistance of the land surface. Drainage densities range from values of 5 km/km^2 for soft slopes with low precipitation and permeable substrates up to numbers higher than 500 km/km^2 for mountainous areas of impermeable rocks, steep slopes, and significant rainfall, such as in badlands (*Fig. 8.5*). Semi-arid zones provide the most spectacular examples; badlands also occur in humid zones with a soft substrate when vegetative cover is removed (deforestation). They also form in the humid climate of New Jersey, as

Figure 8.5 Badlands. Las Bardenas Reales de Navarra. Ebro Depression (Spain).

in Perth Amboy (Schumm, 1956a). Drainage density has also been measured in very different climates and materials in the United States. *Figure 8.6* shows examples.

A problem that arises is the trace of the drainage basin, which can be determined by maps, aerial photographs, and on the ground. Horton (1945) used the blue lines (streams) on maps but not all small streams appear on maps. Another method consists of tracing a blue line and adding crenulations of topographic lines, which are lower-order streams. By comparing this latter method to a network obtained by field surveys, Morisawa (1957) showed no significant difference between the methods. In areas of periglacial and karst dry valleys, the value derived from inflection of the curves provides high values of drainage density (Gregory, 1966). The quality and scale of the maps used have a significant influence because they affect the established network. Also, a fluvial network based on aerial photography provides higher values of drainage density. On the other hand, mapping a drainage network using topographic contour inflections and conducting a survey based on aerial photographs are long and tedious processes. For this reason, rapid calculation methods have been proposed by Carlston and Langbein (1960) who

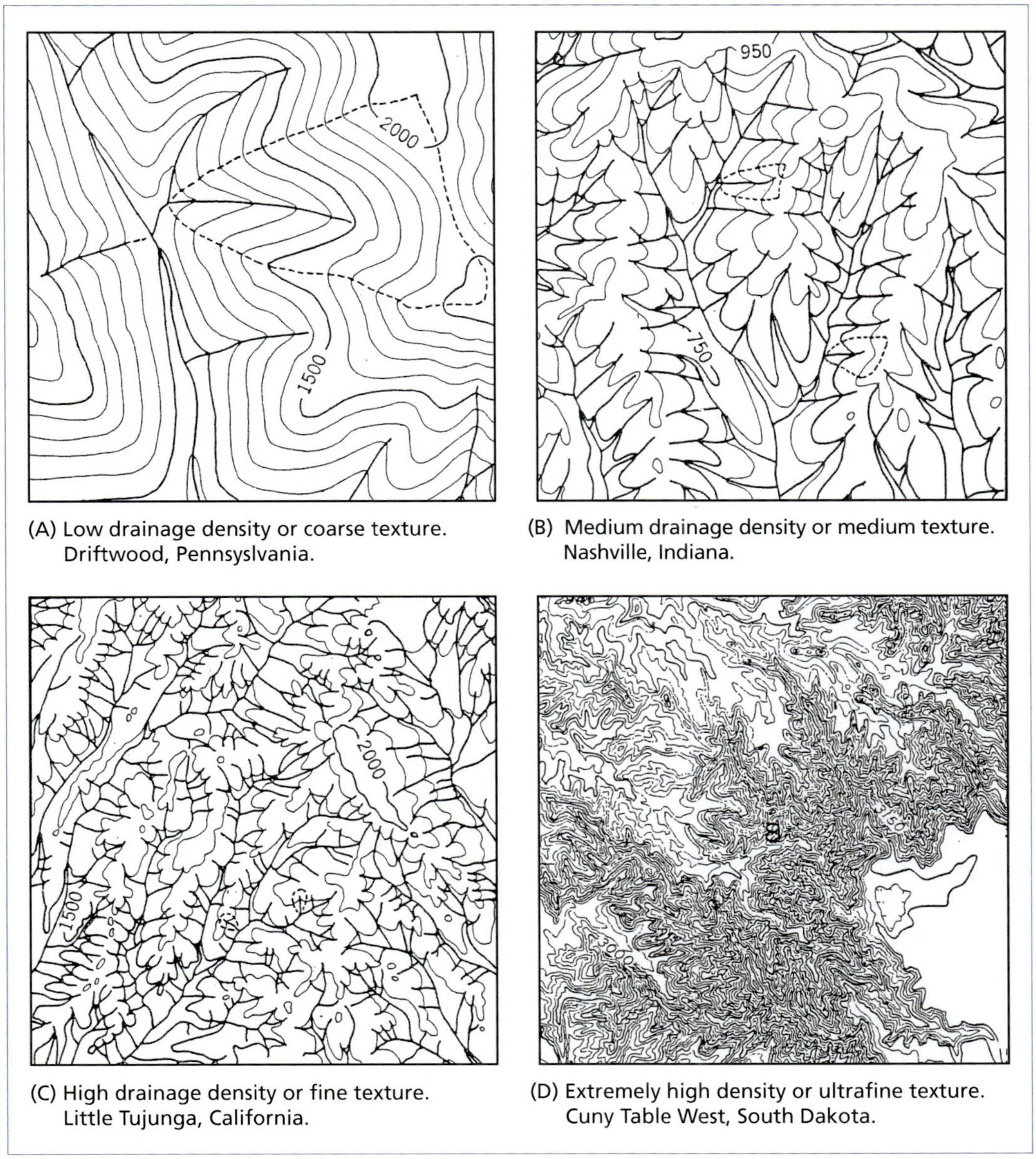

(A) Low drainage density or coarse texture. Driftwood, Pennsyslvania.

(B) Medium drainage density or medium texture. Nashville, Indiana.

(C) High drainage density or fine texture. Little Tujunga, California.

(D) Extremely high density or ultrafine texture. Cuny Table West, South Dakota.

Figure 8.6 Topographic maps of approximately 2.5 km^2 that illustrate the different types of drainage density (Strahler, 1964).

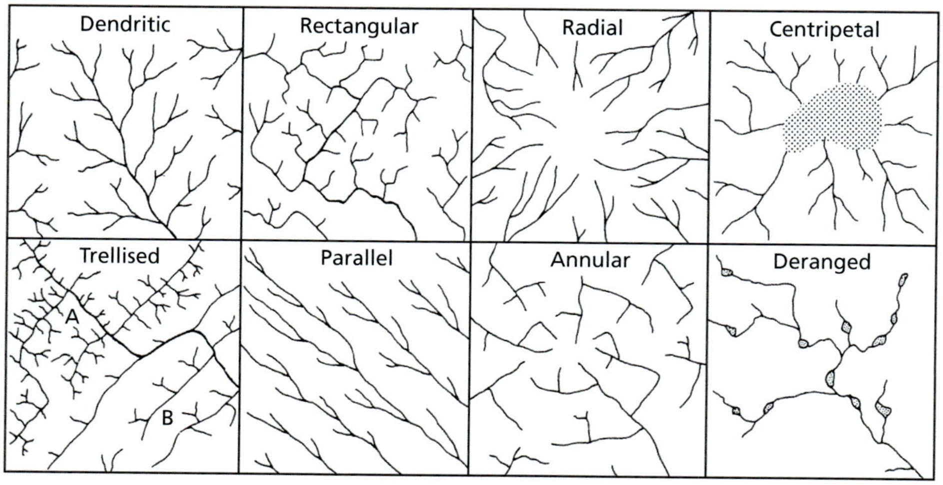

Figure 8.7 Morphological classification of drainage networks (Gregory and Walling, 1973).

calculate the approximate drainage network by tracing a line of known longitude (*L*) on a topographic map and counting the number of streams (*n*) that intersect the line. At least 50 intersections with topographic curves are necessary to provide a suitable value. The drainage density is expressed by $Dd = 1.41\ n/L$.

Until the middle of the last century, the systems used to determine the **network shape** were based on qualitative methods; the network was obtained by comparing it to different types of patterns (*Fig. 8.7*). This process evidently involved significant subjectivity and for this reason it became necessary to express the network shape numerically. One way to express the shape numerically is to measure the different lengths of the streams and their directions and to represent them on a rose diagram (Judson and Andrews, 1955). This method provides the orientation of the entire network, which is usually controlled by a fault system. Another method that takes into account the orientation of the drainage network is based on the measurement of stream junction angles (Lubowe, 1964). This method provides results not very different from the vector method; in both cases, a dendritic network has a unimodal distribution and a system with structural control shows a bimodal distribution.

The **channels system** was also initially classified into qualitative groups and later based on sinuosity, which is the result of the ratio between stream length and valley length. Channels can be straight, braided, or meandering. They will be studied in more detail later in this chapter.

The shape of the drainage basin is important. Elongate basins transmit river discharge over longer distances as a result of rainfall. Basins that are more equilibrated have divides that approximate the shape of a circle. Horton (1932) established a factor of the form $F = S/L^2$, in which S is the basin area and L is the length of the basin. Another method of expressing the shape of a drainage basin is Gravelius' compactness index, in which he provides an idea of the upper and lower proximity of the form of the basin to a circle. This index is based on the ratio between the basin perimeter and the perimeter of the circle equivalent to an area equal to the basin area. Other indices of basin shape are circularity and elongation. A study of the river basin morphometry is further discussed in Strahler (1964), Gregory and Walling (1973), and Zavoianu (1985).

8.5 Hydraulics of fluid flow

The morphology of natural fluvial channels results from the interaction of flow and erodible materials on the channel margin. These interactions vary substantially over different temporal and spatial scales. Due to the variable nature of river flow, the channel adjusts to different forces to maintain a coherent structure. Water flow in a channel is subject to two principal forces, gravity and channel slope; friction opposes water movement (Knighton, 1998).

If we assume uniform and stationary flow, the velocity at which the water passes across successive cross sections of a stream channel is constant or $S_1V_1 = S_2V_2$ (*Fig. 8.8*), where S is the area of the cross section at points 1 and 2, V is the median velocity at the points, and Q is the constant discharge.

Because water is a fluid, it cannot resist these forces. For this reason any force, no matter how small, produces movement, which is either **laminar flow** or turbulent flow.

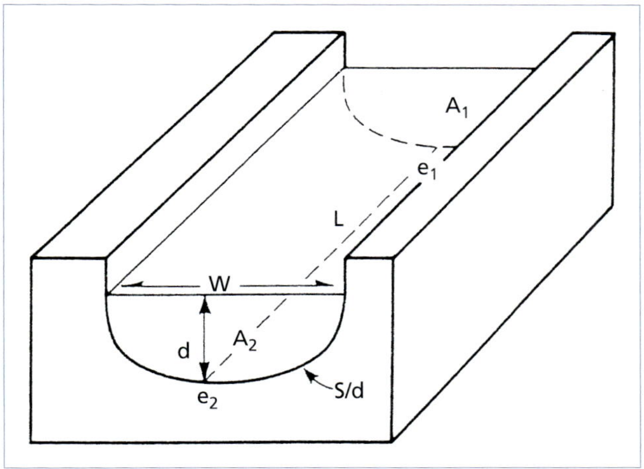

Figure 8.8 Morphometry of a fluvial channel. The width *W* is the width of the channel that contains water. *P* is the wetted perimeter area that is the boundary between the substrate and the water channel. The cross section *S* is the area of river section. The depth *p* is approximately the same as the hydraulic radius *R* which results from dividing the cross sectional area *S* by the wetted perimeter *P* ($R = S/P$). The gradient *g* of the river is the difference in height between two points ($e_1 - e_2$) of the channel bed, divided by the horizontal distance projected between them (*L*). Velocity is the discharge per unit area (Morisawa, 1968).

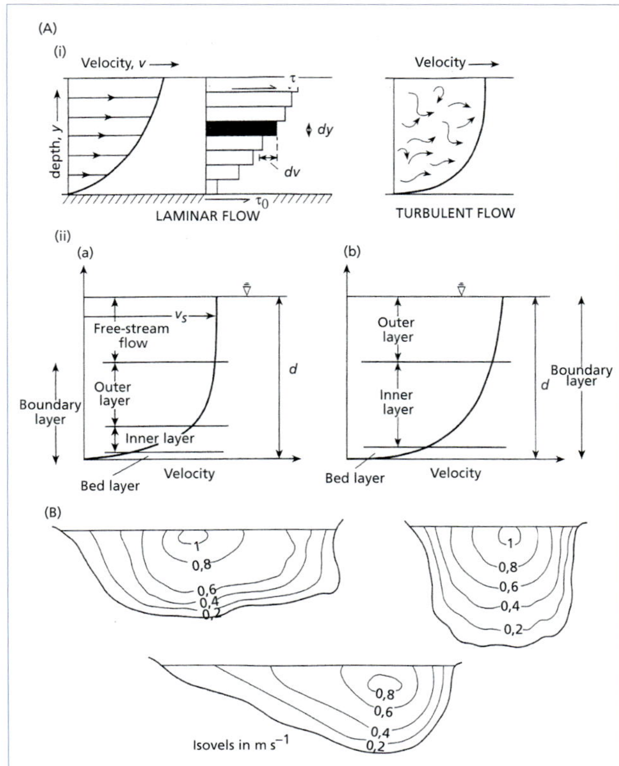

Figure 8.9 Variations in the fluvial flow velocity. (A) With depth: (i) typical velocity profiles for laminar and turbulent flows; (ii) structure of the boundary layer at depth (a) and for shallow flow (b) (B) Cross sections in natural channels (Knighton, 1998).

Water flowing through a straight and smooth channel at very low velocity (fraction of a mm/sec) is moving at laminar flow (*Fig. 8.9Ai*). Each element moves through specific trajectories without significant mixture with adjacent layers. At the contact with the riverbed, the velocity is practically zero and upper layers slide over it, forming a parabolic velocity profile (*Fig. 8.9Ai*). As a result, the shear force is:

$$\tau = \mu \frac{dV}{dy}$$

in which dV/dy is the change in velocity (V) with depth (y). Laminar flow rarely occurs in streams and rivers. It may be found in very tranquil waters.

If the velocity or depth attains a critical value, laminar flow becomes unstable and the parallel layers are destroyed, giving rise to **turbulent flow** (*Fig. 8.9Ai*). This flow is characterized by numerous chaotic movements with secondary eddies superimposed on the frontal flow. Various factors affect the critical velocity at which laminar flow is converted to turbulent flow; these factors include flow viscosity and density, water depth, and rugosity of the channel bed. The most commonly used expression to distinguish between laminar and turbulent flow is the **Reynolds number**:

$$NR = \rho \frac{VR}{\mu}$$

in which ρ is density, V mean velocity, R is the hydraulic radius, and μ is viscosity. Laminar flow has low Reynolds number values and turbulent flow has high values. Stream rugosity depends on the size and shape of materials in the streambed, wide variations in size and shape of the channel, and channel sinuosity. Rugosity is low when the river is near the flood level; it has higher values at low flow. As channel sinuosity increases, rugosity increases; a triangular shaped channel has higher rugosity than a trapezoidal shaped channel. The amount of vegetation and the type of flow also have an influence on rugosity. Hicks and Mason (1991) present a set of data and color plates of New Zealand rivers to visually estimate rugosity coefficients. Another criterion for

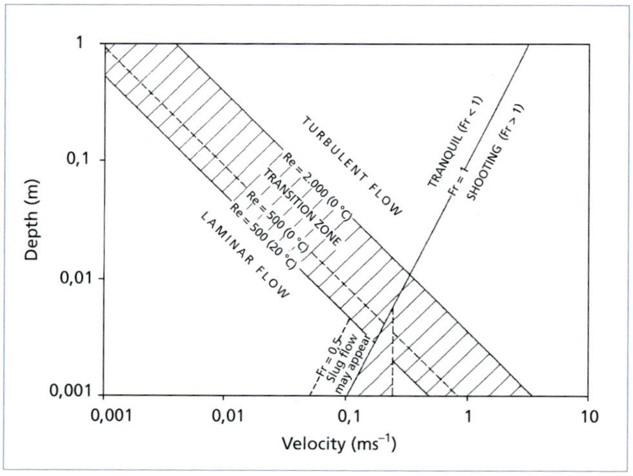

Figure 8.10 Flow regimes in a wide channel, as a function of velocity and depth. The zone of transition between laminar and turbulent flow is defined as the interval of Reynolds number values (*NR*) from 500 to 2000; it is dependent on temperature variations. The Fraude number (F_r) of 1 delimits tranquil turbulent flow and rapid turbulent flow (Sundborg, 1956).

differentiating laminar flow from turbulent flow is the **Froude number** (F_r), which is the ratio between the forces of inertia and forces of gravity:

$$F_r = \frac{V}{gd}$$

where V is velocity, g is gravity, and d is water depth. $F_r = 1$ is equivalent to critical flow, if $F_r < 1$ flow is laminar and if $F_r > 1$ flow is turbulent (*Fig. 8.10*). In most rivers, flow is generally not uniform; it is not permanently turbulent or tranquil, although in rare cases rapid flow occurs (waterfalls, rapids). An example is the lower Mississippi River where F_r is less than 0.20 (Sundborg, 1956).

Velocity is a vector that includes magnitude and direction. It is one of the most sensitive variables. Velocity varies as a function of distance to the riverbed. Velocity increases from zero at the bed to high velocity (*Fig. 8.9Aii*) at the edge of the boundary layer, which is affected by friction with the bed. In shallow flows, the boundary layer may extend to the surface. In theory it can be divided into two sub-layers: (a) an internal layer (or laminar sub-layer) that includes 10–20% of the water thickness above the bed, in which the velocity varies semi-logarithmically with depth; and (b) the external layer, which occupies the upper 80–90%, is highly turbulent, and its profile is far from a semi-logarithmic shape. Velocity increases toward the center of the stream due to decreased friction with the bed and channel walls (*Fig. 8.9B*). Velocity is intimately related to the **flow resistance** of the margins of the channel; there are various equations involving this resistance (Chezy equation, Manning equation and Darcy-Weisbach equation). Of these we will use Manning's:

$$V = \frac{1}{\pi} R^{2/3} S^{1/2}$$

in which π is rugosity, R is hydraulic radius, S gradient of the current, and V is average velocity (m/sec).

A fluvial system has two types of **energy**: potential and kinetic. Potential energy is converted to kinetic energy downstream where it is lost as heat of friction and also due to rugosity, the transverse section of the stream, sinuosity, and the amount of shear generated by eddies. Energy that is not dissipated as heat is used in erosion and transport (Morisawa, 1968). A smaller energy loss results from the internal friction of the transported load but in general, suspended load decreases friction because of less turbulence. **Critical energy** was defined by Bull (1979, 1991) as the energy needed to transport sediment through a channel. Energy variations in riverbeds have been used to explain the sequence of pools and riffles, to determine the shape of the bed for sediment size determinations, and to explain changes in channel systems (Gregory, 2004b). Fluvial systems have been classified according to the quantity and size of the bed load and the energy of the current (Schumm, 1981); sinuosity has been related to current energy (Schumm, 1977). Floodplains are classified on the basis of the energy value of the current: high (>300 W/m^2), medium (100–300 W/m^2), and low energy (10 W/m^2) (Nanson and Croke, 1992).

Once the particles that make up the bottom start moving, the bed is modified and variable landforms develop depending on flow conditions (Allen, 1976). These landforms can be easily observed in sand riverbeds because they follow a sequence that increases with flow intensity (*Fig. 8.11A*); each form has different resistance levels (Simons and Richardson, 1966). These bed modifications are important self-regulating mechanisms of the flow-bed interface. Flow and sediment load increase with the arrival of a flood wave that marks the transition from ripples to dunes (*Fig. 8.12*), which can increase resistance to flow (Richards, 1976). The net resistance to flow remains relatively low until breaking waves arrive and considerable energy is lost (*Fig. 8.11A*). The morphology of bed rugosity for larger sizes is a microtopography of groups of pebbles and channel bars from sequences of pools and riffles (Robert, 1990) (*Fig. 8.11B*).

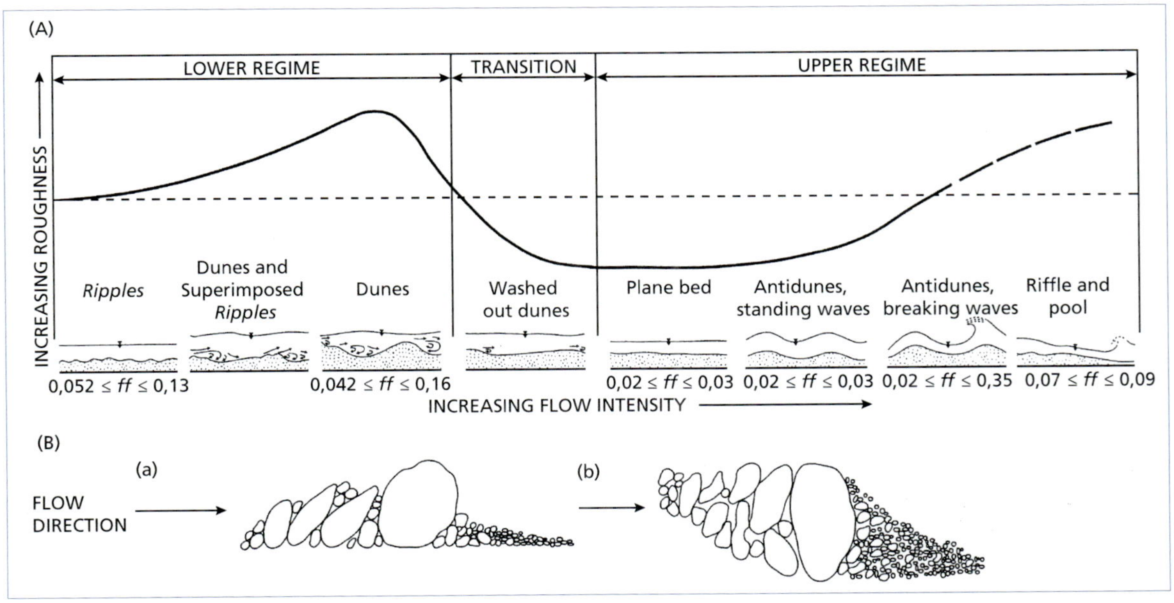

Figure 8.11 Variation in bed shapes as a function of flow intensity and rugosity. (A) The Darcy-Weisbach friction coefficient value is used for sandy beds (*ff*) in laboratory flume experiments (Simons and Richardson, 1966). Gravel riverbeds (A) in profile and (B) in plan view (Knighton, 1998).

Figure 8.12 Linguoid-shaped ripples in an ephemeral stream. Jaisalmer. Thar Desert, India.

The interaction of flow and vegetation is one of the more complex problems in fluvial hydraulics. Green (2005) has stated that resistance to flow is a function of the size of the plant, its structural properties, its location within the channel, local flow conditions, and the percentage of the channel occupied by vegetation. Flume studies show that in braided channels with vegetation, the vegetation directly affects sediment transport and the plan-view shape of the channel. The obstruction causes flow separation and the downstream side of the plant is a preferred place for sedimentation; the number of channels and the degree of braiding increase. In addition, plants stabilize and can even form relatively stable islands downstream of a plant (Coulthard, 2005).

The development of pools and riffles is characteristic of both straight and meandering channels with heterogeneous fluvial bed material (2–256 mm grain size). These channels have undulating beds, with pools and riffles alternating along their length (*Fig. 8.13*), spaced at regular distances of 5 to 7 times the width. This alternation is characteristic of rivers that transport gravel. These gravel accumulations have a lobate shape (Leopold *et al.*, 1964).

8.6 Sediment transport

Discharge information is relatively easy to obtain; it is more difficult to determine a river's sediment load. This load is present in solution, suspension, and at the bottom of the channel. **Dissolved load** consists of ions and molecules produced by chemical weathering; it is generally inorganic material. Its composition depends on a variety of environmental factors (geology, climate, vegetation, and topography). The water of large rivers is similar in chemical composition; the dominant ions are bicarbonates, sulfates, and calcium and sodium chlorides. Water with an acidic pH generally results from oxidation of sulfides or water supplied by swamps and marshes; it may have a higher chemical load. The **suspended load** consists of silt and clay size particles small enough to be transported by turbulent river water. Strong currents carry sand and floods can transport gravel. This supports the observation that the fine fraction moves throughout the channel and is present almost uniformly throughout the water column whereas sand has a tendency to concentrate near the bed (Chorley *et al.*, 1984). The **bedload**, made up of gravel, pebbles, and blocks moves

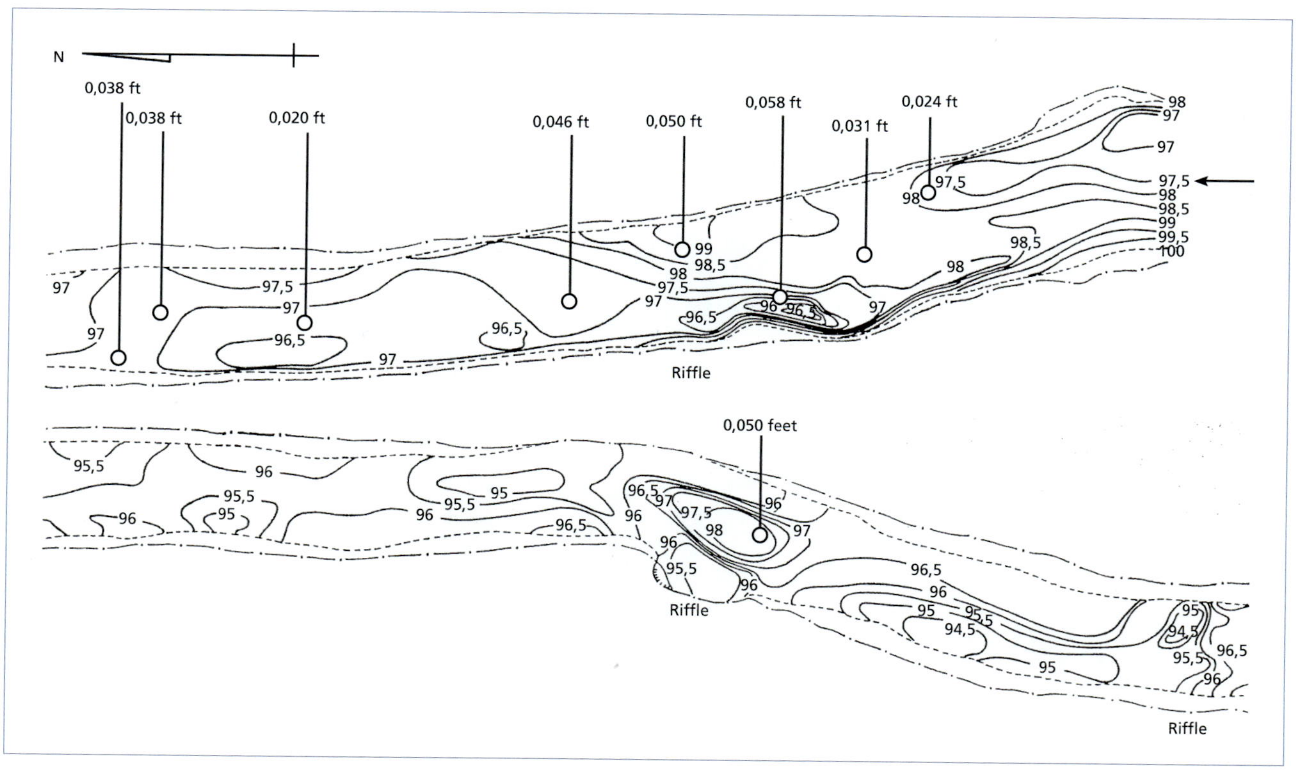

Figure 8.13 Topographic maps of the channel bed of Seneca Creek near Dawsonville, Maryland. The circles show the location of bed material measurements; the numbers associated with the circles indicates the average size in feet (Leopold *et al.*, 1964).

by rolling or dragging along the channel bed. If the current is strong, this material can saltate. The sand bedload moves more slowly than water flow and the grains move in isolation in any direction or as groups by rolling and sliding. Large-size grains move and rounded particles move more rapidly than flat ones. It is easy to collect samples from the suspended load (see Gregory and Walling, 1973) but difficult to sample the bedload because sampling causes flow disturbance (Kondolf *et al.*, 2003). To prevent this, researchers paint pebbles (Laronne and Carson, 1976) or mark them magnetically (Hassan, 1992; Hassan and Ergenziger, 2003) and find them later after flooding to measure their displacement. Experimental studies indicate that particle mobility (for example, transport distance) is greater for small particles than for large ones; particles smaller than 2 cm and lighter than 10 grams have about the same mobility and are preferentially trapped in open vertical structures, which they fill; and finally particles larger than 6 cm are preferentially trapped in imbricated structures (Laronne and Carson, 1976). Catastrophic floods on large rivers such as the Mississippi and the Brahmaputra can build large-scale dunes of pebbles and blocks that can be observed after the flood has passed (Allen, 2004).

Riverbanks are often composed of fine sediments such as fine sand, silt, and clay whereas the bed is composed of sands and gravels with a small quantity of silt and clay. These sediment types differ in both size and physical properties. The finest ones tend to be cohesive and are difficult to erode. The Swedish researcher Hjulström (1935) demonstrated these characteristics in a classic diagram that was later improved by Sunborg (1956) (*Fig. 8.14*). The diagram relates critical velocity to sediment size at the beginning of particle erosion in both water and air. Materials become more cohesive as particles become finer. Thus, as particles less than 0.25 mm in size, they require a stronger current to raise them. But once they are lifted up, the weaker the current needed to transport them. For this reason coarse particles are put into motion when there is enough transport velocity but a much higher velocity is needed to move fine materials (Derruau, 1974).

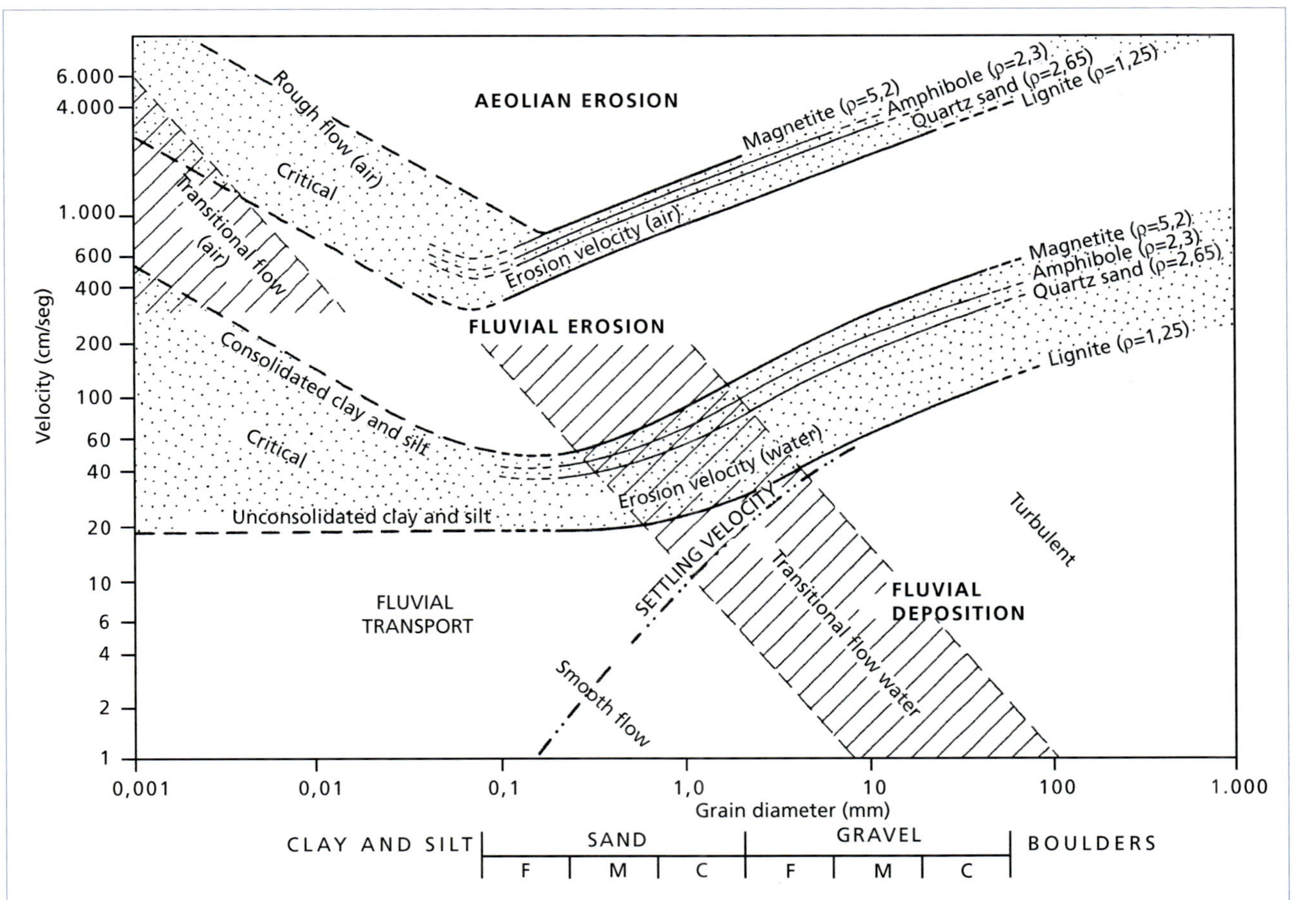

Figure 8.14 Curves showing the relationship between grain size and the critical fluvial and eolian erosion rate for uniform materials with different densities. The critical fluvial erosion rate refers to a height of 1 meter above the riverbed. The two critical zones around these curves and the settling rate curve set the boundaries of the four regimes of fluvial sedimentation, fluvial transport, fluvial erosion, and eolian erosion (Sundborg, 1956).

8.7 Fluvial erosion

Fluvial erosion results from weathering, abrasion, and cavitation processes. **Weathering** is any chemical process that wears down the bed and sides of a fluvial channel. **Abrasion** is mechanical wearing caused by impact of transported particles and mechanical erosion of the bed, which can create potholes (*Fig. 8.15*). **Evorsion** is a type of abrasion whereby the water flow smoothens the riverbed without the intervention of particles. **Cavitation** occurs only at high velocities such as at the base of waterfalls (*Figs. 8.16*, *8.17*) and cataracts (Figs. 18.26, 18.27), in rapids (*Fig. 8.18*), and some artificial conduits (Morisawa, 1968, 1985).

Rivers and streams can erode their beds and walls. **Vertical erosion** in the bed of an alluvial channel occurs when sand and gravel-size particles are mobilized and the load wears down the bed. **Lateral erosion** occurs when the walls of the channel are eroded, normally by undercutting, which leads to landslides or collapse of parts of the channel wall. Continuous lateral erosion widens the channel and vertical erosion causes incision. One of the most spectacular examples of widening and deepening of a valley is the Grand Canyon of the Colorado (USA) (*Fig. 8.19*), which has incised about 1,300 m into a 10-km-wide valley. In regions of massive resistant rocks, retreat of canyon walls is slower than downcutting by rivers, which creates sub-vertical and narrow canyons (*Fig. 16.66*).

Figure 8.16 Kjosfossen waterfall, 93 meters high Norway.

Figure 8.15 Potholes in the granodiorite of Burguillos del Cerro. Badajoz, SW of Spain.

Figure 8.17 Gulfoss Waterfall. Iceland. Photo F. Gutiérrez.

Figure 8.18 Rapids. Gol. Norway.

Figure 8.19 Grand Canyon of the Colorado with stepped margins The Colorado River is at the bottom. Photo F. Gutiérrez.

There are two types of overland flow. Horton's overland flow model is active when rainfall intensity is greater than the infiltration rate. This commonly occurs in arid and semi-arid zones whereas the second type, saturated surface flow, is more common in humid climates (Kirkby, 1969). Flow occurs where permeability decreases downward in the profile so that water is forced to flow laterally. When a horizon that overlies an impermeable horizon becomes saturated, surface flow occurs.

Water flowing downslope becomes concentrated as a result of topographic irregularities; rills and gullies form although the mechanism is not well understood. Most research on rills and gullies has been done in arid and semi-arid regions. For this reason, this topic will be discussed in more detail in Chapter 16, Geomorphology of Arid Regions.

8.8 Longitudinal profile, base level, and stream capture

The longitudinal profile of a river is a line linking the various heights from the river's source to its mouth. Because a river flows toward the sea it slopes toward the sea and downcuts the valley until it reaches the limit, sea level. The longitudinal profile is concave. Many rivers are not obviously concave because parts of them are flat and some are steep. These flat or steep segments are called **knickpoints**, which can result from rocky outcrops, recent active tectonics, and sudden changes in flow. For example, the longitudinal profile of the Rhine River (*Fig. 8.20*) is 1236 km long and has an elevation difference of about 3 km. Its first hundred kilometers have very steep gradients and the rest of the gradient is low. Knickpoints occur at falls on the high Rhine near Schaffhousen and near Bingen in the middle Rhine (Ahnert, 1996).

The general base level is the minimum altitude, which is sea level, although there can be local base levels such as lakes, bands of resistant rock, reservoirs, etc.

The concept of a **graded profile** in a fluvial system refers to a state in which the active variables and the base level are constant (Mackin, 1948). If base level changes, rivers adjust to the equilibrium profile by changes in the channel system, width, and rugosity or by modifying the channel slope through erosion or sedimentation. The variables seen in the channel can change in a short period of time. As a result, it is more appropriate to talk about a quasi-equilibrium state. The more significant the modifications to the profile (for example, neotectonics), the greater the time needed to adjust to the graded profile.

A fluvial valley can elongate by headward erosion. In easily eroded rock such as marls or clays headward erosion is rapid and can reach the watershed divide of another river, finally reaching the other stream and causing **stream capture** (*Fig. 8.21*). Stream capture is often revealed by a capture elbow that leaves the downstream water captured in a dry valley.

8.9 Fluvial channel systems

By observing channel systems, geomorphologists can distinguish **straight**, **meandering**, and **braided** rivers. However there is a wide range of channel systems between straight, meandering, braided, and anabranching (*Fig. 8.22*). Straight and meandering channels are distinguished by their **sinuosity** (*S*) (Schumm, 1963b), which is the ratio between the length of the channel (L_c) and the length of the valley (L_v) or the ratio of valley slope (P_v) to channel gradient (P_c)

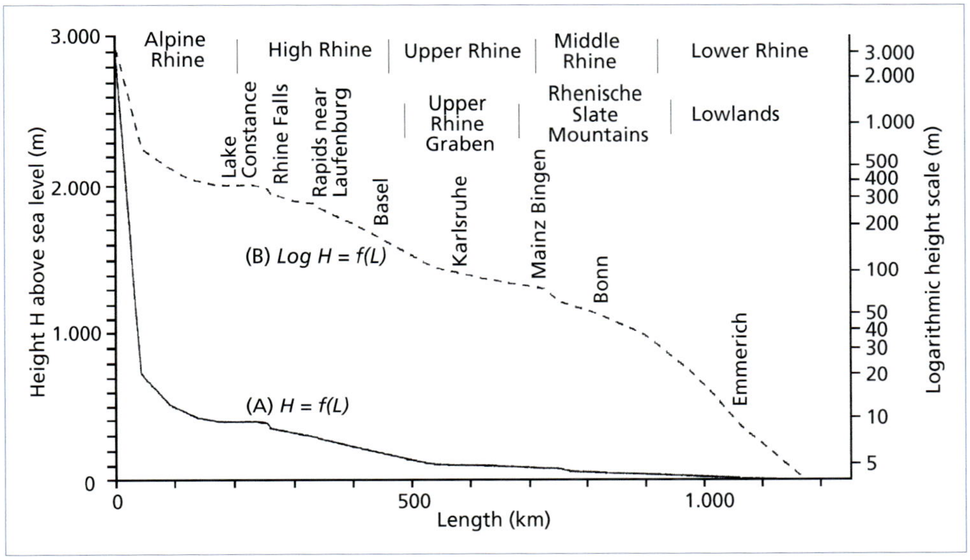

Figure 8.20 Longitudinal profile of the Rhine River. (A) linear height scale; (B) logarithmic height scale (Ahnert, 1996).

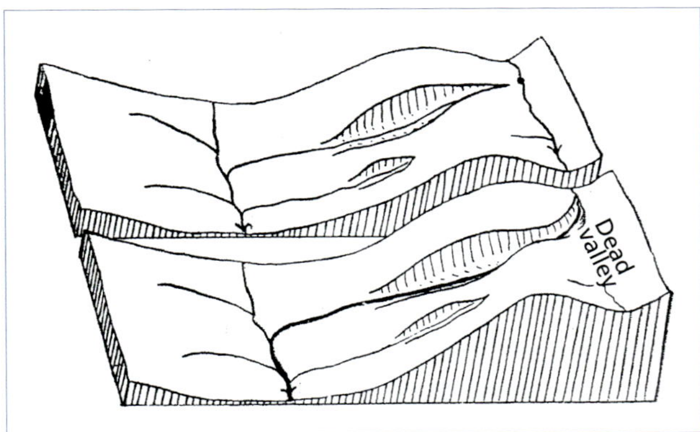

Figure 8.21 Capture by headward erosion (Derruau, 1974).

$$S = \frac{L_C}{L_V} = \frac{P_V}{P_C}$$

In addition to changes in sinuosity, channel width variability and meander belt width can be bimodal; two meander systems of different dimensions can be superimposed (*Fig. 8.23*). A braided channel is made up of bars or islands (*Fig. 8.24*) and the **degree of braiding** (Brice, 1964) is expressed as the percentage of length of the channel divided by the length of one or more bars. It is a variable index because bar length depends on the time of measurement. The **degree of anastomosing** (Brice, 1964) is the percentage of the measured length that is occupied by large islands.

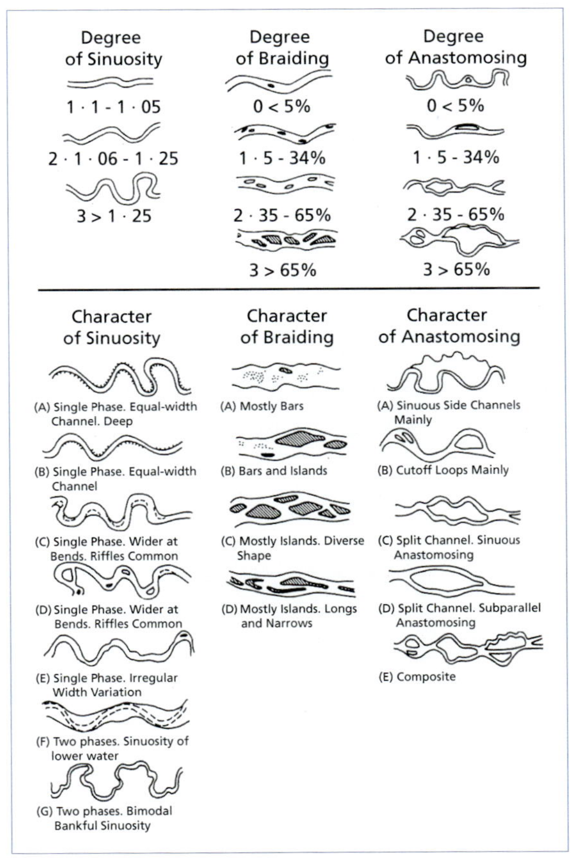

Figure 8.22 Classification of alluvial channels, including the plan-view shape of a single channel and anastomosing channels. The laterally inactive channels are straight and sinuous shapes, while the laterally active channels are meandering and braided (Nanson and Knighton, 1993).

A "confined braided channel" is one in which water covers the channel during floods and a system of submerged bars develops. As discharge decreases, bars emerge, creating a braided channel. In a "free braided channel" the braiding occurs on an unconfined alluvial plain. Kiss and Sipos (2007) found that during the high water of flooding a large volume of sediment erodes and as the flood ebbs, the predominant process is accumulation. In braided rivers that have islands, sediment is deposited upstream of the island. Transport zones form downstream of the end of the island due to the increased current energy. Even when discharge increases, the alluvial plain is never completely submerged. Conditions associated with the development of braided rivers include steep slopes, variable flow, coarse-grained sediments, and significant sediment supply (Miall, 1977). For equality of load with meanders, braided channels develop with greater slopes. The width/depth ratio plays an important role. Channels that are 20 times wider than deep tend to be straight; values between 15 and 150 develop alternating bars that very possibly form meanders; and for numbers on the order of 150, multiple bars develop that are interpreted as the beginning of anastomosing (Paola, 2004).

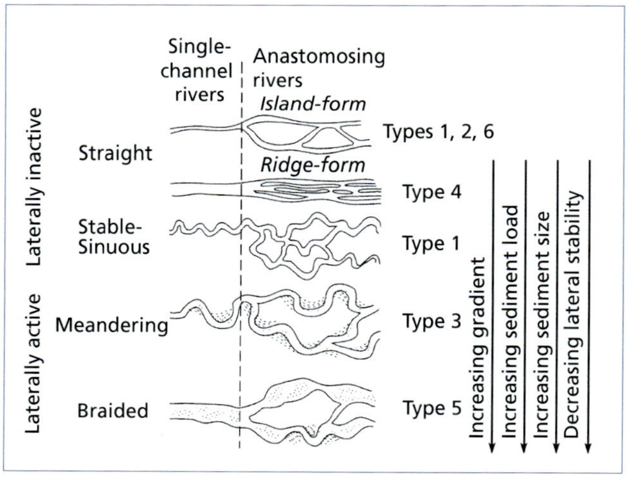

Figure 8.23 Channel systems types (Brice, 1964).

A braided system can form in two ways (Chorley *et al.*, 1984). During active aggradation, the bottom load is deposited as bars, flow is divided, and a typical braided system develops. The other type of braided channel is associated with a steep gradient in which high current energy transports large quantities of bedload. A wide, not-very-deep channel is needed to move this material; it requires a steep slope and large shear forces acting on the riverbed.

Our understanding of braided river dynamics has increased enormously in the last decade with the introduction of methods for measuring flow and bed morphology (Sambrook Smith *et al.*, 2006). Numerous studies have been

Figure 8.24 System of braided channels where the Kotsina River flows into the Cooper River. Southeastern Alaska. Photo P. Lucha.

done on geometry, flow, and sedimentary processes, as well as systems of erosion and sedimentation in the Brahmaputra/ Jamuna River (Bridge and Lunt, 2006, among others). These studies have identified complicated systems of convergence and divergence of flow velocity associated with bars. Bars may originate in the center of a channel or by cutting off meander bars. Once formed, the bars grow by lateral accretion in the downstream direction. Bars in these rivers can be 15 m high, 1.5–3 km long and 0.5–1 km wide. During monsoons, accretion can be on the order of kilometers.

In an anastomosing channel network, the typical channel width is much smaller than the bars whereas in braided rivers these two lengths are comparable. Individual branches of anastomosing channels can be straight, braided, or meandering (Schumm, 1977).

Nanson and Knighton (1996) distinguish six types of **anastomosing channels** based on current energy, sediment size, and morphological characteristics *(Fig. 8.23)*. Types 1 and 3 have low energy and types 4 and 6 have higher energy. The figure shows a plan view of various types of anastomosing rivers. Type 1 is a river with cohesive sediments, channels with a low width/depth ratio, and almost no lateral migration. Type 2 is predominantly composed of sandy islands. Type 3 is a meandering river with mixed bed load that has laterally active meanders. Type 4 is a river with sandy belts parallel to the channel margin. Type 5 is predominantly composed of gravels in laterally active rivers and braided channels in mountainous areas. Type 6 is made up of stable rivers with a predominantly gravel load that migrate within small channels in relatively scabrous basins. Vegetation plays an important role in the development and maintenance of anastomosing rivers. These rivers are not found in formations that predate the Devonian, which is when plants began to develop. Plant roots increase the strength of channel walls (Nanson and Gibling, 2004). The development of anastomosing channels occurs during frequent flooding within channel walls that resist erosion along with mechanisms that block or restrict channel development and the triggering of avulsions (Tooth and Nanson, 1999). Many large rivers have anastomosing channel systems due to a low gradient and a bedload of fine sand (Latrubese, 2008). Latrubese considers the anastomosing system to be the final member of the straight, meandering, and braided system.

Meandering channels have high sinuosity. In a classic meandering channel, channel width remains constant while its sinuosity increases, in such a way that in plan view the channel can be described as a single sinuous line *(Fig. 8.25)*. Richards (1982) has described the development of meanders. Above some discharge thresholds, velocity variations are associated with the formation of riffles and pools, which are spaced at 5–7 *W* (channel width) *(Fig. 8.26)*. At low discharges, water flow erodes the pools more intensely and

Figure 8.25 Valley meanders in the region of Oslo, Norway.

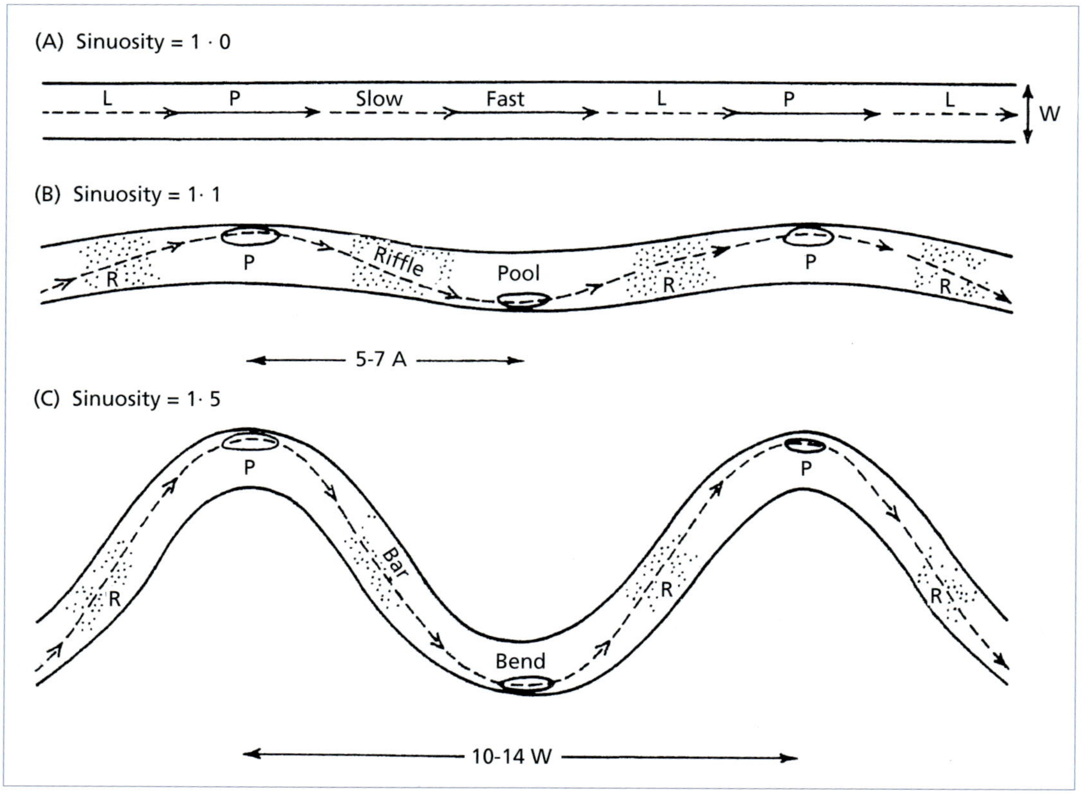

Figure 8.26 Transition from a straight to meandering channel. (A) Includes rapid and slow flows. (B) Development of riffles and pools spaced 5–7 *A*. (C) Development of meanders with a wavelength of 10–14 *A* (Richards, 1982).

causes slow sedimentation at the riffles. In addition to enlarging the transverse sections of streams, average velocity and sedimentation in riffles decreases. All this is associated with the creation of an undulating channel bed. Continued lateral erosion increases sinuosity. Strangulation of a knot channel creates an **oxbow**, forming a lake on the floodplain; in the Ebro Depression these are called "galachos" (Pellicer and Yetano, 1985) (*Figs. 8.27* and *8.29*). The speed of movement in maximum curvature zones can reach 3 m/year according to calculations done over a period of 20 years on 100 curves in the Dane River in northwestern England (Hooke, 2007).

Incised meanders are curved morphologies that deepen substantially as the riverbed erodes and creates a gorge (*Fig. 8.19*). They may be structurally controlled, which causes them to develop a long river in the zone of fracturing. Downcutting and natural leveling are often simultaneous. The incision results from tectonic uplift and is usually a case of **antecedence** in which the river that flows prior to uplift maintains its direction. Incised meanders are a response to a drop in base level (Simon and Darby, 1999; Schumm, 1999).

Khan conducted experimental research in a large flume in 1971 that showed that the progression of fluvial systems

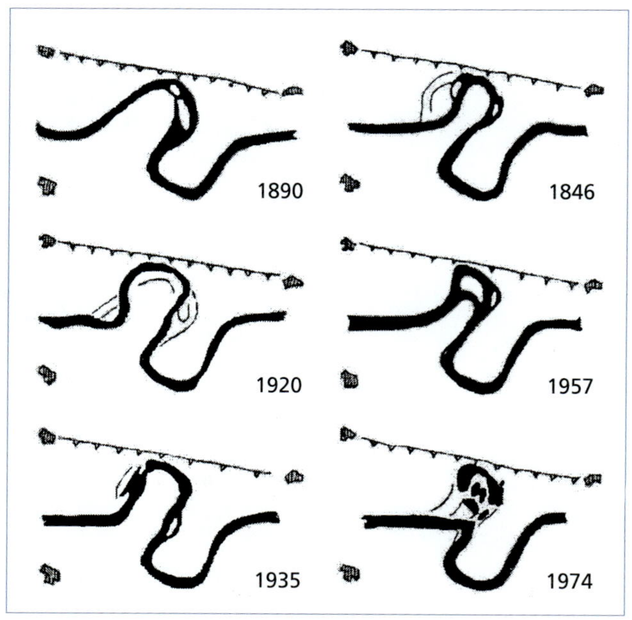

Figure 8.27 Evolution of a meander in Juslibol (Zaragoza) to formation of an abandoned meander (galacho) (Pelllicer and Yetano, 1985).

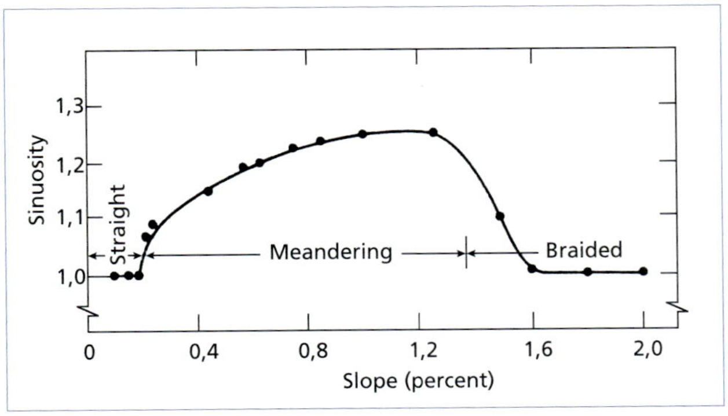

Figure 8.28 Relationship between channel sinuosity and slope of the flume (Schumm and Khan, 1972).

from straight to meandering to braided depends on channel slope or current energy, which reflect sediment load and flow. When channel slope is low, the river is straight but with more inclination, the river exceeds a threshold and develops into a meandering channel. When the slope exceeds 1.6%, the river transforms into a braided channel (*Fig. 8.28*).

8.10 Stability of fluvial channels

An alluvial channel changes over time because it sits atop erodible sediments. Continuous variation of meandering channels over time (*Fig. 8.29*) is a characteristic of alluvial plains that have meanders. Meanders can migrate up to 8 m/year, as the Arkansas River (Nadler and Schumm, 1981) has done for a period of 43 years. As the channel changes through time, its curvature and drift grow and **neck cutoffs** or **chute cutoffs** can occur (*Fig. 8.30*). A channel can also be totally abandoned by **avulsion**. These changes have important implications for political and administrative boundaries.

Schumm (1981) describes a relationship between channel type and relative stability. He distinguishes five systems (*Fig. 8.31*). This classification is not descriptive but is instead based on cause and effect relationships; it illustrates differences that occur in rivers carrying different types of transported loads and having different flow velocities and current energy. It also explains various systems within the same river (Schumm, 1977).

Classifications of alluvial channels must take into account not only the channel type but also the variables that control channel morphology. Channel dimensions basically result from water flow and channel shape, which are related to the quantity and type of sediment transported, which is also a function of sediment type (percentage of gravel, sand, silt and clay) (Schumm, 1981).

When the bedload proportion is small, the channel of suspended load is narrow and deep (depth/width ratio <10). If the channel slope is low, it can be straight (**system 1**) (*Figs. 8.30A* and *8.31*). Channel walls are relatively stable because of their high silt and clay content. As a result, the channel does not drift and bars are able to move down the channel. These channels are not common and present few problems. **System 2** has an intermediate percentage of bedload and the thalweg (line of maximum depth) is sinuous with a width/depth ratio >40 and low sinuosity. Bars drift within the channel (*Fig. 8.30B*) because the thalweg changes over time; sedimentation occurs on one side of the channel and erosion on the other. **System 3** has two types of meandering channels. When the channel of suspended load is highly sinuous (*Figs. 8.31*, *8.32*, and *18.23*) it transports only a small amount of bedload (**System 3a**). The width of the channel is similar and the margins are stable but neck cutoffs can occur. **System 3b** is a less stable type of channel; it transports a mixed load. The channel walls are composed of material that has little cohesion so they are less stable than System 3a channels. The meander increases its radius of curvature and causes typical neck cutoffs (*Figs. 8.33* and *8.30C*). Therefore, the channel is unstable; it erodes in its concave parts and deposits in convex parts. **System 4** is a transition between meandering and braided streams (*Figs. 8.34* and *8.31*); the predominant load is gravel, pebble, and sandy sediments. The channels have variable widths, are shallow, and have a marked gradient. Chute cutoffs of the channel and erosion of the walls are typical. **System 5** (*Fig. 8.31*) has a high bedload and significant inclination and is capable of transporting this load. The channel is unstable because the bars and the thalweg drift throughout the channel. These are typical braided rivers (*Fig. 8.24*), which are common in alluvial plains and alluvial fans; frequent avulsion occurs (*Fig. 8.30*) (Chorley *et al.*, 1984).

The channel system evolves from system 1 to 5 (*Fig. 8.31*) and also shows modification of other morphological aspects. Increased sediment size is accompanied by a change of the system from 1 to 5. Channel stability increases with increased proportion of fine deposits or with a decrease of the bedload/total load ratio. Lower values of flow velocity and of current energy increase stability. The width/depth ratio is high for braided channels, low for straight channels, and intermediate for meandering channels. Braided channels have a high gradient, straight channels have small gradients, and in meandering channels, the slope is intermediate between the two. As a result channel stability decreases from system 1 to 5; systems 4 and 5 are the least stable.

One channel system grades continuously into another. A channel can have changing channel systems along its length. A single channel can meander and with a change in gradient it may behave as a braided channel. As an example,

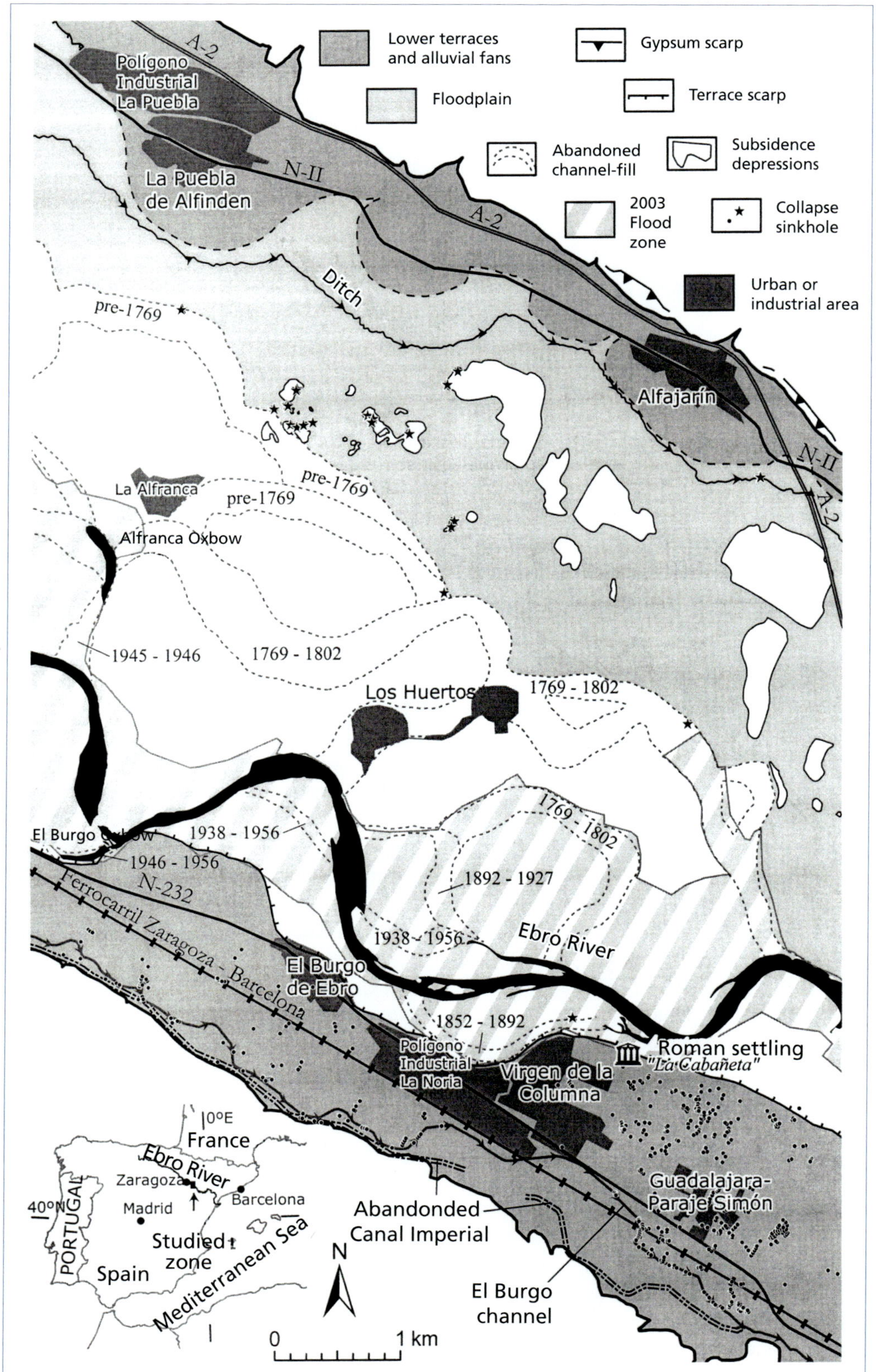

Figure 8.29 Development of meanders in the Ebro River. Zaragoza area. Spain. The dates indicate the time of abandonment time of the channel segments (Gutiérrez *et al.*, 2007).

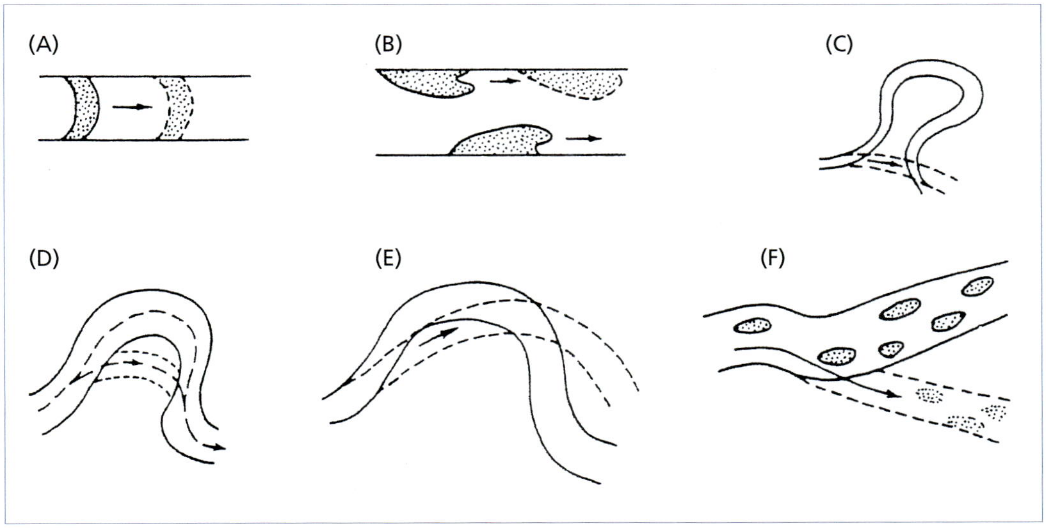

Figure 8.30 Types of channel changes: (A) Drift of a cross bar. (B) Drift of alternating bars. (C) Neck cutoff. (D) Chute cutoff. (E) Meander drift. (F) Avulsion. The continuous lines indicate the current status and those of potential future changes (Shen and Schumm, 1981).

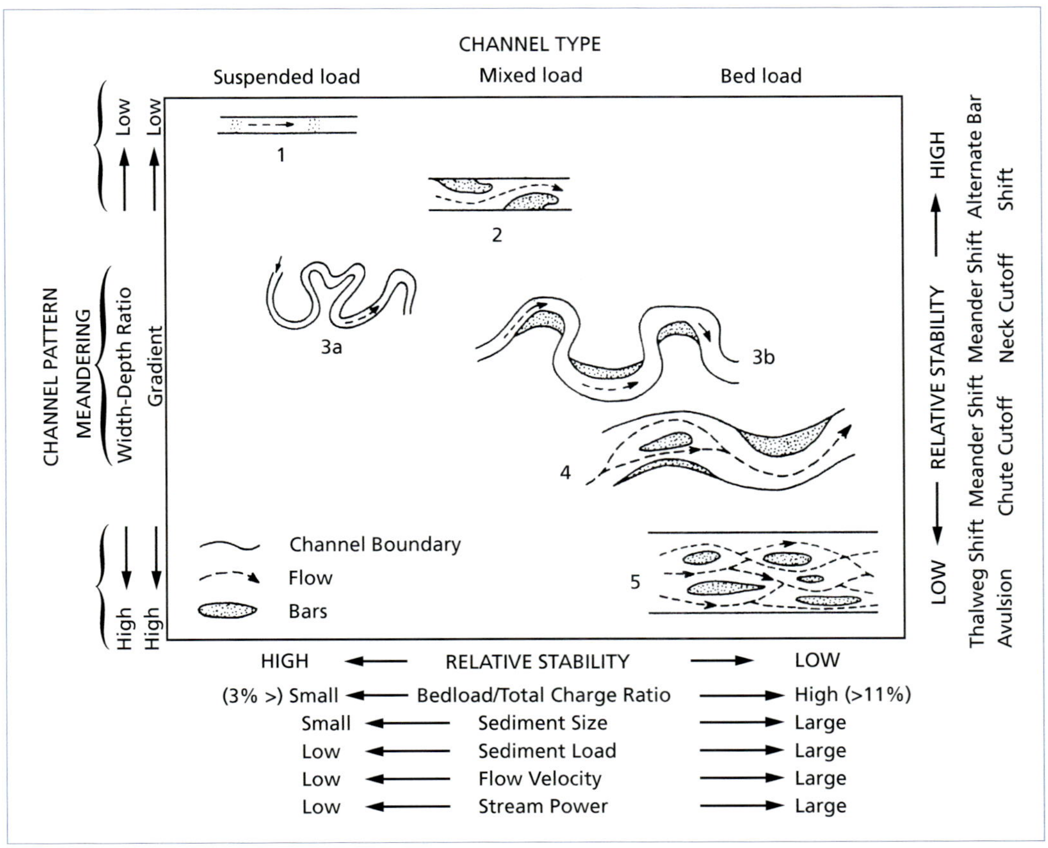

Figure 8.31 Genetic classification of fluvial channels based on the channel system and type of sediment load that indicates the associated variables and relative stability (Schumm, 1981).

Figure 8.32 High sinuosity river (system 3a of Schumm, 1981) at the bottom of the Gregory Rift Valley. Lameluk, Baringo District. Kenya.

Figure 8.33 High sinuosity meandering channel, Copper River Delta. Southeastern Alaska. Photo F. Gutiérrez.

Figure 8.34 Schumm's system 4, with large bars. Yamuna River, Agra, India.

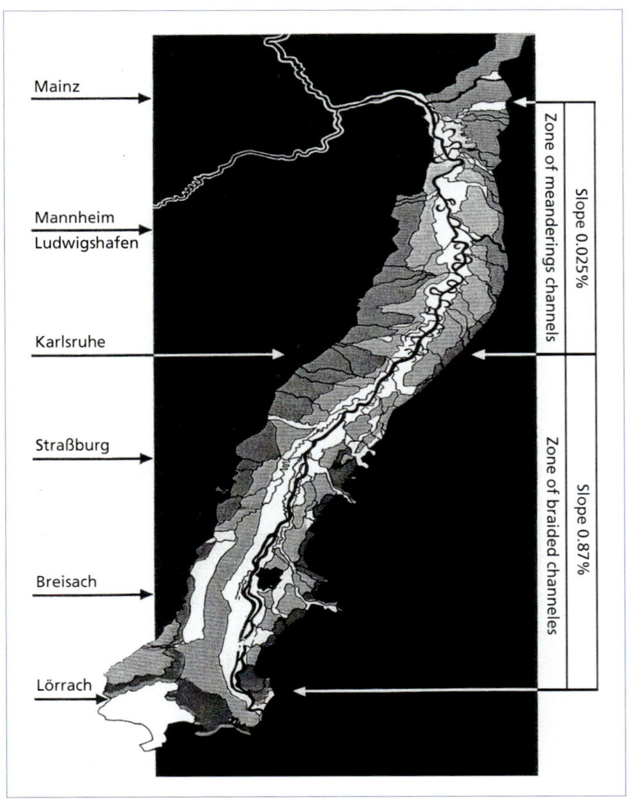

Figure 8.35 Diagram indicating the change in channel system of the Rhine River, as a result of slope change (Schäfer, 1973).

the channel system of the Rhine River (Schäfer, 1973) changes along its length due to a change in gradient (*Fig. 8.35*).

8.11 Metamorphosis of rivers

Rivers change as a result of erosion and sedimentation but these transformations require time. For this reason, flow and sediment modifications do not cause immediate changes in a channel but they initiate changes that can occur over time. Channel changes occur as a result of variation of flow frequency and a change in the amount of material eroded; these modifications may occur at the same time or not. The most important hydrologic problem may be to determine quantitatively how much alteration is needed in the relationship of precipitation-runoff-erosion in order to create a specific type of channel (Dunne and Leopold, 1978). In addition, climate changes during the Pleistocene-Holocene have profoundly affected fluvial systems. Even so, within any region there are sensitive and insensitive fluvial systems and as a result, although extrinsic change may be significant, the response of specific rivers can be considerably delayed (Schumm, 1991).

Fluvial channel changes have provided a complex theme for geomorphic research over the last four decades (Gregory, 1977). In addition, historical analysis of fluvial systems is a new research field for paleoclimatology (Lamb, 1977), paleohydrology (Benito *et al.*, 1998b),

Table 8.2 **CHANGES IN CHANNEL WIDTH ALONG RIVERS IN THE WESTERN UNITED STATES (GRAFF, 1988)**

River	Change	Time
Canadian River, Oklahoma	0.8–3.2 km	Flood in 1906
Río Salado, New Mexico	15–168 m	1882–1918
Red River, Texas-Oklahoma	No change	1874–1937
Red River, Texas-Oklahoma	1.2–0.8 km	1937–1953
Cimarron River, Kansas	15–366 m	1874–1942
Cimarron River, Kansas	366–168 m	1942–1954
Platte River, Nebraska	1161–111 m	1860–1979
South Platte River, Colorado	790–60 m	1897–1959
North Platte River, Wyoming	1200–60 m	1890–1977
Gila River, Arizona	45–90 m 90–610m 610–61 m	1875–1903 1903–1917 1917–1964
Salt River, Arizona	No change	1868–1980
Fremont River, Utah	30–400 m	Flood of 1896

paleoecology (Berglund, 1986) and geoarcheology (Davidson and Schackley, 1976). Historical analysis of a fluvial system requires the reconstruction of systems that date from the pre-industrial era and a determination of the velocity and magnitude of system changes. Historical analysis also makes it possible to predict future impacts and aids in the recommendation of tasks for ecological management (Petts, 1989). Of the numerous examples of channel changes during historical times, some of the most spectacular studies are: in the Vístula valley (Poland), Hooke and Redmond (1989) in the valley of the River Dane (England), Castaldini and Piacente (1995) in the Po River (Italy), and Surian (2006) on Tagliamento, Piave and Brenta rivers (northern Italy), among others.

Changes of channel type are relatively common in the western United States; they have been recognized in numerous rivers (*Table 8.2*) that have changed from meandering to braided (Graff, 1988).

One of the most surprising examples is the metamorphosis of the Cimarron River (Kansas), which was analyzed by Schumm and Lichty (1963). This was a deep meandering river, 15 meters wide; after flood in 1914 it became a straight bedload channel 336 m wide with increased discharge and bedload. This dramatic change is interpreted to be a result of climate fluctuation, agricultural activity, and destruction of natural vegetation.

Another striking example is the Gila River that runs through Arizona and New Mexico (Graf, 1988). Prior to 1890, the channel was narrow and meandering but in 1903 several floods enlarged it into a braided channel more than 600 m wide in some places. Later vegetation growth and sedimentation narrowed the channel and at the beginning of 1980, it returned to its meandering geometry after almost a century. From this we can conclude that braided channels adjust to meandering channels much more slowly than vice versa.

Human activity is quite apparent in the evolution of the South Platte River (Colorado), a braided river that changed in width from 790 m in 1897 to 60 m in 1959, transforming into a narrow channel with a tendency to develop meanders (*Fig. 8.36*). This profound modification is related to regulation of the river; with loss of discharge due to irrigation, vegetation invaded the flood plain and bars, causing the channel to narrow (Nadler and Schumm, 1981).

Direct consequences of human activity are evident in engineering projects (dams, channels, etc.) but other indirect effects can be seen in changes of cross section and shape. The plan-view morphology of the channel, the network, and ecology also change (Gregory, 2006).

Erosion of concave margins of meanders and sedimentation on the convex parts (point bars) cause horizontal instability in this channel system. It also causes meanders to migrate downstream where they may interfere with structures built by humans. This is the case of a bridge over the Cimarron River near Perkins (Oklahoma). When the bridge was built, the river was straight and braided in the bridge area but a large meander a mile upstream shifted and in 1968 affected the structure of the bridge (Chorley *et al.*, 1984).

Although data are not available on sediment load changes, historical examples of channel metamorphosis show

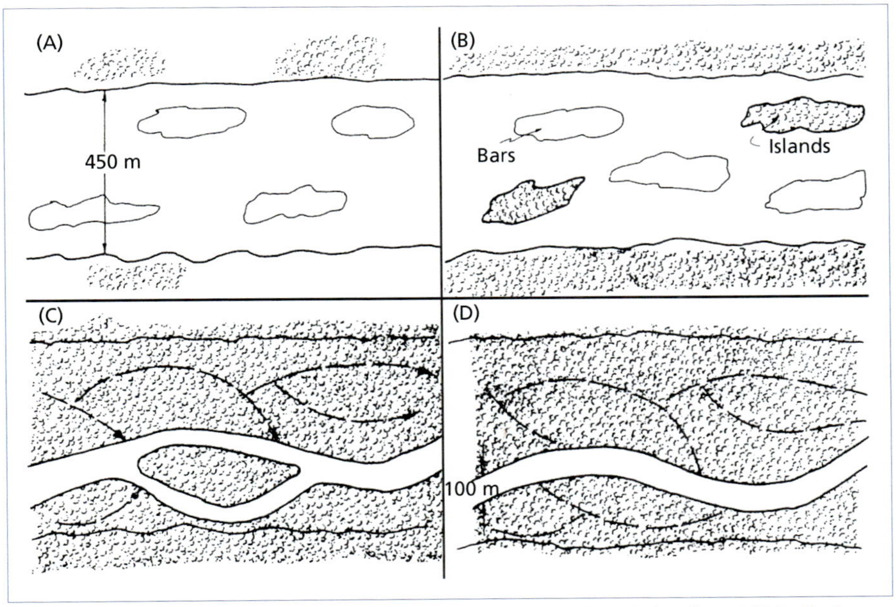

Figure 8.36 Metamorphosis model of the South Platte River: (A) Before 1800, discharge water intermittent and temporary bars; (B) at the end of the century, discharge was perennial and vegetation more luxuriant on the floodplain and islands; (C) at the beginning of 1900, drought caused vegetation to become established below the average annual high water level, converting the bars into islands, with a single channel; (D) modern channel, the islands are created from the floodplain, and vestiges of historical braided channels can be seen on the floodplain (Nadler and Schumm, 1981).

that rivers widen and deepen by destroying their original channels; this occurs when large quantities of bedload material flow down channels during peaks of high floods. If these circumstances persist, channels remain wide and narrow. However, when the flood peak recedes, channel sedimentation and invasive vegetation cause the channel to narrow; this obviously promotes meanders (Schumm, 1977; Chorley *et al.*, 1984).

The examples that we have analyzed are local and the changes have occurred over short time intervals. However, climate and sea level changes during recent geologic epochs have significantly modified channel systems. In Poland, climate change has provided several excellent examples of fluvial metamorphosis on a plain (Starkel, 1995). The Pleistocene river on these plains changed from braided to meandering between 13,000 and 9,000 years ago (*Fig. 8.37*) (Kozarski and Rotnicki, 1977).

One of the most studied rivers is the Mississippi (Schumm, 1977). Fisk's (1944) research provides an excellent example of river metamorphosis related to Quaternary variations of sea level and climate change. Between the glacial maximum that occurred 18,000 BP and the present, two factors have determined the river's behavior. The first is the melting of the Laurentide ice sheet, which provided an immense amount of water and sediments to the upper reaches of the

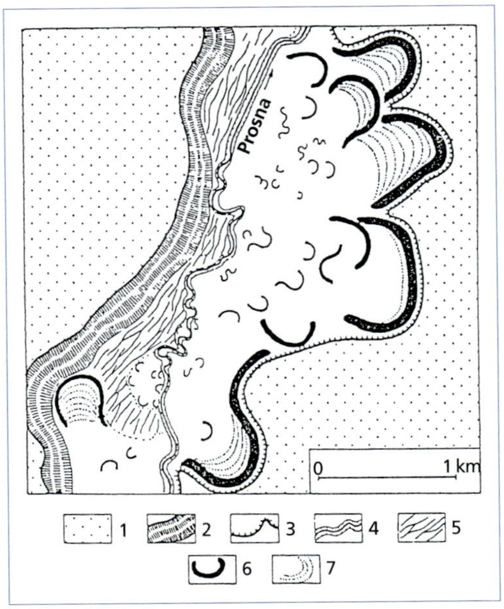

Figure 8.37 Paleochannels in the Prosna en Mirków valley, Poland. (1) High Pleistocene terrace with braided channels; (2 and 3) Slope at the bottom of the valley; (4) Present-day Prosna River; (5) Braided system at the bottom of the valley; (6) Paleomeanders; (7) Meander belts (point bars) (Kozarski and Rotnicki, 1977).

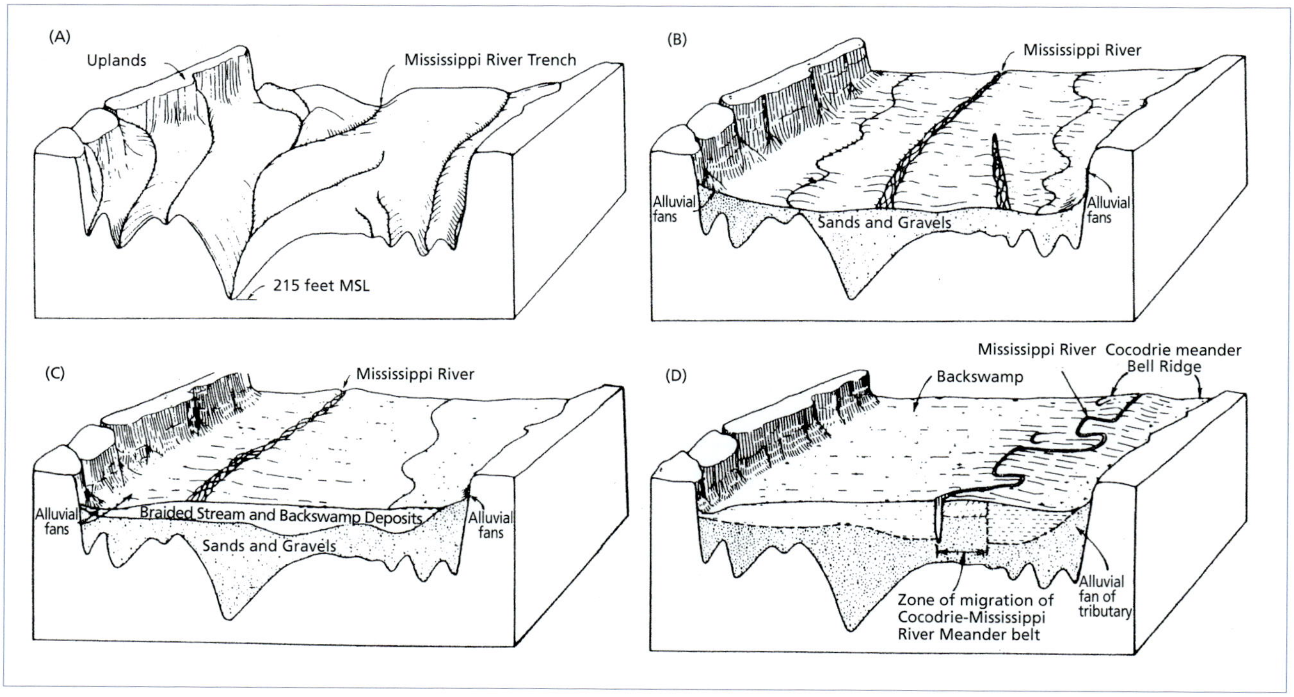

Figure 8.38 Quaternary evolution of the Mississippi River valley. (A) Downcutting stage at the end of the glacial epoch: 18,000–14,000 BP; sea level –130 m; valley slope 0.16 m/km, overloaded river transporting gravel. (B) Stage I of valley aggradation: sea level –30 m; valley slope 0.14 m/km; overloaded braided river transporting sand and gravel. (C) Stage II of valley aggradation: sea level –6 m; valley slope 0.13 m/km, overloaded braided river, transporting silt and sand, creating thick swamp zones of silt and clay. (D) Stage III of valley aggradation: present-day sea level, valley slope 0.11 m/km; development of the Cocodrie meander belt; meandering river with deep channel and wide variation in discharge that gradually replaces the shallow braided channel (Fisk, 1944).

Mississippi. The second is a 130-m drop in sea level in the last 15,000 years. The drop in base level was accompanied by erosion of old alluvial deposits and incision into the valley bottom.

Figure 8.38 shows different stages of the Quaternary history of the Mississippi River: (A)- Maximum sea level drop. The river incised 120–137 m and its tributaries also downcut, producing irregular topography with a slope of 0.16 m km. (B)- Beginning of ice cap melting. Water and sediments (sands and gravel) are introduced into the river. Particle size decreases toward the top as a result of the progressive decrease of slope and retreat of the ice cap. During this time the Mississippi was a braided system. (C)- Continuous but slow rise of sea level. Load of sand, silt, and clay. The braided system continues in spite of decreased gradient and sediment supply to the river. (D)- Constant sea level and reduced load. The river becomes meandering and flows down a single channel due to decreased slope.

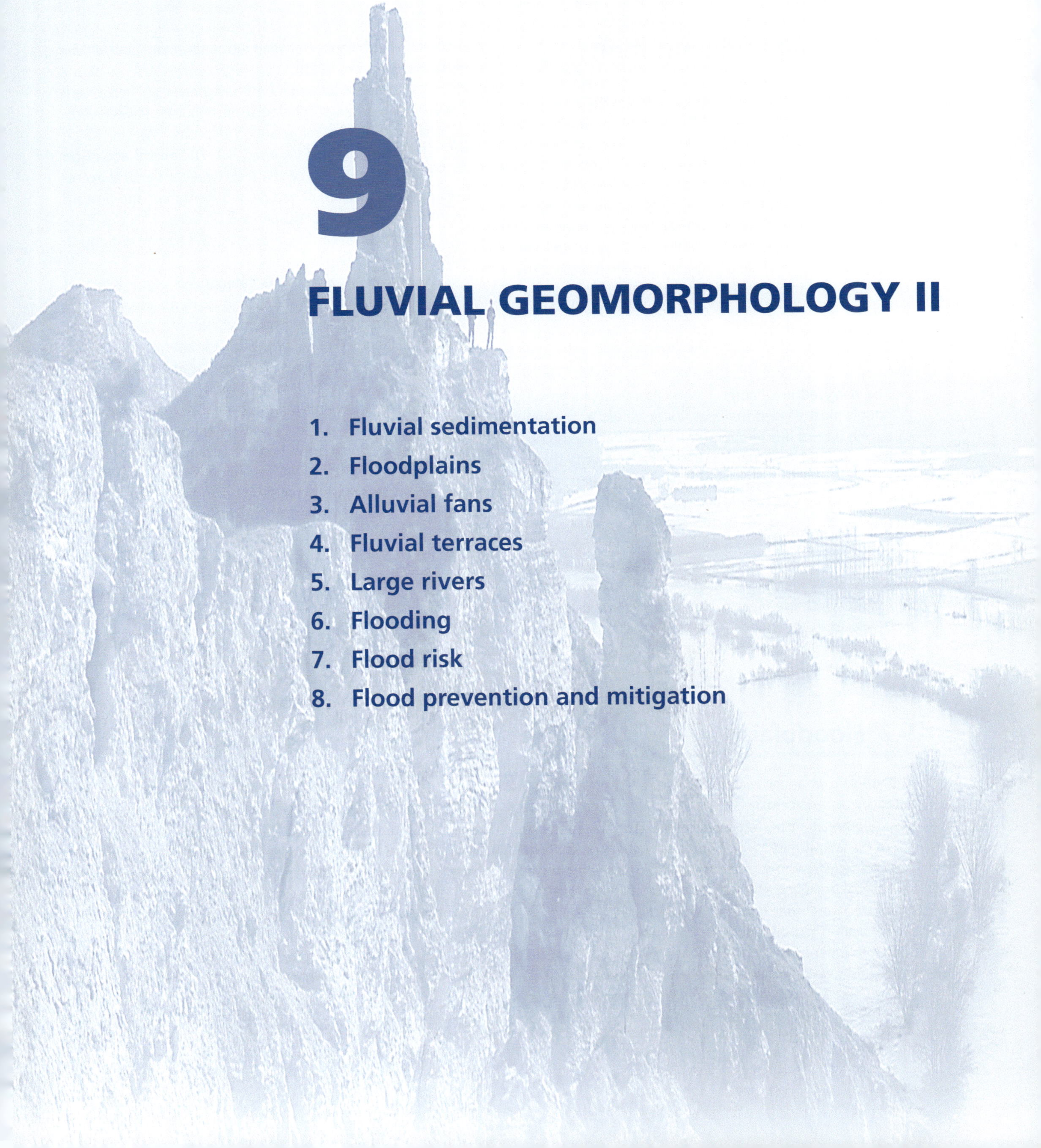

9

FLUVIAL GEOMORPHOLOGY II

1. Fluvial sedimentation
2. Floodplains
3. Alluvial fans
4. Fluvial terraces
5. Large rivers
6. Flooding
7. Flood risk
8. Flood prevention and mitigation

9.1 Fluvial sedimentation

The material that a river picks up and transports is deposited downstream or along valleys as in piedmonts and coastal plains. An analysis of deposits can provide information on velocities and conditions and dynamics of the processes operating at the time of sedimentation (Morisawa, 1985). Fluvial sediments are deposited primarily in topographical discontinuities and as valley fill and marginal deposits. Topographical discontinuities are created by faults, concave crustal structures, glacial scours, marine abrasion, and other factors; these conditions lead to alluvial fans and other landforms. Valley fill involves a set of complex processes that produce flood plains and terraces. Marginal deposits result from sedimentation in still water where transport velocity has decreased such as deltas and playa deposits (Chorley *et al.*, 1984). The latter will be analyzed in Chapter 11 on Littoral Geomorphology. Although most eroded material finally arrives at the mouth of a river, part of it is temporarily deposited along its course. During transport each particle has a "fall" velocity that represents the sedimentation threshold for the particle. If the current velocity is less than the fall velocity, the particle will be deposited.

Alluvial deposits are economically important as groundwater sources and as sources of aggregate for construction. Some alluvium contains heavy minerals, diamonds, and gold, which are actively mined in rivers and terraces in many places in the world. Ancient playa and delta deposits are frequently sources of petroleum and natural gas deposits (Chorley *et al.*, 1984).

9.2 Floodplains

A floodplain is an alluvial surface that is frequently inundated by an adjacent river. The floodplain is composed of river deposits transported by excavation of the river channel or by lateral migration. Some researchers have claimed that braided streams do not have well developed floodplains because of constant channel drift, as compared to a single sinuous channel that migrates slowly within an established alluvial plain. This assumption is not correct; floodplain development is independent of the channel system (Bridge, 2003). Sedimentation during flooding gradually builds a floodplain of sand and clay over periods of hundreds or thousands of years. Floodplains are part of the continuous interaction involved in the balance between erosional energy and alluvial resistance (Nanson, 1986).

River-deposited sediments have been classified in various ways. Most geomorphologists distinguish between vertical accretion and lateral accretion deposits (Mackin, 1937; Fisk, 1947; Leopold and Wolman, 1957). **Lateral accretion** results from lateral shifts of channels whereas **vertical accretion** is a consequence of vertical sedimentation and upward growth by deposition of suspended particles. This classification seems quite simple but it is not always easy to distinguish between the two deposit types.

Leopold *et al.* (1964) categorized eight types of sub-environments in a floodplain and Happ *et al.* (1940) distinguished six types of alluvial sediments. *Figure 9.1* is a schematic drawing of floodplain geometry (Allen, 1964, 1970). The two basic types of fluvial deposits are channel and overbank deposits (Marzo, 1989). Channel deposits include channel lag and point-bar deposits, which are created by lateral accretion; they underlie the second type, overbank deposits, which result from vertical accretion. Examples of overbank deposits are levees, floodplains, and crevasse-splays. Another deposit type is a channel fill deposit, which is intermediate between the first two types.

Channel lag deposits contain coarse sediments at the base of a channel that is later filled with sand and clay. Channel lag deposits accumulate as lenticular bodies in the deepest parts of a channel. They do not form thick layers and are often discontinuous.

Point bars are the most characteristic geomorphic features of meanders and the primary deposit in meandering channels. The form and size of point bars vary according to the size of the river (Reineck and Singh, 1975). In small rivers, they are depositional landforms on top of the convex sides of meanders that dip toward the channel; they are composed of coarse sediments. In large rivers, they take the form of **scroll bars** that alternate with swales, which are elongate depressions, and channels where sedimentation occurs during flooding (Sunborg, 1956; Ollero, 1996) *(Fig. 9.2.)*. Scroll bars are usually several meters taller than swales and are filled with fine stratified sediments. Point bars are generally discontinuous and lenticular in shape. They form during lateral migration of a meandering river during flooding; scroll bars form during successive flooding (Collison, 1996; Miall, 1996).

In *Figure 9.3*, the cross sections show that the channel has moved laterally a distance equal to or somewhat greater than the channel width and thus point bar sedimentation preserves an approximately constant channel width. Sections

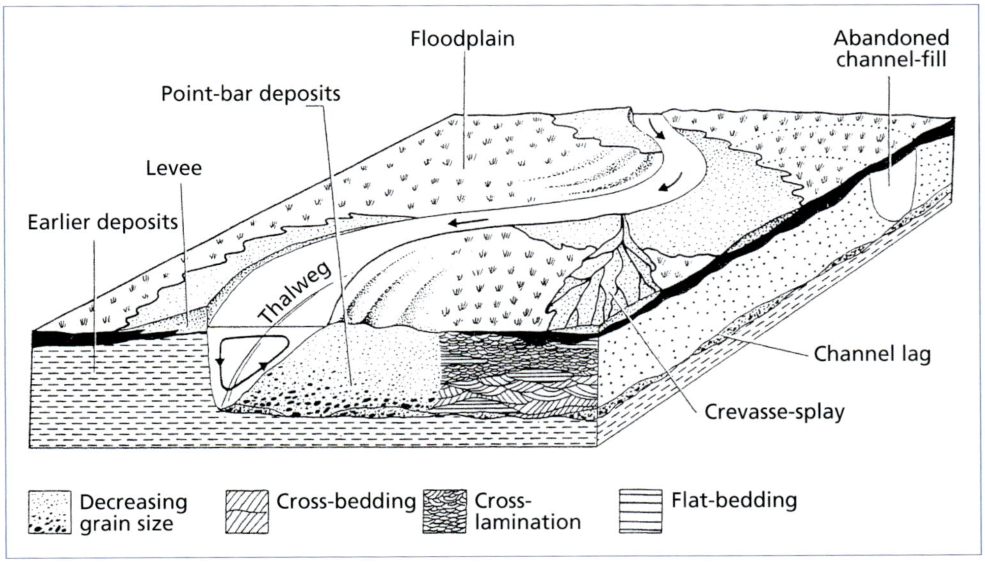

Figure 9.1 Diagram of a meandering river crossing a floodplain (Allen, 1964, 1970).

measured over a period of eight years show that the net volume of erosion was equal to sedimentation. At the point of maximum curvature, erosion slightly exceeded sedimentation; downstream from this point sedimentation exceeded erosion (Leopold *et al.*, 1964).

Levees are wedge-shaped bands of sediments that border fluvial channels *(Fig. 9.1)*. The maximum elevation of a levee is equal to or near that of a channel. Levees dip slightly from the margin of the channel toward the outside edge of the floodplain. Levees reach their greatest thickness on concave sides of the channel; along convex sides they are superimposed on point bars. Along the Mississippi River, they can be 3 km wide and 9 m high (Farrell, 1987; Törnqvist and Bridge, 2002). They form when flooding overbanks the channel margin; sediment grain size decreases away from the channel (Allen, 1965). Levee deposit facies are similar to upper channel fill deposit; they have planar lamination at the base as a result of flooding and superimposed strata that contain small cross stratification formed by ripples, which may be associated with convoluted lamination (Farrell, 2001). Successive sets of flooding generate strata at a single point that can vary in facies and produce vertical sequences that are several meters thick (Bridge, 2003).

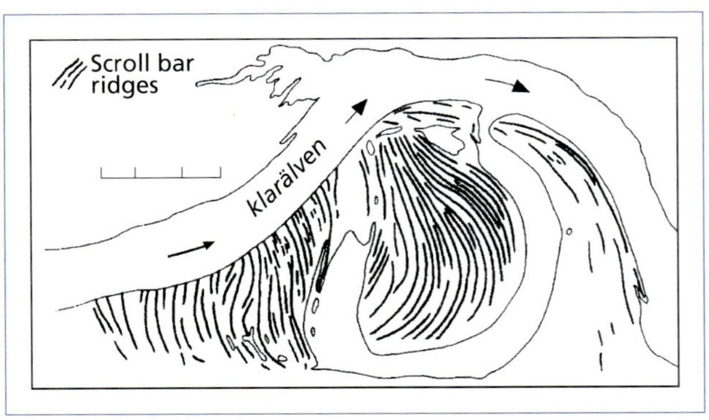

Figure 9.2 A point bar in the Klavälven River (Sweden) showing scrolls bars (Sundborg, 1956).

The scale of **crevasse splays** depends on the dimensions of the river. They can range in width from tens to hundred of meters. They incise into levees and other swamp deposits and can form small deltas (Miall, 1996). They form during large floods; they can traverse the top of levees as sandy tongues *(Fig. 9.1)* and extend 10 km onto the flood plain (Smith, 1986).

Floodplain deposits are found in the lowest parts of these plains. They may extend meters to several kilometers. Their lithology can vary in cross section, showing that they are affected by changes in sedimentary processes. Even so, the accumulations are primarily silt and clay (Aslan, 2007). When the floodplain dries up for a considerable period of time, edafogenetic processes occur; these layers remain intercalated as paleo-soils in the set of floodplain deposits (Miall, 1996).

Abandoned channel fills result from processes of neck cutoff or chute cutoff, creating *oxbows*, which are called galachos in the Ebro River. These channels persist as lakes or ponds, sometimes for long periods. As a result, the water is static and slowly fills with clays deposited from suspension. The lithofacies are fine-grained and contain peat and other organic material intercalated within it (Miall, 1996).

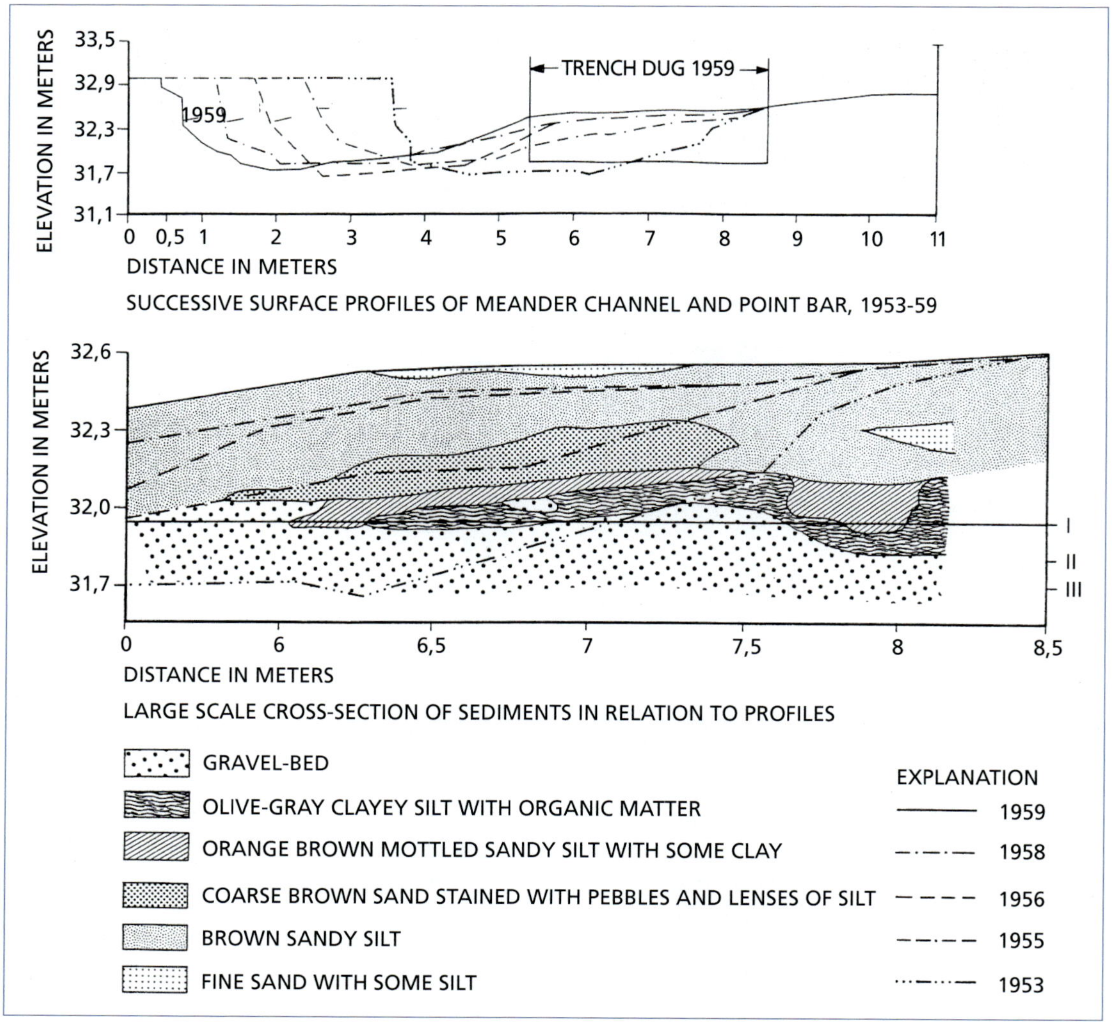

Figure 9.3 Successive cross sections over a period of several years on a meander bar in Watts Branco near Rockville, Maryland. The dimensions of the profiles and distribution of sediments in the section are exaggerated (Leopold *et al.*, 1964; modified in Reineck and Singh, 1975).

9.3 Alluvial fans

9.3.1 Definition, prior work, and terminology

There are numerous definitions of alluvial fans based on geometry, sedimentation area, and processes involved in fan development (Rachocki, 1981). The definition by Bull (1968), one of the best investigators in this field, states that an alluvial fan is a cone-shaped fluvial depositional body radiating down-slope from the point where a stream channel emerges from a mountain area (*Figs. 9.4* and *9.5*). Alluvial fans are found in almost all geomorphic environments: alpine, wet tropical, arid, periglacial and others.

The history of our understanding of alluvial fans, also called dejection cones (Gómez Villar, 1996), goes back to the 19th century when these landforms first received modest attention. It was not until the late 19th century after the exploration of the arid southwestern parts of the United States that excellent papers were published on this topic (Gilbert, 1875; McGee, 1897). During the first half of the 20th century, investigators studied the origin of these fans and sedimentary processes, reworking, and tectonic disturbances of fan surfaces (Johnson, 1932a, b; Blackwelder, 1928, 1931). The major impulse of investigations into these depositional landforms, however, took place early in the second half of the last century with the publication of research by Blissenbach (1952, 1954), Bull (1963, 1964a, b, 1977a, b) and Denny (1967).

Figure 9.4 Alluvial fan at the foot of the Sierra de Alhamilla (Province of Almería). Betic Cordillera. In distal areas of the fan extensive crops are covered by plastic.

Figure 9.5 Alluvial fan and braided network Sinai Desert. Photo J. Rosell.

Alluvial fans were later studied from other points of view. Researchers studied both ancient and modern alluvial fans, their geometry, sedimentary and erosional processes, deposits, and factors involved in their development and evolution. Specific publications resulting from these studies will be cited and discussed in this chapter. Even so, we should highlight the papers and monographs that summarize most of our understanding of alluvial fans (Yazawa *et al.*, 1971; Bull, 1972, 1977b; Schumm, 1977; Colombo, 1979, 1989; Rachocki, 1981; Nilsen and Moore, 1984; Ashida, 1985;

Lecce, 1990; Rachocki and Church, 1990; Cooke *et al.*, 1993; Blair and McPherson, 1994a, b, 2009; Gómez Villar, 1996; Harvey, 1997, 2004b, c; Harvey *et al.*, 2005).

Alluvial fans have also been studied through simulations (Rachocki, 1981) and experimental laboratory models (Schumm, 1977; Schumm *et al.*, 1987). In spite of the considerable problems involved, simulations have provided good results (Blair and McPherson, 1994a). Laboratory models have reproduced the origin of sieve lobes in sandbox analogues (Hooke, 1967). Moreover, the use of sand and silt analogues has led to the conclusion that debris flow deposits predominate in the proximal zone of the fan, channel deposits interfinger in the middle fan position, and fluvial deposits dominate in distal areas (Hooke, 1967, 1987). Other laboratory analogues have been specifically designed to establish relationships among drainage basin evolution, fan development, and fan stratigraphy (Schumm, 1977; Hooke and Rohrer, 1979; Schumm *et al.*, 1987).

The study of alluvial fans has also grown with the application of new dating techniques. Isotopic and physical dating now complement traditional methods of cartography and correlation of fan surfaces based on relative ages recorded by desert pavements and caliche development (Harvey *et al.*, 2005). Cosmogenic, U/Th, and luminescence dating (OSL) can be used.

Now that we have reviewed the early history and trends of alluvial fan research, we will study the relevant features of fans and the terms commonly used in the literature. Alluvial fans are landforms that occur at mountain front locations (*Figs. 9.4* and *9.5*) and in valleys (*Fig. 9.6*) in any morphoclimatic zone. They range in size from small dejection cones less than 50 m long (Harvey and Wells, 2003) to mega-fans that measure 60 km in length such as the classic fans of the Kosi (*Fig. 9.7*) and Gandak Rivers that deposit large volumes of solid load on the Indo-Ganges plain, creating a **mega alluvial fan** characterized by continuous lateral migration of the rivers that flow over them. In these very dynamic rivers, avulsions are common, possibly related to tectonics; it has been calculated that the Kosi River has migrated about 100 km in 200 years (Wells and Dorr, 1987; Agarwal and Bhog, 1992). These mega-fans are made up of sandy layers (generally with gravel at the top) alternating with other silt-clay layers. The sedimentary layers contain intercalated eolian deposits.

An alluvial fan is formed when detrital material is deposited in the dewatering zone of a drainage basin; the detrital material comes from erosion within the drainage basin. Water and sediment fluxes concentrate in a drainage channel or a sediment supply channel in the lowermost part of the basin. When these flows reach the mountain front,

Figure 9.6 Alluvial fans eroded by the Mendoza River. Uspallata (Province of Mendoza). Main cordillera of the Andes.

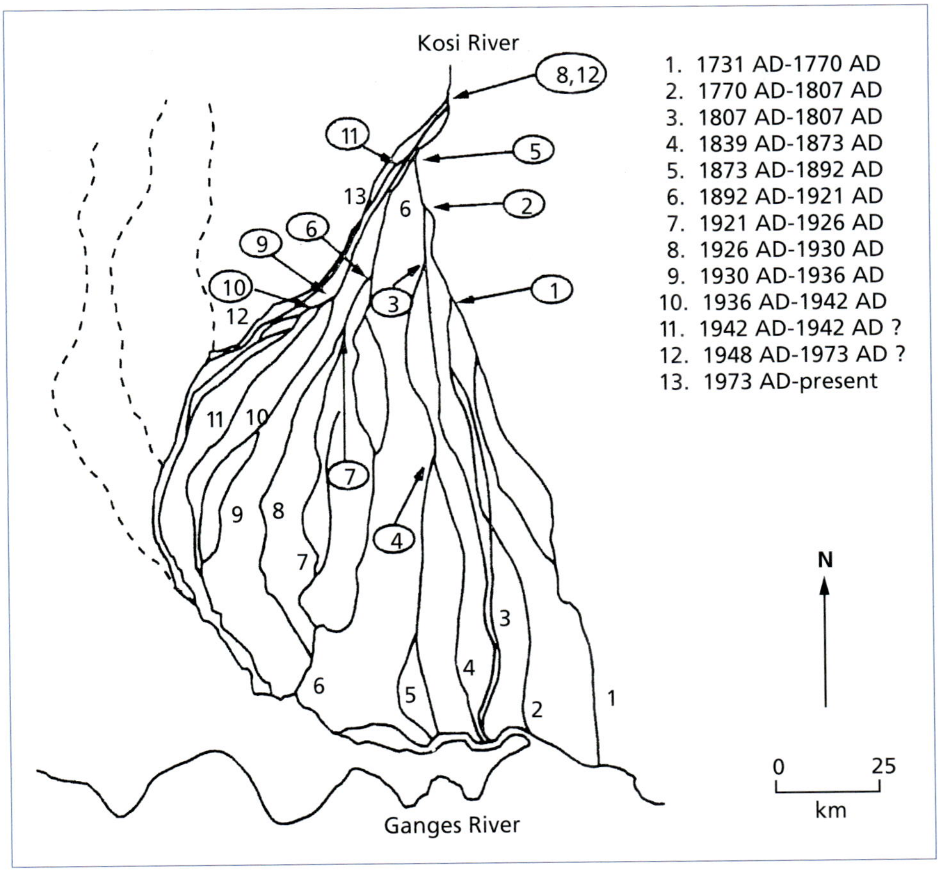

Figure 9.7 Sequence of avulsion by the Kosi River. The numbers with circles indicate avulsion points (AD = after Christ) (Mackey and Bridge, 1995).

they expand into unconfined flow, deposit their load, and construct an alluvial fan due to the significant slope reduction and decreased flow velocity.

In these landforms, individual fans frequently coalesce downslope from the mountain front locations. In arid and semi-arid zones, alluvial fans are the most characteristic landform on piedmonts in basin and range desert areas (*Figs. 9.4* and *9.5*). Distally, the detrital fraction that makes up the main fan body may grade to thinner sediments of playa-lake environments.

Blissenbach (1954) divides alluvial fan bodies into three different zones (*Fig. 9.8*): the **proximal zone**, or fan head, located at the apex; the **middle zone**, between the fan head and the lower parts of the fan; and the **distal zone**, the area farthest from the apex. Finally, he also includes a lowermost area where fans coalesce.

The climatic classification of alluvial fans into 'humid' and 'dry' fans, or in other words, **arid zone fans** formed by intermittent fluvial flows and **humid zone fans** generated by perennial flows (Schumm, 1977) has been widely accepted. However, this classification remains controversial; it was expanded by McGowen (1979) who points out that debris flows are more common in arid zones than in humid areas where fluvial sedimentation prevails. More recently, Nilsen (1993) and Stanistreet and McCarthy (1993) relate fan shape and processes to climate and conclude that their semi-conical shape results from debris flows in desert environments ("dry type") whereas other authors maintain that this geometry also develops in humid temperate regions (Harvey, 1984a; Blair, 1987).

There are various reasons why fans develop in deserts (Harvey, 1997). Vegetation cover is sparse or absent, intense storm rainfall is frequent, and overland flow processes predominate on hill-slopes. As a result, storms mobilize abundant sediments. Steep mountain streams with flashy runoff regimes provide high rates of sediment transport and delivery to mountain front locations. Finally, sediment transfer from the drainage basin to the fan body is sporadic and is basically linked to the aforementioned intense rainfall conditions.

Alluvial fans, which are common in desert mountain areas, are also found in other climatic environments: arctic, alpine, temperate, humid and even tropical (Harvey *et al.*, 2005).

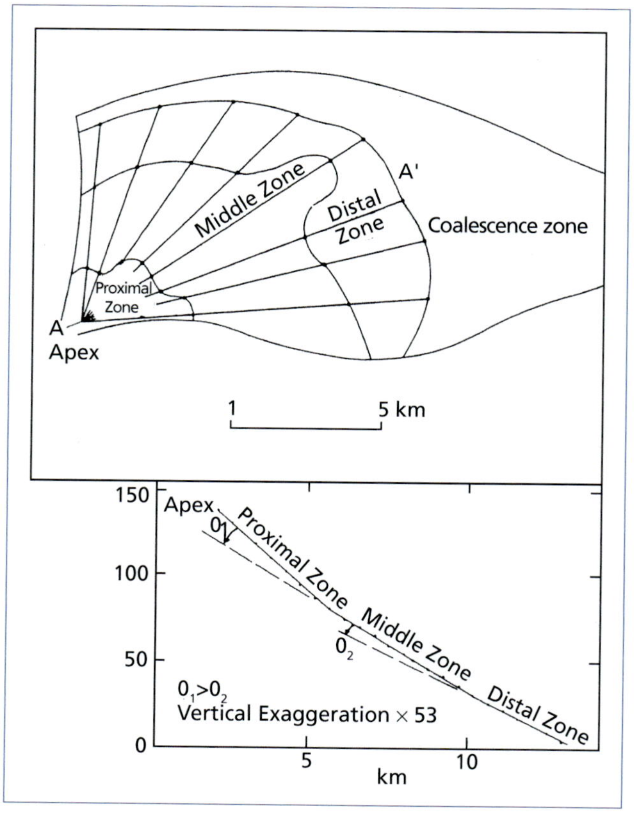

Figure 9.8 Different zones in an alluvial fan and longitudinal profile of an alluvial fan. The figure refers to the Tumey Gulch fan, western Fresno County, California (Bull, 1964a).

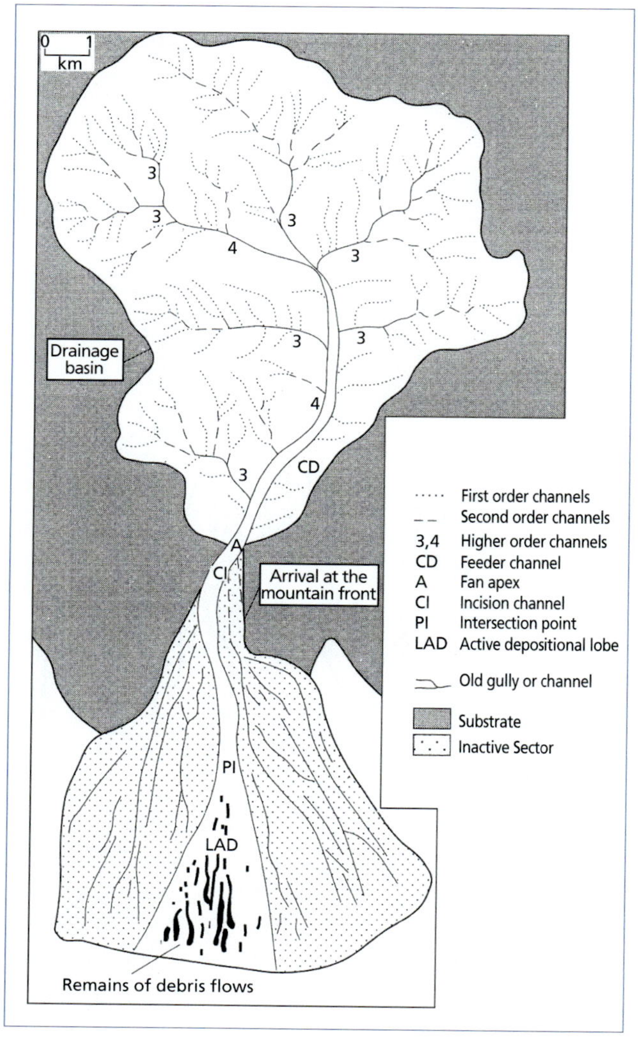

Figure 9.9 Principal morphological features of an alluvial fan system. The example is the Trail Canyon fan in Death Valley, California (Blair and McPherson, 1994a).

9.3.2 Alluvial fan morphology

The main morphological features of alluvial fan systems (Blair and McPherson, 1994a) *(Fig. 9.9)* are listed and defined in the following sentences. The **drainage basin** comprises the upland area; it generally has steep slopes that provide water and sediments. The **feeder channel** transfers the fluid and solid loads to the **fan apex**, which is the discharge point of the drainage basin at the mountain front; if the front is straight, this apex can be seen easily. The **incised channel** is the down-slope prolongation of the feeder channel within the fan, which can divide into several channels. The incised channel usually terminates in the proximal or mid fan areas because this is an **intersection point**. From this point on, flow is no longer confined and it expands down-slope to form an **active depositional lobe**. **Headward-eroding gullies** are common features in the distal portion of the fan. These channels may progress toward the incised channel, causing the active depositional lobe to shift to another part of the fan (Denny, 1967).

The overall morphology of an individual fan should reflect characteristics of its source area because sedimentary processes are controlled by the sediment supply source (Harvey, 1996, 1997, 2002a). Morphology-based analyses of alluvial fans are primarily based on morphometric properties such as shape, area, and slope.

Alluvial fans are semicircular in **shape** or are shaped like a cone segment; they have concave longitudinal profiles *(Fig. 9.8)* and convex traverse profiles. Detailed topographical maps make it possible for geomorphologists to describe alluvial fans in terms of the following mathematical relationship (Troeh, 1965):

$$Z = P + SR + LR^2$$

where Z is the height of a point on the fan surface, P is the height of the fan apex, S is the fan slope at point P, R is the radial distance from P to Z, and L is the average slope along its longitudinal profile.

The **fan area** is the most commonly studied parameter; studies have demonstrated a simple relationship between the fan area and the drainage basin area, as expressed by the following equation:

$$A_f = pA_c^q$$

where A_f is the fan area and A_c^q is the drainage area, both measured in km^2. Studies of different fan groups have indicated similar values for the exponent q, ranging from 0.7 to 1.1, but p values have a wider range, from 0.1 to 2.1. These variations are interpreted to be due to differing fan ages in different regions and also to differences in the bedrock's resistance to erosion in the supply basin (Harvey, 1997). *Figure 9.10* shows the relationship between parameters of different fans in different areas. Fans in the California Coastal Ranges have larger surface areas per unit drainage area than fans in the Death Valley region. The regression line for fans in southeastern Spain was obtained from 68 fans (Harvey, 1987a).

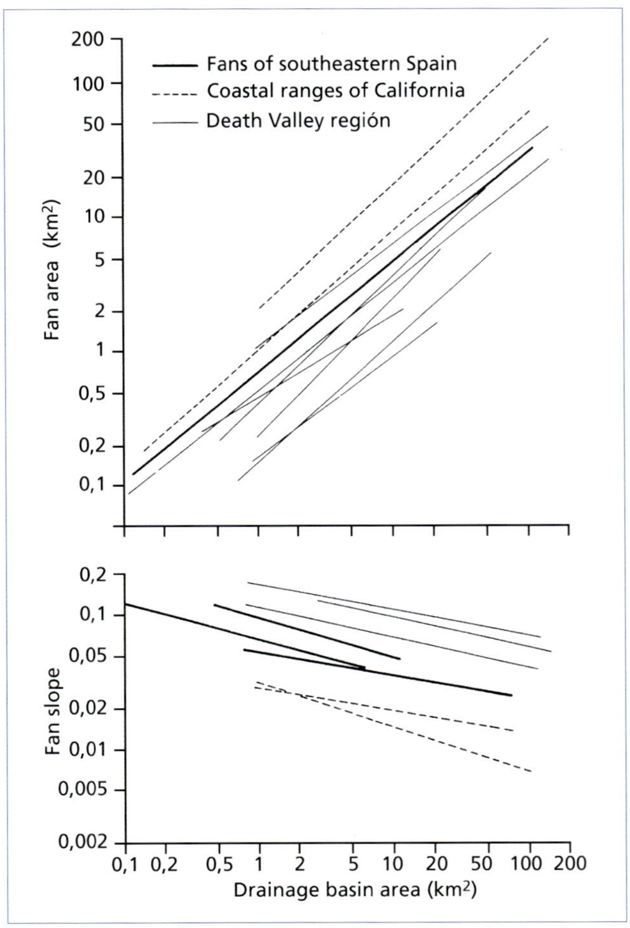

Figure 9.10 Relationship between the drainage basin and the fan slope for three different fans. A clear contrast can be seen between the greater inclination of the *debris flow* fans and the lesser slope of fluvial and intermediate fans (Harvey, 1984a).

The **fan gradient** is usually considered to be the axial fan surface slope in the upper part of the fan; the most characteristic values range from 2° to 12°. Its relation to the fan is:

$$P_a = aA_c^b$$

where P_a is fan gradient. *Figure 9.8* shows that the values of the exponent b range between –0.35 and –0.15; those of the constant a show a greater range, from 0.03 to 0.17, which is interpreted as a result of various sedimentary processes. Fans in the Death Valley region are steeper than those in the California Coastal Ranges; fans in southeastern Spain are intermediate between the two. On the other hand, in southeastern Spain where abundant debris flows occur, fans have steeper slopes than those dominated by fluvial and intermediate fans (*Fig. 9.11*), which shows the importance of sedimentary processes on fan gradients (Harvey, 1984a).

9.3.3 Factors that control alluvial fan development

Fans are deposited when the sediment supply exceeds the transport capacity of the feeder channel. This assumes the presence of a geomorphic threshold defined by Bull (1979) as the **critical stream power threshold**. Even so, fan deposition is also affected by numerous factors that vary from fan to fan.

Topographic factors have a direct influence on sediment supply. Erosion is more pronounced on steep slopes.

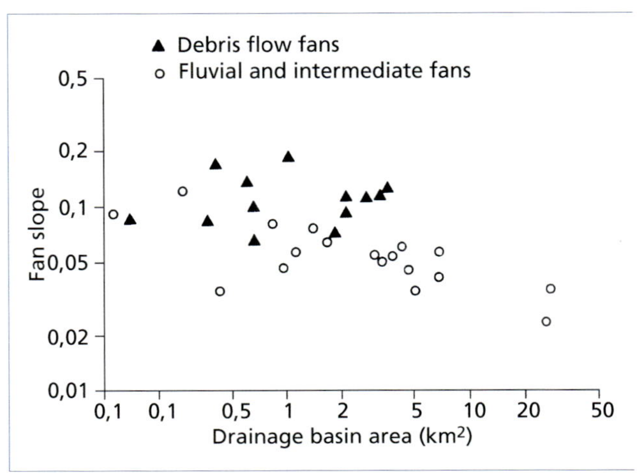

Figure 9.11 Morphometric relationships between the drainage basin and area and slope of the fan in three regions (Harvey, 1987b).

Also, steep gradients are a necessary condition for debris flows, which predominate in fans supplied by small drainage basins with steep slopes. In contrast, fluvial processes predominate in fans supplied by large basins with gentler slopes.

Weathering of the outcropping geologic substrate on slopes causes clasts to move. This **lithologic factor** has a major influence on alluvial fan development. Weathering processes are controlled by rock type; different rock types produce different sediments. The size of the resulting weathered clasts depends mainly on granulometry (in detrital rocks) and the number of discontinuity planes per unit volume in the bedrock. Widely spaced planes can result in large blocks. Silt and clay sized particles, which are formed primarily by chemical weathering, are more abundant in fine-grained lithologies such as shales, siltstones, and tephras. These fine materials are a necessary precondition for the generation of debris flows in drainage basins, as Harvey (1984a, 1987b) pointed out in various alluvial fans of southeastern Spain.

The **climate factor** directly and indirectly controls weathering processes that occur during fan development. Water availability directly affects rock weathering and sediment transport; it also indirectly affects plant growth. The amount, intensity, duration, and frequency, especially of intense rainfall events, are factors that determine the supply of solid load to the fan. The lack or scarcity of precipitation drastically limits weathering and transport processes. Even so, in an area such as Death Valley, with a mean annual precipitation of 43 mm, the solid load that supplies the fans results from rainstorm events (Blair and McPherson, 1994b). Temperature can also affect fan development by increasing chemical weathering rates. However, temperature decreases in higher areas of a system may cause freeze-thaw cycles. The amount of precipitation also increases with altitude. Vegetation affects soil loss and slope stability due to root development; root penetration into the soil varies according to dominant plant types. Consequently, when the percentage of plant cover decreases due to climatic changes or human activity (overgrazing, deforestation, fires, etc.), slopes may become unstable, which can be seen in changes in fan systems (Bull, 1990; Harvey *et al.*, 1999b; Harvey, 2004b).

Fans can develop in areas of both tectonic quiescence and tectonic activity. The latter are characteristic of basin and range deserts such as those in the semiarid southwestern United States and central Asia. Crustal movements (the **tectonic factor**) create suitable topographic settings for fan development such as depressions at the base of active mountain fronts (*Fig. 9.12*). At these locations, thick sequences of alluvial fans of various geological ages may accumulate and record a long depositional history. Neotectonics constitutes the main control of fan location and development in these areas; many old sedimentary sequences of alluvial fans preserved in the geologic record have been interpreted primarily from a tectonic perspective (Silva *et al.*, 1992b; Harvey, 1997; 2002a; Kirkby *et al.*, 2003; Viseras *et al.*, 2003; Jain and Sinha, 2004). However, tectonic activity is not limited to the main range-front fault (*Fig. 9.13*); brittle deformation, fractures, and recent faulting have also been recognized throughout the entire fan system. These recent tectonic movements may affect deposition in alluvial fans (*Fig. 9.14*) and cause significant shifts in channel networks in active depositional lobes of fans, thus modifying fan development.

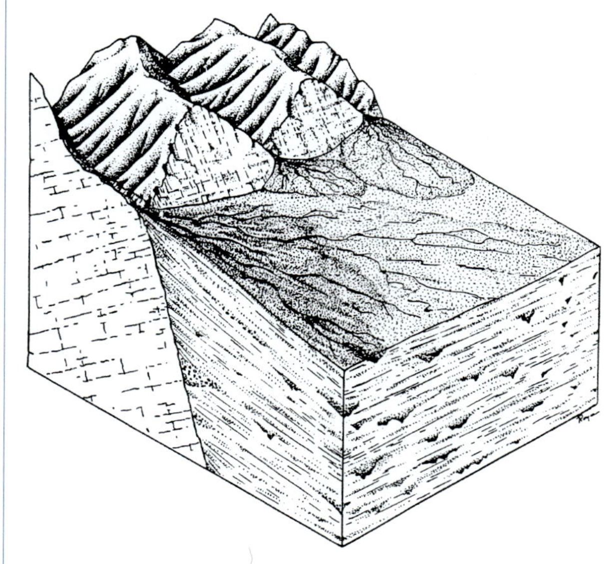

Figure 9.12 Block diagram of a tectonically active front and its piedmont in which significant thicknesses of sediments are deposited at the foot of the mountainous zone. The presence of triangular facets indicates recent activity of the fault along the mountain front (Bull, 1977b).

Tectonic activity can also affect fan development by directly modifying the fan slope. When the mountain front uplift rate exceeds the incision rate of the fan channel (*Fig. 9.15*), fan slope increases and sedimentation occurs at the base of the uplifted block. On the contrary, if the channel incision rate exceeds the uplift rate, active deposition shifts to the mid or distal zones of the fan. *Figure 9.15* illustrates how sedimentation has ceased in the proximal zone and the lower diagram shows that a new apex and another fan segment have been created (telescopic or segmented fan) (Bull, 1964b, 1977b; Harvey 2002a). This process produces fan segments of different ages situated at different topographic positions. These different segments may have caliches, desert varnish and pavements, and highly variable degrees of dissection. The origin of segmented fans (*Fig. 9.16*) is still highly controversial; various hypotheses have been proposed based

Figure 9.13 Materials from the Cordillera of the Andes that develop at the foot of an alluvial fan, marked by an active fault. Cobija (Northern Chile). Photo J. Rodríguez-Vidal.

Figure 9.14 Tectonic tilting and destruction of Pleistocene fluvial deposits related to alluvial fans on the piedmont of the Cordillera of the Andes near the city of Mendoza (Argentina).

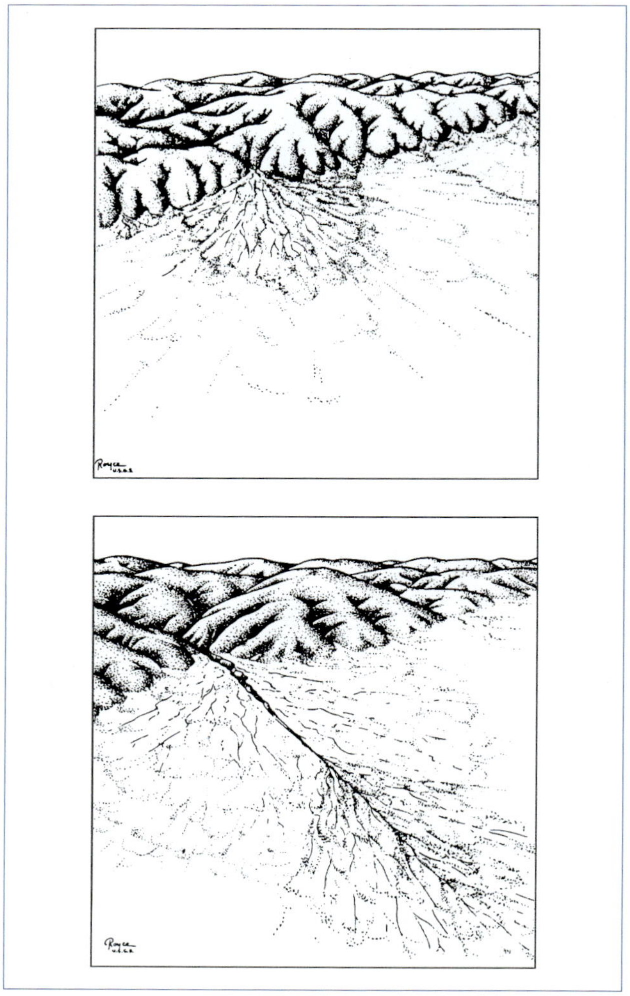

Figure 9.15 Development of alluvial fans as a function of the degree of uplift in the mountainous area. Top: uplift and aggradation exceeds incision of the channel. Bottom: channel incision exceeds uplift; in this case the depositional area is displaced toward the middle zones, creating a new fan segment (telescopic fan) (Bull, 1977b).

Figure 9.16 Incised telescopic fans in which the upper segments are being degraded by rilling and gullying. It has also been affected by fluvial erosion of the Jachal River. The location is between Rodeos and San José de Jachal (Province of San Juan, Argentina).

on recent extreme events, complex responses, and climatic and tectonic changes (see Cooke *et al.*, 1993).

Blair and McPherson (1994b) consider a set of **extrinsic factors** related to alluvial fan systems. Fans can be affected by eolian sedimentation; windblown deposits can modify or even obstruct fan channels. Blocking of channels in distal fan locations can sometimes create ponds. Fluvial channels developed in the lower part of these depressions can incise downward into the distal-fan area sediments and lower the base level. If lakes or seas surround alluvial fans, subaerial deposits may supply these confined water reservoirs and produce fan deltas. Variations in sea or lake levels will directly affect the erosional and sedimentary processes of the fan in the form of base-level changes (Harvey *et al.*, 1999a; Harvey, 2004b). Finally, volcanic activity may generate lava flows that interfere with and modify the internal fluvial network. Volcanic ash falls within the drainage basin can trigger debris flows from steep slopes during storm events.

9.3.4 Sedimentary processes

Two types of sedimentary processes are involved in fan development: primary and secondary (Blair and McPherson, 1994a, b). Primary processes, or **sediment supply processes**, transport sediment from the drainage basin to mountain front locations where alluvial fans develop. The amount of water and sediment and the availability of fine sediments are fundamental to the type of transport (Wells and Harvey, 1987; Harvey, 1997). A low concentration of fines and a high water/sediment ratio favor fluvial type transport of bedload by traction. As the content of fines increases, clasts can be matrix-supported and mobilized as a debris flow by plastic flow in which the solid and liquid phases move as a unit. Secondary processes, or **reworking processes**, erode the fan by remobilizing sediments previously deposited by primary processes. Many of these processes act on fan surfaces over long periods of time whereas sediment is supplied to fans during low frequency and high intensity storm events.

9.3.4.1 Sediment supply processes

Sedimentary supply processes are of two types (Blair and McPherson, 1994a, b). One type moves surface materials or from the rocky substrate of slopes whereas the other transports material in fluvial channels. Mass movements on slopes involve processes that break up rocky material; these processes include congelifraction and avalanches, topples, rock avalanches, landslides, and debris flows (described in Chapter 7).

In **fluvial processes**, sediments and water are separate phases during transport; detrital particles are mobilized during turbulent flow as bed or suspended load. When the solid load makes up a significant fraction, from 20 to 47%, hyper-concentrated flows occur (Costa, 1988a). Two types of fluvial processes are sheetfloods and channel flows (Blair and McPherson, 1994a, b).

Sheetfloods are the most common process of sediment contribution to alluvial fans. These are broad unconfined flows that spread over the piedmont surface when they are no longer confined by the channel. Based on aerial-photo analysis Blair (1987) described the catastrophic sheetflood on the Roaring River fan (Rocky Mountains National Park, USA) as a continuous 0.5 m-deep sheet of water flowing across the 320 m-long active depositional lobe of the fan. Sheetflooding occurs during major low-frequency high-intensity storm events that can cause catastrophic flooding (Gutiérrez *et al.*, 1998a, b). These sporadic events are separated by periods of sparse fan activity and reworking by secondary processes. The deposits are composed of a variety of blocks, gravel, sand, and silt. They usually occur as interbedded laminar couplets, 10–30 cm thick, with synsedimentary slopes of 2°–8° parallel to the fan surface; major axes of the blocks are perpendicular to the gradient although sometimes they are imbricated (*Fig. 9.17*) (Blair and McPherson, 1994b). Sinuous gravel ribbons oriented transverse to the fan slope have been observed on occasion.

Figure 9.17 Imbricated blocks in the Arás alluvial fan (Tena Valley, Central Pyrenees). They were deposited on 7 August 1996 during a catastrophic flash flood.

Incised channels dissect the fan apex or the proximal sector of the fan; these can be several meters deep. They are pathways for sediment transfer from the proximal to the distal zones of the fan and thus are feeder channels for the active depositional lobe. These channels also contribute to fan sediment reworking by eroding the incised channel beds. The resulting sedimentary facies in these channels includes fluvial sand and gravels on the channel bed, which are mobilized during flooding.

A distinctive sedimentary feature of alluvial fans is a **sieve lobe** or **sieve deposit,** which Hooke (1967) was able to reproduce by using sand and gravel in laboratory studies. Lobate deposits accumulate in an area, typically an intersection point, where the flow loses its ability to transport a load and the water completely percolates into the permeable substrate. The coarsest material accumulates at the front of the lobe; upstream deposits are finer grained. Sieve deposits, which commonly develop during floods, occur mainly in mid and distal fan zones. They can also form from previously deposited debris flows when runoff remove finer material. The resulting deposit has a matrix-free, clast-supported fabric. In this last case, the origin of the sieve lobes is secondary.

Three-dimensional analysis of alluvial fans shows numerous and complex facies variations that may result from significant climatic or tectonic changes. Particle size typically decreases down-slope; coarser material is deposited in proximal areas and finer material in distal areas. Fluvial channel deposits (such as blocks and gravels) and debris flow deposits predominate in proximal areas (*Fig. 9.18C, D*) (Harvey, 1997). According to Miall (1978) if the dominant sedimentary facies are fluvial, the result is "Scott" type sedimentary sequences. If debris flows are abundant, "Trollheim-type" sequences will occur (*Fig. 9.18A, B*). Sieve deposits may occur in mid-fan areas but sheetflood deposits are more common; the latter extend towards the distal part of the fan where sands and silt

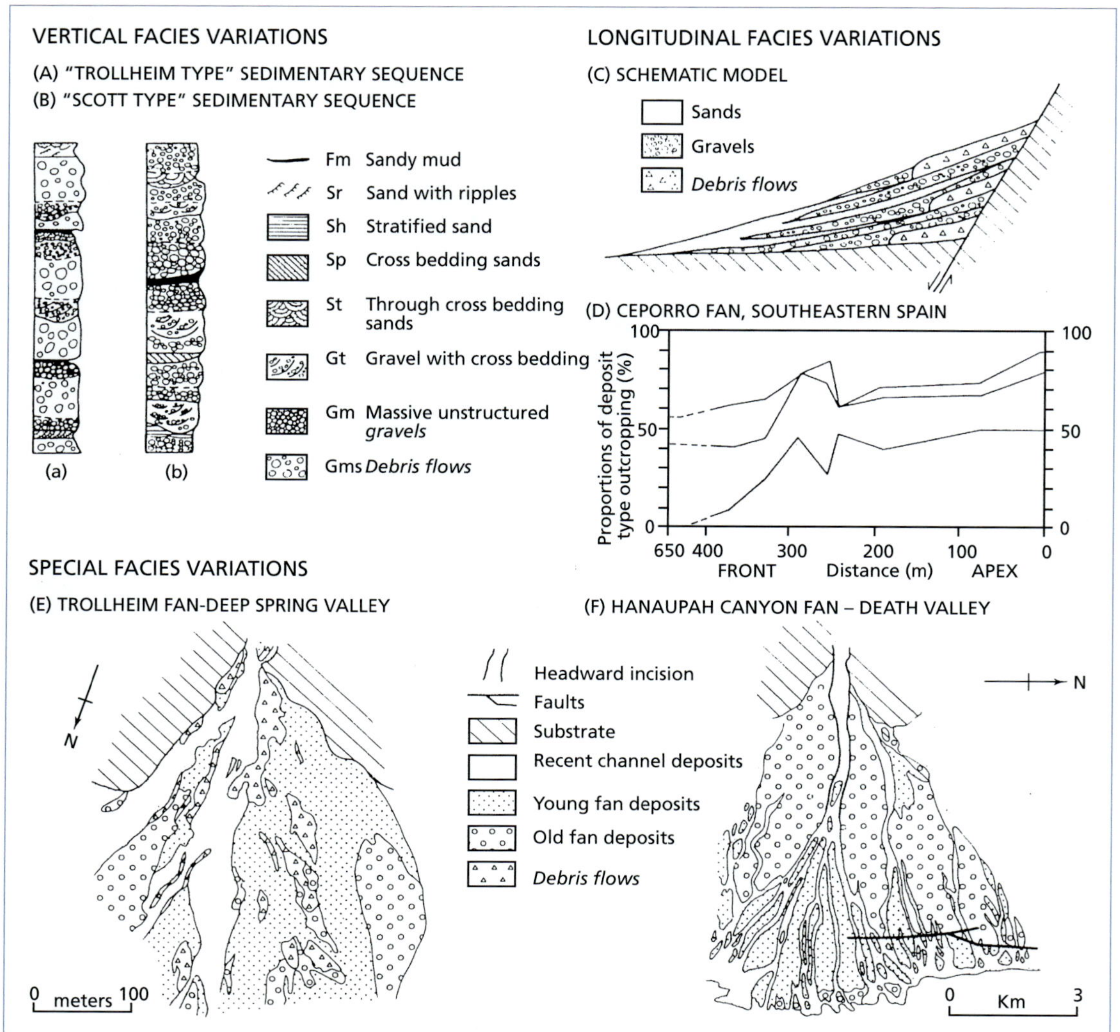

Figure 9.18 Variations in sedimentary facies in alluvial fans. Vertical facies variations: (A) "Trollheim" type and (B) "Scott" type sedimentary sequences (Miall, 1978). Proximal and distal facies relationships: (C) Schematic model (Rust, 1979). (D) Variations along the fan in the proportions of facies exposed in sections of the Ceporro fan, Southeastern Spain (Harvey, 1984b). Spatial facies variations: (E) Zone of a fan apex with sedimentation in the proximal area. Trollheim Fan, Deep Springs Valley (California) (Hooke, 1967). (F) Fan with channel incised into the proximal zone and aggradation in the distal zone. Hanaupah Canyon fan, Death Valley (California) (Hunt and Mahey, 1966). Figure compiled by Harvey (1997).

predominate. Facies distribution is notably different between non-incised fans and fans with an incised channel in the proximal zone; in the latter type, more recent deposits are located towards the distal areas of the fan (Fig. 9.18E, F).

9.3.4.2 Reworking processes

As previously mentioned, alluvial fans develop mainly in response to high intensity storm events but the return period of these high energy events may be long. Because of the rare occurrence of these storm events, a fan surface is subject to a variety of reworking processes over long periods of non-deposition; depositional surfaces created by primary sedimentary processes can be extensively modified. In addition, in areas of active Quaternary tectonics, sediments can be or have been affected by ductile and brittle deformation such as folding, faulting (Fig. 9.18F), and tilting. Various authors have analyzed these diverse reworking processes (Blair and McPherson, 1994a).

Runoff is the most important reworking process on alluvial fans. These flows result from ordinary rainfall episodes that occur in the intervals between intense storm events. Runoff winnows and transfers the fine material present in any type of deposit towards the distal fan area. It also generates a network of rills and distributary channels radiating down-slope from the fan apex to distal areas where wash channels predominate (Denny, 1967). **Groundwater** activity is also important in fans; fans are excellent aquifers. Groundwater availability affects plant growth and erosional processes by runoff. In addition, water movement can cause calcite cementation and if these solutions contain carbonates, they can generate petrocalcic soils (caliche).

Wind action is another factor that reworks fan surfaces in arid regions. The wind picks up and winnows fine particles, creating desert pavements. The exposed clasts are subject to wind abrasion, which polishes and facets them, creating ventifacts that are sometimes covered with desert varnish. Sand-sized particles transported by the wind can also be deposited on the fan surface around obstacles such as plants or can be deposited as continuous sand sheets. These eolian deposits can reach significant thicknesses on the lee-side of slopes and can potentially disrupt fluvial networks.

Weathering of fan sediments includes mechanical, chemical, and biological activity. Mechanical weathering, including salt crystal growth, wetting and drying cycles, and congelifraction can break up rock; these processes are controlled by water and salts. Biological weathering consists of biochemical processes resulting from bioturbation, which is burrowing, and material removal caused by plant and animal activity. In arid zones, calcium carbonate precipitation near the fan surface produces petrocalcic soils (caliche) (Gile and Hawley, 1966; Machette, 1985; Harvey, 1987a). This cementation causes these detrital deposits to be very resistant to erosion and it considerably decreases the infiltration capacity of the coated deposits (Harvey, 1990, 1997). In humid environments chemical weathering continuously breaks down minerals and creates large quantities of fine particles.

9.3.5 Dynamics and evolution of alluvial fan development

Alluvial fans are the fastest-developing landforms in arid areas (Mabbutt, 1977). Average accumulation rates of alluvial fans in Death Valley (USA) are estimated to be one meter per 1,000 years (Hooke, 1972). Alluvial fans are dynamic geomorphic systems that show progressive morphologic changes during their development (Harvey, 1997). Harvey has proposed a fan evolution model that relates fan and channel profiles to incision and aggradation in the medial parts of the fan (Harvey, 1987a, 1997) (Fig. 9.19). Types A and B are

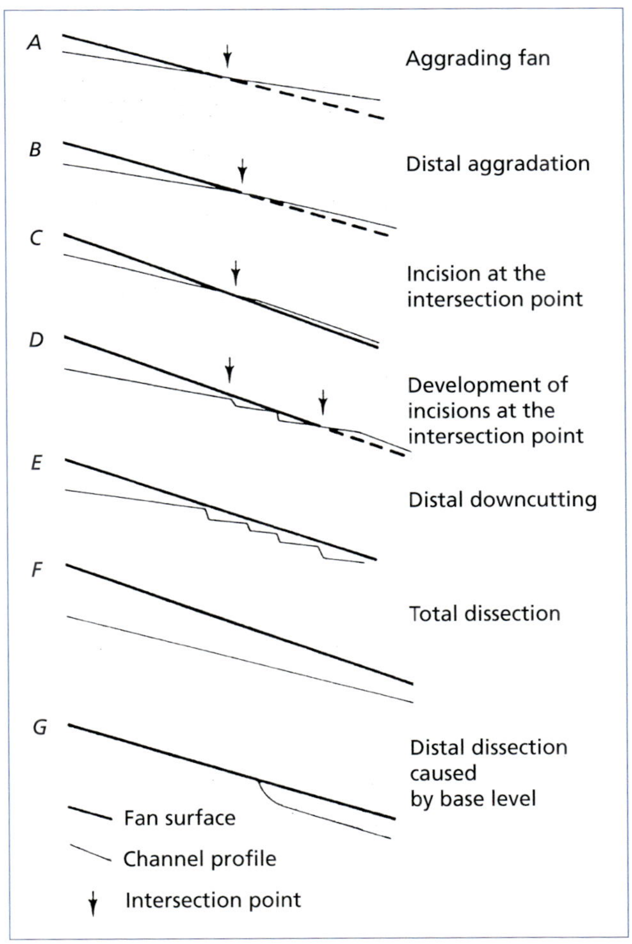

Figure 9.19 Model that establishes the relationships between surface profiles of the fan and the channel with dissection and sedimentation in the middle and distal parts of the fan (Harvey, 1987a).

characterized by the occurrence of headward entrenchment and aggradation in the distal area. In Type C, scour occurs at the intersection point and in Types D and E, this scour is mobilized toward distal areas as a result of the displacement of the intersection point, which finally results in the total fan dissection of type F. Lowering of the base level can cause headward erosion from the front of the fan (Type G). This model helps to explain the dynamics of aggradation and dissection within a fan where tectonics, climate change, human activity, and extreme events all play a role in the modification of erosional and depositional processes.

The Kosi River on the Indo-Ganges plain has constructed a mega-fan (*Fig. 9.7*) (Gole and Chitale, 1966; Wells and Dorr, 1987; Gohain and Parkash, 1990; Singh *et al.*, 1993) as a result of abundant precipitation (1500 mm/year) in the drainage basin; this fan contrasts with arid region fans where annual rainfall is low, riverbeds are ephemeral, and fan sizes are much smaller (Chorley *et al.*, 1984). The Kosi River drainage starts in the heights of the Himalaya; it drops its high load where it enters the Indo-Ganges plain and constructs the Kosi mega-fan. Between the year 1731 and the present (*Fig. 9.7*), it has shifted 113 km to the west by successive avulsions (Short and Blair, 1986, p. 297), which indicate sediment mobilization over an area of more than 9500 km^2. Avulsions have occurred rapidly; the period between them is approximately 18 years (Bridge, 2003). Stevaux and Souza (2004) determined the frequency of avulsions on various rivers; it is 28 years for the Kosi River and 1400 years for the Mississippi. At the fan apex, the channel is composed of blocks and clasts; these coarse materials are rarely transported farther than 15 km from the apex where medium and fine grain sediments predominate. The gradient of the Kosi River decreases gradually from a slope of 20 cm/km at the head to 4 cm/km in the distal areas. In areas of lower slope, the river is composed of several channels no wider than 1.5 km.

The mechanism by which the Kosi River migrates to the west is not well understood. Wells and Dorr (1987) propose a statistical relationship between tectonics and the westward shift of a group of channels; they attribute the shift to topography that is favorable for sedimentation. However, Gohain and Parkash (1990) and Singh *et al.*, (1993) think that the westward shift is due to tectonic tilting.

Other models of fan development are described in the literature. Davis' (1905) proposal is based on his cyclical concept of landscape evolution. He proposes that during a juvenile stage, high relief continuously provides sediments to the piedmont. As the relief is worn down during the final stages of the cycle, less erosion occurs and sheetfloods lead to the predominance of pediments.

Sorriso-Valvo (1988) proposes an evolutionary model for fans based on mass movements in the region of Calabria (Italy). He proposes four principal stages of fan development (*Fig. 9.20*). The first stage, called creep, involves surface landslides on the lower part of the slope and surface deformation on the upper slope. This first stage may last a long time. In stage B, or collapse, landslides and debris flows predominate on the upper part, causing significant amphitheater-shaped gravitational scars, while collapse material accumulates on the piedmont, forming a fan. The river may erode the collapse material during high-intensity low-frequency rainfall events. Subsequent mass movements may remobilize weakened bedrock and debris remaining near the scar, reactivating previous slides and generating new debris flows. These processes can rapidly build a fan. Stage C is the construction of the alluvial

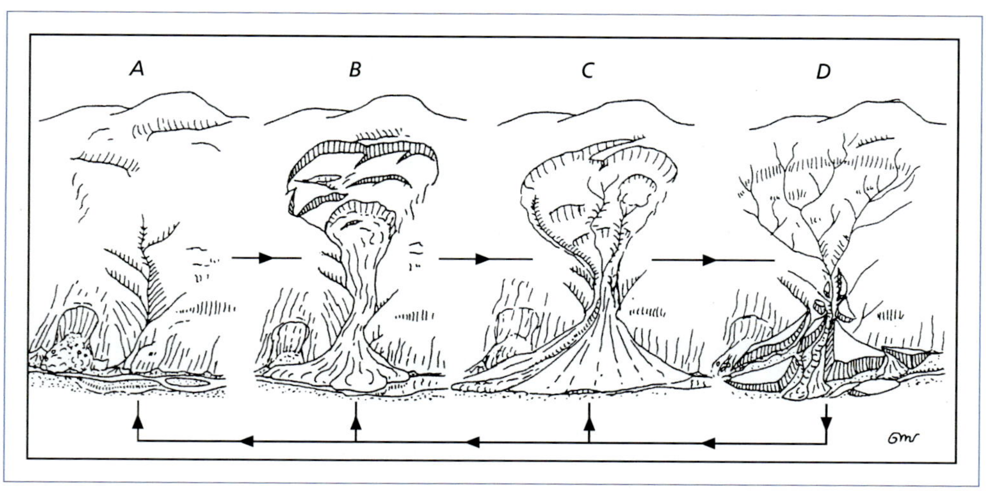

Figure 9.20 Cyclic evolution of a fan as a result of slope movements. The arrows indicate different episodes. The shape is typical for bedrock with ductile behavior. In brittle and jointed rocks the principal flow of stage *B* is caused by high velocity rock avalanches (Sorriso-Valvo, 1988).

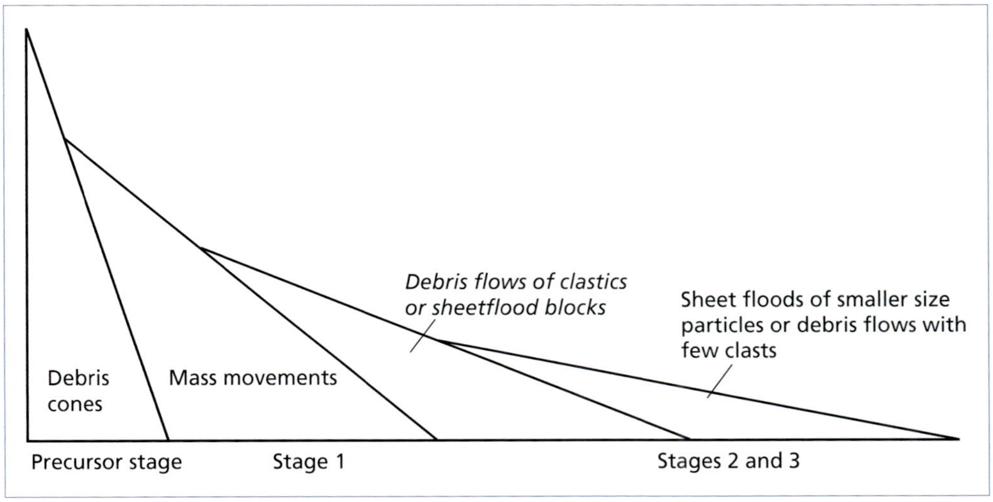

Figure 9.21 Schematic diagram of slopes in the evolutionary stages of an alluvial fan (the 2x vertical exaggeration). The various stages are distinguished by their respective dominant supply processes. The depositional slopes decrease toward the distal sector, the length of the fan radius increases during evolution, and the *knickpoints* decrease in height toward the distal zone (Blair and McPherson, 1994a, b).

fan. Finally, fan destruction, or stage D, occurs when the slide material has been removed and runoff becomes powerful enough to erode the fan from apex to the front. Sorriso-Valvo considers the model to be cyclic and cites the succession of stages A-B-C-D-B-C-D... to be a common evolutionary trend. His model is based primarily on the contribution of mass movements to the generation of alluvial fans and he does not consider the contribution of long-term runoff.

Blair and McPherson (1994a, b) propose a similar four-stage model (*Fig. 9.21*), beginning with mass movements and development of talus cones at the mountain front (*Fig. 9.22*). In Stage One incipient drainage basins have steep relief; first order channels, landslides, and debris flows predominate. During Stage Two, gravitational mobilization of these deposits builds the alluvial fan and enlarges the drainage basin, which takes on the shape of an amphitheater. The fan area also increases during this stage, with its radial length reaching values of 1 to 5 km. During this stage, the fan slope decreases to between 5° and 15°. During Stage Three, the entire alluvial fan system continues to expand, the number of first-order channels in the drainage area increases, and channel orders of 4 or 5 develop. Mass movements continue to occur in the drainage basin; debris flow and sheetflood processes deposit material in the fan (*Fig. 9.23*). The fan radial length may increase from 3 to 10 km or more and fan slope decreases to 2 to 8°. The last stage can continue to develop until rivers or valley glaciers erode the fan although the fan system can also be reactivated by tectonic activity with enough intensity to mobilize the basal fan environments to the distal part.

Figure 9.22 Debris cones in Punta Vacas (Province of Mendoza, Argentina). Main Cordillera of the Andes.

Figure 9.23 To the left of the photo an alluvial fan with its feeder channel and to the right, a debris cone with a drainage basin in the shape of an incipient amphitheater. The location is east of Punta Vacas (Province of Mendoza, Argentina). Cordillera of the Andes.

Studies of Quaternary fans suggest that tectonics may play a fundamental role in controlling the location and position of alluvial fans but that fan sequences respond primarily to climate changes (Harvey *et al.*, 2005).

Large alluvial fan systems, especially those formed at the margins of sedimentary basins, can be preserved in the geological record. Responses to climate and tectonic changes operate at different time scales (Harvey, 2002b). Quaternary fans respond to climate change over periods of 10^2–10^4 years whereas tectonic change occurs over periods of 10^4 years.

In studying the development of alluvial fans we must keep in mind the effects of stream capture on fan deposition; these can be important modifications of the sedimentary sequence of the depositional lobe (Mather, 2000; Mather *et al.*, 2000).

9.4 Fluvial terraces

Fluvial terraces (*Fig. 9.24*) are parts of the floodplain that remain above the bank-full level of the river as a result of the river's downcutting (Leopold *et al.*, 1964). A fluvial terrace is a flat surface and a scarp; it is common to apply also the term "terrace" to the deposit. The terms flat surface and scarp are occasionally used to name the deposit but it is more appropriate to call it an alluvial deposit to distinguish it from its topographic characteristics. The terrace dips slightly downstream unless tectonic movement has affected it (Harden, 2004). Floodplains, which are relatively flat, are composed of alluvium; they develop between the river and the valley margins. Floodplain width (Wolman and Leopold, 1957) (*Fig. 9.25*) depends on the size of the river, the erosion rate, and the strength of the rock material that makes up the valley walls. Floodplains can be found in arroyo valleys that are tens of meters wide (Arauzo and Gutiérrez-Elorza, 1994; Bull, 1997; Harvey and Gutiérrez-Elorza, 2005) but they are generally absent from headwater tributaries because erosion predominates in this environment and there is not enough time for lateral movement of a given magnitude. In humid climates, floodplains are absent from most headwater channels but appear at the point where channel flow changes from ephemeral to perennial, which indicates a groundwater contribution to streamflow. During a period of constant climate without vertical tectonic movement or change of base level, lateral channel movement enlarges the valley (Leopold *et al.*, 1964). If there is a change in tectonic or climatic factors, channel width changes and the channel abandons the floodplain level, causing excavation or aggradation. Downcutting dissects the floodplain, which persists as a continuous or discontinuous nearly flat bench that stands out above the river valley. Thus we can define a terrace as an abandoned floodplain (Leopold *et al.*, 1964). Alternating processes of aggradation and downcutting create terrace sequences (*Figs. 9.26* and *9.27*) that provide a record of changes in river discharges and sediment supply over time (Harden, 2004).

Terraces that erode bedrock (**bedrock or strath terraces**) are different from terraces composed of alluvial material (**alluvial terraces**). These categories correspond to McGee's (1897) destructive and constructive terraces. **Structural terraces** are erosional benches of resistant rocks that are produced by differential erosion (Colorado, Grand Canyon, USA). A floodplain is not a terrace even though it may be affected by fluvial processes. The study of terrace sequences uses a numbering system that starts with the oldest terrace T_1 and the others T_2, T_3, and so on (*Fig. 9.28*).

Figure 9.24 Terraces of the Snake River. Grand Teton. Rocky Mountains. Wyoming, United States. Photo F. Gutiérrez.

Figure 9.25 Alluvial plain and terrace of the Mendoza River in the Andes. Uspallata (Province of Mendoza). Argentina.

Paired terraces form when vertical incision of the river is faster than lateral migration of the fluvial channel (*Fig. 9.28*). **Unpaired terraces** occur when lateral migration of the channel is faster and downcutting is slow (Ritter, 1986).

Field identification and interpretation of a terrace can be a significant problem. Parts of a floodplain near the valley slopes are subject to local sedimentation from slope erosion caused by runoff, mass movements, alluvial fans (*Fig. 9.29*), volcanic ash, and eolian particles. As a result, near valley margins, the floodplain tends to dip slightly from the valley edges toward the valley axis. Colluvium resulting from various processes interfingers with valley alluvium deposited by fluvial aggradation. In some cases colluvium can be the predominant valley fill (*Fig. 9.30*) (Lattman, 1960).

To reconstruct timing and the location of degradation and aggradation in valley fill, geomorphologists describe and

Figure 9.26 Terrace sequence in the Jachal River in the Precordillera of the Andes, Province of San Juan. Argentina.

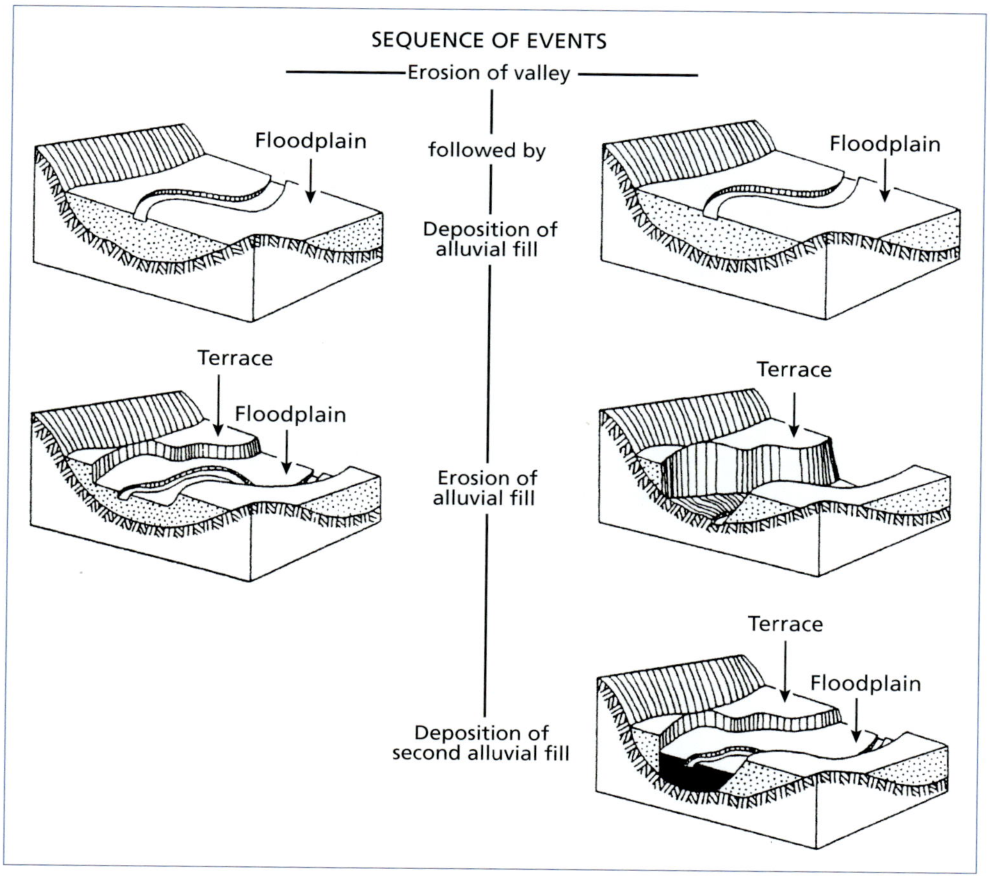

Figure 9.27 Block diagram illustrating the developmental stages of a terrace. The two sequences of events produce the same surface geometry (Leopold *et al.*, 1964).

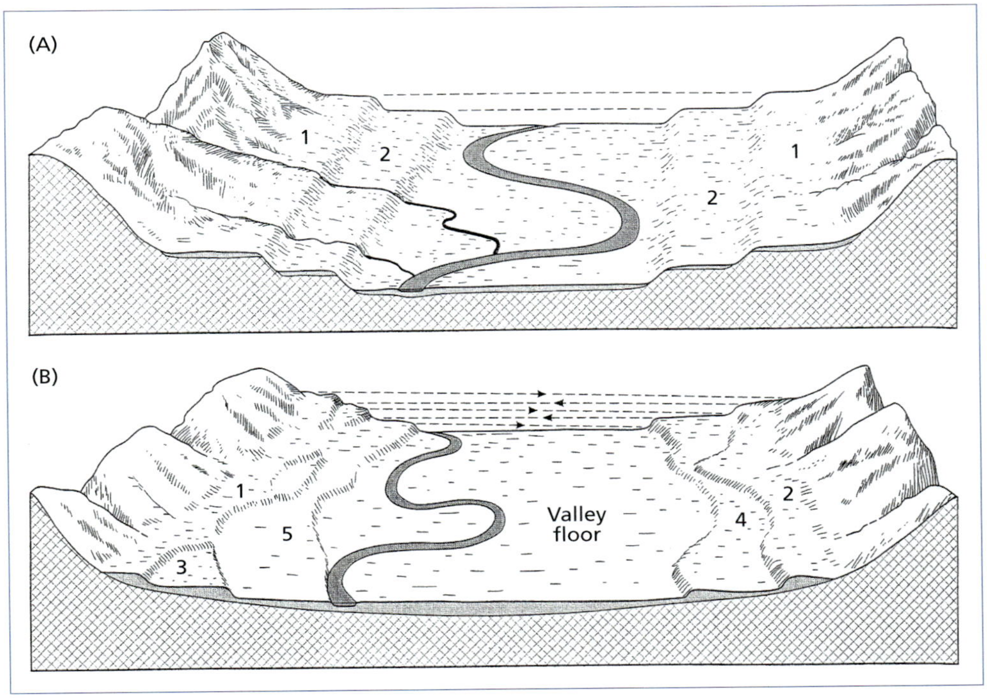

Figure 9.28 Paired and unpaired terraces. (A) Paired, polycyclic terraces. (B) Unpaired, non-cyclic terraces. Terraces are numbered 1, 2, 3 and so on. Terrace 1 (T_1) is the oldest. Adapted from Sparks (1972) and Thornbury (1954).

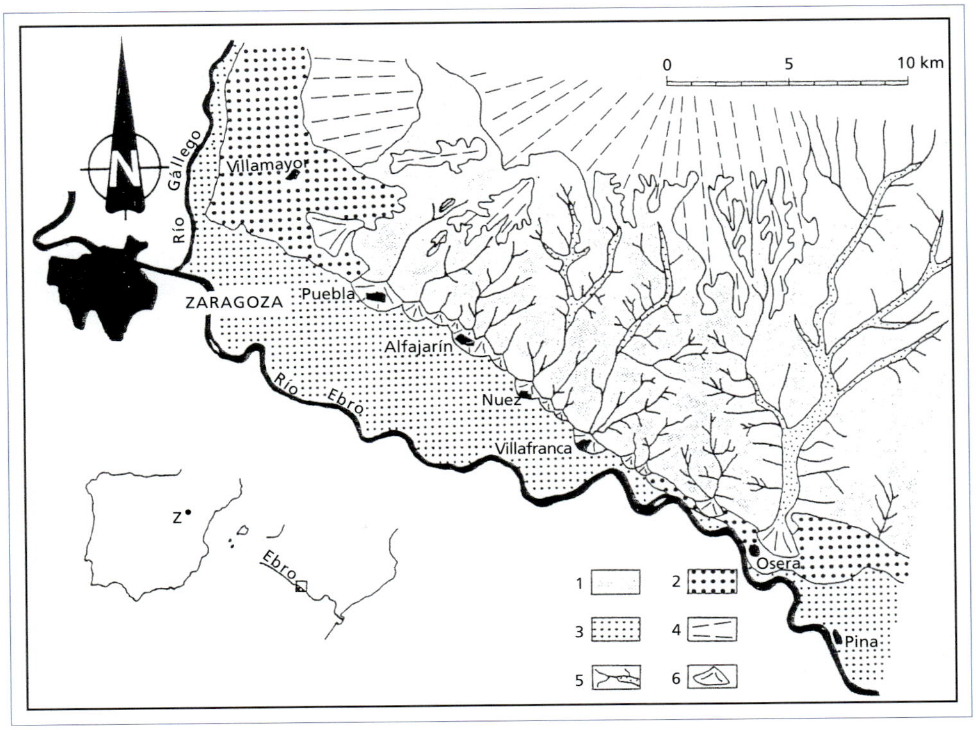

Figure 9.29 Flood plain and alluvial fans on the left margin of the Ebro River, downstream of Zaragoza. 1. Miocene gypsum 2. Terraces 3. Floodplain. 4. Pediment levels 5. Gullies and valleys with flat floors. 6. Alluvial fans (Arauzo and Gutiérrez-Elorza, 1994).

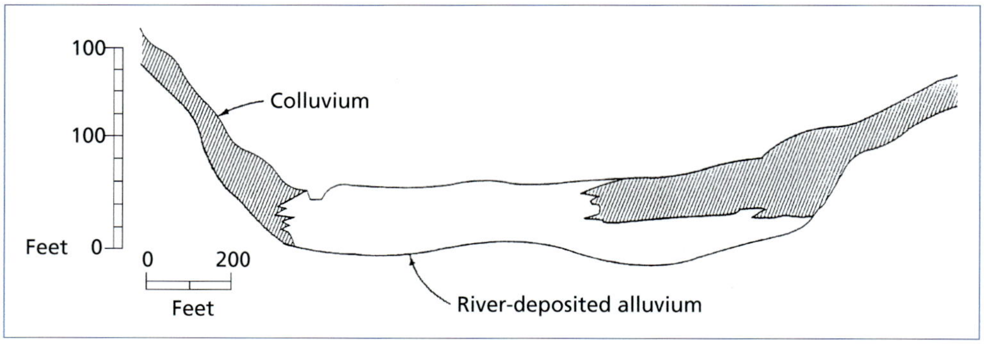

Figure 9.30 Cross section excavated for a dam in the valley of Beaverdam Run, Cambria County. Pennsylvania, United States (Lattman, 1960).

establish the relative ages of terrace surfaces and deposits. They use a precision barometer to measure terrace heights with respect to the present-day floodplain along different sections of the valley. The longitudinal profile of a terrace surface corresponds to the inclination of the ancient floodplain.

The relative age of terrace surfaces is estimated by using the stage of development of paleosoils and by analyzing the degree of degradation of terrace scarps (Bridge, 2003). The age of a terrace is determined through the use of paleontological remains and absolute dating of organic material by ^{14}C (Macklin *et al.*, 1994), dating of sands by simulated optical luminescence (Peña *et al.*, 2004), and dating of clasts on the terrace surface by cosmogenic isotopes such as ^{10}Be and ^{26}Al (Handcock *et al.*, 1999). Also, on some ancient terraces, materials that were not previously consolidated become cemented by carbonates and silica and iron oxides (Costa and Baker, 1981). Scientists can obtain relative ages of terraces and valley floors that contain archeological remains (Gutiérrez and Peña, 1998).

A terrace incises into its own deposits when flow energy is high, when base level decreases, and when its erosional capacity increases with increased load. Current discharge and energy can increase due to climate change or by processes that increase the channel gradient, including a decreased base level. In terms of climate, incision occurs when the climate becomes more humid, with the melting of snow (warmer climate), or when climate-vegetation-soil conditions upstream in the basin create conditions of sudden runoff (Harden, 2004). It can also occur when precipitation decreases in a semi-arid zone and is accompanied by a lower percentage of plant cover (Langbein and Schumm, 1958; Knox, 1984). Anthropic activities such as deforestation, construction of homes, roads and railroads, mining, etc. clearly influence sudden flood.

An increased channel gradient can result from Quaternary tectonics, isostatic rebound, headward propagation of *knickpoints*, faulting, or lowering of the base level (See Chapter 3, Tectonic Geomorphology). During glacial periods, sea level dropped about 130 m.

The sediment supply to a river can decrease as a result of climate changes or agrarian practices that increase vegetative cover and decrease mass movements and eolian erosion. Dams hold back sediments and their reservoirs can eventually fill up; a study of the sedimentary record can provide important data on environmental changes that have taken place during sedimentation. Reservoirs behind Roman masonry dams in Muel and Almonacid de la Cuba, Zaragoza (Pueyo *et al.*, 2001) have filled up. The capture of a river by one of lesser gradient can cause terraces of the captured river to remain as a record of the capture (Rodriguez Vidal, 1986).

Terraces can record tectonic events, climate change, and other environmental changes that modify the erosional capacity and sediment load of a river. Dating terraces makes it possible to calculate incision rates (Harden, 2004).

A large part of geomorphology is stratigraphic geology (Leopold *et al.*, 1964). A detailed sedimentological study is required to correlate terraces. Soils (*Fig. 9.31*) are valuable indicator horizons (Birkeland, 1984; Bull, 1991) but they must be used properly. The terrace unit contains a **chronosequence** of weathering, soils, and terraces. Climate change, a factor that creates terraces, also modifies pedogenic processes. New conditions can gradually erase the original character of the soil but some evidence may remain. For example, the presence of kaolinite in soil indicates humid stages; it can be found in deposits and terrace soils in arid climates. If the climate changes from arid to humid, montmorillonite changes to kaolinite and we cannot detect the climate change. Correlation of terraces with glacial advances provides valuable data on climate change.

William Morris Davis' chronology of denudation is based on the assumption that fluvial terraces, like marine terraces, are **thalassostatic**; they form in relation to the base level. However, terrace development is primarily the result of tectonic and climatic factors. Tectonic deformations can cause differential gradient changes that affect fluvial terraces (Benito-Calvo and Pérez-González, 2007). Records of terraces on the Rhine River show climate change and the erosional history of the region; uplift in the medial parts and subsidence in the lower parts of the valley can also be recognized (Fairbridge, 1968b). **Multiple terraces** can also be recognized. Along the Carrión River (Duero Basin, Spain) 20 terrace levels have been recorded, at least some of them of tectonic origin (Olivé *et al.*, 1982).

Superimposed terraces have also been recognized in fluvial systems; these may be a result of negative crustal warping, or, more commonly, the dissolution of evaporites when rivers encounter these formations. Numerous examples have been published showing the Jalón, Ebro, Gállego rivers (Province of Zaragoza, Spain) and the Noguera-Ribargozana River (Province of Lérida, Spain) where significant subsidence of alluvial deposits has occurred during the Quaternary, resulting in alluvium more than 100 m thick (*Figs. 9.32*

Figure 9.31 Profile of a red fersiallitic soil in a cut in a terrace where gravel is being quarried. San José de la Rinconada (Sevilla, Spain).

Figure 9.32 Superimposed terraces on the Gállego River created by dissolution of underlying evaporites. The Gállego River flows over old terraces. San Mateo de Gállego, Ebro Depression, Spain.

and 9.33) in which striking angular unconformities can be seen (See Chapter 6, *Fig. 6.61*) (Gutiérrez and Gutiérrez, 1998; Benito *et al.*, 1998, 2000; Gutiérrez *et al.*, 2001, 2007; Guerrero *et al.*, 2004, 2006; Lucha *et al.*, 2008a) (*Fig. 9.34*).

The evidence provided by fluvial terraces can be complex and difficult to interpret. Erosion can cause ancient terraces to disappear; rarely is the total terrace sequence preserved (Harden, 2004). Terraces in a drainage basin may respond more to local factors than to climatic and tectonic changes. The complex response (Schumm, 1975) of a fluvial system can lead to the formation of a small terrace that lacks the external variables that produced it. Terraces can also form catastrophically, such as those that developed upstream of Pyramid Lake (California, USA) where six levels of terraces formed in various episodes over 44 years as the lake level began to drop (Born and Ritter, 1970); the study was done by using sequential photographs and historical evidence.

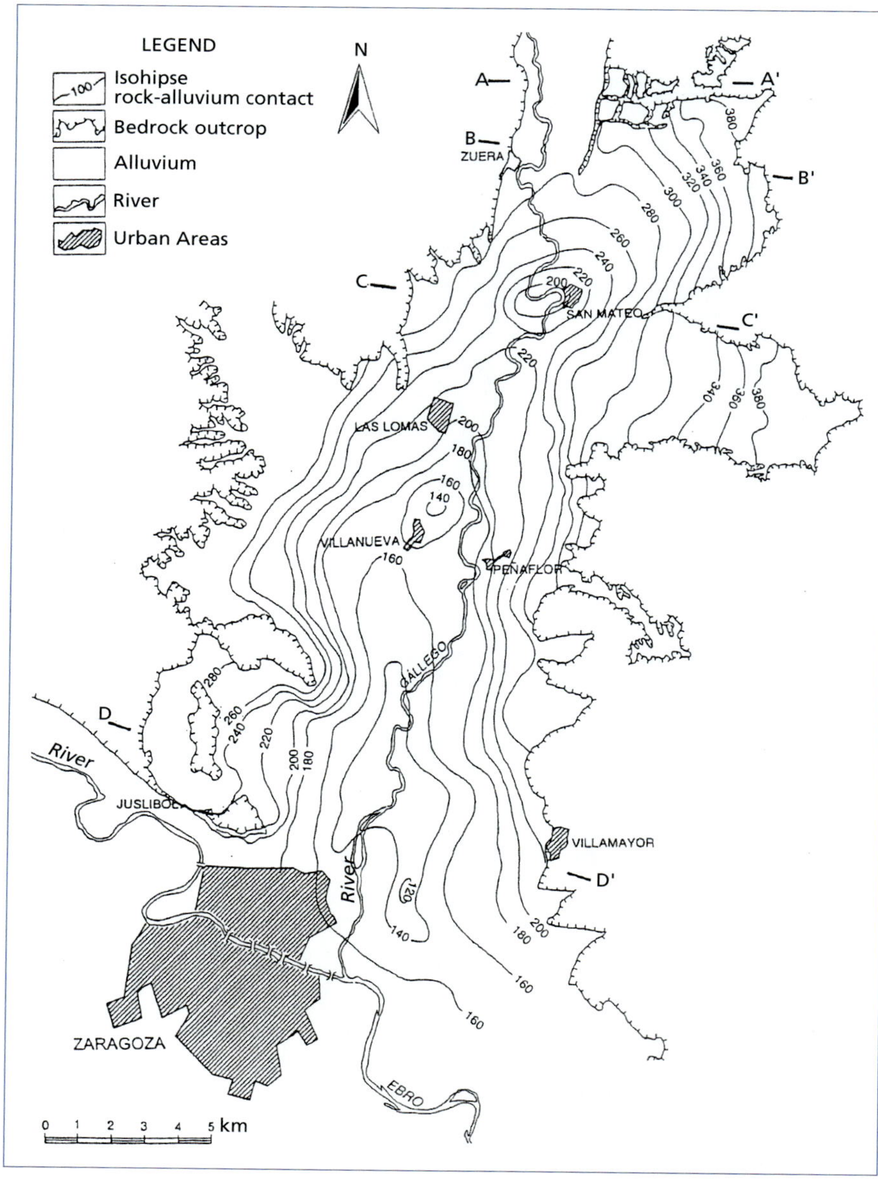

Figure 9.33 Isohypse map of the alluvium-rocky substrate contact in the lower reach of the Gállego River. The depocenters can be seen in San Mateo, Villanueva, and between Zaragoza and Villamayor. The letters A-A', B-B', etc. refer to the cross sections in *Figure 9.34* (Benito *et al.*, 2000).

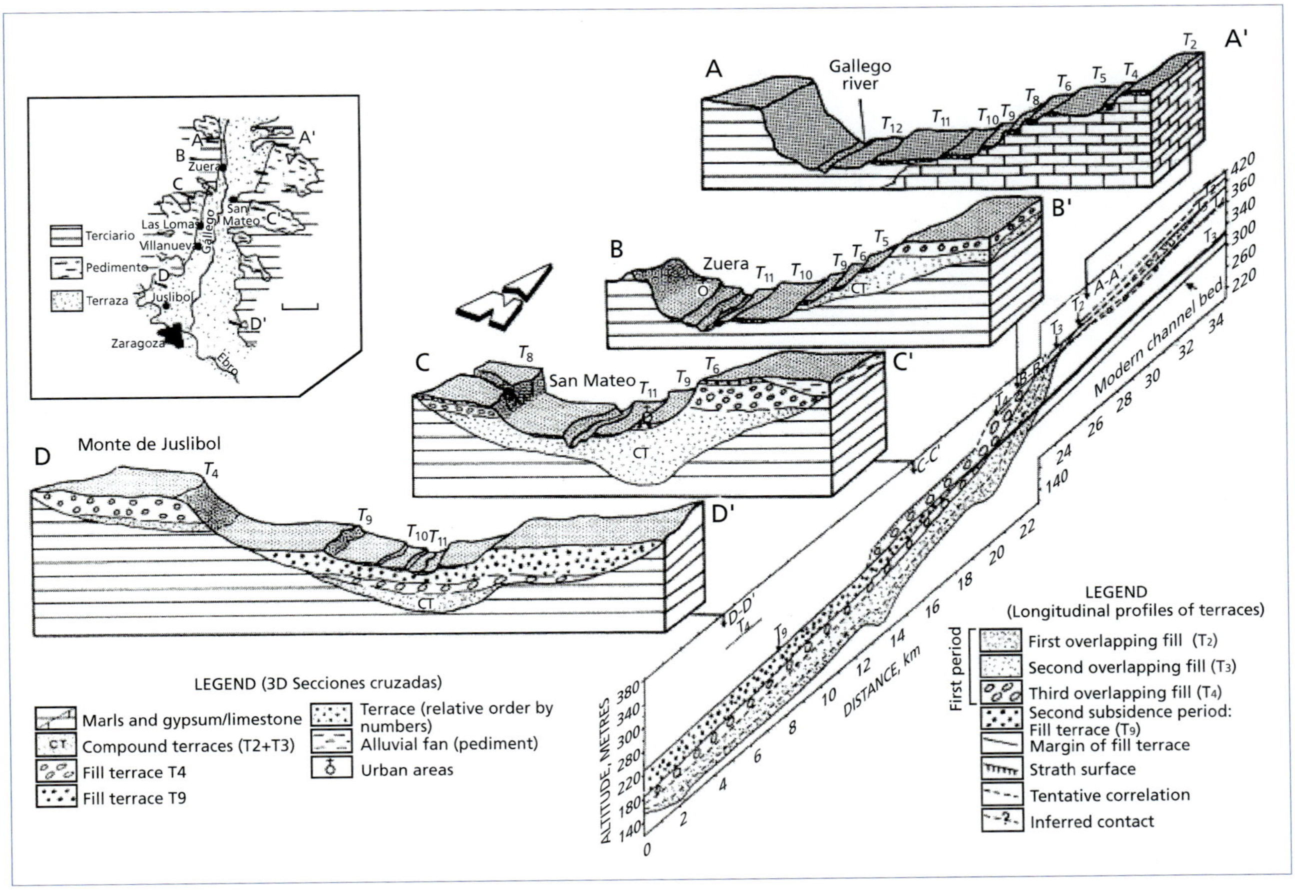

Figure 9.34 Cross sections of the terrace fill along the Gállego River and longitudinal profiles of terrace surfaces and alluvial fill affected by subsidence due to the evaporite dissolution. The terraces in A-A' were formed in Tertiary deposits, forming a typical terraced landform. Downstream of Zuera (B-B', C-C') the terrace alluvium increases in thickness under the present-day floodplain. In the lowest part (D-D') are piled up and superimposed on older alluvium (made of up terraces T_2 and T_3) in angular or parallel unconformity (Benito *et al.*, 2000).

9.5 Large rivers

Large rivers are a major morphology of continents. Some, like the Nile and the Tigris-Euphrates, have been important parts of great civilizations of the past (Schumm, 2005). In spite of their importance, we have limited knowledge about the formation and evolution of large rivers; as a result, the data available to us are not adequate for planning. This situation results from the fact that hydrologic, geomorphic, and sedimentologic studies have focused on smaller drainage basins. As a result, our lack of knowledge prevents us from making predictions of global climate change (Gupta, 2007).

First let us define a large river. By studying the 50 largest rivers in the world, Potter (1978) established four basic properties of large rivers: drainage basin, length, transport of a large volume of detritus, and high discharge (*Table 9.1*). Altogether, the 50 largest rivers of the world drain approximately 47% of the continental surface, excluding Greenland and Antarctica (Potter, 1978). Schumm (1994) compiled a data set on the fifty largest rivers; he considered any river less than 900 km in length to be a small river. He estimates the drainage basin of a small river to have an average surface area of 100,000 km^2, with channels that range up to 20 km in width and 40 m in depth. Small rivers have intense erosion, especially at the confluence of rivers, and large underwater dunes with maximum wavelengths of 100 m and heights of 5 m. Miall (2006) says that the three fundamental parts of a sedimentological study of a large geological river are: 1) prediction of its plate tectonic context, 2) analysis of the scale of depositional elements, and 3) determination of sediment origin. Miall also states that the best way to study a historical large river system is by the size of its depositional elements, which can be observed primarily where large outcrop surfaces are found.

Table 9.1 **SELECTED CHARACTERISTICS OF 24 LARGE RIVERS (COMPILED BY GUPTA, 2007)**

River	Average annual discharge (109 $m^3 \cdot s^{-1}$)	Length (km)	Drainage basin area (106 km^2)	Current average annual load of suspended sediment (10^6t)
1. Amazonas	6300	6000	5.9	1000–1300
2. Congo	1250	4370	3.75	43
3. Orinoco	1200	770	1.1	150
4. Ganges-Brahamaputra	970	2900	1.06	900–1200
5. Changjiang	900	6300	1.9	480
6. Yenisei	630	5940	2.62	5
7. Mississippi	530	6000	3.22	210
8. Lena	510	4300	2.49	11
9. Mekong	470	4880	0.79	150–170
10. Paraná-Uruguay	470	3965	2.6	100
11. St. Lawrence	450	3100	1.02	3
12. Irrawaddy	430	2010	0.41	260
13. Ob	400	>5570	2.77	16
14. Amur	325	4060	2.05	52
15. MacKenzie	310	4200	2.00	100
16. Zhujiang	300	2197	0.41	80
17. Salween	300	2820	0.27	around 100
18. Columbia	250	2200	0.66	8
19. Indo	240	3000	0.97	50
20. Magdalena	240	1540	0.26	220
21. Zambezi	220	2575	1.32	20
22. Danube	210	2860	0.82	40
23. Yukon	195	3200	0.83	60
24. Niger	190	4100	2.27	40

Determination of the total sediment load of a large river is difficult because of the complex measurements required. Dams reduce the sediment load of rivers and also reduce discharge at the mouth (Meade, 1996). In general, a large river has a wide channel, is long, has a large drainage basin, and often transports a large sediment volume annually. The large discharge is probably the most common characteristic of a large river in its natural state (Gupta, 2007). Measurements conducted by various authors who have studied large rivers can vary considerably. There is a relationship between length of the river and the size of the drainage basin (Leopold *et al.*, 1964) but no relationships between other variables.

Some rivers, like the Nile, flow through various climatic zones; the Nile originates in a tropical zone and flows through an arid zone for much of its length. A river can also have hydrologic variations and different types of channels along its course. The Quaternary history of large rivers is often complicated because of different changes observed in the headwaters, along the valley, and near the coast (Gupta, 2007). We should keep in mind that most large rivers have been modified by man by way of dams, channels, and so on, and thus have been modified from their original state.

In this part we will analyze the characteristics of the Mississippi River and in other chapters of Climatic Geomorphology we will discuss the most significant features and problems of the Lena, Colorado, Congo, and Amazon rivers and the deltas of the Nile and Ganges-Brahamaputra rivers through their environmental impacts.

The Mississippi River is not as long as the Amazon or the Nile but its drainage basin is larger than those of the Amazon, Congo, and Nile rivers (Schumm and Winkley, 1994). Glacial deposits are present in the northern part of the basins of the Ohio and Missouri rivers, which are tributaries of the Mississippi. A loess cover one to several meters thick covers a large part of the basin. These layers of sand and complex dunes are common in the Great Plains (Knox, 2007).

Today, the Mississippi River basin is made up of four large tributaries that drain over 90% of the area. These tributaries are the Missouri (~45%), Upper Mississippi (~17%), Ohio (~15%) and Arkansas (~15%). Downstream of the confluence of the Ohio River and the lower Mississippi, the contribution of water and sediments is less than a quarter of the total for the entire Mississippi and all tributaries. Studies conducted in the 1920s prior to the construction of large dams on the Mississippi River show that the average annual discharge of the Ohio River represented 43% of the total Mississippi River basin, in spite of the fact that it drains only 15% of the total basin area (Svenson, 1938). This study also showed that the average annual discharge of the Missouri River basin, which covers 45% of the total area of the Mississippi River drainage basin, contributed only 14% of the average annual discharge. However, during glacial periods, the Ohio, upper Mississippi, and Missouri rivers contributed large quantities of water and sediment to the lower Mississippi (Knox, 2007).

Most of the drainage morphology of the Mississippi River system developed during the recent geologic past. The ages of valley fill and their relationships to regional stratigraphy indicate that valleys incised deeply into the rocky riverbed probably during the Late Pliocene (2–3 million years) (Anderson, 1988). In contrast to the relative geographical stability of the lower Mississippi, the headwaters of the upper Mississippi system seem to have undergone considerable modification over the last 2.5–3 million years in response to advances and retreats of the ice sheets.

Beginning about 14,000 ^{14}C years BP (16,800 calendar years BP) the southern margin of the ice sheet retreated northward during the upper Wisconsin interglacial and large proglacial lakes formed between the ice sheet and the former positions of the ice front. Catastrophic rupture occurred in many of these lakes, causing large floods that eroded downstream valleys (Teller and Clayton, 1983). One of the largest floods was that of Lake Agassiz, a proglacial lake that originally formed shortly after 11,700 ^{14}C years BP (11.670 calendar years BP). The water of Lake Agassiz drained into the Minnesota River and then into the upper Mississippi. It is thought that the magnitude of maximum flooding of Lake Agassiz was about 1 million m^3/second (Matsch, 1983). Maximum incision occurred rapidly after the retreat of ice from the upper Mississippi.

The lowlands of the lower Mississippi extend about 780 km from the confluence of the Mississippi and Ohio rivers to the Gulf of Mexico. Wisconsin glaciation is associated with a wide braided river in the lower Mississippi that transported gravel and sand to the Gulf of Mexico (Saucier, 1994). In this part of the river, Quaternary sea level changes caused stages of accumulation and erosion. The lower Mississippi has been characterized by a meandering morphology during the Holocene *(Fig. 8.38)*. The thickness of deposits increases downstream in the lower Mississippi valley. Alluvial activity in the Mississippi River system during the Holocene has been influenced by fluvial adjustments of sedimentation and erosion during the Late Pleistocene and also by climate and vegetation changes (Knox, 2007).

Unlike most large alluvial rivers, the Mississippi River does not have a concave longitudinal profile. Convexities are present in the profile of the upper Mississippi but they result from alluvial fans or doming, which causes flooding (Schumm, *et al.*, 1994; Schumm, 2005) *(Fig. 3.21)*. The normally straight and relatively deep profile seems to reflect the need for a relatively large energy gradient significant enough to transport large volumes of bed load. Downstream a modest concavity connects with sea level in the Gulf of Mexico (Schumm *et al.*, 1994).

The construction of many large dams on the Mississippi River basin during the 20th century has had little impact on controlling large floods and only a modest influence on sediment supply. In almost all cases, these dams significantly reduce the variability of natural discharge downstream, which also produces significant changes in channel morphology. One of the best-documented examples is that of the Platte River *(Fig. 8.36)* (Schumm, 1977). Dams on the Missouri River were constructed to provide irrigation water, to generate hydroelectricity, and to increase downstream discharge during hot seasons to promote river navigation. Meade *et al.* (1990) show that construction of five large dams on the Missouri River in the Dakotas between 1953 and 1963 virtually stopped the flow of sediments in the Missouri River.

The major reduction of the sediment load by dams and the different channels are well exhibited in the lower Mississippi because the Missouri River has been the principal sediment contributor to the lower Mississippi during the Holocene and recent epochs. The average sediment discharge into the Gulf of Mexico today is estimated to be about half of the amount transported prior to 1953 (Keown *et al.*, 1986).

Prior to about 200 years ago, the drainage basin was only slightly affected by human activity except in localized areas of Native American encampments (Knox, 2002). During the 18th century the eastern drainage basin and during the 19th century the western drainage basin experienced significant modification of woods and grasslands by exploitation, agriculture, and settlements. At the beginning of the 20th century, the Mississippi River was significantly affected by anthropic activity, which caused intense soil erosion in areas of extensive agriculture. In addition, wind action has accelerated erosion and increased the sediment load of runoff (Knox, 2002). In spite of the reductions of runoff and soil erosion related to conservation practices and water retention by dams, floods and pollution by sediments and chemical products used by man are the major problems of the Mississippi River (Knox, 2002). At the end of the 20th century the approximate overall land use in the Mississippi River drainage basin was: 3% crops, 24% grassland, 32% woods, 2% urban and the remainder other categories (U.S. Water Resource Council 1978)

In relation to global warming, the diurnal temperature range has decreased over most of the drainage basin of the Mississippi River and minimum annual temperatures are higher, especially in drainage basins north of the Missouri River and in the upper Mississippi (Easterling *et al.*, 1997). Today there is not yet enough information to speculate on the hydrologic responses to global warming that will affect other parts of the Mississippi River. The large capacity of the lower Mississippi channel normally make it possible to retain the volumes of flood water flow coming from the tributaries of the upper Mississippi and Missouri rivers, but when the two rivers combine with the runoff of the Ohio, significant floods will occur along the lower Mississippi (Knox, 2007).

9.6 Flooding

9.6.1 Introduction

Many definitions have been proposed for floods (Smith and Ward, 1998). Chow's (1956) definition is: "a flood is a relatively high discharge that exceeds the natural channel's capacity to carry away the runoff." Discharge, which can exceed 1000 m^3/sec, can also be described by its period of return; for example, a 100 m^3/second flood occurs once every 100 years (Wohl, 2004). A 10 to 50 km^2 drainage area is usually associated with flash flooding. In alluvial fans, floods result from large discharges through drainage channels that become unconfined at the apex, allowing the water to flow as laminae across the alluvial fan surface. Large floods introduce enormous discharges into the channel-valley system (Baker and Costa, 1987). In some cases, floods produce a surprisingly meager geomorphic response (Costa, 1974) but other floods cause spectacular effects (Baker, 1977; Gupta, 1984). Investigations conducted prior to 1980 dealt with morphological and sedimentological aspects (Wolman and Gerson, 1978) but collected little quantitative data on cataclysmic floods that resulted in a landform response (Baker and Costa, 1987).

Significant documentation is available on major catastrophes caused by flooding (Schumm, 2005). The Bible discusses a flood that occurred 3,000 years prior to Christ, probably in the basins of the Euphrates and Tigris and probably resulting from intense rain for 40 days and 40 nights that covered the world for 150 days. It destroyed all life except for the creatures in Noah's Ark (Costa and Baker, 1981). It is well known that the Egyptians depended on the annual flooding of the Nile River to irrigate and fertilize their land. The Egyptian floods washed away high solubility salts that had accumulated on the surface from capillary rise and also deposited silts and clays that fertilized the soil (Schumm, 2005). Hernando de Soto's exploration in 1543 coincided with a month of flooding on the Mississippi River. Historical documents tell of flooding in India in 1737 that resulted from a tropical cyclone and caused 100,000 deaths. Floods in China in 1887 and 1931 killed 900,000 and 3.7 million people, respectively. These are only approximate estimates (Burton, 1989). *Figure 9.35* shows floods heights in the Ebro River delta since 1617.

Flooding generally affects almost the entire length of a river but there are exceptions. For example, flooding can impact the downstream area of a channel or an upstream flood may dissipate by the time it reaches downstream. However, except for extreme events that modify both the channel and the valley, flood impacts are ephemeral because flood losses by channel readjustment are small in magnitude (Schumm, 2005). Contrary to the rules of hydraulic geometry, river channels in eastern Australia decrease in size downstream. This phenomenon is due to the fact that floods overflow their channels downstream and channels transport only part of the flood (Nanson and Young, 1981). During flooding, erosion predominates during high velocity and current energy. For example, during the 1976 Big Thompson flood in Colorado, the channel reached a width of almost 40 m, had slopes of 0.2–0.4, and caused extensive undermining. However, 80 m wide channels with gradients less than 0.02 caused minimal erosion and extensive sedimentation. These observations can be explained by the fact that the upstream channel is narrow and deep whereas downstream it is a broad braided channel (Wohl, 2000).

In many parts of the world, floods that invade plains along the river and lower coastal areas constitute a major natural risk. Areas prone to flooding are often densely populated and are an economic backbone to numerous countries (Verstappen, 1983).

Floods are the most lethal and destructive natural hazards; they affect industrialized and developing countries alike (McGuire *et al.*, 2004). At least 60 large flood disasters

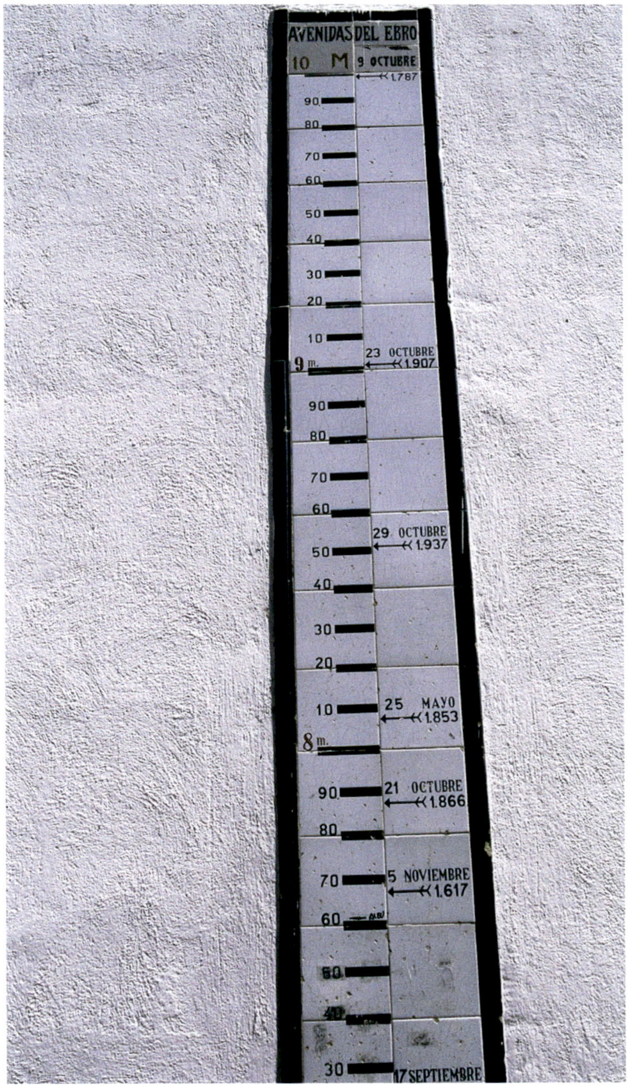

Figure 9.35 Scale of principal floods on the Ebro River in the town near the delta. The maximum flood reached 10 m on 9 October 1787.

occurred during the 1980s, each of which killed more than 100 people; 10 of these disasters took 1000 lives. They occurred in 17 countries, primarily in Asia and South America (UNEP, 1991). It should be pointed out that although most floods are primarily natural phenomena (although they are intensified by human activity such as changes in soil use), the risk of flooding increases when human activity is involved (Smith and Ward, 1998).

A significant part of flood investigations is included in articles and works edited by Baker *et al.*, (1988) and Beven and Carling (1989) and in a book by Smith and Ward (1998). There are also numerous volumes on natural risks that include important geomorphic aspects of flooding and loss of human life and property.

9.6.2 Types of floods

Floods can be classified into two types: those that occur in the interior of continents (fluvial floods) and those that affect coastal areas (coastal floods).

Fluvial floods occur basically in floodplain valleys when water overbanks channel margins or artificial dikes. Under humid conditions, the water table is near the surface and the response to high water is rapid; this is called a **water table flood**. Under dry conditions, the infiltration capacity of the soil is minimal and the response to rainfall is rapid; this type of flooding is called a **flash flood**. In arid zones, when channels are not clearly defined, **sheet flooding** occurs. In urban areas large rainfall events can exceed the capacity of underground pipes and cause flooding (Smith and Ward, 1998).

Coastal floods occur in low coastal areas including estuaries and deltas when seawater advances and overtops artificial dikes. Direct floods occur during storms and high tides when exceptionally large wind-generated waves affect restricted bays. They can also be caused by tsunamis.

9.6.3 Causes and factors that intensify flooding

Table 9.2 shows the causes of floods and the factors that accentuate them. Most fluvial floods are directly or indirectly due to meteorological events such as intense rainfall over a long period.

Among meteorological factors, localized and intense **convective storms** are highly effective at causing floods in small drainage basins. This precipitation can be shown on an isohyet map. Large **cyclonic precipitation** causes low intensity rainfall that produces flooding over large drainage basins. Cyclonic storms of tropical origin cause some of the world's major flooding. Hurricane Agnes, a cyclonic storm, affected the eastern coast of the United States between 14 and 23 June 1972 (Bailey and Patterson, 1975) *(Fig. 9.36)*.

When the duration and intensity of precipitation exceed the infiltration capacity of the ground, runoff increases rapidly. If precipitation is equally distributed throughout the basin, it does not usually exceed the infiltration capacity of the soil but if rainfall is concentrated in a small area of the basin, it can rapidly exceed infiltration capacity and cause high runoff. Additionally, a storm that moves along the basin axis can produce a large flood. If the storm moves from the headwaters to the mouth of the basin, runoff from the headwaters can reach the mouth at the same time as the runoff from around the mouth of the basin. The result is a high flood peak. After days of constant rain prior to a large storm, the infiltration capacity is low and the response is a large flood (Costa and Baker, 1981).

Table 9.2 **CAUSES OF FLOODS AND FLOOD INTENSIFYING FACTORS (SMITH AND WARD, 1998)**

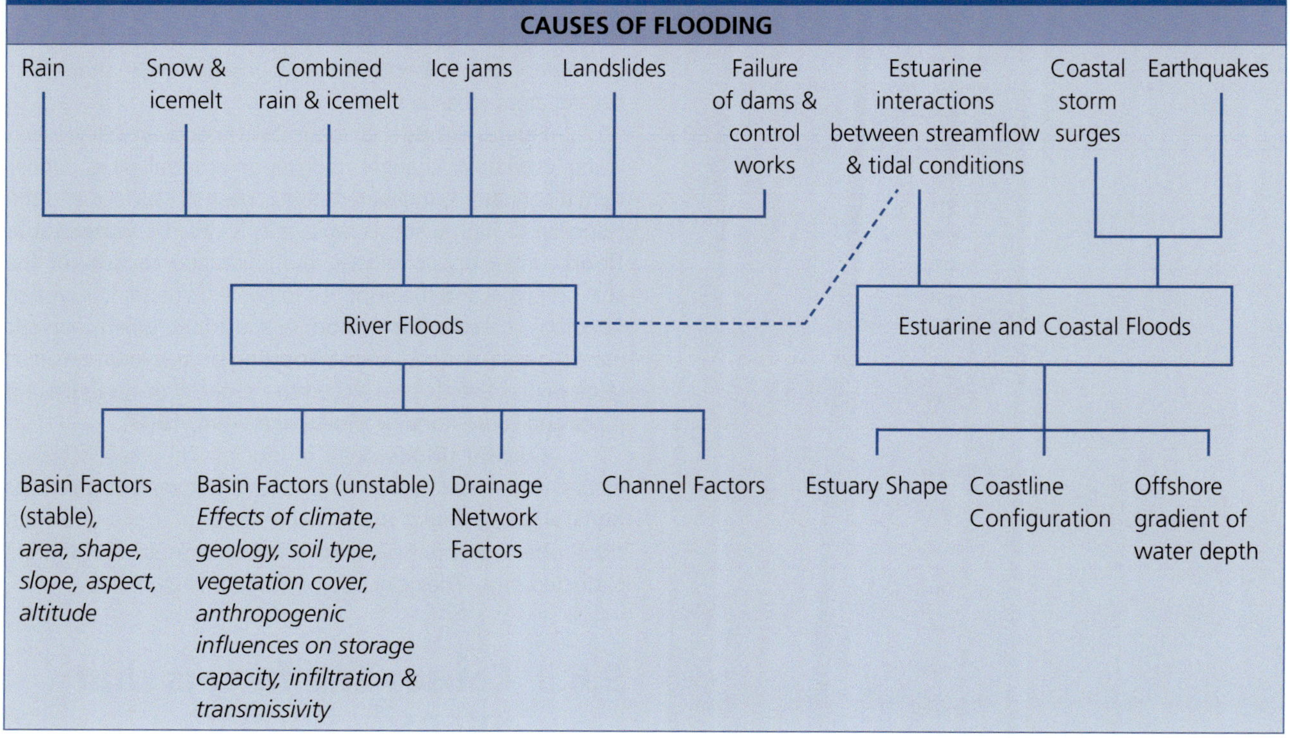

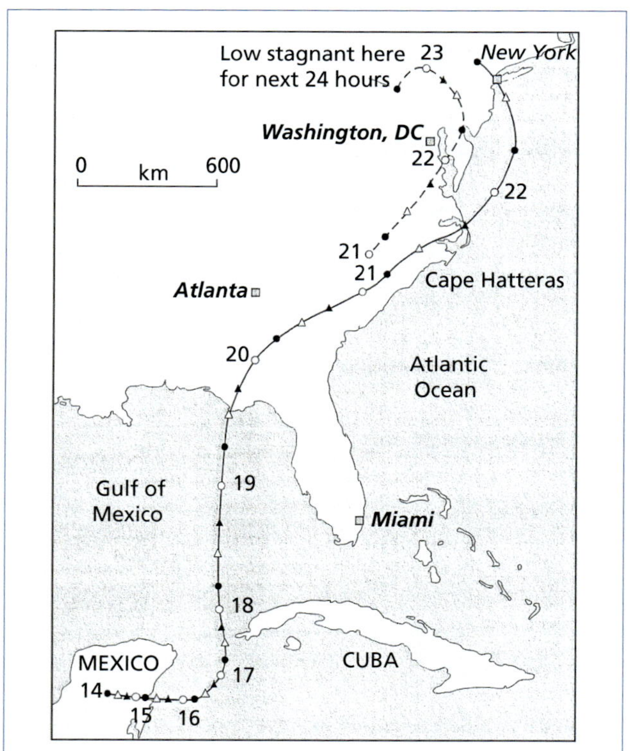

Figure 9.36 Trajectory of Hurricane Agnes (14–23 June 1972) (Bailey and Patterson, 1975).

Melting of snow and ice during spring and summer is another cause of flooding. In areas of cold winters, blocks of ice can dam up rivers and create ice jams (Thomas and Goudie, 2000; p. 264). Large landslides can block rivers; an example is the interglacial landslide in Flims (Switzerland) (See Chapter 7, *Fig. 7.21*). Dam failure can also cause flooding; an example is the Roaring River flood in the Rocky Mountains (Blair, 1987). Costa (1988b) has analyzed the hydrologic and geomorphic characteristics of dam failures.

Climate, as expressed by temperature, precipitation, and plant cover, fundamentally controls most flooding. According to Knox (1983, 1985), a small climate change has been responsible for significant fluvial adjustments during the Holocene. River chronologies and paleohydrologic reconstructions also indicate that rivers respond quite rapidly to changes in climatic conditions.

Large precipitation-induced floods occur primarily between 40°N and 40°S latitude, normally around coastlines where masses of warm air advance over the continent and produce intense and extensive precipitation (Costa, 1987).

Flood characteristics vary somewhat because of the climatic areas where they occur. Ruptures of ice dams holding back glacial melt water (Lake Missoula) from the large continental ice sheets caused catastrophic or cataclysmic flooding (Baker and Costa, 1987) during the Pleistocene that created enormous channel scablands in the northwestern United

States *(Fig. 14.25)* (Bretz, 1923; Baker, 1973). Geomorphologists have analyzed the characteristics and development of fluvial systems in periglacial areas and have determined that they primarily result from the melting of permafrost and its intensity. The water supply of many systems comes from snowmelt, large floods, or a combination of the two. The upper course of the Mississippi River has these characteristics; 75% of all floods occur during the months of March and July (Knox, 1988).

Ancient civilizations in arid zones near the Tigris-Euphrates and Nile rivers understood the power and effects of floods. The most catastrophic documented flood occurred in the spring of 1887 when the Yellow River or Huang Ho (China) overbanked and killed between 1.5 and 7 million people (Davis, 1992). The flood affected 2000 cities and towns and property losses were high. In September 1911 the Yangtze River flooded; it affected four Chinese provinces, killed about 200,000 people, and left half a million homeless. In Farahzad (Iran) on 17 September 1954, 2000 pilgrims who were camped along a narrow gorge died when they were surprised by a flash flood.

In the Finke River basin in the semiarid climate of central Australia, scattered rains caused floods; peak discharges reached 1200 m^3/second (Williams, 1971). These values are similar in size to the 1975 floods of Wadi El Arish in the hyperarid zone of the Sinaí desert (Shick, 1988).

Floods are the most common geomorphic risk in the humid tropics due to the intense and prolonged rains that occur in these areas. The use of floodplains to grow agricultural products considerably increases this risk. Flooding in Asia significantly damages 4 million hectares of land on average every year, adversely affecting 17 million people (Reading *et al.*, 1995).

Convective clouds are the major producers of rain in tropical low-latitude regions. *Figure 18.54* shows the global distribution of storms of different frequency types: the categories are more than 100, between 50 and 100, between 20 and 50, and less than 20 per year (Lamb, 1972). A high storm frequency belt can be seen in areas near the Equator. Precipitation in the tropics results from the convergence of air masses from the two hemispheres (in the Inter Tropical Convergence Zone), from tropical cyclones, and as a result of the orographic effect (Hayden, 1988). In the first type, convergence and rising air generate a band of clouds around the Equator. Seasonal variation of this band of clouds causes monsoonal rains that occur between June and September in Asia, where on average 7 monsoonal depressions occur per year (Gupta, 1988). These rains can sometimes last 9 days. Cyclones move from east to west and maximum rains occur on islands or continental margins where the heated surface of the earth and the existing relief cause ascending movement and intensify the convectivity (Hayden, 1988). In Cilaos, Reunion Island, 1870 mm of rain fell in one day (Gupta, 1988). Tropical cyclones with winds higher than 116 km/hour are called hurricanes; in East Asia they are called typhoons. The orographic effect causes air masses to rise; as the air masses condense, they cause maximum diurnal precipitation during the period of major solar heating.

Gauging stations are rare in the humid tropics and are generally located on large rivers; because of this, records of large floods are also limited. However, the return periods of intense precipitation are well known and these values can be related to those of major floods for which we have approximate dates (Gupta, 1988).

Drainage basin hydrology has changed significantly in urban areas, primarily as a result of impermeable street and roof surfaces that increase flood frequency (Leopold, 1968). These urbanization-induced hydrologic changes affect urban areas when channel and sewer networks are insufficient to evacuate water; this causes flooding in cities such as Kuala Lumpur (Malaysia) and Singapore (Gupta, 1984, 1993).

The frequency of high magnitude events is greater in humid tropical zones. Precipitation occurs in tropical cyclones and during monsoon periods (Gupta, 1988). All tropical rivers have variable discharge peaks during the rainy season although some have two peaks throughout the year, a principal and a secondary peak (Latrubese *et al.*, 2005).

Floods on the Branco River in western Brazil, an area that receives about 200 mm of precipitation per year, has discharge peaks higher than 800 m^3/second every year; a peak of 1914 m^3/second occurred on 17 February 1988. Deforestation, which is not yet very intense, and agricultural use of the alluvial plain have increased flood peaks (Latrubese *et al.*, 2000). Francesco *et al.* (2002) studied the 1982/1983 flooding on the Paraná River in Argentina. Average annual precipitation is about 2000 mm in Paraguay and Brazil whereas in Argentina it is 1200–1400 mm. Discharges reached 60,200 m^3/second in Corrientes (Argentina) in July 1983. The total area affected by flooding was 3,337,000 hectares and economic damages were 1.8 billion dollars.

Floods can be exacerbated by numerous factors *(Table 9.2)*. River floods can intensify as a result of factors associated with the drainage basin, channels, and their network (Patton, 1988). The drainage basin area is important in terms of the amount of precipitation the area has received even though on occasion rain falls everywhere in the basin. The combination of basin shape and drainage system affects the maxima or the flood peaks *(Fig. 9.37)*. Some of the more complex relationships, such as variable basin factors, have a significant influence on three important hydrologic variables: water storage, infiltration, and transmissivity.

Once precipitation reaches the soil, it is converted into fluvial flow depending on the physical characteristics of the drainage basin and fluvial channels. Rodda (1969) separates

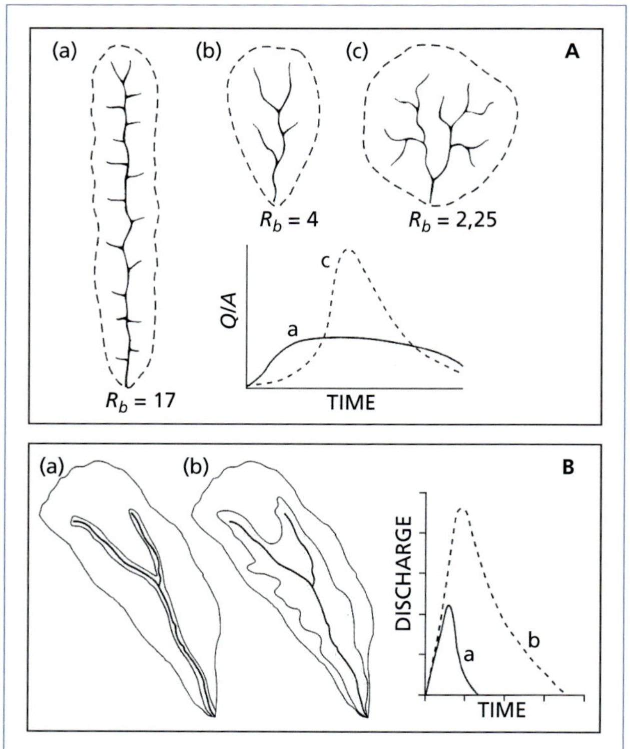

Figure 9.37 Examples of intensification of fluvial erosion: (A) Relationships between basin shape, bifurcation ratio (R_b) and flood maxima (Smith and Ward, 1998). (B) Effects of areal variation on flood maxima (Strahler, 1964).

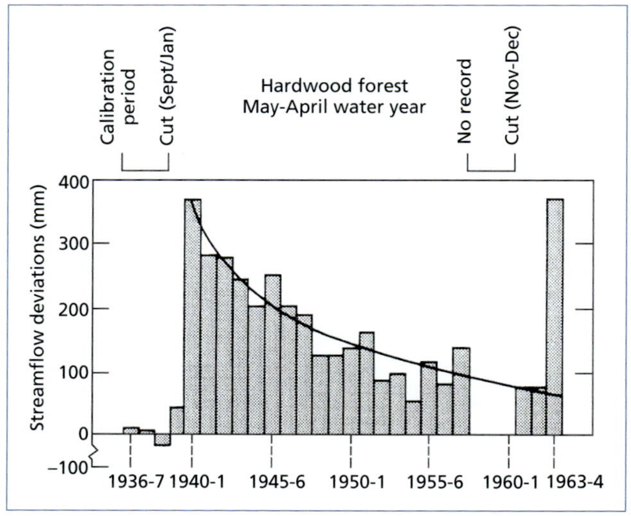

Figure 9.38 Increase in water quantity after deforestation (Hibbert, 1967).

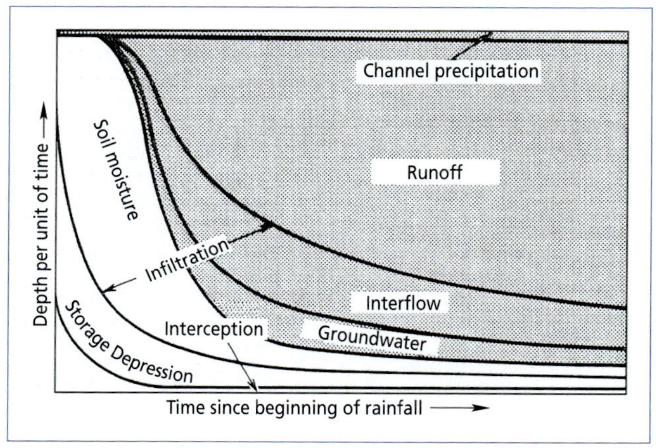

Figure 9.39 Diagram indicating storm rainfall distribution. The shaded zone shows the contribution of water to fluvial flow (Linsley *et al.*, 1975).

the various controls on the flooding response into two principal categories: transitory and permanent. The transitory category includes climatic factors; permanent factors are associated with drainage basin characteristics. It is difficult to quantify each factor because of their strict interdependence.

Soil use in the basin is an important control. Hubbert (1967) documented a change in water production after tree cutting (water increase) followed by rilling (*Fig. 9.38*).

Steep slopes increase the surface flow rate and decrease infiltration. Slope orientation can produce different vegetation densities and colluvium thicknesses (Linsley *et al.*, 1975). A small portion of rain falls directly in the channel. Vegetation and other obstacles intercept another fraction during the first moments of precipitation (10–20% of precipitation in forests). Water then infiltrates and can move laterally into the channel. When it reaches the water table, it forms part of the groundwater supply to river channels, which is essential in maintaining fluvial flow during dry periods. Finally, when the infiltration capacity of the ground has been exceeded, water flows across the surface in fine sheets of up to 1 cm thick at velocities of 10–500 m/hour toward small rills and channels (*Fig. 9.39*) (Costa and Baker, 1981).

In areas with permeable earth materials (glacial deposits, loess, and so on.) such as in the north central United States, flood peaks are small because these earth materials have a high infiltration capacity and water migrates into the vadose zone; in areas of thin or absent surface covers, peaks are higher after the same amount of precipitation.

9.6.4 Magnitude and frequency of flooding

Two of the most interesting factors in flood planning and engineering are water height outside the channel, which determines the extent of the flooded area, and the time that floodwater remains outside the channel. The 1973 Mississippi River flood was outside the channel for weeks whereas floods on other rivers are outside the banks only a few hours.

Channel discharge is the product of the cross section and average velocity; it is expressed in m^3/second:

$$Q = A \cdot V = a \cdot d \cdot v$$

where *a* is channel width and *d* is average depth. It is measured several times every year at discharge gauging stations (*Fig. 9.40*). The water surface height is also recorded each time the discharge is measured; this makes it possible for scientists to obtain a height/flow curve. Because the gauging station keeps a continuous record of height, it is easy to convert it to *Q* by using the curve.

A **hydrograph** shows the variation of discharge over time at a point (Díez Herrero, 2002; 2008). The shape and dimension of a hydrograph are controlled by the spatial and temporal distribution of the storm rainfall and by the physical characteristics of the basin. The flood peak is the time at which the water quantity has reached its maximum; lag time is the interval between the precipitation maximum and the time of the flood peak (*Fig. 9.41*).

Knowledge of flood magnitude and frequency, among other factors, is important in the design of hydraulic structures because the design is based on the flood magnitude and the return period (Costa and Baker, 1981). Catastrophic flooding results from high-intensity low-frequency storms; smaller floods are caused by sudden local convective storms that occur more frequently than high intensity storms.

As we have indicated, we can calculate water discharge on the flow-height curve. A list of annual maximum flows can be represented on a discharge-frequency curve to produce a **recurrence interval** (*R*) for each discharge, using the following ratio:

$$R = N + 1/M$$

in which *R* is the recurrence interval in years (*Fig. 9.42*), *N* is the number of years on record, and *M* is the discharge of each of the recorded years. The reciprocal of the recurrence interval expressed in years is the **return period** ($R = 1/P$) in which *P* is the probability that an equal or higher flood will occur in a year. For example, a flow with a 20% probability of being equaled or exceeded in any year has an *R* of five years (Costa and Baker, 1981; Keller and Blogett, 2006). Once a flood has occurred, its recurrence interval remains the same.

Normally, the return period of a catastrophic flood is long when compared to the historical record. In this case the best method for estimating flood risk is to study the Holocene stratigraphic record. Remains that contain carbon (^{14}C), archeological remains (relative age) and fluvial sands (OSL) provide valuable chronostratigraphy for establishing return periods (Smith and Ward, 1998).

When fluvial data is not available, observations indicate that channel width is less variable than depth or

Figure 9.40 Gauging station in the Cabriel River. El Cañizar, Province of Cuenca (Spain).

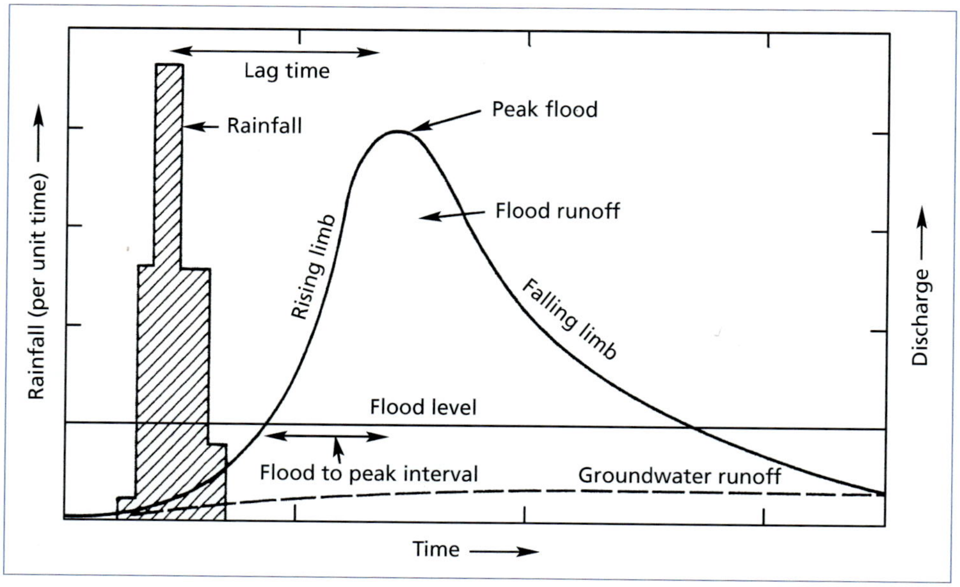

Figure 9.41 Typical flood hydrograph showing the most important characteristics (Costa and Baker, 1981).

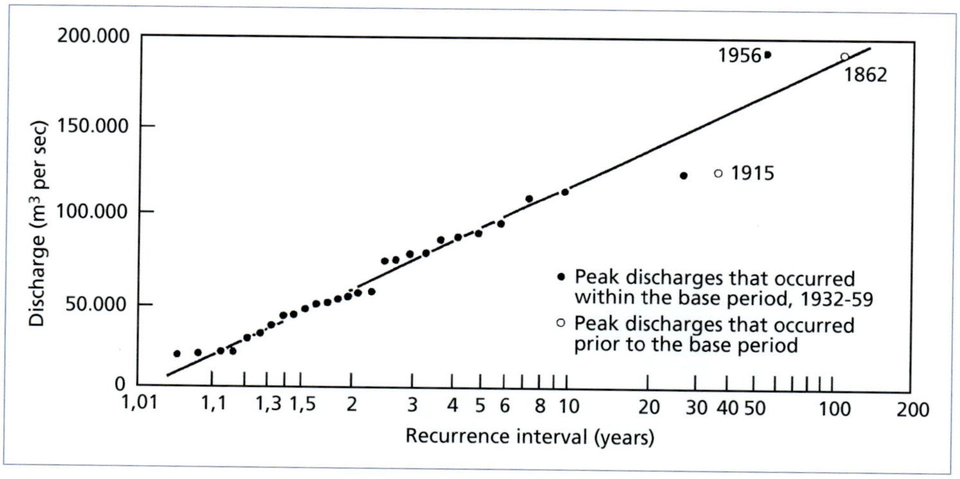

Figure 9.42 Flood frequency curve of the Eelen River, Scotia, California (United States). United States Geological Survey (in Bolt *et al.*, 1975).

cross-sectional area. Thus channel width can be used as an independent variable to determine flood data such as the magnitude of flooding between 10 and 50 years. Good results can be obtained in arid zones (Costa and Baker, 1981).

9.6.5 Mapping of flood risk areas

Flood risk mapping involves multiple aspects because of the need to apply methods and techniques of different matters. Geomorphic mapping of flood-prone areas is a useful exercise because it provides maps of flood areas. Stratigraphic methods, already discussed, provide significant help in determining recurrence intervals. To this list we must add satellite imagery (Brown *et al.*, 1987). Digital simulation models can also be used. Valid dates can also be obtained from archives. Maps at a scale of 1:25,000 are used most often; they are a basic document for understanding flooding (Martinez *et al.*, 1987).

In addition to the analysis of hydrographs, discharge, height and water volume, other flood dimensions such as floodwater velocity and seasonality of flooding are equally important in our understanding of floods. Large rivers that peak during flood stages usually have a long return period.

As a result, these characteristics must be taken into account when mapping flood-prone areas. The United States Geological Survey prepared the first flood limit maps in 1959.

Floods occur primarily in floodplains, which are potentially high value areas for human occupation because they have water, good soils, construction materials, and are areas highly suitable for transport and other activities. In the United States, one billion dollars are spent every year to control flooding. The United States Water Resources Council and the Federal Insurance Administration have adopted a recurrence interval of 100 years as the risk level for the Federal Flood Insurance Program. This number is a compromise between large catastrophic floods that have low probability and more frequent floods that cause continuous property damage. This limit seems moderate, since two-thirds of flood losses have recurrence intervals of 1 to 10 years (Costa and Baker, 1981).

Various methods are used to map the extent of floods. Methods based on highly detailed topography (engineering-hydrologic-hydraulic method) are expensive. Other less onerous methods (physiographic, pedogenic, vegetation, and others) are most suitable when used in combination. Costa and Baker (1981, p. 372–379); Díez-Herrero *et al.*, 2008, p. 109–145) describe them in more detail. Cendrero *et al.* (1987) did detailed cartography of geologic risks for urban and rural planning in Vizcaya (Spain).

9.6.6 Paleohydrology

Paleohydrology studies floods that occurred prior to direct measurement by modern hydrologic methods or documentation in other human records. Direct measurement includes instrumental records; historical flood records are also documented in other human records (Baker, 1989, 1998, 2003, 2008b; Baker *et al.*, 2002; Benito, 2002, 2006; Benito *et al.*, 1998b, 2004, Díez-Herrero, 2002). The increase in paleohydrologic studies stems from the pioneering work of G.A. Dury and S.S Schumm during the 1960s (Williams, 1988).

Sequences of multiple sedimentation units can record various paleofloods, a term coined by Leopold and Miller (1954) to describe floods that are sometimes intercalated with units deposited during non-flood periods (*Figs.* 9.43 and 9.44). This idealized figure shows the number of flood

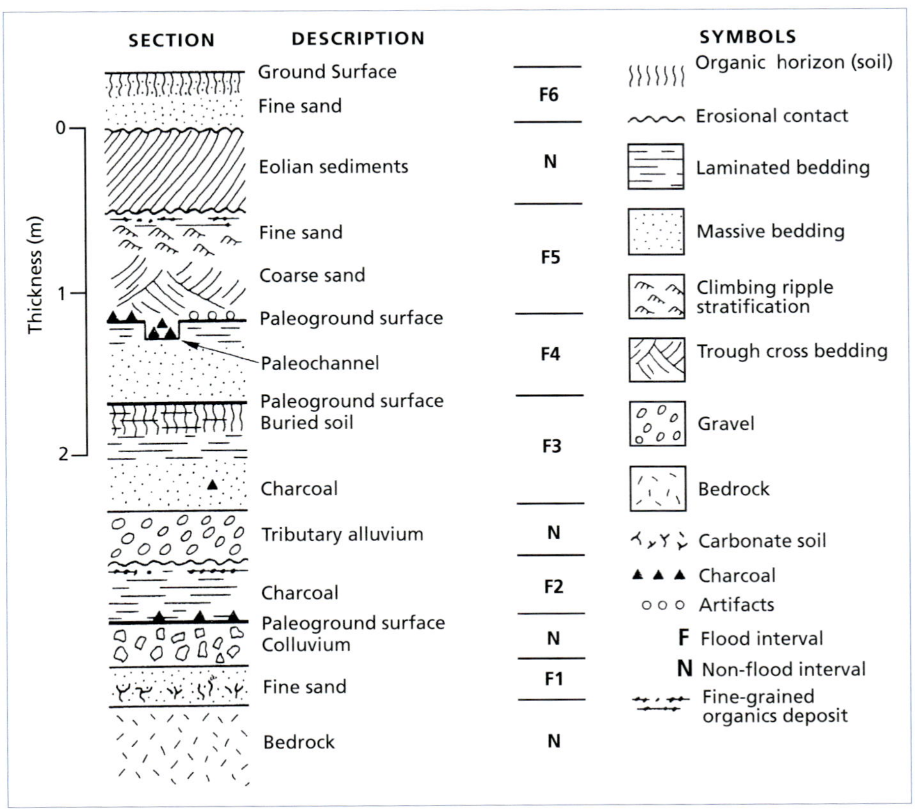

Figure 9.43 Idealized stratigraphic column of a slackwater sequence that contains six flood deposits (numbered from F1 to F6) and various deposits from non-stagnant environments (N). The symbols on the right indicate the complexity of this hypothetical section (Baker, 1989).

Figure 9.44 Fluvial sequence with flooding. Nahal Zin, Negev Desert (Israel).

intervals (Baker, 1989) inferred from a stagnant water environment (Kochel and Baker, 1988). It is fundamentally important to obtain precise dates of paleoflood events. A wide variety of procedures can be used: archeological materials, dendrochronology, thermoluminescence, and radiocarbon dating. Figure 9.45 shows the flood frequency curve for Raccoon Creek in Moffatts, Pennsylvania. There are numerous distributions: Gumbel (G), log. Gumbel (L), and log. Pearson III. The Gumbel distribution is used for data with a 95% confidence level. All these are statistically derived methods that involve the extrapolation of numerous data (Llamas, 1993).

It is important to know the ages of paleofloods in order to compare them with known **climate changes** (Knox, 1985, 2000, 2006). Benito *et al.* (1996, 1998a) and Benito

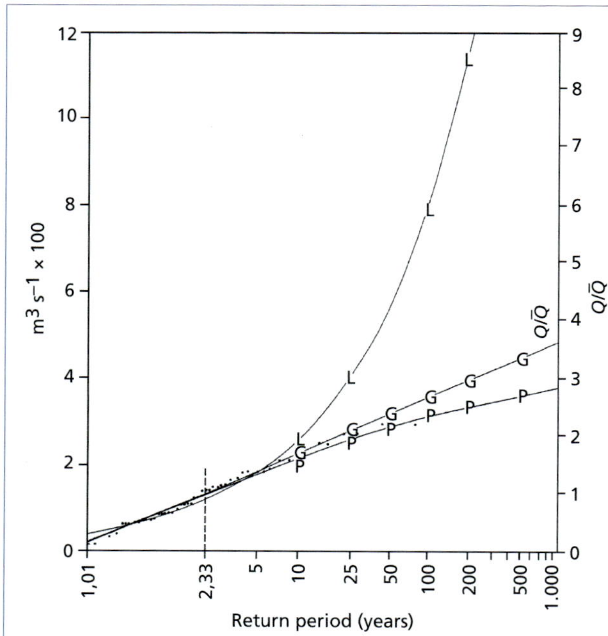

Figure 9.45 Representation of Gumbel (G) points, log. Gumbel (L) and log. Pearson type III, for Raccoon Creek on Moffatts Mill, Pennsylvania. Area of the basin: 461 km^2 (Reich, 1973).

(2002) studied paleofloods in the Tajo River in central Spain. They found floods in the historical and geological record with ages of 1150–1290 AD, 1400–1500, and 1850–1990. The first of these periods corresponds to the climatic transition at the end of the Medieval Warm Period or the Small Climate Optimum (from 900–1000 AD and 1200–1300) (Hughes and Díaz, 1994). Flooding occurred at the end of autumn and winter. The second period is characterized by humid and hot conditions (Font, 1988), which preceded the very cold climate of the Little Ice Age (1600 to 1800–1900 AD) (Grove, 2004). The third period coincides with a variety of temperatures from cold (1820 AD) to hot (1885) and returning to cold (1915 AD).

The science of flood hydrology has matured to the extent that it is considered essential in the understanding of large and rare floods. These floods are increasing and creating risk for the human population. **Globally** over the last three decades considerable advances have been made in determining the ages of paleofloods, quantifying magnitudes and the dynamics of paleoflood processes, and finally in the preparation of risk models. This work also provides valuable data on global climate change (Baker, 2006).

In **Europe** a significant amount of radiocarbon dating has been done on fluvial units in Great Britain, Poland, and Spain; the goal has been to investigate the relationship between climate change, flooding, and the dynamics of Holocene rivers (Benito, 2003; Brown, 2003; Macklin *et al.*, 2006). ^{14}C ages on fluvial sequences coincide with a change in the sedimentation rate. Multiple phases of the frequency of more significant floods are characterized by accelerated erosion and floodplain sedimentation. Fifteen periods of maximum flooding have been identified, eleven of them coinciding with phases of maximum lake levels in central Europe. This relationship shows climatic control, recognized in the response of rivers to climate change, which occurs more rapidly in rivers than in lacustrine systems.

Thorndycraft and Benito (2006) used 74 radiocarbon dates in **Holocene fluvial environments in Spain** to identify various periods of increased fluvial activity; they found a good correspondence to the activity described by Bond *et al.* (2001) in the study of core sampling in the North Atlantic. Maximum flooding in the Holocene coincides with cold climate phases during 9530–9280, 2880–2430, and 520–265 cal. years BP. Another cold phase was identified at 10855–10230 and other warm phases between 9030–8780 and 975–790 cal. years BP.

The **greenhouse effect** theory implies that the hydrologic cycle will accelerate. Recent studies in North America and Europe have not detected evidence of increased flood frequency or magnitude in the 20th century although a low to moderate increase in fluvial channel flow has been observed (Lins and Cohn, 2003). These authors conclude that greenhouse effect warming caused by humans is probably producing significant changes in average fluvial regimes rather than extreme hydrologic changes (Lins and Cohn, 2003).

Results of investigations by six important working groups have led to the establishment of a protocol recommending the use of past hydrological events related to an understanding of global climate change to attempt to determine the relationships between them (Gregory *et al.*, 2006).

Scientific and technological interest in paleoflood studies is evident from the estimation of return periods used for construction design and risk planning (Thorndycraft *et al.*, 2003). Also, in terms of application, hydrologic behavior should be simulated over the long term in order to select sites for underground low and high activity radioactive waste repositories and to estimate flood risk (Baker, 1998).

9.7 Flood risk

Compared to other natural hazards, floods cause the most human fatalities and property losses, which Burton *et al.* (1978) estimate at about 40% of total losses from all natural catastrophes. The 20 most deadly recent major floods killed seven million people (*Table 9.3*); between 1900 and 1999, floods caused 103,870 deaths and affected 1.5 million people with a cost of 243 billion dollars in property damage (McGuire *et al.*, 2004). Floods affect all countries of the world although one third of the major disasters have occurred in Asia. The most destructive recent flood occurred in central Europe in 2000; it killed 230 people and caused economic losses of 55 billion euros (Benito and Thorndycraft, 2004b).

Table 9.3 **MOST CATASTROPHIC FLOODS SINCE 1900 (MCGUIRE *ET AL.*, 2004)**

Year	Country (Location)	Death toll	Cause
1931	China (Yangtze)	3,700,000	River flood
1959	China (north)	2,000,000	River flood
1938	China (Hwangho)	500,000	Dam failure
1970	Bangladesh	300,000	Storm surge
1935	China	142,000	River flood
1991	Bangladesh	138,866	Storm surge
1908	China (Yangtze)	100,000	River flood
1911	China (Yangtze)	100,000	River flood
1949	China	57,000	River flood
1949	Guatemala	40,000	River flood
1954	China (Dongting Lake)	40,000	River flood
1999	Venezuela	30,000	Flash floods and *debris flows*

The most common and highest risk floods occur on floodplains along rivers and low elevation coasts. Risk is high because of large populations and significant resources in these areas (*Table 9.4*). While we expend extraordinary human effort to control the natural environment, risk on a global level is increasing; as a result, risk causes greater loss of human life in less developed countries but the property losses incurred in these extreme events are small. About 95% of deadly disasters affect the two-thirds of the world's inhabitants who live in developing countries. In contrast, three-fourths of property losses occur in rich countries (Burton *et al.*, 1978).

Accurate information on flood losses is generally available only in developed countries. *Figure 9.46* shows flood losses in the United States for the period of 1925–1994. In spite of large inter-annual variations, there is an obvious increasing trend of property losses as a result of the population's high standard of living. The 1993 peak is a result of the midwestern US flooding; significant flooding in 1972 resulted from tropical storm Agnes. However, a number of human fatalities resulted from the wide trajectory of tropical storm Agnes from Florida to New York (Smith and Ward, 1998).

Table 9.4 **MAJOR ECONOMIC AND INSURED LOSSES FROM FLOODS (McGUIRE *ET AL.*, 2004)**

Year	Country	Economic losses (million US$)	Insured losses (million US$)
1998	China (Yangtze River)	30,000	1,000
1996	China (Yangtze River)	26,500	400
1993	USA (Mississippi Basin)	16,000	1,000
2002	Germany, Austria, Czech Republic, Hungary	13,500	3,100
2000	Europe (Italy, France, Switzerland)	8,500	420
2003	China (Yangtze, Huaihe rivers)	8,000	
1993	India and Nepal	7,100	
1995	China (Yangtze River)	6,720	
1997	Central and Eastern Europe	5,900	795
2002	Europe (Italy, Germany, Austria, Czech Republic), Russia, Ukraine	5,100	193
1995	Western Europe (Germany, France, Belgium, Holland)	3,500	950
1953	Holland, United Kingdom	3,000	
1998	Argentina	2,500	
1997	USA (North Dakota) and Canada (Manitoba)	2,000	400
1997	USA (Washington, Oregon, California, Idaho, Nevada)	2,000	400
2000	United Kingdom	1,500	1,090
2002	Southern France	1,200	700

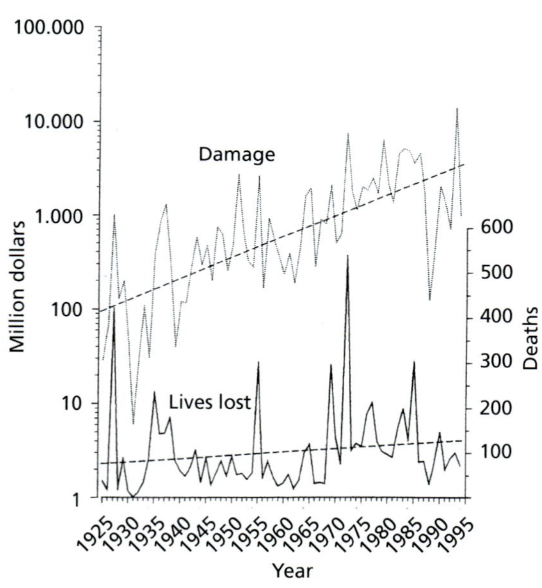

Figure 9.46 Annual fatalities and economic losses caused by flooding in the United States for the period of 1925–1994. Property losses were calculated using the 1990 value of the dollar (United States Department of Commerce, in Smith and Ward, 1998).

Direct effects of flooding include damage to homes and public and industrial buildings. Indirect effects include losses to agriculture (*Fig. 9.47*), industrial and commercial production, and cattle, and destruction of roads (*Fig. 9.48*). Railroads can also be disrupted as a result of deep incision by rivers (*Fig. 9.49*). The high energy of river currents and clasts can cause bridges to collapse partially (*Fig. 9.50*). Significant accretion of sediments can result near bridges built 30–40 years ago and the bed load may almost reach the transit level of the bridge (*Fig. 9.51*). Secondary impacts are caused by contaminated water, scarcity of food, and increased illness, all which may persist many months after a flood.

Alluvial fans characteristic of the piedmont landscape in arid and semi-arid mountainous zones are formed by the discharge of sediments transported by rivers flowing from mountainous areas into areas of lower slope. Alluvial fan areas pose serious problems for human activities (Beaty, 1974; French, 1987) because floods are rare, unpredictable, and vary significantly within the same flow and between successive flows. Water or detritus may predominate in a flow; detritus-dominated flow can obstruct channels. They can also migrate; avulsions are common (Cooke, 1984).

Flood risk is obvious but it does not affect all parts of the fan equally. Risk is extreme along an active channel, moderate at the headwaters and in the distal zones, and low

Figure 9.47 Flooding on the floodplain of the Ebro River on 8 February 2003. Escarpe de Remolinos (Zaragoza). Photo F. Gutiérrez.

Figure 9.48 Medenine-Gaber (Tunisia) highway cut off by waters from a wad. Photo F. Gutiérrez.

in the middle section (*Fig. 9.52*) (Kesseli and Beaty, 1959). A common engineering and design solution is to design and maintain a single flow channel (Graf, 1988). However, a better defense against this risk is to identify danger zones by creating flood risk maps, as was done in the city of Suez and its environs (Cooke *et al.*, 1982) (*Fig. 9.53*). These maps can be used to evaluate construction densities and to site structures far from channels. Engineers also fight against sediment contributions by constructing small dams (*check dams*) and dredging channels in the supply basin. In addition, it is

Figure 9.49 The cone of the Medio Arroyo and the railroad that runs over it have been incised by the Grande River leaving the tracks hanging. Cordillera of the Andes Province of Jujuy, Argentina. Photo J.L. Peña.

Figure 9.50 Bridge destroyed by a flood. Be'er Sheva. Negev Desert. Israel.

possible to install devices that alert the public in the event of flooding (Porath and Shick, 1974; Shick, 1995).

Fan channels are unstable because they fill frequently due to the coarse sediment supply. Sediment blockage creates new channels that can sometimes be part of the abandoned channel system. In southern Jordan in 1966, significant floods partially destroyed the city of Maan, killing 70 people. The fundamental cause was floodwater that reused an abandoned channel (Graf, 1988).

Perhaps the most important problem in the study of risk and its consequences in these areas is that fans can remain dry for significant periods of time, as in Southern California,

Figure 9.51 Significant accretion of fluvial deposits of the León River in the area of a highway bridge. Cordillera of the Andes, Province of Jujuy, Argentina. Photo J.L. Peña.

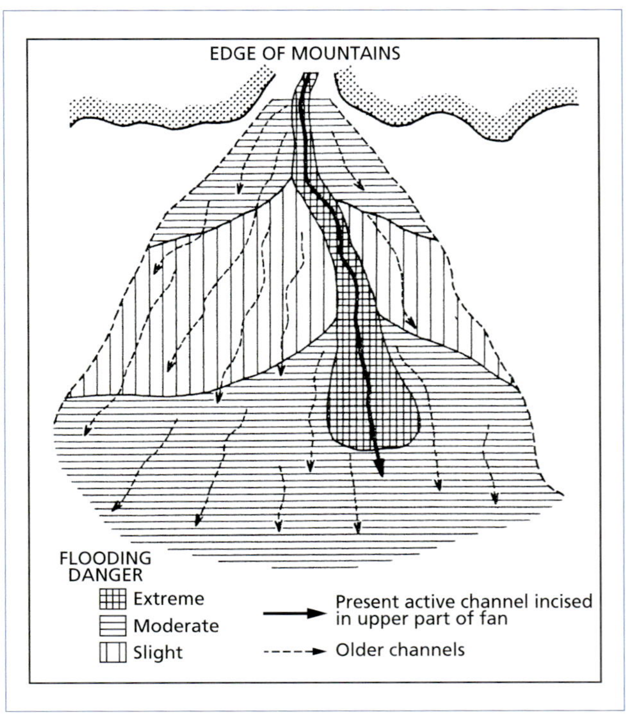

Figure 9.52 Danger related to flooding on an alluvial fan (Kesseli and Beaty, 1959; en Shick, 1971).

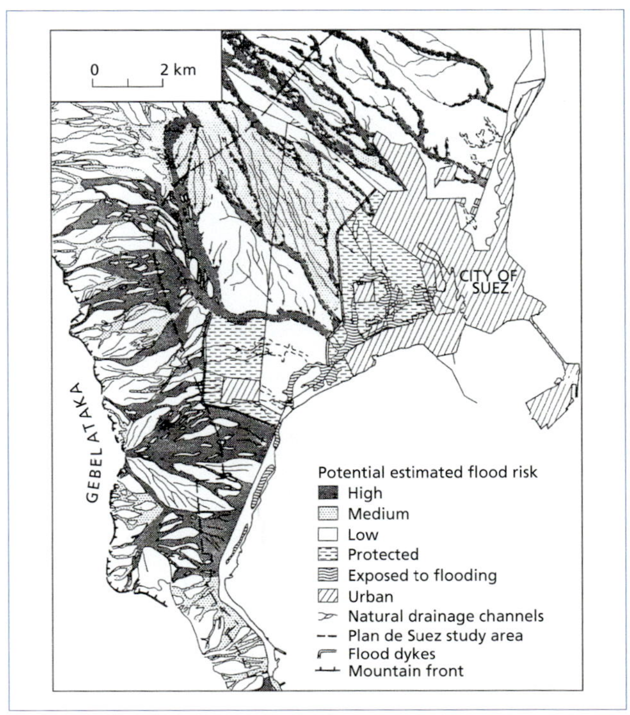

Figure 9.53 Flood risk map for the Suez, Egypt area (Halerow and Partners, 1978; in Cooke *et al.*, 1982).

where the recurrence interval is 70 years. As a result, disasters occur from time to time and only high intensity events gain public attention (Burton *et al.*, 1978).

For inhabitants of arid areas, the struggle against flooding has been constant throughout history. At the beginning of the 8th century, Arabs built the city of Calatayud (Province of Zaragoza) partially on an alluvial fan; it has experienced frequent floods caused by storm events. To alleviate this risk a dam and a channel were built to divert the waters from the Barranco de la Rúa to the Rambla de Ribota (*Fig. 9.54*). This Arabic engineering structure drained a surface of 6.3 km^2 out of the 8.2 km^2 that makes up the basin of the Barranco de la Rúa. As the population grew toward the Jalón River, especially during the 13th and 14th centuries, flood risk increased. During these periods and somewhat afterward, the Balsa Valparaíso was built near the apex of the fan to mitigate flooding. After the 14th century, a tunnel almost a kilometer long was excavated to drain water toward the Barranco de Longia; it diverted 98% of the surface flow of the Barranco de la Rúa from the drainage basin to decrease flood risk. The Barranco de las Pozas, which has a smaller drainage area, continues to pose problems for the urban area of Calatayud (Gutiérrez, 1998). In the neighboring city of Daroca (Province of Zaragoza), the kilometer-long Galería de la Mina was excavated in 1555 to divert water from the principal rambla that circulates along what is today Mayor Street, the location of significant flooding in the past (*Fig. 9.55*).

9.7.1 Flood risk in Spain

Floods are natural catastrophes that cause major economic and social impacts in Spain. These phenomena affect extensive areas of the country simultaneously, but cause severe effects only in very localized areas. In just hours or days, floods can cause significant economic losses and harm a large number of victims (Pujadas, 2002). Of all geomorphic risks, floods are the most significant (Martinez *et al.*, 1987). This is confirmed in a predictive study by Ayala *et al.* (1987a) in which they estimate losses of 28.2 billion dollars for the period from 1986 to 2016. In Spain has four risk areas: 1) sudden floods along the Mediterranean coast; 2) torrential floods in mountain systems; 3) floods along medium and lower sections of large peninsular rivers, and 4) downcutting along flat and endorheic zones in the central sectors of large basins (Díez-Herrero *et al.*, 2008).

Spain has experienced historic floods (*Table 9.5*). Between the years 1407 and 1877, Seville experienced 89 large floods; between 1544 and 1946, 49 floods were recorded in Malaga; and finally the Segura River overflowed its banks 27 times in the period between 1528 and 1983 (Martinez *et al.*, 1987). Due to large floods that affected Valencia, Alicante, and Murcia, and along the southern slope of the Pyrenees, the administration studied flooding problems in the territory, and found a total of 1400 conflict points (Aguilera, 1986) (*Fig. 9.56*) (*Table 9.6*). Martínez *et al.* (1987) prepared a map that establishes distinct zones of potential

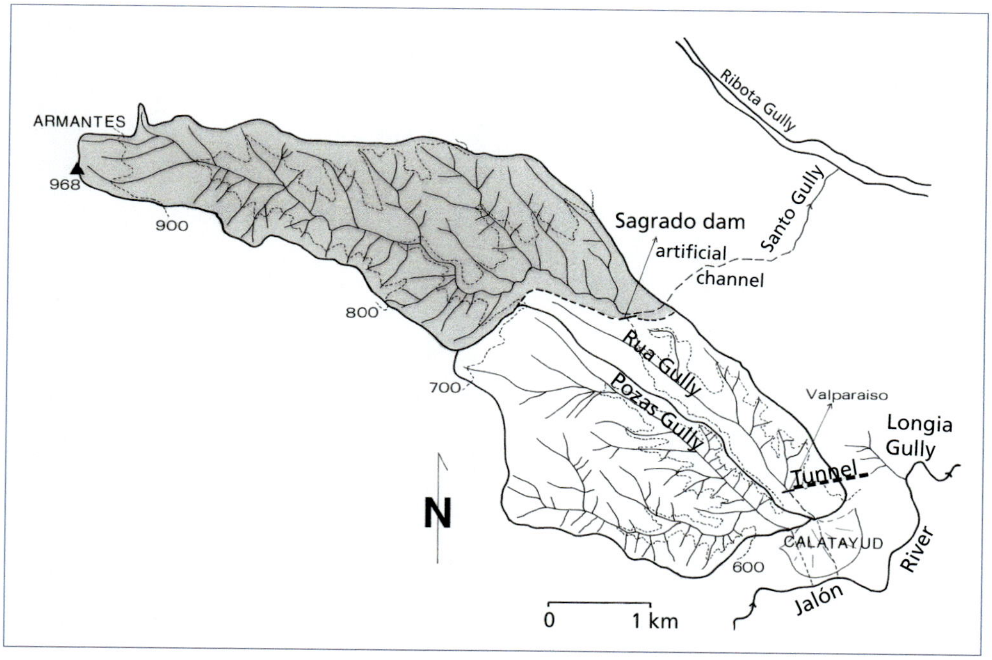

Figure 9.54 Struggle against flooding in the city of Calatayud, Province of Zaragoza. The city is located on an alluvial fan (Gutiérrez, 1998).

Figure 9.55 Gully caused by artificial excavation. Outfall of the Galería de la Mina (Daroca, Province of Zaragoza).

risk (maximum, intermediate, and minimal) as a function of frequency, danger, and magnitude of flooding.

Atlantic rivers now show a decreased frequency of ordinary flooding, although the magnitude of catastrophic floods has remained the same and even increased in spite of the flood reduction effect of reservoirs. In the Mediterranean basin, flooding history indicates that extreme floods have occurred during periods of highly irregular precipitation, both seasonal and annual (Benito, 2006 and Benito *et al*., 2008).

Major floods have occurred in the **Mediterranean region** (Martinez *et al*., 1987; Conesa, 1995; Diaz de Terán *et al*., 1997; López Bermúdez *et al*., 2002; Alonso-Sarriá, *et al*., 2002). They commonly occur between September and November; precipitation between 500–600 mm has been recorded, with peaks of 150 mm/hr. The dry gullies overflow their banks for most of the year (Segura, 1990). The area has been partially deforested, which increases overland flow, and the floods carry a large bedload that increases losses. The Mediterranean area is not well regulated by dams because dams have short lifespans. Numerous significant floods have occurred (*Table 9.5*).

In the **Pyrenees** rainstorm-generated floods are relatively frequent (Comisión Nacional de Protección Civil, 1983; Instituto Geológico y Minero de España, 1985). During the snowmelt period at the end of spring and beginning of summer runoff increases but does not often cause flooding. In addition, floods on the most important rivers are tempered by significant vegetation, higher drainage density, and a greater regulation of rivers by dams (Díaz de Terán *et al*., 1997). Even so, significant flooding occurred in 1982 (*Table 9.5*).

The coastal regions of **Cantabria** and **the Basque Country**, particularly the latter, have short rivers that are poorly hierarchized; in addition, the rivers are narrow and deep. These characteristics, together the presence of urban areas and numerous industries in the valleys, are associated with frequent catastrophic flooding (*Table 9.5*, year 1983) (Martínez *et al*., 1987; Olcina, 1994; Díaz de Terán *et al*., 1997).

Finally, the **large rivers in the interior of Spain** (Ebro, Duero, Tajo, Guadiana and Guadalquivir) are well regulated by dams, channelization, dikes, and trenching. As a result, during large floods they can exceed their banks, as they do in the provinces of Sevilla and Zaragoza (*Fig. 9.47*).

Flood analysis of the Spanish territory showed that most destructive **flash floods** result from high intensity rainfall, with discharges that greatly exceed the infiltration capacity of the watershed (Suarez and Regueiro, 1997). Suarez and Regueiro call the floods that occur on large interior rivers **slow floods**. The SAIH (Automatic Hydrologic Information System), initiated in 1984, was installed in this basin with permanent

Table 9.5 **LARGEST FLOODS IN SPAIN (SUÁREZ AND REGUEIRO, 1997; CENDRERO *ET AL.*, 1997)**

Year	Location	Damages (pesetas)	Death toll
1651	Murcia	Serious	1,000
1779	Ribera Baja del Júcar	Significant	numerous
1802	Lorca (Murcia)	Destruction of the city	700
1874	Cataluña	Major damage, more than seven hundred homes destroyed	600
1879	Murcia	Major damage	800
1891	Consuegra (Toledo)	Partial destruction of the town	359
1940	Gerona		90
1957	Valencia	Major damage, more than three hundred buildings destroyed, 10 billion of 1957	86
1959	Zamora		150
1962	Cataluña	Major damage. Five thousand homes destroyed, 2.7 billion in losses.	1,000
1963	Murcia, Almería	Major damage	300
1965	Cáceres		47
1971	Cataluña	Major damage. 7.0 billion in losses	400
1972	Valdepeñas	Significant damage	22
1973	Southeast	Major damage over an extensive area	300
1979	Ciudad Real		22
1982	Pirineos		30
1982	Levante	Major damage. 300 billion in losses	38
1983	País Vasco and Cantabria	More than 150.0 billion in losses	40
1987	Southeast		7
1989	Málaga and Southeast	200 billion	42
1994	Cataluña	30 billion	9
1995	Yebra (Guadalajara)	Significant damage	10
1996	Biescas (Huesca)	Several billion	87
1997	Alicante and Southeast	10 billion	5

rivers to prevent losses and alert the population about possible flooding. The system consists of control points with sensors (1728 throughout Spain) that analyze and send data on temperatures, levels and water channels, volumes behind dams, and other parameters. This information is collected at 170 concentration points that send it to basin processing centers. With all this data, models of possible flooding are simulated for the purpose of providing alarms and studying potential solutions (Martínez *et al.*, 1987; Pujadas, 2002). Ortiz (1993) conducted a detailed SAIH study of the Ebro Basin.

9.7.2 Two catastrophic floods: Barranco de Arás (Huesca, 1996) and the Riada del Júcar (Valencia, 1982)

These two cases have been chosen for contrast. The first is a case of flash flooding that developed over a very short time period. The Riada del Júcar event resulted from intense precipitation that produced spectacular floods over a wide region of Spain.

Figure 9.56 Maps showing trouble spots in Spain where there have been fatalities and economic losses (Aguilera, 1988).

Table 9.6 **DISTRIBUTION BY HYDROGRAPHIC BASIN OF TROUBLE SPOTS AND ZONES OF POTENTIAL FLOOD RISK IN THE SPANISH PENINSULA (AGUILERA, A. MOPU REVIEW, MARCH, 1986)**

		Zones of potential risk			
Hydrographic basin	**Trouble spots**	**Maximum**	**Intermediate**	**Minimum**	**Total**
Northern Spain	300	16	39	68	123
Duero	72	15	20	48	83
Tajo	25	6	12	70	88
Guadiana	66	4	9	47	60
Guadalquivir	177	15	21	68	104
Southern Spain	121	6	21	60	87
Segura	9	3	5	10	18
Júcar	173	13	28	91	132
Ebro	221	18	45	220	283
Western Pyrenees	172	7	16	36	59
TOTAL	1,336	103	216	718	1,037

On 7 August 1996, an intense short-duration convective storm occurred in the 18.56 km^2 **Arás watershed** (Central Pyrenees); it drained into an alluvial fan where a campground was located (*Figs. 9.57* and *9.58*). Eighty-seven people were killed and the economic losses were high (*Fig. 9.59*). This catastrophe has been studied scientifically and discussed by numerous investigators (Ayala, 2002a; Benito *et al.*, 1998c; Cáncer, 1996; García-Ruiz *et al.*, 1996; Gutiérrez *et al.*, 1998a, b; Riosalido *et al.*, 1996).

This small high relief basin is composed of three sub-basins (Aso, Betés and La Selva) that feed the Barranco de Arás alluvial fan located in the valley of the Gállego River. An artificial channel with a capacity of 150 m^3/second had previously drained the alluvial fan. The Aras gully had more than 30 small check dams that had filled with sediments prior to the event. Precipitation data obtained from the Centro Meteorológico Nacional de la Sierra de Alcubierre radar station indicates that the most intense precipitation occurred on 7 August at a rate of 178.4 mm/day (*Fig. 9.59*) with maximum intensity of 153 mm/hr. falling in a 10-minute interval. Most of the precipitation fell in about 70 minutes. The storm caused immense runoff; it moved a large solid load that caused catastrophic damage (*Fig. 9.60*) and significant geomorphic changes in the drainage basin, especially the Betés sub-basin.

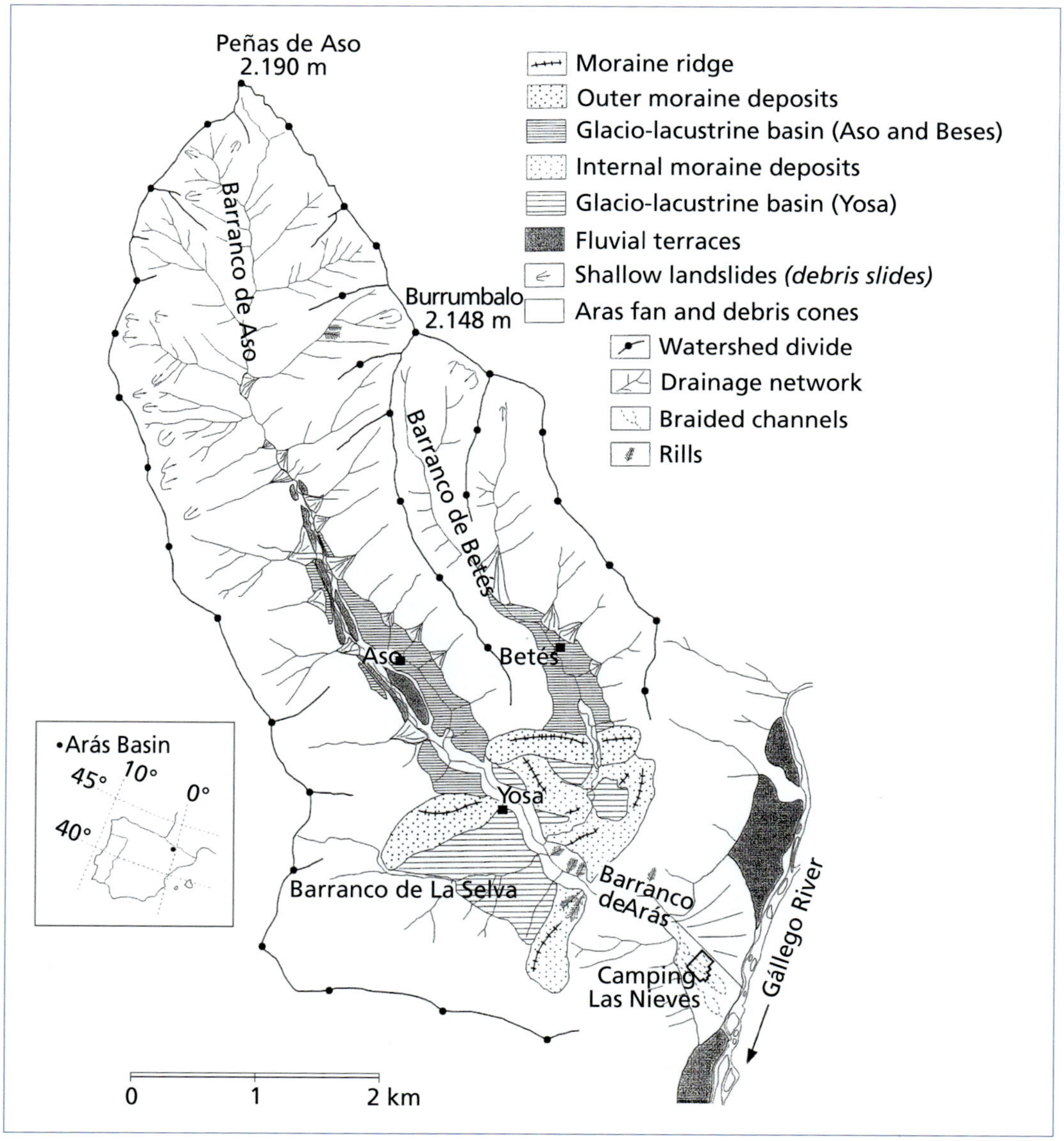

Figure 9.57 Geomorphic map of the basin and alluvial fan of Arás (Gutiérrez *et al.*, 1998a).

Figure 9.58 View of the Arás alluvial fan, two weeks after the catastrophe when the erosional and depositional features had been removed. The artificial straight and stepped channel that discharges into the Gállego can be seen. To the right of it and more to the bottom of the photo are shown the Las Nieves campground, populated by black poplars. The depositional lobe of large blocks at the head of the fan terminated abruptly at the contact with the poplar limit.

The elevated discharge concentrated in the Arás gully destroyed most of the *check dams* (*Fig. 9.61*); the principal channel incised and enlarged, liberating a large quantity of sediments that were deposited in the southern sector of the alluvial fan, forming a depositional lobe of large imbricated blocks (*Figs. 9.17* and *9.62*).

In the Las Nieves campground, which was located on this part of the fan and had opened in 1988, 87 people died and physical damages were estimated at 55 million dollars. This flash flood is the largest disaster in terms of human life that has occurred in Spain in the last 30 years. There were two stages in the development of the flood, according to the sedimentological and geomorphic analysis of the lobe of the flooded fan. The first corresponds to the discharge peak in the unconfined gully that deposited a lobe of large blocks and partially fossilized the artificial channel at the head of the fan, effectively damming the floodwaters. The second stage is associated with decreased discharge, when floodwaters incised into the detritus that had previously accumulated, creating two principal pathways.

On 19, 20, and 21 October 1982, **intense rains occurred over the southeastern region of the Iberian Peninsula** affecting particularly the basins of the Segura, Vinalopó, and Júcar rivers (know the Riada del Júcar). This rainy weather hit the Alicante coast on 19 October and moved northwest and north during the next forty hours, finally reaching the banks of the Júcar between the 20th and 21st (Piqueras, 1983). It caused numerous floods as it moved (*Fig. 9.64*). A map has been prepared on the basis of the 270 data points of this figure. The clouds covered an area of 140,000 km^2 and precipitation penetrated strongly into the interior. Maximum precipitation of 650 mm occurred in the Barranco Salado de Bicorp. Maximum precipitation reached intensities of 180 mm/hour in Alicante (Perez-Cueva and Armengol, 1983).

The Júcar basin, in its middle lower part, is a suitable setting for the concrete mechanism of easterly winds combined with a ("cold drop") at height on the vertical and existing topography that causes a funneling effect when extreme precipitation occurs (Roselló, 1983).

Figure 9.65 shows rain data; three maximum discharges can be seen; they are separated in time but all occurred on 20 October. Precipitation of 500 mm between 7 a.m. and 8 a.m. on 20 October is associated with the

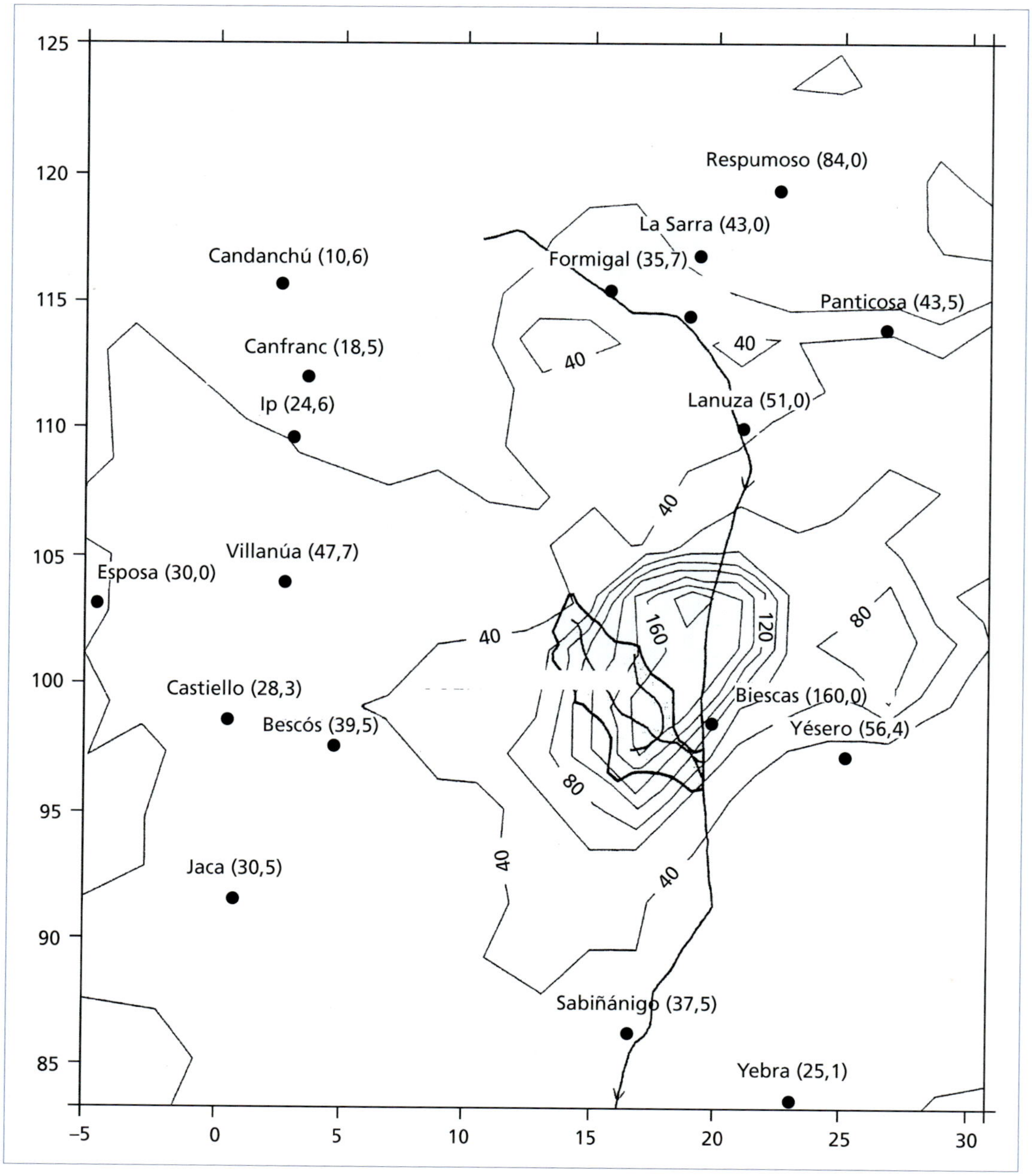

Figure 9.59 Total daily precipitation (mm) from a rain gauge (in boldface) and daily precipitation isohyet map (mm) derived from radar data in the area of the Arás basin for 7 August 1996. Note the restricted spatial distribution of the storm. The figure was re-delineated on the basis of a report from the Instituto Nacional de Meteorológia (Riosalido *et al.*, 1996).

fact that at 9 a.m. flooding began in Cofrentes; maximum discharge was estimated at 4,000 m^3/second in Escalona and 6,000 m^3/second in Millares. At this time, all the gullies covered National Highway 340 (Roselló, 1983). The other peak occurred between 9 and 12 a.m. with rainfall of more than 450 mm; and the third maximum occurred at 4 p.m. after rainfall of more than 400 mm. The catastrophic effects were primarily focused on agriculture, collapse and damage to several houses, and significant impact to communications.

In addition, the **Tous dam** failed; it had been under construction as part of the Large Dams Project of 1967. The dam was seated on limestone and conglomerates and was designed to produce hydroelectric energy, control floods, and

Figure 9.60 Vehicles and campers damaged by the 7 August 1996 flood. The photograph was taken the day after the catastrophe. Photo F. Gutiérrez.

Figure 9.61 *Check dams* destroyed by impact with large blocks transported down the Arás gully; the large blocks came from the erosion of moraines. Photo F. Gutiérrez.

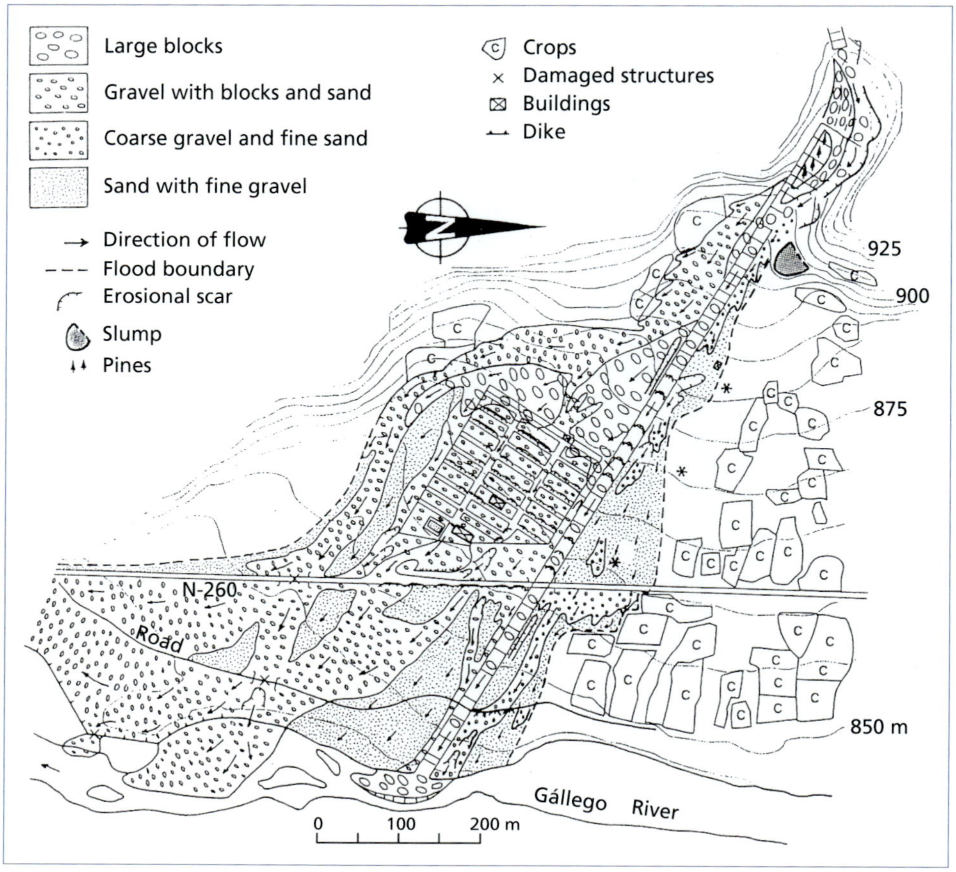

Figure 9.62 Map of the active depositional lobe of the Arás alluvial fan, showing the principal sedimentological (textural facies) and geomorphic features created by the 7 August 1996 flood. The contours correspond to the alluvial fan topography prior to the flood. Points * refer to the location of blocks deposited by previous floods (Gutiérrez *et al.*, 1998a).

Figure 9.63 Depositional lobe of large blocks and tree trunks that did not flow over the camping facilities, indicated by the poplar grove.

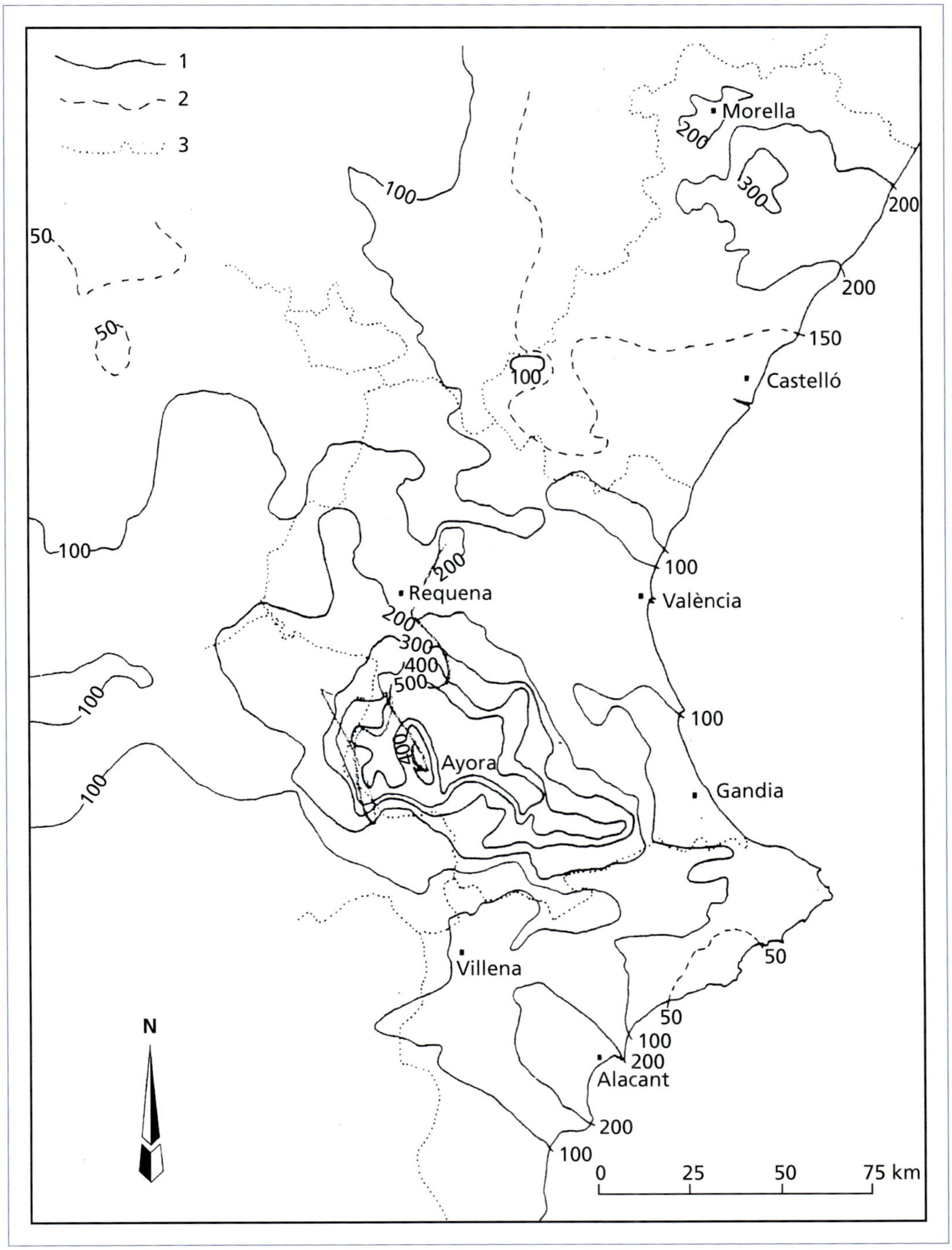

Figure 9.64 Isohyet map of rain that fell on the País Valenciano and contiguous zones from October 18–22. (1) Isohyet. (2) Auxiliary Isohyet. (3) Provincial boundary (Pérez-Cueva and Armengot, 1983).

supply water for irrigation. On the 19th of October at 6 p.m., the volume of water behind the dam was 51 h m^3 with an incoming discharge of 27 m^3/second (Piqueras, 1983). The floodgates of the spillways could not be opened because no electrical energy was available and the electric generator was not working. At 5 p.m. on the 20th, water began to overflow the dam and erode it; at 7:13 p.m. part of the dam gave way, opening a breach about 18 m. The losses from the storm of October 1982 were calculated at about 1800 million euros (Suárez and Regueiro, 1997).

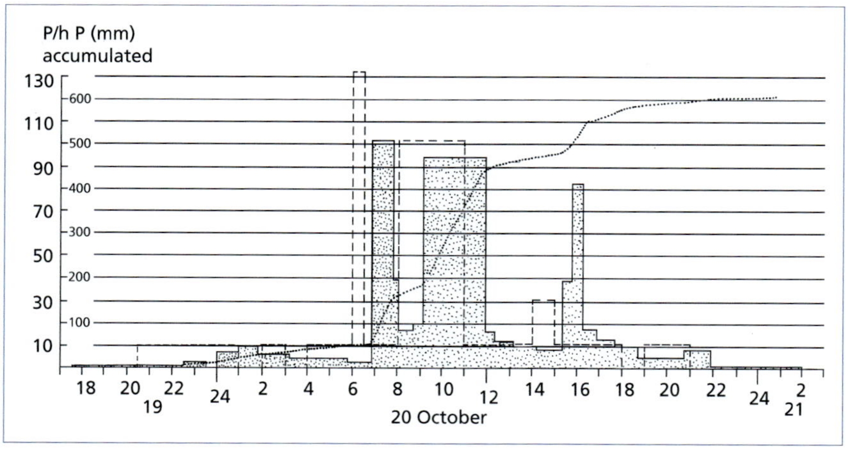

Figure 9.65 Interpretation of the Cofrentes pluviograph according to the Centro de Estudios Hidrográpicos. Hourly precipitation and accumulated rains have different scales. The first interpretation by the Comisión de Aguas del Júcar is shown in dotted lines. (Roselló, 1983)

9.8 Flood prevention and mitigation

The goal of prevention is to reduce risk by modifying the hydrodynamic characteristics of the river (Smith and Ward, 1998). Human activities such as structures in the river or modifications to it can change the fluvial channel. Land use changes within the drainage basin can also modify the channel (Knox, 2001). Human impacts on fluvial channels have increased throughout the 20th century as a result of intensive occupation of floodplains and substantial modification of fluvial channels (Downs and Gregory, 2004).

Dam construction alters flow and sediments. A wedge of sediments is deposited upstream of the dam and the downstream bed is excavated. Humans have constructed dams for several millennia. In 3100 BC, King Menes, founder of Dynasty 1, dammed the Nile near Memphis to protect the city from flooding (Downs and Gregory, 2004). Dams have been built throughout history for various reasons. The Romans built the now-filled Muel and Almonacid de la Cuba dams in the Province of Zaragoza. During the past century humans have put forth significant effort to construct dams throughout the world (*Fig. 9.66*). Spain has participated in this expansion; today there are more than 700 dams in the country (Martínez *et al.*, 1987). Large dams smooth out big floods in the downstream direction (Costa and Baker, 1981). Even so, dams do not provide flood control, but rather a "partial or a specific amount of protection" (Leopold and Maddock, 1964). Dams are also constructed to generate hydroelectric energy, provide water for irrigation, and establish recreational areas. On the other hand, dams retain significant quantities

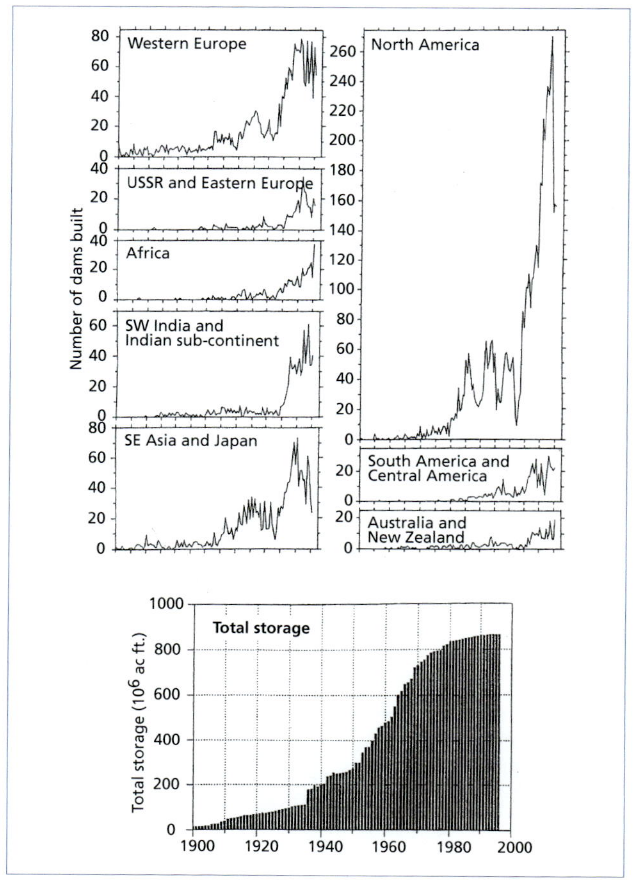

Figure 9.66 Worldwide dam construction. The upper part shows dam construction by regions (Beaumont, 1978), and the lower part show the increased reservoir capacity of dams in the United States (Graf, 2001).

of sediments, as shown by dam construction on the Colorado River (United States) (*Fig. 9.67*). At the headwaters of rivers, numerous small dams (*check dams*) can reduce erosion and control runoff to small floods that cause agricultural damage but they are not effective in controlling runoff caused by large storms (*Fig. 9.61*). Dams, especially earthen dams, may fail because they contain expansive clay and a high sodium content that interacts with construction material. Other causes of failure include failure of the bedrock, inadequate spillways, deficient construction, irregular settling, landslides, earthquakes, and others (Gruner, 1963; González de Vallejo *et al.*, 2002; Butler and Malanson, 2005). Other factors to consider include calculation of a large dam's lifespan and impacts from dam construction. The Nile Delta (Chapter 11, Section 11.12) is an example of this impact.

The construction of concrete or rock-fill **dikes** is the second most important flood protection method. Dikes are built parallel to the riverbed to contain overbanking; other types of dikes form angles (Charlton, 2008). Along the Mississippi River urban dikes have been designed to hold off floods with a return period of 200 years, whereas flood-control dikes in agricultural areas are built for a return period of 50 years (Smith and Ward, 1998). In practice, the dike height rarely exceeds 10 meters and is commonly less than 6 m.

Channelization increases the transport capacity of a natural river channel, and as a result, it contains flood peaks. Channelization involves various processes, such as increasing the river's slope or enlarging and deepening the river to increase its cross-sectional area. **Dredging** is the removal of the riverbed to create or maintain a certain volume of flow capacity. Dredging is most often done to maintain a depth that facilitates navigation on large rivers (Downs and Gregory, 2004). Extraction of minerals of economic interest in the flood plain (for example, diamonds in Tortilla, Ivory Coast) can also affect river characteristics (Schumm, 1977). **Placer** mining in the riverbed degrades the bed; during later storms this degradation can cause a *knickpoint* that moves upstream. The migration of a dredging-induced *knickpoint* can erode bridge footings (Downs and Gregory, 2004). **Enlargement** of the channel can partially prevent instability problems along the banks; these problems result from channel deepening, especially in easily eroded materials. Another important flood-alleviating practice consists of **trenching** (*Fig. 9.68*); it eliminates obstacles to water flow such as vegetation, which causes flooding. Rivers are occasionally channelized, such as the Pico and Vena rivers, which had caused significant flooding in the city of Burgos (Spain) prior to channelization. A more grandiose example is that of the South Solution of Valencia where a wide new riverbed was constructed to prevent flooding of the city. The river had moved away from the ancient riverbed and it was later developed as a park for urban use (Martínez *et al.*, 1987). One direct method is to modify meandering streambeds. Along the lower Mississippi 15 neck cutoffs occurred between 1929 and 1942, reducing the length of the river by 240 km and lowering flood levels by 2–2.5 m (Smith and Ward, 1998).

Floods along large rivers such as the Amazon occur every year; in urban areas like Manaus, parallel dikes are used to protect against high water levels. Small houses on the river bank are elevated about 6 meters above the normal discharge to prevent water from flooding them (*Fig. 9.69*). This is also done in Malaysia (*Fig. 9.70*).

Although urban areas occupy less than 3% of the earth's surface, the effect of **urbanization** on flood hydrology

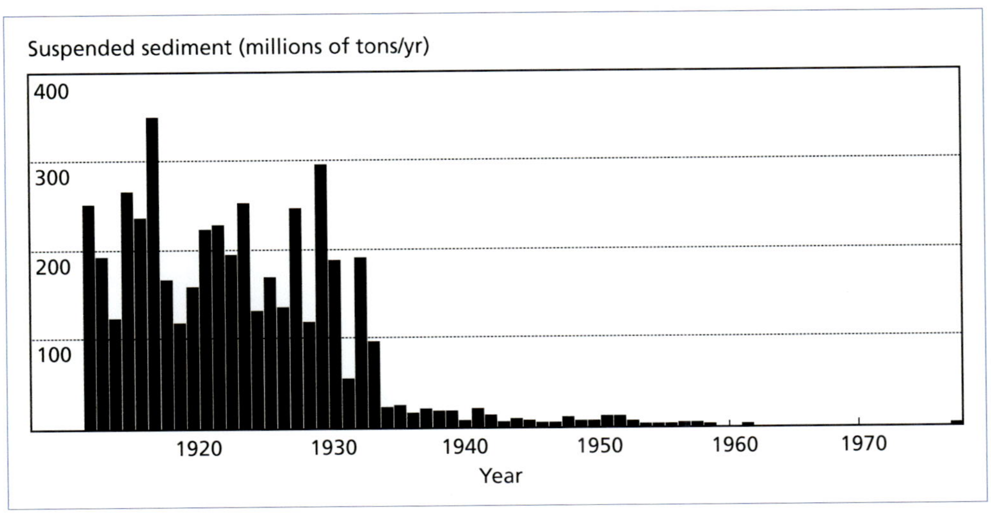

Figure 9.67 Historical sedimentation in the Colorado River (United States), in millions of tons per year (Schwarz *et al.*, 1990).

Figure 9.68 Trenching of the Ribera de Alcazaba (Badajoz, Spain).

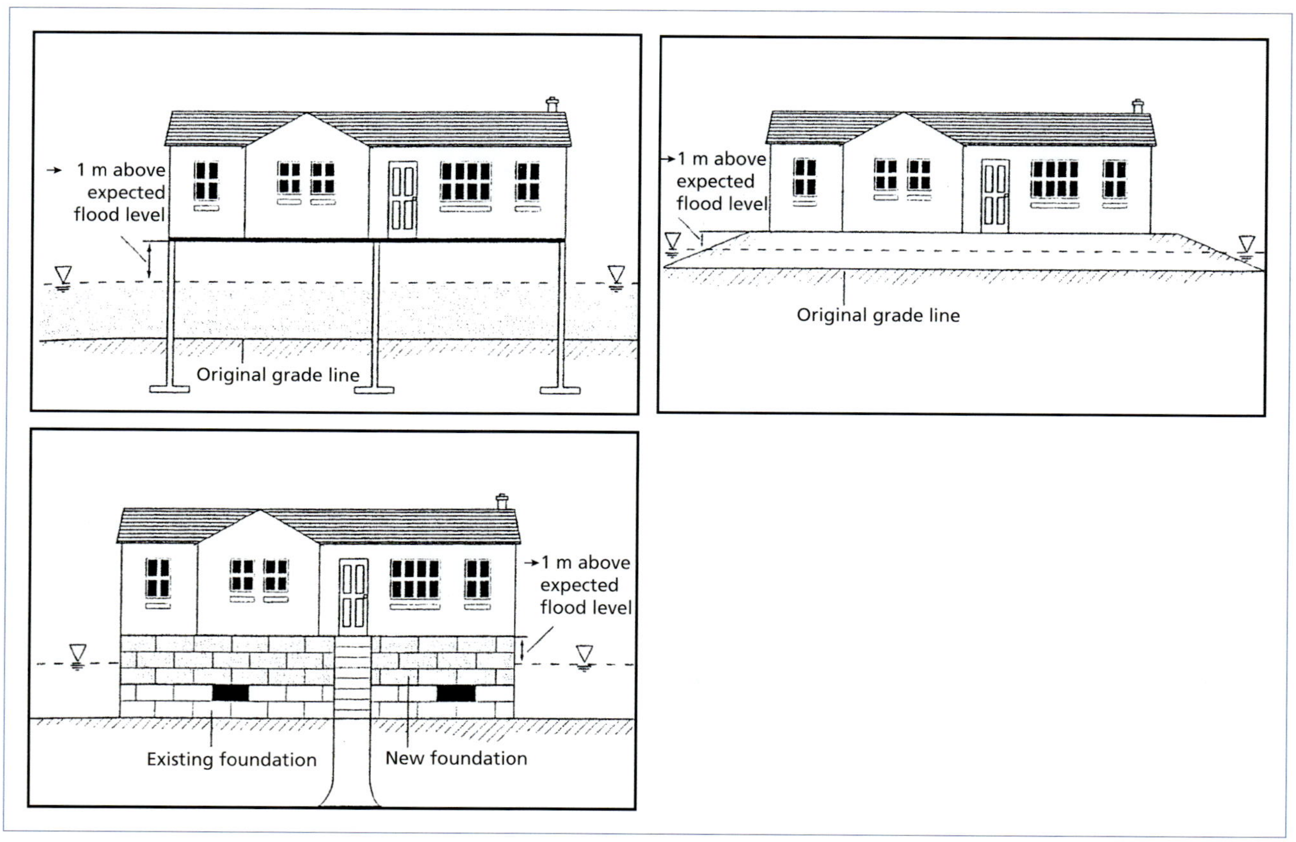

Figure 9.69 Houses protected from flooding by elevation (Dozier and Nancey, 1993).

Figure 9.70 Houses elevated for flood protection. Mulu, Sarawak, Malaysia. Photo F. Gutiérrez.

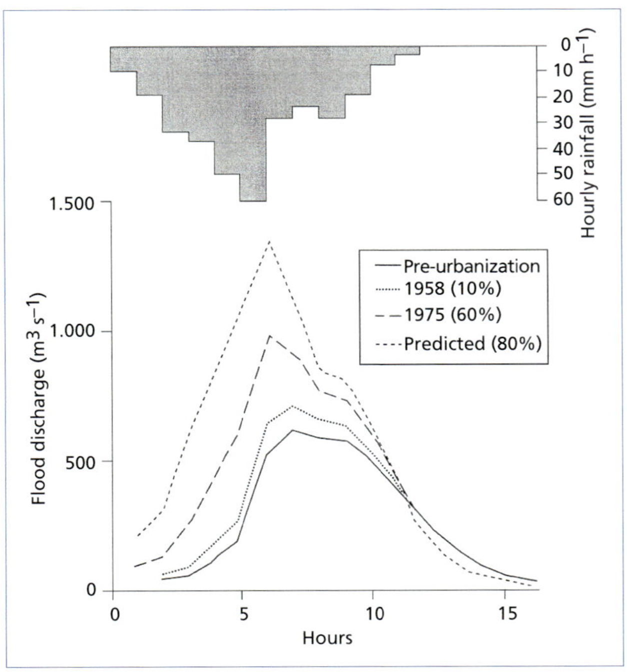

Figure 9.71 Flood hydrographs that show the expansion of urbanization (shown as a percentage of basin without urbanization) in the western part of the Tokyo Japan megalopolis (Yoshimoto and Suetsugi, 1990).

and flood risk is disproportionately large. Urbanized surfaces are less permeable than the surfaces that they replace. As a result water runs off rapidly and hydrographs have more pronounced peaks than they previously had (Leopold, 1968) (*Fig. 9.71*).

A good understanding of flood conditions is essential to developing effective **mitigation**. Mitigation and notification projects are becoming more common; they are integrated into planning and alarm systems that operate in real time through the use of computers. All these efforts make it possible to reduce loss of life and economic damage caused by flooding. Flood situations should be communicated to people so that they take protective measures prior to the arrival of floods (Smith and Ward, 1998).

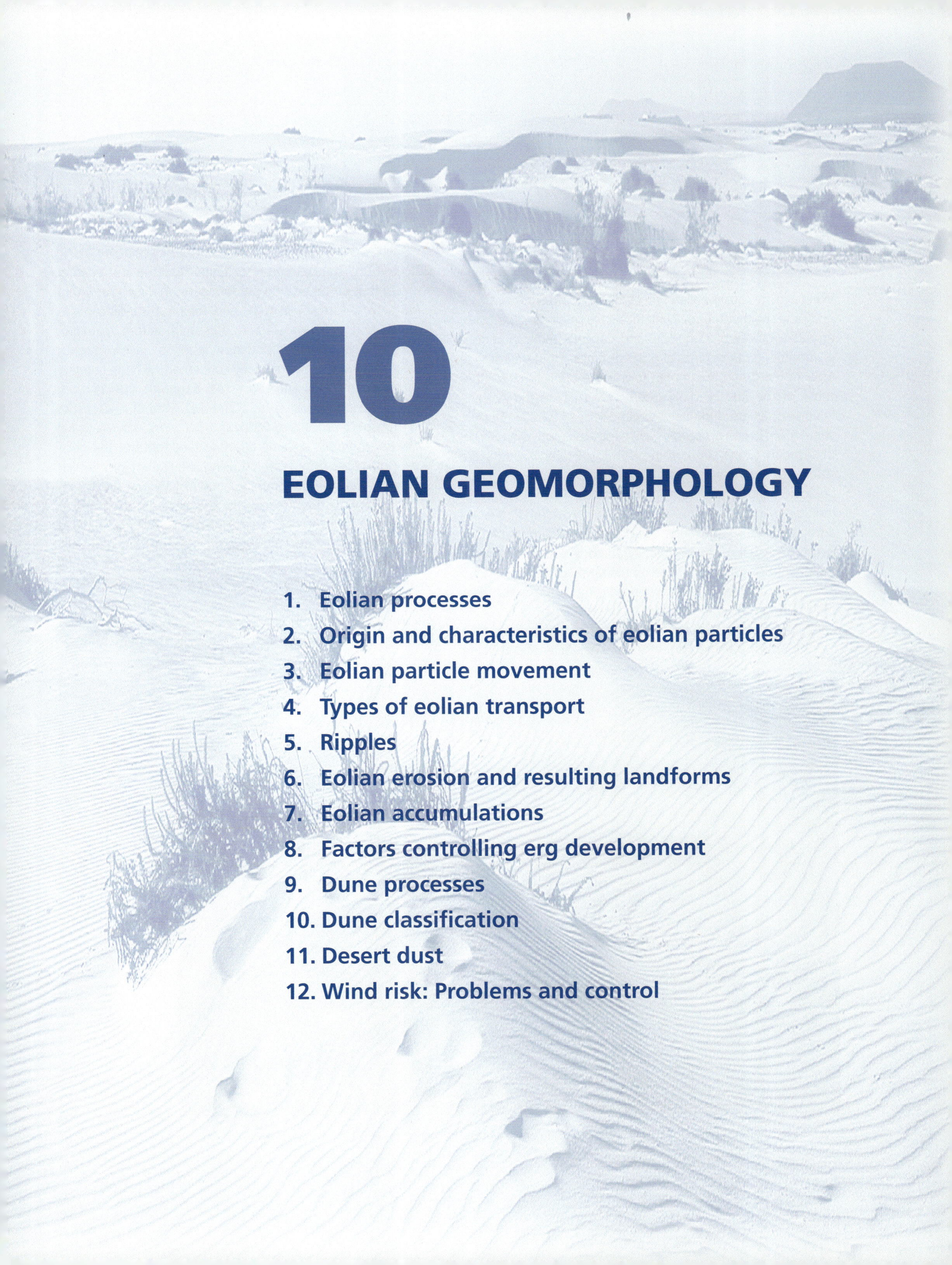

10

EOLIAN GEOMORPHOLOGY

10.1 Eolian processes

Wind activity causes eolian processes including erosion, transport, and sedimentation. Wind mobilizes small-sized particles, erodes a variety of materials, and deposits them in various environments, sometimes at great distances from their source area. In the desert, winds can form large dune fields such as those of the Rub al Khali desert (*Fig. 10.1*) in the Arabian Peninsula, where dune fields cover about 560,000 km^2 (Allison, 1997). Eolian processes have deposited more than 300 m of loess in China (Derbyshire, 1983a). Eolian deposits are studied by a variety of scientific disciplines, including physics, geology, biology, and agronomy (Pye and Tsoar, 1990). This multidisciplinary scientific research has provided significant advances in our knowledge of eolian activity and its effects. Recent research on eolian geomorphology deals mainly with the quantification of processes and analysis of landforms and internal structures of eolian deposits; the information they provide helps us understand past climates.

During the 19th century, wind was considered to be a less important transport agent than water. Scientific thinking changed at the end of the 19th century and beginning of the 20th century. The German researchers Passarge and Walther showed that eolian erosion predominates in deserts and that wind activity forms inselbergs and their surrounding plains (Goudie and Wilkinson, 1977). Research shifted to exploration of new territories and description of landforms but researchers were not particularly concerned with the connection between forms and processes. Bagnold's (1941) work was a milestone because he quantified wind characteristics and transport processes. Following his work, conventional aerial photographs became an important tool but the decisive boost came from research using satellite images, which clearly showed the geomorphic action of the wind (McKee, 1979a). These techniques make it possible to analyze yardang systems, which can cover areas up to 100,000 km^2 (Libyan desert, Tibesti massif, Lut desert in eastern Iran, and others). Recent studies of orbital images from Mars and Venus clearly show the importance of wind processes on these planets (Greely and Iversen, 1985). The search for hydrocarbon discoveries in Mesozoic eolian sandstones in Europe and the USA was also a stimulus for this research. The large number

Figure 10.1 Dune Fields Al Liwa, Rub al Khali Desert (United Arab Emirates).

of absolute dates determined by ^{14}C and luminescence techniques since 1980 has also increased our knowledge of the eolian domain.

During the 1980s and 1990s, research focused primarily on field studies of wind and sand flow in individual dunes; in the mid 1990s researchers analyzed turbulent structures in flow because of their importance in explaining sandy flow (Livingstone *et al.*, 2007) The methods used are explained in Knott and Warren (1981). Geomorphologists increased their understanding by incorporating more complex numerical models and using computers to investigate fluid dynamics (CFD) and by studying radar images to understand the internal structure of dunes.

Wind also acts as a geomorphic agent in cold climates. In these climates, abrasion and deflation are erosional agents that clearly affect the genesis of dune fields and asymmetrical valleys (Seppälä, 2004).

10.2 Origin and characteristics of eolian particles

Materials transported by wind often have a diverse composition that includes mineral and rock fragments and organic particles such as shell remains. The grain sizes of the mobilized particles are usually silt and clay but sometimes clay aggregates can be transported (if they contain salts) (Bowler, 1973; Wasson, 1983). Quartz is the most common material because it is abundant at the Earth's surface and resistant to chemical weathering (Bagnold, 1941). Feldspars and heavy minerals are also present at lesser percentages; they are subject to gradual weathering.

The primary physical characteristics of eolian particles that determine their ability to be eroded and transported by wind are specific weight and particle size and shape (Watson, 1989a). The various specific weights of particles are determined by mineralogical differences. Groups of eolian particles commonly show a bimodal or polymodal grain-size distribution, which indicates the influence of mineralogy on mobility (Willets, 1983). Eolian sand accumulations are usually well sorted although they can vary significantly in both size and sorting in different deserts, within the same erg, and even in individual dunes (Watson, 1989). Grain shape also affects mobility; flat shaped particles are less likely to be mobilized than rounded grains (Willets, 1983). Quartz sand grains from desert dunes are generally sub-rounded in shape; in an eolian accumulation, larger particles are more rounded than smaller ones. Abrasion during transport determines the degree of roundness (Krinsley and Wellendorf, 1980). Eolian particles are rounder than those of glacial, fluvial, and littoral environments.

The origin of particles in desert eolian accumulations is highly variable (Demangeot, 1981; Pye and Tsoar, 1990). Wind-transported particles may be weathered grains entrained from the surface of a rock but most particles come from unconsolidated materials. On continental areas these materials occur in desert river bottoms and their terraces, pediment surfaces and alluvial fans, lacustrine terraces, and sebkha or playa floors, and volcanic ash. Marine beaches are also an important source area for coastal eolian belts (*Fig. 10.2*). During periods of glaciation and the resulting gradual sea-level decline, large submarine surfaces became subaerial regions and some have been affected by eolian activity.

In addition to transporting sand-sized particles, wind can also mobilize silts that can travel as dust storms over distances of thousands of kilometers.

Figure 10.2 Aerial view of the dune belts at the mouth of the Guadalquivir River. Parque Nacional de Doñana (Province of Huelva, Spain).

10.3 Eolian particle movement

Wind transports sands in different ways, some of them complex. Opposing forces such as gravity, friction, and cohesion affect loose particles being moved by the wind (Iversen et al, 1987). Wind speed increases above any ground protuberance and the increase is accompanied by a pressure decrease according to the Bernoulli equation. As a result a loose particle can be **uplifted**; this force is more intense at the ground surface where maximum wind speed occurs (Cooke *et al.*, 1993). In addition, lifting is much more significant in dust turbulences. **Drag** includes surface creeping, which is friction between the particle and the air; it causes rolling. Drag also includes shape drag, which results from the pressure difference between the windward and leeward side of the particle; it causes rolling and sliding (Livingstone and Warren, 1996). Once grains are airborne, **bombardment** occurs between a particle and the substrate and between grains (collisions). These processes are accompanied by drag and lifting of loose particles.

An important parameter to know is the minimum speed required to move particles of a particular size. Entrainment is a function of the average particle size and the wind's shear velocity. Bagnold (1941) studied this relationship and calculated values for critical shear velocity thresholds (U_*) for various particle sizes. This relationship is shown in *Figure 10.3*, which shows that a higher velocity is required to mobilize larger particles. Even so, due to electrostatic cohesive forces, the smallest particles (<0.06 mm) also need high wind velocities to be entrained. Sand size particles (0.04–0.40 mm) are the most easily mobilized. As lifting and drag energy increases, a critical wind shear velocity value is required to move particles. This is the **fluid threshold** (or **static threshold**). When the grains are entrained, ballistic impacts reduce the energy required to keep the particles moving. This energy is the **impact threshold** (or **dynamic**), which requires less velocity to move a stationary grain by fluid pushing; this threshold is equivalent to 80 percent of the static threshold (Anderson and Haff, 1988). These curves apply to flat, homogeneous, dry surfaces with isolated loose clasts (Williams *et al.*, 1994). Sediment variability, surface crusts, slope inclination, humidity, and roughness also affect threshold values (Wiggs, 1997).

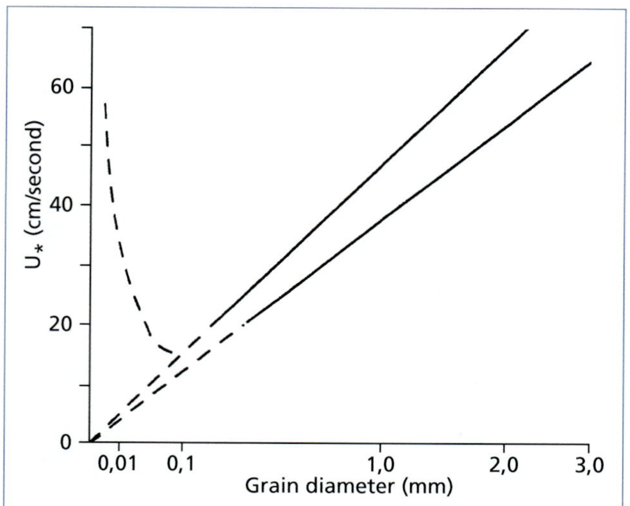

Figure 10.3 Relationship between shear velocity thresholds (U_*), particle size, and fluid and impact threshold (Bagnold, 1941).

10.4 Types of eolian transport

When particles exceed the mobilization threshold, they can be transported in four different ways: creep, reptation, saltation, and suspension (*Fig. 10.4*). The transport method depends mainly on the grain size of available particles (Bagnold, 1941). The boundaries between these different transport types are not sharp and the transition from one to another is not very clear.

Creep involves translation or turning of coarse particles (0.5–2 mm) on the surface; particles are pushed by the force of the wind or by the ballistic impact of grains during saltation. Creep can begin several moments before particle mobilization by saltation (Nickling, 1983). Determination of the amount of material that moves by creeping is difficult because of the difficulty of distinguishing creep from reptation. However, it is estimated that creep represents one quarter of the mobilized bed load.

Recent research has defined **reptation** as a transitional type of movement between creep and saltation; in reptation, particles move in small jumps caused by the impact of high-energy grains (Anderson and Haff, 1988). Grains can lose 40 percent of their velocity as a result of impact and this energy can be transmitted to other grains and can move them from the bed. Reptation differs from creep in that during reptation, particle transport continually changes from reptation to saltation.

Saltation (*Fig. 10.5*) involves the uplift of a particle (approximate size 0.1 mm) that travels a distance along a ballistic trajectory (Bagnold, 1941) and then falls and hits the surface. The impact energy lifts other grains. Therefore, saltation is an important transport mechanism with cascading effects (Nickling, 1988). Uplift angles range from 30° to 50°, and speeds are higher on the side facing the wind. According to different models and experiments, most particles fall at angles between 10° and 15° (Anderson, 1989). Saltation distances are 12 to 15 times the height (Livingstone and Warren,

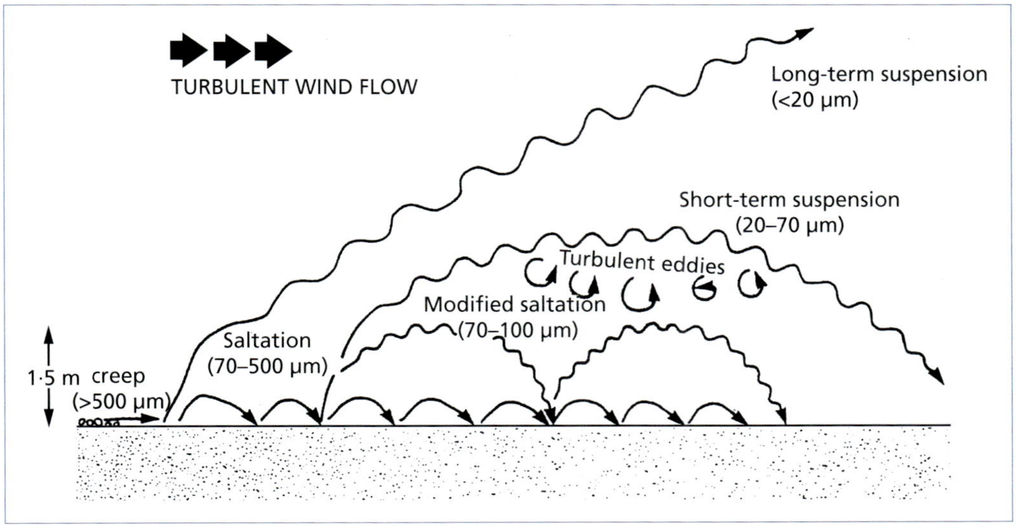

Figure 10.4 Types of Eolian Transport. The range of particles size is shown for the various types of transport (Pye, 1987).

Figure 10.5 Sand grains moved by saltation. Al Liwa, Rub al Khali Desert (United Arab Emirates).

1996); smaller particles, which are entrained at higher velocities, can travel higher and thus their saltation distance is longer (Anderson and Bunas, 1993). Saltation heights can reach 3 m but the average is 0.2 m (Pye and Tsoar, 1990). Bagnold (1941) recognized that saltation is higher over a rocky surface or a surface with pebbles than over a bed of loose sand *(Fig. 10.6)*.

Pure saltation occurs when the turbulent vertical velocity has no significant effect on particle trajectories. Between pure saltation and pure suspension, the particle trajectory is modified due to settlement velocity and inertia. This type of transport has been called "**modified saltation**" *(Fig. 10.7)* (Nalpanis, 1985). This transitional classification shows that no sharp transition exists between saltation and suspension.

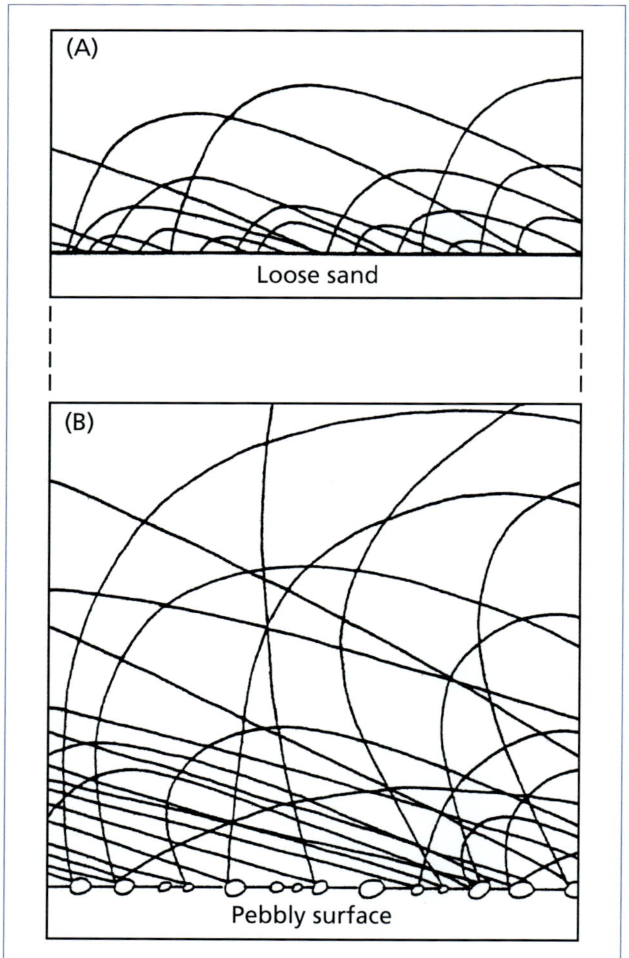

Figure 10.6 Schematic representation of saltation trajectories of (A) loose sand and (B) surface with pebbles (Bagnold, 1941).

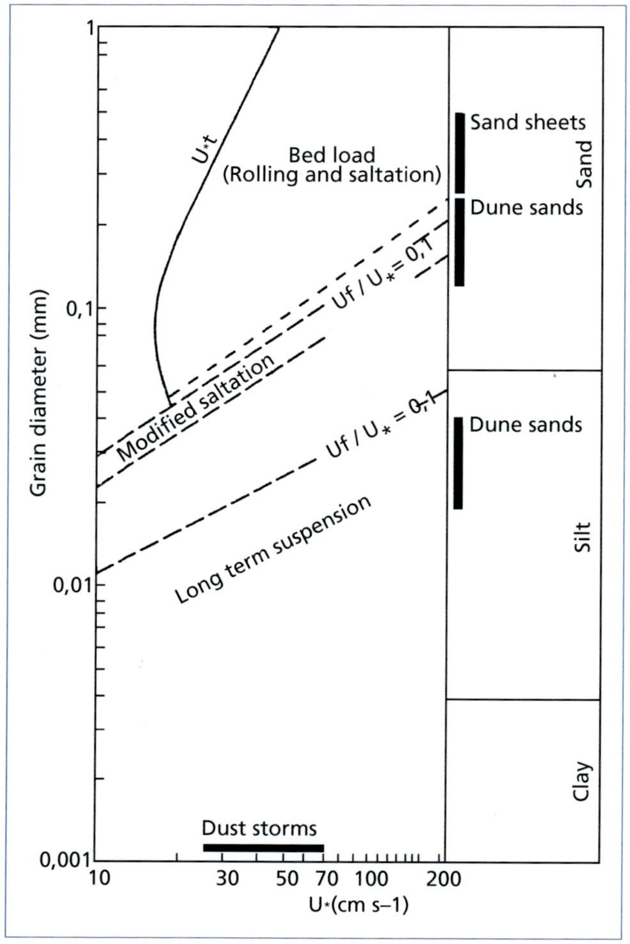

Figure 10.7 Relationships between grain size, shear velocity, and type of sediment transport. U* is the fall velocity of the particle (Tsoar and Pye, 1987).

The smallest particles (<0.06 mm) are transported by suspension within turbulent airflow *(Fig. 10.4)*, which can keep fine sediment at a considerable height for many days; it is finally deposited as dust or loess. This sediment can travel more than 300 km and create a deposit between 30 and 60 mm over large surface areas. Within suspension transport we can distinguish grains that remain airborne for long periods of time (**long term suspension**) from those that fall quickly on the surface (**short term suspension**) (Tsoar and Pye, 1987) *(Figs. 10.4* and *10.7)*.

10.5 Ripples

Ripples, the smallest eolian landforms *(Fig. 10.8)*, cover dunes; they are the initial response of sand surfaces to wind transport. Although they are depositional forms, they will be analyzed in this chapter because they are closely related to the previously described types of transport. Ripples form perpendicularly to the wind transport direction even though they flow obliquely on slopes due to gravity. Most of them are ephemeral and move much more rapidly than dunes (Livingstone and Warren, 1996). Ripples are usually asymmetrical in cross section. The stoss side is convex with angles of 8–10° and the lee side has a slope angle between 20° and 30° (Sharp, 1963). Some ripples have sharp crests while others have rounded ones. Wavelengths range from a few centimeters up to tens of meters and heights range from 1 cm to about 30 cm. Particle size determines wavelength and height; coarse sand ripples have a greater distance between crests than fine sand ripples (Sepälä and Lindé, 1978).

Mega-ripples are remarkable landforms with wavelengths that range from 1 to 25 m. They are created by strong winds and are composed of coarse sand. In cross section they

Figure 10.8 Ripples in Al Liwa, Rub al Khali Desert (United Arab Emirates).

are generally more symmetrical than smaller ripples; the contrast in grain size between crest and trough areas is higher than in smaller ripples. Bagnold (1941) thought several centuries were needed to build mega-ripples but Sharp (1963) estimates that of few hours of a windstorm are sufficient. **Chiflones** are wavy ribbons of coarse and fine sand, 100–150 m long and 1–1.5 m wide (Simons and Eriksen, 1953). They form when coarse sands are channelized between obstacles such as shrubs.

Several models have been proposed to explain ripple formation. The first model (Bagnold, 1941) established a relationship between ripple wavelength as observed in experimental tunnels and the length of sand paths during saltation. According to Bagnold, irregularities on the surface act as erosional and depositional sites. The stoss side is bombarded by particles transported by saltation that transfer energy as they collide; as a result they mobilize new grains and the process continues. Trajectories have the same length as ripples when sand size and wind speed remain the same. This hypothesis is very attractive because it explains many characteristics and variables related to ripple formation. Sharp (1963) thinks that particles in ripples move primarily by reptation. Irregularities on the bottom surface and interactions between moving grains form local highs on the former surface, creating incipient short-wavelength, low-height ripples that grow progressively bigger. A third model, proposed by Anderson (1987), is based on experimental data and numerical simulations. This model establishes differences between low-energy impact reptation and long trajectory high-energy impact saltation. It identifies different saltation paths, some of them longer than ripple wavelengths. This model contrasts with Bagnold's model because saltation impact does not contribute directly to ripple formation. According to Anderson's work, reptation processes are dominant. Because reptation distances increase with wind shear strength, the ripple wavelength also increases. This agrees with data obtained in experimental tunnels (Seppälä and Lindé, 1978).

In conclusion, eolian ripples are closely related to sand saltation although new models suggest that reptation provides the initial conditions for ripple formation, which quickly evolve as a result of saltation impacts and by grains transported by reptation (Lancaster, 1995).

10.6 Eolian erosion and resulting landforms

Wind creates eolian erosional landforms by lifting particles from the surface and causing them to make impact with rocks at significant speeds. These landforms are more common in areas of scarce or absent vegetation, especially in arid zones, but they also occur in periglacial and marine beach environments.

Wind erodes by **deflation**, that is, by entraining loose material from the ground surface and transporting it. The average deflation rate is about 1 mm/year (Cooke *et al.*, 1993). **Abrasion** is the mechanical wearing down of cohesive material. Ballistic impacts by grains causes surface rock to fracture until the surface becomes stabilized and ceases to fracture, which suggests that the abrasion rate reaches a maximum and then decays. Most authors think that only sand grains cause abrasion but others maintain that eolian dust also abrades although to a lesser extent (Greely and Iversen, 1985). Abrasion rates are highly variable, about several millimeters per year; rates depend on rock type, wind speed, and grain supply. The maximum abrasion height ranges between 0.1 and 0.4 m. These figures were obtained from field experiments during intense storms (Sakamoto-Arnold, 1981) and from theoretical models (Anderson, 1986).

10.6.1 Ventifacts

Ventifacts are clasts or blocks with smooth surfaces that are polished and faceted by the impact of wind-transported particles (*Fig. 10.9*). They are associated with pits, flutes, grooves, and helicoidal patterns that develop on both boulders and rock outcrops. These micro-morphologies appear only in rocks that are resistant to weathering. They are found in cold and hot deserts and in coastal environments (Knight, 2008).

Facets develop perpendicularly to wind direction. The number of facets varies; researchers have described up to twenty (Higgins, 1956). More common, however, is the presence of two or three intersecting facets that form edges. The German words **einkanter**, **zweikanter**, and **dreikanter** are commonly used to indicate that they have one, two, or three facets respectively. Ventifact surfaces are often polished and different types of microforms develop on them (Whitney and Dietrich, 1973; Laity, 1995); they have been studied by electron microscope (Whitney, 1979; Sellier, 2006). **Pits** are small closed depressions created by wind action on non-homogeneous rock or on pre-existing holes. Pits are inclined 55° to 90° from the wind direction and as a result they occur on the stoss side. Where the angle decreases considerably, we find a transition to **flutes** (Sharp, 1949). These are spoon-shaped forms with dead ends and U-shaped cross sections. They are commonly open to the stoss direction. Their length, width, and height vary greatly, from millimetric sizes up to centimeters (Cooke *et al.*, 1993). They form on horizontal or slightly sloping surfaces in very different kinds of rocks and are not related to structure (Breed *et al.*, 1989; Laity, 1994). **Grooves**, which are open at both ends, are longer than flutes. Like previously described forms, they indicate the predominant wind direction in the area. If the rock composition is not homogeneous, the wind erodes softer areas more

Figure 10.9 Ventifact. Central sector of the Duero Depression (Spain).

intensely. The forms that result from the presence of inhomogeneities are called **etchings**. Other associated microforms are the **helicoidal forms** developed on marble in the Mojave Desert of California. They are shallow grooves with spiral morphologies related to the wind direction. They range from several millimeters up to several centimeters wide and deep. They occur in areas of very high wind speeds (Laity, 1995).

The origin of ventifacts has been studied in both field research (Sharp, 1964, 1980) and the laboratory (Kuenen, 1960; Whitney and Dietrich, 1973; Whitney and Splettstoessar, 1982; Knight, 2008) for different particle sizes. Ventifacts result from impacts of particles transported by saltation and suspension; sand is the most important factor in their configuration (Laity, 1994). However, ventifacts are not often found in dune fields or surrounding areas although they are common in desert pavements with scarce sand.

Preliminary research indicated that facets were inherited forms or that they at least controlled subsequent abrasion (Higgins, 1956; Kuenen, 1960). Other explanations point to an origin resulting from winds blowing from different directions and making impact with the rock (Kuenen, 1960). This hypothesis can explain some ventifacts but in mountain zones, winds blow primarily from a single direction (Sharp, 1949). One well-accepted theory is based in the rotation of pebbles around a vertical axis, with overturning that results from excavation beneath the pebble by deflation. The overturning can also result from expansion and drying of clays and by animal activity (Livingstone and Warren, 1996). This explains why smaller ventifacts have more facets and vice versa. It also accounts for the occurrence of ventifacts with facets on both the upper part and the bottom. Some authors point out that these micromorphologies (polished surfaces, pits, flutes, grooves, and helicoidal forms) are probably caused by a vortex of eolian dust (wind rotating around an axis) rather than by sand impact (Maxson, 1940; Higgins, 1956). Wind tunnel experiments can reproduce these micromorphologies (Whitney and Dietrich, 1973; Dietrich, 1977; Whitney, 1979). In any case, sand grain impact and eolian dust erosion can explain most shapes related to ventifacts (Breed *et al.*, 1989, 1997).

Wind direction can be determined from ventifacts and their associated forms; for this reason, fossil ventifacts provide good records of wind circulation during past geological periods (Knight, 2008; Gillies *et al.*, 2009). If they are also dated using desert varnish techniques (Dorn, 1995), it is possible to determine paleoenvironmental conditions and the intensity of eolian activity.

10.6.2 Yardangs

The word **yardang** comes from the eastern limit of China's Taklimakan desert. A yardang looks like the hull of a capsized ship *(Fig. 10.10)* although yardangs often have flat tops

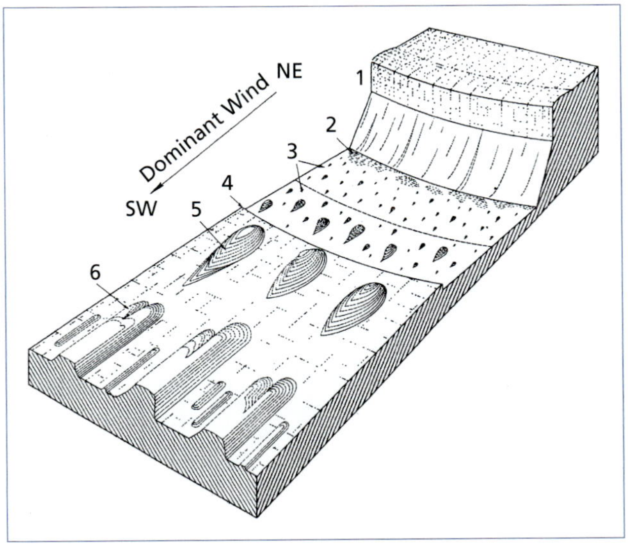

Figure 10.10 Diagram of the general situation of yardangs in the Borkou area, Tibesti Massif (Republic of Chad). 1. Scarp that separates the depression from an upper surface. 2. Lacustrine zone, sometimes subactual. 3. Small yardangs in the depression. 4. Scarp toward the lower surface. 5. Yardangs on the margin of the platform. 6. Crests and sandy corridors (Mainguet, 1972).

and lack a keel. The windward face is wider and higher than the leeside. Yardangs have a variety of forms *(Fig. 10.11)*. Halimov and Fezer (1989) describe eight types of yardangs in the hyperarid Quaidam depression of central China. Their classification is based on morphology and they suggest a temporal succession in the genesis of yardangs. The length/width ratio of a yardang ranges from 3:1 to 10:1 (McCauley *et al.*, 1977). Yardangs usually occur in groups with their long axes parallel to the direction of the most intense winds. They occur in all large deserts of the world except the Australian desert. They have also been identified on Mars (Greeley and Iversen, 1985). Some authors (Cooke *et al.*, 1993) classify yardangs according size: microyardangs (1 m) *(Fig. 10.12)*, mesoyardangs (10–100 m), and megayardangs (1 km). The classification by Livingston and Warren (1996) seems to be more useful: landforms up to 100 m long are called yardangs and longer shapes that form ridges (Mabbutt, 1977, Laity, 1994) are called megayardangs *(Fig. 10.13)*. Cross sections of megayardangs are U-shaped with the bottom as the corridor if there is no fluvial channel (Blackwelder, 1934). Sand, pavements, and sometimes barchan dunes parallel to ridges may cover them (Laity, 1994, 2008). Megayardangs are well developed in the southern part of the Tibesti Massif (central-eastern Sahara) where the ridges can reach a height of 200 m, a length of 4 km, and can be separated from each other by distances of 0.5 to 2 km. They cover a surface area greater

Figure 10.11 Keel-shaped yardangs that have developed on Miocene gypsum-bearing deposits. In the background, La Playa lake, platform of exhumed Bujaraloz and Miocene sediments superimposed on the Sierra of Santa Quiteria. (Ebro Depression, Spain).

Figure 10.12 Microyardangs in a cemented dune. Matti Sebkha (United Arab Emirates).

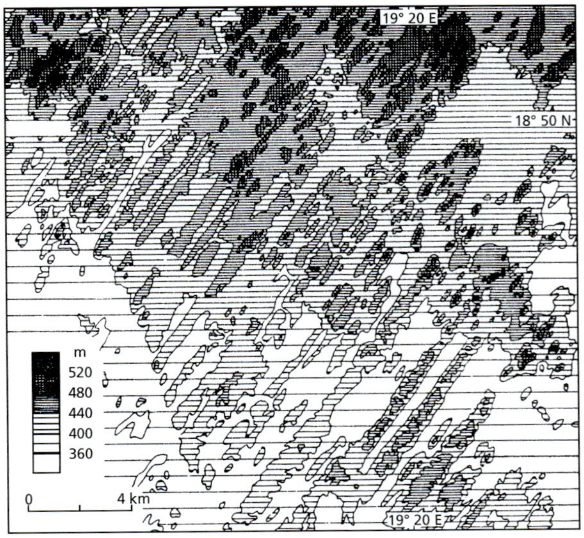

Figure 10.13 Sandstones worked by the wind, resulting in a crest-and-corridor morphology. Lower zones of Borkou, south of the Tibesti Massif (Republic of Chad) (Hagedorn, 1968).

than 600,000 km^2 (Hagedorn, 1968; Mainguet, 1968, 1972). Very large groups also occur in the hyperarid deserts of Egypt, Iran, Arabia, central Asia, Namibia, Perú and the southern Andes (McCauley *et al.*, 1977; Goudie, 2007). Yardangs are generally carved into soft materials such as consolidated eolian sands, fluvial sediments, and Cenozoic lacustrine deposits although many Saharan ridges occur on Cambrian sandstones. They can develop in a wide variety of lithologies from semi-consolidated sediments to granites or even harder rocks (Goudie, 1989b, 2007). The direction, dip, and jointing of the rock also affect the shape and development of yardangs. Yardangs also occur on gypsum and continental Miocene limestone in the central sector of the Ebro Depression (Spain) in a typically semi-arid environment (Gutiérrez *et al.*, 2002). Here, yardangs occur on the lee side of large playas (*Fig. 10.14*) and are oriented N122E, coinciding with the prevalent wind direction ("*cierzo*") (northwest wind). Yardangs also develop in unconsolidated lacustrine deposits (Gutierrez *et al.*, 2002). Yardangs of the ridge-and-swale type composed of Miocene gypsums and argillites are present within these depressions (Quinto de Ebro, province of Zaragoza, Spain).

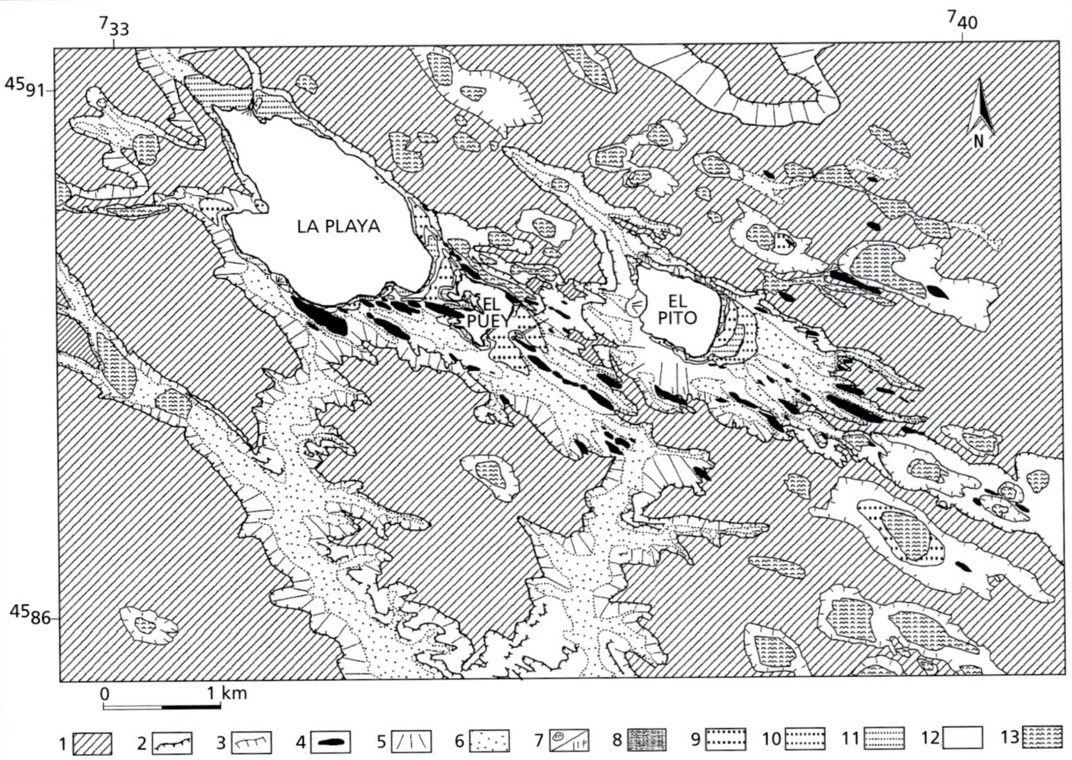

Figure 10.14 Geomorphic map of playa-yardang systems about 5 km south of Bujaraloz (Province of Zaragoza, Spain); (1) structural platform; (2) scarps; (3) margin of the depressions; (4) yardang; (5) slope deposits; (6) corridors and flat-bottom valleys; (7) alluvial fans and gullies; (8) lacustrine terrace T_3; (9) lacustrine terrace T_{2a}; (10) lacustrine terrace T_{2b}; (11) lacustrine terrace T_1; (12) playa bottom; (13) bowl-shaped sinkhole (Gutiérrez *et al.*, 2002).

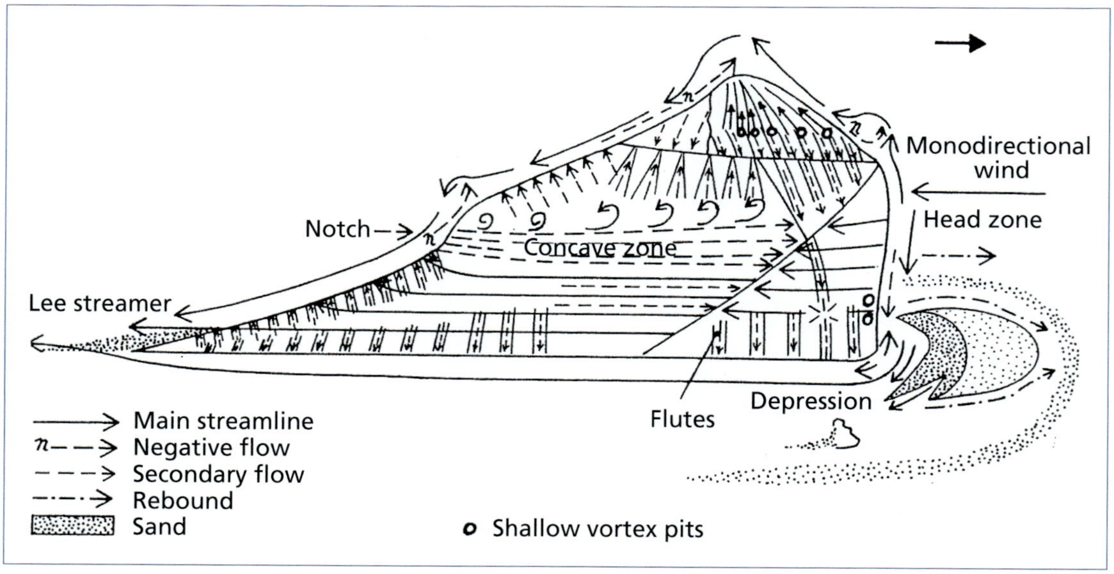

Figure 10.15 Flow line systems derived from a small yardang in Black Hiel southwest of the hyperarid desert in Egypt (Whitney, 1983).

Abrasion and deflation processes probably form yardangs (McCauley *et al.*, 1979; Ward and Greely, 1984) with slight modifications caused by water erosion (Laity, 1994). Abrasion contributes to the polish and causes basal excavation to depths of one or two meters (Grolier *et al.*, 1980). Deflation is important in the development of yardangs composed of poorly consolidated lithologies. Similar to ventifacts, a vortex of dust particles seems to be an important process in shaping yardang morphology (Whitney, 1978, 1983, and 1985) (*Fig. 10.15*). Many yardangs have previously undergone fluvial erosion; they are more elongated if the drainage network is parallel to the prevailing winds (Embabi, 1999; Laity, 1994, 2008) and the direction of fracturing (Gutiérrez *et al.*, 2002). Fluvial-lacustrine activity predominates during some intervals and at other times eolian activity predominates; these are periods of alternating deposition and sediment transport in the troughs (Mainguet, 1972; Gutiérrez *et al.*, 2002). Finally, wind tunnel experiments (McCauley *et al.*, 1979; Ward and Greeley, 1984) using synthetic sediments show that it is possible to identify positive main fluxes and secondary negative return flows along the flanks. These researchers conclude that abrasion and deflation are the fundamental processes in yardang formation (*Fig. 10.15*) (Whitney, 1983).

10.6.3 Deflation basins

Wind excavates closed depressions of various sizes. Chapter 17 discusses the origin of closed depressions and analyzes some fundamental features that characterize so-called erosional basins. In this section we will focus on aspects related to their geometry, distribution, and origin.

Deflation basins are closed depressions that may cover 12 percent of large surfaces (Cooke *et al.*, 1993). They are usually circular, elliptical, or kidney-shaped and have rounded margins. They are abundant in semiarid zones of the world and are well developed in central and southern Africa, the central United States, the Pampas and Pantanal of South America, southern and western Australia, Manchuria, and western Siberia (Goudie, 1991; Goudie and Wells, 1995). They range considerably in size; the surface area of these basins in southwestern Australia ranges from 0.004 to 100 km^2 with an average size of 0.05 km^2 (Killigrew and Gilkes, 1974). In southern Africa they range from 0.05 to 30 km^2 with an average of 0.2 km^2 (Goudie and Thomas, 1985). These authors calculate densities of 100 basins per 100 km^2.

Deflation basins develop preferentially in poorly consolidated sediments. Certain lithologies such as the sandstones of southern Africa are highly suitable for the development of these closed depressions. Certain rocks such as claystones with high salt content disintegrate quickly due to saline weathering; this process promotes deflation (Gutiérrez *et al.*, 2005a). Swelling clays are associated with a clay-rich regolith that cracks deeply during dry periods; these disintegration products can be easily exported by the wind (Price, 1963).

During the Late Pleistocene desiccation of the great pluvial lakes in the western United States triggered a period of intense deflation that excavated lake bottoms, mobilized dune fields, and created lunettes (Goudie and Wells, 1995). Similar effects have been observed on other continents.

Deflation basins commonly form in ancient drainage systems that have developed in large areas of low relief. Excavation of creeks and small rivers by deflation can disconnect the network and generate deflation basins although paleochannels can still be recognized (Bowler, 1986). Dunes can also cut off drainage channels and produce deflation basins.

In sandy deserts, there are large surface areas with interdunal deflation basins. They form as a result of the growth of dune crest fields from parabolic dunes. The development of these dunes involves the formation of elliptical-shaped deflation basins (Lancaster, 1978).

Large fields of deflation basins also occur in coastal areas, which are more humid environments. In the Carolina bays between Maryland and northern Florida, deflation basins occupy an 1100 km-wide strip parallel to the coast; about 500,000 basins have been identified in the area (Prouty, 1952). They apparently formed during the Wisconsin glaciation as elliptical depressions of giant parabolic dunes built by northwest winds that blew along the sandy surfaces of the coastal plain when the water table was low. When the water table rose, the depressions stabilized. Lunettes formed later during dryer periods (Price, 1968; Goudie and Wells, 1995).

Sand-clay dunes often form along the leeward margins of deflation basins. These half-moon-shaped dunes are called **lunettes**. They commonly reach heights of 10 to 50 m. The largest lunette in the world is in the Sebkha Kelbia in central Tunisia; it is 165 m high, 17 km long, and 9 km wide. The surface of a lunette is often vegetated, which promotes accretion. Desiccation causes particles from the bottom of the deflation basin to fragment into micro-polyhedrons. These particles are mobilized by winds with velocities greater than 3 m/second, which transport and deposit the particles on the outer rim of the closed depression (Perthuisot and Jauzein, 1975). The presence of salt prevents vegetative growth in the bottom of the depression and causes the aggregation of clay particles that will also be exported out of the basin. Pellets of salt and clay that form the lunette are leached by rain; the salt partially returns to the sebkha bottom, establishing a salt cycle (Tricart, 1969).

In some cases, multiple lunettes can occur in the leeward area (*Fig. 10.16*). The most external dune has high quartz content; clay content increases towards the interior, resulting in clay dunes (Bowler, 1973, 1986). It is thought that deflation of fluvial sediments builds the external dune. In contrast, internal lunettes develop by deflation of playa deposits that accumulate later after the initial deflation stage (Goudie and Wells, 1995) (*Fig. 10.17*).

Late Pleistocene (25,000 years BP) lunettes are found in the Southern High Plains (Texas and New Mexico). They were produced by episodic sedimentation with brief intervals of stability and soil formation in contrast to an accumulation of thick mid-Holocene deposits followed by a relatively long period of stability and soil formation (Holliday, 1997).

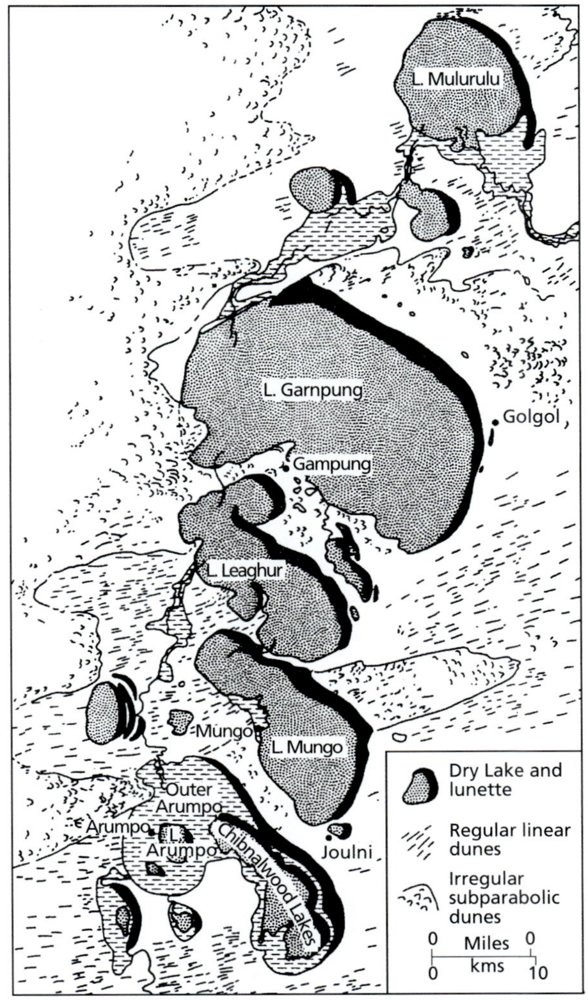

Figure 10.16 Set of deflation basins and small lunettes with dunes that surround them. Northwest of New South Wales, southeastern Australia (Bowler and McGee, 1978).

Large deflation basins (*Fig. 17.52*) such as the Qattara Depression commonly have a polygenic origin although deflation plays a fundamental role. Albritton *et al.* (1990) show that this depression was initially a fluvial valley that was split by karstic processes during the late Miocene and later deepened by deflation. Mass movements and fluvial processes also modified it. In addition, saline weathering prepared materials for subsequent deflation. Quaternary climatic oscillations also play a role in the modification of

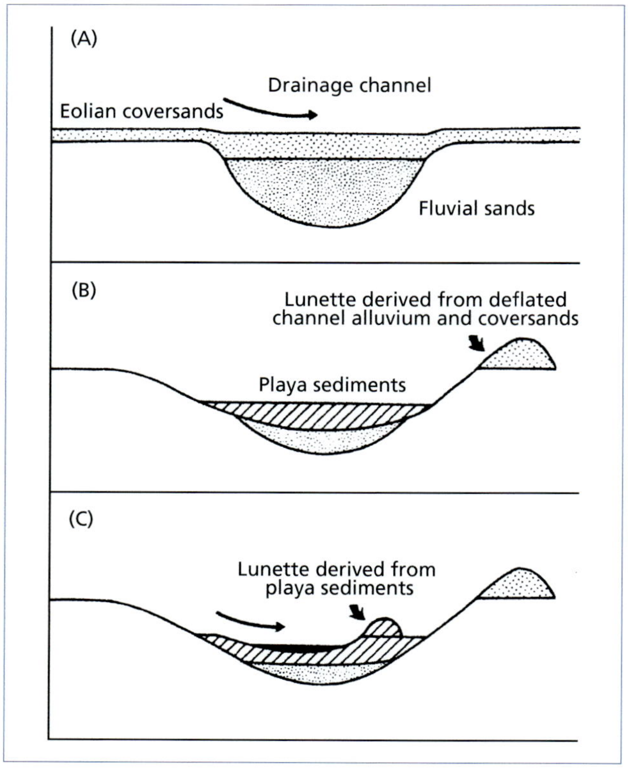

Figure 10.17 Model of small lunette development by deflation of fluvial deposits, eolian sand sheets, and playa deposits (Goudie and Wells, 1995).

geomorphic processes and the genesis of depressions. It is obvious that large deflation basins have long and complex histories. In smaller deflation basins, deflation predominates although other processes are also active, as previously discussed. According to thermoluminescence dates, the lunettes of Witpan (South Africa) originated in the last 2,000 years; most of the sand from the lunettes is older (Telfer and Thomas, 2006). Lunettes occur on the Manchega Plain (Spain); Bernat and Perez-Gonzalez (2008) also call them "lunas de arcilla" (clay moons). These lunettes originate from playa-type depressions with clay and saline sediments that are deflected by the wind like pellets. The ages obtained by OSL and AMS ^{14}C point to the existence of two generations: 29–19 ka BP and 13.5–7 ka BP.

10.7 Eolian accumulations

Eolian accumulations result from sedimentation of wind-deposited particles; they are sand and dust deposits. Beginning with a study of eolian sands, we will analyze their distribution, processes, and types of accumulations.

Dunes are undoubtedly the most remarkable desert landforms (*Fig. 10.18*). Over the last two decades the use of satellite images has significantly advanced research on these landforms. Eolian sands cover about five percent of the Earth's surface (Thomas, 1997b); sand accumulations cover

Figure 10.18 Dune field in Jaisalmer. Thar Desert (India).

20 percent of the world's arid zones. The proportion of this sand cover varies from desert to desert, from less than 1 percent in North and South America (Lancaster, 1995) to 50 percent in Australia (Mabbutt, 1977). The word **erg** or **sand sea** means a large surface of at least 125 km^2 that is covered by sand (Fryberger *et al.*, 1979; Thomas, 1997c). An erg is called a *koum* in Central Asia and *nafud* in Arabia.

Figure 10.19 shows the global distribution of large ergs and distinguishes between active ergs and marginal ergs that are stabilized by plant colonization. Most ergs are active when average annual precipitation is above 150 mm (Wilson, 1973). The largest concentration of these sand seas is in the Northern Hemisphere. Wilson (1973) identified 58 ergs with surface areas greater than 12,000 km^2 in various deserts of the world. Most ergs are in the Sahara desert; the Chech erg in southern Algeria (312,000 km^2) is the largest one in the Sahara. The Rub al Khali erg (*Fig. 10.20*), which occupies one third of the Arabian Peninsula, has a surface area of 560,000 km^2, and the Great Sandy-Gibson Desert in western Australia covers an area of 630,000 km^2. The tallest dunes in the world are found in the Badain Jaran Desert (northern China); they reach heights of 400 m (Dong *et al.*, 2004).

It may be difficult to distinguish between coastal and desert dunes in some areas. This problem occurs in the coastal zones of Namibia (Lancaster, 1982; Lancaster and Ollier, 1983), Perú (Finkel, 1959), and Oman (Goudie *et al.*, 1987). In some zones, continental dunes obtain their sand supply from coastal regions whereas in other areas ergs have prograded towards the sea (Pye and Tsoar, 1990).

A hierarchical classification of ergs (Wilson, 1972; Lancaster, 1994) is based on the identification of three types with distinct spatial patterns: a) eolian ripples (0.1–1 m); b) isolated individual dunes (50–500 m); and c) compound and complex dunes (>500 m). Each element in this hierarchy corresponds to the dynamics of wind patterns in the region.

Eolian sands in western Libya become redder with age and transport distance. The sand pigment is a film composed of iron oxyhydroxides (Walker, 1979). In the northwestern Sahara, sand grain size decreases with the age of the sand and the color of eolian sands indicates that yellow sands turn red with increasing age, which shows that the ergs are derived from each other (Demangeot, 1981). The color intensity is not a real measure of the age of the deposit, only a criterion for correlation (Walter, 1986).

Other large bodies of eolian sands that are not dune shaped are called **sand sheets.** Their surface area ranges from a few square kilometers up to 100,000 km^2 in the eastern Sahara (Breed *et al.*, 1987). Vegetation, grain size, shallow water tables, seasonal floods, and the occurrence of surface crusts control their development (Kocurek and Nielson, 1986).

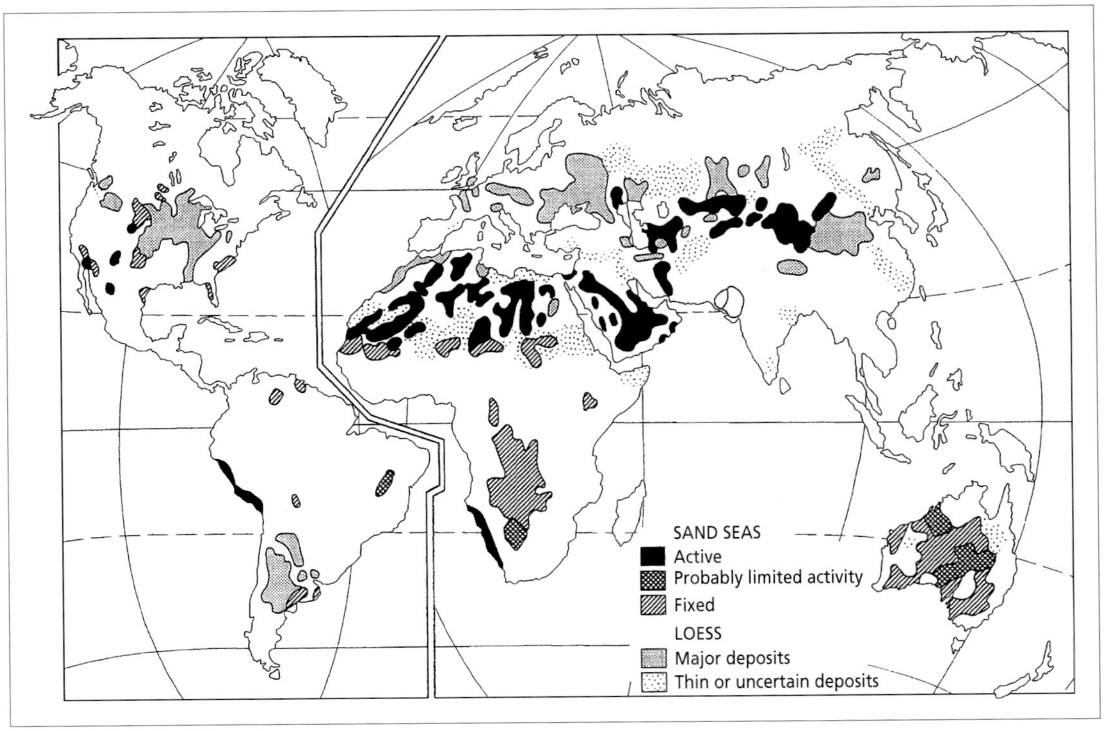

Figure 10.19 Global distribution of eolian deposits (Thomas, 1997).

Figure 10.20 Barchanoid shapes in the southern erg of Rub al Khali (United Arab Emirates).

10.8 Factors controlling erg development

Erg formation has three requirements: a) large supplies of sand, b) suitable climatic and topographic conditions, and c) sufficient wind energy (Pye and Tsoar, 1990).

About 90 percent of the particles in active ergs are sand sized because silt is characteristically transported far away from deserts and clays are mobilized by the wind only as pellets. Most sand comes from fluvial, littoral, and lacustrine deposits. In the Namib (Lancaster and Ollier, 1983) and Sinai deserts eolian transport has formed dunes in fluvial and littoral environments. Moving sand can flow along fixed paths and cover large distances, reaching the distal zones of ergs (Wilson, 1973). The maximum speeds observed in eolian sheets in Mauritania range from 62.5 to 162.5 m^3/s in width per year (Sarnthein and Walger, 1974) whereas the typical speed is about 3.49 m^3/s width per year in barchans. With a high sand supply rate, large dune fields can form in a few decades or centuries. Ergs, however, require much more time to develop. The average erg thickness in the Sahara is 21–43 m although in Simpson's desert (Australia) the average is one meter (Wilson, 1973). These values are low compared to Paleozoic and Mesozoic eolian formations, which are several hundred meters thick. This difference may be related to subsidence in sedimentary basins where more eolian activity may occur during periods of subsidence especially before the appearance of terrestrial plants (Kocurek, 1988).

Eolian processes are usually more significant when vegetative cover is scarce; as a result, extensive ergs occur in zones with <250 mm of rainfall although in southern Africa the limit for active dunes ranges from 100 to 150 mm (Lancaster, 1981). In some ergs, distinct accumulative and non-accumulative phases can provide evidence of past climatic changes as observed in the Gulf of Oman (Cooke *et al.*, 1993).

Ergs can be classified as dynamic or static depending on their mobility over time. Ergs always migrate in the direction of the predominant wind; dynamic ergs are found in flat areas where they do not encounter significant obstacles (Pye and Tsoar, 1990). Static ergs, on the contrary, occur in topographical depressions bordered by steep slopes (*Fig. 10.21A*) (Wilson, 1973). A clear example is the Great Eastern Erg located south of the cliffs of the Tademait plateau (Fryberger *et al.*, 1979). These authors propose various topographic influences for erg development (*Fig. 10.21B*), including (1) the shadow of topographic barriers, (2) shallow desert depressions, and (3) zones with reduced sand-transporting energy. The resulting energy can dissipate, however, (4) when surface winds are deflected to sites where accumulation is likely, and (5) when winds blow from opposite directions.

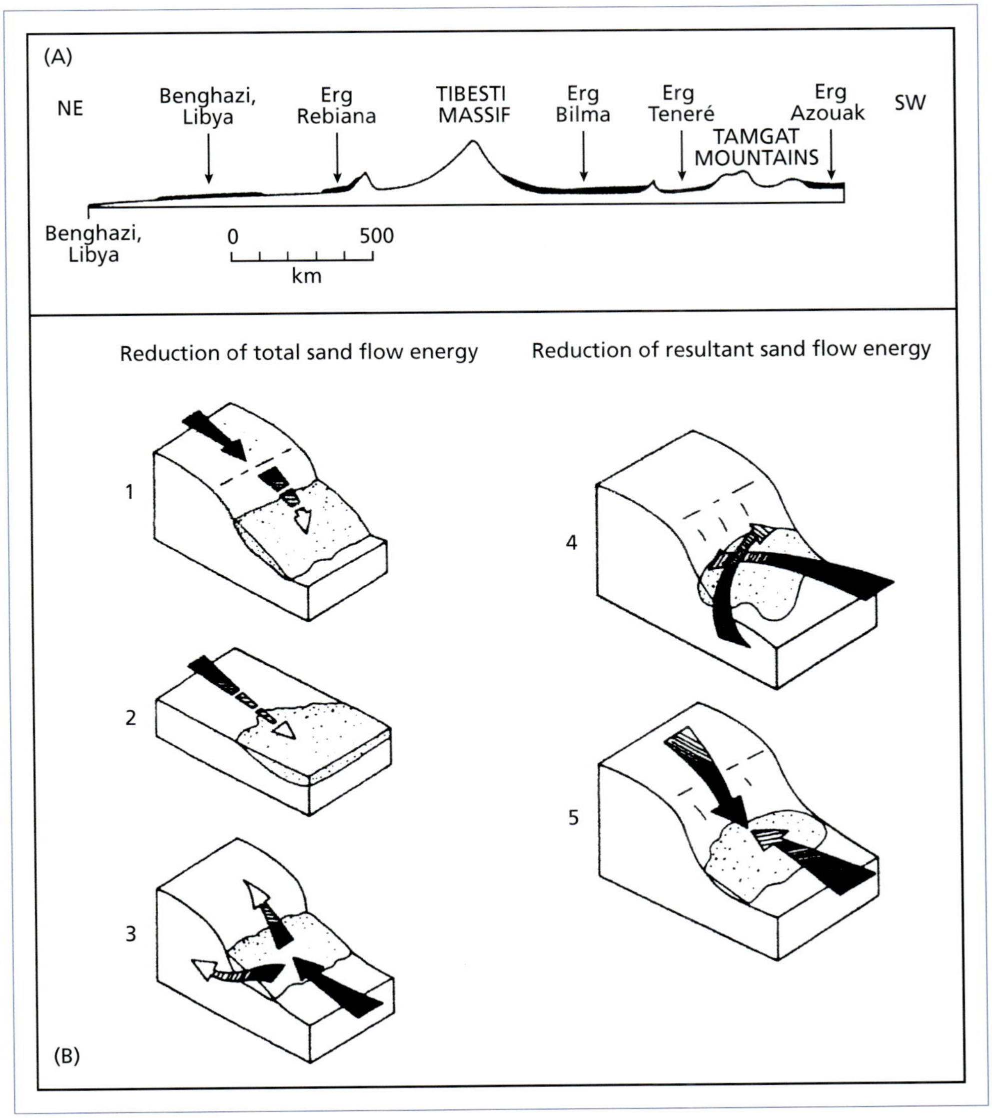

Figure 10.21 (A) Relationships between ergs and topography in northern Africa (Wilson, 1973). (B) Types of topographic influences in erg development (Fryberger *et al.*, 1979).

At high sand transport velocities, flow exceeds sedimentation and eolian sheets and barchans can develop. Ergs form when sand flow is saturated and accumulation exceeds transport (Wilson, 1971). Wilson used sand flow directions recorded by weather stations to draw sand flow maps for the Sahara desert (Livingstone and Warren, 1996) *(Fig. 10.22)*. In this way he discovered the provenance and sedimentation areas of Saharan ergs.

In tropical deserts high eolian energy environments are found in trade wind belts that border anti-cyclonic areas whereas low energy zones are situated near high-pressure centers (Livingstone and Warren, 1996). The Sahel region south of the Sahara desert is a low energy area where particles transported by trade winds from high-energy areas to the northeast can accumulate.

10.9 Dune processes

Onset, development, and achievement of equilibrium dune morphology reflect a balance between erosion and sedimentation, both of which control the shape of a dune (Lancaster, 1995). Decreased sand transport causes an abundance of sand in an area; when the sand supply exceeds erosion,

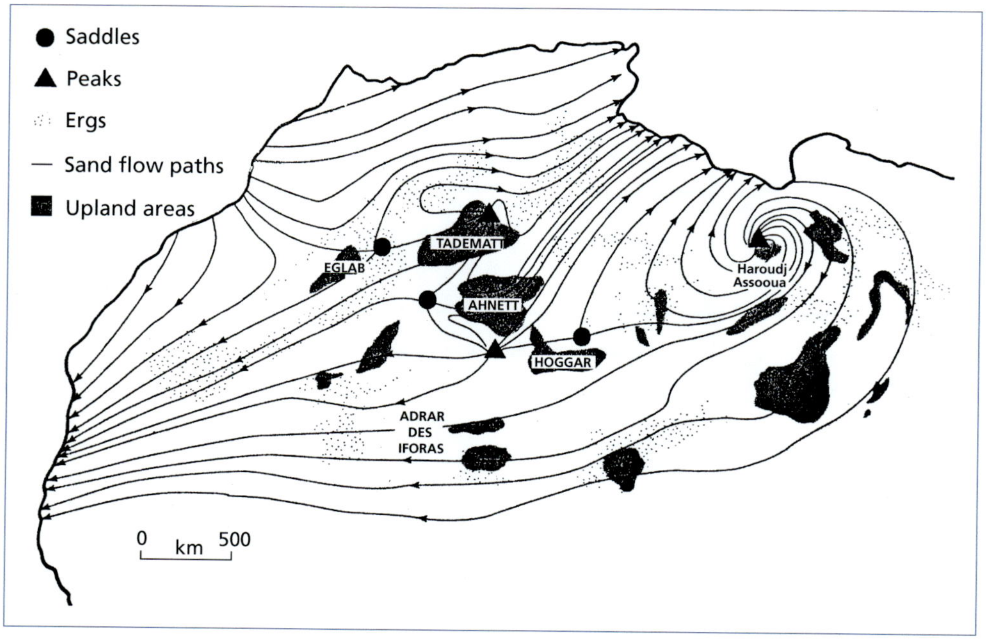

Figure 10.22 Sand flow lines that indicates the principal provenance and sedimentation areas in the Sahara (Wilson, 1971; Mainguet, 1978).

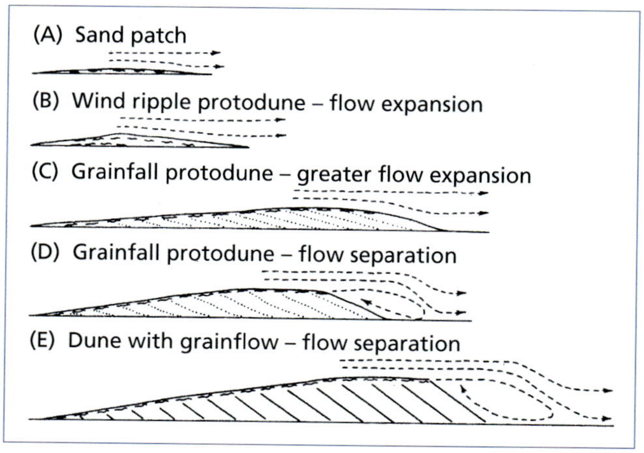

Figure 10.23 Stages in dune initiation on the leeward face. The internal structure is represented by wavy lines for *ripples*, dotted lines for laminae of fallen grains and continuous lines for cross-bedding generated by grain flow (Kocurek *et al.*, 1992).

sedimentation occurs and accumulation increases. Ground surface irregularities also have a considerable impact on transport velocity because they interact with the flow and generate secondary flows (Lancaster, 1994).

The onset of **dune formation** is not a well understood. When sand transport velocity decreases, local sedimentation caused a dune to begin to form. In studies on dune development on Padre Island (Texas), Kocurek *et al.* (1992) indicate that accumulation begins around small surface irregularities and small vegetation obstacles or boulders; this process creates shadow dunes. In their publication, they identify five steps of dune evolution associated with progressive changes on the lee-side slope due to flow separation and expansion (*Fig. 10.23*): (A) small irregular accumulations of dry sand a few centimeters high; (B) proto-dunes 0.1–0.35 m high with ripple-covered surfaces; (C) proto-dunes 0.25–0.40 m high with sand grains rolling down on the leeward slope; (D) barchans 1–1.5 m tall with grain flow, and (E) crescent dunes 1–2 m high. The three first stages can generate embryonic dunes.

Bagnold (1941) pointed out that intense sand transport is accompanied by instability transverse to flow, which causes sand deposition in longitudinal strips 1–3 m wide. He pointed out that these strips could form the core of future longitudinal dunes.

In small sand accumulations, erosion prevails at low and intermediate wind speeds but accretion processes predominate at higher wind speeds (Greely and Iversen, 1985). Dunes can be also built by strong wind gusts that sweep the sand and deposit it when wind intensity decreases; small eolian accumulations initiate dune formation (Warren and Knott, 1983).

Many hypotheses have been proposed to explain the regular repetition of successive dunes that have approximately the same shape and similar spacing. The models are based in dunes that develop under water; their evolution is based on kinematic instability and turbulence (see Cooke *et al.*, 1993; Livingstone and Warren, 1996).

On the other hand, as dunes grow, airflow is compressed towards the crest, which increases the wind speed and causes expansion and divergence on the leeward slope; this creates highly variable secondary flow systems (Lancaster, 1995) (Fig. 10.24).

Windward slopes of dunes are areas where important changes occur including modifications of slope inclination, roughness, shear speed, and sand discharge (Livingstone and Warren, 1996). The entire slope is a zone of maximum erosion (Bagnold, 1941); wind speed increases at the highest point. Differences of wind speed between the slope and the crest measured on barchans can exceed a ratio of 1/42. These conditions occur on straight slopes with inclinations of 5–10°. High wind speeds, however, have been indicated on very steep slopes (Lancaster, 1985) but other authors have noted that slope inclination is low in areas of strong winds (Gaylord and Dawson, 1987). This obvious lack of agreement indicates the need for further research.

Crests can be straight or can occupy a highest point that may be separated from the edge at the top of the leeward face (Fig. 10.25). These variations can be explained by small dome-shaped dunes with a clear separation between crest and rim at the top of the leeward face; they become larger dunes with less separation and finally evolve to a straight slope with a crest (Lancaster, 1987). Reverse evolution has also been suggested (Capot-Rey, 1963). Another interpretation proposed by Bagnold (1941) suggests that the separation between crest and rim results from variations in shear speed and sand transport speed. He also indicates that in dome-shaped dunes the crest may be mobile while the bottom remains static. Lancaster (1985) documented these variations; he also indicated speed ratios of 158:1 and 13:1 between the bottom speed of the dune and the crest for strong and gentle winds, respectively. These observations

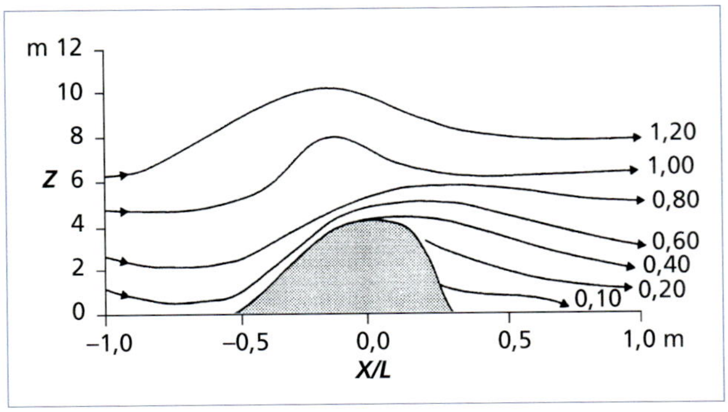

Figure 10.24 Wind flow over an isolated barchan in the Salton Sea, California. Note the flow compression on the stoss side and expansion on the lee side (Lancaster, 1994).

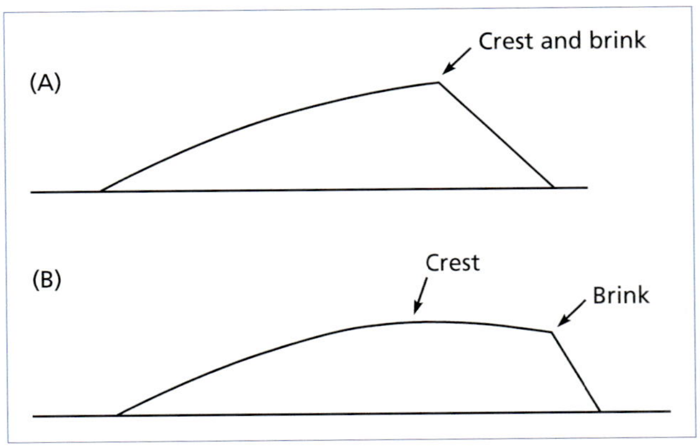

Figure 10.25 Crest of a dune (highest point) and the rim (top of the leeward slope): (A) the crest and the rim are the same and in (B) they are separated (Livingstone and Warren, 1996).

Figure 10.26 Bottleneck shaped flow on the leeward side of a dune. Ythan Stuary and Sands of Forvie. Aberdeenshire (Scotland). Photo F. Gutiérrez.

Figure 10.27 Plates of sand sliding down the leeward slope. Playa del Trabucador, Ebro Delta. Photo F. Gutiérrez.

show that sand transport occurs only in the crest areas of dunes during periods of low wind speeds whereas during strong winds mobilization occurs throughout the dune. Crest height can decrease during periods of low wind speeds and increase at high wind speeds.

On **leeward slopes**, wind speed and sand transport decrease rapidly as flow expands between the crest and the rim and separates on the leeward face. Flow is highly irregular, gusty, and involves flow inversions, which can be detected with smoke. Wind speeds at the base of the slope are low and variable; in crescentic dunes in the Namib Desert they range from 0.4 to 0.8 of the crest values and up to 0.04 of the dunes of Padre Island, Texas (Sweet and Kocurek, 1990). Flow is complex in longitudinal and star dunes, especially if the crest line is sinuous. Flow separation occurs on the crest but when flow is oblique to the crest, whirls appear on the leeward slope and sand is transported along the dune (Tsoar, 1983a). Sand grains can fall from the crest or cross it by saltation. Sand grains can also move en masse from scarps of 5–10 mm, flow as avalanches, and form bottleneck morphologies (*Fig. 10.26*). These avalanche tongues are a few centimeters deep, about 0.5 m wide, and move at speeds of about 0.2 m/second (Hunter, 1985). With morning dew or in saline environments, increased surface cohesion may mobilize sand plates (*Fig. 10.27*) (McKee, 1979b).

10.10 Dune classification

Researchers have described a wide variety of dune types although the main processes as previously described are common to all types. In addition, multiple terms have been applied to a single form (Breed and Grow, 1979), which creates confusion when investigators attempt to classify dunes and their associations.

Existing dune classifications can be grouped into two categories (Lancaster, 1995). The first consists of morphodynamic classifications that relate a dune type to the wind that created it or to the sediment supply. The second type, morphological classifications, considers only the external shape of the dune. Numerous morphodynamic classifications have been proposed (Aufrère, 1928; Clos-Arceduc, 1969; Hunter *et al.*, 1983). The latter authors used the relationship between dune orientation and wind direction to define transverse, oblique, and longitudinal dunes. However, research into dune dynamics shows that different parts of the same dune can be transverse, oblique, and longitudinal at the same time (Tsoar, 1983a; Lancaster, 1989b). Mainguet (1983, 1984) draws a distinction between erosional forms (parabolic dunes and sand ridges) and depositional morphologies (barchans, transverse chains, longitudinal dunes, and pyramidal dunes).

Table 10.1 EOLIAN DUNE CLASSIFICATION (LIVINGSTONE AND WARREN, 1996)

DUNES

FREE				ANCHORED	
Transverse	Linear	Star	(Sheets)	Vegetation	Topography
Transverse Barchan Dome Reversing	Sandridge Seif	Star Network	Zibar Sand sheets	Nebkha Parabolic Coastal Blowout	Echo Climbing Cliff-top Falling Lee Lunette

McKee and Bigarella (1979) promoted a morphological classification based on dune shape and number of faces. Lancaster (1995) and Thomas (1997b) proposed other dune classification based on this classification. In addition, another type of dune classification is based on practical goals, such as the trafficability of dunes. Recently, Tsoar *et al.* (2004) classify dunes as a function of their morphodynamics: migrating dunes such as transverse dunes, elongating dunes, linear dunes and accumulating dunes.

For simplicity we shall use the classification proposed by Livingstone and Warren (1996), which is a modification of a previous one (Cooke *et al.*, 1993). This classification is based on a dune's form and movement characteristics and in some cases is related to the presence of vegetation and topographic obstacles (*Table 10.1*).

10.10.1 Transverse dunes

In transverse dunes, the main flow direction is normal to the crest throughout almost the entire year. As these dunes grow they maintain equilibrium between shape and size. They are longer in the direction perpendicular to the predominant wind and the windward slopes have less inclination than leeward slopes.

The word **barchan** is of Turkish origin; it describes isolated half-moon shaped dunes that have developed on top of a coherent substratum such as a pediment, a desert pavement, or a sebkha. They commonly occur on the margins of ergs and in sand transport corridors associated with depositional zones. As a general rule, they are small in size and their height is generally a tenth of their width. Smaller ones move rapidly, whereas larger ones, up to several decameters high, migrate very slowly. The stoss slope is clearly convex with an average angle of 12° whereas the lee side slope has an inclination of 33–34° (Pye and Tsoar, 1990). The horns point into the direction of the wind (*Fig. 10.28*). Barchan dunes advance at speeds directly related to the sand transport velocity over the crest and inversely related to the crest height (Bagnold, 1941). Speeds are highly variable and can reach 63 m/year (see Thomas, 1997b). Maximum velocity occurs at the crest (Wang *et al.*, 2008). As sand supply increases, barchans join together laterally to form barchanoid ridges (Kocurek *et al.*, 1992), also called megabarchans (Cooke *et al.*, 1993; Lancaster, 1995; Thomas, 1997b). With increasing amounts of sand complex barchanoid ridges can form (*Fig. 10.20*) (Thomas, 1997b), which are also called compound crescent dunes (Breed and Grow, 1979; Lancaster, 1989a). These dunes characteristically have a principal ridge that rises between 20 to 80 m over the area of the interdune with a large number of barchanoids on the stoss slopes and in the crest areas. The spacing of the larger ridges ranges from 700 m to 2000 m. Single barchans and barchan associations cover about 40 percent of sandy deserts (Fryberger and Goudie, 1981).

Erosion and sedimentation in barchans are characterized by erosion of the stoss slope and sedimentation in the lee slope. Howard *et al.* (1978) measured height changes of a barchan surface over a period of two weeks. They found maximum erosion in the center of the stoss side and sedimentation on the majority of the lee areas (*Fig. 10.29*). Barchans can be asymmetrical as a result of transverse winds that transport sand and cause the horn to elongate (Norris, 1966).

Wind tunnel experiments show that shear velocity exercised by the wind is the fundamental force in eolian sand transport (Butterfield, 1993). Two-dimensional field

Figure 10.28 In foreground, a small barchan. In the background are nebkhas. Jaisalmer. Thar Desert (India).

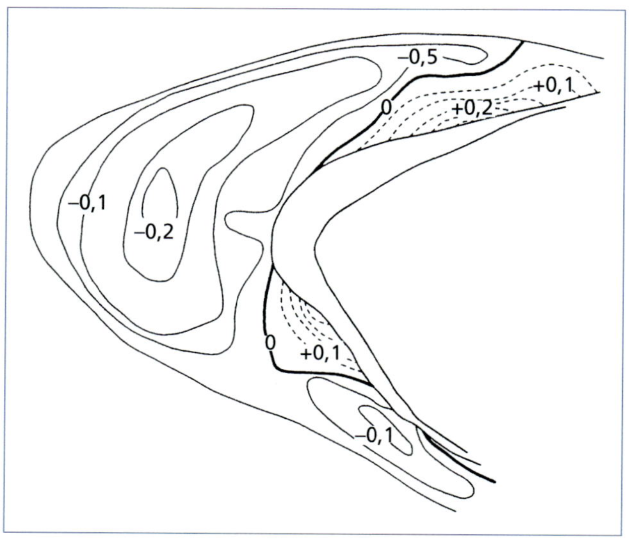

Figure 10.29 Height changes of a barchan surface over a two-week period (Howard *et al.*, 1978).

measurements in barchans indicate that velocity (Frank and Kocurek, 1996, Lancaster *et al.*, 1996) and transported sand collected in traps increases toward the dune crest (Wiggs *et al.*, 1996). The latter authors show the fundamental role of turbulence in wind tunnels. In addition, the results of the application of CFD support the measurement of maximum wind velocities at dune crests (Parsons *et al.*, 2004). The model of Hersen *et al.* (2004) shows that for barchans the quantity of sand lost from a dune is proportional to the width of its horns because this is the only place where sand is lost. However, both physical and numerical models should be taken into consideration along with field investigations (Livingstone *et al.*, 2007). On the coastal plain of southern Israel a barchan changed into a shrub-covered parabolic dune within 60 years as a result of changes in land use. The bushes trapped sand and increased the crest height. This trapped dune did not deposit on the leeward side. In this case the dune became narrower, higher, and concave-shaped during the transformation from barchan to parabolic dune (Ardon *et al.*, 2009).

Transverse dunes are composed of strips that are fundamentally perpendicular to the wind. They have an asymmetrical profile; their stoss side has a flatter slope and the leeward side is steeper (*Fig. 10.30*) (Cooper, 1958; Zazo *et al.*, 2008). The belt undulates, dune height generally does not exceed 10 m, and spacing between dunes ranges from 100 to 600 m (Hesp and Thom, 1990; Gutiérrez *et al.*, 2011a) (*Fig. 10.31*). In explaining the formation of transverse dunes, Warren (1979) proposed an initial phase in which winds create a sinuous system with sedimentation occurring in high areas and erosion in low areas. Only coarse sand is mobilized on the stoss side; it accumulates at the crest where it moves slowly and the dune increases in height in this way and develops the leeward face. Transverse dunes form on flat surfaces (Hesp and Thom, 1990). In the Corralejo (Fuerteventura, Canary Islands) dune complex, Valdemoro *et al.*, (2007) used sequential aerial photographs to estimate dune displacement at 87 m between 1987 and 2002. This rapid displacement does not support experimental field investiga-

Figure 10.30 Active transverse dunes. Corralejo. Fuerteventura. Canary Islands, Spain.

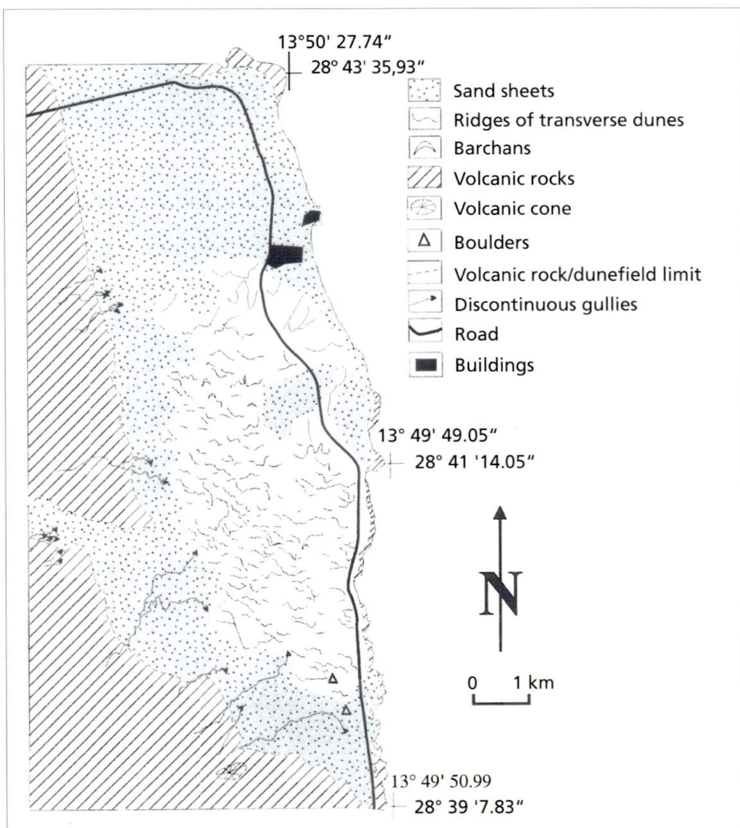

Figure 10.31 Geomorphological map of the Corralejo dune field. Fuerteventura (Canary Islands). (Gutiérrez *et al.*, 2011a).

tions on transverse dunes conducted by Sharp (1966), which detected only a small dune advance over a study period of 15 years. This lack of agreement shows the need for meticulous experimental field studies on the transverse dunes at Corralejo (Gutiérrez *et al.*, 2011a).

Dome dunes are included in the transverse dune group because they have a similar orientation and the same type of sand transport. Dome dunes are usually low and flat-crested. They are circular or elliptical in shape and no slope is steeper than the others. Although not common in most deserts, they are abundant in the Chinese Taklimakan desert (Zhende, 1984) where they can be 40–60 m high and 500–1000 m in diameter, in contrast to other sandy areas where they are quite small. Dome-shaped dunes are common on the margins of dune fields. Glennie (1970) observed that low wind speeds form oval hills whereas high wind speeds build longitudinal forms. Other authors maintain that strong unidirectional wind activity is responsible for dome-shaped dunes because it prevents crest formation (McKee and Bigarella, 1979; Breed and Grow, 1979). Edgell (1990) claims that dome dunes form under conditions of reduced sand supply and variable winds.

Reversing dunes are included in the transverse dune group because sand transport in them is perpendicular to the crests. They are created by winds that flow in opposite directions in different seasons. Thus the reversing dunes of Oman (Cooke *et al.*, 1993) result from strong summer monsoon winds (large dunes) and the opposite weak winter winds (small dunes).

10.10.2 Linear dunes

Linear dunes, also called **longitudinal dunes**, are the most common type; they can be more than 200 km long. They are rectilinear, parallel, and have regular spacing between ridges. In cross section they are more symmetrical than transverse dunes. Numerous crests and lows can be observed in the longitudinal profiles of large and middle-sized dunes. The smaller ones, however, are quite straight. These dunes usually move in the direction of the dominant wind but some move in an oblique direction. A linear dune is also called a *seif, sif,* or a *silk*; the corridors between sand ridges are known as *gassi, goud*, and other names (see Cooke *et al.*, 1993).

Linear dunes occur as simple, composite, and complex varieties (Pye and Tsoar, 1990). Simple linear dunes are narrow ridges (*Fig. 10.32*) that may exceed 200 km in length; they have a straight or wavy crest. The transverse profile is straight or rounded; they have regular interdunal spacing and a high dune-to-interdune area ratio. They are common in the Kalahari and Simpson deserts (Lancaster, 1995). Linear dunes cover approximately 40% of the surface of Australia (Wasson *et al.*, 1988) where they form a counterclockwise spiral (*Fig. 10.33*). In these regions they range from 2 to 35 m in height and 150 to 250 m in width and have interdune spacing of 200 to 450 m. They can be 200 km long. Sometimes the dunes are partly covered by vegetation as in Nizzana on the border between Israel and Egypt (*Fig. 10.34*). A notable characteristic of these dune

Figure 10.32 "El Dragon" seif on the Iquique coast, northern Chile, Photo J. Rodríguez Vidal.

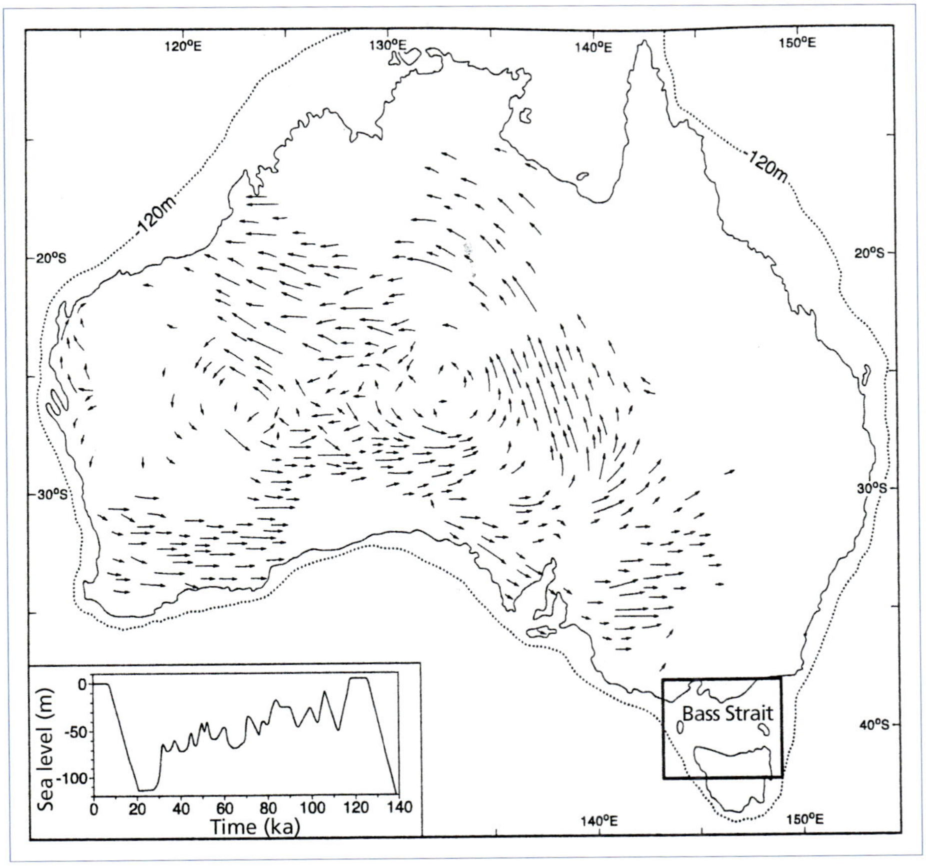

Figure 10.33 Orientation of linear dunes in Australia (Wasson *et al.*, 1988). The lower frame shows sea level over the last 40 ka (Lambeck and Chapell, 2001, in: Duller and Augustinus, 2006).

Figure 10.34 Simple linear dunes partially covered with vegetation. Nizzana experimental station. Negev Desert, Israel.

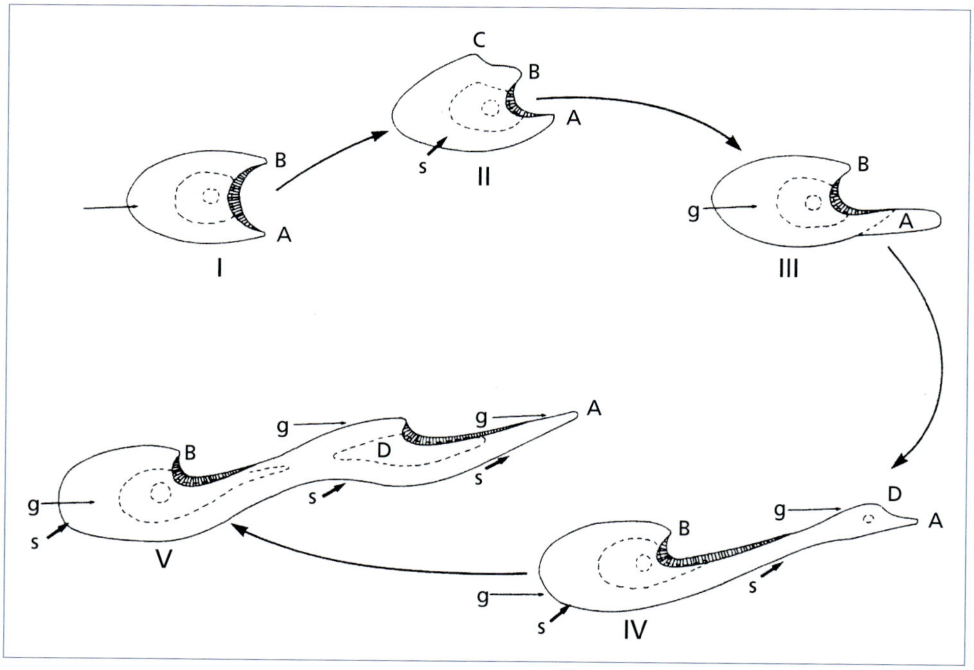

Figure 10.35 Model showing the development of seifs from barchans; s shows strong winds and g shows weak winds (Bagnold, 1941).

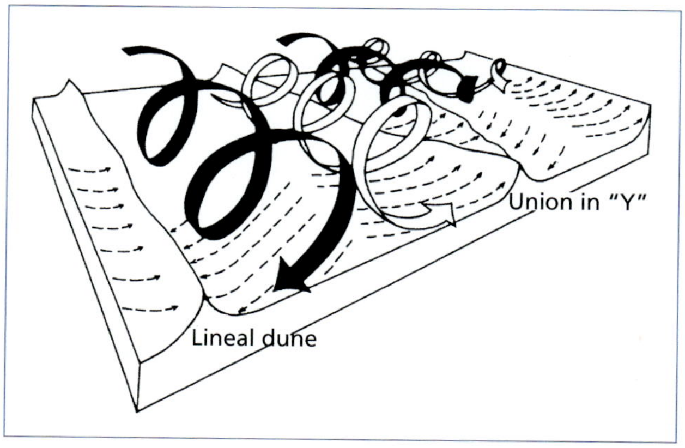

Figure 10.36 Hypothesis on linear dune formation from helicoidal flows (Livingstone and Warren, 1996).

fields is the coalescence of crests to form a "Y" or a tuning-fork shape (*Fig. 10.36*); these are common in areas where dunes are closely spaced. Interdune zones may be covered by sand or may be desert pavements partly covered by vegetation.

Two or more dune ridges very close to each other or superimposed on the crest form a composite linear dune. They have been well studied in the Namib Desert (Lancaster, 1983) where their ridges are asymmetrical, 25 to 40 m high, and are a spaced 1200–2000 m apart; interdune areas are covered by sand. Linear complex dunes such as the large dunes of the Namib are 50–150 m high and spaced 1–2 km apart; they have a wavy main crest and crescentic dunes on the flanks. Interdune areas are covered by sand that is rippled sub-parallel or slightly oblique to the main crest.

Wind patterns of longitudinal dunes are variable. They are commonly unimodal or bidirectional but multiple directions can sometimes be identified. Linear dunes are often primarily associated with permanent winds from a single direction with superimposed oblique winds.

The origin and dynamics of linear dunes have been the focus of many studies as reflected in the summaries of Tsoar (1989), Pye and Tsoar (1990), Cooke *et al.* (1993) and Livingstone and Warren (1996). Bagnold (1941) proposed that seifs are derived from barchans by the elongation of one of the horns in an area of bi-directional winds with different intensities (*Fig. 10.35*). Other researchers think that the elongation occurs at the front of a parabolic dune (Verstappen, 1968). The hypothesis that linear dunes are created by helicoidal flows (Bagnold, 1953) still has its proponents (*Fig. 10.36*); this hypothesis is supported by field investigations and wind tunnel research (Tseo, 1993). Dunes develop in areas of convergence and corridors form between them. This model also seems to explain their regular spacing.

Tsoar (1983a) and Livingstone (1989) think that linear dunes result from wind action in two directions. These dunes migrate laterally as demonstrated by the flume experiments of Rubin and Ikeda (1990). The migration can be identified because the transport direction does not coincide with the strike of the dune and also by studying the internal dune structure (Rubin, 1990). Today, researchers use ground penetrating radar (GPR) to identify these features in the field (Bristow *et al.*, 2000). In an area of the Namib Desert Bristow and his colleagues found that the divergence angle between the two principal wind directions was about 135° and the transport ratio between them was 3:1, which indicates lateral migration. In contrast, Livingstone (2003) did not find perceptible lateral migration in 21 years of observations. Tsoar *et al.* (2004) used aerial photography to study the Sinai but did not observe lateral movements. Perhaps more years of observation are needed to detect lateral displacement.

10.10.3 Pyramidal and network dunes

Pyramidal and network dunes both have numerous slip faces that develop as a consequence of multiple wind directions. **Pyramidal dunes** are also called star dunes, cones, sand mountains, *ghourds*, and other local names (Pye and Tsoar, 1990; Cooke *et al.*, 1993). They are the largest dunes in ergs; in Algeria they can be more than 400 m high (Wilson, 1973). As a result, they support a large amount of sand and are located in zones with high sedimentation rates. In addition to their large size and pyramidal morphology they also have three or four radial arms (*Fig. 10.37*). About 11 percent of dunes are the pyramidal type (Fryberger and Goudie, 1981) and they occupy 40% of the Great Eastern Erg of Algeria (Breed and Grow, 1979). The arms coalesce on the top of the dune where the slope inclination ranges between 15° and 30°. Slope inclinations are lower (5–10°) near the base of the dune and small crescentic and reversing dunes can be seen there. Dune spacing varies from 150 to 6700 m, according to Lancaster's compilation (1995) and dune widths range from 180 to 6000 m. A clear direct relationship generally exists between dune height and spacing in pyramidal dunes (Lancaster, 1989b). These dunes can grow several centimeters per year; they appear to migrate slowly rather than maintaining a compensatory balance (Nielson and Kocurek, 1987).

Wind roses prepared for pyramidal dunes in different deserts indicate that winds are multidirectional and have variable energy (Lancaster, 1995) especially during the months when most sand transport takes place. As a result, net transport is usually low. Interactions between seasonal wind changes concentrate sand sedimentation in the central parts of the dune. The main arms of a pyramidal dune are oriented approximately transverse or slightly oblique to the direction of the predominant local wind. Smaller arms of the pyramidal dune are oriented parallel and transverse to the direction of the secondary winds (*Fig. 10.38*) (Lancaster, 1989b, c). Areas of eolian erosion and sedimentation may alternate with the season (Wang, *et al.*, 2005). Besides this, topographic barriers exert a major influence on pyramidal dunes (Breed and Grow, 1979) because they modify regional winds, increase their variability, or create traps that affect sand transport.

Network dunes form under a continuous sand cover; they are composed of dunes a few meters high. They are inter-

Figure 10.37 Star dune. Erg Chebbi (Morocco). Photo J.L. Peña.

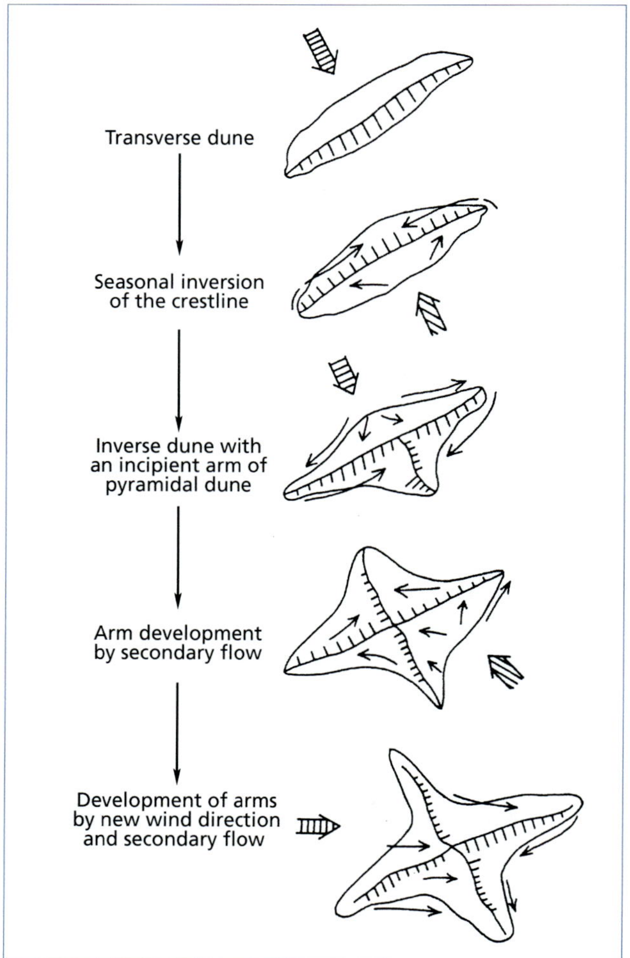

Figure 10.38 Model for the formation of a star dune by development of secondary fluid circulation, so that the dune moves according to multidirectional wind regimes (Lancaster, 1989b).

laced, alveolar, or rhombic-shaped networks that extend several hundred meters. They have been interpreted as groups of transverse dunes that have overlapped as a result of varying seasonal wind directions (Aufrère, 1935). Their complexity depends on the intensity and persistence of the seasonal winds.

10.10.4 Sand sheets and zibars

Sand sheets are accumulations of small dunes that develop on flat terrain and that lack slip faces (*Fig. 10.39*). **Sand strips** and **sand ribbons** are elongate varieties of sand sheets (Breed and Grow, 1979). Their surfaces may be irregular or undulated and they may or may not have *ripples*. Many sand sheets are the foundation or support for other mobile dunes. Worldwide the extent of sand sheets exceeds one and a half million square kilometers and, as mentioned previously, the maximum known sand sheet surface covers an area of 100,000 km^2 along the borders of Egypt, Sudan, and Libya. The thickness of sand sheet deposits ranges from a few centimeters to 10 m (Breed *et al.*, 1987). Grain size ranges from fine sand, sometimes with a significant silt content to coarse poorly sorted sand (Pye and Tsoar, 1990). Egypt consists of sandy plains with residual deposits of fine gravel and pebbles that form the surface layer. With rare exceptions sand sheets are poorly sorted deposits. Underlying levels usually consist of sub-horizontal layered sand beds (Lancaster, 1995).

Sand sheets are affected by a set of factors that control their formation (Kocurek and Nielson, 1986). Vegetation reduces dune movement and growth and causes sand to accrete in horizontal laminae. These sparse vegetation circumstances are common on desert margins. Where a coarse sand layer is present, it is likely that the fine fraction has been exported by deflation. A shallow water table or periodic or seasonal flooding makes dune formation more difficult or impossible. The presence of surface crusts or algal mats also prevents sand transport and dune building.

Like sand sheets, **zibars** are flat areas with thin deposits and dunes that lack slip faces (Kocurek and Nielson, 1986). Zibars form ripples and mega-ripples, are composed of coarse sand, and settle on hard substrates. This explains the term zibar, which is derived from the Arabic word zibara, which means a hard surface where a vehicle can travel. Zibars are common in areas where fine materials have been deflated. They are found in different environments from the central Sahara Desert to zones with sparse vegetation in southern California (Cooke *et al.*, 1993).

10.10.5 Dunes anchored by vegetation

Low-mobility dune areas invite plant growth because sand holds water (Tsoar and Moller, 1986). The vegetative cover modifies wind flow and traps eolian sand.

Dunes anchored by vegetation are also called **phytogenetic dunes** (Cooke *et al.*, 1993). They result from the stabilization of eolian sand around plants (*nebkhas* and coastal dunes) and from the erosion of surfaces covered by vegetation (*blowouts* and parabolic dunes). Although dunes are almost stationary, their shape and size change during seasonal or longer cycles.

The most common type of vegetation-anchored dune is a **nebkha**, which is also called a shrub dune, shadow dune, *hummock* dune, *rebdou*, and other terms (Pye and Tsoar, 1990; Cooke *et al.*, 1993). Nebkhas can cover large areas of lowlands in semiarid and coastal areas (Hesp, 1981). They resemble small ridges trapped by plants, elongated in the direction of the sand transport, and located on the leeward side (Hesp, 1981) (*Figs. 10.40*, and *10.28*). Nebkhas

Figure 10.39 Sand sheet with knobs, El Portillo. Fuerteventura. Canary Islands, Spain.

Figure 10.40 Group of nebkhas on the margin of the Thar Desert. Jaisalmer (India).

consist of sand, silt, and clay aggregates. The nebkha shape is a result of its size, density, and plant growth (Capot-Rey, 1957). Plants need to be at least 10–15 cm high to trap sand effectively. Nebkhas vary considerably in size; they can reach a height of 10 m and a length of one kilometer when trapped by a group of trees (Cooke *et al.*, 1993). The dune reaches its maximum height on the lee side near the plant because reverse flows are higher in this location. Plant longevity is also important because only long-lived species can retain a large quantity of sand. In fact, many nebkhas are ephemeral because they have formed around annual plants. In some cases perennial plants experience changes in the water table, precipitation, and sediment supply that can abruptly modify eolian dynamics (Gile, 1975). Non-eolian processes such as raindrop impact and spray, runoff, and animal activity affect some nebkhas. Because numerous processes are involved, it is difficult to evaluate the individual contribution of each (Cooke *et al.*, 1993).

Blowouts are produced by deflation; they are closed depressions located on eolian sands that are unconsolidated to the depth of plant roots (Livingstone and Warren, 1996). Blowouts can undergo significant changes in short time periods (Gares and Nordstrom, 1987; Jungerius and van der Meulen, 1989; Carter *et al.*, 1990; Gares, 1992). Some authors restrict the term blowout to morphologies generated in vegetated areas whereas others such as Lancaster (1995) extend the term to morphologies that occur on bare dunes (*Fig. 10.41*). Blowouts are found in continental dunes, essentially on hot and cold desert margins (Seppälä, 2004), and in vegetated coastal dunes. One of the most studied groups of blowouts is the Meijendel located in the coastal dunes of Holland (*Fig. 10.42*) (Jungerius, 1984). Two types of blowouts have been identified: saucer and trough (Cooper, 1967) (*Fig. 10.43*). Blowouts elongate in the wind direction; the major axis is 10 to 30 m long and can exceed 100 m in some cases (Hesp, 2002 and 2004; Hesp and Hyde, 1996). Plant cover is critical to the development of these landforms. Blowouts can get started when plant cover is sparse or disappears due to prolonged drought resulting from an arid climate change. Vegetation reduces rugosity, which increases the scour surface and the export of material. Sand can be lost as a result of overgrazing, rodent activity, and fire. In addition, plants at the top of a dune are the first to dry out and thus eolian erosion first occurs in these areas. Vegetation can be sometimes pulled up by hurricanes and unusual storms. Deflation deepens the non-vegetated area. Winds reach their maximum speeds in the bottoms of depressions, generating vertical vortices that throw sand toward the edges of hollows. In areas of abundant precipitation, water erosion also widens the blowout (De Ploey, 1980). Exported sand settles on plants at the margins of the closed depression, forming a blow dune that is part of the blowout (Glenn, 1979; Carter *et al.*, 1990; Hesp and Hiyde, 1996; Hesp, 2004). Eolian erosion reaches

Figure 10.41 Blowouts on the Al Liwa dune chain. Rub al Khali Desert (United Arab Emirates).

Figure 10.42 Blowout with sandy bottom and shrub and arboreal vegetation. Small blowouts can be seen in the foreground. Meijendel. Coastal dunes of Holland.

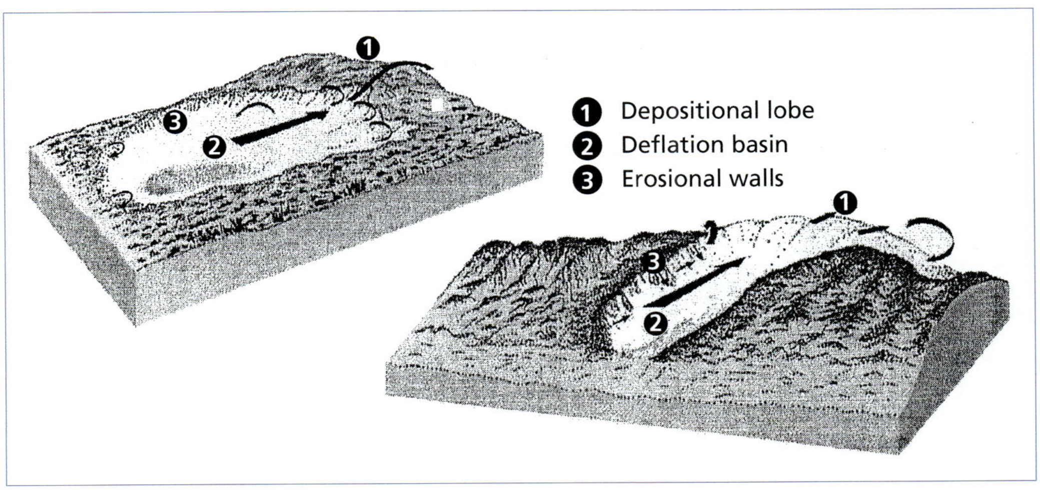

Figure 10.43 Diagram of Saucer-shaped and trough-shaped blowouts that show dissection by the wind (Hesp, 2002).

its maximum during the dry season when the depression widens and may migrate. Even so, most blowouts have a short lifespan because they are quickly covered by vegetation; indeed, half of the blowouts in Holland disappeared within about nine years (Jungerius and Van der Meulen, 1989). Mobilized sand has the form of a fine plume-shaped layer. During intense erosion a dune forms in the leeward area and the dune advances over existing vegetation. As a result, the area of floating sand enlarges and a parabolic dune is formed.

In the area of Tierra de Pinares (Duero Basin, Spain), Gutiérrez *et al.* (2005a) propose an evolutionary model for blowouts studied in a continental area (*Fig. 10.44*). The model begins with deposition of a sheet sand during

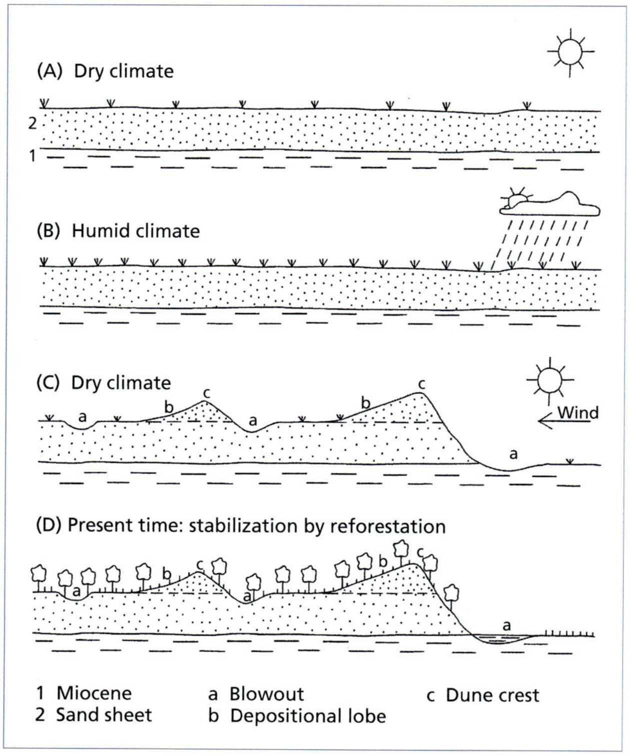

Figure 10.44 Schematic evolution of blowouts on sand sheets in Tierra de Pinares (Duero Depression, Spain) (Gutiérrez *et al.*, 2005).

climatic conditions dryer than the present. During a later humid period vegetation grows, stabilizes, and prevents deflation. Blowouts and dunes form during a later dryer climate. Blowouts deepen into the clay substrate by pelletization of clay particles and later deflation. An intense forest repopulation that occurred later almost totally stabilized the deflation.

Parabolic dunes have a U- or V-shape in plan view and their two arms are partly covered by vegetation on the stoss side. They occur in cold climates, coastal areas, and desert margins. The largest area of parabolic dunes is found in the Thar Desert of India where they cover an area of about 100,000 km^2 (Verstappen, 1970) (*Fig. 10.45*). The arms are 1–2 km long and the fronts are 10–70 m high: they advance essentially by avalanche. Many of these dunes coalesce because they have different movement speeds, creating imbricated or multi-lobed parabolic dunes with multiple crests and slip faces (Wasson *et al.*, 1983). Little is known about the speed at which parabolic dunes advance but it ranges from 0.05 m/year in zones of dense vegetation to 13 m/year in some coastal regions. These dunes seem to be associated with moderately developed plant cover and a unidirectional wind regime. Vegetation plays a fundamental role in the genesis of parabolic dunes because it protects the less mobile arms from eolian activity and allows the central part to advance (Robertson-Rintoul, 1990). It is thought that parabolic dunes derive from blowouts and grow by

Figure 10.45 Front of a parabolic dune slightly covered with vegetation. Tena-Jaisalmer. Thar Desert (India).

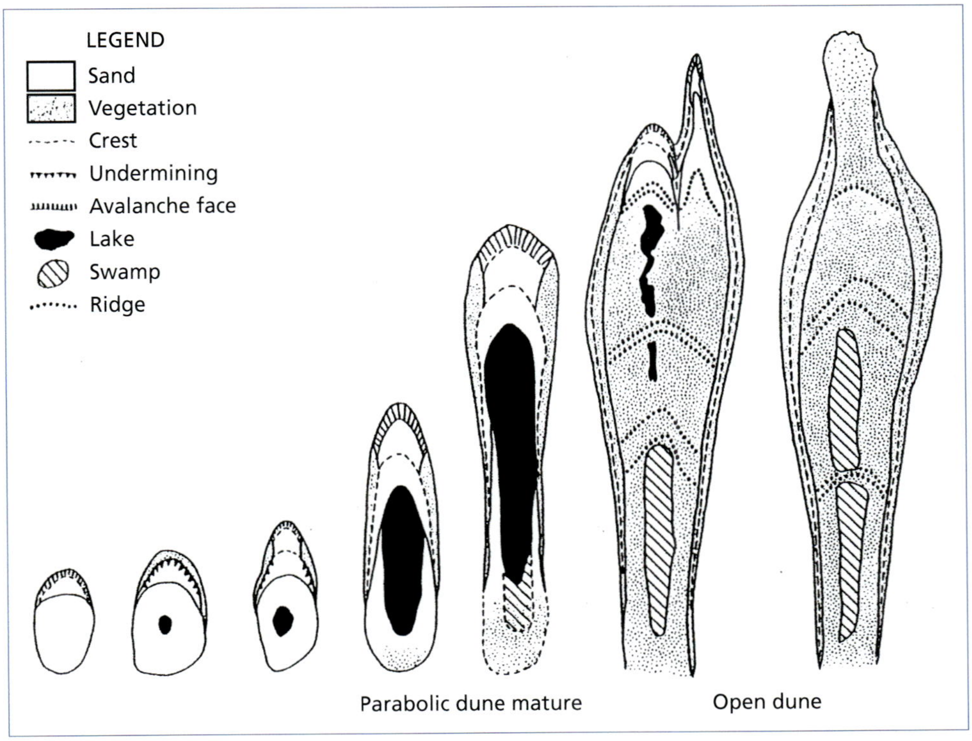

Figure 10.46 Diagrams showing the stages of growth and eventual dissipation of elongate parabolic dunes in the Cape Flattery area (Pye, 1982).

incorporating sand eroded from underlying sediments although this sand supply can cease if erosion reaches an outcrop of cohesive bedrock or the capillarity front where the sand is moist and cannot be entrained. When erosion reaches the water table, it can create swamps and lakes. The arms of parabolic dune then converge and decrease in height and finally a sand tongue passes over the apex (*Fig. 10.46*) (Pye, 1982).

Coastal dunes are formed by a set of eolian processes that interact with lithological, marine, fluvial, slope, soil, and cultural processes (Livingstone and Warren, 1996). The type and extent of these dunes depend on the quantity of available sand and the supply rate. The supply rate is a function of the existence of a sediment source and a strong littoral drift towards land. The sediment source is related to rivers, beaches, coastal erosion, and estuarine sands.

Coastal dunes can be classified simply into primary dunes and secondary dunes. The development of primary dunes is directly related to coastal processes. The sand in secondary dunes comes from the coast but other processes also intervene. Coastal dunes include eolian sheets, blowouts, and parabolic dunes.

A poor sand supply to coastal dunes causes sand to spread inland and form a thin layer. However, an abundant sand supply forms ridges that run parallel to the coastline and move inland slowly. The ridges can be more than 100 m high and have straight slip faces (Goldsmith, 1985). An abundant sand supply to coastal dunes can bury the vegetation; in this case the dunes look very similar to the desert dunes. These dunes are known as transgressive and are usually the transverse type. Abundant vegetation causes a continuous ridge or sometimes a system of ridges to develop parallel to a prograding coast. These ridges can grow at high rates. When vegetation is sparse as on arid coasts, ridges are discontinuous and form linear vegetated sand mounds.

The evolution of coastal dune ridges can give rise to highly complex morphologies with blowouts, transverse dunes, and parabolic dunes through the successive building of parallel ridges that may develop over hundreds of years (Pethick, 1984).

10.10.6 Topographically anchored dunes

When moving eolian sands encounter topographical obstructions such as hills, cliffs, blocks, or buildings, the airflow accelerates and expands around the obstacle. These circumstances cause erosion or sand accumulation and the resulting dunes are static.

Wind tunnel experiments (Tsoar, 1983b) have provided abundant data on the relationship between topographic obstacles and dune generation. If the wind blows perpendicularly on a slope that dips more than 60°, it creates **echo dunes**, which are separated from the obstacle by a distance equal to three times the height of the obstacle. A reverse flow occurs at the base of the cliff that sweeps the sand and creates a corridor between the obstacle and the sand ridge. A small initial echo dune increases in height by this mechanism until it reaches the equilibrium condition, which is 0.3–0.4 times the height of the obstacle. If the scarp is sinuous, the dune shape varies along its length.

Sand is transported over obstacles that have slopes of less than 30°; however, at higher inclinations sand is deposited and forms **climbing dunes** (*Fig. 10.47*). They are common on desert hills and cliffs. Sands that reach the crest of the cliff are deposited when they encounter a zone of reduced speed. These eolian accumulations are called **cliff-top dunes**. Leeward of wide obstacles, the wind decreases, forming **falling dunes**. When the obstacle is narrow, **lee dunes** form; these are usually straight sand ridges that can extend for long distances in the shelter of the obstacle. In the Draa of Malichigdane in Mauritania lee dunes can extend more than 100 km (Breed and Grow, 1979). Some dunes that occur around rounded craters on Mars have been interpreted as lee dunes (Greely and Iversen, 1985). **Lunettes**, half-moon shaped dunes, were described in Section 10.6.3.

10.11 Desert dust

The last two decades of the 20th century saw increased interest in eolian dust deposits as paleoenvironmental records. Desert dust occurs in terrestrial sequences, deep-sea sediments, and cores from polar ice caps. Research provides information about changes on the extent and intensity of aridity, glacial periods and advances, and changes in global atmospheric circulation (Pye, 1995).

Entrainment, transport, and sedimentation of eolian dust are important not only to geomorphologists but also to meteorologists, climatologists, ecologists (McTainsh and Strong, 2007), and environmental scientists. In addition to geomorphic implications that open up a whole range of research possibilities (loess, varnish, crusts, erosion, and so forth), eolian dust is also of interest in applied studies, which are increasingly important because of population growth in arid zones (Péwé, 1981; Middleton, 1997).

10.11.1 General characteristics

Desert dust consists of small particles carried by wind in arid environments. These particles can be suspended in the air or deposited. This atmospheric dust can come from environments other than arid zones (examples are cosmic dust, volcanic dust, marine salts, etc.) and the source areas can be identified by their composition and size (Goudie and Middleton, 2006).

Figure 10.47 Climbing seif. Aswan (Egypt). Photo J.L. Peña.

Inhabitants of western Europe are familiar with desert dust because almost every seven years or so there are "red rains" made up of eolian dust from the Sahara desert; these red rains leave a thin film on the surfaces of automobiles. The sedimentation rate of this dust is 0.25 kg/m^2/year (Goosens and Offer, 1990) on average. Globally it is estimated that the amount of dust removed from deserts is about 1.8–2 billion tons per year (D'Almeida, 1989). A dust storm is defined as visibility conditions of less than 1000 m due to dust. About ten dust storms occur every year in Egypt, about 30 in China, and an average of 60 in the Mexico City and in Kazakhstan (Goudie, 1978).

Desert dust contains two principal particle sizes that are distinguished by their travel distance. Dust that has traveled less than 100 km is between 0.005 to 0.05 mm in size; particles smaller than 0.002 mm travel longer distances (Péwé, 1981). The largest particles are transported by whirlwinds and dust storms and produce extensive loess deposits whereas smaller particles move as aerosols in the troposphere and remain suspended until they are entrained by rain and deposited (*Fig. 10.48*).

Large dust clouds and fronts are called **dust storms** or *haboob*. At the front of a storm (*Fig. 10.49*) hot air rises

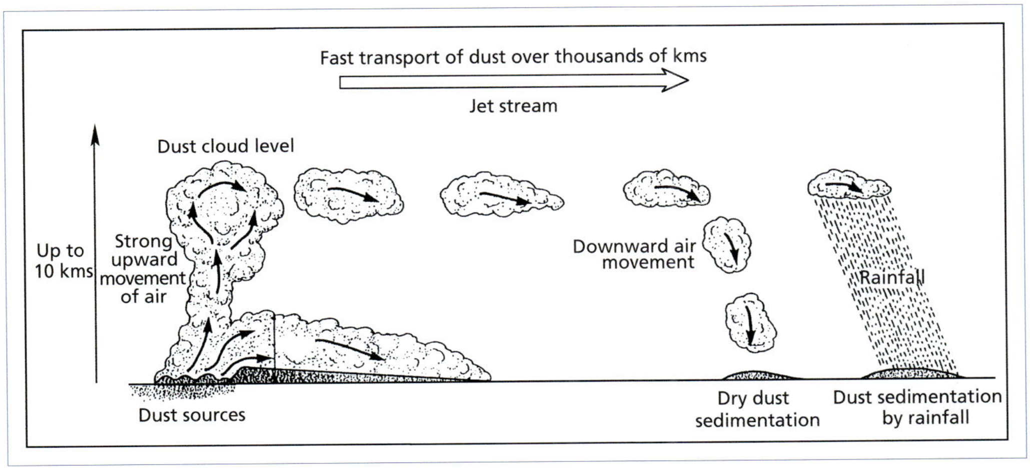

Figure 10.48 Schematic representation of types of transport and sedimentation of eolian dust (Pye, 1995).

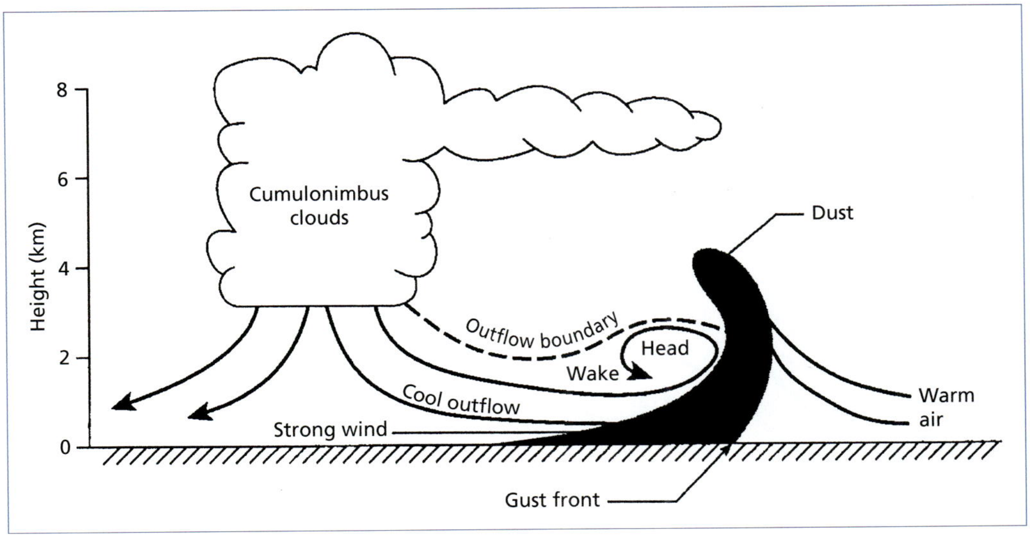

Figure 10.49 Cross section of a dust storm produced by cold flow from a cumulonimbus, with the principal margin propagating along the storm front, called the flow boundary. The strong winds that prevail at the boundary constitute the storm front. The top margin of cold are is called the *nose* and the upper part is known as the head. The rapid movement of cold air and the dust storm front (in shadow) rises into the atmosphere (Warner, 2004).

past the condensation level and produces rain. Dry air that reaches high levels of the storm descends and cools by raindrop evaporation. Descending dry air hits the soil and is deviated to the front of the storm where it moves in large lobes that form a dust front that can rise 1000–2500 m. Observations indicate that these lobes can reach a velocity of 50 m/second (Pye and Tsoar, 1990; Shao, 2008).

The frequency of dust storms varies considerably and it has fluctuated throughout the Quaternary (Goudie and Middleton, 2006). In Nouakchott (Mauritania) there is a big contrast between the variation in the frequency of dust storm days in a year and total annual precipitation. In 1970 precipitation was 48.1 mm whereas in 1979 it was 17.9 mm; these numbers represent 32% and 12%, respectively, of average rainfall for the period of 1949–1967. The number of dust storms increased from 6 in 1960 to 65 in 1974 (Goudie and Middleton, 1992). Storm frequency has increased in recent decades in northern Africa as a result of increased drought. The surface of eolian dust has increased as the Aral Sea has dried up. However, the Great Plains of the United States and in part of China and Australia show a clear decrease in dust storms, perhaps due to improved soil use (Goudie and Middleton, 2006).

10.11.2 Source area and generating processes

Desert dust is composed of inorganic and organic materials. The most abundant mineral is quartz followed by feldspars, calcite, dolomite, mica, clay minerals, oxyhydroxides, heavy minerals, gypsum, halite, and opal. The mineral composition may indicate the source area of the dust. For example, Criado and Porta (2003) found that the dust rain from a January 1999 storm in the Canary Islands came in part from the Precambrian basement of Tiris (south of the old Spanish Sahara) based on its quartz content (Alia, 1945, 1952). The organic part is composed of diatoms, phytoliths, spores, and pollen. As we would expect, composition varies according to source area and distance.

The principal areas that now contribute to eolian dust are subtropical deserts located in a wide belt from western Africa to central Asia and arid and sub-humid regions where dry soils are exposed to strong winds several times during the year (Péwé, 1981; Middleton *et al.*, 1986) (*Fig. 10.51*).

The Sahara Desert is one of the world's most significant exporters of eolian dust (Goudie and Middleton, 2001; Giles, 2005; Engelstaedter *et al.*, 2006). Transport and sedimentation of this dust have environmental effects on climate change, pedogenesis, oceanic sedimentation, and interstratification in glacial ice (Goudie, 1978). Eolian dust also creates problems in terrestrial and air transport, illness related to the transport of pathogenic particles, and respiratory problems for people and cattle among others (Idso, 1967; Goudie, 1978; Pewé, 1981 and others). Eolian dust can be seen on satellite images, which can be used to determine dust direction and speed. Large dust storms can cross the Atlantic Ocean to the Caribbean Sea (*Fig. 10.51*) where they provide nutrients (Idso, 1967; Pewé, 1981).

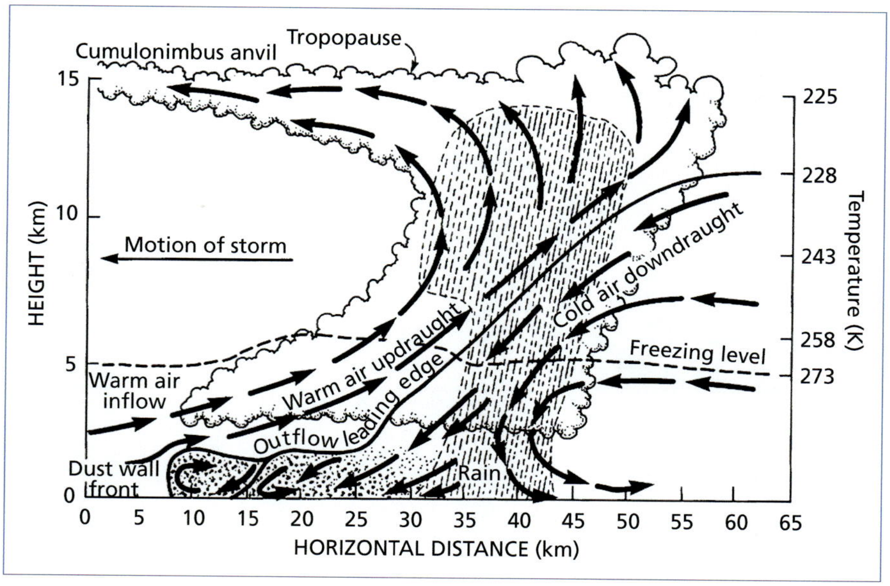

Figure 10.50 Main deserts, fundamental directions and distances of eolian dust transport (Péwé, 1981).

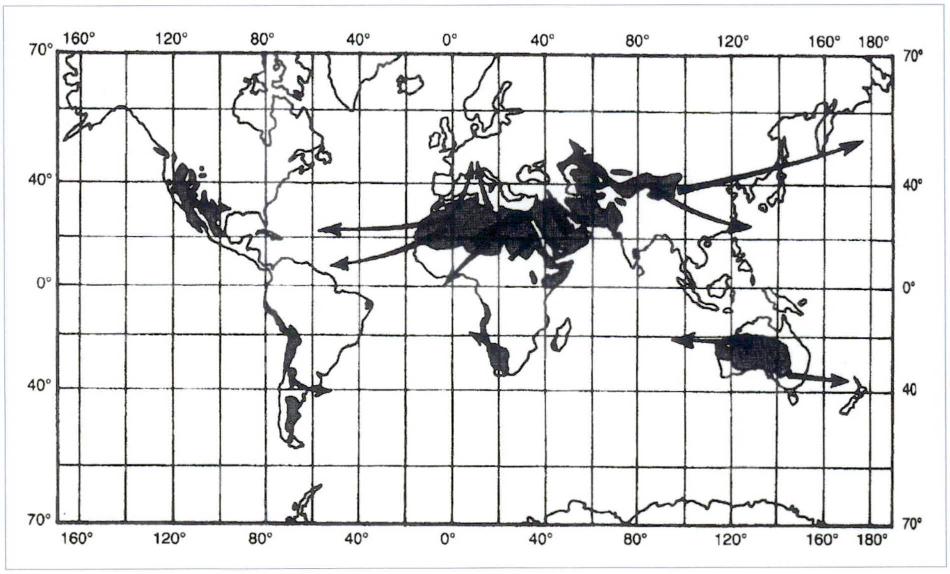

Figure 10.51 Main deserts, fundamental directions and distances of eolian dust transport (Pewé, 1981).

Dust storms were initially studied from data recorded by weather stations (Coudé-Gaussen, 1991) but today they are studied by the IDDI (*Infra-red Dust Index*), which uses decreased temperatures to map the distribution of mineral aerosols in Africa (Brooks and Legrand, 2000). More recently, the TOMS technique (*Total Ozone Mapping Spectrometer*) provides better results (Herman *et al.*, 1997; Goudie and Middleton, 2001, 2006; Prospero *et al.*, 2002; Washington *et al.*, 2003, 2006). Prospero *et al.*, (2002) used the aerosol index from NIMBUS-7 to examine the distribution of dust sources throughout the world over a period of 13 years (1980–1992). The Bodélé Depression (Chad) is the world's most significant source of eolian dust emissions (Giles, 2005; Goudie and Middleton, 2006; Warren *et al.*, 2007) (Fig. 10.52). Farther west another important zone covers eastern Mauritania, western Mali, and south-central Algeria (Goudie and Middleton, 2001, 2006; Prospero *et al.*, 2002; Washington *et al.*, 2003, 2006; Engelstaedter *et al.*, 2006). The Bodélé Depression, associated with Lake Mega-Chad, is exposed to strong winds over an area that lacks vegetation; as a result it offers optimum conditions for deflationary processes (Washington *et al.*, 2006).

Areas that most commonly generate eolian dust are characterized by a cover of loose sediments composed of significant quantities of sand and silt but little clay. *Ouads* are an example of this type of environment: they contain poorly sorted and non-vegetated sediments. They are usually braided channels with extensive lateral migration. Playas, chotts, and littoral sabkhas, although varied in composition, contain zones characterized by the presence of fine-grained material, which is another source of eolian dust. Alluvial fans are supply zones especially their middle and distal areas where the percentage of fine particles is significant. Ancient dry lakes also can supply diatom dust, salt, gypsum, and clay minerals. Large regions covered by loess sediments are important dust source areas particularly where vegetation is lacking. Volcanoes are an important source of dust as shown by the emissions from Mount St. Helens in the United States in 1980 and Pinatubo in the Philippines in 1991. Both volcanoes emitted large quantities of dust into the atmosphere where it was later transported over significant distances (Pye, 1987; Cooke *et al.*, 1993; Livingstone and Warren, 1996).

Various mechanisms can create fine particles. Chemical weathering produces the majority of clay-sized particles; however, the origin of the silt-size particles that constitute loess is poorly understood and subject to controversy. Frost weathering can play a fundamental role in generating fine particles in cold deserts. Processes of thermal weathering, wetting and drying, and salt weathering are the basic mechanisms that produce fine-grained material in deserts. We can add eolian abrasion to this list. In playa-lake environments silt and clay particles form agglomerates that are exported by the wind. Eolian dust can incorporate particles from profiles that have undergone chemical weathering in tropical humid environments. Finally, another source is material of organic origin: phytoliths, diatoms, radiolarians, echinoderm spicules, pollen, spores, and so forth. The generation of small-size particles increases with human activity as the human population increases (El-Askary and Kafatos, 2008).

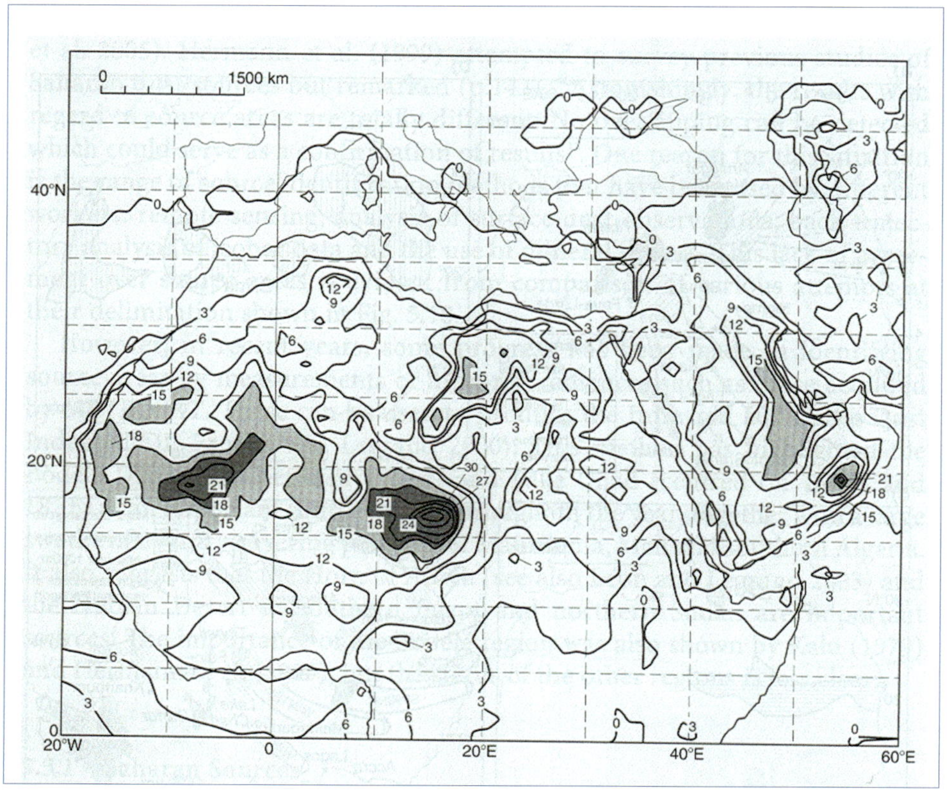

Figure 10.52 Average values of the aerosol index of North Africa and the Middle East from TOMS data (1980–1993; 1997–2000) (Middleton and Goudie, 2006).

10.11.3 Entrainment, transport and sedimentation

The entrainment of dust particles from the soil surface is a function of wind speed; wind type, sediment characteristics, and the presence of obstacles (vegetation cover, topography, and so forth) control the process (Middleton, 1997). Strong winds can mobilize and transport dust particles under different conditions and at different scales (Pye, 1987).

Wind types capable of transporting dust range from simple gusts to dust storms with advance fronts up to 1000 km long (*Fig. 10.48*). **Dust devils** are small convective vortices that occur when the ground surface has been intensely heated, causing upward movement of air and dust. These inverted cone-shaped vortices are visible over long distances. They commonly reach heights of 3 to 100 m although they can exceed 1000 m (Idso, 1967). They are 0.5–3 m in diameter on the ground. They move laterally and can cause significant damage.

Dust storms are called **haboob**, an Arabic word that means violent wind. Air currents descending from large cumulo-nimbus clouds probably generate them. They resemble dust walls with heights that range from 300 m to 3000 m and have speeds as high as 200 m/s. They reduce horizontal visibility to one kilometer or less (Middleton, 2004). On satellite images they measure about 500–600 km wide and form plumes up to 2500 km long (Péwé, 1981).

Other winds that also transport fine particles are **dust plumes**, which move parallel to the soil surface as horizontal vortices. The **shamal** is a wind that blows to the north during the summer in Iraq and the Persian Gulf. Most of these dust-laden winds do not reach one kilometer in height and their ground speed is less than 40 km/hour. Another wind that transports eolian dust is the **katabatic wind**, which blows sediments from the foothills of large mountain ranges, like the **zonda** wind in Argentina and the **Santa Ana** in California. Similar winds occur in the Andean altiplano and on the Tibetan plateau (Pye, 1987; Livingstone and Warren, 1996; Middleton, 1997).

Globally, the quantity of particles moved by dust storms is estimated at a billion tons per year, approximately half of it from the Sahara (Middleton, 2004). Bullard and Livingstone (2009) estimate that each year three billion tons of dust are removed from the Earth's surface. Saharan dust has a significant impact on climate processes, nutrient cycles, soil formation, and sediment cycles. Dust storm activity in the

Sahara occurs in areas with median annual precipitation of 100–200 mm (Goudie and Middleton, 2001).

In semiarid zones of medium latitudes, the passage of cold fronts is probably one of the most important causes of eolian dust mobilization. Like haboobs, they create large dust walls up to three kilometers high and move at maximum speeds of 40 m/second. They have been described in northern China, Israel, Egypt, the High Plains of Texas and New Mexico, and in southern Australia (Pye, 1987).

Sometimes the dust reaches such heights that the jet stream transports it at high speeds over distances of thousands of kilometers. Thus Saharan dust travels to central and northern Europe and crosses the Atlantic Ocean where it has been identified on Caribbean Islands (Prospero *et al.*, 1970) *(Fig. 10.51)*. Eolian dust can move from China to the French Alps (>20,000 km) (Grousset *et al.*, 2003).

Dust particles are deposited as dry and humid sedimentation (Goudie and Middleton, 2006; Shao, 2008). Dry sedimentation consists of a dust flow from the atmosphere to the surface by means of turbulent molecular diffusion and later gravitational fall. Humid sedimentation is based on dust flow to the surface via rainfall.

Pye (1987) shows that sedimentation of eolian dust occurs in several ways: a) reduction of wind speed and turbulence; b) aggregation of electrostatically charged particles; and c) entrainment of atmospheric dust by rain (Pye, 1987). Although sedimentation also occurs in oceans and wet areas, we shall deal essentially with dust deposited in arid zones. The quantities deposited are variable but the average is about 100 t/km^2/year near source areas (Goudie, 1995). This amount decreases quickly with distance from the source area. Coarser sized desert dust travels in the lower parts of the atmosphere and is often deposited when it encounters obstacles perpendicular to wind flow. Nevertheless the finest particles, which travel in the higher parts of the atmosphere, are not affected by roughness of the terrain and can remain suspended for long periods of time (Coudé-Gaussen and Rognon, 1988; Coudé-Gaussen, 1991).

Dust storm activity was at its height during the glacial maximum possibly because of reduced precipitation (Middleton, 2004). Hyperarid zones contain only small quantities of dust because precipitation is rare and few obstacles are present. However, large quantities of eolian dust are deposited on desert margins, which are semiarid zones. These particles are easily trapped by dispersed vegetation. In northern Nigeria, up to three meters of desert dust have been deposited during the last 40,000 years (McTainish, 1987). At present, sedimentation rates range from 0.1 to 0.7 mm/year in different areas of Israel and the Sahel. In the Bodelé Depression (Chad) deflation has removed 4 m of sediments (diatomite) at a rate of 1.6 mm/year; the sediments were deposited during earlier humid phases (Bristow *et al.*, 2009).

Dust particles affect climate directly and indirectly. Indirect influences are seen in the effects of eolian dust on the biogeochemical cycle (Moreno and Canals, 2004). Changes in the quantity of dust in the atmosphere modify the balance of radiation and thus surface temperatures (Goudie and Middleton, 2006). Ridgwell (2003) shows that dust can affect fertilization of oceanic biota, and in turn, take CO_2 out of the atmosphere, reducing the greenhouse effect.

10.11.4 Geomorphic implications

Desert dust has considerable importance in our attempt to understand geomorphic processes that have implications for human and economic activities.

Dust contributes to the formation of various types of surface formations in arid and semi-arid areas. For example, calcium carbonate contributions to soils produce **caliche** in areas that lack calcareous rocks as shown by studies of calcretes in Namibia and southeastern Spain (Blümel, 1982). Silica-rich dust may be important in the formation of **silcretes** in deserts (Summerfield, 1983). Likewise, deposition of gypsiferous dust is the primary cause of **gypsum crust** formation in southern Tunisia (Coque, 1955). **Desert varnish** is a thin manganese and iron rich film that covers rocks in arid zones. As discussed previously, varnish-forming particles come from eolian dust deposition. Laminations in varnish can be carefully analyzed and used for paleoclimatic interpretations based on fluctuations of manganese and iron concentrations (Oberlander, 1994). Desert dust can also affect the **weathering** of buildings. In **dune** areas, sedimentation of dust and salts can form a crust capable of stabilizing the dunes by promoting the formation of biocrusts and establishment of plants (Pye and Tsoar, 1987). Eolian dust has also been found in playas, sebkhas, alluvial fans, and ouads where it causes compositional modifications. This dust is quickly mobilized by water within these depositional environments. Many **soils** in desert and peridesert regions contain significant quantities of eolian dust and some soils are almost totally composed of these allochthonous supplies. According to Rapp (1984) the *terra rossa* of Spain, Italy, and other Mediterranean countries is essentially composed of reworked dust from northern Africa. This desert dust fertilizes many soils including some in the Caribbean Islands. It also affects soil salinization and acidity (Goudie and Middleton, 2006).

Desert dust deposited in oceans and seas constitutes a significant percentage of **oceanic sediments** *(Fig. 21.31)*. The study of these accumulations provides useful information on modifications to wind systems and environmental changes that have occurred on continents during the Quaternary (Middleton, 1997). This research provides information on arid phases and their intensity in terrestrial environments. Ice accumulations, especially in polar **ice caps**, preserve a record of eolian dust that can be retrieved from test drillings. In Greenland during the Last Glacial Maximum (18,000 years ago) the dust concentration was 40 times greater than today. Phases with high dust content are followed by others with small or nonexistent quantities of dust, indicating the alternation of high and low aridity periods.

10.11.5 Loess

Loess, a term derived from the German *löss*, is a clastic terrestrial sediment, essentially composed of accumulations of silt-sized particles; it occurs as massive outcrops *(Fig. 10.52)*. Loess forms from the accumulation of eolian dust (Smalley and Smalley, 1983; Pye, 1995; Pye and Sherwin, 1999). It can have a variety of colors including gray, white, yellow, brown, and red. Loess can contain buried soils or paleosoils *(Fig. 15.22)* that can provide a record of glacial-interglacial cycles when they are dated by thermoluminescence (Muhs, 2007). Primary loess is of eolian origin; it can be distinguished from secondary loess, which has been re-deposited or has originated from non-eolian processes. Loess outcrops show a degree of chemical and biological weathering and pedogenic processes.

Loess accumulations cover about 10 percent of the earth's terrestrial surface (Pecsi, 1968). They occur mainly in Eurasia (Haase *et al.*, 2007) *(Fig. 20.28)*, the United States *(Fig. 15.20)* (Bettis *et al.*, 2003), and Argentina *(Fig. 10.19)*, (Muhs and Bettis, 2003). Loess covers more than a million square kilometers of China. The loess layer has an irregular thickness and covers irregularities of the terrain. It is usually less than 30 m thick although it exceeds 300 m in Lanzhou, China (Derbyshire, 1983b). The largest loess accumulations on the planet are found in China where they cover an area of 440,000 km^2 (Liu, 1985). The size of loess particles ranges from 20 to 40 μm, with a tendency to very fine sizes. When loess contains more than 20 percent sand it is called sandy loess and when the clay content exceeds 20 percent it is called clayed loess (Pye, 1987).

Loess mineralogy is highly variable because it reflects the composition and evolution of the source area. Quartz usually predominates (45–55%) followed by feldspars, carbonates, heavy minerals, volcanic glass, and clay minerals (Mush, 2007). More than 50 percent of loess is composed of silt-size particles but the clay and sand contents are highly variable.

Loess tends to collapse mechanically. When dry, it can support vertical loads but its shear strength decreases considerably when it becomes wet; underlying material flows and slides (Middleton, 1997). Dry loess has vertical fracture systems that can be observed in scarped outcrops.

The formation of a loess deposit can be simple or highly complex. Loess can form as a result of weathering and later deflation. An intermediate phase of fluvial transport usually precedes eolian activity. The initial particles can result from glacial weathering and/or erosion or undergo transport and deposition as a result of fluvioglacial activity; they are finally mobilized by deflation (Smalley, 1972).

Many processes produce silt-sized particles: removal of particles from bedrock, glacial crushing, gelivation, fluvial abrasion, eolian abrasion, haloclastism, chemical weathering, clay pellet aggregation, and biological processes (see Pye, 1995). Because of the processes that create loess, some authors call the extensive loess deposits related to Quaternary continental glaciations **periglacial** or **glacial loess**. Others call loess on the margins of arid zones **peri-desert** loess (Smalley, 1995; Muhs and Bettis, 2003).

In central Asia deserts, particles are weathered in high mountain environments and later transported and deposited by rivers in lower areas where they are entrained by deflation (Dodonov, 2007). During the Quaternary large quantities of sediments, including silts, were produced in these tectonically active areas (Smalley, 1990). They are called **peri-mountain loess** on the basis of the environment where they were generated. In contrast to the deserts of central Asia, deserts located on shields and platforms do not have extensive loess deposits because the latter areas have low relief, are tectonically stable, and, as a result, produce sparse sediments (Smalley, 1995). Iriondo and Krohling (2007) have pointed out the difference between the non-traditional loess in the Ebro Depression (Spain) where the dominant wind ("*cierzo*") deflates gypsum-bearing particles and terrace, alluvial fan, and pediment deposits.

A thick and extensive accumulation of loess deposits requires an abundant dust supply over tens or even hundreds of thousands years and the presence of topographical and vegetation barriers to slow down the dust. In low-relief deserts, sediment production is poor (peri-desert loess) whereas in mountainous neotectonically active deserts, sediments production is very high (peri-mountain loess).

10.12 Wind risk: Problems and control

Wind moves sand and dust in many climate types but wind activity is greatest in arid areas. Because of sparse or absent vegetation in these areas, the soil is subject to wind action. Limited soil development also means that the land surface is easily eroded; these conditions affect 39 percent of the world's arid land (UNEP, 1992).

A **tornado** is a funnel-shaped vortex of air that circulates rapidly from the ground surface up to a cumulonimbus cloud *(Fig. 10.53)*. About 600 tornadoes occur in the

United States every year (Eagleman, 1983). Tornadoes move at speeds of 50–200 km/hour, which is faster than cyclones; they generate internal winds that exceed 400–500 km/hour. They are accompanied by heavy precipitation (Olcina and Miró, 2002).

Tornadoes can be highly destructive (Nalivkin, 1983). The lifting force of tornadoes is significant; they can move objects that weigh 200–300 tons over distances of 10 meters. They are capable of suctioning millions of tons of water and forming marine waterspouts (*Fig. 10.55*). Between the years of 1916–1950, 5,204 tornadoes in the United States caused 7,961 deaths and losses estimated at 500 million dollars (Bryant, 1991). Tornadoes occur primarily in the central and southeastern parts of the United States (*Fig. 10.54*) (Gillette and Hanson, 1989). Most of them occur in springtime and the beginning of summer when warm and humid air from the Gulf of Mexico meets cold dry air heading south from Canada and eastward from the Rocky Mountains. Destructive tornadoes also occur in Russia, central and southern Asia, Africa, and Australia (McGuire *et al.*, 2004).

Tornadoes are classified on the Fujita Scale (*Table 10.2*), which relates wind speed to damages caused. The Pearson Scale, which measures the length and width of the tornado trajectory, is also used.

Tornadoes are rare in Spain; they occur primarily on the Mediterranean coast and a few occur in the interior of the Peninsula (Olcina and Miró, 2002). Gaya (1996, 1999) has shown that about six tornadoes occur every year in Spain including marine waterspouts (*Fig. 10.55*). They occur primarily between May and October. They have intensities of F0 to F3 on the Fujita scale and winds between 182 and 332 km/hour. Locally, tornadoes can cause intensely destruc-

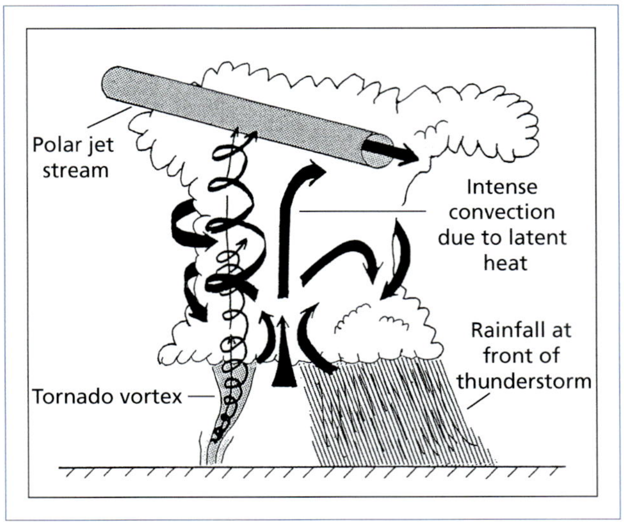

Figure 10.53 Origin of a funnel-cloud tornado with storm that is being traversed by a jet stream (Eagleman, 1983).

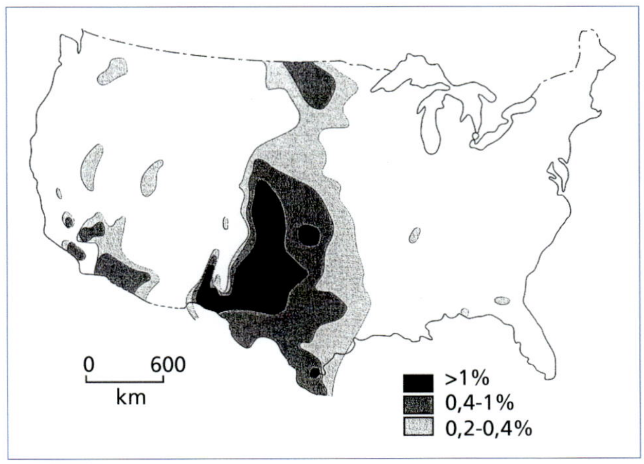

Figure 10.54 Tornadoes in the United States. Annual frequency of tornado hours with visibility of less than 11 Km (Gillette and Hanson, 1989).

Table 10.2 FUJITA TORNADO DAMAGE SCALE (MCGUIRE *ET AL.*, 2004)

F number	Wind speed (ms^{-1})	Damage	Expected effects
F0	18–32	Light	Chimneys and signs damaged; tree branches broken.
F1	33–49	Moderate	Roofs damaged; mobile homes overturned; moving cars pushed off the road.
F2	50–69	Considerable	Roofs torn off; mobile homes demolished; large trees uprooted.
F3	70–92	Severe	Walls torn off buildings; cars lifted; most trees uprooted.
F4	93–116	Devastating	Well constructed homes destroyed; cars thrown.
F5	117–142	Incredible	Buildings lifted off foundations and carried considerable distances to disintegrate; cars thrown more than 100 m; trees debarked.

Figure 10.55 Marine waterspout near the coast of Cádiz (Spain). Photo F.J. Gracia.

tive effects on buildings and agriculture; an example is the event that occurred on 8 August 1992 in the Ejea de los Caballeros (Province of Zaragoza) (*Fig. 10.56*).

In the United States "derechos" or downbursts are related to tornadoes in the sense that they are intense storms. Downbursts contain winds with speeds faster than 25 m/second; they can affect areas that extend 500 km. They are less destructive than tornadoes; in the United States they are two orders of magnitude smaller than tornadoes (McGuire *et al.*, 2004).

The movement of particles driven by tornadoes and dust storms has a considerable effect on human settlements; it causes tremendous damage to buildings, crops, and transport and communications networks (Lockeretz, 1978; Middleton *et al.*, 1986). In addition, wind deposit accumulations can cover existing vegetation and increase the problem of desertification (UNESCO-FAO, 1977). Environmental planning should be used to avoid these high-risk areas (Cooke *et al.*, 1982).

Deflation, transport, and sedimentation can all cause problems. Deflation can reduce the quantity of fine particles and soil nutrients, and cause loss of soil fertility (Middleton, 1990). It can also undermine structures such as lampposts or telephone posts and cause them to collapse. As they are transported, particles cause abrasion that produces hollows and furrows. Dust in the wind can also reduce visibility at airports and suffocate people and animals (Idso, 1967; Pewé, 1981; Pye, 1987; Middleton, 1997; Goudie and Middleton, 2006). For these reasons, Arizona has set up a dust storm alert system that consists of electronic signboards on roads and announcements on public radio to warn its citizens of danger. Dust can also enter houses and cause health problems by contaminating food and drinking water. Saline dust can adversely affect buildings. However, wind-blown dust also has a positive effect: it can provide soil nutrients to new areas.

The amount of dust that moves around the globe is on the same order as the magnitude of sediments transported by rivers (Livingstone and Warren, 1996). A significant part of windblown dust comes from human activities. Agricultural work such as plowing breaks up soil aggregates and in dry periods plumes of dust can be seen behind tractors (Lee *et al.*, 1993). Traffic on unpaved roads is another source of dust. Removal of vegetation by grazing or fire can also increase wind erosion (O'Hara, 1997). Desiccation of the Aral Sea has increased the amount of windblown dust (Goudie, 1994b). Wind sedimentation can bury houses, crops, canals (*Fig. 10.57*) and roads (*Fig. 10.58*). If the wind along a highway always blows sand in the same direction, engineers can slope the highway slightly into the wind to reduce maintenance costs (Kerr and Nigra, 1952) (*Fig. 10.59*). If large areas are affected it can be costly to dig out buildings and land (Livingstone and Warren, 1996).

Control measures should be aimed at reducing the wind's transport capacity, reducing the sand supply, and,

Figure 10.56 Destruction of grain silos as a result of the 8 August 1992 tornado. Ejea de los Caballeros (Province of Zaragoza, Spain). Photo C. Sancho.

Figure 10.57 Acequia partially buried by sand. This is part of the plan to irrigate a portion of the Thar Desert from the Grand Canal of Indira Gandhi. Jaisalmer (Province of Rajastan, India).

Figure 10.58 Dunes advancing toward a highway. Maharaga, south of the Rub al Khali Desert (United Arab Emirates).

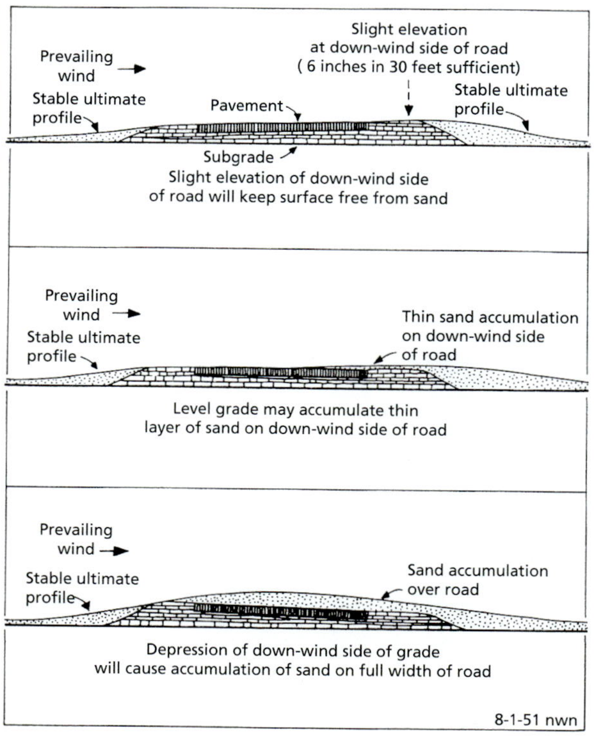

Figure 10.59 Effect of the inclination of a highway surface on sand accumulation (Kerr and Nigra, 1952).

if necessary, deviating windblown sand (Watson, 1990; Pye and Tsoar, 1990; Mainguet, 1991).

Wind control techniques (Cooke *et al.*, 1982; FAO, 1988; Watson, 1990; Nordstrom, 2008) include re-establishing natural vegetation or planting other species to stabilize the surface, depending upon the aridity of the area. In hyper-arid areas, the extreme lack of water may make this impossible. Fast-growing species such as *Tamarix* sp., acacias and eucalyptus should be planted (García Salmerón, 1967; Pye and Tsoar, 1990). Authorities in China have planted vegetation corridors 300–500 m wide next to the railway from Bagoton to Lanchou to slow down erosion (Watson, 1990).

Another control method is surface stabilization, which consists of using stones, oil, asphalt, synthetic latex, polyvinyl, gelatin or cellulose fibers to cover the surface. This method is usually effective for less than five years. Chemical sprays have been widely used to create a chemical crust a few millimeters thick; this method has produced good results when combined with the re-establishment of vegetation (García Salmerón, 1967; Cooke *et al.*, 1982; Livingstone and Warren, 1996).

Palisades are commonly used in desert areas to slow down the wind and trap the sand it carries (*Figs. 10.60* and *10.61*). They are commonly made from palm leaves and from low cost wood from local vegetation (Nordstrom, 2008). Both sides of these structures reduce the transport capacity of the wind. The volume of sand that accumulates is proportional to

Figure 10.60 System of palisades to protect a palm grove. Nefta (Tunisia). Photo J.L. Peña.

Figure 10.61 Group of parallel palisades. Sinai Desert. Photo J. Rosell.

the height of the palisade used. They should be installed at some distance from the object to be protected and perpendicular to the prevailing wind. In the case of newly constructed buildings, prior to construction the builder should investigate the vulnerability of the proposed location to sand movement and dune invasion (Peters, 1964). A series of palisades should be set up at different distances from the object. In this way, as the most distant palisade fills rapidly, the lifespan of the intermediate windbreak will be increased by a factor of four, and the closest by a factor of nine (Fig. 10.62) (Kerr and Nigra, 1952). Some palisades are set up in a zigzag fashion to counteract winds from several directions (Fig. 10.63) (Watson, 1990).

The numerous techniques used to stabilize dunes vary in effectiveness. Sometimes it is enough to change a dune's direction. Small dunes can be removed simply by moving the sand elsewhere; this procedure is cheap if the sand is subse-

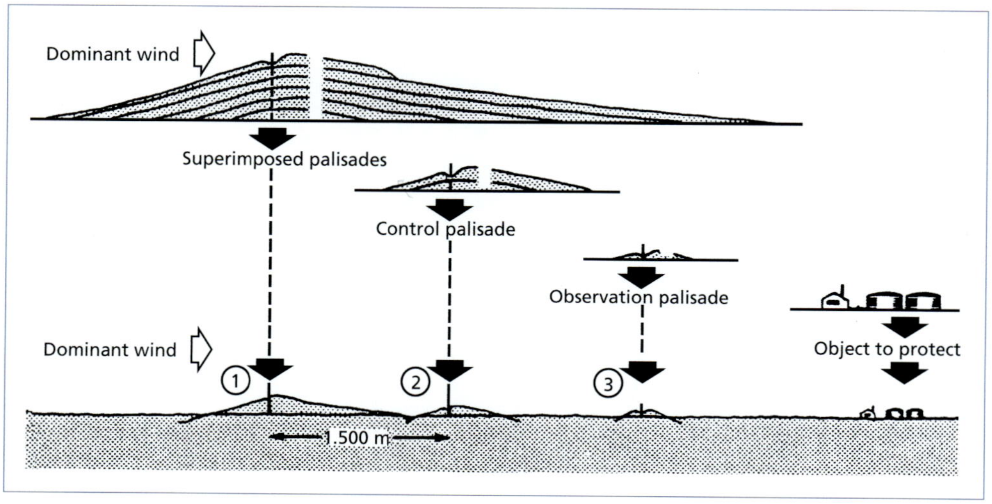

Figure 10.62 Series of parallel palisades to protect extensive areas such as towns, industrial plants, etc. (from Kerr and Nigra, 1952).

Figure 10.63 Palisades installed in zigzag pattern to trap sand transported by winds from varying directions. Photo J.L. Peña.

quently used in construction. In other cases, tree and shrub planting can stabilize a dune. Digging ditches parallel to the dune axis can destroy a dune but this solution is expensive and temporary. Stabilization with vegetation can be very expensive; it is more useful to use palisades or to cover dunes with stones. Applying a surface treatment of oil to the arms of crescent-shaped dunes causes deflation of the central part; a very much larger-than-original mass of sand then accumulates on the windward side (*Fig. 10.64*) (Kerr and Nigra, 1952). During the construction of the Trans Caspian railroad, engineers used water to soak the wheel ruts of the highway to provide consistency; they intercalated clay layers and placed wooden palisades on upwind dunes to stabilize mobile dunes in the first 50 km from the Caspian Sea to Samarkand (Walton, 1969).

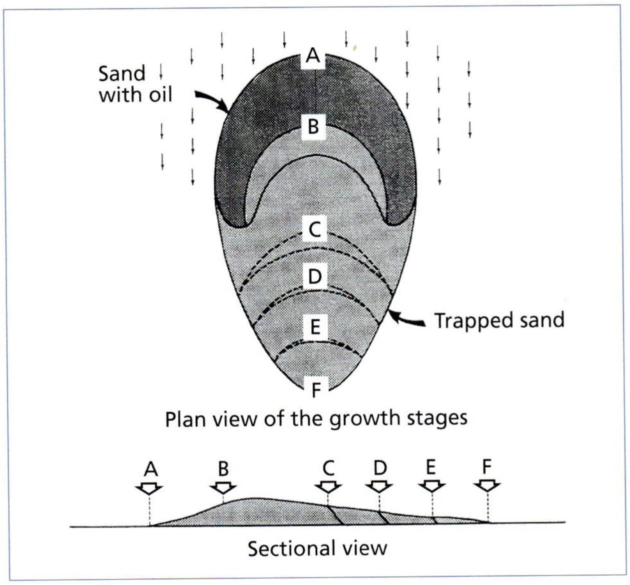

Figure 10.64 Stabilization of a barchan using petroleum as a surface treatment on the arms of the barchan (from Kerr and Nigra, 1952).

In terms of renewable energy, the use of wind is an important technology with a promising future. Wind has a greater motive force in some areas than in others; the former offer the opportunity of **wind energy**. Movement of windmill vanes (*Fig. 10.65*) can be transmitted to turbines to produce electricity. The clustering of large windmill fields leads to electricity production that can be of considerable importance during energy crises as other types of energy (oil, natural gas, nuclear, and solar) become expensive or scarce. As the human population increases almost exponentially our energy needs will increase accordingly. Energy from wind generators produced almost the same amount generated by nuclear power plants in Spain in 2008 (about 23% of Spain's total electricity consumption).

Figure 10.65 Windmills field Villafranca (Province of Navarra, Spain). In the distance, the Moncayo with snow.

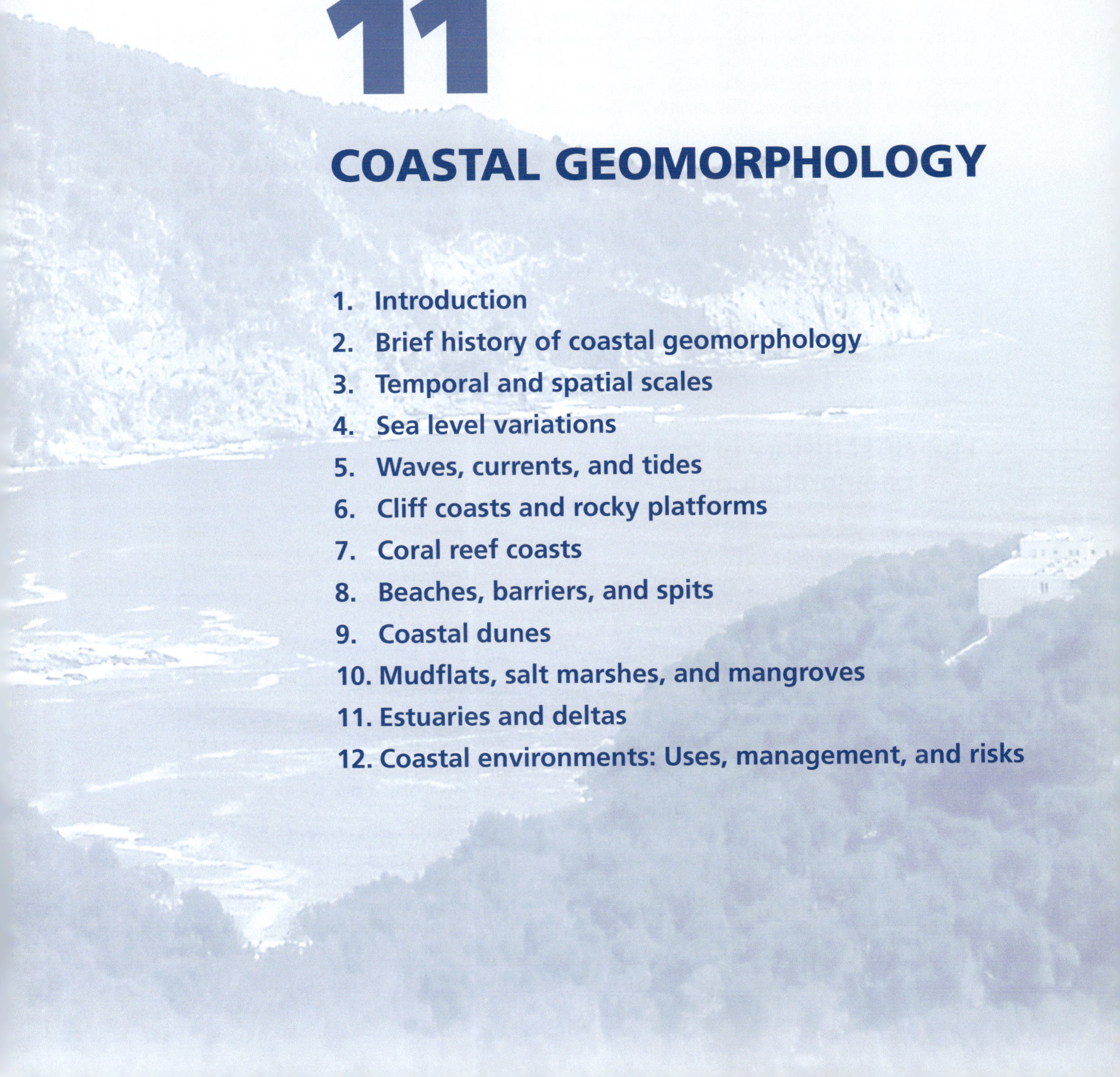

11

COASTAL GEOMORPHOLOGY

11.1 Introduction

Bird (2000) defines coastal geomorphology as the study of the evolution of coastal landforms, the processes that affect them, and the changes they undergo. These landforms include cliffs, rocky coastal margins, beaches, dunes, estuaries, lagoons, and deltas. The world's coastal margins are about 100 m wide and 440,000 km long; humans use them for industrial purposes, transportation, and recreation. These human activities profoundly impact these areas. The United Nations estimates that 66% of the world's population lives within a few kilometers of the coast (Pethick, 1984). As a result, food production, communication, cities, and recreational areas all affect the coast. Coasts have significant problems that include flooding, erosion, pollution, and sea level variations that require constant attention. Interdisciplinary studies by coastal engineers, oceanographers, and geomorphologists have led to significant advances in coastal management (Stephenson and Brander, 2003). In addition, coasts are constantly changing; the change may be catastrophic or imperceptible. Modifications may occur over decades or centuries but they can also take place in a period of hours or minutes.

11.2 Brief history of coastal geomorphology

Prior to the 20th century, research on coastal geomorphology was limited to engineering topics related to coastal erosion (Walker and MacGraw, 2005). Charles Darwin made important observations on coral reefs and the origin of atolls during his famous voyage on the British ship, the Beagle, in 1835 (Shepard, 1959). Suess (1888) in his voluminous book, *The Face of the Earth*, proposed the first tectonic classification of coastlines: a) straight (or Pacific), which are parallel to the coast, and b) irregular (or Atlantic) shown by structures that form angles to the coastline. Based on William Morris Davis' widely accepted ideas on the erosional cycle that were published in 1899, Johnson (1919) applied the cycle to coastline evolution and classified coasts as emersion coasts and immersion coasts.

In the middle of the 20th century, geomorphologists classified coastlines on the basis of coastal structure, described coastlines in stable and mobile regions (Cotton, 1954), and studied stages in their development inspired by the Davisian erosional cycle (Shepard, 1948) and progradational and retreating coasts (Valentin, 1952).

During the 1950s and 1960s, geomorphic process studies transformed geomorphic investigations (Strahler, 1952). These studies focused on research on sediment movement in the coastal environment. In 1970, King published his book *Beaches and Coasts* on the dynamics of coastal processes. Although some investigators conducted process studies, both in the field and in the laboratory, historical geomorphology continued to advance, especially with respect to historical sea level variations, an area in which dating techniques were fundamental (Woodroffe, 2002). In addition, the influence of plate tectonics led to the classification of Pacific coasts (active plate margins) and Atlantic coasts (passive plate margins) (Inman and Nordstrom, 1971). Bloom (1978) distinguished between low coasts and coasts with cliffs.

In the last decades of the 20th century, scientists discovered that during interglacial periods sea level was 80–150 m below the present day sea level. Fairbridge (1961) and many others established eustatic variation curves. These contributions enabled scientists to understand that sea levels had varied along the various coastlines of the world. With this information Pirazzoli (1991, 1996) developed a sea level history along different coasts; one of the objectives of this research was the correlation of various coastlines.

Global warming has generated renewed interest in sea level fluctuation studies because of significant implications that may affect shorelines and their evolution (Barth and Titus, 1984). In Houghton *et al.*, (1996, 2001) and in IPCC (2007) numerous authors have enumerated the factors that contribute to sea level change, past sea level changes, and predictions through the year 2100.

11.3 Temporal and spatial scales

Researchers have investigated the evolution of coastal geomorphology in small and large areas and for highly variable time intervals; in doing so, they have developed a hierarchy of temporal and spatial scales (Cowell and Thom, 1994) (*Fig. 11.1*).

The smallest scale corresponds to the scale of instantaneous time where the principles of fluid dynamics apply. The morphological response to waves and tides, storm waves, tsunamis, etc. occur at this scale (Ingmanson and Wallace, 1989); these are low-frequency high-magnitude events that

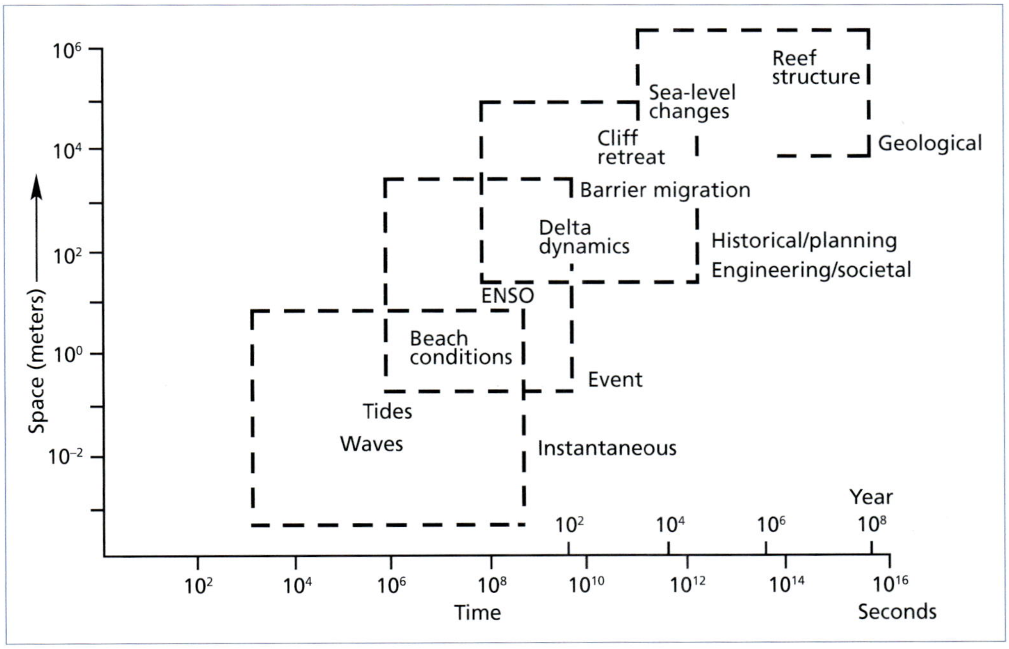

Figure 11.1 Representational of spatial and temporal scales applicable to coastal evolution. Large-size landforms operate at larger time scales, whereas smaller littoral landforms respond to shorter time intervals (modified from Cowell and Thom, 1994, in Woodroffe, 2003).

occupy a higher rank on the scale. The historical approach covers multiple decades and large areas; this approach groups multiple events. Engineering focuses on the coast as modified by humans; planning and social initiatives include the study of future coastal variations and impacts on the human habitat. The geologic time scale takes into account coastal evolution throughout geologic time, particularly during the Quaternary; the geologic scale involves a coastline that is many miles long (Zazo *et al.*, 1994).

11.4 Sea level variations

The coastline is the place where oceans make contact with continents. Oceans cover 70% of the Earth's surface and the sea level at the continental contact defines the coastline. Historical movement of coastlines provides a way to study past sea level changes and crustal movements (Mörner, 1983). It may difficult to distinguish the oceanic and tectonic components of relative sea level changes (Bird, 2000). These oscillations have occurred throughout all geologic time. Indicators of sea level include marine organisms (barnacles, corals, oysters, etc.). Ancient levels are recorded in the coastal stratigraphy and geomorphology along depositional coasts. On rocky coasts, the indicators are emergent abrasion platforms (*Fig. 11.2*), cliffs, and notches (Masselink and Hughes, 2003).

11.4.1 Causes of sea level changes

A primary cause of sea level change is fluctuation of the volume of ocean water. Water is present on the Earth in various forms and its total volume is constant (*Table 11.1*). Most water is found in oceans and seas; snow and groundwater also account for a significant quantity although the amount of groundwater is uncertain (Pirazzoli, 1996).

The oceans are interconnected; thus sea level changes are global or **eustatic** (Suess, 1888). Eustatic changes are of different types (Fairbridge, 1961). **Tectonic-eustatic** sea level changes result from tectonic changes in the shape of ocean basins. The movements may be epeirogenic, orogenic, or isostatic. **Sediment-eustatic** changes are related to the capacity of oceans, which decreases as they fill with sediments transported from land to the sea. Sediment-eustatic changes also a result from sediment compaction. **Glacio-eustatic** changes result from the growth and disintegration of glaciers and ice caps. In addition to these three principal types, **steric** change results from increased water temperature. As the ocean heats up and expands, sea level rises; the opposite occurs as water cools (Warrick, 1993). **Volcanism** can also raise or lower sea level. These changes can also be affected by **hydrostasis,** which occurs as a result of ocean floor subsidence, rapid melting of ice sheets, and increased water volume. In addition,

Figure 11.2 Emergent erosional platform and cliff coast with rocks. Aird Fenish, western part of Lewis Island, Hebrides Islands (United Kingdom).

Table 11.1 **ESTIMATES OF STORAGE VOLUMES AND EQUIVALENT WATER DEPTH. (MASSELINK AND HUGHES, 2003)**

Parameter	Present volume (km^3)	Equivalent water depth
Atmospheric water	13,000	36 mm
Oceans and seas	1,370,000,000	3.8 Km
Lakes and reservoirs	125,000	35 cm
Rivers and channels	1,700	5 mm
Swamps	3,600	10 mm
Biological water	700	2 mm
Moisture in soils and the unsaturated zone	65,000	18 cm
Groundwater	4,000,000 to 60,000,000	11 to 166 m
Frozen water	32,500,000	90 m
Antarctica	29,300,000	81 m
Greenland	3,000,000	8 m
Others	200,000	1 m

geoidal changes can cause sea level changes; the ocean surface is not flat but is determined by the field of gravity. This geoid surface has large undulations and depressions (Mörner, 1976 and 1980a). The following also contribute to sea level changes though to a lesser extent: **ocean currents** in cross gradients that develop when currents are strong and **anthropic activity** in the form of subsidence (extraction of water, gas, and oil), artificial structural loads, discharge of materials, dredging, and others (Bird, 2000).

11.4.2 Holocene eustatic changes and future predictions

About 11,500 years BP (Roberts, 1989) during the Holocene, ice melting caused the **Flandrian Transgression**, which occurred in a discontinuous manner. Sea level has remained stable over the last 6,000 years along all coasts worldwide and, as we have discussed, the present phase is a high sea level (*highstand*).

A number of sea level variation curves have been proposed in association with the Flandrian Transgression (*Fig. 11.3*). These curves indicate a rapid rise until about 6,000 years ago. Fairbridge proposes various fluctuations that are higher than present-day sea level but the proposals of other authors are not higher than this level (Curray *et al.*, 1969). The differences are due to the difficulty of separating eustatic from isostatic effects. In addition, Mörner (1987) maintains that eustatic curves must be defined regionally rather than globally.

Beginning around 6,000 years BP the glacial-eustatic activity ended. Since then sea level variations have been caused primarily by currents that redistribute ocean water masses (Pirazzoli and Morhange, 2005).

Scientists study the present-day sea level through tidal records, geodesic surveys, and biological indicators but it is difficult to measure sea level variation precisely (Bird, 2000).

Because of the connection between global warming and greenhouse gases, as discussed in reports by the IPCC (International Panel of Climatic Change), a major controversy has arisen between two groups. One group has made alarming predictions of rapid glacier and ice sheets melting and accompanying sea level rise; the other group shares the more moderate views of geomorphologists who specialize in field-data based sea level variations.

The IPCC estimates are generated by geophysical load models, which Pirazzoli (2004) claims are based on approximations involving simplistic assumptions that are not supported by precise field-study data. According to Pirazzoli, no proof exists that the Antarctic ice sheet will contribute additional water to the ocean.

Field data and tide gauges show that the mean global sea level rose 1.0–1.1 mm/year between 1850 and 1930. Between 1930 and 1950, sea level dropped. At the end of the 20th century, there is no indication of acceleration. Satellite topographic surveys indicate no changes for the decade of the 1990s (Mörner, 2004). These data do not coincide with the predictions of the IPCC models (Houghton *et al.*, 2001), which estimate a rise of between 11 and 77 cm by the year 2100 based on the IS92a hypothesis (*Fig. 19.10*). This number is somewhat less than the number published in 1996 (Houghton *et al.*, 1996), which ranged from 20 to 86 cm on the basis of the same hypothesis. Field, tidal, and satellite data indicates no need to fear the future flooding predicted by most global warming scenarios (Mörner, 2004). Observations of average ocean temperatures since 1961show that temperatures have increased to depths of about 3,000 m; this causes sea water to expand and thus causes sea level to rise (IPCC, 2007) (*Table 11.2*). Sea level rose 1.8 mm/year for the period from 1961 to 2003. The rise was more rapid during the period of 1993–2003 when it reached 3.1 mm/year. The total rise for the 20th century is estimated to be 0.17 m (IPCC, 2007). However, Rahmstorf *et al.*, (2009) maintain that sea level will rise 0.80 m and perhaps > 1 m by the year 2100, values very much higher than those indicated by the IPCC (2007).

The planet is undoubtedly warming and it is partially due to increased greenhouse gases. In terms of sea level, increased temperature is directly related to sea level variation but the temperatures show no acceleration of sea level rise (Zazo, 2006).

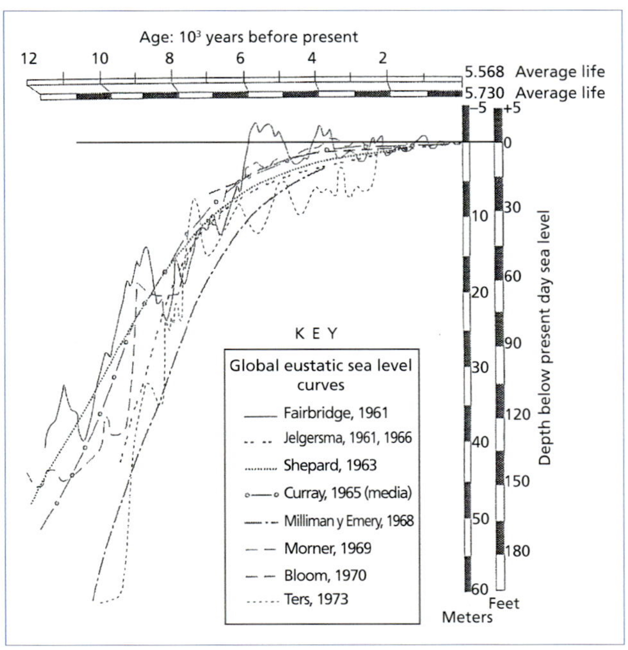

Figure 11.3 Sea level variation curves for the Holocene (Curry, 1969; in Flor, 2004).

Table 11.2 OBSERVED SEA LEVEL RISE AND ESTIMATED CONTRIBUTIONS FROM VARIOUS SOURCES (IPCC, 2007)

Rate of sea level rise (mm/year)		
Beginning of sea level rise	**1961–2003**	**1993–2003**
Thermal Expansion	0.42 ± 0.12	1.6 ± 0.5
Glaciers and small ice caps	0.50 ± 0.18	0.77 ± 0.22
Greenland Ice Sheet	0.05 ± 0.12	0.21 ± 0.07
Antarctic Ice Sheet	0.14 ± 0.41	0.21 ± 0.35
Total climatic contributions to sea level rise	1.1 ± 0.5	2.8 ± 0.7
Total observed sea level rise	1.8 ± 0.5[a]	3.1 ± 0.7[a]
Difference (Observed rise less the sum of observed climatic contributions)	0.7 ± 0.7	0.3 ± 1.0

[a] Data prior to 1993 are tide gauges and after 1993 are satellite altimetry.

11.5 Waves, currents, and tides

Processes that change the shape of coasts include wind, waves, tides, and currents. It is best to study them separately because they interfere with each other. Waves are the principal energy source along most coasts. Tides are important on some coasts; they can also dominate in estuaries and bays.

Waves are wind-generated undulations of water surface. They are actually orbital water motions that diminish rapidly toward the bottom; their motion is very weak at a depth approximately equal to half their **wavelength** (L/2) (Fig. 11.4). The **depth** where wave action is imperceptible is called the **base of the wave**. The **height of the wave** (A) is the elevation difference between the crest and the trough; it is proportional to wind velocity. The **period** (P) is the time it takes a wave to move a distance equal to the wavelength.

Most waves are created by wind (wind waves). In addition, waves can move outward from a storm area; these are called *swells*. Strong storms generate large waves; storms in the center of the ocean can generate 20 m high waves that are capable of traveling more than 500 km at speeds of 80 km/hour. As these waves approach the coast, they become taller or steeper and can become breakers (Bird, 2000).

The largest waves are associated with strong winds in temperate latitudes where west winds blow. These are found in the southern hemisphere in latitudes around 60° S and in the northern hemisphere between 40° and 60° N. Intense tropical storms develop near the Equator; these are called hurricanes in the Atlantic, typhoons in Asia, and cyclones in the western Pacific. In addition, monsoons generate waves along the coasts of India and Southeast Asia.

When a wave approaches the coast, its crest forms an angle with the topographic contours of the seafloor and water depth varies along the crest of the wave. If the wave is in shallow water, the wave velocity also changes along its crest; it moves more slowly than waves in deeper water. This causes the wave crest to rotate in such a way that the wave trajectory also curves. This process is called **wave refraction**. Orthogonal trajectories converge on promontories, which increases wave height and concentrates energy at these points. Orthogonal waves diverge in bays, causing energy to disperse (Woodroffe, 2002) (Fig. 11.5). These wave transformations are important for coastal currents, sediment transport, and coastal morphology.

When the water velocity at the wave crest exceeds the wave velocity and moves ahead of the rest of the wave, the wave breaks. Water falling from breakers is the major erosional force of waves on the seafloor. Breakers are of four types (Fig. 11.6). *Spilling breakers* develop in shallow areas and produce a large amount of foam. *Plunging breakers* are typical of deeper marine floors; the crest curves inward, making a hole underneath and the crest collapses on the water surface. *Surging breakers*, in which the frontal face has only a small amount of foam and breaks at the base, are found on deeper beaches. Lastly, *collapsing breakers* are intermediate between the surging and the plunging types. These different breaker types are related to the beach slope and relative water depth.

Rip currents occur on many beaches; they are hazardous when strong. As the wave-transported water recedes, it produces a channel-shaped return flow. These currents, which

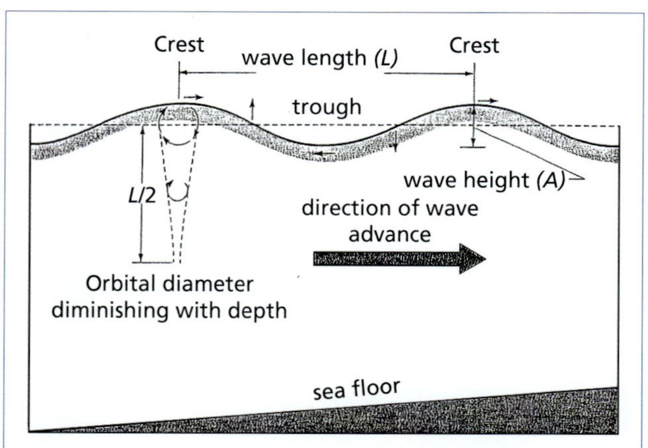

Figure 11.4 Terminology used for waves and wave advance direction (Bird, 2000).

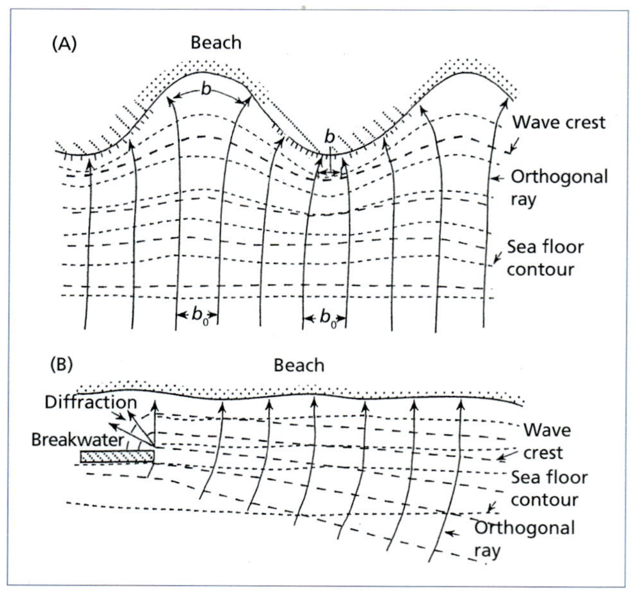

Figure 11.5 Wave refraction (A) Wave trajectories approaching the coast perpendicularly. Divergence of trajectories is indicated by $b > b_0$. (B) Waves approaching the coast obliquely. Diffraction takes place behind breakwaters (Woodroffe, 2002).

are common in areas where large waves occur, disappear into the sea (*Fig. 11.7*). Rip channels can be 30 m wide and currents in them can have velocities of 8 km/hour. To prevent the risk of rip currents, a swimmer should swim parallel to the coast until the flow of the rip current disappears (Shepard, 1959). **Beach cusps** occur in rip currents on coarse-grain beaches; they have a half-moon shape and are regularly spaced along the beach. They are typical of low energy waves that break directly on the beach (Masselink *et al.*, 1997). Large beach cusps can measure 200 m in width along the beach; they are associated with rip currents in Monterrey Bay, California (Thornton *et al.*, 2007). Currents that flow along the coast carry sediments; this process is called **longshore drift**. Their presence can be determined by the direction of movement of spits that form in estuaries and river mouths. Another indicator of longshore drift is sand deposited along jetties, natural promontories, and rocky outcrops. Fluorescent or radioactive tracers and other methods can identify longshore drift.

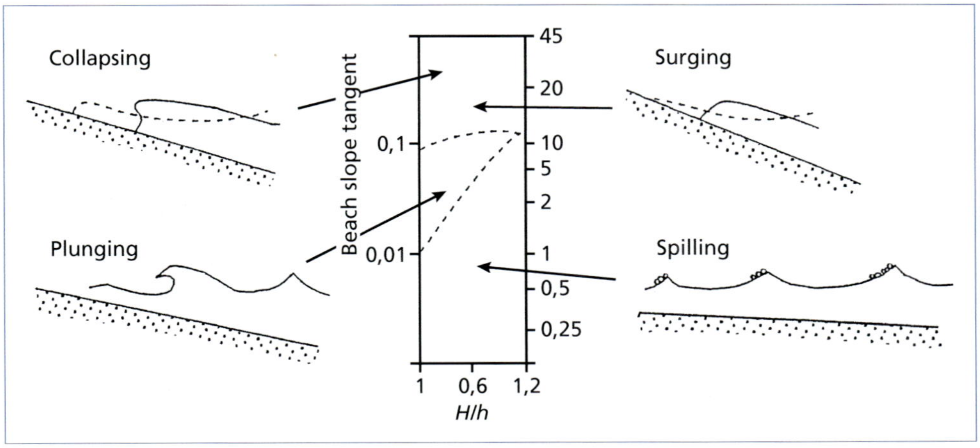

Figure 11.6 Different breaker shapes generated as a function of beach slope and relative height (A/a) where *a* is water depth (Galvin, 1968).

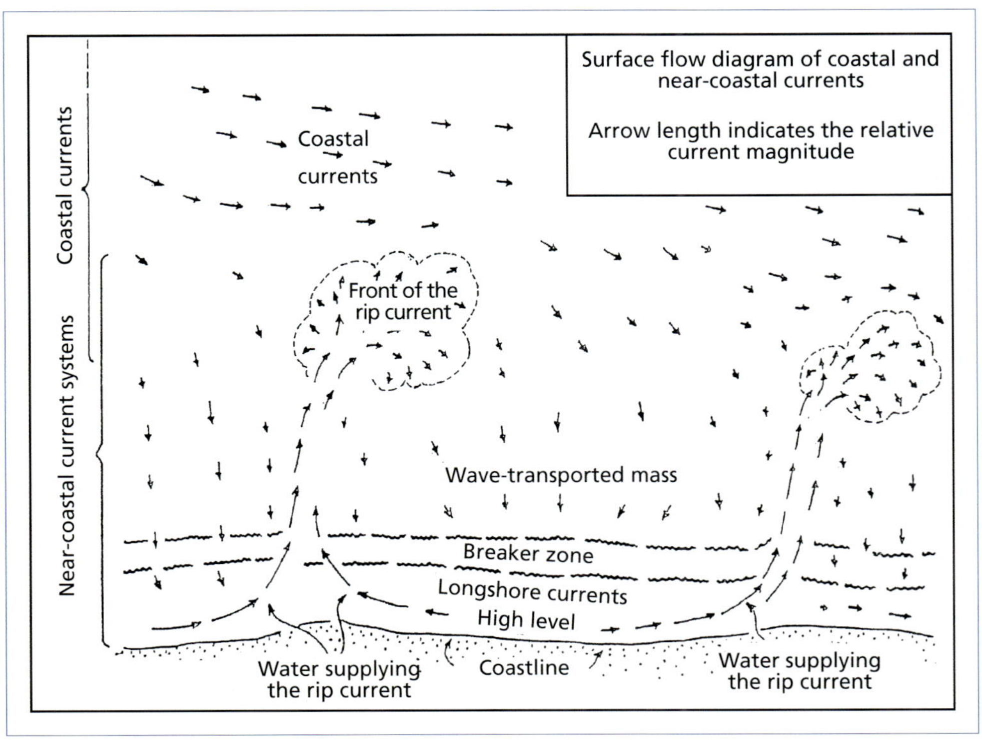

Figure 11.7 Coastal currents and development of rip currents (Shepard, 1959).

Sudden landslides on the seafloor and earthquakes associated with submarine faults can generate **tsunamis** (a Japanese word that means "port wave") or seaquakes, also called (erroneously) sea waves (González, 1999). Other less significant causes are submarine landslides, such as the collapse of the Krakatoa caldera in 1883 that generated waves up to 30 m high. Undersea atomic explosions have caused anthropogenic tsunamis (Bikini Atoll) (Bryant, 2001).

Tsunami wavelengths can exceed 100 km; in high seas their crest heights are less than a meter and thus are not likely to be detected by ships. Their propagation velocity increases with depth; in deep oceans such as the Pacific, which is 5 km deep, tsunami velocity can exceed 500 km/hour. When a tsunami reaches shallow water, its velocity decreases rapidly and the wave height can reach 30 m, which causes intense coastal erosion (Kench *et al.*, 2008). Tsunami waves often occur in a series separated by several minutes up to an hour; any of these waves may do major destruction. The Hawaiian tsunami of 1 April 1946 was the eighth wave in a series of intensely destructive waves (Bolt, *et al.*, 1975).

Tsunamis can be triggered by other causes such as submarine faults with a vertical component; horizontal displacement like that of the 1906 San Francisco earthquake, which moved 6 m along the fault, does not produce tsunamis. They are most common in areas of crustal destruction, such as the Pacific Ocean belt. Tsunamis can also be caused by volcanic activity and associated processes (*Table 11.3*). Beget (2000) discusses the differences between tsunamis produced

Table 11.3 **CHARACTERISTICS OF SELECTED VOLCANO-GENERATED TSUNAMIS (MODIFIED FROM BEGÉT, 2000)**

Location	Age	Type of volcano	Cause of tsunami	Wave height	Distance traveled	Tsunami impact
Santorini Mediterranean Sea	1638 BC ¿?	Island	Explosive eruption, caldera formation. Pyroclastic flow into the sea	10–50 m[a]	150–500 km?[a]	Destruction of the Minoan civilization on Crete[a]
Unzen, Japan Ariake Sea	1782 AD	Coast	Debris avalanche into the sea	10–55 m	20–50 km	More than 10,000 deaths
Tambora, Indonesia Java Sea	1815 AD	Island	Pyroclastic flow into the sea	>10 m	>100 km	More than 10,000 deaths
Krakatoa, Indonesia Sunda Strait	1838 AD	Island	Explosive eruption, caldera formation. Pyroclastic flow into the sea	5–35 m	800 km	More than 36,000 deaths
Augustine, Alaska Cooke Inlet	1883 AD	Island	Debris avalanche into the sea	7–9 m	>100 km	Damage to coastal communities, no known deaths
Ritter Island, Nueva Guinea Bismarck Sea	1888 AD	Island	Debris avalanche into the sea	4–12 m	450 km	Numerous fatalities
Paluweh Island, Indonesia Java Sea	1928 AD	Island	Debris avalanche into the sea	5–10 m	>100 km	More than 150 deaths
Iliwerung, Indonesia Java Sea	1979 AD	Coast	Debris avalanche into the sea	9 m	>100 km	More than 500 deaths
Kick'em Jenny Caribbean Sea	Numerous	Submarine	Submarine eruptions	1–5 m	1–10 km	No known fatalities
Karymskoye, Kamchatka Lake Karymskoye	1996 AD	Caldera lake	Phreatomagmatic eruptions	30 m	5 km[b]	No known fatalities

[a] The size, distanced traveled and impact of a tsunami wave from the Santorini eruption is based on the magnitude of the eruption and height of tsunami deposits.

[b] Tsunami waves in the Karymskoye caldera caused flooding many kilometers downstream of the caldera.

Table 11.4 **RELATIONSHIP BETWEEN EARTHQUAKE MAGNITUDE AND TSUNAMI RUN-UP HEIGHT IN THE SEA OF JAPAN (BRYANT, 2001)**

Earthquake magnitude (Richter scale)	Tsunami magnitude	Maximum wave (meters)
6.0	−2	<0.3
6.5	−1	0.5–0.75
7.0	0	1.0–1.5
7.5	1	2.0–3.0
8.0	2	4.0–6.0
8.2	3	8.0–12.0
8.5	4	16.0–24.0
8.8	5	>32

by submarine and sub-lacustrine eruptions, landslides and rock avalanches, and explosive eruptions. Nott (2006) adds meteorite impacts in oceans to the list of tsunami causes. Scientists have determined a relationship between earthquake magnitude, tsunamis, and wave heights (*Table 11.4*) (McGuire *et al.*, 2004).

As a result of the devastating Hawaiian tsunami of 1 April 1946, rapid communication systems, trajectory maps, and tsunami travel times were established for the Hawaiian Islands (*Fig. 11.8*). Tsunami alerts must be very fast because a tsunami travels from Chile to Hawaii in about 10 hours and from Chile to Japan in 20 hours. Alarm systems are composed of scientific equipment, communication systems, and alarm notices. Tsunami sensors must be able to detect waves just a few decimeters high above the deep ocean (Fernando *et al.*, 2008).

On 17 July 1999, a magnitude 7.1 earthquake occurred in Papua New Guinea; its epicenter was 20 km

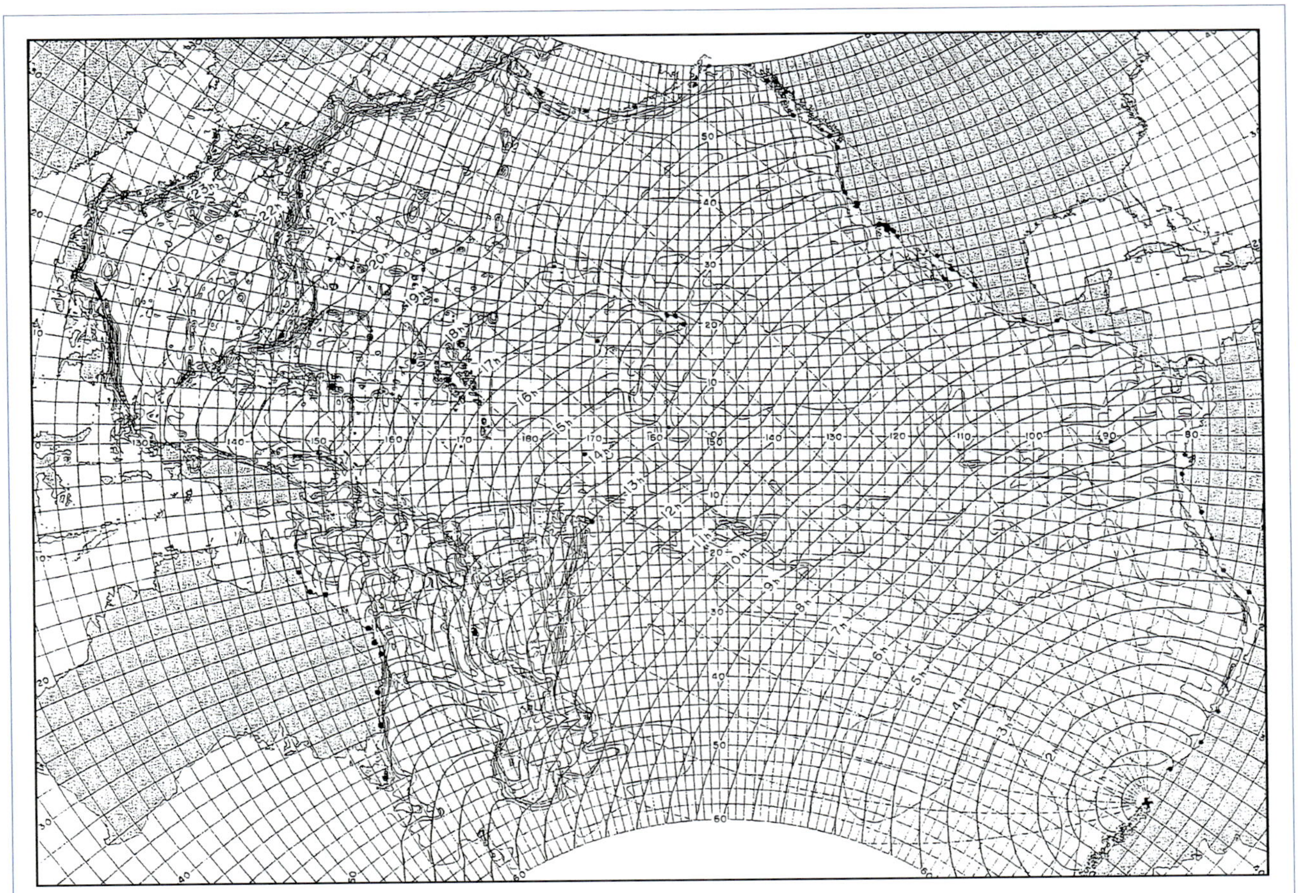

Figure 11.8 Map of the Pacific Ocean indicating tsunami travel times for the 1960 Chile earthquake. The curvature of the wave front shows refraction in response to changes of ocean depth (Japan Meteorological Agency, in Bolt *et al.*, 1975).

Figure 11.9 Large boulders deposited by tsunami waves generated by the 1755 Lisbon earthquake. Trafalgar Beach (Cadiz, Spain).

from the coast and the hypocenter was at a depth of 33 km. The earthquake caused a series of destructive tsunamis; the largest wave was 30 m high. The tsunami wiped out many fishing villages. Tsunamis caused 2200 deaths and left 6000 people homeless (Hugget, 2003).

Geomorphic surveys, isotopic dating of coastal spits, large boulders in beach areas (*Fig. 11.9*), and historical data make it possible to reconstruct the effects of the 1755 tsunami along the coast of the Gulf of Cádiz that resulted from the Lisbon earthquake (Dabrio *et al.*, 1998; Kelletat *et al.*, 2005; Gracia *et al.*, 2006; Morales *et al.*, 2008). Tsunami waves were refracted on the continental platform at Cadiz; waves higher than 10 m in some places caused significant erosion (*Fig. 11.10*).

The tragic tsunami in the Indian Ocean on 26 December 2004 generated by a 9.4 earthquake in Sumatra, Indonesia, led to renewed interest in sedimentological methods of paleo-tsunami investigations to determine the age and return periods and to conduct paleo-environmental reconstructions (Williams *et al.*, 2005). In 2005 the journal Marine Geology published a special issue, Volume 215, dedicated to tsunamis to address this concern. Tsunami records show that larger magnitude tsunamis occurred before historic times (Nott, 2006).

The 2004 Indian Ocean earthquake, known as the Sumatra-Andamza, is one of largest the world has

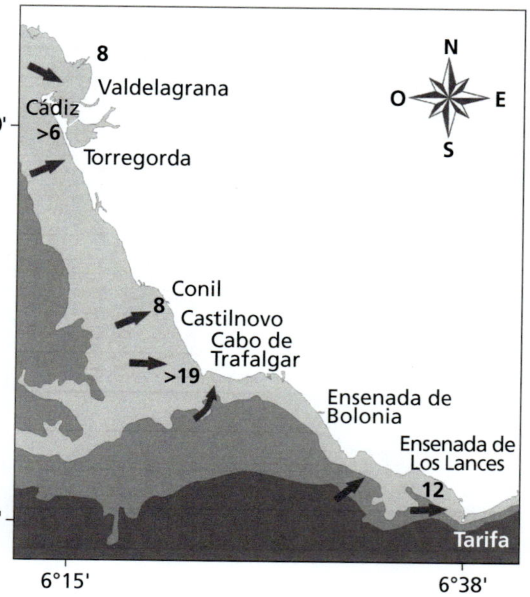

Figure 11.10 Arrival directions of the 1755 tsunami waves along the Cadiz coast, from field data. The numbers indicate the maximum height of inundation by the tsunami in meters, determined from various historical documents (Cadiz coast, Spain) (Gracia *et al.*, 2006).

experienced since 1900. It killed 230,000 people in a number of countries and is without doubt the largest disaster in history (Satake, 2007). The tsunami caused numerous fatalities in the following countries: Indonesia – 167,736 deaths; Sri Lanka – 35,322; India – 18,045; and Thailand – 8,212 (Inoue, 2007).

The Sri Lanka tsunami arrived as a sequence of two or more principal waves whose maximum heights exceeded 10 m. Flooding spread inland about 2 km into the eastern center of the island (Inoue, 2007). In Thailand, modifications of coastal morphology were limited; coral boulders weighing more than 40 tons were transported along with centimeter and decimeter-thick sand sheets, which in some places reached thicknesses of 2 m (Kelletat *et al.*, 2007). Wave heights of 6–10 m reached the coast two hours after the earthquake. Tsunami deposits covered the low coastal plains and penetrated more than 1 km inland (Hori *et al.*, 2007). Satellite imagery studies of the Thailand coast before and after the tsunami have made it possible to calculate the beach area and to locate the position of coastline change. Two years after the tsunami, the beaches have reached equilibrium status (Chowoong *et al.*, 2009).

The geomorphic impact of the tsunami north of Sumatra includes beach erosion, destruction of sand barriers that had protected lagoons and river mouths, numerous erosional scarps 0.5 to 1.5 m high along river margins, and 220 boulders (0.7–1.5 m) transported to 7 m above sea level. Beach erosion is estimated at about 30 m^3/m and locally exceeded 80 m^3/m (Paris *et al.*, 2009).

As previously mentioned, volcanogenic tsunamis can occur in lakes. The two principal lakes in Nicaragua, Lake Managua and Lake Nicaragua, are related to volcanogenic tsunami processes. Tsunami deposits have been identified in the geological record in Lake Managua and massive intrusions of pyroclastic flows have been identified in both lakes along with collapsed volcanic flanks that have generated avalanches. Phreatomagmatic eruptions have also generated maars (Freundt *et al.*, 2007).

As a result of the devastating tsunami in Sumatra, authorities have decided to establish an alarm system for the Indian Ocean (Marris, 2005), similar to the system operating in the Pacific Ocean.

In addition to wave-generated currents, other oceanographic currents affect the coastal domain, the global distribution of organisms, and the climates that affect them. **Ocean currents** are produced by wind action on the surface of the open ocean and are pushed by pressure gradients and density differences resulting from temperature and salinity variations, called thermohalines. In addition, the spatial arrangement of land areas and the Coriolis force both have an affect on currents (*Fig. 11.11*). In the northern hemisphere, circulation rotates in a clockwise direction whereas in the southern hemisphere, circulation is counterclockwise. Warm currents occur in the western part of oceans whereas along eastern coasts circulation is cold and seawater undergoes less evaporation. These factors cause distinctly arid coasts on the western side of large continental masses (Peru-Chile Deserts and the Namibian Desert).

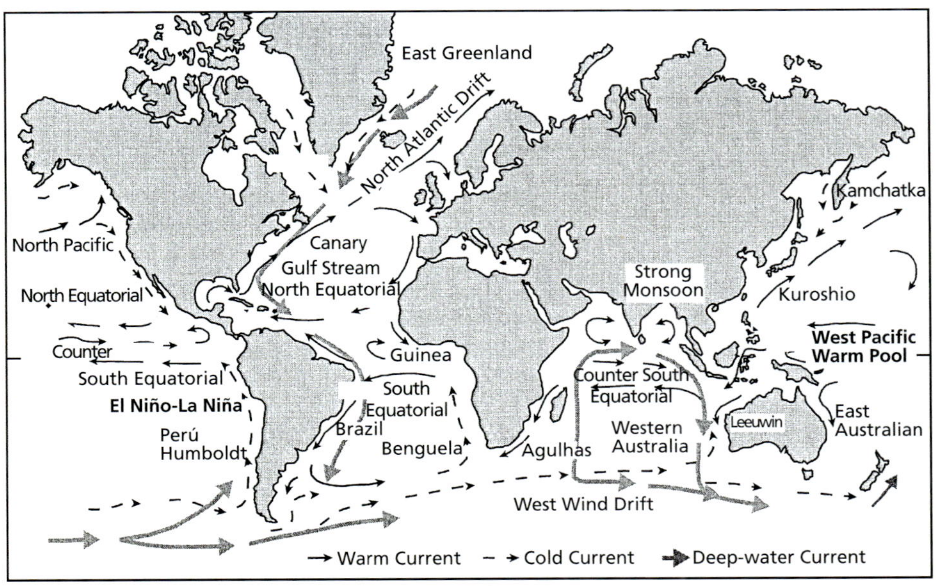

Figure 11.11 Global oceanic currents and deep-water circulation (Woodroffe, 2002).

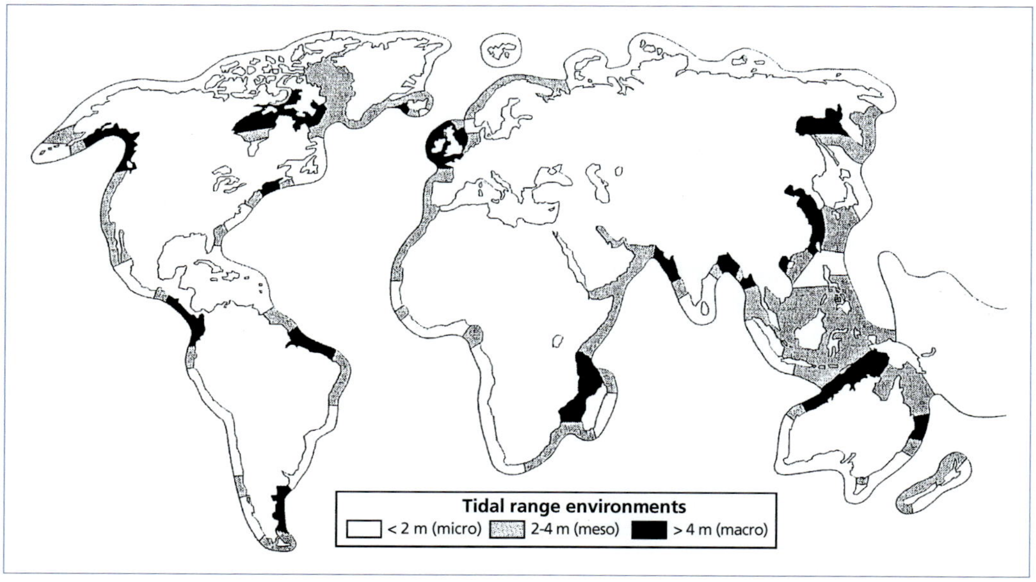

Figure 11.12 Global distribution of tidal intervals (modified from Davies, 1980).

Tides are water movements produced by the gravitational attraction of the moon, and, to a lesser extent, the sun. Almost all coasts experience a rising tide (high tide) and a falling tide (low tide). The range or tidal oscillation in the ocean is very small (<1 m) but increases near coasts where it can exceed 10 m. The tidal range depends on the width and inclination of the continental platform, the location and shape of the continents, and the presence of large bays. Thus the tidal range is controlled mainly by the configuration of the oceans and coasts (Masselink and Hughes, 2003) (*Fig. 11.12*). Macro-tidal ranges exceed 4 m and are found in restricted seas and estuaries. Micro-tidal ranges are less than 2 m; they are primarily found along open seacoasts and in practically all closed seas. A large tidal range can develop along a broad intertidal zone such as in the bay of Mont Saint Michel (France) where about 20 km of sand and silt plains are exposed. The largest tides in the world are located in the Bay of Fundy (Nova Scotia) where the difference between high and low tides is 16 m (Raven *et al.*, 2008). Tidal fluctuations also occur in lakes such as in the Great Lakes of the United States.

Tides reach their maximum values when the sun and the moon have the same orientation with respect to the Earth. These conditions occur at the time of the full moon and the new moon; they produce spring tides. When the moon and the sun are located at right angles with the Earth at the center, the two forces work in opposition, resulting in small tides called neap tides.

Entrances to bays and straits between islands often have strong tidal currents. Tidal channels are the location of significant erosional action as a result of the flow and reflux of tides.

The most important source of inter-annual climate variability on a global scale is called the **ENSO** (El Niño Southern Oscillation). **El Niño** (*Fig. 11.11*) is defined as approximately 2°C or more anomalous warming of surface water temperatures in the central Pacific and the eastern part of the southern hemisphere (Glanz, 2001). Its name comes from "The Christ Child" since it generally surges every year around Christmas time along the western coast of South America. The eastern equatorial Pacific is frequently dominated by rising deep water (*upwelling*) along the coast; these cold waters, which are rich in nutrients, increase the fish population. However, the winds that blow along the coast weaken the upwelling and the surface waters end up being warmer with a resulting decrease in fish (Glanz, 2004).

This warming lasts until March, and during this time warm water accumulates in the equatorial central and eastern Pacific Ocean. The warm mass along the coast is associated with a significant decrease in fish along the Peruvian coast and storms in dry areas of Peru and Chile. El Niño is the warm phase of the major ENSO cycle. The opposite of El Niño is **La Niña**, a cold cycle in the eastern Pacific. It occurs when warm water moves toward the western Pacific and exposes cold water at the surface of the tropical central and eastern Pacific (Mock, 2007).

The consequences of El Niño/La Niña are usually complex and not very well understood. Scientists are beginning to study the effects of El Niño, which include ecosystem changes, primarily in fauna. Attempts have also been made to reconstruct paleo-ENSO. Historical documents, dendrochronology, and coral can provide annual resolution for ENSO reconstructions but studies of El Niño effects for previous

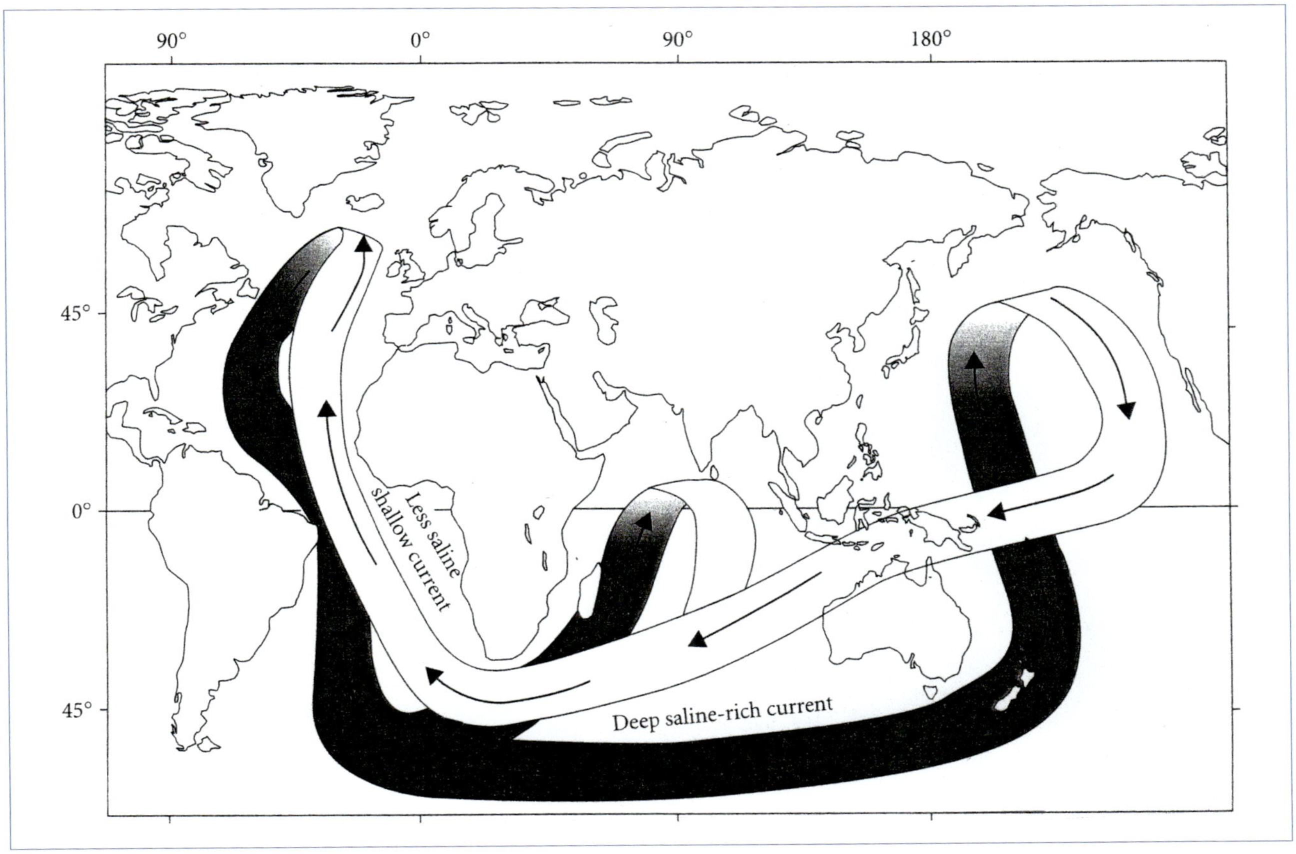

Figure 11.13 Large-scale salt transport system (ocean/global belt) currently operating in the oceans (Broecker and Denton, 1990b).

epochs are very limited. In investigations in southern Peru, Fontugne *et al.* (1999) found less frequent and less intense El Niño events between 9000 and 3380 years BP.

Thermohaline circulation is the transference of warm water from the tropics to higher latitudes (*Fig. 11.13*). Broecker (1987) shows that this thermohaline oceanic transport occurs today as it has in the past. This transfer is the primary conductor of thermal energy, nutrients, and dissolved oxygen between the shallow and deep oceans. This interchange across great distances undoubtedly affects the global climate in a profound way. The present-day global warming of higher latitudes may be due to weaker thermohaline circulation (Häkkinen and Rhines, 2004).

11.6 Cliff coasts and rocky platforms

Cliffs are highly inclined slopes that end at a coast (Emery and Kuhn, 1982). They are a very clear zone of transition between the continent and the sea. Three quarters of the world's coasts are cliffs (Bird, 2000) (*Fig. 11.14*). As waves attack cliffs, they form erosional morphologies. Rocky coasts generally evolve very slowly as in Galicia (Pérez-Alberti *et al.*, 2000). Coasts composed of clastic rocks evolve rapidly; this evolution can be analyzed by using different research techniques (maps, areal photographs, *in situ* measurements, and others) that measure geomorphic changes that have occurred. Books by Trenhaile (1987) and Sunamura (1992) on rocky coast geomorphology provide detailed analyses of processes and landforms in these environments.

Cliff rocks are subject to numerous erosional processes: physical and chemical weathering, bioerosion, and mass movement (Trenhaile, 1987, Masselink and Hughes, 2003). **Mechanical weathering** by waves mobilizes loosened material, abrasion causes mechanical wear, and pounding waves exert hydraulic action that enlarges cracks.

Weathering is probably the principal erosional mechanism (Trenhaile, 1987). Gelifraction and wetting-and-drying cycles weaken both capillaries and cracks. Salt weathering, or haloclastism, increases volume for the growth of crystals, widens planes of weakness, and produces honeycombs (*Fig. 11.15*). Chemical weathering includes numerous reactions

Figure 11.14 High altitude cliff on a rocky coast in a basalt trap with dikes (trap), Acantilado de los Gigantes. Puerto Santiago. (Island of Tenerife, Spain).

Figure 11.15 Alveoles developed on a coastal cliff of Jurassic sandstone. Elgol. Isle of Skye (United Kingdom). Photo F. Gutiérrez.

Figure 11.16 Cliff of soft rocks and loose boulders. Ajim. Southeastern Tunisia.

such as dissolution, hydrolysis, oxidation, and hydration. Bioerosion is the mobilization of rocks by organisms such as lithophagous mollusks; it can also occur as a result of biochemical processes (Fox, 2005). Bioerosion is more significant on tropical coasts due to the abundance of organisms present (Spencer, 1988). When periglacial processes are active on high latitude coasts, rock outcrops on these coasts are subject to gelifraction, resulting in basal fans of angular clasts. With the spring thaw, tundra cliffs degrade rapidly. As a result, landslides deliver tundra and glacial till to the sea where these materials are subject to wave action. This thermoabrasion can cause tens of meters of retreat in a few weeks (Bird, 2008).

Mass movements are common on cliffs partly because of their instability and partly because waves attack the base of cliffs (Trenhaile, 1987). Boulders of various sizes fall frequently especially along cliffs composed of soft rocks (May, 1972), which have a high rate of retreat (*Fig. 11.16*). Falls (*Figs. 11.16* and *11.18*), topples, rotational and planar landslides, earthflows, and mudflows are common on marine cliffs (Bird, 2008)

Cliff morphology is a result of the previously described processes. Sunamura (1992) distinguishes three types of coastal cliff morphologies (*Fig. 11.17*); inclined rocky platforms, horizontal rocky platforms, and vertical cliffs. The first type includes platforms that dip gently without any significant topographic discontinuity. They have many names: coastal platforms, littoral platforms, submarine platforms, benches, and others (Sunamura, 1992). The second type includes horizontal platforms that drop noticeably toward the sea. The third type includes sub-vertical cliffs that have no erosional platform where they enter the sea. Cliff morphology is related to lithological and structural variations.

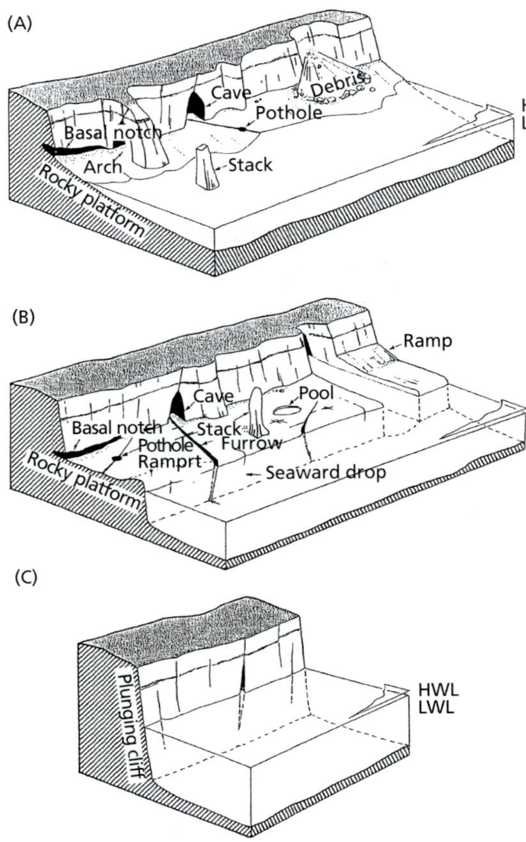

Figure 11.17 Three types of landforms along coasts with cliffs, showing primary erosional features: (A) inclined rocky platform; (B) horizontal rocky platform; (C) vertical cliff (Sunamura, 1992).

A basal **notch** usually marks the high tide level of the cliff base (*Fig. 11.18*). More resistant rocky zones become promontories or capes where columnar **stacks** and **small islands** can be observed (*Fig. 11.19*). **Bays** form where the coast is composed of materials that erode more easily than surrounding materials (*Fig. 11.20*).

Cliffs taller than 300 m are called mega-cliffs (Bird, 2008). They are found along the coasts of Peru and Chile in areas of significant recent tectonic uplift and on volcanic Atlantic islands such as Tenerife (Canary Islands) (*Fig. 11.14*).

Cliff profiles are related to lithological-structural controls. In horizontal formations of alternating hard and soft rocks, differential erosion generates stepped profiles. In massive rocks, profiles are vertical; in soft rocks they may also have vertical slopes due to rapid development by mass movements (*Fig. 11.16*). When marine erosion is interrupted along a cliff and subaerial processes predominate, the escarpment degrades and the profile is predominantly convex (*coastal bluff*) (Bird, 2008) although slopes evolve as a function of rock strength and climatic characteristics. Tropical coasts often have this type of morphology due to intense subaerial weathering and denudation. A long geomorphic history of marine and subaerial processes can frequently be seen on detailed cliff profiles. This type of cliff is called a **compound cliff** (Griggs and Trenhaile, 1994). Convex and straight slopes, erosional terraces, slope breccias, karst deposits and landforms, and dunes are all present along the cliffs of Gibraltar (Rodríguez-Vidal *et al.*, 2004).

As cliffs retreat, erosion penetrates into zones of weakness and forms marine caves. Waves penetrate into caves and if the waves rupture the roof of the cavity, they expel water outside through a hole, forming a **blowhole**. When wave action excavates rocks along weakness planes, large **gorges** or cracks can form. Natural **marine arches** (*Fig. 11.21*) can sometimes develop from marine caves (Trenhaile, 1987; Sunamura, 1992).

Cliff **retreat** is usually measured by taking the top of the cliff as a point of departure although detailed cliff profiles of cliff morphologies can also be used (Bird, 2008). Cliff erosion can be easily observed after large storm waves. Sunamura's (1992) book compiles linear retreats along cliffs in numerous parts of the world; the retreats show the following average values:

- 1 mm/year along granite cliffs
- 1 mm to 1 cm/year in limestones
- 1 cm/year in shales
- 10 cm to 1 m/year in chalk and Tertiary sedimentary rocks
- 1 to 10 m/year in glacial deposits
- At least 10 m/year in volcanic ash

Figure 11.18 Basal notch at the bottom of a cliff of soft rocks and loose boulders. Ajim. Southeastern coast of Tunisia.

Figure 11.19 Small islands, cliffs, and erosion platform in Cretaceous limestones and marls. Santander.

Figure 11.20 "Cala" (bay) in granite along the Costa Brava (Province of Gerona, Spain).

Figure 11.21 Marine arches in magnesium-rich limestones. Marsholen Bay (northwestern England). Photo F. Gutierrez.

Figure 11.22 Abrasional platform at the base of a cliff. Santander. Spain.

Figure 11.23 Jurassic cliffs and Tertiary basic *sills*, hanging valley, and waterfall. The Kilt Rock. Isle of Skye (United Kingdom). Photo F. Gutiérrez.

Many cliffs are surrounded by **rocky or abrasion platforms** (*Fig. 11.22*). Some are horizontal and are located in the tidal zone in micro- and meso-tidal environments (*Fig. 11.17A*). Others dip (*Fig. 11.17B*) slightly toward the sea (1–5°) and extend from the cliff base to below the tidal action; they are common in macro-tidal environments. However, different platform types are more related to the balance of wave action and rock strength. When the balance is tilted in favor of rock strength, vertical cliffs form (*Fig. 11.17C*) without any erosional platform (Sunamura, 1992). Some steep slopes have hanging valleys truncated by cliff retreat; water can flow from the valleys as waterfalls (*Fig. 11.23*). These circumstances show that the cliff retreat rate is much greater than valley incision.

On calcareous coasts, limestone dissolution produces various types of karren with noticeable asperities. At the top of small carbonate cliffs, typical projections can form (*Fig. 11.24*) whereas during low tide on the surface of calcareous erosional platforms, small closed depressions form. They are several centimeters deep and have flat bottoms that are bordered by sinuous carbonate concretions made by calcareous algae and serpulid worms (*Fig. 11.25*). Guilcher (1953, 1954) provides a detailed description of these microforms. In addition, Twidale *et al.* (2005) analyze the role of *etching* on the formation of platforms and accompanying features.

Figure 11.24 Peak landform in Pleistocene calcarenites. Los Escollos. Cabo de Gata (Province of Almería, Spain). Photo F. Gutiérrez.

Figure 11.25 Erosional platform at low tide with "*pools*" bordered by carbonate concretions. Torregorda, San Fernando (Province of Cádiz, Spain) Photo F.J. Gracia.

11.7 Coral reef coasts

Coral reef coasts have rocky mound, platform, or strand morphologies that are slightly elevated above the rocky marine floor; they are composed primarily of the remains of skeletal organisms. Their external structure is essentially composed of branched coral that has grown on top of older skeletons. Waves introduce various types of detritus into the interiors of these structures (Shepard, 1959). Coral reefs live in a sensitive balance between biological construction and destructive physical processes that control the reef morphology. In terms of reef productivity, this complex community has a wide variety of animal life; it has been analyzed in detail by Ingmanson and Wallace, 1989, and Viles and Spencer, 1995.

Coral reefs are located only in areas where water remains warm most of the year, where the *temperature* of the coldest month never dips below 18°C and the hottest month does not exceed 34°C. They are found in tropical water between 30°N and 30°S latitude (Guilcher, 1988). Although coral growth depends on water salinity, corals are highly tolerant because they can withstand salinities of about 30 to 38 and even to 40–41 parts per thousand in the Red Sea. **Turbidity** of the water is also an important concern for reef growth because corals need adequate light levels to grow. In tropical areas with intense weathering, rivers carry abundant suspended load. The absence or lack of coral reefs in the water of southern and eastern Asia can be explained by high turbidity (Guilcher, 1988). Reef distribution is also related to the dispersion of **plankton**. The supply of zooplankton, which is food for the coral, is highly dependent on water circulation; for this reason reefs are more abundant in the windward areas of continents or islands than on the leeward side.

Corals are estimated to grow at a rate of between 1 and 14 mm/year (Spencer, 1994). As they grow upward they may be exposed to low tidal levels and, consequently, wave action. Storm waves break the skeleton of the coral and generate large quantities of coral gravel and sand that lodge in the lower parts of reefs.

The principal types of coral reefs are fringing, barrier, and atoll (Guilcher, 1988) *(Fig. 11.26)*. **Fringing reefs** build both upward and outside in shallow seas on the margins of continents or islands. They form a reef platform at the level of low tide. Some platforms have been shaped by waves; they are rocky erosional platforms. Coastal reefs may have a shallow and narrow channel that communicates with the interior. **Barrier reefs** form at some distance from the coastline and are more or less parallel to it; they often have an internal lagoon in their center that does not exceed 30 m in depth. Along its seaward side the Australian Great Barrier Reef dips 40°–50°; the water is 1,800 m deep in its northern part and about 180 m in the south. The depth of the lagoon oscillates between 18 and 45 m and the bottom in made up of detritus of marine and fluvial origin (Bird, 2008).

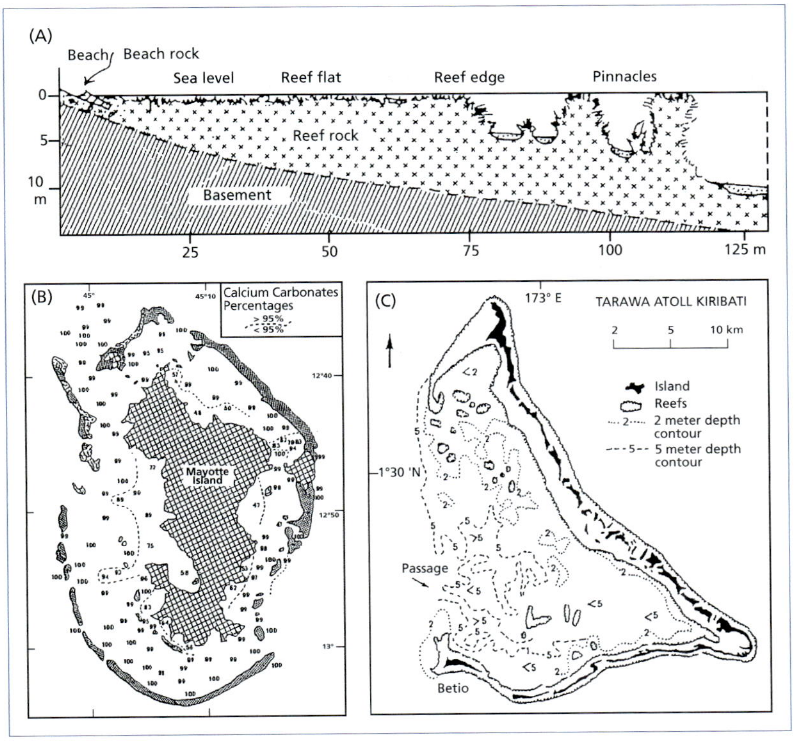

Figure 11.26 (A) Fringing reefs with shallow channel. Aqaba, Jordan. Red Sea. (B) Barrier reef of Mayotte Island, Indian Ocean. (C) Tarawa Atoll, Kiribati, Micronesia. (Compiled by Masselink and Hughes, 2003).

Atolls are reefs that encircle a lagoon in which there is no island or volcanic edifice. Atolls grow above sea level and are generally more or less asymmetrical with differences on the leeward and windward sides. Like barrier reefs, they are exposed to subaerial karstification. Of the 425 atolls in the world, most are located in the Indian and Pacific oceans while a few isolated ones occur in the Caribbean Sea (Stoddart, 1965b). The Maldive Islands are home to some larger atolls; they can reach 75 km in diameter. Transverse channels that drain toward the lagoon are more abundant on the leeward side. Several low islands with tropical vegetation are present in the lagoon.

Darwin's **theory of subsidence** (1842) provides a genetic developmental sequence of the three principal types of coral reefs (*Fig. 11.27*). Since they cannot grow in deep water, atolls begin their growth on a shallow water substrate. Coastal reefs grow around an island that slowly subsides. Over time, a lagoon develops between the island and the reef, creating a barrier reef. The later disappearance of the island by subsidence generates the atoll. Drilling on the Enewetak atoll (Marshall Island) in the Pacific Ocean showed that the atoll is made up of 1,250 m of shallow water coral limestone deposited on top of volcanic rock that is present at depth of 3,200 m on the ocean floor (Ladd *et al.*, 1967).

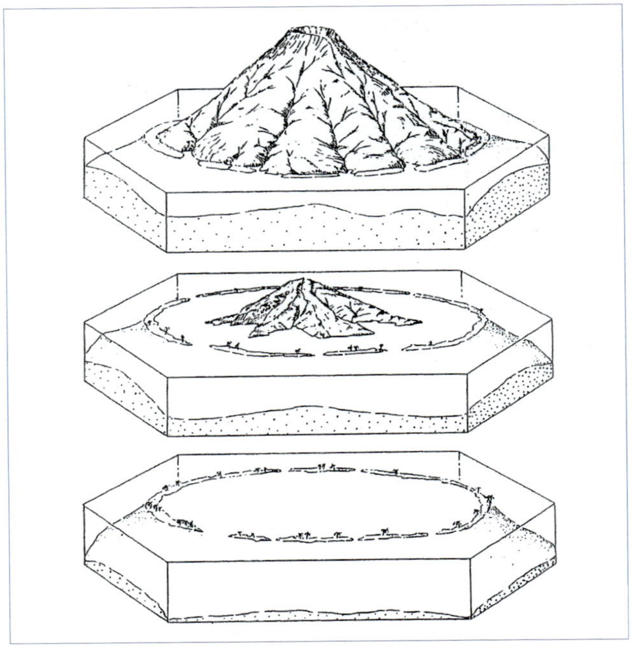

Figure 11.27 Block diagram showing Charles Darwin's hypothesis of atoll development as a result of subsidence of a volcanic island and reef growth (Sherpard, 1959).

11.8 Beaches, barriers, and spits

Beaches are accumulations of sand and gravel or a mixture of the two located at the interface of land and sea where dynamics are primarily controlled by waves (Flor, 2004). Beaches make up approximately 40% of the world's coasts (Bird, 2008). Because they are composed of loose materials, beaches act as dampeners to absorb, reflect, and dissipate energy liberated along the coastal margin (Anthony, 2005). During the last four decades, international tourism has led to concerns about coastal margins. Tourist invasion has eroded about 70% of the world's beaches (Bird, 1996).

Beaches are three-dimensional landforms that undergo significant changes in their **profiles**; these can be studied by field transects and remote sensing. Some beaches, such as those in Rota (Province of Cádiz, Spain), have accreted during the short period between 1996 and 2002 as a result of favorable weather conditions (Anfuso and Gracia, 2005). Other beaches, however, such as those in the southeastern United States, have retreated about 4 m/year; this spectacular rate of retreat has caused significant a loss of land and coastal property (Finkl, 1993).

Beaches vary in size and shape. Some are wide and straight and extend for kilometers such as the beaches on the Gulf of Cádiz (Spain). Others occur as a set of *pocket*

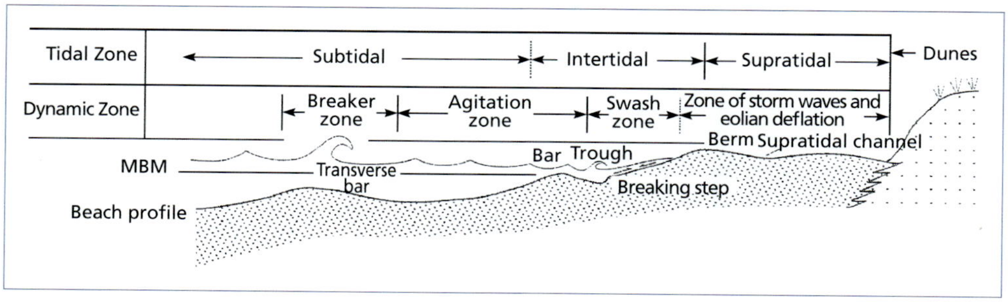

Figure 11.28 Morphological and dynamic zoning of a tidal beach in cross section. The coastal zone is the portion submerged under the average low tides up to the base level of waves, also called the offshore or shoreface zone, depending on the author (Flor, 2004).

Figure 11.29 Bar and intertidal groove on the Playa de Castilla. Parque Nacional de Doñana. In the background, Matalascañas (Province of Huelva, Spain).

beaches separated by promontories along coasts where bays are present. Pocket beaches have a constant sediment volume (Davies, 1980).

Beach **sediment origin** is varied; beach materials come from rivers, cliff erosion, seafloor sediments, wind activity, and even anthropic activity.

The upper limit of a beach is a line indicating the uppermost reach of major storm waves; the lower limit is the depth below which beach processes no longer effect observable changes to the sedimentary floor. Beaches contain various **zones**; zonation of tidal beaches is more complex. Beach zones are belts that run parallel to the coast. They are characteristically dynamic zones. *Figure 11.28* shows the English terms used to distinguish the various parts of a tidal beach (Flor, 2004).

The intertidal and subtidal zones have **bars** (*Fig. 11.29*) that are symmetrical or asymmetrical sedimentary landforms; up to 30 bars can be present. They are parallel or oblique to the coast, continuous or compartmented, linear, sinuous, or crescent shaped. There are various hypotheses regarding the origin of bars, all related to wave action (Woodroffe, 2002; Greenwood, 2005).

A **barrier** is a sand and/or gravel accumulation formed by waves, tides, and wind. Barriers run parallel to the coast, sit above the present-day sea level, and often dam up terrestrial drainage or block coastal lagoons (Horikawa, 1988; Sanjaume and Pardo, 2000) (*Figs. 11.30* and *11.31*). **Spits** (*Fig. 11.32*) are subaerial outcrops of sediments that accumulate as a result of longshore drift currents that transport sands and gravels. Spits are common on points where the coast has an abrupt inflection (Rodríguez *et al.*, 2000). Spits are straight, curved, or hook-shaped. **Cuspate forelands** are sandy promontories that generally have a series of belts. A strand of beach sand that is joined to an island forms a **tombolo** (*Fig. 11.33*). The word tombolo comes from Orbetello (Italy); tombolos result from coastal drift. The Iberian Peninsula has good examples of tombolos in Peñíscola, Ifach, Trafalgar, and Gibraltar.

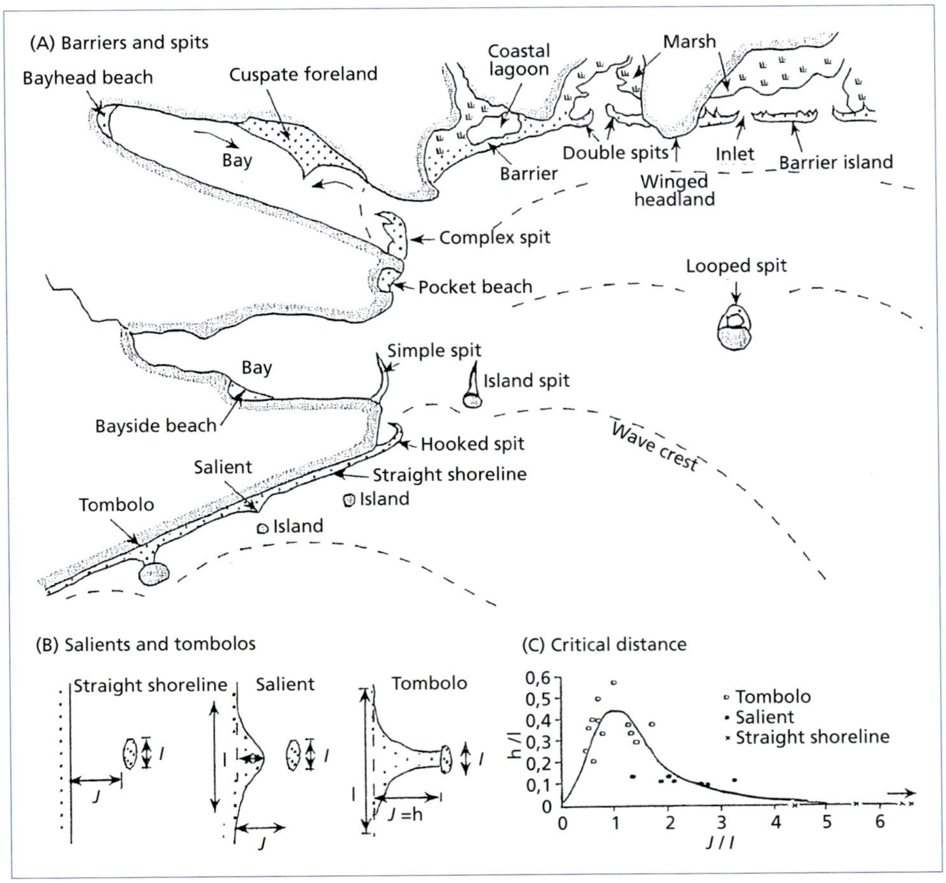

Figure 11.30 Various types of spits, tombolos, and salients or sand capes. (A) Diagram of the arrangement of these landforms with respect to wave trajectory. (B) Definition of the parameters and presence of salients and sandbars related to island diameter. (C) Critical distance from the coast (Horikawa, 1988).

Figure 11.31 Littoral bar. Cies Islands (Province of Pontevedra, Spain).

Figure 11.32 Spits on an island with a semiclosed bay at its center. Northwestern coast of the United States.

Figure 11.33 Tombolo. Eubea Island (Greece) Photo J.L Peña.

11.9 Coastal dunes

Coastal dunes *(Fig. 11.34)* are made of beach sands that accumulate above the tidal level (Nordstrom *et al*., 1990; Nordstrom, 2008). Wind can move dry beach sand grains to build dunes; most of this sand measures between 0.1 and 0.3 mm. Unlike desert dunes, coastal dunes are subject to a variety of processes including as wave action and vegetation (Bagnold, 1941). Coastal dunes require a significant sand supply and abundant wind action. Storm waves can erode dunes but over time and under favorable weather conditions, wind can return sand to the dune system. This shows that maintenance of coastal dunes is important for the protection and use of coastal areas.

Dunes are abundant in middle latitudes (Jennings, 1964) where large tidal intervals are present. They are less well developed in low and high latitudes (Davies, 1980). In low latitudes, abundant coastal vegetation, low energy wave levels, and other less important factors limit dune growth. At high latitudes, the lack of sandy sediments and the presence of ice activity and pack ice prevent wave action, creating obstacles to dune growth.

When vegetation is spaced as it is in arid zones, it can retain sand and form a *nebkha* or **coppice dune**. Also, in areas of scarce vegetation, parabolic dunes and *blowouts* can develop. Dense vegetation can trap sand and cause vertical accretion, which can create a belt of coastal dunes (Arens, 1996). A linear dune (*seif*) *(Fig. 10.32)* and fields of barchan dunes can develop (Inman *et al*., 1966) although they are rare.

Along some coasts, as in the province of Huelva (Spain), numerous dune belts *(Fig. 10.2)* (Flor, 1990; Rodríguez, 1998) form by progradation. They parallel the coast and may reach heights of 30 m. Erosion can modify these belts along a gradation from initially straight dunes to *blowouts* and finally to parabolic dunes (Pethick, 1984) *(Fig. 11.35)*. Advance rates of 1–10 m have been measured in areas where non-vegetated dunes are 10 to 20 m high (Bird, 2000).

When coastal dunes are calcareous (that is, composed of sand grains and shell particles) underlying dunes can lithify as a result of percolating water; this process converts the sand into a rock called calcarenite (Bird, 2008). The most generic term used for them is **eolianite**, which describes their origin. On steep outcrops along coasts various deposits can be seen that show a coastal marine origin, eolianites, and the development of pedogenic processes that create red soils *(Fig. 11.36)*. The overall section represents a complex geomorphic history.

Figure 11.34 Littoral dunes with barchan fronts and blowouts. Maspalomas (Gran Canary, Spain).

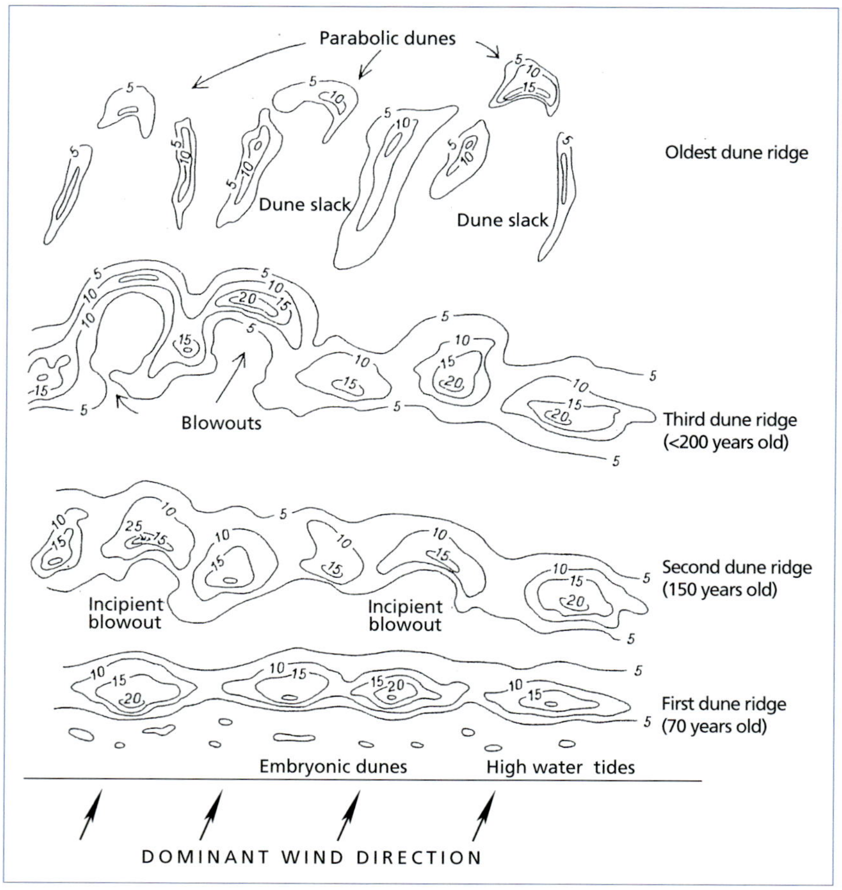

Figure 11.35 Scheme showing the developmental sequence of dune belts. Blowouts in the second and third dune belts producing parabolic dunes in the last stage indicated (Pethick, 1984).

Figure 11.36 Pleistocene coastal deposits in El Aculadero, Puerto de Santa María (Province of Cádiz, Spain). The cliff from bottom to top is the back of the present-day beach, which has loose boulders. At the base are *lagoon* deposits. There is red soil on the second level. Above are eolianites with crossbedding and, at the top, development of a second level of red soil.

11.10 Mudflats, salt marshes, and mangroves

These features develop in intertidal zones primarily on coasts with moderate to low energy wave action (*Fig. 11.37*). Intertidal areas are composed primarily of sand and silt deposits called *sandflats* or *mudflats*; they are also known generically as tidal flats. They occur only on land, are generally slightly inclined toward the sea, and extend from the high tide level to the low tide level. Tidal flats are furrowed by a network of channels. When salt-tolerant plants (halophytes) colonize these intertidal plains they form salt marshes and mangrove swamps especially in higher areas (Bird 2008). Mudflats, salt marshes, and mangrove swamps are laterally extensive; they are associated with wide bays, deltas, estuaries, salt barriers, and areas of open sea. These planar landforms are important because they are home to highly productive ecosystems. In addition, researchers study their sediments for paleoecological and paleoenvironmental reconstructions (Woodroffe, 2002).

The sediments present are sand, mud, clay, and organic material; sources include coastal erosion, the sea floor, and fluvial contributions (Pethick, 1984). Particles move primarily in suspension and are deposited mostly in the intertidal zone. Although very fine-grained deposits are cohesive and consequently erosion resistant, they can be partially removed as tides flow in and out.

Mudflats are not completely flat. A mudflat with a slightly convex profile shows that accretion predominates whereas a slightly concave profile indicates that erosion is the dominant factor. We have noted that mud is more cohesive than sand and thus requires significant flow rates for erosion to occur (Hjulström, 1935). Stages of accretion alternate with phases of erosion on all time scales. These stages depend on sediment supply, tidal interval, and also on the season because mudflats erode in fall and winter and accrete in spring and summer. They also leave records of sediment changes of this type over longer periods of time.

Salt marshes are mudflats vegetated by halophilic plants; dominant processes in salt marshes are different from those in mudflats. Salt marshes occur in temperate zones; they are not present at polar latitudes. Spits and barriers often close them off (*Fig. 11.38*).

Salt marsh surfaces are higher and thus flood less frequently. They flood most commonly as a result of

Figure 11.37 Intertidal plain and tidal channels. Headlands of the Ría de Vigo. Arcade (Province of Pontevedra, Spain).

Figure 11.38 Marshes and tidal channels sheltered by a barrier island. Long Island (New York). North-eastern coast of the United States.

high-amplitude spring tides. The major differences between saltflats and mudflats are vegetative cover (*Salicornia, Juncus*, and others) and the intricate channel network that covers most salt marshes. Plants trap significant sediments and vegetation density dominates water flow (Neumeier and Ciavola, 2004). Vegetation produces organic matter that mixes with inorganic sediments. During spring and summer, water salinity is high and flocculation increases. Small *salt pans* have been identified in salt marshes; these may be areas that lack vegetation or may be the remains of abandoned channels where salts precipitate due to evaporation.

Mangroves are composed of bushes and halophilic trees, of which there are about 70 species in the Indo-Pacific region (Bird, 2008). Mangroves develop at the top of the intertidal zone on coasts where estuaries and lagoons are present and on coasts sheltered from strong wave action such as bays or inlets on the leeward side of promontories, islands, or reefs (Guilcher, 1979). They are found in tropical areas. Their trunks and root systems are intertwined and thus they resist erosion, as shown by the presence of small plants in salt marshes. Mangroves also have a network of channels. Mangroves can extend to a width of several kilometers along macro-tidal coasts; their width increases with the tidal interval.

11.11 Estuaries and deltas

An **estuary** (*Fig. 11.39*) is the seaward part of a submerged valley system that receives sediments from fluvial and marine sources; tidal, wave, and fluvial processes form its facies (Dalrymple *et al*., 1992a). The Flandrian Transgression inundated coastal valleys prior to the stabilization of the sea about 6000 years ago. Since then, sedimentation has been filling estuaries and progradation has been creating *fan deltas*. The Vélez River (Málaga, Spain) became an estuary as the Flandrian Transgression extended inland. Later the estuary began to fill at a rate of 1–2 m/millennium. Archeological remains show Phoenician settlements dating from 650 BC located above the present-day level of the valley. At the end of the 15th century the ships of the Catholic Kings sailed about 5 km up the estuary to surround the town of Vélez Málaga. After the Moors were expelled, sedimentation increased significantly as a result of deforestation and soil erosion. Deposits from the 16th and 17th centuries are 20 m thick. The estuary filled and the Vélez River has built a prograding delta where the town of Torre del Mar is now located (*Fig. 11.40*) (Hoffman and Schulz, 1987).

Figure 11.39 Human impact on an estuary. The outer zone has several spits and pocket beaches. Coast of Connecticut. United States.

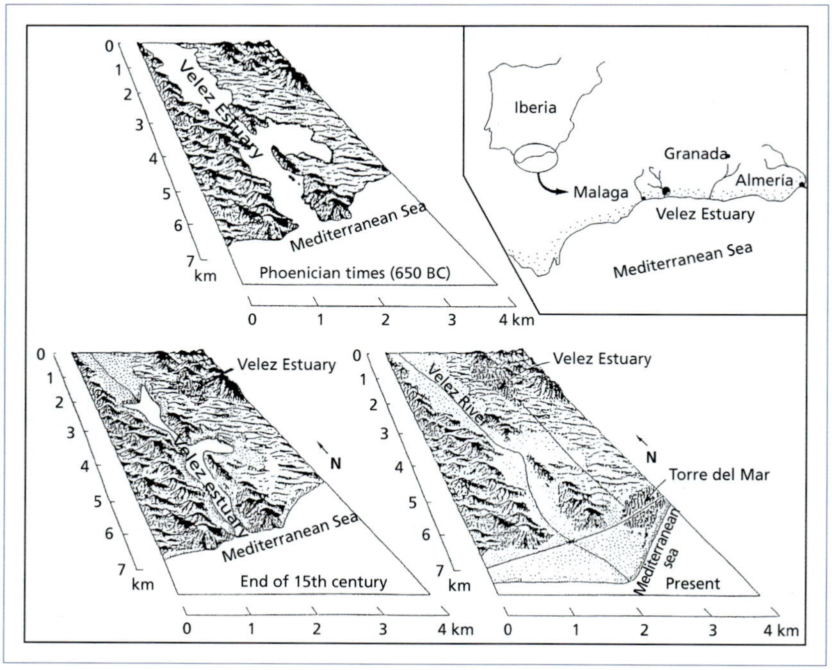

Figure 11.40 Recent evolution of the coast of Vélez Málaga (Province of Málaga, Spain) (Hoffman and Schulz, 1987).

Estuaries are unique ecosystems that provide food for multiple organisms. For humans, estuaries are navigation, settlement, and recreation areas and areas that provide erosion protection and opportunities for mineral extraction and waste disposal (Jackson, 2004).

Estuary morphology results from the constant interplay between fluvial and marine contributions; it is subject to the hydrodynamic processes of fluvial flow, tidal currents, and wave action, biological processes including the growth of salt marshes and mangroves, and processes such as flocculation. Marine water is denser than river water; river water enters an estuary as an overlying wedge that remains at the top of the water column. As the river water moves toward the sea, the line between the two water densities remains very sharp (Bird, 2008).

Dalrymple *et al.* (1992b) divide estuaries into three zones: an internal zone where fluvial processes predominate, an external zone where marine action predominates, and a central zone where tidal and fluvial processes mix. According to Davies (1964), wave dominated estuaries are found in macro-tidal environments (tidal interval > 4 m); these funnel-shaped estuaries have wide mouths and high current velocities. The lower part of the estuary has elongate tidal sandbars oriented parallel to the current direction (Hayes, 2005) (*Fig. 11.41*). Meso-tidal estuaries have a broad intertidal area with mudflats, mangroves, or salt flats, and multiple meandering channels. Micro-tidal estuaries are dominated by wave action and river discharge currents; beaches, spits, and sand banks often develop at their mouths.

At low tide, estuary channels bring water from the river to the sea between sand, mud, and clay banks. Incoming and outgoing tidal currents shape the channels and banks and their shape, position, and dimensions change frequently. Mudflats and salt flats are analogous to the floodplain of a river because they develop above the average tidal level. However, salt flats are covered by water only every two or three years whereas mudflats flood about 700 times a year (Pethick, 1984).

Fjords are ancient glacial troughs where the sea invaded during the Flandrian Transgression. They are analyzed in more detail in Chapter 13 on Glacial Geomorphology.

Rias (*Fig. 11.42*) are similar to fjords but the inundated valley is fluvial. Von Richthofen defined them in 1886 in his study of the rias of Galicia. Vidal Romaní (1984) studied the significance and origin of rias as described by various investigators through 1983. Méndez and Rey (2000) later wrote a historical review and summary of more recent investigations. The word "ría" is Spanish, from Galicia and Asturias; a ría is a long and narrow funnel-shaped inlet created by subaerial erosion. Some rias in Galicia are controlled by fractures (Pagés, 2000). Rias are shallower than fjords, and their inland portions are narrower than the seaward sections. Mudflats and salt flats occur at the headlands of some rias. To understand more about Galician rias, researchers have conducted systematic cartographic studies of submarine sediments (Rías of

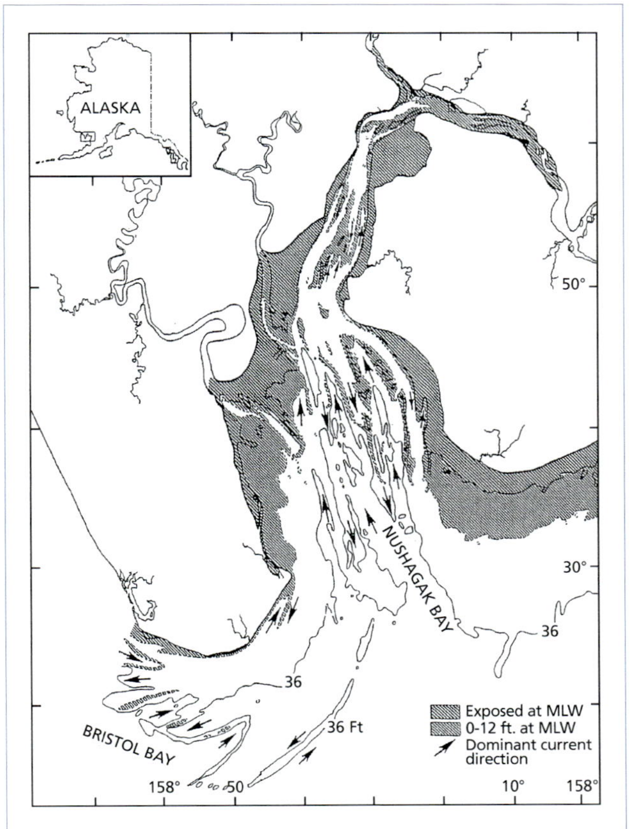

Figure 11.41 Tide dominated bay (tidal range 5.5 m). Note the alignment of the sandbanks parallel to tidal currents. Nushagak Bay, Alaska. United States (Hayes, 2005).

Arousa, Pontevedra, and Vigo) using seismic techniques and analyzing sedimentary fill (Vilas *et al.*, 2005).

A **delta** can be defined as a seaward-protruding convex coastal structure at the mouth of a river. "A delta forms when fluvial sediment contribution exceeds redistribution by marine processes, such as waves, currents, and tides" (Arche, 1989). Deltas formed during late Quaternary marine transgression and significant sediments continue to be deposited during the Holocene. These landforms make up 1% of the world's coastlines (Bird, 2008). Their name comes from the fact that many deltas are shaped like the Greek letter Δ.

Deltas have historically had an important socioeconomic role (Giosan and Goodbred, 2007). Primitive civilizations and the beginnings of agriculture were located on subaerial deltas. Today deltas support large urban population centers such as Shanghai, Bangkok, and Cairo, among others (Wright, 2004). Most deltas have subaqueous and subaerial components both below and above low tide (Coleman, 1981). The subaerial part consists of an upper and lower **delta plain**. The upper delta plain is formed by the contribution of alluvial sediments that remain above high tide. In the lower delta plain, the fluvial channel becomes a tidal channel with a distributary network that may flow to the coast at the **delta front** (*Fig. 11.43*). Fluvial channels in deltas are subject to tidal inflow and outflow (*Fig. 11.44*). Seaward of the delta front is the **prodelta**, which has an abrupt slope that promotes the seaward deposition of clay materials. These highly unstable fine sediments are prone to collapse. A prograding delta creates a **delta lobe** that requires subsidence to grow and a sediment supply to feed it.

Figure 11.42 Ría of Xallas, excavated on top of granite. Note the sandy spit that partially closes the mouth of the estuary. Ezaro (Province of La Coruña, Spain). Photo F.J. Gracia.

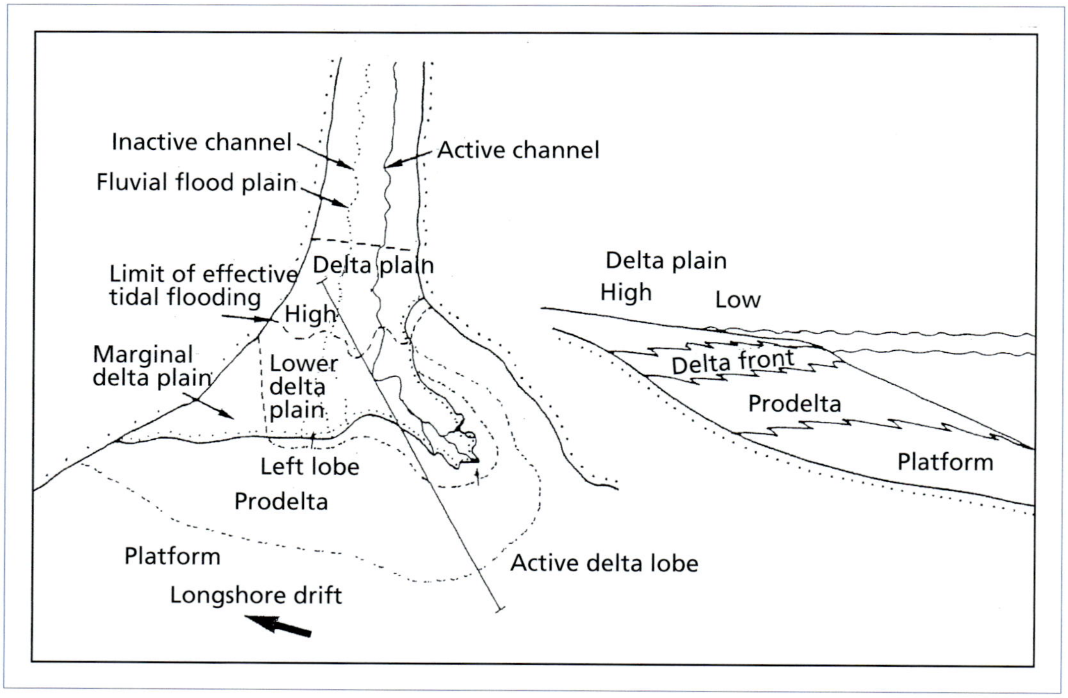

Figure 11.43 Delta system components (modified from Coleman and Prior, 1980) and idealized section (Dabrio and Zazo, 1988).

Figure 11.44 Fluvial channels in the Copper River delta. Southern Alaska. Photo F. Gutiérrez.

Deltas form at the mouths of rivers that supply abundant sediments; deltas need rivers with large drainage basins (Hori and Saito, 2007) where climate, geology, and topography play important roles. However, sediments transported and deposited by coastal currents also feed deltas (Giosan and Goodbre, 2007).

Deltas vary enormously in size and shape depending on sediment supply, coastal configuration, bathymetry in near-shore areas (open sea, gulf, lagoon, lake), the effects of waves and currents acting on accumulated sediments, and the existence of low-gradient continental platforms where sediments accumulate and prograde (Wright, 1985; Bird 2008). For these reasons, large deltas form primarily on passive continental margins (Wright, 1985).

Galloway's (1975) widely accepted delta classification uses a ternary diagram to show delta morphology as a function of fluvial sediment contribution and wave and tidal energy (Dabrio, 1984) (*Fig. 11.45*). **Fluvial-dominated deltas** are formed by large river basins that discharge into protected seas with low energy waves and a small tidal range. The rivers deposit interfingering sand bars that rapidly prograde seaward (such as the Mississippi Delta). Salt flats and swampy areas are well developed in the delta plain of the Mississippi River (Kosters and Sutter, 1993). **Wave-dominated** deltas are shaped by seafloor waves in the open ocean where arcuate bars can form (Senegal Delta). **Tidal dominated deltas** are noted for their wide tidal interval and often a complex system of mangroves, plains, and sandy bays (*Figs. 11.44*, *11.62* and *11.64*) (The Ganges-Brahmaputra Delta). Delta-front platforms in front of tidal-dominated deltas are wider than those of other delta types (Nittouer, 1986). *Figure 11.45* shows different delta types with mixed domains.

The advance or retreat of delta coastlines can be seen on successive maps and aerial photographs over time. The various evolutionary stages of a delta can be traced from the time the sea reached its high level (*highstand*), which is when sediments began to fill the mouth of the river. Deltas develop initially during regressive marine phases (Boyd *et al.*, 1992) as rivers supply sediment during a period of low sea level (Stanley and Warne, 1994). After the initial delta formation, relative sea level change varies considerably from delta to delta due to hydro-isostatic effects (Nakada *et al.*, 1991), tectonic uplift or subsidence, and/or sediment compaction (Roberts *et al.*, 1994).

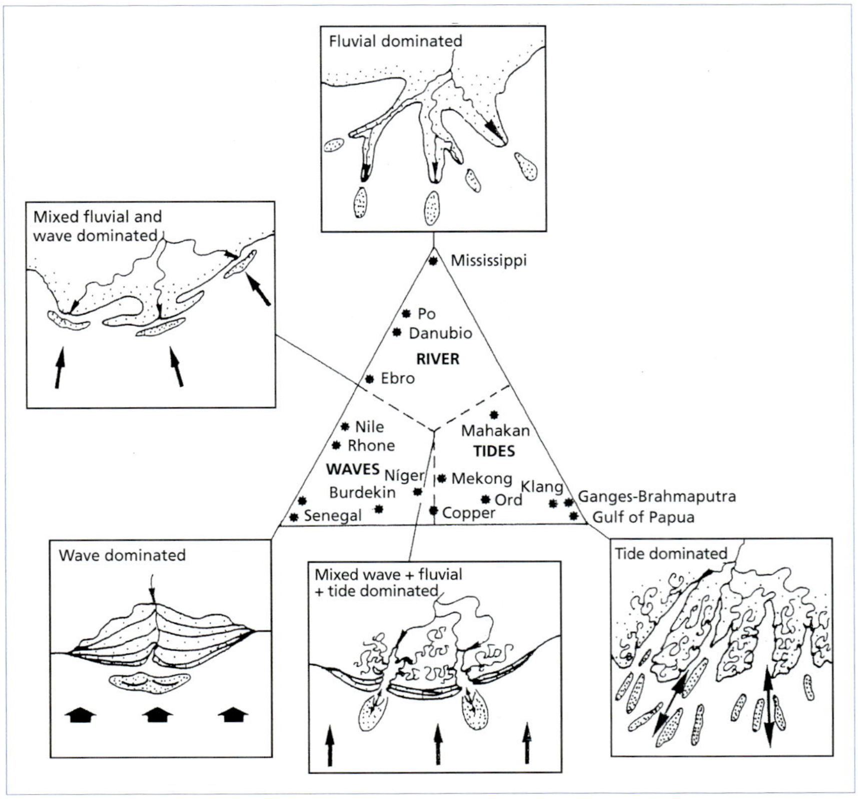

Figure 11.45 Ternary classification of delta types based on the delta front regime (Galloway, 1975, Elliot, 1986) and conceptual models of respective landforms (Dabrio, 1984).

Delta margins result from the interplay between fluvial sediment contributions and processes of marine erosion. Some deltas, such as the Mississippi, undergo significant and rapid progradation (*Fig. 11.46*) (Holmes, 1965) as a result of a favorable balance of fluvial supply. The elongate and digitate shape of the Mississippi delta is rare; it results from a combination of cohesive sediments, low wave height, and weak tidal currents (Wright, 2004). Progradation can be spectacular especially in shallow tropical seas; for example, some deltas in Java prograde more than 200 m/year (Bird, 2008).

Fluvial gravel and sand mobilized by wave and coastal current action supply many deltas that are bordered by beaches and spits; the Ebro Delta is an example (*Fig. 11.47*). Often delta beaches, bars, and spits enclose lagoons and swampy areas that may be incorporated into delta growth (Bird, 2008).

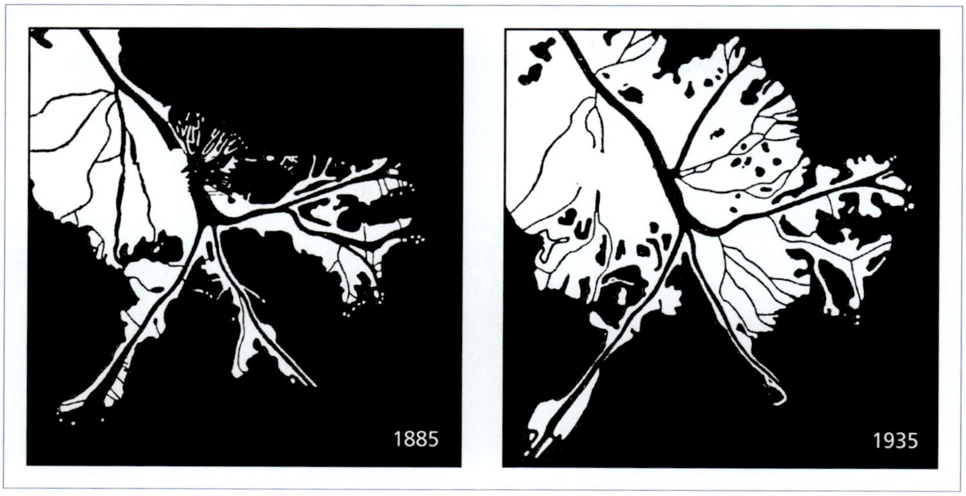

Figure 11.46 Growth of the Mississippi Delta over 50 years. Classic example of a digitate delta (Holmes, 1965).

Figure 11.47 Ebro Delta. Landsat satellite photograph.

11.12 Coastal environments: Uses, management, and risks

Humans have substantially modified most coasts; this has sometimes taken place in a matter of a few years (Fig. 11.48) and the trend is expected to continue in the future (Messerli *et al.*, 2000). By 2020, the human population along the 60 km wide coastal fringe is expected to increase by 75% (Edgren, 1993). Spain has about 7000 km of coastline of the following composition: 57.7% steep coast; 30.1% sandy beaches; 1.1% very fine sediments; 0.9% estuaries; and 10.2% artificial coasts (Andrés and Gracia, 2002). In addition, more than 35% of the country's population lives within 5 km of the coast (Díaz de Terán and Cendrero, 1992).

The interaction between human activities and the coast disrupts the natural system in ways that range from almost imperceptible to catastrophic. Coastal environments vary considerably in their capacity to absorb anthropic pressure (Carter, 1988).

In spite of the benefits of living along the coast, unsuitable coastal management in the past has caused numerous problems (Carter, 1988). For this reason, coastal engineers, oceanographers, and geomorphologists now collaborate closely to confront the complex problems of coastal use (Stephenson and Brander, 2003) and to reach the human goal of living in harmony with nature. Humans use the coast for many purposes: tourism, residence, recreation, industry, commerce, waste disposal, agriculture, aquaculture, fishing, nature preserves, natural resource exploitation, and military and strategic purposes (Carter, 1988). Most coasts support multiple interrelated activities. Most uses affect the natural dynamics of the coastal environment.

These activities require coastal management, which is an approach that incorporates political, economic, and environmental issues. International Programs of the United Nations, such as the 1982 *Law of the Sea, LOS,* and other national coastal management programs focus on the future of coastal environments in recognition of the increased demand for various uses of these areas. To date no unified international program has been established for coastal management although two primary models are available, an American and a British model. It is important to note that coastal management policies can vary from country to country. For example, the primary preoccupation of Holland is coastal management because in the future large storm waves may flood more than 50% of the country. Major efforts have been directed at the recovery of salt marshes for agriculture (rich *polderlands*) and for marine defense (Fig. 11.49). In spite of these efforts, a 1953storm in the North Sea ruptured the dikes; the subsequent flooding killed 1,835 people (McGuire, 2004).

Figure 11.48 Recently built tourist area. Puerto Rico. Gran Canary, Spain.

Coastal **tourism** and residential and recreational activities are major industries along the coast because many people along the coast depend on tourist income. Tourism has expanded significantly in recent decades as reflected in spectacular "*booms*" in the Caribbean, Southeast Asia, and Spain, among other countries. More than 56 million tourists visit Spain every year.

Tourism in coastal zones leads to numerous environmental impacts. Many impacts affect the quality of life of the visitors in a positive way while others, such as excessive building and environmental, biological, physical, and ecological impacts on nature are negative (Hernández *et al.*, 2005). As a result, the key is to protect and preserve many areas, resolve erosion and water pollution problems, and confront the loss of species diversity. Sustainable development is possible through planning, appropriate use, and education (Miller and Hadley, 2005).

The petroleum crisis in the 1970s led to increased research into renewable and non-polluting **energy**. One possible form of renewable energy is the use of wave and

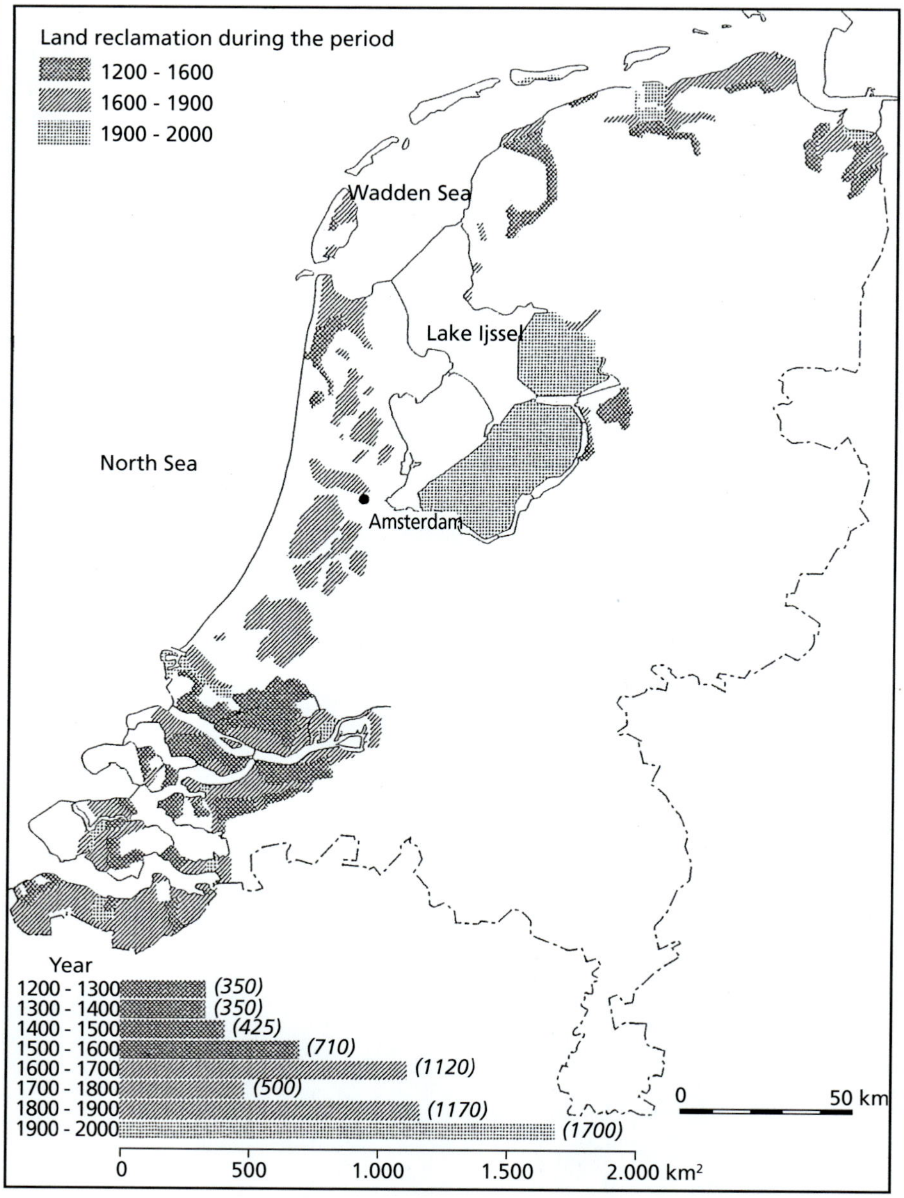

Figure 11.49 Land reclamation in Holland between the years 1200–2000 (Geography of the Netherlands, in Terwindt, 2005).

tide dynamics as energy sources. The main problem associated with these technologies is their high installation cost per amount of energy produced. Today, numerous countries are experimenting with various techniques for producing low-cost energy. Logically the most favorable coasts for these experiments are those with significant wave action. Likewise, tidal plants should be sited on coasts with medium to large tidal ranges because tidal flow channel velocities of 2–3 m/second are needed to generate 3–4 Kw/m^2 of power (Carter, 1988; Flor, 2004). Tidal plants were recently built in the Cantabrian Sea and tidal plants are already producing energy in France, Russia, China, and Canada (Raven *et al.*, 2008).

In addition to wave and tide energy, other possible coastal energy sources include wind, biomass, and differences in water temperature and salinity (Carter, 1988). Thermal and nuclear power plants are also being built; they require water for operation and have their own set of environmental issues.

We must also consider crude oil spills from large oil tankers (more than 100,000 tons) such as the *Torrey Canyon* and the *Amoco Cadiz* in 1967 and 1968. The sinking of the tanker *Prestige* in 2002 caused an even larger crude oil spill and unprecedented damage primarily along the Galician coasts (López-Bedoya and Pérez-Alberti, 2008; Mahmood and Shokr, 2008).

Water, petroleum, and gas are produced along some coasts. Significant extraction can cause surface **subsidence**; the land surface subsides to the greatest extent in the area of maximum pumping. Water extraction has caused subsidence along the margins of the Venice lagoon; this and other factors have caused several towers to lean (*Fig. 11.50*) (Day *et al.*, 1999). Storm waves can flood the subsidence cone, as has occurred in the city of Venice and along other subsiding coasts (Bird, 2008). When water extraction from an aquifer reaches a certain level, seawater intrudes into the aquifer so that subsequent pumping produces saline water, which is potable only up to a salt content of 2%. Saltwater wedges in coastal aquifers have caused major environmental problems particularly in rapidly developing coastal areas of Florida, southern Spain, and Yugoslavia (Carter, 1988).

We should also point out the effects of fecal contamination, industrial wastewaters, and dumps; these can cause illness, toxicity, and oxygen depletion. Dredging, mining activities, mobilization of coastal sediments (*Fig. 11.49*), and desalination plants all lead to serious environmental and ecological consequences.

Dredging is done to facilitate navigation lines, for beach maintenance, to produce construction materials, to extract minerals (gold, rutile, cassiterite, and diamonds), and also to restore certain environments. The problem with dredging is its waste, which affects the environment both directly and indirectly. The remobilization of sediments disturbs sedimentary equilibrium, affects biota, and increases

Figure 11.50 Canal and inclined tower. Venice (Italy).

turbidity. Dredging of mineral fines along the Hallsands coast in southwestern England significantly eroded the coast and destroyed the town of Hallsands (Carter, 1988)

Coastal environments are dynamic; they provide continuous flows of mass, energy, and information. Human intervention in natural processes can accelerate and even double the rate of erosion (Milliman and Syvitski, 1992). This intervention has considerably accelerated during the last century as a result of increased deforestation, expansion and mechanization of agriculture, the proliferation of dams, and increased coastal urbanization. On the other hand, cliff erosion contributes no more than 5% of beach material. Higher retreat rates of up to 1 m/year occur on cliffs composed of unconsolidated sediments (glacial and fluvioglacial deposits on the coast of Lowestoft, Suffolk in eastern England) and dunes on the eastern coast of the United States.

Waves move sand in front of and along the coastal margin. Once an equilibrium beach profile is established, transport occurs primarily along the coast (Inman and Jenkins, 2005). **Coastline erosion** is common today (Paskoff, 1981, 1985; Bird, 1985; Carter, 1988). On some coasts, this tendency goes back to previous centuries as shown by the cartography of Zazo *et al.*, 1986 and Dabrio *et al.*, 1993 (*Figs. 11.51* and *11.52*). Topographic surveys, maps, ancient

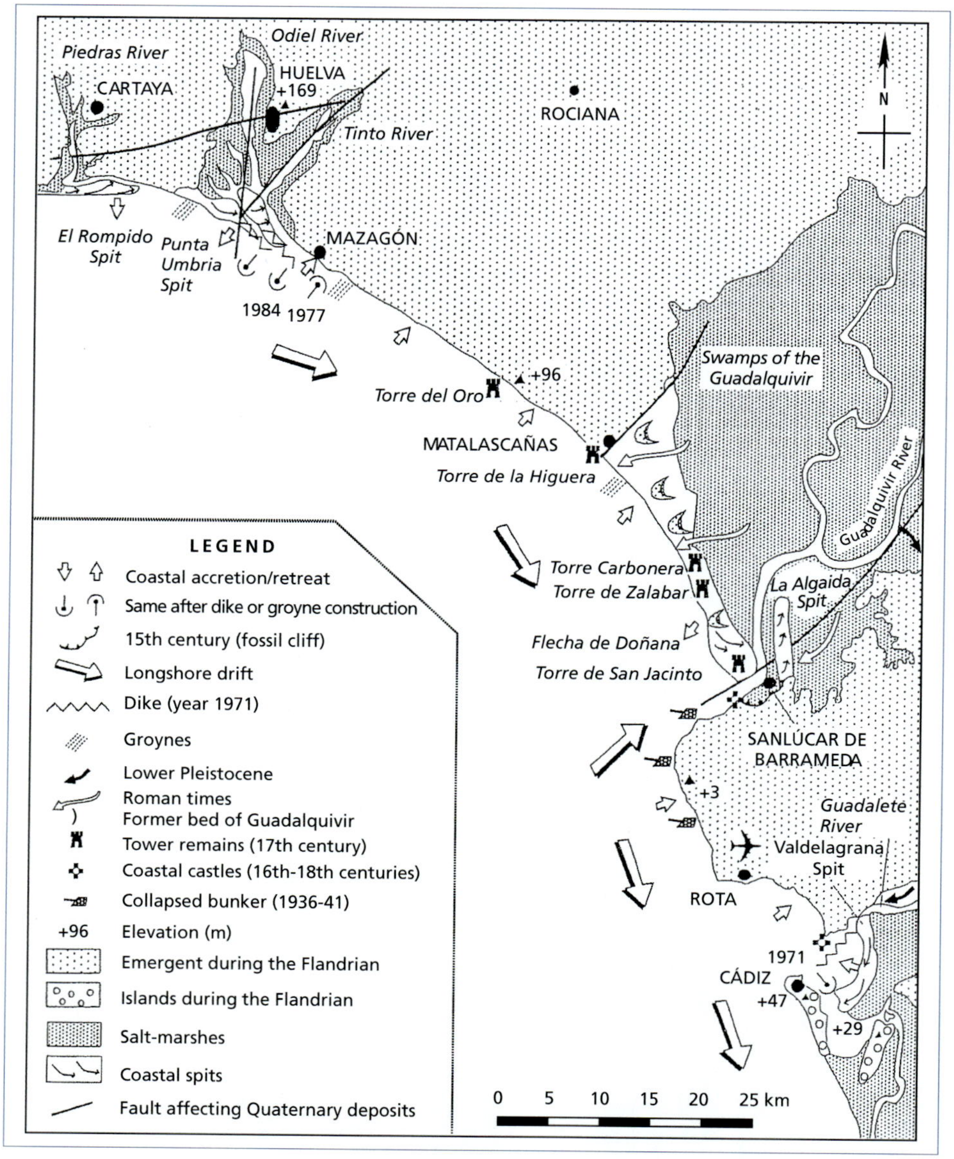

Figure 11.51 Coastal areas of the Gulf of Cadiz (southeastern Spain) with indication of some coastal processes and modifications resulting from human activity. (Simplified by Zazo *et al.*, 1986).

Figure 11.52 Higuera Tower, built in the 17th century, knocked down by coastal dynamics due to coastal retreat. Playa de Castilla, Matalascañas (Province of Huelva, Spain).

maps, and conventional and satellite aerial photography can be used to measure erosion rates. Resolution depends on the scale used (Gracia *et al.*, 2005; Del Río and Gracia, 2009).

Various methods can be used to **control coastal erosion** along cliffs and low-gradient slopes. Engineers have traditionally constructed static structures between the sea and the coast. *Figure 11.53A-E* shows **seawalls**, **dikes** and **protective coverings** (Carter, 1988). To stand up against storm waves, massive barriers are constructed that must be capable of absorbing pressure and enduring wave attacks. A breakwater *(Fig. 11.53D)* can be built with gabions or prefabricated units such as tetrapods (blocks with four arms). Along low-energy coasts, structures are simpler; they include as jetty breakwaters, protective coverings, and dikes *(Fig. 11.53E, F)*.

Groins, also called groynes, are wall-like structures perpendicular to the coast; made of concrete, metal, or wood, they are used to trap sediments carried by longshore drift. They are not a panacea to erosion and the effective distribution of groins is more art than science (Carter, 1988). The intercepted sediment is deposited on the side opposite the direction of the longshore drift current *(Fig. 11.54)*. Groins produce good results in micro-tidal ranges. The placement (length and spacing) of groins is fundamental *(Fig. 11.55)*. In a series of groins, the first is installed to establish a part of the coastline and hold back some of the drift material. Then parallel groins are installed in series, to the last one, called the terminal groin. In this way, *bypassing* allows the drift to continue. If groins are arranged with more spacing, a phenomenon called "*flanking*" can occur; this results in erosion of the landward side of the groin. It sometimes happens that all the sand is exported and the rocky substrate is exposed, as occurred on the Villas de Benicasim beach (Province of Castellón, Spain) during the 1980s.

Periods of storm waves clearly indicate that erosion has affected many beaches worldwide *(Fig. 11.56)*. Erosion can affect beach profiles and, if natural processes do not regenerate them, it may be necessary to add sand from another area to reestablish equilibrium. Miami Beach (Florida) used 13 million m^3 of sand to regenerate its beaches between 1976 and 1981 at a cost of 60 million dollars (Woodroffe, 2002).

Beaches should never be artificially replenished with sand from nearby coastal dunes; this is not an appropriate option because the sediment balance between the beach and the dune are intimately related (Paskoff, 1985; Psuty, 1988). Dunes and the *backshore* constitute a sand resource unit that should be cared for and preserved (Dabrio and Zazo, 1988).

In developing countries the preferred method for protecting coastal margins from beach erosion is by supplying sand artificially. Although this technique has been used in various environmental settings, it remains highly controversial. However, artificially supplying sand to a beach is the

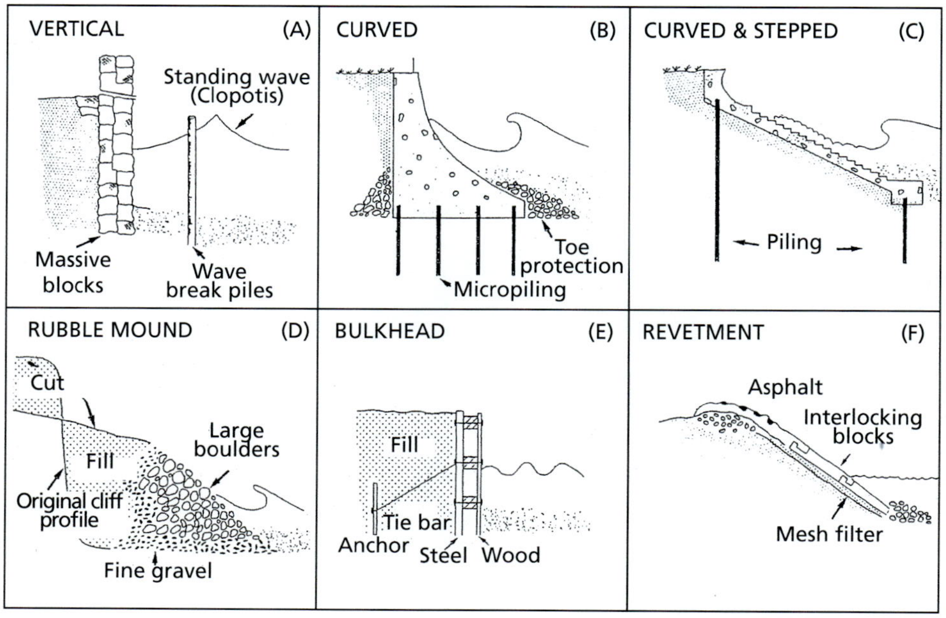

Figure 11.53 Different types of designs for coastal protection based on an energy sequence (high to low, A to F). (A) Vertical wall constructed of strong boulders. (B) Curved protection wall with basal protection (C) Curved and stepped wall reinforced with pilings (D) Jetty with deposits on the top of the cliff (E) Wooden or steel bulkhead. (F) Covering made of boulders, gabions, or asphalt (Carter, 1988).

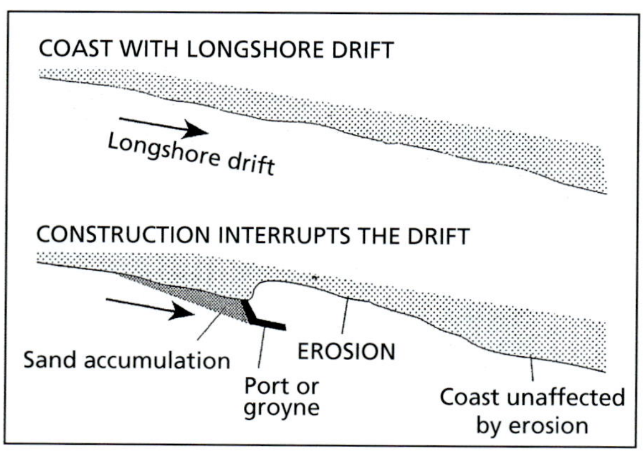

Figure 11.54 Erosion due to the interruption of longshore drift and sand accumulation in the port or groin in the opposite direction of drift (Dabrio *et al.*, 1993).

most practical method for protecting beaches from flooding by strong swells, for extending the coastline toward the sea, and for enlarging recreational beaches (Finkl and Walter, 2005). Sand can also be added to build new beaches. This costly practice is done in tourist areas to attract visitors. An example is the Playa de las Teresitas, which was created north of the capital of Santa Cruz de Tenerife (Canary Islands). At the end of an arcuate coast, two perpendicular groins and a large jetty parallel to the coast were built about a hundred meters from the coastline. A space was left between the groins and the jetty for a seawater connection. Sand was deposited along the coastline to build the beach (Flor, 2004). Given the high cost of beach construction and restoration, it is important to know **the durability of the beach,** which is defined as the time that elapses between the sand addition and the loss of at least 50% of the fill volume; this is the half-life of a regenerated beach. A determination of beach durability requires an understanding of the temporal evolution of beach profiles (Leonard *et al.*, 1990).

In the United States, many scientists approve of the protective measure of artificially adding sand to replenish eroded beaches whereas another smaller group is opposed to it primarily for environmental, economic, sociological, and political reasons (Finkl and Walter, 2005).

Coastal dunes provide natural buffers to coastal erosion and are considered a basic priority in coastal management. For this reason, it is obviously important to preserve coastal dunes. Coastal dunes serve as breakwaters for large waves, often with frequent touchups by humans, as in coastal Denmark and Holland; dunes also prevent flooding of the

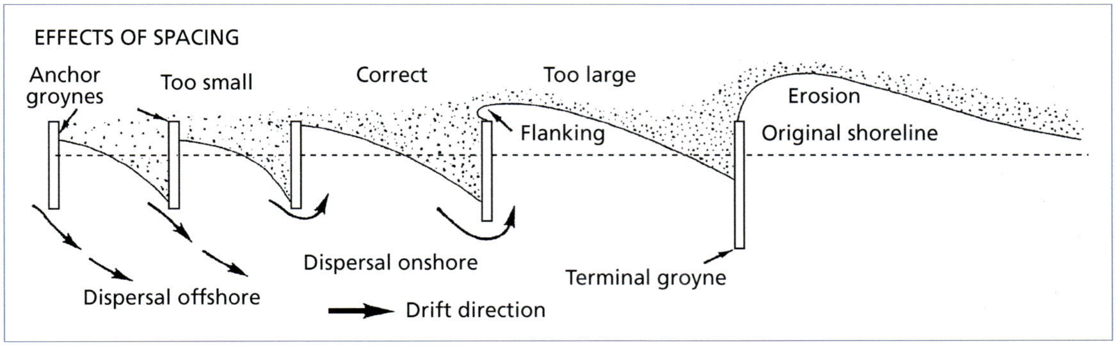

Figure 11.55 Spacing is perhaps the most important of all the factors related to groins. Correct spacing is most likely a function of wave parameters. Closely spaced groins do not retain sediment and widely spaced ones can promote "*flanking*," which is erosion of the area inside the groin. Frequently the end groins are responsible for erosion by the longshore current (Carter, 1988).

Figure 11.56 The Nautical Club of Mazagon was severely damaged by storm waves on 5 June 1996 (Province of Huelva, Spain).

backbeach by tidal seawater (Bakker *et al.*, 1990) (*Figs. 11.57* and *10.40*). Humans have exerted considerable impact on coastal dunes. The type and quantity of vegetation on beaches and dunes is critically important. Diversity decreases near beaches because few species can endure sand movement and salt spray (Moreno-Casasola, 1986). Landward sand accumulation and protection from salt spray promote the growth of woody bushes in seaward parts and tree-like species toward the interior (*Fig. 11.58*). *Figure 11.58* shows the dune types that develop along the coast and different situations indicated for restoration (Nordstrom, 2008).

As a result of various anthropic activities, many dune systems are in advanced states of disintegration; some have been completely destroyed. Dunes are considered to be fragile and sensitive because even minor impacts can modify them. Dune construction and stabilization is a common practice (*Fig. 11.59*) that employs a variety of techniques (FAO, 1988) discussed in Chapter 10 on Eolian Geomorphology.

Figure 11.57 Lighthouse and blowout. Rubjerg Knude Lighthouse, western coast of Jutland. Littoral dunes in Denmark. Photo J. Guerrero.

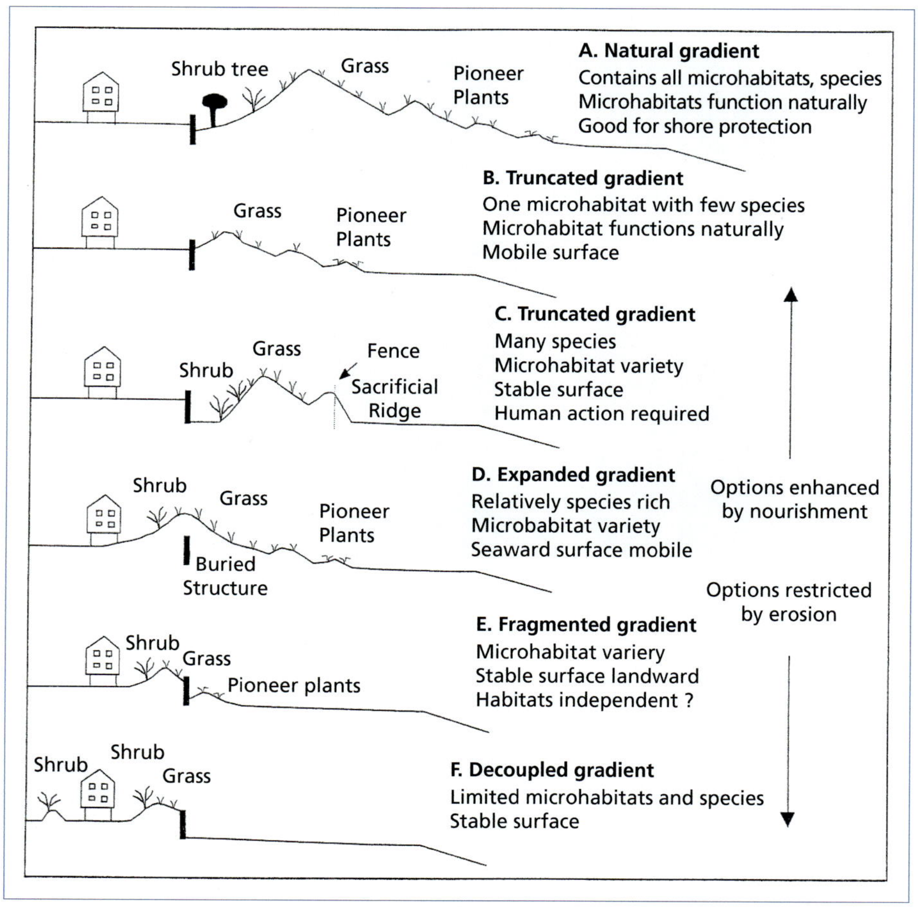

Figure 11.58 Types of dunes that develop on a coast, representing different examples of restoration (Nordstrom, 2008).

Figure 11.59 Littoral dunes with parallel palisade lines used for dune stabilization. Note that the palisade line, located to the left, has little accumulation due to the fact that the other palisades have retained a significant amount of sand and some parts have even been buried. Littoral dunes of Punta Paloma, Tarifa (Province of Cadiz, Spain). Photo F.J. Gracia.

Dams are more obvious forms of anthropic interference in fluvial systems. Thirty percent of the totalload of all rivers is trapped behind large dams (Vörösmarty *et al.*, 1997). Humans have built dams to use and manage rivers for thousands of years. The most common uses are irrigation and hydroelectric energy production. The first large dams in the United States were constructed in the 1930s. Dams to retain water and sediments considerably reduce the solid load that reaches the sea. The reduced solid load affects estuaries and changes the coastline and beach morphology; it also leads to salt water intrusion, inland salinization of deltas, and ecosystem modifications (Chen, 2005; Warrick *et al.*, 2009).

The construction of the Grand Aswan Dam in 1964significantly affected the Nile River (Egypt) (*Fig. 11.60*). Prior to dam construction, the Nile River carried 124 million tons of sediment to the sea every year and deposited about 10 million tons on the delta plain. Today almost 100% of the sediments are retained in the reservoir, Lake Nasser (Woodward *et al.*, 2007). Stanley (1996) has described the sedimentation in Lake Nasser as the "new Nile delta." In the mid 1990s the dimensions of the lake had grown to 250 km in length, 12 km in width, and 40 m in thickness. The principal sources of suspended sediment downstream of the Aswan come from fluvial erosion and eolian dust contributions. Suspended sediments are present in concentrations greater than 850 mg/l in Cairo. Only a small part reaches the coast because of dams

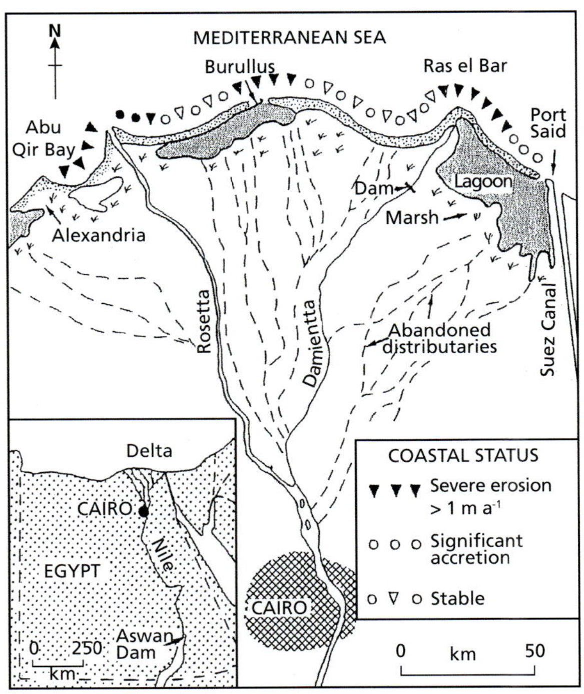

Figure 11.60 Erosion problems on the coastal front of the Nile Delta as a result of sediment retention by the Aswan Dam located 400 km upstream (Carter, 1988).

Table 11.5 **SILT CONCENTRATIONS (IN PARTS PER MILLION) IN THE NILE RIVER AT GAAFRA BEFORE AND AFTER CONSTRUCTION OF THE ASWAN HIGH DAM (ABUL-ATA, 1978)**

BEFORE (averages for the period 1958–63)											
Jan.											Dec.
60	50	45	42	43	85	674	2702	2422	925	124	77
AFTER											
44	47	45	50	51	49	48	45	41	43	48	47
RATIO											
1.5	1.1	1.0	0.8	0.8	1.7	17.0	60.0	59.1	21.5	2.58	1.63

in the delta (Stanley, 1996). Given the negative balance between fluvial contributions and coastal erosion, some authors claim that the Rosseta promontory of the Nile Delta has retreated up to 106 m/year (Frihy *et al.*, 2003). *Table 11.5* compares the silt concentrations (expressed as parts per million) in Gaafra before and after construction of the Aswan Dam. Prior to construction, the period that includes summer and the beginning of autumn was characterized by high silt concentrations but after construction, there is practically no observable variation (Sharaf El Din, 1977; Abul-Ata, 1978). This destructional phase of the Nile Delta is associated with subsidence exceeding 4 mm/year in the eastern part of the delta (Stanley, 1988, 1996; Stanley and Warner, 1993; 1998; Frihy *et al.*, 2003). In some places, the projected retreat is estimated at 30 km for the year 2100 with disastrous consequences for the more than one million people who live in this area. Numerous measures implemented to protect the coast have been quickly destroyed by erosion. In addition, water table decline and salinity increases have affected a large portion of land in the delta (Mainguet, 1991; Beaumont, 1993).

Dam construction has also decreased the bedload of the Ebro River during the 20th century. During the first third of the 20th century, bedload was estimated at one million metric tons/year; between 1950–1970, it decreased to less than 730,000 tons/year and between 1970 and 1990, it was calculated at 300,000 tons/year. During the last decade of the 20th century, the average contribution was less than 100,000 tons/year. By comparing these results to the transport capacity (320,000 tons/year) of the coastal zone of the Ebro Delta, it is obvious that this area has had a negative sedimentary balance since approximately 1970, which has accelerated delta erosion (Guillén, *et al.*, 1992). Until recent times, the Ebro Delta was prograding at a rate of 60 m/year (Somoza *et al.*, 1998) *(Fig. 11.61)*. The construction of large dams especially in the Pyrenees and the Mequinenza and Ribarroja dams in the 1950s, in particular, substantially decreased the terrigenous continental contribution to the delta and transformed it into an erosional delta; some delta locations have retreated 3 km since 1950 (Sánchez-Arcilla *et al.*, 1998).

In contrast, the Huang River (Yellow, China) delta is prograding significantly as a result of increased sediment from land use changes and the accompanying deforestation on loess plateaus (Milliman *et al.*, 1989).

Natural risks are physical phenomena that occur in coastal zones that cause loss of human life, property damage, and environmental degradation. **Tropical cyclones**, also called hurricanes in the Atlantic Ocean, cyclones in the Indian Ocean, and typhoons in the western Pacific are large low pressure systems that develop and intensify in the open ocean between latitudes of ±5° and 25° (Gornitz, 2005).

Bangladesh, previously eastern Pakistan, which is located in the humid tropics, is severely affected by catastrophic floods. It is a small poor country with a surface area of 144, 835 km^2 and a population of 110 million, most of whom live on flood plains with an annual *per capita* income of about 140 € (Brammer, 1990; Khalil, 1990). The Ganges-Brahmaputra delta in the southern part of the country is a tide-dominated delta that is also supplied by the Meghna River *(Fig. 11.62)*. The delta covers an area of 138,000 km^2; its northern part is 10–15 m above sea level (Singh, 2007). Numerous islands in the delta have population densities of up to 1000 people/km^2, most of whom do high-intensity farming (Carter, 1988). About 60% of the country is at risk of flooding *(Fig. 11.63)*. The wide delta and river flood plains are the location of two large closed shallow depressions. Elevated zones consist of Pleistocene terraces and Tertiary hills (Rasid and Pramanik, 1993). Maximum annual precipitation of 11,615 mm has been recorded in Cherrapunji in the northeastern part of the country. Rains diminish from this region to the west where average annual precipitation is 1250 mm (Brammer, 1990). The Ganges, Brahmaputra,

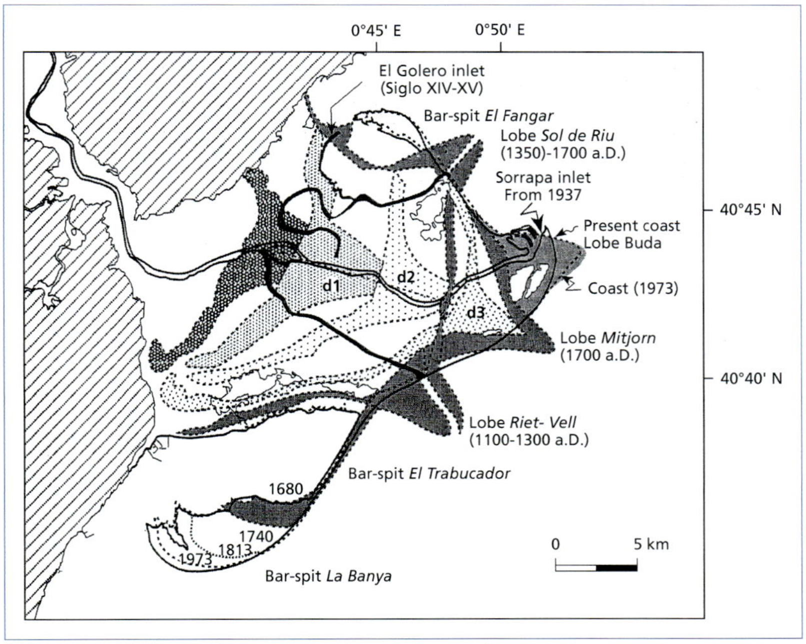

Figure 11.61 Evolution of the Ebro Delta (Spain) during the Holocene. The principal lobes are indicated and are associated with a principal distributary channel. Note how delta fronts have prograded (episodes "*d*") (Somoza *et al.*, 1998).

and Meghna are 10 km wide in their lower sections. The maximum discharge during flooding is 160,000 m^3/sec, three times the normal discharge of the Mississippi. The Ganges, Brahmaputra and Meghna rivers transport a load of 2.5×10^9 tons/year (Coleman, 1969). The three rivers have a noticeable seasonality: a) high discharge levels from the beginning of spring to the end of summer due to snowmelt from the Himalaya and b) monsoon rains in June-July, which can cause significant flooding.

The Bay of Bengal has a tropical cyclone frequency of 20% (Barry and Chorley, 1987). When cyclones reach the coast, strong winds and decreased atmospheric pressure (*storm surges*) cause sea level to rise quickly. These waves can reach heights of 6–7 m; they penetrate deeply into the low-elevation delta areas along the coastal fringe and flood houses and cultivated fields. The effects of the marine advance are accentuated when the cyclone coincides with high tide, which happened during the 12 November 1970 cyclone (Smith and Ward, 1998). Strong destructive winds in excess of 200 km/hour and intense cyclonic precipitation added to the marine flooding (Reading *et al.*, 1995).

Bangladesh has experienced about 20 significant floods since the middle of the 20th century (Brammer, 1990). Some have caused only property damage but most have incurred a high death toll. The destructive cyclone that occurred on

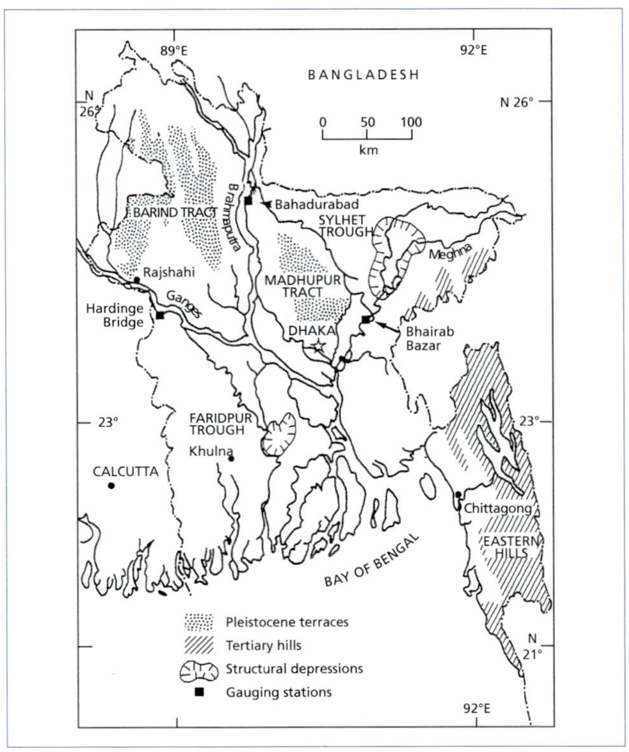

Figure 11.62 Principal physiographic characteristics and rivers of Bangladesh (Rashid and Pramanik, 1993).

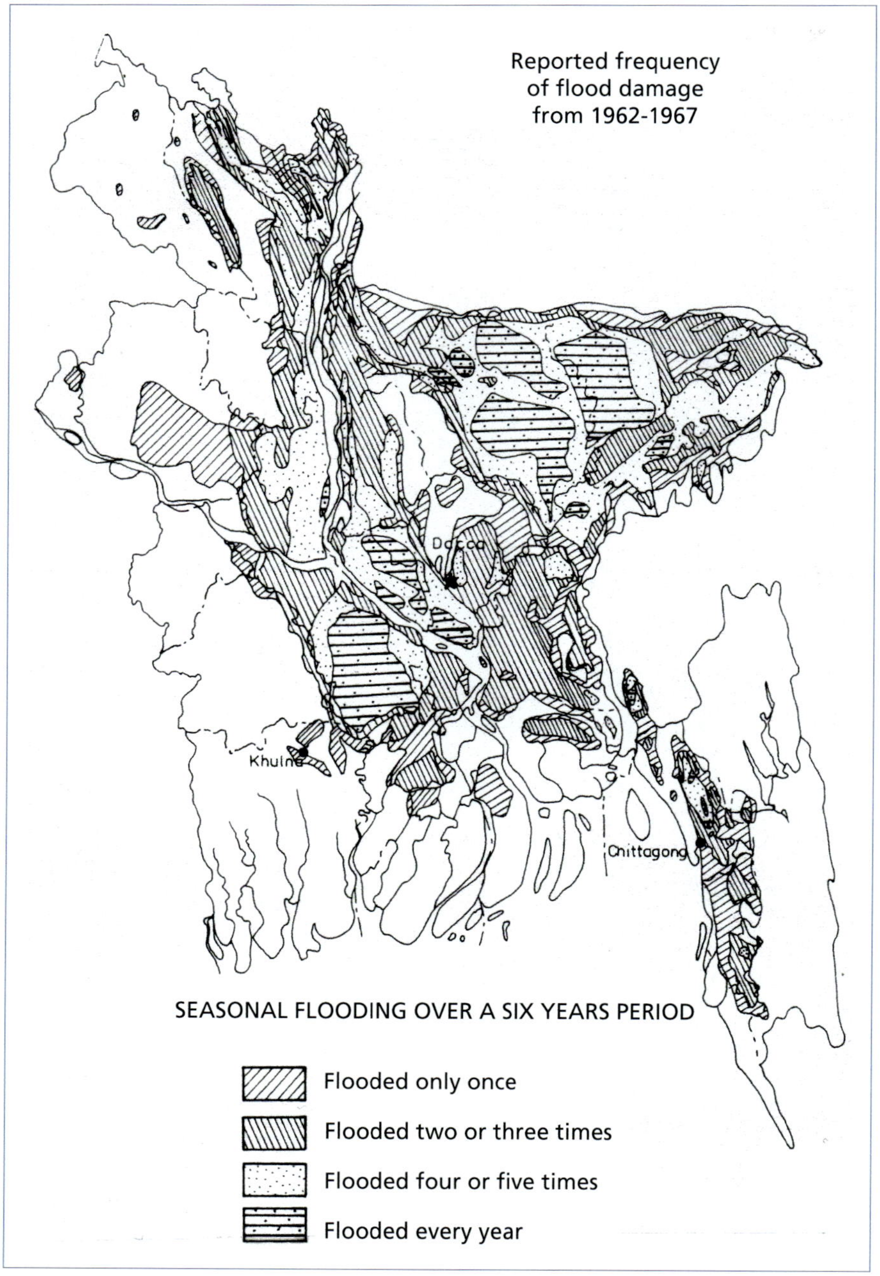

Figure 11.63 Hydrogeologic map of Bangladesh at a scale of 1:500.000, from 1971, based on EP-WAPDA Annual Reports (Verstappen, 1983).

12 November 1970 is considered the worst natural disaster of the 20th century. The cause was a cyclone that hit the coast at 11 o'clock at night; it caused 300,000 deaths (Burton *et al.*, 1978; Carter 1988). Additional catastrophes occurred in subsequent years. Strong monsoonal precipitation in 1987 and 1988 (Blaikie *et al.*, 1994) (*Fig. 11.64*) led to river floods that killed 1657 and 2379 people, respectively, according to official data (Brammer, 1990). On 30 April 1991, another catastrophic cyclone killed 70,000 people in southeastern Bangladesh and destroyed millions of homes (Davis, 1992; Reading *et al.*, 1995).

Few measures can protect against these natural catastrophes because the country is extremely poor. Earthen dikes have been built and trees have been planted although the

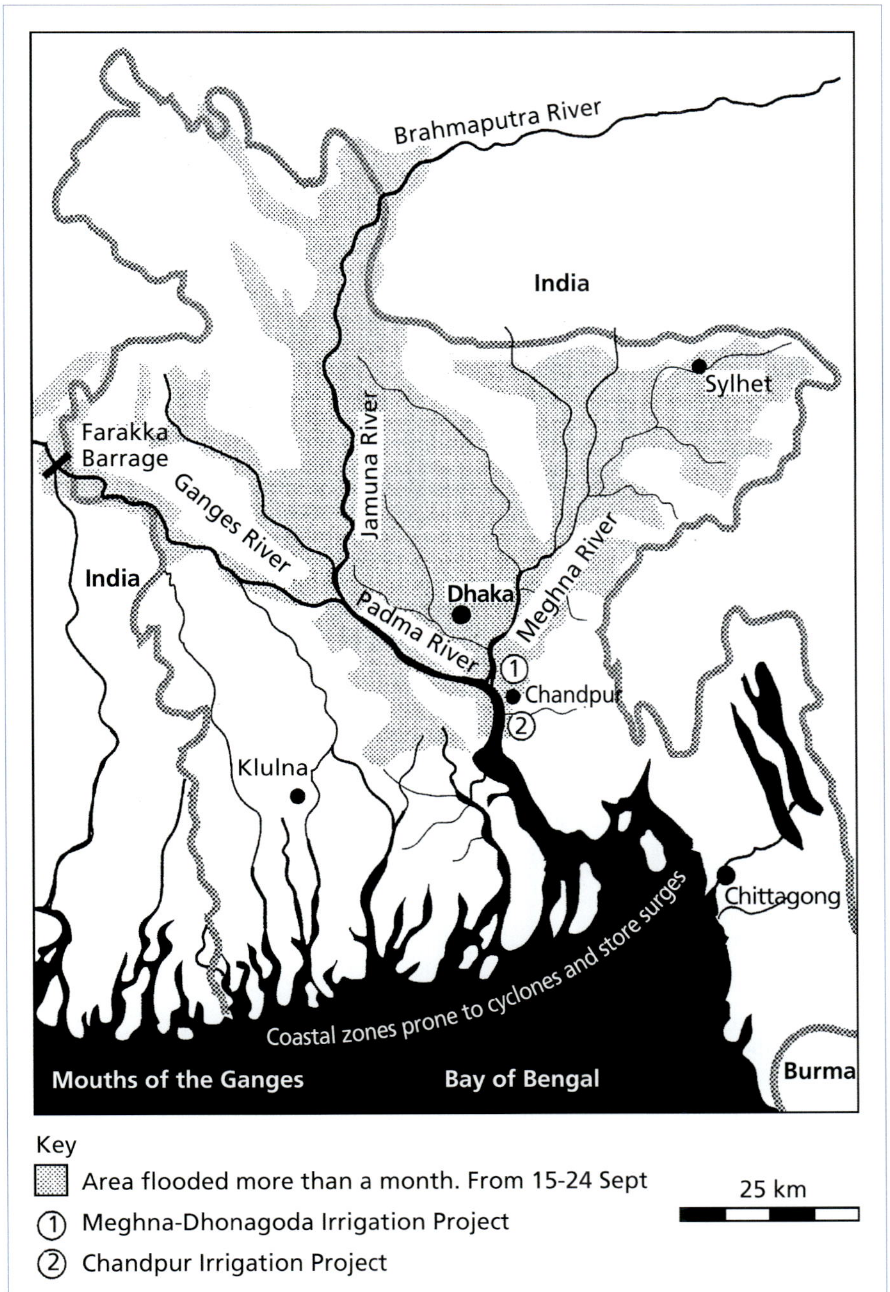

Figure 11.64 Primary rivers and tributaries of Bangladesh, showing the area flooded in September 1988) (Blaikie *et al*., 1994).

primary goal of the former is to prevent salinization. During the 1970 cyclone about 5% of the population used official refuges and 38% climbed into trees. This cyclone was identified three days in advance and was tracked by satellites but its intensity was not calculated (Burton *et al*., 1978). In spite of alarms, very few people evacuated and the majority remained on their land, perhaps for religious reasons (Carter, 1988). However, during the 1991 cyclone, weather forecasts and alarms led to a higher evacuation rate.

Since 1954 up-to date information has become available on flood zones and population growth in parts of the country that are prone to risk (*Fig. 11.65A*). Between 1954 and 1988, floods affected 20% of the nation, which is considered normal. Floods that cover more than 60% are

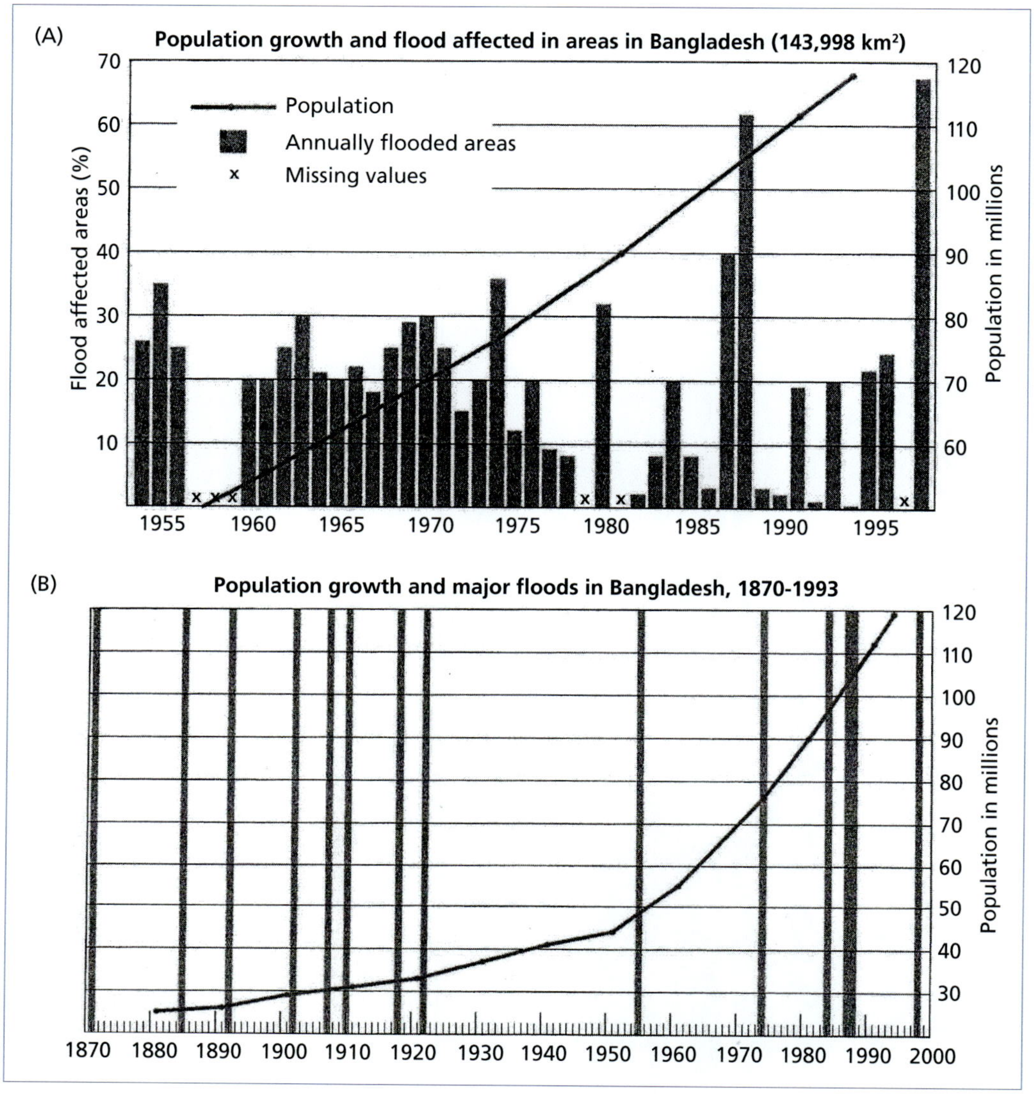

Figure 11.65 (A) Population increase and areas affected by flooding in Bangladesh. (B) Population increase and primary floods (Messerli *et al.*, 2000).

extreme events that have killed more than 2000 people. In addition, the population of Bangladesh continues to grow; it doubled between 1961 (55 million) and 1991 (111 million) (*Fig. 11.65B*). This means an increase of population density from 374 people per km^2 in 1961 to more than 800 in 1994. In spite of catastrophic flooding in 1987 and 1988, there is no evidence of significant tendencies toward increased flooding, either natural or anthropic. However, losses have increased due to the use of new techniques in flood plains (Messerli *et al.*, 2000).

More recently, Hurricane Katrina struck the city of New Orleans (United States) on 29 August 2005, destroying 80% of the city, inundating the principal port on the Mississippi River, and killing about 1300 people.

12

CLIMATIC GEOMORPHOLOGY

1. Introduction
2. The beginnings of climatic geomorphology
3. The structure of climatic geomorphology
4. The evolution of climatic geomorphology
5. Applications and future trends
6. Climatic geomorphology: Processes and morphoclimatic differentiation
7. The concept of zonation in climatic geomorphology
8. Principal morphoclimatic zones

12.1 Introduction

Climatic geomorphology is the study of how climate controls the landscape (Tricart and Cailleux, 1955, 1965). According to Büdel (1968) climatic variation controls all dynamic processes and has priority over all structural components. Cholley (1950) claims that a climate-dependent erosional system shapes all morphology and acts on various landforms according to their structure and tectonics. In reality, each climatic zone has its own climatic processes. Rock strength varies according to climate; Tricart and Cailleux (1965) and Büdel (1968) point out the different behaviors of limestone and granite in different climatic environments. Tricart and Cailleux (1965) also show that weathering of a given lithology and the resulting forms differ depending on the climate in which the morphological processes are operating. They analyzed the behavior of sandstones in very different climates: in Fouta-Djalon (Guinea-Konakry) where rainfall totals about 1500 mm per year; the sandstones of Adrar (Mauritania) with annual rainfall between 100 and 200 mm; and sandstones of the Vosges (France) where annual precipitation is about 1000 mm. In other words, these are wet tropical, arid, and temperate climates, respectively. These regions have similar characteristics; they have significant relief and deeply dissected valleys. Researchers observed variations in the intensity of processes and in the resulting landforms in these areas although structure also affects the landforms (Tricart and Cailleux, 1965). In Chapter 5 of this volume we discussed examples of sandstone weathering in various climatic environments.

Büdel (1968) claims that climate determines the manner in which exogenic processes actively shape the landscape. In contrast, non-climatic structural factors represent only variable passive phenomena. In addition, climatic variance changes systematically when the large climatic zones of the Earth are considered.

Structure is a framework on which climate-controlled processes are superimposed. The nature of the rocks, their structure, and general tectodynamics control the type of dissection that occurs in a climate. Climatic landforms are subordinate to structural relief and this dependency is related to differences of scale. Structural influences generally predominate at a small scale while morphoclimatic influences dominate at a larger scale. Up to a certain point, landforms of the climatic domain can be distinguished from those of the structural domain (Tricart and Cailleux, 1965; Tricart, 1968).

If we add time, we have a four-dimensional synthesis (Büdel, 1968). Over long time periods, existing landforms have experienced climate changes. Consequently, these morphologies tend to adapt to the new climates that are imposed; the result is the coexistence of relict forms and recent forms. Distinguishing between the two can be difficult but it is an important task for a geomorphologist; this is the core of climatogenic geomorphology (Büdel, 1963, 1977, 1980, 1982). These relict, inherited, surviving, or past forms are often good indicators of paleoclimates (Baulig, 1940; Jahn, 1954; Tricart, 1957). A significant example is the detailed cartography that Serrat (1977) did in his study of the eastern Pyrenees; to emphasize the distinction between active and inherited periglacial landforms he prepared two maps, one showing present-day landforms and the other showing relict morphologies.

Cholley (1950) claims that if erosional systems are related to certain climatic conditions, it should be possible to recognize the remains of erosional systems formed by previous systems within the morphology of a region. In summary, because of the short duration of activity in an area, the weakest present-day processes are less evident than the oldest landforms; in addition, climatogenic factors involved in the origin of these forms are more significant than tectogenetic factors.

The existence of a tropical morphological zone is still a subject of debate (Stoddart, 1969a, b: Douglas, 1978; Ollier, 1983; Gupta, 1993; Twidale and Lageat, 1994; Twidale, 2003, 2004). Twidale fundamentally builds on his early criticism of works on tower karst by Brook and Ford (1976, 1978) in which they point out that tower karst morphology can no longer be considered specific to wet tropical climates because it also occurs in cold environments. These authors did their research in the McKenzie Mountains (Canada). The surface area covered by this morphology is not very large especially compared to the 600,000 km^2 of tower karst in southern China and northern Vietnam (Sweeting, 1972). It seems simple to interpret these tower-shaped morphologies as a process of morphological convergence or equifinality (Birot, 1955) caused by the surface dissolution of carbonates, slope evolution, and fluvial action in an area of discontinuous permafrost with a dense conifer forest. The fungiform landform of the Ciudad Encantada (Cuenca, Spain) is interpreted to have resulted from deep differential weathering that favored later fracturation and denudation of the regolith in which the tower forms occur by exhumation (Twidale and Centeno, 1993; Twidale, 2004).

12.2 The beginnings of climatic geomorphology

During the late 19th and early 20th centuries, pioneering scientists explored unknown or poorly known regions of the Earth. Their goals included the discovery of new territories, evaluation of mining and agricultural resources, and colonization and Christianization of the inhabitants. The same goals had led to exploration during previous centuries; numerous examples can be cited on most continents. Early explorers wrote accounts of mining and geology but unlike the excellent works on botany and zoology, there were no detailed observations about landforms.

These expeditions led to the publication of numerous papers that were purely descriptive and superficial; they rarely contributed to rigorous scientific investigation. Some German naturalists, however, started to think about the processes that form the Earth's surface; examples are the works of Von Richtofen (1886) and Albrecht Penck (1894). During this period Martonne (1913) introduced the term '**climatic geomorphology**.'

The Swiss geologist Agassiz proposed his glacial theory and analyzed ice deposits in his seminal work *"Etudes sur les Glaciers"* [Glacier Studies] (1840). It was a groundbreaking publication on **glacial geomorphology**. In 1873 the English geologist Geikie published his book *The Great Ice Age*, in which he analyzed theories of ice movement, causes of glaciation, and the origin of glacial and post-glacial deposits. The first quarter of the 20th century was period of significant progress in the recognition of existing glacial landforms and deposits. The work of Russell (1893) on the Malaspina Glacier, the paper by Garwood (1899) on the Spitsbergen glaciers, and Kendall's (1902) dissertation on glacier lakes are examples of the work of this period. During the second half of the century studies focused on the origin of erosional and depositional landforms of formerly glaciated areas (Price, 1973). The American geologist R.F. Flint's book on *Glacial Geology and the Pleistocene* (1947), and Charlesworth's (1957) two-volume treatise on the *Quaternary Age* summarized the research on relict glacier landforms and deposits.

Although there was some understanding of frost action during the 19th century, the origin of **periglacial geomorphology** dates from the beginning of the 20th century. The Polish scientist Lozinski introduced the term 'periglacial' in 1909 when he showed that periglacial zones are contiguous with Pleistocene ice sheets and are covered with clasts generated by freeze-thaw cycles. Although it was later determined that periglacial processes could occur in areas far from ice caps, the term 'periglacial' continues to be used although with a broader application (Thorn, 1992). At about same time, in 1906, Andersson introduced the concept of solifluction, which he applied to the climatic characteristics and morphologic features of these cold regions. Russian scientists conducted extensive investigations in the expanses of Siberia but the international scientific community paid little attention because they were published in Russian. At the beginning of the 20th century, the settlement of Alaska led to the first well-recognized advances in periglacial research (Caines, Capps, and Eakin). Because the northern territories of Europe were inaccessible, periglacial geomorphology there developed thirty years later (French and Karte, 1988). These advances were mainly related to paleogeographic and environmental reconstructions (i.e. Büdel, Cailleux, Dylik, Edelman, Poser, Tricart, Troll) primarily in central and western Europe.

The most active period in the development of **desert geomorphology** was the end of the 19th century in the arid western United States of America (King, 1976). In the course of several geological expeditions, Powell (1834–1918), Dutton (1841–1912), and especially Gilbert (1843–1918) noticed a new kind of arid landscape that showed the erosional work of water. Powell, in his primary work entitled *Exploration of the Western Colorado River* (1875), introduced the key concept of base level as a prelude to a broader idea of peneplanation. He also established a genetic classification of fluvial networks in rivers as consequent, antecedent, and superimposed drainages. Dutton contributed immensely to the knowledge of isostasy; he also provided detailed descriptions of alluvial fans and established the earliest model of parallel retreat in slope evolution in desert zones. Fisher had discussed this idea (1868) although he used it for the retreat of marine cliffs and for the retreat of cirques and debris slopes (Fisher, 1872). Gilbert was the primary American geomorphologist of the 19th century. His excellent reports on the *Geology of the Henry Mountains* (1877) and on *Debris Transport by Overland Flows* (1914) are significant advances in the mechanics of fluvial processes, sediment transport, lateral river erosion and pediment formation, slope development, and so forth. Gilbert can be considered a pioneer in surface process research. His report on the *History of Lake Bonneville* (1890) is one of the classics of geomorphology. In this work Gilbert emphasized the use of ancient lake-levels and resulting lake terraces for understanding the origin of the present-day Great Salt Lake in Utah and as a method to be applied to other littoral zones. He discussed the isostatic rebound that followed evaporation of the large Pleistocene pluvial lakes (Tinkler, 1985). Gilbert contributed to many branches of geomorphology (Yochelson, 1980).

Scientific contributions from other arid zones of the world were relatively less important (Graf, 1988). French scientists conducted numerous expeditions in the Sahara Desert. They were primarily descriptive works but some of them clearly dealt with climate change (i.e. Chudeau, Flammond, Gautier, Urvoy, among others). Passarge described the Kalahari Desert in 1905, Kaiser analyzed the Namib Desert in 1921, and Little studied part of the Arabian Desert in 1925. The Iranian and Thar (India) deserts were explored in the 1870s (Blanford, Oldham). Research on the Australian desert focused on the age of large erosional surfaces and their relationship to resulting weathering products (Juston, 1934). Following an initially descriptive period in South America, Frengueli conducted desert research on the playa-lakes (salares) of the Atacama in 1928 and Walther Penck did significant research at La Puna in 1920. At the beginning of the 20th century the prevailing idea was that wind was the primary geomorphic agent in the planation of desert surfaces (i.e. Hedin, Passarge, Walther). Among others, Bryan (1922), however, determined that wind causes only small modifications. Wind research done by Bagnold (1941) in the Egyptian Desert was compiled into his classic dissertation entitled *The Physics of Aeolian Sands and Desert Dunes*, which provides a solid basis for understanding eolian processes. This work is one of the fundamentals of eolian geomorphology.

In **tropical geomorphology** the work of the North American scientist Dana (1849) on landforms of central Pacific volcanic islands is notable because it focuses on the influence of a tropical climate on landforms (Chorley *et al.*, 1984). In 1822, Thomson had reported numerous isolated hills standing above the surrounding plains along the Mozambique – Tanzania border (Douglas and Spencer, 1985b). Bornhardt (1900) made similar observations in East Africa and introduced the term *inselberg* to describe this kind of hill. Darwin (1890) and Branner (1896) highlighted the occurrence of thick weathering horizons, and much earlier Buchanan (1807) had made the first scientific descriptions of laterite weathering horizons in India. During the first half of the 20th century numerous works were published on tropical geomorphology as a result of the exploration of new territories in all climatic zones; these works include those of Cushing, Falconer, Grund, Hayes, Hubert, Passarge, and Sapper (Douglas, 1978). In 1926 Thorbecke presided over the first meeting on climatic geomorphology (Düsseldorf, Germany) during which the participants established the bases of morphoclimatic classifications. The most outstanding works of this period are perhaps studies by the German scientist Freise in the 1930s in the State of Rio de Janeiro (Brazil) and by Sapper (1935) on the *Geomorphology of the Humid Tropics,* the most significant summary of knowledge at that time.

12.3 The structure of climatic geomorphology

German researchers held a meeting on 22–23 September 1926 in Düsseldorf (Germany) on the analysis of landforms of various climatic environments and morphoclimatic classification (Thorbecke, 1927). The results of this conference are undoubtedly the most serious attempt to approach the subject of climatic geomorphology. However, the treatment by some authors of certain zones lacked differential morphoclimatic features (Beckinsale and Chorley, 1991).

In about 1950 Büdel's first works (1944, 1948) appeared; he later amplified these ideas in his book on *Climatic Geomorphology* (1977). In a 1980 publication he defines morphoclimatic zones as a set of landforms resulting from processes active in different types of climates. In his 1948 paper he distinguishes eight morphoclimatic zones using morphological and climatic criteria; in his 1977 book *Climatic Geomorphology* he recognized ten zones, excluding high mountains. He also proposed the term climatogenic geomorphology (Büdel, 1963); its fundamental objective is to reconstruct climate and processes active during a particular period. As Bremer (1996) points out, climatic geomorphology leads to climatogenic geomorphology because relict forms of past climates are found in almost all regions of the world.

Tricart and Cailleux (1965) develop the concept of zonation in geomorphology; they prepared a map of morphoclimatic regions, recognizing vegetation as possibly the dominant factor. They distinguish 13 morphoclimatic regions on Earth, including azonal mountain areas. Birot (1949a) analyzes the influence of climate on slope evolution for various types of rocks. During the 1950s a climatic classification was also established for forms resulting from processes that affect limestone (Lehmann *et al.*, 1954) and granite (Wilhelmy, 1958). Peltier (1950) related various exogenic processes to average annual temperatures and total annual precipitation, which he used to propose a classification of nine morphogenetic regions; his system has been slightly modified by Wilson (1968, 1969). This division was criticized for the limitations inherent in using only two variables (Stoddart, 1969a). Birot's book, *The Erosional Cycle in Different Climates* (1960) is an interesting contribution to arid and wet tropical zones for which he analyzes the dominant morphologies.

The papers discussed in this part constitute the framework of climatic geomorphology; they cover the development of fundamental concepts, the effect of climate on landform-generating processes, and the criteria used to distinguish different morphoclimatic zones.

12.4 The evolution of climatic geomorphology

Until approximately the middle of the 20th century, geomorphology was basically concerned with reconstructing different stages of a given age during which certain erosional and tectonic processes actively generated landforms. According to Chorley (1978) this is the '**historical stage**' when geomorphology was establishing erosional models, analyzing the chronology of denudation, and studying structural morphology. As mentioned above, it was also a period of exploration of large territories and description and definition of various forms (**taxonomic stage** described in Chorley, 1978). In addition, researchers were interested in studying processes and quantifying different morphoclimatic zones in laboratory experiments. As a result, climatic geomorphology evolved; the appearance of numerous and extensive monographs on various morphoclimatic zones shows that major progress has been made in recent decades.

In the second half of the 20th century, **glacial geomorphology** made significant advances in the study of ice behavior (Glen, Hambrey, Llibroutry, Nye, Paterson, Weertman, and others). There were also numerous investigations of erosional processes and glacial sedimentation (Boulton, Dreimanis, Drewry, Iverson, Linton, Menzies, Sugden, among others) and work on fluvioglacial environments and glaciotectonics (Price, Menzies, Van der Wateren, etc.). Together with articles that appeared in specific and general journals, numerous volumes were published on topics such as till, moraines, drumlins, glaciotectonics, and fluvioglacial and glaciolacustrine environments. In addition, books of a general nature on glacial geomorphology appeared; these include books by Embleton and King in 1975a, b, Sugden and John in 1976, and the voluminous treatises of Menzies (Ed.) of 1995–1996 and Benn and Evans from 1998.

During this period, researchers in **periglacial geomorphology** continued to investigate processes of ice activity (gelifraction, depth, uplift, cracking, and sorting), mass movements, and fluvial and eolian activity, both in the field and the laboratory. A large research group is dedicated to various subjects (Clark, Corte, French, Lachenbruch, Lautridou, Mackay, Pissart, Washburn, among others). Various existing processes produce forms that are highly variable and their relationship to others is not well understood. They form in specific micro- and meso- environments that lead indirectly to big questions that are common in the periglacial environment (Barsch and Caine, 1984). Many researchers excel in the study of periglacial landforms. Among those who have made the most significant contributions are Barsch, Büdel, Czudek, Demek, French, Mackay, Pissart, Rapp, Tricart, Washburn and others. Numerous books have been published on periglacial geomorphology; among the most important are Washburn (1979), considered by Thorn (1992) as the "Bible" of the discipline; one edited by Clark in 1988, which includes chapters contributed by various authors that cover most of periglacial geomorphology; and finally French's 1996, 2007 book, which provides an update on the knowledge of the discipline.

The extreme conditions of many deserts of the world have led to numerous expeditions and team research over the last fifty years. As a result we are gaining knowledge of the landforms in these inhospitable areas. The **geomorphology of arid zones** has found an excellent tool in satellite imagery, which has been used since the 1970s. Satellite images can be used as maps and their scale makes it possible to analyze large desert environments as a unit, particularly those composed of sand. The growing interest in the knowledge of these desert environments is shown by the recent creation of important research centers to study the problems inherent to these regions. Weathering processes and the resulting small-scale features of landforms have been studied by Cooke, Evans, Goudie, Mabbutt, Twidale and Viles. Dorn and Oberlander have conducted research on varnishes and Goudie, Machette, Marion, Reeves, Summerfield, Watson, Nash, Wright and others have studied crusts. There is abundant recent literature on experimental work on complex processes related to water activity on slopes and micro-basins (Abrahams, Bryan, De Ploey, Parsons, Schumm, Thornes, among others). The primary researchers on the fluvial geomorphology of deserts are Leopold, Wolman, Miller, Schumm and Shick. Research has focused on slope evolution and retreat (Grossman, Howard, King, Mabbutt, Oberlander, Selby, Twidale, etc.). Numerous researchers (Birot, Dresch, King, Mabbutt, Mensching, Tricart, Twidale and others) have covered the controversial theme of pediments and others (Blair, Bull, Denny, Harvey, Hooke, Mcpherson, among others) have studied the problem of alluvial fans. Work on playas and *sebkhas* is also abundant (Eugster, Glennie, Hardie, Krinsley, Goudie, Thomas, etc.). Erosion and eolian sedimentation is described in a broad bibliography of journal-published articles (Kocurek, Lancaster, Livingstone, Mainguet, McCauley, Péwé, Pye, Thomas, Tsoar, Warren, among others). Pye and Tsoar (1990), Lancaster (1995), and Livingstone and Warren (1996) have all published recent books on eolian geomorphology. In addition, Pye (1987) and Goudie and Middleton (2006) have published monographs on eolian dust. In addition, Tricart (1969), Mabbutt (1977), Cooke *et al.* (1993), Abrahams and Parsons (1994), Parsons and Abrahams (2009) and Thomas (1989, 1997, 2011) have published various books on the geomorphology of arid regions. These last two monographs have been produced with the support of numerous researchers who are specialists in the various subjects.

Until recently, **tropical geomorphology** had not been studied systematically (Gupta, 1993). Some authors (Stoddart, 1969 a, b; Twidale and Lageat, 1994) still question its classification as a morphoclimatic zone. Studies on weathering processes and laterites have been the subject of numerous articles by well-known authors such as Berry, Birot, Goudie, Loughnan, McFarlane, Ollier, Ruxton, and Thomas. The study of water erosion has increased over the last three decades especially in agro-forest research where it is conducted primarily on closed parcels. The works of Douglas, Spencer, Gupta, Walling and Latrubese stands out in fluvial transport. Frequent mass movements affect the regolith in particular; articles by Brand, Lal, So, and Tricart provide a representative sample of this research. Büdel (1957) began the study of planation and inselbergs, which predominate in many areas of the tropics. In his theory on double planation processes, he distinguishes the upper denudation surface and the weathering front, which deepens as a result of weathering agents. Continuous erosion causes the appearance of etchsurfaces. Bremer, Ollier, Thomas, and Twidale have discussed the evolution of these ideas in different tropical zones. Balazs, Lehmann, Sweeting, and others have described extensive carbonate outcrops that undergo intense dissolution and give rise to conical and tower karst landforms, specifically in this morphoclimatic zone. Books on tropical geomorphology are not as numerous as those published on other climatic environments. Books by Birot in 1973 and Tricart in 1974a provide important contributions but the most profound and fundamental is without doubt Thomas's 1994a work, which followed up his excellent 1974 book. Following the latter book, Faniran and Jeje in 1983 published their work in which they updated the knowledge of this field, as did Wirthmann in 1987. Books on the geomorphology of granites by Twidale in 1982, Gerrard (1988), Vidal and Twidale (1998), Twidale and Vidal (2005), and Migon (2006) all deal with the issues of tropical geomorphology.

12.5 Applications and future trends

Although climatic geomorphology has always had an important role in applied studies, its practical content has expanded since the 1960s as a result of process geomorphology studies. One of the most influential fields is the investigation of natural risk from external dynamics such as salt weathering, water and eolian erosion, flooding, landslides, subsidence risk, among others. Another field that has evolved considerably in the last three decades is *environmental geomorphology*. As a result of increased human population and resource exploitation, the physical environment is now subject to constant change. Humans have become a significant agent to the point that we now use the term anthropic geomorphology. To live in harmony with nature, man needs to understand the processes that operate in the ecosystem. Climatic geomorphology plays a fundamental role in this area.

Today geomorphology has become an encyclopedic body of knowledge and investigation (Dury, 1978) as reflected in the Encyclopedia of Geomorphology edited by Fairbridge (1968a) and Goudie (2004a) as well as other more specific encyclopedias on deserts and the environment in 1999, caves in 2004, coastal and karst sciences in 2005, and the Quaternary (2007) among others. Fairbridge hopes that by the end of the 20th century 200,000 articles on subjects related to this discipline will be published annually. These data give an idea of the extraordinary development of geomorphology.

In the near future we hope for the following in the field of climatic geomorphology: continuous progress in process studies, a considerable increase in multidisciplinary research, encouraging progress in the analysis of remote sensing data, progressive advances in paleo-morphological studies for climate prediction, and the development of applied research, especially in the fields of environmental geomorphology (Gutiérrez, 1990).

As we have seen, geomorphology has clearly changed direction from a general focus on studies of denudation chronology to a research emphasis on surface processes. This has been accompanied by a metamorphosis or fragmentation of the discipline (Chorley, 1978; Thorn, 1988b). As Chorley indicates, geomorphology has become submerged in a *process syndrome*. The underlying problem is the transition from large spatial and temporal scales, from process studies to the long-term evolution of landforms (Barsch, 1990). For this reason, as Chorley points out, geomorphologists must conduct research at all scales and try to pull together the different mechanisms responsible for landform evolution, whatever their temporal-spatial dimension.

12.6 Climatic geomorphology: Processes and morphoclimatic differentiation

Climatic geomorphology can be defined as the branch of geomorphology that identifies climatic factors such as the intensity, frequency, and duration of rainfall, freeze intensity, wind direction and strength, and explains landform formation under different climatic conditions (Ahnert, 1996). This recent definition takes into account the magnitude and frequency of various processes. Initially the pioneering work of Peltier (1950) considered only climatic parameters, such as annual mean temperature and annual total precipitation to establish a relationship between the regions and six morphological processes: chemical weathering, frost action; total weathering, rainfall erosion, mass movement, and wind action *(Fig. 12.1)*. Peltier considered "these diagrams to represent a simple schematic diagram of a concept." Overall, he established two different morphogenetic elements:

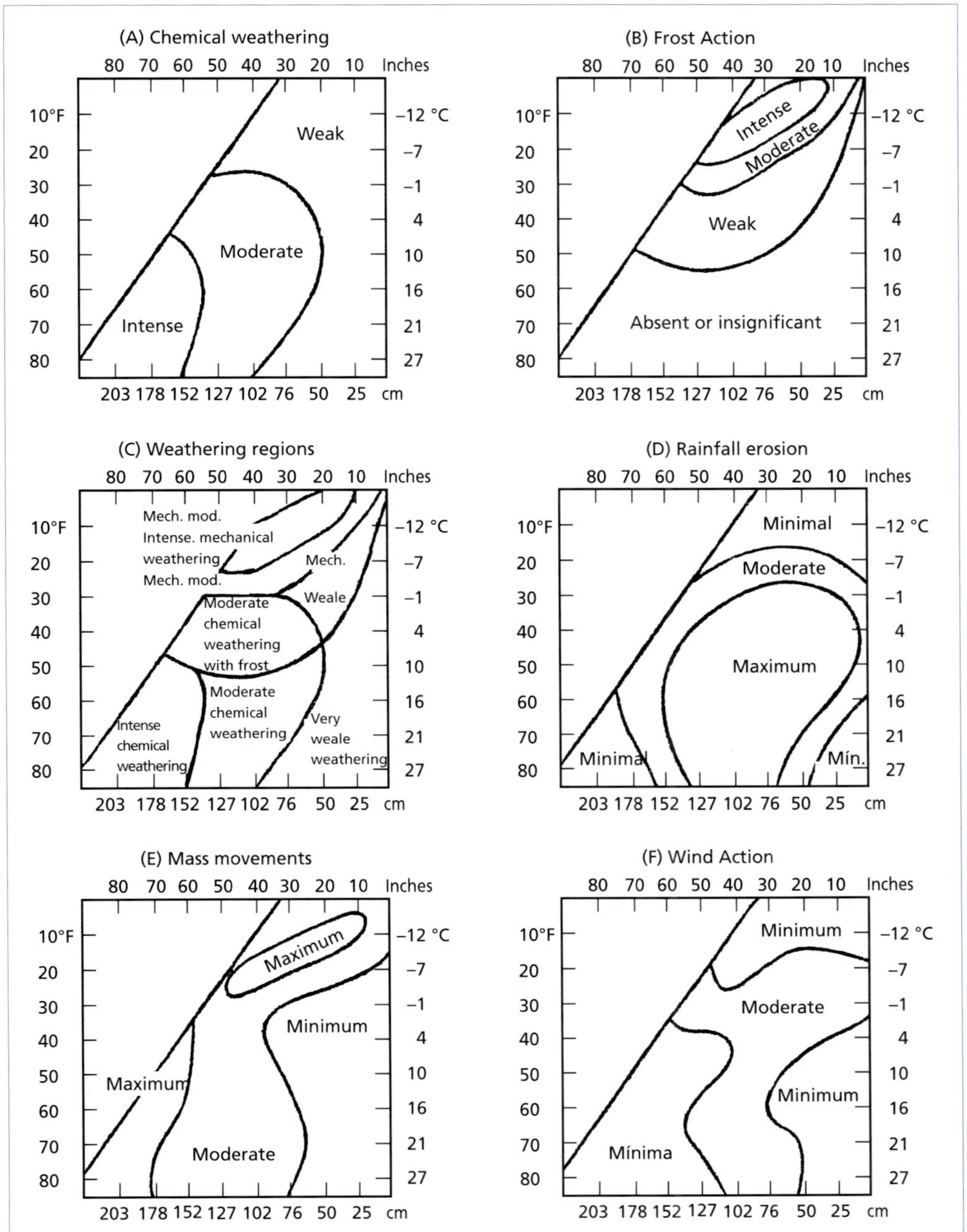

Figure 12.1 Diagrams showing the relative importance of selected processes as a function of average annual temperature and total annual rainfall (Peltier, 1950).

weathering processes and agents that transport weathering products. From prior analysis he proposed nine morphogenetic regions (*Fig. 12.2*) categorized on the basis of a set of characteristic processes. Peltier's classification is very qualitative and highly subjective (Derbyshire, 1973). Leopold *et al.* (1964) conducted a study similar to Peltier's. Wilson (1968, 1969) also conducted a similar analysis; he changed the diagram on frost action to one of mechanical weathering, modified the fields Peltier had established for relative intensities of processes, and set up six zones called climate-process systems. Wilson focused on temperature and rainfall variation over a year, which he calls seasonability, and its influence on processes (*Fig. 12.3*).

Tanner (1961) used potential evapotranspiration in place of temperature because it provides a more realistic idea of water availability and is more closely related to the percentage

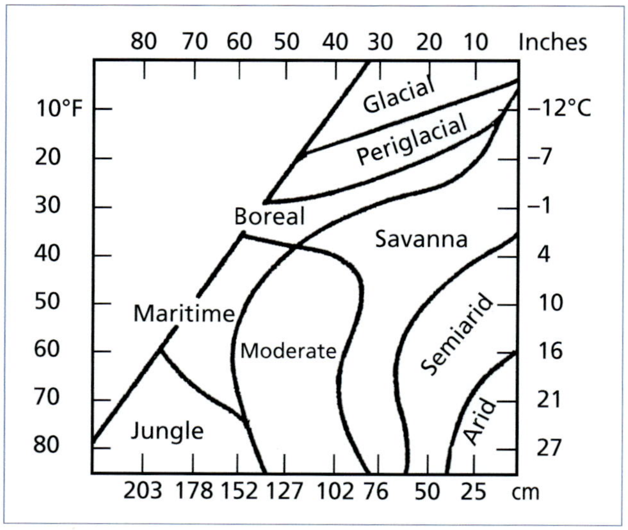

Figure 12.2 Morphogenetic regions based on the relative intensity of various processes (Peltier, 1950).

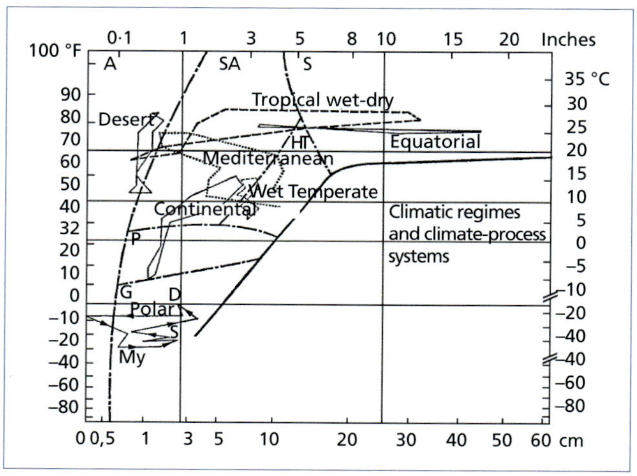

Figure 12.3 Climatic regimes and climate-process systems. The graphs show monthly precipitation and temperature data taken from Strahler (1965). The scales have been modified to highlight arid regions. The limits of climate-process systems are shown by dashed and dotted lines and by abbreviations: A-Arid; SA-Semiarid; J-Jungle; HT-Humid Temperate; P-Periglacial; and G-glacial (Wilson, 1968).

of vegetation. He also established four fundamental morphogenetic regions: glacial, temperate, arid, and jungle, to which tundra, savannah, and semiarid have been added.

An interesting similarity is Common's (1966) proposal in his work on landslides and morphogenetic regions that denudation processes can also be used to identify different landform types. Although numerous variables are involved in the triggering of landslides, Common prepared three global maps based on significant precipitation and temperature characteristics of interest (precipitation regime and total precipitation intervals and temperature).

Global studies have also been conducted on climate and vegetation control related to weathering intensity and the resulting products (Strakhov, 1967). Strakhov recognized a clear-cut climatic zonation with maximum leaching in the tropical jungle zone and in the taigal-podzol, which is where most precipitation occurs. Alteration is most intense and deepest in the tropics due to high temperatures that accelerate chemical reactions. In contrast, weathering in tundra and desert zones is minor due to sparse rainfall. This north-south profile makes it possible to establish a set of climate-related zones (*Fig. 12.4*). Strakhov's book also contains a map showing the geographic distribution of weathering types.

Another estimate of morphoclimatic differences is based on regional differences in water-induced erosion although this parameter is highly controlled by non-environmental factors. Researchers have established various curves based on the relationship of annual precipitation (Langbein and Schumm, 1958; Fournier, 1960; Wilson, 1973) or annual runoff (Douglas, 1977; Dendy and Bolton, 1976) to sediment production (tons/km^2/year). Most authors recognize an erosional maximum around 300 mm of precipitation and others designate a significant denudation zone above 1500 mm. Corbel's (1964) paper summarized erosion data for various climatic zones; he considers the tropics to have the least erosion. Fournier (1960) arrives at very different conclusions by studying 78 river basins ranging in size from 2,460 to 1,060,000 km^2. To do this, he correlated the climatic parameter p^2/P (where p is precipitation in the rainiest month and P is total annual precipitation) with the suspended load carried by rivers. Unlike Corbel, he found that increased erosion is related to increased precipitation. Fournier (1960) prepared a worldwide erosion distribution map and found maximum values in the seasonal wet tropics and a decrease toward equatorial regions and arid zones. The main conclusions of Fournier's book are supported by Strakhov (1967), who proposes a very similar map of global erosion. Milliman and Meader (1983) (*Fig. 12.5*) studied denudation rates from dissolved and suspended loads in rivers from reservoir deposits and geological sedimentation rates. *Figure 12.5A* shows erosion in the world's major river basins; *Figure 12.5B* shows the annual sediment discharge from the world's major rivers to the oceans.

Chorley *et al.* (1984) claim that by conducting a simple morphoclimatic classification, called morphogenetic, they can analyze average annual temperature, total annual rainfall, and seasonality expressed by the temperature of the hottest month and the number of months of precipitation above 50 mm (*Fig. 12.6*). The diagram shows eight regions that can be divided into two large groups: a) First order

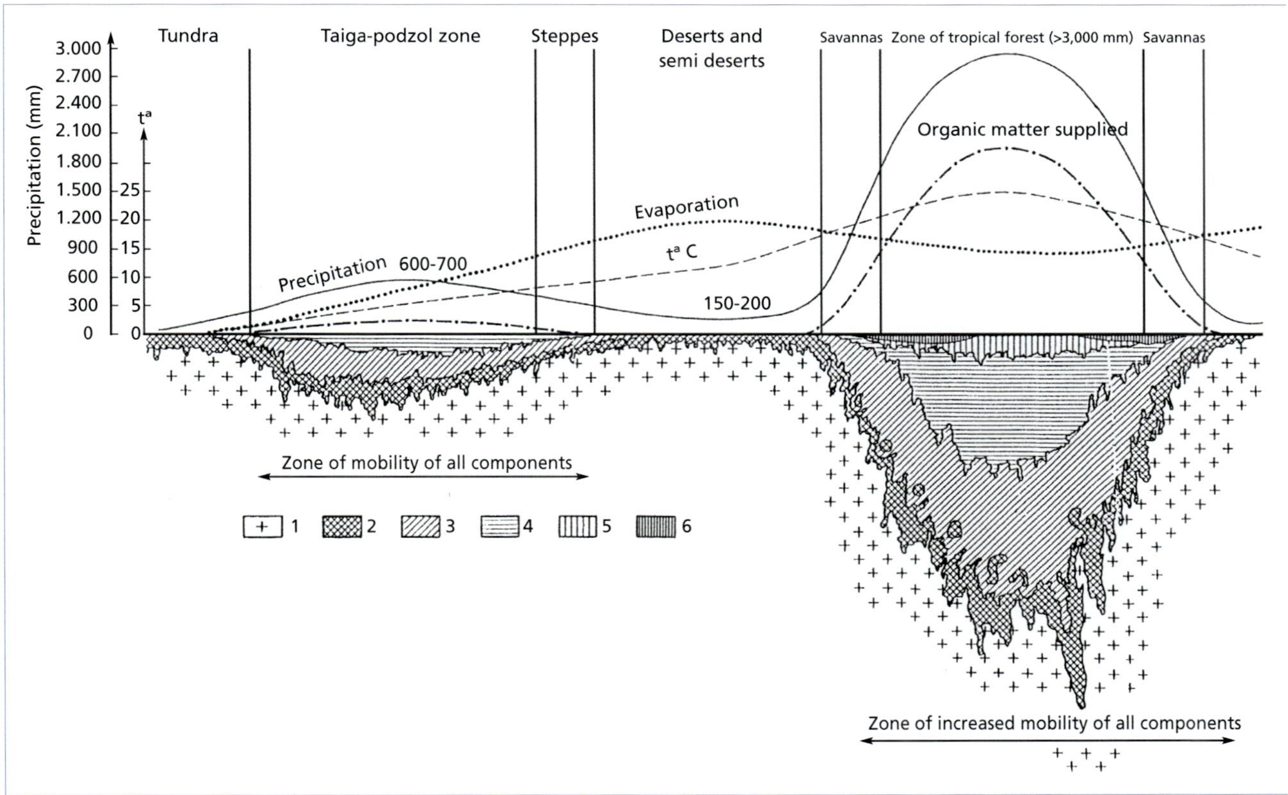

Figure 12.4 Weathering zones related to latitude. 1-Fresh rock. 2-Detritus, slightly chemically altered. 3-Zone of dominant hydrolysis. 4-Zone of kaolinite. 5-Zone of iron and aluminum oxides. 6-Ferricrete (Strakhov, 1967).

morphoclimatic regions, called non-seasonal by Beckinsale and Chorley (1991); these include glacial, arid, and wet tropical regions. Their characteristics include non-seasonal processes and generally low erosion except for sporadic events such as spasmodic glaciers, desert storms, and mass movements on slopes, respectively. The central parts of these non-seasonal regions are only mildly affected by climate change. b) Second order or seasonal morphoclimatic regions include humid-dry tropical semiarid, dry continental, wet mid-latitude, and periglacial regions, according to the previously cited authors. Second order regions also have seasonal processes, occasionally of great intensity, and significant climate changes. In these regions it is possible to distinguish 1) hot climates (wet-dry tropical and semiarid) where processes vary considerably over the duration of the wet season and 2) cold climates (cold continental, humid/wet mid-latitude, and periglacial) where processes vary as a function of summer temperatures and, to some extent, rainfall amounts. Chorley *et al.* provide a map that modifies the map of Tricart and Cailleux (1965) and minutely describes the dominant processes and landforms in the eight zones they establish.

The data normally used in geomorphology are supplied directly by climatology, such as the data used by Peltier, Tanner (1961), Leopold *et al.*, (1964), Wilson (1968, 1969), Chorley *et al.*, (1984), etc. These data may be reasonably adequate for making broad generalizations about processes. However, most of the numbers are averages and for this reason are inadequate because many processes are discontinuous by nature and vary in frequency and magnitude (Ahnert, 1987b, 1996; De Ploey *et al.*, 1991). Consequently, researchers need values for rainfall intensity, wind velocity, and frequency and duration of freezing. These numbers can be used for magnitude-frequency analysis to characterize the components of the morphoclimate. Ahnert (1987b) proposed a magnitude-frequency index for precipitation that can be used to recognize the entire distribution of geomorphically relevant meteorological events. De Ploey *et al.* (1991) also used daily precipitation values to measure cumulative potential erosion, which can provide a basis for estimating erosion on slopes undergoing laminar flow. This analysis is useful for characterizing not only morphoclimates but can also be applied to other branches of science (Ahnert, 1987b). This method cannot identify climatic aspects, such as seasonality, which is not an event. Even so, it is a first step toward a more comprehensive morphoclimatology.

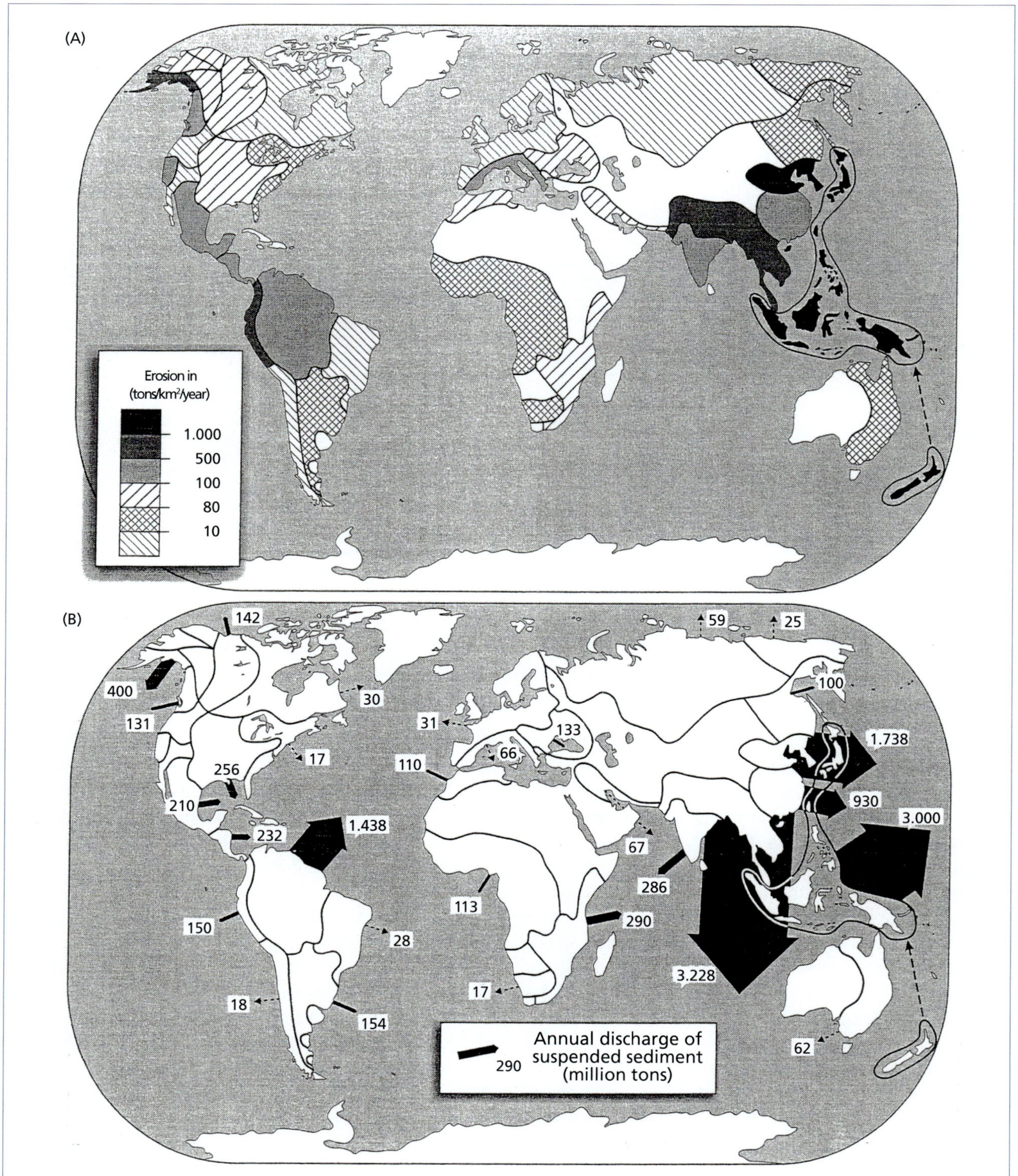

Figure 12.5 (A) Erosion in major river basins of the world. Blank spaces indicate a fundamental lack of discharge to the ocean. (B) Annual suspended sediment discharge into the major drainage basins of the world. The width of the arrows corresponds to the relative discharge. The numbers refer to the average annual amount, in millions of tons. The direction of the arrows does not indicate the sediment direction movement (Milliman and Meade, 1983; adapted by Huggett, 2003).

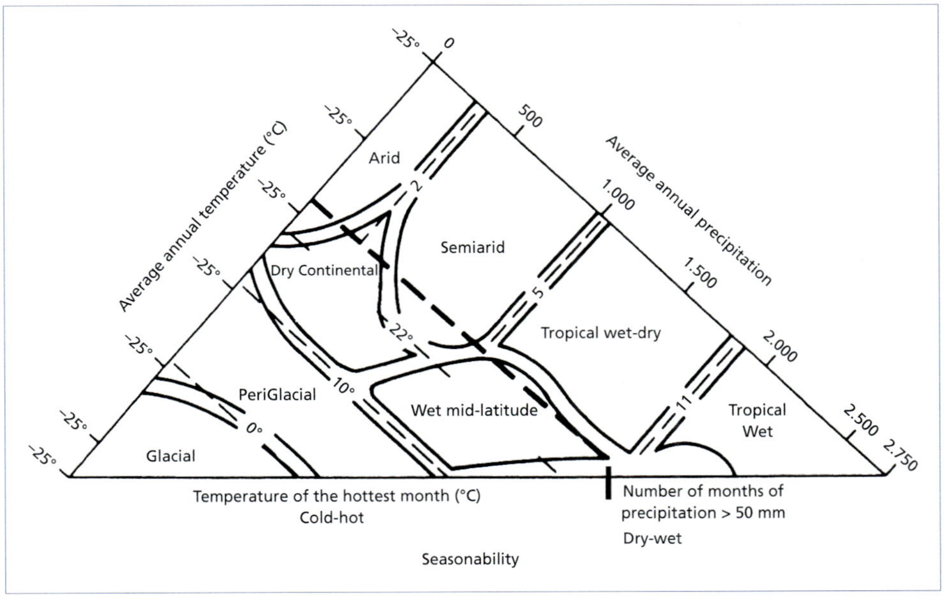

Figure 12.6 Principal present day morphoclimatic regions of the world obtained from the average annual temperature (°C), total annual rainfall (mm), average number of wet months (above 50 mm rainfall), and average temperature of the hottest month (°C) (Chorley *et al.*, 1984).

Although considerable consensus exists on the concept and mission of climatic geomorphology, there are also objections and criticisms of it. Stoddard (1969a) thinks it unrealistic to consider only a group of factors because it leads to distortion. Climatic factors are important but not dominant. Landforms are the result of a complex interaction of climate, lithology, and vegetation; it is impossible to separate climatic geomorphology from other types of geomorphology. Baker and Twidale (1991) maintain that only glacial and arid morphoclimatic regions can be easily recognized whereas the remaining climatic landforms are difficult to identify and define. Twidale and Lageat (1994) point out that glacial, periglacial, and arid morphoclimatic zones cover 50% of the Earth and the other half is made up of wet zones (jungle, savanna, moderate, maritime, and boreal from Peltier's classification). In these areas many existing landforms are of doubtful origin (Stoddard, 1969b). Twidale and Lageat (1994) analyzed another circumstance related to climatically generated processes and common mechanisms that vary over different climatic regions. Landforms associated with erosion and fluvial sedimentation are present in most climatic regions (gullying, gorges, meanders, alluvial fans, braided rivers, etc.). Tafonis and gnammas can also be observed in all climatic environments and patterned ground occurs in both periglacial and arid zones. This shows a problem of morphological convergence or equifinality similar to previous examples (Birot, 1955) in which activities of different processes or different magnitudes generate similar landforms. Even so, these authors recognize that while the influence of climate may be overestimated, there is no doubt that it is important. They also criticize the classification of the wet tropical region on the basis of its abundant mass movements and etched landforms because these forms also occur in other climatic zones.

12.7 The concept of zonation in climatic geomorphology

Variations that occur at the lithospheric-atmospheric contact can be considered zones on a global scale because they occupy wide areas extending from the poles to the equator; they result from differences in rainfall amounts and amounts of solar energy received (Derbyshire, 1976). Prior to its use in geomorphology, the zonal concept was applied to other branches of physical geography such as climatology, pedology, and biogeography. Dokuchayev (1883) first proposed large pedologic regions based on climate differences; he distinguished between zonal, azonal, and intrazonal soils. The first group, zonal soils, includes evolved soils that vary depending on the ecological-climatic zone where they formed. Azonal soils are not highly evolved; they have features similar to the parent rock. Finally, the intrazonal group is well formed but they are affected by poor drainage, salinity, and other factors.

Troll (1944) introduced the concept of zonation to climatic geomorphology in his paper on periglacial microforms. The term "**zonal**" is used for any process or phenomenon that is distributed along lines generally parallel to latitude lines; examples are the equatorial jungle, coral reefs, and continental ice caps. Each of these environments is a morphoclimatic zone (Tricart and Cailleux, 1965) as suggested by Cholley (1950) and Galon (1954). The term **azonal** (or plurizonal or worldwide) indicates processes that occur in various zones such as fluvial, eolian, and littoral activity. Finally the term **extrazonal** describes a phenomenon characteristic of a particular zone that can occur locally in other zones. Examples are glacial and periglacial processes and landforms that exist in high equatorial mountains.

Mountainous areas have a climatic zonation or **altitudinal gradient**. In low-latitude high mountains, climate ranges from temperate at the base to periglacial and finally to glacial at the top. The landforms in each zone are affected by a characteristic climatic control (Büdel, 1948).

The altitudinal zonation is a result of temperature and rainfall variations. Temperature decreases gradually with altitude, air becomes rarified, and solar radiation increases. Under these circumstances rocks are subject to more intense thermal effects (Tricart and Cailleux, 1965). An altitude increase of 1000 m in the Alps is equivalent to a temperature decrease resulting from moving 1000 m closer to the North Pole. This thermal decrease is accompanied by changes in vegetation both in terms of altitude and latitude (Büdel, 1977; 1982). In addition, increased elevation is associated with a characteristic sequence of morphology and size of patterned ground as studied by Furrer (1972) in the Alps and in the Karakorum mountains. These circumstances, which have also been described in other mountainous areas, provide reliable evidence of an altitudinal gradient. Precipitation varies as a function of height and as elevation increases, precipitation increases to a maximum and then diminishes continuously. In the mountains one location is characterized by maximum rainfall and as a result, the most intense clouds. Another important effect is exposure; shady areas in middle latitudes preserve snow much longer than sunny areas. This considerably affects altitude-related variations in periglacial morphogenesis. However, the influence of exposure varies from one zone to another; it is not limited to mountains in medium latitudes. In inter-tropical zones where atmospheric circulation is more constant, windward areas are wet and leeward areas are dry (Tricart and Cailleux, 1965).

We should also mention that climatic differences are primarily responsible for landform variety at scales of 10^6–10^7 km^2; lithological and structural differences predominate at 10^4–10^5 km^2, and landforms are the result of various erosional processes at 10^2 km^2 (Chorley *et al.*, 1984).

12.8 Principal morphoclimatic zones

Classification into different morphoclimatic zones requires a consideration of the dominant present day processes at work in each zone. However, it must also take into account landforms inherited from the Quaternary and Tertiary epochs. This preoccupation with climate change began with Penck and Brückner (1901–1909) in their work on the glacial periods in the Alps. In addition, Passarge (1904), in his book on the Kalahari, noted clear evidence of alternation between arid and savannah conditions. Büdel (1948) insisted on the importance of paleo-forms and in 1963 proposed the term climatogenetic geomorphology for the study of relict landforms, that is, landforms created in other environments that allow us to make inferences about processes active in climates that existed long ago. These climate changes are most obvious in the middle latitudes. Birot (1960) also noted that the influence of Tertiary and Quaternary climate change could be seen from the Arctic to the Equator. The distinction between present day and inherited landforms is fundamental to understanding the evolution of relief and the climate variations that have affected it because otherwise we may incorrectly interpret the genesis of a landform.

Division of the Earth's surface into morphoclimatic zones is a complex task for several reasons. Various parts of the world are not known to have fundamental morphological features (Central Asia, South America, etc.) and yet it is difficult to trace their boundaries (Tricart and Cailleux, 1965) although this problem is somewhat resolved through the use of satellite photography. In addition, we must understand that a morphoclimatic map is not the result of a total or partial superposition of climatic, vegetation, and soil maps although they are clearly related. A map of morphoclimatic zones must be an original map on which the morphoclimate is a particular category of natural phenomena (Tricart and Cailleux, 1965; Büdel, 1977; 1982). We also know that morphoclimatic boundaries are not sharp except along coastlines, although areas such as the Black Forest, the Po plain, and the Appalachians can be clearly distinguished (Büdel, 1977; 1982). In contrast, transition zones between African savannas and semiarid zones can cover very large areas. Only the cores of morphoclimatic zones are easy to recognize.

Büdel (1948; 1963; 1977) and Tricart and Cailleux (1965) wrote major summaries of climatic geomorphology; they prepared separate maps of morphoclimatic zones. Using climatic and morphological criteria, Büdel (1948) proposed seven morphoclimatic zones for the Earth and prepared a map of morphoclimatic zones for the Old World (Europe, Asia, and Africa). Tricart and Cailleux (1965) criticized this global classification for four reasons: its lack of homogeneity, the classification of permafrost into three regions, the consideration of a Mediterranean zone as a unique category, and its confusing terminology. Büdel (1963) considered that non-climatic factors (*petrovarianze* or lithological strength, *epirovarianze* or epeirogenesis, distance to the base level, influence of relief and human activity) would affect the formation of relief if the climate of the Earth were uniform. The influences arise from climatic zones differences, which are more significant than the five non-climatic factors. For Büdel, at each point on the Earth, climate determines the fundamental combination of morphogenetic processes. On the basis of the effects present in each climatic region where these complex active morphogenetic processes operate, the Earth's surface can be classified into five morphoclimatic zones (*Fig. 12.7*) without considering climate variation with elevation. This classification is based primarily on morphological criteria, which correlates to climate. The zones are named for the present-day climate although in middle latitudes the influence of past climates is apparent because the geomorphic evolution of these regions occurs more slowly.

The classification proposed by Tricart and Cailleux (1965) is based on two types of criteria: 1) large climatic and biogeographic zones that provide the principal divisions and 2) subdivisions based on the preceding criteria combined with paleoclimatic differences. They established the following zones, which are applicable only to low- elevation regions where altitudinal gradient is not present.

1. **Cold Zone.**
   a. Glacial domain.
   b. Periglacial domain.

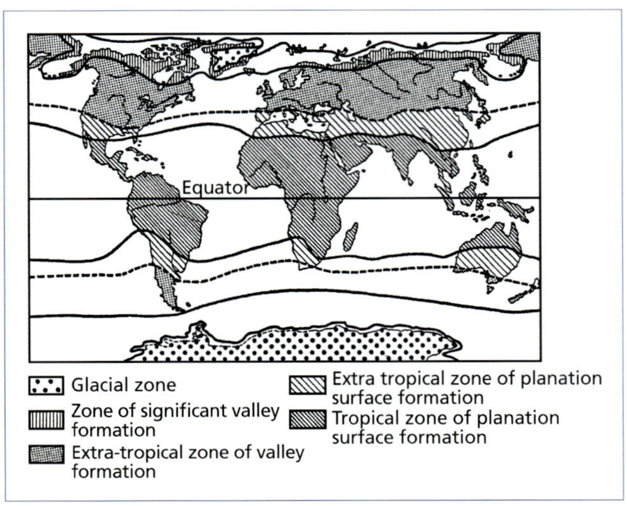

Figure 12.7 The world's present day morphoclimatic zones (Büdel, 1963).

2. **Forest Zone in Medium Latitudes**.
   a. Maritime domain with Quaternary glacial and periglacial heritages.
   b. Continental domain with significant influence of Quaternary and present-day ice. Includes the possible presence of permafrost at depth.
   c. Mediterranean domain with minor influence of relict Quaternary landforms.
3. **Arid and Subarid Zones.** They can be subdivided:
   a. Into steppes and deserts on the basis of the water deficit of xerophitic scrubland.
   b. Into cold and hot zones on the basis of temperature.
4. **Intertropical Wet Zone**.
   a. Savanna domain with less dense plant cover, characterized by deep chemical weathering that is discontinuous in time and by significant overland flow events.
   b. Jungle domain, with thick plant cover and intense chemical weathering.

Based on these zones, Tricart and Cailleux (1965) classify the various morphoclimatic regions present on the Earth (*Fig. 12.8*). They do not clearly distinguish between morphoclimatic and climatogenic influences (Beckinsale and Chorley, 1991). Perhaps the classification of the periglacial domain into five morphoclimatic zones on the basis of present-day or relict forms is excessive. According to Stoddart (1969a) the map emphasizes factors that control relief more than the landforms themselves. Derbyshire (1973) claims that they consider vegetation as a dominant geomorphic factor. He also points out that the classification terminology is more related to pedology and biogeography than to climate and geomorphology.

Büdel's (1977) map of morphoclimatic zones (*Fig. 12.9*) shows morphoclimatic boundaries that encircle areas in which landform-generating processes are active and indicates that they depend on the present-day climate for uniform behavior. In zonation, the central cores must be considered fundamental since the boundaries between them are not clear. Also, the map shows only low and medium altitude zones and omits high mountains. Establishing morphoclimatic zones is highly complex because it requires the distinction of numerous and complex processes; without adequate knowledge it is impossible to distinguish between morphoclimatic zones. In addition, it is sometimes difficult to discern between present-day and relict landforms (that is, landforms from the upper Cretaceous to the Holocene). If a researcher can recognize a landform that originates in active present-day areas, he can then recognize relict forms in other zones. Present-day processes in middle latitudes are exceptionally weak and because of this, about 95% of the relief of these areas is relict forms (Büdel, 1977).

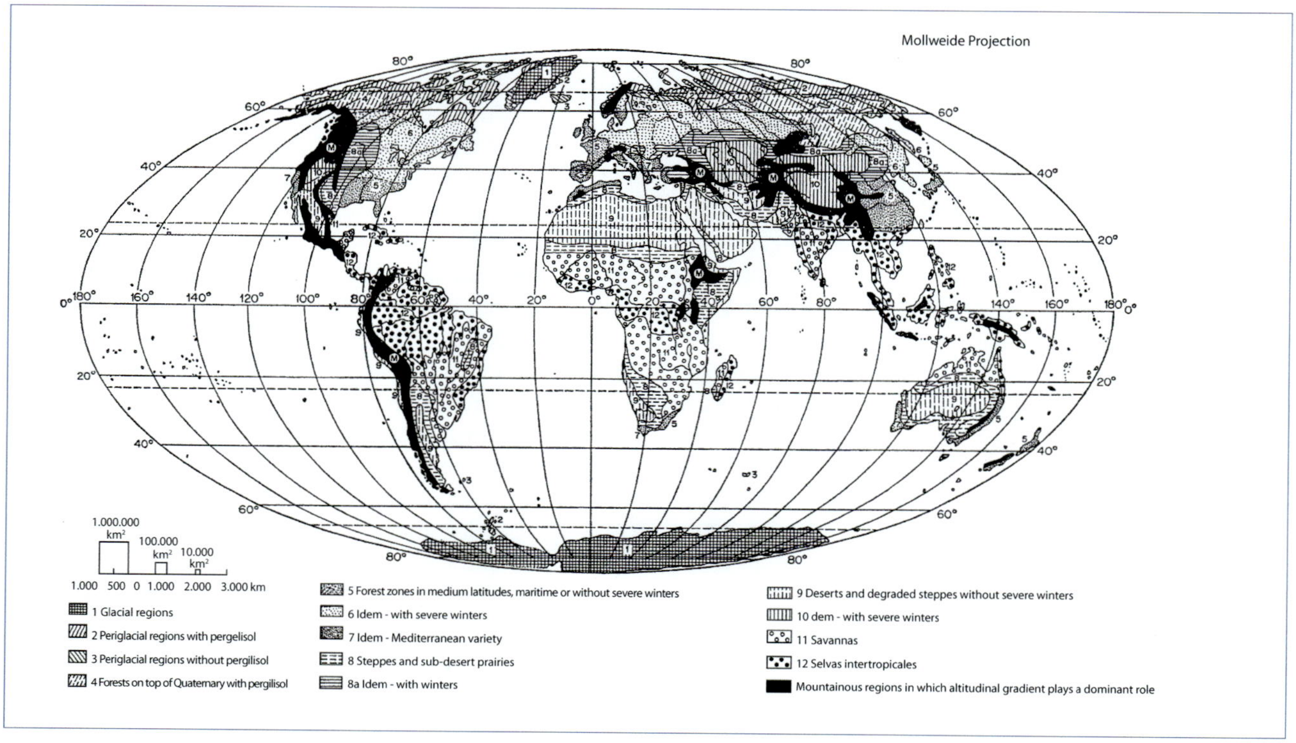

Figure 12.8 The world's morphoclimatic zones (Tricart and Cailleux, 1965).

The most significant features of the principal morphoclimatic zones are taken from Büdel's book and the excellent synthesis of Büdel's ideas written by Kiewietdejonge (1984).

1- **Glacial Zone** to which Büdel gave special attention; the mechanisms that create the relief of this zone are more easily studied in areas of previous glaciation than under present-day glaciers.

2, 3- **Periglacial Zone with formation of deeply incised valleys**. Büdel calls this subpolar and also the solifluction zone (Büdel, 1948) because of its importance in mass movements on periglacial slopes. His northern zone is bounded by glaciers; the southern zone is bounded by solifluction, the margin of arboreal vegetation, and disappearance of permafrost. Forest and solifluction boundaries often coincide whereas permafrost may lie beneath the forest. The surface layer is structured as patterned ground. One of the major problems is the rapid incision of the fluvial network. Slopes can evolve by parallel retreat, leading to tripartite slopes.

4- **Ectropical Zone with formation of smoothed valleys**. It is the most varied zone and possibly the most complex, with numerous and diverse forms of relief. In Europe, 95% of the landforms in this zone are inherited forms. Stepped planation surfaces (W. Penck's *piedmonttreppen*) are common. Pediments, terraces, and wide shallow valleys are present. In areas subject to erosion by ice sheets, extensive landscapes of hills and lakes can form. Finally sheet covers of loess may be present, derived from materials generated during previous cold periods. This and the following zone are the most important zones in the study of climatogenic geomorphology.

5, 6- **Subtropical Zone of mixed morphogenesis**. It includes the Mediterranean and monsoonal regions, which are morphoclimatically different. Büdel did not include descriptions of the monsoonal region in his book and for this reason we will discuss only the Mediterranean region. The evolution of this transition zone has been affected by tropical, periglacial, glacial, and wet and arid temperate climates. To these climate types we must add the anthropic factor, which is extraordinarily important in triggering water-induced erosion. The result is a superposition of relict and present-day forms arranged fundamentally in vertical zonation. There is major Tertiary planation in limestones and accompanying karstic landforms and extensive systems of stepped pediments in valleys and terraces within these intermontane basins.

7, 8- **Arid Zone with preservation of planations and their later transformation or development.** This

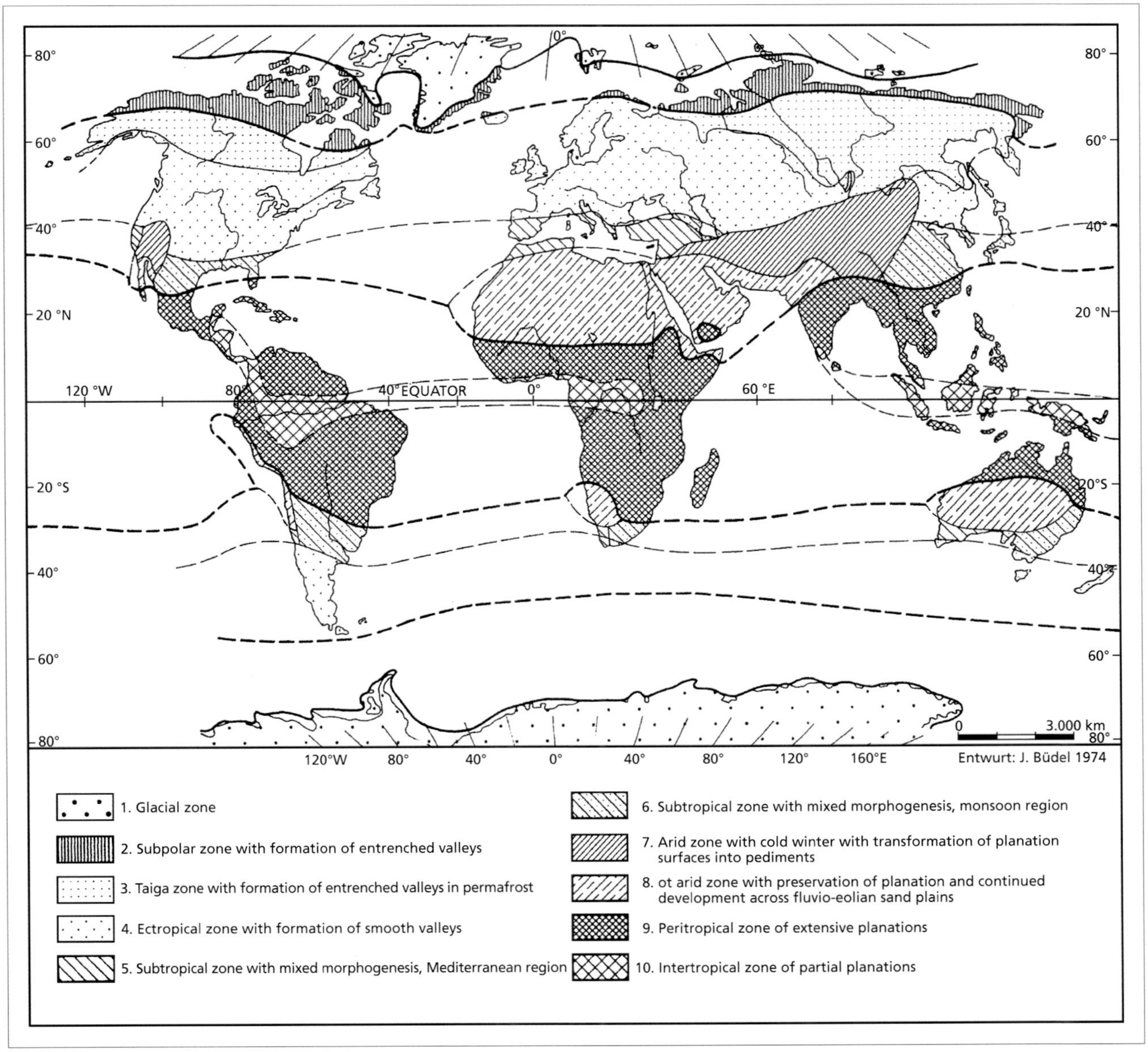

Figure 12.9 Present day morphoclimatic zones, excluding high mountains (Büdel, 1977).

zone includes desert areas with cold winters and tropical deserts. In tropical deserts, wetting and drying processes and saline weathering are intense. These include outstanding areas of planation with inselbergs and extensive hamadas and rock deserts. In zones with less relief, alluvial materials in pediments and terraces usually have caliches at the top. This alluvium is the source area of sand deserts. Arid zones with cold winters are areas of transformation more so than regions of later planation development. Gelifraction is the dominant form of weathering. The most representative forms of topographic landforms are pediment systems.

9- **Peritropical Zone (wet-dry) with extensive planations**. Intense chemical weathering occurs here to depths of several hundred meters and laterites form. The extensive plains can be explained by Büdel's (1957) double planation mechanism for tectonically stable zones on which overland flow and basal weathering surfaces gradually degrade to form etchplains on which inselbergs stand out.

10- **Equatorial Zone with partial planations**. This zone has not been well investigated due to dense vegetation. Chemical weathering is more intense than in the previously discussed region. In regions that are boundaries with the preceding zone, planation is common and inselbergs formed in wet-dry climates during the Tertiary. In internal zones, transformation of planation surfaces is minor because the mechanism has been reduced in both time and space.

In the morphoclimatic zone classifications we have analyzed we see significant variation in the number of zones the authors have set up: Büdel (1948) 7; Peltier (1950) 9; Büdel (1963) 5; Tricart and Cailleux (1965) 13; Wilson (1968) 6; Büdel (1977) 10; and Chorley *et al.* (1984) 8. These modifications respond to the various criteria the authors used to compartmentalize the Earth into morphoclimatic zones.

Monographs published on various morphoclimatic zones have reduced the number of zones to four: glacial, periglacial, arid, and wet tropical. Temperate regions have not been treated together because of the number of relict landforms present. In writing texts on these four zones, the authors have made classifications that may be more precise and better substantiated than those proposed in the previously cited morphoclimatic classifications.

13

GLACIAL GEOMORPHOLOGY I

1. Glaciers
2. Glacial erosion

13.1 Glaciers

13.1.1 Introduction

For many years scientists have recognized glaciation in the erosional and sedimentary features of the Earth's early geologic record; these glacial features date back to the Precambrian (Eyles, 1993), *(Fig. 13.1)*. Quaternary glaciation began during the late Tertiary and the remains of the ice cover approximately 10% of the earth's emergent surface. During the last glacial maximum about 18,000 years ago, ice covered almost one third of the continental area (Flint, 1971, Sugden and John, 1976). Thus the results of glacial activity are visible in numerous erosional and depositional landforms that cover a significant portion of the planet.

Research on ancient and present-day glacial activity is multi-faceted. Works on glacial geology are the core of Quaternary studies, which focus primarily on glacial stratigraphy and chronology. Geomorphologists analyze glacial processes and landforms shaped by ice masses and their evolution. Glaciology is the study of the nature, physical behavior, and work done by glaciers (Lliboutry, 1965; Iverson, 2006). In most cases, glaciological investigations are quite different from those done by glacial geomorphologists although there is some overlap between the two fields of study (Sugden and John, 1976). This intersection is natural because glaciers are rapidly evolving features that undergo modifications over days or months; furthermore, glaciers help us study present-day processes and the resulting forms.

In recent decades a new field of study, planetary geomorphology, has developed within glacial geomorphology. Researchers have found that Martian ice sheets cover about 30% of the planet surface during the winter; these ice sheets shrink to residual ice caps that cover only 1% of the planet in the summer. Some researchers think valley glaciers formed some Martian channels. Others think that Mars has experienced a climate change similar to the Earth's late Cenozoic climate change (Cutts, *et al.*, 1979; Baker, 1981a, and Carr (1981, 2006). Carr's (2006) book describes Martian morphology and speculates on its surface evolution up to the present. Planetary geomorphology research may supply important data leading to better understanding of glacial periods on Earth.

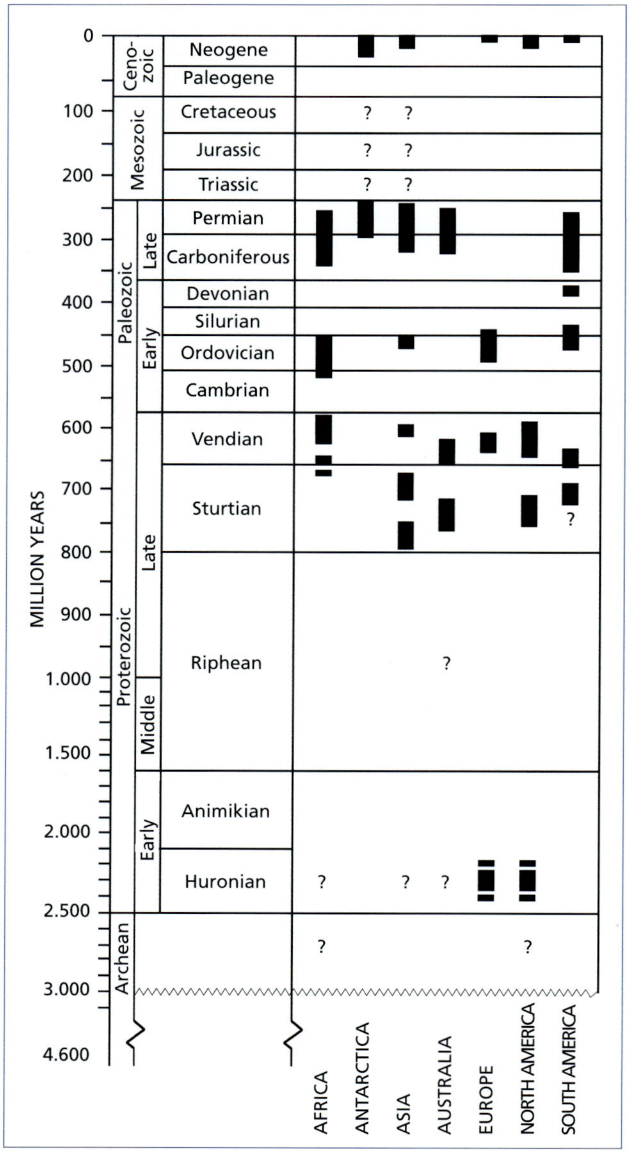

Figure 13.1 Glacial epochs in Earth history (Eyles, 1993).

13.1.2 Present and past extent of glaciers

Glacial ice currently covers approximately 15.8 million km^2 *(Table 13.1)*, about 10% of the Earth's emergent land area. Only 3% of this ice is located outside the large Antarctic and

Table 13.1 **EXTENT OF RECENT GLACIERS (FROM WORLD GLACIER MONITORING SERVICE, 1989) AND MAXIMUM EXTENT DURING THE PLEISTOCENE (ACCORDING TO DATA PREPARED BY FLINT, 1971)**

Region	Present day area (km^2)	Maximum extent during the Pleistocene (km^2)
Antarctica	13,593,310	13,800,000
Greenland	1,726,400	2,295,300
Mexico, United States, Canada and Alaska	276,100	16,217,091
Fennoscandia	51,046	7,169,145
Africa	10	1900
Asia (including Russia)	185,211	3,951,000
Alps, Pyrenees	2,921	39,000
South America	25,908	870,000
Australasia	860	30,000
TOTAL	15,861,766	44,373,436

Greenland ice sheets, which are the major fresh water reserves of the hydrosphere (Lvovitch, 1967). The Antarctica ice sheet covers an area of 13.6 million km^2, has an average ice thickness of 2.2 km, and a maximum thickness of about 5 km. The water equivalent of the ice contained in this ice sheet is about 30.1 million km^3, an amount that would cause sea level to rise 59 m if it melted completely. The Greenland Ice Sheet covers a surface of 1.7 million km^2; it is 1.6 km thick on average and has a maximum thickness of 3.4 km and a volume of 2.4 million km^3 (Donn *et al.*, 1962; Flint, 1971; Sudgen and John, 1976).

The remaining ice accumulations, which measure no more than half a million km^2, are distributed over the entire planet although they are concentrated in the northern hemisphere and in medium and low latitude mountains (Alps, Karakorum, Himalaya). The extent of African glaciers is almost negligible but the Andes have an outstanding extent of ice. In 1989, the World Glacier Monitoring Service calculated that glaciers (see Hambrey, 1994) cover a total surface of 15,861,766 km^2.

During its maximum extent in the Pleistocene, ice covered an area larger than 44 million km^2 (Flint, 1971), approximately 30% of the Earth's emergent land surface. The ice volume on the terrestrial surface is estimated at about 100 million km^3 (Donn *et al.*, 1962). Table 13.1 shows the difference between the extent of today's glaciers and the surface covered during the last glacial maximum during the Quaternary.

The Antarctic and Greenland ice sheets have undergone relatively little fluctuation whereas the Fennoscandian and Laurentide complex have retreated significantly during the Quaternary. The latter ice sheets were almost as large as the present-day Antarctic ice sheet and they extended under of 40°N latitude (Fig. 13.2).

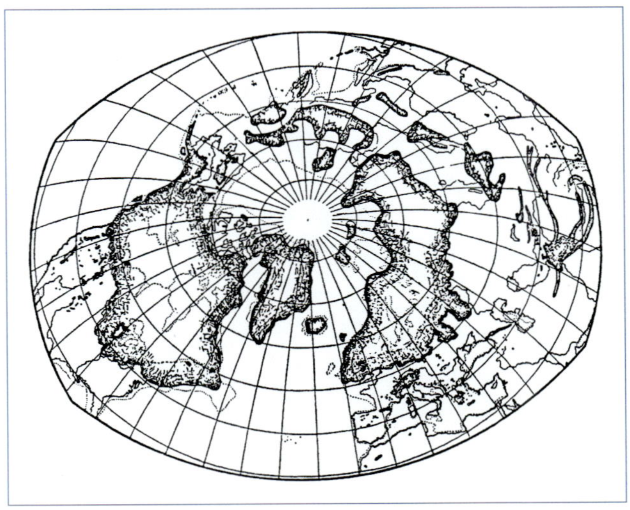

Figure 13.2 Maximum extent of Pleistocene ice sheets in the Northern Hemisphere (from Antevs and Flint, in Holmes, 1944).

13.1.3 Mass balance in glaciers: Accumulation and ablation

Mass balance in a glacier is determined by its gain or loss of snow and ice, which is normally measured in water equivalents (quantity of water formed by melting). The term **accumulation** refers to all processes that add material to a glacier. Precipitation is the most important factor although avalanches can supply significant material to valleys. Eolian activity plays a secondary role because it redistributes snow masses.

Ablation refers to processes that reduce the glacial mass. The primary process is melting, which is controlled by solar radiation, rain, air temperature, cloud cover, amount of detritus, and other factors (Sharp, 1988). Fine rock particles absorb more solar radiation than snow; as a result, they sink into the snow as it melts around clasts. This process generates small cylindrical conduits. In contrast, thick detrital layers isolate the underlying snow from melting, potentially creating gravel mounds with interiors composed of ice. If the detrital layer has an irregular surface, differential melting can generate mounds and depressions. When large boulders are present, the resulting morphology has pedestals or **glacial mesas** (*Fig. 13.3*) consisting of a boulder sitting on top of ice. These mesas are frequently tilted because ice on the sunny side will melt and cause that side of the mesa to collapse. Differential melting occasionally generates numerous and profound depressions, called glacial karst topography because of its similarity to landscapes in limestone areas (Clayton, 1964). Another important ablation process is the generation of large ice blocks or icebergs that float in the water near glaciers that flow into lakes or seas. This process is controlled by the degree of fracturing in the ice mass.

Figure 13.3 Glacial mesa on the Godwin Austen Glacier (Karakorum, Pakistan). In the distance is K2 (8,611 m), the second tallest peak on Earth. Photo J. López-Martínez.

All glaciers have two parts: in the upper part called the accumulation zone, ice and snow accumulate greater than they melt, and in the lower part called the ablation zone, the opposite process occurs. Between these two zones is an **equilibrium line** where the balance is zero (Müller, 1962).

The difference between accumulation and ablation over a period of a year is the **net balance** of the glacier (Andrews, 1975; Paterson, 2004) (*Fig. 13.4*). If the balance is positive, the glacier has gained snow and ice (net annual accumulation) and if the balance is negative, the glacier has lost material (net annual ablation). If the balance is zero, accumulation and ablation are in equilibrium. The winter balance is usually positive and the summer balance is negative. The difference between them is the net balance. These balances are calculated from field measurements or (Hagen and Reeh, 2004), ice on the sea (Wadhams, 2004) and remote sensing techniques (Bamber and Kwok, 2004). There are not many mass balance records for glaciers; one of the most complete is the 28-year record of the Storglaciären in Sweden, which shows a positive balance for five years, a zero balance for one year, and ablation in the remaining years (Ostrem *et al.*, 1973; in Sugden and John, 1976) (*Fig. 13.5*).

Satellite imagery is now used to determine the mass balance of continental ice caps; the satellite detects elevation changes in the depth of the firn layer. For most of Antarctica, the magnitudes of the changes of firn layer depth are comparable to changes observed in the height of the ice cap (Helsen, *et al.*, 2008). Increased snow or lower temperature causes the firn layer to thicken (Cuffey, 2008).

13.1.4 Transformation of snow into ice

Glaciers grow in locations where snow accumulation exceeds snowmelt. This situation requires special climatic conditions and landforms that preserve them (Price, 1973).

Snow that falls in the accumulation zone undergoes profound changes that finally transform it into glacial ice. The ice particles are crystals ranging in size from millimeters to decimeters; they are composed of numerous snowflakes.

Recently fallen snow or powdery snow has very low density (sometimes 0,05) and high porosity (95%); it can be

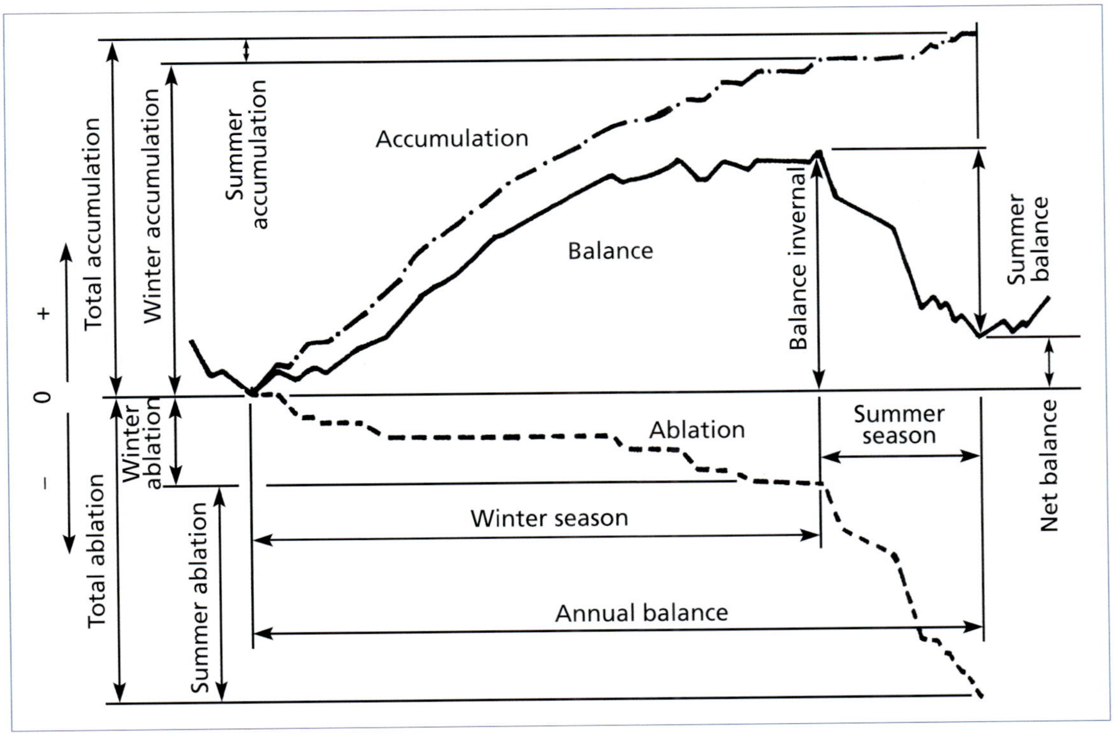

Figure 13.4 Graph showing mass balance in a glacier (in Andrews, 1975).

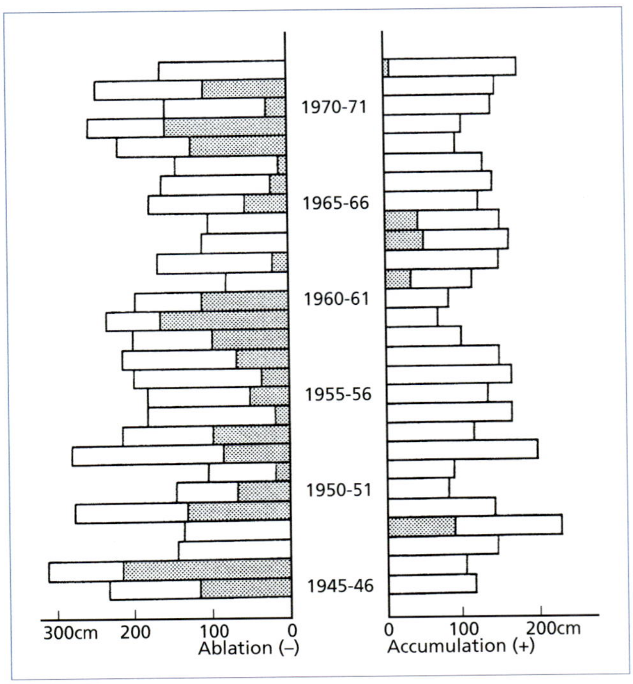

Figure 13.5 Mass balance in Storglaciären (Sweden) for the period 1945–1973 showing accumulation, ablation, and net annual balances (Ostrem *et al*., 1973).

considered an eolian sediment (*Fig. 13.6*). Fresh snow soon changes to granular or old snow with a density of 0.3; this snow consists of spherical granules of loose snow with high permeability. As transformation continues, density increases as grains are modified, pore space is lost, and packing increases. At a density of 0.4, snow becomes **firn**, which has 50% porosity and a grain size ranging from one to several millimeters. In Antarctica firn can be tens of meters thick (Paterson, 2004). As it continues to compact, the pores between grains remain sealed; firn is transformed into glacial ice when it reaches a density of 0.80-0.85. This mass is impermeable and the air inside it is present as bubbles; any increase in density is due to compression of the bubbles. This transformation of snow into ice is accompanied by slippage between crystals, changes in size and shape, and internal deformation (Paterson, 2004).

The snow-to-ice transformation rate varies from place to place because the process is sensitive to temperature and the rate of snow accumulation. The presence of melt water in summer accelerates the transformation. The curves in *Figure 13.7* (Paterson, 1994) show the difference of firn density with depth for a dry snow glacier in Greenland and a wet snow glacier in the Yukon. In the dry snow glacier, firn becomes ice at a depth of 66 m after more than 100 years of transformation whereas in the wet snow glacier the conversion occurred at 13 m over a period of 3–5 years.

Figure 13.6 Windblown snow. Heritage Range, Antarctica, about 80° south latitude. Photo J. López-Martínez.

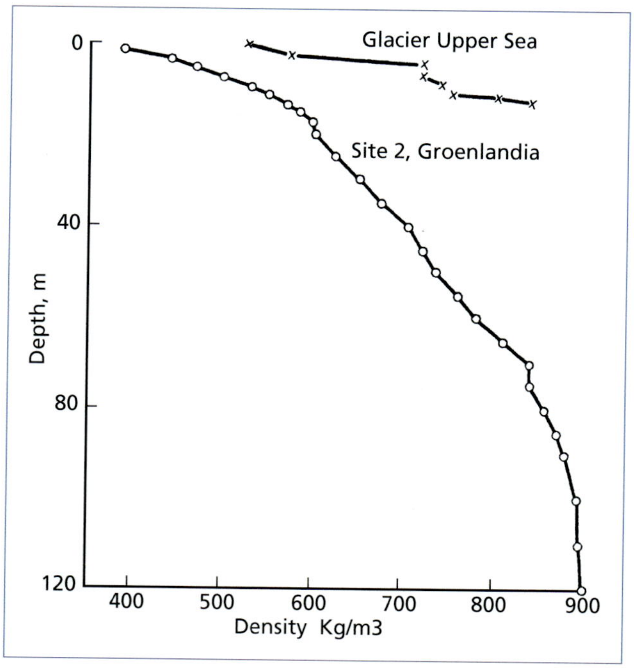

Figure 13.7 Variation in firn density with depth in wet snow, Upper Seward (Yukon) glacier and the Greenland Ice Sheet (Paterson, 2004).

13.1.5 Glacier classification

Of the numerous classifications of ice accumulations that have been proposed, the most commonly used are based on the easiest distinctions and are related to the internal temperature regime and geometry of glaciers.

13.1.5.1 Thermal classification

Water changes to ice at 0°C at one atmosphere of pressure but the process occurs at a slightly lower temperature when pressure is higher. For example, at the base of a 1,500-m-thick glacier, the melting point is –1°C. "**Cold ice**" has a temperature below the melting point and "**warm ice**" is so close to the melting point that it contains water (Ahlmann, 1935).

Cold ice exists when the glacier surface temperature is very low in winter and where there is little or no melting in summer. These conditions exist in Antarctica, Greenland, and the highest areas of large mountains. In Antarctica, researchers have measured air temperatures ranging from –89.2°C to –30°C; temperature increases with depth as a result of geothermal heating and internal deformation of the ice mass during glacial flow (Sugden, 1982).

Warm ice forms when enough heat is present to cause melting. During the summer, glacier surface ice melts. The melt water percolates into the ice mass where it can freeze and liberate latent heat, which increases the temperature of each gram of frozen water 1°C per 160 gr of ice. This liberation of latent heat can be a significant heat source. Under these circumstances, the entire glacier can be composed of ice at the melting point under pressure. At the base of the ice, geothermal heating can be sufficient to raise the ice temperature to the melting point. Warm ice can exist at the bottom of glaciers even though surface layers are cold ice (Budd *et al.*, 1970).

These types of ice help us distinguish between **temperate glaciers** and **polar glaciers**. This classification is highly simplistic because thermal conditions of glaciers can vary in space and time (Sugden, 1977; Paterson, 1994). The Antarctic is predominantly a polar glacier but parts of it have warm ice at the base. Firn transforms to ice much faster in temperate glaciers than in polar glaciers. These ice temperature differences are extremely important in glacial geomorphology especially because a thin water film can lubricate the bed of a temperate glacier and help it move. In addition, this subglacial water plays a significant role in erosion and sedimentation.

13.1.5.2 Morphological classification

The morphological classification of glaciers is based on geometry, position, and size of the ice masses. Some of these ice accumulations are constrained by topography whereas others show no type of confinement.

Examples of **continental ice sheets** are Antarctica (*Fig. 13.8*) and Greenland. Ice sheets are dome-shaped; they have a convex surface in response to ice flow. The ice is thick at the center of the accumulation; the surface dips smoothly, the slope gradually increases, and thickness decreases toward the margins. These sheets fossilize the underlying topography although occasionally the rocky substrate protrudes upward through the ice mass; these outcrops are called **nunataks** (*Fig. 13.9*).

Antarctica is made up of two main ice caps: the Western Antarctic Ice Sheet and the Eastern Antarctic Ice Sheet (*Fig. 13.10*). Smaller ice sheets are present on the Antarctic Peninsula and tens of glaciers are located in mountain areas on the margins of continental masses. The maximum height of ice sheets exceeds 4000 m. Ice flows from domes by sliding and/or basal deformation all the way to the margins. In the Greenland Ice Sheet, also known as the **inlandsis,** ice flows from central regions at rates of several meters per year, while the velocity of the ice in outlet glaciers is hundreds to thousands of meters per year (Sugden, 1982; Slaymaker and Kelley, 2007).

Ice shelves (*Fig. 13.11*) are large masses that cover the ocean surface. Some of them such as the Ross and Filchner-Ronne sheets in the Antarctic cover areas greater than 500,000 km^2; they occupy large bays and cover an area

Figure 13.8 View of the Antarctic ice sheet near the Ellsworth Mountains. Photo J. López-Martínez.

Figure 13.9 Nunatak in the Sentinel Range, Ellsworth Mountains, Antarctica. Photo J. López-Martínez.

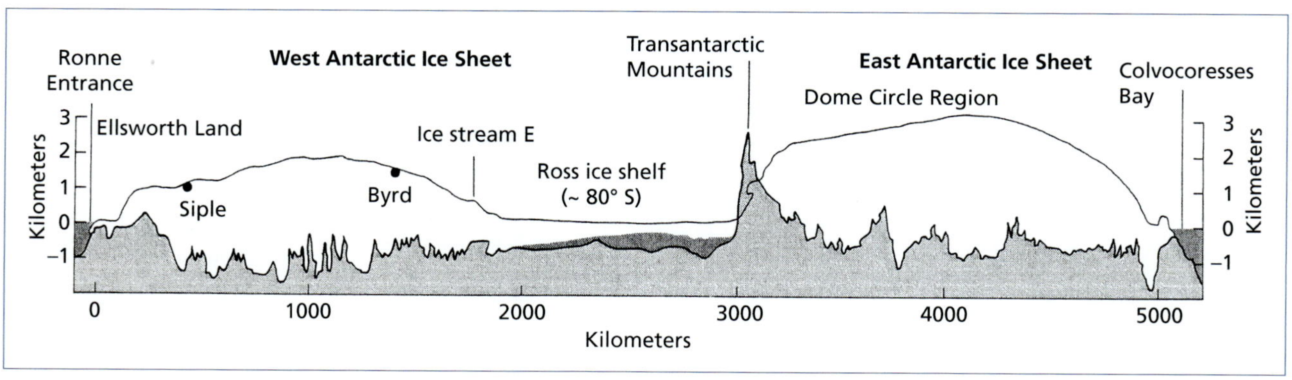

Figure 13.10 Cross section of ice sheets of western and eastern Antarctica, showing the irregular nature of the rocky substrate, ice thickness, and floating ice platforms (Hambrey and Alean, 1992).

corresponding to 7% of the entire Antarctic. Ice in the ice shelves can be 200 m thick and cliffs at shelf margins can be 30 m high, a combination that causes **iceberg** landslides (*Fig. 13.12*) and melting at the base. Mercer (1968, 1978) first proposed that ice shelves are generated by the disintegration as a consequence of industrial air pollution that causes a catastrophic liberation of ice into the sea and produces a sea level rise of about 5 m per century. Another hypothesis is that marine melt water penetrates into crevasse wedges. Water penetration and expansion of refreezing water can help

Figure 13.11 Ice platform and partially frozen sea on the coast of western Antarctica. Photo J. López-Martínez.

break up the ice (Scambos *et al.*, 2000). These floating ice masses contain particles of different sizes that fall to the seafloor (**dropstones**) as the ice melts (*Fig. 13.13*). Ice masses can become thin, split up, and continue to float on the ocean surface; these masses are called **pack ice** (*Fig. 13.14*).

Ice domes that measure less than 50,000 km^2 (Sugden and John, 1976) are called **mountain** or **plains ice caps** depending on the topographic position they occupy. The former sits on top of a mesa-like zone (*Fig. 13.15*); the best example is the Vatnajökull ice cap in southeastern Iceland. Plains ice caps are located in areas of low relief, such as the Barnes ice cap in the Arctic.

Cirque glaciers occupy the heads of valleys in mountainous areas (*Fig. 13.16*). They are small ice masses found in amphitheaters that generally have steep walls. Cirque glaciers can be hundreds of meters wide and tens of meters thick. During glaciation they are the first to develop and the last to disappear.

Valley glaciers are laterally confined by rocky walls. When supplied by cirques, they are called **alpine-type valley glaciers** (Embleton and King, 1975a) (*Fig. 13.17*). They flow radially from large masses and can be fed by secondary valley glaciers. The width of the valley is related to the number of tributaries. The mass of a valley glacier is long and narrow and terminates in an ice tongue, and the longitudinal slopes are variable. Some valley glaciers exceed 100 km in length. They are common today in high mountains. In the peripheral zone of ice sheet, ice masses can escape from the margin and become **outlet glaciers** (*Figs. 13.18* and *13.19*), which are like appendices that protrude from the margin of the ice cap. When the front of a valley glacier is an extensive plain, ice spills on top of it, widening it considerably and creating **piedmont glaciers.** The classic example is the Malaspina Glacier in Alaska, which extends as a wide lobe into the coastal plains of the Pacific. This glacier, which is the ablation zone of the Lower Seward Glacier, is 600 m thick; it occupies a large depression extending 250 m below sea level (Benn and Evans, 1998). Its compressive flow causes intense deformation of the moraines (Sharp, 1958) (*Fig. 13.20*).

Figure 13.12 Iceberg in the Bransfield Strait, western Antarctica. Photo J. López-Martínez.

Figure 13.13 Erratic boulder of metamorphic rock from the Antarctic Peninsula, located in background, transported by an iceberg and released by melting. The photo was taken at high tide. Photo A. Martín-Serrano.

Figure 13.14 Broken up pack ice near the Antarctic Peninsula. Large seals are visible on the ice blocks. Photo J. López-Martínez.

Figure 13.15 Myrdalsjökull ice cap (Iceland). Photo F. Gutiérrez.

Figure 13.16 Monte Perdido cirque glacier. Parque Nacional de Ordesa. Central Pyrenees Sierras (Spain). Photo F. Gutiérrez.

Figure 13.17 Alpine-type valley glacier Spegazzini Glacier Patagonia, Argentina. Photo C. Sancho.

Figure 13.18 Outlet valley glacier. Svinafellsjökull (Iceland). Photo F. Gutiérrez.

Figure 13.19 Outlet valley glacier and pack ice in Whisky Bay. James Ross Island, east of the Antarctic Peninsula Photo A. Martín-Serrano.

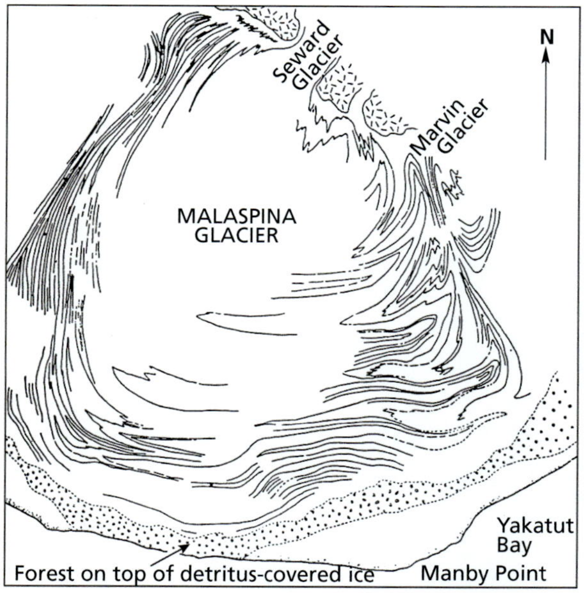

Figure 13.20 Map showing similar folds developed in the ice and moraines of the Malaspina Glacier in Alaska (Sharp, 1958).

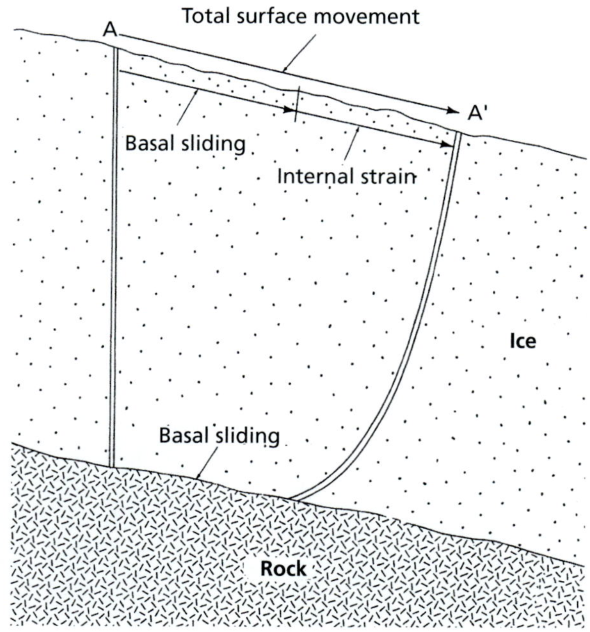

Figure 13.21 Longitudinal section of a glacier showing how a vertical boring deforms over time. The total movement AA' measured at the surface is the sum of internal deformation and basal sliding (Sharp, 1988).

Topography controls other glaciers, such as **ice fields**, which are accumulations on flat surfaces. It can be difficult to distinguish them from ice caps. Other small glaciers occupy small topographic depressions or steep slopes in mountains (**niche** or **slope glaciers**).

13.1.6 Glacier movement

Ice is a crystalline solid that flows easily by gravity. In a glacier, this movement continuously transfers material from the accumulation zone to the ablation zone. This flow occurs as two processes: **internal deformation** and **basal sliding** (Sharp, 1988) *(Fig. 13.21)*. The internal deformation is flow by **creep**; it results from a force applied over a long time period. Creep mechanisms include intergranular slipping, recrystallization, and sliding along lattice planes of ice crystals (Weertman, 1983).

Basal sliding is the displacement of the ice mass along its bed (Weertman, 1957). This movement includes the process of **basal plastic flow** in which the ice flows by rolling large obstacles; these advances can occur at any ice temperature *(Fig. 13.22)*. The other mechanism is called **sliding by refreezing**. The base of the glacier is a rough surface with small ridges and depressions that measure a centimeter to a meter. A temperate glacier, that is, a glacier near its melting point under pressure, can thaw with increased pressure on an upstream ridge and refreeze downstream after the pressure is reduced. If ridge protuberances are small, the transference of latent heat can promote new melting. Layers of refrozen ice, which is clear, can be seen at the base of glaciers; this ice is laminated and several centimeters thick (Kamb and LaChapelle, 1964).

Polar glaciers undergo only minor basal sliding because they are frozen to the bed although relatively rapid sliding can occur near the ice-rock interface. The presence of a film of water at the base of a temperate glacier reduces the friction between ice and rock and promotes basal sliding. If the substrate of a temperate glacier is composed of unconsolidated sediments such as a basal moraine, the underlying material can become water saturated and as a result can considerably increase the rate of basal slipping.

The rates at the surface of a glacier vary spatially and temporally. The rate of movement along a glacier ranges from 3 to 300 m/year and can reach 1–2 km/year in areas with steep walls. Cross-sectional velocities are highest in the center and decrease rapidly toward the margins due to friction with the walls. The velocities along the margins range from 10 to 65% of the maximum at the center. These glacier surface velocity calculations are obtained from equipment that includes stakes driven into the ice and a theodolite located at a fixed point on the rocky margin. To determine velocities at depth, researchers drill cores *(Fig. 13.21)* to the base, install

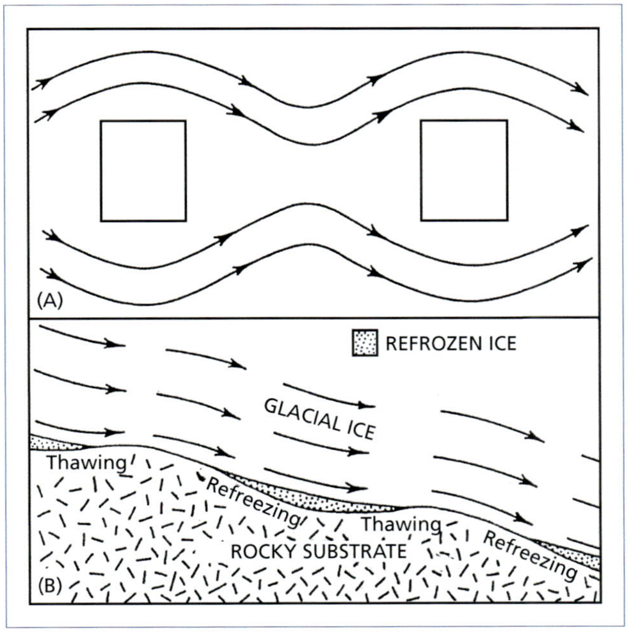

Figure 13.22 Influence of irregularities on the bed under the ice flow. (A) Process of basal plastic flow seen in plan view (Weertman, 1957). (B) Refreezing sliding mechanism, in section (Kamb and LaChapelle, 1964).

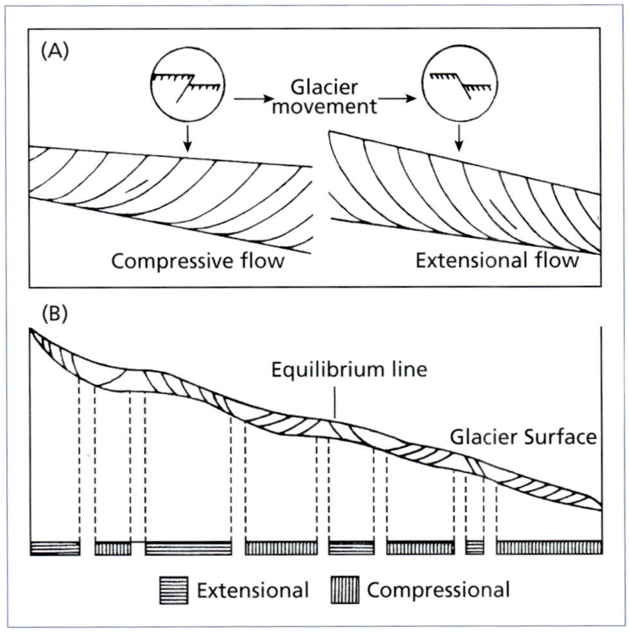

Figure 13.24 (A) Compressional and extensional flow and sliding planes. (B) Distribution of compressional and extensional flow in a glacier (Nye, 1952).

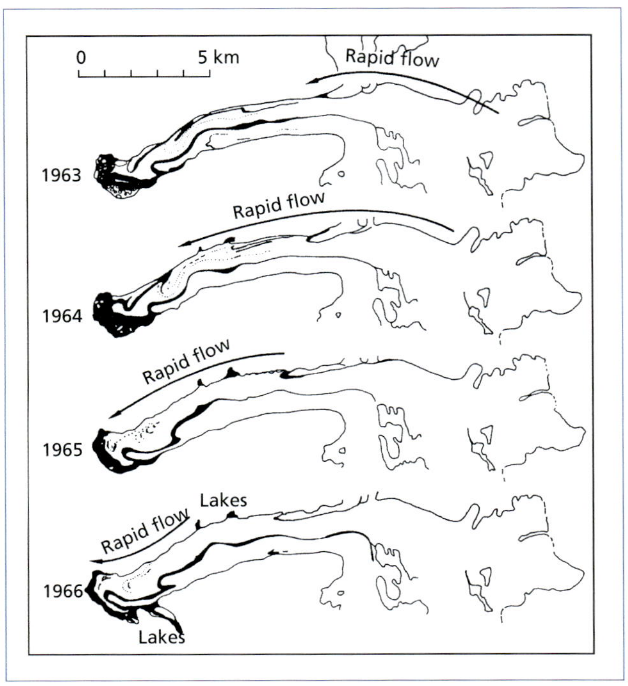

Figure 13.23 Evolution of the Tikke Glacier (British Columbia) with spasmodic flow between 1963 and 1966 and showing deformation of the moraine surfaces (Meier and Post, 1969).

a flexible conduit, and use an inclinometer to monitor movement. The velocity at the top of the ice is greater than the velocity at the bottom.

Some valley glaciers have abnormally rapid flows in which the ice mass moves as a wave at velocities of up to one hundred times the average value and with slipping up to 11 km. These are known as **surging glaciers** (*Fig. 13.23*); vibrations and loud noises accompany these sudden movements (Meier and Post, 1969; Sharp, 1988; Clarke, 1991; Dowdeswell *et al.*, 1991; Evans and Rea, 2003; Evans, 2004a). These events usually have a periodicity of a few years to several centuries. Surging glaciers have features such as chaotically deformed surfaces, intensely sheared margins, significant changes in thickness, and intensely folded central moraines. The intense shearing at the margins of a surging glacier can lead to foliation of the ice (Pfeffer, 1992). High-pressure subglacial water causes these rapid flows, which can lead to catastrophic advances. They can be triggered by seismic activity or significant changes in precipitation.

Flow regimes vary along the length of the glacier (*Fig. 13.24*); one type is **compressional flow**, which reduces the velocity of the glacier (Nye, 1952). In compressional flow, slipping planes that curve upward may transport detritus to the glacier surface. The opposite flow regime, **extensional flow**, occurs in zones where the ice velocity is increasing; here, the shear planes curve downward and become tangent with the bed. This flow predominates above the equilibrium line.

13.1.7 Glacier structures

The deformation of moving ice masses generates structures similar to those observed in deformed rocks. Glacial deformation is easier to analyze than rock deformation because gravity is the only force acting on glaciers (Menzies, 1995a).

Stratification predominates in accumulation areas, as shown by the alternation of winter ice layers and the mud-bearing layers formed during the summer melt. Glacial flow causes **foliation**; this is a planar structure that resembles veining or streaking of the ice. The ice veins are alternating bands of light blue-colored ice and white ice that contains air bubbles; the white ice is more abundant (Sharp, 1988). These bands range in width from millimeters to several meters. Foliation is well developed at valley walls, where it runs parallel to rocky outcrops. Far from the margins and near the glacier tongue, foliation runs transverse to the direction of ice flow (Rutter, 1965). It may be difficult to distinguish stratification from foliation when the latter runs parallel to sedimentary layers.

Glaciers are often folded and faulted (Hambrey, 1977); these structures are observed on glacier surfaces and in marginal cliffs (*Fig. 13.25*). **Folds** range in size and orientation. They are quite obvious in surging glaciers (*Fig. 13.23*) and in some piedmont glaciers such as the Malaspina, which contains large folds similar to those that occur in moraines and ice bands (*Fig. 13.20*).

In areas of high glacial slope, gravity **faults** and rotational landslides occur. Thrust planes develop in the glacier tongue when moraines impede the advance of the ice.

Ojives or **Forbes bands** are alternating layers that extend to the surface of temperate valley glaciers (Paterson, 1994). They are arch-shaped and their convexity points in the flow direction; this curvature is due to faster ice flow in the central parts (*Fig. 13.26*). They are spaced 50 to 200 m apart. The banding is due to light-colored and dark-colored ice. The dark bands are intensely foliated ice and mud generated by the mixing of broken ice, mud, and snow, and subsequent compaction. The less foliated light bands contain white snow with abundant air bubbles. The model that best explains the formation of ojives in the Bas Glacier d'Arolle (Switzerland), as proposed by Posamentier (1978), is slow variation of reverse faults. According to this model, multiple shear zones form and rise from the base of the ice to the glacier surface, creating the dark ojive bands (Goodsell *et al.*, 2002). Some authors think that each pair of bands represents a year, which implies an annual origin. In areas of steep glacier gradient, ice advances longitudinally during winter from areas of steepest slope to lower areas where intense compression occurs;

Figure 13.25 Fold in ice, shown by the alternation of light and dark bands. Livingston Island, Southern Shetland Islands, Antarctica. Photo J. López-Martínez.

Figure 13.26 In the foreground, ogives (Forbes banding) in the Mer de Glace glacier, Mont Blanc massif, French Alps. Photo J. López-Martínez.

the cracks fill with snow, creating the light-colored band. In summer, the cracks are open and fill with fragments of ice and detritus; the dark band forms when they are compressed (Nye, 1958; Sharp, 1988).

Cracks or **crevasses** (*Fig. 13.27*) are the most common structures in glaciers; they are formed by extensional forces (Menzies, 1995a). They are generally straight or slightly arcuate and subvertical. Crevasses can be tens to several thousand meters long and can range in width from a millimeter to several meters. Their maximum depth is about 35 m because behavior becomes plastic beneath the ice. Crevasses are excellent pathways for meltwater infiltration. When crevasse systems meet, the glacier surface is transformed into a broken mass of dentate pinnacles called **séracs**.

Glacial crevasses are classified into systems (*Fig. 13.28*) (Sharp, 1960). **Marginal** or **chevron crevasses** form as a result of extensional forces. They form as a result of faster flow at the center of the glacier than at the margins where friction with the rock wall slows down movement. These crevasses intersect with the glacier margin at angles of about 45°. **Transverse crevasses** (*Fig. 13.29*), which occur in areas of higher velocity, are perpendicular to flow, and are convex upgradient in the glacier. **Extensional crevasses** occur in areas where a valley glacier enlarges and in ridge areas of the bed. These cracks curve upgradient and form angles less than 45° with the margin. **Radially extensional crevasses** form at the end of the glacier tongue. After formation, crevasse

Figure 13.27 Crevasses in the Worthington Glacier (Alaska). In the background, outlet glacier. Photo F. Gutiérrez.

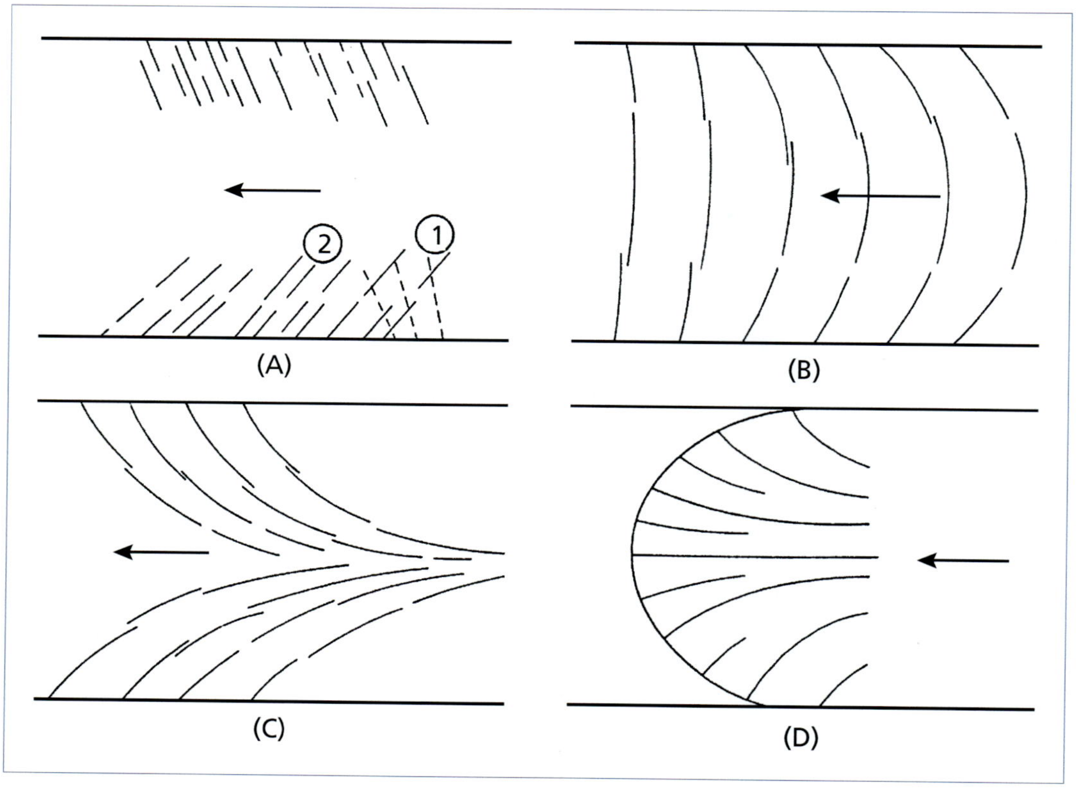

Figure 13.28 Types of crevasses in valley glaciers (A) Marginal (1-old bent cracks, 2-newly formed cracks). (B) Transverse. (C) Extensional. (D) Radially extensional. The arrow indicates the direction of ice movement (Sharp, 1960).

Figure 13.29 Transverse crevasse. Aneto Glacier. Aragon Pyrenees. Photo F. Gutiérrez.

Figure 13.30 Bergschrund in the Aneto Glacier. Aragon Pyrenees. Photo F. Gutiérrez.

systems are modified by glacial flow. Marginal crevasses can rotate and transverse and extensional crevasses can enlarge until they finally close and form a vein of blue ice.

A **bergschrund** (Lliboutry *et al.*, 1976) is a crevasse that separates ice from rock at the top or head of a glacier. The bergschrund is created as ice adheres to the wall while the rest of the glacier advances. They occur near steep walls (*Fig. 13.30*) and can extend for hundreds of meters. Bergschrund systems are more common than a single bergschrund.

13.2 Glacial erosion

The capacity of ice to substantially modify present-day landforms has been recognized for many years. In spite of numerous publications on the subject, our knowledge of ice activity is limited because of the difficulty of studying ice effects on the bed beneath a glacier, which is where most erosion occurs. Researchers have dug tunnels, particularly in cirques, to try to observe this activity. In addition to the difficulty of observing erosional activity, we lack knowledge of pre-glacial relief, which prevents us from adequately estimating the amount of glacial erosion. Another obstacle is the fact that glaciations have been superimposed throughout history in many areas, which compounds the difficulty of interpreting the past (Price, 1973).

Glacial erosional activity occurs at the ice-rock (or ice-glacial deposits) interface. We know that stationary ice or ice that does not contain rocky material (that is, clean ice) cannot erode fresh or slightly jointed rock. Ice of this type will erode fresh or slightly jointed rock only slightly or not as it moves. However, if the material beneath the glacier is unconsolidated, ice can substantially modify its morphology. The most significant erosion occurs beneath thick temperate glaciers that pluck fragments from the glacial bed and move them great distances. These fragments erode rock or loose material at the interface as they are transported.

13.2.1 Erosional processes

As stated previously, our knowledge of glacial erosion is limited. Scientists generally infer processes indirectly through laboratory experiments or by the deductive method used in the earth sciences. From the observed effects we try to determine the processes responsible for the erosional landforms, in this case glacial landforms (Goudie, 2004f).

Numerous **factors** control the intensity of glacial erosional processes. The factors fall into three large categories: those inherent to the glacial system, those related to lithological and structural characteristics of the rock substrate, and those related to the geometry of the rock substrate (Sugden and John, 1976; Drewry, 1986).

The **erosional capacity of the glacial system** partially depends on the basal ice temperature. A temperature near the melting point leads to more significant erosional effects; glaciers with very low basal temperatures erode the substrate only if detritus is present at the interface. In addition to temperature, the basal velocity of the ice mass is also important because it controls the amount of detritus transported and, consequently, the erosional activity of detritus and rock fracturing due to differential pressures. The thickness of the ice mass is fundamentally important in the glacial system because it affects the friction at work at the ice-rock interface.

Characteristics of the rock substrate such as rock hardness and presence of discontinuity planes (stratification, jointing, schistosity, etc.) also affect erosion. The permeability of the rocky bed is important because it determines meltwater infiltration and the erosional capacity of melt water. The absence of melt water limits the amount of water available for future refreezing.

Finally, the **geometry of the rock bed beneath the glacier**, such as its shape, rugosity, and slope, are also factors that control the intensity of glacial erosion.

In addition to the characteristics of the ice and the rocky substrate, changes in glacial erosion over time within a glacial epoch also affect glacial landforms. The greatest erosional intensity occurs in the first stages of a glacial advance (Tricart and Cailleux, 1962); ice masses can easily remove the pre-existing weathering layer during their initial movements. Once the regolith is removed, the ice requires greater force to erode the unaltered substrate.

Researchers have calculated approximate **glacial erosion rates** (Andrews, 1975). They have estimated material transport (bed, suspended, and dissolved load) by glacial melt water near the glacier tongue by assuming that this material originated in the glacier environment. These are therefore estimates of glacial erosion. Average rates range from 1 to 5 m per 1000 years, which is perhaps too high; these numbers should perhaps be divided by a factor of 2 to 10. Theoretical calculations are used to estimate the glacial erosion rate over long time periods. They have been applied to cirques, valley glaciers, and fjords. The variables used in the calculations are volume, area, and age of the landform being analyzed. The erosion rate is obtained by dividing volume by area to obtain erosion or reduction for the time period that the area has been subject to ice activity. For cirques, the rate ranges from 5 cm/1000 years to 40 cm/1000 years; the lowest values are for cirques in polar regions and higher values are for the cirques of western Scotland. In any case, these calculations should be considered as illustrative only.

13.2.1.1 Types of processes

The various mechanisms can be classified as follows:

13.2.1.1.a Abrasion

Ice wears down rocks as it passes over them; glacial erosional features such as striations and grooves, among others, clearly demonstrate this. The abrasion process can be studied directly by digging tunnels to the ice-rock interface and by direct observation of natural subglacial cavities. Laboratory experiments can study it indirectly, for example by analyzing the slipping behavior of ice for different rocks under various circumstances (Embleton, 1979a). Thermal conditions at the ice-substrate interface (with or without melting) and mechanical properties of the glacial substrate (soft or hard) determine processes that may be reactivated (Clarke, 2005).

Ice has a hardness of 1.5 on the Mohs scale at 0°C; thus it is much softer than most minerals. Therefore, ice is unlikely to scratch or wear down rocky surfaces over which it flows. For this reason, researchers think that rock fragments carried by a glacier erode the rock substrate at the ice-rock contact (Sugden and John, 1976).

Calculations of **abrasion rates** in many places indicate a range in values from 0.001 mm/year for small glaciers to 36 mm/year for a crystalline limestone substrate below the Argentière Glacier (Alps) under an ice thickness of 100 m and a glacial rate of 250 m/year (Boulton, 1974). The higher number is perhaps too high because it does not take into account the limestone dissolution process, which is significant in cold water.

The factors that affect the **abrasion process** can be categorized into two large groups: those related to **transported fragments** and **characteristics of the rocky substrate** and those related to particularities of the ice mass (Menzies, 1995b). Clean ice has no abrasive effect on the rocky bed because it lacks hardness but ice at the interface is scratched and pitted, showing that it is abraded by the rock bottom. The abrasion process requires rock fragments at the base of the glacier to do its work. Ice sheets and polar glaciers cause only minor abrasion partly because there are few fragments in the ice mass. In borings more than two thousand meters deep in ice sheets, researchers have found only a few meters that contain significant detrital material although borings in Greenland (Herron and Langway, 1979) and in the Antarctic (Gow *et al.*, 1979) show abundant detritus at the base of these ice sheets. Temperate glaciers contain more detritus and therefore have a higher erosional capacity although the fragment content of these glaciers is not the only factor that affects abrasion intensity. Another important factor is the relative hardness of the fragments and the rock substrate (Drewry, 1986). Harder fragments obviously have a higher abrasive potential. Morphological characteristics are also important; the constant friction between fragments and the rock bed reduces angularity and consequently decreases abrasion. A new supply of particles or breakup of previously abraded fragments can compensate for this reduced abrasion. Erosion at the ice-rock interface produces a fine-grained material which sometimes lodges between the ice mass and the rock base and inhibits erosional activity; when this occurs, the presence of subglacial water can help remove this film of "glacial flour." We must also consider the permeability of the rock substrate because the presence of water at the interface increases flow velocity and is a factor that controls abrasion, as discussed below.

The other group of abrasive processes is called **glaciological factors** (Embleton, 1979a). When the basal ice temperature is low enough, glacier ice can freeze to the bed; in this case, more force is required for sliding to occur. If we also consider the fact that few fragments are present at this temperature, as discussed previously, the abrasive effects will be almost negligible. Basal flow velocity is very important because it controls the quantity of particles that can flow

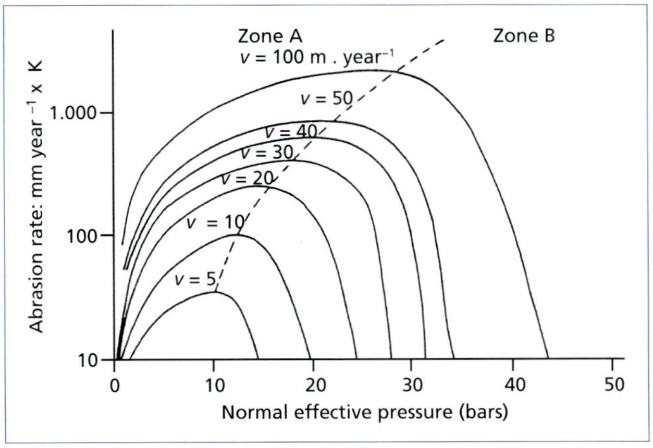

Figure 13.31 Theoretical abrasion rates and normal effective pressure for various ice velocities. In zone A abrasion increases with increasing pressure; in zone B abrasion decreases to zero, causing particles to be deposited. Values of k, c and p depend on the relative hardness of fragments, the rocky substrate, and the amount of detritus (Boulton, 1974).

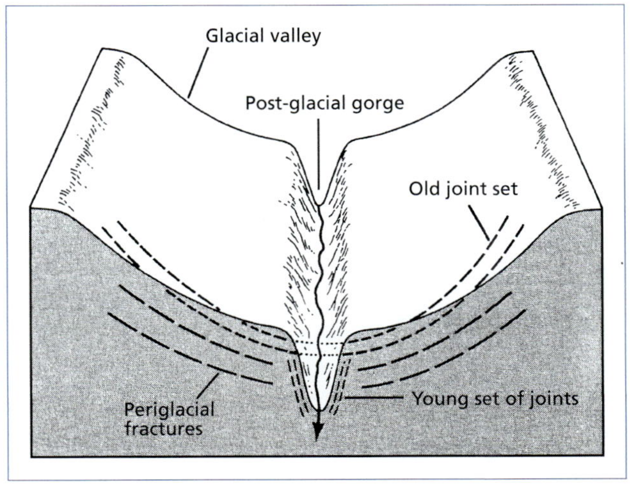

Figure 13.32 Sheeting systems in the rocks of the Vaiont Valley (Italy). The old system developed after the glacial fill was deposited and the most recent originated in a tight gorge incised by the river (Kiersch, 1983).

away from a point as surface flow from the ice-rock contact; a higher velocity is required for more erosion to occur. The ice mass thickness is also significant. A particle at the ice-rock contact is subject to the pressure of the weight of the ice column and the fragments it contains. For a given volume, abrasive intensity increases as pressure increases (Boulton, 1974) or as ice thickness increases. However, at a certain threshold, friction between the fragment and the rocky substrate retards the advance of the fragment so that ice flows over it rather than plucking the particle at the same rate as it moves it *(Fig. 13.31)*. *Figure 13.31* shows two areas: A, where abrasion increases with increased pressure and B, where abrasion decreases with a new pressure increase until it reaches the zero point.

13.2.1.1.b Fracturing

Fractures are of two types: fractures caused by glacier movement and fractures that existed prior to the glacier. It may be difficult to distinguish between the two.

Rock fragments at the base of the ice not only scratch and make grooves in the bed; they can also fracture the substrate and extract splinters of it as the ice exerts pressure on the rock. This process is called ***quarrying*** (Boulton, 1974; Iverson, 1995); it occurs as friction cracks are generated.

The average value of shear at the base of the glacier is approximately one bar (Embleton, 1979a). The movement of rock fragments at the base of the glacier over the underlying rock bed causes fracturing. This basal fracturing is most intense on reverse slopes of glacial flow. Thermal conditions within glaciers control the glacial fracturing process, which requires intense shear forces, particularly in the case of polar glaciers where the ice-rock contact is a unit because it is frozen.

One singularly important factor is the condition of the underlying material prior to passage of the glacier. **Periglacial fracturing of the substrate** is an important process that prepares material for easy exportation during the later ice advance (Boyé, 1949).

In his classic work on Yosemite Valley, Matthes (1930) analyzed the significance of **sheeting** in the plucking and material removal processes. He indicated that the optimum joint spacing for mobilization of underlying rock ranges from 1.5 to 7.5 m. It may be difficult to determine whether topographical jointing is preglacial or whether it results from unloading after the ice disappeared. Two sheeting systems may occur during different epochs; if they intersect (Kiersch, 1983) the zones of intersection are areas of weakness *(Fig. 13.32)*. In any case, sheeting has a considerable control on landforms.

Existing joints of any origin are conduits where subglacial water can infiltrate into the rock bed; if this interstitial water later freezes, **gelifraction** can break the rock. Even minor temperature oscillations around the freezing point can cause gelifraction. Gelifraction is observed in glacial cirques where melt water infiltrates through bergschrunds. On the scale of climate change, this process can be significant during oscillations of the glacier.

Researchers have studied landforms generated by the removal of pre-existing **weathering profiles** (Bakker, 1965; Thomas, 1994a); they have found regoliths of variable thicknesses that were generated prior to glaciations. The presence of this easily eroded material in certain regions must be included in the interpretation of glacial erosional landforms and the quantification of ice erosion. The Central Iberian system, which has thick layers of pre-Quaternary weathering deposits, is an area where this evidence should be part of the study of glacial landforms (Gutiérrez and Rodríguez, 1978).

Fracturing and quarrying of the rock substrate, either inherited or of glacial origin, and the availability of loose pre-glacial material are important factors in glacial erosion; they are more important than abrasion because abrasion accounts for no more than 40% of total erosion (Hallet, 1996).

13.2.1.1.c Removal of detritus

When a glacier slides over its bed, the rate of particle movement depends primarily on the size and shape of the particles and the roughness of the substrate. A traction force that exceeds resistance to friction (Sugden and John, 1976) is necessary to move a particle. Effective erosional activity removes from the substrate fragments that have been generated by the previously discussed processes. In this way, glaciers erode exposed rock. Subglacial melt water removes some of the detritus (Hallet, 1979; Drewry, 1986). This process is significant beneath temperate glaciers but negligible under polar glaciers. Basal material under temperate glaciers is also moved by the pressure of the ice mass on fragments and small sized water-soaked particles at the rock-ice contact; these flow toward areas of lesser pressure (Rea, 2007).

13.2.2 Landforms produced by glacial erosion

As a glacier passes over and erodes a surface, it generates various forms. *Table 13.2* (Sugden and John, 1976) is a classification of landforms generated by glacial activity. This classification, which is necessarily somewhat subjective, has three categories: unconfined ice flow, ice flow confined within rocky channels, and the interaction of glacial and periglacial activity. A second variable used for classification is the position of the landform: elevated or depressed. A third factor in the classification is the morphology of the resulting form: aligned or partially aligned. These criteria are plotted on the y-axis and their sizes are plotted logarithmically on the x-axis.

Some landforms are difficult to interpret because it is uncertain whether they were formed by ice or subglacial water or even by a combination of these media, which are closely related. As a result, it can be difficult to distinguish between glacial and fluvioglacial erosion.

Table 13.2 CLASSIFICATION OF GLACIAL EROSION FEATURES (ADAPTED FROM SUGDEN AND JOHN, 1976)

Process	*Relief Type*	*Relief Shape*	*Micro ← SCALE → Macro* m^{-2} (1 cm) m^{-1} (10 cm) m^{0} (1 m) m^{1} (10 m) m^{2} (100 m) m^{3} (1 km) m^{4} (10 km) m^{5} (100 km) m^{6} (1,000 km) m^{7} (10,000 km)	
Areal ice flow	*Eminence*	*Streamlined*	*← Whaleback →← Streamlined spur →* *← Rock drumlin →*	*Landscape of areal scouring*
		Part-streamlined	*← Roche moutonnée — Flyggberg →*	
	Depression	*Streamlined*	*← Striae →← Groove →* *← P-form →*	
		Part-streamlined	*← Rock basin →*	
Linear flow in rock cannel	*Depression*	*Streamlined*	*← Trough →*	*Landscape of linear erosion on ice sheet*
Interaction of glacial and periglacial	*Depression*		*← Alpine trough →* *← Cirque →*	*Valley glacier landscape*
	Eminence		*Residual summit or ← horn →*	*Nunatak landscape*

Most textbooks classify erosional landforms into two large categories: large and small landforms. This distinction is not realistic because there is broad variation in landform sizes.

13.2.2.1 Striations, grooves, and glacial polish

Friction of moving particles against the rocky bed at the base of the glacier erodes and scratches the bed, creates channels, and leaves a polish on the underlying rock surface. Friction also extracts particles from the rock bottom and incorporates them into the glacier.

Striations are a common glacial feature *(Fig. 13.33)*; these are thin aligned grooves about a meter long and several millimeters wide and deep. Striations develop most easily in fine-grained rock and disappear with exposure to weathering. They weather rapidly in limestone because of its high solubility. For example, striations on a limestone surface in Ingleborough (United Kingdom) were exposed to the atmosphere after glacial till was removed; they disappeared in about ten years (Sweeting, 1966). Striations are not restricted to the rock substrate; they can also occur in glacial till. In some cases, the existence of striated clasts can help in the interpretation of the origin of puzzling deposits (Tricart and Cailleux, 1962). This has to be done cautiously because unrelated processes such as avalanches, mudflows, subglacial currents, and others can also cause striations. Striations are most common in areas of reverse slope but also occur on vertical walls. Under the microscope, striations appear as multiple fractures that are concave in the flow direction (Iverson, 1995; Rea, 2007). They indicate direction but not the sense of glacial flow; in addition, several directions or striation systems can intersect. Morphological irregularities of the substrate can explain directional variations; they can also be used to differentiate glacial stages or phases of ice advance. The study of the large ice sheets requires a statistical analysis of hundreds of striations to obtain reliable data on the direction of glacial flow for a specific epoch.

Grooves are another glacial feature *(Fig. 13.34)*. These furrows of variable dimensions generally measure over a meter in length. Gigantic grooves measuring 12 km long, 30 m deep, and 100 m wide (Smith, 1948) have been found in the Mackenzie River valley in northwestern Canada. Their direction agrees with glacial flow inferred from other features although it is difficult to determine if were formed strictly by glacial processes. Like striations, they seem to have originated from friction of silt- or sand-size particles over the rock; the grooves are formed by the activity of large individual fragments or groups of fragments. But this may not be the only process responsible for these grooves; many authors suggest the action of subglacial currents (Bradwell *et al.*, 2008) or a similar preglacial relict landform that was highlighted by later activity.

Figure 13.33 Glacial striations. Fuerte de Santa Elena. Head of the Gállego River. Aragon Pyrenees (Spain).

Constant abrasion produces **glacial polish** *(Fig. 13.35)*. Observed with a hand lens or microscope, the surface of a glacially polished rock is scored by fine scratches that are similar to small striations. This polish can be easily observed in fine-grained rocks and it disappears rapidly when exposed to weathering (Rea, 2007).

13.2.2.2 Friction cracks and p-shapes

Friction cracks are small glacial features that occur as fractures or concavities in the rock, with or without the extraction of material from the rock substrate.

Works by Gilbert (1906), Harris (1943), and Dreimanis (1953) are the classic papers on the classification and origin of friction cracks. **Friction cracks** occur most often in medium grained rocks where striations and polish are less obvious. They are common in hard and brittle rocks such as

Figure 13.34 Glacial grooves and striations in the Rongbuk Valley, Everest Massif, Tibet. Photo J. López-Martínez.

Figure 13.35 Glacial polish and till. Pordoi Pass (Dolomites, Italy).

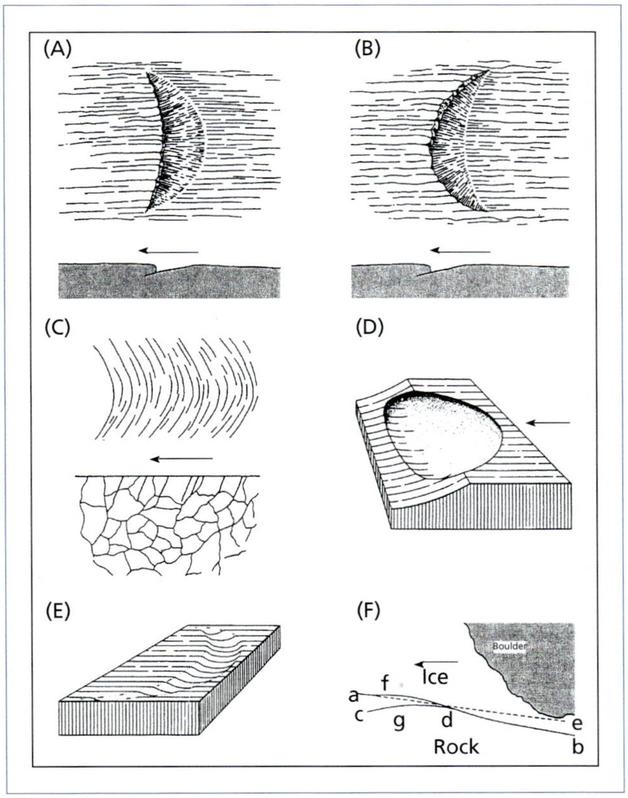

Figure 13.36 Friction cracks and sichelwannen (according to various authors, in Embleton and King, 1975a). The horizontal arrow indicates the direction of ice movement. (A) lunate fracture (plan and section view); (B) crescent gouge (plan and section view); (C) crescentic fractures (plan and section view); (D) conchoidal fracture; (E) sichelwannen; (F) formation of a crescent gouge. ade – original rock surface; afdb – rock surface deformed by pressure transmitted from the block contained in the ice; cgd – conoid fracture; fg – secondary fracture formed when rock wedge fgd breaks.

granite, basalt, quartzite, and others. Embleton and King (1975a) classified them into four types (*Fig. 13.36*):

i. **Lunate fracture** (*Fig. 13.36A*) in which the horns of the fracture point in the direction of the ice movement. This feature is composed of two fractures, a low-angle one, and a vertical fracture; the intersection of the two produces a slab of rock.
ii. **Crescentic gouge** (*Figs. 13.36B, 13.37*) in which the horns point in the direction opposite to ice flow. They range in size from 2 cm to 2 m, commonly occur in groups of two to six, and are generally aligned one behind the other. They also occur with the two fractures previously discussed.
iii. **Crescentic fractures** (*Fig. 13.36C*); these are concave in the direction of glacier flow and are composed of subvertical fractures. Unlike the previous two types, they do not extract slabs.
iv. **Conchoidal fracture** (*Fig. 13.36D*) in which the fracture plane is concave upwards.

These types are more common in reverse slope areas. Friction cracks are commonly preserved because they are larger than striations; thus they can withstand longer periods of weathering and erosion. The line perpendicular to the lunate forms shows the direction of ice flow. Direction of movement can be determined by the inclination of the small angle fracture that dips downward below the glacier (Rea, 2007). However, this observation, which was pointed out by Gilbert (1906) and Harris (1943), is not always true; there are many cases in the literature that contradict it.

Friction cracks generally result from the pressure of a boulder on the rocky substrate (*Fig. 13.36F*). This pressure is greater in areas of reverse slope (Boulton, 1974); most of these features come from such zones. Friction cracks begin with elastic deformation followed by rupture that generates a conoid fracture. The vertical fracture that cuts to the one of lower inclination originates behind the advance of the boulder. This type of crack can be reproduced experimentally with a knife or a steel bowl on glass.

P-forms or plastically sculpted surfaces (Dahl, 1965) occur in areas that have been subjected to ice activity. Kor *et al.*, (1991) claim that these shapes, which occur in hard rock, are produced by subglacial melt water rich in sediments; they call them **S-forms** (sculpted) and classify them as transverse, longitudinal, and non-directional shapes relative to melt water flow. These small forms do not exceed 20 m in size. The most common shape is the **sichelwannen** (*Fig. 13.36E*), which is a lunate shaped depression in hard rock; its horns point in the direction of ice flow. They occur in surfaces with any inclination. Researchers do not agree on their origin. The most accepted hypothesis attributes them to turbulent and sediment-rich low-pressure subglacial water activity (Kor *et al.*, 1991). However, other authors support the idea of abrasion (basal ice loaded with detritus) or erosion caused by the expulsion of water from saturated basal till (Gjessing, 1965; Gray, 1981).

13.2.2.3 Roches moutonnées and other landforms

Roches moutonnées are aligned hills that often occur as groups. They are asymmetrical; the side with lesser slope is frequently polished and striated and the other side, which may be steep, has an irregular and striated surface. This characteristic arrangement indicates the direction of ice movement, which is from the side with lesser slope toward the side with greater dip (*Fig. 13.38*). These forms occur frequently

Figure 13.37 Crescentic gouges. Threshold of the Gerber cirque. Leridan Pyrenees (Spain).

Figure 13.38 Roches moutonnées. Tollstringen. Trolls massif (Norway).

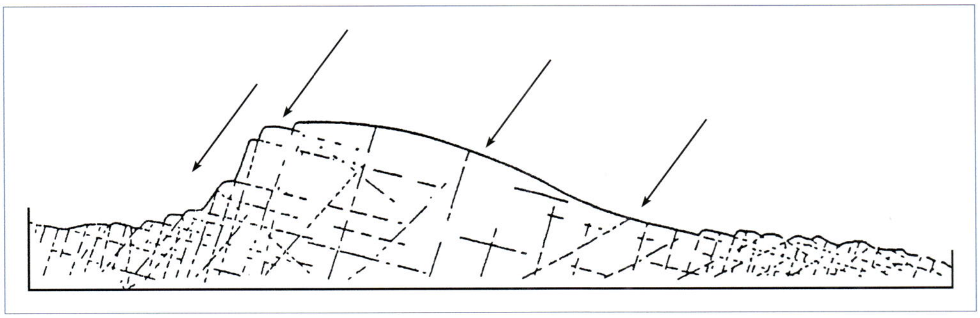

Figure 13.39 Longitudinal profile of a roche moutonnée showing the influence of jointing on its shape. The arrows indicate the direction of the ice push (Matthes, 1930).

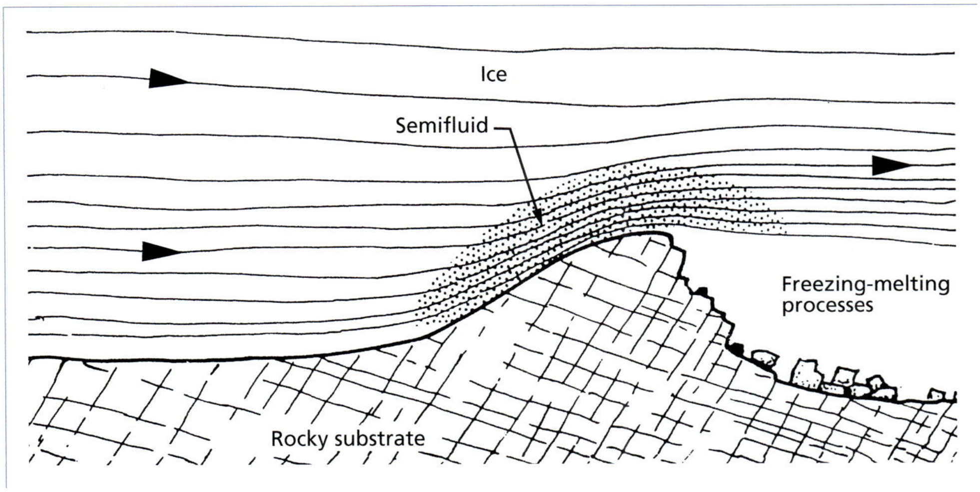

Figure 13.40 Processes that generate a roche moutonnée (adapted from Carol, 1947).

in crystalline rocks; they are very common in areas that have been covered by ice sheets and in thresholds near cirques. They vary in size from less than a meter to several hundreds of meters up to the size of **flyggbergs**, which can extend for kilometers and reach heights of hundreds of meters. Large roches moutonnées can contain smaller ones (Rea, 2007).

Their origin is not well understood. In some cases joint spacing appears to be important in their formation (Matthes, 1930; Sugden *et al.*, 1992) (*Fig. 13.39*) although this is not true in all cases. Other authors, such as Carol (1947), explain the steep side of the roche moutonnée as a product of gelifraction in ice that is at the melting temperature (*Fig. 13.40*). As the glacier passes over the protuberance, pressure increases and the ice becomes semi-plastic; on the steeply inclined side, pressure decreases and the melt water refreezes, breaking the rock. These observations were made in a subglacial cavity. Even so, as Embleton and King (1975a) point out, it is still not clear whether roches moutonnées form only under temperate ice as proposed by the subglacial gelifraction hypothesis.

Another set of landforms related to or associated with roches moutonnées include **whalebacks, rock drumlins,** and **streamline spurs**. These are elongate landforms with smooth slopes; their sizes are shown in *Table 13.2*. Jointing may control the boundary between the hills of whalebacks. Rock drumlins are sometimes associated with drumlin fields in glacial material. The origin of streamline spurs is unknown; they are not necessarily glacial in origin but the smoothness of their slopes and their orientation parallel to the direction of glacial flow implies a glacial origin.

In addition to these landforms with positive relief, depressions and lakes of different sizes, primarily **rock basins**, pit glacial landscapes. These depressions vary greatly in shape; structure plays a significant role in their morphology as well as their genesis. They may result from pre-existing, pre-glacial irregularities that were perpetuated and emphasized by ice activity; in other places, glacial erosion has generated these closed depressions.

13.2.2.4 Cirques

Cirques and glacial valleys are the most spectacular forms of glacial erosion. Glacial geomorphology texts claim that Jean de Charpentier first used the word *cirque* in 1823 to describe landforms in the Pyrenees. Multiple names in various languages and countries have been used for cirques. A glacial cirque is defined as a concave semicircular or semi-elliptical depression delineated by steep slopes that has been occupied by ice (*Fig. 13.41*). Some authors consider the existence of a rocky basin as part of the definition but not all cirques have this characteristic.

If the basin is rocky, the threshold present at the outlet of the cirque can be rocky or composed of glacial till. The walls of the cirque are usually fragmented and broken unlike the basal surface, which is generally smooth and bears small glacial erosion features.

Cirque dimensions vary widely; they range from tens of meters to kilometers in width. The largest known is the Walcott Cirque in Antarctica; it is 16 km wide and its walls are about 3000 m high (Price, 1973). The size depends on many variables, including the duration of glaciation and the structural and lithological characteristics of the rock mass on which the cirque is located (Embleton and King, 1975a). The shape of a cirque is also a function of the underlying rock.

The **morphometry** of a cirque can be described in great detail according to the set of characteristics in *Figure 13.42* (Andrews and Dugdale, 1971). Andrews (1975) claims that another value, the cotangent of the angle between the cirque threshold and the top of the final wall, indicates the intensity of glacial erosion. Another index is the degree of occlusion (Evans, 1969) defined as the number of degrees of the longest topographic contour; 180° shows that the cirque walls are parallel. Researchers have also proposed other parameters as morphometric characteristics of cirques. These values generally lack precision either because they are derived from topographic maps and or because they involve subjective parameters of the researcher. Even so, as Andrews (1975) indicates, "the morphological analysis of cirques can provide significant information on the erosional processes in the cirque, although only of a deductive type: the morphology is a response to the processes." Today specific other techniques are used to define and measure cirques (Evans and Cox, 1995); geographic information systems (GIS) are also used (Federici and Spagnolo, 2004).

To better understand the **processes** that generate cirques, we need to understand the structure and movement of ice within them. The most precise observations to date were made by McCall (1960) in his study of the small cirque glacier Vesl-Skautbreen (Norway). He dug tunnels (*Fig. 13.43*) to measure ice velocities at different points and identified various summer ablation surfaces. He observed that ice flow lines dip 30° in the upper zone of the glacier and 26–28° in the lower part but they dip in the direction opposite the slope

Figure 13.41 Cirque and moraine belt, formed on top of lower Cretaceous materials at the head of the Miera Valley, (Cantabria, Spain).

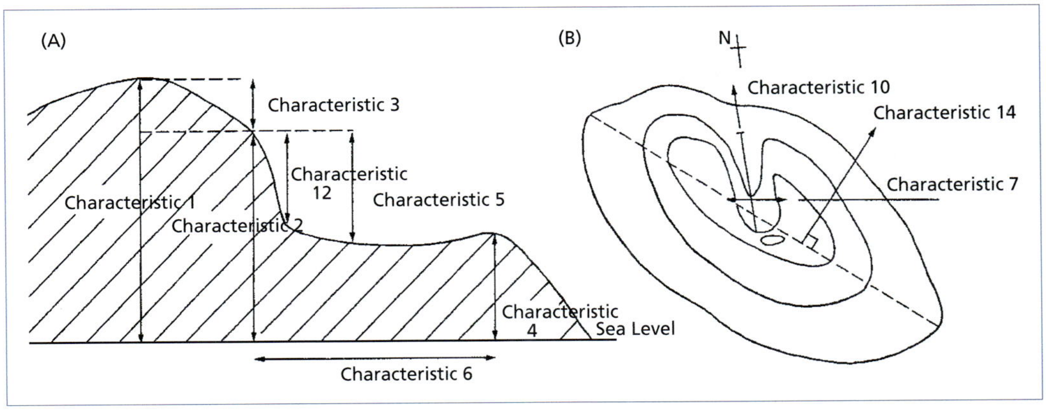

Figure 13.42 Features of a cirque (Andrews and Dugdale, 1971). (A) cross section of a cirque, (B) plan view of a cirque and its summit area. The different characteristics are as follows: 1-Maximum elevation of the mountain where the cirque is located. 2-Maximum elevation of the cirque. 3- Difference between 1 and 2. 4-Elevation of the cirque threshold. 5-Maximum vertical development of the cirque. 6-Length of the major axis of the cirque. 7-Maximum width of the cirque perpendicular to the major axis. 8-Length/width ratio. 9-Length/height ratio. 10-Direction of the major axis. 11-Type of ice mass: a) empty, no ice; b) snow field; c) ice plate; d) cirque glacier. 12-Height of the final cirque wall. 13-Ratio between height of the final wall and the length of the cirque. 14-Regional gradient angle.

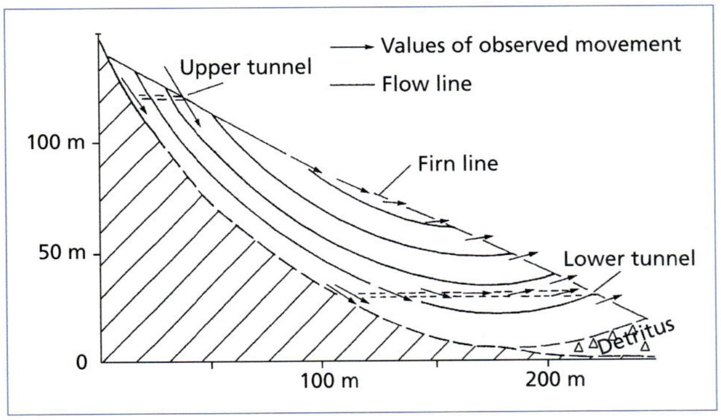

Figure 13.43 Cross-section of the cirque glacier of Vesl-Skautbreen (Norway) showing flow lines and velocities. From McCall (1960), simplified.

of the cirque and the dips decrease downdip. This arrangement suggests a synclinal structure of ice layers. The velocity distribution shows rapid movement in the upper zone of the glacier and slower rates in the lower part; in this area the various velocity vectors indicate updip movement of the ice. All these data show that movement occurs as rotational sliding.

Two erosional processes operate in a cirque; they result from glacial and periglacial activity. At the ice-rock contact, abrasion causes continuous wear on the walls and bottom; this creates the rocky basin present in many cirques (Richardson and Homlund, 1996). Rotational sliding of the ice creates the depression; this movement, in turn, causes ice to surmount the threshold and remove the detritus. The other process, gelifraction, is very effective in rocks that crop out on top of the glacier (Gardner, 1987). Here, freeze-thaw action is more efficient in temperate glaciers than in polar glaciers. This implies that the cirque enlarges fundamentally by gelifraction and deepens as a result of abrasion. Using various methods to determine erosional rates of cirques, researchers have determined erosion to be about 500 mm/ky (Benn and Evans, 1998). Retreat appears to be more rapid than excavation. On the other hand the elevations of the bottoms of cirques do not seem to change much during glaciation. Elevation is generally used to calculate the approximate elevation of perpetual snow, which is the 0°C summer isotherm considering the lowest cirques of equal orientation (Flint, 1971).

Most authors think that cirques form after the first accumulation of snow in a pre-existing depression where gelifraction occurs and particles are evacuated by snow melt in the summer (**nivation**) (Thorn and Hall, 1980; Thorn, 2004). As a result, the depression enlarges and a nivation niche forms. When the excavation reaches a certain depth, snow persists from year to year in the nivation niche and is converted to firn and eventually to ice. This is how an embryonic cirque forms. The snow accumulation stage is called the incipient phase. When snow begins to persist from year to year, the cirque has reached the firn phase (Tricart and Cailleux, 1962).

Once formed, a cirque continues to develop if circumstances are conducive. Its **evolution** is controlled by numerous factors. The duration of glaciation and the number of glaciations play a fundamental role in cirque development.

Figure 13.44 Aerial view of horns, cirques, *arêtes*, and glacial valleys. Anchorage Region (Alaska). Photo P. Lucha.

Also, a glacier evolves differently depending on its location, that is, whether it is in temperate or subpolar latitudes or polar areas. Another important variable is the structure and lithology of the rock mass on which it is situated (Embleton, 1979a). Sometimes a cirque forms on a landform composed of different lithologies; in a monoclinal series this can create **stepped cirques.** The degree of fracturing and the spacing between planes of weakness also affect mechanical disintegration by gelifraction. Sheeting can be an important determinant in the shape of a cirque. Another fundamental factor is the orientation of the cirque. In middle latitudes in the northern hemisphere, cirques develop well if they face north and east because this position protects them from solar radiation and snow can persist more easily. Researchers have analyzed the effect of climate on cirque development (Derbyshire and Evans, 1976). These authors say that in spite of numerous regional studies on the shape and distribution of cirques it is difficult to generalize on this topic.

Cirque development is accompanied by a parallel retreat of cirque walls (Fisher, 1872) that creates *arêtes* at the boundaries with neighboring cirques (*Fig. 13.44*). This type of cirque growth leads to **coalescent cirques** (*Fig. 13.45*). The retreat of the final wall of a cirque forms a **horn** (*Figs. 13.46* and *13.47*), which may have three or four faces.

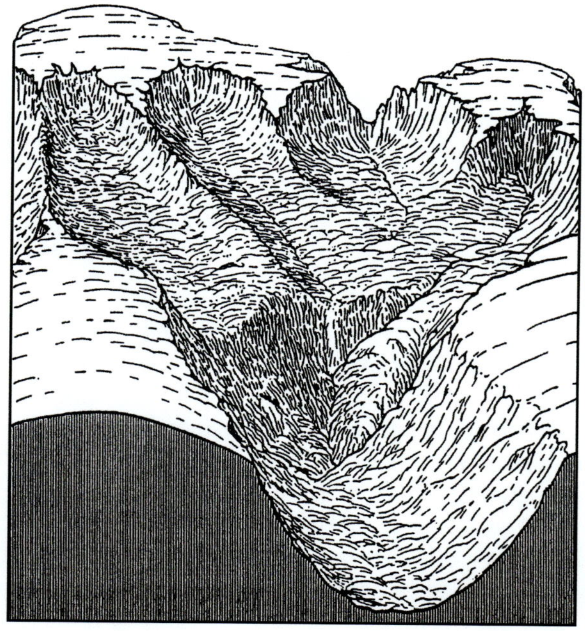

Figure 13.45 Coalescent cirques, *arêtes*, and glacial valley (in Holmes, 1944, modified from Davis, 1906).

Figure 13.46 The Cervino or Matterhorn in the Alps, the boundary between Italy and Switzerland, has a typical horn morphology. Photo J. López-Martínez.

Figure 13.47 Horn in the central Andes. El Portillo (Chile).

13.2.2.5 Glacial valleys

Glacial valleys are a characteristic landform formed by glacial erosion. They differ from fluvial valleys because ice in a glacial valley occupies a large part of the transverse profile of a valley whereas rivers flow over only a small part of a fluvial valley (Price, 1973). In glacial valleys, channelized ice masses excavate troughs that can be hundreds to thousand of meters deep (*Figs. 13.48* and *13.49*). This intense erosion affects the entire windward slope of mountains covered by heavy snowfall. As a result, they can supply thick glaciers that flow down preglacial valleys and modify the relief in considerable and sometimes spectacular ways.

Linton (1963) has classified glacial valleys into four types. The **alpine type** is supplied by a cirque or group of cirques in high areas. The **icelandic type** is supplied by an ice cap that escapes through preglacial valleys. The **compound type** is formed when the glacier does not encounter a previously existing valley where it can unload all its ice or which is the same thing the superposition of preglacial valleys on top of others developed from the partial or complete destruction of the preglacial watershed. Finally, the **intrusive or reverse type** of glacial valley is one in which the ice flows upgradient in the preglacial valley. This type occurs primarily in areas of low relief.

A glacial valley has distinct transverse and longitudinal profiles. The **transverse profile** generally results from ice that erodes old fluvial valleys; this activity widens and deepens old valleys. Glacial valleys are generally U-shaped (*Figs. 13.49* and *13.50*) and have steep walls; they have flatter floors than fluvial valleys because later deposits of colluvium and alluvium level the valley floor. U-shaped valleys exist in other environments such as cradle valleys in periglacial environments (Tricart, 1967) but the slopes of cradle valleys are completely covered by detritus; the walls of glacial valleys are generally bare rock. Glacial valleys can also have V-shaped transverse sections if subglacial channels have eroded the valley (Evans, 2007). In spite of these exceptions, the typical profile of a glacial valley is parabolic or nearly so, possibly because this shape offers the least resistance to friction (Flint, 1971). Sometimes, the walls of a glacial valley have **plains** or **shoulders**; the interpretation of these forms has generated significant controversy (Derruau, 1965). The best explanation seems to be downcutting by a reduced end-of-glaciation glacier in a broad valley that was shaped by a glacier occurring during a time of maximum glaciation.

Field research on transverse profiles is difficult because glacier walls are steep. Investigators generally use maps or photogrammetric surveys and for this reason they do not know if the rock is bare or covered with deposits. Researchers

Figure 13.48 Glacial valley in the central Andes, showing thick accumulations of gelifracts on the slope, which supply the valley floor. Puente del Inca (Argentina).

Figure 13.49 Llanganuco U-shaped valley, of glacial origin, Cordillera Blanca, Peru. Photo J. López-Martínez.

need these data to correctly interpret the evolution of this profile.

Analysis of the **longitudinal profile** reveals a number of irregularities including **basins** and **steps** that distinguish this profile from that of a fluvial valley. When the ice retreats, the basins become lakes or are sometimes filled in by sediments. Thresholds are commonly signs of glacial erosion. After ice disappears from a principal glacial valley that has eroded more deeply than lateral valleys, a series of **hanging valleys** become visible; **faceted spurs**, triangular or trapezoidal, can also be seen among the hanging valleys (*Figs. 13.51* and *13.52*). This postglacial topography creates waterfalls such as those in the Yosemite (California) and Lauterbrunnen (Switzerland) valleys.

The **origin of the steps** has a different explanation. Variations in lithology or different fracture spacing (Matthes, 1930) (*Fig. 13.53*) can generate steps in a profile. Thresholds are also attributed to gradient changes in the preglacial valley that have been preserved or emphasized by ice. Another suggested origin is the coalescence of glacial valleys. As discussed previously, deep differential weathering plays a significant role. Glaciers can easily erode the regolith during early stages and uncover depressions or highs at the basal weathering surface (Bakker, 1965). In any case, basins indicate the capacity of ice to flow against the gradient.

The processes involved in the **genesis of a glacial valley** are generally the same as those that generate other

Figure 13.50 Zezere glacial valley with U-shaped cross section. Sierra de la Estrella (Portugal), the westernmost prolongation of the Central System of the Iberian Peninsula.

Figure 13.51 Block diagram showing hanging valleys, faceted spurs, U-shaped valleys, cirques, *arêtes*, and a horn (Davis, 1906).

Figure 13.52 Hanging valley. West coast of Norway. Photo J. López-Martínez.

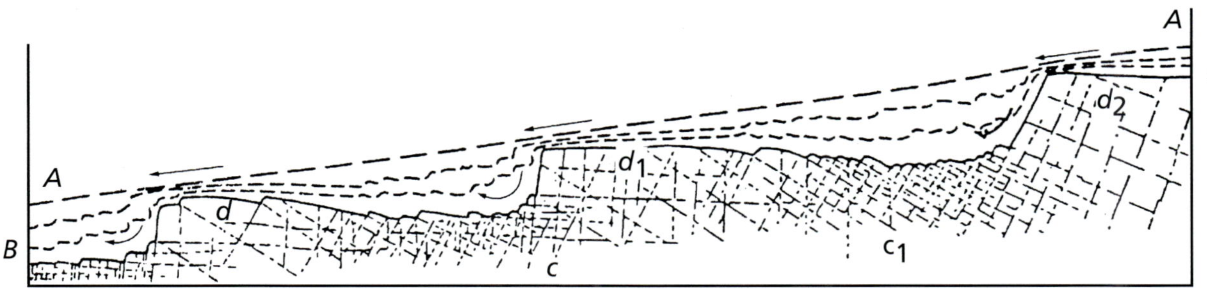

Figure 13.53 Longitudinal profile showing the development of a stepped glacier as a result of different joint spacing. AA represents the preglacial profile. BB is the stepped glacier. The areas d, d1 and d2 are zones of where glacial erosion difficult due to joint spacing; the contrary is true at c and c1. The arrows indicate the direction of the ice movement (Mathes, 1930).

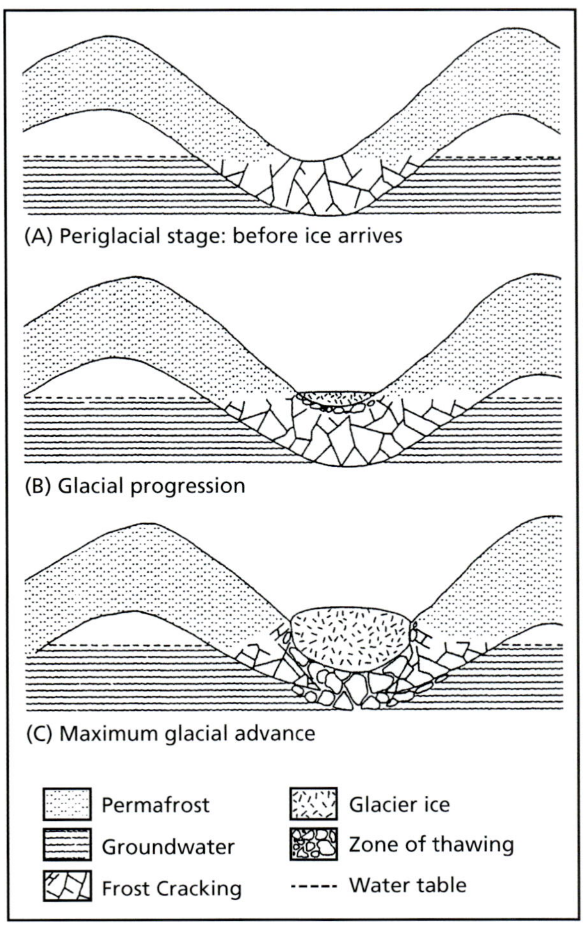

Figure 13.54 Periglacial breaking mechanism in glacial valleys (Cailleux, 1952).

glacial landforms. Abrasion is more effective in temperate glaciers than in polar glaciers. Jointed blocks move when pushed by other blocks transported by the glacier. Another process is the periglacial breaking (Fig. 13.54) first described by Boyé (1949), which Cailleux (1952) used to describe glacial valleys. During the first periglacial stage, gelifraction fractures the substrate of the valley floor. The advancing glacier tongue acts as a thermal layer that causes the permafrost to disappear; the glacier can easily remove gelifracts. The generation of concave sheeting subparallel to the transverse section helps construct the parabolic profile and perpetuate it. Nye's (1952) distribution of extensive and compressional glacial flow partially explains the existence of thresholds and basins (Fig. 13.24). Increased erosion occurs in compressive zones; once irregularities appear, they tend to be preserved or even accentuated.

The elevation of a glacier sometimes rises because it is blocked by another glacier or because it accumulates more material than it loses. In these cases, a preglacial breach can occur in the watershed. The glacier may discharge into an adjacent valley, causing glacial erosion of the breach and the formation of a U shape in the same rocky basin (Penck, 1905). These breaches are called **glacial diffluence**. When the level rises in such a way that all the flows become diffluent ice, the term **transfluence** is used (Embleton and King, 1975a).

13.2.2.6 Glacial landforms in low relief regions

In previous sections we analyzed the erosional activity of ice in areas of highly contrasting relief and their most significant landforms, cirques and glacial valleys. However, much of the emergent land surface previously covered by glacial ice consists of low relief regions. These circumstances are present on elevated plateaus and in low-lying plains (Davies, 1969; Embleton and King, 1975a). Ice sheets covered most of these areas; erosional landforms in these areas contain features that distinguish them from alpine glacial landforms.

Many researchers maintain that ice under ice sheets can protect preglacial relief (Ambrosse, 1964). Many glaciated areas show no or very few signs of glacial erosion and sedimentation (Sugden, 1974). These regions are probably zones where ice in contact with the bed was below the freezing point; thus there was no movement at the ice-rock interface. This protective role of a glacier is shown in various ways. In some places, valleys show no relationship to the glacial flow direction and in other areas they preserve preglacial planation surfaces with only slight modifications.

Despite the evidence of the protective role of ice sheets, there is more abundant proof of glacial erosion in areas of sparse relief. In these regions, glacial erosional processes generate a typical **knock and loch topography** (Linton, 1963) (*Fig. 13.55*). This topography has indistinct relief with hills (tens of meters high) and depressions; lakes or peat bogs often occupy these depressions, which are generally not very deep. The lakes are aligned and elongate due to structural influence (Evans, 2007). The accompanying drainage network is highly irregular. The hills often have striations and glacial polish; they are often roches moutonnées. The hills may also have smooth and elongate shapes reminiscent of drumlins; for these reasons the landform is called a **rock drumlin**. The difference between a rock drumlin and the classic drumlin is compositional: classic drumlins are composed of glacial till and fluvioglacial deposits. The knock and loch topography is typical of northwestern Scotland and the Baltic and Canadian shields (Sugden, 1978; Gordon, 1981; Rea and Evans, 1996). Glacial sedimentation is characterized by small discontinuous thin sheets of moraine. Another form that develops under these circumstances, although not exclusively, is called **crag and tail**, which is an elongate till located in the shelter of a hill. Jahns' (1943) observations on this type of relief show that erosion is more significant on hills in areas protected from ice flow. He draws this conclusion from the erosion of sheeting structures prior to the passage of the glacier.

13.2.2.7 Fjords

Fjord is a Norwegian word for a long and deep arm of the sea that extends landward; fjords have straight segments and steep walls and are very deep (Fairbridge, 1968c). The longest fjords are the Noverstfjord and Sund in Greenland (230 km), Sognefjord in Norway (220 km), and Greepy Fjord/Nansen Sound in the Arctic of Canada (Evans, 2004b). They occur along high latitude coasts (*Figs. 13.56* and *13.57*) and are characterized by a threshold, either totally or partially submerged, at the mouth of the fjord and a deep basin

Figure 13.55 Knock and loch topography South Harris, Hebrides Islands (United Kingdom).

Figure 13.56 Geiranger Fjord (Norway).

Figure 13.57 Seydisfjördur Fjord (Iceland). Photo F. Gutiérrez.

Figure 13.58 Strandflats. Alesund (Norway).

within the fjord (Syvitski *et al.*, 1987). The threshold is generally rocky and the basins can be enormously deep. The largest measured depth of a fjord basin is 1,288 meters in the Messier Channel (Chile) but some Antarctic basins may be 2,000 m deep (Andrews, 1975). As authors Andrews (1975), and Twidale (1976a) point out, some fjord areas (eastern Arctic Canada, Greenland, Norway, and New Zealand) show clear structural control. Some researchers consider this as proof of a structural genesis of fjords but today structural control is considered merely accessory (Powell, 2003).

Most fjords occur within glacial valleys. Fjord walls also have faceted spurs and hanging glacial valleys (*Fig. 13.52*) that pour ice into the principal valley that has been converted into a fjord by postglacial sea level rise. Glacial erosion appears to be responsible for excavation of the basin and the presence of the threshold, which can be explained by thinning and spilling of the glacier as it reached the continental platform.

From a tectonic point of view, fjords are primarily located on convergent and divergent continental margins. In active tectonic environments, uplift and tectonic depressions are important for the accumulation of snow and sediments. Study of the sedimentary record of fjords provides important scientific information for earth science (Powell, 2003).

Steep walls, hanging valleys, and waterfalls characterize the fjords along the southern coast of Norway (Hardanger, Sogner, and Geiranger fjords). Farther north, in the region between Alesund and Trondheim, slopes are much flatter, similar to the *sea lochs* or *firths* of Scotland. In these areas, low-lying rocky islands (*skjaergard*) or partially submerged erosional platforms (*strandflats*) (*Fig. 13.58*) are present. These flat areas are interpreted as having been formed by marine, subaerial, and subglacial erosion, and even as etched surfaces (Benn and Evans, 1998).

Scottish **firths** are long coastal inlets associated with glaciation along low-relief coasts. They are distinguished from fjords by their irregular form, lack of basins, and U-shaped transverse sections. In any case on low-relief coasts, it is difficult to distinguish inlets generated by glacial erosion (Augustinus, 1992).

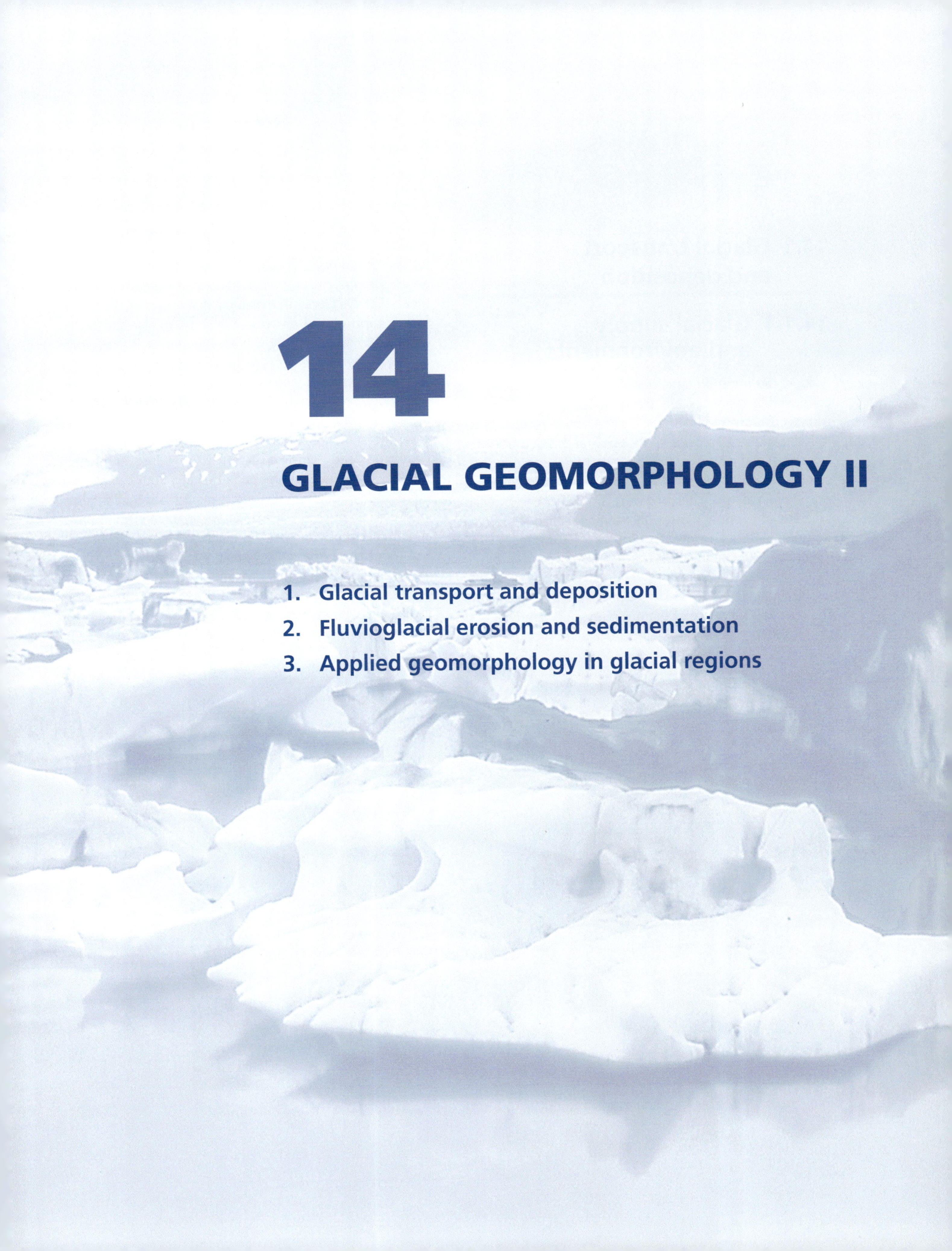

14

GLACIAL GEOMORPHOLOGY II

14.1 Glacial transport and deposition

14.1.1 Glacial supply and environments

Glaciers can transport large quantities of rocky material over considerable distances but glacial erosion is not the only source of transported detritus. Gelifraction of rocks that surround glaciers can provide clasts. Snow and rock avalanches, landslides, and extraglacial streams that flow into the glacial complex also provide detritus. Explosive volcanic activity, as in some Andean volcanic regions, can contribute pyroclastic material to snow masses. In addition, eolian activity can transport particles to glaciers (Boulton, 1978; Small, 1987; Kirkbride, 1995; Benn and Evans, 1998).

In western Greenland, detritus is discharged primarily through the basal ice layer along the ice sheet margin. The relative abundance of detritus is an important control of glacial sedimentary processes and characteristics. Sediment production varies considerably between ice caps and glacial valleys (Knight *et al.*, 2002).

Detritus can be transported in three different environments within a glacier. The material that moves on the ice surface is **supraglacial detritus**, which is angular and slightly modified by glacial activity, if at all. This detritus is more abundant on valley glaciers because the rocky walls provide material to the ice surface whereas in ice caps nunatak walls are the only supply. In the accumulation zone, snow may cover supraglacial detritus; however, in mountainous areas detritus can cover the entire ice surface, creating black glaciers as in the Karakorum and Himalaya (*Fig. 14.1*). This cover protects the snow from melting (Eyles and Lazorek, 2007). In general, the amount of detritus increases at the front of the tongue of most glacial valleys; most material in the ablation zone was in an intraglacial position prior to melting. Supraglacial detritus is transported as if by conveyor belt; this method is capable of moving large boulders great distances.

Englacial detritus can be disseminated throughout the ice mass; it rarely exceeds 15% of the ice volume. It may also be present as planar bands up to 5 m thick separated by

Figure 14.1 Black glacier of Baltoro at about 4,300 m in the Karakorum (Pakistan). Photo J. López Martinez.

clean ice; this detritus may constitute up to 80% of the total mass (Embleton, 1979a). These bands are composed of crack fillings although in most cases they are subglacial debris displaced from the top along flow lines. This transport occurs in areas of compressive flow in the glacial valley; at other times terminal moraines cause compression that obstructs the ice advance (Goldthwait, 1951; Boulton, 1970a).

Subglacial detritus comes from the valley bottom itself or from material that penetrates into cracks and tunnels. This bottom debris is abraded and broken; it is more rounded and has a higher percentage of fine material. Transport occurs by traction (Sharp, 1988) in which particles slide and roll, pushed by the glacier; if the bed is deformable, the particles will press into the bed or become lodged in it. Clasts shapes and orientation affect the manner of sliding and, as a result, elongate particles become oriented in the flow direction (Price, 1973).

14.1.2 Mechanics of glacial deposition

Detritus in the ice mass can be transported along a glacier margin, deposited on the valley bottom, incorporated into glacial meltwater, or carried away by icebergs. Glacially transported sediment is deposited according to a set of primary processes; other processes will later modify the sediments (Whiteman, 1995). Boulton (2006) established three primary shapes of till sedimentation: supraglacial melting or thawing (*melt-out*), subglacial (*lodgement*), and deformation.

Supraglacial melting (*Fig. 14.2*) is the primary form of ablation; it occurs in temperate glaciers but not in cold glaciers where little melting takes place. In temperate glaciers melting can achieve rates of 12 m/year whereas in the Antarctic the rate is limited to a few centimeters a year (Embleton, 1979a). Supraglacial melting is much more significant than melting at the base of the glacier. Melting carries away debris that becomes supraglacial till, which can preserve structures formed during glacial transport (Lawson, 1979). More detritus gradually builds up on the till layer through continuous accretion. The presence of this cover of supraglacial material slows down the melting of underlying ice and creates *hummocky* topography that forms as the underlying ice melts and the overlying sediment collapses (Eyles and Lazorek, 2007). Supraglacial material can remain in place or be moved by meltwater; in glacial tongues, flow and rotational sliding affects most supraglacial till (Dreimanis, 1988). The water-soaked till moves by gravity; movement depends on surface gradient, rugosity, and water content. Flow velocities of a few meters per minute have been recorded and detritus can move distances of tens and even hundreds of meters (Sudgen and John, 1976).

Figure 14.2 Supraglacial melting at the front of the Root Glacier (Alaska). Photo P. Lucha.

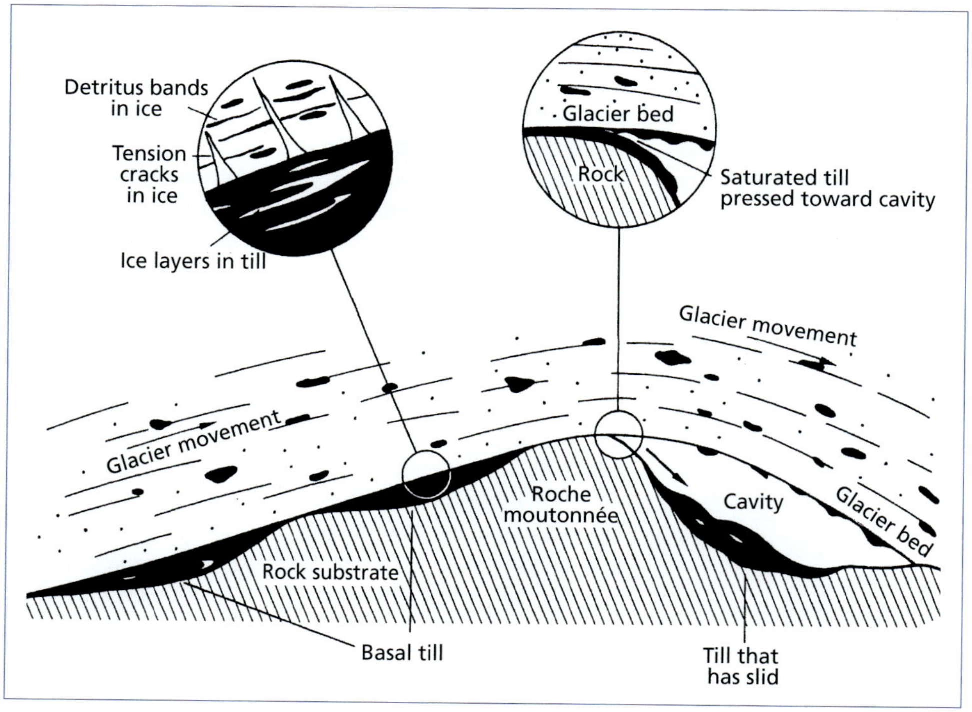

Figure 14.3 Accumulation of subglacial till at the base of the Svalbard glaciers, Spitsbergen. Note the cavity in the shelter of the roche moutonnée, which is filled with till (Boulton, 1971, modified by Sugden and John, 1976).

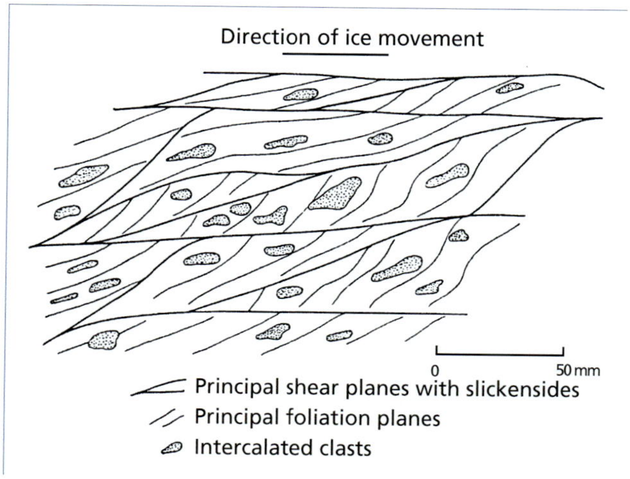

Figure 14.4 Diagram showing shear structures in subglacial till resulting from forces beneath the moving ice (Boulton, 1970a).

Three heat sources can play a role in **subglacial melting**: geothermal flow, frictional heat due to the sliding of ice, and heat generated by pressure increases caused by obstructions on the irregular substrate (Embleton, 1979a). All these heat contributions cause subglacial ice to melt and existing debris to be deposited. Subglacial *lodgement* involves low pressure melting of a moving glacier; the continuous freeing of particles leads to accretion and creates basal till. Till gradually fills in irregularities in the bed and thus smooths the valley floor profile (Boulton, 1971) (*Fig. 14.3*). Subglacial till contains shear structures resulting from forces active during ice movement (Boulton, 1970a, b) (*Fig. 14.4*). Water-saturated subglacial sediments deform under the weight of the ice (Boulton, 1975, 1982). Subglacial flow pushes pressurized till into basal cracks and cavities. This plastic flow of water-saturated till has an annual periodicity in temperate glaciers. In summer, the water infiltrates through cracks, saturates the till, and causes it to flow; in winter, the process stops. Wetting causes the fabric of the till to change. In addition, moving ice erodes the subglacial till and generates a group of furrows and drumlin-like shapes called *fluted moraines* (Sudgen and John, 1976).

Melting along glacier margins causes supraglacial, englacial, and subglacial detritus to be deposited. This **marginal discharge** creates moraine ridges that show the positions of the glacier margin. In general, till has no structure. An advancing glacier can **push** on loose material deposited at the frontal margin and create a push moraine. The till is folded and fractured and forms imbricate structures. The fabric is oriented parallel to the ice movement direction (Croot, 1987; Van der Wateren, 1987; Eyles and Lazorek, 2007).

Figure 14.5 Erratic blocks Markarfljót (Iceland). Photo F. Gutiérrez.

14.1.3 Till characteristics

Till is a poorly sorted sediment composed of a wide range of grain sizes; boulder-size fragments are contained within a fine-grained clay-rich matrix. Till is not stratified, and it is generally composed of a broad variety of rock types; the clasts can be faceted and striated (Sugden and John, 1976). Some investigators use the term "moraine" as an equivalent of "till" but the word "moraine" has a morphological meaning whereas till is a sedimentological term (Price, 1973).

Glaciers have a high transport capacity as shown by **erratic blocks** (*Fig. 14.5*), which are boulders found on rocky surfaces or on top of material deposited by glacial activity. They generally have a different lithology than the material on which they are sitting (Evans, 2007). The dimensions of these blocks can be spectacular, such as the Schollen erratic in Germany, which measures 4 km by 2 km by 120 m (Price, 1973). The Okotoks erratic block near Calgary (Canada) weighs 118,000 tons (Sharp, 1988). We can use lithology to determine provenance of the block and to calculate the distance the block has traveled. Some erratic blocks have traveled up to 1200 km (*Fig. 14.6*) (Flint, 1971).

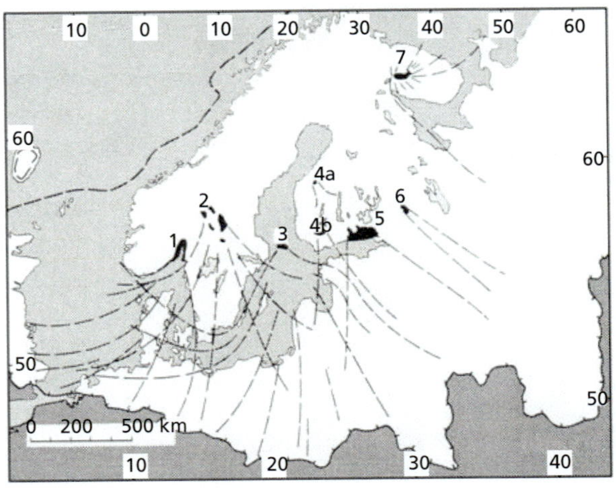

Figure 14.6 Trajectories of erratic blocks shown on the Scandinavian ice cap. 1. Rocks in the Oslo district. 2. Dala porphyries. 3. Rapakivi granites from the Aland Islands and quartz-bearing porphyries. 4. Olivine diabase from Satakunta and uralitic porphyry from Hameenlinna. 4 a. Karnaite from Lappajärvi. 5. Rapakivi granites from Viipuri. 6. Rapakivi granites from Lake Ladoga. 7. Nepheline syenite from Umptek and Lujavr-Urt (in Flint, 1971).

Particle size distribution can provide valuable data on till formation. Benn (2004) has compiled a list of field techniques and analytical methods for the study of glacial sediments. Most tills have a bimodal distribution with one peak associated with block-gravel size fragments and peak for the matrix (Dreimanis and Vagners, 1971) (*Fig. 14.7*). For a given lithology, grain size decreases with distance from the source area; a predominance of mineral grains over rock fragments indicates a higher degree of till maturity. This shows sorting mechanisms in glaciers that are related to the breakup of fragments. If the fragments are calcareous, the till can be substantially modified by carbonate dissolution (Sudgen and John, 1976).

Till **fabric** describes particle arrangement. This fabric is often called glaciotectonic (Elson, 1989). Particle organization may be visible, especially in the orientation of the major axes of large fragments (*Fig. 14.8*). The direction and dip of elongate clasts in till are used to determine the direction of former ice flow (Flint, 1971; Benn, 1995). These studies of till fabric should be done at multiple locations in a study area because a unique data point can lead to an erroneous interpretation. In addition, these investigations can help distinguish two or more types of till in a sequence (Benn, 2007).

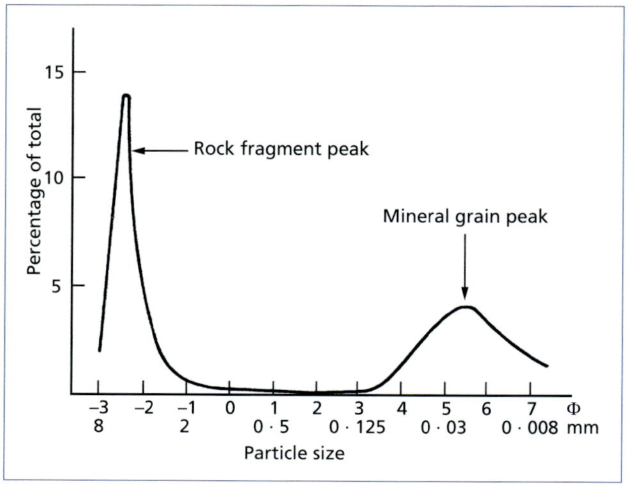

Figure 14.7 Bimodal size distribution of till particles (Dreimanis and Vagners, 1971).

14.1.4 Landforms resulting from glacial deposition

Deposits linked to ice mass activity cover approximately 10% of the continental surface of the Earth (Price, 1973). These deposits can be thick, as in the Spokane Valley (United States) where they measure 400 m in thickness (Flint, 1971). They can also be merely a thin film of deposits that fossilize on the substrate. Moraines are composed of till or a mixture of till and fluvioglacial deposits. These accumulations are different

Figure 14.8 Till deposit showing the distribution of large fragments. Head of the Gállego River (Aragonese Pyrenees).

types of landforms depending on glacial dynamics, their position with respect to the ice, and the processes that formed them.

Depending on their location with respect to the glacier, they are classified as **terminal, lateral, and medial moraines**. The first two types show advance or retreat positions of the glacier; medial moraines *(Fig. 14.9)* occur at the intersection of two valley glaciers. Medial moraines are generally superficial and thin. Lateral moraines can be attached to the rock wall next to the glacier; they sometimes form ridges separated from the valley wall *(Fig. 14.10)*. In both cases, their sediment supply comes primarily from debris along the walls. In some cases, lateral moraines unite with terminal moraines to create moraine ridges or arcs, which can dam up meltwater and create ice-dammed lakes. These moraine ridges are easily eroded by fluvial and fluvioglacial water, leaving a group of small ridges; the remains of these terminal moraine arcs may show vestiges of various stages of glacial retreat.

Determination of moraine types can also be based on ice mass activity and on the subglacial and marginal environments in which they formed (Chorley *et al.*, 1984) *(Fig. 14.11)*. Subglacial landforms form under an active temperate glacier where the melting base of the glacier supplies particles. **Till sheets**, also known as **ground moraines,** are large landforms. They are broad till-covered plains where till thickness ranges from a half meter to tens of meters; they can cover surfaces as large as thousands of km^2 (Kemmis, 1989). Today till covers broad areas of European and North American plains that were covered by ice sheets in the past. Till sheets have an irregular topography that has rock substrate outcrops, small hills, and lakes; moraines also contain lobate landforms. Some till sections are intercalated gravel and sand, indicating alternating glacial and fluvioglacial environments. Thicknesses are highly variable; the presence of a thick section is interpreted as representing a stationary period of the glacier whereas a thin layer may indicate rapid ice retreat. Today, water erosion is destroying till sheets; fluvial incision makes it possible to observe good sections of these sequences. Tills can also be deformed as ice advances over the sediments (Eyles and Lazorek, 2007). Till sheets result from advances and retreats of Pleistocene ice sheets and, as a result, the sedimentological and chronological study of these deposits provides valuable data on the Quaternary geology of these regions.

Other subglacial landforms are aligned parallel to ice flow. **Fluted moraines** are alternations of rectilinear ridges and furrows formed on till; they outcrop on the margins of retreating glaciers (Eyles and Lazorek, 2007). They can be tens to hundreds of meters long but they generally do not exceed 5 m in height. They form in sheltered areas where boulders have been deposited on the ice bed; this creates a

Figure 14.9 Central moraine of the upper Baltoro glacier at 5,000 m in the Karakorum (Pakistan). Photo J. López Martínez.

Figure 14.10 Moraine ridge. Worthington Glacier (Alaska). Photo P. Lucha.

zone of low pressure downstream (Boulton, 1976; Gordon *et al.*, 1992; Benn and Evans, 1996). The longitudinal profile of these forms is similar to that of elongate drumlins; some authors consider them to be linear, very narrow drumlins (Prest, 1968). One unresolved problem in their interpretation is the regular spacing between ridges in some moraines.

Drumlins are elongate hills in which the major axis is parallel to the direction of ice movement. Drumlins are shaped like an inverted spoon or an elongate semi-ellipsoid. They range in length from 100 to 1,000 m. They are about 10 m tall and their width varies directly as the degree of elongation (Menzies, 1979; Chorley *et al.*, 1984; Clark *et al.*, 2009). Their longitudinal profile has an up-gradient side with a steep dip and maximum height; their downstream end is pointed. Drumlins rarely occur in isolation. They normally occur in groups that form drumlin fields and they generally have an en echelon arrangement that can be described as "eggs in a basket" topography (*Fig. 14.12*). Drumlins occur primarily in old ice sheets in areas where thousands of drumlins are present such as the center of Finland, Northern Ireland, New York State, and Ontario (Canada) (Benn and Evans, 1998) (*Fig. 14.13*). They are normally composed of till but sometimes contain intercalated sand and gravel; they can even have a rocky nucleus.

Numerous papers and books have been written on the origin of drumlins; these theories can be summarized as two major ideas. The first theory proposes ice erosion of pre-existing till cover and the second considers drumlins to have formed from subglacial till accumulations that aligned

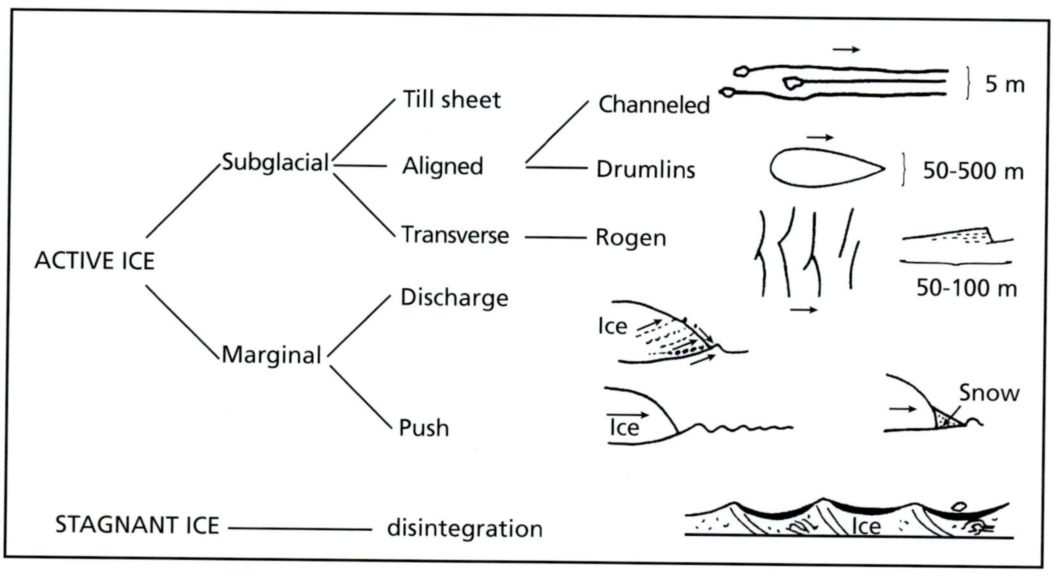

Figure 14.11 Classification of glacial moraines (Chorley *et al.*, 1984).

Figure 14.12 Eggs-in-a-basket topography generated by ice moving from right to left (Holmes, 1965).

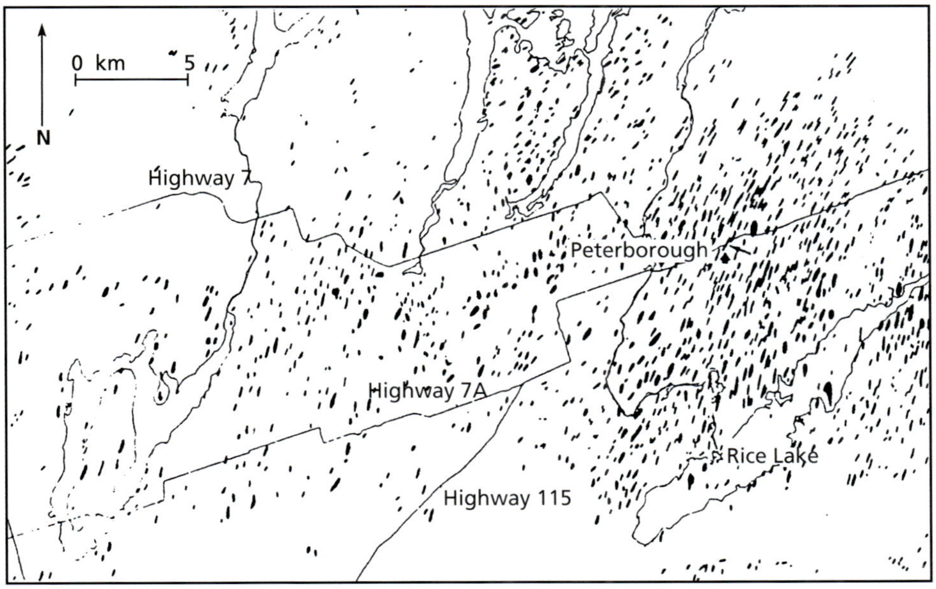

Figure 14.13 Drumlin field in Peterborough (Ontario, Canada) (Sharpe, 1987).

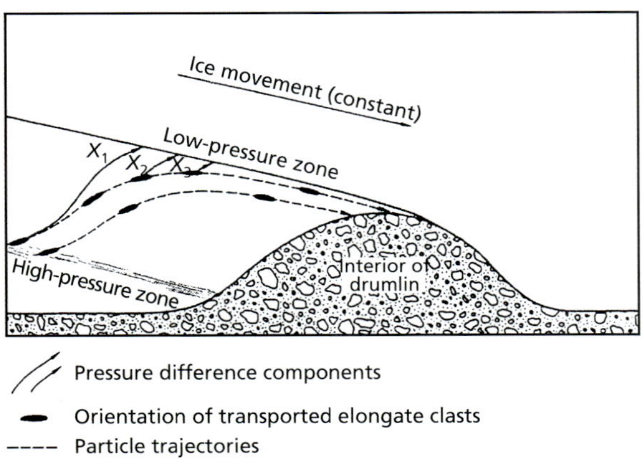

Figure 14.14 Theoretical flow model showing the trajectories of particles on the flanks of a growing drumlin (Evenson, 1971).

as landforms. The first theory involves the action of two glaciations or significant advances of ice masses in which the first glacial advance deposits material and the second erodes it (Embleton and King, 1975a).

The presence of rocky ridges or large boulders suggests that drumlins can form by accretion around boulder ridges as observed in concentrically layered drumlins. Their initial formation stage would be crag and tail morphologies. The drumlin landform is related to basal pressure variations within a moving glacier, as the orientation of clasts in the drumlin till seems to indicate. Elevated parts of the drumlin are low-pressure areas (*Fig. 14.14*) and particles are transported to these higher zones from high-pressure areas at lower elevations (Evenson, 1971). This interpretation corroborates the observation that in drumlin fields till cover is thinner in areas between drumlins. Boulton (1987) explains drumlins as a function of the rheological properties of subglacial sediments and their water content. As a consequence of glacial stresses, thick materials are slightly deformed and fine material flows easily. Some internal areas can come loose and become highly deformed, unlike areas with stable nuclei.

Drumlin concentrations in some areas seem to indicate specific conditions for their formation. Ice thins toward the glacier margin and as a result, subglacial till can reach a critical stage of dilatancy *(Fig. 14.15A)* (Smalley and Unwin, 1968). This dilatancy occurs when till particles are subjected to forces that cause them to move relative to each other and break loose from each other. This process involves expansion or volume increase and greater resistance to shear forces. Not all tills are dilatant materials and a minimum of blocks is required to make them so. This compacted till resists later forces so that the rest of the till flows around it. Stress C is critical to drumlin formation *(Fig. 14.15B)*. If it is greater than A, drumlins do not form and the till is dragged by the glacier. If stress is less than B, the till will deform continuously and finally form a terminal moraine. According to this hypothesis, drumlins originate upstream from terminal moraines.

However, and in spite of numerous mechanisms suggested for drumlin formation, the genesis of these landforms remains problematical (Clark et al., 2009). They are quite possibly polygenetic landforms. Furthermore, researchers have not observed active drumlin formation.

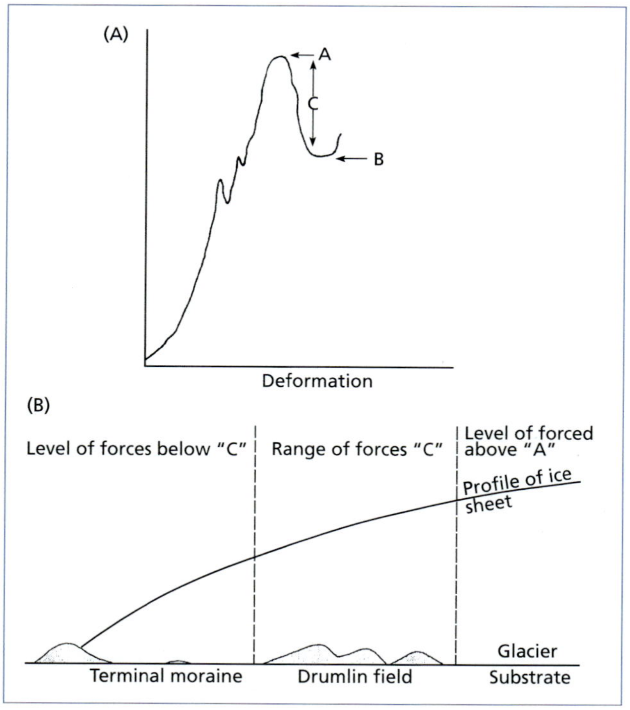

Figure 14.15 (A) Load-deformation curve for glacial till. (B) Cross section of a glacier edge showing forces involved in drumlin formation (Smalley and Unwin, 1968).

Another type of moraine that forms in active ice and in subglacial environments is the **Rogen valley moraine** named for the Swedish lake where they were first described. These are asymmetrical moraine ridges perpendicular to ice flow with a low-gradient slope upstream. They are arc-shaped and slightly concave upwards to ice movement. They range in length from 100 to 3,000 m and from 5 to 10 m high. The glacial till is intensely deformed and has imbricate structures verging downstream (Lundquist, 1989). Rock splinters are also present within this fabric (Moran, 1971). All this indicates intense compressive flow below ice that exerts pressure toward the margin; this pressure intensely and rapidly deforms the basal till and forms transverse belts (Price, 1973; Finlayson and Bradwell, 2008). Another possibility is that till pressure builds a transverse moraine under a water column of a marginal lake (Andrews and Smithson, 1966).

Ice dumped moraines *(Fig. 14.11)* form from material that accumulates from melting ice that has been transported in various environments within the glacier. Consequently, these deposits reflect multiple processes. The material is supraglacial or englacial in origin; it reaches the surface along shear planes and slides along the ice surface or is deposited by melting ice (Sugden and John, 1976). These ridge-shaped moraines are located at glacier margins. Like medial moraines and lateral moraines, marginal moraines contain transported material *(Fig. 14.10)*; they can combine with frontal moraines to generate broad moraine ridges. These moraines generally show no preferred orientation of rock fragments although some inherited fabric can be observed.

In marginal environments, glaciers that end up on terra firma discharge significant volumes of materials at their margins. The amount of particles deposited depends on the volume of solid material transported by the glacier; the size of the moraine accumulation is also a function of the time that the glacier remained in the same place. For example, the moraine ridges of the Franz Josef Glacier (New Zealand), which is located on the coastal plain at the foot of mountains, reached a height of about 430 m during a stationary stage.

Push moraines *(Fig. 14.11)* are located at frontal margins; they are the remains of ice advances that have dragged previously deposited glacial material. This *bulldozer* effect is accompanied by the formation of small asymmetrical moraine ridges that do not exceed two meters in height; the sides facing the ice have a higher gradient. These deposits are composed of glacial till and often fluvioglacial material. Pushing by the ice causes folds and imbricate reverse faults (Sharp, 1988; van der Wateren, 1992) *(Fig. 14.16)*. Glaciers are known to advance in winter, which is a period of minimal ablation; these annual advances of a continuously retreating glacier create small ridges whose spacing is related to the ablation balance (Hewitt, 1967; Worsley, 1974). These ridges are called **annual or De Geer moraines** (Hoppe, 1959). Glaciers

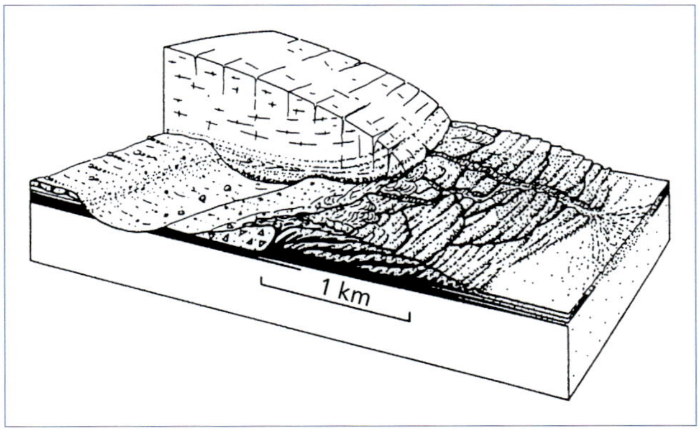

Figure 14.16 Simplified block diagram of the Holmtrömbreen (Spitzbergen, Norway) push moraine showing faults and recumbent folds. The moraine is composed of fine-grained sediments (in black) and tills (triangles). Exaggerated vertical scale (Van der Wateren, 1992, in Ehlers, 1996).

with spasmodic flow create large push moraines during their advances and also much larger moraine ridges with more marked internal deformation (Evans and Rea, 2003).

Melting of inactive ice forms **disintegration moraines** (*Fig. 14.11*). As long as ice is still present in these inactive glaciers, higher zones are covered by debris that protects the underlying snow from melting; particles flow or slide toward depressions. The disappearance of the ice causes inverted relief, with older accumulations settling into depressions (Boulton, 1972). The assemblage consists of a *hummocky* moraine characterized by undulating terrain composed of small hills and closed depressions (lakes).

Figure 14.17 Subglacial melting at the ice front in the Darwin Cordillera, Tierra del Fuego. Photo J. López Martínez.

14.2 Fluvioglacial erosion and sedimentation

14.2.1 Meltwater

Meltwater is part of the glacial system; it is the main product of glacial ablation (*Fig. 14.17*). In addition to the role of meltwater in glacial velocity and particle flow in the subglacial medium, it also plays a significant role in erosion and deposition, particularly outside ice masses, as shown by rapid changes in drainage networks (Price, 1980) (*Fig. 14.18*). Melting increases with distance from the equilibrium line. Maximum melting occurs in distal parts of the glacier as shown by the superposition of a fluvioglacial subsystem of the glacier; in this case, the intensity of glacial geomorphic action becomes subordinate to fluvioglacial activity in the final parts of the ice mass (Sugden and John, 1976).

Meltwater can be surficial or internal; it is much more significant at the surface. Surficial meltwater decreases with altitude. Other meltwater agents are summer rainwater, water from adjacent valleys, and groundwater (Benn and Evans, 1998). Basal and internal water come from two sources. The first is geothermal heating, a medium capable of melting a 6 mm-thick sheet of ice every year. The other source is basal sliding and internal deformation, which produces enough heat to melt a 10–15 m-thick layer per year (Walder, 1982). Maximum melting occurs in temperate maritime environments; it decreases toward high latitudes and toward the interior of continents.

Melting in ablation areas forms channels (*Fig. 14.19*) that occupy various positions inside and outside the ice mass; these channels can travel highly variable distances (Brodzikowki and Van Loon, 1987). Depending on the thermal regime, water can flow over the surface or through the

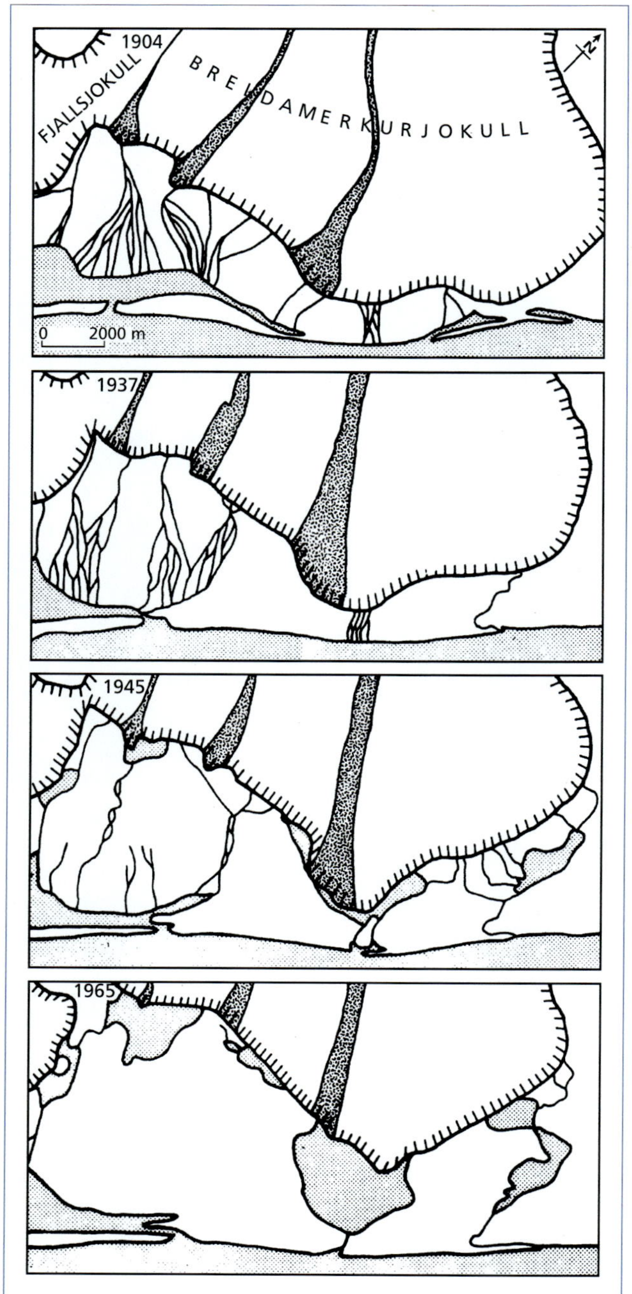

Figure 14.18 Evolution of proglacial drainage in Breioamerkurjökull (Iceland) (Price, 1980).

interior of the glacier. Channel sizes range from meters to kilometers. In **marginal** *(Fig. 14.20)* and **submarginal channels**, meltwater drains along the contact between the ice mass and the rocky slope in a lateral or frontal position relative to the glacier. These channels are short, rarely exceeding 2 km in length. **Supraglacial channels** can develop on top of ice sheets; they generally form dendritic networks (Ferguson, 1973; Dyke, 1993). Channels are straight or meandering and have depths that range from a few centimeters to several meters *(Fig. 14.21)*. Supraglacial channel water can disappear abruptly into cracks, which gradually widen and become cylindrical wells called **moulins** that create glacial karst topography (Stenborg, 1970).

The system of **intraglacial** and **subglacial channels** *(Fig. 14.22)* has internal drainage similar to soluble rocks that host large galleries and lakes *(Fig. 14.23)*. Melting through glacial cracks forms this set of internal conduits in glaciers whereas dissolution generates conduits in karst. If formed by forced or phreatic flow, gallery cross-sections are circular but if flow is vadose or free, the conduits are elongate in the vertical direction. Internal network conduits generally widen in summer as a result of higher temperatures and a greater water supply to the system (Pohjola, 1994). Tracer studies done in Storglaciären (Sweden) show that intraglacial drainage is braided (Hooke *et al.*, 1988). Subglacial cavities sometimes store water in rock basins; geophysical methods have been used to study spectacularly large (180 × 45 km) rock basins in Antarctica (Robin *et al.*, 1977). Meltwater that drains outside an ice mass forms **proglacial channels** *(Fig. 14.24)*.

The **discharge** of these fluvioglacial environments varies enormously over both short and long time periods. Diurnal fluctuations are demonstrated by low flow in the morning and an abrupt increase at the end of the afternoon. Seasonal variations show minima in the winter and maxima in the summer (Menzies, 1995a). Flows from the Argentière Glacier in the Mont Blanc massif are 0.1–1.5 m^3/second in winter and 10–11 m^3/second in summer (Boulton and Vivian, 1973). Surging glaciers occasionally produce spectacular drainage or sudden drainage of subglacial lakes and marginal dammed lakes (Benn and Evans, 1998). These large flows cause significant floods called **jökulhlaup** in Iceland; they are characterized by high discharges for a few hours followed by a rapid decrease. One of the best-known examples is Lake Grimsvötn located on top of a volcano west of the Vatnajökull Ice Cap. About once every ten years it drains suddenly with a flow of 40,000–50,000 m^3/second and floods about 1,000 km^2 of a gravel and sand plain (sandur). Its normal flow is approximately 400 m^3/second (Thoraninsson, 1953a; Björnsson, 1992).

The upper Pleistocene geomorphic record contains surprising cataclysmic **paleo-floods** in the Altay Mountains (Siberia) (Baker *et al.*, 1993; Rudoy, 1998; Carling *et al.*, 2002; Herget, 2005). Some paleofloods were discovered recently and others have been known for years, such as Lake Missoula in the northwestern United States (Bretz, 1923). The Spokane flood is an example of a flood that has been recognized for years (Baker, 2008a,b). Maximum flows in these floods were several million cubic meters per second. Advancing ice cap

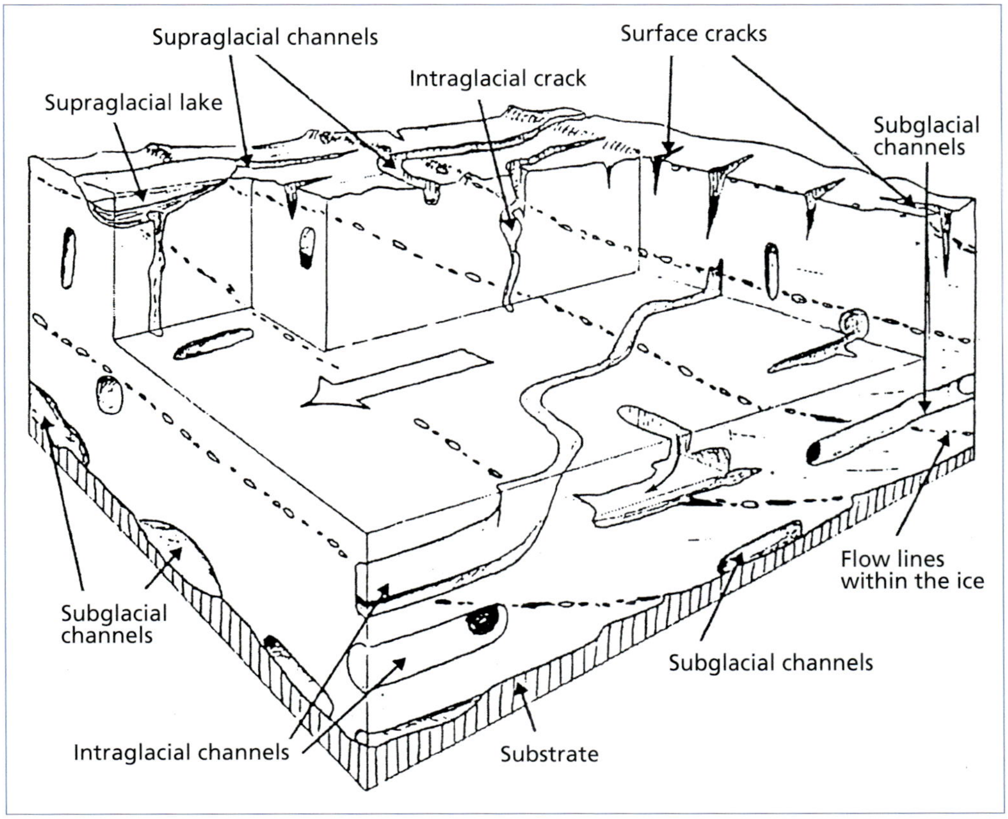

Figure 14.19 Schematic block diagram showing the various types of channels in a temperate glacier (Brodzikowski and Van Loon, 1991).

Figure 14.20 Marginal channel and lateral moraine. Svinafellsjökull (Iceland). Photo F. Gutiérrez.

Figure 14.21 Meandering supraglacial channel in the Southern Shetland Islands, Antarctica. Photo J. López Martínez.

lobes can block drainage and create dammed lakes that lead to flooding. Basal rupture may cause the dam to rupture, which liberates water suddenly and unleashes a gigantic flood.

Lake Missoula (Baker and Bunker, 1985; Baker, 2007) (*Fig. 14.25*) was located in northwestern Montana. The Purcell lobe of the Cordilleran Ice Sheet created it during Wisconsin glaciation. The lake covered a surface area of 7,500 km^2 to a maximum depth of 600 m; it had a water volume of 2,500 km^3. All these data were estimated from coastline research. The ice rupture created a peak discharge of 21.3 million m^3/second, about one hundred times the average discharge of the Amazon or about 20 times the average discharge of all the rivers in the world. Velocities reached 70 km/hour in channels. The study of deposited material has helped researchers identify 40–60 catastrophic floods between 15,300 and 12,700 years ago. The flood extended west as braided channels along the basalt plateau of the Columbia River; water was 300 m deep in the Columbia River gorge. On the plains it eroded most of the loess cover and created a channelized denuded rock landscape (*Channeled Scabland*) (Bretz 1969; Baker, 1981b). In gravels, it produced giant ripples with 100 m wavelengths and heights of 5 meters and large potholes and rock basins up to 30 m deep. The Missoula floodwaters flowed onto the continental platform of the Pacific as channels and associated turbidity currents (Normack and Reid, 2003). In the Columbia River Gorge, discharges calculated at 10 million m^3/second (Benito, 1997) caused spectacular erosional and depositional

Figure 14.22 Subglacial channel in the Biafo Glacier located at 3,200 m. Karakorum Mountains (Pakistan). Photo J. López Martínez.

Figure 14.23 Intraglacial conduit and lake discovered after surface collapse of the glacier. Baltoro Glacier, Karakorum (Pakistan). Photo J. López Martínez.

Figure 14.24 Proglacial channels, marginal lakes, plain and sandar valley in the Skeidararjökull Glacier (Iceland). Photo F. Gutiérrez.

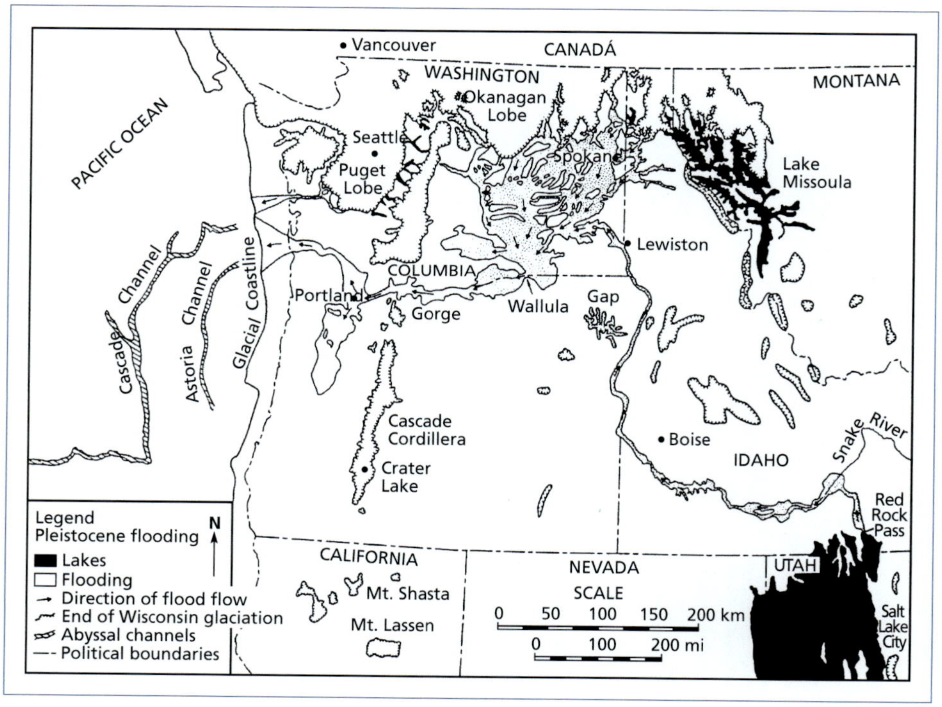

Figure 14.25 Regions of the northwestern United States affected by catastrophic flooding at the end of the Wisconsin glaciation (Baker and Bunker, 1985).

landforms. Other paleo-floods have been studied on the Snake River, a tributary of the Columbia, which overflowed Lake Bonneville, the precursor to the Great Salt Lake of Utah (O'Connor, 1993). Lake Agassiz on the Canadian Shield overflowed repeatedly toward the Gulf of Mexico, the Arctic Ocean, and Hudson Bay (Teller *et al.*, 2002).

14.2.2 Landforms resulting from fluvioglacial erosion

Water in meltwater channels can transport high sediment load and cause significant erosion. This water can move at velocities as high as 8–15 m/second. When subject to high hydrostatic pressure they can cause cavitation of the rocky bed (Drewry, 1986). Meltwater carries a large amount of material in suspension and in bed load. In Norway glacial meltwater carries 3,800 mg/l of suspended load; bed load can be 25% of the total peak flow load in the summer. This water also carries a small dissolved load; in the Argentière glacier (Alps) the dissolved load exceeds the suspended load during the winter (Sugden and John, 1976).

Various features are attributed to meltwater activity, such as **P forms**, or sculpted surfaces (Kor *et al.*, 1991). They are discussed in Section 13.2.2.2 on glacial erosional landforms. Controversy surrounds their interpretation as glacial or fluvioglacial landforms although most researchers consider them fluvioglacial (Allen, 1971). The origin of P-forms can be considered using a number of criteria. (1) If P-forms are on the glacier bed on an arrangement unlikely for the course of subglacial meltwater, they may have a fluvial origin. (2) If the glacier bed is massive and P-forms indicate the flow of ice, then they are generated by ice erosion (Rea et al., 2007). Potholes are also typical forms developed in the bed of these melt channels, which can reach depths of 20 m and widths of 15 m. They are usually deeper than wide and have spiral grooves produced by clasts on a turbine movement (Kor et al., 1991).

The geomorphic features that most reliably indicate fluvioglacial erosion are channels incised into the rocky substrate or on top of loose material produced by glacial activity. **Marginal channels**, lateral or frontal, can incise into the ice or vertically erode into the substrate, which is usually glacial till. These channels are generally short, rarely exceeding 2 km. Channels that are subparallel to topographic contours are thought to have a marginal origin although this is questionable because they can be subglacial channels that drain near the margin of a glacier (Price, 1973). **Subglacial channels** can incise to a depth of 100 m and can be 1–2 km wide and 75 km long (Selby, 1985). They usually have steep walls and flat bases with some irregularities. They often terminate abruptly; it is thought that they originate at the ice

Figure 14.26 Proglacial Lake. Jókulsarlon (Iceland). Photo F. Gutiérrez.

Figure 14.27 Proglacial channels on the Isle of Marambio that discharge water into the Weddell Sea. In the background are several icebergs. Antarctica. Photo A. Martín-Serrano.

outlet that creates an alluvial fan. They are interpreted as subglacial because of their anomalous positions within the drainage system. When the ice margin has enough slope to evacuate meltwater, channels form at greater distances from the glacier margin. These **proglacial channels** direct their water to larger-order collectors or to lakes (*Fig. 14.26*) and seas (*Fig. 14.27*), creating a flow that has significant erosional capacity. **Overflow channels**, which drain excess lacustrine water, are related to glaciolacustrine playa or delta deposits.

14.2.3 Landforms resulting from fluvioglacial deposition

As previously discussed, the fluvioglacial environment is characterized by a wide fluctuation of discharge and, consequently, of transport capacity. This fluctuation can be seen as both abrupt changes in particle size and rapid variations in sedimentary structures, both laterally and vertically. In addition, these variations often result in fluvioglacial landforms that are discontinuous in nature. Continuous erosion that modifies these landforms in areas of glacial retreat can emphasize these discontinuities (Menzies, 1995b, 2002). In addition, fluvioglacial materials often mix with glacial till deposits; fluvioglacial sediments are often reworked glacial till (Flint, 1971). These observations show that the environment is in constant and profound change and that erosion and sedimentation substantially change the morphology of these areas.

Transport and sedimentation mechanisms are essentially identical to those operating in other fluvial, lacustrine, and marine environments. Ripples, dunes, and cross stratification are common in sand deposits in fluvioglacial channels. Sands alternate with gravel when current energy is higher. Sediments are sorted primarily by seasonal and annual variations and to a lesser extent by daily variations. Silt and clay size particles travel greater distances and are generally deposited in glaciolacustrine and glaciomarine environments (Price, 1973).

Lacustrine sedimentation occurs in subglacial, marginal, and proglacial lakes. These contain basal layers representing distal sedimentation of fine particles and frontal and top layers that show the contribution of coarser proximal material.

Fluvioglacial channel deposits usually contain clasts that are more rounded than till clasts but this distinction is not a conclusive criterion because some fluvioglacial deposits have undergone shorter transport and thus less rounding of clasts. In addition, particle size distribution in fluvioglacial channels differs from that of till. The distribution is not bimodal because, unlike tills, silt and clay particles have been exported out of the fluvioglacial channels.

Fluvioglacial sediments generally have a deformed primary fabric called a **glaciotectonic,** or **cryotectonic structure** (Aber and Ber, 2007; Evans, 2007). The melting of ice buried under these deposits causes collapse and subsidence that create secondary sedimentary structures

Figure 14.28 Glaciotectonic structures in fluvioglacial deposits. Isaba (Navarre Pyrenees, Spain).

(Fig. 14.28) such as vertical stratification, folds, and faults. Diapiric structures, dike injections, convolutions, and others are also present; they are related to saturated fine sediments that are affected by overloading (van der Wateren, 1995).

Fluvioglacial deposits can be identified (Price, 1973) on the basis of their sedimentary environment by noting differences between meltwater channel deposits and those formed in confined water such as lakes and seas. They can be subdivided according to their location relative to the ice mass (Table 14.1). Proglacial deposits are deposited at some distance from the ice margin; they have smoother morphologies and the sedimentary structures are not deformed. Sediments deposited in contact with ice have distorted structures; the ice mass or later settling after the melting of buried ice produces these landforms. They can also be classified according to the dominant sediment type and by the shape of the accumulation.

Eskers (*osar* in Scandinavia) are ridges composed of stratified deposits that contain rounded gravel and sand and a few boulders. Eskers are oriented essentially parallel to the regional direction of ice movement. They can be sinuous or straight; they have sharp or planate crests that are sometimes discontinuous. They range in length from a few meters to 400 km, with interruptions (Aylswoth and Shilts, 1989; Punkari, 1997) (Fig. 14.29). They can reach heights of 200 m and widths of 3 km. The latter two dimensions depend on their total length; longer eskers tend to be narrower and taller. They can form single ridges or an interconnected network of confluent and diffluent eskers that come together at a principal ridge. The angles of esker slopes are about 30°, similar to the sedimentary angle of repose although as they weather their slope decreases to 5 to 10° (Embleton and King, 1975a). The slopes dip toward the esker margins and they can be distorted by the melting of underlying ice.

Eskers form in subglacial (Fig. 14.30), englacial, supraglacial, and marginal channels. They form primarily after the significant end-of-summer peak discharge when water velocity in the channels decreases; they can be totally or partially destroyed by new peak discharges (Sugden and John, 1976; Menzies and Shilts, 1996; Boulton *et al.*, 2007). Bottom deposits in subglacial channels can accumulate between ice walls; the cross-sectional shape of the tunnels plays a significant role in the final esker morphology (Price, 1973). However, this hypothesis of subglacial channels does not adequately explain the existence of tunnels that are hundreds of kilometers long. The origin of englacial channels is similar that of supraglacial channels. Research conducted in areas of rapid ice retreat in the Casement Glacier (SE Alaska) and the Breidamerkurjökull Ice Cap (Iceland) used aerial photographs taken in different years; photogrammetric processing of the photos showed that base profiles and esker crests have

Table 14.1 **CLASSIFICATION OF FLUVIOGLACIAL DEPOSITS AND LANDFORMS (PRICE, 1973)**

DOMINANT SEDIMENT	ENVIRONMENT	GENERAL FORM	RELATIONSHIP TO ICE	GENETIC TERM
		ICE-CONTACT	**DEPOSITS**	
Sand and gravel	Fluvial	Ridge	Marginal, subglacial, englacial supraglacial	**Esker**
Sand and gravel	Fluvial	Mound	Marginal, subglacial, englacial, supraglacial	**Kame, Kame complex = Kame and kettle**
Sand and gravel	Fluvial	Spread with depressions	Marginal	**Kettled sandur**
		PROGLACIAL	**DEPOSITS**	
Sand and gravel	Fluvial	Spread	Proglacial	**Sandur**
Silt and clay	Lacustrine	Spread	Proglacial/marginal	**Lake plain Lacustrine**
Sand and gravel	Lacustrine	Terrace, ridge	Proglacial/marginal	**Beach**
Clay, sand and gravel	Lacustrine	Terrace	Proglacial/marginal	**Delta**
Silt and clay	Marine	Spread	Proglacial/marginal	**Raised mud bottom**
Sand and gravel	Marine	Terrace, ridge	Proglacial/marginal	**Raised beach**
Clay, sand and gravel	Marine	Terrace	Proglacial/marginal	**Raised delta**

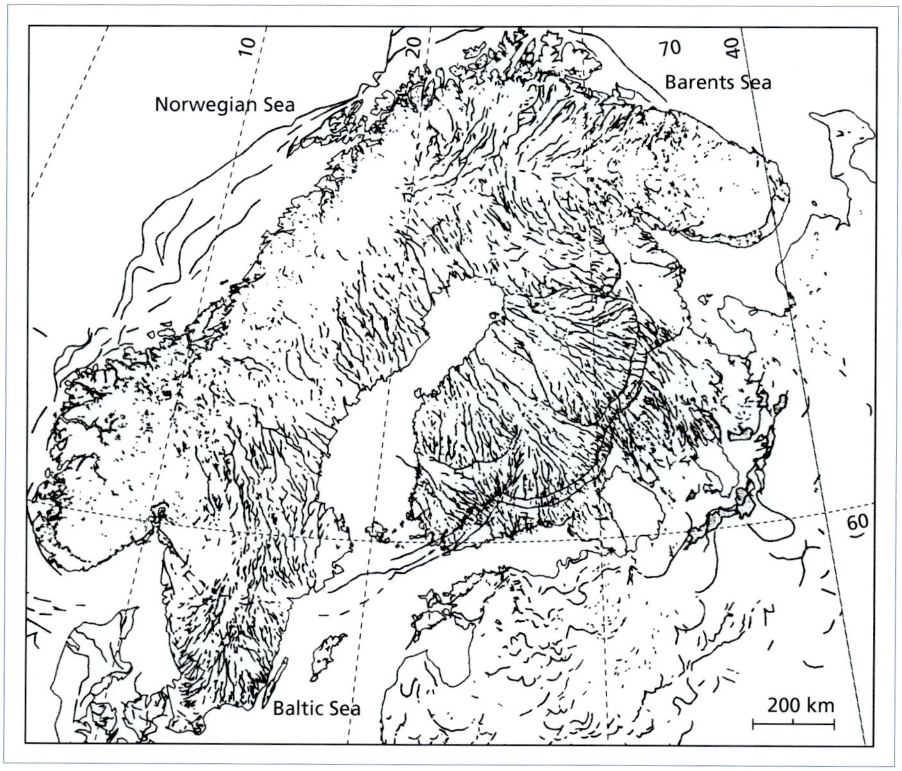

Figure 14.29 Eskers and marginal ice features in the central sector of the Scandinavian ice sheet. The eskers form well-organized systems; they are rare in sedimentary rocks areas where subglacial water can drain through permeable layers (Punkari, 1997).

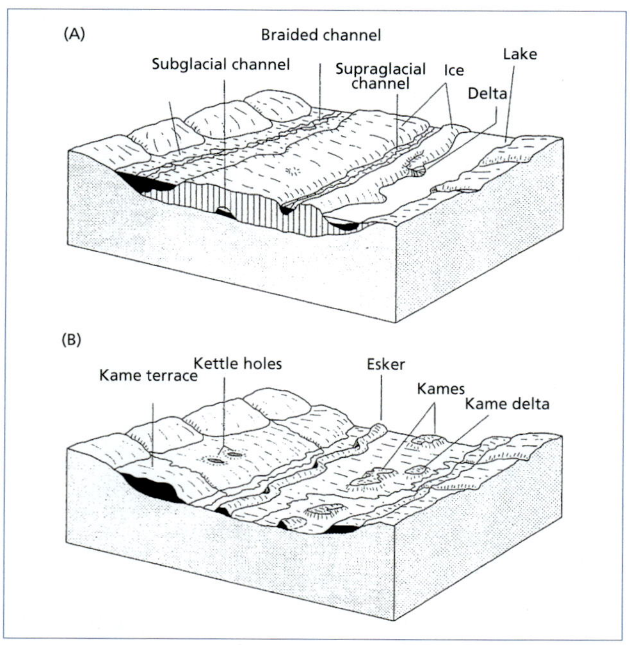

Figure 14.30 Fluvioglacial landforms formed in contact with ice. (A) Melting stage. (B) After glacial retreat (Summerfield, 1991; based on Flint, 1971).

both decreased significantly in height (Price, 1966, 1969) (*Fig. 14.31*). This reduction is due to the melting of buried ice during the period between photographs. This indicates an englacial or supraglacial origin.

Kames are small hills composed of stratified sand and gravel. They range in size from a few meters to conical hills up to 50 m high and 400 m wide at the base. They are formed by sedimentation in holes in the ice and in supraglacial and englacial cavities (Johnson and Menzies, 1996) (*Fig. 14.32*). The term "kame" is also used to describe landforms that occur in particular positions within and along a glacier. When a channel flows into a lake located in an ice mass, the sediments deposited along the margins of the lake form a **kame delta** after the ice melts (*Fig. 14.30*). The accumulation of channel deposits that flow between the valley wall and the lateral margin of the ice form **kame terraces** (*Fig. 14.33*). In kame terraces, the side of the deposit located near the ice has an abrupt margin due to ice melting. **A kame complex** is created by the accumulation of sediments in supraglacial depressions (**glacial karst topography**) and later settling due to ice melting. Kames can be modified by subsidence if they contain underlying ice; they can form subcircular depressions called **kettle holes**. A set of small hills associ-

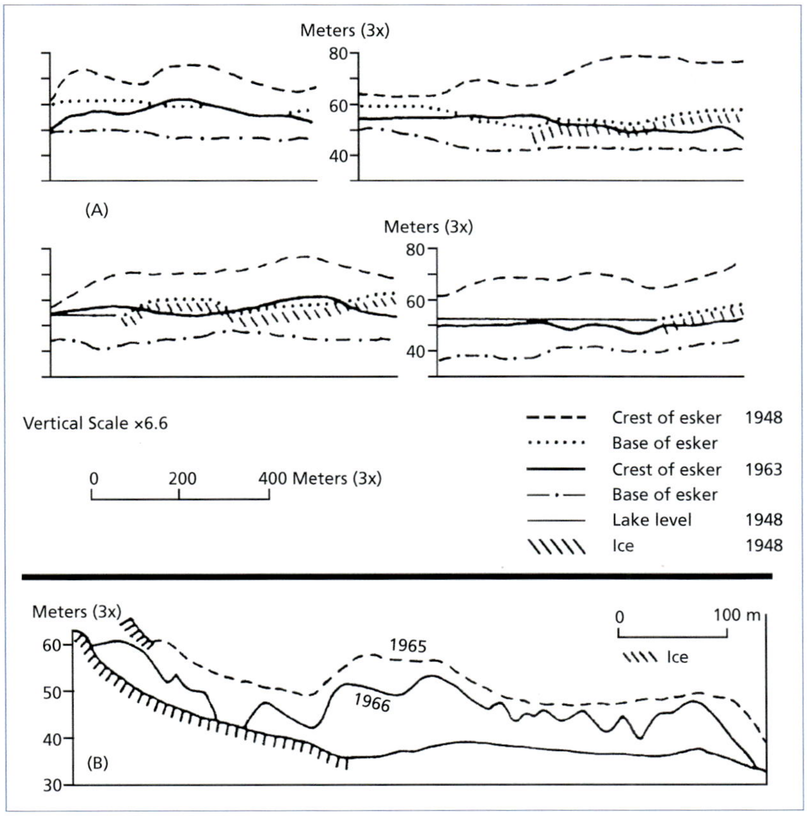

Figure 14.31 Esker profiles in the Casement Glacier (Alaska) (A) and in Breidamerkurjökull (Iceland) (B) (Price, 1966 and 1969).

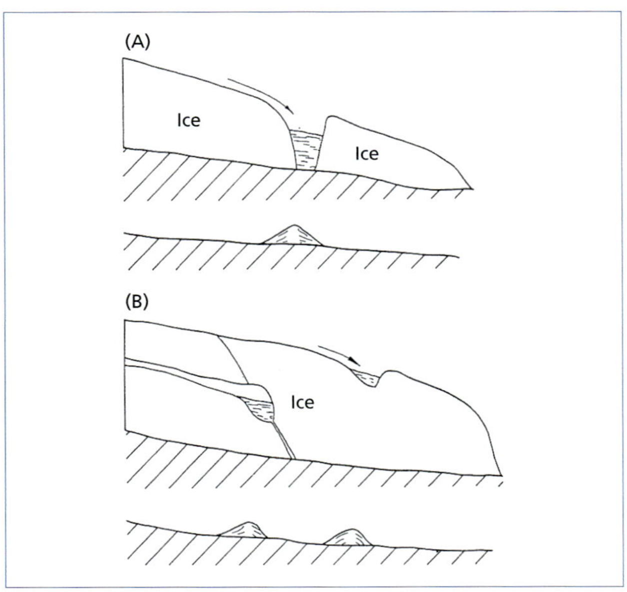

Figure 14.32 Kame formation. A: Crack open to the ice-rock interface and formation of a void in which sediments accumulate. B: Supraglacial and englacial cavities produce shapes similar to the preceding one (Price, 1973).

ated with depressions is called **kame and kettle topography** (Mickelson and Winguth, 2007), which is equivalent to a kame complex (Gray, 1991).

When ice masses discharge large volumes of water from their margins, the water transports and deposits particles far from the ice margins in various environments, creating proglacial deposits (*Table 14.1*). Fluvioglacial sedimentation predominates in these zones because of the decreased transport capacity of meltwater. The water discharges via a braided channel network that is highly mobile laterally and that deposits gravel and sand in broad sedimentary plains called **outwash** or **sandur** (**sandar** in plural) (Maizels, 1995, 1997) (*Fig. 14.24*). Outwash plains are analogous to alluvial fans but have intense seasonal variations. Particle size in sandar decreases from proximal areas of the glacier to distal areas (Boothroyd and Nummedal, 1978) (*Fig. 14.34*). Longitudinal sections are concave whereas cross sections are generally convex. **Sandur valleys** form when they are confined by glacial troughs (*Fig. 14.35*) whereas valleys that form at ice cap margins are called **sandur plains** (*Fig. 14.24*) The proximal zone of a sandur has few melt channels; it is generally a gravel surface with numerous closed depressions resulting from

Figure 14.33 Kame terrace affected by rilling. Talon River, Jura. Photo J.L. Peña.

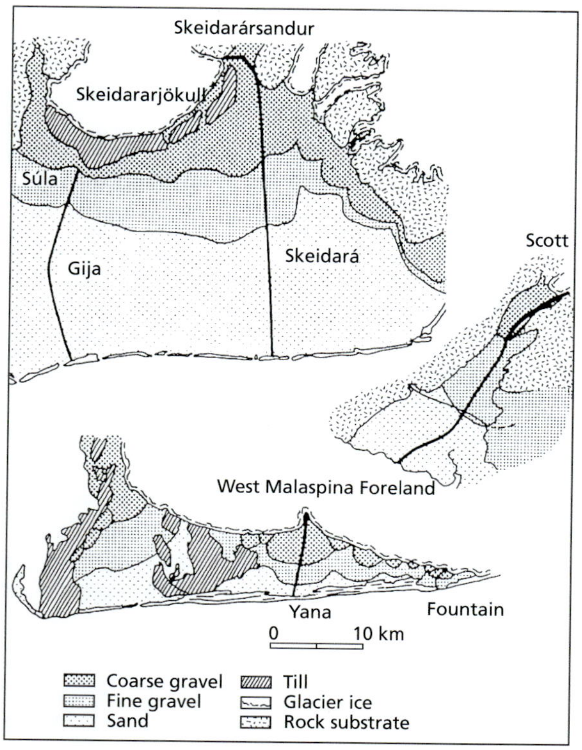

Figure 14.34 Grain size distribution in some sandar. The Skeidararsandar is in Iceland and the Scott and Malaspina systems are in Alaska (Boothroyd and Nummedal, 1978).

the melting of buried ice, which are called **kettled sandar** or **pitted outwash** (Price, 1969) (*Fig. 14.36*). The number of braided channels increases with distance from the ice mass; the channels are shallow (1–2 m), highly mobile, and usually flood during significant discharge. Finally, the distal zone is characterized by the presence of proglacial lakes. During the retreating phases of the glacier, the proximal zones of the sandur erode and new sandar develop from the previous ones.

Lakes are common in fluvioglacial environments especially in marginal and proglacial positions. The discharge of material into these lakes forms **glaciolacustrine playas and deltas**. If water later drains from the lake, the bottom deposits become **plains of lacustrine origin**. These deposits are rhythmic in nature and have seasonal variations. Called **varves** (Zolischka, 2007), they have been used in absolute chronology with very different results. Each varve is composed of a pair of layers, one formed by sand deposited during the summer and the other by silt deposited during the rest of the year. The value of the varve dating method has been criticized because more than one pair can form during a single year and other years may lack deposits due to variations in discharge or deviations of the channels that supply the lake (Smith *et al.*, 1982).

Because tributary glaciers are often dammed up by the principal glacial valley (*Fig. 14.37*), moraine deposits interfinger with marginal fluvial and lacustrine deposits (Bordonau, 1992, 1993) (*Fig. 14.38*).

Figure 14.35 Myrdalsjökull ice cap (Iceland) and sandur valley. Photo F. Gutiérrez.

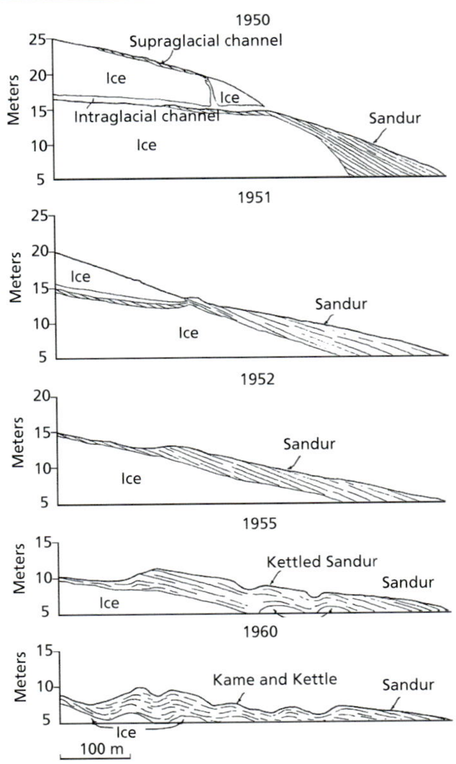

Figure 14.36 Diagram showing forms resulting from sedimentation in supraglacial and intraglacial channels. The sequence was obtained from aerial photographs of Breidamerkurjökull (Iceland) (Price, 1969).

In glaciers near the sea, meltwater discharges into estuaries, fjords, and bays to create large accumulations of glacial and fluvioglacial sediments (*Fig. 14.39*). If these sediments become emergent due to the glacio-isostatic rise that occurs after glaciation, they form **raised playas and deltas and emergent glaciomarine bottom muds** that have morphologies similar to glaciolacustrine deposits. These deposits can be identified as marine on the basis of their faunal content.

14.3 Applied geomorphology in glacial regions

14.3.1 Introduction

Present day ice mass activity and characteristics of materials deposited by glaciers and their meltwater in past epochs cause practical problems in glacial areas. Deglaciation has become a significant problem for human populations and for the sustainable development of high mountain areas, especially in semi-arid areas such as the eastern slopes of the Andes and the Cordilleras of Central Asia. In these areas, glaciers act as reservoirs because the meltwater they store provides water and irrigation. As glaciers have retreated, water has become less available (Etzelmüller and Hagen, 2004).

Some functional processes that take place in ice accumulations are accompanied by significant geomorphic risks

Figure 14.37 Lateral moraine and closed depressions in the valley of the Miera River (Cantabria, Spain).

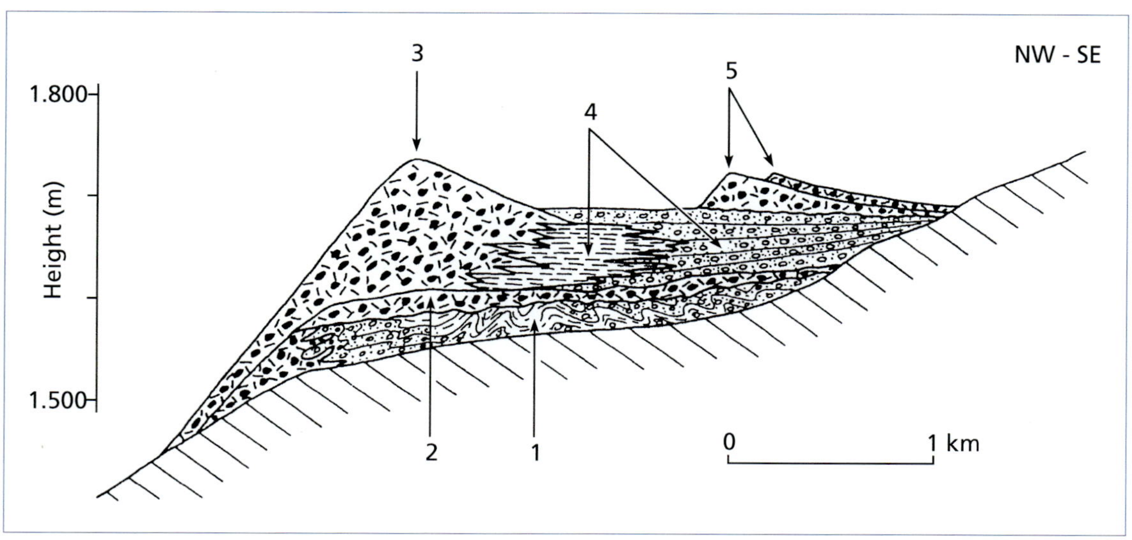

Figure 14.38 Cross-section of the transverse moraine complex at Cerler, Esera River Valley, central southern Pyrenees, Spain. (3). Lateral moraine. (1 and 2). Older deposits. (4). Lacustrine deposits with interdigitation that formed as a result of glacial damming of a tributary valley. A later advance in the tributary valley deposited younger terminal moraine ridges (5). All deposits are related to the last glaciation (Bordonau, 1993).

Figure 14.39 Melting and sediment discharge at the glacier front in the South Shetland Islands, Antarctica. Photo J. López Martínez.

(Catson, 2005) that we will analyze later. However, these areas are also beneficial to humans. Tourist activities such as sightseeing, skiing, mountain climbing, and other sports, have increased considerably in recent decades. The summer thaw in mountainous areas creates excellent waterfalls that can be used to produce hydro-electrical energy. In addition, glacial meltwater is an important source of irrigation water throughout extensive arid zones of the globe such as the Thar Desert in northwestern India, which is irrigated by the large Indira Gandhi and Bhakra canals and their subsidiary network. Other arid areas such as the coastal desert of Peru and Chile and areas of Mendoza and San Juan at the foot of the Pre-Cordillera of the Argentine Andes are irrigated by water from snow and ice melt in adjacent cordilleras (*Fig. 17.74*).

Other important resources come from groundwater found in Quaternary deposits of glacial origin. In North America, for example, a large portion of water used for domestic and industrial purposes comes from fluvioglacial materials (Thornbury, 1954). One of the most significant features of these glacially formed sequences is the wide variation of grain size, thickness, and geometry; as a result, the various layers have very different values of transmissivity, permeability, and storage (Derbyshire and Lowe, 1986). Till is a poor aquifer because of its high clay content whereas gravel and sand deposits are excellent groundwater reservoirs (Lloyd, 1983) (*Fig. 14.40*). In addition to constant lateral variations in these records, barriers to water flow make it difficult to understand their aquifer characteristics. The presence of fossilized pre-glacial valleys in these deposits can occasionally create significant water storage; their identification requires the most precise geomorphic history possible for the region.

14.3.2 Risks caused by ice mass activity

Tufnell (1984) lists three principal types of glacial risks in addition to icebergs: ice avalanches, dammed glacial lakes, and glacial expansion and retreat in mountainous areas. All three affect human populations, highways, towns, hydroelectric plants, irrigation channels, and agricultural fields (Derbyshire *et al.*, 2001).

14.3.2.1 Icebergs

The breakup of ice masses creates icebergs; this is particularly true of ice caps that reach the sea or ocean (*Fig. 14.41*). Icebergs pose a risk for navigation and for petroleum drilling platforms. The latter circumstance has led to predictive studies on iceberg movement along the east coast of Canada (Marko *et al.*, 1988) (*Fig. 14.42*). Icebergs can travels thousands of

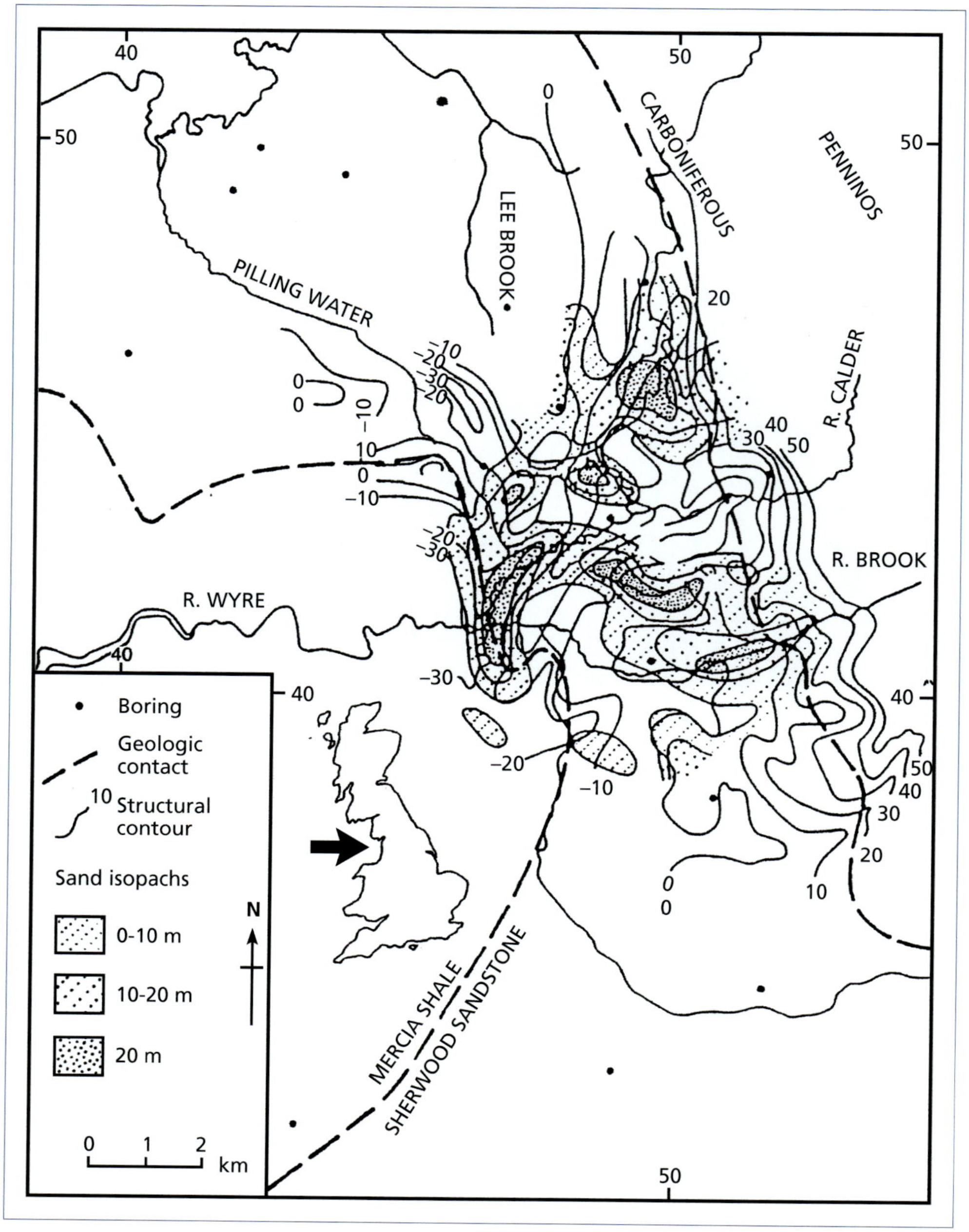

Figure 14.40 Map showing structural contour of the top of the Sherwood Sandstone, with overlying fluvioglacial sand isopachs in the Fylde area of Lancashire, United Kingdom. (Lloyd, 1983).

kilometers before disintegrating. They move by surface ocean currents but they can also be affected by deeper currents and by the wind.

Off the eastern coast of North America, icebergs from the Greenland Ice Sheet move along the Labrador Current. About a hundred or more icebergs a month cross latitude lines during period between March and August; they drift on the order of 14 km/day. They are less common below 50°N latitude. In Antarctica icebergs travel parallel to the continent in a westerly direction until they encounter eastern currents that transport them away from the coast (Hambrey and Alean, 1992).

Figure 14.41 Icebergs between Terranova and Iceland. Photo. F. Pueyo.

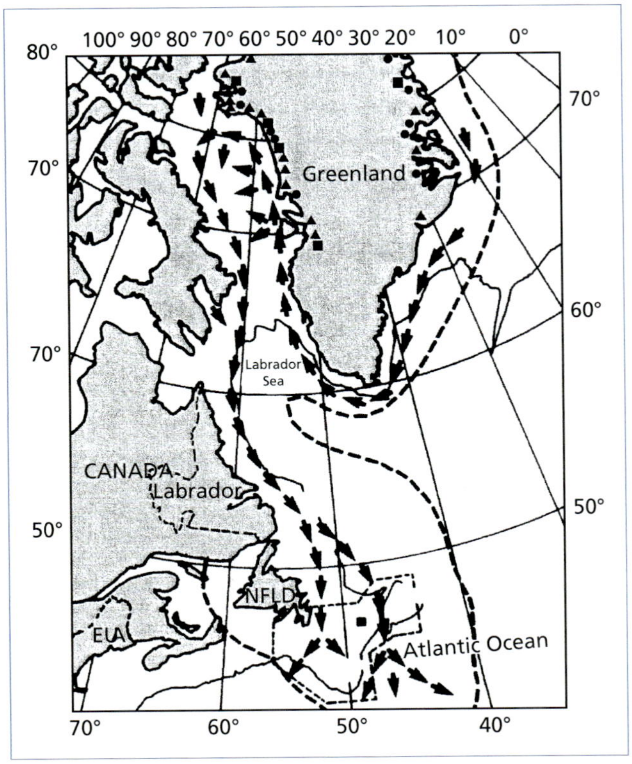

Figure 14.42 Iceberg circulation system off the eastern coast of North America (El-Sabh and Murty, 1988).

14.3.2.2 Ice avalanches

Ice avalanches occur in mountainous areas and affect towns, transportation routes, ski resorts, and hydroelectric plants. They are much less common than snow avalanches although they can occur frequently and become tourist attractions such as the Balmhorngletscher scarp in Switzerland (Hambrey and Alean, 1992). Research using aerial photographs has made it possible to determine their maximum dimensions; this data is valuable for risk determination. The movement of ice avalanches depends on the roughness of the terrain and on their triggering period (McClung and Schaerer, 1993). In winter their velocity is higher because the avalanche slides over a cover of snow that facilitates movement; in summer there is more friction and movement is slower. Ice avalanches begin as a rupture that creates new cracks and ice block slides that often precede the main movement by weeks. Ice avalanches can extend hundred of meters up the opposite valley slope.

The oldest known ice avalanche occurred in the Canton of Valais (Switzerland) on 31 August 1597; an ice mass broke loose from Balmengletscher Glacier and fell on the town of Eggen where it buried all 81 inhabitants (*Table 14.2*). In the same canton an ice mass broke loose from the Mattmarn-valle de Saas on 30 August 1965 during the construction of a hydroelectric dam. The ice avalanche came from the tongue of the Allalingletscher; it moved a million cubic

Table 14.2 **GLACIER RELATED DISASTERS IN SWITZERLAND (EXCLUDING THOSE INVOLVING MOUNTAINEERS) (HAMBREY AND ALEAN, 1992)**

Year	Location	Event	Fatalities
1595	Gietro Glacier	Lake outburst	160
1597	Balmengletscher	Ice avalanche	81
1636	Weisshorn	Ice and snow avalanche	37
1782	Altels	Ice avalanche	4
1818	Gietro Glacier	Lake outburst	40
1819	Weisshorn	Ice and snow avalanche	2
1895	Altels	Ice avalanche	6
1965	Allalingletscher	Ice avalanche	88

meters, buried the construction area, and killed 88 workers (Hambrey and Alean, 1992; McClung and Schaerer, 1993). This catastrophe was the catalyst for research on ice avalanches, especially those related to mechanical causes. Since then researchers have calculated dam strength related to wave trains generated in reservoirs by avalanche activity. Another well known ice avalanche occurred on 14 August 1949 on the Tour Glacier in the French Alps; the lower part of the glacier broke loose on a steep slope; it moved about three million cubic meters and killed six tourists. Causes of ice avalanches are attributed to the weakening of the glacier tongue due to the alternation of high daily temperatures and cold nights over the previous days, which causes significant melting and refreezing (Vivian, 1979).

14.3.2.3 Glacier-dammed lakes

These marginal lakes can form at the junction of two valley glaciers or when a glacial tongue enters a valley that lacks ice where the glacier covers the valley in a cross-wise direction (Costa and Schuster, 1988). The glacier dam causes a lake to fill slowly, especially during the summer. The lake may fill completely and the excess water may drain though supraglacial or marginal overflow channels. Water may also escape rapidly or the ice dam may rupture suddenly, causing abrupt drainage of the water reservoir and potentially catastrophic floods downstream. Water may fill and drain in annual cycles but ruptures are unpredictable (Benn and Evans, 1998).

Dammed lakes are common in polar and subpolar regions; they drain through overflow channels whereas ice dams in temperate glaciers drain water more commonly by rapid rupture of the ice mass (*Fig. 14.43*). Lakes in temperate latitudes are smaller than those in polar regions but because they are near areas of human activity they pose a greater risk; they will also cause more damage when they rupture (Hambrey and Alean, 1992).

These ice-dammed lakes usually drain periodically; this may indicate that water depth and the consequent hydrostatic pressure may be the principal control of lake drainage. Various rupture mechanisms have been proposed for ice dams. When the hydrostatic pressure of the lake water is higher than the ice pressure in the dam, the ice body can rupture at the base and the lake can drain subglacially. Another factor is the increased fracturing that results from ice flow and high hydrostatic pressure. In addition, drainage within the ice mass causes small channels to widen, which decreases the strength of the dam. In addition to these factors, seismic movements destabilize the system (Costa, 1988b).

Sudden discharges of lake water cause enormous floods, called **jökulhlaups** in Iceland; they transport immense amounts of sediments that create sandar or gravel and sand plains (*Table 14.3*) (Costa and Schuster, 1988; Maizels, 1997). This activity is often catastrophic. These floods produce two types of hydrographs (Costa, 1988b). The first is a curve that rises abruptly with a clear peak that is almost impossible to measure correctly, followed by an abruptly descending limb; the time interval between the beginning of the rise and the end of the descent can be minutes or hours. These hydrographs portray sudden ruptures of ice dams. In the other type of hydrograph (*Fig. 14.44*) the rising curve is gradual with a clear peak and an abrupt fall that corresponds to the draining of the reservoir; the time interval between the beginning and the end of the discharge is hours to days. These hydrograph features are characteristic of floods caused by the widening of intraglacial tunnels or the overflowing of lakes (Church, 1988).

Iceland provides the most classic and best-studied examples, even though volcanic activity is superimposed on glacial activity. The most spectacular cases are linked to the

Figure 14.43 Lake dammed by the Perito Moreno Glacier (Argentina) where periodic slides and overflows occur. Photo J. López Martínez.

Table 14.3 **WELL-DOCUMENTED EXAMPLES OF COLLAPSED GLACIAL DAMS THAT HAVE PRODUCED JÖKULHAUPS (COSTA AND SCHUSTER, 1988)**

Name of lake	Location	Year of collapse	Dam height (m)	Lake volume (m^3)	Flood peak (m^3/s)
Missoula	Montana, USA	16,000–12,000 a. B.P.	1,078	$2,184,000 \times 10^6$	21.3×10^6
Vatnsdalur	Iceland	1898	372	120×10^6	3000
Chong Kumdan (Shyok)	India	1929	120	1350×10^6	22650
Demmevam	Norway	1937	406	11.6×10^6	1000
Graenalon	Iceland	1939	535	1500×10^6	5000
Corner	Switzerland	1944	?	$>6 \times 10^6$	200
Gjanupsvatn	Iceland	1951	167	20×10^6	370
Lake George	Alaska, USA	1958	40	1730×10^6	10100
Tulsequah	British Columbia Canada	1958	210	229×10^6	1556
Summit	British Columbia Canada	1965	620	251×10^6	3260
Ekalugad Valley	Baffin Island, Canada	1967	120	4.8×10^6	200
Strupvatnet	Norway	1969	186	2.6×10^6	150
Hazard Lake	Yukon, Canada	1978	300	19.6×10^6	640

coexistence of both environments; these will be analyzed in section 14.3.2.4 of this chapter. However, during the recent glacial retreat following the Little Ice Age (Grove, 1988, 2004), jökulhlaups have been more frequent but have had discharge due to the thinning of ice barriers.

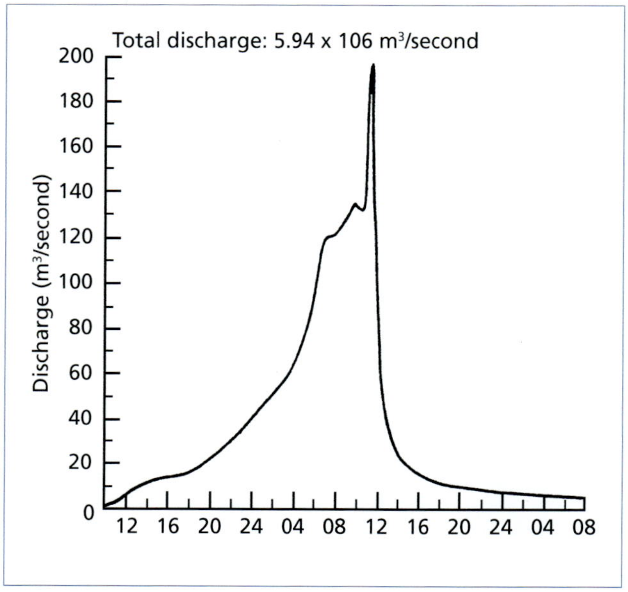

Figure 14.44 Jökulhlaup of July 1967 in Ekalugad Fjord, Baffin Island, caused by overflow of an ice dam (Church, 1988).

Numerous cases of glacial dam ruptures have occurred in the Alps. Floods have been recorded in the Otz Valley in the Alps since the year 1600 (*Fig. 14.45*). The tributary glacier Vernagt flows into the river and creates a dammed lake that has ruptured numerous times (Grove, 1988, 2004; Gerrard, 1990). Other ruptures have occurred in the Allalin Glacier (Switzerland) and the Gietroz Glacier (Switzerland), causing floods that killed 140 in 1595 and 50 in 1818, respectively (Vivian, 1979).

In addition, abundant records describe sudden ruptures of lakes located inside alpine glaciers. The most notorious catastrophe of this type took place on 12 July 1892 as a result of the rupture of a subglacial cavity in the Tête Rousse glacier on the western slope of Mont-Blanc; the flood killed 175 people and caused significant property losses. Glacial ruptures have also occurred on the Miage Glacier (Alps), causing numerous floods during the last two centuries and on the Trient Glacier (Alps) where they occur during the months of July and August and recur every three to five years (Vivian, 1979).

Terminal moraines also create dammed lakes that fill with meltwater and rain (*Fig. 14.46*) (*Table 14.4*) (Costa and Schuster, 1988). The moraines erode easily during rapid thawing of the glacier or during storms. The overflow channel incises rapidly into the deposits, quickly increasing the volume of water to be evacuated. To prevent overflowing and subsequent destruction of the moraine dam, conduits

Figure 14.45 Dates of disasters caused by the Fischbachenchängenfeld stream. Otz Valley. Tyrol Alps (Austria). Photo F. Gutiérrez.

Figure 14.46 Lake of glacial origin, dammed by a terminal moraine ridge. Cordillera Blanca, Peru. Photo J. López Martínez.

Table 14.4 WELL-DOCUMENTED EXAMPLES OF COLLAPSED MORAINE DAMS (COSTA AND SCHUSTER, 1988)

Lake name and place	Location	Date of collapse	Change in lake level (m)	Volume discharged ($m^3 \times 10^4$)	Flood peak (m^3/s)	Rupture mechanism
Madatschferner	Austria	5 August 1874	–	–	–	–
Galrittferner	Austria	7 August 1890	–	–	–	Ice Fall
Cohup	Peru	13 December 1941	–	–	–	–
White Branch	Oregon, USA	July 1942	–	–	360	–
Tempanos	Argentina	1942–1953	–	–	–	Excessive meltwater
Jancarurish	Peru	1950	21	600–1000	7000–8000	Collapse of subglacial cavity
Artesoncoda	Peru	16–17 July	113	–	–	Ice Fall
Broken Top	Oregon, USA	7 October 1966	4.6	18.9	71	Ice Fall
Squaw Creek	Oregon, USA	7 September 1970	25.6	33.3	297	–
Safuna Alta	Peru	1970	38	490 (stored in the lake)	–	Piping induced by earthquakes
Klattasine	British Columbia, Canada	June 1971–September 1973	13	170	>1000	–
Moraine nº13	Soviet Union	3 August 1977	5.2	8.64	210	Thawing of frozen soil
Mingbo Valley, Dudh Kosi River	Nepal	3 September 1977	30	490	1100	Thawing of ice nuclei? Excessive runoff
Nostetuko	British Columbia, Canada	19 July 1983	38.4	650	11000	Ice Fall
Dig Tsho	Nepal	4 August 1985	22.5	800	1600	Ice Fall

can be drilled to the bed to provide drainage. An example of catastrophic destruction of a terminal moraine lake occurred in the Santa Ana River Valley (Peruvian Andes) on 15 March 1941. It caused a major flood that destroyed the town of Huaraz and killed 6,000 people (Hambrey and Alean, 1992).

14.3.2.4 Volcanism and glaciers

Heat liberated by volcanic activity can cause glaciers to melt suddenly. The resulting water mixes with volcanic ash and produces **lahars** or mudflows that flow by gravity; they can affect human activity on the lower parts of the volcanic complex. Abundant meltwater can cause spectacular and dramatic floods.

Three examples in different parts of the world, very different from each other, illustrate the dramatic activity of lahars and the terrible consequences in terms of both human life and economic losses. The crater lake atop the Ruapehu Volcano (2,797 m) on the northern island of New Zealand ruptured suddenly on Christmas Eve in 1953; it melted snow and ice, creating a lahar that destroyed a railroad bridge shortly before an express train passed. The locomotive and five cars fell into the riverbed and 151 people died (Francis, 1993).

Mount St. Helens (2,950 m) in the Cascade Range in the state of Washington (USA) erupted on 18 May 1980; it had begun its activity two months previously with seismic events and small phreatic eruptions. Major damage was averted because the volcano had been carefully monitored (Crandell *et al.*, 1975; Wesson, 1983). The huge eruption caused the north wall of the mountain to rupture, reduced the height of the volcano by 400 m, and triggered ice avalanches. Ash from the explosion covered glaciers near the top of the volcano; in the following years the glaciers advanced because they were protected from solar radiation. Lahars up to 30 m long destroyed houses and forests in their paths. Some dams held back the lahars. Bridges in other rivers held back ripped-up trees and dammed up the water, causing floods and later bridge failure due to the force of the water. The eruption killed 57 people and property losses were calculated at about one billion dollars (Lipman and Mullineaux, 1981; Hickson and Peterson, 1990).

The Nevado del Ruiz volcano (5,400 m), located at 5° north latitude in the Colombian Andes, is topped by an ice cap. On 13 November 1985 it erupted a small amount of pyroclastic material. The eruption, like that of Mount St. Helens, had been predicted a year in advance by seismic activity and a phreatic explosion. In spite of its low energy, the eruption triggered ice and snow avalanches but more importantly, the ash melted the ice mass. The resulting waters carried large quantities of recent and old ash, creating lahars or mudflows that moved by gravity toward the adjacent densely populated valleys (*Fig. 4.64*). The lahars occurred in two principal phases. On the western slope they were channeled by the Lagunillas River and moved at speeds of 30–40 km/hour; they reached the town of Armero, about 60 km distant from the volcano, in the middle of the night. The lahars practically buried the town in a few minutes; 22,000 people died. In the town of Chinchiná on the eastern slope, mudflows killed 2,000 people (Parra and Cepeda, 1990; Voight, 1990).

Geothermal heating can increase temperatures at the base of temperate or polar glaciers, melt the ice mass, and create subglacial lakes as a result. The largest are probably located below the Antarctic Ice Sheet (Hambrey and Alean, 1992). The largest of all known subglacial lakes, Lake Vostok, is located under almost 4 km of ice directly beneath the Russian Antarctic station of the same name. The lake has a surface area of approximately 15,000 km^2, an average water depth of 125 m or more, and a considerable thickness of accumulated sediments at the bottom (Kapitsa *et al.*, 1996). Recent work on Lake Vostok (Siegert, 2005a, b) using reflection radar indicates the presence of about 100 subglacial lakes. Lake Vostok is 250 km long and 50 km wide. Two explanations of its origin have been proposed: melting of underlying ice (Kapitsa *et al.*, 1996) or a proglacial origin related to a lake that survived Antarctic glaciation for about 15 million years (Duxbury *et al.*, 2001).

The first explanation seems more viable because it relies on geothermal heating to melt the ice. Lakes present below small ice caps in Iceland have a well-documented history of volcanic activity below glacial ice and of resulting floods. Ten percent of these ice caps occur in areas of active volcanic rifting. The Grímsvötn subglacial lake near the center of the Vatnajökull ice cap is located in a geothermally active zone that has produced multiple eruptions throughout recorded history (Thorarinson, 1953a). A lake rupture generates enormous floods (jökulhlaups) on the Skeidarársandur plains located south of the ice cap. These plains are the major sandur of Iceland; they cover an area of 1,000 km^2 and have a 1/200 slope. Twelve floods have occurred; on average they happen every ten years. The average flood volume has been 3–3.5 km^3 and the maximum discharge 10,000 m^3/second. Each jökulhlaup has transported about 30 million tons (Bjornsson, 1979). These floods may be caused by increased geothermal heating or by an eruption; the 1934 eruption moved 150 million tons. It is difficult to build a permanent highway network on the Skeidarársandur plain because of the frequent floods.

In southernmost Iceland, the Katla Volcano located beneath the Myrdalsjökull ice cap has erupted since 1625. The last fissural eruption in 1918 had a recorded water discharge of 200,000 m^3/second. As a result of this flooding, the southern coast has advanced 200 m; its advance has been calculated at 2.2–2.5 km since 1660 (Bjornsson, 1979).

14.3.3 Problems caused by glaciotectonics

Glaciotectonics is defined as the glacially induced structural deformation of rocks and sediments as a direct result of glacial movement or glacial loading (Aber *et al.*, 1989). Glacial activity can cause locally severe deformation of existing geological formations (Harris and Murton, 2005; Waller and Tackwell, 2005) including folding, fracturing, and brecciation of stratigraphic sequences, and can reduce the mechanical strength of rock and sediments. As a result, it is important to understand this deformation because it can play an important role in mineral exploration and mining, civil engineering, and soil mapping.

In some western Canadian mines, glaciotectonics (Aber *et al.*, 1989), also called cryotectonic activity, has thrusted up to 12,000,000 m^3 of coal and created a depression at the base that is masked by later sedimentary cover. The coal has oxidized because of its near surface location; this coal is not exploited because of the resulting reduced heat value and its increased water content (Aber and Ber, 2007).

In other places cryotectonic deformation can affect the substrate and decrease its shear strength. Under these circumstances, landslides can occur preferentially in these weak materials, damaging highways and bridges; this has occurred in the upper Cretaceous shales in the South Saskatchewan River Valley (Canada) (Krahn *et al.*, 1979).

14.3.4 Geological engineering considerations for non-glaciated areas

Glacial sediments vary widely in grain size (*Fig. 14.47*), internal structure, thickness, geometry, and surface shape. These changes often occur over short distances and create significant problems in civil engineering studies (Derbyshire and Love, 1986).

In these situations, detailed geomorphic and geological maps and, if possible, maps specifically adapted for the stated purposes are used (Strachan and Dearman, 1983). Borings should also be installed to determine the thickness and geometry of these cryogenic deposits. In addition, mechanical soils tests help us understand the geotechnical properties and the behavior of these materials.

Till and fluvioglacial sand and gravel deposits are often used as construction material for highways and railroads. Glaciomarine and glaciolacustrine silts and clays are low-strength intervals in these glacial sequences; thus they can cause serious slope stability problems (Legget, 1974).

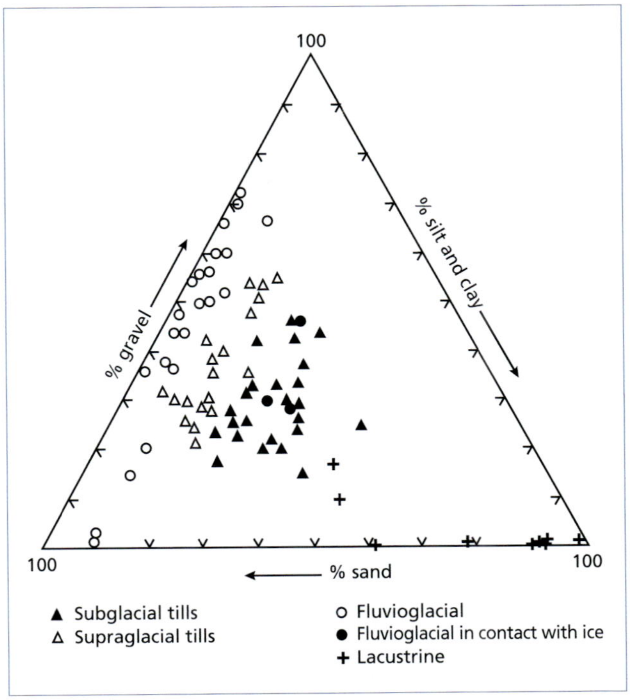

Figure 14.47 Glaciogenic sediments and grain size (Derbyshire and Love, 1986).

Glaciomarine clays that are several thousand years old can be elevated onto the continent during postglacial epochs through glacio-isostasy; their thixotropic behavior makes them extremely dangerous because of their propensity to slide. These marine sediments have a high sodium chloride content that adheres to clay particles and provides cohesion but when fresh water dissolves the salt, they lose cohesion; the clay liquefies and can flow long distances even on very flat slopes. They are called **quick clays** because of this behavior change and later flow; they are common in Canada and Scandinavia (Veder, 1981, Aune, 1983). Their high sensitivity and the fluidity of the materials mean that quick clays are highly susceptible to landslides (Lefebvre, 1996).

One of the most significant landslides linked to quick clay activity occurred in the coastal city of Anchorage (Alaska) on 27 March 1964; largest earthquake ever recorded in North America triggered the landslide. The landslide rupture appears to have been due to the presence of water-saturated glaciomarine clays (*Fig. 14.48*). The slide flowed toward valleys and marine bays and destroyed 75 houses in a residential area. In this type of landslide, damage is commonly most severe in areas near the trench at the top and along the "push ridges" of the frontal zone (Hansen, 1965).

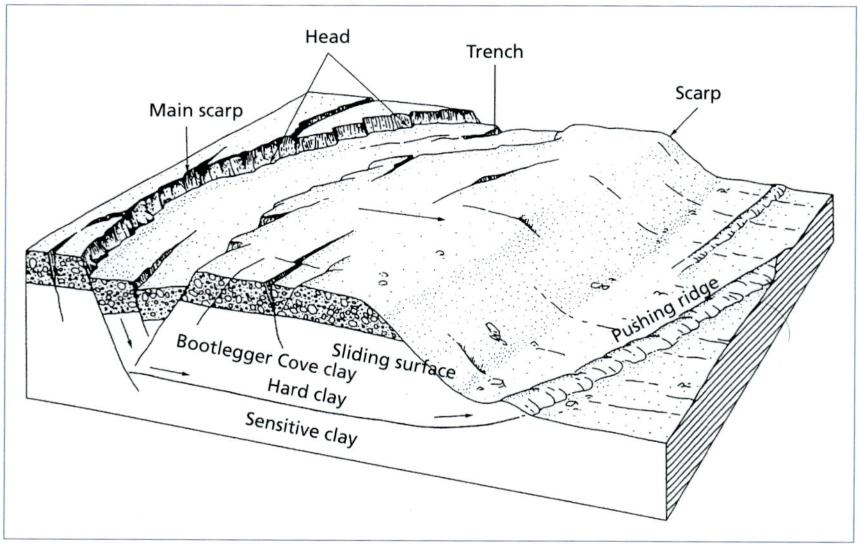

Figure 14.48 Block diagram of the Anchorage (Alaska) landslide triggered by the 27 March 1964 earthquake (Hansen, 1965).

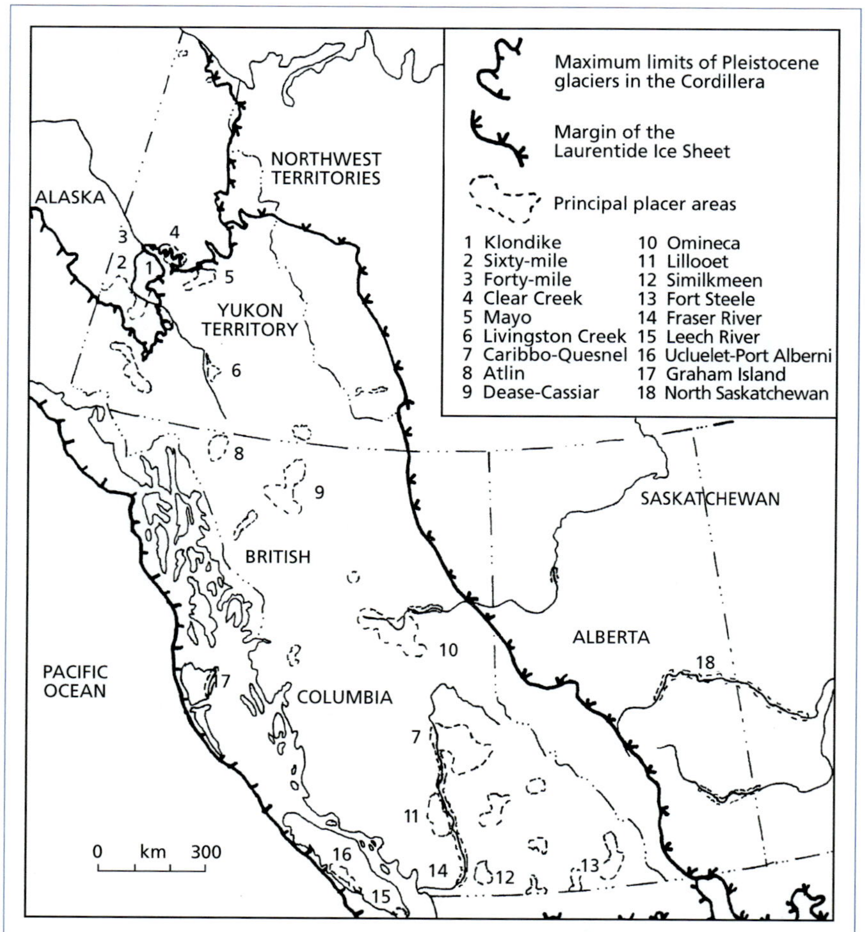

Figure 14.49 Principal areas of western Canadian alluvial placers related to the maximum limits of Pleistocene glaciation (Levson and Morison, 1996).

14.3.5 Placer deposits in glacial environments

Placers are accumulations of heavy minerals that have been weathered out of the source rock by hydrous, eolian, and glacial processes. Placers are significant economic sources of gold, diamonds, platinum, tin, and other mineral ores. Placers produced by ice activity are mined in northern Asia, Canada, and Alaska although most placer mines are found in non-glacial areas (Boyle, 1979).

Placer deposits occur in a broad spectrum of sedimentary environments including alluvial, colluvial, glacial, fluvioglacial, littoral, and marine (Levson and Morison, 1996). Alluvial placers are found in active and abandoned streams, including arroyos, braided and meandering rivers, alluvial fans, and fan deltas. Gold is primarily concentrated with clast sizes and loose blocks and in conglomeratic facies (Fayzullin, 1969).

Fluvial placers are the most frequent type; they occur in recent alluvium and low terraces (*Fig. 14.49*). In meandering rivers, they are found in *point bars* and in the internal curve of meanders. In braided rivers the distribution of heavy minerals is sporadic and discontinuous; they are localized in transverse bars, channel inflections, channel intersections, and segments where flow converges (Smith and Minter, 1980).

In alluvial fans, placers are found in interstratified gravels and debris flow sediments. They form after glaciers have begun to retreat, in places where the lack of vegetation causes slope instability. Once a slide is triggered, the deposits are eroded by glacial meltwater.

15

PERIGLACIAL GEOMORPHOLOGY

1. The periglacial environment
2. Periglacial landforms
3. Applied geomorphology in periglacial regions

15.1 The periglacial environment

15.1.1 Introduction

The Polish geomorphologist W. von Lozinski introduced the term 'periglacial' in 1909 to describe climatic conditions and geomorphic features of areas peripheral to Pleistocene ice-sheets in the Carpathians. The term was later extended to include processes and landforms characteristic of cold climates regardless of their proximity to glaciers or glaciated areas. Periglacial activity predominates in vast zones that are unrelated to glacial areas such as eastern Siberia (Jahn, 1975). The term periglacial has been used broadly because of its imprecise definition. The word 'geocryology,' which is commonly used by Russian researchers, has been suggested as an alternative term (Washburn, 1979). Geocryology is commonly defined as the study of terrestrial materials that have temperatures below 0°C, in other words permanently or seasonally frozen ground (Tricart, 1967; Washburn, 1979; French, 1996, 2007).

Periglacial environments are characterized by freeze-thaw cycles and permafrost or permanently frozen ground. Either or both occur within the periglacial environment although some periglacial landforms are not associated with the presence of permafrost. The periglacial environment is present in polar areas and in middle and low latitude alpine areas of cordilleras throughout the world (Harris, 1986). Today the periglacial domain covers about one-fifth of the global land surface; periglacial conditions covered an additional 20 percent of the Earth's surface during the Pleistocene cold periods (French and Karte, 1988).

Paleoclimatologists recognize the importance of periglacial phenomena in paleoclimatic reconstructions and of permafrost (Isarin, 1997). Ecologists and environmental scientists have pointed out the vulnerability of present-day periglacial regions as a consequence of increased human use and the implications of expected climate change (Vandenberghe and Thorn, 2002).

Periglacial environments have a wide range of climatic conditions with mean annual temperature close to or far below the freezing point and the range between maximum and minimum yearly temperatures is generally large. Total annual precipitation varies considerably from one environment to another with values between 130 and 1400 mm according to Peltier (1950) and between 50 and 1250 mm according to Wilson (1969). Tricart (1967) established a climatic classification based on temperature, precipitation, wind conditions, and their seasonal distributions. He distinguished three main types of periglacial climates:

A) **Dry climates with cold winters.** These conditions occur in subpolar areas of the northern hemisphere. These areas have low-temperature winters, short summers, low precipitation, and strong winds. This climate favors the presence of pergelisol (permafrost). The morphogenetic system of this climate type is characterized by strong frost action, limited influence of runoff, and significant wind activity.

B) **Cold wet climates with extreme winters.** This group includes arctic and mountain types. The **arctic type** is controlled by the ocean and thus has wide climate variations. Temperatures are similar to climate A but with a lower annual range. Precipitation is higher than 300 mm, violent winds are common, and permafrost is present. As a result, frost action is less intense or persistent than in climate A, snow cover reduces eolian processes, and runoff is relatively important. The **mountain type** occurs in alpine meadow areas in temperate zones. Temperatures are similar to the arctic type but with higher annual averages and smaller ranges. Precipitation is more significant than in the arctic climate. Slope and orientation play important roles in these areas. Overall frost action is significant, permafrost is generally absent, wind influence is weak, and runoff-related processes are widespread.

C) **Climates with small annual temperature ranges.** The mean temperature is close to 0°C and the annual temperature range is around 10°C. There are two types of climates with small annual temperature ranges. The climate of **high latitude islands** has significant weather instability, a small temperature range, and snow precipitation above 400 mm that minimizes the effects of wind. These climatic conditions lead to multiple freeze-thaw cycles that have limited penetration into the ground. **Low latitude mountain** climates lack seasonal temperature variations; the daily range is high and larger than the annual range. Precipitation is high except in arid mountains. These climatic conditions prevent permafrost formation and increase numerous freeze-thaw cycles and the weak penetration of frost action; there is insignificant wind activity except in arid mountains.

The periglacial environment has two principal types of vegetation: the subarctic or northern forest and the arctic tundra. There is thus a distinction between forest periglacial environments and woodland-free environments (French, 1996).

15.1.2 Permafrost characteristics

Permafrost, also called pergelisol or permanently frozen ground, is defined as a soil or a surficial deposit or rock mass that remains below 0°C continuously for more than two years; the definition does not consider other properties such as moisture content and lithology (Muller, 1945). Glaciers and ice sheets are not considered to be permafrost. Permafrost covers approximately 22% of the land surface of the Northern Hemisphere and extensive areas in the Andes and the Antarctic (Harry, 1988). Globally, permafrost underlies 14% of the Earth's surface (Washburn, 1979). Half of Canada and 80% of Alaska are underlain by permafrost (Harry, 1988). Permafrost also occurs at the bottom of shallow seas in high latitudes (**submarine permafrost**). Two types of permafrost occur on continents: **polar permafrost** and **alpine permafrost.** Alpine permafrost occurs in mountain environments and is a function of latitude and elevation. The lower altitude limit of permafrost increases from north to south (Harris, 1988).

Continuous permafrost wedges out toward low latitudes and breaks up into patches, giving way to **discontinuous** and **sporadic permafrost** (Harris, 1986) (*Fig. 15.1*). This classification is based on the percentage of underlying permafrost at a particular location. The proportion of frozen ground in continuous permafrost is higher than 80 percent, is between 30 and 80 percent in discontinuous permafrost, and is less than 30 percent in sporadic permafrost.

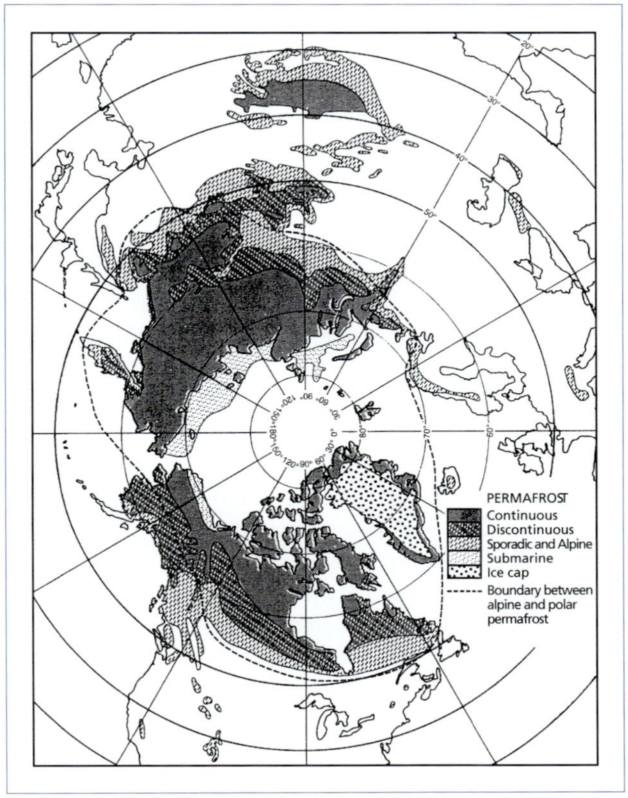

Figure 15.1 Permafrost distribution in the Northern Hemisphere (Harris, 1986).

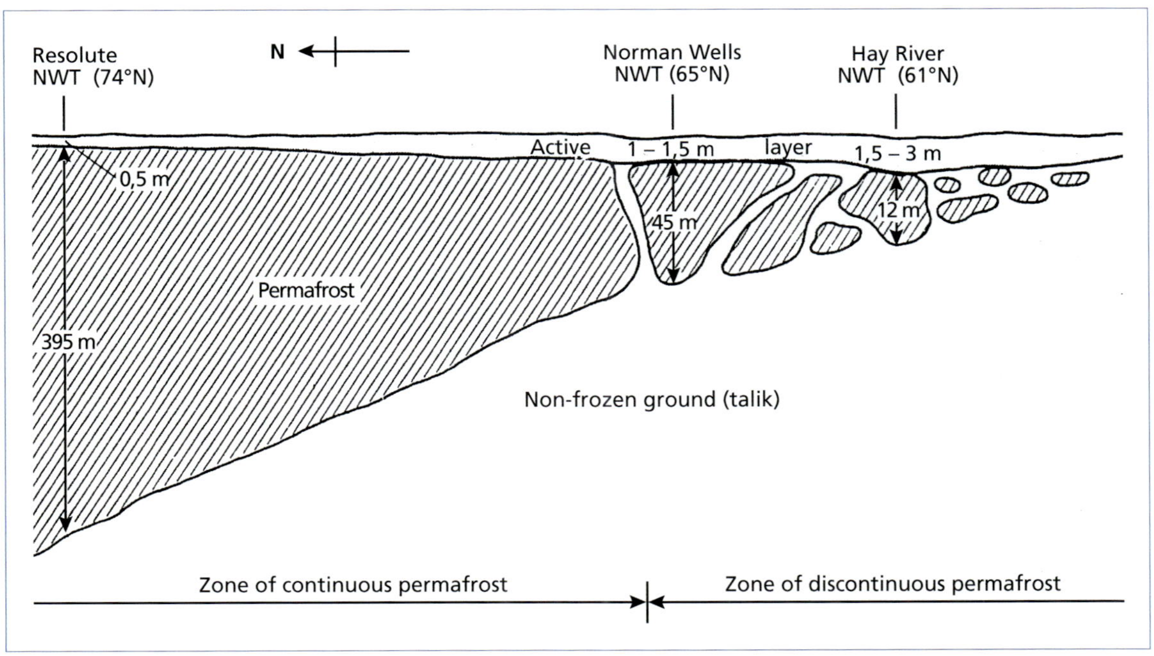

Figure 15.2 North-south cross section of permafrost in Canada showing a decrease in thickness toward the south and the difference between continuous and discontinuous permafrost (Brown, 1970).

Figure 15.2 (Brown, 1970) is a north-south cross section of Canada that shows permafrost increment with increased latitude and increased fragmentation of blocks separated by unfrozen ground, called **talik**, with decreased latitude The maximum documented thickness of permafrost occurs in the River Markha in Siberia where it reaches a depth of 1450 m (Washburn, 1979).

During summer the upper part of the permafrost thaws; this zone, which is subject to continuous cycles of freezing and thawing, is called the **active layer** or **mollisol** although some authors restrict this term to seasonally frozen ground, that is ground that remains below 0°C only in the winter. The thickness of the active layer varies from year to year. The **transition layer** is a near-surface layer that contains abundant ice, indicating a transition toward the upper zone of permafrost (Shur *et al.*, 2005). Overall, the active layer thickens at lower latitudes *(Fig. 15.2)* but does not exceed 3–4 m in depth.

Water that freezes within permafrost comes from the atmosphere, runoff, or underground flow. The frozen water cements mineral and organic particles together; ice may occur as individual crystals, particle coatings, or pore fillings. Ice that formed in pre-existing sedimentary sequences is called epigenetic; ice that forms during sedimentation is called syngenetic (Harry, 1988); an example of the latter is the advance of permafrost in deltaic and fluvial sequences.

Ice bodies in permafrost have variable geometries (Washburn, 1979). Ice can be found in pores, holes, and cavities and in both unconsolidated sediments and rocks (French, 2007). Remains of old **glacial ice** that are found buried by sediments in permafrost can be recognized by their deformation. **Icings** or **aufeis** are tabular or planar-convex bodies of fresh-water ice that form in the winter where groundwater emerges as springs or where floodwaters freeze in river valleys. They may be incorporated into permafrost upon burial. **Ice lenses** are horizontal lenticular layers of clear ice of variable thickness; they form in situ as ice segregates during the freezing process. **Ice veins** and **wedges** are vertical cracks in the ground that fill with ice. Meteoric water supplies the veins, which form wedges as they grow. The V-shaped veins taper downward and the ice crystals have a vertical arrangement (Lachenbruch, 1962) *(Fig. 15.3)*; as they expand, they deform the adjacent sediment. Ice wedges formed by epigenetic freezing are commonly 2–3 m wide and penetrate about 5–8 m. In Siberia, syngenetic ice wedges that form in alluvium can be 5 m wide and 40–50 m deep (Harry, 1988); Bray *et al.*, 2006). They frequently form part of polygonal systems. If ice wedges disappear during warming, the existing void may fill with detritus. Fossil or relict wedges are indicators of the presence of permafrost during previous epochs (Eissmann, 1981; Murton, 2007) *(Fig. 15.4)*. Finally, the ice may become the nucleus of a mound, called **pingo ice**. Pingos are massive

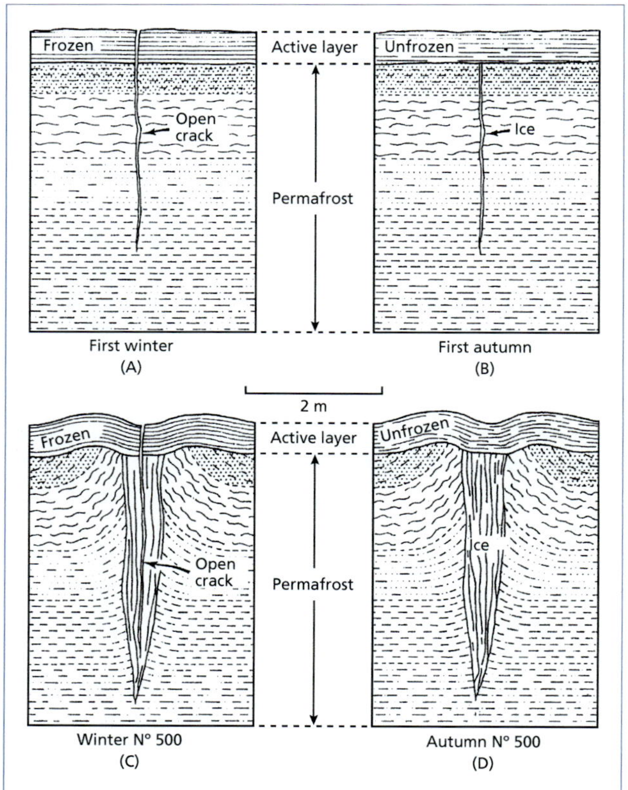

Figure 15.3 Evolution of an ice wedge (Lachenbruch, 1962).

fresh-water ice bodies with a flat convex morphology; they answer to particular hydrological and geothermic local conditions such as the freezing of meteoric water, water-rich sediments, or artesian water intruded through the permafrost.

Permafrost depth is determined by the balance between increasing heat at depth due to the geothermal gradient (1°C every 30–60 m) and heat loss at the surface. The basal limit of permafrost is the depth where the temperature is 0°C as a result of the geothermal gradient (Brown, 1970) *(Fig. 15.5)*. The active layer forms near the surface; its base is at a depth where the maximum annual temperature is 0°C. Temperature fluctuations that occur near the surface due to seasonal fluctuations attenuate at depth to zero annual thermal oscillation; this occurs between 6 and 16 m (French, 1996).

Permafrost is subject to aggradation or increased extent of permafrost or degradation over time due to changes in soil temperature (Williams, and Smith, 1989). Subtle climatic variations may affect permafrost at temperatures near 0°C. In addition climatic changes that occur on the scale of decades or centuries may significantly modify permafrost area and thickness.

Most permafrost is relict; it was formed in past Quaternary periods and is not in equilibrium with present climatic conditions. Abundant evidence corroborates this assertion

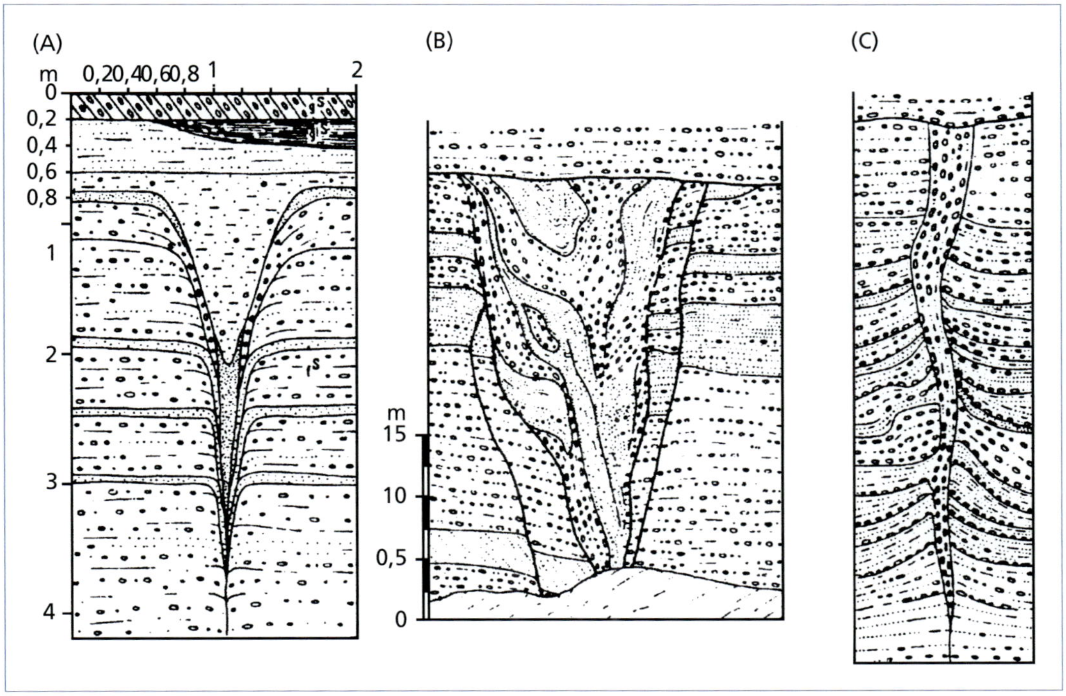

Figure 15.4 Casts of fossil ice wedges in the area of Leipzig, Saxony (Eissmann, 1981).

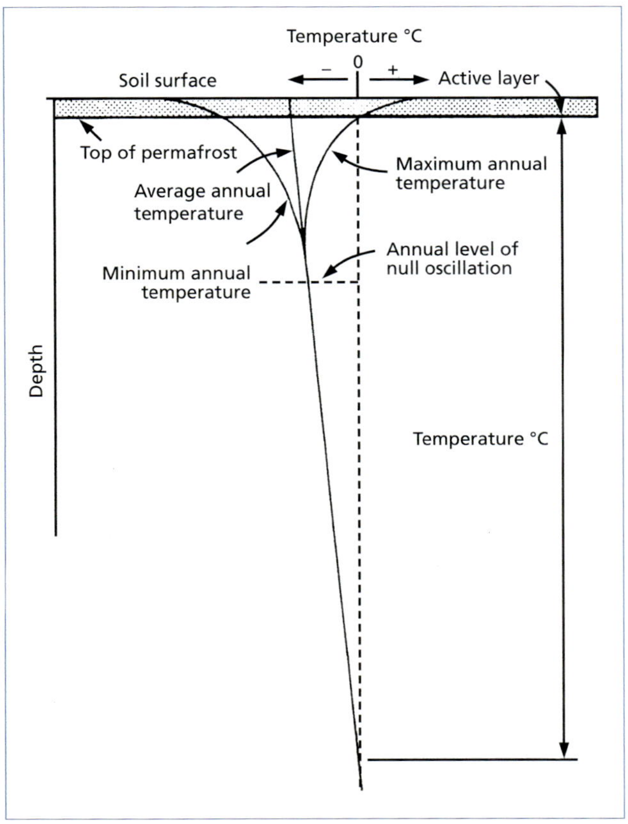

Figure 15.5 Thermal regime in permafrost (Brown, 1970).

but perhaps the most convincing is the discovery of tissues of woolly mammoths (*Mamuthus primigenius*) and other Pleistocene animals preserved in Siberian permafrost (in Washburn, 1979; French, 2007). The presence of tissue shows that permafrost already existed when the animals died; otherwise they would have decomposed.

15.1.3 Periglacial processes

Most morphogenetic processes operating in the periglacial domain are also active in other morphoclimatic zones. However, processes related to freeze-thaw cycles reach their maximum intensity in the periglacial environment. In addition, thawing on slopes promotes mass movements. These processes predominate in cold regions along with surface runoff and eolian activity in cold desert areas (King, 1976).

15.1.3.1 Frost action

Frost action is the term used for processes that result from freezing and thawing. In many situations, multiple processes linked to frost action act jointly but in some cases they may be associated with processes of a different origin. The study of freezing processes can be relatively complex because it requires knowledge of the magnitude and duration of temperatures below 0°C. Furthermore, frost action may affect materials that have different moisture contents and variable heat conductivities (Williams and Smith, 1989). The understanding of frost-related processes requires knowledge of freeze-thaw cycles at the ground surface throughout the year. Daily variations in these zones are due to insolation changes. In some parts of the Andes more than 300 annual freeze-thaw cycles have been recorded (Troll, 1944). An annual average of 80 freeze-thaw cycles has been recorded at the Ransol meteorological station at an altitude of 1640 m in the Pyrenees in Andorra (Raso and García Loureiro, 1998). At the University Hostel (2510 m) in the Sierra Nevada in the Betic Cordillera of Spain, 117 have been recorded in a year (Gómez Ortiz and Salvador Franch, 1997).

Multiple processes are associated with frost action; some of their most significant effects are analyzed below. Gelifraction was discussed in Chapter 5 on Weathering.

15.1.3.1.a Frost heaving

Frost-induced expansion exerts pressures in all directions. The vertical and horizontal components of the resulting displacements are called **heaving** and **translation** (Eakin, 1916). Heaving is more significant than translation. By installing sticks and pins to different depths in the ground, researchers have made field measurements of the magnitude of heaving. Maximum heaving occurs in areas with high moisture content during the spring when the ground has a relatively higher proportion of water from ice melting. Sticks driven into the ground may show different degrees of heaving or may even be expelled. Heaving values of 1–5 cm/year are common although much larger values have been measured (French, 1996). Heaving increases with depth to about 30 cm. It affects mainly the upper part of the active layer; the lower part is passive with respect to frost action. Studies done in the Sierra Nevada of Spain at an altitude of 3200 m indicate that the lower boundary of the horizon affected by frost action is at a depth of 25–30 cm (Gómez Ortiz and Salvador Franch, 1997). The heaving of rounded clasts around finer particles is common in periglacial areas; experiments indicate that larger clasts move upward more rapidly than small ones (Corte, 1969; 1971). Planar particles tend to rotate and rest on edge with the larger faces perpendicular to the cooling surface (*Fig. 15.6*) (Schunke, 1974), as shown in laboratory studies (Kaplar, 1965). Clast lifting causes vertical sorting of soil particles; this is a very common characteristic of periglacial accumulations.

Researchers have explained frost heaving in two ways. One hypothesis proposes that both stones and fine particles rise as the ground expands (*Fig. 15.7*). During thawing the fine grained material settles but ice remains beneath the clasts. Later the ice lens melts and the hole resulting from the melting fills with fine particles (*frost-pull hypothesis*) (French, 2007). This mechanism seems to be the most

Figure 15.6 Tabular clasts arranged vertically due to frost heaving. Sierra de Javalambre, Province of Teruel (Spain).

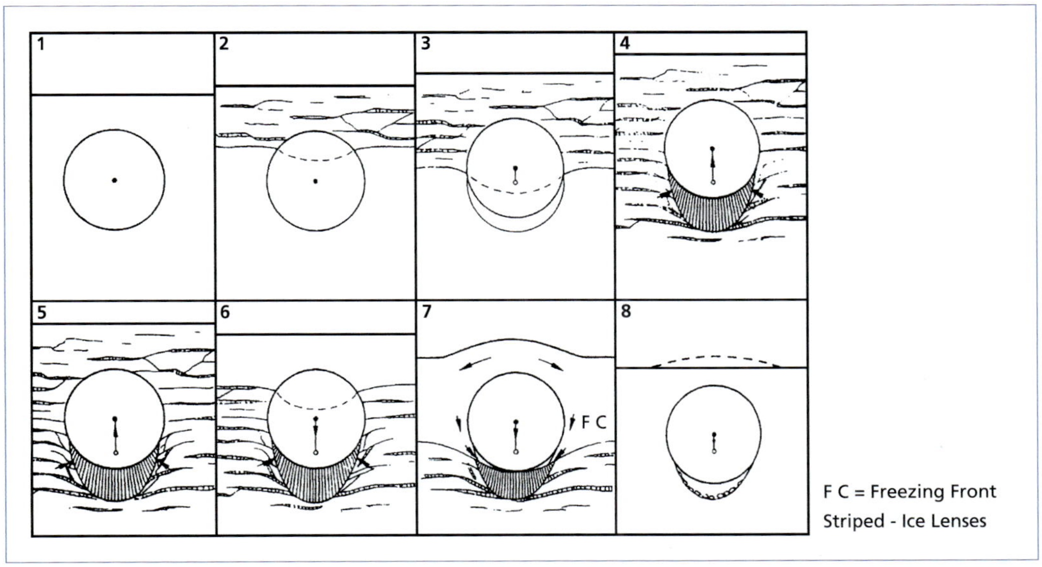

Figure 15.7 Clasts uplifted by frost heaving (Beskow, 1930; in Washburn, 1979).

common. The second hypothesis applies to clasts located near the surface. The water that flows around the clasts collects below them; as it freezes it pushes the clasts upward. The clasts have a greater thermal conductivity than surrounding fine particles so the base of the clast reaches the freezing point first. During thawing, the clast does not return to its initial position because the frost translation component compresses the hole originally occupied by the clast; fine material slides toward the position of the basal ice (*frost-push hypothesis*) (Bowley and Burghardt, 1971).

Another frost heaving mechanism is related to **needle ice** or **pipkrake**, which are ice crystals that form near the surface and perpendicular to it. These crystals, which range in size from a few millimeters to 40 cm, are common in alpine environments. As they grow, they lift clasts that have collapsed during the thaw and in so doing sort the soil (French, 2007). Some features of patterned ground in the French Alps are thought to be related to thawing and collapse of needle ice (Pissart, 1977). **Nubbins** are oval or elongate ground bulges that measure centimeters in size. They are thought to result from needle ice. **Voids around clasts** (*Fig. 15.8*) result from frost heaving and later settling during thawing (Washburn, 1979).

15.1.3.1.b Mass displacement

Mass displacement is movement of unconsolidated material as a result of frost action. Particle movement has a predominantly vertical component although it may also be affected by horizontal displacement. Scientists have suggested numerous mechanisms to explain these internal movements but **cryostatic pressure** seems to be the main cause (Washburn, 1956). Cryostatic pressure increases as freezing-induced pressures propagate to unfrozen parts of the ground located between the freezing front and the top of the permafrost. The ground surface starts to freeze in autumn and the freezing front advances downward during the winter. Because different parts of the ground have variable moisture contents, the ground freezes irregularly, causing differential volumetric expansion. The pressures transmitted to the unfrozen ground may liquefy it or generate dome-shaped bulges at the ground surface. Laboratory experiments have demonstrated the high magnitude of cryostatic pressure (Corte, 1969; Pissart, 1970).

Researchers believe that cryostatic pressures are the primary genetic cause of **cryoturbation** or **periglacial involution** (Sharp, 1942b; Vanderberghe, 2007) although other causes have also been suggested. These chaotic structures are reflected in cross section as disharmonic folds, diapiric injections, and faults that form during freezing and thawing of the soil in seasonally frozen zones (*Fig. 15.9*). Involutions are also observed in other morphoclimatic zones, primarily arid, but the mechanisms that cause differential soil expansion are obviously different in the periglacial environment (French, 1996).

15.1.3.1.c Ice cracking

As the temperature of frozen soils decreases, thermal contraction causes cracking. The soils behave as rigid solids; fissures and cracks form in the ice as the temperature falls. The cracks commonly occur as systems of polygons with four, five, or six sides (Lachenbruch, 1962).

Figure 15.8 Voids around clasts. Sierra Pelarda (Province of Teruel, Spain).

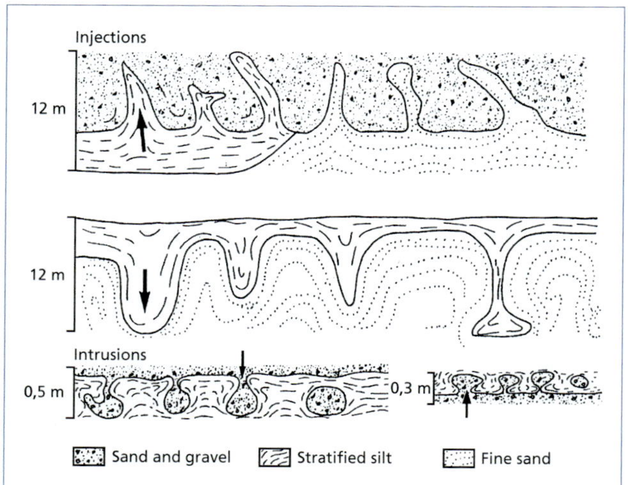

Figure 15.9 Formation of some types of periglacial involutions (Selby, 1985).

Researchers have classified frost cracks into two main types. Cracks form in winter; in the spring, they fill with snow, percolating water, groundwater, and water vapor. The upper part melts during the seasonal thaw whereas cracks in permanently frozen snow are preserved. The cracking is repeated every year; as cracks fill and freeze, they form an **ice wedge** (*Fig. 15.3*) (Lachenbruch, 1962; Yershov, 1998). Ice wedges have vertical banding; the number of bands shows the age of the wedge. They are generally thousands of years old (Yershov, 1998). Ice wedges are typical of humid environments; they require specific temperature conditions for preservation and growth. In Alaska, the southern boundary of active ice wedges corresponds to an average air temperature of 6°C to –8°C (Péwé, 1966). When they thaw, they fill with detritus; preserved ice wedge casts are valuable paleoclimatic indicators (Murton, 2007). In dry environments with less than 100 mm of precipitation, the lack of moisture makes it impossible for ice to fill the wedges; under these circumstances they may fill with eolian sand from winter winds or other detritus as in central Yakutia (Yershov, 1998). This is how **sand wedges** form (Péwé, 1959; Jahn, 1975) (*Fig. 15.10*). Their paleoclimatic significance is similar to ice wedges because they are genetically equivalent.

In addition to ice cracks, other mechanisms are known to cause fissures in periglacial areas. Desiccation cracking is a common process that forms small polygons. Evaporation causes desiccation; in addition, experimental studies show that moisture loss in areas of ice formation also causes desiccation (Washburn, 1973, 1979). Another mechanism is dilatation cracking caused by differential expansion of surface materials. Heaving causes radial cracks and subsequent collapse after thawing creates systems of concentric cracks such as those observed in pingo formation (Benedict, 1970; Jahn, 1975).

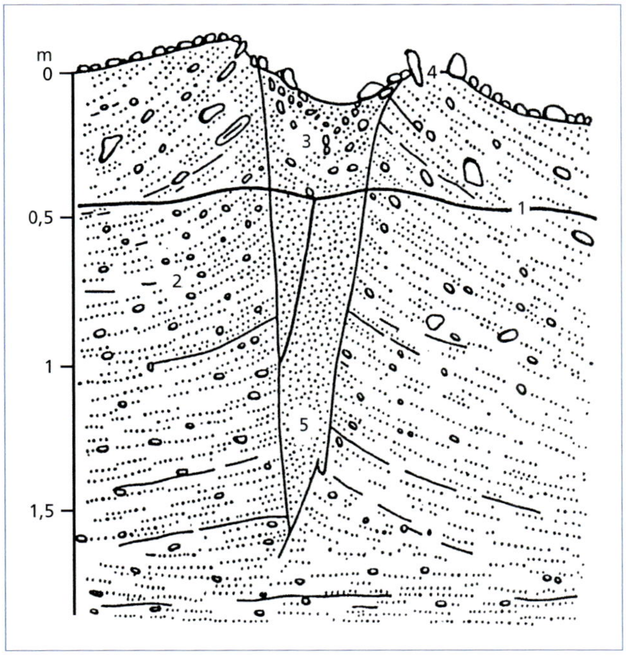

Figure 15.10 Sand wedge. 1- Top of permafrost, below the active layer. 2- Slightly stratified sands and gravels. 3- Collapsed sands and clasts. 4- Clasts. 5- Unstructured sands (Péwé, 1962).

15.1.3.1.d Ice sorting

Ice sorting is a complex mechanism of soil displacement and segregation by particle size. The previously explained processes of clast heaving, needle ice action, and mass displacement all contribute to particle sorting.

Laboratory experiments show three types of sorting (Corte, 1966, 1969; 1971). The first is heaving, which causes larger particles to move upwards, creating horizontal levels of different particle sizes. This is called **vertical sorting**. As the freezing front advances, the finest particles move in parabolic trajectories away from the front, leaving coarser particles behind. This is called **lateral sorting**. Finally, when dome-shaped bulges form at the surface, coarser particles move by gravity toward the margins of the dome, leaving the finer material at the center. This **mechanical sorting** can also be observed in polygons where the clasts located in the center of the bulged cell move toward cracks and fill them. Experimental studies show that sorting increases with increased moisture and when freezing is slow.

15.1.3.2 Chemical weathering

In periglacial environments, frost-shattered clasts commonly accumulate on surfaces of variable slope gradients; this shows that gelifraction is the primary weathering process. Although this process generates primarily large clasts, it can also produce clay-size particles as has been demonstrated in experiments conducted in climate chambers (Guillien and Lautridou, 1970).

Because water is in the solid state except during thawing stages, chemical weathering is generally limited (Hall *et al.*, 2002). In addition, chemical reaction rates slow down at low temperatures. However, rounded blocks of diabase and sandstones that show exfoliation due to hydration and oxidation processes have been observed in some areas (Czeppe, 1964). In some cases authors interpret chemical weathering features as indicators of past non-periglacial conditions. In Mediterranean mountain environments, as in the Sierra Nevada of Spain, chemical weathering of mica-schist during the warm season is thought to be significant in silt production (A. Gómez Ortiz, pers. comm.). In coastal environments, wind-transported seawater spray often leaves a sodium-rich saline efflorescence on rocks. Snowmelt helps the saline water percolate into the rock; later precipitation of dissolved ions may fracture the rocks and cause them to disintegrate. This salt weathering process (haloclasty) is attributed to the formation of honeycomb and tafoni weathering along with wind action that helps transport the particles generated by the breakup of the rock (Washburn, 1969; Selby, 1972; French and Guglielmin, 2002).

Even so, the activity and intensity of chemical weathering in these periglacial areas is not well understood because not much research has been done in this field. Surprising results were obtained from a nine-year study of an area of mica-schist and amphibolite in northern Sweden (Rapp, 1986). This experiment showed that 48% of the weathering products were removed by dissolution in runoff. Other research shows very low values for this type of transport.

Carbonate rock solubility increases with increased carbon dioxide content in water; it reaches maximum solubility in water at 0°C. For this reason, periglacial environmental conditions favor the dissolution of carbonate rock. However, solution rates are lower than in warm areas due to sparse biotic activity, which is generally the main source of CO_2 (Birot, 1960). Kärkebagge conducted field experiments in which he installed limestone and granite disks in the soil; at the end of five years the limestone disks had weathered much more than the granite (Thorn *et al.*, 2002). Ice and snow supply water during the thaw; carbonate dissolution creates karren and doline fields and widens discontinuity planes. Water penetrates into the bottom of the massif and forms shafts and cavities, especially in alpine areas. In areas of continuous permafrost, meltwater cannot penetrate the permafrost so it flows on the surface. Small cavities can form in discontinuous permafrost; they do not exceed 300 m long and stalactites and stalagmites may also form in them (Sweeting, 1972).

15.1.3.3 Mass movement

Mass movement is defined as the movement of detrital material downslope under the influence of gravity. These processes are significant in periglacial environments because the high water content of the active layer reduces friction and soil cohesion. Permafrost is impermeable and the top of the permafrost behaves like a water-lubricated plane to facilitate the movement of the overlying active layer (French, 1996). Different mass movement processes have different intensities due to climatic, topographic, and lithologic characteristics.

In steep areas of glacial origin, marine cliffs, and fluvial scarps, water penetrates into cracks in the rocks; as it freezes it expands and can cause sporadic **rockfalls**. These occur primarily during the thaw and are exacerbated by basal undermining (Harris, 2007). The result is a gelifract-covered slope at the foot of a rocky cliff. Scarp retreat rates calculated by different methods show values of 0.3 to 0.6 mm/year (André, 1993), well below those recorded in temperate and semiarid areas.

Solifluction is one of the most common processes in periglacial areas. This type of movement can occur under a variety of climatic conditions; it is called **gelifluction** in periglacial regions (Baulig, 1957). **Frost creep** is a related process that results from soil displacement as it dilates perpendicularly into the ground during freezing and settles during thawing (Washburn, 1967) (*Fig. 15.11*). Three types of movements have been identified. The first, **potential frost creep (PFC)** is related to frost heaving of the soil (**P1** to **P2**). The soil expands perpendicularly to the slope because the slope is the cooling surface (Taber, 1929). Gelifluction (**G**) is another component and retrograde movement (**R**) works in opposition to the preceding processes; retrograde movement is due to cohesion and interference between particles (Davison, 1889). The particle trajectory during a freeze-thaw cycle is **P1** to **P4**. These movements work together but gelifluction is usually the dominant process and, for this reason the resulting landforms are called gelifluction landforms.

Frost creep increases with the number of freeze-thaw cycles, slope angle, and soil moisture content available for heaving; it decreases with depth (Washburn, 1979). Gelifluction can be active on slopes with 1° gradients. It begins during the spring thaw and reaches its maximum intensity in summer; it decreases in autumn because less soil moisture is present as a result of evaporation and internal flow. Gelifluction also increases with slope and soil water content; it decreases gradually with depth. Experimental studies show that gelifluction is restricted to the top 50 cm of the active layer (French, 1996). Laboratory simulation experiments indicate that gelifluction occurs only as the upper parts of the

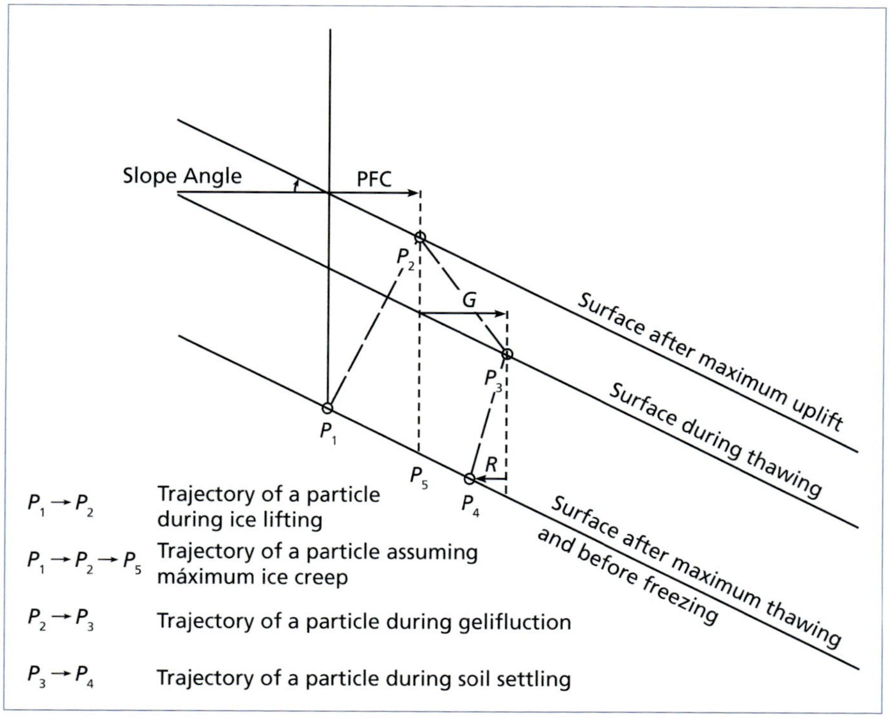

Figure 15.11 Diagram indicating the movement of particles affected by potential ice creep (PIC), gelifluction (G), and retrograde movement (R) (Washburn, 1979).

soil profile, whereas thawing of deeper layers causes very little downslope movement (Harris and Davies, 2000); Other experiments conducted by these investigators (Harris *et al.*, 2003) show that gelifluction is not time-dependent and that its behavior is elasto-plastic. Gelifluction is controlled by grain size; porous sediments provide good drainage and do not favor saturated flow whereas fine particles retain water and lead to saturated flow. Although vegetation retards soil flow, gelifluction measurements conducted in vegetated areas provide values much higher than those in zones that lack vegetation. The explanation for this result may be the higher moisture content of vegetated places, which retain water and reduce runoff (Washburn, 1967). In addition, gelifluction rates are generally higher on shaded slopes than on sunny slopes because of their higher water content. Higher solar radiation, however, accelerates thawing. Numerous field experiments have been conducted to determine and quantify gelifluction. On slopes ranging from 5° to 15°, values between 1 and 12 cm/year have been measured (Washburn, 1979). Soil flow also tends to orient the major axes of clasts parallel to the maximum slope.

When a large block slides on top of a solifluidal mass, its major axis orients parallel to the slope, creating a furrow upslope and a heap downslope as a result of the pushing. This morphology, which is relatively common in periglacial environments, is called a **ploughing block** (Tufnell, 1972) (*Fig. 15.12*).

Slope movement occurs frequently on medial and distal parts of slopes in active gelifluction areas (*Fig. 15.13*). It is confined to the active layer where soil saturation exceeds the shear strength and causes slope instability (Lewkovicz and Harris, 2005). The top of the permafrost determines the position of the sliding surface and also acts as a lubrication zone. The slides that occur tend to be pellicular **mudflows** (*Fig. 15.14*) whose head scars are generally smaller than 5 m in diameter; they rarely exceed lengths of 100 m (French, 1996). Although thawing is the fundamental cause, mudflows can also be triggered by summer rain.

15.1.3.4 Nival processes

Nival activity refers to rapid snow movements called avalanches and to slower movements related to static snow deposits; these processes are called nivation.

Avalanches are rapid downslope movements of snow that commonly occur in mountains within temperate zones but rarely in polar regions. They occur on slopes with gradients from 15 to 50°; they do not commonly occur on steeper slopes because the higher gradient prevents a thick accumulation of snow cover. They can travel at velocities

Figure 15.12 Ploughing block. Puerto de la Bonaigua (Pyrenees, Spain).

Figure 15.13 Multiple landslides affecting debris slides at the base of a slope with active gelifluction. These deposits overlie a terrigenous Cretaceous series. James Ross Island, east of the Antarctic Peninsula. Photo A. Martín-Serrano.

Figure 15.14 Mud flows at the foot of a gelifluction slope. James Ross Island, east of the Antarctic Peninsula. Photo A. Martín-Serrano.

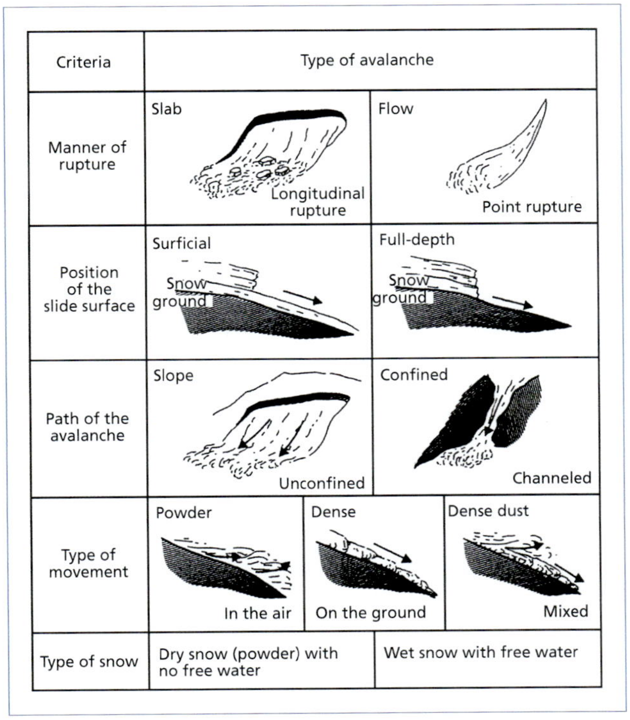

Figure 15.15 Avalanche classification. Compiled and adapted from various authors. (López-Martínez, 1988).

greater than 100 km/hr. Several criteria are used to classify avalanches (Lopez Martínez, 1988) (*Fig. 15.15*). An avalanche can be composed of dry or wet snow, may transport snow or snow and detritus, can slide over the ground, or can move through the air. Some avalanches are contained within a channel while others are not confined at all.

Several factors cause the snow cover to be unstable (Embleton, 1979b). Avalanches are especially common after heavy snow falls that overload previously deposited snow. The presence of frozen ice layers covered by new precipitation creates structurally weak planes that can trigger avalanches. In addition, the percolation of snow meltwater can generate avalanches.

Two fundamental types of avalanches can be distinguished (Furdada, 1996). **Powder avalanches** are an unformed mass with a very limited geomorphic effect although they can destroy trees and houses in their trajectories. **Slab avalanches** are rigid bodies of cohesive snow that move on a sliding surface controlled by an old snow layer (surficial avalanche) or by the snow-ground interface (full-depth avalanche) (*Fig. 15.16*). Full-depth slab avalanches generally incorporate detrital material; they can have significant erosional effects. Avalanches are especially common during the thaw period and in winter (Owens, 2004). Movements of water-saturated snow masses or slush avalanches that flow down fluvial valleys in Arctic areas during the spring should also considered avalanches. They can transport blocks that weight up to 100 tons down slopes that have a 5° gradient (Hestners, 1985).

Figure 15.16 Full-depth avalanche. French Alps. Photo F. Gutiérrez.

Avalanches are highly variable in space and time. They are rare in some zones but occur regularly in others. Sometimes avalanches re-use previous avalanche routes; these are valuable indicators to use in avalanche risk cartography.

As previously stated, whereas the geomorphic activity of powder avalanches is limited, full-depth avalanches can do significant erosion. Many talus slopes and detritus cones are partially produced by avalanches (*Figs. 15.17* and *15.18*) Avalanches remove and transport loose particles; they erode as they move and deepen the channels they travel (Keylock, 1997).

Nivation, first defined by Matthes (1900), designates a group of processes including gelifraction, gelifluction, frost creep, and runoff from snowmelt (Thorn and Hall, 2002). Water from thawing can freeze again where permafrost is

Figure 15.17 Avalanche channels, talus slopes, and detritus cones. Main cordillera of the Andes. Punta Vacas (Province of Mendoza, Argentina).

Figure 15.18 Talus slopes at the foot of acicular landforms composed of basalts. James Ross Island, east of the Antarctic Peninsula. Photo A. Martín-Serrano.

present beneath the snow. In unfrozen soils, a snow thickness of more than a meter has a temperature near the freezing point and the snow layer protects the ground from freeze-thaw cycles (Embleton and King, 1975b). Thin snow cover promotes gelifraction. Along the eastern slopes of Peñalara (Sierra of Guadarrama, Spain) nivation is effective only on outcrops of moraines or the weathering layer. The western slope shows no signs of nivation activity except where catastrophic mass movements have altered the slope profiles (Palacios *et al.*, 2003). Nivation is more active in subarctic and alpine environments; the most notable geomorphic feature is **nivation niches** (Thorn, 1988a).

15.1.3.5 Fluvial activity

Fluvial activity in periglacial environments has not been studied extensively. Fluvial activity was traditionally considered irrelevant because soils are frozen for a significant part of the year. Periglacial area rivers (*Fig. 15.19*) do not have an allogenic supply; therefore they have flowing water only during the thaw period (*Fig. 15.19*). For this reason researchers thought that runoff played a minor role in landform development (French, 1996).

Rivers in periglacial areas are supplied by snow and melted ice, rain, groundwater, a small quantity of frozen soil, and extraperiglacial runoff. Most rivers are strongly affected by snowmelt; their seasonal flow pattern has been described as a nival regime that consists of a high flow period sustained by meltwater with prominent diurnal runoff (Vandenberghe and Woo, 2002).

Snow that falls in winter (Birot, 1960) thaws over a period of two to three weeks in the spring. Snowmelt causes floods whose hydrographs show a steep rising limb and an acute flood peak followed by a prolonged falling limb that lasts two or three months (Arnborg *et al.*, 1967). During the summer, most recharge comes from fields of permanent snow or glaciers. In arid zones, rainstorms may occur during the summer. Although annual precipitation is small, runoff is concentrated over a few days; it is estimated that between 30 and 90% of the annual runoff occurs within two or three weeks of snowmelt flooding (Clark, 1988). Runoff finally ends at the onset of winter when a new ice stage begins.

In periglacial environments, sediment transport is generally low; it may consist partially of glacial sediments. Bedload, which usually predominates, may account for 90 percent of the sediment load, dissolved load is negligible (<1%), and suspended load ranges from 5 to 25 percent (Church, 1972). Various methods have been used to estimate transported volumes in different drainage basins; values of surface lowering

Figure 15.19 Gully created by thawing and slopes covered with heterometric detritus that covers terrigenous Cretaceous materials. James Ross Island, east of the Antarctic Peninsula. Photo A. Martín-Serrano.

are between 0.03 and 0.5 mm/year (Arnborg *et al.*, 1967). Beyrich *et al.*, (2003) reach similar conclusions.

Fluvial channels are predominantly *braided* because of their high bedload; channels are unstable and contain highly mobile bars that may be inundated during high-stage periods. River water causes thermal erosion; it helps thaw frozen riverbanks and undercuts the banks, leading to collapse. This continuous thermal erosion widens the bed and creates U-shaped channels (French, 1996). In addition, large rivers can create a talik under their bed and incise into it.

Large rivers that cross periglacial regions are the MacKenzie in Canada and the Obi, Yenisei, and Lena in Siberia. Costard and Gautier (2007) summarize the hydrology and morphology of the Lena River. The Lena is 4,300 km long and has more than 500 tributaries with a total length of 60,000 km. The surface area of its drainage basin is 2.49×10^{16} km^2. It discharges 525 km^3 of water into the Arctic Ocean every year; it carries between 12 and 17 million tons of suspended load and 49 million tons of dissolved load. At its mouth a large delta has formed; it has a surface area of 30,000 km^2 and supports abundant fauna and flora. Costard and Gautier (2007) collected this data from various Russian authors. The Lena River flows almost north-south and consequently crosses various latitudinal zones inhabited by a wide variety of flora and fauna. Boreal forest (taiga) covers 83% of the basin and 20° of the latitude. Near the Arctic Circle the landscape is arctic tundra.

The Lena River crosses Yakutia where the Earth's lowest temperatures (–72°C) and maximum permafrost thickness (1500 m) have been recorded. Most of central Yakutia is not covered by snow because precipitation is sparse; the lack of snow has helped form and preserve continuous and discontinuous permafrost.

The permafrost along the floodplain of the Lena River contains a large quantity of massive ice and ice wedges. Fluvial thermokarst is present as collapsed pingos and alases (Soloviev, 1973). Alases form as flat bottom landforms and depressions with steep slopes as massive ice layers melt. They may coalescence to produce mature alas valleys. The bottom of an alas has a chaotic topography of collapsed masses and temporary sag ponds. Alases range between 100 m and 15 km in length but are normally about 1 km long. Thermokarst is located primarily on the Pleistocene plains of central Yakutia (Czudek and Demek, 1970b). As a result, alases cover 40–50% of the plains of central Yakutia. At the mouth of the Lena River the subaerial delta covers about an area of about 28,500 km^2 that remains frozen about six months of the year. Its tundra has various types of patterned ground and thermokarst alases.

The fluvial system is composed of anastomosing rivers with multiple channels separated by permanent alluvial islands (Knighton and Nanson, 1993). As a result, the sediment load, which is primarily sandy, does not move large distances; instead, it accumulates in wide bars, long islands, and anastomosing branches (Gautier and Costard, 2000). The headlands of the islands undergo significant erosion, with average values of 12 m/year and maximum values reaching 20 m.

The Lena River is of interest because of its spectacular floods, which are controlled by periglacial dynamics. Yang *et al.* (2002) show that maximum floods are recorded in May in association with positive thermal anomalies. As flooding begins, the river transports large quantities of ice and floating trunks that create local dams. The dams cause the water level to rise, which rapidly floods the floodplain and sometimes even the first flood terrace. In May 2001, the region experienced its worst flood in 100 years, which affected hundreds of thousands of people. In many places, it was necessary to use explosives to destroy the dams. This event was caused by a combination of a severe winter and an abnormally warm spring (Yang *et al.*, 2002).

According to Yang *et al.* (2002), the average temperature in the Lena River basin has increased 1.3°C since 1930. This temperature rise increases the thickness of the active permafrost layer and the extent and subsidence of thermokarst depressions (Pavlov, 1994; Serreze *et al.*, 2002). Climate change is undoubtedly affecting the behavior of the river in a substantial way. In a study of six large Eurasian rivers near the Arctic, Peterson *et al.* (2002) calculated an annual increase in water volume of 2 ± 0.7 km^3, an increase of approximately 7%.

15.1.3.6 Wind action

In Arctic and Antarctic regions where extreme cold and arid conditions prevent vegetation growth, wind can be an important erosional agent because there are few obstacles to impede it (Yershov, 1998). During the Pleistocene, significant transport and deposition of eolian particles occurred over broad areas of Eurasia and the central United States (Embleton and King, 1975a, b) *(Fig. 15.20)*. Silt constitutes an extensive deposit in central Europe but it is patchy in Great Britain *(Fig. 20.28)*. Three phases of eolian sand deposition have been documented in the period between the Last Glacial Maximum and the beginning of the Holocene but they do not strictly coincide with stadial or glacial conditions. Aridity, sparse vegetation cover, and delayed responses of eolian activity to climate change determine the timing of eolian phases (Kasse, 2002). In addition, the wind redistributes snow during the winter. In the summer, snow melts more rapidly on sun-facing slopes. These changes in snow cover are important in nivation processes and the creation of landforms such as asymmetric valleys and oriented lakes, which will be discussed below.

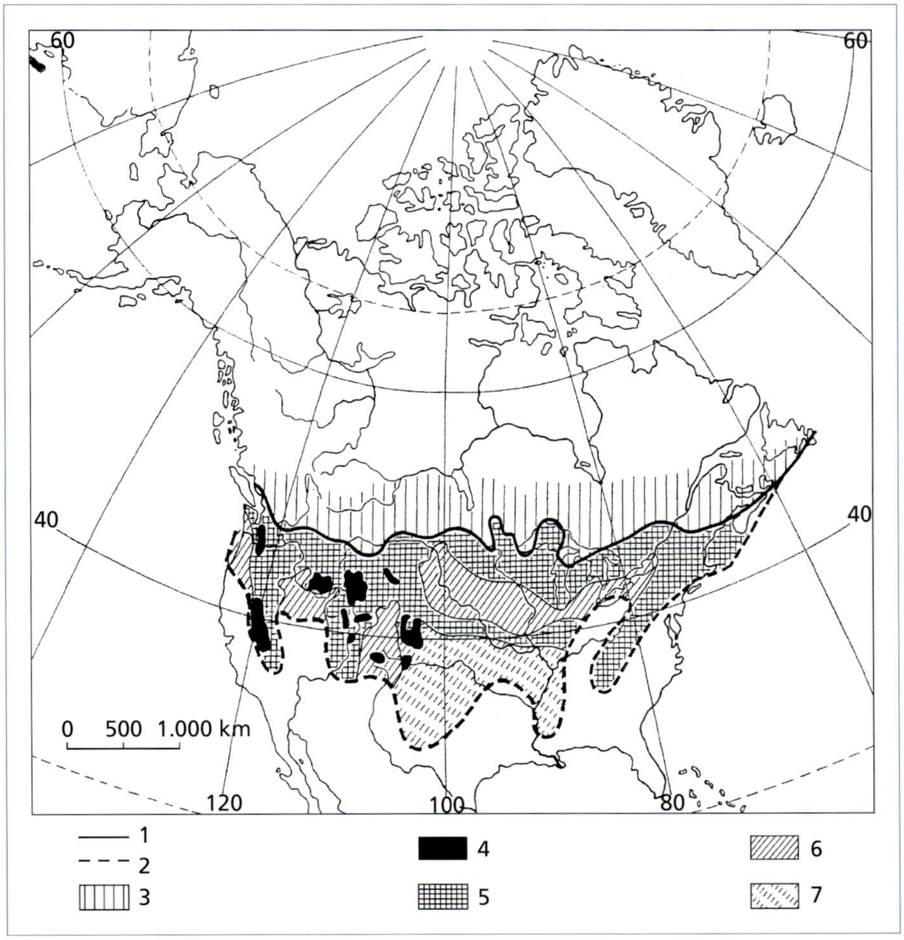

Figure 15.20 Pleistocene periglacial zone of North America according to Brunnschweler's map (interpreted by Jahn, 1975). Boundary of the periglacial zone: 1- northern; 2- southern; 3- Wisconsin glaciation; 4- polar alpine tundra with continuous permafrost; 5- polar alpine tundra with discontinuous permafrost; 6- loess in tundra; 7- loess in steppe-forest.

In cold polar regions wind action predominates in summer, causing the deflation of fine particles from the land surface and creating periglacial pavements on alluvial plains and till piles (*Fig. 15.21*). Eolian abrasion creates ventifacts (Pissart, 1966b; French, 2007). Overexcavation in some places causes deflation blowouts (Seppälä, 2004). During transport, successive impacts create ventifacts or faceted clasts, yardangs in rocky outcrops (elongate mushroom forms), microyardangs in unconsolidated sediments (knife-shaped and aligned features), and tafoni near the base (French, 1996). High latitude dunes supply glacial and fluvioglacial accumulations. They are common in North America, Europe, Asia, Tibet, Argentina, and in areas of Antarctica that lack ice. Several small dune fields have been found in unfrozen areas of Antarctica. In their study of Victoria Valley (Antarctica), Bourke *et al.* (2009) calculated dune field movements at 1.5 m/year.

The combined action of eolian transport and deposition of snow and mineral particles creates nivo-eolian deposits (Pissart, 1966b; Koster and Dijkmans, 1988). As in ice masses, differential melting produces topographies of small mounds and depressions that disappear from one year the next. The most outstanding deposit formed by particle accumulation in sand dunes and particularly as silt layers is called **loess** (Péwé, 1955 and 1959) (*Fig. 15.22*). Loess is transported by dust clouds that deposit sediments far from the source. It is a massive very well sorted deposit; 50 to 60 percent of the particles range from 0.01 to 0.05 mm (silt) in size and the rest are sand and clay size (Muhs, 2007). Loess is gray in color; as it weathers it becomes brownish-gray. It contains ice wedge structures, solifluction, and cryoturbation casts. During the Pleistocene, extensive accumulations of loess were deposited on the southern margins of the North American and Eura-

Figure 15.21 Pavement in the northeastern part of the Isle of Seymour (Antarctica). Photo F. Nozal.

Figure 15.22 Loess deposit with buried soils. Neerijse (Belgium).

sian ice sheets; the study of these accumulations has been of great paleoclimatic importance (Muhs and Bettis, 2003; Muhs, 2007). However, not all loess is periglacial; much of it originates in tropical desert environments.

The dominant wind and atmospheric conditions (isobars) of past periods can be inferred from paleocurrents obtained from internal structures of old eolian deposits (Poser, 1950) (*Fig. 15.23*).

15.2 Periglacial landforms

The previously described periglacial processes are responsible for a variety of periglacial landforms that commonly form in two principal environments (1) high latitudes and (2) medium and low latitudes. They are most extensive at high latitudes with seasonal climatic contrasts whereas climatic variations are primarily daily fluctuations in mountainous areas of medium and low latitudes such as the Andean cordilleras, the Himalaya, and mountains of equatorial Africa. Differences in climatic conditions play an important role in most processes that generate periglacial landforms (King, 1976).

Variations in climatic belts during the Pleistocene changed the distribution of periglacial environments. It is important to differentiate between relict and active periglacial

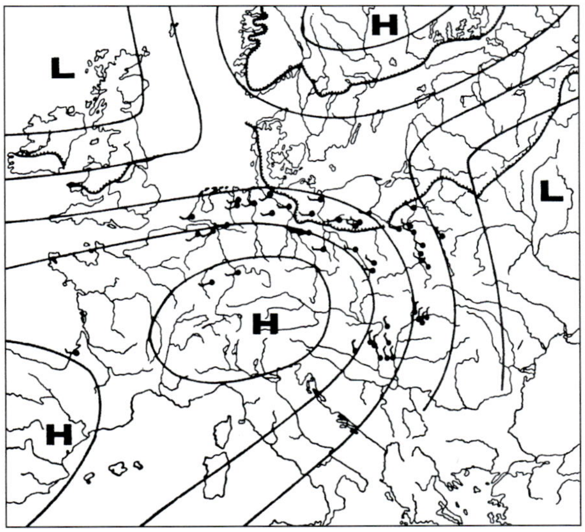

Figure 15.23 Atmospheric pressure in summer and wind directions in Europe during the Late Glacial epoch. H and L: High and low pressures, respectively (Poser, 1950).

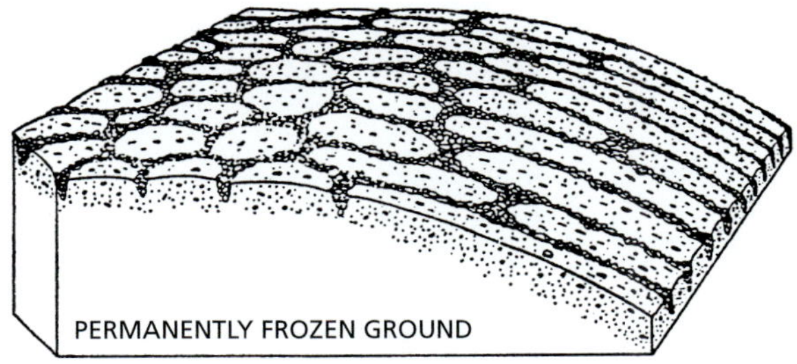

Figure 15.24 Block diagram showing the transformation of sorted polygons that form on very low gradient surfaces to sorted stripes as slope gradient increases (Sharpe, 1938).

landforms because the identification of relict landforms allows us to infer the existence of past periglacial conditions in certain areas (Washburn, 1979).

15.2.1 Patterned ground

Patterned ground consists of small-scale landforms of circular, polygonal, and banded geometries; they attracted the attention of early explorers in periglacial areas to the detriment of other more widespread morphologies (French, 1996). Patterned ground, however, is not specific to periglacial areas; similar forms occur in other environments (Hallet, 1990; Ahnert, 1994), especially in hot deserts (Hunt and Washburn, 1966). Thus we have a problem of morphological convergence or equifinality; different processes generate the same morphologies in response to variations of soil moisture. This is important to keep in mind when making environmental interpretations of relict forms. In the semidesert environment of the Ebro Depression and near Zaragoza (about 200–300 m of elevation), several investigators (Johnson, 1960; Brosche, 1971, 1972) interpreted various structures in terrace deposits as ice wedges and cryoturbation although these forms in fact result from the dissolution of underlying gypsum and swelling of clay-rich formations among other processes (Zuidam, 1976). The only morphologies specific to the periglacial environment are those caused by thermal cracking of frozen ground because these forms provide evidence of permafrost or intense cold.

Periglacial patterned ground forms in the active layer; Washburn's classical morphologic systematization (1956) classifies patterned ground into circles, polygons, nets, steps, and stripes. The classification describes the presence or absence of particle sorting for each type. Circles, polygons, and networks develop primarily on horizontal surfaces. These cells tend to occur on slopes with gradients between 2 to 7° and they elongate to form stripes (Sharpe, 1938; Büdel, 1960) (*Fig. 15.24*).

Circles range in diameter from 0.5 to 3 m. Circles that lack sorting bulge and their vegetation-free center is cracked. Circles usually have a high content of fine material; cross sections show structures that indicate uplift of the central areas. Sorted circles have a rim of stones that surrounds an inner part composed of fine-grained material. The planar stones tend to rest on edge with their large faces oriented vertically (Furrer, 1968). Both sorted and unsorted circles may occur in isolation (*Fig. 15.25*) or grouped (*Fig. 15.26*). Holness (2003) claims that circle formation is due to differential frozen uplift associated with convection cells. The most important processes involved in circle formation are differential frost heaving, mass displacement, and frost sorting (Jahn, 1975).

Polygons with diameters less than one meter and those larger than one meter are classified separately because they may have different origins (Washburn, 1979). Small unsorted polygons may be as small as 5 cm in diameter whereas the large ones can exceed 100 m. They form in soils with variable sized particles where vegetation tends to concentrate along the margins and emphasize the polygonal pattern. Some polygons have ice wedges along their margins. These are domed cells during the growth stage and depressions with ponds during the recession stage (Jahn, 1972). The margins exhibit the opposite behavior. Some polygons may host small pingos and smaller polygonal forms inside the principal cell. In sorted polygons the margins are formed of stones that surround a cell of finer material (*Fig. 15.27*). The

Figure 15.25 Sorted and elongate circle of rocks with relict of a funnel-shaped sinkhole at the center. The margin is composed of clasts that are smaller than the ones in the central zone. Unsorted steps can be seen on the slopes of the dolines. They are relict landforms. Sierra de Javalambre, Province of Teruel (Spain).

minimum size of the smaller features is 10 cm and the maximum size of the large polygons does not exceed 10 m. Clasts decrease in size with depth and/or narrow downward or they may gradually enlarge and merge with a continuous layer of clasts (Jahn, 1975) (*Fig. 15.28*). Cracking is fundamental to polygon formation. Large polygons seem to be produced by frost cracking whereas the smaller forms result from desiccation or dilation cracking (Washburn, 1979).

Nets are intermediate forms between polygons and circles; they can be sorted or unsorted. **Hummocks** are included in this group (*Fig. 15.29*). These are mound-like forms covered by vegetation; their maximum diameters are 1–2 m and they are about 50 cm high. Their internal structure shows cryoturbation. They occur above the tree line in areas of low slope; their origin appears to be related to mass displacement (Schunke and Zoltai, 1988) or cryoturbation (Ballantyne, 2007) although other types of origins have been proposed.

Steps are another type of patterned ground; steps form benches on slopes that dip 3–20°. The margins of unsorted steps are vegetated whereas the borders of sorted steps are composed of imbricated clasts (Sharp, 1942a). Unsorted steps may be derived from hummocks and the sorted variety may evolve from circles and sorted polygons (Washburn, 1979). Their origin may be related to differential mass movement.

Bands or **striated soils** are a set of strips oriented parallel to the steepest slope gradient. Unsorted stripes consist of parallel lines of vegetation-covered ground and intervening strips of relatively bare ground. Sorted stripes are parallel lines of stones and intervening strips of fine material with the latter often several times wider than the coarser lines (*Fig. 15.30*). Stripes are generally straight and may reach 120 m in length (Washburn, 1969). The clasts are usually on edge with the major axis parallel to the stripes. Clast size generally decreases with depth and the sorting penetrates a meter at most. On slopes with gradients greater than 2° sorted polygons grade progressively into sorted stripes (Collard, 1973) (*Fig. 15.24*) because mass movements seal the sides perpendicular to the slope gradient (Furrer, 1968).

Attempts have been made to explain the origin of patterned ground in periglacial environments but it is still not well understood. Patterned ground is undoubtedly polygenic; similar forms may result from different processes. In addition, certain processes may generate different landforms (Washburn, 1979). Washburn discusses 19 hypotheses on

Figure 15.26 Sorted circles. South Shetland Islands (Antarctica). Scale = 1 m. Photo J. López-Martínez.

Figure 15.27 Sorted polygons. South Shetland Islands (Antarctica). Scale = 1 m. Photo J. López-Martínez.

the genesis of patterned ground. This shows that in spite of abundant literature on this theme, the origin of periglacial patterned ground remains unresolved.

15.2.2 Ice-cored mounds

Ice-cored mounds are small hills produced by frost action. Most of them are generated as new permafrost forms in non-frozen floodplain areas and limnic basins (Yershov, 1998). Researchers have proposed so many terms to classify them that significant terminological confusion has resulted. These mounds can be perennial like palsas and pingos or seasonal, such as hydrolaccoliths or blisters. The latter, which are 1–8 m tall and 3–50 cm in diameter, have not been studied extensively because of their ephemeral nature and formation in winter under snow cover. They form as groundwater freezes on top of permafrost (Pollard, 1988).

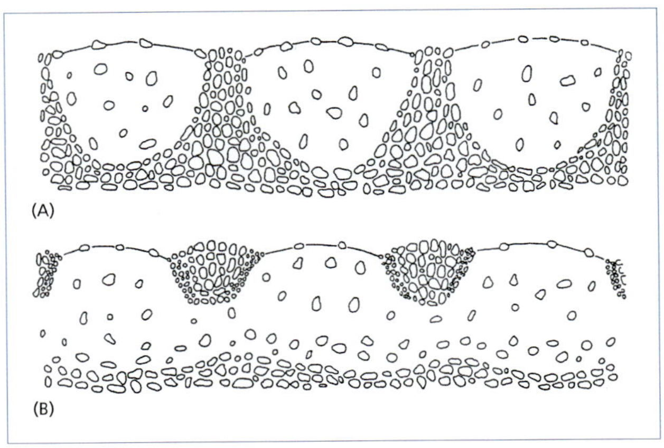

Figure 15.28 Transverse section of sorted polygons. A: Margin of rocks that get coarser with depth. B: Margin of rocks that become finer with depth (Jahn, 1975).

Figure 15.29 Hummock. High valley of the Ter River (East Pyrenees, Spain).

Figure 15.30 Relict sorted striated soils. Sierra de Javalambre, Province of Teruel (Spain).

15.2.2.1 Palsas

Palsas are small hills that form in swampy areas and discontinuous or sporadic permafrost areas (Zoltai, 1971). They are present today in areas with a mean annual temperature below 0°C such as Canada, Iceland, Sweden and alpine regions (Kershaw and Gill, 1979). Palsas occur in isolation or in groups. They are distinguished on the basis of their morphology: dome-shaped, ribbon-shaped, and platform. The dome shapes are between 0.5 and 7 m high, 10–30 m wide, and 15–150 m long. Large size palsas tend to be considerably less conical than smaller ones. Ribbon-shaped palsas are lower in height; they may be parallel or perpendicular to contour lines and can extend several hundred meters in length. Platform palsas rise 1 to 1.5 m above the surrounding bog and may cover several square kilometers. They are related to differential thawing. Finally, palsa complexes are composed of several types of palsas at different growth stages; they contain numerous closed depressions created by thawing (Seppälä, 1988). Some authors use the term *string bogs* to designate a group of different types of palsas. The surface of a palsa is cracked due to dilation and desiccation (Embleton and King, 1975b).

Palsas are classified into two types on the basis of their internal structure. Palsas with a peat nucleus have a surface layer of dry peat that thaws in summer. Between 80 to 90% of the interior is composed of small ice crystals and lenses of segregated ice 5–10 mm in thickness that thickens to 5–10 cm toward the bottom. The base is usually composed of silt-sized mineral particles. Palsas with a silt nucleus or **lithalsas** (Pissart, 2000, 2002) have a surface layer of peat and a nucleus of silt and clay with lenses and veins of ice arranged in a horizontal and vertical reticular fashion (Seppälä, 1988).

Palsas form as a result of different heat conductivities of dry and saturated frozen peat; the heat conductivity of saturated and frozen peat is about twenty times higher than dry peat (Washburn, 1979). As the peat gets wet in winter, its thermal conductivity increases and it penetrates deep into the ice, forming segregated ice and dome-shaped bulges as a result of cryostatic pressures. The outer peat layer dries up in summer and has low heat conductivity; it acts as an insulator that prevents the inner part of the palsa from thawing. Palsas may disintegrate gradually from their margins inward because of rising temperatures; they may turn into small shallow sag ponds where a new generation of peat can begin (Friedman *et al.*, 1971). Seasonal palsas linked to ooze have been reported in the Paramera of Avila Province (Central Spain) at an elevation of 1200 m (Molina and Pellitero, 1982).

Although most palsas in Scandinavian countries formed 1000–3000 years ago, absolute dating indicates that some of them formed recently (Seppälä, 1988).

15.2.2.2 Pingos

Pingo is the Eskimo word for mound. These structures are also called hydrolaccoliths; in Siberia they are called bulgunniaks. Pingos are conical shaped mounds that have an ice nucleus and that generally stand up above extensive plains. Submarine pingos were recently discovered on continental shelf. Scientists do not know if they grew under the sea or if they formed on land and were later invaded by marine waters. The highest documented pingo on land and above sea level is the 48 m high Ibyuk pingo, located near the MacKenzie Delta (Canada). Most pingos range in diameter from 30 to 600 m; their diameter is inversely related to their height. Some are more than a kilometer long and have a maximum height of 9 m. Because they maintain a constant diameter as they grow, their slope gradient increases but never exceeds 45° (Pissart, 1988). Two types of cracks occur at the surface. Dilatation radial cracks converge at the top; they result from the upward push of the ice core as it grows (Washburn, 1979). The opening of the cracks may lead to partial thawing of the underlying ice, leading to a crater-like subsidence depression at the center. Concentric cracks, which are not as conspicuous as radial cracks, result from thawing of the ice core during the negative growth stage of the pingo (Müller, 1959). Unlike palsas, pingos have a massive ice core that penetrates several meters below the ground surface (Tricart, 1967; Lundquist, 1969). The cover material is usually loose sediment such as gravel, sand, and silt but pingos with sandstone and shale covers have also been observed. The cover layer may reach a thickness of 14 m. Their growth rate ranges from very low to 1.5 m/year (Mackay, 1973). All known pingos are younger than 10,000 years; some of them are only a few hundred years old. As temperatures rise, pingos can gradually disintegrate and become thaw lakes.

Two origins of pingo formation have been proposed. The so-called **closed-system or hydrostatic** origin suggests that pingos of the MacKenzie Delta (Canada) and Central Yakutia (Siberia) developed in areas of thick continuous permafrost (Mackay, 1979). They form in relation to lacustrine depressions. As permafrost develops, entrapped water freezes in the lake and creates a massive ice core through downward percolation and aggradation. The volume increase generates cryostatic pressure that domes the lake sediments and may even cause water to be extruded towards the surface. If water reaches the surface, it may form an aufeis; if it contains gas it may cause an explosion. Water does not generally reach the surface; instead it freezes and forms a massive ice core (Mackay, 1979) (*Fig. 15.31*). The cryostatic origin of the MacKenzie Delta pingos is supported by the fact that 98% of the 1380 mapped pingos are located at the edge of or inside contemporaneous or old lakes (Stager, 1956). A number of

open-system or hydraulic pingos are located in areas of relatively thin permafrost in Alaska and Greenland (Mackay, 1979). They form as a result of a pressure gradient due to height differences (Holmes *et al.*, 1968) (*Fig. 15.32*); for this reason, pingos often form in topographic lows such as valley bottoms or distal sectors of low angle slopes (Müller, 1959).

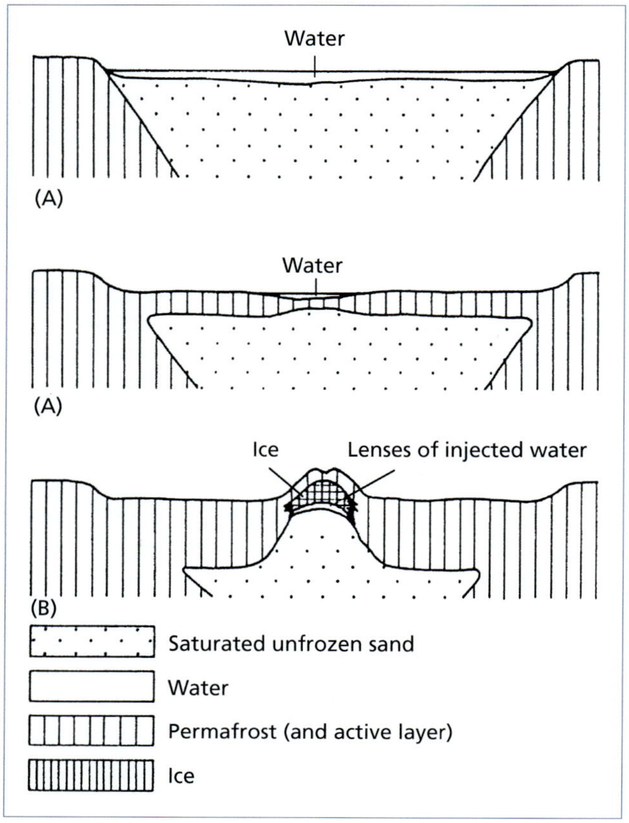

Figure 15.31 Formation of a closed pingo system by cryostatic pressure (MacKay, 1979).

15.2.3 Slope morphology and evolution

Slopes in periglacial environments have a wide range of morphologies because they develop under varying conditions of temperature, moisture, lithology, and vegetation cover. Multiple processes may be involved in their origin such as congelifraction (frost shattering), mass movement, runoff, or nivation. These processes may operate together or separately with variable intensity. Many of the resulting forms are not specific to periglacial regions; they also form in other morphoclimatic zones (French, 2007). In addition, some periglacial slope morphologies formed in the past (relict forms) and may not be in equilibrium with current conditions.

15.2.3.1 Gelifluction slopes

Differential movement and downslope displacement of colluvial deposits due to frost creep and gelifluction create landforms that can be distinguished on the basis of their geometry (Washburn, 1979). **Gelifluction sheets** have a broad lateral extent and a festooned lower edge. They form on slopes with gradients greater than 1 to 3°. **Gelifluction benches** are characterized by a terrace shape; their long dimension tends to run parallel to topographic contours. **Gelifluction lobes** (*Fig. 15.33*) have a tongue-shaped morphology and widths of 30–50 m; in combination with benches they may have gradients of 20–25°. The center of the lobe moves faster than the margins (Harris, 1981; Matsuoka, 2008). Gelifluction

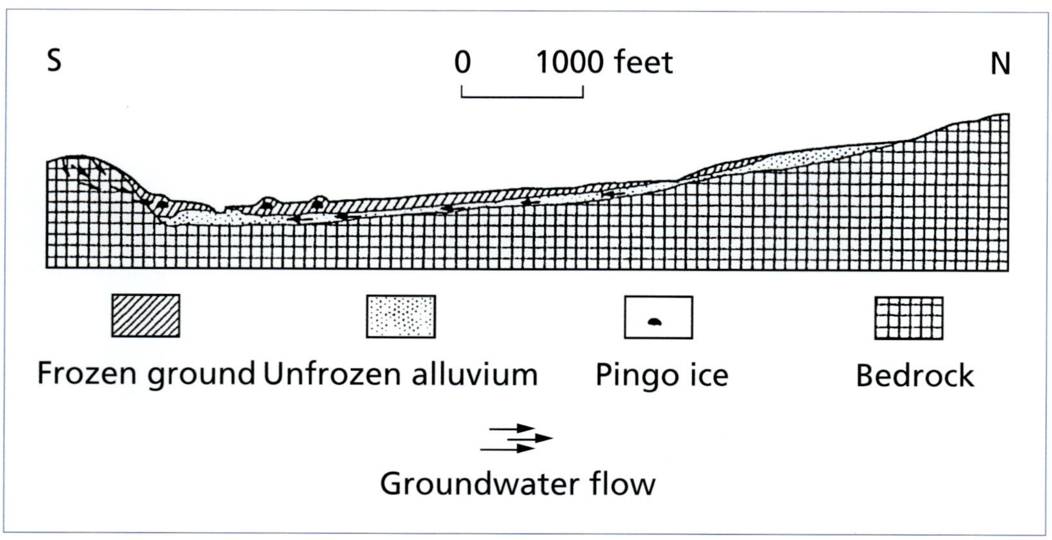

Figure 15.32 Formation of an open pingo system by hydraulic pressure (Holmes *et al.*, 1968).

Figure 15.33 Gelifluction lobes. Martin Lake Area (Alaska). Photo F. Gutiérrez.

Figure 15.34 Materials moved by gelifluction that have reached glacial ice and moved across it. Whiskey Glacier on James Ross Island, east of the Antarctic Peninsula. Photo A. Martín-Serrano.

deposits that elongate considerably in the direction of maximum slope are called **gelifluction streams**. In summer these masses may slide (as *debris flows*) over glacial ice at the surface, which is at the melting point; this promotes basal sliding (*Fig. 15.34*).

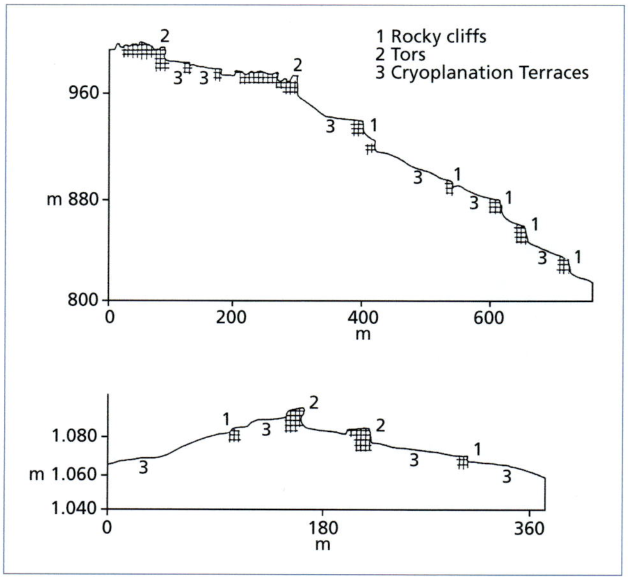

Figure 15.35 Slope profiles with altiplanation terraces and tors in the Hruby Jesenik Mountains (Czechoslovakia). (Czudek, 1964).

All these landforms develop easily on slopes with sparse vegetation and on sun-facing slopes where high solar radiation promotes thawing. Gelifluction deposits are usually heterometric; some have rough stratification. Clasts in these deposits are commonly angular and are oriented with their major axis parallel to the direction of movement (Cailleux and Tricart, 1956). These characteristics may also be present in deposits caused by gelifluction in other morphoclimatic regions but those that form in periglacial conditions are more angular and are not affected by chemical weathering.

15.2.3.2 Cryoplanation terraces and cryopediments

Cryoplanation terraces and cryopediments are low-angle planated surfaces that form in periglacial regions; they have been given multiple descriptive and genetic names in the geomorphic literature. They are intimately related to the periglacial cycle (Peltier, 1950; Birot, 1960). **Cryoplanation terraces** form in the middle and upper parts of hillslopes and mountains up to elevations of 3,000 m. They have a terraced profile and outcrops of bedrock and *tors* (rocky residual relief) (Czudek, 1964) (*Figs. 15.35* and *15.36*). These terraces form on slopes with gradients lower than 25° and at the foot of small scarps. Widening of the highest terrace may eventually form a planation surface at the summit. Terrace widths range from 5 m to more than a kilometer and their lengths range

Figure 15.36 Granite tors east-northeast of Fairbanks (Alaska). Photo F. Gutiérrez.

Figure 15.37 Cryoplanation terrace and outlet glacier at the Hardanger Ice Cap (Norway).

from 30 m to more than 10 km. The gradient ranges from 1 to 14° and the scarp height may reach 50 m. The contact between the scarp and terrace is sharp and usually has an accumulation of snow. Gelifluction deposits up to 3 m thick may cover the terraces (Priesnitz, 1988). Initially, terraces may be structurally controlled although their formation is linked to nivation processes involving material supplied by congelifraction and snowmelt water and mobilization by gelifluction and pellicular runoff. These nivation processes cause scarp retreat, widen the terrace, and increase snow retention capacity (Demek, 1969). Excellent examples of cryoplanation terraces have been documented in the Sierra Nevada (Betic Cordillera, Spain), in the Pyrenees (Gómez Ortiz, 1996), and in Hardanger (Norway) (*Fig. 15.37*).

Cryopediments are slightly dipping erosional surfaces that form at the foot of slopes in valley margins. Generally, only one level is present although in some cases several stepped levels may form. They are larger than altiplanation or cryoplanation terraces; they can extend to several tens of kilometers. Cryopediment slopes range from 1 to 12° and have a straight or slightly concave longitudinal profile. A thin veneer (<2 m) of detrital material often covers them; bedrock may crop out locally. Gelifluction landforms are common in the proximal sectors of cryopediments and patterned ground commonly occurs in distal areas (Priesnitz, 1988). Cryopediments form as a result of frost action in the source area where gelifluction and runoff mobilize detritus; this also occurs on cryoplanation terraces. The latter process is more active in cryopediments and as a result, particles are transported longer distances. In distal sectors, runoff is diffuse and particles are transported in shallow braided fluvial channels. Continuous particle production and transport cause slopes to retreat slowly and cryopediments to enlarge (Czudek and Demek, 1970a). Cryopediments have an obvious morphological resemblance to hot desert pediments although their origin is different (Demek, 1969).

15.2.3.3 Talus slopes and debris cones

Talus slopes and cones are accumulations of angular clasts on slopes. Although talus is common in periglacial environments and especially in alpine areas, it also occurs in other morphoclimatic zones such as deserts. **Talus slopes or scree** are wedge-shaped heaps of fragmented rock that cover the middle and distal part of the slope; clasts come from higher cliff areas composed of resistant rocks (Luckman, 2007). The thickness of talus slopes ranges from very thin to 30 m (Brunner and Scheidegger, 1974) (*Figs. 15.38* and *15.39*). Talus slopes commonly have a concave longitudinal profile with a higher gradient at the top. The talus slope reflects the angle of repose of the coarser particles, which ranges from 25 to 40° (French, 1996). If the scree slope is long, clast sorting may

Figure 15.38 Talus slopes and debris cones in Punta Vacas (Province of Mendoza, Argentina), Main Cordillera of the Andes. Note the significant progradation of the slope debris onto the dry valley floor.

Figure 15.39 Talus slopes and debris cones. In the foreground, fluvioglacial deposits. Punta Vacas area (Province of Mendoza, Argentina). Main Cordillera of the Andes.

Figure 15.40 Large debris cones in Punta Vacas (Province of Mendoza, Argentina), Main Cordillera of the Andes.

occur; the smaller clasts are at the top and the larger ones below, as a result of their greater kinetic energy (Washburn, 1979; Church *et al.*, 1979). The long axis of elongate clasts is usually parallel to the slope gradient.

Multiple processes can fragment rock at the rock scarp near the top of the slope although congelifraction is the dominant process in periglacial regions. The continuous supply of frost-shattered clasts causes rock scarp retreat; this process is largely controlled by rock type and structure. Retreat rates of about one mm/year have been measured in Laponia and Spitzbergen (Rapp, 1957) and one to three mm/year in Great Britain (Ballantyne and Kirkbride, 1987). Clasts move along the channel by rolling, creep, and occasionally by sliding. Movement is usually restricted to the upper half meter of the channel; velocity is highly variable, ranging from 1 to 500 cm/year (Washburn, 1979).

When rock fragments from congelifraction concentrate in a channel, they incise bedrock as they move; these channels may become preferential pathways for rock and snow avalanches. Rock fragments deposited at the foot of these channels create **debris cones**, which are similar to talus slopes although they have convex profiles *(Fig. 15.40)* and usually an obvious basal concavity (Luckman, 2007). Adjacent cones may coalesce and develop morphologies similar to talus slopes.

15.2.3.4 Block fields, block slopes and block streams

These occur on top of or at the foot of hard rocks of various lithologies that have bedding or joint planes spaced decimeters apart. Weathering, which is primarily congelifraction in periglacial environments, acts on these rocky masses and generates block-size clasts that measure decimeters or meters in addition to smaller size particles. These block accumulations may be covered by vegetation; in this case interstitial fill is present within the block framework. When fine particles at the surface are removed by water, the deposit develops an open texture in which the blocks support each other. Depending on their topographic position, these accumulations may be called block fields or plains, block slopes, or block streams, which are located in valley bottoms or in previously existing incisions on slopes (Washburn, 1979). Most of the forms described in the literature are relict (Wilson, 2007). All these morphologies occur in elevated areas at high and alpine latitudes.

Block fields are extensive accumulations of angular clasts located on plateaus or summits at high or medium latitudes (Rea, 2007) where more than half the surface is covered with blocks and smaller size particles. Block fields develop on horizontal or low gradient surfaces; an inclination of <10° is the limit between block fields and block slopes (Rea *et al.*, 1996). They show evidence of intense gelifraction activity (Smith, 1953; Dahl, 1966). Numerous block fields are present in the northwestern Iberian Peninsula (Valcárcel and Pérez Alberti, 2002; García de Celis, 2002).

Block slopes *(Fig. 15.41)* are relatively common accumulations; they occur at the foot of scarps of resistant rocks where frost weathering liberates gelifracts that move downslope. The long axes of clasts are generally oriented in the direction of maximum slope; the deposit is sorted with larger clasts at the surface and finer ones at depth (Gutiérrez and Peña, 1977; Boelhouwers, 1999). A high percentage of fine particles can be seen in distal parts of the slope; clast size decreases progressively from top to bottom. Differential movement of the accumulation causes lobular and bench-like shapes in the middle and lower parts of the slope (Gutiérrez and Peña, 1977) *(Fig. 15.42)*.

Figure 15.41 Slope of quartzite blocks. Paleozoic massif of Tremedal, Sierra de Albarracín (Province of Teruel, Spain).

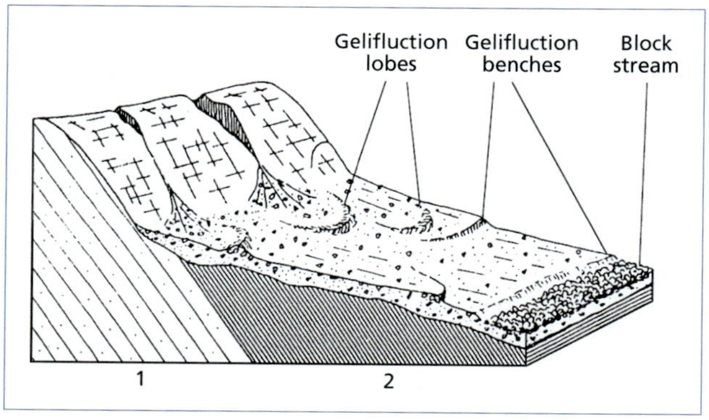

Figure 15.42 Schematic block diagram showing periglacial landforms on slope and valley in the Tremedal Massif, central eastern Iberian Cordillera. 1: Lower Valentian Quartzites 2: Valentian slates (Gutiérrez and Peña, 1977).

Block streams (*Fig. 15.43*), which are made up of large gelifracts, occupy slope gullies or the bottom of principal valleys. The latter are much less common than the former; they are supplied by blocks from the slopes and they end up forming block streams. A spectacular example of a block stream is in the Paleozoic Tremedal massif (*Fig. 15.44*) in the central-eastern sector of the Iberian Cordillera (Gutiérrez and Peña, 1977) where most main and slope valleys contain block streams. The maximum dimensions of block streams are 2.6 km long and 0.25 km wide. Their thickness is difficult to estimate because the bedrock is not visible although thicknesses of about 4 m have been measured. The longitudinal slopes of principal valleys vary widely even within the same valley. There may be transverse steps several meters tall, possibly of structural origin. On the block surface small closed depressions about a meter in diameter result from the illuviation of underlying fine material and subsequent settling.

Beneath the open-texture surface blocks, a heterometric matrix fills in between the blocks. The absence of fines at the surface is attributed to runoff washing away this material, possibly as a result of snow melt (Andersson, 1906; Smith and Smith, 1945; Smith, 1953; Potter and Moss, 1968). This lack of fine material prevents plant growth. No fabric type has observed in the Tremedal block streams, unlike other areas where researchers have observed lobe shapes in the block arrangement and high dips of the long axes in the cross-slope direction (Potter and Moss, 1968; Caine, 1972). Movement in the Tremedal must have been minor because none of the block streams in the principal valleys has reached the lower part of the massif. If this had reached the lower part, they would have generated block pediments or fans of blocks. Today, water circulates under the block accumu-

Figure 15.43 Quartzite block river. Paleozoic massif of Tremedal, Sierra de Albarracín (Province of Teruel, Spain).

lations and flows out at the foot of the transverse steps. Lichens cover the blocks, which indicates lack of movement. Harris *et al.* (2001) calculated velocities of 0.6 to 2.1 m/year in field experiments in block streams in the Qinghai Province of China.

The processes involved in the generation of these morphologies are congelifraction, ice creep, and gelifluction, sorting by ice and by movement of fine particles to the base of the deposit, and mechanical illuviation of the interstitial matrix (Washburn, 1979). Even so, the origin of the Tom Mays Canyon (Texas) block stream studied by Degenhardt *et al.* (2007) has been interpreted as a rock avalanche.

15.2.3.5 Rock glaciers

Rock glaciers are outstanding features of alpine periglacial areas. They are tongue- or lobe-shaped masses of angular clasts; they have ice in the interior and they move downslope (*Fig. 15.45*) (Barsch, 1996; Kääb, 2007). The ice cements the detritus or is an ice nucleus covered by rock fragments (Potter, 1972); it creates a surface depression as it melts. This feature has caused problems in classifying rock glaciers as glacial or periglacial forms (Tricart and Cailleux, 1962). Today, researchers agree that glaciers with an ice nucleus are part of the glacial system and those with interstitial ice are typical periglacial morphologies characteristic of permafrost environments. In addition, confusion in the distinction between block streams and slopes and rock glaciers arises from the transition that sometimes exists between these landforms (Corte, 1976).

Rock glaciers generally occur at the foot of steep walls that may be the margins of cirque amphitheaters or abrupt slopes of trough valleys (Serrat, 1979; Gutiérrez and Peña, 1981; Alonso and Trombotto, 2009) (*Fig. 15.46*). Rock glaciers are supplied by congelifraction clasts that move downslope creating talus slopes, block slopes and streams, and debris cones, and occasionally moraine deposits. They are classified morphologically (Wahrhaftig and Cox, 1959) as **tongue-shaped rock glaciers** if they have a length/width ratio greater than one, **lobate rock glaciers** if their length/width ratio is less than one, and **spatulate rock glaciers** if they are similar to tongue-shaped rock glaciers but have broader distal parts (Barsch, 1996) (*Fig. 15.47*). Some authors add other types: **composite rock glaciers** formed by the superposition of several individual rock glaciers (Washburn, 1979).

Rock glaciers range in width from 100 to 500 m; they can exceed a kilometer in length and range up to 50 m thick. Rock glaciers have scarped fronts due to their movement; the gradient at the front ranges from 35 to 45° (Barsch, 1988). The lateral margins are also abrupt. Their interior has a surface relief of arcuate ridges and grooves that is generally perpendicular to the flow direction (*Figs. 15.45* and *15.48*). The volume transported by a rock glacier normally exceeds a million cubic meters of which 40–50% is composed of clasts

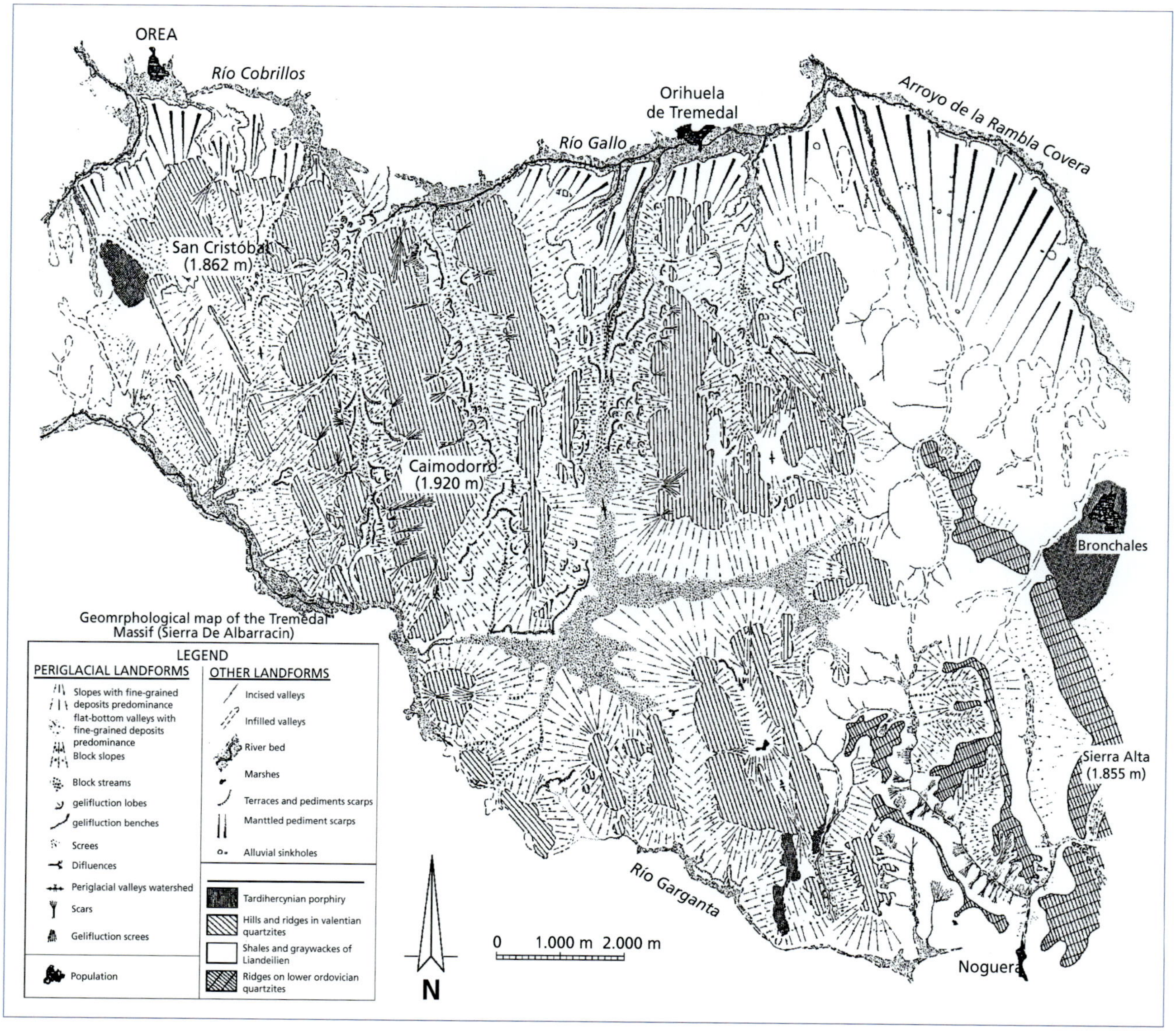

Figure 15.44 Geomorphic map of the Tremedal Massif, Sierra de Albarracín (Provinces of Teruel and Guadalajara, Spain) (Gutiérrez and Peña, 1977).

and 50–60% is interstitial ice and ice lenses. Their movement, which results from the plastic deformation of ice, is controlled by slope, ice content, temperature, and the size of clasts in the interior. Their velocity, which is usually higher in summer than in winter, ranges from several centimeters (Barsch and Hell, 1975) *(Fig. 15.49)* to 5 m/year. The highest velocities have been measured at the rock glacier in Obergurl (Innsbruck, Austrian Alps) at abrupt changes in slope. The weight of the rock glacier and the push of slope deposits on the slope above the entrainment zone help move the glacier. Rock glaciers move by creep and ice deformation processes; maximum surface velocity occurs at their center line and internal velocities decrease parabolically toward the rocky substrate (Kääb, 2007). Rock glaciers move more slowly than ice glaciers and much faster than deposits that move as a result of gelifluction.

Considering their functionality, we can distinguish several types, including **active rock glaciers,** which have an external 2 meter layer of coarse detritus and a nucleus of finer material with ice lenses in which motion can be determined by the advance of the front. Active rock glaciers are part of permafrost; they indicate continuous or discontinuous permafrost on a regional scale (French, 2007). Recently active rock glaciers have been found at 2,600 to 2,950 meters in the central-eastern Pyrenees (Spain); they move 22 cm/year (Serrano and Agudo, 1998; Serrano *et al.*, 2002). Recent

Figure 15.45 Rock glacier and moraine ridge. Breckenridge. Rocky Mountains of Colorado (United States). Photo F. Gutiérrez.

geophysical studies using radar (GPR) on various rock glaciers throughout the world have established relationships between internal structure and surface morphology. Lobate sedimentation is related to catastrophic episodes of rock falls from cirque walls. Longitudinal GPR profiles suggest that the morphological difference between tongue and lobate rock glaciers may be due to cirque geometry, the frequency and location of discharges of detritus, and the magnitude of the valley gradient **Inactive rock glaciers** do not move but still have a frozen nucleus and an external layer that exceeds a thickness of 10 m (Barsch, 1988). The first symptoms of inactivity are plant colonization on the front and the development of a detrital fan at the foot of the frontal slope. Finally, in **relict or fossil rock glaciers** (Barsch, 1977b), the internal ice has melted and thus they may have cryokarst collapse features.

Rock glaciers develop best in continental and semiarid climates because extremely wet conditions generate glaciers (Höllerman, 1983; Kääb, 2007). Research shows that active rock glaciers are less than 10,000 years old; thus they formed after the last glaciation of the Würm, that is, during the Holocene or Postglacial (Barsch, 1996). In the South Shetland Islands, west of Antarctica, Serrano and López-Martínez (2000) studied nine rock glaciers and categorized them into three groups: (a) those that postdate the last major glacial retreat; (b) those that have an age of 2000 year BP but predate the Little Ice Age, and (c) those that formed during the Little Ice Age. The age estimated by OSL (Optical Stimulated Luminescence) for block streams in the Falkland Islands ranges from 54 to 16 Ka, which suggests that they have been active since 16 Ka (Kääb, 2007; Hamson *et al.*, 2008). André *et al.*, (2008) think they formed by denudation of an alteration profile of subtropical or temperate origin, possibly of Tertiary age, and were later reworked by periglacial processes.

15.2.3.6 Grèzes litées

These are slope deposits of stratified angular clasts (*Fig. 15.50*). Some authors call these deposits *éboulis ordonnés* although others use this term for stratified screes formed by gravity. In **grèzes litées** the strata are layers of equigranular pebble-sized clasts alternating with finer material that is predominantly sand, silt, and clay (Tricart, 1952a). If the deposit contains numerous boulder-sized fragments, it is called *groise litée*. In the grèzes, the large clast layer has a grain-supported structure; that is, there is no matrix in the interstices, and as a result, the clasts are loose. When these deposits are composed of limestone components, the layer of fine material is usually cemented because it retains percolation water. The transition from one layer to another is usually clear and channel-shaped structures can sometimes be observed. Unlike stratified screes, stratification in a grèzes litées deposit has nothing to do with the present-day topographic surface. Grèzes litées occasionally fossilize glacial moraines as in the Andes of Ecuador and Bolivia (Heine, 1995) or in Santa Elena at the head of the Tena Valley in the Central Pyrenees (Spain). Grèzes litées occur in any orientation although they are most commonly located on slopes with southern exposure. They are found on slopes that range from 7 to 45°; they can be 40 m thick. The layers tend to be thicker downslope (Guillien, 1951).

Grèzes litées are relatively common in mid latitudes but are unknown in polar zones. Their formation requires thermal oscillations around 0°C that enable congelifraction to generate gelifracts, enough water to sort the deposit, and scant or non-existent vegetation (DeWolf, 1988). Stratification and sorting are interpreted to be due to meltwater from snow accumulations, which is the essential agent for sedimentation of coarse material and the eluviation of the

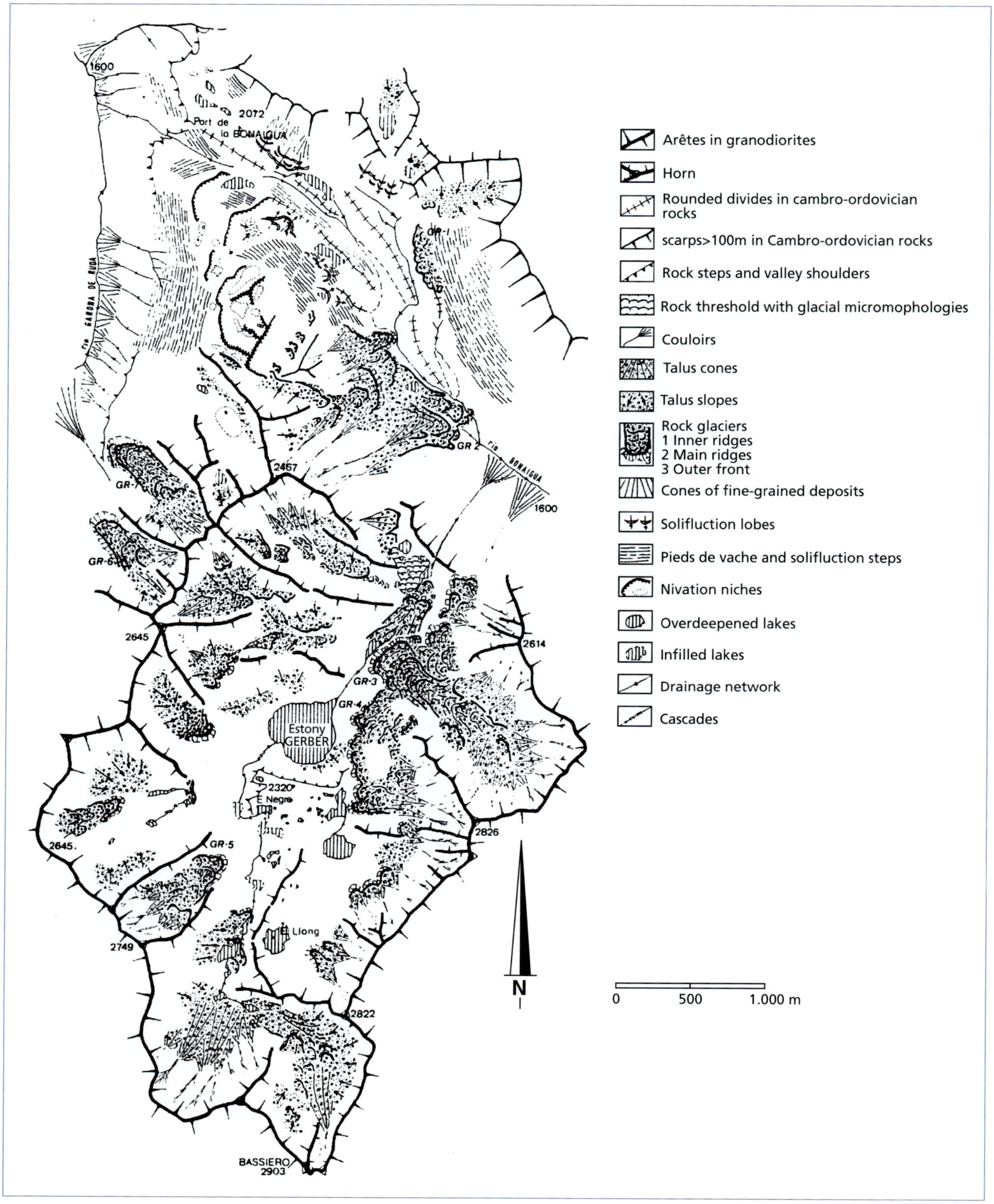

Figure 15.46 Morphological map of the Bonaigua area (Lérida Pyrenees, Spain) (Gutiérrez and Peña, 1981).

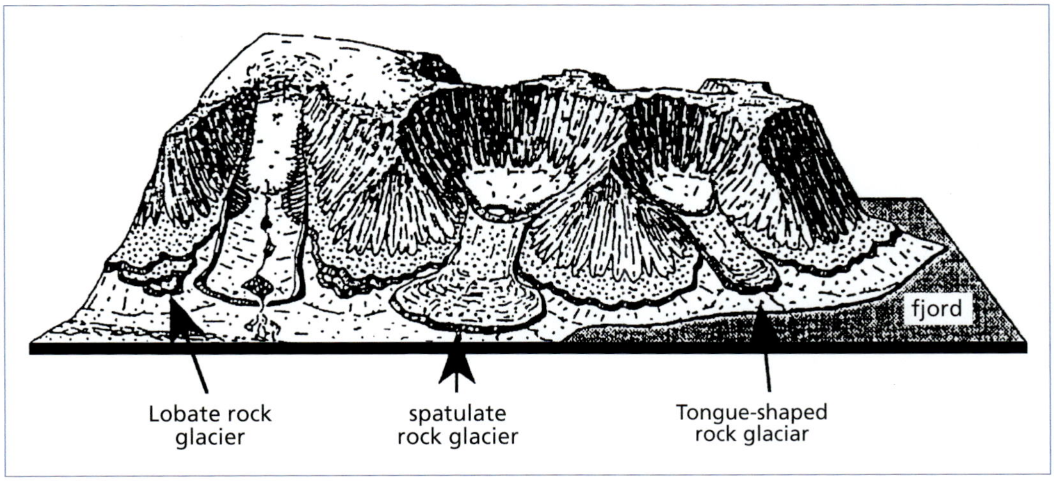

Figure 15.47 Various types of rock glaciers in mountain polar environments (Barsch, 1996; modified by Humlum, 1982).

Figure 15.48 Ridges and arched furrows in a rock glacier situated on the right margin at the head of the Garona River (Pyrenees, Spain).

smallest fines (Guillien, 1951). Francou (1988, 1990) and Lautridou and Francou (1992) observed typical stratified slope deposits in the Andes of Perú at 4,400 to 4,900 m of elevation. These deposits have gradients of 33 to 35°; they are characterized by large solifluction lobes that move several centimeters a year. A cross section of the deposit along the lobe margin reveals its internal structure; the lobes move very slowly and develop like a carpet that buries clasts under the lobes (*Fig. 15.51*). At the same time the ice push/lifting separates fines from clasts, which move by ice *creep* and *pipkrake* at higher velocities than solifluction lobes.

Van Steijn *et al.* (1995) show that research on the morphology and sedimentology of stratified slope deposits can be used to infer the processes that formed them. Vari-

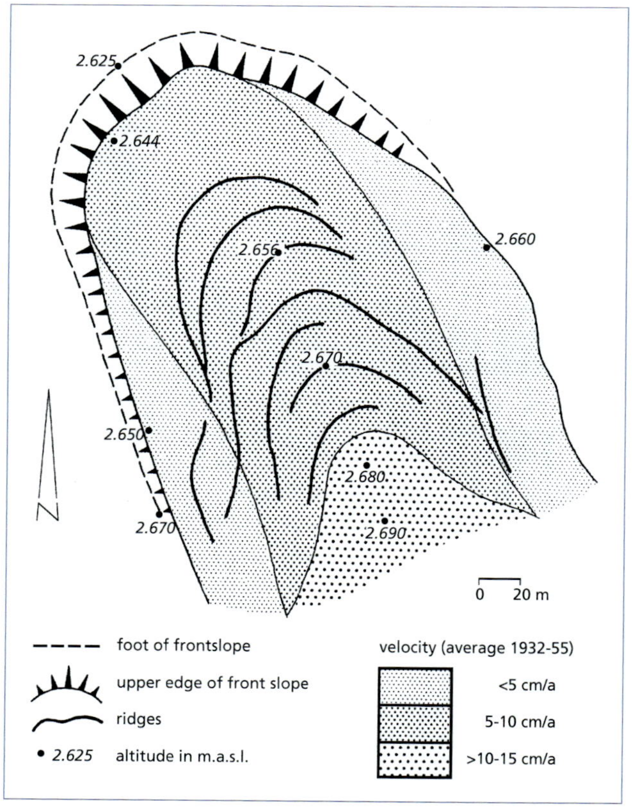

Figure 15.49 Zones of equal velocity of the Murtél I rock glacier, Corvatsch, High Engadin, Swiss Alps. The velocity of each zone is the average for the period of 1932–1955 (23 years); the average for the entire rock glacier is 7.1 cm/year (Barsch and Hell, 1975).

Figure 15.50 Grèzes litées in Entrambasguas, Sierra de Albarracín (Province of Teruel, Spain).

ous origins have been proposed, which may act alone or in combination: solifluction laminae or lobes, flows of dry particles or particles cemented by ice, *debris flows*, and ice-eolian transport. Researchers also discuss the problem of convergence of landforms.

In the central Pyrenees (Spain) a large number of grèzes litées occur primarily at elevations of 600–800 m. Researchers think there may be more than one phase of formation although the principal stage is the Late-Glacial (Peña *et al.*, 1998) date of stratified debris at Eripol, 20,060 ± 180 BP. García Ruiz *et al.* (2000, 2001) think the deposits formed by gelifluction lobes and debris flows during the coldest stage of the upper Pleistocene and during the Late-Glacial.

15.2.3.7 Nivation landforms

Nivation niches are found on slopes beneath a thin film of snow *(Fig. 15.52)*. Meltwater penetrates into bedrock material and at night when temperatures reach the freezing point, frost weathering occurs. The particles generated by this process can move downslope by gelifluction and ice creep aided by runoff; they can be seen as debris accumulations at the foot of the niche (Thorn, 1988b). Particle leaching inside the nivation niche generates periglacial pavements. This evidence has led to the idea that the most important nivation process is the freeze-thaw cycle. In low latitude areas such as the Cordillera of the Andes, more than 300 annual freeze cycles have been recorded (Troll, 1944); these are areas where nivation may be potentially significant. The continuous activity of nivation processes enlarges the niches, which begin as small depressions about a meter across; later they can reach dimensions of almost a kilometer if they coalesce with other niches. They enlarge along margins where melting and subsequent congelifraction are most active because the protective action of the snow in the center prevents the freeze-thaw process. These margins are modified throughout the year as a function of seasonal variations. Niches that reach large dimensions are called **thermocirques** (Selby, 1985). Nivation niches have a

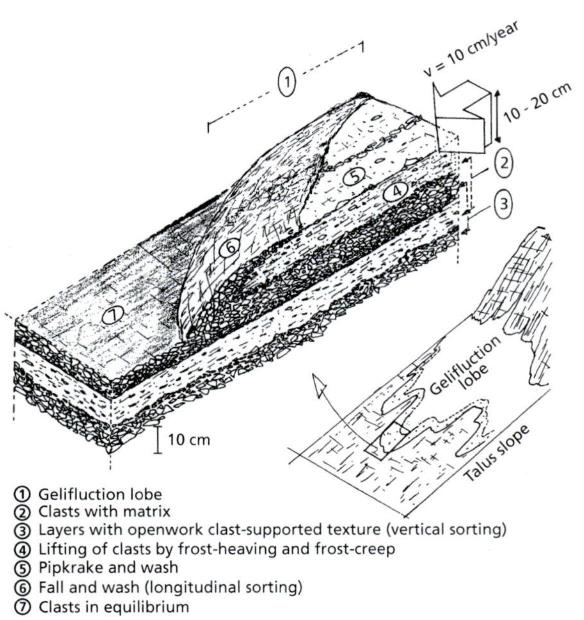

Figure 15.51 Origin of stratified slope deposits (Francou, 1988).

scarp or arched amphitheater and a terrace or low gradient slope at their base, which aids in the removal of detritus (Christiansen, 1998) (*Fig. 15.53*). They form more rapidly on rocky outcrops than in vegetated soils; they have been observed to form in a few years in nivation niches beneath snow patches. The niches are embryonic cirques where glaciers gradually take the place of nivation processes when snow that accumulates in the interior of the niche persists long enough to be converted to ice (Tricart and Cailleux, 1962).

Nivation ridges can form at the foot of snow accumulations that are present at the base of scarped slopes; these are gelifract ridges (Butzer, 1976; Luckman, 2007) or **protalus ramparts** (Washburn, 1979) (*Fig. 15.54*). Clasts from the upper rocky part of the slope slide on top of the snow and accumulate at its margin. When the margin melts, the ridge remains behind and gets disconnected from the slope profile. Ridges that are parallel to each other have occasionally been observed; this indicates parallel slope retreat (Sharpe, 1938) (*Fig. 15.55*).

15.2.3.8 Slope evolution

Slopes that form in periglacial environments are classified according to their shapes and the processes that generate them. The primary types are 1) cliff-talus slopes, which are

Figure 15.52 Nivation niches. Hardanger Massif (Norway).

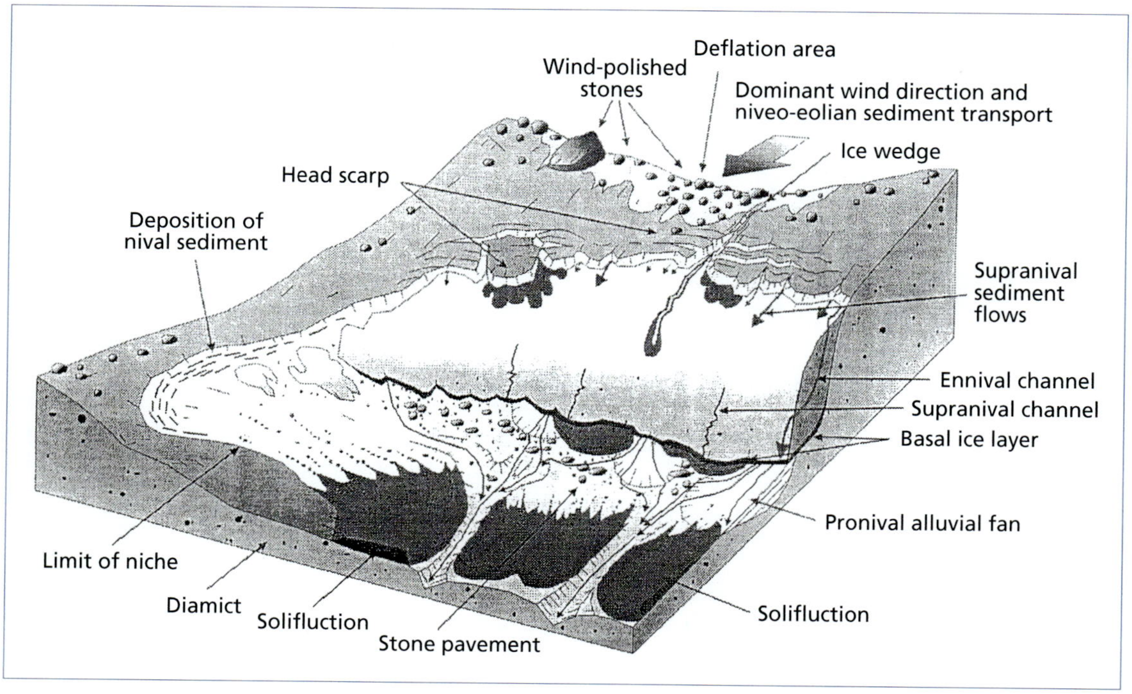

Figure 15.53 Nivation model in the high Arctic for unconsolidated landforms based on observations in the Zackenberg area in northeastern Greenland (Christiansen, 1998).

Figure 15.54 Nivation ridges. Turbón massif (Aragonese Pyrenees). Photo J.L. Peña.

concave and shaped like talus accumulations and detritus cones, 2) benched slopes formed by cryoplanation processes, and 3) convex-concave slopes with continuous cover of solifluidal material (French, 1996) (*Fig. 15.56*).

Cliff-talus slopes evolve by parallel scarp retreat; over time the slope lengthens and its angle and the length of the detritus layer decrease (Carson and Kirkby, 1972). In general, scarps in periglacial environments retreat more slowly than in other morphoclimatic zones although this is a function of lithologic and structural characteristics and microclimate, such as orientation (Souchez, 1966).

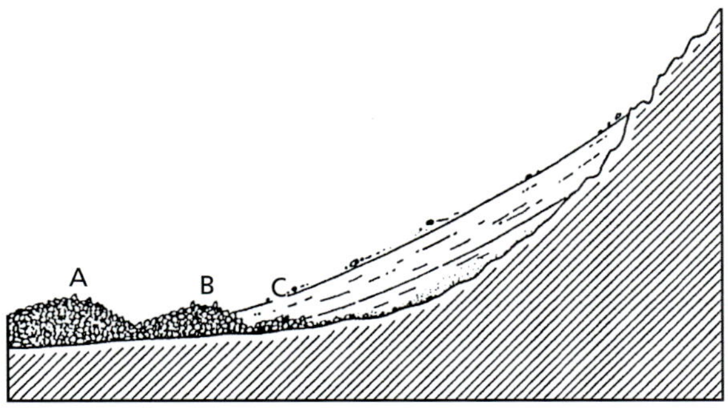

Figure 15.55 Formation of nivation ridges by clastic slip on top of a snow accumulation (Sharpe, 1938).

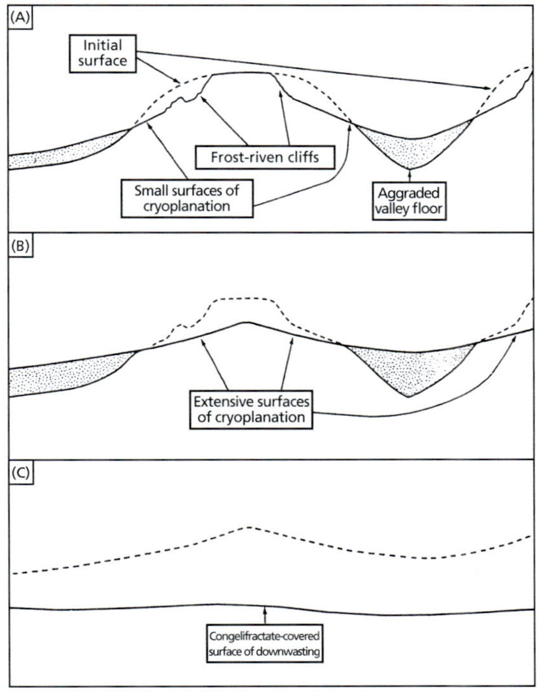

Figure 15.57 The periglaciation cycle (A): youth, (B): maturity, and (C): senility (according to Peltier, 1950, in Small, 1970).

Figure 15.56 Periglacial slopes covered with heterometric material on top of a substrate of detrital Cretaceous lithologies. Fluvial action undermined the base of the slope during the thawing period. James Ross Island, east of the Antarctic Peninsula. Photo A. Martín-Serrano.

Benched slopes result from cryoplanation processes that soften and flatten the profile. Continuous erosion of high areas and sedimentation in low areas, primarily in valley bottoms, involves parallel scarp retreat by congelifraction and enlargement of benched areas. These ideas were expressed by Peltier (1950) in his periglacial "erosion cycle" (*Fig. 15.57*).

Concave-convex slopes have a convex-concave profile that lacks transitions; they are covered by a continuous deposit of gelifracts (*Fig. 15.58*) and are affected by gelifluction, surface washing, and rills in high gradient areas. These slopes may evolve into **tripartite slopes** or **triangular slope facets** (Büdel, 1982) when they are deeply dissected by gullies in the middle part of the slope where tributary gullies flow together (*Figs. 15.59* and *15.60*). Material that has been eroded from the lower parts of slopes is deposited by gullies as small alluvial fans. Once the triangular facet has formed, denudation processes change so that each facet constitutes a relict landform that functions independently of higher parts of the slope. The alternation of accumulation and dissection stages creates a sequence of tripartite slopes separated from each other and in turn parallel to the scarp. Research on triangular facets on Seymour Island (Antarctica) shows that the transition from the accumulation stage to the dissection stage is a geomorphic threshold related to glacio-isostatic adjustment and/or climate change (Gutiérrez *et al.*, 2011b).

15.2.4 Fluvial landforms

Braided channels with significant bed load, high inclination, and high width/depth ratio are the most common channel type in periglacial environments. During summer flood periods, these channels do significant mechanical erosion on channel margins. In addition, runoff water melts the permafrost and undermines river margins; this creates **thermo-erosional niches** (Czudek and Demek, 1970b; Walker, 1973) or thermal-fluvial erosion (French, 2007) (*Fig. 15.61*). This erosion may extend to heights of 3 meters and depths of 1–3 meters. When the niches collapse, they release large masses that provide new load to the river. This explains why thermal erosion is more effective than mechanical erosion but this is true only of non-cohesive sediments (Jahn, 1975). The same erosion occurs along coastal cliffs.

The morphology of valleys that form in periglacial zones has not been well studied; most research is related to the problem of valley asymmetry. The scientific literature on Pleistocene valleys is more abundant; these are relict forms in areas that were subject to periglacial processes such as wide areas of central and western Europe.

Asymmetric valleys are valleys in which one side is notably more inclined than the other. They are common in middle latitude periglacial environments. The asymmetry is not specific to periglacial environments; it can result from

Figure 15.58 Smoothed slopes, possibly relict. Sierra de Javalambre, Province of Teruel (Spain).

various factors, primarily lithological and structural. However, in periglacial regions in the northern hemisphere, the most inclined slope is oriented toward the north. More insolation on sunny slopes causes rapid and long-term thawing, which generates a larger amount of water to trigger gelifluction and mass movement. The eroded materials direct fluvial currents toward shady areas and cause basal undermining; as a result, these slopes will be more steeply inclined (Carson and Kirkby, 1972). In addition, north-oriented slopes experience less thawing during the summer so permafrost helps preserve

Figure 15.59 Triangular slope facets. Northeastern Isle of Seymour. Antarctic Peninsula. Photo F. Nozal.

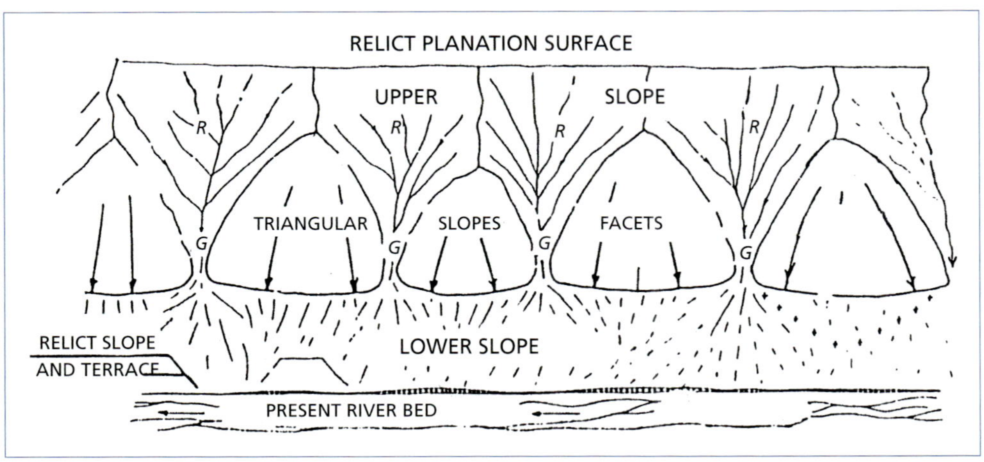

Figure 15.60 Tripartite slope in southeastern Spitsbergen. R: rills on the tops of the slope, C: dissection gullies on the slope (Büdel, 1982).

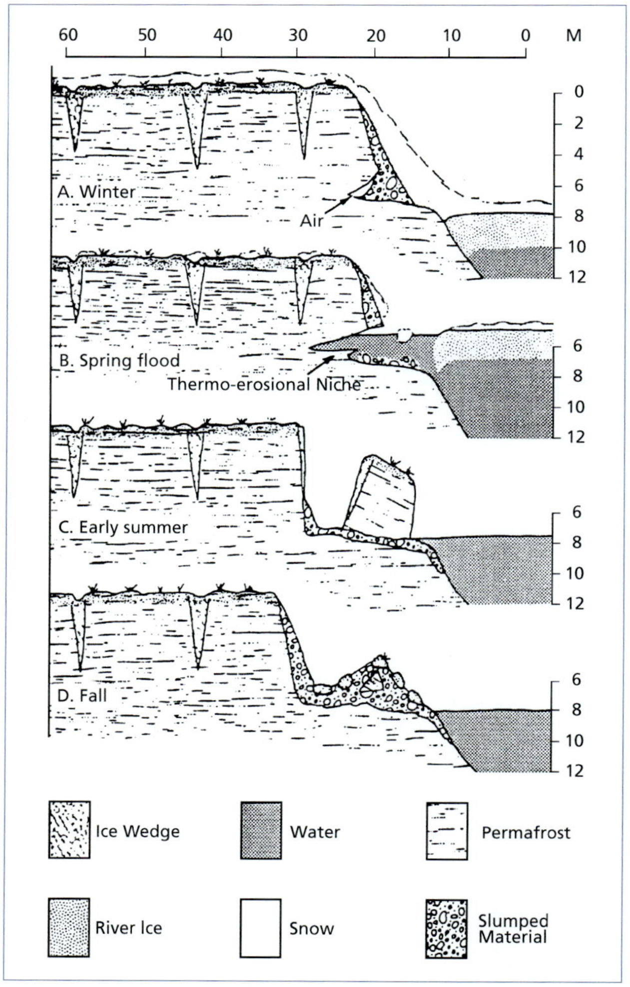

Figure 15.61 Thermal erosion of a fluvial margin (Walker, 1973).

Figure 15.62 Small collapse affecting slope deposits caused by melting of underlying ice during the southern summer. James Ross Island, east of the Antarctic Peninsula. Photo A. Martín-Serrano.

their steeper slope (Gravis, 1969; Czudek, 1973b; Karrasch, 1993). Another reason that may explain valley asymmetry is that dominant wind activity promotes snow accumulation on leeward slopes (French, 1971). During summer, melting provides a large amount of water that accelerates mass movement on slopes.

Low order segments of fluvial basins in non-mountainous regions of middle latitudes have different morphologies. One landform is the **U-shaped valley** (Tricart, 1967), which is U-shaped in cross section and has a longitudinal profile with slight reverse slopes, indicating that runoff plays a minor role. The slopes have a convex-concave morphology because gelifluction is the dominant agent of detrital transport. The valley bottoms are primarily composed of unsorted gelifluction deposits from the slopes. These valleys are common in old periglacial regions of Europe and most are now dry. U-shaped valleys generally do not exceed 3 km in length and toward the bottom they flow into a valley with stream flow or they may become **flat-bottom valleys**, which are the most characteristic landforms in Budel's (1977) ectropical zone. In Iceland they are 5–100 m wide or more, 5–20 m deep and have a gradient of 4–11% (Schunke, 1975). Their bed is primarily composed of non-sorted detritus from slopes and materials that have undergone some longitudinal transport. The connection between the flat bottom and the slopes is concave; it shows the lateral solifluction supply and the inability of the river to remove these deposits. In other places, the junction between the slope and valley bottom is a basal angle that shows significant erosion at the foot of the slopes during flooding (Tricart, 1967). Like U-shaped valleys, these flat-bottom valleys are primarily dry valleys today (Jahn, 1975). Dry valleys are also called *dells*; they have been described in

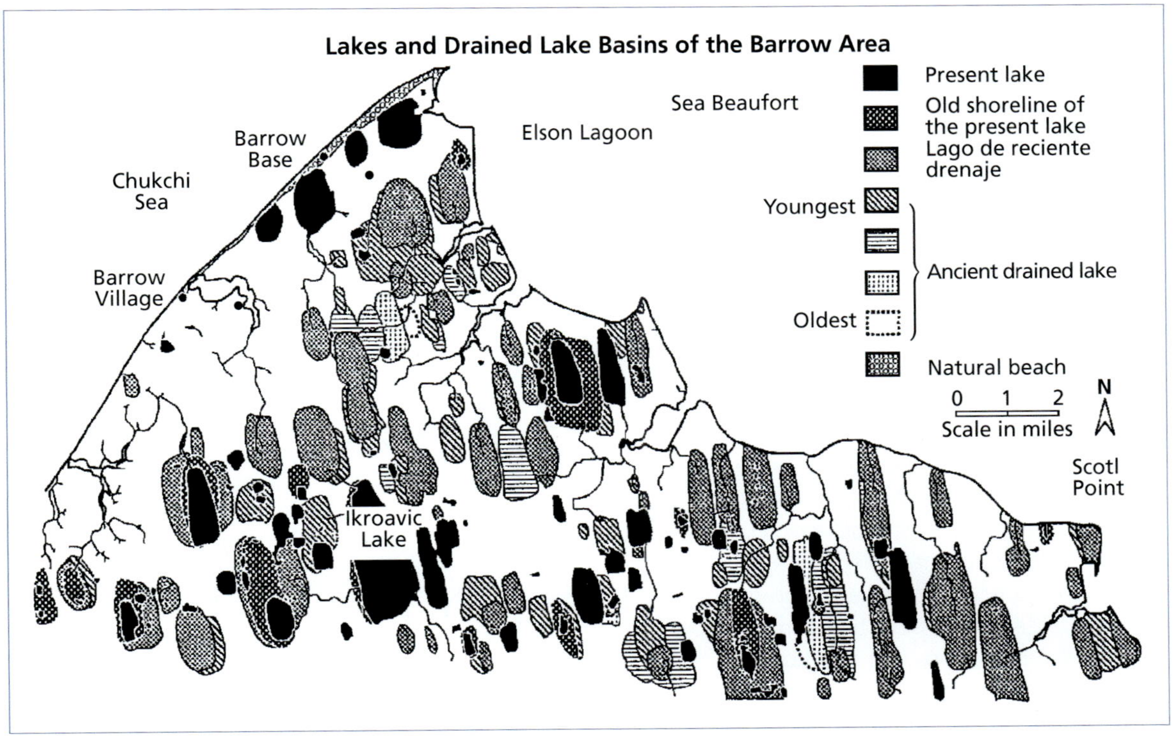

Figure 15.63 Oriented lakes on the Arctic coastal plain, northern Alaska (Carson and Hussey, 1962).

central-western Europe (Washburn, 1979; French, 2007). This dryness may be due to the considerable thickness of the bottom deposits where water can percolate easily by flowing through alluvium. However, dry valleys form in other environments such as karst areas.

15.2.5 Thermokarst or cryokarst

Thermokarst is the term for landforms generated by land surface subsidence (*Fig. 15.62*) resulting from thawing of the upper part of permafrost. The resulting landscape may be highly irregular and even chaotic in places (thermokarst relief in badlands of French, 2007). The term thermokarst is hardly appropriate since the forms produced are not at all similar to exokarst landforms present in carbonate terrains (Pissart, 2004). Thermokarst morphologies occur primarily in low-relief areas such as alluvial sediment plains because the resulting thaw lakes rarely occur in zones of high relief (Williams and Smith, 1989). The landforms are usually active and as a result, periglacial landscapes are rapidly modified in short periods of time (Harry, 1988).

Thawing results from multiple causes. Change of climate toward milder temperatures is the most significant cause affecting large areas. Modifications of plant cover due to fires, deforestation, and construction are locally significant because the loss of cover can trigger thawing due to removal of the insulating vegetation layer. Other geomorphic causes are thermal erosion resulting from standing water, slumping, and mudflows that erode the active layer (French, 1987).

Thaw lakes are the most common forms. They are shallow, less than 4 m deep, and are not wider than 2 km (French, 2007). Thaw lakes form from melting of the permafrost, which triggers ground subsidence that later fills with water. They occur primarily in plains and fill rapidly with sediments and peat. Thaw lakes usually do not persist more than 3000 years. Groups of **oriented lakes** are rare. The best example occurs in the coastal arctic plain of Alaska (Carson and Hussey, 1962) (*Fig. 15.63*). They are elliptical and sometimes rectangular in shape, two or three times longer than wide, with longer axes up to 5 km and depths of 2–6 meters (Black and Barksdale, 1949). In recent work, Hinkel *et al.* (2005) show them to be 15 km long and 6 km wide. They may have complex morphologies due to coalescence of several lakes. Their origin is uncertain; they may be due to the dominant wind activity that deposits sediments on shores perpendicular to the wind directions. The deposits isolate the thawing shores and protect them from wave action; for this reason, beaches have a very smooth gradient. From this time forward, reflux toward the extremities of the lake causes a stronger current in these parts and has an abrasive action that enlarges the lake. Measured retreats are rapid, on the order of 25 m/year (Carson and Hussey, 1962). Even so, considerable controversy surrounds explanations for the orientation of these lakes.

Pingos and collapsed palsas can form closed subcircular depressions; they are shallow and have elevated borders composed of sediments that dip toward the depression (Pissart, 2000, 2002). They fill with lacustrine sediments and peat (De Gans, 1988) (*Fig. 15.64*). These morphologies occur in Europe and Asia in areas that were previously periglacial regions. They are difficult to distinguish from other thermokarst depressions or other fluvioglacial kettles.

When the frozen soil crops out and melts, **thaw collapses** form; these are arc-shaped depressions that open in the downslope direction. They may unite to form a festooned scarp. As they thaw, the surface permafrost becomes saturated with water and flows, leaving a scar several meters high. These cracks retreat at a rate of 7–10 m/year (French and Egginton, 1973), one of the most rapid erosional processes in the periglacial environment. When these collapses occur in areas of ice wedging they create large collapses.

An **alas** is a closed depression that forms in low relief regions; it has abrupt margins and a bottom covered with grass and water. The name comes from central Yakutia (Siberia) (Soloviev, 1962, 1963 in Jahn, 1975; Czudek and Demek, 1970 b; Soloviev, 1973; Yershov, 1998). Alases are rounded or elliptical and often contain a shallow lake at their center. Their diameter ranges from 0.1 to 15 km and their depth from 3 to 40 m. Alases form as a result of permafrost thawing; the coalescence of several depressions may form alas valleys tens of kilometers long. As thawing progresses, different morphologies appear (*Fig. 15.65*). Melting of ice wedges leads to linear depressions separated by bulged cells

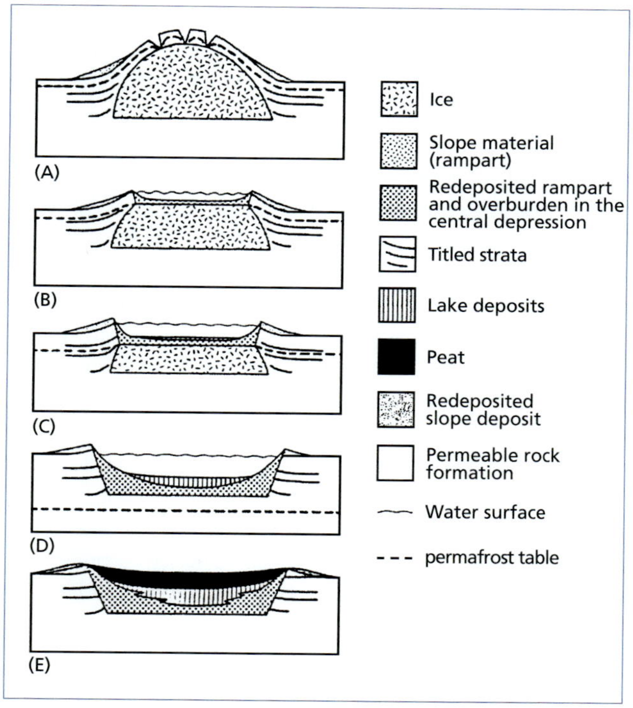

Figure 15.64 Stages of pingo disintegration showing the evolution of the central depression and the elevated margin (de Gans, 1988).

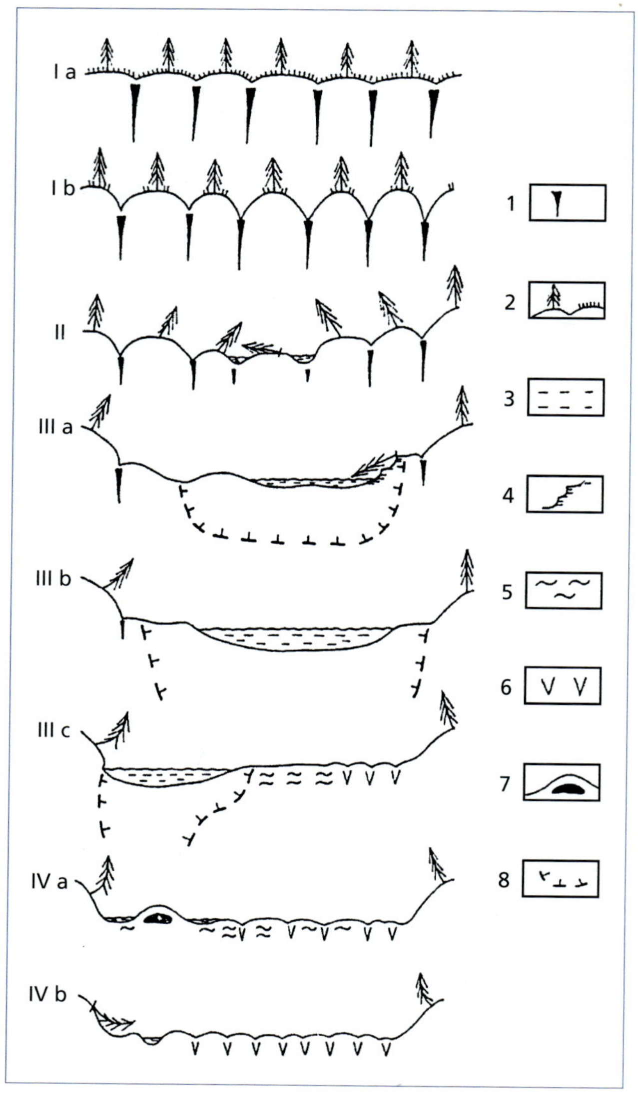

Figure 15.65 Principal stages in the development of alases Ia, original surface with syngenetic ice wedges; Ib, initial stage of thermokarst baydjarakhs; II, duyoda; IIIa, juvenile alas; IIIb, mature alas; IIIc, senile alas; IVa khonu with pingo; IVb Khonu with depression in the place of the collapsed pingo. 1, Wedges that are syngenetic with ice; 2, Pines and grass; 3, Water; 4, Landslides; 5, Alas deposits; 6, Epigenetic ice veins; 7, Pingo; 8, Upper limit of permafrost. (Czudek and Oemek, 1970b; based on publications of Soloviev).

called *baydjarakhs*, which are 3–4 m high and 3–20 m in diameter (stage I). The progressive melting of ice produces a closed depression with disintegrating conical hills known as *duyoda* (stage II). Stage III of alas formation occurs when the bottom is devoid of trees and covered with grass, abrupt margins are present, and lakes perhaps occupy the interior. If the permafrost is reactivated, ice veins and pingos can form at the bottom of the depression (*Khonu*); these features may collapse later (stage IV). In Yakutia, 40–50% of the original surface has been modified by alases; they took several thousand years to form and are related to the Climatic Optimum (2500–9000 BP). At present more than 10% of this territory is active thermokarst.

15.3 Applied geomorphology in periglacial regions

15.3.1 Introduction

Periglacial processes operate in both alpine and high latitude environments. Both are fragile areas where human activity can cause significant disturbances (Ritter, 1986). Permafrost ground poses numerous problems not only for cartographers but also for engineers, builders, miners, oil and gas producers, climate scientists, archeologists, and anyone interested in polar and alpine regions (Heginbottom, 2002). Geocryological engineering is the study of frozen ground and its impact on human life and activities (Yershov, 1998).

During the last few decades many mountainous areas have come under significant pressure as a result of human activities. In developed countries, increasing tourism and other recreational activities are the primary cause of this pressure (David *et al.*, 2009). In developing countries, increased soil use in marginal areas puts pressure on low areas. In some underdeveloped zones, tourism is beginning to be a problem. In addition, deforestation and overgrazing in mountainous areas can modify the activity of geomorphic processes. These factors and human activities have increased natural risks such as avalanches, soil erosion, landslides, flooding, etc. (Gerrard, 1990).

In arctic and subarctic regions the principal problems of economic development center on freezing of the land surface (congelifraction and frost heaving) and thawing of permafrost. These circumstances have limited development in these areas; man has had to resolve these problems with numerous and novel technical solutions (Sudgen, 1982; Walker, 1986).

Humans have occupied high latitudes throughout history (Cooke and Doornkamp, 1990, Harris, 1988) as shown by hunter and gatherer groups whose origins go back some 40,000 years (Sudgen, 1982) but human penetration with advanced technology for exploiting animal, plant, and mining resources is quite recent. This migration comes primarily from circumpolar countries in the northern hemisphere (Soviet Union, United States, Canada, and Scandinavian countries). Russia has a long history dating back to the construction of the trans-Siberian railroad in the 19th century and continuing into the 20th century with the building and installation of modern cities in permafrost areas, such as Norilsk. In North America development began with gold mining in Alaska and the Yukon Territory at the end of the 19th century. After World War II, development increased with the construction of airports and transport routes; it reached its maximum intensity after petroleum was discovered in 1968 in Prudhoe Bay in northern Alaska (Mackay, 1972).

Permafrost areas have few engineering problems if development does not disturb the permafrost (USGS, 1983) but modified permafrost rapidly forms thermokarst and requires many years to re-establish an equilibrium surface (Harry, 1988). Natural causes such atmospheric climate change can cause disturbances that lead to increased freezing or thawing. Little is known about these causes (Williams and Smith, 1989) although researchers think that increased thermokarst topography in some zones of Russia is due to a warmer climate in recent decades. The effects of mild climate variation may significantly affect areas that have median temperatures near 0°C, which includes an extensive area. These modifications cause significant changes in hydrologic regimes, especially runoff. Anthropogenic changes from deforestation, construction of linear projects (highways, railroads, airports, oil pipelines), other construction, and drainage of lakes all lead to thawing. Thawing makes the unstable soil prone to mass movement, subsidence, and ice heaving (USGS, 1983; French, 2007). These facts show that permafrost is highly sensitive to human activity.

To alleviate problems related to anthropic activity in periglacial environments, it is important to understand the processes related to soils that are seasonally or permanently frozen, to adequately evaluate the active forces, and to understand their relationship to environmental variables. This knowledge comes from research on natural landforms and from experience obtained in recent decades on soil behavior related to human activities (Harris, 1986). Another source of knowledge comes from the artificial freezing of soils, which has been done to increase their strength for construction projects (Williams and Smith, 1989). This technique is used to construct mining galleries and tunnels (the metro in various locations in Japan and the tunnel below the Seine in Paris, France) and to provide important data on the behavior of frozen soil. Even so, some processes related to freeze-thawing are very complex and thus are not well understood, such as heat and moisture transfer in frozen soil or the rheological properties of materials. With adequate understanding it is possible to develop construction practices that minimize negative effects. Today **Sensitivity and Land Classification Maps** at scales of 1:250,000 and **Susceptibilities to Land Modification Maps** at a scale of 1:50,000 are available to aid in the determination of areas where land is susceptible to disturbances and thermokarstic subsidence caused by anthropic activity (French, 1996).

15.3.2 Snow avalanches

Avalanches are common in alpine mountains (*Fig. 15.66*); they constitute a significant hazard in regions of developed countries where human activities such as tourism and sports have become popular (López-Martínez, 1989; Vilaplana, 1989; Gerrard, 1990; Ríos and Sáez, 2002). In British Columbia, officials have recorded 25,000 avalanches that have affected highways (McClung, 2003). Snow avalanches may incorporate various types of detritus (rock fragments, soil, trees) that increase their destructive power (Keylok, 1997). They can be triggered by earthquakes (Voight and Pariseau, 1978) (*Fig. 15.67*), thunder, vibrations from human activities, animal movement, ultrasonic blasts from jet engines, activity of skiers, and other activities.

Numerous avalanche disasters have caused significant loss of human life and have affected human activities in various ways (Davis, 1992). At the beginning of October in 218 BC while crossing the Col de la Traversette in the Alps on his march toward Rome, Hannibal and his army were surprised by an avalanche that killed 18,000 soldiers, 2,000 horses, and numerous elephants. During the First World War, between 4,000 and 8,000 soldiers were killed in the Tyrolean Alps, not by war but by avalanches. Well known avalanche disasters have buried mining camps, killing 300 people in Blons in the Austrian Alps (11 November 1954) and 27 miners in Camp Leduc, British Columbia (18 February 1965). In the Carpathians of Czechoslovakia (8 March 1956) strong winds in Tatra caused an avalanche that killed 16 loggers (Bolt, *et al.*, 1975). In 1978 an avalanche in the Col de Messer (Switzerland) covered 60 skiers who were waiting for a ski lift. In 1910 in Wellington (USA) in the Rocky Mountains, an avalanche struck two trains and killed 96 people (Bryant, 1991). Avalanches in Spain killed 45 people between 1978 and 1988 (López-Martínez, 1989) and 43 people between 1988–1995 (Ríos and Sáez, 2002).

Today meticulous research is done to evaluate risk in areas where avalanches interfere with human activities. This research includes measurement of snow layer thickness because avalanche risk increases with snow thickness. Wind can redistribute snow and accumulate it in leeward areas. Temperature also plays an important role because it controls the metamorphism of snow accumulations. Researchers also study snow layer profiles to identify existing discontinuities that are potential sliding planes (McClung and Schaerer,

Figure 15.66 Snow avalanche channelized and deposited as a fan in the distal area. Viso Massif, (Franco-Italian Alps).

1993; Furdada, 1996). Researchers use classifications and scales of snow characteristics such as resistance to penetration, shape and size of crystals, water percentage, surface roughness, and so on (UNESCO, 1981). It is also important to conduct research on slope gradient (maximum risk is between 25 and 50), shape (convexity promotes avalanche formation), and orientation with respect to wind and insolation. Knowledge of the location and trajectories of previous avalanches is critical; it can be obtained by surveys and historical studies. Researchers also calculate the periodicity of avalanches and the distances they travel. All this data is statistically analyzed. This research provides valuable data for avalanche risk evaluation (López-Martínez, 1988; McClung and Schaerer, 1993; Furdada, 1996; Ancey, 1998).

In recent years, fewer avalanche disasters have occurred because of these studies and the application of controls and protective measures to mitigate risk (*Fig. 15.68*). Mitigation methods may be passive: for example, the installation of wind deflectors and snow fences to prevent snow accumulation in leeward areas. Another common technique is the installation of terraces and barriers on slopes (*Figs. 15.69* and *15.70*) to decrease the velocity or alter the path of an

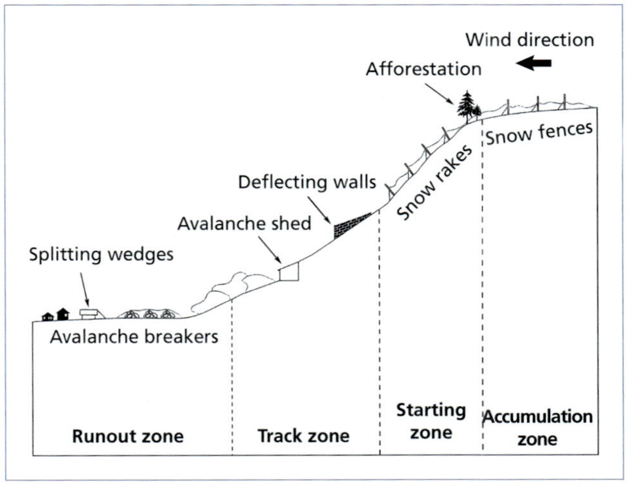

Figure 15.68 Methods to reduce avalanche threats in mountainous areas, including repopulation of forests, rakes, refuges, and deflecting walls (modified from Smith, 2001).

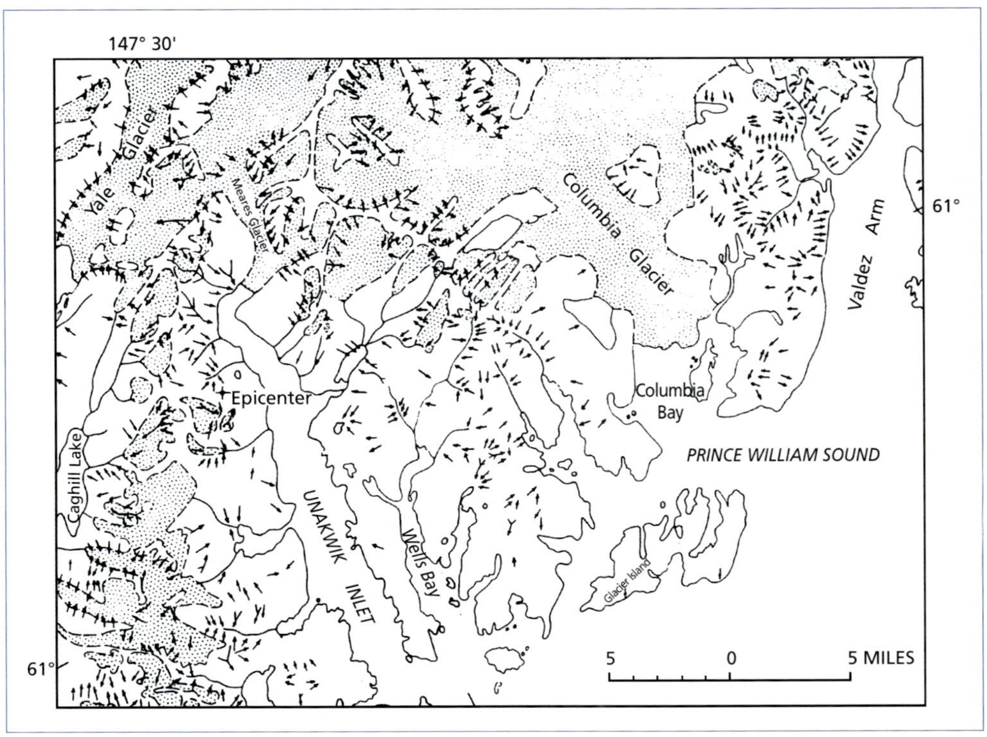

Figure 15.67 Distribution of landslides and avalanches with direction indicators, triggered after the magnitude 8.5 Alaskan earthquake of 27 March 1964. The area is near the epicenter; of the 2036 landslides and avalanches that occurred, 20 were rock avalanches, 58 were snow and rock, and the rest were snow only. The speckled zones are glaciers (Hackman, 1965; in Voight and Pariseau, 1978).

Figure 15.69 Slope rakes to slow down avalanches. Passo de Pordoi. Dolomites, Italy.

Figure 15.70 Palisades and terraces for avalanche protection. Val d'Isère. French Alps. Photo F. Gutiérrez.

avalanche. False tunnels can be constructed along highways and railroads in areas where avalanches are common. An active method involves the artificial triggering of avalanches at an appropriate time before natural causes create one. The most common method is the use of explosives in avalanche prone areas. This technique is commonly used in ski resorts (Bolt *et al.*, 1975; McClung and Schaerer, 1993).

Maps are prepared for planning purposes. **Avalanche Risk Maps** are useful for managing alerts and defense; they are produced at scales of 1:25,000 to 1:100,000. **Avalanche Zone Maps** are fundamental for regional planning because they provide more detail than Avalanche Risk Maps; their scales range from 1:2,000 to 1:5,000 (López-Martínez, 1988) (*Fig. 15.71*). Some maps distinguish the different parts of an avalanche (source, length, and final position). The maps are usually colored; red indicates high risk (construction is not allowed), white indicates no risk, and blue signifies moderate risk (construction is allowed with some restrictions) (Gerrard, 1990).

Furdada (1996) produced a Probable Avalanche Location Map using Geographic Information Systems. The methodologies she used to obtain data include photointerpretation, field studies, and population surveys. She represented this information cartographically and constructed a GIS database. This work is a valuable contribution to natural risk management and territorial planning. Ghinoi and Chung Chang (2005) describe a methodology to locate areas the size of an alpine valley that have high potential for snow instability under certain meteorological conditions.

15.3.3 Mass movement

Mountain slopes are notably unstable due to their high gradients, intensely jointed rocks, and unconsolidated Pleistocene deposits such as glacial till and fluvioglacial and solifluction materials. Mass movement is associated with intense rainfall, rapid thawing of the snow layer, undercutting by rivers, high magnitude earthquakes, deforestation, and other factors. Mass movement includes landslides, rock and detritus avalanches, and mudflows. When these material flows reach

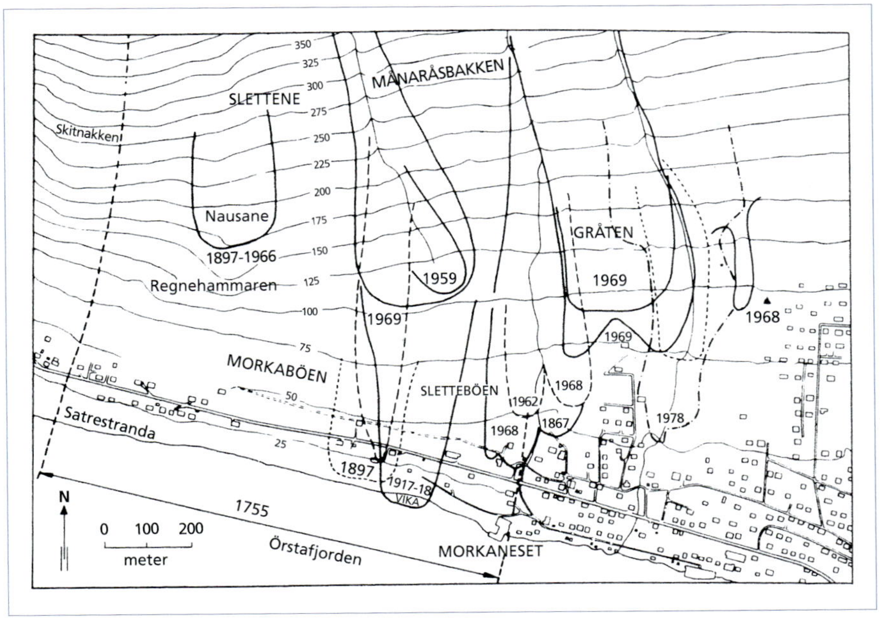

Figure 15.71 Avalanche trajectory map in an area of Norway, simplified from a 1:5,000 map with a contour interval of 5 m (Verstappen, 1983).

riverbeds, they dam up the network and cause upstream flooding (Jones *et al.*, 1983). The classification of landslides in permafrost areas and methods described for locating them, their occurrence, and protective measures were discussed in the chapter on Slopes and Mass Movement (Chapter 7). The characteristic methods in permafrost areas are thermal modifications (Turk, 1996). These circumstances show that in mountainous zones where humans are active, there may be elevated risk due to slope instability. As a result, communication networks, hydroelectric facilities, pipelines and aqueducts are difficult to construct and preserve (Selby, 1993). These projects require the preparation of risk maps of mountain areas for appropriate regional planning.

In mountainous regions, any communications pathway that crosses a **scree** slope must be sited on the lower part of the slope to the extent possible, where the gradient is lower and the risk of falling rock is lower. Vegetation removal in the middle and upper parts may cause instability in the mass of gelifracts due to undercutting. In the Peruvian Andes some slopes are so active that highways must be cleaned daily to remove fallen rock (Gerrard, 1990). As a result, roads in unstable areas require regular maintenance and continuous observation.

There are numerous cases of mass movement in mountain regions. Mass movements of several million cubic meters with velocities higher than 100 km/hour have occurred such as the **rock avalanche** 7 km in length with a drop of 2000 m at Mount Rainier (Washington) and Sherman Glacier (Alaska) triggered by the 1964 earthquake; this last avalanche covered an area of 8.25 km^2. These rock avalanches caused little or no economic damage but the translational rock landslide in Mayunmarca in the central Andes of Peru killed 451 people in April 1974. The cause of the disaster was undercutting by the Mantaro River (Voight and Pariseau, 1978). In other regions, mass movement has affected specific areas repeatedly throughout history. The town of Inzing and various areas of Voralberg in the Austrian Alps (Gerrard, 1990) have suffered numerous catastrophes in recent centuries. In spite of multiple protective measures, on 26 July 1966, a channelized **debris avalanche** struck Inzing; the avalanche moved at 100 km/hour, killed three people, and buried 4 km of highway and 500 m of railroad. Most of the protective structures were destroyed.

Active **rock glaciers** can also be significant geomorphic hazards. They inflict damage in numerous ways: the downslope flow of material, melting of the underlying ice and creation of depressions that measure meters to tens of meters, eluviation of fine interstitial material that results in subsidence, and surface block falls that result from higher velocity on the upper part of the slope, which can free up to a boulder a minute (Giardino and Vick, 1987). Rock glaciers can damage electrical wires, telephone towers, and ski lifts. It makes obvious sense to avoid building in these areas. Places where a highway route, railroad, or pipeline crosses a rock glacier require continuous maintenance; sometimes the route needs to be modified. The state of Colorado provides many examples. Tunnels can be deformed and can even collapse such as the tunnel at Mount Mestas (Colorado), which collapsed due to hydrostatic pressures that built up during thawing (Giardino, 1983).

Landslides are common in the active layer of permafrost regions; these slides are thin and generally travel very fast (Harris, 1981). Collapses and flows with velocities up to 10 m/second occur along the margins of rivers as a result of thermo-erosional niches (Harris, 1988). Pleistocene periglacial zones in medium latitudes commonly contain periglacial deposits characterized by low cohesion and shear strength angles. Building any type of structure that affects these materials causes multiple problems (Derbyshire, 1977).

15.3.4 Engineering problems in frozen soil regions

In these areas humans can alter the thermal equilibrium of permafrost by decreasing or increasing it. Human activities can disrupt the equilibrium in many ways. Forest cutting, fires, and vehicle traffic, and even the creation of agricultural fields destroy vegetation. In the recent past, construction of buildings, highways, railroads, and oil pipelines have been significant causes of infringement onto permafrost. Other impacts involve changes to drainage conditions that result from rerouting rivers, building reservoirs, and drying up peat bogs (Haugen and Brown, 1971). These activities usually reduce permafrost and the changes are generally irreversible. Soil adjusts gradually to a newly imposed thermal regime; in the case of large structures the process takes several years (Cooke and Doornkamp, 1990).

Prior to undertaking any activity in these areas, it is important to obtain reconnaissance information through aerial photography to distinguish and map various types of terrain and plant communities that may indicate moisture conditions. Afterwards, field studies should be conducted to analyze in detail the distribution and nature of the permafrost and the active layer, drainage, surface materials, temperature, and moisture in the atmosphere and soil (Cooke and Doornkamp, 1990).

This toolbox of knowledge makes it possible to choose appropriate construction practices. Methods for reducing geotechnical problems are both passive and active (Muller, 1945). In substrates sensitive to thawing **passive methods** are used to try to maintain materials in their initial thermal state to avoid settling and uplift. **Active methods** involve the replacement of material sensitive to frost action by another material that is more resistant to it. In addition, researchers design structures to support the destructive forces that result from modifications of thermal regimes.

15.3.4.1 Buildings

The first structures in permafrost areas were built on top of basements and wood pillars; they had no significant problems. When concrete began to be used to support structures, significant settling of the structures occurred and cracks appeared in buildings (Harris, 1986).

One characteristic example illustrating construction problems on permanently frozen soils is the town of Atlavik in the MacKenzie River delta (Canada) (Cooke and Doornkamp, 1990; French, 1996). The town was built in 1912 and by 1950 it had a permanent population of 400, which increased to 1500 in the summer with the arrival of Eskimos and Indians. As it grew, inappropriate construction practices caused numerous problems. Atlavik is located on top of fine-grained delta sediments that contain a lot of ice and have poor drainage. As a result, significant uplift and settling occurred and depressed areas formed and later flooded. Authorities decided to build a new town in an area with better geotechnical conditions. Inuvik, a site 48 km from Atlavik, was chosen and construction began in 1955. Today about 1500 people live there. The area was chosen on the basis of detailed studies; it combines the most suitable characteristics possible for the new location. This is a clear example of the necessity of adequately understanding permafrost and local conditions.

The surface capacity that can support a load without deformation depends on the type of material and it can fluctuate enormously between freezing and thawing stages (Swinzow, 1969). Structural deformation occurs during periods when the strength of underlying and adjacent terrains changes in response to seasonal temperature variations. During the frozen period, a building may be uplifted and may even rupture because the structure cannot tolerate movement. During the thaw, the water occupies less volume and the soil loses strength, causing the building to settle. In clay-rich and silty soils that are impermeable and that retain significant meltwater, sediment flows can occur. In addition, because water cannot infiltrate into permafrost, fine sediments in the active layer can become highly plastic, which causes differential settling of buildings. In contrast, coarse grain deposits are very stable since they drain easily. If in addition to seasonal melting of permafrost, progressive warming occurs through the years, deformation increases gradually with increased depth of the active load (Lobacz and Quinn, 1966) *(Fig. 15.72)*. In addition, if ice is present in pores, the substrate is highly stable during the thawing phase but if ice masses are present outside the interstices, the ground is very sensitive to thawing (Harris, 1986). As a result, concrete foundations cause significant problems that are very apparent in large buildings.

For structures in permafrost regions, the heat exchange between the ground, the structure, and the envi-

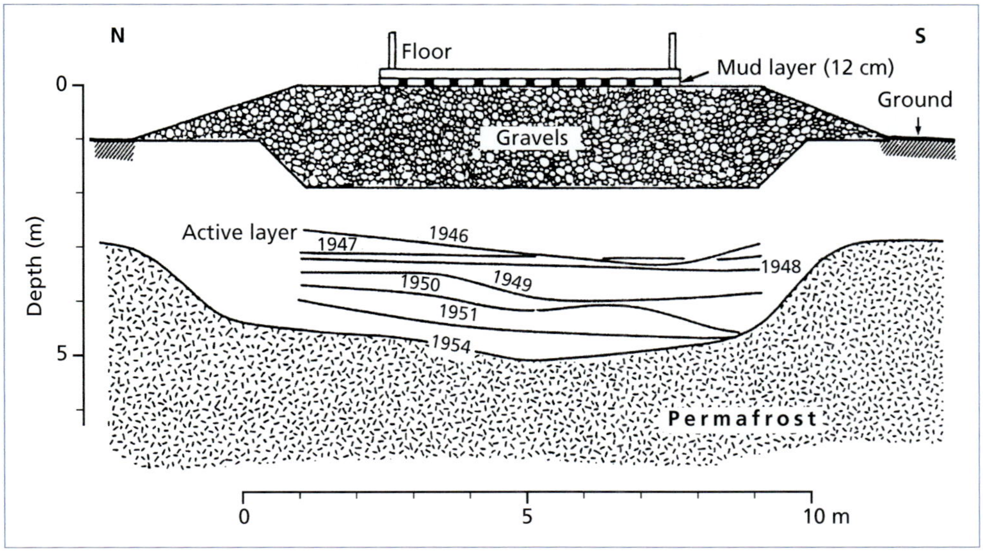

Figure 15.72 Progressive degradation of permafrost under a gravel layer beneath building 4 of the Alaska Field Station, 4.5 km northeast of Fairbanks (Alaska). The active layer was 0.9 m thick prior to construction in 1946 but increased to 1.75 m after tree and brush cover was removed. The lines indicate the surface of the permafrost at the end of the thaw period for different years (Lobacc and Quinn, 1966).

ronment must be regulated (Yershov, 1998). Inhabited buildings isolate the soil from solar radiation and the top of the permafrost can rise due to the resulting shade (Muller, 1945) (*Fig. 15.73*). A building that radiates heat away from its interior to the permafrost is like a thawing bulb below the house; it warms asymmetrically due to differences of solar heating on the various orientations of the building. These variations are accompanied by differential vertical movements that can be observed in the settling and uplift of the corners of the structure (Lobacz and Quinn, 1966) (*Fig. 15.74*). These movements can rupture concrete foundations and crack the walls of buildings. Settling due to thawing is gradual; it stops when it reaches thermal equilibrium, which may take several decades. As an example, of 1230 buildings inspected in Yakutsk, Chita, Vorkuta, and the Republic of Bunyat, 63% were deformed (Yershov, 1998).

Construction techniques vary and in areas of continuous permafrost, the goal is to maintain the frozen layer intact. Building foundations are constructed on top of a framework of wood, steel, or reinforced concrete pilings anchored into the permafrost; they should extend at least 60 cm above the ground surface (Sudgen, 1982; Harris, 1986) (*Fig. 15.75*). In this way, air can circulate freely between the soil and the building foundation; heat radiated by the building can dissipate, preserving the permafrost in the winter. During the summer, thawing is minimal because the building shades the ground surface. Artificial refrigeration can also be installed around the pilings (cryoanchors). Also, ventilation pipes are sometimes installed in the surface layer of the concrete to help cool the permafrost. Finally, buildings are constructed on top of a layer of gravel when possible (*Fig. 15.76*); this facilitates drainage and prevents water stagnation and thermal erosion (Harris, 1986).

15.3.4.2 Linear infrastructure

After World War II numerous highways, railroads, and airports were built in Canada, Alaska, Siberia, and Mongolia (*Fig. 15.77*). At the end of the previous century, the Trans-Siberian Railroad had been built to transport wood to the west. Railroads and highways for heavy use are not generally built in areas of continuous permafrost; they are constructed in areas of discontinuous and sporadic permafrost. In the latter areas, the ground has a temperature of about −2°C; these are the areas that suffer the most damage (Harris, 1986).

Linear structures have varied and complex problems related to differential uplift and settling that result from the removal of plant cover, which modifies the thermal regime and promotes thermokarst formation. These problems can cause spring flooding to destroy bridges. Mass movement can also bury sections of highways and railroads. Another problem is related to spring meltwater that flows over the surface of these linear structures and then freezes again to cause *icings* (Ritter, 1986).

The simplest form of transport in permafrost is a road called a winter highway (Harris, 1986). These are local

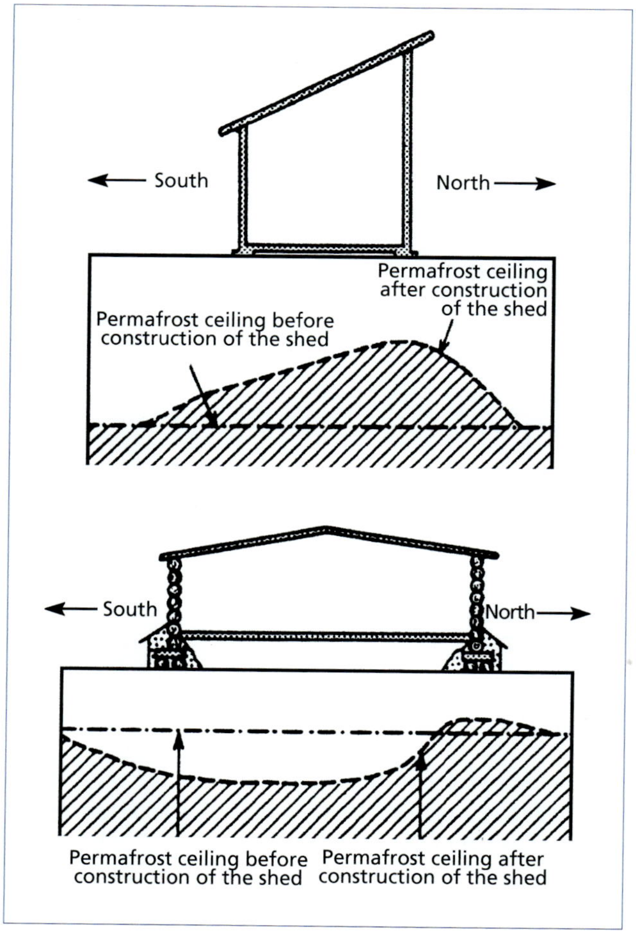

Figure 15.73 Changes to permafrost caused by: (A) inhabited shelter and (B) a house with internal heating (Muller, 1945).

roads built on top of layers of ice or frozen rivers and lakes. Special vehicles are required for driving on them. Railroads and highways that are in continuous use require specific construction techniques, which are costly. These structures are sited in areas of smooth topography to prevent tree removal and to take advantage of good drainage. The fundamental problems are the route and the necessity of using crushed gravel for fill. The fill layer ranges from 0.6 to 1.5 m in thickness; as a result, their construction requires an enormous volume of rock or gravel (Linell and Johnston, 1973). Once the fill is placed where it will support transported loads and vibrations, the permafrost penetrates beneath it (*Fig. 15.78*). A lesser thickness of fill can be used if insulators such as polystyrene plates are used.

In permafrost areas, ferries are commonly used to cross large rivers. Even so, it is occasionally necessary to construct bridges. A talik is generally present beneath river but if the river moves laterally, the talik changes shape and can cause the bridge pilings to subside (*Fig. 15.78*). If the lateral movement is significant, it can cause the permafrost to aggrade, which causes uplift (French, 1996). For this reason, the most important bridge-related problem is stabilization of the river channel.

One of the most significant structures is the 2,451 km Alaskan highway that connects Canada and Alaska; it was completed in 1943 and involved numerous engineering problems (Freedman, 1999) (*Fig. 15.79*).

15.3.4.3 Oil and gas pipelines

These are relatively common in Siberia, Canada, and Alaska. The best known is the Trans-Alaskan Pipeline (*Fig. 15.80*) that

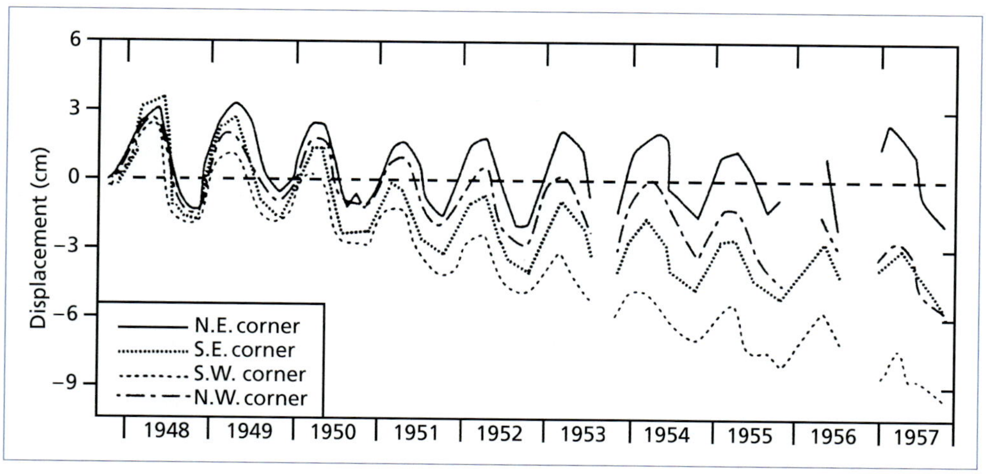

Figure 15.74 Differential uplift and settling at the four corners of Building 9 of the Alaska Field Station, 4.5 km northeast of Fairbanks (Alaska). The movements show uplift and seasonal thawing (up to 5 cm in amplitude) and progressive thawing of the underlying permafrost over the long term (up to 10 cm in 10 years) (Lobacc and Quinn, 1966).

Figure 15.75 Various buildings at the Comodoro Marumbio Base built by Argentina on James Ross Island, east of the Antarctic Peninsula. The buildings are constructed on a set of pilings that allows free air circulation to dissipate heat from radiated heat generated by the building. Photo A. Martín-Serrano.

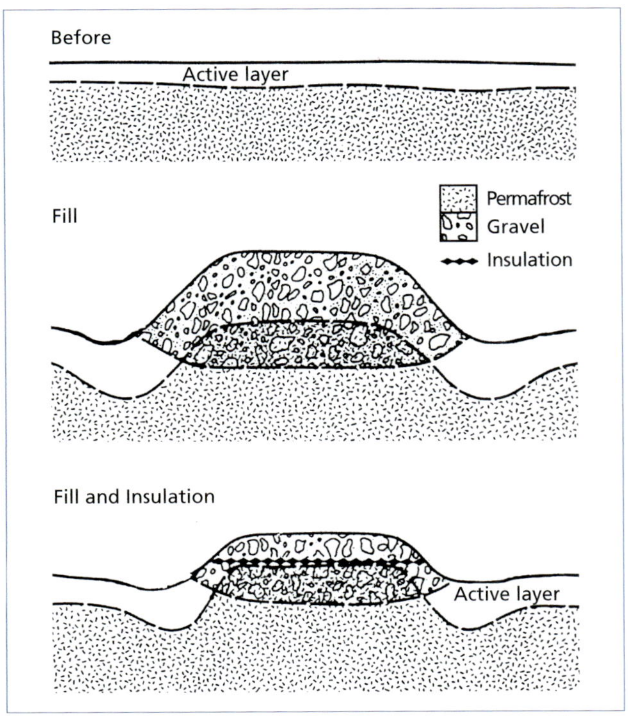

Figure 15.76 Effects on the top of permafrost under the highway as a result of using gravel fill and gravel with insulation. Note the reduction in the volume of gravel necessary when the insulation is used (Harris, 1986).

was built in 1968 after the discovery of a large petroleum deposit (Meleen, 1999). The pipeline starts in Prudhoe Bay on the Arctic coast and terminates in Valdez in the Gulf of Alaska, a distance of 1,286 km. The crude oil is then transported by ship to refineries on the west coast of the United States and to Japan. The pipeline was built between 1969 and 1977 at a cost that totaled eight times its initial budget; the final total was 7.7 billion dollars. Seventy percent of it was built over permafrost (Harris, 1986).

The construction problems involved were numerous and complex. One of the most significant was the temperature of the crude oil; it enters the pipeline at 58°C. A subsurface pipeline transporting warm fluids causes the permafrost to thaw. Without protective measures the area of thawing around the pipe expands with time (Lachenbruch, 1970) (*Fig. 15.81*). The expansion rate decreases but never reaches a state of equilibrium. As a result, the soil loses much of its strength and is incapable of supporting the pipe, causing settling at the ground surface. In addition, differential uplift and settling are common and of different magnitudes because the pipeline crosses areas composed of different materials. Melting occurs more rapidly where the pipeline crosses ice wedges and the effects are more pronounced. Destruction of plant cover during construction also causes problems (Ritter, 1978).

Due to the problems of permafrost zones, in some places the pipeline was built on top of a structure supported

Figure 15.77 Comodoro Marumbio Base built by Argentina on James Ross Island, to the east of the Antarctic Peninsula. The photograph shows the presence of a mire (mollisol) at the airport, which causes significant problems during summer. Photo A. Martín-Serrano.

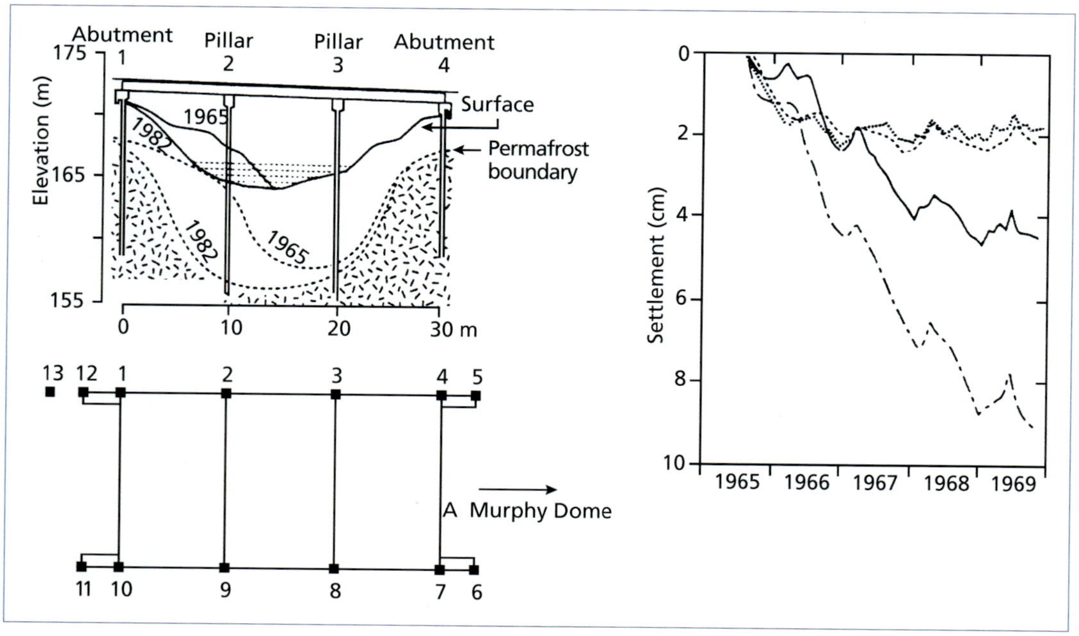

Figure 15.78 Cross-section of the Goldstream Creek bridge and settling of some pilings over time (Crory, 1985; in Harris, 1986).

Figure 15.79 Intersection of the Alaska Highway and the Trans-Alaskan Pipeline. Richardson Highway (Alaska). Photo F. Gutiérrez.

Figure 15.80 Trans-Alaskan Oil Pipeline. Chugach Mountains (Alaska). Photo F. Gutiérrez.

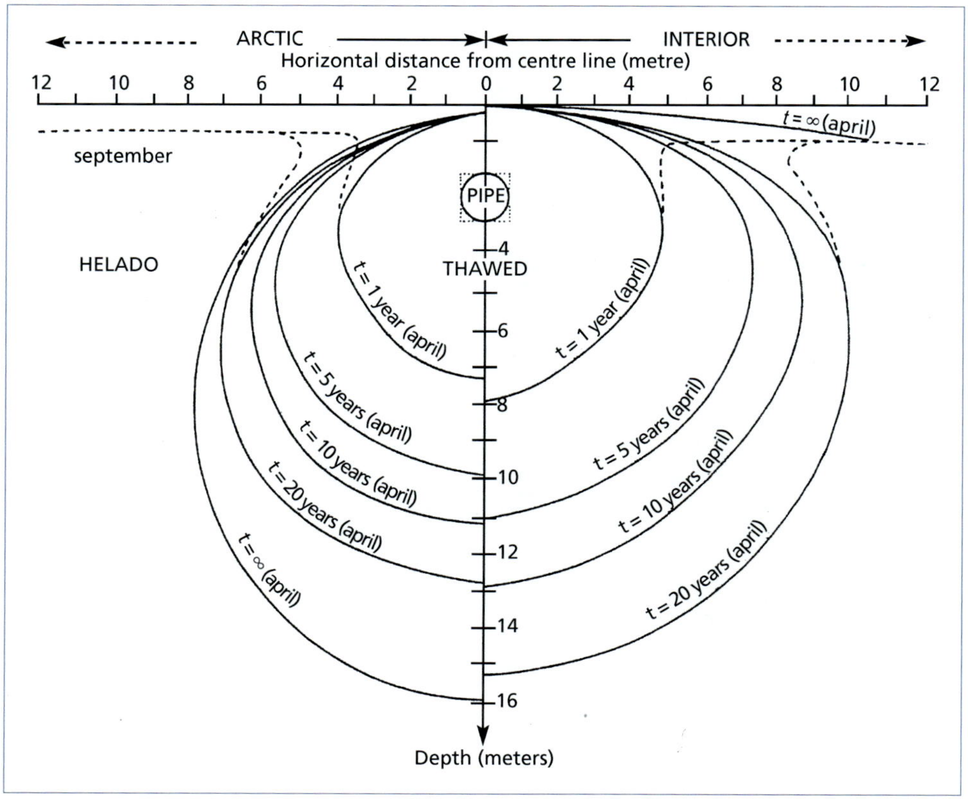

Figure 15.81 Theoretical growth of a thaw cylinder around a heated pipe in silty soil. The pipe is 1.21 m in diameter, was placed at a depth of 2.42 m, and maintain a temperature of 80°C. The curves on the left apply to conditions similar to the Arctic coast of Alaska; those on the right apply to situations near the southern boundary of the permafrost (Lachenbruch, 1970).

by pilings buried to depths of 8 to 20 m (Harris, 1986). In these sections the pilings are equipped with a refrigeration system to prevent permafrost melting (*Fig. 15.80*). In areas that lack permafrost or drainage problems (that is, in sand and gravel zones with good drainage), the pipe was buried.

Similar problems are encountered in the construction of gas pipelines. If possible, gas pipelines are buried to reduce the risk of explosion and to decrease the temperature fluctuation so that the steel will be more resistant to the gas pressure (Harris, 1986).

15.3.4.4 Mining

Mineral resources are abundant in Arctic regions and traditional mining practices have been modified to exploit mines in this region. Modifications address problems of the frozen state of rocks and minerals and the ice content of the ore (French, 2007). In addition, ore can be transported only during the thaw season (Harris, 1986). Underground mining causes subsidence, which in many cases promotes thermokarst processes that substantially transform the topography (Yershov, 1998).

To exploit **placer** deposits, the material must be thawed prior to mechanical treatment. High pressure water blasting is used to melt ice in the ore; in the past, hot water was used but it was costly and inefficient. These practices have a high exploitation cost so they are only used for gold, platinum, cassiterite, and wolframite placer deposits. In Alaska, existing sandur are exploited by using meltwater from Pleistocene glaciers (French *et al.*, 1983).

In **open pit mines,** the rock masses that host the ores contain much less ice than placers. The most important problem is stability of the advancing front, which causes thawing that decreases the rock strength (Harris, 1986). The mine is worked by drilling grids and placing explosives to break loose the ore and surrounding rock. Borehole installation presents problems because water used in drilling can freeze. For this reason, dry boreholes are used. Continuous thawing in these cuts makes it important to remove surface water. These mines, which are very costly, are highly dependent on variations in mineral prices.

The presence of permafrost affects the strength and stability of the surrounding rock and ore mass, and as a result, is important in **underground mining**. Thawing permafrost creates a significant risk of falling material at tunnel entrances (Linell and Lobacz, 1978). Most mines in permafrost areas are ventilated by 5–10°C air, which thaws the roof and walls, affecting their stability. Meltwater causes a continuous flow of water that must be constantly removed by pumping (Harris, 1986). On occasion, the water may burst into a tunnel and cause a flood, which can lead to tragedies.

15.3.4.5 Other activities

Water is necessary for the maintenance of facilities in permafrost areas; for example, pipes have to be heated to 6°–8°C in winter to prevent freezing (Yershov, 1998). Good quality potable water must be provided; water for industrial purposes generally comes from treated wastewater (Harris, 1986). In these regions, it is difficult to obtain low-cost water year round. Groundwater may be present under the permafrost and inside and on top of taliks (French, 1996). The water source must be able to last an entire year and for this reason should be stored in a deep reservoir to prevent a thick ice layer from forming in winter.

Another problem encountered in these regions is related to disposal of **wastewater** and **solid waste**. These substances can transmit disease and damage the environment. Physical-chemical, biological, and crushing and incineration are methods used to dispose of them (Harris, 1986).

In some continuous and discontinuous permafrost zones such as Yakutia (Siberia) there is extensive and significant grain **agriculture**. Here thawing done for agricultural practices causes thermokarst (Czudek and Demek, 1970) and tractors and other types of machinery often get stuck in thawing holes. In any case, the topography formed by alases makes agricultural work difficult.

In recent decades vegetable agriculture has been introduced in these locations to promote self-sufficiency. Vegetables are grown inside buildings in order to avoid ice. The substrate is commonly sand and gravel; it has good drainage and drip systems are used for irrigation (Harris, 1986). Thawing problems are common.

In regions of alpine permafrost, the lower limit of permafrost is generally below the tree line. Even so, exploitation of forest resources can cause thermokarst.

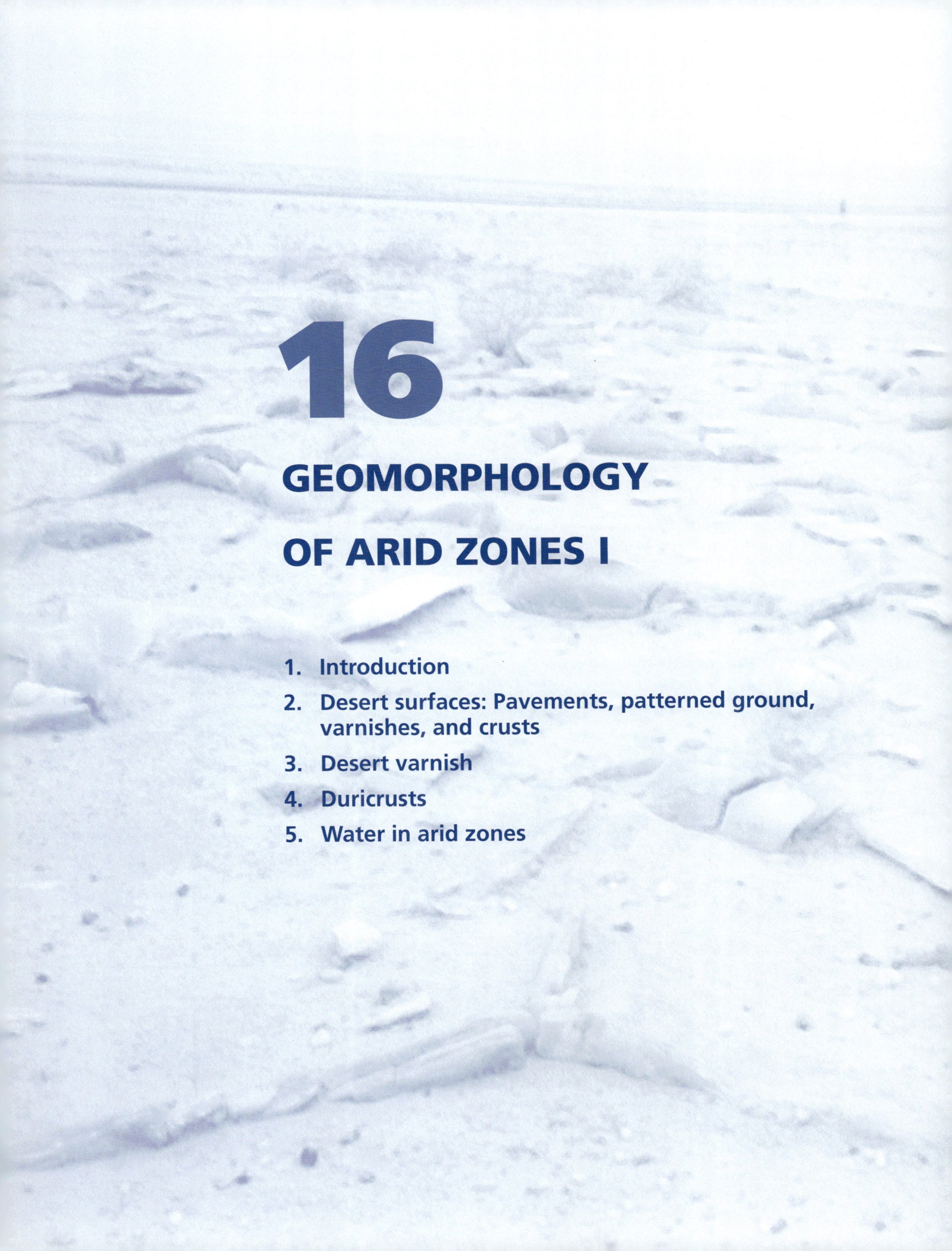

16

GEOMORPHOLOGY OF ARID ZONES I

1. Introduction
2. Desert surfaces: Pavements, patterned ground, varnishes, and crusts
3. Desert varnish
4. Duricrusts
5. Water in arid zones

16.1 Introduction

Arid regions are areas of sparse precipitation and reduced or non-existent plant cover *(Fig. 16.1)*. They are widely distributed at both low and high latitudes, in continental interiors, on coastal margins, and also in high elevation areas. Cold deserts located in medium latitudes have special problems that differ from those of deserts in hot areas; periglacial processes are active in cold deserts.

Many arid regions were explored as Europeans colonized these areas and prospected for new natural resources. This exploration was done throughout history but our knowledge of these areas is primarily limited to modern and contemporaneous times. The first documents written about this exploration are general; they emphasize geographical features and plants and animals. At the end of the 19th century, Gilbert, Powell, and Dutton did the first scientifically rigorous geomorphic work in their study of the deserts of the southwestern United States. However, in spite of these precedents, studies are generally descriptive and use local terminology, which has created confusion among geomorphologists. In recent decades dynamic geomorphology, which applies physical and chemical techniques to explain existing landforms, has made considerable progress in our understanding of arid zone landform evolution.

Geologic, climatic, and vegetation characteristics of hot desert environments vary widely. These overlay geologic formations of different composition and ages in environments of complete crustal stability or significant neotectonic activity. Desert climates are highly variable; they generally lack precipitation and have high temperatures with wide daily variations. Vegetation is scant or non-existent and plant metabolic activity is adapted to the broad temperature variation. Plants obtain water from roots, which sometimes penetrate to a depth of 20 m (Nicholson, 1993). Plants that grow in arid zones, called xerophytes, have mechanisms that adapt to scarce water.

In addition, desert environments are not favorable to man because of their extreme climates and the scarcity or absence of food and water. Human activity in these areas

Figure 16.1 Region of pronounced aridity almost devoid of vegetation. Sinai Desert (Egypt) Photo J. Rosell.

also varies widely; some areas lack population and others are densely inhabited. Worldwide 15% of the population lives in desert regions (Heathcote, 1983). In addition, population is increasing in many arid areas; it has increased 50% or more in the last fifty years. Arid region population exceeded 1.2 billion people by the year 2000. As a result of increasing population, human use of scarce water and plant resources creates numerous environmental problems that require difficult and costly solutions (Cooke *et al.*, 1982; Goudie, 1990a).

Arid regions have been defined in different ways depending on the purpose. Various criteria have been used to categorize arid regions; the most common distinctions are based on vegetation, drainage systems, soil degradation, and water and wind erosion. The most commonly used criterion is water availability because it affects flora and fauna and the dynamics of geomorphic processes. The water balance defines water availability; in hot deserts evaporation and transpiration (ET) and surface and subterranean drainage loss (D) exceed precipitation (P). The difference between the two is the aridity indicator, which is based on ratios of precipitation and temperature (Besler, 1992). Stadler (2004) compiled a list of 41 aridity indicators. Aridity can vary widely even if the amount of precipitation is similar. For example, the Sahel, the Great Plains of the western United States, and the driest areas of Central Europe have annual average precipitation of 400–500 mm; however, the vegetation in these regions ranges from steppe to dense forest (Nicholson, 1993). Thornthwaite (1948) established a humidity index (Ih) that takes evaporation and transpiration into account. Meigs (1953) divided arid areas into groups based on this index. He established three types of arid climates: hyper-arid (Ih < –56), arid (–56 ≤ Ih < –40), semi-arid (–40 ≤ Ih < –20), and a subhumid climate with values between –20 and 0. UNESCO (1979) published a distribution of the world's arid zones based on data from 1600 meteorological stations. The work was later modified in the World Atlas on Desertification (UNEP, 1992; Middleton and Thomas, 1997) prepared for the Program for the Environment of the United Nations. Both are based on the degree of bioclimatic aridity, which is defined as the ratio between precipitation and potential evapotranspiration. Four degrees are aridity are described: hyper-arid (P/ET < 0.05), arid (0.05 ≤ P/ET < 0.20), semi-arid (0.20 ≤ P/ET < 0.50), and dry subhumid between (0.50 ≤ P/ET < 0.65) (FAO-UNESCO, 1971–1979; Mainguet, 1999). Grove (1977) made a simple distinction based on precipitation only. His categories are semi-arid (200–500), arid (25–200 mm), and hyper-arid zones. Hyper-arid zones do not have a seasonal precipitation regime and twelve-month periods without rain have been recorded in these zones.

The amount of the Earth's surface area classified as arid zones depends on the climatic classification used. According to the World Desertification Atlas (1992) and Middleton and Thomas (1997), 37.3% of the global land surface is arid of which 17.7% is semi-arid, 12.1% is arid, and 7.5% is hyper-arid (*Fig. 16.2* and *Table 16.1*). Delineation of the extent of these areas is difficult due to their extreme interannual precipitation variability and their gradual transition from one to another. Boundaries are defined on the basis of vegetation. They are difficult to establish precisely because precipitation, which directly affects plant growth, is variable and anthropogenic activity considerably modifies plant growth in some regions (Thomas, 1989a).

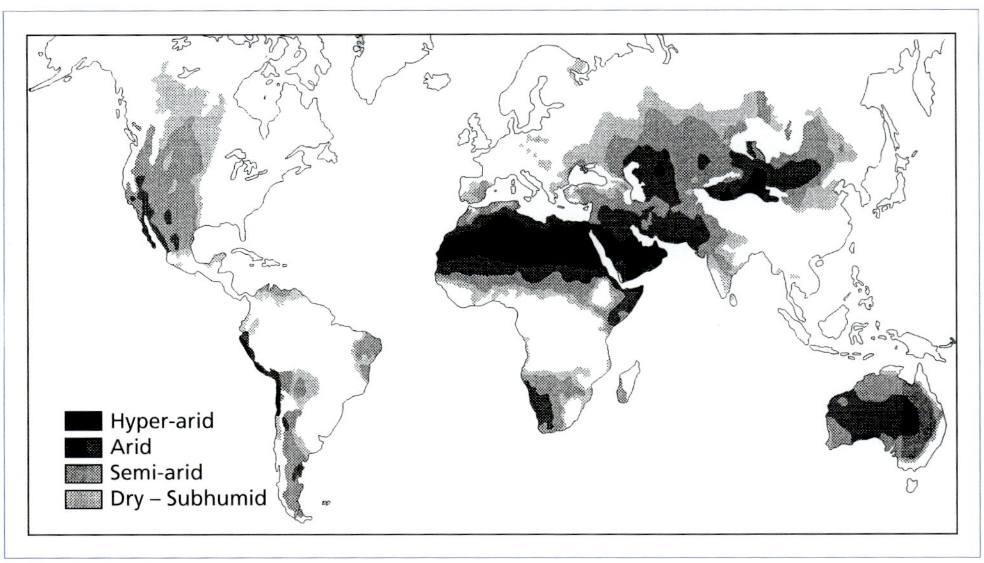

Figure 16.2 Distribution of the world's arid zones (UNEP, 1992).

Table 16.1 **ARIDITY ZONES BY REGION (MILLION HECTARES) (U.N.E.P., 1992)**

	Region							
Zone	**Africa**	**Asia**	**Australasia**	**Europe**	**North America**	**South America**	**Total**	**%**
Cold	0.0	1082.5	0.0	27.9	616.9	37.7	1765.0	13.6
Humid	1007.6	1224.3	218.9	622.9	838.5	1188.1	5100.4	39.2
Dry Subhumid	268.7	352.7	51.3	183.5	231.5	207.0	1294.7	9.9
Semi-arid	513.8	693.4	309.0	105.2	419.4	264.5	2305.3	17.7
Arid	503.5	625.7	303.0	11.0	81.5	44.5	1569.2	12.1
Hyper-arid	672.0	277.3	0.0	0.0	3.1	25.7	978.1	7.5
TOTAL	2965.6	4256.0	882.2	950.5	2190.9	1767.5	13012.7	
% By continent	56.9	37.5	69.4	12.2	23.9	18.9		
% Of global total	34.8	32.9	12.6	2.4	10.4	6.9		

The largest extent of arid zones is a broad belt of northern Africa and Asia *(Fig. 16.2)* including the Sahara, Arabian, Pakistan, Indian and Central Asian deserts. The littoral desert of Namib-Angola and those of Kalahari and Karroo are located in South Africa. The Australian continent does not have hyper-arid areas but 75% of its surface is desert. The western United States and Mexico are classified as arid. Western South America has the coastal desert of Peru and Chile and the arid Andean and Patagonian regions.

16.1.1 Causes of aridity

Climatic, orographic, and oceanic factors control aridity. In some areas, the lack of available rain has one specific cause whereas in others, aridity results from a combination of factors (Walton, 1969; Dresch, 1982; Thomas, 1989, 1997; Cooke *et al.*, 1993; Nicholson, 1993).

The **zonal factor** refers to the location of tropical and subtropical deserts with respect to the trajectories of anticyclonal cells. Areas of low pressure can occasionally affect these deserts although precipitation is low. These areas are characterized by extensive zones of high aridity surrounded by small semi-arid fringes. The most representative example is the Sahara Desert.

Continentality refers to the distance between an ocean and the desert. In temperate latitudes, rain-producing fronts gradually lose their moisture as they move toward continental centers. Cold winters are common in these areas. The more arid zones occupy a small surface that is bordered by extensive semi-arid areas. Deserts in medium latitudes in Central Asia have these characteristics; they are located north of the Himalayas, which rules out the influence of monsoons from the Indian Ocean.

The **orographic effect** creates rain-shadows on the leeward side of mountain chains in regions of dominant winds such as the westerlies and the trade winds. As air masses rise along the cordillera slope, they warm and dry adiabatically. This orographic effect may be accentuated by zonal and continentality factors. Cordilleras located along the western part of the American continent form a rain barrier (for example, Death Valley, California).

The coasts of Peru and Chile *(Fig. 16.3)* and Namib and Angola are deserts because of **cold ocean currents**, the Humboldt and Benguela currents, respectively. Evaporation from the ocean surface is insignificant and fog and dew provide practically no precipitation. Water vapor condenses in air masses over the cold current, creating fog and coastal mists that warm up as they reach the coast; this heating widens the gap between the air temperature and its saturation point and as a result, no rain occurs. These deserts are the most arid in the world.

The **albedo effect**, which refers to the reflectivity of desert surfaces, also causes aridity. In hot deserts, heat generates convective currents. In deserts that are white, reflection can promote daily thermal losses. In contrast, rocky desert environments absorb heat to the maximum extent especially if there is a cover of desert varnish.

16.1.2 Climate features of deserts

Cold and hot deserts are different. Meigs (1953) classifies them according to the average temperature of their coldest months; below 0°C is considered a **cold desert**. Cold deserts occur primarily in middle latitudes. Their most significant

Figure 16.3 Stepped marine terraces up to 300 m above sea level in the Peru-Chile Desert, which is the most arid in the world. San Juan de Marcona (Southern Peru). Photo C. Zazo.

characteristic is their wide annual thermal oscillation. In the Karakun Desert of western Asia, recorded surface soil temperatures range from 79.4°C to –40°C (Nicholson, 1993). Precipitation is controlled by topography; in elevated zones the precipitation is often snowfall. Evaporation is low during cold months; strong winds also occur during these months.

Hot deserts are located in tropical and subtropical zones. These deserts have high temperatures, variable precipitation, high evaporation rates, and significant wind action.

Due to their latitudinal position and the anticyclonal system, they have high insolation; air **temperatures** can reach 58°C in some deserts such as the Sahara (Nicholson, 1993). The sky remains clear at night, allowing temperatures to drop considerably due to radiation; diurnal thermal oscillations are about 22°C in interior deserts at high latitudes although values up of 56°C have been recorded in Tucson (Arizona) (Goudie and Wilkinson, 1977). Annual temperature oscillations are high and seasonal contrasts become more pronounced with increasing latitude. The highest temperature on the Earth's surface, 83.5°C, occurs in the sands of Port Sudan in the Red Sea. In coastal deserts, temperatures show less contrast and have low daily and seasonal variations; landlocked deserts have much higher oscillations. In littoral deserts affected by cold ocean currents, daily variations are about 11°C, which is about half that of interior deserts, and annual ranges are about 18°C (Goudie and Wilkinson, 1977).

Precipitation is extreme; convective-type rains may interrupt dry periods. Temporal variability is accompanied by spatial irregularity; these variations increase as annual precipitation decreases. The variability can be estimated by the ratio of deviation from the average to the average value of annual precipitation; this ratio is expressed as a percentage. Large deserts *(Fig. 16.4)* have values higher than 30%; they are classified as arid and hyper-arid zones (Goudie and Wilkinson, 1977). This spatial and temporal variability is due to the fact that the dominant rains, which result from storms, constitute a significant fraction of the total annual precipitation. A 24-hour desert storm rainfall event can exceed the average annual precipitation recorded over a 30-year period. These rainfall events can produce extraordinary events such as the 1925 storm that discharged 1524 mm on Lima (Peruvian Desert), an area where the average annual precipitation is 46 mm (Nicholson, 1993). In the Atacama, Namib, and Sahara deserts, periods of 10–14 years can go by without any recorded precipitation. In some places houses are constructed of salt blocks *(Fig. 16.5)* as in the arid altiplano, or puna, of Bolivia even though it is not a hyper-arid zone. In littoral deserts, which are near the sea, the air is relatively humid but precipitation is generally less than 50 mm. Fog is

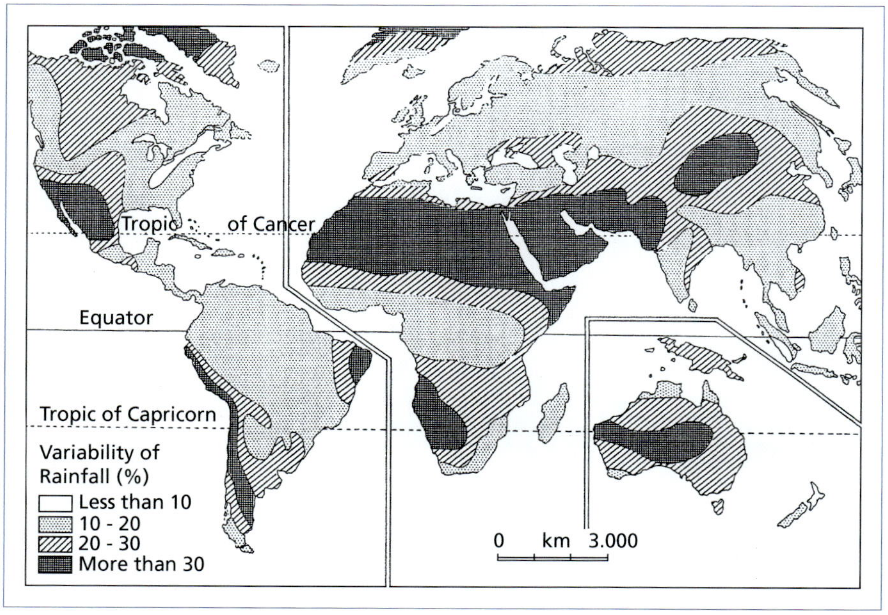

Figure 16.4 Global variability of precipitation (Goudie and Wilkinson, 1977).

Figure 16.5 House built on foundation stones of salt. Uyuni Salar (Bolivia). Photo M.C. Maldonado.

common and it can last more than half the year. It penetrates tens of kilometers inland and provides moisture for plant growth. Fog discharge is an important part of total precipitation (Demangeot, 1981).

Hot **winds** present in desert regions desiccate many areas. These winds blow during the day and diminish at night. They blow from the desert into more humid marginal areas and generally contain dust. Many areas experience these winds and the inhabitants have given them local names such as *trade winds*, *irifi*, *sirocco*, and *harmattan* (Goudie and Wilkinson, 1977). Tornados can also occur; these are masses of hot, turbulent, and particle-laden air. They rise to heights of tens to hundreds of meters but generally last only minutes. Dust storms or *haboob* are more significant; they are walls of particle-charged air that have catastrophic effects on animal life and human activities (Pewé, 1981).

16.1.3 Geomorphic classification of desert areas

We can use geomorphic criteria to classify arid environments. By applying guidelines such as dominant processes and morphostructural types to deserts we can determine the major geomorphic features of these morphoclimatic zones.

Two large geomorphic domains are based on active processes (Cooke, *et al.*, 1993). Wind erosion and deposition predominates in the **eolian domain**. Erosion can produce extensive fields of yardangs as in the deserts of Iran (Lut) and the Sahara (Tibesti, Chad); yardangs are rock outcrops that form in linear arrangements by abrasive wind action. Wind erosion can generate fields of blowout depressions that dot some desert areas (Kalahari desert). The accumulation of wind-transported particles creates extensive sand fields that cover approximately 25% of deserts. A significant portion of sand fields may be stabilized by vegetation (as in the southern Sahara), which indicates that the climate was more arid in the past. The **fluvial domain** includes erosion and accumulation by surface water; water erosion is significant in semi-arid environments. Fluvial activity erodes and deposits part of the bed load in valleys and transports water and the solid load to alluvial fan systems at the foot of mountains; these deposits generally decrease in particle size toward distal areas. The two domains may be superimposed on each other because both eolian and fluvial processes may be active in the same area. The sedimentary record contains numerous intercalations of eolian and fluvial sequences (*Fig. 16.6*). In addition, climate changes that occurred during the Quaternary modified the aridity of numerous areas; as a result, the activity and intensity of eolian and fluvial processes have changed. In other regions such as the Sahel as aridity decreased, fluvial activity increased to the extent of stopping eolian activity by stabilizing and partially destroying the dune systems by water erosion.

The application of morphostructural criteria based on predominant morphology types and geologic structures enables us to distinguish between **shield or platform deserts** and **mountain and depression deserts** (Mabbutt, 1977). Shield or platform deserts include the deserts of Africa, Arabia, India, and Australia (*Fig. 16.7*); they are primarily located in cratonic areas composed of Precambrian rocks and platform areas formed during later periods. Their principal characteristic is flat topography that is interrupted either by recent volcanic massifs such as the mountains of Hoggar and Tibesti in the central Sahara or by large fractures such as those in the east Africa *rift valleys* (*Fig. 16.8*). These large plains are erosional surfaces of various ages (*Fig. 16.9*) or alluvial plains of extensive endorheic basins such as Lake Chad (southern Sahara); they are also associated with eolian landscapes including regs, yardang fields, and ergs. These topographic features, which are common in these deserts, are a function of the well-known long-term crustal stability of these regions.

Mountain and depression deserts are composed of long mountainous lineations separated by depressed areas. This relief formed during the Alpine Orogeny; because of topographic contrast, higher areas have undergone constant denudation and the eroded materials have been deposited in depressions as coalescing alluvial fans. In distal fan areas, runoff collects in closed depressions and forms temporary lakes (*Fig. 16.10*) where evaporite sedimentation occurs (*playa lakes*). The contact between mountains and

Figure 16.6 Eolian deposits fossilized by lower Pleistocene (?) detrital material. Escorihuela. Teruel Province, north of the Teruel semigraben. East central Iberian Cordillera (Spain).

Figure 16.7 Flatness of the central Australian desert from Ayers Rock (Uluru). Gnamma holes can be seen on the sandstones. Photo P. Lucha.

Figure 16.8 Fault scarp in the central sector of the Dead Sea rift valley. Israel.

Figure 16.9 Erosional surfaces and in the foreground, desert pavement. Rann (Great Desert of India or Thar).

Figure 16.10 Uyuni Salar (Bolivia). Bolivian Altiplano. Photo M.C. Maldonado.

depressions is often an active tectonic front where higher areas are uplifted and depressed areas subsequently subside. As a result, intense subsidence occurs in low areas; significant thicknesses of sediments accumulated in these areas during the Tertiary and Quaternary. Significant eolian activity also occurs in depressions. Deserts in the western United States and along the Precordillera and central Cordillera of the Andes have these characteristics. Deserts in Central Asia have similar features although they form at higher elevations and thus their climatic contrasts are more obvious.

16.2 Desert surfaces: Pavements, patterned ground, varnishes, and crusts

Surface and shallow materials in arid zones are subject to processes that act separately or together depending on lithology, climate, geomorphic situation, and other factors. This activity generates desert pavements, patterned ground, varnishes, and crusts.

16.2.1 General characteristics of desert pavements

Desert pavements are composed of angular and rounded rock fragments within a matrix of finer material such as sand, silt, and clay (Mabbutt, 1965a, 1977). They develop on slopes and on low relief surfaces such as alluvial fans, terraces, and pediments. They occur in other morphoclimatic zones although they reach their greatest extent in arid zones, where they may occupy large areas.

They are called **gibber** in Australia, **saï** in Central Asia, and **hammada**, **reg**, and **serir** in the Sahara, the Middle East, and India. Hamadas are tabular structural landforms or pedimentation or pediplanation surfaces with negligible slopes. The surface may have no fragments, in which case it is called a rocky hamada; a boulder hamada is a surface is covered by angular boulders (*Fig. 16.11*) that have not been transported. Boulders in the hamada may have undergone intense weathering or exfoliation that has rounded the fragments (Evanari *et al., 1971*). Regs or serirs can develop on low gradient alluvial surfaces; here the clasts are rounded as a result of transport (*Fig. 16.12*). Today, many regs or serirs are relict forms that are located far from present-day drainage surfaces.

Figure 16.11 Hamada covered by angular residual blocks. A small closed depression (called a *daya* in the Sahara) can be seen in the center; it is full of fine material of a much lighter color. This deposit may have been originally produced by eolian supply and later moved by surface runoff. Rann, Thar Desert (India).

A well-developed pavement is a stable surface but when altered by anthropogenic activity (for example, automobile traffic in the southwestern United States and Saudi Arabia; Webb and Wilshire, 1983), water and wind can substantially erode such pavements. The famous Nazca lines (Fig. 16.13) in southern Peru, which are about 20,000 years old, are desert pavements that moved partially as they were being constructed. In places clasts in the pavement may be prehistoric artifacts; the rocks may be painted or may be petroglyphs (Brackenridge and Shuster, 1986). These circumstances may provide data on the relative age of these desert pavements.

Figure 16.12 Typical reg of rounded clasts developed on an alluvial plain. Bhojka, Thar Desert (India).

Figure 16.13 Nazca lines drawn on desert pavement. Southern desert of Peru. Photo C. Zazo.

16.2.2 Processes involved in the formation of desert pavements

Desert pavements result from one or more processes, which raises questions about convergence of landforms. These processes may be fundamentally extrinsic such as water and eolian activity or intrinsic, which means that the soil itself is modified.

The most common process in pavement formation is **deflation**; the wind removes fine particles and leaves coarse material as a residue that becomes pavement. The removal of fine particles is minimal and as a result, the surface is lowered only slightly. Deflation decreases with increased surface rugosity; it is estimated that exportation is ineffective on a surface that is 50% covered by clasts (Sharon, 1962). A film-like biocrust may occasionally develop on areas occupied by fine material; this prevents deflation and stabilizes a pavement by "sealing" the surface (Campbell *et al.*, 1989).

Some authors (Sharon, 1962) think that **water erosion** is more effective in pavement generation; experimental field studies show that erosion amounts to 5–50 cm on 5% slopes over five years. *Splash* and surface runoff remove fine material although the presence of clasts decreases water flow. The presence of biocrusts considerably reduces *splash*; infiltration capacity decreases significantly (Poesen, 1986) and may be limited to water penetration through cracks.

Another process that generates pavements, although less important than those previously mentioned, is the **upward movement of clasts** from inside the soil and their concentration at the surface (Cooke, 1970a; Mabbutt, 1965a, 1977). In cold deserts, frost heaving causes this soil sorting. In hot deserts, expansive clays may swell as humidity increases; this expansion can lift up the ground. During dry stages, soils contract and crack. Repeated wetting and drying cycles promote pavement formation.

Surface weathering, which is primarily a mechanical process, has been proposed as the cause of boulder hamadas because it fractures the rock and produces a residual pavement of angular blocks. Some authors think **subsurface weathering** (Mabbutt, 1977) may play a significant role in pavement development. Mabbutt says that high subsurface moisture content, especially in the presence of chlorides and sulfates, causes more effective weathering and for this reason, coarse particles gradually disintegrate. Less water is available at the surface so clasts in the pavement weather very slowly. As a result, the pavement is preserved.

16.2.3 Pavement development

From our analysis of pavement formation processes, we see that water and wind erosion work quickly whereas upward particle movement through soils is slower. Subaerial erosion lowers and flattens the surface over the long term (Mabbutt, 1977). Furthermore, as pavement develops, fine material becomes less available. Pavement may evolve slowly over thousands of years; during this time, clasts weather and weathering products are partially removed (Cooke, 1970a). This slow development was observed in a study of a terrace group in the Dead Sea region (Amit and Gerson, 1986). The percentage of clast cover is directly proportional to the age of the terrace. However, the oldest pavements in Death Valley (Hunt and Mabey, 1966) have a higher percentage of fine, more compact particles than more recent pavements. In this case, pavement evolution is contrary to the previous example, perhaps because of more intense surface weathering.

Pavements may increase their salt content during evolution; salts move by capillarity and may be trapped between clasts. Pavements can also continue to receive contributions of eolian material. In any case pavements are relatively stable once they have formed; deflation decreases as clast cover increases and water erosion increases clast cover. These observations show that pavement is a highly stable landform; erosional and depositional processes act on them but with low intensity; they may even counterbalance each other. This shows that pavements are resistant to erosion because they protect the underlying material from water and eolian erosion.

16.2.4 Patterned ground

In the Chapter 15 we discussed how patterned ground (circles, polygons, networks, steps, and striated soils) develops extensively in both high latitude areas and in alpine mountains (Washburn, 1956). Numerous and complex mechanisms generate patterned ground; most are related to freezing and thawing.

Patterned ground can also develop in low gradient areas of arid zones (Hunt and Washburn, 1960; Washburn, 1979) although processes in hot deserts are different from those in periglacial zones (*Fig. 16.14*). Here again we see landform convergence. Patterned ground in arid areas is primarily related to variations in soil moisture content; in environments where salt precipitates, patterned ground is related to periodic flooding and drought. The resulting features may be ephemeral; they may appear and disappear between rain events.

16.2.5 Desiccation cracks

As wet fine-grained sediments dry, they reach their contraction limit and surface cracks form. The cracks may group together to form polygonal systems (*Fig. 16.15*). The shapes of these polygons depend on the physical, chemical, and

Figure 16.14 Desert Pavement in which the varnish-covered clasts appear to be sorted and arranged in circular shapes. Negev Desert (Israel).

Figure 16.15 Polygon development on the margin of a partially inundated sebkha. Sabkhet Bou Jmel. Zarzis region (Southern Tunisia).

mineralogical characteristics of the material and also on environmental factors such as temperature and moisture content. In plan view, cracks may be straight or curved; their length, width, depth, number, and shape are highly variable (Chico, 1963). In profile, polygonal cells may be planar, concave, convex, or irregular.

Most research on crack morphology and genesis comes from experiments on freeze-thaw cycles (Lachenbruch, 1962; Corte and Higashi, 1964; Maizels, 1981); there are few studies on the characteristics and origin of cracks in clay and saline plains in playa environments. Even so, cracks that form in periglacial environments are similar to those in arid environments because the contraction and expansion coefficients of ice and salt are similar. Experiments show that 40–50 times more cracks form in clays than in sands and gravels (*Fig. 16.16*). The length of cracks decrease over time as new ones form and intersect the previous ones. In addition, spacing between cracks is directly proportional to the desiccation rate, clay content, and expansive clay content.

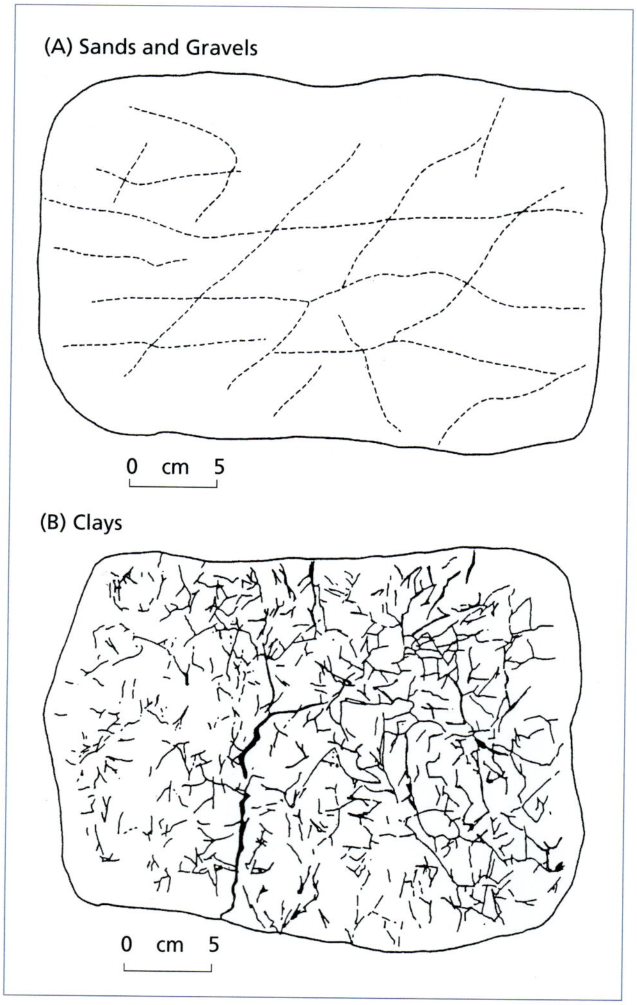

Figure 16.16 System of cracks produced after 12 freeze-thaw cycles under experimental conditions. (A) Sands and gravels. (B) Clays (Maizels, 1981).

Clasts in the interior or on the surface act as departure points for cracks. The shape of the polygonal cell profile also provides information on the origin of these landforms. Concave surfaces indicate rapid drying of the upper material; this process produces **mud curls** on the surface layer of the mud (*Fig. 16.17*). A convex morphology shows that salt is present in the matrix (*Fig. 16.18*); a planar shape indicates slow drying and the absence of salt.

In addition, Lachenbruch (1962) distinguishes between **orthogonal systems** in which cracks intersect at right angles and **non-orthogonal systems** in which the intersection angle is about 120° (*Fig. 16.19*). Orthogonal systems are typical of a non-homogeneous medium in which cracks can express preferred orientations. In contrast, non-orthogonal systems are composed of homogeneous materials that dry uniformly; as a result, cracks form almost instantaneously.

All these cracks form in hot deserts with planar topography, primarily in lacustrine depressions. **Giant desiccation fissures** can also develop locally in these environments (*Fig. 16.20*); the fissures can be kilometers long and wide and more than a meter deep. They have a complex origin; numerous causes have been suggested. Several hypotheses are subsidence-induced cracking due to ground water extraction, fissuring due to hydro-compaction, contraction due to desiccation, and seismic activity. These processes may act separately or together.

In playa environments, non-orthogonal polygonal systems develop in profusion in the saline clays that make up the fringe plains on the margins of evaporite deposition (*Fig. 16.15*). This type of polygonization occurs in gypsum deposits (Tucker, 1978) where the polygons sometimes have thrusted margins (*Fig. 16.21*). In Death Valley (California) sorted polygons develop in gypsum-bearing material (Hunt and Washburn, 1960) (*Fig. 16.22*); drying or cooling causes salt contraction whereas hydration or heating causes expansion. These processes, including salt movement due to capillarity, are thought to be the fundamental causes of the patterning and sorting of these gypsiferous soils. Polygons also develop on salt surfaces in coastal and interior sebkhas (*Fig. 16.23*). These polygons may undergo a complex evolution.

16.2.6 Gilgai

Gilgai is an Australian vernacular term that means a pool or pond; it is pervasive in the scientific literature. Gilgai are small mounds that occur in clay-rich soils and on desert pavements (Hallsworth *et al.*, 1955; Verger, 1964; Harris, 1968; Mabbutt, 1977; Hubble *et al.*, 1983; Dixon, 1994a and b).

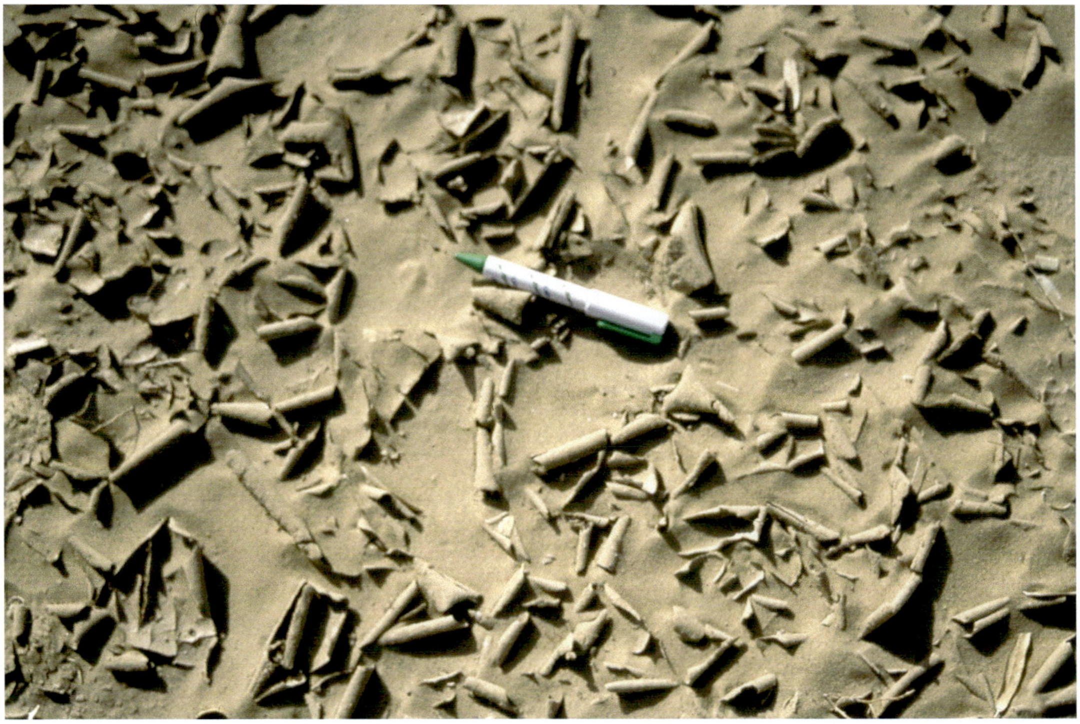

Figure 16.17 Mud curls developed on top of a thin clay layer. Jaisalmer, Thar Desert (India).

Figure 16.18 Non-orthogonal system of convex polygons with traces of raindrops. Closed depression of La Playa (Bujalaroz, Province of Zaragoza, Spain). Ebro Depression.

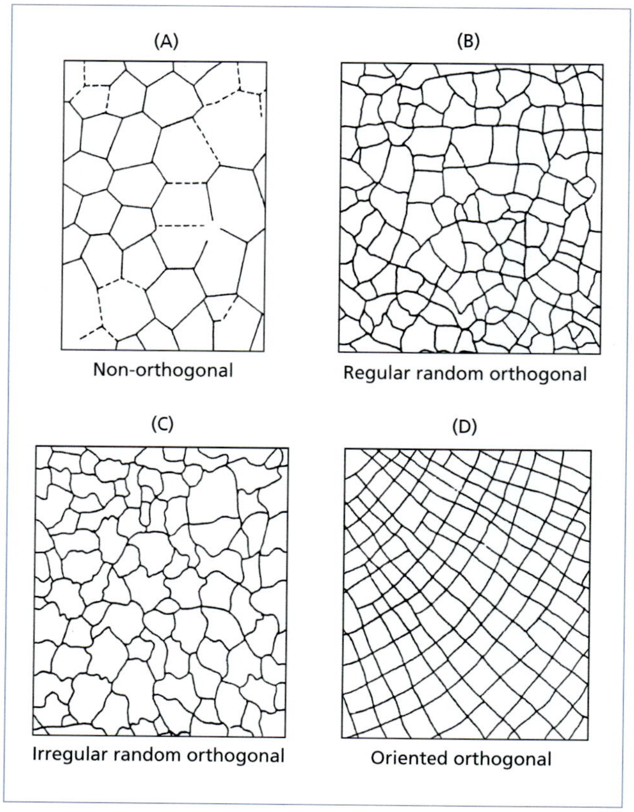

Figure 16.19 Common systems of desiccation cracks (according to Lachenbruch, 1962, and Neal, 1965).

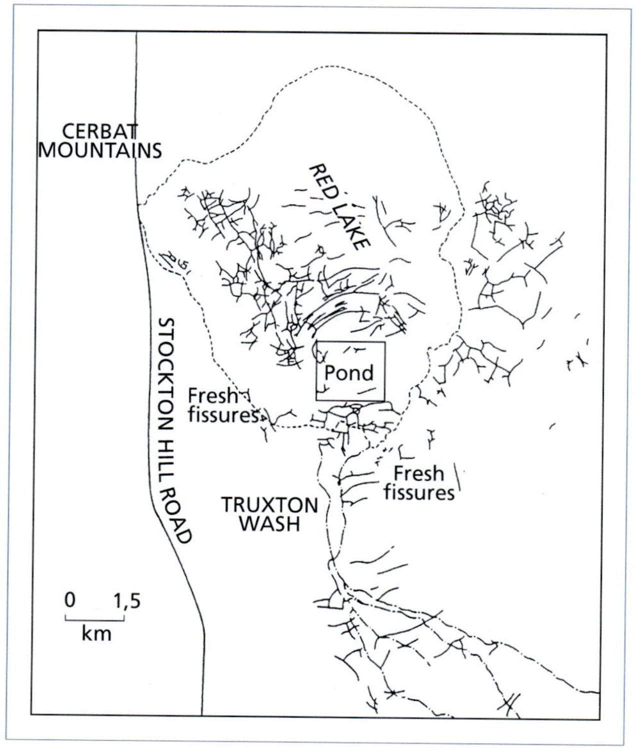

Figure 16.20 Map of giant desiccation fissures with an indication of the most recent, based on aerial photography. Red Lake region (Arizona). (Lister and Secrest, 1985).

Figure 16.21 Gypsum polygons with thrusted edges. Matti Sebkha (United Arab Emirates).

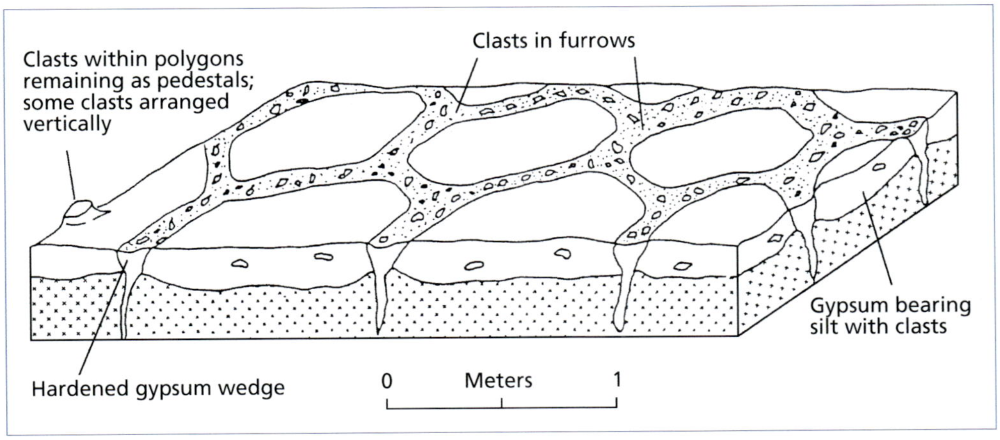

Figure 16.22 Sorted rock polygons in gypsiferous materials. Death Valley (California) (Hunt and Washburn, 1966).

Figure 16.23 Polygons with thrusted margins formed on a saline surface. Sebkha el Melah. Zarzis (Southern Tunisia).

According to Harris (1968) a cross section of a gilgai has three distinct parts (*Fig. 16.24*). One part is a **channel** or depression that may extend 2.5 m below ground level and may be 12 m wide. The other parts are a **platform** and small **mounds** measuring about 50 m in diameter that protrude from the platform. Some authors have categorized gilgai into six types (Dixon, 1994a); these are included in a simple morphological classification by Verger (1964) based on shape and method of grouping (*Fig. 16.24*).

Monographs by Cooke and Warren (1973) and Mabbutt (1977) discuss the processes and factors involved in gilgai formation in areas that have wet and dry seasons. Dry periods crack the surface and wet periods cause swelling. Gilgai occur primarily in tropical and subtropical zones where vertisols (swelling black clay-rich soils) are present. The upward push of wet and swelling subsoil enlarges the gilgai landform. Expansion and retraction of the soil that results from wetting and drying cycles explains the origin of circular and stepped gilgai

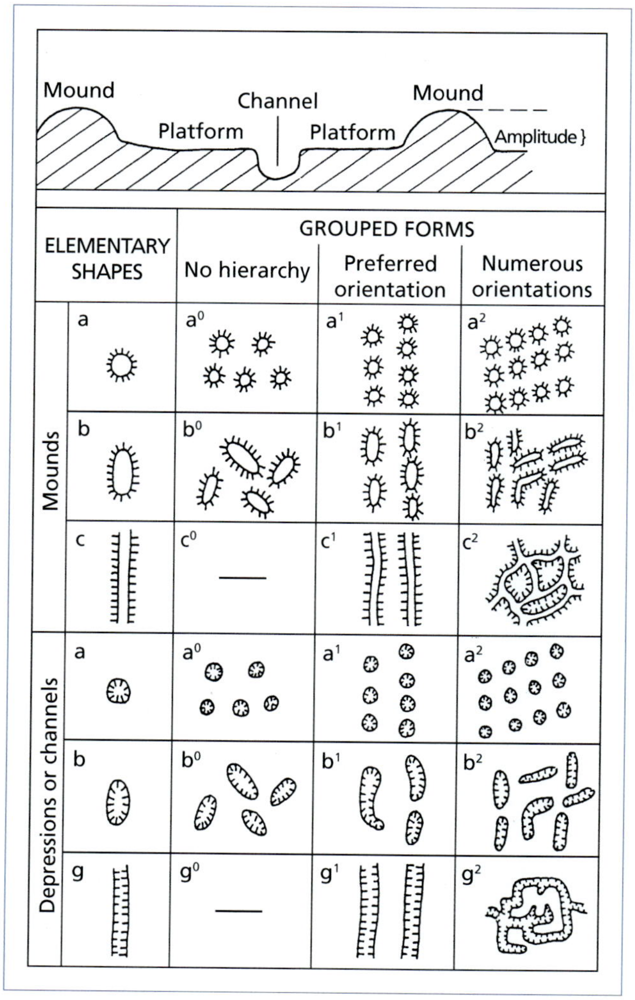

Figure 16.24 In the upper part, a cross section of a gilgai (Harris, 1968). The rest of the figure shows classification of gilgai morphology (Verger, 1964).

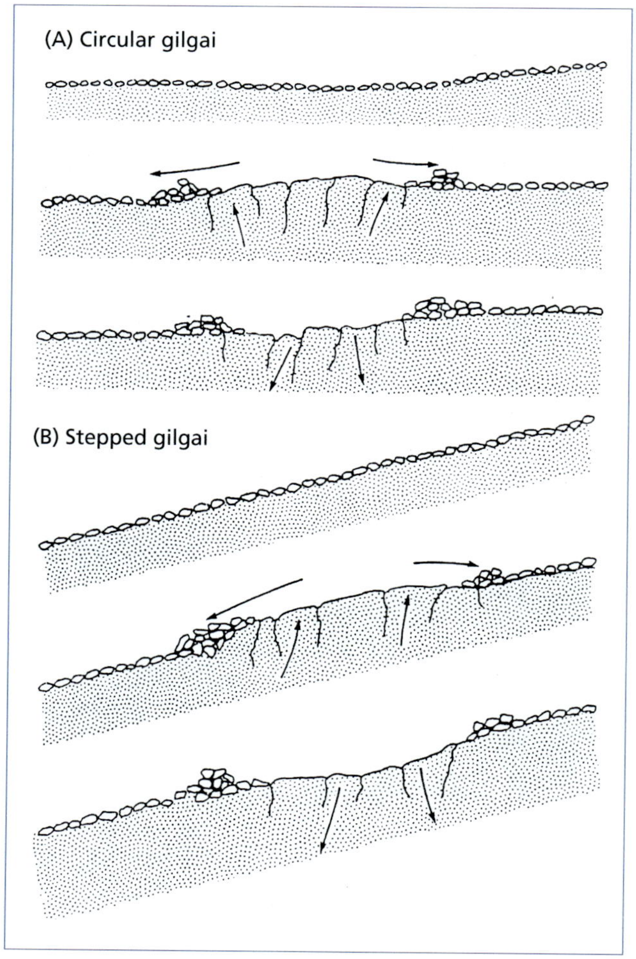

Figure 16.25 Evolution of a circular (A) and stepped (B) gilgai in the central Australian Stony Desert (Ollier, 1966).

(Ollier 1966) (*Fig. 16.25*). Swelling occurs where expansive clays such as montmorillonite and high concentrations of interchangeable sodium are present. Gilgai size increases in direct proportion to swelling clay and sodium content. The expansion that forms gilgai is visible in cross sections of these soils as small fault slickensides and thrust planes. Expansion also causes stakes installed in the soil to be tilted or expelled after a year. Hallsworth *et al.* (1955) also point out another hypothesis and Verger (1964) showed it graphically (*Fig. 16.26*). During the dry season, particles fall into cracks; this added material is thought to cause micro-relief during the wet season because it causes significant pressure at depth. The distinction of diverse mechanisms resulting from the activity of various components of force in different situations is also significant in relation to the cracking and to cycles of expansion-contraction (Knight, 1980), which help explain the genesis of gilgai.

16.3 Desert varnish

Desert varnish is a thin dark-colored film that often covers outcrops in arid regions. It covers up to 75% of rocky surfaces in some deserts of the southwestern United States (Engle and Sharp, 1958). Varnishes are also found in alpine, coastal, fluvial, and subsurface environments and near springs (Dorn and Oberlander, 1982; Dorn, 1998, 2004); these are called rock varnish. They are predominantly black or brown due to their Mn and Fe content.

The origin of varnishes has been the topic of numerous discussions, many of which lack scientific accuracy. The advent of high-resolution techniques (microscopy and electron microprobe) has made detailed observation and exact chemical analysis possible; this technology has considerably improved our understanding of varnishes. This research,

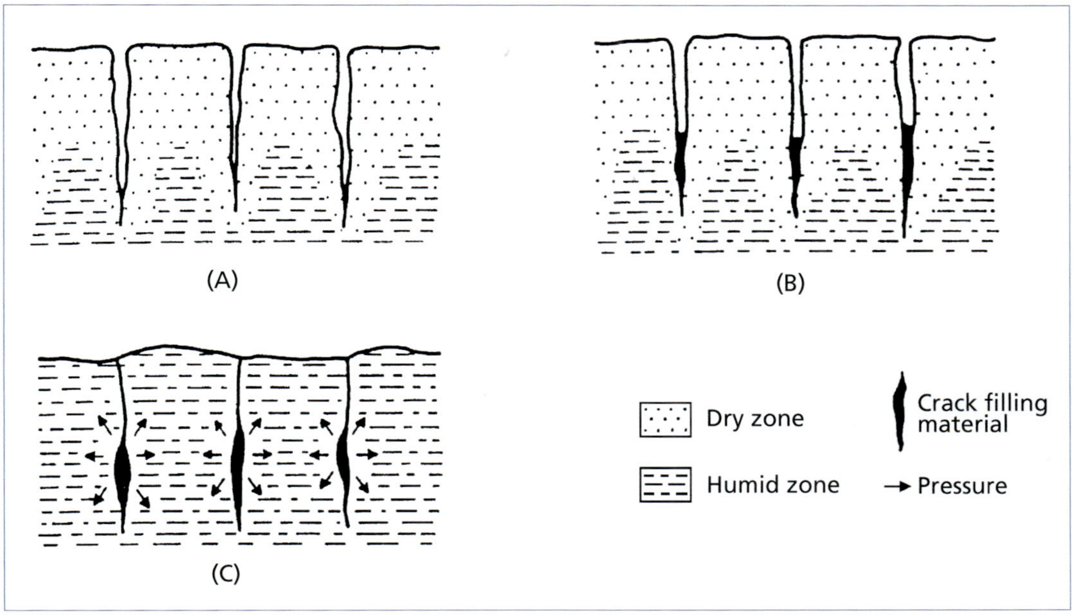

Figure 16.26 Formation of gilgai by wetting and drying (Verger, 1964).

Figure 16.27 Rock surfaces covered with desert varnish. Anti Atlas (Morocco). Photo J.L. Peña.

begun by Engle and Sharp (1958) and continued by numerous researchers, has led to advances in our knowledge of the composition, structure, and origin of varnishes (Dorn and Oberlander, 1981, 1982); significant additional results are expected in the future (Dorn, 2007).

16.3.1 Composition and age

Desert varnish covers rock outcrops (*Fig. 16.27*) and surface clasts (*Fig. 16.28*). It is black if rich in Mn, orange if the Fe content is high, and brown if both elements are present in

Figure 16.28 Varnish film on clasts of a desert pavement. Negev Desert (Israel).

similar proportions. It ranges widely in thickness (0.002–0.5 mm) although it is generally between 0.01 and 0.03 mm thick.

Varnish occurs on all rock types but is more common on silicates. Its primary components are Si, Al, Mn, and Fe, and it also contains lesser amounts of Ca, K, Na, Ba, Ti, Sr, and Cu. All varnishes contain numerous elements in much smaller proportions and locally there may be significant concentrations of particular elements (Dorn, 1994a, 1998, 2007). Varnish is composed of microlaminations of different chemical compositions, primarily variations of Mn content. Many of these elements come from external sources because they are not present in the host rock. Clay minerals are the largest portion, 60–80%; amorphous oxyhydroxides of Fe, Mn, and Si are present in lesser amounts. Some black varnishes have high percentages of Si. They may be 0.01 mm thick and they occur on siliceous rocks. This type of cover is called **desert glaze** or *silica glaze* (Dorn, 1998). Scientists think it originates as chemical precipitation of the monosilicate acid $Si(OH)_4$ as a gel (Krauskopf, 1956) or by the biological intervention of opal phytoliths that dissolve and precipitate (Farr and Adams, 1984).

For decades geomorphic and archeological researchers have used the presence of varnish or degree of patination as an indicator of relative age (Hunt and Mabey, 1966; Tricart, 1969; Demangeot, 1981). Research indicates that varnish does not normally form on Holocene materials, which suggests that at least 10,000 years are required for its formation. Neutron activation, U series, paleomagnetism, and AMS ^{14}C have been used to determine absolute ages of varnishes. These analyses require collection of a 0.5 m^2 surface sample of varnish (Krauskopf, 1956) because of the small amount of organic matter present. Radiocarbon dating has been plagued with contamination problems because some bedrock is incorporated into the varnish sample. Researchers have developed a method to obtain relative ages based on the degree of leaching of varnish constituents: Ca+K/Ti have been considered an indication of leaching but this method also has problems with treatment and analytical determination. All these methods are used in research on fault activity, cliffs retreat rates, ages of Quaternary deposits, and lithic artifacts (Oberlander, 1994; Dorn, 1998).

16.3.2 Formation processes

Researchers used to think that varnish constituents came from weathering of the host rock. Fe- and Mn- rich solutions rise to the surface through capillary processes where they are deposited after being wetted by rain and dew. Today no one doubts that varnish constituents come from external sources either as eolian dust (Dorn, 1998, 2007) or dissolved in water (Perry et al., 2006) (*Fig. 16.29*). Electron microscope research shows that varnish is superimposed and clearly separate from the rock substrate; this supports an external origin for the constituents.

The most accepted hypothesis is biogenic, in which Mn is enriched by fixation by chemotrophic bacteria that oxidize Mn^{2+} to Mn^{4+}, followed by cementation by clays that trap other elements. As this process is repeated the result is varnish accretion and lamination during wetting and drying cycles. Today scientists believe that the bacteria that fix Mn is *Metallogenium personatum*, which is the oldest known living organism (Oberlander, 1994). Other authors propose fungi and cyanobacteria as fixing organisms (Krumbein and Jens, 1981). Some researchers advocate purely physicochemical processes (abiotic hypothesis) (Engle and Sharp, 1958; Elvidge and Moore, 1979). Manganese is more soluble than iron; in desert environments Eh values are high, indicating high oxidation, and pH is alkaline as a result of low leaching rates. According to these authors, variations

Figure 16.29 Varnish developed on sandstone, possibly by precipitation of water on the surface. This was later affected by tafonization processes. Petra (Jordan). Photo J.L. Peña.

in oxidation-reduction and the basicity of the environment produces varnishes. Intense debate continues about varnish generation; it is possible that a combination of biotic activity and variations in the physico-chemistry of the environment may be the path to success in future experiments (polygenetic model) (Krinsley, 1998; Dorn, 2007).

16.3.3 Environmental implications and age of varnishes

The first research on varnishes was focused on their climatic significance; these studies show that their formation required more humid climates than exist at present, meaning that existing varnishes are relict (see Dorn, 1998). They also suggest cycles of patination during which the varnish erodes and new varnish is created.

Today high-resolution techniques make it possible to interpret varnish laminations, which are microlayers of different chemical compositions, as climate-related features. Periods of lower Mn accumulation correspond to arid or interpluvial phases, whereas stages of higher Mn/Fe ratios are indicators of a more humid (Liu, 2003), pluvial, or lacustrine phase. During more arid periods, eolian deflation decreases across lakes surfaces *(Fig. 16.30)*. However, in many places the varnish sections contain records that are not convincing as indicators of paleoclimate oscillations; this shows that varnish formation processes are complex (Oberlander, 1994).

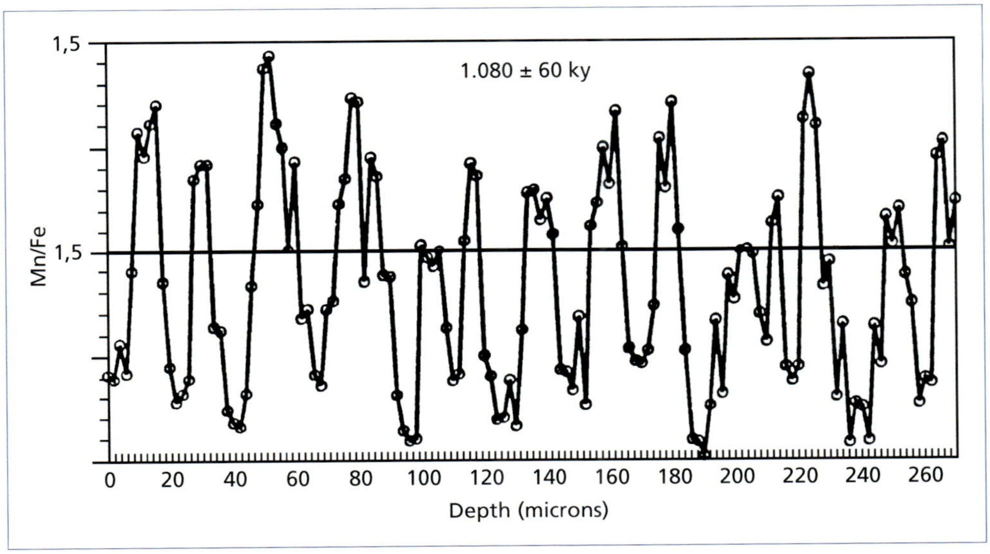

Figure 16.30 Variation in the Mn/Fe ratio with depth in a varnish deposited on a volcanic rock with a K/Ar age of 1080 ± 60 Ky. Coso area (California). The low points are interpreted as arid intervals within the climate oscillation (Oberlander, 1994).

Scientists have studied the microlamination sequence of hundreds of upper Pleistocene/Holocene varnish samples from deserts in the United States. Researchers have recorded 12 dark layers alternating with 13 orange layers deposited in the last 6,000 years (Liu and Broecker, 2007, 2008). Each dark layer represents a wet phase that can be correlated with cold events that persisted on a millennium scale in the North Atlantic by abyssal sedimentation (Bond et al., 1997, 1999).

In a study of rock surfaces in Messak Settafet (southeastern Libya, central Sahara), three microlayers that formed by different accretionary processes record a climatic transition from a humid environment to progressively more arid conditions. The most internal layer contains illuvial clay and calcite; it formed during the early-late Holocene (9200–5500 years BP) under humid environmental conditions. The second microlayer is a typical Mn-rich varnish, biomineralized during a phase of increasing aridity during the middle Holocene (6000–4000 years BP). The most external microlayer was formed by eolian dust; it dates from the last hyper-arid millennia. All this shows a good correlation between desert varnish and environmental changes in the Sahara (Zerboni, 2008).

16.4 Duricrusts

Duricrusts in arid zones have various chemical compositions; these surface and subsurface duricrusts are calcareous, siliceous, and gypsiferous. Laterites form in areas of higher precipitation, primarily on desert margins. They occupy large areas of arid environments *(Fig. 18.3)*; some laterites are so hard that they resist erosion and thus can occur at high elevations (McLaren and Nash, 2007). Laterites are also good indicators of paleoenvironmental conditions; knowledge of Quaternary crusts is useful for interpreting the paleoclimatic record of past geologic epochs (Goudie and Pye, 1983). In addition, they may be of economic interest; calcareous crusts contain uranium and other metals. They can also be used to construct houses and for highway infrastructure (Reeves, 1976).

In this section on crusts we will analyze their nature, properties, significance, the geochemical and biochemical processes involved in their genesis, and finally show the most commonly used models.

16.4.1 Caliches

The term caliche is a Spanish word derived from the Latin *calix*, which means lime or limestone; the term was first used in 1719 (see Reeves, 1976). The term caliche is commonly used in the United States and Mexico (Esteban and Klappa, 1983) but in Spain the term **calcareous crust** is more commonly used, probably influenced by the French (*croute calcaire*) and German (*kalkruste*) terms. Some confusion exists throughout the world because the word has been used for other types of deposits. In addition, numerous local names are used in different areas and countries (Goudie, 1973; Reeves, 1976).

According to Goudie (1972a), "calcrete is a term used for materials of continental origin formed predominantly but not exclusively by calcium carbonate; it ranges from powdery and nodular to very hard and implies cementation of soil, rock, or weathered material within the vadose zone. However, the term does not include speleothems, travertine, cemented littoral deposits, or stromatolites of lacustrine algae". The term "calcrete" is synonymous with caliche (Wright, 2007).

Caliches are whitish in color. They cover about 13% of the Earth's surface and are found in areas with annual precipitation ranging from 400 to 600 mm (Goudie, 1983b). Rutte (1958), in his study on Spanish calcareous crusts, found that they occur in areas where rainfall ranges from 100 to 500 mm; he did not find caliches below or above this rainfall range. However, most caliches occur in areas where annual precipitation is between 100 and 250 mm. Although widely distributed, caliches are characteristic of the world's arid and semi-arid zones *(Fig. 16.31)* (Reeves, 1976). In the geologic record, they are indicators of ancient aridity (Nash, 2004a). They develop on top of surface deposits of pediments, terraces, and eolian accumulations. On the terraces of the Cinca River (Ebro Depression), the degree of calcification of the stepped alluvial levels increases in the oldest deposits and there is no calcification in the modern levels (Sancho and Meléndez, 1992).

Monographs have been published on the specific topic of caliches (Reeves, 1976; Blümel, 1981; Vogt, 1984; Wright and Tucker, 1991); the monographs cover their types, characteristics, and genesis. As Vogt (1984) points out, research on calcareous crusts must include multiple scales. First, it is important to determine their spatial distribution (kilometer scale) and their relationship to the regional geomorphic context, such as their location on top of pediment and terrace systems. Next, research should determine their outcrop level through an analysis of natural cross-sections and excavations in which sedimentary profiles are recorded; this makes it possible to locate the caliches in sequence. Finally, thin section microscopy and electron microscopy is done to analyze textures and micromorphologies that provide data on formation mechanisms. This research should be followed by chemical and mineralogical analysis to aid in the understanding of the processes that generate them.

16.4.1.1 Geochemistry and mineralogy

The chemical composition of caliches calculated from 300 samples collected from all over the world (Goudie, 1972b; 1973;

Figure 16.31 Erosion-resistant caliche on cemented gravel located on the margin of a wadi. Jaisalmer. Thar Desert, India.

1983b) is as follows: $CaCO_3$: 79.28% (CaO^-: 42.62%), SiO_2: 12.30%; MgO: 3.05%; Al_2O_3: 2.12% and Fe_2O_3: 2.03%. Magnesium-rich caliches are called **dolocretes**. The percentages of various components range widely from one location to another as shown by the calcretes of South Australia (Hutton and Dixon, 1981; Dixon, 1994b). A study involving numerous chemical analyses of caliches from Oklahoma and New Mexico (Aristarain, 1970) shows that concentrations of calcium and magnesian carbonate and ferric iron decrease with depth whereas silica increases significantly in the middle of the profile. In addition, Dixon (1994b) examined the role of topography in variations of caliche chemistry and showed a higher percentage of calcium in lower toposequences and higher magnesium in upper ones. Calcium decreases with depth in all of them.

Calcite and dolomite are the dominant carbonate minerals in caliches; they are accompanied by quartz, opal, and clay minerals. Caliches can also contain minor quantities of soluble salts (gypsum), glauconite, phosphates, and heavy minerals, etc. Mineralogical research has been focused particularly on the clay fraction to determine the origin of these minerals and consequently to better understand the processes of caliche formation. The dominant clay minerals are palygorskite and sepiolite although illite, kaolinite, montmorillonite, interstratified illite-montmorillonite, and chlorite are also present (Aristarain 1970; 1971; Gardner, 1972; Reeves, 1976; Hay and Wiggins, 1980; Wright and Tucker, 1991; Wright, 2007). The presence of palygorskite and sepiolite are attributed to neoformation in an environment with sufficient magnesium; they result from the weathering of favorable lithologies (Hay and Wiggins, 1980; Hutton and Dixon, 1981; Sancho *et al.*, 1992). Other researchers maintain that these minerals result from alteration of montmorillonite and interstratified montmorillonite-illite (Watts, 1980).

Radiometric dating has been done on Quaternary caliches. Kelly *et al.* (2000) show in their U/Th dating that the dates obtained are an apparent minimum age for the beginning of caliche formation.

16.4.1.2 Morphology

The literature contains numerous descriptions of caliche morphology. Various classifications have been proposed on the basis of different criteria. Some are based on sequential evolution (Netterberg, 1969; Wright, 2007), others on the degree of hardness of the carbonates (Reeves, 1976), and still others on differences of a pedogenic nature that take into account the degree of calcification (Gile *et al.* 1966; Machette, 1985). Some establish few morphological types

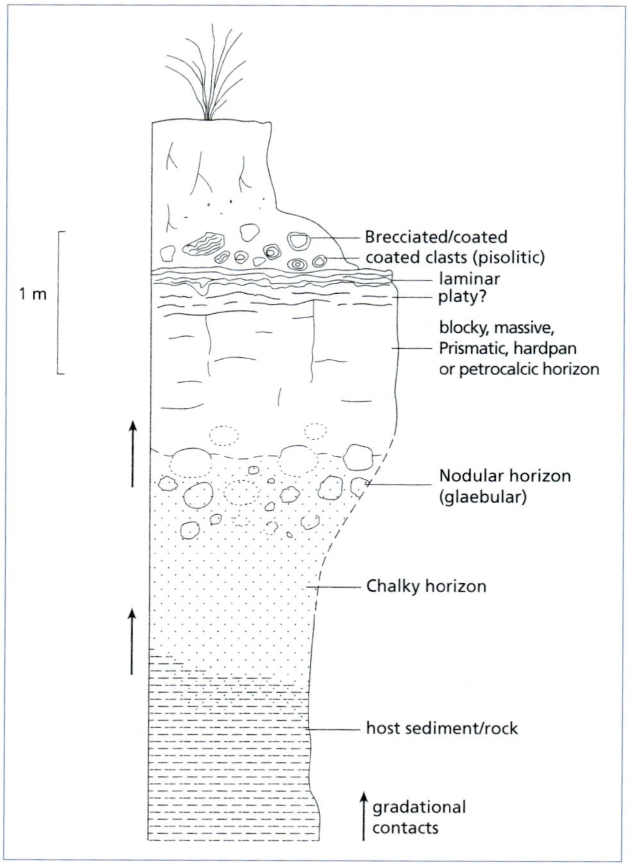

Figure 16.32 Idealized profile of caliche showing different macroscopic morphologies. (Wright, 2007).

whereas others use numerous distinctions (Reeves, 1976). The classic well-developed profile (Alonso-Zarza *et al.*, (1998) (*Fig. 16.32*) has numerous exceptions. Caliches of phreatic and pedogenic origin have very similar morphologies.

The primary macroscopic morphologies (Esteban and Klappa, 1983; Sancho and Meléndez, 1992; Alonso-Zarza, 2003; Wright, 2007) are calcified gravels that normally occur at the base of the caliche profile although small layers can be found within the profile. These are composed of clasts with vadose carbonate envelopes; the matrix may have root molds. **Chalky or powder caliche** (*Fig. 16.33*) is formed from loose silt and sand-sized carbonate particles essentially without material from the enclosing rock. Chalky caliche may be lightly cemented and have a leafy structure. **Hardpan caliche** occurs in the upper part of the profile (*Fig. 16.34*). It is composed of plates that measure centimeters or decimeters and that lock together laterally. Ooids, basal carbonate speleothems, root traces, and desiccation cracks also form. Hardpan caliche can also be deformed. Expansion or contraction can form **pseudoanticlines** (*Fig. 16.34*); the expansion results from hydration of highly soluble salts, alternating periods of wetting followed by drying, calcite crystallization, and subsequent arching. Deformation can also be due to the presence of expansive clays (Reeves, 1976). Finally, **nodular or glaebular caliche** can form on top of fine-grained material in which carbonate cements the particles and generates a large number of nodules (*Fig. 16.35*). It can also come apart as loose or cemented nodules from fragmentation of hardpan caliche

Figure 16.33 Powdery laminated caliche cemented at the top. Alfés (Province of Lérida). Ebro Depression (Spain).

Figure 16.34 Hardpan and nodular caliche affected by a pseudoanticline. La Plana Negra (Province of Zaragoza). Ebro Depression (Spain).

Figure 16.35 Nodular caliche developed on loose fine-grained materials. Bawri. Thar Desert, India.

(*Fig. 16.36*). There is a continuous gradation between the hardpan and nodular varieties.

The macroscopic characteristics of caliche are often apparent on a microscopic scale: nodular texture, oolites, speleothems, carbonate envelopes, and traces of biological activity. All research aimed at a detailed and precise understanding of these micromorphologies and textures leads to a more complete understanding of caliche-generating processes.

16.4.1.3 Origin

Caliche analysis reveals the presence of complex sequences showing the superposition of numerous caliche horizons. In addition, most profiles show various stages of precipitation, dissolution, and reworking.

Caliche formation requires a source of calcium carbonate, which can be bedrock, plant remains, volcanic emissions, or eolian dust (Goudie, 1983b). Many authors consider it likely that carbonates have been transported by wind and dissolved in rainwater (Machette, 1985; Blümel, 1981, 1982) because caliche may develop to great thicknesses on top of rocks that contain almost no calcium.

In semi-arid zones where water is scarce, carbonates do not leach completely but tend to move from one area

Figure 16.36 Nodular caliche with larger nodules at the top. Margalef (Province of Lérida). Ebro Depression (Spain).

to another where they precipitate. The fundamental process implied in caliche genesis is dissolution and precipitation of calcium carbonate in the presence of carbon dioxide:

$$CaCO_3 + CO_2 + H_2O \leftrightarrow Ca^{2+} + 2HCO_3^-$$

Carbonate precipitation involves the loss of CO_2, evaporation, the common ion effect, and biological activity (Wright and Tucker, 1991). Biological mechanisms for fixing carbonates in soils involve lichens, algae, bacteria, and microrhizomes (Klappa, 1979). Biogenic constituents are abundant in Texas caliches (Zhou and Chafetz, 2009); they are generally found in the upper part of the profile. The main biogenic facies of caliche are rhyzolites, stromatolitic laminae, and carbonate-covered particles. Biotic constituents decrease toward more arid regions.

Goudie (1973,1983b), Carlisle (1983), and Wright (2007) suggest two origins for caliches: one pedogenic, which seems to be the most common, and another that does not involve pedogenesis. In either case it is difficult to distinguish one from another because pedogenic processes typically modify non-pedogenic caliches.

Pedogenic caliches form from carbonates that accumulate from top to bottom (Goudie's *descensum model*); they form from leaching of the A horizon and illuviation in the B horizon (petrocalcic horizon). Recent profiles show that pedogenesis begins as powdery filaments and nodules. However, in thousand-year-old profiles, clasts covered with carbonate skins fill the spaces between particles. The presence of calcified filaments, fibrous calcites, and mushroom microfossils support a pedogenic origin of caliches (Vaniman *et al.*, 1994). One difficulty of this hypothesis is that the caliche thickness is generally much greater than the soil; this anomaly can be explained by slow erosion of the soil during caliche production.

Wright's (2007) (*Fig. 16.37*) principal models are pedogenic. The simplest case consists of progressive weathering of a carbonate-rich substrate (*Fig. 16.37B*). During weathering, the caliche develops on top of porous limestones; this implies down-water dissolution as carbonates are accompanied by re-precipitation (Rabenhorst *et al.*, 1991). Using research on caliche profiles in southern Europe as his base, he proposed a rhizogenic origin (Wright *et al.*, 1995) (*Fig. 16.37C*) in which some laminar layers develop from root molds. The progressive illuviation model (Gile et al., 1966; Machette, 1985) (*Fig. 16.37A*) defines a series of time-dependent stages (chronosequences) that are transitional. *Figure 16.37A* shows the progressive accumulation of carbonates (stages 1–4) leading to the paralysis of the profile and the stagnation of water on top of the impermeable petrocalcic profile.

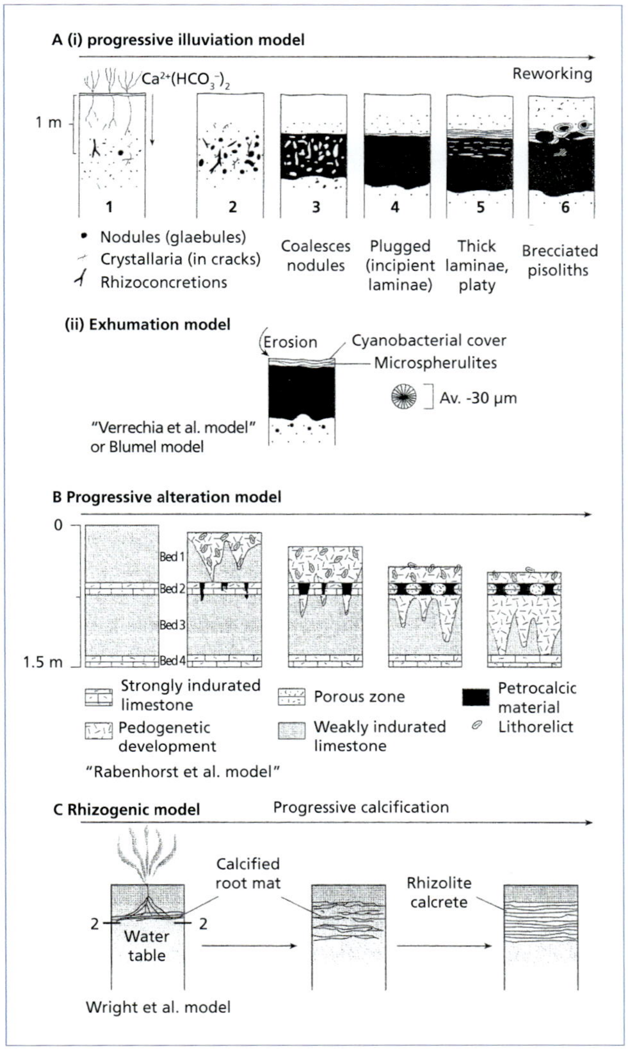

Figure 16.37 Models of pedogenic caliches. (A) (i) after Gile *et al.*, (1966) and Machette (1985); (ii) after Blumel (1982) and Verrechia *et al.*, (1995). (B) after Rabenhorst *et al.*, (1991). (C) after Wright *et al.*, (1995).

Carbonate precipitation in these ponds forms laminar caliches. Carbonate deposited around the grains may generate pisolites. With time, the laminar structures may thicken or the soil may erode. Sometimes the carbonates crack, possibly due to roots, and the fragments may be covered by newly formed carbonates (*Fig. 16.37A(i)*). Blumel (1982) and Verrechia *et al.* (1995) suggest that phototropic microbial communities precipitate laminar caliches and that the precipitation is linked to the presence of microspherulites. Afterwards, an exhumation stage may occur (*Fig. 16.37A(ii)*). Quaternary caliches, including the oldest ones, have been subject to very complex periods of pedogenesis, episodic sedimentation, and erosion (Alonso-Zarza et al., 1998) that make it difficult to decipher their origin.

The ages obtained for caliches in southeastern Spain suggest that the formation of pedogenic caliche horizons fundamentally took place during isotopic warm stages (MIS 1 and 5) and that very little caliche formed during the cold glacial/stadial stages (MIS 2, 3, and 4). They conclude that caliche formation during the Quaternary changed as a result of climate changes but that it occurred during warm periods (Candy and Black, 2009).

When carbonate solutions rise as a result of capillarity and precipitate at the surface by evaporation, **non-pedogenic caliches** form according to Goudie's *per ascensum* model. Caliches can form near the water table by evaporation at the capillary fringe; they can also form below the phreatic level as a result of decreased CO_2 (Land, 1970). Carbonate-rich floodwater can also form this type of caliche, which can reach thicknesses greater than 10 m. They seem to form rapidly because some of the alluvium in them is very young (Machette, 1985) although this origin is a special case. These caliches can be distinguished from pedogenic caliches by the absence of a simple micromorphology in the profile and the presence of calcite cements (Nash and Smith, 1998).

Wright (2007) points out that the thickest phreatic caliches are formed by interstratal cementation, mobilization, and replacement of sediments by carbonate that comes from groundwater. Precipitation occurs in the capillary fringe or below the water table (*Fig. 16.38A*). In fluvial systems, caliches form hardpan bodies tens and even hundreds of kilometers long. Phreatic caliches in alluvial fans are more common in distal zones where the phreatic layer is closer to the surface and prone to evaporation. In distal fan areas where fine-grained playa deposits are found, groundwater is shallow and may overlap vadose and phreatic caliches. Capillary fringe caliches are characterized by fine zones of carbonates associated with the upper part of nodules and tubules (Mack *et al.*, 2000). Phreatic caliches form below the soil and generally lack biogenic characteristics but phreatophyte plants can form laminar caliches and rhizo-concretions in the capillary fringe (Semeniuk and Meagher, 1981). As groundwater flow continues, the Mg/Ca ratio increases as calcite precipitates (*Fig. 16.38B*) (Arakel, 1986). Alonso-Zarza (2003) thinks that phreatic carbonates can rapidly superpose pedogenic caliches and lacustrine-palustrine carbonates. These caliches are generally exhumed during drainage inversion.

Calcretes have numerous paleoenvironmental applications including as temperature measurements and determination of paleoprecipitation. They can also be used to estimate carbon dioxide pressure in the past. Pedogenic caliches can form in various climates including the Arctic but they occur mainly in arid and semi-arid climates (Wright, 2007).

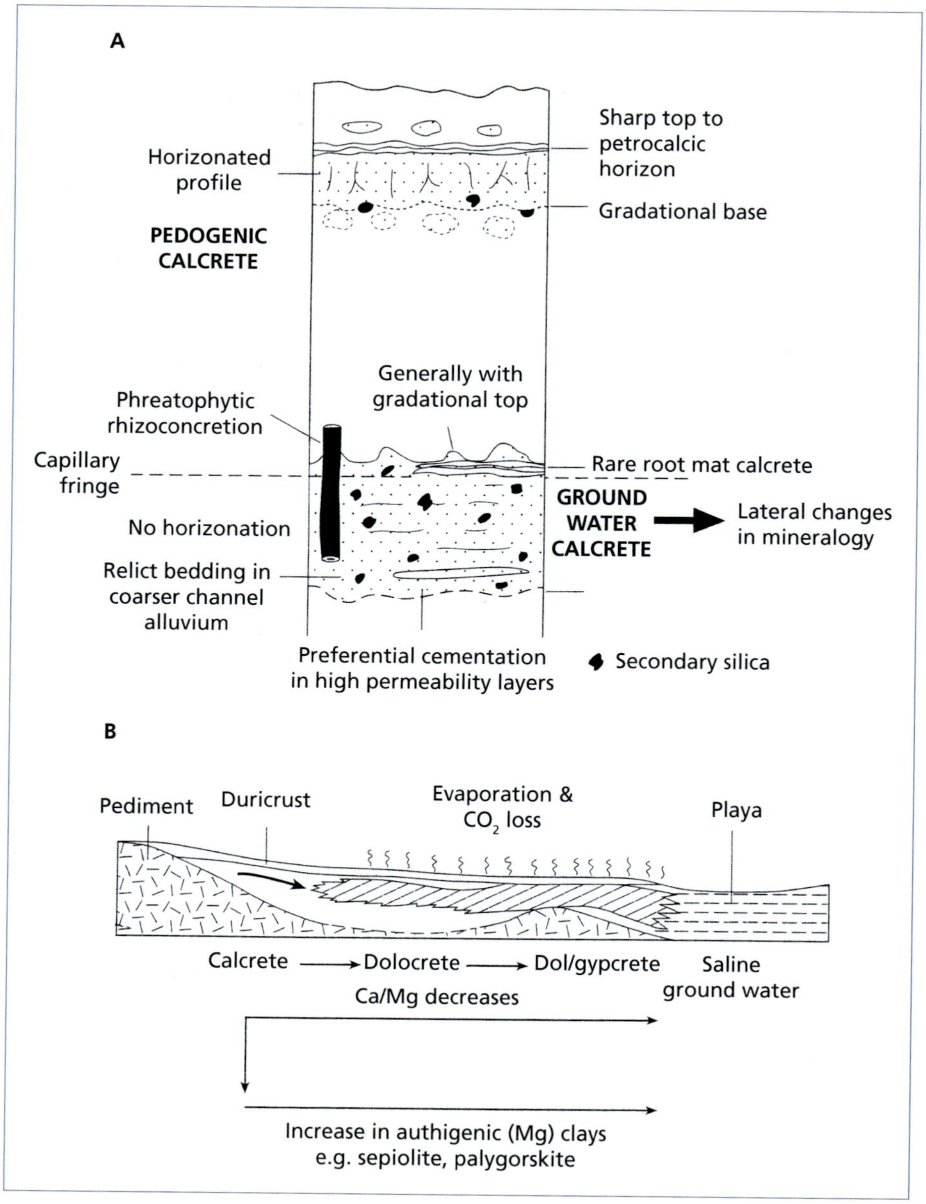

Figure 16.38 Characteristics of phreatic caliches. (A) Macroscopic features, based on various authors. Phreatophyte plants can also produce rhizoconcretions and laminar crusts. (B) Models for development from groundwater and their precipitates in arid and semi-arid alluvial systems (Based on Arakel, 1986; Wright, 2007).

16.4.2 Siliceous crusts

These are also called **silcretes.** Siliceous crusts are gray, brown, or green rocks; they are very hard and brittle, have conchoidal fracture, and are composed of quartz grains cemented by a microcrystalline and amorphous matrix. Their thickness ranges from 1 to 3 m although in some profiles their thickness may exceed 15 m (Nash and Ulliot, 2007). They form primarily in arid environments although they can also form in more humid environments. They are a siliceous replacement product of surficial materials (regolith, sediments, soil); silicification takes place at low temperatures unrelated to volcanism, plutonism, or metamorphism (Summerfield, 1983). They are found in locations that were not glaciated during the late Tertiary and Quaternary (Nash, 2004b). Siliceous crusts are common in Australia

(Fig. 18.3) and South Africa although they are also found in other areas. They typically occupy the crests of hills or mesas because they are resistant to erosion; they may be very extensive when their origin is related to paleosurfaces (Nash and Ulliot, 2007).

In outcrop they have a massive structure and columnar jointing. The crusts also have plate, botryoidal, and pillow structures (Wopfner, 1978). Their micromorphology reflects characteristics of the host rock (grain size and shape, mineralogy, and fabric) and silicification processes. These include micro and cryptocrystalline quartz with indented borders in mineral grains, which support a replacement origin (Milnes *et al.*, 1991). Chemical analysis shows that they contain 85% by weight SiO_2 (Nash, 2004b) and small quantities of aluminum, iron, and titanium oxides and hydroxides (Nash and Ulliot, 2007). The aluminum is associated with clay minerals whereas the titanium appears to be related to the host rock.

The formation of siliceous crusts includes a silica source and dissolution, transport, and precipitation of the silica (Nash, and Ulliot, 2007). The primary source of the silica may be *in situ* solution, weathering of silicates, and eolian dust. In fluvial and lacustrine environments, diatoms may play an important role. Dissolution and transport are active in aqueous systems where pH is clearly alkaline. This transfer must be lateral and vertical (Summerfield, 1983). The lateral model is used for silcretes that form in fluvial and lacustrine environments. The vertical transfer model includes processes similar to the *per descensum* and *per ascensum* models that Goudie (1973) proposed for caliche genesis. Finally, precipitation occurs primarily during evaporation or with a decrease of alkaline pH.

The ages of thick siliceous crusts collected from various places around the world date from the Mesozoic; they are relict accumulations possibly generated in other climatic environments (Cooke *et al.*, 1993).

Pedogenic silcretes (Fig. 16.39A) form over long periods of time (>10^6 years) (Nash and Ulliot, 2007). They create soil profiles on a stable basement and indicate a hiatus or disconformity. Silica may percolate and precipitate in climates that are highly seasonal and that have high evaporation rates (Webb and Holding, 1998). Phreatic silcretes are found in discontinuous lenses (Fig. 16.39B); they are stacked and do not have an organized profile. Their origin is related to actual and ancient phreatic levels and the silicification occurred as a result of oscillations of the water table (Rodas *et al.*, 1994). Lacustrine silcretes commonly develop within or near ephemeral lakes or playas within endorheic basins (Gouide, 1973; Sommerfield, 1982). Their paleoenvironmental significance is controversial; some authors claim they form in arid climates and others maintain that they form in humid climates (see Nash and Ulliot, 2007).

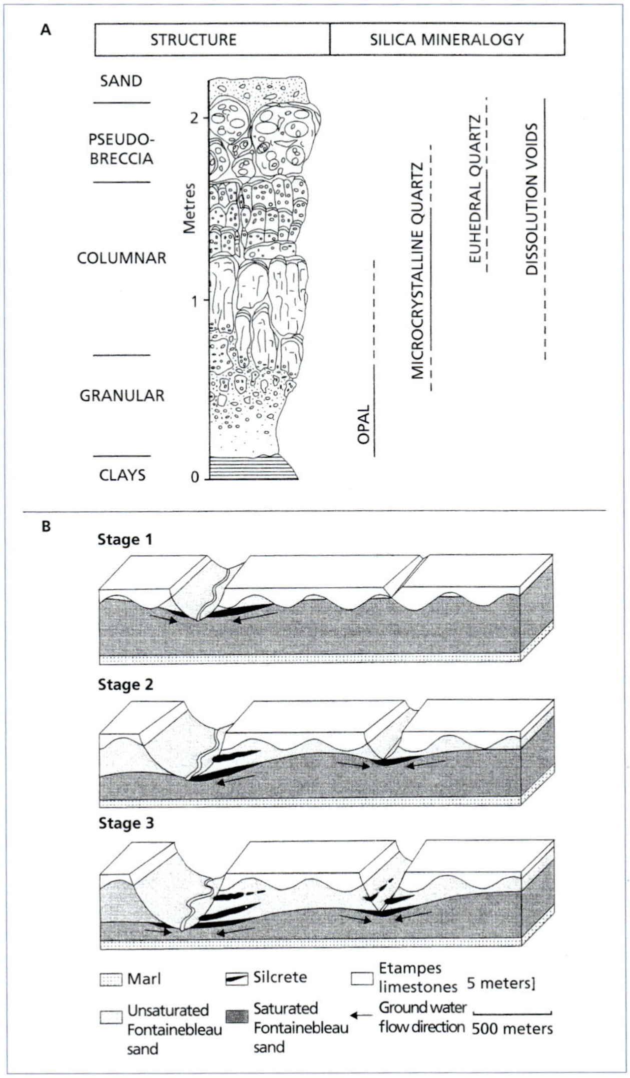

Figure 16.39 (A) Schematic diagram of a typical profile of pedogenic silcrete and vertical arrangement of siliceous cements. (B) Model of phreatic silcrete showing the development of silcrete lenses superimposed on an incised system in a progressive valley (according to various authors, in Wright, 2007).

16.4.3 Gypsiferous crusts

These are found as small areas in numerous arid zones (Coque, 1962). They occur in deserts with annual precipitation below 250 mm (Watson, 1983). Gypsiferous crusts are defined as gypsum accumulations in the top 10 m of soil depth; they are 0.10–5 m thick, their gypsum content is more than 15% by weight, and they have a minimum of 5% more gypsum than the underlying layer (Watson, 1985, 1989b).

They are classified as three types on the basis of their structure and stratigraphic position (Watson, 1979). The first type consists of horizontally stratified crusts of 50–80% gypsum. The second are surface crusts formed by large lenticular crystals from 1 mm to 0.5 m in diameter, known as desert rose crusts or by mesocrystalline material with crystals that measure 0.05 to 1 mm. They can reach thicknesses of 5 m and their gypsum content can range from 50 to 70%. The third type consists of surface crusts formed primarily of alabaster gypsum with crystals <0.05 mm; they are found as columnar crusts that are 1 to 2 m thick and also occur as powdery deposits or surface clasts. They contain about 90% gypsum.

Gypsiferous crusts are formed by capillary action and evaporation of sulfate water or *per descensum* from gypsum-bearing eolian dust (Watson, 1985; Heine and Walter, 1996; Eckardt *et al.*, 2001). These authors claim that the dust in Tunisia comes primarily from sebkhas. In the Namib Desert, H_2S gas from organic-rich marine sediments is released into the atmosphere and carried by the wind in coastal fogs to the interior of the continent (Martin, 1963). Olivier (1995) and Heine and Walter (1996) share this interpretation. For Eckardt and Spiro (1999) the most significant source is the oxidation of marine dimethyl sulfur (CH_3SCH_3). This aerosol is distributed by eolian activity and occasionally by surface runoff. The non-pedogenic gypsiferous crust model is based on gypsum precipitation by groundwater or in lacustrine environments. Because of the purity of these gypsum deposits, they are mined in Namibia and Australia (Eckardt, 2004).

16.5 Water in arid zones

We have already discussed the importance of water in weathering processes and the formation of patterned ground, varnishes, and crusts. In these cases, water forms a thin layer and moves locally. In arid areas water has a larger role because of surface flow and fluvial systems. Schumm (1977) divides fluvial systems into a supply area or drainage basin, which provides the water and sediments, a transfer zone, and distal area or sedimentation zone. A fluvial system can be described by a set of independent and dependent variables (Schumm and Lichty, 1965), which have unequal influence. Time, initial relief, geology (lithology and structure), and climate are the most important independent variables that control hydrology and erosion. Vegetation is a function of climate and lithology. Other dependent variables are relief, hydrology (runoff, discharge, sediment production), drainage network, and slope morphology. In arid zones the first three independent variables vary substantially from desert to desert; however, climate does not vary because it depends on precipitation, which in turn causes runoff and erosion, which is conditioned by plant cover.

16.5.1 Precipitation, vegetation, and evapotranspiration

In most of the world's arid zones **precipitation** data are scarce because the few meteorological stations in these areas are generally distant from each other. Consequently, extrapolation of this sparse data is risky. Precipitation is the product of a variety of atmospheric processes: low pressure fronts, tropical storms, orographic effects, and convective cells (Graf, 1988). Most deserts are located in anticyclonic cells within the global atmospheric circulation system. Precipitation is limited to sporadic penetration of low-pressure fronts into these areas; these are infrequent and generally produce little rain. Tropical storms (cyclones, hurricanes, and typhoons) are less common than low-pressure fronts but they sometimes penetrate into the interior of arid zones such as the Sonoran Desert between Mexico and Arizona. The orographic effect occurs when the pressure and temperature of an air mass decrease as the air mass encounters an obstacle and rises; this process causes condensation and precipitation. As the air moves down the other side of the mountain, it warms; because it has lost most of its moisture, the dry air creates a rain shadow zone. The last process is convective storms that measure a few kilometers in diameter; this process heats the air. If these storms affect small basins, they can cause significant flooding such as the flood on the lower course of the Huerva River, a tributary of the Ebro, on 28 July 1990, which formed a delta at the mouth of the river.

Precipitation in arid zones varies significantly (*Fig. 16.4*); as a result, it can cause major changes in runoff and sediment production. For example, at the experimental station for the study of runoff and erosion in Lanaja (Province of Huesca) in the Ebro Depression (*Fig. 16.40*), the most significant precipitation in a continuous four-year record resulted from storms related to convective cells; these storms occur primarily during the spring and fall, with only a few in the summer. The remainder of the precipitation is rain from small low-pressure fronts (Gutiérrez *et al.*, 1995). For this reason, the highest values of runoff and erosion occur in spring and fall. Intensities vary considerably within rainfall events caused by convective cells. The rain recorded in a single storm may exceed the total annual average precipitation. In addition, experimental basins equipped with multiple rain gauges have abrupt spatial variation and high gradients (*Fig. 16.41*).

The literature contains several cases in which precipitation related to advection of moist air masses provided continuous rainfall on a semi-arid region (Inbar, 1972; in Starkel, 1976). These are Mediterranean winter rains that fall on mountains in Lebanon. Precipitation over four days (20 to 23 of January 1969) ranged from 120 to 360 mm, equivalent to one third of average annual precipitation. In contrast to small

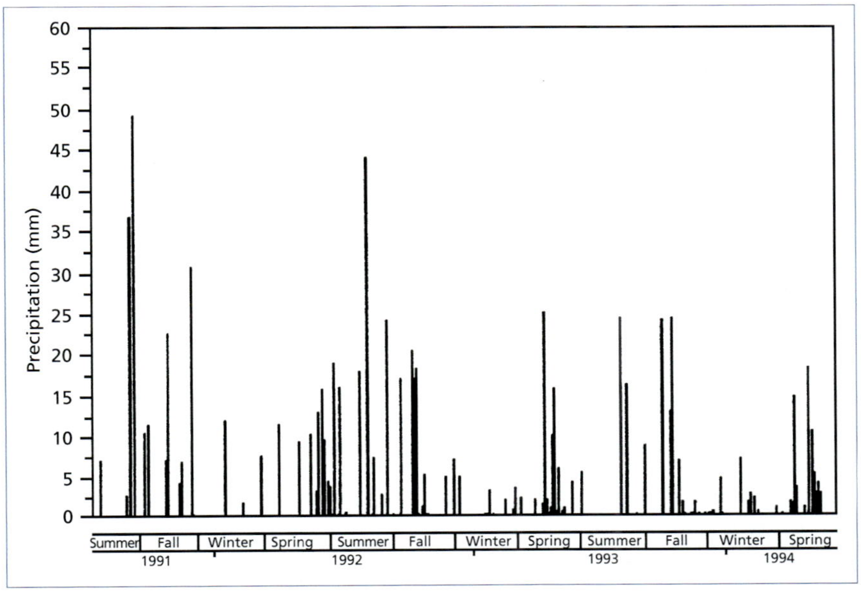

Figure 16.40 Rainfall record at the experimental station for research on water erosion in Lanaja (Province of Huesca, Spain) for the period of July 1991 to May 1994 (Gutiérrez *et al*., 1995).

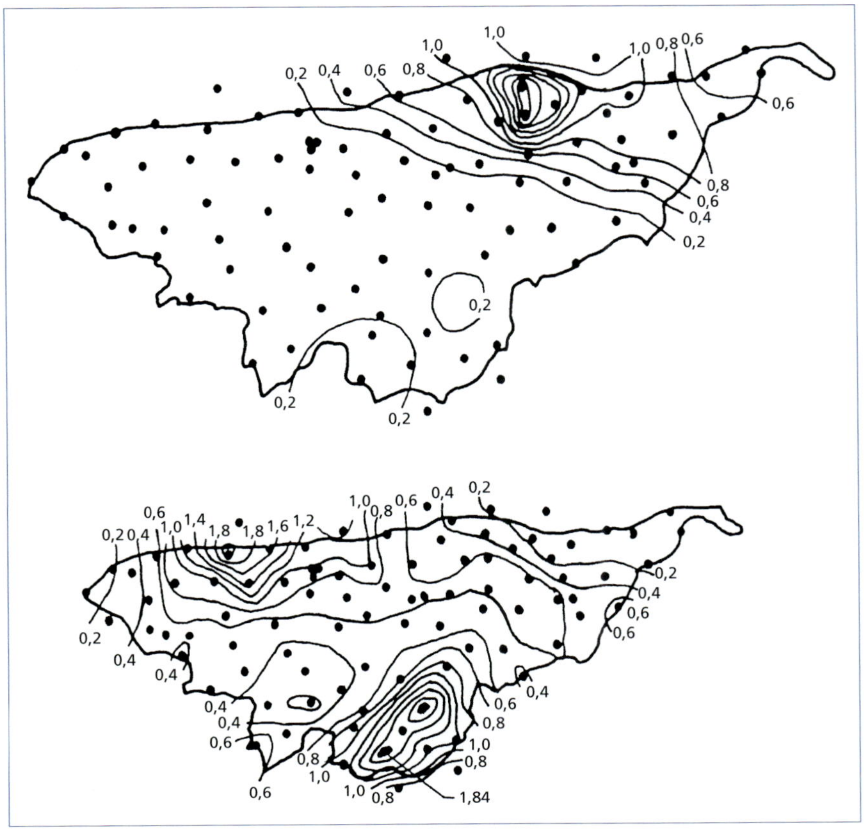

Figure 16.41 Rainfall distribution produced by storm cells in Walnut Gulch, Arizona. Top: Event caused by a single cell. Below: Event with multiple cells. The points represent rain gauges (Renard, 1970).

average intensities of 2–3 mm/hour, rarely exceeding 8 mm/hour, this rainfall has a return period of 100 years. This storm caused intense runoff in the upper basin of the Meshushim; it lowered the surface of this basalt zone 0.04 mm/year, six times the normal. This event shows that low-pressure rainfall can be significant in semi-arid zones.

Vegetation is an essential characteristic of hot arid zone environments; aridity determines whether vegetation is dispersed or not present (Bull and Kirkby, 2002). Vegetation in deserts may be sparse or non-existent. In environments with less than 100 mm of rainfall, vegetation grows only around wadis (*Fig. 16.42*) and in closed depressions. In places where precipitation ranges between 100 and 500 mm natural vegetation is arboreal, which has generally been degraded by humans, shrubs and grasses, which have also been affected by humans and by overgrazing (Francis, 1994). High temperatures and the lack of water force plants to accommodate to these inhospitable environments through physiological, morphological, and behavioral adaptations that become more pronounced as aridity increases (Günster *et al.*, 1993).

We have noted the role of flora in the biological weathering of rocks and the influence of sparse organic material on processes of water erosion. The percentage of plant cover generally correlates well with annual average precipitation. The amount of water held in the leaves of plants is a significant factor in water erosion. In a vegetation study in a semi-arid zone of northeastern Mexico, Návar and Bryan (1990) found that after 230 mm of rainfall fell during 17 storms, 27% of the water was lost to vegetation. Infiltration can also be highly variable; it is considerably higher in zones with vegetation than in denuded soils (López *et al.*, 1984). Some desert environments are more sensitive to vegetation changes, as shown by sediment production. Sediment production increases on slopes with decreased vegetation; it reaches a maximum on denuded soils (*Fig. 16.43A*). However, erosion is hardly affected when the percentage of plant cover is low (*Fig. 16.43B*, *C*) (Schumm, 1977). De Ploey et al., (1976) and Morgan et al., (1986) found that sediment production increased for low values of plant cover (*Fig. 16.43D*). Rogers and Schumm (1991) show experimentally on slopes with 10% gradient that erosion increases rapidly when plant cover decreases from 43% to 15% but erosion decreases below this last number and as a result, at values below 15%, vegetation does not slow down sediment production.

Vegetation affects wind erosion by acting as an obstacle and reducing its velocity; it causes small dunes called *nebkhas* to form on the leeward side of plants. As vegetation increases, dunes become less mobile (Ash and Wasson, 1983) and may stabilize. Biocrusts can also paralyze dune movement as seen in the longitudinal dunes of Nizzana (Negev Desert) (Veste, 1995) (*Fig. 5.18*).

Water in materials at the Earth's surface undergoes **evaporation,** which dries out these materials. In desert regions, desiccation due to long droughts can extend several meters into the ground. Plants absorb water through their roots, which also causes water loss. **Transpiration** occurs in the stomata where water vapor is exchanged between the atmosphere and the leaves. Low humidity, high

Figure 16.42 Shrub vegetation in the bed of the Nahal Zin in the Negev Desert, Israel.

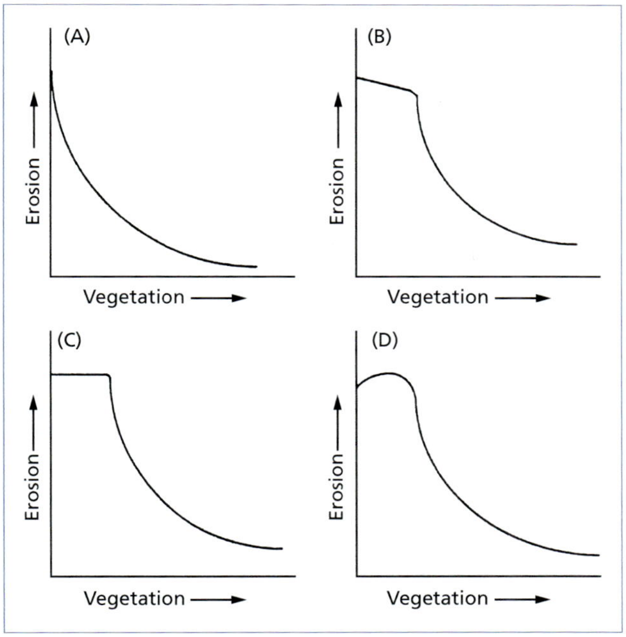

Figure 16.43 Hypothetical relationships between vegetation and erosion. All the curves indicate an exponential increase in erosion for a variable vegetation density, but B, C, and D show a decreasing influence of vegetation for low values plant cover (Rogers and Schumm, 1991).

temperatures, and wind all promote transpiration. In deserts, xerophilic plants adapt to the lack of water by closing their stomata to reduce transpiration and by decreasing leaf size or losing leaves completely, like the cactus plant. In addition to xerophilic plants, phreatophyte plants grow in wadis in arid zones; their roots extend the water table to obtain the water they need (*Fig. 16.42*). Evapotranspiration may account for 85–90% of total precipitation (Renard, 1969) although plant species differ. Most evapotranspiration occurs shortly after rainfall events. Transpiration rates fluctuate between 2.3 and 10 mm/day; during the drier times of the year they may be reduced to 0.1 mm/day (Thornes, 1994).

16.5.2 Desert fluvial networks

Precipitation in arid zones is sparse and variable; this variability is reflected in fluvial systems worldwide. We can distinguish between **endogenous drainage,** which results from precipitation that occurs within the desert area, and **exogenous drainage,** which results from precipitation that originates outside the arid zone but flows through it. The Nile, Indus, Colorado, and Ebro rivers are examples of exogenous drainage. A significant part of desert drainage is **endorheic**, which means it does not flow into the sea (*Fig. 16.44*) because it infiltrates and more importantly, evaporates. Large

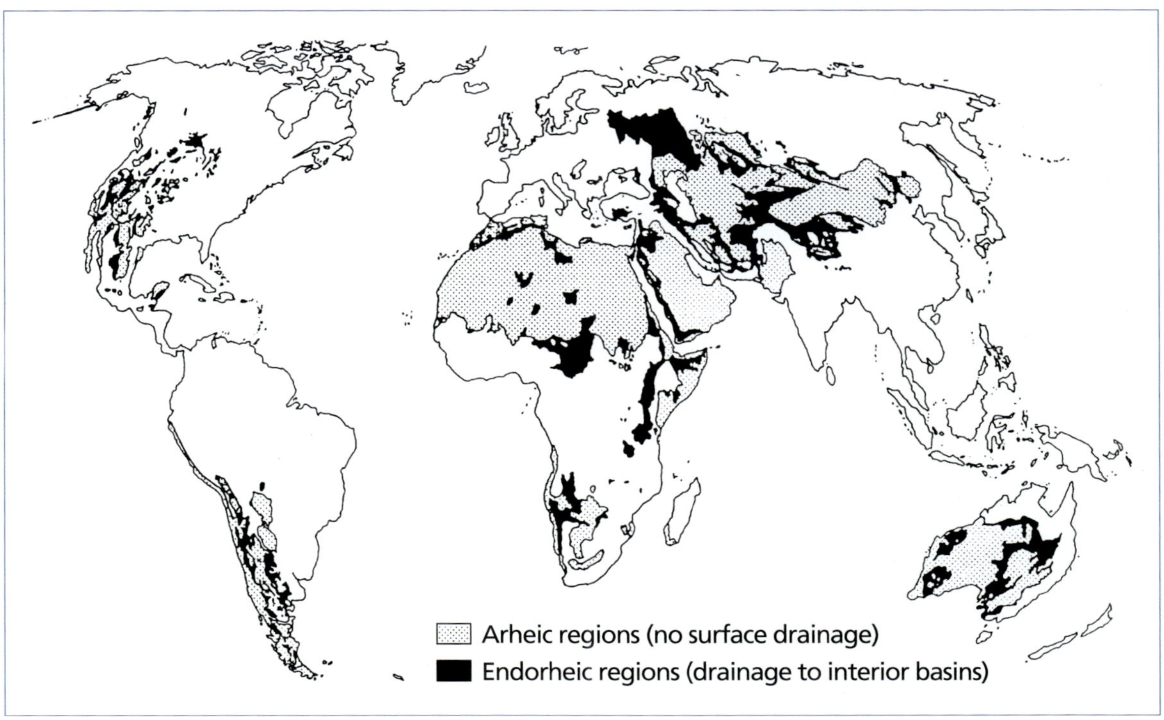

Figure 16.44 Distribution and extent of areas that lack regular drainage to the sea (from Martonne and Aufrère, 1928).

closed depressions also collect surface water; examples are the endorheic basins of Lake Chad in the southern Sahara and Eyre in the central Australian desert. Guilcher (in Demangeot, 1981) estimates that in most Saharan regions, floods flow no farther than 300 km. **Arheic** drainage areas have no regular water circulation; this type of drainage covers significant areas of desert regions. Some deserts have **exorheic** drainage; an example is the southern part of the Atacama Desert (Chile), which receives a large supply of water from the Andes.

Streams in arid zones are usually ephemeral (*Fig. 16.45*) as a result of the sparse water supply they receive; these rivers (wadis = ramblas) remain dry most of the year (*Fig. 16.46*). However, there are also perennial rivers that receive significant allogenic supply, normally from nearby cordilleras that supply snow meltwater (*Fig. 16.47*). In some cases, the more distal drainage no longer functions; this indicates that previous stages (rainy periods) had a higher water contribution. This is the case of the Central Sahara where complex wadi networks receive no water even during major high water periods (Tricart, 1969). The Niger and Senegal rivers, which spring from the tropical mountains of Guinea, encounter extremely dry dune areas as they flow north; the rivers flow across these areas only during rainy periods (Tricart, 1969). Aerial photography and satellite images show that some desert areas previously had fluvial systems and that these riverbeds are now buried under desert plains; an example is the Saraswati River in the Thar Desert (Ghose *et al.*, 1979) (*Fig. 16.48*). Similar circumstances are found in the large endorheic area of Lake Chad, south of the Sahara, where eolian sands have fossilized a significant part of the fluvial network (Grove and Warren, 1968). In the piedmonts of some desert areas, lateral channel migration is common in alluvial fan systems; these channels are reworked by eolian activity during periods when there is no water.

Overland flow in deserts may occur as surface hortonian flow (Horton, 1933), which occurs when rainfall intensity exceeds the infiltration capacity of the soil or rock, or when precipitation saturates the surface material. During rainstorms in arid zones, overland flow occurs by hortonian flow if the downpour exceeds the infiltration rate.

Hydrographs show how discharge varies with time. Hydrographs have a very clear ascending limb and a descending limb that is initially steep and later flattens out (*Fig. 16.49*). Abrupt discharge variations over a short time period are typical of arid zones. The data are from the Las Tosquillas gauging station several kilometers down-

Figure 16.45 Network of ephemeral channels located in the interior of the great Makhtesh Ramon Depression, which is structurally a *combe*. Central Negev Desert, Israel.

Figure 16.46 Crops in the bed of the Sabarmati River (Ahmadabad, northwestern India) showing the sparse overland flow they receive.

Figure 16.47 Streams in the piedmont of the Atlas, showing the importance of snow cover. Assif Mellah, southern Morocco. Photo J.L. Peña.

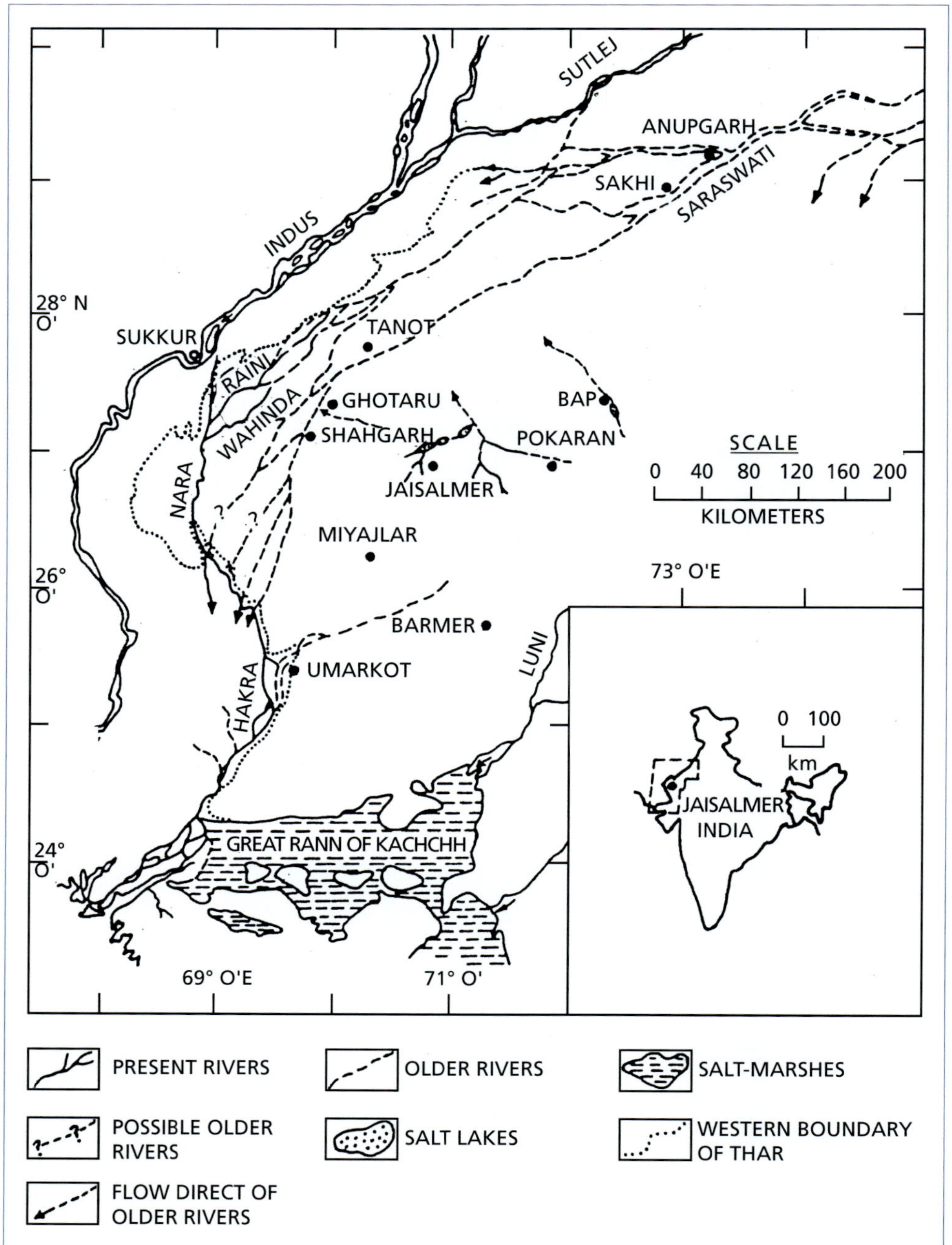

Figure 16.48 Old streambeds of the Saraswati River in the western part of the Thar Desert, India (Ghose *et al.*, 1979).

stream from Ugijar in the Alpujarras of the Province of Granada (Thornes, 1976). The hydrograph is for the period of 24–31 August 1969; it shows a practically instantaneous flood wave followed by an equally sudden decrease. This is typical of flooding in this area caused by heavy rains from summer convection (Thornes, 1976).

River floods in dry zones are classified into four types (Graf, 1988): flash floods, events with a single maximum, events with several maxima, and seasonal floods. Storm rains that fall on small areas of about 8 km in diameter cause flash floods; they are limited to basins of 100 km^2 or less. They have a zero to one maximum in a few minutes or hours and

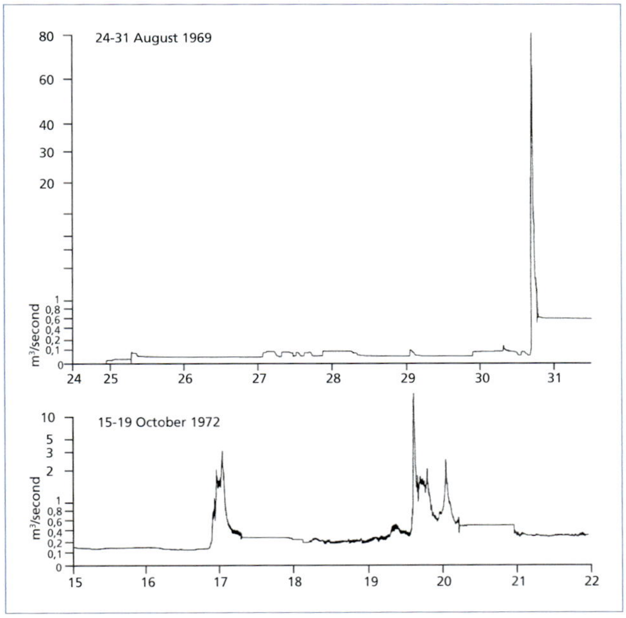

Figure 16.49 Hydrographs recorded at the gauging station at Las Tosquillas, Ugijar River. Alpujarras in the Province of Granada (Spain) (Thornes, 1976).

the response is a wave that advances through the channel as an abrupt turbulent front loaded with detritus. The Ugijar example is this type of flood. Rain events with a single maximum have longer duration than flash floods; they last from several hours to many days (Ward, 1978). Single-maximum floods are caused by storms or low-pressure fronts that affect basins covering thousands of km^2. Floods with multiple maxima are the response to several storms or low-pressure precipitation such as the event recorded on 15–19 October 1972 on the Ugijar River (*Fig. 16.49*). They are caused by several storms or successive low-pressure systems that affect a particular region.

Some water is lost by filtration and evaporation along riverbeds in arid zones. Flooding on small ephemeral streams travels only short distances whereas in larger rivers flood recede gradually, primarily by evaporation, and perhaps even precipitating salts on the margins as they wane (*Fig. 16.50*). This considerably decreases discharge peaks downstream. These transmission losses are shown by hydrographs from two flow-gauging stations 10.9 km apart in Walnut Gulch in Tombstone, Arizona (Renard and Keppel, 1966). Discharge loss is on the order of 57% (*Fig. 16.51*).

Figure 16.50 Evaporation losses at a wadi causing salt precipitation on its margins. Margin of the Matti Sabkha, United Arab Emirates.

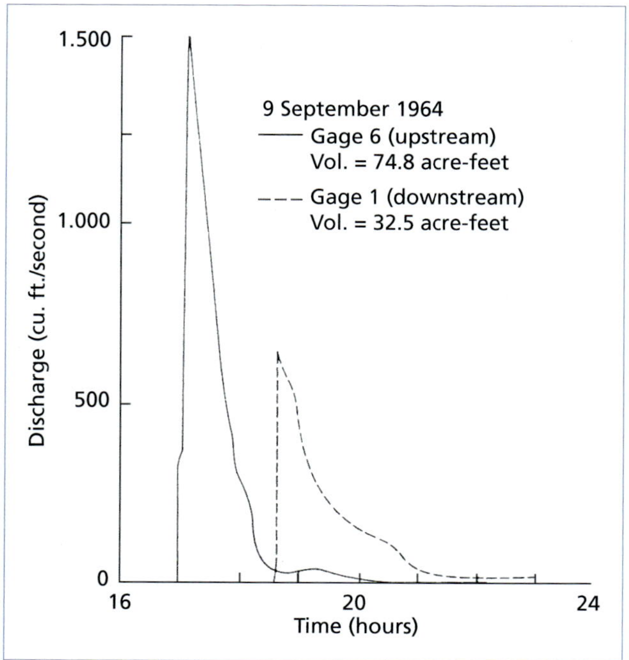

Figure 16.51 Representation of two hydrographs recorded at gauging stations 10.9 km apart during a flood in Walnut Gulch, Arizona. Losses between the two gauging stations were about 57% (Renard and Keppel, 1966).

16.5.3 Runoff and sediment transport on slopes

Rainwater infiltrates into materials at the Earth's surface until the water saturates the soil, that is, until the rainfall exceeds the **infiltration capacity** of the soil. The rate at which water infiltrates is called the infiltration rate; it decreases rapidly at the beginning and later stabilizes. Numerous factors control infiltration: type of rain (intensity and kinetic energy), soil texture and structure, plant cover, and slope of the area. The impact of water drops causes compaction, which reduces infiltration (Morin and Benyami, 1977). Rain simulation research on rock fragments in the laboratory and in the field show that infiltration decreases with increased rock fragment cover (Poesen *et al.*, 1990; Abrahams and Parsons, 1991a). Infiltration also decreases under shrubs where the substantial decrease is due to organic material content and plant remains (Lyford and Qashu, 1969). Biocrusts (Talbot and Williams, 1978) and lichens (Alexander and Calvo, 1990) also decrease infiltration.

Initial **runoff** occurs when rainfall intensity exceeds infiltration capacity. Surface runoff in deserts depends primarily on rainfall intensity because the infiltration rate is a more important factor than the degree of soil saturation. Storm events predominate in arid zones; they rapidly exceed the infiltration rate. In addition, because vegetation is sparse, interception is negligible.

On desert slopes almost all runoff occurs as surface flow although it may rarely occur as subsurface flow in some areas. Foster and Meyer (1975) distinguish between flow and erosion in *interrill* and *rill* zones on slopes. Hydraulics and erosional mechanisms are different in these zones. In **interrill** areas (*Fig. 16.52*) water generally moves as laminar flow that diverges and converges around obstacles (rock fragments, vegetation). As a result, flow rate and depth vary enormously over small distances, which affects the flow type (laminar, turbulent, transitional) (Abrahams *et al.*, 1994). **Rills** are a characteristic feature of arid zone slopes composed of labile materials (*Fig. 16.53*). Longitudinal and transverse rill profiles affect flow velocity and type, which is usually turbulent (Gilley *et al.*, 1990). Velocities range between 1–10 cm/second although they can vary widely along the longitudinal profile (Favis-Mortlock, 2004).

Data related to water erosion on arid zone slopes is limited. Saunders and Young (1983) compiled slope rates. In semi-arid climates, erosion exceeds 2 cm/year (Gutiérrez *et al.*, 1995; Sirvent *et al.*, 1997) whereas in arid regions the values are about 0.1 mm/yr. In semi-arid zones, denudation rates for easily eroded materials are the highest in the world.

Several factors are known to control the intensity of erosion on desert slopes. Abrahams and Parsons (1991b) analyzed the slope factor in their study of simulated rainfall at three experimental stations with different lithologies and vegetation. When the slope is less than 12°, runoff increases slowly; sediment production is directly proportional to slope because of the gravitational component (*Fig. 16.54*). Where slopes exceed 12°, runoff increases rapidly due to the increased slope. This decrease compensates for the increase due to the gravitational component and for this reason sediment production decreases. Simanton et al. (1984) studied the ratio between rocky fragment cover and sediment production on plots; for similar slopes they obtained negative correlations due to numerous factors. However, in laboratory experiments, Poesen and Lavee (1991) find positive correlations when clasts are larger than 5 cm. Erosion also increases after dry periods and decreases during wet periods because plant cover protects the soil from raindrop impacts and also from overland flow.

An important process that affects interrill zones is **rainsplash** because it erodes material that can be moved by laminar flow. When a raindrop hits the soil (*Fig. 16.55*) the impact creates a small compression crater and the dislodged particles are spread on the surface by shear. The rainsplash effect is greater when the kinetic energy of the raindrop is higher. If there is a small film of water on the surface, particle dispersion increases; it reaches its maximum when the thickness of the water layer equals the raindrop diameter (Palmer,

Figure 16.52 Wide zones of interrills separated by small gullies in clay-rich Quaternary materials. The slopes appear to be affected by intense cracking and in turn show have uneven rill development. Las Bardenas Reales (Navarre). Ebro Depression (Spain).

Figure 16.53 Rill development in Miocene clay-rich materials with lenses of sandstones in badland. Las Bardenas Reales (Navarre). Ebro Depression (Spain).

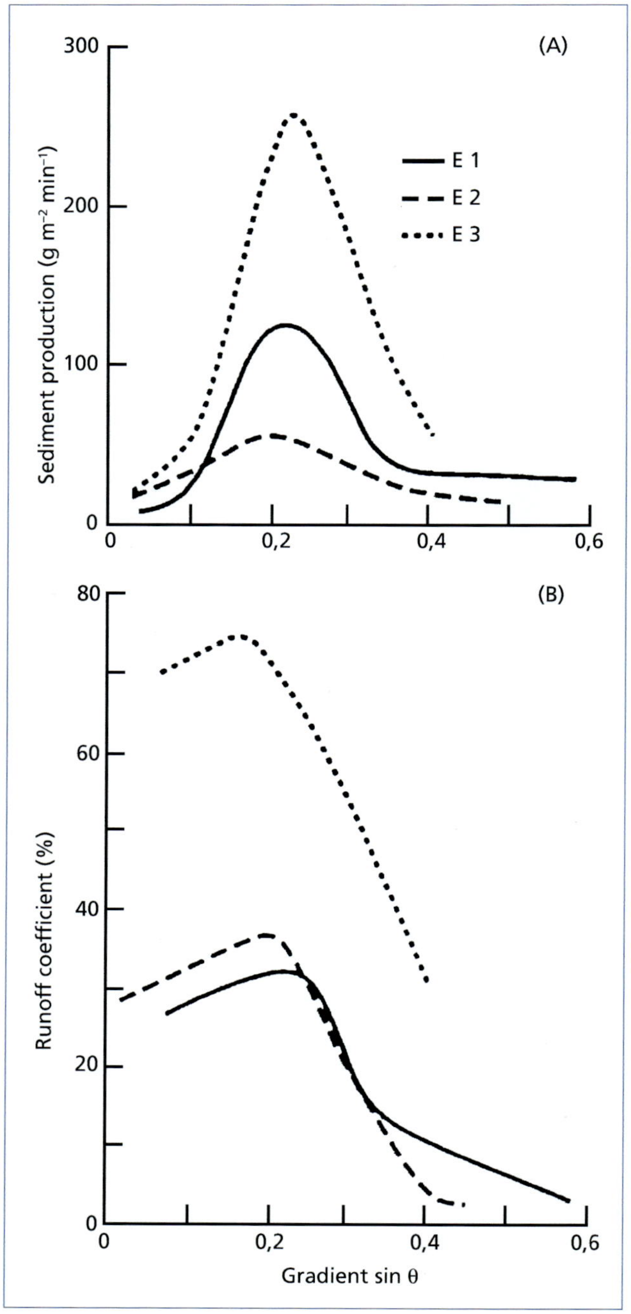

Figure 16.54 Sediment production curves (A) and runoff coefficient (B) related to slope for three sets of experiments called E_1, E_2, and E_3 in Walnut Gulch (Arizona). The experiments on parcels E_1 and E_2 were on Quaternary alluvium and the vegetation was removed on E_1. In E_3 the parcels were located on top of the Bisbee Formation, composed of alternating early Cretaceous limestone and metamorphic slates (Abrahams and Parson, 1991).

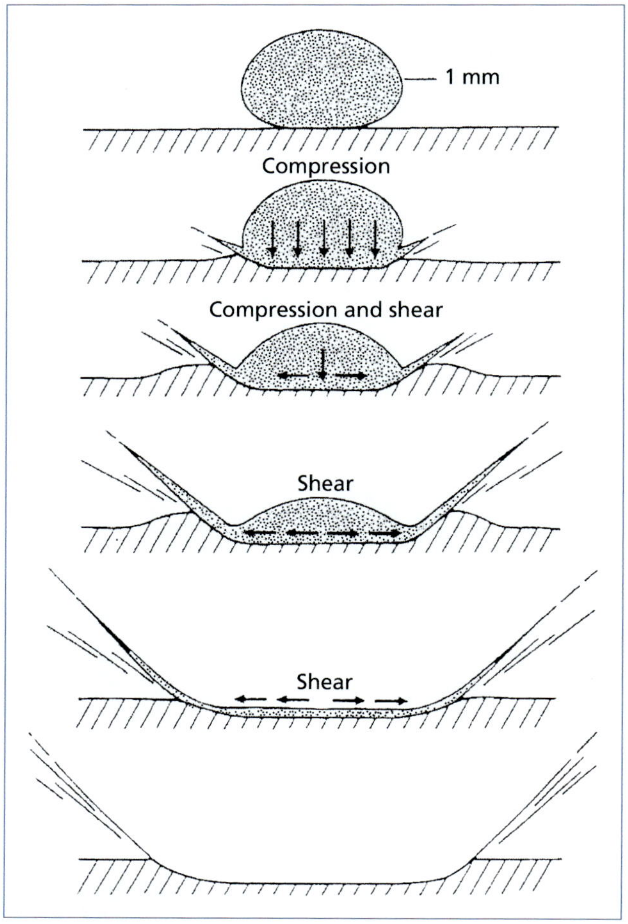

Figure 16.55 Diagram showing rainsplash (Al-Durrah and Bradford, 1982).

1963). On horizontal surfaces, the distance to which the material is ejected decreases exponentially from the point of impact (Savat and Poesen, 1981; Torri *et al.*, 1987). The rainsplash effect is more pronounced as slope and wind velocity increase (Ellison, 1944). This significantly modifies the trajectories of dislodged particles. Vegetative cover, the size and percentage of surface particles, organic material content, soil chemistry, and stability of aggregates also affect these processes (Scoging, 1989). Rainsplash significantly affects particle supply to laminar flow and the transport of particles from interrill areas to rills (Bryan, 1987, 2004).

16.5.4 Water erosion on slopes

Surface flow erosion may occur as laminar flow or by rilling. According to Leopold *et al* (1966), 98% of sediment production in a semi-arid area of New Mexico over a period of ten years came from a slope that lacked rills. In environments with higher precipitation in Kenya, Dunne (1980) did not find rills on long slopes; he confirms this in rainfall simulation experiments (Dunne and Dietrich, 1980). Also, Emmett (1978) shows that some slopes in Wyoming do not have rills and may undergo uniform degradation as a result of laminar erosion. While rill erosion has not received much attention, numerous authors have studied **interrill erosion** and have focused on the interaction of rainsplash and laminar erosion in both field and laboratory experiments. These researchers have taken into account slope and the physical, chemical, and mineralogical properties of soils. They show that interrill processes and erosion rates are extremely complex and varied; they depend on extrinsic factors such as rainfall intensity, raindrop size, and presence or absence of wind, and intrinsic factors including soil texture, aggregate characteristics, surface rugosity, susceptibility to cracking, and the presence and density of organic remains (Bryan, 1987).

Rill erosion occurs primarily as a result of the removal of soil particles by concentrated flow; it affects a small part of the Earth's surface (Bowyer-Bower and Bryan, 1986). Rills can develop when runoff is concentrated on small topographic irregularities. However, rill erosion does not begin until the erosional capacity of the flow exceeds the soils' resistance to particle removal; thus runoff may flow a considerable distance downslope before rills develop (Meyer, 1986). Once a rill has formed, water flow increases rapidly because rill erosion increases with slope length. Rills can also grow by headward erosion; the mechanisms are complex (Bryan and Poesen, 1989).

Dunne and Aubry (1986) explain rill system dynamics as a balance between the intensity of rill and interrill processes or laminar flow processes and rainsplash. Without rainsplash, laminar flow becomes unstable; but if rainsplash occurs, the particles removed from interrills tend to fill and eliminate the rills. This leads to spatial and temporal variations in the effectiveness of these processes. Research conducted on temporal variations in cross-sections of rills with micro-topographic profiles at the Lanaja Experimental Station (Province of Huesca, Spain) found that during times of low precipitation, rills fill with particles and are not capable of concentrating runoff that can transport loose material from the bottom of the rill (*Fig. 16.56*). However, high magnitude and intensity convective rains generate runoff with significant erosive power. In a two-year period, rills that formed in Miocene shale incised 5 cm into their original bed (Sirvent, *et al.*, 1997). This shows the importance of temporal variations, the dependence of rills on interrill processes, and the variability of flow intensity in rills.

Other processes are active in rill formation. Subsurface microconduits (*micropipes*) can form in shales; if they collapse during growth they may crop out as rills (*Fig. 16.57*) (Bryan *et al.*, 1978; Gutiérrez *et al.*, 1988c). They can originate at the intersection of cracks in microrills (*Fig. 16.58*) (Haigh, 1978; Gutiérrez, et al., 1988). In clay-rich lithologies that have alternating layers with different swelling capacities, rills may develop and get lost in strata with a high content of swelling clays. Swelling

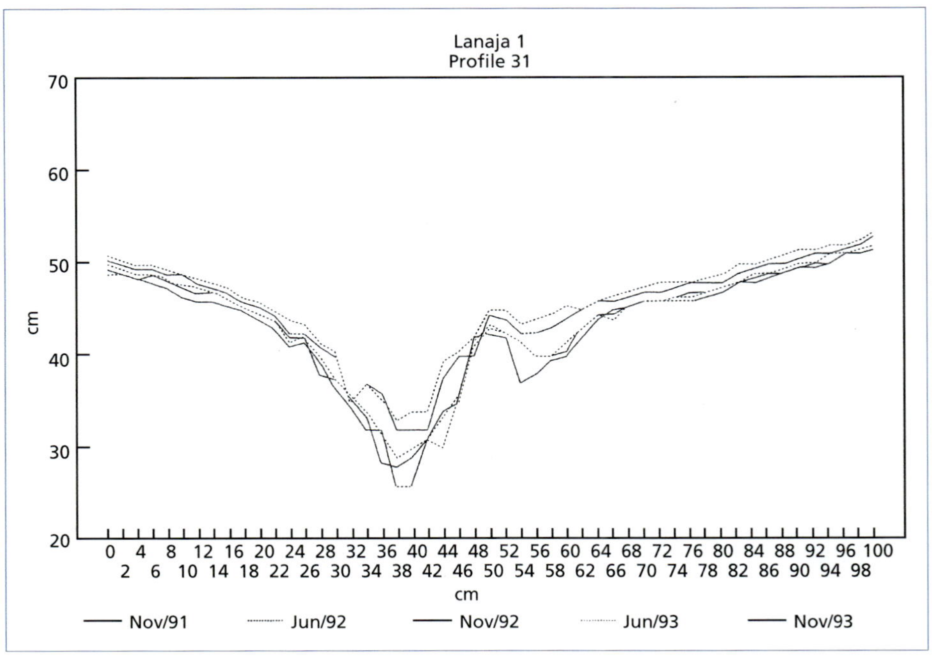

Figure 16.56 Microtopographic profiles over time, for section 31 over time at the Lanaja-1 Experimental Parcel (Province of Huesca) located on a slope composed of Miocene shales in the Ebro Depression (Spain) (Sirvent *et al.*, 1997).

Figure 16.57 Surface conduits (*pipes*) collapses, and bridges in Holocene shales where rills or small gullies form during their evolution. Lupiñén (Province of Huesca). Ebro Depression (Spain).

may be accompanied by the formation of cracks separated by micromounds known as *popcorn* structures, which have a high infiltration capacity (*Fig. 16.59*). In the Dinosaur Badlands, a semi-arid zone in Alberta (Canada), rills form in micropediments in response to hydraulic conditions of surface flow (Hodges and Bryan, 1982). In the same area, Bowyer-Bower and Bryan (1982) observed rills formed as a result of dissolution of resistant siltstones.

Piping or tunneling (*Fig. 16.57*) is common in semiarid environments; they are defined as the natural formation of subsurface drainage in clastic rocks. Piping removes solid particles and transports them in suspension (Parker, 1963). It occurs preferentially in areas where the climate has obvious seasonal contrasts and highly variable precipitation such as Mediterranean climates (Jones, 1981, 2004; Bryan and Yair, 1982). Piping results from a complex combination of factors (Stocking, 1976; Bryan and Yair, 1982; Gutiérrez *et al.*, 1997).

Desiccation cracking is significant in the formation of piping (Parker *et al.*, 1990); desiccation cracking is common in swelling clays (Parker and Jenne, 1967). The materials contract during the summer; storm rains fill the cracks and disperse the clays (Marshall and Workman, 1977). With a sufficient hydraulic gradient, the clays are able to move (Fletcher and Carrol, 1948; Jones, 1981, 2004). Cracks turn into *pipes* as they grow. Pipes are found on flat areas incised by fluvial networks and on slopes. In areas of significant relief, large galleries and collapse pseudodolines can form; examples seen in the Lupiñén area (Province of Huesca, Spain) (*Figs. 16.60* and *16.61*) resemble

Figure 16.58 Small rills formed along a system of cracks in Holocene shales. Lupiñén (Province of Huesca). Ebro Depression (Spain).

Figure 16.59 Rill formation in shales at mid slope that are lost in the stratum with *popcorn* structure. Cuesta del Viento, Colola (Province of San Juan, Argentina). Precordillera of the Andes.

Figure 16.60 Group of funnel-shaped pseudodolines formed by piping on thick slope deposits. Lupiñén (Province of Huesca). Ebro Depression (Spain).

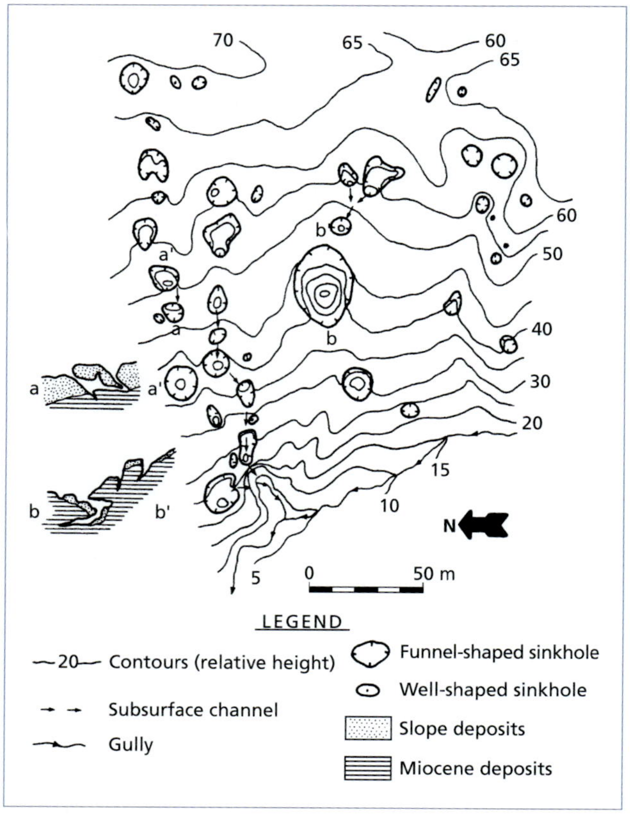

Figure 16.61 Map of pseudodolines formed by piping in slope deposits. Lupiñén (Province of Huesca). Ebro Depression (Spain). (Gutiérrez *et al.*, 1988).

a karst landscape (Gutiérrez *et al.,* 1988). Continuous swelling of gallery roofs caused by piping (Harvey, 2004a) can generate gullies; this process can be seen in the natural arches that form in detrital materials (*Fig. 16.62*). These collapses also increase the rate of headward retreat of gullies (Crouch, 1983). Piping also causes slope instability and contributes to landslide formation (Pierson, 1983).

We have noted that the presence of swelling clays is a significant factor in the formation of piping but clay is not necessary for swelling (Jones, 1981). In sodium-rich environments, illite and other clay minerals may swell (Imeson *et al.*, 1982) 10 to 12% (Benito *et al.*, 1993b and Gutiérrez *et al.*, 1995). The principal factor controlling the susceptibility to piping is the cation content of Na^+ dissolved in the pore water with respect to total Ca^{2+} and Mg^{2+} cations (Sherard and Decker, 1977). These values were obtained by calculating the quantities of cations present in an extract of saturated paste. This information can be used to calculate the percentage of interchangeable sodium (ESP, or Exchangeable Sodium Percentage) or the sodium absorption ratio (SAR). Clays with an ESP higher than 15 (McIntyre, 1979) or a SAR above 5 (Aitchinson and Wood, 1965) have high pH (between 8 and 10) and are very susceptible to piping. In addition, soils that contain high sodium contents promote clay dispersion (Heede, 1971). Sherard's diagram (Sherard *et al.*, 1972a) shows the difference in dispersion fields as a function of sodium content (Gutiérrez *et al.*, 1997) (*Fig. 16.63*). Points located at the dispersion point are susceptible to piping (Benito

Figure 16.62 Natural bridge created by piping. The gully was produced by repeated collapses. Barranco del Tormillo or de la Clamor Vieja. El Tormillo (Province of Huesca). Ebro Depression (Spain).

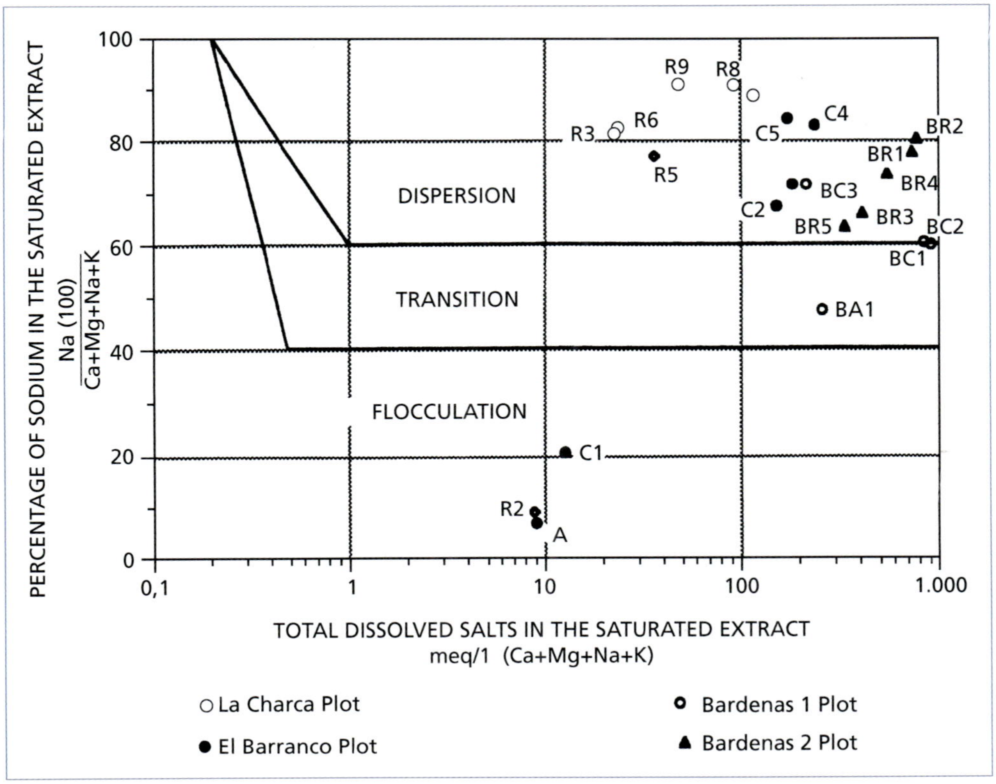

Figure 16.63 Sherard's diagram in which most points indicate dispersion for high percentages of sodium compared to total dissolved salts in the saturated extract (Gutiérrez *et al.*, 1997).

et al., 1993b; Gutiérrez *et al.*, 1997). Research on the physical, chemical, and mineralogical properties of materials is important for civil engineering, especially for the construction of earthen dams because some earthen dams have collapsed as a consequence of piping (Jones, 1981, 2004; Parker *et al.*, 1990). Adding calcium in the form of gypsum or lime (Heede, 1971; Stocking, 1976), compounds that interfere with the piping process, can reduce a high ESP.

16.5.5 Fluvial landforms and processes

Channelized surface runoff erodes the riverbed and transports and deposits particles. Flooding can modify streambed cross sections in spectacular ways (*Fig. 16.64*) by excavation and rapid accretion (Leopold *et al.*, 1964). Semi-arid environments are zones of maximum erosion (Langbein and Schumm, 1958; Fournier, 1960; Corbel, 1964) where there is enough rain and plant cover is scarce.

The stream spends part of its energy on channel **erosion**. Numerous processes erode the stream channel: chemical corrosion or rock dissolution; corrasion or mechanical fracturing of rock by transported fragments (formation of potholes); dragging or hydraulic lifting of boulders; and finally, cavitation or rock fracture that occurs in high velocity currents where vapor-filled bubbles can pound the bedrock and fracture it.

Headward erosion makes channels longer; it has been measured at 0.46 to 0.79 m/year (Leopold *et al.*, 1964). Erosion can also widen channels, particularly by basal undermining and landslide (*Fig. 16.65*) or mass movement of saturated material. In addition to growing in length and width, channels can incise their beds to create deep gorges (*Fig. 16.66*). The longitudinal profiles of some desert rivers have discontinuities (or *knickpoints*) (*Fig. 16.67*) where waterfalls create small closed depressions and undermine the base; this collapses the walls and causes the *knickpoint* to retreat (Bull, 1997). This retreat is intermittent and can be catastrophic depending on sporadic flooding.

As in semi-arid zones, rivers in deserts produce abundant sediment; in their active moments rivers carry a large sediment load, which may be 68% by weight as shown by Bondurant (1951) for the Puerco River, New Mexico. Water

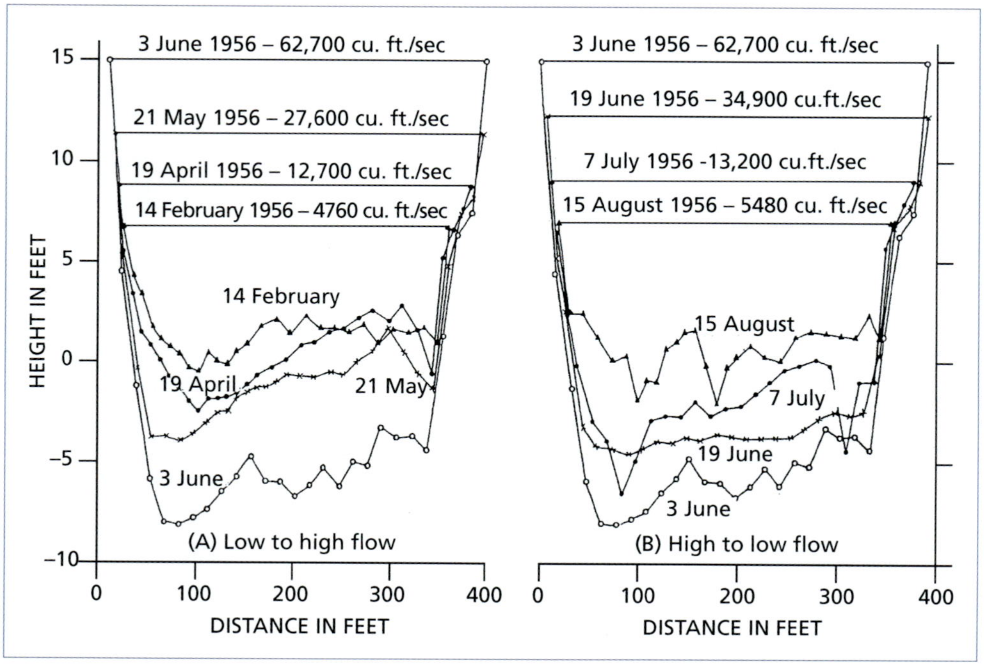

Figure 16.64 Excavation and later filling during the spring snowmelt stage on the Colorado River at Lees Ferry (Arizona) in 1956. (A) Low to high flow and (B) high to low flow (Leopold *et al.*, 1964).

Figure 16.65 Falls of large limestone blocks caused by basal undermining. Canyon of Nahal Zin, Negev Desert (Israel).

Figure 16.66 Todra Gorge. High Central Atlas (Morocco). Photo F. Gutiérrez.

Figure 16.67 Break in the longitudinal profile (*knickpoint*) by differential erosion where the thick layer of Cretaceous limestones is more resistant to erosion. Basal undermining by the waterfall is also visible. Canyon of Nahal Zin, Negev Desert (Israel).

transports sediments in suspension, saltation, and traction in fluvial channels. Bed load moves by saltation and traction. Perennial rivers carry a high suspended load whereas ephemeral streams transport primarily bed load; in arid zones such as the Nahal Yael (southern Israel) bed load may account for 87% of the total load.

After a flood excavates the streambed of an ephemeral stream, the waning flood may deposit an approximately equal amount of sediment into the stream channel. This was demonstrated in the Arroyo de los Frijoles in New Mexico (Leopold *et al.*, 1966) (*Fig. 16.68*). Excavation and filling occur in both sands and gravels; it suggests a stage of dynamic equilibrium (Chorley and Beckinsale, 1980). Excavation also removes detrital material from the bed by runoff. It has been demonstrated that the size distribution of suspended load is proportional to water velocity or flow in desert rivers; the opposite is true of perennial rivers (Frostick *et al.*, 1983) (*Fig. 16.69*). Bed load research is complicated. Traps placed in the riverbed (*Fig. 16.70*) disturb transport processes. The use of painted particles (Leopold *et al.*, 1964) or, even better, magnetized fragments (a magnet is implanted into a clast) makes it possible to use a magnetic locator to find clasts after the flow subsides and study the movement of marked clasts. Leopold *et al.* (1966) recorded that clasts moved 3 km downstream during a flood in a predominantly sandy channel. Schick *et al.* (1987) used marked clasts in a gravel channel and a detector to locate where they were buried. During flooding, excavation and fill moved buried particles toward the surface and vice versa (*Fig. 16.71*).

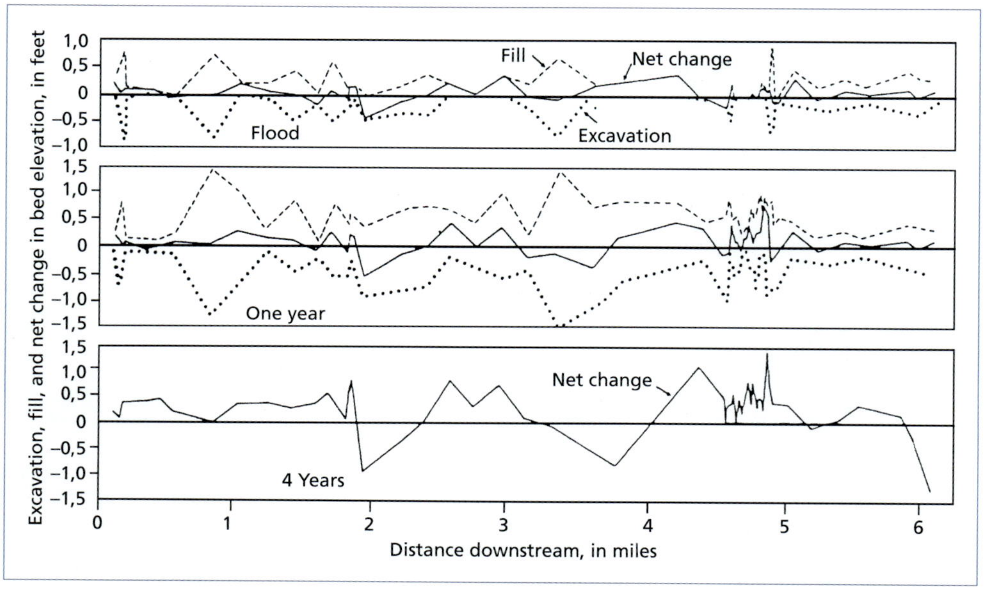

Figure 16.68 Longitudinal profiles showing excavation and fill of various parts of the Arroyo de los Frijoles, New Mexico (Leopold *et al.*, 1966).

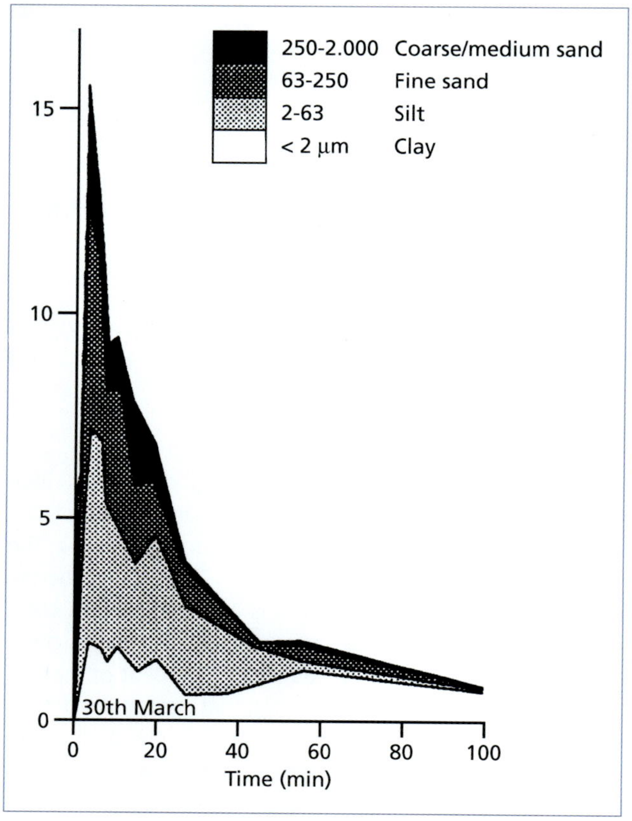

Figure 16.69 Concentration of suspended sediments by grain size, during a flood in the Il Kimere basin, Kenya (Frostick *et al.*, 1983).

Bull (1991, 1997) described a typical ephemeral stream model for arroyos in the southwestern United States. This author documents a repeated sequence of alternating sections of aggradation and degradation along the riverbed. The sequence includes the following: vertical headwaters (*Fig. 16.73*) that concentrate laminar flow, a vertical walled channel that directs flow toward a channel fan apex, braided distribution channels that terminate in a zone of divergent laminar flow, and the sequence ends with convergent laminar flow that drains toward a vertical headwater (*Fig. 16.72*).

As in other morphoclimatic zones, rivers in arid zones deposit transported loads in straight, meandering, and braided channels but deposition also results from unconfined flows in pediments and alluvial fans, which will be analyzed in the next chapter.

16.5.6 Valley fills and incised channels

A characteristic feature of semi-arid zones is alluvial fill and gullies that cut down into them. This combination requires an accumulation stage followed by an excavation stage; this sequence is often interpreted as a result of climate change (Leopold, 1994; Harvey and Gutiérrez, 2005). If alternations persist for a long time period, terrace systems can develop (*Fig. 16.74*). In El Tornillo (Ebro Depression, Spain) the beginning of deposition of the Older Fill

Figure 16.70 Wire mesh trap to catch bed load in an intermittent stream. Nahal Yael Experimental Station, Eilat (Israel).

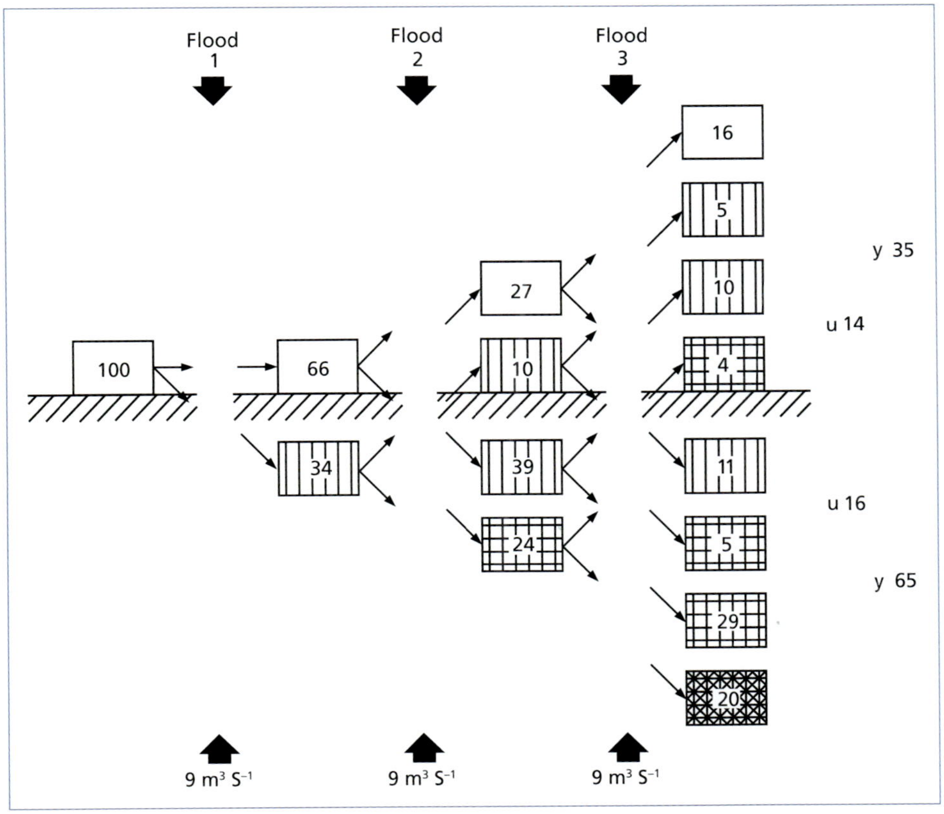

Figure 16.71 Marked gravel clasts buried and exhumed during transport by three successive floods in Nahal Hebron, Israel. Marked clasts were initially located on the surface of the gravel bed (Schick *et al.*, 1987).

Figure 16.72 Vertical headwaters of an ephemeral stream eroded into valley fill on predominantly gypsum-bearing materials. Barranco de Villafranca. Villafranca de Ebro (Province of Zaragoza). Ebro Depression (Spain).

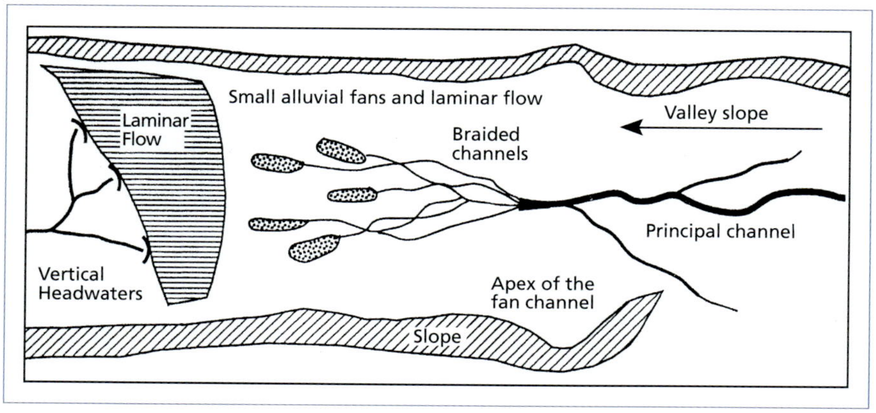

Figure 16.73 Typical erosional and depositional stretches of a fan channel in a discontinuous ephemeral stream (Bull, 1997).

Figure 16.74 Two terrace levels whose age at the most is lower Holocene and continuing until 6.000 cal BP and the earliest Iron Age. El Tormillo (Province of Huesca). Ebro Depression (Spain).

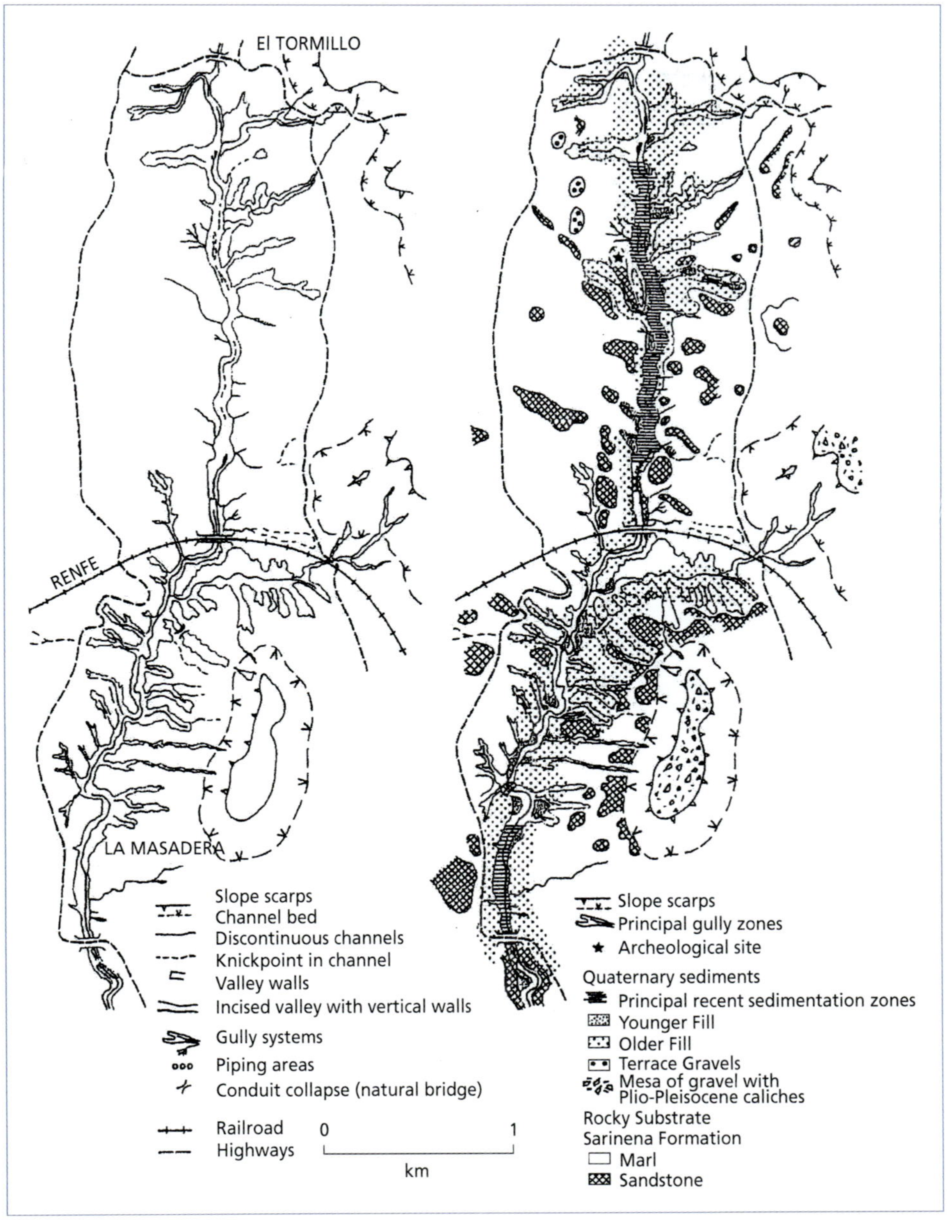

Figure 16.75 Detailed maps of the southern part of El Tormillo (Ebro Depression, Spain). Left: Morphology, drainage and *gully* networks. Right: Substrate and Quaternary geology (Harvey and Gutiérrez-Elorza, 2005).

occurred in the early Holocene at least and continued until 6,000 cal years BP; more recent Younger Fill dates from the Iron Age (Harvey and Gutiérrez, 2005) (*Fig. 16.75*). Incision occurs when the erosional force of the concentrated water flow exceeds the strength of the materials over which they flow (Schumm *et al.*, 1984). This downcutting can form continuous and discontinuous gullies (*Fig. 16.76*). Discontinuous gullies are related to a slope threshold in which the gully lacks sufficient transport capacity to continue (Schumm and Hadley, 1957). In the southwestern United States, gullies with vertical walls are called **arroyos**; in Spain they are called ramblas. These landforms have been studied extensively in the southwestern United States (Cooke and Reeves, 1976) to determine the causes of incision. Researchers have attributed it to changes of climate toward increasing aridity (Antevs, 1952; Leopold, 1976, 1994; Harvey and Gutiérrez, 2005) and high intensity and low frequency precipitation. Other fundamental causes of

Figure 16.76 Small discontinuous gullies on the slope between Rodeos and San José de Jachal (Province of San Juan, Argentina). Precordillera of the Andes.

incision (Chiverrell *et al.*, 2007) are decreased plant cover in semi-arid zones and overgrazing and overfarming in other climatic zones (Chiverrell *et al.*, 2007). The settlement of the western United States is a significant and well-studied example (Cooke and Reeves, 1976). Decreased plant cover and significant storms created gullies that incised more than 50 m. This erosion occurred between 1850 and 1920, especially between 1870 and 1890; the resulting environmental situation forced many settlers to move to more favorable areas. Gully incision and filling may have numerous causes. The timing of these events varies from place to place and each gully has its own history (González, 2001).

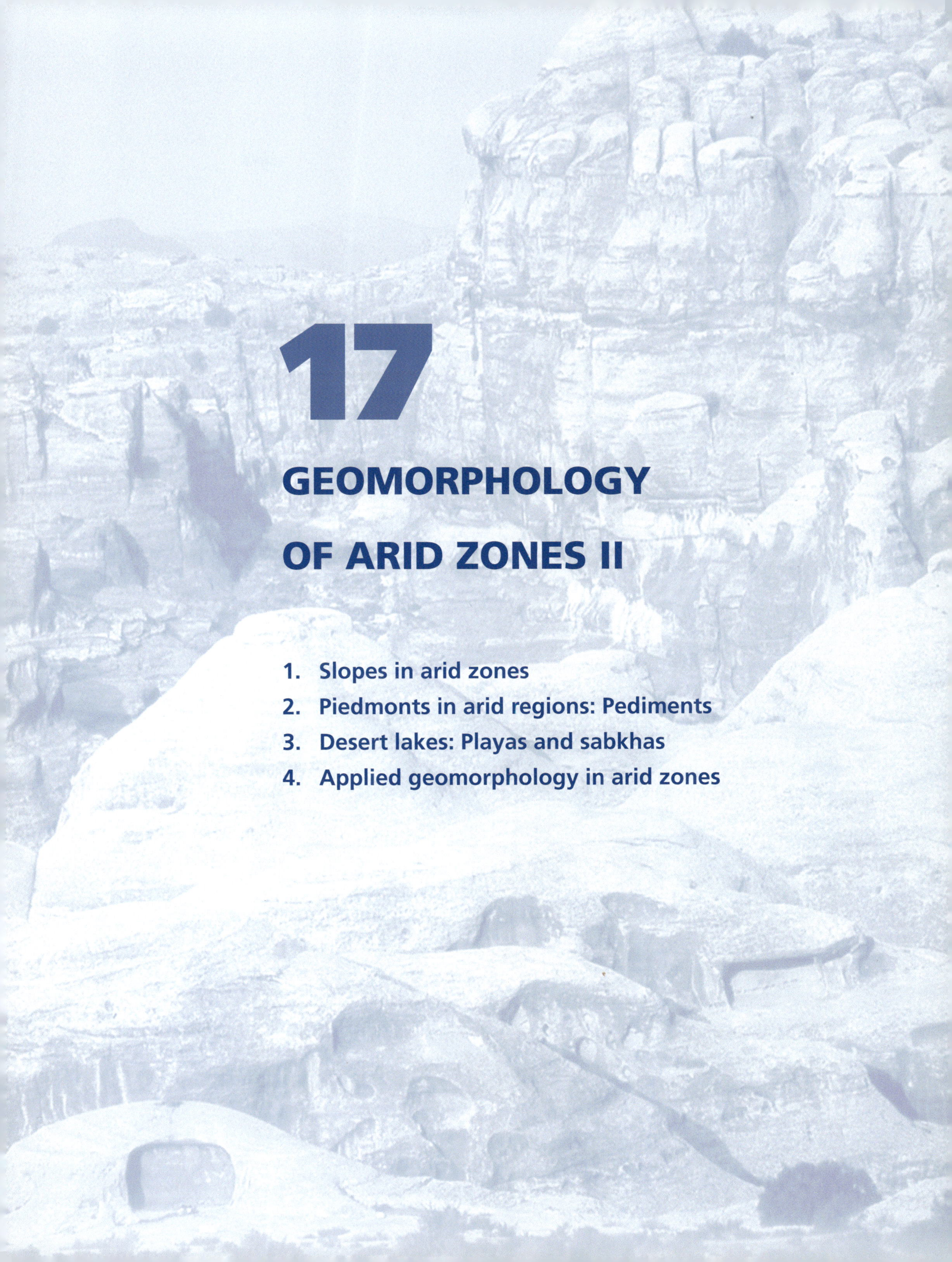

17

GEOMORPHOLOGY OF ARID ZONES II

1. Slopes in arid zones
2. Piedmonts in arid regions: Pediments
3. Desert lakes: Playas and sabkhas
4. Applied geomorphology in arid zones

17.1 Slopes in arid zones

In arid zones, the production of particles by weathering is generally somewhat lower than the mobilization of this debris by runoff or wind. Plant cover, which is closely related to climate, significantly affects this balance. For this reason we say that aridity controls the balance between sediment supply and removal. However, in cold mountainous deserts where congelifraction is the dominant process, gelifract production exceeds removal, especially with extreme aridity; as a result water erosion is insignificant and incapable of moving all the clasts. (*Fig. 15.38*).

Although climate and vegetation affect the development of desert slopes, lithology and bedding plane density are fundamental factors in the development of slope landforms under arid conditions. These geological characteristics directly affect the activity and intensity of weathering. As a result, these factors determine the nature of the slope and also enable us to distinguish three slope types that form on crystalline rocks, stratified formations (*Fig. 17.1*), and easily eroded materials (Mabbutt, 1977). Some authors call the first type massive rock slopes and include sandstones and conglomerates in this group (Howard and Selby, 1994). Others such as Bryan (1922) distinguish cliff slopes, block-dominated slopes, and slopes composed of non-resistant rocks where water can rapidly remove the finest detritus.

17.1.1 Slopes in crystalline rocks

The chemical composition of magma has a direct influence on the morphology of erupted volcanic materials. Basic magma causes extensive basaltic flows that may alternate with pyroclastic layers. Differential erosion of these formations generates stepped slopes or *traps* that make up the large *plateau* basalts of various ages that occur throughout the world. Similar morphologies occur in the *rift valleys* in East Africa. Magmas of acid composition have higher viscosity; they erupt explosively and form domes, plugs, and spines. Examples are the trachytes and phonolites of the Hoggar massif in the central region of the Sahara.

Plutonic rocks form two types of landforms, which are controlled by exfoliation (*sheeting*) and intersecting joint systems. Sheeting, which occurs primarily in massive crystalline rocks, forms by parallel curved jointing at the Earth's surface; for this reason some authors call it jointed topography.

Figure 17.1 Slopes in Precambrian granites and Cretaceous limestones and marls. Negev Desert. Eilat (Israel).

It forms in rocks that were subjected to significant forces within the Earth's crust. As the rock moves toward the surface, its load decreases, allowing the rock mass to relax and decompress (Gilbert, 1904). However, other origins related to different forces have also been proposed (Howard and Selby, 1994). Exfoliation does not occur in rocks with numerous discontinuity planes because the forces are dissipated along the discontinuities.

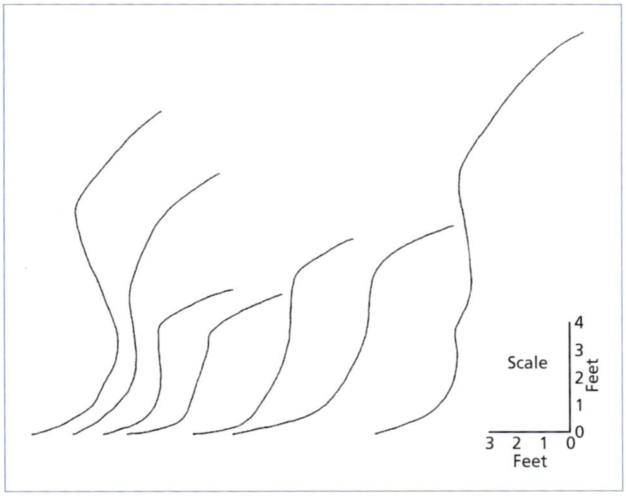

Figure 17.2 Slope profiles from the western end of Pidappa Hill (Twidale, 1962).

Sheeting occurs in almost all climates and is relatively common in crystalline rocks of African and Australian deserts (Twidale, 1981, 1982a, b; Howard and Selby, 1994; Twidale and Vidal Romaní, 2005; Migon, 2006). Sheeting produces domed or hemispheric forms that were first described by the German geologist Bornhardt (1900); Willis (1936) named the domed inselbergs that protrude above plains **bornhardts** (in honor of the discoverer). Many bornhardts result from the denudation of alteration profiles that were formed in previous epochs when the climate was humid and tropical (Büdel, 1957; Ollier, 1988a, 1991; Thomas, 1989a, b, 1994a; Twidale, 1990). Climate change toward arid conditions causes water erosion that creates bedrock outcrops.

A type of mushroom-shaped rocky dome (*Fig. 17.2*) crops out on the northern Eyre Peninsula of southern Australia (Twidale, 1962). The domes are composed of Precambrian granites controlled by curvilinear joints; tafoni and numerous weathering grooves occur on their walls. Their concavity is thought to have originated from subsurface weathering but occasionally the basal profile has double concavities that may be the remains of older levels of subsurface weathering.

When seen from afar, slopes that form in jointed crystalline rocks appear to have rectilinear profiles but a closer look reveals a complex group of rock steps. In his study of igneous rock morphology in southwestern Arizona, Bryan (1927) described a cliff slope composed of blocks or *tors* and castle shapes (*castle kopje*) (*Fig. 17.3*). A slope forms at the base of the cliff; it is covered by debris composed of blocks

Figure 17.3 Tors and castle forms in granite. Phillae, Aswan (Egypt). Photo J.L. Peña.

Figure 17.4 Precambrian granite with folded pegmatites. The high degree of jointing gives the slope profile a rectilinear appearance. Negev Desert. (Israel).

that have weathered in-situ, which is inferred by the lack of blocks on lower parts of the slope. Even so, some slopes are covered with blocks that have moved. If the spacing between joints is obvious, the slope has a stepped profile (Carson, 1971); if not, the morphology is more rectilinear (*Fig. 17.4*). A leaching slope develops at the bottom; it is composed of fine particles produced by weathering of boulders that have moved to the base of the slope. Some slopes form from weathering profiles of plutonic rocks and as a result, their morphological and developmental characteristics will be different (Oberlander, 1972).

17.1.2 Slopes in stratified rocks

In platform deserts (Mabbutt, 1977) in arid zones of cratonic areas, slopes that form on stratified formations of various ages are nearly horizontal; examples include the northern Sahara, eastern Arabia, the Thar in India, and the Colorado Plateau. These slopes are orogenic belts of an intermediate type; examples occur in Lebanon and Israel and in the Tertiary continental basins of arid Spain and the United Arab Emirates.

These slopes form cuesta or mesa morphologies and they are highly varied. In their classic study of the Colorado Plateau, Schumm and Chorley (1966) distinguish various slope types that result from a combination of lithological, climatic, and vegetation characteristics. **Simple slopes** are composed primarily of one rock type. They may be clay-rich materials that form badlands when affected by water erosion. Another subtype is composed of massive rocks such as sandstones and conglomerates that form significant cliffs (*Fig. 17.5*). **Compound slopes** are composed of easily eroded rocks topped by a resistant ledge; these slopes are badlands or debris slopes made up of materials that fall from the cliffs (*Fig. 17.6*). Finally, **complex slopes** are formed from alternating hard and soft layers (*Fig. 17.7*). Simple and compound slopes are analyzed below; complex slopes are considered a hybrid of the two.

17.1.2.1 Simple slopes

In this section we will discuss the morphology and evolution of massive stratified rock slopes, primarily sandstones, and in the next section, we will analyze badlands, which are formed by water erosion of easily eroded materials.

Much of the research on this topic was done in the Colorado Plateau although simple slopes have also been studied in other deserts (Sahara, Australia, Atacama). The evolution of simple slopes is related to rock characteristics and multiple geomorphic processes that act or have acted singly or together in both space and time.

Discontinuities in sandstones such as thin clay intercalations or cross-bedded layers are more easily eroded than massive sandstones; these discontinuities cause noticeable

Figure 17.5 Simple slope composed of spectacular cliffs of massive Neogene conglomerates attached to formations of the southern Pyrenees margin. Mallos de Riglos (Province of Huesca, Spain).

Figure 17.6 Slope composed of a limestone cliff and debris slope that almost totally covers an Oligo-Miocene substrate that is primarily clay. This section is partially affected by rilling and gullying. Ballobar (Province of Huesca). Ebro Depression (Spain).

Figure 17.7 Complex slope corresponding to a cuesta front composed of stratified materials with variable resistance to erosion. Dra Valley, southern Morocco. Photo J.L. Peña.

changes in slope profiles. Vertical joint and fracture systems are zones of weakness that act as conduits for increased moisture; they promote weathering. Weathering products may remain lodged at the bottom of corridors (Doelling, 1985). In areas of high fracture density, the slope retreats rapidly; this process produces successive entrants and salients (Nicholas and Dixon, 1986). Sheeting commonly occurs in the massive rocks that form simple slopes; this can be seen in the arkose of the famous Ayers Rock *(Fig. 18.37)* monolith in the central Australian desert (Twidale, 1978) and also in Colorado Plateau sandstones (Bradley, 1963). Sheeting creates a convex slope and produces slabs from a few centimeters to a meter thick.

The predominant **weathering** processes in sandstones are salt weathering and cycles of wetting and drying although in areas where freezing occurs congelifraction may also be significant. The composition of the cement is also important. In carbonate-cemented sandstones, the carbonate dissolves easily in cold water; siliceous cements require a pH greater than 8 to promote dissolution. The loss of cement causes grain-to-grain disaggregation of the rock. In general, weathering creates convex forms by rounding summits and edges (Baker, 1936) *(Fig. 17.8)*.

In several areas of the Colorado Plateaus (Schumm and Chorley, 1966), weathering causes present-day falls. Even so, some landslides seen in deserts occurred in earlier periods and are thus considered relict forms. Grünert and Busche (1980) studied examples in the central Sahara and DeGraff (1978) studied them in Utah. The landslides they studied are often rotational and may have been triggered in periods of higher precipitation during the Pleistocene; these mass movements can also be reactivated.

Another erosional process observed on simple slopes is **undermining** by groundwater that may spring forth at impermeable levels below massive sandstones. Weathering and water erosion form *alcoves*; as the alcoves evolve, they cause rockfalls and retreat of sandstone slopes (Higgins, 1984; Laity and Malin, 1985; Baker *et al.*, 1990; Higgins and Osterkamp, 1990). These alcoves or cavities co-occur with honeycomb structures and tafonis.

Simple slopes evolve slowly; at the beginning they may behave like compound slopes *(Fig. 17.9)* that retreat parallel to themselves (Fisher, 1868, 1872). In later stages the underlying clay outcrop disappears, the ledges become rounded, and the slope is finally transformed into a simple slope (Schumm and Chorley, 1966). Oberlander (1977) expanded these ideas in his study of sandstone cliffs in Utah. Sandstone cliffs start as compound slopes *(Fig. 17.10)* of easily eroded bedrock; as the slope retreats the compound slope is converted into a simple slope. Slope shape is controlled by the internal structure of the massive sandstones and by thin layers of inter-

Figure 17.8 Rounded shapes and tafoni in Cambrian sandstone of Nubia. Petra (Jordan). Photo J.L. Peña.

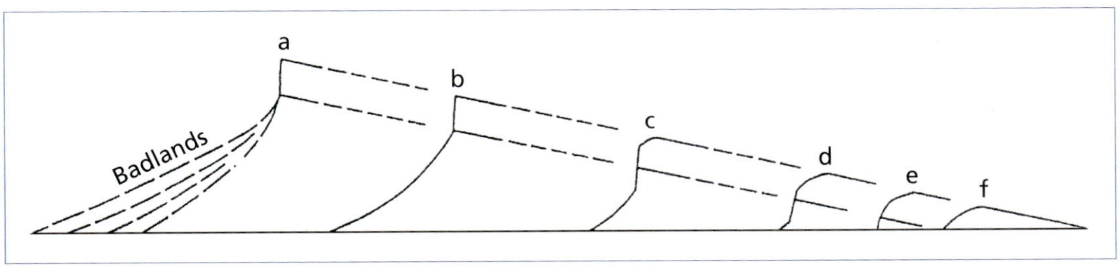

Figure 17.9 Change in slope shape during retreat (Schumm and Chorley, 1966).

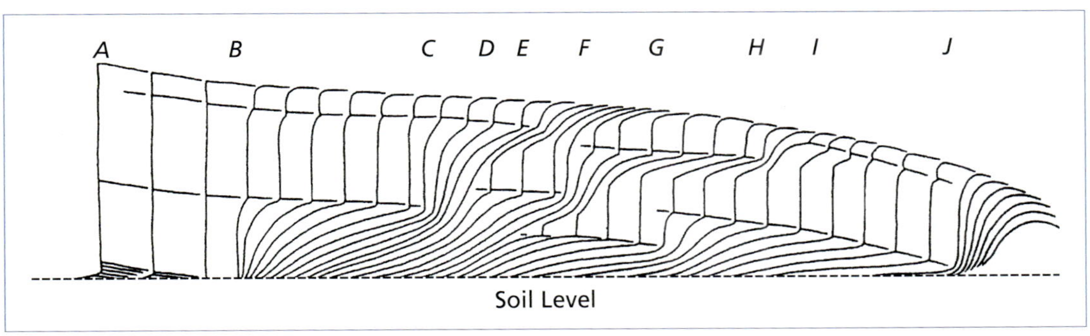

Figure 17.10 Slope development in massive sandstones. *A* shows a cliff on top of a finely stratified bedrock, which is hidden in *B*. The intraformational levels later control the slope landform. Segments located above evolve by undercutting, whereas those located below become rounded (Oberlander, 1977).

stratified clay. These thin beds of clay cause slope changes and at the ends of these layers, the slope profile changes. Cliffs located on top develop by basal undermining, whereas segments located below clay layers are rounded.

17.1.2.2 Compound slopes

Compound slopes have a rocky cliff at the top and a series of easily eroded layers underneath (*Fig. 17.6*). These slope components resemble those described in South Africa in detailed studies by Wood (1942) and Fair (1947, 1948); these slopes were analyzed in depth and redefined by King (1957b). Unlike many researchers in climatic geomorphology, King considers these slopes as "normal." For King, although they are part of a universal model, these slopes are more characteristic of desert environments. The elements that distinguish a slope include the (*Fig. 17.11*) **crest or waning slope**, which is the upper part that sometimes has a convex profile as a result of weathering and creep. The **scarp or free slope** is the outcrop of the most resistant rock; multiple erosional processes act on it to cause retreat. Materials generated by scarp destruction form the **debris slope**, which moves primarily by water erosion. Finally, the **pediment** is a significant concave element that connects the slope to the alluvial plain.

The crest is usually rounded by creep, weathering, and by splash action and laminar erosion. Crest roundness increases with increased dip of the resistant layer of the scarp; a porous or sheeted scarp also increases roundness (Bradley, 1963). When fracturing causes retreat, there is little or no rounding.

Scarps form the edges of mesas, structural platforms, or cuestas. The thickness of the scarp layer determines the height of the cliff and the length of the debris slope. Rock type is also significant in weathering because different processes disaggregate sandstones and limestones. Cementation determines rock resistance; structure is also important. A scarp composed of finely stratified or highly jointed rock

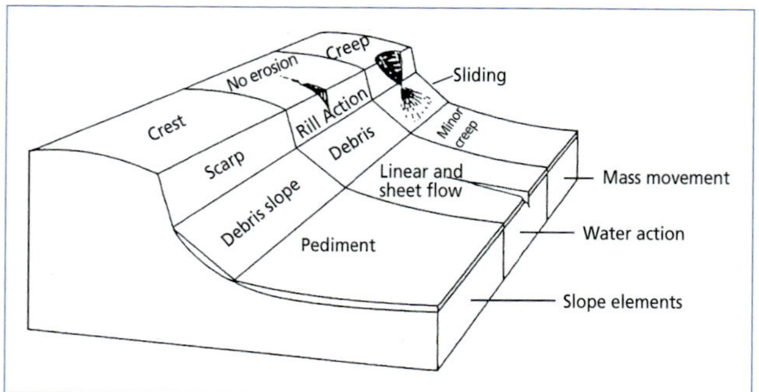

Figure 17.11 Elements of a "normal" slope, water erosion, and mass movement (King, 1957).

has a well-developed debris slope; if these conditions are not present, the slope will be covered with large blocks.

Debris slopes result from the fracturing of resistant scarp rock. This is the source of most material that covers a debris slope although additional material may come from the underlying bedrock. In spite of extensive slope development in many deserts, they have not been as well studied as those in alpine areas (Oberlander, 1997).

Schumm and Chorley (1966) show that the difference between some slopes may be qualitatively expressed by the weathering index (W). The presence or absence of debris at the base of a scarp depends on the rate of fragment generation at the scarp (p) compared to its destruction at the foot of the scarp (d). A ratio greater than one means the slope is accumulating, debris cones form, and debris is abundant. A weathering index W equal to one means signifies equilibrium between p and d. If W is less than one, the boulders that fall disintegrate on impact and the residuum is weathered before the next fall. The latter situation is common in the Colorado Plateau.

This debris can cover all or part of the slope and it may be affected in turn by rilling and gullying that gradually uncover a larger extent of bedrock. The extreme case is the almost total disappearance of the slope cover; when clays are the outcropping material, badlands form. Sometimes the bedrock is composed of hard and soft layers. In this case a stepped topography may form; it results from compound micro-slopes (scarp and debris slope) along the principal slope (*Fig. 17.7*). This type of slope is a complex slope.

Fragments on debris slopes are poorly sorted. They are affected by laminar erosion that removes fine particles and deposits them at the base of the slope, leaving the large fragments to stand out. These fragments can also be significantly weathered, with honeycomb weathering and tafoni in the walls. Finally, small mass movements may affect slope debris.

17.1.2.3 Flatirons

These were originally referred to as *talus flatirons* (Koons, 1955); they have also been called *flatirons* (Everard, 1963); *dreieckshänge* (Wirthmann, 1964; Büdel, 1970); *versants tripartites* (Gossman, 1976); *tripartite slopes* and *triangular slope facets* (Büdel, 1982); and *talus relicts* (Gerson, 1982). They are characteristic landforms of arid and semi-arid zones although as we have noted they also occur in periglacial areas. They have been studied in the deserts of Sahara and Arabia, the southwestern United States, and semi-arid Mediterranean zones.

They begin as a compound slope and are later incised by water erosion (rilling, gullying, and piping) but they retain remnants of individual relict slopes (Koons, 1955) (*Fig. 17.12*).

Howard (1942, p. 24) described their antecedents as: "faceted unions in which a slope intervenes between the mountain slope and the pediment". These paleoslopes have a triangular shape and their apex points toward the scarp (*Fig. 17.13*); trapezoidal morphologies have also been observed. Facets are separated from the scarp and are arranged parallel to it. In their distal parts, they connect with pediments (Blume and Barth, 1972) and finally with fluvial or lacustrine terraces. Flatirons are concave and are covered by debris that is generally poorly sorted and no thicker than 8 m. On the upper part, the angles can reach almost 30° whereas in distal zones, angles range between 2° and 5°. The internal slope of

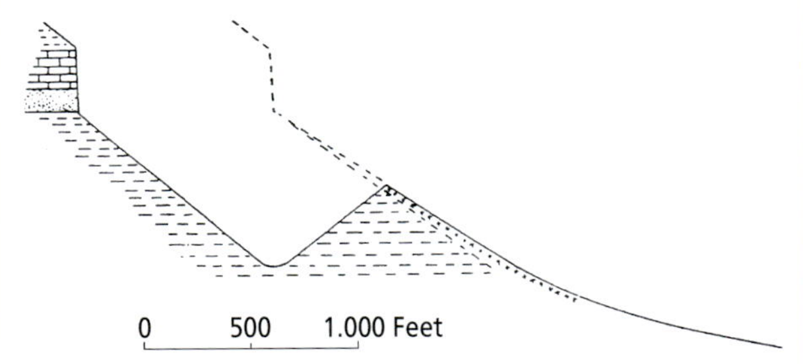

Figure 17.12 Schematic section of a talus flatiron. Western Grand Canyon (United States) (Koons, 1955).

the flatiron dips in the opposite direction than the front and is composed of bedrock. It is evident that flatiron formation requires alternating stages of accumulation and erosion. A series of these alternating stages produces sequences of talus flatirons in which the oldest facets are farthest from the scarp (*Fig. 17.14*). If the erosion persists, inverted relief can form; the flatirons are located in predominantly circular position on top of a closed depression (Morgan *et al.*, 2008).

Flatirons are relict slope accumulations; they become separated by incision of gullies in the middle part of a slope where the rill system converges (Sancho *et al.*, 1988) (*Fig. 17.15*). Once gullies dissect the flatirons, they flow to the lower parts where they deposit their load as alluvial fans. Once a flatiron is isolated, each acts as an independent landform (Büdel, 1970).

If the resistant scarp layer is too thick, accumulation will be excessive (Schmidt, 1987, 1989a); these factors promote rapid slope erosion and scarp retreat. For this to occur, the bedrock must be composed of rocks that are easily eroded. These circumstances occur in compound slopes but not in complex slopes where the alternating resistant and unstable rocks prevent flatiron formation (Gutiérrez *et al.*, 1998a, b).

Koons (1955) provided a model for flatiron formation. First a large rockfall covers the slope; then a dissection

Figure 17.13 Flatiron with a triangular apex, developed at the foot of the Cerro de San Pablo. The area is affected by intense gullying. Villanueva de Huerva (Province of Zaragoza). Ebro Depression, Spain.

Figure 17.14 Sequence of talus flatirons in the Chalamera region (Province of Huesca). Ebro Depression, Spain.

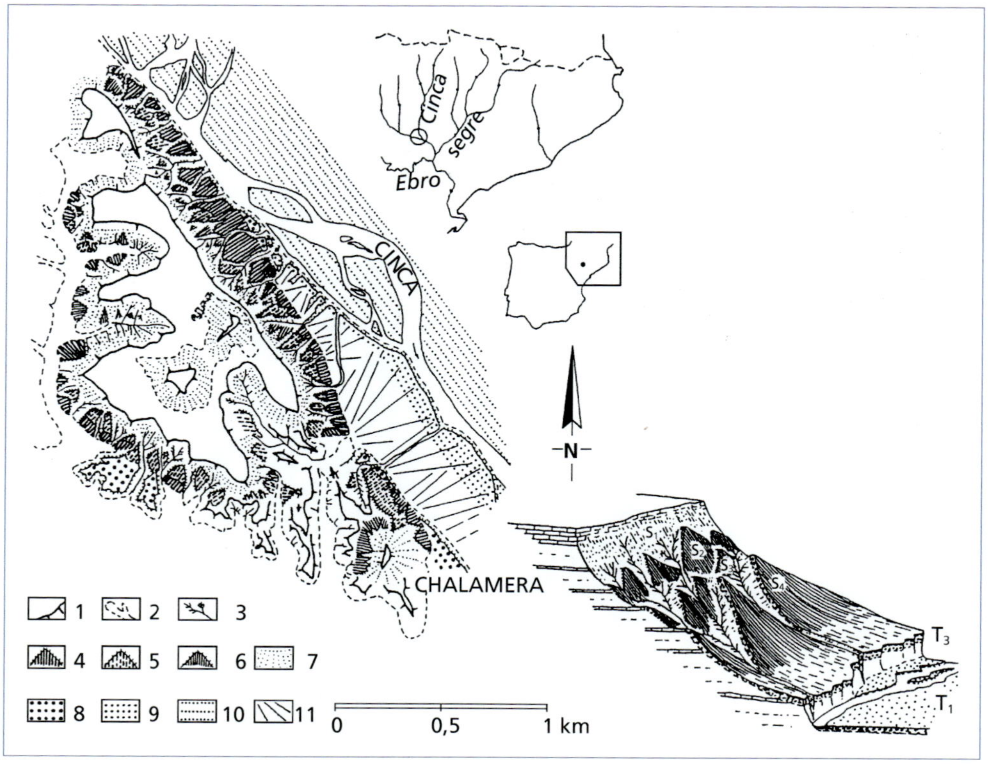

Figure 17.15 Geomorphological map of the Chalamera zone (Province of Huesca, Ebro Depression, Spain) and idealized block diagram of four stages of slope evolution showing three phases of flatirons. 1: Mesas and structural scarps. 2: Flat-floored valleys. 3: Gullies. 4,5,6: Flatirons S_4, S_3 and S_2. 7: Debris slopes S_1. 8,9, and 10: terraces T_3, T_2 and T_1 of the Cinca River. 11: Alluvial fans(Sancho *et al.*, 1988).

stage and individualize the facets (*Fig. 17.16*). Twidale (2007) shows that rock avalanches cover the slopes with deposits and afterwards the base level base decreases and erosion begins. The most accepted hypothesis regarding their origin is related to climate oscillations (Everard, 1963; Gerson, 1982; Gerson and Grossman, 1987; Sancho *et al.*, 1988; Schmidt, 1989b, 1994, 1996; Arauzo *et al.*, 1996a; Gutiérrez *et al.*, 1998a). Slopes with more vegetation generally have more accumulated debris. Decreased plant cover may trigger slope dissection, leading to flatiron formation. Recent periods of human activity that have destroyed vegetation may be triggering factors for the incision of debris slopes (Everard, 1963; Sancho *et al.*, 1988; Gutiérrez and Peña, 1989, 1992, 1998c).

17.1.3 Scarp retreat rates

Slope retreat rate calculation requires two pieces of information: 1) the linear distance between two particular positions of the scarp margin, and 2) the time lapse between these two positions (*Fig. 17.17*). These data are difficult to obtain so geomorphologists use multiple assumptions to estimate the retreat rates. As a result, numerous methodologies have been used to obtain these data (Oberlander, 1997). In the southern part of the Sinai Peninsula, Yair and Gerson (1974) calculate values between 0.1 and 2 mm/year for fault scarp retreat. Luccitta (1975) used K/Ar to date a fine lava cover on top of a retreating scarp in northern Arizona; he obtained values of 6.7 mm/year. In a study of the geometry of consequent

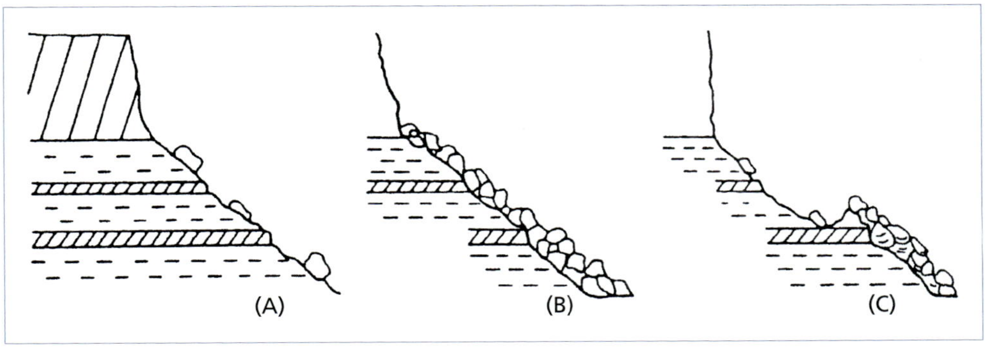

Figure 17.16 Non-cyclical model of flatiron development, according to Koons (1955), in Schmidt (1989a).

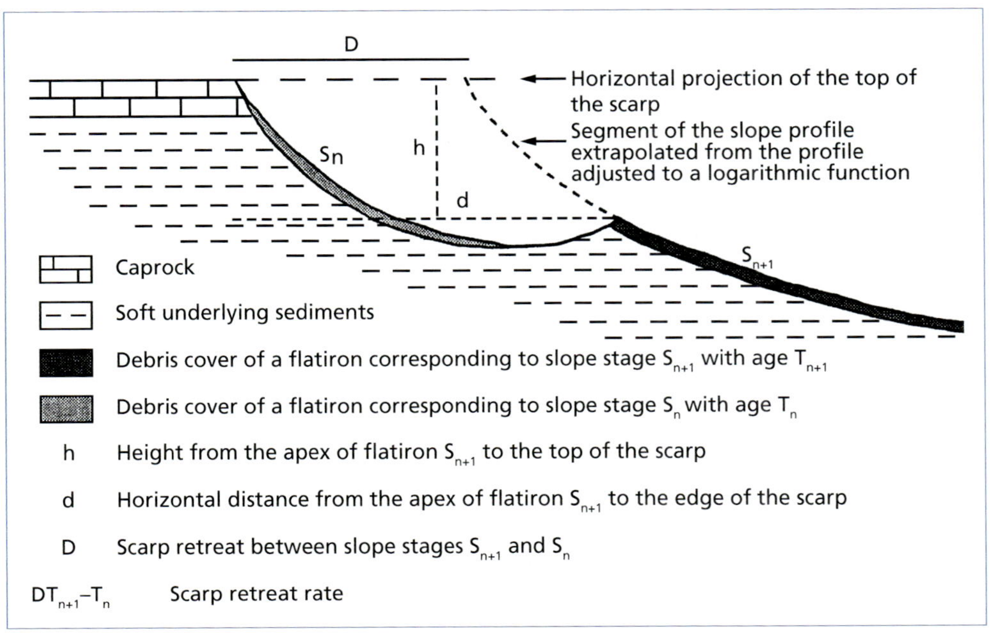

Figure 17.17 Diagram of a talus flatiron indicating the various parameters used to calculate slope retreat rates (Gutiérrez *et al.*, 1998b).

valleys captured on the Colorado Plateaus, Schmidt (1980, 1989b) published values of 0.3 mm/year and 0.5 to 6.7 mm/year in two publications. Another method, highly criticized by some authors, is based on Neotoma accumulations at cave entrances in Arizona; ^{14}C was used to date. Cole and Mayer (1982) calculated rates of 0.45 mm/year. Young (1985) studied scarp retreat on the lower Eocene erosional surface in northwestern Arizona using a scarp position based on the location of straight valleys; he reports values of 0.16 to 0.17 mm/year. Schmidt (1987, 1988, 1996) calculates retreat by graphs based on flatiron profiles. In his 1996 publication, he correlates flatirons with Illinois and Wisconsin glaciation; his results range from 0.2 to 0.35 mm/year. The majority of this research is based on assumed ages. Sancho *et al.* (1988) determine ages from archeological remains on slopes and flatirons; they report retreat values of 0.3 mm/year on the Tertiary formation in the Ebro Depression. Time values are more precise when multiple flatiron slopes are dated by C-14. In the center of the Ebro and Duero depressions, accumulations contain coal and ash remains (Arauzo *et al.*, 1996b, Gutiérrez *et al.*, 1998b, Gutiérrez and Sese, 2001, Gutiérrez *et al.*, 2006). These authors obtain values between 0.9–1 mm/year for the last 35,000 years. These dates show that retreat rates in arid zones around the world are on the order of 0.1 to 7 mm/year. Recently, Gutiérrez *et al.*, (2010) dated 31 samples from flatiron slopes (8 dates by ^{14}C and AMS (Accelerator Mass Spectrometry) and 23 by OSL (Optically Simulation Luminescence)) in large Tertiary depressions of the Ebro, Duero, and Tajo (Spain). The dates obtained by OSL have a large margin of error that increases with the age of the flatiron. As a result, they do not clearly reflect a correlation with climate stages.

We have shown that most of these values were obtained by simplified indirect methods using assumed ages. These factors are a significant disadvantage to using and correlating these numbers; the methods are obviously unreliable and do not provide results that inspire confidence. The sinuosity of the scarp front is another dating problem. The methods generally assume a straight front, which is common but not universally true when the scarp is thick. By extrapolating segments of flatiron slopes to their intersections with extensions of scarp crests, Sancho *et al.* (1988) obtained intersections of varying distances, likely because the scarp front is sinuous (Fig. 17.18). Multiple intersection points can be used to generate an arithmetic mean that is close to the actual retreat value.

Lithological and structural features of scarps are important factors in retreat rates because they determine resistance to erosion (Schumm and Chorley, 1966; Nicholas and Dixon, 1986; Schmidt, 1989a). Rock mineralogy determines its response to dominant weathering processes. The dip of the layers also affects the erosion rate (Howard and Selby, 1994); when dip is greater, less volume of rock is eroded, and vice versa. In addition, the thickness of the scarp rock is inversely related to the retreat rate (Schumm and Chorley, 1966; Schmidt, 1987, 1989b). The

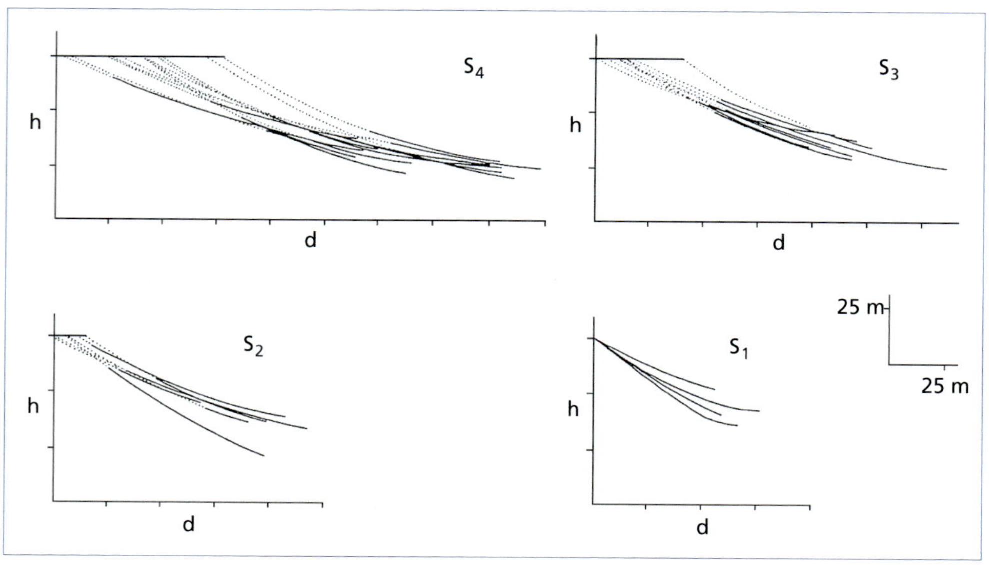

Figure 17.18 Talus flatiron profiles corresponding to various stages (S_1-S_4) drawn on a continuous line and curves extrapolated to points. The height of the flatiron apex with respect to the actual scarp is *h* and *d* is the distance from the apex of the flatiron to the front of the scarp, measured on the basis of aerial photographs (Sancho *et al.*, 1988).

presence of significant scarp rock thicknesses can be used to apply Koons' (1955) acyclic model of flatiron formation, which is accepted by Schipull (1980), Schmidt (1987), and Twidale (2007). According to this model, the alternation of large rockfalls and predominant weathering processes can cause slope retreat. In contrast, a thin scarp retreats more rapidly. The degree of fracturing in a scarp is highly significant to its retreat rate; fracturing largely determines its resistance to erosion. Thus rocks with significant fracture density have higher rates of slope retreat. The intensity of bedrock erosion also obviously affects scarp retreat; when the scarp is composed of easily eroded materials such as clay and marl, rills can reach the scarp base and undermine it, causing rock falls (Schipull, 1980; Gerson and Grossman, 1987).

Another important factor related to scarp retreat is the intensity of physical, chemical, and biological weathering processes, which disintegrate the rock and form debris slopes. These processes are a function of the lithological and structural characteristics of the rock, climate and climate change, and also the period of time the processes are active.

Schumm and Chorley (1966) call the relationship between the degree of particle supply to the slope from the scarp (*p*) and the erosion rate of the accumulations (*d*) the weathering ratio (*W*); Tricart (1957) calls it the morphogenetic equilibrium. Occasionally *p* may greatly exceed *d*; in this case, slope talus can cover the scarp and fossilize it. Under these conditions, slope retreat ceases although chemical weathering may continue (*Fig. 17.19*). This equilibrium between production and destruction can be maintained without much oscillation as long as changes do not significantly affect the system. However, the balance may tip toward greater erosion so that debris slopes eventually remain hanging or isolated. This disruption of equilibrium is due to environmental change; it results from crossing a geomorphic threshold (Schumm, 1979). Most researchers believe that slopes have undergone multiple climate changes during which wet/cold stages are characterized by accumulation and dry/warm stages by erosion. Slope retreat occurs in both types of climate although it may be accentuated in one climate or the other due to variation in weathering intensity, landslide generation, and increased basal undermining.

Finally, we must consider the rate of base level decrease (Howard and Selby, 1994, 2004); it affects the drainage system and may be due to climatic and tectonic causes. Also, the proximity of a slope to a highly erosional river may break the distal part of the slope profile and trigger headward erosion that affects scarp retreat; this appears likely in the Cinca River in the Chalamera area (Province of Huesca, Spain).

Figure 17.19 Slope partially covered with debris that fossilizes the cliff. These are later affected by intense rilling that affects the Miocene red clay substrate. Cerro del Castillo de Alfambra (Province of Teruel). Iberian Cordillera, Spain.

17.1.4 Badlands

Badlands are simple slopes in areas of intense water erosion; they have high drainage density (125–350 km/km^2), sparse vegetation, and steep slopes that are difficult to cross. They form in labile materials primarily in arid and semi-arid environments and often at the foot of scarps *(Fig. 17.20)*. The process that shapes badlands is primarily hortonian flow, which occurs when precipitation intensity exceeds infiltration capacity (Horton, 1945). Hortonian processes require intense storm precipitation, sparse plant cover, low infiltration capacity, easily eroded materials, and relatively steep slopes (Harvey, 2004a).

The interfluves may have a rounded (Davis, 1892; Gilbert, 1909; Penck, 1924; Birot, 1949a; Schumm, 1956b) *(Fig. 16.52)* or sharp *(Fig. 16.53)* morphology. Erosion may generate turret shapes (hoodoos) as interfluves retreat *(Fig. 17.21)* (Scheidegger *et al.*, 1968). Some badlands form in response to anthropogenic activities such as deforestation, construction, mining, and waste disposal (Aghassy, 1973; Harvey, 2004a). Because erosional rates are rapid, badlands provide a small laboratory with complex micro-relief where processes and modifications of landforms occur over brief time periods. We must be cautious in extrapolating these results to larger areas (Campbell, 1989; Campbell and Honsaker, 1982).

Numerous processes are involved in the formation of badlands (Harvey and Calvo, 1991). It is thus important to know the physical, chemical, and mineralogical properties of the **materials** (Gutiérrez *et al.*, 1995) that constitute the zones of intense erosion to understand the nature of the processes involved and their effects on the morphology of the area (Faulkner *et al.*, 2000). Dispersion index and swelling capacity are important physical properties in numerous processes. Chemical analyses provide the sodium absorption ratio (SAR) or the percentage of interchangeable sodium (ESP), which are directly related to whether the material is susceptible to *piping*. X-ray diffraction analysis of the clay fraction on oriented aggregates is used to identify the minerals present. The presence of swelling minerals such as montmorillonite has a major effect on weathering and erosional processes when repeated swelling and contraction occur at the surface.

Badlands generally form in arid and semi-arid **climates** where droughts persist a long time and storms generate the dominant rains. However, badlands can also form in areas with 1075 mm of annual precipitation such as Perth Amboy (New Jersey, although these badlands are anthropogenic in origin (Schumm, 1956a). Precipitation plays an important role in chemistry (Bryan and Yair, 1982); in areas of low precipitation, high solubility salts do not dissolve. On the contrary, if water is readily available, the rocks leach calcium and magne-

Figure 17.20 Rilling zone with sharp interfluves in clays and lenses of Miocene sandstones. The top of the formation is eroded by a cemented cover pediment of Pleistocene age. Las Bardenas Reales (Navarre). Ebro Depression, Spain.

Figure 17.21 Turret-shaped (hoodoo) landform in clays and Miocene sandstones of the Tudela Formation in an intensely water eroded area. Cabezo de Castildetierra (Las Bardenas Reales, Navarre). Ebro Depression, Spain.

sium. With intermediate precipitation, calcium predominates. These variations are reflected in swelling, in the degree of clay dispersion in the regolith, and in infiltration. On a smaller scale, slope orientation affects water availability and as a result, the activity of various geomorphic processes (Yair *et al.*, 1980). In areas of frost action such as western Colorado, permeability varies. Slopes turn into less permeable rill surfaces alternating with highly permeable aggregate-covered slopes (Schumm and Lusby, 1963, Schumm, 1964). In Alberta, (Canada) freeze-thaw cycles affect surface material by preparing it for rapid denudation by later runoff (Campbell, 1974).

Materials that crop out at the surface are subject to multiple physical and chemical processes that destroy the original rock and form a **regolith**. In arid regions regolith is thin (20 to 30 cm); the surface layer has polygonal cracking if the clays lack swelling capacity. If swelling capacity is high, swelling and contraction cycles fracture the surface layer and convert it into fragments with large holes and *popcorn* structures (*Fig. 17.22*). Intense leaching of soluble ions also occurs in this layer. Below is a 5–10 cm thick layer that may be a dense amorphous crust (Gerits *et al.*, 1987) or a granular aggregate (Schumm and Lusby, 1963). The transition to fresh rock occurs in a fragmented, slightly weathered layer. Regoliths vary widely as a function of rock composition, inclination, and slope orientation. During a rainfall event, water infiltrates into the interior of a dry regolith via cracks, conduits, and pores until it saturates the regolith; runoff begins several minutes later. The regolith swells as it becomes wet,

Figure 17.22 *Popcorn* structure showing fragmentation of the surface layer of Tertiary clays. Cuesta del Viento – Colola. Province of San Juan (Argentina). Precordillera of the Andes.

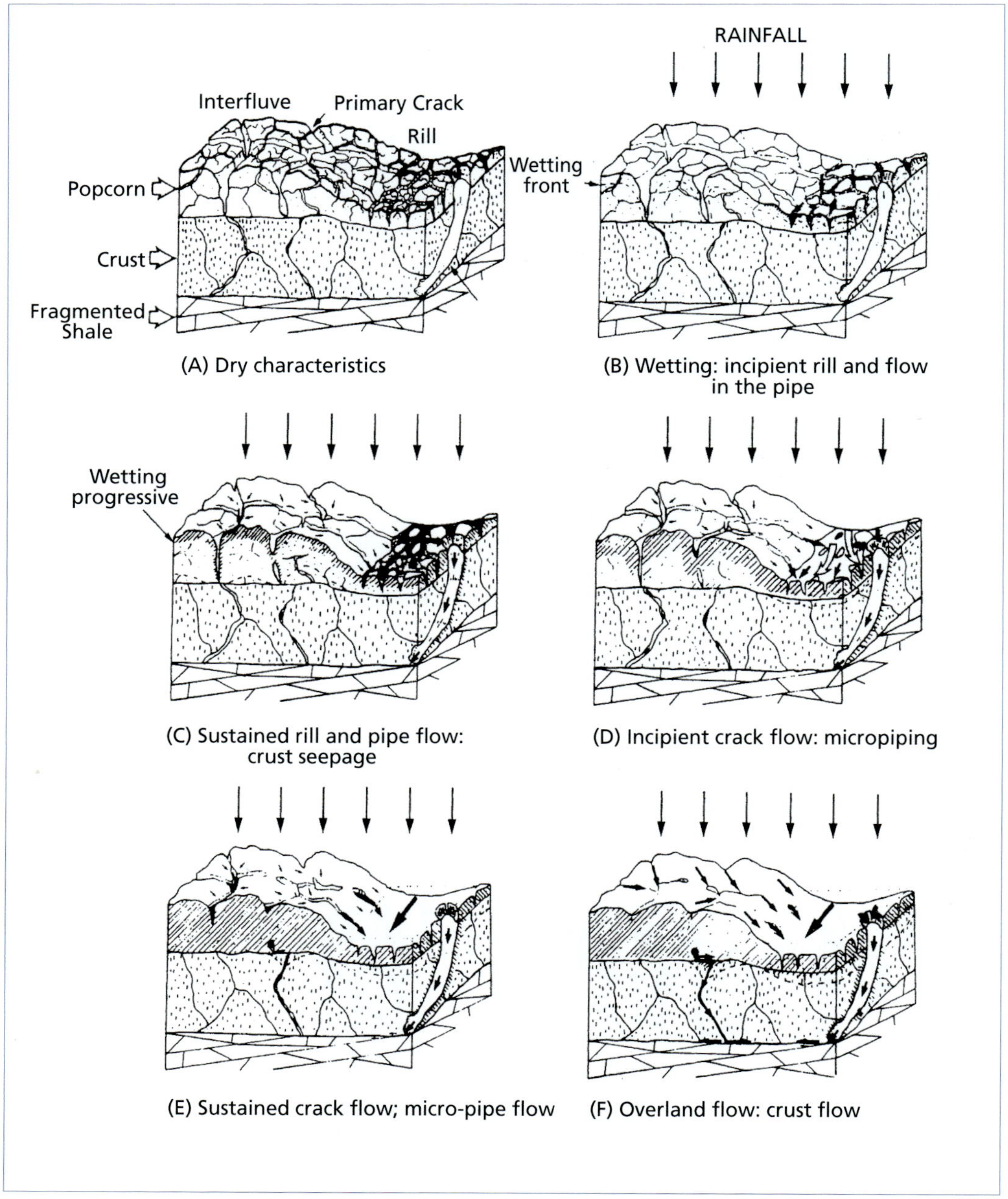

Figure 17.23 Stages of initiation and evolution of runoff from clay slopes (Hodges and Bryan, 1982).

cracks begin to close, and internal flow is limited to large cracks and micropipes (*Fig. 17.23*) (Hodges and Bryan, 1982). The wetting front gradually moves deeper; water infiltrates until surface flow begins and cracks are almost completely closed.

The most significant water erosion processes that affect badlands surfaces are rainsplash, rilling, gullying, and piping, which have already been described. In addition, however, these areas are occasionally subject to mass movement.

Mass movement has a major impact on badlands morphology. If the regolith absorbs water easily, as is the case of the Chadron Formation of South Dakota (Schumm, 1956 a, b), slopes evolve by creep and by small thin landslides that affect primarily the regolith (*Fig. 17.24*). These processes generate rounded interfluves. However, when the surface does not readily absorb water (as in the Brule Formation, also in South Dakota), rilling occurs and the landform develops high-density drainage, steep slopes, and sharp interfluves. Even so, channelized mudflows can follow the paths of existing rills

Figure 17.24 Gullying zone with rounded interfluves showing numerous thin landslides and half-moon shaped head scars. Bedrock in the Guadix-Baza Depression is Miocene marine marl. Dehesas Experimental Station (Granada). Betic Cordillera (Spain).

on these slopes (*Fig. 17.25*). In the Ebro Depression (Spain), mudflows occur during winter when low-pressure fronts bring precipitation. Precipitation and the insignificant evaporation that occurs at low temperatures can saturate the regolith, which may then become liquid and flow; these flows may fill the rills. The mudflows erode during convective rains and a new cycle begins (Gutiérrez *et al.*, 1995; Harvey and Calvo, 1991).

Badlands areas that lack vegetation have extremely high **erosion rates**. As a result, the response of the small-scale features is rapid, which leads to erosion over short time periods. Dynamic and volumetric methods are used to measure soil loss (De Ploey and Gabriels, 1980). Dynamic methods use collectors or Gerlach boxes to measure the quantity of sediment transported by water erosion (*Fig. 17.26*). Volumetric methods measure denudation by erosion pins (*Fig. 17.27*) and bidimensional (*Fig. 17.28*) and tridimensional microprofilers (Campbell, 1974; 1982). Values obtained for soil loss depend on the technique used (Yair *et al.*, 1980); for some authors, erosion values recorded by volumetric techniques (Takei *et al.*, 1981; Rogowski *et al.*, 1985; Sala, 1988) are superior to those obtained by dynamic methods. Sirvent *et al.*, (1997) reach the opposite conclusion, claiming that values obtained by collectors are better than those recorded by erosion pins. Even so, conversion of the volumetric data is a major problem because it requires knowledge of regolith density, which varies through time (Regües *et al.*, 1992).

Figure 17.25 Mudflow contained within a previous rill. Miocene clays in the Hoya de Ayerbe. Los Corrales (Province of Huesca). Ebro Depression, Spain.

Figure 17.26 Bardenas-2 experimental station for the study of water erosion, where collector systems and divisors have been installed in a gullied area of Miocene clays. Las Bardenas Reales (Navarre). Ebro Depression, Spain.

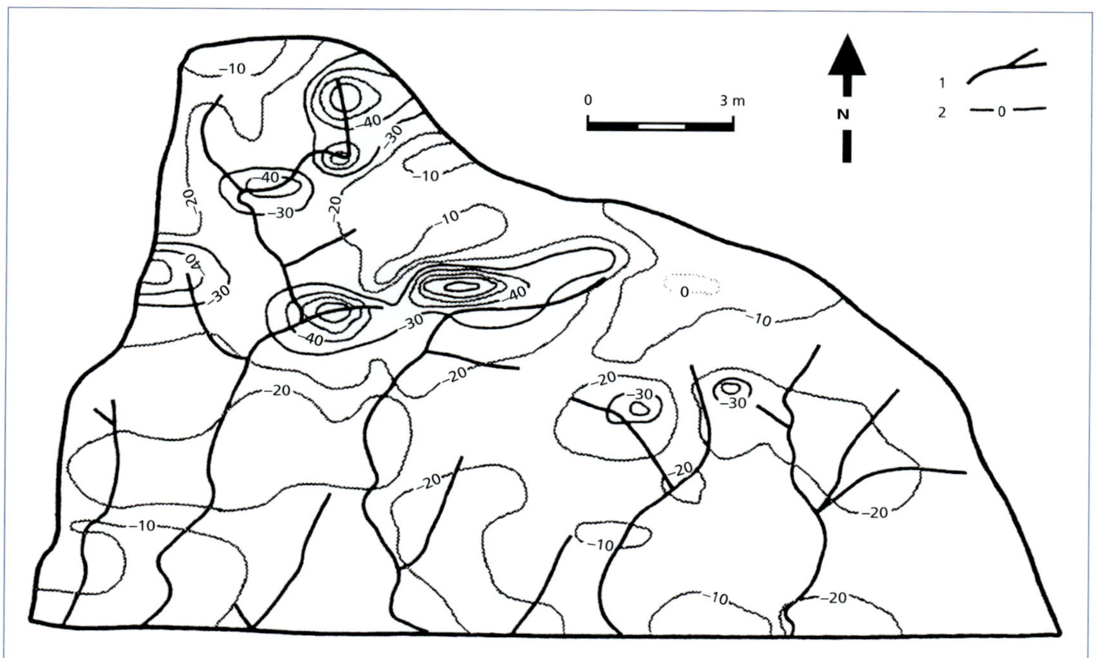

Figure 17.27 Surface lowering contours obtained from the erosion pin technique between November 1991 and November 1993 in the Lanaja-1 experimental plot (Province of Huesca). Ebro Depression, Spain. Legend: 1. Rills. 2. Surface lowering contours (mm). (Gutiérrez *et al*., 1995).

Figure 17.28 Microtopographical profiler composed of 50 pins that takes a profile at fixed points. Profiles are done every six months to study microtopographical variations.

Water erosion causes slope surface lowering that ranges from 2 to 20 mm/year in various experiments conducted in different deserts (Campbell, 1989). Maximum known erosion values were recorded in loess in China: up to 38,000 metric tons/km^2/year. Hadley and Schumm (1961) obtain rates of 13,400 metric tons/km^2/year in the *badlands* of South Dakota. The considerable range of values is due to material type, magnitude and intensity of precipitation, recording time at the experimental stations, and the techniques used.

Because of the rapid erosion rate, we can get at least a partial understanding of the **development and evolution** of this landform. Many aspects of larger morphologies appear in miniature in badlands *(Fig. 17.36)*; observation of small landforms provides information that helps us understand their development (Scheidegger *et al.*, 1968). Research that Schumm (1956a) conducted on the drainage system that developed in the badlands in Perth Amboy shows that its evolution is similar to that of large fluvial systems. Although a dendritic network may develop by chance, channel spacing and distribution are closely related to the erodibility of the material and the amount of runoff generated per unit of surface.

Channels incise rapidly into labile materials and generate steep slopes. As the slopes retreat, micropediments form at their base; they increase in size as slopes retreat. These small pediments can also end up eroding and evolving to slopes (Schumm, 1962).

Research has been conducted in regions with varying degrees of aridity. Yair *et al.*, (1980) found erosion rates of 0.45 mm/year in the northern part of the Negev (Israel), which has 90 mm of average annual precipitation. In the *badlands* of Alberta (Canada) where annual precipitation is 330 mm, rates of 4 mm/year have been recorded (Campbell, 1982). When a particular badlands area is analyzed over a long period interval we see that erosion may have been highly variable, which indicates that these arid and semi-arid areas have undergone multiple climate changes. During these climate changes, precipitation is substantially different during dry periods (interpluvials) and wet periods (pluvials) and as a result, erosion rates are lower or higher, respectively.

The interaction between vegetation and geomorphic processes is a fundamentally important factor that affects runoff and stabilization of eroded slopes (Alexander *et al.*, 1994; Gallart *et al.*, 2002). Even so, the principal factor that controls the formation of badlands is the interaction between base level and surface processes (Harvey, 2004a).

17.2 Piedmonts in arid regions: Pediments

Piedmonts are transitions between high areas where erosion predominates and low areas where transport and deposition are the major processes. In arid zones, aridity controls slope weathering; sediment production is generally minor. Even so, climatic conditions vary considerably in upland areas of mountain deserts and depressions (Mabbutt, 1977); contributions from rainfall or snowmelt may be significant. As a result of allogenic supply, piedmonts will generally receive flowing water with much higher solid loads than are typical in desert areas. Material from higher zones is transported by confined flow within a drainage network hierarchy that becomes unconfined and flows radially onto the piedmont. At the same time, flow velocity and transport capacity diminish abruptly, causing rapid deposition. Erosion-accumulation ramps form on the slopes above the piedmont in arid areas;

Figure 17.29 Detrital piedmont deposits where the distinction between alluvial fans and pediments proposed by Bull (1977b) are not applicable. This is due to minor or no incision of the fluvial system into deposits, which prevents observation of the bedrock. It can be treated as a covered pediment or coalescing alluvial fans. In the distance, the Sierra del Cerro del Coronel (Province of San Juan, Argentina). Precordillera of the Andes.

these are characteristic of the intersection zone. For this reason, piedmont morphologies are usually erosional; they include pediments (glacis) or depositional features such as alluvial fans. Pediments and alluvial fans may be closely related and it may be difficult to distinguish them (*Fig. 17.29*). Bull (1977b) suggests that fans can be distinguished from pediments on the basis of thickness of the deposits; a deposit that exceeds 1/100 of the length of the landform is a fan. In addition, a fan may flatten if the base level is lowered; this creates a pediment that may subsequently be fossilized when later contributions form a fan. In addition, pediments form more readily in areas of relatively flat relief, unlike alluvial fans, which are more common in higher energy topographies.

17.2.1 Glacis and pediments: Definition, terminology and distribution

López-Bermúdez (1973) conducted a historical review of glacis terminology; he determined that Dresch (1938) was the first person to use the term **glacis**. The term originally meant an inclined plain at the foot of a fortification; Dresch used it to describe landforms with a slightly concave longitudinal profile and a slope less than 6°. Tricart (1969) maintains that glacis form vast plains with very gentle slopes in piedmont areas; glacis connect to streambeds and are generally covered with a thin film of debris (*Fig. 17.30*). Significant problems arise when the landform is described in various languages, which has led to extensive terminology and conceptual confusion (Tator, 1952, 1953; Dumas, 1967; Whitaker, 1973, 1979; Tricart, 1969; Pécsi and Szilárd, 1970; Cooke and Warren, 1973; van Zuidam, 1976b; Demangeot, 1981; Cooke *et al.*, 1993). English-speaking researchers use the word **pediment**, which was first used by Gilbert (1877). It is defined by Mabbutt (1977) as a piedmont plain that has been beveled into the bedrock and is separated from the slope by an abrupt slope change (*Fig. 17.31*). A pediment may be covered with debris but its surface is an erosional base; this type of pediment is called a **covered pediment**. In German, the equivalent of the piedmont surface is *fussfläche* (Mensching, 1958).

French researchers use the term glacis; however, for greater clarity and because of its purely descriptive nature, they complement it with other words (Tricart, 1969). For this reason, different authors have used a variety of terms. López-Bermúdez (1973) compiles 28 terms and Whitaker (1973)

Figure 17.30 Pediment connecting to terrace deposits of the Ziz wadi. High Central Atlas (Morocco). Photo F. Gutiérrez.

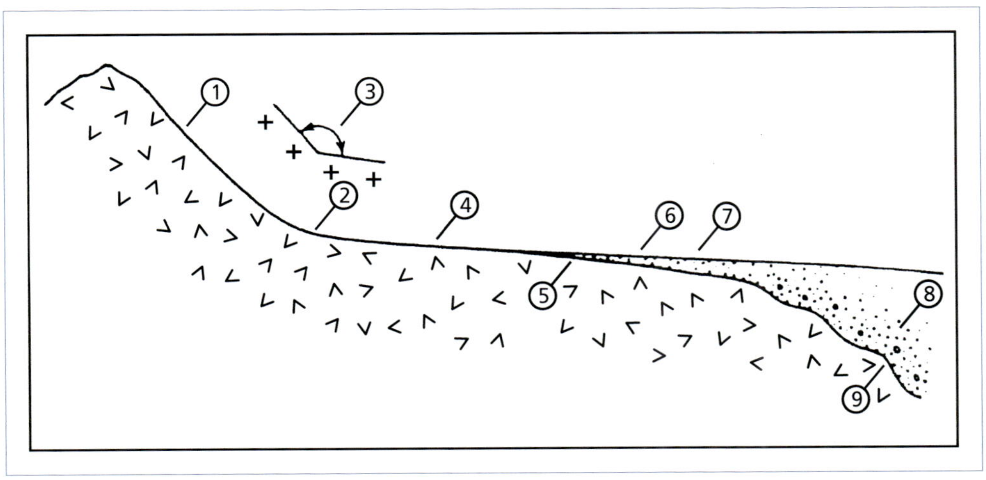

Figure 17.31 Terminology and characteristics of pediments developed in the desert piedmont. 1: Hillslope. 2: Piedmont junction. 3: Piedmont angle. 4: Pediment. 5: Mantled pediment. 6: Mantle. 7: Alluvial plain. 8: Alluvial fill. 9: Suballuvial floor (Mabbutt, 1977).

significantly increased the list. Thus, for example, a glacis that bevels the bedrock and has no deposit (the equivalent of the English pediment) is called a bare glacis (Birot and Dresch, 1966); an ablation glacis (Dresch, 1970); erosional glacis (Dresch, 1938); or a denudation glacis (Tricart, 1969; Tricart *et al.*, 1972). Various terms are also used for a pediment that has an accumulation of detrital material. Thus, Tricart (1969) distinguishes between a *glacis de denudation*, which is practically devoid of deposits, a *glacis d'épandage* that has accumulations of 1 to 3 m, and finally a *glacis d'ennoyage* when the detrital cover is more than 5–10 m thick. Conceptual confusion has added to the terminological profusion. For some French authors, an erosional glacis can form only in soft rocks such as limestones and marls (Dresch, 1957) and if the plain has formed in hard rocks (for example, granite) the term pediment should be used (Birot and Dresch, 1966).

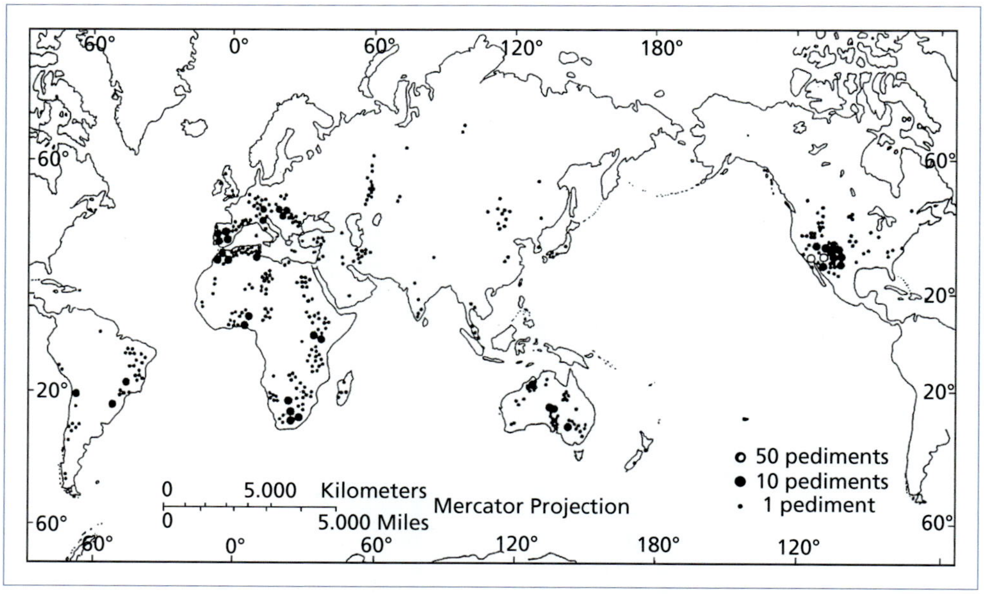

Figure 17.32 Map of pediment distribution indicated in scientific publications (Whitaker, 1973).

To determine whether a rock is soft or hard at simple glance is difficult, which leads to subjectivity. All this has generated numerous studies and controversies. Birot (1960) considers the pediments as the main problem of arid regions. For this reason, to avoid confusion and error we will use the terms **pediment** for erosional surfaces in rock where no deposits present and **mantled or covered pediment** for surfaces that are fossilized by a thin detrital cover.

Pediments develop in a wide variety of climatic, lithologic, and tectonic situations (Dohrenwend, 1994). Given their distribution from subpolar latitudes to arid to humid tropical climates (Whitaker, 1973; 1979), pediments can be considered as azonal morphologies. However, they are much more common in arid and semi-arid environments (*Fig. 17.32*). Whitaker's dictionary (1973) includes cryopediments, which are related to nivation processes in periglacial zones. We believe that most terminologies used in this study are included in arid climates broadly speaking or can be treated as paleo-forms or exhumations of these environments that are now located in humid tropical climates. We maintain that pediment is synonymous with Dresch's (1957) erosional glacis and that these morphologies form primarily in arid and semi-arid zones where sparse vegetation provides few obstacles to runoff.

17.2.1.1 Pediment morphology

Pediments form at the foot of small reliefs or mountain chains. The mountain front, the boundary between the upland area and the piedmont, may be straight or sinuous. Streams that flow to the piedmont cut valleys and form indentations that are called quebradas in South America; they are usually steep-walled valleys. In mountain chains that lack tectonic activity, the front becomes more sinuous over time (Bull and McFadden, 1977). However, the connection to the piedmont may be abrupt, perhaps in response to active faulting or water flowing along the slope base. This type of connection forms an angle between the two surfaces called a **knick**. In other areas the connection may develop over a wide area where there is a significant height contrast; this forms a concave connection between the slope and the pediment. Sometimes it is difficult to distinguish where the pediment begins or where the slope ends.

Few studies have focused on the quantitative analysis of pediments (Corbel, 1963; Mammerickx, 1964; Cooke, 1970b); it is thus essential to discuss the variables involved. Pediment surfaces vary considerably in size; they range from 2.5 to 650 km^2 in the Sonoran Desert (Arizona). In addition, the percentage of total surface area occupied by pediments varies substantially from area to area; pediments cover about 6.7% of the Mohave Desert (California) and 30% of the Sonoran desert. Pediments form between the lower part of the slope and a *stream* or sedimentary plain. Longitudinal profiles range from concave to straight; their slope decreases toward low areas. When pediments surround isolated landforms, or **inselbergs**, their profiles are radial (*Figs. 17.33* and *17.34*). Inselbergs are of two types: **position inselbergs** (*fernlinge* in German), which

Figure 17.33 Inselberg in Cambrian sandstone, with radial low slope pediments at the base. Region between Jodhpur and Jaisalmer. Semi-arid zone of the Thar Desert (India).

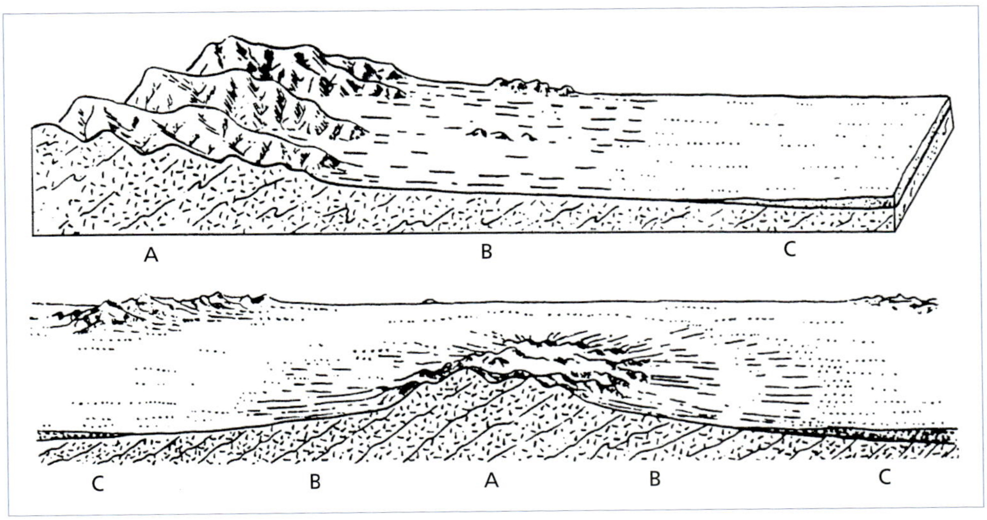

Figure 17.34 Various parts of a pediment. A: Dissected zone (top) and inselberg (bottom) B: Pediment with small residual relief ("*chicots*") (top) C: Mantled or covered pediment (Johnson, 1932a).

are preserved because they are located on the watershed divide, and **resistant inselbergs** (*härtlinge*) composed of rocks that are more resistant to erosion.

Pediment slopes range from 0.5° to 11° (Tator, 1952); their slopes vary along concave segments (*Fig. 17.35*). When they are associated with faults, the slope is more inclined than in areas not affected by these mechanical contacts. Clast or fragment sizes are larger in zones with greater inclination. Some authors think lithology does not affect the gradient (Mammerickx, 1964) whereas others consider it important (Cooke *et al.*, 1993). Climate may also play a significant role because in regions of greater aridity, gradients are smaller (Mabbutt, 1977).

The drainage network is usually incised in proximal areas; it discharges water radially toward the medial and distal parts where it gradually disappears. The most typical

Figure 17.35 Hill with convex and concave segments on the slope and in the front that develop an erosional pediment from a concavity that connects them. Machakos (Kenya).

channel morphology is braided but it changes significantly during flooding. Channels are generally shallow. If the base level decreases, the network incises in distal areas and forms sub-parallel channels (Cooke and Warren, 1973). In most cases, pediments are surfaces of bedrock outcrop; they may be covered by material that is generally alluvial, discontinuous, and very thin. Occasional small elevations (called *chicots* by French authors) may remain as residual landforms (*Fig. 17.34*).

17.2.1.2 Role of geology and climate on pediment formation

As previously stated, French authors make a distinction between pediments in soft versus hard rocks. This distinction emphasizes the major influence of **lithology** and **structure** on pediment location and formation. When these morphologies occur in granite, they generally have a *knick* at the contact between the slope and pediment, which many Australian researchers (Mabbutt, 1966; Twidale, 1967) attribute to more intense weathering at this contact. Although weathering is limited in arid environments, periods of higher rainfall occurred during the Quaternary and even the Tertiary. In southern Arizona, where lithologies are schist and basalt, the slope-pediment angle is considerably higher (Kirkby and Kirkby, 1974). In contrast, in rocks that are less resistant to erosion, the transition between slope and pediment is gradual and is generally concave (Dohrenwend, 1994; Oberlander, 1997) (*Fig. 17.35*); this gradual transition can be observed in pediments in the Ebro Depression (Spain) and in marginal areas of the Mediterranean basin.

The recent tectonic history of a region, or **neotectonics,** has a fundamental influence on pediment formation. Tectonically stable areas are places suitable for the origin and growth of pediments; they are also characterized by highly sinuous mountain fronts. However, when tectonic activity is recent or present, pediments are rare, alluvial fans predominate, and mountain fronts are overwhelmingly straight (Bull, 1977b; Bull and McFadden, 1977). This suggests that crustal stability is an important factor in pediment formation.

In arid morphoclimatic zones, **climate** plays a very different role. In hyper-arid areas (Atacama, central Sahara) the lack of rainfall prevents water erosion (Oberlander 1997); if pediments are present, they result from pluvial periods. However, in arid and semi-arid environments, pediments form where there is more runoff. Climatic variations affect the percentage of plant cover and, as a result, erosional and depositional processes. These changes and their consequences will be analyzed later with pediment evolution.

17.2.1.3 Dominant pediment processes

Surface runoff is perhaps the most important process in desert areas. This seems like a contradiction (Tricart, 1969) because arid regions by definition have a water deficit. Even so, most precipitation comes from storms and runoff encounters few obstacles as it flows over areas with scattered or nonexistent vegetation. Runoff moves in confined flows down rills and gullies and in unconfined flows such as *sheet flow*.

In regions of tectonic stability, the pediment surface can be categorized into an upper erosional zone, an intermediate transport zone, and a lower depositional zone (Johnson, 1932b); the boundaries between them are gradual. When the supply of detritus increases, the zones move upslope; if the supply decreases the zones migrate to distal areas (Cooke and Mason, 1973). Erosion and transport zones are shallow braided channels that move laterally and by sheetflow.

In any of these zones we can find pediment surfaces located higher than the functional pediment; these are relict landforms that are mostly inactive. Surface and subsurface weathering and eolian activity operate on them to create pedocal soils and desert pavements with varnishes.

Australian researchers think that subsurface weathering plays an important role in pediment formation (Ruxton, 1958; Mabbutt, 1966, 1977; Twidale, 1967, 1983), primarily in crystalline rocks. The regolith is easily denuded by runoff, which causes the basal weathering surface to crop out, generating an *etchsurface*. This alteration occurs more easily in savanna areas with seasonal climates and also to a lesser extent in semi-arid environments. Places of maximum alteration are the foot of slopes, the regolith-bedrock contact, and lines of structural weakness.

In soft rocks runoff encounters material that is less resistant to erosion. In carbonate rocks, karst corrosion and mechanical activity of surface water flow causes planation (Nicod, 1992). The Iberian Cordillera has excellent examples of pediments and large plains that reach their maximum development in these calcareous areas.

Research by Schumm (1962) over a period of eight years on the micro-topography of small experimental plots provides valuable data on micro-pediment development. He used erosion pins on siltstone slopes in the Badlands National Monument in South Dakota (USA). Micro-pediments underwent lowering as slope increased and the connection zone with the slope experienced parallel retreat. In addition, in the lower zone some micro-profiles accreted (Fig. 17.36). In his publication Schumm states that "it is highly likely that the formation of these miniature pediments is analogous in general to the formation of pediments," although later (Chorley *et al.*, 1984) indicated that this process may take place on a larger scale at many sites and that *badlands* studies provide a way to understand the development of one type of pediment.

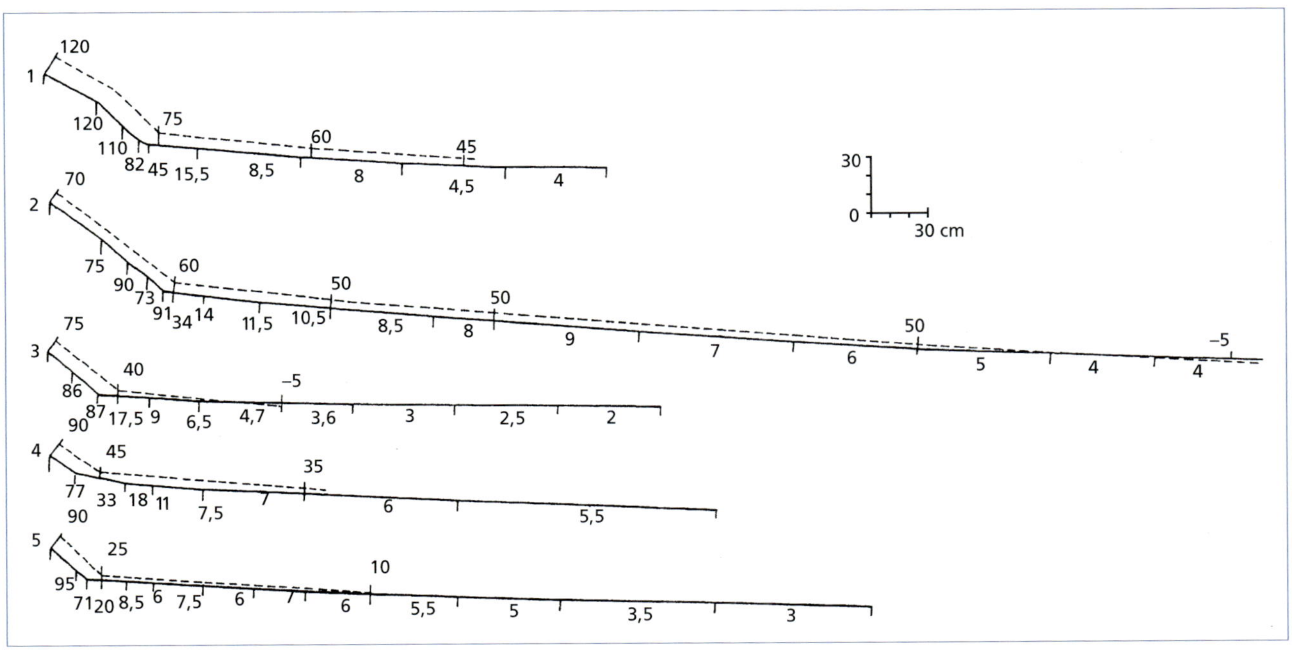

Figure 17.36 Micropediment profiles in the Badlands National Monument (South Dakota). The numbers under the profiles indicate pediment slope as a percent, whereas the upper numbers (at the location points of the erosion pins) show values of lowering by erosion, in millimeters. Profiles in solid lines were done in May 1961 and those in dashed lines were done in July of 1953 (Schumm, 1962).

17.2.1.4 Hypothesis on the origin of pediments

Pediments form in zones where upland areas are eroding. If the upland area is reduced substantially, the pediment surface will be larger. The hypotheses proposed for pediment formation are based on an explanation of mechanisms that degrade topography and promote in piedmont pediments. These hypotheses depend primarily on the nature of runoff that occurs on the piedmont and on preparation of the bedrock by previous water erosion. These hypotheses are not exclusive, but rather complementary.

The classic hypothesis is McGee's (1897) **sheet flooding** in which channelized water from mountainous areas becomes a sheet of water and sediments 0.5 m thick that spills over the area in turbulent unconfined flow. These flows are floods that transport particles of very different sizes and erode the bedrock. Blair (1987) used aerial photography to identify sheet flooding in alluvial fans in the Rocky Mountains. Tricart (1969) shows that it takes a large landform to concentrate water into a channel that can later deposit its sedimentary cargo on the piedmont, which shows that this hypothesis cannot be applied to small landforms. On the other hand, when drainage channels are separated at some distance, the spaces between them are not affected by sheet flooding near the mountain front. In addition, Tricart also suggests that flat or inclined topography is required for sheet flooding to develop. Lawson (1915) introduces weathering processes and mass movement to explain the retreat of slopes whose inclination remains constant during their evolution. He recognizes that sheet flooding is the final process in pediment formation.

The **lateral planation** hypothesis, first proposed by Gilbert (1875, 1877), was accepted by Bryan (1922) and augmented by Johnson (1931, 1932a,b). At the piedmont, the confined channel becomes a group of small braided channels (diffuse runoff). The channels are several decimeters in width and depth. These small channels encounter obstacles in their trajectories such as shrubs and boulders, which make them detour. Their braided morphology, like large rivers, implies that the constant wandering of these small channels causes lateral planation. Erosional activity caused by lateral channel migration slowly destroys small existing landforms and finally forms a pediment. Radial pediments that form at the foot of inselbergs can be explained in this way; they are difficult to explain by sheet flow (Tricart, 1969). Some authors think that lateral planation is more effective when it affects rocks that are more susceptible to erosion (Sharp, 1940; Denny, 1967) whereas in granitic rocks denudation is much less common (Howard, 1942). Although braided runoff activity is verified in field studies, this hypothesis cannot be considered as a general explanation (Mabbutt, 1977).

Twidale (1962) and Corbel (1963) think that chemical weathering processes affect pediment formation. In shield deserts, denudation of weathering profiles that formed during earlier geologic periods may produce etchsurfaces that later become pediments due to planation processes (Mabbutt, 1977). In addition, this weathering is especially intense at the contact with the piedmont (Twidale, 1962, 1967, 1976) where it receives water that runs off the slope. Twidale relates pediments and the origin of overhanging *flared slopes* to this weathering. During humid periods, the weathering front progresses; during dry periods, denudation predominates (Fig. 17.37).

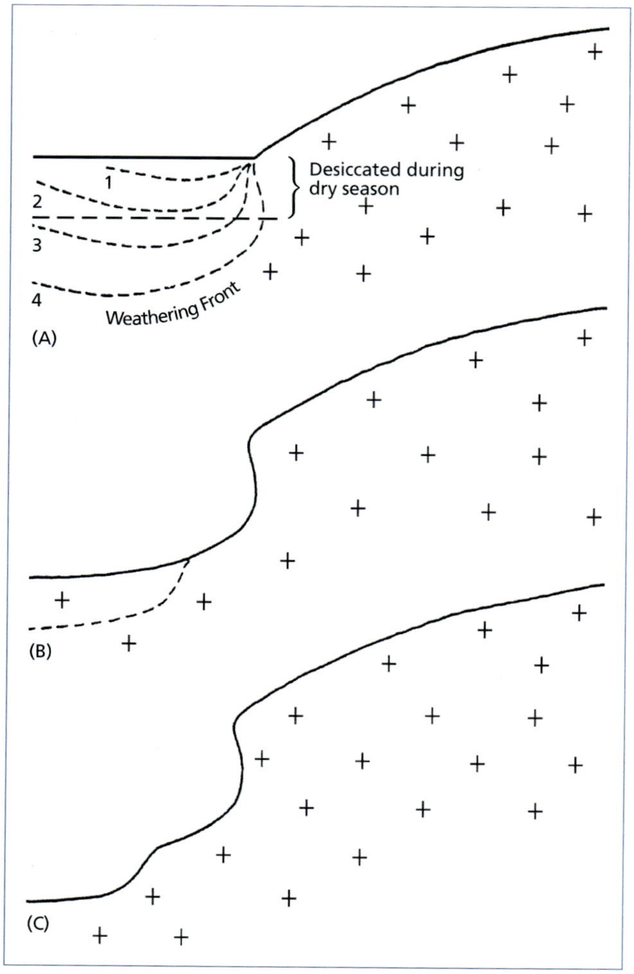

Figure 17.37 Stages in the evolution of overhanging slopes that form by weathering at the slope base and later differential erosion: (A) subsurface infiltration of runoff at the slope base and lowering of the weathering front; (B) lowering of base level and denudation of weathering products, which causes the weathering front to crop out as an overhanging slope; (C) repetition of the process and development of a new overhanging slope (Twidale, 1976).

17.2.1.5 Pediment evolution—pediplains

Runoff that flows across a pediment surface discharges its water in endorheic depressions, in seas and oceans, or more commonly, in desert streams. Base level changes may trigger geomorphic processes that rework the original pediment. In the first case, endorheic basins undergo continuous aggradation as a result of the dismantling of their elevated margins, which decreases the pediment slope. When sea level decreases, as for example during glacial epochs, erosion compartmentalizes and isolates the previous pediment; this can be seen in many places on the Mediterranean shoreline. When sea level rises, the area occupied by pediments decreases and the pediments may be totally covered with marine water.

The most general evolution of pediments is related to *wadis/streams*. Materials transported by surface currents shape an erosional pediment in the proximal zone and a mantled or covered pediment in medial and distal areas. At the juncture, pediment-related clasts may interfinger with alluvium deposited on the alluvial plain by desert rivers; pediment clasts are generally angular and of similar composition because of their lateral supply but clasts from the desert river are polymictic and rounded because of their long longitudinal transport (Gutiérrez and Peña, 1976). When base level decreases due to climate or tectonics, lateral alluvial systems gradually incise into the previous pediment and finally equilibrate to the new base level. The result is erosion and compartmentalization of the previous pediment and formation of a terrace on the *desert stream* margin (Fig. 17.38). Repetition of additional cycles of this type form a set of **stepped pediments** (Fig. 17.39). These are the dissected pediments (Bryan, 1922), *glacis emboîtés* (Joly, 1950; Dresch, 1957) of French authors, and *stepped pediments* or *staged pediments* (Mabbutt, 1977) of English-speaking researchers. This system of stepped pediments shows successive phases of incision by *streams* and stages when pediplanation predominate. This model characterizes pediment development in soft rocks that evolve rapidly in semi-arid climates (for example, in the Ebro Depression and grabens and interior depressions in the Iberian Cordillera, Spain). In contrast, in arid climates where precipitation is rare, planation and downcutting alternate over long periods of time (Figs. 17.40 and 17.41) because large magnitude climatic oscillations are needed to modify the predominant geomorphic processes. In hyperarid areas

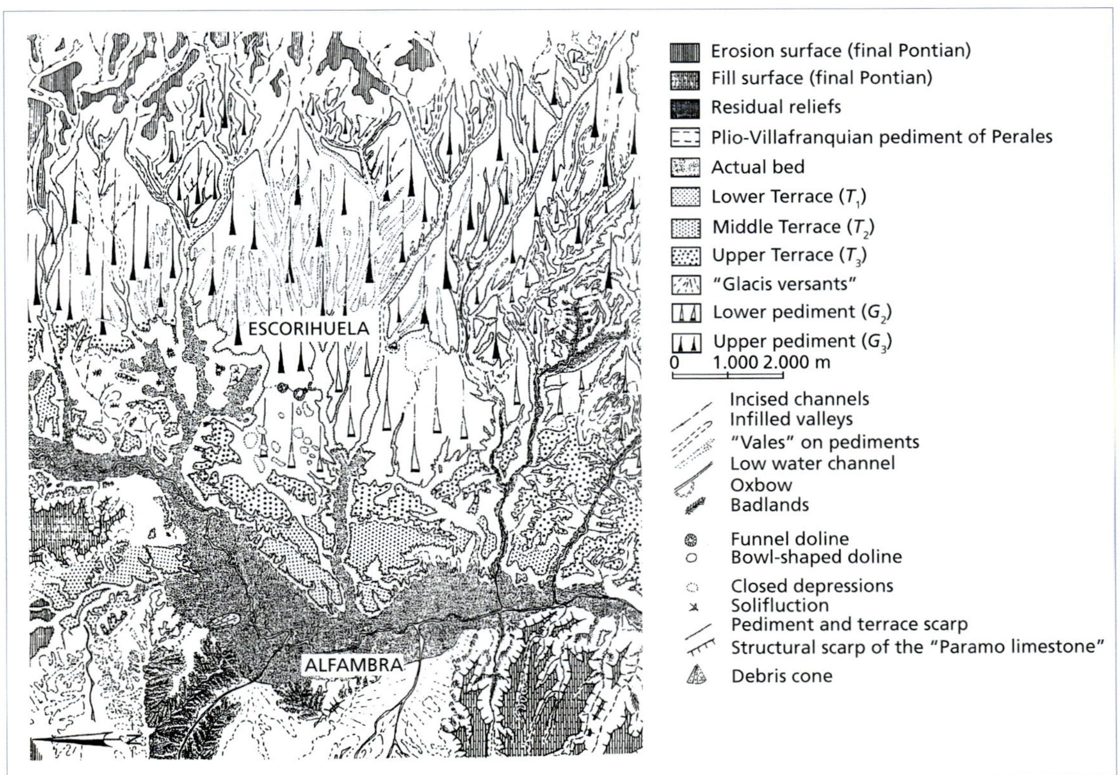

Figure 17.38 Geomorphic map of a sector of the middle branch of the Alfambra River (North of the Teruel graben, Iberian Cordillera, Spain) Upper pediment (G_3) linking with the upper terrace (T_3) and the middle pediment (G_2) with the middle terrace (T_2). Pediment and terrace deposits interfinger in the connection zone (Gutiérrez and Peña, 1976).

Figure 17.39 Various pediment levels that show concavity where they connect to slopes in the Exterior Sierras of the Pyrenees Hoya de Ayerbe (Province of Huesca). Ebro Depression, Spain.

Figure 17.40 Stepped pediment system in the Sinaí Desert. Photo J. Rosell.

Figure 17.41 Complex of stepped pediments with alluvial cover that has evolved to a pavement with desert varnish. Negev Desert (Israel).

fluvial activity is almost non-existent. In hard rocks such as granites, downcutting of lateral tributaries is difficult due to the resistance of the rock; in any case, prior alteration is necessary for pedimentation to occur.

When arid climatic conditions persist over long periods of time in areas of tectonic stability, pediments continue to expand as a result of continuous parallel slope retreat. Inselbergs may form later and they are also finally eroded. The final topography is extensive planation in arid climates, which King (1953) called **pediplains**; these are different from the peneplains described by Davis (1899), which form in humid climates. Pediplains are the final result of a cycle of **pediplanation** that is reactivated by endogenetic movements (*Fig. 17.42*). For King the primary landform of landscapes throughout the world is the pediment. Pediplanation creates pediplains on all continents; they can be incised into each other, generating stepped pediplains (*piedmonttreppen*) (King, 1962, 1976). These pediplains can be fossilized or exhumed (examples are the pre-Triassic and intra-Miocene surfaces of the Iberian Cordillera in Spain) or can crop out (examples are the erosional surfaces that formed later in the same cordillera) (Gutiérrez and Gracia, 1997) (*Fig. 17.43*). By correlating erosional and sedimentary records, King (1976) was able to distinguish six global planation events that occurred between the Jurassic and the Quaternary.

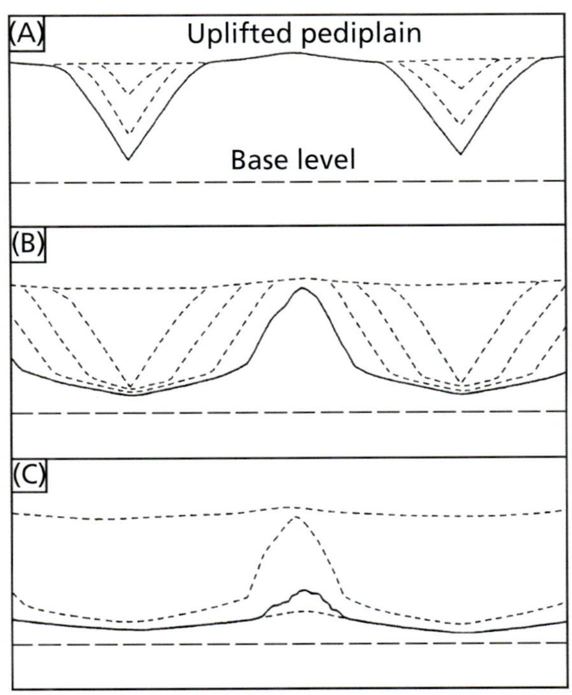

Figure 17.42 Pediplanation cycle (A) youth. (B) maturity. (C) senility (Interpreted by Small, 1970).

Figure 17.43 Erosional surface that bevels folded carbonate materials of Cretaceous age. This surface is a pediplain that formed primarily during mid and late Miocene. Because of its extraordinary development, it is called the Fundamental Erosional Surface of the Iberian Cordillera. The photograph is of the Aliaga region (Province of Teruel). Northeastern Iberian Cordillera, Spain.

17.3 Desert lakes: Playas and sabkhas

Significant portions of desert areas are characterized by endorheism; as a result, water can pond and create lakes of various sizes and origins. These lakes can survive if they are supplied by an amount of surface water and groundwater that exceeds evaporation. As the solute load increases, the lake becomes a salt flat. In arid zones, these lakes may contain no water for long periods of time; sometimes they have an almost seasonal regime and become dry during the summer. Only the Aral Sea, Caspian Sea, and the Dead Sea (*Fig. 17.44*) have a perennial water column supplied by elevated zones or distant areas. The Dead Sea is fed primarily by the Jordan River, which is an overflow channel of the Sea of Galilee located to the north. The high salinity of the Dead Sea (300,000 ppm dissolved solids) prevents the growth of organisms because of its saline supersaturation. Salt lakes, which have a minimum salinity of 5,000 mg/l, are relatively common in arid and semi-arid zones (Shaw and Thomas, 1997). This concentration is the limit between fresh and salt water (Eugster and Hardie, 1978) based primarily on biological tolerance.

There is an abundant nomenclature of closed desert depressions (Tricart, 1969; Cooke and Warren, 1973; Cooke *et al.*, 1993; Currey, 1994; Shaw and Thomas, 1997). Most vocabulary is local but some has attained wider acceptance; for these reasons, the literature is confusing. The term **playa** is Spanish; it was introduced during the exploration of the western United States. Numerous closed depressions are present on the Bujalaroz platform in the center of the Ebro Depression; the largest is La Playa, which is a large salt flat (*Fig. 17.45* and *10.14*) (Gutiérrez *et al.*, 2005b). Bryan (1922) in his study on Papago County (Arizona, United States) shows that the area's only temporary lake is located in a place called Las Playas, a flat area that floods occasionally and later dries and evaporates. In the glossary to his work, Bryan defines a playa as a Spanish word for a place in the lower part of a completely closed basin that is covered by water at irregular intervals. A playa usually consists of layers of clay and silt deposited in a temporary lake that may contain large quantities of soluble salts. The surface generally lacks vegetation. Playas are often called "dry lakes," "alkali flats," or "salinas."

Figure 17.44 The north-central sector of the Dead Sea, seen from the western scarp. Israel.

Figure 17.45 Closed saline depression showing old salt processing facilities in the foreground. Laguna La Playa. Central sector of the Ebro Depression. Bujaraloz (Province of Zaragoza) Spain.

Rosen (1994) defines a playa as a closed depression with a saline surface. Shaw and Thomas (1997) consider the terms *playa* and *pan* as synonyms that define how arid zone basins of variable sizes and origins are affected by flooding of ephemeral surface water of varying extent and periodicity. Other authors use the term *pan* only for small closed depressions. This word is often used in South Africa, Australia, and the United States.

Synonymous terms are *sebkha,* an Arabic word used in North Africa whereas the word *sabkha* is used in countries of the Arabian Peninsula (Fig. 17.46). In Persia, *kavir* is used for salt-rich depressions. *Chott* is an Arabic word used for a saline basin that is fed by groundwater; this word is used in the Saharan area (Chott El Jerid, Tunisia). *Takir,* a term from central Asia, is a depression that is primarily argillaceous; in the Andes of Argentina it is called a *barreal* (Fig. 17.47). In South American deserts, *salar* is used (Chilean altiplano of the Atacama); *salina* and *bolson* are used for depressions with clay and saline bottoms.

Glennie (1970) distinguishes between interior sabkhas and **coastal sabkhas** (Fig. 17.48), which resemble each other; the second are saline plains in arid zones located on the littoral and affected by the penetration of marine water that deposits tidal sediments (Warren, 2006). Because sabkhas are usually located in areas of high evaporation, they precipitate highly soluble salts. There are excellent examples in the Persian Gulf and in southern Tunisia.

Most closed depressions in deserts are small in size; their surfaces range from tens of square meters to almost 10,000 km^2 in the Salar of Uyuni (Bolivia) (Rettig *et al.*, 1980; Risacher and Fritz, 2000), which is located at an elevation of 3,653 m on the altiplano (Fig. 17.49). During the late Pleistocene the Salar of Uyuni covered an area larger than 43,000 km^2 and was called Lake Minchin (Blair, 1986). Altogether these lacustrine areas occupy less than 1% of the desert surface worldwide (Shaw and Thomas, 1997) although in some places in South Africa they may cover 20% of the desert (Goudie and Wells, 1995). The Texas Panhandle has 30,000 to 37,000 lacustrine basins (Osterkamp and Wood, 1987). Except for some lakes of tectonic origin, most basins in arid zones rarely exceed depths of 15 m during their temporary maximum levels (Mabbutt, 1977). Tectonic lakes are elongate and parallel to the dominant regional structure. Deflation, the removal of particles by the wind, can also enlarge the lakes (**oriented lakes**). Closed desert depressions are generally circular in shape. If water and sediment supplies are meager, the floors are irregular. Significant clastic deposition eliminates floor irregularities. The dynamics of lacustrine water, including waves and offshore currents, cause bars, spits, and barrier islands to form; an example is the Laguna de Gallocanta (Provinces of Zaragoza and Teruel, Spain) (Fig. 17.50) (Gracia, 1995).

The lacustrine habitat is ecologically limited to water bodies with depths greater than 2 m; water of shallower

Figure 17.46 Sabkha surrounded by dune fields. Al Liwa, Rub Al Khali desert. United Arab Emirates.

Figure 17.47 Barreal Blanco composed of silts with a high carbonate content that came from the surrounding Cambro-Ordovician calcareous formations. Precordillera of the Andes. Province of San Juan (Argentina).

Figure 17.48 Extensive halite accumulations in the central part of the littoral sabkha of El Melah. Zarzis (Tunisia).

Figure 17.49 Salar of Uyuni from volcanic outcrops on the Isla del Pescado (Bolivia). Photo M.C. Maldonado.

Figure 17.50 Sedimentary littoral landforms during the dry season (September 1996) in the Laguna de Gallocanta (Provinces of Zaragoza and Teruel). Central Iberian Cordillera (Spain). Photo B. Leranoz.

depth is considered palustrine habitat (Cowardin *et al.*, 1979; in Currey, 1994). Lacustrine environments may contain abundant plants from blue-green algae to reeds and also different types of fauna that tolerate saline conditions (Eugster and Hardie, 1978). In addition, the high salinity of some lakes may qualify them as mineral exploitation targets for commodities such as chlorides, nitrates, sodium carbonates, borax, lithium, and potassium. Because of their flat surfaces, they are often used as airstrips for sports and military activities. The littorals of these lakes were places of human habitation throughout prehistory especially during pluvial periods.

17.3.1 Origin of closed depressions in arid environments

Desert lakes can form in various ways. Hutchinson (1957) proposes two primary origins, structural and geomorphic; he sets up 13 large groups of processes that include 75 different types of closed depressions. Mabbutt (1977) proposes four primary origins: tectonic, erosional, and depositional basins and finally volcanic and meteorite craters. Each of them can be subdivided into various types. Shaw and Thomas (1997) generally adopt Mabbutt's classification and expand the number of types.

Tectonic basins are formed by graben and semi-graben generation and by large negative doming. Examples are the East African *rift valleys*, the Dead Sea, Lake Turkana or Rudolph (Kenya), and Natron Lake (Tanzania). Lake Victoria is related to broad arching, similar to Lake Chad in central Africa and Lake Eyre in Australia. In deserts of the western United States, block faulting is common; it creates a system of horsts and grabens (*Fig. 17.51*) similar to the Precordillera of the Andes in Argentina. Small closed depressions sometimes occur along strike-slip faults; they are supplied by groundwater.

Erosional basins in arid zones originate primarily by deflation; they are common in all the world's deserts. The largest known deflation basin is in the Qattara Depression (*Fig. 17.52*) in northwestern Egypt; it covers an area of 18,000 km^2 and extends to a depth of 134 meters below sea level. Deflation areas of this depression form a large dune system located southwest of the depression (Glennie, 1970). The size of this basin is rare but deflation basins are very common; they measure about a hundred square meters, as in the center of the Duero Basin (Spain) (Gutiérrez et al., 2005a). These depressions form where vegetation is absent.

The outcrop of particles too large to be transported by the wind puts an end to the formation of a deflation basin (Tricart, 1969). The same effect occurs in the presence of a thick salt layer. A shallow water table in a playa also limits the depth of deflation (Shaw and Thomas, 1997). Deflation can also create interdune *sebhkas* (*Fig. 17.46*). Hammadas, large flat areas that occur in the Sahara and India, have shallow closed depressions on their surfaces called *dayas*

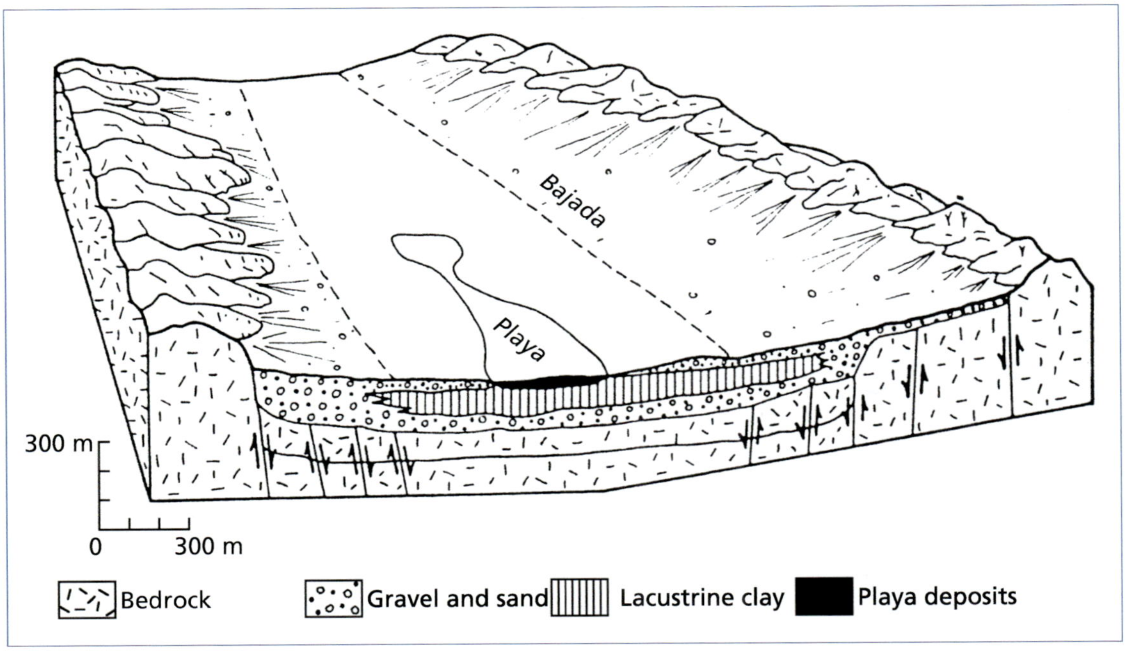

Figure 17.51 Block diagram of a tectonic depression with a playa, typical of the deserts of the southwestern United States. From Motts (1970), in Mabbutt (1977).

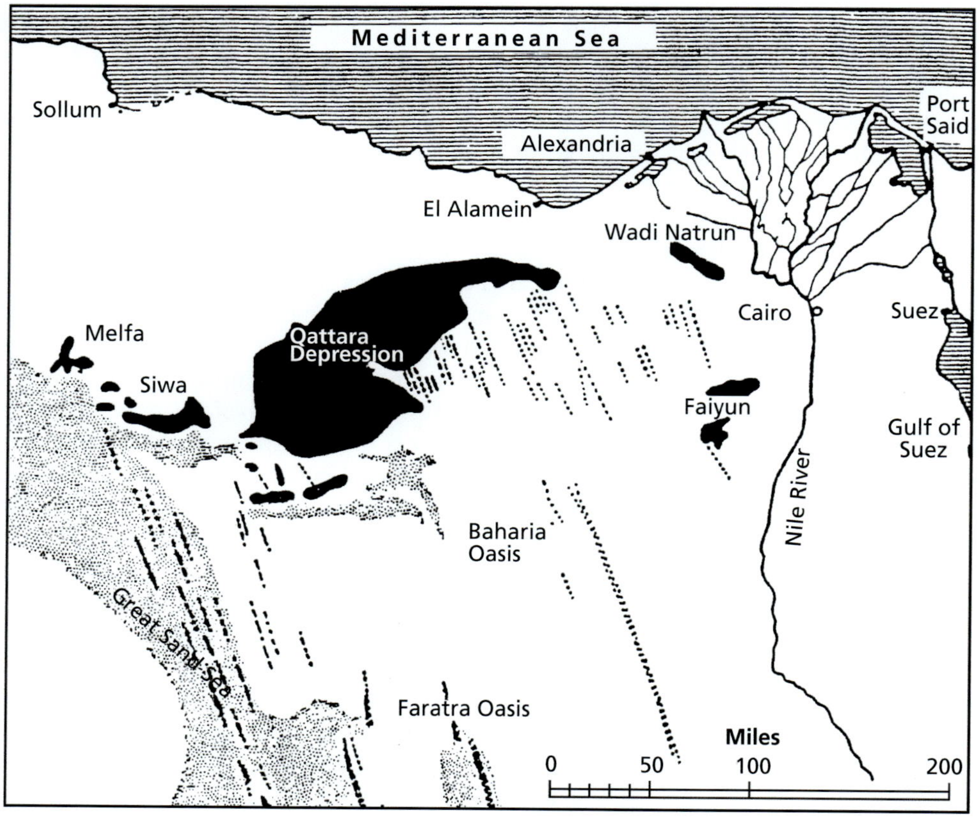

Figure 17.52 Map showing depressions, large sand accumulations, and aligned dunes in the Egyptian desert (Holmes, 1944).

(*Fig. 16.11*). Many geomorphologists think these *dayas* are produced by mechanical disaggregation or eolian activity (Tricart, 1969) although some authors think they result from the karstification of carbonates (see Shaw and Thomas, 1997). As deflation affects the base of closed depressions in arid regions, particles are deposited on the leeward margin, creating a half-moon shaped eolian structure (Hills, 1940); these are called **lunettes** in English (luneta in Spanish and bourrelet in French) (*Fig. 17.53*). These accumulations are formed by particles that range in size from sand to clay. Lunettes form in areas that receive between 100 and 700 mm of precipitation. They can be several kilometers long and can exceed 60 m in height.

In this section on erosional basins, we discuss closed depressions formed by surface and subsurface **karst dissolution,** especially in evaporites. An extensive field of karst depressions is present in the semi-arid zone of Bujaraloz in the center of the Ebro Depression; the depressions are sometimes covered by water (*Figs. 17.45* and *17.54*). In the Laguna de la Playa, the floor is about 4 km long. Dissolution occurs in a Miocene-age formation consisting of alternating limestones and gypsum. The depressions are elongated to the northwest, the direction of the predominant fracture system and the direction of the prevailing wind ("cierzo") (Quirantes, 1965; Pueyo, 1978–79; Sancho and Gutiérrez, 1993: Gutiérrez *et al.*, 2005b).

Piping in semi-arid regions can also generate closed depressions (*Fig. 16.57*). In Lupiñén (Province of Huesca) in the central Ebro Depression, piping occurs on loose debris slopes of variable thickness (up to 15 m) that cover Miocene deposits; a conduit network forms at the contact between the two lithologies due to the high porosity of the slope deposit. Enlargement of the conduits decreases the mechanical resistance of their roofs and triggers collapses. At the surface, they look like pseudodolines up to 35 m wide and 20 m deep with funnel and well morphologies (*Fig. 17.55*) (Gutiérrez *et al.*, 1988).

Animal activity also causes erosion. Large herbivores promote deflationary processes when they consume or trample vegetation (Goudie, 1984). With their feet and trunks, elephants can excavate closed depressions up to 25 m in diameter and 1.5 m deep (Thomas, 1988).

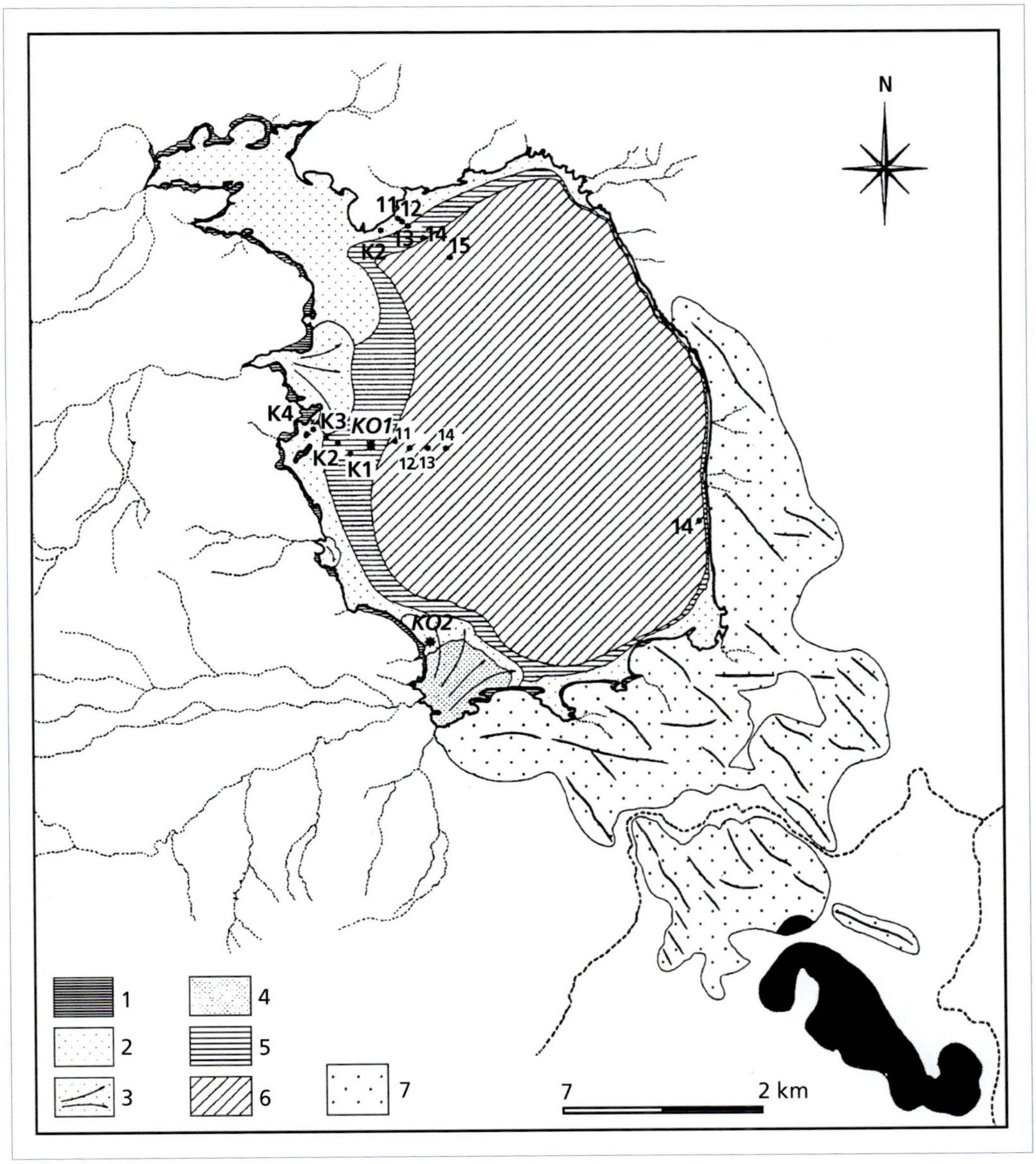

Figure 17.53 The El Kourzia sebkha (Tunisia) and its lunette. 1: Step that encircles the sebkha margin; 2: Zone of marly silts; 3: Alluvial cones of the principal tributaries; 4: Zone of micropolyhedra formation; 5: Zone of marly silts with fine threads of gypsum; 6: Zone of black muds with summer halite crust; 7: lunette of the El Kourzia Sebkha. In black, the old lunette (Perthuisot and Jauzein, 1975).

Sedimentary basins can form when drainage systems are abandoned or replaced by other fluvial networks. Maizels (1987) describes a complex series of paleochannels in Oman that have formed aligned closed depressions over a long period of the Plio-Pleistocene, a time of alternating humid and dry periods. Lacustrine depressions also form in the flood plain. Dunes can also impound streams, as they have south and east of Timbuktu (Mali). Lava flows can also cause the same effects. During coastal evolution, playas and bars form that lead to the formation of littoral sabkhas (Glennie, 1970). Some "albuferas" (coastal lagoons), such as those of Pego and Elche (Spain), form in this way (Sanjaume, 1985).

Volcanic activity also builds **craters** that partially fill with water (*Fig. 4.22*) such as the Zuni Salt Lake (New Mexico, United States). Meteorite impact in arid zones creates **impact craters**; the best-known example is the Meteor Crater (Arizona, United States), which is dry today (*Fig. 2.44*).

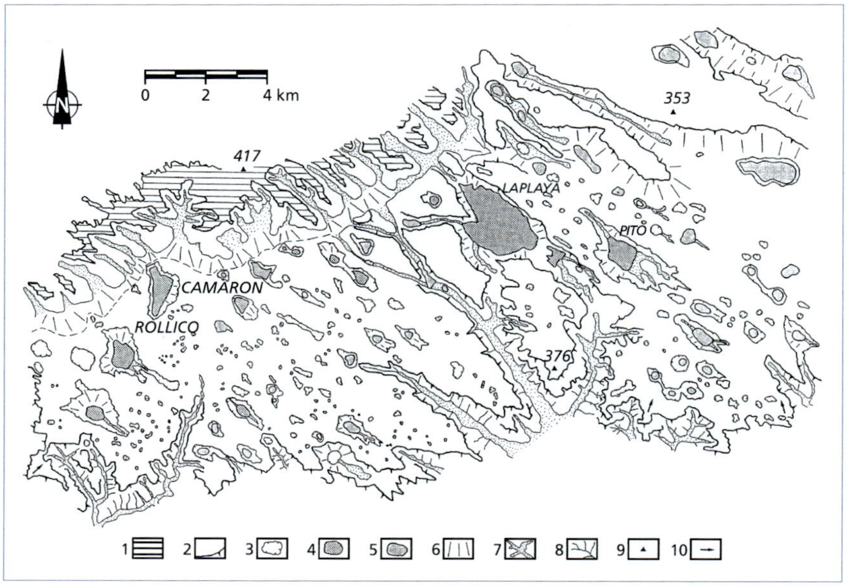

Figure 17.54 Geomorphic map of salt lakes in the Bujalaroz area. 1: Structural landforms on gypsum. 2: Platforms and cuestas on limestones. 3: Karst depressions and dolines. 4: Salt lakes 5: Areas of temporary flooding. 6: Recent pediments and slope deposits. 7: Flat-floored valleys. 8: Gullies. 9: Topographic elevation. 10: Dip (Sancho and Gutiérrez, 1993).

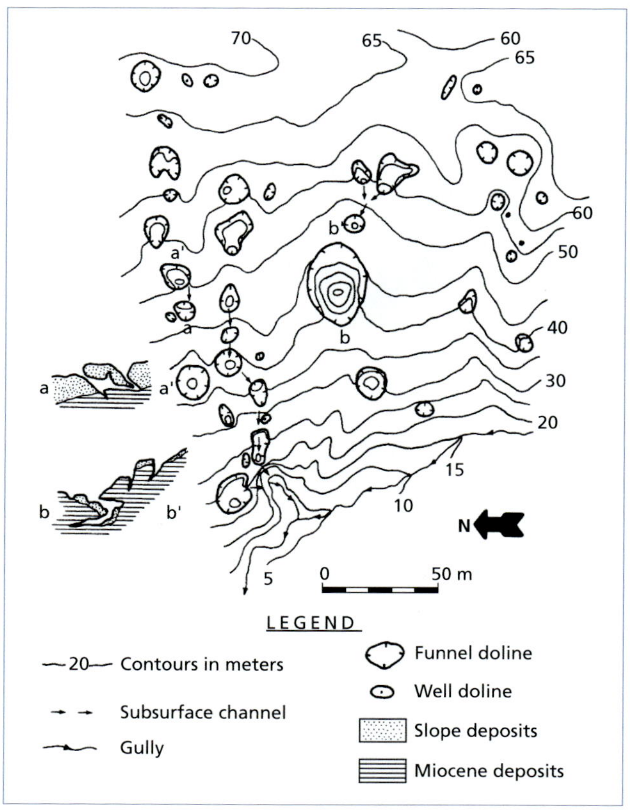

Figure 17.55 Map of pseudodolines formed by piping on slope deposits (Gutiérrez *et al.*, 1988).

17.3.2 General hydrology of arid zones

Lake level fluctuations depend on water supply and losses. Water is supplied by surface runoff, groundwater contributions, and to a lesser extent from direct precipitation on the lake. Water losses are due to evaporation, infiltration, and outflow through an overflow channel if the lacustrine basin fills up. When water supply and water loss are no longer in equilibrium, lake margins expand and contract. In arid regions these modifications occur frequently except in deep lakes such as those of the East Africa *rift valley*. In desert climates sparse precipitation and intense evaporation cause lakes to dry up.

Lake capacity is controlled by the height difference between the watershed boundary and the lake bottom and by the bathymetry of the lakebed. Over time, watershed boundaries gradually erode and sedimentation covers the irregularities of the bed, which decreases the capacity of the lacustrine basin. The height and width can be modified by eolian deflation, which tends to deepen and widen the closed depression. In some regions tectonic activity is common, which also affects lake capacity. Volcanism can also cause significant modifications. During active periods the volcanoes in the East Africa *rift valley* can alter the morphology of lakes located within these large structures.

In some lacustrine basins, connections between different sub-basins can mix water, solutes, and suspended sediments. Channels can transfer these constituents from one lake to another or by overflow of certain morphological subdivisions within a lake (Currey, 1994).

Evaporation lowers desert lake levels; it can easily remove a 2–3 m column of water every year. In the group of diffluent lakes of the Niger near Timbuktu (Mali), evaporation of 2.5 to 3 m was recorded over a period of eight months (Tricart, 1969). Irregular precipitation can cause significant flooding in arid zones; it fills the lake and can remain for several years. These floods are characteristic of large basins; small ones generally dry up in less than a year.

Lake Eyre in the Simpson Desert (Australia) covers a surface area of 8,430 km^2, and is 144 km long and 77 km wide (Warren, 2006). The watershed basin is an area one hundred times larger than the lake surface. The maximum depth of the lake is 15.2 m, which is the lowest point in Australia (Warren, 2006). Annual precipitation ranges from 100 to 500 mm (Bonython and Mason, 1953). Around the middle of the past century, the lake flooded (Fig. 17.56A); rains of 200–400 mm in February and March of 1949 flooded the southern part of the lake. Additional precipitation of about 500 mm fell between January and April of 1950, and later rains filled the lake in September of that year. The desiccation stage began the following month, with increased salinity and a drop in water level, area, and volume (Fig. 17.56B). By 1953 the lake was completely dry. The average salt content

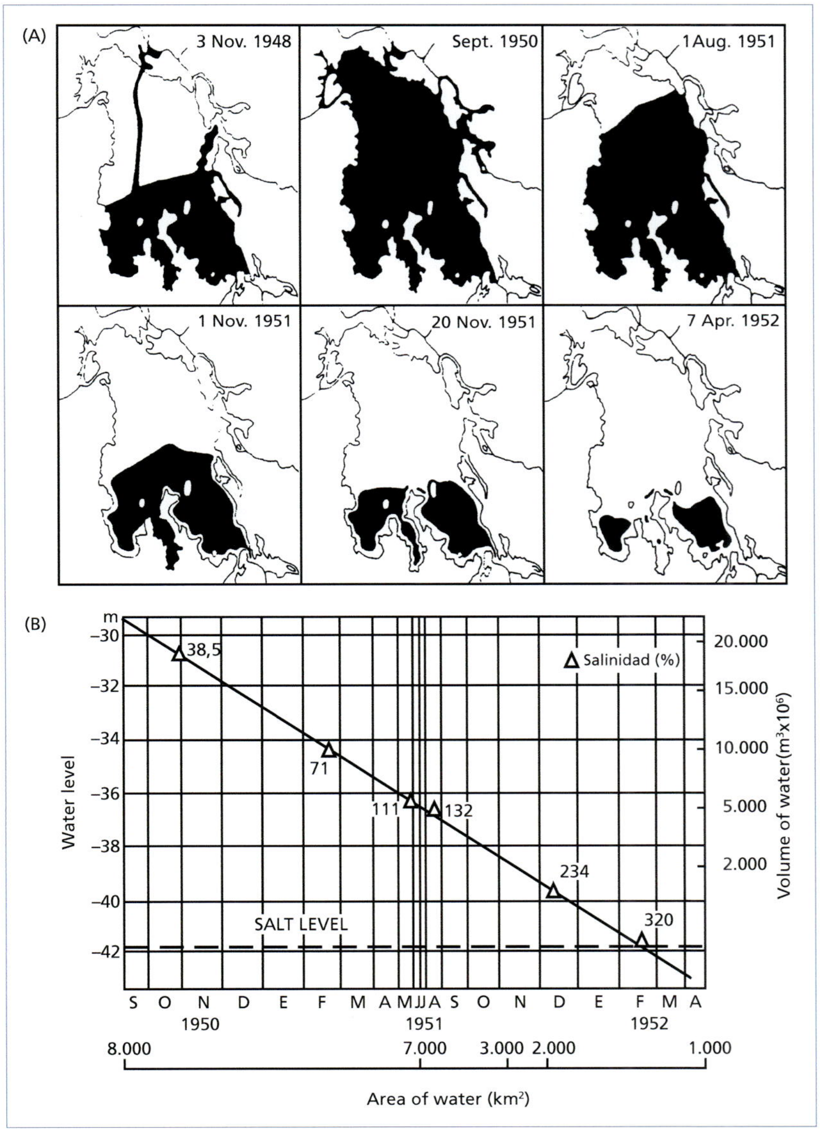

Figure 17.56 Stages of water filling and desiccation in Lake Eyre (Australia). (A) Extent of water at the surface; (B) relationship between level, volume, area, and water salinity (Bonython and Mason, 1953).

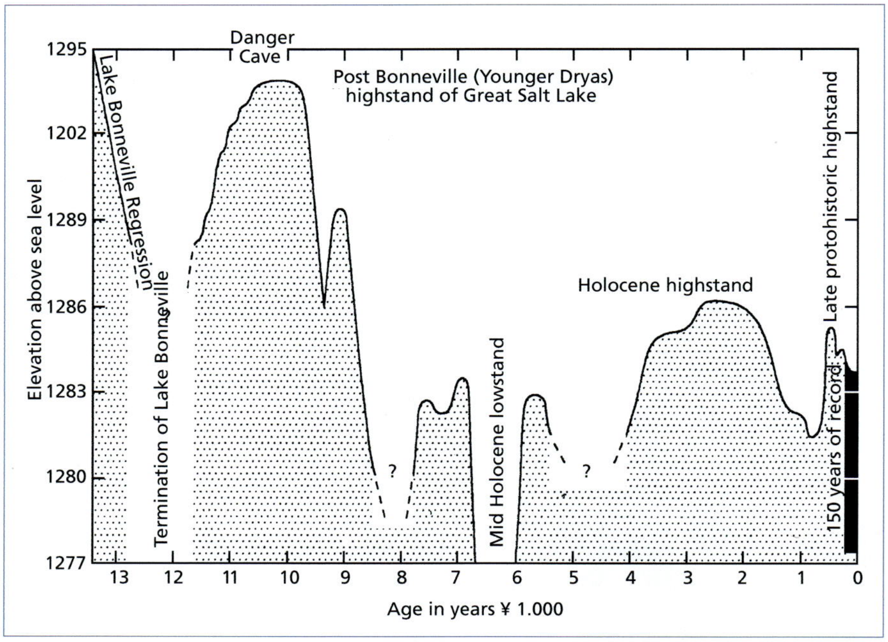

Figure 17.57 Hydrograph of the Great Salt Lake Utah basin for the last 13,000 years (Murchinson, 1989; in Currey, 1994a). The earliest records of human activity, between 11 and 10 Ka, were found in Danger Cave.

of the lake is NaCl (96.14%) and $CaSO_4$ (2.38%). Lake Eyre fills completely about twelve times per century.

In terms of geomorphology, lacustrine processes occur in shallow water where turbulence mixes the water column; these processes generate low energy waves that move toward the coast and along the shoreline where they form beaches and other landforms (Gracia, 1995).

Detailed work on lacustrine margins, including the study of stratigraphic records, have made it possible to determine the rise and fall of Pleistocene and Holocene lake levels, and in this way establish a hydrograph of the lake for these periods (*Fig. 17.57*).

17.3.3 Sedimentation in desert lakes

The nature of sediments deposited in closed depressions in arid zones depends on the hydrologic regime; flooding in dry playas deposits primarily silt and clay, although it can sometimes deposit small channels of fine sand during periods of higher energy. This type of sedimentation occurs primarily on the basin edge where alluvial fans may develop. Water that reaches the central part of the basin is intensely evaporated, which increases salinity and salt precipitation as the salts reach saturation. In humid playas, saline groundwater predominates and deposits range from salt-impregnated sediments to evaporites (Mabbutt, 1977).

In mountain deserts and depressions, the playa landform (*Fig. 17.51*) is characterized by a mountain front, which is often an uplifted block. Alluvial fans develop in the foothills; the proximal facies are composed of sand and gravels that grade into silty and clay-rich plains, often associated with salts in the distal zone. Finally, in topographic lows, a shallow lake develops where evaporites may form (Eugster and Hardie, 1975; Hardie *et al.*, 1978; Anadón, 1989). In basins near tectonically active fronts, these deposits can reach considerable thicknesses. Deserts of the western United States and the Precordillera of the Argentine Andes provide numerous examples.

Ephemeral lakes in shield and platform deserts have very low gradients and contain deposits of sand, silt, and clay with thin evaporites. Bedrock may occasionally crop out. Because of their low relief, these lakes can expand significantly during recharge periods. Eolian activity is the dominant process during dry periods; it can build dune systems in different parts of large closed depressions, as in Lake Chad in central Africa (*Fig. 21.26*).

17.3.3.1 Clay-rich closed depressions

These basins are composed of sand, silt, and clay flats that occupy lake bottoms or form the margins around a central nucleus of evaporites (*Fig. 17.51*). Finer particles remain in suspension and may be deposited slowly whereas channels that travel toward more internal zones transport sands during times of abundant surface water supply. The dissolved load, which is usually minor, mixes with the fine fraction or is transported toward the central salt nucleus. Desiccation of these clay-rich muds generates polygonal systems (*Fig. 16.19*) with concave cells if they lack salts or convex cells when salts are present.

Osterkamp and Wood (1987) describe an interesting example of the formation of small clay basins in the High Plains of Texas and New Mexico. The depressions initially form by deflation and then partially fill with water. The water infiltrates and gradually dissolves the underlying caliches; this causes surface subsidence and enlargement and deepening of the depressions. The depressions subsequently fill with non-salt clay materials.

17.3.3.2 Salt lakes

Salt lakes are natural laboratories for the study of active processes and their evolution. They are also useful environments for interpreting the sedimentary fossil record (Reeves, 1968). Salt lakes have been intensely studied by geologists over the last two decades for their economic importance because they are associated with hydrocarbons and evaporites. In addition, salt lake deposits contain a detailed record of climatic, hydrologic, sedimentological, geochemical, and biochemical conditions (Eugster and Kelts, 1983).

One problem that arises in the study of salt lakes is that of **salt provenance.** In the Senegal Delta, wave-generated marine spray is a source of salt for some *sebkhas*. The disseminated water evaporates and produces fine salt crystals that float in suspension and form a whitish mist. This mist may be pushed toward the land where it precipitates by rainfall or as dew as a result of temperature changes related to nightly increases of relative humidity. Under these circumstances, water adheres to these hygroscopic crystals, enlarges them, and causes them to fall. During intense precipitation, dispersed salt present on the land surface dissolves and is transported to the interior of the *sebkhas*. The water finally evaporates and salt precipitates (Tricart, 1956b). This source of salt is quite significant in arid regions especially where large waves hit the shores. The amount of salt from the sea decreases at a distance of 100 km from the coast (Mabbutt, 1977).

Chemical weathering reactions also produce salt. These reactions dissolve salt-rich geologic formations such as those of Triassic and Tertiary age in Spain. This salt recycling has been common throughout geologic history; it is possible to determine the source of the salt from the isotopic composition of its constituents (Chivas, 2007). As Eugster and Hardie (1978) point out, the final composition of water involved in chemical weathering depends on the minerals present in the rocks, their relative abundance, and the degree of weathering, weathering reactions, and drainage conditions.

Another source of salts is saline crusts present in playas and coastal *sabkhas*; salt becomes available when deflation affects it (Gutiérrez *et al.*, 2002). Salt particles can travel variable distances when pushed by the wind.

Water that covers topographic depressions in desert lakes contains a variable salt content and concentration. The minerals that precipitate are a function of the composition of the water. Precipitation requires a concentrated solution, loss of gases such as CO_2, water mixture, and temperature changes. Depending on the brine type, these processes create different minerals (*Table 17.1*) (Eugster and Hardie, 1978).

The first minerals to form are carbonates, which have an increasing cationic zonation from Ca to Mg. Sulfates form

Table 17.1 **PRINCIPAL SALINE MINERALS IN VARIOUS BRINE TYPES (EUGSTER AND HARDIE, 1978)**

Brine types	Saline Minerals	
Ca-Mg-Na-(K)-Cl	Antarcticite	$CaCl_2$-$6H_2O$
	Bischofite	$MgCl_2$-$6H_2O$
	Carnalite	KCl-$MgCl_2$-$6H_2O$
	Halite	$NaCl$
	Silvite	KCl
	Tachyhydrite	$CaCl_2$-$2MgCl_2$-$12H_2O$
Na-(Ca)-SO_4-Cl	Gypsum	$CaSO_4$-$2H_2O$
	Glauberite	$CaSO_4$-Na_2SO_4
	Halite	$NaCl$
	Mirabilite	Na_2SO_4-$10H_2O$
	Thenardite	Na_2SO_4
Mg-Na-(Ca)-SO_4-Cl	Bischofite	$MgCl_2$-$6H_2O$
	Bloedite	Na_2SO_4-$MgSO_4$-$4H_2O$
	Epsomite	$MgSO_4$-$7H_2O$
	Glauberite	$CaSO_4$-Na_2SO_4
	Gypsum	$CaSO_4$-$2H_2O$
	Halite	$NaCl$
	Hexahydrite	$MgSO_4$-$6H_2O$
	Kieserite	$MgSO_4$-H_2O
	Mirabilite	Na_2SO_4-$10H_2O$
	Thenardite	Na_2SO_4
Na-CO_3-Cl	Halite	$NaCl$
	Nahcolite	$NaHCO_3$
	Natron	Na_2CO_3-$10H_2O$
	Thermonatrite	Na_2CO_3-H_2O
	Trona	$NaHCO_3$-Na_2CO_3-$2H_2O$
Na-CO_3-SO_4-Cl	Burkeite	$NaCO_3$-$2Na_2SO_4$
	Halite	$NaCl$
	Mirabilite	Na_2SO_4-$10H_2O$
	Nahcolite	$NaHCO_3$
	Natron	Na_2CO_3-$10H_2O$
	Thenardite	Na_2SO_4
	Thermonatrite	Na_2CO_3-H_2O

after carbonates; this involves an increase in the Mg/Ca ratio, which may precipitate proto-dolomite or magnesite. After the precipitation of carbonates and sulfates, the brines are considerably more concentrated by evaporation until they reach saturation for the most soluble salts, which are chlorides. As a consequence of the different solubilities of brine components, precipitation within a salt lake follows a zonation (Fig. 17.58) (Hunt *et al.*, 1966); calcium and magnesium carbonates occupy the outermost fringe, sulfates are intermediate, and chlorides fill the central nucleus (Fig. 5.14).

These salt deposits alternate with fine-grained sediments such as fluvial and eolian sands, silts, clays, and levels rich in organic material (kerogens) (Fig. 17.59). This shows the complex history of the sedimentary record that makes it possible to analyze stages of flooding and desiccation and spatial and temporal variations of sedimentary processes (fluvial, eolian).

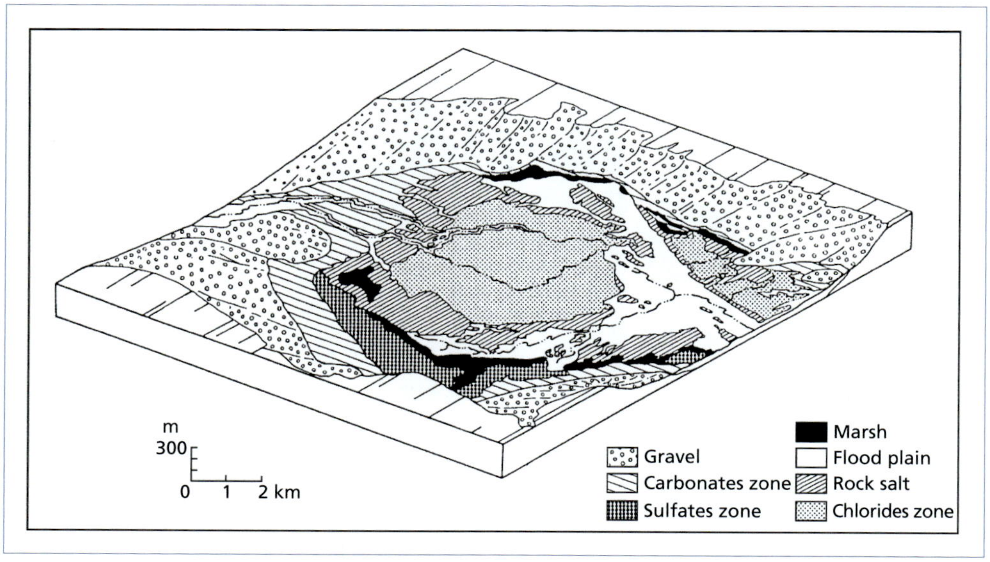

Figure 17.58 Salt zonation on a playa surface. Cottonball Basin, Death Valley (Hunt *et al.*, 1966).

Figure 17.59 Excavation showing a surficial halite layer on top of organic-rich muds. Sabkha el Melah. Zarzis (Tunisia).

Continuous changes in water content make it possible to group playa water into a saline cycle characterized by stages of inundation, concentration by evaporation, and desiccation (Bryant *et al.*, 1994). In addition, playa flooding causes intensive reworking of the surface; salts re-dissolve and fine sediments can move.

Salt lakes have highly varied **surface shapes** that are primarily controlled by groundwater composition and depth. Sedimentation and deflation processes, which are influenced by the water table, also work together to configure this feature (Neal, 1965).

Small mounds develop on the sulfate margins of salt lakes. Called **domes** or **tumuli** (*Figs. 17.60*), they also form in other environments such as flat zones on top of Tertiary gypsum-bearing formations in the Ebro Depression (Artieda, 1993) and the Sorbas region (Almeria) (Pulido-Bosch, 1986) (*Fig. 17.61*). These mounds have dome shapes, measure about 1 m in diameter, and are less than 0.5 m high. The central parts of the dome shapes appear to have collapsed, which reveals the thickness of the layer (about 20 cm) and a void in the center (*Fig. 17.60*). Numerous hypotheses have been proposed for the origin of these landforms. Artieda (2004) attributes mound formation to *in-situ* dissolution and precipitation of surficial gypsum. Secondary gypsum crystals fill the pores; this process increases the volume to the point that crystallization pressures cause the gypsum to expand laterally, which creates the dome morphology. Calaforra (1998) proposes three hypotheses: the first involves compressive tectonic forces; the second is based on volume increase related to the transition from anhydrite to gypsum; and the third proposes that crystallization of the gypsum matrix in the interstices of the macrocrystalline gypsum generates tensions that increase volume and cause uplift. Artieda's hypothesis and Calaforra's third hypothesis share some similarities.

Spring mounds also form near springs or artesian flows (Reeves, 1968; Mabbutt, 1977; Roberts and Mitchell, 1987). They range in size up to 100 m in diameter; they may be related to deep faults. The doming is due to sedimentation around the spring; it may be mechanical or chemical. In the latter case, the resulting hills are composed of travertine of inorganic or organic origin (pinnacles of algae up to 10 m tall) (Neal, 1965). In addition, closed depressions also occur in playas; they form as the underlying salt is dissolved by escaping air or methane gas trapped within the accumulation (Reeves, 1964).

Millington *et al.* (1995) describe the presence of **salt ramps** composed of salt crystals on the crust of Tunisian playas. They occur in groups and have a half-moon shape oriented in the direction of the dominant wind, which concentrates the salt particles into arcuate bands.

In many playas, desiccation after flooding produces a thin smooth salt film that persists as long as it remains above the capillary level or stays wet (Eugster and Kelts,

Figure 17.60 Tumulus formed on the sulfate border of a saline lake. The top has collapsed and the interior is empty. Sabkha el Melah. Zarzis (Tunisia).

Figure 17.61 Gypsum tumulus in the Messinian, partially collapsed. Sorbas (Province of Almería, Spain).

Figure 17.62 Salt blisters composed of fine laminae of saline mud in an interdunal sabkha. Al Liwa. United Arab Emirates.

Figure 17.63 System of saline polygons with raised borders. Pastos Grandes salt flat. Puna, Province of Salta (Argentina). Photo C. Sancho.

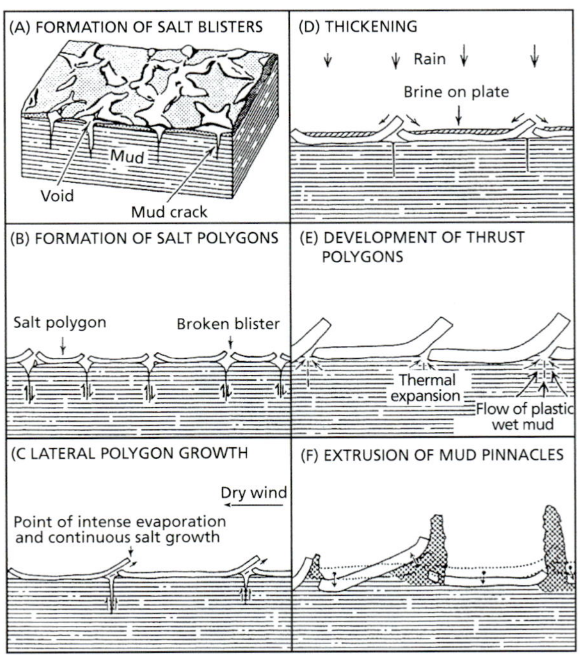

Figure 17.64 Idealized stages in the formation of a saline playa. Gran Kavir (Iran) (Krinsley, 1970).

1983). Desiccation forms **salt blisters** (*Fig. 17.62*) and **salt polygons** up to 10 m in diameter (*Fig. 17.63*). These are the most common morphologies on the surface of salt playas. Cracks form along polygon edges as a result of volume changes caused by wetting and desiccation (Mabbutt, 1977). In his study of the Gran Kavir of Iran, Krinsley (1970) establishes a set of idealized stages in the evolution of the surface morphology of a salt lake (*Fig. 17.64*). The margins of the polygons are points of intense evaporation and precipitation, especially those oriented toward the dominant dry wind. Growth of the salt plate margin causes the saline crust surfaces to override the neighboring polygon (Reeves, 1968) (*Fig. 17.65*). Concave polygons fill with brine that thickens the polygon as the brine precipitates. Extrusion of underlying saline mud can also occur; this process forms pinnacles of dark mud by thermal expansion below the salt crust or by loading exerted by thick saline plates (up to 30 cm thick) above the plastic mud. The morphology produced by salt ramps and mud pinnacles creates an irregular and impassable surface (*Fig. 17.66*).

Patterned ground similar to that of periglacial areas can form in areas of saline sedimentation. This is a clear example of the convergence of landforms. Hunt and Washburn (1960) describe them in Death Valley (California) in salt

Figure 17.65 Plates of halite thrusting over the margins of neighboring polygons. The saline crust is mostly covered with eolian sand. Mattí Sabkha (United Arab Emirates).

Figure 17.66 Saline playa with a very irregular surface due to the formation of thrusting ramps and extrusive pinnacles. Atacama salt flat (Chile). Photo J.L. Peña.

playas near alluvial fans. Polygon, network, stepped, and striated soil morphologies have been recorded. Sorted and unsorted networks have been found on the carbonate border of the playa where debris penetrates into cracks, causing sorting. Polygons develop in the presence of a gypsum or salt layer (*Fig. 17.67*) and on top of detrital material, which fills the cracks when the salts are dry. This filling causes lateral expansion of the polygons when they get wet; this expansion uplifts the polygonal cell.

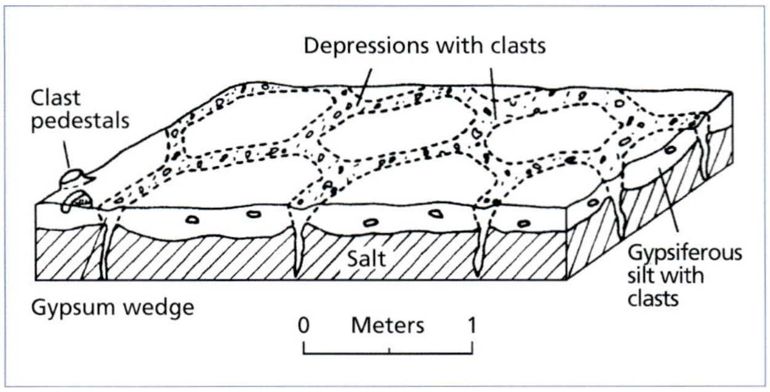

Figure 17.67 Sorted polygons related to cracking of an underlying salt layer (Hunt and Washburn, 1960).

17.3.4 Littoral sabhkas

These supratidal flats have been intensely studied because of their hydrocarbon potential (Chivas, 2007); various processes are involved in their formation. In the last two decades, the littoral sabkhas of the Trucial coast of the Persian Gulf, especially in Abu Dhabi, have been intensely studied (Orti, 1989). Recent and present-day sedimentation taking place in these littoral depressions is being used as a model for interpreting old geologic deposits such as the European Permo-Triassic.

Littoral sabkhas are composed of fragments of shells, ooids, algal mats, stromatolites (*Fig. 17.68*), evaporites, and organic matter (*Fig. 17.69*). Some large littoral sabkhas grade inland (Glennie *et al.*, 1994) to interior sabkhas (Sabkha Matti, United Arab Emirates) where fluvial sedimentation predominates as alluvial fans. These sabkhas are affected by eolian deflation that creates small nebkhas, microyardangs, and dunes. The water that supplies littoral sabkhas comes from spring tides or storm waves although continental water may also contribute. Seaward intratidal plains, lagoons, bars, and islands develop (Cooke *et al*, 1993). The evaporitic surface of the sabkha hosts landforms similar to those present on playas.

Figure 17.68 Dome shaped stromatolites. Sabkha el Melah. Zarzis (Tunisia).

Figure 17.69 Silty-sandy plain with organic content, located near the sea. Mattí Sabkha (United Arab Emirates).

17.4 Applied geomorphology in arid zones

17.4.1 Introduction

Applied geomorphology in desert areas is essentially similar to that of other morphoclimatic zones although certain geomorphic processes common in these areas give rise to specific peculiarities. The lack of water, the scarcity of plant cover, and thin soils cause surface materials to lack effective protection from atmospheric conditions (Tricart, 1969). As a result, the dominant processes in arid regions are wetting and drying, salt weathering, water erosion, channel changes and flooding, and eolian activity.

Practical studies in arid regions are important because of the large expanse of these environments; together with cold regions, they are the two largest morphogenetic systems in the world. In addition, it is difficult to generalize about applied geomorphology in these areas because of the variability of climatic, geologic, and tectonic environments. Furthermore, these regions are generally underdeveloped and relatively unpopulated areas where basic data such as climate and hydrologic records are rare and only locally disseminated. In many cases, the only studies available are isolated and thematic (Gutiérrez and Sancho, 1993).

In 1979 the population living in arid zones (651 million people) was 15% of total global population. Between 1960 and 1974, it increased by 63.5 percent, and reached 1.2 billion people in 2000. This population is now concentrated in 355 arid-area cities that have more than 100,000 inhabitants, mostly in Asia. Numerous geomorphic and environmental problems result from the expansion of cities in arid environments (Cooke *et al.*, 1982; Goudie, 1990a; O'Hara, 1997).

Most landforms in dry regions formed under past climatic circumstances that are very different from current conditions; for this reason detailed studies must be done to distinguish relict forms from landforms generated by present day processes (Gutiérrez, 1986).

Given these considerations it is important to determine the foundation and goal of practical geomorphic studies in desert regions, as clearly defined by Cooke (1977): "Applied geomorphology in arid zones is primarily the recognition and evaluation of landforms, materials, and surface processes, and the management, monitoring, and prediction of landform changes and processes. It is related to practical problems in which geomorphic information usually is only one part of the set of knowledge that is needed. For this reason, an applied geomorphologist in desert regions must be capable of understanding and working with specialists in other disciplines and transmitting his ideas adequately to persons who are not geomorphologists".

17.4.2 Volume changes

Soils change in volume in the presence of swelling clays and high sodium concentrations in the cation exchange capacity. High concentrations of sodium disperse the clays; the forces between particles manifest themselves in the soil as a surficial system of cracks with domed cells. Tests on various samples of Miocene-age mudstones from the Ebro Depression (Spain), which lack swelling clays, show that muds have a swelling capacity of approximately 12% (Gutiérrez *et al.*, 1995).

Volumetric expansion is much higher when significant percentages of swelling clays (smectite, vermiculite, and halloysite) are present; these are also called expansive clays (Ayala, 2002b). This volume-change process is much more common in low flat areas of semi-arid deserts (Cooke *et al.*, 1993). Variations of soil moisture content also cause expansion and contraction of the soil and can generate pressures up to 10 kp/cm^2 (*Fig. 17.70*) (Salinas, 1988). Swelling and retraction form gilgai soils, which are characterized by smooth ground undulations that cover broad semi-arid areas. In this micro-topography, only a small percentage of ground is suitable for cultivation and as a result, these potentially fertile soils must be limited to use as pasture. If machinery is used to level the micro-topography, the undulating topography regenerates within three years. Vertical push movement also lifts stakes that are placed in the ground (Harris, 1968).

In the United States, more than 250,000 houses are built every year on expansive soils and about 10% of them suffer significant damage (Jones and Holtz, 1973). In Spain, 32% of geological formations contain expansive clays and 67% of the territory has climates that can cause swelling (of up to 10%) and contraction (Salinas, 1988). A 3% volume increase can cause damage; therefore, buildings constructed on these soils require special foundations. To build a suitable foundation, it is important to know the depth to which the expansive soil extends. Geotechnical engineers call this zone of volume change the "active layer." According to Salinas (1988), volume changes cause slope deterioration (hillsides and embankments), pipe rupture, destruction of walls, pavement deformation, rupture of curbs and pavements, deformation of foundations, and cracking of structures (*Fig. 17.70*). Rupture of drains may exacerbate some damage. In Chabrillan (southern France) the French high-speed train (TGV) was affected by serious problems related to swelling soils (Tang, *et al.*, 2009).

Another less frequent applied aspect is related to the presence of **large desiccation fissures** in playas of old desert lakes. Dangerous cracks can form quickly and cause traffic problems and serious damage to the infrastructure of linear projects (Cooke *et al.*, 1993).

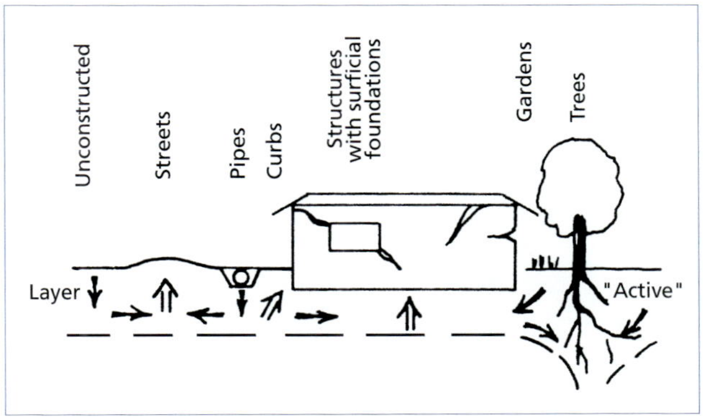

Figure 17.70 Problems caused by the presence of expanding clays. The black arrows point to pathways that transfer humidity and the open arrows show possible damage (Salinas, 1988).

17.4.3 Lacustrine systems

Closed depressions or lakes that lack drainage (Langbein, 1961) are relatively common in arid areas. They are shallow and ephemeral and often occur in groups. Because evaporation exceeds water input, the lakes are dry and saline although the water level may fluctuate as a function of the high variability of recharge.

These arid region lakes may overflow if water input is excessive. In September and October of 1969, rain fell over most of Tunisia for 38 days. Rivers overflowed their banks; lakes filled and spilled their excess water into overflow channels that also caused flooding. The flooding drowned 542 people and destroyed the homes of hundreds of thousands of people; damage to structures and agriculture was significant.

The salinity of these desert lakes may be related to the underground flow of water toward the lake, which occurs in the closed depressions of Bujalaroz (Province of Zaragoza, Spain). The continuous contribution of water provides salt to the system; the salt can be removed only by eolian deflation. These processes make it inadvisable to conduct irrigated agriculture in these areas because the water input from irrigation devices raises the lake water level; thus the lakes become permanent, which affects fauna and flora. The installation of new irrigation systems will cause an environmental change in the region, which may be significant (Sanchez *et al.*, 1993).

Human activities in lakes in arid zones may be highly variable; the consequences often include gradual eutrophication and degradation of lakes (Goudie, 1984). In semi-arid environments, human activity may involve the partial draining of lakes for crop production. This practice is common in small lakes. However, in the Caspian Sea, the largest lake in the world, human modifications have caused a drop in water

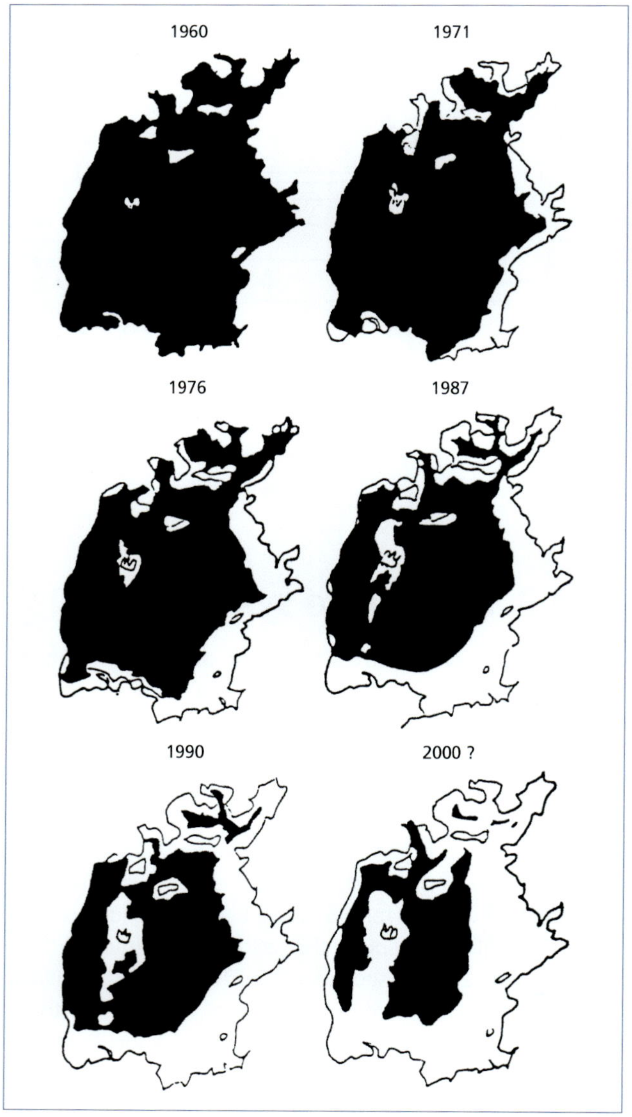

Figure 17.71 Evolution of the Aral Sea during the period 1960–2000 (Letolle and Mainguet, 1993).

level of more than 3 m since 1929 (Goudie, 1992). In Israel, the largest hydrological project is the National Water Carrier, which transports water from the Sea of Galilee along the entire Mediterranean coast, mostly for irrigation purposes (Beaumont, 1993). Salinity of the Dead Sea has increased as a result of decreased fresh water supply from the Jordan River, which originates north of the Sea of Galilee. The increased salinity of the Dead Sea has killed off its animal and plant life although it has promoted salt exploitation in the extreme southern part of the lake.

In the last few decades, the Aral Sea, the fourth largest lake in the world, has experienced one of the greatest ecological disasters ever recorded (Micklin, 1988). The Aral Sea forms part of a NNE-SSW-oriented syneclise (according to Russian authors) that originated at the end of the Pliocene. This arid zone is supplied primarily by the Amu Darya and Sir Darya rivers, which come from the mountains of the Pamir and Tian Shan, respectively. Because of the construction of a gigantic irrigation project for cotton and rice, the perimeter of the Aral Sea has shrunken significantly since 1960 (Létolle and Mainguet, 1993, Mainguet and Létolle, 1994) (*Fig. 17.71*). The sea was originally 16 m deep on average and had a maximum depth of 68 m near the western shore (Cunningham, 1999a). *Table 17.2* shows the principal hydrologic data recorded between 1960 and 1990.

The retreat of the water has caused ports to dry up. Fishing has been reduced from 48,000 metric tons/year in 1957 to zero in 1980. Fishing villages are now deserted. Pesticides have polluted the water and its salinity has increased significantly. Water well levels have decreased 10 m and wells are contaminated. Flora and fauna have been degraded (Létolle and Mainguet, 1993). In addition, a large percentage of irrigated soils have become saline and have been abandoned (Beaumont, 1993). Furthermore, the quality of water available for human consumption is very poor. All these circumstances affect a large portion of the population of about 35 million who suffer from multiple illnesses. The World Bank and the

Table 17.2 **HYDROLOGY OF THE ARAL SEA BETWEEN 1960 AND 1990 (LETOLLE AND MAINGUET, 1993)**

Year	River discharge (km^3)	Water level (m)	Surface (km^2)	Volume (km^3)	Mineralization (g/l)	Fishing (metric tons)
1960	40	53.5	67,900	1090	10	43,430
1965	31	52.5	63,900	1030	10.5	31,040
1970	33	51.6	60,400	970	11.1	17, 460
1975	11	49.4	57,200	840	13.7	2,940
1980	0.5	46.2	52,500	670	16.5	0
1985	0	42	44,200	470	23.5	0
1990	0	39	41,000	330	26.5	0

United Nations Program for the Environment established a grant to alleviate the environmental problems of this region. In 2005, the Republic of Kazakhstan tried to slow down the ecological disaster by constructing a dam in the northern part of the Aral Sea; the dam helped the northern zone but left the southern zone far from any salvation.

17.4.4 Human induced subsidence

Humans cause subsidence in multiple ways: extraction of fluids (water, petroleum, geothermal fluids), hydrocompaction, compaction and lateral expansion by loading, peat drainage, desiccation of clays, and underground mining (Costa and Baker, 1981; Waltham, 1989). Subsidence commonly occurs in arid zones in response to groundwater exploitation. The lack of water in arid areas causes humans to extract it from aquifers for agricultural purposes. Abusive extraction causes induced subsidence due to decreased hydrostatic pressure in the reservoir rock. This subsidence is common in Tertiary and Quaternary unconsolidated detrital deposits, which are formations usually composed of alternating sands and clays. When water is extracted from sand, pore pressure decreases and sand grains have to support the overlying load. Compression of sand is minimal whereas clays are highly compressible. As clays get thinner the surface subsides (Marsden and Davis, 1967). Subsidence usually affects large areas imperceptibly; it creates low-lying cones that damage buildings and structures. These zones of anthropogenic endorheism are subject to flooding; when subsidence occurs along seashores, seawater can invade these areas. In Galveston Bay (Texas) *(Fig. 17.72)* the sea has inundated residential areas (Coates, 1983, 1987).

In the Houston-Galveston (Texas) area, groundwater extraction has caused a broad cone of subsidence and created more than 160 faults with a total length of about 500 km *(Fig. 17.72)* (Holzer, 1984). Of these faults at least 86 have been active in historic times; they are probably reactivations of

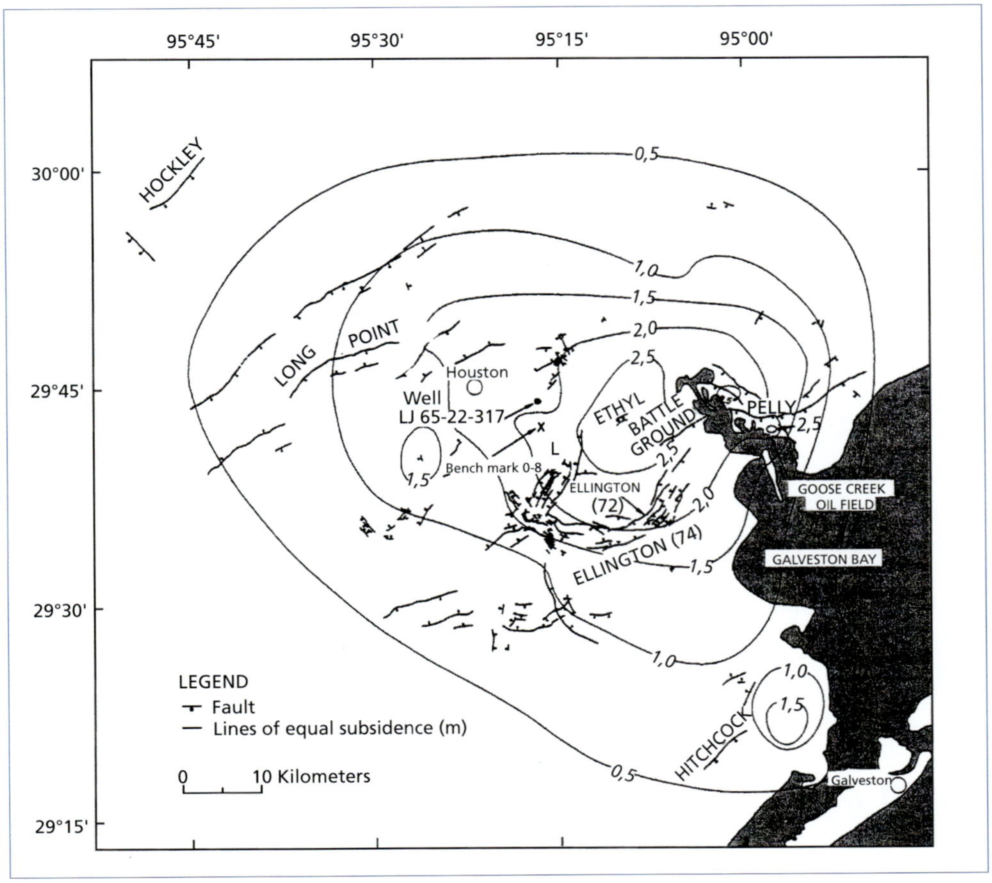

Figure 17.72 Land surface subsidence caused by groundwater pumping between 1906 and 1978 and faults indicating the downdropped side in the Houston-Galveston (Texas) region (in Holzer, 1984; based on data from various authors).

Tertiary faults in overlying Quaternary sediments. Groundwater is extracted primarily from sand and clay lenses (Chicot Aquifer) that form part of a thick sequence of unconsolidated fluvial and marine sediments. Water extraction began at the end of the 19th century and pumping has increased since the 1930s. The response was the creation of a subcircular area of subsidence about 12,200 km^2 in area, with 2.5 m of maximum collapse. The Goose Creek petroleum field is also in this area although its contribution to subsidence appears to be minor. Subsidence was accompanied by faults in the center and on the flanks of the depression; fault density is variable. These faults have been reactivated throughout history and scarps of 1.12 m have been measured on a highway in the area. The faults are easily recognized; they cut across highways, airports and similar structures.

The Central Valley of California holds the record for the magnitude and variety of areas that are subsiding as a result of water extraction. The San Joaquín valley is the most significant area; 13,500 km^2 are affected by settling and a maximum of 9 m of subsidence has occurred. Subsidence began in the 1920s and increased gradually until the middle 1950s, when the rate was 0.55 m/year (Poland and Davis, 1969). Subsidence decreased to 0.33 m/year between 1963 and 1966 and in 1973 it became imperceptible due to the transfer of water from northern California. In 1976 and 1977, subsidence accelerated due to a severe drought that caused renewed use of groundwater.

In south-central Arizona, water exploitation caused significant subsidence in the 1940s. More than 50 zones with a total surface area of 8,000 km^2 are subject to subsidence, cracking, and faults (Holzer *et al.*, 1979). Cracks are usually about 1 m wide and several meters deep. Fault lengths range from 1 to 16.7 km and have scarps of up to 1 m. They have caused significant damage to highways and irrigation systems and the traces of several aqueducts have required modification (Coates, 1983).

In the San Joaquín Valley and in south-central Arizona slow uplift has occurred after the pumping of groundwater. According to the United States Geological Survey about 6 cm of rebound occurred near Casa Grande (Arizona) between 1948 and 1967. The uplift is considered to be the result of isostatic rebound triggered by large water extractions that reduce the load of the underlying materials (Costa and Baker, 1981).

17.4.5 Degradation of arid zones: Desertification

Soil degradation results from decreased soil quality or decreased potential soil productivity. This deterioration can be physical, biological or chemical. Physical degradation results from the loss of fine particles. Biological deterioration is related to decreased organic material. Chemical degradation is related to the concentration of high solubility salts. Soil degradation is the principal factor in desertification (Dregne, 1983; El-Baz and Hassan, 1986; Mainguet, 1991; Chouhan, 1992; Williams and Balling, 1996; Middleton and Thomas, 1997; López Bermúdez, 2002, 2008; Thomas 2004; Glantz, 2005; García-Ruiz and López Bermúdez, 2009). In 2004 desertification affected 9600 million hectares that were previously productive (in López Bermúdez, 2008); the area subject to desertification is home to about 800 million people (*Fig. 17.73*). The World Bank considers desertification and salinity two of the nine main problems facing the world (Goudie, 1990a).

In Spain, about 27 million hectares or 53.4% of the country have significant soil degradation with annual soil losses on the order of a million tons (MOPU, 1987). In south-eastern Spain desertification is aggravated by water erosion, which causes losses of 300 tons/hectare/year (López Bermúdez, 1986). Data on other desertification processes (salinization, water erosion, physical and biological degradation of soil) is not available for all of Spain (López Bermúdez, 1988). Water erosion studies in Spain have expanded recently with the Proyecto Lucdeme and the installation and coordination of the Network of Experimental Stations for Monitoring and Evaluation of Erosion and Desertification (RESEL is the Spanish acronym), which extends over a large part of the national territory (Rojo and Sánchez, 1996).

At the World Conference on Desertification (UNCOD) held in Nairobi in 1977 the United Nations proposed an action plan to combat desertification. **Desertification** was defined as a decrease or destruction of biological potential of land that may lead to desert-like conditions (UNESCO-FAO, 1977). Numerous other definitions emphasize the role of man or climate (Baumer, 1990; Mainguet, 1991; Thomas and Middleton, 1994). The risk of desertification comes from the interaction between human activity and the natural physical and biological system. The human population explosion in the 20th century has aggravated these circumstances (Goudie, 1981b, 1998).

Degradation of deserts involves physical processes such as water and wind erosion and others that have already been discussed. The most important physico-chemical processes in arid areas are **salinization** and **sodification** of soils. Salinization results from the accumulation of soluble salts (chlorides, sulfates and sodium, magnesium, and calcium carbonates) (Rhoades, 1990). Sodification is enrichment in sodium. According to the FAO a soil is classified as saline when its content of soluble salts exceeds 1–2% in the top 20 cm. Salt accumulates in these soils by capillary rise; it creates efflorescences and salt crusts at the surface (*Fig. 17.75*).

Human-induced salinization is a consequence of irrigation in arid and semi-arid areas (*Fig. 17.74*); it affects

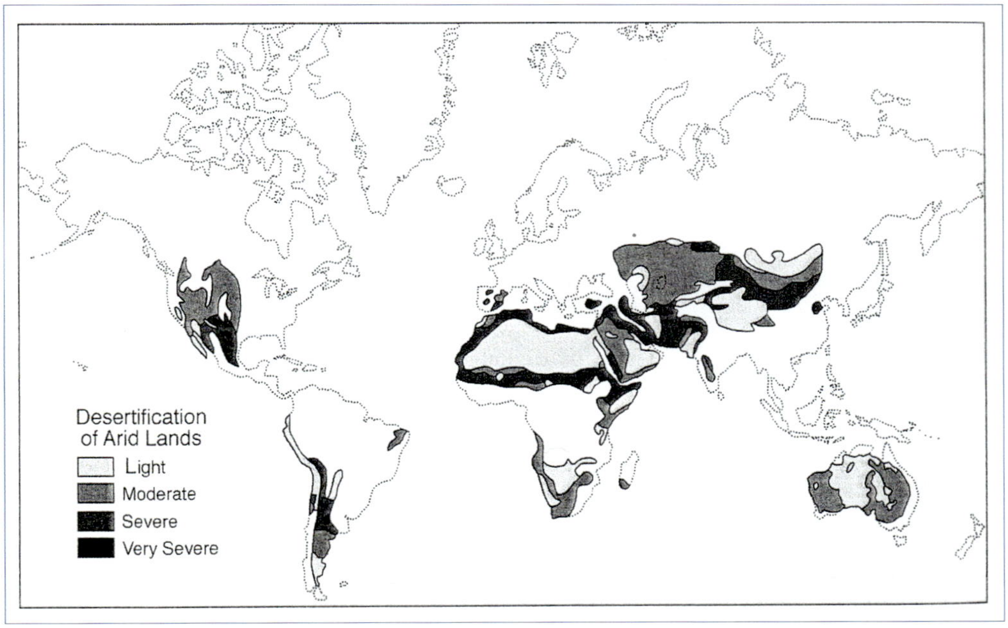

Figure 17.73 Desertification of arid land (Dregne, 1983).

Figure 17.74 Channelized water from the San Juan River flowing from the Andes. San Emiliano dike. San Juan (Argentina). This water is used for multiple purposes on this highly arid land.

one-third of the 200 million irrigated hectares in the world (*Table 17.3*) (Goudie, 1990a). Its effects include modification of soil structure, decreased permeability and porosity, and reduction of natural or cultivated vegetation, including its total disappearance (*Fig. 17.75*). Overexploitation of aquifers in littoral areas may cause salinization because it causes salt-water intrusion. Another cause is the decreased amount of fresh water that reaches delta areas (for example, the Nile delta) as a result of dam construction (the Aswan, in the case of the Nile) (Mainguet, 1991).

Salinization of the Ebro Depression (*Fig. 17.76*) is due to a number of factors: aridity of the area, saline content of outcropping sedimentary rocks, redistribution of salts and their concentration under certain favorable geomorphic circumstances, and diverse intrinsic aspects of soil and agricultural practices (Alberto, 1989; Herrero *et al.*, 1993). In the last category, the predominant salinizing practices are over-irrigation and leveling of land. Sprinklers and drip irrigation can be used to prevent salinization. The problem of salinization linked to irrigation is economically significant (Zekri *et al.*, 1990). The central region of the Ebro Valley is the location of 5,350 km^2 of irrigated land, of which 1,000 km^2 are soils affected by salts; 1,500 km^2 need to be tested to determine their salt levels (Herrero and Aragües, 1988; Herrero *et al.*, 1993).

Table 17.3 **APPROXIMATE PERCENTAGES OF IRRIGATED LAND AFFECTED BY SALINIZATION IN VARIOUS COUNTRIES (GOUDIE, 1990a)**

Algeria	10–15	India	27
Egypt	30–40	Iran	<30
Senegal	10–15	Iraq	50
Sudan	<20	Israel	13
United States	20–25	Jordan	16
Colombia	20	Pakistan	<40
Peru	12	Sri Lanka	13
China	15	Syria	30–35
Cyprus	25	Australia	15–20

In the San Joaquín Valley (California) 160,000 hectares are affected by salt; it is calculated that by the year 2080, 445,000 hectares will be non-productive (Sheridan, 1981).

In desert zones, precipitation varies enormously and consequently, river flow fluctuates significantly. As a result, broad salinity variations can be seen in the Casamance River (Senegal) (*Fig. 17.77*); these variations affect the availability of potable and irrigation water. In October 1968 salinity was low due to a rainy period, whereas in May of 1984 the river as a whole had a salt content higher than seawater (35%) and near the mouth of the river, the salt content was three times higher (Farmer and Wigley, 1985).

As we have discussed, inappropriate agricultural practices together with over-irrigation cause salinization in some cases and leaching in others. For this reason, as Mainguet (1991) states: "It is absolutely vital that irrigation, especially for large projects in arid and semi-arid ecosystems, be considered only as the last option in planning."

Figure 17.75 Grain cultivation affected by salinization in the Hoya de Ayerbe, Province of Huesca. Ebro Depression, Spain.

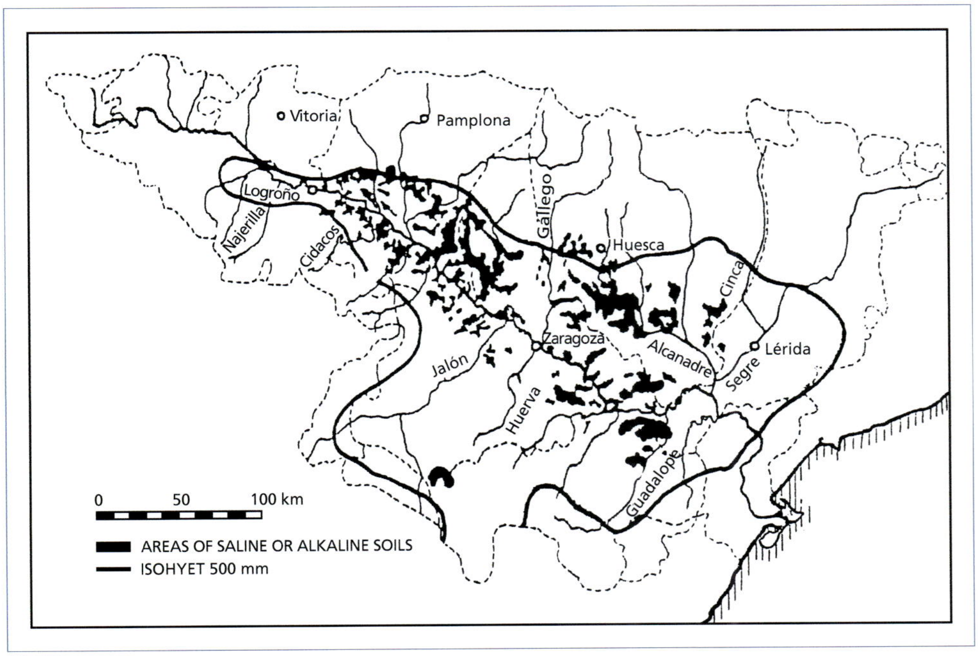

Figure 17.76 Distribution of saline areas in the Ebro Depression (Spain) (Alberto, 1989).

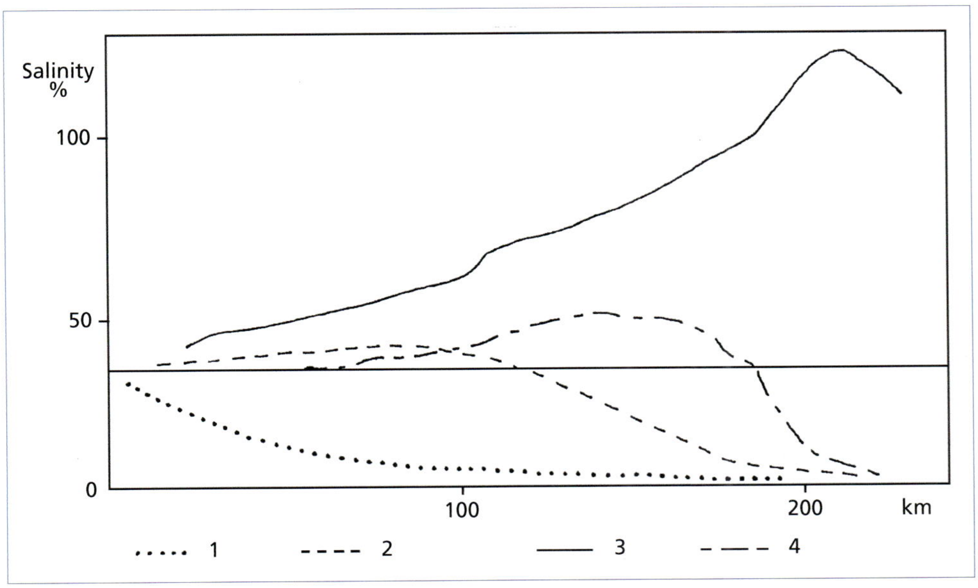

Figure 17.77 Various degrees of salinity of the Casamance River as a function of distance from upstream to the mouth. The horizontal line indicates the salinity of seawater. 1: October 1968. 2: July 1969. 3: May 1984. 4: September 1984 (Farmer and Wigley, 1985; in Mainguet, 1991).

Another cause of degradation in desert areas that is linked to human activity is **abusive agriculture** in zones with poor agricultural potential. It sometimes results from the introduction of inappropriate foreign methods into local agriculture (Thomas and Middleton, 1994). This pressure makes the soil sterile, causes it to be abandoned and to become subject to water and eolian erosion. Numerous examples can be seen in China, the Sahara, and Rajastan (India), etc. **Deforestation** is very important in some areas; it results from the use of wood for domestic purposes because electrical energy and gas are very expensive for most inhabitants of these areas. For this reason, acacias have disappeared from Khartoum (Sudan) over a radius of 100 km and wood now has to be transported 500 km (Goudie, 1990).

Globally the number of beef cattle increased 38% between 1955 and 1976 and sheep and goats increased by 21% (Goudie, 1990). In many areas, the number has exceeded the capacity of the land and has caused **overgrazing** (*Fig. 17.78*). To this we add cultivation of some area and the resulting reduction of the area available for pasture. The loss of plant cover is accompanied by increased water and wind erosion. In northern China, overgrazing has degraded plant cover in the steppe; a series of degradation rings around water wells has resulted (*Fig. 17.79*). They consist of: a central area with a radius of 300 to 500 m where vegetation is sparse and soil is denuded; a second ring with a radius of 500 to 1000 m where grasses and resistant shrubs are present; farther out, at a distance of 1000 to 3000 m from the well, vegetation returns to normal (Mainguet, 1991).

Desertification problems are serious. Desertification can technically be stopped and many degraded areas can be restored for new production but the solutions are costly. Preventive measures to save vegetative cover are based on introducing new species and adopting varieties that are more resistant to water scarcity and salt. It is also important to improve water use, combat water and wind erosion, and

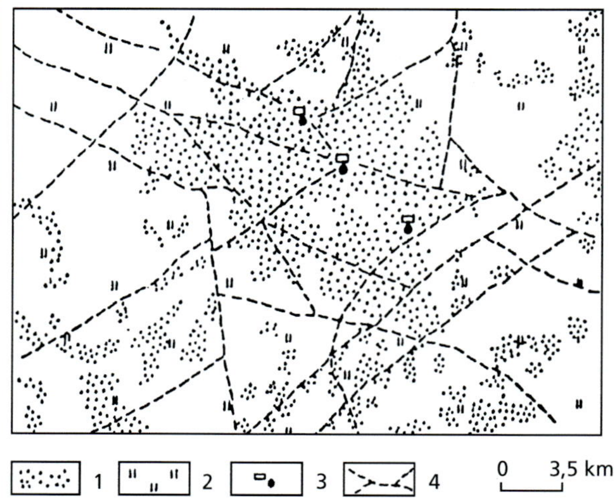

Figure 17.79 Degradation caused by overgrazing. 1: Denuded alluvial surface. 2: Vegetated zone. 3: Shed for cattle and wells. 4: Highways and roads (Mainguet, 1991).

Figure 17.78 Fenced experimental plot for the study of water erosion showing a higher percentage of plant cover than in the area surrounding the plot, four years after its installation. This is due to overgrazing by sheep and goats. Mishor Adumin, Desert of Judea (Israel).

reduce ignorance through social training programs at all levels.

At the request of the General Assembly of the United Nations, the International Year of Deserts and Desertification (IYDD) was commemorated in 2006. Four objectives were established: long term compliance with the United Nations Convention on the Fight against Desertification (UNCCD); increased awareness of the implications of desertification; promotion of exchanges between researchers; and dissemination of knowledge about the UNCCD (Stringer, 2008).

17.4.6 Earth dams and reservoirs

In arid zones the water balance is negative and as a result water is not available for development of these areas. Dams have been a vital part of human civilization for millennia because they have provided sufficient water for human purposes, irrigation networks, and occasionally hydroelectric energy production; the Aswan Dam is an example (Singh, 1966). Dams constructed by rock and/or loose material are called **earth dams**. They are common in arid zones because of their low cost and relatively simple construction. It is estimated that about 2,000 dams have failed through the world since the 12th century (Jansen, 1980). In the United States, 3,604 earth dams have been built, equivalent to about 73% of the total number of dams (Singh, 1996).

The large number of earth dam failures and subsequent flooding show the vulnerability of these structures and the risks they pose to both human life and property (Gutiérrez *et al.*, 2003). Dams collapse for various reasons (Garret and Costa, 1982; Sherard, 1992): piping and karstification are the two of the most common (Lou, 1981, in Singh, 1996). A compilation of various studies conducted throughout the world shows that piping has caused 37% of dam failures. In the western United States, 60% of earth dams higher than 15 m have ruptured due to piping (Von Thun, 1996).

The use of soils that contain dispersive clays in the construction of earth dams can lead to serious geotechnical problems if appropriate measures are not taken (Sherard *et al.*, 1972a; Jones, 1981, 1990; Knodel, 1991). Piping affects earth dams and is a clear risk due to the fact that dispersive clays are widely distributed geographically and occur in a wide variety of sedimentary sequences such as deposits of lacustrine origin, sediments in alluvial fans, and slope deposits or flood plain facies in fluvial sediments (Sherard, *et al.*, 1976). For Colclough (1978), tunnel erosion in Tasmania is the most frustrating and insidious type of erosion. This type of subsurface destruction occurs in the presence of swelling clays or high values of ESP (Exchangeable Sodium Percentage) in the saturated paste (Gutiérrez *et al.*, 1995, 1997; Gutiérrez, 2001, 2005). Water that is active under these circumstances causes what Sherard *et al.* (1972b) call "hydraulic fracturing;" this refers to any form of rupture by piping. It is particularly common in the aridisols of Africa, Australia, and North America (Jones, 1981). These soils are very susceptible to piping but if the clays in them are dispersed in fresh water, the soils can later be used for the construction of earth dams if appropriate precautions are taken (Sherard and Decker, 1977).

Two examples of earth dams in Oman offer interesting data on dam research. Al-Rujeibi *et al.* (2002) indicate that in their country, 17 earth dams have been built to control flooding and to recharge groundwater. The authors studied sediments collected from nine dams. Research shows that water erosion (sediment yield) is directly proportional to discharge and inversely proportional to basin size and amount of precipitation. Also in Omán (Alkhanbashi *et al.*, 2002) where average annual precipitation is 100 mm, rapid development in the study area has increased the demand for water. Excessive pumping has caused a decline in the groundwater level, which has led to the inward migration of the fresh water/salt water interface. An earth dam has stopped the migration of the interface and has improved water quality near the coast.

Structures designed to make water available in arid zones are **reservoirs** of variable sizes depending on the uses for which they are designed. These reservoirs are holes in the ground that fill with water. In order to increase the reservoir's capacity, material extracted from the hole is deposited around it. The San Juan irrigation reservoir located about 2 km north of Altorricón (Province of Huesca, Ebro Depression) (*Fig. 17.80*) is 400 m long and 230 m wide and has a maximum depth of 19 m; its sides are inclined between 24 and 30°. The storage capacity is 850,000 m^3. The area receives an annual average precipitation of 460 mm and the reservoir is excavated primarily in Miocene shales. The horizontal Miocene deposits are in turn fossilized by a covered pediment that is part of the large Barbastro Anticline, the nucleus of which is Oligocene gypsum. As a result, the materials of the 4 m-thick pediment are 40% gypsum. On top of this pediment, the reservoir has been increased over compacted extracted shales.

Miocene clays studied by X-ray diffraction indicate the clay mineralogy is illite and chlorite with some accessory smectite. The exchangeable sodium percentages in the extract of the saturated paste indicate that in some levels clay dispersion (piping) may occur (Gutiérrez *et al.*, 2003).

During fill tests of the San Juan reservoir conducted on 21 January 2001, catastrophic failure occurred; the rupture released 3,000,000 m^3 of water (Batalla and Balasch, 2001), inundated many hectares of agricultural fields, and

Figure 17.80 San Juan reservoir, excavated in Miocene clays, partially filled with water. Altorricon, Province of Huesca (Ebro Depression, Spain).

Figure 17.81 Trapezoidal section break caused by water discharge from the San Juan reservoir. Note the increased surficial deposits. Altorricon, Province of Huesca (Ebro Depression, Spain).

caused damage to the irrigation system. The failure caused a trapezoidal-shaped break (Johnson and Illes, 1976) 10 m deep and 16 × 10 m wide in the reservoir wall (*Fig. 17.81*). The rupture was caused by water circulation though alluvial deposits, which dissolved gypsum particles in the covered pediment materials, promoted by the hydraulic gradient. *Pipes* created during the filling of the reservoir contributed to the failure (Gutiérrez *et al.*, 2003).

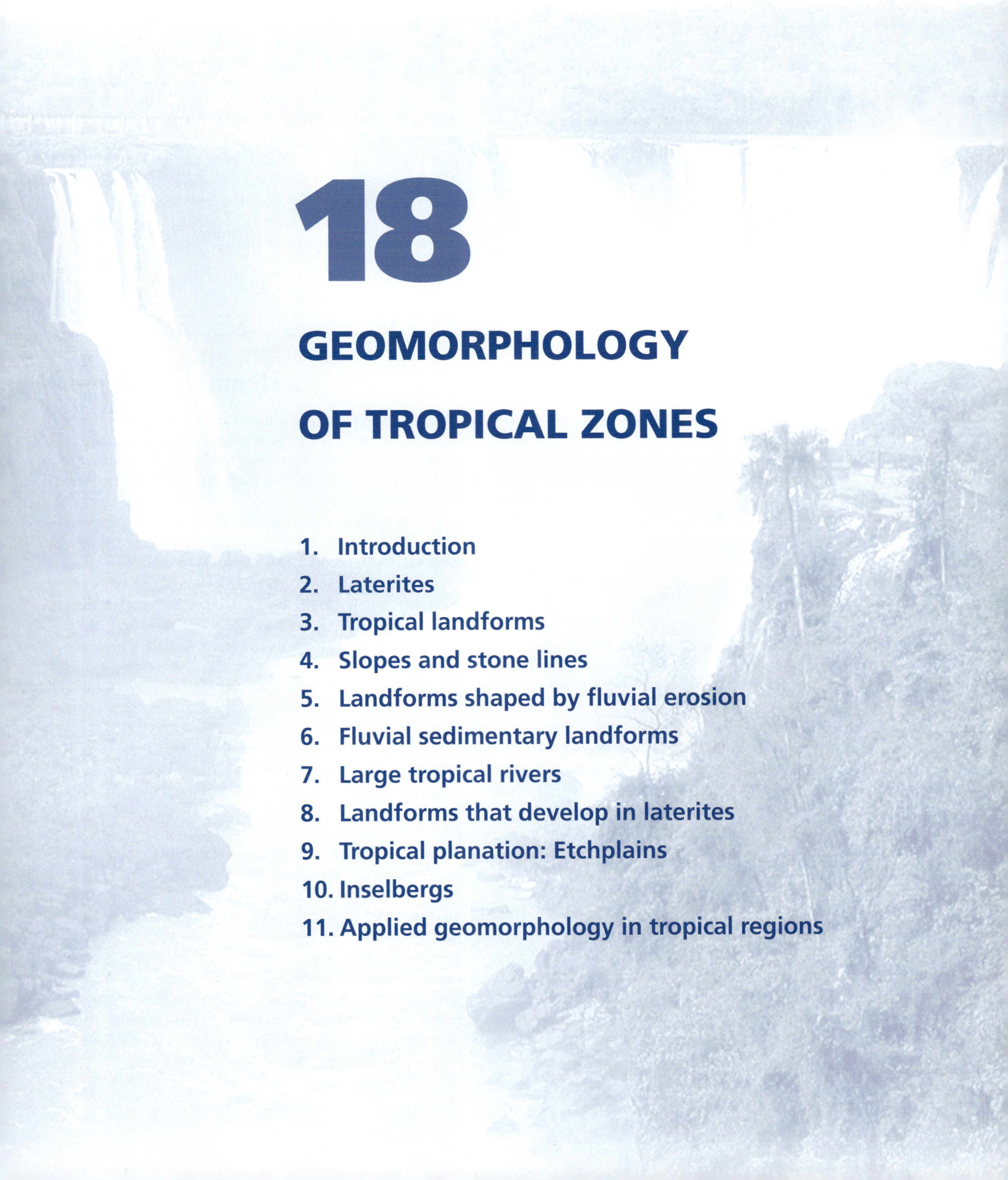

18

GEOMORPHOLOGY OF TROPICAL ZONES

1. Introduction
2. Laterites
3. Tropical landforms
4. Slopes and stone lines
5. Landforms shaped by fluvial erosion
6. Fluvial sedimentary landforms
7. Large tropical rivers
8. Landforms that develop in laterites
9. Tropical planation: Etchplains
10. Inselbergs
11. Applied geomorphology in tropical regions

18.1 Introduction

The tropics include the area between the Tropic of Cancer (23.5°N) and the Tropic of Capricorn (23.5°S). These boundaries mark the area where the sun reaches its zenith position. Thus these low latitude zones receive a large amount of solar radiation. This uninterrupted heat has significant biological consequences because it promotes a thick vegetation cover that has a major effect on morphogenesis in these regions. In these inter-tropical zones we find large hot deserts and extensive areas of high precipitation. These circumstances lead us to classify the different environments on the basis of climatic, biological, and geomorphic criteria, among others. Subdivision of these categories is always difficult especially in continental areas (Reading *et al.*, 1995).

Some authors subdivide the wet tropics on the basis of vegetation; they consider the lack or absence of xeromorphic features in vegetation and the development of rainforest, deciduous forests, and savannahs (Fosberg, *et al.*, 1961). Others consider climatic criteria and use an average coldest month temperature of 18°C as a boundary (Köppen, 1936). Annual average precipitation has also been suggested but researchers disagree on amounts. Köppen (1936) set the boundary at 450–600 mm whereas Tricart (1974) put it between 750 and 800 mm.

About 3.1 billion people, approximately 45% of the world's population, live in tropical zones. Almost all of them live in the wet tropics and about 60% live in east and south Asia. Agriculture, the principal sector of the economy, obviously depends to a great extent on climate variability and catastrophic events (McGregor and Nieuwolt, 1998).

Most of the non-arid tropical area is located in stable cratonic zones, which are composed primarily of Precambrian plutonic and metamorphic rocks; these are essentially central Africa and the Brazilian shield. These areas have undergone long-term denudation accompanied by extensive planation. Due to their flat topographic gradient, the erosion rate is also low but weathering proceeds unobstructed. This situation allows us to study landform evolution over long time periods and under conditions of crustal stability (Thomas, 1974).

However, in regions affected by alpine tectonics where significant relief is present, as in Southeast Asia, Indonesia, Central America, and the Antilles, denudation rates are high and, as a consequence, weathering does not penetrate deeply. In addition, the predominant outcropping rocks are different from those in stable zones; there are large calcareous outcrops where exceptional karst may form.

In previous chapters we said that weathering profiles characteristic of humid tropical zones are commonly found in arid regions, particularly semiarid regions. This shows that in the past much wetter conditions predominated in these dry areas. Thus an understanding of humid tropical processes and deposits help us interpret climate variation that occurred in the past.

18.1.1 Climate characteristics, vegetation, and morphoclimatic domains

Climatic parameters play a fundamental role in the activity and intensity of geomorphic processes and in the nature of resulting forms (Faniran and Jeje, 1983). For this reason it is important to know the magnitude and variability of these parameters.

Temperatures are moderate in the humid tropics; they do not exceed 34°C, unlike tropical deserts where temperatures can rise above 50°C. Mean annual temperatures measured at sea level range between 24 and 30°C. Seasonal and annual variations are less than 2°C. This thermal uniformity breaks down at altitudes where cloudiness increases.

In the humid tropics, humidity is always high and relative humidity generally exceeds 80%. Values throughout the year deviate very little from this number. Potential evapotranspiration is at its maximum in equatorial zones where it ranges from 1000 to 1500 mm/year.

Except in very elevated areas, precipitation consists of rain. Because temperatures are highly uniform, rainfall variability determines the seasons. Rainfall variation is related to rainstorms, which provide 90% of total precipitation in savannah zones. Most storms occur in the afternoon and are of short duration. The record for one-day rainfall is 1870 mm, which was documented on Cilaos (Reunion Island) on 16 March 1952. In humid tropical zones, 40% of precipitation exceeds an intensity of 25 mm/hour and intensities of 340 mm/hour for several minutes have been recorded (Reading *et al.*, 1995). Precipitation regimes are the following (Birot, 1973): 1) Tropical areas of equatorial regions influenced by the Inter-Tropical Convergence Zone, where rains occur throughout the year with a single maximum. Average annual precipitation is on the order of 2000 mm or higher. 2) Areas between 10° north and 10° south of the Equator have two dry and two wet seasons and precipitation ranging from 1000 to 2000 mm. In this humid and dry tropical regime, rainy seasons may be shorter and rain totals may reach 650–1000 mm. 3). A dry regime occurs on desert margins where precipitation ranges from 250 to 650 mm.

Vegetation in the humid tropics is a response to the existing climatic regime. One major category is the equatorial rainforest of 40 m-tall trees with large perennial leaves and thousands of arboreal species (*Fig. 18.1*). The high density of arboreal vegetation casts continuous shade on the soil. Mangrove swamps, which are the characteristic vegetation in coastal areas, occur primarily in basins of the Amazon and the Congo, in Central America, the Antilles, and Southeast Asia and Indonesia. A savanna forms in a tropical humid-dry climate regime. It consists of isolated trees that are sometimes grouped in corridors in valley bottoms and that grow on fields of large grasses that are generally taller than people (*Fig. 18.2*). Summer fires are very common in these dry yellow grasses (*Fig. 18.52*); some authors claim that the fires are the reason that grass predominates over forest. Savannas occur primarily in central and southern Africa where they contain horizontal flat-crowned acacias and baobab; they are also present in South America (*campos cerrados*), India, Southeast Asia, and Australia (Tricart, 1974). Vegetation cover intercepts about 20–30% of rainfall and in this way protects the soil surface. Vegetation also contributes to physical weathering because roots penetrate into joints and promote chemical alteration through the activity of organic acids. Vegetation also protects the soil from water erosion and helps prevent mass movement, which occurs primarily during periods of intense rainfall.

Because temperatures are homogeneous, climatic differences are primarily based on rainfall. An annual precipitation boundary of 750 mm separates the humid tropics from transition zones (Thomas, 1994a). Rainforest and savanna areas include the Af and Am rainforest and Aw and As savanna climates (Köppen, 1936).

Morphoclimatic distinctions are similar because precipitation and the resulting vegetation play a determining role in humid tropical regions. Even so, we must consider multiple geomorphic legacies from other climates present in areas with equatorial and humid and dry climates although these legacies are more common near dry climate zones. Dominant geomorphic processes in the humid tropics are intense chemical weathering, mass movement, and fluvial erosion; the latter is especially significant in humid and dry climate regions (Wilson, 1969).

Deep intense weathering and the evolution of weathering profiles are undoubtedly the most characteristic features of this morphoclimatic zone. Even so, some authors doubt the existence of a tropical climatic geomorphology (Stoddart, 1969a) and others question the designation of a tropical zone in the global morphoclimatic system (Douglas, 1978; Gupta, 1993). The latter author claims that tropical processes and landforms depend only partially on climate and

Figure 18.1 Equatorial rainforest in the Amazon. Manaus (Brazil).

Figure 18.2 Savanna with grass and scattered tree vegetation. Nairobi National Park (Kenya).

that other important factors include neotectonics, lithology, episodic and catastrophic events (cyclones), and anthropogenic activity. This is true, but it is important to remember that these non-climatic processes and factors occur everywhere on the Earth's surface and are superimposed on climatic causes. This point of view refers to heavily populated areas with significant neotectonics. These areas are quite different from African and Brazilian areas where old planations have led to the development of thick weathering deposits and environmental changes have in turn produced lateritic landforms and etched surfaces, which are characteristic of humid tropical zones.

Büdel (1977) claims that humid tropical zones are completely different from other morphoclimatic zones. Terraces, gravel plains, and floodplains do not develop in these zones as they do in ectropic (temperate) zones, nor do loess mantles. The soil is much deeper than in ectropic zones and contains abundant crusts and concretions. Inselbergs of various sizes are common in the tropics; they protrude above extensive erosional surfaces that weather as they form. Mountainous areas are characterized by profound dissection. Büdel assumes that the morphodynamics of low tropical land is totally different and that in these zones, relief is controlled by the creation of **etchsurfaces.**

Tricart and Cailleux (1965) describe an **inter-tropical zone** with warm temperatures and constant fluvial flow. On the basis of rainfall distribution, annual precipitation, and density of plant cover, they also subdivide the zone into a **savanna domain** that has less precipitation and less dense plant cover. Considerable but discontinuous overland flow and intense chemical weathering occur over time. In summary, a savanna is a grassland or dry pasture with dispersed trees. Thick vegetative cover and high precipitation characterize the **rainforest domain**. Chemical and biochemical activity reaches its maximum intensity in this domain. This is undoubtedly the richest and most productive community in the world; the rainforest domain contains 2/3 of the world's plant mass and approximately half of all living species. Rainforests have been reduced to half of their original extent by anthropogenic activity (Cunningham, 1999b).

18.2 Laterites

Various terms are used for laterite in the literature; these include duricrusts, plinthite, ferric laterite, and ferricrete, among others (Thomas, 1974). Laterites are difficult to define although their physical properties, such as resistance and color, have been recognized since the 19th century. During a voyage to western India, Buchanan (1807) described a material that was soft enough to be cut into blocks with an iron instrument; the material hardened as it dried and for this reason it was used for make bricks. In the 20th century, laterite began to be defined on the basis of its chemical composition: its iron and aluminum contents and ratios of SiO_2 to Al_2O_3 and Fe_2O_3. The multiple definitions of laterite (Alexander and Cady, 1962; Sivarajasingham *et al.*, 1962; Norton, 1973; Schellmann, 1981; Widdowson, 2007) are quite similar. Laterites form in an environment of intense chemical alteration where alkaline and alkaline-earth elements are absent; they are enriched in iron and aluminum and contain significant quantities of quartz and kaolinite. They harden by wetting and drying.

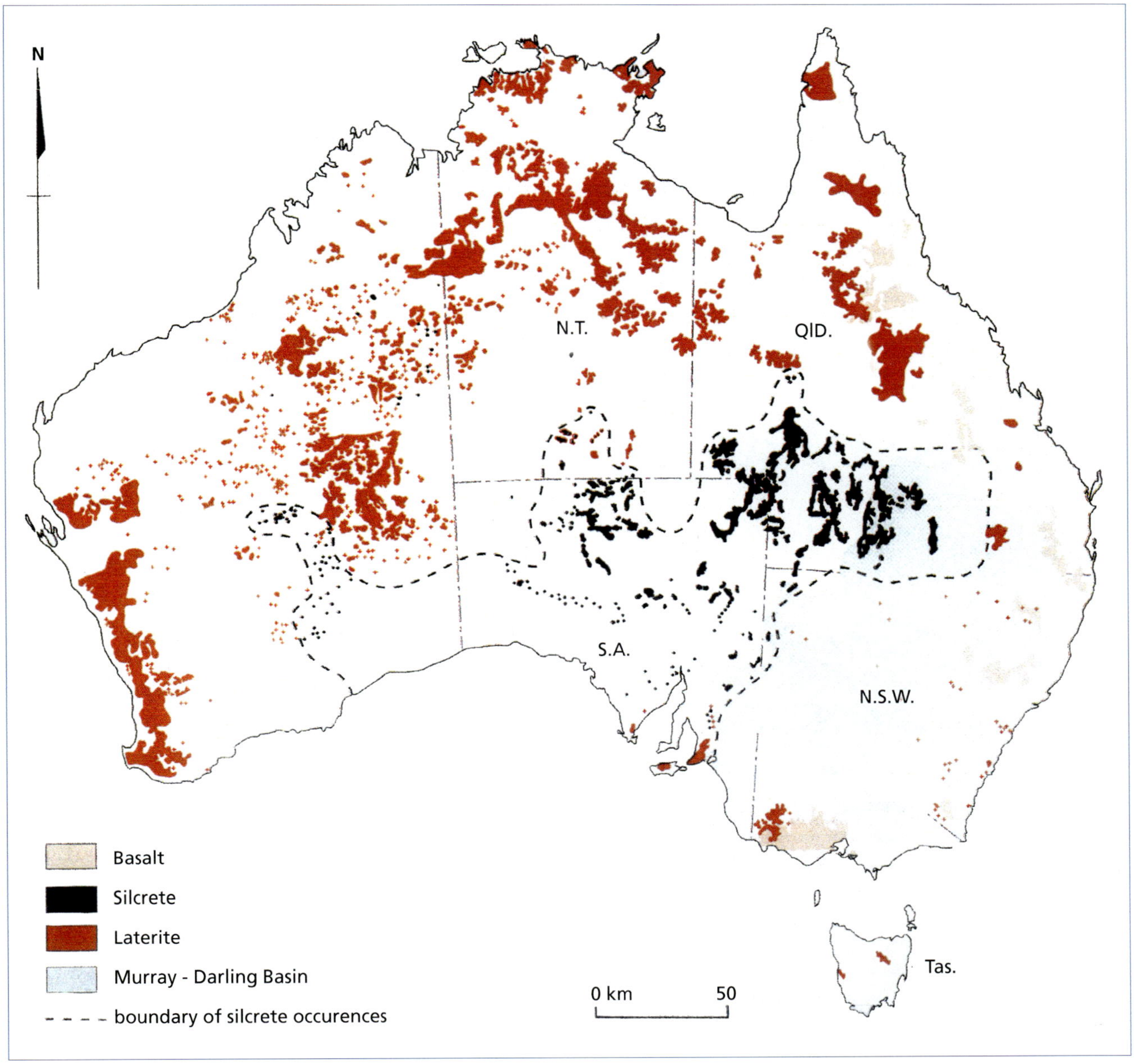

Figure 18.3 Distribution map showing laterites and silcretes in Australia (Twidale, 2007).

Laterites have been described on all continents except Antarctica (Goudie, 1973). Widdowson (2007) describes in detail the various places where laterites have been studied and their ages as determined by various authors. Laterites occupy a large area in northern and western Australia (Twidale, 2007) (*Fig. 18.3*).

Controversy exists regarding the pedological and geomorphic concepts of laterite. Soil scientists see laterite soil as a residual soil enriched in iron on a top surface that has formed under wetting and drying conditions whereas geomorphologists think that laterites form under special geomorphic conditions and that precipitates form in the zone of water table fluctuation. For this reason, soil scientists consider this type of laterite as a tropical podzol soil. This has led to a distinction between pedogenic laterites and laterites generated by water table fluctuation (McFarlane, 1976). This controversy is difficult to understand when layers are 60 m thick or more (Hays, 1967), which are difficult to explain by soil forming processes. According to Goudie (1973) laterites that form by soil processes should be a thin horizon but this is often not the case.

18.2.1 Weathering, composition, and structures

Weathering continuously attacks minerals present in rocks. As weathering increases, the most resistant minerals such as quartz tend to fracture and break up and continuous dissolution occurs along planes of weakness. Cleavage of the more complex silicates provides pathways for alteration. Feldspars, pyroxenes, and amphiboles break into fragments and generate secondary minerals. The crystal structure of micas opens up along cleavage planes (Widdowson, 2007). Figure 18.4 (Borger, 2000) illustrates the physical rupture of the most common rock-forming minerals. The most mobile cations are released during weathering (bases in pedology): Na^+, Ca^{2+}, Mg^{2+} and K^+, although the latter may be retained in authigenic clays. Most primary minerals are destroyed except

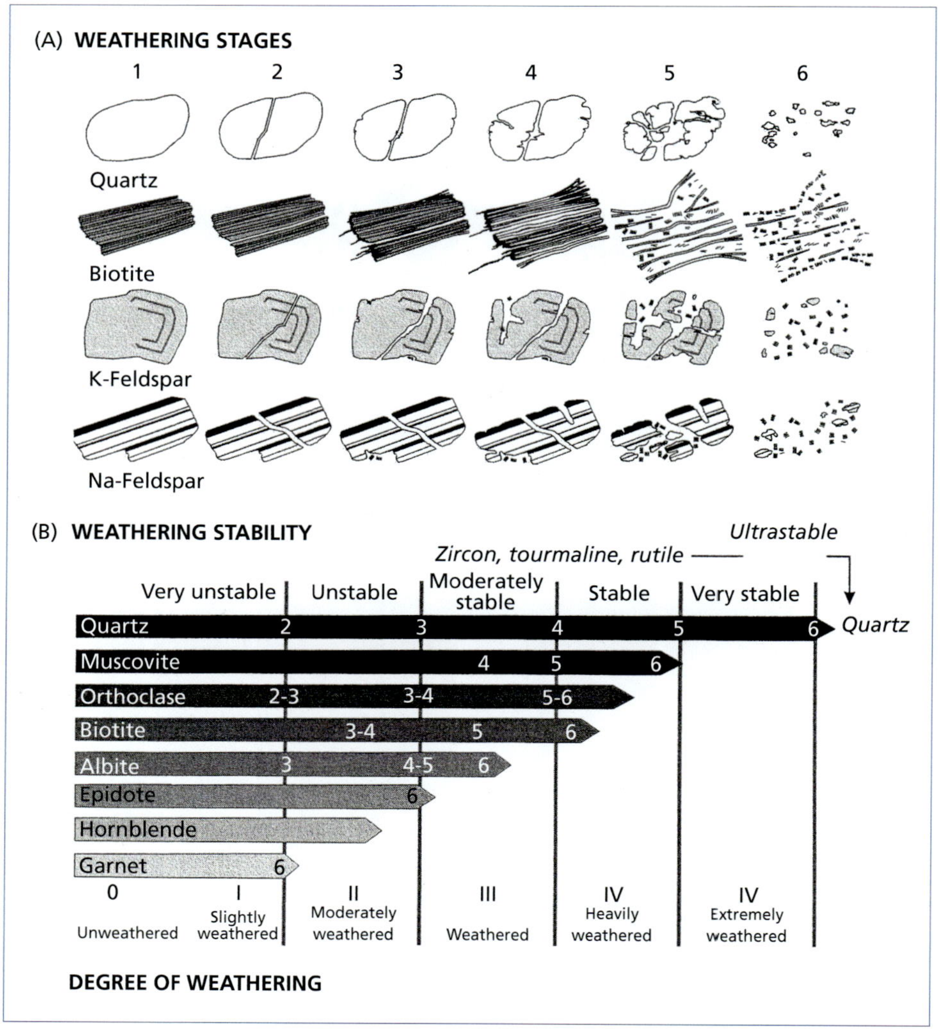

Figure 18.4 (A) Weathering stages in quartz, biotite, K feldspar, and Na feldspar: (1) unaltered mineral; (2) fracturing of the mineral; (3) marginal alteration of Quartz and K feldspar, incipient kaolinization of Na feldspar, mobilization of Fe from biotite which begins to enlarge cleavage planes; (4) corrosion of quartz, K and Na feldspars, discoloration, enlargement and beginning of kaolinization of kaolinite; (5) fragmentation with intense enlarging and kaolinization of biotite; (6) complete dissolution of primary minerals. (B) resistance to weathering and degree of weathering in humid tropical environments: the numbers in the horizontal bars corresponds to stages of quartz weathering shown in (A); the vertical lines show stages of quartz weathering and are also used to classify the resistance to mineralization of other minerals, and as the degree of weathering of any altered mineral (Borger, 2000).

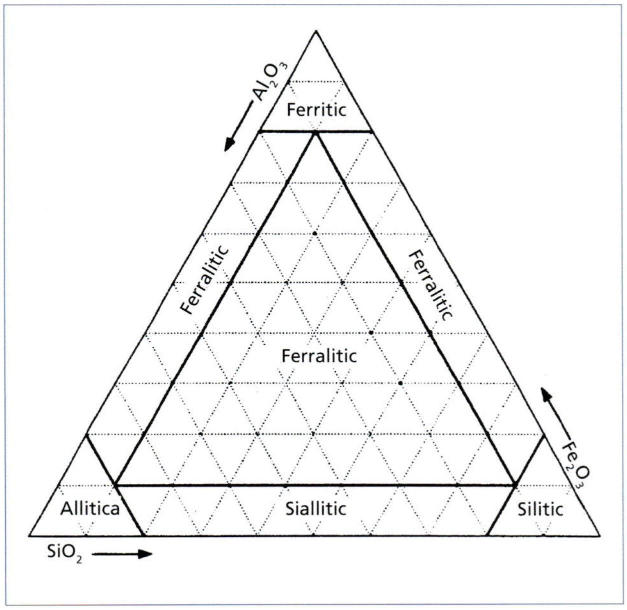

Figure 18.5 Ternary diagram showing the different types of crusts as a function of chemical composition (Dury, 1969).

quartz. As a result, the residuum is enriched in Si, Fe, and Al, in the form of kaolinite, quartz, oxides, and hydroxides of iron and sometimes gibbsite.

Chemical analyses of Al_2O_3, Fe_2O_3, SiO_2, and clay minerals in laterites makes it possible to display them on triangular diagrams and determine their compositions. Some researchers use the percentages of Fe_2O_3, Al_2O_3 and kaolinite (Bardossy and Aleva, 1990) whereas Dury (1969) and Schellmann (1981) substitute kaolinite for SiO_2 in the diagram (*Fig. 18.5*). These studies reflect a wide compositional variety. Thomas (1974) proposed to simplify the broad terminology by using three names: bauxite for an aluminum-rich crust, silcrete when the silica content is high, and laterite for materials of intermediate composition.

Since laterite forms from a residuum resulting from chemical weathering of a rock, we can expect to see a direct relationship to the composition of the parent rock. On a regional scale the iron content may vary whereas on a small scale there may be no clear relationship between the chemistry of the laterite and the parent rock. For example, we cannot generalize that rocks rich in Al_2O_3 favor bauxite formation (McFarlane, 1983).

The principal authigenic minerals that form laterites are oxides and hydroxides of iron and aluminum such as hematite (Fe_2O_3), goethite (FeO·OH), gibbsite Al $(OH)_3$, boehmite (AlO·OH), diaspore (AlO·OH), and clay minerals such as kaolinite and halloysite ($Al_4Si_4O_{10}$ $(OH)_{18}$), which have the same composition (McFarlane, 1983; Bardossy and Aleva, 1990). The greater or lesser abundance of these mineral species is a function of the composition of the parent rock and the various transformations it has undergone (Tardy, 1993). All authigenic minerals move toward stability according to environmental circumstances; in humid tropical environments they are modified primarily by leaching. Kaolinite comes from the alteration of feldspars or authigenic montmorillonite. Goethite comes directly from biotite and hornblende; dehydration of goethite forms hematite. Feldspar weathering may produce gibbsite directly although it is normally formed by the desilication of kaolinite at pH > 8 (see section 5.2.3, Chemical Weathering).

Laterites have different **structures**. **Nodules** that lack internal structure, **pisolites** (diameter > 2 mm), and **oolites** (<2 mm) with concentric layers are common. These occur as loose or packed concretions forming a cemented oolitic or pisolitic laterite. The concentric structure forms by enrichment in the zone of water table fluctuation. Pedogenic pisolites are surrounded by loose material. They have an irregular shape and size and an external layer of manganese (McFarlane, 1976). Other types of lateritic structures are vermicular, vesicular, cellular, and tubular; the classification is based on cavity shape. This terminology is not clearly defined; McFarlane (1976, 1983) uses only the word **vermiform** to characterize laterites with internal tubes and cavities (*Fig. 18.6*). The diameter of these conduits is variable and they are usually filled with kaolin. These structures have a complex origin attributed to termite activity (termite armoring of Erhardt, 1951) and the penetration of plant roots. McFarlane (1976) thinks they result from precipitation, which is found only in laterites formed by water table fluctuation. He also claims that pisolitic laterites are immature laterites whereas the vermiform variety is mature and derived from the former. Because of this relationship, pisolitic and vermiforms are often found together.

18.2.2 Factors that affect formation

Laterites and bauxites occur in a wide variety of rock types (Mabbutt, 1961b; Goudie, 1973; among others) although rock mineralogy substantially influences the composition of lateritic rocks (Thomas, 1994a). Even so, in some zones a particular lithology favors laterite development and in others the same lithology inhibits their formation. This implies that other factors may be more important in laterite formation (McFarlane, 1976).

Laterites form in regions of relative **tectonic stability** characterized by crustal deformation and minimal uplift; these conditions are found on cratons (Australia, Brazil, Africa) (Widdowson, 2007).

Figure 18.6 Vermiform laterite. Eastern Bouaké (Ivory Coast).

Topography is a factor that substantially affects laterite formation. On slopes with significant gradients, abundant runoff carries away weathering products whereas on low-gradient slopes water may penetrate more easily, dissolve the more soluble constituents, and promote accumulation of less mobile constituents. As a result, laterite formation is associated with zones of low topography where overland flow is sparse and aggressive erosion does not occur (Widdowson, 2007). However, laterites are known to be present on the top of mesas and platforms, which are considered to be landforms produced by incision into low relief laterites. Detrital laterites form at the foot of high relief areas (*Fig. 18.7*). Some researchers think laterites form as the ground surface is progressively lowered; the final product is a leveled surface topped by a thick layer of residual material. In this case, the laterite postdates the formation of the planation (McFarlane, 1976, 1983).

The prevailing **climate** in the humid tropics is more suitable for laterite formation but laterites also form in temperate climates (Taylor *et al.*, 1992). Soil temperatures near 27°C promote laterite formation (Thomas, 1994a) because at this temperature percolating water disintegrates rock easily and silica dissolves rapidly. The most favorable rainfall conditions are present in humid and dry tropical climates because humid climates do not favor oxidation/reduction conditions, which mobilize and precipitate iron (Nahon, 1986). In these environments where erosion and overland flow are minor and rainfall continually infiltrates, the water table helps evacuate the more mobile constituents (Widdowson, 2007) and oscillation of the water table may promote the oxidation of ferrous iron. When the system consists of abundant alkaline water, it can practically dissolve quartz (pH > 8).

Biological factors include the effects of vegetation and the activity of organisms and organic compounds. Laterites are commonly associated with zones of savanna vegetation although they also occur in rainforest areas (McFarlane, 1976). Thomas (1994a) points out that bauxites are related to humid conditions of rainforest vegetation whereas more dry conditions tend to be more favorable for laterite formation. The rainforest maintains soil moisture and the plant mass causes significant interception, which reduces water erosion. Plants absorb silica and store it in their leaves. In addition, various microorganisms are involved in the mobilization of aluminum, iron, and silica (McFarlane, 1987). Aluminum and iron can also be leached from chelates as organometallic compounds (Widdowson, 2007).

Most laterites are associated with extensive plains and the time necessary to form plains is measured in millions of years. Ahnert (1970) calculated that several tens of million years are required for the formation of plains. In addition, studies have been done to determine the thickness of the

Figure 18.7 Lateritic breccia on altered gneiss. Machakos (Kenya).

parent rock necessary to generate a determinate thickness of laterite. Considering that granitic rocks contain on average 2% $FeO+Fe_2O_3$, Trendall (1962) calculated that 180 m of weathered granite was required to generate 9 m of laterite in Buganda (Uganda), a process that probably took several million years. Humid tropical zones, especially savanna regions closer to desert margins, have been subject to significant **climate changes**. For this reason extensive lateritic mantles are commonly present in semiarid zones. In addition, numerous lateritic deposits have been found in late Cretaceous and Paleocene rocks in extratropical areas. Basal Tertiary sedimentary deposits in the Duero Depression (Spain) in the Zamora-Salamanca region are composed of transported laterites, kaolinitic sandstones, and silcretes (Martin-Serrano, 1988). Because overlying layers are of Paleocene age, it is thought that prior to the Tertiary the Hesperian Massif (Iberian Massif) had a tropical humid climate. In Portugal, the ferruginous Buçaco sandstone (Birot, 1949b) is likely the sediment that correlates to this tropical climate.

18.2.3 The laterite profile

Laterite is one of a group of weathered materials that organize in zones or horizons called the laterite profile. Millot prepared an ideal profile (1964) (*Fig. 18.8*) inspired by the differentiations of Walther (1915); it is included in the work of Thomas (1974a) and Faniran and Jeje (1983). Loose brown slightly humid material and fragments and ferruginous concretions are found on top of the lateritic horizon. The **laterite**, which is dark or ochre colored, ranges in thickness from a few meters to 60 m and its structures are fundamentally vermiform and pisolitic. Below the lateritic horizon is a zone of **mottled clays** with ferruginous red spots, 1–9 m thick; this layer is not present in all profiles. The transition toward laterite may be sharp or gradual (McFarlane, 1976). The lowest horizon is the **pallid zone**, also called a lithomarge, which is composed of kaolinitic clays and quartz grains; its thickness is variable (5–60 m). Its whitish coloration is due to the leaching of iron and, like the zone of mottled clays, it is not present in all profiles. Generally if the pallid zone is thick the laterite is thin and vice versa (Loughnan *et al.*, 1962). Below is a **transition zone** of variable thickness that is composed of weathered material; it may contain blocks of fresh rock and it may rest on parent rock (weathering front).

This ideal or standard profile serves as a guide for recognizing the various zones that may be present in a lateritic profile but as already discussed not all horizons are present in all profiles. Thick laterites or bauxites sometimes rest almost directly on fresh rock whereas thin laterites sometimes overlie very thick regoliths (McFarlane, 1983).

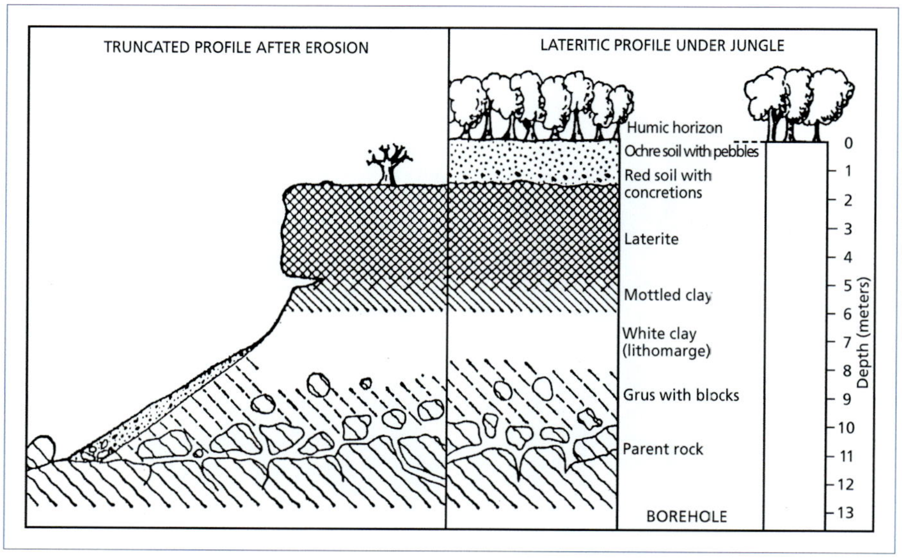

Figure 18.8 Lateritic profile below jungle and its equivalent truncated by erosion. Sudan (Millot, 1964).

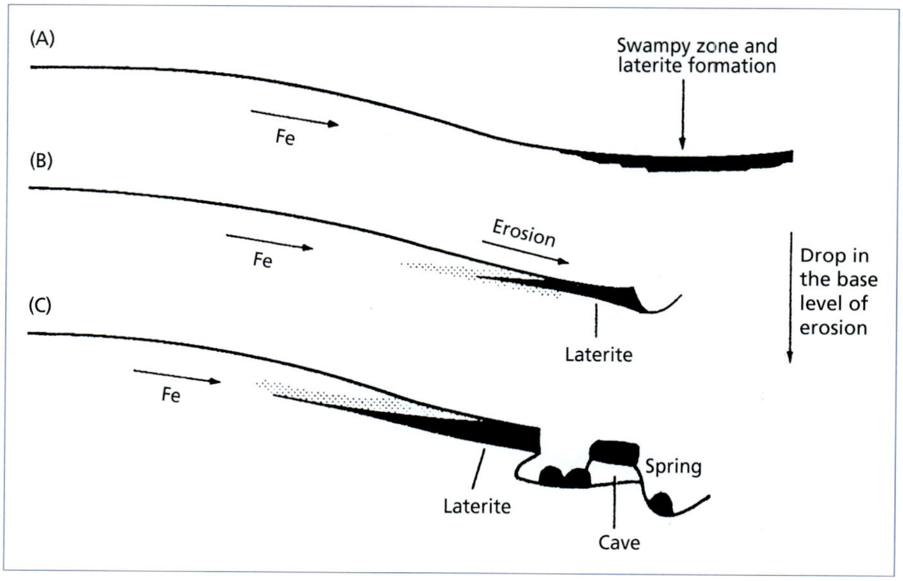

Figure 18.9 Lateral migration of iron oxyhydroxides and laterite formation. A: Fe migration into groundwater by organic acids and accumulation in swampy valley floors. B and C: Drainage incision with falling water table, which causes induration and erosion of laterite (Maignien, 1966).

18.2.4 Laterite formation

An early hypothesis claims that laterites originate from an alteration profile in which the pallid zone releases iron to the laterite. This enrichment results from capillarity and seasonal fluctuations of the water table. Capillary rise does not seem to be very effective and may be overemphasized (Goudie, 1973). In addition, a pallid zone about 60 m thick does not support capillarity as the mechanism. As the water table oscillates, especially in climates with humid and dry seasons, dissolved iron moves upward with the water table rise and can precipitate in that position (McFarlane, 1976; 1983).

Another explanation for iron enrichment is lateral mobilization from high parts of slopes toward valleys (Maignien, 1966) (*Fig. 18.9*) so that iron concentrates in the lower parts of slopes and valley floors. If the erosion indicated in

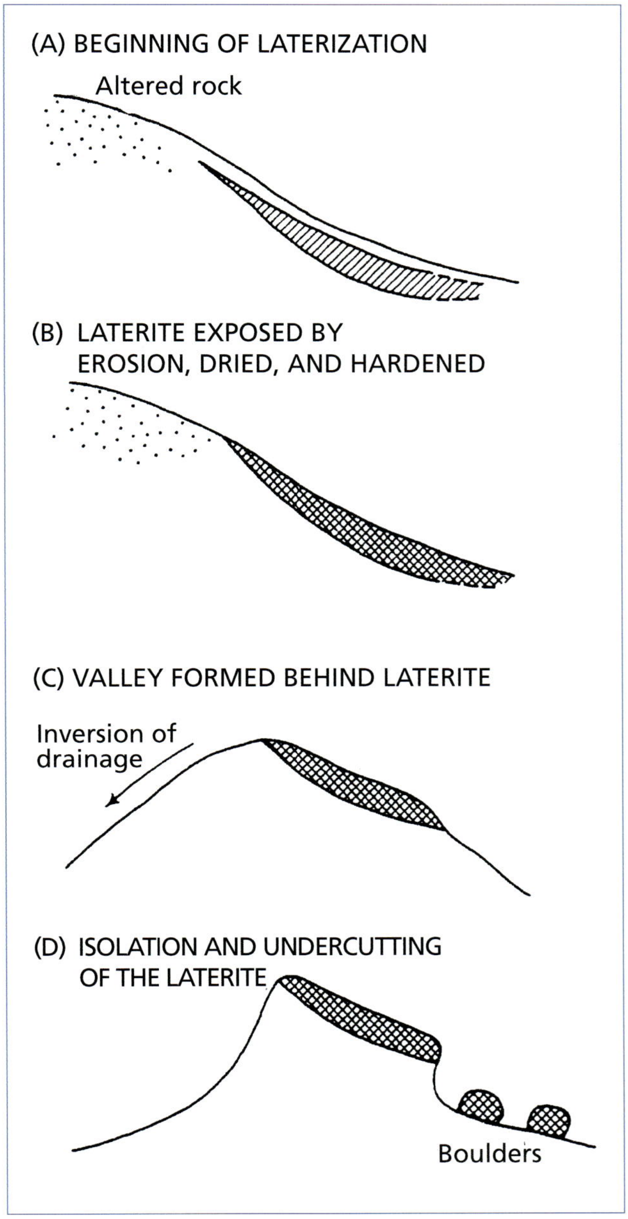

Figure 18.10 Inverted relief from erosion of slope-valley laterites (Clare, 1960; in McFarlane, 1976, *Fig. 20*).

the figure continues for a long time, it can produce inverted relief in which laterites present in low areas come to occupy interfluve positions (McFarlane, 1976) (*Fig. 18.10*).

To explain the great thickness of the pallid zone, Trendall (1962) proposes that as the topographic surface lowers through time, the profile also moves downward. These ideas were developed by McFarlane (1983) and by Tardy and Roquin (1992) to interpret significant thicknesses of laterites that overlie a regolith. Rather than considering the weathering profile as a static column, it is hypothesized that the profile lowers as it loses solutes and fine particles and that the horizons are in constant transformation. As a result, the weathered rock that provides iron occupies increasingly lower topographic positions, similar to the laterite. Research conducted on lower zones in Uganda (McFarlane, 1976; 1983) has made it possible to propose a progression of stages of laterite development that form by oscillation of the water table (*Fig. 18.11*). This evolutionary sequence requires a lowering topographic surface with continual segregation that develops in the interval of water table oscillation and produces a layer of pisolitic laterite. As the relief becomes smoother, these pisolitic laterite layers become massive vermiform laterite. As a result, profile and land surface evolution are interdependent. Erosion of the lateritic profile causes the basal weathering surface or the weathering front to crop out. These outcrops are etch surfaces (Büdel, 1957).

18.2.5 Laterite ages and their paleoenvironmental significance

Laterite formation probably began in the oxygen atmosphere of the early Proterozoic. Intense tropical weathering processes occurred during periods of high CO_2 concentration over a large part of the Earth's surface (Berner, 1991). The best-known episodes are those of the early Cretaceous to the early Eocene during which CO_2 concentrations doubled. Extensive weathering profiles of these ages have been observed around the world (Bardossy and Aleva, 1990; Valeton, 1983).

It is often difficult to determine the age of laterites. Many lateritic crusts are very old and transported laterites sometimes appear in the stratigraphic record. Paleomagnetic dating techniques measure the remanent magnetism preserved in secondary iron minerals in the lateritic crust. This method is useful for places that have changed their latitudinal positions as a result of lithospheric plate motion (Widdowson, 2007). This author also analyzes dating techniques used for laterites and lateritic profiles (K/Ar, $^{40}Ar/^{39}Ar$, uranium series, cosmogenic nuclides, and others).

Although the origin of lateritic profiles seems to depend primarily on the presence of a seasonal tropical climate, the global distribution of laterites of different ages reveals numerous uncertainties involved in various climate interpretations (Bardossy and Aleva, 1990; Tardy, 1993).

18.2.6 Bauxites

Bauxites are also called aluminum laterites and alcretes; they are considered to be the final product of deep weathering because the SiO_2 and Fe^{2+} have been intensely leached

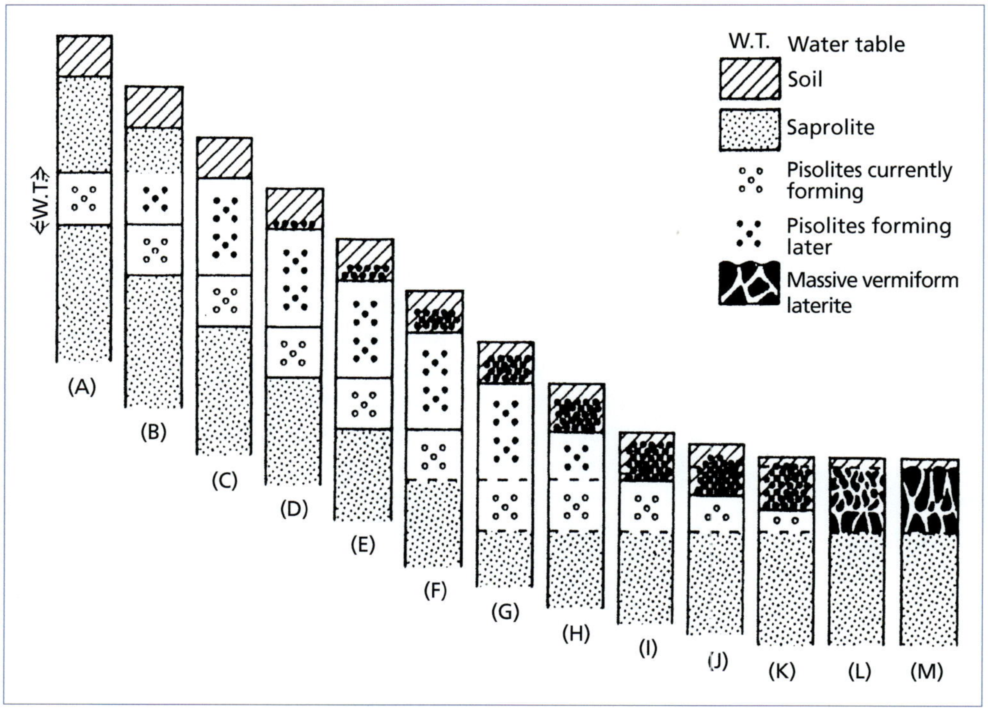

Figure 18.11 Evolution of laterites generated by water table fluctuations (A): segregation of iron to form pisolites within a narrow fringe of water table oscillation; (B) and (C): lowering of the water table causes a lowering of the zone of pisolite formation, with previously formed pisolites remaining in the vadose zone; (D) (E) (F), and (G): pisolites accumulate at the base of the soil; (H) (I) (J) and (K): the water table stabilizes and the continued leaching reduces the vadose zone, which moves the residual pisolite layer to a zone of intermittent saturation; (L) and (M): the pisolitic residuum is converted into massive vermiform laterite rich in goethite (McFarlane, 1983).

(Thomas, 1994a). They are present in the geological record in many places on the planet. Conditions for their development are related to humid tropical climates, porous parent rock, good drainage, and sufficient time for their generation (Ollier and Pain, 1996). Vertical facies distribution in a well-developed bauxitic saprolite depends on the type of parent rock, morphology, and drainage systems (Valeton, 1999). Bauxites can form by alteration of alumino-silicate rocks such as basic igneous rocks (West Africa) (Valeton, 1972; Boulangé and Millot, 1988), metamorphic rocks (Malaysia) (Allen, 1972), and volcanic rocks (Hawaii) (Sherman, 1952). Bauxites can form by direct alteration of alkaline rock with no loss of volume (Boulangé and Millot, 1988) or by indirect weathering in which kaolinite desilication produces gibbsite (Schellmann, 1977). Another type, detrital bauxite, forms breccias or nodular deposits (Allen, 1972). Tropical weathering can affect carbonate rocks and form karstic bauxites (Bardossy, 1981). Bauxites can also form on top of sediments (Tertiary arkoses of Weipa, Queensland, Australia) (Loughnan and Bayliss, 1961) and can form from the deposition of eolian dust (Brimhall *et al.*, 1988). On the western margin of the African rift, two stacked levels of bauxites (Chardon *et al.*, 2006) correspond to King's (1962) post-Gondwana (post-Cretaceous) and African (pre-Miocene) surfaces; this explains the re-bauxitization of bauxite clasts in the lower lateritic level. This implies significant paleoclimatic changes (Tardy, 1993). Tardy and Roquin (1998) found stages of laterite formation in the Triassic, Jurassic, middle Cretaceous, and Paleocene and another more hypothetical one in the Neogene in the Mediterranean basin. They consider the main phase to be the Paleocene.

Many types of bauxite are considered to have formed from the leaching of old laterites. Tardy and Roquin (1992) point out that some aluminum laterites have two layers of gibbsite enrichment: one near the base of the saprolite profile almost at the contact with the parent rock, and the other at the top of the profile near the soil surface.

Bauxitization increases with increased fluvial incision; it has been noted that bauxites are best developed along faults that facilitate leaching. Bauxite often crops out on scarps in lateral connection with the laterite mass, which shows that they do not form a continuous layer and that bauxite is genetically related to scarp evolution (McFarlane, 1983).

Approximately 10% of bauxites are the karst type (Ollier and Pain, 1996). Bauxite fills fissures, depressions, and cavities in karstified carbonate rocks. It was originally thought that bauxites were insoluble residues from the dissolution of calcareous masses but it is difficult to explain the formation of significant quantities of aluminum. According to Bardossy (1981) karstification creates depressed areas by differential dissolution; these closed karst depressions collect aluminum-rich sediments from surrounding areas. In addition, bauxitization is facilitated by the good drainage present in these areas and by protection from erosion. Aluminum enrichment occurs by leaching of foreign contributions (ash and volcanic tephra) transported by water, wind, and gravity. Weathering removes silica from them and enriches them in aluminum (Barany-Kerei, 2004). Carbonate dissolution in turn continues to enlarge the depression.

Bauxites are widely distributed in the geological record around the world (Valeton, 1972). In Spain, bauxites are karstic in origin; they are primarily located in the sub-Pyrenean zone at the lower boundary of the Cretaceous-Jurassic; in the center of the Cadena Costero Catalana in the Infralias and in the lower Cretaceous of northeastern Maestrazgo (NE Spain) (Combes, 1969).

Karstic bauxite formation has increased continuously since the Paleozoic (Bardossy, 1981) with an obvious increase during the Cenozoic, possibly related to increased formation of carbonate rocks. Neogene-age karstic bauxite accounts for one fourth of the world's karstic bauxite.

18.3 Tropical landforms

Tropical climates in areas affected by alpine deformation have high relief and large volcanic cones; examples are the Nevado del Ruiz (Colombia), Cotopaxi (Ecuador), La Soufrière (Guadalupe) *(Fig. 18.12)*, Pinatubo (Philippines), and Merapi (Java). These tropical regions are in mountainous regions of Central America, the Antilles, Southeast Asia, and Indonesia. In the latter three areas, dissolution processes affect large areas of carbonate rocks and have created landforms of conical and turret-shaped hills characteristic of tropical karst. These tropical zones are tectonically controlled and they are still undergoing Quaternary crustal instability. Alteration products generated in these areas are easily removed by fluvial activity. Thus, these areas are subject to vertical movement that generates relief, important explosive volcanic activity, and significant fluvial erosional activity.

In contrast, the shield or cratonic regions of Africa, Brazil-Guyana, India, and Australia are characterized as

Figure 18.12 La Soufrière volcano, with a dome at the summit, located in the southern part of the Island of Guadalupe (Lesser Antilles).

Figure 18.13 The large stratovolcano of Kilimanjaro (5,895 m) in Tanzania, rising above the plains of the Amboseli National Park (Kenya).

tectonically stable, at least during recent geological time. These circumstances have favored the development of extensive planated surfaces on top of primarily Precambrian rocks, which characterize these tropical shields. As a result, the topography is usually smooth and weathering products undergo only minor water erosion. The result is the development of thick weathering profiles that can exceed 100 m. Base-level lowering incises the regolith and creates a set of landforms by differential weathering and erosion (Faniran and Jeje, 1983). In addition, the rift valley system of east Africa system is associated with significant volcanism. The most notable examples of this igneous activity are Mount Kenya (5,194 m) and Kilimanjaro (5,895 m) (*Fig. 18.13*) in Tanzania.

18.3.1 Erosion and sedimentation in fluvial systems

Streams in the humid tropics carry very little bed load. Because of intense alteration the regolith is composed of sand, silt, and clay sized particles, and as a result, the rivers evacuate primarily materials of these dimensions and dissolved load. These sediments have little abrasive power and minimal erosional capacity. Research on bedload in tropical rivers is lacking; little data on this topic is available. More research is focused on mountainous areas where sediments are rapidly mobilized by significant flows that transport the entire solid load (*Fig. 18.29*) to the lower parts of the basin by downcutting into existing rocks. This process is in stark contrast to the morphodynamics of rivers in low-relief areas that remove materials primarily by suspension and dissolution (Thomas, 1994a).

A previous section contained an analysis of the fundamental characteristics of precipitation; we will now examine the hydrological, erosional, and depositional aspects of fluvial systems. During a rainstorm, overland flow occurs when the soil reaches its infiltration capacity. The runoff coefficient is a factor that expresses the percentage of precipitation that flows across the surface; it is an important factor in our understanding of the dynamics of the geomorphic system. Runoff coefficients have been calculated for plots, microbasins, and river basins. The results are highly variable over time because the duration and intensity of storms play a primary role in infiltration.

18.3.1.1 Quantification of soil loss

Soil loss data collected by various authors shows that soil loss varies by study area and depends on multiple factors (Fournier, 1960; Walling, 1984; Douglas and Spencer, 1985c). Even so, small **plots** (*Fig. 18.14*) on slopes with jungle or

Figure 18.14 Experimental plots for the study of water erosion. Katunami Experimental Station. Machakos (Kenya).

savanna vegetation and gradients of 5–15° have similar values of soil loss. Data obtained by Kesel (1977a) in Guyana, by Nortcliff *et al.* (1990) in Brazil, and by Roose (1973, in Thomas, 1994a) in the Ivory Coast range from 0.2 to 8.5 tons/hectare/year. These rates increase substantially when vegetation is cut or burned (Lal, 1986). With this type of vegetation, impact and rainsplash have no effect because a continuous and dense layer of fallen leaves and branches up to 25 cm thick covers the soil (Ruxton, 1967). For this reason, the particles are removed by water that flows under the layer of fallen leaves.

Research on river basins shows that the amount of suspended sediment load depends on the rainfall regime; it also varies from one storm to another. In addition, suspended load increases rapidly with increased discharge (Douglas, 1977). Data collected on suspended loads from experimental plots cannot be compared easily to values obtained in **small basins** because other types of processes are involved in small basins. **Suspended load** values range from 0.4 to 29.5 tons/km^2/year in Nigerian rainforest basins smaller than 20 km^2 (Ogunkoya and Jeje, 1987). In high relief areas such as eastern and southern Africa, sediment loss increases considerably to 100–1000 tons/km^2/year (Walling, 1984). However, in basins larger than 10,000 km^2, the amount of transported sediment decreases substantially (Pilgrim *et al.*, 1982). The Congo River transports only 11.3 tons/km^2/year and the Niger carries 33.1 tons/km^2/year (Milliman and Meade, 1983).

Data compiled by Douglas and Spencer (1985a) on suspended load, runoff, basin area, and percentage of plant cover on tropical rivers in various parts of the world can be used to make some considerations. The source of the Amazon is in the Peruvian Andes, and numerous large tributaries at the head excavate the Andean cordillera. These tributaries have great erosional power and the denudation rates are high whereas erosion is more moderate along middle and lower streams. Suspended load values (*Fig. 18.15*) that range from 0.3 to 115.9 m^3/km^2/year (expressed as volume) have been recorded (Gibbs, 1967). The highest erosion rates are found in islands of Southeast Asia, Indonesia, and Papúa-New Guinea where gradients and runoff are high. The Cilulung River (Indonesia) transports 8,000 m^3/km^2 per year, which is the highest value on record. Low relief areas such as the cratonic regions of Brazil, African, India, and Australia have much lower values, on the order of 30 m^3/km^2/year. In the lower tributaries of the Amazon where relief is almost flat, the load is much smaller, ranging from 0.3 to 3.8 m^3/km^2/year.

Research on plots and small basins in Malaysia show that **dissolved load** constitutes about 27.4% of the total (Douglas and Spencer, 1985a). In fluvial basins, chemical weathering ranges from 143 tons/km^2/year for the Yamuna River, a tributary of the Ganges, to 9 tons/km^2/year for the

Figure 18.15 Confluence of the Negro and Solimoes rivers, tributaries of the Amazon. The Solimoes carries a large suspended load, whereas the Negro transports only a small suspected load. The junction is called the "meeting of the waters" and takes place several kilometers east of Manaus (Brazil).

Congo or Zaire rivers; the Amazon has intermediate load of 35 tons/km^2/year (Sarin *et al.*, 1989). Even so, the average values of dissolved load in tributaries in the low Amazon basin are 5–8 mg/l, which is quite small compared to the average worldwide concentration for rivers, which is 120 mg/l (Walling and Webb, 1986). The highest dissolved concentrations are found in rivers that flow across evaporitic formations where values can reach 60,000 mg/l.

By averaging the load in all the world's rivers researchers calculated the average ratio of suspended load to dissolved load to be 3.5:1 (Meybeck, 1979). The highest values are found in the Ganges-Brahmaputra, which has a ratio of 11.0:1, because these rivers receive a large sediment supply from the Himalayan Cordillera. In contrast, the Congo River basin has little elevation contrast and for this reason its ratio is 0.91:1, which indicates that its dissolved load exceeds its suspended load (Walling and Webb, 1986).

18.4 Slopes and stone lines

In spite of the large surface extent of slopes in these areas (Young, 1972) there are few rigorous studies about them. This may be due to high vegetation density that makes research difficult and to the fact that many observations are based on visual estimation rather than field measurements of angles and landforms (Chorley, 1964).

Long and deep escarpments linked to various factors may interrupt the topography of shield areas. Differential erosion occurs along escarpments that form in sandstone; the most spectacular are found in Roraima (southeastern Venezuela) where vertical walls drop several hundred meters. In Africa, some large scarps are steps on extensive erosional surfaces of various ages (King, 1962). Scarps of tectonic origin created by endogenic activity during the Cenozoic reach their maximum extent along the rift valley system of East Africa (*Fig. 18.16*) where drop-offs may reach 1500 m. They generally occur in basic volcanic rocks; many are thought to be of recent origin because they are not altered by erosion (Nyamweru, 1980). In the Tertiary pre-littoral graben system between Sao Paulo and Río de Janeiro (Brazil) major scarps in the Serras da Mantiqueira and Do Mar are about 1000 m high (Petri and Fúlgaro, 1983).

Bare hills on crystalline rocks are common in cratonic areas; these are commonly inselbergs. They may occur as grouped hemispheric domes called **half oranges** (in Portuguese, *meias laranjas*) common in various states of eastern Brazil. The predominant morphology is convex on the upper

Figure 18.16 View of the bottom and western margin of the Gregory Rift Valley, seen from the Menengai Caldera. Nakuru (Kenya).

slope where creep predominates (Gilbert, 1909; Davis, 1930) and concave profiles in the lower parts. Straight-line faces or occasionally overhanging faces develop between them (Savigear, 1960; Thomas, 1965). Where a hill is composed of blocks, the slope profile is complex and is usually a stepped microrelief (Pugh, 1966).

Slopes on planate areas composed of altered materials are primarily convex, although valley incision changes the profile to convex-concave. Research conducted by Young (1970) in the Mato Grosso (Brazil) using Savigear's (1952) methodology, later amplified by Young (1964), distinguishes five profile types (*Fig. 18.17*): (a) has a continuous convex shape with low angles associated with low order valleys and makes up 44% of all profiles; (b) and (c) have smooth convex crests interrupted by steps and incised by valleys. Profiles (d) and (e) have laterites at their summits and much more obvious concavities. Swan (1972) claims that profiles formed on top of coarse textured regoliths have convex morphologies due to high porosity whereas in fine-grained weathering deposits, slope morphology is concave.

Surface water flow and mass movements are the fundamental processes that build slope accumulations. Vegetation and the rainfall regime control colluvium movement and can trigger landslides during high intensity rainfall. An important problem common to slopes in other morphoclimatic zones is the unknown age of these deposits. Some ages have been obtained from archeological remains; the study of soil sequences superimposed on various slope deposits has established relative dating. In turn, slope deposits may alternate with alluvium or interfinger between them. Climate change toward more humid and dry periods has been recorded in slope deposits in rainforest areas (Thomas and Thorp, 1992) and is common in savanna areas (Watson *et al.*, 1983, among others).

Stone lines (Sharpe, 1938) also called "stone layers" by Stoops (1989) and Twidale (2006) are continuous or discontinuous layers of slightly weathered clasts, normally composed of quartz, quartzite, and laterite fragments. Stone lines are 0.1 to 1 m thick. They are found within slope deposits at depths of 0.5–2 m below the ground surface (*Fig. 18.18*) although in southwestern Nigeria they are 5 m deep. Stone lines are usually straight but on occasion they may have multiple undulations. In general a single stone line develops but Fölster (1969) found three layers on slopes in Nigeria. They are primarily limited to humid tropical zones although they have also been cited in temperate climates (Stoops, 1967; Johnson and Balek, 1991; and Segalen; 1994).

Various hypotheses have been proposed for their origin. One theory suggests that stone lines are residual layers resulting from intense **laminar water erosion** during an

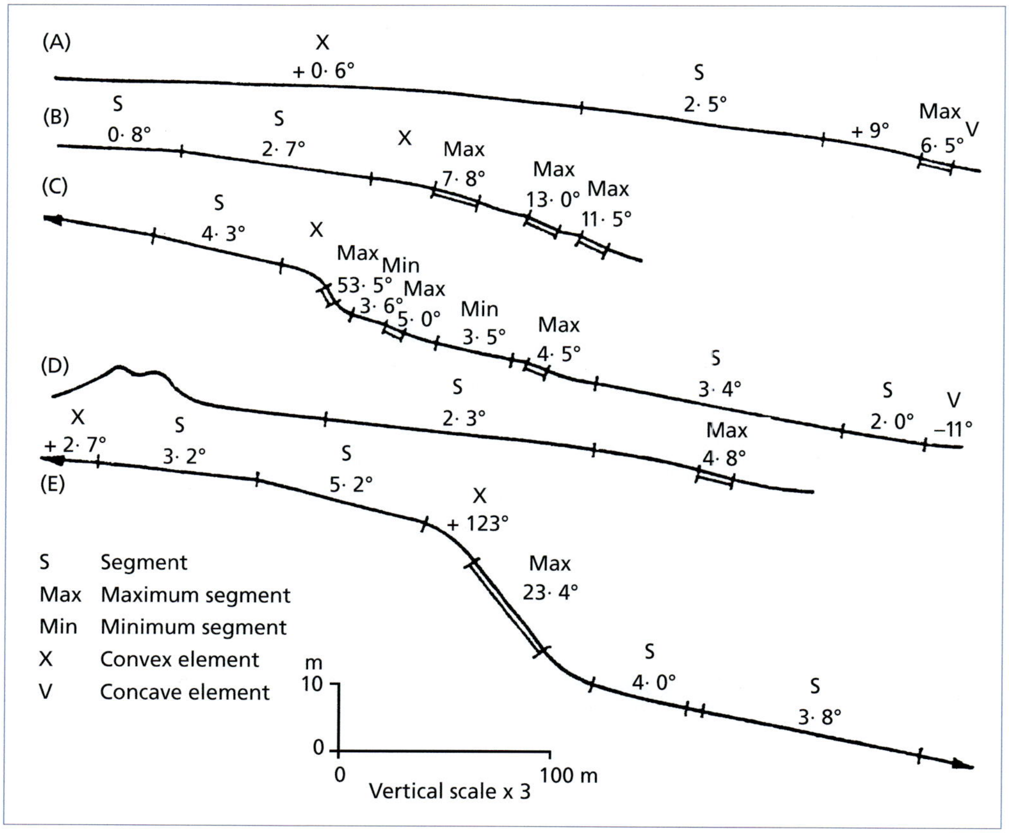

Figure 18.17 Slope profiles in the Mato Grosso region (Brazil) (Young, 1970).

Figure 18.18 Stone lines located 1 m below the surface. The weathered granitic bedrock can be seen in the forefront. Machakos (Kenya).

arid period when the fine material was removed. The stones accumulate on the slope as deep weathering takes place. The stone line is later covered with fine material during a transition period to humid conditions (De Ploey, 1964; Vogt, 1966; De Ploey and Poesen, 1989; Stoops, 1989). It has also been proposed that fine and dissolved material is removed by groundwater below the stone line (Young, 1976; McFarlane and Pollard, 1989).

Other authors propose that **creep** (Sharpe, 1938) affects soil and quartz clasts from small abundant dikes and moves and redistributes the stones. These stones are located at the saprolite-soil contact, creating the stone line (Berry and Ruxton, 1959). Creep is most active at soil discontinuities at depths of 0.5 m; it may be accompanied by small laminar landslides and by interstitial flow that helps remove fine material (Moeyersons, 1989). These processes move stones toward discontinuities and generate stone lines.

Another hypothesis supported by geomorphologists and soil scientists proposes that **termite activity** mobilizes fine fractions of the soil (<2 mm). These small particles are transported from the subsoil to the surface and the remaining coarse fragments form the stone line (Williams, 1978; De Dapper, 1978, 1989). Multiple lines of evidence favor this interpretation: particle size on top of the stone line is the same as that of the termite nest; termite tunnels are usually observed underneath stone lines; these stone lines are synformal below termite nests *(Fig. 18.19)*. Even so, regions like Madagascar have stone lines but rare or no termites (Young, 1972); in other areas rounded stones, allochthonous fragments, or prehistoric artifacts contradict this interpretation. Thomas (1994a) compiled calculations on generation rates by termite bioturbation in the upper part of the soil. It was estimated that a 3–5 m thick layer could form in 6,000–10,000 years although surface erosion was not taken into consideration. In studies of the Northern Territory (Australia) Williams (1978) calculated the creation rate to be 0.03 mm/year, a number much lower when water erosion is considered.

Johnson (1993) and Johnson *et al.* (2005) establish a model called the theory of dynamic denudation in which soil-forming processes in soils and sediments of various size particles are thought to produce the stone layer. The lower layer forms by displacing clasts by biosorting that concentrates a layer of stones whereas faunal activity favors the development of the upper layer and moves large volumes of fine material from depth to the surface.

The interpretation of stone lines as a residual layer requires an earlier climate change toward more arid conditions. Bigarella and Andrade (1965) demonstrated these circumstances in Curitiba (Brazil) where they interpret stone lines as paleopavements. Erosion rates and soil deformation by termites range from thousands to hundreds of thousands of years. During these time intervals significant climate changes are known to have affected geomorphic processes and the flora and fauna of the region. This suggests that processes used to support various hypotheses of stone line origin do not act separately or with the same intensity over long time

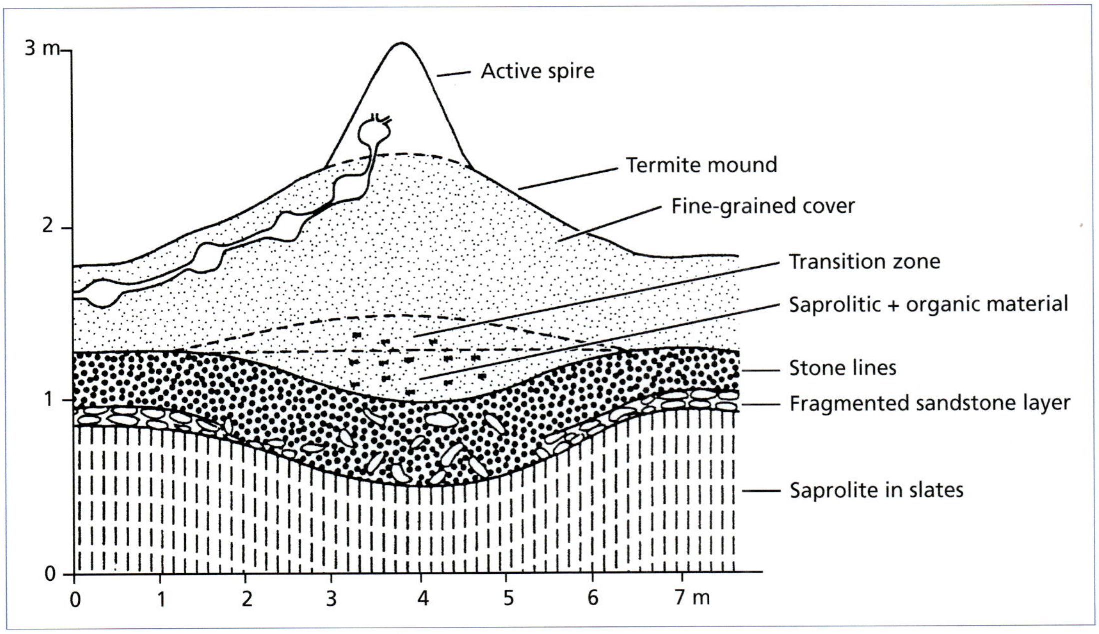

Figure 18.19 Termite activity under a stone line near Lubok Perong, Perak Terap, Malaysia (de Dapper, 1989).

periods. Even so, a sequence of activities can be established beginning with aridification that leads to the removal of fine material and its deposition in valley bottoms. Larger fragments form a stone line that is later buried by fine material of mechanical or biological origin (De Dapper, 1989).

Twidale (2006) thinks that "stone layer" environments occur in climates that range from hyperarid to semiarid and monsoonal. The stone layer is a function of meteorological weather (storms) more than climate. Along with the removal of fine material by water and wind, wetting and drying phenomena in materials that contain swelling clays move the soil, like a gilgai.

18.5 Landforms shaped by fluvial erosion

In some river basin headlands, plant cover may diminish for natural or anthropogenic reasons. Under these circumstances, gullying affects the slope and this process may proceed rapidly (Morgan, 1986) *(Fig. 18.20)*. The **gully head** retreats by **sapping** and the gully walls remain vertical if the gullying erodes materials that are relatively uniform. When the top is resistant rock, the base erodes more rapidly and water that flows over the head has a cascading effect that emphasizes **basal undercutting**. This causes the overlying material to collapse and the head to retreat. The process of piping is relatively common in gully systems and it participates actively in their development (Jones, 1981). The roof of subsurface conduits collapses into the gully head and causes it to retreat (Crouch, 1983) *(Fig. 18.21)*. It can also cause the bed of the gully to collapse, which deepens and widens the gully. Active piping can also generate gullies (Löffler, 1974; Jones, 1990; among others). Numerous factors can cause piping, including cracking, hydraulic gradient, exchangeable sodium percentage (ESP), and the presence of swelling clays. The role of each of these has been discussed in the discussion of piping in the section on arid regions (Section 16.5.4). Headward gully retreat in Nigeria has been studied by comparison of aerial photographs from 1956 to 1972 (Jones, 1975; in Thomas, 1994); researchers obtained rates of 20 m/year for large gullies. Crouch (1990) used microtopographic profilers in New South Wales (Australia) where he calculates a retreat rate of 1 m/year. Oostwoud and Bryan (1994) study gullies by applying a sediment balance and Lameluk (Baringo District, Kenya) obtained retreat rates that range between 0.5 and 10 m/year.

The gullies discussed above are located in cratonic areas where relief is generally low and flat. However, areas shaped by lithospheric plate interaction during alpine orogenesis have irregular topographies. In these areas, fluvial activity has significant erosional capacity and is capable

Figure 18.20 Meandering gully incised into unconsolidated sediments. Kathorin Basin. In the distance, Lake Baringo (Kenya).

Figure 18.21 Collapses due to piping that affects the head, wall, and floor of a gully. Njemp Flats. Baringo District (Kenya).

of transporting large amounts of materials to lower areas. A network of deep gullies with sharp walls illustrates this dynamic (*Fig. 18.22*).

Areas with high relief developed during alpine orogenesis are located primarily in island areas; they have abrupt topographic gradients. Large gullies, which are radial on many volcanic islands, move large quantities of sediments with great speed; they deposit these sediments in areas of where the gradient changes, generally along the coastal margin. This is the case of the coastal fan produced by the Sèche River (Island of Martinique, Lesser Antilles) (*Fig. 18.23*). The natural cross section shows a visible thickness of about 12 m made up of heterometric stratified deposits. Intercalated automobile tires, possibly from the 1960s and 1970s, form a laminated deposit. This leads us to believe that erosional activity is intense and rapid due to high precipitation and frequent cyclones (700 mm/day in August 1970, according to Ercole, 1994) and due to the presence of pyroclastic material and significant chemical weathering in the drainage basin. Deep incision subsequent to fan deposition exposes the deposit. This alternation of intense aggradation and incision has developed over just a few decades.

Rivers that cross extensive plains of tropical cratonic areas flow on top of a weathering layer of variable thickness. For this reason, incision occasionally causes the bedrock to crop out (*Fig. 18.24*). Rivers transport primarily fine particles in suspension, which controls the channel shape. **Straight channels** form, which Schumm (1981) describes as system 1; they have low slopes, relatively uniform width, and minor bed load. Current velocity and energy are reduced, channels are normally straight and deep, and their margins are very stable due to their silty-clay content. As a result, these channels are stable because they do not show much lateral migration (*Fig. 8.31*). Meandering rivers (Schumm's (1981) system 3a) also form when bed load is small compared to total load but they form on a river with a somewhat higher slope; these are sinuous channels (*Fig. 18.25*) with suspended load, similar width, and stable margins due to their constitution although the meanders will finally be cut off.

The longitudinal profile of a river occasionally changes when its slope increases; this causes **rapids** on the rocky bed. Another change occurs at abrupt breaks in the profile; this is illustrated by vertical cascades and cataracts of diverse origins. These abrupt changes in the longitudinal profile of a river are knickpoints, which tend to migrate upstream over time until they disappear (Petts and Foster, 1985). Most rapids are structurally controlled by dominant fracture systems (Howard and Dolan, 1981). Potholes can form in rapids when flow is concentrated on a particular point over a long period of time. They are formed by eddies and can be quite deep; they have steep walls with spiral grooves. Rounded pebbles are often found in the bottom.

A **cascade** is an abrupt interruption of the longitudinal profile of a river where water falls vertically (*Fig. 18.26*); **cataracts** are groups of cascades (*Fig. 18.27*). They are undoubtedly the most spectacular fluvial landforms and are often tourist attractions. In addition, they can provide hydroelectrical energy. Their origin is related to eustatic variations, tectonic uplift, and differential resistance to erosion of the riverbed rocks. Angel Falls (Venezuela), at 980 m the highest cascade in the world, flows over the Roraima sandstones. Iguazú Falls, at the intersection of the borders of Brazil, Paraguay, and Argentina, drops 84 m over the late Jurassic-early Cretaceous Paraná flood basalts. Humid tropical areas also have important falls: Victoria Falls on the Zambezi River on the border between Zambia and Zimbabwe.

Schwarzbach (1967) did a genetic classification of cascades. He distinguishes between destructive cascades

Figure 18.22 Deep gully network dissect acid volcanic materials on the island of Dominica (Lesser Antilles).

Figure 18.23 Alluvial fan deposit incorporating recent tires near its base and later incised. This cut shows the rapidity of geomorphic processes. Sèche River. Island of Martinique (Lesser Antilles).

where water erodes the headlands and constructive cascades, in which water deposits $CaCO_3$ and the cascade grows over time. The cascades and cataracts described above are destructive cascades that have formed in horizontally stacked materials with differential resistance to erosion.

Cascades in tropical rivers erode much less than cascades in higher latitudes (for example, Niágara Falls); tropical rivers have smaller bed loads and as a result, cause less abrasion. In tropical zones with significant topographic contrast, the percentage of bed load increases considerably. Clasts travel only small distances because chemical weathering and abrasion break them down (Douglas, 1977). Evolution of cascades and cataracts depends on knickpoint recession, which is a function of discharge variations and rock strength. Cascades and cataracts may erode more quickly along lines of weakness. Studies on the evolution of Niagara Falls show that the head has gradually attained an arc-shaped morphology, which is thought to be the most stable shape (Philbrick, 1970). Scarp retreat rates for the various cascades and cataracts throughout the world range from 0.25 to 1.65 m/year (Douglass, 1977). Deep closed depressions called plunge basins develop in the rocky bed at the base of cascades; their size is a function of waterfall height, the size of the solid load, and rock strength. The depth of these closed depressions is inversely related to the rate of headward erosion (Philbrick, 1970). As water hits the bottom, it causes retreat because it erodes the base of the wall and may cause it to collapse.

Figure 18.24 The straight channel of the Mara River incised into a weathering mantle on the Loita plains. The amphibolite bedrock can be seen in the foreground. Masai Mara Natural Reserve (Kenya-Tanzania border).

Figure 18.25 Plain where a meandering channel has incised into the regolith. Other tors landforms can be seen nearby; the Tertiary of the southern margin of the Duero Basin is in the background. Campo de Azálvaro (Province of Ávila, Spain).

Figure 18.26 The Thompson Cascade on basalts of the eastern margin of the Gregory Rift Valley. Nyahururu (Kenya).

18.6 Fluvial sedimentary landforms

In the humid tropics, many heads of fluvial networks in low relief areas are elongate depressions with flat bottoms and no clearly defined channel; they are seasonally flooded and covered with herbaceous vegetation (Mäckel, 1974). These areas are called **dambos** in central Africa and other names in various tropical regions of the world (Thomas and Goudie, 1985; Boast, 1990; Thomas, 1994a). They are typical of savanna climates that receive between 600 and 1500 mm of rainfall although they also occur in rainforest areas.

Dambos have concave margins with low gradients, generally between 2° and 6° (Acres *et al.*, 1985). Lines of weakness in the bedrock affect their shape in plan view and they align along the predominant directions of these lines of weakness. Soils that form on dambos are hydromorphic and gley soils (Young, 1976). Their organic content is higher than that of interstream areas and it increases toward the center of the dambo.

No shrub or arboreal vegetation grows on dambos because the water table is very close to the surface. A wide variety of herbaceous vegetation is present and its percentage increases toward the center because of the persistent zone of saturation. Vegetation changes throughout the year because the rainfall regime causes the water table to oscillate (Mäckel, 1974). The vegetation is zoned; arboreal species surround the dambos and end abruptly at

Figure 18.27 Iguazú Fall at low discharge, on the Paraná flood basalts in a place called Garganta del Diablo. Notice the arc shaped margin of the cataract. Garganta del Diablo located on the borders of Brazil, Paraguay, and Argentina.

herbaceous vegetation. Predominantly grasses and other hydrophilic plants grow on top of sandy soils that crop out in the central area where the water table fluctuates (Mäckel, 1985).

Dambos form primarily in central and southern Africa on African and post-African erosion surfaces (King, 1962) on granitic and metamorphic rocks. The flatness of the land promotes dambo formation and allows the water table to flow to the surface as springs or be present at shallow depth (Boast, 1990). In addition, flat relief promotes water infiltration and the development of thick saprolites; according to Raunet (1985) most dambos are located in this environment. Coltrinari and Nogueira (1989) claim that in the Taubaté Basin in southeastern Brazil, the slopes that surround dambos are steps and small circular depressions with or without hydrophilic vegetation. Their origin may be related to landslides of Tertiary sediments.

Some authors think dambos form in relation to drainage system dynamics and evolution whereas others point to differential chemical and biochemical weathering. The first group claims that a stage of valley head incision is followed by alluvium and colluvium fill (*Fig. 18.28*). As fluvial activity diminishes, slope washing transports fine material to the valley bottom. Particle movement seems to coincide with the beginning of the rainy season when less vegetation cover is present (Mäckel, 1974, 1985). Meadows (1985) attributed stages of incision and valley fill in Malawi to climate change.

Another hypothesis holds that dambos form independently of the fluvial network because some are not part of such a network. The principal mechanisms are thought to be chemical and biochemical weathering (McFarlane, 1989; in Boast, 1990). Surface irregularities are interpreted to have formed by differential weathering. Higher fracture density in the bedrock that facilitates water percolation and rock alteration controls these irregularities. Solutes leach toward lower parts of the profile and subsurface flow is directed toward streams. These processes lead to a slow lowering of the land surface and they create areas of highs and lows with very little elevation contrast. In addition, fluvial network rejuvenation affects dambos because incision causes a drop in the water table and increases leaching as the hydraulic gradient

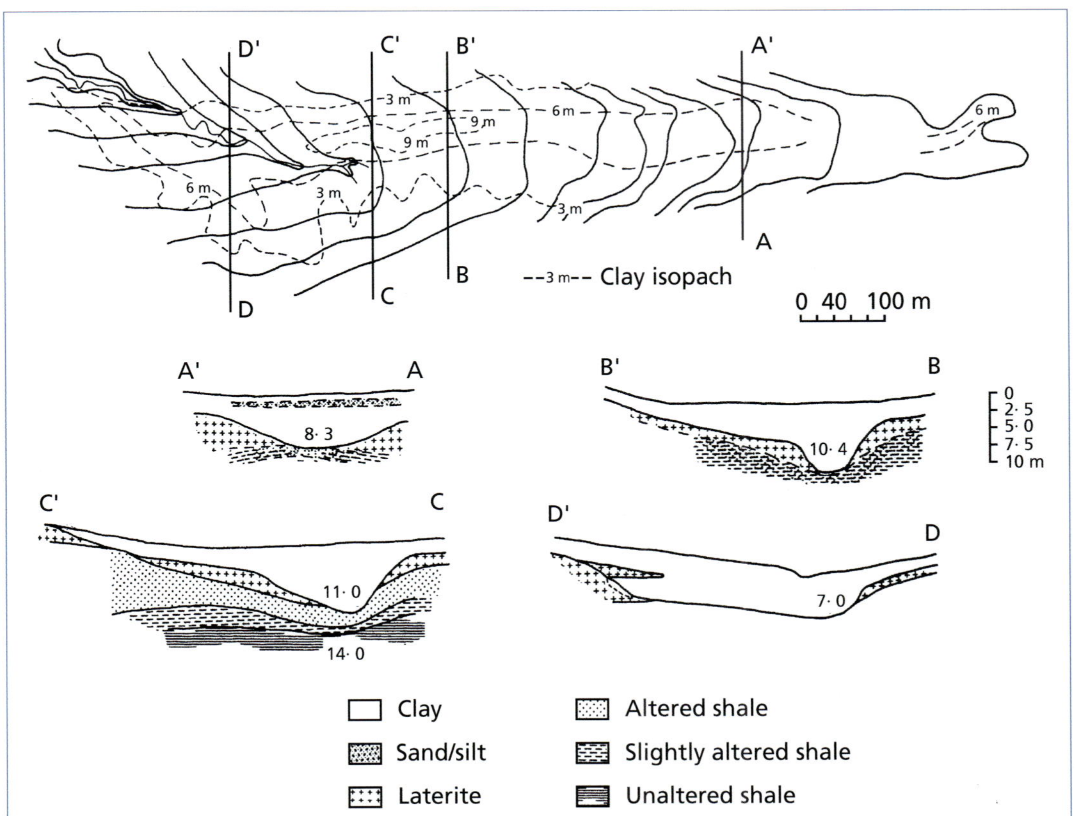

Figure 18.28 Plan-view and cross sectional morphology of a dambo in Kankamo (Zambia) showing the shape of the buried channel and sand, silt, and clay fill (Clark, 1974; in Thomas, 1994a, redrawn by Clarke, 1974 in Fookes and Vaughan, 1986, *Fig. 11.17*).

increases (McFarlane and Whitlow, 1990). McFarlane (1989, in Boast, 1990) proposes an evolutionary sequence that begins with a small circular depression that turns into a large dambo as the surface lowers. Drainage network rejuvenation converts large dambos into smaller linear shapes. Finally, if the water table continues to lower, individual dambos increase in size.

Landforms that form in the **flood plains** of tropical rivers are similar to those that form in other morphoclimatic zones. They are difficult to study because of plant cover and as a result, not much data is available about them. The solid load of tropical rivers is basically sand although when the riverbed slope is higher, the bedload is primarily gravel (*Fig. 18.29*). Fine particles deposited by vertical accretion can form stable channels due to the cohesion of silty-clay particles on the margins. Most of the time the flood plain is not affected by fluvial activity but during big floods, which are common in these areas, intense erosion and deposition occur.

Braided fluvial systems have gravel or sand bars that are separated by low-sinuosity channels. Because the bars are not very high, significant flooding completely covers them and can modify their morphology and position. Meandering rivers (Leopold *et al.*, 1964) develop sand and gravel scroll bars by lateral accretion; the bars are separated from each other by elongate muddy depressions (swales) that fill with well-stratified fine-grained material. Both are components of point bar deposits. When a meandering channel floods, it forms natural levees, which are wedges of fine sand that dip gently from the channel margin toward the floodplain. When a major flood occurs, water flows in small channels that cut into the older sediments, creating sandy tongues called crevasse splay deposits. In the lowest parts of the flood plain, silt and clay particles are deposited in pond and shallow lake zones; these are called backswamp deposits. Figure 18.30 is a map showing the distribution of some of these sub-environments in the Rufiji River (Tanganyika). In terms of muddy zones the most important region is the Tonle Sap or Great Lake in Cambodia, which fills with floodwater from the Mekong River between May and October (Douglas, 1977).

Sedimentary processes are highly complex and change rapidly because of the significant seasonal variations of these rivers. The Auranga River in northeastern India has a monsoonal climate with four months of rainfall (June-September) and storms with intensities up to 200 mm/hour (Gupta and Dutt, 1989). During the dry season, the river is dry and has flat sand bars. During flooding the river destroys the braided bars and reverts to a meandering regime that deposits point bars with sand and gravel bars.

Figure 18.29 Gravel bed load in a braided river coming from the edge of the Irazú volcano (3432 m). Sucio River, Braulio Carrillo National Park (Costa Rica). Photo F. Gutiérrez.

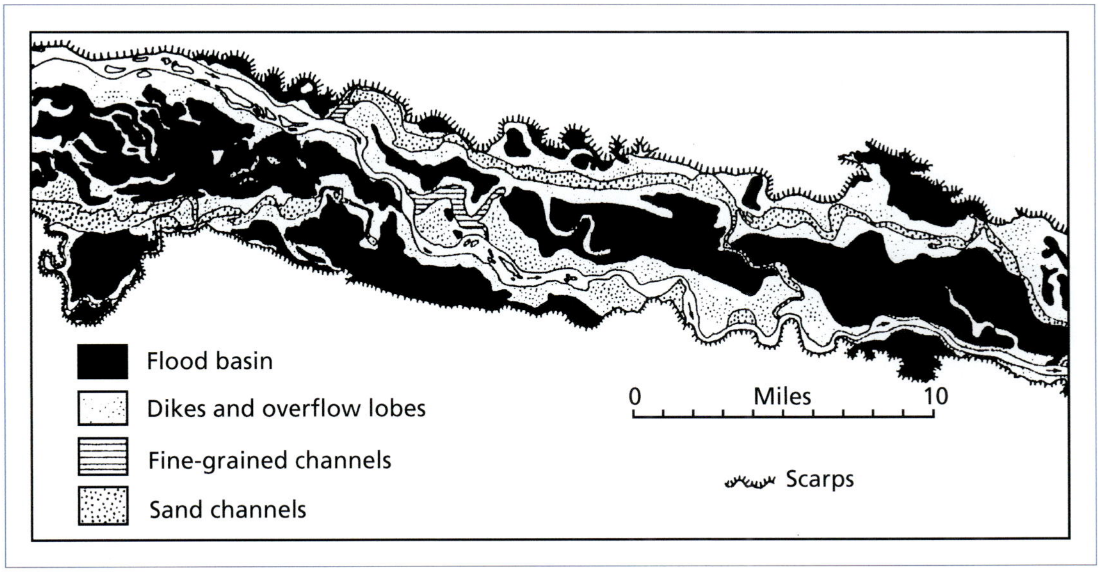

Figure 18.30 Map of part of the Rufiji River (Tanganyika) showing the distribution of various fluvial subenvironments (Anderson, 1961).

18.7 Large tropical rivers

In recent decades researchers have studied the fluvial geomorphology of all morphoclimatic zones. Research has focused on various topics: geomorphology, sedimentology, hydrology of flooding and paleoflooding, and the relationship between tectonic and fluvial processes. Even so, our knowledge of tropical rivers is limited (Labrubesse *et al.*, 2005). Research on tropical rivers describes areas where precipitation is greater than 700 mm/year in river basins that range in size from 10^4 to $6{\cdot}10^6$ km^2.

Hayden (1988) classified the hydroclimatology of flooding in baroclinic climates (which predominate at high latitudes) and barotropic climates (common in tropical low latitudes where atmospheric pressure is constant and thermal gradients are low). In these regions, the dominant mechanisms for rainfall production are related to the Inter-Tropical Convergence Zone where trade winds and cyclonic systems converge (Wohl, 2007). Annual precipitation and rainfall intensity are very high in equatorial and tropical latitudes; these factors combined with orographic effects have produced record rainfall intensities.

Tropical rivers drain a wide variety of geological and geomorphic environments: cordilleras, sedimentary and basaltic platforms, cratonic zones, plains of low ground in sedimentary basins and mixed terrains (Latrubesse *et al.*, 2005). Cordilleras form during alpine orogenies (Andes, Himalaya, and island arcs of Southeast Asia). Platforms contain sedimentary basins and plateau basalts (for example, the basins of north and central Brazil, the Deccan basalts, and the platforms of central Africa). Cratons are low areas that consist of Precambrian rocks (cratons of Brazil-Guyana, Africa, India, and Australia). Finally, lowland plains in Cenozoic sedimentary basins include active foreland basins (basins of the Andes and the Indo-Ganges plains). The diverse basins of the western and eastern Amazon and those of central Africa (Congo Basin) belong in this category. Rivers that flow across the various units of those mentioned above are classified as mixed.

The **hydrology** of rivers shows us the ten largest rivers of the world: the Amazon, Congo, Orinoco, Yangtze, Madeira, Negro, Brahamaputra, Japura, Paraná, and Mississippi. Eight of them are tropical rivers. The Amazon contributes 20% of the world's total river discharge to the ocean (Wohl, 2007). The great complexity of tropical climates makes it impossible to establish a single regime for tropical rivers. Some, like the Congo and the Magdalena, have two flood peaks every year. Two major types can be established on the basis of average monthly discharges: a) rivers with high and low discharges based on rainy periods (Mekong), and b) rivers with two annual flood peaks based on bimodal rainy periods, the principal in the summer and another in the fall (Magdalena and Congo) (Latrubese *et al.*, 2005; Wohl, 2007).

Based on the ratio between daily maximum discharge and average annual discharge (Qmax/Qaverage) several types of regimes can be distinguished (Latrubesse *et al.*, 2005): 1) Rivers that have high discharge variability are high flood regimes; 2) Rainforest river basins have low Qmax/Qaverage ratios and savanna rivers have higher values; 3) Rivers in arid and semiarid environments have very irregular discharges and Qmax/Qaverage is high; 4) Perennial tropical rivers are highly variable, with Qmax/Qaverage ratios up to 190; and 5) The most extreme regimes are the monsoonal rivers of India. During the summer monsoon the rivers have high discharge peaks.

Sediment transport of the largest tropical rivers is highest in high relief areas. Rivers that drain southern and southeastern Asia deposit into the ocean more than 70% of the sedimentary load they receive (Milliman and Meade, 1983). Water erosion is also significant in the Andes; the Magdalena River, which drains the Colombian Andes, contributes 144–200 million tons of suspended load to the Caribbean (Restrepo and Hjerfve, 2000). The Orinoco River transports 150 million tons per year to the sea (Meade, 2007). Rivers that drain tropical rainforest zones carry a large sediment load to internal Tertiary basins. The sediment load of the Amazon measured at Obidos near the point of maximum sediment discharge is about 1,200 million tons per year. Only two rivers in the world, the Ganges-Brahamaputra and the Yellow (Huanghe), have the same order of magnitude (Meade, 2007). Riverbeds that flow over cratons and platforms with savanna and humid tropical climates transport little sediment. The Congo River, the second largest in the world, with an average discharge of 40,000 m^3/second, carries only 40 million tons of sediment per year (Meade, 1996). Laraque *et al.* (1995) think that the total load of sediment and solutes in the Congo near its mouth in Brazzaville is about 91.8 million tons. This annual sediment load is composed of 7.9 tons of sand, 22.8 tons of silt and clay, and 61.1 tons of dissolved material. This number is twice the estimate of other researchers (Runge, 2007).

Channel morphologies of tropical rivers are highly variable. They have the traditional straight, meandering, and braided forms but with numerous transitions among them. For this reason it is helpful to use terminology such as unique and multiple channel systems (Schumm, 1985b) or anabranching systems (Nanson and Knighton, 1996). Rivers that drain mountain ranges contain more than 88% suspended load and 2–15% bed load; the channels alternate between sinuous and straight. Meandering systems, which have a mixed load, are typical of rivers that flow across platforms (Paraguay River, tributary of the Paraná). The Brahmaputra River is a system of multiple channels of moderate sinuosity (Bristow, 1987; Coleman, 1969). Rivers in cratonic zones have minor suspended load compared to bed load, like the Paraná, and have low sinuosity channels.

Additional factors can modify fluvial channels. **Neotectonics** affects fluvial channels; the shape of a river can be used as a geomorphic indicator of active tectonics in large rivers such as the Mississippi (Schumm *et al.*, 2000) and the Amazon (Mertes and Dunne, 2007). Basement topography is another important control since the river conforms to bedrock outcrops that change its longitudinal profile (Latrubesse *et al.*, 2005).

It is important to understand fluvial processes and landforms in order to interpret ancient sedimentary sequences. The analogy of present-day fluvial landforms to understand old sequences is poor and incomplete (Miall, 1996).

18.8 Landforms that develop in laterites

Laterite landforms are found in the humid tropics and in semiarid zones adjacent to savanna regions. Fluvial incision of lateritic deposits is primarily responsible for the formation of tabular deposits topped by laterites. They can persist for long periods because of their resistance to erosion.

These tabular landforms are **platforms** or **mesas** of reduced dimensions that stand out above the landscape; these landforms slope less than 0.5°. Their margins are sharp scarps that retreat as they weather, which reduces the areal extent of the landform. The literature contains many examples of these tabular landforms (*Fig. 18.31*) (Maignien, 1966; Thomas, 1974, 1994a; McFarlane, 1976). Fluvial incision in a weathered area can form a new valley into which Fe^{2+} can flow, causing iron enrichment that leads to the formation of a new level of laterite. Successive downcutting creates a benched or **banked morphology** (*Fig. 18.32*). Destruction of the laterite scarp provides fragments for lower parts and creates **slopes** with loose blocks of laterite (*Fig. 18.33*) and re-cemented colluvium that may connect to pediments (*Fig. 18.7*). Laterites generally harden as a result of fluvial incision. When dry they are very resistant to external agents and after a long period of erosion they occupy dominant positions whereas they had been previously limited to valleys. This variation in their topographic location is called **inverted relief** (*Fig. 18.10*).

Undercutting by mechanical processes and chemical removal by the circulation of underlying groundwater fragments the laterite (*Fig. 18.34*). This causes a slight arching of the rim of the mesa or platform, called **cambering** (Moss, 1965). Underground water flow aided by fractured laterite and porous regolith usually generates tunnels and cavities of various sizes. This subsurface network triggers ductile or brittle subsidence and creates **closed depressions** that range in size from a meter to hundreds of meters and are scattered over the horizontal surface of the tabular landforms. These pseudokarst landforms may have diffuse or sharp margins (De Chetelat, 1938; Goudie, 1973).

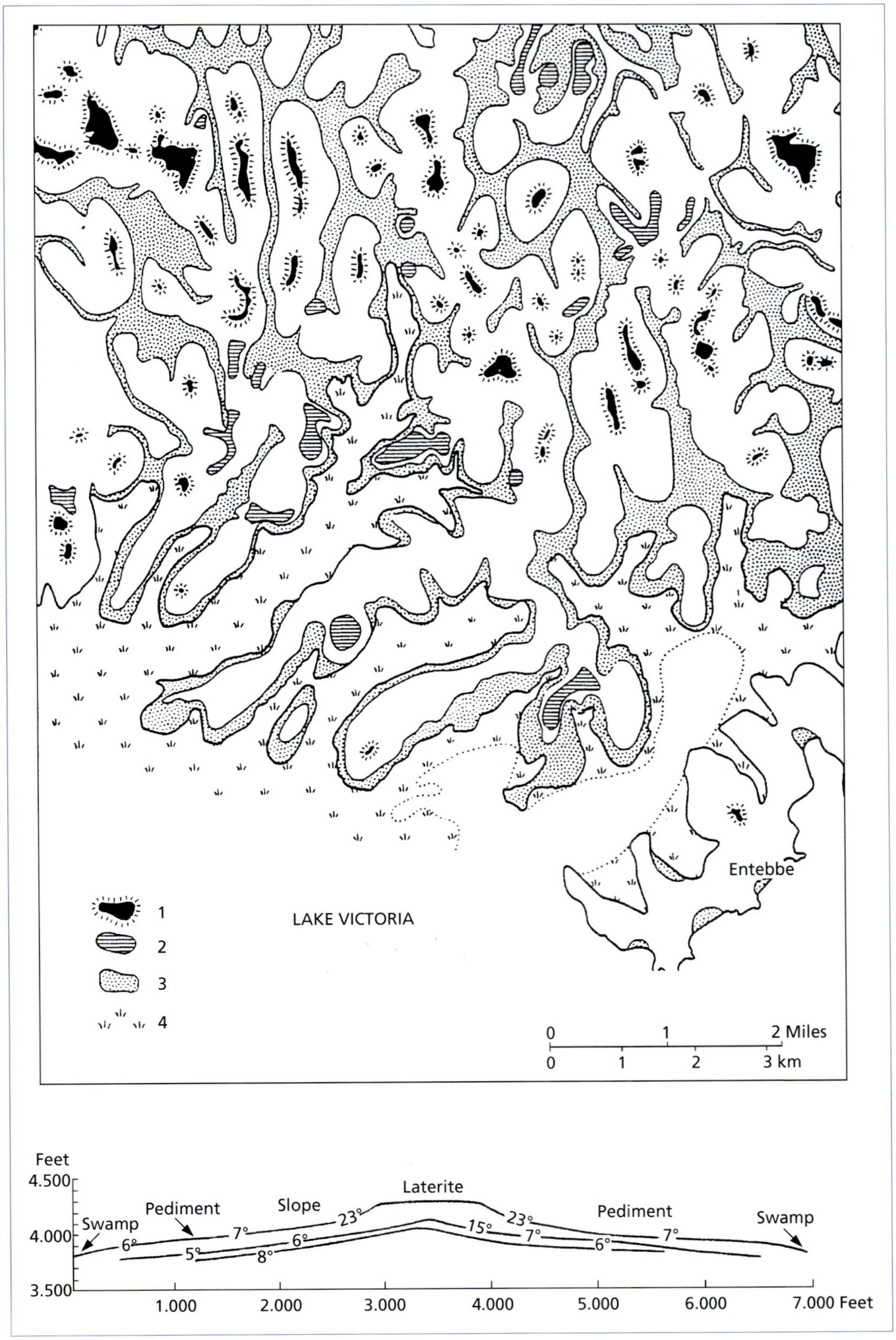

Figure 18.31 Landforms resulting from dissection of ferricretes in Uganda. 1: Flat summits topped by laterites (possible remains of the African surface). 2: Intermediate benches. 3: Alluvial silts and lacustrine layers. 4: Swampy zones with papyrus (Pallister, 1956).

Figure 18.32 Bench morphology in laterites. Nigeria. Photo J. López-Martínez.

Figure 18.33 Slope with loose laterite boulders in an arboreal savanna. Eastern Bouaké (Ivory Coast).

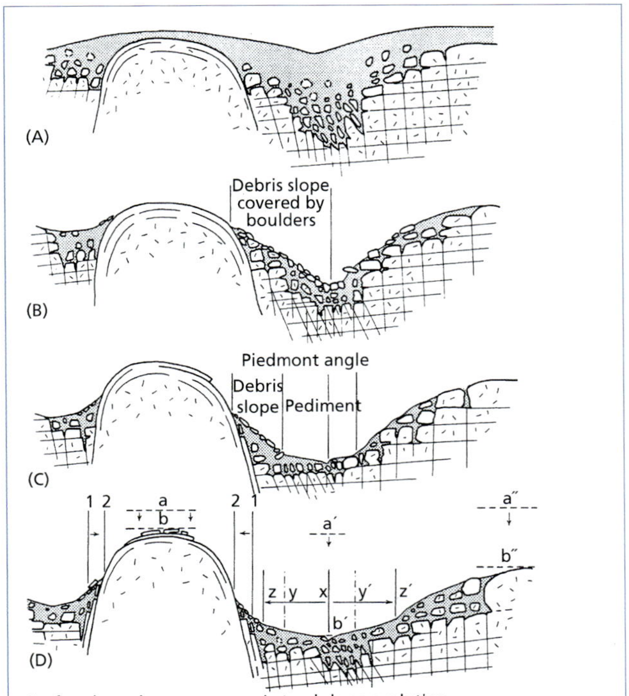

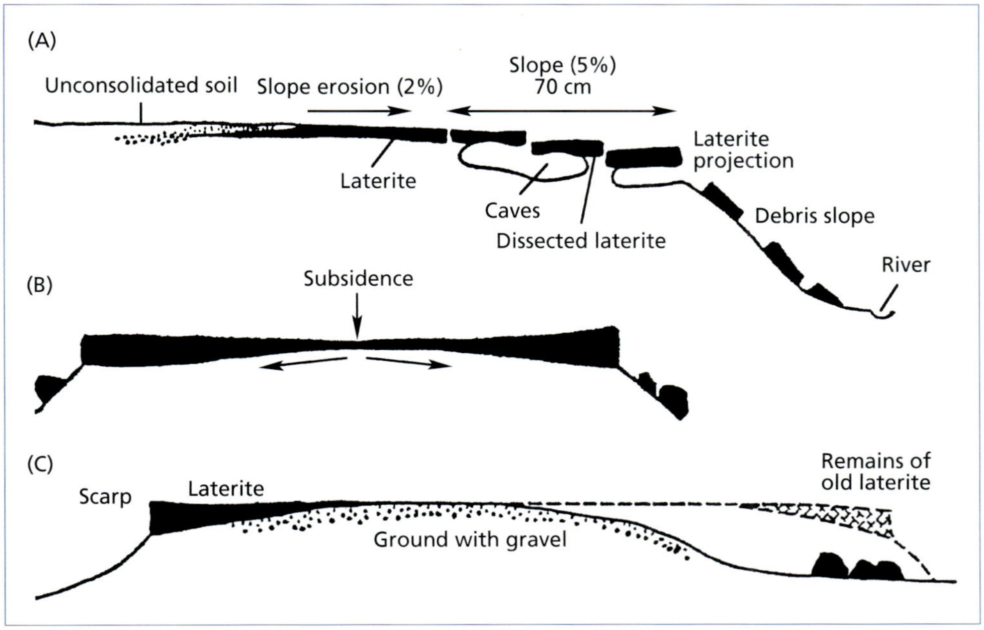

Figure 18.34 Some characteristics of the morphology of lateritic slopes (Maignien, 1966).

18.9 Tropical planation: Etchplains

Broad plains (*Fig. 18.2*) with isolated hills (inselbergs) or mountainous topography cover large areas of tropical zones. The plains are of two types: erosional and sedimentary. Erosional plains develop on variable lithologies and structures whereas sedimentary plains are zones where alluvial and lacustrine sediments accumulate.

Davis (1899) proposed the term **peneplain** for plains in humid areas and King (1942, 1953) proposed the term **pediplain** for arid zones, by extension of pediment; these plains are part of the present-day landscape. These terms and their genetic connotations have been discussed in previous chapters.

Peneplains are generally not perfectly flat areas; instead, they are rounded and undulating plains that exhibit very little elevation contrast (Vidal Romaní and Twidale, 1998; Twidale and Vidal Romaní, 2005). They form on unweathered rocks and are common in South Africa, the Transvaal, Namibia, and Brazil. As we shall see, they are not exclusively limited to temperate zones. Twidale (1983) claims that peneplains are in the process of gradual degradation and should be called **ultiplains**. The change is so slow that old forms survive within the present-day landform (Twidale, 1976a). Some authors think that "pediments and by extension pediplains are not genetically or temporally different from peneplains. They generally coexist, and are simply parts of a larger set called planation surfaces" (Vidal Romaní and Twidale, 1998, p. 101).

Exhumed plains or **surfaces** are also present in the humid tropics; these surfaces are formed by denudation of geological formations that fossilizes erosional surfaces. A classic example is the Valley of the Thousand Hills, Natal (South Africa) where erosion of the horizontal Table Mountain Sandstone uncovers an underlying surface that has domed inselbergs (Thomas, 1978).

When fluvial activity erodes the alteration mantle, the weathering front crops out as an **etchplain** (Wayland, 1933), which other authors call a chemical corrosion plain (Vidal Romaní and Twidale, 1998). Thomas (1989a) has compiled a list of the names proposed for this landform. Wayland calculated the regolith to be 3 m thick; when rapid uplift triggers the next erosional cycle, the regolith will be totally denuded, creating a new surface more or less parallel to the previous one. These cycles of alteration and subsequent erosion can be repeated many times. Wayland's ideas differ primarily on the thickness of weathering, which today is estimated to be tens or hundred of meters.

Büdel (1957) proposed the concept of **double planation surfaces** (runoff surface and basal weathering surface), which are related to the idea of etching. He applied his ideas to tectonically stable areas that have seasonal tropical climates (*Fig. 18.35*). He begins with a plain where the weathering depth ranges from 30 to 60 m; as it erodes, it creates lower plains with scattered small rocky domes (ruwars = knobs)

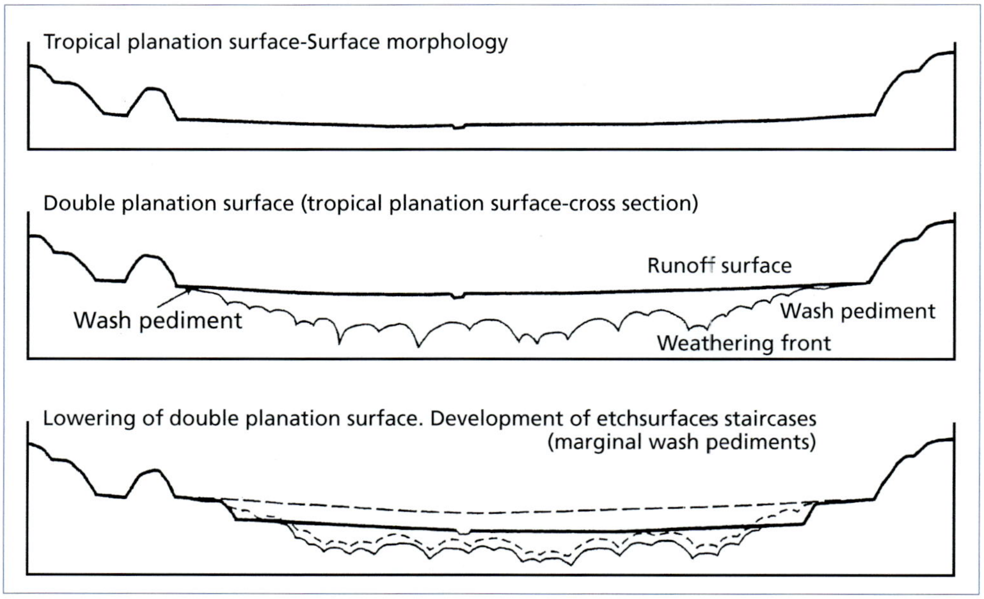

Figure 18.35 Deep weathering and denudation in the humid tropics (Büdel, 1957).

1–3 m high or shield inselbergs. Lowering of the surface can leave isolated round or parallelepiped-shaped blocks, which are low parts of the alteration profile. Valleys with very low-lying slopes or dambos drain the plains. In summary, this theory maintains that soil erosion occurs at the runoff surface whereas saprolite forms and deepens at the basal weathering surface. In seasonal climates, where Büdel (1957, 1965, 1977, 1982) focused his research, plains have almost flat slopes (0.2% in the Tamil Nadu plains of southern India). Water from rainfall cannot organize into a hierarchy of flow paths because a high percentage infiltrates into the regolith. During the dry season, the clay soils crack; roots and other bioturbation conduits provide excellent pathways for water to penetrate when the rainy season comes. This "cycle" promotes weathering throughout the saprolite and basal deepening; in addition, rainfall-generated surface runoff can remove a film of sediments and lower the runoff surface. Büdel's ideas were quickly assimilated by scientists and applied to all humid tropic cratons on the planet (see Thomas, 1974, 1989 a, b; 1994a) including areas outside the tropics with thick alteration profiles in temperate and cold regions (Bakker and Levelt, 1964; Bakker, 1967; Gellert, 1970, Dury, 1971, Büdel, 1977, 1982; Söderman, 1985; Hall, 1986, 1988). This shows how etchplains that range in age from late Mesozoic to Paleocene have survived in Europe (Thomas, 1994a). However, after the Miocene a sandy alteration profile has formed in temperate climates (Bakker and Levelt, 1964; Hall *et al.*, 1989), which may introduce significant confusion into the interpretations.

Ollier (1965, 1983) has pointed out that saprolite varies considerably in depth rather than being a parallel mantle as Wayland and Büdel maintain. Regolith forms deep depressions between protuberances of unaltered rock. Erosion of weathering deposits may create depressed areas surrounded by promontories of fresh rock.

Saprolite plays a significant role in relief configuration because etchplains undergo slow and continuous lowering as the surrounding hills become higher (Bremer, 1971, 1985) as a result of the gradual removal of altered rock. This process of depression formation occurs in highly fractured rocks that allow water to penetrate along discontinuity planes. Büdel (1977, 1982) calls this process divergent weathering.

Thomas (1965, 1974, 1994a) proposed a classification of plains and etch surfaces based on the degree of denudation of lateritic profiles. In the latest version he introduced some modifications but these differences have not been widely accepted.

Thomas (1989b and 1994a) and Pavich (1985) compare basal surface lowering and differential erosion of the regolith to the concept of dynamic equilibrium proposed by Hack (1960, 1975) in which there is a balance between regolith erosion and formation. In contrast, other authors (Ollier, 1991) consider landform formation as a set of events that occur over a long time period. Ollier (1981) considers the Uganda area to have formed as a single etchplain (*Fig. 18.36*). Thick alteration profiles have been preserved on the Buganda surface, which is equivalent to the Gondwana surface. The regolith has been totally denuded in the north; the Acholi Surface formed on top of fresh rock and numerous inselbergs are present on its margins. On the African surface, which is the most extensive one, erosion has removed a deep weathering deposit and caused inselbergs to crop out,

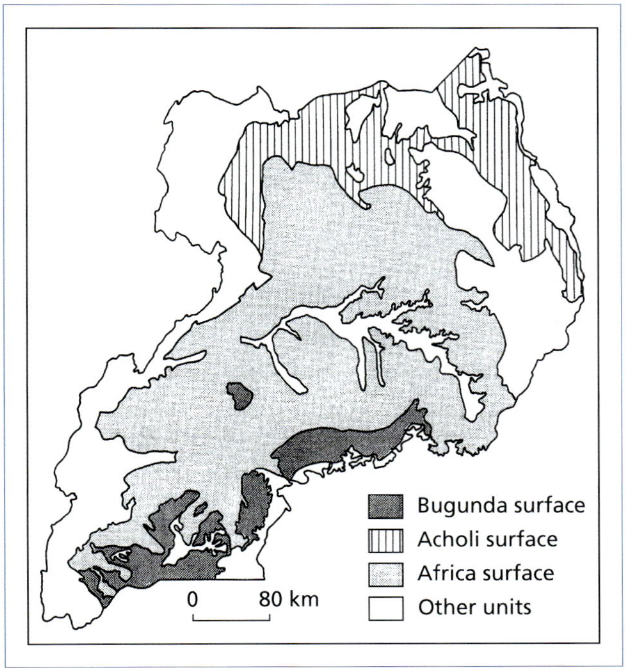

Figure 18.36 Erosion of regolith in Uganda. The Buganda surface is deeply weathered and has remnants of the erosional surface. The African surface is partially eroded. An old regolith has been stripped from the Acholi surface. Recent soils form on fresh rock more readily than on the saprolite (According to Ollier, 1981).

revealing a very irregular basal weathering surface. Miocene volcanics are present in some places on this surface. Ollier thinks that this last surface is at least early Tertiary in age and that the Buganda surface is Mesozoic or older.

18.10 Inselbergs

Inselberg, a German word that means island mountain, was first used by Bornhardt (1900) to describe steep rocky hills that stand out over the tropical plains of Tanganika (today called Tanzania). Inselbergs develop primarily in tropical and subtropical cratonic regions, preferentially on granites and metamorphic rocks but also in sedimentary rocks such as the famous Ayers Rock (Uluru) monolith, which is composed of arkose (*Fig. 18.37*), and in the Olgas (Kata Tjuta) conglomerates, which are both in the central desert of Australia (Twidale, 1978, 1982a, 2007; Vidal Romaní and Twidale, 1998; Twidale and Vidal Romaní, 2005). Research on these landforms is complex because it involves climate, lithology and rock structure, which in turn affect the degree of rock alteration. In addition, it raises fundamental genetic questions such as exhumation of the basal weathering surface and slope retreat. Research on inselbergs must also consider climate changes that have affected these landforms and the persistence or duration of these landforms. As a result many authors have studied different aspects of inselbergs.

Figure 18.37 The Ayers Rock monolith (Uluru) composed of Cambrian arkose stands out on top of the central Australian Desert plain. Photo P. Lucha.

The study of inselbergs has encountered obstacles related to different definitions and terminology. Twidale (1968) defines inselbergs as isolated residual landforms that stand out over tropical plains. Later Twidale (1982a) and Vidal Romaní and Twidale (1998) claimed that inselbergs could occur in any topographical position. Birot (1958) indicated that crystalline domes are located in various positions in the landscape; he based this claim on an analysis of sugar loaves and small rounded hills in the Río de Janeiro area. Thomas (1974, 1978) supports the idea that inselbergs are not necessarily related to plains. This opinion considerably weakens the initial concept of an inselberg. Twidale (1968) points out that inselbergs are present in tropical plains but it has been known for a long time that inselbergs are present in temperate European climates (Penck, 1924). Consequently, either inselbergs form in various types of climate or they are indicators of paleoclimates. In order to determine the global situation of inselbergs, Kesel (1973) analyzed existing publications to determine inselberg distribution. He calculated the following percentages of inselbergs in various climates: 40% in savanna climates; 32% in arid and semiarid zones; 12% in humid and subarctic continental climates; and 6% each in humid tropical, subtropical, and Mediterranean climates. Kesel concluded, like King (1957a), that inselbergs are found in practically all climate types. However, he recognizes that the humid and dry tropics are more favorable regions for inselberg formation, as Cotton (1961, 1962) indicated. Thomas (1978) compiled data on applying numerical indices to identify inselbergs but the indices do not provide differential values and according to the author can be applied to numerous hills.

Young (1972) proposed a general definition; he applies the name inselberg to isolated hills with steep slopes that rise abruptly in zones of moderate slope. He distinguishes several types: mesas with horizontal or lateritic strata, hills with straight slopes (typical of arid zones), and convex-concave slopes covered with regolith, rocky domes or blocks (tors) or castle koppies. Thomas (1976, 1978, 1994a) did a similar classification. German authors distinguish between position inselbergs (fernlinge) and resistant inselbergs (hartlinge). Position inselbergs preserve their location atop the watershed divide where erosion is less intense and resistant inselbergs form because the rock is highly resistant to erosion. Even so, German geomorphologists classify inselbergs into two fundamental types: **boulder inselbergs or tors** and **domed inselbergs or bornhardts**.

18.10.1 Boulder inselbergs (tors)

The morphology of boulder inselbergs varies considerably and is controlled by fracture systems. Boulder inselbergs occur in gneiss, schist, quartzite, diabase, sandstone, and arkose although they form most abundantly in granites (Gerrard, 1988). The boulders (*tors*) can be found in various topographic positions; their sizes and shapes depend on the type of fracturing and its spacing. In Australia, they range in size from 0.25 m to 33 m, although the most common size is 1–2 m. They are spherical or ellipsoidal in shape, principally as a result of exfoliation or spheroidal weathering (*Fig. 18.38*); they occasionally have a parallelepiped shape (*Fig. 18.39*). Some boulders occupy unstable positions; they are called balanced rocks (Vilaplana, 1987; Pedraza *et al*., 1989; Vidal Romaní and Twidale, 1998); they often end up rolling down the slope. Groups of angular blocks that look like old fortresses sometimes occupy summit areas; these are called castle shapes (*castle koppies* or *rock towers*). These boulders are found in all climatic environments.

The origin of the boulders is thought to be deep differential weathering that penetrates more deeply into more intensely fractured zones. Later environmental change erodes the saprolite. As a result, the base of the weathering profile is exposed; it has boulders that show the previously discussed morphology (*Fig. 18.40*) (Linton, 1955). Linton calls this sequence the two-stage hypothesis. Boulder formation can be seen in roadcuts cleared for linear engineering projects. The balance between weathering and erosion must be kept in mind. If weathering is more intense than erosion, the weathering front and accompanying boulders do not crop out; in contrast, if erosion is more active than weathering, the profile breaks down and the basal boulders come loose (Vidal Romaní and Twidale, 1998). Outcropping boulders are subjected to subaerial weathering. Fluvial activity may remove the regolith and boulder channels longer than 10 km may form, such as the Homem River (northern Portugal). Marine, glacial, and eolian erosion can also remove weathering deposits (Vidal Romaní and Twidale, 1998; Twidale and Vidal Romaní, 2005).

The formation of many boulder inselbergs is related to the disintegration of boulders within the inselberg. The fundamental processes involved are boulder weathering, erosion of grus, and lowering of the land surface; these can cause the boulder inselberg to disappear (*Fig. 18.41*) (Thomas, 1978).

As discussed above, boulder inselbergs are widely distributed throughout the world; the deep weathering implies very old alteration in many areas. Linton (1955) describes intense chemical weathering of the Dartmoor granite during the Tertiary. In this case periglacial processes removed the regolith as Demek (1964) proposed for the Bohemian Massif (Czechoslovakia). Czudek and Demek (1971) later interpreted tors as having been formed by weathering and mobilization in a periglacial environment. Mineralogical analysis of the grus at Dartmoor shows a lack of feldspar alteration and low clay content (Doornkamp, 1974). In European latitudes a "sandy weathering product" (Bakker, 1967) may result, forming boulders as it evolves.

King (1948) has proposed that tors, like dome inselbergs, form by slope retreat in arid and semiarid climates and

Figure 18.38 Exfoliation of a granite boulder in the Santa Olalla del Cala batholith (Province of Huelva, Spain).

Figure 18.39 Tors formed in granite on the northern slope of the Serra da Estrêla (Portugal).

that they develop at the foot of a pediment. Czudek (1964) proposed similar ideas, which he had developed for periglacial environments (*Fig. 15.35*).

This discussion shows that more than one hypothesis may explain the origin of these boulders. The two main hypotheses are 1) related to deep weathering and later erosion (Linton's two-stage hypothesis) and 2) related to slope retreat in hot and cold deserts.

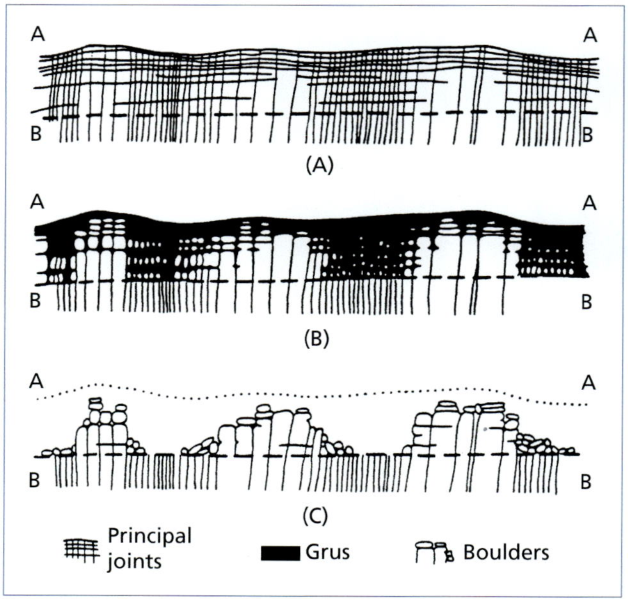

Figure 18.40 Evolution of boulders (*tors*) and granitic landforms based on diagrams by Linton (1955).

18.10.2 Domed inselbergs (bornhardts)

Willis (1936) proposed the term *bornhardt* for crystalline domes in honor of the German explorer Bornhardt (1900) who first described these inselbergs. Bornhardts, which are dome-shaped, occur in massive resistant rocks, primarily granitic, although they can also occur in sandstones and conglomerates (as in central Australia). Twidale (1982b) considers the conglomeratic monoliths of the Mallos de Riglos on the margin of the Southern External Pyrenees of Huesca to be bornhardts (*Fig. 18.42*). This designation can be extended to the conglomeratic hills of Montserrat, Montsant, and Río Iregua that form part of the marginal Tertiary units of the Ebro Depression (Spain).

Bornhardt slopes are usually steep and vary in size from a few meters to several hundred meters (*Fig. 18.43*). In plan view they are circular or elliptical and often fractured (King, 1948; Twidale and Bourne, 1978). When they are elongate and low in height, they are called whalebacks. Some bornhardts are isolated whereas others are grouped into hemispheric shapes called half oranges in Brazil (Birot, 1958). The connection between the rocky monolith and the base of the bornhardt is sharp with a clear *knickpoint* or a transition with no obvious slope modification. Various basic causes explain the abrupt change in slope: contact between rocks with different mechanical properties, contrast between fresh rock and saprolite (Ollier, 1960; Thomas, 1965), and more intense weathering at the slope base (Ruxton, 1958; Twidale, 1962; Mabbut, 1966; Twidale and Bourne, 1978).

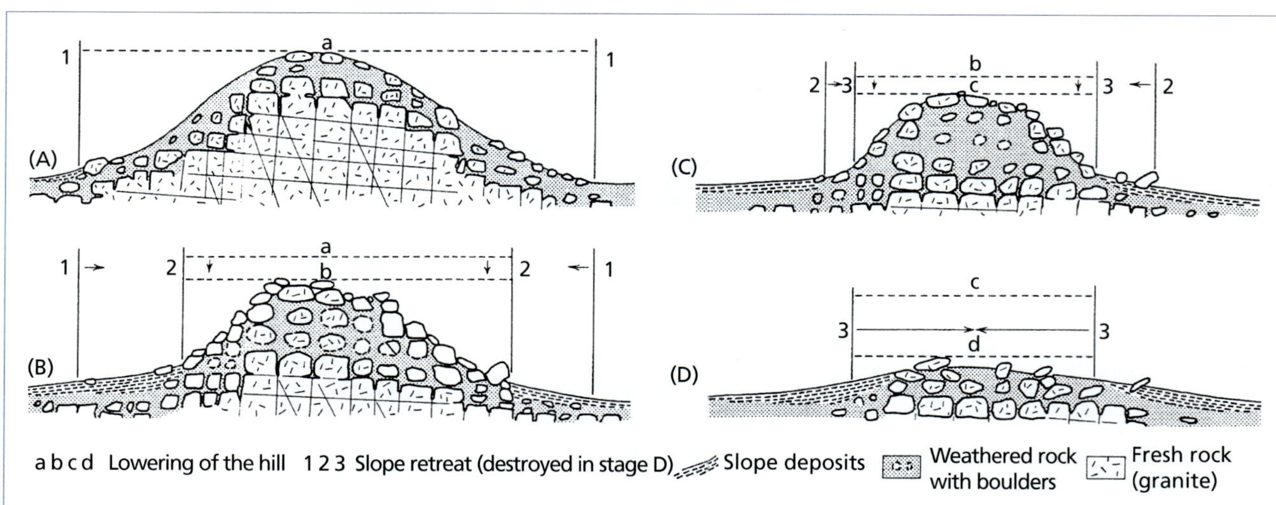

Figure 18.41 Stages of the development of boulder inselbergs, showing the disintegration of interior jointed blocks within the area of vadose water circulation, with the lateral penetration of weathering leading to slope retreat. The first stage (A) can be reached without deep previous weathering, although it does not exclude its occurrence. The denudation implied in the transition from a-b may be related to climate change (according to Thomas, 1978).

Figure 18.42 Monoliths formed on top of massive horizontal conglomerates at the boundary between the Pyrenees Exterior Sierras and the Ebro Depression. Mallos de Riglos (Province of Huesca, Spain).

Figure 18.43 Hundred-meter tall bornhardt. Abuja (central Nigeria). Photo J. López-Martínez.

Bornhardts are found in different climatic environments (Kesel, 1973) from equatorial rainforest zones through medium latitudes (*Figs. 18.44* and *18.45*) to northern Norway (Büdel, 1978). Numerous articles have been written about granitic bornhardts in the Hercynian Massif (Birot, 1949b; Birot and Solé, 1954a, b; Brum Ferreira, 1978; Coudé-Gaussen, 1981; Sanz, 1988; Pedraza *et al.*, 1989; Vidal Romaní and Twidale, 1998; Twidale and Vidal Romaní, 2005; among others). They have also been found in granites of the Catalan Coastal Cordillera (Roque and Pallí, 1998).

Figure 18.44 Dome shapes and other granitic landforms topped by the bell-shaped El Yelmo monolith. La Pedriza de Manzanares (Province of Madrid). Sierra de Guadarrama, Spain.

Figure 18.45 Bornhardt with exfoliation planes at the top. Yosemite National Park (California). Photo J. López-Martínez.

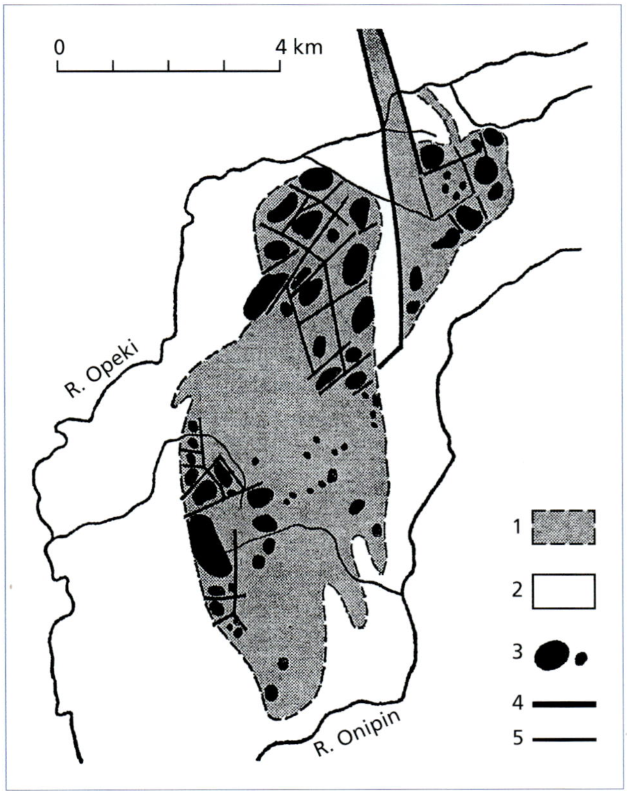

Figure 18.46 Residual hills in southwestern Nigeria. 1: Porphyritic granite with biotite. 2: Gneisses and schists. 3: Inselbergs, small domes, and *koppies*. 4: Fault. 5: Fracture (according to Jeje, 1973).

A group of factors affects the location of bornhardts. Most researchers who have tried to associate bornhardts with specific rock types have concluded that there is no lithological variation in their piedmont. However, some studies have established a relationship between inselberg location and rock type. Jeje (1973) found inselbergs on porphyritic biotitic granite in southwestern Nigeria (*Fig. 18.46*). Brooks (1978) in his research near Johannesburg shows that tors and *castle koppies* occur in a transition zone of gneiss, migmatites, homogeneous granodiorite, and porphyries; in other areas they are preferentially limited to biotitic porphyritic adamellites. In summary, inselbergs are more common in rocks that have undergone potassic metasomatism, in plutons rich in potassium or plutons that contain primary low-temperature quartz. Even so, it is rare to find a relationship between an inselberg and lithology.

Most inselbergs are located in areas of crustal stability (Kesel, 1973) although they are also located in tectonically active mountain ranges. One of the classic examples is Parana Peak in southeastern Brazil, which is part of a horst (Birot, 1958). Twidale (1973, 1982a), Vidal Romaní and Twidale (1998) and Twidale and Vidal Romaní (2005) claim that lateral compressive forces produced by exfoliation exceed those generated by unloading. This hypothesis implies that this mechanism generates domes that are controlled by exfoliation. Numerous measurements of forces in tunnels and wells support this idea (Twidale, 1982b).

At the end of the 19th century, inselbergs were thought to be formed by glacial, marine, and eolian processes (Twidale, 1982b; Vidal Romaní and Twidale, 1998; Twidale and Vidal Romaní, 2005). Fluvial erosion was later considered to be the primary cause. Today two fundamental theories of bornhardt formation are recognized. The first is the **scarp retreat hypothesis**, which was proposed by various authors (Cotton, 1942; Howard, 1942; Pugh, 1956) although it was King (1942, 1948, 1949, 1953, 1962) who championed this idea with the most emphasis. Recently, Kesel (1973, 1977b) has supported King's ideas with studies in the southwestern deserts of the United States; Selby (1977) has done the same with studies in the Namib Desert. King has been criticized because none of his research was done in the humid tropics, which are areas of abundant bornhardts.

They are position inselbergs (*fernlinge*). Over long time periods they undergo parallel scarp retreat. At the same time, the pediment enlarges and finally becomes a pediplain (*Fig. 18.47*). Inselberg development by pediplanation involves progressive size decrease that ends with complete disappearance. Most researchers restrict these processes to arid and semiarid environments where the effects are most pronounced (King, 1957a). King thought that weathering at the scarp base increases its gradient and causes slope recession. However, in hot deserts like the Namib two conflicting interpretations can be proposed. Selby (1977) claims that pediplanation forms inselbergs whereas Ollier (1978) points to their great antiquity and maintains that they are exhumed topography. As a result, a thorough knowledge of past geologic periods is necessary, particularly with respect to climates, to avoid controversial interpretations.

The other theory of bornhardt origin involves **exhumation** of a deeply altered rock massif. When the basal weathering surface is exposed, dome shapes appear; this shows that they formed under an alteration mantle. The hypothesis is similar to Linton's "two stage" theory for boulder inselbergs and tors. This hypothesis has been confirmed in quarry fronts in quartz-diorite in southern Cameroun where a 50 m high dome is surrounded by weathered material (Boye and Frisch, 1973). Incipient dome morphologies can be also observed in highway roadcuts (Thomas, 1978). The domes are located in the less fractured parts of the weathering front. Denudation occurs gradually as a result of fluvial activity and dome-shaped morphologies appear on the ground as small domes that announce an embryonic or incipient bornhardt (Thomas, 1978; Twidale, 1982a) (*Fig. 18.48*).

Figure 18.47 Inselbergs standing out above extensive plains. Narok Plateau (1825 m), southwestern Kenya.

Figure 18.48 Embryonic domal forms in a savanna region. Masai Mara (Kenya).

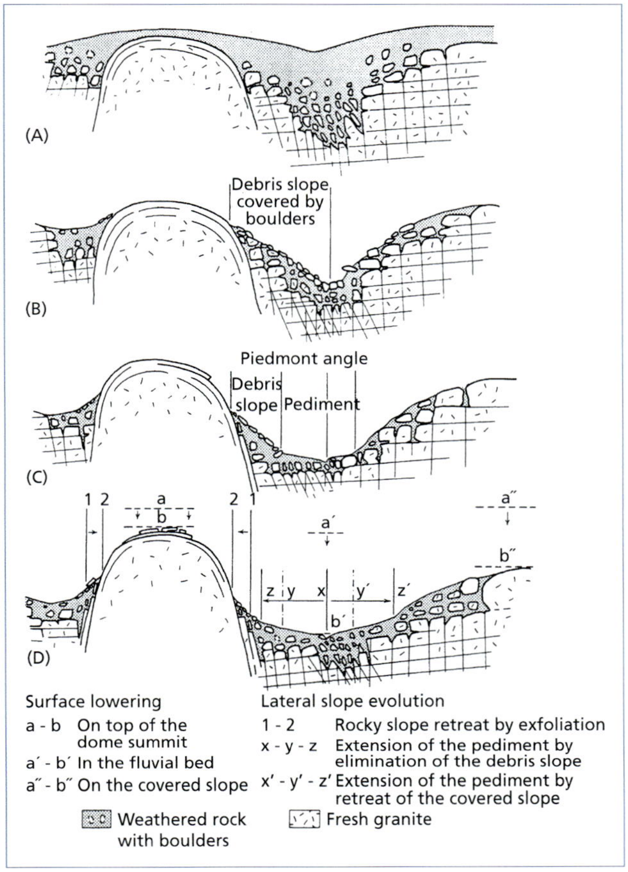

Figure 18.49 Development of bornhardts by continuous weathering of covered slopes during dissection (in Thomas, 1978).

Small bornhardts can form during single periods of deep regolith excavation that may reach depths of 100 m. Large bornhardts are more difficult to explain (Thomas, 1965, 1974). If we accept continuous weathering at the slope base as denudation lowers the surrounding slopes, we can explain bornhardts that are several hundred meters tall *(Fig. 18.49)*. Thus bornhardt formation requires repeated alternations of weathering and denudation over long periods of geologic time (Willis, 1936; Büdel, 1957; Twidale, 1964, 1982b; Thomas, 1978, 1994a). As a result, large bornhardts are quite old geologically.

The scarp retreat theory does not explain landforms related to the rock-plain and rock-saprolite contact. However, the exhumation hypothesis justifies the presence of embryonic domes, small scale features formed along the weathering front (polygonal weathering, tafonis, gnamma pits, weathering flutes, caves, and so on) (Twidale, 1982b; Thomas, 1994a; Vidal Romaní and Twidale, 1998: Twidale and Vidal Romaní, 2005). This hypothesis also supports stepped (Twidale and Bourne, 1975, 1978) and mushroom-shaped slopes (Twidale, 1962) that indicate older subsurface weathering, such as existence of tafoni lines at different heights in Ayers Rock (Twidale, 1978). These forms also show the amplitude of lateral attack on massive bornhardt slopes during each stage of stable base level.

Regardless of whether bornhardts originate by scarp retreat or exhumation, their formation requires a long period of time. Their age is a function of the rate of the geomorphic processes involved and the size of the landform. For this reason we consider bornhardts to have been exposed recently and other inselbergs to have ages of tens of millions of years. Another problem arises when we attempt to explain the survival of large bornhardts over vast geological time. Researchers believe that exfoliation controls the shape of bornhardts and that rainfall flows across their surfaces as it would on an "umbrella," producing minor alteration and erosion (Thomas, 1974). Ages of Mesozoic or early Tertiary have been proposed for many bornhardts. Savigear (1960) thinks that inselbergs in Nigeria were exhumed under a cover of Cretaceous sediments and Barbier (1967) points out that domes (sugar loaves) are exhumed from under the upper Paleozoic sandstone of the Tassili Massif in southern Algeria. In addition, Chaput (1971) shows that the pre-Triassic surface of the eastern Sierra Morena (Spain) was partially exhumed, exposing quartzitic inselbergs. In addition, Twidale and Bourne (1975) and Twidale (1978) claim that the famous Ayers Rock monolith dates back to the Mesozoic stratigraphically and geomorphically.

Because of their age, inselbergs have been subject to endogenetic and exogenetic activity over long time periods. Lithospheric plate movement may have been changed their latitude and their climate; India is an example. In terms of the origin of inselbergs, climate change affects the equilibrium between weathering and erosion rates. Periods when weathering predominates are geomorphically and ecologically stable (stages of biostasy) and when this equilibrium is interrupted, erosion occurs (stages of rhexistasy) (Erhart, 1967). Broad and deep ferralitic weathering occurred in the middle latitudes of Europe, especially during the Eocene (Millot, 1964). This alteration is confirmed in the sedimentary record of Tertiary basins; it provides convincing arguments for ancient deep weathering of areas where inselbergs have formed.

18.11 Applied geomorphology in tropical regions

Tropical zones have a wide variety of topographies and ecosystems that provide important resources but at the same time involve multiple risks (Reading *et al.*, 1995). Tropical climates control the lives of their inhabitants and their economic activities to a greater extent than climates of middle latitudes. The population of tropical regions has exploded in the last 50 years. Today 3.1 billion people live in these areas; this is 45% of the world's population. Approximately 60% of these people live in South and East Asia. Many live in underdeveloped countries with subsistence economies that are primarily dependent on agriculture (McGregor and Nieuwolt, 1998). The need for food has led to intensification and extension of agriculture with a concomitant increase in deforestation. This has led to increased soil erosion (*Fig. 18.50*) and risk of flooding and landslides. These circumstances make it important to use soil conservation practices (Douglas and Spencer, 1982).

Thick regoliths that form in the humid tropics have been thoroughly studied in terms of their implications for engineering projects (Lovegrove and Fookes, 1972). Their depth, structure, and mineral composition vary and the regolith can also vary abruptly (Douglas, 1986). These variations considerably complicate engineering projects.

These humid tropical areas have numerous mineral, petroleum, forest, and agricultural resources; exploitation or management of these resources affects various geomorphic processes, usually triggering or accelerating them. In addition, earthquakes and tsunamis can affect areas located within tectonically active zones. Volcanic risk, especially the explosive variety, is also significant; it may include eruption of nuées ardentes and volcanic ash and the formation of lahars. Finally, tropical cyclones may be the highest impact risk because of their strong destructive winds, flooding, and the landslides that often follow cyclonic precipitation. A recent example is Hurricane Mitch, which caused heavy fatalities and property losses in Nicaragua and Honduras in October and November of 1998.

Humans have reshaped the landscape by terracing slopes (*Fig. 18.51*) and constructing dams and reservoirs of various sizes to modify the fluvial regime and control water erosion on slopes (Wirthmann, 2000). Over the last century, the ecology of the humid tropics has experienced significant crises because of the worldwide population explosion. In zones that are in transition toward semiarid environments, intense desertification is occurring (Mensching, 1983). In many regions around the Mediterranean, cultivation has continued on Tertiary marly, clay-rich, and gypsiferous soils, even after soil loss (an area near Zaragoza, Spain). This is not an option for resistant rocks in the tropics. As a result, as population increases, calamities occur.

In highly vegetated environments, inhabitants set fires and commit similar abuses to provide land for agriculture and cattle production. Fire has been used in the central African savannas (*Fig. 18.52*) since antiquity; it is believed that fires promote the growth of new fresh offshoots of grasses that

Figure 18.50 Gully headland developed in a partially deforested zone, which slightly affects the village and its crops. Muranga (Kenya).

Figure 18.51 Terraces for protection against water erosion and for later use as cropland. Machakos (Kenya).

Figure 18.52 Grasses disturbed and burned, possibly by humans. Masai Mara (Kenya).

feed the cattle and goats of the Masai tribes. Satellite photos shot at night show the lights of fires burning these savannas. In addition, the dense forest mass in the rainforest is subject to abusive tree cutting and fires that humans do for the purpose of selling wood and using the deforested space for cattle production or agriculture. This land later becomes unproductive because humid tropical soils have low fertility; they lack basic elements and humus and they have been partially desilicified. Soil erosion under these conditions is estimated to be ten times greater than under natural conditions (Rapp *et al.*, 1990). In addition, destruction of the tropical rainforest, which is the lung of our planet, increases the carbon dioxide concentration of the atmosphere and thus increases "greenhouse gases."

18.11.1 Catastrophic flooding

Floods are the most common geomorphic risk in the humid tropics because of the intense and prolonged rains that occur in these areas (*Fig. 18.53*). The use of floodplains to grow agricultural products considerably increases this risk. Flooding in Asia on average damages 4 million hectares of land every year and adversely affects 17 million people (Reading *et al.*, 1995).

Convective clouds are the major producers of rain in low-latitude tropical regions. Figure 18.54 shows the global distribution of storms of various frequencies: more than 100 per year, between 50 and 100, between 20 and 50, and less than 20 (Lamb, 1972). A fringe of higher storm frequency occurs around the Equator.

Convergence of air masses from the two hemispheres in the Inter-Tropical Convergence Zone, tropical cyclones, and the orographic effect all generate rainfall in the tropics (Hayden, 1988). The first type involves convergence and rising of air that produces a cloudy band around the Equator. Seasonal variation of this cloud belt causes monsoonal rains from June to September in Asia, where on average seven monsoonal depressions affect the area every year (Gupta, 1988). These rains can sometimes last nine days. Cyclones move from east to west and maximum rains occur on islands or continental margins where the heated land surface and the topography cause the air to rise, thus intensifying convection (Hayden, 1988). For example, a cyclone dropped 1870 mm of rain on Cilaos, Reunion Island, in one day (Gupta, 1988). Tropical cyclones with winds higher than 116 km/hour are called hurricanes; in East Asia they are called typhoons. The orographic effect causes air masses to rise; as water vapor condenses, clouds produce maximum diurnal precipitation during the period of major solar heating.

The few gauging stations in the humid tropics are generally located only on large rivers; as a result, records of large floods are also limited. However, the return periods of

Figure 18.53 Bridge on the Cocos River that was destroyed as a result of flooding by Hurricane Mitch in October–November of 1998. Northern Nicaragua, near the border with Honduras. Photo A. Gómez Sal.

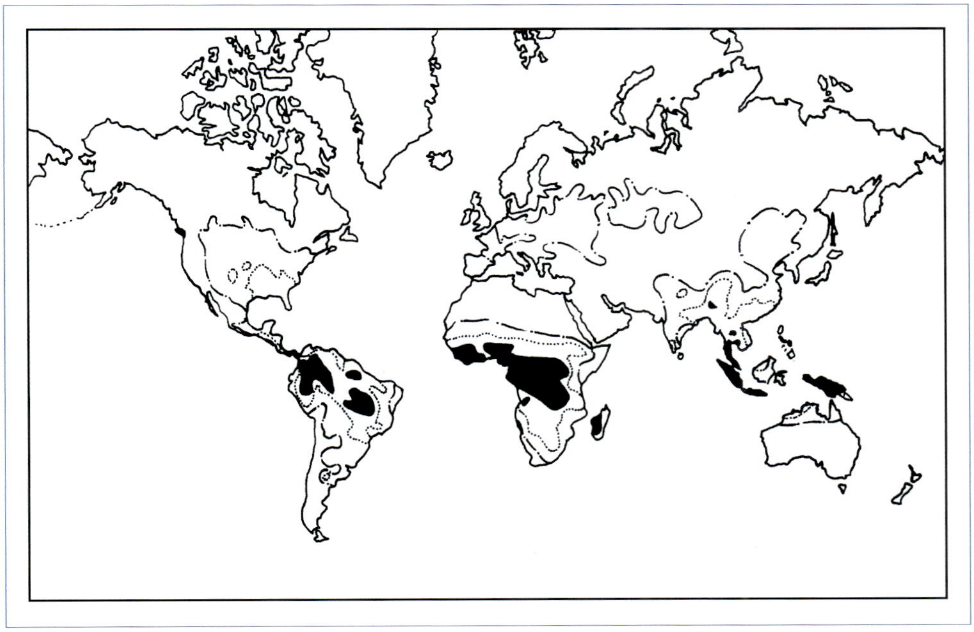

Figure 18.54 Annual storm frequency. Regions with 100 or more storms per year are in black. The dotted line encircles areas with more than 50 storms/year and the dotted dashed line encircles areas with less than 20 storms/year (Lamb, 1972).

intense precipitation are well known and these data can be related to those of major floods for which we have approximate dates (Gupta, 1988).

In urban areas, drainage basin hydrology has been significantly modified, mostly by impermeable street and roof surfaces that increase flood frequency (Leopold, 1968). Urbanization-related hydrologic changes affect these areas when channel and sewer networks are unable to evacuate water; this causes sporadic flooding in cities such as Kuala Lumpur (Malaysia) and Singapore (Gupta, 1984, 1993).

The chapter on Littoral Geomorphology includes a discussion of the numerous floods that have occurred in Bangladesh, which is located in the Ganges Brahamaputra delta.

19

ENVIRONMENTAL CHANGE

1. Introduction
2. The Earth's climate system
3. Climate change and the greenhouse effect
4. Anthropogenic change
5. Toward sustainable development

19.1 Introduction

Climate, sea level, vegetation zones, animal populations, soils, and landforms have all changed throughout Earth history (Goudie, 1977a, 1992) but during the Quaternary continuous environmental changes over very short time periods have become more obvious.

From a geomorphic perspective, these recent environmental changes result from three basic causes: geological, climatic, and anthropogenic. Because the factors that trigger these changes interact or act together, it is it difficult to assess the contribution of each to environmental change.

Changes of the geological environment are primarily due to neotectonic activity (*Fig. 19.1*), which may be significant in areas located along active plate margins where isostatic readjustments and relative lithospheric plate movements cause significant topographic change. Thus in mountain deserts and depressions (Mabbutt, 1977) that have notable crustal instability such in the western United States and central Asia, landforms will be affected by this activity. In contrast, shield deserts and platforms like the Sahara, India, and Australia are generally not affected by this endogenetic activity; it is easier to study of climate change in the latter areas because tectonic activity does not interfere.

Although **humans** have populated the Earth for six or seven million years (Anderson *et al.*, 2007) their interaction with the environment became important only during the agricultural revolution of the last 10,000 years and more so in the recent industrial and medical revolutions (Goudie, 1981b, 2006). Humans affect vegetation, fauna, soil, water, and climate, and thus man is considered a geomorphic agent (Nir, 1983) because human activities modify the global surface (*Fig. 19.2*) and in turn interfere with geomorphic processes.

Climate change has become familiar to us because we have become accustomed to temporal changes in rainfall and temperature on the scale of human lifetimes. These short-term climate variations are followed by other more serious and long-term changes. Goudie (1992) and Boulton (1993) (*Fig. 19.3*) show these circumstances on various

Figure 19.1 Nigüelas fault on the northwestern margin of the Lecrín Valley Padul (Province of Granada). Betic Cordillera, Spain.

climate change scales, over intervals ranging from tens of years to those that occur on a geological time scale.

Climate change (Goodess *et al.*, 1992) that occurs on time scales greater than 20,000 years is considered long-term; Quaternary glaciations are an example. These changes occur as a result of external modifications to the climate system such as changes in the geometry of the Earth's orbit around the sun (Milankovitch cycles). Short-term climate change occurs over periods of 100 to 20,000 years. These changes are partially related to external causes although significant mechanisms may include changes in oceanic circulation, volcanic activity, and the content and type of gases in the atmosphere. The term climate variability is used for changes that occur over periods of less than 100 years; these changes are primarily due to internal mechanisms (Goodess *et al.*, 1992).

In most cases we find a set of hereditary landforms in the landscape that are active under a wide spectrum of environmental conditions different from present-day conditions (*Fig. 19.4*). As discussed above, these processes, which result from tectonic, anthropogenic, and climatic activity, may interfere with the resulting landforms. Consequently, climate change research must determine the contribution of each of them and try to eliminate the "noise" inherent in the tectonic and anthropogenic processes. In recent millennia, humans have significantly modified certain areas such as the countries around the Mediterranean (Goudie, 1981b, 1990b; Mannion, 1997); this anthropogenic activity poses complex challenges to determining the influence of climate change on the origin of existing relief. As a result, the best areas for climate change research are those with prolonged crustal stability and minimal human population throughout history. In such regions, landforms exhibit only the effects of geomorphic processes triggered by climate change.

It is critical for a researcher to have a good understanding of the age of landforms that are related to climate change. The first researchers used relative chronologies of stratigraphic series, morphogenetic sequences, degree of weathering, and soil development (Birkeland, 1984). Recently, especially during the last sixty years, numerous dating methods have been developed that can provide direct or indirect ages for both buried and surface particles. Among the former are: radiocarbon, thermoluminescence, uranium series, cosmogenic isotopes, amino acid racemization, stimulated luminescence, and electron spin resonance. Indirect methods are primarily paleomagnetism and O and C isotopes. The type of sample available determines the method chosen, and precision varies from one technique to another. Some of these methods are under experimental development but their futures promise to be revolutionary for our understanding and interpretation of environmental change (Stokes, 1997).

Figure 19.2 Open pit mining of Cretaceous-age lignites in the Val de Ariño (northern Province of Teruel). Iberian Cordillera, Spain. Photo B. Leranoz.

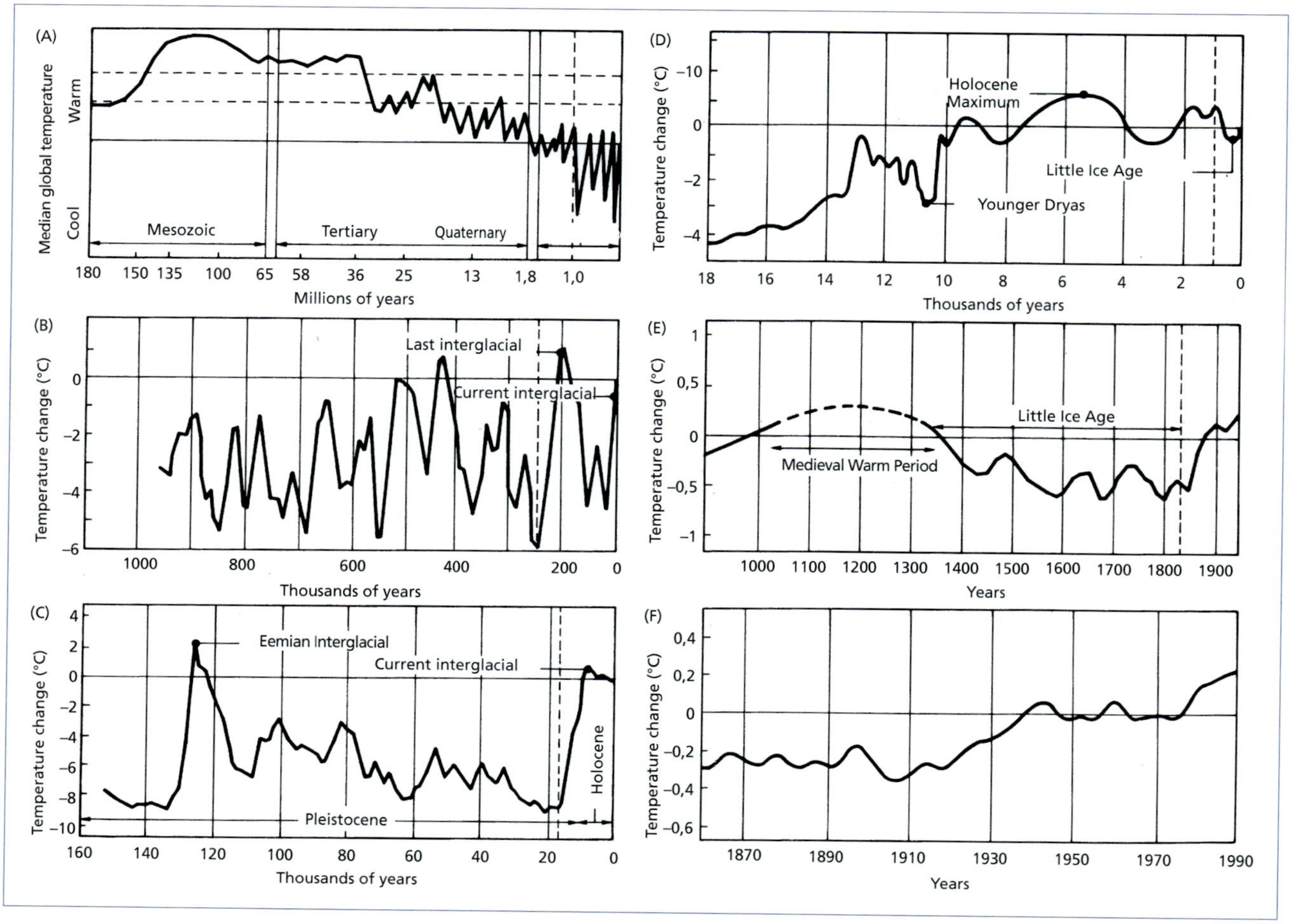

Figure 19.3 Climate change over the last 180 thousand years. (A) Climate change during the period of 10^7–10^8 years, which reflects climatic evolution resulting from plate tectonic activity. (B) Change during the interval of 10^5–10^6 years with glacial and interglacial periods that have cycles of about 100,000 years. (C) Changes during the stage of 10^4–10^5, which shows the structure of change during a glacial/interglacial cycle of 20,000 and 40,000 years, dominated by orbital changes of the Earth. (D) Changes in the last 18,000 years that record a transition from a glacial to interglacial period. (E) Changes during the last millennium in which a Medieval Warm Period and a Little Ice Age can be recognized. (F) Changes during the last century, with possible anthropogenic influence on the climate (Boulton, 1993).

Another significant circumstance stems from our knowledge of the time required for landforms to adjust to new climatic conditions (Trudgill, 1976a). Some authors claim that certain morphologies can persist over long periods of earth history. For example, Twidale (1978) says that the famous inselberg of Ayers Rock (Uluru) in the Australian desert has persisted from the late Mesozoic or early Cenozoic to today. In contrast, badlands areas and active dune fields undergo continuous and rapid evolution. Landform persistence or modification is a function of the spatial location within morphoclimatic zones, among other factors. Landforms located in hyperarid environments have changed only a little or not at all as a result of Quaternary climate change (paralysis of water erosion of Oberlander, 1997). In contrast, landforms in transition areas will obviously be affected by temporary variations of global atmospheric circulation that are accompanied by migrations of climate belt boundaries. These transition areas are more appropriate for attempting to reconstruct climate changes that have taken place during recent periods of Earth history.

As Chorley *et al.* (1984) claim, climate change is one of the fundamental subjects about which geomorphology is profoundly ignorant. These authors claim the causes to be a) continuous climate change through earth history in which the mechanisms of change operate in complex ways, b) the diverse ways in which climate change occurs, and finally c) our lack of understanding of the most significant climate aspects in the generation of erosional and accumulative processes.

Figure 19.4 In the foreground, stepped terraces of the Jachal River and its braided morphology; in the background are structural landscapes of the Precordillera of the Andes where a pediment system has formed. Between Rodeos and San José de Jachal (Province of San Juan, Argentina).

19.2 The Earth's climate system

Earth's temperature conditions and atmospheric composition are unique in the solar system and perhaps in the universe. These conditions are favorable and necessary for life, which has developed over millions of years of biological evolution; the Earth gradually established a set of biogeochemical interactions that operate within the climate system of our planet (Lovelock, 2000). Our planet receives energy from incoming solar radiation and dissipates it into various components of the climatic system. Terrestrial fluids move through convection cells (atmosphere) and currents (oceanic) that carry heat from the equator to the poles. Any alteration of the terrestrial climatic system (*Fig. 19.5*) involves adjustments of its components, which are reciprocally linked through complex interactions that involve a variety of supply mechanisms (Comin and Rodríguez-Arias, 2003).

Because of the dynamic properties of air, atmospheric fluctuations over weeks or years can rapidly mix the entire atmosphere. Ocean processes are slower because marine water is more viscous than air. It is calculated that temporary variability fluctuates from one month to 4000 years. Many atmosphere-ocean interactions vary over decades to centuries, which is the time interval of most interest in climate predictions. An understanding of the mechanisms of these interactions is essential for deciphering past climates and predicting future climate, either from data or through climate models (Comin and Rodríguez-Arias, 2003).

19.3 Climate change and the greenhouse effect

Climate change is familiar to us because we have become accustomed to temporary changes in rainfall and temperature on the scale of a human lifetime. For example, strong torrential rains and flooding occurred in Europe and India in the summer of 2002. Temperatures rose considerably in Europe in the summer of 2003 and caused significant risk to human life. During the summer of 2007, significant flooding affected England whereas southeastern Europe experienced an extreme heat wave with loss of human life and extensive and devastating fires, especially in Greece. These short-term variations are part of deeper and larger amplitude temporary changes.

Climate change researchers have conducted investigations on climate variation over the last 400,000 years and

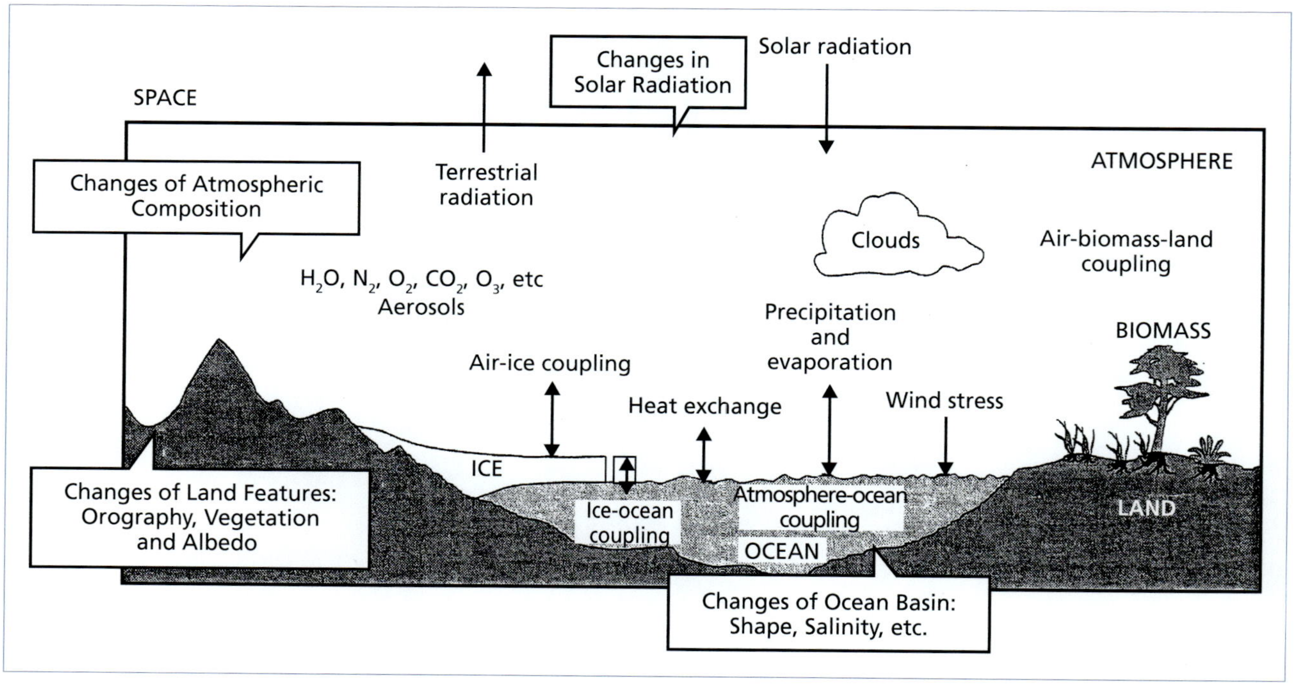

Figure 19.5 The diagram shows the Earth's climate system with examples (boxes) of physical and biological processes related to climate and climate change (Comin and Rodríguez-Arias, 2003).

particularly on the climate record of the last 150 years. They have predicted an average heating of 2°C in the near future and for the year 2100 due to **greenhouse gases**, a group of gases that includes water (H_2O), carbon dioxide (CO_2), nitrous oxide (N_2O), methane (CH_4), chlorofluorocarbons and ozone (O_3). Their concentrations in the atmosphere have increased in recent years; carbon dioxide in particular has contributed to atmospheric heating because it makes up 64% of greenhouse gases. CO_2 is practically transparent to shortwave solar radiation and it absorbs a large part of long-wave radiation. By increasing the temperature of the Earth's surface and the lower atmosphere, it is a major factor responsible for the greenhouse effect. It is clear that an increase in the quantity of CO_2 leads to global warming (Budyko, 1982).

The Earth's greenhouse gas concentration has changed throughout history and climate has changed as a result. The warmest periods correspond to higher greenhouse gas concentrations and cooler periods are related to lower concentrations. This process has been active over tens of thousands of years. Recent human-induced changes in the atmospheric composition have occurred on a scale of decades (Ramanthan, 1988). When scientists use the term climate change as defined by the United Nations Framework Convention on Climate Change, they are referring to changes that have occurred during the post-industrial period. As a result, climate change is attributed to human activities that modify the composition of the atmosphere.

Human activities produce greenhouse gases that contribute to climate change. To determine the effects of these gases on the global climate, we need to know their total amount, their natural and human sources in the atmosphere, natural loss rates, their history and estimates of increases, and their individual and cumulative heating capacities. *Table 19.1*, introduced by Hardy (2003), shows the types of greenhouse gases, the evolution of their concentrations, duration in the atmosphere, anthropogenic sources, and global warming potential.

The most significant contribution to the greenhouse effect occurs when water vapor absorbs atmospheric heat. As a component of the atmosphere, CO_2 participates in most biochemical reactions. *Figure 19.6* is a diagram of the global carbon cycle, which shows reserves of GtC (giga tons of carbon, G = giga = 10^9) in boxes and fluxes (GtC/year) with arrows related to anthropogenic disturbance for the period of 1980 to 1989, expressed in annual averages (Schimel *et al.*, 1996). In the last 150 years human activities have increased the amount of CO_2 in the atmosphere by the combustion of coal, oil, and gas, but the reserves contained in fossil fuels are small compared to the quantities contained soil biota and especially in the great storehouse of oceanic waters. Today approximately 6.5 GtC of CO_2 are emitted to the atmosphere every year and carbon dioxide concentrations continue to increase. On a global scale more than 80% of anthropogenic CO_2 emissions come from transportation and industry. The

Table 19.1 **PRINCIPAL GREENHOUSE GASES (UNEP 2001. *UNITED NATIONS ENVIRONMENT PROGRAMME: INTRODUCTION TO CLIMATE CHANGE*)**

Greenhouse Gas	Chemical Formula	Pre-industrial Concentration (ppbv)	Concentration in 1994 (ppbv)	Duration in atmosphere (years)[a]	Anthropogenic Sources	Potential Global Warming (PGW)[b]
Carbon Dioxide	CO_2	278,000	358,000	Variable	Fossil fuel burning Land Use Changes Cement Production	1
Methane	CH_4	700	1,721	12.2 ± 3	Fossil Fuels Rice paddies Garbage dumps Farms	21[c]
Nitrous Oxide	N_2O	275	311	120	Fertilizers Industrial processes Combustion	310
CFC-12	CCl_2F_2	0	0.503	102	Liquid refrigerants Foams	6,200–7,100[d]
HCFC-22	$CHClF_2$	0	0.105	12.1	Liquid refrigerants	1,300–1,400[d]
Perfluoromethane	CF_4	0	0.070	50,000	Aluminum production	6,500
Sulfur hexafluoride	SF_6	0	0.032	3,200	Dielectric Fluids	23,900

Note: ppbv = parts per billion by volume; 1 ppbv CO_2 in the terrestrial environment is equivalent to 2.13 million tons carbon.
[a] The duration of CO_2 cannot be calculated because of different response rates in various processes of decrease.
[b] Potential Global Warming (PGW) for 100 years.
[c] Includes indirect effects of tropospheric ozone and stratospheric water vapor.
[d] Net PGW (e.g. includes indirect effects due to ozone decrease).

remaining 20% comes from deforestation and biomass burning. Half of the forests of the world were destroyed during the second half of the 20th century. To this global total we must add the combustion of carbonates to produce cement; this process liberates CO_2 to the atmosphere. By the end of the past century the concentration of CO_2 in the atmosphere had increased 31% over the concentration present during the Industrial Revolution; the concentration increases 0.05% each year (Hardy, 2003). There has been more carbon dioxide in the atmosphere in the past than at present; this occurred in the last 420,000 years and also perhaps in the last 20 million years (Houghton *et al.*, 2001).

Methane (CH_4) represents a small portion of greenhouse gases; however, its concentration increases by 1.1% per year. Methane comes from anthropogenic sources such as farms, rice growing, combustion of oil and gas, waste, and coal. Although present at a lesser concentration, nitrous oxide (N_2O) remains in the atmosphere 170 years, increases 0.25% per year, and contributes significantly to global warming. Chlorofluorocarbon (CFC) and hydrochlorofluorocarbons (HCFC) compounds are used in refrigeration and aerosols. In the atmosphere, these greenhouse gases destroy the stratospheric ozone layer that protects the Earth from ultraviolet radiation (Houghton *et al.*, 2001).

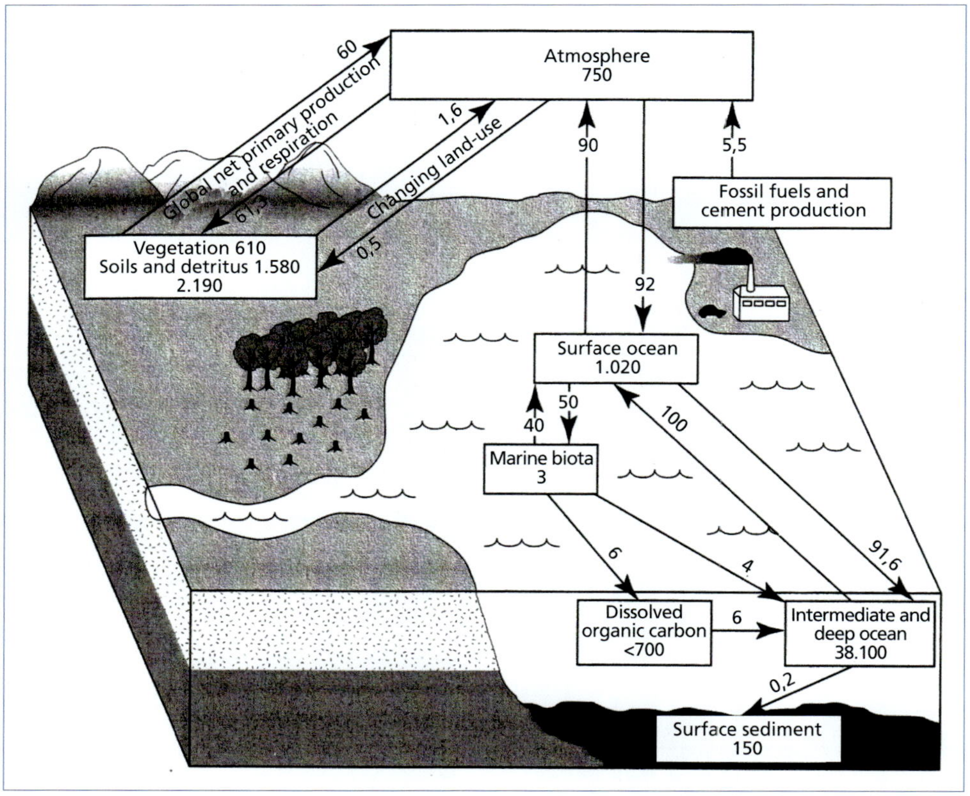

Figure 19.6 Global carbon cycle showing reserves and fluxes (GtC/year) and anthropogenic disturbance over the period of 1980–1989. The numbers show the approximate numbers and fluxes that can fluctuate significantly from one year to the next (Schimel *et al.*, 1996).

19.3.1 Consequences of climate change

The data available show that climate change, primarily temperature change, affects physical and biological systems in many parts of the world. Natural systems can be especially vulnerable to climate change because of their limited capacity for adaptation; other systems may undergo significant and irreversible damage. While some species may increase their population through climate change, others face the risk of extinction, particularly vulnerable species (Thomas *et al.*, 2004). Based on assumptions of average warming for 2050, these authors predict that 15–37% of species will be affected by the danger of extinction. Others ask: What will humans do when this shift occurs? It is evident that the geographical extent of the damage or losses and the number of affected systems will increase with the magnitude and speed of climate change. Sensitivity, capacity to adapt, and vulnerability to climate change all play a role (McCarthy *et al.*, 2001). From this we can infer that we are facing a planetary environmental crisis in which we can consider climate change to be the fever of a sick planet (Delibes and Delibes de Castro, 2005).

As a result of these changes, geomorphic processes and some landforms will change, as shown in *Table 19.2* (Goudie, 2004i). As Goudie indicates, our knowledge of the relationship between geomorphic processes and climatic conditions leads us to the conclusion that some geomorphic environments will respond substantially and all of them will have multiple consequences on human occupation of these environments.

19.3.2 Prediction of climate change

Climate change prediction is based on knowledge of climate change in the recent geologic past and modification of parameters related to the sub-actual climate (*Fig. 19.7*). As a result, climate scientists must extrapolate. Certainly "this extrapolation implies the projections of known information or relationships to the unknown… This involves both extrapolation to the future, which is prediction and extrapolation

Table 19.2 **SOME GEOMORPHIC CONSEQUENCES GLOBAL WARMING (GOUDIE, 2004i)**
Hydrologic
Increased evapotranspiration loss
Increased percentage of precipitation of rainfall as expense of winter snowfall
Increased precipitation as snowfall in very high latitudes
Possible increased risk of cyclones (greater spread, frequency and intensity)
Changes in state of peatbogs and wetlands
Less plant use of water because of increased CO_2 effect on stomatal closure
Vegetation controls
Major changes in latitudinal extent of biomes
Reduction of boreal forest, increase in grassland, etc.
Major changes in altitudinal distribution of vegetation types (c. 500 m for 3°C)
Growth enhancement by CO_2 fertilization
Cryospheric
Permafrost, decay, thermokarst, increased thickness of active layer, instability of slopes, riverbanks, and shorelines
Changes in glacier and ice-sheet rates of ablation and accumulation
Pack ice melting
Coastal
Inundation of low-lying areas (including wetlands, deltas, reefs, lagoons, etc.)
Accelerated coast recession (particularly of sandy beaches)
Changes in rate of reef growth
Spread of mangrove swamp
Eolian
Increased dust storm activity and dune movement in areas of moisture deficit
Soil Erosion
Changes in response to changes in land use, fires, natural vegetation cover, rainfall erosivity, etc.
Changes resulting from soil erodibility modification (e.g. sodium and organic contents)
Subsidence
Desiccation of clays under conditions of summer drought

back in time, which is retrodiction or postdiction… Therefore, when conditions are known and understood and when the history of situations has been established, predictions can be made with some degree of confidence... Earth scientists have difficulty explaining and predicting the behavior of complex open systems" (Schumm, 1991).

To determine the quality of a concept or scientific hypothesis, two possibilities of probability must be kept in mind: P_C-probability that the hypothesis is true and P_F-probability that the hypothesis is false. The sum of these two probabilities is 1; as P_C increases, P_F decreases. Neither can have a value of zero (Strahler, 1987). Strahler thinks that in earth science all hypotheses, except the most simple, have a certain degree of uncertainty. Without doubt, the problem is to correctly determine the probability of extrapolating.

Over the last few decades an enormous amount of technological and research effort has been put forth to explain the possible causes of climate change with the goal of predicting future tendencies, which will aid us in taking mitigation measures. For this reason, physical-mathematical climate models have been developed that simulate the fundamental dynamics of the atmosphere and the ocean (Gutiérrez and Pons, 2006).

To coordinate this research, the World Meteorological Organization and the United Nations Environment Programme (UNEP) set up an Intergovernmental Panel on Climate Change (IPCC) composed of more than 300 experts who work in three groups: basic sciences; impacts, adaptation, and vulnerability; and mitigation. They have published their conclusions in various volumes (Houghton *et al.*, 1990, 1996, 2001) and Solomon *et al.* (2007). After Goudie and Viles (2010) "Geomorphology has played a surprisingly small part in such exercises, which tend to have been dominated by climatologists, oceanographers, and ecologists".

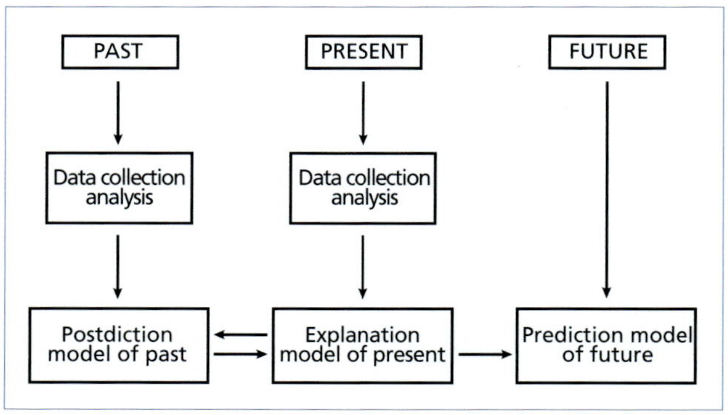

Figure 19.7 Diagram showing how information from the present and the past are used to construct models of the past, present, and future (Schumm, 1985a).

Within the last five hundred years, a cold period called the Little Ice Age affected the northern hemisphere (Grove, 1988, 2004); during this period the climate cooled less than 1°C. It occurred as several oscillations between 1350 and 1830 (*Fig. 19.3E*). The **instrumental record** of weather stations shows that between 1850 and 2000 (*Fig. 19.8*), the temperature rose 0.6°C (Folland *et al.*, 2001). Periods of maximum heating occurred between 1910 and 1945 and from 1976 to 2000. Sea surface temperatures have undergone similar pulsations although the variations and tendencies have been less pronounced (Aguilar *et al.*, 2005). Between 1945 and 1976 the stratosphere cooled more than once. Most models prepared for the 20th century have omitted this anomaly (Weaver, 2003) although recently it has been explained by the presence of sulfur aerosols emitted by explosive volcanic eruptions (see Chapter 4, Section 4.17).

In order to simulate the evolution of the climate system, complex numerical models called **global climate models** have been developed to represent the climate response to different future emission scenarios; this enables scientists estimate future climate change. These models are not perfect. Their capacity to estimate future climate change depends primarily on our knowledge of the physical and chemical processes of the climate system, which are primarily the atmosphere and the ocean. Modelers separate space into three-dimensional cells that are described by complex differential equations. These equations are solved for each variable, starting with initial values. Later approximations are made in relation to the values of each cell. The separation between cells is 2°–5° (250 km) in the atmosphere and 1°–5° (50–250 km) in the ocean; the vertical component is

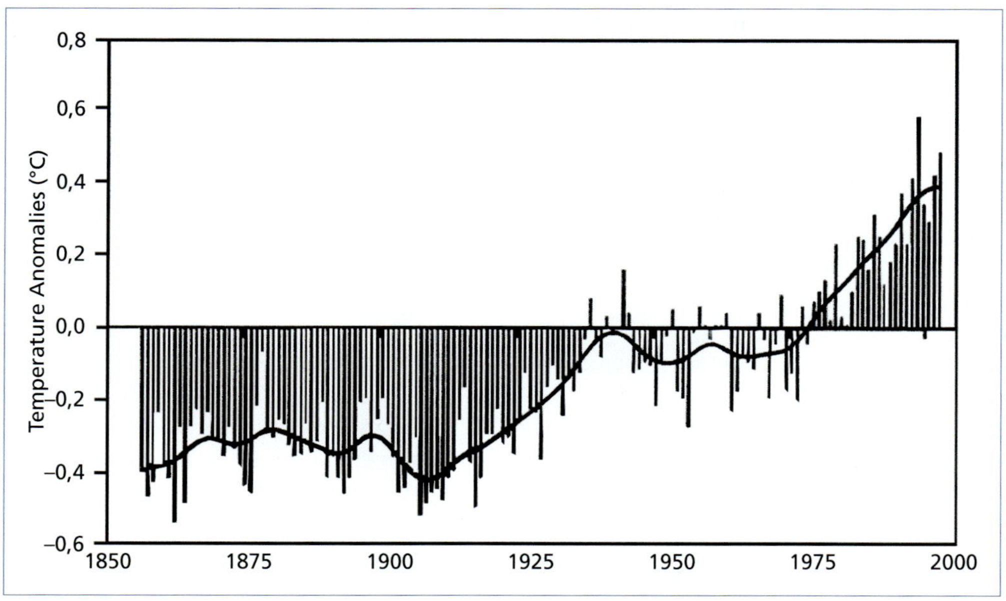

Figure 19.8 Global air temperatures reconstructed from instrumental data for the period 1850–2000. Climate Research Unit (in Aguilar *et al.*, 2005).

1–2 km in the atmosphere and several hundred meters in the ocean. The equations are generally solved for every half hour. In addition, parametrization is used to resolve in an approximate manner physical processes that are active on a smaller scale than cells. Global climate models require powerful computers. The most important centers are the *Hadley Center for Climate Prediction and Research* (London, United Kingdom) and the *Max Planck Institute for Meteorology* (Hamburg, Germany) (Houghton *et al.*, 2001; Bigg, 1985; Hardy, 2003; Gutiérrez and Pons, 2006, among others). Climate models have made it possible to determine the anthropogenic influence of greenhouse gases in the present-day climate.

19.3.3 Evolution of the principal contributions to climate change and uncertainties

Experts with the Intergovernmental Panel on Climate Change (IPCC) presented a report at a 1990 United Nations-sponsored meeting (Houghton *et al.*, 1990) for which they used various models to show that increased greenhouse gases may cause a temperature rise of 0.3°C per decade for most of the scenarios studied. As a result, the temperature rise was predicted to cause ice sheets to partially melt and sea level to rise between 5 and 6 cm per decade.

Between 1990 and 1995 (Houghton *et al.*, 1996) IPCC scientists made considerable progress in the application of simulation models, more powerful statistical methods, studies of sulfate aerosol effects and stratospheric ozone reduction, its anthropogenic nature, and obvious effects on climate change (Santer *et al.*, 1996). This research led the IPCC to develop new scenarios in 1995 for the period of 1990–2100 based on different hypotheses. The 1996 report (Houghton *et al.*, 1996) used only 6 possible scenarios to predict the temperature increase for the year 2100. For the average scenario (IS92a) the authors estimated an increase of about 2°C for air temperature for the year 2100 and 1–2.5°C for the same date for the extreme scenarios (IS92c and IS92e). Sea level rise for the year 2100 due to melting of ice masses and thermal expansion of the oceans was calculated at 50 cm under the IS92a scenario and 15–95 cm for the two extreme scenarios (Houghton *et al.*, 1996). These new numbers considerably reduce the 1990 IPCC's most catastrophic predictions. This significant decrease is due largely to the inclusion of sulfate aerosol prediction and the application of new atmospheric-oceanic models.

Even so, Houghton *et al.* (1996) claim that the predictions contain "significant uncertainties" (p. 5) that arise from lucubrations of future gas emissions, their representation in climate models (such as those related to clouds, oceans, marine ice, and vegetation), and the collected systematics of climate system variables. In addition, short-term and rapid climate changes are difficult to predict because the climate system is not linear and can produce surprises.

Ice cores have revealed rapid climate variations during the last glacial period. Scientists have recognized periods in which temperatures increase 10°C over a few decades and then later return to colder conditions. These sequences, which last between 500 and 200 years, recurred perhaps 20 times during the last glacial period (Bender *et al.*, 1994).

In Scientific Basis of Climate Change 2001, (Houghton *et al.* 2001) propose six different temperature increase curves for six scenarios; they studied 35 scenarios overall (*Fig. 19.9*) (Cubash and Meahl, 2001). The six scenarios show a range of global average temperature fluctuation between 2.0 and 4.5°C for the period between 1990 and 2100 whereas the interval for the 35 scenarios ranges from 1.4 to 5.6°C.

Sea level changes for the period 1990–2100 are predicted to be a rise of 66 cm (Warrick and Oerlemans, 1990). Based on scenario IS92a, sea level rise is estimated to be between 20 and 86 cms for the same period (Warrick *et al.*, 1996). Calculations of future sea-level rise under scenario IS92a, including the effect of sulfate aerosol emissions, range from 11 to 77 cm for the period from 1990 to 2100 (Church and Gregory, 2001); these are somewhat lower numbers than those indicated in the Climate Change 1995 report. *Figure 19.10* shows sea level rise for six scenarios. The gray area shows the range for 35 scenarios (Albrithon and Meira Filho, 2001).

Scientists who study oceanic circulation and climate have united to form the World Ocean Circulation Experiment (WOCE) to conduct the largest and most comprehensive oceanic data-gathering experiment ever done. Satellite measurements, *in situ* observations, and model development will contribute to the first real experiment on the global ocean (Lindstrom and Legler, 2001). WOCE provides data primarily for research on climate variability and prediction and for the development of oceanic models (Smith, 2001).

The 2007 Climate Change Report (Solomon *et al.*, 2007) states that between the pre-industrial period and 2005, carbon dioxide concentrations increased from about 280 ppm to 379 ppm, methane from 715 to 1774 ppb, and nitrous oxide from 270 to 319 ppb. All these gases contribute to warming. Anthropogenic aerosols such as sulfates, organic carbon, nitrate, and dust cause cooling. Observations show that warming of the climate system is indisputable; these observations show an increase of global air and ocean temperatures, extensive snow and ice melting, and increased sea level rise (IPCC, 2007).

Global surface temperatures in the twelve years between 1995 and 2006 were the highest since 1850 (IPCC, 2007); the temperature increased 0.76°C during this interval. In addition, observations since 1961 indicate that the global average ocean temperature has risen to depths of at least 3000 m and the ocean has absorbed more than 80%

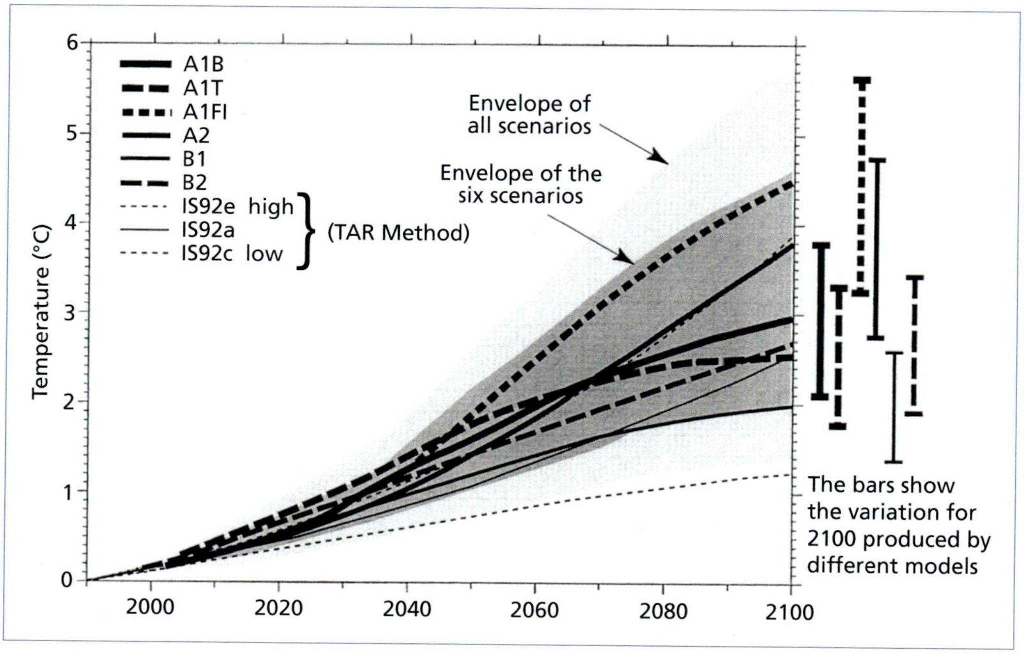

Figure 19.9 Average global temperatures from 1990 for six scenarios. The areas in gray represent the envelope of 35 scenarios (Cubasch and Meehl, 2001).

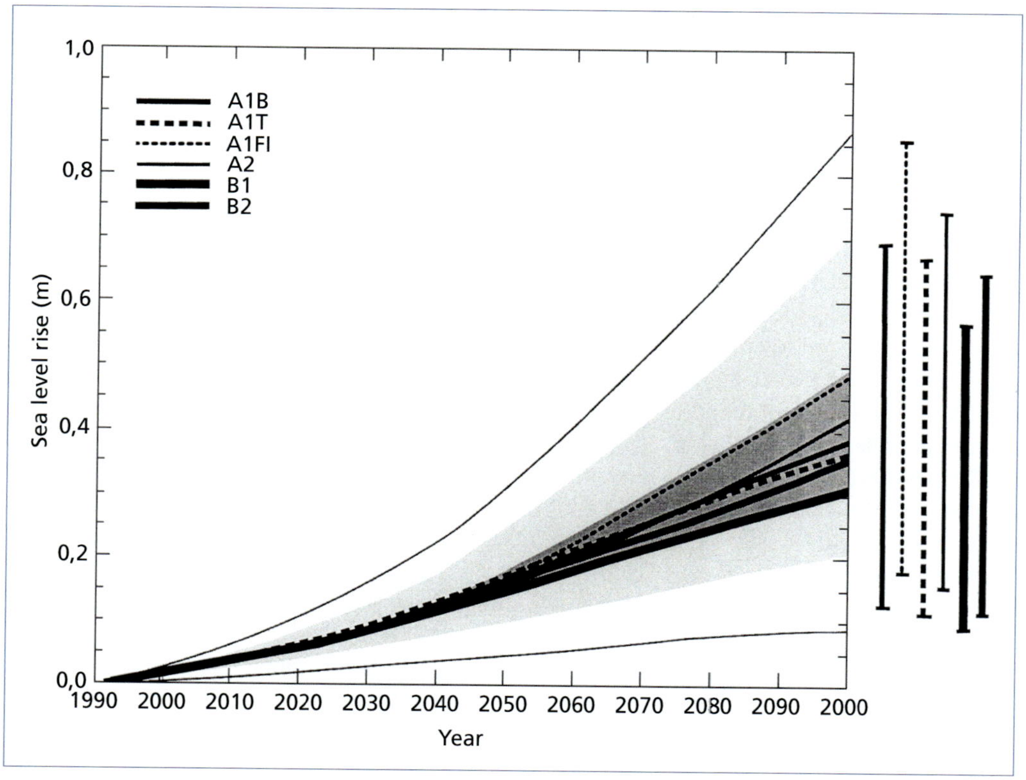

Figure 19.10 Average elevation of the sea level for six scenarios and the area in gray shows the envelope of 35 scenarios (Albritton and Meira Filho, 2001).

of the heat added to the climate system. Partial melting of ice sheets increased between 1993 and 2003 (*Table 11.2*). The most rapid rise occurred between 1993 and 2003. The 20th century total is estimated at 0.17 mm/year. According to the IPCC (2007), temperatures at the top of the permafrost in the Arctic Ocean (up to 3°C) have increased since the 1980s. The surface area covered by seasonally frozen soil has decreased 7% in the northern hemisphere since 1900, with a decrease in the springtime of up to 15%.

Between 1900 and 2005, rainfall increased significantly in the eastern parts of the Americas, northern Europe, and central and northern Asia. Less precipitation has been recorded in the Sahel, the Mediterranean, South Africa, and parts of East Asia. Since 1970, intense and lasting droughts have occurred over large areas. The frequency of intense precipitation has also increased along with warming and increased atmospheric water vapor. During the last 50 years, cold days and nights and freezing are less common than previously; meanwhile increasingly hot days and nights and the number of hot hours have increased. In addition, there is evidence since 1970 of increasingly intense tropical cyclone activity in the North Atlantic, which is correlated with higher temperatures of tropical seas.

The IPCC (2007) summary report predicts 0.2°C warming per decade. If concentrations of all greenhouse gases and aerosols remain at the same levels as the year 2000, warming will be about 0.2°C. If greenhouse gas emissions continue at current rates, warming will increase, which will cause significant changes in the global climate system, that is, changes that are more extreme than those observed in the 20th century.

This brief analysis of the different proposals in climate change publications dating from 1990, 1996, 2001, and 2007 show that prediction efforts contain numerous **successive approximations** that incorporate our continuously increasing knowledge of climate change factors.

Understanding climate change is a complex task because of the number of variables involved. Some variables are known in an approximate way whereas others are difficult to understand; it is also difficult to understand the interactions between these variables. As a result, climate change has been and always will be a subject full of **uncertainty** and controversy (Lomborg, 2008).

Although there is general consensus among scientists about the reality of global warming in the near future, some disagree with the interpretations of temperature increases and express doubts on the subject. Dickinson (1986) points out that general circulation models (GCMs) have many limitations and should be used with caution. In his research on CO_2 releases from ice during the last interglacial and on the balance between CO_2 absorption and loss into the ocean, Sundquist (1993) expresses fundamental uncertainties related to the consequences of anthropogenic CO_2 and thinks that its effects will occur later. Paté-Cornell (1996) maintains that models used to study the effects of global climate change are based on limited knowledge of the primary phenomena (for example, the role of clouds and oceans). Demangeot (1996) finds that tests on the amount of present-day global warming are not convincing; he seeks precision from specialists on ice cap melting, the role of deforestation, and other topics.

Embleton (1989) shows climate change prediction for decades or centuries into the future to be extremely difficult. Data supplied by research on ice cores by Siple (Antarctica) and also measurements of air samples collected at the Mauna Loa observatory (Hawaii) since 1958 show that CO_2 has increased considerably since 1750. There is no doubt that increased CO_2 causes a temperature increase. We have Houghton's (Houghton, *et al.*, 1990) predictions that indicate an increase of up to 3°C per century. If this happens, greenhouse gases will cause a "super-interglacial" (*Fig. 19.11*) that would be the largest interglacial of the entire Quaternary. Embleton (1989) and Goudie (1990b) point to the fundamental role of ocean water, which is a large CO_2 reservoir, and of marine biota in the regulatory dynamics of atmospheric CO_2. The flux of carbon dioxide into the ocean depends on CO_2 pressure and CO_2 concentration in the atmosphere and surface winds. Processes of oceanic transport and geochemical and biological processes in the upper part of the ocean control CO_2 pressure. We must also add incident solar flux

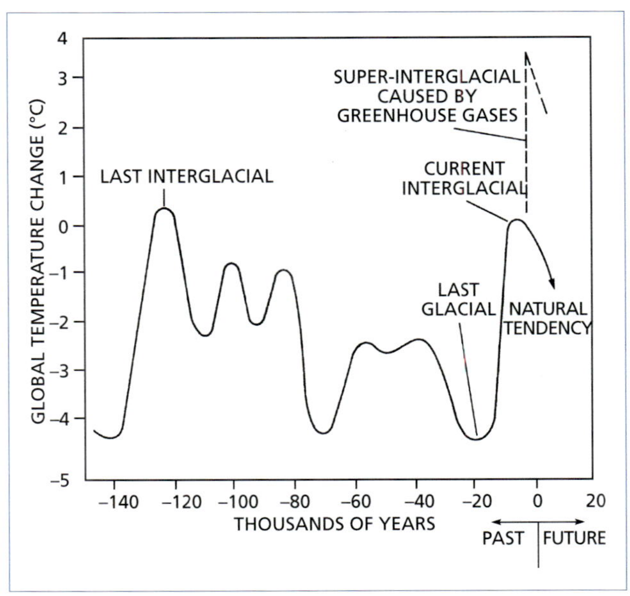

Figure 19.11 Prognosis of a "super-interglacial" produced by greenhouse gas warming, compared to the trends of natural temperatures for the last 140,000 years (Lamont Newsletter 6, in Embleton, 1989).

and latent heat flux, which also affect climate modeling. The system is obviously highly complex; flux modeling of the oceanic-atmospheric interface can be subject to serious errors (Dickinson *et al.*, 1996). In addition, we still do not know the temperature and salinity of deep waters, understand the boundaries between currents and oceanic eddies, or have a deep understanding of thermohaline circulation (McBean *et al.*, 1996).

Statistical analysis of maximum and minimum temperature differences during the day at 2,000 stations around the world over a period of 40 years shows that nights have warmed more than days (nights have warmed 0.84°C whereas days have warmed 0.28°C); this shows that warming is primarily due to increased nocturnal temperatures. A clear statistical relationship shows that **cloud cover** is responsible for these circumstances (Henderson-Sellers, 1992) because clouds regulate nocturnal temperatures by controlling long-wave emissions. During the 20th century, cloudiness has increased 3.4 to 9.4%, possibly because of increased condensation nuclei due to air pollution. This explains the smaller percentage of clouds above oceans than above continents. Understanding variations in cloudiness is obviously of prime importance in adequately analyzing future climate change.

Highly explosive acid **volcanic activity** that occurs sporadically also has an effect on climate. A large eruption can propel ash into the stratosphere where it participates in general atmospheric circulation and prevents the penetration of solar radiation. Surface air temperatures decreased about 2°C around the world after four major volcanic eruptions (Krakatoa, 1883, Santa María, 1962; Agung, 1963, and El Chinchón, 1982). This cooling lasts one or two years after the eruption (Jones and Kelly, 1996) but there is evidence that the effect of the temperature drop continues 5 to 7 years afterward (Hansen *et al.*, 1996). The Pinatubo volcano in the Philippine Islands began erupting in June 1991. Its volcanic aerosols traveled around the world; models indicate that average global air temperatures dropped 0.5–1°C in the northern hemisphere (Self, 2005) and that temperatures during the summer of 1992 were up to 2°C lower than normal (Robock, 2002). The effects of the 1991 eruption were less evident in 1994. Volcanic activity that makes the stratosphere opaque has an obvious influence on climate, even though it is sporadic in nature.

Human industrial and agricultural activity generates considerable **atmospheric dust**. Large quantities of eolian dust are also produced, which can be deposited on distant continents and oceans. Dust rises to the upper part of the troposphere and prevents the penetration of solar radiation; consequently it causes global cooling (Bryant, 1997).

During the 1990s we came to understand the role of anthropogenic **sulfate aerosols** in the set of factors involved in climate change. These aerosols are produced primarily from the burning of coal, oil, and gas; records indicate that during the 20th century the exponential increase of sulfate aerosols in the atmosphere is parallel to that of carbon dioxide. The high reflectivity of sulfate aerosols causes cooling of the troposphere on both clear and cloudy days (Pearce, 1994; Santer *et al.*, 1996).

Houghton *et al.* (1996) show the difficulty of predicting **short-term rapid climate changes**, called "climate surprises" by some authors (Jouzel and Lorius, 1999). Scientists have detected rapid climate variations in the most recent glacial period (Alley, 2004; Broecker, 2006). Boreholes drilled into the Greenland Ice Sheet have indicated periods of warming of about 10°C that occurred over a few decades, after which the climate returned to colder conditions. These sequences, which last between 500 and 2000 years, recurred about 20 times during the last glacial period (Bender *et al.*, 1994); furthermore, they are not limited to Greenland. Deep cores in the Atlantic Ocean show that over the last 70,000 years various sedimentary layers of glacial origin have been deposited, following the melting of a large number of icebergs. These six synchronous stages are called the Heinrich events (Heinrich, 1988; Bond *et al.*, 1993; 1997). These increases of fresh water and decreased surface temperatures of ocean waters cause substantial changes of certain climate parameters. As a result, the fact that rapid climate variations have occurred in the past suggests the possibility of unanticipated triggering of this type of climate change.

Climate change modeling is uncertain and complex (Schiermer, 2007) because of the numerous parameters that must be considered; it is also challenging to understand the interactions among them. An alternative method for the study of climate change is stored in the geologic record (Petit-Maire, 1999); this source provides highly valid data on climate change.

Researchers refer to earlier periods to study the consequences of significant warming on relief. They study the Climate Optimum or Altithermal that occurred about 7,000 BP and compare it to the maximum of the last interglacial that took place about 130,000 BP (Eemian, or Sangamonian in North America) to study the landform response of warmer stages. New climatic circumstances give rise to different processes that modify landforms; these processes can show us the geomorphic panorama that forms them and the consequences derived from them. The capability to make an approximate forecast is a significant contribution of applied geomorphology.

In spite of critics' claims, global warming is real but it remains to be determined the extent to which temperatures increase as a result of natural tendencies or human activities. The two effects are likely superimposed; determining the role of each is a complex task (Embleton, 1989).

The discussion, however, should not be about whether climate change is caused by man; humankind needs to determine is the speed and magnitude of the change and its regional implications (Weaver, 2003).

19.3.4 Mitigation

Mitigation of greenhouse gas effects will require the resolution of numerous technical, economic, political, cultural, and social problems and/or barriers that prevent the full exploitation of technological, economic, and social options aimed at mitigation. Differences in technological, natural, and financial resources among nations or regions and differences in mitigation costs are often key points in the analysis of climate change mitigation (Metz *et al.*, 2001) (*Fig. 19.12*).

Quick decisions are needed to reduce or prevent the impact of human-induced climate change. However, scientists are concerned about the difficulty of determining the collateral effects of some mitigation practices. For this reason, some say "it's better to be safe than sorry" (Hardy, 2003).

CO_2 emissions, which are the primary cause of global warming, have increased 32% since 1750. In addition, the concentration of carbon dioxide in the atmosphere is now at its highest level in the last 420,000 years (McGuire *et al.*, 2004). One possibility under discussion is **CO_2 sequestration** in deep geological formations in abandoned petroleum reservoirs (White *et al.*, 2004). Scientists are currently injecting CO_2 into a sandstone layer 1000 m below the seafloor in the Sleipner oil field in the North Sea. CO_2 capture and storage is a process that involves separating CO_2 from industrial and energy sources, transporting it to a "warehouse," and isolating it from the atmosphere for a long time period. CO_2 can be compressed and stored in geologic formations, in oceans, in carbonate minerals, or can be used in industrial processes (IPCC Working Group III, 2005).

Many countries such as China are making maximum use of hydroelectric energy; the Three Gorges Dam on the Yangtze River (west central China) will be the largest dam in the world. It will produce 18,000 megawatts of electricity, equivalent to one third of the country's consumption. The reservoir will displace 1.3 million people and bury 160 towns and cities, and the dam construction project presents numerous other challenges (Cunningham, 1999).

Alternative energy sources (**non-fossil renewable fuels**) are good resources for mitigating climate change; research in this area is expected to lead to projects of major importance. Wood fiber can also be converted into methyl alcohol for combustion as fuel. Ethyl alcohol derived from fermented corn provides fuel for a considerable percentage of cars in Brazil. These vehicles also have the advantage of not being stolen and exported to other countries because other countries lack this type of fuel.

Wind is another alternative energy source. Many wind energy generators have been set up around the world (*Fig. 10.65*); they still represent a part of total electrical energy but their future is promising. **Solar energy** is also being used increasingly to heat buildings. Solar energy can be used to produce electricity through the use of photovoltaic cells in which semiconductor crystals convert sunlight into electrical energy. Research is also being done to take advantage of **waves** and **tides** as energy sources (see Chapter 11). In addition, **oceanic thermal energy** (OTEC), the potential energy of temperature differences between deep and surface ocean waters, can be used to produce electricity. This process requires a thermal difference of at least 20°C; it may prove useful on islands or coastal areas located near deep water (Hardy, 2003).

Each energy source has advantages and disadvantages (Dresselhaus and Thomas, 2001). Combustion of fossil fuels

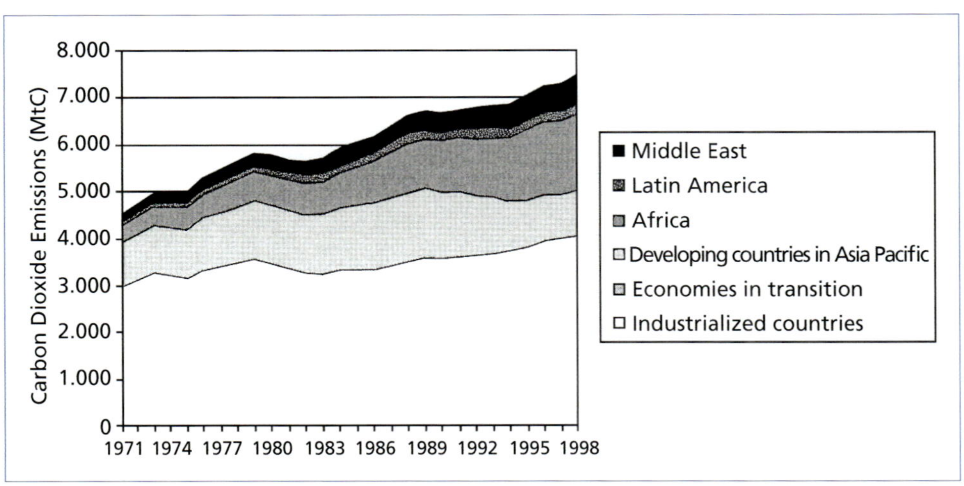

Figure 19.12 Global CO_2 emissions by region for the period 1971–1998 (Metz *et al.*, 2001).

creates greenhouse gases and other contaminants. Hydroelectric plants may interfere with fish migration and flooding over large areas. Generators of solar, eolian, wave and tidal, and ocean heat/thermal energy are geographically limited. Geothermal energy is available only in some volcanic areas. Finally, nuclear plants are costly and produce high activity radioactive waste that is difficult to dispose.

Research is being done on stimulating the growth of phytoplankton to fertilize the oceans with iron. Microscopic floating phytoplankton made of algae and dynoflagellates grow by a process similar to photosynthesis in plants; phytoplankton consumes CO_2 from the atmosphere (Watson *et al.*, 2000).

Nuclear energy, which has numerous proponents and detractors, is an alternative to fossil fuels; it does not emit greenhouse gases. Nuclear energy already provides 17% of the world's electricity. France, the major producer, uses nuclear power to produce 76% of its electricity. But nuclear power has significant accident-related problems such as the Chernobyl accident (Ukraine) on 26 April 1986 and the accumulation of high-activity radioactive waste, which must be buried in repositories at great depths. Currently, no repository in the world can dispose of high-level radioactive waste. One nuclear energy alternative under investigation is nuclear fusion of hydrogen, which produces little waste (Ogden, 1999).

We have stated that sulfur dioxide gases emitted by the Pinatubo volcano in June 1991 caused global cooling. On this basis, Cruzen (2006) has proposed to block solar rays before they reach the Earth and thus control global warming by introducing one or two million tons of sulfur into the stratosphere each year. Sulfate aerosols offer the hope of being able to cool the climate immediately whereas emissions reduction will require decades or generations. Aerosols cool by blocking solar radiation. Therefore, aerosols cool the atmosphere during the day only and cool more during the summer and in the tropics. However, it is ill advisable to apply this or other geo-engineering proposals when general atmospheric circulation models are not adequately known (Morton, 2007).

Climate change is a global problem because greenhouse gas emissions affect all countries; thus the solution must be based on international cooperation. Governments have presented national policies and signed agreements on climate change mitigation.

The 1987 Montreal Protocol adopted an agreement to decrease chlorofluorocarbon gas emissions; these gases adversely affect stratospheric ozone. The Protocol was ratified by 80 countries and significantly reduced the rate of ozone reduction. This Protocol is an excellent example of international effectiveness in problem resolution. At the United Nations-sponsored Conference on Environment and Development in Río de Janeiro in 1992, the fundamental proposal was to "stabilize the concentration of atmospheric greenhouse gases to a level that avoids dangerous anthropogenic interference in the climate system" (UNCED, 1992).

The 1997 Kyoto Protocol was an agreement among about 100 countries to reduce greenhouse gas emissions. The protocol involved the proposal and publication of plans by each nation to alleviate the effects of greenhouse gases. Some countries, including the United States, the largest producer of greenhouse gases, have not ratified the Protocol. The National Academies of Science of 17 nations have said that: "The balance of the scientific evidence requires us to conduct effective measures to prevent changes prejudicial to Earth's climate" (Science, 2002). The Kyoto Protocol gives precedence to research on and the use of non-polluting energy (eolian, solar, geothermal, and so on) and the use of hydrogen-based energy. Industrialized nations must reduce their total CO_2 emissions during the period 2008–2012 (Wigley, 1998), according to the Kyoto Protocol. In December 2009, a climate change summit was held in Copenhagen. Significant discussions and recommendations are expected for the first time because the United States has agreed to participate in the reduction of greenhouse gases.

The climate of Spain is highly variable and its orography and geographic location make it difficult to establish climate forecasts. Temperatures during the 20th century have generally increased; furthermore, a uniform increase throughout the 21st century is expected although predictions vary according to the scenarios contemplated. Rainfall during the 20th century has not been uniform and predictive models show clear discrepancies; even so, all models point to a significant reduction of total annual rainfall. The principal impacts of climate change on Spain are clearly developed in a volume of the Preliminary Evaluation of Impacts in Spain by the Climate Change Effect (Moreno, 2005).

Today, research groups located in different centers and universities study climate change and the various problems it causes. Uriarte (2003) has written a monograph entitled the History of the Earth's Climate that approaches climate study in a global way and focuses on warming of the present day climate. Another monograph, Technologies in the Fight against Climate Change-From Coal to Carbon (Llamas and Romero, 2006), analyzes coal burning and its consequences. Climate change is a concern shared by a wider audience than scientists; all of society is concerned. For this purpose, a meeting entitled Days of Climate Change and Associated Natural Risks was held in Santander (Spain) in 2005. Contributions were published in the Cuaternario y Geomorfología volume 20 (3–4), Zaragoza, 2006. The Real Academia of Exact, Physical, and Natural Sciences, in collaboration with the Areces Foundation, organized an International Symposium on the Critical Evaluation of Climate Change Predictions: A Scientific

Perspective (Madrid, April 2008), coordinated by F. García Novo and A. Cendrero.

The Ministry of the Environment of the Spanish government houses the Spanish Office of Climate Change, which collects and analyzes the research on this topic and also serves as scientific advisor in international meetings and holds conferences on this topic.

19.4 Anthropogenic change

Of the million species that inhabit the Earth, only one-**Homo sapiens**—has a high intellectual capacity (Raven *et al.*, 2008). Man and his activities have modified the face of the planet to obtain resources, occasionally leaving a landscape with scattered hills, gashes, and other degradation that in many cases requires difficult and costly remediation. In addition, humans move large volumes of rock for uses such as the construction of dams and dikes, disposal of mine tailings, and establishment of waste disposal areas. Humans have done most of this activity primarily in the last one hundred years but the pace and scope of these activities has increased in recent decades. Because of these and other activities, man is now considered to be a geomorphic agent (Demek, 1973; Nir, 1983). This last author includes these human manifestations in the term **Anthropogenic Geomorphology**. Haff (2000) considers **Neogeomorphology** as the sub-branch of geomorphology that identifies the principles of landform change caused by anthropogenic forces. The name **Anthropocene** has been proposed as a more appropriate name for the Holocene because of man's erosional activity over the last 10,000 years (Crutzen and Stoemer, 2000). Goudie (2004j) proposes **Anthropogeomorphology** for the study of the human role in landform creation and modification.

Today human activity is transforming the Earth whereas in the past non-human forces shaped the planet. Man emerged as a competitor and today he is the principal agent modifying the land surface (Hooke, 1994; Meyer, 1996). Today "we find ourselves in a situation unprecedented in the four thousand five hundred million years of life on the planet, that for the first time one species has the capacity to affect natural systems on a planetary level with a qualitative and quantitative importance that could equal and even exceed natural agents" (Cendrero, 2003).

Humans have altered **terrestrial ecosystems** enormously through agricultural development, forest modification, exotic species introduction, and changes in biogeochemical cycles (Vitousek *et al.*, 1997). As a result, geomorphic processes have accelerated soil erosion (*Fig. 19.13*). We have made significant advances in understanding these processes and we now know that changes in these processes undoubtedly modify ecosystems (Renschler *et al.*, 2007). Biotic and abiotic processes promotes understanding of many changes to ecosystems, as proven by recent scientific advances. However, we do not understand how these processes are interrelated (Scheffer *et al.*, 2001). Lovelock (1988) claims that the Earth's chemical atmosphere may have led to the initiation of life. This author, in collaboration with the microbiologist Margulis, has promoted the Gaia hypothesis. This hypothesis describes a planet whose inorganic characteristics (fundamentally the atmosphere and oceans) are manipulated by its organisms. Like the Gaia hypothesis, the concept of coevolution of life predicts that the global system evolves toward conditions favorable for organisms (Huggett, 2006).

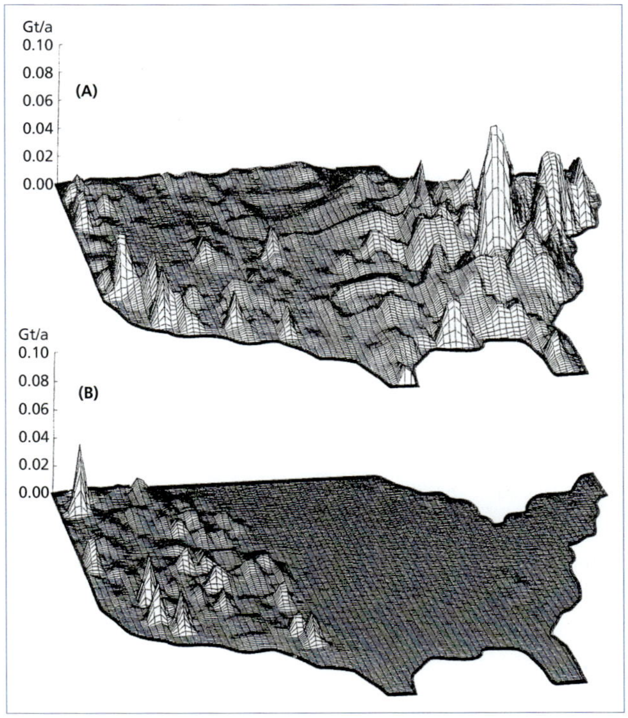

Figure 19.13 Map of the United States showing variations in peak heights for erosion rates in giga tons (giga = 10^9) for a 1° (latitude and longitude) cell. In: (A) human erosion and (B) fluvial erosion (Hooke, 2000).

19.4.1 Changes in populations and society through time

Ehrlisch and Holdren (1971) proposed a formula that is widely used today to reconstruct sources of anthropogenic impact. These authors equate human pressure or impact to the environment (I) to the per capita demand for earth resources (A, by abundance) multiplied by the number of people (P, population) and by impact per unit of resource use (T, technology factor). Their formula is I = PAT, which implies that high levels of impact are associated with large human populations. If consumption and population levels are high (high values of A and P), the formula indicates significant impacts to the environment. Technological (T) changes may independently amplify or diminish the impact by using resources with greater efficiency.

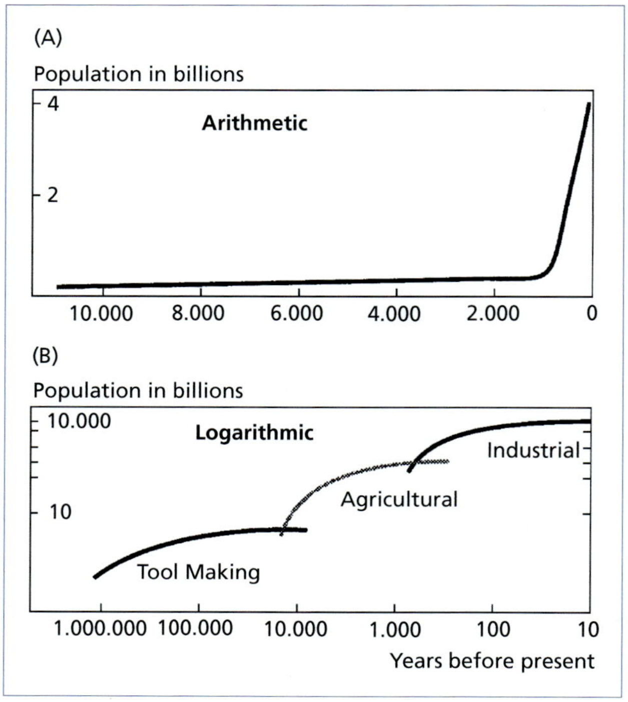

Figure 19.14 Interpretations of global population tendencies. (A) Population growth in the last 11,000 years (arithmetic-exponential curve) (Demeny, 1990). (B) Population increase over the last million years (Whitmore *et al.*, 1990).

Population increase, consumption levels, and technology are the three most influential arguments for social progress. Today, increased population, wealth, and technological capacity are significant advances that produce environmental changes and increased welfare of the population. As Meyer (1996) says, they are the three horsemen of the environmental apocalypse because they have pushed environmental change to its worst levels.

It is only necessary to know concisely the development of human population and of cultural stages with the goal of understanding modifications of the environment and the most discussed landmarks related to human activities throughout history.

Man evolved possibly 6 to 7 million years ago in the rift valleys of East Africa (Kenya, Tanzania, and Ethiopia) and later slowly spread out to cover the surface of the planet. Population growth over the last 10,000 years has been interpreted in various ways (*Fig. 19.14*). Demany (1990) claims that population increased slowly until the Industrial Revolution and that it has increased exponentially since then. Whitmore *et al.*, (1990) describe three revolutions (invention of artifacts, agriculture, and industry) in which humans increased their capacity to change the planet.

One of the most important consequences of the **Agricultural Revolution** (*Fig. 19.15*) was population growth, which is estimated at 5–10 million people during this period. Humans had previously lived from hunting and gathering but at the beginning of the Neolithic (about 10,000 years ago) man began to cultivate and to herd some animals. Communal living arose in association with this change. Over time man constructed dwellings out of adobe and his towns became cities (Hooke, 2000). In other places man found

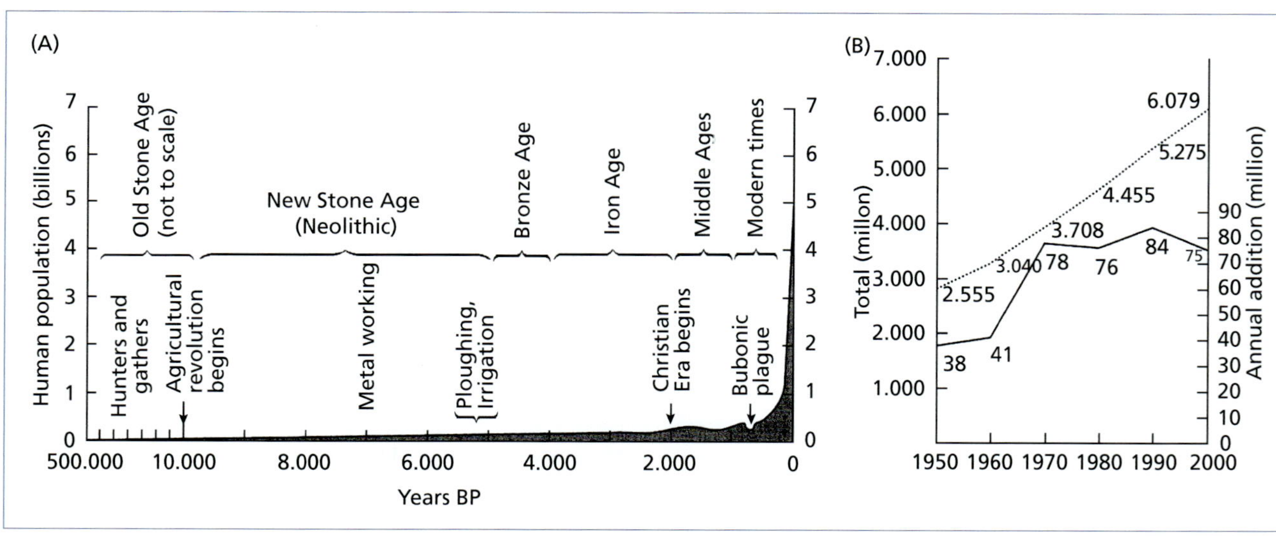

Figure 19.15 (A) Population grown in the last half million years (Ehrlich *et al.*, 1977). (B) Population growth since 1950 (Goudie, 2006).

Figure 19.16 Troglodite dwellings in a limestone massif. Turkey. Photo F. Gutiérrez.

lodging within rocks, a practice that persists today in some areas. Rock is a poor conductor of heat, which makes it a good insulator (*Fig. 19.16*). Population growth creates the need for more cattle to feed the larger population; as a result, the periodic exhaustion of pastures by herds caused seasonal migration. The herders that led flocks to fresh pastures in the mountains left evidence of their daily lives in sculptures and numerous drawings. Non-industrial societies also affect vegetation by burning savanna areas (*Fig. 18.52*) and in middle latitude areas, they cause overgrazing and salinization of irrigated land in arid areas, as in Iraq (Goudie, 2006).

Approximately 4000 years ago humans learned how to melt copper and add small amounts of tin to harden it; this alloy is bronze, which was widely used throughout the ancient world (**Bronze Age**). Many scientists believe that the wheel was invented during this period. The wheel facilitated the transport of geologic materials and man began to construct the first roads at this time. The **Iron Age** began about 2,500 years ago; iron is more abundant than copper and tin and thus it is cheaper. The use of iron had a major impact on the Earth's surface. Concentration of population in cities in arid zones made it necessary to store water for consumption and irrigation. During the Iron Age the Egyptians constructed a 160-km canal to connect the Mediterranean and Red seas. Buildings reached gigantic sizes and roads were considerably expanded. In a period of about 400 years the Romans paved 300,000 km of roads and principal highways. Agriculture also expanded greatly during this time; it was accompanied by significant erosion (Goudie, 1981b, 2006; Hooke, 2000). The first large cities appeared in the Middle East. The great civilizations of Egypt, Mesopotamia, and India collapsed mysteriously around 4,000 years BP. This collapse seems to be related with a dry event around 4.2–4 Ka that affected the monsoon area and the eastern Mediterranean (Gasse, 2000).

Industry, which is a consumer of resources and generator of wealth, has become the principal agent of environmental change, particularly since the **Industrial Revolution** of the 18th century. In the following centuries, industrialization and the highly concentrated human populations that characterize urban-industrial systems have affected terrestrial, aquatic, and atmospheric environments (Mannion, 1997).

The world's **population** reached 1 billion in 1800 and 2 billion in 1930 (*Fig. 19.17*). However, between the 1950s and 2011, the human population has expanded to 7 billion. If fertility levels remain constant, the world's population will be 134 billion in the year 2300. These numbers are not sustainable and thus this scenario is unlikely to occur. A recent report issued by the United Nations Department of Economy and Social Issues forecasts that population will stabilize around 9 billion in the year 2300 (McGuire *et al.*, 2004).

One of the major population problems is the growing concentration of people in large cities known as megalopoli (plural of megalopolis). Today 200 cities have populations of

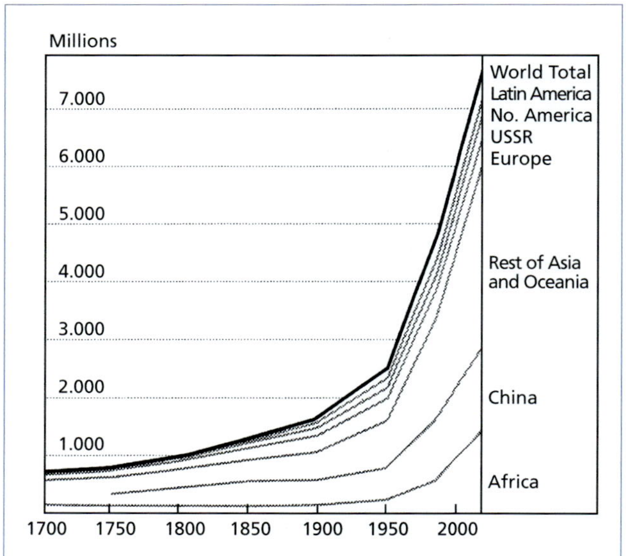

Figure 19.17 Population growth between 1700 and 1985. The graphs predict population up to 2020 (Demany, 1990).

more than two million inhabitants. Many of these megalopoli are located in coastal areas that are exposed to cyclones and flooding; in addition, they are located at lithospheric plate boundaries where earthquakes, tsunamis, volcanic eruptions, and landslides occur. In 1950, nine of the 20 largest cities of the world were located in countries of low and medium per capita income; by 2015 this number is predicted to increase to 17 (*Table 19.3*) (McGuire *et al*., 2004).

Rapid **urbanization** is a primary factor leading to increased vulnerability, particularly for low-income families who live in illegal shantytowns. The urbanization process arises from the pressure of immigrants moving to overpopulated cities where newcomers have few opportunities to occupy safe living spaces. Risks in these communities also include malnutrition and poor health. In addition, cities that have a high proportion of shantytowns are subject to earthquakes (Lima), coastal flooding (Manila), cyclonic flooding (Calcutta), and landslides (Bogota, Río de Janeiro, and Caracas); in landslide areas, people live on steep gradient slopes prone to mass movement (Blaikie *et al*., 1994).

Table 19.3 THE 20 LARGEST CITIES IN THE WORLD IN 1950, 1990, AND PREDICTED FOR 2015 (MCGUIRE *ET AL*., 2004)

1950			1990			2015 (predicted)		
Rank	**City**	**Population (millions)**	**Rank**	**City**	**Population (millions)**	**Rank**	**City**	**Population (millions)**
1	New York	12.3	1	Tokyo	25.1	1	Tokyo	28.9
2	London	8.7	2	New York	16.1	2	Mumbai	26.2
3	Tokyo	6.9	3	Mexico D.F.	15.1	3	Lagos	24.6
4	Moscow	5.4	4	Sao Paulo	15.1	4	Sao Paulo	20.3
5	Paris	5.4	5	Shanghai	13.3	5	Dhaka	19.5
6	Rhine-Ruhr	5.3	6	Mumbai	12.2	6	Karachi	19.4
7	Shanghai	5.3	7	Los Angeles	11.5	7	Mexico D.F.	19.2
8	Buenos Aires	5.0	8	Buenos Aires	11.1	8	Shanghai	18.0
9	Chicago	4.9	9	Calcutta	10.9	9	New York	17.6
10	Calcutta	4.4	10	Beijing	10.8	10	Calcutta	17.3
11	Osaka	4.1	11	Osaka	10.5	11	Delhi	16.9
12	Los Angeles	4.0	12	Seoul	10.5	12	Beijing	15.6
13	Beijing	3.9	13	Río de Janeiro	9.7	13	Metro Manila	14.7
14	Milan	3.6	14	Paris	9.3	14	Cairo	14.4
15	Berlin	3.3	15	Moscow	9.0	15	Los Angeles	14.2
16	Mexico D.F.	2.9	16	Tianjin	8.8	16	Jakarta	13.9
17	Mumbai	2.9	17	Cairo	8.6	17	Buenos Aires	13.9
18	Philadelphia	2.9	18	Delhi	8.2	18	Tianjin	13.5
19	Río de Janeiro	2.9	19	Metro Manila	8.0	19	Seoul	13.0
20	Saint Petersburg	2.9	20	Karachi	7.9	20	Istanbul	12.3

Between 1990 and 1999 about 300,000 people lost their lives and 750 million human beings were affected by hurricanes and flooding, most of them in coastal zones and many on destructive margins. Earthquakes of magnitude 7.5 or higher can occur in the middle of many cities (Jakarta, Tehran, Istanbul, Mexico City). Economic losses are much higher in cities of developed countries where the consequences of natural disasters are more financial than life threatening; the earthquakes of San Francisco (1989), Los Angeles (1994), and Kobe (1995) are well known. Hurricanes in Miami (1992) and Hurricanes Mitch (1998) and Katrina (2005) are examples.

The Munich report contains a proposed danger index. As expected, Tokyo tops the list of high-risk cities, followed at some distance by San Francisco, Los Angeles, Osaka-Kobe-Kyoto, Miami, New York, and Hong Kong. The index provides a useful measure of potential economic losses in developed countries but even the smallest danger indices for these countries indicate that the method has little application. The indices of cities like Paris, London, Istanbul, Tehran and Jakarta do not clearly reflect the true vulnerability of these cities or the serious potential life-threatening dangers due to flooding, hurricanes, or earthquakes (McGuire *et al.*, 2004).

Natural catastrophes occur frequently, primarily as a result of hydrometeorological events. The frequency of these disasters has increased as world population and shantytowns have grown in marginal areas; unfortunately these catastrophes will intensify in the near future. These events have led to a new science, the science of environmental catastrophes, which requires accurate chronologies to establish cause and effect connections and the establishment of links between the geosciences and social sciences. In order to survive these catastrophes, we must learn from the past, adapt to environmental changes, and modify our lifestyles in view of nature, which is casting an ominous shadow (Leroy, 2006).

19.4.2 Human impact on the environment

Previous chapters have included discussions of various Earth processes and resulting landforms; in this chapter we have emphasized the risks inherent in these active processes. Some of these natural and anthropogenic dangers could certainly be treated in more detail but our intention here is to call the reader's attention to their potential consequences. Even so, some of these processes and results will be the focus of this section.

Human activity has created a wide variety of landforms. These landform changes are easily recognized as quarries, mine waste dumps, and rubbish dumps. Other landforms created by indirect anthropogenic processes are less easy to recognize because they are often modifications of natural processes; an example is decreased plant cover (grazing, fire) that accelerates erosion.

Excavations of rocks and minerals are commonplace landforms that have been around for thousands of years. Examples are pits for flint extraction during the Neolithic (2,000 years BC) in Norfolk (England) (Goudie, 1981b) and the deep defensive trenches dug into gypsum by the Iberians (~200 years BC) in the Mediana de Aragón (Province of Zaragoza, Spain) (Burillo *et al.*, 1984); they show the capacity of ancient populations to move earth materials. It may be difficult to determine whether small excavations are natural or of anthropogenic origin.

Extractive power has increased continuously since the Iron Age; it took on enormous dimensions during the Industrial Revolution in both quarry works (*Fig. 19.18*) and mining (*Fig. 19.19*). The world's two largest open pit mines are both copper mines; the Chuquicamata (Chile) mine is more than 700 m deep and the Bingham Canyon Copper Mine (Utah, United States) covers an area of 7.21 km^2 to a depth of 774 m (Goudie, 1981b). These excavations produce gangue that creates large waste heaps (*Fig. 19.20*). These are constructive anthropogenic landforms that severely detract from the original landscape. For this reason, regulations have been established for restoration of the land after mining. In humid areas, grass rapidly colonizes waste dumps but waste heaps can be identified by their slope profiles. In arid areas, however, restoration presents numerous problems. The flat tops of waste dumps can often be used for cereal production but the slopes, which are inclined at the angle of repose, are intensely eroded by water in spite of forest restoration (*Fig. 19.21*). Coalmines contain pyrite; as the pyrite oxidizes, especially from sulfide mines, it produces low pH waters that are highly corrosive and that negatively effect natural ecosystems.

Today numerous waste dumps have obvious negative visual impacts. They are located near large populations; they fill valleys or build landforms, generally as low hills. The wastes sometimes move, as happened in Portiño (Coruña, Spain) (*Fig. 7.39*). Waste is often used to fill estuaries. This generally leads to soil contamination in the area.

Gold extraction by "garimpeiros" (gold miners) causes obvious pollution on riverbeds and terraces in the Amazon basin; this activity reached its height in the 1980s and the beginning of the 1990s. Mining removed a large quantity of sediment. However, the major problem was the mercury that was used as an amalgam; it pollutes river water (Latrubese *et al.*, 2005).

Salt mining that involves water injection into boreholes to dissolve salt and extract brine can affect the land surface. It can affect dwellings, as near the Poza de la Sal (Burgos, Spain) and can modify the landscape by forming groups of hills and closed depressions (Waltham, 1989; Hortvik, 2005).

Figure 19.18 Quarrying of rhyolite porphyry for construction. Albiano near Trento (Italy).

Figure 19.19 Mining of iron oxyhydroxides. Barranco Mine in 1973. Sierra Menera, Ojos Negros (Province of Teruel, Spain).

Figure 19.20 Waste dumps at the potassium salt mines in Suria (Province of Barcelona, Spain). In the foreground is the ore conveyor belt.

Figure 19.21 Waste pile slope from lignite mining is undergoing significant water erosion five years after restoration. Photo taken in 2000. Utrillas (Province of Teruel, Spain).

This process is occasionally accompanied by sudden collapses (Johnson, 2005). Subsidence creates these morphologies; subsidence can be ductile or brittle.

Subsidence can be defined as the gradual or sudden settling (collapse) of the Earth's surface as a result of extraction of materials from the subsurface or by dissolution of soluble underlying rocks. Natural processes and humans can cause subsidence (Costa and Baker, 1981) (*Table 19.4*). Numerous cases of subsidence and their possible causes have been described (Coates, 1987). *Table 19.5* (Coates, 1983) shows cases of maximum subsidence, affected areas, causes, and estimated damages.

Extraction of fluids (water, petroleum, and gas) can cause weak flexure of the ground or ductile subsidence and can generate centripetal drainage in the resulting closed depressions. Gas and oil production have created numerous examples of subsidence; the classic example of subsidence is Long Beach (California) (Mayuga and Allen, 1969) (*Fig. 19.22*) where subsidence exceeded 26 m during the period of 1928–1971. As expected, subsidence rates change as a function of pumping rates. In Ravenna (Italy), water extraction caused subsidence of a few mm/year prior to World War II that increased to 110 mm/year after the war (Teatini *et al.*, 2005) (*Fig. 19.23*). In arid zones, groundwater is commonly extracted for irrigation. Abusive extraction decreases hydrostatic pressure in the reservoir rock, leading to subsidence. Subsidence is more common in unconsolidated detrital deposits, which are typically formations composed of alternating sands and clays.

Mineral extraction from underground tunnels may cause tunnel roofs to collapse and collapses may extend to the land surface. This subsidence occurs on top of shallow mine works. *Figure 19.24* shows the components of mining subsidence, including zones of tension and compression and the angle of rupture (Wohlrab, 1969). In the United States, 200,000 hectares are affected by mining subsidence. Mining subsidence caused 35 m of collapse in the city of Youngtown (Ohio) in 1977. The mine works dated back to 1876; wooden beams in the tunnels had gradually decomposed and lost their mechanical strength. Between 20–30% of the city had to be abandoned because it was built over the mine works.

Differential compaction is one of the most common causes of subsidence; it occurs when recent sediments lose some of the water they contain (Bull, 1974). The Palace of Fine Arts (Walthan, 1989) (*Fig. 19.25*) and the Cathedral in Mexico City (*Fig. 19.26*) are both affected by settlement due to loading. Leaning towers such as those of Pisa, Módena, and Venice (*Fig. 11.50*) result from differential compaction.

Hydrocompaction can also cause subsidence. When water is added to unconsolidated sediments and water-poor sediments, the water forms a film between the grains and decreases the strength of the material. As a result, differential settling occurs at the land surface; it bends the surface and can cause houses to collapse and towers to tilt (Costa and Baker, 1981; Waltham, 1989).

Soil erosion is the movement of surface material as a result of processes of external dynamics. The erosion can be chemical or mechanical. The former, also called chemical weathering, is an attack on soil minerals with the resulting total or partial liberation of ions in solution. Mechanical erosion or physical weathering is the removal of solids from the land surface (*Fig. 19.27*) (Summerfield and Hulton, 1994; Gaillardet *et al.*, 1999). Starting with the Agricultural Revolution, humans became an important factor in the denudation of river basins (Milliman and Syvitski, 1992).

Table 19.4 **SIMPLIFIED CLASSIFICATION OF SUBSIDENCE PHENOMENA (COSTA AND BAKER, 1981)**

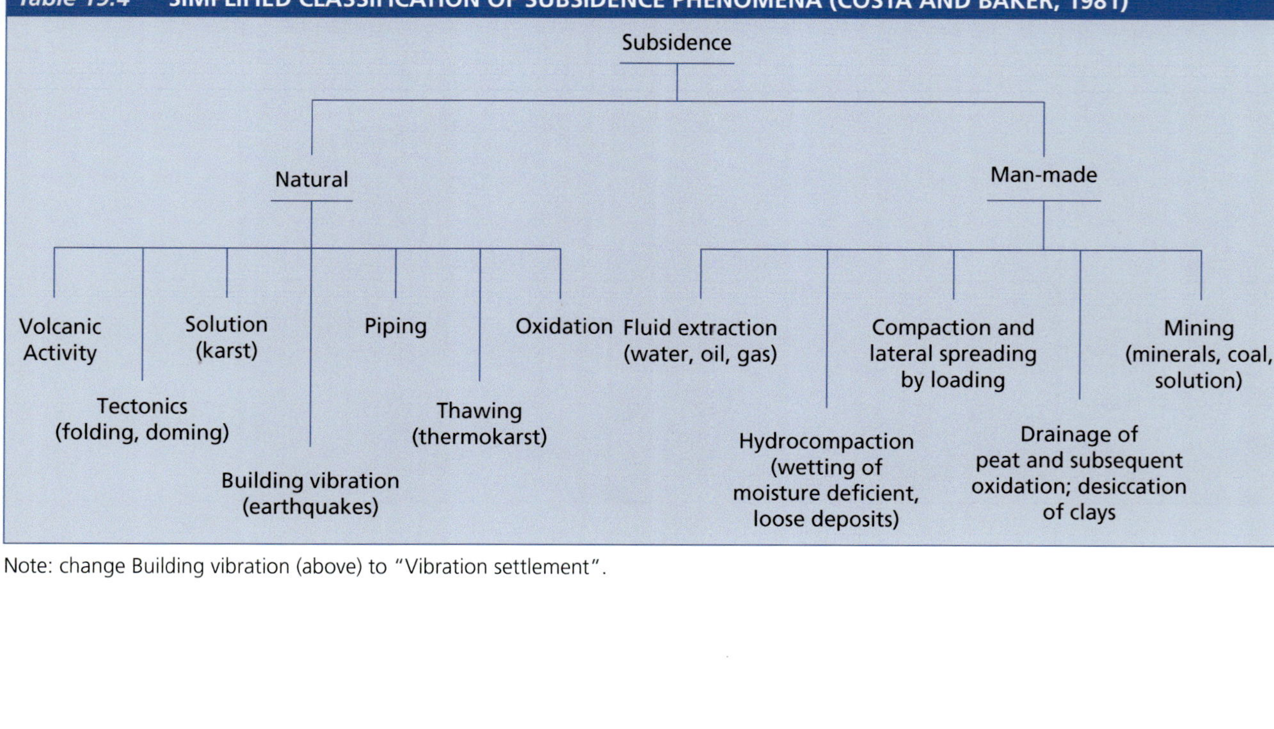

Note: change Building vibration (above) to "Vibration settlement".

Table 19.5 **MAJOR AREAS OF HUMAN-INDUCED LAND SUBSIDENCE (COATES, 1983)**

Name of locality	Maximum subsidence (m)	Affected Area (km^2)	Cause	Damages (millions of dollars)
San Joaquin Valley, California	9.0	13,500	a	100
Houston-Galveston, Texas	2.75	12,170	a	>1,000
Eloy-Pichacho, Arizona and adjacent area	3.6	8,700	a	several million
Tokyo area, Japan	4.6	2,400	a	225 from 1957 to 1970
Nobi Plain, Japan	1.5	800	a	
Po Valley, Italia	3.0	780	d	
Santa Clara Valley, California	3.9	650	a	>25
Baton Rouge, Louisiana	0.5	650	a	>1
		More than 5 cm of subsidence		
Sacramento Valley, California	0.7	500	a	
Osaka, Japan	3.0	500	a	Tens of millions
San Joaquin Valley (southwest), California	5.0	>500	b	
Lake Maracaibo, Venezuela	3.9	450	c	35 up to 1976
London, England	0.35	450	a	
Niigata, Japan	2.6	430	d	Tens of millions
Saga Plain, Japan	1.2	400	a	523 from 1960 to 1979
Venice area, Italia	0.14	400	a	
Debrecen, Hungary	0.42	390	a	
Savannah, Georgia	0.15	330	a	
		More than 2 cm		
Las Vegas, Nevada	1.7	300	a	several million
Valle del Raft, Idaho	2.8	260	a	
Taipei, Taiwan	1.9	230	a	
Ciudad de México, México	8.7	225	a	>500
New Orleans, Louisiana	0.8	150	a	
Victoria-Gippsland, Australia	1.6	102	f	
Saxet oil field, Texas	0.93	92	c	
Wilmington, Long Beach, California	8.8	78	c	200
Chocolate Bayou oil field, Texas	0.53	40	c	
Visonta, Hungary	0.5	40	a	
Haranomachi City, Japan	2.0	25	a	
Goose Creek oil field, Texas	1.0	10	c	
Baldwin Hills, California	3.0	5	c	25
Wairakei, New Zealand	4.8	1,3	e	

Note: a = groundwater well; b = hydrocompaction from surface water; c = oil and gas wells; d = methane in water wells; e = geothermal wells; f = dewatering coalmines.

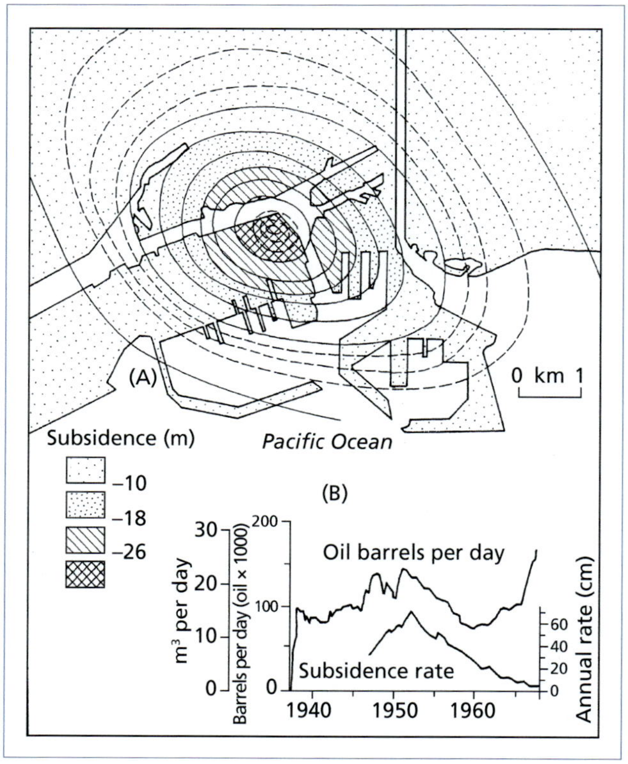

Figure 19.22 (A) Subsidence in the Long Beach, California, area between 1928 and 1971. (B) Petroleum production and subsidence rates in the Wilmington oil field where subsidence has not exceeded 10 m between 1928 and today (Mayuga and Allen, 1969).

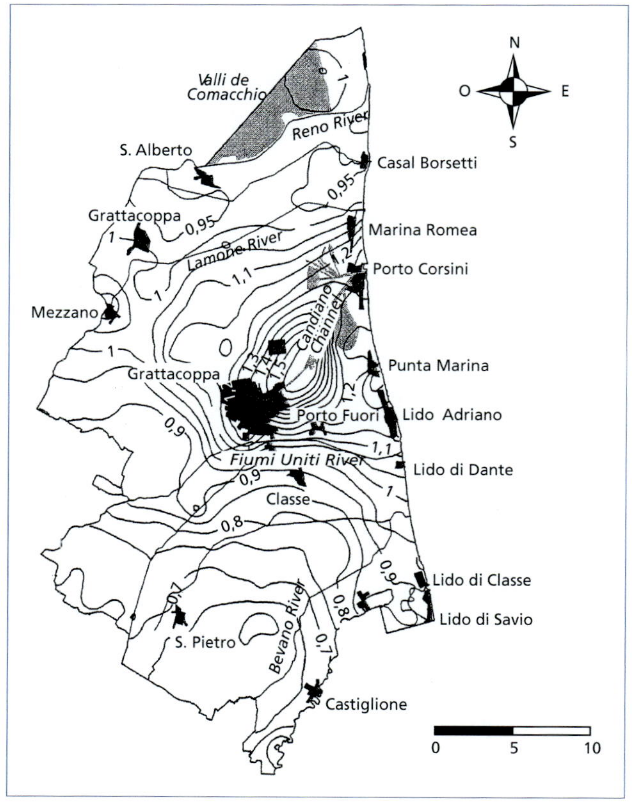

Figure 19.23 Cumulative subsidence (m) in the city of Ravenna (Italy) between 1897 and 2002 (Teatini, *et al.*, 2005).

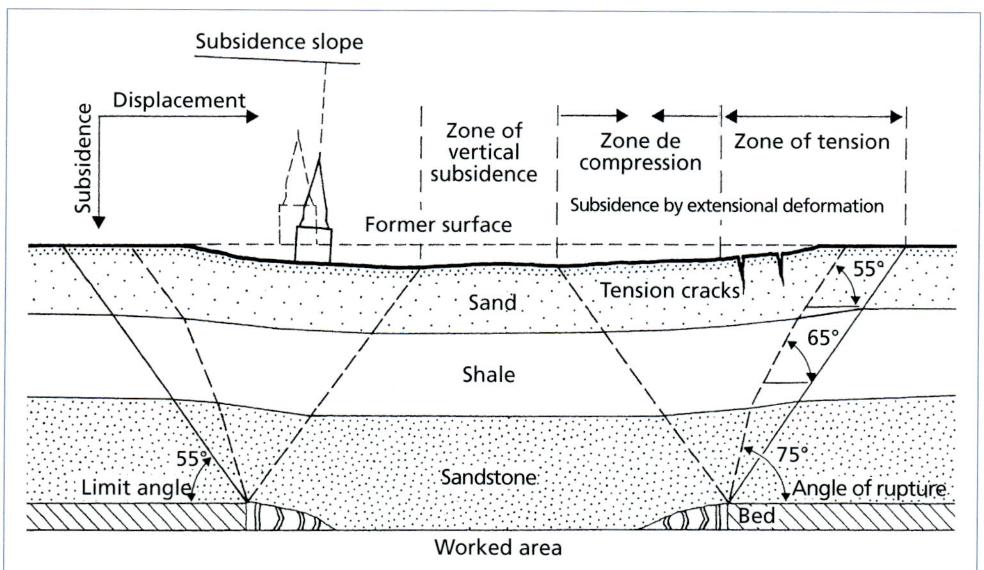

Figure 19.24 Components of mining subsidence. The "angle of rupture" depends on the geologic characteristics of the rock formation. The "limit angle" is defined as the line that connects the collapse limit with the surface subsidence limit (Wohlrab, 1969).

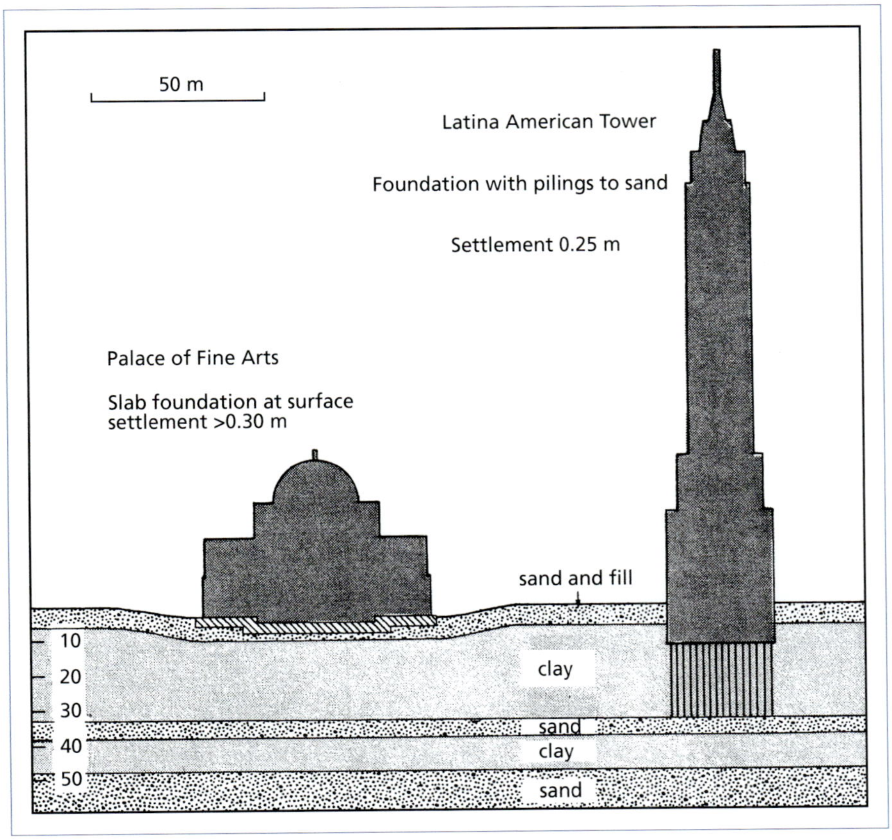

Figure 19.25 Contrast between cementation and the resulting collapse of the Palace of Fine Arts and the Latin American Tower (Mexico City). Both are anchored in montmorillonite-rich clays but the tower is constructed on deep pilings. The drawing is to scale, but the buildings are actually far apart (Waltham, 1989).

Figure 19.26 Tilting of tiles toward the margin of the Cathedral of Mexico City caused by differential consolidation.

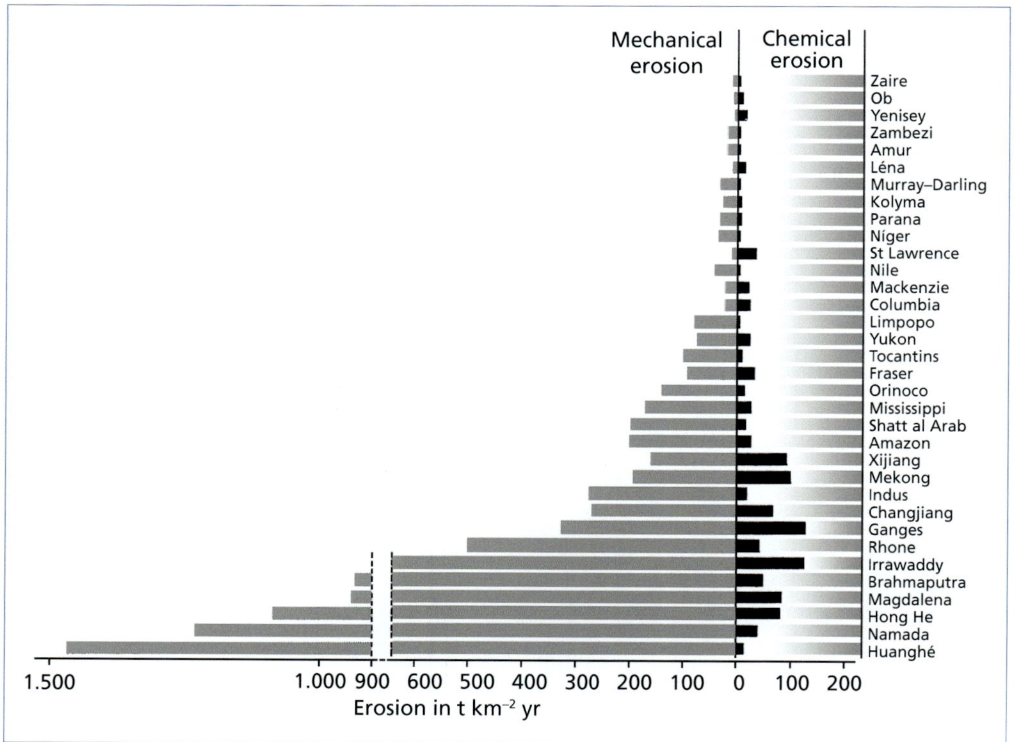

Figure 19.27 Physical and chemical erosion of the largest river basins of the world (based on Summerfield and Hulton, 1994 and Gaillardet *et al.*, 1999).

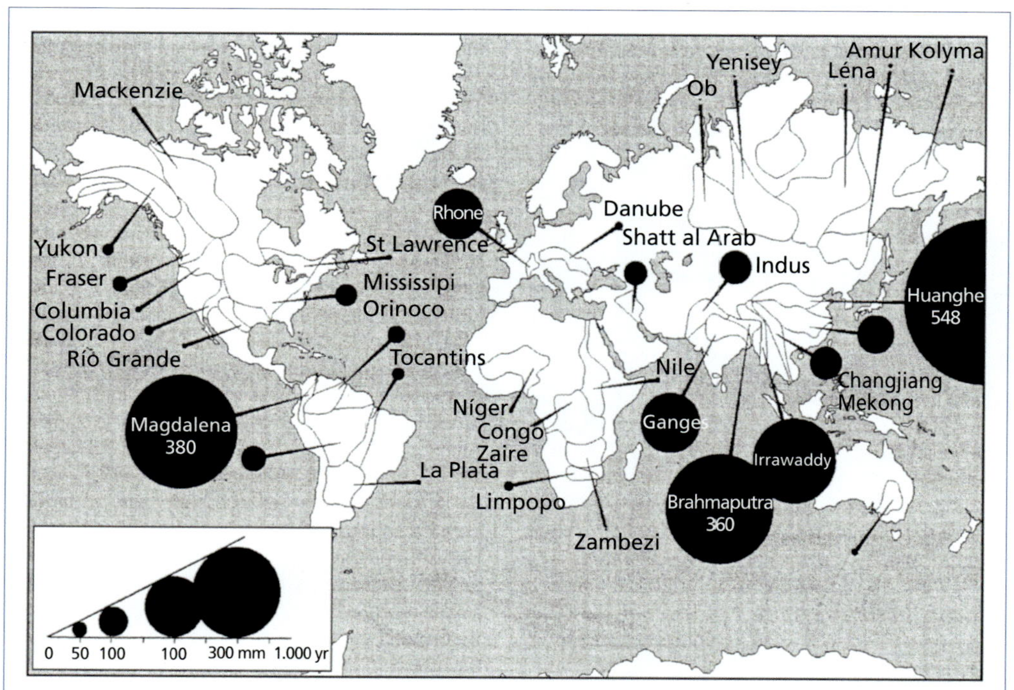

Figure 19.28 Total erosion of the largest drainage basins calculated on the basis of a density of 2.7 g/cm^3 (based on Summerfield and Hulton, 1994 and Gaillardet *et al.*, 1999).

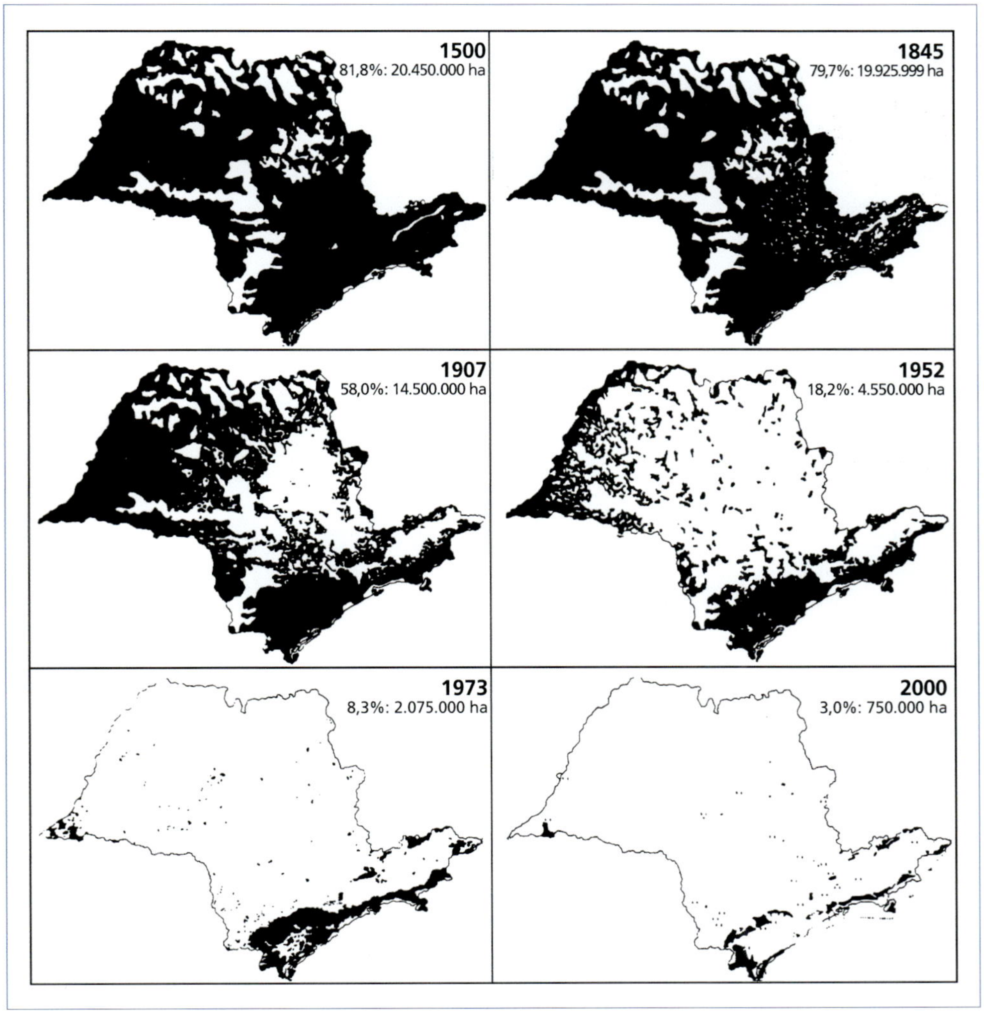

Figure 19.29 Man has practically exhausted the forest area of the State of Sao Paulo (Brazil) (Peters and Loveloy, 1990).

Total erosion, both physical and chemical, is calculated by using an average density of 2.70 gr/cm^3, which can also be expressed in mm/year (*Fig. 19.28*). In large rivers that flow over flat cratonic areas such as the Australian, African, and Brazilian shields, surface lowering is about 10 mm/thousand years. Maximum erosion values occur in the Yellow, Magdalena, and Brahmaputra basins. Average erosion for the land surface is 61 mm/thousand years (Gaillardet *et al.*, 2004).

Soil loss is a significant problem attributed to both natural processes and human intervention (Garcia-Ruiz and López Bermúdez, 2009). The amount and type of vegetation controls erosional processes. Roots have a protective influence that may inhibit soil erosion. Humans have continuously cut down forests (deforestation) (*Fig. 19.29*) primarily to obtain wood for heat and to melt metals, to clear land for farming (Liebault *et al.*, 2005), to expand grazing land, and for other reasons. Throughout history, man has deforested about 8 million km^2 (Meyer, 1996). Forest clearing triggers intense water erosion (*Fig. 19.30*). Pollen research on the upper Paleolithic and Neolithic periods shows a record of the intentional cutting of forests. The land surface of central and northern Europe shows the obvious effects of deforestation to convert land to agricultural use (*Fig. 19.31*). The great stage of deforestation in central and western Europe occurred between the years of 900 and 1900. North America experienced 200 years of brutal deforestation; more wood was cut during this period than in 2000 years in Europe (Goudie, 1981b). Deforestation today affects primarily rainforests and boreal forests (Middleton and Thomas, 1997).

Rainforests have undergone significant annual regression during recent decades. Sixty-two countries contain 78% of the world's rainforests; the FAO estimates that total

Figure 19.30 Rilling in forest areas cleared of vegetation. Barbadillo del Mercado (Province of Burgos, Spain).

Figure 19.31 Deforestation to make space for agriculture. Near Oslo (Norway).

annual deforestation in these countries in 1990 was 16.8 million hectares (Landy *et al.*, 1991). This number is significantly higher than for the period between 1976 and 1980 (9.2 million hectares per year). Rainforest retreat rates around the world vary considerably; Asian rainforests have the highest deforestation rates (World Resources, 1998). Continuous deforestation is a concern because many substances that humans extract from these ecosystems are used for food, drinks, medicines, resins, and dyes. Deforestation also triggers erosion and affects climate change because rainforests are carbon dioxide sinks (Goudie, 2006).

The heavy rains that occur frequently in these ecosystems cause flooding and trigger landslides as a result of deforestation. We must acknowledge that many towns are highly vulnerable because they are located in risk zones that lack appropriate planning.

In 7 BC Strabo described Hispania (Spain) as being rich in forests. Pollen analysis shows that forests were restored during the Arabic period between the 2nd and 11th centuries; the country reached its maximum arboreal coverage during this time. At the end of the 16th century, the Mesta (a Spanish/Castilian institution created to protect cattle) had a major impact on forests because it destroyed them to increase pastureland. During this period significant naval construction also contributed to the destruction of forests. The Disentailment of Mendizábal Law of 1837 encouraged the sale of public woodlands and promoted tree cutting, ploughings, fires and so on (López Cádenas, 1988; García-Ruiz and López Bermúdez, 2009).

Humans began to transform land for agricultural use about 8,000 years BC in the Middle East. These transformations have caused soil degradation and soil erosion since then (Macklin and Lewin, 1997). Table 19.6 summarizes the principal agricultural impacts of erosion and runoff in humid and dry climates (Starkel, 1987). In humid areas where the forest has been cut and in dry regions with seasonal water deficits, overgrazing causes (Fig. 19.32) increased water and eolian erosion. In addition, irrigation of land in arid zones can lead to salinization of cultivated areas.

The agricultural landscape is more sensitive to climate variability than natural landscapes because cultivation and pastures reduce water infiltration and increase the speed and magnitude of surface runoff. Knox (2001) drew these conclusions from his research on the sensitivity of the Upper Mississippi valley to the effects of agriculture.

Soil losses began to increase about 5000 years ago and this trend continues today. Based on natural erosion rates recorded from the Cambrian until today, Ronov (1983) estimated that up to the end of the first millennium, anthropogenic erosion exceeded natural erosion by about 5000 million tons per year (Fig. 19.33). Seasonal soil use practices and the global population increase show that by the end of the

Table 19.6 **IMPACT OF AGRICULTURE ON EROSION AND RUNOFF IN HUMID AND DRY CLIMATES (MODIFIED FROM STARKEL, 1987)**

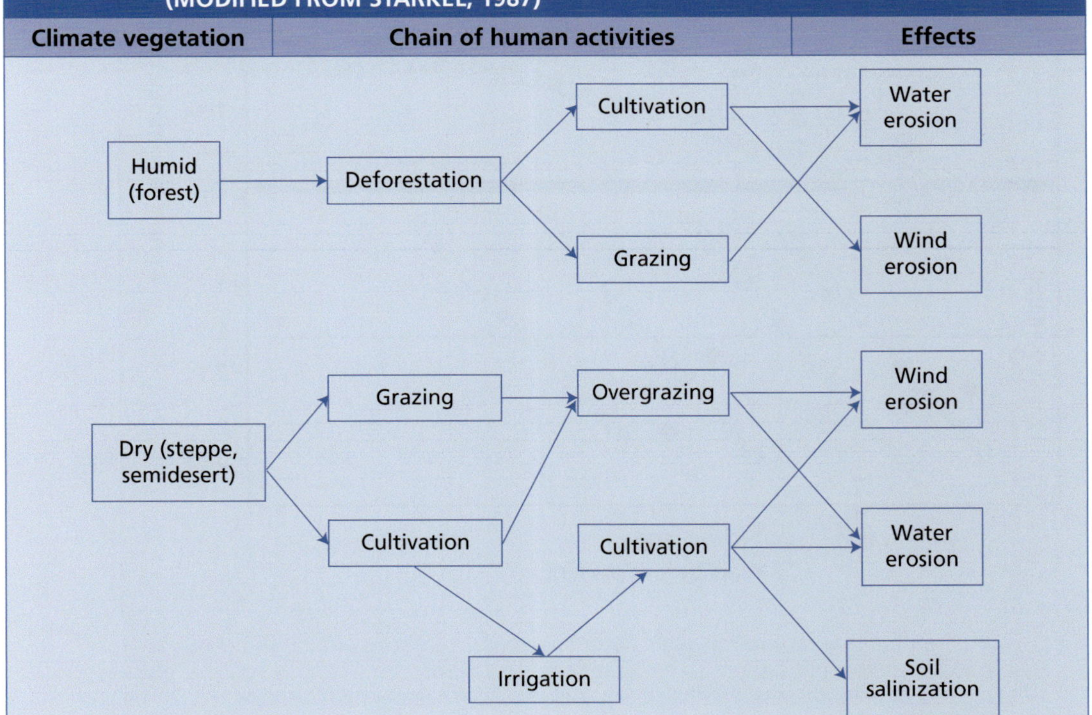

Figure 19.32 Overgrazing by goats. Kalia. Midbar Yehuda. Negev Desert. Israel.

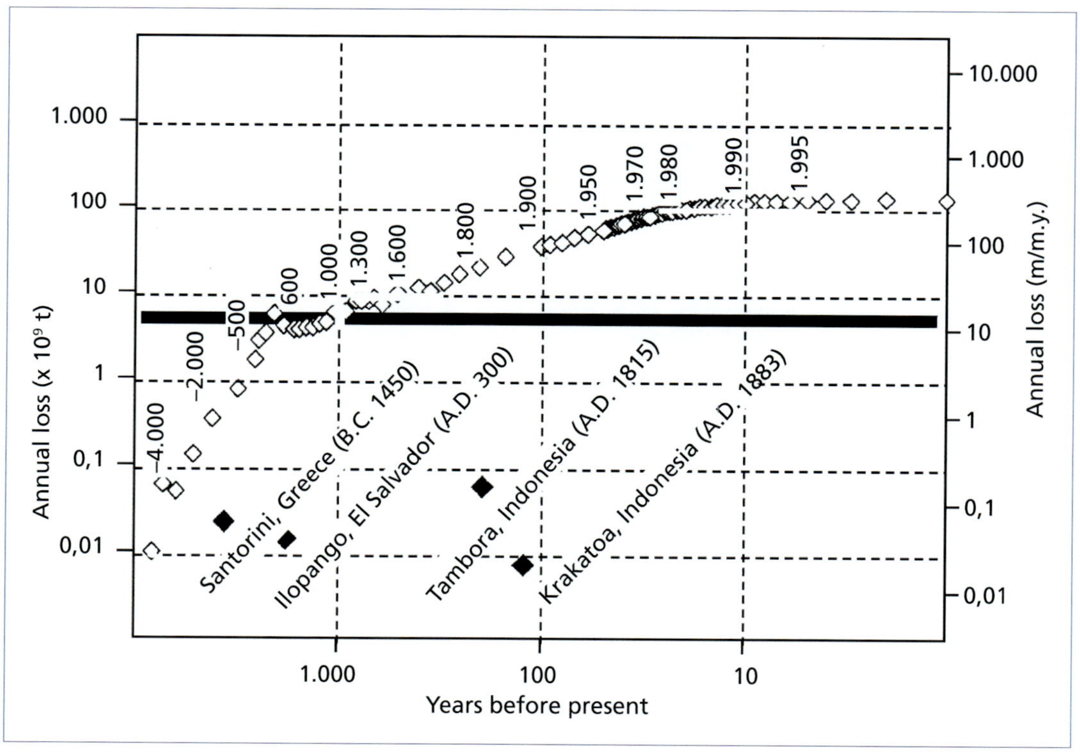

Figure 19.33 Historical anthropogenic erosion rates according to data from Hooke (2000). Anthropogenic erosion rates (black rhombs) are compared to volumes from various volcanic eruptions (dates in parentheses). The black line indicates average denudation, which is 24 m/thousand years (Wilkinson, 2005).

first millennium humans had become the principal agents of erosion (Wilkinson, 2005).

Changes in river basin use and erosion control projects have had significant geomorphic consequences and their effects on the geomorphic evolution of river channels are obvious (Boix-Fayos *et al.*, 2007). These researchers compiled a summary of the dominant processes involved in basin-scale land use changes and the geomorphic and hydrologic responses of various locations as described by numerous authors.

Wolman (1967) wrote the classic work on the relationship between land use changes and soil loss. He studied variation in soil erosion in an area near Washington (United States). *Figure 19.34* shows that erosion is minor under forest conditions, on the order of 0.2 cm/1000 years. The rapid increase in agriculture at the beginning of the 19th century increased soil loss to about 10 cm/1000 years. Returning land to pasture and wood production in the 1940s and 1950s reduced erosion by half. Areas where construction occurred in the 1960s experienced soil loss of almost 10 m/1000 years over very small areas. Finally, in urban areas, erosion is minor, less than 1 cm/1000 years.

The effects of land use changes have also been studied in the central Spanish Pyrenees and in the Ebro Depression. Over the last 50 years, significant land use changes have occurred as a result of land management in the Pyrenees. For centuries grain had been grown on slopes oriented toward the sun; erosion was intense on these slopes. Today this cultivation is being abandoned little by little; shrubs have colonized the slopes and pines have been planted. This change has increased interception, infiltration, and evapotranspiration (Gallart and Llorens, 1996). Runoff and soil loss have decreased as a result of these changes (García-Ruiz *et al.*, 1997). Significant changes have been made in the Ebro Depression during this same period; more recently the agrarian policy of the European Community has introduced changes leading to the abandonment of farming and to other land use changes. These studies show that cultivation is principal factor in triggering erosion in semiarid environments (Navas *et al.*, 1997).

19.5 Toward sustainable development

Activities that directly change relief and indirectly affect greenhouse gases are the two major global problems that threaten biotic and abiotic media. Global warming is the more serious because of the large temperature increase it represents, to which urgent mitigation measures should be applied. Lord Robert May, President of the Royal Society of the Academy of Science of the United Kingdom has claimed that humans have caused climate warming and that humans themselves will have to resolve it (May, 2005).

The introduction to the United Nations Agreement on Climate Change (UNFCC) says that the primary objective of the agreement is to stabilize the concentration of greenhouse gases and that this task will require time to "allow ecosystems to adapt to climate change, ensure that food production is not threatened, and allow economic development

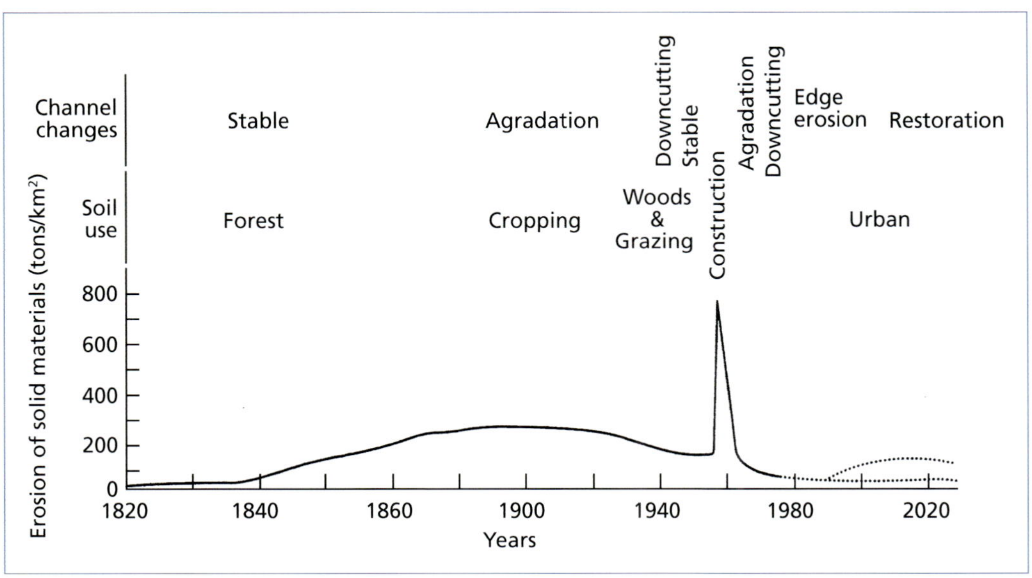

Figure 19.34 Sequence of land use changes and soil losses that began before the advent of extensive agriculture, followed by a period of construction, and later urban development (Wolman, 1967).

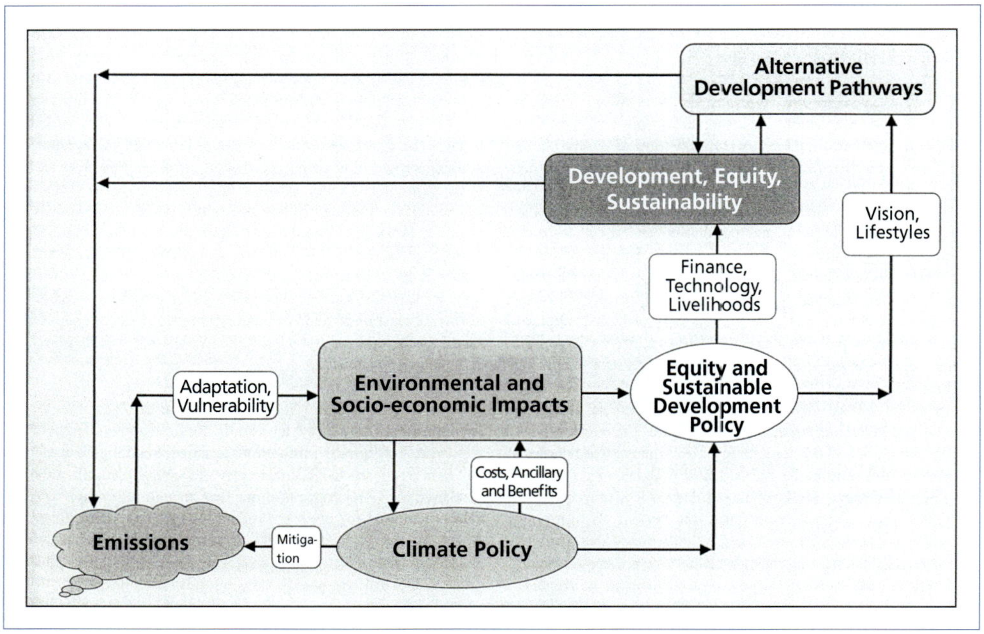

Figure 19.35 The perspective of global sustainability (Banuri and Weyant, 2001).

to occur in a sustainable way" (Article 2). It also specifies various principles to direct this process: equity, less common differentiated responsibilities, caution, profitable cost of the measures, right to a sustainable development and support to an international open economical system Art. 3). People have the right to sustainable development (Art. 4) (UNFCC, 1997). Banuri and Weyant and collaborators (2001) have prepared *Figure 19.35*, which shows the perspective of global change coming from emissions through climate politics, and environmental and socioeconomic impacts to arrive at equitable and sustainable development.

Environmental changes are a complex interaction of geomorphic, climatic, biotic, economic, political, institutional, social, and technological processes. This set of interactions cannot be understood in isolation but rather must be approached through broader social structures that lead to sustainable development. It is obvious that the necessary steps to reach sustainable development will vary from country to country, and for this reason various governments have to develop their own mitigation plans. Developed countries have numerous potential strategies that may build on the past (Banuri and Weyant, 2001).

The FAO claims that in terms of forest restoration as a mitigation measure, Brazil loses an area the size of Cataluña (32,106 km^2) every year and Indonesia loses an area equivalent to Navarra (10,391 km^2). China, which is a leader in reforestation, restores a surface area equivalent to Extremadura (41,635 km^2) each year. However, Chinese businesses obtain large quantities of wood from Latin America, Africa, and Indonesia. An additional factor is the phenomenon of millionaires buying enormous virgin forests or eco-empires to protect them, a practice that is both praised and criticized (Rizzi, 2007). This shows that mitigation practices are beginning to take place in some sectors.

Within the scope of hydro-systems, the Hydrology for the Environment, Life, and Policy (HELP) program seeks to improve the link between hydrology and societal needs to reach sustainable and appropriate water use (Downs and Gregory, 2004). Hydrologists who propose "sustainable" solutions must consider the cost-benefit aspects of their options and the necessities and desires of society. This requires interdisciplinary collaboration between earth scientists and sociologists interested in environmental economy, planning, and social and political strategies, for the purpose of fully evaluating the economic and social validity of different options on various time scales (Piégay *et al.*, 2006).

The cover of the brochure announcing the International Year of Planet Earth 2008, which was published by several Spanish institutions, states: "The human race needs its Planet. We depend on the Earth because we have evolved on it, we will always be a part of it, and we can exist only within a self-sustainable system."

The considerations discussed here lead us to a diagnosis of an ailing planet that cannot heal on its own; it needs healing practices (mitigations) with goals shared by all the countries on Earth.

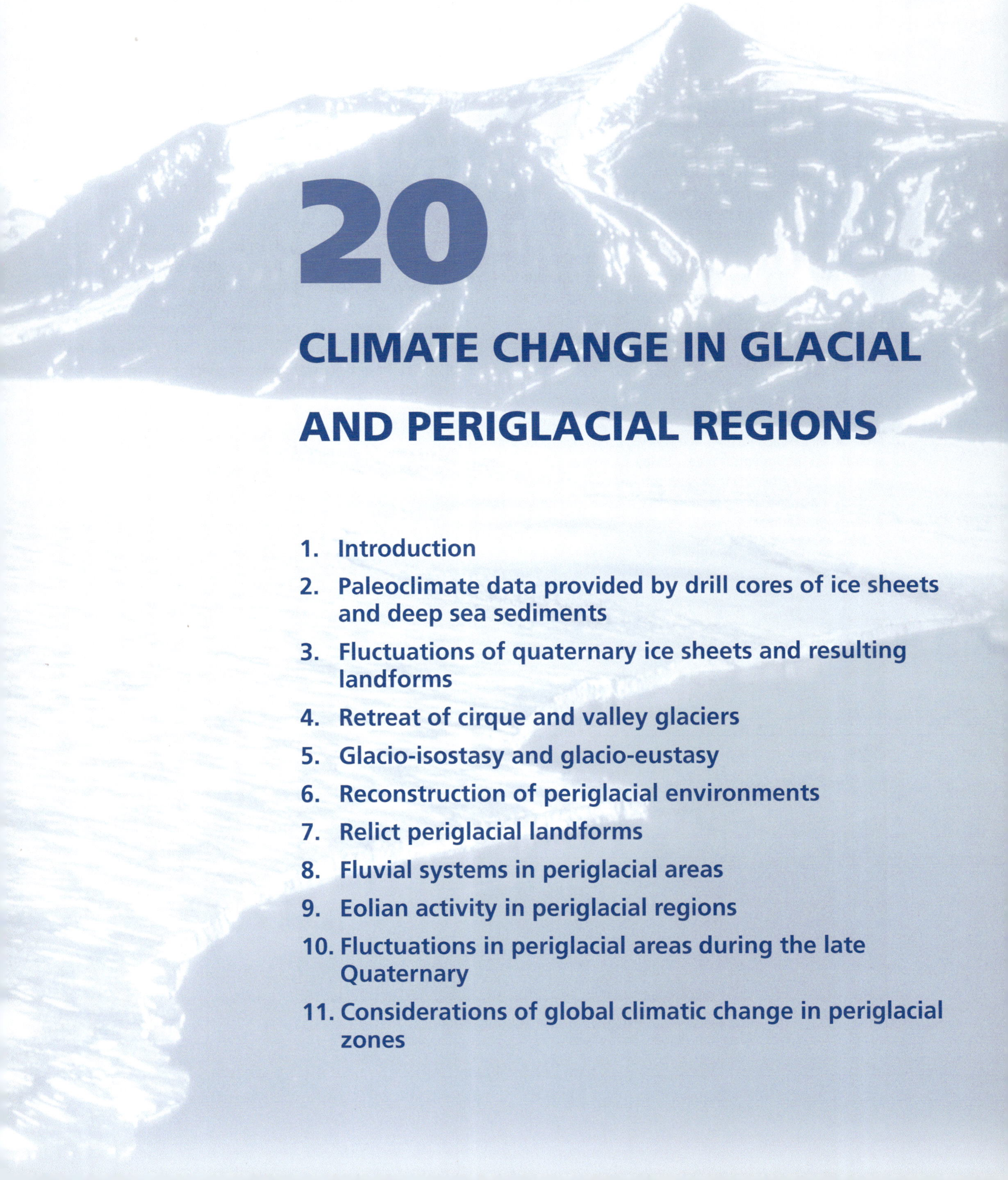

20

CLIMATE CHANGE IN GLACIAL AND PERIGLACIAL REGIONS

20.1 Introduction

Glacial and periglacial regions are morphoclimatic zones where cold conditions predominate. As Tricart (1967) states: "cold countries are those where the geomorphic activity of water is controlled by its existence in the solid state, either permanently or periodically."

Glaciers develop in areas where snow persists from year to year; as a result, it accumulates and turns into ice. Glacier limits are clear-cut; they outline ice accumulations. Periglacial zones are characterized by freezing and thawing but snow does not persist from year to year. Frozen soils (permafrost) may develop under the land surface.

During glacial periods, 30% of the global land surface was covered by ice including extensive areas of North America and Eurasia. Antarctica also expanded during this time. Small ice caps formed and valley glaciers advanced considerably in mountainous areas. Today ice covers 10% of the land surface. Large advances and retreats of ice masses have sculpted extensive areas, eroding rock masses and depositing thick accumulations of glacial material (*Fig. 20.1*).

Movements of glacial fronts are responses to climate change; thus they are a source of information for the study of climate change. In addition, ice sheets and glaciers play an important role in the general atmospheric circulation system.

Responses of continental ice sheets to climate change are difficult to predict because they are a function of reactions between mass balances, geometric changes, and ice fluxes, particularly when outlet glaciers end in deep water or supply ice platforms (Benn, 2006). These ideas emerged from the formation of Larsen Ice Platforms A and B between 1995 and 2002 although these events do not contribute directly to sea level rise. Angelis and Skvarca (2003) analyzed responses to the collapse of the Larsen ice platforms and showed that five of the eight glaciers with greatest supply have accelerated. This indicates that glacial platforms may affect ice sheets dynamics.

As a result of major ice mass advances and retreats during the Quaternary, the periglacial domain has fluctuated widely. During glaciation, the extent and location of periglacial zones were much different from those of today. Extensive areas that were covered by ice are periglacial zones

Figure 20.1 The Root Glacier with extensive glacial deposits Wrangell-St. Elias National Park. Alaska. Photo P. Lucha.

today. Regions around ice margins that were periglacial areas are temperate zones today. As a result, changes from glacial to periglacial and from periglacial to temperate can be recognized.

In other regions, however, the periglacial domain has persisted throughout the Quaternary. One example is a broad region of Siberia where precipitation is so low that it cannot form an ice sheet (King, 1976). The important climatic fluctuations that affect cold areas have shaped a set of new relict forms that persist under new circumstances and others that have formed under existing conditions. Between these two extremes are numerous degraded inherited forms that are difficult to recognize and that raise questions about the identification of relict and modern forms.

20.2 Paleoclimate data provided by drill cores of ice sheets and deep sea sediments

The study of drill cores has revealed past climate changes affecting the entire planet, many of which occurred rapidly. These significant advances in the understanding of climate change have been made primarily in the last six decades.

Cores drilled in ice sheets provide paleoclimatic records (Brook, 2007) that include an annual snowfall record. Data from the upper part of the core show alternating layers resulting from annual precipitation but it is difficult to detect these alternations in deeper layers of the record; for this reason, the deeper layers are dated by indirect methods.

Shallow cores have been drilled in valley glaciers and mountain ice caps in Tibet and the Peruvian Andes but, unlike deep borings into polar ice sheets, shallow cores record recent climate changes. The first deep core was drilled between 1963 and 1966 at Camp Century (Greenland) (Dansgaard *et al.*, 1969). In Antarctica, the 1983 Vostok drill hole (Jouzel *et al.*, 1987) provided a climate record of the last 160,000 years; it goes back farther than the ice of the last interglacial. The basal part of this core reached a depth corresponding to 420,000 years BP (Petit *et al.*, 1999). Recently a consortium called the European Project for Ice Coring in Antarctica (EPICA) drilled into ice at Dome C located 500 km from the Russian Vostok station east of the Antarctic Ice Cap. The basal section of core from this project (EPICA, 2004) may reach 900,000 years or more (Walker, 2004). Between 740,000 BP and 430,000 BP, less pronounced warming occurred during interglacial periods in Antarctica and the warm period occupied a larger interval of each cycle (EPICA, 2004, McManus, 2004) *(Fig. 20.2)*. These research results show that without human intervention a climate similar to our present-day climate can persist into the future (EPICA, 2004).

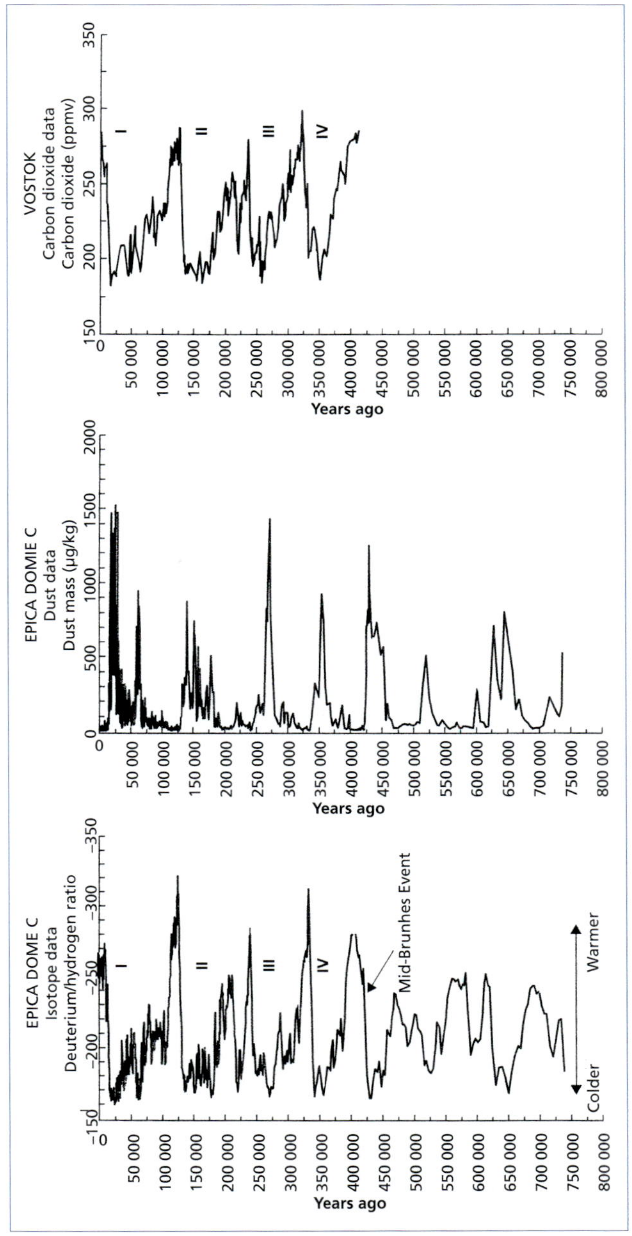

Figure 20.2 Data from drill core in the Antarctic Ice Cap (Petit *et al.*, 1999, EPICA, 2004). The range of glacial/interglacial temperature variations in Antarctica is about 15°C, according to deuterium data. The Roman numerals indicate the ends of glacial periods (Anderson *et al.*, 2007).

The GRIP and GISP2 cores were drilled in central Greenland 28 km from each other. Horizontal ice movements were not recognized in the first core but starting at a depth of 2,750 m in GISP2, discrepancies in interpretation are probably due to ice flow. Ice flow increases toward the margins of the ice cap.

Oxygen atoms that form part of glacial ice contain three isotopes ^{16}O, ^{17}O, and ^{18}O; their relative abundances in nature are 0.2%, 0.04%, and 99.76%, respectively. As water evaporates from the ocean surface, the heaviest isotope tends to remain in the liquid phase and is thus less abundant in the vapor phase; this process is called fractionation. The resulting vapor is poor in ^{18}O compared to the initial water. Condensation is a function of temperature and as a result, the isotopic composition of precipitation shows the temperature at which condensation took place. For this reason, the isotopic composition of glacier ice can be used to reconstruct the temperature of the original precipitation. Researchers determine the isotopes present in ice by melting the ice in the laboratory and analyzing it with a mass spectrometer. The difference between $^{18}O/^{16}O$ and the *standard* water sample (Standard Mean Ocean Water – SMOW) is expressed as $\delta^{18}O$. If the ratio is small, paleo-temperatures were low, and vice versa. Global climate fluctuations can be detected in this way (Dawson, 1992; Ehlers, 1996; Benn and Evans, 1998). *Figure 20.3* is a compilation of data obtained from the GRIP core at a depth of about 3,000 m; it shows interglacial conditions at about 130,000 BP (Eemian, Sangomanian, or the last interglacial). Various interstadials (short periods of warming) and rapid termination of glacial conditions occur around 11,000 years BP. Holocene climate changes are poorly represented according to Dansgaard *et al.*, (1993), but later work by Bond *et al.*, (1997, 1999) has led to some determinations of the Holocene. Broecker *et al.* (1992) were the first to recognize the implications of the discovery of Heinrich events, including the probable links between great waves of icebergs, contribution of melt water to the North Atlantic, interruption of the thermohaline circulation, and cooling at high latitudes (Labeyrie *et al.*, 2007).

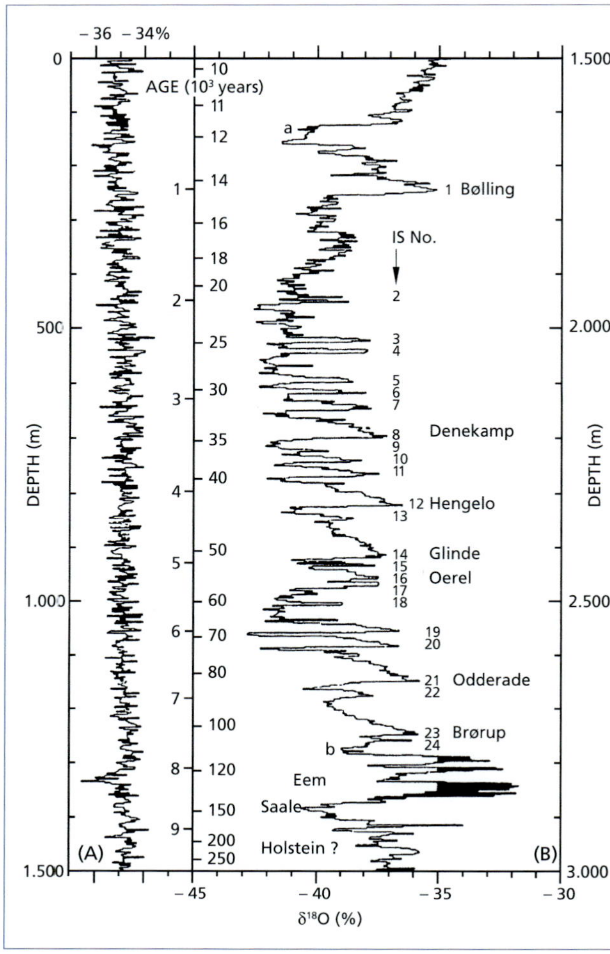

Figure 20.3 Continuous $\delta^{18}O$ record from the GRIP Summit, plotted in two sections on a linear depth scale: (A) from the surface to 1,500 m; (B) from 1,500 to 3,000 m depth. Measurements were made at every 2.2 m of core. The glacial interstadials are numbered to the right of curve B (Dansgaard *et al.*, 1993).

Interstadials are short-duration events that occurred numerous times between 110,000 and 10,000 years. They are called interstadials to distinguish them from cold phases or stadials in the past. *Figure 20.3* shows 24 interstadials, called Dansgaard-Oeschger events" (Bond *et al.*, 1993). Each warm interstadial is linked to a cold interstadial; the pair is called a "**Dansgaard-Oeschger cycle**" (DO cycle), which may last about 1,500 years. Peak deposition of detritus from iceberg melting (IAD) corresponds to the coldest phase of a stadial and it is followed by a warm interstadial.

Research on the Vostok core from eastern Antarctica has provided interesting information on CO_2 variations during the late Pleistocene and Holocene *(Fig. 20.4)* (Lorius *et al.*, 1988). Temporal variation of CO_2 present in air bubbles within the ice correlates well with oxygen isotope fluctuations and, in turn, corresponds to temperature changes.

Research on deep-sea sediments has provided important data related to climate fluctuations during the late Tertiary and Quaternary (Anderson *et al.*, 2007). These sediments contain planktonic foraminifera, which float on the sea while they are alive, and benthic foraminifera, which are limited to the deep ocean. Their shells are composed of $CaCO_3$; the oxygen contains ^{18}O and ^{16}O. These calcareous shells incorporate oxygen isotopes based on the isotopic composition of the seawater. As the water evaporates, the vapor is enriched in ^{16}O and when it precipitates on ice sheets its ^{16}O content is greater than that of the ocean, which is enriched in ^{18}O. As a result, the study of oxygen isotopes aids in the reconstruction of the volume fluctuations of ice sheets, and, in consequence, glacio-eustatic sea level changes. The first person to use this technique was Emiliani (1955) who figured out that the observed fluctuations reflected primarily temperature changes that correspond to glacial and interglacial periods. Analysis of deep cores from the Pacific (Shackleton and Opdyke, 1973, 1976) has shown clearly that the oxygen isotope record reflects the entire Quaternary climatic history (*Fig. 20.5*). Obtaining suitable oxygen isotope records is difficult in oceanic zones where the sedimentation rate is relatively low. In addition, the effects of bioturbation, submarine landslides, and turbidity currents hide original sedimentation and thus complicate interpretations.

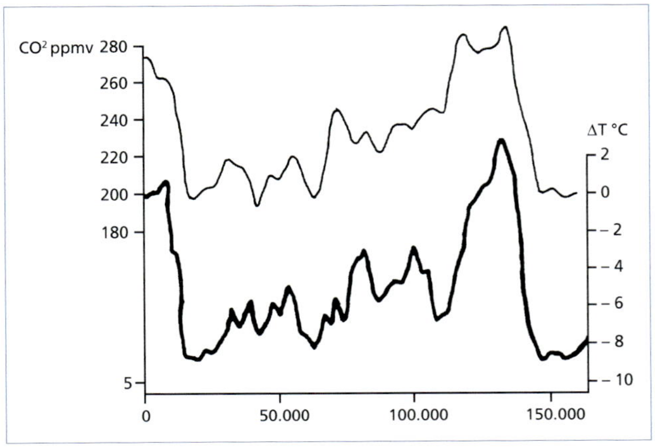

Figure 20.4 Curve showing CO_2 variations over the last 160,000 years (upper part) obtained from drill core in Vostok ice in Antarctica and smooth-curved temperature record from oxygen isotopes from the same core (lower graph) (Lorius *et al.*, 1988).

Deep-sea cores from the North Atlantic as far south as the latitude of the Iberian Peninsula also record significant events of the last 70,000 years, such as **Heinrich events** (Heinrich, 1988; Bond *et al.*, 1993; 1997; Hemming, 2004; Cofaigh, 2007) (*Fig. 20.6*). Unlike interstadials Heinrich events are cool, but like interstadials they are of short duration. Heinrich events appear in the core record as ice-transported clasts (IRD, Ice Rafted Drift) that come from the melting of a large number of icebergs that deposited their solid load on the ocean bottom during separate events (*Table 20.1*). This table shows ages obtained by ^{14}C analyses and expressed in thousands of years. The CALIB 4.1 calibration was used to convert ^{14}C ages into calendar ages (Stuiver and Reimer, 2003; Stuiver *et al.*, 1998) for radiocarbon ages less than 26 Ky. The Meese/Sowers GISP2 time scale was used to calibrate older ages (Meese *et al.*, 1994; Bender *et al.*, 1994) (see Bond *et al.*, 1999). The HO-H2 values are from Bond *et al.* (1997). In addition to the Bond *et al.* (1999) calibration, other calendar age calibrations based on fossil coral ages have been proposed for ages greater than 26 Ky (Fairbanks *et al.*, 2005; Chiu *et al.*, 2005) but Reimer *et al.*, (2006) have criticized them and it seems best not to use them.

Researchers do not know if the processes responsible for Heinrich events result from internal dynamics of ice sheets (MacAyeal, 1993), from periodic external climate changes (Broecker, 1994), or follow the jökulhlaup model, which describes numerous catastrophic water and ice floods that created a large lake at the mouth of Hudson Bay (Lake Zissage) during the Wisconsin glaciation (Johnson and Lauritzen, 1995). Recently Hulbe *et al.*, (2004) show that on the

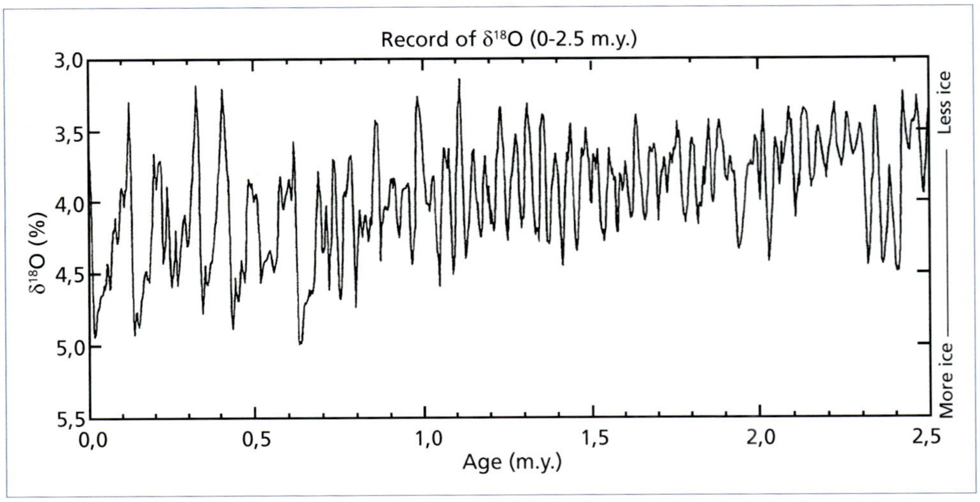

Figure 20.5 Variations in oxygen isotope composition in planktonic foraminifera from the equatorial Indian Ocean during the last 2.5 million years. The global volume of ice has increased, especially during the last 750,000 years and the average duration of cold periods has likewise increased (Crowley and North, 1991; based on data from Raymo *et al.*, 1990).

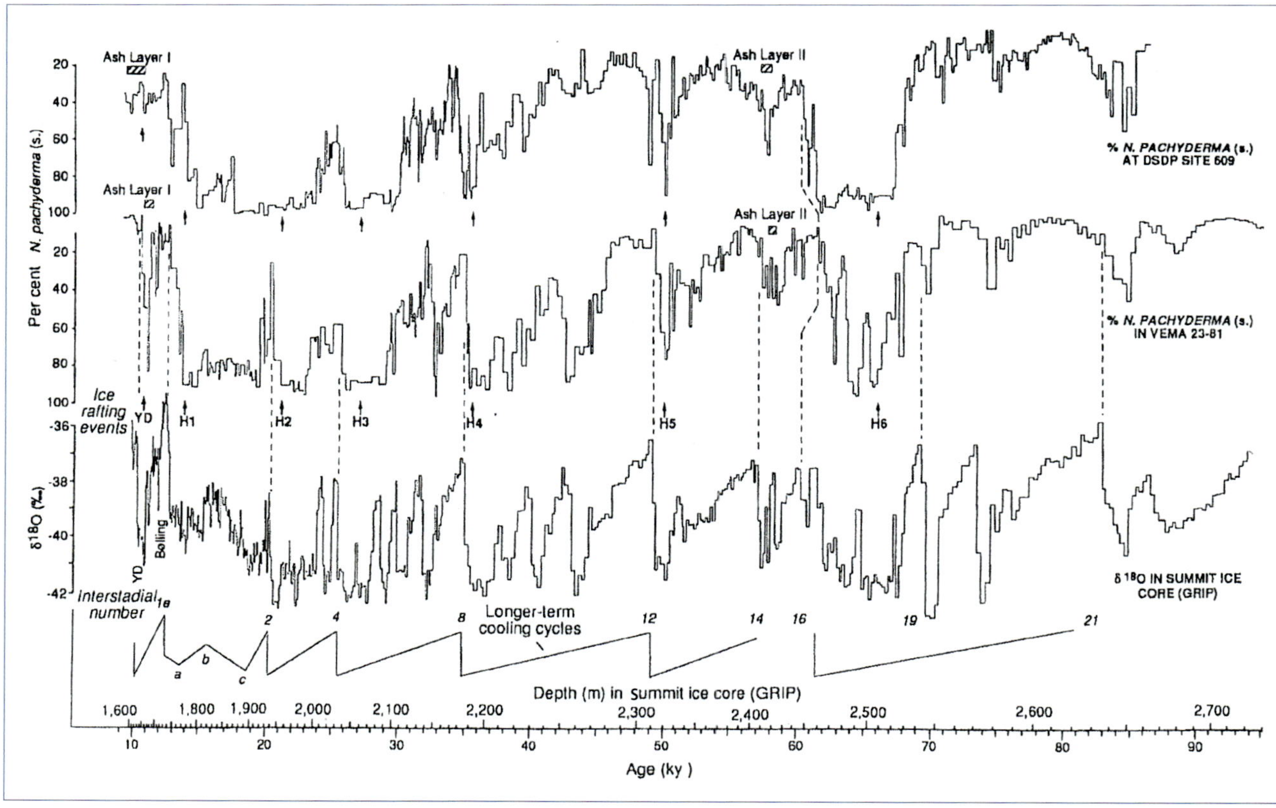

Figure 20.6 Relationship between sea surface temperature, Heinrich events and Dansgaard-Oeschger cycles. Based on various deep-sea cores, the two upper curves show changes in the percent cooling obtained from the foraminifer *N. pachyderma* (a high percentage indicates colder sea surface temperatures). These changes correlate well with temperature oscillations derived from the GRIP ice core. Heinrich events are numbered from H1 to H6; they occur at the end of a cool cycle, before the interstadial (Bond *et al.*, 1993).

Table 20.1 DISTRIBUTION OF THE PRINCIPAL HEINRICH (H) AND YOUNGER DRYAS (YD) EVENTS DURING THE LAST 70,000 YEARS (APPROXIMATE AGES IN THOUSANDS OF YEARS BEFORE PRESENT). H0, H1 AND H2 INDICATE CALENDAR AGES; H3, H4, H5, AND H6 ARE ^{14}C AGES (BOND *ET AL.*, 1993)

Event	Age (Ka)
YD or HO	12.0
H1	15.5
H2	22.0
H3	26.5
H4	35.5
H5	50.1
H6	66.2

periphery of ice platforms that formed on the eastern part of the Laurentide Ice Sheet during conditions of maximum warming, a substantial temperature increase caused surface melting and sudden disintegration of sections of the external ice cap. This creates "armadas" of icebergs that move throughout the North Atlantic *(Fig. 20.7)*. Fresh water from the melted ice sinks into deep water (Rahmstorf, 2002). As these authors point out, the difficulty of this hypothesis is that an explanation of the causes of abrupt climate change does not seem to be a fundamental prerequisite. The disintegration of the Larsen A and B ice platforms in 2002 was related to slow warming that exceeded the threshold at which surface water from melting filled in ice cracks in the platform, which caused the disaggregation of the ice platform margin.

Heinrich events have been correlated with pollen records in Europe and North America (Grimm *et al.*, 1993), loess columns in central China (Porter and Zhisheng, 1995; Porter *et al.*, 2001; Porter, 2007), glacial deposits the western United States (Clark and Bartlein, 1995), and with sets of

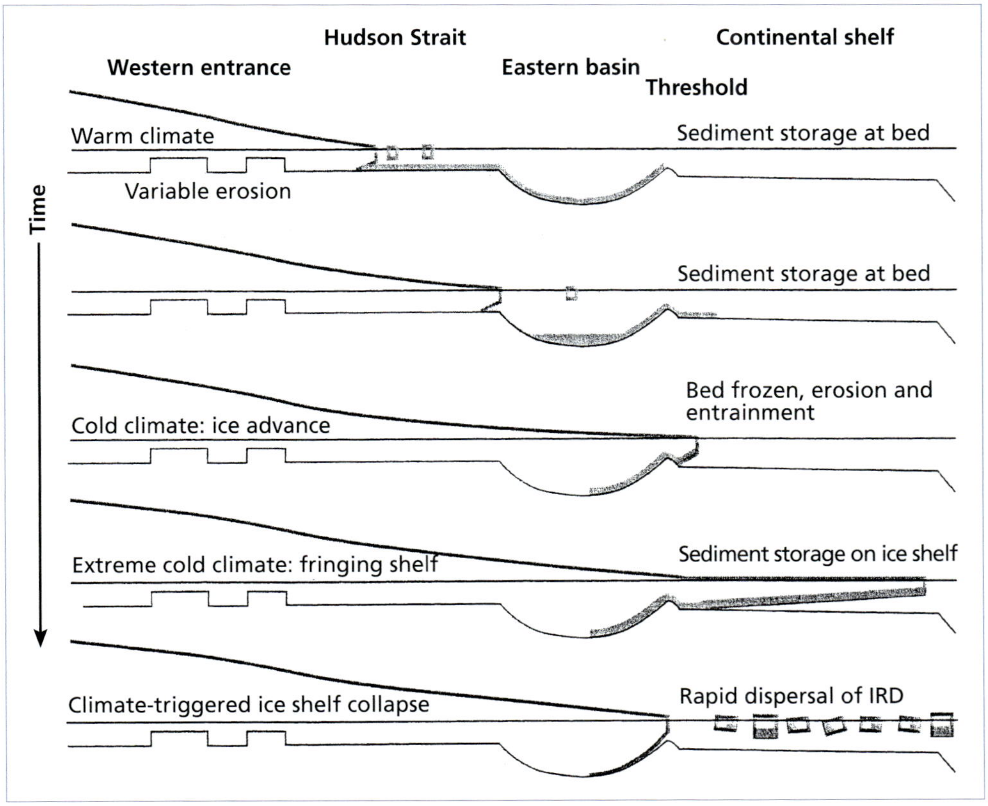

Figure 20.7 During the last glacial maximum, supraglacial streams supplied ice platforms in the Labrador Sea and Baffin Bay. The direction of the streams is shown by the fine lines. The source of sediments and their contribution to platforms are indicated by the thick gray lines (Hulbe *et al.*, 2004).

talus flatirons in northeastern Spain (Gutiérrez *et al.*, 1998b, 2006). This indicates a global climate influence for effects that correspond to Heinrich events, at least in the Northern Hemisphere.

20.3 Fluctuations of quaternary ice sheets and resulting landforms

Glacial and ice sheet fluctuations have been observed and systematically measured for more than a hundred years in various parts of the world (Haeberli, 2004). Glacial advances and retreats have undoubtedly had a bearing on morphogenesis; thus it is important to study past oscillations in great detail to understand the resulting glacial landforms. Glacial landforms are shaped by processes of erosion and deposition; they show up in outcrop after the ice masses have retreated. Existing landforms have been used to calculate the surface and ice flow directions of the ice sheets.

Our understanding of past glacial fluctuations has increased considerably in recent decades due to a better comprehension of the recent relationship between climate and glaciers and due to the rapid development of paleoclimatological technologies and concepts (Solomina *et al.*, 2008).

The maximum extensions of the most recent ice sheets have similar ages. Dawson (1992) compiled global data indicating that the Eurasian Ice Sheet dates from 15,000–20,000 years BP (late Valdai); the Fennoscandian Ice Sheet from 18,000–20,000 years BP (late Weichselian), the Laurentian Ice Sheet from 20,000–25,000 years BP (late Wisconsin), and the Antarctic Ice Sheet from 17,000–21,000 years BP. *Figure 20.8* shows the distribution of glaciers and ice sheets during the last glacial maximum and variation of the equilibrium line as a function of elevation and latitude (Broecker and Denton, 1990a).

Warming that occurred after the glacial maximum melted the ice caps of Eurasia, Fennoscandia, and Laurentia, along with many valley glaciers and small ice caps. The response is seen in changes of the general atmospheric circulation, in sea level rise resulting from an influx of fresh water,

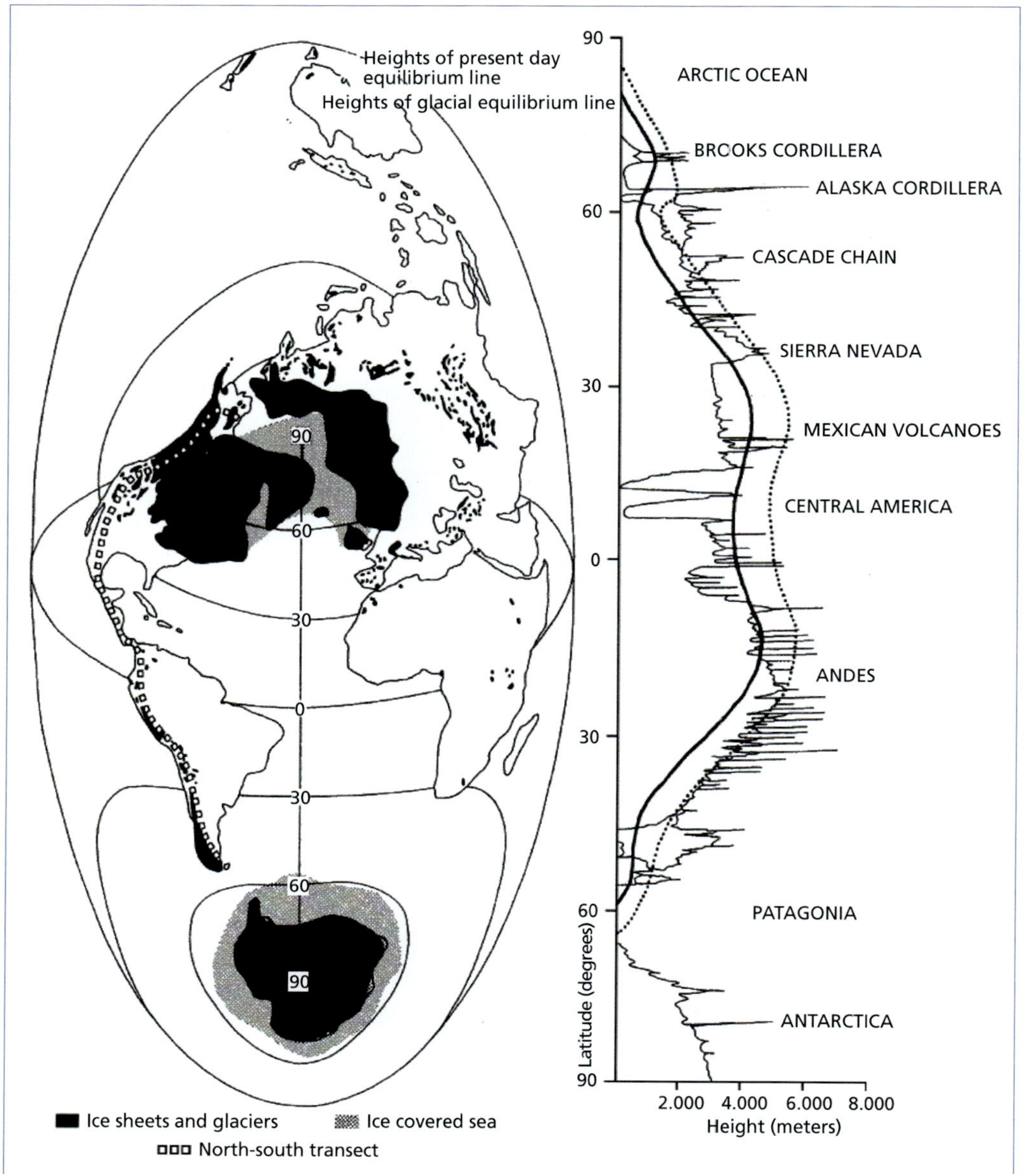

Figure 20.8 Left: Distribution of the principal ice sheets and mountain glacier complexes during the last glacial maximum. Right: North south transect showing the lowering of the height of equilibrium lines [dotted line] (Broecker y Denton, 1990a).

and in turn a modification of oceanic circulation. Melting did not occur at the same time throughout the world: the Cordilleran ice sheet disappeared 10,000 years BP, the Laurentide melted 6,500 years BP, the Fennoscandian had totally melted by about 8,000 years BP, and the Eurasian disappeared 9,000 years BP (according to various authors, in Dawson, 1992).

Determining the retreat chronology of the **Fennoscandian Ice Sheet** is difficult because of discontinuous moraines that are difficult to correlate in fjord areas. After 13,000 years BP, ice sheets thinned rapidly and retreated in response to a warm interstadial that lasted until 11,000 years BP. Between 12,800 and 10,300 years BP, a large lake called the *Baltic* **Ice Lake** formed in southern Scandinavia behind a dam that connected the land of northern Europe and the southern margin of the ice (Fig. 20.9) (Benn and Evans, 1998). About 10,000 years BP the ice retreated, allowing water to flow into the Baltic Sea (*Yoldia Stage*). Later rapid ice retreat between 10,000 and 9,000 years BP reduced the ice sheet to a small expanse north of Sweden and formed a large freshwater lake (*Ancylus Lake*). Finally, glacio-isostatic

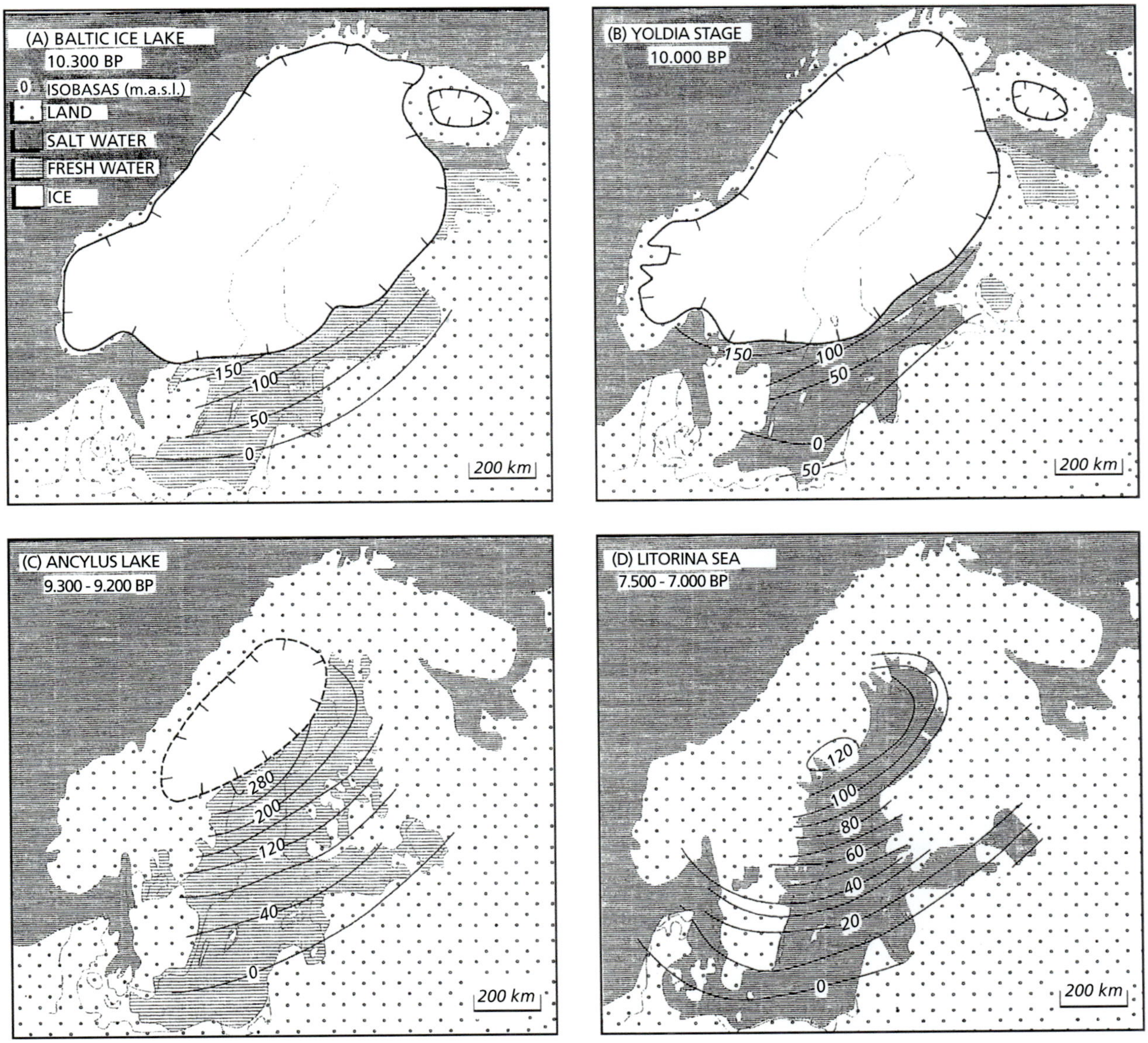

Figure 20.9 Stages in the retreat of the Fennoscandian Ice Sheet and associated lacustrine and marine events. (A) Baltic Ice Lake; (B) Yoldia Stage; (C) Ancylus Lake; (D) Litorina Sea (Benn and Evans, 1998).

rebound and the Flandrian transgression most likely caused marine water to flow into the Baltic Sea and Gulf of Bothnia (*Litorina Sea*).

The **Laurentide Ice Sheet** extended from the Canadian Archipelago in the Arctic Ocean southward to the northern states of the United States. The eastern limit was the Atlantic Ocean and the western margin was the Rocky Mountains. The Cordilleran Ice Sheet formed between this boundary and the Pacific Ocean. As a result of the intense hydrostatic depression, extensive proglacial lakes developed along its southern margin (*Fig. 20.10A*). Ice retreat was characterized by abrupt retreats that caused catastrophic flooding and spectacular changes of relief. The history of ice cap melting is not well understood. The southern margin of the Laurentide Ice Sheet had several ice lobes (*Fig. 20.11A*) whose advances and retreats do not appear to have been synchronous. The first withdrawal began in the eastern part 17,000 years BP whereas to the west the lobe remained active until 14,000 years BP. The Cordilleran Ice Sheet began to retreat after the large *jökulhlaup* that broke the ice dam at Lake Missoula about 13,000 years BP (*Fig. 14.25*) (*Fig. 20.11A*). The ice continued to thin until it disappeared 10,000 years BP. The period

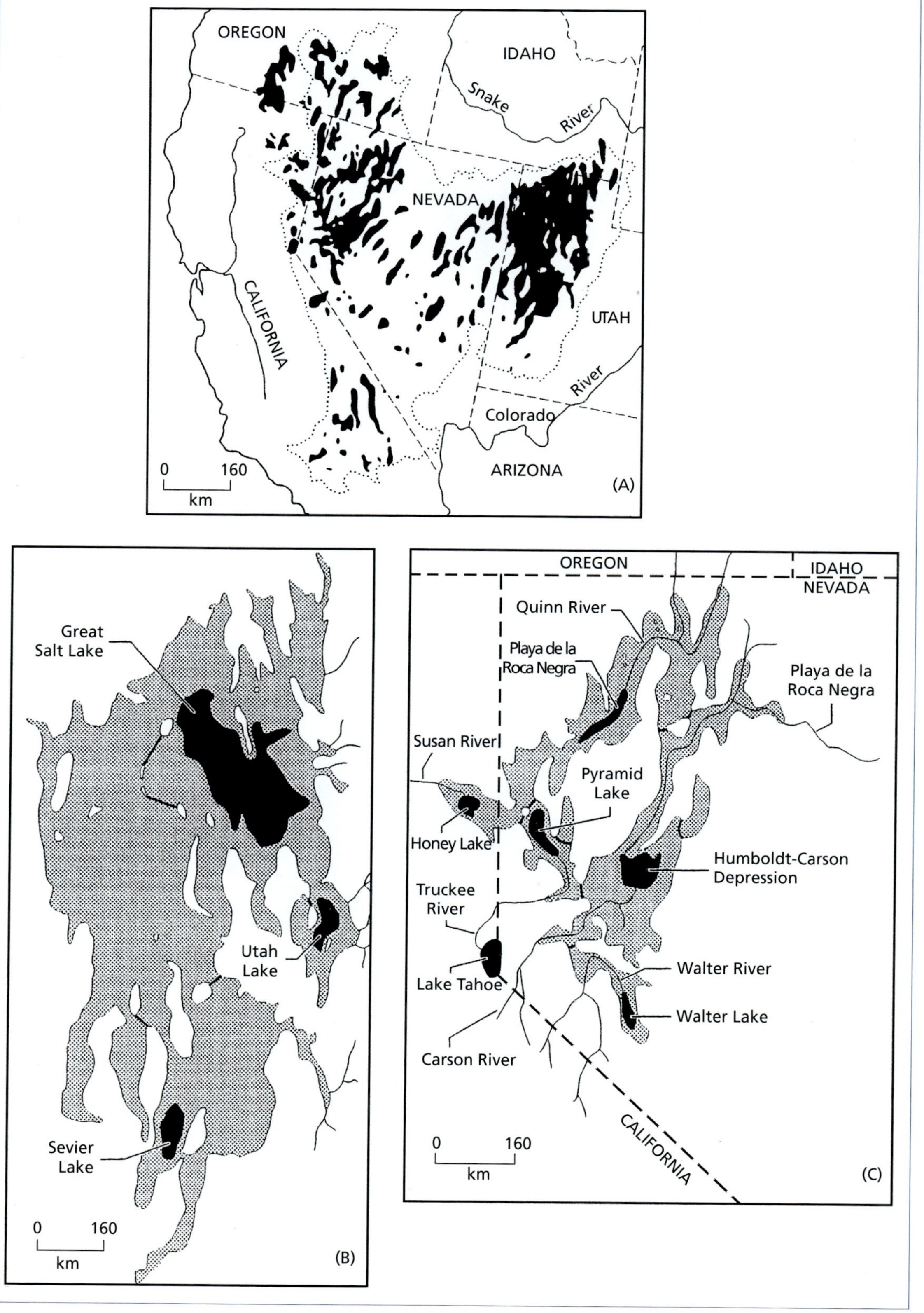

Figure 20.10 (A) Distribution of Pleistocene proglacial lakes in the southwestern United States. (B) Lake Bonneville. (C) Lake Lahontan (Goudie, 2004h).

from 13,000 to 11,000 years BP marked the disappearance of Laurentide Ice Sheet ice from the Great Lakes. The most significant proglacial lake was Lake Agassiz (*Fig. 20.11B*), which drained into the Gulf of Mexico. Later retreat of the ice sheet caused drainage to shift toward the Atlantic Ocean between 11,000 and 10,000 years BP. Lake Agassiz reached its maximum extent between 9,900 and 9,500 years BP when it covered a surface area of 350,000 km^2. Following this period and up until approximately 8,000 years BP, numerous *jökulhlaups* occurred; some volumes were much higher than that released by Lake Missoula. The Laurentide Ice Sheet gradually shrank to the northeast; about 7,000 years BP only small remnants of it remained (in Dawson, 1992).

Research on large groups of erosional and sedimentary landforms and their ages is an important and complex task that may reveal fluctuations and processes resulting from the activity of the large Quaternary ice sheets. We must conduct this research on the basis of precise analogies with present-day environments. In spite of interpretive difficulties, at present we have a highly detailed understanding of large groups of landforms including the processes and geomorphic history of the Laurentide and Fennoscandian ice sheets. Over the last three decades, records obtained from cores into deep marine sediments and ice sheet ice have led to significant advances in our understanding of climate change. This includes substantial scientific progress in the interpretation of landforms created by ice sheets (Hughes, 1992).

Erosional morphologies shaped by ice sheet activity are a function of a number of primary variables. The most important factor is the temperature at the base of the ice because as it reaches the melting point the ice slides and erodes. Geology plays a variable role; although erosion occurs most easily in fractures, the presence of permeable rocks facilitates water drainage and prevents erosion. Topography controls ice thickness, promoting both convergence and divergence of an ice flow (Chorley *et al.*, 1984). Even so, in areas such as Buchan (Scotland) no glacial erosion has occurred because a deep weathering profile remains intact after having been covered by ice (Hall, 1986). This is one of the arguments promoted by scientists who believe in glacial protectionism.

Some authors have tried to calculate the volume of glacial erosion done by ice sheets in medium latitudes. White (1972) thinks their erosional power was significant, and that this erosion moved a thickness of more than 1,000 m into Hudson Bay. The erosional work of ice sheets is facilitated by the presence of thick regoliths, greater than 50 m in northeastern Scotland (Fitzpatrick, 1963). As it advances, the glacier front encounters loose material that it removes easily by pushing it like a "bulldozer" (Bakker, 1965; Thomas, 1994b). Regoliths from pre-glacial chemical weathering, as opposed to those made up of fresh rock, provide good evidence to explain fine-grained tills (Feininger, 1971).

Landforms resulting from areal erosion are evidence of subglacial erosion of the bedrock; this type of erosion shows up as elongate hills, roches moutonnées, and rocky basins. Erosion is more intense along joints, faults, and dikes; this process creates a landform called knock and roch topography, so named by Linton (1963) in Scotland (*Fig. 13.55*). Large areas of these landforms are present on the Canadian Shield and in western Greenland. This topography is characterized by uneven relief controlled by brittle bedrock structures and relief amplitude that does not exceed 100 m.

Glacial deposits cover about 10% of the land surface. **Depositional landforms** related to ice sheets are often difficult to interpret because they are formed by the superposition of materials from various glaciations; in addition, their age is difficult to determine.

Moraines formed by ice sheets have a simple morphology that develops in valley glaciers. At first glance the moraines appear to be straight but their fundamental shape is an arc, as in the Laurentide Ice Sheet; this shape reflects the numerous lobes that make up the frontal margin of the ice cap (*Fig. 20.11*). In many places these large moraines are thrust moraines that are hundreds of kilometers long (*Fig. 20.12*) and up to 400 m thick, with some tills reaching thicknesses of about 400 m. In places they can exceed 200 m in height. Interpretations become difficult when deposits of various ages and similar lithologies are superimposed. ^{14}C dating shows that various lobes of the ice cap have undergone numerous cyclical advances and retreats (Clark, 1994).

Ice sheet retreat forms numerous ice-contact morphologies such as *drumlins*, *kame and kettle*, fluvioglacial shapes such as *eskers* and *pitted sandar,* and glaciolacustrine landforms. These landforms cover most of the area within the successive margins of the ice sheets. Excellent examples in central Canada (*Fig. 20.13*) show that the dividing line of the ice was located to the east where it was surrounded by channeled moraines, drumlins and eskers. Bedrock outcrops and a thin film of glacial material characterize the western margin (Aylsworth and Shilts, 1989). Spectacular examples can be seen in Labrador, east of Canada, in Scandinavia, Finland, and the Kola Peninsula (Russia) (*Fig. 14.29*).

Benn and Evans (1998) attempted to generalize the evolution of the major features of the large continental ice masses during the last glacial retreat, at least for the Laurentide Ice Sheet. They characterized the first period as successive ice advances and retreats that deposited and restructured large quantities of sediments and formed ice dammed lakes. Later, rapidly retreating ice created discontinuous moraines and eskers on top of the bedrock and new dams that formed marginal lakes. During the final retreat stages, moraines were more extensive and a significant network of eskers formed, which indicates the direction of regional water flow.

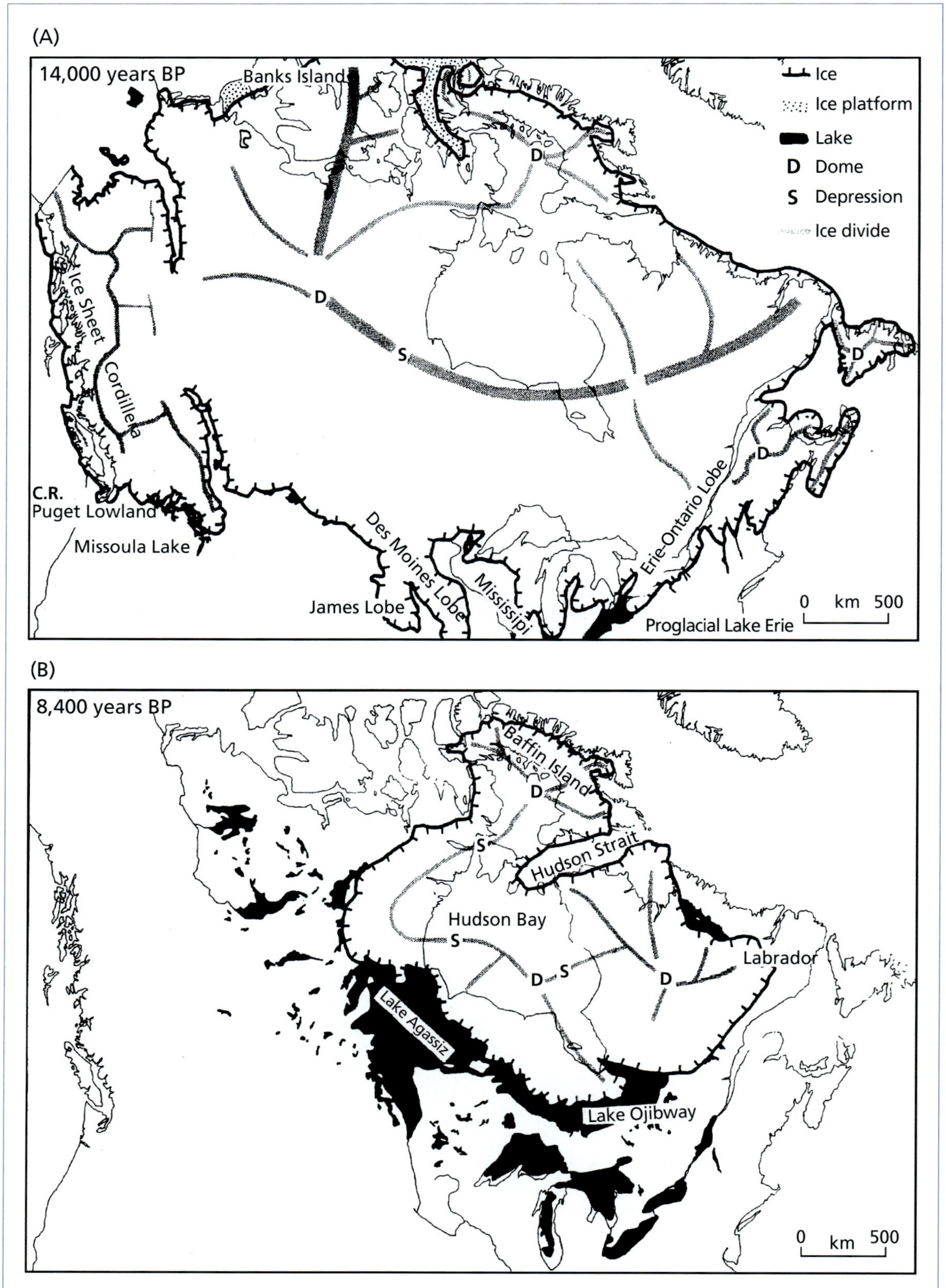

Figure 20.11 Reconstruction of the Laurentide Ice Sheet. (A) Situation 14,000 years BP, indicating the main ice divides. C.R. = Columbia River. (B) Position 8,400 years BP. (Dyke and Prest, 1987).

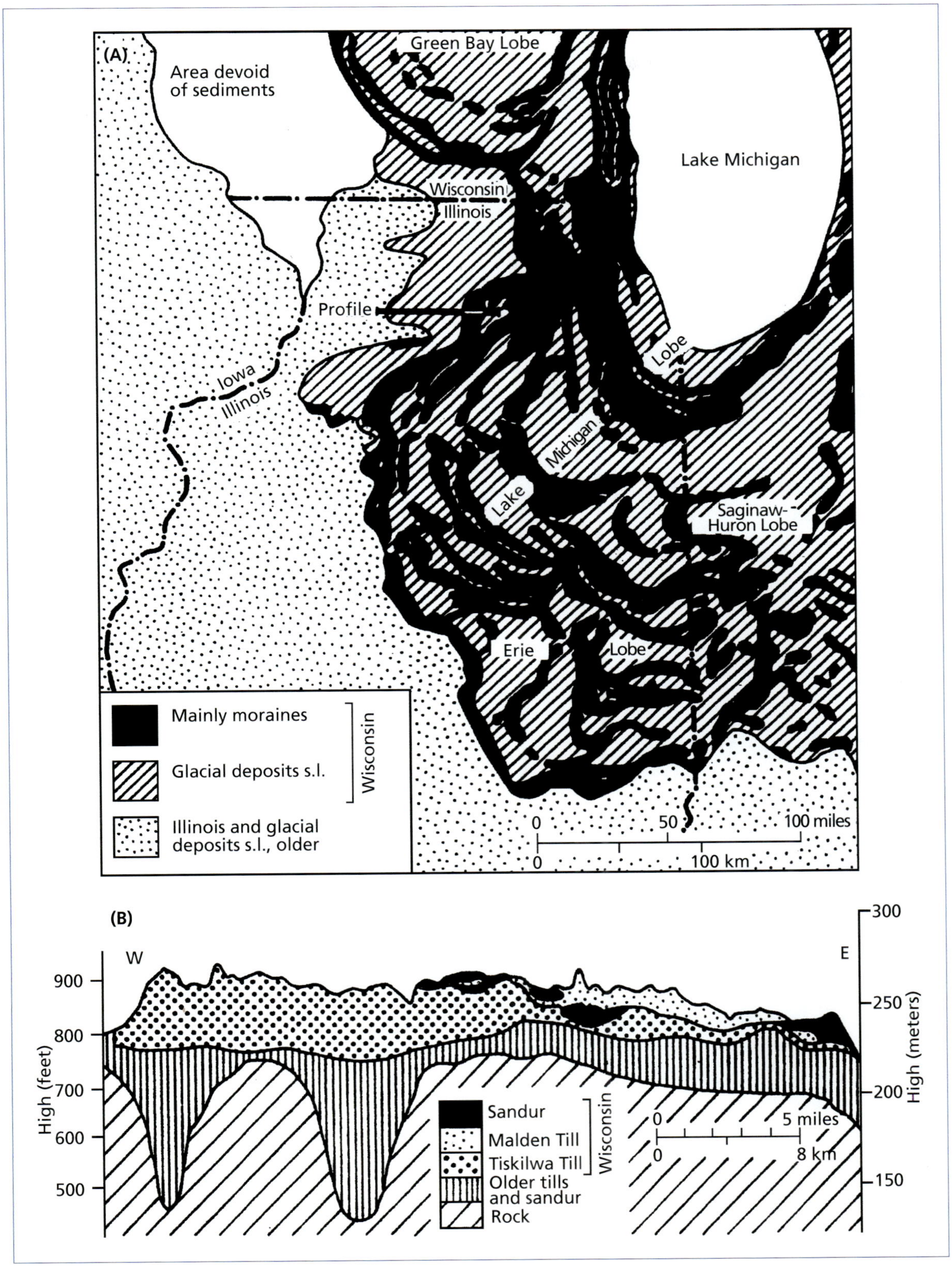

Figure 20. 12 Glacial deposits south of Lake Michigan. (A) At least four lobes of the Laurentide Ice Sheet front formed this configuration (Frye and Willman, 1973). (B) Profile across the till sheets of southern Illinois (Wickham and Johnson, 1981).

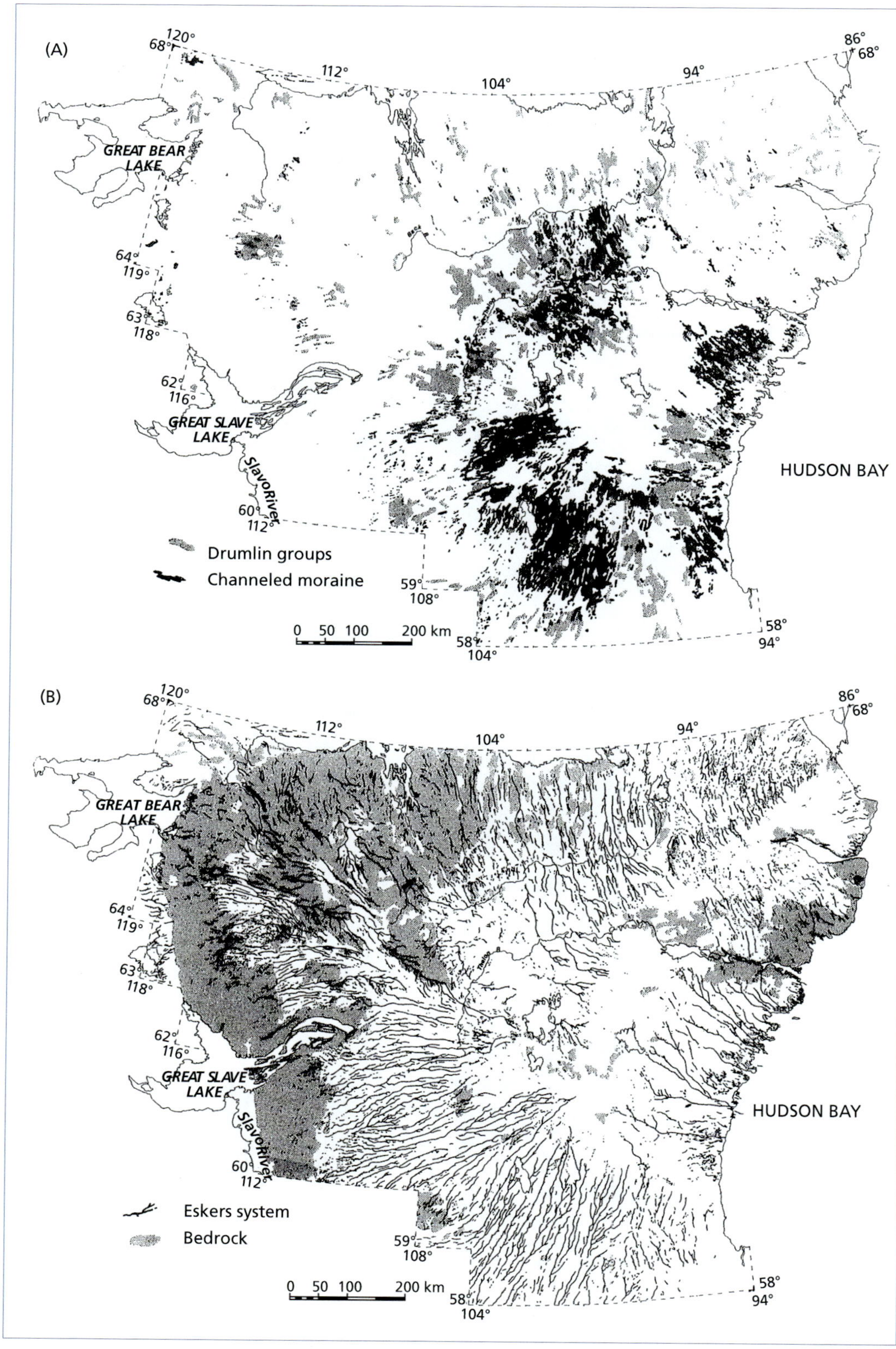

Figure 20.13 Distribution of glacial landforms in Keewatin, central Canada. (A) Location of drumlins and channeled moraines (Rogen moraines). (B) Distribution of eskers and bedrock (Aylsworth and Shilts, 1989).

20.4 Retreat of cirque and valley glaciers

Cirque glaciers are typically indicators of the last glacial stage. As a result, research on these glaciers may reveal the relationship between climate change and glacier fluctuation (Chueca *et al.*, 2002; Grove, 2004).

The landscape of alpine regions is characterized by *horns*, *arêtes* between glaciers, cirques, and longitudinal profiles with frequent overexcavations. In areas where significant retreat has occurred or where snow supply is low the dominant landscape is glacial cirques (*Fig. 20.14*). These landforms develop on all orientations although they form more easily in shady areas protected from insolation and consequently protected from snow melting. As glacial ice disappears, rock glaciers frequently form in cirques.

A good example is the response of the Glaciar de la Maladeta (central Pyrenees) to climate change since the Little Ice Age (Chueca *et al.*, 2005). Research methods used in this study include iconography, topographic maps, aerial photographs, land photographs, and climate data obtained from dendrochronology and the instrument record. The Glaciar de la Maladeta reached its maximum extent in about 1820–1830 according to the age of its moraine deposits (Chueca and Julián, 1996). The glacier covered a surface of 152.3 hectares and gradually thawed to about 54.5 hectares in the year 2000 (*Fig. 20.15*), a reduction of 35.7%. The retreat was not uniformly continuous; instead Chueca *et al.*, (2005) recognized several stages of glacial stabilization, moderate retreat, and marked recession of glacial ice in a group of ten stages. These stages are due to different mass balances in the glacier but no positive balance produced detectable advances.

The temporal comparison of this deglaciation with other Mediterranean alpine zones shows that deglaciation did not occur at the same time because of local factors such as topography, orientation, or the presence or absence of detrital cover or specific topo-climatic conditions. Even so, its area and thickness decreased in all sections, sometimes leading to the disappearance of the glacier Chueca *et al.*, (2005).

Alpine valley glaciers are characterized by their great depth and vertical walls where intense congelifraction occurs and supplies clasts to the margins of the valley floor. These troughs generally form from pre-glacial valleys.

During historic times, glacial fluctuations have caused grave concern to the inhabitants of the valleys because of numerous glacial disasters involving human life and property. Advance of a glacial tongue, rupture of ice dams holding back lakes, and landslides have caused numerous catastrophes (Tufnell, 1984). Information describing these historical events in alpine valleys provides valuable data on glacial evolution (Grove, 2004). Comparison of paintings and engravings from the 18th and 19th centuries with recent or present day photos provides important data on glacier fluctuation (Lamb,

Figure 20.14 Cirque glaciers. Vilcanota Cordillera. Southern Peru. Photo J.L. Peña.

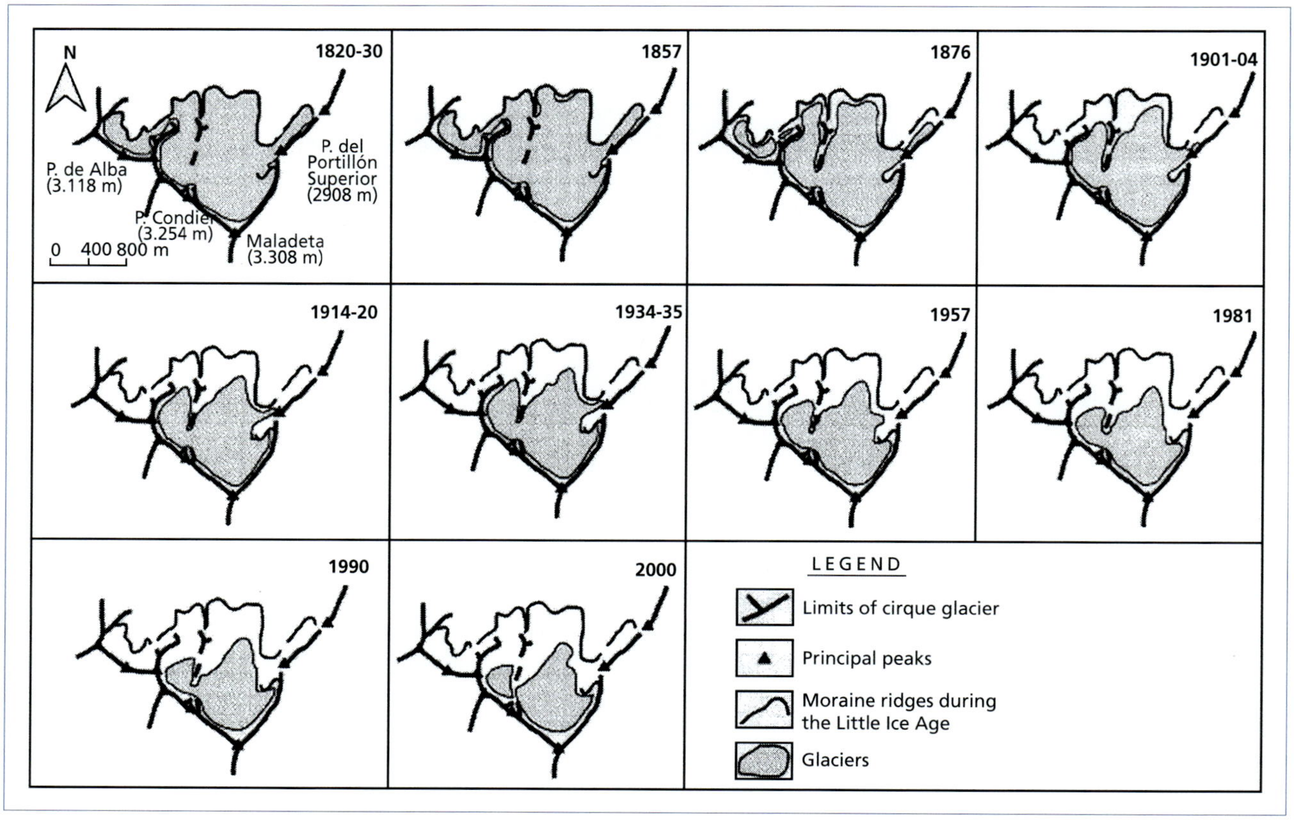

Figure 20.15 Evolution of the Maladeta glacier (central Pyrenees) analyzed through ten stages (Chueca *et al.*, 2005).

1977; Stroeven *et al.*, 1989). In addition, precise research on mass balance has been done in recent decades in relation to hydroelectrical projects; these mass balance studies have linked glacial variation to climate change (Oerlemans, 1989a, b).

A glacial valley is the landform that most spectacularly shows the immense erosional capacity of ice masses. A lateral moraine that unites with a terminal moraine to create a moraine arc is a significant feature. This arc reaches maximum thickness if the glacier remains stationary although glacial melt water usually erodes it gradually. In some valley glaciers, if the terminal parts corresponding to the moraine arcs are well preserved, abundant data can be acquired by lichenometry, which is the case of the Nigardsbreen (Norway) glacial valley (*Fig. 20.16*).

In the last two decades numerous researchers have pointed out the asynchrony of the last glacial maximum in the Pyrenees. Isolated *till* deposits attributed to previous glacial cycles have been discovered in the Spanish Pyrenees. The time of maximum glacial extent is thought to be before 38,000 years BP, prior to the Last Glacial Maximum (LGM) (Vilaplana, 1983; Bordonau, 1992; Montserrat, 1992). A previous glaciation has also been described in various valleys in the French Pyrenees, the Vosges, Alpes, and the Cantabrian Cordillera (northwestern Spain) (see García-Ruiz *et al.*, 2003). Geomorphic and sedimentological research on glacial deposits, palynological analysis of glaciolacustrine sediments, and AMS dating (^{14}C) have been used to decipher the history of the maximum glacial advance and deglaciation in the central Spanish Pyrenees (*Fig. 20.17*). The maximum extent of the Pyrenean glaciers during the last glaciation occurred prior to 30,000 years BP. A later advance occurred during the coldest period (around 20,000 years BP) and synchronous with the maximum global extent of glaciers but in the Pyrenees it was more extensive than the previous advance.

Ice-dammed lakes, which are highly ephemeral, are interesting features of mountain environments. A valley glacier blocks the drainage outlet of secondary valleys, thus forming a lake. Retraction of the ice tongue triggers water drainage from the lake and a new advance may cause another damming event. This enables us to study ice fluctuations related to brief climate changes. Many examples of ice-dammed lakes are present throughout the cordilleras and at high latitudes. Twidale (1976a) discusses numerous examples.

A particular type of glacial sedimentation occurs when the glacier tongue reaches a lake or the sea, an event that is

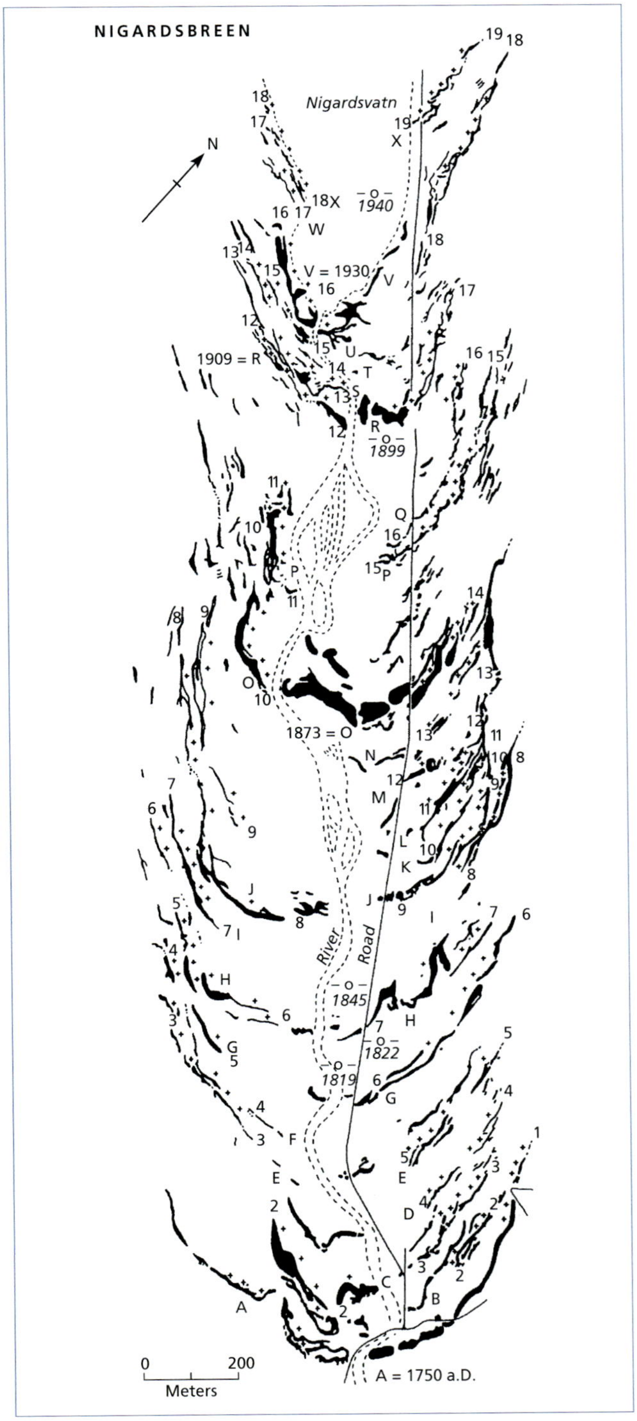

Figure 20.16 Sequence of moraine ridges that show the retreat of the Nigardsbreen (Norway) glacier from its maximum position during the Little Ice Age. The moraines are named with letters (A-X) and numbered from 1 to 19 on the northeastern margin and 1 to 18 on the southwestern margin. The dates were determined by lichenometry and locations are marked (+) (Bickerton and Matthews, 1992).

commonly associated with climate change (*Fig. 20.18*). *Figure 20.19* shows two facies associations proposed by Powell (1981) for the glacier-fjords of Alaska. The association shown in the upper part of the figure is characteristic of stationary glacial margins undergoing slow retraction. These conditions lead to the deposition of large fans or moraines. The lower part shows the margins of a fjord that receives sediments from the margin of the on-land glacier. Meltwater entering the fjord forms deltas with systems of deep, frontal, and top layers (*Fig. 20.20*).

20.5 Glacio-isostasy and glacio-eustasy

Glacio-isostasy accompanies the growth and disappearance of ice sheets. As ice sheets form, the lithosphere deforms under the weight of the ice mass; the size of the depression is a function of the thickness of the ice sheet and the ratio of ice and rock densities, which is approximately 0.3. The resulting depression of the land surface is about one third the thickness of the ice sheet. The depression reaches its greatest depth under the maximum ice thickness and decreases toward the margins. The ice sheet is surrounded by a peripheral depression whose height is a function of the ratio of H/11.5 where H is the height of the ice sheet (for an ice sheet 1,800 m thick, the depression is 155 m). Slight doming occurs outside the depression; its height is H/100, or 18 m in this case (*Fig. 20.21*) (Walcott, 1970). The doming results from lateral transfer of subcrustal material under the ice cap.

When the ice melts, the glacio-isostatic depression recovers as a result of elastic rebound that begins slowly (restricted rebound) but is followed by a period of accelerated uplift (postglacial rebound) and finally a period of residual rebound of lesser magnitude (Andrews, 1970).

Glacio-isostatic rebound is reconstructed by research on uplifted beaches and deltas that contain marine shells and fossiliferous silt and clay deposits that can be dated by ^{14}C. **Isobases** are drawn to represent the surface geometry that forms at sea level during a specific period and that reflects the effects of the ice loading. These isobases have confirmed that during the glacial maximum, the Laurentide Ice Sheet had several domes associated with its central depression (Andrews, 1970, 1975) whereas the Fennoscandian Ice Sheet had only one central dome (Mörner, 1980b) (*Figs. 20.9* and *20.22*) maximum uplift to the north of the Gulf of Bothnia. The isobases have an elliptical morphology that shows that it functioned separately from the dome in the British Isles.

Data obtained from glacio-isostasy have been used to propose models of mantle viscosity. Two models have been put forth: the first states that viscosity is relatively uniform

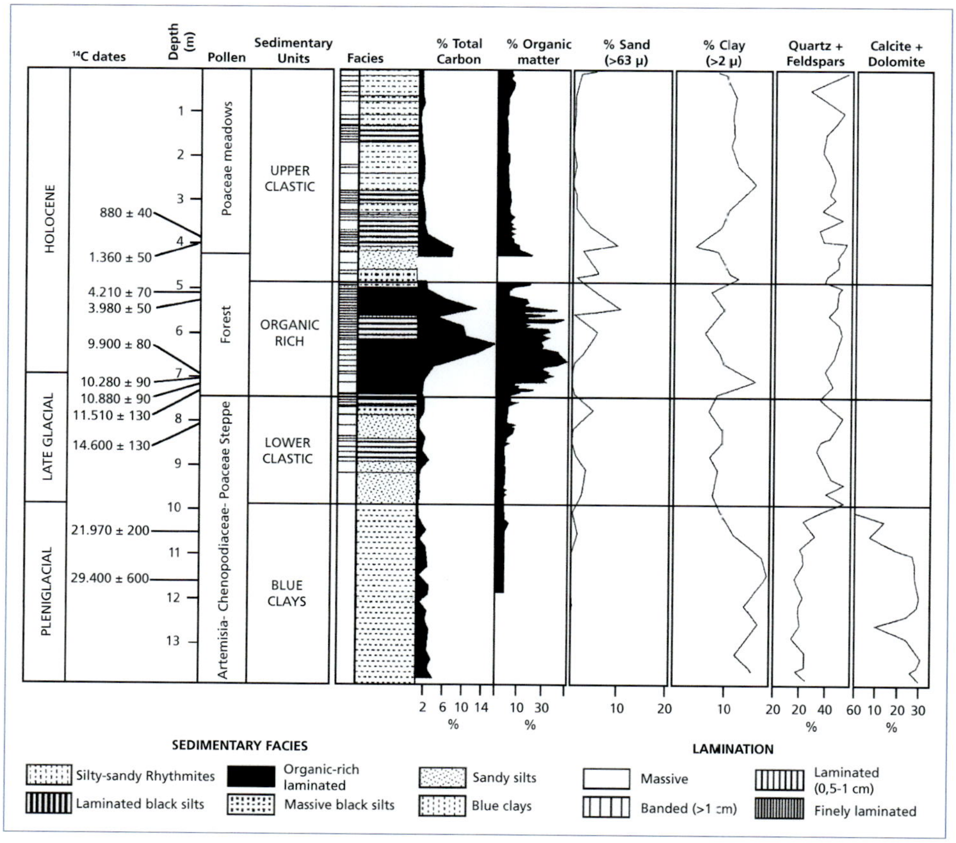

Figure 20.17 Summary of the sedimentary and palynological record of Lake Tramacastilla (central Pyrenees) (Monserrat, 1992; Valero-Garcés *et al.*, 1998; in Garcia-Ruiz *et al.*, 2003).

Figure 20.18 Tongue of an outlet glacier from the Antarctic Ice Sheet, flowing to the sea. Whisky Glacier. James Ross Island (Antarctica). Photo A. Martín-Serrano.

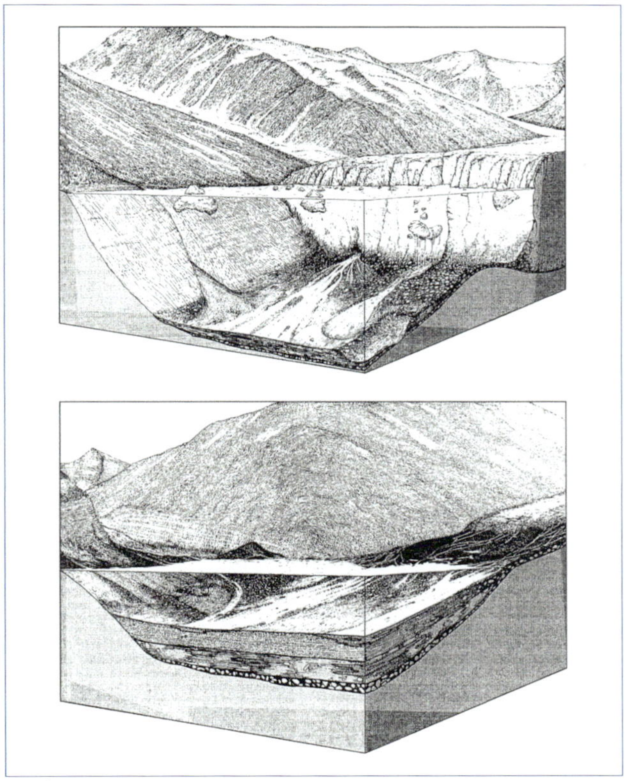

Figure 20.19 Depositional landforms from retreating glaciers based on examples from Alaska. Top: Glacier retreating slowly in shallow water. Below: Margin of a glacier on land that is supplying sediments to a delta (Powell, 1981).

with depth, and the second maintains that viscosity increases at the top of the upper mantle. The values of the second model fit with the unloading of the Laurentide Ice Sheet (Peltier and Andrews, 1983) whereas the first model more adequately describes the Fennoscandian Ice Sheet (Mörner, 1980b). This implies that ice sheet construction and thawing processes are associated with plastic deformation in the asthenosphere.

Increases and decreases of ice masses on continents cause significant sea level changes, which are called **glacio-eustatic**. These fluctuations are identified through the study of oxygen isotopes in foraminifera found in cores from the deep ocean (*Fig. 20.5*).

Other mechanisms are superimposed on glacio-eustatic variations: tectonic causes related to ocean capacity; isostasy; hydrostasy, which is the lithospheric response to water loading by oceans; and the geoidal effect. The most important factors are glaciers and tectonics although glacio-isostatic variations were more significant during the later Quaternary. Reconstructions of ancient sea levels are highly complex in some areas because numerous factors are involved.

Fairnbanks (1989) conducted research on the corals of the Barbados Islands. We know that coral grows when it is exposed above sea level. If sea level rises, coral grows above older corals. Using these facts, radiometric dating can determine the age of the corals and in turn, provide a record of sea level variations in the postglacial period. The record from the Barbados is very similar to one studied in the Indian Ocean (*Fig. 20.5*); they represent sea level fluctuations over much of

Figure 20.20 Bertrand Glacier that is supplying sediments to the Argentine Lake. Patagonia, Argentina. Photo C. Sancho.

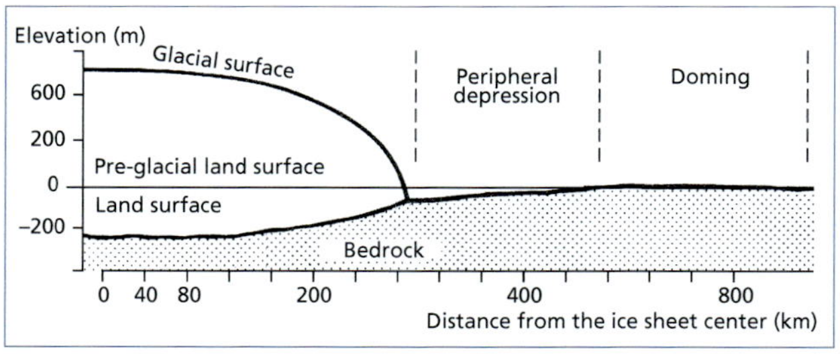

Figure 20.21 Principle of glacio-isostasy showing the depression of the lithosphere beneath the ice sheet (modified, Benn and Evans, 1998).

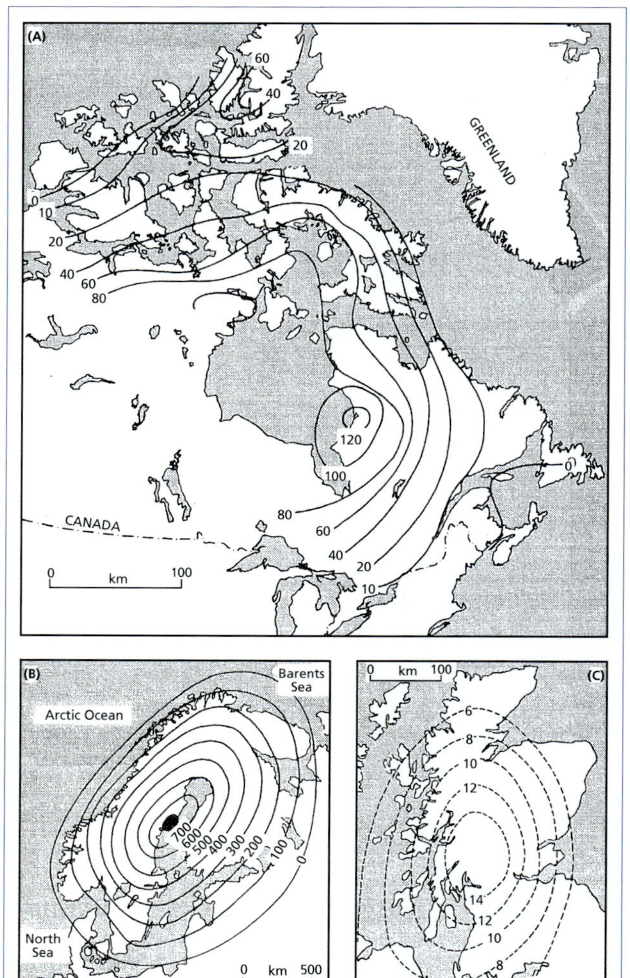

Figure 20.22 Regional isobase map in meters. (A) Isobase map showing the elevation of the eastern Canadian coastline about 6,000 years BP. Note the presence of several domes. (B) Absolute uplift of Scandinavia during the Holocene. (C) Isobases corresponding to maximum uplift of the coastline (7,000–6,000 years BP) in Scotland (Andrews, 1970; Mörner, 1980a).

the Earth's surface but they are not applicable to areas near ice sheets (Boulton, 1993).

The importance of eustasy and isostasy depends on the location relative to glacial ice sheets. On the margins of ice sheets, the isostatic effect is primary. The surface of the Earth has been divided into six **sea level zones** based on postglacial level curve characteristics (Clark *et al.*, 1978; Clark, 1980) *(Fig. 20.23)*. Zone 1 is located on the margins of ice sheets; here the coast shows continuous uplift due to isostatic recovery. Zone 1–2 is a transition zone with initial uplift and later subsidence due to collapse of the marginal doming of the ice sheet. Zone 2 is far from ice sheets; here sea level has undergoes continuous rise. This zone is similar to zone 3 except that for several thousand years it has formed a beach by emergence. Zone 4 includes tropical zones; here the sea rises after glaciation. This zone differs from Zone 5 where sea level rise occurs as the ice sheet melts; when melting stops, emergence occurs as a result of hydrostatic effects. Zone 6 includes all continental platforms except those in Zone 2; it is characterized by weak emergence after melting as a result of isostatic uplift of coastal margins. These sea level curves may be affected by tectonic movement.

20.6 Reconstruction of periglacial environments

Relict periglacial morphologies play an important role in reconstructing the predominant climates of the late Quaternary. The extent to which landforms are modified depends on the location and persistence of periglacial areas, which are common near ice sheets. Lithology has a major influence on the preservation of periglacial landforms. Rocks are susceptible to gelifraction whereas loose materials are easily shaped by periglacial processes. These processes vary according to whether the climate is continental, such as in European Russia, or marine as in Iceland. Paleoclimatic reconstructions should be based on conditions that prevailed during the formation of the landform but this is not always valid because Pleistocene periglacial environments at middle latitudes are very different from those present today at high latitudes (French, 2007). The primary cause is the position of the sun. Both latitudes have summers and winters but there is a major difference and the change between seasons occurs quickly. The rhythm of days and nights is the fundamental difference (arctic night and day). In addition, during the Pleistocene diurnal insolation was more pronounced in middle latitudes than it is today at high latitudes and for this reason congelifraction, ice *creep*, and ice needle activity were possibly several times more intense than in present-day high latitudes. Also, middle latitudes clearly show the effect of the sun's orientation;

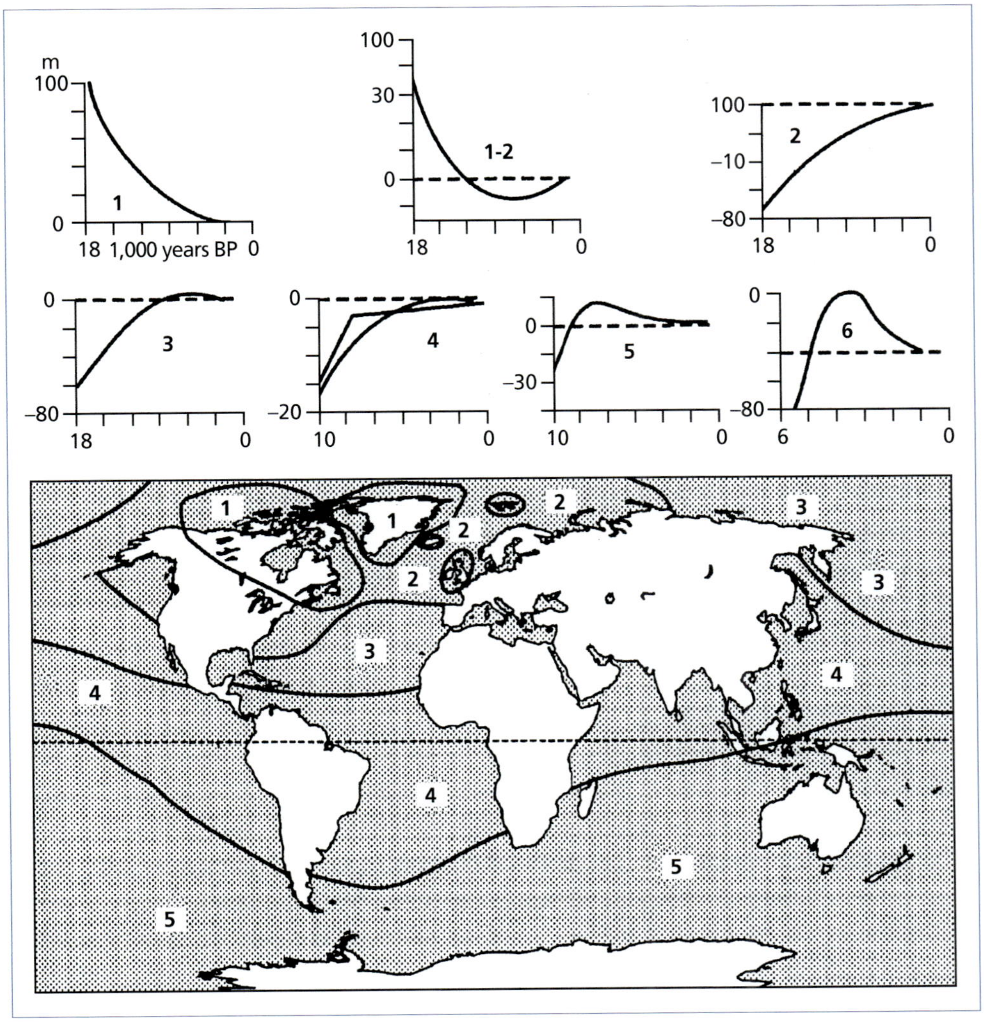

Figure 20.23 Distribution of sea level zones and curves related to typical sea level (Clark *et al.*, 1978).

furthermore, more freeze-and-thaw cycles occurred on sunny slopes than on shady ones. In addition, during the Pleistocene wind action was more significant in middle latitudes than in high latitudes because there were strong gradients along ice sheet margins. Furthermore, extensive plains of *till* and *sandar* developed along the margins, which were the source area of the fine sediments. Finally, fluvial activity must have been more intense in middle latitudes as a result of the greater number of thaw days. In summary, Pleistocene periglacial environments in middle latitudes were substantially different from those of high continental latitudes, for the reasons cited above.

On the basis of these arguments, the principal problem is the need to find precise indicators that testify to past periglacial conditions. The most realistic markers may be casts of ice and sand wedges; their presence indicates the existence of permafrost. Landforms inherited from pingo thermokarst are also considered permafrost indicators, just like fossil rock glaciers. The majority of the remaining periglacial paleoforms are doubtful indicators of periglacial conditions (French, 2007).

20.7 Relict periglacial landforms

French (2007) distinguished between features indicating the development of ancient frozen soils in middle latitudes and those related to the existence of forms that result from ice activity. The knowledge of the latter is subject to various interpretations, whereas features that reveal the presence of ancient permafrost dispel all doubt about their characterization.

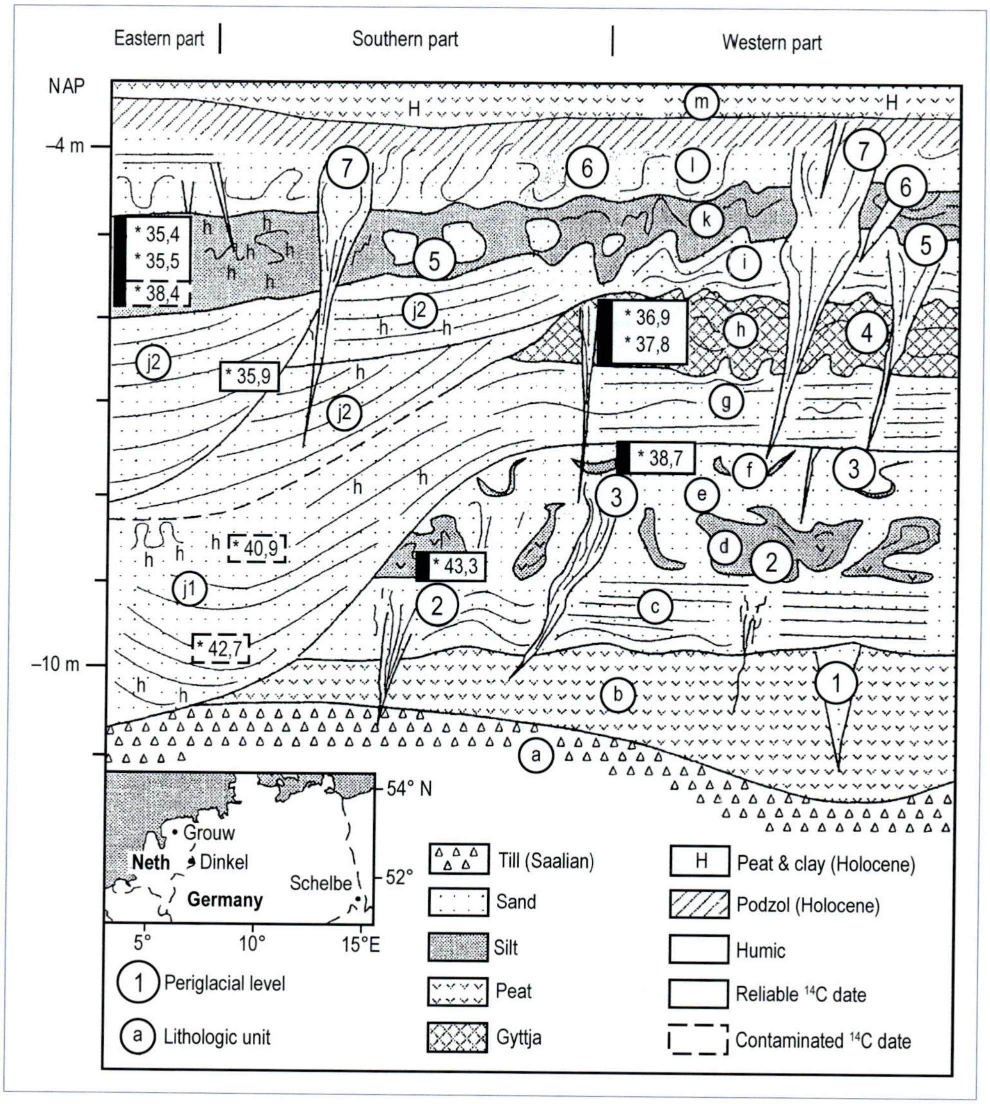

Figure 20.24 Location map of research areas in Holland and Germany and a compound section of Grouw (exaggerated vertical scale). In the eastern part and a portion of the southern part there are piles of channel sequences separated by large unconformities. The western part and part of the southern part contain a record of 8000 years of sedimentation. The southern part is 50 m long (Kasse, 1998).

Fossil wedges are relatively common in middle latitudes; they are the most important feature for identifying ancient permafrost. They may be casts of ice wedges filled with sediments (*Fig. 15.4*), which often form polygonal systems and sand wedges (*Fig. 15.10*). Winter temperatures at the surface of active ice wedges are thought to fluctuate between –15°C and –20°C (Péwé, 1983). Péwé thinks that the median annual air temperatures needed to form ice wedges are on the order of –6°C and –8°C. In central and western Europe (*Fig. 20.24*) (Kasse, 1998) ice wedges are common along a wide belt and their width and depth usually increase toward the southern margin of the ice sheet, which is interpreted to have had longer duration than periglacial conditions and which are also thought to have been more severe (Poser, 1948). In North America, wedge molds located near the Laurentide Ice Sheet along a relatively narrow fringe range in age from 15,000 to 20,000 years BP according to several authors.

Ancient **pingos** can be recognized as a ramp, generally annular in shape, that surrounds a depression. The ramp distinguishes them from other depressions generated by thermokarst, such as palsas. The height of pingo ramps

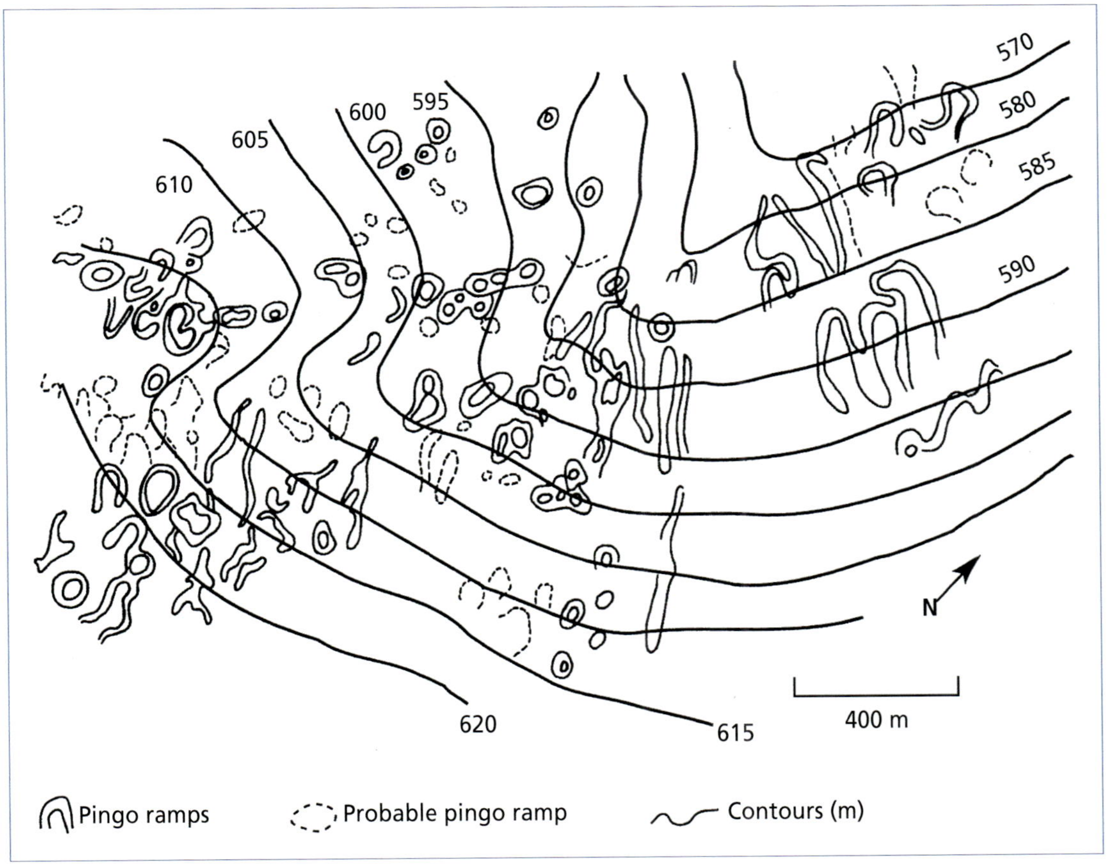

Figure 20.25 Map showing pingo remains in the Hautes Fagnes Moor, Ardennes, near Malmedy, eastern Belgium (Pissart, 1965).

depends on the size of the initial pingo; they range from 0.5 to 5 m. The depression diameter ranges from 200 to 300 m. Pingo remains are found primarily in Western Europe and eastern North America. A classic example is the case Pissart (1965) studied in the Ardennes (Belgium) (*Fig. 20.25*). The irregular shape and the interferences between them have a complex distribution; the pingos are thought to have formed in an open system. The age of most pingos in Europe and North America is Late Glacial, approximately 10,000 to 13,000 years BP.

Fossil **rock glaciers** that are found today below the tree line indicate a periglacial climate with permafrost during their formation, which shows their importance as paleoclimatic indicators (Barsch, 1988, 1996; Serrano and López-Martínez, 2000).

Relict morphologies generated by frost action are common but few provide unequivocal proof of periglacial conditions. Disharmonically deformed soft sediments called **cryoturbations** or **involutions** are present in middle latitudes. Some authors considered them to be permafrost indicators but these deformations are azonal; they can originate from very different processes (see the section on patterned ground in Chapter 15, Section 15.2.1). Periglacial involutions form from ice push and/or from thermokarst.

Block fields, **slopes**, and **streams** are landforms generated by frost cracking in elevated zones. Multiple freeze-thaw cycles favor the development of these landforms. It is necessary to determine whether these block accumulations are active or inactive. Colonization by lichens, as on the block slopes and streams of the Tremedal Massif (Gutiérrez and Peña, 1977), show that the deposit is not moving.

Cryoplanation terraces formed by nivation processes are common in elevated zones of European mountains and especially in the central region of Alaska (Péwé, 1983). Periglacial **tors** and cryoplanation terraces related to them do not unequivocally indicate permafrost conditions (French, 1996) although some authors contend that they do (Dawson, 1992).

According to most authors, **grèzes litées** and ***protalus ramparts*** are related in origin to seasonal snow accumulations. Finally, **solifluction deposits** are poorly sorted and thus very difficult to distinguish from deposits that form in warmer environments.

Also in periglacial areas **triangular slope facets** are located indicative of relict periglacial forms. In the NE of Seymour Island (Antarctica) talus flatirons, have been developed whose origin may be due to a glacioisostatic adjustment or to a climate change (Gutiérrez *et al.*, 2011b)

20.8 Fluvial systems in periglacial areas

The characteristics and evolution of arctic and subarctic rivers are not well understood. French (1996) claimed *braided* channels (*Fig. 20.26*) to be the dominant type in periglacial environments; additional research has been conducted more recently on periglacial hydrology, processes, and landforms (Vandenberghe *et al.*, 1993; Starkel, 1995; Collins *et al.*, 1996; Huisink, 1998, 2000, among others).

Researchers have studied the adaptation of rivers to general climatic conditions for a long time; they have recently analyzed the impact of climate change on fluvial dynamics and morphology. Vegetation, which is a function of climate, is a fundamental factor in water supply and the behavior of fluvial systems. Langbein and Schumm (1958) and Schumm (1977) propose a threshold of 70% plant cover, below which soil loss decreases. Knox (1972, 1984a, 1993) claims that above a threshold of 80% erosion is reduced. In periglacial environments a temperature change of 1°C affects weathering and fluvial morphology; Knox (1984b) and Vanderberghe (2002) find that vegetation can adapt to significant changes in precipitation within 50 to 100 years. The delayed response of vegetation to climate change is accompanied by considerable changes in sediment load and the resulting discharge during brief phases of morphological instability (Mol *et al.*, 2000). Longer lasting climate variations are primarily reflected in changes in plant cover and permafrost conditions or a combination of the two (Vanderberghe and Pissart, 1993). *Table 20.2* (Vanderberghe, 2001, 2002) shows river types as a function of vegetation and permafrost conditions and sediment supply as a function of vegetation. Channels are braided in the absence of vegetation; as plant cover increases, meandering and anastomosing channels predominate. Ephemeral discharge systems develop in all types of frozen soil (Woo and Winter, 1993).

Changes in channel types occur over time and are related to climate changes as Huisink (2000) points out in his study of the late Quaternary of the Vecht River in Holland (*Fig. 20.27*). Huisink *et al.*, (2002) studied periglacial fluvial morphology in a basin with tundra and taiga vegetation in northern European Russia; their results indicate that the morphology of active periglacial channels is influenced by geology, discharge regime, slope, vegetation density, and permafrost conditions. During periods of ice melting there is no significant difference between the hydrologic response of continuous or discontinuous permafrost zones and that of seasonal freezing. The water/sediment ratio is high during large spring floods when the still-frozen surface does

Figure 20.26 Frozen braided river (January 1995). Newfoundland.

not provide much sediment. During these floods incision and downcutting can occur (Huisink, 2000). Runoff between various permafrost zones differs after the spring flood peak when the thickest active layers, which store large quantities of water, begin to melt. This is accentuated during summer rainstorms. Peak runoff decreases as a result of better evaporation and more intense thawing (Woo and Winter, 1993).

Research on the fluvial morphology of the Usa River basin, an area of 93,000 km^2 along this tributary of the Pechora River (Republic of Komi in the Russian Federation), shows that channels are predominantly anastomosing and meandering. This is due to dense vegetation that prevents large quantities of sediment supply from entering the river (Huisink *et al.*, 2002).

Table 20.2 **SUMMARY REPRESENTATION OF PERIGLACIAL RIVER TYPES AS A FUNCTION OF VEGETATION, PERMAFROST CONDITIONS, AND SEDIMENT SUPPLY (VANDENBERGHE, 2001)**

		No vegetation	**Disperse vegetation**	**Continuous vegetation**
Permafrost (partial or continuous)	Low Current energy/ sediment supply High	BRAIDED	BRAIDED MEANDERING BRAIDED	MEANDERING ANASTOMOSING (Anabranching)
Partial permafrost/ deep seasonal freezing	Low Current energy/ sediment supply High		MEANDERING ANASTOMOSING BRAIDED	

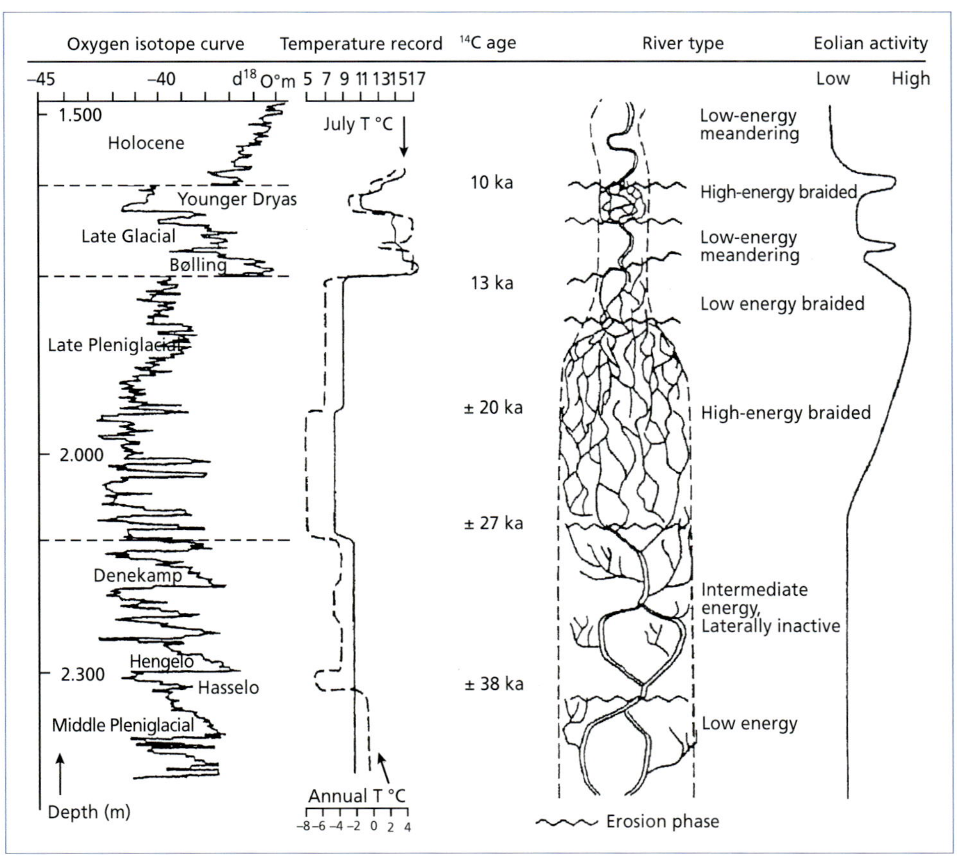

Figure 20.27 Fluvial changes in the Vecht valley (eastern Holland) since the middle Pleniglacial to the Holocene with phases of erosion, eolian activity, average July and annual temperatures, and δ^{18}C curve from Greenland (modified from Huisink, 2000).

Research by Mol *et al.* (2000) on fluvial responses in periglacial climates of various regions of central and western Europe produced ^{14}C and OSL dates of various sedimentary sections. This data shows that climate changes (warm to cool, cool to warm) are generally associated with fluvial instability. Meandering rivers develop during warm phases because vegetation cover is continuous in these rivers and as a result, there is less sediment supply. Braided rivers form during cooler phases when sparse vegetation leads to a high sediment supply, which causes rapid aggradation on an unstable flood plain. Anastomosing ephemeral fluvial systems are an intermediate between the two previous types.

Cycles of incision and aggradation and fluvial modifications that result from climate changes have been established. These results agree with the model proposed by Vanderberghe (1993, 1995) but the response of fluvial systems to climate change does not always coincide with the proposed model, which was applied to different sequences described in periglacial regions in middle latitudes of Europe. Many river basins do not show modifications after climate change. These anomalies can be resolved by introducing the concept of a geomorphic threshold (Schumm, 1973, 1979, 2004), which is a limit a channel has to exceed before it responds to external change. Sometimes the change is not significant enough to exceed the threshold.

Global climate change can considerably affect northern environments, producing changes in permafrost, vegetation, and fluvial morphology (Anisimov *et al.*, 2008). Warm climate conditions favor vegetation growth and permafrost retreat. During the first and long transition from winter to springtime, the snow melts and creates a minor runoff peak. Deeper seasonal thawing increases the loss of permafrost, and as a result, the loss of the soil water storage capacity, which leads to increased base discharge in the summer. Warmer climate conditions, such as those expected during global warming, will involve less seasonal fluctuation of runoff throughout the year (Anisimov and Fitzharris, 2001).

Areas of low relief respond to climate change by brief phases of fluvial instability; these occur both in the transition from a relatively humid (temperate) period to cool (periglacial) one and during relatively cool and warm periods. This instability typically begins with vertical erosion, which is followed by sedimentary filling of this incision. This interpretation, however, presents some problems. First, there are fewer incision phases than climate transitions. Second, erosional remains are rare in the cool-warm transitions; in contrast, obvious erosional remains have been found in warm-cool transitions. In addition, the development of numerous flood plains provides evidence of long and stable phases, both in cold and warm climates. Meandering rivers incise weakly and have a limited lateral extent; as a result, they form narrow floodplains and do not destroy pre-existing braided flood plains. Braided rivers also incise in a minor way but they have considerable lateral development that destroys pre-existing meandering channels and their sediments (Vandenberghe, 2008).

20.9 Eolian activity in periglacial regions

Extensive eolian loess accumulated near the Laurentide Ice Sheet on the Great Plains in the center of the United States and periglacial eolian deposits formed south of the Eurasian Ice Sheet (Koster, 1988). An extensive mantle of eolian sand (loess) covered northwestern Europe during the final glacial period

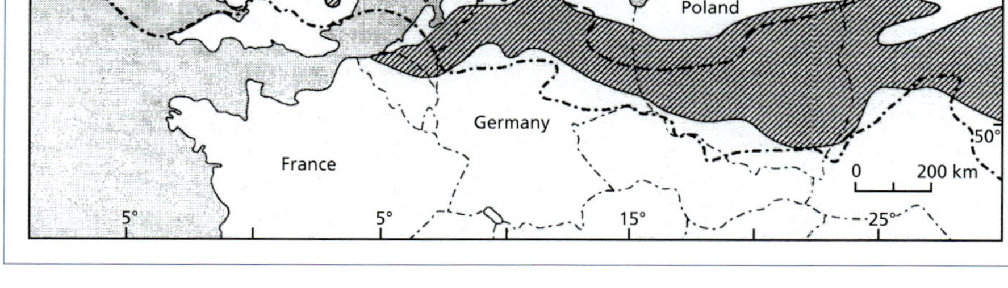

Figure 20.28 Distribution of sand mantles in Europe during the late Weichselian (approximately 14–12.4 ka) (from Koster, 1988; in Kasse, 1997).

Figure 20.29 Loess deposits with horizontal lamination at the base and massive structure at the top. Neerijse (Belgium).

(Weichselian) (approximately 14–12.4 ka) (*Fig. 20.28*). These accumulations are oriented east west, parallel to the margins of the Pleistocene ice sheets. The deposits cover pre-existing morphology and for this reason have been called *coversand*.

The coversand resulted from a group of factors that favored their accumulation (Kasse, 1997). The source area is primarily thick and unconcolidated Quaternary fluvial and fluvioglacial sand deposits (*sandar* or *outwash*). Stratification in the loess is predominantly horizontal (*Fig. 20.29*) and is attributed to poorly drained sedimentary surfaces. Another factor is climate, which was dry and cold with seasonal freezing when the coversand was deposited; the mean annual temperature was between −1°C and −8°C. Previous researchers had characterized the climate as permafrost conditions with temperatures between −6°C and −8°C. Dispersed vegetation cover, low relief (less than 200 m), and the lack of significant topographic barriers are also fundamental factors. Other factors include the occasionally low availability of fine particles due to the presence of frozen or cemented surfaces, degradation of permafrost, and increased aridity. This degradation began with climate warming at the end of the glacial maximum. It was accompanied by increased plant density, increased infiltration, and increased storage capacity of the unconsolidated sandy soil that became a dry surface subject to increased eolian processes. The lack of surface runoff helped preserve the eolian deposits. Increased aridity during the coversand phase is attributed to decreased drainage density and flood plain width. This arid phase is attributed to the principal stage of glacial melt water generation and to southward flow of cold water into the North Atlantic Ocean between 14.2 and 13.2 ka. This arid phase explains the extensive formation of coversands, which was a brief response to climate changes during the retreat of the Weichselian ice sheet (Kasse, 1997).

In central Spain, south of the Duero Depression and north of the Central Cordillera System there are several sand mantles (*Fig. 20.30*) composed of fine to coarse sand with no internal structure or with horizontal or low angle laminations (Gutiérrez-Elorza *et al.*, 2005); the mantles typically have a flat morphology and lack avalanche or *slip faces* (Flyberger *et al.*, 1979; Kocurek and Nielson, 1986). Because of their proximity to the Central Cordillera they can be considered perimontane sand mantles. Parabolic dunes and *blowouts* are present, although vaguely, on top of the sand mantles (Gutiérrez-Elorza *et al.*, 2005). The analysis of these dunes is complex because a dense pine forest planted in the 19th century now covers them; pine growth has distorted the dune features. In addition, natural post-depositional processes that reduce porosity and permeability affect the dunes; these include chemical weathering, dissolution and precipitation by infiltration water, and bioturbation (Pye, 1983). These sand mantles became *blowouts* as a result of climate change (*Fig. 10.44*).

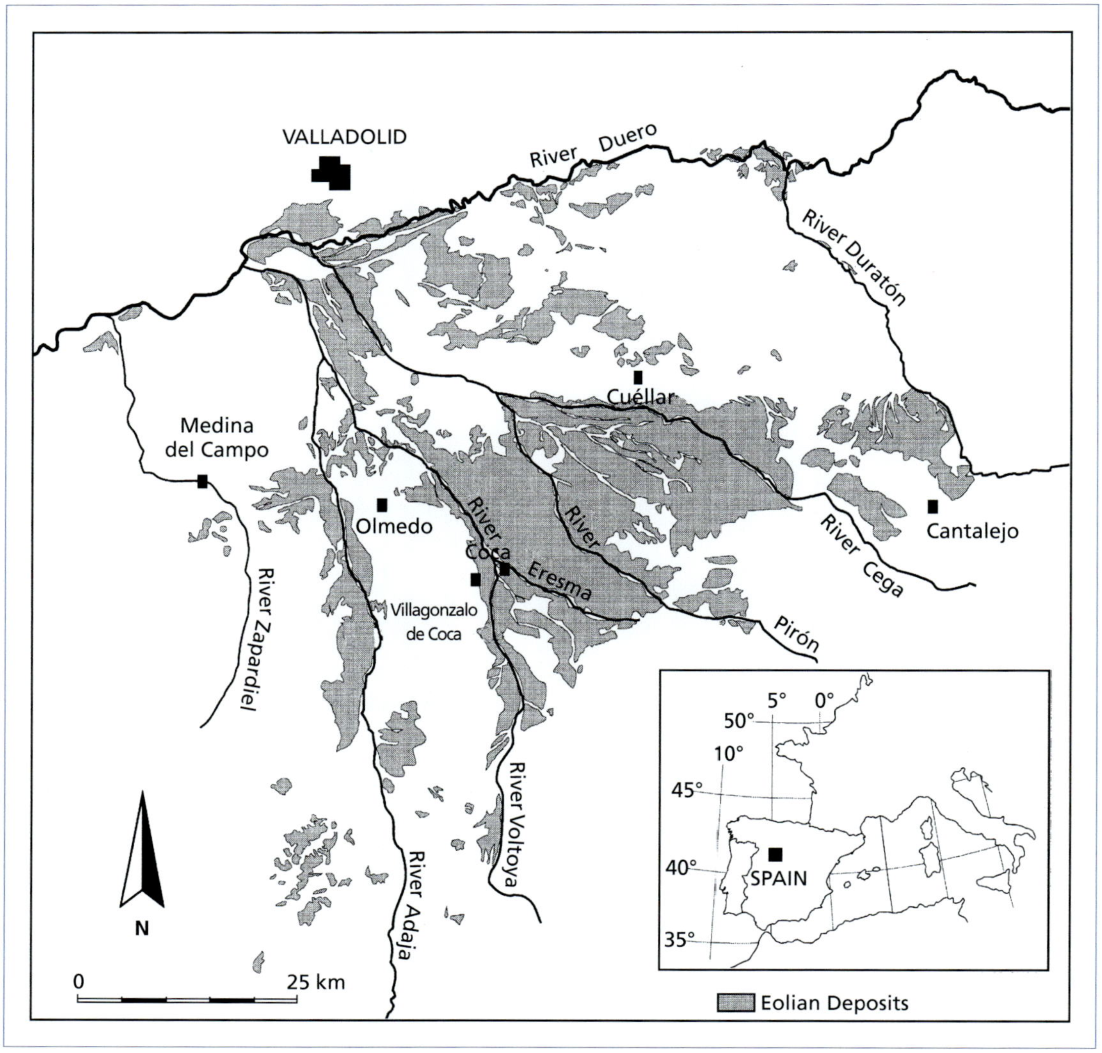

Figure 20.30 Distribution of eolian accumulations in the central-southern sector of the Duero Depression (based on Crespo, 1977; in Gutiérrez *et al.*, 2005a).

Diez *et al.*, (2002) have obtained TL (thermoluminescence) ages of 10.21 ± 2.0 ky for these deposits, which correspond to Younger Dryas. Numerous TL, OSL and ^{14}C dates obtained in this region reveal a cool and dry period in the Younger Dryas and Allerød (12.5 ± 1.0–10.2 ± 2.0 ka) during which reduced vegetation allowed wind to move the sand. Some sediment dates at the top of the mantle correspond to the Atlantic period, a period when deflation was reduced by increased vegetation cover during a gentler climate (Diez and Bateman, 1998; Bateman and Diez, 1999; Diez *et al.*, 2002). These accumulations correlate well with Younger Dryas sediments studied in northwestern Europe (Kasse, 1997) and indicate that climate had a clear influence on eolian processes during this period (Bateman and Diez, 1999; Diez *et al.*, 2002).

20.10 Fluctuations in periglacial areas during the late Quaternary

Most periglacial regions are located on permafrost. Disintegration of permafrost occurs when summer warming exceeds winter cooling and vice versa. Today, periglacial environments cover 25% of the land surface whereas during the last glaciation they may have covered an additional 20–25% of the land surface (French, 2007). Climate change in natural and anthropogenic systems causes ice to melt and increases the thickness of the active layer, which causes subsidence (thermokarst) and surface destabilization (Nelson *et al.*, 2002).

The distribution of plants and the size of eco-zones may be the most reliable criteria for determining the surface area covered by periglacial areas during the late Quaternary. This implies fluctuations of tundra, steppe, and forest zones during these periods. In central Alaska, Jorgenson *et al.*, (2001) describe significant ecological changes such as birch forests, which are found in highlands with lots of ice, being transformed into muddy areas as a result of subsidence caused by ice melting. As a result, palynological research is primary in determining these effects; this work has enabled researchers to establish pollen zones and climatic periods (French, 2007). Faunal remains, such as mollusks and insects, aid in the interpretation of environmental conditions. However, there are few studies aimed at these goals.

Permafrost evolves slowly in response to warming and cooling stages. During the late Pleistocene and Holocene, ice sheet melting and resulting sea level rise occurred much more rapidly than permafrost thinning during these periods (Dawson, 1992). For example, the permafrost of northeastern Siberia is known to have been continuous since the early Pleistocene (Washburn, 1979).

European researchers have the highest degree of knowledge related to climate and vegetation fluctuations that occurred during the late Quaternary. Büdel (1951) distinguished three vegetation zones in **Europe** (tundra, steppe, and forest zones); he also found that most of Europe south of the ice sheets was affected by periglacial conditions except the southern part of the Iberian Peninsula and the Mediterranean coasts. However, the periglacial belt that exists today in central and western Europe has been calculated to be have been about 500 km wide during the last glacial maximum (Weichselian). This estimate was obtained primarily from ice and sand wedge casts and annular pingo ramps. The most important climate record is the Grand Pile pollen diagram (Mook and Woillard, 1982), which shows the first periods of permafrost development about 90,000 years ago. Later permafrost periods are separated by warm interstadials.

In **European** and **Asian Russia** permafrost increased significantly during the late Quaternary, to thicknesses of about 700 m. Continuous permafrost and a belt of lesser width of discontinuous permafrost extends to 50°N latitude, for a total dimension of 400–600 km *(Fig. 20.31)* (Baulin and Danilova, 1984). Permafrost evolution in Russia is similar to that of Europe, with growth periods that coincide with cool stages and alternating warmer periods.

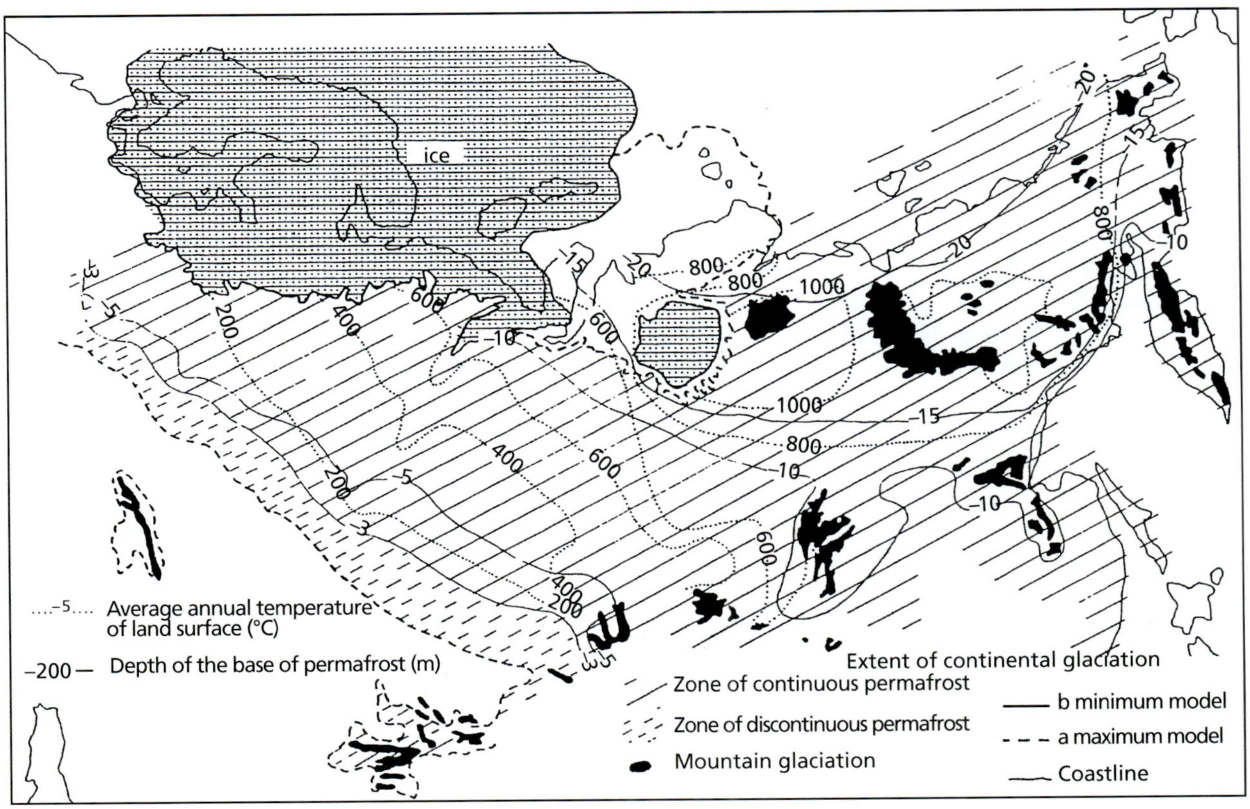

Figure 20.31 Map of permafrost distribution during the last glaciation (late Valdai) in Eurasia (Baulin and Danilova, 1984).

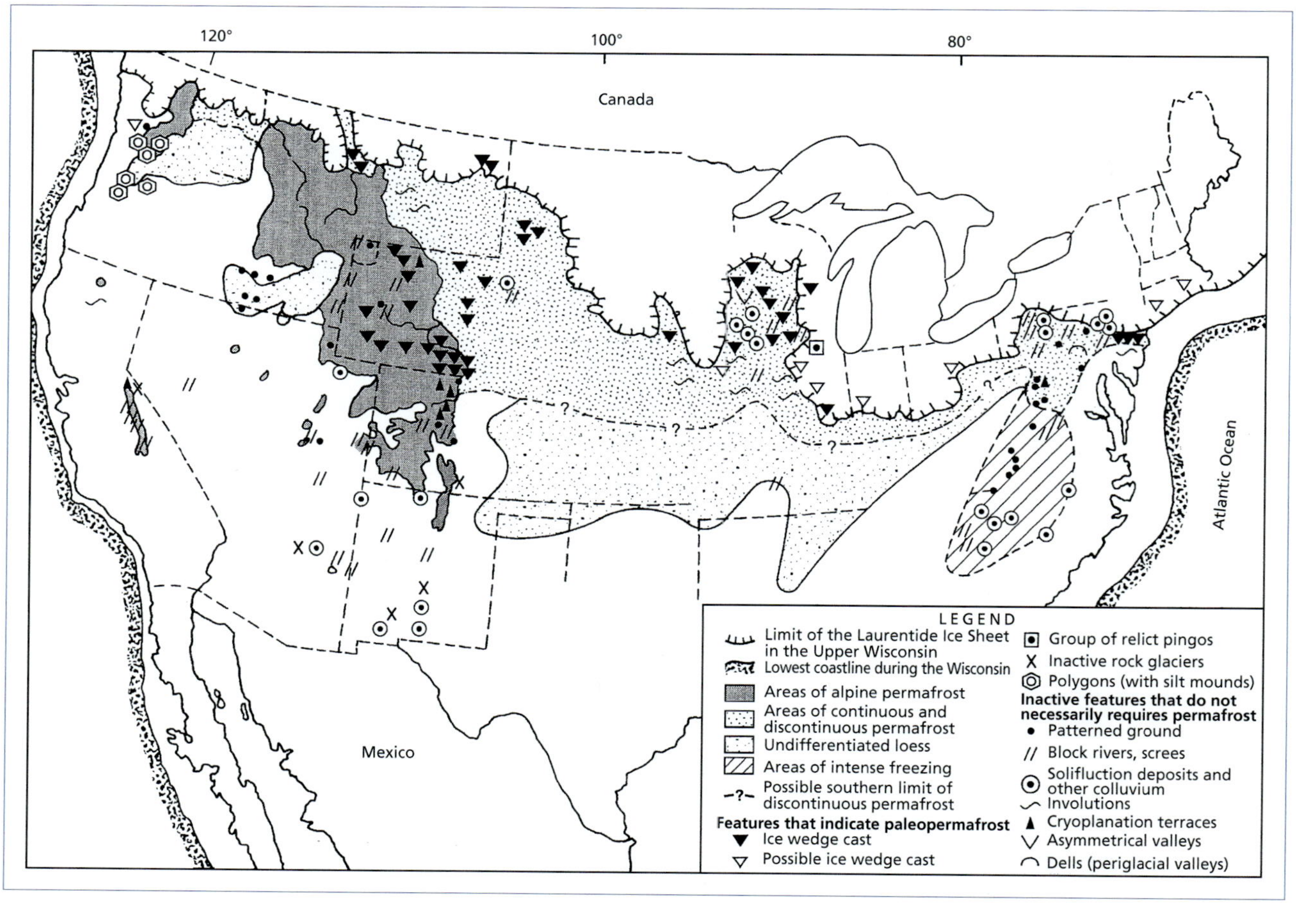

Figure 20.32 Reconstruction of the maximum extent of periglacial conditions south of the ice sheet during the late Wisconsin in the United States, (based on various authors, French, 1996).

The data on relict permafrost in **North America** shows that it dates from the last glacial maximum (late Wisconsin) because information on previous periods is lacking. The permafrost belt south of the Laurentide Ice Sheet is about 100 km wide, much narrower than that of the Eurasian Ice Sheet. The greater advance of the Laurentide Ice Sheet toward more southern latitudes accounts for this; as a result, the zone of intense periglacial conditions is more restricted than in Europe. *Figure 20.32* shows permafrost areas and landforms that indicate fossil permafrost and other inactive periglacial morphologies that do not require permafrost for their generation. Today permafrost is found beneath 82% of Alaska and northwestern Canada (Péwé, 1983). Numerous closed system pingos occur on the Alaska coast and many with open systems are present in the interior; they are all of Holocene age. Extensive cryoplanation surfaces occur in the center of the territory.

Permafrost appears to be less developed in the Southern Hemisphere because the landforms described there do not require permafrost conditions for their formation. Casts of ice wedges have been found only in the southern Patagonia of Argentina (Grosso and Corte, 1989).

20.11 Considerations of global climatic change in periglacial zones

The increased temperatures predicted to result from the greenhouse effect would significantly affect areas occupied by permafrost today. As stated above, permafrost covers about 25% of the land surface. Obviously, the permafrost will thin and its surface area will decrease. A significant increase of the active layer will be accompanied by substantial changes to hydrologic, geomorphic, and biological processes in permafrost regions, to which will be added subsidence processes affecting engineering projects. Based on general circulation models (GCMs), scientific paleo-reconstructions, and the assumption of a global temperature increase of 2°C, scientists conclude that the

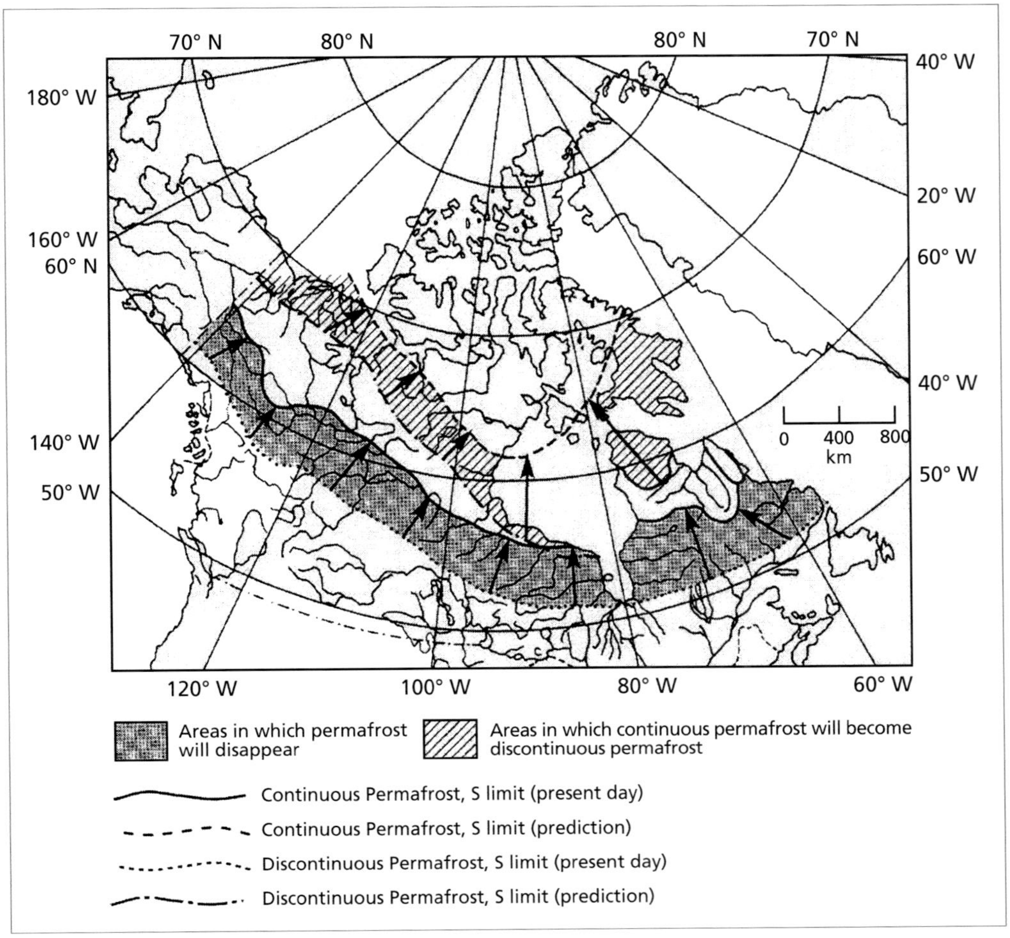

Figure 20.33 Changes in Canadian permafrost as a result of a surface temperature increase of 4°C (French, 1996).

extent of continuous permafrost will decrease by 29%–67% during the next century in the northern hemisphere (Anisimov and Nelson, 1996). In Canada, it is assumed that warming will be +4°C, which will cause the margins of continuous and discontinuous permafrost to shift 300 to 600 km to the north (*Fig. 20.33*). Permafrost areas are highly susceptible to global warming (Boer and de Groot, 1990; Smith, 1993). However, the application of the GCMS models does not lead to precise predictions but rather points to the direction of future change in permafrost areas (Pissart, 1990).

Various studies show that permafrost temperatures have increased over the last 20–30 years. Temperature records from 1970 to 1990 in arctic Europe and subarctic Russia indicate an increase of 0.6°C–0.7°C to three meters of depth (Pavlov, 1994). Similar values are reported in Tibet. As in other areas of the planet, warming is interpreted as being caused by global climate change or by the natural tendency of climate warming.

The general increase of the active layer caused by climate warming leads to additional warming. Large quantities of organic carbon are stored in the surface layers of permafrost. As the active layer thickens, this carbon will move into the atmosphere, basically as carbon dioxide and methane, which will increase greenhouse effect gases (Sturm *et al.*, 2001; Stokstad, 2004).

In addition, permafrost contains valuable information on past climate changes. Cryostratigraphy techniques (French, 1998; Murton, 2007), together with isotopic analysis, provide information on thickness variations of the active layer that have occurred over the last millennia.

21

CLIMATE CHANGE IN ARID AND TROPICAL HUMID REGIONS

1. Arid regions
2. Tropical regions

Rainforest, savanna, and semiarid regions have different morphogenetic processes and we study them separately but they have been affected by significant climate changes during the Cenozoic, as shown by paleoforms that occur anomalously in different environments. For example, eolian accumulations in the Congo rainforest and laterites in arid areas of the southern Niger Republic attest to major climate fluctuations.

This chapter focuses on the late Tertiary and Quaternary evolution of these areas based on an analysis of erosional and depositional landforms. Dating methods (isotopic, paleomagnetic, and thermoluminescence, among others) enable us to establish a sequence of events in time that makes it possible to correlate between different areas. This research has been used to study climate changes that have occurred in recent Earth history. This data also contributes valuable interpretations to future climate prediction (Petit-Maire, 1999).

21.1 Arid regions

The study of landforms in arid zones that are tectonically stable and lack significant anthropogenic activity shows that many existing landforms formed under geomorphic dynamics different from those that predominate today; consequently, we must consider them as paleoforms or inherited forms created under morphoclimatic conditions different from those of today. The impact of Pleistocene climate changes can be seen in all arid zones. Various indicators of these changes are present (Goudie, 2002) *(Table 21.1)*. For example, extensive dune fields covered by steppe vegetation in the Sahel region indicate marked aridity during the period of their formation. In addition, the intricate *wadi* network of the central Sahara no longer contains any runoff during flooding, which indicates that precipitation was higher when they formed (Tricart, 1969, 1977). These landforms indicate profound changes in the distribution of arid zones on the surface of the planet.

This type of evidence led to an old concept of a **pluvial period or lacustrine phase**, which can be defined as a period in which hydrologic and biogeographic conditions were more humid than at present, in contrast to a **interpluvial or arid period** when conditions were drier (Rognon, 1980). This definition accounts for water availability, which is controlled not only by precipitation but also by evapotranspiration. Thus, for the same amount of precipitation, a decrease in temperature makes more water available in the soil. Thus the dual factor of precipitation and temperature make it difficult to define a pluvial period precisely. The alternation of pluvial and interpluvial periods can be recognized in deserts by a set of indirect data of various types: geomorphic, archeological, pedogenic, palynological, hydrologic, and hydrogeologic (Demangeot, 1981; Tricart, 1969, 1977; Goudie, 1992; Thomas, 1997a) although the timing and correlation of these climate changes have occasionally been in dispute (Goudie, 2002).

21.1.1 Fluvial systems

Although precipitation is sparse in arid environments, surface runoff paradoxically plays a fundamental role in the origin of landforms in these areas. This is because sparse vegetation cover and the general lack of soils leaves the geological substrate unprotected and consequently subject to the impact of raindrops. Precipitation, though sparse, is frequently violent and is irregularly distributed over time. Thus semiarid Saharan regions have on average one flood per year but in more arid regions up to ten years can go by without water circulating in the *wadis*.

Many authors focus on the importance of climate change in the modification of geomorphic processes in fluvial systems. Schumm (1977) and Knox (1984b) discuss in detail the direct and indirect climate effects on these systems. In arid regions where plant cover is sparse, vegetation variations resulting from climate change can significantly modify the hydrology. Knox (1984a) set the critical threshold for plant cover at about 70%, which corresponds to mean annual precipitation of 400–500 mm. When plant cover is less than 70% the amount of erodible soil increases rapidly. These circumstances are clearly shown in the curve *(Fig. 21.1)* by Langbein and Schumm (1958) in which they relate mean annual precipitation to annual sediment production. In deserts, runoff increases as precipitation increases; furthermore, erosion increases until it reaches a maximum at about 300 mm of precipitation. When precipitation exceeds this amount, plant cover again predominates and sediment production decreases rapidly. Thus, in extreme deserts water erosion is low but in semiarid regions, storm rainfall frequently produces runoff and sediment production can reach high rates. The upper part of this bell curve is the geomorphologic threshold (Schumm, 1973, 1979, 2004). This graph also shows the importance of plant cover as the most effective way to reduce surface runoff and soil erosion. But these sediment-generating maxima also depend on temperature (Schumm, 1965) *(Fig. 21.2)*; as

Table 21.1 **EVIDENCE OF CLIMATE CHANGE IN DESERTS (GOUDIE, 2002)**

Evidence	Inference
1. Geomorphological	
Fossil dune systems	Past aridity
Breaching of dunes by rivers	Increased humidity
Discordant dune trends	Changed in wind direction
Lake shorelines	Balance of hydrological input and output
Old drainage lines	Integrated hydrological network
Fluvial aggradation and siltation	Dessication
Colluvial Deposition	Reduced vegetation cover and stream flushing
Karstic phenomena (caves, for example)	Increased hydrological activity
Frost screes	Paleotemperature
2. Sedimentological	
Lake floor sediments	Degree of salinity, etc.
Lee dune (lunette) stratigraphy	Hydrological status of lake basin
Spring deposits and tufas	Groundwater activity
Duricrusts and paleosols	Chemical weathering under humid conditions
Dust and river sediments in ocean cores	Amount of eolian and fluvial transport
Loess profiles and paleosols	Aridity and stability
3. Biological and miscellaneous	
Macroplant remains, including charcoal	Vegetation cover
(e.g. in Packrat middens)	
Pollen analysis of sediments	Vegetation cover
Faunal remains	Biomes
Disjunct faunas	Biomes
Isotopic composition of groundwater and speleothems	Paleotemperatures and recharge rates
Distribution of archaeological sites	Availability of water
Drought and famine record	Aridity

temperature increases, erosional maxima occur with higher precipitation. That is, at high temperatures and with greater evapotranspiration, more rainfall is needed to maintain the percentage of plant cover, so there is less runoff and maximum sediment production shifts to the right.

Semiarid regions are areas of intense erosion, as we have seen; floods occur much more frequently because rainwater quickly runs off the surface. In basins with distinct drainage areas *(Fig. 21.3)*, flood magnitude increases considerably as mean annual precipitation decreases from about 500 mm to approximately 375 mm (Knox, 1972). This means that a climate change from relatively humid to dry will cause a channel to enlarge to accommodate larger floods.

In addition, drainage network density tends to increase in semiarid regions and to decrease as precipitation increases (Gregory, 1976) *(Fig. 21.4)*. The graph shows an abrupt decrease in density at 800 mm. Variations are due to various types of rock, relief, and vegetation.

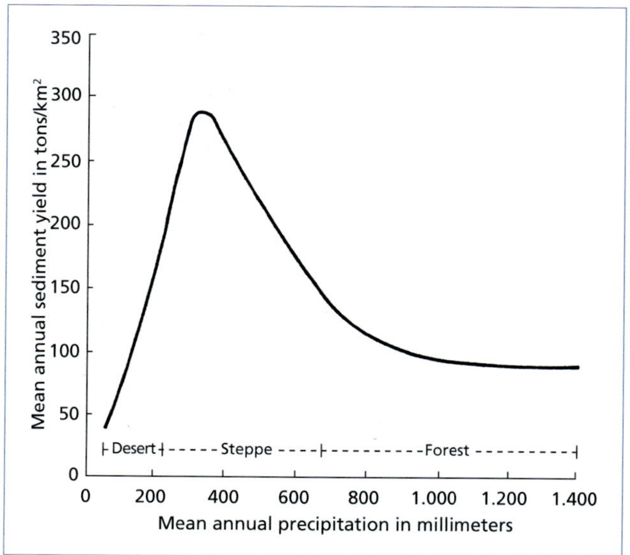

Figure 21.1 Variation of sediment yield in fluvial systems as a function of mean annual precipitation (Langbein and Schumm, 1958).

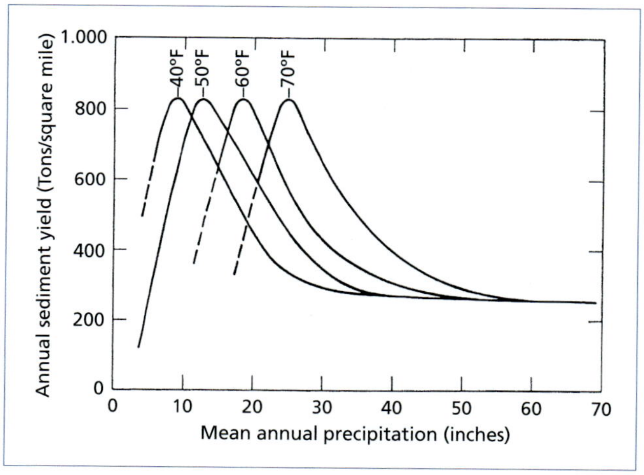

Figure 21.2 Effect of temperature on sediment yield and mean annual precipitation (Schumm, 1965).

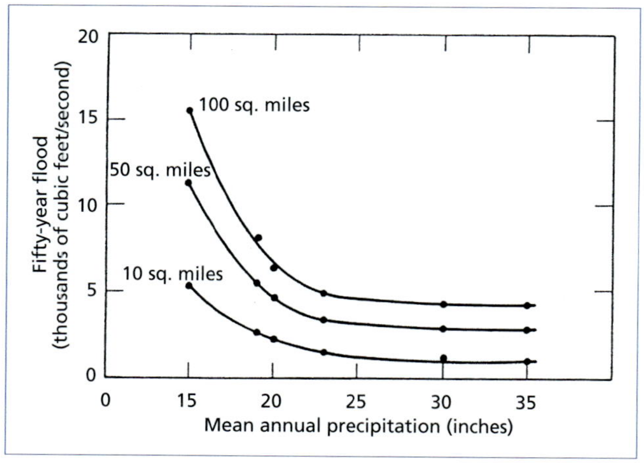

Figure 21.3 Relationship between the magnitude of a flood with a 50 year return period and mean annual precipitation, for three drainage basins of different sizes (Knox, 1972).

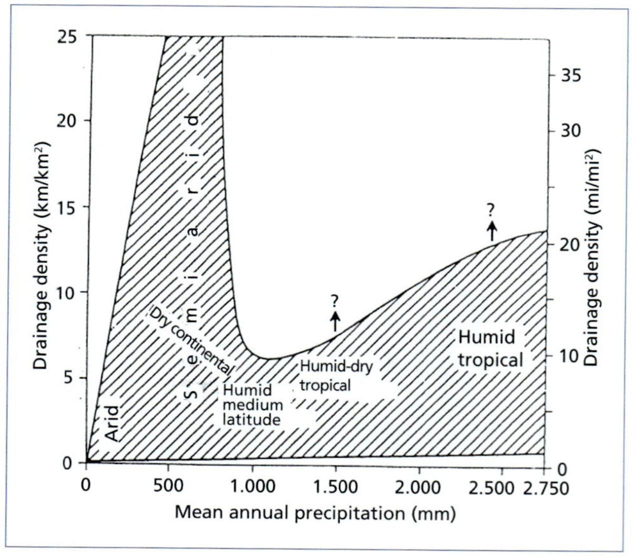

Figure 21.4 Relationship between drainage density and mean annual precipitation (Gregory, 1976, modified by Chorley *et al.*, 1984).

These examples illustrate that landforms in these areas are sensitive to climate change. In arid regions, vegetation is more sensitive to climate and a slight modification of plant cover can cause significant hydrologic changes. Consequently the intensity of erosion can vary substantially in these regions in response to subtle climate changes (Knox, 1993).

Temperature and rainfall oscillations that have occurred throughout the Holocene (the last 10,000 years) have been minor in magnitude (±2°C and ±10–20% of rainfall) (Lamb, 1977) but they have been sufficiently important to affect the intensity of many geomorphic processes that determine the morphological stability of landforms, especially fluvial systems (Knox, 1984b).

21.1.1.1 Slopes and their evolution

The slope profile characteristic of arid areas is most easily recognized in stratified rocks. The profile contains an upper part that is a free slope, a middle part that consists of a slope partially or completely covered by debris, and at the foot is a slight basal concavity where a pediment develops *(Fig. 17.11)* (King, 1957b, 1962).

Slope research throughout most of the United States shows the influence of climate on the evolution and morphology of these profiles (Toy, 1977). To minimize the influence of geology and structure, Toy focused his research on marine shales with dips less than 5°. From this research he determined that slopes in arid zones are shorter, more abrupt, and have a smaller radius of curvature than those in humid regions (Fig. 21.5).

Peña *et al.* (1996) discuss an example of slope evolution during the late Holocene in a semiarid area in the center of the Ebro Depression. They established a series of stages based on periods of accumulation and incision that are supported by archeological research and dating (Fig. 21.6). At the end of the Bronze Age, humans established settlements at high elevations. Many have been excavated and the slopes are covered with deposits. In terms of climate, this stage corresponds to the end of the Subboreal, a period that was cooler and more humid than today (Lamb, 1977; Bintliff, 1982). Settlement on top of hills and slope accumulations continued during the Iberian-Iron stage, and slope incision began at the end of this stage (Burillo *et al.*, 1981a, b, 1983, 1986; Gutiérrez and Peña, 1989, 1992, 1998); rilling became widespread during the Iberian period. The Iron Age climate was humid and cold. Gribbin and Lamb (1978) call it the Cold Epoch of the Ice Age; from then on the climate began to change, initiating the Subatlantic (Wenland and Brison, 1974; Lumley *et al.*, 1976; Lamb, 1977).

During the Roman period the type of settlement changed and scattered houses or villas were built near cultivated land; in slope morphogenesis incision predominated, creating gullies and paleoslopes or triangular slope facets (flatirons) that acted as isolated landforms. In this way, slope erosion was accompanied by the reduction of triangular facets. Arabs, who tended to settle on protected peaks, arrived during the medieval period. The Subatlantic Period includes the maximum warming of medieval times (Gribbin and Lamb, 1978) called the Small Optimum.

These forms and the resulting processes can be seen on slopes and valley fill in both the Ebro Depression (van Zuidam, 1976a, 1976b; Burillo *et al.*, 1985, 1986; Pellicer *et al.*, 1986; Peña and González, 1992; Soriano and Calvo, 1987; Soriano, 1989; Peña *et al.*, 1993; Arauzo and Gutiérrez, 1994; Harvey and Gutiérrez, 2005) and in the Iberian Cordillera (Burillo *et al.*, 1981a, b, 1983). This research shows a good climatic correlation between slope accumulations and the end of Bronze-Iron period. In contrast, although climate correlation can be made locally in valley fill, no correlation exists in many valleys (diachronism), possibly due to intense anthropogenic pressure on the environment.

On the other hand, the lesser solubility of limestones generally prevents the development of karst morphologies. Investigators do not think that karst landforms develop in areas with than 250–300 mm of precipitation (Sweeting, 1972). Studies on present-day dissolution rates of limestone gravestones in Haifa (Klein, 1984) and Jerusalem (Danin, 1983) indicate a weathering rate of about 0.005 mm/year although this dissolution may be influenced by urban pollution and microbiological activity. These results show that dissolution is very low in spite of about 600 mm of precipitation in both cities. More to the east in the Judean Desert (Israel), slopes developed on upper Cretaceous limestones and marls are convex-concave profiles, dissected by fluvial erosion at their base, and covered with a thin layer of debris (Fig. 21.7). Below them are well-developed karren of the *kluftkarren* type (structural karren) (Fig. 21.8), *hohlkarren*, and *kamenitzas* in areas where

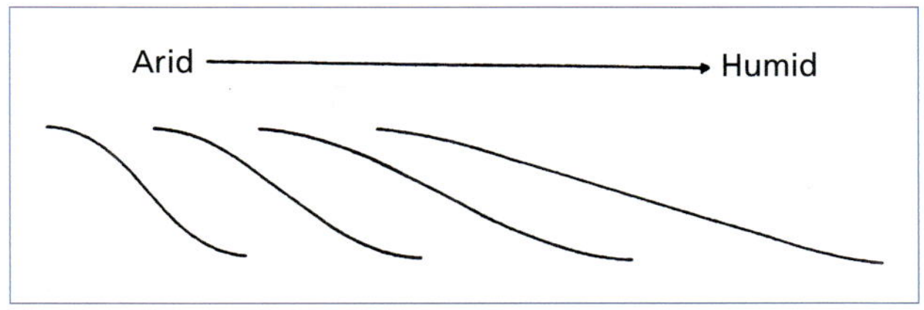

Figure 21.5 Schematic slope profiles for different climates (Toy, 1977).

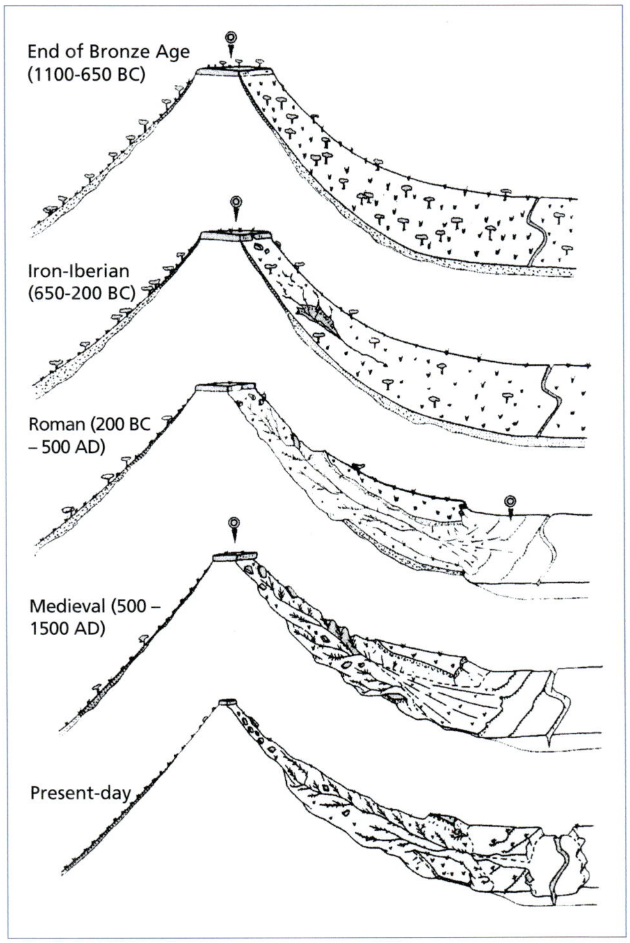

Figure 21.6 Diagram of slope evolution and valley fill from the Bronze Age to today. The symbols indicate the most common movements for human settlements during various periods (modified from Peña *et al.*, 1996).

Figure 21.7 Smoothed slopes on upper Cretaceous limestones and marls that crop out as a result of gully incision. The convex-concave profile shows a previous stage that was more humid than the present (260 mm). Judea Desert. Israel.

Figure 21.8 Structural karren (*kluftkarren*) exhumed from detrital slope cover with cracks greatly widened by dissolution in an area with 100 mm of present day precipitation. Kalia, Midbar Yehuda. Judea Desert (Israel).

Figure 21.9 Cracks in structural karren filled with secondary $CaCO_3$ with a leafy structure. Kalia. Midbar Yehuda. Judea Desert (Israel).

precipitation is less than 100 mm. In many places, cracks in the structural karren appear to be totally filled with secondary $CaCO_3$ that has a leafy structure (*Fig. 21.9*) These circumstances seem to indicate an older stage with much higher precipitation than today and another later, more arid period of carbonate precipitation between the fractures in the karren.

21.1.1.2 Talus flatirons

Our knowledge of the processes that act on slopes in arid and semiarid environments has increased considerably in recent decades. However, research on slope evolution is limited because of the difficulty of dating that temporally pinpoints different relict slopes. However, in semiarid environments with a greater percentage of plant cover and where human occupation has continued since antiquity, it is possible to find carbonaceous remains and archeological materials in deposits that make an age determination possible. In addition the OSL technique is being used more often in dating.

Slope accumulations can be affected after formation by a predominantly incision stage, which may destroy the entire accumulation or produce a relict slope called a *talus flatiron* (Koons, 1955). These are also called *tripartite slopes* (Gossmann, 1976) and *triangular slope facets* (Büdel, 1977, 1982). These landforms are triangular or trapezoidal in plan view and are located near but separated from a scarp (*Figs. 17.13, 17.14* and *21.10*). As a result, facet formation requires an alternation of accumulative and erosional stages during slope evolution. The occurrence of several of these stages can create a sequence of *talus flatirons*; in these sequences the most recent flatirons are located near the scarp. These relict forms are not common throughout the world because they occur preferentially in subhorizontal stratified formations where the scarp overlies easily erodible materials that can be rapidly removed by entrenching processes (Koons, 1955; Everard, 1963; Büdel, 1970; Blume and Barth, 1972; Gerson, 1982; Gerson and Grossman, 1987; Schmidt, 1989a, 1994; Arauzo *et al.*, 1996a). In addition, they occur where the scarp is thin and thus the accumulation of debris is not excessive (Schmidt, 1987); thus, destruction and scarp retreat are rapid. In addition, alternation of hard and soft layers under the scarp does not promote talus flatiron formation.

With this knowledge of factors that promote talus flatiron formation, we can discuss the causes that generate them. Koons (1955) interpreted talus flatirons as an uneven balance between detrital supply to a slope and removal from it (*Fig. 17.16*). An alternative interpretation to this acyclic origin is based on modification of the dominant morphogenetic processes as a result of climate change (Everard, 1963; Gerson, 1982; Gerson and Grossman, 1987; Sancho *et al.*, 1988; Schmidt, 1989, 1994; Arauzo *et al.*, 1996a, b; Gutiérrez *et al.*, 1998a, b; Gutiérrez, 2004; Gutiérrez *et al.*, 2006; Gutiérrez

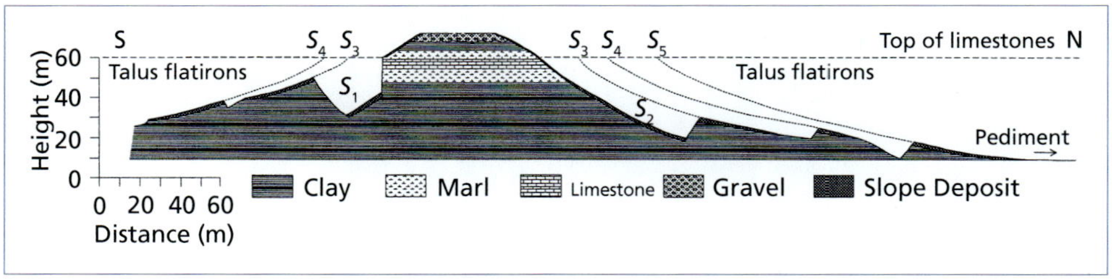

Figure 21.10 Schematic section of Miocene materials and a sequence of five stages of slope evolution. Cerro de San Pablo, Villanueva de Huerva (Province of Zaragoza). Ebro Depression, Spain (Gutiérrez *et al.*, 1998a).

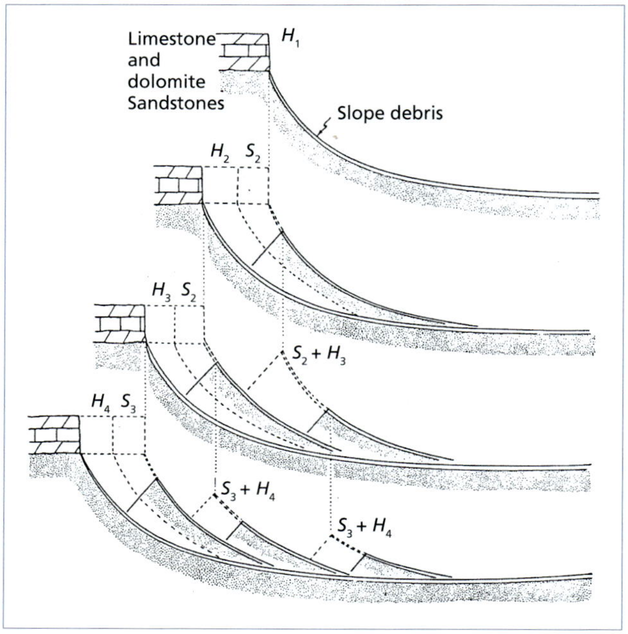

Figure 21.11 Evolution of a sequence of talus flatirons. H = Pluvial regime. S = Interpluvial regime. The flatiron shows the end of pluvial periods (Gerson, 1982).

et al., 2010) (*Fig. 21.11*). During recent periods the impact of anthropogenic activity may be significant in the modification of geomorphic processes (Everard, 1963; Sancho *et al.*, 1988; Gutiérrez and Peña, 1989, 1992, 1998; Arauzo *et al.*, 1996a, b; Gutiérrez *et al.*, 1998a, b).

The importance of climate change has been amply discussed for various types of landforms. In desert regions with sparse plant cover, variations of the cover due to climate change can lead to significant modifications of hydrologic conditions. Climate characteristics in the center of the Ebro Depression coincide approximately with the points of maximum sediment production on Schumm's (1965) curve. According to this diagram, slight variations of temperature and annual precipitation can modify the percentage of plant cover, and as a result, can cause changes in sediment production.

In addition, during more humid periods (pluvials) sediment production processes are more active in the scarp, especially process of undercutting and slumping. This results from increased infiltration and underground flow on top of impermeable layers (Ahnert, 1960). When temperature decreases, the annual number of frozen days increases and congelifraction increases sediment production on the scarp.

A climatic interpretation of accumulation and incision stages leads us to establish possible correlations with known climate sequences. The two most recent stages of slope accumulation are widely represented throughout northeastern Spain (Gutiérrez and Peña, 1989, 1998) and archeological and radiometric techniques have been used to date them at multiple locations. The most recent accumulation, the Post-Medieval, correlates with the Little Iron Age (1,450–1,850 AD) (Grove, 2004). The previous accumulation stage occurred during the Cool Period of the Ice Age (2.900–2.300 BP) (Lamb, 1977; Gribbin and Lamb, 1978; Grove, 1979; Harding, 1982). Gutiérrez and Peña (1989, 1992, 1998) and Gutiérrez *et al.* (2003, 2006) correlated two accumulation stages with two cool periods in the northeastern part of the Iberian Peninsula (*Fig. 21.12*).

Researchers used C^{14} analyses to determine the first dates of the oldest of these accumulations in Mezalocha and Chalamera (Province of Zaragoza) (Gutiérrez *et al.*, 1998a, 2006). These dates correlate well with the Heinrich events established through research on deep ocean floor cores from the North Atlantic. The ages of the accumulations may correspond to H3 (27.8–28.5 ka BP) and H4 (35 ka BP) events (Heinrich, 1988; Bond *et al*, 1993, 1997, 1999). A Heinrich event is the presence of glacial origin sediments from the melting of icebergs that contained sparse foraminifera. The formation stages of these events are related to an enormous profusion of icebergs resulting from rapid breakup of northern ice sheets. These circumstances are accompanied by overall cooling of ocean surface water. As a result, cool water from the Atlantic entered the Ebro Depression. The temperature drop caused evapotranspiration to decrease; this increased

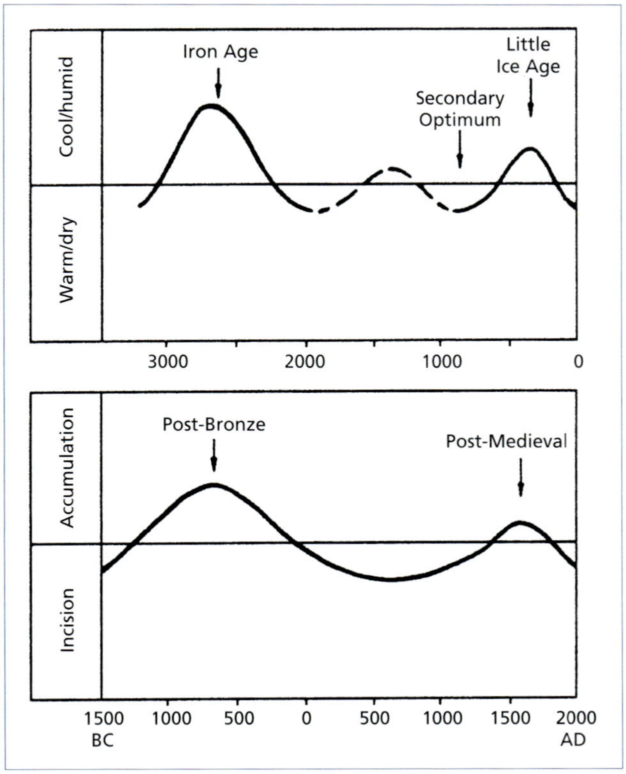

Figure 21.12 Relationship between accumulative and erosional stages in slope deposits in northeastern Spain and the climate periods of the late Holocene (Gutiérrez and Peña, 1989).

the water available in the soil, which led to increased plant cover. These conditions seem to have been suitable for the sediment accumulation on slopes. Heinrich events occur during periods of extreme atmospheric cooling and are followed by clear but short interstadials (Dansgaard *et al.*, 1993).

This implies that accumulation stages coincide with cool climatic periods, whereas incision stages correspond to warm climatic periods, which are periods of talus flatiron formation. For this reason, talus flatirons seem to record climate changes.

In these semiarid environments, slope system changes that cause *talus flatirons* occur over relatively brief time periods. However, areas with more pronounced aridity (*Fig. 21.13*) require climate changes on the order of 100 ka to generate these relict slopes (Gerson, 1982; Gerson and Grossman, 1987; Bull, 1991; Schmidt, 1994, 1996). Research on talus flatirons in Fuerteventura (Canary Islands) where precipitation is 150 mm does not corroborate the previous assertion (Gutiérrez *et al.*, 2011a) because OSL ages of flatirons are 29778 ± 1984 BP and 51842 ± 3874 BP.

Gutiérrez *et al.* (2009a) dated 31 samples from flatiron slopes (8 dates by ^{14}C and AMS and 23 by OSL on talus flatirons in the large Tertiary depressions of the Ebro, Duero, and Tajo (Spain). They interpret them to have a climatic origin although the groups do not coincide exactly with Heinrich events (H_3, H_4 and H_5), quite possibly due to the duration of each Heinrich event and especially to the large margin of error of the OSL dates, which increases their age.

Figure 21.13 Sequence of talus flatirons. Makhtésh Ramon. Negev Desert (Israel).

21.1.1.3 Ramblas and arroyos

Rambla is a term used primarily for ephemeral streams in semiarid zones in the Spanish Mediterranean basin. These streams have flat floors and vertical walls; they fill with water only during major rain events that generate floods that move significant solid loads (Segura, 1990). **Arroyo** is a word introduced by the Spanish colonists of the western United States deserts; it is used for streams 5 to 200 km long with deep cross sections and flat floors. Arroyos are intermittent streams that fill as a result of sporadic storms; they incise into material that is generally unconsolidated (American Geological Institute, 1972; Bull, 1997; Tchakerian, 1994; Goudie, 2004g). Different stages of accumulation and incision can be seen in these alluvial systems (*Figs. 21.14* and *21.15*).

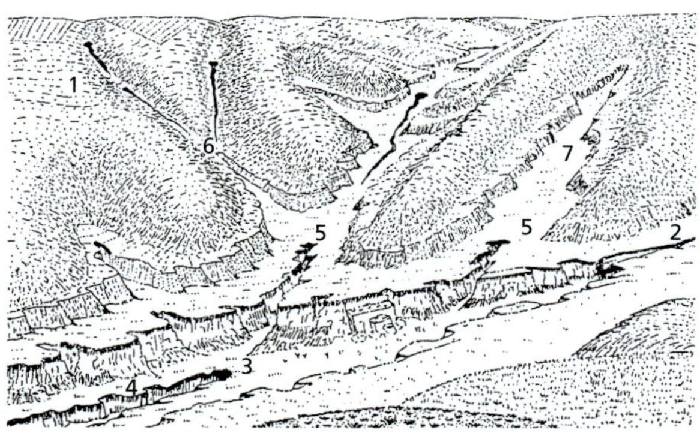

Figure 21.14 Scheme obtained from various photographs showing discontinuous gullies (1) formed in tributary valleys that coalesce by headward erosion (6). There are two *knickpoints* in these valleys (2,3). The entrenching of the valley floor produced terraces (4,5). Valley fill (7) formed/resulting from processes of lateral and longitudinal erosion (Brice, 1966).

These geomorphic systems have been studied most intensely in North American deserts; these studies have produced advanced knowledge of hydrology, paleo-hydrology, geomorphology, and climate change. Discontinuous ephemeral streams deposit valley fill in areas where precipitation ranges from 100 to 500 mm (Bull, 1997). Temperature controls runoff by its influence on evapotranspiration and on vegetation density and type. These effects are more pronounced in arid zones than in subhumid and humid areas (Langbein *et al.*, 1949). The material deposited in ephemeral streams consists primarily of sand, silt, and clay that come from slope erosion. Later, these particles become mobile except clay layers that decrease infiltration and increase cohesion. For a given tractional force, there is a minimum biomass value below which the valley floor becomes unstable and begins to be incised (Schumm, 1980).

Figure 21.15 Terrace formed in valley fill by entrenching of the Villafranca Gully. Villafranca de Ebro (Province of Zaragoza). Ebro Depression, Spain.

In addition, anthropogenic activity is significant because humans modify vegetation cover in many ways: overgrazing, overfarming, deforestation, and burning, among others. Increased biomass affects a channel's resistance to erosion (Graf, 1979). In this regard, Cooke and Reeves (1976) show how the introduction of herbivorous animals into New Mexico and Arizona triggered significant soil erosion.

Multiple stepped levels can be recognized (*Fig. 16.74*) within the sediments that fill ramblas or arroyos; these result from alternating periods of aggradation and incision (Harvey and Gutiérrez-Elorza, 2005). In the central Ebro Depression, an intricate dendritic network of flat-floored valleys (locally called "vales") have formed on, and appear to have incised into, Miocene gypsiferous marls (*Fig. 21.16*); the downcutting shows various levels of Holocene age sediments (Peña *et al.*, 1993; Arauzo and Gutiérrez, 1994). Periods that lack sedimentation can also be identified in these ephemeral streams; they appear in the record as incipient soil profiles. As a result, ephemeral streams are characterized by periods of prolonged sedimentation that alternate with brief intervals of shallow incised channels and the development of entisols (Leopold, 1994).

The aggradation process is followed by a degradation stage that involves incision of the rambla; this begins with rapid incision that decreases exponentially as it approaches a new base level of erosion (Bull, 1997). Cooke and Reeves (1976) proposed a model for southern Arizona based on multiple causes for the initiation of arroyo incision; the model involves the interaction of numerous variables. The principal variables are land use changes, random variations in the frequency and magnitude of processes, and secular climate change. The first and last seem to be the principal causes that initiate arroyos. However, their origin is a classic example of equifinality or convergence of landforms; it is difficult to pinpoint a single mechanism (Cooke and Reeves, 1976).

Research on the origin of arroyo incision is complicated by the difficulty of separating anthropogenic activity from secular or short-term climate changes. An easier way to approach the problem is to point out that erosion increases with more runoff and in arroyo sediments that have less resistance to erosion. In either case vegetation seems to be the key in the initiation and development of arroyos. Plant cover on slopes and riverbanks (*Fig. 21.17*) responds rapidly to temperature changes, precipitation, and anthropogenic interference. Changes in **land use** include overgrazing in arid zones with silty soils, which decreases plant cover, intense trampling, increased impact and splashing by raindrops, and the resulting decrease in infiltration capacity. All of these, together with construction practices (roads, railroads, bridges), deforestation, and fires are the principal triggering modifications caused by human activity. Two examples illustrate these anthropogenic causes.

Figure 21.16 Aerial view of the dendritic network carved into the Miocene gypsum and marls on the flat valley floor on cultivated anthropic terraces. Center of the Ebro Depression, Spain.

Figure 21.17 Rambla or arroyo with channel bed covered by shrub vegetation. The walls are vertical and are composed of clayed silt. Barranco Grande. Las Bardenas Reales (Navarre, Spain).

Thornes (1976) studied the Almazán Basin (Soria, Spain) and showed that overgrazing due either to ranches owned by the local population or as a result of being located on a migration route toward summer pasture in the Urbión Mountains "is a significant factor in the intense erosion of these areas" that caused much of the steppe vegetation to disappear. This disappearance has triggered erosion and the formation of gullies that incise into Tertiary deposits composed of easily erodible materials.

A similar problem arose during the colonization of the southwestern United States. Colonists arrived in great numbers, established farms, built roads, and moved tens of thousands of head of cattle to the land. These circumstances caused much of the existing vegetation to disappear and initiated intense erosion that incised gullies more than 50 m deep within a few years. This erosive process took place between 1850 and 1920 and especially between 1870 and 1890. The colonists were forced to move again and migrate to more favorable areas. This environmental change of anthropogenic origin has been recognized for many years; Cooke and Reeves (1976) wrote an excellent monograph about it.

Other authors (Bryan, 1941; Antevs, 1952) propose a mixed origin for incision; they point out that vegetation was in bad shape at the end of the last century due to climate change and that overgrazing is added. This hypothesis is supported by dendrochronological research that provides evidence for a period of low precipitation (Stockton and Fritts, 1971).

Another model proposed by Cooke and Reeves (1976) is that of **random variation in the frequency and magnitude of processes**. One cause of incision is tectonic lowering of the base level (Bull and Pearthree, 1988). Another involves erosional capacity, which increases exponentially with increased discharge (Leopold *et al.*, 1964). This model proposes that flooding by storms causes incision whereas aggradation predominates during low-flow periods. Schumm and Hadley (1957) suggest that they are due to internal adjustments in the fluvial system. In small basins less than about 50 km^2, sedimentation can result from the dissipation of flow caused by infiltration, evaporation, or other causes. These authors generalize these changes and propose a semiarid erosional cycle in which they contemplate the evolution of aggradation and incision in small drainage basins (*Fig. 21.18*).

The third model proposed as a cause of incision is **minor and secular climate changes**. Ample evidence shows that erosion and deposition in these systems respond over short time periods to climate change and human activity. Leopold (1976) used instruments to analyze modifications to ephemeral streams in New Mexico over a period of 15 years; he determined that arroyos that had eroded for an entire century

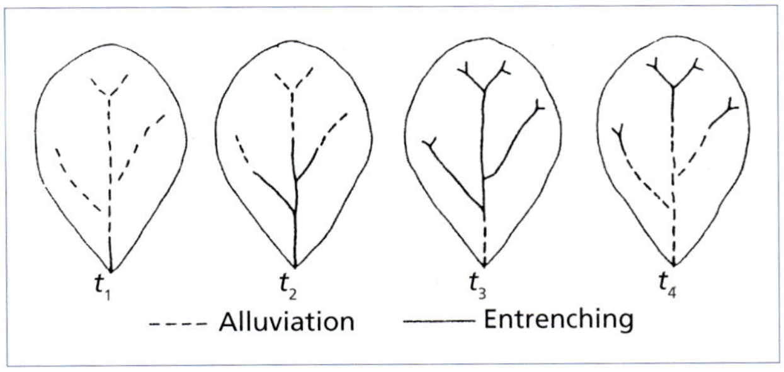

Figure 21.18 Cycle of alluviation and entrenching in small semiarid drainage basins (Schumm and Hadley, 1957).

began to aggrade in 1961. He attributed the change to cooling after 1940. Hereford (1984) describes a similar situation in Arizona. In addition, hydrologic and geomorphic responses to El Niño oscillations may cause droughts (Andrade and Sellers, 1988) or major floods in the southwestern United States. We must keep in mind that this climate disturbance is continuous. Harvey and Gutiérrez-Elorza (2005) studied Holocene valley fill in the eastern Ebro Depression; they identified stages of accumulation and incision (*Fig. 16.74*). They distinguish two principal filling events that represent periods of aggradation; they dated them at 6000 cal years BP and from the Iron Age, with a dominant period of incision between them. These variations are interpreted as resulting from climate change.

There is no doubt that incision of streams increased significantly during the colonization of semiarid zones of the southwestern United States, South Africa, India, and Australia (Cooke *et al.*, 1993). However, in prehistoric periods the alternation of aggradation and incision seems to have been due to climate change (Antevs, 1952). Other researchers, using dendrochronological research on records of thousands of years, have identified the alternation of dry and humid periods related to stages of entrenching and deposition. ^{14}C dating and archeological remains have produced similar conclusions.

On the basis of numerous investigations of alluvial deposits in the southwestern United States, Leopold (1994) compiled a summary of climate history for the period after the retreat of the Wisconsin ice. He identified a period of aggradation starting at the last glaciation to the Altithermal Period or Climate Maximum of about 4000–6000 years ago. Arid conditions during this period triggered broad erosion of valley fill that formed terraces up to 10 m above the riverbed. This was followed by another period of aggradation that ended about 1200–1400 AD (Medieval Warm Period or Second Climate Optimum). Incision of previously deposited fluvial sediments recurred during this period. During the humid and cool stage of the Little Ice Age (1450–1850 AD), aggradation resumed. Between 1880 and 1920 increasing aridity eroded the valleys.

This proposal is general in nature and follows existing dates but is not universally applicable. The sequence of aggradation and erosional stages may be different for headward tributaries than for main channels (Balling and Wells, 1990). In addition, degradation stages probably do not occur simultaneously across a broad region because each fluvial system has a different sensitivity to climatic disturbances (Bull, 1997). Bull calculates that incision requires at least 100 years but that aggradation in incised channels takes more than 500 years.

In summary, it is highly simplistic to attribute the beginning of channel incision to a single cause such as a temperature increase, for example. Modification of vegetation density on slopes plays an important role in incision. Vegetation can be modified by changes in precipitation or temperature, overgrazing, fires, and other human activities. Incision begins when these variables cause sufficient disequilibrium in a system to trigger and maintain the downcutting process. However, some explanations such as climate change are regional in scope whereas others are only locally significant. Due to the complexity of the problems involving ramblas or arroyos, there are defenders and detractors of each hypothesis; however, we must keep in mind that different changes in a single zone may lead to similar geomorphic results.

21.1.1.4 Alluvial fans

Alluvial fans are semiconical landforms created when confined water and sediment flows reach the foot of a mountain; here the current loses its transport power, flow becomes unconfined, and the current deposits its predominantly coarse sediment load. Although alluvial fans form in many climate environments (Rachocki, 1981) our purpose is to discuss climate changes that have affected these fans, which are now located in arid zones.

The study of alluvial fans advanced greatly with the application of sedimentological techniques (Mayer *et al.*, 1984; Grossman and Gerson, 1987), paleosoil analysis (Bull, 1991), charcoal dating, and rock varnish methods (Dorn, 1994a). These methods allow us to infer climate features from alluvial fan deposits and paleosoils and to correlate within the fan and regionally. We can use this information to determine the influence of climate on alluvial fan formation.

Some authors consider the fundamental stages of fan aggradation to have occurred during humid periods (Glennie, 1970; Dorn *et al.*, 1987; Maizels, 1987) whereas others think that deposition increases during more arid stages (Wells *et al.*, 1987; Bull, 1991). Other researchers take an intermediate position and point out that both climate conditions cause aggradation (Williams, 1973; Ponti, 1985). All this points to the complexity of interpretation and the extreme difficulty of correlation.

Four conceptual models have been proposed to relate climate change to alluvial fan development (Dorn, 1994b). A **transition to a drier climate** causes major denudation of deposits accumulated on slopes and valleys in the drainage basin and deposits that were previously held in place by more abundant plant cover during an aggradation stage (Knox, 1984b; Wells *et al.*, 1987; Harvey, 1990, 1997; Bull, 1991). The **paraglacial model** focuses on the large amount of easily erodible material and the large volumes of water liberated during a melt period. This material affects fans because glaciation produces large amounts of sediments that supply alluvial fans; furthermore, these circumstances can persist for several thousand years after the glacial stage. The deserts of the western United States and Central Asia contain numerous examples of alluvial fan drainage basins that have been subject to glacial activity. Another alternative model, which was discussed previously, is the direct relationship between **a humid stage** and **aggradation**. Some authors relate it to glacial conditions in which water is more available (Bull, 1991). Finally, **the periglacial model** proposes significant production of gelifracts that erode and cause aggradation in an alluvial fan (Williams, 1973; Wasson, 1977). These periglacial processes are quite obvious in some mountains where alluvial fan drainage basins develop. In southeastern Spain these fundamental aggradation periods coincide with Quaternary cool phases and dissection coincides with periods of minor sediment supply, which correspond to warmer phases (Harvey, 1978a, 1990, 1996, 1997).

An alluvial fan forms over a long period of time and as a result, climate variables can change during its formation. In spite of the variety of the proposed models, it remains difficult to separate climatic influences from other "noise" such as tectonics or factors intrinsic to the system. However, analysis of the paleovegetation record in the Mojave Desert has resulted in climatic correlations between different areas of alluvial fans (Spaulding, 1990). This type of correlation was done in areas that may have highly variable recent tectonic histories.

Today, researchers agree that alluvial fans form in arid areas as a result of high magnitude low frequency storms (Beaty, 1974; Baker, 1977; Dorn, 1994b; Blair and McPherson, 1994a, b; Harvey, 1997) that have occurred throughout all climatic periods of history. For this reason, it is difficult to predict the effects of climate change (Harvey, 1996). As this author points out, threshold conditions and sensitivity to environmental change (Brunsden and Thornes, 1979) vary considerably from one fan to another. In the semiarid zone of southeastern Spain, storm frequency increased significantly during the period of 1985–1990. Erosion was limited to the apex of the fans where marked entrenching took place; however, the eroded sediment was deposited downstream without any evidence of new incision in the medial and distal zones of the alluvial fans (Harvey, 1996). For these reasons, reconstruction of evolutionary history or data on climate changes that have affected alluvial fans is highly complex (Blair and McPherson, 1994a, b).

21.1.1.5 Desert rivers

Changes in channel shape and deposit type are the primary tools used to determine environmental changes. During historic periods numerous examples of human intervention are known to have triggered important variations in both morphology and fluvial sediments (see Chapter 9). Neotectonic activity is another factor that may affect some zones through the modification and intensity of fluvial processes. To avoid the "noise" of such activity, geomorphic investigations are conducted in cratonic areas of geologic regions that are tectonically stable. Rivers are relatively insensitive to climate change unless the changes are sufficiently notable (Reid, 1994). However, rivers in arid and humid regions are clearly different, not only in their shape but also in the solid load they transport (Wolman and Gerson, 1978). In arid zones, few hydrologic records are available and on occasion they are totally lacking; for this reason recent variations and their relationship to climate are not known. Some researchers, such as Schumm and Lichty (1963) infer climate changes from studies of temporal channel variations. One problem with drainage basins is the response of fluvial system variables to climate change (Schumm and Lichty, 1965). These variables are interrelated; climate affects primarily weathering, pedogenesis, vegetation, and hydrology (Schumm, 1977). In addition variables that are susceptible to change can be more active in one basin than in another.

Modifications to the fluvial system can result from changes in average climate characteristics and also from modifications of the ordinary magnitude-frequency of various events (Knighton, 1998). Small climate changes have also been responsible for variations in fluvial systems (Knox, 1984b). Research on Mississippi River floods over the last 7,000 years shows a warm and dry stage between 5,000 and 3,000 years BP with lesser floods and an abrupt change to a cooler and more humid stage and a response involving larger floods (Knox, 1993). In addition, Knox claims that the most significant floods of the last 9,000 years result from periodic changes in the atmospheric circulation system (Knox, 1995).

When climate change causes discharge to increase, channel width, depth, and slope increase; the opposite happens when discharge decreases. If sediment load increases, erosion increases in the drainage basin; increased sediment can result from climate change toward greater aridity or from deforestation. Decreased solid load can result from climate change to more humid circumstances or from a change in soil conservation practices. In general, any change in discharge is accompanied by a change of sediment type. When climate changes toward more humidity, the percentage of plant cover increases and soil loss decreases as a result. Increased precipitation changes ephemeral streams to perennial, increases drainage density, and promotes erosion on tributaries to the principal river with a resulting increase of their solid load especially during storms. If the climate returns to semiarid conditions, soil loss increases and slope and channel erosion decrease as runoff decreases. These variations of discharge and solid load, which in this case can be considered climate changes from subhumid to semiarid and vice versa, translate into significant hydrologic modifications and changes in channel morphology, which together are called **channel metamorphosis** (Schumm, 1977; Chorley *et al.*, 1984).

Many large drainage basins flow across different climatic zones. The Nile River in central Africa, which has significant discharge, transports a small sediment load. In its middle and lower reaches, discharge decreases and sediment load increases as it crosses the arid countries of Sudan and Egypt. The history of Quaternary climate variations along its course is highly complex and difficult to study (Williams and Williams, 1980).

One of the best-known examples of fluvial metamorphosis is the meandering Murrumbidgee River, which flows across the low-slope Riverine Plains in New South Wales, Australia (Schumm, 1968, 1969). The 60 m-wide channel is entrenched by a flood plain in which several abandoned meanders can be recognized (*Fig. 21.19*). At the bottom of the figure, an old low-sinuosity channel can be seen. Various parameters of present-day, recent, and ancient channels (width, depth, sinuosity, slope, meander wavelength) have been measured, along with solid load and discharge, for the purpose of comparing hydrologic variations and load transport of the channels when they were active. Geomorphic and pedologic observations indicate that the climate was drier when the oldest paleochannels were functioning, and more humid during the activity of the youngest paleochannels.

During the dry stage, river discharge decreases, plant cover is reduced, and more bedload is transported. Change toward more humid conditions is associated with increased discharge, and greater width, depth, and meander wavelength. This result is a complete transformation of the fluvial system. An example of an abandoned fluvial system and its replacement by another is found in Oman, west of the Wahiba sands (Maizels, 1987). A system of five paleochannels form part of a fan composed of cemented gravel (*Fig. 21.20*). Fan erosion caused lowering that was accentuated by eolian deflation; exhumed and inverted relief persists in the form of a ridge that projects about 30 m above the plain. The oldest channels may correspond to humid stages of the lower Plio-Pleistocene whereas the exhumation must be related to Pleistocene semiarid periods with lower sea levels.

Superposition or interferences of eolian and fluvial regimes are common in arid environments. For example, the large Selima sand sheet near the borders of Sudan and Algeria dominates southwestern Egypt. In this hyperarid zone the study of radar images has revealed the presence of a large fluvial channel system buried beneath the sand sheet, which may possibly be early Miocene in age (McCauley, 1982).

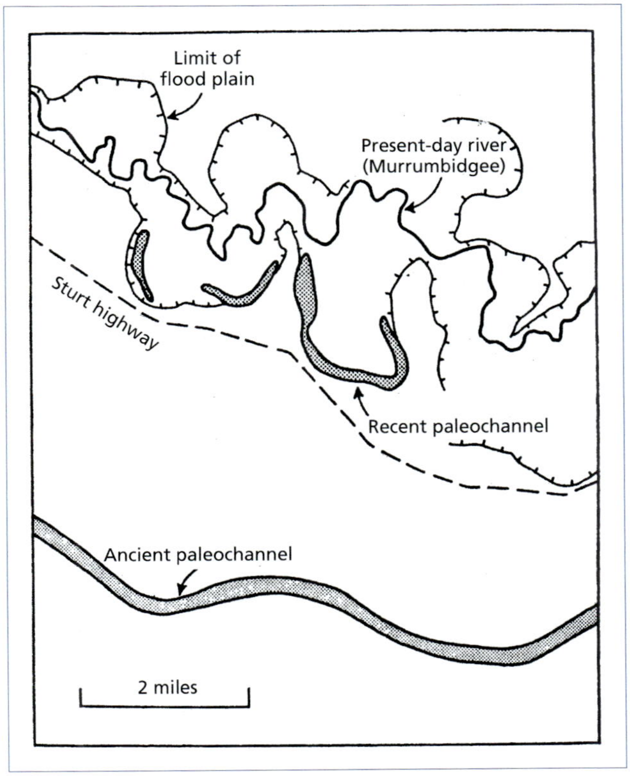

Figure 21.19 The Murrumbidgee meandering river flows over the Riverine plain of minor slope, confined to the floodplain, where various abandoned meanders (recent paleochannels) can be observed. In the lower part of the figure is an ancient paleochannel (Schumm, 1969).

The classic and complex example of alternating eolian and fluvial stages is recognized in the middle reach of the Niger River in Mali (*Fig. 21.21*). Fluvial accumulations were deposited during the humid early Quaternary period; this deposition totally ceased during a later stage of maximum aridity when longitudinal dune or trade wind dunes (ENE-SSW) formed and river water disappeared into them. Later, during a more humid period than today, the rivers crossed the dune fields and supplied numerous closed depressions that still exist; the lakes expanded as shown by the large quantity of Neolithic ceramic remains. Today, water discharge is sufficient for the Niger River to be exorheic in nature although today's precipitation is less than that of the Neolithic period and much higher than that of the dune formation stage (Tricart, 1969; 1977). This theme will be the focus of the dune system section of this chapter.

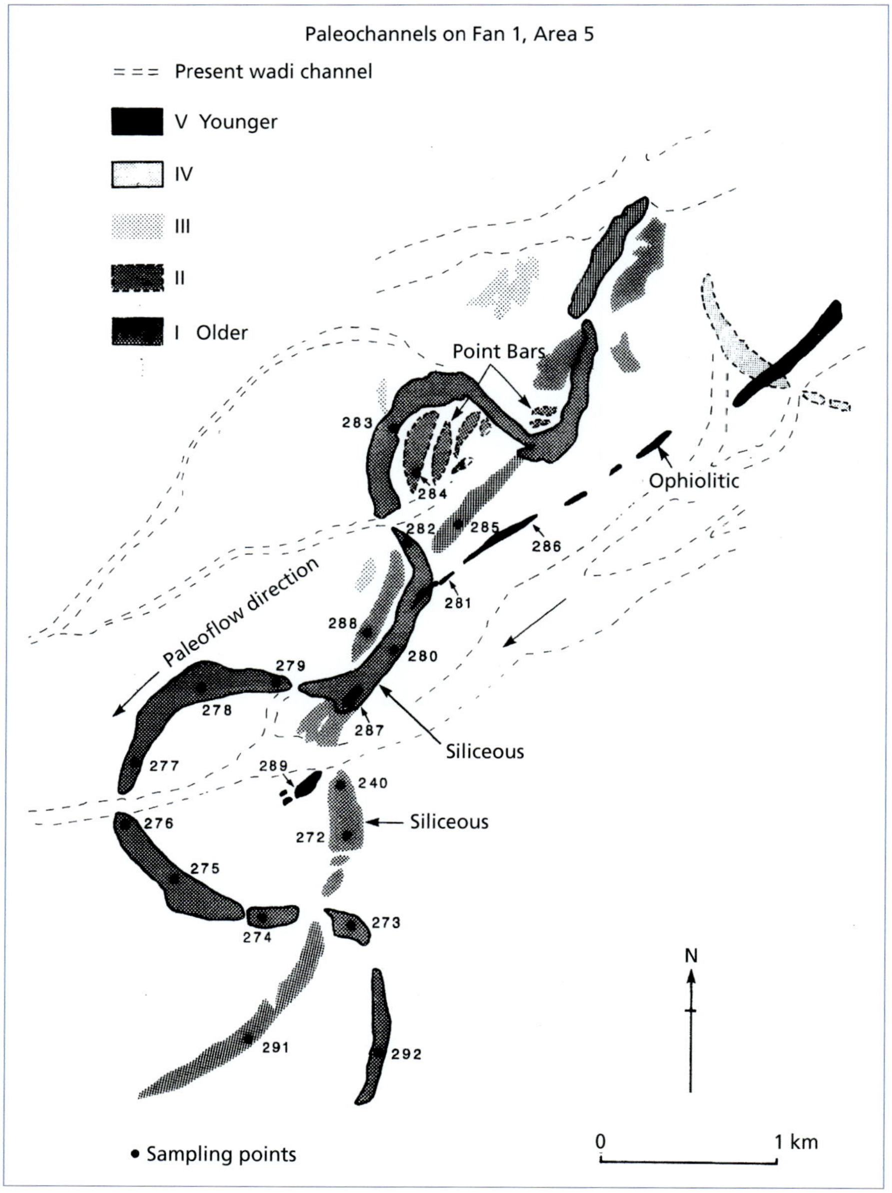

Figure 21.20 System of channels and lithologic changes associated with successive generations of paleochannels in Fan 1 to the west of the Wahiba Sands (Oman). The sinuosity and average width of the channels diminishes progressively over time (Maizels, 1987).

21.1.2 Paleolakes

Lake levels are a good archive of paleohydrology. During the Quaternary, lake sizes fluctuated in response to climate changes that modified the hydrologic balance of lacustrine systems (Jones and Jordan, 2007). In addition, recent tectonic and volcanic activity can modify the capacity of lakes.

Knowledge of ancient lake coastlines and studies of deposited sediments provide valuable data on hydrologic modifications that have affected these coastlines. Many desert lakes show clear geomorphic and stratigraphic evidence of having been much larger than today during late Quaternary periods when effective humidity was high. Continental scale modifications are a point of departure for explaining climate change mechanisms (Street and Grove, 1976). In addition, human activities are focused around lakes; Holocene changes in society appear to be related to changes in lake dynamics (Hoelzmann *et al.*, 2001).

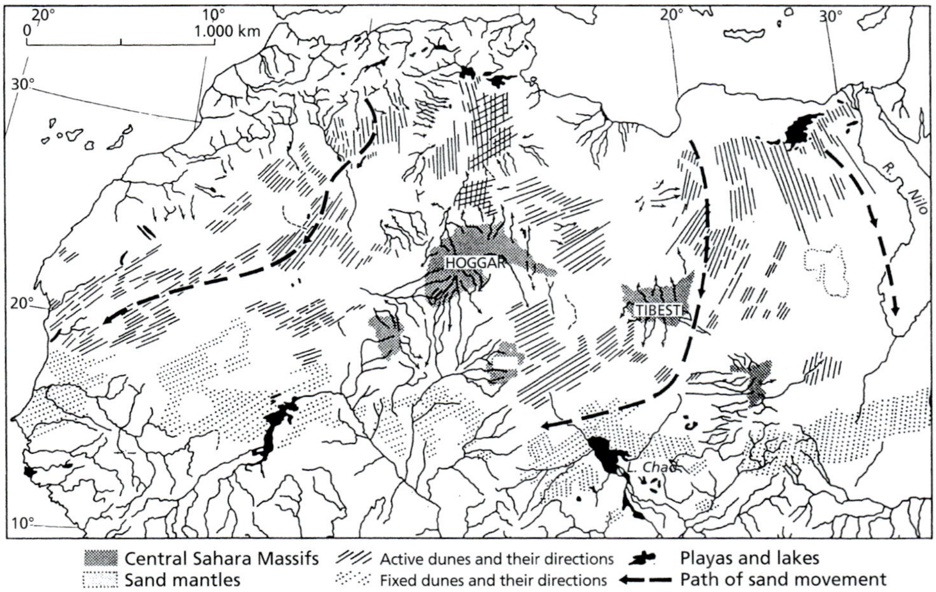

Figure 21.21 Dune fields or *ergs* in the Sahara that tend to develop in endorheic zones and possibly have formed from alluvial deposits. A band of inactive dunes extend from the Atlantic coast of Senegal-Mauritania to the Nile River. Active dunes with the principal trajectories of sand movement are also shown (Mainguet, 1975).

Cartography of lake coastlines requires detailed photogeomorphic interpretation and minute field study (Sack, 1994). Measurements at different points in a lacustrine basin usually show different heights along the same coastline. This may be due to coastal geomorphic processes or to hydro-isostatic rebound in deep lakes (Gilbert, 1890). We can use this data to determine the area and depth of a paleolake at different times. Paleolake levels (Benson, 1978; Benson and Paillet, 1989) are controlled by the presence of sub-basins separated from each other by ridges, as in Lake Bonneville (Gilbert, 1890); the complex dynamics of these sub-basins complicate their study. Another problem arises from paleohydrologic variations that may have affected fluvial basins that supply the paleolake. Any substantial variation that changes water supply can be erroneously interpreted as climate change.

Study of the stratigraphic record within lacustrine depressions provides fundamental data for the understanding of this history (*Fig. 21.22*). Geomorphic, sedimentologic, mineralogical, geochemical, paleontological, and palynological research provides paleoenvironmental information on the various events that have affected the lacustrine system (*Fig. 21.23*) (Valero *et al.*, 2003).

Lakes expand in response to increased precipitation or decreased temperature, which is usually accompanied by decreased evapotranspiration. These circumstances correspond to the old concept of a pluvial period or lacustrine phase (Rognon, 1980). However, the inferences obtained from lake studies are not as trustworthy as those obtained from dune systems because the relationship between precipitation and lake levels is complicated by temperature and other non-climatic factors (Goudie, 1992). Some lakes, such as those related to the east Africa *Rift Valley* (Gasse, 2000) (*Fig. 21.24*), are obviously affected by Quaternary tectonics that can modify the geometry and capacity of the lacustrine basin. Volcanic activity, particularly the extrusive type, also causes pronounced modifications. In spite of these drawbacks, research conducted by numerous authors on East African and southern Sahara lakes shows that they could be excellent paleoclimatic indicators (*Fig. 21.25*).

On shield and platform deserts (Mabbutt, 1977) large lakes form in shallow structural depressions during relatively more humid periods. The largest endorheic basin known is Lake Chad (*Fig. 21.26*), which is bounded by the Guinea Mountains, Congo watershed, and the Central Sahara massifs. Numerous researchers have studied Lake Chad (Barbeau, 1961; Grove and Warren, 1968; Pias, 1970; Servant and Servan-Vildary 1980; Nicholson, 1981; Street-Perrot *et al.*, 1985; Thomas and Thorp, 2003); they have analyzed the various humid and dry phases identified in this basin, both in space and time. During various humid periods, the lake was much larger than at present. The depth of this flat lacustrine basin ranges from 3 to 7 m and today it covers an area of

Figure 21.22 Small excavation formed on the margin of the Melah Sabkha. The base corresponds to gray rhythmite sediments and the roof is composed of a layer of fine grain sand. Zarzis (Southern Tunisia).

Figure 21.23 Extensive saline plain corresponding to the Melah Sabkha and, in the foreground, saline polygons. Zarzis (Tunisia).

Figure 21.24 Lake Nakuru, located in the Gregory Rift Valley, and thus of tectonic origin and affected by present-day tectonic activity. This alkaline lake is famous for the presence of the largest number of pink flamingos in the world. Nakuru (Kenya).

20,000 km^2. During an expansive stage (7,000–5,400 years BP) (Pias, 1970) the lake level was 320 m above sea level and the lake covered an area of 350,000 km^2 ("Megachad"); its water flowed to the Atlantic Ocean and the Nile River. During dry periods it was partially covered by ENE-trending dunes (trade wind dunes). This type of lake is characterized by overlapping lacustrine and eolian landforms as shown in *Figure 21.26*.

Mountain deserts and depressions contain numerous lakes that are commonly deep; they form by stages of fracturing. These lakes often have lacustrine levels at various elevations, beach ridges, spits, small deltas, and plains of saline muds (Gracia, 1995). The Great Basin, located in the northern part of the Basin and Range Province of the United States, has the largest number of "pluvial" lakes (Morrison, 1965); there are about 110–120 of them, formed by extensional block tectonics during the late Pliocene and early Pleistocene (*Fig. 21.27*). Lake Bonneville formed during the maximum extent of the Pleistocene; Gilbert, one of the first American geologists, studied the lake in detail (1890). Lake Bonneville covered an area of 51,640 km^2 and was 304 m deep. Today the Great Salt Lake of Utah, the largest lake remaining from desiccation of Lake Bonneville, covers an area that fluctuates between 2,600 and 6,500 km^2. Lake Lahontan, which has an area of 22,900 km^2, formed to the west of Lake Bonneville. Overflow channels linked some basins and others flowed to the ocean via the Snake and Colorado rivers.

These North American deserts owe their aridity to the orographic effect. High cordilleras (Coastal Range, Sierra Nevada, and Cascade Range) rim the Pacific Ocean and slow down cyclonic circulation coming from the west. As a result, deserts form in the rain shadow. In this area, temperatures were very low, primarily during glacial periods, because the Laurentide Ice Sheet was located nearby, and furthermore, numerous glaciers developed in the high areas of the cordilleras that bordered the coast. As temperatures decreased, evaporation decreased and the depressed areas of the Great Basin received enormous quantities of meltwater from the glaciers. Thus large lakes developed in endorheic zones and spectacular present-day lacustrine terraces provide evidence of them. This area is an ideal region to study the relationship between mountain glaciation and lacustrine phases. The work of Smith and Street-Perrott (1983) analyzes many of these basins and discusses the chronology of these lacustrine phases. They identify a period of expansion between 24,000 and 14,000 years BP; another between 14,000 and 10,000 years BP with broad fluctuations that may or may not be synchronous, and finally an arid or dry phase between 10,000 and 5,000 years BP, when lake levels dropped. Oviatt (1997) points out different stages of lake decline at dates of

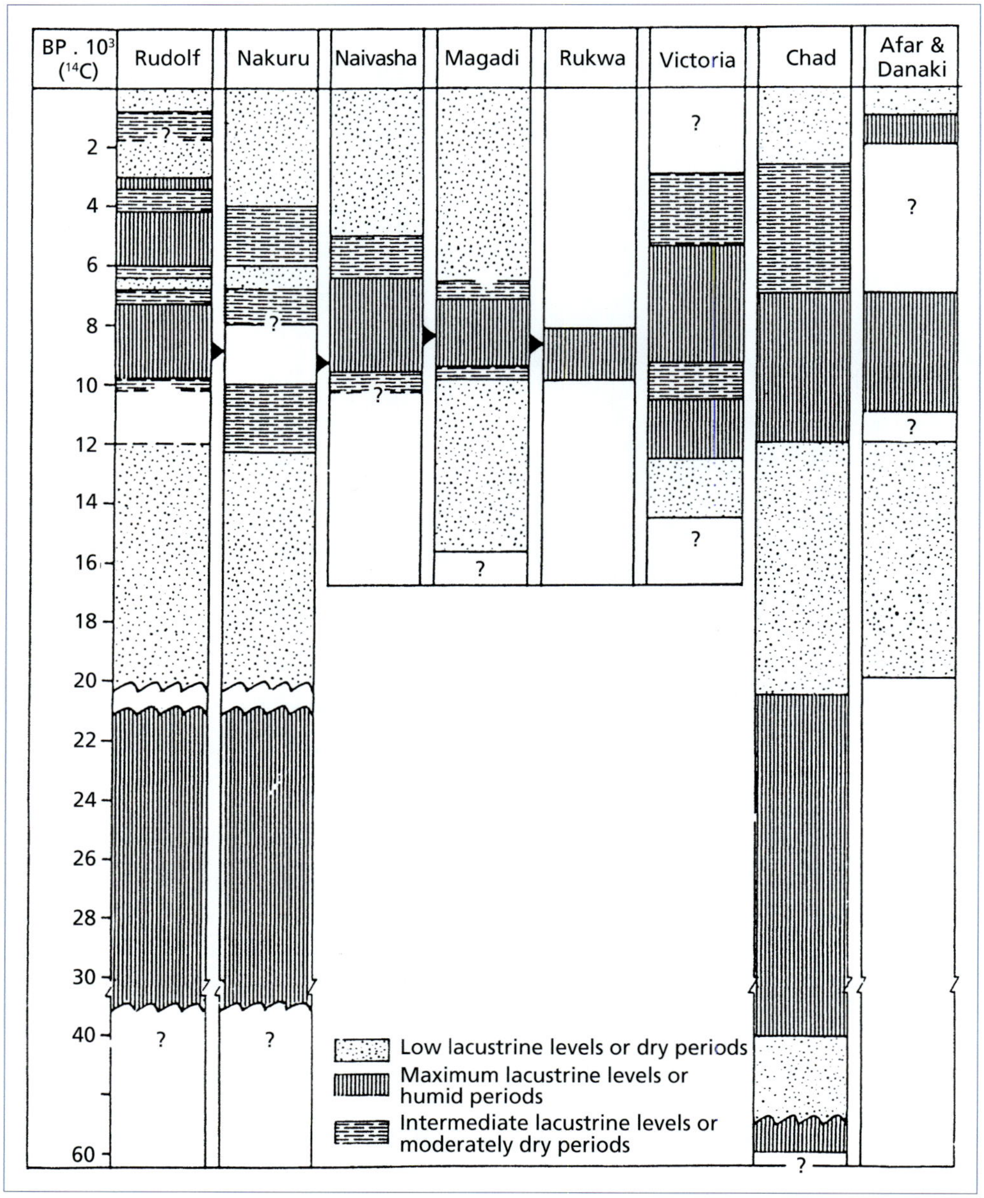

Figure 21.25 Fluctuations in the lacustrine levels of east Africa and the southern center of the Sahara. According to various authors (compiled by Selby, 1985).

21, 18.5–19, 17.5, 16–15.5, 14–13, and 10 ka, all obtained by ^{14}C dating. They are synchronous with the terminations of the H1 and H2 Heinrich events and other smaller stages of iceberg drift (a, b, c and Younger Dryas) in the North Atlantic. *Figure 17.57* shows lake level changes in the Great Salt Lake basin for Holocene periods; they show the first record of human activities between 11 and 10 ka in Danger Cave.

In deep lakes, the study and interpretation of coastlines are complicated because of deformation of the lacustrine basin and hydrostatic rise. However, the study of lacustrine levels is fundamental to determining the magnitude of subsidence in certain areas as a result of the load exerted by the water volume and the subsequent rebound that occurred when it was unloaded. The left side of *Figure 21.28* shows isobaths of Lake Bonneville during the period of 25,000–11,000 years BP and the right side shows isostatic rebound resulting from desiccation of the lake, which is equivalent to about 20% of the water column (Crittenden, 1963). The highest values of isostatic rebound generally correspond to the maximum lake depths.

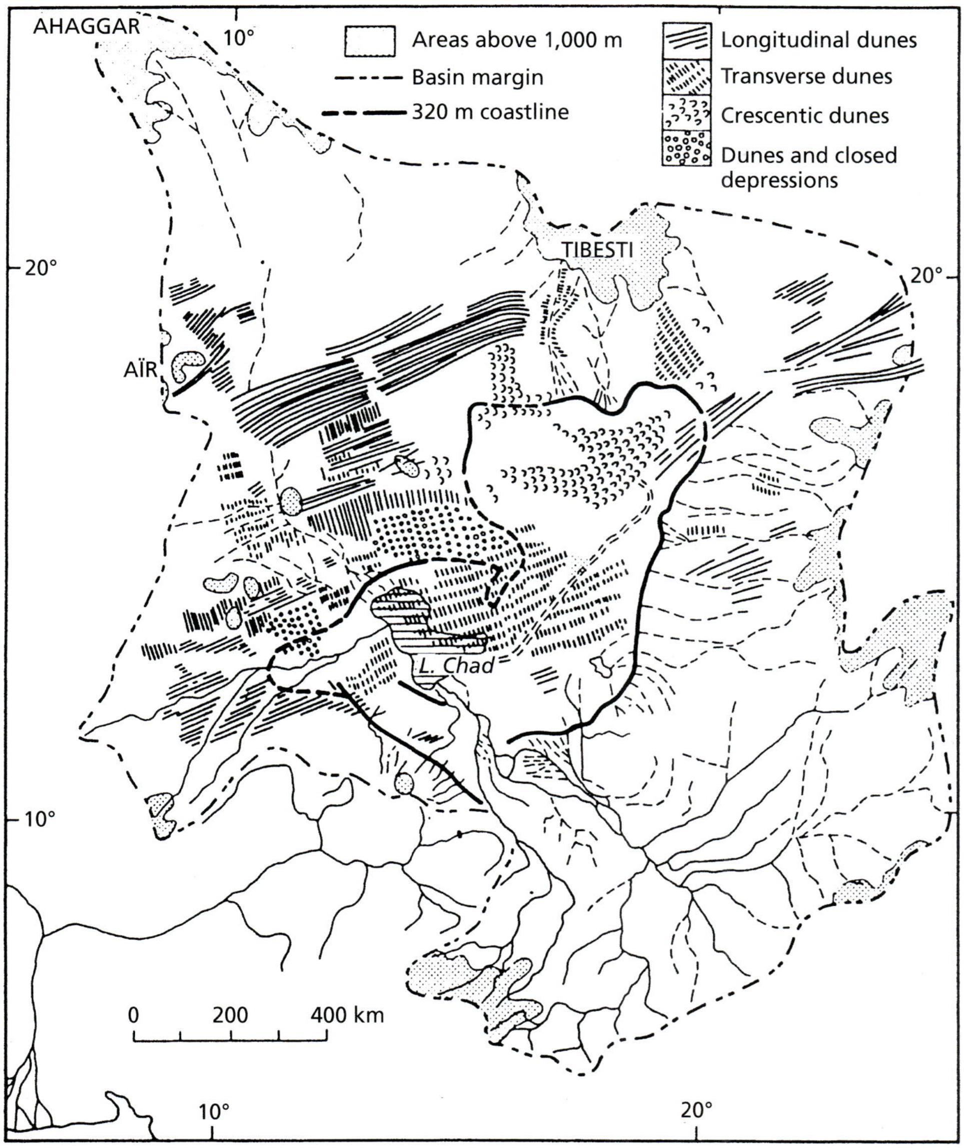

Figure 21.26 The Chad basin with its dune systems and the Megachad coastline at an elevation of 320 m (Grove and Warren, 1968).

One of the most spectacular cases of Pleistocene lake expansion is the Aral-Caspian system, which formed as a result of broad shallow doming. The Sirdaria and Amudaria rivers provided large amounts of glacial melt water to the Aral Sea and the Volga and Ural rivers supplied similar amounts to the Caspian Sea. When the lake level rose 76 m above the Caspian Sea, it flowed into the Aral Sea, forming the largest "pluvial" lake in the world and inundating 1,100,000 km^2 including flowing some 1300 km up the Volga River (Goudie, 1992). The Caspian Sea also connected to the Black Sea through the Mantych Depression. In Asia, lake level changes have been very sensitive to climate change especially during intense droughts in arid and semiarid zones such as the Aral basin and the saline lakes of the Tibetan Plateau, among others (Yu and Ke, 2007). In the Mideast there are records of 30 lakes between 10 ka BP and 30 ka BP. In Central Asia, 34 records contain four lake maxima (25–30 ka BP, 15–17 ka BP, 6–7 ka BP, and 3–4 ka BP). On the Tibetan Plateau and in surrounding areas, 44 records indicate three periods of high lake levels (30–24 ka BP, 22–20 ka BP and 8–5 ka BP). Finally

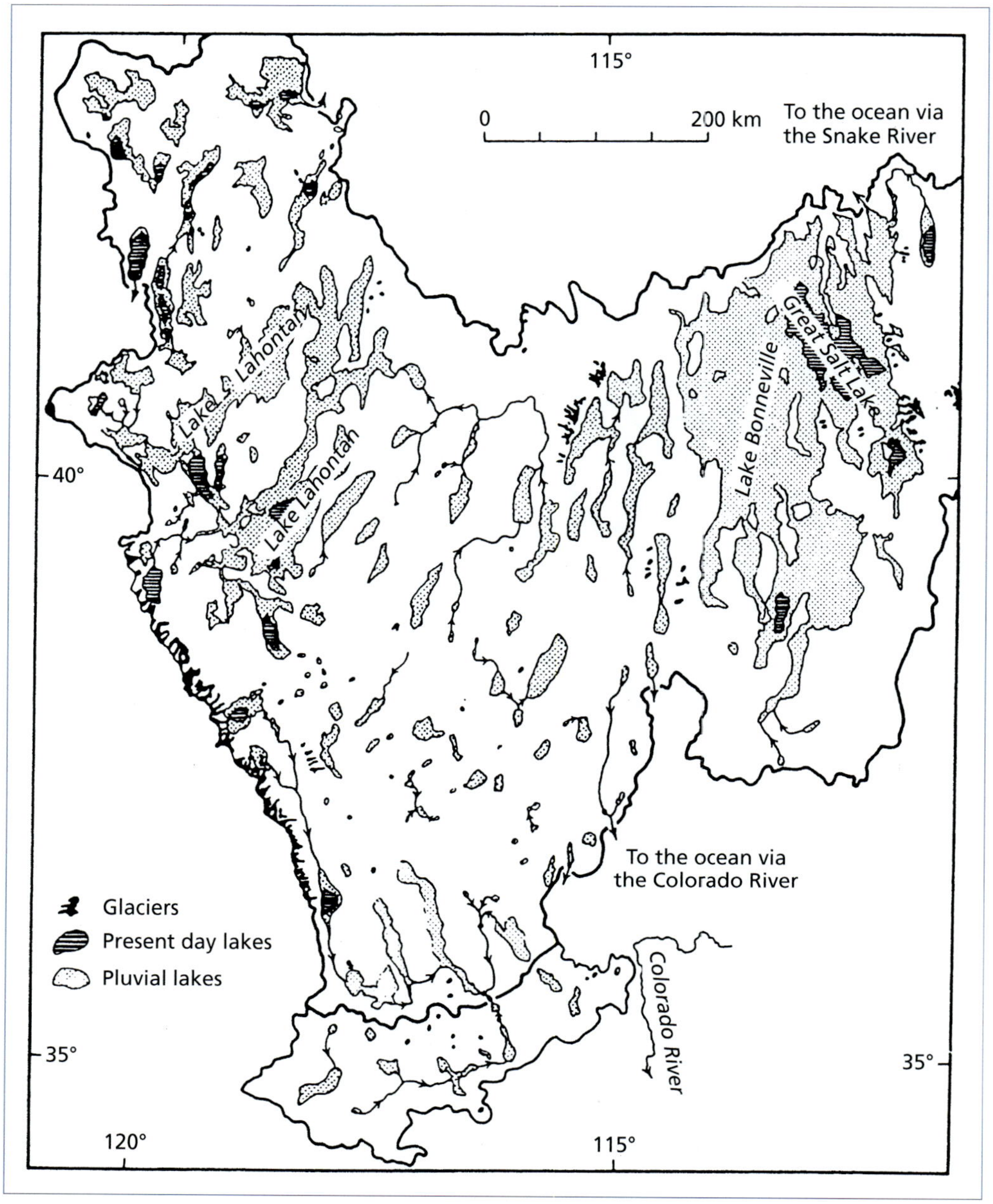

Figure 21.27 Maximum extent of the Great Basin lakes during the post-Sangamon period. The arrows indicate overflow and connections. The thick line indicates the present-day drainage basin of the Great Basin (Morrison, 1965).

in East Asia, high lacustrine levels have been identified in the early and mid Holocene but there are intercalated loess deposits, which contradict dry periods (Yu and Ke, 2007).

Australia has predominantly arid and semiarid environments in its interior; zones of higher precipitation are present in the south, east, and north of the country. Lake Eyre (*Fig. 17.56*), located in south-central Australia, has a large drainage basin with an area of about 2 million km^2. Lake level records show that prior to 130 ky BP the lake was dry and affected by deflation. High lake levels occurred in the following periods: 130–120 ky BP, 95–85 ky BP, 70–60 ky BP, 45–35 ky BP, and 14–3 ky BP. These periods are separated by dry stages of unequal duration (Magee *et al.*, 2004).

In spite of problems with climate interpretations inferred from geomorphic studies of paleolakes, this research is fundamental to paleoclimate interpretation, especially when based on multidisciplinary paleoenvironmental studies.

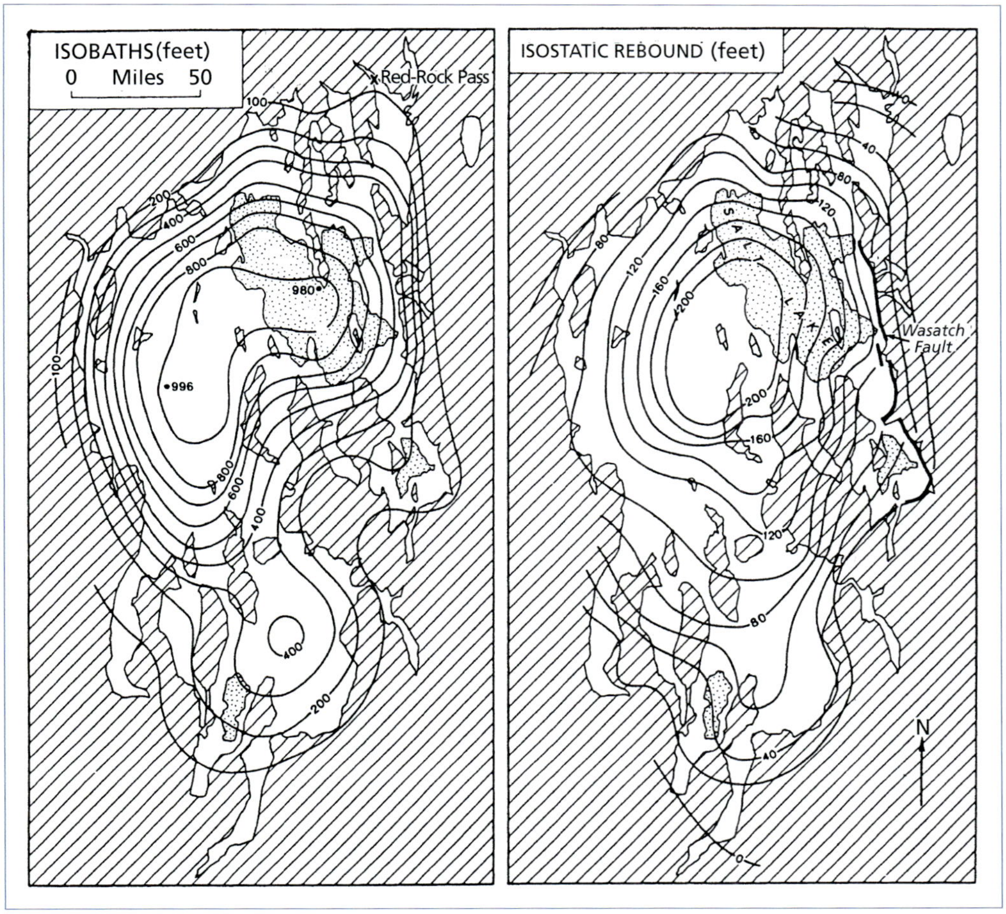

Figure 21.28 Pleistocene Lake Bonneville, Utah. Left, water depth in feet, in which the white area indicates the extent of Lake Bonneville and the dotted zone represents present day lakes. The lake overflowed through Red Rock Pass during the lacustrine phase. Right: Isostatic rebound in feet as a result of decreased water volume, measured by the deformation of the Lake Bonneville coastline. Recent movements along the Wasatch Fault are indicated (Crittenden, 1963).

21.1.3 Dune systems

21.1.3.1 Introduction

Large eolian accumulations or *ergs*, together with erosional landforms in intertropical deserts cover 20–25% of the Earth's surface (Livingstone and Warren, 1996). We know the areal extent of present-day dune fields for most of the world's deserts *(Fig. 10.19)* (Thomas, 1997b), thanks to aerial photography and satellite images. Dunes develop over periods of 1,000 to 100,000 years (Lancaster, 1995); during their formation the ice caps expanded and contracted, which greatly affected lower latitudes through climate and sea level changes. The most significant changes can be seen in fluctuations of large sand desert margins and in changes of sand particle supply and dune mobility. These sand deserts were not formed exclusively during the Pleistocene but many of them increased in size during global climate cooling *(Fig. 21.29)* (Shackelton and Kennet, 1975; Goudie, 1992). Williams (1994) calls this the "late Cenozoic desiccation".

Researchers have identified numerous droughts during the Quaternary. The study of dated ice cores from the upper part of the Greenland Ice Sheet (GISP 2 – Greenland Ice Sheet Project 2) shows fluctuations in ice conductivity on a scale of <5–20 years, which reflects rapid changes in the atmospheric dust content *(Fig. 21.30)* including periods in which the atmosphere was very dusty alternating with periods that lacked dust. These phenomena indicate significant variations in the intensity of eolian activity (Taylor *et al.*, 1993) and rapid changes of atmospheric circulation. Analysis of terrestrial sediments in cores from ocean drill holes helps us interpret the extent, intensity, and duration of arid zones during

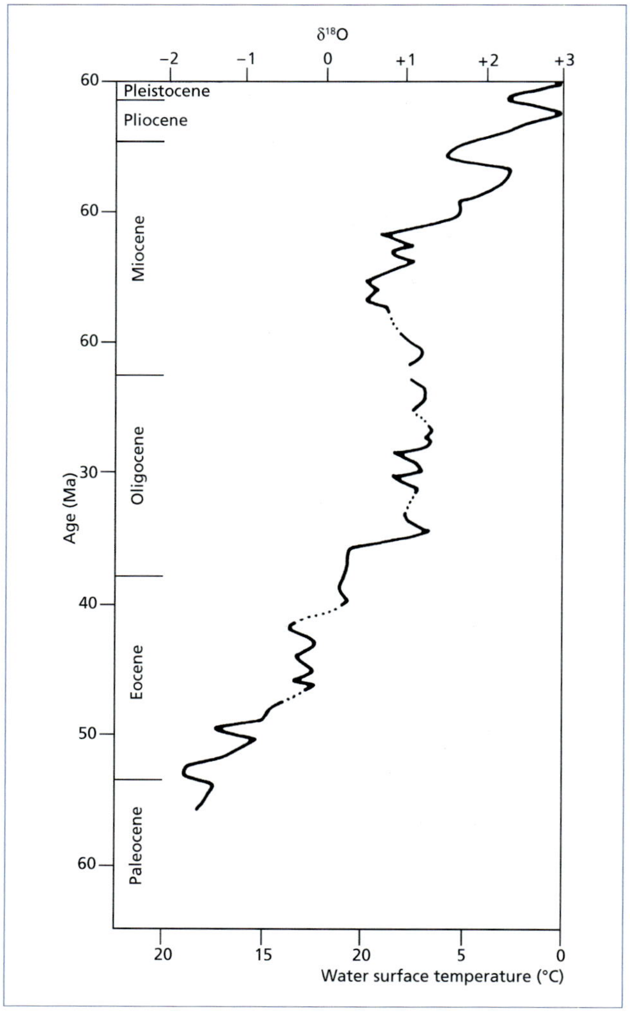

Figure 21.29 Temperatures of the sea surface during the Cenozoic in the southern ocean, obtained from oxygen isotope composition changes in planktonic foraminifera in locations 277, 279, and 281 of the DSDP (Deep Sea Drilling Project) (Schackeaton and Kennett, 1975).

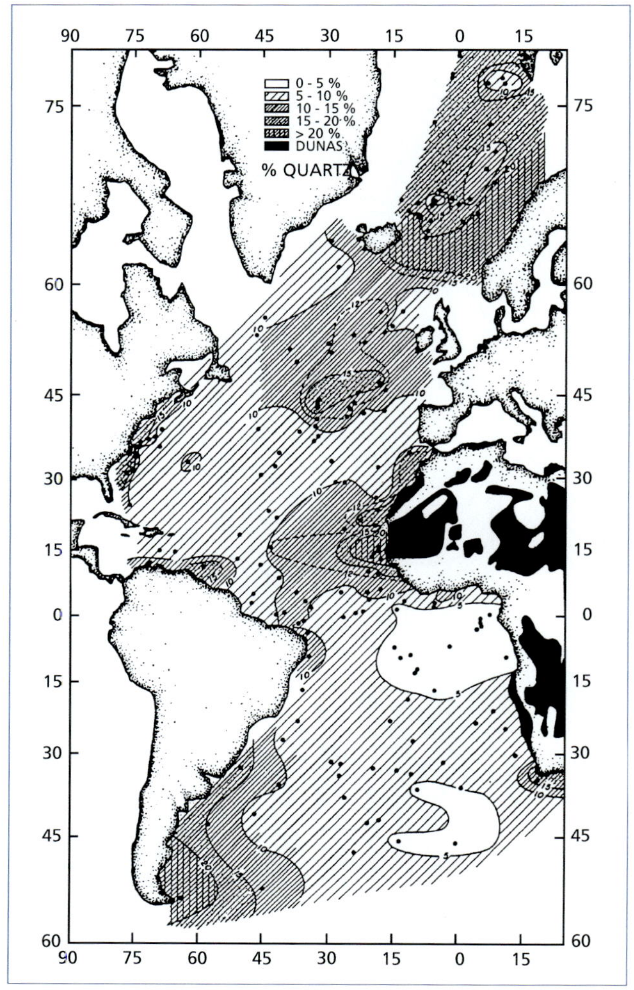

Figure 21.31 Distribution of the percentage weight of quartz (free of carbonates) in sediments that correspond to the last glaciation (18,000 years BP). The sand deserts of continental Africa are shown in black (Kolla *et al.*, 1979).

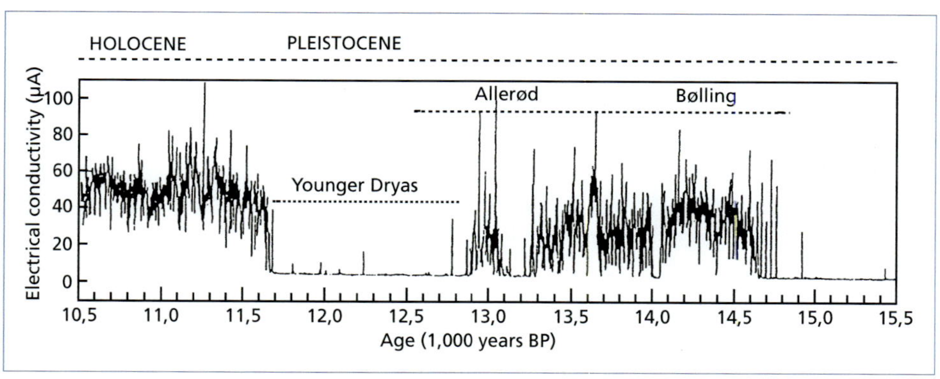

Figure 21.30 Variations in electrical conductivity (related to dust content) of the upper part of the GISP2 core in the Greenland Ice Sheet (Taylor *et al.*, 1993).

the Quaternary. The continents supply fluvial and eolian sediments to the marine floor in intertropical areas. The supply of organic carbonates is relatively constant and some carbonate-rich cores contain only minor terrestrial contributions. High percentages of silt-size quartz indicate eolian contributions whereas fluvial sand is identified by its reddish color (Diester-Haas, 1976). Eolian dust increased during Quaternary cool periods; during this time, the quartz content maximum at the bottom of the Atlantic Ocean moved southward to about 8° north latitude, which indicates an expansion of the arid area and of the Saharan sand deserts (*Fig. 21.31*) (Kolla *et al.*, 1979). The presence of fresh water diatoms in dust in cores corresponding to the last glacial maximum is interpreted as particles mobilized by wind and transported to the ocean from desiccated desert lakes, which also supports the expansion of the arid zone (Parmenter and Folger, 1974; Pokras and Mix, 1985). Important data on the investigation of cores from the ocean bottom can also be obtained by studying the degree of feldspar alteration, pollen and phytoliths, and salinity and temperature of marine waters from oxygen isotope ratios in foraminifera.

This research provides indirect data on major dry periods that can be recognized in the recent geologic record. However, the detailed study of active and inactive dune fields in deserts around the world provides valuable information on their extent, duration, and distribution during periods of extreme aridity and on changes in wind direction and circulation systems throughout time (Lancaster, 1995).

One important problem has been the dating of sand accumulations, which until recently was difficult and relative. For example, ^{14}C dating of organic matter in paleosoils that develop on dunes indicates a pre- or post- date. Another difficulty is the availability of reliable dates because many dune fields form from multiple episodes. At the beginning of the 1980s extraordinary advances in dating resulted from two techniques, thermoluminescence (TL) and optically stimulated luminescence (OSL); these techniques make it possible to date sands and loess (Wintle, 1993) back about 500,000 years. These dates in turn make it possible to correlate paleoclimatic data obtained from other techniques. These data can be used for paleoclimatic correlations and interpretations. As a result of these advances, researchers are now able to identify various dry periods in deserts throughout the world (*Fig. 21.32*) where highly variable chronological histories are observed.

Dunes are sensitive to modifications in atmospheric parameters such as wind direction and intensity and changes in precipitation, which affect evapotranspiration, soil moisture,

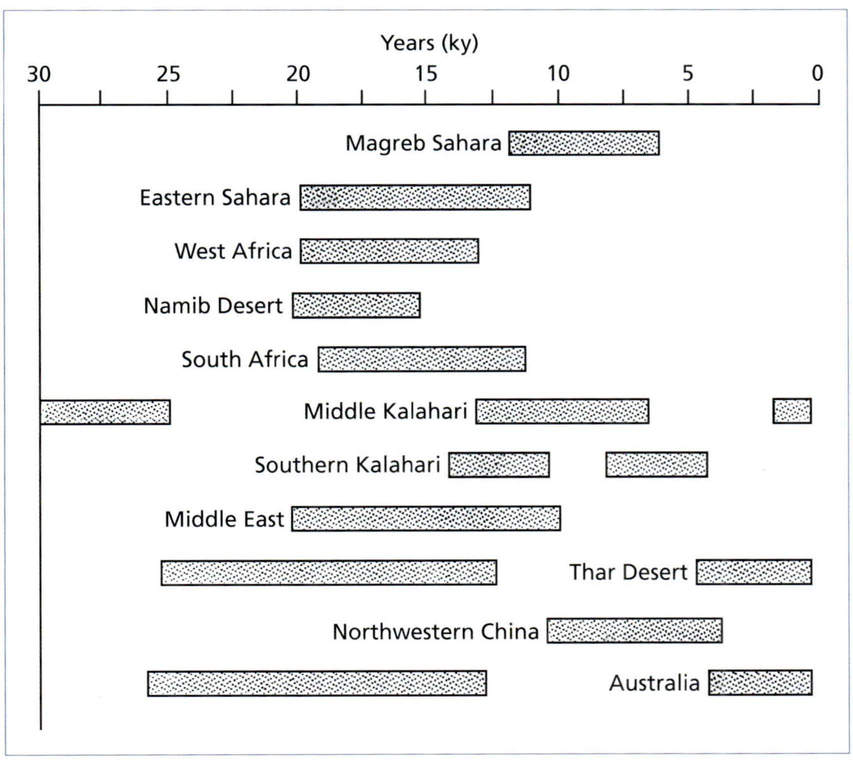

Figure 21.32 Dry periods of dune formation during the last 30,000 years, for tropical and subtropical continental deserts of Africa, Asia, and Australia (compiled from various authors by Tchakerian, 1995).

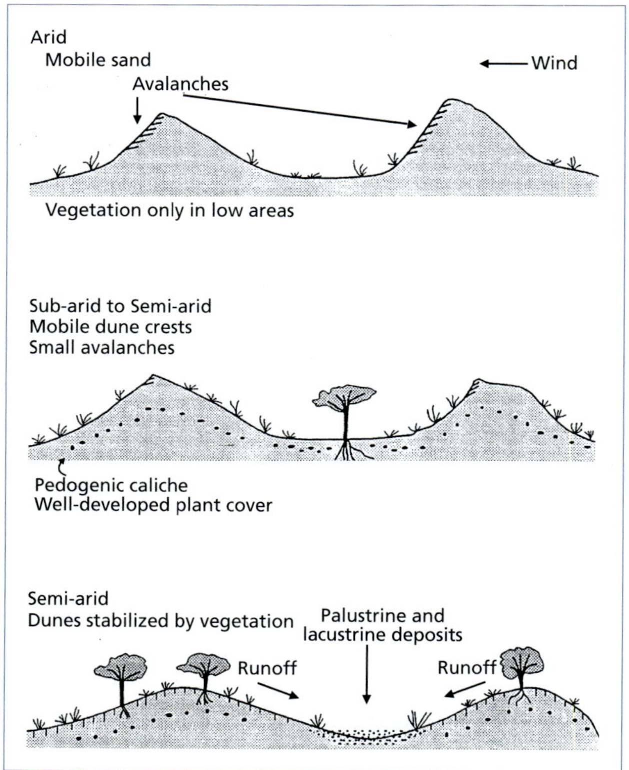

Figure 21.33 Possible dune responses to climate changes (Rognon, 1980).

vegetation percentage, and particle mobility. Dune areas tend to accentuate the effects of dry and humid phases (Rognon, 1980) and respond rapidly to climate changes. If precipitation increases slightly, vegetation grows primarily in the lower parts of dunes and sand moves as avalanches down the leeward side of the upper part of the dune (*Fig. 21.33*). As precipitation increases, vegetation colonizes the entire dune except the crests and may begin to cause calcification of the sandy accumulations. Finally, when precipitation is significant and long term, the water table may crop out in the lower part of the dune (Swezey, 2001) and form palustrine deposits in interdune areas; in addition hardier vegetation becomes established, leading to dune stability and soil formation. Intense precipitation can erode dunes (Talbot and Williams, 1978). In contrast, during dry periods when the water table is low, the lack of water prevents plant growth and allows wind to predominate.

21.1.3.2 Active and inactive dunes

Our understanding of the extent of dune activity is incomplete. On the basis of numerous data, researchers estimate that when mean precipitation exceeds 100–300 mm (Goudie, 1992) vegetation is effective in restricting dune movement. The degree of activity can be obtained from climate indices discussed by Cooke *et al.* (1993), and amplified by Chepil *et al.* (1962), Wasson (1984), and Lancaster (1988). Climatic indicators are imprecise because of the lack of meteorological stations and the wide interannual variability of semiarid climates (Livingstone and Warren, 1996); in some years dunes are active and in years of higher precipitation, dunes lose much of their mobility and even begin to stabilize.

Transport and sedimentation on **active dunes** (*Fig. 21.34*) is shown by *rippled surfaces* and leeward avalanches. Today, overgrazing and other human activities on desert margins can reactivate dunes as in the highly populated area of Rajastan (India) in the Thar Desert (Goudie, 1992). **Dormant dunes** (Lancaster, 1995) or **episodically active dunes** (Livingstone and Warren, 1996) are dunes in which sand movement is minor or absent over long periods of time (*Fig. 21.35*). In general, they are found in areas where mean annual precipitation is between 100–300 mm and where interannual precipitation varies to the extent that plant cover decreases and sand moves during dry periods. In humid years the opposite occurs. Bioturbation can affect primary sedimentary structures in these dunes. Vegetation is well developed with high percentages of perennial plants. **Inactive, fixed or relict dunes** are vestiges of previous drier climates and thus are indicators of paleodeserts (*Fig. 21.36*). Various indicators of relict dunes have been identified (*Table 21.2*) (Goudie, 2002). Inactive dune systems are present in areas where precipitation today ranges from 250 mm to 2,000 mm such as in Zaire (Thomas and Goudie, 1984), the Orinoco Basin (Tricart, 1985), and the swamps of Brazil (Klammer, 1982). These systems include dunes and sand mantles that are stabilized by structured soils where increased precipitation has formed petrocalcic horizons and ferralitic soils (Tricart, 1977; Völkel and Grünert, 1990). In other cases the dunes are cemented, generally in an unequal way, to produce eolianite, which often contains alteration honeycombs and karst features (*Fig. 21.37*). In addition, dunes can be modified by non-eolian processes and eluviation. The crests become rounder, water erosion affects their flanks, and the resulting products fill interdunal depressions (Mabbutt, 1977). Today the flanks of longitudinal dunes in the southern Sahara do not exceed 5°. The initial stages of dune degradation in humid seasons may involve processes of rilling, gullying, and piping that may destroy the original eolian structure. Another cause of dune inactivity is the flooding of dunes during humid stages, as has occurred in the Thar Desert (Singh, 1971) and in the Chad Basin (Grove, 1958); in these places, dunes are found beneath lacustrine deposits. In these relict dune areas, a return to active status is possible only through major environmental changes.

A controversial topic common to other branches of geomorphology is the determination of the amount of time

Figure 21.34 Large barchanoid that show sand movement (in the foreground) Al Liwa, Rub Al Khali Desert. United Arab Emirates.

Figure 21.35 Dormant longitudinal dunes with well developed shrub vegetation. Sand mobility is limited to dune crests whereas the flanks are appear to be stabilized by biocrusts and shrub vegetation. Nizzana experimental station, western Negev Desert. Israel.

Figure 21.36 Inactive dunes covered with shrub and arboreal vegetation. Beer Sheva region, Negev Desert. Israel.

Table 21.2 INDICATORS OF RELICT DUNES (GOUDIE, 2002)

Geomorphological	Pedological	Biological	Archaeological
Degraded slopes	Calcification	Mantled in vegetation	Covered by artifacts of known age
Gully development	Soil horizons		
Mantled in colluvium	Development of clays and silts		
Flooded by lakes			
Submerged in alluvium			

is required to build a dune. Some investigators have indicated that large present-day dunes must be relict (Glennie, 1970; Besler, 1982) because only high intensity winds such as those of the late Pleistocene are capable of generating these landforms. Other researchers have assumed long time periods for their formation. Based on experimental observations Wilson (1972) estimates that the 100 m-tall dunes of the Great Eastern Erg are 10,000 years old; Lancaster (1989a) estimates 42,000 years as the age of the large dunes of the Namib Desert. Dunes 70 m high have formed on the coast of the Province of Huelva (Spain) in the Asperillo; according to archeological and historical dates they formed sometime between the 17th and 20th centuries (Zazo *et al.*, 1999a; Borja *et al.*, 1999).

Extensive areas of stabilized dunes are present today on all continents both in tropical regions and in high latitude areas where they are related to periglacial climates (Sarnthein, 1978) (*Fig. 21.38*). Dune complexes covered an area of 2.5 million km^2 of Africa during the last glacial maximum; it was the largest desert in the world (Thomas and Shaw,

Figure 21.37 Eolianite with cross-stratification affected by alveolization along laminations. Sila. United Arab Emirates.

1991). By comparing the extent of these dune fields to the surface area of today's active dunes, we can infer that major changes in precipitation and the percentage of plant cover have taken place. Today, deserts with active dunes cover about 10% of the land surface between 30°N and 30°S whereas 18,000 years ago, their extent was 50% (Goudie, 1992).

Stabilized longitudinal dunes are present on the margins of most zonal deserts. The most extensive are found in the Sahara north of 5°N (Mainguet *et al.*, 1980) (*Fig. 21.39*); they extend from the Atlantic coast of Senegal to the Sudan and the Arabian Peninsula. The southern margin of the Sahara is a transition zone where desert vegetation gives way to the Sahelian steppe; this change is marked by a precipitation increase of about 1 mm/year/km, calculated by serial satellite images (Tucker *et al.*, 1991). Active processes along the southern margin of the Sahara are highly complex; they involve climatic and anthropogenic changes. Lack of precipitation over the last 30 years has a precedent in droughts recorded during the 19th century. These droughts extended 500 km to the south into areas that today receive 500 mm of precipitation. Three principal generations of dunes have been identified (Grove and Warren, 1968; Talbot, 1980). The oldest is more 20,000 years old; the second between 20,000 and 13,000 years; and the most recent has an age of 5,000 years. Most of them stabilized between 11,000 and 7,000 years. The most recent droughts and deforestation have reactivated some dunes and created new ones (Nickling and Gillies, 1993). Other examples of relict dunes can be found in the Kalahari, northern Arabia, the Thar Desert, and large areas of Australia, Brazil, and Venezuela. These dunes are completely colonized by plants and are covered by soils that have developed on them.

21.1.3.3 Paleoclimatic information from dune systems

Active dunes develop on continental areas where precipitation is than 150 mm (Mainguet *et al.*, 1980); the southern Sahara is an excellent example (*Fig. 21.39*). The presence of relict dunes is irrefutable proof that more arid conditions existed in the past. The presence of these dunes in areas of high precipitation as for example in the Congo rainforest shows the dramatic change that has affected the area. We must also consider that dune reactivation or formation may result from wind velocity variations and their interactions with moisture content (Thomas, 1992).

During the Pleistocene and Holocene, dunes were repeatedly reactivated and stabilized, lakes oscillated over short time periods, and desert dust storms often reached the oceans. The timing and correlation of these events may be subject to controversy but information on climatic tendencies is well known (Goudie, 2002). In particular, dry conditions during and after the Last Glacial Maximum and humid periods during the early and middle Holocene seem to be characteristic of shield

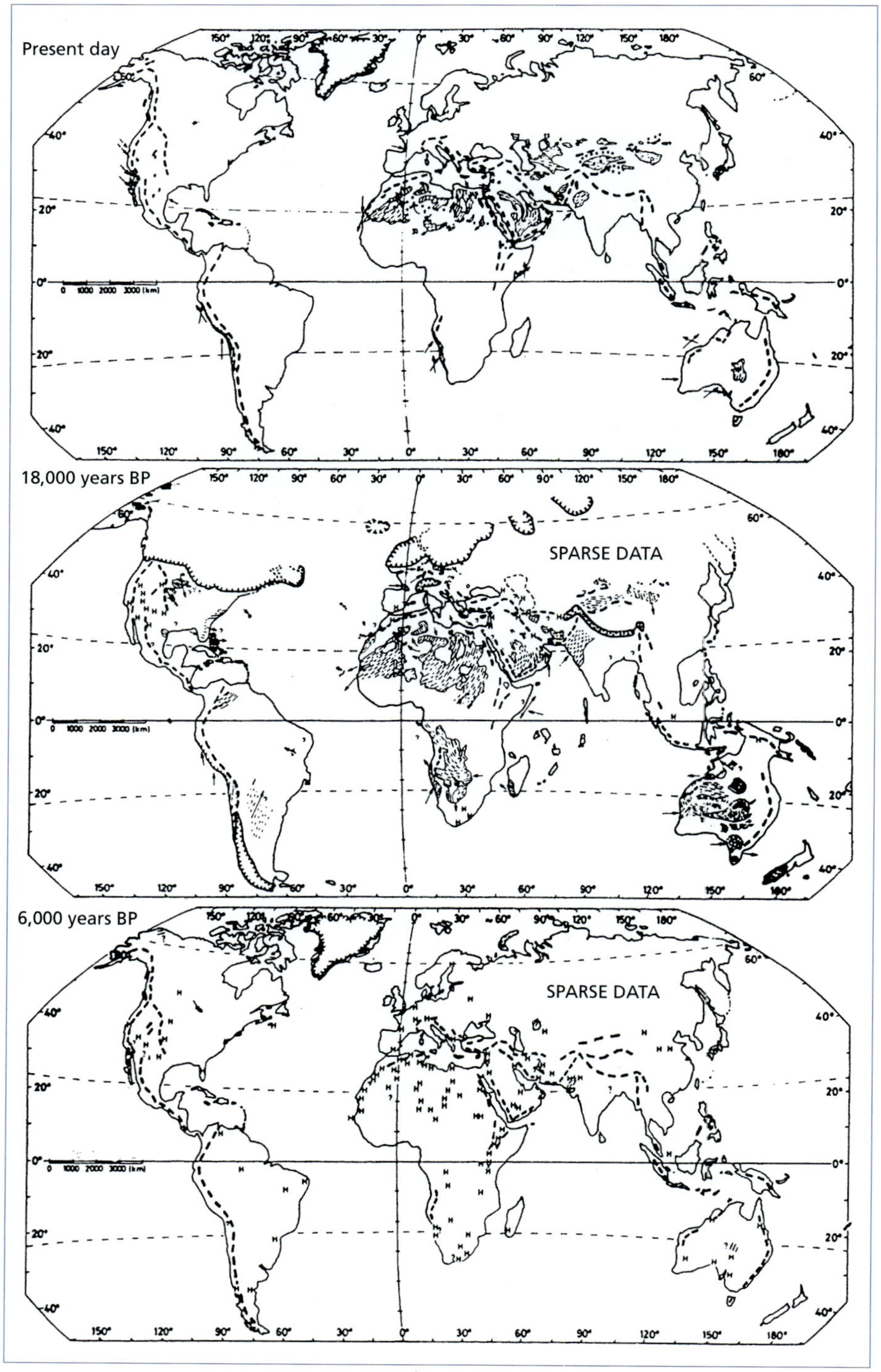

Figure 21.38 Extent of dune fields today, 18,000 years ago (glacial maximum), and 6,000 (Climate Optimum). H indicates humid conditions (Sarnthein, 1978).

and tropical platform deserts but not of arid zones in the southwestern United States. Differences among arid zones in the Sahara, the Arabian Peninsula, and Asia 18,000 years BP, 8,000 years BP, and 2,000 AD are undoubtedly significant (Yan and Petit-Maire, 1994) (*Fig. 21.40*).

As discussed above, the climate of the Sahara was very different from its climate today. Giraffe and elephant bones indicate that they lived in eolian sand areas that lacked vegetation and water. People in these zones painted herbivorous animals and hunting scenes on rock walls. This indicates that the climate

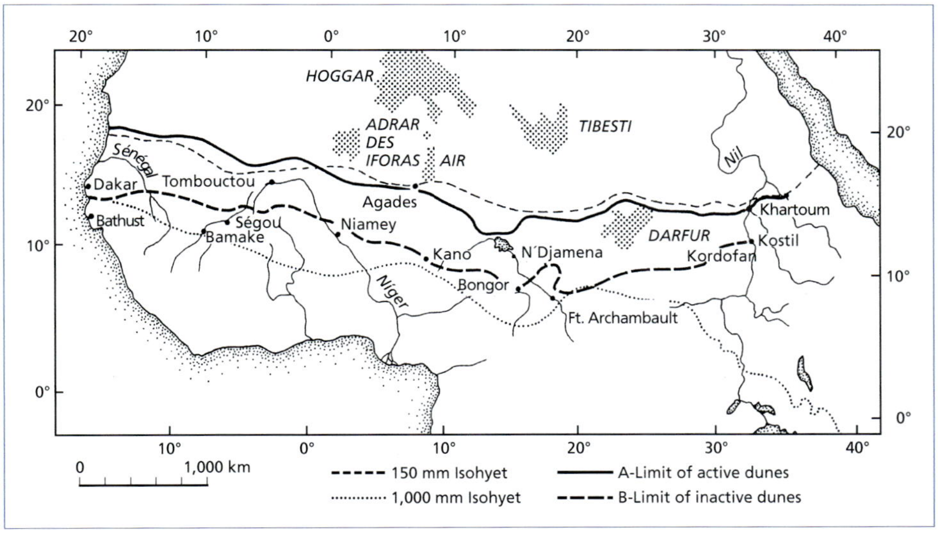

Figure 21.39 Limit of active and relict dunes in the southern regions of the Sahara (Mainguet *et al.*, 1980).

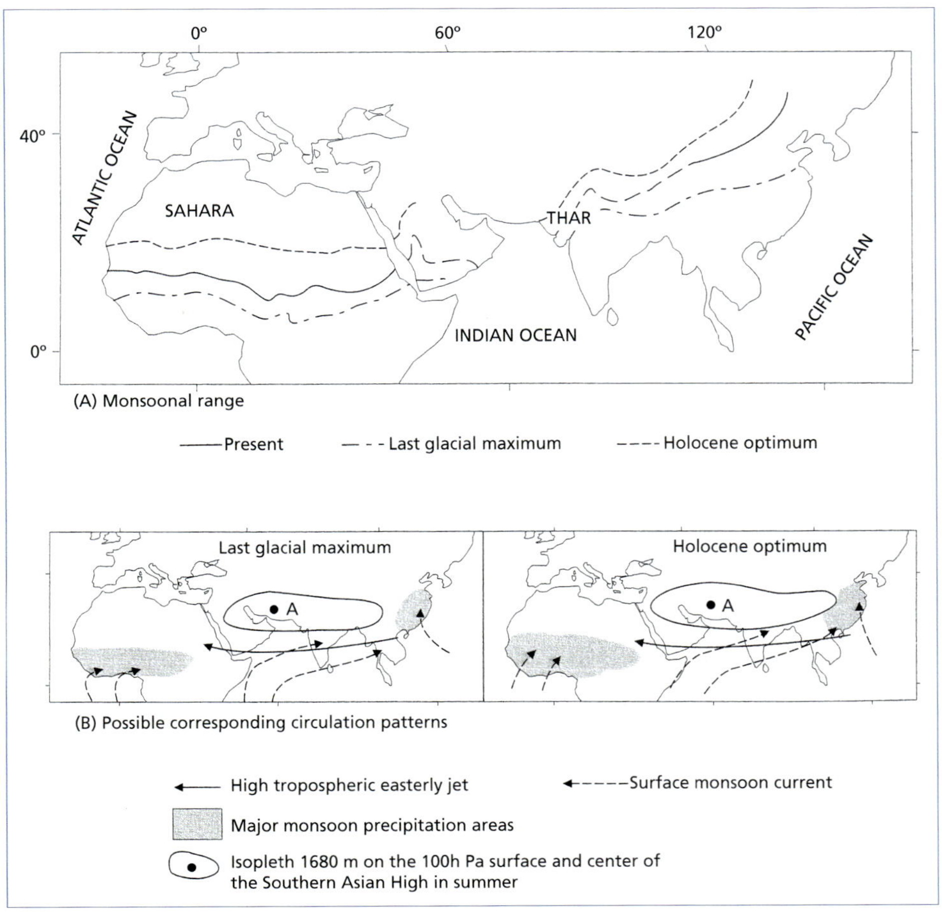

Figure 21.40 (A) Comparison of climate changes throughout the Afro-asiatic transition zone between the Last Glacial Maximum and the Holocene. (B) Possible circulation system (Yan and Petit-Maire, 1994).

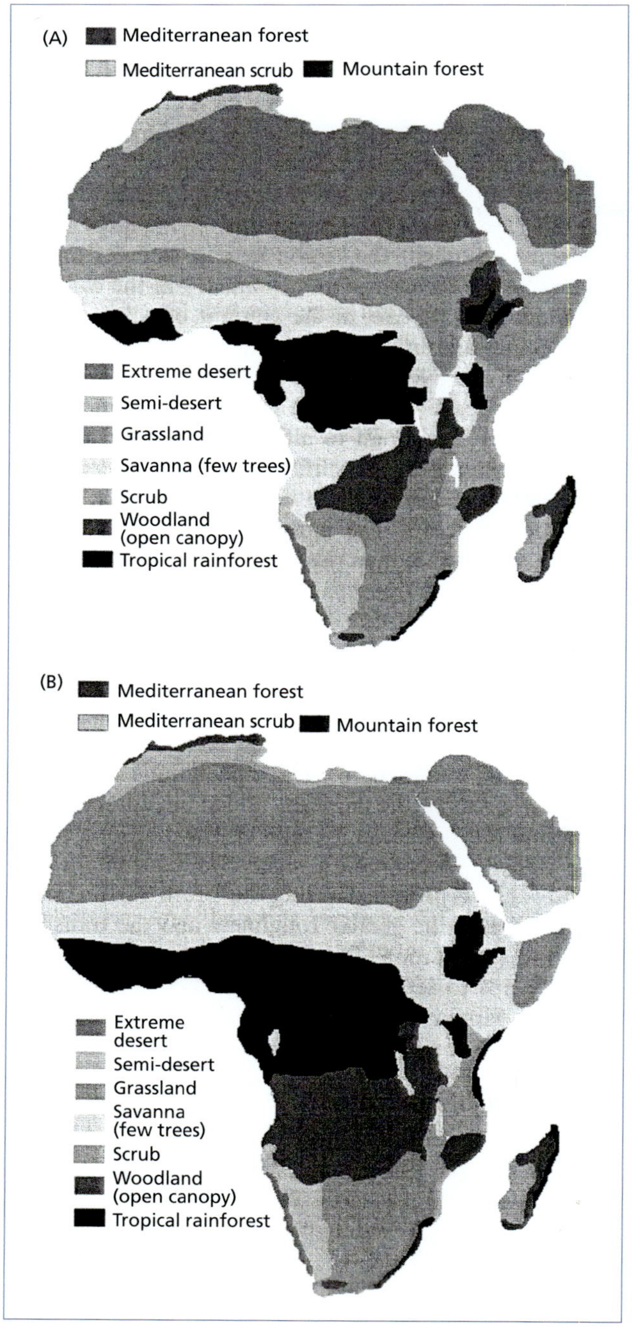

Figure 21.41 Distribution of vegetation into (A) present day and (B) during the "Green Sahara" (8000–7400 years BP). Grasslands (mixed with shrubs) seem to cover the whole Sahara during this period (Adams, 2007).

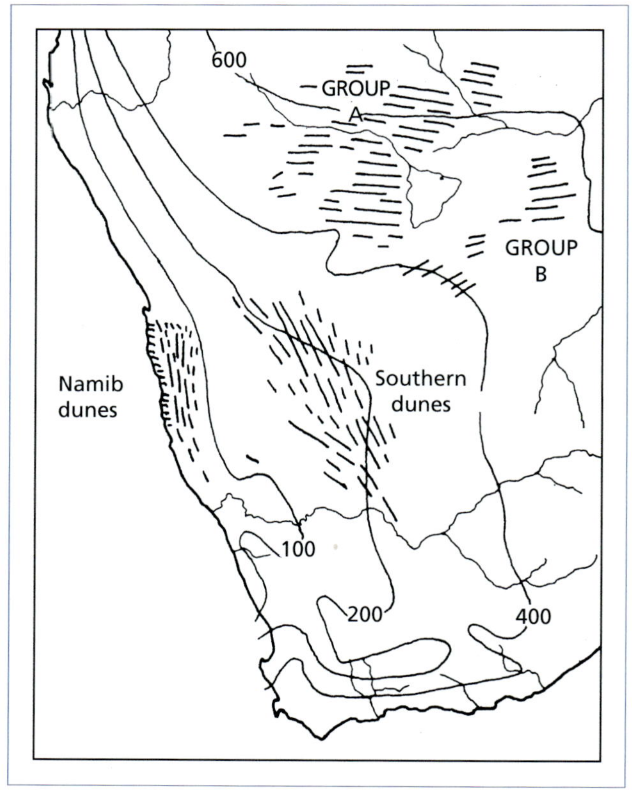

Figure 21.42 Isohyets and relict and dormant dune systems in the Kalahari region. Group A has an age prior to 21,000–30,000 years, group B (18,000), and the southern dunes are the dormant type and their sporadic activity begins in the early Holocene (Lancaster, 1981).

changed several thousand years ago. Studies of dune systems together with pollen, animal fossils, and sediments from paleolakes show that the ancient "Green Sahara" era occurred between 9,000 and 6,000 years BP (*Fig. 21.41*) (Adams, 2007).

Valuable paleoclimatic information can be obtained from a comparison of dune alignment and wind directions responsible for moving the sand. In the southern Sahara, winds (trade winds) blow in the same direction today as they did in the late Pleistocene. However, some regions show a clear contrast between present day and ancient wind directions. In the Kalahari Desert, dunes are primarily inactive (Stone and Thomas, 2008); they are mostly covered by acacia forests, shrubs, and grass. The three dune systems that have been identified in the Kalahari were formed by winds that blew from different paleodirections (*Fig. 21.42*) (Lancaster, 1981, 1989a).

Most present-day sand deserts formed slowly during Quaternary climate changes and subsequent sea level variations. As a result, researchers can identify distinct surfaces of regional extent (Kocurek, 1988) that show different stages of accumulation. These surfaces beveled the eolian deposits during periods when accumulation ceased. Pedogenic processes, water erosion, palustrine and lacustrine sedimentation can be recognized on top of these surfaces.

Deflation of marine, lacustrine, and fluvial sands (*Fig. 21.43*) supplies most sand deserts. A larger and more constant supply area through time leads to more transport, accumulation, and migration of dunes. As the sea level dropped during glacial periods, the Persian Gulf remained emergent for the most part (Kassler, 1973) and acted as a supply area for the Rub al Khali Desert, as indicated by the direction of dune systems (Glennie *et al.*, 1994). These dunes formed between 20,000 and 9,000 years BP (McClure, 1978). Sources of dune sand in the Mojave Desert (California) were fluvial systems and the floors of desiccated lacustrine basins. Aridity increased substantially at the Pleistocene-Holocene boundary and the sand supply abruptly decreased. Eolian activity was limited to dune remobilization (Lancaster, 1995).

Interdune deposits are common in many sand deserts such as the Al Liwa region in the Rub al Khali Desert (*Fig. 21.44*). Interdune deposits form in shallow lakes supplied by springs, rising water tables, or runoff. Their fundamental characteristic is their ephemeral nature, which is sometimes seasonal; short periods of expansion alternate with desiccation. Interdune deposits are important because they may contain fossils that enable us to determine ages and make paleoenvironmental interpretations. For example, interdune lacustrine deposits in the Rub al Khali Desert provide evidence for humid conditions between 2,400 and 3,200 years BP and also between 5,000 and 8,500 years BP (Lancaster, 1995).

The best place to study climate change is an area that has thick eolian accumulations indicating obvious aridity and sequences of predominantly palustrine and lacustrine deposition that indicate substantially more humidity. The Erg de Akchar (western Mauritania) is an excellent example of periodicity caused by climate change. Kocurek *et al.* (1991) (*Fig. 21.45*) conducted research using stratigraphy, ^{14}C dating, and prehistoric artifacts. The substrate consists of longitudinal dunes whose age ranges from 13,000 to 20,000 years BP. Two other stages of eolian reactivation were recognized, separated by surfaces that are identified stratigraphically by their specific characteristics. Finally, the other two periods correspond to phases of paralyzation of eolian activity due to implantation of plant cover. In these periods palustrine and lacustrine systems develop between the dunes with sandy, calcareous and gypsiferous deposition. In interdunal areas near the ocean, marine transgressions convert these depressions into littoral *sabhkas*.

Tributaries from the Hoggar Massif supplied the Niger River along its southeastern reach during the late Pleistocene (*Fig. 21.21*) but these tributaries are practically inactive today. In other tributaries discharge increases downstream to where the river flows into the delta. The upper Niger flows northeast and is supplied by the Guinea Mountains but during the late Pliocene and early Pleistocene it flowed into the Gulf of Senegal (Goudie, 1992). During a later dry phase, extensive

Figure 21.43 Star dune whose base is supported on laminated palustrine deposits. Maharaga. Al Liwa, Rub Al Khali Desert. United Arab Emirates.

Figure 21.44 Interdunal playa with a floor covered mostly by shrub vegetation. These playas can be flooded by precipitation or artesian springs. Al Liwa. Rub Al Khali Desert. United Arab Emirates.

dune fields became established and acted as a barrier to block the earlier river flow (Tricart, 1977). Numerous ephemeral lakes formed as a result. Through later capture or overflow, water flowed toward the Niger, forming a wide elbow in the region of Timbuktu (Mali). The capture took place about 5,000–6,000 years BP. During dry phases the Niger River was not able to cross the extensive dunes and became lost in the lacustrine zones of the dunes located south of Timbuktu. During "pluvial" stages, most of the river discharge can penetrate the eolian accumulations and change the nature of the river basin from endorheic to exorheic. All the lakes in this area expand significantly during "pluvial" phases; however, during dry periods their salt content increases by evaporation, including Lake Fagnibine, which dried up in 1957 (Tricart, 1977). Today the climate allows water from the Niger to flow into the Gulf of Guinea, which shows that precipitation is higher than during the endorheic periods of the river.

The chronological information available shows that most ergs have experienced episodic accumulations during the last 120 ka, with long periods of stability when sedimentation did not occur separated by relatively brief periods of dune formation. Luminescence dating shows widespread dune activity in low-latitude deserts during periods of 120–110 ka BP, 35–10 ka BP, and since 5 ka BP (Lancaster, 2007) (*Fig. 21.46*).

In the Sahel (western Africa), Talbot (1980) showed the existence of a fluvial network tributary to the Niger River in its northern area (*Fig. 21.47*); however, it completely lacked functionality. From this we can infer that in the past the Dallol Bosso River (the present-day relict) may have been the principal trunk of the river. To the east, the enormous surface area of Lake Megachad overflowed from the lake into the hydrologic basin of the Niger River during the humid phase of the transition from Pleistocene to Holocene.

Also in East Africa, Lake Turkana expanded during the humid phase and part of its water overflowed into the Nile River (Harvey and Grove, 1982; Frostick and Reid, 1989) (*Fig. 21.48*).

21.1.4 Climate change in desert regions and uncertainties

The interpretation of some paleoclimatic data suggests that global climate change may cause much higher precipitation in many arid regions and may thus make ecosystems more productive. However, it is difficult to make analogies between past climate changes and those caused by greenhouse gases (Lioubimtseva, 2004).

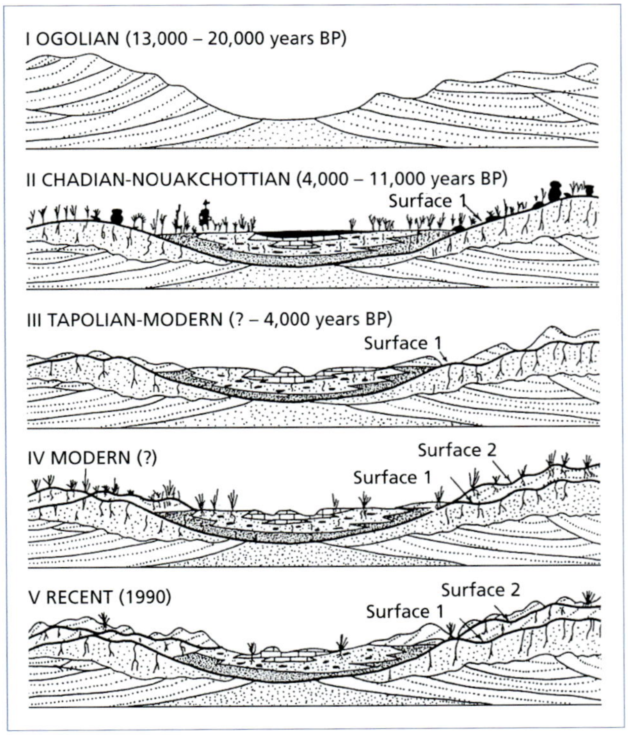

Figure 21.45 Evolutionary sequence of stages of eolian activity and periods of stabilization in the Erg Akchar (Mauritania). (I) Longitudinal dunes formed during the Ogolian constructive phase. (II) Stabilization of dunes by vegetation and formation of lakes between dunes during the Chadian and Novakchottian. (III) Reactivation of dunes and desiccation of lakes during the Tapolian. (IV) Formation of a new surface 2 vegetated during the most recent period. (V) Modern reactivation of dunes with remobilization of ancient ones (Kocurek *et al.*, 1991).

The results of general circulation climate models for arid areas are complex and contradictory for a future climate responding to the greenhouse effect (Hulme, 2001; IPCC, 2001). In spite of large disparities in the responses of arid ecosystems to climate change, the "scenarios" predict increased precipitation in present-day deserts as a result of increased CO_2.

Drastic changes in desert climates during the recent geologic past are well documented in most of the world's arid regions (Nanson *et al.*, 1992; Yang and Petit-Maire, 1994; Petit-Maire and Guo, 1998; Kershaw *et al.*, 2000; Hetzel *et al.*, 2002). Variations of Pleistocene climates can be observed in many present day arid zones. But can paleo-reconstructions provide an understanding of present-day tendencies of arid climates? (Lioubimtseva, 2004).

Abundant information is available on paleo-environmental changes in the **Saharan** and **Arabian** deserts during the late Pleistocene and Holocene. Our knowledge of climate and vegetation changes in these regions has led to various paleo-climatic reconstructions (Sanlaville, 1992; Petit-Maire *et al.*, 1995; Lioubimtseva, 1995; Guo *et al.*, 2000; Swezey, 2001). During the Glacial Maximum, about 21,000 years BP, the 100 mm isohyet moved south to 13–14° north latitude. During the Climate Optimum of the Sahara (8,500–6,500 years BP) precipitation increased significantly from 200–300 mm compared to present-day precipitation data (Petit-Maire *et al.*, 1995). During this time the Sahara-Sahel boundary moved to 23°N, which is 500 km north of its present position and about 1,000 km from its position during the Glacial Maximum (Yan and Petit-Maire, 1994). The expansion of deserts at the expense of arboreal and herbaceous vegetation during the Glacial Maximum may have been caused not only by cooling or drier conditions but also by decreased atmospheric CO_2 (Lioubimtseva, 2004).

Most data from **Australia** suggests that the mid Holocene was generally warmer during winter but cooler in summer than today (Kershaw *et al.*, 2000; Moss and Kershaw, 2000). In general, Australia received less precipitation during every season except spring. As a result, throughout the year the modeled Australian surface had a lower moisture content than at present. In contrast, in northeastern Queensland, in eastern Australia, which is an area of intense paleo-environmental research, the climate was more humid because of increased regional precipitation on an annual cycle.

Biostratigraphic, geomorphic, and archeological data suggest a cold arid stage in **Asia** during the Glacial Maximum. Moraine ages in the Tian Shan mountains record maximum glacial advances between 19,000 and 20,000 years BP and a more recent advance during the Younger Dryas around 15,000 years BP (Hetzel *et al.*, 2002). During the mid Holocene, precipitation increased and a humid phase occurred during the Climate Optimum around 6,000 years BP. The atmospheric circulation mechanism in these deserts is quite different from controls on precipitation in tropical deserts. Atmospheric processes in Russia and Siberia seem to the principal factors that control precipitation in these desert plains.

As a result of the mosaic of climate and vegetation zones in the western **United States** caused by high elevation and complex orographic conditions, the resolution of climate models can indicate with some exactitude only the spatial variability of climatic tendencies in this region. Paleo-reconstructions show very complex spatial changes of vegetation systems in this region, which are reflected by topographic effects. During the Glacial Maximum, the southwestern United States was much more humid and cooler than it is today. These more humid conditions persisted until

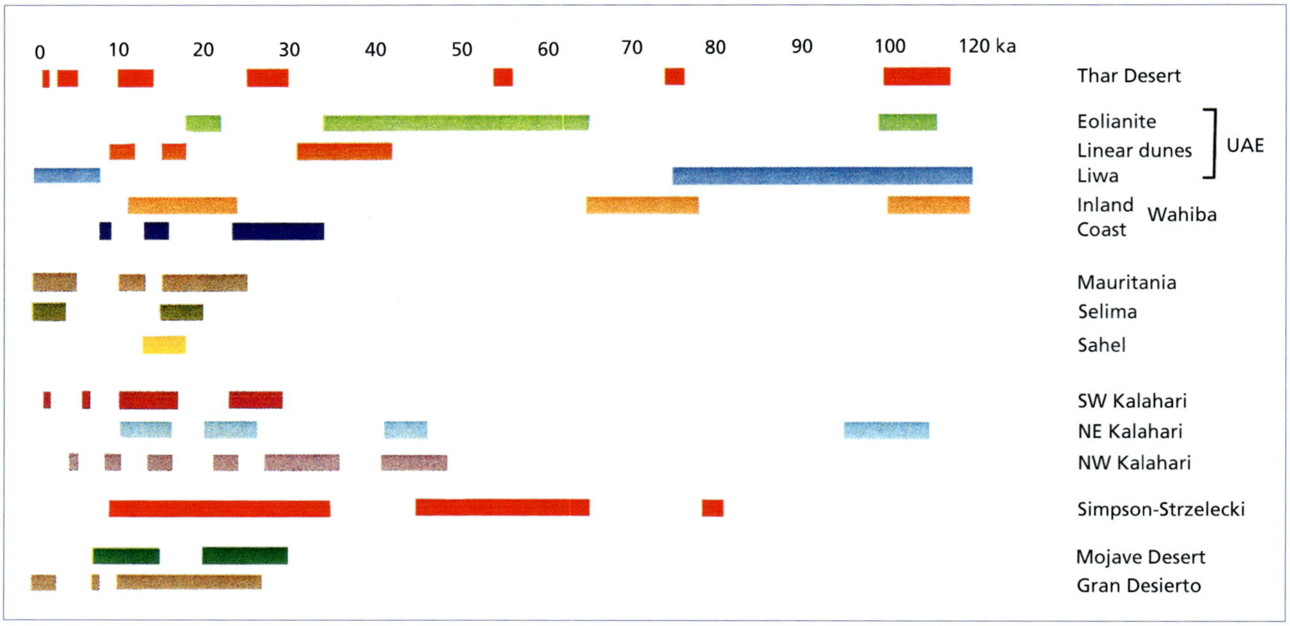

Figure 21.46 Chronological summary of periods of dune construction and/or activity during the last 120 ky, using data from luminescence methods (Compiled by Lancaster, 2007).

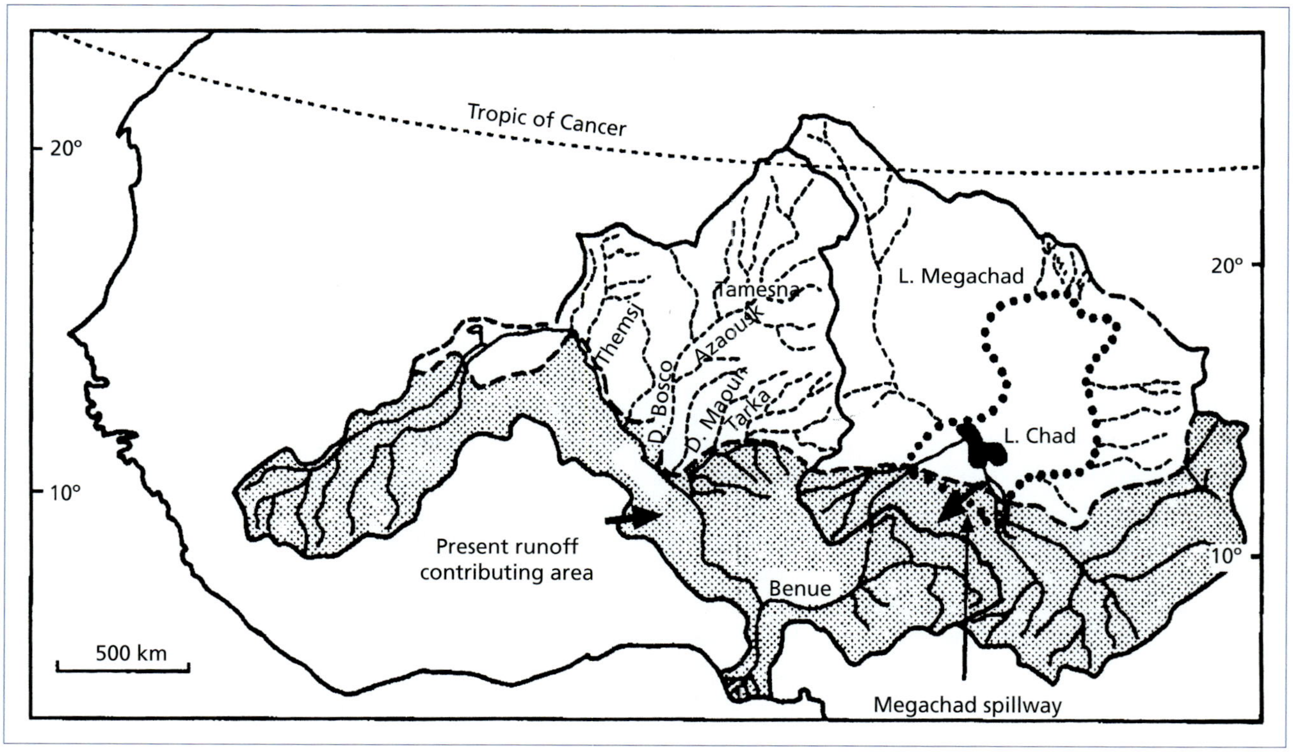

Figure 21.47 Hydrologic basin of the Niger and Benue rivers since 9,000 years BP and the actual area that contributes runoff to the Niger Delta and Lake Chad (dotted). The perimeter of the Lake Megachad basin and relict tributaries of the Benue River (Talbot, 1980).

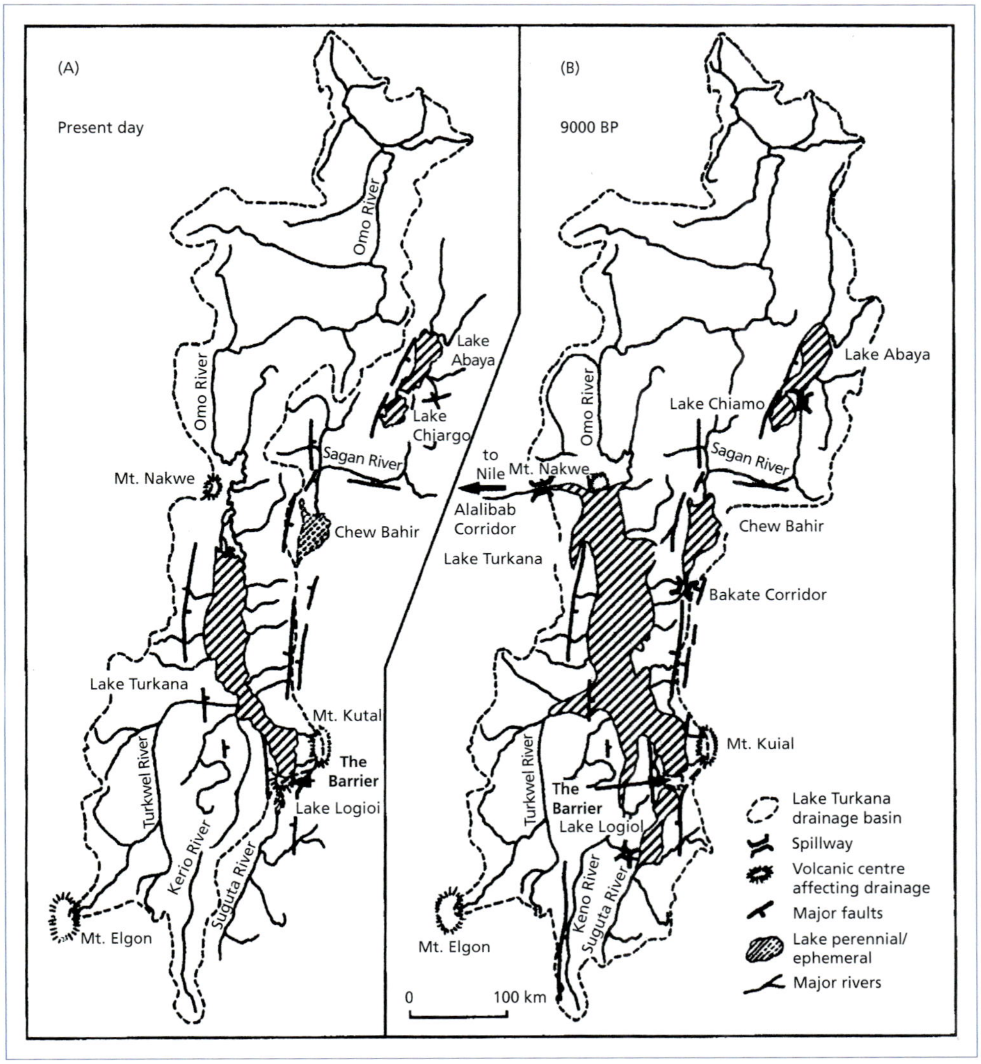

Figure 21.48 Hydrologic basin of Lake Turkana in the northern section of the East Africa Gregory Rift: (A) during the present day of relative desiccation, and (B) during the humid phase of early Holocene about 9,000 years BP, that shows the overflowing to the Nile River (Frostick and Reid, 1989; Harvey and Grove, 1982).

the beginning of the Holocene. The desert climates of the southwestern United States were warmer and drier during the mid Holocene (Thompson *et al.*, 1993). Data from the mid Holocene suggests small climate changes around 6,000 years BP; these have led to diverse interpretations depending on the area studied. Some dates show considerable temporal climate variability during the Holocene in the southwestern United States (Polyak and Asmeron, 2001). These authors point to signs of major climate variability during the second half of the Holocene in New Mexico; climate conditions around 4,000 years BP comparable to today's more humid climate predominated in these deserts until 3,000 years BP. This higher humidity lasted until about 800 years ago. Dean (1996) obtained dendrochronological and archeological ages that suggest two very dry episodes between 1130–1180 AD and 1275–1300 AD.

Arid climates show significant variability over time. Paleodates provide valuable information on global and regional climate change mechanisms over relatively long time periods (centuries and millennia). However, these data cannot be used to provide smaller fluctuations of temperature and precipitation. The interpretation of recent climates is complex due to interannual climate variations and observations from the relatively short period of instrument records (Lioubimtseva, 2004).

The temperature predicted for the year 2050 for the Saharan and Arabian deserts is 1.6°C higher than today's

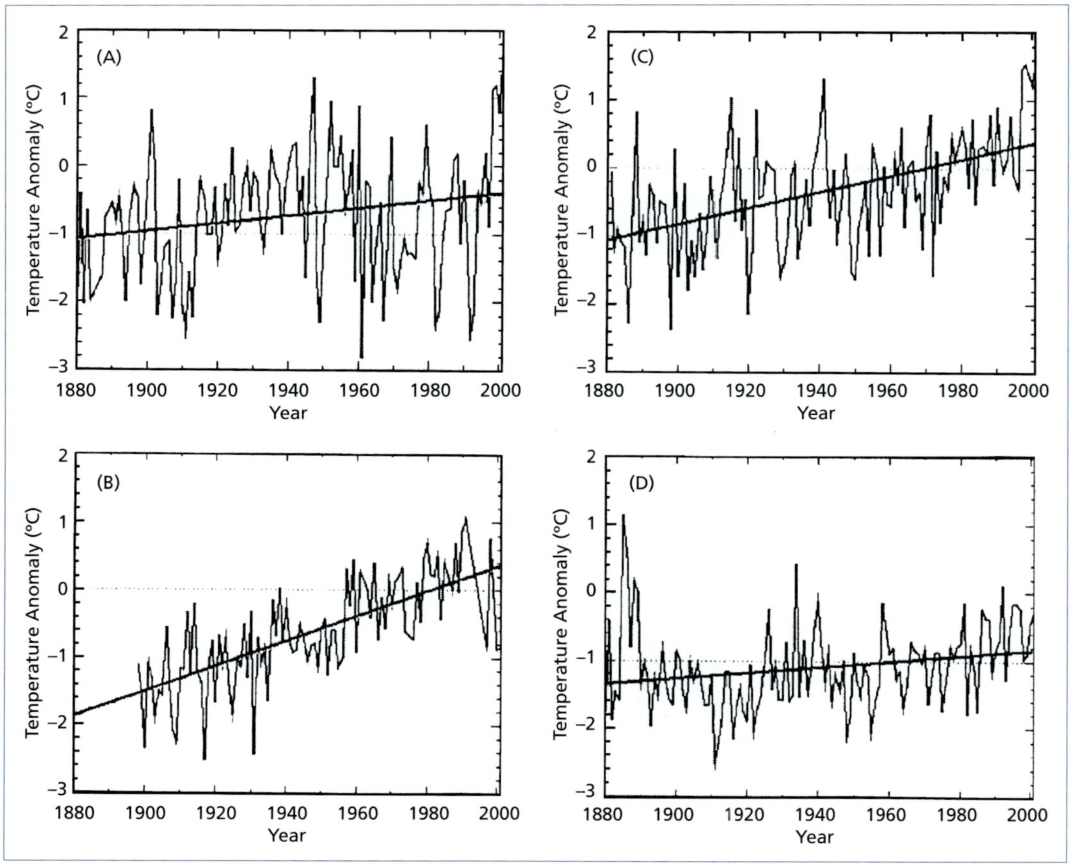

Figure 21.49 Annual temperature trends since 1900. (A) The Sahara and Arabia region; (B) central and western Australia; (C) central Asian deserts; and (D) Southwestern United States (Peterson and Vose, 1997).

temperature (IPCC, 2001). Global temperature tendencies for these areas are expressed as anomalies based on the set of Global Historical Climate Network data (Peterson and Vose, 1997) (*Fig. 21.49*). However, some measurements in the southern and eastern Mediterranean do not reflect these tendencies. A tendency toward cooling of about −0.5°C/year has been detected in North Africa, the Negev Desert, and the Arabian Peninsula (Kutiel *et al.*, 2000). There is also major uncertainty about regional scale precipitation changes simulated by the General Circulation Models (GCM) (Hulme, 2001). Some models state that precipitation increased 15% in the period of 1961–1990 (Joubert and Hewitson, 1997); other researchers forecast less precipitation in the same region (Hulme, 2001). We must remember that the models do not incorporate the effect of dust fluxes from the Saharan and Arabian deserts and urban aerosol pollution along the Mediterranean coast. As a result of these shortcomings, the GCM scenarios have a relatively little predictive power.

It is generally accepted that computer-based modeling offers the most effective method of predicting future global climate and its impacts (Peterson and Vose, 1997; McGuffie and Henderson-Sellers, 2001). Significant advances have recently been achieved in the development of the Earth Model of Intermediate Complexity (EMIC). This model describes most processes implicit in the comprehensive models and also simulates interactions between various components of the climate system, including biogeochemical cycles. EMIC makes it possible to do long-term climate simulation over tens of thousands of years or to conduct a wide range of sensitive experiments over several millennia. In spite of the progress of climate models in recent decades, considerable discrepancies between the different models remain. The principal reason for the high degree of uncertainty is the modeling of arid climates, primarily due to the wide natural variation of desert climates in both time and space. Arid climates have different interannual, interdecadal, and seasonal variations, principally in precipitation. Although temporal precipitation variability is taken into account in many studies, spatial variability is also an important variable in ecosystem processes (Whitford, 2002).

Increased atmospheric CO_2 has a direct and relatively immediate effect on physiological processes in plants;

it increases photosynthesis rates but decreases stomatal apertures and thus reduces a plant's capacity to lose water. This combination of factors leads to a significant increase of effective water in desert vegetation as a result of the high atmospheric CO_2. Projections based on biogeographic models suggest considerable changes to vegetation in arid and semiarid zones as a result of the combination of greenhouse gases and the effects of CO_2 on plant physiology. Even CO_2 experiments on desert vegetation show some very complex responses (Lioubimtseva, 2004).

21.1.5 Causes of paleoclimatic oscillations during the late quaternary

This topic has ignited research on arid zones; this research has focused primarily on the African continent perhaps because it is a zonal desert or because it has been studied and is well understood by geomorphologists from many countries. However, Tricart (1969) claims that reconstruction of paleoclimatic oscillation mechanisms is different for zonal deserts and extrazonal deserts. Zonal deserts owe their existence to their location in subtropical high-pressure areas whereas extrazonal deserts form as a result of particular geographic configurations such as continentality, orographic effect, or cool ocean currents. Extrazonal deserts are usually less sensitive to variations of terrestrial thermal balance and their evolution is governed by more local factors.

Nicholson and Flohn (1980) have compiled theories on general circulation during the Pleistocene in an excellent paper in which numerous authors theorize on the drift of mid latitude cyclone trajectories, thermal contrasts imposed by the presence of large ice masses with respect to pluvial-glacial synchrony, and other subjects.

Nicholson and Flohn (1980) have conducted the most relevant summary on climate and environmental change in Africa and on general atmospheric circulation during the late Pleistocene and Holocene. Seasonal variations were minimal during the maximum ice expansion (18,000 years BP) in the Northern Hemisphere. As a result, cyclonic zones displaced southward and the temperature gradient increased. The Southern Hemisphere, in contrast, had characteristics very similar to those of today.

These authors think that during this period and for a large part of the interval from 20,000 to 12,000 years BP, aridity increased in the southern Sahara and in East Africa. This contradicts data by Sarnthein (1978) who claims that the aridity interval lasted 2,000 to 3,000 years around 18,000 years BP. But this assumption does not explain that from this time until 12,000 years BP, humid conditions prevailed in North Africa, in the massifs within the Sahara, and in southern

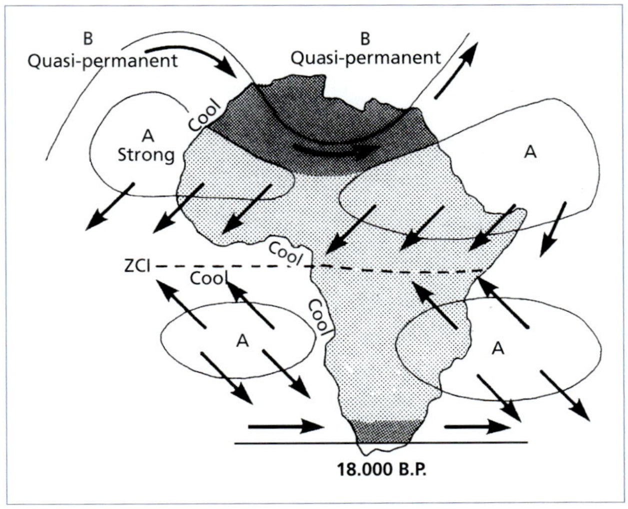

Figure 21.50 Schematic model of atmospheric circulation in Africa for the period of 21,000–12,000 years BP. Dark shading: areas of higher humidity than today. Light shading: zones that were drier than today (Nicholson and Flohn, 1980).

Egypt. Nicholson and Flohn show that major environmental change in tropical Africa coincides with rapid ice sheet growth (pre-22,000 years BP) and their disappearance soon thereafter (12,000 years BP). For most of this period, circulation systems assume aridity in tropical Africa and relatively humid conditions in North Africa (*Fig. 21.50*) as a result of the southward displacement of the atmospheric circulation system.

Approximately 10,000–8,000 years BP, a significant change in the circulation system caused the first subtropical lacustrine episode as a result of increased precipitation; this caused lakes to expand in the southern Sahara from east Africa to Senegal and Mauritania. In turn, aridity became established in northwestern Africa. These circumstances are related to the gradual warming of the northern Atlantic Ocean, a decrease in the temperature gradient, and a maximum drift to the north, which displaced aridity toward the northwest, as shown by dunes advancing into Morocco and western Algeria.

Later, around 7,000 years BP, the ice of the Fennoscandian Ice sheet melted; northern Europe experienced the Climate Optimum around 6,000 years BP while the Southern Hemisphere remained cool. During the period from 6,500–4,500 years BP, both the northern and southern Sahara were more humid than today but with less intensity than during the period of 10,000–8,000 years BP perhaps due to fact that the Northern Hemisphere was much warmer during this third interval.

The Nicholson and Flohn (1980) model is the most complete of the existing models although it contains various contradictions, in particular the interpretation of the southern part of Africa (Thomas, 1997c). The problem becomes highly complicated when interpretations are not limited to the African continent but are made on a global scale. Difficult and major work remains to be done for which it will be necessary to gather new data, reinterpret the past, and use the largest possible number of dates. We must also understand Quaternary climate evolution in the Mediterranean area because it is a transition zone between Europe and warm Africa. This work will require a detailed exploration of arid zones.

21.2 Tropical regions

21.2.1 Introduction

Tropical climates include a wide variety of precipitation regimes and plant groups, from the rainforest to the savanna and transitions to semiarid zones. For this reason, the activity and intensity of geomorphic processes vary according to the zone where we find ourselves. Variations of these morphoclimatic limits throughout time are accompanied by changes in geomorphic processes and, as a result, of existing landforms. Even so, for Douglas (1969) the nuclei of humid tropics are examples of the few stable ecosystems on Earth where plant groups use energy and where there is no water erosion on rainforest plains.

The most significant studies on climate change in the humid tropics were done after 1950. In their research on surface deposits, Brückner (1955) in Ghana and De Ploey (1965) in Zaire found clear evidence of Pleistocene aridity. In South America, ancient dunes found in the Llanos (Venezuela), in the State of Bahía (Brazil), and in the central Amazon (Tricart, 1974a, 1975, 1985) are all examples of evidence of arid conditions in humid tropical areas. Bigarella and Andrade (1965) and Bigarella *et al.* (1969) developed their hypothesis of pediplanation in southeastern Brazil. The work of Aubreville (1962) and Butzer (1978) focus on spatial variations of savanna and rainforest boundaries as a result of climate change. Over the last twenty years numerous new publications on this topic have appeared but research is hampered by a lack of long-term chronologic records (Thomas, 1994a); researchers are forced to resort to the temporal records of lakes in arid zones of Africa and Central America (Street-Perrott *et al.*, 1985) because lakes in tropical regions are rare. Even so, Talbot *et al.* (1984) and Shanahan *et al.* (2006) studied Lake Bosumtwi (Ghana), which is lodged in an impact crater in the humid tropics. The most significant examples are found in the East Africa Rift Valley (*Fig. 21.25*).

Research on the impact of late Quaternary climate change on tropical African river systems focuses primarily on the Nile River (Butzer, 1980; Williams and Faure, 1980; Williams and Williams, 1980; Williams *et al.*, 2000). Activity increased after 15 ky and lake levels have risen (Thomas, 2008).

Records from semiarid lakes and data from ice sheet cores and from deep ocean borings are also available for climate study (Thomas, 2008). ^{14}C dating and AMS ^{14}C are limited to a range of about 60,000 years.

Research on cores from equatorial Atlantic boreholes has focused on the contents of freshwater diatoms, opal phytoliths, pollen, and sapropelitic muds of the last 150,000 years. Depending on the method used, various arid phases have been identified and up to seven dry phases have been recognized (Jausen and Van Iperen, 1991).

21.2.2 Biological modifications

Changes of sea and land configurations in atmospheric circulation or in relief can cause significant modifications in

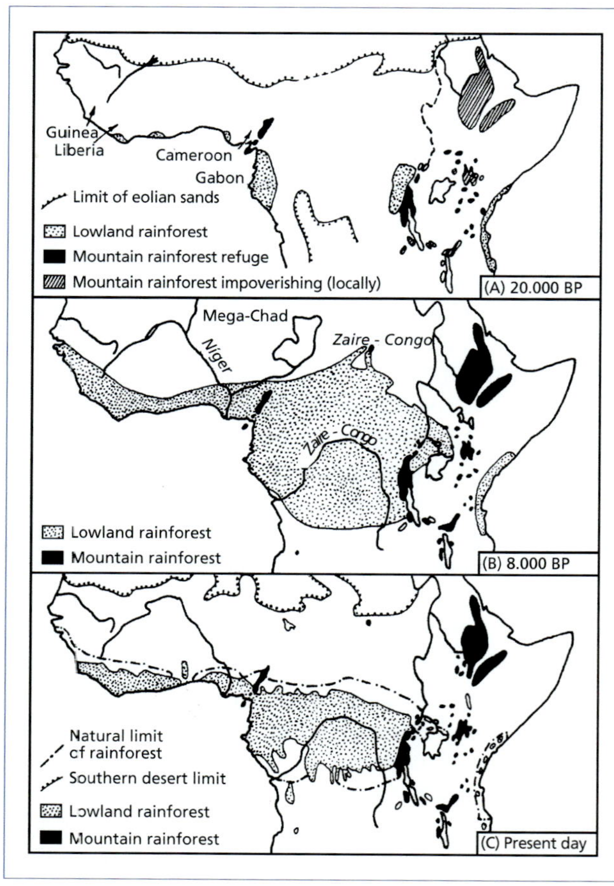

Figure 21.51 Probable distribution of rainforest in central Africa 21,000 years BP and 8,000 years BP compared to today. Note the presence of Great Lake Chad 8,000 years BP (Hamilton, 1976).

moisture content to the extent that they can cause savannas to replace rainforests and vice versa, with the resulting modification of morphogenesis (Tricart, 1974b). In the lower Congo and in southern Ivory Coast, rainforests tend to replace the savanna (Meulenberg, 1949). The opposite effect is observed in Senegal.

In his review of rainforest and savanna areas, Aubreville (1962) found anomalous ecological evidence in both the African humid tropics and in Guyana. These anomalies are interpreted to mean that the rainforest did not recover after the drier conditions that prevailed during the last glacial period. This indicates a southward drift of rainforest regions with the main nucleus in Angola; there are some remaining riverside regions in the Gulf of Guinea and the Congo. This interpretation contradicts the idea of a stable rainforest. Butzer (1978), who studied similar characteristics of the glacial optimum (18,000 BP), proposed an extensive reduction of the tropical rainforest that limited it to three bastions: Guinea-Sierra Leone-Liberia, Nigeria-Cameroon, and Zaire. Research on the distribution of African flora and fauna in rainforest areas conducted by Hamilton (1976) provides important data on the considerable variations that have affected the central-African rainforest during the last 20,000 years (Fig. 21.51). Figure 21.52 shows two global maps of forest and desert distribution for use in

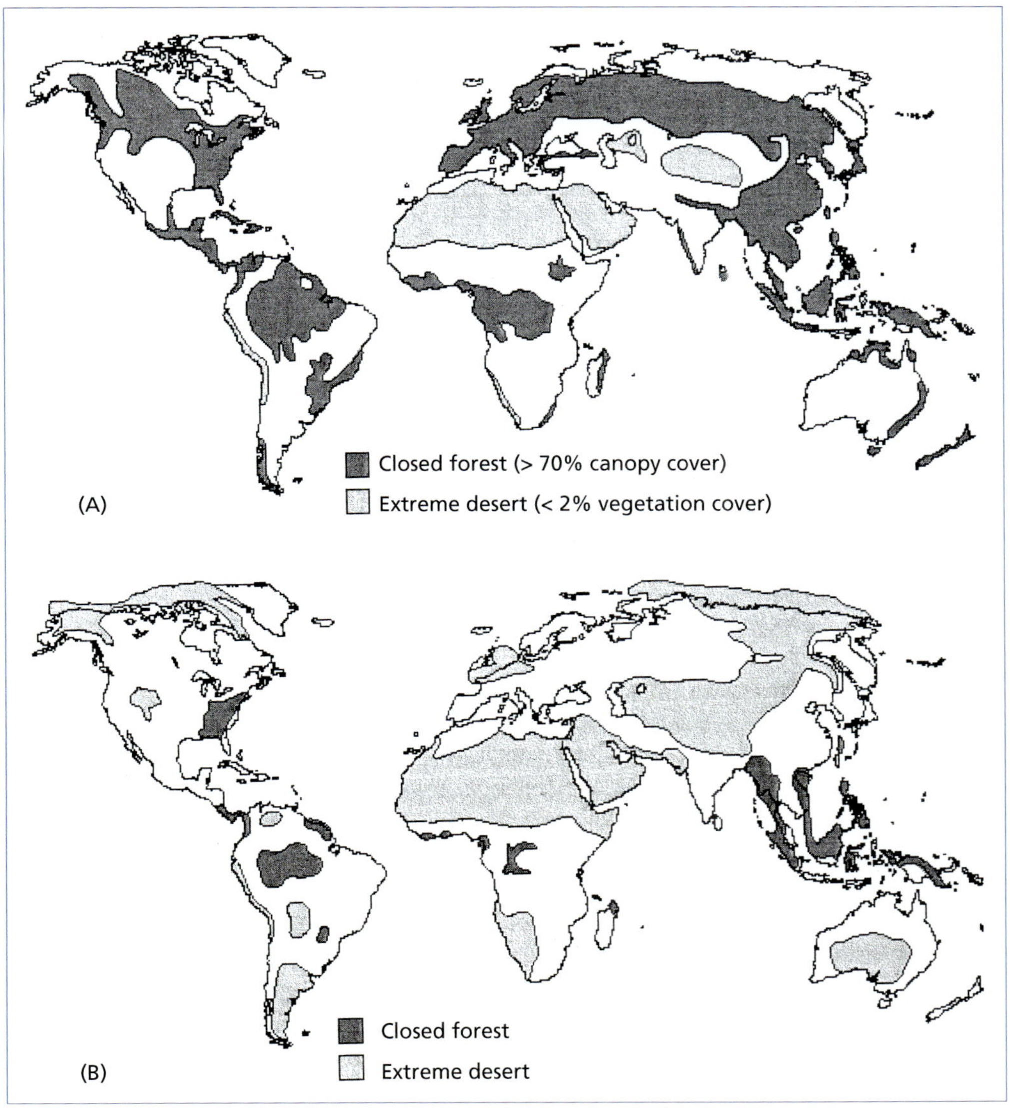

Figure 21.52 Distribution of forests and deserts in: (A) in a present day natural world and (B) during the Last Glacial Maximum (18,000 years BP). The Last Glacial Maximum was much colder and drier and as a result there was much less forest and more desert (Adams, 2007).

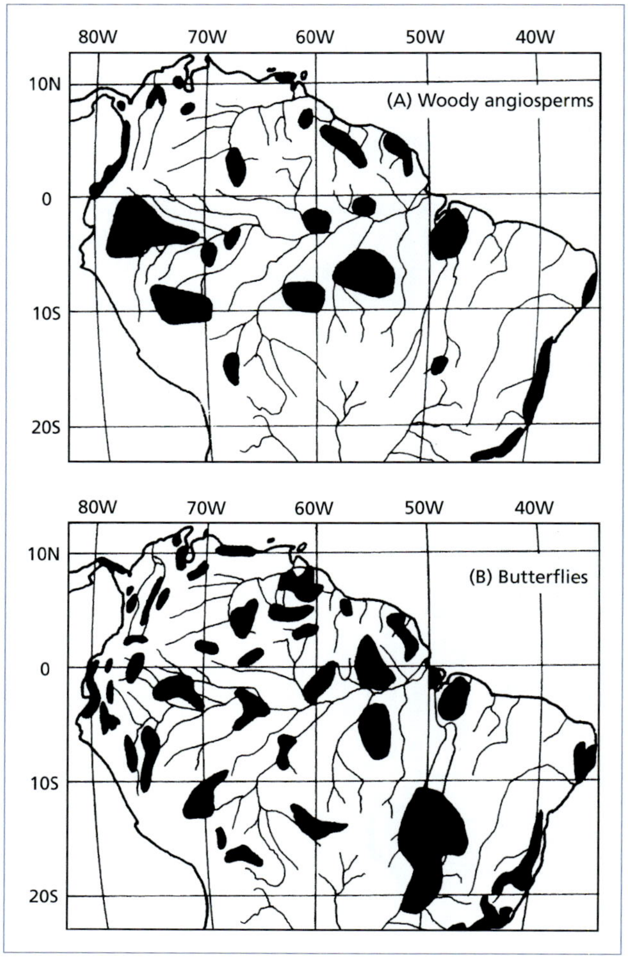

Figure 21.53 Refuge areas in the Amazon Basin for some species of (A) woody angiosperms and (B) butterflies, during dry phases of the Pleistocene (Bradley, 1985).

comparing present day vegetation with that of the Last Glacial Maximum. Areas that are dominated by forests today were shrubs and grasslands, and those that are pasture areas today used to be desert and semideserts (Adams, 2007).

In the South American rainforest, there are curious speciation systems of various animals that seem to have been caused by changes in the areal extent of the rainforest during the late Pleistocene. The rainforest was fragmented in small isolated compartments in which speciation occurred (*Fig. 21.53*) (Goudie, 1983a; Bradley, 1985; Crowley and North, 1991). In Brazil paleogeographic evidence shows the presence of xeromorphic plants in *cerrados and tabuleiros* in central Brazil and western Amazonia. The Barreiras Formation, composed of quartz-bearing and kaolinitic sands, is the predominant outcrop; it was intensely dissected during the last glacial period due to sea level drop resulting from marine regression. Presence of a drier climate facilitated incision, which caused the rainforest to disappear from lowlands along the Amazon River. This implies a reduction of rainforest area during dry periods; remnants of rainforest persist only in the highest areas of Guyana and the Brazilian Shield where precipitation is higher. During the Holocene, the climate became more humid and the Flandrian transgression caused a rise in sea level (Dias de Avila, 1974; Tricart, 1985).

21.2.3 Geomorphological evidence

21.2.3.1 Regolith and relict weathering profiles

The deepest weathering occurs in tropical regions; normally these weathering profiles developed during previous periods although the present-day climate characteristics of the tropics accentuate weathering effects (Ollier and Pain, 1996). Some authors claim that regolith thickness in elevated areas is much less than in areas of low relief; however, others claim that in some cases the thickest weathering deposits are found in high areas and that fresh rock crops out at the base of valleys due to fluvial incision (Pain and Ollier, 1981). From this we infer that weathering depth is a function of the intensity of alteration and the amount of erosion. In addition, extremely thick profiles need deep groundwater for an active weathering front (Ollier, 1988a).

Numerous cases throughout the world show weathering profiles that attained thicknesses of more than 100 m during the Tertiary and Cretaceous. Most of these profiles are not in equilibrium with today's climate conditions; because they have resisted numerous climate changes, it is difficult to interpret them. Chapter 18 contains numerous examples of deep weathering profiles in both the humid tropics and extra-tropical regions. The latter are located in Europe, Australia, India, and the United States and the weathering may date back to the Jurassic. From these investigations we can infer that principal periods of deep weathering took place during the Mesozoic or early Tertiary when thick regoliths developed that were later eroded, and at the same time alteration decreased in intensity (Valeton, 1994). During the Eocene and at various times during the Cretaceous, the land was warmer and more humid than it is in the present-day geographic position of these weathered areas. The poles were warmer and temperatures in the tropics were lower (Walker and Sloan, 1992).

Dating of weathering profiles is generally difficult but several well-known examples are can be cited. In the Mojave Desert (California) upper Miocene basalts (8–9 million years old) overlie weathering profiles that developed in granite. A later more arid stage eroded part of the basalt and the

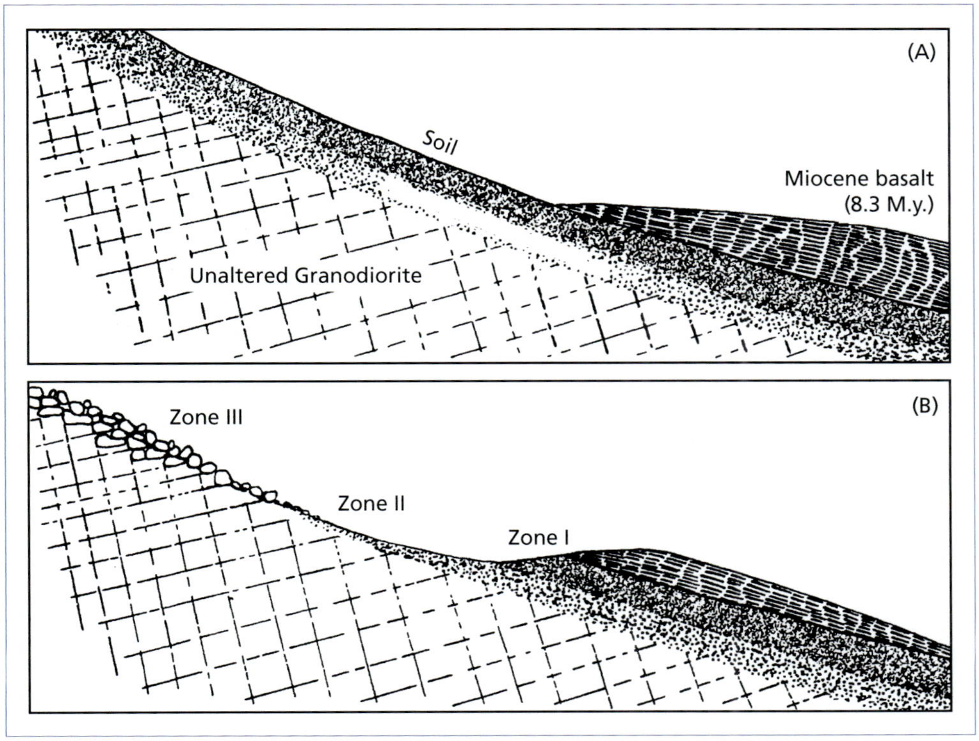

Figure 21.54 Basalts and alteration profiles in the Mojave Desert (California). (A) Immediately after the emplacement of the basalt. (B) Present-day situation. Zone I: relict weathering preserved beneath the basalt. Zone II: partially eroded surface. Zone III: outcrop of residual blocks (Oberlander, 1972).

outcropping profile (Oberlander, 1972) *(Fig. 21.54)*. Oxygen isotope studies of deep regoliths in Australia have made it possible to distinguish profiles that formed during the late Mesozoic and early Tertiary from those generated later than the middle Tertiary. These profile differences are thought to be due to migration of the Australian continent from high to low latitudes during the Mesozoic and Tertiary (Bird and Chivas, 1988).

21.2.3.2 Inselbergs

Inselbergs are spectacular landforms that present major interpretation problems. The principal controversy in the study of inselbergs is their origin. On one side, authors such as King (1948; 1975) and Kesel (1973) consider inselbergs as subaerial residuals generated from scarp retreat processes. Other researchers, among them Büdel (1957); Ollier (1960); Thomas (1965; 1994a); Twidale (1964; 1982a), Vidal and Twidale (1998) and Twidale and Vidal (2005) claim that inselbergs are landforms exhumed from a thick chemical alteration mantle.

According to the first hypothesis, the pediplanation that frequently accompanies these landforms is related to the intrinsic development of the landform and weathering is superimposed on the inselberg-pediment system after its formation. According to the exhumation hypothesis, regolith development is the dominant process in inselberg formation and the accompanying pedimentation plays a later and minor role.

Development of inselberg landforms according to the slope recession theory occurs predominantly in semiarid morphoclimatic zones. These landforms may be formed by exhumation in humid tropical morphogenetic systems. In the latter case, an explanation of small dome inselbergs (*bornhardts*) up to 50 m tall by simple denudation of a thick weathering deposit presents no problems. When this hypothesis is applied to the formation of large *bornhardts* significant difficulties arise but the difficulties decrease when we consider continuous weathering at the foot of a slope as denudation lowers the surrounding plains.

In savanna areas, morphoclimatic histories are lacking (Birot, 1978), at least during the Quaternary, and inselbergs proliferate in these areas (Kesel, 1973). The alternation of humid and dry climate stages promotes the development of these landforms. In addition, we should note that once a bornhardt-plain system is established, differential weathering rates between the two ensure that the system persists (Thomas, 1978).

In some cases, inselbergs have formed in previous periods of the geological past and have persisted until today. Twidale (1978) applied this interpretation to the famous Ayers Rock monolith, which is located in the desert plains of central Australia; from an analysis of correlative sediments and geomorphic interpretations, he concluded that it began to develop during the Mesozoic-Cenozoic transition.

21.2.3.3 Fluvial systems

Channel and fluvial sediment morphology in the humid tropics records the effects of environmental fluctuations, data on fluvial regimes, and their temporal evolution because many of these sediments can be dated (Thomas, 1994a; Thomas and Thorp, 1995).

The Amazon River, which has a discharge equal to 15% of the combined discharge of the world's rivers, has a gradient of about 0.1 m/km near the Peruvian Andes; it decreases drastically to 0.03 m/km along the stretch from Manaus near the basin center to the mouth 1,400 km downstream. These elevation and inclination differences clearly affect channel morphology. Rivers that drain the Andes, such as the Solimoes, the principal trunk of the Amazon, transport bedload and especially suspended load (*Fig. 18.15*); its channel floodplains have low sinuosity. In contrast, rivers in the interior of the river basin such as the Juruá River transport practically no bedload or suspended load and the channels are highly sinuous. This is interpreted to be a function of the river's capacity to move loads during dry periods. These coincide with glacial periods in which channels had low sinuosity (Baker, 1978).

It was previously thought that tropical rivers did not have terraces but exploration of alluvial placers put an end to this idea (Tricart, 1974b). Most rivers that drain the Ivory Coast have a terrace composed of rounded quartz alluvium covered by 2 m of unctuous reddish clay. It is situated 3–4 m above the present day riverbed where gold and diamonds are mined (*Fig. 21.55*) (Vogt, 1959).

In diamond-bearing alluvial placer mines in the center of Sierra Leone, floodplains and channel fill have been studied by stratigraphic sections and drill cores. ^{14}C dating of the materials shows a long history over the last 30,000 years. In addition, older large ferruginous river terraces are present but have not been dated. Along the floodplain margins, black clays from a swampy environment have been identified as dating from a later period (36,000 and 20,500 years BP) and gravel channel fill also dates from a later period. These last two groups seem to have been formed during arid phases whereas excavation and Holocene fill are related to humid periods (Thomas and Thorp, 1985). A fluvial substrate of weathered slate with polygonal cracks filled to a depth of 2 m has been identified in southern Ghana. This intense

Figure 21.55 Meander of the gold-bearing Nzi River that flows across an arboreal savanna and carries a high suspended load. Eastern Bouaké (Ivory Coast).

cracking results from desiccation in dry environments practically devoid of vegetation. A group of diamond-bearing deposits overlie this bed; because they are dated at various levels, it is possible to interpret various dry and humid stages (Hall *et al.*, 1985).

In the Juruá River in the southwestern Amazon basin, two terraces 5–6 m and 12 m above the water level are composed of sands and clays on which a reddish-brown soil has developed. The valley floor is also argillaceous sand. This assemblage shows that present-day observers are looking at an incision phase and that two stages of alluviation occurred in the past. This alternation seems to be related to climate change. In the Solimoes River, more to the northwest, the same sequence of accumulation and incision has been observed and some abandoned meanders have been identified on top of the terrace. Here, significant regression during glacial periods triggered dry climates in which intense entrenching took place. For this reason, terrace formation seems to be eustatic rather than climatic (Tricart, 1985).

The classic works of Bigarella and Andrade (1965) and Bigarella *et al.* (1969) are a clear exponent for the inference of Pleistocene arid stages from the study of pediments. Their study focused on the Serra do Mar and the southward prolongation of littoral regions in the states of Paraná and Santa Catarina in Brazil. They show that stepped valley margins result from an alternation of arid stages when pediments formed and humid phases when weathering and incision predominated (*Fig. 21.56*). Arid phases correspond to glacial stages whereas humid periods correlate with interglacial stages.

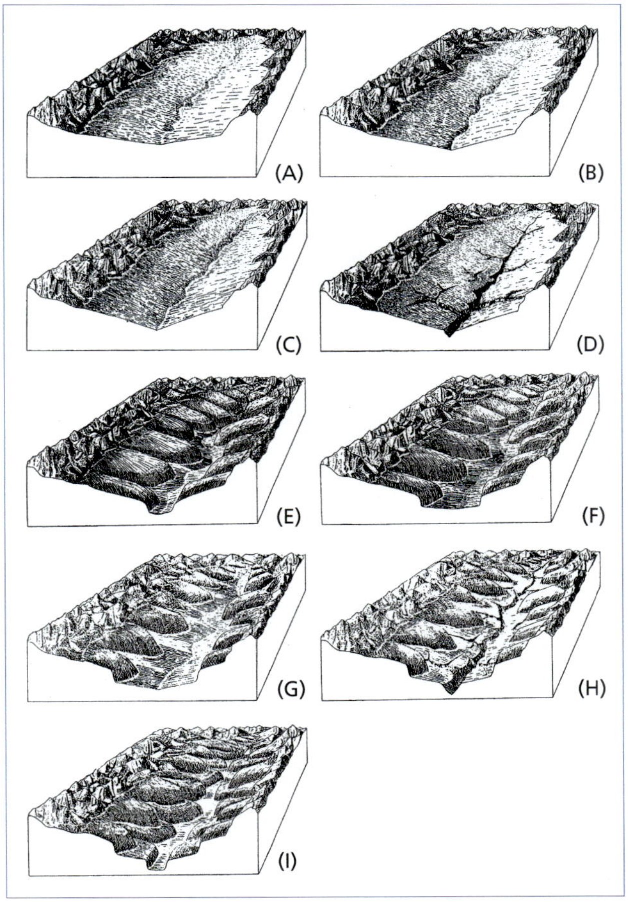

Figure 21.56 Basic scheme of slope evolution. (A) Intramontane surface extent formed by pediplanation of an arid climate. (B) and (C) Lowering of the planate surface caused by a slight drop of local base level, as a result of small climatic fluctuations toward humid conditions during an arid epoch. (D) Generalized dissection of the planate surface due to increased humidity. (E) Widening of the valley, alluviation, and colluviation during brief arid episodes within the humid stage. (F) Scarp retreat and pediment formation under arid conditions. (G) Lowering of pediment slopes during weak humid fluctuations within the arid period. (H) General incision below a new humid period. (I) Widening and alluviation of valleys caused by climate fluctuations tending to aridity (Bigarella *et al.*, 1969).

21.2.3.4 Eolian activity

Eolian accumulations and deflation-generated landforms found in the humid tropics are excellent indicators of prior arid stages. Most research on this subject has been done in central-southern Africa and in northern and central South America.

In southwestern Africa, the cool Benguela current controls aridity; the current reached its greatest energy and produced its greatest aridity during glacial periods. It is also thought to be responsible for the northward shift of the Angola Basin and deep cores from the Atlantic Ocean seem to support this idea. Thus, it reaffirms the ideas of De Ploey (1965) related to the presence of eolian sand from the Kalahari in western Congo (*Fig. 21.57*). This extension is probably explained by the expansion of the Benguela Current (Selby *et al.*, 1979).

The work of Tricart (1985) is a good update of the dispersed knowledge of dry climates in central and northern South America. Extensive dune fields have been identified on the left margin of the river in the Llanos of the Orinoco (Tricart, 1974a). Landsat images show that they are longitudinal NE-SW oriented dunes and that parabolic dunes are also present. These eolian accumulations come from alluvial mantles deposited by rivers from the Andean piedmont. Their age is related to the last glacial period when the glacio-eustatic level dropped. During the Holocene, the climate was warmer

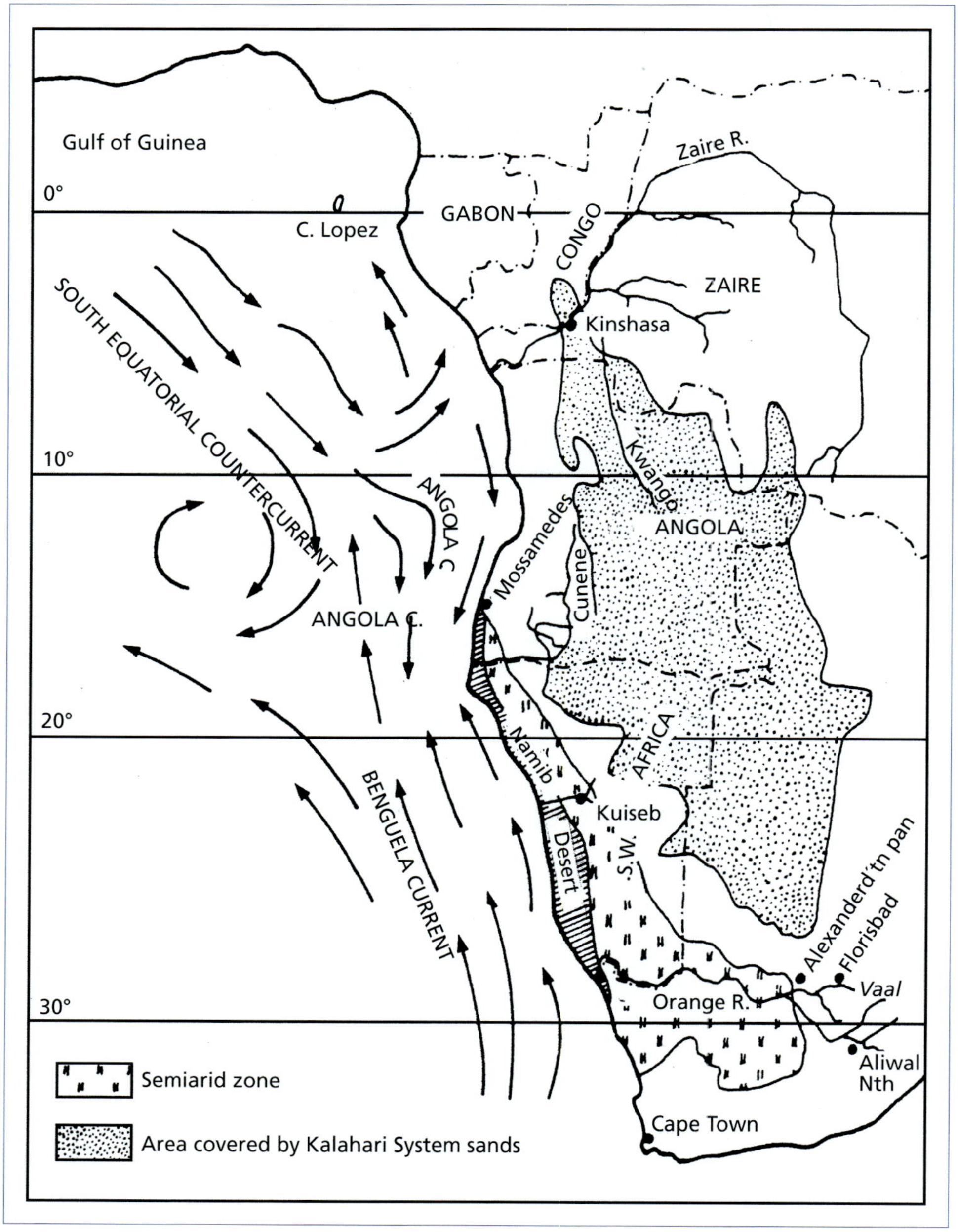

Figure 21.57 Extent of the southwestern African desert. Sands of the Kalahari Desert reached the Congo rainforest as far as the Zaire River (Selby et al., 1979).

and more humid. The dunes have been partially fossilized by alluvium and are flooded during rainy periods. Dunes and deflationary landforms partially fossilized by sediments of the Flandrian transgression have also been identified near the Orinoco delta (Tricart and Alfonsi, 1981).

The Pantanal del Mato Grosso depression (southwestern Brazil) where present-day annual precipitation is about 1,200 mm is the location of multiple large alluvial fans that came from Paleozoic sierras. These forms seem to indicate a period of dry climate with sparse vegetation and dune formation. In the Taquari fan, the largest one, deflation basins up to 3 km in diameter have been identified. Today the Pantanal lives up to its name; it is flooded and colonized by swamp vegetation (Tricart, 1982; Klammer, 1982).

REFERENCES

Aber, J.S. and Ber, A. (2007). *Glaciotectonism*. Elsevier. Amsterdam, 246 p.

Aber, J.S., Croot, D.G. and Fenton, M.M. (1989). *Glaciotectonic Landforms and Structures*. Kluwer. Dordrecht, 200 p.

Abrahams, A.D. and Parsons, A.J. (1991a). Relation between infiltration and stone cover on a semiarid hillslope, southern Arizona. *Journal of Hydrology*, 122, 49–59.

Abrahams, A.D. and Parsons, A.J. (1991b). Relation between sediment yield and gradient on debris-covered hillslopes, Walnut Gulch, Arizona. *Bulletin of the Geological Society of America*, 103, 1109–1113.

Abrahams, A.D. and Parsons, A.J. (Eds.) (1994). *Geomorphology of Desert Environments*. Chapman and Hall. London, 674 p.

Abrahams, A.D., Howard, A.D. and Parsons, A.J. (1994). Rock mantled slopes. In: Abrahams, A.D. and Parsons, A.J. (Eds.). *Geomorphology of Desert Environments*. Chapman and Hall. London, 173–212.

Abul-Ata, A.A. (1978). *Egypt and the Nile after the construction of the High Aswan Dam*. Ministry of Irrigation and Land Reclamation. Cairo.

Acres, B.D., Blaire Rain, A., King, R.B., Lawton, R.M., Mitchell, A.J.B. and Rackham, L.J. (1985). African dambos: their distribution, characteristics and use. *Zeitschrift für Geomorphologie Supplementband,* 52, 63–86.

Adams, J. (1980). Active tilting of the United States mid-continent: geodetic and geomorphic evidence. *Geology*, 8, 442–446.

Adams, J. (2007). *Vegetation-Climate Interaction. How Vegetation Makes the Global Environment*. Springer. Berlin, 232 p.

AE-88 (1996). Norma de Construcción Sismo-Resistente NCS-94. In: Norma Básica de la Edificación (NBE): *Acciones en la Edificación*. Serie Normativas (NBE). Centro de Publicaciones Ministerio de Fomento. Madrid, 235 p.

Agarwald, R.P. and Bhoj, R. (1992). Evolution of Kosi River fan, India: structure implication and geomorphic significance. *International Journal of Remote Sensing*, 13, 1891–1901.

Aghassy, J. (1973). Man-induced badlands topography. In: Coates, D.R. (Ed.). *Environmental Geomorphology and Landscape Conservation. Volume III. Non-urban*. Dowden, Hutchinson and Ross. Stroudsburg, 124–136.

Agliardi, F., Crosta, G. and Zamchi, A. (2001). Structural constraints on deep-seated slope deformation kinematics. *Engineering Geology*, 59, 83–102.

Aguilar, E., Quayle, R.G. and Murray Mitchell, J. (2005). Climate variation: instrumental data. In: Oliver, J.E. (Ed.). *Encyclopedia of World Climatology*. Springer. Berlin, 263–266.

Aguilera, A. (1986). *Revista MOPU, marzo*. Madrid.

Ahlmann, H.W. (1935). Contribution to the physics of glaciers. *Geographical Journal*, 86, 97–113.

Ahnert, F. (1960). The influence of Pleistocene climates upon the morphology of cuesta scarps on the Colorado Plateau. *Annals of Association American Geographes,* 50, 139–156.

Ahnert, F. (1970). Functional relationships betweeen denudation relief and uplift in large, mid-latitude drainage basins. *American Journal of Science*, 268, 243–263.

Ahnert, F. (1987a). Approaches to dynamic equilibrium in theoretical simulations of slope development. *Earth Surface Processes and Landforms*, 12, 3–15.

Ahnert, F. (1987b). An approach to identification of morphoclimates. In: Gardiner, V. (Ed.). *International Geomorphology 1986*, Part II. Wiley. Chichester, 159–188.

Ahnert, F. (1994). Modelling the development of non-periglacial sorted nets. *Catena*, 23, 43–63.

Ahnert, F. (1996). *Introdución to Geomorphology*. Arnold, London, 352 p.

Aires-Barros, L. (1977). Experiments on thermal fatigue of non-igneous rocks. *Engineering Geology*, 11, 227–238.

Aires-Barros, L., Graça, R.C. and Vélez, A. (1975). Dry and wet laboratory tests and thermal fatigue of rocks. *Engineering Geology*, 9, 249–265.

Aitchison, G.D. and Wood, C.C. (1965). Some interactions of compaction, permeability, and post-construction deflocculation affecting the probability of piping failure in small earth dams. *Proceedings 6th International Conference on Soil Mechanics and Foundation Engineering*, 2, 442–446. Montreal.

Aitken, M.J. (1998). *An introduction to optical dating: the dating of Quaternary sediments by the use of photonstimulated luminescence*. Oxford University Press. Oxford, 267 p.

Alberto, F. (1989). La desertización por salinización en el Valle del Ebro. *Azara*, 1, 45–53.

Albritton, C.C., Brooks, J.E., Issawi, B. and Swedan, A. (1990). Origin of the Qattara Depression, Egypt. *Bulletin of the Geological Society of America*, 102, 952–960.

Albritton, D.L. and Meira Filho, L.G. (Coords.) (2001). Technical Sumamry. In: Houghton, J.T., Ding, Y., Griggs, D.J., Noguer, M., Liden, P.J. van der, Dai, X., Maskell, K. and Johnson, C.A. (Eds.). *Climate Change 2001: The Scientific Basis*. Cambridge University Press. Cambridge, 21–83.

Al-Durrah, M. and Bradford, J.M. (1982). New methods of studying soil detachment due of raindrop impact. *Journal of the Soil Science Society of America*, 45, 949–953.

Al-Rujeibi, A., Cheema, T., Al-Harthy, A. and Ahmad, I. (2002). In: Alawaji, H.A. (Ed.). *Geotechnical and Geoenvironmental Engineering in Arid Lands*. Balkema. Lisse, 267–274.

Aleva, G.J.J. (1985). Indonesian fluvial casiterite placers and their genetic environments. *Journal of the Geological Society of London,* 142, 815–836.

Aleva, G.J.J. (1991). Tropical weathering, denudation and mineral accumulation. *Geologie in Mijnbouw*, 70, 35–38.

Alexander, D. (1992). On the causes of landslides: human activities, perceptions and natural processes. *Environmental Geology and Water Sciences*, 20, 165–179.

Alexander, D. (1995). *Natural Disasters*. UCL Press. London, 632 p.

Alexander, R.W. and Calvo, A. (1990). The influence of lichens on slope processes in some spanish badlands. In: Thornes, J.B. (Ed.). *Vegetation and Erosion*. Wiley. Chichester, 385–398.

Alexander, R.W., Harvey, A.M., Calvo, A., James, P.A. and Cerdá, A. (1994). Natural stabilisation mechanism on badland slopes, Tabernas, Almería, Spain. In: Millington, A.C. and Pye, K. (Eds.). *Environmental Change in Drylands: Biogeographical and Geomorphological Perspectives.* Wiley. Chichester, 85–111.

Alexandre, L.T. and Cady, J.G. (1962). *Genesis and Hardening of Laterite in Soils*. Technical Bulletin, 1282. United States Department of Agriculture, Soil Conservation Service, 90 p.

Algermissen, S.T., Perkins, D.M., Thenhaus, P.C., Hanson, S.L. and Bender, B.L. (1982). Probabilistic estimates of maximum acceleration and velocity in rocks in the contiguous United States. *United States Geological Survey Open File Report*, 82–1033, 1–107.

Alía Medina, M. (1945). *Características morfográficas y geológicas de la zona septentrional del Sahara Español*. Instituto José de Acosta, C.S.I.C. Madrid, 260 p.

Alía Medina, M. (1952). *Bosquejo Geológico del Sahara Español*. Mapa Geológico E:1:2.000.000. Dirección General de Marruecos y Colonias. Instituto de Estudios Africanos. Servicio Geológico del África occidental española. Madrid.

Alkattan, M., Oelkers, E.H., Dandurand, J.L. and Schott, J. (1997). Experimental studies of halite dissolution kinetics. 1. The effect of saturation rate and the presence of trace metals. *Chemical Geology*, 137, 201–219.

Alkhanbashi, I., El-Hussain, I. and Cheema, T. (2002). The effect of dam on ground water quality of Wadi Hawasina. In: Alawaji, H.A. (Ed.). *Geotechnical and Geoenvironmental Engineering in Arid Lands*. Balkema. Lisse, 297–304.

Allen, J.G. (1972). Aluminium ore deposits. In: Fairbridge, R.W. (Ed.). *The Encyclopedia of Geochemistry and Environmental Sciences*. Van Nostrand Reinhold. New York, 23–27.

Allen, J.R.L. (1964). Studies in fluviatile sedimentation. Six cyclothems from the Lower Old Red Sandstone, Anglo-Welsh Basin. *Sedimentology*, 3, 163–198.

Allen, J.R.L. (1965). Later Quaternary Niger delta and adjacent areas: sedimentary environments and lithofacies. *Bulletin American Association Petroleum Geologists*, 49, 547–600.

Allen, J.R.L. (1970). A quantitative model of grain size and sedimentary structures in lateral deposits. *Geological Journal*, 7, 129–146.

Allen, J.R.L. (1971). Transverse erosional marks of mud and rock: their physical basis and geological significance. *Sedimentary Geology*, 5, 1–38.

Allen, J.R.L. (1976). Bedforms and unsteady processes: some concepts of classification and response illustrated by common one-way types. *Earth Surface Processes*, 1, 361–374.

Allen, J.R.L. (1986). Earthquake magnitude-frecuency, epicentral distance, and soft-sediment deformation in sedimentary basins. *Sedimentary Geology*, 46, 67–75.

Allen, J.R.L. (2004). Bedform. In: Goudie, A.S. (Ed.). *Encyclopedia of Geomorphology*. Routledge. London, 76–79.

Alley, R.B. (2004). Abrupt climate change. *Scientific American*, 291, 62–69.

Allison, R.J. (1997). Middle East and Arabia. In: Thomas, D.S.G. (Ed.). *Arid Zone Geomorphology: Process, Form and Change in Drylands*. John Wiley. Chichester, 507–521.

Allison, R.J. (2002) (Ed.). *Applied Geomorphology*. Wiley. Chichester, 480 p.

Allison, R.J. and Bristow, G.E. (1999). The effects on the fire on rock weathering: Some further considerations of laboratory experimental simulation. *Earth Surface Processes and Landforms*, 24, 707–713.

Allison, R.J. and Goudie, A.S. (1994). The effects of fire on rock weathering: an experimental study. In: Robinson, D.A. and Williams, R.B.G. (Eds.). *Rock Weathering and Landform Evolution*. Wiley. Chichester, 41–56.

Allison, R.J. and Thomas, D.G. (1993). The sensivity of landscapes. In: Thomas, D.S.G. and Allison, R.J. (Eds.). *Landscape Sensivity*. Wiley. Chichester, 1–5.

Alonso, V. and Tromobotto, D. (2009). Periglacial geomorphology of El Miro area, Cantabrian Mountains, NW Spain. *Zeitschrift für Geomorphologie*, 53, 335–357.

Alonso-Sarriá, F., López-Bermúdez, F. and Conesa-García, C. (2002). Synoptic conditions producing extreme rainfall events along the Mediterranean coast of the Iberian Peninsula. In: Bull, J.L. and Kirkby, M.J. (Ed.). *Dryland Rivers: Hydrology and Geomorphology of Semi-arid Channel*. Wiley. Chichester, 351–372.

Alonso-Zarza, A.M. (2003). Palaeoenvironmental significance of palustrine carbonates and calcretes in the geological record. *Earth-Science Reviews*, 60, 261–298.

Alonso-Zarza, A.M., Silva, P.G., Goy, J.L. and Zazo, C. (1998). Fan surface dynamics and biogenic calcrete development: interactions during ultimate phases of fan evolution in the semi-arid SE Spain (Murcia). *Geomorphology*, 24, 147–167.

Alsharhan, A.S., Glennie, K.W., Whittle, G.L. and Kendall, C.G.St. (Eds.) (1998). *Quaternary Deserts and Climatic Change*. Balkema. Rotterdam, 621 p.

Ambrose, J.W. (1964). Exhumed paleoplains of the Precambrian shield of North America. *American Journal of Science*, 262, 815–857.

American Geological Institute (1972). *Glossary of Geology*. Washington, 805 p.

Amit, R. and Gerson, R. (1986). The evolution of Holocene reg (gravelly) soils in deserts-an example from the Dead Sea region. *Catena*, 13, 59–79.

Amoros, C. and Petts, G.E. (1993). *Hydrosisthémes Fluviaux*. Masson. Paris, 300 p.

Anadón, P. (1989). Lagos. In: Arche, A. (Ed.). *Sedimentología*. Consejo Superior de Investigaciones Científicas. Madrid, 219–276.

Ancey, C. (Ed.) (1998). Guide Beige et Avalanches. Connaissanes, Prafiques, Securité. Edisuol. Aix-en.Provence, 335 p.

Ancochea, E. (2004). Los grandes episodios de deslizamientos. In: Vera, J.A. (Ed.). *Geología de España*. Sociedad Geológica de España e Instituto Geológico y Minero de España. Madrid, 641–642.

Ancochea, E., Fuster, J.M., Ibarrola, E., Cendrero, A., Coello, J., Hernán, F., Cantagrel, J.M. and Jamond, C. (1990). Volcanic evolution of the island of Tenerife (Canary Islands) in the light of new K-Ar data. *Journal of Volcanology and Geothermal Research*, 44, 231–249.

Ancochea, E. and Barrera, J.L. (2002). *La peligrosidad volcánica*. In: Ayala, F.J. and Olcina, J. (Eds.). Riesgos Naturales. Ariel Ciencia. Barcelona. 265–285.

Ancochea, E., Hernán, F., Cendrero, A., Cantagrel, J.M., Fúster, J.M., Ibarrola, E. and Coello, J. (1994). Constructive and destructive episodes in the building of a young oceanic island: La Palma, Canary Islands, and genesis of the Caldera de Taburiente. *Journal of Volcanology and Geothermal Research*, 60, 243–262.

Ancochea, E., Huertas, M.J., Cantagrel, J.M., Fúster, J.M. and Arnaud, N. (2000). Cronología y evolución del edificio Cañadas: Tenerife, Islas Canarias. *Boletín Geológico y Minero*, 111, 3–16.

Anderson, B. (1961). *The Rufiji Basin, Tanganika*. 7. Soils of main irrigatable areas. Rept. Gout. Tanganika, Prelim. Reconnaissance Surv. Rufiji Basin, 125 p.

Anderson, D.E., Goudie, A.S. and Parker, A.G. (2007). *Global Environments through the Quaternary. Exploring Environmental Change*. Oxford University Press. Oxford, 359 p.

Anderson, J.F., Wesnousky, S.G. and Stirling, M.W. (1996). Earthquake size as function of fault slip rate. *Bulletin of the Seismological Society of America*, 86, 683–690.

Anderson, R.C. (1988). Reconstruction of preglacial drainage and its diversion by earliest glacial forebulge in the upper Mississippi Valley region. *Geology*, 16, 254–257.

Anderson, R.S. (1986). Erosion profiles due to particles entrained by wind: application of an eolian sediment transport model. *Bulletin of the Geological Society of America*, 97, 1270–1278.

Anderson, R.S. (1987). A theoretical model for aeolian impact ripples. *Sedimentology*, 34, 943–956.

Anderson, R.S. (1989). Saltation of sand: a qualitative review with biological analogy. In: Gimminghang, C.H.,

Richtie, W., Willets, B.B. and Willis, A.J. (Eds.). *Symposium: Coastal Sand Dunes*, Royal Society of Edinburgh, Proceeding, B96, 149–165.

Anderson. R.S. and Bunas, K.L. (1993). Grain size segregation and stratigraphy in aeolian ripples modelled with a cellular automaton. *Nature*, 365, 740–743.

Anderson, R.S. and Haff, P.K. (1988). Simulation of aeolian saltation. *Science*, 241, 820–823.

Andersson, J.G. (1906). Solifuction; a component of subaerial denudation. *Journal of Geology*, 14, 91–112.

Andrade, E.R. and Sellers, W.D. (1988). El Niño and its effect on precipitation in Arizona and western New Mexico. *Journal of Climatology*, 8, 403–410.

André, M.F. (1993). *Les Versants du Spitsberg*. Presses Universitaires de Nancy. Nancy, 361 p.

André, M.F., Hall, K., Bertran, P. and Arocena, J. (2008). Stone runs in Fakland Islands: Periglacial or tropical?. *Geomorphology*, 95, 524–543.

Andrejchuk, K.V. and Klimchouk, A. (2000). Mechanisms of karst breakdown formation of the gypsum karst of the Fore-Ural Region, Russia (from observations in the Kungurskaja Cave). *Speleogenesis and Evolution of Karst Aquifers*, 2, 1–17.

Andrejchuk, K.V. and Klimchouk, A. (2002). Mechanisms of karst breakdown formation in the gypsum karst of the Fore-Ural region, Russia (from observations in the Kungurskaja Cave). Implication of Speleological Studies from karst subsidence hazard assessment. *International Journal of Speleology*, 31, 89–114.

Andrés de, J.R. and Gracia, F.J. (2002). La franja litoral: riesgos y protección. In: Ayala-Carcedo, F.J. and Olcina, J. (Eds.). *Riesgos Naturales*. Ariel. Barcelona, 1023–1057.

Andrews, J.T. (1970). A geomorphological studies of post-glacial uplift with particular reference to Arctic Canada. *Institute of British Geographers Special Publication*, 2, 156 p.

Andrews, J.T. (1975). *Glacial Systems. An Approach to Glaciers and their Environments*. Duxbury Press. North Scituate, Massachussets, 191 p.

Andrews, J.T. and Dugdale, R.E. (1971). Quaternary history of Northern Cumberland Peninsula, Baffin Island, N.W.T.: Part V: Factors affecting corrie glacierization in Okoa Bay. *Quaternary Research*, 1, 532–551.

Andrews, J.T. and Smithson, B.B. (1966). Till fabric of the cross-valley moraines of north-central Baffin Island. *Bulletin of the Geological Society of America*, 77, 271–290.

Anfuso, G. and Gracia, F.J. (2005). Morphodynamic characteristics and short-term evolution of coastal sector in SW Spain: Implications for coastal erosion management. *Journal of Coastal Research*, 21, 1139–1153.

Angeli, M.G., Gasparetto, P., Menotti, R.M., Pasuto, A., Silvano, S. and Soldati, M. (1996). Rock Avalanche. In: Dikau, R., Brunsden, D., Schrott, R. and Ibsen, M.L. (Eds.). *Landslide Recognition. Identification, Movement and Causes. Wiley*. Chichester, 190–201.

Angelis, H. de, and Skvarca, P. (2003). Glacier surge after ice shelf collapse. *Science*, 299, 1560–1562.

Anisimov, O.A. and Fitzharris, B. (2001). Polar regions (Arctic and Antarctic). In: McCarthy, O.F.C., Leary, N.A., Dokken, D.J. and White, F.S. (Eds.). *Climate Change 2001*. Impacts, Adaptation and Vulnerability Contribution of Working Group II to the Third Assessment Report of the Intergovernamentla Panel of Climate Change. Cambridge University Press. Cambridge, 801–841.

Anisimov, O.A. and Nelson, F.E. (1996). Permafrost distribution in the Northern Hemisphere under scenarios of climatic change. *Global and Planetary Change*, 14, 59–72.

Anisimov, O.A., Vanderberghe, J., Lobanov, V. and Kontratiev, A. (2008). Predicting changes in alluvial channel patterns in North-European Russia under conditions of global warming. *Geomorphology*, 98, 262–274.

Anon (1966). "Ordered Water" molecular pressures offer new clues to rock failure. *Engineering Mining Journal*, 167, 92–94.

Antevs, E. (1952). Arroyo cutting and filling. *Journal of Geology*, 60, 375–385.

Anthony, E.J. (2005). Beach erosion. In: Schwartz, M.L. (Ed). *Encyclopedia of Coastal Science*. Springer. Dordrecht, 140–145.

Arakel, A.V. (1986). Evolution of calcrete in palaeodrainages of the Lake Napperby area, central Australia. *Palaeogeography, Palaeoclimatology, Palaeoecology*, 54, 282–303.

Araña, V. (1988). Geología y Volcanismo. In: Ayala, F.G. and Durán, J.J. (Eds.). *Riesgos Geológicos*. Instituto Geológico y Minero de España. Madrid, 45–82.

Araña, V. y Ortiz, R. (1993). Riesgo Volcánico. In: Martí, J. and Araña, V. (Eds.). *La Volcanología Actual*. Consejo Superior de Investigaciones Científicas. Madrid, 277–385.

Arauzo, T. and Gutiérrez, M. (1994). Evolución de los valles de fondo plano del centro de la Depresión del Ebro. In: Arnáez, J., García-Ruiz, J.M. and Gómez Villar, A. (Eds.). *Geomorfología en España*, vol. 1. Logroño, 277–290.

Arauzo, T., Gutiérrez, M. and Sancho, C. (1996a). Facetas triangulares de ladera como indicadores paleoclimáticos en ambientes semiáridos (Depresión del Ebro). *Geogaceta*, 20, 1093–1095.

Arauzo, T., Gutiérrez-Elorza, M. and Sancho, C. (1996b). Retroceso de escarpes en ambientes semiáridos a partir de facetas triangulares de ladera (Depresión del Ebro). *Cadernos do Laboratorio Xeolóxico de Laxe*, 21, 405–416. Sada (A Coruña).

Arche, A. (1989). Deltas. In: Arche, A (Ed.). *Sedimentología*, vol.1, Consejo Superior de Investigación Científicas. Madrid, 395–449.

Ardon, K., Tsoar, H. and Blumberg, D.G. (2009). Dynamics of nebkhas superimposed on a parabolic dune and their effect on the dune dynamics. *Journal of Arid Environments*, 73, 1014–1022.

Arens, S. (1996). Patterns of sand transport on vegetated dunes. *Geomorphology*, 17, 339–350.

Aristarain, L.F. (1970). Chemical analysis of caliche profiles from High Plains, New Mexico. *Journal of Geology*, 78, 201–12.

Aristarain, L.F. (1971). On the definition of caliche deposits. *Zeitschrift für Geomorphologie*, 15, 274–289.

Arlegui, L.E. and Simón, J.L. (2000). Fracturación y campos de esfuerzos en el Cuaternario del sector central de la Cuenca del Ebro (NE España). *Cuaternario y Geomorfología*, 14, 11–20.

Arnborg, L.B., Walker, H.J. and Peippo, J. (1967). Suspended load in the Colville River, Alaska, 1962. *Geografiska Annaler*, 49, 131–144.

Arribas, A., Polo, V. and Jiménez, E. (1984). *La "Enfermedad de la Piedra" en la Arenisca de Villamayor. Diagnóstico, Tratamiento y Conservación*. Caja de Ahorros y M.P. de Salamanca. Salamanca, 6–169.

Artieda, O. (1993). *Materiales Parentales y Geomorfología en la Génesis de Aridisoles en un sector del centro del Valle del Ebro*. Tesis Doctoral. Universidad de Zaragoza, 567 p. (Unpublished).

Ash, J.E. and Wasson, R.J. (1983). Vegetation and sand mobility in the Australian desert dunefield. *Zeitschrift für Geomorphologie, Supplementband*, 45, 7–25.

Ashida, K. (1985). *Debris Flow. Disaster on Alluvial Fans*. Kokon-shoin. Tokyo, 224 p.

Aslan, R. (2007). Fluvial environments. In: Elías, S.A. (Ed.). *Encyclopedia of Quaternary Science*. Elsevier, Ámsterdam, 672–685.

Attewell, P.B. and Taylor, D. (1988). Time-dependent atmospheric degradation of building stone in a polluting environment. In: Marinos, P.G. and Koukis, G.C. (Eds.). *Engineering Geology of Ancient Works, Monuments and Historical Sites*. Balkema. Rotterdam, 739–753.

Auboin, J. (1961). Propos sur l'orogénèse. *Bulletin Bureau Recherche Geologique et Minière*, 52, 1–24.

Aubreville, A. (1962). Savanisation tropicale et glaciation quaternaires. *Adansonia*, 2, 16–84.

Aufrére, M.L. (1928). L'orientation des dunes et la direction des vents. *Comptes Rendus de l'Academie Scientifique à Paris*, 187, 833–835.

Aufrère, M.L. (1935). Essai sur les dunes du Sahara algérien. *Geografiska Annaler*, 18, 481–550.

Augustinus, P.C. (1992). Outlet glaciar trough size-drainage area relationships. Fiordland, New Zealand. *Geomorphology*, 4, 347–361.

Aune, Q.A. (1983). Quick clays and California's clays: no quick solution. In: Tank, R.W. (Ed.). *Environmental Geology*, Oxford University Press. New York, 145–150.

Avouac, J.P., Tapponnier, P., Bai, M. You, H. and Wang, G. (1993). Active thrusting and folding along the northern Tien Shan and late Cenozoic rotation of the Tarim relative to Daungaria and Kazathstan. *Journal of Geophysical Research*, 98, 6755–6804.

Ayala, F.J. (1994). Socioeconomic impacts and vulnerability resulting from slope movements. In: Corominas, J. and Georgakakos (Eds). *Proceedings of the U.S.-Spain Workshop of Natural Hazards.* Iowa University, 235–254.

Ayala, F.J. (2002a). La inundación torrencial catastrófica del camping "Las Nieves" del 7 de agosto de 1996 en el cono de deyección del Arás (Biescas, Pirineo aragonés). In: Ayala, F.J. and Olcina, J. (Eds.). *Riesgos Naturales*. Ariel. Barcelona, 889–912.

Ayala, F.J. (2002b). Expansividad y colapso de suelos. In: Ayala, F.J. and Olcina, J. (Eds.). *Riesgos Naturales*. Ariel Ciencia. Barcelona, 475–482.

Ayala, F.J., Elizaga, E., González de Vallejo, L., Durán, J.J., Beltrán, F., Oliveros, M.A., Carbó, A., Guillamont, M.L. and Capote, R. (1987a). *Impacto económico y social de los riesgos geológicos en España*. Instituto Geológico y Minero de España. Madrid, 86 p.

Ayala, F.J., Andreu Fe, M., Ferrer, M., de Simón, A. Fernández, I., Olalla, C., Gómez, J., Sanpedro, J. and Cienfuegos. F.J. (1987b). *Manual de Taludes*. Instituto Geológico y Minero de España. Madrid, 456 p.

Aylsworth, J.M. and Shilts, W.W. (1989). Bedforms of the Keewatin Ice Sheet, Canada. *Sedimentary Geology*, 62, 407–428.

Azañón, J.M., Azur, A., Bouth-Rea, G. and Torcal, F. (2004). Small-scale faulting topographic steps and seismic ruptures in the Alhambra (Granada, southeast Spain). *Journal of Quaternary Science*, 19, 219–227.

Baena, J., Bardají, T., Goy, J.L., Silva, P.G., Somoza, L. and Zazo, C. (1993). An attempt of fault activity data file and map modellization. *Bulletin INQUA*, NC 16, 1993–1998.

Baena, J., Barranco, L.M., Zazo, C., Goy, J.L., Estévez, A., Sanz de Galdeano, A., Bardají, T., Silva, P.G., Somoza, L. and Rodríguez-Estrella, T. (1994). *Mapas Neotectónico,*

Sismotectónico y de Actividad de Fallas de la Región de Murcia. Madrid, España. ITGE-CARM, 101 p.

Bagnold, R.A. (1941). *The Physics of Blown Sand and Desert Dunes*. Methuen. London, 265 p.

Bagnold, R.A. (1953). The surface movement of blown sand in relation to meteorology. *Research Council of Israel, Special Publication*, 2, 89–96.

Bahn, P.G. (2004). Atapuerca Caves, Spain: Archaeology. In: Gunn, J. (Ed.). *Encyclopedia of Caves and Karst Science*. Fitzroy Dearborn. New York, 119–120.

Bailey, J.F. and Patterson, J.L. (1975). Hurricane Agnes Rainfall and Floods, June–July, 1972. *United States Geological Survey Professional Paper*, 924.

Baker, A.A. (1936). Geology of the Monument Valley-Navajo Mountain region, San Juan County, Utah. *United States Geological Survey*, 865, 106 p.

Baker, V.R. (1973). Paleohydrology and sedimentology of Lake Missoula flooding in eastern Washington. *Geological Society of America, Special Paper*, 144, 79 p.

Baker, V.R. (1977). Stream channel response to floods with examples from Central Texas. *Bulletin of the Geological Society of America*, 88, 1057–1071.

Baker, V.R. (1978). Adjustment of fluvial system to climate and source terrain in tropical y subtropical environments. In: Miall, A.D. (Ed.). *Fluvial Sedimentology*. Canadian Society of Petroleum Geologists. Memoir 5, 211–230.

Baker, V.R. (1981a). The geomorphology of Mars. *Progress in Physical Geography*, 5, 473–513.

Baker, V.R. (Ed.) (1981b). *Catastrophic Flooding. The Origin of the Chanelled Scabland*. Dowden, Hutchinson and Ross. Stroudsburg, 360 p.

Baker, V.R. (1988). Overview. In: Baker, V.R., Craig Kochel, R. and Patton, P.C. (Eds.). *Flood Geomorphology*. Wiley. New York, 1–8.

Baker, V.R. (1989). Magnitude and frecuency of palaeofloods. In: Beven, K. and Carling, P. (Eds.). *Floods: Hydrological, Sedimentological and Geomorphological Implications*. Wiley. Chichester, 171–183.

Baker, V.R. (1993). Extraterrestrial geomorphology: science and phylosophy of Earthlike planetary landscapes. *Geomorphology*, 7, 9–35.

Baker, V.R. (1998). Palaeohydrology and Hydrological Sciences. In: Benito, G., Baker, V.R. and Gregory, K.J. (Eds.). *Palaeohydrology and Environmental Change*. Wiley. Chichester, 1–10.

Baker, V.R. (Ed.) (2001a). Extraterrestrial geomorphology. *Geomorphology*, 37, 175–352.

Baker, V.R. (2001b). Water and Martian landscapes. *Nature*, 412, 228–236.

Baker, V.R. (2003). A bright future for old flows: Origins, status and future of paleoflood hydrology. In: Thorndycraft, V.A., Benito, G., Barriendos, M. and Llasat, M.C. (Eds.). *Paleofloods, Historical Data and Climatic Variability. Applications on Flood Risk Assessment*. Consejo Superior de Investigaciones Científicas. Madrid, 13–24.

Baker, V.R. (2004a). Extraterrestrial Geomorphology. In: Goudie, A.S. (Ed.). *Encyclopedia of Geomorphology*. Routledge. London, 353–358.

Baker, V.R. (2004b). Megageomorphology. In: Goudie, A.S. (Ed.). *Encyclopedia of Geomorphology*. Routledge. London, 662–665.

Baker, V.R. (2006). Paleoflood hydrology in a global context. *Catena*, 66, 161–168.

Baker, V.R. (2007). Greatest floods and largest rivers. In: Gupta, A. (Ed.). *Large Rivers. Geomorphology and Management*. Wiley. Chichester, 65–74.

Baker, V.R. (2008a). The Spokane flood debates: historical background and philosophical perspective. In: Grafes, R.H., Oldroyd, D. and Grigel, S.A. (Eds.). *History of Geomorphology and Quaternary Geology*. Geological Society of London, Special Publication, 301, 33–50.

Baker, V.R. (2008b). Paleoflood hydrology: Origin, progress, prospects. *Geomorphology*, 101, 1–13.

Baker, V.R. and Bunker, R.C. (1985). Cataclysmic late Pleistocene flooding from glacial Lake Missoula: A review. *Quaternary Science Reviews*, 4, 1–41.

Baker, V.R. and Costa, J.E. (1987). Flood power. In: Mayer, L. and Nash, D. (Eds.). *Catastrophic Flooding*. Allen and Unwin. London, 1–24.

Baker, V.R. and Twidale, C.R. (1991). The reenchantment of geomorphology. *Geomorphology*, 4, 73–100.

Baker, V.R., Benito, G. and Rudey, A.N. (1993). Paleohydrology of Late Pleistocene superflooding, Altai Mountains, Siberia. *Science*, 259, 348–350.

Baker, V.R., Craig Kochel, R. and Patton, P.C. (1988) (Eds.). *Flood Geomorphology*. Wiley. New York, 503 p.

Baker, V.R., Kochel, R.C., Laity, J.E. and Howard, A.D. (1990). Spring sapping and valley network development. In: Higgins, C.G. and Coates, D.R. (Eds.). *Groundwater Geomorphology*. Geological Society of America Special Paper 252, 235–265.

Baker, V.R., Webb, R.H. and Kyle House, P. (2002). The scientific and societal value of paleoflood hydrology. In: Kyle House, P., Webb, R.W., Baker, V.R. and Levish, D.R. (Eds.). *Ancient Floods, Modern Hazards. Principles and Applications of Paleoflood Hydrology*. American Geophysical Union. Washington, 1–19.

Bakker, J.P. (1965). A forgotten factor in the interpretation of glacial stairways (An old question; new problems). *Zeitschrift für Geomorphologie, NF*, 9, 18–34.

Bakker, J.P. (1967). Weathering of granites in different climates. In: Macar, P. (Ed.). *L'Evolution des Versants*. Congrès et Colloque Université Liége, 40, 51–68.

Bakker, J.P. and Levelt, Th.W.M. (1964). An inquiry into the probability of a polyclimatic development of peneplains and pediments (etchplains) in Europe during the Senonian and Tertiary Period. *Publication Service Carte Géologique, Luxembourg*, 14, 27–75.

Bakker, Th.W., Jungerius, P.D. and Klinj, J.A. (Eds.) (1990). *Dunes of the European Coast: Geomorphology-Hydrology-Soils*. Catena Supplement 19, 223 p.

Balazs, D. (1973). Relief types de tropical karst areas. In: Jakucs, L. (Ed.). *Symposium of Karst-Morphogenesis*. Attila Joseph University. Szeged, 16–32.

Ballantyne, C.K. (2002). Glacial Geomorphology. *Earth-Science Reviews*, 21, 1935–2017.

Ballantyne, C.K. (2007). Periglacial Landforms. Patterned ground. In: Elias, S.A. (Ed.). *Encyclopedia of Quaternary Science*. Elsevier. Amsterdam, 2182–2191.

Ballantyne, C.K. and Kirkbride, M.P. (1987). Rockfall activity in upland Britain during the Loch Lomond stadial. *Geographical Journal,* 153, 153, 86–92.

Balling, R.C. and Wells, S.G. (1990). Historical rainfall patterns and arroyo activity within the Zuni River drainage basin, New Mexico. *Annals Association American Geographers*, 80, 603–617.

Banner, J.L. (2004). Radiogenic isotopes: systematics and applications to earth surface processes and chemical stratigraphy. *Earth-Science Reviews*, 65, 141–194.

Banuri, T. and Weyant, J. (Coord.) (2001). Setting the Stage: Climate Change and Sustainable Development. *Climate Change 2001. Mitigation*. Cambridge University Press. Cambridge, 73–114.

Baños, C.J. (2002). El desastre del huracán *Mitch* en Centroamérica. In: Ayala, F.J. and Olcina, J. (Eds.). *Riesgos Naturales*. Ariel. Barcelona, 797–799.

Bárany-Kerei, I. (2004). Bauxite deposits in karst. In: Gunn, J. (Ed.). *Encyclopedia of Caves and Karst Science*. Fitzroy Dearborn. New York, 135–137.

Barazangi, M. and Dorman, J. (1969). World seismicity maps compiled from ESSA, Coast and Geodetic Survey, epicenter data, 1961–1967. *Bulletin Seismological Society of America*, 59, 369–380.

Barbary, J.P., Maire, R. and Shouyuz, Z. (1991). *Grottes et Karsts Tropicaux de Chine Méridionale*. Karstologia, Mémoire nº 4. Venissieux, 232 p.

Barbeau, J. (1961). Morphologie du Quaternaire des abords orientaux du Lac Tchad. *Bulletin Institute Equatoriale Recherches Geologiques et Minières*, 14, 73–82.

Barber, A.J. (1982). Interpretations of the tectonic environment of southwest Japan. *Procedings of Geological Association*, 93, 131–145.

Barberi, F., Carapezza, M.L., Valenza, M. and Villari. L. (1993). The control of lava flow during 1991–1992 eruption of Mt. Etna. *Journal Volcanology Research*, 56, 1–34.

Barbier, R. (1967). Nouvelles réflexions sur le problème des "pains de sucre" à propos d'observations dans le Tassili N'Ajjer (Algérie). *Travaux Laboratoire de Géologie. Faculté des Sciences de Grenoble*, 43, 15–21.

Barch, D. (1977). Eine abschätaung von schottproduktion und schutttransport in bereich activer blockgletscher der schiveizer Alpen. *Zeitschrift für Geomorphologie Supplementband*, 28, 148–160.

Bardossy, G. (1981). *Karst Bauxites*. Akadémiai Kiadó. Budapest, 480 p.

Bardossy, G. and Aleva, G.J.J. (1990). *Lateritic Bauxites*. Elsevier. Amsterdam, 624 p.

Barrère, P. (1964). Le relief karstique dans l'ouest des Pynénées centrals. *Revue Belge de Géographie*. Special Publication, Karst and Climats Froids, 88, 9–62.

Barry, R.G. (1997). Paleoclimatology, climate system processes and the geomorphic record. In: Stoddart, D.R. (Ed.). *Process and Form in Geomorphology*. Routledge. London, 187–214.

Barry, R.G. and Chorley, R.J. (1987). *Atmosphere, Weather and Climate*. Methuen. London.

Barsch, D. (1977a). Eine abschätaung von schottproduktion und schutttransport in bereich activer blockgletscher der schiveizer Alpen. *Zeitschrift für Geomorphologie Supplementband*, 28, 148–160.

Barsch, D. (1977b). Nature and importance of mass-wasting by rock glaciers in the alpine permafrost environments. *Earth Surface Processes*, 2, 231–245.

Barsch, D. (1988). Rockglaciers. In: Clark, M.J. (Ed.). *Advances in Periglacial Geomorphology.* Wiley. Chichester, 69–90.

Barsch, D. (1990). Geomorphology and geoecology. *Zeitschrift für Geomorphologie Supplemenband*, 79, 39–49.

Barsch, D. (1996). *Rock Glaciers*. Springer. Berlín, 331 p.

Barsch, D. and Caine, N. (1984). The nature of mountain geomorphology. *Mountain Research and Development*, 4, 287–298.

Barsch, D. and Hell, G. (1975). Photogrammetrische bewegungsmessungen am blockgletscher Murtèl Oberengadin, Schweizer Alpen. *Zeitschrift für Gletscherkunde und Glazialgeologie*, 9, 111–142.

Barsch, D. and Liedtke, H. (Eds.) (1985) *Geomorphological Mapping in the Federal Republic of Germany Berliner*. Geographische Abhandlungen, H. 39, Berlín, 89 p.

Bashenina, N.W. *et al.*, (1971). *Legend to the International Geomorphological Map of Europe 1:2.500.000*. Institute of Geography. Brno, 30 p.

Bashenina, N.W., Leontiev, O.K., Simonov, J.G., Wyskrebenceva, W.S. and Zaruckaja, J.P. (1960). *Legend of the Geomorhological Map of the Soviet Union a 1:50.000–1:25.000.* Dpto. Geography. University of Moscow (reeditado en 1962 con resumen en francés).

Bashenina, N.W., Gellert, J.F., Joly, F., Klimaszewski, M. and Scholz, E. (1968). Project of the unified key to the detailed geomorphological map of the world. *Folia Geographica*, II, 7–40.

Batalla, R.J. and Balasch, J.C. (2001). Interpretación hidrodinámica y sedimentaria de la rotura de la balsa de San Juan (Altorricón, Huesca). *Cuaternario y Geomorfología*, 15, 109–123.

Bateman, A.M. (1950). *Economic Mineral Deposits*. Wiley. New York.

Bateman, M.D. and Díez, A. (1999). Thermoluminiscence dates and palaeoenvironmental information of the Quaternary sand deposits, Tierra de Pinares. *Catena*, 34, 277–291.

Bauer, F. (1962). Nacheiszeitliche karsformen in de Österreichischen kalkhochalpen. *Proceedings 2nd International Congress of Speleology*. Bari-Lecce-Salerno, 1, 299–329.

Baulig, H. (1938). Questions de terminologie. Consécuent, subsécuent, obséquent. *Journal of Geomorphology*, 224–229 p.

Baulig, H. (1940). Le profile d'équilibre des versants. *Annales de Géographie*, 49, 81–97.

Baulig, H. (1957). Peneplains and pediplains. *Bulletin of the Geological Society of America*, 68, 913–920.

Baulin, V.V. and Danilova, N.S. (1984). Dynamics of Late Quaternary permafrost in Siberia. In: Velichko, A.A. (Ed.). *Late Quaternary Environments of the Soviet Union*. Longman. London, 69–78.

Baumer, M. (1990). *Agroforestry and Desertification. Technical Centre for Agricultural and Rural Cooperation*. Wageningen, 249 p.

Beach, T., Dunning, N. and Doyle, N. (2008). Geoarchaeology and Geomorphology: Soils, sediments and societies. A special issue of Geomorphology. *Geomorphology*, 101, 413–415.

Beaty, C.B. (1974). Debris flow, alluvial fans and a revitalized catastrophism. *Zeitschrift für Geomorphologie Supplement Band,* 21, 39–51.

Beaumont, P. (1978). Man's impact on river systems: a worldwide review. *Area*, 10, 38–41.

Beaumont, P. (1993). *Drylands. Environmental Management and Development.* Routledge. London, 536 p.

Beck, B.F. (2005). Soil piping and sinkhole failures. In: Culver, D.C. and White, B.W. (Eds.). *Encyclopedia of Caves*. Elsevier. Amsterdam, 521–526.

Beckinsale, R.F. and Chorley, R.J. (1968). Geomorphology, History of. In: Fairbridge, R.W. (Ed.). *The Encyclpedia of Geomorphology*. Dowden, Hutchinson and Ross. Stroudsburg, 410–415.

Beckinsale, R.F. and Chorley, R.J. (1991). *The History of the Study of Landforms on the Development of Geomorphology. Vol. 3: Historical and Regional Geomorphology 1890–1950*. Routledge. London, 496 p.

Bedcows, P.A. (2004). Yucatán phreas, Mexico. In: Gunn, J. (Ed.). *Encyclopedia of Caves and Karst Science*. Fitzroy Dearborn. New York, 786–788.

Begét, J.E. (2000). Volcanic Tsunamis. In: Sigurdsson, H. (Ed.). *Encyclopedia of Volcanoes*. Academic Press. San Diego, 1005–1013.

Begin, A.B. and Schumm, S.A. (1984). Gradational thresholds and landforms singularity: significance for Quaternary studies. *Quaternary Research*, 21, 267–274.

Belikov, B.P., Zalesski, B.V., Rozanov, Y.A., Sanina, E.A. and Timchenko, I.P. (1967). Methods of studying the physicochemical properties of rocks. In: *Physical and Mechanical Properties of rocks. Israel Program for Scientific Translations*. Jerusalem, 1–58.

Bender, M., Sowers, T., Dickson, M.L., Orchado, J., Grooters, P., Mayewski, P.A. and Meese, D.A. (1994). Climate connection between Greenland and Antarctica during the last 100.000 yr. *Nature*, 372, 663–666.

Benedict, J.B. (1970). Frost cracking in the Colorado Front Range. *Geografiska Annaler*, 52, 87–93.

Benito, G. (1987). Karstificación y colapsos kársticos en los yesos del sector de la Depresión del Ebro. *Cuaternario y Geomorfología*, 1, 61–76.

Benito, G. (1997). Energy expenditure and geomorphic work of the cataclysmic Missoula flooding in the Columbia River George, USA. *Earth Surface Processes and Landforms*, 22, 457–472.

Benito, G. (2002). La paleohidrología en el análisis de inundaciones. In: Ayala, F.J. and Olcina, J. (Coor.). *Riesgos Naturales*. Ariel Ciencia. Barcelona, 953–967.

Benito, G. (2003). Paleoflood in Europe. In: Thorndycraft, V.R., Benito, G., Barriendos, M. and Llasat, M.C. (Eds.). *Paleofloods, Historical Data and Climatic Variability. Applications on Flood Risk Assessment*. Consejo Superior de Investigaciones Científicas. Madrid, 19–24.

Benito, G. (2006). Riesgo de inundaciones: tendencias históricas y perspectivas de acuerdo con el cambio climático. *Cuaternario y Geomorfología*, 20, 29–44.

Benito, G. and Gutiérrez, M. (1988). Karst in gypsum and its enviromental impact on the Middle Ebro Basin, Spain. *Enviromental Geology and Water Sciences*, 12, 107–111.

Benito, G. and Pérez del Campo, P. (1991). Sinkhole evolution in alluvial deposits within the Central Ebro Basin, Northeast Spain. In: Johnson, A.I. (Ed.). *Land Subsidence*. International Association of Hydrological Sciences, 200, 323–331.

Benito, G. and Thorndycraft, V.R. (Eds.) (2004a). *Systematic, Palaeoflood and Historical Data for the Improvement of Flood Risk Estimation: Methodological Guidelines*. Centro de Ciencias Ambientales. Consejo Superior de Investigaciones Científicas. Madrid, 115 p.

Benito, G. and Thorndycraft, V.R. (2004b). Use of systematic palaeoflood and historical data for the improvement of flood risk estimation: an introduction. In: Benito, G. and Thorndycraft, V.A. (Eds.). *Systematic, Palaeoflood and Historical Data for the Improvement of Flood Risk Estimation*. Consejo Superior de Investigaciones Científicas. Madrid, 5–14.

Benito, G., Machado, M.J. and Sancho, C. (1993a). Sandstone weathering processes damaging prehistoric rock paintings at the Albarracín Cultural Park, NE Spain. *Environmental Geology,* 22, 71–79.

Benito, G., Gutiérrez, M. and Sancho, C. (1993b). The influence of physico-chemical properties on erosion processes in badland areas, Ebro basin, NE-Spain. *Zeitschrift für Geomorphologie*, 37, 199–214.

Benito, G., Pérez del Campo, P., Gutiérrez-Elorza, M. and Sancho, C. (1995). Natural and human induced sinkholes in gypsum terrain and associated environmental problems in NE Spain. *Environmental Geology*, 25, 156–164.

Benito, G., Machado, M.J., Pérez-González, A. and Sopeña, A. (1996). Paleoflood analysis of the Tagus River in the El Puente del Arzobispo. In: Benito, G., Pérez-González, A., Machado, M.J. and Alba, S. de (Eds.). *Paleohydrology in Spain*. Consejo Superior de Investigaciones Científicas. Madrid, 335–347.

Benito, G., Pérez-González, A., Gutiérrez, F. and Machado, M.J. (1998a). River response to Quaternary subsidence due to evaporite solution (Gállego River, Ebro Basin, Spain). *Geomorphology*, 22, 243–263.

Benito, G., Baker, V. and Gregory, K. (Eds.) (1998b). *Palaeohydrology and Environmental Change*. Wiley. Chichester, 353 p.

Benito, G., Gordeck, T. and Enzel, Y. (1998c). The geomorphic and hydrologic impacts in the catastrophic failure of flood-control-dams during the 1996-Biescas flood (Central Pyrenees, Spain). *Zeitscrift für Geomorphologie*, 42, 417–437.

Benito, G., Pérez-González, A., Gutiérrez, F. and Machado, M.J. (2000). Geomorphological and sedimentological features in Quaternary fluvial system affected by solution-induced subsidence in the Ebro Basin, NE Spain. *Geomorphology*, 33, 209–224.

Benito, G., Thorndycraft, V.R., Enzel, Y., Sheffer, N.A., Rico, M., Sopeña, A. and Suárez-Moya, Y. (2004). Paleoflood data collection and analysis. In: Benito, G. and Thorndycraft, V.A. (Eds.). *Systematic Palaeoflood and for the improvement of flood risk estimation: Methodological Guidelines*. Consejo Superior de Investigaciones Científicas. Madrid, 15–27.

Benito, G., Thorndycraft, V.R., Rico, M., Sanchez-Moya, Y. and Sopeña, A. (2008). Palaeoflood and flood plain records from Spain: evidence for long-term climate variability and environmental changes. *Geomorphology*, 101, 68–77.

Benito-Calvo, A. and Pérez-González, A. (2007). Erosion surfaces and Neogene landscape evolution in the NE Duero Basin (north-central Spain). *Geomorphology*, 88, 226–241.

Benn, D.I. (1995). Fabric signature of subglacial till deformation Breidamerkurjökull, Iceland. *Sedimentology*, 42, 735–747.

Benn, D.I. (2004). Clast morphology. In: Evans, D.J.A. and Benn, D.I. (Eds.). *A Practical Guide to the Study of Glacial Sediments*. Arnold. London, 77–92.

Benn, D.I. (2006). Glaciers. *Progress in Physical Geography*, 30, 432–442.

Benn, D.I. (2007). Glacial Landforms, Sediments. Till fabrics analysis. In: Elias, S.A. (Ed.). *Encyclopedia of Quaternary Science*. Elsevier. Amsterdam, 954–959.

Benn, D.J. and Evans, D.J.A. (1996). The interpretation and classification of subglacially-deformed materials. *Quaternary Science Reviews*, 8, 279–291.

Benn, D.I. and Evans, D.J.A. (1998). *Glaciers and Glaciation*. Arnold. London, 734 p.

Bennet, R.B., Huddart, D. and González, S. (2009). Glaciovolcanic landsystems and large-scale glaciotectonic deformation along the Brekkna fjöll-Jarlettur, Iceland. *Quaternary Science Reviews*, 28, 647–676.

Benson, L.V. (1978). Fluctuation in the level of pluvial Lake Lahotan during the last 40.000 years. *Quaternary Research*, 9, 300–318.

Benson, L.V. and Paillet, F.L. (1989). The use of total lake-surface area as an indicator of climatic change: examples from the Lahotan Basin. *Quaternary Research*, 32, 262–275.

Benson, R.C. and Kaufmann, R.D. (2001). Characterization of a highway sinkhole within the gypsum karst of Michigan. In: Beck, B.F. and Herring, J.G. (Eds.). *Geotechnical and Environmental Applications of Karst Geology and Hydrology*. Balkema. Lisse, 103–109.

Berglund, B.E. (Ed.) (1986). *Handbook of Holocene Palaeoecology and Palaeohydrology*. Wiley. Chichester, 869 p.

Bernat, M. and Pérez-González, A. (2008). Inland aeolian deposists of the Iberian Peninsula: Sands dunes and clay dunes of the Duero Basin and the Manchega Plain. Palaeoclimatic considerations. *Geomorphology*, 102, 207–220.

Berner, R.A. (2001). A model for atmospheric CO_2 over Phanerozoic time. *American Journal of Science*, 291, 339–376.

Berry, L. and Ruxton, B.P. (1959). Notes on weathering zones and soils on granite rocks in two tropical regions. *Journal Soil Science*, 10, 54–73.

Berti, M. and Simoni, A. (2005). Experimental evidences and numerical modelling of debris flows intiated by channel runoff. *Landslides*, 2, 171–182.

Besler, H. (1982). The north-eastern Rub' al Khäli within the borders of the United Arab Emirates. *Zeitschrift für Geomorphologie NF*, 26, 495–505.

Besler, H. (1992). *Geomorphologie der Ariden Gebiete*. Wissenschaftliche Bughgesellschaft. Darmstadt, 180 p.

Bettis, E.A., Muhs, D.R., Roberts, H.M. and Wintle, A.G. (2003). Last Glacial loess in the conterminous USA. *Quaternary Science Reviews*, 22, 1907–1946.

Beven, K. and Carling, P. (1989). *Floods. Hydrological, Sedimentological and Geomorphological Implications*. Wiley. Chichester, 290 p.

Beyrich, A.A., Kolstrup, E., Linde, N., Pedersen, L.B., Thrysted, T., Gintz, T. and Dynesius, L. (2003). Assessment of chemical denudation rates using hydrological measurements, water chemistry analysis and electromagnetic geophysical data. *Permafrost and Periglacial Processes*, 14, 387–398.

Bezuidenhout, C.A. and Enslin, J.F. (1970). Surface subsidence and sinkholes in the dolomitic area of the Far West Rand, Transvaal. Republic of South Africa. *Land Subsidence*. International Association of Hydrological Sciences, 49, 482–495.

Bickerton, R.W. and Matthews, J.A. (1992). On the accuracy of lichenometric dates: an assessment based on the "little ice age" moraine sequence of Nigardsbreen, southern Norway. *Holocene*, 2, 227–237.

Bierman, J. Caffree (2001). Slow rates of rock surface erosion and sediment production across the Namib Desert and escarpment, southern Africa. *American Journal of Science*, 301, 326–358.

Bierman, P.R. (1994). Using *in situ* produced cosmogenic isotopes to estimate rates of landscape evolution: a review from the geomorphic perspective. *Journal of Geophysical Research,* 99, 13885–13896.

Biese, W. (1931). Über höhlenbildung: I. Teil Entstehung der gypshöhlen am südl. Harzrand und am kyffhäuser. *Abh. Preuss. Geol. Landsamt*, N.F.H. 137.

Bigarella, J.J. and Andrade, G.O. de (1965). Contribution to the study of the Brazilian Quaternary. *Geological Society of America,* Special Paper, 84, 435–451.

Bigarella, J.J. and Becker, R.D. (Eds.) (1975). International Symposium on the Quaternary. *Boletim Paranaense de Geociências*, 33.

Bigarella, J.J., Mousinho, M.M. and Da Silva, J.X. (1969). Processes environments of the Brazilian Quaternary. In: Péwé, T.L. (Ed.). *The Periglacial Environment*. McGill-Queen's University Press. Montreal, 417–487.

Bigg, G. (1985). *The Oceans and Climate*. Cambridge University Press. Cambridge, 273 p.

Bintliff, J.L. (1982). Paleoclimatic modelling of environmental changes in the East mediterranean region since the last glaciation. In: Bintliff, J.L. and Van Zeist, W. (Eds.). *Paleoclimates, Paleoenvironments and Human Comunities in the Eastern Mediterranean Region Later Prehistoty.* B.A.R. 485–527.

Bird, E.C.F. (1985). *Coasline changes*. Willey-Interscience. Chichester.

Bird, E.C.F. (1996). *Beach Management*. Wiley. Chichester, 281 p.

Bird, E. (2000). *Coastal Geomorphology. An Introduction*. John Wiley. Chichester, 322 p.

Bird, E. (2008). *Coastal Geomorphology. An Introduction*. John Wiley. Chichester, 411 p.

Bird, M.I. and Chivas, A.R. (1988). Oxygen isotope dating of the Australian regolith. *Nature*, 331, 513–516.

Birkeland, P.W. (1974). *Pedology, Weathering and Geomorphology*. Oxford University Press. New York, 285 p.

Birkeland, P.W. (1984). *Soils and Geomorphology*. Oxford University Press. New York, 372 p.

Birot, P. (1949a). *Essai sur Quelques Problèmes de Morphologie Génerale*. Instituto para a Alta Cultura. Centro de Estudos Geográficos. Lisboa, 176 p.

Birot, P. (1949b). *Les surfaces d'erosion du Portugal central et septentrional*. Rapport de la Commision pour la Cartographie des Surfaces d'Aplanissement, Union Géographique International. Congrès International de Géographie. Lisboa, 9–116.

Birot, P. (1954). Problèmes de morphologie karstique. *Annales de Géographie*, 63, 161–192.

Birot, P. (1955). *Les Méthodes de la Morphologie*. Presses Universitaires de France. Paris, 177 p.

Birot, P. (1958). Les dômes cristallines. *Mémoires et Documents*, CNRS, 6, 8–34.

Birot, P. (1959). *Precis de Géographie Physique Generale*. Lib. Armand Colin. Paris. (Trad. Esp. A. Gómiz, Ed. Vicens-Vives, Barcelona, 475 p.).

Birot, P. (1960). *Le Cycle d'Erosion sous les Différents Climats*. Centro de Pesquisas de Geografía do Brasil. Faculdade Nacional de Filosofia. Rio de Janeiro, 137 p.

Birot, P. (1966). *Le Relief Calcaire*. Centre de Documentation Universitaire. Paris, 238 p.

Birot, P. (1968). *Contribution a l'Étude de la Désagrégation des Roches.* Centre de Documentation Universitaire. Paris, 232 p.

Birot, P. (1973). *Géographie Physique Générale de la Zone Intertropicale.* Centre de Documentation Universitaire. Paris, 280 p.

Birot, P. (1978). Evolution des conceptions sur la genèse des inselbergs. *Zeitschrift für Geomorphologie, Supplement Band*, 31, 42–63.

Birot, P. and Dresch, J. (1966). Pédiments et glacis l'Ouest des États-Unis. *Annales de Géographie*, 411, 513–552.

Birot, P. and Solé, L. (1954a). Recherches morphologiques dans le Nord-Ouest de la Peninsule Iberique. *Memoires et Documents du Centre National de la Recherche Scientifique*, 4. Paris, 9–61.

Birot, P. and Solé, L. (1954b). *Investigaciones sobre la Morfología de la Cordillera Central Española*. Instituto Juan Sebastián Elcano. Consejo Superior de Investigaciones Científicas. Madrid, 87 p.

Birot, P., Henin, S., Guillien, J. and Delvert, J. (1968). *Contributions à l'étude de la désagrégation des roches cristallines*. Centre de Documentation Universitaire. Paris, 232 p.

Bisci, C., Dramis, F. and Sorriso-Valvo, M. (1996). Rock Flow (Sackung). In: Dikau, R., Brunsden, D., Schrott, R. and Ibsen, M.L. (Eds.). *Landslide Recognition. Identification, Movement and Causes*. Wiley. Chichester, 150–160.

Bishop, A.W. (1973). The stability of tips and spoil heaps. *Quaterly Journal of Engineering Geology*, 6, 335–376.

Bishop, A.W., Hutchinson, J.N., Penman, A.D.M. and Evans, H.E. (1969). Geotechnical investigations into the causes and circumstances of the disaster of 21 October 1960. A selection of technical reports submitted to the Aberfan Tribunal, Welsh Office. London HMSO.

Bishop, P. (2007). Long-term landscape evolution: linking tectonics and surface processes. *Earth Surface Processes and Landforms*, 32, 329–365.

Bjornsson, H. (1979). Glaciers in Iceland, *Jökull*, 29, 74–80.

Björnsson, H. (1992). Jökulhlaups in Iceland: Prediction, characteristics and simulation. *Annals of Glaciology*, 18, 95–106.

Black, R.F. and Barksdale, W.L. (1949). Oriented lakes of Northern Alaska. *Journal of Geology*, 57, 105–118.

Blackwelder, E. (1925). Exfoliation as a phase of rock weathering. *Journal of Geology*, 33, 793–806.

Blackwelder, E. (1928). Mudflow as a geologic agent in semiarid mountains. *Bulletin of the Geological Society of America*, 39, 465–484.

Blackwelder, E. (1931). Desert plains. *Journal of Geology*, 39, 133–140.

Blackwelder, E. (1933). The insolation hypothesis of rock weathering. *American Journal of Science*, 152, 97–113.

Blackwelder, E. (1934). Yardangs. *Bulletin of the Geological Society of America*, 45, 159–165.

Blackwell, P.G., Back, C.E. and Reimer, P.J. (2006). Important features of the new radiocarbon calibration curves. *Quaternary Science Reviews*, 25, 408–413.

Blaikie, P., Cannon, T., Davis, I. and Wisner, B. (1994). *At Risk Natural Hazards, People's Vulnerability and Disaster*. Routledge. London, 284 p.

Blair, R.W. (1986). Karst landforms and lakes. In: Short, N.M. and Blair, R.W. (Eds.). *Geomorphology from Space. A Global Overview of Regional Landforms*, NASA. Washington, 407–446.

Blair, T.C. (1987). Sedimentary processes, vertical stratification sequences, and geomorphology of the Roaring River alluvial fan, Rocky Mountain National Park, Colorado. *Journal of Sedimentary Petrology*, 57, 1–18.

Blair, T.C. and McPherson, J.G. (1994a). Alluvial fans and their natural distintion from rivers based on morphology, hydraulic processes, sedimentary processes, and facies assemblages. *Journal of Sedimentary Research*, A64, 450–489.

Blair, T.C. and McPherson, J.G. (1994b). Alluvial fans processes and forms. In: Abrahams, A.D. and Parsons, A.J. (Eds.). *Geomorphology of Desert Environments*, Chapman and Hall. London, 354–402.

Blair, T.C. and McPherson, J.G. (2009). Processes and forms of alluvial fans. In: Parsons, A.J. and Abrahams, A.D. (Eds.). *Geomorphology of Desert Environments* (2ª Edición). Springer. Berlín, 413–467.

Blake, D.H. and Paijmans, J.K. (1973). Landforms types of eastern Papua and their associated characteristics. Land Res. Ser. CSIRO Aust., 32, 1–21.

Bland, W. and Rolls, D. (1998). *Weathering: An Introduction to the Scientific Principles*. Arnold. London, 271 p.

Blázquez, R. (2001). Diseño sismorresistente de edificaciones y obras públicas. Normas sísmicas. In: Capote, R. and Martínez-Díaz, J.J. (Eds.). *El Riesgo Sísmico. Prevención y Seguro*. Consorcio de Compensación de Seguros. Madrid, 179–197.

Blissenbach, E. (1952). Relation of surface angle distribution to particle size distribution on alluvial fans. *Journal of Sedimentary Petrology*, 22, 25–28.

Blissenbach, E. (1954). Geology of alluvial fans in semiarid regions. *Bulletin of Geological Society of America*, 65, 175–190.

Bloom, A.L. (1978). *Geomorphology. A Sistematic Analisis of Late Cenozoic*. Prentice Hall. New Jersey, 510 p.

Blount, C.W. and Dickson, F.W. (1973). Gypsum-anhydrite equilibria in systems $CaSO_4 \cdot 2H_2O$ y $CaCO_3$–$NaCl$–H_2O. *American Mineralogist*, 58, 323–331.

Blume, H. and Barth, H.K. (1972). Rampenstufen und Schuttrampen als Abtragungsformen in ariden Schichtstufenlandschaften. *Erkunde*, 26, 108–116.

Blümel, W.D. (1981). *Pedologische und geomorphologische aspekte der kalkkrustenbildung in südwesafrika und Südtospanien*. Karlsruher Geographische Hefte, 10, 228 p.

Blümel, W.D. (1982). Calcretes in Namibia and SE Spain-relations to substratum, soil formation and geomorphic factors. *Catena Supplement*, 1, 67–95.

Boast, R. (1990). Dambos: a review. *Progress in Physical Geography*, 14, 153–177.

Boelhouwers, J.C. (1999). Relict periglacial slope deposits in the Hex River Mountains, South Africa: observations and palaeoenvironmental implications. *Geomorphology*, 30, 345–358.

Boer, M.M. and de Groot, R.S. (1990). *Landscape-ecological impact of climatic change*. IOS Press. Amsterdam, 429 p.

Bögli, A. (1956). Grundformen von Karsthöhlenquerschnitten. *Stalactite*, 6, 56–62.

Bögli, A. (1960). Kalklösung und Karrenbildung. *Zeitscrift für Geomorphologie*, Supplementband 2, 4–21.

Bögli, A. (1964). Mischungskorrosion-ein beitrag zur verkarstungsproblem. *Erdkunde*, 18, 83–92.

Bögli, A. (1980). *Karst Hydrology and Physical Speleology*. Springer. Berlin, 270 p.

Boix-Fayos, C., Barberá, G.G., López-Bermúdez, F. and Castillo, V.M. (2007). Effects of check dams, reforestation and land-use changes on river channel morphology: Case study of the Rogativa catchment (Murcia, Spain). *Geomorphology*, 91, 103–123.

Bolt, A.B. (1999). *Earthquakes*. Freeman. New York, 366 p.

Bolt, B.A., Horn, W.L., Macdonald, G.A. and Scott, R.F. (1975). *Geological Hazards. Earthquakes, Tsunamis, Volcanoes, Avalanches, Landslides, Floods*. Springer, Berlin, 328 p.

Bonacci, O. (1987). *Karst Hydrology with Special Reference to the Dinamic Karst*. Springer. Berlin, 184 p.

Bonacci, O. (2004). Poljes. In: Gunn, J. (Ed.). *Encyclopedia of Caves and Karst Science*. Fitzroy Dearborn. New York, 599–560.

Bonacci, O. and Zivaljevic, R. (1993). Hydrological explanation of the flow in the karst: example of the Crnojevica spring. *Journal of Hydrology*, 146, 405–419.

Bond, G., Broecker, W., Johnsen, S., McManus, J., Labeyrie, L., Jozuel, J. and Bonani, G. (1993). Correlations between climate records from North Atlantic sediments and Greenland ice. *Nature*, 365, 143–147.

Bond, G.C., Showers, W., Cheseby, M., Lotti, R., Almasi, P., de Menocal, P., Priore, P., Cullen, P., Hajdas, I. and Bonani, G. (1997). A pervasive millennial-scale in North Atlantic Holocene and glacial climates. *Nature*, 278, 1257–1266.

Bond, G.C., Showers, W., Elliot, M., Evans, M., Lotti, R., Hajdas, I., Bonani, G. and Johnson, S. (1999). The North Atlantic's 1–2 kgr climate rhythm, relation to Heinrich Events. Dansgaard/Oescher cycles and Little Ice Age. In: Clerk, P.U., Webb, R.S. and Keigwin, L.D. (Eds.). *Mechanisms of Global Climate Change at Millenial Time Scales*. American Geophysical Union. Washington, 35–58.

Bond, G., Kromer, B., Beer, J., Muscheler, R., Evans, M.N., Showers, W., Hoffman, S., Lotti-Bond, R., Hafdas, I. and Bonani, I. (2001). Persistent solar influence on North Atlantic climate during the Holocene. *Science*, 294, 2130–2136.

Bondurant, D.C. (1951). Sedimentation studies at Conchas Reservoir in New Mexico. *Transactions American Society Civil Engineers*, 116, 1292–1295.

Bonython, C.W. and Mason, R. (1953). The filling and drying of Lake Eyre. *Geographical Journal*, 119, 321–330.

Boothroyd, J.C. and Nummedal, D. (1978). Proglacial braided outwash: a model for humid alluvial-fan deposits. In: Miall, A.D. (Ed.). *Fluvial Sedimentology*. Canada Society Petroleum Geology, Memoire 5. Calgary, 641–668.

Borcherdt, R.D. (1975) (Ed.). Studies of seismic zonation of the San Francisco Bay region. *United States Geological Survey Professional Paper*, 941A, 102 p.

Bordonau, J. (1992). *Els complexos glacio-lacustres relacionats amb el darrer cicle glacial als Pirineus*. Geoforma Ediciones. Logroño, 251 p.

Bordonau, J. (1993). The Upper Pleistocene ice-lateral till complex of Cerler (Ésera Valley, central Southern Pyrenees: Spain). *Quaternary International*, 18, 5–14.

Bordonau, J. and Vilaplana, J.M. (1986). Géomorphologie et tectonique récente dans le Val d'Aran (zone axiale des Pyrénées Centrales (Espagne)). *Revue de Géologie Dynamique et de Géographie Physique*, 27, 303–310.

Borgati, L. and Soldati, M. (2005). Fenomeni Franosi. In: Panizza, M. (Ed.). *Manuale de Geomorfologia Applicata*. FrancoAngeli. Milano, 88–123.

Borger, H. (2000). *Mikromorphologie und Paläoenvironment: Die Mineralverwitterung als Zeugnis der cretazisch-tertiären Umvelt in Süddeutschland*. Relief, Boden, Paläoklima 15. Schweizerbart'sche. Stuttgart.

Borja, F., Zazo, C., Dabrio, C.J., Díaz del Olmo, F., Goy, J.L. and Lario, J. (1999). Holocene eolian facies and human sttlements along the Atlantic coast of southern Spain. *The Holocene*, 9, 333–339.

Born, S.M. and Ritter, D.F. (1970). Modern terrace development near Pyramide Lake, Nevada, and its geologic implications. *Geological Society of America Bulletin*, 81, 1233–1242.

Bornand, M. (1979). Application de la méthode des bilans à l'étude de l'évolution pédologique des terrasses alluviales. Las des cailloutis fluvioglaciaires de la moyenne vallée du Rhône. *Bulletin de l'Association française pour l'étude du sol*. Science du sol. Versailles, 2–3, 281–299.

Bornhardt, W. (1900). *Zur Oberflächengestaultung und Geologie Deutsch-Afrikas.* Reimer, Berlin.

Bott, M.H.P. (1982). *The Interior of the Earth, its Structure, Constitution and Evolution*. Arnold. London.

Boulangé, B. and Millot, G. (1988). La distribution des bauxites sur le craton Ouest-Africain. *Bulletin des Sciences Géologiques*, 41, 113–123.

Boulton, G.S. (1970a). On the origin and transport of englacial debris in Svalbard Glaciers. *Journal of Glaciology*, 9, 213–229.

Boulton, G.S. (1970b). On the depositon of subglacial and melt-out tills at the margins of certain Svalbard glaciers. *Journal of Glaciology*, 9, 231–245.

Boulton, G.S. (1971). Till genesis and fabric in Svalbarg, Spitsbergen. In: Goldthwait, R.P. (Ed.). *Till: A Symposium*, Ohio State University Press. Columbus, 41–72.

Boulton, G.S. (1972). Modern arctic glaciers as depositional models for former ice sheets. *Journal of the Geological Society of London*, 128, 361–393.

Boulton, G.S. (1974). Processes and patterns of glacial erosion. In: Coates, D.R. (Ed.). *Glacial geomorphology*, State University of New York. Binghamton, 41–87.

Boulton, G.S. (1975). Processes and patterns of subglacial sedimentation: a theoretical approach. In: Wright, A.E. and Moseley, E. (Eds.). *Ice Ages: Ancient and Modern*. Seel House Press. Liverpool, 7–42.

Boulton, G.S. (1976). The origin of glacially fluted surfaces-observation and theory. *Journal of Glaciology*, 17, 287–309.

Boulton, G.S. (1978). Boulder shapes and grain-size distributions of debris as indicators of transport paths through a glacier and till genesis. *Sedimentology*, 23, 15–37.

Boulton, G.S. (1982). Subglacial processes and the development of glacial bedforms. In: Davidson-Arnott, R.; Nickling, W. and Fahey, B.D. (Eds.). *Research in Glacial, Glacio-Fluvial and Glacio-Lacustrine Systems*. Geobooks, 1–31. Norwich.

Boulton, G.S. (1987). A theory of drumlin formation by subglacial sediment deformation. In: Menzies, J. and Rose, J. (Eds.). *Drumlin Symposium*. Balkema. Rotterdam, 25–80.

Boulton, G.S. (1993). Ice ages and climatic change. In: Duff, D. (Ed.). *Holmes' Principles of Physical Geology*. Chapman and Hall. London, 439–469.

Boulton, G.S. (2006). Glaciers and their coupling with hydraulic and sedimentary processes. En: Knight, P.G. (Ed.). *Glacier Science and Environmental Change*. Blackwell. Maldon, 3–22.

Boulton, G.S. and Vivian, R. (1973). Underneath the glaciers. *Geographical Magazine*, 45, 311–319.

Boulton, G.S., Lunn, R., Vidstrand, P. and Latsepin. (2007). Subglacial drainage by groundwater-channel coupling, and the origin of esker systems: part II-theory and simulation of a modern system. *Quaternary Science Reviews*, 26, 1091–1105.

Bourke, M.C. Ewing, R.C., Finnegan, D. and McGowan, H.A. (2009). Sand dune movement in the Victoria Valley, Antarctica. *Geomorphology*, 109, 148–160.

Bousquet, J.C. (1979). Quaternary strike-slip faults in southern Spain. *Tectonophysics*, 52, 277–286.

Bovis, M.J. (1982). Uphill-facing (antislope) scarps in the Coast Mountains, southwest British Columbia. *Geological Society of America Bulletin*, 93, 804–812.

Bowen, N.L. (1928). *The Evolution of Igneous Rocks*. Dover Publications. New York, 332 p.

Bowler, J.M. (1973). Clay dunes: their occurrence, formation and environmental significance. *Earth Science Reviews*, 9, 315–338.

Bowler, J.M. (1986). Spatial variability and hydrologic evolution of Australian lake basins: analogue for Pleistocene hydrologic change and evaporite formation. *Palaeogeography, Palaeoclimatology, Palaeoecology*, 54, 21–41.

Bowler, J.M. and McGee, J.W. (1978). Geomorphology of the Malle region in semiarid northern Victoria and Western New South Wales. *Proceedings of the Royal Society of Victoria*, 90, 5–20.

Bowley, W.W. and Burghardt, M.D. (1971). Thermodynamics and stones, EOS. *American Geophysical Union Transactions*, 52, 4–7.

Bowyer-Bower, T.A.S. and Bryan, R.B. (1986). Rill initiation: concepts and evaluation on badland slopes. *Zeitschrift für Geomorphologie, Supplementband*, 59, 161–175.

Boyd, R., Dalrymple, R. and Zaitlin, B.A. (1992). Classification of clastic coastal depositional environments. *Geology*, 17, 926–929.

Boyé, M. (1949). Importance du défongage périglaciaire dans l'elaboration des formes glaciaires. *Compte Rendus de l'Académie des Sciences de Paris*, 229, 723–724.

Boyé, M. and Frisch, P. (1973). Dégagement artificiel d'un dóme crystallin au Sud-Cameroun. *Travaux et Documents du Géographie Tropicale, Bordeaux*, 8, 31–62.

Boyle, R.W. (1979). The geochemistry of gold and its deposits. *Bulletin of the Geological Survey of Canada*, 280, 584 p.

Brabb, E.E. (1989). Landslides: Extent and economic significance in the United States. In: Brabb. E.E. and

Harrod, B.L. (Eds). *Landslide: Extent and Economic Significance*. Balkema. Amsterdam, 25–50.

Bradley, R.S. (1985). *Quaternary Paleoclimatology: Methods of Paleoclimatic Reconstruction*. Allen and Unwin. Boston. Massachussets.

Bradley, W.C. (1963). Large-scale exfoliation in massive sandstones of the Colorado Plateau. *Bulletin of the Geological Society of America*, 74, 519–528.

Bradley, W.C., Hutton, J.T. and Twidale, C.R. (1978). Role of salts in development of granitic tafoni, South Australia. *Journal of Geology*, 86, 647–654.

Bradwell, T., Stoker, M. and Krabbendam, M. (2008). Mega grooves and streamlined bedrock in NW Scotland: The role of ice streams in landscape evolution. *Geomorphology*, 97, 135–156.

Brakenridge, G.R. and Shuster, J. (1986). Late Quaternary geology and geomorphology in relation to archaeological site locations, southern Arizona. *Journal of Arid Environments,* 10, 225–239.

Brammer, H. (1990). Floods in Bangladesh: I. Geographical background to the 1987 and 1988 floods. *The Geographical Journal*, 156, 12–22.

Brand, E.W. (1989). Occurrence and significance of landslides in southeast Asia. In: Brabb, E.E. and Harrod, B.L. (Eds.). *Landslides: Extent and Economic Significance*. Balkema. Rotterdam, 303–324.

Brand, E.W., Premchitt, J. and Phillipson, H.B. (1985). Relationship between rainfall and landslides in Hong Kong. En: *IV International Symposium on Landslides*. Toronto, 1984. Vol. 1, 377–384.

Bravo, T. (1964a). El volcán y el país de la Corona. La "Cueva de los Verdes" y los "Jameos". *Publicaciones del Cabildo Insular de Lanzarote*. Arrecife (Lanzarote). (Islas Canarias), 31 p.

Bravo, T. (1964b). Estudio geológico y petrográfico de la isla de La Gomera. *Estudios Geológicos*, 20, 1–57.

Bray, M.T., French, H. and Shur, Y. (2006). Further cryogenic observations in the CRREL permafrost tunnel, Fox, Alaska. *Permafrost and Periglacial Processes*, 17, 233–234.

Breed, C.S. and Grow, T. (1979). Morphology and distribution of dunes in sand seas observed by remote sensing. In: McKee, E.D. (Ed.). A *Study of Global Sand Seas*. United States Geological Survey, Professional Paper, 1052, 253–303. Washington.

Breed, C.S., McCauley, J.F. and Davis, P.A. (1987). Sand sheets of the eastern Sahara and ripple blankets on Mars. In: Frostick, L.E. and Reid, I. (Eds.). *Desert Sediments, Ancient and Modern*. Blackwell. Oxford, 337–360.

Breed, C.S., McCauley, J.F. and Whitney, M.I. (1989). Wind erosion forms. In: Thomas, D.S.G. (Ed.). *Arid Zone Geomorphology.* Belhaven Press. London, 284–307.

Breed, C.S., McCauley, J.F., Whitney, M.I., Tchakarian, V.P. and Laity, J.E. (1997). Wind erosion in drylands. In: Thomas, D.S.G. (Ed.). *Arid Zone Geomorphology: Process, Form and Change in Drylands*. John Wiley. Chichester, 437–464.

Bremer, H. (1971). Flüsse, Flächen- und Stufenbildung in den feuchten Tropen. *Wuzburger Geographische Arbeiten*, 35, 194 p.

Bremer, H. (1985). Soil and slope development in the wet zone of Sri Lanka. In: Douglas, I. and Spencer, T. (Eds.). *Environmental Change and Tropical Geomorphology*. Allen and Unwin. London, 295–302.

Bremer, H. (1996). Climatic, climatogenetic and tectonic geomorphology. In: McCann, S.B. and Ford, D.C. (Eds.). *Geomorphology sans Frontières.* Wiley. Chichester, 34–39.

Bremmer, H. (2002). Tropical weathering, landforms and geomorphological processes: field work and laboratory analysis. *Zeitschrift für Geomorphologie*, 46, 273–291.

Bretz, J.H. (1923). The channelled scablands of the Columbia Plateau. *Journal of Geology*, 31, 617–649.

Bretz, J.H. (1942). Vadose and phreatic features of limestone caves. *Journal of Geology*, 50, 675–811.

Bretz, J.H. (1969). The Lake Missoula floods and the Channeled Scablands. *Journal of Geology*, 72, 505–543.

Brice, J.C. (1964). Channel patterns and terraces of the Loup Rivers in Nebraska. *United States Geological Survey Professional Paper*, 422D.

Brice, J.C. (1975). Airphoto interpretation of the form and behaviour of alluvial rivers. *Final Report for United States Army Research Office*.

Brice, J.L. (1966). Erosion and deposition in the loess mantled Great Plains, Medicine Creek Drainage Basin, Nebraska. *United States Geological Survey*, Professional Paper 350 H, 255–339.

Brice, J.S. and Lunt, I.A. (2006). Depositional models of braided rivers. In: Sambrook Smith, G.H., Best, J.L., Bristow, C.S. and Petts, G.E. (Eds.). *Braided Rivers. Process, Deposits, Ecology and Management*. Special Publication Number 36. International Association of Sedimentologist. Blackwell. Malden, 11–50.

Bridge, J.S. (2003). *Rivers and Floodplains. Forms, Processes and Sedimentary Records*. Blackwell. Oxford, 491 p.

Briersley, G.J. y Fryirs, R.A. (2005). *Geomorphology of River Management: Applications of the River Styles Framework*. Blacwell. Oxford.

Brimhall, G.H., Lewis, C.J., Ague, J.J., Dietrich, W.E., Hampel, J., Teague, T. and Rix, C. (1988). Metal enrichment in bauxites by deposition of chemically nature aeolian dust. *Nature*, 333, 819–824.

Bristow, C.S., Bailey, S.D. and Lancaster, N. (2000). The sedimentary structure of linear sand dunes. *Nature*, 406, 56–59.

Bristow, C.S., Drake, N. and Armitage, S. (2009). Deflation in the dustiest place on the Earth: The Bodélé Depression, Chad. *Geomorphology*, 105, 50–58.

Bristow, Ch. (1987). Anabranching river: channel migration and deposition. In: Ethridge, F., Flores, R.M. and Harvey, M. (Eds.). *Recent Development in Fluvial Sedimentology*. Special Publication-Society of Economic Paleontologists and Mineralogists, 39, 63–74.

Brodzikowski, K. and Van Loon, A.J. (1991). *Glacigenic Sediments*. Elsevier. Amsterdam, 674 p.

Broecker, W.S. (1987). The biggest chill. *Natural History*, 96, 74–82.

Broecker, W.S. (1994). Massive iceberg discharges as triggers for global climate change. *Nature*, 372, 421–424.

Broecker, W.S. (2006). Abrupt climate change revisited. *Global Planetary Change*, 54, 211–215.

Broecker, W.S. and Denton, G.H. (1990a). What drives glacial cycles? *Scientific American*, January 1990, 43–50.

Broecker, W.S. and Denton, G.H. (1990b). The role of ocean-atmosphere reorganizations in glacial cycles. *Quaternary Science Reviews*, 9, 305–341.

Broecker, W.S., Bond, G., Klas, M., Clark, E. and McManus, G. (1992). Origin of the northern Atlantic's Heinrich events. *Climate Dynamics*, 6, 265–273.

Bromhead, E.N. (1979). Factors affecting the transition between the various types of mass movement in coastal cliffs consisting largely of overconsolidated clay, with special reference to Southern England. *Quaterly Journal Engineering Geology*, 12, 291–309.

Bromhead, E.N. (1986). *The Stability of Slopes*. Surrey University Press. New York, 374 p.

Brook, E. (2007). Ice core methods. Overview. In: Elias, S.A. (Ed.). *Encyclopedia of Quaternary Science*. Elsevier. Amsterdam, 1145–1156.

Brook, G.A. and Ford, D.C. (1976). The Nahanni North karst: a question mark of the validity of the morphoclimatic concept of karst development. *Proceedings 6th International Congress of Speleology*, 2, 43–57.

Brook, G.A. and Ford, D.C. (1978). The origin of labyrinth and tower karst and the climatic conditions necessary for their development. *Nature*, 275, 493–496.

Brookfield, M.E. and Ahlbrandt, T.S. (Eds.) (1983). *Eolian Sediments and Processes*, Elsevier. Amsterdam, 660 p.

Brooks, G.A. (1978). A new approach to the study of inselberg landscapes. *Zeitschrift für Geomorphologie Supplementband*, 31, 138–160.

Brooks, N. and Legrand, M. (2000). Dust variability over Northern Africa and rainfall in the Sahel. In: McLaren, S.J. and Kniveton, D.R. (Eds.). *Linking Climate Change to the Land Surface Change*. Kluver. Dordrecht, 1–25.

Brosche, K.U. (1971). Neue Beobacktungen zu vorzeitslichen periglaziaterscheinungen in Ebrobecken. *Zeitschrift für Geomorphologie*, 15, 107–114.

Brosche, K.U. (1972). Vorzeitliche Periglazialerscheinunger im Ebrobecken in der Ungebung voz Zaragoza sowie ein Beitrag zur Ausdehbung von Schuttund Bloakdecken in Zentral-und-W-Teil der Iberische Halbinsel. *Göttingen Geographische Abh.*, 60, 293–316.

Brown, R.J.E. (1970). *Permafrost in Canada; its Influence on Northern Development*. University of Toronto Press. Toronto, 234 p.

Brown, A.G. (2003). Global environmental change and palaeohydrology of Western Europe: A review. In: Gregory, K.J. and Benito, G. (Eds.). *Palaeohydrology: Understanding Global Change*. Wiley. Chichester, 105–121.

Brown, A.G. (2008). Geoarchaeology, the four dimensional (4D) fluvial matrix and climatic causality. *Geomorphology*, 101, 278–297.

Brown, A.G., Gregory, K.J. and Milton, E.J. (1987). The use of Landsat multispectral scanner data for the analysis and management of flooding on the River Severn, England. *Environmental Management*, 11, 695–701.

Bruce, D.A. (2003). Sealing of massive water inflows through karst by grouting. *American Society Civil Engineers Geotechnical Special Publication*, 122, 615–626.

Brückner, W.D. (1955). The mantle rock (laterite) of the Gold Coast and its origin. *Geologische Rundschau*, 43, 307–327.

Brum Ferreira, A. de (1978). *Planaltos e Montanhas do Norte da Beira. Estudo de Geomorfología*. Memórias do Centro de Estudos Geográficos, 4. Lisboa, 374 p.

Brunner, F.K. and Scheidegger, A.E. (1974). Kinematics of a scree slope. *Revista Italiana di Geofisica*, 23, 89–94.

Brunsden, D. (1979). Mass movement. In: Embleton, C. and Thornes, J. (Eds.). *Process in Geomorphology*. Arnold. London, 130–186.

Brunsden, D. (1980). Aplicable models of long term landform evolution. *Zeitschrift für Geomorphologie*, Supplementband 36, 16–26.

Brunsden, D. (1984). Mudslides. In: Brunsden, D. and Prior, B. (Eds.). *Slope Instability*. Wiley. Chichester, 363–418.

Brunsden, D. (1990). Tablets of stone: towards the Ten Commandments of geomorphology. *Zeitschrift für Geomorphologie Supplementband* 79, 1–37.

Brunsden, D. (1993). Mass movement; the research frontier and beyond: a geomorphological approach. *Geomorphology*, 7, 85–128.

Brunsden, D. (2004a). Relaxation time. In: Goudie, A.S. (Ed.). *Encyclopedia of Geomorphology*. Routledge. London, 839–840.

Brunsden, D. (2004b). A critical assessment of the sensivity concept on Geomorphology. *Catena*, 42, 99–123.

Brunsden, D. and Ibsen, M.L. (1996). Mudslide. In: Dikau, A. Brunsden, D. Shortt, L. and Ibsen, M.L. (Eds.). *Landslide Recognition. Identification Movement and Causes* Wiley. Chichester, 103–119.

Brunsden, D. and Thornes, J.B. (1979). Landscape sensivity and change. *Transactions of the Institute of British Geographers*, 4, 463–484.

Bryan, K. (1922). Erosion and sedimentation in the Papago County, Arizona. *United States Geological Survey Bulletin*, 730, 19–90.

Bryan, K. (1927). Pedestal rocks formed by differential erosion. *United States Geological Survey Bulletin*, 790A, 1–15.

Bryan, K. (1928). Niches and other cavities in sandstones at Chaco Canyon, New Mexico. *Zeitschrift für Geomorphology*, 3, 125–140.

Bryan, K. (1941). Pre-Columbian agriculture in the southwest as conditioned by periods of alluviation. *Association American Geographers Annals*, 31, 219–242.

Bryan, R. (2004). Raindrop impact, splash and wash. In: Goudie, A.S. (Ed.). *Encyclopedia of Geomorphology*. Routledge. London, 823–826.

Bryan, R.B. (1987). Processes and significance of rill development. *Catena Supplement*, 8, 1–15.

Bryan, R.B and Yair, A. (1982). Perspectives on studies of badland geomorphology. In: Bryan, R.B. and Yair, A. (Eds.). *Badland geomorphology and piping*. Geobooks. Norwich, 1–12.

Bryan, R.B. and Poesen, J. (1989). Laboratory experiments on the influence of slope length on runoff, percolation and rill development. *Earth Surface Processes and Landforms*, 14, 211–231.

Bryan, R.B., Yair, A. and Hodges, W.K. (1978). Factors controlling the initiation of runoff and piping in Dinosaur Provincial Park badlands, Alberta, Canada. *Zeitschrift für Geomorphologie Supplementband*, 29, 151–168.

Bryant, E. (1991). *Natural Hazards*. Cambridge University Press. Cambridge, 294 p.

Bryant, E. (1997). *Climate, Process and Change*. Cambridge University Press. Cambridge, 209 p.

Bryant, E.A. (2001). *Tsunami: Underreated Hazard*. Cambridge University Press. Cambridge, 350 p.

Bryant, R.G., Sellwood, B.W., Millington, A.C. and Drake, N.A. (1994). Marine-like potash evaporite formation on a continental playa: case study from Chott el Djerid, southern Tunisia. *Sedimentary Geology*, 90, 269–291.

Bryscn, R.A. and Goodman, B.M. (1980). Volcanic activity and climatic change. *Science*, 207, 1041–1044.

Buchanan, F. (1807). *A Journey from Madras through the Countries of Mysore, Kahara and Malabar*. East Indian Company, Vol. 2, 436–461; 559; vol. 3, 66, 89, 251, 258, 378. London.

Bucher, W.H. (1963). Cryptoexplosion structures caused from without or within the Earth (Astroproblemes or Geoproblemes). *American Journal of Science*, 261, 597–619.

Buckman, R.C. and Anderson, R.E. (1979). Estimation of fault scarp ages from scarp-height-slope-angle relationship. *Geology*, 7, 11–14.

Budd, W.F., Jenssen, D. and Radok, U. (1970). The extent of basal melting in Antarctica. *Polarforschung*, 6, 293–306.

Büdel, J. (1944). Die morphologischen wirkungen des eiszeitcklimas im glitocherfreien gebiet. *Geologische Rundschau*, 34, 482–519.

Büdel, J. (1948). Das system der klimatischen Geomorphologie. *Verhandlungen der Deutschen Geographentag*, 27, 69–100 (Traducción inglesa en Derbyshire, E. (Ed.) (1973). *Climatic Geomorphology*. Macmillan. London, 104–130).

Büdel, J. (1951). Die klimazonen des Eiszeitalters. *Eiszeitalter und Gevenwart*, 1, 16–26. (Trad. Inglesa: *International Geology Review*, 1(9), 72–79, 1959).

Büdel, J. (1957). Die "Doppelten Einebnungsflächen" in den feuchten Tropen *Zeitschrift für Geomorphologie,* 1, 201–228.

Büdel, J. (1960). *Die Frostschott zone Südorst Spitsbergen*. Colloquium Geographica, 6. Bonn, 105 p.

Büdel, J. (1963). Klima-genetische Geomorphologie. *Geographische Rundschau*, 15, 269–285. (Traducción inglesa en Derbyshire, E. (Ed.) (1973). *Climatic Geomorphology*. Macmillan. London, 202–227.

Büdel, J. (1965). Die Relieftypen de Flachenspül-Zone Sud-Indiens am Ostabfall Dekans gegen Madras. *Colloquium Geographicum*, 8. Bonn, 100 p.

Büdel, J. (1968). Geomorphology principles. In: Fairbridge, R.W (Ed.). *The Enciclopedia of Geomorphology*. Dowden, Hutchinson and Ross. Stroudsburg, 416–422.

Büdel, J. (1970). Pedimente, Rumpfflächen und Rückland-Steilhänge. *Zeitschrift für Geomorphologie*, 14, 1–57.

Büdel, J. (1977). *Klima-Geomorphologie*. Borntraeger. Berlín, 304 p.

Büdel, J. (1978). Das Inselberg-Rumpfflächenrelief des heutigen Tropen und das Schicksal seiner fossilen Altformen in anderen Klimazonen. *Zeitschrift für Geomorphologie Supplementband,* 31, 79–110.

Büdel, J. (1980). Climatic and climatomorphic geomorphology. *Zeitschrift für Geomorphologie Supplemenband*, 36, 1–8.

Büdel, J. (1982). *Climatic Geomorphology* (transl. Fischer, L. y Busche, D.). Princeton University Press. Princeton, 443 p.

Budyko, M.I. (1982). *The Earth's Climate: Past and Future*. Academic Press. Orlando, 307 p.

Bull, L.J. and Kirkby, M.J. (2002). Dryland river characteristics and concepts. In: Bull, L.J. and Kikrby, M.J. (Eds.). *Dryland Rivers: Hydrology and Geomorphology of Semi-arid Channels*. Wiley. Chichester, 3–15.

Bull, W.B. (1963). Alluvial fan deposits in Western Fresno County, California. *Journal of Geology*, 71, 243–251.

Bull, W.B. (1964a). Geomorphology of segmented alluvial fans in Western Fresno County, California. *U.S. Geological Survey Professional Paper*, 352 E, 89–125.

Bull, W.B. (1964b). History and causes of channel trenching in Western Fresno County, California. *American Journal of Science*, 262, 249–258.

Bull, W.B. (1968). Alluvial fan cone. En: Fairbridge, R.W. (Ed.). *Enclycopedia of Geomorphology*. Dowden, Hutchinson and Ross. Stroudsburg, 7–10.

Bull, W.B. (1972). Recognition of alluvial fan deposits in the stratigraphic record. In: Rigby, J.K. and Hamblin, W.K. (Eds.). *Recognition of Ancient Sedimentary Environments*. Society of Economic Paleontologist and Mineralogists. Special Publication, 16, 63–83.

Bull, W.B. (1974). Geologic factors affecting compactation of deposits in a land subsidence area. *Geological Society of America Bulletin*, 84, 3783–3802.

Bull, W.B. (1977a). Tectonic Geomorphology of the Mojave Desert. *United States Geological Survey*, Contract Report 14-08-001-G-394. Office of Earthquakes, Volcanoes and Engineering. Menlo Park, California, 188 p.

Bull, W.B. (1977b). The alluvial fan environment. *Progress in Physical Geography*, 1, 222–270.

Bull, W.B. (1978). Geomorphic tectonic classes of the south front of the San Gabriel Mountains, California. *United States Geological Survey*, Contract Report 14-08-001-G-394. Office of Earthquakes, Volcanoes and Engineering. Menlo Park, California, 59 p.

Bull, W.B. (1979). Threshold of critical stream power. *Bulletin of the Geological Society of American*, 90, 453–464.

Bull, W.B. (1990). Stream-terrace genesis: implications for soil development. *Geomorphology*, 3, 351–367.

Bull, W.B. (1991). *Geomorphic Responses to Climate Change*. Oxford University Press. Oxford. 326 p.

Bull, W.B. (1997). Discontinous ephemeral streams. *Geomorphology*, 19, 227–276.

Bull, W.B. (2007). *Tectonic Geomorphology of Mountains. A New Approach to Paleosismology*. Blackwell. Oxford, 320 p.

Bull, W.B. (2009). *Tectonically Active Landscapes*. Wiley-Blackwell. Chichester, 326 p.

Bull, W.B. and McFadden, L.D. (1977). Tectonic geomorphology north and south of the Garlock fault, California. In: Doehring, D.O. (Ed.). *Geomorphology in Arid Regions*. Allen and Unwin. London, 115–138.

Bull, W.B. and Pearthree, P.A. (1988). Frequency and size of Quaternary surface ruptures of the Pilaycachi fault, northeastern Sonora, Mexico. *Bulletin Seismological Society America*, 78, 956–978.

Bullard, F.M. (1976). *Volcanoes of the Earth*. University of Texas Press. Austin, 579 p.

Bullard, J.E and Livingstone, I. (2009). Dust. In: Parsons, A.J. and Abrahams, A.B. (Eds.). *Geomorphology of Desert Environments* (2ª Edición). Springer. Berlin, 629–654.

Bullard, T.F. and Lettis, W.R. (1993). Quaternary fault deformation associated with blind thrust faulting, Los Angeles Basin, California. *Journal of Geophysical Research*, 98, 8349–8369.

Bullen, K.E. and Bolt, B.A. (1985). *An Introduction to the Theory of Seismology*. Cambridge University Press. Cambridge.

Buma, J. and van Asch, T. (1996a). Slide (rotational). In: Dikau, R., Brunsden, D., Schrott, L. and Ibsen. M.L. (Eds.). *Landslide Recognition. Identification, Movement and Causes*. Wiley, Chichester, 43–61.

Buma, J. and van Asch, T. (1996b). Soil (debris) spreading. In: Dikau, R., Brunsden, D., Schrott, L. and Ibsen, M.L. (Eds.). *Landslide Recognition. Identification, Movement and Causes*. Wiley, Chichester, 137–148.

Burbank, D.W. and Anderson, R.S. (2001). *Tectonic Geomorphology*. Blackwell. Oxford, 274 p.

Burillo, F., Gutiérrez, M. and Peña, J.L. (1981a). El cerro del castillo de Alfambra (Teruel). Estudio interdisciplinar de geomorfología y arqueología. *Kalathos*, 1, 1–61.

Burillo, F., Gutiérrez, M. and Peña, J.L. (1981b). Las vertientes del Cerro del Castillo de Alfambra (Teruel). In: Díaz, F., Márquez, D. and Rubio, J.M. (Eds.). *V Reunión del GETC*. Sevilla, 231–244.

Burillo, F., Gutiérrez, M. and Peña, J.L. (1983). La geoarqueología como ciencia auxiliar: una aplicación a la Cordillera Ibérica turolense. *Revista de Arqueología*, 26, 6–13.

Burillo, F., Gutiérrez, M. and Peña, J.L. (1984). Las acumulaciones holocenas y su datación arqueológica en Mediana de Aragón (Zaragoza). *Cuadernos de Investigación Geográfica*, 11, 193–207.

Burillo, F., Gutiérrez, M and Peña, J.L. (1985). Datación arqueológica de deformaciones tectónicas en vertientes holocenas de Sierra Palomera (Cordillera Ibérica centroriental). *I Reunión de Cuaternario Ibérico*, Vol. 2, Lisboa, 355–366.

Burillo, F., Gutiérrez, M., Peña, J.L. and Sancho, C. (1986). Geomorphological processes as indicators of climatic changes during the Holocene in the North-East Spain. In: López Vera, F. (Ed.). *Quaternary Climate in Western Mediterranean*. Universidad Autónoma de Madrid. Madrid, 37–44.

Burton, I. (1989). Natural environmental hazards. In: Henry, J.G. and Heinke, G.W. (Eds.). *Environmental Science and Engineering*. Prentice Hall. Englewood Cliffs, 86–113.

Burton, I., Kates, R.W. and White, G.F. (1978). *The Environmental as Hazard*. Oxford University Press. New York, 239 p.

Busenberg, E. and Plummer, L.N. (1982). The kinetics of dissolution of dolomite in CO_2-H_2O systems at 1,5° to 65°C and o to 1 atm P_{CO2}. *American Journal of Science*, 282, 45–78.

Buss, E. and Heim, A. (1981). *Der Bergsturz von Elm, den 11 September 1981*. Zurich, 163 p.

Butler, D.R. (1995). *Zoogeomorphology. Animals as Geomorphic Agents*. Cambridge University Press. Cambridge, 231 p.

Butler, D.R. and Malanson, G.P. (2005). The geomorphic influences of beaver dams and failures of beaver dams. *Geomorphology*, 71, 48–60.

Butterfield, G.R. (1993). Sand transport response to fluctuating wind velocity. In: Clifford, N.J., French, J.R. and Hardisty, J. (Eds.). *Turbulence: Perspectives on Sediment Transport*. Wiley. Chichester, 305–334.

Butzer, K.W. (1965). *Environment and Archaeology*. Methuen. London, 703 p.

Butzer, K.W. (1976). *Geomorphology from the Earth*. Harper and Row. New York, 463 p.

Butzer, K.W. (1978). Climate patterns in an unglaciated continents. *Geographical Magazine*, 51, 201–208.

Butzer, K.W. (1980). Pleistocene history of the Nile Valley in Egypt and Lower Nubia. In: Williams, M.A.J. and Faure, H. (Eds.). *The Sahara and the Nile*. Balkema. Roterdam, 253–280.

Cailleux, A. (1952). Polissage et surcreusement glaciaires dans l'hypothèse de Boyé. *Revue de Géomorphologie Dynamique*, 5, 247–257.

Cailleux, A. (1961). *Histoire de la Géologie*. Presses Universitaires de France. Paris, 128 p.

Cailleux, A. and Tricart, J. (1956). Le probléme de la classification des faits géomorphologiques. *Annales de Géographie*, 65, 162–186.

Caine, N. (1972). Air photo analysis of blockfield fabrics in Talus Valley, Tasmania. *Journal Sedimentary Petrology*, 42, 33–48.

Caine, N. (1980). The rainfall intensity duration control of shallow landslides and debris flow. *Geografiska Annaler*, 62 A, 23–27.

Calaforra, J.M. (1998). *Karstología en yesos*. Instituto de Estudios Almeriense. Almería, 389 p.

Calaforra, J.M. and Pulido-Bosch, A. (1996). Some examples of gypsum karsts and the more important gypsum caves in Spain. *International Journal of Speleology*, 25, 225–237.

Calaforra, J.M. and Pulido-Bosch, A. (1997). Peculiar landforms in the gypsum karst of Sorbas (Southeastern Spain). *Carbonates and Evaporites*, 12, 110–116.

Calaforra, J.M., Forti, P. and Pulido-Bosch, A. (1992). Nota preliminar sobre la influencia en la evolución espeleogenética de los yesos con especial referencia a los afloramientos kársticos de Sorbas (España) y de Emilia.Romagna (Italia). *Espeleotemas*, 2, 9–18.

Calamber, L. and Dantinne, R. (1964). The avalanche of ash at Jupille (Liège) on February 3rd, 1961. *Vol. commemorative to Prof. F. Campus*. Liège. Belgium, 41–57.

Calvin, P. and Cailleux, A. (1962). A quantitative study of cavernous weathering (taffonis) and its application to glacial chronology in Victoria Valley, Antarctica. *Zeitscrift für Geomorphologie*, 6, 317–324.

Calvo, J.P., Alcalá, L., Alonso-Zarza, A.M., van Dam, J. and Gutiérrez-Santolalla, F. (1999). Estratigrafía y estructura del área de Los Mansuetos (Cuenca de Teruel). Precisiones para la definición del estratotipo del Turoliense. *Geogaceta*, 25, 55–58.

Campbell, I.A. (1974). Measurements of erosion on badlands surfaces. *Zeitschrift für Geomorphologie Supplement Band*, 21, 122–137.

Campbell, I.A. (1982). Surface morphology and rates of change during a ten-year period in the Alberta badlands. In: Bryan, R. and Yair, A. (Eds.). *Badland Geomorphology and Piping*. GeoBooks. Norwich, 221–236.

Campbell, I.A. (1989). Badlands and badland gullies. In: Thomas, D.S.G. (Ed.). *Arid Zone Geomorphology*. Halstead Press. New York, 159–193.

Campbell, I.A. and Honsaker, J.L. (1982). Variability in badlands erosion; problems of scale and threshold identification. In: Thorn, C.E. (Ed.). *Space and Time in Geomorphology*. George Allen & Unwin. London, 59–79.

Campbell, R.H. (1975). Soil Slips, Debris Flows, and Rainstorms in the Santa Monica Mountains and Vicinity, Southern California. *United States Geological Survey Professional Paper*, 851, 51 p.

Campbell, S.E., Seeler, J.S. and Golubic, S. (1989). Desert crust formation and soil stabilization. *Arid Soil Research and Rehabilitation*, 3, 217–228.

Cancelli, A., Pellegrini, M. and Tosatti G. (1987). Alcuni esempi di deformazioni gravitative profonde di versante nell'Appennino settrionale. *Memorias Societá Geologica di Italia*, 39, 447–466.

Cáncer, L. (1996). La catastrophe del Barranco de Arás (7/8/1996): procesos naturales e hipótesis explicativa. *Geographicalia*, 33, 51–71.

Candy, J. and Black, S. (2009). The timing of Quaternary calcrete development in semi-arid southeast Spain: Investigating the role of climate on calcrete genesis. *Sedimentary Geology*, 218, 6–15.

Cantagrel, J.M., Arnaud, N., Ancochea, E., Fúster, J.M. and Huertas, M.J. (1999). Repeated debris avalanches on Tenerife and genesis of Las Cañadas caldera wall (Canary Islands). *Geology*, 27, 739–742.

Capote, R. (1997). Terremotos. (Adaptación a la edición española). In: Suárez, L. and Regueiro, M. (Eds.). *Guía Ciudadana de los Riesgos Geológicos*. The American Institute Professional Geologist and Ilustre Colegio Oficial de Geólogos de España. Madrid, 61–78.

Capote, R. and Martínez-Díaz, J.J. (2002). Estado actual de la previsión sísmica. Nuevas perspectivas y metodologías. In: Capote, R. and Martínez-Díaz, J.J. (Eds.). *El Riesgo Sísmico. Prevención y Seguro*. Consorcio de Compensación de Seguros. Madrid, 1–26.

Capote, R., Gutiérrez, M., Hernández, A. and Olivé. A. (1981). Movimientos recientes en la Fosa del Jiloca (Cordillera Ibérica) *V Reunión del Grupo Español de Trabajo del Cuaternario*. Sevilla, 245–257.

Capot-Rey, R. (1957). Le vent et le modelé éolien au Borkou. *Travaux de l'Institut de Recherches Sahariennes*, 15, 155–157.

Capot-Rey, R. (1963). Contribution à l'ètude et la reprèsentation des barkhanes. *Travaux de l'Institut de Recherches Sahariennes*, 22, 37–60.

Caputo, R. and Helly, B. (2008). The use of distinct disciplines to investigate past earthquakes. *Tectonophysics*, 453, 7–19.

Caputo, R. and Pavlides, S.B. (2008). Earthquake geology: Methods and Applications. *Tectonophysics*, 453, 1–6.

Carbonell, E., Bermúdez de Castro, J.M., Arsuaga, J.L. Díez, J.C., Rosas, A., Cuenca-Bescós, G., Sala, R., Mosquera, M. and Rodríguez, X.P. (1995). Lower Pleistocene hominids and artifacts from Atapuerca TD-6 (Spain). *Science*, 269, 826–830.

Cardona, F. (1989–90). *Grans cavitats de Catalunya*. Espeleo Club de Gràcia. Barcelona, 488 p.

Cardona, F. and Viver, J. (2002). *Sota la sal de Cardona*. Espeleo Club de Gràcia. Barcelona, 128 p.

Carey, S.W. (1962). Scale of geotectonic phenomena. *Journal of the Geological Society of India*, 3, 97–105. (Trad. esp. J.M. Ríos: Notas y Comunicaciones Instituto Geológico y Minero de España, 72, 277–288).

Carling, P.A.., Kirkbride, A.D., Parnachov, S., Boronavko, P.S. and Berget, G.V. (2002). Late Quaternary catastrophic flooding in the Altai Mountains of south-central Siberia: a synoptic overview and introduction to flood deposit sedimentology. In: Martini, J.P., Baker, V.R. and Garzón, G. (Eds.). *Floods and Megafloods Processes and Deposits: Recent and Ancient Examples*. International Association of Sedimentologists. Special Publication 32. Oxford, 17–35.

Carlisle, D. (1983). Concentration of uranium and vanadium in calcretes and gypcretes. In: Wilson, R.C.L. (Ed.). *Residual Deposits: Surface Related Weathering Processes and Materials*. Geological Society of London. London, 185–195.

Carlston, C.W. and Langbein, W.B. (1960). Rapid approximation of drainage density: line intersection method. *United States Geological Survey Water Resources Division Bulletin*, 11.

Carol, H. (1947). The formation of roches moutonnées. *Journal of Glaciology*, 1, 57–59.

Carr, M.H. (1981). *The Surface of Mars*. Yale University Press. New Haven.

Carr, M.H. (2006). *The Surface of Mars*. Cambridge University Press. Cambridge, 307 p.

Carracedo, J.C. (1988). El riesgo volcánico. In: Ayala, F.J. and Durán, J.J. (Eds.). *Riesgos Geológicos*. Instituto Geológico y Minero de España, Madrid, 83–97.

Carracedo, J.C. (2006). *Los Volcanes del Parque Nacional del Teide*. Ministerio de Medio Ambiente. Madrid.

Carroll, D. (1970). *Rock Weathering*. Plenum Press. New York, 203 p.

Carson, C.E. and Hussey, K.M. (1962). The oriented lakes of Arctic Alaska. *Journal of Geology*, 70, 417–439.

Carson, M.A. (1971). An application of the concept of threshold slopes to the Laramie Mountains, Wyoming. In: Brunsden, D. (Ed.). *Slope Form and Process*. Institute of British Geographers. Special Publication 3, 31–47.

Carson, M.A. and Kirkby, M.J. (1972). *Hillslope Forms and Processes*. Cambridge University Press. Cambridge, 475 p.

Carter, R.W.G. (1988). *Coastal Environments. An Introduction to the Psysical, Ecological and Cultural Systems of Coastlines*. Academic Press. London, 617 p.

Carter, R.W.G., Hesp, P.A. and Nordstrom, K.F. (1990). Erosional landforms in coastal dunes. In: Nordstrom, K.F., Psuty, N.P. and Carter, R.W.G. (Eds.). *Coastal Dunes: Form and Process*. Wiley. Chichester, 217–249.

Carton, A. (2005). La pericolositá nell'ambiente morfoclimatico glaciale. In: Panizza, M. (Ed.). *Manuale di Geomorfologia Applicata*. FrancoAngeli. Milano, 186–207.

Cashman, K.V., Sturtevant, B., Papale, P. and Navon, O. (2000). Magmatic fragmentation. In: Sigurdsson, H. (Ed). *Encyclopedia of Volcanoes*. Academic Press, San Diego, 421–430.

Castaldini, D. and Panizza, M. (2005). Geomorfologia y rischio sismico. In: Panizza, M. (Ed.). *Manuale di Geomorfologia Applicata*. FrancoAngeli. Milano, 250–301.

Castaldini, D. and Piacente, S. (1995). Channel changes of the Po River Mantova Province, Northern Italy. In: Hickin, E.J. (Ed.). *River Geomorphology*. Wiley. Chichester, 193–207.

Cendrero, A. (1970). The volcano-plutonic complex of La Gomera, Canary Islands. *Bulletin Volcanologique*, 34-II, 537–561.

Cendrero, A. (1971). Estudio geológico y petrológico del complejo basal de la isla de La Gomera. *Estudios Geológicos*, 27, 3–73.

Cendrero, A. (1980). Bases doctrinales y metodológicas. *1ª Reunión de Geología Ambiental y Ordenación del Territorio*. Ponencias. Santander, 1–62.

Cendrero, A. (1992). Planning the Use of the Earth's Surface: an overview. In: Cendrero, A., Lüttig, G. and Wolff, F.C. (Eds.). *Planning the Use of the Earth's Surface*. Springer, Berlin, 3–22.

Cendrero, A. (1997). Riesgos naturales e impacto ambiental. In: Novo, M. and Lara, R. (Eds.). *La Interpretación de la Problemática Ambiental: Enfoques Básicos*. Fundación Universidad-Empresa. Madrid, 21–90.

Cendrero, A. (2003). *De la comprensión de la historia de la Tierra al análisis y predicción de las interacciones entre seres humanos y medio natural*. Real Academia de Ciencias Exactas, Físicas y Naturales. Madrid, 60 p.

Cendrero, A. and Díaz de Terán, J.R. (1985). Caracterización cuantitativa de los procesos de erosión en las series volcánicas de la isla de La Gomera (Canarias). *I Reunião do Quaternario Ibérico*. Actas, vol. 1. Lisboa, 531–543.

Cendrero, A. and Dramis, F. (1996). The contribution of landslides to landscape evolution in Europe. *Geomorphology*, 15, 191–211.

Cendrero, A., Díaz de Terán, J.R., Fernández, E., Garrote, R., González Lastra, J.R., Inoriza, I., Lütting, G., Onamendi, J., Pérez, M. and Serrano, A. (1987). Detailed geological hazard mapping for urban and rural planning in Vizcaya, northern Spain. In: Wolff, F.C. (Ed.). *Geology for environmental planning*. Geological Survey of Norway, Special Publication, 2, 25–41.

Cendrero, A., Díaz de Terán, J.R., Araña, V., Badiola, E.R., Berga, L., Cariacedo, J.C., Chacón, J., Coello, J., Corominas, J., Dabrio, C.J., Durán, J.J., Elízaga, E., Ferrer, M., García Fernández, M., Garzón, M.G., González Díez, A., Goy, J.L., López Martíonez, J., Martínez Goytre, J., Mescua, J., de la Nuez, J., Salinas, J.L., Soler, V., del Val, J. and Zazo, C. (1997). Geomorphological hazards in Spain. In: Embleton, C. and Embleton-Hamman, C. (Eds.). *Geomorphological Hazards of Europe*. Elsevier, Amsterdam, 429–456.

Cendrero, A., Remondo, J., Bonachea, J., Rivas, V. and Soto, J. (2006). Acción humana y procesos geológicos superficiales. ¿Qué le estamos haciendo a la faz de la Tierra?. *Revista Real Academia Ciencias Exactas, Físicas y Naturales*, 100, 187–209. Madrid.

Chacón, J., Irigaray, C., Fernández, T. and El Hamdouni, R. (2006). Engineering geology maps: landslides and Geographical Information Systems. *Bulletin Engineering Geology Environmental*, 65, 341–411.

Chapman, C.A. (1956). The control of jointing by topography. *Journal of Geology*, 66, 552–558.

Chappell, J. (1974), Geology of coral terraces, Huon Peninsula, New Guinea: a study of Quaternary tectonic movements and sea-leved changes. *Geological Society of America Bulletin*, 85, 555–570.

Chapron, E., Ariztegui, D., Mulsow, S., Villarosa, G., Pino, M., Outes, V., Juvignié, E. and Crivelli, E. (2006). Impact of the 1960 major sudduction earthquake in Northern Patagonia (Chile, Argentina). *Quaternary International*, 158, 58–71.

Chaput, J.L. (1971). Aspects morphologiques du sud-est de la Sierra Morena (Espagne). *Revue de Géographie Physique et de Géologie Dynamique*, 13, 55–66.

Chardon, D., Chevillotte, V., Beauvais, A., Grandin, G., Boulangé, B. (2006). Planation, bauxites and epeirogeny: One or two paleosurfaces on the West African margin. *Geomorphology*, 82, 273–282.

Charlton, R.O. (2008). *Fundamentals of Fluvial Geomorphology*. Routledge. London, 234 p.

Chávez, F., Goula, X., Roca, A., María, F., Presmanes, J.A. and López-Arroyo, A. (1999). Escenarios de daños sísmicos en Cataluña. *I Congreso Nacional de Ingeniería Sísmica*. Memorias, T. I, 299–307. Murcia.

Chen, J. (2005). Dams, Effects on Coasts. In: Schwartz, M.L. (Ed*). Encyclopedia of Coastal Science*. Springer. Dordrecht, 357–359.

Chepil, W.S., Siddoway, F.H. and Armbrust, D.V. (1962). Climatic factor form estimating wind erodibility of farm fields. *Journal of Soil and Water Conservation*, 17, 162–165.

Chester, D. (1993). *Volcanoes and Society*. Arnold. London, 351 p.

Chester, D.K., Duncan, A.M., Guest, J.E and Kilburn, C.R.J (1985). *Mount Etna: The Anatomy of a Volcano*. Chapman and Hall, London, 405 p.

Chico, R.J. (1963). *Playa mud cracks: regular and king size*. Geological Society of America, Special Paper 76.

Chigira, M. (2009). September 2005 rain-induced catastrophic rockslides on slopes affected by deep-seated gravitational deformations, Kyushu, southern Japan. *Engineering Geology*, 108, 1–15.

Chiu, T., Fairbanks, R.G., Mortlock, R-A. and Bloom, A.L. (2005). Extending the radiocarbon calibration beyond 26,000 years before present using fossil corals. *Quaternary Science Review*, 24, 1797–1808.

Chivas, A.R. (2007). Terrestrial evaporites. In: Nash, D.J. and McLaren, S.J. (Eds.). *Geochemical Sediments and Landscapes*. Blackwell. Malden, 330–364.

Chiverrell, R.C., Harvey, A.M. and Foster, G.C. (2007). Hillslope gullying in the Solvay Firth-Morlcambe Bay region, Great Britain: Responses to human impact and/or climatic deterioration. *Geomorphology*, 84, 317–343.

Cholley, A. (1950). Morphologie structurale et morphologie climatique. *Annales de Géographie*, 317, 325–335.

Chorley, R.J. (1962). Geomorphology and general systems theory. *United States Geological Survey Professional Paper*, 500B, 10 p.

Chorley, R.J. (1963). The diastrophism background of twentieth century geomorphological thought. *Bulletin of the Geological Society of America,* 74, 953–970.

Chorley, R.J. (1964). The nodal position and anomalous character of slope studies in geomorphological research. *Geographical Journal*, 130, 503–506.

Chorley, R.J. (1969). The drainage basin as the fundamental geomorphic unit. In: Chorley, R.J. (Ed.). *Introduction to Physical Hydrology*. Methuen. London, 37–59.

Chorley, R.J. (1978). Bases for theory in geomophology. In: Embleton, C., Brunsden, D. and Jones, D.K.C. (Eds.). *Geomorphology: Present Problems and Future Prospects*. Oxford University Press. Oxford, 1–13.

Chorley, R.J. and Beckinsale, R.P. (1980). G.K. Gilbert's geomorphology. *Geological Society of America*. Special Paper 183, 129–142.

Chorley, R.J. and Kennedy, B.A. (1971). *Physical Geography: A System Approach*. Prentice-Hall. London, 370 p.

Chorley, R.J., Beckinsale, R.P. and Dunn, A.J. (1964). *The History of the Study of Landforms or the Development of Geomorphology. Vol. 1: Geomorphology Before Davis*. Methuen. London, 678 p.

Chorley, R.J., Beckinsale, R.P. and Dunn, A.J. (1973). *The History of the Study of Landforms or the Development of Geomorphology. Vol. 2: The Life and Work of William Morris Davis*. Methuen. London, 874 p.

Chorley, R.J., Schumm, S.A. and Sudgen, D.E. (1984). *Geomorphology. Earth Sciences and the Past*. Methuen. London, 605 p.

Chou, L., Garrels, R.M. and Wollast, R. (1989). Comparative study of the kinetics and mechanisms of dissolution of carbonate minerals. *Chemical Geology*, 78, 269–282.

Chouhan, T.S. (1992). *Desertification in the World and its Control*. Scientific Publishers. Jodhpur, 172 p.

Chow, V.T. (1956). Hydrologic Studies of Floods in the United States. *International Association Science Hydrological*, 42, 134–170.

Chowoong, M., Phantuwongraj, S., Charoentitirat, T., Chutakkosikanon, V., Yumuang, S. and Charusiri, P. (2009). Beach recovery after 2004 tsunami from Phang-nga, Thailand. *Geomorphology*, 104, 134–142.

Christiansen, H. (1998). Nivation form and processes in unconsolidated processes and landforms, NE Greenland. *Earth Surface Processes and Landforms*, 23, 751–760.

Chueca, J. (1994). *Liquenometría*. Cuadernos Técnicos de la Sociedad Española de Geomorfología. Geoforma Ediciones. Logroño, 34 p.

Chueca, J. and Julián, A. (1996). Datación de depósitos morrénicos de la Pequeña Edad del Hielo: Macizo de la Maladeta. In: Pérez, A., Martín, P., Chesworth, W. and Martínez, A. (Eds.). *Dinámica y Evolución de Medios Cuaternarios*. Xunta de Galicia. Santiago de Compostela, 171–182.

Chueca, J., Julián, A. and Peña, J.L. (2002). Comparación de la situación de los glaciares del Pirineo español entre el final de la Pequeña Edad del Hielo y la actualidad. *Boletín Glaciológico Aragonés*, 3, 13–41.

Chueca, J., Julián, A., Saz, M.A., Crens, J. and López, J.I. (2005). Responses to climatic changes since the Little Ice Age on Maladeta Glaciar (Central Pyrenees). *Geomorphology*, 68, 167–182.

Church, J.A. and Gregory, J.M. (Coords.) (2001). Changes in sea level. In: Houghton, J.T., Ding, Y., Griggs, D.J., Noguer, M., Liden, P.J. van der, Dai, X., Maskell, K. and Johnson, C.A. (Eds.). *Climate Change 2001: The Scientific Basis*. Cambridge University Press. Cambridge, 639–693.

Church, M. (1972). *Baffin Island Sandurs; a Study of Arctic Fluvial Processes*. Geological Survey of Canada. Bulletin 216, 208 p.

Church, M. (1988). Floods in cold climates. In: Baker, V.R., Craig Kochel, R. and Patton, P.C. (Eds.). *Flood Geomorphology*. Wiley. New York, 205–229.

Church, M. (2003). What is the geomorphological prediction?. *Geophysical monograph*, 135, 133–194. American Geophysical Union.

Church, M., Stock, R.F. and Ryder, J.M. (1979). Contemporany sedimentary environments of Baffin Island, N.W.T., Canada: Debris slope accumulations. *Arctic and Alpine Research*, 11, 371–402.

Clapperton, C.M. (1977). Volcanoes in space in time. *Progress in Physical Geography*, 1, 375–411.

Clark, C.D., Hughes, A.L.C., Greenwood, S.L., Spagnolo, M. and Ng, F.S.L. (2009). Size and shape caracteristics of drumlins, derived from a large sample, and associated scaling laws. *Quaternary Science Reviews*, 28, 677–690.

Clark, J.A. (1980). A numerical model of worldwide sea level changes on a viscoelastic earth. In: Mörner, N.A. (Ed.). *Earth Rheology, Isostasy and Eustasy*. Wiley. Chichester, 525–534.

Clark, J.A., Farrell, W.E. and Peltier, W.R. (1978). Global changes in postglacial sea level, a numerical calculation. *Quaternary Research*, 9, 265–287.

Clark, M. y Small, J. (1982). *Slopes and Weathering*. Cambridge University Press. Cambridge, 190 p.

Clark, M.J. (1988) (Ed.). *Advances in Periglacial Geomorphology*. Wiley, Chichester. 481 p.

Clark, M.J. (1988). Periglacial hydrology. En: Clark, M.J. (Ed.). *Advances in Periglacial Geomorphology*. Wiley. Chichester, 415–462.

Clark, P.U. (1994). Unstable behaviour on the Laurentide ice sheet over deforming sediment and its implications for climate change. *Quaternary Research*, 41, 19–25.

Clark, P.U. and Bartlein, C.J. (1995). Correlation of the late Pleistocene glaciation in the western United States with North Atlantic Heinrich events. *Geology*, 23, 483–486.

Clarke, G.K.C. (1991). Length, width and slope influences on glacier surging. *Journal of Glaciology*, 37, 236–246.

Clarke, G.K.C. (2005). Subglacial processes. *Annual Review of the Earth and Planetary Sciences*, 33, 247–276.

Clayton, K.M. (1966). The origin of the landforms of the Malham area. *Field Studies*, 2, 359–384.

Clayton, L. (1964). Karst topography on stangnant glaciers. *Journal of Glaciology*, 5, 107–112.

Cleaves, E.T., Godfrey, A.E. and Bricker, C.P. (1970). Geochemical balance of a small watershed and its geomorphic implications. *Geological Society of America Bulletin*, 81, 3015–3032.

Cloos, H. (1939). Hebung-Spaltung-Volcanismus. *Geologische Rundschau*, 30, 405–527.

Clos-Arceduc, A. (1969). *Essai d'Explication des Formes Dunaires Sahariennes*. Institut Géographique National. Paris, 66 p.

Clowes, A. and Comfort, P. (1982). *Process and Landform*. Oliver and Boyd, 289 p.

Coates, D.R. (1983). Large-scale and land subsidence. In: Gardner, R. and Scoging, H. (Eds.). *Mega-Geomorphology*. Claredon Press. Oxford, 212–240.

Coates, D.R. (1987). Subsurface impacts. In: Gregory, K.J. and Walling, D.E. (Eds.). *Human Activity and Environmental Processes*. Wiley. Chichester, 271–304.

Coe, J.A., Cannon, S.H. and Santi, P.M. (2008). Introduction to special issue on debris flow initiated by runoff, erosion and sediment entrainment in western North America. *Geomorphology*, 96, 247–249.

Cofaigh, C.O. (2007). Glaciamarine sediments and ice-rafted debris. In: Elias, S.A. (Ed.). *Encyclopedia of Quaternary Science*. Elsevier. Amsterdam, 932–945.

Colclough, J.D. (1978). Soil conservation in Tasmania. *Journal of Soil Conservations Service, New South Wales*, 34, 63–72.

Cole, K.L. and Mayer, L. (1982). Use of packrat middens to determine rates of cliff retreat in the Eastern Grand Canyon, Arizona. *Geology*, 10, 597–599.

Coleman, J.M. (1969). Brahmaputra River: Channel processes and sedimentation. *Sedimentary Geology*, 3, 129–239.

Coleman, J.M. (1981) *Deltas: Processes of Deposition and Models for Exploration* Continuing Education Publising. Minneapolis, 102 p.

Coleman, J.M. and Prior, D.B. (1980). Deltaic sand bodies. *American Associattion of Petroleum Geologist*. Continuing Education Course, 15. Tulsa, 171 p.

Collard, M. (1973). Cryoturbations en forme de striès enracinées paralèls à la pente. *Biuletyn Peryglacjalny*, 22, 323–327.

Collins, P.E.F., Fenwick, I.M., Keith-Lucas, D.M. and Worshley, P. (1996). Late Devensian river and floodplain dynamics and related environmental change in Northwestern Europe, with particular reference to a site at Woolhampton, Berkshire, England. *Journal of Quaternary Science*, 11, 357–375.

Collison, J.D. (1996). Alluvial sediments. In: Reading, H.G. (Ed.). *Sedimentary Environments. Processes, Facies and Stratigraphy*. Blackwell. Oxford, 37–82.

Colman, S.M. and Dethier, S.P. (1986). An overview of chemical weathering. In: Colman, S.M. and Dethier, S.P. (Eds.). *Rates of Chemical Weathering of Rocks and Minerals*. Academic Press. Orlando, 1–18.

Colombo, F. (1979). *Introducció a l'Estudi Sedimentologic dels Cons de Deyecció (Alluvial Fans) i dels Sediments Associats*. Seminaris d'Estudis Universitaris, 5. Barcelona, 163 p.

Colombo, F. (1989). Abanicos aluviales. In: Arche, A. (Ed.). *Sedimentología*. Consejo Superior de Investigaciones Científicas, Vol. 1. Madrid, 143–218.

Coltrinari, L. and Nogueira, F. (1989). Dambo-like landforms in southeastern Brazil. *Abstracts of Papers, Second International Conference on Geomorphology, Frankfurt/Main*, 1989. Geöko-plus, 1, 58.

Combes, P.J. (1969). *Recherches sur la Genèse des Bauxites dans le Nord-Est de l'Espagne, le Languedoc et l'Ariège (France)*. Mémoires du Centre d'Études et de Recherches Géologiques et Hydrogéologiques. Montpellier, 342 p.

Comín, F.A. and Rodríguez-Arias, M.A. (2003). What we know about the Climate System? A brief review of current research. In: Rodó, X. and Comín, F.A. (Eds.). *Global Climate. Current Research and Uncertainties in the Climate System*. Springer. Berlin, 3–24.

Comisión Nacional de Protección Civil-Comisión Técnica de Inundaciones (1983). *Estudio de Inundaciones Históricas. Mapa de Riesgos Potenciales*. Servicio de Publicaciones del Ministerio de Obras Públicas y Urbanismo. Madrid.

Common, R. (1966). Slope failure and morphogenetic regions. In: Dury, G.H. (Ed.). *Essays in Geomorphology*. Heineman. London, 53–81.

Conesa, C. (1995). Torrential flow frequency and morphological adjustments of ephemeral channels in south-east Spain. In: Hickin, E.J. (Ed.). *River Morphology*. Wiley. Chichester, 169–192.

Cooke, R.U. (1970a). Stone pavements in deserts. *Annals of the American Asoociation Geographers*, 60, 560–77.

Cooke, R.U. (1970b). Morphometric analysis of pediments and associated landforms in the western Mojave Desert, California. *American Journal of Science*, 269, 26–38.

Cooke, R.U. (1977). Applied geomorphological studies in deserts: a review of examples. In: Hails, J.R. (Ed.) *Applied Geomorphology*. Elsevier. Amsterdam, 183–225.

Cooke, R.U. (1979). Laboratory simulation of salt weathering processes in arid environments. *Earth Surface Processes*, 4, 347–359.

Cooke, R.U. (1981). Salt weathering in deserts. *Proceedings of Geologists Association*, 92, 1–16.

Cooke, R.U. (1984). *Geomorphological Hazards in Los Angeles*. Allen and Unwin. London, 206 p.

Cooke, R.U. (1986). Hot drylands. In: Fookes, P.G. and Vaughan, P.R. (Eds.). *A Handbook of Engineering Geomorphology*. Surrey University Press. New York, 109–124.

Cooke, R.U. and Smalley, I.J. (1968). Salt weathering in deserts. *Nature*, 220, 1226–1227.

Cooke, R.U. and Mason, P. (1973). Desert Knolls pediment and associated landforms in the western Mojave Desert, California. *Revue de Geomorphologie Dynamique*, 20, 71–78.

Cooke, R.U. and Warren, A. (1973). *Geomorphology in Deserts*. Batsford. London, 394 p.

Cooke, R.U. and Reeves, R.W. (1976). *Arroyos and environmental change in the American South-west*. Clarendon Press. Oxford, 213 p.

Cooke, R.U. and Doornkamp, J.C. (1990). *Geomorphology in Environmental Management. An Introduccion*. Clarendon Press. Oxford, 410 p.

Cooke, R.U., Warren, A. and Goudie, A. (1993). *Desert Geomorphology*. UCL Press. London, 526 p.

Cooke, R.U., Brunsden, D., Doornkamp, J.C. and Jones, D.K. (1982). *Urban Geomorphology in Drylands*. Oxford University Press. Oxford, 324 p.

Cooper, A.H. (1988). Subsidence resulting from the dissolution of Permian gypsum in Ripon area; its relevante to mining and water abstraction. In: Culshaw, M.G., Cripps, J.C. and Lovell, M.A. (Eds.). *Engineering Geology of Underground Movements*. Geological Society Engineering Geology, Special Publication, 5, 387–390.

Cooper, A.H. (1995). Subsidence hazards due to the dissolution of Permian gypsum in England: Investigation and remediation. In: Beck, B.F. (Ed.). *Karst Geohazards*. Balkema. Rotterdam, 23–29.

Cooper, A.H. (1998). Subsidence hazards caused by the dissolution of Permian gupsum in England: geology, investigation and remediation. In: Maund, J.G. and Eddleston, M. (Eds.). *Geohazards in Engineering Geology*. Geological Society of London, V.5. London, 265–275.

Cooper, A.H. and Waltham, A.C. (1999). Subsidence caused by gypsum dissolution at Ripon, North Yorkshire. *Quaterly Journal of Engineering Geology*, 32, 305–310.

Cooper, W.S. (1958). *Coastal Sand Dunes of Oregon and Washington*. Geological Society of America, Mem. 72, 164 p.

Cooper, W.S. (1967). Coastal Sand Dunes of California. *Geological Society of America Memoir*, vol. 101, 131 p.

Copons, R., Vilaplana, J.M., Corominas, J., Altimir, J. and Amigó, J. (2005). Rockfall risk management in high-density urban areas. The Andorran Experience. In: Glade, T., Anderson, M. and Crozier, M.J. (Eds.). *Landslide Hazard and Risk*. Wiley. Chichester, 675–698.

Coque, R. (1955). Morphologie et croût dans le sud-Tunesien. *Annales de Géographie*, 64, 359–370.

Coque, R. (1962). La Tunisie présaharienne. Etude Géomorphologique. Impremerie Oberthur, 476 p.

Corbel, J. (1956). Le Karst du Vercours. *Revue de Géographie de Lyon*, 31, 221–241.

Corbel, J. (1957). Karsts hauts-Alpins. *Revue de Géographie de Lyon*, 32, 135–158.

Corbel, J. (1959). Erosion en terrain calcaire (vitesse d'érosion et morphologie). *Annales Géographie*, 67, 97–120.

Corbel, J. (1963). Pédiments d'Arizona. *Centre de Documentation Cartographique et Géographique*. *Memoires et Documents*, 9, 33–95.

Corbel, J. (1964). L'érosion terrestre, étude quantitative (Méthodes-Techniques-Résultats). *Annales de Géographie*, 398, 385–412.

Corominas, J. (1996). Debris slide. In: Dikau, A., Brunsden, D., Shortt, L. and Ibsen, M.C. (Eds.). *Landslide*

Recognition. Identification, Movement and Causes. Wiley. Chichester, 97–102.

Corominas, J. (2006). El clima y sus consecuencias sobre la actividad de los movimientos de ladera en España. *Cuaternario y Geomorfología*, 20, 89–113.

Corominas, J. and Moya, J. (1996). Historical landslides in the Eastern Pyrenees and their relation to rainy events. In: Chacón, J., Irigaray, C. and Fernández, T. (Eds.). *Landslides*. Balkema. Rotterdam, 125–139.

Corominas, J., Remondo, J., Farias, P., Estevao, M., Zézere, J., Díaz de Terán, J., Dikau, R., Schrott, L., Moya, J. and González, A. (1996). Debris flow. In: Dikau, R., Brunsden, D., Schrott, R. and Ibsen, M.L. (Eds.). *Landslide Recognition. Identification, Movement and Causes. Wiley*. Chichester, 161–180.

Corrêa Neto, A.V. (2000). Speleogenesis in quartzite in Southeastern Minas Gerais, Brazil. In: Klimchouk, A.B., Ford, D.C., Palmer, A.N. and Dreybrodt, W. (Eds.). *Speleogenesis Evolution of Karst Aquifers*. National Speleological Society. Huntsville, 452–457.

Corte, A.E. (1966). Particle sorting by repeated freezing and thawing. *Biuletyn Periglacjalny*, 15, 175–240.

Corte, A.E. (1969). Formación en el laboratorio de estructuras como pliegues por congelamiento y descogelamiento múltiple. *Cuartas Jornadas Geológicas Argentinas*, 1, 215–227.

Corte, A.E. (1971). Particle sorting by repeated freezing and thawing. *Biuletyn Peryglacjalny*, 15, 175–240.

Corte, A.E. (1976). Rock glaciers. *Biuletyn Peryglacjalny*, 26, 157–197.

Corte, A.E. and Higashi, A. (1964). Experimental research on desiccation cracks in soil. *U.S. Army Material Command Region Research and Engineering Laboratory*. Research Report 66, 72 p.

Costa, J.E. (1974). Response and recovery of a Piedmot watershed from tropical store Aghes, June 1972. *Water Resources Research*, 10, 106–112.

Costa, J.E. (1984). Physical geomorphology of debris flow. In: Costa, J.E., Flaischer, P.J. (Eds.). *Development and Applications of Geomorphology*. Springer. Berlin, 268–317.

Costa, J.E. (1987). A comparison of the largest rainfall-runoff floods in the United States with those of the Peoples Republic of China and the world. *Journal of Hydrology*, 96, 101–115.

Costa, J.E. (1988a). Rheologic, geomorphologic and sedimentologic differentiation of water floods, hyperconcentrated flows and debris flows. In: Baker, V.R., Craig Kochel, R. and Patton, P.C. (Eds.). *Flood Geomorphology*. Wiley. New York, 113–122.

Costa, J.E. (1988b). Floods from dam failures. In: Baker, V.R., Craig Kochel, R. and Patton, P.C. (Eds.). *Flood Geomorphology*. Wiley. New York, 439–463.

Costa, J.E. and Baker, V. (1981). *Surficial Geology. Building with the Earth*. Wiley. New York, 498 p.

Costa, J.E. and Baker, V. (1987). Flood power. In: Mayer, L. and Nash, D. (Eds.). *Catastrophic Flooding*. Allen and Unwin. Boston, 1–21.

Costa, J.E. and Schuster, R.L. (1988). The formation and failure of natural dams. *Geological Society of America Bulletin*, 100, 1054–1068.

Costard, F. and Gautier, E. (2007). The Lena river: Hydromorphodynamic features in a deep permafrost zone. In: Gupta, A. (Ed.). *Large Rivers. Geomorphology and Management*. Wiley. Chichester, 225–233.

Cotton, B., Degraff, J. and Hart, M. (1996). *Recognition, Investigation and Mitigation of Landslides*. Geological Society of America. Denver.

Cotton, C.A. (1942). *Climatic Accidents in Landscape Making*. Whitcomb & Tombs, Wellington.

Cotton, C.A. (1954). *Deductive morphology and genetic classification of coast*. Science Monthly, 78, 163–181.

Cotton, C.A. (1958). Alternating pleistocene morphogenetic systems. *Geological Magazine*, 95, 125–137.

Cotton, C.A. (1961). Theory of savanna planation. *Geography*, 46, 89–96.

Cotton, C.A. (1962). Plains and inselbergs of the humid tropics. *Royal Society of New Zealand, Transactions (Geology)*, 1, 269–277.

Coudé-Gaussen, G. (1981). *Les Serras da Peneda et do Gerês. Etude Géomorphologique*. Memórias do Centro de Estudos Geográficos, 5, 254 p.

Coudé-Gaussen, G. (1991). *Les Poussières Sahariennes: Cycle Sédimentaire et Place dans les Environments et Paléoenvironments Désertiques.* John Libby Eurotext. Montrouge, 485 p.

Coudé-Gaussen, G. and Rognon, P. (1988). Charactérisation sédimentologique et conditions paléoclimatiques de la mise en place de loess and Nord du Sahara á partir de l'example du sud-Tunesien. *Bulletin de la Societé Géologique de France*, 4, 1081–1090.

Coulomb, J. (1969). *L'expansion des fonds océaniques et la dérive des continents*. Presses Universitaires de France. Paris, 222 p.

Coulthard, T.J. (2005). Effects of vegetation of braided stream pattern and dynamics. *Water Resources Research*, 41, WO4003.

Coussot, P. and Meunier, M. (1996). Recognition, classification and mechanical description of debris flow. *Earth-Science Reviews*, 40, 209–227.

Cowardin, L.M., Carter, V., Golet, F.C. and La Roe, E.T. (1979). Classification of wetlands and deepwater habitats of the United States. U.S. Fish and Wildlife Service. FWS/OBS-79/31.

Cowell, P.J. and Thom, B.J. (1994). Morphodynamics of coastal evolutrion. In: Carter, R.W.G. and Woodroffe, C.D. (Eds.). *Coastal Evolution, Late Quaternary Shoreline Morphodynamics*. Cambridge University Press. Cambridge, 33–86.

Cox, A., Doell, R.R. and Dalrymple, J.B. (1964) Geomagnetic polarity epochs. *Science*, 143, 351–352.

Cox, N.J. (2007). Kinds and problems of geomorphological explanation. *Geomorphology*, 88, 46–56.

Cox, R.T. (1994). Análysis of drainage basin symetry as a rapid technique to identify areas of possible Quaternary tilt-block tectonics: an example from de Misissippi. Embayment. *Geological Society of America Bulletin*, 106, 571–581.

Cramer, H. (1941). Die Systematik des karstdolinen. *Neues Jahrbuch für Mineralogie, Geologie, Paläontologie*, 85, 293–382.

Crandell, D.R., Mullineaux, D.R. and Rubin, M. (1975). Mount St. Helens Volcano: recent and future behaviour. *Science*, 187, 438–441.

Crespo, J. (1997). *Mapa Geológico y Minero de Castilla y León, Escala 1:400.000*. Junta de Castilla y León, 459 p.

Criado, C. (2006). El Teide: formas de modelado. In: Carracedo, J.C. (Coord.). *Los Volcanes del Parque Nacional del Teide*. Ministerio de Medio Ambiente. Madrid. 259–275.

Criado, C. and Porta, P. (2003). An inusual "blood rain" over the Canary Islands (Spain). The storm of January 1999. *Journal of Arid Environments*, 55, 765–783.

Crittenden, M.D. (1963). New data on the isostatic deformation of Lake Bonneville. *United States Geological Survey Professional Paper* 454-E. 31 p.

Crook, R. (1986). Relative dating of Quaternary deposits based on P-waves velocities in weathered granitic clasts. *Quaternary Research*, 25, 281–292.

Croot, D.G. (1987). Glacio-tectonic structures: A mesoscale model of thin-skinned thrust sheets?. *Journal of Structural Geology*, 9, 797–808.

Croista, G.B., and Clagne, J.J. (2009). Dating, triggering, modelling and hazard assessment. *Geomorphology*, 103, 1–4.

Crouch, R.J. (1983). The role of tunnel erosion in gully head progression. *Journal of Soil Conservation, New South Wales*, 39, 148–155.

Crouch, R.J. (1990). Rates and mechanisms of discontinous gully erosion in a red-brown earth catchment, New South Wales, Australia. *Earth Surface Processes and Landforms*, 15, 277–282.

Crowley, T.J. and North, G.R. (1991). *Paleoclimatology*. Oxford University Press. New York, 339 p.

Crozier, M.J. (1975). Techniques for the morphometric analysis of landslips. *Zeitscrift für Geomorphologie*, 17, 78–101.

Crozier, M.J. (1986). *Landslides: Causes, Consequences and Environment*. Croom Helm. London, 252 p.

Crozier, M.J. (2004a). Landslides. In: Goudie, A.S. (Ed.). *Encyclopedia of Geomorphology*. Routledge. London, 605–608.

Crozier, M.J. (2004b). Slope Stability. In: Goudie, A.S. (Ed.). *Encyclopedia of Geomorphology*. Routledge. London, 969–970.

Crozier, M.J. and Glade, T. (2005). Landslide hazard and risk: issues, concepts and approach. In: Glade, T., Anderson, M. and Crozier, M.J. (Eds.). *Landslides Hazard and Risk*. Wiley. Chichester, 1–40.

Cruden, D.M. (1991). A simple definition of a landslide. *Bulletin of the International Association of Engineering Geology*, 43, 27–29.

Cruden, M.C. and Varnes, D.J. (1996). Landslide types and proceses. In: Turner, A.K. and Schuster, R.L. (Eds.). *Landslides: Investigation and Mitigation*. Special Report 247. National Academy Press. Washington, 36–75.

Crutzen, P.J. (2006). Albedo enhancement by stratospheric sulphur injections: a contribution to resolve a policy dilemma?. *Climatic Change*, 77, 211–219.

Crutzen, P.J. and Stoermer, E.F. (2000). The "Antropocene". *Global Change Newsletter*, 14, 12–13.

Cubasch, U. and Meahl, G.A. (Coords.) (2001). Projections of future climate change. In: Houghton, J.T., Ding, Y., Griggs, D.J., Noguer, M., Liden, P.J. van der, Dai, X., Maskell, K. and Johnson, C.A. (Eds.). *Climate Change 2001: The Scientific Basis*. Cambridge University Press. Cambridge, 525–582.

Cuffey, K.M. (2008). A matter of firn. *Science*, 320, 1596–1597.

Cunningham, M.A. (1999). Three Gorges dam (China). In: Cunningham, W.P., Cooper, T.H., Gorham, E. and Hepworth, M.T. (Eds.). *Environmental Encyclopedia*. Jaico Publ. Mumbai, 1024–1025.

Cunningham, W.P. (1999a). Aral Sea. In: Cunningham, W.P., Cooper, T.H., Gorham, E. and Hepworth, M.T. (Eds.). *Environmental Encyclopedia*. Jaico Publ. Mumbai, 58–59.

Cunninghan, W.P. (1999b). Tropical rain forest. In: Cunninghan, W.P., Cooper, T.H., Gorham, E. and Hepworth,

M.T. (Eds.). *Environmental Encyclopedia*. Jaico Publ. Mumbai, 1047–1048.

Curl, R.L. (1966a). Scallops and flutes. *Transactions Cave Research Group of Great Britain*, 7, 121–160.

Curl, R.L. (1966b). Caves as a measure of karst. *Journal Geology*, 74, 798–829.

Curl, R.L. (1974). Deducing flow velocity in caves conduits for scallops. *National Speleological Society America Bulletin*, 36, 1–5.

Curray, J.R. (1969). History of the continental shelves. *The NEW concepts of Continental Margin Sedimentations. Short Course Lecture Notes-JC-VI*, 1–18.

Curray, J.R., Emmel, R.J. and Crampton, P.J.S. (1969). Holocene history of a strand plain, lagoonal coast, Nayarit, México. In: Castañares, A.A. and Phleger, F.B. (Eds). *Coastal Lagoons- A Symposium*. Universidad Nacional Autónoma de Méjico/UNESCO. México City, 35–43.

Currey, D.R. (1994). Hemiarid lake basins: hydrographic patterns. In: Abrahams, A.D. and Parsons, A.J. (Eds.). *Geomorphology of Desert Environments*. Chapman and Hall. London, 405–421.

Curtis, C.D. (1976). Chemistry of rock weathering: fundamental reactions and controls. In: Derbyshire, E. (Ed.). *Geomorphology and Climate*. Wiley. New York, 25–57.

Curtis, L.F., Doornkamp, J.C. and Gregory, K.Y. (1965). The description of relief in field studies of soils. *Journal Soil Science*, 16, 16–30.

Cutts, J.A., Blasius, K.R. and Roberts, W.J. (1979). Evolution of Martian polar landscapes: interplay of long-term variations in perennial ice cover and dust storm intensity. *Journal of Geophysical Research*, 84, 2975–2994.

Cvijic, J. (1893). Das Karstphänomen. *Geogr. Abhand.*, 5, 215–319.

Cvijic, J. (1918). L'hydrographie souterraine et l'évolution morphologique du karst. *Revue de Géographie Alpine*, 6, 375–426.

Cvijic, J. (1924a). The evolution of lapiés. *Geographical Review*, 14, 26–49.

Cvijic, J. (1924b). Types morphologiques de terrains calcaires. *Glasnik Geogr. Drustva*, 10, 1–7. Beograd.

Cvijic, J. (1925). Types morphologiques du terrains calcaires. Le holokarst. Le Merokarst. Types karstiques de transition. *Compte Rendus de la Academie de Sciences de Paris*, 180, 592–594, 757–758, 1038–1040.

Czeppe, Z. (1964). Exfoliation in a periglacial climate. *Geographia Polonica*, 2, 5–10.

Czerwinski, J. and Migon, P. (1993). Microformy wietrzenia granito w masywie Karkonosko-Izerskim. *Czasopismo Geograficzne*, 64, 265–284.

Czudek, T. (1964). Periglacial slope development in the area of the Bohemian Massif in Northern Moravia. *Biuletyn Peryglacjalny*, 14, 169–194.

Czudek, T. (Ed.) (1973a). *Typologické Cleneni Reliéfn CSR (Typological Classification of the relief of the Czech Socialist Republic) 1:500.000*. Institute of Geography. Czeschoslovak Academy of Science. Brno.

Czudek, T. (1973b). Zur Klimatischen Talasymmetrie des Westeiles der Tschechoslowakei. *Zeitschrift für Geomorphologie,* 17, 49–57.

Czudek, T. and Demek, J. (1970a). Pleistocene cryopedimentation in Czechoslovakia. *Acta Geographica Lodziensia*, 24, 101–108.

Czudek, T. and Demek, J. (1970b). Thermokarst in Siberia and its influence on the development of lowland relief. *Quaternary Research,* 1, 103–120.

Czudek, T. and Demek, J. (1971). Pleistocene cryoplanation in the Ceska vysocina highlands, Czechoslovakia. *Transactions, Institue British Geographers,* 52, 95–112.

Czudek, T., Marvan, P., Panos, V. and Rauser, J. (1964). Verwitterung und Abtragunsformen des granits in den Bohemische Masse. *Petermanns Geographische Mitteilungen*, 108, 182–192.

D'Almeida, G.A. (1989). Desert aerosol: characteristics and effects of climate. In: Leinen, M. and Sarnthein. M. (Eds.). *Paleoclimatology and palaeometereology: modern and past patterns of global atmospheric transport*. Kluwer Academic, 311–337. Dordrecht.

D'Ercole, R. (1994). Natural hazards in the French West Indies. An overall view. *Third European Intensive Course on Applied Geomorphology. Tropical Regions*. Université Louis Pasteur. Strasbourg. 10 p.

Dabrio, C.J. (1984). Sedimentación siliciclástica en costas, deltas y mares someros. Ciclo de Seminarios de Sedimentología, 2. *Instituto Geológico y Minero de España*. Madrid, 131–190.

Dabrio, C.J. and Zazo, C. (1988). Riesgos geológicos en zonas litorales. In: *Riesgos Geológicos*. Instituto Geológico y Minero de España. Madrid, 227–250.

Dabrio, C.J., Zazo, C. and Goy, J.L. (1993). Litoral y riesgos geológicos. Ponencia in: Ortiz, R. (Ed.). *Problemática geoambiental y desarrollo*. V Reunión Nacional de Geología Ambiental y Ordenación del Territorio. Murcia, 93–117.

Dabrio, C.J., Goy, J.L. and Zazo, C. (1998). The record of the tsunami produced by the 1755 Lisbon earthquake in Valdelagrana spit (Gulf of Cádiz, Southern Spain). *Geogaceta*, 23, 31–34.

Dahl, R. (1965). Plastically scultured detail forms on rock surfaces in Northern Nordland, Norway. *Geografiska Annaler*, 47, 83–140.

Dahl, R. (1966). Blockfields and other weathering forms in the Narvik Mountains. *Geografiska Annaler*, 48 A, 55–85.

Dalrymple, J.B., Blong, R.J. and Conacher, A.J. (1968). A hypothetical nine unit landsurface form. *Zeitschrift für Geomorphologie*, 12, 60–76.

Dalrymple, G.B., Silver, E.A. and Jackson, E.D. (1973). Origin of Hawaiian Islands. *American Scientific*, p. 306.

Dalrymple, R.W., Zaitlin, B.A. and Boyd, R. (1992a). Estuarine facies models: Conceptual basis and stratigraphic implications. *Journal of Sedimentary Petrology*, 62, 1130–1146.

Dalrymple, R.W., Kight, R.J., Zaitlin, B.A. and Middleton, G.V. (1992b). Dynamics and facies model of a macrotidal sand-bar complex, Cobequid Bay-Salmon River estuary (Bar of Fundy). *Sedimentology*, 37, 577–612.

Danin, A. (1983). Weathering of limestone in Jerusalem by cyanobacteria. *Zeitscrift für Geomorphologie*, 27, 413–421.

Dansgaard, W., Johnsen, S.J., Clausen, H.B., Dahl-Jensen, D., Gundestrup, N.S., Hammer, C.U., Huidberg, C.S., Steffensen, J.P., Sveinbjörnsdottir, A.E., Jouzel, J. and Bond, G. (1993). Evidence for general instability of past climate from a 250 kyr ice-cored record. *Nature*, 364, 218–220.

Dansgaard, W., Johnsen, S.J., Möller, J. and Langway Jr, C.C. (1999). One thousand centuries of climatic record from Camp Century on the Greeland ice sheet. *Science*, 166, 377–381.

Darwin, L. (1842). *The structure and distribution of coral reefs*. Smith, Elder and Co. London.

David, G.C.L., Bledsoe, B.P., Merritt, D.M. and Wohl, E. (2009). The impacts of ski slopes development on stream channel morphology in the White River National Forest, Colorado, USA. *Geomorphology*, 103, 375–388.

Davidson, D.A. and Shackley, M.L. (Eds.) (1976). *Geoarcheology*. Duckworth. London, 408 p.

Davies, J.L. (1964). A morphogenetic approach to the Worlds´shorelines. *Zeitscrift für Geomorphologie*, 8, 127–142.

Davies, J.L. (1969). *Landforms of Cold Climates*. The M.I.T. Press. Cambridge, 200 p.

Davies, J.L. (1980). *Geographical Variation in Coastal Development*. Longman. London, 212 p.

Davis, D.G. (2005). Spelothems: Helictites and Related Forms. In: Culver, D.C. and White, W.B. (Eds.). *Encyclopedia of Caves*. Elsevier. Amsterdam, 549–554.

Davis, L. (1992). *Encyclopedia of Natural Disasters*. Headline, London. 433 p.

Davis, W.M. (1892). The convex profile of bad-land divides. *Science*, 29, 245.

Davis, W.M. (1899). The geographical cycle. *Geographical Journal*, 14, 481–504.

Davis, W.M. (1905). The geographycal cycle in arid climate. *Journal of Geology*, 13, 381–407.

Davis, W.M. (1906). The sculpture of mountains by glaciers. *Scottish Geographical Magazine*, 22, 76–89.

Davis, W.M. (1913). Nomenclature of surface forms on faulted structures. Bulletin *Geological Society of America*, 24, 187–216.

Davis, W.M. (1930). Rock floors in arid and humid climates. *Journal of Geology*, 38, 1–27.

Davis, W.M. (1932). Piedmont bechlands and primärümpfe. *Bulletin of Geological Society of America*, 43, 399–440.

Davis, W.M. (1954). *Geographical Essays*. Dover. New York, 777 p.

Davison, C. (1889). On the creeping on the soilcap through the action of frost. *Geological Magazine*, 6, 255–261.

Dawson, A.G. (1992). *Ice Age Earth. Late Quaternary Geology and Climate*. Routledge. London, 293 p.

Day, J.W., Rybczy, J., Scarton. F., Rismondo, A. Are, D. and Cecconi, G. (1999). Soil accrecionary dynamics, sea-level rise and the survival of wetlands in Venice Lagoon: a field and modelling approach, *Estuarine, Coastal and Shelf Science*, 49, 607–628.

Day, M. (2004). Cone Karst. In: Gunn, J. (Ed.). *Encyclopedia of Caves and Karst Science*. Fitzroy Dearborn. New York, 241–243.

Day, M.J. (2004b). Cenotes. In: Goudie, A.S. (Ed.). *Encyclopedia of Geomorphology*. Routledge. London, 132–133.

Day, M. and Tang, T. (2004). Tower Karst. In: Gunn, J. (Ed.). *Encyclopedia of Caves and Karst Science*. Fitzroy Dearborn. New York, 734–736.

De Chetelat, E. (1938). Le modelé latérique de l'ouest de la Guinée francaise. *Revue de Géographie Physique et Géologie Dynamique*, 11, 5–12.

De Dapper, M. (1978). Couvertures limono-sableuses, stone-line, indurations ferrugineuses et action des termites sur le plateau de La Manika (Kolwezi, Shaba, Zaire). *Géo-Eco-Trop*, 2, 265–278.

De Dapper, M. (1989). Pedisediments and stone-line complexes in Peninsular Malaysia. In: Alexandre, J. and Symoens, J.J. (Eds.). *Stone-Line*. Académie Royale des Sciences D'Outre-Mer, Geo-Eco-Trop, 11, 37–59.

De Gans, W. (1988). Pingo scars and their identification. In: Clark, M.J. (Ed.). *Advances in Periglacial Geomorphology*, Wiley. Chichester, 299–322.

De Ploey, J. (1964). Stone-lines and clayey-sandy mantles in Lower Congo: their formation and the effect of termites. In: Bouillon, A. (Ed.). *Etudes sur les Termites*

Africains. Université de Louvanium. Leopoldville, 399–414.

De Ploey, J. (1965). Position géomorphologique, génèse et chronologie de certains dépots superficiels au Congo occidentale. *Quaternaria*, 7, 131–154.

De Ploey, J. (1980). Some field measurements and experimental data on wind-blown sands. In: Boodt, M. and Gabriels, D. (Eds.). *Assessment of erosion*. Wiley. New York, 541–552.

De Ploey, J. and Gabriels, D. (1980). Measuring soil loss and experimental studies. In: Kirkby, M.J. and Morgan, R.P.C. (Eds.). *Soil Erosion*, Wiley, 63–108.

De Ploey, J. and Poesen, J. (1989). Experimental geomorphology and the interpretation of stone-lines. In: Alexandre, J. and Symoens, J.J. (Ed.). *Stone Lines*. Académie Royale des Sciences D'Outre-Mer, Geo-Eco-Trop, 11, 75–82.

De Ploey, J., Savat, J. and Moeyerons, D. (1976). The differential impact of some soil loss factors on flow, runoff, creep, and rainwash. *Earth Surface Processes*, 1, 151–161.

De Ploey, J., Kirkby, M.J. and Ahnert, F. (1991). Hillslope erosion by rainstorms-a magnitude—frecuency analysis. *Earth Surface Processes and Landforms*, 16, 399–409.

De Waele, J., Plan, L. and Audra, P. (2009). Recent developments in surface and subsurface karst geomorphology. *Geomorphology*, 106-1-8.

Dean, J.S. (1996). Demography, environment, and subsitence stress. In: Tainer, J.A. and Tainter, B.B. (Eds.). *Evolving complexity and enviromental rosk in the prehistoric Southwest*. Addison Wesley, Reading, 25–56.

Dearman, W.R. and Shibakova, V.S. (1989). The tropical environment. In: Dearman, W.R., Sergeev, E.M. and Shibakoya, V.S. (Eds.). *Engineering Geology of the Earth*. Nauka Publishers. Moscow, 128–141.

Dearman, W.R., Baynes, F.J. and Irfan, T.Y. (1978). Engineering grading of weathered granite. *Engineering Geology*, 12, 345–374.

Decker, R. and Decker, B. (1998). *Volcanoes. Academic Version*. W.H.Freeman, New York, 319 p.

Degenhardt, J., Giardino, J., Marston, R. and Pitty, A. (2007). Possible origin of the block stream in Tom Mays Canyon, Franklin Mountains, TX. *Zeitschrift für Geomorphologie*, 51, 377–396.

DeGraff, J.V. (1978). Regional landslide evaluation: two Utah examples. *Environmental Geology*, 2, 203–214.

Delannoy, J.J. (1998). Contribución al conocimiento de los macizos kársticos de las Sierras de Grazalema y de Ronda. In: Durán, J.J. and López Martínez, J. (Eds.). *Karst en Andalucía*. Instituto Tecnológico y Geominero de España. Madrid, 93–129.

Delibes, M. and Delibes de Castro, M. (2005). *La Tierra herida. ¿Qué mundo heredarán nuestros hijos?* Ediciones Destino. Barcelona, 173 p.

Delvert, C. (1968). L'erosion de grès d'Angkor. In: Birot, P., Henin, S., Guilluien, Y. and Delvert, J. (Eds.). *Contribution a l'Etude de la Désagrégation des Roches*. Centre de Documentation Universitaire. Paris, 49–56.

Del Río, L. and Gracia, F.J. (2009). Erosion risks assesment of active coastal cliffs in temperate environments. *Geomorphology*, 112, 82–95.

Demangeot, J. (1976). *Les Espaces Naturels Tropicaux*. Masson. Paris, 190 p.

Demangeot, J. (1981). *Les Milieux Naturels Désertiques*. SEDES. Paris, 261 p.

Demangeot, J. (1996). Certitudes et incertitudes du "global change". *Bulletin Association de Géographers Francaises*, 4, 278–282.

Demany, P. (1990). Population. In: Turner, B.L., Clark, W.C., Kates, R.W., Richards, J.F., Matthews, J.T. and Meyer, W.D. (Eds.). *The Earth as Transformed by Human Action*. Cambridge University Press. Cambridge, 41–54.

Demek, J. (1964). Castle koppies and tors in the Bohemian highland (Czechoslovakia). *Biuletyn Peryglacjalny*, 14, 195–216.

Demek, J. (1965). Slope development in granite areas of Bohemian massif (Czechoslovakia). *Zeitscrift für Geomorphologie, Supplement Band*, 5, 82–106.

Demek, J. (1969). Cryogene processes and the development of cryoplanation terraces. *Biuletyn Peryglacjalny*, 18, 115–125.

Demek, J. (Ed.). (1972). *Manual of Detailed Geomorphological Mapping*. Czechoslovak Academy of Sciences. Akademia. Praga, 344 p.

Demek, J. (1973). Quaternary relief development and man. *Geoforum*, 15, 68–71.

Demek, J. and Embleton, C. (Eds.) (1978). *Guide to Medium-Scale Geomorphological Mapping*. I.G.U. Commision on Geomorphological Survey and Mapping. Akademia. Praga. 348 p.

Dendy, F.E. and Bolton, G.C. (1976). Sediment yield-runoff drainage area relationships in the United States. *Journal of Soil and Water Conservation*, 31, 264–266.

Deng, O., Fergmin, S., Shilong, Z., Mengluan, L. Tielin, W., Weiqi, Z., Burchfel, B.S., Molnar, P. and Peizhen, Z. (1984). Active faulting and tectonics of the Ningxia-Hui Autonomous Region, China. *Journal of Geophysical Research*, 89, 4427–4445.

Denny, C.S. (1967). Fans and pediments. *Americal Journal of Science*, 265, 81–105.

Derbyshire, E. (1973). Introduction. In: Derbyshire, E. (Ed.). *Climatic Geomorphology*. Macmillan. London, 11–18.

Derbyshire, E. (1976). Geomorphology and climate: background. In: Derbyshire, E. (Ed.). *Geomorphology and Climate*. Wiley. London, 1–24.

Derbyshire, E. (1977). Periglacial environments. In: Hails, J.R. (Ed.). *Applied Geomorphology*, Elsevier. Amsterdam, 227–276.

Derbyshire, E. (1983a). On the morphology, sediments and origin of the Loess Plateau of central China. In: Gardner, R.A.M. and Scoging, H. (Eds.). *Megageomorphology*, Oxford University Press. Oxford, 172–194.

Derbyshire, E. (1983b). Origin and characteristics of some Chinese loess at two locations in China. In: Brookfield, M.E. and Ahlbrandt, T.S. (Eds.). *Eolian Sediments and Processes*, Elsevier. Amsterdam, 69–90.

Derbyshire, E. and Evans, I.S. (1976). The climatic factor in cirque variation. In: Derbyshire, E. (Ed.). *Geomorphology and Climate*, John Wiley, 447–494.

Derbyshire, E. and Love, M.A. (1986). Glacial environments. In: Fookes, P.G. and Vaughan (Eds.). A *Handbook of Engineering Geomorphology*. Chapman and Hall. New York, 66–81.

Derbyshire, E., Fort, M. and Owen, L.A. (2001). Geomorphological hazards along the Karakoran Highway: Khumjerab Pass to the Gilgit River, Northernmost Pakistan. *Erdkunde*, 55, 49–71.

Derruau, M. (1965). *Précis de Géomorphologie*. Masson et Cie. Paris. (Trad. esp. L. Solé Sugrañes, Ediciones Ariel, Barcelona, 442 p.).

Derruau, M. (1974). *Précis de Géomorphologie*. Masson. Paris, 453 p.

Dewey, F.F. and Bird, J.M. (1970). Mountain belts and the new global tectonics. *Journal of Geophysical Research*, 75, 2625–2647.

Dewey, J.F. and Burke, K.C. (1974). Hot sports and continental break-up: implications for collisional orogeny. *Geology*, 2, 57–60.

DeWolf, Y. (1988). Stratified slope deposits. In: Clark, M.J. (Ed.). *Advances in Periglacial Geomorphology*, Wiley. Chichester, 91–110.

Dhir, R.P., Kar, A., Wadhawan, S.K., Rajaguru, S.N., Misra, V.N., Singhvi, A.K. and Sharma, S.B. (1992). *Thar Desert in Rajasthan. Land, Man and Environment*. Geological Society of India. Bangalore, 191 p.

Dias de Avila-Pires, F. (1974). Caracterizao zoogeográfica de Provincia Amazónica. *Annales Academia de Ciencias do Brasil*, 46, 133–181.

Díaz de Terán, J.R. and Cendrero, A. (1992). Riesgos geológicos en España. *XXVI Curso de Geología Práctica*. Universidad de Verano. Teruel, 179–231.

Díaz de Terán, J.R., Cendrero, A., Araña, V., Badiola, E.R., Berga, L., Carracedo, J.C., Chacón, J., Coello, J., Corominas, J., Dabrio, C., Durán, J.J., Elízaga, E., Ferrer, M., García Fernández, M., Garzón, M.G., González Díez, A., Goy, J.L., López Martínez, J., Martínez-Goytre, J., Mercua, J., Nuez, J. de la, Salinas, J.L., Soler, V., Val del, J. and Zazo, C. (1997). *Spain*. In: *Geomorphological Hazards of Europe*. Elsevier. Ámsterdam, 429–456.

Dickinson, R.E. (1986). Impact of human activities on climate a framework. In: Clark, W.C. and Munn, R.E. (Eds.). *Sustainable Development of the Biosphere*. Cambridge University Press. Cambridge 252–289.

Dickinson, R.E., Meleshko, V., Randall, D., Sarashik, E., Silva-Dias, P. and Slingo, A. (1996). Climate Processes. In: Houghton, J.T. *et al.*, (Eds.). *Climate Change. The Science of Climate Change*. Cambridge University Press. Cambridge, 193–227.

Diester-Haas, L. (1976). Late Quaternary climatic variations in Northwest Africa deduced from east Atlantic sediment cores. *Quaternary Research*, 6, 299–314.

Dietrich, R.V. (1977). Impact abrasion of harder by softer materials. *Journal of Geology*, 85, 242–246.

Dietz, R.S. (1961). Continental and ocean basin evolution by spreading of the sea floor. *Nature*, 190, 854–857.

Díez, A. and Bateman, M.D. (1998). Interpretación paleoambiental y datación mediante luminiscencia del manto arenoso de la de Tierra de Pinares oriental (Segovia). *Geogaceta*, 24, 107–110.

Díez, A., Bateman, M.D., López Sáez, J.A. and Vegas, J. (2002). Procesos eólicos tardiglaciares en la Submeseta Septentrional: cronología del manto arenoso de la Tierra de Pinares. In: Pérez González, A., Vegas, J. and Machado, M.J. (Eds.). *Aportaciones a la Geomorfología de España en el inicio del tercer milenio*. Instituto Geológico y Minero de España. Madrid, 167–175.

Díez-Herrero, A. (2002). Condicionantes geomorfológicas de las avenidas y cálculo de caudales y calados. In: Ayala-Carcedo, F.J. and Olcina, J. (Coor.). *Riesgos Naturales*. Ariel Ciencia. Barcelona, 921–952.

Díez-Herrero, A., Benito, G. and Laín-Huerta, L. (1992). Regional palaeoflood databases applied to flood hazards and palaeoclimatic analysis. In: Benito, G., Baker, V.R. and Gregory, K.J. (Eds.). *Palaeohydrology and Environmental Change*. Wiley. Chichester, 335–347.

Díez-Herrero, A., Laín-Huerta, L. and Llorente-Isidro, M. (2008). *Mapas de peligrosidad por avenidas e inundaciones. Guia metodológica para su elaboración*. Instituto Geológico y Minero de España. Madrid, 190 p.

Dikau, R. (2004). Mass Movement. In: Goudie, A.S. (Ed.). *Encyclopedia of Geomorphology*. Routledge, London, 644–652.

Dikau, R., Brunsden, D., Schrott, L. and Ibsen M.L. (1996a). Introduction. In Dikau, R., Brunsden, D., Schrott, L. and Ibsen, M.L. (Eds.). *Landslide Recognition.*

Identification, Movement and Causes. Wiley. Chichester. 1–12.

Dikau, R., Schrott, L. and Dehn, M. (1996b). Topple. In: Dikau, R., Brunsden, D., Schrott, L. and Ibsen, M.L. (Eds.). *Landslide Recognition. Identification, Movement and Causes*. Wiley. Chichester, 29–41.

Dixon, J.C. (1994a). Aridic soils, patterned ground and desert paviments. In: Abrahams, A.L. and Parsons, A.J. (Eds.). *Geomorphology of Desert Environments*, Chapman and Hall. London, 64–81.

Dixon, J.C. (1994b). *Duricruts.* In: Abrahams, A.D. and Parson, A.J. (Eds.). *Geomorphology of Desert Environments*. Chapman and Hall. London, 82–105.

Dodonov, A.E. (2007). Loess records: Central Asia. In: Elías, S.A. (Ed.). *Encyclopedia of Quaternary Science*. Elsevier. Amsterdam, 1418–1429.

Doehring, D.O. (Ed.) (1977). *Geomorphology of Arid Regions*. Allen and Unwin. London, 271 p.

Doelling, H.H. (1985). *Geology of Arches National Park*. Utah Geological and Mineral Survey, Map 74.

Dohrenwend, J.C. (1994). Pediments in arid environments. In: Abrahams, A.D. and Parsons, A.J. (Eds.). *Geomorphology of Desert Environments*. Chapman and Hall. London, 321–353.

Dokuchayev, V.V. (1883). *The Russian Chernozem*. St. Petersburg.

Dong, Z., Wang, T. and Wang, X. (2004). Geomorphology og mega dunes in the Badain Jaran Desert. *Geomorphology*, 60, 191–203.

Donn, W.L., Farrand, W.R. and Ewing, M. (1962). Pleistocene ice volumes and sea lowering. *Journal of Geology*, 70, 206–214.

Doorkamp, J.C. (1974). Tropical weathering and the ultramicroscopic characteristics of regolith quartz on Dartmoor. *Geografiska Annaler*, 56, 73–82.

Doornkamp, J.C. and Ibrahim, H.A.M. (1990). Salt weathering. *Progress in Physical Geography*, 14, 335–348.

Dorn, R.I. (1994a). Rock varnish as evidence of climatic change. In: Abrahams, A.D. and Parsons, A.J. (Eds.). *Geomorphology of Desert Environments*. Chapman and Hall. London, 539–552.

Dorn, R.I. (1994b). The role of climatic change in alluvial fan development. In: Abrahams, A.D. and Parsons, A.J. (Eds.). *Geomorphology of Desert Environments*. Chapman and Hall. London, 593–615.

Dorn, R.I. (1995). Alterations of ventifat surfaces at the glacier/desert interface. In: Tchakerian, V.P. (Ed.). *Desert Aeolian Processes.* Chapman and Hall. London, 199–217.

Dorn, R.I. (1998). *Rock Coatings*. Elsevier. Amsterdam, 429 p.

Dorn, R.I. (2003). Boulder weathering and erosion associated with a wildfire. Sierra Ancha Mountains, Arizona. *Geomorphology*, 55, 155–171.

Dorn, R.I. (2004). Desert varnish. In: Goudie, A.S. (Ed.). *Encyclopedia of Geomorphology*. Routledge. London, 251–254.

Dorn, R.I. (2007). Rock varnish. In: Nash, D.L. and McLaren, S.J. (Eds.). *Geochemical Sediments and Landscapes*. Blackwell. Malden, 246–297.

Dorn, R.I. and Oberlander, T.M. (1981). Rock varnish origin, characteristics, and usage. *Zeitschrift für Geomorphologie*, 25, 420–436.

Dorn, R.I. and Oberlander, T.M. (1982). Rock varnish. *Progress in Physical Geography*, 6, 317–367.

Dorn, R.I., Tanner, D.L., Turrin, B.D. and Dohrenwend, J.C. (1987). Cation-ratio dating of Quaternary materials in the east-central Mojave Desert, California. *Physical Geography,* 8, 72–81.

Douglas, I. (1967). Man, vegetation and the sediment yield of rivers. *Nature*, 215, 925–928.

Douglas, I. (1968). Some hydrological factors in the denudation of limestone terrains. *Zeitscrift für Geomorphologie*, 12, 241–255.

Douglas, I. (1969). The efficiency of humid tropical denudation system. *Transactions Institute British Geographers*, 46, 1–16.

Douglas, I. (1977). *Humid Landforms*. The MIT Press. Cambridge, 288 p.

Douglas, I. (1978). Tropical geomorphology: present problems and future prospects. In: Embleton, C., Brunsden, D. and Jones, D.K.C. (Eds.). *Geomorphology. Present Problems and Future Prospects*. Oxford University Press. Oxford, 162–184.

Douglas, I. (1986). Hot wetlands. In: Fookes, P.G. and Vaughan, P.R. (Eds.). *A Handbook of Engineering Geomorphology*. Surrey University Press. Glasgow, 137–149.

Douglas, I. and Spencer, T. (1982). Introduction: Applied Geomorphology in the Tropics. *Zeitschrift für Geomorphologie Supplementband*, 44, 1–3.

Douglas, I. and Spencer, T. (Eds.) (1985a). *Environmental Change and Tropical Geomorphology.* Allen and Unwin. London, 378 p.

Douglas, I. and Spencer, T. (1985b). The history of geomorphology in low latitudes. In: Douglas, I. and Spencer, T. (Eds.). *Environmental Change and Tropical Geomorphology.* Allen and Unwin. London, 3–11.

Douglas, I. and Spencer, T. (1985c). Present day processes as a key to the effects of environmental change. In: Douglas, I. and Spencer, T. (Eds.). *Environmental Change and Tropical Geomorphology.* Allen and Unwin. London, 39–73.

Dowdeswell, J.A., Hamilton, G.S. and Hagen, J.O. (1991). The duration of the active phase on surge-type glaciers: contrasts between Svalbard and other regiones. *Journal of Glaciology*, 37, 388–400.

Downs, P.W. and Gregory, K.J. (2004). *River Channel Management. Towards Sustainable Catchment Hydrosystems*. Arnold. London, 395 p.

Dozier, E.F. and Yancey, T.N. (1993). *Floodproofing Options for Virginia Homeowners*. United States Army Corps of Engineers and Commonwealth of Virginia. Norfolk, Virginia.

Dragovich, D. (1969). The origin of cavernous surfaces (tafoni) in granitic rocks of Southern Australia. *Zeitscrift für Geomorphologie*, 13, 163–181.

Drake, J.J. (1984). Theory and model for global carbonate solution by groundwater. In: Lafleur, R.G. (Ed.). *Groundwater as a geomorphic agent*. Allen and Unwin. London, 210–226.

Dramis, F. and Sorriso-Valvo, M. (1994). Deep-seated gravitacional slope deformations, related landslides and tectonics. *Engineering Geology*, 38, 231–243.

Dregne, H.E. (1983). *Desertification of Arid Lands*. Harwood Academic. New York.

Dreimanis, A. (1953). Studies of friction cracks along shores of Cirrus Lake and Kasakokwog, Ontario. *American Journal of Science*, 251, 769–783.

Dreimanis, A. (1988). Tills: their genetic terminology and classification. In: Goldthwait, R.P. and Matsch, C.L. (Eds.). *Classification of Glacigenic Deposits*, Balkema. Rotterdam, 17–83.

Dreimanis, A. and Vagners, J.J. (1971). Bimodal distribution of rock and mineral fragments in basal tills. In: Goldthwait, R.P. (Ed.). *Till: A Symposium*, Ohio State University Press. Columbus, 237–250.

Dresch, J. (1938). Les surfaces de piémont dans les Djebilet et le Massif Central du Grand Atlas. *Compte Rendu Congress Internationale de Géographie*, Vol. 2, section 2a. Amsterdam, p. 135.

Dresch, J. (1957). Pédiments et glacis d'erosion, pediplains et inselbergs. *Information Géographique*, 21, 183–196.

Dresch, J. (1970). A propos du desert de Chihuahua. *Annales de Géogrpahie*, 79, 51–57.

Dresch, J. (1982). *Géographie des Régions Arides*. Presses Universitaires de France. Paris, 277 p.

Dresselhaus, M.S. and Thomas, I.L. (2001). Alternative energy technologies. *Nature*, 414, 332–337.

Drewry, D. (1986). *Glacial Geologic Processes*. Edward Arnold. London, 276 p.

Dreybrodt, W. (1988). *Processes in Karst Systems. Physics, Chemistry and Geology*. Springer. Berlin, 288 p.

Dreybrodt, W. (2004). Dissolution: Carbonate Rocks. In: Gunn, J. (Ed.). *Encyclopedia of Caves and Karst Science*. Fitzroy Dearborn. New York, 295–298.

Dreybrodt, W. (2005). Speleothem Deposition. In: Culver, D.C. and White, W.B. (Eds.). *Encyclopedia of Caves*. Elsevier. Amsterdam, 543–549.

Drogue, C. and Bidaux, P. (1992), Structural and hydrogeological origin of the tower karst in southern China. *Zeitscrift für Geomorphologie*, 36, 25–36.

Droppa, A. (1966). The correlation of some horizontal caves with river terraces. *Studies in Speleology, Association of the Pengelly Cave Research Centre*, 1, 186–192.

Duchaufour, P. (1977). *Pédologie. 1. Pédogènese et Classification*. Masson. Paris, 477 p.

Duchaufour, P. (1979). Introduction générale. In: Pedro, M.J. (Ed.). Alteration des roches cristallines en milieu superficiel. *Bulletin de l'Association Francaise por l'Etude du Sol*, 2 y 3, 1–3.

Dueñas, C. (2001). La mitigación de los efectos de los terremotos desde el punto de vista de la protección civil. In: Capote, R. and Martínez-Díaz, J.J. (Eds.). *El Riesgo Sísmico. Prevención y Seguro*. Consorcio de Compensación de Seguros. Madrid, 199–219.

Duller, G.A.T. and Augustius, P.C. (2006). Reassessment of the record of linear dune activity in Tasmania using optical dating. *Quaternary Science Reviews*, 25, 2608–2618.

Dumas, B. (1967). Place et signification des glacis dan le Quaternaire. *Bulletin Association Francaise pour l'Etude du Quaternaire*, 3, 223–244.

Dumitrashko, N.V. and Scholz, E. (1978). Classification of geomorphological maps according to scale. In: Demek, J. and Embleton, C. (Eds.). *Guide to Medium-scale Geomorphological Mapping*. E. Schweizerbartsehe. Stuttgart, p. 40.

Dunne, T. (1980). Formation and control of channel networks. *Progress in Physical Geography*, 4, 211–239.

Dunne, T. and Aubry, B.F. (1986). Evaluation of Horton's Theory of sheetwash and rill erosion on basis of field experiments. In: Abrahams, A.D. (Ed.). *Hillslope Processes*. Allen and Unwin. Boston, 31–53.

Dunne, T. and Dietrich, W.E. (1980). Experimental study of Horton overland flow on tropical hillslopes. 1. Soil conditions, infiltration and frequency of runnoff. *Zeitschrift für Geomorphologie Supplementband*, 35, 40–59.

Dunne, T. and Leopold. L.V. (1978). *Water in Environmental Planning*. Freeman. San Francisco, 818 p.

Durán, J.J. (1988). Riesgos asociados al karst. In: *Riesgos geológicos*. Instituto Geológico y Minero de España. Madrid, 261–283.

Durán, J.J. (2002). Riesgos en terrenos kársticos. In: Ayala-Carcedo, F.J. and Olcina, J. (Eds.). *Riesgos naturales*. Ariel Ciencia. Barcelona, 365–378.

Durán, J.J. and Burillo, F.J. (1989). Riesgos geológicos ligados al karst de España. In: Durán, J.J. and López-Martínez, J. (Eds.). *El Karst en España*. Monografía 4, Sociedad Española de Geomorfología. Madrid, 347–359.

Durgin, P.B. (1977). Landslides and the weathering of granite rocks. In: Coates, D.R. (Ed.). *Landslide perspectives*. Reviews in Engineering Geology, vol. 3. Geological Society of America, 127–131.

Dury, G.H. (1969). Rational descriptive classification of duricrusts. *Earth Science Journal*, 3, 77–86.

Dury, G.H. (1971). Relict deep weathering and duricrusting in relation to the palaeoenvironments of middle latitudes. *Geographical Journal*, 137, 511–522.

Dury, G.H. (1978). The future of geomorphology. In: Embleton, C., Brunsden, D. and Jones, D.K.C. (Eds.). *Geomorphology: Present Problems and Future Prospects*. Oxford University Press. Oxford, 263–274.

Duxbury, N.S., Zotikov, A., Nealson, K.H., Romanovsky, V.E. and Carsey, F.D. (2001). A numerical model for an alternative origin of Lake Vostok and its exobiological implications. *Journal of Geophysical Research*, 106, 1453–1462.

Dyke, A.S. (1993). Landscapes of cold-centered Late Wisconsin ice caps, Arctic Canada. *Progress in Physical Geography*, 17, 223–247.

Dyke, A.S. and Prest, V.K. (1987). Late Wisconsinan and Holocene history of the Laurentide Ice Sheet. *Géographie Physique et Quaternaire*, 41, 237–264.

Eagleman, J.R. (1983). *Severe and Unusual Weather*. Van Nostrand Reinhold. New York, 393 p.

Eakin, H.M. (1916). The Yukon-Koyukuk region, Alaska. *United States Geological Survey Bulletin*, 631, 88 p.

Eastearling, D.R., Horton, B., Jones, P.D., Peterson, T.C., Karl, T.R., Parker, D.E., Sallinger, M.J., Razubayev, V., Plunner, N., Jamason, P. and Folland, C.K. (1997). Maximum and minimum temperatures trends for the globe. *Science*, 277, 364–367.

Eckardt, F. (1999). The origin of sulphur and dissolved sulphate in the Central Namib Desert, Namibia. *Sedimentary Geology*, 123, 255–273.

Eckardt, F. (2004). Gypcrete. In: Goudie, A.S. (Ed.). *Encyclopedia of Gemorphology*. Routledge. London, 507–509.

Eckardt, F. and Spiro, B. (1999). The origin of sulphur in gypsum and dissolved sulphate in the Central Namib Desert, Namibia. *Sedimentary Geology*, 123, 255–273.

Eckardt, F., Drake, N.A., Goudie, A.S.,White, K. and Viles, H. (2001). The role of playas in the formation of pedogenic gypsum crusts of the Central Namib Desert. *Earth Surface Processes and Landforms*, 26, 177–193.

Edgell, H.S. (1990). Evolution of the Rub'al Khali Desert. *Journal of King Abdulaziz University. Earth Science*, 3, 109–126.

Edgren, G. (1993). Expected economic and demographic developments in coastal zones word-wide. *World Cost'93*, National Institute for Coastal and Marine Management, Coastal Zone Management Centre. Noordwijk, The Netherlands, 367–370.

Ehlen, J. and Wohl, F. (2002). Joints and landform evolution in bedrock canyons. *Transactions, Japanese Geomorphological Union*, 23, 237–255.

Ehler, J. (2005). Above the weathering front: contrasting approaches to the study and classification of weathered mantle. *Geomorphology*, 67, 7–21.

Ehlers, J. (1996). *Quaternary and Glacial Geology*. John Wiley. Chichester, 578 p.

Ehrlich, P.R. and Holdren, J.P. (1971). Impact of population growth. *Science*, 171, 1212–1217.

Ehrlich, P.R., Ehrlich, A.H. and Holdren, J.P. (1977). *Ecoscience population, resources, environment*. Freeman. San Francisco.

Eisbacher, G.H. and Clague, J.J. (1984). *Destructive Mass Movement in Mountains: Hazard and Management*. Paper 84–16. Geological Survey of Canada. Ontario, Ottawa, 230 p.

Eissmann, L. (1981). Periglaciäre prozesse und permafroststrukturen aus sechs kaltzeiten des quartäs. Ein beitrag zur periglazialgeologie aus des sicht des Saale-Elbe-Gebiets. *Altenburger Naturwissenschaftliche Forschunge*, 1, 171 p.

El-Askary, H. and Kafatos, M. (2008). Natural and anthropogenic aerosol-related hazards affecting megacities. In: Gad-el Hak, M. (Ed.). *Large-scale Disasters. Prediction, Control and Mitigation*. Cambridge University Press. Cambridge, 218–257.

El-Baz, F. (Ed.) (1984). *Desert and Arid Lands*. Martinus Nijhof. The Hague, 222 p.

El-Baz, F. and Hassan, M.H.A. (Eds.) (1986). *Physics of Desertification*. Martinus Nijhoff. The Hague, 473 p.

El-Hamdouni, R., Irigaray, C., Fernández, T., Chacón, J. and Keller, E.A. (2008). Assessment of relative active tectonics, southwest border of Sierra Nevada (southern Spain). *Geomorphology*, 93, 150–173.

Ellenberger, F. (1988). *Histoire de la Géologie. Tome I. Des Anciens à la première moitié du XVIIe Siècle*. Tecnique et Documentation-Lavoisier. Paris, 352 p.

Ellenberger, F. (1994). *Histoire de la Géologie. La grande éclosion et ses premices (1660–1810)*. Technique et Documentation-Lavoisier. Paris, 381 p.

Elliot, T. (1986). Deltas. In: Reading. H.G. (Ed.). *Sedimentary Environments and Facies*. Blackwell. Oxford, 113–154.

Ellison, W.D. (1944). Studies of raindrop erosion. *Agricultural Engineering*, 25, 131–136, 181–182.

El-Sabh, M.I. and Murty, T.S. (1988). *Natural and Man-Made Hazards*. Reidel Pub. Co. 894 p.

Elson, J.A. (1989). Comment on glacitectonite, deformation till and comminution till. In: Goldthwait, R.P. and Matsch, C.L. (Eds.). *Genetic Classification of Glacigenic Deposits*. Balkema. Rotterdam, 85–88.

Elvidge, C.D. and Moore, C.B. (1979). A model for desert varnish formation. *Geological Society of America Abstracts with Programs*, 11, 271.

Embabi, N.S. (1999). Playas of Western Desert, Egypt. *Annales Academiae Scientarum Fenicæ Geologica-Geographica*, 160, 5–47.

Embleton, C. (1979a). Glacial Processes. In: Embleton, C. and Thornes, J. (Eds.). *Process in Geomorphology*. Arnold. London, 272–306.

Embleton, C. (1979b). Nival processes. In: Embleton, C. and Thornes, J. (Eds.). *Process in Geomorphology*. Arnold. Londres, 307–314.

Embleton, C. (1989). Natural hazards and global change. *ITC Journal*, 1989, 169–178.

Embleton, C. and King, C.A.M. (1975a). *Glacial Geomorphology*. Arnold. London, 573 p.

Embleton, C. and King, C.A.M. (1975b). *Periglacial Geomorphology*. Arnold. Londres, 203 p.

Emery, K.O. (1960). Weathering of the Great Pyramide. *Journal Sedimentary Petrology*, 30, 140–143.

Emery, K.O. and Kuhn, G.G. (1982). Sea cliffs: Their processes, profiles and classification. *Geological Society of America Bulletin*, 93, 644–654.

Emery, K.O. and Aubrey. D.F. (1991). *Sea levels, Land Levels and Tide Gauges*. Springer Verlag. New York.

Emiliani, C. (1955). Pleistocene paleotemperatures. *Journal of Geology*, 63, 539–578.

Emmett, W.W. (1978). Overland flow. In: Kirkby, M.J. (Ed.). *Hillslope Hydrology*, Wiley. Chichester, 145–176.

Engelstaedter, S., Tegen, I. and Washington, R. (2006). North African dust emissions and transport. *Earth-Science Reviews*, 79, 73–100.

Engle, C.E. and Sharp, R.P. (1958). Chemical data on desert varnish. *Bulletin of the Geological Society of America*, 69, 487–518.

EPICA (2004). Eight glacial cycles from an Antarctic ice core. *Nature*, 429, 623–628.

EPOCH (European Community Programme) (1993). In: Flageollet, J.C. (Ed.). *Temporal Ocurrence and Forecasting of Landslides in the European Community,* 3 vols.

Eraso, A., Trzhtsinskij, Y. and Castrillo, A. (1995). Dolinas de colapso y karst en yeso en la plataforma cámbrica del este de Siberia. *Boletín Geológico y Minero*, 106, 373–378. Madrid.

Erhardt, H. (1951). Sur l'importance des phénomènes biologiques dans la formation des cuirasses ferrugineuses en zone tropicale. *Campte Rendus de la Académie des Sciences de Paris*, 223, 805–806.

Erhart, H. (1955). "Biostasie" et "Rhexistasie". Esquise d'une théorie sur le rôle de la pedogenése en tant que phénomène géologique. *Compte Rendus de la Academie de Sciences de Paris*, 241, 1218–1220.

Erhart, H. (1967). *La Genèse des Sols en tant que Phénomène Géologique. Esquisse d'une Théorie Géologique et Géochimique. Biostasie et Rhexistasie*. Masson. Paris, 177 p.

Erickson, G.E., Ramírez, C.F., Concha, J.F., Tisnado, G.M. and Urquidi, F.B. (1989). Landslide hazards in the central and Southern Andes. In: Brabb, E.E. and Harrod, B.L. (Eds.). *Landslides: Extent and Economic Significance*. Balkema. Rotterdam, 111–118.

Esbert, R.M., Marcos, R.M. and Montoto, M. (1983). *Las piedras de la Catedral de Oviedo y su Deteriorización*. Colegio Oficial de Aparejadores y Arquitectos Técnicos de Asturias. Oviedo, 145 p.

Esbert, R.M., Gutiérrez, M., Ordaz, J., Meléndez, A., Sancho, C., Montoto, M., Suárez del Río, L.M., Calleja, L., Gómez, V., Alonso, F.J. and Rodríguez, A. (1987). *Estudio Geológico y Alterológico de los Materiales Pétreos de San Juan de la Peña: Especial Referencia a sus Capiteles Románicos*. Departamento de Cultura. Diputación General de Aragón. Zaragoza, 90 p.

Esbert, R.M., Ordaz, J., Alonso, F.J. and Montoto, M. (1997). *Manual de Diagnosis y Tratamiento de Materiales Pétreos y Cerámicos*. Col-legi d'Aparelladors i Arquitectes Técnics de Barcelona. Barcelona, 126 p.

Esbert, R.M., Díaz-Pache, F., Grossi, C.M., Alonso, F.J. and Ordaz, F. (2001). Airbone particulate matter around the Catedral of Burgos (Castilla y León, Spain). *Atmospheric Environmental*, 35, 441–452.

Espíndola, J.M., Macías, J.L., Godínez, M.L. and Jiménez, Z. (2002). La erupción del 1982 del volcán Chichonal, Chiapas, México. In: Hubp, J.L. and Inbar, M. (Eds.). *Desastres naturales en América Latina*. Fondo de Cultura Económica. México, 37–65.

Esteban, M. and Klappa, C.F. (1983). Subaerial exposure environment. In: Scholle, A., Bebout, D.G. and Moore, C.H. (Eds.). *Carbonate Depositional Environments*. American Association Petroleum Geologists. Tulsa, 1–55.

Etzelmüller, B. and Hagen, J.D. (2004). Deglaciation. In: Goudie, A.S. (Ed.). *Encyclopedia of Geomorphology*. Routledge. London, 233–237.

Eugster, H.P. and Hardie, L.A. (1975). Sedimentation in an ancient playa-lake complex: the William Peak Member of the Gran River Formation of Wyoming. *Bulletin of the Geological Society of America*, 86, 319–334.

Eugster, H.P. and Hardie, L.A. (1978). Saline lakes. In: Lerman, A. (Ed.). *Lakes: Chemistry, Geology, Physics*, Springer-Verlag. New York, 237–293.

Eugster, H.P. and Kelts, K. (1983). Lacustrine chemical sediments. In: Goudie, A.S. and Pye, K. (Eds.). *Chemical Sediments and Geomorphology*, Academic Press. London, 321–367.

Evans, D.J.A. (2004a). Surging glaciers. In: Goudie, A.S. (Ed.). *Encyclopedia of Geomorphology*. Routledge, 1028–1029.

Evans, D.J.A. (2004b). Fjord. In: Goudie, A.S. (Ed.). *Encyclopedia of Geomorphology*. Routledge, 374–376.

Evans, D.J.A. (2007a). Glacial Landforms, Sediments. Glacial erractics and till dispersal indicators. In: Elias, S.A. (Ed.). *Encyclopedia of Quaternary Science*. Elsevier. Amsterdam, 975–978.

Evans, D.J.A. (2007b). Glacial Landforms. Glaciotectonic structures and landforms. In: Elias, S.A. (Ed.). *Encyclopedia of Quaternary Science*. Elsevier. Amsterdam, 831–838.

Evans, D.J.A. and Rea, B.R. (2003). Surging glacier landsystems. In: Evans, J.A. (Ed.). *Glacial Landsystems*. Arnold. London, 259–288.

Evans, D.M. (1966). Man-made earthquakes in Denver. *Geotimes*, 10(9), 11–18.

Evans, I.S. (1969). The geomorphology and morphometry of glacial and nival areas. In: Chorley, R.J. (Ed.). *Water, Earth and Man*, Methuen, 369–380.

Evans, I.S. (1969–70). Salt crystallization and rock weathering: a review. *Revue de Géomorphologie Dynamique*, 19, 153–177.

Evans, I.S. (2007). Glacial landforms, erosinal features. Major scale forms. In: Elias, S.A. (Ed.). *Encyclopedia of Quaternary Science*. Elsevier. Amsterdam, 838–852.

Evans, I.S. and Cox, N.J. (1995). The form of glacial cirques in the English Lake District, Cambridge. *Zeitschrift für Geomorphologie*, 39, 175–202.

Evans, S.G., Bishop, N.E., Smoll, L.F., Valderrama Murillo, P., Delaney, K.E and Oliver-Smith, A. (2009). A re-examination of the mechanics and human impact of catstrophic mass flows originating on Nevado Huascarán, Cordillera Blanca, Perú in 1962 and 1970. *Engineering Geology*, 108, 96–118.

Evenari, M.L., Shanan, N. and Tadmor, N.H. (1971). *The Negev: the Challenge of a Desert*. Harvard University Press. Cambridge.

Evenson, E.B. (1971). The relationship of macro and microfabric of till and the genesis of glacial landforms in Jefferson County, Wisconsin. In: Goldthwait, R.P. (Ed.). *Till: A Symposium*, Ohio State University Press. Columbus, 345–364.

Everard, C.E. (1963). Contrats in the form and evolution of hill-side slopes in central Cyprus. *Transaction Institute British Geographers*, 32, 31–47.

Eybergen, F.A. and Imeson, A.C. (1989). Geomorphological processes and climatic change. *Catena*, 16, 307–319.

Eyles, N. (1993). Earth's glacial record and its tectonic setting. *Earth-Science Reviews*, 35, 1–248.

Eyles, N. and Lazorek, M. (2007). Glacial Landform, Sediments. Glacigenic Lithofacies. In: Elias, S.A. (Ed.). *Encyclopedia of Quaternary Science*. Elsevier. Ámsterdam, 920–932.

Faci, E., Rodríguez-Avial, J.I. and Jugo, J. (1988). Estabilización y establecimiento de medidas preventivas en un talud rocoso en Azagra (Navarra*). II Simposio sobre Taludes y Laderas Inestables*. Andorra La Vella, 485–496.

Fahey, B.D. (1986). A comparative laboratory study of salt crystallisation and salt hydration as potential weathering agents in deserts. *Geographiska Annaler*, 68 A, 107–111.

Fair, T.J.D. (1947). Slope form and development in the interior of Natal. South Africa. *Transaction Geological Society of South Africa*, 50, 105–120.

Fair, T.J.D. (1948). Hill-slopes and pediments of the semi-arid Karroo. *South Africa Geographical Journal*, 30, 71–79.

Fairbanks, R.G. (1989). A 17,000-year glacio-eustatic sea level record: influence of glacial melting rates on the Younger Dryas event and deep-ocean circulation. *Nature*, 342, 637–642.

Fairbanks, R.G., Mortlock, R.A., Chiu, T.C., Cao, L., Kaplan, A., Guilderson, T.P., Fairbanks, T.W., Bloom, A.L., Grootes, P.M. and Nadeau, M.J. (2005). Radiocarbon calibration curve spanning 0 to 50,000 years and ^{14}C dates in pristine corals. *Quaternary Science Reviews*, 24, 1781–1796.

Fairbridge, R.W. (1961). Eustatic changes in sea level. *Physics and Chemistry of the Earth*, 4, 99–185.

Fairbridge, R.W. (1968a). *The Encyclopedia of Geomorphology*. Dowden, Hutchinson and Ross. Strousburg, 1295 p.

Fairbridge, R.W. (1968b). Terraces, fluvial-environmental controls. In: Fairbridge, R.W. (Ed.). *Encyclopedia of Geomorphology*. Reinhold. New York, 1124–1138.

Fairbridge, R.W. (1968c). Fjord, fiord. In: Fairbridge, R.W. (Ed.). *The Encyclopedia of Geomorphology*, Dowden, Hutchinson and Ross. Stroudsburg, 358–359.

Fairbirdge, R.W. (1981). The concept of neotectonics. An introduction. *Zeitschrift für Geomorphologie Supplementband* 40, 7–12.

Faniran, A. and Jeje, L.K. (1983). *Humid Tropical Geomorphology*. Longman. London, 414 p.

FAO (1988). *Manuel de fixation des dunes*. Danida. Roma, 57 p.

FAO-UNESCO (1971–1979). *Carte des sols du monde*, vols. 1–10. Paris.

Farmer, G. and Wigley, T.M.L. (1985). *Climatic Trends for Tropical Africa. A research report for overseas development administration*. Climatic Research Unit. School of Environmental Sicences. University East Anglia. Norwich, 136 p.

Farr, T.G. and Adams, J.B. (1984). Rock coatings in Hawai. *Bulletin of the Geological Society of America*, 95, 1077–1083.

Farrant, A. (2004). Paleoenvironments: Clastic Cave Sediments. In: Gunn, J. (Ed.) *Encyclopedia of Caves and Karst Science*. Fitzroy Dearborn. New York, 553–555.

Farrell, K.M. (1987). Sedimentology and facies architecture of overbank deposits of the Mississippi River, False River region, Lousiana. In: Erhridge, F.E., Flores, R.M. and Harvey, M.D. (Eds.). *Recent Development in Fluvial Sedimentology*. Society Economic Paleontology Mineralogy, Publ. 19, 111–120.

Farrell, K.M. (2001). Geomorphology, facies architecture, and high resolution, non-marine sequence stratigraphy in avulsion deposits, Cumberland marshes, Saskatchewan. *Sedimentary Geology*, 139, 93–150.

Faulkner, H., Spirey, D. and Alexander, R. (2000). The role of some sites of geochemical processes in the development and stabilisation of three badland sites in Almería, Southern Spain. *Geomorphology*, 35, 87–99.

Favis-Mortlock, D. (2004). Rill. In: Goudie, A.S. (Ed.). *Encyclopedia of Geomorphology*. Routledge. London, 852–853.

Fayzullin, R.M. (1969). Lithological-facies characteristics and the gold contents of a series of conglomerates of the Kamensk deposit. *Chemical Abstracts*, 70, 108202b.

Federici, P.R. and Spagnolo, M. (2004). Morphometric análisis on the size, shape and areal distribution of glacial cirques in the Maritim Alps (Western French-Italian Alps). *Geografiska Annaler*, 86 A, 235–248.

Feigl, K.L., Agnew, D.C., Bock, Y., Dong, D., Donellan, A., Hager, B.H., Herring, T.A., Jackson, D.D., Jordan, T.H., King, R.W., Larsen, S., Larson, K.M., Murray, M.H., Shen, Z. and Webb, F.H. (1993). Space geodetic measurement of crustal deformation in central and southern California, 1984–1992. *Journal of Geophysical Research*, 98, 21677–21712.

Feininger, T. (1971). Chemical weathering and glacial erosion of crytalline rocks and the origin of till. *United States Geological Survey Professional Paper*, 750 C, 65–81.

Fenelon, P. (1974). Karst de type tropical sous climat tempéré. Phénomènes karstiques, II. *Memoires et Documents du CNRS*, 15, 95–103.

Ferguson, R.I. (1973). Sinuosity of supraglacial streams. *Bulletin of the Geological Society of America*, 84, 251–256.

Fernández-Nieto, C., Fernández, R., Gutiérrez Elorza, M. and Arrese, F. (1981). Papel de la karstificación en la génesis de los yacimientos de Sierra Menera (Teruel y Guadalajara). *Boletín Geológico y Minero*, 92, 127–140.

Fernández-Rubio, R. (1976). Hidrogeología aplicada a la minería. *Industria Minera*. Madrid, 170, 15–36.

Fernández-Rubio, R. (1989). Minería y Karst en España. In: Durán, J.J. and López-Martínez, J. (Eds.). *El Karst en España*. Monografía 4, Sociedad Española de Geomorfología. Madrid, 373–380.

Fernández-Rubio, R. and Benavente, J. (1984). Estudio hidroquímico de los aportes de agua a la mina de Reocín (Cantabria). *VII Congreso Internacional de Minería Metálica*. Barcelona, 2, 257–268.

Fernando, H.J.S., Braun, A., Galappatti, R., Ruwanpura, J. and Wirasinghe, S.C. (2008). Tsunamis: manifestation and aftermath. In: Gad-el-Hak (Ed.). *Large-Scale Disasters. Prediction, Control, and Mitigation*. Cambridge University Press. Cambridge, 258–292.

Finkel, H.J. (1959). The barchans of southern Peru. *Journal of Geology,* 67, 614–647.

Finkl, C.J. (1993). Pre-emptive strategies for enhanced sand bypassing and beach replenishment activities in southeast Florida beaches. *Journal of Coastal Research*, 18, 59–89.

Finkl, C.W. and Walker, H.J. (2005). Beach Nourisment. In: Schwartz, M.L. (Ed). *Encyclopedia of Coastal Science*. Springer. Dordrecht, 147–161.

Finlayson, A.G., and Bradwell, T. (2008). Morphological characteristics, formation and glaciological significance of Rogen morraine in northern Scotland. *Geomorphology*, 101, 607–617.

Finlayson, C., Giles, F., Rodríguez-Vidal, J., Fa, D.A., Gutiérrez, J.M., Santiago, A., Finlayson, G., Allue, E., Baena, J., Cáceres, I., Carrión, J.S., Fernández, Y., Gleed-Owen, C.P., Jimenez, F.J., López, P., López, J.A., Riquelme, J.A., Sánchez, A., Giles, F., Brown, K., Fuentoes, N., Valarino, C.A., Villalpando, A., Stringer, C.B., Martínez, F. and Sakamoto, T. (2006). Late survival of Neardenthals at the southernmost extreme of Europe. *Nature*, 443, 850–853.

Fioll, L.L., Fornós, J.J. and Ginés, A. (1996). Effects of biokarstic processes on the development of solutional rillenkarren in limestones rocks. *Earth Surface Processes and Landforms*, 21, 447–452.

Fisher, M.A. (1868). On the disintegration of a chalk cliff. *Geological Magazine*, 3, 354–356.

Fisher, M.A. (1872). On cirques and taluses. *Geological Magazine*, 9, 10–12.

Fisher, R.V. and Schmincke, H.V. (1984). *Pyroclastic rocks.* Springer, Berlín, 472 p.

Fisher, R.V., Heiken, J. and Hulen, J.B. (1997). *Volcanoes. Crucibles of Change*, Princeton, New Yersey, 315 p.

Fisk, H.N. (1944). *Geological investigation of the alluvial valley of the lower Mississippi River*. River Commision. Viksburg, 78 p.

Fisk, H.N. (1947). *Fine-grained alluvial deposits and their effects on Mississippi River activity*. Mississippi River Commision. Vicksburg. Mississippi, 78 p.

Fitzpatrick, E.A. (1963). Deeply weathered rock in northeast Scotland, its ocurrence, age and contribution to the soils. *Journal of Soil Science*, 14, 33–42.

Flageollet, J.C. (1989). *Les Mouvements de Terrain et leur Prevention*. Masson. Paris, 224 p.

Flageollet, J.C. and Weber, D. (1996). Fall. In: Dikau, R. Brunsden, D., Schrott, L. and Ibsen, M.L. (Eds.). *Landslide Recognition. Identification, Movement and Causes*. Wiley. Chichester, 13–28.

Fletcher, J.E. and Carroll, P.H. (1948). Some properties of soils associated with piping in southeastern Arizona. *Soil Science Society of America Proceedings*, 13, 545–547.

Flint, R.F. (1971). *Glacial and Quaternary Geology*. Wiley. New York, 892 p.

Flor, G. (1990). Tipología de dunas eólicas, procesos de erosión-sedimentación y evolución costera de la Provincia de Huelva (Golfo de Cádiz occidental, S de España). *Estudios Geológicos*, 46, 91–109.

Flor, G. (2004). *Geología Marina*. Libr. SERVITEC. Oviedo. 644 p.

Florsheim, J.L., Keller, E.A. and Best, D.W. (1991). Fluvial sediment transport in response to moderate storm flows following Chaparral Wildlife, Ventura County, Southern California. *Geological Society of America Bulletin*, 103, 504–511.

Flyberger, S.F., Ahlbrandt, T.S. and Andrews, S. (1979). Origin, sedimentary features and significance of low-angle eolian "sand sheet" deposits, Great Sand Dunes National Monument and vicinity, Colorado. *Journal of Sedimentary Petrology*, 49, 733–746.

Folland, C.K., Karl, T.R., Christy, J.R. et al., (2001). Observed climate variability and change. In: Houghton, J.T., Ding, Y., Griggs, D.J., Noguer, M., Liden, P.J. van der, Dai, X., Maskell, K. and Johnson, C.A. (Eds.) (2000). *Climate Change 2001: The Scientific Basis*. Cambridge University Press. Cambridge, 99–181.

Fölster, H. (1969). Slope development in SW Nigeria during the late Pleistocene and Holocene. *Göttinger Bodenkundliche Berichte*, 10, 3–56.

Font Tullot, I. (1988). *Historia del Clima en España. Cambios Climáticos y sus Causas*. Instituto Nacional de Meteorología. Madrid, 296 p.

Fontugne, M., Usselman, P., Lavallée, D. and Julien, M. (1999). El Niño variability in the coastal desert of southern Peru during the mid-Holocene. *Quaternary Research*, 52, 171–179.

Fookes, P.G. (1977). First Glossop Lecture: Geology for Engineeres: the geological model, prediction and performance. *Quaterly Journal of Engineering Geology*, 30, 290–342.

Fookes, P.G. (Ed.) (1997). *Tropical Residual Soils*. Geological Society Engineering Group Working Party Revised Report. Geological Society Publishing. London.

Fookes, P.G. (2005). Hot Wetland. En: Fookes, P.G., Lee, E.M. and Milligan, G. (Eds.). *Geomorphology for Engineers*. Whittles Publ., Caithnees, United Kingdom, 473–500.

Fookes, P.G. and Vaughan, P.R. (1986) (Eds.). *A Handbook of Engineering Geomorphology*. Chapman and Hall. New York, 343 p.

Fookes, P.G., Lee, E.M. and Milligan, G. (Eds.) (2005). *Geomorphology for Engineers*. Whittles. Caithness, 851 p.

Foose, R.M. (1953). Ground water behaviour in the Hershey Valley, Pennsylvania. *Geological Society American Bulletin*, 64, 623–645.

Ford, D.C. (1965). The origin of limestone caverns: a model from the central Mendip Hills, England. *National Speleological América Society Bulletin*, 27, 109–132.

Ford, D.C. (1979). A review of alpine karst in the Southern Rocky Mountains of Canada. *National Speleological Society of America*, Bulletin, 41, 53–65.

Ford, D.C. (1983). Effects of glaciations upon karst aquifers in Canada. *Journal of Hydrology*, 61, 149–158.

Ford, D.C. (1997). Principal features of the evaporites in Canada. *Carbonates and Evaporites*, 12, 15–23.

Ford, D.C. (2004). Karst. In: Gunn, J. (Ed.). *Encyclopedia of Caves and Karst Science*. Fitzroy Dearborn. New York, 473–475.

Ford, D.C. and Williams, P. (1989). *Karst Geomorphology and Hydrology*. Chapman and Hall. London, 601 p.

Ford, D.C. and Williams, P. (2007). *Karst Hydrology and Geomorphology*. Wiley. Chichester, 562 p.

Forman, S.L., Nelson. A.R. and McCalpin (1991). Thermoluminiscence dating of fault scarp derived colluvium: Decephering the timing of paleoearthquakes on the Wasatch fault zone, nort-central Utah. *Journal Geophysical Research*, 96, 595–605.

Fornós, J.J. and Ginés, A. (1996). *Karren Landforms*. Universitat de las Illes Balears. Palma de Mallorca, 450 p.

Forti, P. (1996). Speleothems and cave minerals in gypsum caves. *International Journal of Speleology*, 25, 91–104.

Forti, P. and Hill, C.A. (2004). Spelothems: Evaporite. In: Gunn, J. (Ed.). *Encyclopedia of Caves and Karst Science*. Fitzroy Dearborn. New York, 692–695.

Fosberg, F.R., Garnier, B.J. and Kuchler, A.W. (1961). Delimitation of the humid tropics. *Geographical Review*, 51, 333–347.

Foster, G.R. and Meyer, L.D. (1975). Mathematical simulation of upland erosion using fundamental erosion mechanics. In: *Present and prospective technology for predicting sediment yields and sources*, U.S. Agricultural Research Service Report ARS-S-40. MS: U.S. Department of Agriculture, 190–207.

Fouqué, F., Lèvy, M., Bertrand, M., Barrois, Offret, Kilian, Bergeron and Bréon (1889). Etudes relatives au temblement de terre du 25 décember 1884 et à la constitution géologique du sol ébranlé par les secousses. *Memoires de la Academie de Sciences, Institut National de France*, 30, 772 p.

Fournier, F. (1960). *Climat et Erosion: La Relation entre l'Erosion du Sol par l'Eau et les Précipitations Atmosphériques*. Presses Universitaires de France. Paris, 201 p.

Fowler, C.M.R. (2005). *The Solid Earth. An Introduction to Global Geophysics*. Cambridge University Press. Cambridge, 685 p.

Fox, W.T. (2005). Bioerosion. In: Schwartz, M.T. (Ed). *Encyclopedia of Coastal Science*. Springer. Dordrecht, 191–192.

Francesco, F. de, Colado, V., Schnack, E., Schnack, J. and García-Lozano, L. (2002). La inundación de 1982/1983 en la porción argentina de la cuenca del Río Paraná. In: Hubp, J.L. and Inbar., M. (Eds.). *Desastres Naturales en América Latina*. Fondo de Cultura Económica. México, 317–333.

Francis, C. (1994). Plants on desert hillslopes. In: Abrahams, A.D. and Parsons, A.J. (Eds.). *Geomorphology of Desert Enviroments*. Chapman and Hall. London, 243–254.

Francis, P. (1993). *Volcanoes. A. Planetary Perspective.* Oxford University Press, New York, 443 p.

Francis, P.W., Gardeweg, M., O'Callaghan, L.J., Ramírez, P.F. and Rothery, D.A. (1985). Catastrophic debris avalanche deposit of Socampa volcano, north Chile. *Geology*, 13, 600–603.

Franckel, H. (1988). From continental drift to plate tectonics. *Nature*, 335-127-130.

Francou, B. (1988). Eboulis stratifiés dans les Hautes Andes Centrales du Pérou. *Zeitschrift für Geomorphologie,* 32, 47–76.

Francou, B. (1990). Stratification mechanisms in slope deposits in high subequatorial mountains. *Permafrost and Periglacial Processes*, 1, 249–263.

Frank, A. and Kocurek, G. (1996). Airflow up the stoss slope of sand dunes: limitations of current understanding. *Geomorphology*, 17, 47–54.

Franke, M.W. (1965). The theory behind stalagmite shapes. *Studia Speleologica*, 1, 89–95.

Freedman, B. (1999). Alaska Highway. In: Cunningham, W.P., Cooper, T.H., Gordham, E. and Hepworth, M.T. (Eds.). *Environmental Encyclopedia*. Jaico. Mumbai, 29–30.

French, B.M. (1998). *Traces of Catastrophe: A Handbook of Shock-Metamorphic Effects in Terrestrial Meteorite Impact Structures* (LP5 Contribution n° 954). Houston, Texas: Lunar and Planetary Institute.

French, H.M. (1971). Slope asymmetry of the Beaufort Plain, northwest Banks Island, N.W.T. Canada. *Canadian Journal of Earth Sciences*, 8, 717–731.

French, H.M. (1987). Permafrost and ground ice. In: Gregory, K.J. and Walling. D.E. (Eds.). *Human activity and environmental processes.* Wiley. Chichester, 237–269.

French, H.M. (1996). *The Periglacial Environment*. Longman. Edimburgh, 341 p.

French, H.M. (1998). An appraisal of cryostratigraphy in northwest arctic Canada. *Permafrost and Periglacial Processes*, 9, 297–312.

French, H.M. (2007). *The Periglacial Environment*. Wiley. Chichester, 458 p.

French, H.M. and Egginton, P. (1973). Thermokarst development, Banks Island, Western Canadian Arctic. In: *Permafrost: North American Contribution, Second International Permafrost Conference, Yakutsk,* USSR. National Academy of Science, Publication 2115, 203–212. Washington, DC.

French, H.M. and Guglielmin, M. (2002). Observations con granite weathering phenomena, Mount Keinath, Northern Victoria Land, Antarctica. *Permafrost and Periglacial Processes*, 13, 231–236.

French, H.M. and Karte, J. (1988). A periglacial overview. In: Clark, M.J. (Ed.). *Advances in Periglacial Geomorphology*. Wiley. Chichester, 463–473.

French, H.M., Harris, S.A. and Van Everdingen, R.O. (1983). The Klondike and Dawson. In: French, H.M. and Heginbottom, J.A. (Eds.). *Guidebook to Permafrost and Related Features of the Northern Yukon Territoy and Mackenzie Delta, Canada*. 4th International Conference of Permafrost. Division of Geological and Geophysical Surveys, State of Alaska, 35–63. Fairbanks.

Frenzel, G. (1989). The tafoni weathering of granites. In: *Weathering; its Products and Deposits. Vol. 1: Processes*. Theophrastus. Atenas, 421–435.

Freundt, A., Strauch, W., Kutterolf, S. and Schmincke, H.U. (2007). Volcanogenic tsunamis in lakes: Examples from Nicaragua and general implications. *Pure and Applied Geophysics*, 164, 527–545.

Friedman, J.D., Johansson, C.E., Oskarsson, N., Svensson, H., Thorarinsson, S. and Williams, R.S. (1971). Observations on icelandic polygon and palsa areas. Photointerpretation and field studies. *Geografiska Annaler*, 53, 115–145.

Friend, S. (2002). *Sinkholes*. Pineapple Press. Sarasota, Florida, 96 p.

Frihy, O.E., Debes, E.A. and El Sabed, W.R. (2003). Processes reshaping the Nile delta promontories of Egypt: pre and post protection. *Geomorphology*, 53, 263–279.

Froese, D.G., Lowe, D.L., Knott, J.R. and Slate, J.R. (2008). Global tephra studies: John Westgate and Andrei Savna-Wojcick commemorative volume. *Quaternary International*, 178, 1–3.

Frostick, L.E. and Reid, I. (Eds.) (1987). *Desert Sediments, Ancient and Modern, Special Publication*, 35. Geological Society of London. Blackwell. Oxford, 401 p.

Frostick, L.E. and Reid, I. (1989). Is structure the main control of river drainage and sedimentation in rifts? *Journal Africa Earth Science*, 8, 165–182.

Frostick, L.E., Reid, I. and Layman, J.T. (1983). Changing size distribution of suspended sediment in arid-zone flash floods. *Special Publication International Association of Sedimentologists*, 6, 97–106.

Frumkin, A. (1994). Hydrology and denudation rates of halite karst. *Journal of Hydrology*, 162, 171–189.

Frumkin, A. (1995). Morphology and development of salt caves. *Bulletin of the National Speleological Society*, 56, 82–95.

Frumkin, A. (1996). Determining the exposure age of a karst landscape. *Quaternary Research*, 46, 99–106.

Frumkin, A. (2000a). Dissolution of salt. In: Klimchouk, A.B., Ford, D.C., Palmer, A.N. and Dreybrodt, W. (Eds.). *Speleogenesis. Evolution of Karst Aquifers*. National Speleological Society. Huntsville, 169–170.

Frumkin, A. (2000b). Speleogenesis in salt. The Mount Sedom Area, Israel. In: Klimchouk, A.B., Ford, D.C., Palmer, A.N. and Dreybrodt, W. (Eds.). *Speleogenesis. Evolution of Karst Aquifers*. National Speleological Society. Huntsville, 443–451.

Frumkin, A. (2004). Sedom Salt Karst, Israel. In: Gunn, J. (Ed.). *Encyclopedia of Caves and Karst Science*. Fitzroy Dearborn. New York, 637–639.

Frumkin, A. and Ford, D.C. (1995). Rapid entrenchment of stream profiles in the salt caves of Mount Sedom, Israel. *Earth Surface Processes and Landforms*, 20, 139–152.

Frumkin, A. and Forti, P. (1997). Liquid Crystal Cave. In: Hill, C.A. and Forti, P. (Eds.). *Cave Minerals of the World*. National Speleological Society. Huntsville, Alabama.

Frumkin, A., Magaritz, M., Carmi, I. and Zak, I. (1991). Holocene climatic record of the salt caves of Mount Sedom, Israel. *The Holocene*, 1, 191–200.

Frumkin, A., Schwarcz, P. and Ford, D.C. (1992). Evidence for isotopic equilibrium in stalagmites from caves in a dry region: Jerusalem, Israel. *Israel Journal of Earth Sciences*, 43, 221–230.

Fryberger, S.G. and Ahlbrandt, T.S. (1979). Mechanism for the formation of aeolian sand seas. *Zeitschrift für Geomorphologie*, 23, 440–460.

Fryberger, S.G. and Goudie, A.S. (1981). Arid geomorphology. *Progress of Physical Geography*, 5, 420–428.

Frye, J.C. and Willman, H.B. (1973). Wisconsinian climatic history interpreted from Lake Michigan lobe deposits and soils. *Geological Society of America*, Memoir 136, 135–152.

Furdada, G. (1996). *Estudi de les Allaus al Pirineu Occidental de Catalunya: Predicció Espacial i Aplications de la Cartografía*. Geoforma. Logroño, 315 p.

Furrer, G. (1968). Unterswchungen an Strukturböden in Ostspitbergen, ihre Bedentung für die Erforchung rezenter und fossiler Frostmusterformen in den Alpen baw im Alpen vorland. *Polarforchung*, 6, 202–206.

Furrer, G. (1972). Bewegungsmessungen auf solifluktionsdecken. *Zeitschrift für Geomorphologie Supplemenband*, 13, 87–101.

Fuster, J.M. (1992). *Vulcanismo, riesgos cercanos y remotos*. Discurso inaugural del Año Académico 1992–1993. Real Academia de Ciencias Exactas, Físicas y Naturales. Madrid, 23 p.

Fuster, J.M, Hernan, H., Cendrero, A., Coello, J., Cantagrel, J.H., Ancochea, J.M. and Ibarrola, E. (1993). Geocronología de la isla del Hierro (Islas Canarias). *Boletín Real Sociedad Española de Historia Natural (Geol.)*, 88, 86–97.

Gaál, L. and Bello, P. (2008). Granites and granite caves in western Carpathians. *Cadernos Laboratorio Xeoloxico de Laxe*, 33, 11–18.

Gaillardet, J., Dupré, B., Louvat, P. and Allègre, C. (1999). Global silicate weathering of silicates estimated from large river geochemistry. *Chemical Geology*, 159, 3–30.

Gaillardet, J. (2004). Denudation. In: Goudie, A.S. (Ed.). *Encyclopedia of Geomorphology*. Routledge. London, 240–244.

Galit, K., Yehouda, E. and Yehuda, E. (195). Lake-level variations and active tectonism in the holocene fan-delta of Nahal Darga, Dead Sea region, Israel. *Intenational Conference on Geomorphic Response of Mediterranean and Arid Areas to Climate Change. Post- Conference Negev Fieldtrip*, 56–67.

Gallart, F. and Llorens, P. (1996). Los efectos hidrológicos de la recuperación del bosque en áreas de montaña. In: Lasanta, T. and García-Ruiz, J.M. (Eds.). *Erosión y Recuperación de Tierras en Áreas Marginales*. Sociedad Española de Geomorfología-Logroño, 73–78.

Gallart, F., Solé, A., Puigdefábregas, J. and Lázaro, R. (2002). Badland systems in the Mediterranean. In: Bull, L.J. and Kirkby, M.J. (Eds.). *Dryland Rivers: Hydrology and Geomorphology and Semiarid Channels*. Wiley. Chichester, 299–326.

Galloway, W.E. (1975). Process framework for describing the morphologic and stratigraphic evolution of deltaic depositional system. In: Broussard, M.L. (Ed.). *Deltas: Models for Exploration*. Houston Geological Society. Houston, 87–98.

Galon, R. (1954). Les principaux paysages morphologiques du monde au pont de vue des profiles synthétiques qui les caractérisent. *Czasopismo Geogr.*, 25, 26–37.

Galve, J.P. (2009). *Desarrollo y evolución de técnicas para el estudio de la peligrosidad de dolinas*. Tesis doctoral. Facultad de Ciencias. Universidad de Zaragoza, 189 p. (Unpublished).

Galve, J.P., Gutiérrez, F.. Lucha, P., Guerrero, J., Bonachea, J., Redondo, J. and Cendrero, A. (2009a). Probabilistic sinkhole modelling for hazard assessment. *Earth Surface Processes and Landforms*, 34, 437–452.

Galve, J.P., Gutiérrez, F., Redondo, J., Bonachea, J.,. Lucha, P. and Cendrero, A. (2009b). Evaluating and comparing methods of sinkhole susceptibility mapping in the Ebro Valley evaporite karst (NE Spain). *Geomorphology*, 111, 160–172.

Galve, J.P., Gutiérrez, F., Lucha, P., Bonachea, J., Cendrero, A., Gutiérrez, M., Gimeno, M.J., Pardo, G. and Sánchez, J.A. (2009c). Sinkhole and salt-weathering evaporite karst of the Ebro river valley upstream of Zaragoza city (NE Spain). Geomorphological mapping and analysis as a basis for risk management. *Geomorphology*, 108, 145–158.

Galvin, C.J. (1968). Breaker type classification on three laboratory beaches. *Journal of Geophysical Research*, 73, 3651–3659.

Gams, I. (1978). The polje: the problem of definition. *Zeitscrift für Geomorphologie*, 22, 170–181.

Gams, I. (1992). Les influences des climates dans les régions karstiques. In: *Karst et évolutions climatiques*. Presses Universitaries de Bordeaux, 520 p.

Gams, I. (2005). Tectonic impact on poljes and minor basins (case studies of Dinaric Karst). *Acta Carsologica*, 34, 25–29.

Gams, I., Nicod, J., Julia, M., Anthony, E. and Sauro, U. (1993). Environmental change and human impacts on the mediterranean karst of France, Italy and Dinaric region. In: Williams, P.W. (Ed.). *Karst terrains: Environmental Changes and Human Impact*. Catena. Supplement 25. Cremlingen, 59–98.

Garay, P. (1986). Informe geológico sobre la sima de hundimiento de Pedreguer (Provincia de Alicante). *Jornadas sobre el Karst en Euskadi*. San Sebastián, 323–333.

García de Celis, A. (2002). Formas periglaciares relictas en la Sierra del Suspirón (Cordillera Cantábrica, León): Campos de bloques. In: Serrano, E. and García, A. (Eds.). *Periglaciarismo en Montaña y Altas Latitudes*. Departamento de Geografía. Universidad de Valladolid, 37–52.

García-Ruiz, J.M. and López Bermúdez, F. (2009). *La Erosión del Suelo en España*. Sociedad Española de Geomorfología. Zaragoza, 441 p.

García Ruiz, J.M., White, S., Martí, C., Valero, B., Errea, M.P. and Gómez, A. (1996). *La Catástrofe del Barranco de Arás (Biescas, Pirineo Aragonés)*. Consejo Superior de Investigaciones Científicas. Zaragoza, 54 p.

García-Ruiz, J.M., White, S.M., Lasanta, T., Martí, C., González, C., Errea, M.P., Valero, B. and Ortigosa, L. (1997). Assessing the effect of land-use changes of sediment yield and channel dynamics in the Central Spanish Pyrenees. *Human Impact on Erosion and Sedimentation*. International Association Hydrological Sciences, 245, 151–158.

García Ruiz, J.M., Martí Bono, C., Valero Garcés, B., González Sampériz, P., Lorente, A., Beguería, S. y Edwards, L. (2000). Derrubios de ladera en el Pirineo central español: Significación cronológica y paleoclimática. In: Peña, J.L., Sánchez-Fabre, M. and Lozano, M.V. (Eds.). *Procesos y Formas Periglaciares en la Montaña Mediterránea*. Instituto de Estudios Turolenses. Teruel, 63–79.

García-Ruiz, J.M., Valero-Garcés, B., González-Sampériz, P., Lorente, A., Martí-Bono, C., Beguería, S. and Edwards, L. (2001). Stratified settle in the Central Spanish Pyrenees: paleoenvironmental implications. *Permafrost and Periglacial Processes*, 12, 233–242.

García-Ruiz, J.M., Valero-Garcés, B.L., Martí-Bono, C. and González-Sampériz, P. (2003). Asynchroneity of maximum glaciar advances in the central Spanish Pyrenees. *Journal of Quaternary Science*, 18, 61–72.

García Salmerón, J. (1967). *Erosión Eólica*. Ministerio de Agricultura. Madrid, 579 p.

Gardiner, V. and Dackombre, R. (1983). *Geomorphological Field Manual*. George Allen and Unwin. London, 254 p.

Gardner, J.S. (1987). Evidence for headwall weathering zones, Boundary Glacier, Canadian Rocky Mountains. *Journal of Glaciology*, 33, 60–67.

Gardner, L.R. (1972). Origin of the Mormon Mesa caliche, Clark County, Nevada. *Bulletin of the Geological Society of America*, 83, 143–156.

Gardner, R. (1983). Introduction. In: Gardner, R. and Scoging, H. (Eds.). *Mega-Geomorphology*. Clarendon Press. Oxford, X–XIII.

Gares, P.A. (1992). Topographic changes associated with coastal dune blowouts at island beach State Park, N.J. *Earth Surface Processes and Landforms*, 17, 589–604.

Gares, P.A. and Nordstrom, K.F. (1987). Dynamics of coastal foredune blowout at Island Beach State Park, N.J. In: Graus, N.E. (Ed.). *Coastal Sediments*, 87, ASCE, 213–221.

Garner, H.F. (1966). Derangement of Rio Caroni, Venezuela. *Revue de Géomorphologie Dynamique*, 16, 54–83.

Garret, R.D. and Costa, J.E. (1982). Hydrology, geomorphology and dam-break modelling of the July 15, 1982, Lawn lake and Cascade Lake Dam failures, Larimer County, Colorado. *United States Geological Survey*, Open File, Report, 84–162.

Gascogne, M. (1981). Rates of cave passage entrechment and valley lowering determined from speleothem age measurements. In: Beck, B. (Ed.). *Proceedings of the 8th International Speleological Congress*. Bowling Green, Kentucky, 99–100.

Gasse, F. (2000). Hydrological changes in the Africa tropics since the last glacial maximum. *Quaternary Science Reviews*, 19, 189–211.

Gautier, E. and Costard, F. (2000). Les systèmes fluviaux à chenaux anastomosés en milleu periglaciar: la Léna et ses principaux affluents en Sibérie Centrale. *Geographie Physique et Quaternaire*, 54, 327–342.

Gaya, M. (1996). Caps de Fibló (trombas o tornados). Algunas observaciones recientes. En: *III Simposio Nacional de Predicción del Instituto Nacional de Meteorología*. Ministerio de Medio Ambiente. Madrid, 19–25.

Gaya, M. (1999). Fenómenos severos en las Islas Baleares. Parte I: frentes de racha. Parte II: tornados. *IV Simposio Nacional de Predicción del Instituto Nacional de Meteorología (Memorial Alfonso Ascaso)*. Ministerio de Medio Ambiente. Madrid, 225–233.

Gaylord, D.R. and Dawson, P.J. (1987). Airflow-terrain interactions through a mountain gap, with an example of eolian activity beneath an atmospheric hydraulic jump. *Geology*, 15, 789–792.

Gee, M.D. (1991). Classification on landslide hazard zonation methods and a test of the predictive capability. En: *Landslides-Glissement de Terrain*. VI International Symposium on Landslides. Balkema. Rotterdam, vol. 2, 947–952.

Geikie, A. (1905). *Founders of Geology*. Macmillan. London, 486 p.

Gellert, J.F. (1970). Climatomorphology and palaeoclimates of the central European Tertiary. In: Pecsi, M. (Ed.). *Problems of Relief Planation*. Akadémiai Kaidó. Budapest, 107–112.

Geological Society (1990). Engineering Group Working Party Report: Tropical residual soils. *Quarterly Journal of Engineering Geology*, 23, 1–101.

Gerits, J., Imeson, A.C., Verstraten, J.M. and Bryan, R.B. (1987). Rill development and badland regolith properties. *Catena Supplement*, 8, 141–160.

Gerrard, A.J. (1988). *Rocks and Landforms*. Unwin Hyman. London, 319 p.

Gerrard, A.J. (1990). *Mountain Environments: An examination of the Physical Geography of the Mountains*. Belhaven Press. London, 317 p.

Gerson, R. (1977). Sediment transport for desert watershed in erodible materials. *Earth Surface Processes*, 2, 343–361.

Gerson, R. (1982). Talus relicts in deserts: a key to major climatic fluctuations. *Israel Journal Earth Science*, 31, 123–132.

Gerson, R. and Grossman, S. (1987). Geomorphic activity on escarpments and associated fluvial systems in hot deserts. In: Rampino, M.R., Sanders. J.E., Newman, W.S. and Königsson, L.K. (Eds.). *Climate, History, Periodicity, Predictability*. Van Nostrand Reinhold. New York, 300–322.

Gerstenhauer, A. (1960). Der tropische kelgelkarst in Tabasco (Mexico). *Zeitschrift für Geomorphologie Supplementband*, 2, 22–48.

Gerstenhauer, A. (1968). Ein Karstmorphologischer vergleich zwischen Florida and Yucatan. *Deuscher Geographische Bad Godesberg Wiss. Abh.*, 332–344.

Gèze, B. (1965). *La Spéléologie Scientifique*. Editions de Seuil. Paris, 190 p.

Ghinoi, A. and Chung Chang, J. (2005). STARTER: A statistical GIS-based model for the prediction of snow avalanche susceptibility using terrain features-application to Alta Val Badia, Italian Dolimites. *Geomorphology*, 66, 305–325.

Ghose, B., Kar, A. and Hussain, Z. (1979). The lost courses of the Sarawasti river in the Great Indian Desert—new evidence from Landsat imagery. *Geographical Journal*, 145, 446–451.

Giannini, E. (1951). Alloctonia del macigno nella regione Bagni di Casciana-Chianni in provincia di Pisa (Allochthonous position of the "macigno"-type sandstone in the region of Bagni di Casciana-Chianni in the Pisa province). *Bolletin Societá Geologica Italia*, 70.

Giardino, J.R. (1983). Movement of ice-cemented rock glaciers by hydrostatic pressure: an example from Mt. Mestas, Colorado. *Zeitschrift für Geomorphologie*, 27, 297–310.

Giardino, J.R. and Vitek, S.G. (1987). Geologic engineering aspects of rock glaciers. In: Giardino, J.R., Schroeder,

J.F.Jr. and Vitek, J.D. (Eds.). *Rock Glaciers*. Allen and Unwin. London, 265–287.

Gibbs, S.J. (1967). The geochemistry of Amazon River System. Part I. The factors that control the salinity and composition of the suspended solids. *Bulletin Geological Society of America*, 78, 203–232.

Gilbert, G.K. (1875). Report on the geology of portions of Nevada, Utah, California and Arizona, 1871–1872. In: *Report upon Geographical and Geological Explorations and Surveys West of the One Hundredth Meridian*, 3, 21–187.

Gilbert, G.K. (1877). *Report on the Geology of the Henry Mountains*. United States Geographical and Geological Survey of the Rocky Mountain Region. Washington, DC: U.S. Department of the Interior.

Gilbert, G.K. (1890). Lake Bonneville. *United States Geological Survey, Monograph 1*.

Gilbert, G.K. (1904). Domes and dome structure of the High Sierra. *Bulletin of the Geological Society of America*, 15, 29–36.

Gilbert, G.K. (1906). Crescentic gouges on glaciated surfaces. *Bulletin of the Geological Society of America*, 17, 303–313.

Gilbert, G.K. (1909). The convexity of hilltops. *Journal of Geology*, 17, 344–350.

Gilbert, G.K. (1914). The Transportation of Debris by Running Water. *United States Geological Survey, Professional Paper* nº 86B.

Gile, L.H. (1975). Holocene soils and soil-geomorphic relations in an arid region of southern New Mexico. *Quaternary Research*, 5, 321–360.

Gile, L.H. and Hawley, J.W. (1966). Periodic sedimentation and soil formation on an alluvial-fan piedmont in southern New Mexico. *Soil Science Society of America Proceedings*, 30, 261–268.

Gile, L.H., Peterson, F.F. and Grossman, R.B. (1965). The K horizon: A master soil horizon of carbonate accumulation. *Soil Science*, 99, 74–82.

Gile, L.H., Peterson, F.F. and Grossman, R.B. (1966). Morphological and genetic sequences of carbonate accumulation in desert soils. *Soil Science,* 101, 347–360.

Giles, J. (2005). The dustiest place on the Earth. *Science*, 434, 816–819.

Gillete, D.A. and Hanson, K.J. (1989). Spatial and temporal variability of dust production caused by wind erosion in the United States. *Journal of Geophysical Research*, 94 (D2), 2197–2206.

Gilley, J.E., Kottwitz, E.R. and Simanton, J.R. (1990). Hydraulic characteristics of rills. *Transactions of the American Society of Agricultural Engineers*, 33, 1900–1906.

Gillies, J.A., Nickling, W.G. and Tilson, M. (2009). Ventifacts and wind-abraded rock features in the Taylor Valley, Antarctica. *Geomorphology*, 107, 149–168.

Gillieson, D. (1996). *Caves. Processes, Development, Management*. Blackwell. Oxford, 324 p.

Ginés, A. (2004). Karrens. In: Gunn, J. (Ed.). *Encyclopedia of Caves and Karst Science*. Fitzroy Dearborn. New York, 470–473.

Giosan, L. and Goodbred, S.L. (2007). Deltaic environments. In: Elias, S.A. (Ed.). *Encyclopedia of Quaternary Science*. Elsevier. Amsterdam, 704–716.

Gjessing, J. (1965). On "plastic scouring" and "subglacial erosion". *Norsk Geografisk Tidskrift*, 20, 1–37.

Glade, T. and Crozier, M.J. (2005). The nature of landslide hazard impact. In: Glade, T., Anderson, M. and Crozier, M.J. (Eds.). *Landslides Hazard and Risk*. Wiley. Chichester, 43–74.

Glantz, M.H. (2005). Desertification. In: Oliver, J.E. (Ed.). *Encyclopedia of World Climatology*. Springer. Berlin, 318–324.

Glanz, N.H. (2001). *Currents of Change: El Niño's Impact on Climate and Society*. Cambridge University Press. Cambridge.

Glanz, N.H. (2004). El Niño. In: Oliver, J.E. (Ed.). *Encyclopedia of World Climatology*. Springer. Berlin, 349–351.

Glass, B. (1982). *Introduction to Planetary Geology*. Cambridge University Press, Cambridge.

Glenn, M. (1979). Glossary. In: McKee, E.D. (Ed.). *A Study of Global Sand Seas*. United States Geological Survey Professional Paper, 1052, 399–407.

Glennie, K.W. (1970). *Desert Sedimentary Environments*. Elsevier. Amsterdam, 222 p.

Glennie, K.W. (1972). Permian Rotliegendes of northwest Europe interpreted in the light of modern desert sedimentation studies. *Bulletin of the American Association of Petroleum Geologists*, 56, 1048–1071.

Glennie, K.W., Pugh, J.M. and Goodall, T.M. (1994). Late Quaternary Arabian desert models of Permian Rotliegend reservoirs. *Shell Exploration Bulletin*, 274, 1–19.

Godard, A. (1977). *Pays et Paysages du Granite*. Presses Universitaires de France. Paris, 232 p.

Godt, J.W. and Coe, J.A. (2007). Alpine debris flow triggered by a 28 July 1999 thunderstorm in the central Front Range, Colorado. *Geomorphology*, 84, 80–97.

Goguel, J. (1963). A hypotheses on the origin of the cryptovolcanic structures of the central Platform of North America. *American Journal of Science*, 251, 665–667.

Goguel, J. (1965). *Traité de Tectonique*. Masson, París, 457 p.

Gohain, K. and Parkarsh, B. (1990). Morphology of the Kosi megafan. In: Rachocki, A.H. and Church, M. (Eds.). *Alluvial Fans: A Field Approach*. Wiley. Chichester, 151–178.

Goldich, S.S. (1938). A study of rock weathering. *Journal of Geology*, 46, 17–58.

Goldsmith, V. (1985). Coastal dunes. In: Davis, R.A. (Ed.). *Coastal Sedimentary Environments*. Springer-Verlag. New York, 171–236.

Goldthwait, R.P. (1951). Development of end moraines in east-central Baffin Island. *Journal of Geology*, 59, 567–577.

Gole, C.V. and Chitale, S.V. (1966). Inland delta building activity of the Kosi River. *American Society of Civil Engineering Journal Hydraulic Division* HY2, 111–126.

Gombert, P. (1995). Approche théorique simplifiée de la dissolution karstique. *Karstología*, 24, 41–51.

Gómez Ortiz, A. (1996). *El Relleu d'Andorra. Morfologia glacial i periglacial*. Govern d'Andorra. Ministeri d'Educació, Joventut i Esports, 129 p. Andorra.

Gómez Ortiz, A. and Salvador Franch, F. (1997). Procesos geomórficos periglaciares en el Pandero del Mulhacén (Sierra Nevada). Experimentación de campo sobre la efectividad de la geliturbación y del hielo en el suelo. *Cuaternario y Geomorfología*, 11, 81–97.

Gómez Ortiz, A. and Salvador Franch, F. (1998). Procesos periglaciaress actuales en montaña mediterranea. Ideas clave, trabajos de campo y resultados en Sierra Nevada. In: Gómez Ortiz, A., Salvador Franch, F., Schulte, L. and García Navarro, A. (Eds.). *Procesos biofísicos actuales en medios frios*. Publicacions de la Universitat de Barcelona, Barcelona. 217–234.

Gómez Villar, A. (1996). Abanicos aluviales: aportación teórica a sus aspectos más significativos. *Cuaternario y Geomorfología*, 10, 77–124.

Gómez, A. and Zamorano, J.J. (2002). El Popocatépetl ("cerro que humea"), pasado y presente. In: Hubp, J.L. and Inbar, M. (Eds.). *Desastres naturales en América Latina*. Fondo de Cultura Económica. México, 103–146.

Gómez-Heras, M., Smith, B.J. and Fort, R. (2006). Surface temperatures differences between minerals and crystalline rocks: Implications for granular desaggregation of granites through thermal fatigue. *Geomorphology*, 78, 236–247.

González, Díez, A. (1995). *Cartografía de movimientos de ladera y su aplicación al análisis del desarrollo temporal de los mismos y de la evolución del paisaje*. Tesis Doctoral. Universidad de Cantabria, 2 vols., 415 p. Unpublished.

González Díez, A., Salas, L., Díaz de Terán, J.R. and Cendrero, A. (1996). Late Quaternary climatic changes and mass movement frequency and magnitude in the Cantabrian region, Spain. *Geomorphology*, 15, 291–309.

González Díez, A., Remondo, J., Díaz de Terán, J.R. and Cendrero, A. (1999). A methodological approach for the analysis of the temporal occurence and triggering factors of landslides. *Geomorphology*, 30, 95–113.

González, E. (1970). *Informe sobre los hundimientos existentes en los perfiles P-112 y P-116 (margen derecha de la CN232). Desdoblamiento Zaragoza-Alagón*. Jefatura Regional de Carreteras (Inédito).

González, F.I. (1999). Tsunami. *Scientific American*, 280, 44–55.

González, M.A. (2001). Recent formation of arroyos in the Little Missouri Badlands of southwestern Dakota. *Geomorphology*, 38, 63–84.

González de Vallejo, L.I., Ferrer, M., Ortuño, L. and Oteo, C. (2002). *Ingeniería Geológica*. Pearson Prentice Hall. Madrid, 715 p.

Goodess, C.M., Palutikof, J.P. and Davies, T.D. (1992). *The Nature and Causes of Climatic Change. Assesing the Long Term Future*. Belhaven Press. London, 248 p.

Goodman, R.E. and Bray. J.W. (1976). Toppling of rock slopes. In: *Proceedings, Special Conference on Rock Engineering for Foundations and Slopes*. American Society of Civil Engineers. Boulder, Colorado, vol. 2, 201–234.

Goodsell, B., Hambrey, M.J. and Glasser, N.F. (2002). Formation of band ogives and associated structures at Bas Glacier d'Arolla, Valais, Switzerland. *Journal of Glaciology*, 48, 207–300.

Goosens, D. and Offer, Z.I. (1990). A wind-tunnel simulation and field verification of desert dust deposition. *Sedimentology*, 37, 7–22.

Gordon, J.E. (1981). Ice-scoured topography and its relation ships to bedrock structure and ice movement in parts of northern Scotland and west Greenland. *Geografiska Annaler*, 63 A, 55–65.

Gordon, J.E., Whalley, W.B., Gellatly, A.F. and Vere, D.M. (1992). The formation of glacial flutes: assesment of models with evidence from Lyngsdalen, north Norway. *Quaternary Science Reviews*, 11, 709–731.

Gordon, S.J. and Dorn, R.I. (2005). In situ weathering rind erosion. *Geomorphology*, 67, 97–113.

Gornitz, V. (2005). Natural hazards. In: Schwartz, M.L. (Ed.). *Encyclopedia of Coastal Science*. Springer. Dordrecht, 678–684.

Gospodaric, R. and Habic, P. (1979). Karst phenomena of Cerknisko polje. *Acta Carsologica*, 8, 150–162.

Gossmann, H. (1976). L'importance des processus se déroulant à la ligne de partage locale des eaux por l'évolution des versants sous la dominance du ruissellement pluvial (á l'aide de formules mathématiques élémentaires). *Actes du Symposium sur les versants en pays méditerranées, Aix en Provence, 28–30 Avril 1975*. Aix en Provence: Centre d'Etudes Géographiques et de Recherches Méditerranéennes vol. V, 139–143.

Goudie, A.S. (1972a). On the definition of calcrete deposits. *Zeitschrift für Geomorphologie*, 16, 464–468.

Goudie, A.S. (1972b). The chemistry of world calcrete deposits. *Journal Geology*, 80, 449–463.

Goudie, A.S. (1973). *Duricrusts in Tropical and Sub-Tropical Landscapes.* Clarendon Press. Oxford, 174 p.

Goudie, A.S. (1974). Further experimental investigation of rock weathering by salt and other mechanical processes. *Zeitscrift für Geomorphologie*, 21, 1–12.

Goudie, A.S. (1977a). *Environmental Change*. Clarendon Press. Oxford, 244 p.

Goudie, A.S. (1977b). Sodium sulphate weathering and the desintegration of Mohenjo-Daro, Pakistan. *Earth Surface Processes*, 2, 75–86.

Goudie, A.S. (1978). Dust storms and their geomorphological implications. *Journal of Arid Environments*, 1, 291–310.

Goudie, A.S. (1981a). *Geomorphological Techniques*. George Allen and Unwin. London, 395 p.

Goudie, A.S. (1981b). *The Human Impact. Man's Role in Environmental Change*. Blackwell. Oxford, 326 p.

Goudie, A.S. (1983a). The arid earth. In: Gardner, R. and Scoging, H. (Eds.). *Megageomorphology*. Clarendon Press. Oxford, 152–171.

Goudie, A.S. (1983b). Calcrete. In: Goudie, A.S. and Pye, K. (Eds) *Chemical sediments and geomorphology: precipitates and residua in the near surface environment*. Academic Press. New York, 93–131.

Goudie, A.S. (1984). *The Nature of the Environment*. Blackwell. Oxford, 397 p.

Goudie, A.S. (1988). The geomorphological role of termites and earthworms in the tropics. In: Viles, H. (Ed.). *Biogeomorphology*. Blackwell. Oxford, 365 p.

Goudie, A.S. (1989a). Weathering processes. In: Thomas, D.S.G. (Ed.). *Arid Zone Geomorphology*. Belhaven Press. London, 11–24.

Goudie, A.S. (1989b). Wind erosion in deserts. *Proceedings of the Geologists Association*, 100, 83–92.

Goudie, A.S. (1990a). Desert degradation. In: Goudie, A. (Ed.). *Techniques for Desert Reclamation*. Wiley. Chichester, 1–33.

Goudie, A.S. (1990b). The global geomorphological future. *Zeitschrift für Geomorphologie Supplementband*, 79, 51–62.

Goudie, A.S. (1991). Pans. *Progress in Physical Geography*, 15, 221–237.

Goudie, A.S. (1992). *Environmental Change*. Clarendon Press. Oxford, 329 p.

Goudie, A.S. (1994a). Salt attack on buildings and other structures in arid lands. ín: Fookes, P.G. and Parry, R.H.G. (Eds.). *Engineering Characteristics of Arid Soils*. Balkema. Rotterdam, 15–28.

Goudie, A.S. (1994b). Deserts in a warmer world. In: Millington, A.C. and Pye, K. (Eds.). *Environmental Change in Drylands: Biogeographical and Biogeomorphological Perspectives*. Wiley. Chichester, 1–24.

Goudie, A.S. (1995). *The Changing Earth. Rates of Geomorphological Processes*. Blackwell. Oxford, 302 p.

Goudie, A.S. (1998). The salt weathering hazards in deserts. In: Kalvoda, J. and Rosenfeld, C.L. (Eds.). *Geomorphological Hazards in High Mountain Areas*. Kluwer. Dordrecht, 107–120.

Goudie, A.S. (2002). *Great Warm Deserts of the World. Landscape and Evolution*. Oxford University Press. Oxford, 444 p.

Goudie, A.S. (Ed.) (2004a). *Encyclopedia of Geomorphology*. Routledge. London, 1156 p.

Goudie, A.S. (2004b). Fold. In: Goudie, A. (Ed.). *Encyclopedia of Geomorphology*. Routledge London, 398–401.

Goudie, A.S. (2004c). Tectonic Geomorphology. In: Goudie, A.S. (Ed.) *Encyclopedia of Geomorphology*. Routledge, London, p 1037.

Goudie, A.S. (2004d). Salt Weathering. In: Goudie, A.S. (Ed.). *Encyclopedia of Geomorphology*. Routledge. London, 894–897.

Goudie, A.S. (2004e). Tafoni. In: Goudie, A.S. (Ed.). *Encyclopedia of Geomorphology*. Routledge. London, 1034–1035.

Goudie, A.S. (2004f). Glacial erosion. In: Goudie, A.S. (Ed.) *Encyclopedia of Geomorphology*. Routledge. London, 447–448.

Goudie, A.S. (2004g). Arroyo. In: Goudie, A.S. (Ed.). *Encyclopedia of Geomorphology*. Routledge. London, 32–34.

Goudie, A.S. (2004h). Pluvial lake. In: Goudie, A.S. (Ed.). *Encyclopedia of Geomorphology*. Routledge. London, 797–800.

Goudie, A.S. (2004i). Global Warming. In: Goudie, A.S. (Ed.). *Encyclopedia of Geomorphology*. Routledge. London, 479–485.

Goudie, A.S. (2004j). Anthropogeomorphology. In: Goudie, A.S. (Ed.). *Encyclopedia of Geomorphology*. Routledge. London, 25–27.

Goudie, A.S. (2006). *The Human Impact on the Natural Environment*. Blackwell. Oxford, 357 p.

Goudie, A.S. (2007). Mega-Yardangs: A global analysis. *Geography Compass*, 1/1, 65–81.

Goudie, A.S. and Middleton, N.J. (1992). The changing frequency of dust storm through time. *Climatic Change*, 20, 197–225.

Goudie, A.S. and Middleton, N.J. (2001). Sahara dust storms: nature and consequences. *Earth-Science Reviews*, 56, 179–204.

Goudie, A.S. and Middleton, N.J. (2006). *Desert Dust in the Global System*. Springer. Berlin, 287 p.

Goudie, A.S. and Pye, K. (Eds.) (1983). *Chemical Sediments and Geomorphology*. Academic Press. London, 439 p.

Goudie, A.S. and Thomas, D.S.G. (1985). Pans in southern Africa with particular reference to South Africa and Zimbabwe. *Zeitschrift für Geomorphologie* NF, 29, 1–19.

Goudie, A.S. and Viles, H. (1995). The nature and pattern of debris liberation by salt weathering: a laboratory study. *Earth Surface Processes and Landforms*, 20, 437–449.

Goudie, A.S. and Viles, H. (1997). *Salt Weathering Hazards*. Wiley. Chichester, 241p.

Goudie, A.S. and Viles, H. (2010). *Landscapes and Geomorphology*. Oxford University Press. Oxford, 137 p.

Goudie, A.S. and Wells, G.L. (1995). The nature, distribution and formation of pans in arid zones. *Earth Science Reviews*, 38, 1–69.

Goudie, A.S. and Wilkinson, J. (1977). *The Warm Desert Environment*. Cambridge University Press. Cambridge, 88 p.

Goudie, A.S., Allison, R.J. and McLaren, S.J. (1992). The relation between Modulus of Elasticity and temperature in the context of the experimental simulation of rock weathering. *Earth Surface Processes and Landforms*, 17, 605–615.

Goudie, A.S., Cooke, R.V. and Evans, I. (1970). Experimental investigation of rock weathering by salts. *Area 1970*, 42–48.

Goudie, A.S., Warren, A., Jones, D.K.C. and Cooke, R.U. (1987). The character and possible origins of the aeolian sediments of the Wahiba Sand Sea, Oman. *Geographical Journal*, 153, 231–256.

Gow, A.J., Epstein, S. and Stteeby, W. (1979). On the origin of stratified debris in ice cores from the bottom of the Antarctic ice sheet. *Journal Glaciology*, 23, 185–192.

Goy, J.L. and Zazo, C. (1984). Los piedemontes cuaternarios de Campo Dalias y Campo de Níjar (Almería). *Cadernos del Laboratorio Xeolóxico de Laxe*, 5, 40–53.

Goy, J.L. and Zazo, C. (1986). Synthesis of the Quaternary in the Almería littoral, neotectonic activity and its morphologic features (Eastern Betics, Spain). *Tectonophysics*, 130, 259–270.

Goy, J.L., Simón, J.L. and Zazo, C. (1988). Criterios geomorfológicos aplicados a la neotectónica. In: Gutiérrez, M and Peña, J.L. (Eds). *Perspectivas en Geomorfología*. Monografía nº 2 de la Sociedad Española de Geomorfología, Zaragoza, 4–27.

Goy, J.L., Silva, P.G., Zazo, C., Bardají, T. and Somoza, L. (1991). Model of morphotectonic map and legend. *Bulletin of the INQUA. Neotectonics Commision*, 14, 19–31.

Gracia, F.J. (1985). *Geomorfología de las Bardenas Orientales*. Facultad de Ciencias. Zaragoza, 172 p. Tesis de Licenciatura. Unpublished.

Gracia, F.J. (1995). Shoreline forms and deposits in Gallocanta Lake (NE Spain). *Geomorphology*, 11, 323–335.

Gracia, F.J., Gutiérrez, F. and Gutiérrez, M. (1996). Los poljes de la región de Layna (Cordillera Ibérica Noroccidental). *Cuaternario y Geomorfología*, 10, 33–45.

Gracia, F.J., Gutiérrez, F. and Gutiérrez, M. (2002). Origin and evolution of the Gallocanta polje. *Zeitscrift für Geomorphologie*, 46, 245–262.

Gracia, F.J., Gutiérrez, F. and Gutiérrez, M. (2003). The Jiloca karst polje-tectonic graben (Iberian Range, NE Spain). *Geomorphology*, 52, 215–231.

Gracia, F.J., Anfuso, G., Benavente, J., Del Río, L., Domínguez, L. and Martínez, J.A. (2005). Monitoring coastal erosion at different temporal scales on sandy beaches: application to the Cadiz Coast (SW Spain). *Journal of Coastal Research*, Special Issue 49, 22–27.

Gracia, F.J., Alonso, C., Benavente, J., Anfuso, G. and Del Río, L. (2006). The Different Coastal Records of the 1755 Tsunami Waves along the south Atlantic Spanish Coast. *Zeitschrift für Geomorphologie N.F., Suppl.* Vol, 146, 195–220.

Graf, W.L. (1979). The development of montane arroyos and gullies. *Earth Surface Processes*, 4, 1–14.

Graf, W.L. (1988). *Fluvial Processes in Dryland Rivers*. Springer. Berlin, 346 p.

Graf, W.L. (2001). Damage control: restoring the physical integrity of America's rivers. *Annals of the Association of American Geographers*, 91, 1–27.

Grandstein, F.M., Ogg, J.G. and Smith, A.G. (2004). *A Geological Time Scale*. Cambridge University Press, Cambridge.

Granger, D.E., Fabel, D. and Palmer, A.N. (2001). Pliocene-Pleistocene of the Green River, Kentucky, determined from radioactive decay of cosmogenic ^{26} Al y ^{10}Be in Mammoth Cave sediments. *Geological Society of America Bulletin*, 113, 825–836.

Grant, W.H. (1969). Abrasion pH, and index of weathering. *Clays and Clay Minerals*, 17, 151–155.

Gravis, G.F. (1969). Fossil slope deposits in the northern Arctic asymmetrical valleys. *Biuletyn Peryglacjalny*, 20, 239–257.

Gray, J.M. (1981). P-forms from the Isle of Mull. *Scottish Journal of Geology*, 17, 39–47.

Gray, J.M. (1991). Glaciofluvial landforms. In: Ehlers, J., Gibbard, P.L. and Rose, J. (Eds.). *Glacial Deposits in Britain and Ireland*. Balkema. Rotterdam, 443–453.

Gray, R.E. and Bruhn, R.W. (1984). Coal mine subsidence-Eastern United States. In: Holzer, T.L. (Ed.). *Man-induced land subsidence*. Geological Society of America Reviews in Engineering Geology, 6, 123–149.

Greeley, R. (1985). *Planetary Landscapes*. Allen and Unwin. London, 265 p.

Greeley, R. and Iversen, J.D. (1985). *Wind as a Geological Process on Earth, Mars, Venus and Titan*. Cambridge University Press. Cambridge, 333 p.

Green, D.H. and Ringwood, A.E. (1967). An experimental investigation of the gabbro to eclogite transformation and its petrological applications. *Acta Geochemical and Cosmochemical*, 31, 767–833.

Green, J. and Short, N.M. (1971). *Volcanic landforms and surface features. A volcanic atlas and glossary*. Springer. New York, 522 p.

Green, J.L. (2005). Modelling flow resistance in vegetated streams: review and development of new theory. *Hydrological Processes*, 19, 1245–1259.

Greenway, D.R. (1987). Vegetation and slope stability. In: Anderson, M.G. and Richards, K.S. (Eds.). *Slope Stability. Geotechnical Engineering and Geomorphology*. Wiley. Chichester, 187–230.

Greenwood, B. (2005). Bars. In: Schwartz, M.L. (Ed). *Encyclopedia of Coastal Science*. Springer. Dordrecht, 120–129.

Gregory, K.J. (1966). Dry valleys and the composition of the drainage net. *Journal of Hydrology*, 4, 327–340.

Gregory, K.J. (1976). Drainage networks and climate. In: Derbyshire, E. (Ed.). *Geomorphology and Climate*, Wiley. London, 289–315.

Gregory, K.J. (Ed.) (1977). *River Channel Changes*. Wiley. Chichester, 448 p.

Gregory, K.J. (2004a). Fluvial Geomorphology. In: Goudie, A.S. (Ed.). *Encyclopedia of Geomorphology*. Routledge. London, 392–398.

Gregory, K.J. (2004b). Stream power. In: Goudie, A.S. (Ed.). *Encyclopedia of Geomorphology*. Routledge. London, 1002–1004.

Gregory, K.J. (2006). The human role in changing river channels. *Geomorphology*, 79, 172–191.

Gregory, K.J. and Brown, E.H. (1966). Data processing and the study of landform. *Zeitschrift für Geomorphologie*, 10, 257–263.

Gregory, K.J. and Walling, D.E. (1973). *Drainage Basin Form and Process*. Arnold. London, 456 p.

Gregory, K.J., Benito, G., Dikau, R., Golosov, V., Jones, A.J.J., Macklin, M.G., Parsons, A.J. Passmore, D.G., Poesen, J., Starkel, L. and Walling, D.E. (2006). Past hydrological events related to understanding global change: An ICSU research Project. *Catena*, 66, 2–13.

Gribbin, J. and Lamb, H.H. (1978). Climatic change in historical times. In: Gribbin, J. (Ed.). *Climatic Change*. Cambridge University Press. Cambridge, 68–82.

Griggs, D.T. (1936). The factor of fatigue in rock exfoliation. *Journal of Geology*, 44, 783–796.

Griggs, J.G. and Trenhaile, A.S. (1994). Coastal cliffs and platforms. In: Carter, R.W.G. and Woodroffe, C.D. (Eds). *Coastal evolutions. Late Quaternary shoreline morphodynamics*. Cambridge University Press. Cambridge, 425–450.

Grimm, E.C., Jacobson, G.L., Watts, W.A., Hansen, B.C.S. and Maasch, K. (1993). A 50.000 year record of climate oscillations from Florida and its temporal correlation with the Heinrich events. *Science*, 261, 198–200.

Grolier, M.J., McCauley, J.F., Breed, C.S. and Embabi, N.S. (1980). Yardangs of the Western Desert. *Geographical Journal*, 135, 191–212.

Grossman, S. and Gerson, R. (1987). Fluviatile deposits and morphology of alluvial surfaces as indicators of Quaternary environmental changes in the southern Negev, Israel. In: Frostick, L. and Reid, Y. (Eds.). *Desert sediments: ancient and modern*. Geological Society of London Special Publication, 35, 17–29.

Grosso, S.A. and Corte, A.E. (1989). Pleistocene ice-wedge coasts at 34°S, eastern Andes piedmont, south-west of Sourth America. *Geografiska Annaler*, 71 A, 125–136.

Grousset, F.E., Ginoux, P., Bory, A. (2003). Case study of a Chinese dust plume reaching the French Alps. *Geophysical Science Review*, 30, 1277-

Grove, A.T. (1958). The ancient ergs of Hausaland and similar formations of the south side of the Sahara. *Geographical Journal*, 124, 528–533.

Grove, A.T. (1977). The geography of semi-arid lands. *Phylosophical Transactions of the Royal Society of London*, Serie B, 278, 457–475.

Grove, J.M. (1979). The glacial history of the Holocene. *Progress in Physical Geography*, 3, 1–54.

Grove, J.M. (1988). *The Little Ice Age*. Routledge. London, 498 p.

Grove, J.M. (2004). *Little Ice Age: Ancient and Modern*. Routledge. London, 718 p.

Grove, A.T. and Warren, A. (1968). Quaternary landforms and climate on the south side of the Sahara. *Geographical Journal*, 134, 194–208.

Grund, A. (1914). Der geographische zjdus im karst. *Zeitscrift Gesselchaft Erdkund*, 52, 621–740.

Gruner, E. (1963). Dam disasters. *Proceedings Institute Civil Engineers*, London, 24, 47–60.

Grünert, J. and Busche, D. (1980). Large-scale fossil landslides at the Msak Mallat and Hamadat Manghini escarpement. In: Salem, M.J. and Busrewil, M.T. (Eds.). *Geology of Libya*, Academic Press, Vol 3. London, 849–860.

GSL (Geomorphological Services Ltd.) (1987). *Coastal landslide potential assessment, Isla of Wight Undercliff, Ventnor*. Technical Report Research for the Department of Environment, United Kingdon.

Guadagno, F.M., Forte, R., Revellino, P., Fiorillo, F. and Focareta, M. (2005). Some aspects of the initiation of debris avalanches in the Campania Region: The role of morphological slope discontinuities and the development of failure. *Geomorphology*, 66, 237–254.

Guerra, A.J.T. and Favis Mortlock, D. (2002). Movimentos de mass em Petrópolis, Rio de Janeiro. In: Lugo Hubp, J. and Inbar., M. (Eds.). *Desastres naturales en América Latina*. Fondo de Cultura Económica. México, 447–460.

Guerrero, J. (2008). *Dinámica fluvial y riesgos naturales derivados de la subsidencia kárstica en los valles del Huerva y Ebro*. Tesis doctoral. Facultad de Ciencias. Universidad de Zaragoza, 348 p. (Unpublished).

Guerrero, J., Gutiérrez, F. and Lucha, P. (2004). Paleosubsidence and active subsidence due to evaporate dissolution in Zaragoza city area (Huerva River valley, NE Spain). Processes, spatial dissolution and protection measures for linear infrastructures. *Engineering Geology*, 72, 309–329.

Guerrero, J., Gutiérrez, F. and Lucha, P. (2006). The impact of halite dissolution subsidence on fluvial terrace development. The case study of the Huerva River in the Ebro Basin (NE Spain). *Geomorphology* (in press).

Guerrero, J., Gutiérrez, F., Bonachea, J and Lucha, P. (2008). A sinkhole susceptibility zonation based on paleokarst análisis along a strecht of the Madrid-Barcelona high-speed railway built over gypsum-salt bearing. *Engineering Geology*, 102, 62–73.

Guilcher, A. (1953). Essai sur la zonation et la distribution des formes littorales de dissolution du calcaire. *Annales de Géographie*, 62, 161–179.

Guilcher, A. (1954). *Morphologie littorale et sous-marine*. Presses Universitaires de France. París, 216 p.

Guilcher, A. (1979). Marshes and estuaries in different latitudes. *Interdisciplinary Science Reviews*, 4, 158–168.

Guilcher, A. (1988). *Coral Reef Geomorphology*. Wiley. Chichester, 228 p.

Guillén, J, Díaz, J.I., and Palanques, A. (1992). Cuantificación y evolución durante el siglo XX de los aportes de sedimentos transportados como carga de fondo por el río Ebro al medio marino. *Revista de la Sociedad Geológica de España*, 5, 27–37.

Guillien, Y. (1951). Les grèzes litèes de Charente. *Revue Geographique de Pyrénées et de Sud-Ouest*, 22, 154–162.

Guillien, Y. and Lautridou, J.P. (1970). Recherches de gélifraccion experimentale du Centre de Géomorphologie. I. Calcaires de Charentes. *Bulletin du Centre de Géomorphologie de Caen*, 5. C.N.R.S, 45 p.

Gunn, J. (1981). Hydrological processes in karst depression. *Zeitscrift für Geomorphologie*, 25, 313–331.

Gunn, J. (2004a). Dissolution. In: Goudie, A.S. (Ed.). *Encyclopedia of Geomorphology*. Routledge. London, 263–265.

Gunn, J. (2004b). Valleys in Karst. In: Gunn, J. (Ed.). *Encyclopedia of Caves and Karst Science*. Fitzroy Dearborn. New York, 753–754.

Günster, A., Hadley, N.F., Mitchell, D., Nobel, P.S. and Seely, M. (1993). Form and function. In: Seely, M. (Ed.). *Deserts*. Weldon Ower Limited. Australia, 66–81.

Guo, Z., Petit-Maire, N. and Kroepelin, S. (2000). Holocene non-orbital climatic events in present-day arid areas of northern Africa and China. *Global and Planetary Change*, 26, 97–103.

Gupta, A. (1984). Urban hydrology and sedimentation in the humid tropics. In: Costa, J.E. and Fleischer, P.J. (Eds.). *Developments and Applications in Geomorphology*. Wiley. New York, 240–267.

Gupta, A. (1988). Large floods as geomorphic events in the humid tropics. In: Baker, V.R., Craig Kockel, R. and Patton, P.C. (Eds.). *Flood Geomorphology*, Wiley. New York, 301–315.

Gupta, A. (1993). The changing geomorphology of the humid tropics. *Geomorphology*, 7, 165–186.

Gupta, A. (2007). Introduction. In: Gupta, A. (Ed.). *Large Rivers. Geomorphology and Management*. Wiley. Chichester, 1–5.

Gupta, A. and Dutt, A. (1989). The Auranga: description of a tropical mosnoon river. *Zeitschrift für Geomorphologie*, 33, 73–92.

Gupta, A., Rahman, A., Poh Poh, W. and Pitts, J. (1987). The Old Alluvium of Singapore and the extinct drainage system of the South China Sea. *Earth Surface Processes and Landforms*, 12, 259–275.

Gupta, W.L. (1983). Downstream changes in stream power in the Henry Mountains, Utah. *Annals Association of American Geographers*, 73, 373–387.

Gutiérrez, F. (1996). Gypsum karstification induced subsidence (Calatayud Graben, Iberian Range, Spain). *Geomorphology*, 16, 277–293.

Gutiérrez, F. (1998). *Fenómenos de subsidencia por disolución de formaciones evaporíticas en las fosas neógenas de Teruel y Calatayud (Cordillera Ibérica)*. Tesis Doctoral. Facultad de Ciencias. Zaragoza. Unpublished.

Gutiérrez, F. (2010). Hazards associated with karst. In: Alcántara-Ayala, I. and Goudie, A.S. (Eds.). *Geomorpho-*

logical Hazard and Disaster Prevention. Cambridge University Press. Cambridge, 161–175.

Gutiérrez, F. and Cooper, A. (2002). Evaporite dissolution subsidence in the historical city of Calatayud, Spain; damage appraisal and prevention. *Natural Hazards*, 25, 259–288.

Gutiérrez, F. and Gutiérrez, M. (1998). Geomorphology of the tertiary gypsum formations in the Ebro Depression (Spain). *Geoderma*, 87, 1–29.

Gutiérrez, F., Arauzo, T. and Desir, G. (1994). Deslizamientos en el escarpe en yesos de Aljafarín (Zaragoza). *Cuaternario y Geomorfología*., 8, 54–68.

Gutiérrez, F., Arauzo, T., Sancho, C. and Gutiérrez, M. (1995). Factores desencadenantes de los deslizamientos de la Cantera del Barrio de La Corona (Ejea de los Caballeros, Depresión del Ebro). *Cuaternario y Geomorfología*, 9, 33–50.

Gutiérrez, F., Cooper, A.H. and Johnson, K.S. (2008). Identification, prediction and mitigation of sinkhole hazards in evaporite karst. *Environmental Geology* 53, 1007–1022.

Gutiérrez, F., Desir, G. and Gutiérrez, M. (2003). Causes of the catastrophic failure of an earth dam built of gypsiferous alluvium and dispersive clays. *Environmental Geology*, 43, 842–851.

Gutiérrez, F., Guerrero, J. and Lucha, P. (2007). A classification of sinkholes illustrated from evaporite paleokarst exposures in Spain. *Environmental Geology* (aceptada).

Gutiérrez, F., Gutiérrez, M. and Gracia, F.J. (2005). *Karst, Neotectonics and Periglacial Features in the Iberian Range*. Field Trip Guide C-5. Sixth International Conference on Geomorphology. Zaragoza. 58 p.

Gutiérrez, F., Gutiérrez, M. and Sancho, C. (1998a). La avenida del 7 de Agosto de 1996 en la cuenca y abanico aluvial de Arás (Valle de Tena, Pirineos centrales). Aspectos geomorfológicos y sedimentológicos. *Revista Sociedad Geológica de España*, 11, 71–85.

Gutiérrez, F., Gutiérrez, M. y Sancho, C. (1998b). Geomorphological and sedimentological analysis of a catastrophic flash flood in the Arás drainage basin (Central Pyrenees, Spain). *Geomorphology*, 22, 265–283.

Gutiérrez, F., Gutiérrez, M. and Sancho, C. (1998c). Sedimentological features of the 7th August 1996 catastrophic flash flood in the Arás Mountain torrent and alluvial fan (Central Pyrenees, Spain). *15th International Sedimentological Congres*. Alicante, 401–402.

Gutiérrez, F., Ortí, F., Gutiérrez-Elorza, M., Pérez-González, A., Benito, G., Gracia-Prieto, J. y Durán, J.J. (2001). The stratigraphical record and activity of evaporite dissolution subsidence in Spain. *Carbonates and Evaporites*, 16, 46–70.

Gutiérrez, F., Calaforra, J.M., Cardona, F., Ortí, F., Durán, J.J. and Garay, P. (2004). El karst en las formaciones evaporíticas españolas. In: Andreu, B. and Durán, J.J. (Eds.). *Investigaciones en sistemas kársticos españoles*. Instituto Geológico y Minero de España. Madrid, 49–87.

Gutiérrez-Santolalla, F., Acosta, E., Ríos, S., Guerrero, J. and Lucha, P. (2005). Geomorphology and geochronology of sackung features (up-hill-facing scarps) in the Central Spanish Pyrenees. *Geomorphology*, 69, 298–314.

Gutiérrez, F., Galve, J.P., Guerrero, J., Lucha, P., Cendrero, A., Remondo, J., Bonaechea, J., Gutiérrez, M. and Sánchez, J.A. (2007). The origin, typology, spatial distribution and detrimental effects of the sinkholes developed in the alluvial evaporate karst of the Ebro River valley downstream of Zaragoza city (NE Spain). *Earth Surface Processes and Landforms*, 32, 912–918.

Gutiérrez, F., Johnson, K.S. and Cooper, A. (2008a). Evaporite karst processes landform and environmental problems. *Environmental Geology* 53, 935–936.

Gutiérrez, F., Calaforra, J.M., Cardona, F., Ortí, F., Durán, J.J. and Garay, P. (2008b). Geological and environmental implications of evaporite karst in Spain. *Environmental Geology* 53, 951–965.

Gutiérrez, F., Ortuño, M., Lucha, P., Guerrero,J., Acosta, E., Coratza, P., Picentini, D., Soldati, M. and Beguería, S. (2008c). Episodic displacement on a sackung scarp in Benasque Valley (Central Spanish Pyrenees). Paleosismic record?. *Geo-Temas*, 10, 1011–1014.

Gutiérrez, F., Galve J.P., Lucha, P., Bonachea, J., Jordá, L and Jordá, R. (2009). Investigation of a large collpase sinkhole affecting a multi-store building by means of geophysics and the trenching technique. *Environmental Geology*, 58, 1107–1122.

Gutiérrez, J.M. and Pons, R.M. (2006). Modelización numérica del cambio climático: Bases científicas. Incertidumbres y proyecciones para la Península Ibérica. *Cuaternario y Geomorfología*, 20, 15–28.

Gutiérrez, M. (1986). Some remarks on the problem of climatic changes and geomorphological processes in arid zones. In: López-Vera, F. (Ed.). *Quaternary Climate in Western Mediterranean*. Madrid, 3–29.

Gutiérrez, M. (1990). Historia de la Geomorfología. In: *Historia de la Geología*. Real Academia de Ciencias Exactas, Físicas y Naturales. Madrid, 115–131.

Gutiérrez, M. (1994) (Coord.). *Geomorfología de España*. Rueda. Madrid, 526 p.

Gutiérrez, M. (2001). Geomorfología Climática. Omega. Barcelona, 642 p.

Gutiérrez, M. (2004). Flatiron. In: Goudie, A.S. (Ed.). *Encyclopedia of Geomorphology*. Routledge. London, p. 378.

Gutiérrez, M. (2005). *Climatic Geomorphology*. Elsevier. Amsterdam, 760 p.

Gutiérrez, M. (2006). Erosión e influencia del cambio climático. *Cuaternario y Geomorfología*, 20, 45–59.

Gutiérrez, M. (2007). El papel del hombre en la creación y destrucción del relieve. *Revista de la Real Academia de Ciencias Exactas, Físcas y Naturales*, 101, 211–226.

Gutiérrez, M. and Ibáñez, M.J. (1979). Las "gnammas" de la región de Alcañiz. *Estudios Geológicos*, 35, 193–198.

Gutiérrez, M. and Gracia, F.J. (1997). Environmental interpretation and evolution of the Tertiary erosion surfaces in the Iberian Range (Spain). In: Widdowson, M. (Ed.). *Paleosurfaces: Recognition, Reconstruction and Palaeoenvironmental Interpretation*. The Geological Society. Special Publication, 120. London, 147–158.

Gutiérrez, M. and Gutiérrez, F. (1998). Geomorphology of the Tertiary gypsum formations in the Ebro Depression (Spain). *Geoderma*, 87, 1–29.

Gutiérrez, M. and Peña, J.L. (1976). Glacis y terrazas en el curso medio del río Alfambra. *Boletín Geológico y Minero*, 87, 561–570. Madrid.

Gutiérrez, M. and Peña, J.L. (1977). Las acumulaciones periglaciares del Macizo del Tremedal (Sierra de Albarracín). *Boletín Geológico y Minero*, 88, 109–115.

Gutiérrez, M. and Peña, J.L. (1979a). El Karst de Villar del Cobo (Sierra de Albarracín). *Estudios Geológicos*, 35, 651–654.

Gutiérrez, M. and Peña, J.L. (1979b). El karst de los Llanos de Pozondón (Sierra de Albarracín). *Teruel*, 61–62, 39–46.

Gutiérrez, M. and Peña, J.L. (1979c). Deslizamientos intracuaternario de bloques en la región de Villel (Provincia de Teruel). *Estudios Geológicos*, 35, 299–303.

Gutiérrez, M. and Peña, J.L. (1981). Los glaciares rocosos y el modelado acompañante en el área de la Bonaigua (Pirineo de Lérida). *Boletín Geológico y Minero*, 92, 101–110.

Gutiérrez, M. and Peña, J.L. (1989). Upper Holocene climatic change and geomorphological processes on slopes and infilled valleys from archaeological dating (NE-Spain). *European Conference on Landscape Ecological Impact of Climate Change Lunteren, Netherlands, 3–7 December 1989*, 21p. Lunteren: Dutch Ministry of the Environment.

Gutiérrez, M. y Peña, J.L. (1992). Evolución climática y geomorfológica del Holoceno Superior (Cordillera Ibérica, Depresión del Ebro y Pre-Pirineo). In: Cearreta, A. and Ugarte, F.M. (Eds.). *The Late Quaternary in the Western Pyrenean Region*. Servicio Editorial de la Universidad del País Vasco. Bilbao, 109–124.

Gutiérrez, M. and Peña, J.L. (1994a). Cordillera Ibérica. In: Gutiérrez, M. (Coord.). *Geomorfología de España*, Rueda, Madrid, 251–286.

Gutiérrez, M. and Peña, J.L. (1994b). Depresión del Ebro. In: Gutiérrez, M. (Coord.). *Geomorfología de España*. Rueda. Madrid, 305–349.

Gutiérrez, M. and Peña, J.L. (1998). Geomorphology and Upper Holocene climatic change in northeastern Spain. *Geomorphology*, 23, 205–217.

Gutiérrez, M. and Rodríguez, J. (1978). Consideraciones sobre la morfogénesis del Sistema Central. *Boletín Geológico y Minero*, 89, 109–113.

Gutiérrez, M. and Sancho, C. (1993). Applied geomorphology in arid and semiarid regions. *Second Intensive Course of Applied Geomorphology: Arid Regions*. Universidad de Zaragoza-Erasmus. Zaragoza, 3–31.

Gutiérrez, M. and Sesé, V.H. (2001). Multiple talus flatirons. Variations of scarp retreat rates and the evolution of slopes in Almazán Basin (semi-arid central Spain). *Geomorphology*, 38, 19–29.

Gutiérrez, M. and Valverde, M. (1994). El sistema de poljes del río Guadazaón (Cordillera Ibérica, prov. de Cuenca). *Cuaternario y Geomorfología*, 8, 87–95.

Gutiérrez, M., Benito, G. and Rodríguez, J. (1988). Piping in badlands areas of the middle Ebro basin, Spain. *Catena Supplement*, 13, 49–60.

Gutiérrez, M., Desir, G. and Gutiérrez, F. (2002). Yardangs in semiarid central sector of the Ebro Depression. *Geomorphology*, 44, 155–176.

Gutiérrez, M., Desir, G., Gutiérrez, F. and Marín, C. (2005a). Origin and evolution of playas and blowouts in the semiarid zone of Tierra de Pinares (Duero Basin, Spain). *Geomorphology*, 72, 177–192.

Gutiérrez, M., Desir, G., Gutiérrez, F., Sánchez, J.A., Castañeda, C. and Lucha, P. (2005b). Aeolian Landforms and Saline Lakes (Central Ebro Basin). *Sixth International Conference on Geomorphology*. Field Trip Guide B-7. Zaragoza, 47 p.

Gutiérrez, M., Gutiérrez, F. and Desir, G. (2003). Are talus flatirons from central Spain related with Heinrich events?. *Suppl. Geografía Física y Dinámica Cuaternaria*, 6, 59–63.

Gutiérrez, M., Gutiérrez, F., and Desir, G. (2006). Considerations on the chronological and causal relationship between talus flatirons and palaeoclimatic changes in central and northeastern Spain. *Geomorphology*, 73, 50–63.

Gutiérrez, M., Peña, J.L. and Simón. J.L. (1982). El polje de Vistabella del Maestrazgo (Provincia de Castellón). *Actas Reunión Monográfica sobre el Karst*. Larra, 95–104.

Gutiérrez, M., Peña, J.L. and Simón, J.L. (1983a). Los valles tectónicos recientes de Rubielos de la Cérida (Teruel). *Actas VI Reunión Nacional Grupo Español de Trabajo del Cuaternario*. Vigo- Santiago de Compostela, 449–459.

Gutiérrez, M., Sancho, C. and Arauzo, T. (1998a). Scarp retreat in semiarid environments from talus flatirons (Ebro Basin). *Geomorphology*, 25, 111–121.

Gutiérrez, M., Ibáñez, M.J., Peña, J.L., Rodríguez, J. and Soriano, M.A. (1985). Quelques exemples de karst sur gypse dans la Dépresion de l'Ebre. *Karstologia*, 6, 29–36.

Gutiérrez, M., Peña, J.L., Rodríguez, J. and Simón, J.L. (1983b). Criterios geomorfológicos aplicados al estudio de la neotectónica en áreas continentales (ejemplos en la Cadena Ibérica, Depresión del Ebro y Pirineos). *Primeras Jornadas sobre neotectónica y su aplicación al análisis de riesgos de emplazamientos energéticos e industriales*, J.E.N. Madrid, 158–213.

Gutiérrez, M., Sancho, C., Desir, G., Sirvent, J., Benito, G. and Calvo, A. (1995). *Erosión hídrica en terrenos arcillosos y yesíferos de la Depresión del Ebro*. Proyecto Lucdeme. Ministerio de Agricultura, Pesca y Alimentación. Zaragoza, 389 p.

Gutiérrez, M., Sancho, C., Benito, G., Sirvent, J. and Desir, G. (1997). Quantitative study of piping processes in badland areas of the Ebro Basin. *Geomorphology*, 20, 237–253.

Gutiérrez, M., Sancho, C., Arauzo, T. and Peña, J.L. (1998b). Evolution and paleoclimatic meaning of the talus flatirons in the Ebro Basin (NE of Spain). In: Alsharhan, A.S., Glennie, K.W. and Whittle, G.L. (Eds.). *Quaternary Deserts and Climatic Change*. Balkema. Amsterdam, 593–599.

Gutiérrez, M., Lucha, P., Gutiérrez, F., Moreno, A., Guerrero, J., Martín-Serrano, A., Nozal, F., Desir, G., Marín, C. and Bonaechea, J. (2010). Are talus flatiron sequences in Spain climate-controlled landforms? *Zeitschrift für Geomorphologie* 54 (2), 243–252.

Gutiérrez, M., Lucha, P., Gracia, F.J., Desir, G., Marín, C. and Petit-Maire, N. (2011a). Talus flatirons and aeolian deposits in northern Fuerteventura volcanic island (Canary Islands, Spain). Paleoclimatic considerations. *Geomorphology* doi: 10.1016/j.geomorph.2011.09.020

Gutiérrez, M., Nozal, F. and Montes, M. (2011b). Talus flatirons on Seymour Island (Weddell Sea, Antarctic). Considerations about their genesis. *Cuaternario y Geomorfología* 25 (1–2), 9–19.

Haase, D., Fink, J., Haase, G., Ruske, R., Pécsi, M., Richter, H., Altermann, M. and Jäger, K.D. (2007). Loess in Europe-it's spatial distribution based on a European loess map, scale 1:2,500,000. *Quaternary Science Reviews* 26, 1301–1312.

Hack, G.T. (1960). Interpretation of erosional topography in humid temperate regions. *American Journal of Science*, 252, 80–97.

Hack, J.T. (1973). Stream-profile analysis and stream-gradient index. *United States Geological Survey Journal of Research*, 1, 421–429.

Hack, J.T. (1975). Dynamic equilibrium and landscape evolution. In: Melhorn, W.C. and Flemal, R.C. (Eds.). *Theories of Landform Development*. George Allen and Unwin. London, 87–102.

Hadley, R.F. and Schumm, S.A. (1961). Sediment sources and drainage basin characteristics in upper Cheyenne River basin. *United States Geological Survey Water Supply*, Paper 1531 B, 137–198.

Haeberli, W. (2004). Glaciers and ice caps: historical background and strategies of world-wide monitoring. In: Bamber, J.L. and Payne, A.J. (Eds.). *Mass Balance of the Cryosphere*. Cambridge University Press. Cambridge, 559–578.

Haff, R.K. (2000). Neomorphology, Prediction and Anthropic Landscape. In: Wilcock, P.R. and Iverson, R.M. (Eds.). *Prediction in Geomorphology*. American Geophysical Union. Washington, 15–26.

Hagedorn, H. (1968). Über äolische Abtragung und Formung in der Südest-Sahara. *Erdekunde*, 22, 257–269.

Haigh, M.J. (1978). Microrills and dessication cracks: some observations. *Zeitschrift für Geomorphologie*, 12, 457–461.

Hail, J.R. (Ed.) *Applied Geomorphology*. Elsevier. Amsterdam, 418 p.

Häkkinen, S. and Rhines, P.B. (2004). Decline of subpolar North Atlantic circulation during the 1990s. *Science*, 304, 555–559.

Halimov, M. and Fezer, F. (1989). Eight yardang types in central Asia. *Zeitschrift für Geomorphologie* NF, 33, 205–217.

Hall, A.M. (1986). Deep weathering patterns in north-east Scotland and their geomorphological significance. *Zeitschrift für Geomorphologie*, 30, 407–422.

Hall, A.M. (1988). The characteristics and significance of deep weathering in the Gaick area, Grampian Highlands, Scotland. *Geografiska Annaler*, 70 A, 309–314.

Hall, A.M., Thomas, M.F. and Thorp, M.B. (1985). Late Quaternary alluvial placer development in the humid tropics: the case of the Birim Diamond Placer, Ghana. *Journal of the Geological Society*, 142, 777–787.

Hall, A.M., Mellor, A.M. and Wilson, M.J. (1989). The clay mineralogy and age of deeply weathered rock in north-east Scotland. *Zeitschrift für Geomorphologie Supplementband*, 72, 97–108.

Hall, K., Thorn, C.E., Matsouke, N., and Prick, H. (2002). Weathering in cold regions: some thoughts and perspectives. *Progress in Physical Geography*, 26, 577–603.

Hallam, A. (1973). *A Revolution in the Earth Sciences. From Continental Drift to Place Tectonics*. Clarendon Press, Oxford.

Hallbauer, D.K.J. and Jahns, N.M. (1977). Attack of lichens on quartzite rock surfaces. *Lichenologist*, 9, 119–122.

Hallet, B. (1979). Subglacial regelation water film. *Journal of Glaciology*, 23, 321–334.

Hallet, B. (1990). Spatial self-organization in geomorphology: From periodic bedforms and patterned ground to scale invariant topography. *Earth Science Reviews*, 29, 57–75.

Hallet, B. (1996). Glacial quarrying. A simple theorical model. *Annals of Glaciology*, 22, 1–8.

Halliday, W.R. (2004). Pseudokarst. In: *Encyclopedia of Caves and Karst Science*. Fitzroy Dearborn. New York, 604–608.

Hallsworth, E.G., Robertson, G.K. and Gibbons, F.R. (1955). Studies of pedogenesis in New South Wales, VII: the "gilgai" soils. *Journal of Soil Science, 6*, 1–31.

Hambrey, M.J. (1977). Foliation, minor folds and strain in glacier ice. *Tectonophysics*, 39, 397–416.

Hambrey, M.J. (1994). *Glacial Environments*. UCL Press. London, 296 p.

Hambrey, M. and Alean, J.C. (1992). *Glaciers*. Cambridge University Press. Cambridge, 207 p.

Hamilton, A. (1976). The significance of patterns of distribution shown by forest plants and animals in tropical Africa for the reconstruction of Upper Pleistocene palaeoenvironments: a review. In: Van Zinderen Baker, E.M. (Ed.). *Palaeoecology of Africa*, vol. 9. Balkema. Rotterdam, 63–97.

Hamson, J.D., Evans, D.J.A., Sanderson, D.C.W., Bingham, R.G. and Bentley, M.J. (2008). Contraining the age and formation of stone runs in the Fakland Islands using optically stimulated luminescence. *Geomorphology*, 94, 117–130.

Handcock, G.S., Anderson, R., Chadwick, O. and Finkel, R. (1999). Dating fluvial terraces with ^{10}Be y ^{26}Al profiles: application to the Wind River, Wyoming. *Geomorphology*, 27, 41–60.

Hanks, T.C. and Wallace, R.E. (1985). Morphological analysis of the Lake Lahontan shoreline and beach front fault scarps, Pershing County, Nevada. *Bulletin of the Seismological Society of America*, 75, 835–846.

Hansen, M.J. (1984a). Strategies for classification of landslides. In: Brunsden, D. and Prior, D.B. (Eds.). *Slope Instability*. Wiley. Chichester, 1–25.

Hansen, M.J. (1984b). Landslides hazard analysis. In: Brunsden, D. and Prior, D.B. (Eds). *Slope Instability*. Wiley. Chichester, 1–25.

Hansen, J., Sato, M., Ruedy, R., Lacis, A., Asamaon, K., Borestein, S., Brown, N., Cairns, B., Caliri, G., Campbell, M., Curran, B., Castro, S. de, Druyan, L., Fox, M., Johnson, C., Lerner, J., McCormick, M.P., Miller, R., Minnis, P., Morrison, A., Pandolfo, L., Ramborran, I., Zaucker, F., Robinson, M., Russell, P., Shad, K., Stone, P., Tegen, I., Thomason, L., Wilder, J. and Wilson, H. (1996). A Pinatubo climate modelling investigation. In: Fiocco, G., Fuà, D. and Visconti, J. (Eds.). *The Mount Pinatubo Eruption. Effects on the Atmosphere and Climate*. NATO A51 Series, vol. 42, 233–272.

Hansen, W.R. (1965). Effects of the earthquake of March 27, 1964, at Anchorage, Alaska. *United States Geological Survey Profesional Paper*, 542-A, 64 p.

Happ, S.C., Rittenhouse, G. and Dobson, G.C. (1940). Some principles of accelerated streams and valley sedimentation. *United States Department of Agriculture Technical Bulletin*, 695, 22–31.

Harden, C. (2004). Terrace, river. In: Goudie, A.S. (Ed.). *Encyclopedia of Geomorphology*. Routledge. London, 1039–1043.

Hardie, L.A., Smoot, J.P. and Eugster, H.P. (1978). Saline lakes and their deposits: a sedimentological approach. In: Matter, A. and Tucker, M. (Eds.). *Modern and Ancient Lake Sediments*. International Association of Sedimentologists, Special Publication, 2, 7–41.

Harding, A.F. (1982). Introduction: Climate change and Archaeology. In: Harding, A.F. (Ed.). *Climatic change in Later Prehistory*: Edingburgh University Press. Edinburgh, 1–10.

Hardy, J.T. (2003). *Climate Change. Causes, Effects and Solutions*. Wiley. Chichester, 247 p.

Hare, P.W. and Gardner, T.W. (1985). Geomorphic indicator of vertical neotectonism along converging plate margins, Nicoya, Peninsula, Costa Rica. In: Morisawa, M. and Hack, J.K. (Eds.). *Tectonic Geomorphology*. Allen and Unwin. Boston, 75–104.

Harris, C. (1981). *Periglacial Mass-Wasting, a Review of Research*. Geobooks, Norwich, 204 p.

Harris, C. (2007). Periglacial Landforms. Slope Deposits and Forms. In: Elias, S.a. (Ed.). *Encyclopedia of Quaternary Science*. Elsevier. Amsterdam, 2207–2217.

Harris, C. and Davies, M.C.R. (2000). Gelifluction; observations from large-scale laboratory simulations. *Arctic, Antarctic and Alpine Research*, 32, 202–207.

Harris, C. and Murton, J.B. (2005). Interactions between glaciers and permafrost: an introduction. In: Harris, C. and Murton, J.B. (Eds.). *Cryosphere Systems: Glaciers and Permafrost*. Geological Society of London, Special Publication 242, 1–9.

Harris, C., Haeberli, W., Vonder Muhll, D. and King, L. (2001). Permafrost monitoring in the high mountains of Europe: the PACE Project in its global context. *Permafrost and Periglacial Processes*, 12, 3–12.

Harris, C., Davies, M. and Rea, B.R. (2003). Gelifluction: viscous flow or plastic creep? *Earth Surface Processes and Landforms*, 28, 1289–1301.

Harris, S.A. (1968). Gilgai. In: Fairbridge, R.W. (Ed.) *The Encyclopedia of Geomorphology*. Reinhold. New York, 425–426.

Harris, S.A. (1986). *The Permafrost Environment*. Croom Helm. Kent, 276 p.

Harris, S.A. (1988). The Alpine Periglacial zone. In: Clark, M.J. (Ed.). *Advances in Periglacial Geomorphology*. Wiley. Chichester, 369–413.

Harris, S.E. (1943). Friction cracks and the direction of glacial movement. *Journal of Geology*, 51, 244–258.

Harrison, C.G.A. and Bonatti, E. (1981). The oceanic lithosphere. In: Emiliani, E. (Ed.). *The Oceanic Lithosphere*. The Sea 7, Wiley. New York, 333 p.

Harrison, J.V. and Falcon, N.L. (1934). Collapse Structures. *Geological Magazine*, 71, 529–539.

Harrison, J.V. and Falcon, N.L. (1936). Gravity collapse structures in mountain ranges, as exemplified in southwestern Persia. *Quaterly Journal Geological Society of London*, 92, 91–102.

Harry, D.G. (1988). Ground ice and permafrost. In: Clark, M.J. (Ed.). *Advances in Periglacial Geomorphology.* Wiley, Chichester. 113–149.

Hart, M.G. (1986). *Geomorphology Pure and Applied*. Allen and Unwin. London, 228 p.

Hart, R.S., Anderson, D.I. and Kanamori. M. (1977). The effect of attenuation on gross Earth models. *Journal of Geophysical Research,* 82, 1647–1654.

Härtel, H., Cilek, V., Herben, T., Jackson, A. and Williams, R. (Eds.) (2007). *Sandstone Landscapes*. Academic Press. Praha, 493 p.

Hartlen, J. and Viberg, L. (1988). General Report: Evaluation of landslide hazard. In: Bonnard, C. (Ed.). *Landslides-Glissements de Terrain*. Proceedings V International Symposium on Landslides, vol. 2, 1037–1057. Lausanne.

Harvey, A.M. (1978). Dissected alluvial fans in southeast Spain. *Catena*, 5, 177–211.

Harvey, A.M. (1982). The role of piping in the development of badlands and gully systems in southeast Spain. In: Bryan, R. and Yair, A. (Eds.). *Badland Geomorphology and Piping*. Geobooks. Norwich, 317–335.

Harvey, A.M. (1984a). Debris flows and fluvial deposits in Spanish Quaternary alluvial fans: implications for fan geomorphology. In: Koster, E.H. and Steel, R.J. (Eds.): *Sedimentology of Gravels and Conglomerates*. Canadian Society Petroleum Geologist, Memoir 10, 123–132.

Harvey, A.M. (1984b). Aggradation and disecction sequences on spanish alluvial fans: influence on morphological development. *Catena*, 11, 289–304.

Harvey, A.M. (1987a). Alluvial fans dissection: relationship between morphology and sedimentation. In: Frostik, L. and Reid, I. (Eds.). *Desert Sediments, Ancient and Modern.* Geological Society of London, 35, 87–103.

Harvey, A.M. (1987b). Patterns of Quaternary aggradational and dissectional landform development in the Almería region, southeast Spain: a dry region, tectonically active landscape. *Die Erde*, 118, 193–215.

Harvey, A.M. (1990). Factors influencing Quaternary alluvial fan development in southeast Spain. In: Rachocki and Church, M. (Eds.). *Alluvial Fans: a Field Approach*. Wiley. New York, 247–269.

Harvey, A.M. (1996). The role of alluvial fans in the mountain fluvial systems of southeast Spain: implications of climatic change. *Earth Surface Processes and Landforms*, 21, 543–553.

Harvey, A.M. (1997). The role of alluvial fans in arid zone fluvial systems. In: Thomas, D.S.G. (Ed.). *Arid Zone Geomorphology*. Wiley. Chichester, 231–259.

Harvey, A.M. (2002a). Factors influencing the geomorphology of dry-region alluvial fans. In: Pérez-González, A., Vegas, J. and Machado, M.J. (Eds.). *Aportaciones a la Geomorfología Española en el inicio del Tercer Milenio*. Instituto Geológico y Minero de España. Madrid, 59–75.

Harvey, A.M. (2002b). Effective timescales of coupling in fluvial systems. *Geomorphology,* 44, 175–201.

Harvey, A.M. (2002c). The relationships between alluvial fans and fan channels within mediterranean mountain fluvial systems. In: Bull, L.J. and Kirkby, M.J. (Eds.). *Dryland Rivers: Hydrology and Geomorphology of Semi-arid Channels*. Wiley. Chichester, 205–226.

Harvey, A.M. (2004a). Badland. In: Goudie, A.S. (Ed.). *Encyclopedia of Geomorphology*. Routledge. London, 45–48.

Harvey, A.M. (2004b). The response of dry-region alluvial fans to late Quaternary climatic change. In: Alsharshan, A.S., Wood, W.W., Goudie, A.S., Fowler, A. and Abdellatif, E.M. (Eds.). *Desertification in the Third Millenium*. Balkema. Rotterdam, 83–98.

Harvey, A.M. (2004c). Alluvial fan. In: Goudie, A.S. (Ed.). *Encyclopedia of Geomorphology*. Routledge. London, 15–19.

Harvey, A.M. (2007). Geomorphic instability and change-Introduction: implications of temporal and spatial scale. *Geomorphology*, 84, 153–158.

Harvey, A.M. and Calvo, A. (1991). Process interactions and rill development on badland and gully slopes. *Zeitschrift für Geomorphologie, Supplementband*, 83, 175–194.

Harvey, A.M. and Gutiérrez, M. (2005). Repeated patterns of Quaternary discontinuous gullying at El Tormillo, Ebro Basin, Spain. In: Garcia, C. and Batalla, R. (Eds.). *Catchment Dynamics and River Processes: Mediterranean and Other Climate Regions*. Elsevier. Amsterdam, 53–67.

Harvey, A.M. and Wells, S.G. (2003). Late Quaternary alluvial fan developments relations to climatic change, Soda Mountains, Mojave Desert, California. In: Lancaster, N., Enzel, Y. and Wells, S.G. (Eds.). *Environmental Change in the Mojave Desert*. Geological Society of America, Special Paper, 368, 207–230.

Harvey, A.M. Silva, P.G., Mather, A.E., Goy, J.L. Stokes, M. and Zazo, C. (1999a). The impact of Quaternary sea-level and climatic change of coastal alluvial fans in the Cabo de Gata ranges, southeast Spain. *Geomorphology*, 28, 1–22.

Harvey, A.M., Wigand, P.E. and Wells, S.G. (1999b). Response of alluvial fan systems to the late Pleistocene to Holocene climatic transition: contrasts between the margins of alluvial Lakes Lahontan and Mojave, Nevada and California, USA. *Catena*, 36, 255–281.

Harvey, A.M., Mather, A.E. and Stokes, M. (2005). Alluvial fans: geomorphology, sedimentology, dynamics-introduction. A review of alluvial-fan research. In: Harvey, A.M., Mather, A.E. and Stokes, M. (Eds.). *Alluvial Fans: Geomorphology, Sedimentology, Dynamics*. Geological Society of London, Special Publications, 251, 1–7.

Harvey, C.P.D. and Grove, A.T. (1982). A prehistoric source of the Nile. *Geographical Journal*, 148, 327–336.

Hassan, M.A. (1992). Structural controls of the mobility of coarse material in gravel-bed channels. *Israel Journal of Earth Sciences*, 41, 105–122.

Hassan, M.A. and Ergenzinger, P. (2003). Use of tracers in fluvial geomorphology. In: Kondolf, G.M. and Piégay, H. (Eds.). *Tools in Fluvial Geomorphology*. Wiley. Chichester, 397–423.

Haugen, R.K. and Brown, J. (1971). Natural and man-induced disturbances of permafrost terrane. In: Coates, D.R. (Ed.). *Environmental Geomorphology*. Geom. State University, 139–149. Binghamton, Nueva York.

Hawkins, J. (1995). Evoultion of the Lau Basin-insights from ODP Leg 135. In: *Active Margins and Marginal Basins of the Western Pacific*. Geophysical monograph 88. American Geophysical Union. Washington DC.

Hay, R.L. and Wiggins, B. (1980). Pellets, ooids, sepiolite and silica in three calcretes of the southwest United States. *Sedimentology*, 27, 559–576.

Hayden, B.P. (1988). Flood climates. In: Baker, V.R., Craig Kockel, R. and Patton, P.C. (Eds.). *Flood Geomorphology*, Wiley. New York, 13–26.

Hayes, M.O. (2005). Tide-dominated coasts. In: Schwartz, M.L. (Ed). *Encyclopedia of Coastal Science*. Springer. Dordrecht, 982–984.

Hays, J. (1967). Land surfaces and laterites of the Northern Territory. In: Jennings, J.N. and Mabbutt, J.A. (Eds.). *Landform Studies from Australasia and New Guinea*. Cambridge University Press. Cambridge, 182–210.

Hays, W.W. (1981). Facing geologic and hydrologic hazards: Earth-science considerations. *United States Geological Survey Professional Paper*, 1240-B, 54–85.

Heathcote, R.L. (1983). *The Arid Lands: Their Use and Abuse*. Longman, London.

Heede, B.H. (1971). Characteristics and processes of soil piping in gullies. *United States Department of Agriculture Forest Service, Rocky Mountain Forest and Range Experiment Station Research Paper* RM-68, 15 p.

Heginbottom, J.A. (2002). Permafrost mapping: a review. *Progress in Physical Geography*, 26, 623–642.

Heim, A. (1882). Der Bergsturz von Elm. *Zeitschrift Deutsche Geologische Gessellshaft*, 34, 74–115.

Heim, A. (1932). Bergstruz und Menschenleben. *Beiblat zur Vierteljahrsschrift Gesellchaft in Zurich*, 77, 1–217. Translated by N. Skermer bajo el título *Landslides and Human Lives*. Bitech Publishers, Vancouver, British Columbia, 1989, 195 p.

Heine, K. (1995). Bedded slope deposits with respect to the Late Quaternary glacial sequence in the High Andes of Ecuador and Bolivia. In: Slaymaker, O. (Ed.). *Steepland Geomorphology*. Wiley. Chichester, 257–278.

Heine, K. and Walter, R. (1996). Gypcretes of the central Namib Desert, Namibia. In: Heine, K. (Ed.). *Palaecology of Africa and the Surroundings Islands*. Balkema, vol. 24. Rotterdam, 173–201.

Heinrich, H. (1988). Origin and consequences of cyclic ice rafting in the Northeast Atlantic ocean during the past 130,000 years. *Quaternary Research*, 29, 142–152.

Heirtzler, J.R., LePichon, S. and Baron J.F. (1966). Magnetic anomalies over the Reykjanes. *Deep Sea Research*, 13, 424–443.

Hellsen, M.M., Broecke, M.R. van den, Wal, A.S.W. van de, Berg, W.J. van de, Meijgaard, E. van, Davis, C.H., Li, Y. and Goodwin, I. (2008). Elevation changes in Antarctica mainly determined by accumulation variability. *Science*, 320, 1626–1629.

Hemming, S.R. (2004). Heinrich events: massive late Pleistocene detritus layers of the North Atlantic and their global climate imprint. *Review du Geophysics*, 42, 1–48.

Henderson, E.P. (1956). Large nivation hollows near Knob Lake, Quebec. *Journal Geology*, 64, 607–616.

Henderson-Sellers, A. (1992). Continental cloudiness changes this century. *GeoJournal*, 27, 255–262.

Henscher, S.R. (1987). The implications of joints and structures for slope stability. In: Anderson, M.G. and Richards, K.S. (Eds.). *Slope Stability. Geotechnical Engineering and Geomorphology*. Wiley. Chichester, 145–186.

Herath, J.W. and Grimshaw, R.W. (1971). A general evaluation of the frequency distribution of clay and associated minerals in the alluvial soils of Ceylon. *Geoderma*, 5, 119–130.

Hereford, R. (1984). Climate and ephemeral-stream processes: twentieth-century geomorphology in alluvial stratigraphy of the Little Colorado River, Arizona. *Bulletin of the Geological Society of America*, 95, 654–668.

Herget, J. (2005). *Reconstruction of Pleistocene Ice-Dammed Lake Outburst floods in the Altai Mountain, Siberia*. Geological Society of America, Special Paper 386, 117 p.

Herman, J.R., Bhartia, P.K., Torres, O., Hsu, C., Setfor, C. and Celavier, E. (1997). Global distribution of UV-Absorbing Aerosols from Nimbus 7/TOMS Data. *Journal of Geophysical Research*, 102, 16889–16909.

Hernández Calvento, L., Ruiz Flaño, P., Pérez-Chacón, E., Alonso Bilbao, I., Alcántara-Carrió, J. and Suárez Rodríguez, E. (2005). Transformaciones ambientales en ecosistemas dunares inducidas por el desarrollo turístico: el ejemplo de Maspalomas (Gran Canaria, Islas Canarias). In: Sanjuame, E. and Mateu, F. (Eds.). *Geomorfología litoral i Quaternari*. PUV. Valencia, 203–214.

Hernández, J., Carabias, R., Alonso, A., Jiménez, O., Sánchez, E., Corrochano, A., Carballeira, J. and Martín, J.M. (1984). Alteraciones de las areniscas de Villamayor: causas y soluciones. In: *Estudio sobre las Alteraciones y Tratamiento de la Piedra de Villamayor*. Ediciones Caja de Ahorros y M.P. de Salamanca. Salamanca, 171–314.

Hernández-Enrile, J.L., Carreño, M., Santanach, P. and Martínez, J.M. (2001). ¿Es España un país de actividad sísmica moderada? (Mesa Redonda). In: Capote, R. and Martínez, J.J. (Eds.). *El Riesgo Sísmico. Prevención y Seguro*. Consorcio de Compensación de Seguros. Madrid, 139–146.

Hernández-Pacheco, E. (1950). Las rasas litorales de la costa cantábrica en su segmento asturiano. C.R. Congrès Internationnal Géographie de Lisbonne, 2, 29–86.

Herrero, J. and Aragüés, R. (1988). Suelos afectados por salinidad en Aragón. *Surcos de Aragón*, 9, 5–10.

Herrero, J., Aragüés, R. and Amezketa, E. (1993). Salt-affected soils and agriculture in the Ebro Basin. In: Gutiérrez, M., Sancho, C. and Benito, G. (Eds.). *Second Intensive Course of Applied Geomorphology: Arid Regions*. Universidad de Zaragoza-Erasmus. Zaragoza, 139–150.

Herron, S. and Langway, C.C. (1979). The debris-laden ice at the bottom of the Greenland ice sheet. *Journal Glaciology*, 23, 193–207.

Hersen, P., Andersen, K.H., Elbelrhti, H., Andreotti, B., Claudin, P. and Douady, S. (2004). Corridors of barchan dunes: stability and size selection. *Physical Review*, E69, c11304, 12 p.

Hesp, P.A. (1981). The formation of shadow dunes. *Journal of Sedimentary Petrology*, 51, 101, 111.

Hesp, P.A. (2002). Foredunes and blowouts: initiation, geomorphology and dynamics. *Geomorphology*, 48, 245–268.

Hesp, P.A. (2004). Dune coastal. In: Goudie, A.S. (Ed.). *Encyclopedia of Geomorphology*. Routledge. London, 291–295.

Hesp, P.A. and Hyde, R. (1996). Flow dynamics and geomorphology of a through blowout. *Sedimentology*, 43, 505–525.

Hesp, P.A. and Thom, B.C. (1990). Geomorphology and evolution of active transgressive dune fields. In: Nordstrom, K.F., Psuty, N.P. and Carter, R.W.G. (Eds.). *Coastal Dunes: Form and Process*. Wiley. Chichester, 253–288.

Hess, H.H. (1962). History of ocean basins. In: Engle, A.E.J. (Ed.) Petrologic Studies. A Volume in Honor of A.F. Buddington, *Geological Society of America*, Boulder, 599–620.

Hestners, E. (1985). A contribution to the prediction of slush avalanches. *Annals of Glaciology*, 6, 1–4.

Hetzel, R., Niedermann, S., Tao, M., Kubik, P.W., Ivy-Ochs, S., Gao, B. and Strecker, M.R. (2002). Low slip rates and long-term preservation of geomorphic features in Central Asia. *Nature*, 417, 428.

Hewitt, K.J. (1967). Ice-front deposition and the seasonal affect, a Himalayan example. *Transactions of the Institute of British Geographers*, 42, 93–106.

Hewitt, K.J., Clague, J.J. and Orwin, J.F. (2008). Legacies of catastrophic rock slope failures in mountain landscapes. *Earth-Science Reviews*, 87, 1–38.

Hibbert, A.R. (1967). Forest treatment effects on water yield. In: Sopper, W.E. and Luel, H.W. (Eds.). *Forest Hydrology*. Pergamon, Dordrech, 527–543.

Hicks, D.M. and Mason, P.D. (1991). *Roughness Characteristics of New Zeland Rivers*. Water Resources Survey, 329 p.

Hickson, C.J. (1990). The May 18, 1980, Eruption of Mount St. Helens, Washington State: A synopsis of events and review of Phase I from an eyewitness perspective. *Geoscience Canada*, 17, 127–131.

Hickson, C.J. and Peterson, D.W. (Eds.) (1990). Special Symposium Commemoranting the 10th Anniversary of the Eruption of Mount St. Helens, May 18, 1980. *Geoscience Canada*, 17, 125–187.

Higgins, C.G. (1956). Formation of small ventifacts. *Journal of Geology*, 64, 506–516.

Higgins, C.G. (1975). Theories of landscape development: a perspective. In: Melhorn, W.N. and Flemal, R.C. (Eds.). *Theories of Landform Development*. Allen and Unwin. Boston, 1–28.

Higgins, C.G. (1982). Grazing-step terracettes and their significance. *Zeitschrift für Geomorphologie*, 26, 459–472.

Higgins, C.G. (1984). Piping and sapping: development of landforms by groundwater flow. In: LaFleur, R.G. (Ed.). *Groundwater as a Geomorphic Agent*. Allen & Unwin. Boston, 18–58.

Higgins, C.G. and Osterkamp, W.R. (1990). Seepage-induced cliff recession and regional denudation. In: Higgins, C.G. and Coates, D.R. (Eds.). *Groundwater Geomorphology*. Geological Society of America Special Paper, 252, 291–317.

Hilger, A. (1897). Uber Verwiterunfsvorgänge bei Kristalinischen und Sedimentär-gesteinen. *Landw. Jahrb.*, 8, 1–11.

Hill, C. (2003). Intrastratal karst at the waste isolation pilot plant side, Southeastern New Mexico. In: Johnson, K.S. and Neal, J.T. (Eds.). *Evaporite karst and engineering/environmental problems in the United States.* Oklahoma Geological Survey. Circular 109, 197–209.

Hill, I.D. and Rackham, I.J. (1978). Indications of mass movement on the Jos Plateau, Nigeria. *Zeitschrift für Geomorphologie,* 22, 258–274.

Hills, E.S. (1940). The lunette, a new landform of aeolian origin. *The Australian Geographer*, 3, 15–21.

Hills, E.S. (1963). *Elements of Structural Geology*. Methuen. London, 483 p.

Hinkel, K.M., Frohn, R.C., Nelson, F.E., Eisner, W.R. and Beck, R.A. (2005), Morphometric and spatial analysis of thaw lakes and drained thaw-lake basins in the Western Arctic Coastal Plain, Alaska. *Permafrost and Periglacial Processes*, 16, 327–342.

Hjulström, F. (1935). Studies of the geomorphological activity of rivers as illustrated by the River Fyris. *Bulletin of the Geological Institute of Uppsala*, 25, 221–527.

Hodges, W.K. and Bryan, R. (1982). The influence of material behaviour on runoff initiation in the Dinosaur badlands, Canada. In: Bryan, R. and Yair, A. (Eds.). *Badland Geomorphology and Piping.* GeoBooks. Norwich, 13–46.

Hoelzmann, P.B., Kedig, H., Borke, S., Kropelin, A. and Kruse, H.J. (2001). Environmental change and archaeology: lake evolution and human occupation in the Eastern Sahara during the Holocene. *Palaeogeography, Palaeoclimatology, Palaeoecology*, 169, 193–217.

Hoffmann, G. and Schulz, H.D. (1987). Holocene stratigraphy and changing coastlines at the Mediterranean coast of Andalucia (SE Spain). In: Zazo, C. (Ed.). *Late Quaternary Sea-Level Changes in Spain. Trabajos Neógeno-cuaternario.* Museo Nacional de Ciencias Naturales, C.S.I.C, 10, 153–158. Madrid.

Höllermann, P. (1983). Blockgletscher als mesoformen der periglazialstufe. *Bonner Geographische Abhandlungen*, 67, 1–73.

Holliday, V.T. (1997). Origin and evolution of lunettes on the High Plains of Texas and New Mexico. *Quaternary Research*, 47, 54–69.

Hollinworth, S.E., Taylor, J.H. and Kellaway, G.A. (1944). Large-scale superficial structures in the Northampton Irostone Field. *Quaterly Juornal Geological Society of London*, 100, 1–44

Holmes, A. (1928). Radioactivity and continental drift. *Transaction Geological Society of Glasgow*, 18, 559–606.

Holmes, A. (1944). *Principles of Physical Geology*. Nelson, London.

Holmes, A. (1965). *Principles of Physical Geology*. Nelson. London, 1288 p.

Holmes, D.A. (1968). The recent history of the Indus. *Geographical Journal*, 134, 367–382.

Holmes, G.W., Hopkins, D.M. and Foster, H.L. (1968). Pingos in central Alaska. *Geographic Survey Bulletin*, 1341 H. Washington, 40 p.

Holness, S.D. (2003). Sorted circle in the maritime subantarctic. Marion Island. *Earth Surface Processes and Landforms*, 28, 337–347.

Holzer, T.L. (1984). Ground failure induced by ground-water withdrawal from unconsolidated sediment. In: Holzer, T.L. (Ed.). *Man-Induced Land Subsidence*. Geological Society of America. Reviews in Engineering Geology, 6, 67–105.

Holzer, T.L., Davis, S.N. and Lofgren, B.E. (1979). Faulting caused by groundwater extraction in south-central Arizona. *Journal of Geophysical Research*, 84, 603–612.

Holzhausen, G.R. (1989). Origin of sheet structure.1. Morphology and boundary conditions. *Engineering Geology*, 27, 225–278.

Hooke, J.M. (2007). Spatial variability, mechanisms and propagation of change in an active meandering river. *Geomorphology*, 84, 277–296.

Hooke, J.M. and Redmond, C.E. (1989). Use of cartographic sources for analysing river channel change with examples from Britain. In: Petts, G.E. (Ed.). *Historical Change of Large Alluvial Rivers: Western Europe*. Wiley. Chichester, 79–93.

Hooke, R. leB. (1967). Processes on arid-region alluvial fans. *Journal of Geology*, 75, 438–460.

Hooke, R. leB (1972). Geomorphic evidence of late-Wisconsin and Holocene tectonic deformation, Death Valley. California. *Geological Society of America Bulletin*, 83, 2073–2098.

Hooke, R. leB. (1987). Mass movements in semi-arid environments and the morphology of alluvial fans. En: Anderson, M.G. y Richards, K.S. *Slope Stability*. Wiley. Chichester, 505–529.

Hooke, R. leB. (1994). On the efficacy of humans as geomorphic agents. *GSA Today*, 4, 217 y 224–225.

Hooke, R. leB. (2000). On the history of humans as geomorphic agents. *Geology*, 28, 843–846.

Hooke, R. leB. and Rohrer, W.L. (1979). Geometry of alluvial fans: Effect of discharge and sediment size. *Earth Surface Processes*, 4, 147–166.

Hooke, R. leB. and Lee, B. (1987). Mass movement in semi-arid environments and the morphology of alluvial fans. In: Anderson, M.G. and Richards, K.S. (Eds.). *Slope Stability*. Wiley. Chichester, 505–529.

Hooke, R. leB., Miller, S.B. and Kohler, J. (1988). Character of englacial and subglacial drainage system in the upper part of the ablation area of Storglaciären, Sweden. *Journal of Glaciology*, 34, 228–231.

Hoppe, G. (1959). Glacial morphology and inland ice recession in northern Sweden. *Geografiska Annaler*, 41, 193–212.

Hori, K. and Saito, G. (2007). Classification, Architecture, and evolution of large-rivers deltas. En: Gupta, A. (Ed.). *Large Rivers. Geomorphology and Management*. Wiley. Chichester, 75–96.

Hori, K., Uzumoto, R., Hirouchi, D., Umitsn, M., Janjirawuttukul, N. and Patanakanog, B. (2007). Horizontal and vertical variation of 2004 Indian tsunami depostis: A example of two transects along the western coasts of Thailand. *Marine Geology*, 239, 163–172.

Horikawa, K. (1988). *Nearshore Dynamics and Coastal Processes: Theory, Measurement, and Predictive Models*. University of Tokyo Press. Tokyo, 522 p.

Horton, R.E. (1932). Drainage basin characteristics. *Transactions American Geophysical Union*, 13, 350–361.

Horton, R.E. (1933). The role of infiltration in the hydrologic cycle. *Transactions of America Geophysical Union*, 14, 446–460.

Horton, R.E. (1945). Erosional development of streams and their drainage basins: hydrophysical approach to quantitative morphology. *Bulletin of the Geological Society of America*, 56, 275–370.

Hortvik, K. (2005). Deformation of the surface ground induced by process of inundation of worked out areas after termination of underground mining exploitation. In: Konecny, (Ed.). *Eurorock 2005. Impact of Human Activity on the Geological Environment*. Balkema. Leiden, 197–203.

Hotz, R.D. and Schuster, R.L. (1996). Stabilization of soil slopes. In: Turner, A.K. and Schuster, R.L. (Eds.). *Landslides: Investigation and Mitigation*. Special Report, 247. National Academic Press. Washington, 439–473.

Houghton, J.T., Jenkins, Y.C. and Ephraums, J.J. (Eds.) (1990). *Climatic Change. The IPCC Scientific Assessment*. Cambridge University Press. Cambridge, 365 p.

Houghton, J.T., Meira Filho, L.G, Callander,.B.A., Harris, N., Kattenberg, A. and Masskell, K. (1996). Climatic Change 1995. *The Science of Climatic Change*. Published for the IPCC. Cambridge University Press. Cambridge, 572 p.

Houghton, J.T., Ding, Y., Griggs, D.J., Noguer, M., Liden, P.J. van der, Dai, X., Maskell, K. and Johnson, C.A. (Eds.) (2001). *Climate Change 2001: The Scientific Basis*. Cambridge University Press. Cambridge, 881 p.

Howard, A.D. (1942). Pediment passes and the pediment problem. *Journal of Geomorphology*, 5, 3–31.

Howard, A.D. and Dolan, R. (1981). Geomorphology of the Colorado River in the Grand Canyon. *Journal of Geology*, 89, 269–298.

Howard, A.D. and Selby, M.J. (1994). Rock slopes. In: Abrahams, A.D. and Parsons, A.J. (Eds.). *Geomorphology of Desert Environments*. Chapman and Hall. London, 123–172.

Howard, A.D., Morton, J.B., Gad-el-Hak, M. and Pierce, D.B. (1978). Sand transport model of barchan dune equilibrium. *Sedimentology*, 25, 307–338.

Howard, G.A. and Mitchell, C.W. (1985). *Phytogeomorphology*. Wiley Interscience. New York, 222 p.

Hoyos, M. and Jordá, J.F. (1989). Los yacimientos arqueológicos y paleontológicos asociados a los depósitos kársticos cuaternarios en España. In: Durán, J.J. and López-Martínez, J. (Eds.). *El Karst en España*. Monografía 4, Sociedad Española de Geomorfología. Madrid, 381–390.

Hsu, K.J. (1978). Alber Heim: Observations on landslides. In: Voight, B. (Ed.). *Rockslides and Avalanches*. Elsevier. Amsterdam, 70–93.

Hsu, K.Y. (1975). Catastrophic debris streams (Sturzstroms) generated by rockfalls. *Geological Society of America Bulletin*, 86, 129–140.

Hubbert, M.K. (1967). The strength of earth. *Bulletin of the American Association of Petroleum Geologists*, 29, 1630–1653.

Hubble, G.D., Isbell, R.F. and Northcote, K.H. (1983). Features of Australian soils. In: *Soils: an Australian viewpoint*. Academic Press. Melbourne, 17–47.

Huggett, R.J. (2003). *Fundamentals of Geomorphology*. Routledge. London, 386 p.

Huggett, R.J. (2006). *The Natural History of the Earth*. Routledge. London, 203 p.

Hughes, M.K. y Díaz, H.P. (Eds.). *The Medieval Warm Period*. Kluver. London, 342 p.

Hughes, T.J. (1992). Abrupt climatic change related to unstable ice-sheet dynamics: toward a new paradigm. *Palaeogeography, Palaeoclimatology, Palaeoecology*, 97, 203–234.

Hubp, J.L. and Inbar, M. (2002). Desastres Naturales en América Latina. In: Hubp, J.L. and Inbar., M. (Eds.). *Desastres naturales en América Latina*. Fondo de Cultura Económica, México, 9–33.

Huisink, M. (1998). Tectonic versus climatic controls on the River Maas dynamics the Late Glacial. In: Benito, G., Baker, V.R. and Gregory, K.J. (Eds.). *Palaeohydrology and Environmental Change*. Wiley. Chichester, 99–109.

Huisink, M. (2000). Changing river styles in response to Weichcelian climate changes in the Vecht valley, eastern Netherlands. *Sedimentary Geology*, 133,115–134.

Huisink, M., Moor, J.J., Kasse, C. and Virtanen, T. (2002). Factors influencing periglacial fluvial morphology in the northern European Russian tundra and taiga. *Earth Surface Processes and Landforms*, 27, 1223–1235.

Hulbe, C.L., MacAyeal, D.R., Denton, G.H., Kleman, J. and Lowell, T.V. (2004). Catastrophic ice shelf breakup as the source of Heinrich event icebergs. *Paleoceanography*, 19, PA1004, doi: 10.1024/2003PA000890.

Hulme, M. (2001). Classics in physical geography revisited. *Progress in Physical Geography* 25, 385–387.

Humlum, O. (1982). Rock glacier types on Disko, central West Greenland. *Geog. Tidsskr.*, 82, 59–66.

Hunt, C.B. and Mabey, D.R. (1966). Stratigraphy and structure, Death Valley, California. *United States Geological Survey Professional Paper* 496 A.

Hunt, C.B. and Washburn, A.L. (1960). Salt features that simulate ground patterns formed in cold climates. *United States Geological Survey*, Professional Paper, 494-B, 104–133.

Hunt, C.B., Robinson, T.W., Bowles, W.A. and Washburn, A.I. (1966). Hydrologic basin. Death Valley. California. *United States Geological Survey, Professional Paper*, 494-B.

Hunter, R.E. (1985). A kinematic model for the structure of lee-side deposits. *Sedimentology*, 32, 409–422.

Hunter, R.E., Richmond, B.M. and Alpha, T.R. (1983). Storm-controlled oblique dunes of the Oregon coast. *Bulletin of the Geological Society of America,* 94, 1450–1465.

Hutchinson, G.E. (1957). *A Treatise on Limnology: Geography, Physics and Chemistry*, vol. 1. Wiley. New York, 1015 p.

Hutchinson, J.N. (1968). Mass movement. In: Fairbridge, R.W. (Ed.). *The Encyclopedia of Geomorphology*. Reinhold. New York, 688–695.

Hutchinson, J.N. (1988). Morphological geotechnical parametres of landslides in relation to geology and hydrogeology. In: Bonnard, C. (Ed.) *Landslides*. Proceedings of 5th International Symposium on Landslide, vol. 1, 3–35.

Hutchinson, J.N. (1995). Keynote paper: landslide hazard assessment. In: Bell, D.H. (Ed.). *Landslides*. Balkema. Rotterdam, 1805–1841.

Hutton, J.T. and Dixon, J.C. (1981). The chemistry and mineralogy of some South Australian calcretes and associated soft carbonates and their dolomitization. *Journal of the Geological Society of Australia*, 28, 71–79.

I.S.R.M. (1979). Suggested methods for determining: Water content, porosity, density, absortion and related properties. Swelling and slake-durability index properties. *International Journal of Rock Mechanics and Mineral Science*, 16, 143–156.

Ibsen, M.L., Brunsden, D. Bromhead, E. and Collison, A. (1996a). Block Slide. In: Dikau, R., Brunsden, D., Shortt, L. and Ibsen, M.L. (Eds). *Landslide Recognition. Identification, Movement and Causes*. Wiley. Chichester, 64–77.

Ibsen M.L., Brunsden, D., Bromhead, E. and Collison, A. (1996b). Slab Slide. In: Dikau, A., Brunsden, D., Shortt, L. and Ibsen, M.L. (Eds.). *Landslide Recognition. Identification, Movement and Causes*. Wiley Chichester, 78–84.

Ibsen, M.L., Brunsden, D., Bromhead, E. and Collison, A. (1996c). Flow slide. In: Dikau, R., Brunsden, D., Schrott, R. and Ibsen, M.L. (Eds.). *Landslide Recognition. Identification, Movement and Causes. Wiley*. Chichester, 202–211.

ICOD (1997). (Eds.). Guía Ciudadana de los Riesgos Geológicos. Ilustre Colegio Oficial de Geólogos, 196 p. Traducción español Regueiro, M., Suárez, L. *et al.* (ampliada) de *The Citizens´ Guide to Geologic Hazards.* The American Institut of Professional Geologists, 134 p.

Idso, S.B. (1967). Dust storms. *Scientific American*, 235, 108–114.

Idso, S.B. (1974). Tornado or dust devil: enigma of desert whirlwinds. *American Scientist*, 62, 530–541.

Imeson, A.C., Kwaad, F.J.P.M., and Verstratten, J.M. (1982). The relationship of soil physical and chemical properties to the development of badlands in Morocco. In: Bryan, R.B. and Yair, A. (Eds.). *Badland geomorphology and piping*. Geobooks. Norwich, 47–70.

Ingmanson, D.E. and Wallace, W.J. (1989). *Oceanography. An Introduction*. Wadsworth. Belmont, California, 511 p.

Inman, D.L. and Jenkins, S.A. (2005). Accretion and erosion waves on beaches. In: Schwartz, M.L. (Ed.). *Encyclopedia of Coastal Science*. Springer. Dordrecht, 1–4.

Inman, D.L. and Nordstrom, C.E. (1971). On the tectonic and morphologic classifications of coasts. *Journal of Geology*, 79, 1–21.

Inman, D.L., Ewing, G.C. and Corliss, J.B. (1966). Coastal sand dunes of Guerrero Negro, Baja California. Méjico. *Bulletin of the Geological Society of America*, 77, 787–802.

Innes, J.L. (1983). Lichenometric dating of debris-flow deposits in the Scottish Highlands. *Earth Surface Processes and Landforms*, 8, 579–588.

Innes, J.L. (1985). Lichenometry. *Progress of Physical Geography*, 9, 187–254.

Inoue, S., Wijeyewickrema, A.C., Matsuimoto, H., Miura, H., Guranatna, P., Madurapperuma, M. and Sekiguchi, T. (2007). Field survey of tsunami effects in Sri Lanka due to the Sumatra-Andaman earthquake of December 26, 2004. *Pure and Applied Geophysics*, 164, 395–411.

Instituto Geológico y Minero de España (1985). *Geología y Prevención de Daños para Inundaciones*. Instituto Geológico y Minero de España. Madrid, 412 p.

International Society for Rock Mechanics (ISRM) (1972). Suggested methods for determining water content, porosity, density, absortion and related properties and swelling and slake-durability indes properties (Doc. nº 2). Commission on Standardization of Laboratory and Field Tests, 36 p.

IPCC (2001). *Climate Change 2001. The Scientific Basis*. Contribution of Working Group I to the Thris assemsnet Report on the Intergovernmental Panel on Climate Change. In: Houghton, J.T., Ding, Y., Griggs, D.J., Noguer, M., van der Linde, P.J. and Xiaosu, D. (Eds.). Cambridge University Press. Cambridge, 881 p.

IPCC Workin Group III (2005). *Carbon Dioxide Capture and Storage. Summary for Policymakers*. Cambridge University Press. Cambridge, 431 p.

IPCC (2007). Summary of Policymakers. In: *Climate Change 2007. The Physical Science Basis. Contribution on Working Group I to Fourth Assessment Report of the Intergovernamental Panel of Climate Change* (Solomon, S.D., Quin, P., Manning, M., Chen, Z., Markis, M., Averyt, K.B., Tignarr, M. and Miller, H.L.). Cambridge University Press. Cambridge, 1–18.

Iriondo, M.H. and Kröhling, D.M. (2007). Non-classical types of loess. *Sedimentary Geology*, 202, 352–368.

Isacks, B., Oliver, J. and Sykes, L.R. (1968). Seismology and the new global tectonics. *Journal of Geophysical Research*, 73, 5855–5899.

Isarin, R.F.B. (1997). Permafrost distribution and temperatures in Europe during the Youngest Dryass. *Permafrost and Periglacial Processes*, 8, 313–333.

Ishihara, K. (1989). Liquefaction-induced landslides and debris flow in Tajikistan, URSS. *Landslides News*, 3, 6–7.

ISSMFE (1985). Progress Report: Peculiarities of Geotechnical behaviour of tropical lateritic and saprolitic soils. *First International Conference and Saprolitic Soils*. Committee on Tropical Soils of International Society of Soil Mechanics and Foundation Engineering. Sao Paulo, 449 p.

Iversen, J.D., Greeley, R., Marshall, J.R. and Pollack, J. (1987). Aeolian saltation threshold: effect of density ratio. *Sedimentology*, 34, 699–706.

Iverson, N.R. (1995). Processes of erosion. In: Menzies, J. (Ed.). *Modern Glacial Environments: Processes, Dynamics and Sediments*. Butterworth-Heineman. Oxford, 241–260.

Iverson, N.R. (2006). Laboratory experiments in glaciology. In: Knight, P.G. (Ed.). *Glacier Science and Environmental Change*. Blackwell. London, 449–456.

Iverson, R.M., Logan, M., Denlinger, R.P. and Lahusen, R.J. (2005). Tansformation of valley floods to debris flows large-scale experiments. *Geological Society of America Abstract with Programms*, 37, 35.

Ivy-Ochs, S., Poschinger, H.v., Synal, H.A. and Maisch, M. (2009). Surface exposure dating of the Flims landslide, Graubünden, Switzerland. *Geomorphology*, 103, 104–112.

Jackson, N.L. (2004). Estuary. In: Goudie, A.S. (Ed). *Encyclopedia of Geomorphology*. Routledge. London, 341–344.

Jacobs, J.A., Russell, R.D. and Tuzo Wilson, J. (1959). *Physics and Geology*. McGraw. Hill, New York, 424 p.

Jahn, A. (1954). Balance de denudation de versant. *Czasopismo Geogr.*, 25, 38–64.

Jahn, A. (1964). Slopes morphological features resulting from gravitation. *Zeitscrift für Geomorphologie Supplement Band*, 5, 59–72.

Jahn, A. (1972). Tundra polygons in the Mackenzie Delta. *Göttinger Geographishe Abhandlungen*, 60, 285–292.

Jahn, A. (1975). *Problems of Periglacial Zone*. PWN Polish Scientific Publishers. Varsovia, 220 p.

Jahn, A. (1976). Geomorphological modeling and nature protection in Arctic and Subartic environments. *Geoforum*, 7, 121–137.

Jahns, R.H. (1943). Sheet structure in granites: its origin and use as a mesure of glacial erosion in New England. *Journal of Geology*, 60, 7–98.

Jain, V. and Sinha, R. (2004). Fluvial dynamics of an anabranching river system in Himalaya foreland basin, Bahma river, north Viras plains, India. *Geomorphology*, 60, 147–170.

Jakucs, L. (1977). *Morphogenetic of Karst Regions*. Adam Hilger. Bristol, 284 p.

James, A.N. (1992). *Soluble materials in civil engineering*. Ellis Horwood. Chichester, 439 p.

Jansen, R.W. (1980). *Dams and Public Safety. A Water Resources Technical Publication.* Water and Power Resources Service, United States Department of the Interior. Denver.

Jausen, J.H.F. and Van Iperen, J.M. (1991). A 220.000 years climatic record for the east equatorial Atlantic Ocean and equatorial Africa: evidence from diatoms and opal phytoliths in the Zaire (Congo) deep sea fan. *Palaeoceanography*, 6, 573–591.

Jauzein, A. (1974). Les données sur le système CaSO4.H2O et leurs implications géologiques. *Revue de Géographie Physique et de Géologie Dynamique*, 16, 151–159.

Jeffreys, H. (1929). *The Earth; its origin, history and physical constitution*. Cambridge University Press, Cambridge.

Jeje, L.K. (1973). Inselberg evolution in a humid tropical environment: the example of south-western Nigeria. *Zeitschrift für Geomorphologie,* 17, 194–225.

Jennings, J.N. (1964). The question of coastal dunes humid tropical climates. *Zeitschrift für Geomorphologie*, 8, 150–154.

Jennings, J.N. (1971). *Karst*. The M.I.T. Press. Cambridge, 252 p.

Jennings, J.N. (1973). "Any Milleniums today, lady?". The geomorphic bandwaggon parade. *Australian Geographical Studies*, 11, 115–133.

Jennings, J.N. (1983). The disregarded karst of the arid and semiarid domain. *Karstología*, 1, 61–73.

Jennings, J.N. (1985). *Karst Geomorphology*. Blackwell. Oxford, 293 p.

Jenny, H. (1941). *Factors of Soil Formation*. McGraw-Hill. New York, 281 p.

Jenny, H. (1950). Origin of soils. In: Trask, P.D. (Ed.). *Applied Sedimentation*. Wiley. New York, 41–61.

Jenny, H. (1980). *The Soil Resource*. Springer, New York, 377 p.

Jeschke, A.A., Vosbeck, K. and Dreybrodt, W. (2001). Surface controlled dissolution rates of gypsum in aequeous solutions exhibit nonlinear dissolution kinetics. *Geochimica et Cosmochimica Acta*, 65, 27–34.

Jibson, R.W. (1996). Using landslides for paleoseismic análisis. In: McCalpin, J. (Ed.). *Paleosismology*. Academic Press. London, 397–438.

Jiménez-Salas, J.A., de Justo, J.L. and Serrano, A.A. (1981). *Geotecnia y Cimientos II. Mecánica del Suelo y de las Rocas*. Rueda. Madrid, 1188 p.

Jiskoot, H., Murray, T. and Boyle, P. (2000). Controls on the distribution of surge-type glaciers in Svalbard. *Journal of Glaciology*, 46, 412–422.

Johannessen, C.L., Feireisen, J.J. and Wells, A.K. (1982). Weathering of ocean cliffs by salt expansion in midlatitude coastal environment. *Shore and Beach*, 50, 26–34.

Johnson, A.M. (1970). *Physical Processes in Geology*. Freeman. San Francisco, 577 p.

Johnson, A.M. and Rodine, J.R. (1984). Debris Flow. In: Brunsden, D. and Prior, D.V. (Eds.). *Slope Instability*. Wiley. New York, 257–361.

Johnson, D. (1919). *Shore Processes and Sholine Development*. Prentice Hall. New York, 584 p.

Johnson, D.W. (1931). Planes of lateral corrosion. *Science*, 83, 174–177.

Johnson, D.W. (1932a). Rock planes of arid regions. *Geographical Review*, 22, 656–665.

Johnson, D.W. (1932b). Rock fans of arid regions. *American Journal of Science*, 23, 389–416.

Johnson, D. (1993). Dynamic denudation evolution of tropical, subtropical and temperates landscapes with three tiered soils: toward a general theory of landscape evolution. *Quaternary International* 17, 67–78.

Johnson, D. and Balek, C. (1991). The genesis of quaternary landscapes with stone-lines. *Physical Geography*, 12, 385–395.

Johnson, D., Domier, J. and Johnson, D.N. (2005). Animating the biodynamics of soil thickness using process vector analysis: a dynamic denudation approach to soil formation. *Geomorphology* 67, 23–46.

Johnson, F.A. and Illes, P. (1976). A classification of dam failures. *Water Power Dam Cosntruct*, 28, 43–45.

Johnson, G. (1960). Cryoturbartion at Zaragoza, Northern Spain. *Zeitschrift für Geomorphologie,* 4, 74–80.

Johnson, K.S. (1989). Salt dissolution, interstratal karst, and ground subsidence in the northern part of Texas Panhandle. In: Beck, B.F. (Ed.). *Engineering and Environmental Impacts of Sinkholes and Karst.* Proceedings of the third multiplinary conference on sinkholes and the engineering and environmental impacts of karst. St. Peterburg Beach, Florida, 115–121.

Johnson, K.S. (2005). Salt dissolution and subsidence or collapse caused by human activities. *Geological Society of America. Reviews in Engineering Geology*, 16, 101–110.

Johnson, K.S. (2006). Evaporite-karst problems and studies of United States. *Environmental Geology* (en prensa).

Johnson, K.S. and Neal, J.T. (2003). *Evaporite karst and engineering environmental problems in the United Sates*. Oklahoma Geological Survey. Circular 109, 353 p.

Johnson, R.G. and Lauritzen S.E. (1995). Hudson-Bay-Hudson Strait jökulhlaups and Heinrich events: a hypothesis. *Palaeogeography, Palaeoclimatology, Palaeoecology*, 117, 123–137.

Johnson, W.H. and Menzies, J. (1996). Pleistocene supraglacial and ice-marginal deposits and landforms. In: Menzies, J. (Ed.). *Past Glacial Environments. Sediments, Forms and Techniques*. Butterworth-Heinemann. Oxford, 137–160.

Joly, F. (1950). Pédiments et glacis d'érosion dans le Sud-Est du Maroc. *Congres Internationale de Géographie*, 2, Lisboa. 110–125.

Jones, D.E. and Holtz, W.G. (1973). Expansive soils-the hidden disaster: Civil Engineering. *American Society of Civil Engineerings*, 43, 49–51.

Jones, D.K.C. (1980). British applied geomorphology: an appraisal. *Zeitschrift für Geomorphologie*. Supplementband, 36, 48–73.

Jones, D.K.C., Brunsden, D. and Goudie, A.S. (1983). A preliminary geomorphological assesment of the Karakoram Highway. *Quaterly Journal of Engineering Geology*, 16, 331–355.

Jones, F.O. (1973). Landslides of Rio de Janeiro and the Serra das Araras Scarpment, Brazil. *United States Geological Survey, Professional Paper*, 697, 42 p.

Jones, J.A.A. (1981). *The Nature of Soil Piping. A Review of Research*. Geobooks. Norwich, England, 301 p.

Jones, J.A.A. (1990). Piping effects of humid lands. In: Higgins, C. and Coates, D. (Eds.). *Ground-water Geomorphology. The Role of Subsurface Water in Earth-surface Processes and Landforms*. Geological Society of America, Special Paper, 111–138.

Jones, J.A.A. (2004). Pipe and Piping. In: Goudie, A.S. (Ed.). *Encyclopedia of Geomorphology*. Routledge. London, 784–788.

Jones, P.D. and Kelly, P.M. (1996). The effect of tropical explosive volcanic eruption on surface air temperature. In: Fiocco, G., Fuà, D. and Visconti, G. (Eds.). *The Mount Pinatubo Eruption. Effects on the Atmosphere and Climate*. NATO A51 Series, vol. 42, 95–111.

Jones, R.T. and Jordan, J.T. (2007). Lake Level Studies. In: Elías, S.A. (Ed.). *Encyclopedia of Quaternary Science*. Elsevier. Ámsterdam, 1319–1326.

Jordá, J.F. (1998). *Tectónica placas. Evolución de las ideas sobre la dinámica interna de la Tierra*. Santillana, Madrid, 112 p.

Jordan, T.H. and Minster, R.G. (1988). Measuring crustal deformation in the American West. *Scientific American*, 259, 48–56.

Jorgensen, D.W., Harvey, M.D., Schumm, S.A. and Flam, L. (1993). Morphology and dynamics of Indus River: implications for the Mohen jo Daro site. In: Shroder, J.F. Jr. (Ed.). *Himalaya to the sea*. Routledge. London, 288–326.

Jorgenson, M.T., Racine, C.H., Walters, J.C. and Osterkamp, T.E. (2001). Permafrost degradation and ecological changes associated with a warming climate in central Alaska. *Climatic Change*, 48, 551–571.

Joubert, A.M. and Hewitson, B.C. (1997). Simulating present and future climates of southern Africa using general circulation models. *Progress in Physical Geography*, 21, 51–78.

Jouzel, J. and Lorius, C. (1999). Evolution du climat: du passé récent vers le futur. *Compte Rendus de la Académie des Sciences de Paris*, 328, 229–239.

Jouzel, J. Lorius, C., Petit, J.R., Barkev, N.I., Kotlyakov, V.M. and Petrov, V.M. (1987). Vostok ice core. A continous isotopic temperature record over the last climatic cycle (160.000 years). *Nature*, 329, 403–408.

Judson, S. and Andrews, G.W. (1955). Pattern and form of some valleys in the Drift Area, Wisconsin. *Journal of Geology*, 63, 328–336.

Jungerius, P.D. (1984). A simulation model of blowout development. *Earth Surface Processes and Landforms*, 9, 509–512.

Jungerius, P.D. and van der Meulen, F. (1989). The development of dune blowouts, as measured with erosion pins and sequential air photos. *Catena*, 16, 369–376.

Jurasz, K. and Niewiarowski, W. (1962). *Instruction to the Detailed Geomorphological Map of the Polish Lowland 1:50.000*. Polish Acad. Sciences Geographical Institute. Torún.

Justo, J.L., Azañón, J.M., Azor, A., Saura, J., Durán, P., Villalobos, M., Morales, A. and Justo, E. (2008). Neotectonics and slope stabilization at the Alhambra, Granada, Spain. *Engineering Geology*, 100, 101–119.

Kääb, A. (2007). Periglacial Landforms. Rock Glaciers and Protalus Forms. In: Elias, S.A. (Ed.). *Encyclopedia of Quaternary Science*. Elsevier. Amsterdam, 2236–2242.

Kamb, B. and LaChapelle, E.R. (1964). Direct observations of the mechanism of glacier sliding over bedrock. *Journal of Glaciology*, 6, 159–172.

Kapitsa, A.P., Ridley, J.K., Robin, G. de Q., Siegert, M.J. and Zotikov, I.A. (1996). A large deep freshwater lake beneath the ice of central East Antarctica. *Nature*, 381, 684–686.

Kaplar, C.W. (1965). Stone migration by freezing of soils. *Science*, 149, 1520–1521.

Karrasch, H. (1983). Die Periglaziale tal und reliefa symmetrie. *Abhandlungen der Akademie der Wissenschaften*, 35, 310–327.

Kasse, C. (1997). Cold-climate sand-sheet formation in North-Western Europe (C^{14-}12,4 Ka); a response to permafrost degradation and increased aridity. *Permafrost and Periglacial Processes*, 8, 295–311.

Kasse, C. (1998). Depositional model for cold-climate tundra rivers. In: Benito, G., Baker, V.R. and Gregory, K.J. (Eds.). *Palaeohydrology and Environmental Change*. Wiley. Chichester, 83–97.

Kasse, C. (2002). Sandy eolian deposits and environments and their relation to climate during the Last Glacial Maximum and Late glacial in northwest and centre Europe. *Progress in Physical Geography*, 26, 507–532.

Kassler, P. (1973). The structural and geomorphic evolution of the Persian Gulf. In: Purser, B.H. (Ed.). *The Persian Gulf*. Springer Verlag. New York, 11–32.

Kaufman, D.S., Miller, G.H. and Andrews, J.T. (1992). Aminoacid composition as taxonomic tool for molluscan fossils: an example from Pliocene-Pleistocene Arctic marine deposits. *Acta Geochemica and Cosmochimica*, 56, 2445–2453.

Kaye, C.A. (1967). Kaolinisation of bedrock of the Boston, Massachusetts area. *United States Geological Survey Professional Paper*, 575-C, C165-C172.

Kearey, P. and Vine, F.J. (1996). *Global Tectonics*. Blackwell, Oxford, 333p.

Keller, C. (1946). *El Departamento de Arica*. Ministerio de Economía y Comercio. Santiago de Chile.

Keller, E.A. and Blogget, R.H. (2006). *Natural Hazards: Earth Processes as Hazards, Disasters and Catastrophes*. Prentice Hall. (Trad. esp. P. Gil. Riesgos Naturales, Procesos de la Tierra como riesgos, desastres y catástrofes. Pearson-Prentice Hall. Madrid, 422 p.).

Keller, E.A. and Pinter, N. (1996). *Active Tectonics. Earthquakes, Uphift and Landscape*. Prentice Hall. News Jersey, 338 p.

Keller, W.D. (1957). *The Principles of Chemical Weathering*. Lucas Brothers. Columbia, Missouri, 111 p.

Kelletat, D., Whelan, F., Bartel, P. and Scheffers, A. (2005). New tsunami evidences in southern Spain–Cabo Trafalgar and Mallorca Island. In: Sanjaume, E. and Mateu, J.F. (Eds.). *Geomorfología litoral i Quaternari*. PUV. Valencia, 215–222.

Kelletat, D., Scheffers, S.R. and Scheffers, A. (2007). Field signatures of the SE-Asia mega-tsunami along west coast of Thailand compared to Holocene paleo-tsunami from the Atlantic region. *Pure and Applied Geophysics*, 164, 413–431.

Kelly, M., Black, S. and Rowna, J.S. (2000). A calcrete-based U/Th chronology for landform evolution in the Sorbas Basin, south east Spain. *Quaternary Science Reviews*, 19, 995–1010.

Kemmis, T.L. (1989). Importance of the vegetation process to certain properties of basal tills deposited by the Laurentide ice sheet in Iowa and Illinois, USA. *Annals of Glaciology*, 2, 147–152.

Kench, P.S., Nichol, S.L., Smithers, S.G., McLean, R.F. and Brander, R.W. (2008): Tsunami as agents of geomorphic change in mid-ocean reef islands. *Geomorphology*, 95, 361–383.

Kennedy, B.A. (2004a). Systems in Geomorphology. In: Goudie, A.S. (Ed.). *Encyclopedia of Geomorphology*. Routledge. London, 1031–1033.

Kennedy, B.A. (2004b). *Inventing the Earth. Ideas on landscape development since 1740*. Blackwell, Oxford, 160 p.

Keown, M.B., Dardeau, E.A.J. and Causey, E.M. (1986). Historic trends in the sediment flow regime of the Mississippi river. *Water Reosurces Research*, 22, 1555–1564.

Kerans, Ch. (1988). Karst-controlled reservoir heterogeneity in Ellenberger Group Carbonates of West Texas. *American Association Petroleum Geology Bulletin*, 72, 1160–1183.

Kerr, R.C. and Nigra, J.O. (1952). Eolian sand control. *Bulletin American Associattion Petroleum Geologists*, 36, 1541–1573.

Kershaw, A.P., D'Costa, D.M., McEwan Mason, J.R.C. and Wagstaff, B.E. (1991). Palynological evidence for Quaternary vegetation and environments of mainland southwestern Australia. *Quaternary Science Reviews*, 10, 391–404.

Kershaw, A.P., Quilty, P.G., David, B., Van Huet, S. and McMinn, A. (2000). Paleobiogeography of the Quaternary of Australia. *Memoirs of the Association of Australian Paleontologists*, 23, 471–515.

Kershaw, G.P. and Gill, D. (1979). Growth and decay of palsas and peat plateaux in the Macmillan Pass - Tsichu River area, Northwest Territories, Canada. *Canadian Journal of Earth Sciences*, 16, 1362–1367.

Kesel, R.H. (1973). Inselberg landform elements: definition and synthesis. *Revue de Géomorphologique Dynamique*, 22, 97–108.

Kesel, R.H. (1977a). Slope runoff and denudation in the Rupununi savanna, Guyana. *Journal of Tropical Geography*, 44, 33–42.

Kesel, R.H. (1977b). Some aspects of the geomorphology of inselbergs in central Arizona, U.S.A.*Zeitschrift für Geomorphologie*, 21, 119–146.

Kesseli, J.E. and Beaty, C.B. (1959). *Desert flat conditions in the White mountains of California and Nevada*. United States Army Quatermarter Research and Engineering Centre. Technical Report, EP-108.

Keylock, C. (1997). Snow avalanches. *Progress in Physical Geography*, 21, 481–500.

Khalil, G.M. (1990). Floods in Bangladesh: a question of disciplining the rivers. *Natural Hazards*, 3, 379–401.

Khan, H.R. (1971). *Laboratory study of river morphology*. Ph. D. Dissertation. Colorado State University, 189 p.

Kiersch, G.A. (1964). Vaiont reservoir disaster. *Civil Engineering*, 34, 32–39.

Kiersch, G.A. (1983). The Vaiont reservoir disaster. In: Tank, R.W (Ed). *Enviromental Geology*. Oxford University Press. Oxford, 151–162.

Kiewietdejonge, C.J. (1984). Büdel's geomorphology II. *Progress in Physical Geography*, 8, 365–397.

Kilburn, C.R.J. and Petley, D.N. (2003). Forecasting giant, catastrophic slope collapse: lessons from Vaiont, Northern Italy. *Geomorphology*, 54, 21–32.

Killigrew, L.P. and Gilkes, R.J. (1974). Development of playa lakes in south western Australia. *Nature*, 247, 454–455.

King, C.A.M. (1966). *Techniques in Geomorphology*. Arnold. London, 342 p.

King, C.A.M. (1970). *Beaches and Coasts*. Arnold. London, 570 p.

King, C.A.M. (1976). Introduction. In: King, C.A.M. (Ed.). *Landforms and Geomorphology. Concepts and History.* Dowden, Hutchinson and Ross. Stroudsburg, 1–10.

King, L.C. (1942). *South African Scenery. A Texbook of Geomorphology*. Oliver and Boyd. Edinburgh, 308 p.

King, L.C. (1948). A theory of bornhardts. *Geographical Journal,* 112, 83–87.

King, L.C. (1949). The pediment problem: some current problems. *Geological Magazine*, 86, 245–250.

King, L.C. (1953). Canons of landscape evolution. *Bulletin of Geological Society of America*, 64, 721–752.

King, L.C. (1957a). Landscape study in southern Africa. *Proceedings Geological Society of South Africa*, 50, 22–52.

King, L.C. (1957b). The uniformitarian nature of hillslopes. *Transactions of the Edinburgh Geological Society*, 17, 81–102.

King, L.C. (1961). Cymatogeny. *Transactions Geological Society of South Africa*, 64.

King, L.C. (1962). *The Morphology of the Earth. A Study and Synthesis of World Scenery.* Oliver and Boyd. Edinburgh, 726 p.

King, L.C. (1975). Bornhardt landforms and what they teach. *Zeitschrift für Geomorphologie,* 19, 299–318.

King, L.C. (1976). Planation remnants upon high lands. *Zeitschrift für Geomorphologie N.F.*, 20, 133–148.

King, P.B. and Schumm, S.A. (1980). *The Physical Geography (Geomorphology) of William Morris Davis*. Geo Books. Norwich, 217 p.

Kirkbride, M.P. (1995). Processes of transportation. In: Menzies, J. (Ed.). *Modern Glacial Environments: Processes, Dynamics and Sediments.* Butterworth-Heineman. Oxford, 261–292.

Kirkby, A.T.V. and Kirkby, M.J. (1974). Surface wash at the semiarid break in slope. *Zeitschrift für Geomorphologie,* Supplementband, 21, 151–171.

Kirkby, E., Whipple, K.X., Tang, W.Q. and Chen, Z.L. (2003). Distribution of active rock uplift along the eastern margin of the betan Plate: interference from bedrock channel longitudinal profiles. *Journal of Geophysical Research-Solid Earth*, 108(B4), 2217.

Kirkby, M.J. (1969). Erosion by water on hillslopes. In: Chorley, R.J. (Ed.). *Water, Earth and Man*. Methuen. London, 229–238.

Kirkby, M.J. (1976). Hydrological slope models: the influence of climate. In: Derbyshire, E. (Ed.). *Geomorphology and Climate*. Wiley. Chichester, 247–267.

Kirkby, M.J. (1987). Models in Physics Geography. In: Clark, M.J., Gregory, K.J. and Gurnell, A.M. (Eds.). *Horizons in Physical Geography*. Macmillan. London, 47–61.

Kirkby, M.J. (1994). Thresholds and instability in stream head hollow: a model of magnitude and frequency for wash processes. In: Kirkby, M.J. (Ed.). *Process, Models and Theoretical Geomorphology*. Wiley. Chichester, 215–314.

Kiss, T. and Sipos, G. (2007). Braid-scale channel geometry changes in a sand-bedded river: Significance of low stages. *Geomorphology*, 84, 209–221.

Klammer, C. (1982). Die Palaeowüste des Pantanal von Mato Grosso und die Pleistozane Klimageschichte der Brazilianischen Randtropen. *Zeitschrift für Geomorphologie*, 26, 393–416.

Klappa, C.F. (1979). Lichen stromatolites: criterion for subaerial exposure and a mechanism for the formation of laminar calcretes (caliche). *Journal of Sedimentary Petrology,* 49, 387–400.

Klein, C. (1993). Du Dynamisme des Processus à la Dynamique des Formes en Géomorphologie. Ed. Ophrys. Gap, 299 p.

Klein, M. (1984). Weathering rates of limestones tombstones measured in Haifa, Israel. *Zeitscrift für Geomorphologie*, 28, 105–111.

Klimaszewski, M. (1956). The principles of geomorphological survey in Poland. *Przeglad Geograficzny*, 28, suppl., 32–40.

Klimaszewski, M. (1963). The principles of geomorphological mapping in Poland+annex: Landform list and signs used in the detailed geomorphological map. Problems of geomorphological mapping. *Geogr. Studies*, 46, 67–71 y 139–177.

Klimchouk, A. (2000a). The formation of epikarst and its role in the vadose speleogenesis. In: Klimchouk, A.B., Ford, D.C., Palmer, A.N. and Dreybrodt, W. (Eds.). *Speleogenesis Evolution of Karst Aquifers*. National Speleological Society. Huntsville, 91–99.

Klimchouk, A. (2000b). Speleogenesis in gypsum. In: Klimchouk, A.B, Ford, D.C., Palmer, A.N. and Dreybrodt, W. (Eds.). *Speleogenesis. Evolution of Karst Aquifers*. National Speleological Society. Huntsville. Alabama, 431–442.

Klimchouk, A. (2004a). Caves. In: Gunn, J. (Ed.). *Encyclopedia of Caves and Karst Science*. Fitzroy Dearborn. New York, 203–205.

Klimchouk, A. (2004b). Evaporite karst. In: Gunn, J. (Ed.). *Encyclopedia of Caves and Karst Science*. Fitzroy Dearborn. New York, 343–347.

Klimchouk, A. (2004c). Ukraine gypsum caves and karst. In: Gunn, J. (Ed.) *Encyclopedia of Caves and Karst Science*. Fitzroy Dearborn. New York, 745–748.

Klimchouk, A. (2005). Gypsum Caves. In: Culver, C.C. and White, W.D. (Eds.). *Encyclopedia of Caves*. Elsevier. Ámsterdam, 283–288.

Klimchouk, A. (2009). Morphogenesis of hypogenic caves. *Geomorphology*, 106, 100–117.

Klimchouk, A. and Andrejchuk, K.V. (1996a). Breakdown development in cover beds and landscape features induced by intrastratal gypsum karst. *International Journal of Speleology*, 25, 127–144.

Klimchouk, A. and Andrejchuk, K.V. (1996b). Environmental problems in gypsum karst terrains. *International Journal of Speleology*, 25, 145–156.

Klimchouk, A. and Andrejchuk, K.V. (2005). Karst breakdown mechanisms from observations in the gypsum caves of the Western Ukranie: implications for subsidence hazard assessment. *Environmental Geology*, 48, 336–359.

Knez, M. and Slabe, T. (2004). Highways on Karst. In: Gunn, J. (Ed.). *Encyclopedia of Caves and Karst Science*. Fitzroy Dearborn. New York, 419–420.

Knight, J. (2008). The environmental significance of ventifacts: A critical review. *Earth-Science Reviews*, 86, 89–105.

Knight, M.J. (1980). Structural analysis and mechanical origins of gilgai at Boorook, Victoria, Australia. *Geoderma*, 23, 245–283.

Knight, P-G., Waler, R.I., Patterson, C.J., Jones, A.P. and Robinson, Z.P. (2002). Discharge of debris flow ice at the margin of the Greenland ice sheet. *Journal of Glaciology*, 48, 192–200.

Knighton, D. (1984). *Fluvial Form and Processes*. Arnold. London, 218 p.

Knighton, D. (1998). *Fluvial Form and Processes. A New Perspective*. Arnold. London, 383 p.

Knighton, D. and Nanson, G.C. (1993). Anastomosis and the continuum of channel pattern. *Earth Surface Processes and Landforms*, 18, 613–625.

Knodel, P.C. (1991). *Characteristics and problems of dispersive clay soils*. Research and Laboratory Services Division Materials Engineering Branch, United States Department Interior.Denver.

Knott, P. and Warren, A. (1981). Aeolian processes. In: Goudie, A.S. (Ed.). *Geomorphological Techniques*. Allen and Unwin. London, 226–246.

Knox, J.C. (1972). Valley alluviation in south western Wisconsin. *Annals Association American Geographers*, 62, 401–410.

Knox, J.C. (1984a). Responses of river systems to Holocene climates. In: Wright Jr, H.E. (Ed.). *Late Quaternary Environments of the United States, Volume 2. The Holocene*. University of Minnesota Press. Minneapolis, 26–41.

Knox, J.C. (1984b). Fluvial response to small scale climate changes. In: Costa, J.E. and Fleisher, P.J. (Eds.). *Development and Applications of Geomorphology*. Springer, Berlin, 318–342.

Knox, J.C. (1985). Responses of floods to Holocene climatic change in the Upper Mississippi Valley. *Quaternary Research*, 23, 287–300.

Knox, J.C. (1988). Climatic influence on upper Mississippi valley flood. In: Baker, V.R., Craig Kochel, R. and Patton, P.C. (Eds.). *Flood Geomorphology*. Wiley. New York, 279–300.

Knox, J.C. (1993). Large increases in flood magnitude in response to modest changes in climate. *Nature*, 361, 430–432.

Knox, J.C. (1995). Fluvial systems since 20.000 years BP. In: Gregory, K.J., Starkel, L. and Baker, V.R. (Eds.). *Global Continental Palaeohydrology*. Wiley. Chichester, 87–108.

Knox, J.C. (2000). Sensivity of model and Holocene flood to climate change. *Quaternary Science Reviews*, 19, 439–457.

Knox, J.C. (2001). Agricultural influence on landscape sensivity in the Upper Mississippi River Valley. *Catena*, 42, 193–204.

Knox, J.C. (2002). Agriculture, erosion and sediment yields. In: Orme, A.R. (Ed.). *The Physical Geography of North America*. Oxford University Press. Oxford, 482–500.

Knox, J.C. (2006). Floodplain sedimentation in the Upper Mississippi Valley: natural versus human accelerated. *Geomorphology*, 97, 286–310.

Knox, J.C. (2007). The Mississippi river system. In: Gupta, A. (Ed.). *Large Rivers. Geomorphology and Mangement*. Wiley. Chichester, 145–182.

Kochel, R.C. and Baker, V.R. (1988). Paleoflood analysis using slackwater deposits. In: Baker, V.R., Craig Kochel, R. and Patton, R.C. (Eds.). *Flood Geomorphology*. Wiley, New York, 357–376.

Kocurek, G. (1988). First order and super bounding surfaces in eolian sequences-Bounding surfaces revisited. *Sedimentary Geology*, 56, 193–206.

Kocurek, G. and Nielson, J. (1986). Conditions favourable for the formation of warm-climate aeolian sand sheets. *Sedimentology*, 33, 795–816.

Kocurek, G., Harholm, K.G., Deynoux, M. and Blakey, R.C. (1991). Amalgamated accumulations resulting from climatic and eustatic changes, Akchar Erg, Mauritania. *Sedimentology*, 38, 751–772.

Kocurek, G., Townsley, M., Yeh, E., Sweet, M. and Havholm, K. (1992). Dune and dune-field development stages on Padre Island, Texas: effects of lee airflow and sand saturation levels and implications for interdune deposition. *Journal of Sedimentary Petrology*, 62, 622–635.

Kolla, V., Biscaye, P.E. and Hanley, A.F. (1979). Distribution of quartz in Late Quaternary Atlantic sediments in relation to climate. *Quaternary Research*, 11, 261–277.

Kondolf, G.M., Lisle, T.E. and Wolman, G.M. (2003). Bed sediment measurement. In: Kondolf, G.M. and Piégay, H. (Eds.). *Tools in Fluvial Geomorphology*. Wiley. Chichester, 347–395.

Koons, D. (1955). Cliff retreat in the southwestern United States. *American Journal of Science*, 253; 44–52.

Köppen, W. (1936). Das geographische system der klimate. In: *Handbuch der Klimatologie*, 1, Gerbruder Borntraeger. Berlin.

Kor, P.S.G., Shaw, J. and Sharpe, D.R. (1991). Erosion of bedrock by subglacial meltwater, Georgian Bay, Ontario: a regional review. *Canadian Journal of Earth Sciences*, 28, 623–642.

Koster, E.A. (1988). Ancient and modern cold climate aeolian sand deposition: a review. *Journal of Quaternary Science*, 3, 69–83.

Koster, E.A. and Dijkmans, J.W.A. (1988). Niveo-aeolian deposits and denivation forms, with special reference to the Great Kobuk Sand Dunes, Northwestern Alaska. *Earth Surface Processes and Landforms*, 13, 153–170.

Kosters, E.C. and Sutters, J.R. (1993). Facies relationships and system tracts in the late Holocene Mississippi delta plain. *Journal Sedimentary Petrology*, 63, 727–733.

Kozarski, S. and Rotnick, K. (1977). Valley floods and changes of river channel pattern in the north Polish plain during the late Wurm and Holocene. *Questiones Geographicae*, 4, 51–93.

Krahn, J., Johnson, R.F., Fredlund, D.G. and Clifton, A.W. (1979). A highway failure in Cretaceous sediments at Maymount, Saskatchewan. *Canadian Geotechnical Journal*, 16, 703–715.

Kranjc, A. (2004). Kras, Eslovenia. In: Gunn, J. (Ed.). *Encyclopedia of Caves and Karst Science*. Fitzroy Dearborn. New York, 485–486.

Krauskopf, K.B. (1956). Dissolution and precipitation of silica at low temperatures. *Acta Geochimica et Cosmochimica*, 10, 1–26.

Krinsley, D. (1970). *A Geomorphological and Paleoclimatological Study of the Playas of Iran*. United States Geological Survey, 2 vols. Washington, 486 p.

Krinsley, D. (1998). Models of rock varnish formation constrained by high resolution transmission electron microscopy. *Sedimentology*, 45, 711–725.

Krinsley, D. and Wellendorf, W. (1980). Wind velocities determined from the surface textures of sand grains. *Nature*, 283, 372–373.

Krohn, J.P. (1992). Landslide mitigation using horizontal drainages, Pacific Palisades area, Los Angeles, California. *Reviews in Engineering Geology*, vol. 10, 63–68. Geological Society of America.

Krumbein, W.E. and Jens, K. (1981). Biogenic rock varnishes of the Negev Desert (Israel): an ecological study of iron and manganese transformation by cyano-bacteria and fungi. *Oecologia*, 50, 25–28.

Kuenen, Ph. (1960). Experimental abrasion, 4: eolian action. *Journal of Geology*, 68, 427–449.

Kühnel, R.A. (1987). The role of cationic and anionic scavengers in laterites. *Chemical Geology*, 60, 31–40.

Kumar, R.D. and Hasegawa, S. (2008). Representative rainfall for landslides in the Nepal Himalaya. *Geomorphology*, 100, 429–443.

Kutiel, P., Kutiel, H. and Lavee, H. (2000). Vegetation response to possible scenarios of rainfall variations along a Mediterranean-extreme arid system transect. *Journal of Arid Environments*, 44, 277–290.

Labeyrie, L., Skinner, L. and Cortijo, E. (2007). Sub-Milankovich (D/O/Heinrich) Events. In: Elias, S.A. (Ed.). *Encyclopedia of Quaternary Science*. Elsevier. Amsterdam, 1964–1974.

Lachenbruch, A.H. (1962). *Mechanisms of thermal contraction cracks and ice-wedge polygons in permafrost*. Geological Society of America, Special Paper 70, 65 p.

Lachenbruch, A.H. (1970). Thermal considerations in permafrost. In: Adkison, W.L. and Borsge, M.M. (Eds.).

Geological Seminar on the North Slope of Alaska Proceedings American Association Petroleum Geologists, J1–2 y discussion J2–5. Los Angeles.

Lacroix, A. (1904). *La Montagne Pelée et ses éruptions*. Masson. Paris, 662 p.

Ladd, H.S., Tracey, J.I. and Gross, M.C (1967). Drilling on Midway Atoll, Hawaii. *Science*, 156, 1088–1094.

Lain, R. (2002). Técnicas de estabilización de laderas y taludes. In: Ayala, F.J. and Olcina, J. (Eds.). *Riesgos Naturales*. Ariel Ciencia. Barcelona, 421–444.

Laity, J.E. (1994). Landforms of aeolian erosion. In: Abrahams, A.D. and Parsons, A.J. (Eds.). *Geomorphology of Desert Environments*. Chapman and Hall. London, 506–537.

Laity, J.E. (1995). Wind abrasion and ventifact formation in California. In: Tchakerian, V.P. (Ed.). *Desert Aeolian Processes*. Chapman and Hall. London, 295–321.

Laity, J.E. (2008). *Deserts and desert environments*. Wiley-Blackwell. Chichester, 342 p.

Laity, J.E. and Malin, M.C. (1985). Sapping processes and the development of theater-headed valley networks in the Colorado Plateau. *Bulletin of the Geological Society of America*, 96, 203–217.

Lal, R. (1986). Deforestation and soil erosion. In: Lal, R., Sánchez, P.A. and Cummings, R.W. (Eds.). *Land Clearing and Development in the Tropics*. Balkema. Rotterdam, 299–315.

Lal, R., Lawson, T.L. and Anastase, A.H. (1981). Erosivity of tropical rains. In: De Broodt, M. and Gabriels, D. (Eds.). *Assessment of Erosion*. Wiley. Chichester, 143–151.

LaMarche, V.C. and Hirschboek, K.K. (1984). Frost rings in trees as records of major volcanic eruptions. *Nature*, 307, 121–126.

Lamb, H.H. (1970). Volcanic dust in the atmosphere, with a chronology and assessment of its meteorological significance. *Phylosophical Transactions of the Royal Society of London*, A226, 425–533.

Lamb, H.H. (1972). *Climates: Present, Past and Future*. Vol I, Methuen. London, 613 p.

Lamb, H.H. (1977). *Climates: Present, Past and Future*, Vol. II. Methuen. London, 835 p.

Lamb, H.H. (1982). *Climate, History and the Modern World*. Methuen. London, 387 p.

Lambeck, K. and Chapell, J. (2001). Sea level change through the last glacial cycle. *Science*, 292, 679–686.

LaMoreaux, P.E. (1985). *Panel for the Study of Management of Radioactive Waste at the Oak Ridge NationalLaboratory: A Technical Review*. Board of Radioactive Waste Management, Commision of Physical Sciences, Mathematics and Resources. National Research Council. National Academy Press. Washington, 176 p.

LaMoreaux, P.E., Powell, W.J. and Lefrand, A.E. (1997). Environmental and legal aspects of karst areas. *Environmental Geology*, 29, 23–36.

Lancaster, N. (1978). The pans of the Southern Kalahari, Botswana. *Geographical Journal*, 144, 80–98.

Lancaster, N. (1981). Palaeoenvironmental implications of fixed dune systems in southern Africa. *Palaeogeography, Palaeoclimatology, Palaeoecology*, 33, 327–346.

Lancaster, N. (1982). Dunes on the Skeleton Coast, Namibia (South West Africa): Geomorphology and grain size relationships. *Earth Surface Processes and Landforms*, 7, 575–587.

Lancaster, N. (1983). Linear dunes of the Namib sand sea. *Zeitschrift für Geomorphologie Supplementband*, 45, 27–49.

Lancaster, N. (1985). Variations in wind velocity and sand transport on the windward flanks of desert sand dunes. *Sedimentology*, 32, 581–593.

Lancaster, N. (1987). Variations in wind velocity and sand transport on the windward flanks of desert sand dunes-reply. *Sedimentology*, 43, 516–520.

Lancaster, N. (1988). Development of linear dunes in the southwestern Kalahari, southern Africa. *Journal of Arid Environments*, 14, 233–244.

Lancaster, N. (1989a). *The Namib Sand Sea: Dune Forms, Processes and Sediments*. Balkema. Rotterdam, 180 p.

Lancaster, N. (1989b). Star dunes. *Progress in Physical Geography*, 13, 67–91.

Lancaster, N. (1989c). The dynamics of star dunes: an example from the Gran Desierto, Mexico. *Sedimentology*, 36, 273–289.

Lancaster, N. (1994). Dune morphology and dynamics. In: Abrahams, A.D. and Parsons, A.J. (Eds.). *Geomorphology of Desert Environments*. Chapman and Hall. London, 474–505.

Lancaster, N. (1995). *Geomorphology of Desert Dunes*. Routledge. London, 290 p.

Lancaster, N. (2007). Dune Fields. Low Latitudes. In: Elias, S.A. (Ed.). *Encyclopedia of Quaternary Science*. Elsevier. Amsterdam, 626–642.

Lancaster, N. and Ollier, C.D. (1983). Sources of sand for the Namib Sand Sea. *Zeitschrift für Geomorphologie Supplementband*, 45, 71–83.

Lancaster, N., Nickling, W.G., McKenna, C. and Wyatt, W.E. (1996). Sediment flux and airflow on the stoss slope of a barchan dune. *Geomorphology*, 17, 55–62.

Land, L.S. (1970). Phreatic versus vadose meteoric diagenesis of limestones: evidence from a fossil water table. *Sedimentology* 14, 175–185.

Landy, J.P., Singh, K.D. and Janz, K. (1991). FAO's 1990 reassesment of tropical forest cover. *Nature and Resources*, 27, 21–26.

Langbein, W.B and Schumm, S.S. (1958). Yield of sediment in relation to mean annual precipitation. *Transactions American Geophysical Union*, 39, 1076–1084.

Langbein, W.B. (1961). *Salinity and hydrology of closed lakes*. United States Geological Survey, Professional Paper, 412, 1–20.

Langbein, W.B. *et al.*, (1949). Annual runoff in the United States. *United States Geological Survey*, Circular 52, 14 p.

Langbein, W.B. and Schumm, S.A. (1958). Yield of sediment in relation to mean annual precipitation. *Transactions of the American Geophysical Union*, 39, 1076–1084.

Laraque, A., Brisquet, J.P., Olivry, J.C. and Berthelot, M. (1995). Transport solides et disouss du fleuve Congo (Bilan de six annèes d'observations). In: Olivry, J.C. and Boulegue, J. (Eds.). *Grandes bassis fluviaux péri-atlantiques: Congo, Niger, Amazone*. ORSTOM. Paris, 133–145.

Laronne, J.B. and Carson, M.A. (1976). Interrelationships between bed morphology and bed-material transport for a small, gravel-bed channel. *Sedimentology*, 23, 67–85.

Larsen, M.C., Wieczorek, G.F., Eaton, S. and Sierra, H.T. (2001). The Venezuela landslide and flash flood disaster of December 1999. In: Mungai, A. (Ed.). *2nd Plinius Conference on Mediterranean Storms*, 16–18 October 2000. Siena, Italia.

Latrubesse, E.M. (2008). Patterns of anabranching channels: the ultimate end-member adjustment of mega rivers. *Geomorphology*, 101, 130–145.

Latrubesse, E.M., Aquino da Silva, S. and Moraes, M.J. (2000). Floods in the city of Rio Branco, Brazil: A case study of the impacts of human activities on flood dynamics and effects. In: Marchetti, M. and Rivas, V. (Eds.). *Geomorphology and Environmental Assesment*. Balkema. Lisse, 163–175.

Latrubesse, E.M., Stevaux, J.C. and Sinha, R. (2005). Tropical rivers. *Geomorphology*, 70, 187–206.

Lattman, L.H. (1960). Cross section of flood plain in a moist region of moderate relief. *Journal Sedimentary Petrology*, 30, 275–282.

Lautridou, J.P. (1988). Recent advances in cryogenic weathering. In: Clark, M.J. (Ed.). *Advances in Periglacial Geomorphology.* Wiley. Chichester, 33–47.

Lautridou, J.P. and Francou, B. (1992). Present-day periglacial processes and landforms in mountain areas. *Permafrost and Periglacial Processes*, 3, 93–101.

Lautridou, J.P. and Ozouf, J.C. (1982). Experimental frost shattering 15 years of research at the Centre de Geomorphology du CNRS. *Progress in Physical Geography*, 5, 215–232.

Lawson, A.C. (1915). The epigene profiles of the desert. *University of California Department of Geology Bulletin*, 9, 23–48.

Lawson, D.E. (1979). A comparison of the pebble orientations in ice and deposits of the Matanuska glacier, Alaska. *Journal of Geology*, 87, 629–645.

Lazaridis, G.I. and Pipera, K. (2008). Preliminary Report on granite caves in Greece. *Cadernos Laboratorio Xeoloxico de Laxe*, 33, 101–113.

Lázaro, R., Cantón, Y., Solé-Benet, A., Bevan, J., Alexander, R., Sancho, L. and Puigdefábregas, J. (2008). The influence of competition between lichen colonization and erosion on the evolution surfaces in the Tabernas badlands (SE Spain) and its landscape effects. *Geomorphology*, 102, 252–266.

Le Pichon, X. (1968). Sea floor spreading and continental drift. *Journal of Geophysical Research*, 73, 3661–3697.

Lecce, S.A. (1990). The alluvial fan problem. In: Rachocki, A.H. and Church, M. (Eds.). *Alluvial Fans. A Field Approach*. Wiley. Chichester, 3–24.

Lee, J.A., Wigner, K.A. and Gregory, J.M. (1993). Drought, wind and blowing dust on the southern High Plains of the United States. *Physical Geography*, 14, 56–67.

Lee, M. and Fookes, P. (2005). Climate and weathering. In: Fookes, P.G., Lee, E.M. and Milligan, G. (Eds.). *Geomorphology for Engineers*. C.R. Press. Boca Raton (USA), 1–56.

Lee, S.W. (1981). Landslides in Taiwan. *Proceedings South East Asian Regional Symposium on Problems on Soil Erosion and Sedimentation*. Bangkok, 195–206.

Leeder, M.R. and Alexander, J. (1987). The origin and tectonic significance of asymetrical meander-belts. *Sedimentology*, 34, 217–226.

Leeder, M.R. and Gawthorpe, R.L. (1987). Sedimentary models for extensional tilt block/half-graben basins. In: Coward, M.P., Dewey, J.F. and Hancock, P.L. (Eds.). *Continental Extensional Tectonics*. Geological Society of London, Special Publication 28, 139–152.

Lefebvre, G. (1996). Soft sensitive clays. In: Turner, H.K. and Schuster, R.L. (Eds.). *Landslides. Investigation and Mitigation*. National Academic Press. Washington, 607–619.

Legget, R.F. (1974). Glacial landforms and civil engineering. In: Coates, D.R. (Eds.). *Glacial Geomorphology*. State University of New York. Binghamton, 351–374.

Lehmann, H. (1936). Morphologische Studien auf Java. *Geogr. Abhanld*, Heft 9. Stuttgart, 114 p.

Lehmann, H. (1954). Das Karphaenomen in den verschiedenen Klimazonen. *Erdkunde*, 8, 112–139.

Lehmann, H. (1955). Der tropische kegelkarst in westindien. *Tagungsbericht und wissenschaftliche Abhandlungen*. Wiesbaden, 126–131.

Lehmann, H., Roglic, J., Rathjens, C., Lasserre, G., Harrassowitz, H., Corbel, J. and Birot, P. (1954). Karstphänomen in den verschiedenen klimazonen. *Erdkunde*, 8, 112–122.

Leick, A. (1990). *Satellite Surveying*. John Wiley. New York.

Leinen, M. and Sarnthein, M. (Eds.) (1989). *Palaeoclimatology and Palaeometeorology: Modern and Past Patterns of Global Atmospheric Transport*. Kluwer Academic. Dordrecht.

Lelong, F. (1966). Régimes des nappes phreatiques contenues dans las formations d'altération tropicale. Conséquences pour la pédogenèse. *Sciences de la Terre*, 11, 203–244.

Leonard, L.A., Clayton, T. and Pilkey, D.H. (1990). An analysis of beach design parameters on U.S. East Coast barrier islands. *Journal of Coastal Research*, 6, 15–36.

Leopold, L.B. (1968). Hydrology for urban planning - a guidebook on the hydrologic effects of urban land use. *United States Geological Survey Circular* 554, 18 p.

Leopold, L.B. (1976). Reversal of erosion cycle and climatic change. *Quaternary Research*, 6, 557–562.

Leopold, L.B. (1994). *A View of the River*. Harvard University Press. Cambridge, Massachusetts, 298 p.

Leopold, L.B. and Maddock, T. (1953). Hydraulic Geometry of stream channels and some physiographic implications. *United States Geological Survey Profesional Paper*, 252, 57 p.

Leopold, L.B. and Maddock, T. (1964). *The Flood Control Controversy*. Ronald Press. New York, 176 p.

Leopold, L.B. and Miller, J.P. (1954). Postglacial chronology for alluvial valleys in Wyoming. *United States Geological Survey Supply Paper* 1261, 61–85.

Leopold, L.B. and Wolman, M.G. (1957). River meanders patterns: braided, meandering and straight. *United States Geological Survey Professional Papers*, 282B, 39–85.

Leopold, L.B., Emmet, W.W. and Myrick, R.M. (1966). Channel and hillslope processes in a semiarid area of New Mexico. *United States Geological Survey Professional Paper*, 352G.

Leopold, L.B., Wolman, M.G. and Miller, J.P. (1964). *Fluvial Processes in Geomorphology*. Freeman. San Francisco, 522 p.

Leroy, S.A.G. (2006). From natural hazard to environmental catastrophe: Past and present. *Quaternary International*, 158, 4–12.

Leser, H. and Stäblein, G. (1985). Legend of the geomorphological map 1:25.000 (GMK25)-fifth version in the GMK priority program of the Deutsch Forschungsgemainschaft. *Berliner Geographische Abhandlungen*, 39, 61–89.

Létolle, R. and Mainguet, M. (1993). *Aral*. Springer - Verlag. Paris, 357 p.

Levson, V.M. and Morison, S.R. (1996). Geology of placer deposits in glaciated environments. In: Menzies, J. (Ed.). *Past Glacial Environments: Sediments, Forms and Techniques*. Butterworth-Heinemann. Oxford, 441–478.

Lewis, W.V. (1954). Pressure release and glacial erosion. *Journal of Glaciology*, 2, 417–422.

Lewkovicz, A.G. and Harris, C. (2005). Morphology and geotechnique of active-layer detachment failures in discontinuous and continuous permafrost,northern Canada. *Geomorphology*, 69, 275–297.

Lhénaff, R. (1968). Le poljé de Zafarraya (Province de Granade). *Mélangues de la Casa de Velázquez,* 4, 5–25.

Lièbault, F., Gómez, B., Page, M., Marden, M., Peacock, D., Richard, D. and Trotter, C.M. (2005). Land-use change, sediment production and channel response in upland region. *Rivers Research and Applications*, 21, 739–756.

Lindstrom, E.J. and Legler, D.M. (2001). Developing the WOCE global data system. In: Siedler, G., Church, J. and Goued, J. (Eds.). *Ocean Circulation and Climate*. Academic Press. San Diego, 181–190.

Linell, K.A. and Johnston, G.H. (1973). Engineering design and construction in permafrost regions. *Proceedings of Permafrost 2 st International Conference*. Natural Research Council, 553–575. Yakutsk, U.R.S.S.

Linell, K.A. and Lobacz, E.F. (1978). Some experiences with tunnel entrances in permafrost. *Proceedings 3rd International Conference of Permafrost*. Ottawa, 813–819.

Lins, H.F. and Cohn, T.A. (2003). Floods in the greenhouse: spinning the right tale. In: Thorndycraft, V.R., Benito, G., Barriendos, M. and Llasat, M.C. (Eds.). *Palaeofloods, Historical Floods and Climatic Variability: Applications in Flood Risk Assesment*. Consejo Superior de Investigaciones Científicas. Madrid, 263–268.

Linsley, R.K., Kohler, L.A. and Paulhus, J.L.H. (1975). *Hydrology for Engineers*. Mac-Graw Hill. New York, 482 p.

Linton, D.L. (1955). The problem of tors. *Geographical Journal*, 121, 470–487.

Linton, D.L. (1963). The forms of glacial erosion. *Transaction of Institute British Geographers*, 33, 1–27.

Linton, D.L. (1964). The origin of Pennine tors- an essay in analysis. *Zeitscrift für Geomorphologie*, 8, 5–24.

Liñán, C., Calaforra, J.M., Cañaveras, J.C., Carrasco, F., Fernández Cortés, A., Jiménez Sánchez, M., Martín Rosales, W., Sánchez Martos, F., Soler, V. and Vadillo, I. (2004). Experiencias de monitorización medioambiental en cavidades turísticas. In: Andreo, B. and Durán, J.J. (Eds.). *Investigaciones en Sistemas Kársticos Españoles*. Instituto Geológico y Minero de España. Madrid, 385–42.

Lioubimtseva, E. (1995). Landscape evolution of the Saharo-Arabian area during the last glacial cycle. *Journal of Arid Environments*, 30, 1–17.

Lioubimtseva, E. (2004). Climate change in arid enviroments: revisiting the past to understand the future. *Progress in Physical Geography* 28, 502–530.

Lipman, P.W. (2000). Calderas. In: Sigurdsson, H. (Ed.). *Encyclopedia of Volcanoes*. Academic Press, San Diego, 643–662.

Lipman, P.W. and Mullineaux, D.R. (Eds.) (1981). *The 1980 Eruptions of Mount St. Helens, Washington*. United States Geological Survey Professional Paper, 1250.

Lister, L.A. and Secrest, C.D. (1985). Giant dessication cracks and differential surface subsidence, Red Lake Playa, Mohave County, Arizona. *Bulletin of Association of Engineering Geology*, 22, 299–314.

Liu, T. (1985*). Loess and the Environment*. China Ocean Press. Beijing.

Liu, T. (2003). Blind testing of rock varnish microstratigraphy as a chronometric indicator: results on late Quaternary lava flows in the Mojave Desert, California. *Geomorphology*, 53, 209–234.

Liu, T. and Broecker, W.S. (2007). Holocene rock varnish microstratigraphy and its chronometric application in the drylands of the Western USA. *Geomorphology*, 84, 1–21.

Liu, T. and Broecker, W.S. (2008). Rock varnish microlamination dating of late Quaternary geomorphic features in the drylands of western USA. *Geomorphology*, 93, 501–523.

Livingstone, I. (1989). Monitoring surface change on a Namib linear dune. *Earth Surface Processes and Landforms*, 14, 317–332.

Livingstone, I. (2003). A twenty-one year record of surface change on a Namib linear dune. *Earth Surface Processes and Landforms*, 28, 1025–1031.

Livingstone, I. y Warren, A. (1996). *Aeolian Geomorphology: An Introduction*. Longman. Essex, 211 p.

Livingstone, I., Wiggs, G.F.S. and Weaver, C.M. (2007). Geomorphology of desert sand dunes: A review of recent progress. *Earth-Science Reviews*, 80, 239–257.

Llamas, B. and Romero, E. (2006). *Tecnologías de lucha contra el cambio climático. Del carbón al carbono*. Servicio de Publicaciones de la Universidad de Huelva-ENDESA, 347 p.

Llamas, J. (1993). *Hidrologie Générale: Principes et Applicacions*. Gastan Morin. Québec, 635 p.

Lliboutry, L. (1965). *Tratité de Glaciologie. Glaciers. Variations du climat. Sols gêles*. Masson. Paris, 1040 p.

Lliboutry, L., Briat, M., Cresoveur, M. and Fourchet, M. (1976). 15 m deep temperatures in the glaciers of Mont Blanc (French Alps). *Journal of Glaciology*, 16, 197–203.

Lloyd, J.W. (1983). Hydrogeology in glaciated terrains. In: Eyles, N. (Ed.). *Glacial Geology. An Introduction for Engineers and Earth Scientists*. Pergamon Press. Oxford, 349–368.

Lobacz, E.F. and Quinn, W.F. (1966). Thermal regime beneath buildings constructed on permafrost. *Proceedings 1st International Conference of Permafrost*. National Academy of Sciences, Publication 1287. Washington, 447–491.

Lockeretz, W. (1978). The lessons of the dust bowl. In: Skinner, B.J. (Ed.). *Use and Misuse of Earth's Surface*. Kaufmann. Los Altos, California, 140–149.

Löffler, E. (1974). Piping and pseudokarst features in the tropical lowlands of New Guinea. *Erdkunde*, 28, 13–18.

Lomborg, B. (2008). *En Frío. La Guía del Ecologista Escéptico para el Cambio Climático*. Espasa. Madrid, 284 p.

López Arroyo, A., Martín Marín, A.J. and Mézcua Rodríguez, J.M. (1981). *El terremoto de Andalucía del 25 de diciembre de 1884*. Instituto Geográfico Nacional. Madrid, 139 p.

López-Bedoya, J. and Pérez-Alberti, A. (2008). *El control geomorfológico en la distribución de la contaminación por hidrocarburos. Costa de Galicia (España)*. X Reunión Nacional de Geomorfología. Cádiz. 359–362.

López Bermúdez, F. (1973). *La Vega Alta del Segura*. Departamento de Geografía. Universidad de Murcia. Murcia, 288 p.

López Bermúdez, F. (1986). Evaluación de la erosión hídrica en las áreas receptoras de los embalses de la Cuenca del Segura. Aplicación de la USLE. In: López Bermúdez, F. and Thornes, J. (Eds.). *Estudios sobre Geomorfología del Sur de España*. Universidad de Murcia, 93–101.

López Bermúdez, F. (1988). Desertificación: magnitud del problema y estado actual de las investigaciones. In: Gutiérrez, M. and Peña, J.L. (Eds.). *Perspectivas en Geomorfología*. Monografía de la Sociedad Española de Geomorfología, 2. Zaragoza, 155–170.

López Bermudez, F. (2002). *Erosión y Desertificación. Heridas de la Tierra*. Nívola. Madrid, 190 p.

López Bermúdez, F. (2008). *Desertificación: Preguntas y Respuestas de un Desafío Económico, Social y Ambiental*. Ministerio de Medio Ambiente, Medio Rural y Marino. Fundación Biodiversidad. Madrid, 131 p.

López-Bermúdez, F., Conesa-García, C. and Alonso-Sarriá, F. (2002). Floods: Magnitude and frequency in ephemeral streams of the Spanish Mediterranean region. In: Bull, J.L. and Kirkby, M.J. (Ed.). *Dryland Rivers: Hydrology and Geomorphology of Semi-arid Channel*. Wiley. Chichester, 329–350.

López Cádenas, F. (1988). La erosión en España. In: *Riesgos Geológicos*. Instituto Geológico y Minero de España. Madrid, 163–173.

López, F., Romero, M.A., Ruiz, A., Fisher, G.C., Francis, C. and Thornes, J.B. (1984). Erosión y ecología en la España semiárida. *Cuadernos de Investigación Geográfica*, 10, 113–126.

López-Chicano, M., Calvache, M.L., Martín-Rosales, W. and Gisbert, J. (2002). Conditioning factors in flooding of karstic poljes- the case of the Zafarraya polje (South Spain). *Catena*, 49, 331–352.

López-Martínez, J. (1986). Geomorfología del Macizo Kárstico de la Piedra de San Martín (Pirineo Occidental). Tesis Doctoral. Facultad de Ciencias, 529 p. Zaragoza (Unpublished).

López-Martínez, J. (1988). El riesgo debido a los aludes. In: Ayala, F.J., Durán, J.J. and Peinado,T. (Eds.). *Riesgos geológicos*. Instituto Geológico y Minero de España, 215–225. Madrid.

López-Martínez, J. (1989). Análisis del riesgo y de los accidentes por aludes de nieve. *Encuentro Internacional "Catástrofe y Sociedad"*. Fundación MAPFRE. Madrid, 217–232.

López-Martínez, J. and Durán, J.J. (1989). Usos y aprovechamientos de cavidades y paisajes kársticos. In: Durán, J.J and López-Martínez, J. (Eds.). *El Karst en España*. Monografía 4, Sociedad Española de Geomorfología. Madrid, 391–402.

Lorius, C., Barkov, N.I., Jouzel, J., Korotkevich, Y.S., Kotlyakov, V.M. and Raynaud, D. (1988). Antarctic ice core: CO_2 and change over the last climatic cycle. *EOS*, 69, 681–684.

Loucks, R.G. (1999). Paleocave carbonate reservoirs: origins, burial-depth modifications, spatial complexity and reservoirs implications. *American Association of Petroleum Geologists*, 83, 1795–1834.

Loughnan, F.C. (1969). *Chemical Weathering of the Silicate Minerals*. Elsevier. New York, 154 p.

Loughnan, F.C. and Bayliss, P. (1961). The mineralogy of the bauxitic deposits near Weipa, Queensland. *American Mineralogist*, 46, 209–217.

Loughnan, F.C., Grim, R.W. and Vernet, J. (1962). Weathering of some Triassic shales in the Sydney area. *Journal of the Geological Society of Australia,* 8, 245–257.

Lovegrove, G.W. and Fookes, P.G. (1972). The planning and implementation of a site investigation for a highway in tropical conditions in Fiji. *Quaterly Journal Engineering Geology*, 5, 43–68.

Lovelock, J.E. (2000). *The Ages of Gaia: A Biography of Our Living Earth*. Oxford University Press, Oxford, 255 p.

Lowe, D.J. (2000). Role of speleogenesis in the development of hidrocarbon and mineral deposits. In: Klimchouk, A; Ford, D.C., Palmer, A.M. and Dreybrodt, W. (Eds.). *Speleogenesis. Evolution of Karst Aquifers*. National Speleological Society. Huntsville, Alabama, 473–480.

Lowry, D.C. and Jennings, J.N. (1974). The Nullarbor Karst Australia. *Zeitscrift für Geomorphologie*, 18, 35–81.

Lozinski, W. von (1909). Über die mechanische Verwitterung der Sandsteine im gemässigten Klima. *Bulletin International de l'Academie des Sciences de Cracovie class des Sciences Mathematiques et Naturelles,* 1, 1–25. (Traducción inglesa por T. Mrozek (1992). On the mechanical weathering in sandstones in temperate climates. In: Evans, D.J.E. (Ed.) (1994). Cold Climate Landform. Wiley. Chichester, 119–134.

Lu, Y. and Cooper, A.H. (1997). Gypsum Karst geohazards in China. In: Beck, B.F. and Stephenson (Eds.). *The Engineering Geology and Hydrogeology of Karst Terranes*. Balkema. Rotterdam, 117–125.

Lubowe, J.K. (1964). Stream junction angles in the dendritic drainage pattern. *American Journal of Science*, 262, 325–339.

Lucchitta, I. (1975). Application of ERTS images and image processing to regional geologic problems and geologic mapping in northern Arizona-Part IV B, the Shivwits Plateau. *National Aeronautics and Space Administration Technical Report*, 32–1597, 41–72.

Lucha, P. (2009). *Subsidencia por disolución de evaporitas y halocinesis del anticlinal de Barbastro y el valle del río Cardener (Cuenca Terciaria)*. Tesis doctoral. Facultad de Ciencias. Universidad de Zaragoza, 548 p. (Unpublished).

Lucha, P., Cardona, F., Gutiérrez, F. and Guerrero, J. (2008a). Natural and human-induced dissolution and subsidence processes in salt outcrop of the Cardona Diapir (NE Spain). *Environmental Geology* 53, 1023–1035.

Lucha, P., Gutiérrez, F. and Guerrero, J. (2008b). Environmental problems derived for evaporitic dissolution in the Barbastro-Balaguer anticline (Ebro Basin, NE Spain). *Environmental Geology* 53, 1045–1055.

Luino, F. (2005). Sequence of instability processes triggered by heavy rainfall in the northern Italy. *Geomorphology*, 66, 13–39.

Lumb, P. (1962). The properties of the composed granite. *Geotechnique*, 12, 226–243.

Lumb, P. (1983). Engineering properties of fresh and decomposed igneous rocks form Hong Kong. *Engineering Geology*, 27, 287–296.

Luckman, B.H. (2007). Periglacial Landforms. Talus Slopes. In: Elias, S.A. (Ed.). *Encyclopedia of Quaternary Science*. Elsevier. Amsterdam, 2242–2248.

Lumb, P. (1962). The properties of decomposed granite. *Geotechnique*, 12, 226–243.

Lumb, P. (1983). Engineering properties of fresh and decomposed igneous rocks from Hong Kong. *Engineering Geology*, 27, 287–298.

Lumberg, J. (2005). Karren. In: Culver, D.C. and White, W.B. (Eds.). *Encyclopedia of Caves*. Elsevier. Amsterdam, 315–321.

Lumley, H. de, Renault-Miskovsky, J.C. and Guilaine, J. (1976). Le cadre chronologique et sequènces culturelles. En: *La Préhistoire Française, II*. CNRS, 3–16.

Lundquist, G. (1969). Earth and ice mounds: a terminological discussion. In: Péwé, T.L. (Ed.). *The Periglacial Environment*. McGilt-Queen's University Press. Montreal, 203–215.

Lundqvist, J. (1989). Rogen (ribbed) moraine: identification and possible origin. *Sedimentary Geology*, 62, 281–292.

Lutgens and Tarbuck (1992). *Essentials of Geology*. McMillan. New York.

Lvovitch, M.I. (1967). Water resources of the world and their future. General Assembly of Bern. *International Association of Scientific Hydrologists*, 317–322.

Lyford, F.P. and Qashu, H.K. (1969). Infiltration rates as affected by desert vegetation. *Water Resources Research*, 5, 1373–1376.

Mabbutt, J.A. (1961a). Basal surface or weathering front. *Proceedings of the Geological Association*, 72, 357–358.

Mabbutt, J.A. (1961b). A stripped landsurface in Western Australia. *Transactions of the Institute of British Geographers*, 29, 101–114.

Mabbutt, J.A. (1965a). Stone distribution in a Stony Tableland soil. *Australian Journal of Soil Research*, 3, 131–142.

Mabbutt, J.A. (1965b). The weathered landsurface of central Australia. *Zeitschrift für Geomorphologie*, 9, 82–114.

Mabbutt, J.A. (1966). Mantle-controlled planation of pediments. *American Journal of Science*, 264, 78–91.

Mabbutt, J.A. (1977). *Desert Landforms*. The MIT Press. Cambridge, 340 p.

Macaluso, T. and Sauro, U. (1996). Weathering crust and exposed gypsum surfaces. *International Journal of Speleology*, 25, 115–126.

MacAyeal, D.R. (1993). A low-order model of the Heinrich event cycle. *Paleoceanography*, 8, 767–773.

Macdonald, F.A. (1972). *Volcanoes*. Prentice Hall, New York, 510 p.

Macdougall, J.D. (Ed.) (1988). *Continental flood basalts*. Kluver Academic Publishers. Dordrecht.

Machette, M.N. (1985). Calcic soils of the Southwestern United States. In: Wide, D.L. (Ed.). *Soils and Quaternary Geology of the Southwestern United States*. Geological Society of America, Special Paper 203, 1–21.

Mack, G.H., Cole, D.R. and Trevino, L. (2000). The distribution and discrimination of shallow, authigenic carbonates in the Pliocene-Pleistocene Palomas Basin, southern Rio Grande Rift. *Bulletin of the Geological Society of America*, 112, 643–656.

Mackay, J.R. (1972). The world of the underground ice. *Annals Association American Geographers*, 62, 1–22.

Mackay, J.R. (1973). The growth of pingos, western Arctic coast, Canada. *Canadian Journal of Earth Sciences*, 10, 979–1004.

Mackay, J.R. (1979). Pingos of the Tuktoyaktuk Peninsula area, Northwest Territories. *Géographie Physique et Quaternaire*, 33, 3–61.

Mäckel, R. (1974). Dambos. A study in morphodynamic activity on the plateau regions of Zambia. *Catena*, 1, 327–365.

Mäckel, R. (1985). Dambos and related landforms in Africa-an example for the ecological approach to tropical geomorphology. *Zeitschrift für Geomorphologie Supplementband*, 52, 1–23.

Mackey, S.D. and Bridge, J.S. (1992). A revised FORTRAN program to simulated alluvial stratigraphy. *Comp. Geoscience*, 18, 119–181.

Mackin, J.H. (1937). Erosional history of the Big Horn Basin, Wyoming. *Geological Society of America Bulletin*, 48, 813–894.

Mackin, J.H. (1948). The concept of the graded river. *Geological Society of America Bulletin*, 59, 463–512.

Mackling, M.G. and Lewin, J. (1997). Channel floodplain and drainage basin response to environmental change. In: Thorne, C.R., Hey, R.D. and Newson, M.D. (Eds.). *Applied Fluvial Geomorphology for River Engineering and Management*. Wiley. Chichester, 15–45.

Mackling, M.G., Benito, G., Gregory, K.J., Johnstone, E., Lewin, J. Michczyñska, D.J., Soja, R., Starkel, L. and Thorndycraft, V.R. (2006). Past hydrological events reflected in the Holocene fluvial record of Europe. *Catena*, 145–154.

Mackling, M.G., Passmore, D.G., Stevenson, A.C., Davis, D.A. and Benavente, J.A. (1994). Responses of the rivers and lakes to Holocene environmental change in the Alcañiz region, Teruel, North-east Spain. In: Millington, A.C. and Pye, K. (Eds.). *Environmental Change in Drylands: Biogeographical and Geomorphological Perspectives*. Wiley. Chichester, 114–130.

Magee, J.W., Miller, G.H., Spooner, N.A. and Questiaux, D. (2004). Continuous 150 ky monsoon record for Lake Eyre, Asutralia: Insolation-forcong implications and unexpected Holocene failure. *Geology* 32, 885–888.

Mahaney, W.C. (Ed.). (1984). *Quaternary Dating Methods*. Elsevier, Amsterdam, 431 p.

Mahmood, A. and Shokr, M. (2008). Space measurements for disaster response: the International Charter. In: Gad-el-Hak, M. (Ed.). *Large-scale Disasters. Predic-*

tion, Control and Mitigation. Cambridge University Press. Cambridge, 453–541.

Maignien, R. (1966). *Review of Research on Laterites*. UNESCO Natural Resources Research, vol. 4. Paris.

Mainguet, M. (1968). Le Borkou, aspect d'une modelé éolien. *Annales de Géographie*, 77, 296–322.

Mainguet, M. (1972). *Le Modelé des Grés. Problemes Généraux*. Institut Géographique National. Paris, 657 p.

Mainguet, M. (1975). Etude comparée des ergs à l'échelle continentale (Sahara et déserts d'Australie). *Bulletin de la Association de Géographers Françaises*, 52, 135–140.

Mainguet, M. (1978). The influence of trade winds, local air masses and topographic obstacles on the aeolian movement of sand particles and the origin and distribution of ergs in the Sahara and Australia. *Geoforum*, 9, 17–28.

Mainguet, M. (1983). Dunes vives, dunes fixées, dunes vêtues: une classification selon le bilan d'alimentation, le règime èolien et la dynamique des édifices sableux. *Zeitschrift für Geomorphologie Supplementband*, 45, 265–285.

Mainguet, M. (1984). A classification of dunes based on aeolian dynamics and the sand budget. In: El Baz, F. (Ed.). *Deserts and arid lands*. Martinus Nijhof. The Hague, 59–77.

Mainguet, M. (1991) *Desertification. Natural Background and Human. Mismanagement*. Springer, Berlin, 306 p.

Mainguet, M. (1999). *Aridity. Droughts and Human Development*. Springer. Berlin, 302 p.

Mainguet, M. and Létolle, R. (1994). Le Lac d'Aral est-il un lac hydro-éolian. *Revue de Géomorphologie Dynamique*, 43, 26–34.

Mainguet, M., Canon, N. and Chemin, M. (1980). Le Sahara: géomorphologie et paléogéomorphologie éoliennes. In: Williams, M.A.J. and Faure, H. (Eds.). *The Sahara and the Nile*, Balkema. Rotterdam, 17–35.

Maizels, J.K. (1981). Freeze/thaw experiments in the simulation of sediment cracking patterns. Bedford College London. *Papers in Geography*, 13.

Maizels, J.K. (1987). Plio-Pleistocene raised channel systems of the western Sharqiya (Wahiba), Oman. In: Frostick, L. and Reid, Y. (Eds.). *Desert Sediments: Ancient and Modern*. Geological Society of London Special Publication, 35, 31–50.

Maizels, J.K. (1995). Sediments and landforms of modern proglacial terrestrial environments. In: Menzies, J. (Ed.). *Modern Glacial Environments. Processes, Dynamics and Sediments*. Butterworth-Heinemann. Oxford, 365–416.

Maizels, J.K. (1997). Jökulhlaup deposits in proglacial areas. *Quaternary Science Reviews*, 16, 783–819.

Malin, M.L. (1974). Salt weathering of Mars. *Journal of Geophysical Research*, 79, 3888–3894.

Mammerickx, J. (1964). Quantitative observations on pediments in the Mojave and Sonoran deserts (Southwestern United States). *American Journal of Science*, 262, 417–435.

Mann, A.W. (1998). Oxidised gold deposits: relationships between oxidation and relative position of water-table. *American Journal of Earth Sciences*, 45, 97–108.

Mannion, A.M. (1997). *Global Environmental Change*. Longman. Essex, 387 p.

Marker, M.E. (1976). Cenotes: a class of enclosed karst hollows. *Zeitscrift für Geomorphologie*, Supplement Band, 26, 104–123.

Marko, J.R., Fissel, D.B. and Miller, J.D. (1988). Iceberg movement prediction off the Canadian East Coast. In: El-Sabh, M.I. and Murty, T.S. (Eds.). *Natural and Man-Made Hazards*, D. Reidel Publ., 435–462.

Marris, E. (2005). Sri Lanka sings warn coral thieves of natural wrath. *Nature*, 433, p. 3.

Marrs, R.W. and Kolm, K.E. (Eds.) (1982). *Interpretation of Windflow Characteristics from Eolian Landforms*. The Geological Society of America, Special Paper 192. Boulder, 109 p.

Marsden, S.S. and Davies, S.N. (1967). Geological subsidence. *Scientific America*, 216, 93–100.

Marshall, A.F. and Workman, J.P. (1977). Identification of dispersive clays in the Texas Gulf Coast Area. In: Sherard, J.L. and Decker, R.S. (Eds.). *Dispersive Clays, Related Piping, and Erosion in Geotechnical Projects*. American Society for Testing and Materials, 623, 274–286. Philadelphia.

Martí, J. and Folch, A. (2005). Anticipating volcanic eruptions. In: Martí, J. and Ernst, G.J. (Eds.). *Volcanoes and the Environment*. Cambridge University Press. Cambridge, 90–120.

Martí, J., Mitjavila, J. and Araña, V. (1994). Stratigraphy, structure and geochronology of the Las Cañadas Caldera (Tenerife, Canary Islands). *Geological Magazine*, 131, 715–727.

Martín, A.J. (1984). *Riesgo Sísmico en la Península Ibérica*. Instituto Geográfico Nacional. Madrid, 235 p.

Martín, A.J. (2002). Elementos de vulnerabilidad sísmica y de diseño sismorresistente. In: Ayala, F.J. and Olcina, J. (Eds.). *Riesgos Naturales*. Ariel Ciencia. Barcelona, 328–339.

Martin, H. (1963). A suggested theory for the origin and a brief description of some gypsum deposits of south west Africa. *Transactions Geological Society of South Africa*, 66, 343–350.

Martínez Solares, J.M. and Mezcua, J. (2002). *Catálogo Sísmico de la Península Ibérica*. Monografías IGN, 18. IGN. Madrid, 253 p.

Martínez, J., Garzón, M.G. and Arche, A. (1987). *Avenidas e Inundaciones*. Ministerio de Obras Públicas y Urbanismo. Madrid, 67 p.

Martínez, J.M. (1997). Los terremotos históricos destructivos y la sismicidad actual en España. In: Capote, R. and Martínez, J.J. (Eds.). *El Riesgo Sísmico. Prevención y Seguro*. Consorcio de Compensación de Seguros. Madrid, 103–118.

Martini, J.E.J. (2000). Quarzite caves in Southern Africa. In: Klimchouk, A.B., Ford, D.C., Palmer, A.N. and Dreybrodt, W. (Eds.). *Speleogenesis Evolution of Karst Aquifers*. National Speleological Society. Huntsville, 458–461.

Martini, J.P. (1978). Tafoni weathering, with examples from Toscany, Italy. *Zeitscrift für Geomorphologie*, 22, 44–67.

Martín-Serrano, A. (1988). *El Relieve de la Región Occidental Zamorana. La Evolución Geomorfológica de un Borde del Macizo Hes périco*. Instituto de Estudios Zamoranos Florián de Ocampo. Zamora, 311 p.

Martín-Serrano, A. (Ed.) (2005). *Mapa Geomorfológico de España y del Margen Continental*. Instituto Geológico y Minero de España. Madrid, 232 p.

Martín-Serrano, A., Nozal, F., Salazar, A. and Suárez, A. (2004). *Mapa Geomorfológico de España a Escala 1:50.000.Guía para su Elaboración*. Instituto Geológico y Minero de España. Madrid, 122 p.

Martín-Serrano, A., Vegas, J., García-Cortés, A., Galán. L., Gallardo-Millán, J.L., Martín-Alfageme, S., Rubio, F.M., Ibarra, P.I., Granda, A., Pérez-González, A. and García-Lobón, J.L. (2009). Morphotectonic setting of maar lakes in the Campo de Calatrava Volcanic Field (Central Spain, SW Europe). *Sedimentary Geology*, 222, 52–63.

Martonne, E. de. (1909). *Traité de Geographie. Physique*. A. Colin. Paris (Trad. esp. R. Candel y J. Comas. Ed. Juventud, 1135 p. Barcelona).

Martonne, E. de (1913). Le climat facteur du relief. *Scientia*, 13, 339–355.

Martonne, E. de (1924). *Traité de Géographie Physique*. vol. 2. Le Relief du Sol. Colin. Paris, 499–1057.

Martonne, E. de and Aufrère, L. (1928). L'extension des régions privées d'ecoulement vers l'ocean. *Annales de Géographie*, 38, 1–24.

Marzo, M. (1989). Sistemas fluviales de alta sinuosidad. In: Arche, A. (coord.). *Sedimentología*. Consejo Superior de Investigaciones Científicas. Madrid, 107–141.

Mason, B. (1958). *Principles of Geochemistry*. Wiley. New York, 310 p.

Masselink, G. and Hughes, M.G (2003). *Introduction to Coastal Processes and Geomorphology*. Arnold. London, 354 p.

Masselink, G., Hegge, B.J. and Pattiaraechi, C.B. (1997). Beach cusp morphodynamics. *Earth Surface Processes and Landforms*, 22, 1139–1155.

Masson, D.G., Watts, A.B., Gee, M.J.R., Urgeles, R., Mitchell, N.C., LeBas, T.P. and Canals, M. (2002). Slope failures on the flanks of the western Canary Islands. *Earth-Science Reviews*, 57, 1–35.

Massonnet, D., Feigl, K., Rossi, M and Adragna, F. (1994). Radar interferometric mapping of deformation in the year after the year after the Landers earthquake. *Nature*, 369, 227–230.

Mather, A.E. (2000). Impact of headwater river capture of alluvial system development: an example from SE of Spain. *Journal of the Geological Society, London*, 157, 957–966.

Mather, A.E., Harvey, A.E. and Stokes, M. (2000). Quantifying long term catchment changes of alluvial fan systems. *Bulletin of the Geological Society of America*, 112, 1825–1833.

Mathews, W.H. (1947). "Tuyas" flat topped volcanoes in northern British Columbia. *American Journal of Science*, 245, 560–570.

Matsch, C.L. (1983). River Warren, the southern outlet of Glacial Lake Agassiz. In: Teller, J.T. and Clayton, L. (Eds.). *Glacial Lake Agassiz*. Geological Association of Canada. St. Johns, New Foundland, 231–244.

Matsuoka, N. (2008). Frost weathering and rockwall erosion in the southwestern Swis Alps: Long termed (1994–2006) observations. *Geomorphology*, 99, 353–368.

Matthes, F.E. (1900). Glacial sculpture of the Bighorn Mountains, Wyoming, United States. *Geological Survey 21st Annual Report 1899–1900,* 167–190.

Matthes, F.E. (1930). Geologic History of the Yosemite Valley. *United States Geological Survey Professional Paper*, 160, 54–103.

Maxson, J.H. (1940). Fluting and faceting of rock fragments. *Journal of Geology*, 48, 717–751.

May, R. (2005). Periódico *"El País"*, 12 de Junio de 2005, p. 41.

May, V.J. (1972). Earth cliffs. In: Barnes, R.S.K. (Ed). *The Coastline*. Wiley. New York, 215–235.

Mayer, L., Gerson, R. and Bull, W.B. (1984). Alluvial gravel production and deposition- a useful indicator of Quaternary climatic changes in deserts (a case study in south-western Arizona). *Catena Supplement,* 5, 137–51.

Mayuga, M.N. and Allen, D.R. (1969). Subsidence in the Wilmington oil field. Long Beach, California, USA. *Publications of Institute Sciences Hydrologies*, 88, 66–79.

Mc Gregor, G.R. and Nieuwolt, S. (1998). *Tropical Climatology. An Introduction to the Climates of the Low Latitudes*. Wiley. Chichester, 339 p.

McBean, G.A., Liss, P.S. and Schneider, S.H. (1996). Advancing our understanding. In: Houghton, J.T. *et al.* (Eds.). *Climate Change 1995. The Science of Climatic Change*. Cambridge University Press. Cambridge, 517–531.

McCall, J.G. (1960). The flow characteristics of a cirque glacier and their effect on cirque formation. In: Lewis, W.V. (Ed.). *Investigations on Norvegian Cirque Glaciers*, Royal Geographers Society, Research Series, 4, 39–62.

McCalpin, J.P. (1996a). Field Techniques in Paleosismology. In: McCalpin, J.P. (Ed.). *Paleosismology*. Academic Press. San Diego, 33–83.

McCalpin, J.P. (1996b). Paleoseismology in extensional tectonic enviroments. In: McCalpin, J.P. (Ed.). *Paleoseismology*. Academic Press. San Diego, 85–146.

McCalpin, J.P. (1999). Criteria for determining the seismic significance of sackungen and other scarplike landforms in mountainous regions. Techniques for identifying Faults and Determining their Origins. *United States Nuclear Regulatory Commission*. Washington, 2.55–2.59.

McCalpin, J.P. and Irvine, (1995). Sackungen at the Aspen Highlands ski area, Pitkin County, Colorado. *Environmental and Engineering Geoscience*, 1, 277–290.

McCalpin, J.P. and Nelson, A.R. (1996). Introduction to Paleoseismology. In: McCalpin, J. (Ed.) *Paleosismology.* Academic Press. San Diego, 1–32.

McCarthy, J.J., Canziani, O.F., Leary, N.A., Dokken, D.J. and White, K.S. (Eds.) (2001). *Climate Change 2001: Impacts, Adaptation and Vulnerability*. Cambridge University Press. Cambridge, 1032 p.

McCauley, J.F., Grolier, M.J. and Breed, C.S. (1977). Yardangs. In: Doehring, D.O. (Ed.). *Geomorphology in Arid Regions*. Allen and Unwin. London, 233–269.

McCauley, J.F., Breed,C.S., El-Baz, F., Whitney, M.I., Grolier, M.J. and Ward, A.W. (1979). Pitted and fluted rocks in the Western Desert of Egypt-Viking comparisons. *Journal of Geophysical Research*, 84, 8222–8232.

McCauley, J.F. *et al.*, (1982). Subsurface valleys and geoarchaeology of the eastern Sahara revealed by Shuttle Radar. *Science*, 218, 1004–1020.

McClung, D. (2003). Magnitude and frequency of avalanches in relation to terrain and forest cover. *Arctic, Antarctic and Alpine Research*, 35, 82–90.

McClung, D. and Shaerer, F. (1993). *Avalanche Handbook*. The Mountainers, 272 p.

McClure, H.A. (1978). ArRub' Alkhali. In: Al-Sayari, S.S. and Zötl, J.G. (Eds.). *Quaternary Period in Saudi Arabia*. Springer-Verlag. Viena, 252–263.

McCracken, M.C. (2005). Climate change and global warming. In: Oliver, J.E. (Ed.). *Encyclopedia of World Climatology*. Springer. Berlin, 199–209.

McFadden, L.D., Eppes, M.C., Gillespie, A.R. and Hallet, B. (2005). Physical weathering in arid landscapes due to diurnal variation in the direction of solar heating. *Geological Society of America Bulletin*, 117, 161–173.

McFarlane, M.J. (1976). *Laterite and Landscape*. Academic Press. London, 151 p.

McFarlane, M.J. (1983). Laterites. In: Goudie, A.S. and Pye, K. (Eds.). *Chemical Sediments and Geomorphology*. Academic Press. London, 7–58.

McFarlane, M.J. (1987). The key role of microorganisms in the process of bauxitisation. *Modern Geology*, 11, 325–344.

McFarlane, M.J. (1989). Dambos- their characteristics and geomorphological evolution in parts of Malawi and Zimbawbe, with particular reference to their role in the hydrogeological regime of surviving areas of Africa surface. *Proceedings of the groundwater exploration*. Harare. Zimbawbe, vol. 1, 254–308.

McFarlane, M.J. and Pollard, S. (1989). Some aspects of stone-lines and dissolution fronts associated with regolith and dambo profiles in parts of Malawi and Zimbabwe. In: Alexandre, J. and Symoens, J.J. (Eds.). *Stone-Lines*. Académie Royale des Sciences D'Outre-Mer, Geo-Eco-Trop, 11, 23–35.

McFarlane, M.J. and Whitlow, R. (1990). Key factors affecting the initiation and progress of gullying in dambos in parts of Zimbabwe and Malawi. *Land Degradation & Rehabilitation*, 2, 215–235.

McGee, W.J. (1897). Sheetflood erosion. *Bulletin of the Geological Society of America*, 8, 87–112.

McGill, G.E. and Stromquist, A.W. (1979). The grabens of Canyonlands National Park, Utah: geometry, mechanics and kinematics. *Journal of Geophysical Research*, 84, 4547–4563.

McGinnies, W.G. *et al.*, (1968). *Deserts of the World*. University Arizona Press. Tucson, 788 p.

McGinnies, W.G. and Golman, B.J. (Eds.) (1969). *Arid Lands in Perspective*. University Arizona Press, Tucson.

McGowen, J.H. (1979). Alluvial fan systems. In: Galloway, W.E., Kreitler, C.V. and McGowen, J.H. (Eds.). *Depositional and Ground Water Flow Systems in the Exploration for Uranium*. Texas Bureau of Economic Geology Research Colloquium, 43–79.

McGreevy, J.P. (1982). Frost and salt weathering: further experimental results. *Earth Surface Processes and Landforms*, 7, 475–488.

McGregor, G.R. and Nieuwolt, S. (1998). *Tropical Climatology. An Introduction to Climates of Low Latitudes.* Wiley. Chichester, 339 p.

McGuffie, K. and Henderson-Sellers, A. (2001). *A Climate Modelling Primer*. Wiley. Chichester, 253 p.

McGuire, B., Burton, P., Kilburn, C. and Willets, D. (2004). *World Atlas of Natural Hazards*, Arnold. London, 120 p.

McIntyre, D.S. (1979). Exchangeable sodium, subplasticity, and hydraulic conductivity of some Australian soils. *Australian Journal of Soil Research*, 17, 115–120.

McKee, E.D. (Ed.) (1979a). *A Study of Global Sand Seas*. United States Geological Survey, Professional Paper 1052. Washington, 429 p.

McKee, E.D. (1979b). Introduction to a study of global sand seas. In: McKee, E.D. (Ed.). *A Study of Global Sand Seas*. United States Geological Survey, Professional Paper 1052. Washington, 1–19.

McKee, E.D. and Bigarella, J.J. (1979). Sedimentary structures in dunes. In: McKee, E.D. (Ed.). *A Study of Global Sand Seas*. United States Geological Survey, Professional Paper 1052. Washington, 83–134.

Mckenzie, D.P. and Parker, R.L. (1967). The Pacific: an example of tectonics on a sphere. *Nature*, 216, 1276–1280.

McLaren, S.J. and Nash, D.J. (2007). Geochemical sediments and landscapes: General summary. In: Nash, D.J. and McLaren, S.J. (Eds.). *Geochemical Sediments and Landscapes*. Blackwell. Malden, 443–446.

McManus, J.F. (2004). A great grand-daddy of ice cores. *Nature*, 429, 611–612.

McTainish, G.H. (1987). Desert loess in northern Nigeria. *Zeitschrift für Geomorphologie NF*, 31, 145–165.

McTainish, G.H. and Strong, C. (2007). The role of aeolian dust in ecosystems. *Geomorphology*, 89, 39–54.

Meade, R. (1996). River-sediment inputs to major deltas. In: Milliman, J., Haq, B.U. (Eds.). *Sea Level Rise and Coastal Subsidence: Causes, Consequences and Strategies*. Kluwer. Dordrecht, 63–85.

Meade, R.H. (2007). Transcontinental moving and storage: the Orinoco and Amazon rivers transfer the Andes to the Atlantic. In: Gupta. A. (Ed.). *Large Rivers. Geomorphology and Management*. Wiley. Chichester, 45–63.

Meade, R.H., Yuzyk, T.R. and Day, T.J. (1990). Movement and storage of sediment rivers of the United States and Canada. In: Wolman, M.G. and Riggs, H.C. (Eds.). *Surface Water Hydrology*. Geological Society of America. Boulder, 255–280.

Meadows, M.E. (1985). Dambos and environmental change in Malawi, central Africa. *Zeitschrift für Geomorphologie Supplementband,* 52, 147–169.

Meese, D.A., Alley, R.B., Fiasco, R.J., Germani, M.S., Gow, A.J., Grootes, P.M., Illing, M., Mayewski, P.A., Morrison, M.C., Ram, M., Taylor, K.C., Yang, Q. and Zielinski, G.A. (1994). Preliminary depth-age scale of the GISP2 ice core. *Special CRREL Report* 94–1. United States.

Mehta, P.K. (1983). Mechanism of sulphate attack on Portland cement concrete-another look. *Cement and Concrete Research*, 13, 401–406.

Meigs, P. (1953). The world distribution of arid and semiarid homoclimates. In: *Reviews of Research on Arid Zone Hydrology*, UNESCO. Paris, 203–209.

Meir, M.F. and Post, A. (1969). What are glacier surges?. *Canadian Journal of Earth Sciences*, 6, 807–817.

Meleen, N.H. (1999). Trans-Alaska pipeline. In: Cunninghan, W.P., Cooper, T.H., Gordham, E. and Hepworth, M.T. (Eds.). *Environmental Encyclopedia*. Jaico. Mumbai, 1041–1042.

Meli, R. (2002). El sismo de 1985 en México. In: Hubp, J.L. and Inbar, M. (Eds.). *Desastres naturales en América Latina*. Fondo de Cultura Económica. México, 125–146.

Mellet, J.S. (1990). Ground-penetrating radar enhances knowledge of Earth's surface layer. *Geotimes*, 35, 12–14.

Mellor, A., Short, J. and Kirkby, S.J. (1997). Tafoni in the El Chorro area, Andalucía, southern Spain. *Earth Surface Processes and Landforms*, 22, 817–833.

Melton, F.A. (1959). Aerial photographs and structural geology. *Journal Geology*, 67, 351–370.

Menard, H.W. (1971). The bottom of the oceans. In: Wilson, J.T. (Ed). *Continents adrift*. Scientific American. Freeman, San Francisco.

Méndez, J. and Rey, D. (2000). Perspectiva histórica del conocimiento geológico rías gallegas. *Journal of Iberian Geology*, 26, 21–44.

Mensching, H. (1958). Glacis-Fussfläche-Pediment. *Zeitschrift für Geomorphologie N.F.*, 2, 165–186.

Mensching, H. (1973). Karstscheinungen in Trokengebieten. *Geographische Zeitschrift Beihefte*, 32, 47–53.

Mensching, H. (1983). Die globale desertification als Umweltproblen. *Geographische Rundschau*, 45, 360–365.

Menzies, J. (1979). A review of the literature on the formation and location of drumlins. *Earth Science Reviews*, 14, 315–359.

Menzies, J. (1995a). Hydrology of glaciers. In: Menzies, J. (Ed.). *Modern Glacial Environments. Processes, Dynamics and Sediments.* Butterworth-Heinemann. Oxford, 197–239.

Menzies, J. (1995b). Glaciers and ice sheets. In: Menzies, J. (Ed.). *Modern Glacial Environments. Processes, Dynamics and Sediments*. Butterworth-Heinemann. Oxford, 101–138.

Menzies, J. (1995c) (Ed.). *Modern Glacial Environments. Processes, Dynamics and Sediments*. Butterworth-Heiemann. Oxford, 621 p.

Menzies, J. (1996) (Ed.). *Past Glacial Environments. Sdiments, Forms and Techniques*. Butterworth-Heinemann. Oxford, 598 p.

Menzies, J. (2002) (Ed.). *Modern and Past Glacial Environments*. Revised Student Edition. Butterworth-Heineman. Oxford, 543 p.

Menzies, J. and Shilts, W.W. (1996). Subglacial environments. In: Menzies, J. (Ed.). *Past Glacial Environments. Sediments, Forms and Techniques*. Butterworth-Heinemann. Oxford, 15–136.

Mercer, J.H. (1968). Antartic ice and Sangano Sea Level. *International Association Science Hydrologist Symp*. 79, 217–225.

Mercer, J.H. (1978). West Antartic ice sheet and CO_2 greenhouse effect: A threat of disaster. *Nature*, 271, 321–325.

Mertes, L.A.K. and Dunne, T. (2007). Effects of tectonism, climate change and sea-level change of the modern Amazon River and its floodplain. In: Gupta. A. (Ed.). *Large Rivers. Geomorphology and Management*. Wiley. Chichester, 115–144.

Mescherikov, Y.A. (1968). Neotectonics. In: Fairbridge, R.W. (Ed.). *The Encyclopedia of Geomorphology*. Dowden, Hutchinson and Ross. Stroudsburg, 768–772.

Messerli, B., Grosjean, M., Hofer. T., Núñez. L. and Pfister. C. (2000). From nature-dominated to human-dominated environmental changes. *Quaternary Science Reviews*, 19, 459–479.

Metz, B., Davidson, O., Swart, R. and Pan, J. (2001). Summary for policy makers. *Climate Change 2001. Mitigation*. Cambridge University Press. Cambridge, 1–71.

Meulenberg, O. (1949). *Introduction à l'étude pédologique des sols du territoire du Bas-Fleuve*. Institute Royal Col. Belge. Bruxelles, 133 p.

Meybeck, M. (1979). Concentrations des eaux fluviales éléments majeurs et apports en solution aux océans. *Revue de Géologie Dynamique et de Géographie Physique*, 21, 215–246.

Meyer, L.D. (1986). Erosion processes and sediment properties for agricultural cropland. In: Abrahams, A.D. (Ed.). *Hillslope Processes*. Allen and Unwin. Boston, 55–76.

Meyer, W.B. (1996). *Human Impact of the Earth*. Cambridge University Press. Cambridge, 253 p.

Meyerhoff, H.A. (1940). Migration of erosional surfaces. *Association American Geographers Annals*, 30, 247–254.

Mézcua, J. (1982). *Catálogo General de las Isosistas de la Península Ibérica*. Instituto Geográfico Nacional. Madrid, Publ. 220, 237 p.

Mézcua, J. (2002). Peligrosidad sísmica. In: Ayala, F.J. and Olcina, J. (Eds.). *Riesgos Naturales*. Ariel Ciencia. Barcelona, 313–328.

Mézcua, J. and Martínez, J.M. (1983). *Sismicidad del Area Iberomagrebí*. Instituto Geográfico Nacional. Madrid, Publ. 203, 301 p.

Miall, A.D. (1977). A review of braided-river depositional environment. *Earth-Science Reviews*, 13, 1–62.

Miall, A.D. (1978). Lithofacies types and vertical profile models in braided river deposits: a summary. In: Miall, A.D. (Ed.). *Fluvial Sedimentology*. Canadian Society of Petroleum Geologists, Memoir 5. Calgary, 597–604.

Miall, A.D. (1996). *The Geology of Fluvial Deposits. Sedimentary Facies, Basin Analysis, and Petroleum Geology*. Springer. Berlin, 582 p.

Miall, A.D. (2006). How do we identify big rivers? And how big is big?. *Sedimentary Geology* 186, 39–50.

Mickelson, D.M. and Winguth, C. (2007). Glacial Landforms, Ice Sheets. Evidence of glacier and ice sheet extent. In: Elias, S.A. (Ed.). *Encylopedia of Quaternary Science*. Elsevier. Amsterdam, 871–882.

Micklin, P.P. (1988). Dessication of Aral Sea: A water management disaster. *Science*, 241, 1170–1176.

Middleton, N. (1990). Wind erosion and dust-storm control. In: Goudie, A. (Ed.). *Techniques for Desert Reclamation*. Wiley. Chichester, 87–108.

Middleton, N. (1997). Desert dust. In: Thomas, D.S.G. (Ed.). *Arid Zone Geomorphology. Process, Form in Change in Drylands*. Wiley. Chichester, 413–436.

Middleton, N. (2004). Dust storm. In: Goudie, A.S. (Ed.). *Encyclopedia of Geomorphology*. Routledge. London, 301–304.

Middleton, N. and Thomas, D. (Eds.) (1997). *World Atlas of Desertification*. Arnold. London, 182 p.

Middleton, N., Goudie, A.S. and Wells, G.L. (1986). The frequency and source areas of dust storms. In: Nickling, W.G. (Ed.). *Aeolian Geomorphology*. Allen and Unwin. Boston, 237–268.

Migon, P. (2004). Structural Landform. In: Goudie, A.S. (Ed.). *Encyclopedia of Geomorphology*. Routledge, London, 1007–1008.

Migon, P. (2006). *Granite Landscapes of the World*. Oxford University Press. Oxford, 384 p.

Migon, P. and Lidmar-Bergström, K. (2002). Deep weathering through time in central and northeastern Europe: problems of dating and interpretation of geological record. *Catena*, 49, 25–40.

Migon, P. and Thomas, M.F. (2002). Grus weathering mantles-problems of interpretation. *Catena*, 49, 5–24.

Mijatovic, B.F. (1984). Karst poljes in Dinarides. In: Mijatovic, B.F. (Ed.). *Hydrogeology of Dinaric Karst*. International Associattion of Hydrologists. Hannover, vol. 4, 87–109.

Milanovic, P. (1981). *Karst Hydrogeology*. Water Resources Publications. Litleton, Colorado, 434 p.

Milanovic, P.T. (2000). *Geological Engineering in Karst*. Zebra. Belgrade, 347 p.

Milanovic, P.T. (2004). Dams and Reservoirs on karst. In: Gunn, J. (Ed.). *Encyclopedia Caves of Karst Sciences*. Fitzroy Dearborn. New York, 277–279.

Miller, D.J. and Dunne, T. (1996). Topographic perturbations of regional stresses and consequences bedrock fracturing. *Journal of Geophysical Research*, 101 (B12), 25523–25536.

Miller, M.L. and Hadley, N.P. (2005). Tourism and coastal development. In: Schwartz, M.L. (Ed.). *Encyclopedia of Coastal Science*. Springer. Dordrecht, 1002–1008.

Milliman, J.D., Broadus, J.M. and Gable, F. (1989). Environmental and economics implications of rising sea level and subsiding deltas: the Nile and Bengal examples. *Ambio*, 18, 340–345.

Milliman, J.D. and Meade, R.H. (1983). World-wide delivery of river sediment to the oceans. *Journal of Geology*, 91, 1–21.

Milliman, J.D. and Syvitski, J.P.M. (1992). Geomorphic/tectonic control of sediment discharge to the ocean: The importance of small mountainous rivers. *Journal of Geology*, 100, 525–544.

Millington, A.C. and Pye, K. (Eds.) (1994). *Environmental Change in Drylands: Biogeographical and Geomorphological Perspectives*. Wiley. Chichester, 456 p.

Millington, A.C., Drake, N.A., White, K. and Bryant, R.G. (1995). Salt ramps: wind-induced depositional features on Tunisian playas. *Earth Surface Processes and Landforms*, 20, 105–113.

Millot, G. (1964). *Géologie des Argiles*. Masson. Paris, 499 p.

Millot, G. (1979). Essai de conclusions. *Bulletin de l'Association française pour l'étude du sol*, 2–3, 301–303.

Mills, H.H. (1990). Thickness and character of regolith on mountain slopes in the vicinity of Mountain Lake, Virginia, as indicated by seismic refraction, and implications for hillslope evolution. *Geomorphology*, 3, 143–157.

Milnes, A.R., Thiry, M. and Wright. (1991). Silica accumulations in saprolite and soils in South Australia. In: Nettleton, W.D. (Ed.). *Occurrence, characteristics, and genesis of carbonate, gypsum and silica accumulation in soils*. Soil Science Society of America, Special Publication, 26,121–149.

Miötke, F.D. (1968). Karstmorphologische studien in der glazial-überformten höhenstufe der Picos de Europa, Nordspanien. *Jahrbuch der Geographischen Gesellschaft zu Hannover*, Sonderheft, 4, 161 p.

Mitchell, P. (1988). The influences of vegetation, animals and micro-organism on soil processes. In: Viles, H. (Ed.). *Biogeomorphology*. Blackwell. Oxford, 43–82.

Mock, C.J. (2007). Paleo-ENSO. In: Elias, S.A. (Ed.). *Encyclopedia of Quaternary Science*. Elsevier. Amsterdam, 1934–1941.

Moeyersons, J. (1989). The concentration of stones into a stone-line, as a result from subsurface movements in fine and loose soils in the tropics. In: Alexandre, J. and Symoens, J.J. (Eds.). *Stone-Lines*. Académie Royale des Sciences D'Outre-Mer, Geo-Eco-Trop, 11, 11–22.

Mol, J., Vanderberghe, J. and Kasse, C. (2000). River response to variations of periglacial climate in mid-latitude Europe. *Geomorphology*, 33, 131–148.

Molina, E. and Blanco, J.A. (1980). Quelques précisions sur l'altération du Massif Hercynien espagnol. *Compte Rendus de la Académie des Sciences de Paris*, 290, 1293–1296.

Molina, E. and Pellitero, E. (1982). Formas periglaciares actuales en la Paramera de Avila. Hipótesis sobre su origen. *Boletín de la Real Sociedad Española de Historia Natural*, 80, 49–56.

Molnar, P. and Gibson, J.M. (1994). Very long baseline interferometry and active rotations of crustal blocks in the Western Transverse Ranges, California. *Geological Society of America Bulletin*, 106, 594–606.

Monroe, W.H. (1964a). The zanjón, a solution feature of karst topography in Puerto Rico. *United States Geological Survey, Professional Paper*, 501B, 126–129.

Monroe, W.H. (1964b). Lithological control in the formation of the towers of Puerto Rico. *20th International Geographical Conference Karst Symposium*.

Monroe, W.H. (1966). Formation of tropical karst topography by limestone solution and reprecipitation. *Caribbean Journal Science*, 6, 1–7.

Montserrat, J. (1992). *Evolución glaciar y postglaciar del clima y la vegetación en la vertiente sur del Pirineo: Estudio palinológico*. Instituto Pirenaico de Ecología. Zaragoza, 147 p.

Mook, W. and Woillard, G. (1982). Carbon-14 dates at Gran Pile. Correlation of land and sea chronologies. *Science*, 215, 159–161.

MOPU (1987). *Medio Ambiente en España, 1986*. Dirección General de Medio Ambiente. Ministerio de Obras Públicas y Urbanismo. Madrid, 368 p.

Morales, C. (Ed.) (1979). *Saharan Dust - Mobilization, Transport, Deposition*. John Wiley. New York, 316 p.

Morales, J.A., Borrego, J., San Miguel, E.G., López-González, N. and Carro, D. (2008). Sedimentary record of recent tsunamis in the Huelva Estuary (southwestern Spain). *Quaternary Science Reviews*, 27, 734–746.

Moran, S. (1971). Glaciotectonic structures in drift. In: Goldthwait, R.P. (Ed.). *Till: A Symposium*, Ohio State University Press. Columbus, 127–148.

Moreno-Casasola, P. (1986). Sand movement as a factor in the distribution of plant communities in a coastal dune system. *Plant Ecology*, 65, 67–76.

Moreno, A. and Canals, M. (2004). The role of dust in abrupt climate change: insights from the offshore northwest

Africa and Alborán Sea sediment records. *Contribution Science*, 2, 485–498.

Moreno, J.M. (Coor.) (2005). *Evaluación Preliminar de los Impactos en España por Efecto del Cambio Climático*. Ministerio de Medio Ambiente. Madrid, 822 p.

Morgan, M.L., Matthwes, V., Gutiérrez, F., Thorson, J.P., Madole, R.F. and Hanson, P.R. (2008). From buttes to bowls: Repeated relief inversion in the lanscape of the Colorado Piedmont. *The Geological Society of America, Field Guide* 10, 203–215.

Morgan, R.P.C. (1986). *Soil Erosion and Conservation*. Longman. Harlow, 298 p.

Morgan, R.P.C., Finney, H.J., Lavee, H., Merrit, E. and Noble, C.A. (1986). Plant cover effects on hillslope runoff and erosion: evidence from two laboratory experiments. In: Abrahams, A.D. (Ed.). *Hillslope Processes*. Binghampton Symposia in Geomorphology. Unwin. London, 77–90.

Morgan, W.J. (1968). Rises, trenches, great faults and crustal blocks. *Journal of Geophysical Research*, 73, 1959–1982.

Morin, J. and Benyami, Y. (1977). Rainfall infiltration into bare soils. *Water Resources Research*, 13, 813–817.

Morisawa, M. (1957). Accuracy determination of stream lengths from topographic maps. *Transactions American Geophysical Union*, 38, 86–88.

Morisawa, M. (1968). *Streams: their dynamics and morphology*. McGraw Hill Book Co. New York, 175 p.

Morisawa, M. (1985). *Rivers: Form and Processes*. Longman. London, 222 p.

Mörner, N.A. (1976). Eustasy and geoid changes. *Journal Geology*, 84, 123–152.

Mörner, N.A. (1980a). Eustasy and geoid changes as a function of core/mantle change. In: Mörner, N.A. (Ed.). *Earth Rheology, Isostasy and Eustasy.* Wiley, 535–553.

Mörner, N.A. (1980b). The Fennoscandian uplift: geological data and their geodinamical implication. In: Mörner, N.A. (Ed.). *Earth Rheology, Isostasy and Eustasy*. Wiley. Chichester, 251–284.

Mörner, N.A. (1983). Sea levels. In: Gardner, R. and Scoging, H. (Eds.). *Mega-Geomorphology*. Oxford University Press. Oxford, 73–91.

Mörner, N.A. (1987). Models of global sea-level changes. In: Tooley, M.J. and Shennan, I. (Eds). *Sea-level changes*. Blackwell. Oxford, 332–355.

Mörner, N.A. (2004). Estimating future sea level changes from past records. *Global and Planetary Change*, 40, 49–54.

Morrisey, M.M. and Martin, L.G. (2000). Vulcanian Eruptions. In: Sigurdsson, H. (Ed). *Encyclopedia of Volcanoes*. Academic Press, San Diego, 463–475.

Morrison, R.B. (1965). Quaternary Geology of the Great Basin. In: Wright, H.E. and Frey, D.G. (Eds.). *The Quaternary of the United States*. Princeton University Press. Princeton, New Jersey, 265–285.

Mortensen, H. (1933). Die Salzprengung und ihre Bedeutung für die regional klimatische Gliederung der Wüsten. *Petermanns Geographische Mitteilungen*, 79, 130–135.

Morton, O. (2007). Is this what it takes to save the world? *Nature*, 447, 132–136.

Moss, P.T. and Kershaw, A.P. (2000). The last glacial cycle from the humid tropics of northeastern Austrlia: comparison of a terrestrial and a marine record. *Palaeogeography, Palaeoclimatology, Palaeoecology*, 155, 155–176.

Moss, R.P. (1965). Slope development and soil morphology in a part of south west Nigeria. *Journal of Soil Science*, 16, 192–209.

Mottershead, D.N. (1994). Spatial variations in intensity of alveolar weathering of a dated sandstone structure in a coastal environment, Weston-super-Mare, UK. In: Robinson, D.A. and Williams, R.B.G. (Eds.). *Rock Weathering and Landform Evolution*. Wiley. Chichester, 151–174.

Muhs, D.R. (2007). Loess deposits, origin and properties. In: Elías, S.A. (Ed.). *Encyclopedia of Quaternary Science*. Elsevier. Amsterdam, 1405–1418.

Muhs, D.R. and Bettis III, E.A. (2003). Quaternary loess-paleosol sequences as examples of climate-driven sedimentary extremes. *Geological Society of America*. Special Paper 370, 53–74.

Müller, F. (1959). Beobachtung uber pingos. *Meddeleser om Gronland,* 153, 127 p.

Müller, F. (1962). Zonation in the accumulation area of the glaciers of Axel Heiberg Island, N.W.T., Canada. *Journal of Glaciology*, 4, 302–318.

Müller, L. (1964a). Application of rock mechanics in the design of rock slopes. In: Judd, W.R. (Ed.). *The State of Stresses in the Earth's Crust*. Elsevier. New York, 575–598.

Müller, L. (1964b). The rock slide in the Vaiont Valley. *Rock Mechanics and Engineering Geology,* 2, 3–4.

Muller, R.D., Roest, W.R., Royer, J.Y., Gahagan, L.M. *et al.* (1997). Digital isochrons of the world's ocean floor. *Journal of Geophysical Research*, 102, 3211–8214.

Muller, S.W. (1945). *Permafrost or Permanently Frozen Ground and Related Engineering Problems*. Ann Arbor, J.W. Edwards. Michigan, 223 p.

Munuera, J.M. (1969). *Mapa de Zonas Sísmicas Generalizadas de la Península Ibérica*. Instituto Geográfico Nacional. Madrid, 50 p.

Murton, J. (2007). Periglacial Landforms. Ice Wedges and Ice Wedge Casts. In: Elias, S.A. (Ed.). *Encyclopedia of Quaternary Science*. Elsevier. Amsterdam, 2153–2170.

Murty, T.S. and El-Sabh, M.I. (1988). Edge waves and ice ride-up on shorelines. In: El-Sabh, M.I. and Murty, T.S. (Eds.). *Natural and Man-Made Hazards*, D. Reidel Publ., 429–434.

Mustoe, G.E. (1982). Origin of honeycomb weathering. *Bulletin of the Geological Society of America*, 93, 108–115.

Mustoe, G.E. (1983). Cavernous weathering in the Capitol Reef Desert, Utah. *Earth Surface Processes and Landforms*, 8, 517–526.

Muxart, T. and Birot, P. (1977). *L'Alteration Meteorique des Roches*. Université de Paris-Sorbonne, nº 4. Paris, 279 p.

Mycielska-Dowgiallo, E. (1977). Channel pattern changes during the last glaciation and Holocene in the northern part of the Sandomierz basin and the middle part of the Vistula valley, Poland. In: Gregory, K.J. (Ed.). *River Channel Changes*. Wiley. Chichester, 75–87.

Nadler, C.T. and Schumm, S.A. (1981). Metamorphosis of South Platte and Arkansas Rivers, eastern Colorado. *Physical Geography*, 2, 95–115.

Nagel, R.H. (1962). Geology of the Serra do Navio manganese district. *Economic Geology*, 57, 481–498.

Nahon, D. (1986). Evolution of iron crusts in tropical landscapes. In: Colman, S.M. and Dethier, D.P. (Eds.). *Rates of Chemical Weathering of Rocks and Minerals*. Academic Press. New York, 169–191.

Nahon, D., Beauvales, A. and Trescases, J.J. (1985). Manganese concentration through chemical weathering of metamorphic rocks under lateritic conditions. In: Drever, J.I. (Ed.). *The Chemistry of Weathering*. Reidel. Boston, 277–291.

Nahon, D.B. (1991). *Introduction to the Petrology of Soils and Chemical Weathering*. John Wiley. New York, 313 p.

Nakada, M., Yonekura, N. and Lambeck, K. (1991). Late Pleistocene and Holocene sea-level changes in Japan: implications for tectonic histories and mantle rheology. *Palaeogeography, Palaeoclimatology, Palaeoecology*, 85, 107–122.

Nalivkin, D.V. (1983). *Hurricanes, storms and tornadoes*. Balkema. Rotterdam.

Nalpanis, P. (1985). Saltating and suspended particles over flat and sloping surfaces. II. Experiments and numerical simulations. In: Barndorff-Nielsen, O.E., Moller, J.T., Romer-Rasmussen, K. and Willets, B.B. (Eds.). *Proceedings of International Workshop on the Physics of Blown Sand*. Institute of Mathematics, University of Aarhus, 8, 37–66.

Nanson, G.C. (1986). Episodes of vertical accretion and catastrophic stripping: A model of disequilibrium floodplain development. *Geological Society of America Bulletin*, 97, 1467–1475.

Nanson, G.C. and Croke, J.C. (1992). A genetic classifications of floodplains. *Geomorphology*, 4, 459–486.

Nanson, G.C. and Gibling, M. (2004). Anabranching and anostomosing river. In: Goudie, A.S. (Ed.). *Encyclopedia of Geomorphology*. Routledge. London, 21–25.

Nanson, G. and Knighton, D. (1996). Anabranching rivers: their cause, character and classifications. *Earth Surface Processes and Landforms*, 21, 211–239.

Nanson, G.C. and Young, R.W. (1981). Overbank deposition and floodplain formation on small coastal streams of New South Wales. *Zeitschrift für Geomorphologie*, 25, 332–347.

Nanson, G.C., Price, D.M. and Short, S.A. (1992). Wetting and drying of Australia over the past 300 ka. *Geology*, 20, 791–794.

Nash, D.J. (2004a). Calcrete. In: Goudie, A.S. (Ed.). *Encyclopedia of Geomorphology*. Routledge. London, 108–111.

Nash, D.J. (2004b). Silcrete. In: Goudie, A.S. (Ed.). *Encyclopedia of Geomorphology*. Routledge. London, 959–960.

Nash, D.J. and Smith, R.F. (1998). Multiple calcrete profiles in the Tabernas Basin, southeast Spain: their origins and geomorphic implications. *Earth Surface Processes and Landforms*, 23, 1009–1029.

Nash, D.J. and Ulliot, S. (2007). Silcrete. In: Nash, D.J. and McLaren, S.J. (Eds.). *Geochemical Sediments and Landscapes*. Blackwell. Malden, 95–143.

Nash, L.L. and Gleick, P.H. (1991). Sensitivity of stream-flow in the Colorado Basin to climatic changes. *Journal of Hydrology*, 125, 221–241.

Návar, J. and Bryan, R. (1990). Interception loss and rainfall redestribution by three semiarid growing shrubs in northeastern Mexico. *Journal of Hydrology*, 115, 51–63.

Navas, A. (1990). The effect of hydrochemical factors on the dissolution rate of gypsiferous rocks in flowing water. *Earth Surface Processes and Landforms*, 15, 709–715.

Navas, A., García-Ruiz, J.M., Machín, J., Lasanta, T., Valero, B., Walling, D.E. and Quine, T.A. (1997). Soil erosion of dry farming land in two changing environments of the Central Ebro valley, Spain. In: *Human Impact on Erosion and Sedimentation*. International Association Hydrological Sciences, 245, 13–20.

Neal, J.T. (1965). Environmental setting and general surface characteristics of playas. In: Neal, J.T. (Ed.). *Geology,*

Mineralogy and Hydrology of United States Playas. Air Force Cambridge Research Laboratory, Environmental Research Paper, 96, 1–29.

Neall, V.E. (1996). Hydrological disasters associated with volcanoes. In: Singh, V.P. (Ed.). *Hydrology of Disasters*. Kluwer. Dordrecht, 395–345.

Neall, V.E. (2004). Lahar. In: Goudie, A.S. (Ed.). *Encyclopedia of Geomorphology*. Routledge. London, 597–599.

Nelson, A.R. (1992). Lithofacies analysis of coluvial sediments-an aid in interpreting the recent history of Quaternary normal faults in the Basin and Range province, Western United States. *Journal Sedimentary Petrology*, 62, 607–621.

Nelson, A.R. (2007). Tectonics Locations. In: Elías, S.A. (Ed.). *Encyclopedia of Quaternary Science*. Elsevier, Amsterdam, 3072–3087.

Nelson, F.E., Anisimov, O.A. and Shiklomanov, N.I. (2002). Climatic change and hazard zonation in the circum-Arctic permafrost regions. *Natural Hazards*, 26, 203–225.

Netterberg, F. (1969). Ages of calcretes in southern Africa. *South African Archeological Bulletin*, 24, 117–122.

Neumeier, U. and Ciavola, P. (2004). Flow resistance and associated sedimentary processes in *Spartina maritime* salt-marsh. *Journal of Coastal Research*, 20, 435–447.

Newhall, C.H. and Self, S. (1982). The volcanity explosivity index (V E I): an estimate of explosive magnitude for historical volcanism. *Journal Geophysical Research*, 87, 231–238.

Newton, J.G. (1986). Development of sinkholes resulting from man's activities in the Eastern United States. *United States Geological Survey*. Denver, 54 p.

Nguetnkam, J.P., Kamga, R., Villéras, F., Ekodek, J.D. and Yven, J. (2008). Altérations diffèrentielle du granite in zone tropicale. Exemple de deux sequences etudieé au Cameroun (Afrique centrale). *Compte Rendus Geoscience*, 340, 451–461.

Nicholas, R.M. and Dixon, J.C. (1986). Sandstone scarp form and retreat in the Land of Standing Rocks, Canyonlands National Park, Utah. *Zeitschrift für Geomorphologie*, 30, 167–187.

Nicholson, S.E. (1981). Saharan climates in historical times. In: Allan, J.A. (Ed.). *The Sahara: Ecological Change and Early Economic History*. Menas Press. London, 35–59.

Nicholson, S.E. (1993). What is a desert? In: Seely, M. (Ed.). *Deserts*. Weldon Ower Limited. Australia, 14–25.

Nicholson, S.E. and Flohn, H. (1980). African environmental and climatic changes and the general atmospheric circulation in Late Pleistocene and Holocene. *Climatic Change*, 2, 313–348.

Nickling, W.G. (1983). Grain size characteristics of sediment transported during dust storms. *Journal of Sedimentary Petrology*, 53, 1011–1024.

Nickling, W.G. (Ed.) (1986). *Aeolian Geomorphology*. Allen and Unwin. Boston, 311 p.

Nickling, W.G. (1988). The initiation of particle movement by wind. *Sedimentology*, 35, 499–511.

Nickling, W.G. and Gillies, J.A. (1993). Dust emission and transport in Mali, West Africa. *Sedimentology*, 40, 859–868.

Nicod, J. (1972). *Pays and Paysages du Calcaire*. Presses Universitaires de France. Paris, 244 p.

Nicod, J. (1976). Corrosion du type crypto-karstique dans les karsts mediterrannéens. In: Gams, I. (Ed.). *Karst Processes and Relevant Landforms*. Departement of Geography. Ljubljana University. Ljubljana, 171–180.

Nicod, J. (1979). Sur le rôle de la tectonique et des variations des circulations karstiques et specialement de leurs bourdures. *Annales de la Societé Géologique de Belgique*, 102, 87–93.

Nicod, J. (1992). Formes d'aplanissement et de regularisation des versants dans les roches carbonatées: position des problèmes et eléments de discusion. *Tübinger Geographische Studien*, 109, 1–22. Tübingen.

Nielson, J. and Kocurek, G. (1987). Surface processes, deposits, and development of star dunes, Dumont dune field, California. *Bulletin of the Geological Society of America*, 99, 177–186.

Nilsen, T.H. (1993). What is an alluvial fan?. *International Working Group on Fan Deltas Newsletter*, 1, 3–4.

Nilsen, T.H. and Moore, T.E. (1984). *Bibliography of Alluvial Fan Deposits*. GeoBooks. Norwich, 96 p.

Nir, D. (1983). *Man, a Geomorphological Agent. An Introduction to Anthropic Geomorphology*. Keter Publication. Jerusalem, 165 p.

Nittouer, C.A., Kuehl, S.A., DeMaster, D.J. and Kowssmann, R.O. (1986). The deltaic nature of Amazon shelf sedimentation. *Geological Society of America Bulletin*, 97, 444–458.

Nocera, S. di and Matano, F. (2002). Weathering Surveys in Geomorphology: mapping examples from Sila Massif. In: Allison, R.J. (Ed.). *Applied Geomorphology: Theory and Practice*. Wiley. Chichester, 205–225.

Noguchi, K. and Kamiya, H. (1963). Prediction of volcanic eruption by measuring the chemical composition and amounts of gases. *Bulletin Volcanologique*, 26, p. 367.

Noller, J.S., Sower, J.M. and Lettis, W.R (2000). *Quaternary Geochronology: methods and applications*. AGU Reference Shelf 4. American Geophysical Union, Washington, 582 p.

Nordstrom, K.F. (2008). *Beach and Dune Restoration*. Cambridge University Press. Cambridge, 187 p.

Nordstrom, K., Psuty, N. and Carter, B. (Eds.) (1990). *Coastal Dunes: Form and Process*. Wiley. Chichester, 392 p.

Norris, R.M. (1966). Barchan dunes of Imperial Valley, California. *Journal of Geology*, 74, 292–308.

Normack, V.R and Reid, J.A. (2003). Extensive deposits on the Pacific Plate from Late Pleistocene North-America glacial lake bursts. *Journal of Geology*, 111, 617–637.

Nortcliff, S., Ross, S.M. and Thornes, J.B. (1990). Soil moisture, runoff and sediment yield from differentially cleared tropical rainforest plots. In: Thornes, J.B. (Ed.). *Vegetation and Erosion*, Wiley. Chichester, 420–436.

Norton, S.A. (1973). Laterite and bauxite formation. *Economic Geology*, 68, 353–361.

Norwick, S.A. and Dexter, L.A. (2002). Rates of development of tafoni in the Moenkopi and Kaibab formations in Meteor Crater and on the Colorado Plateau, northeastern Arizona. *Earth Surface Processes and Landforms*, 27, 11–26.

Nott, J. (2006). *Extreme Events. A Physical Reconstruction and Risk Assessment*. Cambridge University Press. Cambridge, 297 p.

Nussbaumer, J. (1998). *Die Gewalt der Natur. Eine Chronic der Naturkatastrophen von 1500 bis heute*. Grundback Sandkorn.

Nyamweru, C. (1980). *Rifts and Volcanoes. A Study of the East African Rift System*. Nelson Africa. Nairobi, Kenia, 128 p.

Nye, J.F. (1952). The mechanics of glacier flow. *Journal Glaciology*, 2, 82–93.

Nye, J.F. (1958). Surges in glaciers. *Nature*, 181, 1450–1451.

O'Connor, G.J.E. (1993). *Hydrology, Hydraulics and Sediment Transport of Pleistocene Lake Bonneville Flooding on the Snake River, Idaho*. Geological Society of America, Special Paper 274, 83 p.

O'Hara, S.L. (1997). Human impacts on drylands geomorphic processes. In: Thomas, D.S.G. (Ed.). *Arid Zone Geomorphology. Process, Form in Change in Drylands*. Wiley. Chichester, 639–658.

Oberlander, T.M. (1972). Morphogenesis of granitic boulder slopes in the Mojave desert, California. *Journal of Geology*, 80, 1–20.

Oberlander, T.M. (1977). Origin of segmented cliffs in massive sandstones of southeastern Utah. In: Doehring, D.O. (Ed.). *Geomorphology in Arid Regions*. George Allen and Unwin. London, 79–114.

Oberlander, T.M. (1994). Rock varnish in deserts. In: Abrahams, A.D. and Parsons, A.J. (Eds.). *Geomorphology of Desert Environments*. Chapman and Hall. London, 106–119.

Oberlander, T.M. (1997). Slope and pediment systems. In: Thomas, D.S.G. (Ed.). *Arid Zone Geomorphology*. Wiley. Chichester, 135–163.

Oerlemans, J. (1989a). On the response of valleys glaciers to climatic change. In: Oerlemans, J. (Ed.). *Glacier Fluctuations and Climatic Change*. Kluwer. Dordrecht, 353–371.

Oerlemans, J. (1989b). *Glacier Fluctuations and Climatic Changes*. Kluwert. Dordrecht, 417 p.

Ogden, J.M. (1999). Prospects for building a hydrogen energy infrastructure. *Annual Review of Energy and the Environment*, 24, 227–279.

Oguchi, C.T., Hatta, T. and Matsuhura, Y. (1999). Weathering rates over 40.000 years based on changes in rock properties of porous rhyolite. *Physics and Chemistry of the Earth* (A), 24, 861–870.

Ogunkoya, O.O. and Jeje, L.K. (1987). Sediment yield from some third order basins of the basement complex rocks in southwestern Nigeria. *Catena,* 14, 383–397.

Okada, A., Takeuchi, A., Tsukuda, T., Ikeda, Y., Watanabe, M., Hirano, S., Masumoto, S., Takehana. Y., Okumura, K., Kamishima, T., Kobayashi, T. and Ando, M. (1989). Trenching study of the Atotsugawa Fault at Nokobi, Miyagawa Village, Gifa Prefecture, central Japan. *Journal Geography*, 98, 62–85.

Olcina, J. (1994). *Riesgos climáticos en la Península Ibérica*. Acción Divulgativa. Madrid, 440 p.

Olcina, J. and Miró, J. (2002). Tormentas de arena y tornados. In: Ayala, F.J. and Olcina, J. (Eds.). *Riesgos naturales*. Ariel Ciencia. Barcelona, 681–702.

Olivé, A., Portero, J.M., Olmo, P. del, Aragonés, E., Carreras, F., Molina, E. and Gutiérrez-Elorza, M. (1982). El sistema de terrazas del río Carrión. *I Reunión sobre la Geología de la Cuenca del Duero. Salamanca, 1979*. Instituto Geológico y Minero de España. Madrid, 451–459.

Oliver-Smith, A. (2002). El gran terremoto del Perú, 1970: el concepto de la vulnerabilidad y la gestión de los desastres en América Latina. In: Hubp, J.L. and Inbar, M. (Eds.). *Desastres naturales en América Latina*. Fondo de Cultura Económica. México, 147–169.

Olivier, J. (1995). Spatial distribution of fog in the Namib. *Journal of Arid Environments*, 29, 129–138.

Ollero, A. (1996). *El Curso Medio del Ebro*. Publicaciones del Consejo de Protección de la Naturaleza de Aragón, 4. Zaragoza, 311 p.

Ollier, C.D. (1960). The inselbergs of Uganda. *Zeitschrift für Geomorphologie*, 4, 43–52.

Ollier, C.D. (1965). Some features of granite weathering in Australia. *Zeitschrift für Geomorphologie*, 9, 285–304.

Ollier, C.D. (1966). Desert gilgai. *Nature*, 212, 581–583.

Ollier, C.D. (1969a). *Volcanoes*. The MIT Press. London, 177 p.

Ollier, C.D. (1969b). *Weathering*. Longman. London, 304 p.

Ollier, C.D. (1977). Applications of weathering studies. In: Hails, J.R. (Ed.) *Applied Geomorphology*. Elsevier. Amsterdam, 9–50.

Ollier, C.D. (1978). Inselbergs of the Namib Desert, processes and history. *Zeitschrift für Geomorphologie Supplementband*, 31, 161–176.

Ollier, C.D. (1981). *Tectonics and Landforms*. Longman, London, 324 p.

Ollier, C.D. (1983). Tropical geomorphology and long-term landform evolution. *Finisterra*, 36–203–221.

Ollier, C.D. (1988a). Deep weathering, groundwater and climate. *Geografiska Annaler*, 70 A, 285–290.

Ollier, C.D. (1988b). The regolith in Australia. *Earth Science Reviews*, 25, 355–362.

Ollier, C.D. (1991). *Ancient Landforms*. Belhaven Press. London, 233 p.

Ollier, C.D. and Ash, J.E. (1983). Fire and rock breakdown. *Zeitscrift für Geomorphologie*, 27, 363–374.

Ollier, C.D. and Pain, C. (1996). *Regolith, Soils and Landforms.* Wiley. Chichester, 316 p.

Ollier, C.D. and Tratman, E.K. (1969). Geomorphology of the caves. In: Tratman, E.K. (Ed.). *The Caves of North-west Clare Ireland*. David and Charles. Newton Abbot, Devon, 59–95.

ONU (1984). *Disaster Prevention and Mitigation. A Compendium of Current Knowledge*. United Nations. New York, 218 p.

Oostwoud, D.J.W. and Bryan, R.R. (1994). Gully head cuts as sediment sources on the Njemps Flats and initial low-cost gully control measures. In: Bryan, R.R. (Ed.). *Soil Erosion, Land Degradation and Social Transition*. Advances in GeoEcology, 27, 205–229.

Opdyke, N.D., Flass, B., Hays, J.D. and Foster, J. (1966). Palaeomagnetic study of Antarctic deep sea cores. *Science*, 154, 349–357.

Ordaz, J., Esbert, R.M., Montoto, M., de Caso, F. and Alonso, F.J. (1984). Estado de alteración y alterabilidad futura de la piedra de Villamayor de los monumentos salmantinos: interpretación petrográfica. In: *Estudio Sobre las Alteraciones y Tratamiento de la Piedra de Villamayor*. Ediciones Caja de Ahorros y M.P. de Salamanca. Salamanca, 215–419.

Ordoñez, S., García del Cura, A., López de Azcona, M.C. and González Martín, J.A. (1983). Contribución al conocimiento de la hidroquímica de Las Lagunas deRuidera (Cuenca alta del río Guadiana). *I Congreso de Geoquímica*, 167–168.

Ortí, F. (1989). Evaporitas marinas. In: Arche, A. (Ed.). *Sedimentología, vol. II*. Consejo Superior de Investigaciones Científicas. Madrid, 89–177.

Ortiz, A. (1993). Hydrology information and voice communication automatic system in the Ebro basin. In: Gutiérrez, M., Sancho, C. and Benito, G. (Eds). *Second Intensive Course on Applied Geomorphology: Arid Regions*. Universidad de Zaragoza. Zaragoza, 89–105.

Osborne, A. (2004). Paleokarst. In: Gunn, J. (Ed.). *Encyclopedia of Caves and Karst Science*. Fitzroy Dearborn. New York, 559–561.

Osterkamp, W.R. and Wood, W.W. (1987). Playa-lake basins on the Southern High Plains and New Mexico: Part 1- Hydrologic, geomorphic and geologic evidence for their development. *Bulletin of the Geological Society of America*, 99, 215–223.

Østrem, G., Haakensen, N. and Melander, O. (1973). *Atlas over Breer i Nord-Skandinavia*. Norges Vassdrags og Elektrisitetsvesen og Stockholm Univ.

Oteo, C. (1984). Potencial de alterabilidad de la piedra de Villamayor. In: *Estudio sobre las Alteraciones y Tratamiento de la Piedra de Villamayor*. Ediciones Caja de Ahorros y M.P. de Salamanca. Salamanca, 477–565.

Ouchi, S. (1985). Response of alluvial rivers to slow active tectonic movement. *Geological Society of America Bulletin*, 96, 504–515.

Oviatt, C.G. (1997). Lake Bonneville fluctuations and global climate change. *Geology*, 25, 155–158.

Owen L.A., Kemp, U., Khattak, G.A., Harp, E.L., Keefer, D.K. and Bauer, M.A. (2008). Landslide triggered by the 8 October 2005 Kashmir earthquake. *Geomorphology*, 94, 1–9.

Owens, I. (2004). Avalanche, snow. In: Goudie, A.S. (Ed.). *Encyclopedia of Geomorphology*. Routledge. Londres, 41–44.

Owens, L.B. and Watson, J.P. (1979). Landscape reduction by weathering in small rhodesian watersheds. *Geology*, 7, 281–284.

Oyagi, N. (1989). Geological and economic extent of landslides in Japan and Korea. In: Brabb, E.E. and Ardo, B.L. (Eds.). *Landslides: Extent and Economic Significance*. Balkema. Rotterdam, 289–302.

Pagés, J.L. (2000). Origen y evolución geomorfológica de las rías atlánticas de Galicia. *Revista de la Sociedad Geológica de España*, 13, 393–403.

Pain, C.F. and Ollier, C.D. (1981). Geomorphology of a Pliocene granite in Papua New Guinea. *Zeitschrift für Geomorphologie*, 25, 249–258.

Palacios, D., de Andrés, N. and Luengo, E. (2003). Distribution and effectiveness of nivation mediterranean: Peñalara (Spain). *Geomorphology*, 54, 157–178.

Pallister, J.W. (1956). Slope development in Buganda. *Geographical Journal*, 122, 80–87.

Palmer, A.N. (1991). Origin and morphology of limestones caves. *Geological Society of America Bulletin*, 103, 1–21.

Palmer, A.N. (2000). Hydrological control of cave patterns. In: Klimchouk, A.B., Ford, D.C., Palmer, A.N. and Dreybrodt, W. (Eds.). *Speleogenesis: Evolution of Karst Aquifers*. National Speleological Society. Huntsville, Alabama, 77–90.

Palmer, A.N. (2004). Carlsbad Cavern and Lechugilla Cave. In: Gunn, J. (Ed.) *Encyclopedia of Caves and Karst Science*. Fitzroy Dearborn. New York, 192–195.

Palmer, R.S. (1963). The influence of a thin water layer on water drop impact forces. *International Association Scientific Hydrology*, 65, 141–148.

Pánek, T., Hradecký, J., Šilhan, K., Smolková, V. and Altová, V. (2009). Time constrains for the evolution of a large slope collapse in karstified mountainous terrains of the southwestern Crimea Mountains, Ukranie. *Geomorphology*, 108, 171–181.

Panizza, M. (1972). Schema di legenda per carte geomorfologiche di dettaglio. *Boletin Societa Geologica de Italia*, 91, 207–237.

Panizza, M. (1973). Glacio pressure implications in the production of landslides in the Dolomitic area. *Geologia Applicata e Idrogeologia*, 8, 289–297.

Panizza, M. (1988). *Geomorfologia Applicata. Metodi di Applicazione alla Planificazione Territoriale e alla Valutazione d'Impatto Ambientale*. La Nuova Italia Scientifica. Roma, 342 p.

Panizza, M. (1991). Geomorphology and seismic risk. *Earth-Science Reviews*, 31, 11–20.

Panizza, M. (1992). *Geomorfología*. Pitagora Editrice. Bologna, 397 p.

Panizza, M. (2005a). Pericolosità e rischio geomorfologici. In: Panizza, M. (Ed.). *Manuale di Geomorfologia Applicata*. FrancoAngeli. Milano, 31–43.

Panizza, M. (2005b). Procesi superficiali di pericolosità. In: Panizza, M. (Ed.). *Manuale di Geomorfologia Applicata*. FrancoAngeli. Milano, 44–87.

Panizza, M. and Piacente, S. (2003). *Geomorfología Cultural*. Pitagora Editrice. Bologna, 350 p.

Paola, C. (2004). Braided River. In: Goudie, A.S. (2004). *Encyclopedia of Geomorphology*. Routledge. London, 98–101.

Parés, J.M. and Pérez- González, A. (1995). Paleomagnetic age for hominid fossils at Atapuerca archaeological site, Spain. *Science*, 269, 830–832.

Paris, R., Carracedo, J.C. and Pérez Torrado, F.J. (2005). Massive flank failures and tsunamis in the Canary Islands. *Zeitschrift für Geomorphology Supplement Band*, 140, 37–54.

Paris, R., Wassmer, P., Sartohadi, J., Lavigne, F., Barthomeuf, B., Desgajes, E., Grancher, D., Beaumont, P., Veautier, F., Brunstein, D. and Gómez, C. (2009). Tsunamis at geomorphic crises: Lessons from the December 26, 2004 tsunami in Lhok Nga, West Band Aceh (Sumatra, Indonesia). *Geomorphology*, 104, 59–72.

Park, C.C. (1981). Man, river system and environments impacts. *Progress in Physical Geography*, 5, 1–31.

Parker, G.G. (1963). Piping, a geomorphic agent in landform development of the drylands. *International Association of Scientific Hydrologists Publication*, 65, 103–113.

Parker, G.G. and Jenne, E.A. (1967). Structural failure of Western U.S. highways caused by piping. *United States Geological Survey. Water Resources Division*, 27p.

Parker, G.G., Higgins, C.G. and Wood, W.W. (1990). Piping and pseudokarst in drylands. *Geological Society of America Special Paper*, 252, 77–110.

Parmenter, C. and Folger, D.W. (1974). Eolian biogenic detritus in deep sea sediments: a possible index of equatorial Ice Age aridity. *Science*, 185, 695–698.

Parra, E. and Cepeda, H. (1990). Volcanic hazards maps of the Nevado del Ruiz volcano, Columbia. *Journal of Volcanology and Geothermal Research,* 42, 117–127.

Parsons, A.J. (1988). *Hillslope form*. Routledge. London. 212 p.

Parsons, A.J. and Abrahams, A.D. (2009) (Eds.). *Geomorphology of Desert Environments*. Springer, 831 p.

Parsons, D.R., Walter, I.J. and Wiggs, G.F.S. (2004). Numerical modelling of flow structures over idealised transverse aeolian dunes of varying geometry. *Geomorphology*, 59, 149–164.

Paskoff, P. (1981). *L'érosion de côtes*. Presses Universitaires de France. París, 127 p.

Paskoff. R. (1985). *Les littoraux. Impact des aménagements sur leur évolution*. Masson. París, 188 p.

Passarge, S. (1904). *Die Kalahari*. Reimer (2 vols.). Berlín.

Passarge, S. (1931). *Geomorfología*. Ed. Labor. Barcelona, 189 p.

Pasuto, A. and Soldati, M. (1996). Rock Spreading. In: Dikau, A., Brunsden, D., Schrott, L. and Ibsen, M.L. (Eds.). *Landslide Recognition. Identification, Movement and Causes*. Wiley. Chichester, 122–136.

Pasuto, A., Silvano, S., Soldati, M. and Tecca, P.R. (1994). Deep-seated gravitational deformations in northeastern Italy. *Special volume for the International Congress IAEG*. Lisboa, 23–27.

Paté-Cornell, E. (1996). Uncertainties in global climate change estimates. An Editorial Essay. *Climatic Change*, 33, 145–149.

Paterson, W.S.B. (2004). *The Physics of Glaciers*. Pergamon. New York, 480 p.

Patton, P.C. (1988). Drainage basin morphometry and floods. In: Baker, V.R., Craig Kocher, R. and Patton, P.C. (Eds.). *Flood Geomorphology*. Wiley. New York, 51–64.

Paukstys, W.J. and Narbutas, V. (1996). Gypsum karst of the Baltic republics. In: Klimchouk, A., Lowe, A., Cooper, A. and Sauro, U. (Eds.). Gypsum karst of the world. *International Journal of Speleology*, 25, 279–284.

Paul, E.A. and Clark, F.E. (1989). *Soil Microbiology and Biochemistry*. Academia Press. London, 273 p.

Pavich, M.J. (1985). Appalachian piedmont morphogenesis: weathering, erosion, and Cenozoic uplift. In: Morisawa, M. and Hack, J.T. (Eds.). *Tectonic Geomorphology*. George Allen and Unwin. London, 27–51.

Pavlov, A.V. (1994). Current changes of climate and permafrost in the Arctic and Sub-Arctic of Russia. *Permafrost and Periglacial Processes*, 5, 101–110.

Pearce, F. (1994). Not warming, but cooling. *New Scientifist*, 9 July, 37–41.

Peck, D.L., Troester, J.W. and Moore, J.E. (1988). Karst hydrogeology in the United States of America. En: *Proceedingsof the 21st Congress of the International Association of Hydrologists*. Karst Hydrology and Karst Environmental Protection. Guilin, China. Open-file Report 88–476. United States Geological Survey. Washington, 19 p.

Pécsi, M. (1968). Loess. In: Fairbridge, R.W. (Ed.). *The Encyclopedia of Geomorphology*. Reinold. New York, 674–678.

Pécsi, M. (Ed.) (1987). *Loess and Environment*. Catena Supplement. Brockenblick, 114 p.

Pécsi, M. and Szilárd, J. (1970). Planated Surfaces: principal problems of research and terminology. In: Pécsi, M. (Ed.). *Problems of Relief Planation*. Akadémiai Kiadó, 13–27. Budapest.

Pedraza, J., Sanz, M.A. and Martín, A. (1989). *Formas Graníticas de La Pedriza*. Agencia de Medio Ambiente. Madrid, 205 p.

Pedro, G. (1966). Essai sur le caractérisation géochimique des différents processus zonaux résultant de l'altération superficielle. *Compte Rendu de la Academie des Sciences de Paris*, 262, 1828–1831.

Pedro, G. (1968). Distributions des principaux types d'altération chimique à la surface du globe. *Revue de Géographie Physique et Géologie Dynamique*, 10, 457–470.

Pedro, G., Jamague, M. and Bejon, J.C. (1969). Mineral interactions and transformations in relation to pedogenesis during the Quaternary. *Soil Science*, 107, 462–469.

Pedro, J. (1993). Un phenomene normal de la surface de la planete: l'alteration des roches. In: Vicente, M.A., Molina, E. and Rives, V. (Eds.). *Alteración de Granitos y Rocas Afines*. Consejo Superior de Investigaciones Científicas. Madrid, 17–50.

Peel, R.F. (1974). Insolation weathering: some measurements of diurnal temperature changes in exposed rocks in the Tibesti region, central Sahara. *Zeitschrift für Geomorphologie, Supplementband* 21, 19–28.

Peeters, Y, van Ost, K., Govers, G., Verstraeten, G., Rommens, T. and Poesen, J. (2008). The compatibility of erosion data at different temporal scales. *Earth Planetary Science Letters*, 265, 138–152.

Pellerin, F.M. (1980). La porosimétrie au mercure apliquée a la l'étude géotechnique des sols et des roches. *Bulletin Liaison Laboratoire des Ponts et Chaussées*, 106, 105–116.

Pellicer, F. and Yetano, L.M. (1985). El galacho de Juslibol: un ejemplo de meandro abandonado. *Cuadernos de Investigación Geográfica*, 11, 193–124.

Pellicer, F., Peña, J.L. and Ibáñez, M.J. (1986). Estudio geomorfológico del yacimiento de Burrén y Burrena (Depresión del Ebro): génesis del relieve y evolución holocena. *Estudios Hom. Dr. Antonio Beltrán*. Zaragoza, 33–45.

Peltier, L.C. (1950). The geographic cycle in periglacial regions as it is related to climatic geomorphology. *Annals of the Association of American Geographers*, 40, 214–236.

Peltier, W.R. and Andrews, J.T. (1983). Glacial geology and glacial isostasy of the Hudson Bay region. In: Smith, D.E. and Dawson, A.G. (Eds.). *Shorelines and Isostasy*. Academic Press. London, 285–320.

Penck, A. (1894). *Morphologie der Erdoberfläcke*. Engelhotn (2 vols.). Stuttgart.

Penck, A. (1905). Glacial features in the surface of the Alps. *Journal Geology*, 13, 1–19.

Penck, W. (1924). *Die Morphologische Analyse*. Geographische Abhandlungen, 2. Stuttgart, 1–283. Traducción inglesa por H. Czech y K.C. Boswell (1953). *Morphological Analysis of Landforms. A Contribution to Physical Geology*. Macmillan. London, 429 p.

Pentecost, A. (2005). *Travertine*. Springer. Berlin, 445 p.

Peña. J.L. (Ed). (1997). *Cartografía Geomorfológica Básica y Aplicada*. Geoforma. Logroño, 227 p.

Peña, J.L. (1997a). Los mapas geomorfológicos: características y tipos. In: Peña, J.L. (Ed.). *Cartografía Geomorfológica Básica y Aplicada*. Geoforma Ediciones. Logroño, 13–24.

Peña, J.L. (1997b). Los sistemas de cartografía geomorfológica. In: Peña, J.L. (Ed.). *Cartografía Geomorfológica Básica y Aplicada*. Geoforma Ediciones. Logroño, 85–102.

Peña, J.L. and González, J.R. (1992). Modelo evolutivo de los cambios en la dinámica geomorfológica del Baix Cinca y Segre (Depresión del Ebro) durante el Pleistoceno superior-Holoceno a partir de los datos geoarqueológicos. *Cuaternario y Geomorfología*, 6, 103–110.

Peña, J.L., Jiménez, A. and Echevarría, M.T. (1989). Geomorphological cartography and evolutionary aspects of the Sierra de Albarracín poljes. Eastern Iberian Ranges, Teruel, Spain. *Geografía Física e Dinámica Quaternaria*, 12, 51–57.

Peña, J.L., Julián, A. and Chueca, J. (1991). Séquences évolutives des accumulations holocènes à la Hoya de Huesca dans le contexte general du bassin de l'Ebre (Espagne). *Physio-Géo*, 22/23, 55–60.

Peña, J.L., Echeverría, M.T., Petit-Maire, N. and Lafont. R. (1993). Cronología e interpretación de las acumulaciones holocenas de la Val de las Lenas (Depresión del Ebro, Zaragoza). *Geographicalia*, 30, 321–332. Zaragoza.

Peña, J.L., Chueca, J. and Julián, A. (1998). Los derrubios estratificados del sector central pirenaico: cronología y límites altitudinales. In: Gómez Ortiz, A., Salvador Franch, F., Schulte, L. and García Navarro, A. (Eds.). *Procesos biofísicos actuales en medios fríos*. Publicaciones de la Universitat de Barcelona. Barcelona, 205–216.

Peña, J.L., González, J.R. and Rodríguez, J.I. (1996). Paleoambientes y evolución geomorfológica en yacimientos arqueológicos del sector oriental de la Depresión del Ebro durante el Holoceno Superior. In: Pérez-Alberti, A., Martín, P., Chesworth, W. and Martínez, A. (Eds.). *Dinámica y Evolución de Medios Cuaternarios*. Xunta de Galicia, 63–80.

Peña, J.L., Pellicer, F., Julián, A., Chueca, J., Echeverría, M.T., Lozano, M.V. and Sánchez, M. (2002). *Mapa Geomorfológico de Aragón*. Consejo de Protección de la Naturaleza. Zaragoza, 54 p.

Peña, J.L., Lewis, C, McDonald, E., Rhodes, E. and Sancho, C. (2004). Ensayo cronológico del Pleistoceno medio-superior de la cuenca del río Cinca (Pirineos y Depresión del Ebro). In: Benito, G. and Diez Herrero, A. (Eds.). *Contribuciones Recientes sobre Geomorfología* (Actas de la VIII Reunión Nacional de Geomorfología, Toledo). Consejo Superior de Investigaciones Científicas. Vol. I. Madrid, 165–172.

Pérez-Alberti, A., Blanco, R., Costa, M., Vázquez, M. and Valcárcel, M. (2000). Dinámica y evolución de las costas rocosas. El ejemplo de Galicia. In: Andrés, J.R. and Gracia, F.J. (Eds.). *Geomorfología Litoral. Procesos Activos*. Monografía nº 7 de la Sociedad Española de Geomorfología. Instituto Geológico Minero de España. Madrid, 175–185.

Pérez-Cueva, A. and Armengot, R. (1983). El temporal de octubre de 1982 en el marco de las lluvias torrenciales en la cuenca baja del Júcar. *Cuadernos de Geografía*, 32–33, 61–86.

Perfit, M.R. and Davidson, J P. (2000). Plate tectonics and volcanism. In: Sigurdsson, H. (Ed). *Encyclopedia of Volcanoes*. Academic Press, San Diego, 89–113.

Perry, R.S., Lynne, B.Y., Sephton, M.A., Kolb, V.M., Perry, C.C. and Staley, J. (2006). Backing black opal in the desert sun: The importance of silica in desert varnish. *Geology*, 34, 737–754.

Persons, B.S. (1970). *Laterite: Genesis, Location, Use*. Plenum Press. New York, 103 p.

Persons, B.S. (1984). Laterite, Engineering Geology. In: Finkl, C.W. (Ed.). *The Encyclopedia of Applied Geology*. Van Nostrand Reinhold. New York, 320–326.

Perthuisot, J.P. and Jauzein, A. (1975). Sebkhas et dunes d'argile: l'enclave endoreique de Pont du Fars, Tunisie. *Revue de Géographie Physique et de Géologie Dynamique*, 17, 295–306.

Perucca, L.P. and Moreiras, S.M. (2006). Liquefaction phenomena associated with historical earthquakes in San Juan and Mendoza Provinces, Argentina. *Quaternary International*, 158, 96–109.

Peters, J. (1964). Stabilization of sand dunes at Vanderberg Air Force Base. *Journal Soil Mechanics and Foundations Division American Society Civil Engineering*, 90, 97–106.

Peters, R.L. and Lovejoy, T.E. (1990). Terrestrial fauna. In: Turner, B.L. Clark, W.C., Kates, R.W., Richards, J.F., Mattews, J.T. and Meyer, W.B. (Eds.). *The Earth as Transformed by Human Action*. Cambridge University Press. Cambridge, 353–369.

Peterson, B.J., Holmes, R.M., McClellan, J.W., Vörösmarty, C.J., Lammers, R.B., Shiklomanov, A.I. and Rahmstortf, S. (2002). Increasing river discharge to the Arctic Ocean. *Science*, 298, 2171–2173.

Peterson, T.C and Vose, R.S. (1997). An overview of the Global Historical CLimatology Network temperatura data base. *Bulletin of the American Meteorological Society*, 78, 2837–2849.

Pethick, J. (1984). *An Introduction to Coastal Geomorphology*. Arnold, London, 260 p.

Petit, J.R., Jouzel, J., Raynud, D., Barkov, N.I., Barnola, J.M., Basile, I., Benders, M., Chappellaz, J., Davis, M., Delaygne, G., Delmotte, M., Kotlyakov, V.M., Legrand, J.M., Lipenkov, V.I., Lorius, C., Pépin, L., Ritz, C., Saltzman, E. and Stievenard, M. (1999). Climate and atmospheric history of the past 420,000 years from the Vostok ice core, Antarctica. *Nature*, 399, 429–436.

Petit-Maire, N. (1999). Variabilité naturelle des environments terrestres: les deux dernieres extrêmes climatiques (18.000 ± 2.000 et 8.000 ± 1.000 ans BP). *Compte Rendus de la Académie des Sciences de Paris,* 328, 273–279.

Petit-Maire, N. and Guo, Z.T. (1998). Mid-Holocene climatic change and Man in the present-day Sahara desert. In: Alsharhan, A.S., Glennie, K.W., Whittle, G.L. and

Kendall, C.G. (Eds.). *Quaternary Deserts and Climatic Change*. Balkema. Rotterdam, 351–356.

Petit-Maire, N., Sanlaville, P. and Yan, Z. (1995). Oscillations de la limite nord du domaine des moussons africaine, indienne et asiatique, au cours du dernier cycle climatiue. *Bulletin Societé Géologique de France*, 166, 213–220.

Petri, S. and Fúlgaro, V.F. (1983). *Geología do Brasil*. Editora da Universidade de Sao Paulo. Sao Paulo, 631 p.

Petts, G. (1989). Historical análisis of fluvial hydrosystems. In: Petts, G. (Ed.). *Historical Change of Large Alluvial Rivers*. Wiley. Chichester, 1–18.

Petts, G.E. and Amoros, C. (1996). The Fluvial Hydrosystem. In: Petts, G.E. and Amoros, C. (Eds.). *Fluvial Hydrosystems*. Chapman and Hall, London, 1–12.

Petts, G. y Foster, I. (1985). *Rivers and Landscape*. Arnold. London, 274 p.

Peulvast, J.P. and Vanney, J.R. (2001). Géomorphologie Structurale, Terre, Corps Planétaires Solides. Gordon and Breach. Paris, 501 p.

Péwé, T.L. (1955). Origin of the upland silt near Fairbanks, Alaska. *Bulletin of the Geological Society of America*, 66, 699–724.

Péwé, T.L. (1959). Sand wedge polygons (tesselations) in the McMurdo Sound region, Antarctica. *American Journal of Science*, 257, 545–552.

Péwé, T.L. (1966). Paleoclimatic significance of fossil wedges. *Biuletyn Peryglacjalny*, 15, 65–73.

Pewé, T.L. (1981). Desert dust: an overview. In: Pewé, T.L. (Ed.). *Desert dust: origin, characteristics and effect on man*. Geological Society of America, Special Paper 186, 1–10. Boulder, Colorado.

Péwé, T.L. (Ed.) (1981). *Desert Dust: Origin, Characteristics and Effect of Man.* Geological Society of America, Special Paper, 186. Boulder, 303 p.

Péwé, T.L. (1983). The periglacial environment in North America during Wisconsin time. In: Porter, S.C. (Ed.). *Late Quaternary of the United States, vol. I, The Late Pleistocene*. Longman. London, 157–189.

Pfeffer, K.H. (1973). Flachenbildung in den kalkgebieren. In: Semmel, A. (Ed.). *Neur Erbegnisse der Karstforschung in dem Tropen und in Mittelmrellzum*. Geogr. Zeit. Beiheft.

Pfeffer, W.T. (1992). Stress-induced foliation in the terminus of variegated Glacier, Alaska, USA, formed during the 1982–83 surge. *Journal of Glaciology*, 38, 213–222.

Philbrick, S.S. (1970). Horizontal configuration and the rate of erosion of Niagara Falls. *Bulletin of the Geological Society of America,* 81, 3723–3732.

Phillips, J.D. (2005). Weathering instability and landscape evolution. *Geomorphology*, 67, 255–272.

Pias, J. (1970). *Les Formations Sédimentaires Tertiaires et Quaternaires de la Cuvette Tchadienne et les Sols qui en Dérivent*. ORSTOM, Mem. 43.

Pichler, E. (1958). Aspectos geológicos dos escorregamentos de Santos. *Noticia Geomorfológica*, 2, 40–44.

Piccini, L. (1995). Karst in siliceous rocks: Karst landforms and caves in the Auyán-tepui, Bolivar, Venezuela. *International Journal of Speleology* 24, 2–13.

Piégay, H. and Schumm, S.A. (2003). Systems approaches in fluvial geomorphology. In: Kondolf, G.M. and Piégay, H. (Eds.). *Tools in Fluvial Geomorphology*. Wiley. Chichester, 105–134.

Piégay, H., Salvador, P.G. and Astrade, L. (2000). Réflexions relatives à la variabilité spatiale de la mosaique fluviale à l'echelle d'un tronçon. *Zeitschrift für Geomorphologie*, 44, 317–342.

Piégay, H., Grant, G., Nakamura, F. and Trustum, N. (2006). Braided river management: for assessment of river behaviour to improved sustainable development. In: Sambrook Smith, G.H., Best, J.L., Bristow, C.S. and Petts, G.E. (Eds.). *Braided Rivers: Process, Deposits, Ecology and Management*. Blacwell. Malden, 257–275.

Pierce, K.L. (1986). Dating methods. In: *Active Tectonics*. National Academic Press. Washington, 195–214.

Pierson, T.C. (1983). Soil pipes and slope stability. *Quaterly Journal of Engineering Geology*, 16, 1–11. London.

Pilgrim, D., Cordery, I. and Baron, B. (1982). Effects of catchment size on runoff relationships. *Journal of Hydrology*, 58, 205–221.

Piotrovski, M.V., Simonov, Y.C. and Aristarkhova, L.B. (1972). Detailed geomorphological mapping in mapping prospecting. In: Demek, J. (Ed.). *Manual of Detailed Geomorphological Mapping*. Czechoslovak Academy of Sciences.Akademia. Praga, 267–277.

Piqueras, J. (1983). Crónica de la riada del Xúquer (20/21-X-1982). *Cuadernos de Geografía*, 32–33, 39–58.

Pirazzoli, P.A. (1991). *World Atlas of Holocene Sea-Level Changes*. Elsevier. Amsterdam, 300 p.

Pirazzoli, P.A. (1996). *Sea-level changes: the last 20.000 years*. John Wiley. Chichester, 211 p.

Pirazzoli, P.A. (2004). A review of posible eustatic, isostatic and tectonic contributions in eight late-Holocene relative sea-level histories from the Mediterranean area. *Quaternary Science Reviews*, 24, 1989–2001.

Pirazzoli, P.A. (2005). A review of possible eustatic, isostatic and tectonic contributions in eight late-holocene relative sea-level histories from the Mediterrean area. *Quaternary Science Reviews*, 24, 1989–2001.

Pirazzoli, P.A. and Morhange, C. (2005). Interpretation des variations du niveau de la mer a la fin de l'Holocene le long d'une transversale nord-sud en Méditerranée centrale. In: Sanjaume, E. and Mateu, J.F. (Eds.). *Geomorfología litoral i Quaternari*. PUV. Valencia, 263–371.

Pirazzoli, P.A., Laborel, J. and Stiros, S.C. (1996). Earthquake clustering in the Eastern Mediterranean during historical times. *Journal of Geophysical Research*, 101 (B3), 6083–6097.

Pissart, A. (1965). Les pingos des Hautes Fagnes: le problème de leur genèse. *Annales Societé Géologique de Belgique*, 89, 377–402.

Pissart, A. (1966a). Etude de quelques pentes de l'île Prince Patrick. *Annales, Societé Géologique de Belgique*, 89, 377–402.

Pissart, A. (1966b). Le rôle geomorphologique du vent dans la région de Mould Bay (Ile Prince Patrick, N.W.T., Canada). *Zeitschrift für Geomorphologie*, 10, 226–236.

Pissart, A. (1970). Les phénoménes physiques éssentiels liés au gel; les structures périglaciaires qui en résultent et leur signification climatique. *Annales, Société Géologique de Belgique*, 93, 7–49.

Pissart, A. (1977). Apparition and evolution des sols structuraux périglaciares de haute montagne. Expériences de terrain au Chambeyron (Alpes, France). *Abhandlungen der Akademie der Wissenschaften in Göttingen, Math. Physik Klasse*, 31, 142–156.

Pissart, A. (1988). Pingos: An overview of the present state of knowledge. In: Clark, M.J. (Ed.). *Advances in Periglacial Geomorphology*, Wiley. Chichester, 279–297.

Pissart, A. (1990). Advances in periglacial geomorphology. *Zeitschrift für Geomorphologie Supplementband*, 79, 119–131.

Pissart, A. (2000). Remnants of lithalsas of the Hautes Fagnes, Belgium: a summary of present-day knowledge. *Permafrost and Periglacial Processes*, 11, 327–356.

Pissart, A. (2002). Palsas, lithalsas and remnants of these periglacial mounds: a progress report. *Progress in Physical Geography*, 26, 605–621.

Pissart, A. (2004). Thermokarst. In: Goudie, A.S. (Ed.). *Encyclopedia of Geomorphology*. Routledge. Londres, 1049–1051.

Plafker, G. and Ericksen, G.E. (1978). Nevados Huascarán avalanches, Perú. In: Voight, B. (Ed.). *Rockslides and Avalanches*, 1. Elsevier. Amsterdam, 277–314.

Plafker, G., Hudson, T., Bruns, T. and Rubin, M. (1978). Late Quaternary offsets along the Fairweather fault and crustal plate interactions in southern Alaska. *Canadian Journal of Earth Science*, 15, 805–816.

Ploey, J. de (1980). Some field measurements and experimental data on wind-blown sands. In: Boodt, M. de and Gabriels, D. (Eds.). *Assessment of erosion*. John Wiley. New York, 541–552.

Ploey, J. de and Cruz, O. (1979). Landslides in the Serra do Mar, Brazil. *Catena*, 6, 111–122.

Poessen, J. de (1986). Surface sealing as influenced by slope angle and position of simulated stones in the top layer of loose sediments. *Earth Surface Processes and Landforms,* 11, 1–10.

Poesen, J. and Lavee, H. (1991). Effects of size and incorporation of synthetic mulch on runoff and sediment yield from interrills in a laboratory study with simulated rainfall. *Soil and Tillage Research*, 21, 209–223.

Poesen, J., Ingelmo-Sánchez, F. and Mucher, H. (1990). The hydrological response of soil surfaces to rainfall as affected by cover and position of rock fragments in the top layer. *Earth Surface Processes and Landforms*, 15, 653–672.

Pohjola, V.A. (1994). TV-video observations of bed and basal sliding on Storglaciären, Sweden. *Journal of Glaciology*, 39, 111–118.

Pokras, E.M. and Mix, A.C. (1985). Eolian evidence for spatial variability of Late Quaternary climates in tropical Africa. *Quaternary Research*, 24, 137–149.

Poland, J.F. and Davis, G.H. (1969). Land subsidence due to withdrawal of fluids. En: Varnes, O.J. y Kierch, G. (Eds.). *Reviews in Engineering Geology. Geological Society of America*, 2, 187–269.

Pollard, W.E. (1988). Seasonal frost mounds. In: Clark, M.J. (Ed.). *Advances in Periglacial Geomorphology*, Wiley. Chichester, 201–229.

Polyak, V.K. and Asmerom, Y (2001). Late Holocene climate and cultural changes in the southwestern United Status. *Science*, 294, 148–151.

Polynov, B. (1937). *The Cycle of Weathering*. Murby. London, 220 p.

Ponti, D.J. (1985). The Quaternary alluvial sequence of the Antelope Valley, California. *Geological Society America Special Paper,* 203, 79–96.

Porath, A. and Schick, A.P. (1974). The use of remote sensing systems in monitoring desert floods. *International Association of Hydrological Sciences*. Publ. 112, 133–138.

Porter, S.C: (1981). Recent glacier variations and volcanic eruptions. *Nature*, 291, 139–141.

Porter, S.C. (2007). Loess Record/China. In: Elias, S.A. (Ed.). *Encyclopedia of Quaternary Science*. Elsevier. Amsterdam, 1429–1440.

Porter, S.C. and Zhisheng, A. (1995). Correlation between climate events in the North Atlantic and China during the last glaciation. *Nature*, 375, 305–308.

Porter, S.C., Hallet, B., Wu, X.H. and An, Z.S. (2001). Dependence of near-surface magnetic susceptibility on dust accumulation rate and precipitation of the Chinese Loess Plateau. *Quaternary Research*, 55, 271–283.

Posamentier, H.W. (1978). Thoughts on ogive formation. *Journal of Glaciology*, 20, 218–220.

Poser, H. (1948). Boden-und klimaverhältnisse in Mittel und Westeuropa während der Würmmeiszeit. *Erdkunde*, 2, 53–68.

Poser, H. (1950). Zur rekonstruktion der spätglazialen luftdruckverhältnisse in Mittel-und Westeuropa auf grund der vorzeitlichen binnendünen. *Erdkunde*, 4, 81–88.

Potter, D.B. and McGill, F.E. (1978). Valley anticlines of the Needles District, Canyonlands National Park. *Bulletin Geological Society of America*, 89, 952–960.

Potter, N. (1972). Ice-cored rock glacier, Galena Creek, Northern Absaroka Mountains, Wyoming. *Bulletin of the Geological Society of America*, 83, 3025–3057.

Potter, N. and Moss, J.H. (1968). Origin of the Blue Rocks block field and adjacent deposits, Berks County, Pennsylvania. *Bulletin of the Geological Society of America*, 79, 255–262.

Potter, P.E. (1978). Significance and origin of big rivers. *Journal of Geology*, 86, 13–33.

Powell, R.D. (1981). A model for sedimentation by tidewater glaciers. *Annals of Glaciology*, 2, 129–134.

Powell, R.D. (2003). Subacuatic landsystem: fjords. In: Evans, D.J.A. (Ed.). *Glacial Landsystems*. Arnold. London, 313–347.

Powell, W.L. and LaMoreaux, P.E. (1969). A problem of subsidence in a limestone terrane at Columbia, Alabama. *Alabama Geological Survey*, Cir. 56.

Prada, J.L., Valenciano, A. and Navarro, A. (1996). Procesos de alteración de materiales pétreos en edificios de interés histórico. *Acta Geológica Hispánica*, 30, 97–110.

Prest, N.K. (1968). Nomenclature of moraines and ice-flow features as applied to the glacial map of Canada. *Geological Survey of Canada*, 66–57.

Price, R.J. (1966). Eskers near the Casement glacier, Alaska. *Geografiska Annaler*, 48, 111–125.

Price, R.J. (1969). Moraines, sandar, kames and eskers near Breidamerkurjökull, Iceland. *Transactions of Institute British Geographers*, 46, 17–43.

Price, R.J. (1973). *Glacial and Fluvioglacial Landforms*. Longman. Edinburgh, 242 p.

Price, R.J. (1980). Rates of geomorphological changes in proglacial areas. In: Cullinford, R.A., Davidson, D.A. and Lewin, J. (Eds.). *Timescales in Geomorphology*, John Wiley. Chichester, 79–93.

Price, W.A. (1963). Physico-chemical and environmental factors in clay-dune genesis. *Journal of Sedimentary Petrology*, 33, 766–778.

Price, W.A. (1968). Carolina Bays. In: Fairbridge, R.W. (Ed.). *The Encyclopaedia of Geomorphology*. Reinhold. New York, 102–109.

Price. R.J. (1970). Moraines at Fjallsjökull, Iceland. *Artic and Alpine Research*, 2, 27–42.

Priesnitz, K. (1988). Cryoplanation. In: Clark, M.J. (Ed.). *Advances in Periglacial Geomorphology*, Wiley. Chichester, 49–67.

Prospero, J.M., Bonatti, E., Schubert, C. and Carlson, T.N. (1970). Dust in the Caribbean atmosphere traced to an African dust storm. *Earth and Planetary Science Letters*, 9, 287–293.

Prospero, J.M., Ginoux, J.P., Torres, O., Nicholson, S.E. and Gill, T.E. (2002). Environmental characterization of global sources of atmospheric soil dust identified by the Nimbus 7 Total Ozone Mapping Spectometer (TOMS) absorbing aerosol product. *Review of Geophysics*, 40, 2–1, 2–31.

Prouty, W.F. (1952). Carolina Bays and their origin. *Bulletin of the Geological Society of America*, 63, 167–224.

Psuty, N.P. (1988). Sediment budget and dune beach interaction. *Journal of Coastal Research*, 3. Special Issue, 1–4.

Pueyo, E.L., Mauritsch, H.G., Pocoví, A., Scholger, R., Millán, H. and Valero, B. (2001). Archeomagnetostratigraphy of the filling of the Roman dam. Preliminary report on Almonacid de la Cuba (NE Spain). *AARCH 2001. Arqueomagnetism Applied for Rescue Cultural Heritages*. Abstract. Leuven.

Pueyo, J.J. (1978–79). La precipitación evaporítica actual en las lagunas del área: Bujaraloz, Sástago, Caspe, Alcañiz y Calanda (Provs. de Zaragoza y Teruel). *Revista del Instituto de Investigaciones Geológicas*, 33, 5–56. Barcelona.

Pueyo-Anchuela, O., Pocoví, A., Soriano, M.A. and Casas-Sainz, A.M. (2009a). Characterization of karst hazards from the perspective of the doline triangle using GPR-Examples from central Ebro Basin (Spain). *Engineering Geology*, 108, 225–236.

Pueyo-Anchuela, O., Pocoví, A., Casas-Sainz, A.M., Soriano, M.A. and Ansón, D. (2009b). Radarfacies and sinkhole hazards. *Geophysical Research Abstracts*, 11, 8116.

Pugh, J.C. (1956). Fringing pediments and marginal depressions in the inselberg landscape of Nigeria. *Transactions of the Institute of British Geographers*, 22, 15–31.

Pugh, J.C. (1966). Landforms in low latitudes. In: Dury, G.H. (Ed.). *Essays in Geomorphology*. American Elsevier. New York, 121–138.

Pujadas, J. (2002). Las inundaciones en España: impacto económico y gestión de riesgo. In: Ayala-Carcedo, F.J. and Olcina, J. (Eds.). *Riesgos Naturales*. Ariel Ciencia. Barcelona, 879–888.

Pulido- Bosch, A. (1986). Le karst dans les gypses de Sorbas (Almería). Aspects morphologiques et hydrogéologiques. *Karstologia*, Memoire 1, 27–35.

Pulina, M. (1974). Chemical denudation on the carbonate karst areas (Summary). *Geographical Studies*, 105, p. 159. Warszawa.

Pulina, M. (2005). Le karst et les phénomènes karstiques similaires des régions froides. In: Salomon, J.N. and Pulina, M. (Eds.). *Les Karsts des Regions Climatiques Extrêmes*. Karstología, Memoires, 14, 11–100.

Pulina, M. and Trzcinski, J.B. (1996). Remarques sur les phénomenes karstiques en Sibérie Orientale. *Guide des Terrains Karstiques Choisis de la Sibérie Orientales et de l'Oural*, 7–11. Université de Silésie. Sosnovic.

Punkari, M. (1997). Subglacial processes of the Scandinavian Ice Sheet in Fennoscandia inferred from flow-parallel features and lithostratigraphy. *Sedimentary Geology*, 111, 263–283.

Pye, K. (1982). Morphological development of coastal dunes in a humid tropical environment, Cape Bedford and Cape Flattery, North Queensland. *Geografiska Annaler*, A64, 212–227.

Pye, K. (1983). Early postdepositional modification of aeolian dune sands. In: Brookfield, M.E. and Ahlbrandt, T.S. (Eds.). *Eolian Sediments and Processes*. Elsevier. Amsterdam, 197–219.

Pye, K. (1987). *Aeolian Dust and Dust Deposits*. Academic Press. London, 334 p.

Pye, K. (Ed.) (1993). *The Dynamics and Environmental Context of Aeolian Sedimentary Systems*. Special Publication 72, Geological Society of London.

Pye, K. (Ed.) (1994). *Sediment Transport and Depositional Processes*. Blackwell, Oxford.

Pye, K. (1995). The nature, origin and accumulation of loess. *Quaternary Science Reviews*, 14, 653–667.

Pye, K. and Lancaster, N. (Eds.) (1993). *Aeolian Sediments: Ancient and Modern*. International Association of Sedimentologists. Blackwell. Oxford, 167 p.

Pye, K. and Sharwin, D. (1999). In: Goudie, A.S., Livingstone, I. and Stokes, S. (Eds.). *Aeolian Environments, Sediments and Landforms*. Wiley. Chichester, 214–238.

Pye, K. and Tsoar, H. (1987). The mechanics and geological implications of dust transport and deposition in deserts, with particular reference to loess formation and dune sand diagenesis in the northern Negev, Israel. In: Frostick, L.E. and Reid, I. (Eds.). *Desert Sediments, Ancient and Modern*. Special Publication 35, Geological Society of London. Blackwell. Oxford, 139–156.

Pye, K. and Tsoar, H. (1990). *Aeolian Sand and Sand Dunes*. Unwin Hyman. London, 396 p.

Quinlan, J.F. (1967). Classification of karst types: A review and synthesis emphasizing the North America literature, 1941–1966. *Natural Speleological Society Bulletin*, 29, 107–109.

Quirantes, J. (1965). Notas sobre las lagunas de Bujalaroz-Sástago. *Geographica*, 12, 30–34.

Rabenhorst, M.C., West, L.T. and Wilding, L.P. (1991). Genesis of calcic and petrocalcic horizons in soils over carbonate rocks. In: Nettelton, V.D. (Ed.). *Occurrence, Characteristics and Genesis of Carbonate and Silica Accumulations in Soils.*, Soil Science Society of America, Special Publication 26, 61–74.

Rachocki, A.H. (1981). *Alluvial Fans: An Attemp at an Empirical Aproach*. Wiley. New York, 161 p.

Rachocki, A.H. and Church, M. (Eds.) (1990). *Alluvial Fans. A Field Approach*. Wiley. Chichester, 391 p.

Radbruch-Hall, D.H. (1978). Gravitational creep of rocks masses on slopes. In: Voight, B. (Ed.). *Rockslides and Avalanches. Natural Phenomena*. Elsevier. Amsterdam, 607–657.

Rahmstorf, S. (2002). Ocean circulation and climate during the past 120,000 years. *Nature*, 419, 207–214.

Rahmstorf, S., Cazenave, A., Church, J.A., Hansen, J.E., Keeling, R.F., Parker, D.E. and Sommerville, R.C.J (2009). Recent climate observations compared to projections. *Science*, 316, 709.

Ramanthan, V. (1988). The greenhouse theory of climate change: a test by an inadvertent global experiment. *Science*, 240, 293–299.

Rampino, M.R. and Self, S. (1982). Historic eruptions of Tambora (1815), Krakatau (1833) y Agung (1963). *Quaternary Research*, 18, 127–143.

Ramsay, J.G. and Huber, M.I. (1987). *Modern Structural Geology*. Vol.2. Academic Press. New York.

Rantz, S.E. (1970). Urban sprawl and flooding in Southern California. *United States Geological Survey*. Circular 601B.

Rapp, A. (1957). Studien über Schutthalden in Lappland und auf Spitzbergen. *Zeitschrift für Geomorphologie*, 1, 179–200.

Rapp, A. (1984). Are terra rossa soils in Europa eolian deposits from Africa?. *Geologiska Föreningens I Stocklom Förhendlingar*, 105, 161–168.

Rapp, A. (1986). Slope processes in high latitude mountains. *Progress in Physical Geography*, 10, 53–68.

Rapp, A. and Hastean-Dahlin, A. (1990). Improved management of drylands by water harvesting in third world countries. In: Boardman, J., Foster, I.D.L. and Dearing, J.A. (Eds.). *Soil Erosion and Agricultural Land*. Wiley. Chichester, 241–251.

Rasid, H. and Pramanik, M.A.H. (1993). Areal extent of the 1988 flood in Bangladesh: How much did the satellite imagery show? *Natural Hazards*, 8, 189–200.

Raso, J.M. and García Loureiro, M.C. (1998). Oscillacions termomètiques a l'entorn de 0°C a Andorra. In: Gómez Ortiz, A., Salvador Franch, F., Schulte, L. and García Navarro, A. (Eds.). *Procesos biofísicos actuales en medios frios*. Publicacions de la Universitat de Barcelona. Barcelona, 217–234.

Raunet, M. (1985). Les bas-fonds en Afrique et à Madagascar. *Zeitschrift für Geomorphologie Supplementband,* 52, 25–62.

Raven, P.H., Berg, L.R. and Hassenzahl, D.M. (2008). *Environment* (Sixth Edition). Wiley. Chichester, 509 p.

Raymo, M.E., Ruddiman, W.F., Shackleton, N.J. and Oppo, D.W. (1990). Evolution of Atlantic-Pacific $\delta^{13}C$ gradients over the last 2.5 m.y. *Earth Planetary Science Letters*, 97, 353–368.

Rea, B.R. (2007). Glacial landforms, erosional features. Micro to macro scale forms. In: Elias, S.A. (Ed.). *Encyclopedia of Quaternary Science*. Elsevier. Amsterdam, 853–864.

Rea, B. (2007). Periglacial Landforms. Block fields (Felsenmeer). In: Elias, S.A. (Ed.). *Encyclopedia of Quaternary Science*. Elsevier. Amsterdam, 2225–2236.

Rea, B.R. and Evans, D.J.A. (1996). Landscapes of areal scouring in NW Scotland. *Scottish Geographical Magazine*, 112, 47–50.

Rea, B.R., Whalley, W.B. and Porter, E.M. (1996). Rock weathering deformation of summit blockfield slopes in Norway: Examples and implications. In: Anderson, M.G. and Brooks, S.M. (Eds.). *Advances in Hillslopes Processes*. Wiley. Chichester, 1257–1275.

Reading, A.J., Thompson, R.D. and Millington, A.C. (1995). *Humid Tropical Environments.* Blackwell. Oxford, 429 p.

Reeves, C.C. (1964). Gas rings from Terry County, Texas. *Journal Sedimentary Petrology*, 34, 190–193.

Reeves, C.C. (1968). *Introduction to Paleolimnology*. Elsevier. Amsterdam, 228 p.

Reeves, C.C. (1976). *Caliche: Origin, Classification, Morphology and Uses*. Estacado Books. Lubbock. Texas, 233 p.

Regües, D., Pardini, G. and Pini, R. (1992). Estudio del comportamiento de regolitos arcillosos en zonas abarrancadas frente a variaciones de la temperatura y la humedad. In: López, F., Conesa, C. and Romero, M.A. (Eds.). *Estudios de Geomorfología en España*. Murcia, 171–182.

Reich, B.M. (1973). *Effects on Agnes Floods on Annual Series in Pennsylvania*. Pennsylvania State University. Institute for Research on Land and Water Resources. Res. Publication 74.

Reiche, P. (1950). *A Survey of Weathering Processes and Products.* New Mexico University. Publications in Geology n° 3, 95 p.

Reicherter, K.R., Jabaloy, A., Galindo-Zaldívar, J., Ruano, P., Becker-Heidmann, P., Morales, J., Reiss, S. and González-Lodeiro, F. (2003). Repeated palaeoseismic activity of the Ventas de Zafarraya fault (S Spain) and its relation with the 1884 Andalusian earthquake. *International Journal of Earth Sciences*, 92, 912–922.

Reid, I. (1994). River landforms and sediments: evidence of climatic change. In: Abrahams, A.D. and Parsons, A.J: (Eds.). *Geomorphology of Desert Environments*. Chapmand and Hall. London, 571–592.

Reimer, P.J., Baillie, G.L., McCormac, G., Reimer, R.W., Bard, E., Beck, J.W., Blacwell, P.G., Buck, C.E., Burr, G.S., Edwards, R.L., Friedrich, M., Guilderson, T.P., Hogg, A.G., Hughen, K.A., Kromer, B., Manning, S., Southon, J.R., Stuiver, M., van der Plicht, J. and Weyhenmeyer, C.E. (2006). Comment on 'Radiocarbon calibration curve spanning 0 to 50,000 years BP based on paired $^{230}Th/^{234}U/^{238}U$ and ^{14}C dates on pristine corals' by R.J. Fairbanks et al. (Quaternary Science Reviews 24 (2005), 1781–1796) and 'Extending the radiocarbon calibration beyond 26,000 years before present using fossil corals' by T.C.Chiu et al. (Quaternary Science Reviews, 24 (2005), 1797–1808). *Quaternary Science Reviews*, 25, 855–862.

Reineck, H.E. and Singh, I.B. (1975). *Depositional Sedimentary Environments with Reference to Terrigenous Clastics*. Springer. Berlin, 439 p.

Remondo, J. (2001). *Elaboración y validación de mapas de susceptibilidad de deslizamientos mediante técnicas de análisis espacial*. Tesis Doctoral. Universidad de Oviedo. Unpublished.

Remondo, J., González, A., Díaz de Terán, J.R. and Cendrero, A. (2003a). Landslide susceptibility models utilising spatial data analysis techniques. A case study from the lower Deba Valley, Guipúzcoa (Spain). *Natural Hazards*, 30, 267–279.

Remondo, J., González, A., Díaz de Terán, J.R., Fabbri, A., Cendrero, A. and Chung, C.F. (2003b). Validation of landslide susceptibility maps; examples and applications from a case study in northern Spain. *Natural Hazards*, 30, 437–449.

Remondo, J., Soto, J., González-Díez, A., Díaz de Terán, J.R. and Cendrero, A. (2005a). Human impact on geomorphic processes and hazards in mountain areas in northern Spain. *Geomorphology*, 66, 69–84.

Remondo, J., Bonachea, J. and Cendrero, A. (2005b). A statistical approach to landslide risk modelling at a basin scale: from landslide susceptibility to quantitative assessment. *Landslides*, 2, 321–328.

Rémy, F. (2007). *Histoire de la Glaciologie*. Vuivert. Adapt-Snes, 170 p.

Renard, K.G. (1969). Evaporation from an ephemeral stream bed: discussion. *Journal of Hydraulics Division, Proceedings of the American Society of Civil Engineers*, 95, 2200–2204.

Renard, K.G. (1970). The hydrology of semiarid rangeland watersheds. *United States Deparment Agriculture Research Service Publishing*, 41–162.

Renard, K.G. and Keppel, R.V. (1966). Hydrographs of ephemeral streams in the Southwest. *Proceedings of the American Society Civil Engineers. Journal of the Hydrology Division*, 92, 33–52.

Renault, P. (1953). Caràcteres généraux des grottes gréuseses du Sahara meridional. *Premier Congrès Internationale de Spéléologie*. Paris, T 2, 1–15.

Renault, P. (1958). Eléments de spéléologie karstique. *Annales de Spéléologie*, 13, 23–48.

Renschler, C.S., Doyle, M.W. and Thoms, M. (2007). Geomorphology and ecosystems: Challenges and keys for success in bridging disciplines. *Geomorphology*, 89, 1–8.

Restrepo, J.D. and Kjerfve, B. (2000). Magdalena river: interanual variability (1975–1995) and revised water discharge and sediment load estimates. *Journal of Hydrology*, 235, 137–149.

Rettig, S.L., Jones, B.F. and Risacher, F. (1980). Geochemical evolution of brines in the Salar of Uyuni, Bolivia. *Chemical Geology*, 30, 57–79.

Reynaud, A. (1971). *Epistémologie de la Géomorphologie*. Masson. Paris, 128 p.

Reyne, A., Jamveit, B., Mathiesen, J. and Malte-Sorenssen (2008). Controls on rock weathering rates by reaction-induced hierarchical fracturing. *Earth and Planetary Science Letters*, 275, 364–369.

Rhoades, J.D. (1990). Soil salinity-Causes and controls. In: Goudie, A.S. (Ed.). *Techniques for Desert Reclamation*. Wiley. Chichester, 109–134.

Riba, O. and Llamas, M.R. (1962). Canales de la Violada, Monegros y Flumen. *I Coloquio Internacional sobre las Obras Públicas en terrenos yesíferos*. Servicio Geológico de Obras Públicas. Madrid, 23 p.

Richards, K.S. (1976). Channel and flow geometry. *Progress in Physical Geography*, 1, 65–102.

Richards, K.S. (1982). *Rivers: Form and Process in Alluvial Channels*. Methuen. London, 258 p.

Richards, K.S. (1987). Fluvial Geomorphology. *Progress in Physical Geography*, 11, 432–457.

Richardson, C. and Holmund, P. (1996). Glacial cirque formation in northern Scandinavia. *Annals of Glaciology*, 22, 102–106.

Ridgwell, A.J. (2003). Dust in Earth system: the biochemical linking of land, air and sea. *Philosophical Transaction Royal Society*, 360 A, 2905–2924.

Ringwood, A.E. (1974). The petrological evolution of island arc systems. *Journal of the Geological Society of London*, 130, 183–204.

Ríos, S. and Sáez, M.T. (2002). Aludes. In: Ayala, F.J. and Olcina, J. (Eds.). *Riesgos Naturales*. Ariel Ciencia. Barcelona, 411–420.

Riosalido, R. (coord.) (1996). *Estudio Meteorológico de la Situación del 7 de Agosto de 1996 (Biescas)*. Ministerio de Medio Ambiente. Madrid, 90 p.

Risacher, F. and Fritz, B. (2000). Bromine geochemistry of the salar de Uyuni and deeper salt crust, Central Altiplano, Bolivia. *Chemical Geology*, 167, 373–392.

Ristic, D.M. (1976). Water regime of flooded karst poljes. In: Yeryevich, V. (Ed.). *Karst Hydrology and Water Resources Publications*. Fort Collins,C.O, 301–318.

Ritter, D.E. (1986). *Process Geomorphology*. W.M.C. Brown, Dubuque. Iowa, 603 p.

Rittmann, A. (1962). *Volcanoes and their Activity*. Interscience. New York.

Rivas, A., Cabezas, J., Carrasco, F., Durán, J.J. and González-Ríos, (2004). Las cuevas turísticas españolas: un recurso natural de gran interés ecológico, económico y social. In: Andeso, B. and Durán, J.J. (Eds.). *Investigaciones en Sistemas Kársticos Españoles*. Instituto Geológico y Minero de España. Madrid, 367–384.

Rives, V. and Vicente, M.A. (1993). Formas de alteración de la arenisca de Villamayor en distintos microambientes de edificios salmantinos. In: Vicente, M.A., Molina, E. and Rives, V. (Eds.). *Alteración de Granitos y Rocas Afines*. Consejo Superior de Investigaciones Científicas. Madrid, 75–82.

Rizzi, A. (2007). Periódico *"El País"*, 22 de Abril de 2007, p. 36–37.

Robert, A. (1990). Boundary roughness in coarse-grained channels. *Progress in Physical Geography*, 14, 42–70.

Robert, M. (1993). Role du facteur biologique dans la degradation des roches et des monuments. In: Vicente, M.A., Molina, E. and Rivas, V. (Eds.). *Alteración de Granitos y Rocas Afines*. Consejo Superior de Investigaciones Científicas. Madrid, 103–11.

Robert, M., Razzaghe, M.K., Vicente, M.A. and Veneau, G. (1979). Rôle du facteur biochimique dans l'altération des minéraux silicatés. *Bulletin de l'Association Francaise pour l'Etude du Sol*, 2 y 3, 153–17.

Roberts, C.R. and Mitchell, C.W. (1987). Spring mounds in southern Tunisia. In: Frostick, L.E. and Reid, I. (Eds.). *Desert Sediments: Ancient and Modern*. Geological Society of London, Special Publication, 35, 321–336.

Roberts, H.H., Bailey, A. and Chuecher, G.J. (1994). Subsidence in the Mississippi River delta-important influences of valley filling by cyclic deposition, primary consolidation, phenomena and early diagenesis. *Gulf Coast Association Geological Society Transactions*, 44, 619–629.

Roberts, N. (1989). *The Holocene*. Blackwell. Oxford, 227 p.

Robertson-Rintoul, M.J. (1990). A quantitative analysis of the near-surface wind-flow pattern over coastal parabolic dunes. In: Nordstrom, K.E., Psuty, N. and Carter, B. (Eds.). *Coastal Dunes: Form and Process*. Wiley. Chichester, 57–78.

Robin, G. De Q., Drewry, D.J. and Meldrum, D.T. (1977). International studies of ice sheet and bedrock. *Phylosophical Transactions of the Royal Society of London*, 279, 185–196.

Robinson, D.A. and Williams, R.B.G. (1992). Sandstone weathering in the High Atlas, Morocco. *Zeitscrift für Geomorphologie*, 4, 413–429.

Robinson, D.A. and Williams, R.B.G. (1994). Introduction: Advances in rock weathering studies. In: Robinson, D.A. and Williams, R.B.G. (Eds.). *Rock Weathering and Landform Evolution*. Wiley. Chichester, XV-XXI.

Robock, A. (2002). Pinatubo eruption: the climatic aftermath. *Science*, 295, 1242–1244.

Roca, A. and Udías, A. (1976). Sismicidad y riesgo sísmico de la zona de Cataluña y Pirineos. *Revista de Geofísica*, 35, 183–207.

Rock-Color Chart. (1979). *The Geological Society of America*. Boulder. Colorado.

Rockwell, T.K., Keller, E.A. and Johnson, D.L. (1984). Tectonic geomorphology of alluvial fans and mountain fronts near Ventura, California. In: Morisawa, M. and Hack, T.J. (Eds.). *Tectonic Geomorphology*. Allen and Unwin. Boston, 183–207.

Rodas, M., Luque, F.M., Mas, R. and Garzón, M.G. (1994). Calcretes, palycretes and silcretes in the paleogene detrital sediments of the Duero and Tajo Basisn, central Spain. *Clay Minerals*, 29, 273–285.

Rodda, J.C. (1969). The flood hydrograph. In: Chorley, R.J. (Ed.). *Water, Earth and Man*. Methuen. London, 405–418.

Rodríguez de la Torre, F. (1884). *Los terremotos alicantinos de 1829*. Publicaciones del Instituto de Estudios Alicantinos. Alicante, 323 p.

Rodríguez Vidal, J. (1986). *Geomorfología de las Sierras Exteriores Oscenses y su Piedemonte*. Colección de Estudios Altoaragoneses. Huesca, 172 p.

Rodríguez, A. (1998). *Geomorfología del Parque Nacional de Doñana y su Entorno*. Ministerio de Medio Ambiente. Madrid, 148 p.

Rodríguez, A., Cáceres, L.M. and Rodríguez, J. (2000). Dinámica y evolución de flechas litorales: el litoral onubense. In: Andrés, J.R. and Gracia, F.J. (Eds.). *Geomorfología Litoral. Procesos Activos*. Monografía n° 7 de la Sociedad Española de Geomorfología. Instituto Tecnológico y Geominero de España. Madrid, 101–113.

Rodríguez, J. and Navascués, L. (1982). La tafonización de las areniscas miocenas en los alrededores de Huesca. *Tecniterrae*, 19, 7–12.

Rodríguez-Vidal, J., Cáceres, L.M., Finlayson, J.C., Gracia, F.J. and Martínez-Aguirre, A. (2004). Neotectonics and shoreline history of the Rock of Gibraltar, Southern Iberia. *Quaternary Science Reviews*, 23, 2017–2019.

Rogers, J.D. (1992). Recent developments in landslides mitigation techniques. *Reviews in Engineering Geology*, vol. 10, 95–118. Geological Society of America.

Rogers, J.V. (2002). Recent developments in landslide mitigation techniques. In: Slosson, J.E., Keene, A.G. and Johnson, J.A. (Eds.). *Landslide/Landslide Mitigation*. Geological Society of America. Reviews in Engineering Geology, vol. 9, 95–118.

Rogers, R.D. and Schumm, S.A. (1991). The effect of sparse vegetative cover on erosion and sediment yield. *Journal of Hydrology*, 123, 19–24.

Roglic, J. (1940). Geomorphologische studien äber das Duvangsku Polje in Bosnien. *Mitt. Geogr. Ges. Wien*, 83, 152–177.

Roglic, J. (1952). Les surfaces de corrosion dans le karst dinarique. *Proceedings Gen. Assembly, 8th International Congres 17th*. Washington, 366–369.

Roglic, J. (1960). Das Verhältnis der Flusserosion zum karstprozess. *Zeitscrift für Geomorphologie*, 4, 116–128.

Roglic, J. (1964). "Karst valleys" in the Dinaric karst. *Erdkunde*, 18, 113–116.

Roglic, J. (1972). Historical review of morphological concepts. In: Herak, M. and Springfield, V.T. (Eds.). Karst: Important Karst Regions of the Northern Hemisphere. Elsevier. Amsterdam, 1–18.

Roglic, J. (1974). Les caràcteres spéciphiques du Karst Dinarique. *Memoires et Documents*, C.N.R.S., 15, 269–278.

Roglic, J. (1979). Les carácteres specifiques du karst dinarique. *Memoires et Documents*, 15. Phénomenes Karstiques. C.N.R.S., Paris, 269–278.

Rognon, P. (1980). Pluvial and arid phases in the Sahara: The role of nonclimatic factors. In: Sarnthein, M.; Seibold, E. and Rognon, P. (Eds.). *Palaeoecology of Africa. Sahara and Surrounding Seas*, Balkema. Rotterdam, 45–62.

Rogowski, A.S., Khanbilvardi, R.M. and Deangelis, R.J. (1985). Estimating erosion on plot, field and watershed scales. In: El-Swaifi, S.A., Molldenhaner, W.C. and Andrew, L.

(Eds.). *Soil Erosion and Conservation*. Soil Conservation Society of America, 149–166.

Rojo, L. and Sánchez, M.C. (1996). *Red de Estaciones Experimentales de Seguimiento y Evaluación de la Erosión y Desertificación (RESEL)*. Catálogo de Estaciones, Proyecto Lucdeme. Dirección General de Conservación de la Naturaleza. Ministerio de Medio Ambiente. Madrid, 121 p.

Romagnoli, C., Casalbore, D., Chiacci, F.L. and Bosman, A. (2009). Offshore evidence of large-scale lateral collapses on the eastern flank of Stromboli, Italy, due to structurally-controlled, bilateral flank instability. *Marine Geology*, 262, 1–13.

Romero, A. (1990). Itinerario: Vuelta a la Isla. Mirador del Tanque. *Monografías de la Sociedad Española de Geomorfología*, 5, 323–324.

Ronov, A.B. (1983). The Earth's sedimentary shell: Quantitative patterns of its structure, compositions and evolution. *American Geological Institute Reprint Series*, 5, 1–73.

Roose, E.J. and Lelong, F. (1976). Les facteurs de l'érosion hydrique en Afrique tropicale. Etudes sur petites parcelles expérimentales du sol. *Revue de Géographie Physique et de Géologie Dynamique*, 18, 365–374.

Roque, C. and Pallí, L. (1998). Geología de l'Ardenya i Formes Granítiques Associades. *Publicacions de l'Institut d'Estudis del Baix Empordá*, 17, 5–42.

Rosell, J. and Linares, R. (2001). Grandes deslizamientos en el frente de la lámina cabalgante del Montsec (Sierra del Montsec, Prepirineo central). *Revista Sociedad Geológica de España*, 14, 255–268.

Roselló, V.M. (1996). La revinguda del Xúquer i el desastre de la Ribera (20–21 Octubre, 1982). Una perspectiva geográfica. *Cuadernos de Geografía*, 32–33, 3–38.

Rosen, M.R. (1994). The importance of groundwater in playas: a review of playa classifications and the sedimentology and hydrology of playas. In: Rosen, M.R. (Ed.). *Paleoclimate and Basin Evolution of Playa Systems*. Geological Society of America, Special Paper, 289, 1–18.

Rouse, W.C. (1984). Flowslides. In: Brunsden, D. and Prior, D.B. (Eds.). *Slope Instability*. Wiley. Chichester, 491–522.

Roux, A.L. (1982). *Cartographie polythématique appliqué à la gestion écologique des eaux: étude d'un hydrosistème fluvial: le Aut. Rhône francais*. Report CNRS. Lyon, 116 p.

Rowland, S.K. and Walker, G.P.L. (1990). Pahoehoe and aa in Hawaii. Volumetric flow rate controls the lava structure. *Bulletin Volcanologie*, 52, 615–628.

Rubiera, J. (2002). El huracán Mitch: el más desastroso en la historia de centroamérica. In: Lugo Hubp, J. and Inbar, M. (Eds.). *Desastres naturales en América Latina*. México, 289–314.

Rubin, D.M. (1990). Lateral migration on linear dunes in the Strzeleck; Desert, Australia. *Earth Surface Processes and Landforms*, 15, 1–14.

Rubin, D.M. and Ikeda, H. (1990). Flume experiments on the alignment of transverse, oblique and longitudinal dunes indirectionally varying flows. *Sedimentology*, 37, 673–684.

Rudoy, A. (1998). Mountain ice-dammed lakes of southern Siberia and their influence on the development and regime of the intracontinental runoff systems of North Asia in the Late Pleistocene. In: Benito, G., Baker, V.R. and Gregory, K.J. (Eds.). *Palaeohydrology and Environmental Change*. Wiley. Chichester, 215–234.

Ruhe, R.V. (1975). Climatic geomorphology and fully developed slopes. *Catena*, 2, 309–320.

Runge, J. (2007). The Congo river, central Africa. In: Gupta. A. (Ed.). *Large Rivers. Geomorphology and Management*. Wiley. Chichester, 293–309.

Rupke, N.A. (1970). Continental drift before 1900. *Nature*, 227, 349–350.

Russ, D.R. (1982). Style and significance of surface deformation in the vicinity of New Madrid, MO. *United States Geological Survey Professional Paper*, 1236-H, 95–114.

Rust, B.R. (1979). Facies models 2: Coarse alluvial deposits. In: Walker, R.G. (Ed.). *Facies Models*. Geoscience Reprint, Series 1, Kitchener, Ontario, Canada, 9–21.

Rutte, E. (1958). Kalkruste in Spanien. *Neues Jahrbuch Geologische und Paläeontologische*, 106, 52–138.

Rutter, N.W. (1965). Foliation pattern of Gulkana glacier, Alaska Range, Alaska. *Journal of Glaciology*, 5, 711–718.

Rutter, N.W. and Catto, N.R. (Ed.) (1995). *Dating Methods Quaternary Deposits*. Geological Association of Canada. Geotext 2. St. John´s. New foundland, 308 p.

Ruxton, B.P. (1958). Weathering and sub-surface erosion in granite at the piedmont angle, Balos Sudan. *Geological Magazine*, 95, 353–377.

Ruxton, B.P. (1967). Slope wash under nature primary rainforest in northern Papua. In: Jennings, J.N. and Mabbutt, J.A. (Eds.). *Landform Studies from Australia and New Guinea*. Cambridge University Press. London, 85–94.

Ruxton, B.P. and Berry, L. (1957). Weathering of granite and associated erosional features in Hong Kong. *Bulletin of the Geological Society*, 68, 1263–1292.

Ruxton, B.P. and Berry, L. (1961). Weathering profiles and geomorphic position on granite in two tropical regions. *Revue de Géomorphologie Dynamique*, 12, 16–31.

Ryan, J.W. anf Ma, C. (1998). NASA-ESFC Geodetic VLBI Program, a twenty-year retrospective: Geodesy and

Global change. *Physics and Chemistry of the Earth*, 23, 1041–1052.

Sack, D. (1994). Geomorphic evidence of climate change from desert-basin paleolakes. In: Abrahams, A.D. and Parsons, A.J. (Eds.). *Geomorphology of Desert Environments*. Chapman and Hall. London, 616–630.

Sakamoto-Arnold, C.M. (1981). Eolian features produced by the December, 1977 windstorm, southern San Joaquin Valley, California. *Journal of Geology*, 89, 129–137.

Sala, M. (1988). Slope runoff and sediment production in two mediterranean mountain environments. *Catena Supplement*, 12, 13–29.

Salas, R. (1993). *Análisis de las variaciones climáticas holocenas en la Región Cantábrica, a partir de estudios palinológicos; influencia de la degradación diferencial del polen en las interpretaciones paleoclimáticas*. Tesis Doctoral. Universidad de Zaragoza, 583 p. Unpublished.

Salas, R. (1995). Los estudios polínicos en España, utilizados en la reconstrucción cllimática de los últimos 10.000 años. *Cuadernos Laboratorio Xeológico de Laxe*, 20, 67–98.

Salinas, J.L. (1988). Riesgos ligados a arcillas expansivas. In: Ayala, F.J.; Durán, J.J. and Peinado, T. (Eds.). *Riesgos Geológicos*. Instituto Geológico y Minero de España. Madrid, 295–304.

Salomon, J.N. (2000). *Précis de Karstologie*. Presses Universitaires de Bordeaux. Bordeaux, 250 p.

Salomon, J.N. (2005). La spécifité des karsts tropicaux. In: Salomon, J.N. and Pulina, M. (Eds.). *Les Karsts des Regions Climatiques Extrêmes*. Karstología, Memoires, 14, 103–156.

Sambrook Smith, G., Best, J.L., Bristow, C.S. and Petts, G.E. (2006). Braided rivers: where have we come in 10 years? Progress and future needs. In: Sambrook Smith, G., Best, J.L., Bristow, C.S. and Petts, G.E. (Eds.). *Braided Rivers. Process, Deposits, Ecology and Management*. Special Publication Number 36. International Association of Sedimentologists. Blackwell. Malden, 1–10.

Sanches Furtado, A.F.A. (1968). Altération des granites dans les régions intertropicales sous différents climats. *9th International Congress of Soil Science*, 4, 403–409.

Sánchez, J.A., Martínez, F.J., Floría, E. and Schumann, S. (1993). Hydrogeological characterization of the Monegros Lakes. In: Gutiérrez, M.; Sancho, C. and Benito, G. (Eds.). *Second Intensive Course on Applied Geomorphology: Arid Regions*. Zaragoza, 245–252.

Sánchez-Arcilla, A., Jiménez, J.A. and Valdemoro, H. (1998). The Ebro Delta: morphodynamics and vulnerability. *Journal of Coastal Research*, 14, 754–752.

Sancho, C. and Benito, G. (1990). Factors controlling tafoni weathering in the Ebro Basin (NE Spain). *Zeitscrift für Geomorphologie*, 34, 165–177.

Sancho, C. and Gutiérrez, M. (1990). Análisis morfométrico de la tafonización de la arenisca de Peraltilla (Anticlinal de Barbastro, Depresión del Ebro): Influencia de los factores mineralógicos-texturales. *Cuaternario y Geomorfología*, 4, 131–145.

Sancho, C. and Gutiérrez, M. (1993). Geomorphological features of the Bujalaroz salt lakes. *Second Intensive Course of Applied Geomorphology: Arid Regions*, 241–243. Universidad de Zaragoza.

Sancho, C. and Meléndez, A. (1992). Génesis y significado ambiental de los caliches pleistocenos de la región del Cinca (Depresión del Ebro). *Revista de la Sociedad Geológica de España*, 5, 81–93.

Sancho, C., Gutiérrez, M., Peña, J.L. and Burillo, F. (1988). A quantitative approach to scarp retreat starting from triangular slope facets (Central Ebro Basin, Spain). *Catena Supplement*, 13, 139–146.

Sancho, C., Gutiérrez, M. and Peña, J.L. (1991). Erosion and sedimentation during the upper Holocene in the Ebro Depression: Quantification and environmental significance. In: Sala, M., Rubio, J.L. and García Ruiz, J.M. (Eds.). *Soil Erosion Studies in Spain*. Geoforma. Logroño, 219–228.

Sancho, C., Meléndez, A., Signes, M. and Bastida, J. (1992). Chemical and mineralogical characteristics of Pleistocene caliche deposits from the Central Ebro basin. *Clay Minerals*, 27, 293–308.

Sanjaume, E. (1985). *Las Costas Valencianas. Sedimentología y Geomorfología*. Sección de Geografía. Universidad de Valencia. Valencia, 505 p.

Sanjaume, E. and Pardo, J.E. (2000). Albuferas litorales: estado de la cuestión. In: Andrés, J.R. de and Gracia, F.J. (Eds.). *Geomorfología litoral y procesos activos*. Monografía nº 7 de la Sociedad Española de Geomorfología. Instituto Geológico y Minero de España. Madrid, 115–139.

Sanjurjo, J. and Alves, C.A.S. (2006). Degradación de rocas graníticas empleadas como material de construcción en edificios de A Coruña (NW España). *Cuadernos Laboratorio Xeolóxico de Laxe*, 31, 11–28.

Sanjurjo, J., Vidal Romaní, J.R., Pallí, L. and Roque, C. (2003). Espelotemas de ópalo y pseudokarst granítico. *Cuatenario y Geomorfología*, 21, 123–134.

Sanlaville, P. (1992). Science del terre et archéologie: l'evolution de la Basse Mésopotamie à l'Holocene. In: Miskovsky, J.C. (Ed.). Les applications de la géologie à la connaissance de l'environment de l'homme. *Bulletin de la Société Géologique de France*, 160, 11–18.

Santanach, P. (2001). Fallas activas en España: tasas de movimiento y recurrencia sísmica. In: Capote, R. and Martínez, J.J. (Eds.). *El Riesgo Sísmico. Prevención y Seguro*. Consorcio de Compensación de Seguros. Madrid, 119–137.

Santer, B.D., Wigley, T.M.L., Barnett, T.P. and Anyamba, B. (1996). Detection of climatic change and attribution of causes. In: Houghton, J.T. *et al.* (Eds.). *Climate Change 1995. The Science of Climatic Change*. Cambridge University Press. Cambridge, 406–443.

Sanz, C. (1988). *El Relieve del Guadarrama Oriental*. Comunidad de Madrid. Consejería de Política Territorial. Madrid, 547 p.

Sarin, M.M., Krishnaswami, S., Dilli, K., Somayajulu, B.L.K. and Moore, W.S. (1989). Major ion chemistry of the Ganga-Brahmaputra river system: weathering processes and fluxes to the Bay of Bengal. *Acta Geochimica et Cosmochimica*, 53, 997–1009.

Sarnthein, M. (1978). Sand deserts during the last glacial maximum and climatic optimum. *Nature*, 272, 43–46.

Sarnthein, M. and Walger, K. (1974). Der äolische Sandstrom aus der W-Sahara zur Atlantikküste. *Geologische Rundschau*, 63, 1065–1087.

Satake, K., Oklal., E.A. and Borrero, J.C. (2007). Tsunami and its hazard in the Indian and Pacific Oceans: Introduction. *Pure and Applied Geophysics*, 164, 249–259.

Saucier, R.T. (1994). Evidence of late glacial runoff in the Lower Mississippi Valley. *Quaternary Science Reviews*, 13, 973–981.

Saunders, I. and Young, A. (1983). Rates of surface processes on slopes, slope retreat and denudation. *Earth Surface Processes and Landforms,* 8, 473–501.

Sauro, U. (1996). Geomorphological aspects of gypsum karst with special emphasisis on exposed karst. *International Journal of Speleology*, 25, 105–114.

Sauro, U. (2005). Closed depressions. In: Culver, D.C. and White, W.B. (Eds.). *Encyclopedia of Caves*. Elsevier. Amsterdam, 108–127.

Savat, J. and Poesen, J. (1981). Detachment and transportation of loose sediments by raindrop splash. Part I. The calculation of absolute data on detachability and transportability. *Catena*, 8, 1–17.

Savigear, R.A.G. (1952). Some observations on slope development in South Wales. *Transactions of the Institute of British Geographers*, 18, 31–51.

Savigear, R.A.G. (1956). Technique and terminology in the investigation of slope forms. *Slopes Commision Report*, 1, 66–75.

Savigear, R.A.G. (1960). Slopes and hills in West Africa. *Zeitschrift für Geomorphologie Supplementband*,1, 156–171.

Scambos, T.A., Hulbe, C., Fahnestock, M. and Bohlander, J. (2000). The link between climate warming and break-up of ice shelves in the Antartic Peninsula. *Journal of Glaciology*, 46,516–530.

Scarth, A. (1994). *Volcanoes*. UCL Press, London, 273 p.

Schackelton, N.J. and Opdyke, N.D. (1973). Oxygen isotope and paleomagnetic stratigraphy of equatorial Pacific core V28–238: oxygen isotope temperatures and ice volumes on a 10^5 and 10^6 year scale. *Quaternary Research*, 3, 39–55.

Schackleton, N.J. and Kennett, J.P. (1975). Paleotemperature history of the Cenozoic and initiation of Antarctic glaciation: oxygen and carbon isotope analyses in DSDP Sites 277, 279 and 281. In: Kennett, J.P., Houtz, R.E., Andrews, P.B., Edwards, A.R. *et al.* (Eds.). *Initial Reports of the Deep Sea Drilling Project N° 29*. U.S. Government Printing Office. Washington, D.C.

Schackelton, N.J. and Opdyke, N.D. (1976). Oxygen isotope and paleomagnetic stratigraphy of equatorial Pacific core V28–239, late Pliocene to latest Pleistocene. In: Cline, R.M. and Hays, R.D. (Eds.). *Investigation of Late Quaternary Paleoceanography and Paleoclimatology*. Memoirs of the Geological Society of America, 145, 449–464.

Schäfer, W. (1973). Der Oberhein, sterbende Landschaft?. *Natur Museum*, 103, 1–29.

Scheffer, M. Carpenter, S., Foley, J.A., Folke, C. and Walker, B. (2001). Catastrophic shifts in ecosystems. *Nature*, 413, 591–596.

Scheidegger, A.E. (1965). The algebra of stream-orden numbers. *United States Geological Survey Professional Paper*, 525B, B-187–189.

Scheidegger, A.E. (1991). *Theorical Geomorphology*. Springer. Berlin.

Scheidegger, A.E., Schumm, S.A. and Fairbridge, R.W. (1968). Badlands. In: Fairbridge, R.W. (Ed.). *The Encyclopedia of Geomorphology*. Dowden, Hutchinson and Ross. Strousburg, 43–48.

Schellmann, W. (1977). The formation of lateritic silicate bauxites and criteria for their exploration and assessment of deposits. *Natural Resources Development*, 5, 119–134.

Schellmann, W. (1981). Considerations on the definition and classification of laterites. In: Proceedings International Seminar on Lateritisation Processes, Trivandrum. *Lateritisation Processes*. Balkema. Rotterdam, 1–10.

Schick, A.P., Lekach, J. and Hassan, M.A. (1987). Vertical exchange of coarse bedload in desert streams. In: Frostick, L.E. and Reid, I. (Eds.). *Desert Sediments: ancient and modern*. Geological Society of London, Special Publication 35, Blackwell. Oxford, 7–16.

Schiermeir, Q. (2007). What we don't know about climate change? *Nature*, 445, 580–581.

Schimel, D., Alves, D., Enting, I., Heimann, M. et al., (1996). Radiative forcing of climate change. In: Houghton, J.T., Meira Filho, L.G., Callander, B.A., Harris, N., Kattenberg, A. and Maskell, K. (Eds.). Climate *Change 1995. The Science of Climatic Change*. Cambridge University Press. Cambridge, 65–131.

Schipull, K. (1980). Die Cedar Mesa -Schichtstufe aud dem Colorado Plateau-ein Beispiel für die Morphodynamik arider Schichtstufen. *Zeitschrift für Geomorphologie*, 24, 318–331.

Schmidt, G.A., Schindell, D.T., Miller, R.L., Mann, M.E. and Rind, D. (2004). General circulation modelling of Holocene climate variability. *Quaternary Science Reviews*, 23, 2167–2181.

Schmidt, K.H. (1980). Eine neue Metode zur Ermittlung von Stufenruckwanderungsrate, dargestellt am Beispiel der Black Mesa Schichtstufen, Colorado Plateau, USA. *Zeitschrift für Geomorphologie*, 24, 180–191.

Schmidt, K.H. (1987). Factors influencing structural landform dynamics on the Colorado Plateau about the necessity of calibrating theoretical models by empirical data. *Catena Supplment*, 10, 51–66.

Schmidt, K.H. (1988). Rates of scarp retreat: a means of dating neotectonic activity. In: Jacobshagen, V.H. (Ed.). *The Atlas System of Morocco Studies on its Geodynamic Evolution*. Lecture Notes in Earth Science, 15. Berlin, 445–462.

Schmidt, K.H. (1989a). Talus and pediment flatirons - erosional and depositional features of dryland cuesta scarps. *Catena Supplement*, 14, 107–118.

Schmidt, K.H. (1989b). The significance of scarp retreat for Cenozoic landform evolution on the Colorado Plateau, USA. *Earth Surface Processes and Landforms*, 14, 93–105.

Schmidt, K.H. (1994). Hillslopes as evidence of climatic change. In: Abrahams, A.D. and Parsons, J. (Eds.). *Geomorphology of Deserts Environments.* Chapman & Hall. London, 553–570.

Schmidt, K.H. (1996). Talus and pediment flatirons-indicators of climatic change on scarp slope on the Colorado Plateau, USA. *Zeitschrift für Geomórphologie Supplementband* 103, 135–158.

Schmincke, H.W. (2004). *Volcanism*. Springer, Berlín, 324 p.

Schumm, S.A. (1956a). Evolution of drainage systems and slopes in badlands at Perth Amboy, New Jersey. *Bulletin of the Geological Society of America*, 67, 597–646.

Schumm, S.A. (1956b). The role of creep and rainwash on the retreat of badland slopes. *American Journal of Science*, 254, 693–706.

Schumm, S.A. (1962). Erosion on miniature pediments in Badlands National Monument, South Dakota. *Bulletin of the Geological Society of America*, 73, 719–724.

Schumm, S.A. (1963a). Rates of surficial rock creep on hillslopes in western Colorado. *Science*, 155, 560–561.

Schumm, S.A. (1963b). Sinuosity of alluvial rivers of the Great Plains. *Bulletin of the Geological Society of America*, 74, 1089–1100.

Schumm, S.A. (1964). Seasonal variations of erosion rates and processes on hillslopes in western Colorado. *Zeitschrift für Geomorphologie Supplement Band*, 5, 215–238.

Schumm, S.A. (1965). Quaternary paleohydrology. In: Wright, H.E. and Frey, D.G. (Eds.). *The Quaternary of the United States*. Princeton University Press. Princeton, 783–794.

Schumm, S.A. (1968). River adjustement to altered hydrologic regime, Murrumbidgee River and paleochannels. *United States Geological Survey Professional Paper*, 598.

Schumm, S.A. (1969). Geomorphic implications of climatic changes. In: Chorley, R.J. (Ed.). *Water, Earth and Man*. Methuen. London, 525–534.

Schumm, S.A. (1973). Geomorphic thresholds and complex response of drainage systems. In: Morisawa, M. (Ed.). *Fluvial Geomorphology*, George Allen and Unwin. London, 299–310.

Schumm, S.A. (1975). Episodic erosion: a modification of the geomorphic cycle. In: Molhorn, W. and Flemal, R. (Eds.). *Theories of Landform Development*. Allen and Unwin. London, 69–85.

Schumm, S.A. (1977). *The Fluvial System*. Wiley. NewYork, 338 p.

Schumm, S.A. (1979). Geomorphic thresholds: the concept and its applications. *Transactions of Institute of British Geographers*, 4, 485–515.

Schumm, S.A. (1980). Some applications of the concept of geomorphic thresholds. In: Coates, D.R. and Vitek, J.D. (Eds.). *Thresholds in Geomorphology*. Georges Allen and Unwin. London, 473–485.

Schumm, S.A. (1981). Evolution and response of the fluvial system, sedimentologic implications. *Society of Economic, Palaeontologists and Mineralogists. Special Publication* 31, 19–29.

Schumm, S.A. (1985a). Explanation and extrapolation in geomorphology: seven reasons for geologic uncertainty. *Transactions of the Japanese Geomorphological Union*, 6, 1–18.

Schumm, S.A. (1985b). Patterns of alluvial rivers. *Annual Review of Earth and Planetary Sciences*, 13, 5–27.

Schumm, S.A. (1986). Aluvial river response to active tectonics. In: *Active Tectonics: Studies in Geophysics*. National Academic Press. Washington, 80–94.

Schumm, S.A. (1991). *To Interprete the Earth: Ten Ways to be Wrong*. Cambridge University Press. Cambridge, 133 p.

Schumm, S.A. (1994). *The Variability of Large Alluvial Rivers*. American Society of Civil Engineers. New York, 467 p.

Schumm, S.A. (1999). Causes and Controls of Channel Incision. In: Darby, S.E. and Simon, A. (Eds.). *Incised River Channels*. Wiley. Chichester, 19–23.

Schumm, S.A. (2004). Threshold, Geomorphic. In: Goudie, A.S. (Ed.). *Encyclopedia of Geomorphology*. Routledge. London, 1051–1052.

Schumm, S.A. (2005). *River Variability and Complexity*. Cambridge University Press. Cambridge, 220 p.

Schumm, S.A. and Chorley, R.J. (1966). Talus weathering and scarp recession in the Colorado Plateau. *Zeitschrift für Geomorphologie*, 10, 11–36.

Schumm, S.A. and Hadley, R.F. (1957). Arroyos and the semiarid cycle of erosion. *American Journal of Science*, 255, 161–174.

Schumm, S.A. and Khan, H.R. (1972). Experimental study of channel patterns. *Geological Society of America Bulletin*, 83, 1755–1770.

Schumm, S.A. and Lichty, R.W. (1963). Channel widening and flood plain construction along Cimarron river in south-western Kansas. *United States Geological Survey Professional Paper*, 352D, 71–88.

Schumm, S.A. and Lichty, R.W. (1965). Time, space and causality in geomorphology. *American Journal of Science*, 263, 110–119.

Schumm, S.A. and Lusby, G.C. (1963). Seasonal variation of infiltration capacity and runoff on hillslopes in western Colorado. *Journal of Geophysical Research*, 68, 3655–3666.

Schumm, S.A. and Winkley, B.R. (1994). The character of large alluvial rivers. In: Schumm, S.A. and Winkley, B.R. (Eds.). *The Variability of Large Alluvial Rivers*. American Society of Civil Engineers. New York, 1–9.

Schumm, S.A. Harvey, M.D. and Watson, C.C. (1984). *Incised Channels: Morphology, Dynamics and Control*. Water Resources Publications. Littleton, Colorado, 200 p.

Schumm, S.A. Mosley, M.P. and Weaver, W.E. (1987). *Experimental Fluvial Geomorphology*. Wiley. Chichester, 413 p.

Schumm, S.A., Rutherford, I.D and Brooks, J. (1994). Precutoff morphology of the lower Mississippi River. In: Schumm, S.A. and Winkley, B.R. (Eds.) *The Variability of Large Alluvial Rivers*. American Society of Civil Engineers. New York, 13–44.

Schumm, S.A., Dumont, J.F. and Holbrook, J.M. (2000). *Active Tectonics and Alluvial Rivers*. Cambridge University Press. Cambridge, 276 p.

Schunke, E. (1974). Formungsvorgäuge an schneeflecken im isländischen Hochland. In: Poser, H. (Ed.). *Geomorphologische Prozesse und Prozesskombinationen in der Gegenwart unter Verschiedenen Klimabedingungen*. Akademie Wissenschaften zu Göttingen Abh., Math.-Phys., 29, 274–286.

Schunke, E. (1975). *Die periglazialerscheinungen Islands in Abhängigkeit von Klima und substrat*. Akademie Wissenchaften zu Göttingen, Math-Phys., Kl. Folge 3, 30, 273 p.

Schunke, E. and Zoltai, S.C. (1988).-Earth hummocks (thufur). In: Clark, M.J. (Ed.). *Advances in Periglacial Geomorphology*, Wiley. Chichester, 231–245.

Schuster, R.L. (1996). The 25 most catastrophic landslides of the 20th century. In: Chacón, J., Irigaray, C. and Fernández, F. (Eds.). *Landslides*. Balkema. Rótterdam, 53–62.

Schuster, R.L. and Fleming, R.W. (1996). Socioeconomic Significance of landslides and mudflows. In: Kozlouskii, E.A. (Ed). *Landslides and Mudflows*. UNESCO/UNEP. Moscow, 131–141.

Schwarz, H.E., Enel, W.J., Dickens, P., Rogers, P. and Thompson, J. (1990). Water quality and flows. In: Turner, B.L., Clark, W.C., Kates, R.W., Matthews, J.T. and Meyer, W.B. (Eds.). *The Earth as Transformed by Human Action*. Cambridge University Press. Cambridge, 253–270.

Schwarzbach, M. (1967). Islandische wasserfalle und eire genetische systematik des wasserfalles überhaupt. *Zeitschrift für Geomorphologie,* 11, 377–417.

Science (2002). The science of climate change. *Editorial Science*, 292, 1261.

Scoging, H. (1989). Run-off generation and sediment mobilization by water. In: Thomas, D.S.G. (Ed.). *Arid Zone Geomorphology*, Belhaven Press. London, 87–116.

Ségalen, P. (1994). Les sols ferrallitiques et leur répartition géographique. ORSTOM. *Office de la Recherche Scientifique et Technique d'Outre-Mer*. Paris, 198 p.

Segura, F.S. (1990). *Las Ramblas Valencianas*. Departamento de Geografía. Universidad de Valencia, 229 p.

Seijmonsbergen, A.C., Woning, M.P., Verhoef, P.N.W. and de Graaff, L.W.S. (2005). The failure mechanism of a Late Glacial Sturzstrom in the Subalpine Molasse (Leckner Valley, Voralberg, Austria). *Geomorphology*, 66, 277–286.

Selby, M.J. (1972). Antarctic tors. *Zeitschrift für Geomorphologie, Supplementband*, 13, 73–86.

Selby, M.J. (1977). Bornhardts of the Namib Desert. *Zeitschrift für Geomorphologie,* 21, 1–13.

Selby, M.J. (1982). Controls on the stability and inclinations of hillslopes formed on hard rock. *Earth Surface Processes and Landforms*, 7, 449–467.

Selby, M.J. (1985). *Earth's Changing Surface*. Clarendon Press. Oxford, 607 p.

Selby, M.J. (1993). *Hillslope Materials and Processes*. Oxford University Press. Oxford, 451 p.

Selby, M.J., Hendy, C.H. and Seely, M.K. (1979). A late Quaternary lake in the central Namib Desert, southern Africa, and some implications. *Palaeogeography, Palaeoclimatology, Palaeoecology*, 26, 37–41.

Self, S. (2005). Effects of volcanic eruptions on the atmosphere and climate. In: Martí, J. and Ernst, G.J. (Eds.). *Volcanoes and the Environment*. Cambridge University Press. Cambridge, 152–174.

Sellier, D. (2006). Contribution à la caractérisation des marques de la corrasion éolienne par l'analyse exoscopique de ventifacts du Massif Armoricaine. *Zeitschrfit für Geomorphologie*, 50, 393–420.

Semeniuk, V. and Meagher, T.D. (1981). Calcretes of Quaternary coastal dunes in south western Australia: a capillary-rise phenomenon associated with plants. *Journal of Sedimentary Petrology*, 51, 47–68.

Seppälä, M. (1988). Palsas and related forms. In: Clark, M.J. (Ed.). *Advances in Periglacial Geomorphology*, Wiley. Chichester, 247–278.

Seppälä, M. (2004). *Wind as a Geomorphic Agent in Cold Climates*. Cambridge University Press. Cambridge, 358 p.

Seppälä, M. and Lindé, K. (1978). Wind tunnel studies of ripple formation. *Geografiska Annaler*, 60 A, 29–40.

Serrano, E. and Agudo, C. (1998). Glaciares rocosos activos de los Pirineos. In: Gómez Ortiz, A., Salvador Franch, L., Schulte, A. and García Navarro, A. (Eds.). *Procesos biofísicos actuales en medios fríos*. Publicaciones de la Universitat de Barcelona. Barcelona, 133–154.

Serrano, E. and López-Martínez, J. (2000). Rock Glaciers in the south Shetland Islands, Western Antarctica. *Geomorphology*, 35, 145–162.

Serrano, E., Angulo, C. and González Trueba, J.J. (2002). El dominio de alta montaña periglaciar en el Macizo del Posets (Pirineo aragonés). in: Serrano, E. and García, A. (Eds.). *Periglaciarismo en Montaña y Altas Latitudes*. Departamento de Geografía. Universidad de Valladolid, 163–184.

Serrat, D. (1977). *Estudio Geomorfológico del Pirineo Oriental (Puigmal-Costabona)*. Tesis Doctoral. Departamento de Geomorfología y Tectónica. Universitat de Barcelona, 222 p (Unpublished).

Serrat, D. (1979). Rock glacier morainic deposits in the eastern Pyrenees. In: Schlüchter, Ch. (Ed.). *Moraines and Varves. Origin, Genesis, Classification*. Balkema. Rotterdam, 93–100.

Serreze. M.C., Bromnwich, D.H., Clark, M.P., Etringer, A.J., Zhang, T. and Lammers, R. (2002). The large scale hydroclimatic of the terrestrial Arctic drainage. *Journal Geophysical Research*, 108 (D2).

Servant, M. and Servant-Vildary, S. (1980). L'environnement Quaternaire du bassin du Tchad. In: Williams, M.A.J. and Faure, H. (Eds.). *The Sahara and the Nile*, Balkema. Rotterdam, 133–162.

Shakesby, R.A. and Doerr, S.H. (2006). Wildfire as a hydrological and geomorphological agent. *Earth-Science Reviews*, 74, 269–307.

Shalem, N. (1962). The karst in the salt mountain of Sodom (Dead Sea). *Proceedings and International Congres Speleology*, 1, 353–359.

Shanahan, P.M., Overpick, J.P., Wheeler, C.W., Beck, J.W., Pigarti, J.S., Talbot, M.R., Scholz, C.A., Peck, J. and King, J.W. (2006). Paleoclimatic variations in West Africa from a record of late Pleistocene and Holocene lake level stands of Lake Bosumtwi, Ghana. *Palaeogeography, Palaeoclimatology, Palaeoecology*, 242, 287–302.

Shao, Y. (2008). *Physics and Modelling of Wind Erosion*. Springer. Berlin., 452 p.

Sharaf El Din, S.H. (1977). Effect of the Aswan High Dam on the Nile flood and on the estuarine and coastal circulation pattern along the Mediterranean Egyptian coast. *Limnology and Oceanography*, 22, 194–207.

Sharma, P.R. (Ed.). *Applied Geomorphology in the Tropics*. Rishi Publications. Varanasi, India, 182 p.

Sharon, D. (1962). On the nature of hamadas in Israel. *Zeitschrift für Geomorphologie*, 6, 129–147.

Sharp, A.D., Trudgill, S.T., Cooke, R.U., Price, C.A., Crabtree, R.W., Pickle, A.M. and Smith, D.I. (1982). Weathering of balustrade on St. Paul's Cathedral, London. *Earth Surface Processes and Landforms*, 7, 387–389.

Sharp, M.J. (1982). Modification of clasts in lodgement tills by glacial erosion. *Journal of Glaciology*, 28, 475–481.

Sharp, R.P. (1940). Geomorphology of the Ruby-East Humboldt Range, Nevada. *Bulletin of the Geological Society of America*, 51, 337–372.

Sharp, R.P. (1942a). Soil structures in the St. Elias Range, Yukon Territory. *Journal Geomorphology*, 5, 274–304.

Sharp, R.P. (1942b). Periglacial involutions in northeastern Illinois. *Journal of Geology*, 50, 113–133.

Sharp, R.P. (1949). Pleistocene ventifacts east of the Bighorn Mountains, Wyoming. *Journal of Geology*, 57, 175–195.

Sharp, R.P. (1958). Malaspina Glacier, Alaska. *Bulletin of the Geological Society of America*, 69, 617–646.

Sharp, R.P. (1960). *Glaciers*. Condon Lecture Publishing, University Oregon Press. Eugene, 78 p.

Sharp, R.P. (1963). Wind ripples. *Journal of Geology*, 71, 617–636.

Sharp, R.P. (1964). Wind-driven sand in the Coachella valley, California. *Bulletin of the Geological Society of America*, 75, 785–804.

Sharp, R.P. (1966). Kelso dunes, Mojave Desert, California. *Geological Society of America Bulletin*, 77, 1045–1074.

Sharp, R.P. (1980). Wind-driven sand in Coachella Valley, California: further data. *Bulletin of the Geological Society of America*, 91, 724–730.

Sharp, R.P. (1988). *Living Ice*. Cambridge University Press. Cambridge, 225 p.

Sharpe, C.F.S. (1938). *Landslides and Related Phenomena*. Columbia University Press. New York, 138 p.

Sharpe, D.R. (1987). The stratified nature of drumlins from Victoria Island and southern Ontario. In: Menzies, J. y Rose, J. (Eds.). *Drumlin Symposium,* Balkema. Rotterdam, 185–214.

Shaw, P.A. and Thomas, D.S.G. (1993). Geomorphological processes, environmental change and landscape sensivity in the Kalahari Region of Southern Africa. In: Thomas, D.S.G. and Allison, R.J. (Eds.). *Landscape Sensivity*, Wiley. Chichester, 83–96.

Shaw, P.A. and Thomas, D.S.G. (1997). Pans, playas and salt lakes. In: Thomas, D.S.G. (Ed.). *Arid Zone Geomorphology. Process, Form and Change in Drylands*. Wiley. Chichester, 293–317.

Shen, H.W. and Schumm, S.A. (1981). Methods for assessment of stream-related hazards to highways and bridges. *Federal Highway Administration, Office of Research and Development Report* FHWA/RD'80/160.

Shepard, F.P. (1948). *Submarine Geology*. Harper, New York, 348 p.

Shepard, F.P. (1959). *The Earth Beneath the Sea*. Oxford University Press. London, 275 p.

Sherard, J.L. (1992). Embankment dam craking. In: Sherard, J.L., Sukhmander, J. and Singh, V.P. (Eds.). Embankment Dams: James L.Sherard contributions. *America Society Safety Engineering Geotechnical*, Special Publication, 32, 271–353.

Sherard, J.L. and Decker, R.S. (1977). Summary- Evaluation of Symposium on dispersive clays. In: Sherard, J.L. and Decker, R.S. (Eds.). *Dispersive clays, related piping, and erosion in geotechnical projects*. American Society for Testing and Materials, 623, 467–479.

Sherard, J.L., Decker, R.S. and Ryker, N.L. (1972a). Piping in earth dams of dispersive clay. *Proceedings of American Society Civil Engineering. Special Conference on Performance of Earth and Earth-supported Structures*, 1, 589–626.

Sherard, J.L., Decker, R.S. and Ryker, N.L. (1972b). Hydraulic fracturing in low dams of dispersive clay. *Proceedings of American Society Civil Engineering. Special Conference on Performance of Earth and Earth-supported Structures*, 653–689.

Sherard, J.L., Duningan, L.P. and Decker, R.S. (1976). Identification and nature of dispersive soils. American Society Civil Engineering. *Journal Geotechnical Engineering Division Proceedings*, 102, 287–301.

Sheridan, D. (1981). *Problems of Desertification of the United States*. Council Environmental Quality, 142 p.

Sheridan, M.F. and Wohletz, K.H. (1981). Hydrovolcanic explosions: The systematics of water pyroclast equilibrium. *Science*, 212, 1387–1384.

Sheridan, M.F. and Wohletz, K.H. (1983). Hydrovolcanism: Basic considerations and review. *Journal of Volcanology and Geothermal Research,* 17, 1–29.

Sherman, G.D. (1952). The genesis and morphology of the alumina-rich laterite clays. In: *Problems of Clay and Laterite Genesis*. American Institute of Mining and Metallurgical. New York, 154–161.

Shick, A.P. (1971). A desert flood: physical characteristics; effects of man, geomorphic significances, human adaptation-a case study of the southern Arava watershed. *Jerusalem Studies in Geography*, 2, 91–155.

Shick, A.P. (1988). Hydrologic aspects of floods in extreme arid environments. In: Baker, V.R., Craig Kochel, R. and Patton, P.C. (Eds.). *Flood Geomorphology*. Wiley. New York, 189–203.

Shick, A.P. (Ed.) (1992). *Surficial Processes and Landscape Evolution: Rift Valley and Arid Terrains*. Laser Pages Publishing Ltd.. Jerusalem, 191 p.

Shick, A.P. (1995). Fluvial processes on an urbanizing alluvial fan: Eilat, Israel. *Geophysical Monograph*, 89, 209–218.

Shopov, Y.Y. (2004). Dating of karst landforms. In: Gunn, J. (Ed.). *Encyclopedia of Caves and Karst Science*. Fitzroy Dearborn. New York, 281–283.

Short, N.M. and Blair, R.W. (1986). *Geomorphology from Space. A Global Overview of Regional Landforms*. National Aeronautics and Space Administration. Washington, 717 p.

Shreve, R.L. (1975). The probabilistic-topology approach to drainage basin geomorphology. *Geology*, 3, 527–529.

Shroder, J.F.Jr. (1980). Dendrogeomorphology. Review and new techniques of the tree-ring dating. *Progress in Physical Geology*, 4, 161–188.

Shur, Y., Hinkel, K.M. and Nelson, F.E. (2005). The transient layer: implications for geocryology and climate-science change. *Permafrost and Periglacial Processes*, 16, 5–18.

Siegert, M.J. (2005a). Reviewing the origin of subglacial Lake Vostok and its sensivity to ice sheet changes. *Progress in Physical Geography*, 29,156–170.

Siegert, M.J. (2005b). Lakes beneath the ice sheet: The ocurrence, analysis and future exploration of Lake Vostok and of the Antartic subglacial lakes. *Annual Review of Earth ad Planetary Sciences*, 33, 215–245.

Sigurdsson, H. (2000). Indroduction. In: Sigurdsson. H. (Ed.). *Encyclopedia of Volcanoes*. Academic Press. San Diego, 1–13.

Sigurdsson, H., Carey, S., Cornell, W. and Pescatore, T. (1985). The eruption of Vesubius in AD 79. *National Geographic Research,* 1, 332–387.

Sillières, P. (1997). Baelo Claudia: Una ciudad romana de la Bética. Junta de Andalucía-Casa de Velásquez. Madrid.

Silva, P.G. (1994*) Evolución Geodinámica de la Depresión del Guadalentín desde el Mioceno Superior hasta la Actualidad: Geotectónica y Geomorfología*. Facultad de Ciencias Geológicas. Universidad Complutense-Tesis Doctoral (Unpublished). Madrid., 642 p.

Silva, P.G. and Harvey. A.M. (1991). Actividad tectónica cuaternaria y desarrollo geomorfológico de abanicos aluviales en la Depresión del Guadalentín (Murcia). *Abstracts VIII Reunión del Cuaternario Ibérico*. Valencia, 17–19.

Silva, P.G., Harvey, A., Zazo, C. and Goy, J.L. (1992a). Geomorphology, depositional style and morphometric relationships of Quaternary alluvial fans in the Guadalentin Depresion (Murcia, SE España). *Zeitscrift für Geomorphologie*, 36, 325–341.

Silva, P.G., Goy, J.L., Zazo, C., Bardají, T., Somoza, L. and Dabrio, C.J. (1992b). Evaluación geomorfológica de la actividad tectónica cuaternaria a lo largo de frentes montañosos de falla en el SE de España. *Actas III Congreso Geológico de España*, T.2, 96–100.

Silva, P.G., Goy, J.L. and Zazo, C. (1992c). Características estructurales y geométricas de la zona de la Falla de Lorca-Alhama. *Geogaceta*, 12, 7–11.

Silva, P.G., Goy, J.L., Somoza, L., Zazo,C. and Bardají, T. (1993). Lanscape responses to strike-slip faulting linked to collisional setting: Quaternary tectonics and basin formation in the Eastern Betics (SE Spain). *Tectonophysics*, 224, 289–303.

Silva, P.G., González-Hernández, F.M., Goy, J.L., Zazo, C. and Carrasco, P. (2001). Paleoseismicity and historical seismicity in the Mallorca Island (Baleares, Spain): A preliminary approach. *Acta Geológica Hispánica*, 36, 245–266.

Silva, P.G., Goy, J.L., Zazo, C. and Bardají, T. (2003). Fault generated mountain fronts in southeast Spain: geomorphologic assessment of tectonic and seismic activity. *Geomorphology*, 250, 203–226.

Silva, P.G., Borja, F., Zazo, C., Goy, J.L., Bardají, T., De Luque, L., Lario, J. and Dario, C.J. (2005). Archaeoseismic record at the ancient Roman City of Baelo Claudia (Cádiz, south Spain). *Tectonophysics*, 408, 129–146.

Silva, P.G., Goy, J.L., Zazo, C., Bardají, T., Lario, J., Somoza, L., Luque, L. and González-Hernández, F.M. (2006). Neotectonic fault mapping at the Gibraltar Strait Tunnel area, Bolonia Bay (South Spain). *Engineering Geology*, 84, 31–47.

Simanton, J.R. Rawitz, E. and Shirley, E. (1984). Efects of rock fragments on erosion of semiarid rangeland soils. *Soil Science Society of America Special Publication*, 13, 65–72.

Simkin, T. and Siebert, L. (1996). *Volcanoes of the World*. Smithsonian Institution and Geoscience Press. New York.

Simkin, T., Sieber, L. and Blong, R. (2001). Policy form: disaster-lessons from the historical record. *Science*, 291, 255.

Simon, A. and Darby, S. (1999). The nature and significance of incised river channels. In: Darby, S. and Simon, A. (Eds.). *Incised River Channels*. Wiley. Chichester, 3–18.

Simón, J.L. (1983). La distención plio-cuaternaria de la Cadena Ibérica oriental. Mapa de contornos estructurales de la penillanura fundamental deformada. *Boletín Geológico y Minero*, 94, 496–502.

Simonett, D.S. (1967). Landslides distribution and earthquakes in the Bewanni and Toricelli Mountains, New Guinea, a statistical analysis. In: Jennings, J.N. and Mabbutt, J.A. (Eds.). *Landform studies from Australia and New Guinea*. Cambridge University Press. London, 64–84.

Simonett, D.S. (1968). Cuesta. In: Fairbridge, R.W. (Ed.). *The Enciclopedia of Geomorphology*. Halsted Press. Stroudsburg, Pennsylvania, p. 233.

Simons, D.B. and Richardson, E.V. (1966). Resitance flow in Alluvial Channels. *United States Geological Survey Professional Paper* 422 J, 61 p.

Simons, F.S. and Eriksen, G.E. (1953). Some desert features of northwest central Perú. *Boletín de la Sociedad Geológica del Perú*, 26, 229–245.

Sinclair, W.C. (1982). Sinkhole development resulting from ground-water withdrawal in the Tampa Area, Florida. *United States Geological Survey*. Water Resources Invest. 81–150.

Singer, A. (1979/1980). The paleoclimatic interpretation of clay minerals in soils and weathering profiles. *Earth-Science Reviews,* 15, 303–326.

Singh, G. (1971). The Indus valley culture seen in context of post-glacial climate and ecological studies in north-west India. *Archaeology and Anthropology in Oceania*, 6, 177–189.

Singh, H., Parkash, B. and Gohain, K. (1993). Facies and analysis of the Kosi megafans deposits. *Sedimentary Geology*, 85, 87–113.

Singh, I.B. (2007). The Ganga River. In: Gupta, A. (Ed.). *Large Rivers. Geomorphology and Management*. Wiley. Chichester, 347–371.

Singh, V.P. (1996). *Dam Break Modelling Technology*. Kluwer. Dordrecht, 242 p.

Sirvent, J., Desir, G., Gutiérrez, M., Sancho, C. and Benito, G. (1997). Erosion rates in badland areas recorded by collectors, erosion pins and profilometer techniques (Ebro Basin, NE-Spain). *Geomorphology*, 18, 61–75.

Sivarajasingham, S., Alexander, L.T., Cady, J.G. and Cline, M.G. (1962). Laterite. *Advances in Agronomy*, 14, 1–60.

Skempton, A.W. and Hutchinson, J.N. (1969). Stability of natural slopes and embankment foundations. *Proceedings 7th International Conference of Soil Mechanics and Foundation Engineering*. State of Art Volume. Mexico City, 291–340.

Skina, R. and Friend, P.J. (1994). River systems and their sediment flux, Indo-Gengetic Plains, northern Bihar, India. *Sedimentology,* 41, 27–52.

Slaymaker, O. (2000) (Ed.). *Geomorphology, Human Activity and Global Enviromental Change*. Wiley. Chichester, 322 p.

Slaymaker, O. and Kelly, R.E.J. (2007). *The Cryosphere and Global Environmental Change*. Blackwell. Malden, 261 p.

Small, R.J. (1970). *The Study of Landforms. A Textbook of Geomorphology*. Cambridge University Press. Cambridge, 502 p.

Small, R.J. (1987). Englacial and supraglacial sediment: transport and deposition. In: Gurnell, A.M. and Clark, M.J. (Eds.). *Glacio-Fluvial Sediment Transfer-an Alpine Perspective*, John Wiley. Chichester, 111–145.

Smalley, I.J. (1972). The interaction of great rivers and large deposits of primary loess. *Transactions of the New York Academy of Sciences*, 34, 534–542.

Smalley, I.J. (Ed.) (1975). *Loess. Lithology and Genesis*. Dowden, Hutchinson and Ross. Stroudsburg, Pennsylvania, 429 p.

Smalley, I.J. (1990). Possible formation mechanisms for the modal coarse-silt quartz particles in loess. *Quaternary International*, 7/8, 23–27.

Smalley, I.J. (1995). Maxing the material: the formation of silt-sized primary mineral particles for loess deposits. *Quaternary Science Reviews,* 14, 645–651.

Smalley, I.J. and Piotrowski, J.A. (1987). Critical strength/stress ratios at the ice-bed interface in the drumlin forming process: From "dilatancy" to cross-over. In: Menzies, J. and Rose, J. (Eds.). *Drumlin Symposium,* Balkema. Rotterdam, 81–86.

Smalley, I.J. and Smalley, V. (1983). Loess material and loess deposits: Formation, distribution and consequeneces. In: Brookfield, M.E. and Ahlbrandt, T.S. (Eds.). *Eolian Sediments and Processes*. Elsevier. Amsterdam, 51–68.

Smalley, I.J. and Unwin, D.J. (1968). The formation and shape of drumlins and their distribution and orientation in drumlin fields. *Journal of Glaciology*, 7, 377–390.

Smart, C. and Worthington, S.R.H. (2004b). Water tracing. In: Gunn, J. (Ed.). *Encyclopedia of Caves and Karst Science*. Fitzroy Dearborn. New York, 769–771.

Smart, C. and Worthington, S.R.H. (2004a). Springs. In: Gunn, J. (Ed.). *Encyclopedia of Caves and Karst Science*. Fitzroy Dearborn. New York, 699–703.

Smellie, J.L. (2007). Glacial Landform. Quaternary vulcanism, subglacial landforms. In: Elias, S.A. (Ed.). *Encyclopedia of Quaternary Science*. Elsevier. Amsterdam, 784–798.

Smith, B.J. (1978). The origin and geomorphic implications of cliff foot recesses and tafoni on limestone hamadas in the northwest Sahara. *Zeitscrift für Geomorphologie*, 22, 21–43.

Smith, B.J. (1994). Weathering processes and forms. In: Abrahams, A.D. and Parson, J.P. (Eds.). *Geomorphology of Desert Environments*, Chapman and Hall. London, 39–63.

Smith, B.J. and Warke, P. (1997). Controls and uncertainties in the weathering environment. In: Thomas, D.S.G. (Ed.). *Arid Zone Geomorphology.* Wiley. Chichester, 41–54.

Smith, B.J., Magee, R.W. and Whalley, W.B. (1994). Breakdown patterns of quartz sandstone in a polluted urban environment, Belfast, Northern Ireland. In: Robinson, D.A. and Williams, R.B.G. (Eds.). *Rock Weathering and Landform Evolution*. Wiley. Chichester, 131–150.

Smith, D.G. (1986). Anastomosing river deposits, sedimentation rates and basin subsidence, Magdalena River, northwestern Colombia, South America. *Sedimentary Geology*, 46, 177–196.

Smith, D.I. (1969). The solution erosion of limestone in an arctic morphogenetic region. In: Stelel, O. (Ed.). *Problems in Karst Denudation*. Institute of Geography. Brno, 99–110.

Smith, D.I., Greenaway, M.A. and Spate, A.P. (1989). Measure of limestone erosion in Australia using the micro-erosion meter. In: Seuffert, O. (Ed.). Abstracts of papers and posters. *Second International Conference on Geomorphology*. Frankfurt, 272 p.

Smith, E.A. (1986). The structure of the Arabian Heat Low. *Monthly Weather Review*, 114, 1067–1083.

Smith, G.I. and Streat-Perrott, F.A. (1983). Pluvial lakes of the Western United States. In: Porter, S.C. (Ed.). *Late-Quaternary Environments of the United States*, University of Minnesota Press. Minneapolis, 190–212.

Smith, H.T.U. (1948). Giant glacial groves in Nothwest Canada. *American Journal of Science*, 246, 503–514.

Smith, H.T.U. (1953). The Hickory Run boulderfield, Carbon County, Pennsylvania. *American Journal of Science*, 25, 625–642.

Smith, H.T.U. and Smith, A.P. (1945). Periglacial rock streams in the Blue Ridge area. *Bulletin of the Geological Society of America*, 56, 1198.

Smith, K. (2001). *Environmental Hazards: Assesing Risk and Reducing Disaster*. Routledge. Londres, 389 p.

Smith, K. and de Sánchez, B.A. (1992). Erosion hazards in a Brazilian suburb. *Geography Review*, 6, 37–41.

Smith, K. and Ward, R. (1998). *Floods: Physical Processes and Human Impacts*. Wiley. Chichester, 382 p.

Smith, M.W. (1993). Climatic change and permafrost. In: French, H.M. and Slaymaker, O. (Eds.). *Canada's Cold Environments*. McGill-Queen's University Press. Montreal. 291–311.

Smith, N. (2001). Ocean and climate prediction-the WOCE legacy. In: Siedler, G., Church, J. and Gould, G. (Eds.). *Ocean Circulation and Climate*. Academic Press.San Diego, 585–602.

Smith, N.D. and Minter,W.E.C. (1980). Sedimentologic central of gold and uranium and two Writwatersrand palaeoplacers. *Economic Geology*, 75, 1–14.

Smith, N.D., Vendl, M.A. and Kennedy, S.K. (1982). Comparison of sedimentation regimes in four glacier-fed lakes of western Alberta. In: Davidson-Arnott, R., Nickling, W. and Fahey, B.D. (Eds.). *Research in Glacial, Glaciofluvial and Glaciolacustrine Systems*. Geo Books. Norwich, 203–238.

Smith, R.L. (1960). Ash flows. *Geological Society American Bulletin*, 71, 795–842.

So, C.L. (1971). Mass movement associated with the rainstorm of June 1966 in Hong Kong. *Transactions Institute British Geographers*, 53, 55–66.

Söderman, G. (1985). Planation and weathering in eastern Fennoscandia. *Fennia*, 163, 347–352.

Soldati, M., Corsini, A. and Pasuto, A. (2004). Landslides and climate change in the Italian Dolomites since the Late glacial. *Catena*, 55, 141–161.

Solomina, O., Haeberli, W., Kull, C. and Wiles, G. (2008). Historical and Holocene glacier-climate variations. *Global and Planetary Change*, 60, 1–9.

Solomon, S., Qin, D., Manning, M., Marquis, M., Averyt, K., Tingor, M.M.B., LeRoy, H. and Chen, Z. (2007). *Climate Change 2007. The Physical Science Basis*. Cambridge University Press. Cambridge, 996 p.

Soloviev, P.A. (1973). Thermokarst phenomena and landforms due to frost heaving in Central Yakutia. *Biuletyn Peryglacjalny*, 23, 135–155.

Somoza, L., Barnolas, A., Arasa, A., Maestro, A., Rees, J.F. and Hernández-Molina, F.J. (1998). Architectural stacking patterns of the Ebro delta controlled by Holocene high-frequency eustatic fluctuations, delta-lobe switching and subsidence processes. *Sedimentary Geology*, 117, 11–32.

Soriano, M.A. (1989). Infilled valleys in the central Ebro Basin (Spain). *Catena*, 16, 357–367.

Soriano, M.A. and Calvo, J.M. (1987). Características, datación y evolución de los valles de fondo plano em las inmediaciones de Zaragoza. *Cuaternario y Geomorfología*, 1, 283–293.

Sorriso-Valvo, M. (1988). Landslide-related fans in Calabria. *Catena Supplement*, 13, 109–121.

Sorriso-Valvo, M. and Gullá, G. (1996). Rock Slide. In: Dikau, R., Brunsden, D. Shortt,L and Ibsen, M.L. (Eds.). *Landslide Recognition. Identification, Movement and Causes*, Wiley. Chichester, 85–96.

Souchez, P. (1966). Gélivation et évolution des versants en bordure de l'Islandsis d'Antartides orientale. In: Macar, P. (Ed.). *L'Evolution des Versants. Les Congrès et Colloques de L'Univesité de Liége*, 40, 291–298.

Sowers, G.F. (1996). *Building of Sinkholes*. ASCE Press. New York, 202 p.

Sparks, B.W. (1972). *Geomorphology. Geographies for advanced study*. Longman. London, 530 p.

Sparks, R.S.J. (1993). Génesis y evolución de magmas. In: Martí, J. and Araña, V. (Eds.). *La Volcanología Actual*. Consejo Superior de Investigaciones Científicas. Madrid, p. 1–44.

Sparks, R.S.J. and Walker, G.P.L. (1973). The ground surge deposit: a third type of pyroclastic rock. *Nature*, 241, 62–64.

Sparks, R.S.J., Self, S. and Walker, G.P.L. (1973). Products of ignimbrite eruptions. *Geology*, 1, 115–118.

Sparks, R.S.J., Bursik, M.I., Carey, S.N., Gilbert, J.S., Glaze, L.S., Sigurdsson, H. and Woods, A.W. (1997). *Volcanic Plumes*, Wiley, 574 p.

Spate, A.P., Jennings, J.N., Smith, D.I. and Greenaway, M.A. (1985). The micro-erosion meter: use and limitations. *Earth Surface Processes and Landforms*, 10, 427–440.

Spaulding, W.G. (1990). Vegetation and climatic development of the Mojave Desert: The last glacial maximum to the present. In: Betancourt, J.L., Van Devender, T.R. and Martin, P.S. (Eds.). *Packrat Middens, the Last 40,000 Years of Biotic Change*. University of Arizona Press. Tucson, 166–199.

Spencer, T. (1988). Coastal biogeography. In: Viles, H.A. (Ed). *Biogeography*. Blackwell. Oxford, 255–318.

Spencer, T. (1994). Tropical coral islands-an uncertain future?. In: Roberts, N. (Ed). *The Changing Global Environment*. Blackwell. Oxford, 190–209.

Sperling, C.H.B. and Cooke, R.U. (1985). Laboratory simulation of rock weathering by salt crystallization and hydration processes in hot, arid environments. *Earth Surface Processes and Landforms*, 10, 541–555.

St. Onge, D.A. (1968). Geomorphologic Maps. In: Fairbridge,R.W. (Ed.). *The Encyclopedia of Geomorphology*. Reinhold Book Corp. New York, 388–403.

Stadler, S.J. (2004). Aridity Indexes. In: Oliver, J.E. (Ed.). *Encyclopedia of World Climatology*. Springer. Berlin, 89–94.

Stager, J.K. (1956). Progress report on the analysis on the characteristics and distribution of pingos east of the Mackenzie Delta. *Canadian Geographer*, 7, 13–20.

Stanistreet, I.G. and McCarthy, T.S. (1993). The Okavango Fan and the classification of subaerial fan systems. *Sedimentary Geology*, 85, 115–133.

Stanley, D.J. (1988). Subsidence in the Northeastern Nile Delta: rapid rates, possible causes, and consequences. *Science*, 240, 497–500.

Stanley, D.J. (1996). Nile delta: extreme case of sediment entrapment on a delta plain and consequent coastal land loss. *Marine Geology*, 129, 189–195.

Stanley, D.J. and Warne, A.G. (1993). Nile delta: recent geological evolution and human impact. *Science*, 260, 628–634.

Stanley, D.J. and Warne, A.G. (1994). Worldwide initiation of Holocene marine deltas by deceleration of sea-level rise. *Science*, 265, 228–231.

Stanley, D.J. and Warne, A.G. (1998). Nile delta in its distruction phase. *Journal of Coastal Research*, 14, 794–823.

Starkel, L. (1976). The role of extreme (catastrophic) meteorological events in contemporary evolution of slopes. In: Derbyshire, E. (Ed.). *Geomorphology and Climate*. Wiley. London, 203–246.

Starkel, L. (1987). Man as a cause sedimentologic changes in the Holocene. *Striae*, 26, 5–12.

Starkel, L. (1995). The place of the Vistula river valley in the late Vistulian-early Holocene evolution of the European valleys. In: Frenzel, B. (Ed.). European river activity and climate change during the Lateglacial and Holocene. *Paläoklimaforschung*, 14, 75–88.

Steijn, H. van, Bertran, P., Francou, B. and Teixier, J.P. (1995). Models for the genetic and environmental interpretation of stratified slope deposits: Review. *Permafrost and Periglacial Processes*, 6, 125–146.

Stenborg, T. (1970). Delay of run-off from a glacier basin. *Geografiska Annaler*, 52, 1–30.

Stephenson, W.J. and Finlayson, B.L. (2009). Measuring erosion with the micro-erosion meter-Contribution to understanding landform evolution. *Earth-Science Reviews*, 95, 53–62.

Stephenson, W.J. and Brander, R.W. (2003). Coastal Geomorphlogy into the 21st century. *Progress in Physical Geography*, 27, 607–623.

Stevaux, J.C. and Souza, I.A. (2004). Floodplain construction in an anastomotised river. *Quaternary International*, 114, 55–65.

Steven T.A. and Lipman, P.W. (1976). Calderas of the San Juan Volcanic field, south-western Colorado. *United States Geological Survey Professional Paper*, 958, 35 p.

Stevens, F.R. (1974). *Rugged Landscape, the Geology of Central New Zealand*. A.H. and A.W. Reed, Wellington, 286 p.

Stille, H. (1924). *Grundfragen der Vergleichender Tectonik*. Borntraeger. Berlin.

Stock, J.D. and Dietricht, W.E. (2006). Erosion of steepland valleys by debris flows. *Geological Society of America Bulletin*, 118, 1125–1148.

Stocking, M. (1976). Tunnel erosion. *Rhodesia Agricultural Journal*, 73, 35–39.

Stockton, C.W. and Fritts, H.C. (1971). Conditional probability of ocurrence for variations in climate based on width of annual tree-rings in Arizona. *Tree-Ring Bulletin*, 31, 3–24.

Stoddart, D.R. (1965a). Geography and the ecological approach. The ecosystem as a geographical principle and method. *Geography*, 50, 242–251.

Stoddart, D.R. (1965b). The shape of atolls. *Marine Geology*, 3, 369–383.

Stoddart, D.R. (1969a). Climatic Geomorphology: review and reassesment. In: Board, C. *et al.*, (Eds.). *Progress in Geography*. Arnold. London, 160–222.

Stoddart, D.R. (1969b). Climatic Geomorphology. In: Chorley, R.J. (Ed.). *Introduction to fluvial processes*. Methuen. Suffolk, 189–201.

Stokes, S. (1997). Dating of desert sequences. In: Thomas, D.S.G. (Ed.). *Arid Zone Geomorphology*, Wiley, 607–637. Chichester.

Stokstad, E. (2004). Defrosting the carbon freezer of the north. *Science*, 304, 1618–1620.

Stone, A.E.C. and Thomas, D.S.G. (2008). Linear dune accumulation chronologies from the southwest Kalahari, Namibia: challenges of reconstructing late Quaternary palaeoenvironments from eolian landforms. *Quaternary Science Reviews*, 27, 1667–1681.

St-Onge, D.A. (1968). Geomorphologic maps. In: Fairbridge, R.W. (Ed.). *The Encyclopedia of Geomorphology*. Reinhold. New York, 388–403.

Stoops, G. (1967). Le profil d'alteration au Bas-Congo (Kinshasa). Sa description et sa genèse. *Pedologie* 17, 60–105.

Stoops, G. (1989). Contribution of in situ transformations to the formation of the stone-layers complexes in Central Africa. *Geo-Eco- Trop* 11, 139–149.

Stovel, H. (1998). *Risk preparedness: a management manual for world cultural heritage*. ICEROM. Roma.

Strachan, A.D. and Dearman, W.R. (1983). Engineering geological mapping in glaciated terrains. In: Eyles, N. (Ed.). *Glacial Geology: An Introduction for Engineerings and Earth Scientists.* Pergamon Press. Oxford, 229–246.

Strahler, A.N. (1950a). Davis' concept of slope development viewed in the light of recent quantitative investigations. *American Association of Geographers Annals*, 40, 209–213.

Strahler, A.N. (1950b). Equilibrium theory of erosional slopes approached by frequency distribution analysis. *American Journal of Science*, 248, 673–696.

Strahler, A.N. (1952). Dynamic basis of geomorphology. *Bulletin of the Geological Society of America*, 66, 923–938.

Strahler, A.N. (1956a). Quantitative slope analysis. *Bulletin of the Geological Society of America*, 67, 571–596.

Strahler, A.N. (1956b). The nature of induced erosion and aggradation. In: Thomas, W.L. (Ed.). *Man's Role in Changing the Face of the Earth*. University of Chicago Press. Chicago, 621–638.

Strahler, A.N. (1964). Quantitative geomorphology of drainage basins and channels networks. In: Chow, V.T. (Ed.). *Handbook of Applied Hydrology*. McGraw Hill. New York, 4–39; 4–76.

Strahler, A.N. (1968). Quantitative Geomorphology. In: Fairbridge, R.W. (Ed.). *The Encyclopedia of Geomorphology*. Dowden, Hutchinson and Ross. Strousburg, 898–912.

Strahler, A.N. (1980). Systems theory in Physical Geography. *Physical Geography*, 1, 1–27.

Strahler, A.N. (1987). *Science and Earth History. The Evolution-Creation Controversy*. Prometheus Books. Buffalo, New York, 552 p.

Strahler, A.N. and Strahler, A.H. (1974). *Introduction to Environmental Science*. Hamilton Publishing. Santa Bárbara, California.

Strakhov, N.M. (1967). *Principles of Lithogenesis*. Oliver and Boyd, vol. 1. Edinburgh, 245 p.

Street, F.A. and Grove, A.T. (1976). Environmental and climatic implications of late Quaternary lake-level fluctuations in Africa. *Nature*, 261, 385–390.

Street, G.J. and Anderson, A. (1993). Airborne electromagnetic surveys of the regolith. *Exploration Geophysics*, 24, 795–800.

Street-Perrott, F.A., Roberto, N. and Metcalfe, S.L. (1985). Geomorphic implications of late Quaternary hydrological and climatic changes in the Northen Hemisphere tropics. In: Douglas, I. and Spencer, T. (Eds.). *Environmental Change and Tropical Geomorphology*. Allen and Unwin. London, 165–183.

Street-Perrot, F.A., Mitchell, J.F.B., Marchand, D.S. and Brunner, J.S. (1990). Milankovitch and albedo forcing of the tropical monsoon: a comparison of geological evidence and numerical simulations for 9000 yrs. BP. *Transctions of the Royal Society of Edinburgh (Earth Science*), 81, 407–427.

Striebel, T. (2008). Granite caves in the Fichtelgebirge Mountains. *Cadernos Laboratorio Xeolóxico de Laxe*, 33, 167–174.

Stringer, L.C. (2008). Reviewing the International Year of Deserts and Desertification 2006: What contribution towards combating global desertification and implementing the United Nations Convention to Combat Desertification?. *Journal of Arid Environments*, 72, 2065–2074.

Stroeven, A., Wal, R. van de and Oerlemans, J. (1989). Historic front variations of the Rhone Glacier: simulation with an ice flow model. In: Oerlemans, J. (Ed.). *Glacier Fluctuations and Climatic Change*. Kluwer. Dordrecht, 391–405.

Stuiver, M. (1970). Tree rings, varve and carbon-14 chronologies. *Nature*, 228, 454–455.

Stuiver, M. and Kra, R. (1986). Calibration issues. *Radiocarbon*, 28, 805–1030. United States Department of Commerce (1972). *News*. September 22, 1972, NOAA 72–112, 3 p.

Stuiver, M. and Reimer, P.J. (2003). Extended ^{14}C data base and revised CALIB 3.0 ^{14}C age calibration program. In: Stuiver, M. and Kra, R.S. (Eds.). *Calibration 1993. Radiocarbon*, 35, 215–230.

Stuiver, M., Reimer, P.J., Bard, J.W., Beck, G.S., Burr, K.A., Hughen, B., Kromer, G., McComac, J., van der Plicht, J. and Spurk, M. (1998). INTCAL98. Radiocarbon age calibration, 24.000-0 cal BP. *Radiocarbon*, 40, 1041–1063.

Sturm, M., Racine, C. and Tape, K. (2001). Climate change: increasing shrub abundance in the Arctic. *Nature*, 411, 546–547.

Suárez, L. and Regueiro, M. (1997). *Guía ciudadana de los riesgos geológicos. Deslizamientos y aludes*. Ilustre Colegio Oficial de Geólogos de España. Madrid, 196 p.

Sueoka, T. (1988). Identification and classification of granitic residual soils using chemical weathering index. En: *Geomechannics in Tropical Soils*. Balkema, Vol.1, 55–61. Rotterdam.

Suess. F. (1888). *Das Antlitz der Erde*. 3 vols. F. Tempsky. Viena.

Sugden, D.E. (1974). Landscapes of glacial erosion in Greenland and their relationship to ice, topographic and bedrock conditions. In: Brown, E.H. and Waters, R.S. (Eds.). *Progress in Geomorphology*, Institute British Geographers, 7, 177–195.

Sugden, D.E. (1977). Reconstruction of the morphology, dinamics and thermal characteristics of the Laurentide ice sheet at its maximum. *Arctic and Alpine Research*, 9, 27–47.

Sugden, D.E. (1978). Glacial erosion by the Laurentide ice shect. *Journal of Glaciology*, 20, 367–391.

Sugden, D.E. (1982). *Arctic and Antarctic. A Modern Geographical Systems*. Blackwell. Oxford, 472 p.

Sugden, D.E. and John, B.S. (1976). *Glaciers and Landscape*. Edward Arnold. London, 376 p.

Sugden, D.E., Glasser, N.F. and Clapperton, C.M. (1992). Evolution of large roches moutonnées. *Geografiska Annaler*, 74 A, 253–264.

Summerfield, M.A. (1982). Distribution, nature and genesis of silcrete in arid and semi-arid southern Africa. *Catena Supplement* 1, 37–65.

Summerfield, M.A. (1983). Silcrete. In: Goudie, A.S. and Pye, K. (Eds). *Chemical sediments and geomorphology: precipitates and residua in the near surface environment*. Academic Press. New York, 59–91.

Summerfield, M.A. (1991). *Global Geomorphology*. Longman. Exxex, 537 p.

Summerfield, M.A. (2005a). A tale of two scales, or the two geomorphologies. *Transaction, Institute British Geographers*, 30, 402–415.

Summerfield, M.A. (2005b). The changing landscape of geomorphology. *Earth Surface Processes and Landforms*, 30, 779–781.

Summerfield, M.A. and Hulton, N.J. (1994). Natural controls of fluvial denudation rates in major world drainage basins. *Journal of Geophysical Research*, 99, NO B7, 13871–13883.

Sunamura, T. (1992). *Geomorphology of Rocky Coasts*. Wiley, Chichester, 302 p.

Sunborg, A. (1956). The River Klarälven, a study of fluvial processes. *Geografiska Annaler*, 38, 127–316.

Sundquist, E.T. (1993). The global carbon dioxide budget. *Science*, 259, 934–941.

Surian, N. (2006). Effects of human impact on braided river morphology: examples from northern Italy. In: Sambrook Smith, G.H., Best, J.L., Bristow, C.S. and Petts, G.E. (Eds.). *Braided Rivers. Process, Deposits, Ecology and Management*. Special Publication Number 36. International Association of Sedimentologists. Blackwell. Malden, 325–338.

Sutherland, D.G. (1984). Geomorphology and mineral exploration: some samples for exploration for diamondiferous placer deposits. *Zeitschrift für Geomorphologie, Supplementband*, 51, 95–108.

Sutherland, D.G. (1985). Geomorphological controls on the distribution of placer deposits. *Journal of the Geological Society of London,* 142, 727–737.

Svenson, B. (1938). The Ohio and Mississippi River floods of January-February 1937. *Monthly Weather Review*, Suppl 37, 1–55.

Swan, S.B.St C. (1972). Land surface evoution and related problems with reference to a humid tropical region: Johor, Malaya. *Zeitschrift für Geomorphologie,* 16, 160–181.

Sweet, M.L. and Kocurek, G. (1990). An empirical model of aeolian dune lee-face airflow. *Sedimentology*, 37, 1023–1038.

Sweeting, M.M. (1958). The karstlands of Jamaica. *Geographical Journal*, 124, 184–199.

Sweeting, M.M. (1966). The weathering of limestones, with particular references to the carboniferous limestone in Northern England. In: Dury, G.H. (Ed.). *Essays in Geomorphology*. American Elsevier. New York, 177–210.

Sweeting, M.M. (1972). *Karst Landforms*. Macmillan, London, 362 p.

Sweeting, M.M. (1981). Introduction. In: Sweeting, M.M. (Ed.). *Karst Geomorphology*. Hutchinson Ross. Stroudsburg, Pennsylvania, 1–8.

Sweeting, M.M. (1995). *Karst in China. Its Geomorphology and Environment.* Springer. Berlin, 265 p.

Swezey, C. (2001). Eolian sediment resposnse to late Quaternary climate change: temporal and spatial patterns in the Sahara. *Palaeogeography, Palaeoclimatology, Palaeoecology*, 167, 119–155.

Swinzow, G.K. (1969). Certain aspects of engineering geology in permafrost. *Engeneering Geology*, 3, 177–215.

Swoboda-Colberg, N.G. and Drever, J. (1993). Mineral dissolution rates in plot-scale field and laboratory experiments. *Chemical Geology*, 105, 51–69.

Syvinski, J.P.M., Burrell, D.C. and Skei, J.M. (1987). *Fjords: Processes and Products*. Springer. New York, 379 p.

Szunyogh, G. (2005). Theoretical investigation on the duration of karstic denudation on bare, sloping limestone surface. *Acta Carsologica*, 34, 9–23.

Taber, S. (1929). Frost heaving. *Journal of Geology*, 37, 428–461.

Tabor, R.W. (1971). Origin of the ridge-top depressions by large scale creep in the Olympic Mountains,

Washington. *Geological Society of America Bulletin*, 82, 1811–1822.

Takahashi, T. (1991). *Debris Flow*. Balkema. Rotterdam, 165 p.

Takei, A., Kobashi, S. and Fukushima, Y. (1981). Erosion and sediment transport measurement in a weathered granite mountain areas. In: *Symposium on Erosion and Sediment Transport Measurement, Florence*. International Association Hydrological Sciences, 133, 493–502.

Takeuchi, H., Uyeda, S. and Kanamori, H. (1970). *Debate about the earth*. Freeman, Cooper and Co. San Francisco, 281 p.

Talbot, M.R. (1980). Environmental responses to climatic change in the West African Sahel over the past 20,000 years. In: Williams, M.A.J. and Faure, H. (Eds.). *The Sahara and the Nile*. Balkema. Rotterdam, 37–62.

Talbot, M.R. and Williams, M.A.J. (1978). Erosion of fixed dunes in the Sahel, Central Niger. *Earth Surface Processess*, 3, 107–113.

Talbot, M.r., Livingstone, D.A., Plamer, P.G., Malley, J., Melark, J.N., Delibrias, G. and Gulliksen, S. (1984). Preliminary results from sediment cores from Lake Bosumtwi, Ghana. *Palaeocology of Africa*, 16, 173–192.

Tang, A.M., Cui, Y.J., Trinh, V.N., Saerman, Y. and Marchadier, G. (2009). Analisis of the railway heave induced by soil swelling at a site in southern France. *Engineering Geology*, 106, 68–77.

Tanner, W.F. (1961). An alternate approach to morphogenetic climates. *Southeastern Geology*, 2, 251–257.

Tardy, Y. (1993). *Pétrologie des Laterites et des Sols Tropicaux.* Masson. Paris, 459 p.

Tardy, Y. and Roquin, C. (1992). Geochemistry and evolution of lateritic landscapes. In: Martini, I.P. and Chesworth, W. (Eds.). *Weathering, Soils and Paleosols.* Elsevier. Amsterdam, 407–443.

Tardy, Y. and Roquin, C. (1998). *Dérive des Continents- Paléoclimats et Alterations Tropicales*. Ed. BRGM. Orleans, 473 p.

Tarling, D.H. and Tarling, M.P. (1971). *Continental Drift. A study of the Earth`s Moving Surface*. Penguin Books, Harmondsworth, 142 p.

Tator, B.A. (1952). Pediment characteristics and terminology (part I). *Annals of the Association of American Geographers*, 42, 295–317.

Tator, B.A. (1953). Pediment characteristics and terminology (part II). *Annals of the Association of American Geographers,* 43, 47–53.

Taylor, G. and Eggleton, R.A. (2001). *Regolith Geology and Geomorphology*. Wiley. Chichester, 325 p.

Taylor, G.R., Eggleton, R.A., Holzhauer, C.C., Maconachie, L.A., Gordon, M., Brown, M.C. and McQueen, K.G. (1992). Cool climate lateritic and bauxitic weathering. *Journal of Geology*, 100, 669–677.

Taylor, K.C., Lamorey, G.W., Doyle, G.A., Alley, R.B., Grootes, P.M., Mayewskill, P.A., White, J.W.C. and Barlow, L.K. (1993). The 'flickering switch' of late Pleistocene climate change. *Nature*, 361, 432–436.

Taylor, R.K. and Cripps, J.C. (1987). Weathering effects: Slopes in mudrocks and over-consolidated clays. In: Anderson, M.G. and Richards, K.S. (Eds.). *Slope Stability. Geotechnical Engineering and Geomorphology*. Wiley. Chichester, 405–445.

Taylor, R.T. and Scott, M.M. (1985). *The Continental Crust: its Composition and Evolution*. Blackwell. Oxford.

Tazieff, H. (1979). *Nyiragongo: the forbidden volcano*. Casell. London, 287 p.

Tazieff, H. and Derruau, M. (1990). *Le volcanisme et sa prévention. Masson*, París, 256 p.

Tchakerian, V.P. (1994). Palaeoclimatic interpretations from desert dunes and sediments. In: Abrahams, A.D. and Parsons, A.J. (Eds.). *Geomorphology of Desert Environments*.Chapman and Hall. London, 631–643.

Tchakerian, V.P. (Ed.) (1995). *Desert Aeolian Processes*. Chapman and Hall, London.

Tchakerian, V.P. (1999). Arroyo. In: Mares, M.A. (Ed.). *Encyclopedia of Deserts*. University of Oklahoma Press. Noman, p. 41.

Teatini, P., Ferronato, M., Gambolati, G., Bartoni, W. and Gonella, M. (2005). A century of land subsidence in Ravenna (Italy). *Environmental Geology*, 47, 831–846.

Teixell, A., Bertotti, G., Frizon de Lamotte, D. and Charroud, M. (2009). The geology of vertical movements of the lithosphere: An overview. *Tectonophysics*, 475, 1–8.

Telfer, M.W. and Thomas, D.S.G. (2006). Complex Holocene lunette dune development, South Africa: Implications for paleoclimate and models of pan development in arid regions. *Geology*, 34, 853–856.

Teller, J.T. and Clayton, L. (1983). *Glacial Lake Agassiz*. Geological Association of Canada, Special Paper 26. St. Johns. New Foundland, 451 p.

Teller, J.T., Leverington, D.V. and Mann, J.D. (2002). Freshwater outburst to the ocean from the glacial Lake Agassiz and their role in climatic change during the last deglaciation. *Quaternary Science Reviews*, 21, 879–887.

Temple, P.H. and Rapp, A. (1972). Landslides in the Mgeta area, western Uluguru Mountains, Tanzania. *Geografiska Annaler*, 54 A, 157–193.

Terakawa, A. and Matsuo, O. (1996). Earthquakes. In: Singh, V.P. (Ed.). *Hydrology of Disasters*. Kluwer. Dordrecht, 427–434.

Terwindt, J.H.T. (2005). Polders. In: Schwartz, M.L. (Ed.). *Encyclopedia of Coastal Science*. Springer. Dordrecht, 778–780.

Terzaghi, K. (1943). *Theoretical Soil Mechanics.* Wiley, New York, 510 p.

Terzaghi, K. (1950). Mechanism of landslides. In: Paige, S. (Ed.). *Application of Geology to Engineering Practice*. Geological Society of America. New York, 83–123.

Terzaghi, K. and Peck, R.B. (1948). *Soil Mechanics and Engineering Practice*. Wiley. New York, 566 p.

Tharp, T.M. (1995). Mechanisms of upward propagation of cover-collapse sinkholes. *Engineering Geology*, 52, 23–33.

The Rock-Color Chart (1979). The Geological Society of America. Boulder, Colorado. United States.

The Tanna Fault Trenching Research Group (1983). Trenching Study for Tanna Fault at Myoga, Shizuoka Prefecture, Japón. *Bulletin Earthquakes Institute University of Tokyo*, 58, 797–830.

Thomas, C.D., Cameron, A., Green, R.E., Bakkenes, M., Beaumont, L.J., Collingham, Y.C., Erasmus, B.E.M., Martínez, F., Grainger, A., Hannah, L., Hughes, L., Huntley, B., van Jaarsveld, A.S., Midley, G.F., Miles, L., Ortega-Huerta, M.A., Townsend, A., Phillips, O.L. and Williams, S.E. (2004). Extinction risk for climate change. *Nature*, 427, 145–158.

Thomas, D.S. (1988). The biogeomorphology of arid and semi-arid environments. In: Viles, H. (Ed.). *Biogeomorphology*, Blackwell. Oxford, 193–221.

Thomas, D.S.G. (1989a). The nature of arid environments. In: Thomas, D.S.G. (Ed.). *Arid Zone Geomorphology*. Belhaven Press. London, 1–8.

Thomas, D.S.G. (1992). Desert dune activity: concepts and significance. *Journal of Arid Environments*, 22, 31–38.

Thomas, D.S.G. (Ed.) (1989). *Arid Zone Geomorphology*. Belhaven Press. London, 372 p.

Thomas, D.S.G. (Ed.) (1997). *Arid Zone Geomorphology*. Wiley. Chichester, 713 p.

Thomas, D.S.G. (1997a). Arid environments: their nature and extent. In: Thomas, D.S.G. (Ed.). *Arid Zone Geomorphology. Process, Form and Change in Drylands.* Wiley. Chichester, 3–12.

Thomas, D.S.G. (1997b). Aeolian sand deposits. In: Thomas, D.S.G. (Ed.). *Arid Zone Geomorphology*, Wiley. Chichester, 372–412.

Thomas, D.S.G. (1997c). Reconstructing ancient arid environments. In: Thomas, D.S.G. (Ed.). *Arid Zone Geomorphology*, Wiley. Chichester, 577–605.

Thomas, D.S.G. (2004). Desertification. In: Goudie, A.S. (Ed.). *Encyclopedia of Geomorphology*. Routledge. London, 254–257.

Thomas, D.S.G. and Goudie, A.S. (1984). Ancient ergs of the Southern Hemisphere. In: Vogel, J.C. (Ed.). *Late Cenozoic Palaeoclimates of the Southern Hemisphere*. Balkema. Rotterdam, 407–418.

Thomas, D.S.G. and Goudie, A.S. (2000). *The Dictionary of Physical Geography*. Third Edition. Blackwell. Oxford, 610 p.

Thomas, D.S.G. and Middleton, N.J. (1994). *Desertification: Exploding the Myth.* Wiley. Chichester, 194 p.

Thomas, D.S.G. and Shaw, P.A. (1991). *The Kalahari Environment.* Cambridge University Press. Cambridge, 284 p.

Thomas, M.F. (1965). Some aspects of the geomorphology of domes and tors in Nigeria. *Zeitschrift für Geomorphologie*, 9, 63–81.

Thomas, M.F. (1966). Some geomorphological implications of deep weathering patterns in crystalline rocks in Nigeria. *Transactions of the Institute of British Geographers*, 40, 173–193.

Thomas, M.F. (1974). *Tropical Geomorphology. A Study of Weathering and Landform Development in Warm Climates*. Macmillan. London, 329 p.

Thomas, M.F. (1976). Criteria for the recognition of climatically induced variations in granite landforms. In: Derbyshire, E. (Ed.). *Geomorphology and Climate*. Wiley. London, 411–445.

Thomas, M.F. (1978). The study of inselbergs. *Zeitschrift für Geomorphologie Supplementband,* 31, 1–41.

Thomas, M.F. (1983). Contemporary denudation systems and the effects of climatic change in the humid tropics-some problems from Sierra Leone. In: Briggs, D.J. and Waters, R.S. (Eds.). *Studies in Quaternary Geomorphology.* GeoBooks. Norwich, 195–214.

Thomas, M.F. (1986). Savanna. In: Fookes, P.G. and Vaughan, P.R. (Eds.). *A Handbook of Engineering Geology*. Surrey University Press. Glasgow, 125–136.

Thomas, M.F. (1989a). The role of etch processes in landform development. I. Etching concepts and their application. *Zeitscrift für Geomorphologie*, 33, 129–142.

Thomas, M.F. (1989b). The role of etch processes in landform development. II. Etching and the formation of relief. *Zeitscrift für Geomorphologie*, 33, 157–167.

Thomas, M.F. (1994a). *Geomorphology in the Tropics. A study of Weathering and Denudation in Low Latitudes*. Wiley. Chichester, 460 p.

Thomas, M.F. (1994b). Ages and geomorphic relationships of saprolite mantles. In: Robinson, D.A. and Williams, R.B.G. (Eds.). *Rock Weathering and Landform Evolution*. Wiley. Chichester, 287–301.

Thomas, M.F. (2001). Landscape sensivity in time and space-an introduction. *Catena*, 42, 83–98.

Thomas, M.F. (2004). Landscape sensitivity to rapid environmental change- a Quaternary perspective with examples from tropical areas. *Catena*, 55, 107–124.

Thomas, M.F. (2005). Savanna. In: Fookes, P.G., Lee, E.M. and Milligan, G. (Eds.). *Geomorphology for Engineers*. Whittles Publ., Caithnees, United Kingdom, 454–472.

Thomas, M.F. (2008). Understanding the impacts of Late Quaternary climate change in tropical and subtropical regions. *Geomorphology*, 101, 146–158.

Thomas, M.F. and Goudie, A.S. (Eds.) (1985). Dambos: small channells valleys in the tropics. Characteristics, formation, utilisation. *Zeitschrift für Geomorphologie Supplementband*, 52, 228 p.

Thomas, M.F. and Thorp, M.B. (1985). Environmental change and episodic explanation in the humid tropics of Sierra Leona: the Koidn etchplain. In: Douglas, I. and Spencer, T. (Eds.). *Environmental Change and Tropical Geomorphology*. Allen and Unwin. London, 239–267.

Thomas, M.F. and Thorp, M.B. (1992). Landscape dynamics and surface deposits arising from late Quaternary fluctuations in the forest-savanna boundary. In: Furley, P.A., Proctor, J. and Ratter, J.A. (Eds.). *Nature and Dynamics of Forest-Savanna Boundaries*. Chapman & Hall. London, 215–253.

Thomas, M.F. and Thorp, M.B. (1993). The geomorphology of some Quaternary placer deposits. *Zeitschrift fúr Geomorphologie Supplementband*, 87, 183–194.

Thomas, M.F. and Thorp, M.B. (1995). Geomorphic response to rapid climatic and hydrologic change during the Late Pleistocene and Early Holecene in the humid and sub-humid tropics. *Quaternary Science Reviews*, 14, 193–207.

Thomas, M.F. and Thorp, M.B. (2003). Palaeohydrological reconstructions for tropical Africa since the last Glacial Maximum-evidence and problems. In: Gregory, K.J. and Benito, G. (Eds.). *Palaeohydrology. A Contribution to Global Change*. Wiley. Chichester, 167–192.

Thompson, D.M. (2005). The history of the use and effectiveness of instream structures in the United States. Geological Society of America. *Reviews in Engineering Geology*, 16, 35–50.

Thomspson, R.S., Whitlock, C., Bartlein, P.J., Harrison, S.P. and Spaulding, W.G. (1993). Climatic changes in the western United States since 18,000 yr. BP. In: Wright, H.E.., Kutzbach, J.E., Webb, T. Ruddiman, W.F., Street-Perrot, F.A. and Bartlein, P. (Eds.). *Global climates since the last glacial maximum*. University of Minnesota Press. Minneapolis, 468–513.

Thompson, S.C., Clague, J.J. and Evans, S.G. (1997). Holocene activity of the Mt. Currie scarp, Coast Mountain, Bristish Comlumbia, and implications for its origin. *Environmental Engineering Geoscience*, 3, 329–348.

Thoraninsson, S. (1953a). Some new aspects of the Grimsvöth problem. *Journal of Glaciology*, 2, 267–274.

Thoraninsson, S. (1953b). The crater groups of Iceland. *Bulletin Volcanologie*, 14, 3–44.

Thorbecke, F. (1927). Morphologie der klimazonen. *Düsseldorfer Geographische Vorträge und Erörterungen*, Part 3, 1–100.

Thorn, C. (1988a). Nivation: a geomorphic chimera. In: Clark, M.J. (Ed.). *Advances in Periglacial Geomorphology*. Wiley. Chichester, 3–31.

Thorn, C.E. (1988b). *Introduction to Theoretical Geomorphology*. Unwin Hyman. Boston, 247 p.

Thorn, C.E. (1992). Periglacial geomorphology: what, where, when?. In: Dixon, J.C. and Abrahams, A.D. (Eds.). *Periglacial Geomorphology*. Wiley. Chichester, 1–30.

Thorn, C.E. (2004). Nivation. In: Goudie, A.S. (Ed.). *Encyclopedia of Geomorphology*. Routledge. London, 718–721.

Thorn, C.E. and Hall, K. (1980). Nivation: an arctic-alpine comparison and reappraisal. *Journal of Glaciology*, 25, 109–124.

Thorn, C.E. and Hall, K. (2002). Nivation and cryoplanation: the case for scruting and integration. *Progress in Physical Geography*, 26, 533–550.

Thorn, C.E., Darmody, R.G., Dixon, J.C. and Schlyter, P. (2002). Weathering rates of buried machine-polished rock diks, Karkevagge, Swedish Lapand. *Earth Surface Processes and Landforms*, 27, 831–845.

Thornbury, W.D. (1954). *Principles of Geomorphology*. Wiley. New York, 618 p.

Thorndycraft, V.R. and Benito, G. (2006). The Holocene fluvial chronology of Spain: evidence from a newly compiled radiocarbon database. *Quaternary Science Reviews*, 25, 234.

Thorndycraft, V.R., Benito, G., Llasat, C. and Barriendos, M. (2003). Palaeofloods, Historical Data and Climatic Variability: Applications in Flood Risk Assessement. In: Thorndycraft, V.R., Benito, G., Barriendos, M. and Llasat, C. (Eds.). *Palaeofloods, Historical Floods and Climatic Variability: Applications in Flood Risk Assesement*. Consejo Superior de Investigaciones Científicas. Madrid, 1–9.

Thornes, J. (1976). *Semi-arid Erosional Systems: Case Studies from Spain*. The London School of Economics and Political Sciences. London, 79 p.

Thornes, J. (1994). Catchment and channel hydrology. In: Abrahams, A.P. and Parsons, A.J. (Eds.). *Geomorphology of Desert Environments*. Chapman and Hall. London, 257–287.

Thornes, J. and Brunsden, D. (1977). *Geomorphology and Time*. Methuen. London, 208 p.

Thornthwaite, C.W. (1948). An approach towards a rational classification of climate. *Geographical Review*, 38, 55–94.

Thornton, E.B., Macmahan, J. and Sallenger, A.H. (2007). Rip currents, mega-cusps, and eroding dunes. *Marine Geology*, 240, 151–167.

Thouret, J.C. (2004). Geomorphic processes and hazards of volcanic mountains. In: Owens, P.N. and Slaymaker, O. (eds.). *Mountain Geomorphology*. Arnold. London, 242–273.

Tianchi, L.C. (1989). Landslides: extent and economic significance in China. In: Brabb, E.E. and Harrod, B.L. (Eds.). *Landslides: Extent and Economic Significance*. Balkema. Rotterdam, 271, 288.

Tilling, R. (1993). Volcanic hazards and their mitigation: progress and problems. *Reviews of Geophysics*, 27.

Tilling, R. (2005). Volcano hazards. In: Martí, J. and Ernst, G.J. (Eds.). *Volcanoes and the Environment*. Cambridge University Press. Cambridge, 55–89.

Timms, B.V. (1992). *Lake Geomorphology*. Gleneagles Publishing. Adelaide, 180 p.

Tinkler, K.J. (1985). *A Short History of Geomorphology*. Barnes and Noble. Tokowa. New Jersey, 317 p.

Tooth, S. and Nanson, G.C. (1999). Anabranching rivers of the Northern Plains of arid central Australia. *Geomorphology*, 29, 211–233.

Törnqvist, B.E. and Bridge, J.S. (2002). Testing a 3-D process-based alluvial architecture model with new field data: the control of spatial variation in floodplain deposition rate on avulsion frequency. *Sedimentology*, 49, 891–905.

Torraca, G. (1981). *Porous Building Materials: Materials Science for Architectural Conservation*. International Centre for the Study of Preservation and Restoration of Cultural Property. Rome, 141 p.

Torri, D., Sfalanga, M. and Del Sette, M. (1987). Splash detachment: runoff depth and soil cohesion. *Catena*, 14, 149–155.

Toy. T.J. (1977). Hillslope form and climate. *Bulletin of Geological Society of America*, 88, 16–22.

Trendall, A.F. (1962). The formation of "apparent peneplains" by a process of combined lateritisation and surface wash. *Zeitschrift für Geomorphologie*, 6, 183–197.

Trenhaile, A.S. (1987). *The Geomorphology of Rock Coasts*. Clarendon Press. Oxford, 384 p.

Tricart, J. (1952a). *Cours de Géomorphologie. Géomorphologie Climatique*. C.D.U.. Paris, 193 p.

Tricart, J. (1952b). La géomorphologie et le notion d'échelle. *Revue de Géomorphologie Dynamique*, 3, 213–218.

Tricart, J. (1956a). Etude expérimentale du problème de la gélivation. *Biuletyn Peryglacialny*, 4, 285–318.

Tricart, J. (1956b). Geomorphologic dynamique du Delta du Senegal. *Revue de Géomorphologie Dynamique*, 7, 65–86.

Tricart, J. (1956c). La Géomorphologie et la pensée marxiste. *La Pensée*, 69, 55–76.

Tricart, J. (1957). L'evolution de versants. *L'Information Géographique*, 21, 108–115.

Tricart, J. (1962). *L'épiderme de la terre*. Masson. Paris, 167 p.

Tricart, J. (1965). *Principes et Méthodes de la Geomorphologie*. Masson, París, 496 p.

Tricart, J. (1967). *Le Modelé des Régions Périglaciaires*. SEDES. Paris, 512 p.

Tricart, J. (1968). *Precis de Géomorphologie. I. Geomorphologie Structurale*. Societé d'Edition d'Enseignement Supérieur. Paris, 322 p.

Tricart, J. (1969). *Le Modelé des Régions Séches*. SEDES. Paris, 472 p.

Tricart, J. (1972). Normes pour l'etablissement de la carte geomorphologique détaillée de la France. *Memoires et Documents*. Centre Nationale de la Reserche Scientiphique. Paris, 37–105 (Trad. esp. D. Serrat. Instituto Jaime Almera. Barcelona, 1976).

Tricart, J. (1974a). *Le Modelé des Régions Chaudes: Forêts et Savanes*. SEDES. Paris, 345 p.

Tricart, J. (1974b). Existence de périodes sèches au Quaternaire en Amazonie et dans le régions voisines. *Revue de Géomorphologic Dynamique*, 23, 145–158.

Tricart, J. (1975). Influence des oscillations climatiques récents sur le modelé en Amazonie Orientale (région de Santarém) d'aprés des images de radar latéral. *Zeitschrift für Geomorphologie*, 19, 140–163.

Tricart, J. (1977). El concepto de "pluvial". *Actas de la III Reunión Nacional del Grupo Español de Trabajo del Cuaternario*. Zaragoza, 7–20.

Tricart, J. (1978). *Géomorphologie aplicable*. Masson. Paris, 204 p.

Tricart, J. (1979). Mapas geomorfológicas, ¿por qué?.*Acta Geol. Hispánica*, 14, 416–420.

Tricart, J. (1982). El Pantanal: Un ejemplo del impacto geomorfológico sobre el ambiente. *Informaciones Geográficas*, 29, 81–97. Universidad de Chile. Santiago.

Tricart, J. (1985). Evidence of Upper Pleistocene dry climates in northern South America. In: Douglas, I. and Spencer, T. (Eds.). *Environmental Change and Tropical Geomorphology*, Allen and Unwin. London, 197–217.

Tricart, J. and Alfonsi, P. (1981). Action éoliennes récents aux abords du delta de l'Orénoque. Bulletin AGG, 476, 75–82.

Tricart, J. and Cailleux, A. (1955). *Introduction a la Géomorphologie Climatique*. Centre de Documentation Universitaire. Paris, 228 p.

Tricart, J. and Cailleux, A. (1962). *Le Modelé Glaciaire et Nival*. SEDES. Paris, 508 p.

Tricart, J. and Cailleux, A. (1965). *Introduction a la Géomorphologie Climatique*. SEDES. Paris, 306 p.

Tricart, J. and Demek, J. (1972). Concept and content of applied and special detailed geomorphological maps. En: Demek, J. (Ed.). *Manual of Detailed Geomorphological Mapping*. Akademia. Praga, 261–267.

Tricart, J., Raynal, R. and Besancon, J. (1972). Cônes rocheux, pediments, glacis. *Annales de Géographie*, 43, 1–24.

Troeh, F.R. (1965). Landform equations fitted to topographic maps. *American Journal of Science*, 263, 616–627.

Troll, C. (1944). Strukturböden, solifluktion und Frosklimate der Erde. *Geologische Rundschau*, 34, 545–694.

Trombe, F. (1952). *Traité de Spéléologie*. Payot. Paris, 376 p.

Trombe, F. (1974). *La Spéléologie*. Presses Universitaires de France. Paris, 124 p.

Trudgill, S.T. (1976a). Rock weathering in climate: quantitative and experimental aspects. In: Derbyshire, E. (Ed.). *Geomorphology and Climate*, Wiley. London, 59–99.

Trudgill, S.T. (1976b). The subaerial and subsoil erosion of limestones on Aldabra Atoll, Indian Ocean. *Zeitschrif für Geomorphology, Supplementband* 26, 201–210.

Trudgill, S.T. (1985). *Limestone Geomorphology*. Longman. London, 196 p.

Trudgill, S.T., High, C.J. and Hanna, F.K. (1981). Improvements of the micro-erosion meter. *British Geomorphological Research Group, Technical Bulletin*, 29, 3–17.

Tseo, G. (1993). Two types of longitudinal dune fields and possible mechanisms for their development. *Earth Surface Processes and Landforms,* 18, 627–643.

Tsoar, H. (1983a). Dynamic processes acting on a longitudinal (seif) dune. *Sedimentology*, 30, 567–578.

Tsoar, H. (1983b). Wind tunnel modelling of echo and climbing dunes. In: Brookfield, M.E. and Ahlbrandt, T.S. (Eds.). *Eolian Sediments and Processes*, Elsevier. Amsterdam, 247–259.

Tsoar, H. (1989). Linear dune-forms and formation. *Progress of Physical Geography*, 13, 507–528.

Tsoar, H. and Moller, J.T. (1986). The role of vegetation in the formation of linear dunes. In: Nickling, W.G. (Ed.). *Aeolian Geomorphology.* Allen and Unwin. Boston, 75–95.

Tsoar, H. and Pye, K. (1987). Dust transport and the question of desert loess formation. *Sedimentology*, 34, 139–154.

Tsoar, H., Blumberg, D.G. and Stoler, Y. (2004). Elongation and migration of sand dunes. *Geomorphology*, 57, 293–302.

Tsukamoto, Y. and Kusakabe D. (1984). Vegetative influences on debris slide occurrences on steep slopes in Japan. *Proceedings of Symposium on Effects Forest Land use on Erosion and Slope Stability*. Environment and Policy Institute. Honolulu, Hawaii.

Tsuya, H. and Morimoto, R. (1963). Types of volcanic eruptions in Japan. *Bulletin Volcanologie*, 26, 209–222.

Tucker, C.J., Dregne, H.E. and Newcomb, W.W. (1991). Expansion and contraction of the Sahara Desert form 1980 to 1990. *Science*, 253, 299–301.

Tucker, M.E. (1978). Gypsum crust (gypcrete) and patterned ground from northern Iraq. *Zeitschrift für Geomorphologie*, 22, 89–100.

Tufnell, L. (1972). Ploughing blocks with special reference to north-west England. *Biuletyn Peryglacjalny*, 21, 237–270.

Tufnell, L. (1984). *Glacier Hazards*. Longman. London, 97 p.

Turcotte, D.L. and Schubert, F. (2002). *Geodinamics*. Cambridge University Press, Cambridge, 456 p.

Turk, R.G.Jr. (1996). Permafrost. In: Turner, H.K. and Schuster, R.L. (Eds.). *Landslides. Investigation and Mitigation*. National Academic Press. Washington, 620–645.

Turkington, A.V., Philips, J.O. and Campbell, S.W. (2005). Weathering and landscape evolution. *Geomorphology*, 67, 1–6.

Turner, A.K. and Jayaprakash, G.P. (1996). Introduction. In: Turner, A.K. and Schuster, R.L. (Eds.). *Landslides: Investigation and Mitigation*. Special Report 247. National Academy Press. Washington, 3–11.

Twidale, C.R. (1962). Steepened margins of inselbergs from north-western Eyre Peninsula, South Australia. *Zeitschrift für Geomorphologie*, 6, 51–69.

Twidale, C.R. (1964). A contribution to the general theory of domed inselbergs. *Transactions of the Institute of British Geographers*, 34, 91–113.

Twidale, C.R. (1967). Hillslopes and pediments in the Flinders Ranges, South Australia. In: Jennings, J.N. and Mabbutt, J.A. (Eds.). *Landform Studies from Australia and New Guinea*. Cambridge University Press. Cambridge, 95–117.

Twidale, C.R. (1968). Inselberg. In: Fairbridge, R.W. (Ed.). *Encyclopedia of Geomorphology*. Reinhold. New York, 556–559.

Twidale, C.R. (1971). *Structural Landforms.* The MIT Press, Cambridge, 247 p.

Twidale, C.R. (1973). On the origin of the sheet jointing. *Rock Mechanics*, 5, 163–187.

Twidale, C.R. (1976a). *Analysis of Landforms*. Wiley. Sydney, 572 p.

Twidale, C.R. (1976b). On the survival of palaeoforms. *American Journal of Science*, 276, 77–94.

Twidale, C.R. (1978). On the origin of Ayers Rock, Central Australia. *Zeitschrift für Geomorphologie, Supplement Band*, 31, 177–206.

Twidale, C.R. (1981). Granite Inselbergs. *Geographical Journal*, 147, 54–71.

Twidale, C.R. (1982a). The evolution of bornhardts. *American Scientist*, 70, 268–76.

Twidale, C.R. (1982b). *Granite Landforms*. Elsevier. Amsterdam, 372p.

Twidale, C.R. (1983). Pediments, peneplains and ultiplains. *Revue de Géomorphologie Dynamique,* 32, 1–35.

Twidale, C.R. (1990). The origin and implications of some erosional landforms. *Journal of Geology*, 98, 343–364.

Twidale, C.R. (2003). "Canons" revisited and reviewed: Lester King's views of landscape evolution considered 50 years later. *Geological Society of America Bulletin*, 115, 1155–1172.

Twidale, C.R. (2004). Karst Towers: some questions and answers. *Studia Geologica Salmanticensia*, 40, 57–84.

Twidale, C.R. (2006). Notes on the origin and significance of stone layers. *Cadernos Laboratorio Xeolóxico de Laxe*, 31, 127–141.

Twidale, C.R. (2007). *Ancient Australian Landscapes*. Rossenberg. Dural Delivery, 144 p.

Twidale, C.R. and Bourne, J.A. (1975). Episodic exposure of inselbergs. *Bulletin of the Geological Society of America*, 86, 1473–1481.

Twidale, C.R. and Bourne, J.A. (1978). Bornhardts. *Zeitschrift für Geomorphologie Supplementband,* 31, 111–137.

Twidale, C.R. and Bourne, J.A. (2008). Caves in granitic rocks: types, terminology and origins. *Cadernos de Laboratorio Xeolóxico de Laxe*, 33, 35–57.

Twidale, C.R. and Centeno, J.D. (1993). Landform development at the Ciudad Encantada, Cuenca, Spain. *Cuadernos de Laboratorio Xeolóxico de Laxe*, 18, 257–269.

Twidale, C.R. and Corbin, E.M. (1963). Gnammas. *Revue de Géomorphologie Dynamique*, 14, 1–20.

Twidale, C.R. and Lageat, Y. (1994). Climatic Geomorphology: a critique. *Progress in Physical Geography*, 18, 319–334.

Twidale, C.R. and Vidal Romaní, J.R. (2005). *Landforms and Geology of Granite Terrains.* Balkema. Leiden, 351 p.

Twidale, C.R., Bourne, J.A. and Vidal Romaní, J.R. (2005). Beach etching and shore platforms. *Geomorphology*, 67, 47–61.

U.S.G.S. (1983). Permafrost. In: Tank, R.W. (Ed.). *Environmental Geology*. Oxford University Press. Oxford, 163–170.

Ugolini, F.C. (1986). Processes and rates of weathering in cold and polar desert environments. In: Colman, S.M. and Dethier, D.P. (Eds.). *Rates of Chemical Weathering of Rocks and Minerals*. Academic Press. Orlando, 193–235.

UNCED (1992). *Conferencia Medio Ambiente y Desarrollo*. U.N.C.E.D. Brasil-92. Ministerio de Obras Públicas y Transportes. Madrid, 101 p.

UNDRO (1991). *Mitigating natural disasters: Phenomena, Effects and Actions-A Manual for Policy Makers and Planners*. Office of the UN Relief Coordinator. Geneva.

UNEP (1991). *Environmental Data Report, 1991/1992*. United Nations Environments. Programm. Blackwell. Oxford, 363 p.

UNEP (1992). *World Atlas of Desertification*. Arnold. London, 69 p.

UNEP (2001). *United Nations Environment Programme: Introduction to Climate Change*.

UNESCO (1979). *Map of the World Distribution of Arid Regions*. MAB Technical Note 7.

UNESCO (1981). *Avalanche Atlas*. UNESCO. Paris, 265 p.

UNESCO-FAO (1977). Desertification: an overview. *Desertification. Its Causes and Consequences*. Pergamon Press. Oxford, 1–61.

UNFCC (1997). United Nations Framework Convention on Climate Change. Report of the Ad-hoc Group of the Berlin. Mandate on the Work of its Sixth Session. Bonn, March 1997.

USGS (1974). *Earthquake Information Bulletin*, 6, 14.

USGS (1983). Permafrost. In: Tank, R.W. (Ed.). *Environmental Geology*. Oxford University Press. Oxford, 163–170.

U.S. Water Resources Council (1978). *The Nation's Water Resources- The Second National Water Assesment*. Vol. A-4, Streamflow Conditions. United States Water Resource Council. Washington DC, 255 p.

Uriarte, A. (2003). *Historia del Clima de la Tierra*. Servicio Central de Publicaciones del Gobierno Vasco. San Sebastián, 306 p.

Valcárcel, M. and Pérez Alberti, A. (2002). Los campos de bloques en montañas del noroeste de la Península Ibérica: génesis y significado paleoambiental. In: Serrano, E. and García, A. (Eds.). *Periglaciarismo en Montaña y Altas Latitudes*. Departamento de Geografía. Universidad de Valladolid, 13–26.

Valdemoro, H.I., Jiménez, J.A., Alonso, I. Lorente, P. and Rodríguez-Herrerías, M. (2007). Coastal dune field evolution in conditions of limited sediment availability. *Coastal Sediments'07, ASCE*. New Orleans; May, 2007.

Valentin, H. (1952). *Die Küsten der Erde*. Pertemans Geographische. Mitteilungen. Gotha J. Perthes, 118 p.

Valentine, F.K. and Fisher, R.V. (2000). Pyroclastic surges and blasts. In: Sigurdsson, H. (Ed.): *Encyclopedia of Volcanoes*. Academic Press. San Diego, 571–580.

Valero-Garcés, B., Zeroual, E. and Kelts, K. (1998). Arid phases in the western Mediterranean region during the Last Glacial Cycle reconstructed from lacustrine records. In: Benito, G., Baker, V. and Gregory, K.J. (Eds.). *Paleohydrology and Environmental Change*. Wiley. Chichester, 67–80.

Valero-Garcés, B., Navas, A., Mata, P., Delgado-Huertas, A., Peña Moné, J., Machín, J., González Sampériz, P., Schwald, A., Ariztegui, D., Schellmann, M., Bao, R. and González-Barrios, A. (2003). Sedimentary facies analyses in lacustrine cores: from inicial core descriptions to detailed paleoenvironmental reconstructions. A case study from Zoñar lake (Cordoba Province, Spain). In: Valero-Garcés, B. (Ed.). *Limnogeology in Spain: A tribute to Kerry Kelts*. Biblioteca de Ciencias del Consejo Superior de Investigaciones Científicas. Madrid, vol. 3, 385–414.

Valeton, I. (1972). *Bauxites*. Elsevier. Amsterdam, 226 p.

Valeton, I. (1983). Palaeoenvironment of lateritic bauxites with vertical and lateral differentiation. In: Wilson, R.C.L. (Ed.). *Residual Deposits: Surface Related Weathering Processes and Materials*. Geological Society of London, Special Publication 11, 77–90.

Valeton, I. (1994). Element concentration and formation of ore deposits by weathering. *Catena*, 21, 99–129.

Valeton, I. (1999). Saprolite-bauxite facies of ferralitic duricrust on paleosurfaces off former Pangea. *Special Publication, International Association of Sedimentologists*, 27, 153–158.

Van Burkolow, A. (1945). Angle of repose and angle of sliding friction: an experimental study. *Geological Society American Bulletin*, 56, 669–707.

Van der Wateren, D.F.M. (1987). Structural geology and sedimentology of the Dammer Berge push moraine, FRG. In: Van der Meer, J.J.M. (Ed.). *Tills and Glaciotectonics*, Balkema. Rotterdam, 157–182.

Van der Wateren, D.F.M. (1992). *Structural Geology and Sedimentology of Push Moraines. Processes of Soft Sediment Deformation in a Glacial Environment and the Distribution of Glaciotectonic Styles*. Dissertation. Amsterdam, 230 p.

Van der Wateren, F.M. (1995). Processes of glaciotectonism. In: Menzies, J. (Ed.). *Modern Glacial Environments. Processes, Dynamics and Sediments*. Butterworth-Heinemann. Oxford, 307–335.

Van Zuidam, R.A. (1975). Geomorphology and Archaeology-Evidences of interrelation at historical sites in the Zaragoza region, Spain. *Zeitschrift für Geomorphologie N.F.*, 19, 319–328.

Van Zuidam, R.A. (1976a). *Geomorphological Development of the Zaragoza Region, Spain*. I.T.C. Enschede, 221 p.

Van Zuidam, R.A. (1976b). Periglacial-like feature in the Zaragoza region, Spain. *Zeitschrift für Geomorphologie,* 20, 227–234.

Van Zuidam, R.A. (1982). Considerations on systematic medium-scale geomorphological mapping. *Zeitschrift für Geomorphologie*, N.F., 26, 473–480.

Vanderberghe, J. (1993). Changing fluvial processes under changing periglacial conditions. *Zeitschrift für Geomorphologie. Supplement Band*, 88, 17–28.

Vanderberghe, J. (1995). Timescales, climate and river development. *Quaternary Science Reviews*, 14, 631–638.

Vanderberghe, J. (2001). A typology of Pleistocene cold-based rivers. *Quaternary International*, 79, 111–121.

Vanderberghe, J. (2002). "Climate geomorphology under fire". *Terra Nostra*, 2002/6, 382–385.

Vanderberghe, J. (2007). Periglacial Landform. Cryoturbation Structures. In: Elias, S.A. (Ed.). *Encyclopedia of Quaternary Science*. Elsevier. Amsterdam, 2147–2158.

Vanderberghe, J. (2008). The fluvial cycle at cold-warm-cold transitions in lowland regions. A refinement of theory. *Geomorphology*, 38, 275–284.

Vanderberghe, J. and Pissart, A. (1993). Permafrost change in Europe during the last glacial. *Permafrost and Periglacial Processes*, 4, 121–135.

Vanderberghe, J. and Thorn, C.E. (2002). Progress in periglacial research. *Progress in Physical Geography*, 26, 479–506.

Vanderberghe, J. and Woo, M. (2002). Modern and ancient periglacial river types. *Progress in Physical Geography*, 26, 533–550.

Vaniman, D.T., Chipera, S.J. and Bish, D.L. (1994). Pedogenesis of siliceous calcretes at Yucca Mountains, Nevada. *Geoderma*, 63, 1–17.

Vaqueiro, M., Barreiro, B., Costas, R., Suárez, R. and Groba, X. (2006). Relation between structure and morphology in the development of the granite cave of "O Folón" (Vigo, Galicia, Spain). *Cadernos Laboratorio Xeolóxico de Laxe*, 31, 87–103.

Varnes, D.J. (1958). Landslide types and processes. In: Eckel, E.B. (Ed.). Landslides and Engineering Practice. *Highway Research Board Special Report* 29. NAS-NCR Publication 544, 20–47.

Varnes, D.J. (1978). *Slope movement: types and processes*. In: Schuster, R.L. and Krizek, R.J. (Eds.). Landslides: Analysis and Control. *Transportation Research Board Special Report* 176. National Academy of Sciences. Washington, 11–33.

Varnes, D.J. (1984). *Landslide, Hazard Zonation: A Review of Principles and Practice*. UNESCO. Paris.

Vasconcelos, P.M., Becker, T.A., Renne, P.R. and Brimhall, G.H. (1992). Age and duration of weathering by $^{40}K/^{40}Ar$

y $^{40}Ar/^{39}Ar$ analysis of potassium manganese oxides. *Science*, 258, 451–455.

Veder, C. (1981). *Landslides and Their Stabilization*. Springer-Verlag. New York, 247 p.

Vegas, J., Galán, L., Pérez-González, A. and García-Cortés, A. (2004). El archivo lacustre del mar de Fuentillejo (Campo de Calatrava). Primeros protocolos científicos y estrategia de trabajo para el estudio paleoclimático del Cuaternario en el centro de España. *Boletín Geológico y Minero*, 115, 641–650.

Verger, F. (1964). Mottereaux et gilgais. *Annales de Géographie,* 73, 413–430.

Verrechia, E.P., Freytet, P., Verrechia, K.E. and Dumon, J.L. (1995). Spherulites in calcrete laminar crust: biogenic $CaCO_3$ precipitation as a major contributor to crust formation. *Journal of Sedimentary Research*, 65, 690–700.

Verstappen, H.Th. (1964). Karst morphology of the Star Mountains (Central New Guinea) and its relation to lithology and climate. *Zeitschrift für Geomorphologie*, 8, 40–49.

Verstappen, H.Th. (1968). On the origin of longitudinal (seif) dunes. *Zeitschrift für Geomorphologie NF*, 12, 200–220.

Verstappen, H. Th. (1970a). Introduction to ITC systems of geomorphological survey. KNAG Geografrisch Tijdschrift, 4, 85–90.

Verstappen, H.Th. (1970b). Aeolian geomorphology of the Thar Desert and palaeo-climates. *Zeitschrift für Geomorphologie Supplementband,* 10, 104–120.

Verstappen, H.Th (1977). *Remotre Sensing in Geomorphology*. Elsevier. Amsterdam, 214 p.

Verstappen, H.Th (1983). *Applied Geomorphology. Geomorphological Surveys for Environmental Developments*. Elsevier. Amsterdam, 437 p.

Verstappen, H.Th. and Van Zuidam, R. (1968). ITC System of Geomorphological Survey ITC. *Textbook of Photointerpretation*, VII-2. 52 p.

Verstappen, H.Th. and Van Zuidam, R. (1991). *The ITC System of Geomorphologic Survey*. ITC Publ. 10. Enschede, 89 p.

Veste, M. (1995). Structures of geomorphological and ecological units and ecosystems processes in the linear dune system near Nizzana/Negev. *Bielefelder Ökologische Beitrage*, 8, 85–96.

Vicente, G. de, Crone, A.J., Berryman, K.R., Wenowsky, S.G. and Villamor, P. (2001). Catálogo sísmico y registro geológico: El necesario encuentro de dos enfoques del problema. In: Capote, R. and Martínez-Díaz, J.J. (Eds.). *El Riesgo Sísmico. Prevención y Seguro*. Consorcio de Compensación de Seguros. Madrid, 89–101.

Vicente, M.A., Sánchez, M., Lombardía, F., Vicente, I., Saavedra, J., Brufau, A., Rodríguez, P. and Nieves, V. (1984). In: *Estudio sobre las Alteraciones y Tratamiento de la Piedra de Villamayor*. Ediciones Caja de Ahorros y M.P. de Salamanca. Salamanca, 421–475.

Vidal Romaní, J.R. (1984). A orixe das Rías Galegas. Estado da cuestión. *Cuadernos da Area de Ciencias Mariñas.* Seminario do Estudos Galegos, 1, 13–25.

Vidal Romaní, J.R. (1991). Kind of plane fabric and their relation to the generation of granitic forms. *Caderno do Laboratorio Xeolóxico de Laxe*, 16, 301–312.

Vidal Romaní, J.R. and Gracia Prieto, F.J. (1987). Formación de cavidades en rocas graníticas bajo condiciones no epigénicas. *Cuadernos do Laboratorio Xeológico de Laxe*, 12, 47–57.

Vidal Romaní, J.R. and Twidale, C.R. (1998). *Formas y Paisajes Graníticos.* Universidade da Coruña. Servicio de Publicacións. A Coruña, 411 p.

Vidal Romaní, J.R., Twidale, C.R., Campbell, E.M. and Centeno, J.D. (1995). Pruebas morfológicas y estructurales sobre el origen de las fracturas por descamación. *Cadernos Laboratorio Xeolóxico de Laxe*, 20, 307–346.

Vidal Romaní, J.R., Twidale, C.R. and Bourne, J.A. (2003). Siliceous cylindrical speleothems in granitoids in warm semiarid and humid climates. *Zeitschrift für Geomorphologie*, 47, 417–437.

Viers, F. (1974). *Eléments de Géomorphologie*. Fernand Nathan, Paris. (Trad. Esp. L. Casassas, Oikos-Tau, Barcelona, 320 p.).

Vilaplana, J.M. (1983). *Estudi del Glaciarisme Quaternary de les Altes Valls de la Ribagorça*. Tesi Doctoral. Dpto. de Geomorfología i Tectónica. Universidad de Barcelona, 322 p (Unpublished).

Vilaplana, J.M. (1987). *Guia dels Paisatges Granítics dels Països Catalans*. Kapel S.A. Barcelona., 182 p.

Vilaplana, J.M. (1989). Los aludes: un riesgo mal conocido. *Encuentro Internacional "Catástrofes y Sociedad"*. Fundación MAPFRE. Madrid, 201–216.

Vilas, F., Bernabeu, A.M. and Méndez, G. (2005). Sediment distribution pattern in the Rias Baixas (NW Spain): main facies and hydrodynamic dependence. *Journal Marine Systems*, 54, 216–276.

Viles, H.A. (1984). Progress report: Biokarst. *Progress in Physical Geography*, 8, 523–543.

Viles, H.A. (1988). Introduction. In: Viles, H.A. (Ed.). *Biogeomorphology*. Blackwell. Oxford, 1–8.

Viles, H.A. (1995). Ecological perspectives on rock surface weathering: Towards a conceptual model. *Geomorphology*, 13, 21–35.

Viles, H.A. (2001). Scale issues in weathering studies. *Geomorphology*, 41, 63–71.

Viles, H.A. (2004). Biokarstificación. In: Gunn, J. (Ed.). Encyclopedia of Caves and Karst Science. Fiztroy Dearborn. New York, 147–148.

Viles, H.A. (2005). Microclimate and weathering in the Central Namib Desert, Namibia. *Geomorphology*, 67, 189–209.

Viles, H.A. and Goudie, A.S. (2007). Rapid saltweathering in the coastal Namib desert: implications for landscape development. *Geomorphology*, 85, 49–62.

Viles, H.A. and Pentecost, A. (1994). Problem in assessing the weathering action with an example of epiliths on sandstone. In: Robinson, D.A. and Williams, R.B.G. (Eds.). *Rock Weathering and Landform Evolution*. Wiley. Chichester, 99–116.

Viles, H.A. and Spencer, T. (1995). *Coastal Problems*. Arnold. London, 350 p.

Villamayor, P. and Berryman, K.R. (1999). La tasa de desplazamiento de una falla como aproximación de primer orden en las estimaciones de peligrosidad sísmica. *Primer Congreso Nacional de Ingeniería Sísmica*. Asociación Española de Ingeniería Sísmica. Murcia, vol.1, 153–163.

Vincent, P. and Clarke, V. (1980). Terracette morphology and soil properties: a note on a canonical correlation study. *Earth Surface Processes*, 5, 291–295.

Vine, F.J. (1966). Spreading of the ocean floor: New evidence. *Science*, 154, 1405–1415.

Vine, F.J. (1977). The continental drift debate. *Nature*, 266, 19–22.

Vine, F.J. and Matthews, D.H. (1963). Magnetic anomalies over oceanic ridges. *Nature*, 199, 947–949.

Vischer, S.S. (1945). Climatic maps of geological interest. *Bulletin of the Geological Society of America*, 56, 713–736.

Viseras, C., Calvache, M.L. Soria, J.M. and Fernández, J. (2003). Differential features of alluvial fans controlled by tectonic or eustatic accomodation space. Examples from the Betic Cordillera, Spain. *Geomorphology*, 50, 181–202.

Vita-Finzi, C. (1986). *Recent Earth Movementes: An Introduction to Neotectonics*. Academic Press. London, 226 p.

Vitousek, P.M., Mooney, H.A., Lubehenco, J. and Melillo, J.M. (1997). Human domination of earth's ecosystems. *Science*, 277, 494–499.

Vittori, E., Labini, S.S. and Serva, L. (1991). Palaeosismology: review of the state of the art. *Tectonophysics*, 193, 9–32.

Vivian, R. (1979). *Les Glaciers sont Vivants*. Ed. Denoëll. Paris, 240 p.

Vogt, J. (1959). Aspects de l'évolution morphologique récente de l'Ouest Africain. *Annales de Géographie*, 68, 193–206.

Vogt, J. (1966). Le complexe de la stone-line. Mise au point. *Bulletin du Bureau de Recherches Géologiques et Minières*, 4, 3–51.

Vogt, T. (1984). *Croûtes Calcaires: Types et Genèse*. Université Louis Pasteur. Institut de Géographie. Strasbourg, 239 p.

Voight, B. (1978). Lower Gross Ventre Slide, Wyoming, U.S.A. In: Voight, B. (Ed.). *Rockslides and Avalanches*. Elsevier. Amsterdam, 113–166.

Voight, B. (1990). The 1985 Nevado del Ruiz Volcano catastrophe: Anatomy and retrospection. *Journal of Volcanology and Geothermal Research*, 42, 151–188.

Voight, B. and Pariseau, W.G. (1978). Rockslides and avalanches: an introduction. In: Voight, B. (Ed.). *Rockslides and Avalanches, 1. Natural Phenomena*. Elsevier. Amsterdam, 1–67.

Völkel, J. and Grunert, J. (1990). To the problem of dune formation and dune weathering during the Late Pleistocene and Holocene in the southern Sahara and Sahel. *Zeitschrift für Geomorphologie*, 34, 117.

Von Thun, J.L. (1996). Understanding seepage and piping failures-the number one dam safety problem in the west. *ASDSO Western Regional Conference*. Lake Tahoe. Nevada.

Von Richtofen, F. (1886). *Führer für Forchungsreisende*. Oppenheim. Berlín.

Vörösmarty, C.J., Meybeck, M., Fekote, B. and Sharma, K. (1997). The potential impact of neo-Castorization on sediment transport by the global network of rivers. Human impact on erosion arid sedimentation. *IAHS Publication*, 245, 261–273.

Vravec, M. (1994). Some thoughts on the pull-apart origin of karst poljes along the Idrija strike slip fault zone. *Acta Carsologica*, 23, 155–167.

Wahrhaftig, C. and Cox, A. (1959). Rock Glaciers in the Alaska Range. *Bulletin of the Geological Society of America*, 70, 383–436.

Walder, J.S. (1982). Stability of sheet flow of water beneath temperate glaciers and implications for glacier surging. *Journal of Glaciology*, 28, 273–293.

Walcott, R.I. (1970). Isostatic response to loading of the crust in Canada. *Canadian Journal Earth Sciences*, 70, 383–436.

Walker, G. (2004). Frozen time. *Nature*, 429, 596–597.

Walker, G.P.L. (1980). The Taupo pumice; product of the most powerful known (ultraplinian) eruption. *Journal of Volcanology and Geothermal Research*, 8, 69–94.

Walker, H.J. (1973). The morphology of the North Slope. In: Britton, M. (Ed.). *Alaska Arctic Tundra*. AINA. Washington, 49–92.

Walker, H.J. (1986). Periglacial environments. In: Fookes, P.G. and Vaughan, P.R. (Eds.). *A Handbook of Engineering*

Geomorphology, Surrey University Press. London, 82–96.

Walker, H.J. and McGraw, M. (2005). History. Coastal Geomorphology. In: Schwartz, M.L. (Ed). *Encyclopedia of Coastal Science*. Springer. Dordrecht, 519–526.

Walker, J.G.C. and Sloan, L.C. (1992). Something is wrong with climate theory. *Geotimes*, June, 16–18.

Walker, T.R. (1979). Red color in dune sand. In: McKee, E.D. (Ed.). *A Study of Global Sand Seas*. United States Goverment Printing Office. Washington, 61–81.

Wallace, R.E. (1977). Profiles and ages of young fault scarps, north-central Nevada. *Geological Society of American Bulletin*, 88, 1267–1281.

Waller, R.J. and Tackwell, G.W. (2005). Glacier-permafrost interactions and glaciotectonic landform generation at the margin of the Leverett Glacier, West Greenland. In: Harris, C. and Murton, J.B. (Eds.). *Cryosphere Systems: Glaciers and Permafrost*. Geological Society of London. Special Publication, 242, 39–50.

Walling, D.E. (1984). The sediment yields of African rivers. In: Walling, D.E., Foster, S.S.D. and Wurzel, P. (Eds.). *Challenges in African Hydrology and Water Resources*. International Association of Hydrological Sciences, N° 144, 265–283.

Walling, D.E. (1987). Rainfall, runoff and erosion of the land: a global view. In: Gregory, K.J. (Ed.). *Energetics of Physical Environment*. Wiley. Chichester, 89–117.

Walling, D.E. and Webb, B.W. (1986). Solutes in river systems. In: Trudgill, S.T. (Ed.). *Solute Processes*. Wiley. Chichester, 251–327.

Walter, H. (1986). The Namib Desert. In: Evenari, M., Noy-Meir, L. and Goodall, D.W. (Eds.). *Hot Deserts and Arid Shrublands*. Elsevier. Amsterdam, 245–282.

Waltham, A.C. (1989). *Ground Subsidence*. Blackie. Glasgow, 202 p.

Waltham, A.C. and Fookes, P.G. (2003). Engineering classification of karst ground conditions. *Quaterly Journal of Engineering Geology and Hydrogeology*, 36, 101–118.

Waltham, T., Bell, F. and Culshaw, M. (2005). *Sinkholes and Subsidence. Karst and Cavernous Rock in Engineering and Construction*. Springer-Praxis, Chichester, 382 p.

Walther, J. (1915). Laterite in Westaustralien. *Zeitschrift der Deustschen Geologischen Gessellschaft*, 62, 1–7, 46–53.

Walton, K. (1969). *The Arid Zones*. Aldine. London, 175 p.

Wang, X., Dong, Z., Yang, P., Zhang, J. and Quian, G. (2005). Wind energy environments and dunefield activity in the Chinese deserts. *Geomorphology*, 65, 33–48.

Wang, Z.T., Zhang, J.W., Zang, Q.M., Qiang, M.R., Chen, F.H. and Ling, Y.Q. (2008). Barchans of Minqin: Sediment Transpor. *Geomorphology*, 96, 233–238.

Ward, A.W. and Greeley, R. (1984). Evolution of yardangs at Rogers Lake, California. *Bulletin of the Geological Society of America*, 95, 829–837.

Ward, R. (1978). *Floods: a Geographical Perspective*. Macmillan. London.

Ward, S. (2004). Suffosion. In: Goudie, A.S. (Ed.). *Encyclopedia of Geomorphology*. Routledge. London, p. 1025.

Ward, W.H. (1945). The stability of natural slopes. *Geographical Journal*, 105, 170–197.

Warke, P.A. and Smith, B.J. (1994). Short-term rock temperature fluctuations under simulated hot desert conditions: some preliminary data. In: Robinson, D.A. and Williams, R.B.G. (Eds.). *Rock Weathering and Landform Evolution*. Wiley. Chichester, 57–70.

Warner, T.T. (2004). *Desert Meteorology*. Cambridge University press. Cambridge.

Warren, A. (1979). Aeolian processes. In: Embleton, C. and Thornes, J. (Eds.). *Process in Geomorphology*. Edward Arnold. London, 325–351.

Warren, A. and Knott, P. (1983). Desert dunes: a short review of needs in desert dune research and a recent study of micrometeorological dune-initiation mechanisms. In: Brookfield, M.E. and Ahlbrandt, T.S. (Eds.). *Eolian Sediments and Processes*, Elsevier. Amsterdam, 343–352.

Warren, A., Chappell, A., Todd, M.C., Bristow, C., Drake, N., Engelstaedter, S., Martins, V., M'bainayel, S. and Washington, R. (2007). Dust-raising in the dustiest place on earth. *Geomorphology*, 92, 25–37.

Warren, J. (1999). *Evaporites*. Blackwell. Oxford, 438 p.

Warren, J. (2006). *Evaporites: Sediments, Resources and Hydrocarbons*. Springer. Berlin, 1035 p.

Warren, J. (2007). *Evaporites*. Blacwell. Oxford, 1050 p.

Warrick, J.A., George, D.A., Gelfenbaum, G., Ruggiero, P., Kaminsky, G.M. and Beirne, M. (2009). Beach morphology and change along the mixed grain-size delta of the dammed Elwha River, Washington. *Geomorphology*, 111, 136–148.

Warrick, R.A. (1993). Climate and Sea Level Change: a synthesis. In: Warrick, R.A., Barrow, E.M. and Wigley, T.M. (Eds). *Climate and Sea Level Change: Observations, Projections and Implications*. Cambridge University Press. Cambridge, 3–21.

Warrick, R.A. and Oerlemans, J. (1990). Sea level rise. In: Houghton, J.T., Jenkins, G.J. and Ephraums, J.J. (Eds.). *Climate Change. The IPCC Assessment. Cambridge University Press*. Cambridge, 257–281.

Warrick, R.A., Le Provost, C., Meier, M.F., Oerlemans, J. and Woodworth, P.L. (1996). Changes in sea level. In: Houghton, J.T., Meira Filho, L.G., Callander, B.A., Harris, N., Kattenberg, A. and Maskell, K. (Eds.). Climate *Change 1995. The Science of Climatic Change*. Cambridge University Press. Cambridge, 358–405.

Warwick, G.T. (1964). Dry valleys of the southern Pennines, England. *Erdkunde*, 18, 116–123.

Washburn, A.L. (1956). Classification of patterned ground and review of suggested origins. *Bulletin of the Geological Society of America*, 67, 823–865.

Washburn, A.L. (1967). Instrumental observations on mass wasting in the Mesters Vig District, Northeast Greenland. *Meddelelser om Gronland*, 166, 318 p.

Washburn, A.L. (1969). Weathering, frost action and patterned ground in the Mesters Vig District, Northeast Greenland. *Meddelelser om Gronland*, 176, 303 p.

Washburn, A.L. (1973). *Periglacial Processes and Environments*. Arnold. London, 320 p.

Washburn, A.L. (1979). *Geocryology. A Survey of Periglacial Processes and Environments*. Arnold. Londres, 406 p.

Washington, R., Todd, M., Middleton, M.J. and Goudie, A.S. (2003). Dust-storm areas determined by Total Ozone Monitoring Spectometer and surface observations. *Annal of the Association Geographers*, 93, 297–313.

Washington, R., Todd, M.C., Lizcano, I. Tegen, C., Flamant, C., Koren, I., Ginoux, P., Engelstaedter, S., Bristow, C.S., Zender, C.S., Goudie, A.S., Warren, A. and Propsero, J.M. (2006). Links between topography, wind, deflation, lakes and dust: The case of the Bodélé Depression. *Geophysical Research Letters*, vol. 33, LO9401, doi: 10.1029/2006GL 25827.

Wasson, R.J. (1977). Last-glacial alluvial fan sedimentation in the lower Derwent Valley, Tasmania. *Sedimentology* 24, 781–799.

Wasson, R.J. (1983). Dune sediment types, sand colour, sediment provenance and hydrology in the Strzelecki-Simpson dunefield, Australia. In: Brookfield, M.E. and Ahlbrandt, T.S. (Eds.). *Eolian Sediments and Processes*, Elsevier. Amsterdam, 165–195.

Wasson, R.J. (1984). Late Quaternary palaeo-environments in the desert dunnefields of Australia. In: Vogel, J.C. (Ed.). *Late Cenozoic Palaeoenvironmental of the Southern Hemisphere*, Balkema. Rotterdam, 419–432.

Wasson, R.J., Rajaguru, S.N., Misra, V.N., Agarwal, D.P., Dhir, R.P., Singhvi, A.K. and Kameswara Rao, K. (1983). Geomorphology, Late Quaternary stratigraphy and palaeoclimatology of the Thar dune field. *Zeitschrift für Geomorphologie Supplementband,* 45, 117–151.

Wasson, R.J., Fitchett, K., Mackey, B. and Hyde, R. (1988). Large-scale patterns of dune type, spacing and orientation in the Australian Continental Dunefield. *Australian Geographer*, 19, 89–104.

Wataron, R.T., Rodhe, A., Deschger, H. and Siegenthaler, U. (1990). Greenhouse gases and aerosols. In: Houghton, J.T., Jenkins, Y.C. and Ephraums, J.J. (Eds.). *Climatic Change. The IPCC Scientific Assessment*. Cambridge University Press. Cambridge, 1–40.

Watson, A. (1979). Gypsum crusts in deserts. *Journal of Arid Environments*, 2, 3–20.

Watson, A. (1983). Gypsum crusts. In: Goudie, A.S. and Pye, K. (Eds). *Chemical Sediments and Geomorphology: Precipitates and Residua in the near Surface Environment*. Academic Press. New York, 133–161.

Watson, A. (1985). Structure, chemistry and origins of gypsum crusts in southern Tunisia and the central Namib Desert. *Sedimentology*, 32, 855–875.

Watson, A. (1989a). Windflow characteristics and aeolian entrainment. In: Thomas, D.S.G. (Ed.). *Arid Zone Geomorphology*. Belhaven Press. London, 209–231.

Watson, A. (1989b). Desert crusts and varnishes. In: Thomas, D.S.G. (Ed.). *Arid Zone Geomorphology*. Hallstead Press. New York, 25–55.

Watson, A. (1990). The control of blowing sand and mobile desert dunes. In: Goudie, A. (Ed.). *Techniques for Desert Reclamation*. Wiley, 35–85.

Watson, A., Price-Williams, D. and Goudie, A.S. (1983). Palaeoenvironmental interpretation of colluvial sediments and palaeosols of the Late Pleistocene hypothermal in southern Africa. *Palaeogeography, Palaeoclimatology and Palaeoecology*, 45, 225–250.

Watson, A.J., Bakker, D.C.E., Ringwell, A.J., Boyd, P.W. and Law, C.S. (2000). Effect of iron supply on Southern Ocean CO_2 uptake and implications for glacial atmospheric CO_2. *Nature*, 407, 730–733.

Watts, A.B. and Masson, D.G. (1995). A giant landslide on the north flank of Tenerife, Canary Islands. *Journal Geophysical Research*, 100, 24487–24498.

Watts, S.H. (1980). Quaternary pedogenic calcretes from the Kalahari (southern Africa): mineralogy, genesis and diagenesis. *Sedimentology*, 27, 661–686.

Watts, S.H. (1983). Weathering processes and products under arid Arctic conditions. *Geografiska Annaler*, 65 A, 85–98.

Watts, W.A., Allen, J.R.M. and Huntley, B. (1996). Vegetation history and palaeoclimate of the last glacial period at Lago Grande di Montichio, Southern Italy. *Quaternary Science Reviews*, 15, 133–153.

Wayland, E.J. (1933). Peneplains and some other erosional platforms. *Annual Report and Bulletin*, Protectorate of Uganda Geological Survey, Department of Mines, Note 1, 77–79.

Weaver, A.J. (2003). The science of climate change. *Geoscience Canada*, 30, 91–109.

Webb, J.A. and Golding, S.D. (1998). Geochemical mass-balance and oxygen-isotope constraints on silcrete formation and its palaeoclimatic implications in southern Australia. *Journal of Sedimentary Research*, 68, 981–993.

Webb, R.H. and Wilshire, H.G. (Eds.) (1983). *Environmental Effects of Off-road Vehicles*. Springer. New York.

Weertman, J. (1957). On the sliding of glaciers. *Journal of Glaciology*, 3, 33–38.

Weertman, J. (1983). Creep deformation of ice. *Annual Review of Earth and Planetary Sciences*, 11, 215–240.

Wegener, A. (1915). *Die. Entstehung der Kontinente und Ozeane*. Sammlung Vieweg, Braunschweig, 94 p.

Wellman, H.W. and Wilson, A.T. (1965). Salt weathering a neglected geological erosive agent in coastal arid environments. *Nature*, 205, 1097–1098.

Wells, D.L. and Coppersmith, K.J. (1994). New empirical relationships among magnitude, rupture length, rupture width, rupture area and surface displacement. *Bulletin of the Seismological Society of America*, 84, 974–1002.

Wells, N. and Dorr, J.A. (1987). A reconnaissance in sedimentation of the Kosi alluvial fan on India. In: Ethridge, f., Flores, R.M. and Harvey, M. (Eds.). *Recent Development in Fluvial Sedimentology*. Special Publication-Society of Economic Paleontologists and Mineralogists, 39, 51–61.

Wells, S.G. and Dorr, J.A. (1987). Shifting of the Kosi River, northern India. *Geology*, 15, 204–207.

Wells, S.G. and Harvey, A.M. (1987). Sedimentologic and geomorphic variations in storm generated alluvial fans, Howgill Fells, Northwestern England. *Bulletin Geological Society of America,* 98, 182–198.

Wells, S.G., Metadden, L.D. and Dohrenwend, J.C. (1987). Influence of late Quaternary climatic changes on geomorphic and pedogenic processes on a desert piedmont, eastern Mojave Desert, California. *Quaternary Research*, 27, 130–146.

Wenland, W.M. and Bryson, R.A. (1974). Dating climate episodes of the Holocene. *Quaternary Research*, 4, 9–24.

Wesson, R.L. (1983). The eruption of Mount St. Helens - Entering the era of real-time geology. In: Tank, R.W. (Ed.). *Environmental Geology*, Oxford University Press. New York, 50–62.

Wesson. R.L., Helley, E.J., Lajoie, K.R. and Wentworth, C.M. (1975). Faults and future earthquakes. In: Borcherdt, R.D. (Ed.). *Studies for seismic zonation in the San Francisco Bay region*. United States Geological Survey Professional Paper 941-A, A5–A30.

Westgate, J.A. and Naeser, N.C. (1998). Tephrocronology and fission-track dating. In: Rutter, N.W. and Catto, N.R. (Eds.). *Dating Methods for Quaternary Deposits*, Geological Association of Canada, Geotext. St. John's, 15–28.

Whirthmann, A. (2000). *Geomorphology of the Tropics*. Springer. Berlin, 314 p.

Whitaker, C.R. (1973). *A Bibliography of Pediments*. GeoAbstracts. Norwich, 95 p.

Whitaker, C.R. (1979). The use of the term 'pediment' and related terminology. *Zeitschrift für Geomorphologie*, 23, 427–39.

White, D.J., Burroves, G., Davis, T., Hajmal, Z., Hirsche, K., Hutcheon, I., Majer, E., Rostron, B. and Whittaker, S. (2004). Greenhouse gas sequestration in abandoned oil reservoirs. The International Energy Agency Weyburn pilot project. *GSA Today*, 14, 4–10.

White, E.L. and White, B.D. (1969). Processes of cavern breakdown. *National Speleological Society Bulletin*, 31, 83–96.

White, E.L. and White, B.D. (2000). Breakdown morphology. In: Klimchouk, A.B., Ford, D.C., Palmer, A.N. and Dreybrodt, W. (Eds.). *Speleogenesis Evolution of Karst Aquifers*. National Speleological Society, 427–429.

White, W.A. (1972). Deep erosion by continental ice sheets. *Bulletin of the Geological Society of America*, 83, 1037–1056.

White, W.B. (1988). *Geomorphology and Hydrology of Karst Terrains*. Oxford University Press. Oxford, 464 p.

White, W.B. (2005). Gypsum flowers and related speleothems. In: Culver, C.C. and White, W.D. (Eds.). *Encyclopedia of Caves*. Elsevier. Ámsterdam, 288–291.

White, W.B. and Culver, D.C. (2005). In: Culver, D.C. and White, W.B. (Eds.). *Encyclopedia of Caves*. Elsevier. Amsterdam, 81–85.

Whiteman, C.A. (1995). Processes of terrestrial deposition. In: Menzies, J. (Ed.). *Modern Glacial Environments: Processes, Dynamics and Sediments*. Butterworth-Heineman. Oxford, 293–308.

Whitford, W. (2002). *Ecology of Desert Systems*. Academic Press. London, 343 p.

Whitmore, T.M., Tuner, B.L., Johnson, D.L., Kates, R.W. and Gottschan, T.R. (1990). Long-term population change. In: Turner, B.L.; Clark, W.C., Kates, R.W., Matthews, J.T. and Meyer, W.B. (Eds.). *The Earth as Transformed by Human Action*. Cambridge University Press. Cambridge, 26–39.

Whitney, M.I. (1978). The role of vorticity in developing lineation by wind erosion. *Geological Society of America Bulletin*, 89, 1–18.

Whitney, M.I. (1979). Electron micrography of mineral surfaces subject to wind-blast erosion. *Bulletin of the Geological Society of America*, 90, 917–934.

Whitney, M.I. (1983). Eolian features shaped by aerodynamic and vorticity processes. In: Brookfield, M.E. and Ahlbrandt, T.S. (Eds.). *Eolian Sediments and Processes*. Elsevier, Amsterdam, 223–245.

Whitney, M.I. (1985). Yardangs. *Journal of Geological Education*, 33, 93–96.

Whitney, M.I. and Dietrich, R.V. (1973). Ventifact sculpture by windblown dust. *Bulletin of the Geological Society of America*, 84, 2561–2581.

Whitney, M.I. and Splettstoesser, J.F. (1982). Ventifacts and their formation: Darwin Mountains, Antarctica. In: Yaalon, D.H. (Ed.). *Aridic Soils and Geomorphic Processes*. Catena Supplement 1. Braunsehweig, 175–194.

Wickham, S.S. and Johnson, W.H. (1981). The Tiskilwa till, a regional view of this origin and depositional processes. *Annals of Glaciology*, 2, 176–182.

Widdowson, M. (2007). Laterite and Ferricrete. In: Nash, D.J. and McLaren, S.J. (Eds.). *Geochemical Sediments and Landscapes*. Blackwell. Malden, 46–94.

Wieczorek, G.F. (1996). Landslides triggering mechanisms. In: Turner, A.K. and Schuster, R.L. (Eds.). *Landslides: Investigation and Mitigation*. Special Report, 247. National Academic Press. Washington, 76–90.

Wiggs, G.F.S. (1997). Sediment mobilisation by the wind. In: Thomas, D.S.G. (Ed.). *Arid Zone Geomorphology: Process, Form and Change in Drylands*. John Wiley. Chichester, 351–372.

Wiggs, G.F.S., Linvingstone, I. and Warren, A. (1996). The role of streamline curvature in sand dune dynamics: evidence from field and wind tunnel measurements. *Geomorphology*, 17, 29–46.

Wigley, T.M.L. (1998). The Kyoto Protocol: CO_2, CH_4 and climate implications. *Geophysical Research Letters*, 25, 2285–2288.

Wilcock, P.R. and Iverson, R.M. (2003). Prediction in Geomorphology. *Geophysical Monograph* 135, 3–11. American Geophysical Union.

Wilford, G.E. and Wall, J.A.D. (1965). Karst topography in Sarawak. *Journal Tropical Geography*, 21, 44–70.

Wilhelmy, H. (1958). *Klimamorphologie der Massengesteine*. Westerman. Brunswick, 238 p.

Wilkinson, B.H. (2005). Humans as geologic agents: A deep-time perspective. *Geology*, 33, 161–164.

Willets, B. (1983). Transport by wind of granular materials of different grain shapes and densities. *Sedimentology*, 30, 669–679.

Williams, G.E. (1973). Late Quaternary piedmont sedimentation, soil formation and paleoclimates in arid south Australia. *Zeitschrift für Geomorphologie*, 17, 102–125.

Williams, G.P. (1971). Flood deposits of the sand-bad ephemeral streams of central Australia. *Sedimentology*, 17, 1–40.

Williams, G.P. (1988). Paleofluvial estimates from dimensions of former channel and meanders. In: Baker, V.R., Craig Kocher, R. and Patton, P.C. (Eds.). *Flood Geomorphology*. Wiley. New York, 321–334.

Williams. H. (1941). Calderas and their origins. *University of California Publication, Department of Geological Sciences*, 25, 239–246.

Williams, H. and McBirney, A.R. (1979). *Volcanology*. Freeman and Cooper. San Francisco, 391 p.

Williams, H.F.L., Hutchinson, A. and Nelson, A.R. (2005). Multiple sources for late-Holocene tsunamis at Discovery Bay, Washington State, USA. *The Holocene*, 15, 60–73.

Williams, J.H. and Vineyard, J.M. (1976). Geologic indicators of subsidence and collapse in karst terrain in Missouri. *55th Annual Meeting Transportation Research Board*. Washington.

Williams, J.J., Butterfield, G.R. and Clark, D.G. (1994). Aerodynamic entrainment thresholds: effect of boundary layer flow conditions. *Sedimentology*, 41, 309–328.

Williams, M.A.J. (1978). Termites, soils and landscape equilibrium in the Northern Territory of Australia. In: Davis, J.L. and Williams, M.A.J. (Eds.). *Landform Evolution in Australasia*. Australian National University Press. Camberra, 128–141.

Williams, M.A.J. (1994). Cenozoic climatic changes in deserts: a synthesis. In: Abrahams, A.D. and Parsons, A.J. (Eds.). *Geomorphology of Desert Environments*, Chapman and Hall. London, 644–670.

Williams, M.A. and Balling, R.C. (1996). Interactions of Desertification and Climate. Arnold. London, 270 p.

Williams, M.A. and Faure, H. (Eds.) (1980). *The Sahara and the Nile*. Balkema. Rotterdam, 607 p.

Williams, M.A.J. and Williams, F.M. (1980). Evolution of the Nile basin. In: Williams, M.A.J. and Faure, H. (Eds,). *The Sahara and the Nile*. Balkema. Rotterdam, 207–224.

Williams, M.A.J., Adamson, D., Cock, B. and McEvedy, R. (2000). Late Quaternary environment of the White Nile region, Sudan. *Global and Planetary Change*, 26, 305–316.

Williams, P.J. and Smith, M.W. (1989). *The Frozen Earth. Fundamental in Geocryology*. Cambridge University Press. Cambridge, 306 p.

Williams, P.W. (1966). Limestone pavements with special reference to West Ireland. *Transactions of the Institute of British Geographers*, 40, 155–172.

Williams, P.W. (1969). The geometric effect of groundwater. In: Chorley, R.J. (Ed.).*Water, Earth and Man*. Methuen. London.

Williams, P.W. (1971). Illustrating morphometric analysis of karst with examples from New Guinea. *Zeitschrift für Geomorphologie*, 15, 40–61.

Williams, P.W. (1972). Morphometric analysis of polygonal karst in New Guinea. *Bulletin of the Geological Society of America*, 83, 761–796.

Williams, P.W. (1978). Interpretations of Australasian karst. In: Davies, J.L. and Williams, M.A.J. (Eds.). *Landform Evolution of Australasia*. Australian National University Press. Camberra, 259–288.

Williams, P.W. (1983). The role of subcutaneous zone in karst hydrology. *Journal Hydrology*, 61, 45–67.

Williams, P.W. (1993). Environmental change and human impact of karst terrains: an introduction. In: Williams, P.W. (Ed.). *Karst Terrains: Environmental Changes and Human Impact. Catena,* Supplement, 25, Cremlingen, 1–20.

Williams, P.W. (2004a). Dolines. In: Gunn, J. (Ed.). *Encyclopedia of Caves and Karst Science*. Fitzroy Dearborn. New York, 304–310.

Williams, P.W. (2004b). Dolines. In: Goudie, A.S. (Ed.) *Encyclopedia of Geomorphology*. Routledge. London, 266–270.

Williams, R.B.G. and Robinson, D.A. (1981). Weathering of sandstone by the combined action of frost and salt. *Earth Surface Processes and Landforms*, 6, 1–9.

Williams, R.B.G. and Robinson, D.A. (1994). Weathering flutes on siliceous rocks in Britain and Europe. In: Robinson, D.A. and Williams, R.B.G. (Eds.). *Rock Weathering and Landform Evolution*. Wiley. Chichester, 413–432.

Williams, S.N. and Self, S. (1983). The October 1902 Plinian eruption of Santa María Volcano, Guatemala. *Journal of Volcanology and Geothermal Research*, 16, 33–56.

Willis, B. (1936). *East African Plateaus and Rift Valleys- Studies in Comparative Seismology*. Carnegie Institute, Publication. Washington, 470 p.

Wilson, A.F. (1984). Origin of quartz-free gold nuggets and supergene gold found in laterites and soils - a review and some new observations. *Australian Journal of Earth Science*, 31, 303–316.

Wilson, C.J.N. (1993). Ignimbritas y calderas: perspectivas históricas, ideas actuales y desarrollos futuros. In: Martí, J. and Araña, V. (eds.). *La Volcanología Actual*. Consejo Superior de Investigaciones Científicas. Madrid, 197–275.

Wilson, C.J.N. and Houghton, B.F. (2000). Pyroclast transport and deposition. In: Sigurdsson, H. (Ed.). *Encyclopedia of Volcanoes*. Academic Press. San Diego, 545–554.

Wilson, I.G. (1971). Desert sandflow basins and a model for the development of ergs. *Geographical Journal*, 137, 180–199.

Wilson, I.G. (1972). Aeolian bedforms-their development and origins. *Sedimentology*, 19, 173–210.

Wilson, I.G. (1973). Ergs. *Sedimentary Geology*, 10, 77–106.

Wilson, J.T. (1963). A possible origin of Hawaiian islands. *Canadian Journal of Physics*, 41, 863–870.

Wilson, J.T. (1965). A new class of faults and their bearing on continental drift. *Nature*, 207, 343–347.

Wilson, J.T. (1966). Did the Atlantic close and then reopen?. *Nature*, 211, 676–681.

Wilson, L. (1968). Morphogenetic classification. In: Fairbridge, R.W. (Ed.). *The Encyclopedia of Geomorphology*. Dowden, Hutchinson and Ross. Stroudsburg, 717–729.

Wilson, L. (1969). Les relations entre les processus géomorphologiques et le climat moderne comme méthode de paleoclimatologie. *Revue de Géographie Physique et de Géologie Dynamique*, 11, 301–314.

Wilson, L. (1973). Variations in mean annual sediment yield as a function of mean annual precipitation. *American Journal of Science*, 273, 335–349.

Wilson, P. (2007). Periglacial Landforms. Block/rock Streams. In: Elias, S.A. (Ed.). *Encyclopedia of Quaternary Science*. Elsevier. Amsterdam, 2217–2225.

Winker, C.D. and Howard, J.D. (1977). Correlation of tectonically deformed shorelines on the southern Atlantic Coastal Plain. *Geology* 5, 123–127.

Winkler, E.M. (1973). *Stone: Properties, Durability in Man's Environment*. Springer. New York, 230 p.

Winkler, E.M. and Singer, P.C. (1972). Crystallisation pressure of salts in stone and concrete. *Bulletin of the Geological Society of America*, 83, 3509–3514.

Wintle, A.G. (1993). Luminescence dating of aeolian sands-an overview. In: Pye, K. (Ed.). *Dynamics and Environmental Context of Aeolian Sedimentary Systems*. Geological Society of London, Special Paper 72. London, 49–58.

Wirthmann, A. (1964). Die Landformen der Edge-Inseö in Südost-Spitzbergen. *Ergebnisse der Stanforland-Expedition*, V.2, 53 p.

Wohl, E.E. (2000). Geomorphic effects of flood. In: Wohl, E.E. (Ed.). *Inland Flood Geomorphology*. Cambridge University Press. Cambridge, 167–193.

Wohl, E.E. (2004). Flood. In: Goudie, A.S. (Ed.). *Encyclopedia of Geomorphology*. Routledge. London, 378–380.

Wohl, E.E. (2007). Hydrology and discharge. In: Gupta. A. (Ed.). *Large Rivers. Geomorphology and Management*. Wiley. Chichester, 29–44.

Wohlrab, B. (1969). Effects of mining subsidences on the ground water and remedial measures. *Publications of International Association of Sciences Hydrological*, 89, 52–512.

Wolfe, S.A. (2007). Dune Fields. High Latitudes. In: Elias, S.A. (Ed.). *Encyclopedia of Quaternary Science*. Elsevier. Amsterdam, 599–607.

Wolman, M.G. (1967). A cycle of sedimentation and erosion in urban river channels. *Geografiska Annales*, 49-A, 385–395.

Wolman, M.G. and Gerson, R. (1978). Relative scales of time and effectiveness of climate in watershed geomorphology. *Earth Surface Processes*, 3, 189–208.

Wolman, M.G. and Leopold, L.B. (1957). River flood plains: some observations on their formation. *United States Geological Survey Professional Paper*, 282C, 87–107.

Woo, M.K. and Winter, T.C. (1993). The role of permafrost and seasonal frost in the hydrology of northern wetlands in North America. *Journal of Hydrology*, 141, 5–31.

Wood, A. (1942). The development of hillside slopes. *Proceeding of Geological Association*, 53, 128–140.

Wood, H.O. and Neumann, F. (1931). Modified Mercalli intensity scale of 1931. *Bulletin of the Seismological Society of America*, 21, 277–283.

Woodroffe, C.D. (2002). *Coasts. Form, Process and Evolution.* Cambridge University Press, Cambridge, 623 p.

Woodward, J.C., Macklin, M.C. and Lewin, J. (1994). Pedogenetic Weathering and relative-age dating of Quaternary alluvial sediments in the Pindus Mountains of Northwest Greece. In: Robinson, D.A. and Williams, R.B.G. (Eds.). *Rock Weathering and Landform Evolution*. Wiley. Chichester, 259–283.

Woodward, J.C., Macklin, M.C., Krom, M.D. and Williams, M.A.J. (2007). *The Nile: Evolution, Quaternary River environments and material fluxes*. Wiley. Chichester, 261–292.

Wopfner, H. (1978). Silcretes of northern South Australia and adjacent regions. In: Langford Smith, T. (Ed.). *Silcrete in Australia*. Department of Geography, University of New England, 93–141.

World Glacier Monitoring Service (1989). *World Glacier Inventory*, IAHS, (ICSI). UNEP-UNESCO.

World Resources (1998). *World resources 1996–1997*. Oxford University Press. Oxford.

Worsley, P. (1974). Recent "annual" moraine ridges at Austre Okstindebreen, North Norway. *Journal of Glaciology*, 13, 265–277.

Wright, H.E. Jr. (1984). Sensivity in response time of natural systems to climatic change in the late Quaternary. *Quaternary Science Reviews*, 3, 91–131.

Wright, L.D. (1985). River deltas. In: Davis, R.A. (Ed.). *Coastal Sedimentary Environments*. Springer. New Cork, 1–76.

Wright, L.D. (2004). River Delta. In: Goudie, A.S. (Ed.). *Encyclopedia of Geomorphology*. Routledge. London, 862–866.

Wright, V.P. (2007). Calcrete. In: Nash, D.L. and McLaren, S.J. (Eds.). *Geochemical Sediments and Landscapes*. Blackwell. Malden, 10–45.

Wright, V.P. and Tucker, M.E. (1991). Calcretes: An Introduction. In: Wright, V.P and Tucker, M.E. (Eds.). *Calcretes*. Series 2. International Association of Sedimentologists. Blackwell. Oxford, 1–22.

Wright, V.P., Platt, N.H., Marriot, S.B. and Beck, V.H. (1995). A classification of rhizogenic (root-formed) calcretes, with examples from the Upper Jurasic-Lower Cretaceous of Spain and Upper Cretaceous of southern France. *Sedimentary Geology*, 100, 143–158.

Wyllie, D.C. and Norrish, N.I. (1996). Stabilization of rock slopes. In: Turner, A.K. and Schuster, R.L. (Eds.). *Landslides: Investigation and Mitigation*. Special Report, 247. National Academic Press. Washington, 474–504.

Yair, A. and Gerson, R. (1974). Mode and rate of escarpment retreat in an extremely arid environment (Sharm el Sheikh, southern Sinai Peninsula). *Zeitschrift für Geomorphologie*, 21, 106–121.

Yair, A., Lavee, H., Bryan, R.B. and Adar, E. (1980). Runoff and erosion processes and rates in the Zian Valley badlands, northern Negev, Israel. *Earth Surface Processes*, 5, 205–225.

Yan, Z. and Petit-Maire, N. (1994). The last 140 ka in the Afro-Asian arid/semiarid transitional zone. *Palaeogeography, Palaeoclimatology, Palaeoecology*, 110, 217–233.

Yanes, A. (1990). Configuración litoral de Tenerife y La Gomera. *Monografías de la Sociedad Española de Geomorfología*, 5, 197–215.

Yang, D., Kane, D.L., Hinzman, L.D., Zhang, X., Zhang, T. and Ye, H. (2002). Siberain Lena River hydrology regime and recent change. *Journal Geophysical Research*, 107 (D23).

Yatsu, E. (1966). *Rock Control in Geomorphology*. Sozosha. Japan.

Yatsu, E. (1988). *The Nature of Weathering: An Introduction*. Sozosha. Tokyo, 624 p.

Yazawa, D., Toya, H. and Kaizuka, S. (1971). *Alluvial Fans*. Kokon Shoin. Tokyo, 318 p.

Yeats, R.S., Sieh, K. and Allen, C.R. (1997). *The Geology of Earthquakes*. Oxford University Press. New York, 568 p.

Yershov, E.D. (1998). *General Geocryology*. (Translated from Russian, Obschchaya Geokriologiya, Nedra, 1990). Cambridge University Press. Cambridge, 580 p.

Yochelson, E.L. (1980) (Ed.). The scientific ideas of G.K. Gilbert. *Geological Survey of America Special Paper 183*.

Yokohama, I. (1993). *Vigilancia de Volcanes Activos: Métodos Geofísicos y Geodésicos*. In: Martí, J. and Araña. V. (Eds.). *La Volcanología Actual*. Consejo Superior de Investigaciones Científicas. Madrid, 445–530.

Yoshikawa, T. (1970). On the relations between Quaternary tectonic movement and seismic crustal deformation in Japan. *Bulletin of the Departament of Geography. University of Tokyo*, nº 2.

Yoshimoto, T. and Suetsugi, T. (1990). Comprehensive flood disaster prevention measures in Japan. In: Massing, H., Packman, J. and Zuidema, F.C. (Eds.). *Hydrological Processes and Water Management in Urban Areas*. International Association Hydrological Sciences, Publ. 198, 175–181.

Young, A. (1964). Slope profile analysis. *Zeitschrift für Geomorphologie Supplementband,* 5, 17–27.

Young, A. (1970). Slope form in part of the Mato Grosso, Brasil. *Geographical Journal*, 136, 383–392.

Young, A. (1972). *Slopes*. Longman. Edingurgh, 278 p.

Young, A. (1976). *Tropical Soils and Soil Survey*, Cambridge University Press. London, 468 p.

Young, R. y Young, A. (1992). *Sandstone Landforms*. Springer. Berlin, 163 p.

Young, R.A. (1985). Geomorphic evolution of the Colorado Plateau margin in West-Central Arizona: a tectonic model to distinguish between the causes of rapid, symmetrical scarp retreat and scarp dissection. In: Morisawa, M. and Hack, J.T. (Eds.). *Tectonic Geomorphology*, Allen and Unwin. Boston, 261–278.

Young, R.M. (1987). Salt as an agent in the development of cavernous weathering. *Geology*, 15, 962–966.

Young, R.W. (2004). Sandstones Geomorphology. In: Goudie, A.S. (Ed.). *Encyclopedia of Geomorphology*. Routledge. London, 906–908.

Yu, G. and Ke, X. (2007). Lake Level Studies/Asia. In: Elias, S.A. (Ed.). *Encyclopedia of Quaternary Science*. Elsevier. Amsterdam, 1343–1359.

Yuan, D. (1987). New observations on tower karst. In: Gardiner, V. (Ed.). *International Geomorphology 1986*. Proceedings of 1st International Conference on Geomorphology, Part II. Wiley. Chichester, 1109–1123.

Zaruba, Q. and Mencl, V. (1969). *Landslides and Their Control*. Elsevier. Amsterdam, 205 p.

Zaruba, Q. and Mencl, V. (1982). *Landslides and Their Control*. Elsevier, Amsterdam, 324 p.

Zavoianu, I. (1985). *Morphometry of Drainage Basin*. Elsevier. Amsterdam, 238 p.

Zazo, C. (2006). Cambio climático y nivel del mar: La Península Ibérica en un contexto global. *Cuaternario y Geomorfología*, 20, 115–130.

Zazo, C., Goy, J.L. and Dabrio, C.J. (1986). Late Quaternary and recent evolution of coastal morphology of the Gulf of Cádiz (Huelva-Cádiz), Soutwestern Spain. *First International Symposia on Harbours, Port Cities and Coastal Topography*. Haifa, Israel. Proceedings 200–203.

Zazo, C., Goy, J.L., Somoza, L., Dabrio, C.J., Belluomini, G., Improta, S., Lario, J., Bardají, T. and Silva, P.G. (1994). Holocene sequence of sea-level fluctuations in relation to climatic trends in the Atlantic-Mediterranean linkage coast. *Journal of Coastal Research*, 10, 933–945.

Zazo, C., Dabrio, C.J., Borja, F., Goy, J.L., Lezine, A.M., Lario, J., Polo, M.D., Hoyos, M. and Boersma, J.R. (1999a). Pleistocene and Holocene aeolian facies along the Huelva Coast (southern Spain): climatic and neotectonic implications. *Tectonophysics*, 77, 209–224.

Zazo, C., Silva, P.G., Goy, J.L., Hillaire-Marcel, C., Lario, J., Bardají, T. and González, A. (1999b). Coastal uplift in continental collision plate boundaries: Data from the Last interglacial marine terraces of the Gibraltar Strait area (South Spain). *Tectonophysics*, 301, 95–109.

Zazo, C., Hansen, A., Goy, J.L., Criado, C., González, J.A., Dabrio, C.J., Hillaire-Marcel, C., Bardají, J., Silva, P.G., Cabero, G., Civis, J., Soler, V. and Ghaleb, B. (2008). *Geomorfología volcánica y variaciones del nivel del mar en Lanzarote y Fuerteventura (Plio-Cuaternario)*. Field Trip Guide. VII Congreso Geológico de España. Las Palmas de Gran Canaria.

Zekri, S., Albisu, L.M., Aragües, R. and Herrero, J. (1990). *Impacto económico de la salinidad de los suelos en la agricultura de Bárdenas I*. Comunicaciones Instituto Nacional de Investigaciones Agrarias, Serie Economía, 36. Madrid, 129 p.

Zerboni, A. (2008). Holocene rock varnish on the Messak Plateau (Lybian Sahara): Chronology of the weathering processes. *Geomorphology*, 102, 640–651.

Zhang, C., Li, W. and Day, M. (2005). Towards establishing effective protective boundaries for the Lunan Stone Forest using an online spatial decision support system. *Acta Carsologica*, 34, 151–187.

Zhende, Z. (1984). Aeolian landforms in the Taklimakan Desert. In: El-Baz, F. (Ed.). *Desert and Arid Lands*. Martinus Nijhoff. The Hague, 133–144.

Zhou, J. and Chafetz, H.S. (2009). Biogenic caliches in Texas: The role of organisms and effect of climate. *Sedimentary Geology*, 222, 207–225.

Zielinski, G.A. (1995). Stratospheric loading and optical depth estimates of explosive volcanism over the last 2100 years derived from the GISP2 Greenland ice core. *Journal of Geophysical Research*, 100, D10: 20937–20955.

Zischinky, U. (1966). On the deformation of high slopes. *Proceedings of First Congress on the International Society of Rock Mechanics*, vol. 2, 179–185.

Zolischka, B. (2007). Varved Lake Sediments. In: Elias, S.A. (Ed.). *Encyclopedia of Quaternary Science*. Elsevier. Amsterdam, 3105–3114.

Zoltai, S.C. (1971). Southern limit of permafrost features in peat landforms, Manitoba and Saskatchewan. *Geological Association of Canada, Special Paper,* 9, 305–310.

Zuchiewicz, W. (2004). Fault and fault scarp. In: Goudie, A. (Ed). *Encyclopedia of Geomorphology*. Routledge, London, 362–364.

INDEX

For Product Safety Concerns and Information please contact our EU representative GPSR@taylorandfrancis.com Taylor & Francis Verlag GmbH, Kaufingerstraße 24, 80331 München, Germany

Batch number: 10733584

Printed by Printforce, the Netherlands